Bovine Reproduction

"Let our cattle bear, without mishap and without loss"
　　　　Psalms 144: 14, *New American Standard Bible*

Bovine Reproduction

Edited by

Richard M. Hopper DVM, Diplomate ACT
Department of Pathobiology and Population Medicine
College of Veterinary Medicine
Mississippi State University
Starkville, Mississippi, USA

WILEY Blackwell

This edition first published 2015
© 2015 by John Wiley & Sons, Inc.

Editorial Offices
1606 Golden Aspen Drive, Suites 103 and 104, Ames, Iowa 50014-8300, USA
The Atrium, Southern Gate, Chichester, West Sussex, PO19 8SQ, UK
9600 Garsington Road, Oxford, OX4 2DQ, UK

For details of our global editorial offices, for customer services and for information about how to apply for permission to reuse the copyright material in this book please see our website at www.wiley.com/wiley-blackwell.

Authorization to photocopy items for internal or personal use, or the internal or personal use of specific clients, is granted by Blackwell Publishing, provided that the base fee is paid directly to the Copyright Clearance Center, 222 Rosewood Drive, Danvers, MA 01923. For those organizations that have been granted a photocopy license by CCC, a separate system of payments has been arranged. The fee codes for users of the Transactional Reporting Service are ISBN-13: 978-1-1184-7083-1/2015.

Designations used by companies to distinguish their products are often claimed as trademarks. All brand names and product names used in this book are trade names, service marks, trademarks or registered trademarks of their respective owners. The publisher is not associated with any product or vendor mentioned in this book.

The contents of this work are intended to further general scientific research, understanding, and discussion only and are not intended and should not be relied upon as recommending or promoting a specific method, diagnosis, or treatment by health science practitioners for any particular patient. The publisher and the author make no representations or warranties with respect to the accuracy or completeness of the contents of this work and specifically disclaim all warranties, including without limitation any implied warranties of fitness for a particular purpose. In view of ongoing research, equipment modifications, changes in governmental regulations, and the constant flow of information relating to the use of medicines, equipment, and devices, the reader is urged to review and evaluate the information provided in the package insert or instructions for each medicine, equipment, or device for, among other things, any changes in the instructions or indication of usage and for added warnings and precautions. Readers should consult with a specialist where appropriate. The fact that an organization or Website is referred to in this work as a citation and/or a potential source of further information does not mean that the author or the publisher endorses the information the organization or Website may provide or recommendations it may make. Further, readers should be aware that Internet Websites listed in this work may have changed or disappeared between when this work was written and when it is read. No warranty may be created or extended by any promotional statements for this work. Neither the publisher nor the author shall be liable for any damages arising herefrom.

Library of Congress Cataloging-in-Publication Data

Bovine reproduction / [edited by] Richard McRae Hopper.
    p. ; cm.
  Includes bibliographical references and index.
  ISBN 978-1-118-47083-1 (cloth)
  1. Cattle–Reproduction. 2. Cattle–Breeding. I. Hopper, Richard M., 1954– editor.
  [DNLM: 1. Cattle–physiology. 2. Reproduction. 3. Reproductive Medicine–methods.
4. Reproductive Techniques–veterinary. SF 768.2.C3]
  SF768.2.C3B67 2015
  636.2′1–dc23
                              2014013293

A catalogue record for this book is available from the British Library.

Wiley also publishes its books in a variety of electronic formats. Some content that appears in print may not be available in electronic books.

Cover image: courtesy of the Editor
Cover design by: Modern Alchemy LLC

Set in 9/11pt Palatino by SPi Publisher Services, Pondicherry, India
Printed and bound in Singapore by Markono Print Media Pte Ltd

1   2015

To Donna

Contents

Contributors, x

Preface, xvi

Section I: The Bull

Anatomy and Physiology

1. Anatomy of the Reproductive System of the Bull, 5
   Ben Nabors and Robert Linford

2. Endocrine and Exocrine Function of the Bovine Testes, 11
   Peter L. Ryan

3. Thermoregulation of the Testes, 26
   John P. Kastelic

4. Endocrine Control of Testicular Development and Initiation of Spermatogenesis in Bulls, 30
   Leonardo F.C. Brito

Breeding and Health Management

5. Bull Development: Sexual Development and Puberty in Bulls, 41
   Leonardo F.C. Brito

6. Breeding Soundness Examination in the Bull: Concepts and Historical Perspective, 58
   Richard M. Hopper

7. Evaluation of Breeding Soundness: The Physical Examination, 64
   James Alexander

8. Evaluation of Breeding Soundness: Basic Examination of the Semen, 68
   Richard M. Hopper and E. Heath King

9. Ultrasound Evaluation of the Reproductive Tract of the Bull, 79
   Harry Momont and Celina Checura

10. Management of Breeding Bull Batteries, 92
    E. Heath King

11. Management of Bulls at Custom Collection Studs, 97
    Gary Warner

12. Testicular Degeneration, 103
    Albert Barth

13. Vesicular Adenitis, 109
    Albert Barth

14. Inability to Breed due to Injury or Abnormality of the External Genitalia of Bulls, 113
    Herris Maxwell

Reproductive Surgery

15. Local and Regional Anesthesia for Urogenital Surgery, 131
    Misty A. Edmondson

16. Surgery of the Scrotum and its Contents, 136
    Dwight F. Wolfe

17. Restorative Surgery of the Prepuce, 142
    Dwight F. Wolfe

18. Restorative Surgery of the Penis, 155
    Dwight F. Wolfe

19. Bovine Urolithiasis, 172
    Katharine M. Simpson and Robert N. Streeter

20. Preparation of Teaser Bulls, 181
    Gretchen Grissett

Section II: The Cow

Anatomy and Physiology

21. Anatomy of the Reproductive System of the Cow, 191
    Ben Nabors and Robert Linford

22. Initiation of Puberty in Heifers, 195
    Charles T. Estill

23. Neuroendocrine Control of Estrus and Ovulation, 203
    Marcel Amstalden and Gary L. Williams

24 **Ovarian Follicular and Luteal Dynamics in Cattle,** 219
    Gregg P. Adams and Jaswant Singh

25 **Maternal Recognition and Physiology of Pregnancy,** 245
    Caleb O. Lemley, Leticia E. Camacho and Kimberly A. Vonnahme

Breeding and Health Management

26 **Biosecurity and Biocontainment for Reproductive Pathogens,** 259
    Carla Huston

27 **Beef Replacement Heifer Development,** 267
    Terry J. Engelken

28 **Heifer Development: From Weaning to Calving,** 272
    Ricardo Stockler

29 **Interaction of Nutrition and Reproduction in the Beef Cow,** 276
    William S. Swecker Jr

30 **Interaction of Nutrition and Reproduction in the Dairy Cow,** 283
    Butch Cargile and Dan Tracy

31 **Estrus Detection,** 290
    Rhonda C. Vann

32 **Artificial Insemination,** 295
    Ram Kasimanickam

33 **Pharmacological Intervention of Estrous Cycles,** 304
    Ram Kasimanickam

34 **Pregnancy Diagnosis: Rectal Palpation,** 314
    David Christiansen

35 **Biochemical Pregnancy Diagnosis,** 320
    Amanda J. Cain and David Christiansen

36 **Reproductive Ultrasound of Female Cattle,** 326
    Jill Colloton

37 **Beef Herd Health for Optimum Reproduction,** 347
    Terry J. Engelken and Tyler M. Dohlman

38 **Dairy Herd Health for Optimal Reproduction,** 353
    Carlos A. Risco

39 **Herd Diagnostic Testing Strategies,** 359
    Robert L. Larson

40 **Beef Herd Record Analysis: Reproductive Profiling,** 364
    Brad J. White

41 **Evaluating Reproductive Performance on Dairy Farms,** 370
    James A. Brett and Richard W. Meiring

42 **Marketing the Bovine Reproductive Practice: Devising a Plan or Tolerating a System,** 374
    John L. Myers

Obstetrics and Reproductive Surgery

43 **Vaginal, Cervical, and Uterine Prolapse,** 383
    Augustine T. Peter

44 **Inducing Parturition or Abortion in Cattle,** 396
    Albert Barth

45 **Management to Prevent Dystocia,** 404
    W. Mark Hilton and Danielle Glynn

46 **Dystocia and Accidents of Gestation,** 409
    Maarten Drost

47 **Obstetrics: Mutation, Forced Extraction, Fetotomy,** 416
    Kevin Walters

48 **Obstetrics: Cesarean Section,** 424
    Cathleen Mochal-King

49 **Retained Fetal Membranes,** 431
    Augustine T. Peter

50 **Postpartum Uterine Infection,** 440
    Colin Palmer

51 **Cystic Ovarian Follicles,** 449
    Jack D. Smith

52 **Postpartum Anestrus and its Management in Dairy Cattle,** 456
    Divakar J. Ambrose

53 **Surgery to Restore Fertility,** 471
    Richard M. Hopper

Pregnancy Wastage

54 **Fetal Disease and Abortion: Diagnosis and Causes,** 481
    Wes Baumgartner

55 **Infectious Agents: *Campylobacter*,** 518
    Misty A. Edmondson

56 **Infectious Agents: *Trichomonas*,** 524
    Mike Thompson

57 **Infectious Agents: Leptospirosis,** 529
    Daniel L. Grooms

58 **Infectious Agents: Brucellosis,** 533
    Sue D. Hagius, Quinesha P. Morgan and Philip H. Elzer

59 **Infectious Agents: Infectious Bovine Rhinotracheitis,** 541
    Ahmed Tibary

60 **Infectious Agents: Bovine Viral Diarrhea Virus,** 545
    Thomas Passler

61 **Infectious Agents: Epizootic Bovine Abortion (Foothill Abortion),** 562
    Jeffrey L. Stott, Myra T. Blanchard and Mark L. Anderson

62 **Infectious Agents: *Neospora*,** 567
    Charles T. Estill and Clare M. Scully

63 **Infectious Agents: Mycotic Abortion,** 575
    Frank W. Austin

64 **Early Embryonic Loss Due to Heat Stress,** 580
    Peter J. Hansen

65 **Bovine Abortifacient and Teratogenic Toxins,** 589
    Brittany Baughman

66	**Heritable Congenital Defects in Cattle, 609** *Brian K. Whitlock and Elizabeth A. Coffman*		75	**Superovulation in Cattle, 696** *Reuben J. Mapletoft and Gabriel A. Bó*
67	**Abnormal Offspring Syndrome, 620** *Charlotte E. Farin, Callie V. Barnwell and William T. Farmer*		76	**Embryo Collection and Transfer, 703** *Edwin Robertson*
68	**Strategies to Decrease Neonatal Calf Loss in Beef Herds, 639** *David R. Smith*		77	**Cryopreservation of Bovine Embryos, 718** *Kenneth Bondioli*
69	**Management to Decrease Neonatal Loss of Dairy Heifers, 646** *Ricardo Stockler*		78	**Selection and Management of the Embryo Recipient Herd for Embryo Transfer, 723** *G. Cliff Lamb and Vitor R.G. Mercadante*
			79	**Evaluation of *In Vivo*-Derived Bovine Embryos, 733** *Marianna M. Jahnke, James K. West and Curtis R. Youngs*

Section III: Assisted and Advanced Reproductive Technologies

70	**Use of Technology in Controlling Estrus in Cattle, 655** *Fred Lehman and Jim W. Lauderdale*		80	**Control of Embryo-borne Pathogens, 749** *Julie Gard*
71	**Cryopreservation of Semen, 662** *Swanand Sathe and Clifford F. Shipley*		81	***In Vitro* Fertilization, 758** *John F. Hasler and Jennifer P. Barfield*
72	**Utilization of Sex-sorted Semen, 671** *Ram Kasimanickam*		82	**Cloning by Somatic Cell Nuclear Transfer, 771** *J. Lannett Edwards and F. Neal Schrick*
73	**Control of Semen-borne Pathogens, 679** *Rory Meyer*		83	**The Computer-generated Bull Breeding Soundness Evaluation Form, 784** *John L. Myers*
74	**Bovine Semen Quality Control in Artificial Insemination Centers, 685** *Patrick Vincent, Shelley L. Underwood, Catherine Dolbec, Nadine Bouchard, Tom Kroetsch and Patrick Blondin*			

Glossary, 790

Index, 795

Contributors

Gregg P. Adams DVM, MS, PhD, Diplomate ACT
Professor
Veterinary Biomedical Sciences
Western College of Veterinary Medicine
University of Saskatchewan
Saskatoon, Saskatchewan, Canada

James Alexander DVM
Alexander Veterinary Services
Bentonia, Mississippi, USA

Divakar J. Ambrose MVSc, PhD, PAg, PAS
Dairy Research Scientist, Livestock Research Branch
Alberta Agriculture and Rural Development
ARD Professor, University of Alberta
Edmonton, Alberta, Canada

Marcel Amstalden DVM, PhD
Associate Professor
Department of Animal Science
Texas A&M University
College Station, Texas, USA

Mark L. Anderson DVM, PhD, Diplomate ACVP
Professor
School of Veterinary Medicine
University of California
Davis, California, USA

Frank W. Austin DVM, PhD
Professor
Department of Pathobiology and Population Medicine
College of Veterinary Medicine
Mississippi State University
Starkville, Mississippi, USA

Jennifer P. Barfield PhD
Department of Biomedical Sciences
Animal Reproduction and Biotechnology Laboratory
Colorado State University
Fort Collins, Colorado, USA

Callie V. Barnwell PhD
Department of Animal Science
North Carolina State University
Raleigh, North Carolina, USA

Albert Barth DVM, MS, Diplomate ACT
Large Animal Clinical Sciences
Western College of Veterinary Medicine
University of Saskatchewan
Saskatoon, Saskatchewan, Canada

Brittany Baughman DVM, MS, Diplomate ACVP
Mississippi Veterinary Diagnostic Laboratory
Jackson, Mississippi, USA

Wes Baumgartner DVM, PhD, Diplomate ACVP
Assistant Professor
Department of Pathobiology and Population Medicine
College of Veterinary Medicine
Mississippi State University
Starkville, Mississippi, USA

Myra T. Blanchard CLS, MS
School of Veterinary Medicine
University of California
Davis, California, USA

Patrick Blondin PhD
L'Alliance Boviteq Inc.
St-Hyacinthe, Quebec, Canada

Gabriel A. Bó DVM, PhD
Instituto de Reproducción Animal
  Córdoba (IRAC)
Cno. General Paz – Paraje Pozo del Tigre-
  Estación General Paz
CP 5145 Córdoba, Argentina

Kenneth Bondioli PhD
Associate Professor
School of Animal Sciences
Louisiana State University
Baton Rouge, Louisiana, USA

Nadine Bouchard BSc
L'Alliance Boviteq Inc.
St-Hyacinthe, Quebec, Canada

Contributors

James A. Brett DVM
Associate Clinical Professor
Department of Pathobiology and
 Population Medicine
College of Veterinary Medicine
Mississippi State University
Starkville, Mississippi, USA

Leonardo F.C. Brito DVM, PhD,
Diplomate ACT
ABS Global Inc.
DeForest, Wisconsin, USA

Amanda J. Cain BS
Graduate student DVM/PhD
Department of Pathobiology and
 Population Medicine
College of Veterinary Medicine
Mississippi State University
Starkville, Mississippi, USA

Leticia E. Camacho MS
Graduate Research Assistant
Department of Animal Sciences
North Dakota State University
Fargo, North Dakota, USA

Butch Cargile DVM, MS
Nutrition and Management Consultant
Progressive Dairy Solutions
Twin Falls, Idaho, USA

Celina Checura MV, MS, PhD
Department of Medical Sciences
School of Veterinary Medicine
University of Wisconsin-Madison
Madison, Wisconsin, USA

David Christiansen DVM
Assistant Clinical Professor
Department of Pathobiology and
 Population Medicine
College of Veterinary Medicine
Mississippi State University
Starkville, Mississippi, USA

Elizabeth A. Coffman DVM,
Diplomate ACT
Theriogenology Resident
Department of Veterinary Clinical Sciences
College of Veterinary Medicine
Ohio State University
Columbus, Ohio, USA

Jill Colloton DVM
Bovine Services, LLC
Edgar, Wisconsin, USA

Catherine Dolbec MSc
L'Alliance Boviteq Inc.
St-Hyacinthe, Quebec, Canada

Tyler M. Dohlman DVM
Resident, Theriogenology
Department of Veterinary Diagnostics and
 Production Animal Medicine
Lloyd Veterinary Medicine Center
College of Veterinary Medicine
Ames, Iowa, USA

Maarten Drost DVM,
Diplomate ACT
Professor Emeritus
College of Veterinary Medicine
University of Florida
Gainesville, Florida, USA

Misty A. Edmondson DVM, MS,
Diplomate ACT
Associate Professor
Department of Clinical Sciences
College of Veterinary Medicine
Auburn University
Auburn, Alabama, USA

J. Lannett Edwards PhD
Professor and Graduate Director
Department of Animal Science
University of Tennessee
Knoxville, Tennessee, USA

Phillip H. Elzer PhD
Professor and Head
Department of Veterinary Science
Louisiana State University
Baton Rouge, Louisiana, USA

Terry J. Engelken DVM, MS
Associate Professor
Department of Veterinary Diagnostics and
 Production Animal Medicine
Lloyd Veterinary Medicine Center
College of Veterinary Medicine
Ames, Iowa, USA

Charles T. Estill VMD, PhD, Diplomate ACT
Associate Professor
Department of Clinical Sciences
College of Veterinary Medicine
Oregon State University
Corvallis, Oregon, USA

Charlotte E. Farin PhD
Professor
Department of Animal Science
North Carolina State University
Raleigh, North Carolina, USA

William T. Farmer PhD
Department of Animal Science
North Carolina State University
Raleigh, North Carolina, USA

Contributors

Julie Gard DVM, PhD, Diplomate ACT
Associate Professor
Department of Clinical Sciences
College of Veterinary Medicine
Auburn University
Auburn, Alabama, USA

Danielle Glynn RVT
College of Veterinary Medicine
Purdue University
West Lafayette, Indiana, USA

Gretchen Grissett DVM
Clinical Resident
Department of Clinical Sciences
College of Veterinary Medicine
Kansas State University
Manhattan, Kansas, USA

Daniel L. Grooms DVM, PhD, Diplomate ACVM
Professor
College of Veterinary Medicine
Department of Large Animal Clinical Sciences
Michigan State University
East Lansing, Michigan, USA

Sue D. Hagius BS
Department of Veterinary Science
Louisiana State University
Baton Rouge, Louisiana, USA

Peter J. Hansen PhD
Distinguished Professor and L.E. "Red" Larson Professor
Department of Animal Sciences
University of Florida
Gainesville, Florida, USA

John F. Hasler PhD
Bioniche Animal Health, Inc.
Laporte, Colorado, USA

W. Mark Hilton DVM, Diplomate ABVP – Beef Cattle Practice
Clinical Professor, Beef Production Medicine
College of Veterinary Medicine
Purdue University
West Lafayette, Indiana, USA

Richard M. Hopper DVM, Diplomate ACT
Professor
Department of Pathobiology and Population Medicine
College of Veterinary Medicine
Mississippi State University
Starkville, Mississippi, USA

Carla Huston DVM, PhD, Diplomate ACVPM
Department of Pathobiology and Population Medicine
College of Veterinary Medicine
Mississippi State University
Starkville, Mississippi, USA

Marianna M. Jahnke MS
Lecturer
Veterinary Diagnostic and Production Animal Medicine Department
College of Veterinary Medicine
Iowa State University
Ames, Iowa, USA

Ram Kasimanickam BVSc, DVSc, Diplomate ACT
Associate Professor
Department of Veterinary Clinical Sciences
College of Veterinary Medicine
Washington State University
Pullman, Washington, USA

John P. Kastelic DVM, PhD, Diplomate ACT
Professor, Cattle Reproductive Health – Theriogenology
Head, Department of Production Animal Health
University of Calgary, Faculty of Veterinary Medicine
Calgary, Alberta, Canada

E. Heath King DVM, Diplomate ACT
Assistant Clinical Professor
Department of Pathobiology and Population Medicine
College of Veterinary Medicine
Mississippi State University
Starkville, Mississippi, USA

Tom Kroetsch MSc
The Semex Alliance
Guelph, Ontario, Canada

G. Cliff Lamb PhD
Assistant Director and Professor
North Florida Research and Education Center
University of Florida
Marianna, Florida, USA

Robert L. Larson DVM, PhD, Diplomate ACT, ACN, ACVPM (Epidemiology)
Professor, Production Medicine
Clinical Sciences
College of Veterinary Medicine
Kansas State University
Manhattan, Kansas, USA

Jim W. Lauderdale PhD
Lauderdale Enterprises
Augusta, Michigan, USA

Fred Lehman DVM, MABA, Diplomate ACT, PMP
Overland Park, Kansas, USA

Caleb O. Lemley PhD
Assistant Professor
Department of Animal and Dairy Sciences
Mississippi State University
Starkville, Mississippi, USA

Robert Linford DVM, PhD, Diplomate ACVS
Professor
Department of Clinical Sciences
College of Veterinary Medicine
Mississippi State University
Starkville, Mississippi, USA

Herris Maxwell DVM, Diplomate ACT
Clinical Professor
Department of Clinical Sciences
College of Veterinary Medicine
Auburn University
Auburn, Alabama, USA

Reuben J. Mapletoft DVM, PhD, Diplomate ACT
Distinguished Professor
Department of Large Animal Clinical Sciences
Western College of Veterinary Medicine
University of Saskatchewan
Saskatoon, Saskatchewan, Canada

Richard W. Meiring DVM, Diplomate ACVPM
Clinical Professor
Department of Pathobiology and Population Medicine
College of Veterinary Medicine
Mississippi State University
Starkville, Mississippi, USA

Vitor R.G. Mercadante DVM, MS
Research Assistant
North Florida Research and Education Center
University of Florida
Marianna, Florida, USA

Rory Meyer DVM, Diplomate ACT
Staff Veterinarian
Alta Genetics
Watertown, Wisconsin, USA

Cathleen Mochal-King DVM, MS, Diplomate ACVS
Assistant Clinical Professor
Department of Clinical Sciences
College of Veterinary Medicine
Mississippi State University
Starkville, Mississippi, USA

Harry Momont DVM, PhD, Diplomate ACT
Professor
Department of Medical Sciences
School of Veterinary Medicine
University of Wisconsin-Madison
Madison, Wisconsin, USA

Quinesha P. Morgan PhD
Department of Veterinary Science
Louisiana State University
Baton Rouge, Louisiana, USA

John L. Myers DVM, Diplomate ACT
Pecan Drive Veterinary Services
Vinita, Oklahoma, USA

Ben Nabors DVM
Department of Clinical Sciences
College of Veterinary Medicine
Mississippi State University
Starkville, Mississippi, USA

Colin Palmer DVM, MVetSc, Diplomate ACT
Professor, Theriogenology
Department of Large Animal Clinical Sciences
Western College of Veterinary Medicine
University of Saskatchewan
Saskatoon, Saskatchewan, Canada

Thomas Passler DVM, PhD, Diplomate ACVIM
Assistant Professor
Departments of Clinical Sciences and Pathobiology
College of Veterinary Medicine
Auburn University
Auburn, Alabama, USA

Augustine T. Peter BVSc, MVSc, MSc, PhD, MBA, Diplomate ACT
Professor
Veterinary Clinical Sciences
College of Veterinary Medicine
Purdue University
West Lafayette, Indiana, USA

Carlos A. Risco DVM, Diplomate ACT
Professor and Chair
Large Animal Clinical Sciences
College of Veterinary Medicine
University of Florida
Gainesville, Florida, USA

Edwin G. Robertson DVM
Harrogate Genetics International
Harrogate, Tennessee, USA

Peter L. Ryan PhD, Diplomate ACT (honorary)
Associate Provost
College of Veterinary Medicine
Mississippi State University
Starkville, Mississippi, USA

Swanand Sathe BVSc, MVSc, MS, Diplomate ACT
Assistant Professor, Theriogenology
Lloyd Veterinary Medical Center
College of Veterinary Medicine
Iowa State University
Ames, Iowa, USA

F. Neal Schrick PhD
Professor and Chair
Department of Animal Science
University of Tennessee
Knoxville, Tennessee, USA

Contributors

Clare M. Scully DVM, MA
Theriogenology Resident
Department of Clinical Sciences
College of Veterinary Medicine
Oregon State University
Corvallis, Oregon, USA

Clifford F. Shipley DVM, Diplomate ACT
Attending Veterinarian for Agricultural Animals
Agricultural Animal Care and Use Program
College of Veterinary Medicine
University of Illinois
Urbana, Illinois, USA

Katharine M. Simpson DVM, MS, Diplomate ACVIM
Clinical Assistant Professor
College of Veterinary Medicine
Ohio State University
Columbus, Ohio, USA

Jaswant Singh BVSc, MVSc, PhD
Professor
Veterinary Biomedical Sciences
Western College of Veterinary Medicine
University of Saskatchewan
Saskatoon, Saskatchewan, Canada

David R. Smith, DVM, PhD, Diplomate ACVPM (Epidemiology)
Mikell and Mary Cheek Hall Davis Endowed Professor
Department of Pathobiology and Population Medicine
College of Veterinary Medicine
Mississippi State University
Starkville, Mississippi, USA

Jack D. Smith DVM, Diplomate ACT
Associate Professor
Department of Pathobiology and Population Medicine
College of Veterinary Medicine
Mississippi State University
Starkville, Mississippi, USA

Ricardo Stockler DVM, MS, Diplomate ABVP (Dairy)
Dairy Production Medicine Clinician
Veterinary Medicine Teaching and Research Center (VMTRC)
University of California Davis
Tulare, California, USA

Jeffrey L. Stott MS, PhD
Professor
School of Veterinary Medicine
University of California
Davis, California, USA

Robert N. Streeter DVM, MS, Diplomate ACVIM
Associate Professor
Department of Veterinary Clinical Sciences
Center for Veterinary Health Sciences
Oklahoma State University
Stillwater, Oklahoma, USA

William S. Swecker Jr DVM, PhD, Diplomate ACVN
Professor and Associate Department Head
Large Animal Clinical Sciences
Virginia-Maryland Regional College of Veterinary Medicine
Blacksburg, Virginia, USA

Ahmed Tibary DVM, MS, PhD, Diplomate ACT
Professor
Department of Clinical Sciences
College of Veterinary Medicine
Washington State University
Pullman, Washington, USA

Mike Thompson DVM, Diplomate ACT
Willow Bend Animal Clinic
Holly Springs, Mississippi, USA

Dan Tracy DVM, MS
Technical Services Veterinarian
Multimin USA
Auburn, Kentucky, USA

Shelley L. Underwood PhD
L'Alliance Boviteq Inc.
St-Hyacinthe, Quebec, Canada

Rhonda C. Vann PhD
Research Professor
Brown-Loan Experiment Station
Raymond, Mississippi, USA

Patrick Vincent PhD
L'Alliance Boviteq Inc.
St-Hyacinthe, Quebec, Canada

Kimberly A. Vonnahme PhD
Associate Professor
Department of Animal Sciences
North Dakota State University
Fargo, North Dakota, USA

Kevin Walters DVM, Diplomate ACT
Assistant Clinical Professor
Department of Pathobiology and Population Medicine
College of Veterinary Medicine
Mississippi State University
Starkville, Mississippi, USA

Gary Warner DVM
Bovine Division
Elgin Veterinary Hospital
Elgin, Texas, USA

James K. West DVM, MS
Armbrust Professor of Clinical Medicine
Director of Embryo Transfer Services
Iowa State University
Ames, Iowa, USA

Brad J. White DVM, MS
Associate Professor
Department of Clinical Sciences
College of Veterinary Medicine
Kansas State University
Manhattan, Kansas, USA

Brian K. Whitlock DVM, PhD, Diplomate ACT
Associate Professor
Department of Large Animal Clinical Sciences
College of Veterinary Medicine
University of Tennessee
Knoxville, Tennessee, USA

Gary L. Williams PhD
Regents Fellow, Faculty Fellow and Professor
Animal Reproduction Laboratory
Department of Animal Science
Texas A&M University
Beeville, Texas, USA

Dwight F. Wolfe DVM, MS, Diplomate ACT
Professor
Department of Clinical Sciences
College of Veterinary Medicine
Auburn University
Auburn, Alabama, USA

Curtis R. Youngs PhD
Department of Animal Science
Iowa State University
Ames, Iowa, USA

Illustrators

Alison Anderson
College of Veterinary Medicine
Mississippi State University
Starkville, Mississippi, USA
Chapter 76

Barbara DeGraves
Bowling Green, Kentucky, USA
Chapters 16, 17, and 18

Rachel Fishman
Class of 2017
College of Veterinary Medicine
Mississippi State University
Starkville, Mississippi, USA
Chapter 54

Mal Hoover
Medical Illustrator/Graphic Design Specialist
Kansas State University
Manhattan, Kansas, USA
Chapter 20

McRae Hopper
Starkville, Mississippi, USA
Chapters 47 and 53

Kathleen June Mullins
Germinal Dimensions
915 Allendale Ct
Blacksburg, Virginia, USA
Chapter 2

Rachel Oman DVM
Food Animal Medicine Resident
Oklahoma State University
Stillwater, Oklahoma, USA
Chapter 19

Tyla Barkley DVM
Ardmore Animal Hospital
Ardmore, Oklahoma, USA
Chapter 43

Preface

There is an old fable in which three penniless and hungry travelers come to a small town. Unsuccessful in finding work or even a handout, one concocts a novel plan. He goes to the middle of the village carrying three fist-sized rocks and announces with great aplomb that he is planning to make his famous "stone soup." The skeptical but curious villagers gather. Well of course he needed a kettle and some water. The inquisitive villagers wondered if that was all. "Yes," he replied, "but it is better with a little garnish to improve the flavor." One villager thought that he could spare some carrots, another some potatoes, and a third some meat. This continued with virtually everyone in the village contributing. The result of course was a wonderful soup and everyone enjoyed a fine meal, while experiencing an object lesson in cooperation.

The story bears an ironic resemblance to the development of this text. The editor, like the plucky traveler, personally short on ability and resources but acutely aware of a need, enlisted the assistance of those who possessed both. Excellent reference texts were available on equine and small animal theriogenology, but a current bovine text was much needed. The goal was to produce a text that would service the needs of the veterinary student and bovine practitioner, as well as the graduate student and resident.

While I would readily admit that this text could be improved with respect to the choices made vis-à-vis the organization of the book or the order of some chapters, I honestly do not believe that I could have done any better than the contributors selected. The authors of this text represent a wide array of specialties and educational and experiential backgrounds. I will forever be grateful for their assistance and immensely proud of their individual contributions. I would also like to acknowledge the efforts of my graduate assistant, Amanda Cain, who in addition to contributing a chapter, prepared the glossary of terms and index. Likewise, I would like to thank everyone at Wiley for their help. Erica Judisch, the commissioning editor, was so very helpful in guiding a novice through the early phases of this book. Susan Engelken, the managing editor for this book, was incredible to work with, always patient, always competently and quickly responding to any issue or concern. Dr Joe Phillips, the copy editor Wiley enlisted, deserves the credit for identifying errors that I missed and enhancing the readability of this text.

Additionally, I would like to acknowledge on a personal level those who have been so important to me from the standpoint of my life and career. First of all I would thank my parents, Lewis and Barbara Hopper, who were always supportive of my goals and aspirations, and my family, wife Donna and children Tricia (her husband Caleb), McRae, and Molly, who I will always consider to be my greatest accomplishments. Also, as this goes to press I can announce a wonderful addition, a granddaughter by the name of Abigail Betty Butts.

I would also be remiss to not use this opportunity to thank some of my professors and instructors at Auburn who were so influential to me professionally and important to me personally. First on this list would be Dr Robert Hudson, but also Drs Bob Carson, Howard Jones, Donald Walker, Ram Purohit, John Winkler, and Howard Kjar. Likewise, I need to thank my colleagues at Mississippi State who have alternatively both encouraged and tolerated me through this long process.

I sincerely hope the reader finds this text useful.

Richard M. Hopper
Starkville, Mississippi

SECTION I

The Bull

ANATOMY AND PHYSIOLOGY

1. Anatomy of the Reproductive System of the Bull 5
   Ben Nabors and Robert Linford

2. Endocrine and Exocrine Function of the Bovine Testes 11
   Peter L. Ryan

3. Thermoregulation of the Testes 26
   John P. Kastelic

4. Endocrine Control of Testicular Development and Initiation of Spermatogenesis in Bulls 30
   Leonardo F.C. Brito

Chapter 1

Anatomy of the Reproductive System of the Bull

Ben Nabors and Robert Linford

Department of Clinical Sciences, College of Veterinary Medicine, Mississippi State University, Starkville, Mississippi, USA

Introduction

The anatomy of the reproductive system of the bull can be grouped functionally into the components of production, transport, and transfer of spermatozoa (Figure 1.1).

Production

The testicular parenchyma contains the cellular machinery for spermatogenesis and steroid production (Figure 1.2). The parenchyma is arranged in indistinct lobules of convoluted tubules called seminiferous tubules. The seminiferous tubules contain the spermatogonia from which the mature sperm cells develop. Sertoli cells are also located within the lumen of the seminiferous tubules. The Leydig cells that are responsible for the production of the male hormone testosterone are located between the seminiferous tubules in the interstitial space.[1]

Testes

The testes are housed in the scrotum. The scrotum is suspended between the thighs in the inguinal region. The scrotum consists of external and internal layers. The external layer is made up of the skin, tunica dartos, superficial perineal fascia, external spermatic fascia, cremasteric fascia, internal spermatic fascia, and parietal vaginal tunic. The skin of the scrotum and tunica dartos muscle are closely adhered whereas the fascial layers are easily separated from the skin and the parietal vaginal tunic as in a closed castration technique. The coverings of the testicle itself consist of the visceral vaginal tunic and the tunica albuginea.[2] The visceral vaginal tunic is the innermost layer of the vaginal tunic, an outpouching of abdominal peritoneum that passes through the inguinal canal into the scrotal sac. The potential space between the parietal and visceral vaginal tunic is the vaginal cavity (Figure 1.3). The purpose of the vaginal cavity is for temperature regulation of the testicle by raising it closer to the body through contraction of the tunica dartos and cremaster muscles. The tunica albuginea is a thick fibrous capsule that covers the testicle and maintains the testicular contents under pressure.[3] Internally the tunica albuginea forms the axially positioned mediastinum testis from which connective tissue septa divide the testis into indistinct lobules. This connective tissue framework supports the vasculature, nerves, parenchyma, and tubular system of the testicle. The scrotum of the bull is pendulous due to the dorsoventral orientation of the testes contained within.[1]

Spermatic cord

The spermatic cord includes the ductus deferens, vasculature, lymphatic vessels, and nerves of the testicle and epididymis.[4] Essentially the spermatic cord consists of all the tissue within the vaginal tunic so it extends from the vaginal ring within the abdominal cavity to the testicle.[5]

Transport

Spermatozoa are transported from the testicles through a tubular system consisting of the convoluted seminiferous tubules, straight seminiferous tubules, rete testis, efferent ductules, epididymis, ductus deferens, and urethra (Figure 1.4). The tubular system allows for maturation and storage of spermatozoa and provides fluid to ease movement of the spermatozoa.

Tubular transport system

The convoluted seminiferous tubules are the location of the spermatogenic process: the development of spermatogonia to primary spermatocytes, to spermatids, and finally to spermatozoa.[1] This process occurs within the wall of the seminiferous tubule. Specific regions of the

6   The Bull: Anatomy and Physiology

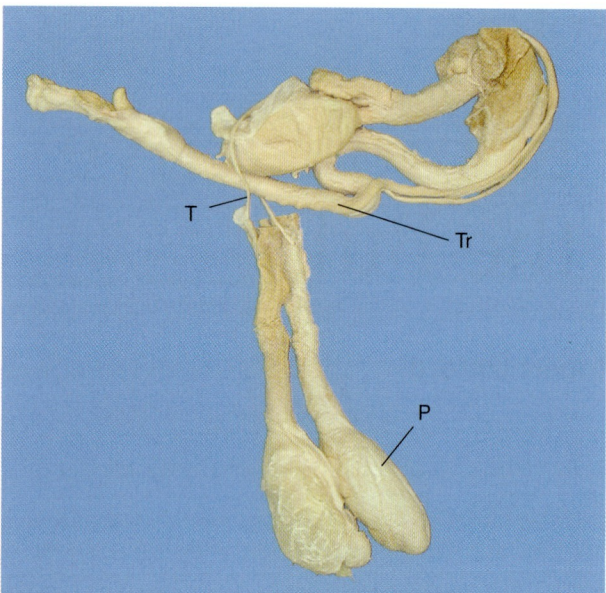

Figure 1.1 Reproductive system of the bull. P, production; T, transport; Tr, transfer.

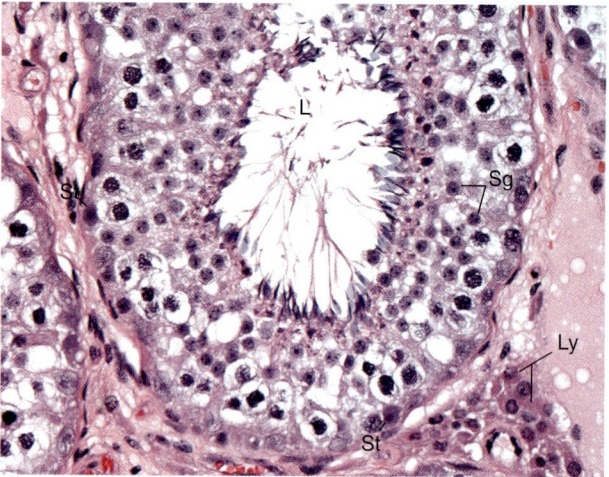

Figure 1.2 St, Sertoli cell; Ly, Leydig cells; Sg, spermatogonia; L, lumen of seminiferous tubule.

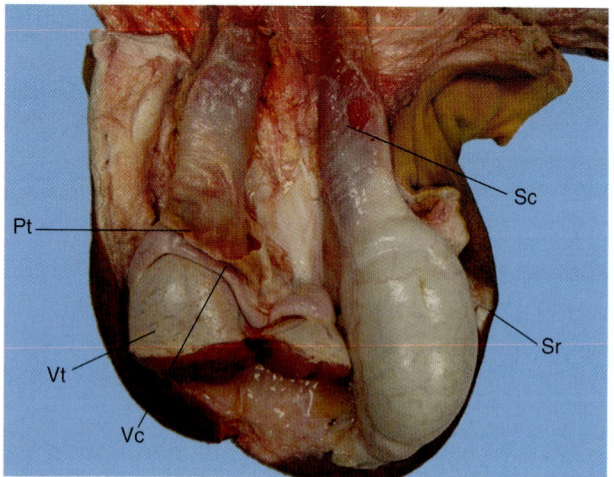

Figure 1.3 Vt, visceral vaginal tunic; Pt, parietal vaginal tunic; Vc, vaginal cavity; Sc, spermatic cord.

Figure 1.4 Ep-h, head epididymis; Et, efferent tubules; Mt, mediastinum testis; Rt, rete testis; Tp, testicular parenchyma; Ep-t, tail of epididymis.

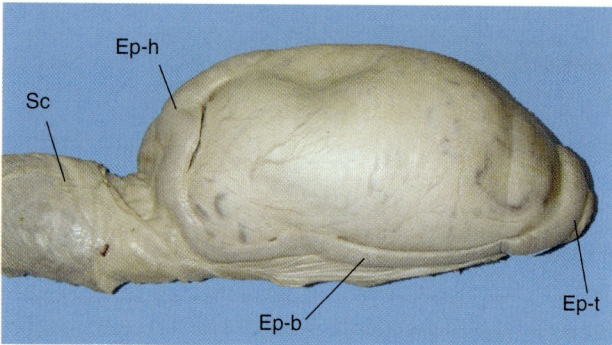

Figure 1.5 Sc, spermatic cord; Ep-h, head of epididymis; Ep-b, body of epididymis; Ep-t, tail of epididymis.

tubule are devoted to a particular stage of development, so that each stage can be identified by specific histological techniques.[6] Upon the completion of spermiogenesis, the spermatozoa are released into the lumen of the convoluted seminiferous tubule to begin transit through the straight seminiferous tubule. The straight seminiferous tubule is simply the connection between the convoluted seminiferous tubule and the rete testis. The rete testis is a "network of irregular labyrinth spaces and interconnected tubules."[2] The rete testes are located within the mediastinum testis connecting the seminiferous tubules to the efferent ducts that exit the testicle at the extremitas capitata (head). The efferent tubular system continues as the epididymis on the external surface of the testis (Figure 1.5). The epididymis is divided into a head, a body located on the medial surface, and a tail located at the distal extremitas caudate.

Ductus deferens

The ductus deferens is attached to the medial side of the testicle by the mesoductus.[5] The ductus deferens is the continuation of the tail of the epididymis (Figure 1.6). The ductus deferens enters the abdominal cavity through the inguinal canal, crosses the lateral ligament of the bladder, and before it ends at the colliculus seminalis in the urethra it widens into the ampulla.[5]

Anatomy of the Reproductive System of the Bull

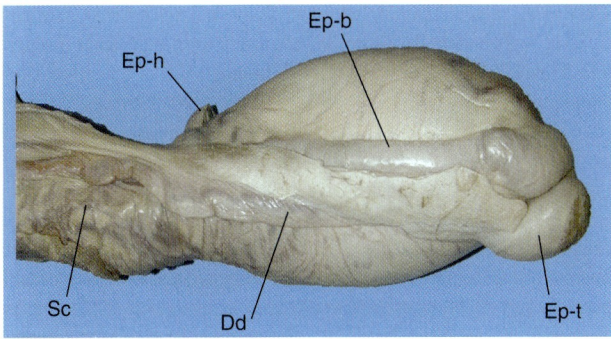

Figure 1.6 Sc, spermatic cord; Ep-h, head of epididymis; Ep-b, body of epididymis; Ep-t, tail of epididymis; Dd, ductus deferens.

Transfer

The transfer of spermatozoa from the bull to the cow is achieved by the process of intromission, which requires erection of the penis and ejaculation of sperm. The pertinent anatomy for these processes to occur includes the penis, the musculature of the penis, the vasculature, and the innervations.

Penis

The penis of the bull can be divided into a root, body, and glans penis (Figure 1.7). The root of the penis can be defined as the origin of the erectile tissue that comprises the penis as well as the origin of the muscles of the penis. The erectile tissue that makes up the bulk of the penis is the corpus cavernosum. The paired corpora cavernosa originate separately on each side of the ischiatic arch medial to the ischiatic tuberosity. These individual limbs are termed the crura of the penis. The crura pass ventromedially until they join to form the body of the penis. The corpus spongiosum is the erectile tissue that surrounds the urethra. The origin of the corpus spongiosum, called the bulb of the penis, originates between the crura along the midline of the ischiatic arch. Therefore the root of the penis is composed of the crura (corpus cavernosum) and the bulb (corpus spongiosum).

The erectile tissue is enclosed in the dense outer covering of the tunica albuginea. The tunica albuginea is a dense covering that consists of an inner circular layer and outer longitudinal layer of fibers. The inner circular layer sends trabecular scaffolds throughout the corpus cavernosum for the attachment of the cavernous endothelium.

Located caudal to the root of the penis are the muscles of the penis: the ischiocavernosus, bulbospongiosus, and retractor penis muscles (Figure 1.8). The paired ischiocavernosus muscles originate on the medial surfaces of the ischiatic tuberosities overlying the crura; the muscle fibers pass ventromedially in a "V" fashion until ending a short distance on the body of the penis.[1] During erection the ischiocavernosus muscle contracts pushing blood from the cavernous spaces of the crura into the body of the penis.[7] The bulbospongiosus muscle lies caudal to the bulb of the penis, originating along the ischiatic arch and continuing until the junction of the crura.[1] The bulbospongiosus muscle fibers run transversely across the bulb of the penis and contraction of this muscle results in propulsion of the ejaculate through the urethra.[7] The retractor penis muscle extends from the caudal vertebrae and internal anal sphincter to insert distal to the sigmoid flexure.[8] These paired muscles relax during erection allowing the penis to extend from the prepuce and contract during quiescence, retracting the penis into the sheath.[8]

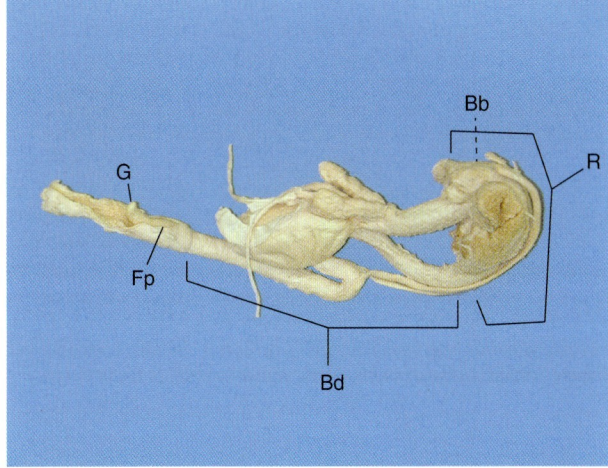

Figure 1.7 R, root of penis; Bb, bulb of penis; Bd, body of penis; Fp, free part of penis; G, glans penis.

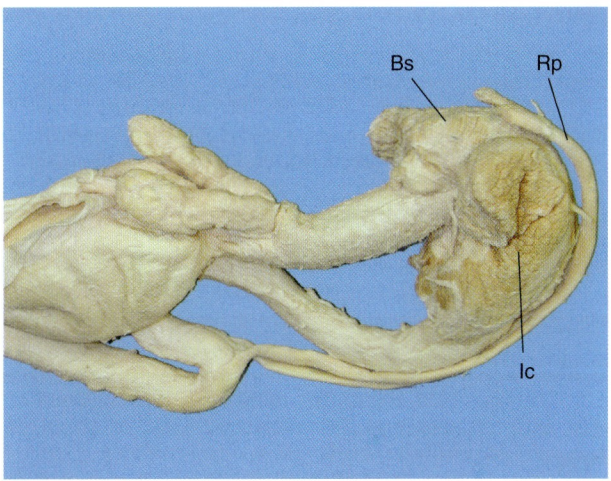

Figure 1.8 Bs, bulbospongiosus muscle; Ic, ischiocavernosus muscle; Rp, retractor penis muscle.

The body of the penis begins where the two crura meet distally to the ischiatic arch; it extends craniad, along the ventral body wall to become at the mid-ventral abdomen the free part of the penis (Figure 1.9). The body of the penis is bent in an "S" shape called the sigmoid flexure. The proximal bend of the sigmoid flexure opens caudally and is located near the scrotum. The distal bend is opened cranially and the short suspensory ligaments of the penis attach the penis to the ventral surface of the ischiatic arch.

The glans penis is a small restricted region at the tip of the free part of the penis[8] (Figure 1.10). The free part of the penis is the distal extent from the attachment of the internal lamina of the prepuce to the glans penis.[8] The free end of the penis is twisted in a counterclockwise direction as viewed from the right side, illustrated by the

8   The Bull: Anatomy and Physiology

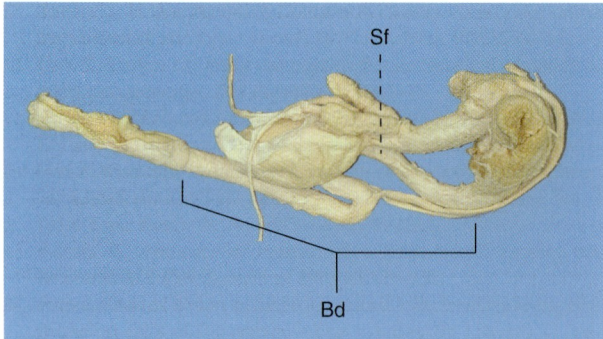

Figure 1.9 Bd, body of penis; Sf, sigmoid flexure.

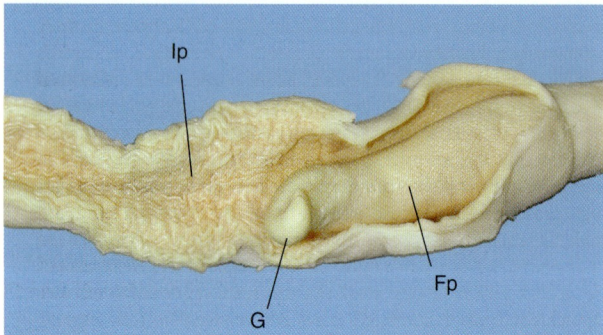

Figure 1.10 G, glans penis; Fp, free part; Ip, internal lamina of prepuce.

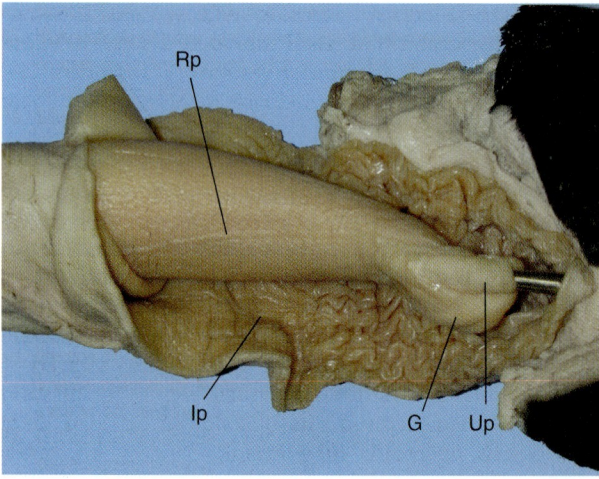

Figure 1.11 Rp, raphe of penis; Ip, internal lamina of prepuce; G, glans penis; Up, urethral process.

oblique direction of the raphe of prepuce continued as the raphe of the penis to the urethral process (Figure 1.11). The twist of the free end of the penis is due to the attachment of the apical ligament. The apical ligament of the penis is formed by the longitudinal fibers of the tunica albuginea leaving the body of the penis just distal to the sigmoid flexure and reattaching near the apex of the penis.[9]

The prepuce of the penis is composed of an external and internal fold or lamina[8] (Figure 1.12). The external lamina is the haired outer fold of skin attached to the ventral abdomen.

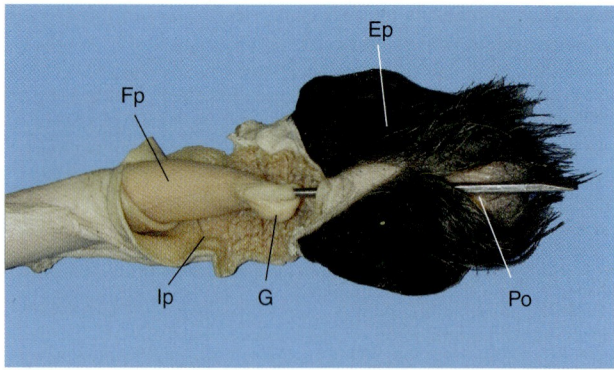

Figure 1.12 Fp, free part penis; G, glans penis; Ip, internal lamina of prepuce; Ep, external lamina of prepuce; Po, preputial orifice.

The haired skin terminates at the preputial orifice where the external fold turns inward to line the preputial cavity as the internal lamina. The internal lamina serves to attach the external lamina to the penile epithelium.

Blood supply

Before ejaculation can occur the testis must produce spermatozoa. This requires an adequate blood supply for the metabolic demands of cellular division for spermatogenesis and steroidogenesis. The arterial blood supply to each testis is provided by a testicular artery, a direct branch of the abdominal aorta arising caudal to the renal arteries. The testicular artery crosses the lateral abdominal wall and then passes ventrally through the inguinal canal.[10] As the testicular artery approaches the testis it begins to spiral with the nearby tortuous pampiniform plexus of the testicular vein forming a vascular cone. This arterial/venous arrangement is an effective thermoregulatory apparatus.[11]

An adequate blood supply to the penis and associated muscles is required for the processes of erection, ejaculation, and tissue maintenance. This comes by way of the internal iliac artery. The internal iliac artery is a direct continuation of the abdominal aorta at the entrance to the pelvic cavity. The umbilical artery, a branch of the internal iliac, supplies the ductus deferens and the bladder.[4] The prostatic artery leaves the internal iliac and supplies the prostate, vesicular glands, ductus deferens, ureter, and urethra.[4] As the internal iliac continues through the pelvic cavity it divides into the caudal gluteal and internal pudendal.[10] The internal pudendal gives off the ventral perineal artery, urethralis artery, and continues as the artery of the penis.[10] The artery of the penis gives off the artery of the bulb of the penis, which supplies the bulbospongiosus muscle and the cavernous spaces of the corpus spongiosum[12] (Figure 1.13). The deep artery of the penis is another branch of the artery of the penis that enters the crus of the penis and supplies the erectile tissue, the corpus cavernosum.[12] After the deep artery branches off, the artery of the penis continues as the dorsal artery of the penis which passes along the dorsal aspect of the penis toward the glans penis and prepuce. It is responsible for maintenance of penile tissue during quiescence.[13]

Anatomy of the Reproductive System of the Bull

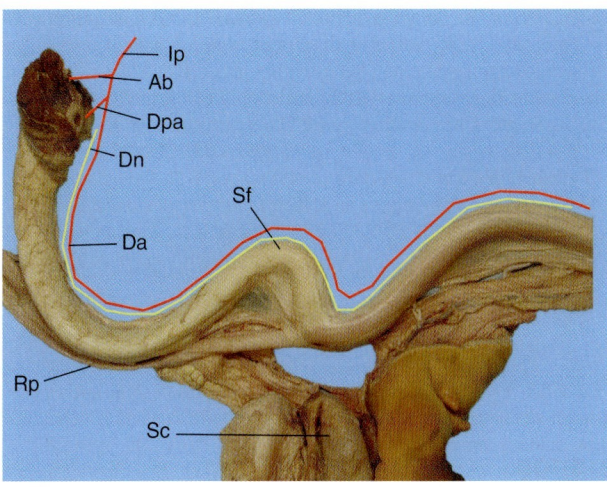

Figure 1.13 Ip, internal pudendal artery; Ab, artery of the bulb of the penis; Dpa, deep artery of the penis; Da, dorsal artery of the penis; Dn, dorsal nerve of the penis; Rp, retractor penis muscle; Sc, spermatic cord; Sf, sigmoid flexure.

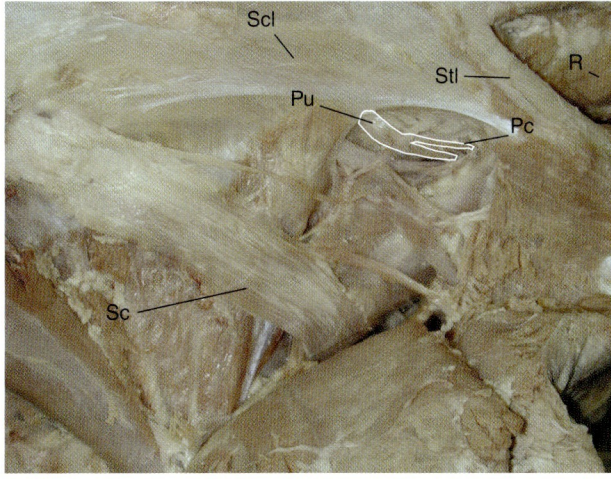

Figure 1.14 Sc, sciatic nerve; Pu, pudendal nerve; Pc, proximal cutaneous branch of pudendal nerve; Scl, sacrosciatic ligament; Stl, sacrotuberous ligament; R, rectum.

Nervous supply

The innervation of the external genitalia of the bull consists of the pudendal nerve and its branches. The pudendal nerve carries motor, sensory, and parasympathetic nerve fibers.[4] The pudendal nerve passes through the pelvic cavity medial to the sacrosciatic ligament and divides as it approaches the lesser ischiatic notch of the pelvis into proximal and distal cutaneous branches supplying the skin of the caudal hip and thigh.[4,8] The pudendal nerve continues through the ischiorectal fossa, terminating in a preputial branch, a scrotal branch, and finally the dorsal nerve of the penis.[6] The pelvic nerve provides parasympathetic innervations from the sacral plexus.[1] The hypogastric nerve contributes sympathetic fibers from the caudal mesenteric plexus to the genital system[1] (Figure 1.14).

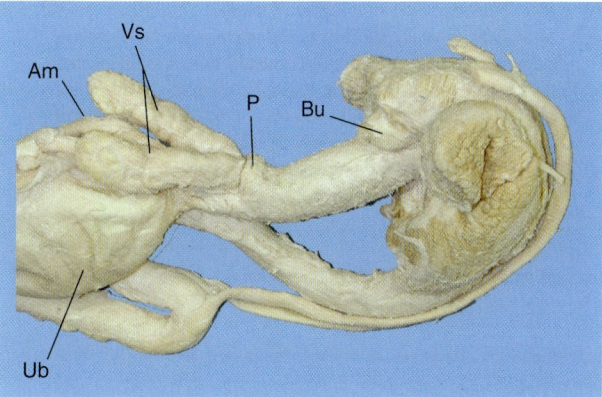

Figure 1.15 Am, ampulla; Vs, vesicular glands; P, prostate; Bu, bulbourethral gland; Ub, urinary bladder.

Accessory glands

The accessory genital glands of the bull include the vesicular gland, ampulla of the ductus deferens, and the prostate and bulbourethral glands (Figure 1.15). The bilateral vesicular gland is the largest accessory gland in the bull and contributes the greatest volume to the ejaculate. It is a lobated gland of firm consistency. It lies dorsal to the bladder and lateral to the ureter and ampulla of the ductus deferens.[1] The body of the prostate lies dorsal to the urethra between and caudal to the vesicular glands. The disseminate part of the prostate is concealed in the wall of the urethra and covered by the urethral muscle.[1] The ampulla, vesicular glands, and prostate all empty their contents into the urethra through the colliculus seminalis. The bilateral bulbourethral gland lies on each side of the median plane dorsal to the urethra; it is mostly covered by the bulbospongiosus muscle. Its duct opens into the urethral recess[1] (Figure 1.15). The urethral recess is a blind pouch that exits dorsally into the penile urethra at the level of the ishiatic arch. The presence of this structure makes it difficult to pass a catheter retrograde into the bladder.

References

1. Nickel R, Schummer, A, Seiferle E, Sack WO (eds). *The Viscera of the Domestic Mammals*. Berlin: Springer-Verlag, 1973, xiv, p. 401.
2. Johnson A, Gomes W, Vandemark N. *The Testis*. New York: Academic Press, 1970, v, illus. 3.
3. McGavin M, Zachary J. *Pathologic Basis of Veterinary Disease*, 4th edn. St Louis, MO: Mosby Elsevier, 2007, p. 1476.
4. Schaller O, Constantinescu G. *Illustrated Veterinary Anatomical Nomenclature*. Stuttgart: F. Enke Verlag, 1992, vi, p. 614.
5. Ross M, Kaye G, Pawlina W. *Histology: A Text and Atlas*, 4th edn. Philadelphia: Lippincott Williams & Wilkins, 2003, p. 875.
6. Mullins KJ, Saacke R. *Illustrated Anatomy of the Bovine Male and Female Reproductive Tracts: From Gross to Microscopic*. Columbia, MO: National Association of Animal Breeders, 2003, p. 79.
7. Watson J. Mechanism of erection and ejaculation in the bull and ram. *Nature* 1964;204:95–6.
8. Budras K-D. *Bovine Anatomy: An Illustrated Text*. Hannover, Germany: Schlütersche, 2003, p. 138.
9. Ashdown R, Smith J. The anatomy of the corpus cavernosum penis of the bull and its relationship to spiral deviation of the penis. *J Anat* 1969;104:153–160.

10. Schummer A, Wilkens H, Vollmerhaus B, Habermehl K-H. *The Circulatory System, the Skin, and the Cutaneous Organs of the Domestic Mammals*. Berlin: Springer-Verlag, 1981, p. 158.
11. Brito L, Silva A, Barbosa R, Kastelic J. Testicular thermoregulation in *Bos indicus*, crossbred and *Bos taurus* bulls: relationship with scrotal, testicular vascular cone and testicular morphology, and effects on semen quality and sperm production. *Theriogenology* 2004;61:511–528.
12. Ashdown R, Gilanpour H, David J, Gibbs C. Impotence in the bull. (2) Occlusion of the longitudinal canals of the corpus cavernosum penis. *Vet Rec* 1979;104:598–603.
13. Beckett S, Wolfe D, Bartels J, Purohit R, Garrett P, Fazeli M. Blood flow to the corpus cavernosum penis in the bull and goat buck during penile quiescence. *Theriogenology* 1997;48:1061–1069.

Chapter 2

Endocrine and Exocrine Function of the Bovine Testes

Peter L. Ryan

College of Veterinary Medicine, Mississippi State University, Starkville, Mississippi, USA

Introduction

The normal bovine male reproductive system consists of paired testes retained within a sac or purse-like structure known as the scrotum, which is formed from the outpouching of skin from the abdomen and consists of complex layers of tissue. The testes are accompanied by a number of supporting structures including spermatic cords, accessory sex glands (prostate, bulbourethral, paired vesicular glands), penis, prepuce, and the male ductal system. The testicular duct system is extensive and comprises the vas efferentia found within the testes, the epididymis, vas deferens, and the urethra, all of which are located external to the testes. The reader is referred to the excellent chapter on the anatomy of the reproductive system of the bull in this book (Chapter 1). The primary functions of the testes are to produce male gametes (spermatozoa) and the endocrine factors, such as steroid (testosterone) and protein hormones (inhibin, insulin-like peptide 3), that help regulate reproductive function of the bull in concert with hormonal secretions from the hypothalamus (gonadotropin-releasing hormone) and pituitary glands (luteinizing hormone, follicle-stimulating hormone). The testes consist of parenchymal tissue that supports the interstitial tissue and includes the steroid-producing Leydig cells, vascular and lymphatic system, and the seminiferous tubules within which the germinal tissue develops with the support of the nurse cells more commonly known as Sertoli cells. Chapter 4 discusses in detail the endocrine factors responsible for testicular development and initiation of spermatogenesis in the bull, and thus this chapter focuses more on the regulation and function of the adult testes. This chapter will not undertake a treatise of those conditions that disrupt testicular function but rather will focus, as practically as is possible, on what is known of the endocrine and exocrine function of the bovine testes. Much of the endocrine and exocrine function of the testes is similar across mammalian species, and where specific information is absent for the bovine, examples will be given from other domestic species when possible. It has not been possible to cite the many significant contributions to the field of endocrine and exocrine function of the testes. Thus, where and when possible, the reader is referred to selected citations for additional reading.

Historical perspective

It has been evident for many centuries that the testes exercise control over the characteristics of the male body. The results of castration in domestic animals and human males made this very clear, but provided no clues as to the mechanism of control. Pritchard[1] noted from Assyrian records dating some 15 centuries BC that the castration of men was used as punishment for sexual offenders, which suggests that the effect of castration on fertility and behavior was recognized at that time. Knowledge of the effects of castration of livestock dates back to the Neolithic Age (*c.* 7000 BC) when animals were first thought to have been domesticated.[2] The effects of castration were understood by Aristotle (300 BC) who provided very detailed and clear descriptions of testicular anatomy and function.[3] It was not until the seventeenth century that a detailed account of testicular and penile anatomy was presented by Regnier de Graaf[4] in a treatise on the male reproductive organs. De Graaf indicated the existence of the seminiferous tubules and suggested that the production of the fertile portion of the semen occurred in the testes. The first microscopic examination of the testes was undertaken by Antonie van Leeuwenhoek in 1667 where he demonstrated and reported the presence of germ cells in the seminal fluid.[5]

Detailed study of the testis began in the mid-nineteenth century. In 1840, Albert von Kölliker discovered that spermatozoa develop from cells residing in the testicular (seminiferous) tubules. This major discovery was followed by Franz Leydig's[6] description of the microscopic characteristics of the interstitial cells. Later, Enrico Sertoli,[7] an Italian scientist, correctly described the columnar cells running from the basement membrane to the lumen of the *tubuli seminiferi contorti* (seminiferous tubules) of the testes, and Anton von Ebner is credited with introducing the concept of the symbiotic relationship between Sertoli cells and the developing germinal cells.[8,9]

The first clear demonstration that the testes are involved in an endocrine role was made by Arnold Berthold in 1849, while studying the testes of the rooster. He concluded that the regulation of male characteristics was brought about by way of blood-borne factors. The most compelling evidence for an endocrine function of the testes being associated with Leydig cells was presented by two French scientists, Bouin and Ancel in 1903. They reported that ligation of the vas deferens in dogs, rabbits, and guinea pigs was followed by degeneration of the seminiferous tubules, but no castration effects were observed and no degenerative changes of the interstitial cells, and thus concluded that internal secretions of the testes were synthesized by the Leydig cells.[10] By the mid 1930s it was clear that the male hormone emanating from the testes was testosterone[11] and that the function of the testes was controlled by pituitary hormones.[12,13] Smith[12] demonstrated that the pituitary gland must secrete substances (now known as gonadotropins) responsible for the stimulation of testicular growth and maintenance of function in the rat. Greep and Fevold[14] restored Leydig cell function in hypophysectomized rats with crude preparations of luteinizing hormone (LH) and reestablished the male secondary sex characteristics. Further evidence was elucidated in favor of a steroid secretory function for Leydig cells from a study on postnatal development in bulls where changes in testicular androgen levels paralleled the differentiation of these cells.[15] Using cell culture techniques, Steinberger *et al.*[16] provided direct evidence that the Leydig cells are the primary source of steroid hormone synthesis in the testes. Later, it became apparent that the Leydig cells are essential for providing the androgenic stimuli that are required for the maintenance of spermatogenesis in the germinal epithelium.[16,17]

The testis

The bovine testes are paired, capsulated, ovoid-like structures located in the inguinal region and suspended in a pendulous scrotum away from the abdominal wall. The proximal relationship of the testes to the abdominal wall varies and may depend on season and ambient temperatures. The cremaster muscle plays an important role in thermal regulation of the testis. The size of the testis varies with breed, but typically the adult testis weighs 300–400 g and is about 10–13 cm long and 5–6.5 cm wide.[18] The tough fibrous capsule covering each testis consists of three tissue layers: the outer layer, the tunica vaginalis; the tunica albuginea, which consists of connective tissue composed of fibroblasts and collagen bundles; and the inner layer, the tunica vaginalis, which supports the vascular and lymphatic systems.[19] The capsule is the main structure that supports the testicular parenchyma, the functional layer of the testes, which consists of the interstitial tissue and seminiferous tubules. The interstitial tissue is found in the spaces between the seminiferous tubules and consists of clusters of Leydig cells, which are primarily responsible for steroid hormone biosynthesis and secretion, along with vascular and lymph vessels that supply the testicular parenchyma. The seminiferous tubules originate from the primary sex cords and contain the germinal tissue (spermatogonia, the male germ cell) and a population of specialized cells, the Sertoli cells, which not only support the production of spermatozoa but also form tight junctions with each other, creating one of the most important components of the blood–testis barrier.[20] This structure prevents the entry of most large molecules and foreign material into the seminiferous tubules that may disrupt normal spermatogenesis. The most important substances synthesized by the testes and released into the vascular system are peptide and steroid hormones. However, fluids from the seminiferous tubules may pass into the interstitial tissue via the basal lamina, where they may enter the testicular lymphatic and vascular systems, or into the tubule lumen via the apical surface of the Sertoli cells.[19]

The scrotum and spermatic cords

The scrotum is composed of an outer layer of thick skin and three underlying layers, the tunica dartos, the scrotal fascia, and the parietal vaginal tunica. The scrotal skin is extensively populated with numerous large adrenergic sweat and sebaceous glands that are highly endowed with thermal receptors and nerve fibers. Neural stimulation from the thermal receptors enables the tunica dartos, which consists of smooth muscle fibers and lies just beneath the scrotal skin, to contract and relax in response to changes in temperature gradients and facilitates the cooling of the scrotal surface via scrotal glandular sweating.[19] Thus the scrotum plays an important role not only in housing and protecting the testes but also has a role in thermoregulation of the testes. The spermatic cord connects the testes to the body and provides access to and from the body cavity for vascular, neural, and lymphatic systems that support the testes. In addition, the spermatic cord accommodates the cremaster muscle, the primary muscle supporting the testes, and the pampiniform plexus, a complex and specialized venous network that wraps around the convoluted testicular artery.[21] This vascular arrangement is very important in temperature regulation of the testicular environment. The plexus consists of a coil of testicular veins that provide a counter-current temperature exchange system: this is an effective mechanism whereby warm arterial blood entering the testes from the abdomen is cooled by the venous blood leaving the testes. Testicular arteries originate from the abdominal aorta and elongate as the testis migrates into the scrotum.[19] In cattle and other large domestic ruminants these arteries are highly coiled, reducing several meters of vessel into as little as 10 cm of spermatic cord.[19] The arterial coils and venous plexus are complex structures that form during fetal life in cattle.[19,22] Because of the pendulous nature of the bovine scrotum, testicular cooling is facilitated by the contraction and relaxation of the cremaster muscle, which draws the testes closer to the abdominal wall during cooler ambient temperatures and vice versa during warmer temperatures. Figure 2.1 shows bright-field and thermal images of the bovine testes that demonstrate the change in temperature from the neck to the tip of the scrotum as the testes thermoregulate during elevated environmental temperatures. Scrotal and testicular thermoregulation is a complex process

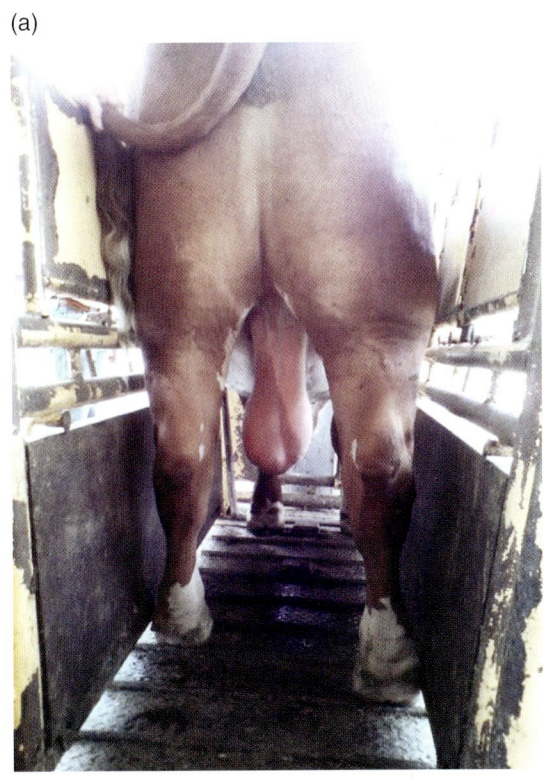
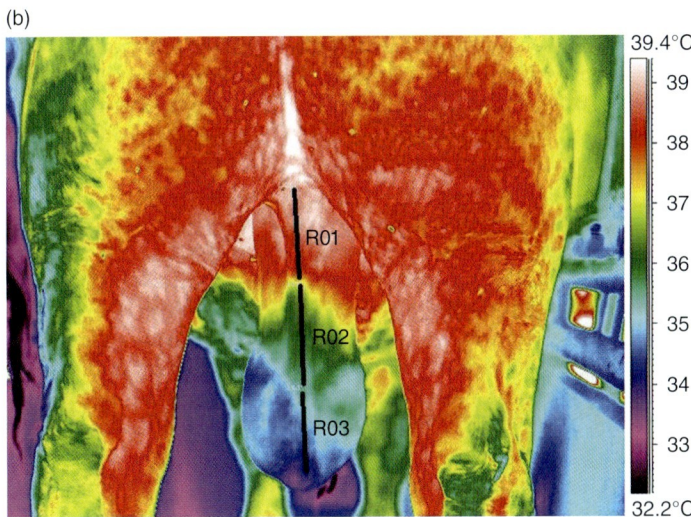

Temperature [°C]	Min	Max	Max–Min	Avg	Stdev
R01	37.7	39.1	1.4	38.6	0.3
R02	35.5	36.9	1.4	36.1	0.3
R03	34.0	35.5	1.5	34.7	0.4

Figure 2.1 Bright-field and thermal images of 2-year-old Hereford bull testes. The images were taken using an FLIR SC600 thermography unit (FLIR Systems Sweden, AB) on June 25, 2013 at midday with atmospheric temperature of 29°C and humidity of 75%. (a) Bright-field image of the testes. (b) Thermal image of the testes as seen in (a), identifying three regions of interest (RO1, RO2 and RO3) as indicated by the vertical bars along which temperature values were obtained by means of the software ThermaCAM Research Pro 2.7. The image is pseudo-colorized with temperature scale bar to help visualize the change in temperature gradient from scrotal neck to scrotal tip. (c) Temperature values, minimum (Min), maximum (Max), the difference between Min and Max and the average (Avg) temperature for each region of interest (RO), and the standard deviation (st dev) which assess the variation of temperature in each region. Note the almost 4°C change in average temperature from RO1 to RO3.

involving a number of local mechanisms that strive to maintain the testes at environmental and physiological conditions conducive for normal spermatogenesis. For additional reading on testicular thermoregulation in the bull the reader is referred to the review by Kastelic et al.[23] and Chapter 3 of this book.

Interstitial tissue (Leydig cells)

Franz Leydig, a German zoologist, first described the interstitial cells of the testes in 1850 and these cells have since been known as Leydig cells. The Leydig cells reside in the interstitial tissue of the testis, a meshwork of loose connective tissue filling the spaces between the seminiferous tubules and blood vessels. In mammalian testes, the Leydig cells occur mainly as clusters in the angular interstices between the seminiferous tubules and are closely associated with the walls of small arterioles.[8,24] The Leydig cell content of testes varies from species to species. The Leydig cells are thought to be the principal source of androgens in the testis. The development of the Leydig cells, via metamorphosis of mesenchymal precursor cells, has been observed to be continuous throughout life after the time of puberty in the bull.[25] Christensen[26] provides a very detailed and interesting review of the history of Leydig cell research dating from Leydig's description of the cells in the 1850s to the confirmation provided in the mid-1960s that these cells were indeed primarily responsible for testicular androgen synthesis and secretion. There is extensive evidence to suggest that early fetal Leydig cells are steroidogenically active in some mammalian species including the pig[27] and sheep.[28]

The Leydig cells of most mammalian species studied are basically similar, with some minor variations in appearance, size, and the relation of Leydig cell clusters to the lymph or blood vessels of the interstitial tissue. Some variation in the extent of cytoplasmic structures exists, but ultrastructurally Leydig cells show considerable overall similarity.[8] Fawcett et al.[29] described in detail the morphology of interstitial tissue of several mammalian species, and categorized three groups based on the abundance of Leydig cells and the relationship between volume of intertubular lymph structures and connective tissue. In the first group are the guinea-pig and rodents (rat, mouse). In these species only 1–5% of the testicular volume is occupied by Leydig cells, for example 2.8% in the rat.[30] The bull, monkey, elephant, and human fall into the second group. In these species, the connective tissue of the interstitium is very loose and the Leydig cells are scattered throughout the interstitium and are closely associated with a well-developed lymph system. The Leydig cells comprise only a small portion of the testicular volume (~15%).[8] In the third group are the domestic boar and horse. In these animals there is abundant interstitial tissue packed with Leydig cells (20–60% of testicular volume).[31] The reason for the high density of Leydig cells in these species is not known, but Parkes[10] and Fawcett et al.[29] have attributed this phenomenon to the vast amounts of estrogens produced by the boar and stallion and the large quantities of musk-smelling 16-androstenes secreted by the boar testes.[32] For more detailed discussions on the cytology of Leydig cells the reader is referred to an excellent chapter by de Kretser and Kerr.[33]

Endocrine function of the testis

Hypothalamic–pituitary hormone regulation of the testis

Hormone action depends on the release of hormones from the appropriate endocrine gland and transportation via the vascular circulatory system to the target tissue where the hormone binds to cellular receptors, thus inducing a physiological response. In some cases these receptors are very hormone-specific. The response at the target tissue may depend on the level of receptor expression and concentration of hormone. Some hormones regulate their own receptors, others may require synergism between two hormones, and others still may have their receptors regulated by other hormones.[34] The general characteristics of neuroendocrine regulation of the mammalian testis by the hypothalamic–pituitary axis are well established,[35] but in some species this may be seasonally regulated. Domestic cattle are considered to be continuous or nonseasonal breeding species.[36] However, there is information in the literature to indicate that domestic beef and dairy bulls have a functional hypothalamic–pituitary–gonadal axis that is seasonally regulated and thus may influence levels of gonadal steroidal and germ cell production. In a study using composite breeds of mature bulls, Stumpf et al.[37] observed that season of the year influenced the profile of gonadotropins in the circulation of both gonadectomized and intact males, and that the greatest secretion of gonadotropins occurred at the spring equinox. In addition, these authors observed that season of the year also influenced secretion of testosterone in intact males, but that more testosterone was released in response to LH during the time of the summer solstice. Others have reported similar effects, where location affected reproductive traits of bulls including semen quality and blood concentrations of LH and testosterone.[38,39]

In addition, sensitivity in responsiveness to exogenous gonadotropin administration and testosterone secretion in bulls was observed to be seasonally influenced.[40]

The regulation of testicular function by hormonal mechanisms depends on the integrated actions of gonadotropins, such as LH, follicle-stimulating hormone (FSH) and prolactin, and steroids (androgens and estrogens) on the Leydig cell.[41] Gonadotropin-releasing hormone is the primary hypothalamic hormone governing the regulation of the synthesis and release of the gonadotropins LH and FSH by the anterior portion of the pituitary gland. LH is primarily responsible for testosterone production by the Leydig cells while FSH facilitates Sertoli cell proliferation and support of the germinal cells. Although it has been well established that gonadotropins stimulate testicular function, it was during the 1970s that the basic mechanism and site of action in the testis were identified. LH has been shown to be the gonadotropin essential for the maintenance of testicular testosterone production.[41,42] Catt and Dufau[43] have reviewed the mechanisms of action of the gonadotropins and concluded that they all elicit target cell response by similar mechanisms. In fact, LH provides the most important physiological regulation of the production of androgens by the Leydig cells of the testis.[44,45]

Leydig cells contain plasma membrane-bound receptors that specifically bind LH. The binding of LH stimulates adenyl cyclase to produce cyclic adenosine monophosphate (cyclic AMP), a second messenger in the cell cytoplasm, which in turn activates cyclic AMP-dependent protein kinase thereby increasing the conversion of cholesterol to pregnenolone. The action of LH can be readily demonstrated in hypophysectomized and intact males. Removal of the pituitary gland is followed by rapid cessation of testosterone production, loss of enzymes involved in steroidogenesis, and testicular atrophy.[46] There is little direct evidence to support a role of FSH in Leydig cell steroidogenesis, but there is some evidence indicating that this gonadotropin may play a function in the conversion of androgens to estrogens in the Sertoli cells.[47] Bartke et al.[42] have suggested that FSH may act on the aromatizing enzyme system of testosterone biosynthesis, but little evidence exists for action on other steps of the steroidogenesis pathway in the testis. In bulls, testosterone secretion is not tonic, but is characterized by episodic pulses dictated by the release of luteinizing hormone releasing hormone (LHRH) from the hypothalamus and LH from the anterior pituitary gland.[48] Schanbacher[48] reports that a temporal relationship exists between concentrations of LH and testosterone in the blood, and that there is evidence that episodic secretion depends on discrete episodes of LHRH discharge from the hypothalamus.

Prolactin

Prolactin (PRL) is known to enhance the effect of LH on spermatogenesis. The physiological importance of PRL in the regulation of Leydig cell function was first determined using three animal models:[41,42] the golden hamster, where testicular regression can be induced by short photoperiod; hypophysectomized mice; and the hereditary dwarf mouse with congenital PRL deficiency and infertility. In all three models there is a deficiency in plasma concentrations of PRL and testosterone and testicular atrophy is evident. PRL therapy, on the other hand, stimulates spermatogenesis in the dwarf mouse and restores testicular function in the golden hamster; in hypophysectomized animals, PRL, in the presence of LH, induces spermatogenesis and restores plasma testosterone. In men, PRL has an important role in the control of testosterone and reproductive function.[49] At very high concentrations (hyperprolactinemia), PRL has an antigonadal effect in men, inhibiting testicular function, and is associated with hypogonadism.[50]

Other regulatory factors

The effects of other regulatory factors, such as growth hormone (GH), on testicular steroidogenesis have not been fully clarified. However, in view of the structural similarities between PRL and GH, it is thought that both hormones may have comparable effects on Leydig cell function.[42] The effect of GH may be more significant during puberty (see Chapter 4). However, administration of gonadal steroids to intact animals is known to cause reduced androgen production, an effect that has been attributed to inhibition of gonadotropin secretion with secondary effects on testicular endocrine function. Estrogen receptors have also been detected in interstitial tissue of the testis and since decreased testosterone levels are not correlated with a corresponding change in plasma LH, some authors suggest that estrogen may exert a direct inhibitory action on Leydig cell function.[51,52] The effects of corticosteroids on Leydig cell function have been noted. Boars treated with adrenocorticotropin (ACTH) experienced increased testicular testosterone simultaneously with increased secretion of adrenal corticosteroids.[53] A transient increase in peripheral circulation of both corticosteroids and testosterone was first observed in boars following acute treatment with ACTH while a decrease in testosterone occurred following chronic treatment.[54] These authors concluded that prolonged stressful conditions may lead to chronic elevation in ACTH levels, thus suppressing testosterone production and inducing poor breeding activity.

In reviewing the literature, Tilbrook et al.[55] have determined that stress-related influences on reproduction is complicated and not fully understood, but the process may involve a number of endocrine, paracrine, and neural systems. The significance of stress-induced secretion of cortisol varies with species. In some instances, there appears to be little impact of short-term increases in cortisol concentrations and protracted increases in plasma concentration seem to be required before any deleterious effect on reproduction is apparent. In a comprehensive review, Moberg[56] describes the influence of the adrenal axis on gonadal function in mammals. Subsequently, the same author published a detailed review[57] of the impact of behavior on domestic animal reproductive performance and placed particular emphasis on how intensive livestock management practices can contribute to stress-induced disruption of normal reproductive function. However, the emphasis has primarily been on studies investigating the effects of stress-induced ACTH release in females.

Elevated plasma ACTH results in increased adrenal corticosteroid synthesis and under chronic conditions may inhibit gonadotropin-releasing hormone and gonadotropin hormone production, thereby reducing gonadotropin (LH) release from the pituitary gland and thus impacting normal ovarian function in females. Nevertheless, studies have shown that stress-induced elevation in ACTH affects males in a physiologically similar manner by impacting gonadotropin regulation of testicular function. Liptrap and Raeside[58] observed the effects of cortisol on gonadotropin-releasing hormone in boars, while Matteri et al.[59] observed that stress or acute ACTH treatment suppresses gonadotropin-releasing hormone-induced LH release in the ram. In Holstein bulls treated with a bolus of ACTH (0.45 IU/kg), plasma LH or FSH concentrations did not appear to be negatively affected, but it did appear to reduce plasma testosterone concentrations.[60] Pharmacological concentrations of ACTH (200 IU every 8 hours) over a 6-day period resulted in reduced plasma testosterone in yearling and mature bulls within 8 hours and 4 days of initial treatment, respectively, and this decrease persisted in both age groups for an additional 24 hours after the last ACTH injection.[61] While semen viability, concentration, and sperm output were unaffected by the prolonged ACTH treatment concomitant with a subsequent marked increase in glucocorticoids and decrease in testosterone, a small increase in semen content of immature sperm or sperm with abnormal heads was observed.[61] Thus, it appears that under prolonged stressful conditions, gonadotropin secretion is more likely to be suppressed thus inhibiting reproductive performance. In addition, studies have demonstrated that glucocorticoids can inhibit gonadotropin secretion in some circumstances. Tilbrook et al.[55] conclude that suppression of reproduction is more likely to occur under conditions of chronic stress and may involve actions at the level of the hypothalamus or pituitary. In addition, they indicate that there are likely to be species differences in the effect of glucocorticoids on gonadotropin secretion, and the presence of sex steroids and the sex of an individual are also likely to be factors.

Cytokines are members of the family of growth factors, and the interleukins in particular may act at the level of the hypothalamus where they may control the release of gonadotropin-releasing hormone, thus influencing the release of pituitary gonadotropins and ultimately gonadal steroid synthesis and release. Svechniko et al.[62] have reported that testicular interleukin (IL)-1 may play a role in the paracrine regulation of Leydig cell steroidogenesis in rats. Others have reported that testicular interleukins may play an important functional role in both normal testicular function and under pathological conditions.[63] There is some evidence in the bovine that cytokines such as tumor necrosis factor (TNF)-α and interferon (IFN)-γ may play a role in nitric oxide regulation in luteal endothelial cells by increasing inducible nitric oxide synthase (iNOS) activity, thereby accelerating luteolysis, while progesterone is thought to suppress iNOS expression in bovine luteal cells.[64] Whether these or other cytokines have similar effects on testicular cells of the bull has yet to be reported.

Steroid synthesis by Leydig cells

Since the beginning of the twentieth century, Leydig cells have been considered the probable source of testicular androgens.[65] Berthold[66] was the first to observe from experimentation with the rooster that the testes produced a substance that influenced secondary sex organ development and maintenance.[26] The first isolation of an androgen, androsterone, from human urine[67] and the crystallization of testosterone from bull testes[11] established the major site of testosterone production as the testis. Extensive literature has emerged over the past 50 years on the function of the Leydig cell. These studies, employing a variety of techniques, have identified and confirmed the Leydig cells as the primary source of testicular androgens and elucidated the important pathways in androgen biosynthesis.[5, 26, 52, 68] LH has also been confirmed as the major pituitary hormonal stimulus on the Leydig cells.[46] Ewing et al.[68] have suggested that because Leydig cells are concentrated in clusters in the interstitial tissue of the testis, they must influence seminiferous tubular and peripheral androgen-dependent functions (accessory sex glands) by hormonal signals rather than by cell-to-cell interaction.

The primary steroids produced and secreted by the bull testis are shown in Table 2.1. The conversion of the precursor cholesterol to androgens and estrogen is facilitated by a series of enzymatic reactions involving hydroxylation, dehydrogenation, isomerization, C–C side-chain cleavage (lyase), and aromatase activity. Testosterone is now recognized as the principal steroid responsible for the endocrine functions of the testis since it is synthesized in copious amounts by mammalian Leydig cells. The Leydig cell has also been identified as a major source of other androgens and estrogens.[5,69] In addition, the Leydig cells of the boar testis produce large amounts of the musk-smelling steroids, Δ^{16}-androstenes.[70] It has become apparent that the primary function of the Leydig cell is the provision of the androgen stimulus required for the initiation and maintenance of spermatogenesis in the germinal epithelium within the seminiferous tubules.[71] Some of the earliest functions of the Leydig cell are associated with regulation of the male reproductive organs, the organization of parts of the brain, pituitary secretion of gonadotropins, and accessory sex organ development of the neonate to ensure the appropriate response by these tissues to testicular steroids in the adult male.[69] The dependence of male sex accessory organs on testicular hormones is not restricted to the early fetal period but occurs also during puberty and throughout male adult life.[72]

Cholesterol has been described as an obligatory intermediate in testosterone synthesis.[5] Testosterone is synthesized from a pool of metabolically active cholesterol, which is derived from either *de novo* biosynthesis of cholesterol, cholesterol esters stored as lipid droplets in the cell cytoplasm, or from blood plasma.[68] The conversion of cholesterol to testosterone involves five main enzymatic steps that include 20,22-lyase, 3β-dehydrogenase isomerase, 17α-hydroxylase, 17,20-lyase, and 17β-hydroxysteroid dehydrogenase, and in some species the conversion of androgens to estrogens via the aromatase enzyme system.[73] Conversion of cholesterol to pregnenolone is the initial step in the pathway and is catalyzed by the cholesterol side-chain cleavage enzyme complex, a three-step process that takes place in the mitochondria of the cell.[45] Briefly, pregnenolone is formed from cholesterol (C_{27} sterol) by cleavage of the bond between C-20 and C-22 catalyzed by the multienzyme complex of the side-chain cleavage system in the mitochondria, and

Table 2.1 Primary steroid and peptide hormones synthesized and/or secreted by the bull testis.

	Hormone	Site of synthesis
Steroid family		
Steroid precursors	Cholesterol	*De novo* biosynthesis, fat deposits, or from blood
Cholesterol (27 carbons)	22-Hydroxycholesterol	
	20,22-Dihydroxycholesterol	
Progestins (21 carbons)	$\Delta^{[5]}$-Pregnenolone	Leydig cells (mitochondria)
	17α-Hydroxypregnenolone	
	17α-Hydroxyprogesterone	
	Progesterone	
Androgens (19 carbons)	Dehydroepiandrosterone	Leydig cells (microsomal compartments)
	$\Delta^{[4]}$-Androstenedione	
	$\Delta^{[5]}$-Androstenediol	
	Testosterone	
	Dihydrostestosterone	
Estrogens (18 carbons)	Estrone	Leydig cells
	Estradiol-17β	
Peptide family		
Relaxin-like peptides	Relaxin/insulin-like peptide-3	Leydig cells
Neuropeptides	Oxytocin	Leydig cells
	Glial cell-derived factor	Sertoli cells
Cytokine family	Activin	Sertoli cells
	Inhibin	Sertoli cells
Glycoproteins	Androgen-binding protein	Sertoli cells
	Testicular transferrin	Sertoli cells

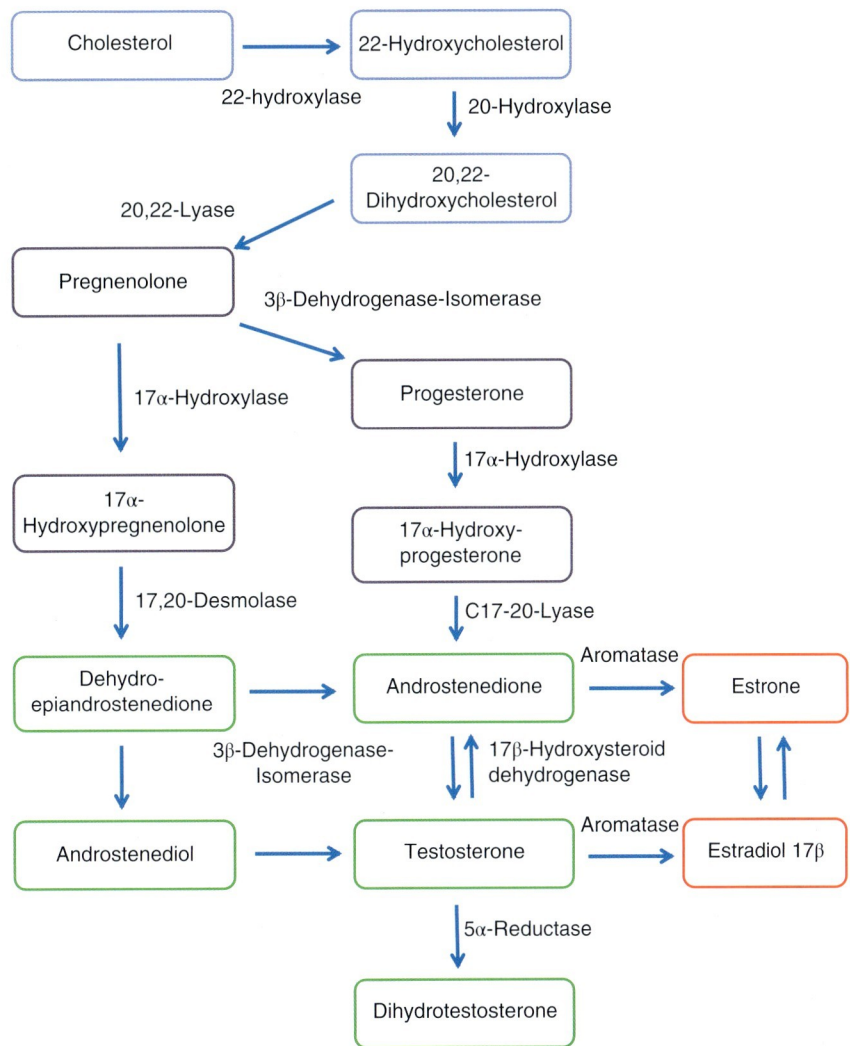

Figure 2.2 Synthetic pathway of testosterone and conversion to active androgen and estrogen metabolites in the bull testis. Relevant enzyme systems involved in the synthesis are shown; in some instances the enzyme reactions are reversible. Color code: blue, C_{27} cholesterol steroid precursors; purple, C_{21} progestin steroids; green, C_{19} androgen steroids; red, C_{18} estrogen steroids.

metabolized in the microsomes by the microsomal enzyme complex 3β-hydroxysteroid dehydrogenase/isomerase. Pregnenolone is also an obligatory intermediate in Leydig cell steroid synthesis. Van der Mollen and Rommerts[52] have indicated that the conversion of C_{21} steroids (i.e., pregnenolone) to C_{19} steroids (testosterone) may occur in the mammalian testis through two biosynthetic pathways. The crucial step in the biosynthesis of pregnenolone to androgens is the cleavage of the two-carbon side-chain of 17α-hydroxyprogesterone or 17α-hydroxypregnenolone by the 17,21-lyase enzyme complex.[74] This is regarded as an irreversible reaction producing "weak" androgens (androstenedione, dehydroepiandrosterone, respectively). Through the action of 17β-hydroxysteroid dehydrogenase, androstenedione is converted to the more potent androgen testosterone, which occurs in the microsomal compartments of the cell. However, testosterone is converted to dihydrotestosterone (DHT) by the 5α-reductase enzyme system, and is regarded as the more biologically active androgen produced by the testis. A more detailed account of the mechanisms involved in testosterone biosynthesis can be found in the review by Hall,[45] but a schematic overview of steroid synthesis in the bovine testis is presented in Figure 2.2.

The unusual abundance of Leydig cells in domestic boars and stallions promoted the hypothesis that this may be related to the fact that both species secrete significant amounts of estrogens,[75] but the significance of the vast quantities of estrogens produced by these two species is unexplained. It has been suggested that estrogens act synergistically with testosterone to enhance both secretory activity of accessory sex organs and sexual behavior in boars castrated after puberty.[76] Estrogens are C_{18} steroids and are formed by the conversion of androgens by the aromatase enzyme system to produce estrone and estradiol from androstenedione and testosterone, respectively. Of interest in the boar are the musk-smelling Δ^{16}-androstene steroids (pheromones) that are regarded quantitatively as the most abundant steroids produced by the boar testis and contribute to the familiar "boar taint" of pork.[77] However, there is insufficient evidence to demonstrate that bull testis produces estrogens in the quantities found in the boar and stallion, nor is there evidence that the bull secretes much in the way of the Δ[16]-androstene steroids. However, what is now well documented is that testosterone is the most potent androgen produced by Leydig cells in mammalian testes, and the site of action is primarily on seminiferous tubule target cells, thus influencing the reduction division of the spermatogenic cells.[78] Androgens stimulate production of androgen-binding protein (ABP) by the Sertoli cells,[79] and this acts as an intracellular carrier of testosterone and DHT within the Sertoli cells. Testosterone is also the most important determinant of the rate of formation of fructose by the seminal vesicles of human, bull, and ram, and of citric acid by the prostate and seminal vesicle glands of the bull, ram, and human.[78]

Oxytocin is a nine amino acid neuropeptide hormone normally associated with the hypothalamic–posterior pituitary system and the regulation of parturition and lactation in the female, but has also been shown to have an endocrine and paracrine role in male reproduction.[19,80] There is evidence reported in the literature that oxytocin is produced and secreted by the male reproductive tract including the testis.[81,82] Moreover, there is now evidence to show that oxytocin is produced locally by the testis and that it has a paracrine role in modulating testicular steroidogenesis and contractility of the male reproductive tract.[83] In addition, it has been shown that the Leydig cells are the testicular site of production of this hormone, and that oxytocin acts as a paracrine hormone influencing the contractility of the peritubular myoid cells.[19] The contraction of myoid cells in the seminiferous tubule epithelium is thought to facilitate sperm transport through the testicular parenchyma emptying into the rete testis and on into the epididymal system. It has been shown that, within the prostate, testosterone is converted by 5α-reductase to DHT, which stimulates growth of the prostate gland. Nicholson[83] has postulated that oxytocin increases the activity of 5α-reductase resulting in increased concentrations of DHT and growth of the prostate, but that androgen feedback reduces oxytocin concentrations in the prostate thereby modulating prostate gland growth. Definitive evidence of oxytocin synthesis within the bovine testis has come from studies on oxytocin gene expression in the seminiferous tubules.[84,85]

Relaxin and INSL-3

Insulin-like peptide 3 (INSL3; formerly known as relaxin-like factor) is a peptide hormone belonging to the relaxin–insulin family of peptide hormones.[86,87] It was first identified in the testes of pigs[88] and is now thought to be an important factor in regulating normal testicular decent,[87] particularly during the second phase where it acts on the gubernaculum as demonstrated using knockout mice. The peptide hormone has been identified in male and female tissues of other species, including human, marmoset monkey, sheep, goat, bovine, as well as deer, dog and other species.[89] It is produced in large quantities by the Leydig cells of both the fetal and adult testes, and circulating INSL3 concentrations have been measured in the blood of adult male mammals including the rat (5 ng/mL),[90] mouse (2 ng/mL),[90] and human (0.8–2.5 ng/mL).[91,92] Although INSL3 has been successfully extracted from bovine testis[93] and found to be present in amniotic fluid from human male fetuses,[94] it is only recently that Anand-Ivell et al.[95] using new time-resolved fluorescence immunoassay to directly detect INSL3 in the blood and body fluids of ruminants reported that mid-gestation (day 153) cows carrying a male fetus showed significantly higher maternal blood concentrations of INSL3 compared with cows carrying a female fetus. The authors speculate that INSL3 provides the first example of a gender-specific fetal hormone with the potential to influence both placental and maternal physiology. In a recent review, it is reported that INSL3 is a major secreted product of the interstitial Leydig cells of the mature testes in all male mammals.[96] Moreover, current evidence points to autocrine, paracrine, and endocrine roles, acting through the G-protein-coupled relaxin family receptor 2 (RXFP2), although more research is required to characterize these functions in detail. Indeed, recent studies have provided evidence of the presence of both RXFP1 and RXFP2 receptors in porcine spermatozoa,[97] which would suggest that relaxin may play an important role in sperm motility.[98] However, the presence of relaxin receptors, relaxin-binding activity, and enhanced sperm motility has yet to be demonstrated in the bovine species, but with the advent of new technologies this may be possible.[99]

Seminiferous tubules, Sertoli cells, germinal tissue

Within segments of the testicular parenchyma lie the seminiferous tubules, originally formed from the primary sex cords. The seminiferous tubules contain the Sertoli cells, also known as the nurse cells of the developing germinal cells, the spermatogonia, which are elongated cells found within the tubules. The Sertoli cells were first described by the Italian physiologist Enrico Sertoli in the early 1860s. Structurally, these cells are very important in that they form the blood–testis barrier, which separates the interstitial blood compartment of the testes from the adluminal compartment of the seminiferous tubules. In addition to Sertoli cells nurturing the development and maturation of the germinal cells through the various stages of spermatogenesis, these cells act as phagocytes by removing or consuming residual cytoplasm and surplus spermatozoa material. Sertoli cells, which are also very active secretory cells, produce anti-Müllerian hormone (AMH), a glycoprotein secreted during early fetal life by the testis to direct the appropriate development of the male reproductive tract by causing the regression of the Müllerian ducts.[100,101] Mutations of AMH gene lead to persistent Müllerian duct syndrome in male human patients;[102] whether the same observation is true in cattle has not been reported. However, recent studies successfully employed the use of AMH profiles as a novel biomarker to evaluate the existence of functional cryptorchid testis in Japanese black calves.[103] A more comprehensive review of Sertoli cell cytology may be found in the review by de Kretser and Kerr.[33]

In addition, Sertoli cells secrete two closely related protein complexes of the transforming growth factor cytokine superfamily (activin and inhibin) that have opposite regulatory effects on the synthesis and regulation of pituitary FSH secretion. Activin upregulates while inhibin downregulates FSH synthesis and secretion.[104,105] Both of these hormones are secreted after puberty and play an important role in mammalian reproduction, which is coordinated by assorted neural, neuroendocrine, endocrine, and paracrine cell–cell communication pathways.[106] Activin is also thought to facilitate androgen synthesis in the testis by enhancing LH, thereby promoting spermatogenesis. In male mammals, inhibin production is thought to be promoted by testicular androgens and may function locally in the testis to help regulate spermatogenesis. Much less is known about inhibin in relation to its mechanism of action, but it is thought that it may compete with activin for the binding of the activin receptor and/or binding to the inhibin-specific receptor.[105] However, Phillips[107] in reviewing the literature notes that inhibin not only acts as a feedback regulator of FSH in the male, but it may also be an important paracrine and autocrine regulator of testis function. It has been shown from molecular studies that both forms of inhibin (A and B) are produced by the bovine testis and that the proportion of the mature forms of inhibin A and inhibin B increases as bulls age, but that total inhibin production by Sertoli cells decreases.[108] In a later study, Kaneko et al.[109] reported that inhibin A and inhibin B increased in the testis of bulls during postnatal development, and that immunoreactive inhibin A could be detected in the plasma of these bulls. Others have attempted to use blood concentrations of inhibin and gene expression along with other endocrine and genetic markers as a predictor of fertility in Brahman bulls.[110] Unfortunately, there appears to be little information in the current literature on the role activin plays in the bovine male. Its role as a paracrine regulator in the testis is presumed but has not been conclusively demonstrated in domestic species, although *in vitro* studies in porcine Leydig cells would suggest a role in androgen synthesis.[107] However, it is thought to be an important intrafollicular factor in the regulation of follicle selection in the bovine female.[111,112] A role in inflammatory processes for activin has been postulated.[107]

Additional factors secreted by the Sertoli cell include ABP, estradiol (by conversion of testosterone via the aromatase enzyme complex), glial cell-derived neurotrophic factor, transferrin,[113] and the more recently identified ERM transcription factor. Testicular transferrin and ABP are two glycoproteins produced and secreted into the lumen of the tubule by Sertoli cells, and which have important biological effects on differentiation and maturity of sperm.[114,115] ABP is a glycoprotein that binds specifically to testosterone, DHT, and 17β-estradiol. By binding to ABP, the steroid hormones testosterone and DHT are less lipophilic and become more concentrated within the lumen of the seminiferous tubules, thereby helping to facilitate spermatogenesis. ABP synthesis is regulated by FSH activity on the Sertoli cells, which may be enhanced by insulin and testosterone. ABP can especially combine with testosterone to maintain a high concentration of androgen in the seminiferous tubule, which facilitates the development and maturation of spermatogenic cells. Androgen is transferred to the target cells and via the nuclear transfer mechanism into the nucleus to combine with androgen receptors, which regulate translation of the target genes. The characteristic of ABP combining with androgen shows that ABP participates in sexual differentiation and maturation of germ cells.[113]

Sertoli cell gene products include growth factors, metabolic enzymes, transport proteins (e.g., transferrin, ABP), inhibin, proteases, antiproteases, energy metabolites, and structural components.[116] Transferrin is a major secretory product of differentiated Sertoli cells and is postulated to transport iron sequestered by the blood–testis barrier to the developing germ cells. Fe^{3+} in blood is transferred to Sertoli cells, combined with transferrin synthesized by Sertoli cells, and further transferred to spermatogenic cells at the development stage to promote germ cell growth and maturation. It is well known that the synthesis and secretion of transferrin and ABP are regulated by FSH and androgen. The gonadotropin FSH is an important endocrine hormone required for the regulation of Sertoli cell function.[117] FSH has been shown to regulate the expression of most Sertoli cell genes, including the FSH receptor, ABP, transferrin, plasminogen activator, and aromatase.[118–121]

Glial cell-derived neurotrophic factor (GDNF) is a small protein normally associated with the promotion of neuron survival.[122] Its role in testicular function is not fully understood, but there is some evidence to suggest that it may be involved in promoting undifferentiating spermatogonia, which ensures stem cell self-renewal during the perinatal period.[123] Moreover, Johnston et al.[124] demonstrated in the adult rat that cyclic changes in GDNF expression by Sertoli cells are responsible for the stage-specific replication and differentiation of stem spermatogonia, the foundational cells of spermatogenesis. Recent reports in the literature have demonstrated that GDNF, rather than induce proliferation of spermatogonia stem cells, enhances self-renewal and increases survival rate of bovine spermatogonia in culture.[125] An interesting observation reported by Harikae et al.[126] is a rare case of a freemartin calf exhibiting the transdifferentiation of ovarian somatic cells into testicular somatic cells including Sertoli cells, Leydig cells, and peritubular myoid cells, and that the Sertoli cells stained positive for GDNF protein. They speculate that these observations suggest that mammalian XX ovaries may have a high potential for sexual plasticity. For a more comprehensive review of the function and regulation of Sertoli cells, the reader is referred to the excellent review by Russell and Griswold.[127]

Spermatogenesis and maturation of spermatozoa

The first studies on the process of spermatogenesis and the spermatogonia of the domestic bull were reported in the latter part of the nineteenth century,[128] but it was not until 1931 that the chromosome number of spermatogonia and spermatocytes was determined and that the diploid number was 60.[129] Subsequent studies in the era after the Second World War described the germ cells and their cytoplasmic structures in more detail.[128] It is now well established that spermatogenesis is a complex cellular process whereby spermatozoa, the male haploid germ cell or gamete, are formed from the diploid spermatogonia stem cells through a series of cellular transformations. These complex transformations occur in the seminiferous tubules of the mammalian testes and may proceed over an extended period of time, which is species dependent. Spermatogenesis has been described morphologically in distinct and recognizable cellular "stages" or "phases" that progress through highly organized and precisely timed cycles.[35] During fetal development primordial germ cells migrate to the embryonic testes where they undergo mitotic division to form gonocytes. Just prior to puberty, gonocytes differentiate into the primary pool of A0 spermatogonia, the stem cells from which all subsequent classes of spermatogonia arise. Spermatogenesis proceeds through three distinct stages within the seminiferous tubules of the testis. The first stage is spermatocytogenesis, a proliferative phase where spermatogonia undergo a series of mitotic divisions to form primary spermatocytes. The second stage is meiosis where the primary spermatocytes undergo reduction division of the chromosomal number, from primary spermatocytes (4n) to secondary spermatocytes which are diploid (2n) to the final division that produces round haploid (1n) spermatids. The final stage is differentiation of the haploid spermatids in a process referred to as spermiogenesis, where the rounded spermatids undergo a series of metamorphic changes to form the elongated and flagellated spermatozoa.[35,130]

Spermatocytogenesis

The initial proliferation occurs at the basal membrane compartment of the seminiferous tubules where the germ cells reside. Spermatogonia emerge from the basal germ cell layer, where spermatogonia progress through the differentiation cycle and undergo a series of mitotic divisions, the number of which may be species dependent, forming A1 to A4, I and B spermatogonia. Spermatogonia move forward from the basal lamina of the seminiferous tubules as a cohort connected to each other by intercellular bridges, thereby allowing interconnection of cytoplasm between cells. These cytoplasmic bridges may facilitate intercellular communication and thus support the synchronized development of the cohort of spermatogonia. The bull and ram undergo four mitotic divisions during spermatocytogenesis, yielding 16 primary spermatocytes from each active spermatogonium.[18] During this process, a pool of stem germ cells (dormant spermatogonia A1 and A2) is maintained that provides for new generations of spermatogonia (stem cell renewal). This allows the continual production of spermatozoa in the adult male, but the precise mechanism of this regeneration process or renewal of germinal stem cells is not completely understood.[131] During this process, there may be a level of cell loss resulting from cellular degeneration. These cells are resorbed by Sertoli cell cellular phagocytosis.

Meiosis

Mitotic division provides a continuous supply of spermatogonia, while meiotic division is a process whereby B spermatogonia undergo reduction division to form haploid spermatocytes.[35] This meiotic process not only halves the number of chromosomes in spermatocytes, but also ensures genetic diversity by DNA replication and crossing-over, random events that produce genetically unique individual spermatozoa. This replication and crossing-over of DNA material occurs during the first meiotic division, while the second meiotic division results in the formation of the haploid spermatocytes. Spermatocytes may be found at all stages of development in seminiferous epithelium due to the prolonged nature of the meiotic period of spermatogenesis, which ranges from 18 to 21 days in the bull (Figure 2.3a). At the completion of this meiotic process in bulls, each B spermatogonium will result in the production of four haploid spermatids, with a total of 64 spermatids emerging from a single active (A3) spermatogonium.[132–134]

Spermiogenesis

The final phase is the differentiation or transformation of the round spermatids into elongated, flagellated, and highly condensed mature spermatozoa that are released into the seminiferous tubule lumen.[35] This process is remarkably similar across domestic species and involves a complex series of events whereby the spermatid undergoes metamorphosis into a highly organized motile cellular structure.[33] Differentiation consists of four main phases that occur in the adluminal compartment between adjacent Sertoli cells: the Golgi phase, the cap phase, the acrosomal phase, and the maturation phase. During the Golgi phase, small Golgi vesicles within the spermatid cytoplasm fuse to form a larger complex structure, the acrosomic vesicle or acrosome. The acrosome is a membrane-bound vesicle or lysosome that contains several enzymes including acid hydrolase, acrosin, esterases, hyaluronidase, and zona lysine. As differentiation progresses, the acrosome begins to migrate to form a cap (capping) over the nucleus of the cell. During capping, the acrosomic vesicle flattens and covers approximately one-third of the nucleus.[35] At the same time the nucleus and cytoplasm undergo elongation, and the nucleus begins to occupy the head region of the spermatid. While the head region of the spermatid is undergoing elongation, the midpiece and tail are being formed, thus initiating the maturation phase. Mitochondria migrate in the cytoplasm to form a spiral assembly around the flagellum posterior to the nucleus, which defines the midpiece of the spermatid. The flagellum originates from the distal centriole that gives rise to the axoneme, which is composed of nine pairs of microtubules arranged radially around two central filaments.[21] The axoneme is connected to the base of the nucleus and extends through the midpiece and continues on to form the principal piece or "tail" of the spermatid. As the cytoplasm from the spermatid is shed during the formation of the tail, a cytoplasmic droplet is formed on the neck of the spermatozoon. Once the metamorphosis is complete, the flagellated and elongated spermatid is now referred to as a spermatozoon (Figure 2.3b). These metamorphic changes are all important developmental steps to ensure that the spermatozoa have the ability to be not only motile but also capable of fertilization and delivery of nuclear material when they come in contact with the female gamete (oocyte).

Spermiation

The final stage of development results in the release of the spermatozoa into the lumen of the seminiferous tubules by a process referred to as spermiation.[33] As the final stages of

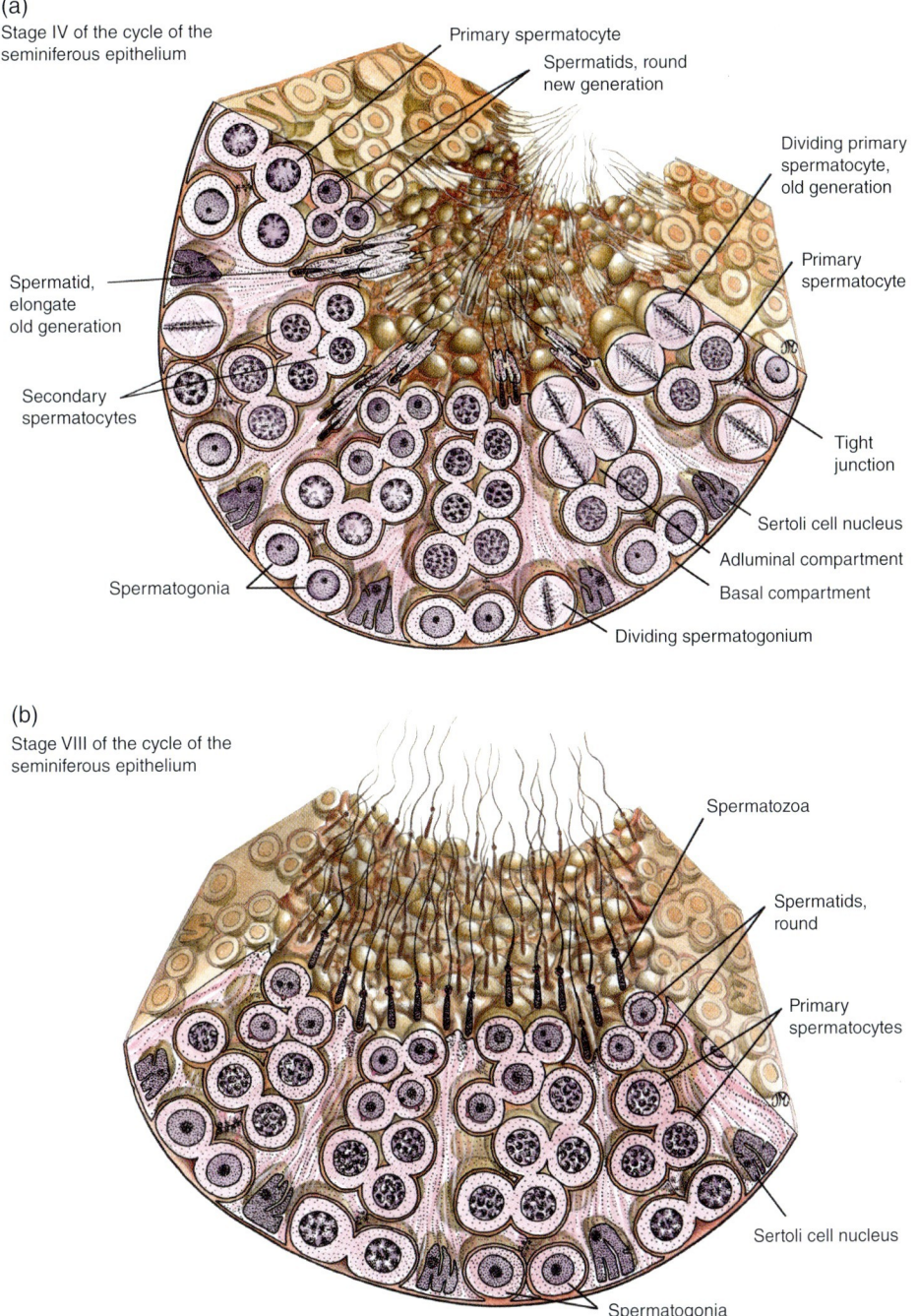

Figure 2.3 Schematic representations of the germinal epithelium of the bull testis during stage IV and stage VIII of the seminiferous epithelium cycle. Stage IV of the cycle (a) is characterized by the very active generation of primary spermatocytes into secondary spermatocytes during Meiosis I and progress on with the completion of the meiotic division of Meiosis II and the formation of spermatids. Stage VIII of the cycle (b) is characterized by the emerging free spermatozoa that have undergone differentiation from the round spermatids. Images reproduced with permission from K. June Mullins and Richard G. Saacke *Illustrated Anatomy of the Bovine Male and Female Reproductive Tract*. Germinal Dimensions Inc., Cadmus Professional Communications, Science Press Division, Ephrata, PA, 2003.

spermiogenesis are completed, the spermatozoa emerge from the adluminal compartment of the seminiferous tubules tail toward the apex of the Sertoli cells and the tubule lumen. During the process of migration from the rete testis through the epididymis and ejaculate, the spermatozoa shed their cytoplasmic droplets, but not all emerge as normal spermatozoa in the ejaculate (Figure 2.4). Before spermatozoa are capable of fertilization they must undergo additional maturation in the epididymis. Capacitation, the final stage of maturation where spermatozoa gain the ability to penetrate the zona pellucida (the outer membrane of the oocyte), occurs in the female reproductive tract. For more specific descriptions of the process of spermatogenesis in the bull the reader is referred to a series of articles by Amann,[132] Berndtson and Desjardins,[133] and Johnson *et al*.[134]

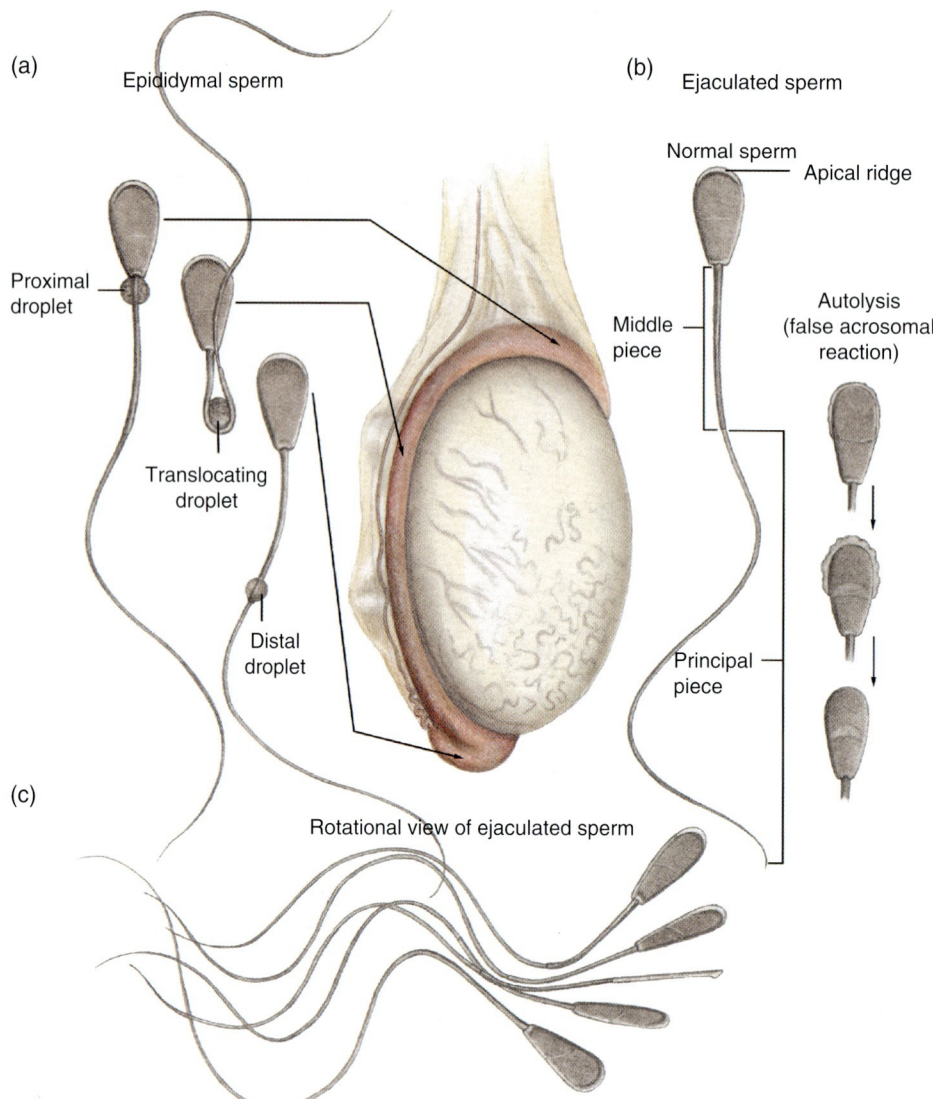

Figure 2.4 Schematic representation of spermatozoa reflecting visible changes as they migrate from the rete testis through the caput (head), corpus (body), and cauda (tail) regions of the epididymis to the ejaculate. Sperm found in the rete testis, efferent ducts, and caput epididymis are identified by the proximal nature of the protoplasmic droplet, while in sperm in the distal caput through to the corpus epididymis the protoplasmic droplet has translocated to the bend of the flagellum; by the time the spermatozoa reach the cauda epididymis the droplet has located at the distal portion of the middle piece of the flagellum (a). Schematic representations of ejaculated normal and autolysis spermatozoa are shown in (b). Normal sperm show an intact acrosome in the form of the apical ridge along the apical edge of the spermatozoan head, while the spermatozoan head images to the right show examples of false or premature acrosomal reactions typically associated with dead or dying spermatozoa. The paddle-shaped nature of the bovine spermatozoon is shown in (c). Images reproduced with permission from K. June Mullins and Richard G. Saacke *Illustrated Anatomy of the Bovine Male and Female Reproductive Tract*. Germinal Dimensions Inc., Cadmus Professional Communications, Science Press Division, Ephrata, PA, 2003.

Seminiferous epithelial cycle and wave of spermatogenesis

The cycle of the seminiferous epithelium refers to a cohort of germ cells that progress by clonal expansion within a segment of the seminiferous tubule and change in synchrony with other layers over time through a series of well-characterized cellular associations referred to as stages.[35,130] Cross-sections along the length of the seminiferous tubules will have different morphological appearances, but within a given section the cellular activity will be similar. Successive sections of the seminiferous tubules will have cohorts of germ cells at sequential stages of development with repetition, thus producing a "wave" effect along the seminiferous tubules.[35] Germ cell development progresses from the basal lamina of the seminiferous tubule, and cells migrate toward the tubule lumen and not in a lateral direction. This synchronous behavior is thought to be associated with biochemical signaling facilitated by the intercellular bridges, but the process is not fully understood. In the bull, it takes about 4.5 cycles of the seminiferous epithelium to complete spermatogenesis, and each cycle is estimated to take approximately 13.5 days, thus requiring 61 days to complete the process.[132,134] Spermatogenesis continues throughout the natural lifespan of the postpubertal bull. The domestic bull attains puberty at approximately 10–12 months of age when mature sperm begin to appear in the ejaculate. It is estimated that the daily sperm output of the dairy bull is approximately 8 billion and approximately 7 billion in the beef bull, with the level of production of spermatozoa increasing from puberty to maximum potential at approximately 3 years of age. The productivity potential of a bull is determined by the size of the testes, which may be breed dependent, and is thought to be related to being a component of Sertoli cell volume within the seminiferous tubules. Hence, breeding soundness examinations of bulls to be used as sires is dependent not only on sperm quality but also on testicular size and volume.[135]

References

1. Pritchard J. *Ancient Near Eastern Texts (relating to the Old Testament)*, 2nd edn. Princeton: Princeton University Press, 1955, pp. 159–222.
2. Steinach E. *Sex and Life: Forty Years of Biological and Medical Experiments*. New York: Viking Press, 1940.
3. Bremner W. Historical aspects of the study of the testis. In: Burger HG, de Kretser DM (eds) *The Testis*. New York: Raven Press, 1981, pp. 1–8.
4. de Graaf R. Treatise on the human reproductive organs (translated by Jocelyn and Setchell). *J Reprod Fertil Suppl* 1972;17:1–64.
5. Eik-Nes K. Biosynthesis and secretion of testicular steroids. In: Hamilton DW, Greep RO (eds) *Handbook of Physiology*, Section 7, Vol. V. Washington, DC: American Physiological Society, 1975, pp. 467–483.
6. Leydig F. Zur anatomie der manlichen geslechtsargane und analdrusen de saugethiere. *Z Wiss Zool* 1850;2:1.
7. Sertoli E. De l'esistenza di particulari cellulae ramificatai new canalicoli seminiferi del l'testicolo umano. *Morgagni* 1865;7:31–40.
8. Christensen A. Leydig cells. In: Hamilton DW, Greep RO (eds) *Handbook of Physiology*, Section 7, Vol. V. Washington, DC: American Physiological Society, 1975, pp. 57–94.
9. Fawcett D. Ultrastructure and function of the Sertoli cell. In: Hamilton DW, Greep RO (eds) *Handbook of Physiology*, Section 7, Vol. V. Washington, DC: American Physiological Society, 1975, pp. 21–55.
10. Parkes A. The internal secretions of the testis. In: Parkes A (ed.) *Marshall's Physiology of Reproduction*, Vol. III. London: Longmans, 1966, pp. 412–569.
11. David K, Dingemanse E, Freud J, Laque F. Uber krystellinishces manliches hormone aus hoden (testosteron), wirksamer als aus harn oder uas cholesterin bereitestes androsteron. *Z Physiol Chem* 1935;233:281.
12. Smith P. Hypophysectomy and replacement therapy in the rat. *Am J Anat* 1930;45:205–273.
13. Greep R, Fevold H, Hisaw F. Effects of two hypophyseal hormones on the reproductive system of the mare rat. *Anat Rec* 1936;65:261–271.
14. Greep R, Fevold H. The spermatogenic and secretory functions of the gonads of hypophysectomized adult rats tested with pituitary FSH and LH. *Endocrinology* 1937;21:611–618.
15. Hooker C. The postnatal history and function of the interstitial cell of the testis of the bull. *Am J Anat* 1994;74:1–37.
16. Steinberger A, Walther J, Heindel K, Sanbourn B, Tsai Y, Steinberger E. Hormone interactions in Sertoli cells. *In Vitro* 1979;15:23–31.
17. de Kretser D, Kerr J, Rich K, Risbridger F, Dobos M. Hormonal factors involved in normal spermatogenesis and following the disruption of spermatogenesis. In: Steinberger A, Steinberger S (eds) *Testicular Development, Structure and Function*. New York: Raven Press, 1980, pp. 107–115.
18. Bearden H, Fuquay J, Willard S. The male reproductive system. In: *Applied Animal Reproduction*, 6th edn. Upper Saddle River, NJ: Pearson Prentice Hall, 2004, pp. 22–35.
19. Setchell B, Maddocks S, Brooks D. Anatomy, vasculature, innervation and fluids of the male reproductive tract. In: Knobil E, Neill JD (eds) *Physiology of Reproduction*, Vol. I. New York: Raven Press, 1993, pp. 1063–1176.
20. Bardin C, Cheng C, Mustow N, Gunsalus G. The Sertoli cell. In: Knobil E, Neill JD (eds) *Physiology of Reproduction*, Vol. I. New York: Raven Press, 1993, pp. 1291–1336.
21. Senger P. The organization and function of the male reproductive system. In: *Pathways to Pregnancy and Parturition*, 2nd edn. Redmond, OR: Current Conceptions Inc., 2003, pp. 44–79.
22. Wyrost P, Radek J, Radek T. Morphology and development of the bovine testicular artery during fetal and neonatal periods [in Polish]. *Pol Arch Weter* 1990;30:39–56.
23. Kastelic J, Cook R, Coulter G. Scrotal/testicular thermoregulation and the effects of increased testicular temperature in the bull. *Vet Clin North Am: Food Anim Pract* 1997;13:271–282.
24. Fawcett D, Long J, Jones A. The ultrastructure of the endocrine glands. *Recent Prog Horm Res* 1969;25:314–368.
25. Hooker C. The intertubular tissue of the testis. In: Johnson AD, Gomes WR, VanDemark NL (eds) *The Testis*, Vol. I. New York: Academic Press, 1970, pp. 483–550.
26. Christensen A. A history of the Leydig cell. In: Payne AH, Hardy MP (eds) *The Leydig Cell in Health and Disease*. Totowa, NJ: Humana Press, 2007, pp. 3–30.
27. Stewart D, Raeside J. Testosterone secretion by the early fetal pig testes in organ culture. *Biol Reprod* 1976;15:25–28.
28. Attal J. Levels of testosterone, androstenedione, estrone and estradiol-17β in testes of fetal sheep. *Endocrinology* 1969;85:280–289.
29. Fawcett D, Neaves W, Flores M. Observations on intertubular lymphatics and the organization of the interstitial tissue of the mammalian testis. *Biol Reprod* 1973;9:500–532.
30. Mori H, Christensen A. Morphometric analysis of Leydig cells in the normal rat testis. *J Cell Biol* 1980;84:340–354.
31. Belt W, Cavazos L. Fine structure of the interstitial cells of Leydig in the boar. *Anat Rec* 1967;158:335–350.
32. Booth W. Testicular steroids and boar taint. In: Cole DJA, Foxcroft GR (eds) *Control of Pig Reproduction*. London: Butterworth Scientific, 1982, pp. 25–44.
33. de Kretser D, Kerr J. The cytology of the testis. In: Knobil E, Neill JD (eds) *Physiology of Reproduction*, Vol. I. New York: Raven Press, 1993, pp. 1177–1290.
34. Bearden H, Fuquay J, Willard S. Neuroendocrine and endocrine regulation of reproduction. In: *Applied Animal Reproduction*, 6th edn. Upper Saddle River, NJ: Pearson Prentice Hall, 2004, pp. 36–60.
35. Hess R, de Franco L. Spermatogenesis and cycle of the seminiferous epithelium. *Adv Exp Med Biol* 2008;636:1–15.
36. Tucker H. Seasonality in cattle. *Theriogenology* 1982;17:53–59.
37. Stumpf T, Wolfe M, Roberson M, Kittok R, Kinder J. Season of the year influences concentration and pattern of gonadotropins and testosterone in circulation of the bovine male. *Biol Reprod* 1993;49:1089–1095.
38. Godfrey R, Lunstra D, Jenkins T et al. Effect of season and location on semen quality and serum concentrations of luteinizing hormone and testosterone in Brahman and Hereford bulls. *J Anim Sci* 1990;68:734–749.
39. Godfrey R, Lunstra D, Jenkins T et al. Effect of location and season on body and testicular growth in Brahman and Hereford bulls. *J Anim Sci* 1990;68:1520–1529.
40. Jiménez-Severiano H, Quintal-Franco J, Vega-Murillo V et al. Season of the year influences testosterone secretion in bulls administered luteinizing hormone. *J Anim Sci* 2003;81:1023–1029.
41. Dufau H, Hsueh A, Cigorraga D, Baukal A, Catt K. Inhibition of Leydig cell function through hormonal regulatory mechanisms. *Int J Androl* 1978;1(Suppl. s2a):193–239.

42. Bartke A, Hafies A, Bex F, Dalterio S. Hormonal interactions in regulation of androgen secretion. *Biol Reprod* 1978;18:44–54.
43. Catt K, Dufau M. Basic concept of the mechanism of action of peptide hormones. *Biol Reprod* 1976;14:1–15.
44. Dufau M, Veldhuis J, Fraioli F, Johnson M, Catt K. Mode of bioactive LH secretion in man. *J Clin Endcocrinol Metab* 1983;57:993–1003.
45. Hall P. Testicular steroid synthesis: organization and regulation. In: Knobil E, Neill JD (eds) *Physiology of Reproduction*, Vol. I. New York: Raven Press, 1993, pp. 1335–1362.
46. Purvis K, Hanson V. Hormonal regulation of Leydig cell function. *Mol Cell Endocrinol* 1978;12:123–138.
47. Fritz I. Sites of action of androgens and follicle stimulating hormone on cells of the seminiferous tubules. In: Litwack G (ed.) *Biochemical Actions of Hormones*, Vol. V. New York: Academic Press, 1978, pp. 367–382.
48. Schanbacher B. Hormonal interrelationships between hypothalamus, pituitary and testis in rams and bull. *J Anim Sci* 1982;55(Suppl. 2):56–67.
49. Rubin R, Poland R, Tower B. Prolactin-related testosterone secretion in normal adult men. *J Clin Endocrinol Metab* 1978;42:112–116.
50. Risbridger G, Hodgson Y, de Krestser D. Mechanisms of action of gonadotrophins on the testis. In: Burger HG, de Kretser DM (eds) *The Testis*. New York: Raven Press, 1981, pp. 195–211.
51. Aoki J, Fawcett D. Is there a local feedback from the seminiferous tubules affecting activity of Leydig cells? *Biol Reprod* 1978;19:144–158.
52. van der Mollen H, Rommerts F. Testicular steroidogenesis. In: Burger HG, de Kretser DM (eds) *The Testis*. New York: Raven Press, 1981, pp. 213–238.
53. Juniewicz P, Johnson B. Influence of adrenal steroids upon testosterone secretion by the boar testis. *Biol Reprod* 1981;20:409–422.
54. Liptrap R, Raeside J. Increase in plasma testosterone concentration after injection of adrenocorticotrophin into the boar. *J Endocrinol* 1975;66:123–131.
55. Tilbrook A, Turner A, Clarke I. Effects of stress on reproduction in non-rodent mammals: the role of glucocorticoids and sex differences. *Rev Reprod* 2000;5:105–113.
56. Moberg G. Influence of the adrenal axis upon the gonads. *Oxford Rev Reprod Biol* 1987;9:456–496.
57. Moberg G. How behavioral stress disrupts the endocrine control of reproduction in domestic animals. *J Dairy Sci* 1991;74:304–311.
58. Liptrap R, Raeside J. Effect of cortisol on the response to gonadotrophin releasing hormone in the boar. *J Endocrinol* 1983;97:75–81.
59. Matteri R, Watson J, Moberg G. Stress or acute adrenocorticotrophin treatment suppresses LHRH-induced LH release in the ram. *J Reprod Fertil* 1984;72:385–393.
60. Barnes M, Kazmer G, Birrenkott G, Grimes L. Induced gonadotropin release in adrenocorticotropin-treated bulls and steers. *J Anim Sci* 1983;56:155–161.
61. O'Connor M, Gwazdauskas F, McGilliard M, Saacke R. Effect of adrenocorticotropic hormone and associated hormonal responses on semen quality and sperm output of bulls. *J Dairy Sci* 1985;68:151–157.
62. Svechnikov K, Sultana T, Söder O. Age-dependent stimulation of Leydig cell steroidogenesis by interleukin-1 isoforms. *Mol Cell Endocrinol* 2001;182:193–201.
63. Amjad AI, Söder O, Sultana T. Role of testicular interleukin-1alpha tIL-1alpha in testicular physiology and disease. *J Coll Physicians Surg Pak* 2006;16:55–60.
64. Yoshioka S, Acosta T, Okuda K. Roles of cytokines and progesterone in the regulation of the nitric oxide generating system in bovine luteal endothelial cells. *Mol Reprod Dev* 2012;79:689–696.
65. Bouin P, Ancel P. Recherches sur les celles interstitielles du testicles des mammiferes. *Arch Zool Exptl Gen* 1903;1:437–523.
66. Berthold AA. Transplantation der hoden. *Arch Anat Physiol Wiss Med* 1849;16:42.
67. Butendant A. Uber die chemie der sexual hormon. *Z Angew Chem* 1932;45:655.
68. Ewing L, Davis J, Zirkin B. Regulation of testicular function: a spatial and temporal view. In: Greep RO (ed.) *Reproduction of Physiology III*. International Review of Physiology, Vol. XXII. Baltimore: University Park Press, 1980, pp. 41–115.
69. Ewing L, Brown S. Testicular steroidogenesis. In: Johnson AD, Gomes WR (eds) *The Testis*, Vol. IV. New York: Academic Press, 1977, pp. 239–287.
70. Prelog V, Ruzicka L. F. Untersuchungen uber organextrakte uber zwei moschsartig riechende steroide aus schweintetesextrakten. *Helv Chim Acta* 1944;27:61.
71. Steinberger E. Hormonal control of mammalian spermatogenesis. *Physiol Rev* 1971;51:1–22.
72. Orgenbin-Critz M. The influence of testicular function on related reproductive organs. In: Burger HG, de Kretser DM (eds) *The Testis*. New York: Raven Press, 1981, pp. 239–254.
73. Griffin J. Male reproductive function. In: Griffin JE, Ojeda SR (eds) *Textbook of Endocrine Physiology*. New York: Oxford University Press, 2000, pp. 243–264.
74. Ewing L, Zirkin B, Chubb C. Assessment of testicular testosterone production and Leydig cell structure. *Environ Health Perspect* 1981;38:19–27.
75. Velle W. Urinary oestrogens in the male. *J Reprod Fertil* 1966;12:65–73.
76. Joshi H, Raeside J. Synergistic effects of testosterone and oestrogens on accessory sex glands and sexual behavior of the boar. *J Reprod Fertil* 1973;33:411–423.
77. Booth W. Changes with age in the occurrence of C19 steroids in the testis and submaxillary gland of the boar. *J Reprod Fertil* 1975;42:459–472.
78. Setchell B. Endocrinology of the testis. In: *The Mammalian Testes*. New York: Cornell University Press, 1978, pp. 109–180.
79. Ritzen E, Hansson V, Frenchn F. The Sertoli cell. In: Burger HG, de Kretser DM (eds) *The Testis*. New York: Raven Press, 1981, pp. 171–194.
80. Ivell R, Balvers M, Rust W, Bathgate R, Einspanier A. Oxytocin and male reproductive function. *Adv Exp Med Biol* 1997;424:253–264.
81. Wathes D. Possible actions of gonadal oxytocin and vasopressin. *J Reprod Fertil* 1984;71:315–345.
82. Wathes D. Oxytocin and vasopressin in the gonads. *Oxford Rev Reprod Biol* 1989;11:87–99.
83. Nicholson H. Oxytocin: a paracrine regulator of prostatic function. *Rev Reprod* 1996;1:69–72.
84. Ang H, Ungofroren H, De Bree F et al. Testicular oxytocin gene expression in seminiferous tubules of cattle and transgenic mice. *Endocrinology* 1991;128:2110–2117.
85. Ungefroren H, Davidoff M, Ivell R. Post transcriptional block in oxytocin gene expression within the seminiferous tubules of the bovine testis. *J Endocrinol* 1994;140:63–72.
86. Ivell R, Bathgate R. Insulin-like peptide 3 in Leydig cells. In: Payne AH, Hardy MP (eds) *The Leydig Cell in Health and Disease*. Totowa, NJ: Humana Press, 2002, pp. 279–290.
87. Ivell R, Anand-Ivell R. The biology of insulin-like factor 3 (INSL3) in human reproduction. *Human Reproduction Update* 2009;15:463–476.
88. Adham I, Burkhardt E, Banahmed M, Engel W. Cloning of a cDNA for a novel insulin-like peptide of the testicular Leydig cells. *J Biol Chem* 1993;268:26668–26672.
89. Ivell R, Bathgate RA. Reproductive biology of the relaxin-like factor (RLF/INSL3). *Biol Reprod* 2002;67:699–705.
90. Anand-Ivell R, Heng K, Hafen B, Setchell B, Ivell R. Dynamics of INSL3 peptide expression in the rodent testis. *Biol Reprod* 2009;81:480–487.
91. Bay K, Hartung S, Ivell R et al. Insulinlike factor 3 serum levels in 135 normal men and 85 men with testicular disorders: relationship to the luteinizing hormone–testosterone axis. *J Clin Endocrinol Metab* 2005;90:3410–3418.
92. Anand-Ivell R, Wohlgemuth J, Haren M et al. Peripheral INSL3 concentrations decline with age in a large population of Australian men. *Int J Androl* 2006;29:618–626.

93. Bullesbach E, Schwabe C. The primary structure and disulfide links of the bovine relaxin-like factor (RLF). *Biochemistry* 2002;41:274–281.
94. Anand-Ivell R, Ivell R, Driscoll D, Manson J. Insulin-like factor 3 levels in amniotic fluid from human male fetuses. *Hum Reprod* 2008;23:1180–1186.
95. Anand-Ivell R, Hiendleder S, Vinoles C et al. INSL3 in the ruminant: a powerful indicator of gender- and genetic-specific feto-maternal dialogue. *PLoS ONE* 2011;6(5):e19821.
96. Ivell R, Wade J, Anand-Ivell R. INSL3 as a biomarker of Leydig cell functionality. *Biol Reprod* 2013;88:1–8.
97. Feugang J, Rodriguez-Munoz J, Willard S, Bathgate R, Ryan P. Examination of relaxin and its receptors expression in pig gametes and embryos. *Reprod Biol Endocrinol* 2011;9:10.
98. Feugang J, Rodríguez-Munoz J, Willard S, Ryan P. Effects of relaxin on motility characteristics of boar spermatozoa during storage. In: Sixth International Symposium on Relaxin and Related Peptides 2012.
99. Feugang J, Youngblood R, Greene J et al. Application of quantum dot nanoparticles for potential non-invasive bio-imaging of mammalian spermatozoa. *J Nanobiotechnol* 2012;10:45.
100. Behringer R. The in vivo roles of Müllerian-inhibiting substance. *Curr Top Dev Biol* 1994;29:171–187.
101. Cate R, Mattaliano R, Hession C et al. Isolation of the bovine and human genes for Müllerian inhibiting substance and expression of the human gene in animal cells. *Cell* 1986;45:685–698.
102. Belville C, Van Vlijmen H, Ehrenfels C et al. Mutations of the anti-Müllerian hormone gene in patients with persistent mullerian duct syndrome: biosynthesis, secretion, and processing of the abnormal proteins and analysis using a three-dimensional model. *Mol Endocrinol* 2004;18:708–21.
103. Kitahara G, El-Sheikh Ali H, Sato T et al. Anti-Müllerian hormone (AMH) profiles as a novel biomarker to evaluate the existence of a functional cryptorchid testis in Japanese Black calves. *J Reprod Dev* 2012;58:310–315.
104. Ying S. Inhibins and activins: chemical properties and biological activity. *Proc Soc Exp Biol Med* 1987;186:253–264.
105. Robertson D, Burger H, Fuller P. Inhibin/activin and ovarian cancer. *Endocrinol Related Cancer* 2004;11:35–49.
106. Suresh P, Rajan T, Tsutsumi R. New targets for old hormones: inhibins clinical role revisited. *Endocrinol J* 2011;58:223–235.
107. Phillips D. Activins, inhibins and follistatins in the large domestic species. *Domest Anim Endocrinol* 2005;28:1–16.
108. Kaneko H, Noguchi J, Kikuchi K, Hasegawa Y. Molecular weight forms of inhibin A and inhibin B in the bovine testis change with age. *Biol Reprod* 2003;68:1918–1925.
109. Kaneko H, Matsuzaki M, Noguchi J, Kikuchi K, Ohnuma K, Ozawa M. Changes in circulating and testicular levels of inhibin A and B during postnatal development in bulls. *J Reprod Dev* 2006;52:741–749.
110. Fortes M, Reverter A, Hawken R, Bolormaa S, Lehnert S. Candidate genes associated with testicular development, sperm quality, and hormone levels of inhibin, luteinizing hormone, and insulin-like growth factor 1 in Brahman bulls. *Biol Reprod* 2012;87:58.
111. Ginther O, Beg M, Bergfelt D, Kot K. Activin A, estradiol, and free insulin-like growth factor I in follicular fluid preceding the experimental assumption of follicle dominance in cattle. *Biol Reprod* 2002;67:14–19.
112. Beg M, Ginther O. Follicle selection in cattle and horses: role of intrafollicular factors. *Reproduction* 2006;132:365–377.
113. Xiong X, Wang A, Liub G et al. Effects of p,p'-dichlorodiphenyldichloroethylene on the expressions of transferrin and androgen-binding protein in rat Sertoli cells. *Environ Res* 2006;101:334–339.
114. Griswold M. Protein secretion by Sertoli cells: general considerations. In: Russell LD, Griswold MD (eds) *The Sertoli Cell*. Clearwater, FL: Cache River Press, 1993, pp. 195–200.
115. Kelce W, Stone C, Laws S, Gray L, Kemppainen J, Wilson E. Persistent DDT metabolite p,p'-DDE is a potent androgen receptor antagonist. *Nature* 1995;375:581–585.
116. Griswold M. Protein secretions of Sertoli cells. *Int Rev Cytol* 1988;110:133–156.
117. Tapanainen J, Aittomaki K, Huhtaniemi LT. New insights into the role of follicle-stimulating hormone in reproduction. *Ann Med* 1997;29:265–266.
118. Fritz I, Rommerts F, Louis B, Dorrington J. Regulation by FSH and dibutyryl cyclic AMP of the formation of androgen-binding protein in Sertoli cell-enriched cultures. *J Reprod Fertil* 1976;46:17–24.
119. Suire S, Fontaine I, Guillou F. Follicle stimulating hormone (FSH) stimulates transferrin gene transcription in rat Sertoli cells: cis and trans-acting elements involved in FSH action via cyclic 3′,5′-monophosphate on the transferrin gene. *Mol Endocrinol* 1995;9:756–766.
120. Schteingart H, Meroni S, Pellizzari E, Perez A, Cigorraga S. Regulation of Sertoli cell aromatase activity by cell density and prolonged stimulation with FSH, EGF, insulin and IGF-1 at different moments of pubertal development. *J Steroid Biochem Mol Biol* 1995;52:375–381.
121. Skinner M, Griswold M. Secretion of testicular transferrin by cultured Sertoli cells is regulated by hormones and retinoids. *Biol Reprod* 1982;27:211–221.
122. Lin L, Doherty D, Lile J, Bektesh S, Collins F. GDNF: a glial cell line-derived neurotrophic factor for midbrain dopaminergic neurons. *Science* 1993;260:1130–1132.
123. Liu T, Yu B, Luo F et al. Gene expression profiling of rat testis development during the early post-natal stages. *Reprod Domest Anim* 2012;47:724–731.
124. Johnston D, Olivas E, DiCandeloro P, Wright W. Stage-specific changes in GDNF expression by rat Sertoli cells: a possible regulator of the replication and differentiation of stem spermatogonia. *Biol Reprod* 2011;85:763–769.
125. Aponte P, Soda T, van de Kant H, de Rooij D. Basic features of bovine spermatogonial culture and effects of glial cell line-derived neurotrophic factor. *Theriogenology* 2006;65:1828–1847.
126. Harikae K, Tsunekawa N, Hiramatsu R, Toda S, Kurohmaru M, Kanai Y. Evidence for almost complete sex-reversal in bovine freemartin gonads: formation of seminiferous tubule-like structures and transdifferentiation into typical testicular cell types. *J Reprod Dev* 2012;58:654–660.
127. Russell L, Griswold M (eds). *The Sertoli Cell*. Clearwater, FL: Cache River Press, 1993.
128. Kramer M, de Lange A, Visser M. Spermatogonia in the bull. *Z Zellforschung* 1964;63:735–758.
129. Krallinger H. Cytologische studien an einigen haussäugetieren. *Arch Tierernahr Tierz* 1931;5:127–187.
130. Parks J, Lee D, Huang S, Kaproth M. Prospects for spermatogenesis in vitro. *Theriogenology* 2003;59:73–86.
131. Senger P. Endocrinology of the male and spermatogenesis. In: *Pathways to Pregnancy and Parturition*, 2nd revised edn. Redmond, OR: Current Conceptions Inc., 2005, pp. 214–239.
132. Amann R. Reproductive capacity of dairy bulls. IV. Spermatogenesis and testicular germ cell degeneration. *Am J Anat* 1962;110:69–78.
133. Berndtson W, Desjardins C. The cycle of the seminiferous epithelium and spermatogenesis in the bovine testis. *Am J Anat* 1974;140:167–180.
134. Johnson L, Wilker C, Cerelli J. Spermatogenesis in the bull. In: *Proceedings of the Fifteenth Technical Conference on AI and Reproduction*. Columbia, MO: National Association of Animal Breeders, 1994, pp. 9–27.
135. Johnson L, Varner D, Roberts M, Smith T, Keillor G, Scrutchfield W. Efficiency of spermatogenesis: a comparative approach. *Anim Reprod Sci* 2000;60–61:471–480.

Chapter 3

Thermoregulation of the Testes

John P. Kastelic

Department of Production Animal Health, Faculty of Veterinary Medicine, University of Calgary, Calgary, Alberta, Canada

Introduction

Since one bull may be responsible for breeding 20 (natural service) to thousands (artificial insemination) of females, bull fertility is critically important. Although sterile bulls (total inability to reproduce) are uncommon, there can be a wide range in bull fertility, particularly in the absence of selection for fertility.[1] It is well established that a bull's testes must be 2–6°C cooler than core body temperature for fertile sperm to be produced; consequently, increased testicular temperature, regardless of cause, reduces semen quality.[2] Although the underlying cause of infertility in bulls is frequently unknown, I speculate that it is often increased testicular temperature.

Anatomy and physiology

Regulation of testicular temperature is dependent on several features. Scrotal skin is typically thin, with minimal hair and an extensive subcutaneous vasculature, facilitating heat loss by radiation.[3] The scrotal neck is the warmest part of the scrotum; a long distinct scrotal neck (and pendulous scrotum) reduces testicular temperature by increasing the area for radiation and enabling the testes to move away from the body. The tunica dartos, a thin sheet of smooth muscle under the scrotal skin, is controlled by sympathetic nerves and contracts and relaxes in cold and warm environments, respectively.[4] The cremaster muscle also contracts to draw the testes closer to the body under cold ambient conditions.[4] Dorsal to the testis is the testicular vascular cone,[5] comprising the highly coiled testicular artery surrounded by the pampiniform plexus, a complex venous network. The testicular vascular cone functions as a classic countercurrent heat exchanger, transferring heat from the artery to the vein, contributing to testicular cooling. Characteristics of scrotal surface temperatures and the testicular vascular cone in bulls aged 0.5–3 years have been reported.[6] In a recent study in bulls,[7] testicular vascular cone diameter increased with age; furthermore, increased testicular vascular cone diameter and a decreased distance between arterial and venous blood in this structure were associated with increased percentage of normal sperm and fewer sperm with defects.[7] In a comparative study of semen quality and scrotal/testicular thermoregulation in *Bos indicus*, *Bos taurus*, and *B. indicus/B. taurus* crossbred bulls, there were significant differences among these three genotypes in the vascular arrangement, characteristics of the testicular artery (e.g., wall thickness), and thickness of the tunica albuginea; overall, *B. indicus* bulls had the best thermoregulatory capacity whereas *B. taurus* bulls had the worst, with crossbred bulls intermediate.[8]

Sweating and whole-body responses also contribute to testicular cooling. In bulls, sweat gland density is highest in the scrotal skin.[9] In rams, apocrine sweat glands in the scrotum discharge simultaneously (up to 10 times per hour) when scrotal surface temperature is about 35.5°C.[10] In these animals, respiration rate increases in association with scrotal surface temperature, reaching 200 breaths per minute when scrotal surface temperature is 38–40°C.[11]

Surface and internal temperatures

In beef bulls, average temperatures at the top, middle, and bottom were 30.4, 29.8, and 28.8°C (scrotal surface); 33.3, 33.0, and 32.9°C (scrotal subcutaneous); and 34.3, 34.3, and 34.5°C (intratesticular).[12] Therefore, top-to-bottom differences (gradients) in temperature were 1.6, 0.4, and –0.2°C for scrotal surface, scrotal subcutaneous, and intratesticular temperatures, respectively. Moving dorsal to ventral, the scrotum gets cooler, whereas the testis (independent of the scrotum) gets warmer. These temperature gradients are consistent with their vasculature, as the scrotum is vascularized from top to bottom, whereas the testis is essentially vascularized from bottom to top. In that regard, the testicular artery exits the bottom of the testicular vascular cone, courses the length of the testis (under the corpus epididymis), and at the bottom of the testis ramifies into multiple branches that spread dorsally and laterally across the surface of the testis before entering the testicular parenchyma.[13] Blood within the testicular artery was a similar temperature at the top of the testis compared with the bottom of the testis, but was significantly cooler at the point of entry into the testicular parenchyma (intra-arterial temperatures at these locations were 34.3, 33.4, and 31.7 °C, respectively[14]). Therefore, both the scrotum and the testis are warmest at the origin of their blood supply (top of scrotum and bottom of testis), but they both get cooler distal to that point (i.e., bottom of scrotum, top of testis). Remarkably, these opposing temperature gradients collectively result in a nearly uniform intratesticular temperature *in situ*.[15]

In beef bulls, internal temperatures of the caput, corpus, and cauda epididymis averaged 35.6, 34.6, and 33.1 °C (gradient, 2.5 °C).[12] That the caput was warmer than the testicular parenchyma at the top of the testis was attributed to the proximity of the caput to the testicular vascular cone. Furthermore, it was noteworthy that the cauda, critical in sperm storage and maturation, was cooler than the testicular parenchyma.

It has been well established that bulls fed moderate-energy diets after weaning have better semen quality than those fed high-energy diets. In one study, beef bulls fed a moderate- versus high-energy diet for 168 days after weaning had a larger scrotal surface temperature gradient (3.9 vs. 3.4 °C, $P<0.02$), more morphologically normal sperm (68.8% vs. 62.5%, $P<0.01$), and a higher proportion of progressively motile sperm (53.4% vs. 44.5%, $P<0.006$).[16] Perhaps increased dietary energy reduced heat loss, thereby increasing the temperature of the testes and scrotum.

Sources of testicular heat

Testicular blood flow and oxygen uptake were measured in eight Angus bulls to determine the relative importance of blood flow versus metabolism as sources of testicular heat.[17] Blood flow in the testicular artery averaged 12.4 mL/min. Arterial blood was warmer (39.2 vs. 36.9 °C, $P<0.001$) and had more hemoglobin saturated with oxygen than blood in the testicular vein (95.3 vs. 42.0%, $P<0.001$). Based on blood flow and hemoglobin saturation, the oxygen used by one testis (1.2 mL/min) was calculated to produce 5.8 calories of heat per minute, compared with 28.3 calories per minute attributed to blood flow. Therefore, the major source of testicular heat is blood flow not metabolism.

The testis usually operates on the brink of hypoxia.[4] Increased temperature increases metabolism, with a concurrent need for increased oxygen to sustain aerobic metabolism. However, in rams,[4] blood flow changes little in response to increases in testicular temperature and consequently the testes become hypoxic. Increasing blood oxygen saturation is not practical, since the blood is nearly completely saturated under normal conditions. Although increasing blood flow would increase the delivery of oxygen, it would also bring considerable additional heat into the testes. Therefore, increasing heat loss from the scrotum would appear to be the most appropriate response to increased testicular temperature.

Pathogenesis of heat-induced changes in sperm morphology

The testis operates on the brink of hypoxia under physiological conditions,[18] whereas in situations of increased scrotal/testicular temperatures, metabolism and oxygen utilization increase but blood flow to the testis remains constant, resulting in frank hypoxia.[19] To determine the relative contributions of hyperthermia and hypoxia to heat-induced changes in sperm morphology, mice were subjected to ambient temperatures of 20 and 36 °C and concurrently exposed to hyperoxic conditions to prevent the hypoxia induced by elevated testicular temperature.[20] Hyperoxia (known to increase testicular oxygen saturation) did not protect against hyperthermia-induced deterioration in sperm quality. Furthermore, since sperm characteristics were not significantly different between mice exposed to hypoxic versus normoxic or hyperoxic conditions at 20 °C, it appeared that the effects of increased testicular temperature on semen quality were due to hyperthermia per se and not hypoxia.[20] Similar results were obtained with a scrotal insulation model in rams (Kastelic *et al.*, unpublished results). Although further work is needed to verify these findings, they do call into question the long-standing dogma that the effects of increased testicular temperature on sperm morphology are due to hypoxia.

Evaluation of scrotal surface temperature with infrared thermography

Infrared thermograms of the scrotum of bulls with apparently normal scrotal thermoregulation were symmetrical left to right, with the temperature at the top 4–6 °C warmer than at the bottom.[21,22] More random temperature patterns, including a lack of horizontal symmetry and areas of increased scrotal surface temperature, were interpreted as abnormal thermoregulation of the testes or epididymides. Nearly all bulls with an abnormal thermogram had reduced semen quality;[21,22] conversely, not every bull with poor-quality semen had an abnormal thermogram. Consequently, infrared thermography is a useful tool for breeding soundness evaluation of bulls, although it does not replace collection and evaluation of semen. In one study, 30 yearling beef bulls, all deemed breeding sound on a standard breeding soundness examination, were individually exposed to about 18 heifers for 45 days.[23] Pregnancy rates 80 days after the end of the breeding season were similar (83 vs. 85%) for bulls with a normal or questionable scrotal surface temperature pattern, respectively, but were higher ($P<0.01$) than pregnancy rates for bulls with an abnormal scrotal surface temperature pattern (68%).

Effects of increased testicular temperature

Increased ambient temperature

The effect of increased ambient temperature on semen quality has been widely reported. In one study, ambient temperatures of 40 °C at a relative humidity of 35–45% for 12 hours reduced semen quality.[24] Furthermore, *Bos taurus* bulls are more susceptible than *Bos indicus* bulls to high ambient temperatures.[24] In that regard, decreases in semen quality were less severe, occurred later, and recovered more rapidly in crossbred (*Bos indicus* × *Bos taurus*) bulls than in purebred *Bos taurus* bulls exposed to high ambient temperatures.[25]

Scrotal insulation as a model of increased testicular temperature

Scrotal insulation is frequently used to increase testicular temperature. In one study,[26] the scrotum of *Bos indicus* × *Bos taurus* bulls was insulated for 48 hours. The nature and time (day 0, start of insulation) of the morphologically abnormal sperm that resulted were as follows: decapitated, days 6–14; abnormal acrosomes, days 12–23; abnormal tails, days 12–23; and protoplasmic droplets, days 17–23. Therefore, scrotal heating affected sperm in the caput epididymis as well as spermatids. Although daily sperm production was not affected, epididymal sperm reserves were reduced by nearly 50% (9.2 vs. 17.4 billion), particularly in the caput (3.8 vs. 6.6 billion) and cauda (3.7 vs. 9.5 billion), perhaps due to selective resorption of abnormal sperm in the rete testis and excurrent ducts. In another study,[27,28] the scrota of six Holstein bulls were insulated for 48 hours (day 0, initiation of insulation). The number of sperm collected was not significantly different, but the proportion of progressively motile sperm decreased from 69% (prior to insulation) to 42% on day 15. The proportion of normal sperm was not significantly different from day −6 to day 9 (80%), decreased abruptly on day 12 (53%), and reached a nadir on day 18 (14%). Although there was considerable variation among bulls in the type and proportion of abnormal sperm, specific abnormalities appeared in a consistent chronological sequence: tailless, days 12–15; diadem, day 18; pyriform and nuclear vacuoles, day 21; knobbed acrosome, day 27; and Dag defect, day 30. When sperm were collected 3–9 days after insulation and examined immediately, their motility and morphology were similar to pre-insulation values.[27] Compared with semen collected prior to insulation, following freezing, thawing and incubation at 37 °C for 3 hours,[27] there were significant reductions in the proportion of progressively motile sperm (46 vs. 31%, respectively) and the proportion of sperm with intact acrosomes (73 vs. 63%). Freezing plus post-thaw incubation manifested changes that had occurred in sperm that were in the epididymis at the time of scrotal insulation.

In another study,[29] scrotal insulation (4 days) and dexamethasone treatment (20 mg/day for 7 days) were used as models of testicular heating and stress, respectively. Some bulls seemed predisposed to produce sperm with a particular abnormality. Pyriform heads, nuclear vacuoles, microcephalic sperm, and abnormal DNA condensation were more common in insulated than dexamethasone-treated bulls. Conversely, dexamethasone treatment resulted in an earlier and more severe effect on epididymal sperm, an earlier and greater increase in distal midpiece reflexes, and an earlier increase in proximal and distal droplets. Overall, the types of defective sperm and the time of their detection were similar for the two treatments.

Insulation of the scrotal neck

The scrotal neck of five bulls was insulated for 7 days (days 1–8) as a model of bulls with excessive body condition (which typically have considerable fat in the scrotal neck). Sperm within the epididymis or at the acrosome phase during insulation appeared to be most affected.[30] Insulated bulls had twice as many sperm with midpiece defects and four times as many with droplets on day 5, fewer normal sperm and three times as many with midpiece defects and droplets on day 8, fewer normal sperm on days 15 and 18, and more sperm with head defects on days 18 and 21. Semen quality in insulated bulls had nearly returned to pre-insulation values by day 35. In a second experiment,[30] scrotal subcutaneous temperature increased 2.0, 1.5, and 0.5 °C at the top, middle, and bottom of the testis, respectively, and intratesticular temperature was 0.9 °C higher at the corresponding three locations 48 hours after scrotal neck insulation compared with before insulation. Clearly the scrotal neck is an important site of heat loss.

Increased epididymal temperature

In most mammals, the cauda epididymis is cooler than the testes,[31] facilitating its sperm storage function. Increasing cauda temperature disrupts absorptive and secretory functions, changes the composition (ions and proteins) of the cauda fluid, and increases (approximately threefold) the rate of sperm passage through the cauda.[31] Consequently, the number of sperm in the first ejaculate declines, with an even more dramatic decline in sperm number in successive ejaculates. In addition, increased temperature seems to hasten sperm maturation.[31]

Effects of increased temperature on testicular cells

Although heating seems to affect Sertoli and Leydig cell function, germ cells are the most sensitive.[32] All stages of spermatogenesis are susceptible, with the degree of damage related to the extent and duration of the increased temperature.[32] Spermatocytes in meiotic prophase are killed by heat, whereas sperm that are more mature usually have metabolic and structural abnormalities.[18] Heating the testis usually decreases the proportion of progressively motile and live sperm, and increases the incidence of morphologically abnormal sperm, especially those with defective heads.[33] In a recent study in bulls,[34] increased testicular temperature caused a lack of chromatin protamination and subtle changes in sperm head shape. Although there is considerable variation among bulls in the nature and proportion of defective sperm, the order of appearance of specific defects is relatively consistent.[28,29] Unless spermatogonia are affected, the interval from cessation of heating to restoration of normal sperm in the ejaculate corresponds to the interval from the beginning of differentiation to ejaculation.[30] Following scrotal insulation in bulls, blastocyst rate was more sensitive than cleavage rate (in an *in vitro* system[35]). Even though sperm morphology has returned to normal, their utilization may result in decreased fertilization rates and an increased incidence of embryonic death.[36]

Summary of increased testicular temperature

When scrotal/testicular temperature is increased (regardless of the cause), sperm morphology is generally unaffected initially (for an interval corresponding to epididymal transit time) but subsequently declines.[33] In some studies,[26,30] sperm in the epididymis at the time of scrotal heating were morphologically abnormal when collected soon after heating. In another study,[27] changes in these sperm were manifest only after they were frozen, thawed, and incubated. Sperm morphology usually returns to pretreatment values within approximately 6 weeks of the thermal insult. However, a prolonged and/or severe increase in testicular temperature will increase the interval for recovery. In general, the decrease in semen quality following increased testicular temperature is related to the severity and the duration of the thermal insult.

References

1. Cates W. Observations on scrotal circumference and its relationship to classification of bulls. In: *Proceedings of the Annual Meeting of the Society for Theriogenology, Cheyenne, WY*, 1975, pp. 1–15.
2. Waites G. Temperature regulation and the testis. In: Johnson AD, Gomes WR, VanDemark NL (eds) *The Testis*, Vol. I. New York: Academic Press, 1970, pp. 241–279.
3. Dahl E, Herrick J. A vascular mechanism of maintaining testicular temperature by counter-current exchange. *Surg Gynecol Obstet* 1959;108:697–705.
4. Setchell B. The scrotum and thermoregulation. In: Setchell B (ed.) *The Mammalian Testis*. Ithaca: Cornell University Press, 1978, pp. 90–108.
5. Hees H, Leiser R, Kohler T, Wrobel K. Vascular morphology of the bovine spermatic cord and testis. Light and scanning electron microscopic studies on the testicular artery and pampiniform plexus. *Cell Tissue Res* 1984;237:31–38.
6. Cook R, Coulter G, Kastelic J. The testicular vascular cone, scrotal thermoregulation, and their relationship to sperm production and seminal quality in beef bulls. *Theriogenology* 1994;41:653–671.
7. Brito L, Barth A, Wilde R, Kastelic J. Testicular vascular cone development and its association with scrotal temperature, semen quality, and sperm production in beef bulls. *Anim Reprod Sci* 2012;134:135–140.
8. Brito L, Silva A, Barbosa R, Kastelic J. Testicular thermoregulation in *Bos indicus*, crossbred and *Bos taurus* bulls: relationship with scrotal, testicular vascular cone and testicular morphology, and effects on semen quality and sperm production. *Theriogenology* 2004;61:511–528.
9. Blazquez N, Mallard G, Wedd S. Sweat glands of the scrotum of the bull. *J Reprod Fertil* 1988;83:673–677.
10. Waites G, Voglmayr J. The functional activity and control of the apocrine sweat glands of the scrotum of the ram. *Aust J Agric Res* 1963;14:839–851.
11. Waites G. The effect of heating the scrotum of the ram on respiration and body temperature. *Q J Exp Physiol* 1962;47:314–323.
12. Kastelic J, Coulter G, Cook R. Scrotal surface, subcutaneous, intratesticular, and intraepididymal temperatures in bulls. *Theriogenology* 1995;44:147–152.
13. Gunn S, Gould T. Vasculature of the testes and adnexa. In: Greep RO (ed.) *Handbook of Physiology*, Section 7, Vol. 5. Washington, DC: American Physiological Society, 1975, pp. 117–142.
14. Kastelic J, Cook R, Coulter G. Contribution of the scrotum, testes, and testicular artery to scrotal/testicular thermoregulation in bulls at two ambient temperatures. *Anim Reprod Sci* 1997;45:255–261.
15. Kastelic J, Cook R, Coulter G. Contribution of the scrotum and testes to scrotal and testicular thermoregulation in bulls and rams. *J Reprod Fertil* 1996;108:81–85.
16. Coulter G, Cook R, Kastelic J. Effects of dietary energy on scrotal surface temperature, seminal quality, and sperm production in young beef bulls. *J Anim Sci* 1997;75:1048–1052.
17. Barros C, Oba E, Brito L et al. Testicular blood flow and oxygen evaluation in Aberdeen Angus bulls. *Rev Bras Reprod Anim* 1999;23:218–220.
18. Setchell B, Voglmayr J, Hinks N. The effect of local heating on the flow and composition of rete testis fluid in the conscious ram. *J Reprod Fertil* 1971;24:81–89.
19. Setchell B. The Parkes Lecture. Heat and the testis. *J Reprod Fertil* 1998;114;179–194.
20. Kastelic J, Wilde R, Bielli A, Genovese P, Bilodeau-Goeseels S, Thundathil J. Hyperthermia is more important than hypoxia as a cause of disrupted spermatogenesis. *Reprod Domest Anim* 2008;43(Suppl. 3):166.
21. Coulter G. Thermography of bull testes. In: *Proceedings of the 12th Technical Conference on Artificial Insemination and Reproduction*. Columbia, MO: National Association of Animal Breeders, 1988, pp. 58–63.
22. Purohit R, Hudson R, Riddell M, Carson R, Wolfe D, Walker D. Thermography of the bovine scrotum. *Am J Vet Res* 1985; 46:2388–2392.
23. Lunstra D, Coulter, G. Relationship between scrotal infrared temperature patterns and natural-mating fertility in beef bulls. *J Anim Sci* 1997;75:767–774.
24. Skinner J, Louw G. Heat stress and spermatogenesis in *Bos indicus* and *Bos taurus* cattle. *J Appl Physiol* 1966;21:1784–1790.
25. Johnston J, Naelapaa H, Frye J. Physiological responses of Holstein, Brown Swiss and Red Sindhi crossbred bulls exposed to high temperatures and humidities. *J Anim Sci* 1963;22:432–436.
26. Wildeus S, Entwistle K. Spermiogram and sperm reserves in hybrid *Bos indicus* × *Bos taurus* bulls after scrotal insulation. *J Reprod Fertil* 1983;69:711–716.
27. Vogler C, Saacke R, Bame J, DeJarnette J, McGilliard M. Effects of scrotal insulation on viability characteristics of cryopreserved bovine semen. *J Dairy Sci* 1991;74:3827–3835.
28. Vogler C, Bame J, DeJarnette J, McGilliard M, Saacke R. Effects of elevated testicular temperature on morphology characteristics of ejaculated spermatozoa in the bovine. *Theriogenology* 1993;40:1207–1219.
29. Barth A, Bowman P. The sequential appearance of sperm abnormalities after scrotal insulation or dexamethasone treatment in bulls. *Can Vet J* 1994;35:93–102.
30. Kastelic J, Cook R, Coulter G, Saacke R. Insulating the scrotal neck affects semen quality and scrotal/testicular temperatures in the bull. *Theriogenology* 1996;45:935–942.
31. Bedford J. Effects of elevated temperature on the epididymis and testis: experimental studies. In: Zorgniotti AW (ed.) *Temperature and Environmental Effects on the Testis*. New York: Plenum Press, 1991, pp. 19–32.
32. Waites G, Setchell B. Physiology of the mammalian testis. In: Lamming CE (ed.) *Marshall's Physiology of Reproduction*, 4th edn, Vol. 2. Edinburgh: Churchill Livingstone, 1990, pp. 1–105.
33. Barth A, Oko R. *Abnormal Morphology of Bovine Spermatozoa*. Ames, IA: Iowa State University Press, 1989, p. 139.
34. Rahman M, Vandaele L, Rijsselaere T et al. Scrotal insulation and its relationship to abnormal morphology, chromatin protamination and nuclear shape of spermatozoa in Holstein-Friesian and Belgian Blue bulls. *Theriogenology* 2011;76:1246–1257.
35. Fernandes C, Dode M, Pereira D, Silva A. Effects of scrotal insulation in Nellore bulls (*Bos taurus indicus*) on seminal quality and its relationship with in vitro fertilizing ability. *Theriogenology* 2008;70:1560–1568.
36. Burfening P, Ulberg L. Embryonic survival subsequent to culture of rabbit spermatozoa at 38° and 40°. *J Reprod Fertil* 1968;15:87–92.

Chapter 4

Endocrine Control of Testicular Development and Initiation of Spermatogenesis in Bulls

Leonardo F.C. Brito

ABS Global, DeForest, Wisconsin, USA

Introduction

The process of testicular development that leads to initiation of spermatogenesis in bulls involves complex maturation mechanisms of the hypothalamus–pituitary–testes axis. Sexual development can be divided into three periods according to changes in gonadotropins and testosterone concentrations, namely the infantile, prepubertal, and pubertal periods. These changes are accompanied by changes in testicular cell proliferation and differentiation (Figure 4.1).

Infantile period

The infantile period is characterized by low gonadotropin and testosterone secretion and relatively few changes in testicular cellular composition. This period extends from birth until approximately 2 months of age in *Bos taurus* bulls.

Gonadotropin secretion during the infantile period is low due to reduced gonadotropin-releasing hormone (GnRH) secretion; maturation changes within the hypothalamus result in increased GnRH pulse secretion and drive the transition from the infantile period. Increased GnRH secretion is dependent on either the development of central stimulatory inputs or removal of inhibitory inputs. Hypothalamus weight and GnRH content do not increase during the infantile period, but hypothalamic concentrations of estradiol receptors decrease after 1 month of age.[1] However, the hypothesis that GnRH secretion is low during infancy due to elevated sensitivity of the hypothalamus to the negative feedback of sex steroids (gonadostat hypothesis) has been questioned in bulls, since castration does not alter luteinizing hormone (LH) pulse frequency or mean concentrations before 2 months of age.[2] Nonetheless, since GnRH secretion into hypophyseal portal blood is not necessarily accompanied by LH secretion during the infantile period, experiments that use LH concentrations to infer GnRH secretion patterns during this period need to be interpreted with caution.[3] Another possibility is that removal of opioidergic inhibition and/or increased dopaminergic activity may be involved in triggering the increase in GnRH secretion during the infantile period. Opioidergic inhibition of LH pulse frequency during the infantile period has been demonstrated by increased LH secretion between 1 and 4 months of age in bulls treated with naloxone, an opioid competitive receptor antagonist,[4] whereas concentrations of norepinephrine, dopamine, and dopamine metabolites increased twofold to threefold in the anterior hypothalamic–preoptic area in bulls aged 0.5–2.5 months.[5]

Direct evaluation of blood samples from the hypophyseal portal system has demonstrated that GnRH pulsatile secretion increases linearly from age 2 weeks (3.5 pulses per 10 hours) to 12 weeks (8.9 pulses per 10 hours) in bulls. Although GnRH secretion into hypophyseal portal blood was detected at 2 weeks, pulsatile LH secretion was not detected in jugular blood samples before 8 weeks of age. In addition, GnRH pulses are not necessarily accompanied by LH secretion until 8–12 weeks of age, when all GnRH pulses result in LH pulses. The increase in pulsatile GnRH release from 2 to 8 weeks of age without a concomitant increase in LH secretion may represent a reduced ability of the pituitary gland to respond to GnRH stimulus.[3] The period in which GnRH pulses do not stimulate LH secretion correspond to a period during which there is an increase in pituitary weight, GnRH receptor concentration, and LH content.[1] Moreover, frequent GnRH treatments during the infantile period in calves increases pituitary LH-β mRNA, LH content, and GnRH receptors, with resulting increases in LH pulse frequency and mean concentrations,[6] indicating that increased GnRH pulse frequency results in increased pituitary sensitivity to GnRH. With time, the increased GnRH secretion results in the increased LH pulse frequency observed during the prepubertal period.

Bovine Reproduction, First Edition. Edited by Richard M. Hopper.
© 2015 John Wiley & Sons, Inc. Published 2015 by John Wiley & Sons, Inc.

Endocrine Control of Testicular Development and Initiation of Spermatogenesis in Bulls

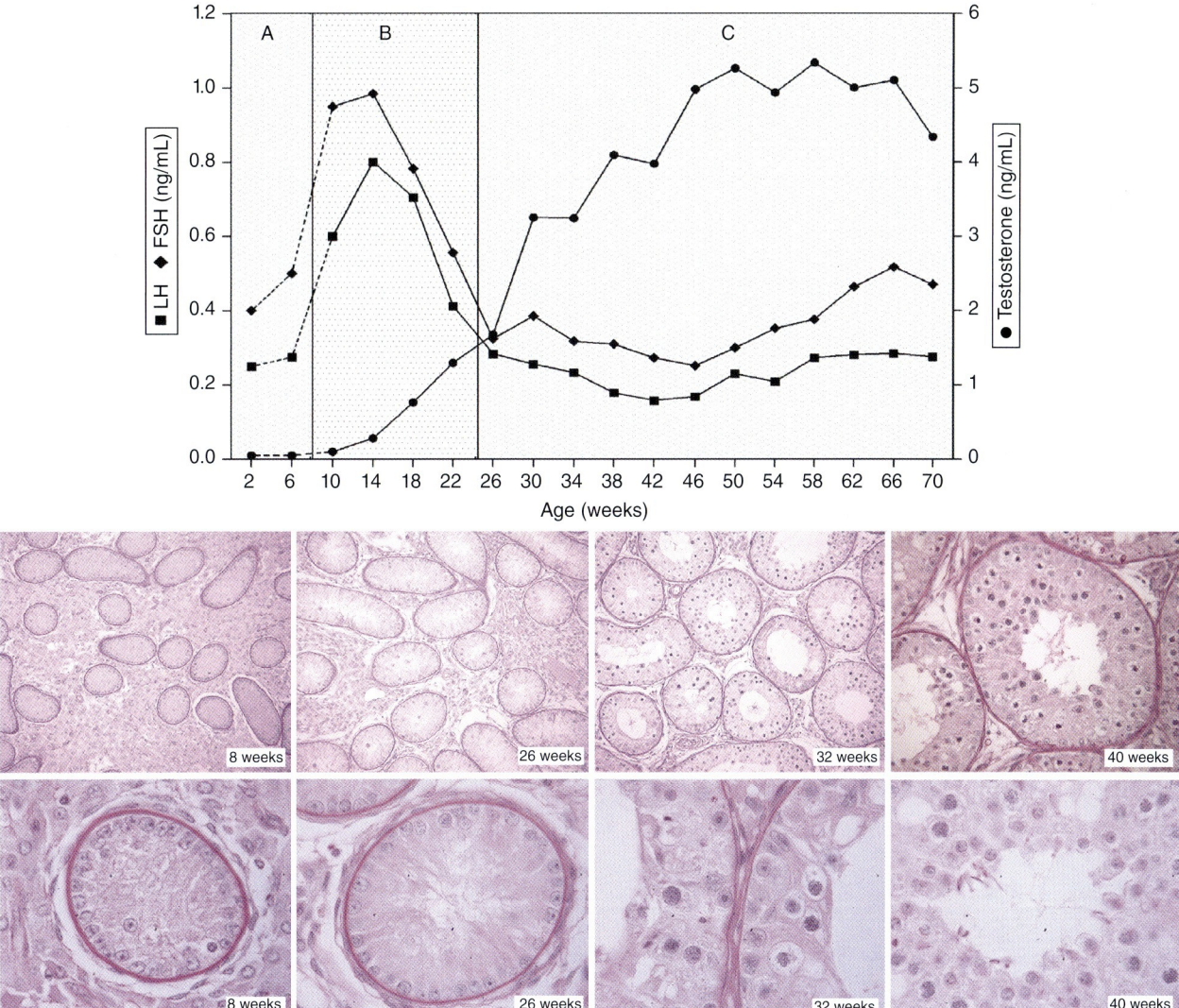

Figure 4.1 Mean serum LH, FSH, and testosterone concentrations in bulls during the infantile (A), prepubertal (B), and pubertal/postpubertal periods (C). Data from 2 to 6 weeks are adapted from Hereford × Charolais bulls.[15] Data from 10 to 70 weeks are from Angus and Angus × Charolais bulls receiving adequate nutrition.[16,26,27,34] Infancy is the period that extends from birth until approximately 8 weeks of age. During this period gonadotropin and testosterone concentrations are low, the testicular parenchyma is occupied mostly by interstitial tissue, seminiferous cords are lined by undifferentiated Sertoli cells, and centrally located gonocytes with large nuclei can be observed. The prepubertal period is characterized by a dramatic increase in gonadotropin concentrations (the early gonadotropin rise) and by slowly increasing testosterone secretion. The prepubertal period extends from approximately 10 to 26 weeks of age and during this period a cord lumen begins to develop and gonocytes migrate toward the basement membrane differentiating into spermatogonia. Gonadotropin concentrations decrease concomitantly with a rapid increase in testosterone concentration during the pubertal period, which also coincides with the start of a phase of rapid testicular growth. Formation of the tubular lumen is evidence of Sertoli cell differentiation and development of a functional blood–testis barrier that precedes the appearance of primary spermatocytes and spermatids around 32 weeks of age. With continuous increase in diameter, seminiferous tubules occupy most of the testicular parenchyma and start to produce mature sperm at approximately 40 weeks of age.

From birth until approximately 2 months of age, mesenchymal-like cells comprise the majority of the cells in the testicular interstitial tissue. Typical Leydig cells constitute about 6% of all intertubular cells at 1 month of age and a number of these cells are found in an advanced degenerative state, probably as remnants of the fetal Leydig cell population. Degenerating fetal and newly formed Leydig cells coexist until 2 months of age, but only Leydig cells formed postnatally are observed thereafter.[7,8] The diameter of the seminiferous tubules is approximately 50 μm during the infantile period; tubule is actually a misnomer, since these are in fact solid cords with no lumen at this stage of development. Undifferentiated Sertoli cells (or undifferentiated supporting cells) are the predominant intratubular cells from birth until approximately 4 months of age. The number of undifferentiated Sertoli cells remains constant until 1 month of age, but cell multiplication is maximal between 1 and 2 months of age, decreasing thereafter until approximately 4 months of age. During the infantile period, the membranes of neighboring undifferentiated Sertoli cells contain few interdigitations and no special junctional complexes.[9,10] The germ cell population is composed solely of gonocytes (or prespermatogonia) at birth. Gonocytes are usually centrally located and have a large nucleus (~12 μm in diameter) with a well-developed nucleolus. Gonocyte proliferation slowly resumes between 1 and 2 months of age.[9,11]

Prepubertal period

The prepubertal period is characterized by a temporary increase in gonadotropin secretion, the so-called early gonadotropin rise. The early gonadotropin rise is a critical event in the sexual development of bulls. It is not only associated with dramatic changes in testicular cellular composition, initial increase in testosterone secretion and timing of attainment of puberty, but also has long-lasting effects on testicular growth and sperm production. This period extends from approximately 2 to 6 months of age in *Bos taurus* bulls.

The early gonadotropin rise is driven by increased GnRH pulse secretion, as demonstrated by a dramatic increase in LH pulse frequency (Figure 4.2); pulsatile discharges of follicle-stimulating hormone (FSH) have been observed in bulls, but are much less evident than those of LH. The number of LH pulses increases from less than one per day at 1 month to approximately 12–16 per day (one or more pulse every 2 hours) at approximately 4 months of age. Changes in pulse amplitude during this period are not consistent among reports; amplitude may be reduced, unchanged, or augmented.[1,12–16] LH-binding sites in testicular interstitial tissue have been demonstrated in bulls at birth and at 4 months of age and pulsatile LH secretion is an essential requirement for Leydig cell proliferation and differentiation and for maintenance of fully differentiated structure and function.[17] Mesenchymal-like cells in the testes cease to proliferate around 4 months of age and start to differentiate into contractile myofibroblasts and Leydig cells. Differentiating, mitotic, and degenerating Leydig cells are observed in close proximity from 4 to 7 months of age. Leydig cell numbers and mass per testis increase from 1 month (0.42 billion and 0.15 g/testis, respectively) to 7 months of age (6 billion and 5.8 g/testis, respectively), but mitosis after this age is rare.[7,8]

The characteristic pulsatile nature of LH secretion is important for testosterone production, since continuous exposure of Leydig cells to LH results in reduced steroidogenic responsiveness due to downregulation of LH receptors.[18]

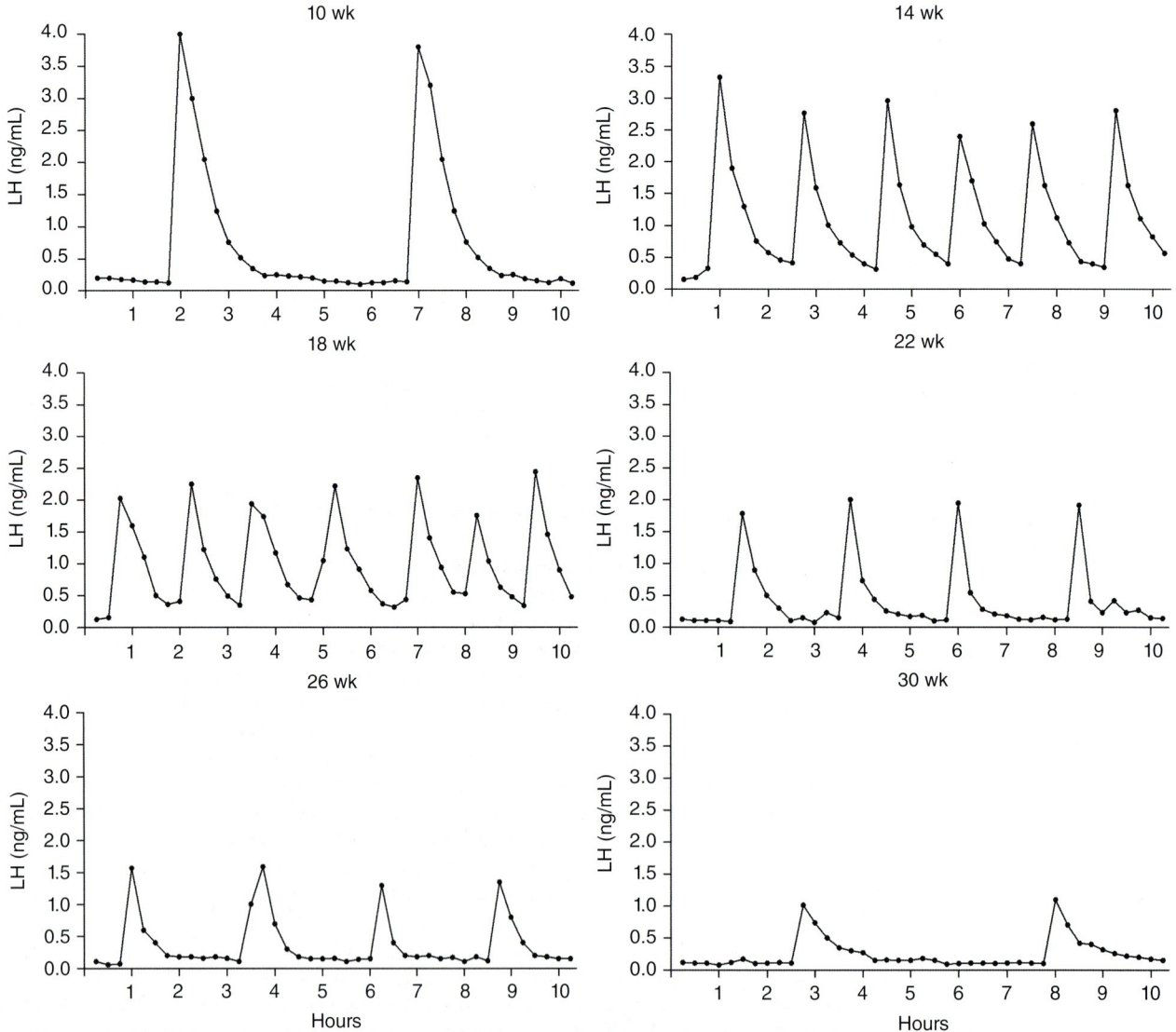

Figure 4.2 Serum LH concentrations between 10 and 30 week of age in Angus and Angus × Charolais bulls. Graphs closely exemplify the mean pulse frequency and pulse amplitude observed in bulls receiving adequate nutrition.[16, 26, 27]

Initiation of Leydig cell steroidogenesis is characterized by increased androstenedione secretion, which decreases as the cells complete maturation and begin secreting testosterone. During the first 3–4 months of age, testosterone concentrations are low and secretion does not necessarily accompany LH pulses. After this age, LH pulses are followed by testosterone pulses and mean testosterone concentrations begin to increase. The number of testosterone pulses increases from 0.3–2.3 pulses per 24 hours at 1–4 months of age to 7.5–9 pulses per 24 hours at 5 months of age.[19–22]

The crucial role of the early gonadotropin rise (especially LH secretion pattern) in regulating sexual development in bulls has been demonstrated in several studies using a variety of approaches. Prolonged treatment with a GnRH agonist in calves aged 1.5–3.5 months decreased LH and FSH pulse frequency, pulse amplitude and mean concentrations at 3 months of age, delayed the peak mean LH concentration from 5 to 6 months of age, and reduced FSH and testosterone concentrations between 3.5 and 4.5 months of age. These hormonal alterations were associated with delayed puberty and reduced testes weight and number of germ cells in tubular cross-sections at 11.5 months of age. On the other hand, treatment with GnRH every 2 hours to mimic pulsatile secretion from 1 to 1.5–2 months of age increased LH pulse frequency and mean concentration during the treatment period and resulted in greater scrotal circumference, testes weight, seminiferous tubules diameter, and number of germ and Sertoli cells in tubular cross-sections at 12 months of age.[23–25] The LH secretion pattern during the prepubertal period is also associated with age at puberty in bulls raised in contemporary groups, suggesting that this is the physiological mechanism by which genetics affect sexual development. Studies have shown that LH pulse frequency was greater around 2.5–5 months of age and that mean LH concentrations increased earlier and reached greater maximum levels in early- than in late-maturing Hereford bulls (age at puberty 9.5 and 11 months, respectively).[13,15]

Additional support for the crucial role that the early gonadotropin rise plays in sexual development in bulls has been provided by studies demonstrating that nutrition during the prepubertal period affects LH secretion pattern, age at puberty, and testicular development. In one study in

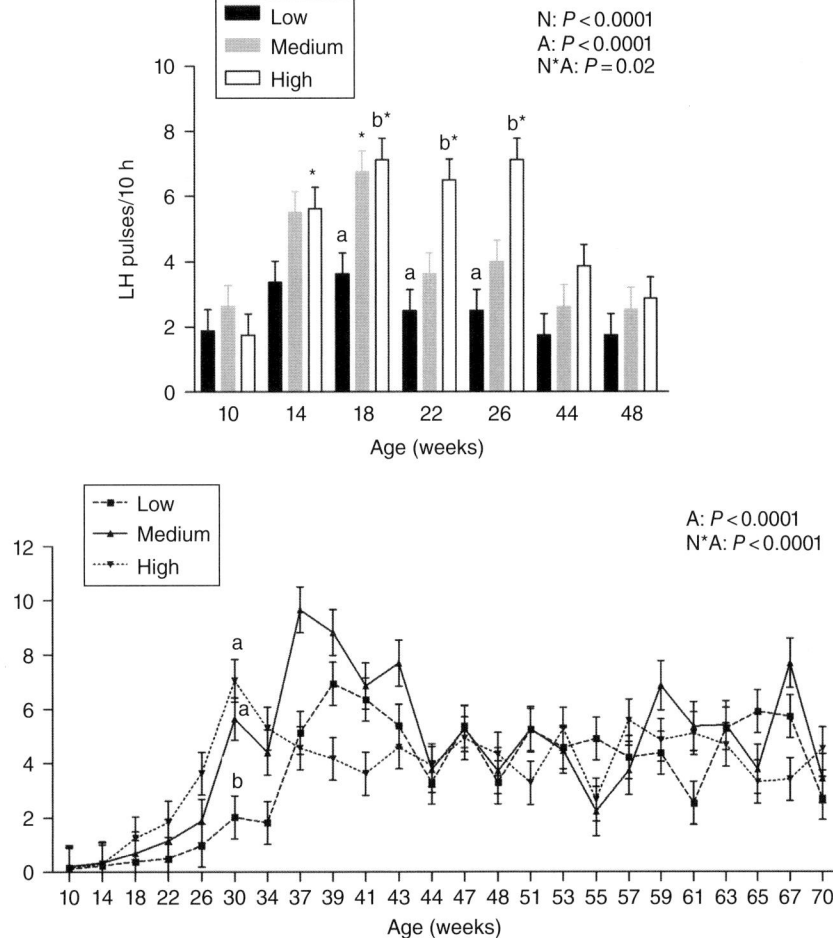

Figure 4.3 Mean (± SEM) number of LH pulses and serum testosterone concentrations in Angus and Angus × Charolais bulls receiving low, medium (control), or high nutrition from 10 to 70 weeks of age. N, A and N*A indicate nutrition, age, and nutrition-by-age interaction effects, respectively. Superscripts indicate differences ($P < 0.05$): a,b, group differences within age; *, age differences within group. Bulls in the low nutrition group were older at puberty (321 days) than bulls in the medium and high nutrition groups (299 and 288 days, respectively), whereas bulls in the high nutrition group had greater paired-testes weight at 70 weeks of age (655 g) than bulls in the low and medium nutrition groups (520 and 549 g). Adapted from Brito L, Barth A, Rawlings N *et al*. Effect of nutrition during calfhood and peripubertal period on serum metabolic hormones, gonadotropins and testosterone concentrations, and on sexual development in bulls. *Domest Anim Endocrinol* 2007;33:1–18 with permission from Elsevier.

which bulls received different nutrition from 2 to 16 months of age, reduced LH pulse frequency during the prepubertal period resulted in delayed puberty in bulls receiving low nutrition, while a more sustained increase in LH pulse frequency in bulls receiving high nutrition was associated with hastened testosterone production and greater testes weight at 16 months of age when compared with bulls receiving low and medium (control) nutrition (Figure 4.3).[16]

These observations were corroborated by additional studies designed to investigate the effects of nutrition specifically during the prepubertal period. Bulls that received high nutrition only from 2 to 7 months of age also had a more sustained increase in LH pulse frequency, had greater testosterone secretion, were 2 weeks younger at puberty, and had greater testes weight at 16.5 months of age when compared with bulls receiving control nutrition (Figure 4.4).[26] On the other hand, reduced LH secretion resulting from low nutrition from 2 to 7 months of age was associated with increased age at puberty and smaller testes weight at 16 months of age even when these bulls received control or high nutrition after 7 months of age and LH and testosterone secretion were not different after the change in nutrition (Figure 4.5).[27]

During the prepubertal period there is a progressive increase in the proportion of testicular parenchyma occupied by seminiferous tubules, and seminiferous tubule diameter increases to approximately 125 μm at 6 months of age.[14,28] FSH-binding sites can be observed in seminiferous tubules of bull calves at birth and at 4 months age and increased FSH concentrations stimulate the proliferation of undifferentiated Sertoli cells.[17] Although there is considerable evidence that FSH is essential for normal Sertoli cell function, the period of Sertoli cell differentiation coincides with the initiation of testosterone secretion by the Leydig cells, indicating that testosterone may also be involved in promoting maturation of undifferentiated Sertoli cells. At approximately 4 months of age, undifferentiated Sertoli cells enter the G_0 phase of the cell cycle for the rest of the bull's life. With the end of the proliferative phase, undifferentiated Sertoli cells begin to transform into adult-type Sertoli cells. Opposing cell membranes of adjacent undifferentiated Sertoli cells start to develop extended

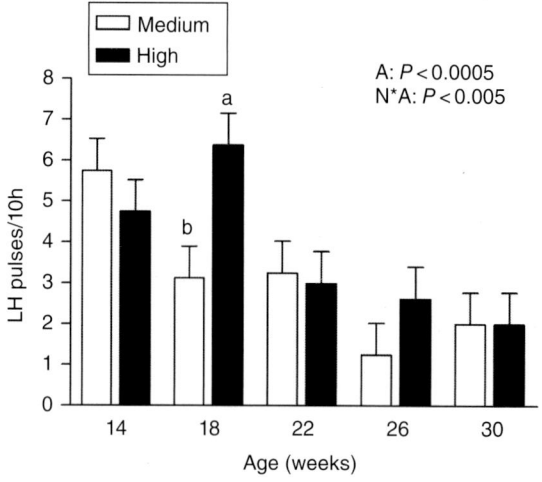

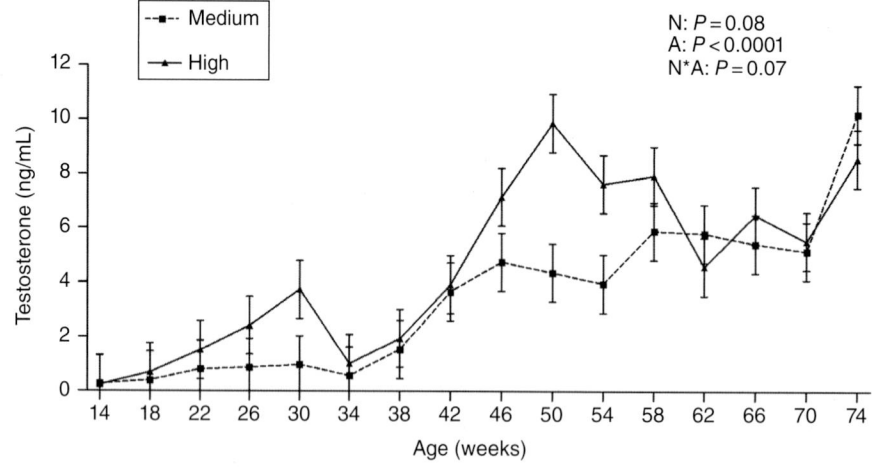

Figure 4.4 Mean (± SEM) number of LH pulses and serum testosterone concentrations in Angus and Angus × Charolais bulls receiving medium (control) or high nutrition from 10 to 30 weeks of age and the same medium nutrition from 31 to 74 weeks. N, A and N*A indicate nutrition, age, and nutrition-by-age interaction effects, respectively. Superscript a and b indicate differences ($P = 0.09$) between groups within age. Bulls in the high nutrition group were younger at puberty (314 days) and had greater paired-testes weight at 74 weeks of age (610 g) than bulls in the medium group (327 days and 531 g, respectively). Adapted from Brito L, Barth A, Rawlings N et al. Effect of improved nutrition during calfhood on serum metabolic hormones, gonadotropins, and testosterone concentrations, and on testicular development in bulls. *Domest Anim Endocrinol* 2007;33:460–469 with permission from Elsevier.

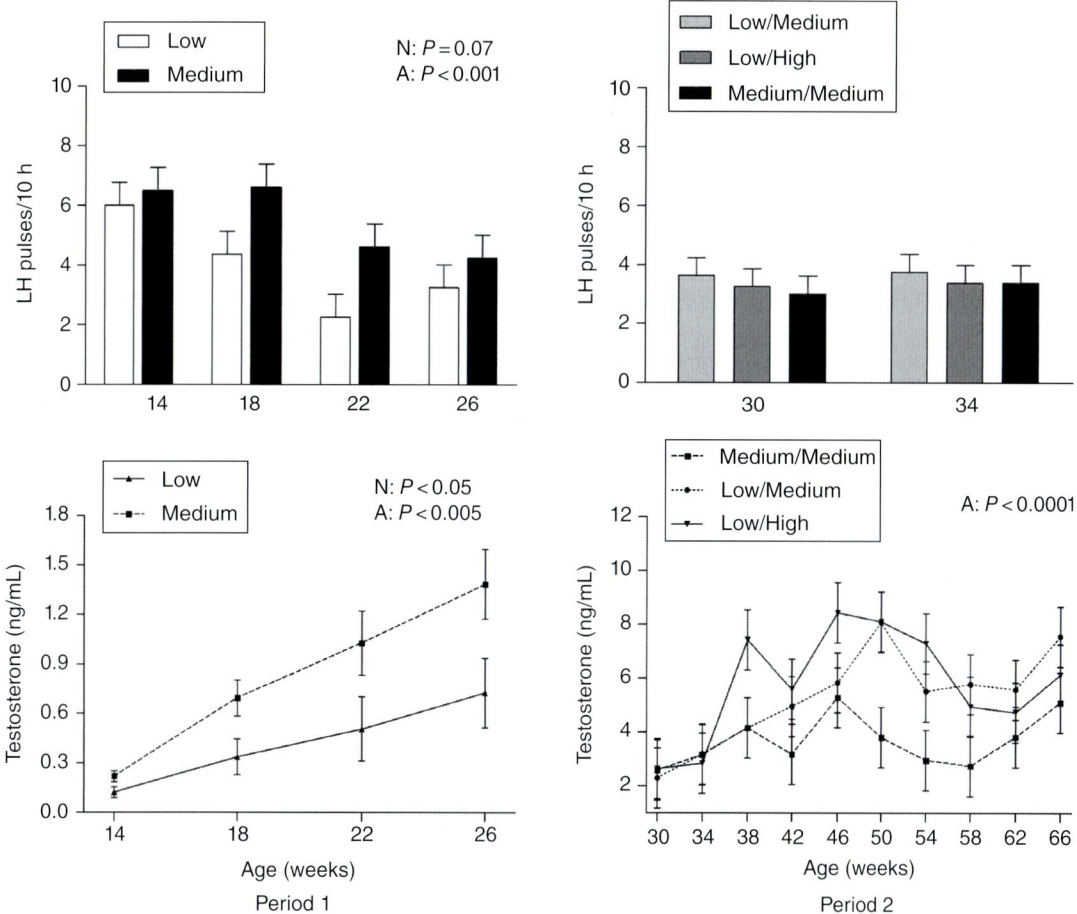

Figure 4.5 Mean (± SEM) number of LH pulses and serum testosterone concentrations in Angus and Angus × Charolais bulls receiving medium (control) from 10 to 70 weeks of age or low nutrition from 10 to 26 weeks of age and either medium or high nutrition from 27 to 70 weeks of age. N, A and N*A indicate nutrition, age, and nutrition-by-age interaction effects, respectively. Bulls in the medium/medium nutrition group were younger at puberty and had greater paired-testes weight at 70 weeks of age than those in the low/medium nutrition group (293 vs. 331 days and 600 vs. 528 g, respectively). Age at puberty and paired-testes weight for bulls in the low/high nutrition group were intermediate (313 days and 553 g, respectively). Adapted from Brito L, Barth A, Rawlings N et al. Effect of feed restriction during calfhood on serum concentrations of metabolic hormones, gonadotropins, testosterone, and on sexual development in bulls. *Reproduction* 2007;134:171–181 with permission from the Society for Reproduction and Fertility.

junctional complexes above the spermatogonia and in the basal portion of the tubules; "cracking" of the tubular cytoplasm is first detected around 6 months of age.[9–11] FSH secretion, maturation of Sertoli cells, and increased testosterone secretion are probably also involved in the differentiation of gonocytes into spermatogonia. Gonocytes are gradually displaced to a position close to the basal lamina and divide by mitosis, originating A-spermatogonia. Differentiation and degeneration result in the complete disappearance of gonocytes from the seminiferous tubules by 5 months of age. A-spermatogonia divide mitotically to form In-spermatogonia and B-spermatogonia, that in turn enter meiosis around 4–5 months of age.[9,11,14,28,29]

Pubertal period

The pubertal period is characterized by reduced gonadotropin secretion, increased testosterone secretion, initiation of spermatogenesis, and the eventual appearance of sperm in the ejaculate. This period also coincides with the start of a phase of rapid testicular growth (see Chapter 5 for testicular growth charts) and extends from approximately 6 to 12 months of age in *Bos taurus* bulls.

The rapidly increasing testosterone secretion and possibly increased hypothalamic sensitivity to negative feedback from androgens are likely responsible for the decrease in LH secretion during the pubertal period, whereas inhibin produced by Sertoli cells may act on the gonadotrophs to limit FSH secretion, since immunization with inhibin antiserum results in a marked increase in FSH concentrations in prepubertal bulls.[30,31] After 7 months of age, Leydig cell mass increases slowly but continuously to reach about 10 g in the young adult testis at 24 months of age as a result of considerable increase in Leydig cell volume (hypertrophy); Leydig cell mitochondrial mass more than doubles from 10 to 24 months of age.[8] Testosterone pulse frequency does not increase after the peripubertal period and remains at approximately 4.5–6.8 pulses per 24 hours from 6 to 10 months of age. However, pulse amplitude increases during the pubertal period with consequent increase in testosterone mean concentrations until approximately 12 months of age. Elevated testosterone secretion is essential for initiation of spermatogenesis.[6,19–21,32]

Seminiferous tubule diameter increases to approximately 200 μm at 8 months of age and reaches 240 μm by 16 months.[14, 28, 33] Total seminiferous tubule length increases from 830 m per testis at 3 months of age to 2010 m per testis at 8 months of age in Holstein bulls.[28] Most Sertoli cells complete their morphological differentiation and attain adult structure after 6–7 months of age. Junctional complexes consisting of many serially arranged points or lines of fusion involving neighboring Sertoli cell membranes can be observed. These junctions form a functional blood–testis barrier and divide the tubular epithelium into a basal compartment containing spermatogonia and an adluminal compartment containing germ cells at later stages of spermatogenesis; formation of a functional blood–testis barrier is accompanied by formation of the tubular lumen and precedes the appearance of primary spermatocytes and more advanced germ cells.[9–11] In Holstein bulls, the number of adult-type Sertoli cells increases dramatically from 202 to 8862 million cells per testis between 5 and 8 months of age, respectively.[28]

Germ cell proliferation is maximal between 4 and 8 months of age and represents the expansion of the spermatogonial stem cell. In Holstein bulls, the number of spermatogonia increases from 181 million cells per testis at 4 months of age to 3773 million cells per testis at 8 months of age; the number of spermatogonia continues to increase until approximately 12 months of age.[28] Primary spermatocyte numbers increase slowly until 8 months of age, when the numbers exceed the number of spermatogonia. Secondary spermatocytes and round spermatids first appear at approximately 6–7 months of age, whereas elongated spermatids appear around 8 months of age. The number of spermatids increases rapidly after 10 months of age when spermatid numbers exceed the numbers of any other germ cell. Mature sperm appear in the seminiferous tubules at approximately 8–10 months of age. Testes weighing more than 100 g in Swedish Red-and-White bulls or more than 80 g in Holstein bulls are likely to be producing sperm.[9, 11, 14, 28, 29] Spermatogenesis eventually reaches a level of efficiency (i.e., increasing number of more advanced germ cells resulting from individual precursor cells) that results in the production of a number of sperm sufficient for those to appear in the ejaculate, the physiological event that characterizes puberty (see Chapter 5).

Metabolic hormones during the prepubertal and pubertal periods

The mechanisms controlling reproduction and energy balance are intrinsically related and have evolved to confer reproductive advantages and guarantee the survival of species. The neural apparatus designed to gauge metabolic rate and energy balance has been denominated the body "metabolic sensor." This sensor translates signals provided by circulating (peripheral) concentrations of specific hormones into neuronal signals that ultimately regulate the GnRH pulse generator and control the reproductive process. Metabolic indicator hormones may serve as signs to the hypothalamus–pituitary–gonadal axis and affect sexual development. The patterns of some of these hormones have been studied in growing beef bulls (Figure 4.6). In contrast to species in which circulating growth hormone (GH) concentrations continue to increase until after puberty, GH concentrations decrease during the pubertal period in bulls.[16,34] Differences in the stage of body development at which each species attains puberty are likely responsible for the different GH profiles among species. Accordingly, the GH profile in bulls seems to indicate that a relatively advanced stage of body development must be attained before the gonads are efficiently producing sperm. The differences in GH secretion among species may be due to the regulatory role of steroids on GH secretion. In other species steroids stimulate GH secretion, but GH concentrations do not differ between intact bulls and castrated steers.[35] Moreover, decreasing GH concentrations during sexual development are observed along with increasing testosterone concentrations, indicating that steroids do not have a positive feedback on GH secretion in bulls as in other species.[16,34]

Circulating insulin-like growth factor (IGF)-I concentrations in bulls increase continuously and only reach a plateau

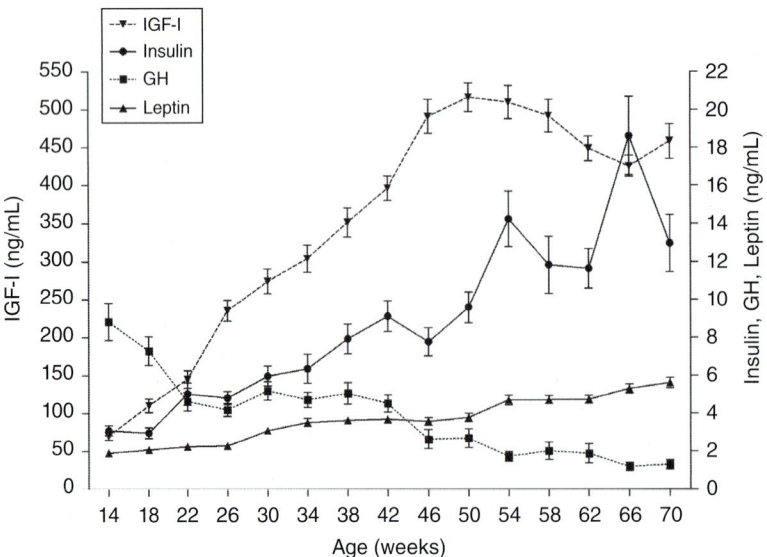

Figure 4.6 Mean (± SEM) serum IGF-I, insulin, GH, and leptin concentrations during sexual development in Angus and Angus × Charolais bulls receiving adequate nutrition.[16, 26, 27, 34]

Endocrine Control of Testicular Development and Initiation of Spermatogenesis in Bulls

(or decrease slightly) after sexual development is mostly completed after 12–14 months of age; increasing circulating concentrations of IGF-binding protein 3 and decreasing concentrations of IGF-binding protein 2 are also observed during sexual development.[16,26,27,34,36–38] The concomitant decrease in circulating GH concentrations with the increase in IGF-I concentrations during sexual development in bulls indicates that there are either drastic changes in liver sensitivity to GH or that other sources are responsible for IGF-I production. A possible IGF-I source might be the testes, since Leydig cells are capable of secreting this hormone in other species. Observations that intact bulls tend to have greater IGF-I concentrations than castrated steers at 12 months of age further support the hypothesis that the testes might contribute substantially to circulating IGF-I concentrations during the prepubertal and pubertal periods in bulls.[35] Close temporal associations observed in a series of nutrition studies strongly suggest that circulating IGF-I might be involved in regulating the GnRH pulse generator and the magnitude and duration of the early gonadotropin rise in beef bulls.[16,26,27]

A possible effect of IGF-I on testicular steroidogenesis in bulls has also been suggested. Leydig and Sertoli cells produce IGF-I, indicating the existence of paracrine/autocrine mechanisms of testicular regulation involving IGF-I.[39,40] It is assumed that most of the IGF-I in the testes is produced locally and that circulating IGF-I may play a secondary role in regulating testicular development and function. However, the temporal patterns and strong associations among circulating IGF-I concentrations, testicular size, and testosterone secretion observed in bulls receiving different nutrition argue for a primary role for this hormone.[16,26,27] The primary role of increased circulating IGF-I during the pubertal period may be to promote the increase in testosterone concentrations by regulating Leydig cell multiplication, differentiation, and maturation. Since testosterone upregulates IGF-I production and IGF-I receptor expression by Leydig and Sertoli cells,[41] the establishment of a positive feedback loop between IGF-I secretion and testosterone production may be important for sexual development.

Circulating leptin and insulin concentrations also increase during the pubertal period in bulls. However, developmental and nutritional differences in LH pulse frequency are not related to differences in leptin or insulin concentrations in beef bulls.[16,27] Other studies have also demonstrated that leptin does not stimulate *in vitro* GnRH secretion from hypothalamic explants or gonadotropin secretion from adenohypophyseal cells collected from bulls and steers maintained at an adequate level of nutrition.[42] These results indicate that the role of these hormones in regulating GnRH secretion, if any, might be purely permissive in bulls.

GnRH-independent testicular development

The rapid testicular growth observed after 6 months of age in bulls occurs when circulating gonadotropin concentrations are decreasing, which points to the existence of important GnRH-independent mechanisms regulating testicular development. The period of accelerated testicular growth coincides with increasing circulating IGF-I and leptin concentrations and strong associations between these hormones and testicular size have been observed in growing beef bulls,[16,34] indicating that metabolic hormones may be involved in regulating GnRH-independent testicular development. Since there was no association between circulating metabolic hormones and gonadotropin concentrations in these studies, the possible effects of metabolic hormones on testicular growth are likely direct and independent of the hypothalamus and pituitary. Accelerated testicular growth in bulls involves increases in seminiferous tubule diameter and length, volume of testicular parenchyma occupied by seminiferous tubules, and total number of germinal cells.[9,28] Although IGF-I and leptin concentrations are associated with testicular size, there are no associations between these hormones and seminiferous tubule diameter and area, seminiferous epithelium area, or volume occupied by seminiferous tubule (L.F.C. Brito, unpublished results). These observations suggest that increased circulating IGF-I and leptin concentrations are associated with increased length of the seminiferous tubules and likely with overall increases in the total number of testicular cells. Considering the cellular events in the testis during the pubertal period, the temporal patterns of metabolic hormone concentrations indicate that circulating IGF-I and leptin could be involved in regulating Leydig cell multiplication and maturation, Sertoli cell maturation, and germ cell multiplication during the period of accelerated GnRH-independent testicular growth in bulls.

Testicular concentrations of LH and FSH receptors in beef bulls decrease around 5–6 months of age, but increase thereafter until at least approximately 13 months of age, which might increase the sensitivity of Leydig and Sertoli cells to the low concentrations of gonadotropins during the rapid testicular growth phase.[29] Other mechanisms that might be associated with GnRH-independent testicular growth include changes in testicular concentrations and bioavailability of growth factors such as transforming growth factor (TGF)-α and TGF-β1, TGF-β2 and TGF-β3 and interleukin (IL)-1α, IL-1β and IL-6.[43,44] In addition, experiments evaluating the effect of nutrition on sexual development have demonstrated that the impact of gonadotropins on target tissues during the prepubertal period have long-term effects on testicular development in bulls. Bulls with either greater LH pulse frequency or more sustained increase in LH pulse frequency during the early gonadotropin rise had a more prolonged period of increased testicular growth and greater testicular size at 15–16 months of age, even when no differences in metabolic hormones or testosterone concentrations were observed after 6 months of age. These results indicate that the putative effects of circulating metabolic hormones, gonadotropins, local growth factors, and other unknown factors during the period of rapid testicular growth might be dependent on the previous LH exposure during the prepubertal period. The LH secretion pattern during the early gonadotropin rise seems to "prime" testicular development and dictates maximum testicular size in bulls.[16,26,27]

References

1. Amann R, Wise M, Glass J, Nett T. Prepubertal changes in the hypothalamic–pituitary axis of Holstein bulls. *Biol Reprod* 1986; 34:71–80.
2. Wise M, Rodriguez R, Kelly C. Gonadal regulation of LH secretion in prepubertal bull calves. *Domest Anim Endocrinol* 1987;4: 175–181.

3. Rodriguez R, Wise M. Ontogeny of pulsatile secretion of gonadotropin-releasing hormone in the bull calf during infantile and pubertal development. *Endocrinology* 1989;124:248–256.
4. Evans A, Currie W, Rawlings N. Opioidergic regulation of gonadotrophin secretion in the early prepubertal bull calf. *J Reprod Fertil* 1993;99:45–51.
5. Rodriguez R, Benson B, Dunn A, Wise M. Age-related changes in biogenic amines, opiate, and steroid receptors in the prepubertal bull calf. *Biol Reprod* 1993;48:371–376.
6. Rodriguez R, Wise M. Advancement of postnatal pulsatile luteinizing hormone secretion in the bull calf by pulsatile administration of gonadotropin-releasing hormone during infantile development. *Biol Reprod* 1991;44:432–439.
7. Wrobel K, Dostal S, Schimmel M. Postnatal development of the tubular lamina propria and the intertubular tissue in the bovine testis. *Cell Tissue Res* 1988;252:639–653.
8. Wrobel K. The postnatal development of the bovine Leydig cell population. *Reprod Domest Anim* 1990;25:51–60.
9. Abdel-Raouf M. The postnatal development of the reproductive organs in bulls with special reference to puberty (including growth of the hypophysis and the adrenals). *Acta Endocrinol* 1960;Suppl. 49:1–109.
10. Sinowatz F, Amselgruber W. Postnatal development of bovine Sertoli cells. *Anat Embryol (Berl)* 1986;174:413–423.
11. Wrobel K. Prespermatogenesis and spermatogoniogenesis in the bovine testis. *Anat Embryol (Berl)* 2000;202:209–222.
12. Amann R, Walker O. Changes in the pituitary–gonadal axis associated with puberty in Holstein bulls. *J Anim Sci* 1983;57:433–442.
13. Evans A, Davies F, Nasser L, Bowman P, Rawlings N. Differences in early patterns of gonadotrophin secretion between early and late maturing bulls, and changes in semen characteristics at puberty. *Theriogenology* 1995;43:569–578.
14. Evans A, Pierson R, Garcia A, McDougall L, Hrudka F, Rawlings N. Changes in circulating hormone concentrations, testes histology and testes ultrasonography during sexual maturation in beef bulls. *Theriogenology* 1996;46:345–357.
15. Aravindakshan J, Honaramooz A, Bartlewski P, Beard A, Pierson R, Rawlings N. Pattern of gonadotropin secretion and ultrasonographic evaluation of developmental changes in the testis of early and late maturing bull calves. *Theriogenology* 2000;54:339–354.
16. Brito L, Barth A, Rawlings N et al. Effect of nutrition during calfhood and peripubertal period on serum metabolic hormones, gonadotropins and testosterone concentrations, and on sexual development in bulls. *Domest Anim Endocrinol* 2007;33:1–18.
17. Schanbacher B. Relationship of in vitro gonadotropin binding to bovine testes and the onset of spermatogenesis. *J Anim Sci* 1979;48:591–597.
18. Saez J. Leydig cells: endocrine, paracrine, and autocrine regulation. *Endocrinol Rev* 1994;15:574–626.
19. Rawlings N, Hafs H, Swanson L. Testicular and blood plasma androgens in Holstein bulls from birth through puberty. *J Anim Sci* 1972;34:435–440.
20. Rawlings N, Fletcher P, Henricks D, Hill J. Plasma luteinizing hormone (LH) and testosterone levels during sexual maturation in beef bull calves. *Biol Reprod* 1978;19:1108–1112.
21. McCarthy M, Hafs H, Convey E. Serum hormone patterns associated with growth and sexual development in bulls. *J Anim Sci* 1979;49:1012–1020.
22. Lacroix A, Pelletier J. Short-term variations in plasma LH and testosterone in bull calves from birth to 1 year of age. *J Reprod Fertil* 1979;55:81–85.
23. Chandolia R, Evans A, Rawlings N. Effect of inhibition of increased gonadotrophin secretion before 20 weeks of age in bull calves on testicular development. *J Reprod Fertil* 1997;109:65–71.
24. Chandolia R, Honaramooz A, Bartlewski P, Beard A, Rawlings N. Effects of treatment with LH releasing hormone before the early increase in LH secretion on endocrine and reproductive development in bull calves. *J Reprod Fertil* 1997;111:41–50.
25. Madgwick S, Bagu E, Duggavathi R et al. Effects of treatment with GnRH from 4 to 8 weeks of age on the attainment of sexual maturity in bull calves. *Anim Reprod Sci* 2008;104:177–188.
26. Brito L, Barth A, Rawlings N et al. Effect of improved nutrition during calfhood on serum metabolic hormones, gonadotropins, and testosterone concentrations, and on testicular development in bulls. *Domest Anim Endocrinol* 2007;33:460–469.
27. Brito L, Barth A, Rawlings N et al. Effect of feed restriction during calfhood on serum concentrations of metabolic hormones, gonadotropins, testosterone, and on sexual development in bulls. *Reproduction* 2007;134:171–181.
28. Curtis S, Amann R. Testicular development and establishment of spermatogenesis in Holstein bulls. *J Anim Sci* 1981;53:1645–1657.
29. Bagu E, Cook S, Gratton C, Rawlings N. Postnatal changes in testicular gonadotropin receptors, serum gonadotropin, and testosterone concentrations and functional development of the testes in bulls. *Reproduction* 2006;132:403–411.
30. Kaneko H, Yoshida M, Hara Y et al. Involvement of inhibin in the regulation of FSH secretion in prepubertal bulls. *J Endocrinol* 1993;137:15–19.
31. Rawlings N, Evans A. Androgen negative feedback during the early rise in LH secretion in bull calves. *J Endocrinol* 1995;145:243–249.
32. McCarthy M, Convey E, Hafs H. Serum hormonal changes and testicular response to LH during puberty in bulls. *Biol Reprod* 1979;20:1221–1227.
33. Brito L, Barth A, Wilde R, Kastelic J. Effect of growth rate from 6 to 16 months of age on sexual development and reproductive function in beef bulls. *Theriogenology* 2012;77:1398–1405.
34. Brito L, Barth A, Rawlings N et al. Circulating metabolic hormones during the peripubertal period and their association with testicular development in bulls. *Reprod Domest Anim* 2007;42:502–508.
35. Lee C, Hunt D, Gray S, Henricks D. Secretory patterns of growth hormone and insulin-like growth factor-I during peripubertal period in intact and castrate male cattle. *Domest Anim Endocrinol* 1991;8:481–489.
36. Renaville R, Devolder A, Massart S, Sneyers M, Burny A, Portetelle D. Changes in the hypophysial–gonadal axis during the onset of puberty in young bulls. *J Reprod Fertil* 1993;99:443–449.
37. Renaville R, Massart S, Sneyers M et al. Dissociation of increases in plasma insulin-like growth factor I and testosterone during the onset of puberty in bulls. *J Reprod Fertil* 1996;106:79–86.
38. Renaville R, Van Eenaeme C, Breier B et al. Feed restriction in young bulls alters the onset of puberty in relationship with plasma insulin-like growth factor-I (IGF-I) and IGF-binding proteins. *Domest Anim Endocrinol* 2000;18:165–176.
39. Spiteri-Grech J, Nieschlag E. The role of growth hormone and insulin-like growth factor I in the regulation of male reproductive function. *Horm Res* 1992;38(Suppl. 1):22–27.
40. Bellve A, Zheng W. Growth factors as autocrine and paracrine modulators of male gonadal functions. *J Reprod Fertil* 1989;85:771–793.
41. Cailleau J, Vermeire S, Verhoeven G. Independent control of the production of insulin-like growth factor I and its binding protein by cultured testicular cells. *Mol Cell Endocrinol* 1990;69:79–89.
42. Amstalden M, Harms P, Welsh T, Randel R, Williams G. Effects of leptin on gonadotropin-releasing hormone release from hypothalamic–infundibular explants and gonadotropin release from adenohypophyseal primary cell cultures: further evidence that fully nourished cattle are resistant to leptin. *Anim Reprod Sci* 2005;85:41–52.
43. Bagu E, Gordon J, Rawlings N. Postnatal changes in testicular concentrations of transforming growth factors-alpha and -beta 1, 2 and 3 and serum concentrations of insulin like growth factor I in bulls. *Reprod Domest Anim* 2010;45:348–353.
44. Bagu E, Gordon J, Rawlings N. Post-natal changes in testicular concentrations of interleukin-1 alpha and beta and interleukin-6 during sexual maturation in bulls. *Reprod Domest Anim* 2010;45:336–341.

BREEDING AND HEALTH MANAGEMENT

5. Bull Development: Sexual Development and Puberty in Bulls — 41
   Leonardo F.C. Brito

6. Breeding Soundness Examination in the Bull: Concepts and Historical Perspective — 58
   Richard M. Hopper

7. Evaluation of Breeding Soundness: The Physical Examination — 64
   James Alexander

8. Evaluation of Breeding Soundness: Basic Examination of the Semen — 68
   Richard M. Hopper and E. Heath King

9. Ultrasound Evaluation of the Reproductive Tract of the Bull — 79
   Harry Momont and Celina Checura

10. Management of Breeding Bull Batteries — 92
    E. Heath King

11. Management of Bulls at Custom Collection Studs — 97
    Gary Warner

12. Testicular Degeneration — 103
    Albert Barth

13. Vesicular Adenitis — 109
    Albert Barth

14. Inability to Breed due to Injury or Abnormality of the External Genitalia of Bulls — 113
    Herris Maxwell

Chapter 5

Bull Development: Sexual Development and Puberty in Bulls

Leonardo F.C. Brito

ABS Global, DeForest, Wisconsin, USA

Introduction

Age at puberty is a major determinant of cattle production efficiency. The ability to breed animals at younger ages reduces generation intervals and increases genetic gains. However, reduced sperm production and poor semen quality due to immaturity are common causes of poor reproductive performance of young bulls and represent a serious loss of superior genetic stock. The ability to collect and freeze semen from younger bulls is also desired to reduce the time required for progeny testing and to accelerate the process of artificial insemination and sire selection. Therefore, an understanding of pubertal changes and the factors that affect sexual development is required in order to promote the successful use of young bulls for reproductive purposes.

Testicular development

Sexual development is associated with marked gonadal growth. Scrotal circumference (SC) is highly correlated with testicular weight (Figure 5.1) and is the most common endpoint evaluated to determine testicular development. The testicular growth curve in bulls shows an initial period of little growth followed by a rapid growth phase and then by a plateau (Figure 5.2). Although the overall pattern of testicular growth is fairly similar in all breeds, the characteristics of the growth curve are greatly affected by genetics. In general, the rapid growth phase is shorter and testicular growth plateaus sooner in bulls from breeds that mature faster (reach puberty earlier) than in bulls from late-maturing breeds, resulting in marked differences in the curve slope. This is especially evident when *Bos taurus* bulls are compared with *Bos indicus* bulls, which in general reach puberty later than the former. The asymptotic value of the testicular growth curve, namely adult testicular size (Figure 5.3), also differs considerably among breeds.[1–7] These same differences can be observed within breeds between early- and late-maturing bulls (Figure 5.4), emphasizing the effects of genetics on testicular growth.[8–10]

SC is a moderately heritable trait in cattle; yearling heritability estimates are 0.36–0.55 in Angus,[11–15] 0.28 in Brahman,[16] 0.40–0.71 in Hereford,[13,14,17–23] 0.67 in Holstein,[24] 0.46 in Limousin,[25] 0.39–0.60 in Nelore,[26–29] 0.32 in Red Angus,[30] and 0.48 in Simmental bulls.[13] Therefore, direct selection can have a very significant impact on SC. For example, selection of Santa Gertrudis bulls based on minimum SC over a 10-year period resulted in significant changes in average SC in one herd,[31] whereas testicular weight at weaning was greater in the progeny sired by Limousin bulls with high expected progeny difference (EPD) for SC compared with progeny sired by bulls with average or low EPD.[32] Several studies have also demonstrated moderate to high phenotypic correlations between SC and growth traits and estimates of the genetic correlations with growth traits are generally positive (Table 5.1). Therefore, either the combination of direct selection for SC and/or indirect selection for growth traits is likely responsible for the general trend of increasing SC over the years in certain breeds (Figure 5.5).

Heritability estimates for SC vary according to age. Studies have demonstrated that heritability estimates increase with age until approximately 1 year of age (or 15–18 months of age in *Bos indicus* bulls), whereas estimates for 2-year-old bulls are lower.[4,14,17,24,26,27,29,33] Therefore, selection based on yearling SC is recommended over selection based on measurements obtained at other ages. Yearling SC is commonly recorded in performance evaluation programs for beef bulls, but age effects are very pronounced around this age since testicular growth is rapid during this stage of development (Figure 5.6). In order to adjust SC measurements to age 365 days,

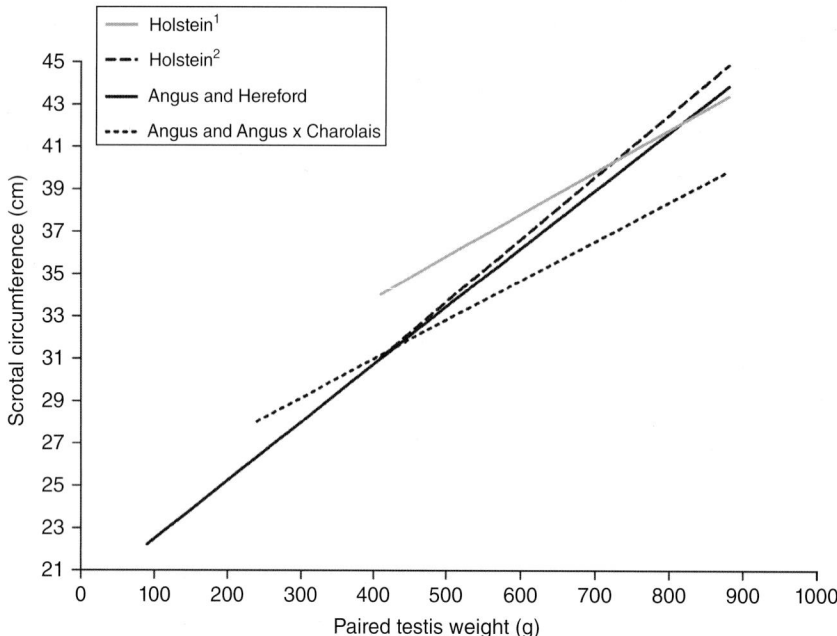

Figure 5.1 Regression lines for paired-testes weight (PTW) according to scrotal circumference (SC). Holstein[1], measurements obtained from mature Holstein bulls ($n=35$); PTW = $-1298.5 + (50.2 \times SC)$.[116] Holstein[2], measurements obtained from Holstein bulls ($n=47$) 19–184 months old; PTW = $-654.4 + (34 \times SC)$.[117] Angus and Hereford SC measurements obtained from Hereford ($n=199$) and Angus ($n=136$) bulls 11–30 months old; PTW = $-722.28 + (36.53 \times SC)$.[3] Angus and Angus × Charolais SC measurements obtained from Angus and Angus × Charolais bulls ($n=111$) 14–16 months old; PTW = $-1274 + (54.04 \times SC)$.

the Beef Improvement Program recommends use of the adjustment factors described in Table 5.2. Correlation coefficients between SC at 1 year of age and SC and paired-testes weight at 2 years of age in Angus and Hereford and bulls were 0.76 and 0.65, respectively, demonstrating that a bull with relatively small or large testes as a yearling will generally have comparable testes size as a 2 year old.[3]

Attempts to establish guidelines for selection of bulls at weaning based on the likelihood of attainment of certain minimum yearling SC have produced mixed results. In one study, it was recommended that the minimum SC in Angus and Simmental bulls 198–291 days old should be 23 or 25 cm to ensure an SC of 30 or 32 cm at 365 days of age, respectively; the same recommendations for Hereford bulls were 26 and 28 cm.[7] In another study, differences between bulls that attained a minimum yearling SC of 34 cm and bulls that did not were observed for adjusted SC at 200 days of age (23.3 vs. 20.5 cm, respectively). Based on these results, it was suggested that SC at weaning could be used to select bulls for breeding and 23 cm was proposed as the minimum SC standard at 200 days.[34] However, this study included bulls from several breeds with known differences in patterns of testicular growth and mature size, while using a singular and very strict yearling SC minimum. SC at 240 days of age could be used as a tool to select bulls with a high probability of meeting the minimum requirements for SC at 365 days of age (i.e., Simmental 32 cm; Angus, Charolais, and Red Poll 31 cm; Hereford 30 cm; Limousin 29 cm); sensitivity and specificity analysis for determining cutoff values indicated that the probability of Charolais bulls with SC ≥24 cm, Simmental and Limousin bulls with SC ≥22 cm, and Angus, Hereford and Red Poll bulls with SC ≥21 cm attaining minimum requirements was greater than 80%. However, SC at weaning was not useful as a culling tool, since a large portion of bulls, irrespective of breed, met the minimum requirements at 365 days of age even when SC was below 21 cm at 240 days of age.[35]

Although the heritability of semen traits is generally low, SC is positively associated with sperm production and semen quality and genetic correlations between SC and semen traits are generally favorable (Table 5.1). This suggests that direct selection for SC would be more effective in bringing about sperm production and semen quality improvement than direct selection pressure on semen traits themselves. In addition, several studies have reported an association between sire SC and daughter puberty. In Brahman and Hereford cattle, genetic correlations between SC and heifer ages at first detected ovulatory estrus, first breeding, and first calving were −0.32, −0.39 and −0.38, respectively.[16,36] In another study with beef cattle, favorable relationships between greater sire SC and ages at puberty and at first calving were demonstrated by negative correlation coefficients between the two traits.[37] In a population of composite beef cattle, the correlation coefficient among parental breed group means for SC and percentage of pubertal females at 452 days of age was 0.95, whereas the correlation with female age at puberty was −0.91.[5] A significantly greater proportion of females had reached puberty at 11 and 13

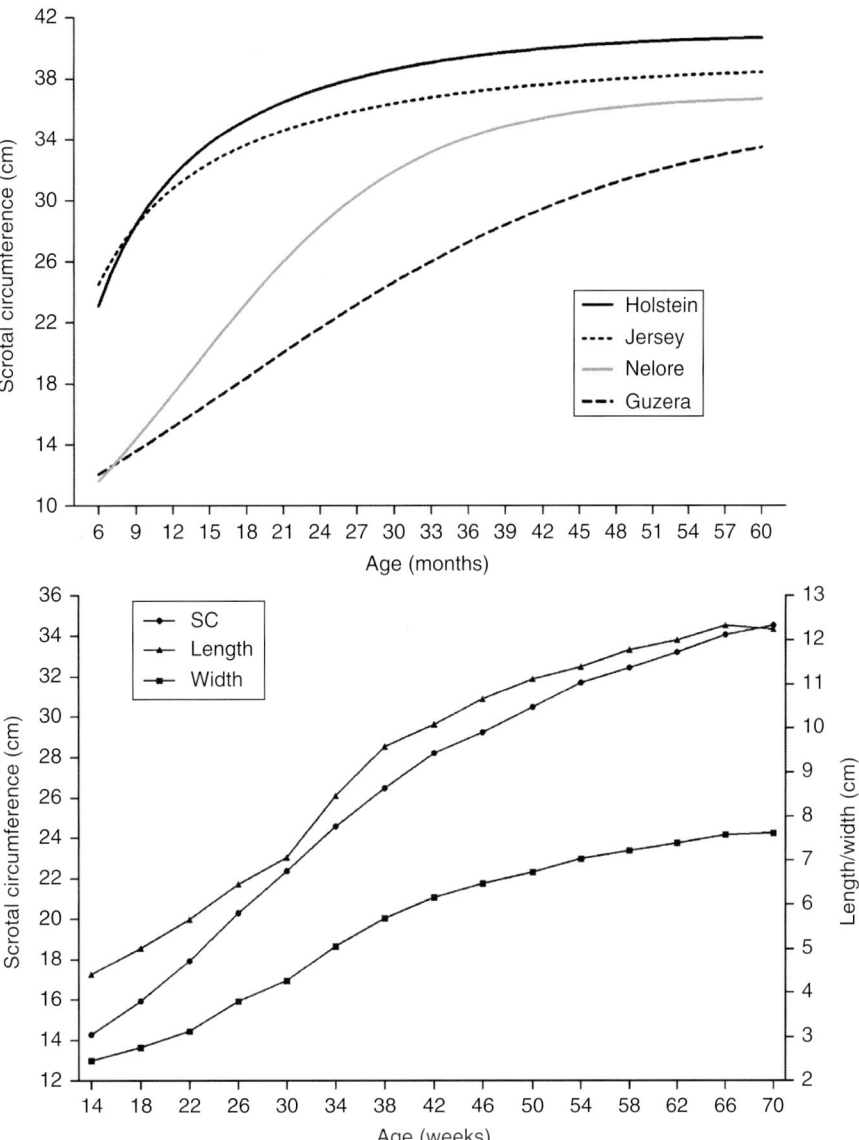

Figure 5.2 *Top*: Regression curves for scrotal circumference (SC) according to age in Holstein, Jersey, Nelore, and Guzera bulls. Holstein measurements ($n = 9614$) obtained from bulls 6–77 months old; SC = $-11.75 + [56.7 \times (\log Age)] - [15.3 \times (\log Age)^2]$ (ASB Global Inc., unpublished results). Jersey measurements ($n = 1038$) obtained from bulls 7–75 months old; SC = $-0.6814 + [40.26 \times (\log Age)] - [10.27 \times (\log Age)^2]$ (ABS Global Inc., unpublished results). Nelore measurements obtained from bulls ($n = 532$) 7–43 months old; SC = $36.9/1 + [4.22^{-(0.11 \times age)}]$.[118] Guzera measurements ($n = 7410$) obtained from bulls 2–69 months old; SC = $35.96/1 + [2.86^{-(0.002 \times age \text{ in days})}]$.[119] *Bottom*: Testicular length and width associations with SC between 3 and 16 months of age in Angus and Angus × Charolais bulls ($n = 111$) receiving adequate nutrition.

months of age when sired by Limousin bulls with high SC EPD compared with females sired by bulls with low or average EPD.[32]

Although sire SC is associated with daughter puberty, evaluation of the genetic correlation between SC and pregnancy rates has produced low estimates that in some cases are not different from zero.[28,30,38] A possible explanation for these observations is a nonlinear relationship between the traits. One study in Hereford cattle indicated that the effect of SC breeding values on heifer pregnancy exhibits a threshold relationship. As SC increases in value, there is a diminishing return for improved heifer pregnancy, suggesting that selection for a high SC breeding value may not be an advantage for increased heifer pregnancy over selection for a moderate SC breeding value.[22] Although it would seem that the favorable genetic relationship between SC and age at puberty does not completely translate to heifer pregnancy, it is important to note that the experimental design might have confounded some of the referred results, since it is obvious that when the entire group of heifers reach puberty before exposure to breeding, those heifers reaching puberty at younger ages would have no advantage in conception over those reaching puberty at older ages. Moreover, end-of-season pregnancy rates were used in these studies as opposed to per-cycle pregnancy rates and the value of having heifers conceiving early rather than late in the season might have been lost.

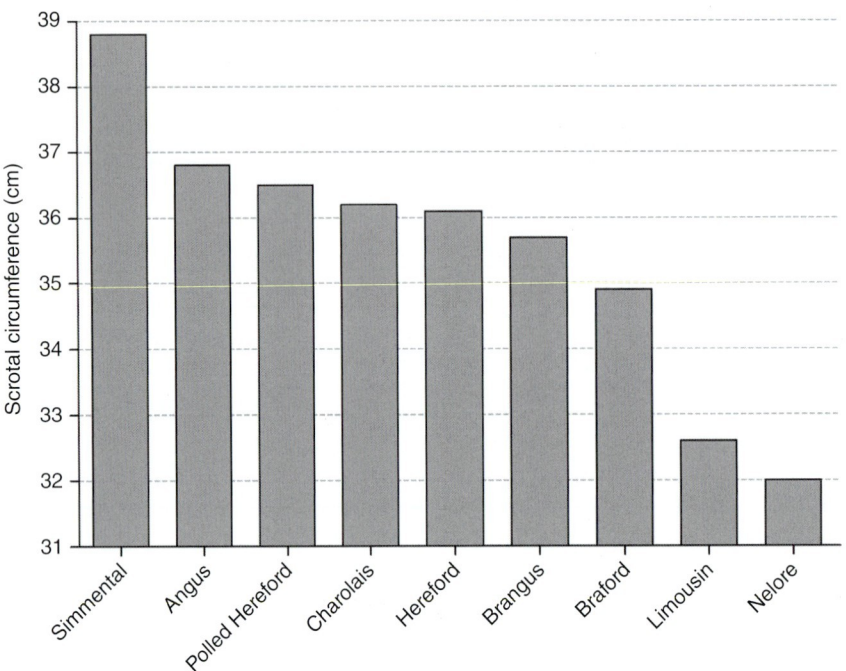

Figure 5.3 Weighted mean scrotal circumference (SC) reported for 2-year-old bulls of various breeds: Simmental, $n=540$; Angus, $n=2310$; Polled Hereford, $n=3546$; Charolais, $n=1015$; Hereford, $n=4369$; Brangus, $n=1312$; Braford, $n=1210$; Limousin, $n=149$; Nelore, $n=5903$.[4, 120, 121]

Puberty

After spermatogenesis is established, there is a gradual increase in the number of testicular germ cells supported by each Sertoli cell and an increase in the efficiency of the spermatogenesis, i.e., an increase in the number of more advanced germ cells resulting from the division of precursor cells. The yields of different germ cell divisions, low during the onset of spermatogenesis, increases progressively to the adult level.[39–41] Testicular histological changes and increasing efficiency of spermatogenesis are accompanied by increasing testicular echogenicity. Testicular ultrasonogram pixel intensity starts to increase approximately 12–16 weeks before puberty, and reaches maximum values right around puberty[42] (Figure 5.7). If the initial changes in testicular echogenicity are associated with Sertoli cell differentiation and meiosis is not completed until formation of a functional blood–testis barrier, then 12–16 weeks seems to be the interval required for the gradual increase in the efficiency of spermatogenesis that eventually leads to the appearance of sperm in the ejaculate. That testicular echogenicity does not change significantly after puberty indicates that a certain developmental stage of the testicular parenchyma must be reached before puberty, a conclusion corroborated by the observation that testicular echotexture at puberty did not differ between early- and late-maturing bulls.[9] In addition, testicular echogenicity did not change with age in mature bulls,[43] suggesting that the composition of testicular parenchyma remained relatively consistent after puberty.

In general terms, puberty is defined as the process by which a bull becomes capable of reproducing. This process involves development of the gonads and secondary sexual organs, and development of the ability to breed. For research purposes, however, puberty in bulls is usually defined as an event instead of a process. Most researchers define attainment of puberty by the production of an ejaculate containing 50 million or more sperm with 10% or more motile sperm.[44] The interval between the first observation of sperm in the ejaculate and puberty as defined by these criteria is approximately 30–40 days in *Bos taurus* bulls.[45,46] Age at puberty determined experimentally can be affected by the age that semen collection attempts are performed, the interval between attempted collections, the method of semen collection (artificial vagina or electroejaculator), the response of the bull to the specific semen collection method, and the experience of the collector(s). Moreover, age at puberty is affected by management, nutrition (see below), and genetics. Table 5.3 describes weight, SC, and age at puberty in different breeds. Although data from large trials comparing bulls of different breeds raised as contemporary groups are scarce, some liberties could be taken to make some generalizations. Dairy bulls usually mature faster and attain puberty earlier than beef bulls. Bulls from continental beef breeds (with the exception of Charolais) usually attain puberty later than bulls from British beef breeds, especially Angus bulls. Bulls from double-muscled breeds are notorious for being late-maturing. Puberty is delayed in bulls from tropically adapted *Bos taurus* breeds and in nonadapted bulls raised in the tropics. In general, *Bos indicus* bulls attain puberty at considerably older ages than *Bos taurus* bulls.

There is large variation in age and body weight at puberty across breeds and within breeds. Although on average *Bos taurus* bulls attain puberty with SC between 28 and 30 cm regardless of the breed, the fact that there is still considerable variation in SC at puberty is sometimes

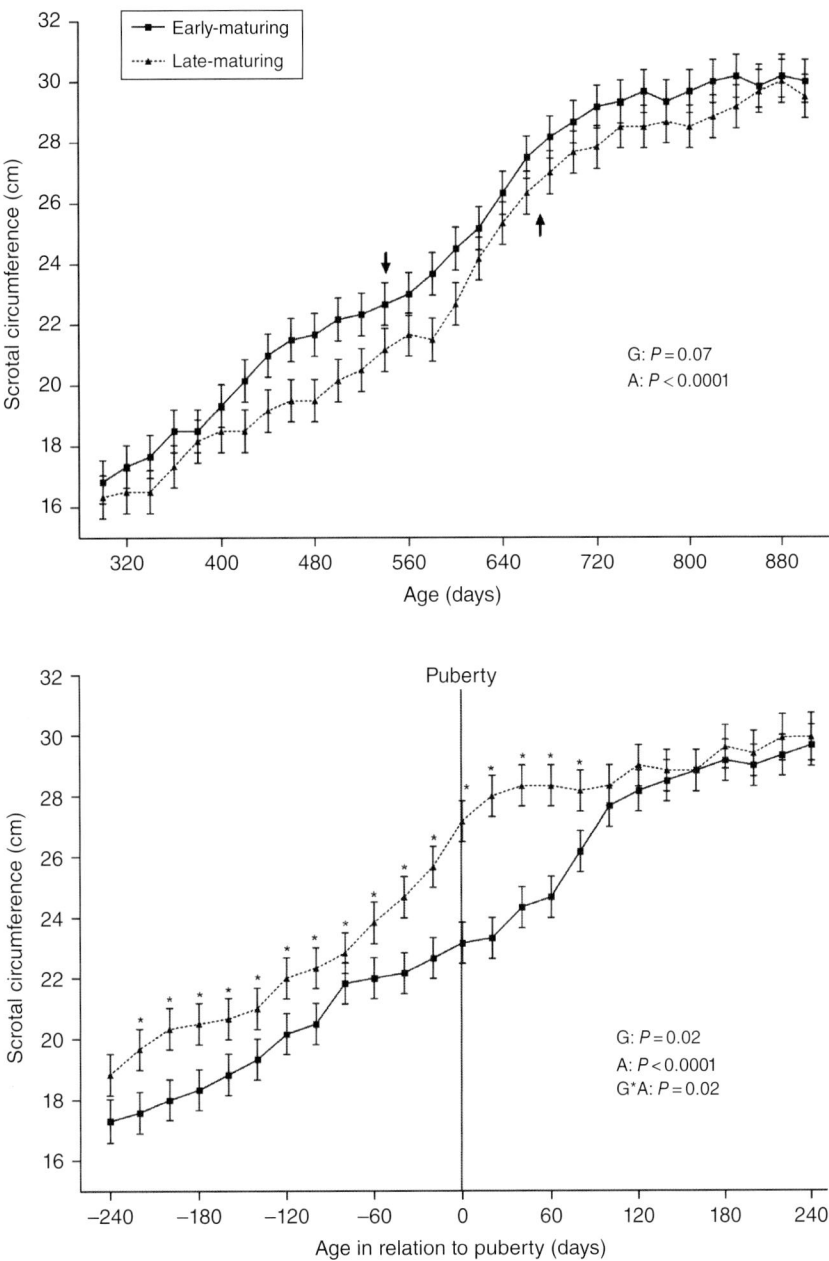

Figure 5.4 Mean (± SEM) scrotal circumference (SC) according to age (*top*) and age at puberty (*bottom*) in early- and late-maturing Nelore bulls (*n* = 6 per group). Arrows indicate mean age at puberty (ejaculate containing ≥50 × 10⁶ sperm with ≥10% motile sperm; downward arrow indicates early; upward arrow indicates late). Early-maturing bulls had greater body weight (not shown) and SC than late-maturing bulls during the entire experimental period. In addition, early-maturing bulls were lighter and had smaller SC at puberty than late-maturing bulls, indicating that sexual precocity is not related to attainment of a threshold body or testicular development earlier, but that these thresholds are lower in early-maturing bulls. G, A and G × A indicate group, age, and group-by-age effects, respectively. Means with superscripts indicate differences between groups within age (a, $P<0.05$; b, $P<0.005$; c, $P<0.01$). Adapted from Brito L, Silva A, Unanian M, Dode M, Barbosa R, Kastelic J. Sexual development in early- and late-maturing *Bos indicus* and *Bos indicus* × *Bos taurus* crossbred bulls in Brazil. *Theriogenology* 2004;62:1198–1217 with permission from Elsevier.

overlooked. Interesting observations have been reported in studies evaluating differences between early- and late-maturing bulls. Bulls that attain puberty earlier were generally heavier and had greater SC than bulls that attained puberty later; however, both weight and SC were smaller at puberty in early-maturing bulls[8,9,47] (Figure 5.4). These observations not only indicate that precocious bulls develop faster, but also suggest that sexual precocity is not simply related to earlier attainment of a threshold body or testicular development. In fact, these thresholds seem to be lower in early-maturing bulls, and late-maturing bulls must reach a more advanced stage of body and testicular development before puberty is attained.

Table 5.1 Genetic correlations (r_g) between scrotal circumference and growth traits in bulls.

Breed	Growth trait	r_g	Reference
Angus	Yearling weight	0.24–0.68	Knights et al.[12]; Meyer et al.[14]; Garmyn et al.[15]
	Sperm concentration	0.54	
	Sperm motility	0.36	
	Total sperm defects	−0.23	
Composite	Yearling weight	0.40–0.43	Mwansa et al.[33]
Hereford	Weaning weight	0.08–0.86	Meyer et al.[14]; Neely et al.[17]; Nelsen et al.[18]; Bourdon & Brinks[19]; Crews & Porteous[20]; Kriese et al.[21]; Kealey et al.[23]
	Yearling weight	0.30–0.52	
	Weaning–yearling ADG	0.22–0.35	
	Sperm concentration	0.77	
	Sperm motility	0.34	
	Normal sperm	0.33	
Hereford/Simmental	Sperm concentration	0.20	Gipson et al.[13]
	Sperm motility	0.11	
	Total sperm number	0.19	
Limousin	Weaning weight	0.14	Keeton et al.[25]
Nelore	Weaning weight	0.36	Yokoo et al.[26]; Boligon et al.[27]
	Yearling weight	0.34	
	Longissimus muscle area	0.28	
	Backfat thickness	0.17	
Red Angus	Yearling intramuscular fat	0.05	McAllister et al.[30]
	Yearling carcass marbling score	0.01	
Various breeds	Birth–weaning ADG	0.02	Lunstra et al.[85]; Smith et al.[107]
	Yearling weight	0.10–0.63	
	Weaning weight	0.56	
	Weaning–yearling ADG	0.59	

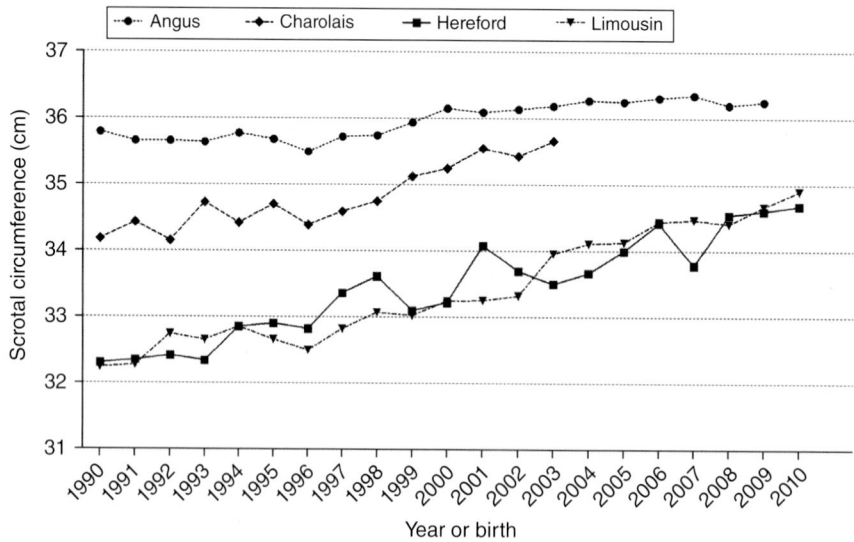

Figure 5.5 Mean scrotal circumference (SC) in registered yearling bulls according to year of birth. Measurements for Angus (unspecified number of bulls) and Limousin ($n = 73\,757$; 1184–5200/year) bulls are adjusted to 365 days of age (sources: North American Limousin Foundation and American Angus Association). Measurements for Charolais ($n = 6984$; 121–997/year) and Hereford ($n = 5553$; 360–536/year) bulls are unadjusted measurements obtained between 321 and 421 days of age (sources: Canadian Hereford Association and Canadian Charolais Association).

Bull Development: Sexual Development and Puberty in Bulls

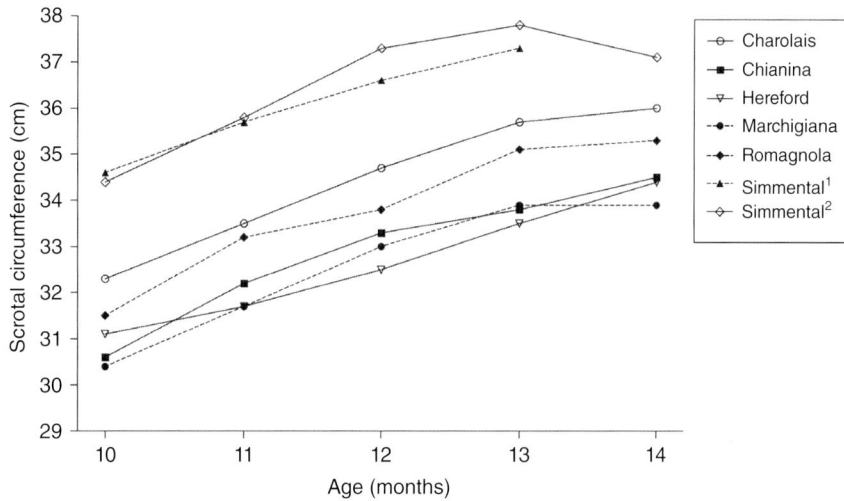

Figure 5.6 Mean scrotal circumference (SC) in yearling bulls. Charolais: $n = 246$ to 2622/age (Canadian Charolais Association). Chianina, Marchigiana, and Romagnola: $n = 455, 415$, and 425/age, respectively.[122] Hereford: $n = 77$ to 2510/age (Canadian Hereford Association). Holstein: $n = 162$ to 1004/age (ABS Global Inc.). Simmental[1]: $n = 129$ to 2276/age (Canadian Simmental Association). Simmental[2]: $n = 120$ to 2022/age (American Simmental Association). Standard deviations are all between 2 and 3 cm.

Table 5.2 Age adjustment factors for scrotal circumference (SC) at 365 days of age according to breed.

Breed	Age adjustment factor
Angus	0.0374
Charolais	0.0505
Gelbvieh	0.0505
Hereford	0.0425
Limousin	0.0590
Red Angus	0.0324
Simmental	0.0543

365-day SC = actual SC + [(365 − age) × age adjustment factor].[108]

Spermatogenesis efficiency reaches adult levels at approximately 12 months of age in Holstein bulls[39,48] and 2.5–3.5 years of age in *Bos indicus* bulls.[49] Individual variation in spermatogenesis efficiency is relatively small and is not affected by ejaculation frequency; values between 10 and 14 million sperm per gram of testicular parenchyma have been reported for bulls.[1,6,39,48–54] Since spermatogenesis efficiency is somewhat constant among bulls, daily sperm production of a bull is largely dependent on the weight of the testes. Considering testicular weight at different ages, yearling *Bos taurus* bulls are expected to produce around 4–5 billion sperm per day, whereas adult bulls are expected to produce around 7–9 billion sperm per day. Sperm output (number of sperm in the ejaculate) in bulls ejaculated frequently is essentially the same as sperm production.[51] One important difference between young and older bulls is the capacity of the epididymis to store sperm. Evaluation of sperm numbers in the tail of the epididymis in 15- to 17- month-old Holstein bulls demonstrated that sperm available for ejaculation corresponded to approximately 1.5–2 days of sperm production, whereas in 2- to 12-year-old bulls stored sperm numbers corresponded to approximately 3.5–5 days of sperm production.[55,56] These observations are especially important for artificial insemination centers and indicate that more frequent semen collection is necessary to maximize sperm harvest from young bulls, whereas semen collection intervals of less than 3 days have smaller effects on increasing sperm harvest from older bulls. Sperm output increases with increased ejaculation interval up to the number of days required for epididymal storage capacity to reach its limit. Sperm that are not ejaculated are eliminated with urine or during masturbation.

Semen quality in peripubertal bulls is poor and a gradual improvement characterized by increase in sperm motility and reduction in morphological sperm abnormalities is observed after puberty. The most prevalent sperm defects observed in peripubertal bulls are proximal cytoplasmic droplets and abnormal sperm heads (approximately 30–60% and 30–40% at puberty, respectively; Figure 5.8).[8,47,57] The difference between age at puberty and age at satisfactory semen quality (≥30% sperm motility, ≥70% morphologically normal sperm) was 110 days in *Bos indicus* bulls[9] and 50 days in *Bos taurus* beef bulls; 10% of the latter did not have satisfactory semen quality by 16 months of age[58] (Figure 5.8). In western Canada, the proportions of *Bos taurus* beef bulls with satisfactory sperm morphology (≥70% morphologically normal sperm) at 11, 12, 13, and 14 months of age were approximately 40, 50, 60, and 70%, respectively.[59] Similarly, only 48% of *Bos taurus* beef bulls 11–13 months old in Sweden had less than 15% proximal cytoplasmic droplets and less than 15% abnormal sperm heads.[60]

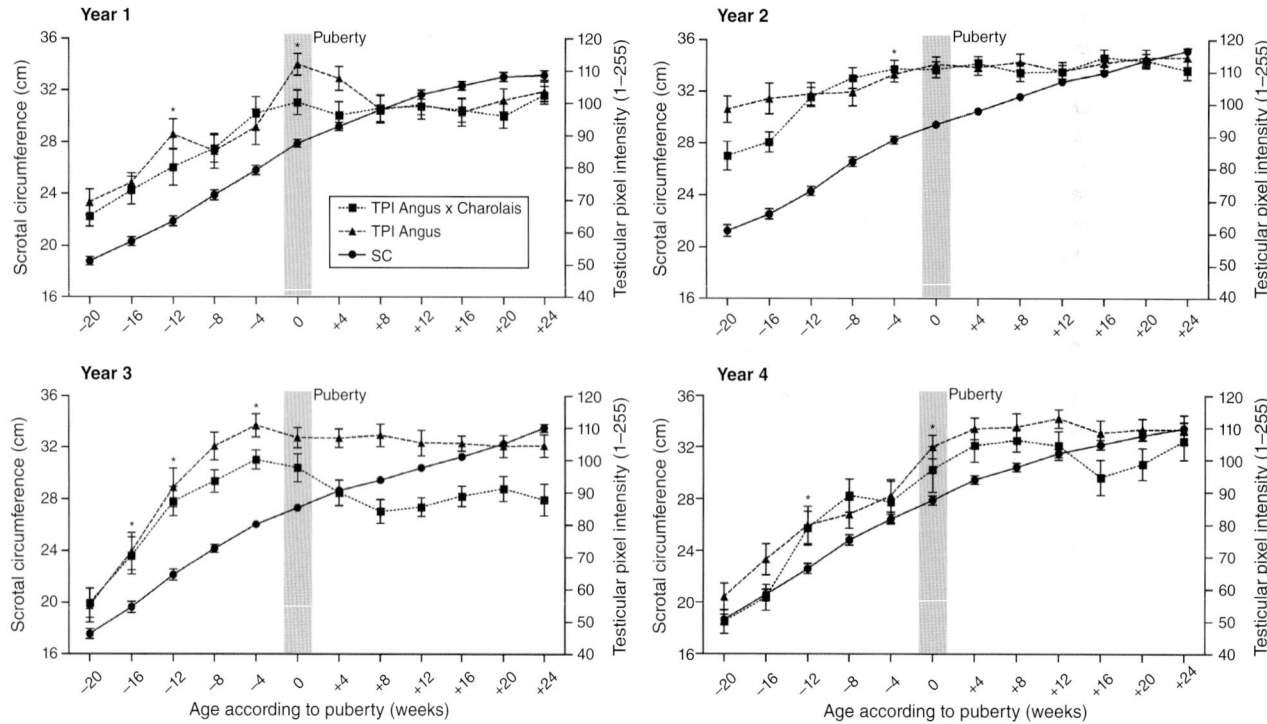

Figure 5.7 Mean (± SEM) scrotal circumference (SC) and testicular ultrasonogram pixel intensity (TPI) according to age at puberty in Angus and Angus × Charolais bulls (Year 1, $n=37$; Year 2, $n=39$; Year 3, $n=43$; Year 4, $n=33$). TPI, determined on a scale of 1 (black) to 255 (white), started to increase 16–12 weeks before puberty and reached maximum values 4 weeks before or at puberty. These results indicate that a certain developmental stage of the testicular parenchyma must be reached before puberty and that the composition of the parenchyma remains consistent after puberty. Overall, TPI was greater ($P<0.0001$) in Angus × Charolais than in Angus bulls. TPI means with superscript asterisks indicate overall change ($P<0.05$) with age. Adapted from Brito L, Barth A, Wilde R, Kastelic J. Testicular ultrasonogram pixel intensity during sexual development and its relationship with semen quality, sperm production, and quantitative testicular histology in beef bulls. *Theriogenology* 2012;78:69–76 with permission from Elsevier.

Table 5.3 Age, weight, and scrotal circumference (SC) at puberty (ejaculate with ≥50 million sperm and ≥10% sperm motility) in different breeds.

Breed	Age (months)[a]	Weight (kg)	SC (cm)	References
Angus	10.1	309	30.0	Wolf et al.[44]
Bos taurus beef crosses	7.8–9.7	272–339	27.9–28.3	Lunstra & Cundiff[6]; Lunstra et al.[45]; Casas et al.[109]
Brahman	15.9–17.0	350–430	28.2–33.0	Chase et al.[110]; Fields et al.[111]; Rocha et al.[112]; Silva-Mena[113]
Brown Swiss	8.7–10.2	233–295	25.9–27.2	Lunstra et al.[45]; Jimenez-Severiano[46]
Charolais	9.4	396	28.8	Barber & Almquist[70]
Guzera	18.2	310	25.6	Troconiz et al.[114]
Gyr	17–19.2	315–346	26.2–27.9	Martins et al.[10]
Hereford	9.6–11.7	261–391	27.9–32.0	Evans et al.[8]; Wolf et al.[44]; Lunstra et al.[45]; Pruitt et al.[91]
Holstein	9.4–10.9	276–303	28.4	Jimenez-Severiano[46]; Killian & Amann[48]
Nelore	14.8–19.7	232–298	21.7–24.3	Brito et al.[9]; Troconiz et al.[114]; Freneau et al.[115]
Red Poll	9.3	258	27.5	Lunstra et al.[45]
Romosinuano	14.2	340	28.8	Chase et al.[110]
Simmental	10.6–11.4	328–419	30.6–34.0	Pruitt et al.[91]

[a] Transformed from days or weeks from original reports.

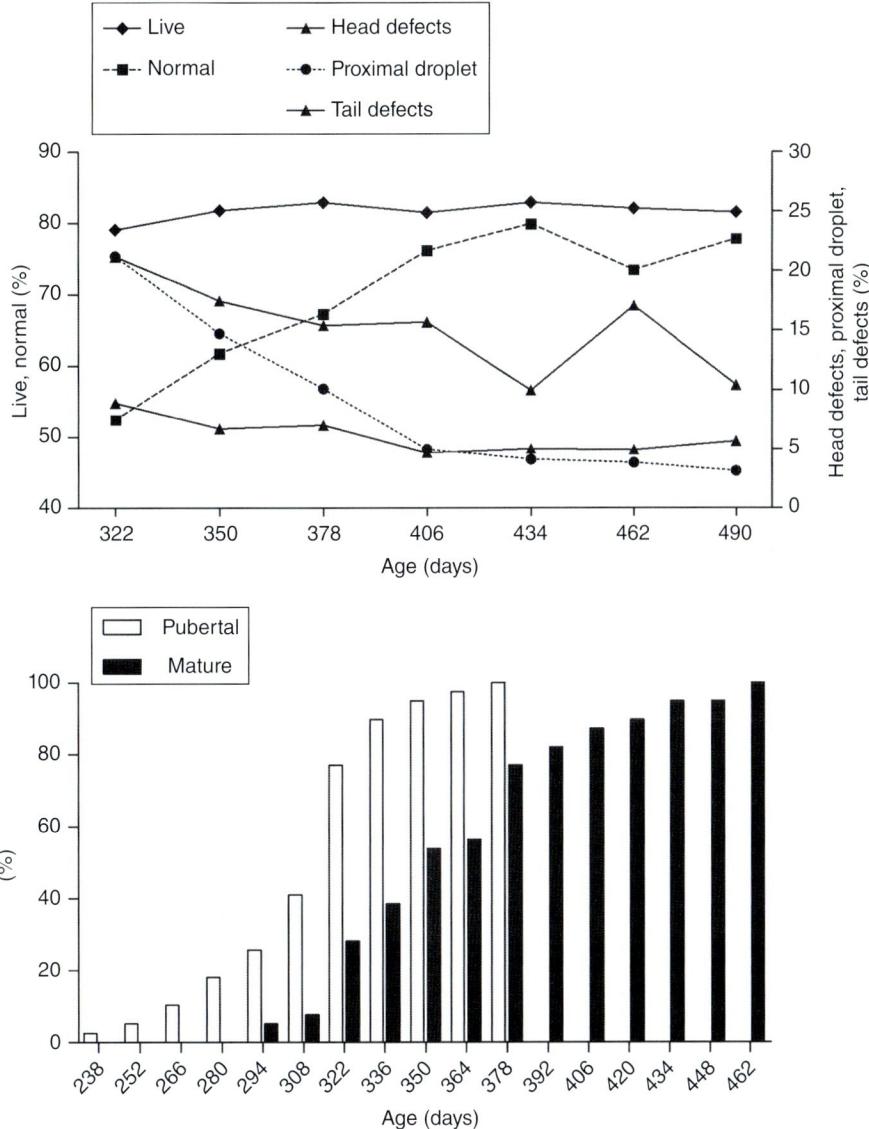

Figure 5.8 *Top*: Mean sperm viability and morphology according to age in Angus and Angus × Charolais bulls (*n* = 39). *Bottom*: Proportion of pubertal (ejaculate containing ≥50 × 10⁶ sperm with ≥10% motile sperm) and mature (ejaculate containing ≥30% motile and ≥70% morphologically normal sperm) bulls according to age. The interval between puberty and maturity was approximately 50 days. Adapted from Brito L, Barth A, Wilde R, Kastelic J. Effect of growth rate from 6 to 16 months of age on sexual development and reproductive function in beef bulls. *Theriogenology* 2012;77:1398–1405 with permission from Elsevier.

These observations have profound implications on the ability of producers to use yearling bulls and the ability of artificial insemination centers to produce semen for progeny testing at the youngest possible age.

Development of accessory sex organs

Anatomical changes of the testicular vascular cone as the bull ages and starts to produce sperm are indications that the efficiency of the counter-current heat exchange mechanism between the testicular artery and veins in the pampiniform plexus needs to improve in order to cope with increasing testicular metabolism. In *Bos taurus* beef bulls the testicular vascular cone diameter measured by ultrasonography increases until approximately 13.5 months of age, or until 1–8 weeks before SC reaches a plateau[61] (Figure 5.9). In crossbred beef bulls, the length and diameter of the testicular artery in the vascular cone increased from 6 to 12 months of age (1.8 m and 1.9 mm vs. 3.1 m and 3.5 mm, respectively), but did not increase significantly thereafter.[62] The testicular artery length and volume in the vascular cone were 1.6 m and 6 mL in 15-month-old Angus bulls and 2.2 m and 11.4 mL in 28–month-old crossbred bulls, respectively.[54] Other studies have reported that the testicular vascular cone length is approximately 10–15 cm and that testicular artery length varies from 1.2 to 4.5 m in adult beef bulls of several breeds.[63–65] In addition to the lengthening of the testicular artery, the distance between the arterial and venous blood in the testicular vascular cone also decreases with age as

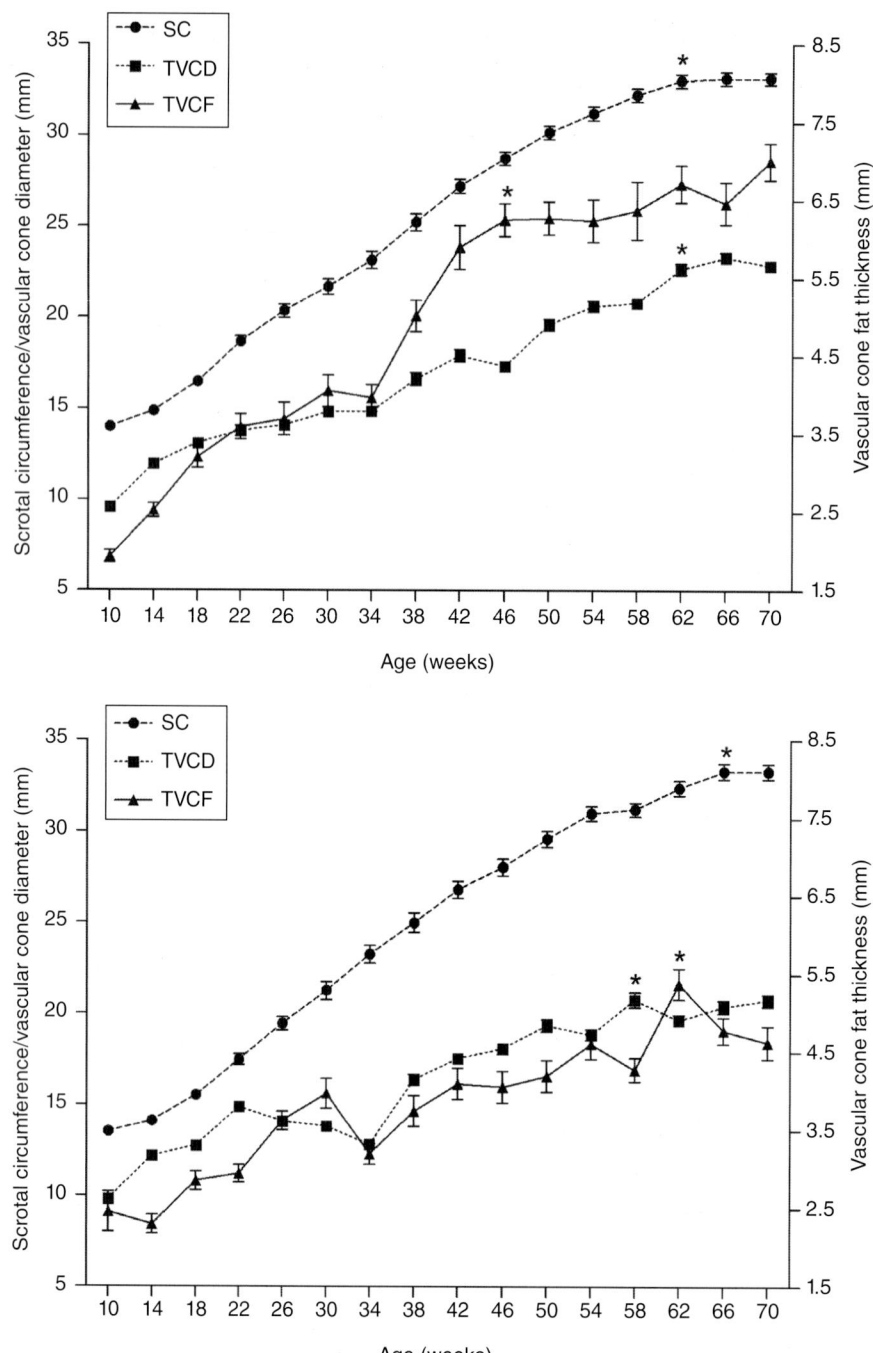

Figure 5.9 Mean (± SEM) scrotal circumference (SC), testicular vascular cone diameter (TVCD), and fat thickness (TVCF) in Angus and Angus × Charolais bulls in 2 years (*top*: n=37; *bottom*: n=33). Testicular vascular cone diameter increased with age following testicular development, whereas vascular cone fat thickness increased similar to a pattern observed for body backfat. Means with superscript asterisks indicate last significant ($P<0.05$) change with age. Adapted from Brito L, Barth A, Wilde R, Kastelic J. Testicular vascular cone development and its association with scrotal temperature, semen quality, and sperm production in beef bulls. *Anim Reprod Sci* 2012;134:135–140 with permission from Elsevier.

a result of thinning of the artery wall (317 and 195 μm at 6 and 36 months of age, respectively) and reduction of the distance between the artery and the closest veins.[62]

The epididymis continues to grow until at least 6 years of age in Holstein bulls and epididymal weight increases from 9 g at 8 months of age to 15, 23, 27, and 38 g at 12, 18, 25–48, and 73–96 months of age, respectively.[48,66] The greatest dimension of the epididymis tail measured by ultrasonography from an oblique plane near the distal pole of the testis in Friesian bulls increased from 0.9 cm at 3 months of age to 1.8 and 2.7 cm at 12 and 24 months of age, respectively.[67] The tube-like vesicular glands in newborn calves increase in length and become lobulated during development. The weight of the vesicular glands increased until approximately 4 years of age in Holstein bulls, from 13 g at 8 months of age to 26, 35, 54, and 78 g at 12, 18, 25–48, and 73–96 months of age, respectively.[48,66] The maximum diameter of vesicular glands in Friesian bulls increased significantly only from age 3 months (1.1 cm) to 9 months (1.6 cm), suggesting that further increase in gland weight is a result of increase in length.[67] Prostate diameter also only increases significantly until 12 months of age (0.85 cm from 0.49 cm at 3 months), whereas maximum diameter of the bulbourethral glands increased from 1.2 cm at 3 months of age to 1.5 and 2.1 cm at 12 and 24 months of age, respectively.[67]

In *Bos taurus* bulls, the sigmoid flexure of the penis begins to develop at about 3 months of age, penis length increases by up to five times by the onset of puberty, and length continues to increase until sexual maturity.[68] The penis in Friesian bulls 13–19 months old measured 73–89 cm,[69] whereas the penis in Holstein bulls aged 25 months or older measured 95–106 cm.[66] First protrusion of the penis during mounting was observed at approximately 8 months of age, whereas complete separation of penis and sheath was observed at approximately 8.5 months of age in Angus, Charolais, and Hereford bulls.[44,70] Complete sheath–penile detachment evaluated during electroejaculation was observed around the same time of puberty, whereas first completed service evaluated during libido testing was only observed approximately 1 month after puberty.[45]

Development of sexual behavior

Development of sexual behavior has been studied in Hereford bulls. Mounting in response to an estrual female was first observed by 3, 6, and 9 months of age in 18.5, 26, and 48% of the bulls, respectively. By 12 months of age 59% of the bulls had their first ejaculations, whereas by 15 months 78% of the bulls registered a complete service. The number of services increased with age until 18 months of age.[71] In Angus, Brown Swiss, Hereford, Angus × Hereford, and Red Poll the first completed service was observed around 11 months of age.[45] Rearing seems to affect development of sexual behavior, since Hereford calves raised in individual pens had a greater number of services when tested for the first time than bulls raised in groups, although the differences quickly disappeared once these bulls were grouped together.[72] In another study, the influence of the presence of females during bull rearing was evaluated. In the first 2 hours of being exposed to females in estrus, males raised with females had 73% more services than bulls raised in isolation. However, bulls reared in isolation compared favorably in further tests, illustrating the fact that a learning process occurred rather quickly after exposure to females.[73]

Age and experience have major effects on libido and serving capacity test results. An overall increase in test scores have been observed in several studies with yearling bulls when tests were conducted repeatedly over relatively short intervals, indicating that a maturing and/or learning process occurs rapidly after exposure of bulls to females. Exposure of yearling bulls classified as low serving capacity to estrual females for 4 days resulted in increased serving capacity score when bulls were tested within a week later; 85% of these bulls moved into the medium- or high-serving capacity category.[74] No correlations between the numbers of services at 12 months of age with sexual performance at older ages were observed in Hereford bulls. Only when bulls reached 18 months did individual differences in serving capacity remain consistent from one age to another, i.e., 18, 21, and 24 months.[75] In one study, only 53% of 113 yearling bulls completed a service during 10 min of libido testing,[76] whereas in a different study with yearling Angus and Hereford bulls the variance of the score obtained during eight libido tests conducted over a 2-month period ranged between 69 and 73%.[77] These observations suggest that attempts to predict sexual performance are likely to be unproductive until bulls can fully express their inherent sexual behavior and serving ability. When adult Angus bulls were evaluated, the time required to complete six services when exposed individually to 10 estrous females increased with age from 31 min in 2-year-old bulls to 43, 55, and 67 min in 3-, 4-, and 5-year-old bulls, respectively.[78] However, in another study no difference in libido score or servicing capacity test results between 2- and 3-year-old Angus and Hereford bulls was observed.[79]

Effects of nutrition on sexual development

Very few studies have evaluated the effect of nutrition from birth to maturity on sexual development and reproductive function in bulls. Holstein bulls receiving low nutrition (approximately 60–70% of requirements) from birth were older at the time the first ejaculates containing motile sperm were collected and had smaller testes, whereas bulls receiving high nutrition (approximately 160% of requirements) had earlier puberty and larger testes compared with bulls receiving control nutrition (100% of requirements).[80,81] These observations have been corroborated and expanded in a series of more recent experiments that have also shown that the most pronounced effects of nutrition occur during the prepubertal period. These studies have demonstrated that the adverse effects of low nutrition during the prepubertal period cannot be compensated with improved nutrition during the pubertal period and that the beneficial effects of high nutrition during the prepubertal period are sustained even if maintenance diets are fed thereafter. Low nutrition during the peripubertal period in beef bulls reduced gonadotropin secretion, delayed the increase in circulating testosterone concentrations, delayed puberty, and resulted in decreased testicular size at 16 months of age, whereas high nutrition produced the opposite results.[82–84]

The effects of nutrition on sexual development and reproductive function in bulls are mediated through the hypothalamic–pituitary–testes axis. Nutrition affects the gonadotropin-releasing hormone (GnRH) pulse generator in the hypothalamus, since differences in luteinizing hormone (LH) pulse secretion in bulls receiving different nutrition can be observed even in the absence of differences in pituitary LH secretion capability as determined by GnRH challenge.[83,84] Interestingly, though, when low nutrition was imposed on bulls by limiting the amount of nutrients in a ration fed *ad libitum*, only reduced LH pulse frequency was observed, whereas reduced LH pulse frequency, mean and peak concentrations, and secretion after GnRH challenge were observed when nutrition was restricted by restricting the availability of food.[82] These results seem to indicate that the inhibitory effects of limited availability of nutrients on LH secretion appeared to be exerted only on the hypothalamus, whereas the combination of limited availability of nutrients and the sensation of hunger experienced by bulls with restricted intake affected both hypothalamic and pituitary function, producing a much more severe inhibition of LH secretion. The effect of nutrition on Leydig cell number and/or function in bulls receiving different diets was demonstrated by differences in testosterone secretion after GnRH challenge even in the absence of differences in LH secretion after the challenge.[83,84]

Differences in yearling SC due to age of the dam in beef bulls could also be interpreted as an indication that nutrition during the pre-weaning period affects sexual development, although possible effects *in utero* cannot be completely ruled out. SC in *Bos taurus* beef bulls increases as age of the dam increases until 5–9 years of age and decreases as dams get older. Adjustment factors of 0.7–1.4, 0.2–1.0, 0.1–1.0, and 0.3–0.75 cm for yearling SC have been suggested for bulls raised by 2-, 3-, 4-, and ≥10-year-old dams, respectively.[18–22,85] In these studies, the inclusion of weight as a covariate in the models describing SC resulted in decreased effects of age of the dam, indicating that the effect of age of the dam on testicular growth seems to be primarily the result of age of the dam effects on bull's body weight, likely related to differences in milk production. This theory is also supported by reports that, similarly to that observed in bulls receiving low nutrition, LH secretion after GnRH challenge was greater from 3.5 to 6 months of age in bulls raised by multiparous than in bulls raised by primiparous females.[86]

Several studies reported in the literature describe the effects of nutrition only during the pubertal period, in other words after the initial hormonal changes that regulate sexual development have occurred. In general, these studies indicate that low nutrition has adverse effects on growth and sexual development. In one study, bulls receiving one-third of the amount supplied to their twin controls had lower body and vesicular gland weights, vesicular gland fructose and citric acid contents, and circulating and testicular testosterone concentrations, whereas circulating androstenedione concentrations were increased.[87] In another series of experiments, beef bulls 8–12 months old receiving diets with low levels of crude protein (8, 5, and 1.5%) for periods of 3–6 months had markedly reduced testes, epididymis, and seminal glands weights compared with control bulls fed diets containing 14% crude protein. Moreover, seminiferous tubule diameter and seminiferous epithelium thickness were smaller in bulls with a restricted protein intake.[88,89]

Although low nutrition during the pubertal period has adverse effects on reproductive function, the potential beneficial effects of high nutrition after weaning are questionable at best. Effects of energy on sexual development were not consistent in a study with Simmental and Hereford bulls fed diets with a low, medium, or high energy content (approximately 14, 18, and 23 Mcal/day, respectively) from 7 to 14 months of age. Dietary energy affected sexual development in Simmental bulls, but not in Hereford bulls. Simmental bulls in the high-energy group were heavier and had greater SC and testosterone concentrations than bulls in the low-energy group (in general, the medium-energy group was intermediate). However, increased dietary energy did not hasten age at puberty. The only semen trait affected by dietary energy was semen volume, which was depressed in Simmental bulls in the medium-energy group. Serving capacity was greater for Hereford bulls on the high-energy diet, but medium- and high-energy diets were associated with a decrease in the number of services between two testing periods in Simmental bulls. There was a trend for lower sperm motility and proportion of normal sperm in Simmental bulls fed the low-energy diet.[90,91]

In Holstein bulls producing semen for artificial insemination, high energy intake was associated with visual evidence of weakness of the feet and legs and increased reaction time after 3 years of age.[81] Under field conditions, post-weaning high-energy diets are frequently associated with impaired reproductive function in bulls, likely related to altered testicular thermoregulation due to excessive fat deposition above and around the testes in the scrotum. In one report, sperm motility decreased and the proportion of sperm defects increased with age in Hereford bulls fed to gain more than 1.75 kg/day, which was significantly different from bulls fed to gain approximately 1 kg/day (control). Even after the high-nutrition diet was changed to a control diet, bulls previously receiving high nutrition continued to have lower semen quality. There was greater deposition of fat around the testicular vascular cone in the scrotal neck in bulls in the high-nutrition group and the difference between body and testes temperature was reduced in this group compared with bulls in the control group. This difference was still present after the high-energy diet was changed and the bulls had lost a considerable amount of weight, indicating that fat accumulated in the scrotum is more difficult to lose than other body fat.[92]

In another series of experiments, Angus, Hereford, and Simmental bulls were fed high nutrition (80% grain and 20% forage) or medium nutrition (forage only) from approximately 6.5 until 12–24 months of age. In general, bulls receiving high nutrition had greater body weight and backfat, but paired-testes weight was not affected by diet. Moreover, bulls receiving high nutrition had lower daily sperm production and epididymal sperm reserves, and a greater proportion of sperm abnormalities. The authors indicated that increased dietary energy may adversely affect sperm production and semen quality due to fat deposition in the scrotum, which reduces the amount of heat that can be radiated from the scrotal skin, thereby increasing the temperature of the testes and scrotum.[93–96] Observations from a different study indicated that bulls fed high-nutrition diets had greater SC and scrotal weight than bulls fed medium-nutrition diets, but paired-testes weight was not different between the two groups.[97] Growth rate between 6 and 16 months of age did not affect sexual development and reproductive function in Angus and Angus × Charolais bulls. However, greater body weight at various ages was associated with reduced age at puberty and maturity, and with larger testes at 16 months of age, indicating that improved nutrition might be beneficial, but only when offered before 6 months of age. Average daily gains of 1–1.6 kg/day did not result in excessive fat accumulation in the scrotum, increased scrotal temperature, or reduction in sperm production and semen quality, and could be considered "safe" targets for growing beef bulls.[58]

This summary of the literature supports the intuitive assumption that low nutrition has adverse effects on sexual development and reproductive function regardless of the bull's age. However, most research seems to indicate that high nutrition is only beneficial during the first 6 months of life, which presents a challenge to bull producers. Beef bull calves are usually nursing until 6–8 months of age and very little attention is paid to their nutrition, whereas nutrition offered to dairy bull calves is often suboptimal. Efforts to obtain maximum weight gain during the first months after birth by offering high-nutrition diets and adopting management practices like creep-feeding will be compensated by reduced age at puberty and greater sperm production capacity in adult bulls. It is also clear that although high-nutrition diets after 6 months of age might be associated with greater SC, this effect is likely the result of fat accumulation in the scrotum and not actually greater testicular size. Moreover, sperm production, semen quality, and serving capacity are all compromised in bulls receiving excessive nutrition after this age. Adjusting diets accordingly to maximize growth but to prevent overconditioning after the peripubertal period is advisable.

Implications of sexual development and puberty for breeding soundness evaluation

In general terms, evaluation of bulls for breeding soundness involves a general physical examination, a specific examination of the reproductive system (including measurement of SC), and examination of the semen. According to the guidelines of the Society for Theriogenology (SFT), bulls must be free of any physical abnormalities, have a minimum SC of 30–34 cm depending on age, have 30% or more progressively motile sperm, and have 70% or more morphologically normal sperm in order to be classified as satisfactory potential breeders.[98] Since developing bulls experience significant changes in SC and semen quality over relatively short periods of time, the effect of age and the interaction of age with breed need to be considered for breeding soundness evaluation. The SFT guidelines list the breeding soundness category "classification deferred," which is used for bulls that the veterinarian considers likely to improve with time after reflecting on the bull's signalment, history, test results, and nature of the deficiencies. Most bulls classified in this category are young animals that may not have reached the suggested minimum for SC and semen quality by the time of the examination.

The proportion of yearling *Bos taurus* beef bulls classified as satisfactory potential breeders increases from approximately 55% at 10 months of age, to 72% at 11 months, and 78% at 12 months of age as more bulls reach minimum requirements for SC and semen quality.[99,100] Since there are breed differences in sexual development, a larger proportion of bulls from early-maturing breeds are expected to obtain satisfactory classification at a younger age than bulls from late-maturing breeds. An extreme example was found with tropically adapted Senepol bulls; the proportion of satisfactory yearling bulls was only 25–50%, but this increased to 70–85% at approximately 24 months of age.[101,102] It has also been reported that the proportion of satisfactory bulls decreases after 5–6 years of age as bulls become more prone to develop physical and reproductive problems.[103,104]

Special consideration must also be given to *Bos indicus* bulls, since these animals in general attain puberty at considerably older ages than *Bos taurus* bulls. Although SC in selected mature *Bos indicus* and *Bos taurus* bulls does not differ,[43] using SC guidelines for young bulls devised for the latter unjustifiably penalizes the former. Satisfactory semen quality is also observed at older ages in *Bos indicus* bulls. In Florida, Brahman and Brahman × Nelore bulls only had greater than 70% normal sperm after 24 months of age; these observations are consistent with those in Nelore bulls in Brazil.[9] In addition, the proportion of total sperm defects was approximately 57% in Nelore bulls 12–18 months old[105] and 20% of Nelore bulls were still considered immature at 2 years of age based on the spermiogram.[106] Less than 5% of *Bos indicus* bulls were considered satisfactory potential breeders before 15 months of age and the proportion increased to only 20–50% between 18 and 22 months of age.[101] Accordingly, it might be prudent to delay breeding soundness evaluation for *Bos indicus* bulls until 2 years of age.

References

1. Johnson W, Thompson J, Kumi-Diaka J, Wilton J, Mandell I. The determination and correlation of reproductive parameters of performance-tested Hereford and Simmental bulls. *Theriogenology* 1995;44:973–982.
2. Bell D, Spitzer J, Bridges W Jr, Olson L. Methodology for adjusting scrotal circumference to 365 or 452 days of age and correlations of scrotal circumference with growth traits in beef bulls. *Theriogenology* 1996;46:659–669.
3. Coulter G, Keller D. Scrotal circumference of young beef bulls: relationship to paired testes weight, effect of breed, and predictability. *Can J Anim Sci* 1982;62:133–139.
4. Coulter G, Mapletoft R, Kozub G, Cates W. Scrotal circumference of two-year-old bulls of several beef breeds. *Theriogenology* 1987;27:485–491.

5. Gregory K, Lunstra D, Cundiff L, Koch R. Breed effects and heterosis in advanced generations of composite populations for puberty and scrotal traits of beef cattle. *J Anim Sci* 1991;69: 2795–2807.
6. Lunstra D, Cundiff L. Growth and pubertal development in Brahman-, Boran-, Tuli-, Belgian Blue-, Hereford- and Angus-sired F1 bulls. *J Anim Sci* 2003;81:1414–1426.
7. Pratt S, Spitzer J, Webster H, Hupp H, Bridges W Jr. Comparison of methods for predicting yearling scrotal circumference and correlations of scrotal circumference to growth traits in beef bulls. *J Anim Sci* 1991;69:2711–2720.
8. Evans A, Davies F, Nasser L, Bowman P, Rawlings N. Differences in early patterns of gonadotrophin secretion between early and late maturing bulls, and changes in semen characteristics at puberty. *Theriogenology* 1995;43:569–578.
9. Brito L, Silva A, Unanian M, Dode M, Barbosa R, Kastelic J. Sexual development in early- and late-maturing *Bos indicus* and *Bos indicus* × *Bos taurus* crossbred bulls in Brazil. *Theriogenology* 2004;62:1198–1217.
10. Martins J, Souza F, Ferreira M *et al*. Desenvolvimento reprodutivo de tourinhos Gir selecionados para produção de leite. *Arq Bras Med Vet Zootec* 2011;63:1277–1286.
11. Latimer F, Wilson L, Cain M, Stricklin W. Scrotal measurements in beef bulls: heritability estimates, breed and test station effects. *J Anim Sci* 1982;54:473–479.
12. Knights S, Baker R, Gianola D, Gibb J. Estimates of heritabilities and of genetic and phenotypic correlations among growth and reproductive traits in yearling Angus bulls. *J Anim Sci* 1984;58: 887–893.
13. Gipson T, Vogt D, Ellersieck M, Massey J. Genetic and phenotypic parameter estimates for scrotal circumference and semen traits in young beef bulls. *Theriogenology* 1987;28: 547–555.
14. Meyer K, Hammond K, Mackinnon M, Parnell P. Estimates of covariances between reproduction and growth in Australian beef cattle. *J Anim Sci* 1991;69:3533–3543.
15. Garmyn A, Moser D, Christmas R, Minick Bormann J. Estimation of genetic parameters and effects of cytoplasmic line on scrotal circumference and semen quality traits in Angus bulls. *J Anim Sci* 2011;89:693–698.
16. Vargas C, Elzo M, Chase C Jr, Chenoweth P, Olson T. Estimation of genetic parameters for scrotal circumference, age at puberty in heifers, and hip height in Brahman cattle. *J Anim Sci* 1998;76: 2536–2541.
17. Neely J, Johnson B, Dillard E, Robison O. Genetic parameters for testes size and sperm number in Hereford bulls. *J Anim Sci* 1982;55:1033–1040.
18. Nelsen T, Short R, Urick J, Reynolds W. Heritabilities and genetic correlations of growth and reproductive measurements in Hereford bulls. *J Anim Sci* 1986;63:409–417.
19. Bourdon R, Brinks J. Scrotal circumference in yearling Hereford bulls: adjustment factors, heritabilities and genetic, environmental and phenotypic relationships with growth traits. *J Anim Sci* 1986;62:958–967.
20. Crews D Jr, Porteous D. Age of dam and age at measurement adjustments and genetic parameters for scrotal circumference of Canadian Hereford bulls. *Can J Anim Sci* 2003;83: 183–188.
21. Kriese L, Bertrand J, Benyshek L. Age adjustment factors, heritabilities and genetic correlations for scrotal circumference and related growth traits in Hereford and Brangus bulls. *J Anim Sci* 1991;69:478–489.
22. Evans J, Golden B, Bourdon R, Long K. Additive genetic relationships between heifer pregnancy and scrotal circumference in Hereford cattle. *J Anim Sci* 1999;77:2621–2628.
23. Kealey C, MacNeil M, Tess M, Geary T, Bellows R. Genetic parameter estimates for scrotal circumference and semen characteristics of Line 1 Hereford bulls. *J Anim Sci* 2006;84: 283–290.
24. Coulter G, Rounsaville T, Foote R. Heritability of testicular size and consistency in Holstein bulls. *J Anim Sci* 1976;43: 9–12.
25. Keeton L, Green R, Golden B, Anderson K. Estimation of variance components and prediction of breeding values for scrotal circumference and weaning weight in Limousin cattle. *J Anim Sci* 1996;74:31–36.
26. Yokoo M, Lobo R, Araujo F, Bezerra L, Sainz R, Albuquerque L. Genetic associations between carcass traits measured by real-time ultrasound and scrotal circumference and growth traits in Nelore cattle. *J Anim Sci* 2010;88:52–58.
27. Boligon A, Silva J, Sesana R, Sesana J, Junqueira J, Albuquerque L. Estimation of genetic parameters for body weights, scrotal circumference, and testicular volume measured at different ages in Nellore cattle. *J Anim Sci* 2010;88:1215–1219.
28. Eler J, Silva J, Evans J, Ferraz J, Dias F, Golden B. Additive genetic relationships between heifer pregnancy and scrotal circumference in Nellore cattle. *J Anim Sci* 2004;82:2519–2527.
29. Quirino C, Bergmann J. Heritability of scrotal circumference adjusted and unadjusted for body weight in Nellore bulls, using univariate and bivariate animal models. *Theriogenology* 1998;49:1389–1396.
30. McAllister C, Speidel S, Crews D Jr, Enns R. Genetic parameters for intramuscular fat percentage, marbling score, scrotal circumference, and heifer pregnancy in Red Angus cattle. *J Anim Sci* 2011;89:2068–2072.
31. Godfrey R, Randel R, Parish N. The effect of using the Breeding Soundness Evaluation as a selection criterion for Santa Gertrudis bulls on bulls in subsequent generations. *Theriogenology* 1988; 30:1059–1068.
32. Moser D, Bertrand J, Benyshek L, McCann M, Kiser T. Effects of selection for scrotal circumference in Limousin bulls on reproductive and growth traits of progeny. *J Anim Sci* 1996;74: 2052–2057.
33. Mwansa P, Kemp R, Crews D Jr, Kastelic J, Bailey D, Coulter G. Comparison of models for genetic evaluation of scrotal circumference in crossbred bulls. *J Anim Sci* 2000;78:275–282.
34. Coe P, Gibson C. Adjusted 200-day scrotal size as a predictor of 365-day scrotal circumference. *Theriogenology* 1993;40: 1065–1072.
35. Barth A, Ominski K. The relationship between scrotal circumference at weaning and at one year of age in beef bulls. *Can Vet J* 2000;41:541–546.
36. Toelle V, Robison O. Estimates of genetic correlations between testicular measurements and female reproductive traits in cattle. *J Anim Sci* 1985;60:89–100.
37. Smith B, Brinks J, Richardson G. Relationships of sire scrotal circumference to offspring reproduction and growth. *J Anim Sci* 1989;67:2881–2885.
38. Martinez-Velazquez G, Gregory K, Bennett G, Van Vleck L. Genetic relationships between scrotal circumference and female reproductive traits. *J Anim Sci* 2003;81:395–401.
39. Macmillan K, Hafs H. Gonadal and extra gonadal sperm numbers during reproductive development of Holstein bulls. *J Anim Sci* 1968;27:697–700.
40. Curtis S, Amann R. Testicular development and establishment of spermatogenesis in Holstein bulls. *J Anim Sci* 1981;53: 1645–1657.
41. Aponte P, de Rooij D, Bastidas P. Testicular development in Brahman bulls. *Theriogenology* 2005;64:1440–1455.
42. Brito L, Barth A, Wilde R, Kastelic J. Testicular ultrasonogram pixel intensity during sexual development and its relationship with semen quality, sperm production, and quantitative testicular histology in beef bulls. *Theriogenology* 2012;78:69–76.

43. Brito L, Silva A, Rodrigues L, Vieira F, Deragon L, Kastelic J. Effect of age and genetic group on characteristics of the scrotum, testes and testicular vascular cones, and on sperm production and semen quality in AI bulls in Brazil. *Theriogenology* 2002;58:1175–1186.
44. Wolf F, Almquist J, Hale E. Prepuberal behavior and puberal characteristics of beef bulls on high nutrient allowance. *J Anim Sci* 1965;24:761–765.
45. Lunstra D, Ford J, Echternkamp S. Puberty in beef bulls: hormone concentrations, growth, testicular development, sperm production and sexual aggressiveness in bulls of different breeds. *J Anim Sci* 1978;46:1054–1062.
46. Jimenez-Severiano H. Sexual development of dairy bulls in the Mexican tropics. *Theriogenology* 2002;58:921–932.
47. Aravindakshan J, Honaramooz A, Bartlewski P, Beard A, Pierson R, Rawlings N. Pattern of gonadotropin secretion and ultrasonographic evaluation of developmental changes in the testis of early and late maturing bull calves. *Theriogenology* 2000;54:339–354.
48. Killian G, Amann R. Reproductive capacity of dairy bulls. IX. Changes in reproductive organ weights and semen characteristics of Holstein bulls during the first thirty weeks after puberty. *J Dairy Sci* 1972;55:1631–1635.
49. Wildeus S, Entwistle K. Postpubertal changes in gonadal and extragonadal sperm reserves in *Bos indicus* strain bulls. *Theriogenology* 1982;17:655–667.
50. Weisgold A, Almquist J. Reproductive capacity of beef bulls. VI. Daily spermatozoal production, spermatozoal reserves and dimensions and weight of reproductive organs. *J Anim Sci* 1979;48:351–358.
51. Amann R, Kavanaugh J, Griel L Jr, Voglmayr J. Sperm production of Holstein bulls determined from testicular spermatid reserves, after cannulation of rete testis or vas deferens, and by daily ejaculation. *J Dairy Sci* 1974;57:93–99.
52. Almquist J. Effect of long term ejaculation at high frequency on output of sperm, sexual behavior, and fertility of Holstein bulls; relation of reproductive capacity to high nutrient allowance. *J Dairy Sci* 1982;65:814–823.
53. Godinho H, Cardoso F. Gonadal and extragonadal sperm reserves of the Brazilian Nelore zebu (*Bos indicus*). *Andrologia* 1984;16:131–134.
54. Brito L, Silva A, Barbosa R, Kastelic J. Testicular thermoregulation in *Bos indicus*, crossbred and *Bos taurus* bulls: relationship with scrotal, testicular vascular cone and testicular morphology, and effects on semen quality and sperm production. *Theriogenology* 2004;61:511–528.
55. Amann R, Almquist J. Bull management to maximize sperm output. In: *Proceedings of the 6th Technical Conference on Artificial Insemination and Reproduction*. Columbia, MO: National Association of Animal Breeders, 1976, pp. 1–10.
56. Amann R. Management of bulls to maximize sperm output. In: *Proceedings of the 13th Technical Conference on Artificial Insemination and Reproduction*. Columbia, MO: National Association of Animal Breeders, 1990, pp. 84–91.
57. Lunstra D, Echternkamp S. Puberty in beef bulls: acrosome morphology and semen quality in bulls of different breeds. *J Anim Sci* 1982;55:638–648.
58. Brito L, Barth A, Wilde R, Kastelic J. Effect of growth rate from 6 to 16 months of age on sexual development and reproductive function in beef bulls. *Theriogenology* 2012;77:1398–1405.
59. Arteaga A, Baracaldo M, Barth A. The proportion of beef bulls in western Canada with mature spermiograms at 11 to 15 months of age. *Can Vet J* 2001;42:783–787.
60. Persson Y, Soderquist L. The proportion of beef bulls in Sweden with mature spermiograms at 11–13 months of age. *Reprod Domest Anim* 2005;40:131–135.
61. Brito L, Barth A, Wilde R, Kastelic J. Testicular vascular cone development and its association with scrotal temperature, semen quality, and sperm production in beef bulls. *Anim Reprod Sci* 2012;134:135–140.
62. Cook R, Coulter G, Kastelic J. The testicular vascular cone, scrotal thermoregulation, and their relationship to sperm production and seminal quality in beef bulls. *Theriogenology* 1994;41:653–671.
63. Kirby A. Observations on the blood supply of the bull testis. *Br Vet J* 1953;109:464–472.
64. Kirby A, Harrison R. A comparison of the vascularization of the testis in Afrikaner and English breeds of bull. *Proc Soc Study Fertil* 1954;6:129–139.
65. Hofmann R. Die Gefäbarchitektur des bullenhodens, Zugleich eim Versuch ihrer funktionellen Deutung. *Zentralblatt für Vetrinarmedizin* 1960;7:59–93.
66. Almquist J, Amann R. Reproductive capacity of dairy bulls. II. Gonadal and extra-gonadal sperm reserves as determined by direct counts and depletion trials: dimensions and weight of genitalia. *J Dairy Sci* 1961;44:1668–1678.
67. Abdel-Razek A, Ali A. Developmental changes of bull (*Bos taurus*) genitalia as evaluated by caliper and ultrasonography. *Reprod Domest Anim* 2005;40:23–27.
68. Coulter G. Puberty and postpubertal development of beef bulls. In: Morrow AD (ed.) *Current Therapy in Theriogenology*, 2nd edn. Philadelphia: Saunders, 1986, pp. 142–148.
69. Ashdown R, David J, Gibbs C. Impotence in the bull: 1. Abnormal venous drainage of the corpus cavernosum penis. *Vet Rec* 1979;104:423.
70. Barber K, Almquist J. Growth and feed efficiency and their relationship to puberal traits of Charolais bulls. *J Anim Sci* 1975;40:288–301.
71. Price E, Wallach S. Development of sexual and aggressive behaviors in Hereford bulls. *J Anim Sci* 1991;69:1019–1027.
72. Lane S, Kiracofe G, Craig J, Schalles R. The effect of rearing environment on sexual behavior of young beef bulls. *J Anim Sci* 1983;57:1084–1089.
73. Price E, Wallach S. Short-term individual housing temporarily reduces the libido of bulls. *J Anim Sci* 1990;68:3572–3577.
74. Boyd G, Corah L. Effect of sire and sexual experience on serving capacity of yearling beef bulls. *Theriogenology* 1988;29:779–790.
75. Price E, Wallach S. Inability to predict the adult sexual performance of bulls by prepuberal sexual behaviors. *J Anim Sci* 1991;69:1041–1046.
76. Chenoweth P, Brinks J, Nett T. A comparison of three methods of assessing sex-drive in yearling beef bulls and relationships with testosterone and LH levels. *Theriogenology* 1979;12:223–233.
77. Landaeta-Hernandez A, Chenoweth P, Berndtson W. Assessing sex-drive in young *Bos taurus* bulls. *Anim Reprod Sci* 2001;66:151–160.
78. de Araujo J, Borgwardt R, Sween M, Yelich J, Price E. Incidence of repeat-breeding among Angus bulls (*Bos taurus*) differing in sexual performance. *Appl Anim Behav Sci* 2003;81:89–98.
79. Chenoweth P, Farin P, Mateos E, Rupp G, Pexton J. Breeding soundness and sex drive by breed and age in beef bulls used for natural mating. *Theriogenology* 1984;22:341–349.
80. Bratton R, Musgrave S, Dunn H, Foote R. Causes and prevention of reproductive failure in dairy cattle: II. Influence of underfeeding and overfeeding from birth to 80 weeks of age on growth, sexual development, and semen production in Holstein bulls. *Cornell University Agricultural Experimental Station Bulletin* 1959;940:1–45.
81. Flipse R, Almquist J. Effect of total digestible nutrient intake from birth to four years of age on growth and reproductive development and performance of dairy bulls. *J Dairy Sci* 1961;44:905–914.

82. Brito L, Barth A, Rawlings N et al. Effect of feed restriction during calfhood on serum concentrations of metabolic hormones, gonadotropins, testosterone, and on sexual development in bulls. *Reproduction* 2007;134:171–181.
83. Brito L, Barth A, Rawlings N et al. Effect of improved nutrition during calfhood on serum metabolic hormones, gonadotropins, and testosterone concentrations, and on testicular development in bulls. *Domest Anim Endocrinol* 2007;33:460–469.
84. Brito L, Barth A, Rawlings N et al. Effect of nutrition during calfhood and peripubertal period on serum metabolic hormones, gonadotropins and testosterone concentrations, and on sexual development in bulls. *Domest Anim Endocrinol* 2007;33:1–18.
85. Lunstra D, Gregory K, Cundiff L. Heritability estimates and adjustment factors for the effects of bull age and age of dam on yearling testicular size in breeds of bulls. *Theriogenology* 1988;30:127–136.
86. Bagu E, Davies K, Epp T et al. The effect of parity of the dam on sexual maturation, serum concentrations of metabolic hormones and the response to luteinizing hormone releasing hormone in bull calves. *Reprod Domest Anim* 2010;45:803–810.
87. Mann T, Rowson L, Short R, Skinner J. The relationship between nutrition and androgenic activity in pubescent twin calves, and the effect of orchitis. *J Endocrinol* 1967;38:455–468.
88. Meacham T, Cunha T, Warnick A, Hentges J Jr, Hargrove D. Influence of low protein rations on growth and semen characteristics of young beef bulls. *J Anim Sci* 1963;22:115–120.
89. Meacham T, Warnick A, Cunha T, Hentges J Jr, Shirley R. Hematological and histological changes in young beef bulls fed low protein rations. *J Anim Sci* 1964;23:380–384.
90. Pruitt R, Corah L. Effect of energy intake after weaning on the sexual development of beef bulls. I. Semen characteristics and serving capacity. *J Anim Sci* 1985;61:1186–1193.
91. Pruitt R, Corah L, Stevenson J, Kiracofe G. Effect of energy intake after weaning on the sexual development of beef bulls. II. Age at first mating, age at puberty, testosterone and scrotal circumference. *J Anim Sci* 1986;63:579–585.
92. Skinner J. Nutrition and fertility in pedigree bulls. In: Gilmore D, Cook B (eds) *Environmental Factors in Mammal Reproduction*. London: MacMillan, 1981, pp. 160–168.
93. Coulter G, Kozub G. Testicular development, epididymal sperm reserves and seminal quality in two-year-old Hereford and Angus bulls: effects of two levels of dietary energy. *J Anim Sci* 1984;59:432–440.
94. Coulter G, Carruthers T, Amann R, Kozub G. Testicular development, daily sperm production and epididymal sperm reserves in 15-mo-old Angus and Hereford bulls: effects of bull strain plus dietary energy. *J Anim Sci* 1987;64:254–260.
95. Coulter G, Bailey D. Epididymal sperm reserves in 12-month-old Angus and Hereford bulls: effects of bull strain plus dietary energy. *Anim Reprod Sci* 1988;16:169–175.
96. Coulter G, Cook R, Kastelic J. Effects of dietary energy on scrotal surface temperature, seminal quality, and sperm production in young beef bulls. *J Anim Sci* 1997;75:1048–1052.
97. Seidel G Jr, Pickett B, Wilsey C, Seidel S. Effect of high level of nutrition on reproductive characteristics of Angus bulls. In: *Proceedings of the 9th International Congress on Animal Reproduction and Artificial Insemination*. Madrid: Editorial Garsi, 1980, p. 359.
98. Hopkins F, Spitzer J. The new Society for Theriogenology breeding soundness evaluation system. *Vet Clin North Am Food Anim Pract* 1997;13:283–293.
99. Kennedy S, Spitzer J, Hopkins F, Higdon H, Bridges W Jr. Breeding soundness evaluations of 3,648 yearling beef bulls using the 1993 Society for Theriogenology guidelines. *Theriogenology* 2002;58:947–961.
100. Higdon H, Spitzer J, Hopkins F, Bridges W Jr. Outcomes of breeding soundness evaluation of 2,898 yearling bulls subjected to different classification systems. *Theriogenology* 2000;53:1321–1332.
101. Chenoweth P, Chase C Jr, Thatcher M, Wilcox C, Larsen R. Breed and other effects on reproductive traits and breeding soundness categorization in young beef bulls in Florida. *Theriogenology* 1996;46:1159–1170.
102. Godfrey R, Dodson R. Breeding soundness evaluations of Senepol bulls in the US Virgin Islands. *Theriogenology* 2005;63:831–840.
103. Carson R, Wenzel J. Observations using the new bull-breeding soundness evaluation forms in adult and young bulls. *Vet Clin North Am Food Anim Pract* 1997;13:305–311.
104. Barth A, Waldner C. Factors affecting breeding soundness classification of beef bulls examined at the Western College of Veterinary Medicine. *Can Vet J* 2002;43:274–284.
105. Fonseca V, Santos N, Malinski P. Andrological classification of zebu bulls based on scrotal perimeter and morpho-physic of the semen characteristics. *Rev Bras Reprod Anim* 1997;21:36–39.
106. Vale Filho V, Bergmann J, Andrade V, Quirino C, Reis S, Mendonca R. Andrologic characterization of Nelore bulls selected for the first breed season. *Rev Bras Reprod Anim* 1997;21:42–44.
107. Smith B, Brinks J, Richardson G. Estimation of genetic parameters among breeding soundness examination components and growth traits in yearling bulls. *J Anim Sci* 1989;67:2892–2896.
108. Beef Improvement Federation. *Guidelines for Uniform Beef Improvement Programs*, 9th edn, 2010. Available at http://www.beefimprovement.org
109. Casas E, Lunstra D, Cundiff L, Ford J. Growth and pubertal development of F1 bulls from Hereford, Angus, Norwegian Red, Swedish Red and White, Friesian, and Wagyu sires. *J Anim Sci* 2007;85:2904–2909.
110. Chase C Jr, Chenoweth P, Larsen R et al. Growth and reproductive development from weaning through 20 months of age among breeds of bulls in subtropical Florida. *Theriogenology* 1997;47:723–745.
111. Fields M, Hentges J Jr, Cornelisse K. Aspects of the sexual development of Brahman versus Angus bulls in Florida. *Theriogenology* 1982;18:17–31.
112. Rocha A, Carpena M, Triplett B, Forrest D, Randel R. Effect of ruminally undegradable protein from fish meal on growth and reproduction of peripuberal Brahman bulls. *J Anim Sci* 1995;73:947–953.
113. Silva-Mena C. Peripubertal traits of Brahman bulls in Yucatan. *Theriogenology* 1997;48:675–685.
114. Troconiz J, Beltran J, Bastidas H, Larreal H, Bastidas P. Testicular development, body weight changes, puberty and semen traits of growing guzerat and Nellore bulls. *Theriogenology* 1991;35:815–826.
115. Freneau G, Vale Filho V, Marques A Jr, Maria W. Puberdade em touros Nelore criados em pasto no Brasil: características corporais, testiculares e seminais e de índice de capacidade andrológica por pontos. *Arq Bras Med Vet Zootec* 2006;58:1107–1115.
116. Hahn J, Foote R, Seidel G Jr. Testicular growth and related sperm output in dairy bulls. *J Anim Sci* 1969;29:41–47.
117. Coulter G, Foote R. Relationship of testicular weight to age and scrotal circumference of Holstein bulls. *J Dairy Sci* 1976;59:730–732.
118. Quirino C, Bergmann J, Vale Filho V, Andrade V, Pereira J. Evaluation of four mathematical functions to describe scrotal circumference maturation in Nellore bulls. *Theriogenology* 1999;52:25–34.

119. Loaiza-Echeverri A, Bergmann J, Toral F *et al*. Use of nonlinear models for describing scrotal circumference growth in Guzerat bulls raised under grazing conditions. *Theriogenology* 2013; 79:751–759.
120. Menegassi S, Barcellos J, Peripolli V, Pereira P, Borges J, Lampert V. Measurement of scrotal circumference in beef bulls in Rio Grande do Sul. *Arq Bras Med Vet Zootec* 2011;63:87–93.
121. Silveira T, Siqueira J, Guimaraes S, Paula T, Miranda Neto T, Guimaraes J. Maturação sexual e parâmetros reprodutivos em touros da raça Nelore criados em sistema extensivo. *Rev Bras Zootec* 2010;39:503–511.
122. Sylla L, Stradaioli G, Borgami S, Monaci M. Breeding soundness examination of Chianina, Marchigiana, and Romagnola yearling bulls in performance tests over a 10-year period. *Theriogenology* 2007;67:1351–1358.

Chapter 6

Breeding Soundness Examination in the Bull: Concepts and Historical Perspective

Richard M. Hopper

Department of Pathobiology and Population Medicine, College of Veterinary Medicine, Mississippi State University, Starkville, Mississippi, USA

Introduction

Evaluation for breeding soundness of the bull provides information that is vital for the cowman and in turn the beef cattle industry. In fact information from the National Animal Health Monitoring System (NAHMS) reports on utilization of various management practices by cattlemen reveals that the bull breeding soundness examination (BBSE) is a valued and much-utilized procedure. At each category of operation based on herd size, the percentage of cattlemen who utilized a BBSE was equal to or greater than those that utilized palpation for pregnancy and several other veterinary services.[1] Additionally, numerous studies reveal the advantage to cattlemen of this management tool, when either a herd fertility standard as measured through pregnancy rate or an economic value is used to compare BBSE-evaluated and -passed bulls versus general population bulls.[2]

A correctly performed BBSE provides an assessment of a bull's breeding potential but only for the date performed. For a bull to successfully breed cows in a pasture environment he must be able to survive (maintain body condition), identify females in estrus, and deliver fertile semen into their reproductive tract. Thus a BBSE should include thorough physical, reproductive tract, and semen examinations.

The evolution of the BBSE that is currently in use is a testimony to the foresight and persistence of a group of veterinarians (both practitioners and academicians) and animal scientists who identified a problem, specifically subfertility in bulls, and worked to continually improve the metrics that could be used to validate the fertility or subfertility of a bull.

What is a BBSE?

While it is true that very few bulls are in fact completely sterile, the incidence of subfertility can be 20–40% in groups of bulls.[3] Thus the goal of a BBSE is to identify subfertile bulls so that they may be removed from the breeding population. Therefore major emphasis must be placed on physical attributes that affect a bull's ability to breed cows and metrics of semen production and quality. Over the years, increased understanding of the factors possessed by a bull that impact overall herd fertility, as measured by both pregnancy rate and percentages of cows impregnated at each time period within the calving season, has led to increased emphasis on measurement of scrotal circumference (SC) and microscopic evaluation of sperm morphology. SC, specifically in bulls under the age of 4 years, is highly correlated to sperm production. Evaluation of sperm morphology is likely the single most important criterion to be evaluated with respect to identifying fertile versus subfertile bulls.

Historically, the need to identify sterile or marginally fertile bulls in North America came to a point of importance in the winter of 1949 after a brutal snowstorm that occurred throughout the northern section of the mid-western and Rocky Mountain states of the United States. Many bulls suffered severe frostbite of the scrotum and so their subsequent fertility was an obvious concern. In response veterinarians initiated attempts at evaluating the semen of these bulls. Semen samples were obtained by manual deviation of the penis into an artificial vagina as the bull mounted a restrained cow.[4] While this is commonly performed at bull studs, the practice did not lend itself to range cattle, so this was both labor-intensive and dangerous to the operator. Once a semen sample was obtained, a microscopic examination of motility was performed. While this event and the ensuing activity viewed through the lens of our current standards for evaluation of bull fertility might seem crude, this was indeed the impetus for the development of the electroejaculator, the early organization (Rocky Mountain Society for the Study of Breeding Soundness) that subsequently became today's Society for Theriogenology (SFT), the professional organization that sets the standards for breeding soundness examinations for all domestic species of both genders, an American Veterinary Medical Association approved specialty, the American College of Theriogenologists (ACT), and likely much of the literally thousands of publications resulting from the last 60 years of research in bovine fertility.

So, finally, to answer the question: What is a BBSE? Basically stated, a BBSE must identify whether or not a bull so examined "is able to produce adequate numbers of normal spermatozoa and possess the ability and desire to deposit these sperm into a cow."[5] This statement, made by Dr J.N. Wiltbank at the Annual Meeting of the SFT in 1982, continues to adequately describe the mission of a BBSE.

Overview of the current BBSE

The standards for the BBSE that are currently in use were adapted in 1992 by a select committee of the SFT and in fact the form developed and utilized is under copyright by that organization. This form has been effectively constructed to include all of the components of a BBSE in an organized format and on a single page (Figure 6.1). A useful aspect of the format is that a veterinarian new to the procedure or an infrequent user can easily utilize the form as a template, ensuring that all aspects of the examination are completed.

A previous evaluation system followed the premise that once the physical attributes of a bull had been examined and found satisfactory, the rest of the examination (SC, sperm motility, and sperm morphology) could be evaluated with a scoring system. Thus the SC would be measured and a score applied, as well as an individual score for motility and one for morphology. A bull's "score" was a total of the three individual scores. It was soon determined that this system was flawed in that a very good score in one area could help a bull pass, even when he might have a substandard score in another area. The current system is an improvement in that a bull must meet minimum standards in each area in order to be classified "satisfactory." Additionally, there is evidence that the current standards put in place in 1992 are more stringent and have resulted in fewer marginal bulls passing.[6–8]

Another aspect of this evaluation system is the acknowledgment that the fertility of some bulls, typically young peripubertal bulls, as well as bulls that have experienced a transient illness resulting in a raised body temperature or a bull that has been exposed to extremes in environmental temperature (high or low) or to stress, will improve over time, once the inciting cause has been removed. If in the opinion of the veterinarian performing the BBSE, a bull does not currently meet standards but is likely to improve, the bull can be classified as "deferred." This bull should then be retested at a later date, typically 60 days later, but often earlier based on the judgment of the veterinarian performing the examination. A bull that for any reason does not meet the standards of the BBSE and is unlikely to meet them at a later date is classified "unsatisfactory."

Thus the BBSE begins with a history, which in fact may be a herd history. A general physical examination, followed by an examination of the urogenital system, and measurement of the SC all precede the collection of semen. Microscopic examination of sperm motility and morphology follow.

Because of the importance of this procedure, separate chapters of this book have been devoted specifically to the physical and urogenital examination and the collection and evaluation of semen (Chapters 7 and 8). Additional chapters cover the use of the electronic BBSE (Chapter 83), a recently developed software program that allows information derived from the BBSE to be captured and stored digitally, and the enhanced diagnostic evaluation of semen (Chapter 74).

Limitations and concerns

Despite the scientific advances in both our understanding of the most crucial parameters of bull fertility and the implementation of improvements in evaluation and classification of these parameters, there appears to be a perception that there are some potential limitations to this procedure. The first and most obvious is that the BBSE in its current format does not include the assessment of libido. Another limitation could be inconsistency among different veterinarians performing the examination, due to either equipment differences or differences in the level of competence or attention to detail.

Concerns about the current standards center on the minimum "threshold" metrics, particularly the minimum motility level of 30%.[9] Additionally, the minimum SC of 30 cm for a bull 12–15 months old is thought to be too small by some, but a barrier for certain breeds that mature more slowly. Finally, each advance in our understanding of the etiology and/or the actual effect on fertility of a specific sperm abnormality tends to raise new questions with respect to what might be acceptable or unacceptable limits for numbers of abnormal sperm.

Evaluation of libido

Libido can be defined as the willingness and eagerness of a male animal to mount and to attempt service of a female.[10] While it is intuitive that libido is an important criterion in the assessment of fertility, the development of a repeatable and predictive test may be unachievable.[11] The expression of libido by an individual is indeed subject to a great number of factors, such as bull to cow ratio, number of bulls in a bull battery, behavioral tendencies (suppression of mating behavior of a subordinate bull in the presence of a dominant bull), breeding experience, being overweight, climatic effects, and breed.[11] All these issues are relevant to attempts at evaluating libido through testing, in addition to the presence of people performing the test and environmental factors.

Most of the tests utilized to evaluate libido are categorized more accurately as serving capacity tests. These tests appear to be very good at identifying bulls with defects that preclude breeding. However, differences in testing, for example

Figure 6.1 Bull Breeding Soundness Evaluation form. Society for Theriogenology, 1992. Used with permission.

test duration or the number of females utilized and whether they are restrained or in estrus, potentially affects the results.[11] Additionally, breed differences are such that *Bos taurus* and *Bos indicus* bulls would require different customized testing protocols. Most *Bos taurus* bulls will mount any female that is restrained regardless of stage of estrus, while most *Bos indicus* bulls hesitate to mount any female not in estrus.

The greatest value for a test that would accurately evaluate or predict the potential level of libido in bulls would be for use in young virgin bulls. However, this category of bull displays the most variation in response and research has shown that serving capacity tests may not accurately predict subsequent fertility. In a study confirming this, young bulls were categorized as having either high or low serving capacity based on the results of a serving capacity test and then placed in multi-sire herds (3 bulls to 100 heifers). The breeding season was 44 days. The heifers exposed to the low-capacity bulls had a first cycle 60% pregnancy rate and an overall pregnancy rate of 84%, compared with 50% and 71% for the high-capacity bulls. Each treatment had a replicate; additionally, there was a third treatment utilizing synchronized heifers and only high-capacity bulls, which displayed 59% and 77% for first service and overall pregnancy rates.[12] This study leads one to suspect that there are other aspects of mating behavior in bulls at play that impact potential libido and, more importantly, fertility as measured by seasonal pregnancy rate. For example, the high-capacity bulls might breed a single heifer or cow several times, while leaving other in-heat females alone. Also, most bulls that had been identified as low capacity had improved performance on an identical serving capacity test conducted after the season. This verified previous work showing that experience and attaining physiologic maturity enhances libido.[13] Other work comparing serving capacity test results with BBSE results showed that the BBSE score (this project was completed prior to the implementation of the 1992 standards) was more useful in identifying groups of bulls that impregnated more cows.[14]

The use of an internal artificial vagina (IAV), a device that can be placed in the vagina of a cow in estrus, has been evaluated as a means both to evaluate libido and as an alternative to the use of electroejaculation for semen collection.[15] Although the semen samples collected by this method were of comparable quality to those collected by electroejaculation, samples were not obtained from all bulls with the IAV. Additionally, the IAV has been used to compare breeding behavior of young *Bos taurus* versus *Bos indicus* bulls.[16] However, *Bos indicus* bulls, specifically the Nelore bulls used in this study, did not display sufficient response for this to be a useful way to collect semen, although their lack of sexual activity likely confirmed its use as a method of libido evaluation.

Serving capacity tests do help identify bulls that have physical defects; however, most of these defects should be identified by careful physical examination, which is already part of the standard BBSE. An important exception is the use of a mating test in the diagnostic work-up of a bull believed to have a breeding injury, but this is obviously different from the serving capacity tests that involve the evaluation of groups of young bulls.

Given the current lack of a serving capacity test that consistently and correctly differentiates the potential fertility of young bulls, plus the time-consuming nature of this type of test, and with the caveat that this procedure does engender the ire of animal welfare groups, it is not likely to become a part of the standard BBSE as performed in North America. However, it is an extremely useful diagnostic tool for young bulls suspected of having insufficient libido or the inability to physically breed a cow and indeed it is still currently a part of the standard BBSE as utilized in Australia.[17]

Inconsistency of results

Another area that has been cited as a potential limitation of the BBSE is the inconsistent results among veterinarians performing this procedure. The two most commonly expressed reasons are differences in equipment (microscope quality or type) and differences among operators. Both issues center on consistency in the evaluation of sperm morphology, although in the past there have been concerns about measurement of SC as well.

With respect to the issue of differences in equipment, specifically the type of microscope, there seems to exist among theriogenologists a strong preference for the use of phase contrast microscopy for the evaluation of morphology. A study comparing the evaluation of sperm morphology with a wet mount (fixed with isotonic formol saline) sample utilizing differential interference phase contrast (DIC) or eosin–nigrosin (EN) stained dry-mount bright-field (BF) microscopy provides insight on this issue.[18] In this study sample slides (either wet mount for the DIC or stained dry mount for BF) were examined at 1000× magnification. DIC was superior in identifying morphologic abnormalities of the sperm head such as the presence of vacuoles and this is of importance as these abnormalities are categorized as uncompensable. However, the percentage normal (i.e., the ability to identify normal sperm) was the same with both types of microscopy. Thus, based on this study, the use of BF microscopy and the EN stain is adequate for the routine BBSE. Another study comparing the use of the EN stain with BF microscopy, Feulgen stain with BF microscopy, or phase contrast microscopy revealed that the Feulgen stain identified more head defects and the phase contrast revealed more distal cytoplasmic droplets.[19] However, the authors felt that the differences were not enough to be clinically important.

Of more concern is that there is inconsistency among veterinarians due to differences in their ability to evaluate a spermogram.[6,20] Additionally, this does not take into account veterinarians or others who perform BBSEs or the euphemistic "semen check" but do not include the evaluation of sperm morphology. This underscores the need for adequate training for veterinarians who wish to provide this service and the protection of the public (cattlemen) from those who pass off a substandard service as a BBSE.

Concerns with consistency with respect to the measurement of SC has been, or could be, alleviated by the use of currently available scrotal tapes that have meters that allow different users to apply similar levels of tension when adjusting the tape around the scrotum for measurement. Observation of different veterinary students measuring SC for the first time and then comparing their measurement to those of experienced clinicians validates this.

Minimum threshold standards

The SFT BBSE standard sets minimum threshold metrics for each of the fertility categories deemed to be most important: motility, SC, and morphology. Approaching this issue, one should remember that the goal of the examination has never been to differentiate levels of fertility, but instead to set a standard that separates out the subfertile individual that should be culled.

The motility threshold of 30% (or "fair") motility reflects a concern that often a bull's semen motility cannot be fairly evaluated in field conditions, as well as a lack of evidence that it is predictive of natural service fertility.[21] It should be noted that evaluation of motility and concern even over the term "progressive" as a descriptor for motility is controversial among those evaluating stallion fertility.[22] Thus, for now, a low threshold for motility is likely adequate.

The morphology threshold of 70% normal with no distinction with regard to classification of abnormalities can be justified by a study in which cows and heifers exposed to bulls with either 70% or 80% morphologically normal sperm had similar pregnancy rates. Additionally, these pregnancy rates were statistically higher than cows and heifers exposed to untested bulls,[23]

The SC threshold of a minimum of 30 cm for a 12–15 month old bull can be questioned on the basis of whether this is optimum for the breeder who wishes to improve cow fertility in light of what is known about the correlation between SC and the subsequent fertility of offspring, but not from a basis of the specific bull meeting a baseline standard of fertility.[24] Conversely, there are breeds, notably the Texas Longhorn, Brahman, and Limousin, that tend to have smaller SC and often breeders of these cattle feel that the standards discriminate against their bulls, especially in light of their experience that many of these bulls display reasonable fertility as measured by the pregnancy rate of cows they are exposed to. Thus the question: Should there be breed-by-breed standards?

The minimum threshold values in general should be considered just that – minimums – and breeders who wish to improve can of course set standards for their bulls that exceed these metrics, particularly with respect to SC.

Conclusion

It is the intention of this chapter to provide a perspective of this important management tool, while explaining aspects of the examination that are critical in nature to both its integrity and usefulness. Additionally, it is important to raise concerns, as these provide an impetus for continual improvement.

Other evaluation standards for a BBSE exist besides the one set forth by the SFT. Veterinarians in Canada (Western Canadian Association of Bovine Practitioners) and Australia (Australian Cattle Veterinarians) have developed and use their own BBSE protocols.[25] However, with the exception of the aforementioned inclusion of a serving capacity test in the Australian standard, similarities among the three standards seem to outweigh differences.

Continuous changes and improvements of this examination are likely, based on the history of these groups and the efforts of other groups such as the Association for Applied Animal Andrology. As always researchers, both veterinarians and animal scientists, continue to add to our understanding of bull fertility and its assessment.

References

1. National Animal Health Monitoring System. Bull management. In: *Beef 2007–08, Part 2 Reference of Beef Cow-calf Management Practices in the United States.* Fort Collins, CO: USDA, 2009, pp. 26–31.
2. Wiltbank J, Parish N. Pregnancy rates in cows and heifers bred to bulls selected for semen quality. *Theriogenology* 1986;25:779–783.
3. Kastelic J, Thundathil J, Brito L. Bull BSE and semen analysis for predicting bull fertility. *Clin Theriogenol* 2012;4:277–287.
4. Walker D. The history of bovine penile surgery (Bartlett Lecture). In: *Proceedings of the Annual Meeting of the Society for Theriogenology*, 1994, pp. 28–37.
5. Wiltbank J, Parish N. Evaluation of bulls for potential fertility. In: *Proceedings of the Annual Meeting of the Society for Theriogenology*, 1982, pp. 141–154.
6. Johnson K. An observational study of breeding soundness examinations in beef bulls: effects of the revised standards and the evaluating veterinarian (a preliminary analysis). In: *Proceedings of the Annual Meeting of the Society for Theriogenology*, 1996, pp. 298–303.
7. Higdon H, Spitzer J, Hopkins F, Bridges W. Outcomes of breeding soundness evaluation of 2898 yearling bulls subjected to different classification systems. *Theriogenology* 2000;53:1321–1332.
8. Kennedy S, Spitzer J, Hopkins F, Higdon H, Bridges W. Breeding soundness evaluations of 3648 yearling bulls using the 1993 Society for *Theriogenology* guidelines. *Theriogenology* 2002;58:947–961.
9. Barth A. Evaluation of sperm motility in bull breeding soundness evaluations. *Society for Theriogenology News* 2001;24(3):4–5, 12.
10. Chenoweth P. Libido and mating behavior in bulls, boars, rams: a review. *Theriogenology* 1981;16:155–177.
11. Petherick J. A review of some factors affecting the expression of libido in beef cattle, and individual bull and herd fertility. *Appl Anim Behav Sci* 2005;90:185–205.
12. Boyd G, Healy V, Mortimer R, Piotrowski J. Serving capacity tests are unable to predict the fertility of yearling bulls. *Theriogenology* 1991;36:1015–1025.
13. Price E, Wallach S. Development of sexual and aggressive behavior in Hereford bulls. *J Anim Sci* 1991;69:1019–1027.
14. Farin P, Chenoweth P, Tomky D, Ball L, Pexton J. Breeding soundness, libido, and performance of beef bulls mated to estrus synchronized females. *Theriogenology* 1989;32:717–725.
15. Barth A, Arteaga A, Brito L, Palmer C. Use of internal artificial vaginas for breeding soundness evaluation in range bulls: an alternative for electroejaculation allowing observation of sex drive and mating ability. *Anim Reprod Sci* 2004;84:315–325.
16. Cruz F, Lohn L, Marinho L et al. Internal artificial vagina (IAV) to assess breeding behavior of young *Bos taurus* and *Bos indicus* bulls. *Anim Reprod Sci* 2011;126:157–161.
17. Fordyce G, Entwistle K, Norman S, Perry V, Gardiner B, Fordyce P. Standardizing bull breeding soundness evaluations and reporting in Australia. *Theriogenology* 2006;66:1140–1148.
18. Freneau G, Chenoweth P, Ellis R, Rupp G. Sperm morphology of beef bulls evaluated by two different methods. *Anim Reprod Sci* 2010;118:176–181.
19. Sprecher D, Coe P. Differences in bull spermiograms using eosin–nigrosin stain, Feulgen stain, and phase contrast microscopy methods. *Theriogenology* 1996;45:757–764.
20. Brito L, Greene L, Kelleman A, Knobbe M, Turner R. Effect of method and clinician on stallion sperm morphology evaluation. *Theriogenology* 2011;76:745–750.

21. Ott R. Breeding soundness examination in bulls. In: Morrow AD (ed.) *Current Therapy in Theriogenology*, 2nd edn. Philadelphia: Saunders, 1986, pp. 125–136.
22. Love C. The stallion breeding soundness evaluation: revisited. *Clin Theriogenol* 2011;3:294–300.
23. Wiltbank J, Parrish N. Pregnancy rates in cows and heifers bred to bulls selected for semen quality. *Theriogenology* 1986;25:779–783.
24. Smith M, Morris D, Amoss M, Parish J, Williams J, Wiltbank J. Relationships among fertility, scrotal circumference, seminal quality, and libido in Santa Gertrudis bulls. *Theriogenology* 1981;16:379–395.
25. Chenoweth P. Evaluation of natural service bulls: the "other" BSE [guest editorial]. *Vet J* 2004;168:211–212.

Web resources

American College of Theriogenologists: www.theriogenology.org
American Association of Bovine Practitioners: www.aabp.org
Association for Applied Animal Andrology: www.animal-andrology.org
Australian Cattle Veterinarians: www.aacv.com.au
Society for Theriogenology: www.therio.org
Western Canadian Association of Bovine Practitioners: www.wcabp.com

Chapter 7

Evaluation of Breeding Soundness: The Physical Examination

James Alexander

Alexander Veterinary Services, Bentonia, Mississippi, USA

Introduction

When a bull breeding soundness examination (BBSE) is performed for the purpose of evaluating a bull's potential to breed in a natural service situation, the physical component takes on paramount significance. In fact when a physical abnormality is observed or the bull does not meet the standards for scrotal circumference (SC), the examination need not proceed to evaluation of semen characteristics. Thus, goals of the physical examination portion of a BBSE are the identification of undesirable genetic traits, structural or physical impediments to breeding, and pathology of the reproductive tract. Even though it is common for inadequacies to exist in this area of the BBSE, it is often ignored or rushed through without attention to detail or in a consistent systematic manner. Therefore it is of utmost importance that the physical examination be done using a consistent protocol. Additionally, the SC, which is an indirect metric for testicle size and thus sperm production potential, must be measured and meet minimum standards.

History

The history can be addressed prior to or during the physical examination. The history can provide the examiner with insight into genetics, condition, management, vaccination, any disease testing programs, and history of disease in the herd. The breed or breed composition should be obtained to help establish what genetic abnormalities or which conformation traits may be common to that breed. The ration fed previously and at present should be established as well as the level of each ingredient in the rations and the poundage fed to each bull. The veterinarian needs to determine what body condition scores were attained at different stages of development. Management practices should be established with regard to large or small group facilities, parasite control, pasture or dry lot, through space, and water availability. These management practices can influence the level of foot problems and increased incidence of diseases such as vesiculitis. The herd vaccination program should be ascertained, if for no other reason than for buyer information. Information on disease control programs should be established for buyer information and the reason for those programs. The history of disease in the herd may explain the results of the BBSE examination and again be useful information to the herd owner and/or buyer.

Basic physical examination

For a bull to survive and function as a natural service sire (pasture or range environment), he must be able to walk, eat, and see. Thus a basic examination should begin with a history, examination at a distance, and an assessment of a bull's conformation, gait and overall appearance.

Examination at a distance

If possible, the bulls should be observed from the fence of the holding pen while they are standing quietly. This is a good time to observe their conformation, and overall appearance. Bull identification numbers are collected of those that will need closer examination for any problems that may be seen. Move through the bulls slowly while noting any other problems. This is a good time to observe each bull's gait, possible foot issues, sight deficiencies, and bilateral symmetry. The absence of bilateral symmetry is the examiner's guide to areas of potential abnormalities.

Evaluation of Breeding Soundness: The Physical Examination

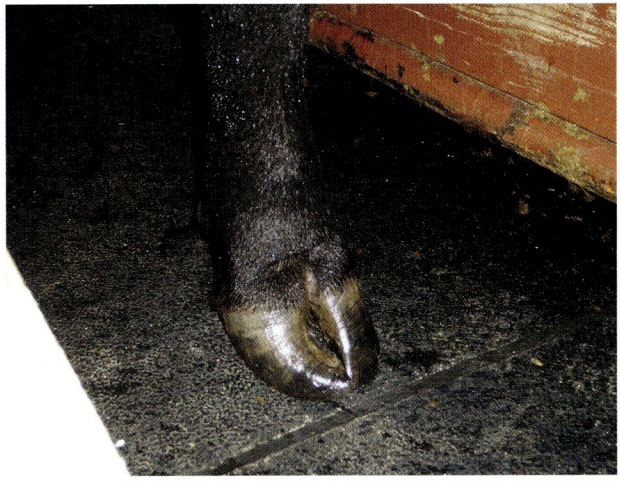

Figure 7.1 A 3-year-old Angus bull that exhibits the screw claw abnormality. This bull will require annual hoof trimming to remain sound.

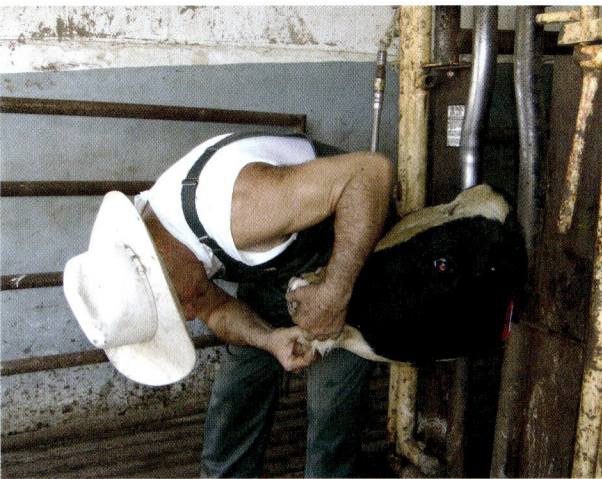

Figure 7.2 Examination of the oral cavity, dentition, and aging.

Conformation

Selection for conformation has been influenced by judges for show purposes and is not always the conformation most desirable for durability of the bull in a pasture or range environment. Because a bull is required to cover large areas, mount and lunge multiple times in a single day, feet and legs are often the trait that determines his ability to service cattle and his longevity as a sire. Hoof and hock abnormalities lead to lameness, which affects his ability to accomplish this purpose as well as lowering sperm quality. Screw claw, chronic laminitis, and interdigital fibromas are common. The incidence of screw claw appears to be increasing in beef breeds (Figure 7.1). These abnormalities should be recorded in the feet and legs section of the BBSE form. Defects that may be inherited should be listed in the comment block with a statement that recommends use in a terminal cross only. Research has shown that animals with shorter hoof length and greater dorsal hoof angle develop less hoof lameness.[1] Dorsal hoof angles of 50 and 55° in the rear and front hooves, respectively, appear to be the more desirable. The angle of the hock should be between 155 and 170°, based on information derived from the Conformation Determination System (CDS), which was a method of relating hock angles of dairy heifers with other anatomical landmarks and then correlating those metrics with the incidence of future unsoundness.[2]

Examination close up

Once the bull enters the chute, the examination should begin with the head. The areas on and around the eyes should be examined for squamous cell carcinomas, corneal damage, and lymphomas. The nasal passages are checked for even and equal air flow, the oral cavity is examined for abnormalities, and the age of the bull is confirmed (Figure 7.2). This is also a good vantage from which to observe the front feet for interdigital fibromas and the screw claw abnormality.

Examination of the reproductive system

Scrotum and testicles

The scrotum, testes, and spermatic cord should be examined carefully for fluid, fibrotic tissue, size, shape, and texture. The scrotum should be examined for scars and its ability to extend or stretch to allow the testicles to cool. The testes should be palpated and have the texture of meat. Any fibrotic or swollen areas should be noted. The epididymides should be palpated for abnormalities such as epididymitis or fibrosis. The spermatic cord is palpated from the external inguinal ring distal to the testicle for abnormalities such as hernias, hematomas, fibrosis, or fluid. The external prepuce should be observed and palpated for abscesses, swellings, as well as hematoma of the penis directly anterior to the scrotum.

Internal reproductive genitalia

A transrectal examination is utilized to evaluate the internal reproductive genitalia and can be aided or enhanced with the use of ultrasound. Each of the secondary sex organs should be carefully identified and palpated for any changes from the norm. The urethralis muscle is the first to be encountered and should be palpated for abscesses and tumors. As you palpate forward, the seminal vesicles (vesicular glands) are encountered. Palpation of both seminal vesicles should begin at the bifurcation with the vesicle surrounded by the hand as it is palpated toward and to each point. Texture, size, and distinction of the nodules should be noted. Vesiculitis is one of the most common abnormalities found during the internal reproductive examination. Evidence of previous infection or injuries to the seminal vesicles can also be diagnosed by the presence of fibrosis and adhesions. The ampullae should the examined next for abnormalities as well as gently massaged to aid in semen collection. Approaches to the treatment of seminal vesiculitis are discussed in Chapter 13.

Penis and internal prepuce

The penis has already been palpated during the examination of the external prepuce but must be carefully examined and palpated while extended during the collection stage of the BBSE. Problems such as warts, hair rings, lacerations, and persistent frenulum will be discovered at this point. The internal prepuce presents itself for thorough examination during erection, allowing the opportunity to observe lacerations, warts, and fibrotic areas from old injuries that may cause deviations of the penis preventing entry. The necessity of completely extending the penis at some point of the examination cannot be overemphasized as the previously mentioned conditions will be missed if this is not done. In-depth coverage of urogenital injuries and their diagnosis, management, and correction are covered in Chapters 14 and 16–18.

Example of an examination protocol

The examination begins with a history and examination at a distance. Evaluate conformation and be certain to record observations and identification of those bulls that have issues. Watching each bull approach the chute is a good time to observe the gait of the bull and overall appearance. Once in the chute, the bull's identification (tag, tattoo, or brand ID), eyes, head, and front claws are observed and observations recorded.[3] From the side of the bull, the testicles are palpated and SC measured, the prepuce is palpated for abnormalities, and the hind feet are observed (Figure 7.3). At this point the scrotal measurement and any observations are recorded. The rear of the bull is then approached and the back, hindlegs, and especially the hock joints are observed and/or palpated. Transrectal palpation is then performed, again using a constant sequence. In the pelvic area, the urethralis muscle is followed forward to the seminal vesicles, which are palpated and massaged, the fornix of the seminal vesicles is located and the ampullae are palpated and massaged, as are the inguinal rings, pelvic lymph nodes, the kidney, and any viscera within reach. The ampullae and the seminal vesicles are then massaged again. The urethralis muscle and the prostate are identified and massaged while progressing toward the anus. In addition to identifying abnormalities or eliciting a painful response, which may indicate a problem, palpation and massage serves to stimulate the bull and facilitate ejaculation (Figure 7.4). Thereafter, the electroejaculator is placed on the standard program and the probe is inserted into the rectum. At this point any observation or abnormalities from palpation are recorded. As erection and protrusion occur, the internal prepuce and the penis are examined by observation and palpation for abnormalities. After microscopic examination of the semen has been done, the bull is then released from the chute. The bull is observed leaving the chute and the chute area for lameness. It is of utmost importance that a protocol be developed to insure that all areas of the physical examination and SC are checked and recorded.

Figure 7.4 Transrectal palpation to evaluate the internal reproductive tract and prepare the bull for electroejaculation.

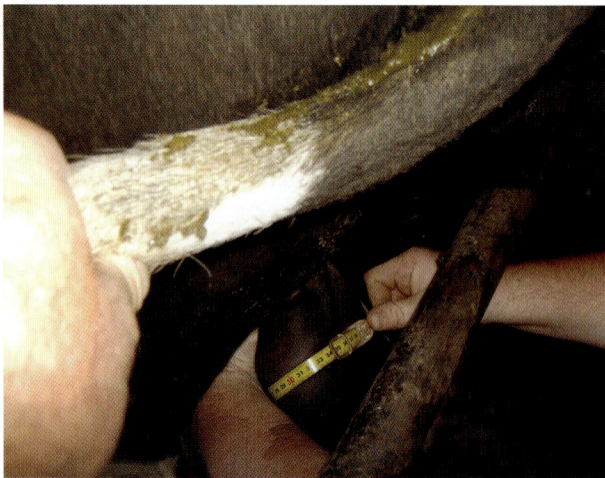

Figure 7.3 The correct way to measure the scrotal circumference. The testes are pushed toward the bottom of the scrotum and the scrotal tape is placed at the widest area. The scrotal tape should be pulled tightly.

Scrotal circumference

The SC measurement is obtained while doing the general and reproductive examination.[3] It can be obtained from the side or the rear, depending on preference, facilities, and the bull's temperament. After gently forcing both testes to the bottom of the scrotum, the primary considerations are to avoid spreading the testes apart and to ensure sufficient pressure that the top surface of the measure tape is level with the skin (see Figure 7.3). Record the measurement at this point. If the circumference is below the minimum requirement, there is no need to continue the examination. Young bulls or emaciated bulls may receive a deferred classification in some cases. The SFT minimum thresholds[4] for SC are listed below:

- 30 cm at <15 months
- 31 cm at >15 to ≤18 months
- 32 cm at >18 to ≤21 months
- 33 cm at >21 to ≤24 months
- 34 cm at >24 months.

References

1. Vermunt J, Greenough P. Structural characteristics of the bovine claw: horn growth and wear, horn hardness and claw conformation. *Br Vet J* 1995;151:157–180.
2. Vermunt J, Greenough P. Hock angles of dairy heifers in two management systems. *Br Vet J* 1996;152:237–242.
3. Alexander J. Bull breeding soundness evaluation: a practitioner's perspective. *Theriogenology* 2008;70:469–472.
4. Chenoweth P. A new bull breeding soundness form. In: *Proceedings of the Annual Meeting of the Society for Theriogenology*, 1993, pp. 63–70.

Chapter 8

Evaluation of Breeding Soundness: Basic Examination of the Semen

Richard M. Hopper and E. Heath King

Department of Pathobiology and Population Medicine, College of Veterinary Medicine, Mississippi State University, Starkville, Mississippi, USA

Introduction

An evaluation of semen is performed for several reasons, most commonly following the physical examination as part of a standard bull breeding soundness examination (BBSE). The BBSE in turn is performed for pre-purchase, annually or semi-annually prior to breeding season turnout, or for diagnostic purposes. A semen evaluation alone can be performed for prognostic reasons following an illness or injury to the reproductive system or in the case of an older or injured bull of high genetic merit that might be a candidate for use by artificial insemination. This chapter primarily deals with microscopic examination of semen as a part of a BBSE.

A basic assessment of semen consists of the microscopic evaluation of motility and morphology. An assessment of sperm numbers or sperm production is determined by measurement of the scrotal circumference (covered in Chapter 7). The assessment of morphology has been shown to be of critical importance and therefore the proper evaluation of sperm morphology, as well as the significance of certain sperm abnormalities, will be emphasized.

Spermatogenesis is a dynamic process and so dramatic changes in the quality of semen produced by bulls can occur within days. This basic fact renders the results of this examination historic in nature within days.

Semen evaluation as part of a standard BBSE

Only after successfully meeting the standards of the physical examination is a bull subjected to collection and evaluation of his semen. To proceed with the evaluation only negates the all-important minimum standard placed on scrotal circumference or the importance of structural soundness. In order to evaluate the semen a sample must be collected in a safe and efficient manner. Semen motility should be assessed immediately but a slide for examination of cell morphology can be prepared and analyzed immediately or at a later date.

Semen collection

Semen can be collected by electroejaculation, manual (per rectal) stimulation of the internal genitalia, or with an artificial vagina utilizing a jump animal. Additionally, an internal artificial vagina placed within the vagina of an estrual female can be utilized.[1,2] Collection via manual stimulation provides samples of inconsistent quality among bulls,[3] but can be useful in young bulls that respond negatively to electroejaculation. Use of an artificial vagina and jump animal is typically used only at bull studs for collection of semen slated for freezing. The internally placed artificial vagina provides the advantage of allowing the assessment of libido. However, of these, EE is the most consistent and reliable method of obtaining a diagnostic sample.

Observation of bulls and specifically the vocalization of some bulls during electroejaculation have led to concerns that this procedure may be painful or at least stressful. A study was performed in which serum cortisol and progesterone levels, relative aversion, and the degree of vocalization were all quantified.[4] While the results of the aversion metric appeared to be equivocal, there were increases in serum cortisol and progesterone, along with increased incidence of vocalization, tending to substantiate these concerns to some degree. However, with the recent elucidation of substance P as a neuropeptide involved in the integration of pain as well as stress and anxiety, it has been shown that electroejaculation does not specifically result in pain.[5]

Bovine Reproduction, First Edition. Edited by Richard M. Hopper.
© 2015 John Wiley & Sons, Inc. Published 2015 by John Wiley & Sons, Inc.

Evaluation of Breeding Soundness: Basic Examination of the Semen

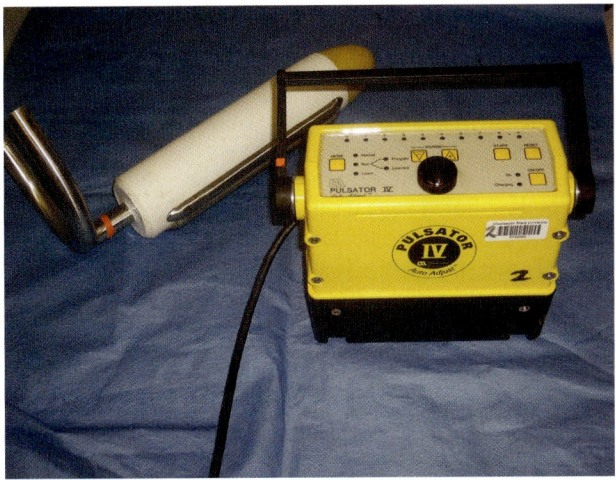

Figure 8.1 Pulsator IV Electroejaculator and probe. Courtesy of Lane Manufacturing, Inc.

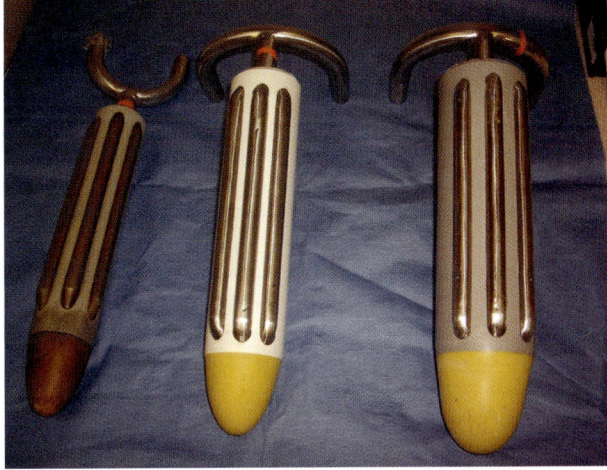

Figure 8.2 Electroejaculator probes (60, 75, and 90 mm). Utilization of the largest diameter probe that can easily be inserted facilitates semen collection. Note that the two larger-diameter models have an upright yoke and are "weighted" (heavier than standard models).

All the currently used electroejaculator models are similar in function, although there are differences in operation and in bull response. Some models utilize a rheostat that allows manual increases in voltage, others a preprogrammed increase, and at least one model allows the selection of either. Thus the type of model selected is typically based on personal preference (Figure 8.1).

From a technical standpoint little has changed in the voltage delivery of an ejaculator in the more than 50 years since the first patent was issued (Marden-US #2,808,834). However, continual progress has been made in probe design, with respect to electrode placement, size, and the yoke. In general the largest diameter probe that can easily be placed in the rectum of a bull should be the size used and heavier probes seem to ensure better contact than lighter probes. A general guideline for probe size is the use of a 7.5-cm diameter probe for bulls weighing 544–907 kg and a 9-cm probe for larger as well as older bulls.[6] The larger (9 cm) also seems to work best in Brahman (Figure 8.2) and Brahman cross bulls. Smaller probes (6–6.5 cm) are available and are often necessary for yearling bulls. The probe yoke is the U-shaped extension on the back of the probe which fits around the tail and serves to prevent probe displacement from the rectum and maintain proper electrode alignment. A yoke that is oriented upward is preferred as a "butt bar" is often used in the chute to facilitate restraint. If a bull attempts to lie down in the chute, the horizontally oriented yoke will hang on butt bars forcing the anterior aspect of the probe down. While this rarely results in serious damage to the rectum, it should be avoided (Figure 8.3).

Efficient collection of semen from a bull by electroejaculation can be facilitated by a couple of techniques employed during the physical examination. During per rectal palpation of the bull's internal genitalia, spend a few extra minutes massaging the ampullae and applying digital pressure to the urethralis muscle and bulbourethral gland. This will stimulate relaxation of the penis and often the bull will begin to dribble pre-seminal and seminal fluid. Next, before introducing the electroejaculator probe, close your hand into a loose fist and thrust your arm back and forth, fatiguing and relaxing the bull's rectum. This action facilitates entry of the probe. The aforementioned massage and fatiguing of

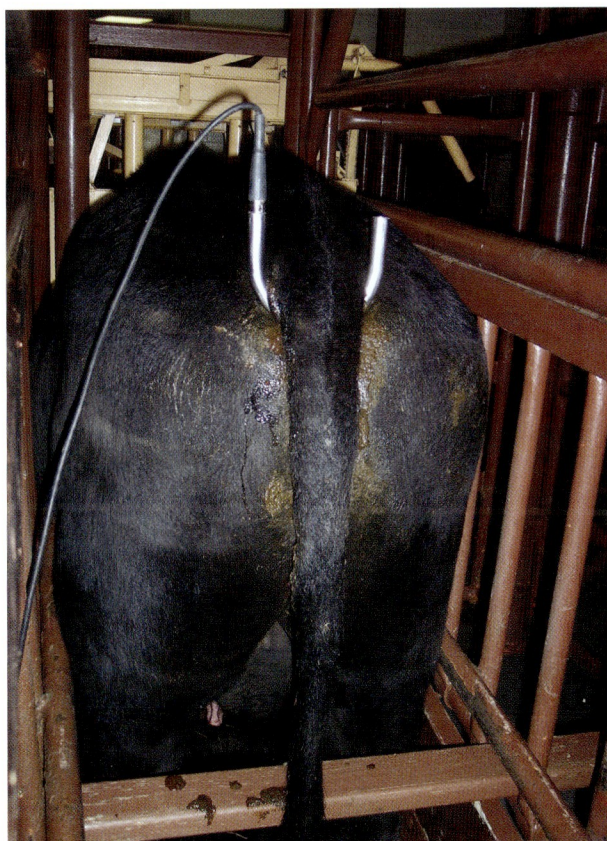

Figure 8.3 Adequate restraint for a bull to be examined for soundness and utilization of upward oriented yoke.

the rectum take only a few minutes and seem to both decrease the bull's reaction to electrostimulation and speed the process of collection. Then, as seamlessly as possible, insert the probe into the rectum as the arm is being removed.

After the probe has been inserted into the rectum, either a preprogramed cycle of stimulation can be used or a manually controlled stimulation cycle. The preprogramed cycle has the advantage of freeing up the operator to perform other tasks, and also provides stimulation in a controlled cyclical frequency that through experience has been shown to be effective in a majority of bulls. Manual control provides the operator with the flexibility to provide less stimulation if the bull seems sensitive or, conversely, to both speed up the frequency and increase the intensity of stimulation if necessary. Again, selection of a certain model of ejaculator is largely based on personal preference. To summarize, when utilizing manual control, provide the first two to three rounds of stimulus slowly, carefully observing the bull. Depending on his response, increase the amount of stimulus with each successive attempt, pausing for a "thousand-one, thousand-two, thousand-three count" at the peak of stimulus, then turning stimulus down for a one-count rest and repeat. Do not collect the first emissions; wait until the seminal fluid becomes cloudy. Since the penis should be extended at this time, it is also a good time to evaluate the penis for defects. Likewise the penis can be grasped with a gauze pad as this will allow better visualization, ensure a cleaner semen sample, and prevent retraction (Figure 8.4).

Occasionally, ejaculation will not have occurred even with the full utilization of electrical stimulation. At this point stop stimulation and allow a rest period. Usually within 2–3 min of rest, semen will spontaneously dribble from the penis. If semen is not obtained following this "rest" period, reinitiate stimulation with two to four rapidly increasing cycles.

In general bulls that have been out with cows and actively breeding within the last 24 hours will be harder to collect. Also, nervous and fractious bulls are often harder to collect. If sedation is necessary, xylazine (20–50 mg i.v.) can be used. Acepromazine should *not* be used as it hampers collection and increases the incidence of urine contamination (urospermia).

Semen can be collected into a cone attached to a collection handle and this in turn can be enclosed within a warm water "bath" or likewise insulated. Alternatively, semen can be collected into a Styrofoam coffee cup. These cups seem to adequately protect the sample from moderate temperature stress and as these cups are used in the food service industry, cleanliness (sterility?) is assumed.

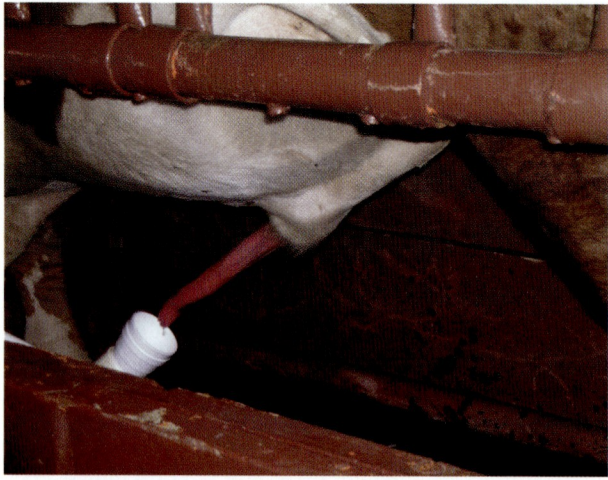

Figure 8.4 Bull with penis extended allowing proper visualization.

Gross evaluation of semen

After collection the semen should first be observed grossly. A rough estimation of the concentration can be made based on the opacity (or lack of) and the color of the semen. Very concentrated samples look like heavy cream while very dilute samples have the appearance of watered-down skim milk. Yellow-tinted semen can result from urine contamination and this can be substantiated by smell or by use of a blood urea nitrogen test strip. Additionally, semen contaminated with urine will have rapidly declining or often no motility when examined microscopically. Conversely, a light-yellow or gold appearance is also associated with very highly concentrated semen and the presence of riboflavin,[7] which is a common finding in many Jersey and some Angus bulls. Red- or brown-colored semen is due to the presence of blood or blood pigments and the source of this contamination must be determined.

Evaluation of semen motility

After quickly evaluating the semen grossly, a small "standing" drop is placed on a prewarmed slide and evaluated under low-power microscopy (40–200×) for gross motility. Thick, dark, rapidly oscillating swirls are indicative of excellent motility (defined as high-velocity or high-speed motility), a high percentage of sperm that are progressively motile, and a sample of high concentration. This type of sample would typically be classified as "very good." A sample that displays slower moving swirls is classified as "good." A "fair" sample displays no swirls, but significant individual sperm movement. A "poor" sample has no or very little movement/oscillation. Because the concentration of a sample impacts the gross motility designation, individual motility should be assessed if there is any question about the validity of a motility rating based on gross motility (i.e., low-power evaluation of the standing drop). Individual motility can be assessed using 200–400× microscopy and, depending on the concentration, a coverslip over either the previously examined droplet or a diluted droplet (diluted with warmed sodium citrate solution). Individual motility is classed "very good" if greater than 70%, "good" if 50–69%, "fair if 30–49%, and "poor" if less than 30%.[8]

Based on the current standards set forth by the Society for Theriogenology (SFT), bulls must have a minimum of "fair" sperm motility based on either individual or gross assessment.[9] While this may seem to be low, it is a minimum threshold and while there is a positive correlation between motility and fertility with the use of artificial insemination, this does not appear to be the case with natural service bull fertility.[10]

Evaluation of sperm morphology

Accurate evaluation of sperm morphology begins with the preparation of a stained sample of diagnostic quality and the use of a bright-field microscope that has at least 100× oil immersion objective (1000× magnification) or the use of a high-quality phase-contrast microscope. Additionally, one must make a commitment to the careful examination of at least 100 sperm cells. Although, strictly speaking, one who could differentiate normal from abnormal sperm could perform this procedure adequately, the ability to identify the various commonly occurring sperm abnormalities and an understanding of their significance is important.

Preparation of a slide

Evaluation of sperm morphology depends on the preparation of a good semen smear. Begin with a clean warm slide, realizing that some "new" slides may be contaminated with detergents, etc. that interfere with staining. For years India ink was utilized as a semen stain. It does not actually stain the sperm cells, but provides a dark background to the white unstained cells. A vital stain, eosin–nigrosin, is currently recommended for field use, due to its ease and consistent staining properties. The eosin portion will penetrate dead sperm cells, staining them pink (red is dead) and leaving live cells unstained (white) against the dark background provided by the nigrosin component. A smear stained with Diff Quik will allow better visualization of white cells and is also an adequate stain for visualizing sperm morphology. Another useful staining procedure is the Feulgen staining method. This technique begins with preparation of a smear and then allowing it to dry for an hour. Next, place the slide in 5N HCl for 30 min. Wash the slide by running water into the corner of a staining dish containing the slide for 2 min. Then place the slide in Schiff's reagent for 30 min. Wash again as before and air dry. The Feulgen technique is superior for identifying the nuclear vacuole (crater) defect and because the process removes fat globules, it is an excellent staining technique for smears from extended semen.

To stain a semen sample, first place a small droplet of stain on the slide, then add a drop of semen and mix. Placing the semen drop first followed by the stain can result in contamination of the stain solution if the tip of the bottle or dropper inadvertently touches the semen. Once the stain and semen are mixed, a second slide is used to push (spread) the semen across the slide in the same manner as a blood smear is prepared. Because the feathered edge is the best area for evaluation, an alternative method is to create several thickness gradients by stopping and starting as the mixture is spread. It is also often a good idea to make a second slide at the same time, as it is faster to make two than to come back later and make another slide if it turns out that the first slide was not of diagnostic quality (Figure 8.5).

Once the slide is dry, the smear can be examined either with phase-contrast or bright-field microscopy at 1000× (i.e., the oil immersion objective for the bright-field microscope).

Categorization of sperm abnormalities

In an attempt to better quantify sperm abnormalities, several classification systems have been developed.[6,11] The commonly employed classifications are "primary/secondary," "major/minor," and "compensable/uncompensable." Additionally, abnormalities can simply be counted by each specific defect, which is often done when a BBSE is performed for diagnostic reasons. Quantifying each defect can be cumbersome as there are at least 25 currently recognized sperm defects.[12]

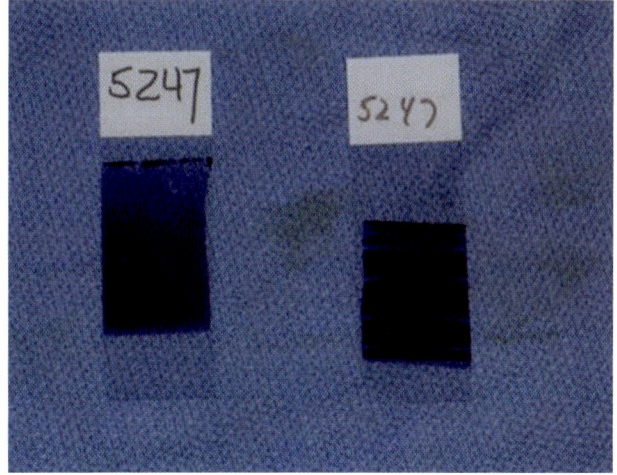

Figure 8.5 Two techniques for preparing a semen smear. The stop/start smear provides multiple thickness gradients.

The primary/secondary sperm classification system, which is still utilized by the SFT, categorizes an abnormality based on the suspected origin of the defect. Primary abnormalities are testicular in origin or, more specifically, occur during spermatogenesis,[13] whereas secondary abnormalities are epididymal in origin. However, as advances are made abnormalities are continually reclassified and in fact some abnormalities can be caused during either phase.[6] There is also a tendency to ascribe more importance or significance to those abnormalities designated as primary and this may not be appropriate for each individual abnormality (Table 8.1).

The major/minor sperm classification system is based on the suspected significance on fertility.[14] Major defects are obviously those whose presence in the ejaculate decreases fertility. Minor defects, which at the time of the original work were not thought to be especially significant, have been subsequently shown to decrease fertility but only when in significant numbers. Interestingly, the primary/secondary categories that came along later and were designated based on presumed origin differ only slightly from the major/minor categories defined by Blom.

Categorizing semen defects that result in reduced pregnancy rates into a classification of compensable/uncompensable was introduced by Saacke et al.[15] Simply defined, compensable defects in an ejaculate, or more specifically an insemination dose, are those that will not result in a lowered fertility rate if there are adequate additional normal sperm. Thus these defects can be "compensated for" in natural breeding and also in artificial insemination providing additional normal sperm are present in adequate numbers. Examples of these are defects where motility or the ability to traverse the barriers of the female genital tract is compromised, rendering the sperm unable to reach the ovum or able to reach the ovum but unable to initiate the fertilization process.[16]

Conversely, additional normal sperm will not compensate for the presence of some abnormalities and these defects are categorized as uncompensable. Sperm with these defects can reach the ovum and initiate the fertilization process but are unable to sustain it or result in poor embryonic viability.[15,16] These uncompensable sperm

Table 8.1 Sperm abnormalities as categorized by Society for Theriogenology.

Primary abnormalities
Underdeveloped
Double forms
Acrosome defect (knobbed acrosome)
Narrow head
Crater/diadem defect
Pear-shaped defect
Abnormal contour
Small abnormal head
Free abnormal head
Proximal droplet
Strongly folded or coiled tail (Dag)
Accessory tail

Secondary abnormalities
Small normal heads
Giant and short broad heads
Free normal heads
Detached, folded, loose acrosomal membranes
Abaxial implantation
Distal droplet
Simple bent tail
Terminally coiled tail

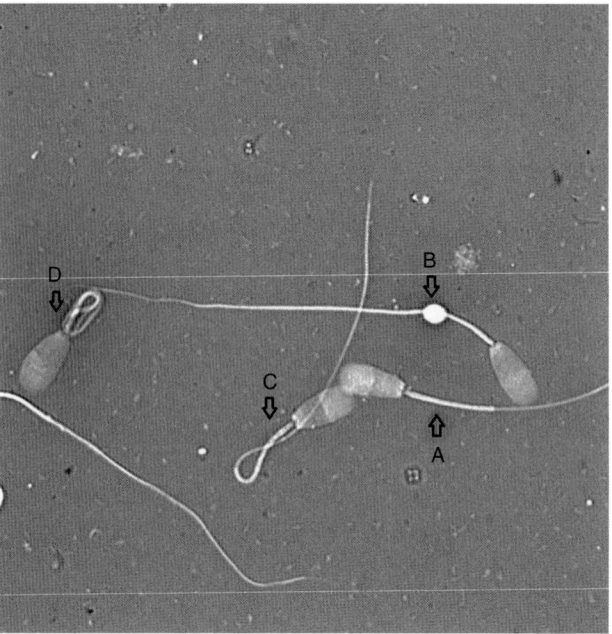

Figure 8.6 (a) Normal sperm; (b) distal droplet defect; (c) bent tail defect; (d) Dag defect.

compete with and often preempt a normal sperm's opportunity to fertilize an ovum. Thus, an uncompensable defect will theoretically decrease fertility by the percentage rate that it is present in an ejaculate or insemination dose.

Complicating the designation of sperm defects as compensable/uncompensable is the fact that, like the other classification schemes, new advances are likely to result in changes to a category of defects and the complex nature of male infertility in general. However, approaching evaluation of abnormalities with this approach is of tremendous value in helping the veterinarian determine the prognostic implications of a prevalent defect in an ejaculate.

Further complicating the issue of evaluation of sperm morphology is the reality that within a specific ejaculate there may be morphologically normal sperm that have chromatin damage. Sperm with chromatin damage are uncompensable in that they can achieve fertilization but the resulting embryo is often of poor quality. However, chromatin-damaged but normally appearing sperm can likely be predicted by the predominant presence of morphologically abnormal cells in an ejaculate. This has been demonstrated by scrotal insulation studies which revealed that the resultant manifestation of defects is dependent on the stage of spermatogenesis at the time of an insult[17] and that both the number and type of defects seen reflect a larger number of defective but normally appearing sperm.[15]

The best approach therefore is to realize that abnormal sperm in an ejaculate are either a reflection of testicular health as it relates to spermatogenesis or a genetic etiology. Carefully evaluate a minimum of 100 sperm at 1000× and "pass" only those bulls that have a minimum of 70% normal.

Note prevalent or predominant sperm abnormalities, since the occurrence of certain defects make the presence of "normally appearing" but defective sperm more likely. Bulls with over 70% normal but with 20–25% nuclear or other uncompensable defects should be closely scrutinized. Thus, it is prudent when evaluating the bull with a borderline spermiogram to consider the type of defects and their prevalence and perhaps err on the side of caution.

Normal sperm

Since there may be some confusion in the categorization of abnormal sperm by veterinarians performing a basic BBSE, the ability to accurately ascertain the percent normal may be the most important aspect of evaluating the spermiogram. In fact some veterinary andrologists have suggested that classification of abnormal sperm as part of the routine BBSE should, or could, be discontinued (6).

Most of our clients and virtually all veterinary students when looking at a magnified display of a stained population of sperm can intuitively identify a normal sperm cell. However, subtle abnormalities, or what we as veterinarians would consider not so subtle such as the nuclear vacuole defect, will be easily missed by the novice or casual observer. Thus a complete description of the normal is warranted.

Basically, the normal sperm cell (Figure 8.6) can be divided simply into the head and tail for the purposes of description and also as a way to better define the location of defects. The head, which is 8–10 μm long, 4–4.5 μm wide, and 1–1.5 μm thick,[7] is composed of nuclear material and the acrosome. The tail, which is approximately 40–45 μm in length, can be further divided into the neck, midpiece, principal piece, and endpiece.

As previously stated, when stained with a vital stain such as eosin–nigrosin, dead cells will take up the stain and appear pink, with the acrosome staining much lighter and thus differentiating the head. As the head of the sperm is virtually completely composed of chromatin, the nuclear shape is correlated to fertility[18] and indeed the sperm heads of highly fertile bulls are very consistent in size.[13]

Sperm abnormalities: classification, etiology and prognostic implications

A BBSE is in its essence a prognostic evaluation. Therefore an understanding of the etiology and significance of sperm abnormalities is useful to the veterinarian performing the examination. The goal of this section is to provide a reference for each of the more common sperm defects with respect to classification, etiology, and prognostic significance.

Cytoplasmic droplet

Small numbers of these defects may be found in the ejaculate of normal bulls and especially peripubertal bulls. Normally, droplets should be "shed" during epididymal transit. Proximally located cytoplasmic droplets or simply proximal droplets are due to abnormal spermiogenesis,[13] are testicular in origin, and appear to be a manifestation of sperm incompetence. Indeed, in men decreased fertility due to the increased presence of cytoplasmic droplets has been shown to result from higher levels of sperm DNA denaturation.[11] Distally located cytoplasmic droplets (distal droplets) are not believed to be of much significance and are thought to be epididymal in origin. Thus the proximal droplet is classified as either a primary or major defect, while the distal droplet is classified as either a secondary or minor defect.

The distal droplet defect is believed to be compensable and the proximal droplet defect uncompensable.[6,13] However, the possibility that the proximal droplet defect is compensable may be supported by work that used ejaculates of either high or low numbers of sperm with this defect for *in vitro* fertilization.[19] In this study, fertilization was decreased as the percentage of proximal droplets in an ejaculate increased, but of the ova that were fertilized, cleavage rates were similar (note that in this study there were normal sperm in each population of sperm cells exposed to the ova).

This is a good example of the complexity of categorizing sperm defects. Using the basic metric of a defect that lowers fertility at the level of its expression in the ejaculate, then the proximal droplet defect is clearly uncompensable. However, another standard definition for an uncompensable defect is one that can initiate fertilization but which does not lead to normal embryonic development. Thus the placement of the proximal droplet defect in the uncompensable category is problematic, since the aforementioned study[19] validated the literature[13] in asserting that a sperm with a proximal droplet is unable to fertilize an ovum (Figure 8.7).

Perhaps the best approach is to treat this defect (proximal droplet) as a clinical sign or "marker" for abnormal spermiogenesis, with the potential underlying cause either immaturity or conversely testicular degeneration. So young bulls can be placed in the deferred category and retested since improvement of the spermiogram is likely, but older bulls with this defect have a more guarded to poor prognosis for return to fertility.

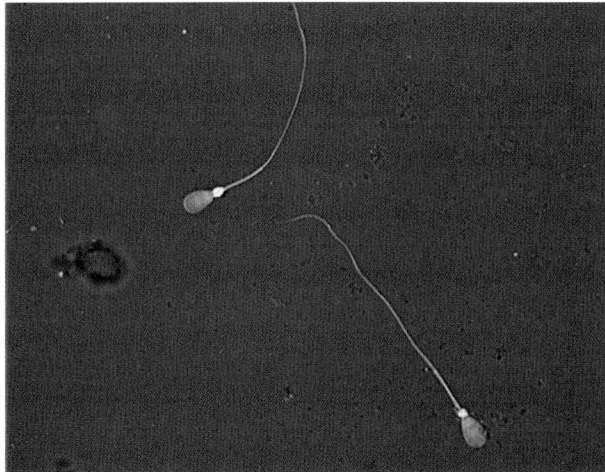

Figure 8.7 Proximal droplet defect.

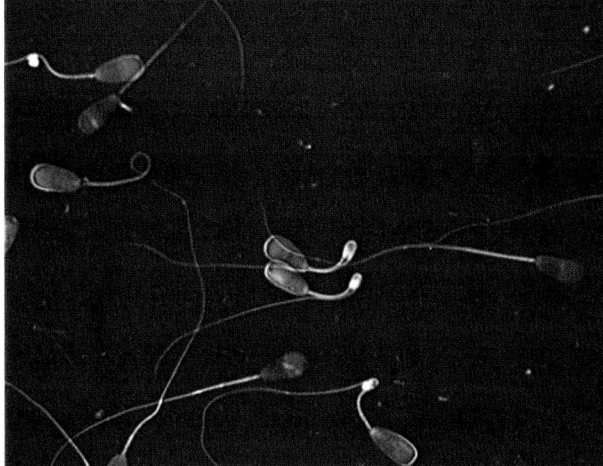

Figure 8.8 Distal midpiece reflex defect.

Distal midpiece reflex

The distal midpiece reflex (DMR) is the most common abnormality of the sperm tail.[13] Considered to be epididymal in origin and therefore a secondary defect, it is also categorized as a minor defect and compensable. This defect is compensable due to the lack of forward progressive motility. Evidence for its origin is based on its rapid appearance in the ejaculate of bulls within a few days of a thermal insult. This defect appears as a sharp hairpin bend at the distal midpiece,[20] with a cytoplasmic droplet within the bend. If there is no droplet present, it is likely that the "bend" is due to contact with a hypotonic solution, presumably the stain that was used. It can also be identified during the evaluation of motility as these sperm will be swimming backwards (Figure 8.8).

The primary etiology is a negative effect on epididymal function due to depressed testosterone levels, which can in turn be caused by stress, thermal stress (either high or low), exogenous estrogen, or induced hypothyroidism, although normal fertile bulls can have up to 25% of this defect in an ejaculate[13] due presumably to its compensable nature. However, when this defect is present at a prevalence of 20–25% in the ejaculate of a bull that meets the standards

74 The Bull: Breeding and Health Management

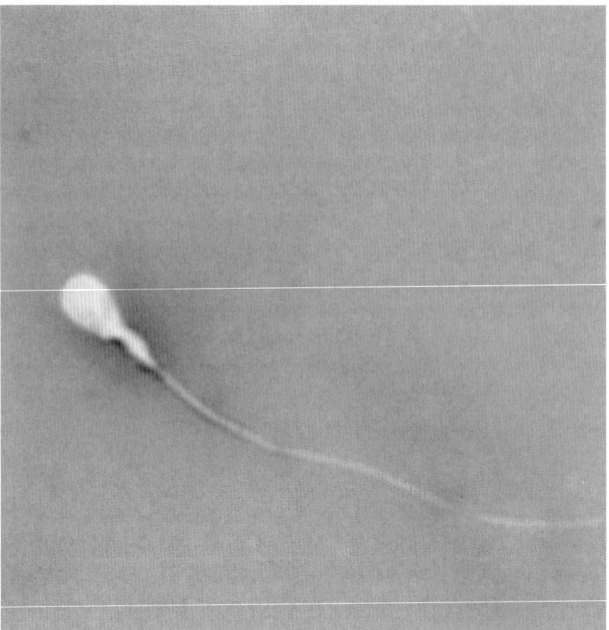

Figure 8.9 Pseudodroplet defect.

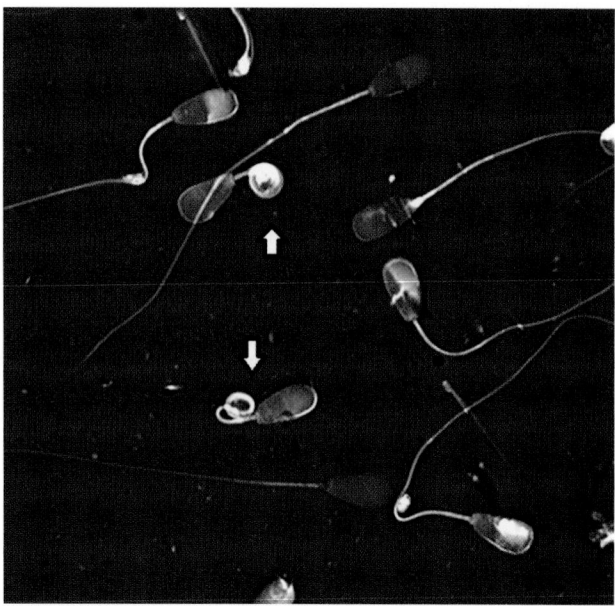

Figure 8.10 Dag defect.

(>70% normal) when tested at a time of moderate weather and absence of stress, it must be realized that although the bull is fertile now levels of this defect can increase dramatically during times of environmental temperature extremes. Additionally, it has been shown to be heritable in Jersey bulls, some of which would have up to 100% DMR defective sperm in an ejaculate,[6] and the possibility of this defect being heritable in other breeds is something that should be considered.

Abnormal midpiece

This category of defects includes the "pseudodroplet" defect, the mitochondrial sheath defect, and segmental aplasia. Additionally, the midpiece may appear swollen, "corkscrew," bent, or asymmetric (Figure 8.9). Compensable because of the impact this defect has on motility, it is classified as a primary defect in the SFT system. Since the development of this sperm region occurs almost completely during spermiogenesis,[13] the specific origin for most of these defects is undoubtedly testicular. It has been shown that some forms of this group of defects can be caused by increased levels of gossypol,[21] a compound found in the cotton plant, and specifically cottonseed, in the diet of bulls. Bulls fed diets high in gossypol appear to be especially sensitive to this compound during puberty.[22] The etiology of defects caused by gossypol appears to result from damage to sperm structure during spermiogenesis with further damage occurring during epididymal transit.[21]

Strongly folded or coiled tail

Also referred to as the "Dag" defect, this is also an abnormality of the midpiece (Figure 8.10). Although small numbers of this defect appear in the ejaculate of normal bulls, it can be found in high numbers (>50%) in some individuals. This defect has been classified as primary and of epididymal origin,[23] specifically the cauda epididymis. Named for the Jersey bull from whose ejaculate this defect was first identified, it was immediately believed to have a genetic basis because the same defect was identified in a full brother.[24] A breeding trial utilizing a normal bull that had produced offspring with the defect confirmed this suspicion when he was bred to 120 of his daughters and produced sons with the defect. At least in the Danish Jersey, the Dag defect can be transmitted to offspring as a recessive trait.[25]

Additionally, there are increased zinc levels in the plasma and sperm of bulls affected with this defect (bulls with >50% defects). This could presumably be due to either dietary imbalance or a genetically predisposed sensitivity and thus it is not known whether zinc plays a causal or contributory role.[25]

Abaxial placement

The abaxial placement sperm type is listed as a secondary defect in the SFT classification scheme. In a series of breeding trials consisting of superovulated heifers that were slaughtered 7 days after breeding for embryo recovery, the artificial breeding of synchronized heifers, and finally a competitive breeding situation, the presence of high numbers (50, 88, and 100%) of abaxial sperm did not negatively impact fertility by either decreasing conception or embryonic survivability.[26] Thus abaxial implantation should be considered as simply a variant of the normal as it is in horses and swine.

Acrosome defect: knobbed acrosome

The knobbed acrosome defect can be identified as an apical swelling that may protrude from, or fold over, the head[27] and appears most often as a flattening or indention of the apex.[13] Early on this defect was identified as having a genetic etiology, specifically an autosomal sex-linked recessive trait

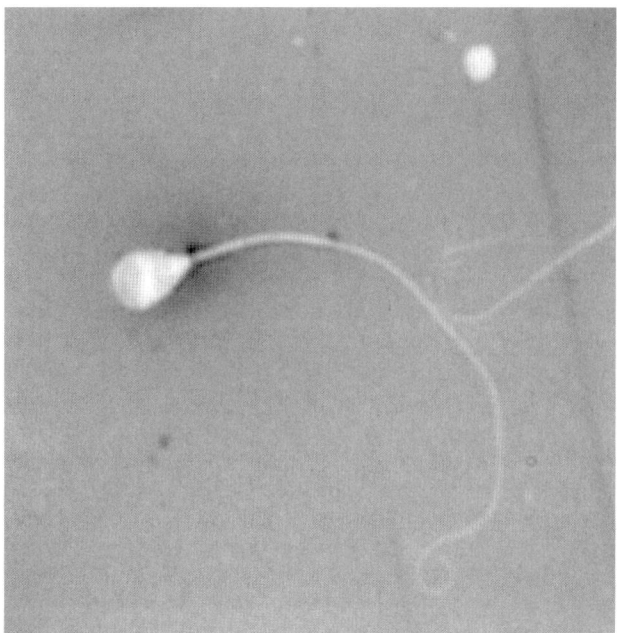

Figure 8.11 Pyriform defect.

in the Friesian breed.[24,27] A genetic etiology should be considered when this defect is prominent in the ejaculate over time, but when identified with several other defects an environmental (temperature-related) cause is likely.[13,17,28] It is considered to be both a major and primary defect, and the best current evidence is that it is uncompensable.[28] The uncompensable nature of this defect is not straightforward as it actually appears to be compensable based on the fact that sperm with this defect do not traverse the reproductive tract of cows efficiently,[29] and those that do are unable to penetrate the zona pellucida.[28] However, this defect is an example of those imperfections whose presence denotes the occurrence of normal-appearing but defective cohorts.[17,28,30] These defective but morphologically normal cells, though able to penetrate ova, yielded lower rates of fertilization and reduced cleavage by zygotes.[28,30] Therefore from a practical perspective this defect should be deemed uncompensable.

When this defect is encountered in large numbers with other head defects, it is prudent to defer and recheck the bull in 60–90 days. Also, scrutinize more closely those bulls whose ejaculate displays this defect predominantly in large numbers (>20%) as it is known it can have a genetic basis and expresses infertility at levels higher than its occurrence within an ejaculate.

Pyriform-shaped heads

These include "pear-shaped" and tapered heads and are often referred to as such (Figure 8.11). This is the most common defect of the sperm head[13] and is commonly found in low numbers in the ejaculates of fertile bulls.[20] Because there are bulls of normal fertility that have narrowed sperm and because there appear to be variations in the range of "taperedness," it can be hard to distinguish at what point a designation is made between normal and pyriform.[13] In fact, in the human, sperm categorized as pyriform or pear-shaped are no longer considered abnormal.[31]

This abnormality is currently categorized as both a primary and major defect. The evidence for whether this abnormality is compensable is equivocal. In general, sperm with misshapen heads do not traverse the reproductive tract, but sperm with this defect apparently do,[29] although this appears to depend on the level of deformity.[32] The level of deformity also apparently impacts fertilization rates as trials evaluating this defect reveal decreased levels of zona penetration, fertilization, and cleavage rates.[32,33] For example, the previously cited work revealed that semen containing this defect at high percentages (85% pyriform heads) had zona penetration at about half the rate of control (90% normal) semen. Considering that semen containing a high percentage of pyriform head defects still resulted in some, albeit much lower, fertilization, these authors came to the conclusion that this could be due to the presence of a small number of normal sperm as well as a percentage of less affected pyriform sperm that may be capable of successful fertilization, suggesting that this abnormality could be partially compensable.[33]

With respect to etiology, this defect is seen following environmental heat stress, validated by scrotal insulation studies,[17] and also from bulls with testicular hypoplasia.[13] In addition to environmental causes of heat stress, the scrotal insulation effects of fat deposition around the scrotum that results from heavy feeding during gain tests have the same deleterious effect. Bulls examined after recently coming off a gain test or experiencing adverse environmental extremes that have this abnormality in numbers that contribute to not meeting the metric for percent normal sperm should be deferred. In the case of bulls with testicular hypoplasia, they typically do not meet BBSE standards for scrotal circumference.

Other head size abnormalities

Small (microcephalic) and giant (macrocephalic) heads are categorized as secondary and minor defects. They are commonly found in very small numbers in the ejaculates of bulls of normal fertility. These defects can be observed with a myriad of other defects (pyriform, vacuoles, etc.) following a disturbance in spermatogenesis, but still rarely exceed 5–7% of the ejaculate.[13] Misshapen heads are generally excluded as these sperm traverse the reproductive tract and therefore would be considered compensable.

Narrow heads, categorized as primary by the SFT classification and minor according to Blom, can be normal as there are fertile bulls that consistently produce sperm with a narrow head profile.[13,32] However, in these bulls all the sperm heads in an ejaculate were very consistent in size. Therefore any sperm with a head size or profile that is observably narrower than the others should be counted as a defect.

Nuclear vacuole defect

The nuclear vacuole defect is also termed a "crater" and includes the diadem defect, which is a string or line of vacuoles around the acrosome–nuclear cap junction[34] (Figure 8.12). This abnormality is categorized as both a primary and major defect and is uncompensable.

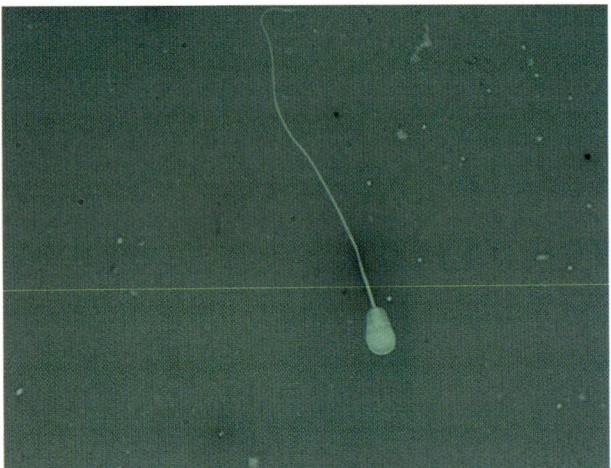

Figure 8.12 Nuclear vacuole defect. Courtesy of Dr Robert Carson.

While small numbers (<15%) of this defect in an ejaculate can be compatible with fertility, larger numbers suggest a disturbance in spermatogenesis and in most instances where this defect is present at 10% or greater it is accompanied by other defects that reduce semen quality.[35] Because zona binding and perhaps zona penetration (research is equivocal on this) is similar between normal sperm and sperm with this defect, but fertilization failure is higher, this defect meets the criteria for its categorization as uncompensable.[32,35,36]

The cause of this defect is undoubtedly environmental (temperature-related) effects on spermatogenesis, as the appearance of this sperm abnormality follows within days of the administration of dexamethasone or the application of scrotal insulation.[37] The possibility of a genetic etiology has been ruled out.[38] The prognosis for recovery is good if the inciting cause is eliminated.[37,39]

Incidentally, this defect can be found more efficiently with smears stained with the Feulgen stain[13] but can be identified using eosin–nigrosin stained smears at 1000× magnification. Routine evaluation of spermiograms that does not distinguish this defect on a frequent basis is likely an indication to upgrade the microscope.

Detached head defect

A detached normal head is a commonly encountered defect that is categorized as both a secondary and minor abnormality (Figure 8.13). It is compensable due to the obvious lack of motility. Although found in small numbers in virtually all ejaculates, it is seen in high numbers in bulls following sexual rest ("rusty load" scenario), in peripubertal bulls, and in bulls that have experienced a recent stress with or without a high fever. It is also found in young bulls with testicular hypoplasia, but these should have already been excluded from further testing following an inadequate scrotal circumference measurement. Additionally, there are reportedly bulls that accumulate senescent sperm[13] similar to stallions that are diagnosed as sperm accumulators. In addition to the absence of motility, and presence of numerous dead cells, a high percentage of cells will have detached heads. In a case report that included a breeding trial, bulls

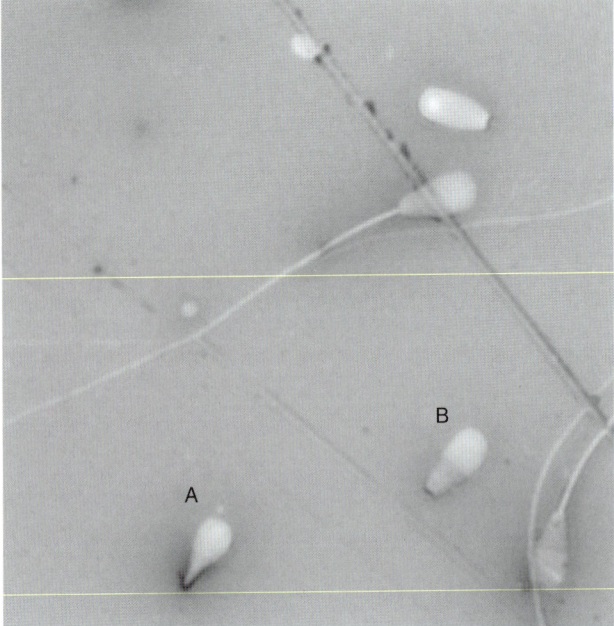

Figure 8.13 Detached head defects: (a) pyriform head; (b) normal head.

with this condition had normal a spermiogram following multiple collections and normal fertility after the first week of the breeding season.[40]

When this defect is found at threshold levels, often the bull can be recollected or at worst placed in the deferred category. Because it is an abnormality of epididymal origin and epididymal transit is around 11 days, these bulls can be retested as early as 2 weeks later.

Abnormal detached heads should be categorized based on the nuclear defect observed. These are usually major and uncompensable. Detached heads in the presence of separated and motile tails is a different, distinct abnormality. This defect has a genetic basis and, when observed, 80–100% of sperm will be affected.[24] Affected bulls will be sterile.

Stump tail defect

The stump tail defect (Figure 8.14), which should not be confused with the detached head, is a defect of the midpiece, but unlike most of the other midpiece abnormalities it likely has a genetic basis,[13,24,41] although it may be seen in a spermiogram with other midpiece defects in cases of gossypol toxicity.[23] Like other midpiece defects it is categorized as both primary and major. It is compensable, but because of the likelihood of a genetic origin its presence in the ejaculate in significant numbers (>20–25%) justifies close scrutiny. This defect appears as a head with a small tail stump attached, with or without a retained cytoplasmic droplet.

Terminally coiled tail defect

The terminally coiled tail defect, also termed a coiled principal piece,[13] is categorized as secondary and minor (Figure 8.15). Because of poor motility it is compensable. This defect is not commonly found, although it can be seen

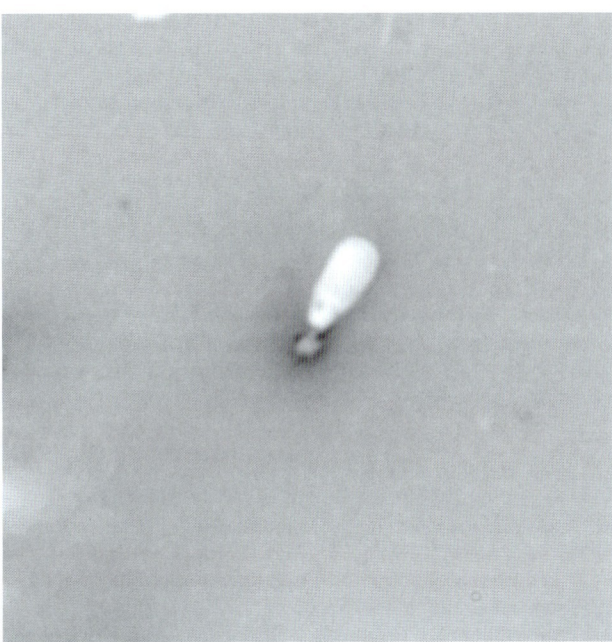

Figure 8.14 Stump tail defect.

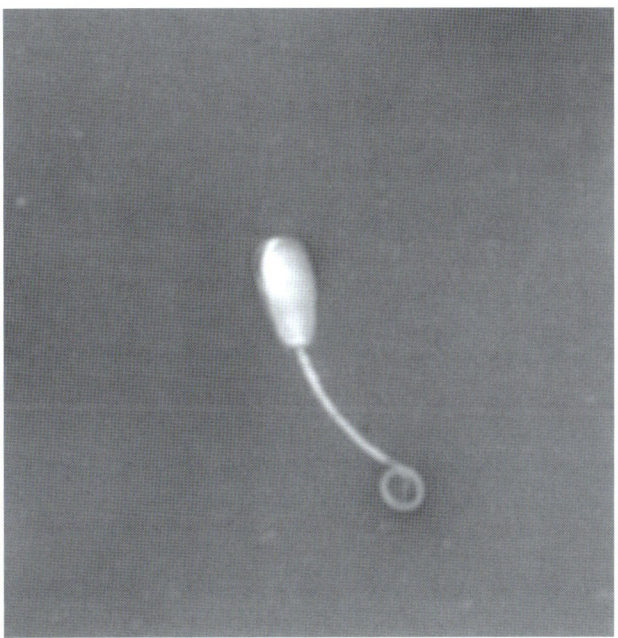

Figure 8.15 Terminally coiled tail defect.

along with other defects following environmental temperature stress, as documented by a scrotal insulation trial.[13] It is also a defect that is increased proportionally following gossypol toxicity.[42]

Simple bent tail defect

A sperm with a bent tail defect is categorized as secondary and minor. It is a compensable defect. The primary concern, when observing this abnormality, is determining whether the defect is real or was induced during staining. The use of old stain that has become hypertonic will often cause a "bowing" or slight bending of sperm tails. In the absence of other defects, you should closely observe these cells or possibly prepare another smear.

References

1. Barth A, Arteaga A, Brito L, Palmer C. Use of internal artificial vaginas for breeding soundness evaluation in range bulls: an alternative for electroejaculation allowing observation of sex drive and mating ability. *Anim Reprod Sci* 2004;84:315–325.
2. Cruz F, Lohn L, Marinho L *et al*. Internal artificial vagina (IAV) to assess breeding behavior of young *Bos taurus* and *Bos indicus* bulls. *Anim Reprod Sci* 2011;126:157–161.
3. Palmer C, Brito L, Arteaga A, Soderquist L, Persson Y, Barth A. Comparison of electroejaculation and transrectal massage for semen collection in range and yearling feedlot beef bulls. *Anim Reprod Sci* 2005;87:25–31.
4. Palmer C. Welfare aspects of theriogenology: investigating alternatives to electroejaculation in bulls. *Theriogenology* 2005;64:469–479.
5. Whitlock B, Coffman E, Coetzee J, Daniel J. Electroejaculation increased vocalization and plasma concentrations of cortisol and progesterone, but not substance P, in beef bulls. *Theriogenology* 2012;78:737–746.
6. Barth A. Evaluation of potential breeding soundness in the bull. In: Youngquist RS, Threlfall W (eds) *Current Therapy in Large Animal Theriogenology*, 2nd edn. St Louis: Saunders, 2007, pp. 228–240.
7. Roberts S. *Veterinary Obstetrics and Genital Diseases (Theriogenology)*, 3rd edn. Woodstock, VT: published by author, 1986.
8. Reference table of the Bull Breeding Soundness Evaluation Form. Society for Theriogenology, Association Office, PO Box 3007, Montgomery, AL 36109, USA. Available at www.therio.org
9. Chenoweth P, Spitzer J, Hopkins F. A new breeding soundness evaluation form. In: *Proceedings of the Annual Meeting of the Society for Theriogenology*, 1992, pp. 63–70.
10. Ott R. Breeding soundness examination in bulls. In: Morrow AD (ed.) *Current Therapy in Theriogenology*, 2nd edn. Philadelphia: Saunders, 1986, pp. 125–136.
11. Chenoweth P. Interpreting the spermiogram. In: *Proceedings of the Annual Meeting of the Association of Applied Animal Andrology*, 2006, Paper 1. Available at www.animalandrology.org
12. Enciso M, Cisale H, Johnston S, Sarasa J, Fernandez J, Gosalvez J. Major morphological sperm abnormalities in the bull are related to sperm DNA damage. *Theriogenology* 2011;76:23–32.
13. Barth A, Oko R. *Abnormal Morphology of Bovine Spermatozoa*. Ames, IA: Iowa State University Press, 1989.
14. Blom E. The ultrastructure of some characteristic sperm defects and a proposal for a new classification of the bull spermiogram. *Nord Vet Med* 1973;25:383–391.
15. Saacke R, Dalton J, Nadir R, Nebel J, Bame J. Relationship of seminal traits and insemination time to fertilization rate and embryo quality. *Anim Reprod Sci* 2000;60–61:663–677.
16. Saacke R. Sperm morphology: its relevance to compensable and uncompensable traits in semen. *Theriogenology* 2008;70:473–478.
17. Volger C, Bame J, DeJarnette M, McGilliard M, Saacke R. Effects of elevated testicular temperature on morphologic characteristics of ejaculated spermatozoa in the bovine. *Theriogenology* 1993;40:1207–1219.
18. Ostermeier C, Sargeant G, Yandell B, Evenson D, Parrish J. Relationship of bull fertility to sperm nuclear shape. *J Androl* 2001;22:595–603.
19. Amann R, Seidel G, Mortimer R. Fertilizing potential in vitro of semen from young beef bulls containing a high or low percentage of sperm with a proximal droplet. *Theriogenology* 2000;54:1499–1515.

20. Smith J. What's known about selected sperm abnormalities in the bull? *Clin Theriogenol* 2009;1:141–146.
21. Chenoweth P, Chase C, Risco C, Larsen R. Characterization of gossypol-induced sperm abnormalities in bulls. *Theriogenology* 2000;53:1193–1203.
22. Chenoweth P, Risco C, Larsen R, Velez J, Tran T, Chase C. Effects of dietary gossypol on aspects of sperm quality, sperm morphology and sperm production in young Brahman bulls. *Theriogenology* 1994;72:445–452.
23. Wenkoff M. A sperm mid-piece defect of epididymal origin in two Hereford bulls. *Theriogenology* 1978;4:275–282.
24. Chenoweth P. Genetic sperm defects. *Theriogenology* 2005;64:457–468.
25. Koefoed-Johnsen H, Anderson J, Andresen E, Blom E, Philipsen H. The Dag defect of the tail of the bull sperm. Studies on inheritance and pathogenesis. *Theriogenology* 1980;14:471–475.
26. Barth A. Abaxial tail attachment of bovine spermatozoa and its effect on fertility. *Can Vet J* 1989;30:656–662.
27. Cran D, Dott H. The ultrastructure of knobbed bull spermatozoa. *J Reprod Fertil* 1976;47:407–408.
28. Thundathil J, Meyer R, Palasz A, Barth A, Mapletoft R. Effect of the knobbed acrosome defect in bovine sperm on IVF and embryo production. *Theriogenology* 2000;54:921–934.
29. Mitchell J, Senger P, Rosenberger J. Distribution and retention of spermatozoa with acrosomal and nuclear abnormalities in the cow genital tract. *J Anim Sci* 1985;61:956–967.
30. Thundathil J, Palomino J, Barth A, Mapletoft R, Barros C. Fertilizing characteristics of bovine sperm with flattened or indented acrosomes. *Anim Reprod Sci* 2001;67:231–243.
31. Dubin N, Calapano C, Berger N. The significance of pyriform (pear-shaped) sperm heads in sperm morphology. *Fertil Steril* 2001;76(Suppl.):S243–S244.
32. Saacke R, DeJarnette J, Bame J, Karabinus D, Whitman S. Can spermatozoa with abnormal heads gain access to the ovum in artificially inseminated super- and single-ovulating cattle? *Theriogenology* 1998;50:117–128.
33. Thundathil J, Palasz A, Mapletoft R, Barth A. An investigation of the fertilizing characteristics of pyriform shaped bovine spermatozoa. *Anim Reprod Sci* 1999;57:35–50.
34. Coulter G, Oko R, Costerton J. Incidence and ultrastructure of "crater" defect of bovine spermatozoa. *Theriogenology* 1978;9:165–173.
35. Foote R. Bull sperm surface "craters" and other aspects of semen quality. *Theriogenology* 1999;51:767–775.
36. Pilip R, Del Campo M, Barth A, Mapletoft R. In vitro characteristics of bovine spermatozoa with multiple nuclear vacuoles: a case study. *Theriogenology* 1996;46:1–12.
37. Barth A, Bowman P. The sequential appearance of sperm abnormalities after scrotal insulation or dexamethasone treatment in bulls. *Can Vet J* 1994;34:93–101.
38. Miller D, Heudka F, Cates W, Mapletoft R. Infertility in a bull with a nuclear defect: a case report. *Theriogenology* 1982;17:611–621.
39. Heath E, Ott R. Diadem/crater defect in spermatozoa of a bull. *Vet Rec* 1982;110:5–6.
40. Barth A. case-based studies of infertility in bulls. In: *Proceedings of the Annual Meeting of the American Association of Bovine Practitioners*. Auburn, AL: American Association of Bovine Practitioners, 2012.
41. Arriola J, Johnson L, Kaproth M, Foote R. A specific oligoteratozoospermia in a bull: the sperm tail stump defect. *Theriogenology* 1985;23:899–913.
42. Hassan M, Smith G, Ott R et al. Reversibility of the reproductive toxicity of gossypol in peripubertal bulls. *Theriogenology* 2004;61:1171–1179.

Web resources

Lane Manufacturing, source of ejaculator referenced in figures: www.lane-mfg.com

Chapter 9

Ultrasound Evaluation of the Reproductive Tract of the Bull

Harry Momont and Celina Checura

Department of Medical Sciences, School of Veterinary Medicine, University of Wisconsin-Madison, Madison, Wisconsin, USA

Introduction

Diagnostic ultrasound has become the standard for medical imaging of soft tissue. The genital tract is an obvious target for this diagnostic modality and while ultrasound is routinely employed for evaluation of the female genital system, its use in male domestic animals remains much less common. In the bull, for example, the routine breeding soundness evaluation (BSE) as recommended by the Society for Theriogenology does not include ultrasonographic assessment of the genital system.[1] In most cases of infertility in the bull, adequate diagnostic and prognostic information is obtained without the use of ultrasonography. A major advantage of ultrasonography, however, is the ability to localize and objectively assess morphologic changes in tissue.

A thorough history, complete physical examination, and a BSE should precede an ultrasound examination of the bull's genital system. We routinely perform ultrasound evaluations for bulls with palpable or visible abnormalities of the genital system; those with pyospermia, hemospermia, azoospermia, or oligospermia; those with low semen volume; and bulls that persistently produce high numbers of morphologically abnormal or immotile sperm. As with any medical technology, the use of ultrasonography for examination of the bull will expand as veterinarians become more familiar with the modality and its application.

Equipment and general methodology

A wide variety of ultrasound equipment is available for diagnostic imaging. Most veterinary clinics providing service to cattle owners will already have a portable, real-time, B-mode system with a linear intrarectal probe that is used for reproductive examination of cows. All illustrations in this chapter were obtained with this type of system. Typical probe frequencies range from 5 to 7.5 MHz and are more than adequate for imaging the bull's reproductive tract. More detailed images of the epididymis and penile tissues can be obtained using probes with a smaller footprint and higher frequency. Representative images may be captured, labeled, measured, and saved for inclusion in the medical record. These applications are available with most ultrasound systems. In order to adequately assess a system you intend to purchase, it is advisable to use it in an actual clinical setting so you can critically evaluate functionality and image quality under real-world conditions. Beyond the primary issue of image quality, durability and resistance to damage from dirt and moisture are important considerations when purchasing an ultrasound unit to be used in cattle environments.

Operator safety is always a concern when working with bulls. A chute or stock with a head catch and squeeze option is ideal. For added security, the bull's head should be secured with a halter. Most bulls will tolerate scrotal, perineal, and rectal examinations with little restraint beyond confinement in the chute. If a kick bar is used behind the bull, extreme caution is needed to avoid placing the arms and hands in a dangerous position. For fractious bulls or for examination of the prepuce and penis cranial to the scrotum, we prefer sedation (xylazine 0.01–0.02 mg/kg i.v.).

While it is beyond the scope of this chapter, a fundamental understanding of ultrasound principles and an appreciation for common image artifacts are essential for anyone offering diagnostic ultrasonography services. For those requiring a more expansive treatment of the basic principles of veterinary ultrasonography, excellent texts are available.[2,3] A bovine reproductive anatomy monograph is available from the National Association of Animal Breeders[4] and McEntee[5] is an excellent resource for those seeking a review of the pathology of the male genital system.

Images are obtained by either a transcutaneous or transrectal scan depending on the anatomical location of the tissue to be examined. If long or dense hair covers the area, the quality of the transcutaneous image is improved by first shaving the skin. Shaving is usually not required for examination of the scrotum and testes. Image quality is enhanced when the scrotal skin is stretched tightly over the testes. To do this, firmly grasp the scrotal neck dorsal to the testes and pull the testes ventrally while conducting the examination. In all cases of transcutaneous ultrasonography, a coupling agent is required to achieve good contact between the probe and skin. Any air or gas between the probe and the tissue to be examined will interfere with the acquisition of an image. The use of ultrasonic coupling gel as a coupling agent has the advantage of long duration, since it does not evaporate. Additionally, it is approved by most manufacturers for contact with the probe surface. Alcohol (70% isopropyl) also makes a good coupling agent, with the advantage that it does not have to be cleaned from the probe or patient after the examination is completed. A disadvantage of alcohol is that it evaporates and often needs to be reapplied to complete the examination. Additionally, not all probe surfaces are approved for contact with alcohol or other solvents. Placing a protective cover containing a small amount of coupling gel over the probe will prevent alcohol from contacting the probe surface. A disposable examination glove works well for this purpose.

For transrectal examination of the pelvic organs, the probe is covered with an examination sleeve containing just enough coupling gel to provide good contact between the probe and sleeve. Most probes designed for transrectal use in horses and cattle will fit snugly within a finger of a glove or sleeve. The feces are evacuated from the rectum and a manual examination is conducted before the covered probe is inserted. In our experience, no coupling agent is necessary to maintain contact between the sleeve and rectal mucosa beyond the lubricant normally used for palpation per rectum. If a protective cover is not used, the probe should be cleaned and disinfected between examinations following the recommendations of the probe manufacturer.

While diagnostic ultrasonography is generally regarded as safe for examination of reproductive tissues and fetuses,[3] few published data exist in the veterinary literature. Coulter and Bailey[6] exposed yearling beef bulls to ultrasound (3 min for each testis, one time, using a 5-MHz linear transducer) and found no discernible effects on sperm numbers, morphology, or motility during a 70-days period after the examination. They concluded that diagnostic ultrasonography should be safe for examination of the bull scrotum and testes. A reasonable approach is to use the minimum power and contact time necessary to complete your examination.

Examination of the scrotum

B-mode ultrasonography of bull testes was first described by Pechman and Eilts[7] in 1987. A thorough visual and manual examination of the scrotum and its contents should always precede the ultrasound examination. The scrotum and testes are then examined by ultrasound in both a sagittal and transverse plane with the probe applied to the cranial, lateral, or caudal surface of the testis depending on the examiner's preference. A complete examination includes the spermatic cord, the entirety of the testicular parenchyma, the epididymis, and the scrotal wall. The body of the epididymis and the ductus deferens are difficult to detect unless they are grossly abnormal.

Normal ultrasonic appearance of the testis is shown in Figure 9.1. The testis parenchyma appears as a homogeneous tissue with a stippled medium echogenicity that surrounds the centrally located and more hyperechoic (whiter) mediastinum testis. The mediastinum testis is normally less than

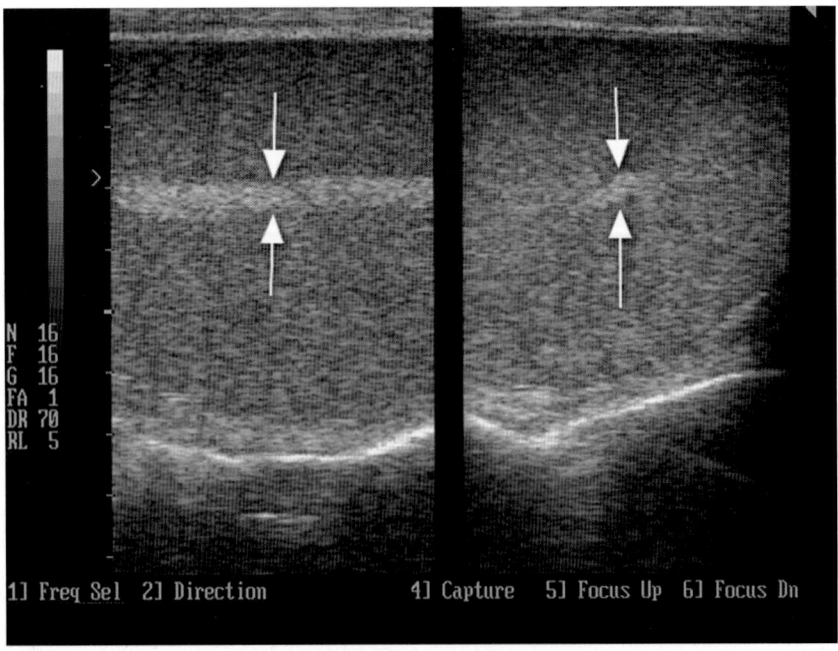

Figure 9.1 Caudal ultrasonographic views of the left testis of a yearling dairy bull. Caudal is at the top and cranial at the bottom of each ultrasound image. On the left is a sagittal view with the mediastinum testes (between arrows in each image) appearing as a more hyperechoic linear structure running through the approximate center of the less echoic testis parenchyma. On the right side of the image is a transverse view of the same testis with the mediastinum in cross-section, again in the middle of the less echoic parenchyma. The scrotal wall is visible adjacent to the probe on the top (caudal) and on the bottom (cranial) of the testis in each image. The bright white line at the bottom of each image is a gas artifact that marks the far side of the scrotal skin.

5mm wide. The parenchyma consists primarily of convoluted seminiferous tubules that contain the sustentacular (Sertoli) and germinal cells but also includes the adjacent interstitial (Leydig) cells along with associated vascular, neural, and stromal tissues. The seminiferous tubules are connected by the straight tubules to the rete testis that comprises a series of interconnected channels within the mediastinum testis. Sperm pass dorsally through the rete testis to the efferent ducts that in turn penetrate the capsule of the testis before uniting as the epididymal duct.

Changes in the ultrasonic appearance of the testis parenchyma can reflect normal events around puberty as well as pathology. An increase in echogenicity (increased pixel intensity, or a whiter parenchyma) has been reported to occur in association with sexual development around the time of puberty,[8,9] but ultrasonography was no better at predicting puberty than was measurement of scrotal circumference.[9] While pixel intensity has been correlated with attainment of sexual maturity as assessed by semen quality,[10] it seems to be a better predictor of future semen quality than the present status of the bull.[11] In any event, objective assessment of these subtle changes in pixel intensity is not possible by real-time visual interpretation of an ultrasound image. To accurately assess pixel intensity, images must be obtained and digitally preserved using rigorously standardized methods. The digital files are then sampled and evaluated using computer software that can objectively assess the gray scale of each pixel in the sampled area, providing both an average intensity and measure of variation. Given the limited clinical utility and the great effort required, Brito et al.[9] concluded that "though testicular ultrasonogram pixel intensity might be useful for research purposes, clinical application of this technology in the present form for bull breeding soundness evaluation is not justifiable."

The majority of the discrete ultrasonographic changes detected in the testis are more hyperechoic than normal tissue. These may appear as small foci scattered throughout the testis (Figure 9.2) or they may be more localized (Figure 9.3). Larger hyperechoic lesions are often seen to radiate from the rete to the periphery of the testis, suggesting the complete involvement of one or more seminiferous tubules (Figure 9.4). The histologic basis for the increase in echogenicity is usually fibrosis[12] but the echogenicity itself does not confirm histology or etiology. Tumors are generally more hyperechoic than the normal testis parenchyma (Figure 9.5) but mixed or hypoechoic echogenicity is also possible.[3] Mineralized lesions are extremely hyperechoic and are always accompanied by intense or complete shadowing causing a loss of image distal to the lesion (Figure 9.6).

Barth et al.[12] have proposed a scoring system for fibrotic lesions in the bull testis based on the size and frequency of hyperechoic lesions detected by ultrasonography in a single transmediastinal plane. They found that the occurrence of fibrotic lesions in young bulls aged between 3 and 20 months was not associated with decreased semen quality. This is consistent with their histologic findings, where seminiferous tubules within the area of fibrosis were generally affected but immediately adjacent germinal tissue was quite often completely normal. However, bulls that experienced a dramatic increase in fibrosis score at a second examination (≥4 points, 4–13 months later) were more likely to have poor semen quality than bulls that did not experience this degree of change.[12] While bulls in this report with up to 50% of their parenchyma affected were able to produce mostly normal sperm, it is reasonable to suspect that they would have reduced production capacity.

The cause of fibrotic lesions in the testis is unknown. Possible etiologies include infectious or inflammatory conditions, developmental defects of the seminiferous tubules or their connecting ducts, autoimmunity, degenerative conditions, or aging. Ram testes inoculated with *Trueperella pyogenes* (formerly *Arcanobacterium pyogenes*) eventually developed hyperechoic lesions that consisted of fibrotic tissue.[13] A severe outbreak of bovine respiratory syncytial virus (BRSV) in a group of bulls was associated with an increased prevalence of fibrotic lesions in one report, though a cause-and-effect relationship could not be confirmed.[12] On the other hand, scrotal insulation resulting in a dramatic decrease in semen quality did not cause fibrosis or any other ultrasonically detectable change in the testis within 4 months[14] or 6 months[11] after the insulation event. A progressive fibrosis that is more marked in the ventral testis has been described as a histologic feature of aging in the bull.[15]

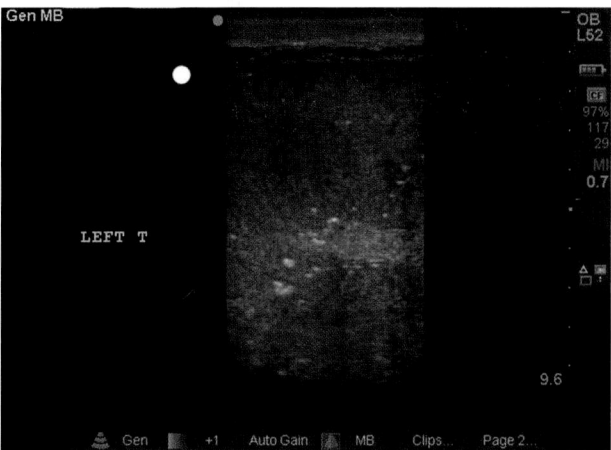

Figure 9.2 Sagittal view of the left testis of a mature bull with small hyperechoic foci distributed throughout the testis parenchyma.

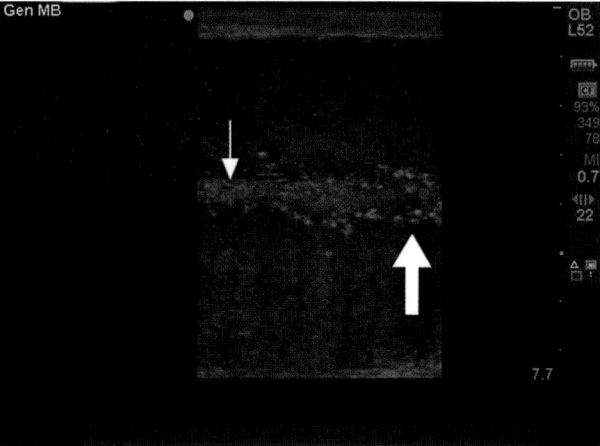

Figure 9.3 Sagittal view of a testis with small hyperechoic foci (large arrow) surrounding the mediastinum (small arrow) suggestive of lesions in the straight ducts that drain the seminiferous tubules into the rete testis.

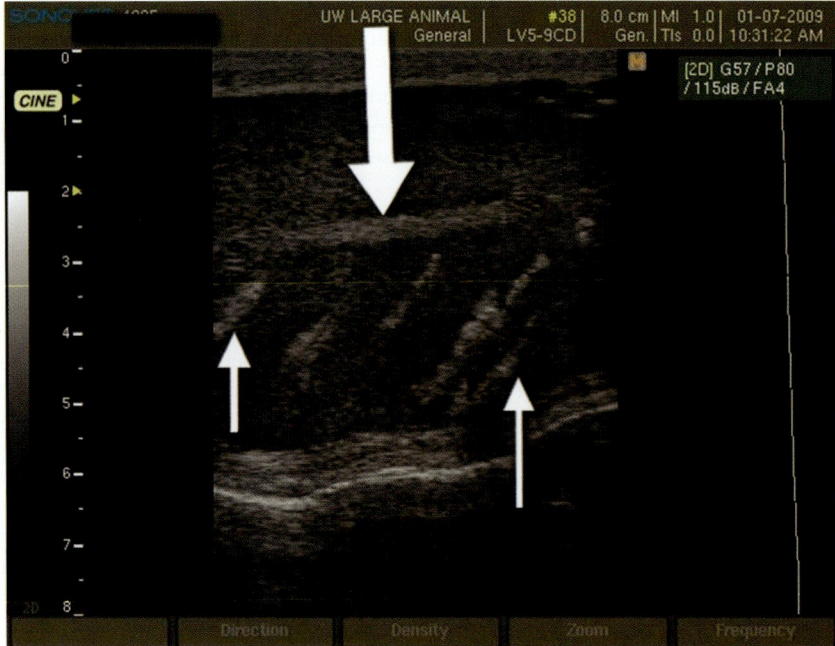

Figure 9.4 Sagittal view of a testis with multiple linear radiating hyperechoic lesions, two of which are highlighted by the smaller arrows. A normal mediastinum testis is shown (large arrow).

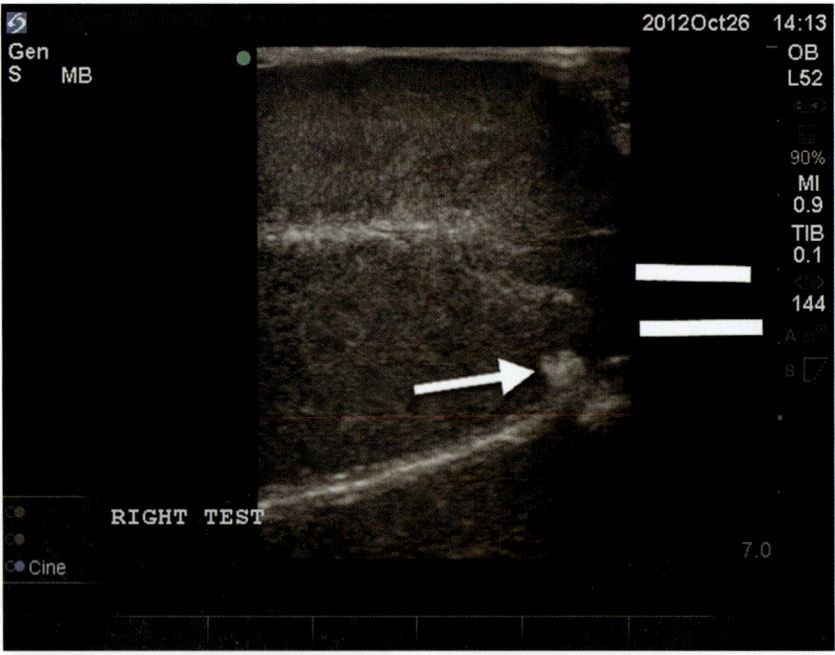

Figure 9.5 Sagittal view of the right ventral testis of an aged beef bull. Histologically, the hyperechoic lesion (arrow) was a Sertoli cell tumor.

Distinct hypoechoic lesions in the testis parenchyma are seen less commonly (Figure 9.7). Intratesticular inoculation of *T. pyogenes* in rams caused an initial hypoechoic change associated with edema, neutrophilic infiltration, and degenerative changes in the seminiferous tubules.[13] Since an increase in echogenicity is seen at puberty in association with proliferation of the spermatogenic tissue, it seems reasonable to speculate that areas affected by a degenerative process would be more hypoechoic by contrast.

A hypoechoic distension of the rete testis has been reported in Ayrshire cattle.[16] Most of the bulls in this report had bilateral distension and were azoospermic by 16 months of age, suggesting a block to sperm outflow. A genetic predisposition to the condition was suspected. A similar unilateral lesion was reported in a mature Holstein bull in conjunction with agenesis of the ipsilateral seminal vesicle, epididymal body and tail as well as low semen volume.[17] Figure 9.8 shoes an ultrasonogram from a yearling Holstein bull with a history of oligospermia and low semen volume.

Ultrasound Evaluation of the Reproductive Tract of the Bull 83

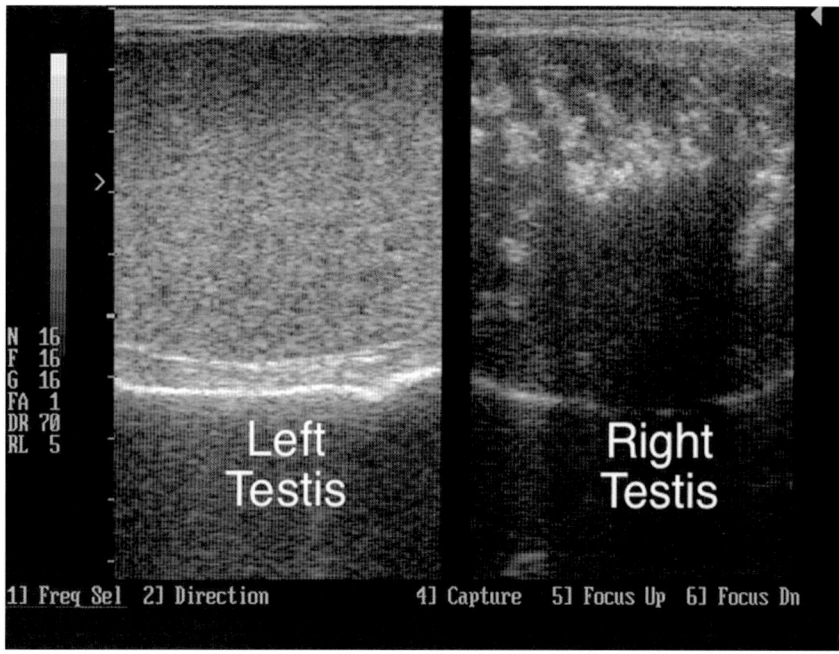

Figure 9.6 Sagittal views of the left and right testes of a yearling dairy bull. The left testis is normal while the hyperechoic areas of the right testis cast prominent dark shadows in the distal tissue and represent areas of mineralization.

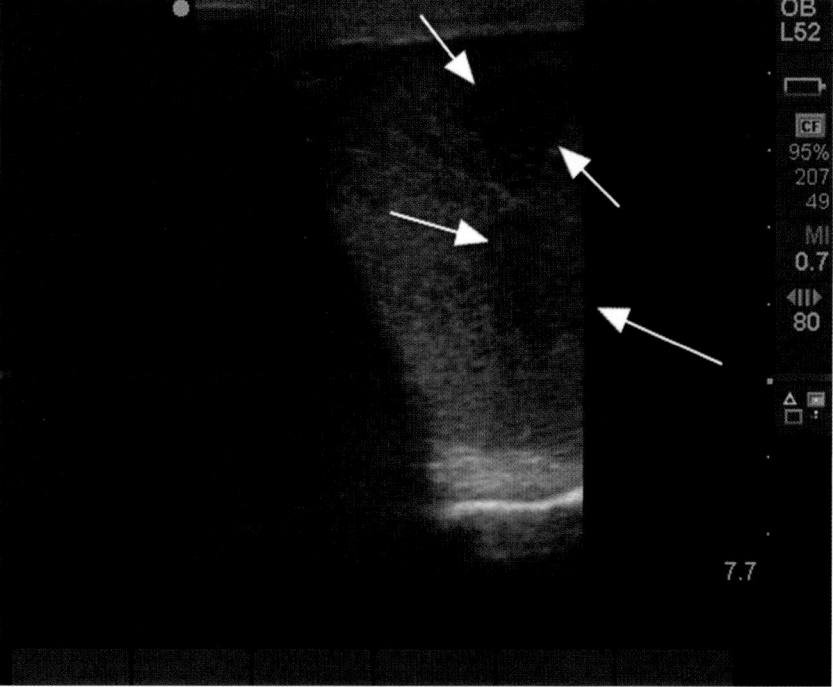

Figure 9.7 Transverse view of the testis containing two discrete hypoechoic wedge-shaped areas (between arrows) of unknown etiology or significance.

Note the enlarged hypoechoic appearance of the mediastinum. The contralateral testis appeared to be normal (not shown). Surgical removal of the affected testicle confirmed a diagnosis of agenesis of the body of the epididymis and fluid distension of the rete testis. The distension can also affect the efferent ducts. In our experience, extensive and uniform hypoechoic distension of the rete testis (Figure 9.9) is pathognomonic for sperm outflow obstruction, most often congenital. In unilateral cases, the presenting complaint is usually low sperm numbers and semen volume in an otherwise normal young bull. Congenital cases may be associated with aplasia affecting the epididymis, deferent duct, ampulla, or seminal vesicle.

The head of the epididymis is best seen in a sagittal view and is found on the dorsocranial aspect of the testicle. It is usually less echoic than the parenchyma of the testis. The tail of the epididymis can be readily located by palpation and is found on the ventral aspect of the testis. Acquired

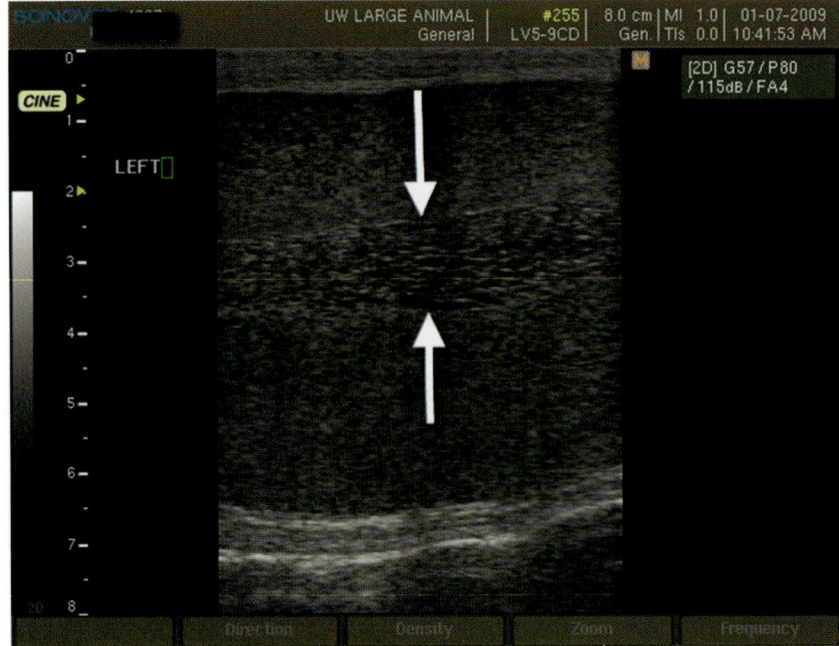

Figure 9.8 Sagittal view of the left testis of a mature Holstein bull that presented with a complaint of low semen volume and oligospermia. Note the hypoechoic distension of the rete testis (dark linear area between arrows). The distension was secondary to a complete outflow obstruction resulting from the absence of the epididymal body on this side.

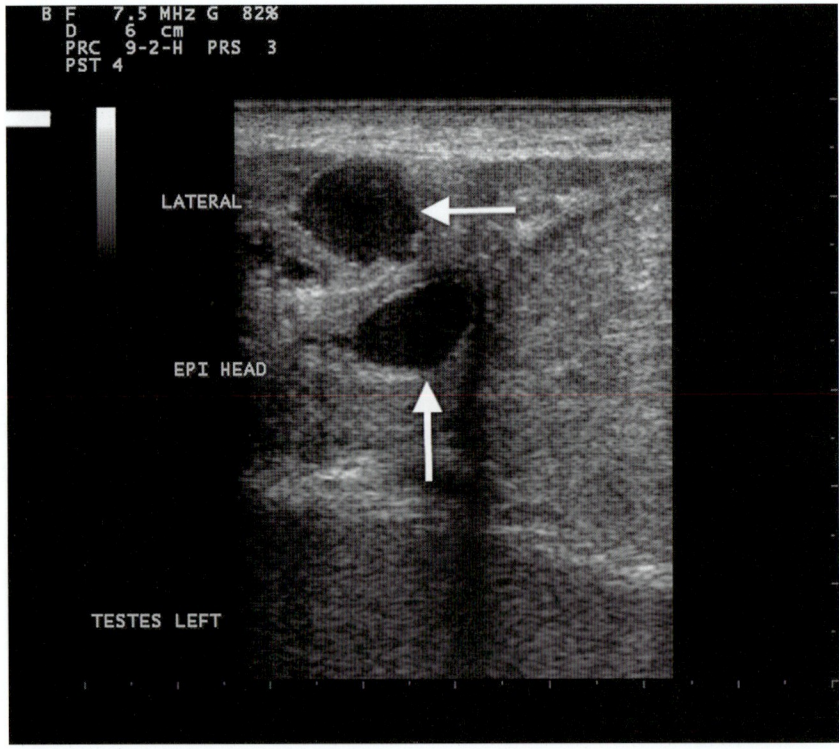

Figure 9.9 Sagittal view of the testis. Note the hypoechoic dilations (arrows) in or near the head of the epididymis. Normal testis parenchyma is seen to the right of the cysts.

lesions of the epididymides can lead to outflow obstruction. A cystic distension of the head of the epididymis is shown in Figure 9.9. It is worth noting that cysts of the appendix epididymis (proximal mesonephric duct remnant) are also found in this area and even those as large as a few centimeters in diameter usually do not interfere with normal epididymal function.[5] The epididymal tail (Figure 9.10) can be difficult to image with a transrectal probe and for fine detail one should consider using a higher-frequency probe with a smaller footprint.

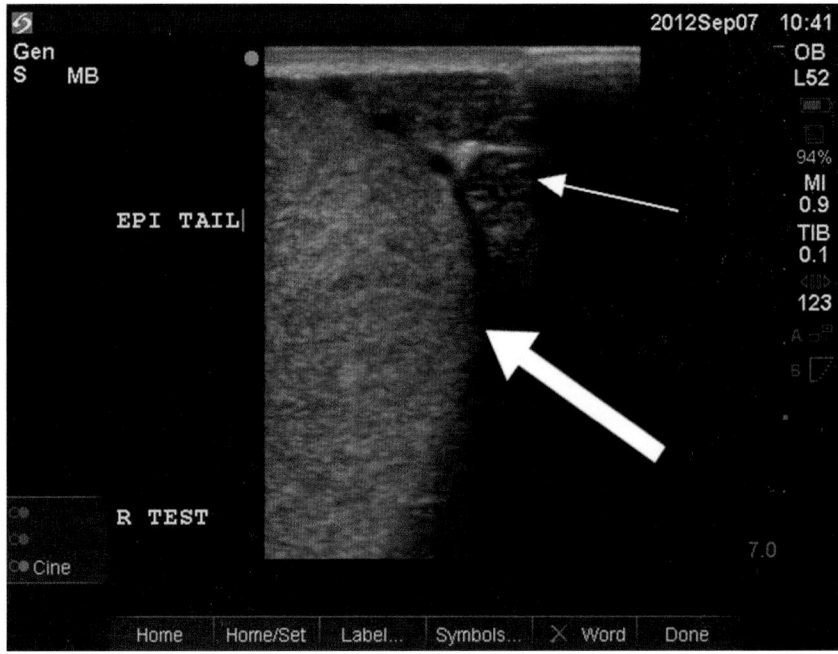

Figure 9.10 Sagittal view of the ventral testis (large arrow) and tail of the epididymis (small arrow) of a normal bull.

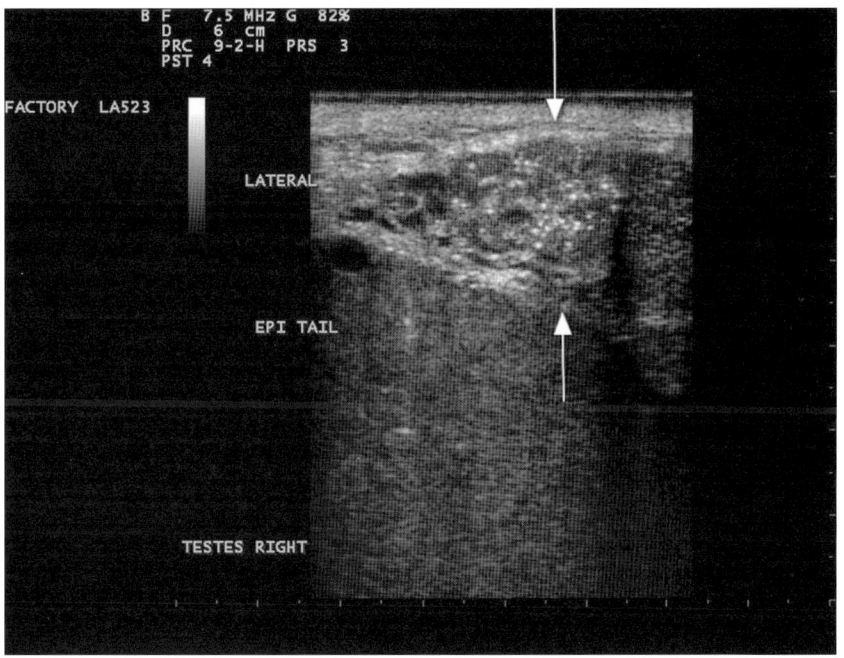

Figure 9.11 Transverse view of the ventral aspect of the right testis of an aged beef bull. The tail of the epididymis is seen between the arrows. The small white foci seen in the epididymal tissue between the arrows are sperm granulomas.

Congenital, traumatic, and inflammatory lesions can be seen here with sperm stasis and granuloma formation as possible sequelae. Alterations of size and echogenicity are characteristic of disease in this tissue. Multiple small sperm granulomas are visible as focal hyperechoic lesions in the tail of the epididymis (Figure 9.11).

Ultrasonography can provide more specific information about the extent, location, and resolution of scrotal pathology. Examples include inflammation, sperm granuloma, spermatocele, fibrosis, mineralization, herniation, scrotal hydrocele, tumors, abscesses, trauma, hemorrhage, outflow obstruction, and lesions of the scrotal wall and vasculature. The echogenic pattern seen is variable and often quite heterogeneous. In cases of testicular trauma (Figure 9.12), ultrasound can confirm the actual degree of damage to the testis and provide a more accurate prognosis for recovery.

The final element of the scrotal ultrasound evaluation is an assessment of the scrotal neck and spermatic cord (Figure 9.13). While herniation, trauma, hemorrhage, and scrotal hydrocele can affect this region, they are usually

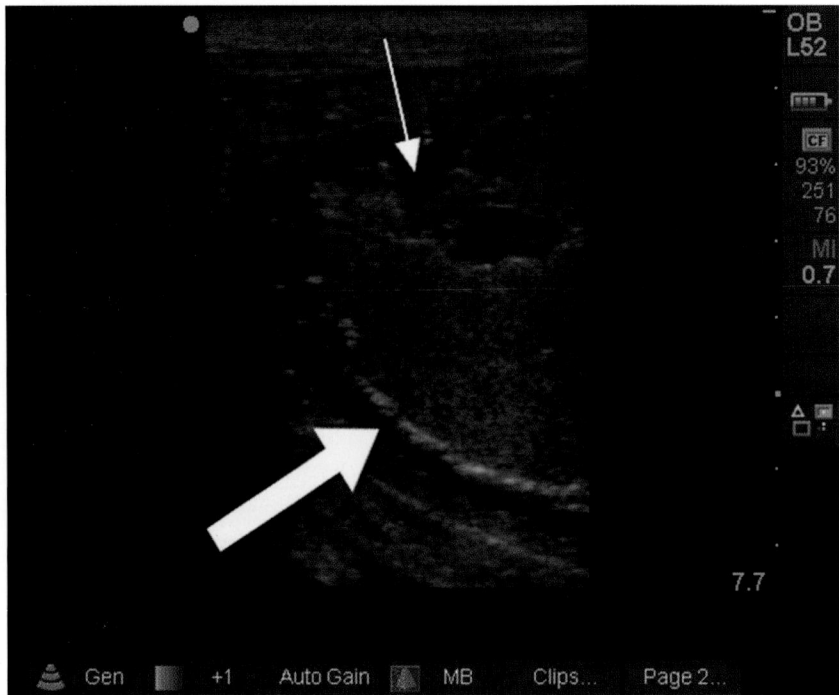

Figure 9.12 Sagittal view of the dorsal testis. The small arrow points to a traumatic lesion of the testis capsule and visceral vaginal tunic. The hypoechoic fluid surrounding the testicle is a persistent scrotal hydrocele (large arrow).

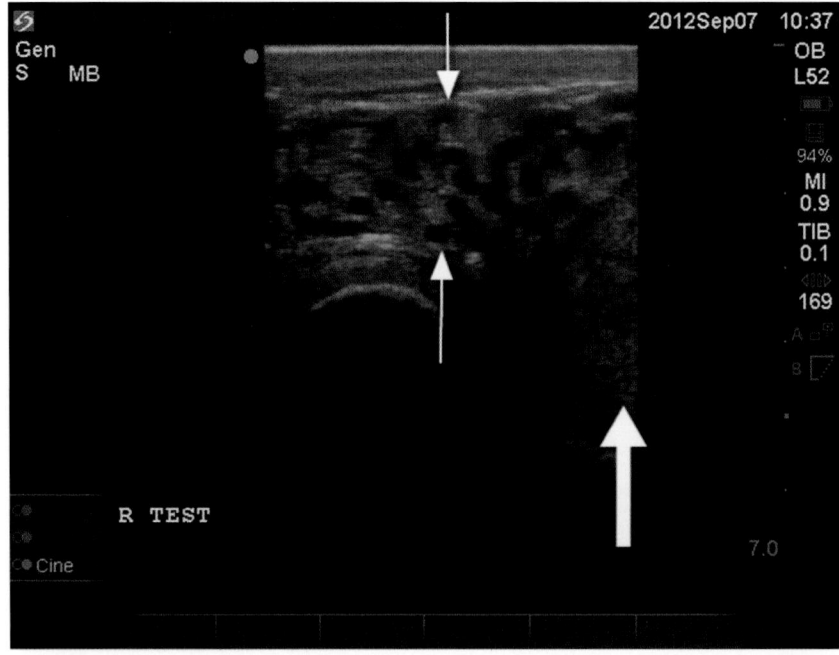

Figure 9.13 Sagittal view of the vascular cone (between small arrows) just dorsal to the testis (large arrow). The lumen of the testicular artery is represented by the black irregular areas in the vascular cone. The network of veins forming the pampiniform plexus is not usually visible due to their small size.

more evident in the scrotum proper. The vas deferens cannot be reliably imaged in normal bulls. The primary ultrasonic feature of this region is the vascular cone consisting of the prominent tortuous testicular artery overlain by a fine network of very small veins called the pampiniform plexus. Brito et al.[18] reported an increase in vascular cone diameter up to 13.5 months of age in beef bulls and increased vascular cone diameter was positively correlated with an increased percentage of normal sperm.

The direction, relative velocity, and volume of blood flow in arteries can be assessed by use of Doppler ultrasonography.[3] A color Doppler ultrasonogram of the testicular artery is shown in Figure 9.14. In camelids, pulsed-wave Doppler ultrasonography detected a difference in mean peak

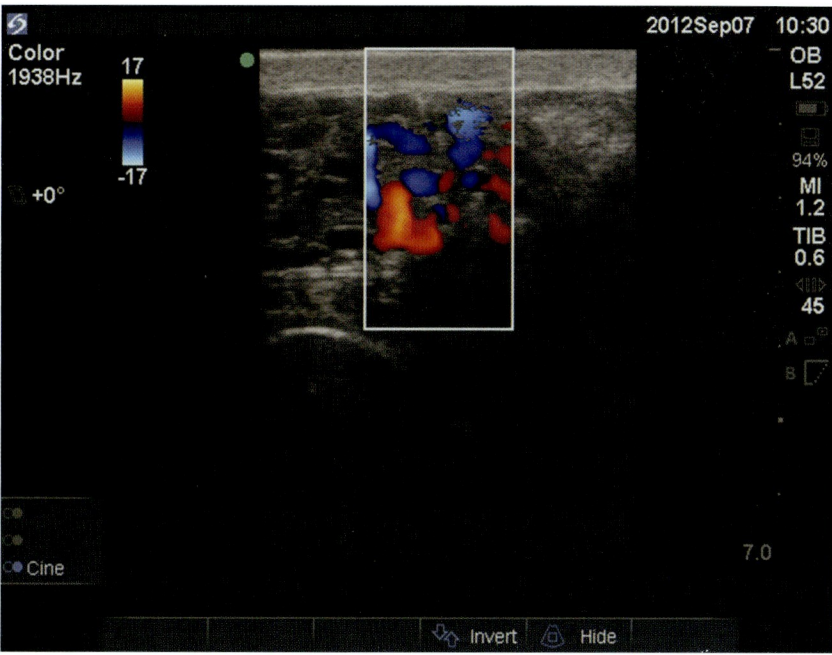

Figure 9.14 Sagittal view of the vascular cone using color Doppler mode. The direction of blood flow in the vessels is indicated by color and the velocity by shade within color.

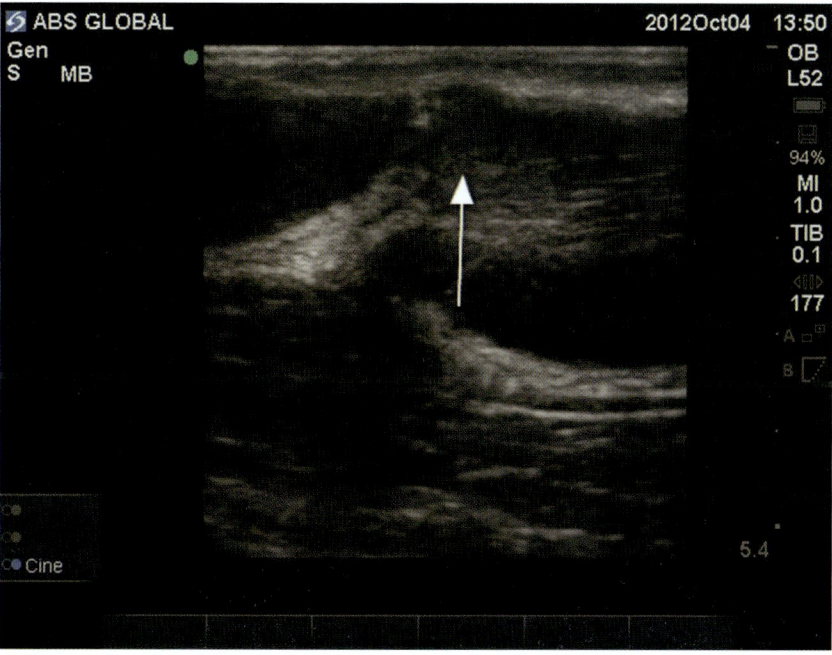

Figure 9.15 Midline sagittal view of the body of the prostate gland (ovoid tissue above arrow). Dorsal is on top, ventral on the bottom, cranial to the left, and caudal to the right.

systolic and end-diastolic velocities in the testicular arteries of fertile compared with infertile males.[19] Similar vascular studies may provide a fertile direction for research into the causes and diagnosis of idiopathic infertility in the bull.

Examination of the pelvic organs

The anatomy of the pelvic accessory sex organs is reviewed in Chapter 1. Ultrasonic assessment of the accessory glands of the bull was first described in 1987 by Weber et al.[20]

Transrectal ultrasound images are generally sagittal but may also be oblique to the long axis. An ultrasonic evaluation of the tissues is performed after first evacuating fecal material and manually palpating the reproductive tract. The transverse band of prostatic tissue on the dorsal aspect of the cranial extent of the urethralis is a convenient landmark (Figure 9.15). The urethralis is then imaged in a caudal direction (Figure 9.16) until the paired bulbourethral glands are detected (Figure 9.17). Moving cranially, the urethralis is followed until the prostate is encountered again. The paired

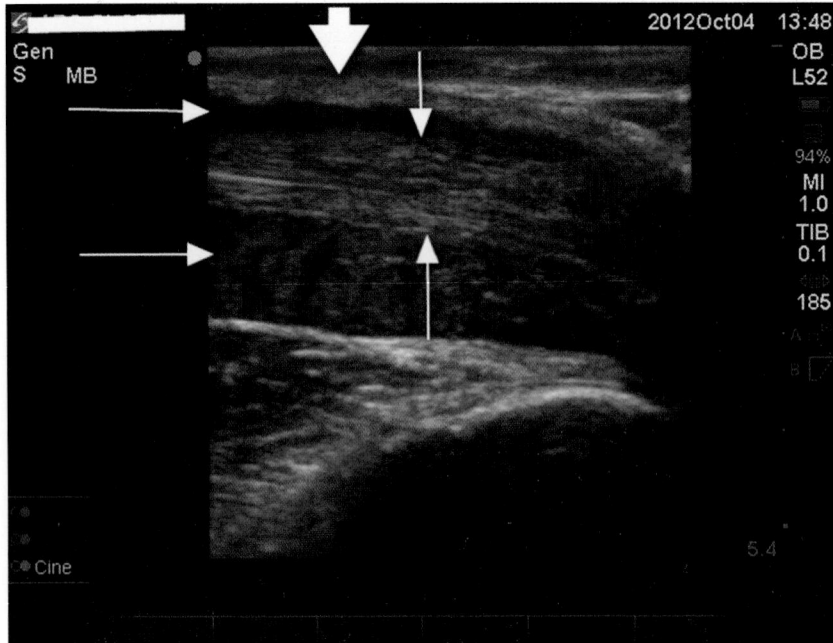

Figure 9.16 Sagittal view of the urethralis just caudal to the body of the prostate. The horizontal arrows on the left point to the more hypoechoic urethral muscle above and below the urethra. The more hyperechoic tissue between the vertical arrows is the cavernous tissue immediately surrounding the pelvic urethra. The disseminate prostate is seen as a hyperechoic linear structure on the dorsum of the urethralis (large vertical arrow). Dorsal is on top, ventral on the bottom, cranial to the left, and caudal to the right.

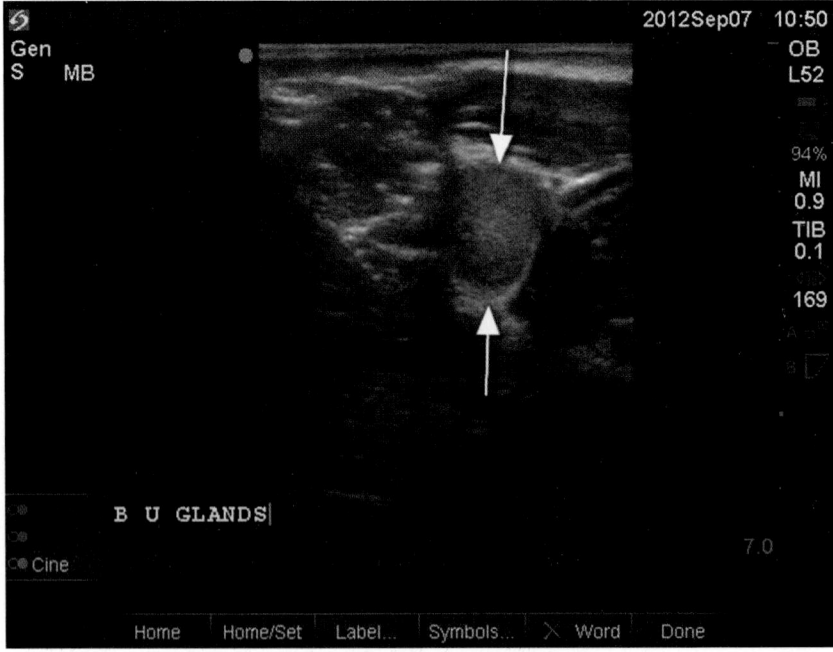

Figure 9.17 Sagittal view of the bulbourethral (Cowper's) gland (between arrows). Dorsal is on top, ventral on the bottom, cranial to the left, and caudal to the right.

ampullae are found just cranial to the prostate, on or near the midline (Figure 9.18). The paired vesicular glands are found just lateral to the ampullae (Figure 9.19). An ultrasonic assessment of the bladder should also be done at this time to check for anomalies, cystitis, or calculi.

Pathology of the bulbourethral glands and prostate is uncommon. Aplasia[5] and sperm accumulation[21] have been reported in the ampullae. Inflammation of the vesicular glands is relatively common in bulls and should be suspected whenever an asymmetry in size or echotexture is encountered. Inflamed glands are larger and often have a mixed echogenicity while abscessed glands show a loss of lobar architecture and often have a globoid area of variable echogenicity.[22] A bacterial infection is most commonly

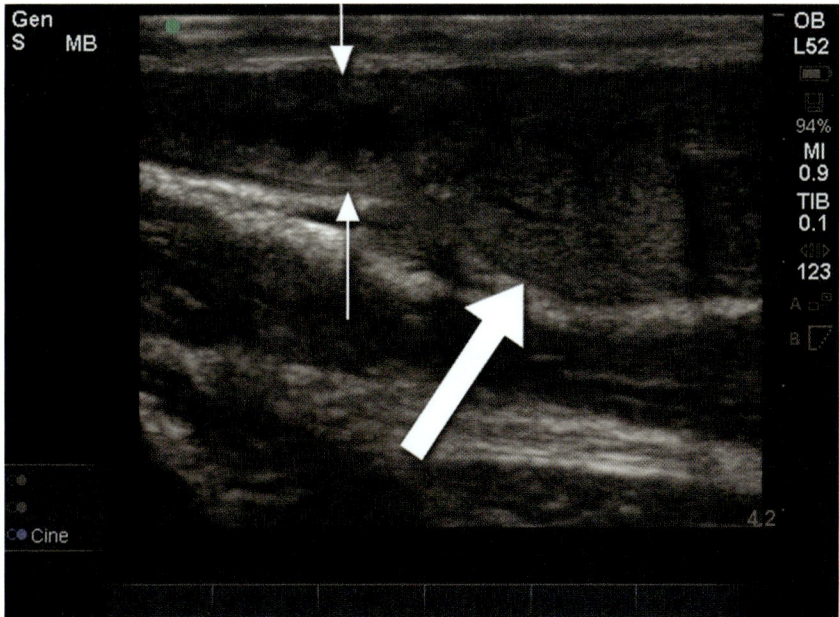

Figure 9.18 Sagittal view of the ampulla (between small arrows). Note the distended hypoechoic lumen surrounded by gland tissue. Luminal contents may or may not be seen during examination of normal bulls. The lobar tissue above and to the right of the large arrow is an adjacent seminal vesicle. Dorsal is on top, ventral on the bottom, cranial to the left, and caudal to the right.

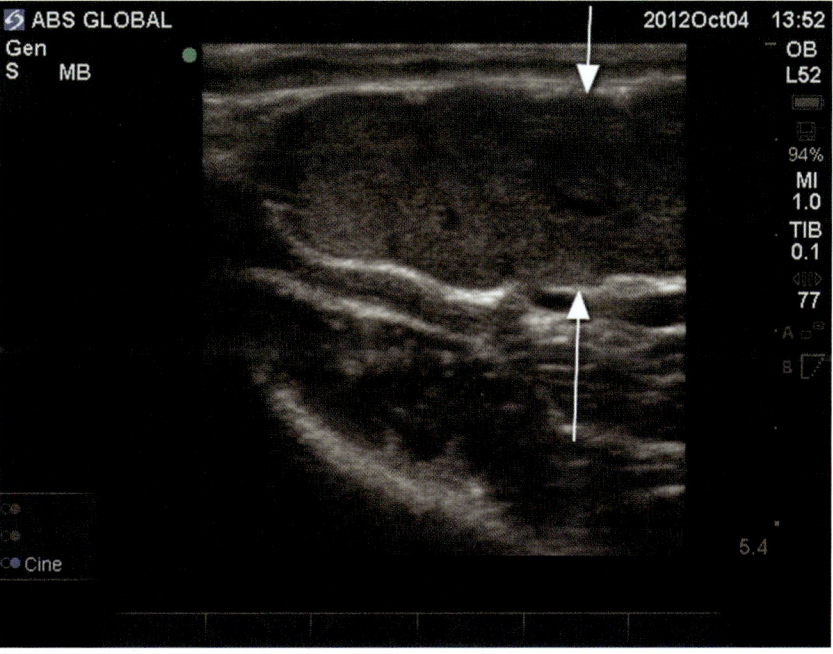

Figure 9.19 Oblique view of the seminal vesicle (horizontal tissue between arrows). Note the lobar architecture. A short segment of the hypoechoic central duct is seen in the center of the gland between the two arrows.

incriminated as the cause of seminal vesiculitis but sterile inflammation resulting from urine or sperm leaking into the glands is also a potential etiology. Figure 9.20 is an ultrasonogram from a 2-year-old beef bull with segmental aplasia of the right vesicular gland. Note the cystic dilation of the occluded central duct. The gland was much larger compared with the gland on the left. This bull also had aplasia of the right epididymal tail and a grossly distended hypoechoic right rete testis.

Examination of the penis and prepuce

Ultrasound can be used to examine the penis, prepuce, and penile urethra. Ultrasonography of the perineal area caudal and dorsal to the scrotum will generally require shaving the hair to obtain good images. The risks associated with ultrasound examination of the penis and prepuce cranial to the scrotum should be balanced with the lower probability of the examination providing valuable diagnostic and prognostic

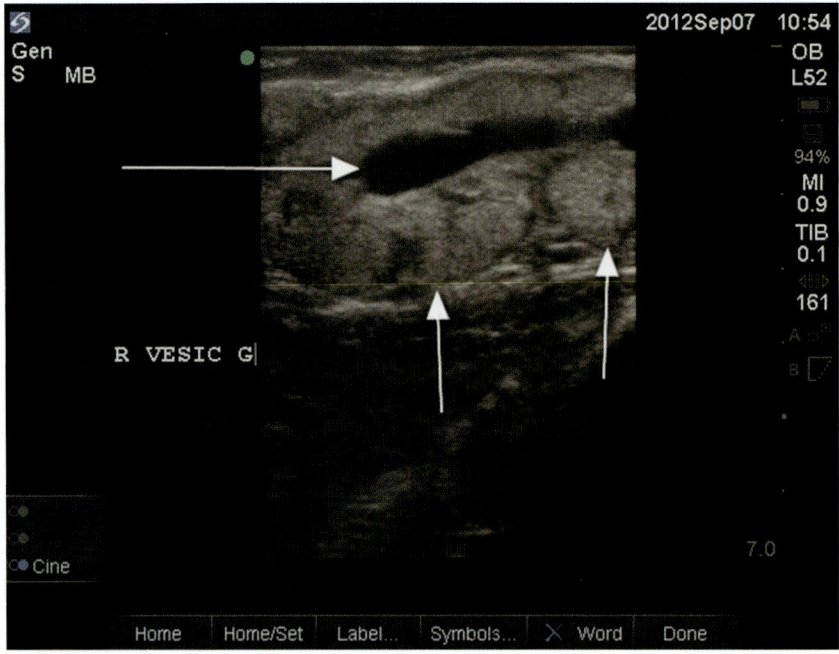

Figure 9.20 Oblique view of the seminal vesicle (above vertical arrows). This bull had a segmental aplasia of the seminal vesicle on this side resulting in the cystic distension of the central duct (horizontal arrow).

information beyond that obtained by visual inspection and palpation. Direct observation of the penis and prepuce during mating or semen collection is necessary to detect many conditions, such as penile hypoplasia, genital warts, persistent frenulum, hair rings, or balanoposthitis. Ultrasonically relevant lesions affecting the penis and prepuce include trauma, neoplasia, hematoma, abscess, calculi, urethral dilation and rupture, and vascular shunts.

Summary

Diagnostic ultrasound is a safe noninvasive method for acquiring additional diagnostic and prognostic information in cases of infertility in the bull. Clinical competence with bull genital ultrasonography can be readily achieved by routinely imaging the genital tract of normal healthy bulls. This will provide the frame of reference necessary to identify and interpret pathologic changes as they are encountered. The common lesions are seldom subtle and are readily detected by systems providing average or better image quality. We recommend ultrasonography as an adjunct to the BSE for all cases of subfertility in AI sires and valuable natural service sires. Routine screening of young sires as part of a pre-purchase evaluation should be done with caution as the most common ultrasound change encountered is a fibrotic lesion or foci in the testes. These lesions have been shown to have little or no effect on semen quality in young bulls. Advances in ultrasound diagnostic capability, especially as it relates to assessment of blood flow in the scrotum, may expand our ability to identify more subtle causes of infertility in the bull in the future.

References

1. Chenowith P, Hopkins F, Spitzer J, Larsen R. Guidelines for using the bull breeding soundness examination form. *Clin Theriogenol* 2010;2:43–50.
2. Ginther O. *Ultrasonic Imaging and Animal Reproduction. Fundamentals*. Cross Plains, WI: Equiservices Publishing, 1995.
3. Nyland T, Mattoon J, Wisner E. Physical principles, instrumentation and safety of diagnostic ultrasound. In: Nyland T, Mattoon J (eds) *Veterinary Diagnostic Ultrasound*. Philadelphia: WB Saunders, 1995, p. 17.
4. Mullins K, Saacke R. *Illustrated Anatomy of the Bovine Male and Female Reproductive Tracts. From Gross to Microscopic*. Blacksburg, VA: Germinal Dimensions, 2003.
5. McEntee K. *Reproductive Pathology of Domestic Animals*. San Diego: Academic Press, 1990, pp. 224–383.
6. Coulter G, Bailey D. Effects of ultrasonography on the bovine testis and semen quality. *Theriogenology* 1988;30:743–749.
7. Pechman R, Eilts B. B-mode ultrasonography of the bull testicle. *Theriogenology* 1987;27:431–441.
8. Chandolia R, Honaramooz A, Omeke B, Pierson R, Beard A, Rawlings N. Assessment of development of the testes and accessory glands by ultrasonography in bull calves and associated endocrine changes. *Theriogenology* 1997;48:119–132.
9. Brito L, Barth A, Wilde R, Kastelic J. Testicular ultrasonogram pixel intensity during sexual development and its relationship with semen quality, sperm production, and quantitative testicular histology in beef bulls. *Theriogenology* 2012;78:69–76.
10. Brito L, Silva A, Unanian M, Dode M. Sexual development in early and late maturing *Bos indicus* and *Bos indicus × Bos taurus* crossbred bulls in Brazil. *Theriogenology* 2004;62:1198–1217.
11. Arteaga A, Barth A, Brito L. Relationship between semen quality and pixel-intensity of testicular ultrasonograms after scrotal insulation in beef bulls. *Theriogenology* 2005;64:408–415.
12. Barth A, Alisio L, Aviles A, Arteaga A, Campbell J, Hendrick S. Fibrotic lesions in the testis of bulls and relationship to semen quality. *Anim Reprod Sci* 2008;106:274–288.
13. Gouletsou P, Fthenakis G, Cripps P, Papaioannou N, Lainas D, Amiridis G. Experimentally induced orchitis associated with *Arcanobacter pyogenes*: clinical, ultrasonographic, seminological and pathological features. *Theriogenology* 2004;62:1307–1328.
14. Sidibe M, Franco L, Fredriksson G, Madej A, Malgrem L. Effects on testosterone, LH and cortisol concentrations, and on ultrasonographic appearance of induced testicular degeneration in bulls. *Acta Vet Scand* 1992;33:191–196.

15. Humphrey J, Ladds P. A quantitative histological study of changes in the bovine testis and epididymis associated with age. *Res Vet Sci* 1975;19:135–141.
16. Andersson M, Alanko M. Ultrasonography revealing the accumulation of rete testis fluid in bull testicles. *Andrologia* 1991;23:75–78.
17. Williams H, Revell S, Scholes S, Courtenay A, Smith R. Clinical, ultrasonographic and pathological findings in a bull with segmental aplasia of the mesonephric duct. *Reprod Domest Anim* 2010;45:212–216.
18. Brito L, Barth A, Wilde R, Kastelic J. Testicular vascular cone development and its association with scrotal temperature, semen quality, and sperm production in beef bulls. *Anim Reprod Sci* 2012;134:135–140.
19. Kutzler M, Tyson R, Grimes M, Timm K. Determination of testicular blood flow in camelids using vascular casting and color pulsed-wave Doppler ultrasonography. *Vet Med International* 2011;Article ID638602.
20. Weber J, Hilt C, Woods G. Ultrasonographic appearance of bull accessory sex glands. *Theriogenology* 1988;29:1347–1355.
21. Barth A. Sperm accumulation in the ampullae and cauda epididymides of bulls. *Anim Reprod Sci* 2007;102:238–246.
22. Momont H, Meronek J. Seminal vesiculitis. In: Haskell SR (ed.) *Blackwell's Five-Minute Veterinary Consult: Ruminant*. Baltimore: Lippincott Williams & Wilkins, 2008, pp. 822–825.

Chapter 10

Management of Breeding Bull Batteries

E. Heath King

*Department of Pathobiology and Population Medicine, College of Veterinary Medicine,
Mississippi State University, Starkville, Mississippi, USA*

Introduction

The purchase and maintenance cost of the bull battery represents a significant expense to the cow calf producer. Improper management of this investment can have negative economic consequences through both a reduction in herd productivity and the loss of bulls due to injury or death. A reduction in the size of the calf crop or loss of the calf crop is easily noticed by the producer, but many subfertile bulls are capable of inefficiently producing offspring and this may go unnoticed in some herds. This inefficiency leads to delayed conception, which has been estimated to cost 23–27 kg of weaning weight for every 21 days a cow remains open during the breeding season.[1] In an attempt to compensate for poor bull management, producers will often increase the stocking density of bulls within their herd. The expense of purchasing and maintaining excess bulls can also limit the profitability of a herd.

The management of breeding bulls will vary somewhat between herds, but the overall goal should be to provide a group of disease-free, structurally sound and physically fit, fertile bulls with good libido. Achievement of this goal requires an understanding of both the general and reproductive health requirements of breeding bulls and how these requirements change during the production phase of the bull. The production phase of a breeding bull can be divided into three time periods: the pre-breeding period, breeding period, nonbreeding period.

Pre-breeding period

The pre-breeding period begins 60–90 days prior to the start of the breeding season. Breeding soundness evaluations, vaccinations, and diagnostic testing for reproductive pathogens should all occur at the beginning of this period. Performing these procedures early gives the producer time to find replacements if necessary. Vaccination prior to the start of the breeding season should provide protection to the bull while reducing or eliminating his significance as a disease reservoir or vector. Vaccinating bulls at least 60 days prior to the breeding season should allow any viral shedding or transient negative effects on spermatogenesis to subside before breeding. Body condition can also be assessed at this time and minor adjustments to the diet made if necessary.

Breeding soundness evaluation

Breeding soundness evaluations should be performed by a veterinarian on all bulls prior to the breeding season. Any bulls deemed unsatisfactory potential breeders, by the standards for breeding soundness of the Society for Theriogenology, should not be used for breeding and bulls placed in the deferred category should be reevaluated prior to use. Herds with two breeding seasons should have their bulls examined before each season. The complete breeding soundness evaluation of breeding bulls is covered in Chapters 7 and 8. Lameness and other foot problems are often first identified during the breeding soundness evaluation. Performing these examinations early will allow for treatment and corrective trimming prior to breeding, possibly salvaging bulls that may have been culled otherwise.

Vaccination

As a general recommendation, breeding bulls should be vaccinated for the same pathogens as the mature cow herd. Vaccination should include protection against clostridial diseases, leptospirosis, campylobacteriosis, bovine viral diarrhea virus, and bovine herpesvirus 1. Vaccine efficacy and safety varies between vaccine types and products. The goal of this section is not to endorse specific products but to demonstrate the role vaccination plays in control of these pathogens and to address any negative impacts vaccination may have on bull fertility.

Bovine Reproduction, First Edition. Edited by Richard M. Hopper.
© 2015 John Wiley & Sons, Inc. Published 2015 by John Wiley & Sons, Inc.

Clostridial disease

Clostridial diseases are not transmitted directly between cattle and are not considered reproductive pathogens, but they are invariably fatal. Therefore, vaccination is solely aimed at protection of the bull. All clostridial vaccines are killed products and should be boostered according to the label directions. Annual revaccination is recommended especially in all cattle 2 years of age and younger.[2]

Leptospirosis

Leptospirosis is a common cause of abortion and infertility of cattle worldwide. More than 200 serovars of *Leptospira* have been identified each of which is supported within its own maintenance host. The severity and chronicity of disease caused by organisms of the genus *Leptospira* depends on whether infection is taking place in the maintenance host (host-adapted) for that specific serovar or an incidental host (non-host-adapted).[3] Cattle are the maintenance host for serovar *hardjo*, which contains two genetically distinct types: *Leptospira interrogans* serovar *hardjo* type *hardjoprajitno* found primarily in the British Isles and *Leptospira borgpetersenii* serovar *hardjo* type *hardjo-bovis* found throughout the world including North America.[4,5] Infection of cattle with a non-host-adapted serovar is characterized by acute disease, late term abortion, and transient renal shedding of the organism. Infection with the serovar *hardjo* is characterized by chronic infections, prolonged renal shedding, abortion, and infertility.[4] *Leptospira* species are primarily shed in the urine of infected cattle. Transmission is possible via semen and venereal transmission of serovar *hardjo* is thought to be common.[3,6]

Vaccination of cattle against leptospirosis is a complex and controversial subject. Many pentavalent vaccines marked for prevention and control of leptospirosis contain antigens for *hardjoprajitno* not *hardjo-bovis*, the type most commonly isolated in North America.[5] Challenge studies evaluating the efficacy of pentavalent vaccines for protection against *hardjo-bovis* have yielded conflicting data; however, these studies utilized different products and the time period from initial vaccination until administration of the booster differed greatly.[7,8] Administration of a commercial monovalent *hardjo-bovis* vaccine has been shown to prevent renal colonization and shedding of *hardjo-bovis* following challenge.[9] The subject is further complicated by the true incidence of infertility that can be attributed to *hardjo-bovis*. A study in California involving 207 first-lactation cows found that cows seropositive for serovar *hardjo* had a mean time from calving to conception 34 days longer than seronegative cattle.[10] However, when beef cattle were vaccinated with a commercially available monovalent *hardjo-bovis* vaccine and administered oxytetracycline to eliminated the carrier state of *hardjo-bovis*, no improvements in pregnancy and calving rates were observed.[11] The decision to vaccinate and the product used should be based on the prevalence of leptospirosis in the area and its potential impact on reproductive performance in the herd. All the commercial vaccines currently available are killed products which should be boostered according to the label and revaccination annually is recommended.

Campylobacteriosis

Bovine venereal campylobacteriosis caused by *Campylobacter fetus* subsp. *venerealis* results in transient infertility in cattle and occasional abortion. The bull serves as an asymptomatic carrier transmitting the organism to cows during coitus. Vaccination of bulls with an oil adjuvanted killed bacterin can be both protective and curative.[12–14] Exposure of 4- and 5-year-old vaccinated bulls to *C. fetus*-infected heifers resulted in transmission to only 1 of 17 unvaccinated naive heifers and carrier status was not established in any of the five bulls exposed.[12] Therapeutic vaccination of experimentally infected 5-year-old bulls with a two-dose series of an oil adjuvanted killed bacterin was curative in 8 of 10 bulls.[13] While prophylactic vaccination will not clear infection in all bulls, it is an important and inexpensive method of control. The efficacy of vaccines containing an aluminum hydroxide adjuvant is questionable.[14,15] All bulls should be vaccinated and boostered according to the vaccine label and annual revaccination is recommended.

Bovine viral diarrhea

Clinical manifestation of bovine viral diarrhea virus (BVDV) infection in cattle can vary from subclinical to fatal.[16] BVDV is transmitted primarily by direct contact with bodily fluids of infected cattle. Venereal transmission via semen from acutely and persistently infected bulls is also well documented.[17,18] Cattle that are born persistently infected with BVDV shed large amounts of the virus and are considered the major reservoir. Control of BVDV centers around elimination of these persistently infected individuals combined with vaccination and biosecurity measures to prevent exposure to BVDV.[16]

The bull is also capable of maintaining a prolonged testicular infection that persists beyond the initial viremia.[19,20] These localized prolonged testicular infections can persist for at least 2.75 years and have been induced by vaccination of seronegative peripubertal bulls with a modified live vaccine containing noncytopathic BVDV.[20,21] Semen from one bull with prolonged testicular BVDV infection was capable of infecting heifers through artificial insemination; however, others studies evaluating transmission from bulls with prolonged testicular infection have not resulted in transmission.[20,22] While the exact risk of transmission from these bulls appears to be low, further studies are needed before the use of vaccines containing noncytopathic BVDV can recommended in breeding bulls.[14,20] Current recommendation are to vaccinate all bulls with a vaccine containing cytopathic modified live BVDV at least 28 days prior to breeding and booster according to the product label recommendations.[14]

Infectious bovine rhinotracheitis

Bovine herpesvirus (BHV)-1 is a common cause of abortion, conjunctivitis, and respiratory disease of cattle. Genital infections also occur resulting in balanoposthitis of bulls and pustular vulvovaginitis of cows. Cattle that become infected often develop latent infection that can reactivate later, complicating BHV-1 control in the herd. The virus is shed by the mucous membranes of the respiratory tract, conjunctiva, and genital tract and transmission occurs primarily through direct contact. Venereal transmission occurs during natural mating and through artificial insemination with contaminated semen.[23]

Vaccine products containing killed or modified live BHV-1 are available for cattle. In general, the protection induced by modified live viral products is more rapid and of longer duration than killed viral products. However, modified live BHV-1 vaccine products are capable of establishing latency that can be reactivated under stress and have been reported to cause abortion if used inappropriately in pregnant cows.[24-26] Most modified live products available for use in pregnant cows carry a label which states that the vaccine may be used in pregnant cows and calves nursing pregnant cows only if those cattle were vaccinated with the same product within the last 12 months. In order to err on the side of caution, it is my opinion that bulls should be vaccinated with the same BHV-1 vaccine product as the mature cow herd. Vaccine products containing modified live BHV-1 should not be administered to bulls unless the cow herd received the same vaccine product in the previous 12 months.

Trichomoniasis

Trichomoniasis caused by the protozoan *Tritrichomonas foetus* subsp. *venerealis* is a venereal disease resulting in endometritis, embryonic death, abortion, and pyometra in infected cattle. The bull serves as an asymptomatic carrier.[27] Prevention of bovine trichomoniasis is centered on biosecurity practices to avoid introduction of infected individuals into the herd. Only one *T. foetus* vaccine is currently available in the United States. Administration of this killed vaccine to bulls aged 55–72 months prevented colonization of the preputial epithelium following intrapreputial challenge.[28] Within affected herds, vaccination of bulls and cows may reduce the economic impact of this pathogen.[14]

Breeding period

The length of the breeding period is generally determined by the producer and can vary from 65 to 365 days. A controlled calving/breeding season has obvious economic advantages and simplifies the overall health management of the herd; however, in many herds especially in the southern United States the bull is housed with the cow herd year round. The following discussion on the stocking rate of bulls will assume that the herd has a controlled breeding season of 65–90 days.

Stocking rate

The correct stocking rate of bulls, or more specifically the bull to cow ratio that will result in the most efficient and successful servicing of females, is not known. Typical bull to cow ratios in the United States ranges from 1 : 20 to 1 : 30 for mature bulls.[29] Young bulls less than 3 years of age can be placed with one cow per month of age, for example a 15-month-old bull can be expected to service 15 cows during the breeding season.[30] Assuming that a bull is fertile and has passed a breeding soundness examination, other factors that will influence the number of females he can successfully service during the breeding season include his libido or "sex drive" and his social ranking, both of which are difficult to quantify and predict. The assessment and measurement of bull libido can be attempted by performing a servicing capacity test; however, no standardized testing method exists and results vary depending on age of the bulls tested, breed, and the methodology of the test.[31] An Australian trial of 1100 servicing capacity tests found that measurements of sexual behavior were not consistently correlated to the calf output of the tested bulls and that the main value of a servicing capacity test may be to identify bulls that are unable to mount and/or gain intromission.[32]

Social dominance occurs in multi-sire herds where the presence of a dominant bull suppresses mating activity of other bulls. Dominance is generally expressed by older, more senior bulls and the effects may become more pronounced as the number of females assigned to each bull is reduced.[29,33] The impact of social dominance on individual bull performance was demonstrated by a series of trials from northern Australia which examined the effect of the bull to cow ratio in multi-sire breeding groups on reproductive performance. Reducing the percentage of bulls to cows from 6% to 2.5% had no impact on conception patterns between groups. Variation in calf output between bulls was noted in both groups but was significantly reduced when fewer bulls were utilized. The area of movement of individual bulls within the pasture expanded as the percentage of bulls to cows decreased, suggesting that the effects of social dominance were reduced by increasing the breeding pressure on the bull battery. Interestingly, bull attrition due to injury was also reduced.[33]

Recently workers from Auburn University published the following formula for bull to cow ratio:

$$B = 1 + \left[(T - N)/1/2N \right]$$

where B is number of bulls required to breed the given number (T) of cows in the breeding group and N is the individual bull's fair complement of cows as a single-sire breeding group. N is calculated for bulls less than 3 years of age by assigning one cow per month of age. Scrotal circumference is used to calculate N for bulls greater than 3 years of age, where one cow is assigned for each centimeter of scrotal circumference.

This formula assumes that after the first bull each additional bull added to a multi-sire herd can only be expected to reliably service half the number of females that he would service in a single-sire breeding group.[30] Utilizing this formula for a 200-cow breeding group will demonstrate the inefficiency that many people believe exists in a multi-sire breeding group. The calculated requirements for this group would be nine adult bulls with 40-cm scrotal circumferences; however, if this same group was divided into single-sire breeding groups of 40 cows each, only five bulls would be needed to service the entire herd. Clearly, single-sire breeding groups would be a more efficient use of bulls in this situation, but the added expense of fencing required for single-sire breeding groups and the risk of bull injury/illness going unnoticed makes producers hesitant to adopt this management strategy.

Scientific recommendations on bull stocking rates are difficult because variations in libido exist between bulls but are hard to predict and the effects social dominance on breeding performance is not well understood. While the effects of social dominance and breeding overlap may reduce the efficiency of multi-sire breeding groups in terms of the number of bulls needed compared with single-sire breeding groups, these effects have been reportedly reduced by increasing the breeding pressure on the bulls.[33] Clearly, more research is needed in this area so that more accurate recommendations can be made to producers, allowing them to avoid purchasing and maintaining excess bulls. Until further research is published, the Auburn formula provides a good guide when making recommendations to producers and is not likely to result in inadequate bull power.

Producer observation

Libido is not assessed during the routine breeding soundness examinations of bulls, so it is up to the producer to ensure that the bull is actively servicing cows. Producers should take care to observe that intromission is achieved when a female is mounted and make note of the date when the individual was serviced. Breeding injuries and lameness are best identified early so that treatment can be initiated and adjustments made before the herd's reproductive performance is reduced. Routine observation of the herd and documentation of breeding activity during the breeding season is especially important in the single-sire breeding group to prevent loss or reduction in the calf crop.

Nonbreeding period

The nonbreeding period provides an opportunity for the bulls to rest and replace any weight that was lost during the breeding season. Body condition scoring or weighing bulls at this time will allow producers to make the necessary dietary changes to insure that the bull battery enters the next breeding season at an appropriate body condition. The effect of body condition on breeding performance of mature bulls is not well documented; however, logic would suggest that avoiding overconditioning will reduce the effort to mount and the force on the hindlimbs during mating. Commonly, breeding bulls are fed to reach a body condition score of 6–7 prior to entering the breeding season.[34] Dietary energy excess has been shown to have detrimental effects on semen quality in yearling bulls. These effects are thought to be related to impaired thermoregulation of the testes due to increased scrotal fat deposition.[35]

Housing during the nonbreeding period

Group housing is generally recommended for bulls used in multi-sire breeding groups to reduce bull attrition due to injury during the breeding season by allowing the social hierarchy to be established and maintained in the absence of females. Some separation of bulls may be necessary depending on the uniformity of the bull battery. Young bulls less than 2 years of age may need to be grouped separately to avoid competition for feedstuffs and allow for continued growth and development. Bulls finishing the breeding season in poor body condition may likewise be grouped together or fed separately until their body condition improves. Housing recommendations in cold climates with significant snowfall may also include the addition of a shelter with bedding to avoid scrotal frostbite.

Parasite control

Specific recommendations on parasite control will vary depending on the geographical region. Most bulls are dewormed once yearly prior to the breeding season. In warmer more humid climates, twice-yearly deworming may be necessary especially in younger bulls. Yearling bulls are at the highest risk for internal parasitism and may need to be dewormed immediately following the breeding season to remove the parasite burden accumulated while in the cow herd. Fly control is also important in bulls to eliminate fly strike in the scrotal region, which can lead to inflammation and impaired thermoregulation of the testes.[34]

Biosecurity concerns for replacement bulls

The introduction of replacement bulls into the cow herd represents a significant risk for disease introduction. The primary pathogens of interest in the United States include *Brucella abortus*, *Campylobacter fetus* subsp. *venerealis*, *Tritrichomonas foetus* subsp. *venerealis*, *Leptospira* species, *Mycobacterium avium* subsp. *paratuberculosis*, *Mycobacterium bovis*, BVDV, and BHV-1.

In addition to diagnostic screening for potential pathogens, all replacement bulls should be quarantined for a minimum of 30 days before exposure to the cow herd. This period will allow viral shedding to cease in bulls with acute BVDV and BHV-1. Producers should also be aware that no diagnostic test is 100% sensitive and therefore bulls should be purchased from herds with good health programs. Purchasing bulls from herds that are certified free for a specific pathogen provides additional security. Likewise, purchasing virgin bulls eliminates the risk for *C. fetus* and *T. foetus*.

Chapter 26 provides a complete overview of biosecurity measures for the beef herd.

References

1. Barth A. Evaluation of potential breeding soundness of the bull. In: Youngquist RS, Threlfall WR (eds) *Current Therapy in Large Animal Theriogenology*, 2nd edn. Philadelphia: Saunders, 2007, pp. 228–240.
2. Staempfli H, Oliver O. Diseases caused by *Clostridium* species. In: Howard J (ed.) *Current Veterinary Therapy in Food Animal Practice*, 3rd edn. Philadelphia: Saunders, 1993, pp. 567–573.
3. Yaeger M, Holler L. Bacterial causes of bovine infertility and abortion. In: Youngquist RS, Threlfall WR (eds) *Current Therapy in Large Animal Theriogenology*, 2nd edn. Philadelphia: Saunders, 2007, pp. 389–399.
4. Grooms D. Reproductive losses caused by bovine viral diarrhea virus and leptospirosis. *Theriogenology* 2006;66:624–628.

5. BonDurant R. Selected diseases and conditions associated with bovine conceptus loss in the first trimester. *Theriogenology* 2007;68:461–473.
6. Wentink G, Frankena K, Bosch J, Vandehoek J, van den Berg T. Prevention of disease transmission by semen in cattle. *Livest Prod Sci* 2000;62:207–220.
7. Rhinehart C, Zimmerman A, Buterbaugh R, Jolie R, Chase C. Efficacy of vaccination of cattle with the *Leptospira interrogans* serovar *hardjo* type *hardjoprajitno* component of a pentavalent *Leptospira* bacterin against experimental challenge with *Leptospira borgpetersenii* serovar *hardjo* type *hardjo-bovis*. *Am J Vet Res* 2012;73:735–740.
8. Bolin C, Thiermann B, Handsaker A, Foley J. Effect of vaccination with a pentavalent leptospiral vaccine on *Leptospira interrogans* serovar *hardjo* type *hardjo-bovis* infection of pregnant cattle. *Am J Vet Res* 1989;50:161–165.
9. Bolin C, Alt D. Use of a monovalent leptospiral vaccine to prevent renal colonization and urinary shedding in cattle exposed to *Leptospira borgpetersennii* serovar *hardjo*. *Am J Vet Res* 2001;62:995–1000.
10. Guitian J, Thurmond M, Hietala S. Infertility and abortion among first lactation dairy cows seropositive or seronegative for *Leptospira interrogans* serovar *hardjo*. *J Am Vet Med Assoc* 1999;215:515–518.
11. Kasimanickam R, Whittier W, Collins J et al. A field study of the effects of a monovalent *Leptospira borgpetersenii* serovar *hardjo* strain *hardjo-bovis* vaccine administered with oxytetracycline on reproductive performance in beef cattle. *J Am Vet Med Assoc* 2007;231:1709–1714.
12. Clark B, Dufty J, Monsbourgh M, Parsonson I. Studies of venereal transmission of *Campylobacter fetus* by immunised bulls. *Aust Vet J* 1975;51:531–532.
13. Vasquez L, Ball L, Bennett B et al. Bovine genital campylobacteriosis (vibriosis): vaccination of experimentally infected bulls. *Am J Vet Res* 1983;44:1553–1557.
14. Givens M. Assessment of available vaccines for bulls to prevent transmission of reproductive pathogens. *Clinical Theriogenology* 2012;4:308–313.
15. Cobo E, Cipolla A, Morsella C, Cano D, Campero C. Effect of two commercial vaccines to *Campylobacter fetus* subspecies on heifers naturally challenged. *J Vet Med B Infect Dis Vet Public Health* 2003;50:75–80.
16. Walz P, Grooms D, Passler T et al. Control of bovine viral diarrhea virus in ruminants. *J Vet Intern Med* 2010;24:476–486.
17. Kirkland P, McGowan M, Mackintosh S. The outocome of widespread use of semen from a bull persistently infected with pestivirus. *Vet Rec* 1994;135:527–529.
18. Kirkland P, McGowan M, Mackintosh S, Moyle A. Insemination of cattle with semen from a bull transiently infected with pestivirus. *Vet Rec* 1997;140:124–127.
19. Voges H, Horner G, Rowe S, Wellenberg G. Persistent bovine pestivirus infection localized in the testes of an immune-competent, non-viraemic bull. *Vet Microbiol* 1998;61:165–175.
20. Givens M, Riddell K, Edmondson M et al. Epidemiology of prolonged testicular infections with bovine viral diarrhea virus. *Vet Microbiol* 2009;139:42–51.
21. Givens M, Riddell K, Walz P et al. Noncytopathic bovine viral diarrhea virus can persist in testicular tissue after vaccination of peri-pubertal bulls but prevents subsequent infection. *Vaccine* 2007;25:867–876.
22. Niskanen R, Alenius S, Baule C, Belak S, Voges H, Gustafsson H. Insemination of susceptible heifers with semen from a non-viremic bull with persistent bovine virus diarrhea virus infection localized in the testes. *Reprod Domest Anim* 2002;37:171–175.
23. Kelling C. Viral diseases of the fetus. In: Youngquist RS, Threlfall WR (eds) *Current Therapy in Large Animal Theriogenology*, 2nd edn. Philadelphia: Saunders, 2007, pp. 399–408.
24. Pastoret P, Babiuk L, Misra V, Griebel P. Reactivation of temperature-sensitive infectious bovine rhinotracheitis vaccine virus with dexamethasone. *Infect Immun* 1980;29:483–488.
25. Mitchell A. An outbreak of abortion in a dairy herd following inoculation with an intramuscular infectious bovine rhinotracheitis vaccine. *Can Vet J* 1974;15:149.
26. O'Toole D, Miller M, Cavender J, Cornish T. Pathology in practice. *J Am Vet Med Assoc* 2012;241:189–191.
27. Strickland L. Bovine trichomoniasis: a review. *Clin Theriogenol* 2009;1:289–306.
28. Cobo E, Corbeil L, Gershwin L, BonDurant R. Preputial cellular and antibody responses of bulls vaccinated and/or challenged with *Tritrichomonas foetus*. *Vaccine* 2010;28:361–370.
29. Chenoweth PJ. Bull drive and reproductive behavior. In: Chenoweth PJ (ed.) *Topics in Bull Fertility*. Ithaca, NY: International Veterinary Information Service, 2000. Available at www.ivis.org
30. Wenzel J, Carlson R, Wolfe D. Bull-to-cow ratios: practical formulae for estimating the number of bulls suggested for successful pasture breeding of female cattle. *Clin Theriogenol* 2012;4:477–479.
31. Petherick J. A review of some of the factors affecting expression of libido in beef cattle, and individual bull and herd fertility. *Appl Anim Behav* 2005;90:185–205.
32. Bertram J, Fordyce G, McGowan M et al. Bull selection and use in northern Australia 3. Servicing capacity test. *Anim Reprod Sci* 2002;71:51–66.
33. Fordyce G, Fitzpatrick L, Copper N, Doogan V, De Faveri J, Holroyd R. Bull selection and use in northern Australia 5. Social behavior and management. *Anim Reprod Sci* 2002;71:81–99.
34. Christmas R. Management and evaluation considerations for range beef bulls. In: Chenoweth PJ (ed.) *Topics in Bull Fertility*. Ithaca, NY: International Veterinary Information Service, 2000. Available at www.ivis.org
35. Barth A, Brito L, Kastelic J. The effect on nutrition on sexual development of bulls. *Theriogenology* 2009;70:485–494.

Chapter 11

Management of Bulls at Custom Collection Studs

Gary Warner

Elgin Veterinary Hospital, Elgin, Texas, USA

Veterinary services for the commercial bull stud

The semen collection – or "custom bull stud" – industry has markedly evolved over the last 50 years. Initially, a large majority of product was provided by large cooperatives or studs that maintained a large resident bull herd, catering primarily to the dairy industry and its needs. Most bulls in these studs were either purchased outright by stud management or secured by a long-term lease, and thus breeders/owners very rarely realized the full rewards of their production successes. Beef cattle were processed as an aside, as the beef industry's use of frozen semen products was not widespread. As technology and breeding synchronization systems have evolved, the demand for semen in the beef industry has skyrocketed and custom bull studs have developed throughout the United States. These custom studs allow bull owners the opportunity to present herd sires with proven genetic merit, the opportunity to process and store these quality genetics, and to retain ownership and possession of the bull. Also, it gives bull owners the option to access the international marketplace by allowing them to transport the frozen semen product anywhere in the world.

To assure health quality of the frozen semen products, any semen produced for export must meet standards set by Certified Semen Services, Inc. (CSS). Recommendations for housing, collection, handling, and health testing are made and supervised by administrators within this organization. Recommendations for health standards for residents at the bull studs are taken from guidelines published by the World Organization for Animal Health (OIE). This group concerns itself with advising importing and exporting countries on protecting themselves from introducing disease into their countries from foreign sources. Many countries will take the guidelines of the OIE and formulate entry requirements for cattle as well as bovine germplasm, both semen and embryos. Guidelines are specific and must be followed as entry is forbidden unless standards are met.

Veterinarian's role in the bull stud

As previously mentioned, the primary role of a veterinarian in an artificial insemination center (AIC) is to insure biosecurity and animal welfare for the bovine population housed there. Standards for testing protocols are established and readily available on the website of the National Association of Animal Breeders (NAAB) at www.naab-css.org. These are standard recommendations for the industry that are accepted by USDA/APHIS but certainly may be augmented by any further testing considered appropriate by the center's veterinarian.

The ultimate responsibility of the veterinarian is to assure the welfare of the animals under his or her care. Since the stud is usually compensated by the units of semen provided, it is in the stud's best interest to ensure that the cattle in its charge are adequately cared for. This means that feed, water, forage, and facilities meet the best of standards for quality. Although some facilities may retain the services of a nutritionist, others may rely on their veterinarian for suggestions on ration formulation or forage sampling and testing. Supplementation of minerals should not be forgotten for long-term resident herd members. Quality control of commercial feed rations should be observed and the resident veterinarian should periodically check feed and storage facilities to insure proper care and quality.

Since the focus of this chapter is the private veterinarian consulting for the bull stud, we should concentrate on the start of the process: entry into the bull stud. There are three periods at which health testing requirements must be performed: pre-isolation, isolation, and routine testing that will take place at regular intervals after entry into the resident herd.

Bovine Reproduction, First Edition. Edited by Richard M. Hopper.
© 2015 John Wiley & Sons, Inc. Published 2015 by John Wiley & Sons, Inc.

Table 11.1 Basic AI center testing protocol per CSS testing requirements.

	Testing environment		
	Pre-entry to isolation	**Isolation**	**Resident herd**
Physical examination	Conducted by accredited veterinarian	Conducted by accredited veterinarian	Conducted by accredited veterinarian
Tuberculosis	Negative intradermal tuberculin test (within 60 days prior to entry)	Negative intradermal tuberculin test at least 60 days after pre-entry test	Negative intradermal tuberculin test at 6-month intervals
Brucellosis	Official test of state where bull is located. Blood serum test (CF, BAPA or Card)	CF and one BAPA or Card test at least 30 days after pre-entry testing	CF and one BAPA or Card test at 6-month intervals
BVDV	One negative virus isolation test performed on either whole blood (animals less than 6 months of age) or serum	Virus isolation and serologic testing at least 10 days after entry into the isolation facility. If seropositive, virus isolation of semen required	Not required
Leptospirosis	Blood test for 5 serotypes important in USA[a]	Blood test for 5 serotypes important in USA[a]. At least 30 days after pre-entry test	Blood test for 5 serotypes important in USA[a] at 6-month intervals
Campylobacteriosis	Not required	Series of negative culture tests of preputial material or screening by FA with any positive FA tested by culture for final determination. 1, 3 or 6 consecutive weekly tests; number of necessary tests dependent on age	Negative single culture test of preputial material or FA for screening test at 6-month intervals
Trichomoniasis	Not required	Series of negative microscopic examinations of cultured preputial material. 1, 3 or 6 consecutive weekly tests; number of necessary tests dependent on age	Negative single microscopic test of cultured preputial material at 6-month intervals

[a]L. pomona, L. hardjo, L. canicola, L. icterohaemorrhagiae, L. grippotyphosa.
BAPA, buffered acidified plate antigen; BVDV, bovine viral diarrhea virus; CF, complement fixation; FA, fluorescent antibody.
Source: adapted from Certified Semen Services, Inc. For additional information on CSS, see http://www.naab-css.org/about_css/

The first round of testing, pre-isolation testing, may be done at the farm of origin by the resident veterinarian. Pre-entry requirements should be met within 30 days prior to entering the isolation facility. If bull owners do not wish to trouble themselves with testing, many custom studs offer pre-isolation facilities to house the bull while testing is conducted. These pre-entry requirements also apply to any mount or jump animals to be used in isolation or the resident herd. The isolation and resident testing requirements will be conducted upon arrival at the collection facility and at prescribed intervals thereafter. All testing phases require a complete physical examination, a negative intradermal test for tuberculosis, a negative test for bovine brucellosis, and a negative test for bovine leptospirosis. Those with titers of 1 : 100 or greater for leptospirosis may be reevaluated in 2–4 weeks. Those with titer no greater than 1 : 400 will be considered stabilized and allowed entry into isolation. Testing for bovine viral diarrhea virus (BVDV) should be performed before entry and a negative result obtained; the testing must only be conducted by virus isolation from whole blood or serum performed in cell culture, followed by evaluation of cell cultures with immunoperoxidase (IP), fluorescent antibody (FA), enzyme-linked immunosorbent assay or polymerase chain reaction (PCR). Lastly, testing for campylobacteriosis and trichomoniasis is required during the isolation period and at 6-month intervals thereafter. For both diseases, bulls are required to undergo a series of weekly tests, with the required number varying depending on the age of the bull. Table 11.1 outlines a basic AI center testing protocol. Please refer to www.naab-css.org for complete details and the most recent updates regarding testing requirements.

Examination of entry-level animals should be conducted as soon as possible after admission to pre-isolation. A good physical should include observation of ambulation on a surface that provides good footing, preferably loose soil or grass. At this time all physical features should be assessed from the muzzle to the tail, paying particular attention to the feet (Figure 11.1). After this initial examination, the bull should be properly restrained in a squeeze chute, head gate, or lock stanchion. For beef bulls, a squeeze chute would be most desirable for restraint. Starting at the head, the oral cavity should be examined, making sure that all teeth are present and showing even wear (particularly important in more aged herd sires). The eyes should be evaluated for clarity, any presence of neoplasia, or previous injury. Make sure both eyes respond to movement and function normally. The ears are checked for any unusual discharge; in some areas of the country they should also be checked for ear ticks. The bull's chest should be auscultated and any abnormalities noted, as should the entire abdominal cavity. Particular attention should be paid to rumenal contractions, both in strength and number. Several peripheral regional lymph nodes are palpated for

Figure 11.1 Bull going into chute.

Figure 11.2 Ideal conditions for housing bulls in central Texas, Elgin Breeding Service.

any evidence of lymphadenitis. Next, the underline of the bull should be evaluated from the bottom of the chute. The preputial orifice is palpated and examined as well as the scrotum, with palpation of the testes and all epididymal structures. Record the scrotal circumference with a scrotal measuring tape. Evaluation of the lower limb joints and hooves is performed at this time as well. Lastly, rectal palpation should be performed to evaluate abdominal structure, with a rectal temperature being taken and recorded. It is vital to pay attention to accessory reproductive organs, with specific interest in the health of the seminal vesicles. During the rectal examination, it is a good idea to massage the crus of the penis to allow an assistant the opportunity to fully extend the penis. Both vaginal rings are palpated and any abnormalities noted. After examination and pre-entry testing, follow the recommendation of the NAAB/CSS for entry into isolation after presentation at the stud. Some producers may not be interested in submitting a bull for further testing as they have no desire to export semen. It is always recommended to follow CSS guidelines on any mature bull that has been breeding cows before presentation for collection.

At the time of entry into isolation at the bull stud, or AIC as some like to be called, the bull is assigned a unique sire code number per the NAAB uniform coding system, if the stud is a NAAB participant. This code is separate and apart from any tattoo or brand used in a breed registry. The bull may display this code in a number of ways but most AIC's use ear tags with the code plainly marked. This code can be used by the center to identify product produced by the bull, although other permanent identification is used for health testing purposes (i.e., hide brands or tattoos).

After admission the bull entering the isolation facility must go into a pen that has been properly cleaned of any residual organic material deposited by the former occupant and disinfected (Figure 11.2). For stalls with dirt floors, a disinfectant applied as a mist over the area should be sufficient to meet NAAB standards. The veterinarian should be aware of the facilities and any damaged portions of housing that could injure an animal. The isolation facility must be separate and apart from the resident housing. All equipment used for cleaning, feeding, watering, and semen collection of the isolation bulls must remain separate from the resident herd. This includes feed storage and any mechanical means of transport. Workers must change boots or protective cover before entering or leaving an isolation area and going to other quarantine areas. Testing on the isolation bulls starts with an intradermal test for tuberculosis performed at least 60 days after the pre-entry test; a test for bovine brucellosis not sooner than 30 days after the pre-entry test, with negative result; and a test for bovine leptospirosis to be conducted no sooner than 30 days after pre-entry test, with negative result or subject to stabilized titer of less than 1 : 400. Bovine campylobacteriosis requires up to six cultures and examination of preputial wash material collected at weekly intervals. Bovine trichomoniasis has the same weekly culture requirement during the isolation period. Testing for BVDV is performed no longer than 10 days after entry via whole blood or serum, with evaluation by FA, IP or PCR on cell culture. All bulls having semen collected will have future testing for BVDV before semen is released. Semen testing by PCR is preferred, with at least one passage through cell line culture.

Veterinarians should also familiarize themselves with the diagnostic laboratory they will be working with, particularly with regard to sample submission requirements, the intricacies of testing procedures, and especially testing protocols and the reporting of test results, as understanding results is very important. Any positive test results, particularly on pre-entry tests, should be evaluated in conjunction with previous vaccination history. Frequently, certain vaccines produce a prolonged immune response that may interfere later on with attempts to approve semen for export. Leptospirosis vaccination produces a persistent titer for prolonged periods in certain cases.

Once the bull meets all the requirements of the isolation protocol, he is allowed to enter the resident herd where testing is continued to assure compliance with international protocols. The required testing is preformed semi-annually on each bull occupying the resident herd. Previous testing is continued at a less frequent pace (every 6 months), with tuberculosis, brucellosis, leptospirosis, trichomoniasis, and campylobacteriosis all evaluated using the same testing protocols as discussed earlier. Bulls must be restricted from any

100 The Bull: Breeding and Health Management

Figure 11.3 Semen canisters for storage of frozen semen and embryos. Note the cleanliness and organization of the facility.

Figure 11.4 Collection area.

contact with any other cloven hoofed animal; if such contact occurs, the bull must be returned to isolation and the testing protocol for isolated bulls must be followed once again.

Facility overview is another responsibility of the veterinarian. Supervision of the artificial vagina (AV) preparation room, collection area, bull housing, semen lab and processing area should all be evaluated for cleanliness, absence of any cross-contamination, and ability to disinfect contact surface areas. Semen processing equipment should be clean and serviceable and packing and shipping areas clean and organized (Figure 11.3). The veterinarian is responsible for all these areas, as noted when signing off on any federal export certificate. Accuracy is the goal of the veterinarian to assure that standards set by the OIE and NAAB/CSS are met to insure continued access to all markets.

The collection process

The collection of semen for cryopreservation is a fairly simple and straightforward process, except for those with no experience. Therefore the procedures should only be administered by someone who is competent and experienced at handling bulls and very familiar with the collection process. Good facilities are paramount to a successful collection attempt, particularly when dealing with beef bulls that have primarily had pasture breeding experience. Collection of semen from bulls with an AV is not for the faint of heart. The procedure basically involves allowing the donor bull to mount or "jump" another steer, bull or cow and diverting the penis laterally in order to introduce the extended penis into the AV. Usually, once bulls are familiar with the technique, they will readily mount and serve the AV. Training them to do this is an art, particularly when dealing with less docile mature beef bulls and bucking bulls.

CSS mandates guidelines to be followed for bull health and cleanliness. Steers are the most common mount animal unless the facility prefers the use of a phantom (dummy structure designed for supporting a mounting bull). Using cows as mount animals is discouraged because of the potential for penetration and contamination by the bull. Mount animals should have their hair closely clipped from the hindquarters, allowing for ease of cleaning and disinfection, which should be accomplished after every use. The process usually requires at least two people and preferably three. One person is assigned the mount animal, which is usually haltered or handled by nose ring. This person's responsibility is to position the mount animal for mounting and to assist in keeping it in position during collection. In large facilities there is also a bull handler who may have the bull haltered or tethered to control him during the mounting process. Most beef bulls in custom studs are free to mount the jump animal. Lastly, the semen collector manages the AV and is in charge of the entire collection area (Figure 11.4). It is this person who ensures the sample is properly identified and collected hygienically and is responsible for the safety and welfare of the team as well as the animals.

The procedure of semen collection is rather straightforward. Several bulls are usually penned separately along a "bull alley" facing the collection area. Mount animals are allowed to move along in front of the bulls in order to initiate some arousal. The bulls are allowed to observe other bulls mounting and being collected from this vantage point. It can be very evident that the "bull gallery" is aroused by the amount of vocalization that occurs as a bull is being collected.

Once it is time for a bull to be collected, he is allowed to leave his pen and enter the mount area. At this time, the mount handler will move the mount animal to prevent the bull from mounting. This may continue for several minutes until such a time as the semen collector is satisfied the bull is properly aroused. The semen collector may further determine that the bull needs to "false mount" to further his arousal and thus increase the quality and quantity of semen he produces. This is accomplished by the bull handler allowing the bull to mount but the semen handler diverting the penis away from the mount animal as the mount is withdrawn from under the bull. As discussed previously, when the semen collector believes the bull is properly stimulated, the bull is allowed to mount for collection (Figure 11.5). After collection the bull is either returned to housing or is placed back in the bull alley for another collection attempt.

Management of Bulls at Custom Collection Studs

Figure 11.5 Proper technique for collection of bull.

Figure 11.6 A well-designed work area for performing lameness examinations at a custom stud. Note the hydraulic tilt table and footbath in foreground.

The preparation of the AV is important as the tactile sensation received by the bull is paramount to his semen production. It should simulate the natural breeding act as closely as possible. The AV is constricted, with a ridged tube that has a port allowing water to be added between the rigid structure and a latex liner. Within this liner another is placed to facilitate collection of the semen from the bull. This liner has a tapered end that allows the inclusion of a glass or plastic receptacle that will serve to hold the collected semen sample. Water added to the water jacket should be in the temperature range 40–60 °C and varies according to the preference of the bull; this has to be determined by the semen collector. Also, the amount of water within the jacket determines the rigidity of the liner, which can impact the bull's reaction to the AV. It is very important that a sterile nonspermicidal lubricant be applied to the upper one-third of the AV. This improves sensation and reduces the possibility of abrasive injury to the bull's penis.

For those bulls which are too fractious or are incapacitated and unable to jump a mount animal, the collector has the option to use an electroejaculator. Although it is discouraged in the semen production industry, it may be used when all else fails. Usually the bull will produce a slightly less concentrated ejaculate and with the potential for a somewhat higher abnormal sperm count. However, with proper stimulus (rectal massage of the ampulla and seminal vesicles) adequate semen may be collected for processing. Massage of the pelvic genitalia alone usually does not produce adequate semen product for processing and is discouraged in the industry.

Common problems of bulls at stud

Lameness

Lameness is the most common problem in bulls at stud. The primary complaint from the stud manager will be lameness of acute onset. Some of the more common insults are hoof or foot rot, hairy heel wart, (bovine dermatitis), hoof abscess, hoof lacerations, broken toes, or overgrown toes. Hoof abscesses due to wall separations ("white line disease") is most often identified as the problem in these lame bulls. Most white line disease can be traced back to subclinical

Figure 11.7 Assisting in preparation of hoof block application.

laminitis occurring some time in the bull's yearling development. The best curative approach is to access a hydraulic tilt table and to place the bull in lateral recumbency in order to remove the damaged sole and/or wall (Figures 11.6 and 11.7). Feed and water are withheld for 24 hours before table placement to reduce the chance of rumenal tympany while in lateral recumbency. If over 1 cm of sole or a hoof wall is removed, it is recommended that an elevated block be placed on the opposing toe in order to reduce contact with the ground by the injured toe. Post-procedure antibiotics and analgesics are necessary to prevent infection and pain due to the lesion. Other injuries causing lameness are usually accidents that can occur in the collection areas, for example shoulder "sprains" when bulls get their forelimb hung up on the back of the mount animal and stifle injuries when mounting or dismounting the mount animal.

Sometimes the only method available to properly diagnose the problem is diagnostic anesthesia, which often requires the use of a tilt table. Superficial injuries may be treated with a few days of nonsteroidal anti-inflammatory drugs. The more serious may require more aggressive measures.

Treatment of issues such as foot rot or hairy heel wart may best be treated via footbath or topical therapy. The more stubborn infections may require antibiotics and periodic topical antibiotic bandages. If lameness persists after a few days of therapy, it is always recommended to question the original diagnosis and reinvestigate the problem.

Occasionally, a diffusely swollen limb is observed associated with a joint (most often a tarsus) with accompanying pyrexia, which can signal initiation of septic arthritis secondary to a generalized bacteremia. These are usually consequences of some kind of stress, most often change of feed or feeding methods. Any bull presenting with generalized cellulitis in a limb should be considered a medical emergency. Most bulls are not weight-bearing and can be difficult to deal with in local settings, so it is often better to refer these cases to a facility capable of handling this kind of care.

Penile injuries

One of the more common penile injuries observed in the bull stud is preputial laceration/avulsion at the site of the preputial reflection. This injury occurs as a friction "tear" due to misapplication of the AV as it is placed on the bull to collect the semen sample. Most often, the only evidence of injury initially will be drops of blood in the AV or in the ejaculate, or by blood dripping from the preputial orifice. This is to be considered a medical emergency and should be addressed as soon as possible after injury. After a pudendal nerve block and/or local anesthesia, the wound should be debrided with copious lavage of the site. Debridement should be followed by suturing with simple interrupted sutures or by surgical stapling. Bandaging of the orifice to maintain the penis in the sheath, with antibiotics and anti-inflammatory drugs, will be required for several days. After recovery, the bull should only be collected with an electroejaculator or rested from sexual activity for about 4 weeks.

Occasional preputial lacerations from self-trauma may occur. It is not uncommon to see bulls "catch" the prepuce with a dew claw as they start to rise. These may only require antibiotics and topical medication instead of debridement and surgery. Each case should be independently assessed. Occasionally, the injury may scar sufficiently to reduce the diameter of the prepuce and impede full extension of the penis. If this occurs a penile reefing or circumcision may be required in the future.

Gastrointestinal tract problems

AICs will need veterinary services to treat a wide variety of gastrointestinal problems. Most are simple rumenal tympanys or bloat, followed by nonspecific and transient diarrheas. Most are thought to be due to mistakes in feeding protocols but on occasion may be caused by changes in gastrointestinal motility brought on by a host of other primary problems. These incidents will respond to systemic treatment of laxatives, oral fluids, and analgesics if warranted. More serious gastrointestinal problems encountered include hemorrhagic bowel disease, perforating abomasal ulcers, intestinal intussusception, or intestinal volvulus. Once diagnosed, these are best referred to those most experienced in dealing with such problems as these bulls will need surgery and intensive aftercare.

Occasionally, an acute disease due to a foreign body may occur in a bull at an AIC. Of course, the penetrating foreign body could have been present for years; such subtle problems should not be overlooked when making a differential list. Advising a custom stud owner on per-os administration of rumen magnets should be considered when creating entry requirements.

Urinary tract problems

Urinary tract problems are rare at stud but do occasionally occur. Initial presentation of a bull with urinary obstruction is much like any other abdominal problem: colic-like symptoms, kicking at abdomen, elevated heart rate, increased salivation, straining attempts to defecate, urinate, etc. Most often bulls may respond to analgesic therapy along with smooth muscle relaxants and urinary acidification. It is to be remembered that if the urinary obstruction is relieved, there are probably still many calculi left in the bladder. When urinary calculi are involved it is wise to diagnose the type of calculi present and adjust therapy accordingly, maintaining the bull on acidification until the calculi significantly decrease in a urine sample. In some areas calculi do not respond to urinary acidification, presenting the clinician with a whole new set of problems. Most bulls that have calculi problems also have an ongoing cystitis requiring bacterial culture and sensitivity testing and prolonged antibiotic therapy.

Conclusion

Consulting for an AIC can be rewarding both financially and professionally. Many breeders working with an AIC are the elite within their breed and by exposure to the veterinarian through association with the bull stud new client relationships can develop. Consultation services, from disease prevention to improving reproductive efficiency and development, can occur through contacts made with the AIC.

Chapter 12

Testicular Degeneration

Albert Barth

*Large Animal Clinical Sciences, Western College of Veterinary Medicine,
University of Saskatchewan, Saskatoon, Saskatchewan, Canada*

Introduction

Degeneration of the testis parenchyma may be temporary or permanent. Temporary degeneration is by far the most common form of degeneration and, with removal of the cause, the potential for recovery of testis function is good. It may be difficult to distinguish between a severe disturbance of spermatogenesis, with a marked decline in semen quality, and degeneration. The use of the term "degeneration" is arbitrary, but usually means there has been a loss of scrotal circumference, the testes may feel soft, and the percentage of normal sperm has declined to extremely low values, usually less than 20%. With temporary degeneration there is loss of germinal cell layers near the lumen in many or all of the seminiferous tubules. Sertoli cells and spermatogonia, which are more resistant to damage, are likely to be retained in the tubules, and after the cause of degeneration is removed these cells serve to repopulate the tubules with spermatocytes and spermatids with an accompanying increase in testis size and semen quality. Advanced degenerative changes resulting in permanent damage include spermiostasis, tubular mineralization, granuloma formation, thickened basement membranes, and fibrosis in focal or diffuse areas of the parenchyma.

Pathogenesis

Factors involved in the pathogenesis of testis degeneration include abnormal thermoregulation of the testes, nutritional excesses or deficiency, toxicity, inheritance, congenital blockage of sperm outflow from the testes, infectious disease, severe trauma, and senile atrophy.

The testes of scrotal mammals function normally when testis temperature is maintained at 2–6 °C less than body temperature.[1] Scrotal insulation has been used in many experiments to produce temporary testicular degeneration.[2,3] An experiment involving scrotal insulation for up to 10 days was conducted to determine whether abnormal thermoregulation might induce fibrotic testes lesions.[4] Scrotal insulation resulted in severe changes in semen quality, but did not result in any fibrotic lesions within 6 months after insulation. Therefore, abnormal thermoregulation of the testes that is reversible (e.g., obesity) may not be a common cause of permanent damage. Other less common causes of chronic abnormal thermoregulation that may lead to temporary or permanent testicular degeneration include severe chronic scrotal dermatitis, heat and swelling due to orchitis, trauma, chronic illness which may involve prolonged fever and endotoxins that are potent inhibitors of gonadotropin secretion, a congenitally short scrotum, severe scrotal frostbite leading to loss of testis mobility within the scrotum, and incomplete descent of a testis into the scrotum due to hereditary or congenital factors.

Obesity is a leading cause of testis degeneration in bulls. The prevailing feeding and husbandry practices throughout North and South America and in many countries of the world fatten bulls for shows and sales. Many bulls become obese and fat accumulates in the scrotum leading to chronic abnormal thermoregulation with increased testicular temperatures.[2] After loss of fat from the scrotum, regeneration of seminiferous tubules may occur with return to normal sperm production.[5]

Nutritional deficiencies are seldom to blame for testis degeneration in bulls. Naturally occurring deficiencies are usually multiple and deficiency of any single nutrient as a cause of bull infertility has rarely been reported. Vitamin deficiencies are likewise rarely observed as a cause of infertility in the larger male domestic animals if forages are normal in quality or quantity. Poor feed quality resulting in loss of body weight may depress spermatogenesis via an endocrine effect rather than by a direct effect of specific vitamin or mineral deficiencies on developing germ cells. One study showed that protein supplementation of bulls on poor-quality forage significantly increased dry matter intake, enabling maintenance of live weight. In unsupplemented bulls, weight loss was accompanied by a significant decrease in scrotal circumference and daily sperm production per gram of testis. However, histological sections did not indicate any apparent differences in seminiferous tubule activity between bulls that received protein supplementation and bulls that did not.[6] Vitamin A deficiency can cause testicular degeneration in laboratory rats,[7] but in

Bovine Reproduction, First Edition. Edited by Richard M. Hopper.
© 2015 John Wiley & Sons, Inc. Published 2015 by John Wiley & Sons, Inc.

bulls under practical feeding conditions, diets that result in long-term marginal vitamin A deficiency or a relatively short-term absence of vitamin A intake probably have minimal effects on spermatogenesis.[8] There is little evidence that deficiencies of vitamins B, C, D, or E are even occasional causes of infertility in domestic animals.[9] Although vitamin E is essential for reproduction in the rat, naturally occurring deficiencies resulting in testis degeneration in livestock have not been reported. Mineral deficiencies may affect reproduction in male animals, but reports are rare. Deficiencies of calcium, manganese, zinc, iodine, potassium, and selenium have not been proven to be causes of testis degeneration. Deficiencies of phosphorus, cobalt, iron, zinc, and copper may cause anemia, lack of appetite, and loss of weight and thus adversely influence reproduction. Deficiencies of these minerals are often associated with a lack of protein and low levels of vitamin A.

The potential of many drugs, chemicals, and heavy metals to cause testis degeneration has been shown by intentional experimental exposure (mainly of rats and mice); however, there are very few reports of naturally occurring cases in bulls.[9,10] Ingestion of plant toxins may cause testis degeneration. *Fusarium* mold in grain crops is common and one of the toxins produced is zearalenone. Zearalenone, which has an estrogenic effect, has the potential for detrimental effects on semen quality and testis degeneration. In one report, 23 young rams became infertile on a farm where the animals were fed grain containing zearalenone 5–20 mg/kg for several months. Two bulls fed maize containing zearalenone 20 mg/kg for 72 days had poor semen quality after 21 days. Histological examination of the rams' testicles showed complete destruction of the germinal epithelium and aspermia. In the bulls, degeneration of the germinal epithelium was marked only in certain areas of the testicles, but more than 75% of sperm were in a degenerated form.[11] Cottonseed meal in the diet may expose bulls to the effects of gossypol toxicity. Gossypol causes alterations in mitochondrial structure and function. This results in abnormalities of spermatocytes, spermatids, and mature sperm.[12] Although gossypol affected semen quality, testicular degeneration did not appear to be associated with gossypol.[13,14] Locoweed (*Astragalus lentiginosus*) has the potential to cause temporary testis degeneration as was shown by experimentally feeding locoweed to mature rams.[15] The use of growth-promoting implants has the potential to impair testicular development.[16,17] When bulls were implanted with zeranol (Ralgro) at birth and at 3 and 6 months of age, or every 3 months from birth through 18 months of age, scrotal circumference was reduced but tended to recover with increasing age. There was little effect on the reproductive organs when bulls were implanted with zeranol after 7 months of age.

Congenital and inherited disorders of the testes and epididymis may be involved in testis degeneration. Young bulls with testicular hypoplasia may produce semen with satisfactory quality, but are more prone to develop testicular degeneration at 2–3 years of age.[10] In young bulls a small scrotal circumference has been correlated with a lack of germinal epithelium within the seminiferous tubules.[18,19] Hypoplasia may be unilateral or bilateral. Some authors report that most cases are unilateral and that the left side is more frequently affected,[10] but there appears to be no precise definition for testicular hypoplasia based on physical measurement. Histologically, testicular hypoplasia is defined based on cell populations in the seminiferous tubules. Testicular hypoplasia is congenital and possibly hereditary in origin. It has been studied in Swedish Highland cattle in which it appears to be caused by an autosomal recessive gene with incomplete penetrance.[20] The condition exists in many breeds today, although at a very low frequency.[21] Double muscling (myofiber hyperplasia), which is inherited as an autosomal recessive trait, is associated with a high incidence of bilateral testicular hypoplasia;[22] however, it appears there is no information available to compare the incidence of testis degeneration in double-muscled breeds and non-double-muscled breeds.

The embryo needs to connect its gonads with the mesonephric urinary system in order to develop its reproductive excurrent duct system. The connection comes via the rete tubules that develop between the fetal gonad and the mesonephros. The ureter for the mesonephros, the mesonephric duct, will become the epididymis. The efferent ductules develop from the mesonephric tubules and join to the mesonephric duct via the rete tubules during embryonic development. Failure of some of the efferent ductules to join with the rete may later (at puberty) result in sperm impaction of the blind-ending ductules. Rupture of sperm-impacted ductuli would lead to formation of sperm granulomas. If all the efferent ductules are obstructed, the seminiferous tubule fluid and sperm that are produced cannot leave the testes. The testes may become enlarged and edematous and then degenerate.

Infectious organisms may be involved in testis degeneration. *Eperythrozoon* infection in a group of young bulls led to anemia, scrotal and hindlimb edema, and soft testes. Loss of scrotal thermoregulation was likely the main cause of testicular degeneration and poor semen quality.[23] In another report, testicular degeneration and loss of libido occurred when beef bulls were experimentally inoculated with *Anaplasma marginale*.[24] Testicular degeneration was confirmed by histopathology and semen evaluation. Picornavirus and bovine enterovirus isolated from semen and feces of a bull were implicated as a cause of orchitis, testicular degeneration, aspermatogenesis, and loss of libido in a bull.[25]

Diagnosis

During clinical evaluations it may be difficult to draw a line between disturbances of testis function that lead to reduced sperm output and increased sperm abnormalities and more severe disturbances of testis function that might be called degeneration. A clinical diagnosis of degeneration must be decided on an arbitrary basis. What some would call degeneration others might call a severe disturbance of spermatogenesis. A semen sample is not necessarily of low concentration in cases of degeneration since markedly reduced sperm output can still result in filling of the ampullae over periods of sexual rest.

The use of ultrasonography has not proven to be clinically useful for distinguishing between normal testis tissue and tissue that has lost germinal epithelium.[26] When a great deal of fibrous tissue is present, testes may lose their firm resilient feel on palpation and take on a harder, more wooden feel; however, the diagnosis of fibrosis is best accomplished with ultrasonography. The normal echographic anatomy of slaughterhouse specimens of bull testes has been studied[27] and conducting ultrasonography on testes has been shown to be safe.[28] Ultrasonography has been used to examine the effects of testis degeneration induced by scrotal insulation in bulls;[4,29,30] however, there were no evident visible changes in ultrasonograms after induced testicular degeneration. Furthermore, some studies have failed to demonstrate significant correlations between computer analysis of ultrasonogram pixel intensity and semen quality, either during breeding soundness evaluations or after scrotal insulation.[29,31] However, one study showed that echogenic changes induced by scrotal insulation preceded an increase in sperm abnormalities.[4] In the latter study, changes in pixel intensity accounted for 13–25% of the variation in semen quality of ejaculates collected 2–4 weeks after the ultrasonographic examination.

Prognosis

Recovery to normal testis tone and spermatogenesis is possible in most cases of testis degeneration if the cause can be removed. In cases of obesity, it might be possible to turn bulls out to pasture for weight loss. When breeding seasons are short and the bull to female ratio is about 1 : 35, there is sufficient breeding overlap to prevent an infertile dominant bull from making a significant impact on herd fertility.[32,33] In my experience, it takes 3–4 months for sufficient weight loss and for seminiferous epithelium regeneration before semen production returns to normal.

Testis fibrosis

Focal lesions of fibrosis may be found quite commonly in some groups of bulls after the post-weaning period. Diffuse fibrosis is much less common and in aged bulls it occurs more commonly in the ventral areas of the testes.[10] Routine ultrasonography of the testes of bulls may reveal the presence of few to many foci of fibrosis, with or without the production of elevated numbers of abnormal sperm. In one study,[34] ultrasonography of the testes was done in bulls at three locations in western Canada ($n = 325$) and one in Argentina ($n = 387$) to determine the prevalence of fibrotic lesions and to examine the relationship between fibrotic lesions and location, age, breed, testis size, and semen quality. Bulls in western Canada are typically fed forage and grain-based rations after weaning, whereas in Argentina weaned calves in the study did not receive grain supplementation until 18 months of age. Bulls used in the study ranged in age from 3 to 20 months and ultrasonography was done at different ages within groups of bulls.

Bulls in all groups in Canada and Argentina had been vaccinated at weaning with clostridial vaccines and vaccines for bovine viral diarrhea virus (BVDV), infectious bovine rhinotracheitis, and parainfluenza 3. In some cases bulls also received *Mannheimia* and *Histophilus somni* vaccines. In all groups there were no unexpected health issues except in Group 1 in Argentina. In that group there was a severe outbreak of respiratory disease with high morbidity and 6% mortality. The primary cause of the disease outbreak appeared to be due to bovine respiratory syncytial virus (BRSV). In the following year bulls at this location were vaccinated for BRSV and no further problems associated with BRSV occurred.

Ultrasonographic examinations in some groups of bulls in western Canada were done monthly from 3 to 15 months of age. In other groups in Canada and Argentina, ultrasonography was done after weaning at 5–8 months of age and repeated at 12–14 months of age in Canada and 18–20 months of age in Argentina. Testis assessment was done by orienting the transducer vertically on the surface of the scrotum and moving it 180° around the testis on the dorsal and ventral half of the testes. Testes were scored for fibrosis from none to very severe based on the number and size of fibrotic lesions (Figures 12.1 and 12.2).

Immunohistochemistry was performed on testis tissues of 10 bulls from the group that was affected by BRSV. Only one of these bulls had fibrotic lesions in the testes. Testis tissue was submitted to a diagnostic laboratory for immunohistochemistry for BVDV and BRSV.

Fibrotic lesions scored as slight or mild were very common (31.3%) in the testes of a group of bulls raised under intensive rearing conditions in western Canada. Only 2.5% had moderate scores and none had severe scores for fibrosis. In the more extensive rearing conditions of Argentina, in a group that was not affected by respiratory disease associated with BRSV, fibrotic lesions were also mainly slight or mild but a higher incidence was observed (48.6%). In that group, 1.9% scored as moderate and 0.95% as severe. In the group in Argentina that was affected by BRSV, fibrotic lesion scores

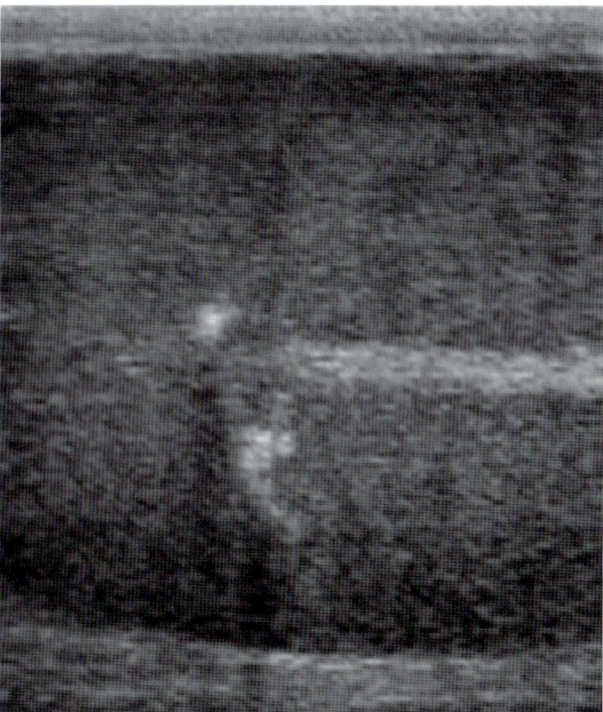

Figure 12.1 An ultrasonogram of testis scored as having mild fibrosis.

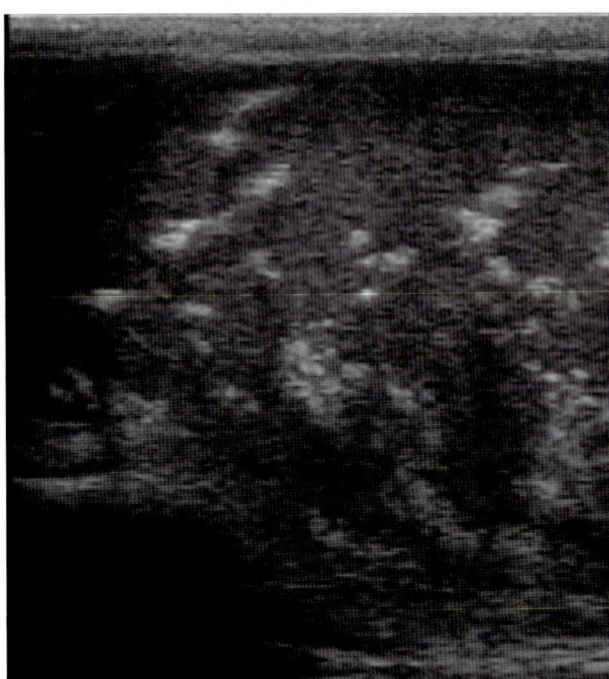

Figure 12.2 An ultrasonogram of testis scored as having very severe fibrosis.

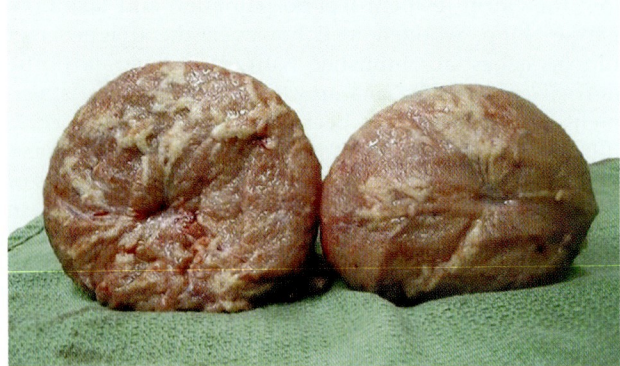

Figure 12.3 Cross-section of a testis affected by fibrotic lesions. The lesions appear to radiate from the rete testis toward the periphery suggesting that individual tubles are destroyed and replaced by scar tissue.

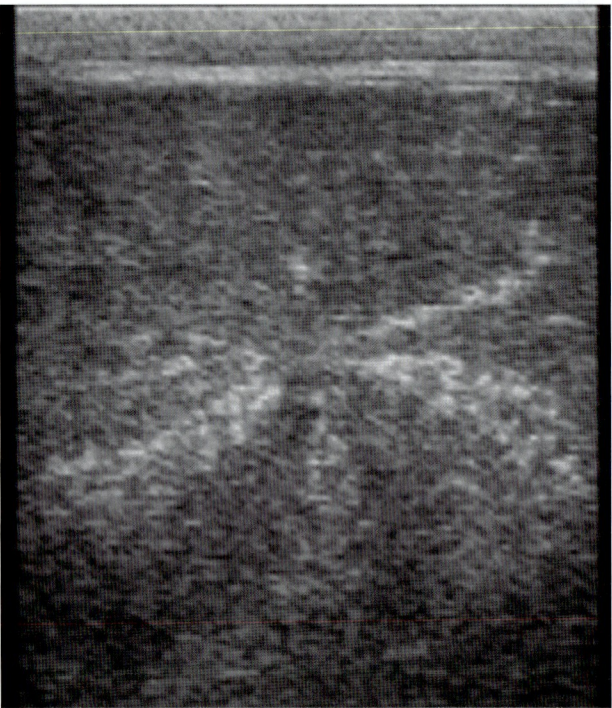

Figure 12.4 Ultrasonogram of a testis with fibrotic lesions radiating from the rete testis toward the periphery.

were higher and were present more frequently. Scores for slight to mild, moderate and severe to very severe, were 58.3, 17.1, and 5.7%, respectively.

Fibrotic lesions appeared as early as 5–6 months of age and the number of cases continued to increase until at least 12–14 months of age. The severity of lesions increased in some cases during this period, although it appears that the development of lesions occurred during a finite period of pubertal development. There was some indication that during the active process that leads to fibrosis, spermatogenesis is adversely affected; however, the presence of a large number of fibrotic lesions that may occupy as much as 50% of the testis parenchyma did not preclude the production of a high percentage of morphologically normal sperm. In bulls that scored as none, mild, moderate and severe for testis fibrosis, the median percentage of normal sperm was 88.5, 87, 86 and 90% respectively. The prevalence of lesions was not influenced by breed or testis size.

Histological evaluation showed that the lesions consisted of masses of fibrous tissue with fine fibrillation, or heavy bundles of wavy tissue situated among the seminiferous tubules. Usually there was a reduction in germinal cells in seminiferous tubules within the masses of fibrous tissue. In some tubules, the germinal cells and Sertoli cells were missing entirely and hyaline material was present in the lumen. Similar abnormalities could be seen in seminiferous tubules adjacent to fibrous tissue masses; however, completely normal tubules were often seen immediately beside or very close to fibrous tissue masses. Inflammatory cells were not seen in or beside the lesions in any of the cases. This suggests that neither the insult that caused the fibrosis nor the proximity of fibrous tissue to seminiferous tubules had a permanent effect on the function of adjacent tubules.

With real-time ultrasonography, many of the lesions appeared to extend from the rete testis toward the periphery of the testis parenchyma and appeared to involve the entirety of one or more tubules at specific sites within the testis. This pattern was seen in both gross sections of the testis (Figure 12.3) and in ultrasonograms (Figure 12.4). It appeared that entire tubules may be destroyed with replacement by fibrotic tissue. In other cases, isolated lesions were found in central or peripheral parts of the testis parenchyma without extension to the rete testis. In such cases, tubular destruction may have occurred at a specific site with part of the tubule spared from destruction.

The cause of the fibrotic change of testis tissue observed is speculative. Maldevelopment of embryonic tubule-to-rete connections, an infectious disease process affecting tubule health and development, abnormal testes thermoregulation, and trauma are possibilities in the etiology of tubule destruction and fibrosis.

Developmental changes of the testes might be involved in the etiology of fibrotic lesions. Tubularization of the seminiferous cords within the testis tissue begins at about 4 months of age and is completed in nearly all bulls by 6 months of age.[35] Sperm release into the tubules occurs as early as 8 months of age and occurs in most bulls by 10 months of age. Therefore, tubule impaction with sperm and tubule rupture with escape of sperm into the peritubular space of the testis tissue would not be expected to occur much before 8 months of age. Because sperm are antigenically foreign, a tissue reaction against escaped sperm could result in fibrosis. Impaction of the seminiferous tubules might be due to congenital failure to develop a tubular connection to the rete tubules. Tubule rupture and a tissue reaction against sperm should not occur until the testes are capable of producing sperm. As sperm production increases toward the time of sexual maturity, testes lesions would be expected to increase in number and severity.

Bacterial and viral infections occur commonly in the early post-weaning period when maternal immunity has waned and animals from different backgrounds are mixed and held together in close quarters.[36] Thus an infectious process, as a cause of fibrotic lesions, would fit the time period when testis lesions are appearing in young bulls. Bacterial or viral infections would be expected to cause swelling and pain of the testes; however, it is possible that the inflammation is insufficient to be detected by personnel involved in maintaining bulls. Infectious agents could cause inflammation of the arterioles and capillaries in the testes leading to local tissue necrosis and the formation of foci of fibrosis.[10] Viral infections may also target specific cells within the seminiferous tubules, such as the Sertoli cells or individual germinal cells, resulting in tubule damage and fibrous tissue infiltration.[37,38]

Two common viral diseases of cattle, bovine herpesvirus (BHV)-1 and BVDV, have been investigated for their role in male reproductive infections. Both viruses have been isolated from semen[39,40] and BHV-1 has been associated with degenerative changes in the seminiferous epithelium, perhaps due to the stress of illness and fever;[41] however, there are no reports that associate BHV-1 or BVDV with lesions in the bovine testis.

The association of a greater prevalence of testis fibrosis with the occurrence of a severe outbreak of BRSV respiratory disease in the Group 1 bulls in Argentina, combined with a lesser prevalence of testis lesions in Group 2 that was vaccinated against BRSV, suggests that BRSV may be involved in the etiology of testis fibrosis. However, immunohistochemistry of testis tissues from 10 bulls that were culled and slaughtered from Group 1 failed to reveal the presence of BRSV antigen. The failure to find evidence of BRSV antigen might indicate that BRSV does not multiply in testis tissue or it might be due to the lack of fibrotic lesions in 9 of 10 of the testes from which samples were submitted and because the samples were obtained about 12 months after the BRSV incident.

Viral agents multiply within the testes of several species. In one experiment, mature male domestic European rabbits were infected with an attenuated strain of myxoma virus. The animals developed symptoms of myxomatosis 7–10 days after infection and had high titers of the virus in the testes. After 20–30 days the testes were 50% of normal size and affected animals had interstitial orchitis and epididymitis.[42] Protein and genomic studies of a porcine paramyxovirus showed a close relationship to the human mumps virus,[43,44] which is known to multiply in the human testis resulting in orchitis in postpubertal men.[45] Furthermore, when 9-month-old boars were experimentally infected with porcine paramyxovirus, histopathological epididymal alterations and testicular atrophy associated with degeneration of seminiferous tubules occurred.[38] In another study, colostrum-deprived pigs were inoculated with porcine circovirus type 2 alone, porcine parvovirus alone, or with both viruses and were examined at necropsy between 21 and 26 days after infection. All pigs that received both viruses became ill. Hepatomegaly and enlarged kidneys were prominent findings in these animals; however, granulomatous lesions were apparent in many types of tissues including the testis.[46]

Obesity is associated with testis degeneration and scrotal insulation has been used to produce testicular degeneration; however, scrotal insulation did not result in fibrotic lesions.[4] In one study, deficient, normal, and excessive dietary intakes did not have an effect on the prevalence of fibrotic lesions. However, in cases of prolonged abnormal thermoregulation, especially where semen quantity and quality are very severely reduced, with loss of testis tone and size, and with severe seminiferous tubular degeneration, it is conceivable that scar tissue infiltration might occur and result in testis fibrosis.

Trauma to the testes has also been proposed as a cause of testis fibrosis.[10] Trauma caused by a blow to the testes from kicking or butting could occur at any age, but might be more frequent during crowding in pens or frequent handling in chute systems. Therefore, the appearance of fibrosis in conjunction with post-weaning penning and feeding could be considered in the etiology of testis fibrosis. However, in this regard, I once observed a bull receive an acute blow to the testes with a piece of metal pipe. Subsequent ultrasonography the following day and several times over the course of a month did not reveal any damage to the testes. This suggests that testis trauma of the sort that might be experienced commonly in penned bulls would not cause fibrotic lesions.

Interestingly, fibrotic lesions in the testes were not associated with poor semen quality. Even bulls with very severe fibrosis produced semen with up to 94% morphologically normal sperm. These results indicate that the presence of relatively large amounts of scar tissue within the testis parenchyma did not prevent the remaining unaffected parenchyma from producing normal sperm. Large amounts of scar tissue would be expected to reduce sperm production; however, the amount of semen produced by bulls with different degrees of severity of fibrosis has not been investigated.

References

1. Waites G. Temperature regulation and the testis. In: Johnson AD, Gomes WR, VanDemark NL (eds) *The Testis*, Vol. I. New York: Academic Press, 1970, pp. 241–279.
2. Setchell B. The scrotum and thermoregulation. In: *The Mammalian Testis*. Ithaca, NY: Cornell University Press, 1978, pp. 50–69, 90–103.
3. Kastelic J, Cook R, Coulter G. Scrotal/testicular thermoregulation and the effects of increased testicular temperature in the bull. *Vet Clin North Am Food Anim Pract* 1997;13:271–282.

4. Arteaga A, Barth A, Brito L. Relationship between semen quality and pixel-intensity of testicular ultrasound images after scrotal insulation in beef bulls. *Theriogenology* 2005;64:408–415.
5. Barth A, Oko R. *Abnormal Morphology of Bovine Spermatozoa*. Ames, IA: Iowa State University Press, 1989, pp. 142–143.
6. Ndama P, Entwhistle K, Lindsay J. Effect of protected protein supplements on some testicular traits in Brahman cross bulls. *Theriogenology* 1983;20:639–650.
7. Evans H. Testicular degeneration due to inadequate vitamin A in cases where E is adequate. *Am J Physiol* 1932;99:477–486.
8. Rhode L, Coulter G, Kastelic J, Bailey D. Seminal quality and sperm production in beef bulls with chronic dietary vitamin A deficiency and subsequent realimentation. *Theriogenology* 1995;43:1269–1277.
9. Gomes W. Chemical agents affecting testicular function and male fertility. In: Johnson AD, Gomes WR, VanDemark NL (eds) *The Testis*, Vol. III. New York: Academic Press, 1977, pp. 241–279.
10. McEntee K. *Reproductive Pathology of Domestic Mammals*. San Diego, CA: Academic Press, 1990, pp. 260–261.
11. Vanyi A, Timar I, Szeky A, Fusariotoxicoses IX. The effect of F-2 fusariotoxin (zearalenone) on the spermatogenesis of rams and bulls. *Magyar Allatorvosok Laja* 1980;35:777–780.
12. Robinson J, Tanphaichitr N, Bellve A. Gossypol-induced damage to mitochondria of transformed Sertoli cells. *Am J Pathol* 1986;125:484–492.
13. Chenoweth P, Chase C Jr, Risco C, Larson R. Characterization of gossypol-induced sperm abnormalities in bulls. *Theriogenology* 2000;53:1193–1203.
14. Velasquez-Pereira J, Chenoweth P, McDowell L et al. Reproductive effects of feeding gossypol and vitamin E to bulls. *J Anim Sci* 1998;76:2894–2904.
15. Panter K, James L, Hartley W. Transient testicular degeneration in rams fed locoweed (*Astragalus lentiginosus*). *Vet Hum Toxicol* 1989;31:42–46.
16. Deschamps J, Ott R, McEntee K, Heath E, Henri R. Effects of zeronal on reproduction in beef bulls: scrotal circumference, serving ability, semen characteristics, and pathological changes in the reproductive organs. *Am J Vet Res* 1987;48:137–147.
17. Newman J, Tennessen T, Tong K, Coulter G, Mears G, Doornenbal H. Effects of zeronal implantation on growth, feed conversion, testicular development and behavioral traits of young bulls fed for slaughter. *Can J Anim Sci* 1990;70:1005–1016.
18. Veeramachaneni D, Ott R, Heath E, McEntee K, Bolt D, Hixon J. Pathophysiology of small testes in beef bulls: relationship between scrotal circumference, histopathologic features of testes and epididymides, seminal characteristics and endocrine profiles. *Am J Vet Res* 1986;47:1988–1999.
19. Settergren I, McEntee K. Germ cell weakness as a cause of testicular hypoplasia in bulls. *Acta Vet Scand* 1992;33:273–282.
20. Eriksson K. Heritability of reproduction disturbances in bulls of Swedish red and white cattle. *J Nordisk Veterinaermed* 1950;2:943–966.
21. Humphrey J, Ladds P. Pathology of the bovine testis and epididymis. *Vet Bull* 1975;45:787–797.
22. Michaux C, Van Sichem-Reynaert R, Beckers J, De Fonseca M, Hanset R. Endocrinological studies on double muscled cattle: LH, GH, testosterone and insulin plasma levels in the first year of life. In: King JWB, Ménissier F (eds) *Muscle Hypertrophy of Genetic Origin and its Use to Improve Beef Production*. London: Martinus Nijhoff, 1982, pp. 350–359.
23. Welles E, Tyler J, Wolfe D, Moore A. *Eperythrozoon* infection in young bulls with scrotal and hindlimb edema, a herd outbreak. *Theriogenology* 1995;43:557–567.
24. Swift B, Reeves J, Thomas G. Testicular degeneration and libido loss in beef bulls experimentally inoculated with *Anaplasma marginale*. *Theriogenology* 1979;11:277–290.
25. Weldon S, Blue J, Wooley R, Lukert P. Isolation of picornavirus from feces and semen from an infertile bull. *J Am Vet Med Assoc* 1979;174:168–169.
26. Eilts B, Pechman R. B-mode ultrasound observations of bull testes during breeding soundness examinations. *Theriogenology* 1988;30:1169–1176.
27. Pechman R, Eilts B. B-mode ultrasonography of the bull testicle. *Theriogenology* 1987;27:431–441.
28. Coulter G, Bailey D. Effects of ultrasonography on the bovine testis and semen quality. *Theriogenology* 1988;30:743–749.
29. Brito L, Silva A, Barbosa R, Unanian M, Kastelic J. Effects of scrotal insulation on sperm production, semen quality, and testicular echotexture in *Bos indicus* and *Bos indicus* × *Bos taurus* bulls. *Anim Reprod Sci* 2003;79:1–15.
30. Sidibe M, Franco L, Fredriksson G, Madej A, Malmgren L. Effects on testosterone, LH and cortisol concentrations, and on testicular ultrasonographic appearance of induced testicular degeneration in bulls. *Acta Vet Scand* 1992;33:191–196.
31. Kastelic J, Cook R, Pierson R, Coulter G. Relationship among scrotal and testicular characteristics, sperm production, and seminal quality in 129 beef bulls. *Can J Vet Res* 2001;65:111–115.
32. Barth A. *Bull Breeding Soundness Evaluation*, 2nd edn. Saskatoon: Western Canada Association of Bovine Practitioners, 2000, pp. 6–7.
33. Blockey MdeB. Observations on group mating of bulls at pasture. *Appl Anim Ethol* 1979;5:15–34.
34. Barth A, Alisio L, Aviles M, Arteaga A, Campbell J, Hendrick S. Fibrotic lesions in the testis of bulls and relationship to semen quality. *Anim Reprod Sci* 2008;106:274–288.
35. Curtis S, Amann R. Testicular development and establishment of spermatogenesis in Holstein bulls. *J Anim Sci* 1981;53:1645–1657.
36. Nagaraja T, Galyean M, Cole N. Nutrition and disease. *Vet Clin North Am Food Anim Pract* 1998;14:257–277.
37. Bouters R. A virus with enterogenic properties causing degeneration of the germinal epithelium in bulls. *Nature* 1964;201:217–218.
38. Ramirez-Mendoza H, Hernandez-Jauregui P, Reyes-Leyva J, Zenteno E, Moreno-Lopez J, Kennedy S. Lesions in the reproductive tract of boars experimentally infected with porcine rubulavirus. *J Comp Pathol* 1997;117:237–252.
39. Kupferschied H, Kihm U, Bachmann P, Muller K, Ackerman M. Transmission of IBR/IPV virus in bovine semen: a case report. *Theriogenology* 1986;25:439–443.
40. McClurkin A, Coria M, Cutlip R. Reproductive performance of apparently healthy cattle persistently infected with bovine viral diarrhea virus. *J Am Vet Med Assoc* 1979;174:1116–1119.
41. French E. Relationship between infectious bovine rhinotracheitis (I.B.R.) virus and a virus isolated from calves with encephalitis. *Aust Vet J* 1962;38:555.
42. Fountain S, Holland M, Hinds L, Janssens P, Kerr P. Interstitial orchitis with impaired steroidogenesis and spermatogenesis in the testes of rabbits infected with an attenuated strain of myxoma virus. *J Reprod Fertil* 1997;110:161–169.
43. Berg M, Bergvall A, Svenda M, Sundqvist A, Moreno-Lopez J, Linne T. Analysis of the fusion protein gene of the porcine rubulavirus LPMV. Comparative analysis of paramyxovirus F proteins. *Virus Genes* 1997;14:55–61.
44. Linne T, Berg M, Bergvall A, Hjertner B, Moreno-Lopez J. The molecular biology of the porcine paramyxovirus LPMV. *Vet Microbiol* 1992;33:263–273.
45. Ray C. Mumps. In: Wilson JD, Braunwald E, Isselbacher KJ, Petersdorf RG, Martin JB, Fauci AS, Root RK (eds) *Harrison's Principles of Internal Medicine*, 12th edn. New York: McGraw-Hill, 1991, p. 718.
46. Kennedy S, Moffett D, McNeilly F et al. Reproduction of lesions of postweaning multisystemic wasting syndrome by infection of conventional pigs with porcine circovirus type 2 alone or in combination with porcine parvovirus. *J Comp Pathol* 2000;122:9–24.

Chapter 13

Vesicular Adenitis

Albert Barth

Large Animal Clinical Sciences, Western College of Veterinary Medicine, University of Saskatchewan, Saskatoon, Saskatchewan, Canada

Prevalence

The prevalence of vesicular adenitis (vesiculitis) is highest in yearling bulls, but the incidence may vary greatly and depends partly on the criteria used for making a diagnosis. Reported incidences may be higher if enlargement and increased firmness of the vesicular glands without pus in the semen is sufficient for a positive diagnosis. In addition, abattoir findings and histological examination of tissues may elevate the incidence above clinical diagnosis made on transrectal palpation and semen sampling. In general, clinical vesiculitis, with enlarged glands and pus in the semen, has an incidence of 1–5% in populations of yearling bulls;[1–3] however, much higher incidences might be found in individual groups of bulls.[4]

Cause

A large variety of organisms has been implicated in the cause of vesicular adenitis, including numerous types of bacteria, viruses, mycoplasma, ureaplasma, and chlamydia.[5,6] Where brucellosis has not been controlled, *Brucella abortus* is the primary cause of vesicular adenitis.[7] *Trueperella pyogenes* (formerly *Arcanobacterium pyogenes*) is one of the most frequent isolates in North America.[8,9] *Histophilus somni* is also a common isolate from bulls with vesicular adenitis during BSE.[4]

Pathogenesis

A definitive pathogenesis for vesiculitis has not been determined, although proposed routes of infection include infectious agents ascending the genitourinary tract, agents descending from the upper urinary or reproductive tracts, hematogenous invasion, or direct invasion from local sources.[8] Ascending infections, perhaps acquired from bulls riding each other, seem unlikely since repeated infusion of a culture of *T. pyogenes* into the distal penile urethra of a bull failed to produce vesiculitis.[5] A hematogenous source of infection has often been favored as an explanation, since vesiculitis has been associated with high-energy diets.[6] High-energy diets predispose to the development of rumenal acidosis, leading to rumenitis followed by bacteremia. *Trueperella pyogenes* and Gram-negative anaerobic bacteria, the most common isolates from liver abscesses of feedlot animals, are also commonly isolated from vesicular gland infections.[6,8] However, in Brazil and Argentina where bulls are raised primarily at pasture, incidences of vesiculitis as high as 17.7 and 18.6% have been reported.[10,11] In a report from Argentina, of 489 young bulls on five farms, 18.6% had vesiculitis diagnosed by rectal palpation; 6% also had epididymitis, orchitis or erosions on the penile mucosa. This suggests a relationship between vesiculitis and infections in other parts of the reproductive tract (descending infection). Bacterial and viral cultures of inflamed vesicular glands were often negative,[12] suggesting the possibility of chemical irritation. Congenital abnormalities of the ducts and vessels opening into the urethra at the colliculus seminalis[13] or a lack of synchrony in the ejaculatory process[12] may lead to reflux of semen and urine into the vesicular glands, causing inflammation but not necessarily bacterial infection. Vesiculitis has not been reported in feedlot steers that commonly suffer from rumenal acidosis and liver abscesses. Rapid development of the duct system of vesicular glands during puberty may allow reflux of semen and urine to occur in young bulls; conversely, lack of development of the vesicular glands may spare steers from vesiculitis.

Effect on fertility

The effect of vesiculitis on sperm viability is variable and is likely influenced by changes in pH and in the components of seminal plasma. Cytokines produced by white blood cells can adversely affect sperm motility and have been shown to reduce fertilizing ability of human sperm in the hamster ovum penetration test.[14] Activated granulocytes can produce large amounts of reactive oxygen species that are damaging to sperm, although seminal plasma contains antioxidant substances that protect sperm.[15] Usually, sperm motility is reduced in the presence of purulent material;

however, in some cases sperm motility and morphology are good despite the presence of large numbers of leukocytes and flocculi of pus. This may explain why there are reports of both poor and good fertility when affected bulls have been used in natural service.[3,7]

Diagnosis

The great majority of bulls with vesiculitis show no outward symptoms of disease. Usually vesiculitis is discovered during transrectal palpation of the internal reproductive tract during breeding soundness evaluations. Most commonly, one or both vesicular glands feel enlarged and indurated with loss of lobulation. In a small proportion of cases abscessation may occur and rarely there may be adhesions between the glands and the surrounding tissues; even more rarely, an abscessed gland may fistulate to the rectum. Ultrasonography is the best way to establish whether abscessation has occurred and this information is valuable in formulating a prognosis for recovery after treatment. Inflammation in other parts of the reproductive tract may be associated with vesiculitis (e.g., ampullitis, urethritis); however, inflammation in these sites is not easily clinically detected. The finding of enlarged indurated glands may or may not be associated with the presence of pus in semen samples. Many bulls recover spontaneously from vesiculitis and may retain enlarged indurated glands without excreting purulent material. The presence of leukocytes in semen also does not necessarily indicate the presence of vesiculitis since leukocytes might originate from other sites of inflammation in the reproductive tract or the urinary bladder. A common source of leukocytes in semen samples is the surface of the penis and prepuce that may be irritated from riding of pen mates or breeding females.

Special techniques are required to obtain useful samples for culture of causative agents. Collection of jets of semen into sterile culture tubes after disinfection of the glans penis nearly always results in the growth of many environmental contaminants, including staphylococci, streptococci, *Escherichia coli*, *Pseudomonas*, *Histophilus*, and many others. A method for collection of vesicular gland secretions for bacterial culture has been described as follows.[16] The bull may be blindfolded and tranquilized for restraint and for ease of penile protrusion. The preputial hair is clipped and the penis is extended by transrectal massage of the urethralis. The glans penis is held with sterile gloves and surgical sponges. The end of the penis is washed and disinfected and the urethra is irrigated with sterile saline administered through a sterile teat cannula. Then a 25–30 cm sterile Silastic tube is passed up the urethra leaving 2.5–5 cm protruding from the penis. The vesicular glands are then massaged to propel secretions into the urethra to be collected into a sterile vial.

Treatment

Vesicular adenitis is commonly treated with parenteral antibiotics, although until recently there have been no publications of data on treatment outcomes. Although vesicular gland bacterial isolates are usually sensitive to most antibiotics, the success rate of treatment has been poor.[7,9] In one case, (unpublished) a 2-year-old Hereford bull with clinical vesiculitis was treated with four different antibiotics (penicillin, oxytetracycline, chloramphenicol, and sulfamethazine), each one for a period of 1 month in succession. *Trueperella pyogenes* had been isolated and found to be sensitive to these antibiotics. After 4 months of treatment, the bull continued to produce purulent semen. This case confirmed the premise that many commonly used antibiotics are unable to penetrate the infected vesicular gland tissue in sufficient concentrations to inhibit bacterial growth.

An experiment was reported in which clinically normal bulls were treated with antibiotics followed by biopsy of the vesicular glands.[17] (At this point it is important to note that there may be differences in the pharmacokinetics of antibiotics in normal gland tissue compared with infected tissue.) The biopsies were homogenized and the tissue extracts were placed within wells on agar plates inoculated with *T. pyogenes* and *Histophilus somni*, both known to be sensitive to the antibiotics used. Tissue concentrations of penicillin, ceftiofur, or florfenicol were insufficient to inhibit growth of *H. somni* in any of the bulls (five bulls per antibiotic). Two macrolide antibiotics, tilmicosin and tulathromide, achieved sufficient concentrations in vesicular gland tissue to inhibit growth of *T. pyogenes* in four of five bulls and two of five bulls, respectively. In the second part of this experiment, some antibiotics were administered at twice the recommended dosage. With twice the recommended dose of penicillin, ceftiofur, or florfenicol, only ceftiofur achieved sufficient tissue concentrations to inhibit *H. somni* (three of five cases). Inhibitory tissue concentrations for *T. pyogenes* were achieved in some cases by all antibiotics: two of five, one of five, and one of five cases for penicillin, ceftiofur, and florfenicol, respectively. It appears that even at twice the recommended dose, often these antibiotics do not reach inhibitory concentrations in vesicular gland tissue.

Based on the results of previous experiments, two antibiotics, tilmicosin and tulathromycin, were used in a field trial to test their effectiveness in the treatment of clinical vesiculitis. In addition, there was evidence that tilmicosin and tulathromycin accumulated in macrophages and neutrophils and were subsequently released slowly from such cell types.[18] Thus immune defense cells would carry these antibiotics into sites of infection.[19] A single tulathromycin treatment, or two tilmicosin treatments 72 hours apart, at a dose indicated by the label information, were administered to bulls with clinical vesiculitis. Response to treatment was determined 21–28 days after the beginning of treatment and response was considered to be positive if there was no evidence of pus in semen samples and there were one or less white blood cells per microscope field at 1000× magnification. The recovery rate was higher for bulls treated with tulathromycin (22 of 25, 88%) than for tilmicosin (11 of 23, 48%) and both antibiotics resulted in higher recovery rates than occurred in the untreated control group (0 of 17). In the tulathromycin group, six bulls were mature (2 and 3 years old) and all recovered after treatment. In the tilmicosin group, five bulls were mature (2–6 years of age). Two of the 2-year-old bulls treated with tilmicosin recovered, but the other mature bulls did not. In the untreated control group, all bulls were yearlings and none recovered. Many of the semen culture and sensitivity tests indicated that the organisms isolated were resistant to tulathromycin and tilmicosin.

This included organisms considered likely to be pathogens; however, a large proportion of isolated organisms were considered contaminants. Nevertheless, recovery occurred after treatment in a large proportion of the cases. Both antibiotics resulted in cures of vesiculitis, including some bulls that were aged 2 years or older. Although the time of onset of vesiculitis in these bulls could not be determined, it is likely that the infections were of a chronic nature. If this is so, then a 6-day course of antibiotics was apparently able to overcome chronic infections in many mature bulls. This study indicates that tulathromycin appears to be an antibiotic of first choice for the treatment of vesiculitis. However, other antibiotics that were not tested in this experiment may also be useful. When bacterial resistance to tulathromycin is encountered, other antibiotics might be useful if administered at two to three times the recommended dose.

Failure of some bulls to respond to antibiotic treatment has led to a study of the feasibility of chemical ablation by intraglandular injection of formalin.[20] Although the preliminary results were successful, the method appears not to have found wide advocacy at this time. Iodine has also been injected into the vesicular gland, but not as a treatment for vesiculitis and the effect was not described.[21] Intraglandular injection of antibiotics has also been investigated.[17] Intraglandular injection of ceftiofur or procaine penicillin G through a pararectal cannula was done in 14 bulls with clinical vesiculitis. Of 13 bulls, seven recovered after one intraglandular injection of ceftiofur. Three of the six bulls that did not recover following one injection of ceftiofur recovered after a second intraglandular injection with penicillin. Another bull received an initial treatment of intraglandular penicillin and recovered. Three bulls that did not recover after ceftiofur and penicillin intraglandular treatment recovered after receiving three treatments of tilmicosin by subcutaneous injection. Semen samples were obtained by electroejaculation every 4 days for 6 weeks after intraglandular antibiotic injection. Intraglandular injection usually resulted in blood-tinged ejaculates in the following 2 weeks. The amount of blood discoloration and leukocytes in semen decreased with successive ejaculates and, in some bulls, the affected vesicular glands became softer on palpation. Intraglandular injection of ceftiofur or penicillin via the ischiorectal fossa appeared to be safe and effective in the treatment of vesicular adenitis and could be considered for cases that do not respond to parenteral treatment with antibiotics. No treatments have been described for vesicular glands that have abscessation.

Surgical removal of infected glands might be the final remedy provided that adhesions are not extensive and the size of the gland is moderate.[3] Libido, ability to serve, and emission of semen were normal after unilateral or bilateral vesiculectomy; however, complications from surgery may arise. In one case,[22] retarded emptying of the ampullae, perhaps due to nerve damage due to the vesiculectomy procedure, resulted in dilated ampullae, and a high level of detached heads and dead sperm. Lateral and ventral pararectal surgical approaches have been described. A ventral pararectal approach to the vesicular glands reduces the risk of intraoperative hemorrhage and postoperative hematoma formation. Additionally, the potential to damage nerves that lie adjacent to the rectum and pelvic urethra is reduced. Consequently, bladder atony and ejaculatory failure are less likely to develop after vesiculectomy by a ventral approach.[23] After seminal vesiculectomy semen volume, total sperm output, and percent motile sperm decreased, while pH increased; however, post-thaw survival of cryopreserved sperm did not change significantly after vesiculectomy.[24]

Prevention of vesiculitis

There are several main hypotheses for the pathogenesis of vesiculitis, but none have been verified through research. Without a clear indication of the causes of vesiculitis it is difficult to design preventative programs. Vaccination protocols to maintain health and resistance of bulls to infections may play an important role in reducing vesiculitis. Feed bunk management to prevent rumenal acidosis may be important considering the premise that bacteria may reach the vesicular glands by a hematogenous route. A possible genetic basis has also been proposed,[25] and as such veterinarians involved in herd management programs should closely monitor the genetic lines of bulls that have vesiculitis. Early detection and treatment should improve the success rate of managing vesiculitis. However, in one study, early detection did not reduce the incidence of clinical vesiculitis in yearling bulls. In that study, nine veterinary practitioners examined 2207 bulls by transrectal palpation at 9–12 months of age and assigned them to three groups: (i) positive for vesiculitis and receiving tilmicosin at 1 mL per 30 kg body weight every second day for three treatments; (ii) positive for vesiculitis, but not treated; and (iii) negative for vesiculitis and not treated. Bulls were considered to be positive for vesiculitis if one or both glands were enlarged and hardened. Transrectal palpation of the glands was done again at a pre-sale evaluation of semen 28–70 (mean 42.8) days after the first examination. Semen was collected by electroejaculation and evaluated for the presence of pus and/or leukocytes by light microscopy. The proportion of bulls with vesiculitis at the initial examination was 4.4%. At the second evaluation, the total number of bulls with vesiculitis decreased to 1.3%. There was no difference in the proportion of bulls with enlarged glands between the positive treated group (15 of 66, 23%) and the positive untreated group (7 of 31, 23%). Furthermore, there was no difference in the proportion of bulls with pus in their semen between the positive treated group (4 of 66, 6.1%) and the positive untreated group (2 of 31, 6.5%). Early detection and treatment appeared not to be efficacious; however, the overall incidence of the disease was quite low in this group of bulls. In herds where vesiculitis has a very high incidence, early transrectal examination and treatment may be worthy of consideration.

References

1. Carroll E, Ball L, Young S. Seminal vesiculitis in young beef bulls. *J Am Vet Med Assoc* 1968;152:1749–1757.
2. Ball L, Griner L, Carrol E. The bovine seminal vesiculitis syndrome. *Am J Vet Res* 1964;25:291–302.
3. Roberts S. *Veterinary Obstetrics and Genital Diseases, Theriogenology*, 3rd edn. Woodstock, VT: published by author, 1986, p. 847.

4. Grotelueschen D, Mortimer R, Ellis R. Vesicular adenitis syndrome in beef bulls. *J Am Vet Med Assoc* 1994;205:874–877.
5. Galloway DB. A study of bulls with the clinical signs of seminal vesiculitis: clinical, bacteriological and pathological aspects. *Acta Vet Scand* 1964;5(Suppl. 2):1–122.
6. Cavalieri J, Van Camp S. Bovine seminal vesiculitis. A review and update. *Vet Clin North Am Food Anim Pract* 1997;13:233–241.
7. McCauley A. Seminal vesiculitis in bulls. In: Morrow AD (ed.) *Current Therapy in Theriogenology*. Philadelphia: WB Saunders, 1980, p. 402.
8. Dargatz D, Mortimer R, Ball L. Vesicular adenitis of bulls: a review. *Theriogenology* 1987;28:513–521.
9. Phillips P. Seminal vesiculitis: new strategies for an old problem. In: *Proceedings of the Annual Meeting of the Society for Theriogenology*, 1993, pp. 59–66.
10. Gomes M, Wald V, Correa M, Machado R. Prevalence of bovine seminal vesiculitis (BSV) in the Province of Rio Grande do Sul, Brazil. *A Hora Veterinária* 2000;19(114):45–48.
11. Villar J, Pereira J, Vautier R, Smitsaart E. Seminovesiculitis. 1. Outbreak in young bulls. Preliminary report. *Veterinaria Argentina* 1987;4(33):255–264.
12. Linhart R, Parker W. Seminal vesiculitis in bulls. *Comp Cont Educ Pract Vet* 1988;10:1429–1432.
13. Blom E. Malformation of the pelvic genital organs as a possible predisposing factor in the pathogenesis of vesiculitis in the bull. In: *9th International Congress on Animal Reproduction and Artificial Insemination*. Madrid: Editorial Garsi, 1980, Vol. III, p. 217.
14. Hill J, Haimovici F, Politch J, Anderson D. Effects of soluble products of activated lymphocytes and macrophages (lymphokines and monokines) on human sperm motion parameters. *Fertil Steril* 1987;47:460–465.
15. Schopf R, Schramm P, Benes P, Morsches B. Seminal plasma-induced suppression of the respiratory burst of polymorphonuclear leukocytes and monocytes. *Andrologia* 1984;16:124–128.
16. Parsonson I, Hall C, Settergren I. A method for the collection of bovine seminal vesicle secretions for microbiologic examination. *J Am Vet Med Assoc* 1971;158:175–177.
17. Rovay H, Barth A, Chirino-Trejo M, Martinez M. Update on treatment of vesiculitis in bulls. *Theriogenology* 2008;70:495–503.
18. Giguere S. Macrolides, azalides, and ketolides. In: Giguere S, Prescott JF, Baggot JD, Walker RD, Dowling PM (eds) *Antimicrobial Therapy in Veterinary Medicine*, 4th edn. Oxford: Wiley-Blackwell, 2006, pp. 191–205.
19. Siegel T, Earley D, Smothers C, Sun F, Ricketts A. Cellular uptake of the triamilide tulathromycin by bovine and porcine phagocytic cells in vitro. *J Anim Sci* 2004;82(Suppl. 1):186.
20. Waguespack R, Shumacher J, Wolfe D, Sartin E, Heath A, Smith R. Preliminary study to evaluate the feasibility of chemical ablation of the seminal vesicles in the bull. In: *Proceedings of the 37th Annual Conference of the American Association of Bovine Practitioners*, 2004, pp. 295–296.
21. Hoover T. Bacterial seminal vesiculitis in bulls. In: *Proceedings of the Annual Meeting of the Society for Theriogenology*, 1974, pp. 92–98.
22. Nothling J, Volkmann D. Dilation of the ampullae and an increased incidence of loose sperm heads after bilateral vesiculectomy in a bull. *Reprod Domestic Anim* 1997;32:321–324.
23. Hooper R, Taylor T, Blanchard T, Schumacher J, Edwards J. Ventral pararectal approach to the seminal vesicles of bulls. *J Am Vet Med Assoc* 1994;205:596–599.
24. Alexander F, Zemjanis R, Graham E, Schmeh M. Semen characteristics and chemistry from bulls before and after vesiculectomy and after vasectomy. *J Dairy Sci* 1971;54:1530–1535.
25. Blom E. Studies on seminal vesiculitis in the bull. II. Malformations of the pelvic genital organs as a possible predisposing factor in the pathogenesis of seminal vesiculitis. *Nord Vet Med* 1979;31:241–250.

Chapter 14

Inability to Breed due to Injury or Abnormality of the External Genitalia of Bulls

Herris Maxwell

Department of Clinical Sciences, College of Veterinary Medicine, Auburn University, Auburn, Alabama, USA

Introduction

Impotence, the inability to sire offspring, may be due to abnormalities in sperm production and/or inability to copulate. Inability to copulate is further divided into inability to perform the copulatory act due to physical limitations or inability to copulate due to erection failure. This chapter reviews important causes of impotence due to abnormalities of the genitalia of the bull. Abnormal spermatogenesis is covered in other chapters of this book.

Impact

Natural service typically utilizes bull to cow ratios of 1 : 15 to 1 : 50.[1] In single-sire units inability of the bull to breed will have a devastating effect on reproductive efficiency. Similarly, when a socially dominant bull in a multi-sire unit is unable to complete the copulatory act, the decrease in breeding efficiency may not be masked by the presence of additional subservient bulls. Although artificial insemination eliminates the requirement for coitus, semen must be collected from the bull and most semen processed for the artificial insemination industry is collected by methods that utilize an artificial vagina (AV) as the bull mounts a teaser or phantom to mimic the copulatory act. While some impediments to the copulatory act may be overcome by the collection of semen using electroejaculation, others may not be easily managed.

Physiology of erection in the ruminant

The unique and interesting features contributing to extension and erection of the penis of the bull have been described.[2,3] A brief summary of the process follows.

The fibroelastic ruminant penis is encased in the rigid tunica albuginea. The erectile tissues of the corpus cavernosum penis (CCP) and corpus spongiosum penis (CSP) are contained within this strong, relatively inelastic case. In the nonerect state the penis is maintained within the prepuce and sheath and assumes a sigmoid shape due to the traction of the retractor penis muscles, which insert on the ventral surface of the mid-shaft of the penis at the distal bend of the sigmoid flexure. The proximal terminus of the CCP is the bulb of the penis, which divides into two crura that attach to the ischiatic arch on either side of the pelvis and are covered by the ischiocavernosus muscles.

At the initiation of sexual arousal the retractor penis muscles relax in synchrony with increased blood flow from the deep arteries of the penis into the crura and the CCP fills with blood. The lack of tension from the retractor muscles coupled with complete filling of the CCP initiates partial straightening of the sigmoid flexure. Contractions of the ischiocavernosus muscles force the crura against the pelvis and occlude all vascular flow to and from the CCP resulting in a closed hydraulic system. Further rhythmic contractions of the ischiocavernosus muscles apply pressure to the blood contained within the crura and this pressure is transferred to the blood contained within the CCP to generate the pressure necessary for full erection and extension of the penis. The pressure generated in the CCP far exceeds the arterial pressure of the circulatory system, reaching 14 000 mmHg (1.87 MPa) or higher at the time of peak erection.[4]

Coitus

Although brief, the breeding act requires orchestration of interrelated simple and complex events. The bull must identify the female in estrus, approach, extend the penis,

---

Bovine Reproduction, First Edition. Edited by Richard M. Hopper.
© 2015 John Wiley & Sons, Inc. Published 2015 by John Wiley & Sons, Inc.

mount appropriately, achieve full erection, adjust his stance and posture while locating the vulva with the penis, achieve intromission, make the breeding lunge, ejaculate, and dismount. The physical requirements are obvious and structural soundness is a necessary component of the successful breeding act. In cattle the coital act occurs rapidly, with time from mounting to intromission and ejaculation generally less than 10s and the time from intromission to ejaculation 1–2s.[5,6] Although some false mounts are expected, excessive numbers may be associated with pathology of the axial or appendicular skeleton or may indicate problems with the erection process, impediments to penile protrusion and extension, or lack of innervation of the glans penis.

Following a successful breeding attempt the bull has a brief refractory period, the length of which varies among bulls and is likely affected by age, physical condition, and the presence of other bulls and females exhibiting estrus. For young structurally sound bulls the refractory period may be only a few minutes and many bulls can be expected to copulate multiple times per day.

Abnormalities of the prepuce

The prepuce is an invagination of abdominal skin that contains the free portion of the nonerect penis and covers the body of the penis behind the glans. The penile attachment of this invagination of skin occurs caudal to the glans penis and in the nonaroused state results in the formation of a fornix in the preputial cavity adjacent to the free portion of the penis. The preputial orifice is located caudal to the umbilicus and is surrounded by a tuft of long hair that serves to divert urine away from the skin surrounding the orifice following micturition. When erect, the free portion of the penis extends beyond the preputial orifice and the hairless reflection of skin within the preputial cavity extends with the penis and covers the shaft. The subcutaneous tissues of the prepuce must be freely moveable to allow extension and retraction of the penis and the preputial reflection and these specialized connective tissues are referred to as the elastic layers of the prepuce.

Congenital and developmental anomalies of the prepuce

Congenital preputial stenosis

Although congenital preputial stenosis could prevent extension of the penis, stenosis secondary to injury or laceration of the preputial tissues is more common.

Incomplete separation of the penile and preputial epithelium

Prior to puberty, the epithelium of the free portion of the juvenile penis and the integumentary epithelium of the penile portion of internal lamina of the prepuce are joined by an interdigitating attachment.[7–9] Exposure to the trophic effects of androgens as bulls approach puberty leads to an increase in penile size and to development of the sigmoid flexure characteristic of ruminants. Simultaneous with the increase in penile size, the epithelial attachments between the prepuce and the free portion of the penis weaken and begin separating. Separation may be facilitated by mounting activity and early attempts at erection during sexual role-playing in young bulls. Until this separation is complete at 8–11 months of age, extension of the penis is not possible.[7,8,10]

Normal separation is sometimes delayed and may result in an inability to fully extend the penis at the time of a breeding soundness examination or during an observed breeding attempt (Figure 14.1). The attachment will spontaneously regress in most bulls given adequate time to mature but assisted separation by gentle application of manual traction to the penis is sometimes advocated in cases where partial separation has already occurred (Figure 14.1b,c).

Young bulls sometimes tear the epithelial attachment prematurely and hemorrhage into the epithelium, resulting in formation of a localized hematoma. Such injuries may go unnoticed and are often self-limiting.[10] Should bacteria gain access to the damaged tissues and hematoma, the formation of an abscess may result in phimosis if the bull's ability to extend the penis is compromised. Case management should emphasize isolation from other animals to discourage sexual role-play accompanied by broad-spectrum antibiotic treatment to minimize complications.

Persistent frenulum

The frenulum is a thin band of tissue found in juvenile bulls that extends from the midline of the ventrum of the tip of the free portion of the penis to the attachment of the preputial epithelium near the base of the free portion. Similar to the epithelial connection between the penis and prepuce, this band of tissue normally separates as the bull matures. Unlike the epithelial tissues of the penis and prepuce, the frenulum often contains a large vein. When the process of normal separation fails, the frenulum may persist and the result is a band of epithelium and connective tissue joining the tip of the penis with the preputial epithelium near the junction of the prepuce and free portion of the penis (Figure 14.2). A persistent frenulum results in a sharp ventral bend of the distal penis as it becomes erect. Persistent frenulum is often detected in young bulls during a routine pre-service breeding soundness examination,[11] but occasionally is not detected until the bull attempts natural service.

The cause or causes of failure of separation of the preputial epithelium and/or frenulum are unknown. Although unproven, a genetic association is suspected in some breeds.[12,13] Surgical transection of the frenulum is curative[13,14] but use of affected animals only as terminal sires is recommended.

Injuries to the prepuce

Bulls sometimes sustain injuries to the prepuce severe enough to interfere with breeding. Bulls from breeds with substantial *Bos indicus* influence have a pendulous sheath and excessive preputial skin accompanied by a large preputial orifice and bulls with this phenotype are predisposed to preputial trauma at the time of breeding. *Bos taurus* breeds

Inability to Breed due to Injury or Abnormality of the External Genitalia of Bulls

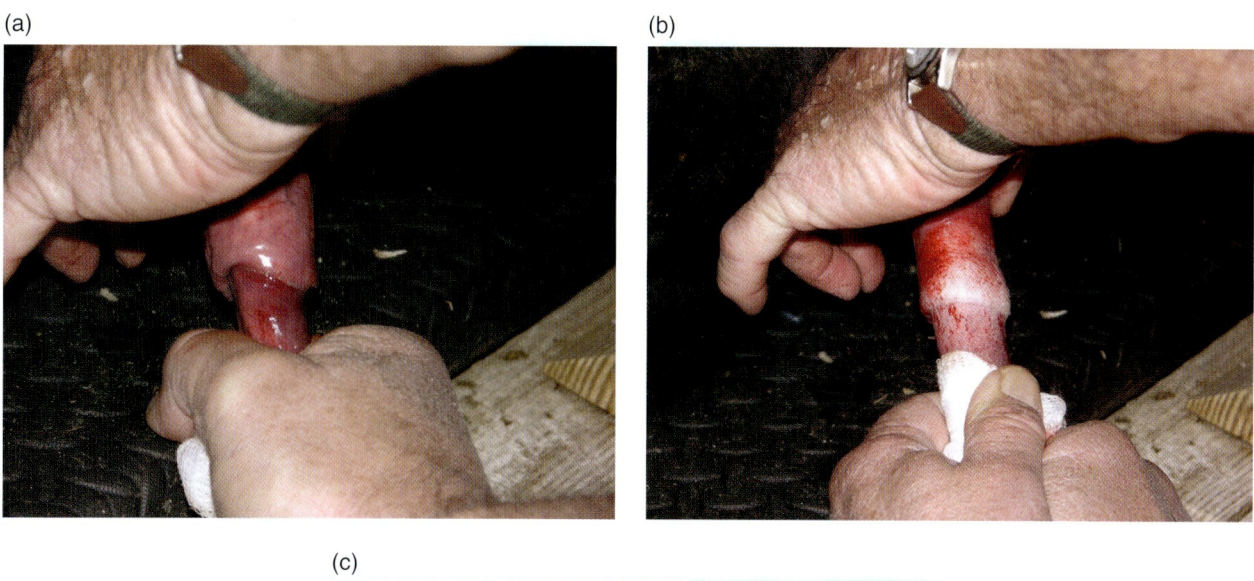

Figure 14.1 (a) Failure of separation of the epithelium of the free portion of the penis and the penile portion of the prepuce. (b) Note the adherence between the preputial and penile skin. Application of gentle traction to the penis to facilitate adherence breakdown. (c) Penile and preputial epithelium following breakdown of the epithelial adherence. Courtesy of Robert L. Carson and Dwight Wolfe.

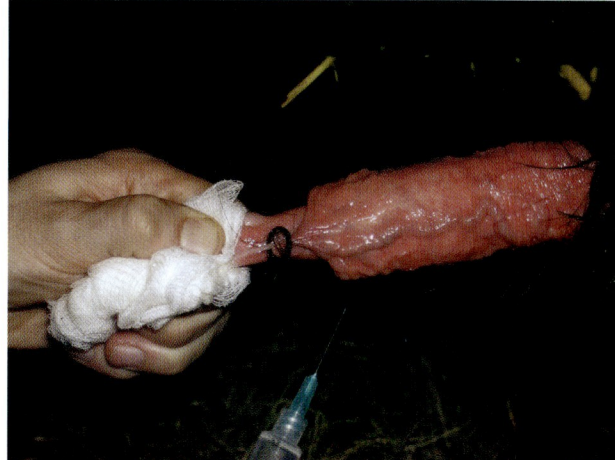

Figure 14.2 Persistent frenulum. The location on the ventral surface of the penis adjacent to the median raphe is typical. Note the accumulation of hair between the penis and the frenulum. Courtesy of Jack Smith.

are less predisposed to preputial trauma but preputial injury at the time of the ejaculatory lunge may occur in any breed.

Preputial laceration

Although bulls with a pendulous sheath and excessive preputial skin may traumatize the preputial tissues independent of the breeding act, most serious preputial injuries occur at the time of the ejaculatory lunge. As the free portion of the penis enters the vagina during coitus, preputial skin slides caudally up the shaft of the penis toward the abdomen of the bull and folds of redundant skin gather at the preputial orifice. This "bunching" of preputial skin usually occurs without incident, but when preputial tissue is trapped between the abdomen of the bull and the bony pelvis of the female at the time of intromission compressive forces generated by the ejaculatory lunge can be focused on the entrapped tissues. In mild cases the preputial epithelium remains intact and the accumulation of edema in the damaged tissues results in an uncomplicated preputial prolapse. More serious injury occurs when compression of the entrapped prepuce disrupts the epithelium, with subsequent exposure and damage of the underlying elastic tissues. Although commonly referred to as preputial laceration, the injury is actually the result of bursting of preputial tissues in response to compressive force.[15]

Preputial lacerations initiated during the breeding act predictably occur on the ventrum of the prepuce, with the initial injury in the preputial tissues oriented longitudinally, parallel to the long axis of the bull's body. Following the injury, as the penis is retracted the damaged preputial tissues are drawn toward the preputial orifice and the disrupted tissues at the site of the laceration fail to maintain normal alignment. As a result, the defect becomes oriented transversely (Figure 14.3). If the preputial tissues cannot be retracted through the preputial orifice, the transverse orientation of the lacerated tissues results in shortening of the caudal aspect of the exposed prepuce and the characteristic "elephant trunk" appearance of the prolapsed tissues (Figure 14.4).

Trauma and disruption of the preputial epithelium and underlying elastic tissues result in inflammation and edema and the open wound inevitably becomes septic. As dependent edema in the prolapsed tissues increases the size and weight of the externalized prepuce, traction on the prepuce results in increasing amounts of preputial tissue becoming exposed. Additional trauma, mutilation, or desiccation of the unprotected preputial tissues complicates wound management and treatment. Wound contracture at the site of the injury further distorts the tissues as the reoriented wound undergoes fibrosis (Figure 14.5).

The preputial retractor muscles serve to elevate the prepuce and this elevation can minimize edema formation in damaged tissues. Because many polled bulls lack retractor prepuce muscles, preputial prolapse following laceration in naturally hornless bulls tends to become more severe.[16]

Wolfe and Carson have constructed a four-point classification scheme that incorporates the severity of the preputial injury to estimate the prognosis for return to function and guide treatment decisions (Table 14.1).

Medical management of preputial laceration and prolapse is aimed at control of sepsis, reduction of edema, and eventual return of the damaged tissues to the preputial cavity. Application of emollients to prevent desiccation and topical antibiotics should be combined with light bandaging. Careful cleansing and flushing of the wound with dilute antiseptic solutions combined with debridement of devitalized tissues is necessary. Topical antibiotic therapy is usually sufficient if wound management is adequate. Bandaging of the exposed preputial tissues must be done carefully to avoid restricting blood flow and devitalizing tissues but proper application of a circumferential bandage to prevent desiccation and to apply mild compression is useful. To apply a bandage following cleansing and application of an emollient antibiotic ointment, first place a short length of latex tubing into the preputial orifice and position it with one end in the preputial cavity proximal to the torn epithelial tissues and the other end exiting the preputial orifice to allow urine egress from the prepuce. Then place a piece of clean 5-cm orthopedic stockinette over the exposed preputial tissues and snugly apply

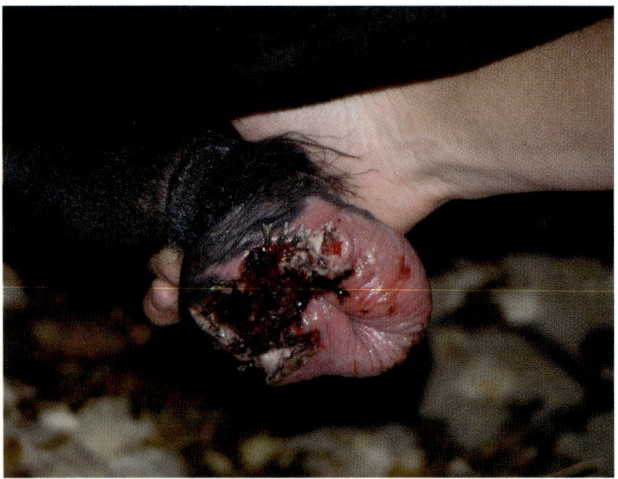

Figure 14.3 Preputial laceration in a breeding bull. The laceration assumed a transverse orientation as the preputial tissues were retracted toward the preputial orifice.

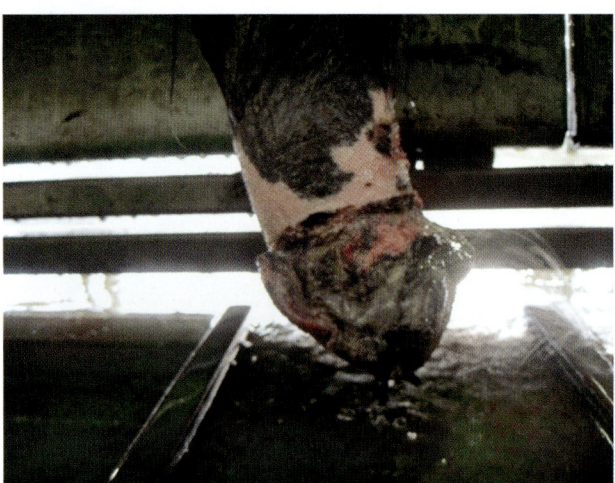

Figure 14.4 Preputial laceration with prolapse of the damaged preputial tissues. Reorientation of the wound results in the elephant trunk appearance of the prolapsed preputial tissue.

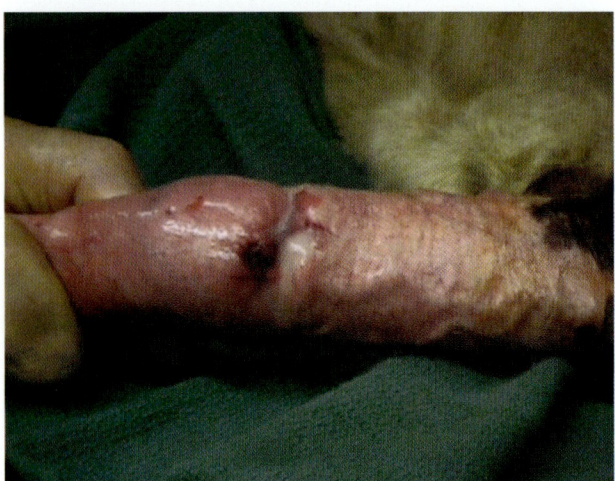

Figure 14.5 Wound contracture and fibrosis at the site of a preputial laceration. Courtesy of Richard Hopper.

Inability to Breed due to Injury or Abnormality of the External Genitalia of Bulls

Table 14.1 Classification of preputial prolapse.

Category	Description	Treatment and prognosis
I	Simple preputial prolapse with slight to moderate edema without laceration, necrosis, or fibrosis	Either conservative or surgical treatment with good prognosis
II	The prolapsed prepuce has moderate to severe edema, may have superficial lacerations or slight necrosis but has no evidence of fibrosis	Surgery is the usual course of therapy with a good to guarded prognosis
III	There is severe edema of the prolapsed prepuce with deep lacerations, moderate necrosis, and slight fibrosis	Surgery is indicated and the prognosis is guarded
IV	The prolapsed prepuce has been exposed for quite some time and has severe edema, deep lacerations, deep necrosis, fibrosis, and often abscess	Surgery and salvage by slaughter are the only options, and a guarded to poor prognosis follows surgery

Source: Wolfe D, Beckett S, Carson R. Acquired conditions of the penis and prepuce (bulls, rams, and bucks). In: Wolfe DF, Moll HD (eds) *Large Animal Urogenital Surgery*. Baltimore: Williams and Wilkins, 1998, p. 258.

(a) 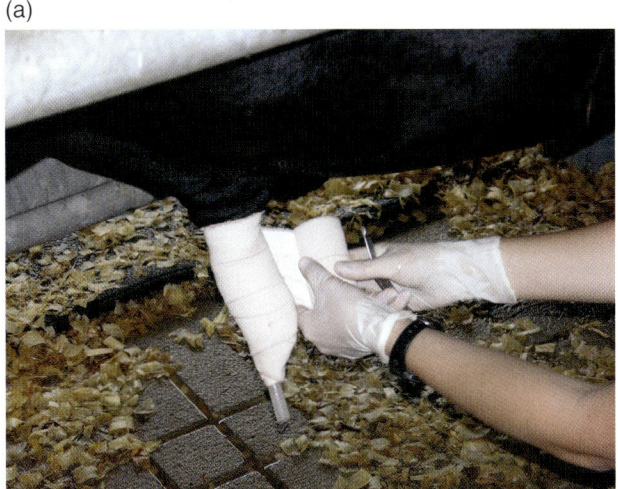 (b)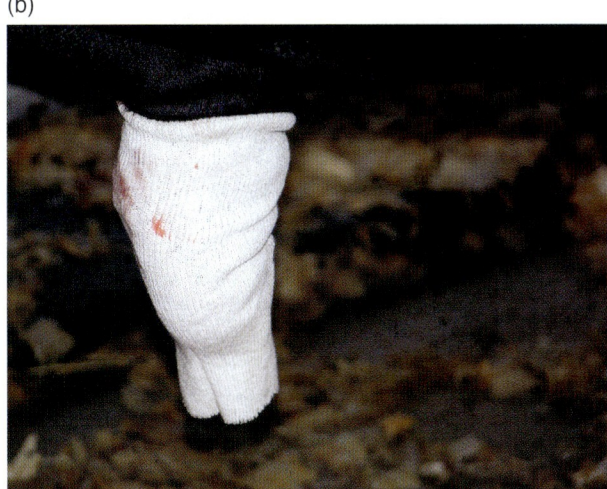

Figure 14.6 Bandaging of the prolapsed preputial tissues following application of an emollient and topical antibiotics. Placement of a urine egress tube to evacuate urine from the prepuce, a light stockinette to protect the exposed tissues (a), and an overlapping elastic tape pressure bandage secured to the preputial hairs and urine egress tube distally and to the skin of the haired sheath proximally (b).

an elastic tape bandage over the stockinette, prepuce, and urine egress tube beginning at the distal end of the prolapsed tissue, overlapping the tape as it advances up the prepuce to the preputial orifice where it can be secured to the haired skin of the sheath (Figure 14.6). Following bandaging, the edematous prepuce may be suspended by application of a bib or sling made of net material or burlap supported by straps encircling the bull's abdomen (Figure 14.7). Frequent bandage changes are necessary and the wound should be treated locally each time it is exposed. Cold water hosing for 10–15 min at each bandage change will reduce edema and remove necrotic debris. With diligent treatment many bulls may be returned to service without surgery[17] but repeat injury is common.

Surgical treatment following preputial laceration can improve outcome[18] and is indicated when the bull's value and remaining breeding life is sufficient to justify the expense.[18,19] Surgery must always be preceded by preoperative wound management. Excellent descriptions of the surgical options appear elsewhere in this book (see Chapter 17).

Retropreputial abscess

Preputial injury and laceration are not limited to *Bos indicus* influenced breeds. In *Bos taurus* bulls preputial injury may occur at the time of breeding in a manner identical to that described for *Bos indicus* bulls but the outcome is often altered by phenotype. *Bos taurus* breeds are more likely to retract all the damaged tissues into the preputial cavity following injury and as a result the wound is less likely to be noticed early. The compromised elastic tissues within the preputial cavity are contaminated with bacteria and cellulitis and phlegmon rapidly develop and often progress to abscess formation. The preputial swelling may be confined to a well-defined area adjacent to the sheath, or may be more diffuse and occasionally extend from the preputial orifice caudally toward the scrotum.

Retropreputial abscess formation is more likely in bulls of *Bos taurus* than *Bos indicus* influenced breeds due to lack of redundant skin and generally tighter sheath conformation. Affected bulls present with an obvious swelling visible through the overlying skin of the sheath that may be accompanied by the presence of pus or blood at the preputial

Figure 14.7 Burlap "bib" applied to the bull's abdomen to suspend the edematous preputial tissues.

orifice. Diagnosis is based on physical examination and palpation, sometimes augmented by ultrasound imaging of the tissues. This condition must be differentiated from the enlargement of the elastic tissues seen following rupture of the tunica albuginea of the penis. In contrast to the lesion seen with rupture of the tunica albuginea, retropreputial abscesses are generally nonsymmetrical and located distal to the sigmoid flexure nearer the level of the preputial fornix (Figure 14.8). Retropreputial abscess formation carries poor prognosis for future breeding. Destruction and impairment of the elastic tissues frequently result in adhesion formation within the elastic tissues of the prepuce and the overlying skin or in compromise of the diameter of the preputial lumen, either of which may prevent extension of the penis.[20]

Therapy for retropreputial cellulitis, phlegmon, and abscessation relies on systemic antibiotic administration and local wound management. Daily flushing of the preputial tissues with dilute antiseptic solutions and cold water hosing of the sheath aids in resolution of cellulitis. Drainage of a retropreputial abscess into the preputial lumen at the site of the original injury is difficult but if possible may facilitate recovery. No attempt should be made to drain a retropreputial abscess through the overlying skin of the sheath as damage and sepsis of the underlying elastic tissues is inevitable and formation of peri-penile adhesions will decrease the chance of a successful outcome.[17,20] Even with aggressive therapy the prognosis is guarded to poor and many affected bulls will never return to service.[20]

Phimosis

Phimosis, the inability to extend the penis, effectively prevents the bull from breeding and may be diagnosed at the time of an observed breeding or by induction of erection with an electroejaculator. Phimosis may be due to stenosis of the preputial opening or lumen, adhesions within the elastic layers of the prepuce and surrounding skin, or occasionally to abnormalities of the distal penis including the presence of large penile fibropapillomas.

Stenosis or stricture of the preputial lumen can occur following preputial injury despite appropriate and apparently successful medical or surgical management. Scar tissue replaces damaged elastic tissues and contracture and cicatrix

Figure 14.8 Retropreputial abscess following preputial laceration in a young bull. Note location of the swollen tissues in the distal sheath.

formation may constrict the preputial lumen and result in preputial stenosis sufficient to prevent extension of the penis (Figure 14.9). If circumferential constriction of the preputial cavity occurs distal to the end of the tip of the nonerect penis, sexual arousal and engorgement of the penis will force the distal portion of the penis down the prepuce until the constriction is encountered and the preputial lamina will be forced out the preputial orifice without exposure of the free portion of the penis. Strictures sometimes interfere with the evacuation of urine from the preputial cavity. Noncircumferential scar formation at the site of a healed preputial laceration may also compromise the preputial lumen enough to prevent penile extension.

Extension of the penis is dependent on appropriate function of the elastic tissues and severe damage can prevent the gliding action necessary to allow the free portion of the penis and internal lamina of the prepuce to exit the preputial orifice. Formation of adhesions between the penis, elastic tissues and skin following trauma can effectively limit the movement of the penis and result in partial or complete failure of extension. The site of the adhesion may sometimes be identifiable by a deformation of the contour of the overlying skin at the time of attempted erection.

Phimosis sometimes occurs due to the presence of a penile wart (papilloma) larger than the lumen of the prepuce or the opening of the preputial orifice. Incision of the internal lamina of the prepuce sufficiently to allow extension of the penis and removal of the wart followed by closure of the incised preputial tissue can be curative if the value of the bull is sufficient to justify the expense (Figure 14.10). Restraint on a tilt table and administration of local analgesia or an internal pudendal nerve block facilitates this procedure.

Inability to Breed due to Injury or Abnormality of the External Genitalia of Bulls

Many bulls affected by phimosis are culled. No treatment is effective if adhesions of the elastic tissues are severe. In cases of preputial stenosis, restoration of the preputial lumen may be accomplished by resection of the compromised tissues and anastomosis of the remaining prepuce if sufficient healthy preputial tissues remain. Similarly, surgical scar revision can be useful for restoring the integrity of the preputial lumen when the amount of healthy preputial skin is insufficient for preputial resection. Both techniques are described in Chapter 17.

Avulsion of the prepuce

The attachment of the reflection of preputial skin to the free portion of the penis is susceptible to avulsion injury at the time of breeding but this injury occurs most often when semen is collected with an AV. If the AV is inappropriately sized or not properly lubricated, preputial skin may adhere to the latex liner and be avulsed or torn from its attachment to the penis at the time of the ejaculatory lunge. This injury is often recognized immediately following service of the AV when blood is seen at the external preputial orifice. Extension of the penis allows observation of the avulsion injury, typically located on the ventrum of the penis.[21] Unlike preputial lacerations which occur at the time of breeding, the underlying tissues are not severely traumatized and immediate repair by direct suture of the avulsed tissues is recommended. Prognosis following early detection and repair is good.[21] If not noticed early and repaired immediately, hemorrhage and wound sepsis may complicate management but careful dissection and reattachment of the avulsed tissue can still be successful.

When collecting semen with an AV the person collecting the semen should be appropriately trained, a suitable AV and liner should be selected and properly prepared, and bulls should always be observed closely following collection.[21]

Abnormalities of the penis

Congenital abnormalities

Congenital abnormalities of the penis occur sporadically.

Hypospadias

Hypospadias, an abnormal opening in the urethra that occurs when the embryonic urethral groove fails to close or closes incompletely anywhere along its length, has been reported in bulls and other species. In some cases the entire length of the penile urethra may be involved and be easily observed at birth.[22] Cases which involve only the distal urethra may go unobserved until the bull is examined during a breeding soundness examination.[23] Slight caudal relocation of the urethral meatus may not significantly affect fertility.

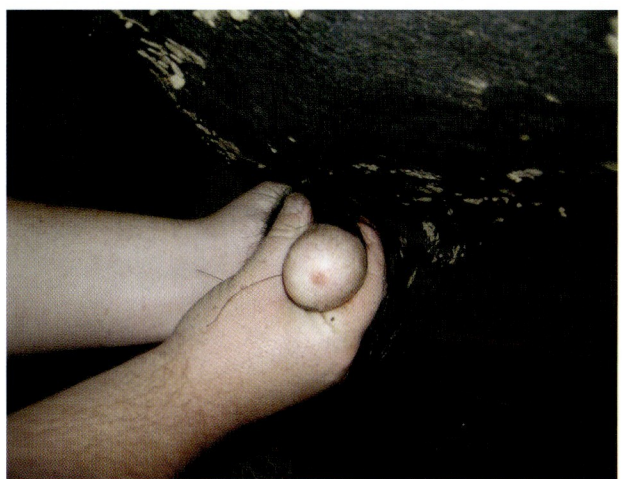

Figure 14.9 Phimosis due to circumferential stricture of the preputial cavity following preputial laceration. Courtesy of Craig Easley.

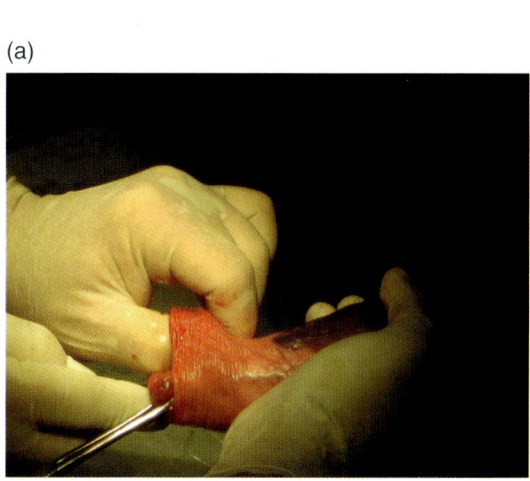
(a)

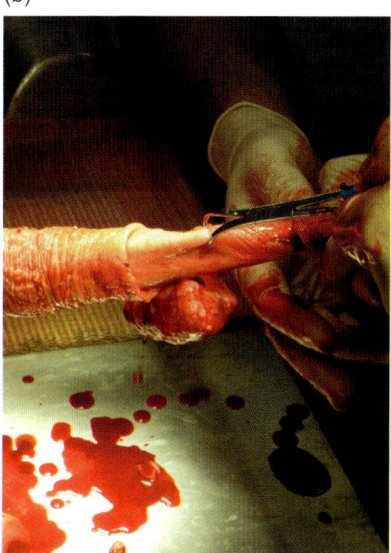
(b)

Figure 14.10 Phimosis due to the presence of a penile fibropapilloma larger than the preputial orifice. The tip of the free portion of the penis has been grasped with sponge forceps (a). Following incision of the internal lamina of the prepuce to enlarge the preputial orifice, the penis was extended and the fibropapilloma surgically excised (b).

Congenital short penis

Congenital short penis is seen uncommonly and must be differentiated from a partial phimosis secondary to preputial injury or fibrosis of the retractor penis muscles. Some affected bulls may sire calves by natural service during the first season or two of their career but copulation becomes impossible as they reach mature body size. Diagnosis is made by observation of breeding and measurement of the distance from the preputial orifice to the tip of the extended penis. The etiology of congenital short penis is unknown.[24] Affected bulls are unable to breed naturally and should be culled.

Diphallus

Diphallus, or duplication of the penis, has been reported sporadically.[25,26] Duplication may involve only glans penis, or the free portion of the penis may be involved.

Congenital vascular shunts

Anomalous vascular anastomoses between the peri-penile circulation and the CCP result in shunts that allow blood from the CCP to exit the erectile tissues and destroy the integrity of the closed hydraulic system necessary for complete erection.[27] Vascular shunts may occur as a congenital anomaly in young bulls or form following an injury which disrupts the integrity of the tunica albuginea of the penis. Regardless of etiology, when erection is stimulated at a test breeding or with an electroejaculator, partial erection and protrusion of the glans penis is initiated followed by loss of intrapenile pressure and failure to achieve full erection.

Vascular anastomoses between the CCP and extracorporeal vasculature found in the distal free portion of the penis are usually multiple and thought to result from congenital flaws in the integrity of the tunica albuginea.[28] As blood escapes through these distally located shunts at the time of sexual stimulation, "blushing" of the preputial tissues may be observed as blood from the CCP enters the venous circulation beneath the preputial skin. A presumptive diagnosis may be made following observation of a test mating or attempts to induce erection with an electroejaculator. Confirmation requires demonstration of the shunts with radiographic contrast studies in which radiopaque contrast media is injected directly into the CCP as serial radiographs are taken. In the normal penis the contrast media remains within the CCP until it eventually exits at the crura of the penis (Figure 14.11). Visualization of the contrast media within the peri-penile vasculature outside the CCP is diagnostic (Figure 14.12). Demonstration of anomalous vessels in the distal portion of the penis may be enhanced by occlusion of the distal peri-penile vessels by application of a tourniquet to the penis proximal to the location of the shunt(s).[29] Unlike acquired vascular shunts of the CCP, congenital shunts are seldom amenable to surgical correction.

Anatomic and developmental abnormalities

At the time of puberty androgens initiate growth of the juvenile penis and initiate separation of the penile and preputial epithelium,[7] as discussed under abnormalities of

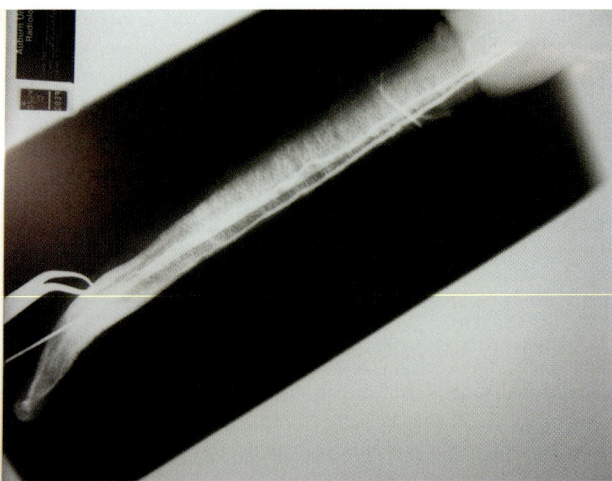

Figure 14.11 Normal cavernosogram. Contrast media injected into the CCP at the level of the free portion of the penis remains within the CCP. The ventral canals and cavernous spaces are clearly outlined.

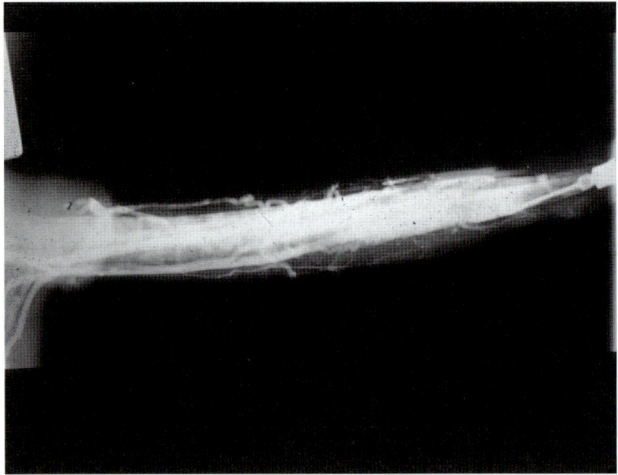

Figure 14.12 Cavernosogram demonstrating multiple shunts from distal cavernous spaces to the peri-penile vasculature. Courtesy of Robert L. Carson and Dwight Wolfe.

the prepuce. The sigmoid flexure develops as the penis enlarges and the retractor penis muscles attached to the proximal flexure of the sigmoid maintain the nonerect penis within the sheath. The rigid tunica albuginea of the penis contains the pressures generated at the time of erection and gives the erect penis it tubular shape. Uniform filling of the cavernous spaces of the CCP is required for complete erection, and the dorsal apical ligament of the penis elevates the free portion of the erect penis to allow successful intromission.

Penile deviation

Deviation of the penis results in inability to copulate. Affected bulls may have a history of one or more successful breeding seasons and there is often no history of trauma or penile injury. In the bull the dorsal apical ligament of the penis originates from the tunica albuginea proximal to the free portion of the penis, runs along the dorsum of the free portion beneath the penile skin, and rejoins the tunica albuginea near the distal end of the CCP. The ligament inserts on the distal penis centrally with a broad set of fibers and on the left lateral aspect of the free portion of the penis with a narrower and better defined set of fibers.[9,30] This ligament gives support to the erect penis and maintains the normal alignment of the penis as the bull positions himself for coitus and uses the tip of the penis to search for the vulva. Following intromission and achievement of peak erectile pressure the ligament may "slip" to the left and the distal penis will sometimes assume a corkscrew shape and spiral within the vagina at the time of the ejaculatory lunge.[6] When erection is induced with the artificial stimulus of an electroejaculator, the penis will frequently assume this same spiral orientation, often prior to ejaculation.

Spiral deviation of the penis

Pathologic spiral deviation of the penis occurs when the dorsal apical ligament slips laterally prior to intromission (Figure 14.13). The role of the dorsal apical ligament in the development of spiral deviation is well described but the factor or factors leading to the premature occurrence of this otherwise normal phenomenon are poorly understood.[9,30,31] Malfunction of the dorsal apical ligament due to a shortening of the ligament or lengthening of the penis as the bull ages was once commonly felt to be the cause but remains unproven. More recent speculation suggests that intravaginal spiral deviation occurs normally in many bulls associated with peak erectile pressure and that spiral formation occurs prematurely in bulls in which peak erectile pressure occurs prior to intromission. In either case, bulls with premature spiral deviation of the penis are unable to complete the copulatory act.

A diagnosis of spiral deviation is suspected based on history and a description of the penis during the breeding attempt but diagnosis can only be confirmed by an observed test mating. Because the penis of many normal bulls will spiral under the artificial stimulus of electroejaculation, diagnosis following observation of spiraling during stimulation with an electroejaculator is insufficient. Deviation may be intermittent, especially early in the development of the condition, and repeated observations of test breedings may be required to confirm the diagnosis.

No medical therapy is available to correct spiral deviation of the penis. Surgical correction of spiral deviation relies on induction of fibrous tissue to strengthen the attachments of the dorsal apical ligament to the penis. The available surgical techniques are described in Chapter 18.

Ventral deviation of the penis

Ventral deviation of the penis is less common than spiral deviation and the etiology is uncertain. The penis assumes a ventral curvature as erection progresses and has been referred to as a "rainbow" due to the arc formed by the erect penis. Like spiral deviation, ventral deviation is best diagnosed with an observed test mating but observation under stimulation with an electroejaculator can be useful because, unlike spiral deviation, ventral deviation is not a normal phenomenon. Most commonly the area of ventral deviation originates in the shaft of the penis proximal to the origin of the dorsal apical ligament (Figure 14.14). Surgical correction of ventral deviation of the penis is less likely to be successful than correction of spiral deviation. Occasional reports of successful correction by pexy or supplementation of the dorsal apical ligament are likely limited to cases in which the deviation is restricted to the free portion of the penis[32] (Figure 14.15).

S-shaped deviation of the penis

The S-shaped deviation of the penis is the least common type of penile deviation. It develops in mature bulls and the serpentine curvature of the penis appears to result from a mismatch of the length of the penis and dorsal apical ligament. The penis may appear excessively long or the apical ligament may have undergone contracture following repeated injury. The dorsal apical ligament prevents the fully

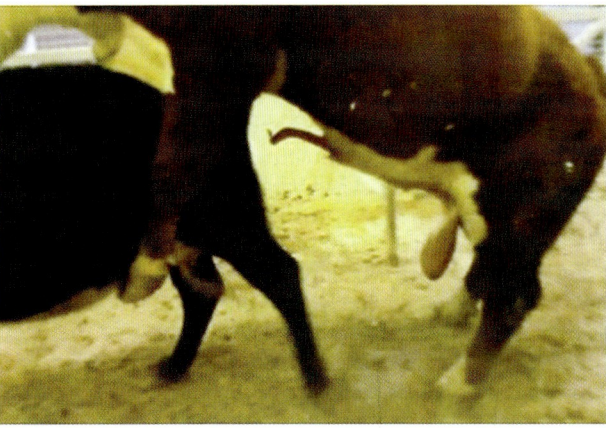

Figure 14.13 Spiral deviation of the penis demonstrated during a test mating.

Figure 14.14 Ventral deviation of the penis demonstrated during stimulation with an electroejaculator. Note that the deviation begins proximal to the origin of the dorsal apical ligament of the penis. Surgical repair has been unsuccessful in such cases. Courtesy of Robert L. Carson and Dwight Wolfe.

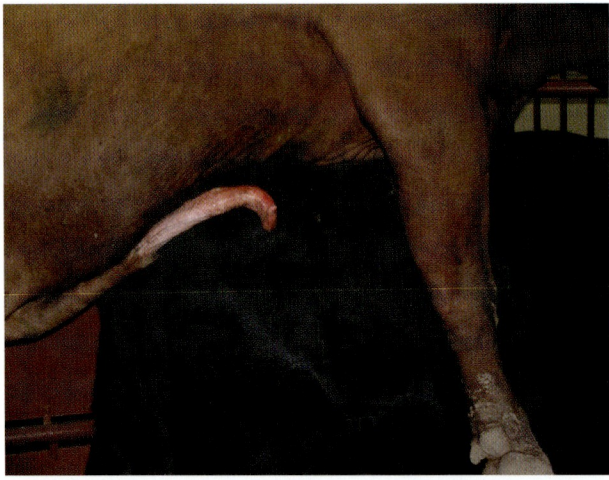

Figure 14.15 Ventral deviation of the penis demonstrated during a test mating. Note that in this case the deviation begins distal to the origin of the dorsal apical ligament of the penis. Surgical supplementation of the dorsal apical ligament may be attempted in these cases. Courtesy of Richard Hopper.

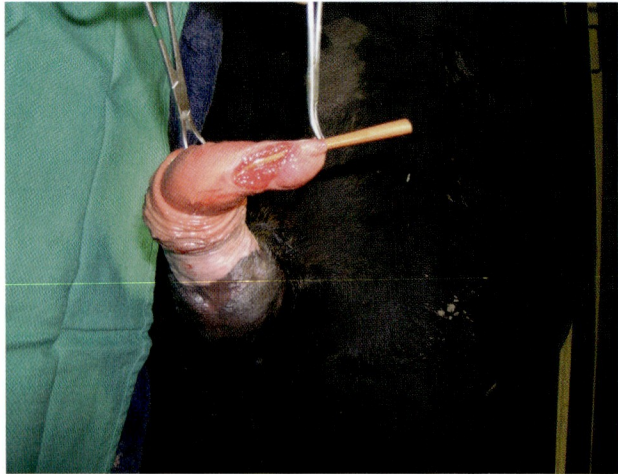

Figure 14.16 Urethral fistula. Courtesy of Richard Hopper.

erect penis from assuming its normal straight orientation and the bull may be unable to control the penis sufficiently to locate the vulva of the cow. Semen from affected bulls could be collected for use in artificial insemination, but no effective treatment for the condition exists.

Persistent frenulum

Persistent frenulum has been described under abnormalities of the prepuce. Diagnosis by observation is obvious in most cases but may sometimes be confused with phimosis in bulls with excessive redundant preputial skin if the ventral bending of the prepuce prevents extension of the glans beyond the preputial orifice (Robert L. Carson, personal communication).

Penile injury

Urethral fistula

Following laceration of the urethra, constrictive injury from a penile hair ring, or urethral necrosis associated with the presence of a urethral calculus, a urethral fistula may develop (Figure 14.16). The bull's ability to deposit semen properly in the cranial vagina may be compromised depending on the location of the fistula.[33]

Paraphimosis

Paraphimosis, the inability to retract the penis into the preputial cavity, may occur following penile laceration or preputial trauma. Edema effectively reduces the diameter of the preputial orifice and the nonerect penis remains exposed beyond the preputial orifice. The exposed preputial and penile epithelium desiccates rapidly and the superficial layers become necrotic and slough. The exposed penis is typically discolored and assumes a mild corkscrew orientation (Figure 14.17).

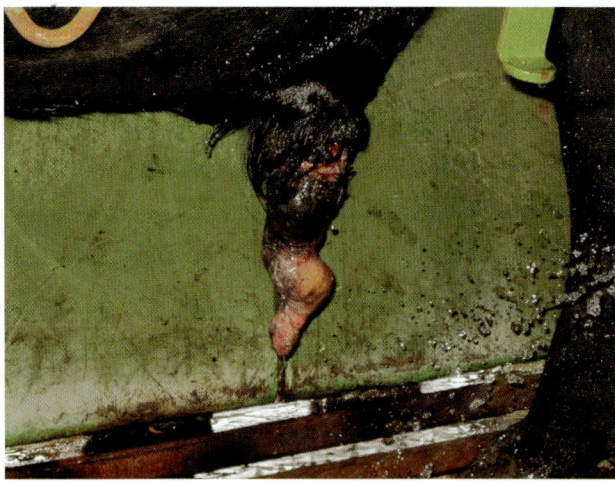

Figure 14.17 Paraphimosis following traumatic injury to the preputial trauma. The exposed penile and preputial epithelium desiccates rapidly and the free portion of the penis forms a spiral.

Paraphimosis following breeding injury warrants a grave prognosis and treatment must be initiated early to be successful. Apply emollient ointments and humectants to protect the exposed skin and cover the damaged tissues with a length of orthopedic stockinette or other light bandage material. Frequent bandage changes and fresh application of dressings combined with daily cold water hosing should continue until the penis can be retracted into the prepuce. Continue preputial lavage with antiseptics and application of antibiotic ointments or emollients for at least a week after the penis is returned to the sheath. Even with aggressive early treatment return to service is unlikely and the chance for a successful outcome decreases the longer treatment is delayed following injury.

Paraphimosis is sometimes associated with the presence of a penile papilloma on the distal penis large enough to prohibit retraction through the preputial orifice. Surgical removal of the wart is curative.

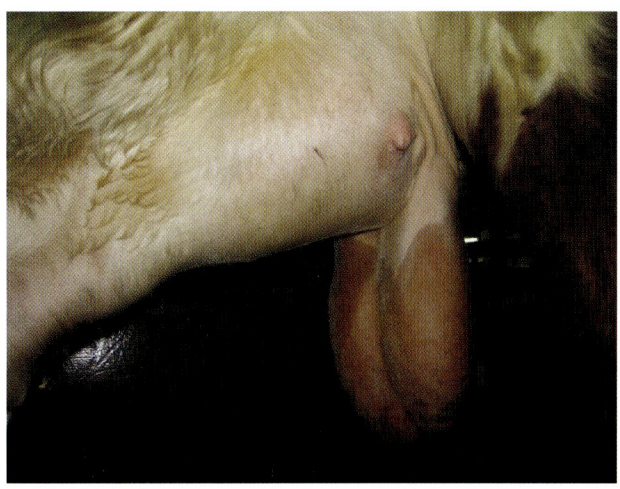

Figure 14.18 Hematoma of the penis (rupture of tunica albuginea). Note the location of the swelling, dorsal to the penis and cranial to the scrotal neck. Courtesy of Richard Hopper.

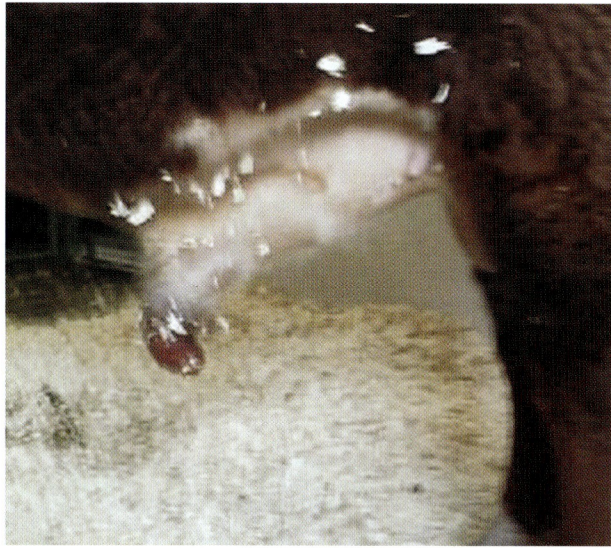

Figure 14.19 Preputial prolapse associated with penile hematoma due to rupture of tunica albuginea of the penis.

Penile hematoma (rupture of the tunica albuginea of the penis)

While any extravascular accumulation of blood in the vicinity of the penis could technically be termed a penile hematoma, in the bull the terms "hematoma of the penis" and "penile hematoma" are usually reserved for the breeding injury that results in rupture of the tunica albuginea. The penis of ruminants is well equipped to withstand high intrapenile pressure created within the CCP. At the time of erection, venous outflow to the CCP is obstructed and contraction of the ischiocavernosus muscles increases pressure within the penis to 14 000 mmHg (1.87 MPa) or greater.[3] Pressures of this magnitude are easily contained by the thick fibrous tunica albuginea of the penis which encompasses the CCP. However, if a cow or heifer collapses during coitus or an ill-timed breeding lunge accidently forces the erect penis against the escutcheon, sudden angulation of the penile shaft may increase intrapenile pressure to greater than the 70 000 mmHg (9.3 MPa) required to rupture the tunica albuginea.[34] Although occasionally seen at other sites,[35,36] rupture of the tunica albuginea consistently and almost inevitably occurs on the dorsum of the penis opposite the attachment of the retractor penis muscles on the distal bend of the sigmoid flexure.

Rupture of the tunica albuginea is accompanied by extravasation of the blood contained in the CCP and formation of a hematoma visible as a symmetrical enlargement at the site of the rupture. While the erect penis contains a relatively small quantity of blood (about 150–250 mL), the hematoma may grow from comparatively small to quite large if repeated sexual stimulation results in additional attempts at erection and continued leakage of blood through the rent. In some cases several liters of blood may accumulate in the peri-penile tissues and form a large symmetrical swelling dorsal to the penis near the neck of the scrotum (Figure 14.18). Hematoma of the penis is diagnosed based on physical examination findings and may be differentiated from retropreputial abscess by the typical location and symmetry of the swelling.

Following rupture of the tunica albuginea, ventral migration of blood and edema within the peri-penile elastic tissues often results in a secondary prolapse of preputial tissues through the preputial orifice. These mild to moderate preputial prolapses may have a distinct bluish tinge as a result of subcutaneous blood and may be the first sign noticed by the owner or manager of the bull (Figure 14.19).

Rupture of the tunica albuginea is seldom life-threatening but the injury and the complications which follow can result in permanent loss of reproductive function. Potential complications following penile hematoma include abscess formation at the site of the hematoma, formation of adhesions between the penis and peri-penile tissues, development of vascular shunts between the CCP and the surrounding vasculature, injury to the prolapsed preputial tissues, and destruction of the dorsal nerves of the penis. Injury to the dorsal nerves of the penis at the time of injury or by entrapment injury as scar tissues remodel can result in loss of sensation to the distal penis, rendering the bull unable to breed by natural service. Even following apparently successful management and resolution, recurrence of injury may occur during subsequent attempts at breeding.[20]

Case management options following rupture of the tunica albuginea include salvage for slaughter, surgical removal of the blood clot coupled with repair of the rent, or medical management. If the injury is recognized early, surgical removal of the hematoma and closure of the rent increases the likelihood for restoration of breeding ability and decreases the incidence of other post-injury sequelae. Surgical repair of hematoma of the penis and postoperative care is covered in Chapter 18. Nonsurgical management includes broad-spectrum antibiotic coverage to decrease the likelihood of abscess formation, twice-daily cold water hydrotherapy of the affected area, local treatment of the secondary preputial prolapse, and strict enforcement of sexual rest for 60–90 days. Medical management is advocated when diagnosis has been delayed or when the economic value of the bull will not justify the expense of surgery and aftercare.

Erection failure

Inability to achieve or maintain penile erection (impotentia erigendi) precludes natural service. A history of failure to impregnate females in the breeding pasture or observation of unsuccessful breeding is often the presenting complaint. A well-taken history including previous breeding performance, breeding injuries, and the owner's description of the appearance of the bull at the time of attempted coitus are valuable but observation of the penis during an attempt at erection is a required element for diagnostic evaluation. Use of the electroejaculator to induce erection may be useful but a controlled test mating is preferred.

Because painful stimuli from the spine, rear limbs or pelvis may interfere with the willingness or ability of the bull to achieve erection and complete the breeding act, a physical examination of the bull at rest and in motion is mandatory. Appropriate management or correction of painful musculoskeletal and spinal conditions may be useful and return some bulls to breeding soundness. True erection failure may involve disruption of vascular components of the erection mechanism or failure of the corpus cavernosum to fill completely.

Erection failure due to vascular shunts

Sexual stimulation of the bull is followed by increased blood flow through the crura of the penis and into the CCP. This mechanism may be mimicked by stimulation with an electroejaculator. As discussed previously, the tunica albuginea encases the erectile tissues in the CCP and there are normally no functional venous outlets along the body or shaft of the penis at the time of erection, allowing the intact tunica albuginea to effectively maintain a closed hydraulic system and contain the suprapysiologic pressures generated as the ischiocavernosus muscles rhythmically contract against the blood-filled crura of the penis.[2,4] If the integrity of the fibrous tunica albuginea is compromised, anastomoses between the CCP and the surrounding peripenile vasculature may form and provide an escape route for blood contained in the CCP. Should this occur, pressure sufficient to achieve or maintain erection cannot develop.[28,37] Communication of the CCP with the CSP will produce a similar result because venous drainage of the CSP is not occluded during erection.[29,38] Formation of vascular shunts may follow traumatic disruption of the tunica albuginea or be associated with a congenital weakness of the tissues of the tunica albuginea.

Diagnosis of vascular shunts as a cause of erection failure is suspected based on findings of an observed test mating and confirmation depends on demonstration of the vascular communication of the CCP and surrounding vasculature or CSP using radiographic contrast studies. Cavernosography utilizes water-soluble radiographic contrast media (Renografin 76, Squibb Diagnostics, New Brunswick, NJ). Best results are obtained with the bull restrained in lateral recumbency on a tilt table. Extend the penis and place towel clamps under the dorsal apical ligament and apply sufficient traction on the towel clamps to maintain extension of the penis. A length of suture or umbilical tape should be attached to the towel clamps to keep the hands of the assistant out of the radiograph beam. At the same time, place a 30-cm loop of suture through the skin under the retractor penis muscles and apply traction to pull the more proximal portion of the penis away from the abdomen for better radiographic visualization. Make an initial scout film without contrast media to establish appropriate radiographic settings. While utilizing the towel clamps to extend the penis and the loop of suture under the retractor penis muscles to pull the penis from the abdomen, insert a 16-gauge needle into the CCP on the dorsum of the free portion of the penis and attach an infusion set of sufficient length to keep the operator's hands out of the radiographic field (Figure 14.20). Inject saline

(a) 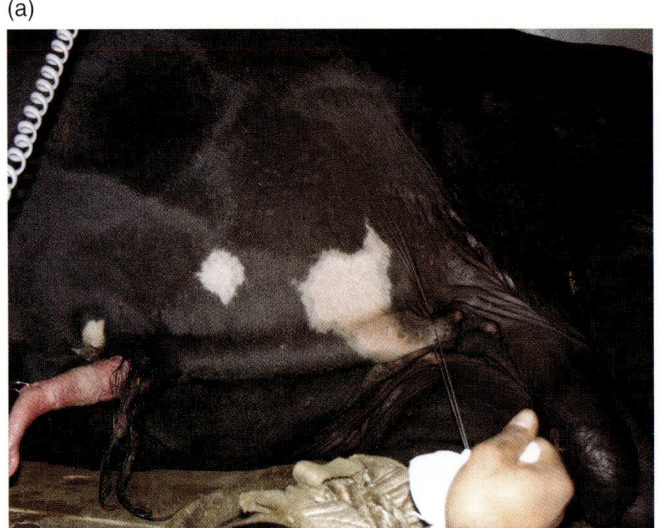 (b)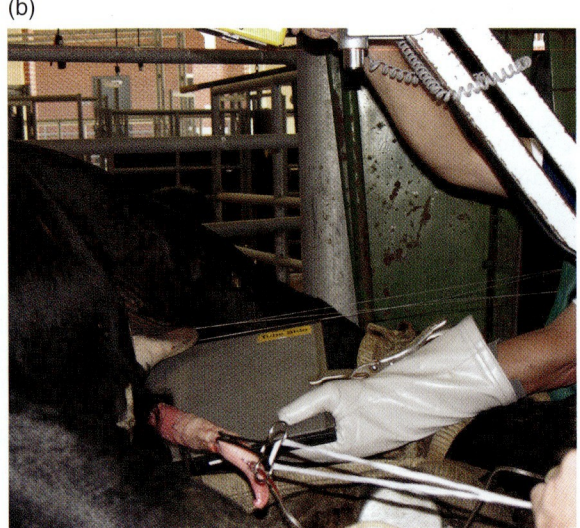

Figure 14.20 Bull prepared for cavernosography. Note suture placed through the abdominal skin and under retractor penis muscles to pull penis away from abdomen (a). As the bull is positioned, an extension set is attached to the hypodermic needle placed into the CCP of the free portion of the penis, the penis is held in extension with a towel clamp placed under the dorsal apical ligament, and traction is applied to the preplaced sutures under the retractor penis muscles (b).

into the CCP to insure proper placement, and once proper needle placement is ascertained, inject 15 mL of the radiographic contrast media and begin the radiographic series. Have the assistant slowly inject additional contrast media over the next 60 s as serial radiographs are taken of as much of the penis as possible. Take enough exposures to allow visualization of the free portion of the penis and penile shaft to the level of the distal sigmoid flexure.[39]

Formation of vascular shunts is recognized as a potential complication following rupture of the tunica albuginea as described under the section on penile hematoma. Anastomoses which develop following disruption of the integrity of the tunica albuginea and subsequent penile hematoma formation are located at the site of injury, the dorsum of the penis at the level of the distal bend of the sigmoid flexure (Figure 14.21). Penetrating injury to the tunica albuginea at sites other than the location of a penile hematoma may also be followed by shunt formation. Surgical correction of vascular shunts that occur following trauma may be successful (see Chapter 18 for details).

Cases of erection failure due to multiple vascular shunts involving multiple defects in the tunica albuginea of the free portion of the penis are thought to be the result of a congenital weakness in the structural integrity of the tunica. In such cases there is no history of penile trauma or injury. Affected bulls fail to achieve erection following sexual stimulation and may have bluish discoloration of the penile or preputial skin when attempting to breed or when attempts are made to induce erection with an electroejaculator. Discoloration is the result of blood exiting the CCP through the peri-penile vasculature. Multiple distal vascular shunts are most often diagnosed in bulls during the first breeding season. Vascular shunts in the distal penis are readily demonstrable with cavernosography (see Figure 14.12). Unlike vascular shunts that form secondary to traumatic disruption of the tunica albuginea, surgical correction of multiple congenital shunts is seldom possible.

Filling defects of the CCP

Erection and maintenance of the normal penile form is dependent on complete distension of the unobstructed cavernous spaces with blood under pressure. If the cavernous spaces are blocked sufficiently to prevent blood from completely filling the CCP, engorgement of the penis can only progress from the proximal portion of the penis to the area of the filling defect.[40] The clinical presentation depends on the location of the obstruction. Complete obstruction of more distal portions of the CCP may result in extension of the penis without erection of the more distal portions (Figure 14.22). Partial blockade of the cavernous spaces in the distal free portion of the penis may cause the erect penis to deviate ventrally or laterally.

Cavernosography can demonstrate filling defects of the penis located distal to the sigmoid flexure. The filling defects may be congenital, due to fibrosis following trauma, or subsequent to cavernositis.

Denervation injury

When mounting an estrus female the bull must position himself to make searching motions using the penis to locate the vulva and make intromission. Sensory innervation to the glans and free portion of the penis is necessary for the bull to align the penis, achieve intromission, and successfully complete the breeding act. Without sensory innervation of the distal penis the bull is unable to locate the vagina and the coital act cannot be completed.[41]

Sensory input from the distal penis is transmitted by branches of the paired dorsal nerves of the penis through the pudendal nerves to reach the spinal cord and brain. Although disruption of any portion of the neurologic pathway could result in loss of sensation, damage to the dorsal nerves of the penis is the most likely etiology of penile desensitization. Denervation injury does not interfere with erection but affected bulls are unable to breed by natural service and usually cannot be successfully collected with an AV. Semen collection with an electroejaculator for artificial insemination is possible.

An observed test mating utilizing an estrus female in a confined area with adequate footing remains the best method to evaluate function of the dorsal nerves of the penis. Successful intromission and ejaculation definitively demonstrate normal nerve function.[20] In bulls with impairment of sensation to the distal penis, erection is not

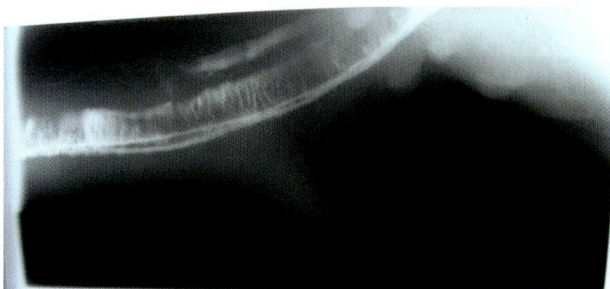

Figure 14.21 Cavernosogram of the penis of a bull demonstrating escape of contrast media into the peri-penile circulation at the level of the distal sigmoid flexure. Courtesy of Dwight Wolfe.

Figure 14.22 Failure of erection of the distal portion of the penis due to occlusion of the cavernous spaces of the CCP.

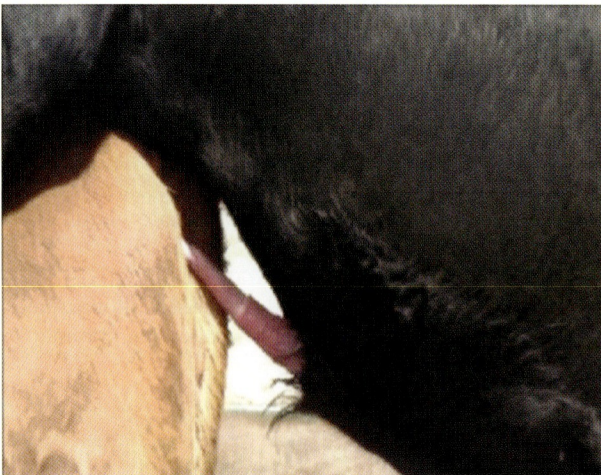

Figure 14.23 Failed attempt at copulation due to denervation injury of the penis. During the test mating, the bull was unsuccessful in locating the vulva with the penis. Note the location of the penis beside the tail head of the cow.

Figure 14.24 Electrodes and stimulator in place for nerve conduction study of dorsal penile nerves. Electrical stimulation applied through the electrode on the free portion of the penis is recorded as the signal passes the proximally placed electrodes.

impaired but the searching motions made with the penis will be ineffective and the bull will be unable to achieve intromission and complete the breeding act. Affected bulls often place the penis to the side of the tail head, on the cow's hip, or below the vulva near the udder (Figure 14.23). Observation of typical signs at the time of test mating coupled with a history of reproductive failure is usually sufficient for diagnosis but definitive electrodiagnostic testing of nerve function[39,42] is available at some veterinary teaching hospitals (Figure 14.24).

References

1. Rupp G, Ball L, Shoop M, Chenoweth P. Reproductive efficiency of bulls in natural service: effects of male to female ratio and single- vs multiple-sire breeding groups. *J Am Vet Med Assoc* 1977;171:639–642.
2. Beckett S, Hudson R, Walker D, Vachon R, Reynolds T. Corpus cavernosum penis pressure and external penile muscle activity during erection in the goat. *Biol Reprod* 1972;7:359–364.
3. Lewis J, Walker D, Beckett S, Vachon R. Blood pressure within the corpus cavernosum penis of the bull. *J Reprod Fertil* 1968;17:155–156.
4. Beckett S. Physiology of erection and ejaculation In: Wolfe DF, Moll HD (eds) *Large Animal Urogenital Surgery*. Baltimore: Williams and Wilkins, 1998, p. 211.
5. Roberts S. Coition or copulation. In: *Veterinary Obstetrics and Gynecology: Theriogenology*, 3rd edn. Woodstock, VT: published by author, 1986, p. 416.
6. Seidel G Jr, Foote R. Motion picture analysis of ejaculation in the bull. *J Reprod Fertil* 1969;20:313–317.
7. Ashdown R. The adherence between the free end of the bovine penis and its sheath. *J Anat* 1960;94:198–204.
8. Ashdown R. Development of penis and sheath in the bull calf. *J Agric Sci* 1960;54:348–352.
9. Ashdown R, Ricketts S, Wardley R. The fibrous architecture of the integumentary coverings of the bovine penis. *J Anat* 1968;103:567–572.
10. Wolfe D, Carson R. Juvenile anomalies of the penis and prepuce: bulls. In: Wolfe DF, Moll HD (eds) *Large Animal Urogenital Surgery*. Baltimore: Williams and Wilkins, 1998, p. 233.
11. Spitzer J, Hopkins F, Webster H, Kirkpatrick F, Hill H. Breeding soundness examination of yearling beef bulls. *J Am Vet Med Assoc* 1988;193:1075–1079.
12. Roberts S. Deviation of the penis. In: *Veterinary Obstetrics and Gynecology: Theriogenology*, 3rd edn. Woodstock, VT: published by author, 1986, p. 795.
13. Elmore R. Surgical repair of bovine persistent penile frenulum. *Vet Med Small Anim Clin* 1981;76:701–704.
14. Anderson D. Surgery of the prepuce and penis. *Vet Clin North Am Food Anim Pract* 2008;24:245–251.
15. Wolfe D, Hudson R, Walker D. Common penile and preputial problems of bulls. *Comp Cont Educ Pract Vet* 1983;5:S447–S456.
16. Long S, Hignett P, Lee R. Preputial eversion in the bull: relationship to penile movement. *Vet Rec* 1970;86:192–194.
17. Memon M, Dawson L, Usenik E, Rice L. Preputial injuries in beef bulls: 172 cases (1980–1985). *J Am Vet Med Assoc* 1988;193:481–485.
18. Desrochers A, St-Jean G, Anderson D. Surgical management of preputial injuries in bulls: 51 cases (1986–1994). *Can Vet J* 1995;36:553–556.
19. Kasari T, McGrann J, Hooper R. Cost-effectiveness analysis of treatment alternatives for beef bulls with preputial prolapse. *J Am Vet Med Assoc* 1997;211:856–859.
20. Wolfe D, Beckett S, Carson R. Acquired conditions of the penis and prepuce (bulls, rams, and bucks). In: Wolfe DF, Moll HD (eds) *Large Animal Urogenital Surgery*. Baltimore: Williams and Wilkins, 1998, pp. 237–272.
21. Parker W, Braun R, Bean B, Hillman R, Larson L, Wilcox C. Avulsion of the bovine prepuce from its attachment to the penile integument during semen collection with an artificial vagina. *Theriogenology* 1987;28:237–256.
22. Kumi-Diaka J, Osori D. Perineal hypospadias in two related bull calves, a case report. *Theriogenology* 1979;11:163–164.
23. Vidal G, Traslavina R, Cabrera C, Gaffney P, Lane V. Theriogenology question of the month. Hypospadias. *J Am Vet Med Assoc* 2011;239:1295–1297.

24. Gilbert R. The diagnosis of short penis as a cause of impotentia coeundi in bulls. *Theriogenology* 1989;32:805–815.
25. Roberts S. *Veterinary Obstetrics and Genital Diseases: Theriogenology*, 3rd edn. Woodstock, VT: published by author, 1986.
26. Bähr C, Distl O. Case report. Diphallus in a German Holstein calf. *Dtsch Tierarztl Wochenschr* 2004;111:85–86.
27. Ashdown R, Gilanpour H. Venous drainage of the corpus cavernosum penis in impotent and normal bulls. *J Anat* 1974;117:159–170.
28. Young S, Hudson R, Walker D. Impotence in bulls due to vascular shunts from the corpus cavernosum penis. *J Am Vet Med Assoc* 1977;171:643–648.
29. Gilbert R, van den Berg S. Communication between the corpus cavernosum penis and the corpus spongiosum penis in a bull diagnosed by modified contrast cavernosography: a case report. *Theriogenology* 1990;33:577–581.
30. Ashdown R, Smith J. The anatomy of the corpus cavernosum penis of the bull and its relationship to spiral deviation of the penis. *J Anat* 1969;104:153–160.
31. Ashdown R, Coombs M. Spiral deviation of the bovine penis. *Vet Rec* 1967;80:737–738.
32. Hopper R, King H, Walters K, Christiansen D. Selected surgical conditions of the bovine penis and prepuce. *Clin Theriogenol* 2012;4:339–348.
33. Roberts S. Miscellaneous causes for loss of libido or inability to copulate. In: *Veterinary Obstetrics and Gynecology: Theriogenology*, 3rd edn. Woodstock, VT: published by author, 1986, pp. 807–808.
34. Beckett S, Reynolds T, Walker D, Hudson R, Purohit R. Experimentally induced rupture of corpus cavernosum penis of the bull. *Am J Vet Res* 1974;35:765–767.
35. Wolfe D, Mysinger P, Hudson R, Carson R. Ventral rupture of the penile tunica albuginea and urethra distal to the sigmoid flexure in a bull. *J Am Vet Med Assoc* 1987;190:1313–1314.
36. Ashdown R, Glossop C. Impotence in the bull: (3) Rupture of the corpus cavernosum penis proximal to the sigmoid flexure. *Vet Rec* 1983;113:30–37.
37. Hudson R, Beckett S, Walker D. Pathophysiology of impotence in the bull. *J Am Vet Med Assoc* 1972;161:1345–1347.
38. Wolfe D, Hudson R, Walker D, Carson R, Powe T. Failure of penile erection due to vascular shunt from corpus cavernosum penis to the corpus spongiosum penis in a bull. *J Am Vet Med Assoc* 1984;184:1511–1512.
39. Wolfe DF, Moll HD. Examination and special diagnostic procedures of the penis and prepuce: bulls, rams, and bucks. In: Wolfe DF, Moll HD (eds) *Large Animal Urogenital Surgery*. Baltimore: Williams and Wilkins, 1998, pp. 225–227.
40. Ashdown R, Gilanpour H, David J, Gibbs C. Impotence in the bull: (2) Occlusion of the longitudinal canals of the corpus cavernosum penis. *Vet Rec* 1979;104:598–603.
41. Beckett S, Hudson R, Walker D. Effect of local anesthesia of the penis and dorsal penile neurectomy on the mating ability of bulls. *J Am Vet Med Assoc* 1978;173:838–839.
42. Mysinger P, Wolfe D, Redding R *et al.* Sensory nerve conduction velocity of the dorsal penile nerves of bulls. *Am J Vet Res* 1994;55:898–900.

REPRODUCTIVE SURGERY

15. Local and Regional Anesthesia for Urogenital Surgery — 131
    Misty A. Edmondson

16. Surgery of the Scrotum and its Contents — 136
    Dwight F. Wolfe

17. Restorative Surgery of the Prepuce — 142
    Dwight F. Wolfe

18. Restorative Surgery of the Penis — 155
    Dwight F. Wolfe

19. Bovine Urolithiasis — 172
    Katharine M. Simpson and Robert N. Streeter

20. Preparation of Teaser Bulls — 181
    Gretchen Grissett

Chapter 15

Local and Regional Anesthesia for Urogenital Surgery

Misty A. Edmondson

Department of Clinical Sciences, College of Veterinary Medicine, Auburn University, Auburn, Alabama, USA

Introduction

Local and regional analgesia are routinely used in cattle because they are considered safe and effective. There are many advantages of local and regional analgesia over general anesthesia. These benefits include the need for minimal equipment (syringes, needles, and anesthetic drug), lower risk of toxic effects especially in debilitated animals, and reduction in the risk of bloat, regurgitation, and nerve or muscle damage that may associated with animals placed in recumbency. Many surgical procedures can be performed safely and humanely using a combination of physical restraint, mild sedation or tranquilization, and local or regional anesthesia. Local anesthetic techniques are usually simple, inexpensive, and provide a reversible loss of sensation to a relatively well-defined area of the body with minimal effects on the rest of the body. The most common techniques used in bovine reproduction include infiltration anesthesia, nerve block anesthesia, and epidural anesthesia.

Local anesthetics

Many local or regional anesthetic drugs are available that can produce reversible loss of autonomic, motor, and sensory function with acceptable onset times and predictable duration.[1] These drugs vary in their potency, toxicity, and cost.[2] Lidocaine hydrochloride 2% and mepivacaine hydrochloride 2% have become two of the most widely used local anesthetic agents in cattle due to limited toxicity and low cost. Lidocaine hydrochloride has a duration of 90–180 min, is three times more potent that procaine, and diffuses into tissues more widely.[1,3] The addition of a vasoconstrictor such as epinephrine (5 μg/mL) to the local anesthetic solution (0.1 mL epinephrine 1 in 1000 added to 20 mL of local anesthetic) increases the potency and duration of activity of both regional and epidural anesthesia. However, local anesthetics containing epinephrine 1 in 200 000 should not be used in wound edges or in the subarachnoid space due to the risk of causing tissue necrosis and spinal cord ischemia.[1]

Anesthesia for laparotomy

Anesthesia of the paralumbar fossa and abdominal wall can be achieved by several techniques, including infusion of the incision or line block, the inverted L block, the proximal paravertebral nerve block, and the distal paravertebral nerve block. These anesthetic techniques are commonly used for reproductive procedures such as cesarean section, ovariectomy, and cervicopexy.

Line block

Infusion of local anesthetic into the incision site or a line block may be used to desensitize a selected area of the paralumbar fossa. An 18-gauge 3.8-cm needle is used to infuse multiple small injections of 10 mL of local anesthetic solution subcutaneously and into the deep muscle layers and peritoneum. Pain of successive injections may be alleviated by placing the edge of the needle into the edge of the previously desensitized area at an angle of approximately 20–30°.[2] In heavily muscled or overweight cattle, it may be necessary to use an 18-gauge 7.5-cm needle to penetrate through the large amount of subcutaneous fat to reach the deep muscle layers. The amount of local anesthetic needed to acquire adequate anesthesia depends on the size of the area to be desensitized. Adult cattle weighing 450 kg can safely tolerate 250 mL of a 2% lidocaine hydrochloride solution.[2] Delayed healing of the incision site is a possible complication of infiltration of local anesthetic at the surgical site.

Bovine Reproduction, First Edition. Edited by Richard M. Hopper.
© 2015 John Wiley & Sons, Inc. Published 2015 by John Wiley & Sons, Inc.

Inverted L

The inverted L block is a nonspecific regional block that locally blocks the tissue bordering the caudal aspect of the thirteenth rib and the ventral aspect of the transverse processes of the lumbar vertebrae.[4] An 18-gauge 3.8-cm needle is used to inject up to a total of 100 mL of local anesthetic solution in multiple small injection sites into the tissues bordering the dorsocaudal aspect of the thirteenth rib and ventrolateral aspect of the transverse processes of the lumbar vertebrae (Figure 15.1). This creates an area of anesthesia under the inverted L block. Advantages of the inverted L block are that it is simple to perform, does not interfere with ambulation, and deposition of anesthetic away from the incision site minimizes incisional edema and hematoma.[1] Disadvantages include incomplete analgesia and muscle relaxation of the deeper layers of the abdominal wall (particularly in obese animals), possible toxicity after larger doses of anesthetic, and increased cost because of larger doses of local anesthetic.[1]

Proximal paravertebral

The proximal paravertebral nerve block desensitizes the dorsal and ventral nerve roots of the last thoracic (T13) and first and second lumbar (L1 and L2) spinal nerves as they emerge from the intervertebral foramina. To facilitate proper needle placement of anesthetic, the skin at the cranial edges of the transverse processes of L1, L2, and L3, and at a point 2.5–5 cm off the dorsal midline can be desensitized by injecting 2–3 mL of local anesthetic using an 18-gauge 2.5-cm needle. A 14-gauge 2.5-cm needle is used as a cannula or guide needle to minimize skin resistance during insertion of an 18-gauge 10- to 15-cm spinal needle. Approximately 5 mL of local anesthetic may be placed through the cannula to anesthetize the needle tract for further needle placement.

To desensitize T13, the cannula needle is placed through the skin at the anterior edge of the transverse process of L1 at approximately 4–5 cm lateral to the dorsal midline. The 18-gauge 10- to 15-cm spinal needle is passed ventrally until it contacts the transverse process of L1. The needle is then walked off the cranial edge of the transverse process of L1 and advanced approximately 1 cm to pass slightly ventral to the process and into the intertransverse ligament. A total of 6–8 mL of local anesthetic is injected with little resistance to desensitize the ventral branch of T13. The needle is then withdrawn 1–2.5 cm above the fascia or just dorsal to the transverse process and 6–8 mL of local anesthetic is infused to desensitize the dorsal branch of the nerve.

To desensitize L1 and L2, the needle is inserted just caudal to the transverse processes of L1 and L2. The needle is walked off the caudal edges of the transverse processes of L1 and L2, at a depth similar to the injection site for T13, and advanced approximately 1 cm to pass slightly ventral to the process and into the intertransverse ligament. A total of 6–8 mL of local anesthetic is injected with little resistance to desensitize the ventral branches of the nerves. The needle is then withdrawn 1–2.5 cm above the fascia or just dorsal to the transverse processes and 6–8 mL of local anesthetic is infused to desensitize the dorsal branch of the nerves (Figure 15.2).

Evidence of a successful proximal paravertebral nerve block includes increased temperature of the skin; analgesia of the skin, muscles, and peritoneum of the abdominal wall of the paralumbar fossa; and scoliosis of the spine toward the desensitized side. Advantages of the proximal paravertebral nerve block include small doses of anesthetic, wide

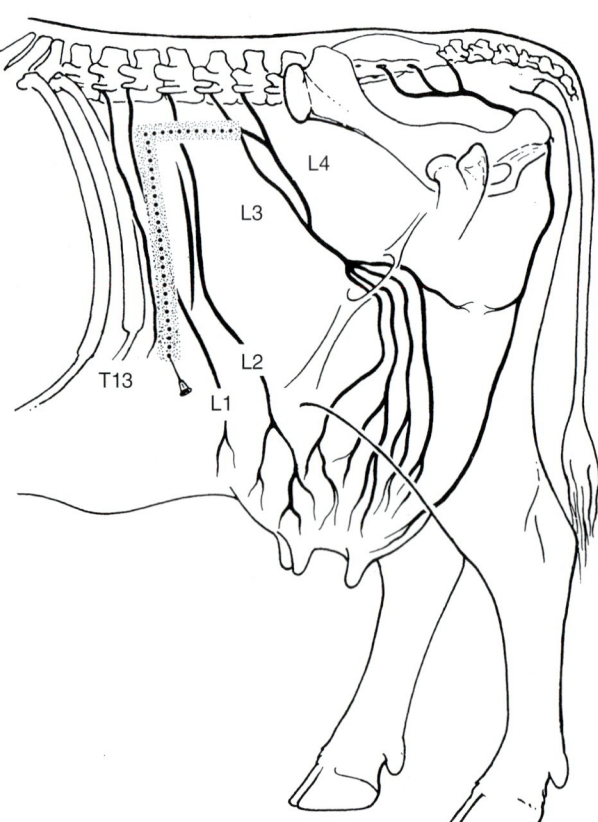

Figure 15.1 Inverted L. With permission from *Lumb and Jones' Veterinary Anesthesia*.

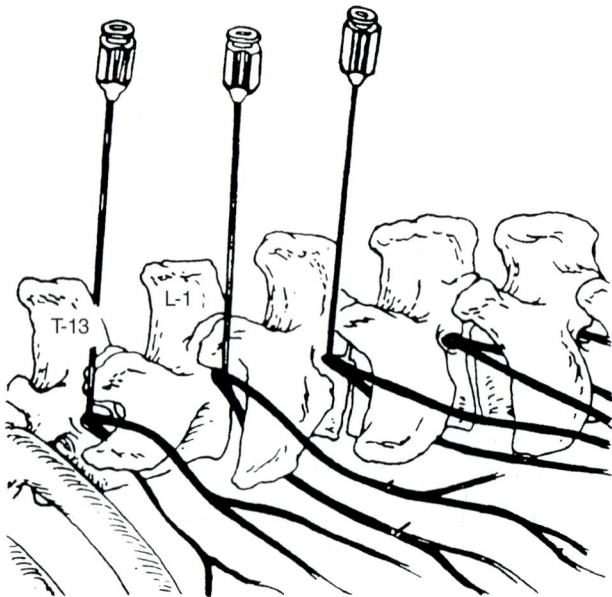

Figure 15.2 Proximal paravertebral. With permission from Skarda R. Techniques of local analgesia in ruminants and swine. *Vet Clin North Am Food Anim Pract* 1986;2:631.

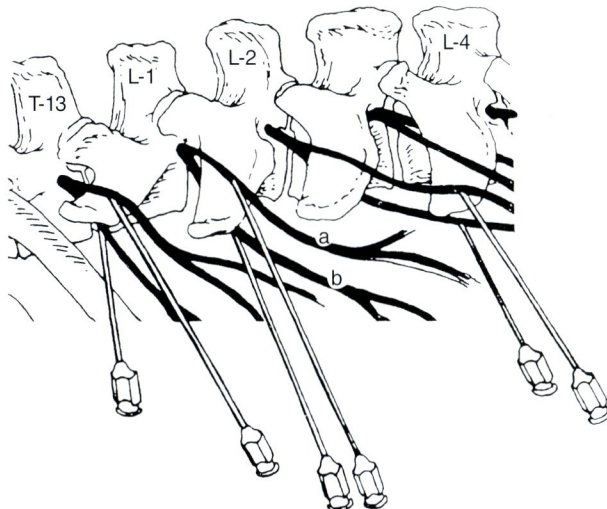

Figure 15.3 Distal paravertebral. With permission from Skarda R. Techniques of local analgesia in ruminants and swine. *Vet Clin North Am Food Anim Pract* 1986;2:635.

and uniform area of analgesia and muscle relaxation, decreased intra-abdominal pressure, and absence of local anesthetic at the margins of the surgical site. Disadvantages of the proximal paravertebral nerve block include scoliosis of the spine, which may make closure of the incision more difficult, difficulty in identifying landmarks in obese and heavily muscled animals, and more skill or practice required for consistent results.[2-4]

Distal paravertebral

The distal paravertebral nerve block desensitizes the dorsal and ventral rami of the spinal nerves T13, L1, and L2 at the distal ends of the transverse processes of L1, L2, and L4, respectively. An 18-gauge 3.5- to 5.5-cm needle is inserted ventral to the transverse process and 10 mL of local anesthetic is infused in a fan-shaped pattern. The needle can then be removed completely and reinserted or redirected dorsal to the transverse process, in a caudal direction, where 10 mL of local anesthetic is again infused in a fan-shaped pattern. This procedure is repeated for the transverse processes of the second and fourth lumbar vertebrae (Figure 15.3). Advantages of the distal paravertebral nerve block compared with the proximal paravertebral nerve block include lack of scoliosis, easier performance, and more consistent results. Disadvantages of the distal paravertebral nerve block compared with the proximal paravertebral nerve block include larger doses of anesthetic and variations in efficiency caused by differences in anatomic pathways of the nerves.[2-4]

Anesthesia of the perineum

Anesthesia of the perineum is routinely performed for obstetric procedures and for relief of rectal tenesmus in the cow. Many urogenital surgeries in cows (i.e., replacement of vaginal/cervical prolapse) and bulls/steers (ischial urethrostomy, penile amputation) also require anesthesia of the perineum.

Caudal epidural anesthesia

Caudal epidural anesthesia is an easy and inexpensive method of analgesia that is commonly used in cattle. A high caudal epidural at the sacrococcygeal space (S5–Co1) desensitizes sacral nerves S2, S3, S4, and S5. The low caudal epidural at first coccygeal space (Co1–Co2) desensitizes sacral nerves S3, S4, and S5; as the anesthetic dose increases, nerves cranial to S2 may also become affected.[5] If possible, the hair should be clipped and the skin scrubbed and disinfected.

Standing alongside the cow, the tail should be moved up and down to locate the fossa between the last sacral vertebra and the first coccygeal vertebra (first freely moveable space) or between the first and second coccygeal vertebrae. An 18-gauge 3.8-cm needle (with no syringe attached) is directed perpendicular to the skin surface. Once the skin is penetrated, place a drop of local anesthetic solution in the hub of the needle (hanging drop technique). The needle should then be advanced slowly until the anesthetic solution is drawn into the epidural space by negative pressure. The syringe may then be attached to the needle and anesthetic solution slowly injected with no resistance (Figure 15.4a). The dose of local anesthetic to be used is 0.5 mL per 45 kg body weight.

Continuous caudal epidural anesthesia

Continuous caudal epidural anesthesia is used in cattle with chronic rectal and vaginal prolapse that experience continuous straining after the initial epidural. This procedure is performed by placing a catheter into the epidural space for intermittent administration of local anesthetic. A 17-gauge 5-cm spinal needle (Tuohy needle) with stylet in place is inserted into the epidural space at Co1 to Co2 with the bevel directed craniad. The stylet is removed, and 2 mL of local anesthetic is injected to determine if the needle is in the epidural space. A catheter is inserted into the needle and advanced cranially for 2–4 cm beyond the needle tip. The needle is then withdrawn while the catheter remains in place (Figure 15.4b). An adapter is placed on the end of the catheter and the catheter secured to the skin on the dorsum. Local anesthetic solution may then be administered as needed.[2]

More recently, α_2-adrenergic agonists and opioids either alone or in combination with local anesthetic solutions have been used for epidural anesthesia. Epidural administration of the α_2-agonist xylazine hydrochloride (0.05 mg/kg) diluted in 5–12 mL of sterile saline or xylazine hydrochloride (0.3 mg/kg) added to 5 mL of 2% lidocaine hydrochloride offer similar anesthesia to lidocaine. Although the duration of anesthesia is prolonged (4–5 hours) using these combinations, systemic effects (sedation, salivation, ataxia) may also occur.[1] Epidural administration of opioids, such as morphine (0.1 mg/kg) diluted in 20 mL of sterile saline, is used to provide analgesia for a prolonged period (approximately 12 hours) without interfering with motor function. Disadvantages of using opioids for epidural anesthesia are that the analgesia is not as potent as lidocaine and the maximum effect of a morphine epidural may not occur for 2–3 hours. Caudal epidural administration of morphine may be used to help alleviate pain in the perineal area and straining.[6]

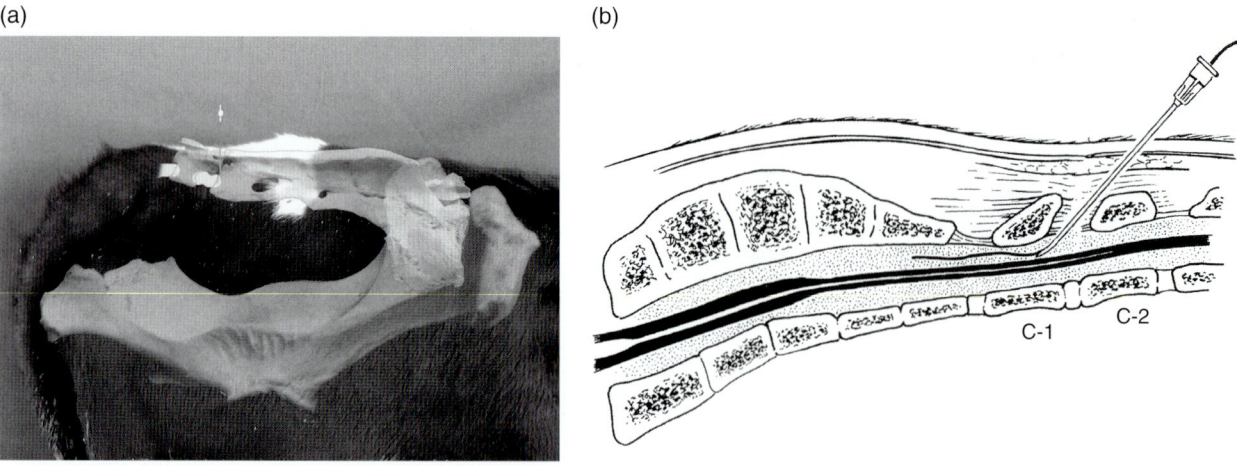

Figure 15.4 (a) Needle placement for epidural. Image courtesy of Dr Douglas Hostetler. (b) Catheter placement for continuous flow epidural. Adapted with permission from *Lumb and Jones' Veterinary Anesthesia*.

Sacral paravertebral anesthesia

Sacral paravertebral anesthesia is used to relieve rectal tenesmus associated with rectal prolapse without affecting the sciatic nerve and function of the tail or the animal's ability to stand. The sacral paravertebral nerve block is used to provide analgesia to the pudendal nerve (pudic nerve), medial hemorrhoidal nerve (pelvic splanchnic nerve), and caudal hemorrhoidal nerve (caudal rectal nerve) by blocking S3, S4, and S5 as they branch off the spinal cord thereby providing analgesia to the anus, vulva, and vagina.[1,7] In bulls, S3 supplies motor function to the retractor penis muscles. Physical restraint in a squeeze chute and/or sedation may be beneficial in order to prevent lateral movement of the animal during the procedure. In addition, a caudal epidural may be helpful if the animal is fractious. The skin over the dorsal sacrum should be clipped of hair and surgically prepared for the procedure. The paired S5 foramina are 1–2 cm lateral to the sacral coccygeal joint. The S4 foramina are about 3–4 cm cranial and more lateral to the S5 foramina. The S3 foramina are an additional 3–4 cm cranial to the S4 foramina. A stab incision can be made dorsal to each foramen to aid in the introduction of a 5- to 7-cm 18-gauge needle. The foramina can be palpated rectally with a finger placed in or over the ring which allows for identification of the foramen and ensures correct needle placement. Once the needle has entered the osseous ring, inject 2–3 mL of lidocaine hydrochloride; this should be repeated for each foramen.[7] The use of a lidocaine/alcohol mixture has also been described to manage tenesmus following chronic cervicovaginal prolapse or rectal prolapse. A mixture of 1 mL of 2% lidocaine hydrochloride and 2 mL of 95% ethyl alcohol has been used effectively.[7]

Anesthesia of the penis and prepuce

Desensitization of the internal pudendal nerve block

The procedure for bilateral internal pudendal (pudic) nerve block was first described by Larson[8] to facilitate relaxation of the bull's penis without causing locomotor impairment. The internal pudendal nerve block can be used in the

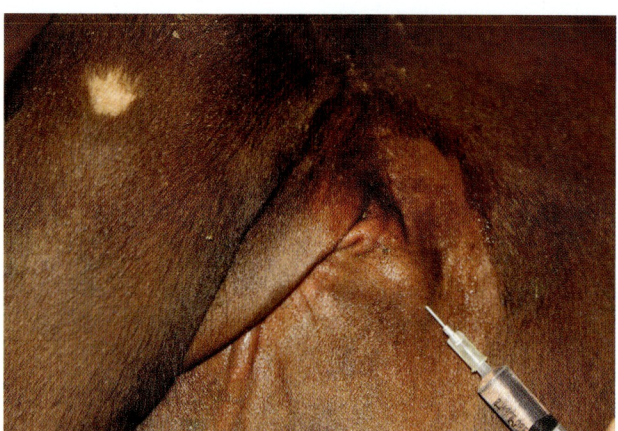

Figure 15.5 Injection of 2 mL local anesthetic in the skin at the ischiorectal fossa.

standing bull for penile relaxation and analgesia distal to the sigmoid flexure and examination of the penis. In the standing female, the internal pudendal nerve block can be used to relieve straining caused by chronic vaginal prolapse. This technique may also be used for surgical procedures of the penis, such as repair of prolapses, removal of peri-penile tumors, removal of penile papillomas or warts, and other minor surgeries of the penis and prepuce.

This procedure involves desensitizing the internal pudendal nerve and the anastomotic branch of the middle hemorrhoidal nerve using an ischiorectal approach. The internal pudendal nerve consists of fibers originating from the ventral branches of the third and fourth sacral nerves (S3 and S4) and the pelvic splanchnic nerves. The skin at the ischiorectal fossa on both sides is clipped, disinfected, and desensitized with approximately 2 mL of local anesthetic (Figure 15.5). A 14-gauge 1.25-cm needle is inserted through the desensitized skin at the ischiorectal fossa to serve as a cannula. An 18-gauge 10- to 15-cm spinal needle is then directed through the cannula to the pudendal nerve. The operator's left hand is placed into the rectum to the level of the wrist and the fingers directed laterally and ventrally to identify the lesser sacrosciatic foramen. The lesser sciatic

Local and Regional Anesthesia for Urogenital Surgery

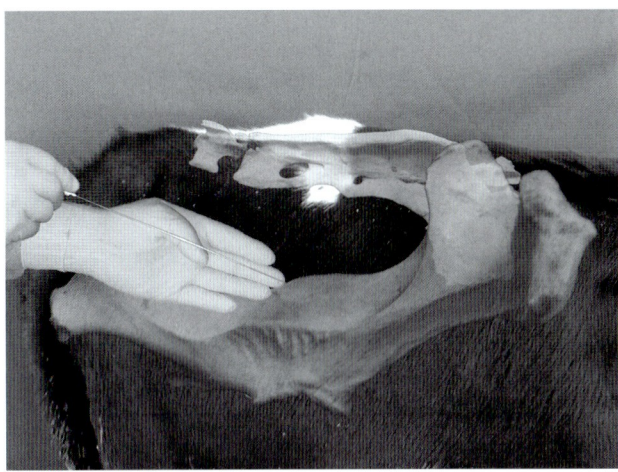

Figure 15.6 Needle placement for pudendal nerve block. Image courtesy of Dr Douglas Hostetler.

foramen is first identified by rectal palpation as a soft depression in the sacrosciatic ligament. The internal pudendal nerve can be readily identified lying on the ligament immediately cranial and dorsal to the foramen and approximately one finger's width dorsal to the pudendal artery passing through the foramen. The internal pudendal artery can be readily palpated a finger's width ventral to the nerve. The spinal needle is held in the operator's right hand and introduced through the cannula in the ischiorectal fossa. The spinal needle is directed medial to the sacrosciatic ligament and directed cranioventrally (Figure 15.6). The needle is not felt until it has been introduced approximately 5–7 cm and can then be repositioned to the nerve. Once at the pudendal nerve, 20 mL of local anesthetic is deposited at the nerve. The needle is then partially withdrawn and redirected 2–3 cm more caudodorsally where an additional 10 mL of local anesthetic is deposited at the cranial aspect of the foramen to desensitize the muscular branches and the middle hemorrhoidal nerve. The needle is then removed and the sites of deposition are massaged to aid in dispersal of the local anesthetic. This procedure is then repeated on the opposite side of the pelvis. Relaxation of the penis varies and may take as long as 30–40 min for full effect. Effectiveness of the block can be assessed by firmly squeezing the tail of the epididymis of each testicle. The bull's inability to lift or retract the testicle signifies adequate analgesia. The duration of the internal pudendal nerve block lasts 2–4 hours.[7]

Desensitization of the dorsal nerve of the penis

The dorsal nerve of the penis may be desensitized at a location just proximal to the surgical site. With the bull restrained, the penis should be manually extended and a towel clamp should then be placed under the dorsal apical ligament. Alternatively, a gauze tourniquet may be placed around the free portion of the penis to aid in penile extension. With the dorsal aspect of the penis thoroughly cleansed, 2–4 mL of 2% lidocaine hydrochloride should be infused subcutaneously across the dorsum of the penis proximal to the lesion.[9]

Alternatively, the dorsal nerve of the penis may also be desensitized as it passes over the ischial arch for penile anesthesia and relaxation. The skin associated with the penile body and located 10 cm ventral to the anus and 2.5 cm from midline is infiltrated with 2–4 mL of 2% lidocaine hydrochloride using a small gauge needle (22–25 gauge). A 20-gauge 4-cm needle is then inserted through the desensitized skin and advanced for 5–7 cm to contact the pelvic floor. Aspiration ensures that the needle is not in the dorsal artery of the penis. The needle is then withdrawn approximately 1 cm and the area infiltrated with 20–30 mL of 2% lidocaine hydrochloride. The procedure is then repeated on the opposite side of the penis. Analgesia and paralysis of the penis will occur within 20 min and should last for 1–2 hours.[1]

Anesthesia for castration

Castration of bulls is a very common surgical procedure in general practice. Historically, castration was often performed with minimal or no anesthesia. However, anesthesia for castration is more commonly practiced because calves benefit from anesthesia with improved feed consumption and rate of gain. Depending on the age and size of the animal, the surgery is usually performed with chemical and/or regional anesthesia (scrotum and testicles). Depending on the size of the calf, the proposed line of incision for removal of the distal aspect of the scrotum should be subcutaneously infiltrated with 5–10 mL of 2% lidocaine hydrochloride. In bulls, an 18-gauge 3.8-cm needle is inserted at an angle (30–45°) into the center of the testicle and 10–15 mL of local anesthetic per 200 kg body weight is injected into the parenchyma of each testicle. The anesthetic quickly enters the lymphatics and desensitizes the sensory fibers in the spermatic cord. For smaller animals or calves, a smaller needle (20 gauge, 2.5 cm) may be used to administer 2–10 mL of 2% lidocaine hydrochloride.[1]

References

1. Skarda R. Local and regional anesthetic techniques: ruminants and swine. In: Tranquilli WJ, Thurmon JC, Grimm KA (eds) *Lumb and Jones' Veterinary Anesthesia and Analgesia*, 4th edn. Oxford: Wiley-Blackwell, 2007, pp. 731–746.
2. Skarda R. Techniques of local analgesia in ruminants and swine. *Vet Clin North Am Food Anim Pract* 1986;2:621–663.
3. Edwards B. Regional anesthesia techniques in cattle. *In Practice* 2001;23:142–149.
4. Noordsy J, Ames N. Local and regional anesthesia. In: Noordsy J, Ames N (eds) *Food Animal Surgery*, 4th edn. Yardley, PA: Veterinary Learning Systems, 2006, pp. 21–42.
5. Noordsy J, Ames N. Epidural anesthesia. In: Noordsy J, Ames N (eds) *Food Animal Surgery*, 4th edn. Yardley, PA: Veterinary Learning Systems, 2006, pp. 43–55.
6. Navarre C. Numbing: nose to tail. In: *Proceedings from the 39th Annual Convention of the American Association of Bovine Practitioners*, 2006, pp. 53–55.
7. Hopper R, King H, Walters K, Christiansen D. Management of urogenital surgery and disease in the bull: the scrotum and its contents. *Clin Theriogenol* 2012;4:332–338.
8. Larson L. The internal pudendal (pudic) nerve block. *J Am Vet Med Assoc* 1953;123:18–27.
9. Wolfe D, Beckett S, Carson R. Acquired conditions of the penis and prepuce. In: Wolfe DF, Moll HD (eds) *Large Animal Urogenital Surgery*, 2nd edn. Baltimore: Williams and Wilkins, 1998, pp. 237–272.

Chapter 16

Surgery of the Scrotum and its Contents

Dwight F. Wolfe

Department of Clinical Sciences, College of Veterinary Medicine, Auburn University, Auburn, Alabama, USA

Normal scrotum and testes

The scrotum of the bull is a dependent appendage of the ventral abdominal skin that supports and protects the testicles and helps regulate testicular temperature. The testis is an abdominal organ maintained 2–7 °C cooler than body temperature.[1] Testicular thermoregulation is a complex process that includes contraction of the tunica dartos within the scrotal wall to alter scrotal surface area in concert with the cremaster muscles to regulate the distance of the testicles from the body. The scrotum is one of the few places in the bovine where sweat glands are found. Blood temperature is regulated by counter-current heat exchange in the pampiniform plexus just proximal to the testicle between the testicular artery and vein. These mechanisms collectively function to maintain testicular temperature 4–6 °C below core body temperature, which is optimal for normal semen production.[1,2]

Diagnosis of testicular disease or injury

Bulls with scrotal or testicular disease or injury frequently have scrotal enlargement, which may be unilateral or bilateral and can be caused by a variety of conditions.[2,3] It is important to use both visual and palpable information to determine the source of the swelling and make a definitive diagnosis.

The most common cause of scrotal enlargement in the bull is fluid accumulation within the vaginal cavity.[3] Fluid accumulation is usually unilateral and may be due to periorchitis, hydrocele, or hematocele. Appreciable fluid accumulation is readily detectable by palpation of the testis; the consistency of the fluid may be thin and easily displaced by palpation, or thicker (e.g., purulent material or clotted blood) which may be confirmed by ultrasound. Additionally, fibrinous or fibrous adhesions may form between the testis or epididymis and parietal vaginal tunic and these may be detected by palpation as the testis is moved within the vaginal cavity.

Orchitis is usually subclinical but rarely may range to severe and perhaps suppurative.[4,5] The dense fibrous tunica albuginea surrounding the testicle limits this organ's potential for enlargement and therefore there is usually not significant testicular swelling. Although orchitis may affect one or both testes, greater than 25% size difference between the two testes should be considered abnormal. Orchitis in the bovine is usually due to hematogenous spread of bacteria but in rare cases may be due to puncture wounds through the scrotum. Occasionally, traumatic testicular rupture may occur and is diagnosed by an amorphous clot-like consistency by palpation and confirmed by ultrasound.

Epididymitis occurs more commonly than orchitis in bulls and may be diagnosed by thorough examination of the testes and scrotal contents.[6-8] The condition is often unilateral in the bull and may present as a swollen painful epididymal tail in the acute phase of the disease. Chronic epididymitis usually results in epididymal tails that are small and firm, and infertility is often caused by their eventual obstruction. Periorchitis – inflammation of the peritesticular tissues with fluid accumulation and adhesions – frequently accompanies epididymitis or orchitis.

Trauma is a common cause of pathological changes in scrotal contents. Trauma to the scrotum may produce hematocele (hemorrhage into the vaginal cavity) or hydrocele (accumulation of inflammatory exudate or transudate in this cavity).[3] Trauma severe enough to damage the tubular integrity of the testicle or epididymis may lead to an autoimmune reaction due to the extravasated haploid sperm that produce granulomatous reactions.

Swelling of the wall of the scrotum can usually be differentiated from other causes of scrotal enlargement by palpation of the thickened scrotal skin, presence of pitting edema, with confirmation by ultrasound (Figure 16.1). Generalized swelling of the scrotal wall may be caused by infection with

---

Bovine Reproduction, First Edition. Edited by Richard M. Hopper.
© 2015 John Wiley & Sons, Inc. Published 2015 by John Wiley & Sons, Inc.

Surgery of the Scrotum and its Contents 137

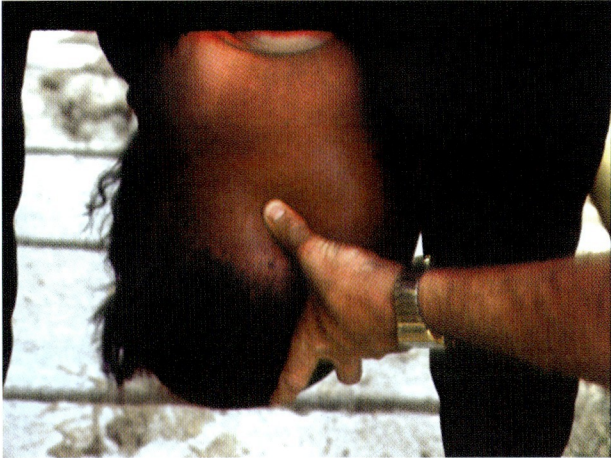

Figure 16.1 Pitting edema of the scrotal wall.

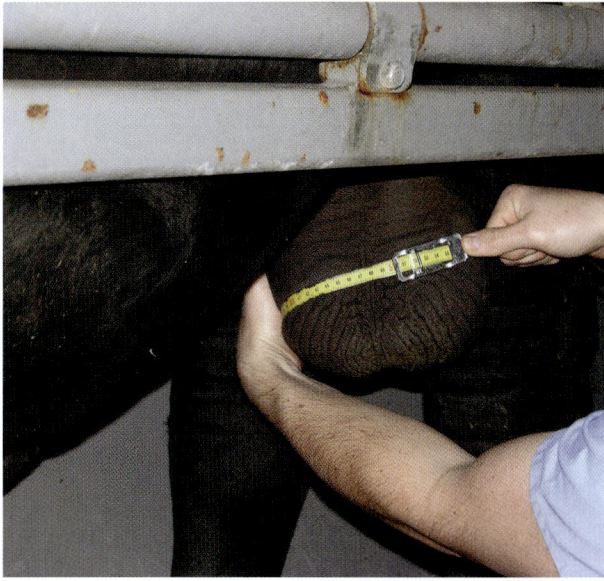

Figure 16.3 Scrotal circumference measurement.

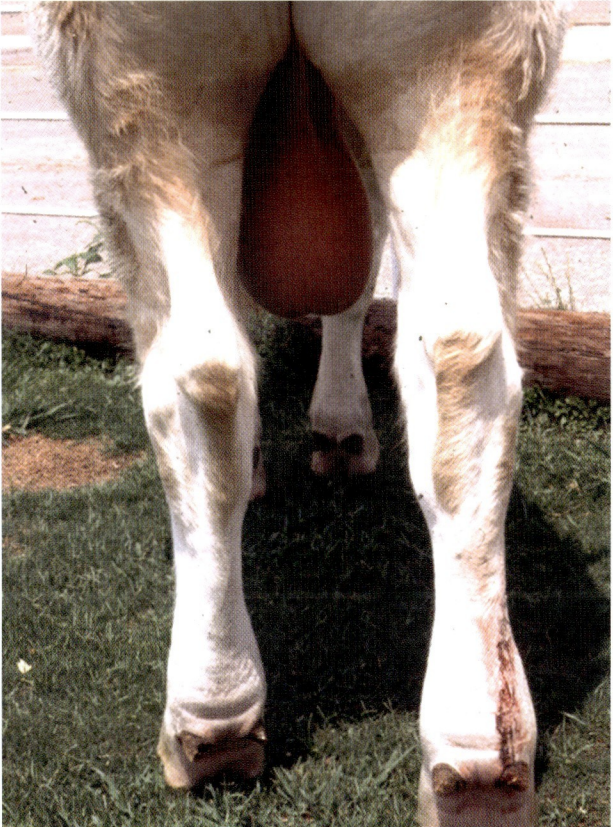

Figure 16.2 Generalized edema of scrotal wall due to *Mycoplasma weyenoii*.

Physical examination of the scrotum and testicles

Carefully examine the scrotum for dermatitis, edema, scar tissue, and symmetry.[2] Palpate the testicles for relative size, firmness, symmetry, evidence of pain or swelling, presence of fluid in the vaginal cavity, and the ability of the testicles to move freely within the vaginal cavity. There should be no more than 10% difference in the size of the testes and normal testicular tone approximates that of liver. Scrotal circumference is heritable and highly correlated with daily sperm output, sperm reserves, serving capacity, and age of puberty of the bull's offspring. Measure the scrotal circumference with a nonelastic tape at the widest circumference of the scrotum (Figure 16.3). This measurement can be compared against normal values that are readily available in tables for scrotal circumference in different age bulls. However, the recommended minimum scrotal circumferences from the Society for Theriogenology bull breeding soundness evaluation serve as an excellent reference:[12]

- 30 cm at >12 to <15 months of age
- 31 cm at >15 to <18 months of age
- 32 cm at >18 to <21 months of age
- 33 cm at >21 to <24 months of age
- 34 cm at >24 months of age.

Ultrasound of the testicles

Diagnosis of scrotal or testicular disease may be aided by B-mode real-time ultrasound using a 5-MHz probe. Normal testicles are homogeneous and moderately echogenic[3] (Figure 16.4). The mediastinum testis is a readily identifiable hyperechoic area in the center of the testicle when viewed in the transverse plane or a hyperechoic line when viewed in the sagittal plane (Figure 16.5). The head, body and tail of the epididymis are less echogenic than the testicle

Mycoplasma wenyonii[9,10] (Figure 16.2). This condition is not treated surgically but should be considered when managing scrotal wall thickness. Crushing or blunt trauma to the scrotal wall may disrupt the normal muscular and vascular architecture of the scrotal wall with resultant increased wall thickness. If these injuries are severe and unilateral, removal of the affected half of the scrotum and its associated testicle may be warranted.

Primary testicular tumors are not common in bulls but unilateral or bilateral scrotal enlargement has been reported in bulls with mesothelioma.[4,5,9,11]

and are readily identified as they course along the testicle. Thickness of the scrotal skin and vaginal tunics and the presence of fluid within the vaginal cavity are readily determined.

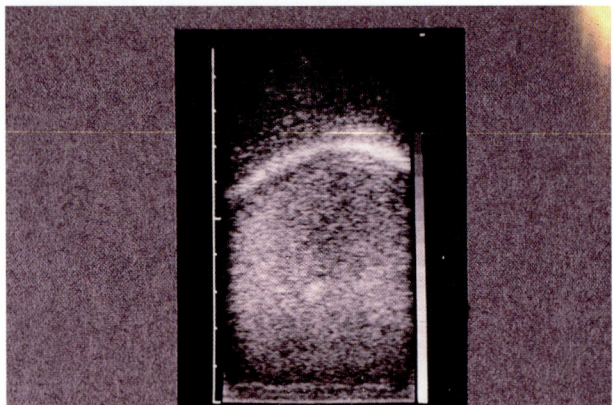

Figure 16.4 Ultrasound of normal testis in transverse plane. Mediastinum testis is hyperechoic area in center of testis.

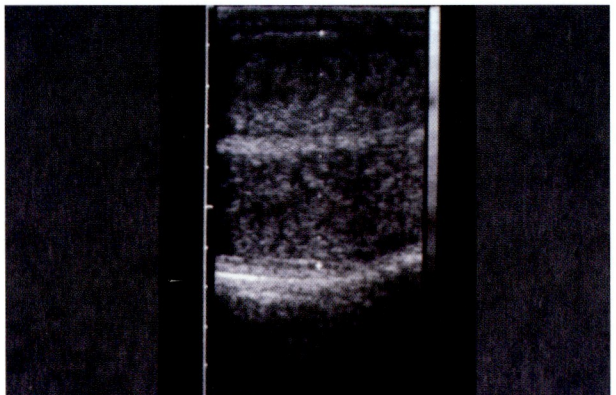

Figure 16.5 Sagittal ultrasound of normal bovine testicle. Mediastinum testis is hyperechoic line in center of testis.

Thermography of the scrotum

Testicular temperature must be below body temperature for normal spermatogenesis in bulls. Infrared thermography provides a noninvasive noncontact imaging technique to determine normal and abnormal thermal patterns in bulls and other animals.[11] Thermography utilizes sensitive infrared imaging to produce a color photograph that portrays variations in surface temperature that reflect the temperature of tissues immediately beneath the skin. The normal thermogram of the scrotum in all species is a left and right symmetric pattern with a constant decrease in temperature gradient from the base to the apex (Figure 16.6). In bulls a temperature gradient of 4–6 °C from base to apex is considered normal. A gradual decrease from base to apex with concentric color bands is consistent with normal function of the vascular counter-current heat-exchange mechanism of the testes. An excessively cool or excessively warm area as evidenced by thermographic color is consistent with testicular disease or injury[13,14] (Figure 16.7).

Semen evaluation

It is beyond the scope of this chapter to thoroughly discuss collection and evaluation of bull semen. Semen from bulls with significant scrotal or testicular pathology generally does not meet the minimum 70% morphologically normal sperm of which more than 30% are progressively motile, and these bulls are thus considered subfertile.

Surgical repair of testicular problems

The only conditions of the bull scrotum readily amenable to surgical repair are those that involve only one testicle.[13] When one testicle is inflamed, the heat thus generated affects the contralateral testicle and the reversibility of degenerative change depends on the severity and duration of the insult.[4] Valuable animals with unilateral orchitis,

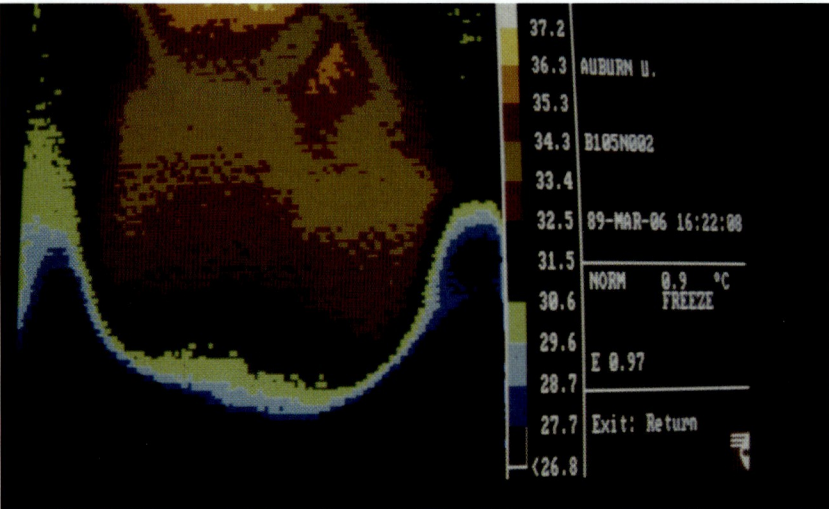

Figure 16.6 Normal thermographic pattern of bull scrotum.

Surgery of the Scrotum and its Contents 139

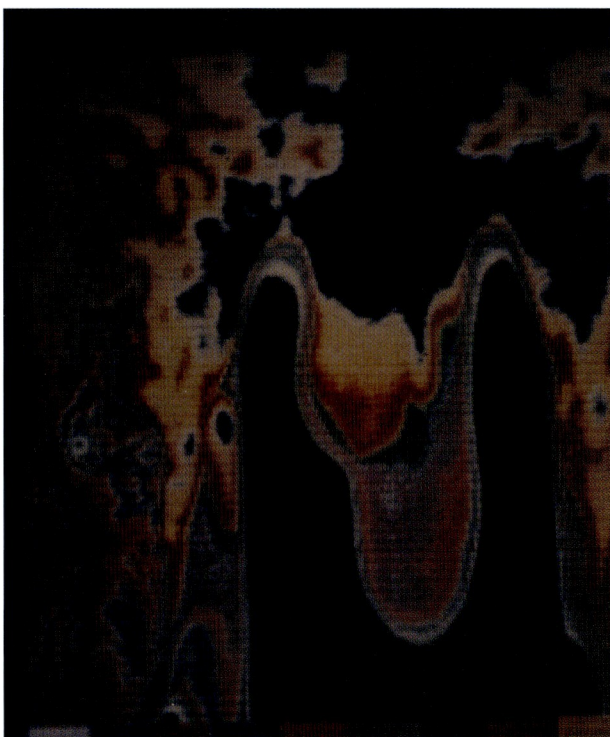

Figure 16.7 Thermograph of scrotum of bull with left inguinal hernia. Left side of scrotum is markedly warmer than right.

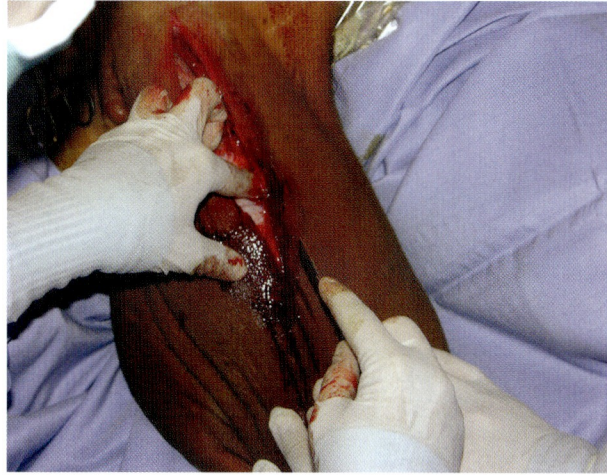

Figure 16.8 Vertical skin incision on lateral aspect of scrotum.

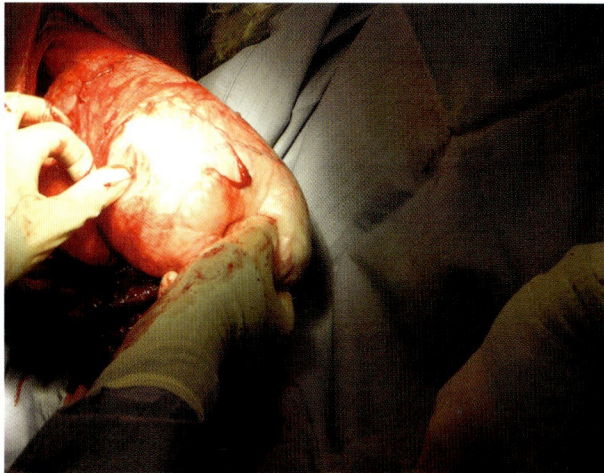

Figure 16.9 Testicle within parietal vaginal tunic bluntly dissected from scrotal fascia.

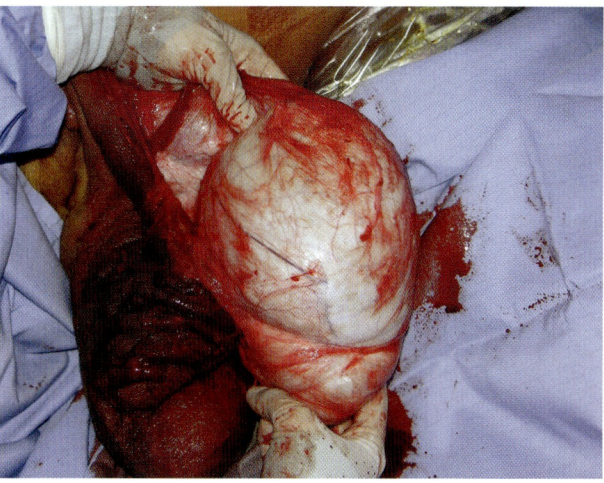

Figure 16.10 Parietal tunic incised to expose testicle within vaginal tunic.

periorchitis, testicular rupture, or crushing of the scrotal wall may be able to resume breeding soundness with unilateral castration.

Although unilateral castration is not an emergency procedure, the prognosis for return to fertility improves if surgery is performed early in the disease process. Early intervention can reduce further insult to the contralateral testicle generated by inflammation, with less likelihood that the testicle will undergo irreversible degeneration. Most bulls will return to fertility following surgery if the remaining testicle is not severely compromised.[15–17] Compensatory hypertrophy in the normal testis should allow the bull to subsequently produce up to 75% of normal sperm capacity. Therefore, advise owners that restrictions may be necessary following surgery. Additionally, although a bull may be productive in a herd situation, a unilaterally castrated bull will not pass a standardized breeding soundness evaluation and is therefore ineligible for certain shows or sales.

Restrain the bull in lateral recumbency under anesthesia or heavy sedation and local anesthesia of the scrotum and spermatic cord. Clip and prepare the scrotum for aseptic surgery.[15,18] On the lateral aspect of the scrotum make a vertical skin incision approximately the length of the testicle, preserving the parietal vaginal tunic (Figure 16.8). Bluntly dissect the affected testicle within the parietal vaginal tunic from the surrounding scrotal fascia (Figure 16.9). If the condition is chronic, variable amounts of fibrous tissue and adhesions of the parietal and visceral vaginal tunics may be encountered. Make a vertical incision through the parietal tunic sufficiently long to remove the testis, and exteriorize the testicle and spermatic cord (Figures 16.10 and 16.11).

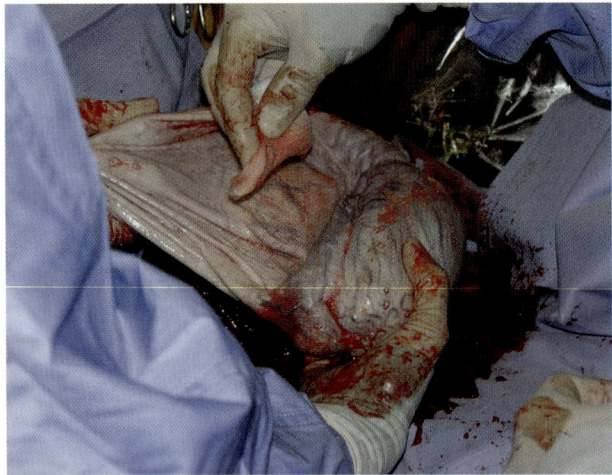

Figure 16.11 Parietal vaginal tunic incised and testicle exposed.

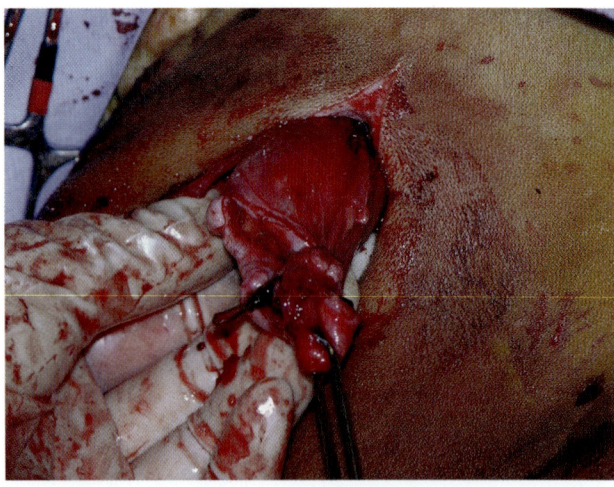

Figure 16.13 Transected stump of vaginal tunic following inverting suture closure.

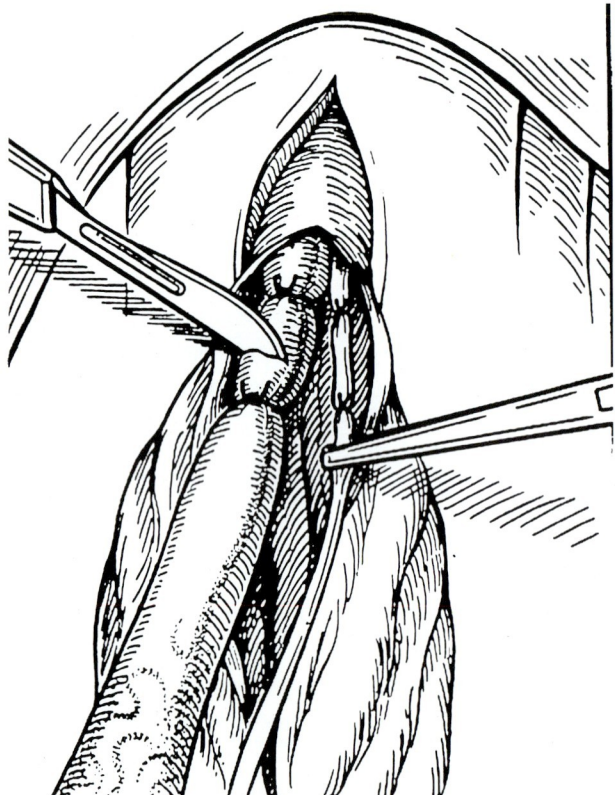

Figure 16.12 Schematic of double ligation and transection of spermatic cord and vessels.

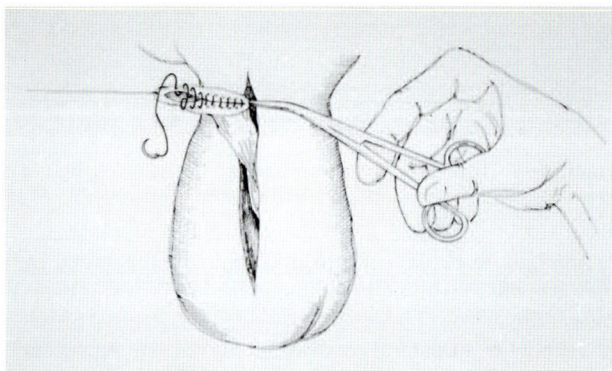

Figure 16.14 Schematic of inverting closure of vaginal tunic.

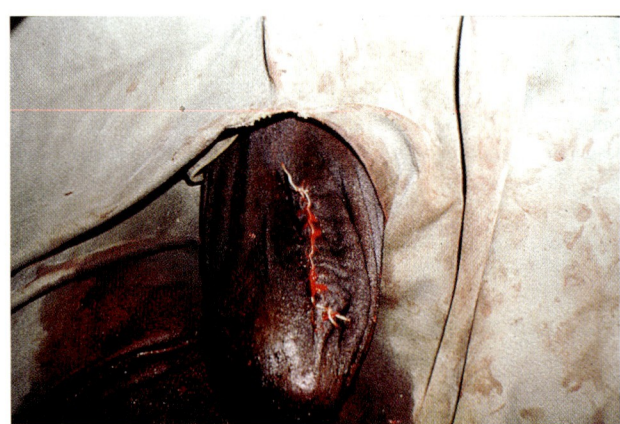

Figure 16.15 Scrotal skin incision closed with continuous interlocking pattern.

Isolate and doubly ligate the spermatic artery, vein, and ductus deferens with #0 chromic gut 8 cm proximal to the pampiniform plexus and transect between the two ligatures (Figure 16.12). Transect the vaginal tunic, then ligate and transect the external cremaster muscle distal to the stump of the spermatic cord (Figure 16.13). Close the vaginal tunic with an inverting pattern such as a Connell or Parker–Kerr using #0 chromic gut (Figure 16.14).

Excise excessive scrotal skin to sufficiently minimize dead space within the surgical site and close the tunica dartos in a simple continuous pattern with #0 chromic gut. Close the skin with a continuous pattern of the surgeon's choice (Figure 16.15).

Administer systemic antibiotics for 5–7 days postoperatively and observe the bull daily for swelling or drainage from the surgical site. Remove sutures 10 days postoperatively and the bull should be ready to return to service when the spermiogram returns to normal. Normal bulls undergoing unilateral orchiectomy had normal spermiograms by 14 days after surgery.[16] Bulls with unilateral scrotal pathology and transient testicular degeneration of the remaining testicle should be expected to return to normal sperm production approximately 60 days after removal of the diseased or injured testicle.[17] However, if the surgery is performed during the warmer months of the year, normal testicular function may not resume until ambient temperatures moderate.

References

1. Setchell B. *The Mammalian Testis*. Ithaca, NY: Cornell University Press, 1978.
2. Waites G. Temperature regulation and the testis. In: Johnson AD, Gomes WR, VanDemark NL (eds) *The Testis*, Vol. I. New York: Academic Press, 1970, pp. 241–280.
3. Heath A, Purohit R, Powe T, Pugh D. Anatomy of the scrotum, testes, epididymis and spermatic cord. In: Wolfe DF, Moll HD (eds) *Large Animal Urogenital Surgery*, 2nd edn. Baltimore: Williams and Wilkins, 1998, pp. 213–220.
4. Wolfe D, Hudson R. Diseases of the testicle and epididymis. In: Howard J (ed.) *Current Veterinary Therapy. Food Animal Practice*, 3rd edn. Philadelphia: WB Saunders, 1993, pp. 794–796.
5. Wolfe D, Wilborn R. Diagnosis and management of conditions of the scrotum and testis. In: Anderson DE, Rings DM (eds) *Current Veterinary Therapy. Food Animal Practice*, 5th edn. St Louis: Saunders, 2009, pp. 360–363.
6. Humphrery J, Ladd P. Pathology of the bovine testis and epididymis. *Vet Bull* 1975;45:787–797.
7. McEntee K. Pathology of the testis of the bull and stallion. In: *Proceedings of the Annual Meeting of the Society for Theriogenology*, 1977, pp. 80–91.
8. McEntee K. Pathology of the epididymis of the bull and stallion. In: *Proceedings of the Annual Meeting of the Society for Theriogenology*, 1979, pp. 103–109.
9. Welles E, Tyler J, Wolfe D, Moore A. *Eperythrozoon* infection in young bulls with scrotal and hindlimb edema, a herd outbreak. *Theriogenology* 1995;43:557–568.
10. Niemark H, Johansson K, Rikihisa Y, Tully J. Proposal to transfer some members of the genera *Haemobartonella* and *Eperythrozoon* to the genus *Mycoplasma* with descriptions of *Candidatus Mycoplasma haemofelis*, *Candidatus Mycoplasma haemomuris*, *Candidatus Mycoplasma haemosus*, and *Candidatus Mycoplasma wenyonii*. *Int J Syst Evol Microbiol* 2001;5:891–899.
11. Wolfe D, Carson R, Hudson R *et al*. Mesothelioma in cattle: 8 cases (1970–1988). *J Am Vet Med Assoc* 1991;199:486–491.
12. Chenoweth P, Spitzer J, Hopkins F. A new bull breeding soundness evaluation form. In: *Proceedings of the Annual Meeting of the Society for Theriogenology*, 1993, pp. 63–70.
13. Purohit R, Hudson R, Riddell M, Carson R, Wolfe D. Thermography of the bovine scrotum. *Am J Vet Res* 1985;46:2388–2392.
14. Heath A, Carson R, Purohit R, Sartin E, Wenzel J, Wolfe D. Effects of testicular biopsy in clinically normal bulls. *J Am Vet Med Assoc* 2001;220:507–512.
15. Wolfe D. Unilateral castration for acquired conditions of the scrotum. In: Wolfe DF, Moll HD (eds) *Large Animal Urogenital Surgery*, 2nd edn. Baltimore: Williams and Wilkins, 1998, pp. 313–320.
16. Wolfe D, Hudson R, Carson R, Purohit R. Effect of unilateral orchiectomy on semen quality in bulls. *J Am Vet Med Assoc* 1985;186:1291–1293.
17. Heath A, Baird A, Wolfe D. Unilateral orchiectomy in bulls: a review of eight cases. *Vet Med* 1996;8:786–792.
18. Wolfe D, Carson R. Unilateral orchiectomy in bulls. In: Dziuk PJ, Wheeler M (eds) *Handbook for Methods of Study of Reproductive Physiology in Domestic Animals*. Urbana, IL: University of Illinois Animal Science Department, 1992, pp. 270–272.

Chapter 17

Restorative Surgery of the Prepuce

Dwight F. Wolfe

Department of Clinical Sciences, College of Veterinary Medicine, Auburn University, Auburn, Alabama, USA

Introduction

Bulls commonly sustain preputial injuries during natural breeding activity. The injuries vary in severity and the therapeutic approach necessary to return the bull to breeding soundness. The veterinarian should thoroughly examine the injured tissues to establish a definitive diagnosis of the cause and extent of preputial injury and economic and therapeutic options for clinical management of the bull.

Examination of the prepuce

The nonerect penis is contained within the sheath, a double invagination of skin along the ventral abdominal wall. The skin of the sheath is covered with hair and joins the hairless prepuce at the preputial orifice.[1] The preputial orifice is highly vascular with gradual transition from haired skin to nonhaired epithelium of the prepuce. The prepuce terminates at the free portion of the penis several centimeters proximal to the glans penis at a junction known as the preputial ring.[2] This area is occasionally injured by avulsion of the epithelium between the prepuce and free portion of the penis during artificial insemination and more rarely during natural mating.

The length and diameter of the prepuce varies considerably among bulls of varying breeds and ages but the prepuce of an adult bull is 35–40 cm long and approximately 4 cm in diameter. The prepuce of *Bos indicus* breeds averages 5.5 cm longer than bulls of *Bos taurus* breeds. *Bos indicus* breeds have a more pendulous sheath and the preputial orifice may exceed 10 cm in diameter in some *indicus* bulls, compared with 2–4 cm in *B. taurus* breeds.[3]

There are multiple interdigitating layers of elastic tissue between the preputial skin and the tunica albuginea of the penis. These elastic layers allow the penis to slide within the sheath from full retraction to full extension. There is wide variation among bulls such that the penis extends 25–60 cm beyond the sheath during full erection and therefore full excursion of the glans penis may be greater than 1 m.[1,3]

Bulls with preputial injury may prolapse the preputial epithelium distal to the end of the sheath or develop phimosis whereby the bull is unable to freely extend the penis and prepuce through the end of the sheath.

It is best to examine bulls from a distance for preputial prolapse or for swelling within the sheath. The conformation of the sheath is important and the distal end of the sheath should be no lower than a line drawn from the hock to the carpus. The distal end of the sheath, the preputial orifice, should not be excessively large and the angle of the sheath should roughly approximate a line drawn along the ventral aspect of the sheath that intersects the lower front leg or foot (Figure 17.1).

Restrain the bull in a chute with moderate squeeze pressure on his sides and place a sturdy bar behind him to limit his ability to kick during the examination. The preputial hairs should be free of calculi, exudate, or hemorrhage. Since bulls do not extend the penis during urination, urine courses down the preputial epithelium and off the preputial hairs. These hairs assist with removal of urine from the preputial orifice and help prevent maceration of the epithelium due to chronic exposure to urine.

In the normal relaxed bull the penis and prepuce should be completely withdrawn within the sheath, although naturally polled bulls may have a slight prolapse of the prepuce when they are relaxed. Palpate the entire penis through the sheath for symmetry and presence of swelling or fibrous tissue. Preputial abscesses are usually circumscribed swellings along the midportion of the sheath, while penile hematoma produces swelling on the dorsum of the penis at the distal bend of the sigmoid flexure (Figure 17.2). Swelling due to penile hematoma is usually symmetrical along the long axis of the penis, while retropreputial abscess is usually located more along one side of the penis (Figure 17.3). Generalized swelling within the sheath along the penis is due to cellulitis from preputial

Restorative Surgery of the Prepuce 143

Figure 17.1 Normal length and angle of sheath of a bull indicated by hashed lines.

laceration or from urine contamination of the peri-penile elastic tissue (Figure 17.4).

An assistant performing gentle massage of the accessory sex glands via rectal palpation facilitates manual extension of the penis. Manually extend the penis and when visible grasp the free portion of the penis with a dry surgical sponge and complete penile extension. The skin surface of the penis should be moist and pink with no evidence of swelling, vesicles, pustules, papillomas, lacerations, or scar tissue.

Congenital and juvenile conditions of the prepuce

As young bulls complete puberty the penis develops a sigmoid flexure as the penis grows in length and diameter. The surface epithelium of the free portion of the penis is firmly attached to the epithelium of the prepuce at birth. Separation of these interdigitating tissues begins at approximately 4 weeks of age and proceeds caudally until

(a)  (b)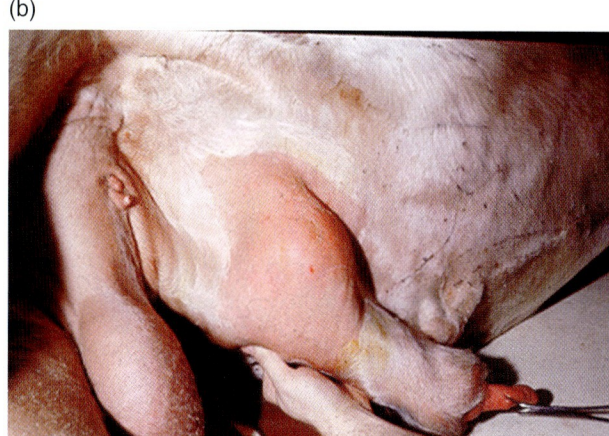

Figure 17.2 (a) Swelling due to retropreputial abscess in a bull. (b) Organized swelling in sheath due to retropreputial abscess in a bull.

(a) 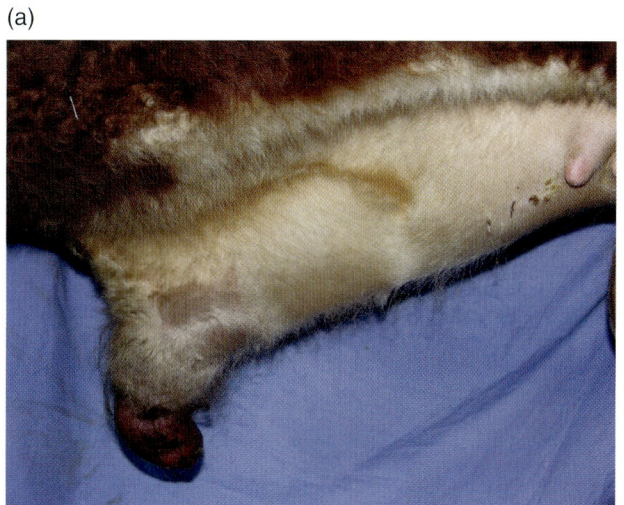 (b)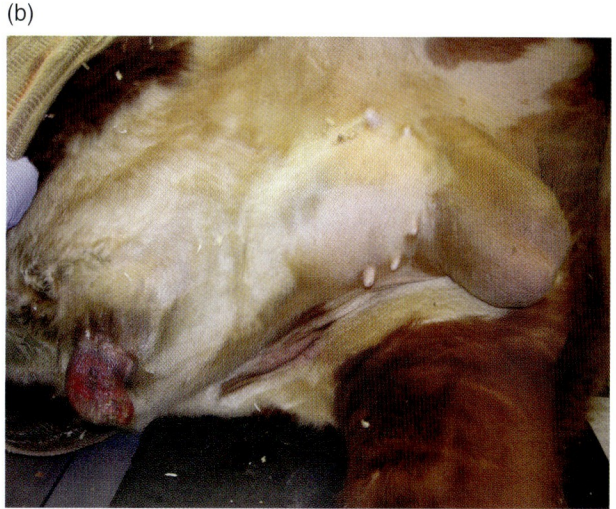

Figure 17.3 (a) Swelling in sheath near the scrotum due to penile hematoma in a bull. Note bruised appearance of preputial prolapse. (b) Symmetrical swelling in sheath due to penile hematoma in a bull. Note bruised appearance of sureputial prolapse.

complete separation occurs between 8 and 11 months of age. The separation occasionally occurs prematurely in young bulls, perhaps caused by juvenile attempts at mounting and penile extension, resulting in hematoma formation from hemorrhage of the surface epithelial layers (Figure 17.5).

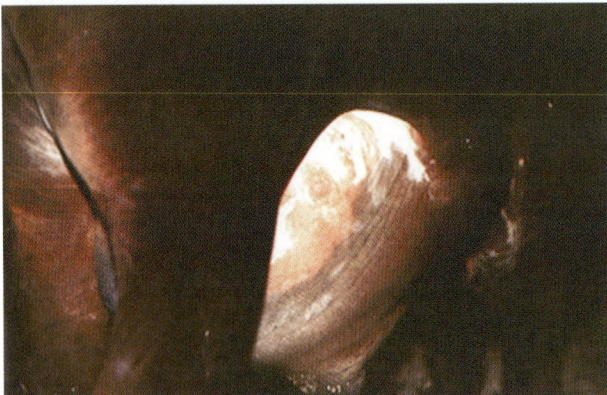

Figure 17.4 Severe generalized swelling in sheath due to urethral rupture in a bull.

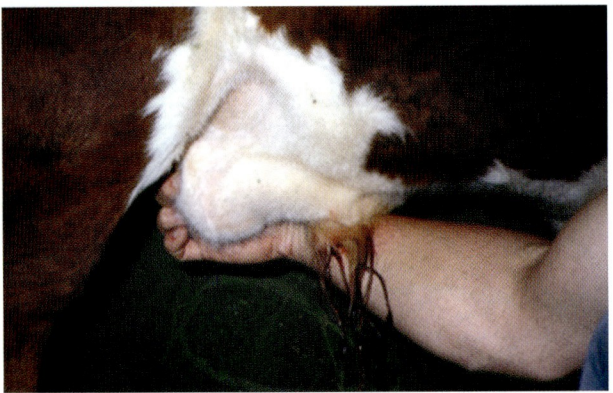

Figure 17.5 Prepubertal calf with swelling in sheath due to premature separation of skin of prepuce and penis.

Treatment is symptomatic and no surgery is indicated. These young bulls may have swelling along the distal sheath with evidence of hemorrhage on the preputial hairs. Full recovery may be expected unless excessive fibrosis develops between the surface epithelial layers, which may permanently prevent penile extension.

The penile frenulum is a thin band of collagenous connective tissue on the ventral midline that extends over the basal 80% of the free end of the penis. During normal separation of the epithelium of the penis and prepuce of young bulls, the frenulum ruptures allowing complete separation of the glans penis and prepuce. When this band of tissue does not rupture the penis can extend but the persistent frenulum interferes with straightening of the tip of the penis so that intromission may be impaired. The persistent frenulum is easily diagnosed by physical examination as a band of tissue from the median raphe at the posterior of the glans penis to the prepuce. This epithelium-covered band may be thin or broad and usually contains one or more blood vessels. Persistent frenulum is easily repaired surgically[4] (see Chapter 18) (Figure 17.6).

Preputial trauma, frostbite and balanoposthitis

Bulls commonly sustain injuries to the prepuce during breeding. The extent of disruption of the surface epithelium and peri-penile elastic tissue determines the prognosis and therapeutic approach for returning the bull to breeding soundness. The veterinarian should understand the etiology of preputial injury and the therapeutic options for the animal.[5,6]

Etiology

Primary preputial prolapse in the bull is usually a sequela to breeding injury or frostbite or to balanoposthitis caused by herpesvirus infection (infectious bovine rhinotracheitis/

(a)
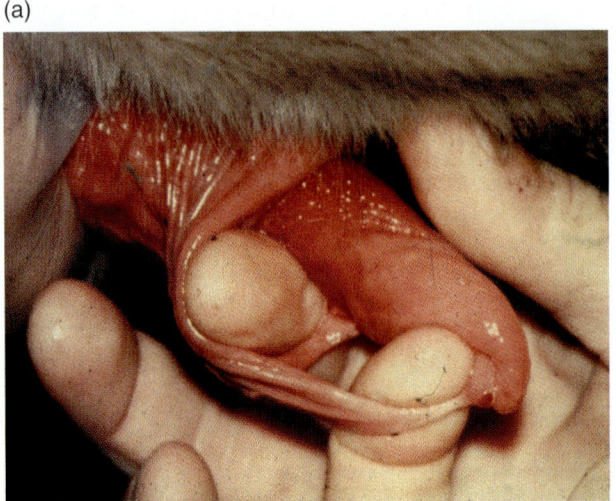

(b)
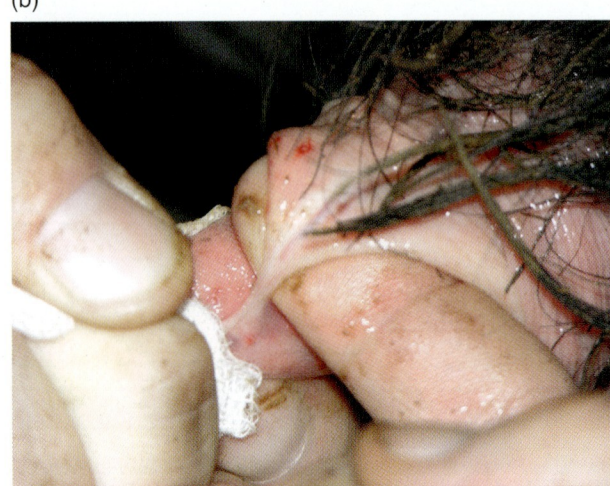

Figure 17.6 (a) Persistent frenulum with two epithelial bands in a bull. (b) Short persistent frenulum in a bull.

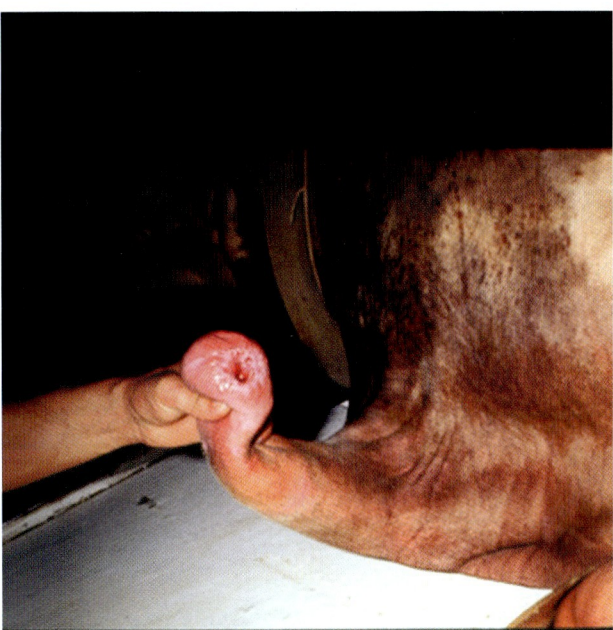

Figure 17.7 Severe preputial stenosis due to frostbite.

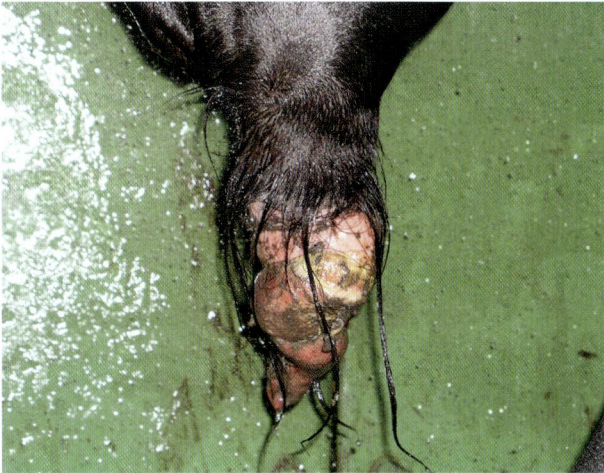

Figure 17.8 Paraphimosis secondary to preputial laceration.

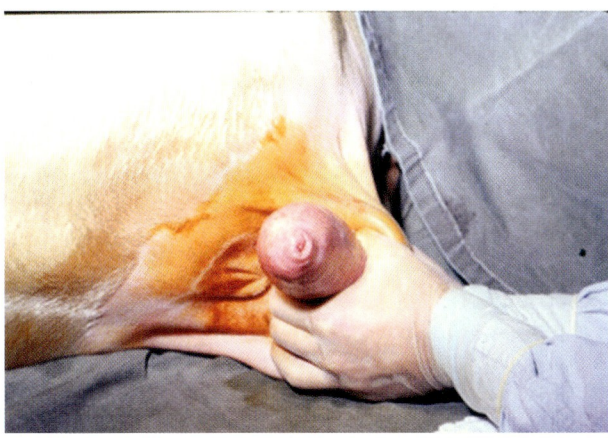

Figure 17.9 Phimosis secondary to preputial laceration.

infectious pustular vulvovaginitis, or IBR-IPV). Secondary preputial prolapse is often seen with penile hematoma or urethral rupture. Bulls suffering preputial frostbite may undergo considerable necrosis of the preputial epithelium and heal with mild to severe preputial stenosis (Figure 17.7). IBR-IPV most commonly affects young bulls and the prepuce may be extremely edematous with vesicles or pustules on the preputial epithelium. These bulls usually respond well to sexual rest, application of emollient ointments and conservative therapy, which may include hydrotherapy and support bandaging.

Preputial lacerations

Laceration of the prepuce of *Bos taurus* breeds usually does not lead to preputial prolapse. The damaged prepuce is typically withdrawn into the sheath and swelling of the sheath may be observable. Minor injuries are often unnoticed and heal without complication with only minor superficial scarring visible on the surface epithelium of the prepuce.[5-8] Some *Bos taurus* bulls with preputial laceration develop paraphimosis, where the damaged tissues will not allow retraction of the penis into the sheath (Figure 17.8). Alternatively, some bulls that do retract the penis into the sheath develop phimosis due to stricture of the injured prepuce (Figure 17.9).

Laceration of the prepuce with subsequent preputial prolapse occurs more commonly in *Bos indicus* breeds and their crosses due to the pendulous sheath and longer prepuce. During breeding the excess prepuce is forced caudally and forms a collar at the preputial orifice when intromission is achieved. This collar of prepuce becomes forcefully entrapped between the bull's abdomen and the vulva and pelvis of the cow during the ejaculatory lunge, with subsequent contusion and occasionally laceration or bursting of the skin on the longitudinal axis of the ventral aspect of the prepuce. Edema quickly develops in the traumatized skin and underlying elastic tissue leading to prolapse of the prepuce. As the penis is withdrawn into the preputial cavity the longitudinal tear assumes a transverse orientation that effectively shortens the ventral aspect of the prepuce. As edema accumulates in the damaged tissues the prolapsed prepuce increases in size and assumes the appearance of an elephant's trunk with the lumen of the prepuce directed caudally. The laceration is evident as a transverse wound on the caudal aspect of the prolapsed tissues (Figure 17.10).

If the injured bull continues to attempt breeding, the prolapsed tissues suffer additional trauma and perhaps worsen the prognosis for return to breeding soundness. As the dependent edema increases, the size and weight of the prolapsed tissues also increases thereby leading to further swelling and prolapse. These conditions seem to become more severe in naturally polled bulls due to lack of the retractor preputial muscle in homozygous polled bulls. In naturally polled bulls retraction of the prepuce passively follows the penis which is retracted by the paired retractor penis muscles. The edematous prolapsed prepuce frequently cannot be withdrawn into the preputial cavity and often sustains additional external trauma and drying, with subsequent extreme mutilation, cellulitis, necrosis, fibrosis, and risk of frostbite in colder climates.

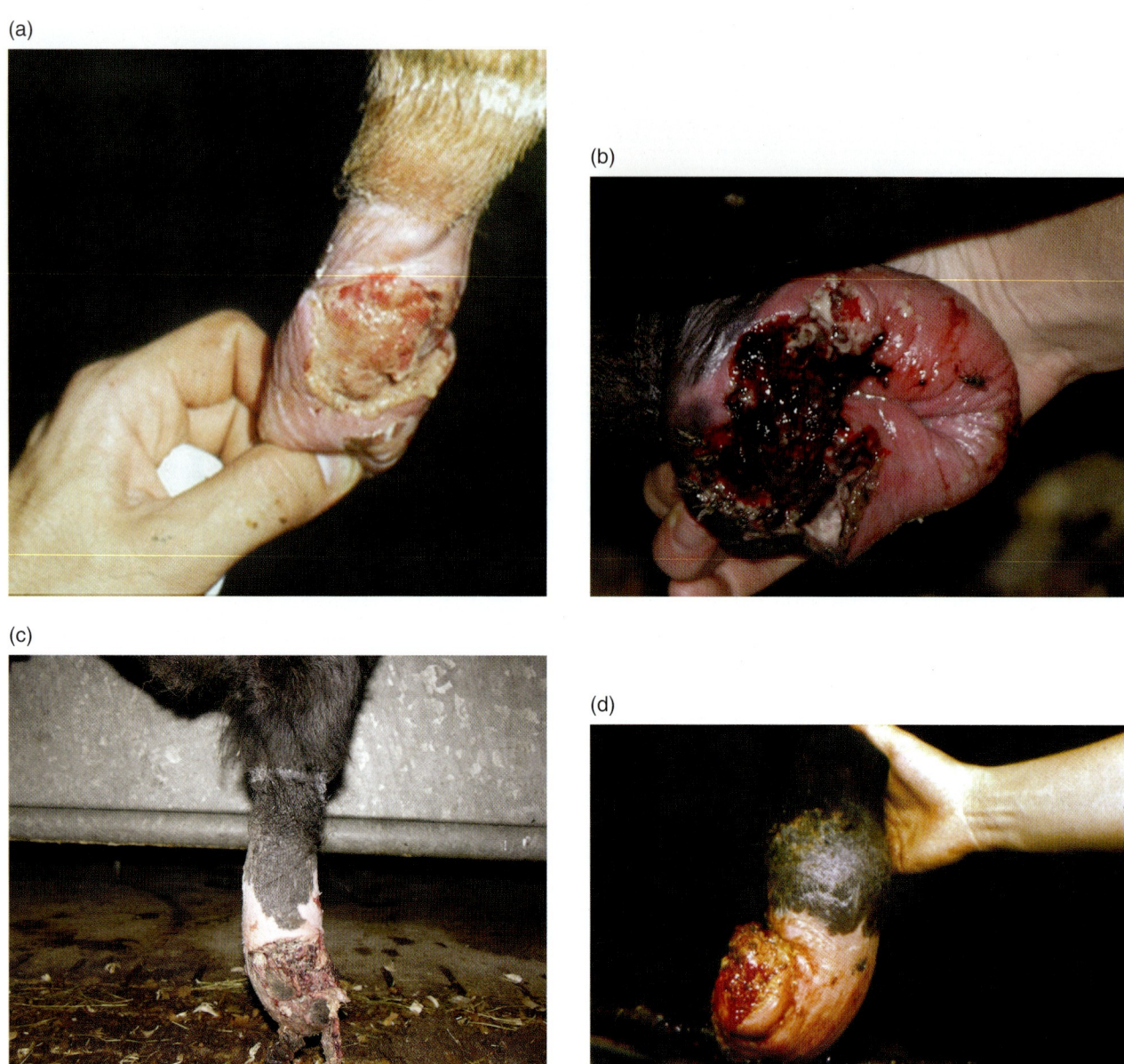

Figure 17.10 (a) Moderately severe preputial prolapse. (b) Fresh preputial laceration with minimal necrosis. (c) Severe preputial laceration with minimal edema. (d) Preputial prolapse with laceration and severe edema.

The most critical factors for determining the potential for return to breeding soundness are the length of the preputial skin that is lacerated and the extent of damage to the underlying peri-penile elastic tissue. If surgical repair of the damaged prepuce is indicated, the remaining prepuce must be at least 1.5 times the length of the free portion of the penis in order to allow sufficient penile and preputial extension for breeding. The peri-penile elastic tissue must also be free of scar tissue to allow extension and retraction of the penis and prepuce.[5]

Medical management of preputial prolapse

The prolapsed tissues should be cleaned with surgical antiseptic scrub and an emollient antiseptic ointment should be applied. Effective ointments include Wound Wonder® (Dr J.P. Brendemeuhl, woundwonder.com) or 2 g tetracycline powder and 60 mL scarlet oil mixed in 500 g anhydrous lanolin, or commercially available human hemorrhoidal ointments. Frequently this symptomatic treatment allows bulls with minor laceration and prolapse to resume breeding soundness following at least 60 days of sexual rest.[5–7]

If the tissue damage is more extensive and edema prevents return of the prepuce into the preputial cavity, compression bandaging is indicated. Following cleansing of the tissues with a surgical scrub, apply the emollient ointment and cover the prolapsed prepuce with a clean 5-cm diameter orthopedic stockinette (Figure 17.11). Insert a minimum 6.45-mm diameter latex tube into the lumen of the prepuce for urine drainage (Figure 17.12) and snugly apply elastic tape from the distal end of the prepuce proximally ensuring that the tape is anchored to the haired skin of the sheath (Figure 17.13). Change the bandage

Restorative Surgery of the Prepuce 147

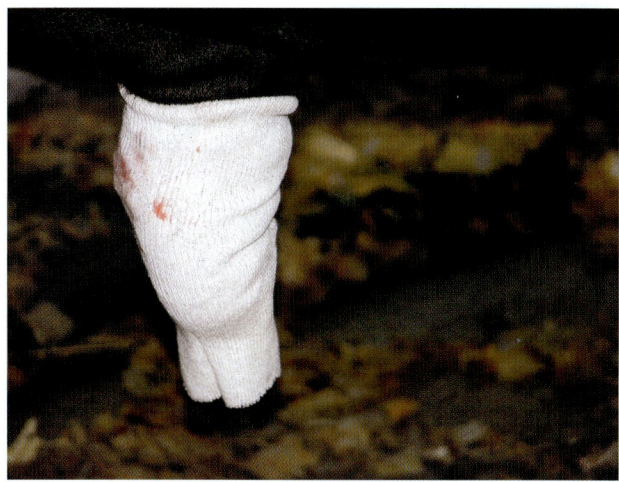

Figure 17.11 Orthopedic stockinette applied over prolapsed prepuce.

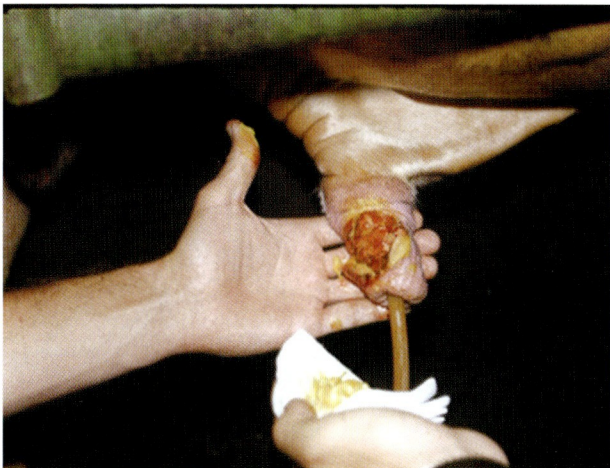

Figure 17.12 Latex tube inserted into preputial lumen for urine drainage.

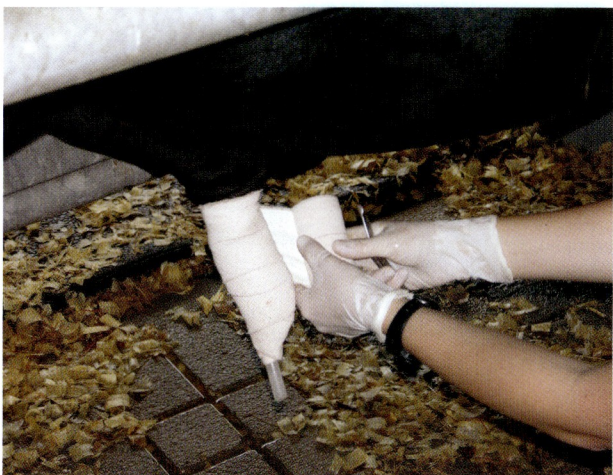

Figure 17.13 Prepuce bandaged with latex tube in place for urine drainage.

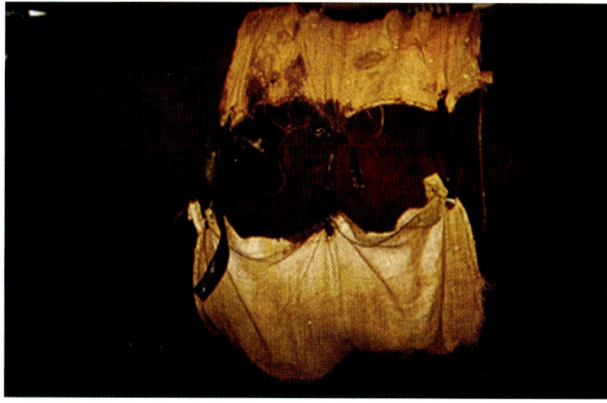

Figure 17.14 Support sling for bull with pendulous preputial prolapse.

every second or third day or sooner if it becomes loose or extremely soiled.

Bulls with a particularly pendulous preputial prolapse may benefit from a support sling to lift the sheath and prolapsed tissues close to the body. A sling fashioned by cradling the prepuce and sheath in a large piece of burlap or other loosely woven material may be held in place with bungee cords across the bull's back (Figure 17.14). The support sling may be preferred to avoid adhesive scald from repeated bandaging in bulls with especially pendulous sheath or severe prolapse with deep laceration where prolonged treatment is necessary.

When the initial prolapse is reduced in bulls with mild to moderate preputial prolapse, the prepuce should be reverted into the preputial cavity and held in place by elastic bandaging with a urinary drainage tube incorporated into the bandage. I adamantly oppose placing a pursestring suture in the preputial orifice because of the risk of abscess formation and subsequent stenosis of the preputial orifice. Reduction of the prolapse and use of a pursestring retaining suture may be acceptable to protect the injured tissues of bulls destined for immediate slaughter.

Bulls with preputial laceration and prolapse may sustain superficial and possibly deep necrosis and slough skin and elastic tissues after the first few bandage changes. These areas must be covered by healthy granulation tissue before reversion into the preputial cavity or before surgery is considered.

Because they do not usually prolapse the prepuce after injury, bulls of *Bos taurus* breeds are more likely to develop retropreputial abscess and cellulitis within the sheath. Administration of systemic antibiotics, irrigation of the preputial cavity with mild antiseptic solutions, and hydrotherapy may reduce the risk of severe complications. Allow at least 60 days of forced sexual rest and evaluate the prepuce for scar formation that restricts penile extension prior to resumption of breeding. Impaired extension is likely since the original wound was longitudinal but healed transversely.

Acute lacerations of the prepuce should not be sutured as the peri-penile elastic tissue is contaminated and suturing the wound is likely to lead to abscess formation. These lacerations should be managed medically as previously described until the wound granulates.

148   The Bull: Reproductive Surgery

Surgical management of preputial injury

Avulsion of the prepuce

Avulsion of the distal prepuce from its attachment on the dorsum of the free portion of the penis occurs most commonly in bulls that serve an artificial vagina (AV) for semen collection for artificial insemination. This injury rarely occurs during natural breeding. The prepuce adheres to the latex liner of the AV and the prepuce is torn loose from the penis when the bull makes the ejaculatory lunge. Contrary to other types of lacerations, those that occur in an AV should be sutured immediately. The wounds are quickly diagnosed and comparatively clean and excellent healing is expected with immediate primary closure of the wound. If the injury is not immediately closed by suture then surgical repair should be delayed until the wound is completely covered by healthy granulation tissue. With local anesthesia of the penis, prepare the surgical field for aseptic surgery, undermine the retracted prepuce and slide it distally to the normal position and appose the prepuce to the free portion of the penis with absorbable suture.

Circumcision

Resection and anastomosis of the prepuce, frequently called circumcision, provides the best prognosis for return to breeding soundness when scar tissue in the prepuce prevents normal penile extension and coitus. Circumcision entails full extension of the penis and prepuce, which may require incision of fibrous stenotic areas of the prepuce. The guideline for the maximum amount of prepuce that may be removed is that the remaining prepuce must be at least 1.5 times the length of the free portion of the penis (Figure 17.15). If the prepuce is left excessively long, repeat of the prolapse is likely when the bull resumes breeding. However, if the prepuce is shortened excessively, the bull will be unable to completely extend the penis.

Following aseptic preparation of the surgical field apply a 2.5-cm Penrose drain around the penis and prepuce at the preputial orifice as a tourniquet and make circumferential incisions through the preputial epithelium proximal and distal to the area to be removed. Make a longitudinal incision through the epithelium to join the circumferential incisions (Figure 17.16). Some surgeons prefer use of a CO_2

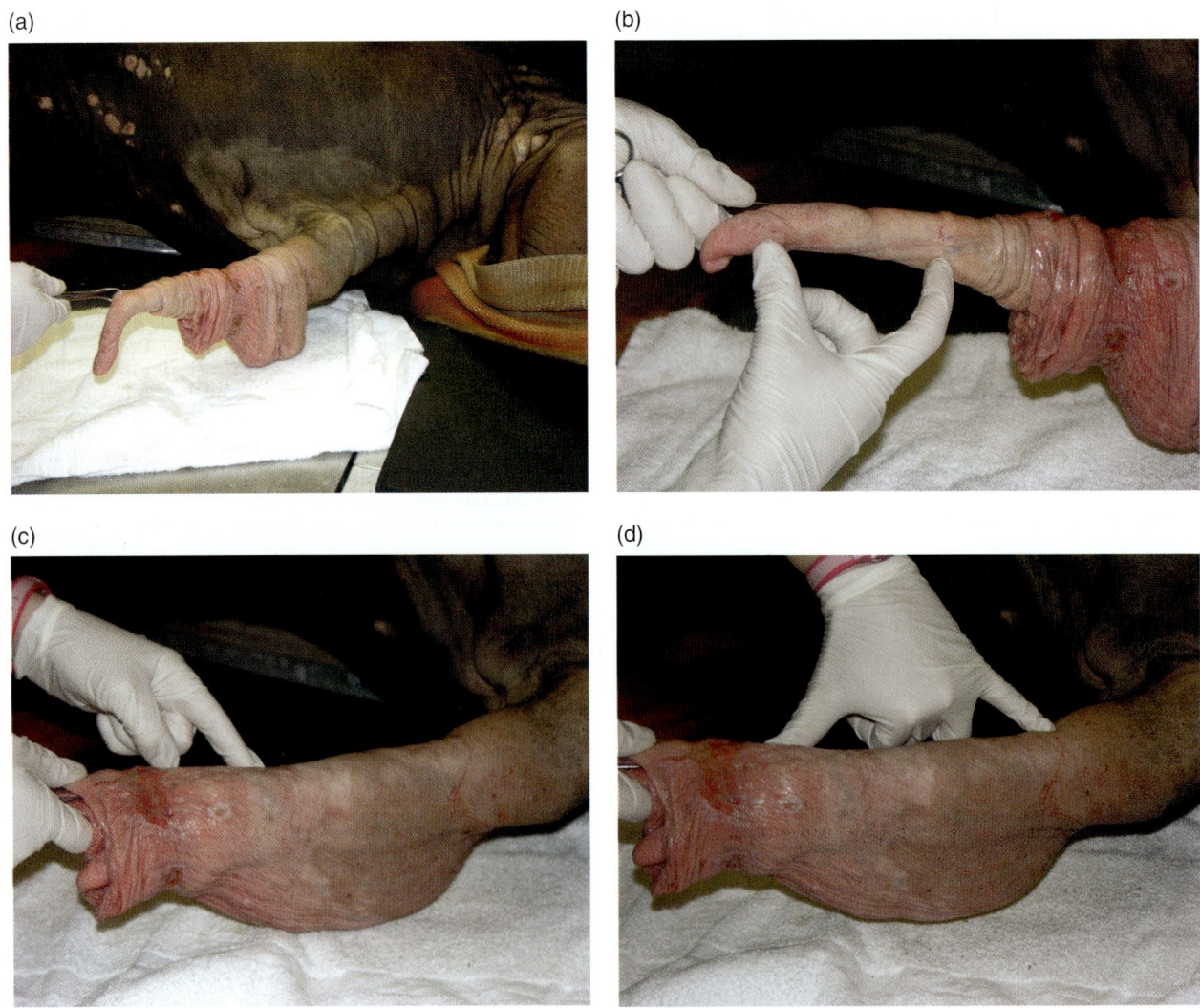

Figure 17.15 (a) Penis extended in preparation for circumcision. Note loose excessive prepuce. (b) Determination of length of free portion of prepuce. (c) Proximal end of free portion of prepuce indicated by surgeon's left forefinger. (d) Length of excess prepuce measured from end of sheath (surgeon's small finger) to preputial ring (surgeon's thumb).

Restorative Surgery of the Prepuce 149

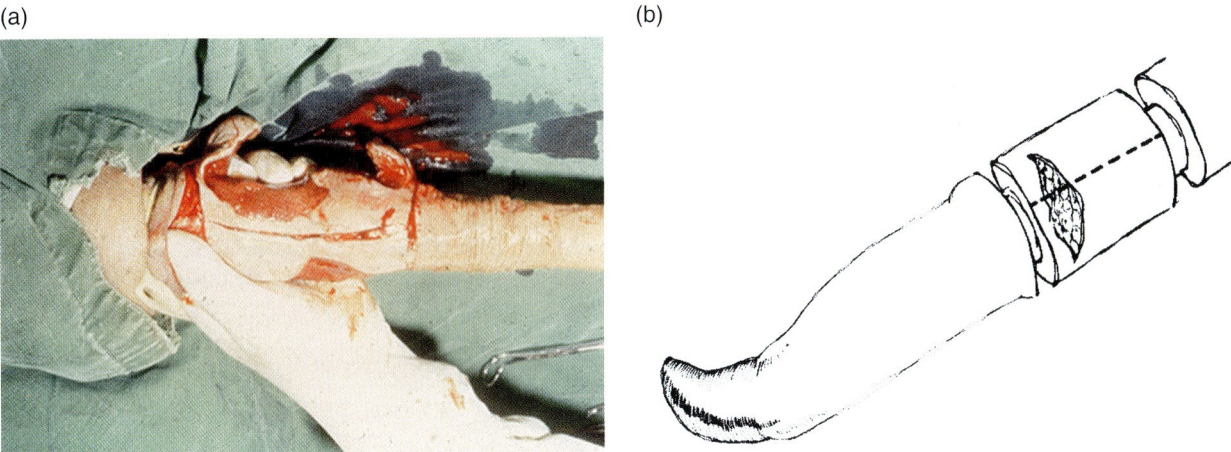

Figure 17.16 (a) Circumferential and longitudinal skin incisions in the preputial epithelium. Note Penrose drain (arrow) placed as a tourniquet around the prepuce. (b) Schematic of circumferential and longitudinal skin incisions for circumcision of a bull.

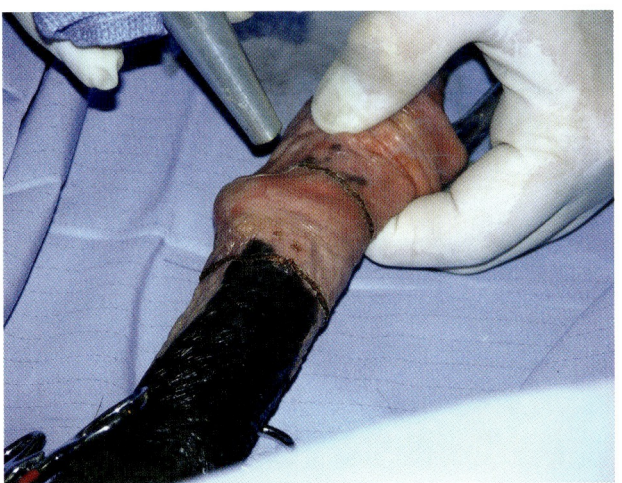

Figure 17.17 Circumferential laser incision in preputial epithelium for circumcision.

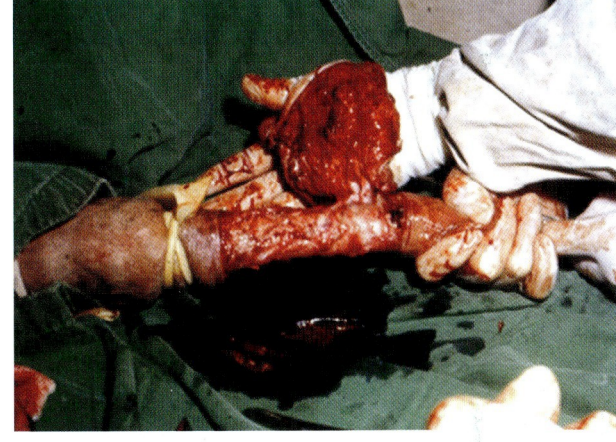

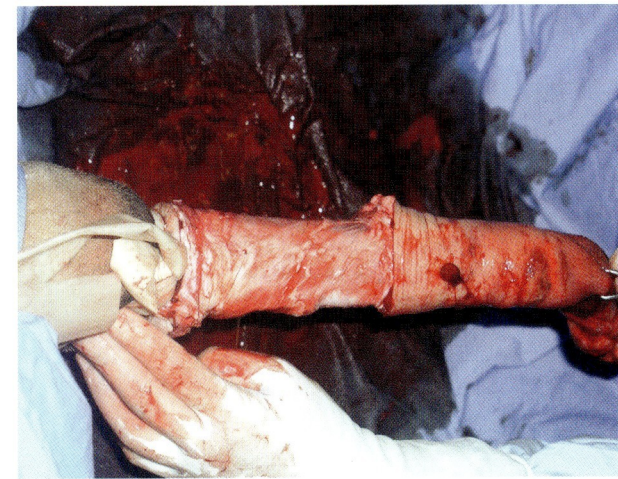

Figure 17.18 (a) Removal of preputial skin and underlying scar tissue. (b) Preputial epithelium and scar tissue excised.

laser to minimize hemorrhage during this procedure (Figure 17.17). Remove the skin of the injured area by blunt and sharp dissection, taking care to only dissect to the depth necessary to remove fibrotic tissues (Figure 17.18).

Following removal of the excessive preputial skin and scar tissue ligate or seal all visible elastic tissue vessels by judicious electrocautery. Remove the tourniquet and ligate any remaining vessels to ensure complete hemostasis. Close the subcutaneous elastic tissue with #0 chromic gut in a simple interrupted pattern or in a simple continuous pattern tied at each quadrant of the circumference of the prepuce (Figure 17.19). Close the skin of the prepuce in similar fashion ensuring end-to-end apposition of the skin edges (Figure 17.20). Place a 2.5-cm Penrose drain over the free portion of the penis to provide urine drainage away from the surgical site. When securing the Penrose, place three or four absorbable sutures through the latex tubing and the

150 The Bull: Reproductive Surgery

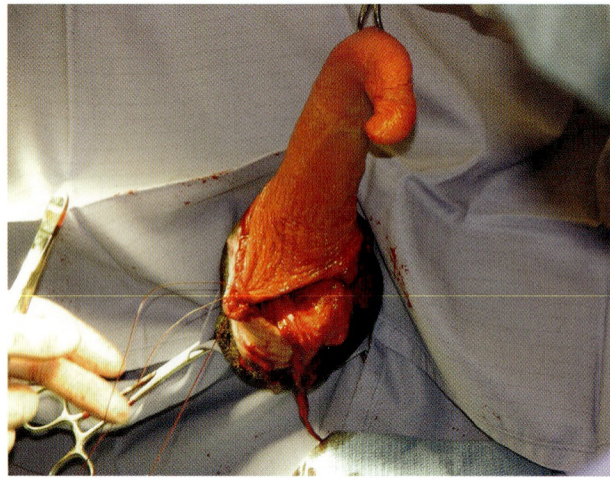

Figure 17.19 Subcutaneous elastic tissue of prepuce closed with simple continuous sutures.

skin of the penis (Figure 17.21). Revert the penis and prepuce into the preputial cavity and place a firm but flexible rubber tube of maximum diameter into the lumen of the prepuce (Figure 17.22). Firmly wrap the distal portion of the haired sheath with elastic tape to serve as a pressure bandage on the incision (Figure 17.23). Remove the pressure bandage and rubber tubing in 3–5 days and leave the Penrose drain on the penis until skin sutures are removed 10 days postoperatively. Ensure a minimum of 60 days sexual rest and evaluate the bull prior to resumption of breeding.

Amputation of the prepuce

Amputation of the prepuce is occasionally performed on young bulls of *Bos indicus* breeds to prophylactically shorten an excessively long prepuce to reduce the likelihood of preputial laceration when the bulls begin breeding. The advantage of this procedure is that it is quick to perform compared with circumcision. The disadvantage is that there is more elastic tissue removed with this procedure and likely edge-to-edge

(a)
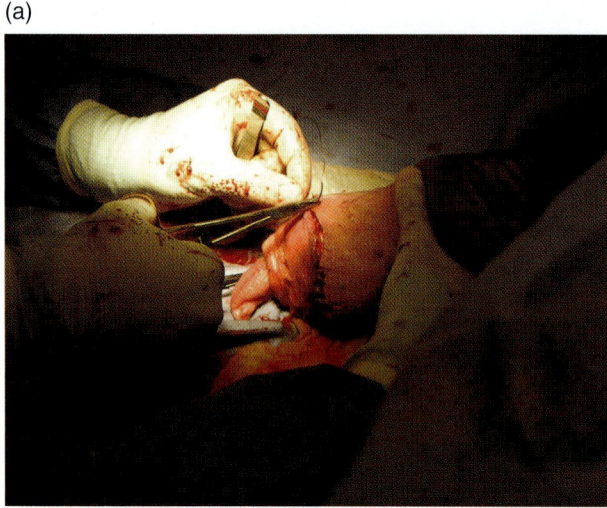

(b)
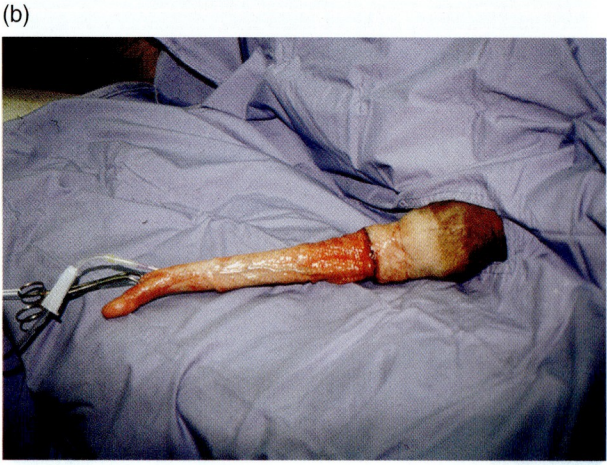

Figure 17.20 (a) Closure of preputial epithelium. (b) Closure of the preputial skin with simple continuous suture pattern ensuring edge-to-edge skin apposition.

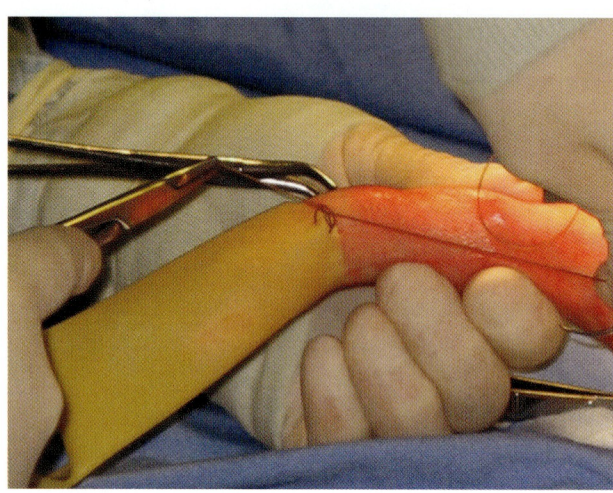

Figure 17.21 Penrose tube sutured over free portion of the penis for urine drainage.

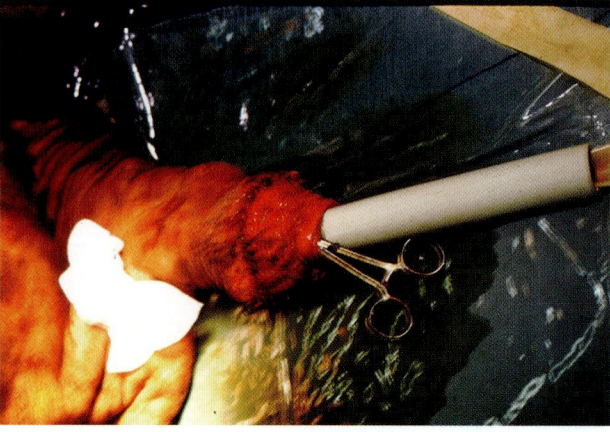

Figure 17.22 Firm flexible rubber tube placed over Penrose drain in preparation to revert penis to retracted position.

Restorative Surgery of the Prepuce

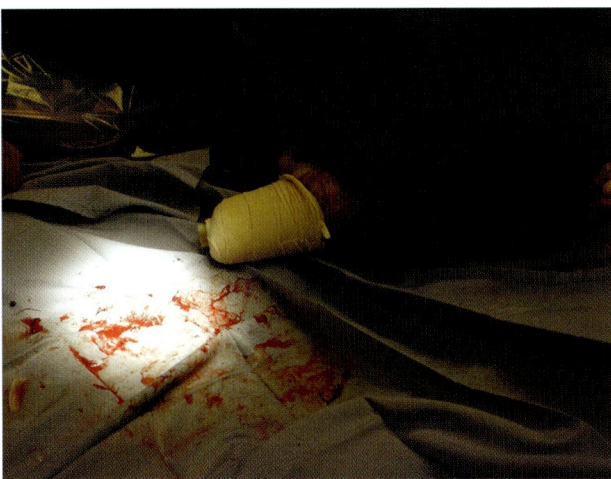

Figure 17.23 Elastic bandage applied over sheath and firm rubber tube in lumen of prepuce.

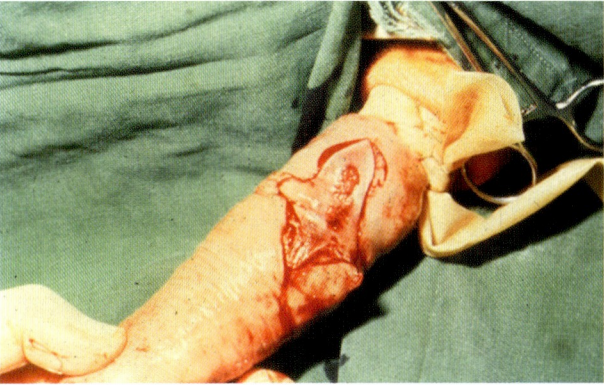

Figure 17.24 Excise scar tissue from preputial epithelium.

apposition of skin edges creating a higher incidence of preputial stenosis compared with circumcision. The surgeon should keep in mind that the prepuce is a tubular structure and that with wound contracture stenosis is more likely.

With the bull adequately restrained for aseptic surgery, infiltrate local anesthetic in the skin of the sheath just proximal to the preputial orifice. Incise the prolapsed prepuce into the preputial lumen approximately one-third of its circumference and ligate bleeders. Suture the internal and external layers of skin with #0 absorbable suture using a simple continuous pattern. Repeat the procedure for the remaining two-thirds of the circumference of the preputial prolapse. Suture a Penrose drain over the glans penis as previously described and revert the prepuce into the preputial cavity. Place an elastic bandage over the sheath as previously described and manage postoperatively as for circumcision.

Preputial reconstruction by scar revision

Following healing of preputial laceration, if the length of the healthy prepuce is insufficient to allow resection and anastomosis the bull may be returned to breeding soundness with scar revision and preputial reconstruction. Restrain the bull and prepare the prepuce for aseptic surgery. With the penis extended excise only the superficial transverse scar tissue (Figure 17.24). Return the edges of the preputial laceration to their original longitudinal orientation. Loosely place an absorbable bootlace suture in a longitudinal plane such that the free ends are toward the sheath (Figure 17.25). Place a 2.5-cm Penrose drain over the free potion of the penis and revert the penis into the preputial cavity. Tighten the sutures to appose the skin edges (Figure 17.26). The purpose of this procedure is to allow the surgical wound to heal in a longitudinal plane and to maintain normal elastic tissue function. Since no elastic tissue or preputial skin is removed with this procedure, healing should reduce the risk of preputial stenosis and allow subsequent penile extension. Most bulls require 60–120 days for the penis to freely extend following this procedure. Do not forcefully extend the penis during the convalescent period as the potential trauma increases the likelihood of

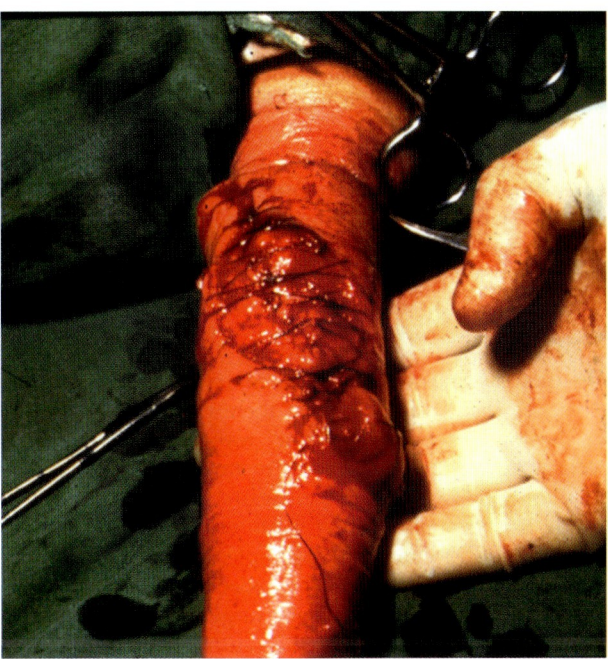

Figure 17.25 Placement of bootlace suture in longitudinal orientation of prepuce.

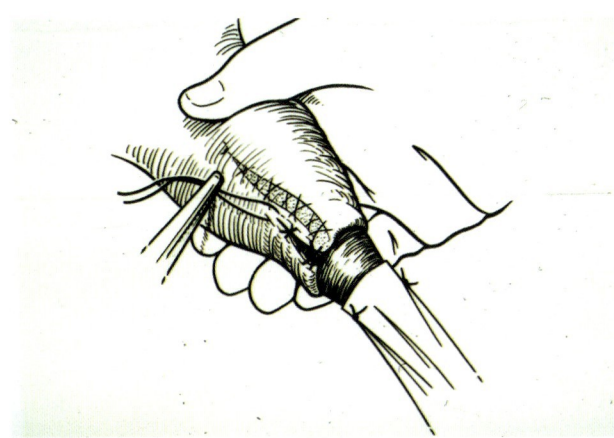

Figure 17.26 Schematic of closing incision with bootlace suture when penis returned to retracted position.

excessive scar formation. Most bulls begin to masturbate within a few weeks following surgery and will stretch contracted tissues without permanent damage.

Inability to extend the penis (phimosis)

The most common cause of phimosis in adult bulls is preputial stenosis resulting from scar tissue. The stenosis is most often a sequela of either breeding injury laceration or frostbite. Either case may be corrected surgically by circumcision if removal of the damaged tissues leaves sufficient healthy prepuce and elastic tissue for normal penile extension (Figure 17.27).

Healed preputial laceration may lead to preputial stenosis and affected bulls exhibit phimosis, the inability to extend the penis. These injuries may occur in any breed but preputial lacerations are more common in *Bos indicus* breeds and their crosses than in *Bos taurus* breeds. *Bos indicus* bulls usually prolapse the injured prepuce, while preputial injuries in *Bos taurus* bulls usually do not lead to prolapse. However, *Bos taurus* bulls are much more likely to develop preputial stenosis with the resultant inability to extend the penis.

The most limiting factors for the likelihood of returning to breeding soundness for a bull with preputial laceration are the extent of damage to the peri-penile elastic tissue and the length of preputial skin that is lacerated.

Creation of preputial stoma

Occasionally bulls with preputial injury or frostbite develop extensive scar tissue such that there is insufficient length of healthy prepuce for scar revision and preputial reconstruction or circumcision. Creation of a stoma from the preputial cavity through the sheath may allow sufficient penile protrusion for semen collection. In my experience two bulls were able to successfully complete coitus following this surgical procedure. With the bull adequately restrained and following preparation for aseptic surgery, make an elliptical incision through the skin on the ventral sheath sufficiently proximal to the scar tissue to penetrate the lumen of the preputial cavity (Figure 17.28). The diameter of the stoma should be sufficiently large to allow the free portion of the penis to extend but not sufficiently large that remaining prepuce may prolapse. Following removal of the elliptical skin, continue dissection through the peri-penile elastic layers and elevate the prepuce through the opening in the sheath (Figure 17.29). Incise the prepuce (Figure 17.30) and exteriorize the free

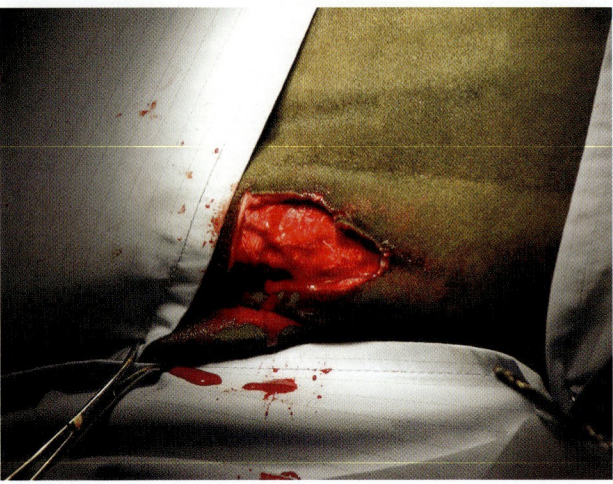

Figure 17.28 Triangular flap of skin removed from sheath (apex of triangle oriented toward scrotum).

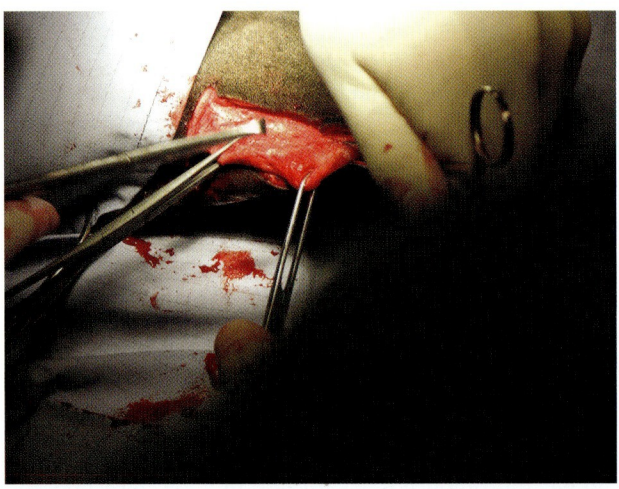

Figure 17.29 Prepuce elevated through incision in sheath.

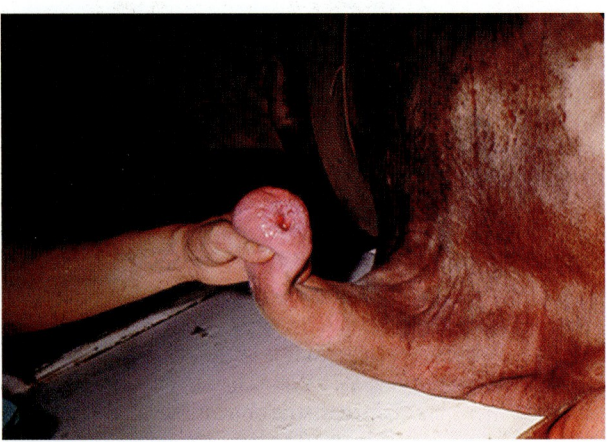

Figure 17.27 Paraphimosis due to stenosis from preputial laceration.

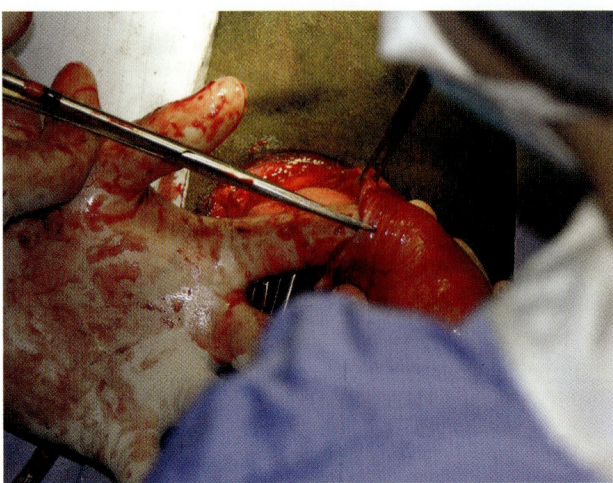

Figure 17.30 Incision into preputial cavity to create stoma.

Restorative Surgery of the Prepuce 153

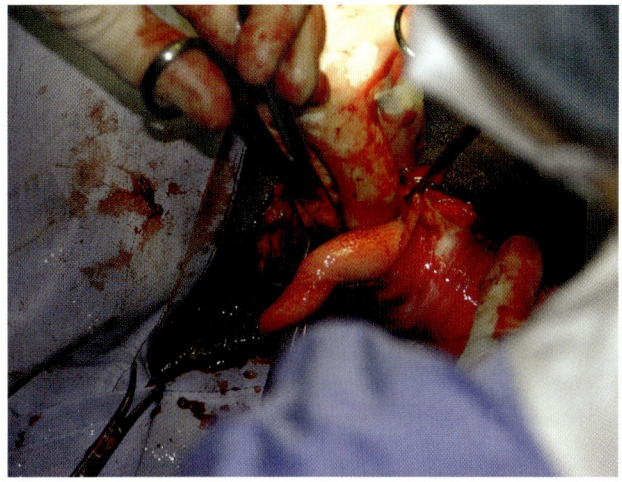

Figure 17.31 Free portion of penis exteriorized through preputial stoma.

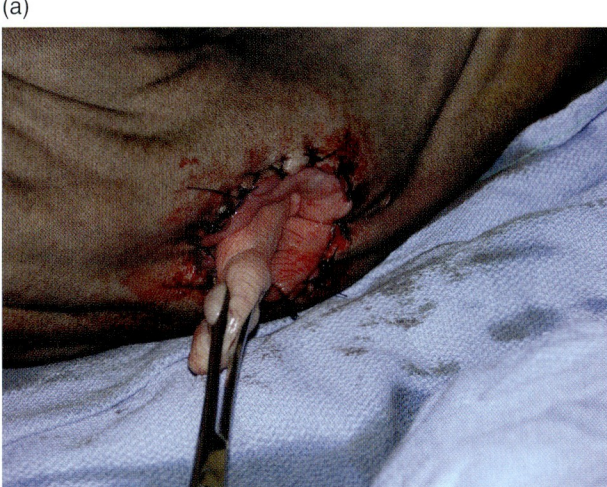

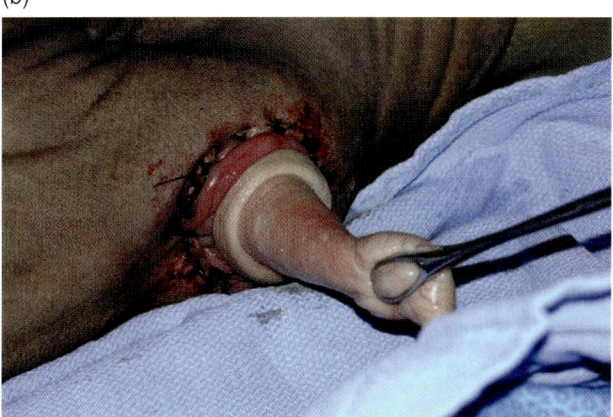

Figure 17.32 (a) Preputial epithelium apposed to skin of the sheath with interrupted absorbable sutures. (b) Completed suturing of prepuce to skin of sheath to create stoma.

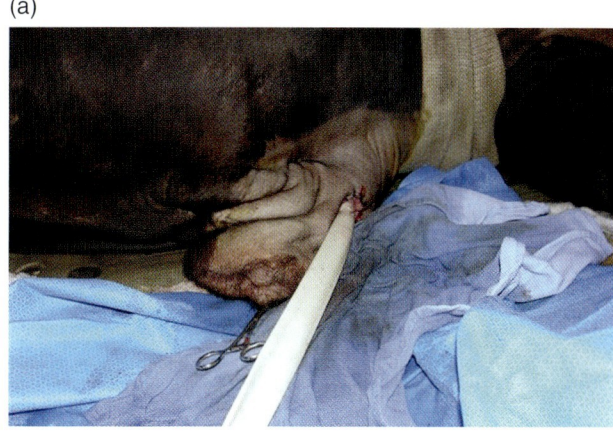

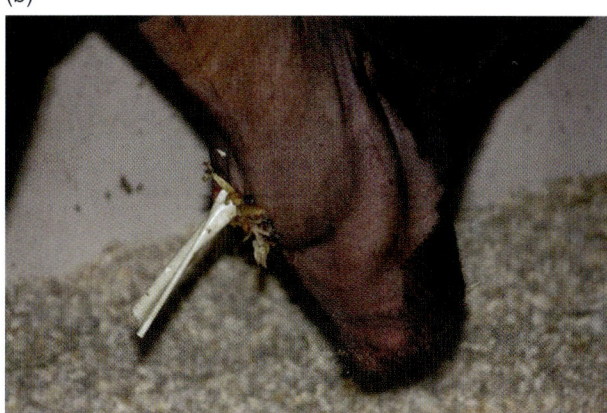

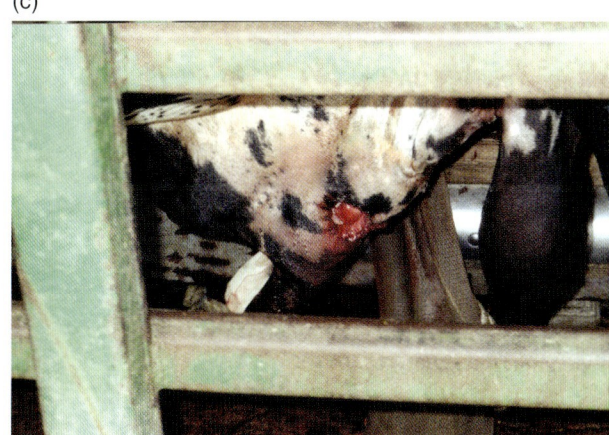

Figure 17.33 (a) Penrose drain sutured over free portion of penis. (b) Penrose drain over penis through preputial stoma for urine drainage. (c) Postoperative bull with preputial stoma (note Penrose drain through natural sheath opening).

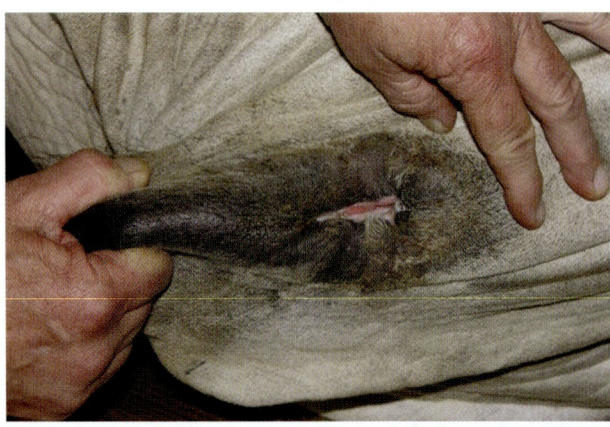

Figure 17.34 Preputial stoma completely healed ready for attempted semen collection.

portion of the penis through the incision (Figure 17.31). Using #0 absorbable suture appose the skin of the prepuce to the skin of the sheath with a simple interrupted pattern (Figure 17.32). Suture a 2.5-cm Penrose drain over the free portion of the penis to allow postoperative urine drainage away from the incision (Figure 17.33). Remove the Penrose drain 10 days postoperatively and allow 60 days sexual rest prior to attempted semen collection (Figure 17.34).

References

1. Ashdown R, Pearson H. Anatomical and experimental studies on eversion of the sheath and protrusion of the penis in the bull. *Res Vet Sci* 1973;15:13–24.
2. Ashdown R, Rickets S, Wardley R. The fibrous architecture of the integumentary coverings of the bovine penis. *J Anat* 1968;103:576–578.
3. Bellenger C. A comparison of certain parameters of the penis and prepuce in various breeds of beef cattle. *Res Vet Sci* 1971;12:299–304.
4. Ashdown R, Pearson H. The functional significance of the dorsal apical ligament of the bovine penis. *Res Vet Sci* 1971;12:183–184.
5. Wolfe D, Beckett S, Carson R. Acquired conditions of the penis and prepuce. In: Wolfe DF, Moll HD (eds) *Large Animal Urogenital Surgery*, 2nd edn. Baltimore: Williams and Wilkins, 1998, pp. 237–272.
6. Memon M, Dawson L, Usenik E, Rice L. Preputial injuries in bulls. *J Am Vet Med Assoc* 1988;193;484–485.
7. Dawson L, Rice L, Morgan G. Management of preputial prolapse in the bull. Fact sheet, Society for Theriogenology, 1989.
8. Wolfe D. Surgical procedures of the reproductive system of the bull. In: Morrow DA (ed.) *Current Therapy in Theriogenology*, 2nd edn. Philadelphia: WB Saunders, 1986, pp. 353–380.

Chapter 18

Restorative Surgery of the Penis

Dwight F. Wolfe

Department of Clinical Sciences, College of Veterinary Medicine, Auburn University, Auburn, Alabama, USA

Examination of the penis

Examine the penis and prepuce of bulls from a distance followed by manual examination with the bull safely restrained in a chute with a drop side. The conformation of the sheath is important and the distal end of the sheath should be no longer than a line drawn from the hock to the carpus. The distal end of the sheath, the preputial orifice, should not be excessively large and the angle of the sheath should roughly approximate a line drawn along the ventral aspect of the sheath that intersects the lower front leg or foot[1] (see Figure 17.1).

With the bull in a chute, apply moderate pressure on the bull's sides with the chute squeeze mechanism and place a sturdy bar behind him to limit his ability to kick during the examination. The preputial hairs should be free of calculi, exudate, or hemorrhage. The penis and prepuce should be contained within the sheath, although naturally polled bulls may have a slight prolapse of the prepuce when they are relaxed. Palpate the entire penis through the sheath for swelling or fibrous tissue.[1]

With the aid of an assistant manually extend the penis, grasp the free portion of the penis with a dry surgical sponge, and complete penile extension. The skin of the penis should be moist and pink with no evidence of swelling, vesicles, pustules, papillomas, lacerations, or scar tissue.

Juvenile penile conditions

Penile fibropapilloma

Etiology

Fibropapilloma or warts caused by bovine papilloma virus are fairly common in young bulls. The virus is believed to enter the penile skin through wounds or abrasions sustained during homosexual activity among young bulls. The virus causes neoplastic growth of fibroblasts that is not locally invasive or metastatic. Often several bulls in a group will develop penile fibropapillomas and affected bulls frequently do not have obvious lesions on other parts of the body. Other neoplastic growth on the bovine penis is extremely rare.[2]

Diagnosis

Bulls with penile fibropapilloma are usually 1–3 years of age and frequently the earliest sign of the lesion is scant hemorrhage from the preputial cavity following attempted coitus. Often the lesion is first noted when the bull is observed masturbating or is presented for semen collection or breeding soundness examination. The growths are typically single pedunculated masses near the glans penis (Figure 18.1), although multiple or sessile growths occasionally develop (Figure 18.2). Very large lesions may prevent complete retraction of the penis into the preputial cavity. Rarely, continued growth of the lesion with the penis retracted into the preputial cavity may prevent penile extension resulting in phimosis.

Penile fibropapillomas are usually easily removed with the bull restrained on a tilt table or in a squeeze chute.[1] Manually extend the penis and either place a towel clamp under the dorsal apical ligament (Figure 18.3) or a gauze tourniquet around the penis proximal to the growth to aid in holding the extended penis. Prepare the surgical field and infiltrate 2–4 mL of 2% lidocaine subcutaneously across the dorsum of the penis proximal to the lesion (Figure 18.4). Repeat the surgical preparation and carefully identify the urethra to avoid incising this tissue during excision of the growth. Catheterize the urethra with a 10-French male dog urinary catheter to help identify the urethra and thus avoid this structure (Figure 18.5). Dissect the skin of the penis at the base of the lesion until the growth is completely removed. In lieu of sharp dissection with a scalpel, judicious dissection with a CO_2 laser may assist hemorrhage control. Large growths are more easily removed by gradually debulking the lesion until the entire mass is removed. Ligate any small vessels and close the skin with #0 absorbable suture (Figure 18.6). Very small growths may be excised without anesthesia or suture closure. Remove the towel clamp and return the penis to the preputial cavity. Application of topical or systemic antimicrobials is optional.

Bovine Reproduction, First Edition. Edited by Richard M. Hopper.
© 2015 John Wiley & Sons, Inc. Published 2015 by John Wiley & Sons, Inc.

156 The Bull: Reproductive Surgery

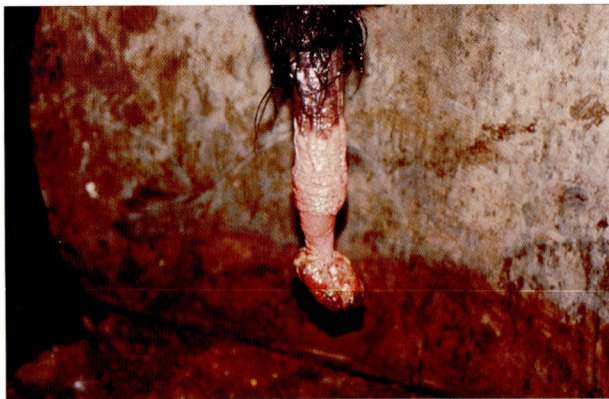

Figure 18.1 Penile fibropapilloma encompassing glans penis.

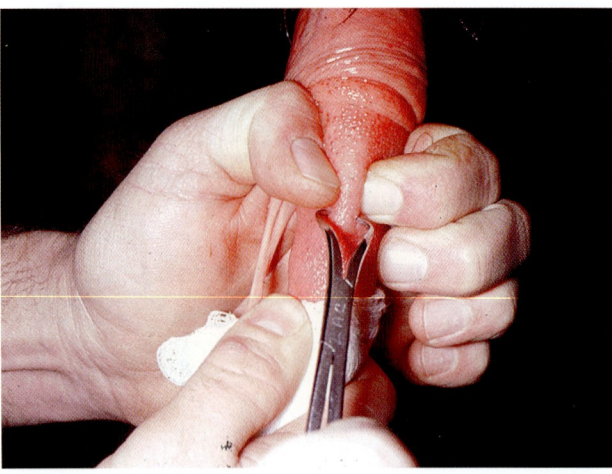

Figure 18.3 Towel clamp placed under dorsal apical ligament to hold penis in extension.

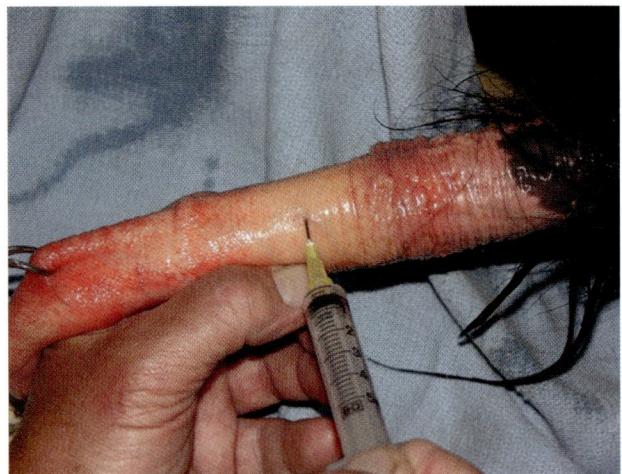

Figure 18.4 Local anesthetic injected subcutaneously over dorsum of penis for anesthesia of distal penis.

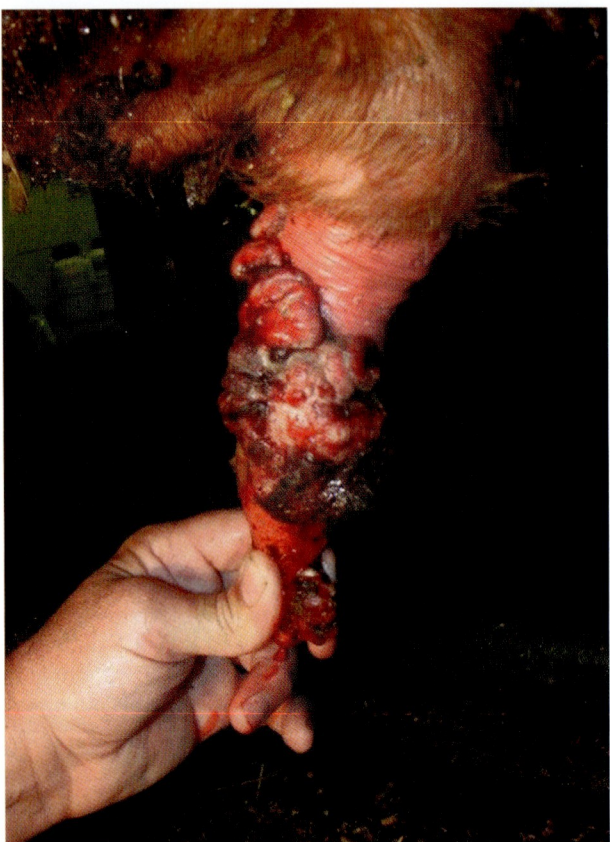

Figure 18.2 Multiple sessile penile fibropapillomas on penis and prepuce.

Postoperative care

Penile fibropapilloma may recur since new growth can occur if the bull is an active state of disease. Complete removal of the growth with an adjacent margin of unaffected penile surface epithelium lessens the likelihood of recurrence. Commercial or autogenous wart vaccine has been suggested to prevent or reduce recurrence of the lesions. Anecdotal evidence suggests the use of Immunoboost (Bioniche Animal Health, USA) may help prevent recurrence of penile lesions. Bulls treated for penile fibropapilloma should be examined for healing or regrowth 4 weeks following surgery before entering breeding service.

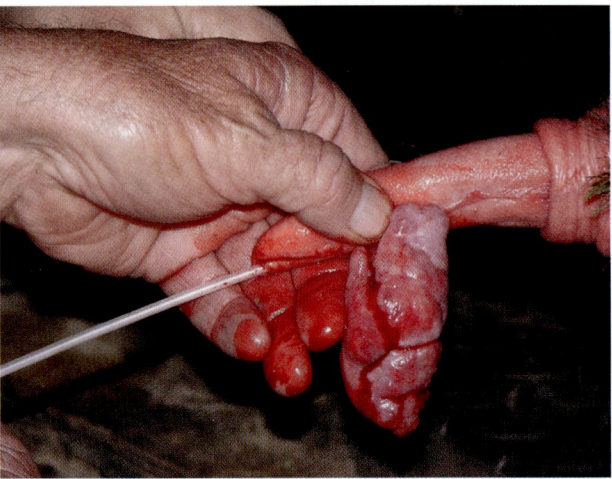

Figure 18.5 Male dog urinary catheter inserted into urethra of bull penis.

Restorative Surgery of the Penis   157

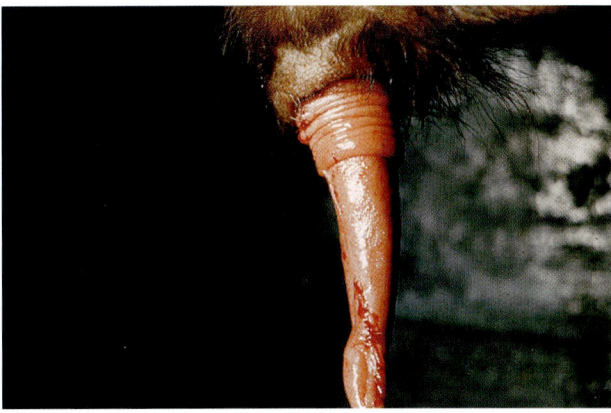

Figure 18.6 Closure of epithelium of penis following removal of fibropapilloma.

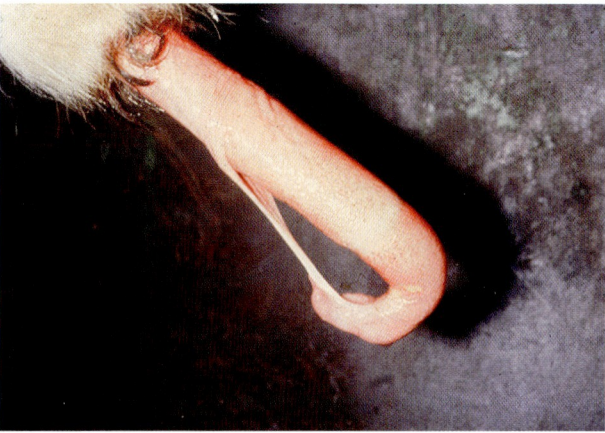

Figure 18.8 Persistent frenulum preventing complete straightening of erect penis.

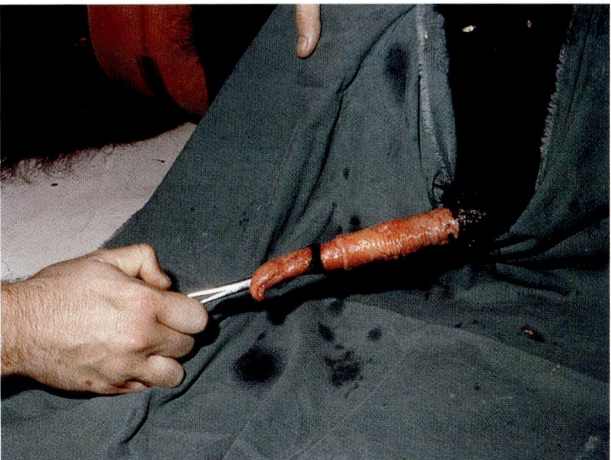

Figure 18.7 Penile hair ring.

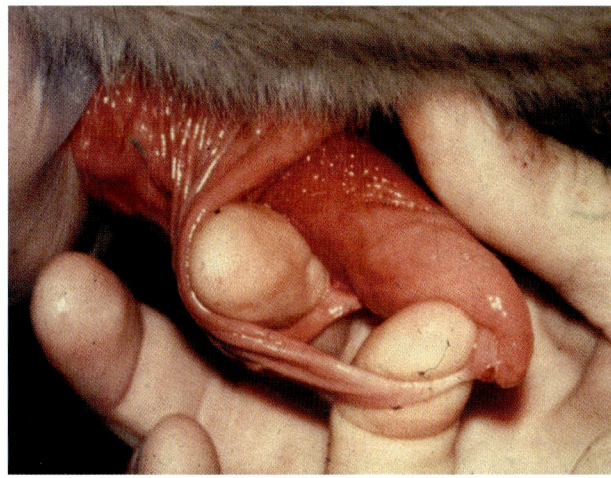

Figure 18.9 Persistent frenulum with two branches attached from prepuce to penis.

Penile hair rings

Body hair may accumulate on the penis of young bulls during homosexual riding and may develop an encircling ring on the penis of the more aggressive bull[1] (Figure 18.7). The encircling ring may become sufficiently tight to cause pressure necrosis with damage to the urethra causing urethral fistula formation. In severe cases the constricting ring may cause avascular necrosis of the surface epithelium of the penis and in the most severe cases damage the dorsal nerves of the penis or slough the entire glans penis. Treatment involves removal of the encircling ring of body hair and topical application of an emollient antibacterial agent. If a urethral fistula has formed, surgical repair may be necessary to restore breeding soundness.

Persistent frenulum: delayed preputial–penile separation

The penis of the bull calf cannot be extended prior to puberty due to interdigitating attachment of the skin of the penis and prepuce and lack of a sigmoid flexure. During puberty androgen production shifts from androstenedione to testosterone, and the attachment of the penis and prepuce begins and should be complete between 8 and 11 months of age. Young bulls are occasionally presented with incomplete separation at 12–14 months of age.[1,3] In these bulls separation can be completed by pulling the prepuce back from the free portion of the penis. These tissues should separate easily and hemorrhage is seldom a problem.

The penile frenulum is a thin band of connective tissue on the ventral midline of the free portion of the penis which adjoins the prepuce. Normally the frenulum ruptures during penile separation from the prepuce. When the frenulum does not rupture, the penis extends but the frenulum causes ventral bending of the distal penis during extension (Figures 18.8, 18.9 and 18.10). Surgical repair is relatively simple. I recommend ligating each end of the frenulum and transecting the tissue to reduce the possibility of hemorrhage, although not all practitioners adhere to this practice (Figures 18.11, 18.12 and 18.13) The owner should be advised that this condition is considered to be heritable and retaining the bull's sons as sires is not recommended (see also Chapter 17).

Anatomy and physiology of erection

Erection in the bull occurs when blood flow increases in the deep artery of the penis and into the crus penis and subsequently into the corpus cavernosum penis (CCP) following

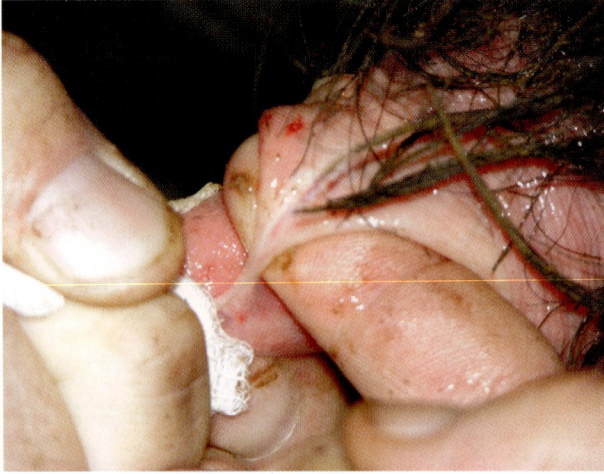

Figure 18.10 Very short persistent frenulum preventing separation of penis and prepuce.

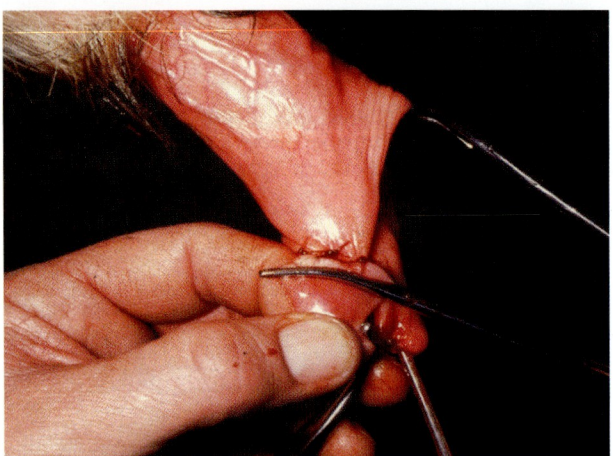

Figure 18.11 Transfixation ligatures on each end of persistent frenulum.

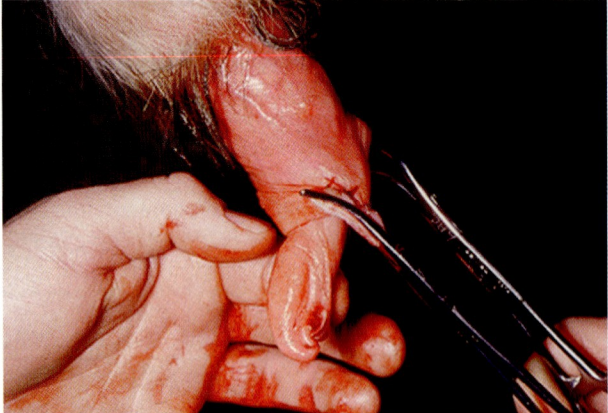

Figure 18.12 Excision of frenulum adjacent to transfixation ligatures.

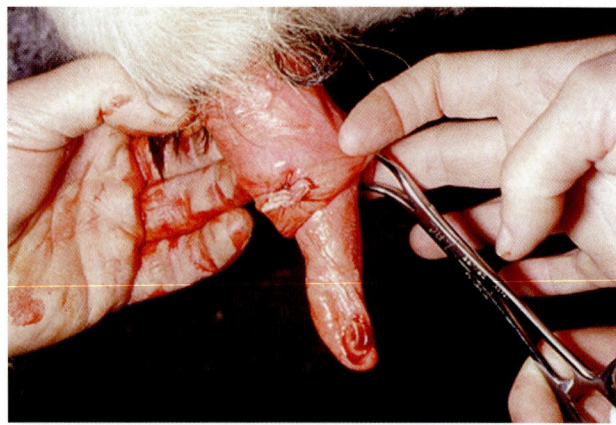

Figure 18.13 Completion of excision of persistent frenulum.

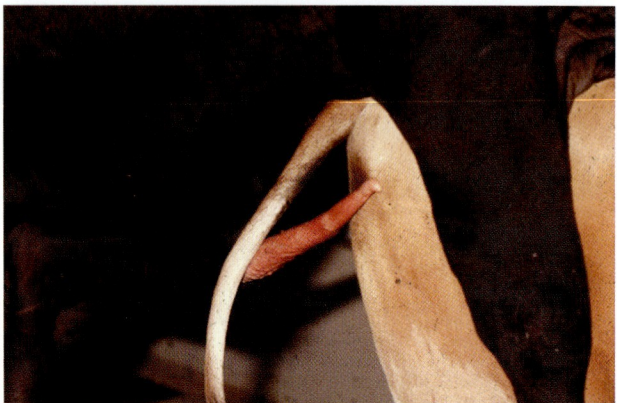

Figure 18.14 Bull with lack of innervation cannot achieve intromission.

olfactory or visual sexual stimulation.[4–6] The CCP in the bull is a closed system in that erectile blood flows into the penis from the crus and leaves this same area during detumescence following erection. The stimulation that causes this reflex dilation of the deep artery of the penis also causes relaxation of the retractor penis muscles which hold the penis in the preputial cavity. As the retractor penis muscles relax, the sigmoid flexure relaxes and the mildly engorged penis protrudes from the sheath.[6–8] With continued sexual stimulation the ischiocavernosus muscles begin rhythmic contraction that raises blood pressure from the normal resting state of 15 mmHg within the CCP. Peak pressure within the CCP may be greater than 14 000 mmHg (1.87 MPa). This rapid increase in blood pressure within the CCP causes complete penile extension and erection. Following ejaculation the ischiocavernosus muscles relax, detumescence occurs as blood pressure within the CCP decreases, and the penis is withdrawn back into the preputial cavity.[8]

Erection may be induced in the bull with an ejaculator, although the optimal method for evaluating erection is with observed test mating. Normal function of the penile nerves is essential for coitus and is most accurately assessed by observed test mating or by semen collection using an artificial vagina.[9,10]

Test mating

Bulls with erectile dysfunction do not achieve sufficient erection pressure to complete coitus.[11,12] Bulls with nerve dysfunction mount the cow but there are no penile searching motions near the vulva and the bull fails to make intromission.[9,10] Usually the penis is placed along the cow's hip or below the vulva in the escutcheon area above the cow's udder (Figure 18.14).

Diagnosis of CCP fistulas or vascular shunts (cavernosography)

Contrast radiography of the CCP, known as cavernosography, may confirm vascular defects in the penis.[11–14] Although the procedure may be performed with the bull in a squeeze chute, the technique is more easily accomplished with the bull restrained on a table in lateral recumbency. Manually extend the penis and place a towel clamp under the dorsal apical ligament approximately 5 cm from the distal end of the penis to aid in manipulation of the penis during the procedure (Figure 18.15). Place a length of umbilical tape through the rings of the towel clamp to allow complete removal of the hands from the radiographic field during the procedure (Figure 18.16). Place a double strand of heavy suture (0.6 mm) through the skin of the sheath, between the retractor penis muscles and the penis, and through the skin on the opposite side of the sheath (Figure 18.17). This suture serves to retract the penis away from the abdominal wall to enhance visualization of the sigmoid flexure of the penis.

With the penis fully extended, insert a 16-gauge 3.8-cm needle at a 45° angle proximally through the skin and tunica albuginea and into the CCP. Place the needle on the dorsum of the penis near the towel clamp. To ascertain that the needle has penetrated the tunica albuginea and the tip is within the CCP inject 10 mL sterile saline which should flow into the CCP with ease. Attach a sterile extension set to the needle for ease of injection and to position the hands away from the radiographic field (Figure 18.18). Place a radiographic cassette under the penis, then inject 15 mL of water-soluble radiographic contrast media (Renograffin 76, Squibb Diagnostic, New Brunswick, NJ) and expose the film. Slowly inject an additional 15–30 mL of media as the radiographic series is performed. Remove the cassette and quickly place another cassette more proximal under the penis. By using 43-cm cassettes the entire penis up to the sigmoid flexure may be radiographed with two or three exposures (Figure 18.19). Ideally all radiographic exposures should be completed within 60 s.

There are no vascular communications from the CCP to peri-penile vasculature in the normal penis, and there

(a)

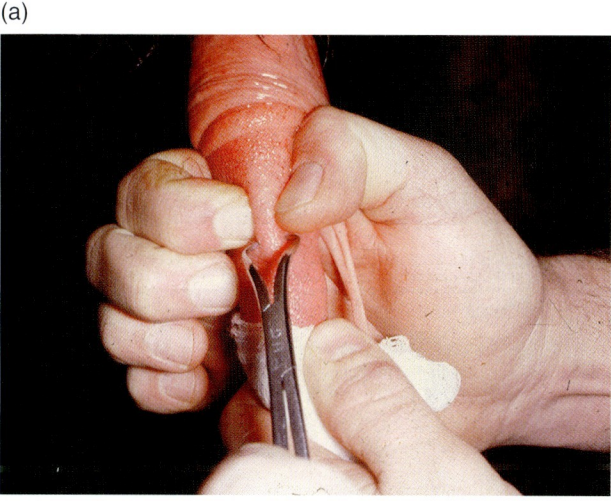

(b)

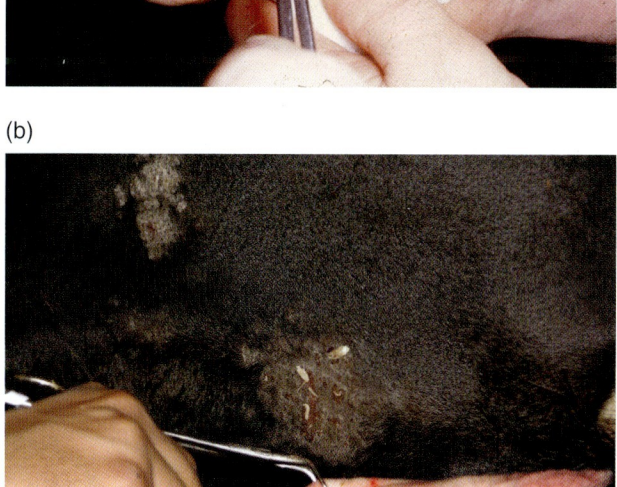

Figure 18.15 (a) Correct placement of towel forceps under dorsal apical ligament to hold penis. (b) Penis held in extension by towel forceps under dorsal apical ligament.

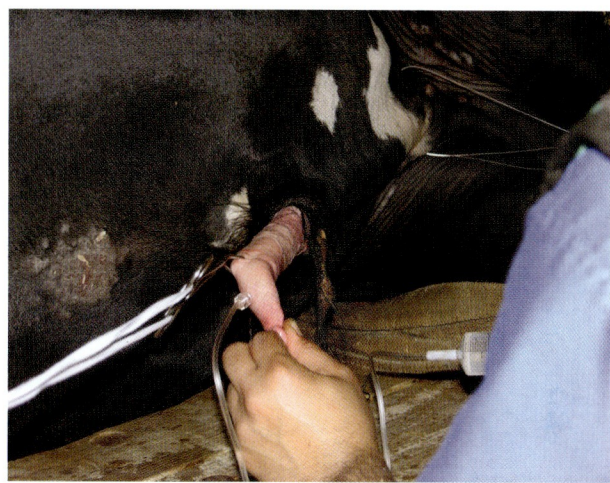

Figure 18.16 Umbilical tape passed through rings of towel forceps to allow hands to be clear of radiographic field.

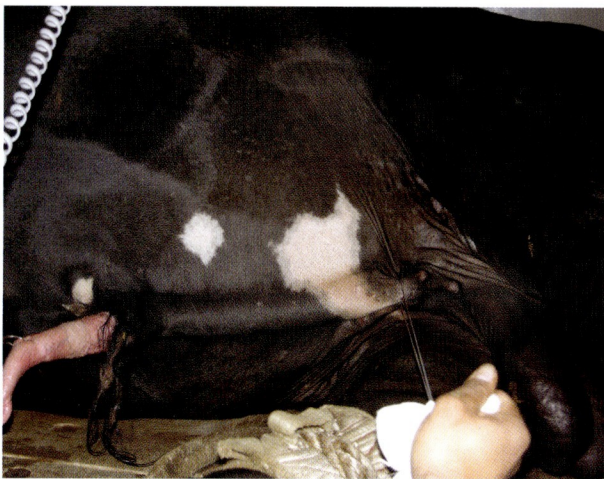

Figure 18.17 Double strand of heavy suture placed through sheath between retractor penis muscles and penis.

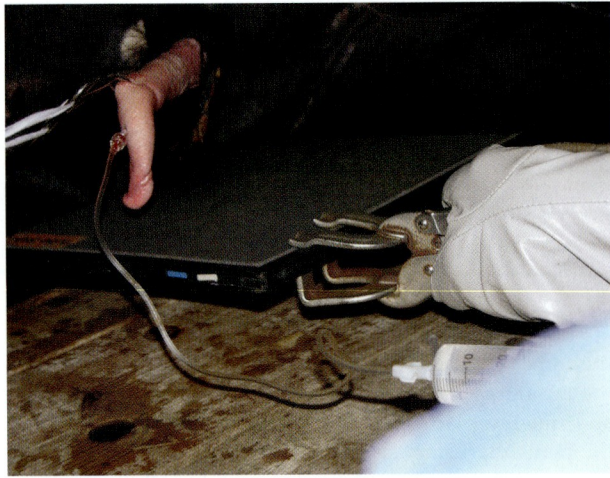

Figure 18.18 Needle inserted into CCP and attached to sterile intravenous extension set to allow positioning of hands away from radiographic field.

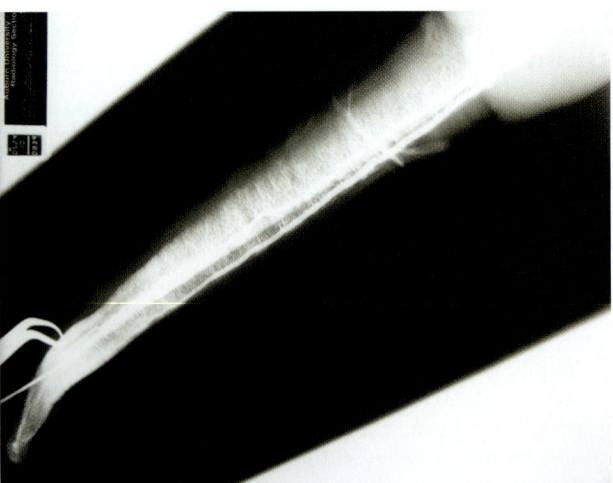

Figure 18.20 Contrast radiograph of normal bovine penis. All contrast media is within CCP.

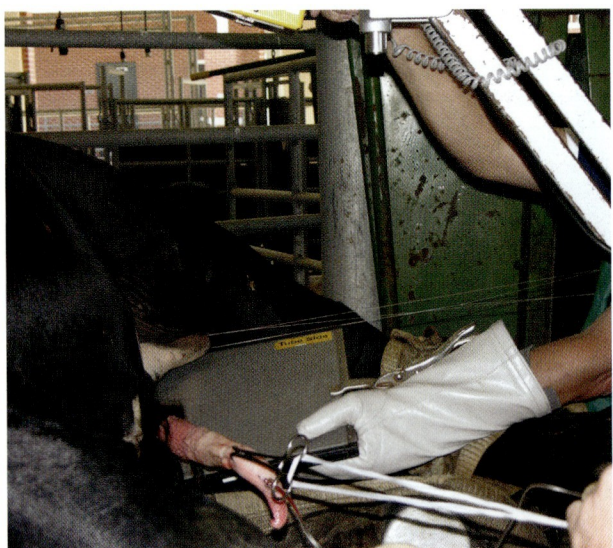

Figure 18.19 Radiographic cassette placed under penis for contrast radiography.

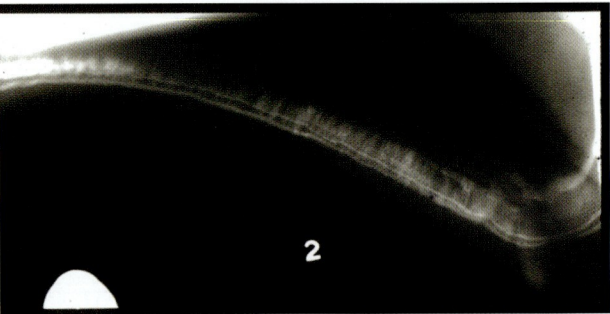

Figure 18.21 Post-hematoma CCP shunt at dorsum of distal bend of sigmoid flexure of bull penis.

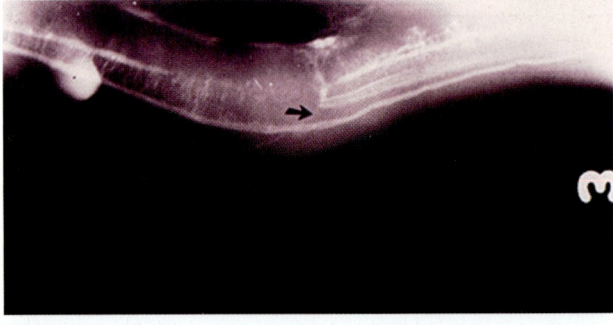

Figure 18.22 Post-hematoma shunt from CCP to corpus spongiosum penis (arrow) at distal bend of sigmoid flexure of penis.

should be no contrast media outside the CCP (Figure 18.20). A vascular shunt is identifiable as contrast media exiting the CCP (Figures 18.21 and 18.22).

Congenital vascular shunts: erection failure due to corpus cavernosal shunts (impotentia erigendi)

Occasionally young bulls fail to achieve intromission due to congenital corpus cavernosal vascular shunts. These bulls usually are normal on physical examination but fail to achieve adequate intracorporeal pressure for erection. When observed during erection, either by test mating or with electroejaculation, the free portion of the penis becomes noticeably bluish during attempted erection. The bluish discoloration is due to blood from a relatively porous tunica albuginea of the penis exiting the CCP and being removed by subcutaneous capillaries and veins. These shunts may be confirmed by cavernosography (Figure 18.23). Typically the shunts are multiple and not considered repairable.

Hematoma of the penis

The term "hematoma of the penis" is used to indicate rupture of the tunica albuginea of the penis, an injury that happens much more frequently in bulls than other farm animals and may lead to permanent infertility. Common terms for this condition are broken or fractured penis.[1,15–17] Hematoma of

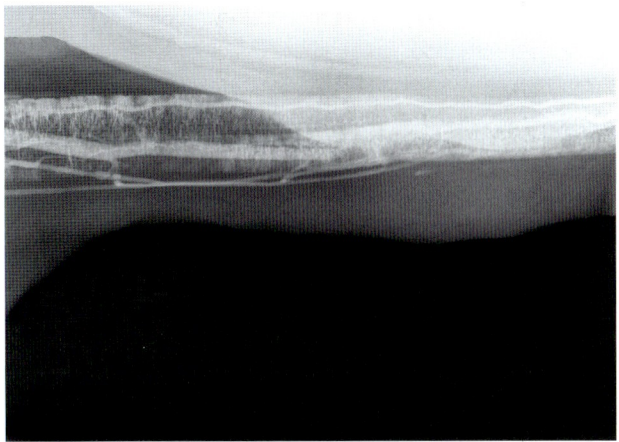

Figure 18.23 Contrast media outside CCP revealing multiple vascular shunts.

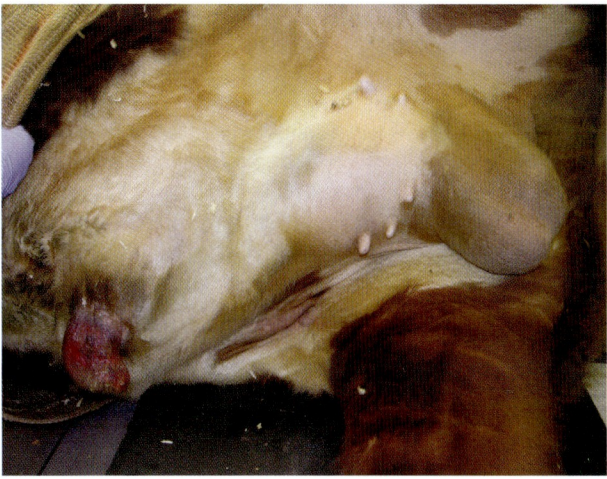

Figure 18.24 Symmetrical swelling in sheath immediately cranial to scrotum due to penile hematoma.

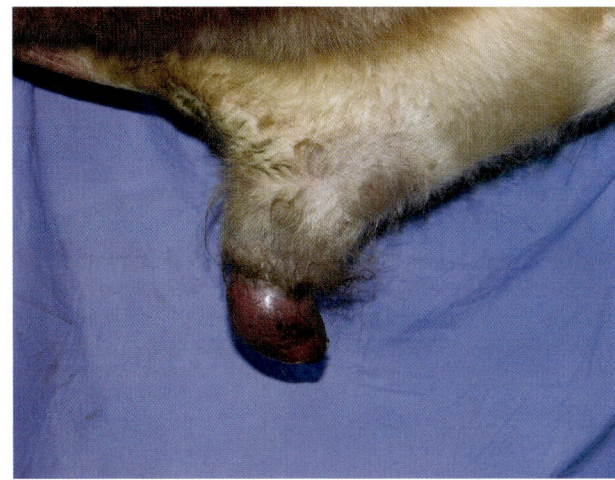

Figure 18.25 Bruised appearing preputial prolapse due to hematoma of the penis.

the penis due to rupture of the tunica albuginea is much more serious than small hematomas that may rarely develop due to rupture of superficial vessels of the prepuce.

A bull normally approaches the cow from the side, then moves to her hindquarters to smell the vulva and confirm that she is in estrus. The bulbospongiosus muscles begin rhythmic contractions that are visible as pulsations just ventral to the anus. As the retractor penis muscles relax the penis begins to protrude from the sheath and the bull prepares to mount. As he mounts the penis becomes engorged and the free portion should visibly extend from the sheath. When the bull fully mounts the erection becomes complete and the glans penis makes two or three searching motions near the vulva. The bull achieves intromission and ejaculates in one forceful thrust and then dismounts the cow.

If the bull does not achieve intromission, is misaligned with the vagina, or the female slips or goes down during the breeding thrust, the penis may abruptly bend. The penis is a closed vascular system during erection and when the penis bends the effective volume of the CCP is reduced. Blood pressure within the CCP increases drastically due to the suddenly reduced volume and the tunica albuginea of the penis ruptures.[17] Experimental studies have shown that the average rupture pressure is 75 000 mmHg (10 MPa) and that the penis consistently ruptures transverse to the long axis on the dorsum of the distal bend of the sigmoid flexure.[18] The rupture of the tunica albuginea is predictably 2–7.5 cm in length. The volume of blood in the CCP at the time of rupture is less than 250 mL. This blood under pressure forcefully enters the peri-penile elastic tissue, creating swelling immediately cranial to the base of the scrotum.

Diagnosis

This is easily diagnosed by history and physical examination.[1, 15–17] The injury usually occurs early in the breeding season when the bull is an aggressive breeder and a large number of females may be cycling. Occasionally the injury happens late in the breeding season when the bull may be fatigued and perhaps has lost body condition and is not as athletic or becomes awkward or clumsy during his breeding attempts.

Immediately after rupture of the tunica albuginea, the bull retracts the penis into the sheath and blood and edema within the peri-penile elastic tissue cause visible swelling in the sheath immediately cranial to the base of the scrotum (Figure 18.24). The size of the swelling is variable depending on how many breeding attempts the bull makes following the injury. Some bulls continue to attempt erection and breeding and therefore the volume of blood entering the peri-penile elastic increases.[16,19,20] The volume of blood within the CCP is less than 250 mL during erection but repeated attempts at breeding may cause significant swelling at the base of the scrotum, ranging in size from as small as a softball to larger than a volleyball.

Blood and edema within the peri-penile elastic tissue migrates ventrally and may lead to mild to severe prolapse of the prepuce. The prolapsed tissue appears swollen and has a distinct bluish or bruised appearance due to subcutaneous blood accumulation (Figure 18.25).

Observe the bull from a distance to determine the conformation of the sheath and then restrain the bull in a chute for physical examination. Palpate the distal bend of the sigmoid flexure to confirm rupture of the tunica albuginea. Swelling from penile hematoma is symmetrical on the dorsum of the penis at the neck of the scrotum. This swelling must be differentiated from swelling from urethral rupture or from retropreputial abscess or cellulitis associated with preputial laceration. Urethral rupture produces swelling ventral to the sigmoid flexure. Retropreputial abscess is usually located more distally along the shaft of the penis and is seldom symmetric along the dorsum of the penis[1,15,20] (Figure 18.26). Ultrasound examination of the swollen area may aid in diagnosis. Occasionally, hemorrhage from the tunica albuginea rupture will extend caudally along the retractor penis muscles causing these muscles to be palpably enlarged. The skin of light-colored bulls may show bruising from extravasated blood at the time of penile rupture (Figure 18.27). Swelling within the peri-penile elastic tissues prevents penile extension and the veterinarian should avoid manual attempts to extend the penis as such efforts may tear already injured elastic tissues.

Conservative therapy

Approximately 50% of bulls with rupture of the tunica albuginea return to breeding soundness without surgical repair of the injured penis. Conservative therapy of absolute sexual rest for a minimum of 60 days with systemic antibiotic therapy for the first 7–10 days is recommended. Daily hydrotherapy of the swollen sheath may stimulate circulation to the damaged tissues and assist with resorption of the edema and blood clot within the elastic tissues. If preputial prolapse is present, an emollient ointment should be applied and the protruding tissues protected with a support bandage or sling to prevent secondary trauma (see Chapter 17). Avoid extending the penis for at least 60 days and the bull should have a complete breeding soundness examination prior to returning to service.

Surgical procedure

Approximately 75% of bulls treated with surgical repair of penile hematoma are expected to return to breeding soundness if the surgery is performed quickly after the injury and the prepuce is not seriously injured.[1,15,19,20] Surgery should be performed as quickly as practical after the injury; I prefer to perform surgery between 3 and 7 days after the injury.

Following a 24–48 hour fast place the bull on a surgical tilt table in right lateral recumbency and induce general anesthesia. Secure the left rear leg in flexed abduction to provide optimal surgical access and clip and prepare the sheath and neck of the scrotum for aseptic surgery (Figure 18.28).

Make a 20-cm skin incision just cranial and parallel to the rudimentary teats over the swelling on the left lateral aspect of the sheath (Figure 18.29). Extend the incision through the subcutaneous tissue until the clotted blood of the hematoma is encountered in the peri-penile elastic tissue (Figure 18.30). Ligate bleeding vessels as necessary and manually evacuate the majority of clotted blood to visualize the penis within the peri-penile elastic tissue. Copiously lavage the cavity with warm 3% povidone-iodine in sterile saline. Grasp the penis and manually exteriorize the distal bend of the sigmoid flexure through the incision site (Figure 18.31).

Locate the distal bend of the sigmoid flexure by identifying the attachment of the retractor penis muscles. Identify the ventrum of the penis by palpating the longitudinal urethral groove and using the urethral groove as a landmark make a 20-cm longitudinal incision through the elastic tissue on the lateral aspect of the penis to expose the tunica albuginea of the penis at the level of attachment of the retractor penis muscles

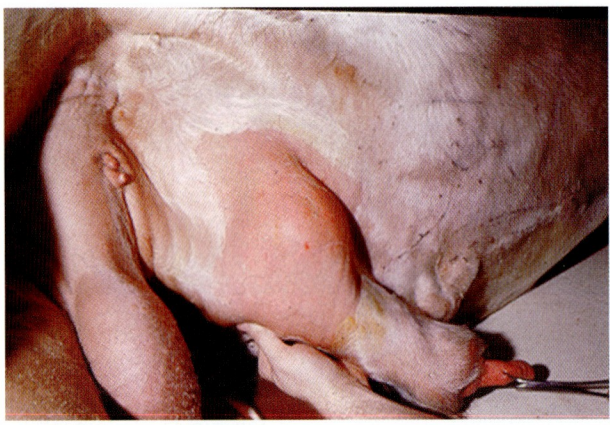

Figure 18.26 Swelling in sheath due to retropreputial abscess.

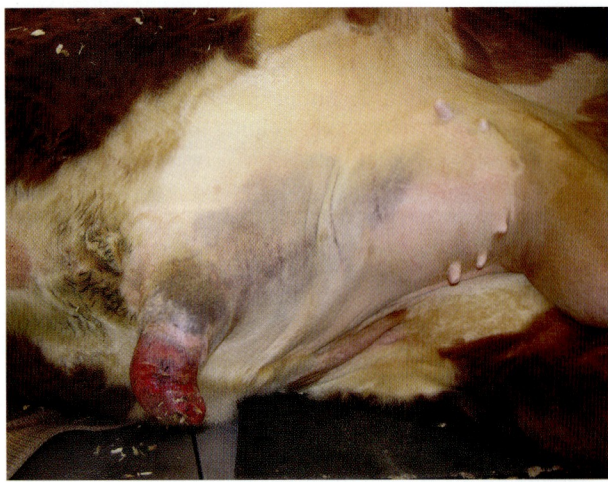

Figure 18.27 Bruising of skin of sheath due to penile hematoma.

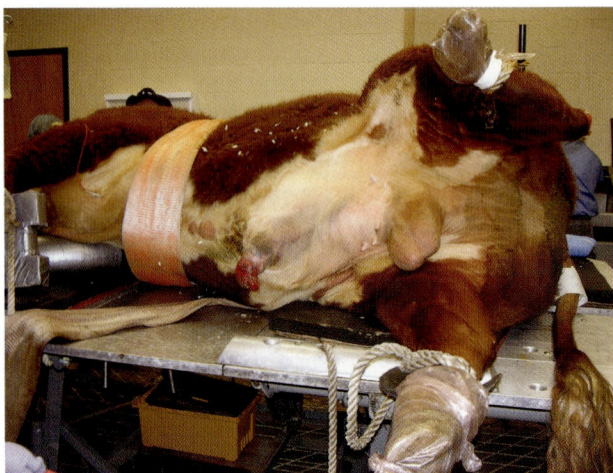

Figure 18.28 Bull positioned in right lateral recumbency with left rear leg secured in flexed abduction to expose surgical site.

Restorative Surgery of the Penis

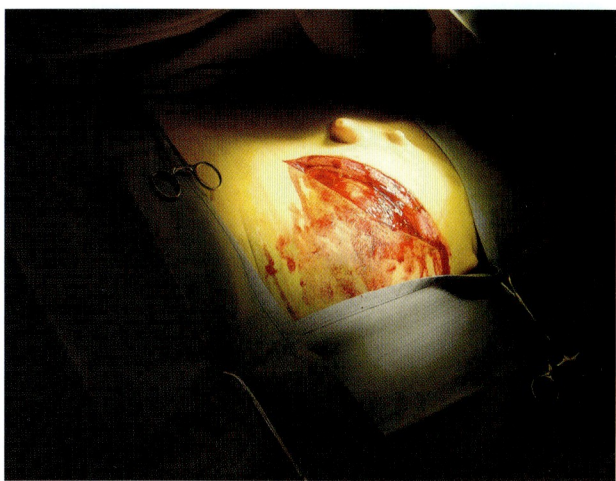

Figure 18.29 Skin incision cranial and parallel to teats for hematoma surgery.

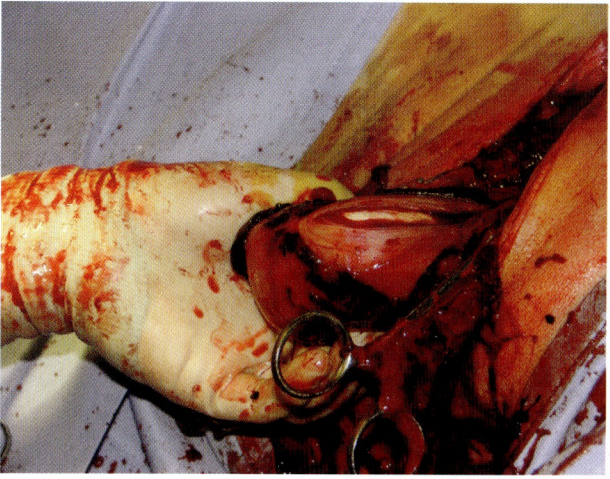

Figure 18.32 Lateral longitudinal incision in peri-penile elastic tissue to expose tunica albuginea of penis.

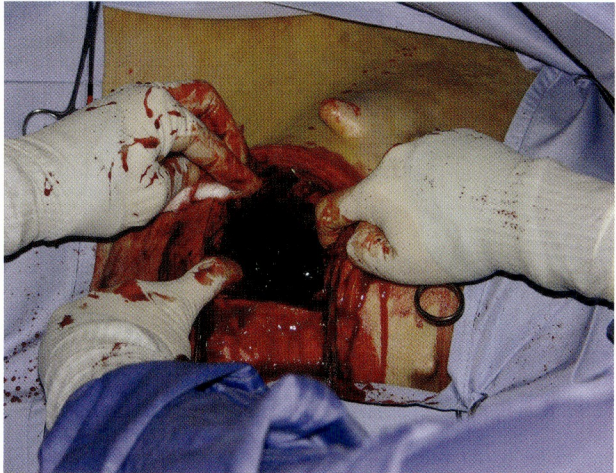

Figure 18.30 Blood clot from penile hematoma exposed through skin incision.

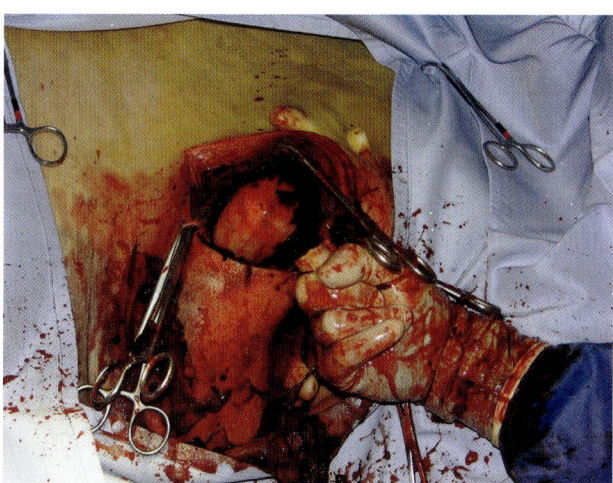

Figure 18.31 Exteriorize distal bend of sigmoid flexure of penis through incision.

(Figure 18.32). Carefully bluntly elevate the elastic tissue dorsally to expose the transverse tear in the dorsum of the tunica albuginea (Figure 18.33). This approach avoids damaging the dorsal arteries, veins and nerves of the penis.

Remove as little tissue as possible while carefully trimming the frayed edges of the wound to ensure smooth apposition of the wound edges (Figure 18.34). Removal of excessive tissue may sufficiently shorten the dorsum of the erect penis, which may prevent complete straightening of the sigmoid flexure in subsequent breeding attempts.

Firmly appose the trimmed edges of the tunica albuginea with #1 polyglycolic acid suture in a bootlace pattern (Figure 18.35). Apposition of the edges of the dense fibrous tunica albuginea is easier if the sutures are loosely placed and traction is then applied from either end of the sutures to achieve edge-to-edge apposition of the tissues (Figure 18.36). Appose the elastic tissues over the penis and close the longitudinal incision with simple continuous #2-0 chromic gut sutures in the superficial layers. Return the penis to its normal position within the sheath and irrigate the surgical field with warm 3% povidone-iodine in saline. Gently remove remaining blood clots and close the subcutaneous tissues with #0 chromic gut using a simple continuous pattern and the skin with 0.6-mm synthetic nonabsorbable suture using a continuous interlocking pattern (Figure 18.37).

Postoperative care

Administer systemic antibiotics for 5–7 days postoperatively. A small seroma normally forms in the hematoma cavity and should begin to subside within 10 days following surgery. Remove skin sutures 10–14 days following surgery and ensure sexual rest for at least 60 days. Continue treatment of preputial prolapse as needed but this condition usually resolves spontaneously within 1 week after surgery in *Bos taurus* bulls. In *Bos indicus* bulls the preputial prolapse may persist for longer than 2 weeks unless there is serious secondary damage to the prolapsed tissues. Occasionally bulls of *Bos indicus* breeds sustain sufficient preputial damage prior to hospitalization that circumcision may be necessary to return the bull to breeding soundness.

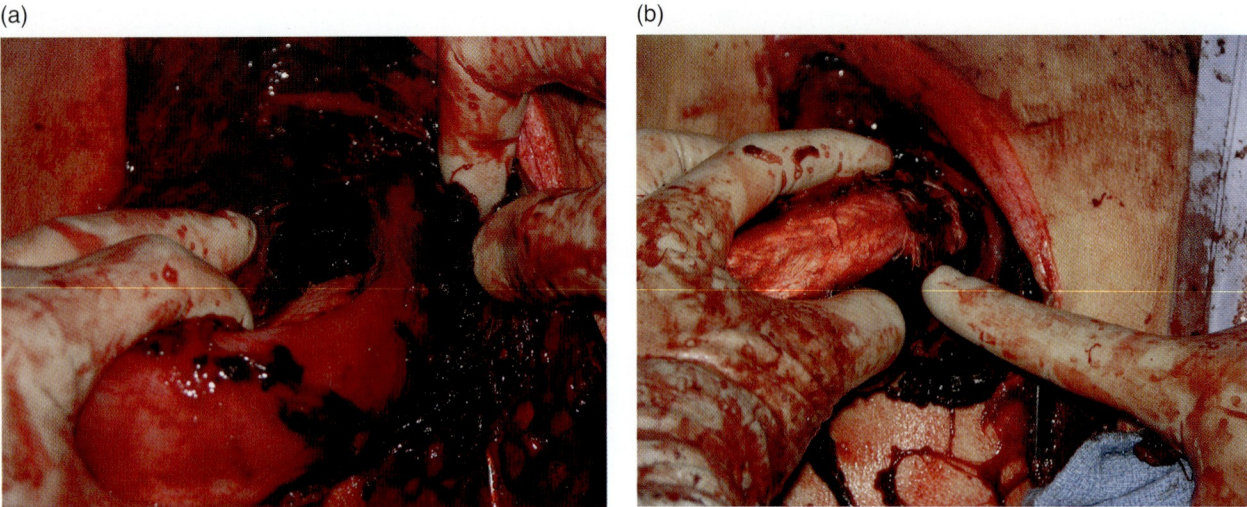

Figure 18.33 (a) Peri-penile elastic tissue dissected from dorsum of penis to expose rupture in tunica albuginea. (b) Transverse tear in dorsum of tunica albuginea of penis.

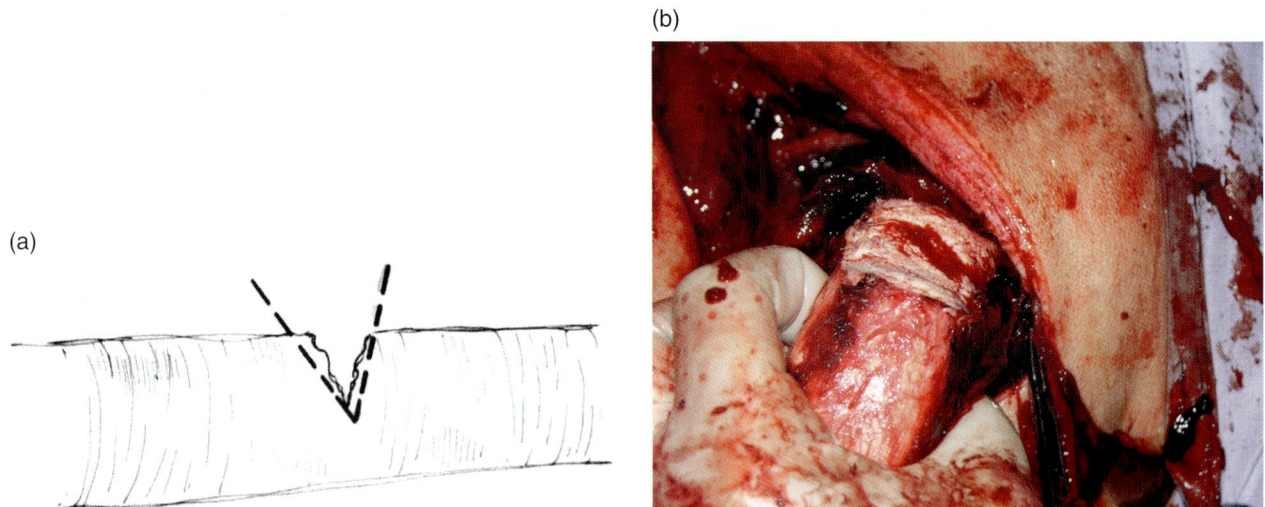

Figure 18.34 (a) Schematic of trimming of frayed edges of tear in dorsum of tunica albuginea of penis. (b) Frayed edges of tunic albuginea trimmed.

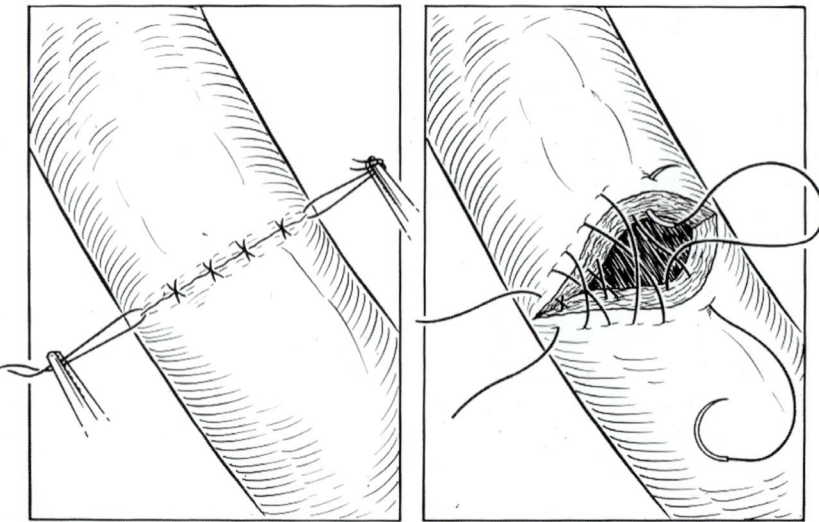

Figure 18.35 Schematic of bootlace suture closure of tunica albuginea.

Restorative Surgery of the Penis

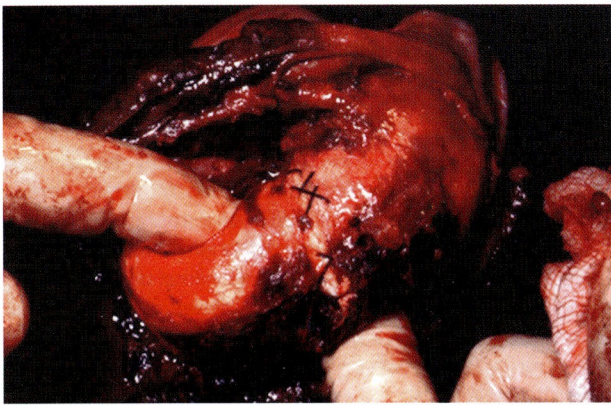

Figure 18.36 Bootlace suture closure of tunica albuginea of penis completed.

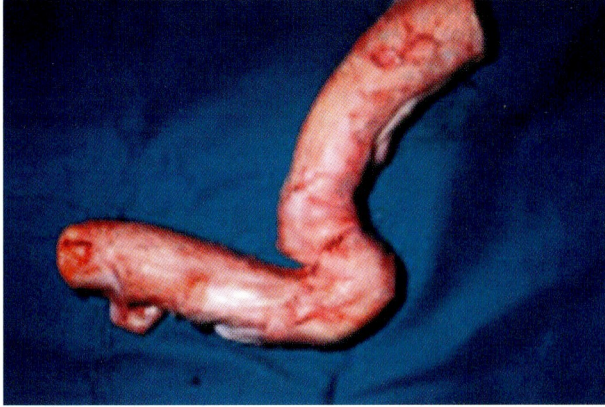

Figure 18.38 Fibrosis from penile hematoma that prevents straightening of distal bend of sigmoid flexure of penis.

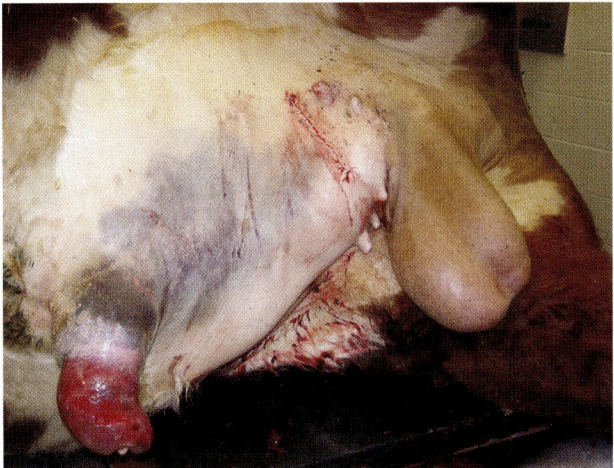

Figure 18.37 Completed suture closure of sheath following repair of tunic albuginea.

Complications

Several potential complications of rupture of the tunica albuginea of the penis should be considered when selecting a therapeutic plan for an injured bull. These complications include development of vascular shunts that prevent erection, loss of innervation of the free portion and glans penis, abscess of the hematoma, adhesions that prevent complete extension of the sigmoid flexure (Figure 18.38) or of the peri-penile elastic tissue to the sheath which prevent normal penile extension, recurrence of the injury, and damage to the prepuce caused by prolapse secondary to the penile hematoma.

Erection failure caused by vascular shunt formation

During tunica albuginea healing following penile hematoma, vascular shunts may develop between the CCP and the peri-penile vasculature that prevents the CCP from being a closed system and therefore pressure within the CCP adequate for erection cannot be achieved[1,11,12,20] Affected bulls typically display normal libido and penile engorgement but do not develop sufficient blood pressure within the penis to achieve intromission. Cavernosography for diagnosis of vascular shunt has been discussed earlier in this chapter.

Repair of vascular shunts

If the CCP shunt is confirmed by contrast cavernosography, surgical repair is similar to repair of the ruptured tunica albuginea following the initial injury.[1] Table the bull in right lateral recumbency and following induction of anesthesia and surgical preparation of the sheath adjacent to the neck of the scrotum make a 20-cm skin incision cranial and parallel to the teats. Exteriorize the penis within its elastic tissue through the skin incision and identify the attachment of the retractor penis muscles on the ventrolateral aspect of the distal bend of the sigmoid flexure. Palpate the urethral groove on the ventral aspect of the penis and using this landmark make a longitudinal incision through the elastic tissue on the left lateral aspect of the penis to expose the tunica albuginea. Carefully bluntly dissect and reflect the elastic tissues dorsally to expose the dorsum of the tunica albuginea of the penis. This approach reduces the risk of surgical trauma to the dorsal vessels and nerves of the penis.

The shunt is usually identified as a small adhesion on the dorsum of the tunica albuginea. Remove the shunt by careful full-thickness transverse wedge excision of the dorsum of the tunica albuginea of the penis. Carefully appose the edges of the tunica albuginea with one or two simple interrupted #1 polyglycolic acid sutures, ligate the excised wedge of tunica albuginea from the peri-penile elastic tissues, and appose the elastic tissues over the lateral aspect of the penis. Close the surgical site as described for hematoma of the penis.

Administer prophylactic antibiotics for 5 days and remove skin sutures 10 days after surgery. Ensure the bull has 60 days sexual rest and observe for mating ability when breeding resumes.

Loss of nerve innervation

The stimuli for intromission and ejaculation are generated primarily in the glans penis and free portion as the distal penis contacts the vagina.[7,9,10] If the dorsal nerves are damaged the nerve stimuli cannot reach the brain and the bull may not be able to achieve coitus even though he is capable of erection. If 75% of the dorsal nerves are damaged, the bull usually cannot ejaculate.

When the tunica albuginea ruptures, the jet of blood under high pressure causes significant damage to the peripenile elastic tissues and perhaps to the dorsal nerves of the penis, which lie in close proximity to the tunica albuginea. This pressure can impair the nerves, including the sensory fibers associated with ejaculation. Loss of sensation may be immediate if the dorsal nerves are irreparably damaged at the time of penile rupture or the nerves may become incorporated in fibrous tissue during healing of the hematoma site and become stretched or torn during subsequent penile extension when the bull masturbates or resumes breeding. The typical history with delayed denervation is that the bull serves a few cows and then is no longer able to attain intromission. Although nerve conduction velocity studies can confirm nerve injury, observed test mating with failure of intromission or failure to ejaculate into a normally prepared artificial vagina is usually conclusive evidence of loss of nerve sensation.[21] Attempted surgical repair or spontaneous healing of dorsal penile nerves is usually unsuccessful, although the bull may have semen collected by electroejaculation for freezing and subsequent artificial insemination.

Abscessation of the hematoma

Clotted blood accumulated within the tissues following rupture of the tunica albuginea may become infected resulting in abscess formation. This likely occurs due to depressed natural defenses in the injured tissues and the pooled blood, which is an excellent growth media for bacteria. Bacteria may gain entry to the injured tissues via the hematogenous route. Subsequent scar tissue that develops with resolution of the abscess and healing of the damaged tissues is often so extensive that penile extension is impaired. Bacteria from the abscess may additionally gain access to the cavernous spaces within the CCP, with eventual fibrosis of the spaces such that penile erection is not possible.[1,18]

Adhesions that prevent penile extension

Bulls recovering from penile hematoma may not be able to complete coitus due to excessive fibrous tissue that adheres from the penis to the sheath, or along the dorsum of the penis that prevents straightening of the distal bend of the sigmoid flexure. In either case the scar tissue may prevent adequate penile extension and these bulls are usually not able to return to breeding soundness.

Recurrence of penile hematoma

Occasionally a bull recovered from penile hematoma will again rupture the tunica albuginea in a future breeding season. These bulls may have innate weakness of the tunica albuginea but reinjury is more likely when the initial injury is not surgically repaired. Perhaps repeat injury occurs more frequently in bulls that are either overly aggressive breeders or that lack athletic agility. The majority of bulls that repeat this injury are less than 5 years of age, indicating that perhaps

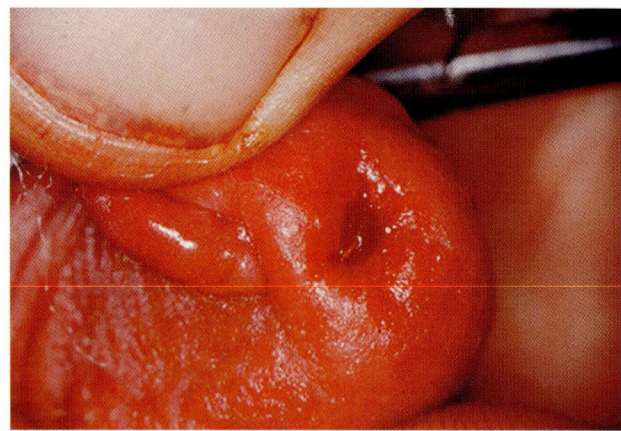

Figure 18.39 Fistula from CCP near tip of glans penis of a bull.

bulls predisposed to repeat injury leave the herd during the first few breeding seasons or that older bulls develop more refined breeding behavior.

Penile lacerations

Lacerations of the penis are rare in bulls and often only involve the skin of the penis. Minor scars are frequently found on the glans or free portion of the penis suggesting that the majority of lacerations are minor and heal spontaneously without complication.

Deep lacerations or punctures of the bull penis are rare but may penetrate the tunica albuginea of the penis.[1] These wounds may heal spontaneously with a fistula that allows hemorrhage when a bull attempts erection (Figure 18.39). Even though the bull may achieve sufficient pressure within the CCP to allow intromission and ejaculation, the fistula may allow blood contamination of the ejaculate preventing fertilization of the ova. Also, rarely, the puncture may heal with neovascularization and vascular shunt formation that prevents development of sufficient pressures within the CCP to complete coitus.

Penile deviations

Deviations of the penis may occasionally prevent intromission and coitus. Bovine penile deviations are classified as either spiral (corkscrew), ventral, or S-shaped.[1,22–26] Typically, bulls have had one or more successful breeding seasons prior to development of penile deviations with usually no known history of previous penile injury. Most bulls with these conditions are 3 or 4 years old and occasionally a bull over that age will develop penile deviation.

Spiral deviation

Spiral deviation of the penis occurs more commonly than either ventral or S-shaped deviations. One potential etiology involves malfunction of the dorsal apical ligament. The apical ligament, a thick collagen band, arises from the dorsum of the tunica albuginea about 2.5 cm proximal to the distal end of the prepuce and inserts into the tunica albuginea near the distal end of the CCP. The function of this structure is to help

Restorative Surgery of the Penis 167

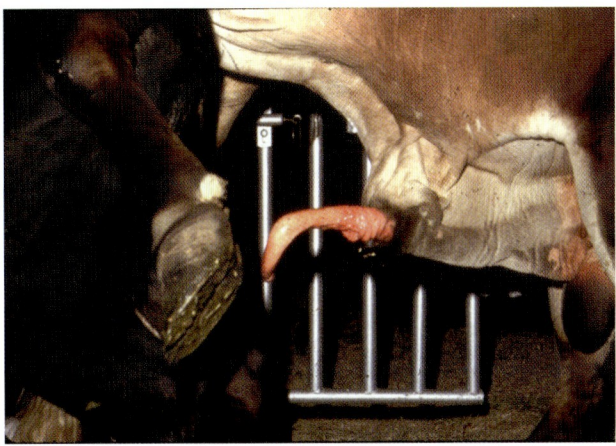

Figure 18.40 Spiral deviation of penis during attempted breeding.

maintain the shape of the penis during erection.[24] Historically, spiral deviations were considered to result from a short or inadequate apical ligament that slipped off to the left side of the penis at peak erection allowing the penis to spiral and therefore preventing intromission.[23–25]

However, the rapidly coordinated events leading to erection and the high maximum pressures within the CCP at ejaculation suggest that affected bulls may prematurely achieve maximum erection pressures prior to intromission. This suggestion may be supported by the fact that the penis of many normal bulls likely develops a spiral orientation in a cow's vagina during ejaculation.[26] Although the definitive cause of spiral deviations remains uncertain, apical ligament malfunction and premature achievement of high CCP pressure may be involved.

Many bulls develop spiraling of the distal penis during electroejaculation and masturbation.[27] Therefore, spiral deviation should only be diagnosed as a pathologic condition when it consistently develops during natural mating[1,18,24] (Figure 18.40).

Bulls with spiral deviation of the penis are most often 3 years old and have had a successful previous breeding season. Development of this condition may be insidious, occurring intermittently at first then more consistently as breeding efforts continue. Therefore, several test matings may be necessary to confirm the diagnosis. In *Bos indicus* bulls the deviation may occur prior to penile extension such that the deviation remains hidden within the bull's pendulous sheath and excessive prepuce. For bulls with this sheath conformation manual retraction of the prepuce during attempted breeding may be necessary to differentiate from erection or other causes of penile extension failure.

Ventral penile deviation

Ventral penile deviation occurs much less commonly than spiral deviation and the etiology is uncertain. Ventral deviations may occur as a result of altered blood flow through the ventral portion of the CCP or due to stretching or injury of the apical ligament, both of which probably result from chronic traumatic injury.[24,25]

Ventral penis deviation is obvious as long gradual curvatures of the erect penis (Figure 18.41). The curvature frequently originates proximal to the junction of the sheath and

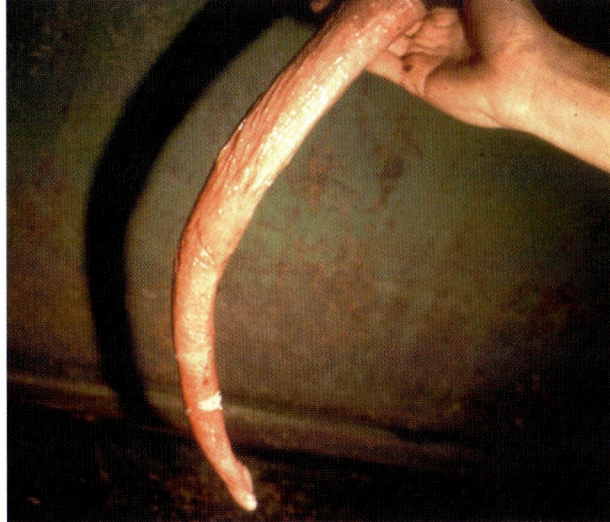

Figure 18.41 Ventral deviation of the bull penis.

prepuce.[24,25] These conditions become more apparent as erection pressure increases. Although ventral penile deviation can be induced during electroejaculation, observation during natural breeding is recommended, especially for less obvious cases. These conditions are only considered to be pathologic when they consistently prevent intromission.

Repair of spiral and ventral penile deviations

The prognosis for return to breeding soundness is greater following surgical repair of spiral than for ventral deviations. I recommend only attempting surgical repair of ventral deviations when the deviation is limited to the free portion of the penis. Spiral and ventral penile deviations are both repaired with a fascia lata graft. A narrow strip of fascia lata is sutured between the dorsal apical ligament and the dorsum of the tunica albuginea to serve as a fibroblast lattice to strengthen adjacent structures and stabilize the apical ligament on the dorsum of the penis.[28]

Fascia lata graft technique

Withhold feed for 24–48 hours and water for a minimum of 24 hours prior to general anesthesia. Place the bull in right lateral recumbency and clip and aseptically prepare a 20 cm wide by 40 cm long area dorsal to the left patella and extending toward the tuber coxae (Figure 18.42). Clip the hair around the preputial orifice, manually extend the penis and place towel forceps under the distal portion of the apical ligament to aid with penile extension.

Aseptically prepare the penis and prepuce and leave the towel forceps in place and allow the penis to return to the sheath while the fascia lata graft is harvested. The preplaced towel forceps will facilitate penile extension once the fascia lata harvest is complete.

Beginning 10 cm dorsal to the anterior border of the patella, extend a 20-cm skin incision toward the tuber coxae. Continue the incision through the superficial fascia to expose the deeper, thicker layer of deep fascia lata (Figure 18.43).

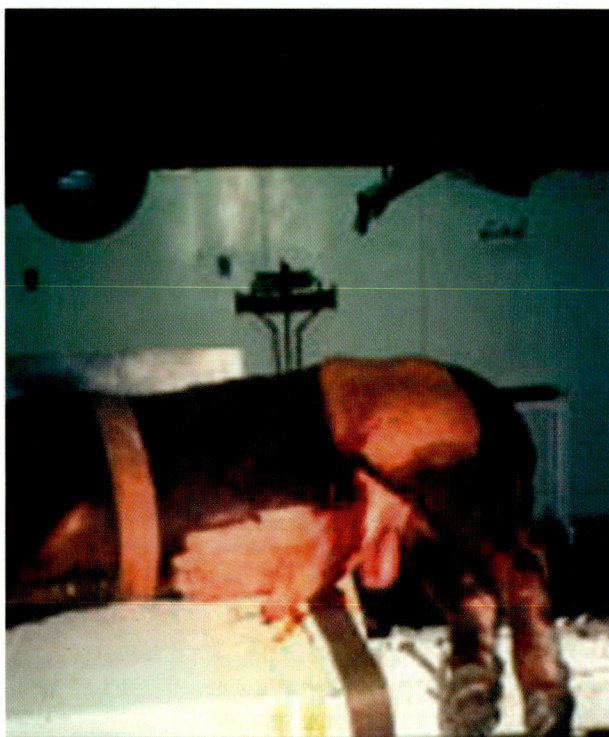

Figure 18.42 Bull positioned and clipped for fascia lata graft surgery.

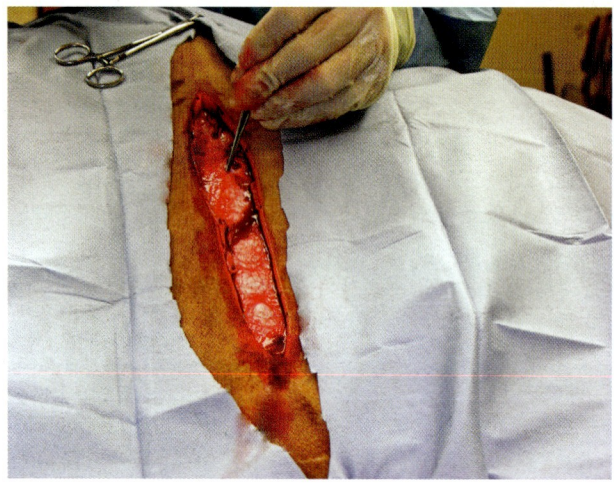

Figure 18.43 Incision through skin and superficial fascia lata to expose deep fascia lata.

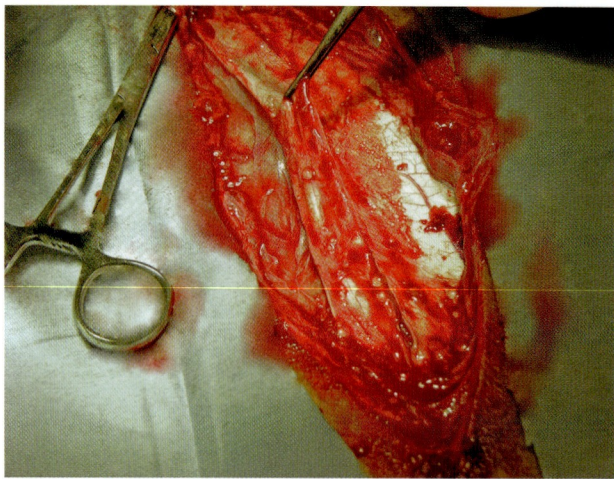

Figure 18.44 Incise 2-cm wide strip of deep fascia lata in preparation for excising graft.

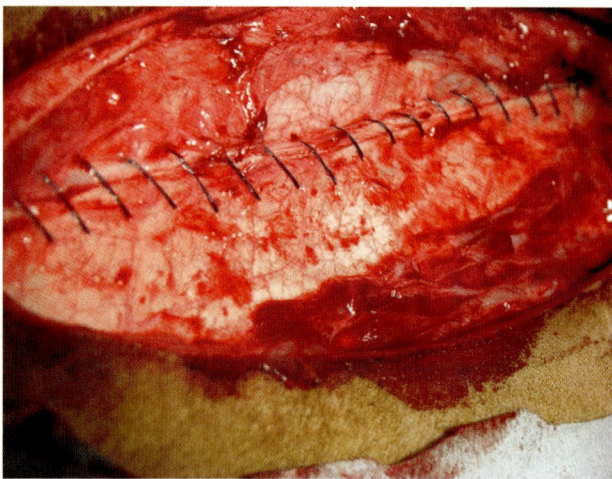

Figure 18.45 Suture closure of deep fascia lata.

Excise a 2 × 15 cm section of the deep fascia lata and place in warm saline while closing the incision (Figure 18.44). Close the deep layer of fascia lata with #1 polyglycolic acid suture in a bootlace or simple continuous pattern (Figure 18.45). Secure closure is imperative to prevent muscle herniation that could result in lameness. Close the superficial layer of fascia lata and subcutaneous tissues in two layers with #1 absorbable suture. Close the skin with nonabsorbable 0.6-mm synthetic suture in a continuous interlocking pattern. Remove all loose connective tissue and fat from the fascia lata strip to prepare the graft and then replace it in warm saline (Figure 18.46).

Extend the penis and place a 2.5-cm Penrose drain around the penis and prepuce near the distal end of the sheath as a tourniquet. Make a 20-cm skin incision along the dorsum of the penis beginning just caudal to the glans penis. Continue the incision through the thin fascial layers in the free portion of the penis and through the loose elastic layers in the preputial portion of the incision to expose the apical ligament (Figure 18.47).

Divide the apical ligament by making a midline incision through it thickest portion along its entire length. Reflect the two edges laterally to expose the tunica albuginea of the penis, which will serve as a bed for the fascia lata graft (Figures 18.48 and 18.49). Carefully avoid the two veins that drain the corpus spongiosum penis. These large exhaust veins from the corpus spongiosum penis are located deep to the apical ligament on the right ventral aspect of the penis near the glans penis.

Place the fascia lata graft on the tunica albuginea under the apical ligament as far proximally as possible. Ensure that the graft lays flat on the tunica albuginea and with #0 polyglycolic acid suture material place three or four simple interrupted sutures through the proximal border of the fascia lata graft and into the tunica albuginea (Figure 18.50). Avoid completely penetrating the entire thickness of the tunica albuginea into the CCP as this could result in vascular shunt formation and erection failure. Similarly, place interrupted sutures at 10-mm

Restorative Surgery of the Penis

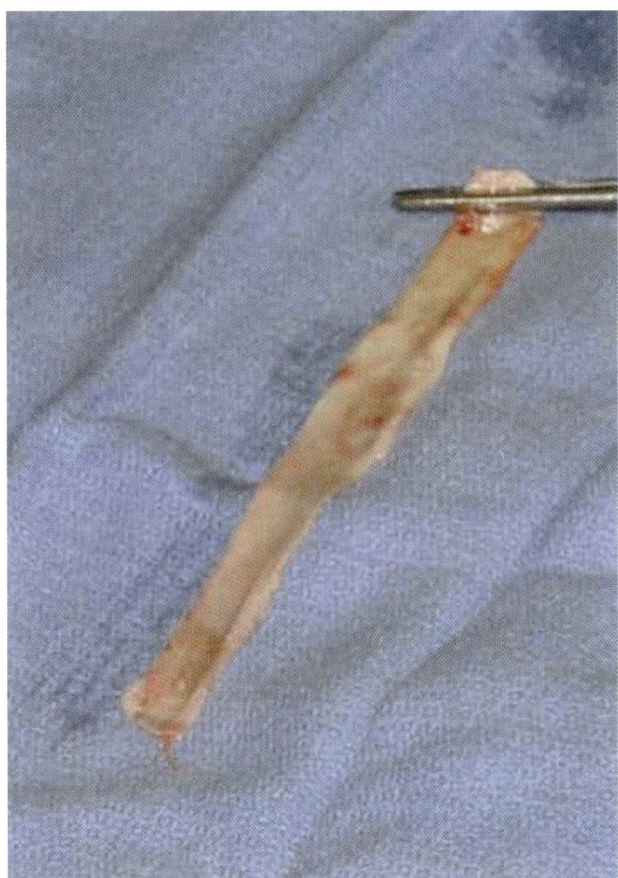

Figure 18.46 Fascia lata graft with loose areolar tissue removed.

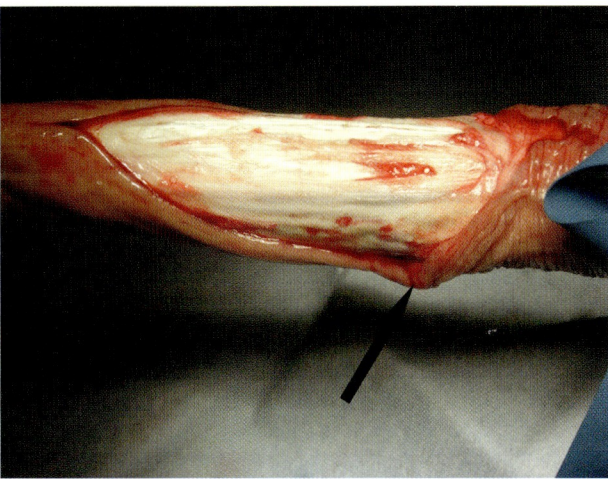

Figure 18.47 Dorsal longitudinal incision through penile surface epithelium (arrow denotes distal end of prepuce).

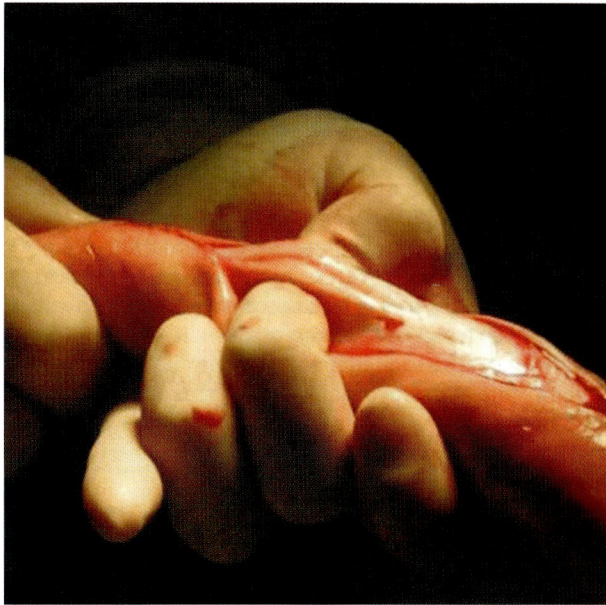

Figure 18.48 Beginning incision to divide dorsal apical ligament.

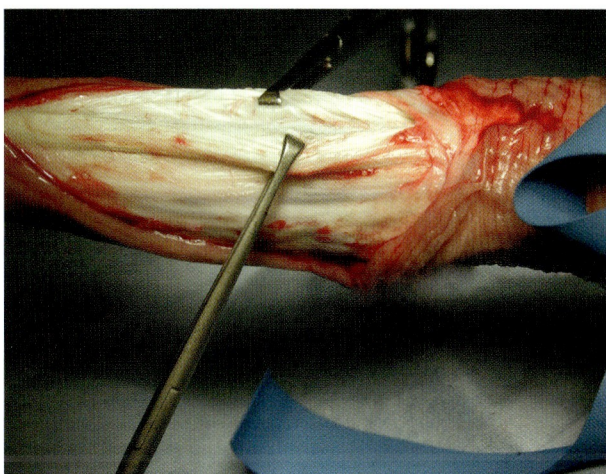

Figure 18.49 Dorsal apical ligament divided along its length and retracted with Allis tissue forceps.

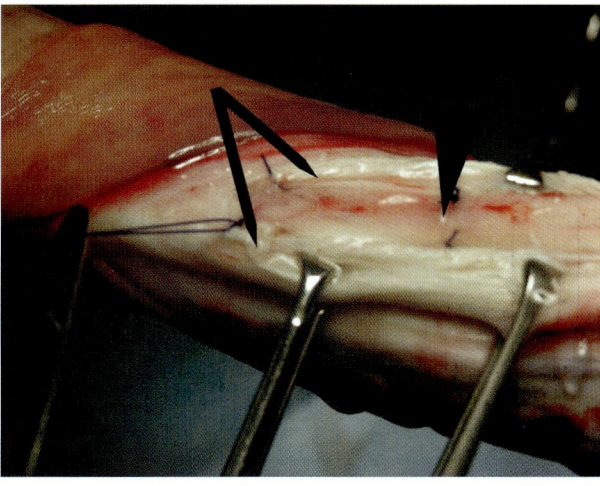

Figure 18.50 Fascia lata graft (heavy arrow) secured with simple interrupted sutures (small arrows indicate divided apical ligament).

intervals along the lateral border of the graft while stretching the fascia lata in both directions (Figure 18.51). Trim the graft to fit at the distal end of the apical ligament and suture under mild tension with three interrupted sutures.

Remove the tourniquet and ensure hemostasis for the surgical field. Appose the edges of the apical ligament over the fascia lata graft with #1 polyglycolic acid sutures placed 15mm apart in a simple interrupted pattern that engages the underlying fascia lata (Figures 18.52 and 18.53).

170 The Bull: Reproductive Surgery

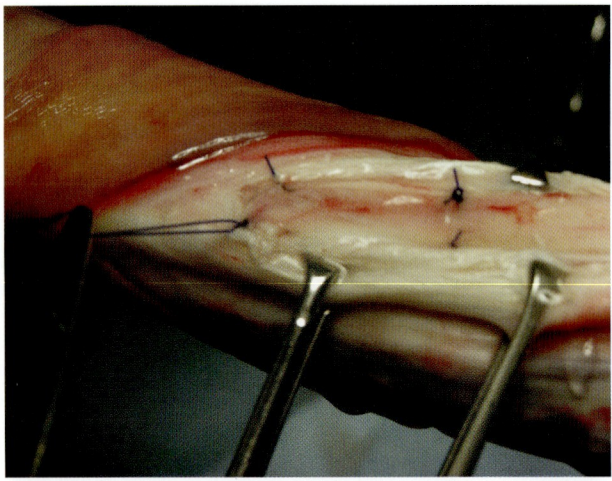

Figure 18.51 Fascia lata graft secured with simple interrupted sutures.

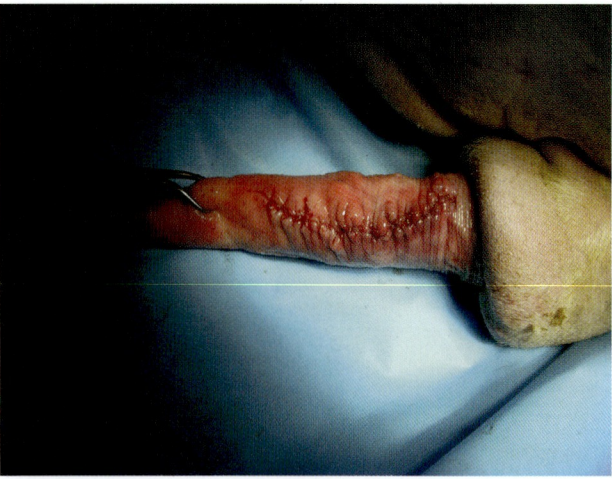

Figure 18.54 Interrupted suture closure of penile and preputial epithelium.

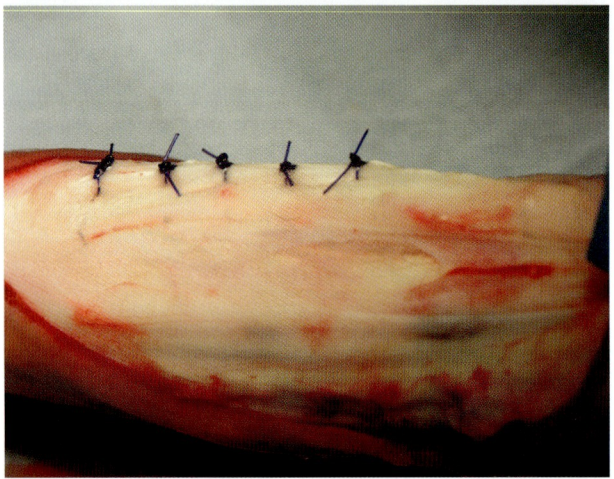

Figure 18.52 Apical ligament closed with simple interrupted sutures.

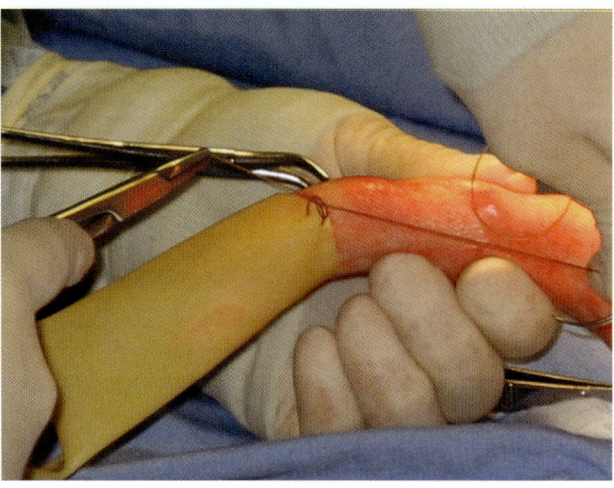

Figure 18.55 Penrose drain sutured over free portion of penis.

Appose the elastic tissue over the apical ligament with #3-0 chromic gut in a closely spaced simple continuous pattern, then close the skin with a simple interrupted suture pattern using #0 absorbable suture (Figure 18.54). Place a 2.5-cm Penrose drain over the glans penis to ensure urine egress from the incision and secure with three evenly spaced absorbable sutures avoiding the penile urethra (Figure 18.55). Apply antibacterial ointment to the incision before removing the towel forceps and allowing the penis to retract into the sheath.

Administer systemic antibiotics and remove the Penrose drain and penile skin sutures 10 days postoperatively. Do not remove the thigh skin sutures for 3 weeks to reduce the risk of muscle herniation. Ensure complete sexual rest for a minimum of 60 days and the bull should undergo a breeding soundness examination prior to resumption of breeding.

References

1. Wolfe D, Beckett S, Carson R. Acquired conditions of the penis and prepuce. In: Wolfe DF, Moll HD (eds) *Large Animal Urogenital Surgery*, 2nd edn. Baltimore: Williams and Wilkins, 1998, pp. 237–272.

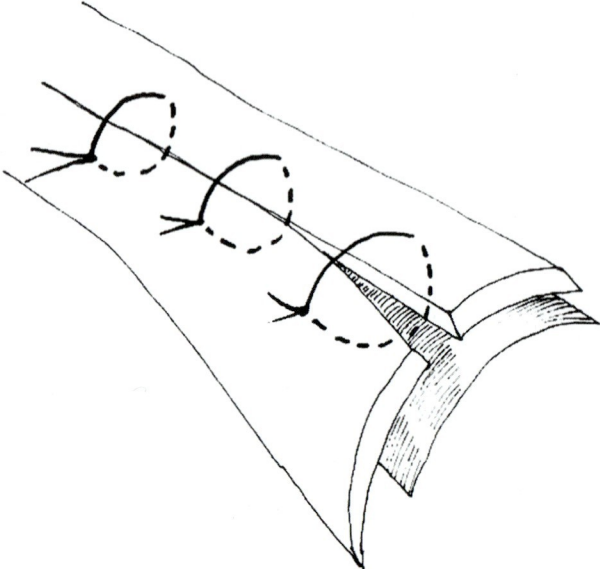

Figure 18.53 Schematic of fascia lata graft suture closure.

2. McEntee K. Fibropapilloma of the external genitalia of cattle. *Cornell Vet* 1950;40:304–312.
3. Wolfe D, Rodning S. Diagnosis and management of juvenile anomalies of the penis and prepuce. In: Anderson DE, Rings DM (eds) *Current Veterinary Therapy. Food Animal Practice*, 5th edn. St Louis: Saunders, 2009, pp. 340–341.
4. Watson J. Mechanism of erection and ejaculation in the bull and ram. *Nature* 1964;204:95–96.
5. Ashdown R. The angioarchitecture of the sigmoid flexure of the bovine corpus cavernosum penis and its significance in erection. *J Anat* 1970;106:403–404.
6. Ashdown R. Functional anatomy of the penis in ruminants. *Vet Anat* 1973;14:22–25.
7. Larson L, Kitchell R. Neural mechanisms in sexual behavior. II. Gross neuroanatomical and correlative neurophysiological studies of the external genitalia of the bull and ram. *Am J Vet Res* 1958;19:853–865.
8. Becket S, Walker D, Hudson R, Reynolds T, Vachon R. Corpus cavernosum penis pressure and penile muscle activity in the bull during coitus. *Am J Vet Res* 1974;35:761–764.
9. Izumi T. Neuro-anatomical studies on the mechanism of ejaculatory reflexes in the bull. *Special Bulletin of Ishikawa Prefecture College of Agriculture* 1980;9:45–85.
10. Beckett S, Hudson R, Walker D, Purohit R. Effect of local anesthesia of the penis and dorsal penile neurectomy on the mating ability of bulls. *J Am Vet Med Assoc* 1978;173:838–839.
11. Hudson R, Walker D, Young S, Bartels J. Impotentia erigendi in bulls due to vascular shunts from the corpus cavernosum penis. In: *Proceedings of the 10th World Congress on Buiatrics*, 1978, pp. 35–43.
12. Young S, Hudson R, Walker D. Impotence in bulls due to vascular shunts form the corpus cavernosum penis. *J Am Vet Med Assoc* 1977;171:643–648.
13. Glossup C, Ashdown R. Cavernosography and differential diagnosis of impotence in the bull. *Vet Rec* 1986;118:357–360.
14. Moll H, Wolfe D, Hathcock J. Cavernosography for diagnosis of erection failure in bulls. *Comp Contin Educ Prac Vet* 1993;15:1160–1164.
15. Cardwell W. The surgical correction of preputial and penile disorders of the bull. *Southwest Vet* 1961;(Summer):270–273.
16. Farquahasson J. Fracture of the penis in the bull. *Vet Med* 1952;47:175–176.
17. Ashdown R, Glossup C. Impotence in the bull. III. Rupture of the corpus cavernosum penis proximal to the sigmoid flexure. *Vet Rec* 1983;11:30–37.
18. Beckett S, Reynolds T, Walker D, Hudson R, Purohit R. Experimentally induced rupture of the corpus cavernosum penis of the bull. *Am J Vet Res* 1974;35:765–767.
19. Hudson R. Surgical correction of hematoma of the penis in bulls. In: Williams EI (ed.) *Proceedings of the 9th Annual Convention of the American Association of Bovine Practitioners*, 1976, pp. 67–69.
20. Wolfe D, Hudson R, Walker D. Common penile and preputial problems in bulls. *Comp Cont Educ Prac Vet* 1983;5;S447–S455.
21. Mysinger P, Wolfe D, Redding R, Hudson R, Purohit R, Steiss J. Sensory nerve conduction velocity of the dorsal penile nerves of the bull. *Am J Vet Res* 1994;55:898–900.
22. Carson R. Corpus cavernosal filling defects. In: *Proceedings of the First Hudson-Walker Theriogenology Symposium, Auburn University, Auburn, Alabama*, 1990, pp. 47–49.
23. Ashdown R, Pearson H. The functional significance of the dorsal apical ligament of the bovine penis. *Res Vet Sci* 1971;12:183–184.
24. Walker D. Deviations of the bovine penis. *J Am Vet Med Assoc* 1972;147:677–682.
25. Hanselka D. Bovine penile deviations: a review. *Southwest Vet* 1973;3:265–271.
26. Ashdown R, Smith J. The anatomy of the corpus cavernosum penis of the bull and its relationship to spiral deviation of the penis. *J Anat* 1969;104:153–159.
27. Seidel G, Foote R. Motion picture analysis of ejaculation in the bull. *J Reprod Fertil* 1969;20:313–317.
28. Walker D, Young S. The fascia lata implant technique for correcting bovine penile deviations. In: *Proceedings of the Annual Meeting of the Society for Theriogenology*, 1979, pp. 99–102.

Chapter 19

Bovine Urolithiasis

Katharine M. Simpson[1] and Robert N. Streeter[2]

[1]*College of Veterinary Medicine, Ohio State University, Columbus, Ohio, USA*
[2]*Department of Veterinary Clinical Sciences, Center for Veterinary Health Sciences, Oklahoma State University, Stillwater, Oklahoma, USA*

Introduction

Obstructive urolithiasis is a costly and challenging disorder of male ruminants that can be associated with significant morbidity and mortality in affected animals and can have devastating effects on breeding soundness in bulls. The variety of clinical presentations and diverse classes of livestock affected allow for many treatment options. The most appropriate means of therapy will depend on the location and duration of obstruction, intended purpose of the animal, and economic considerations.

Pathophysiology

Uroliths are composed of an organic matrix or nidus combined with mineral salts.[1,2] Initially, soluble minerals precipitate into an insoluble crystallized matrix, and ultimately form calculi.[3] The composition and concentration of minerals available to contribute to urolith formation are a product of diet, environment, and management. Inflammatory, cellular, or necrotic debris may provide a nidus for calculi development,[4] but is uncommon in ruminants.

High-concentrate diets often contain low calcium to phosphorus ratios and frequently result in phosphatic calculi (calcium, ammonium, magnesium),[3] most commonly struvite.[5] Feedlot and show animals on grain-based rations are therefore at risk.[3,4] Pastured cattle grazing western grasses with high levels of silica are at risk of developing silicate uroliths, even after they have stopped consuming the offending forages. Oxalate- and carbonate-containing calculi are less commonly encountered in bovines.[3]

Inadequate water intake also predisposes to development of urolithiasis. Clinical disease may be precipitated by decreased water availability or consumption, and a seasonal influence has been reported.[6] In the winter, decreased voluntary intake or frozen water sources may be problematic.[4] Dehydration following high ambient temperatures and lack of available palatable water sources may occur in the summer, particularly in drought conditions.[3]

Normal pH of ruminant urine is 7 or greater, a physiologic factor that can contribute to calculogenesis as some types of uroliths precipitate in alkaline urine (struvite, calcium phosphate, calcium carbonate).[3] Urine pH does not appear to affect silicate calculi.[1,3]

Both steers and intact bulls are predisposed to urinary obstruction from urolithiasis because of their long and relatively narrow urethra compared with females. Castration at an early age results in a decreased urethral diameter[2,7] and increased likelihood of obstructive urolithiasis; steers 8–12 months old are reported to be at highest risk of clinical disease.[4] Delaying castration until the animal is at least 6 months of age is therefore recommended by some authors,[2,4,7] but welfare concerns regarding late castration should be taken into consideration before implementing this practice. In cattle, urinary obstruction most commonly occurs at or near the sigmoid flexure, although more proximal obstructions are occasionally seen. Multiple uroliths and cystoliths may be present, particularly with phosphatic calculi.[3,8] Silica-containing uroliths are usually singular.[3] Ureteroliths and nephroliths can also occur but may be less common in cattle than small ruminants; however, one report in grass-fed dairy cows describes ultrasonographic detection of renal calculi in 63% of pastured cattle and 93% of cows in tie stalls in Japan.[9]

While the prevalence of obstructive urolithiasis is relatively low in mature breeding bulls,[4] urothelial injury secondary to obstruction or surgical treatment can be catastrophic to breeding soundness in these animals. Additionally, erection failure from vascular obstruction of the corpus cavernosum following obstructive urolithiasis has been reported in ruminants.[10]

Clinical findings

Clinical signs vary depending on the stage of the disease process. Frequent posturing to urinate, tail swishing, dysuria, stranguria, hematuria, and partial to complete anuria may be present initially. Blood or calculi are sometimes visible on the preputial hairs (Figure 19.1). Careful palpation of the

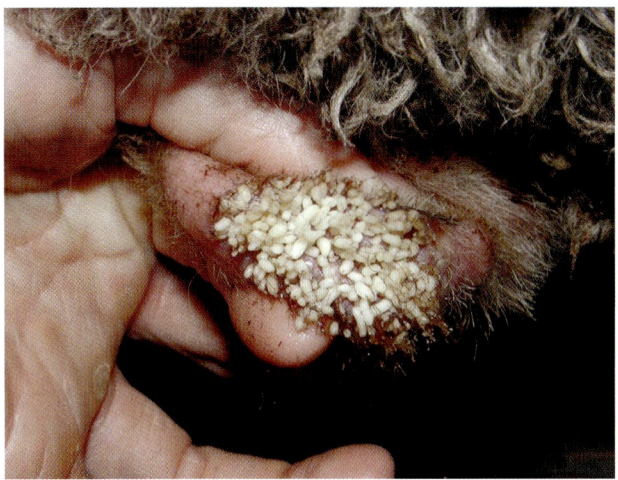

Figure 19.1 Urethral calculi visible on preputial hairs.

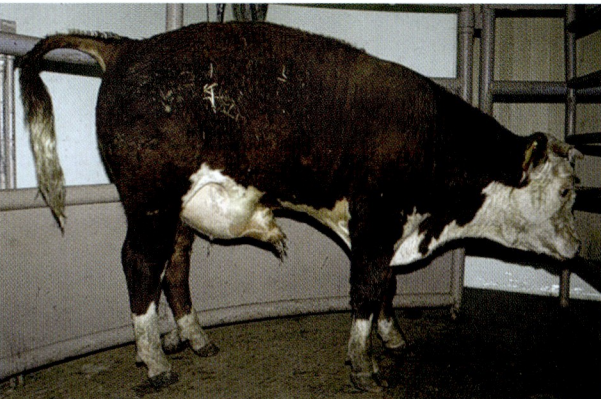

Figure 19.2 Subcutaneous peri-penile swelling typically extends from the base of the scrotum cranially and involves the entire ventrum.

penis immediately caudal to the scrotum or its remnant may reveal the location of large urethral calculi. Pain may be elicited when the site of obstruction is palpated, but is not consistent. Nonspecific signs including decreased appetite to complete anorexia, hypodipsia, depression, and lethargy may also be apparent or reported by the owner.

Acute urinary obstruction can result in colic signs including bruxism, stretching, treading the hind feet, or kicking at the abdomen. Cattle with acute urethral obstruction may display urethral pulsations ventral to the anus without concurrent urination; these may be palpable per rectum along with a distended urinary bladder. Signs of colic or dysuria typically resolve following urethral or urinary bladder rupture, and clinical improvement may be seen initially followed by signs consistent with uremia.[11] Rupture of the urinary tract usually occurs within 48 hours of complete obstruction.

Ruptured urethra results in peri-penile subcutaneous swelling that is cool to the touch and pits on digital pressure; this can extend from the base of the scrotum cranially to involve the entire ventrum (Figure 19.2). If left untreated, necrosis of the affected tissues occurs followed by sloughing.[12] Bilateral ventral abdominal distension with a fluid wave on ballottement is suggestive of ruptured urinary bladder[11] and resultant uroabdomen. Rectal examination usually reveals a small, partially filled urinary bladder.[13] Intact males more commonly rupture the bladder rather than the urethra because of a more substantial tunica albuginea; scrotal hydrocele may result from uroabdomen. Cattle with ruptured urinary bladders can survive for 2 weeks or more following cystorrhexis.[14] Although rare in ruminants, ureteral calculi may be palpable per rectum.[15]

Diagnostics

A presumptive diagnosis can usually be made based on history and physical examination findings, including palpation of the penis and rectal examination when possible. If urine can be obtained, urinalysis with examination of sediment should be performed to confirm the diagnosis. Hematuria, proteinuria, and alkalinuria are expected findings. Crystalluria is variably present; determination of crystal type assists with formulation of a treatment plan and prognosis. Concurrent bacterial cystitis or pyelonephritis can contribute to the development of urolithiasis and should be ruled out. In valuable animals, a serum biochemistry profile is recommended and may reveal azotemia, hyperphosphatemia, hypocalcemia, or metabolic acidosis.[14] Hyponatremia and hypochloremia are usually present following urinary bladder rupture[14,16] but hyperkalemia is not common in cattle.[14,17]

Transcutaneous and transrectal ultrasonography are useful for assessing the integrity of the urinary tract, the location of obstructing uroliths and the presence of cystoliths, ureteroliths, or nephroliths in the proximal urinary tract, and for indicating the existence of uroabdomen, hydroureter or hydronephrosis. Peri-penile edema may suggest early urethral rupture and signify a lower prognosis. Ultrasonography of bovine urinary tract disorders has recently been reviewed.[18] If uroabdomen is suspected, abdominocentesis and comparison of abdominal fluid creatinine to serum creatinine should be performed. A ratio of 2 : 1 or greater confirms uroabdomen.[11,19]

If uroliths can be acquired, analysis allows determination of calculi type; although not immediately available, this information can guide management strategies to minimize likelihood of recurrence and to prevent additional cases in the herd. In the short-term, placing one or two uroliths in an acidifying agent to establish whether they will dissolve allows the clinician to determine if chemolysis would be an effective adjunctive therapy. Survey or contrast radiography is unlikely to be diagnostic unless the patient is a young calf. However, urethroscopy may be useful for assessing the location of the obstruction, extent of urothelial damage and presence and location of strictures, and for providing additional avenues of treatment.

Treatment

When formulating a treatment plan in cases of obstructive urolithiasis, the clinician should take into consideration the intended use of the animal, integrity of the urinary tract, location of the obstruction or obstructions, and economic constraints. If severe electrolyte, acid–base, or fluid imbalances exist, these should be corrected prior to sedation or anesthesia. Analgesia and antimicrobial and anti-inflammatory drugs are indicated. Meat withdrawal times and degree of renal compromise should be considered prior to drug administration. In cases where the urinary tract is still intact, caution should be exercised during restraint as fractious or struggling animals are at risk of urinary bladder rupture.

Medical treatment

In cases of acute urethral obstruction, relaxation of the retractor penis muscles may allow extrusion of calculi by straightening the sigmoid flexure. This may be accomplished by local anesthetic blockade via a pudendal or ischial penile nerve block.[20] Phenothiazine tranquilizers may decrease urethral spasm[21] or induce relaxation of the retractor penis muscles,[22,23] allowing the calculi to pass. The α_2-adrenergic agonists including xylazine are not recommended due to marked diuretic effects.

If penile extrusion can be accomplished with these procedures, retrograde urethral catheterization can be attempted. This is more easily done standing in mature bulls rather than castrated or prepubertal animals and may be hastened by massage per rectum. Infusion of a small volume of lidocaine may be used to decrease urethral spasm following successful catheterization. Gentle urohydropulsion can be used to attempt to dislodge calculi; this may be facilitated by having an assistant manually occlude the pelvic urethra during hydropulsion to maximize urethral expansion at the level of the obstructing urolith. In our experience hydropulsion in cattle is usually unsuccessful, likely due to the relative chronicity of most cases when presented and the resultant periurethral swelling and necrosis. Multiple or forceful attempts at either catheterization or hydropulsion should be avoided due to the risk of urothelial trauma or rupture and resultant stricture or fistula formation, respectively. If the obstruction is not rapidly relieved, surgical intervention is necessary.

Surgical treatment

Acute urethral obstruction in salvage animals

In a commercial or salvage animal, perineal urethrostomy is the most cost-effective method of reestablishing urinary patency. This surgery can be performed low (immediately caudal to the sigmoid flexure) or high (ventral to the rectum at the level of the tuber ischii). It is of paramount importance that the procedure be performed proximal to the level of the urethral obstruction. Careful palpation along the penis may reveal the location of the obstruction; alternatively, ultrasonography is very useful in determining the site of obstruction. The choice of technique is generally determined by the size of the animal and the length of time before the animal is marketable.

Perineal urethrostomy

In young or lightweight animals, a low perineal urethrostomy with penile amputation is commonly performed as it is a more permanent technique that allows for resolution of azotemia and continued growth of the animal prior to slaughter. This is sometimes referred to as "penile transposition" and various methods of performing this technique have been described.[13,15,24,25]

Surgery can be performed standing with caudal epidural anesthesia and local infiltration of lidocaine, or in dorsal recumbency with lumbosacral epidural and heavy sedation or general anesthesia. The region from the scrotum to the anus is aseptically prepared. A 10–15 cm incision is made on the midline immediately caudal to the scrotum and extended through the fascia between the semimembranosus muscles.[15] The paired retractor penis muscles are encountered and can be misidentified as the penis. The penis is grasped and elevated to the level of the skin with a combination of blunt dissection and traction, and the retractor penis muscles may be retracted or be ligated proximally and removed. The dorsal vessels of the penis are ligated, and penile amputation is performed 5 cm distal to the dorsal apex of the skin incision.[13] The penis distal to the amputation may be removed or left in place. Removal of the distal penis is facilitated in cases of urethral rupture with subsequent peripenile tissue necrosis and its removal eliminates a potential source of infection.[25] In cases involving an obstructed but not ruptured urethra, sharp dissection of the preputial attachments is required to extirpate the distal penis. If the distal penis is to be left in place, the dorsal penile artery and vein should be reflected from the penile stump rather than ligated. The stump is sutured to the skin using nonabsorbable monofilament suture material in a horizontal mattress pattern through the corpus cavernosum, taking care not to incorporate the urethra. The penis should not be rotated when it is exteriorized and secured to the skin so that the urethra remains on the dorsal surface.[15] Spatulation of the urethra is optional. If elected, the urethra is incised to the apex of the skin incision, and the urethral mucosa and tunica albuginea are sutured to the skin using 2-0 monofilament suture in a simple continuous or interrupted pattern. Depending on the procedure, the skin is routinely closed distal and/or proximal to the urethrostomy. In animals that will not be salvaged for several months, amputation and removal of the penis with urethral spatulation affords better apposition of the urethral mucosa to the skin[15] and is ideal for decreasing the likelihood and rapidity of stricture formation.

In mature animals and bulls, hemorrhage following this procedure may be substantial but can be controlled by placement of a catheter into the urethra that fits securely and compresses the corpus spongiosum. This should be removed in 2–3 days. Low perineal urethrostomy should not be used in animals intended for breeding.

Ischial urethrostomy

In heavier animals that are near salvage weight (≥ 318 kg), high perineal urethrostomy is ideal to allow the uremia to resolve over approximately 30 days until the animal is suitable for slaughter.[12] This procedure is done standing with a caudal epidural for local analgesia. The perineal region from the anus to the base of the scrotum is aseptically prepared.[3] A 10-cm midline incision is made starting 5 cm below the anus, over the ischial arch.[3] The dense layer of fascia overlying the retractor penis muscles is incised and the muscles exposed.[3] Blunt dissection between the retractor penis muscles reveals the bulbospongiosus muscle.[3] The urethrotomy is made through the bulbospongiosus muscle or just distal to its attachment.[15] When performing this surgery, it is critical to ensure the incision into the urethra is made on the midline in the event that bulging of the urethra is not palpable following dissection down to the penis. Hemorrhage from the corpus spongiosum in the region of the urethral diverticulum is expected and may appear excessive to individuals performing this procedure for the first time. Following high perineal urethrostomy, a 20–28 French Foley catheter is advanced into the bladder and sutured in place where it exits the skin. This can be facilitated by inserting a rigid stylet with a slight curve on the end into the lumen of the Foley. The balloon is distended with saline and retracted until it is seated within the bladder trigone. No sutures are placed into the urethra with this technique. A one-way valve utilizing the fingertip of a surgical or examination glove with a slit cut into the end should be affixed to the external end of the catheter to minimize bacterial contamination and air aspiration into the bladder. This surgery can also be utilized in cases where a previous low perineal urethrostomy has strictured or reobstructed.

Following surgery, a section of inner tubing or impervious plastic should be adhered to the skin just ventral to the anus to prevent fecal contamination of the surgical site.

Urethrotomy

Urethrotomy can be used in cases of acute urethral obstruction (without urethral rupture) provided the location of the calculus is identified via palpation, catheterization, or ultrasound. Following caudal epidural anesthesia, aseptic preparation and local infiltration of 2% lidocaine, an incision is made immediately cranial or caudal to the scrotum and the penis is visualized. The urolith is identified and the affected segment of the penis is exteriorized. Towel clamps or tissue forceps may be used to attempt to crush the calculus. Phosphatic calculi tend to be soft and less amenable to fracture.[15] If unsuccessful after two attempts, urethrotomy should be performed immediately distal to the obstruction through relatively healthy urothelium. Most authors recommend primary closure of the urethra,[13,15,25] usually with 3-0 absorbable swaged-on suture in a simple interrupted pattern. Urethral closure can be facilitated by suturing over a catheter. Routine closure of the subcutaneous tissues and skin are performed. Urethral dehiscence, stricture, and peri-penile adhesions are common following this procedure, particularly if urethral mucosa is devitalized. Postoperative urinary diversion via cystostomy tube or ischial urethrostomy is ideal for reducing the likelihood of dehiscence or fistulation at the urethral incision site. Alternatively, short-term (24–48 hour) postoperative urethral catheterization may be utilized.

Acute urethral obstruction in breeding animals

If medical therapy and hydropulsion are not successful at relieving the urinary obstruction, certain forms of lithotripsy (including Ho:YAG laser,[26] extracorporeal shockwave,[27] electrohydraulic shockwave[28]) or basket catheter retrieval[29,30] are less invasive methods of resolving obstructive urolithiasis that may be available at referral centers. If these modalities are not available, a number of surgical options exist. Tube cystostomy with or without cystotomy would be considered the ideal surgical treatment in breeding animals. Urethral surgery is ideally avoided in bulls intended for breeding because of the risk of stricture or fistula formation at the incision site or adhesions of the penis if performed distally.

Laser lithotripsy

Laser lithotripsy has been successful in establishing urethral patency in a steer, goats, and pot-bellied pigs.[26,31] Retrograde passage of an endoscope and laser fiber is possible in standing bulls with pudendal nerve block or in recumbent animals under general anesthesia. Normograde urethroscopy and laser lithotripsy may also be performed through the fistula created by tube cystostomy if the endoscope is long enough. Alternatively, the endoscope can be advanced through an ischial urethrotomy site distally to the level of the calculi.

Laser lithotripsy requires an endoscope with at least one instrument port. We have used a similar technique as previously described.[26] Prior to urethroscopy, a catheter extension set can be attached to the instrument port. A needle is used to puncture the rubber port that feeds directly into the instrument channel, and a silica optical fiber is fed into the channel until it is only a few millimeters from the channel exit. The fiber is then connected to a holmium:yttrium–aluminum–garnet (Ho:YAG) laser. A 60-mL syringe filled initially with 1% lidocaine is attached to the extension port to allow for local anesthesia, flushing of urethral debris, and hydropulsion via the same instrument channel that the laser fiber occupies. The fiber is then advanced and visualized until it comes into contact with the urolith; it should be centered directly on the stone (Figure 19.3). The laser is fired in a pulsatile manner, with urethral flushing between firing, until the calculus fractures into fragments small enough to dislodge with hydropulsion or retrieve with a wire basket catheter.[32] Calculi that are firmly wedged into the urethra are most amenable to laser lithotripsy because they do not move when contacted by the laser fiber.[32] Additionally, calculi with irregular surfaces are reportedly less likely to deflect the laser fiber into the surrounding soft tissues than smooth spherical calculi.[32] Clinical experience suggests that previous chemolytic treatment in animals with surgical urinary diversion (tube cystostomy, ischial urethrotomy) results in uroliths with uneven surfaces that fracture readily.

Following lithotripsy in animals with an ischial urethrotomy, the endoscope should be passed to the external urethral orifice and the laser fiber or a long guidewire (~280 cm) advanced past the level of the preputial orifice. An assistant holds this in place as the endoscope is removed from the urethra. A transurethral 8–10 French 150-cm silicone Foley catheter (with the distal 2 mm removed) or sterilized polyethylene tubing (200 cm in an adult[13]) can then be advanced retrograde to the level of the ischial urethrotomy. The fiber or guidewire is retracted until the end is 2–3 mm distal to the

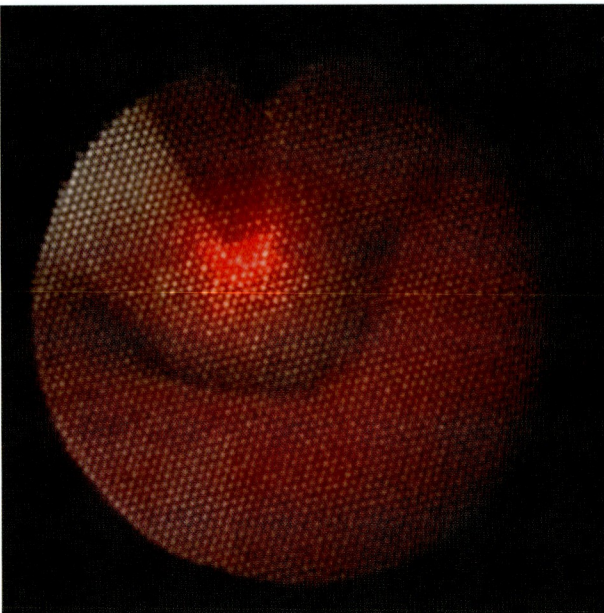

Figure 19.3 Endoscopic image of urolith within the urethra.

end of the catheter and the catheter is advanced into the bladder. The balloon of the catheter is inflated with saline and the catheter secured to the skin of the ventral abdomen with a Chinese finger cuff ligature. A one-way valve is attached to the external end of the catheter. This is left in place for 7–10 days to allow second-intention healing of the urethra and is particularly important if urothelial compromise is present near the level of the previous obstruction. Laser damage to the urothelium can also result in urethral stricture;[32] direct mucosal contact or prolonged firing times should thus be avoided. Careful positioning of the laser fiber, good visualization, and adequate irrigation are essential. Cost of equipment usually limits the use of this technology to referral institutions.

Basket catheter and urolith retrieval

Basket catheter retrieval of uroliths has been used to relieve urinary obstruction in bulls while maintaining reproductive capability.[30] The procedure can be done standing or in lateral recumbency with sedation or anesthesia. After exteriorization of the penis, the instrument is passed into the external urethral orifice and advanced beyond the level of the obstruction with gentle probing and rotation. The basket is opened and the catheter withdrawn until the calculus is encompassed within.[30] The wire basket is closed and the catheter is removed from the urethra.[30] If initially unsuccessful, the process can be repeated. Urethral calculi were easily removed in approximately 55% of male calves, via retrograde catheterization with a 5.5 French 70-cm basket stone dislodger and radiography to determine proper catheter placement.[29] Surgery was necessary in the remainder of cases. Alternatively, urethroscopy may facilitate appropriate placement of the wire basket. It is suggested that subsequent inflammation in the region of the sigmoid flexure could lead to penile adhesions, which may be circumvented by continued breeding on a limited basis.[30]

Cystotomy

Cystotomy[33] and tube cystostomy[34,35] are associated with the best long-term prognosis for survival and reproduction in small ruminants. Urethral surgery and its related complications are avoided, and removal of cystic calculi is facilitated. Disadvantages include the necessity of general anesthesia, cost and duration of surgery, and potential for temporary urethral obstruction from inflammation postoperatively.[15]

The urinary bladder is exteriorized via a paramedian skin incision adjacent to the prepuce, and either paramedian or midline celiotomy.[36] Cystorrhaphy is performed in cases of ruptured bladder. Resection of necrotic regions of the bladder wall may be necessary in cases of rupture following chronic distension. Routine cystotomy is followed by aspiration and lavage of the trigone, bladder neck, and most proximal urethra to dislodge cystoliths. Uroliths adhered to the urinary bladder wall with fibrin or cellular debris can be removed with a soft-edged surgical spoon. Normograde urethral catheterization with an 8–10 French polypropylene catheter and hydropulsion with saline may relieve urethral obstructions but should be done with caution to avoid initiating or exacerbating urethral rupture.[34] Retrograde hydropulsion may also be required and can be performed by an assistant; concurrent occlusion of the distal urethra prevents loss of flush solution.[36] Occlusion of the proximal urethra by the surgeon maximizes urethral dilation and probability of dislodging uroliths.[37] However, excessive time and effort at relieving urethral obstructions via hydropulsion are contraindicated because of the risk of urothelial trauma. If urethral patency is not readily established following cystotomy, tube cystostomy should be performed. The cystotomy is closed in one or two layers with inverting pattern(s) using absorbable suture.[36] Celiotomy and skin incisions are routinely closed. Prognosis for survival of cattle undergoing cystotomy and indwelling urethral catheterization is 70%.[38]

Tube cystostomy

Tube cystostomy provides urinary diversion to allow for relief of urethral spasm and swelling and healing of the urothelium, and offers a direct route for administering chemolytic agents into the urinary bladder[15,39,40] (Figure 19.4). Although typically performed with cystotomy, tube cystostomy alone has been successful and shortens the length of the procedure.[36] In cattle, this surgery can be performed standing through a laparotomy incision, although optimal access to the urinary bladder is obtained through a ventral celiotomy as previously described. The urinary bladder is visualized and exteriorized. A Foley catheter or similar retention catheter (mare uterine lavage tube, gastrostomy tube) is advanced through the skin via a stab incision opposite the side of the initial skin incision and 3 cm from the prepuce. Use of Carmalt or hemostatic forceps clamped but not locked onto the end of the catheter facilitates placement. The end of the Foley is advanced into the urinary bladder through a stab incision on the cranioventral aspect near the apex; this can be done through a preplaced purse-string suture.[15] The balloon is distended with saline to the maximum level recommended. A pursestring suture (if not preplaced) is placed into the seromuscular layer of the urinary bladder around the catheter,[15] but should not

Bovine Urolithiasis 177

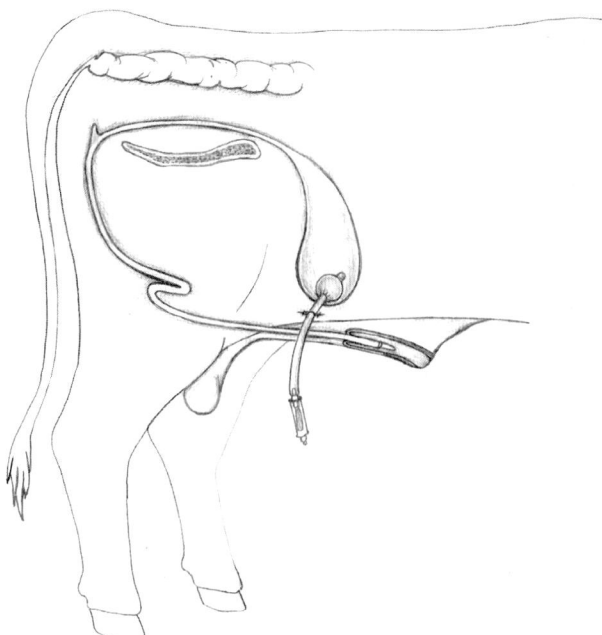

Figure 19.4 Placement of a Foley catheter in the tube cystotomy procedure. Courtesy of Rachel Oman.

penetrate the lumen. Tension is placed on the Foley catheter and the urinary bladder is pulled snugly against the abdominal wall.[15] Cystopexy to the internal rectus sheath using two to three interrupted sutures can be performed[15] if excessive tension is encountered during apposition. Abdominal and skin incisions are routinely closed. A Chinese finger cuff ligature should be placed to secure the Foley catheter to the ventral abdomen, but a pursestring suture should not be placed around the catheter's exit from the skin, to allow for drainage. A one-way valve is affixed to the end of the catheter. The tube should be left in place for a minimum of 7–10 days. After penile inflammation has subsided, the catheter is intermittently occluded to allow normograde voiding and relief of urethral obstruction. An average of 10–12 days is required for this to occur in small ruminants.[34] The tube should remain in place for 24–48 hours following normal urination in the event that reobstruction occurs.[15,34]

If the obstruction does not pass, cystic irrigation with up to 250 mL of an acidifying agent[41] for 30 min twice a day can be implemented starting 5–7 days postoperatively. Use of high-volume commercially available buffered acetic acid solution (Walpole's) or hemiacidrin (Renacidin®) may be cost-prohibitive in the long term; use of ammonium chloride solution[41] or anecdotally use of dilute acetic acid (vinegar) is reported. Antimicrobial therapy should be implemented preoperatively and continued for the duration of catheter placement.[34] Complications have included physical obstruction of the catheter postoperatively (uroliths, blood clots, cellular debris),[34] tube kinking, dislodgement or loss of the catheter,[42] balloon deflation,[34] cystitis,[34] uroperitoneum following tube removal,[34] adhesions of the urinary bladder to the abdominal wall,[15] and failure to resolve urethral obstructions.[43] Use of appropriate catheter size, single use of catheters, and concurrent cystotomy with cystolith removal may alleviate some of these problems.[34] Minimally invasive tube cystostomy[44] and laparoscopic-assisted placement of cystic catheters[45] have also been described in small ruminants. Prognosis for survival in cattle treated with tube cystostomy is reported to be as high as 90%.[38,42] In small ruminants without urethral rupture, prognosis for breeding ability is good; similar results would be anticipated in cattle.

Ischial urethrotomy

Ischial urethrotomy as described previously (see section Ischial urethrostomy) may provide temporary urinary diversion as well as access to the distal urethra to allow for laser lithotripsy or chemolysis in cases where the obstruction remains unrelieved. The urethral diameter is larger in this region, which may explain our experience that clinically relevant urethral stricture is not common following this surgery. If distal urethral patency is immediately obtained, primary closure of the urethra and passage of a transurethral catheter (either ~200-cm polyethylene tubing with rounded ends in an adult[13] or a 150-cm silicone Foley catheter) proximally into the urinary bladder and exiting the distal urethra allows urinary diversion and healing of the urethral incision. The catheter should be sutured to the ventral abdomen with a Chinese finger cuff ligature.

If the obstruction persists, a 20–28 French Foley catheter is advanced into the urinary bladder. If relief of urethral spasm and swelling fails to allow the calculi to pass within 3–5 days, other methods of calculi removal are necessary. Retrograde urethral catheterization and urohydropulsion may flush the urolith to the ischial urethrotomy site for removal.[13] Urethroscopy of the distal urethra via the ischial urethrotomy site allows visual confirmation of remaining uroliths and assessment of urothelial damage (Figure 19.5). Up to a 4.0 mm (12 French) flexible endoscope at least 90–100 cm long is adequate for reaching the sigmoid flexure in most mature bulls. Young animals with a smaller urethral diameter may require a 6 French (2.0 mm) endoscope. If phosphatic uroliths (struvite or apatite) are present, chemolysis may be effective provided the urothelium does not appear compromised. To facilitate this, a 6–12 French silicone Foley catheter can be advanced into the distal urethra to within 1–2 cm of the urolith. A Christmas tree adapter with a female Luer connector is placed into the end of the catheter and an injection cap is attached; this apparatus is sutured to the skin adjacent to the urethrotomy site using a Chinese finger cuff ligature. The catheter's balloon is dilated with 1–3 mL saline until slight resistance is met, and 5 mL of an acidifying solution (acetic acid or hemiacidrin) is injected into the catheter via the port. The infusion should be stopped immediately if resistance is encountered or the animal shows signs of acute discomfort. After 15–30 min, the balloon is deflated and 10 mL of saline is injected to dilute the acidifying agent. This procedure can be repeated two to three times daily until dissolution of the urolith is adequate to allow extrusion. We have used this combination of procedures successfully in a valuable breeding bull. However, the complication rate and long-term breeding soundness of bulls with ischial urethrotomies has not been evaluated. Reversal of a strictured perineal urethrostomy site using buccal mucosal graft urethroplasty was successful in a goat.[46]

Maintaining a sterile bandage over the incision site with stay sutures and laparotomy sponges covered with a

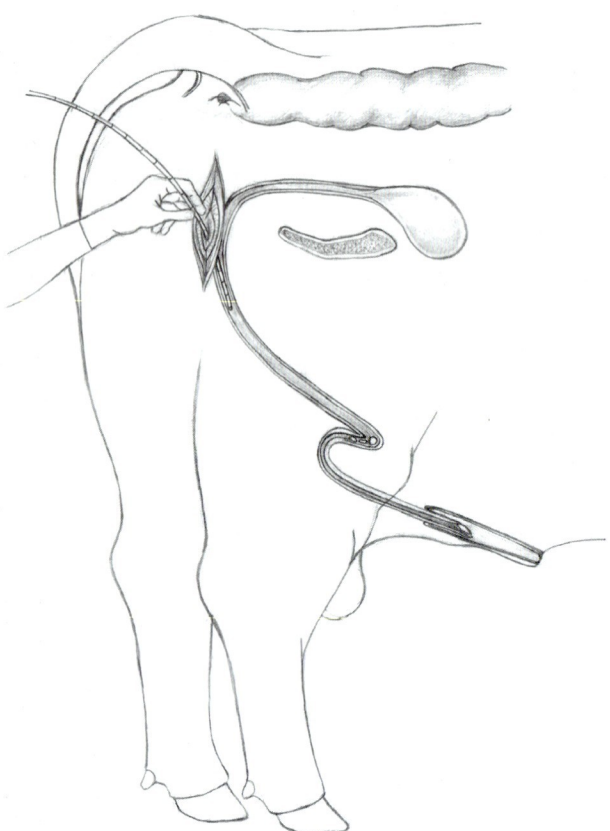

Figure 19.5 Passage of a flexible endoscope for assessment of the distal urethra via an ischial urethrostomy. Courtesy of Rachel Oman.

surgical towel is recommended (Figure 19.6), as is daily surgical scrubbing of the site. This should be covered by inner tubing as previously described.

Urethrotomy

Urethrotomy to remove calculi can also be performed as previously described for salvage animals. The potential complications associated with urethral incisions may be reduced if alternative methods of urinary diversion are concurrently employed. This can be accomplished with an additional ischial urethrotomy and passage of either polyethylene tubing or a silicone Foley catheter proximally into the urinary bladder and then via the urethra to the external urethral orifice as previously described (see section Laser lithotripsy). The catheter can be maintained for 7–10 days to allow the distal urethrotomy to heal. Alternatively, tube cystostomy may be utilized. Prognosis for return to breeding is guarded.[24]

Urinary bladder marsupialization

Urinary bladder marsupialization can be performed in cases with unresolved or recurrent urethral obstructions, severe urothelial damage, or detrusor areflexia/atonicity of the urinary bladder. In the latter, breeding ability may be preserved assuming the urethra is patent. Laparoscopically assisted urinary bladder marsupialization has been described in a goat.[47] Urine scald, ascending urinary tract infections, stoma stricture, and mucosal prolapse are

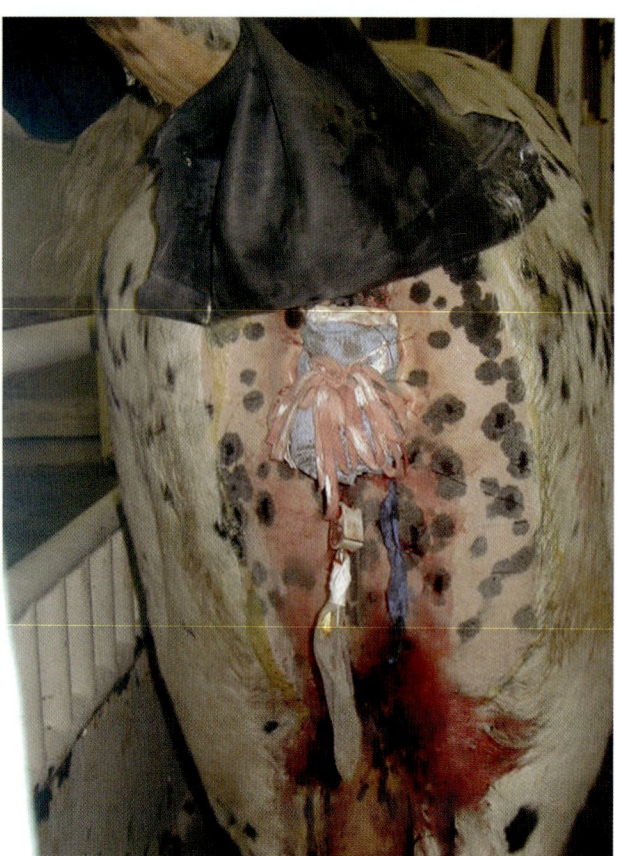

Figure 19.6 Maintenance of a sterile bandage over the urethrotomy site with stay sutures and laparotomy sponges covered with a surgical towel is recommended.

potential complications.[48–51] Some authors do not recommend the procedure in bovines, as high complication and mortality rates are reported (up to 77%).[22,51]

Urethral rupture

By the time the urethra ruptures, significant necrosis has occurred in the penis making direct urethral surgery in the region of the obstruction unrewarding. Most cases should be managed by a more proximal perineal urethrostomy. In breeding animals, urethral rupture and peri-penile accumulation of urine induces severe inflammation and will likely result in severe peri-penile adhesions that eliminates use of the animal in natural service. Urethral strictures or fistula may also result. Prognosis for return to natural breeding is poor. Most authors therefore recommend permanent urethrostomy.[13,15,24] The potential for semen collection following perineal urethrostomy has not been described but could be attempted in valuable breeding animals. Alternatively, tube cystostomy for urinary diversion and second-intention healing of the urethra can be performed but prognosis for return to breeding remains guarded.[36] Ischial urethrotomy and laser lithotripsy have resulted in normal urination following acute urethral rupture with mild peri-penile urine accumulation in a steer.[26]

In cases with long-standing urethral rupture, a significant amount of urine can accumulate in the peri-penile soft tissues of the ventral abdomen. If left in place, the caustic nature of the urine will result in widespread necrosis of the ventral abdominal tissues and predispose the animal to localized infection. Several drainage incisions should be made in the ventral abdomen over the accumulated urine to allow for escape from the tissue. Orienting the stab incisions parallel to the subcutaneous abdominal veins minimizes the risk of hemorrhage. Additionally, distal penile amputation and removal facilitates drainage through the preputial orifice.[15] Survival rate in animals treated for ruptured urethra is reported to be 93%.[23]

Ruptured bladder

In valuable breeding or pet animals, primary suture closure of the bladder via celiotomy is indicated. A standing laparotomy may give adequate exposure for craniodorsal tears, but ventral paramedian incision affords the best exposure to the bladder. The rupture in the bladder may be a solitary rent in the craniodorsal or cranioventral bladder, or it may be multiple small holes secondary to chronic overdistension and necrosis. Contrast cystography in juveniles or exploratory surgery in adults is required to differentiate the location and type of rupture. In salvage animals, the rupture will often seal with fibrin if the tear is dorsal and urinary tract patency can be achieved via perineal or ischial urethrostomy. Our preference is placement of a Foley catheter directly into the bladder lumen via ischial urethrostomy. The catheter is sutured in place and left until the animal is salvaged.

For valuable animals where the urinary obstruction remains unrelieved, moderate to severe urethral trauma is present, or the defect in the bladder is inaccessible for primary closure, tube cystostomy is the best option. Even in salvage animals, this tube can be left in place for 4 weeks or more.[52] Some animals will start to urinate normally without additional intervention.[52]

Severe uremia and electrolyte disturbances must be alleviated prior to performing general anesthesia, and the uroabdomen must be drained regardless. This can be achieved with a rumen trocar and sleeve or a fenestrated thoracic catheter. Abdominal drainage should be performed gradually (over 30 min or more), preferably with intravenous fluid support. This should be performed prior to positioning the animal in recumbency for any surgery as the excess peritoneal fluid will compress the diaphragm and caudal vena cava, compromising the patient. Rapid removal of uroabdomen at surgery can result in hypotensive crisis and death.

The survival rate in steers with urinary bladder rupture is reportedly 90% in animals treated with tube cystostomy[52] compared with 55%[53] to 86%[54] in animals treated with cystorrhaphy, with or without urethrotomy, and transurethral catheterization. Spontaneous healing of the bladder occurs in approximately 50% of animals following urethrotomy and drainage of the abdomen.[23]

Urethral stricture

Urethral strictures may occur in cattle subsequent to urolithiasis or in response to its treatment. Bougienage (balloon dilation) can be performed with an esophageal balloon dilator (up to an 8 French) at 207 kPa, using 1 min of inflation followed by 45 s deflation. This is done three times in succession. The success of this procedure depends on the number of strictures, their length, and location. In our experience, resolution of strictures of the distal penis and particularly the glans is moderately successful. Alternatively, one of us (R.N.S.) has utilized laser ablation via ischial urethrotomy to treat a focal circumferential urethral stricture in a bull. No clinical evidence of stricture was subsequently seen. Transurethral stent placement may also be an option in valuable breeding animals.

Postoperative care

Intravenous fluids for postobstruction diuresis are recommended in azotemic animals, and may be advantageous in preventing urethral or catheter obstruction with cellular debris.[36] Physiologic saline is the fluid of choice, with supplementation of other electrolytes as needed. Antimicrobials should be continued a minimum of 5–7 days following surgery, or longer if a catheter is left in place. Beta-lactams are ideal, and cephalosporins used per label tend to have comparatively short withdrawal times. Ammonium chloride is recommended at 200–400 mg/kg daily by mouth until urinary acidification is achieved. Nonabsorbable sutures should be removed 10–14 days after surgery. Dietary, environmental, and management factors that predispose to urolith development should be addressed to avoid reobstruction.

Prevention

Ideally, urolith composition should be determined to allow for specific prevention and control measures to be accurately established. Diets should be adjusted to contain a calcium to phosphorus ratio of 1.5–2.0 : 1. Dietary magnesium should be limited to less than 0.2%. Adequate water intake is critical. The availability and quality of water sources should be examined. Water intake can be enhanced by addition of sodium chloride to the diet. Ammonium chloride can be added to the diet to reduce urinary pH to assist in the control of apatite, struvite, and carbonate stones. Females can be grazed on siliceous forages in problem areas to reduce morbidity from silicate urolithiasis.

References

1. Udall R, Chow F. The etiology and control of urolithiasis. *Adv Vet Sci Comp Med* 1969;13:29–57.
2. Hawkins W. Experimental production and control of urolithiasis. *J Am Vet Med Assoc* 1965;147:1321–1323.
3. Larson B. Identifying, treating, and preventing bovine urolithiasis. *Vet Med* 1996;91:366–377.
4. Hardisty J, Dillman R. Factors predisposing to urolithiasis in feedlot cattle. *Iowa State University Veterinarian* 1971;33:77–81.
5. Parrah J, Hussain S, Moulvi B, Singh M, Athar H. Bovine uroliths analysis: a review of 30 cases. *Isr J Vet Med* 2010;65:103–107.
6. Sharma A, Mogha I, Singh G, Amarpal, Aithal H. Incidence of urethral obstruction in animals. *Indian J Anim Sci* 2007;77:455–456.
7. Marsh H, Safford JW. Effect of deferred castration on urethral development in calves. *J Am Vet Med Assoc* 1957;130:342–344.
8. Frank F, Meinershagen W, Scrivner L, Keith T, Baron R. Urolithiasis. 1. Incidence of bladder calculi, urine properties, and urethral diameters of feedlot steers. *Am J Vet Res* 1961;22:899–901.

9. Yamada H, Abe N, Sota K, Kameya T. Ultrasonic detection of renal pelvic calculi in healthy dairy cows. *J Japan Vet Med Assoc* 1991;44:108–111.
10. Todhunter P, Baird A, Wolfe D. Erection failure as a sequela to obstructive urolithiasis in a male goat. *J Am Vet Med Assoc* 1996;209:650–652.
11. Divers T. Urinary tract disorders in cattle. *Bov Pract* 1989;24:150–153.
12. Winter R, Hawkins L, Holterman D, Jones S. Catheterization: an effective method of treating bovine urethral calculi. *Vet Med* 1987;82:1261–1268.
13. Wolfe D. Urolithiasis. In: Wolfe DF, Moll HD (eds) *Large Animal Urogenital Surgery*, 2nd edn. Baltimore: Williams & Wilkins, 1998, pp. 349–354.
14. Sockett D, Knight A, Fettman M, Kiehl A, Smith J, Arnold S. Metabolic changes due to experimentally induced rupture of the bovine urinary bladder. *Cornell Vet* 1986;76:198–212.
15. Hooper R, Taylor T. Urinary surgery. *Vet Clin North Am Food Anim Pract* 1995;11:95–121.
16. Donecker J, Bellamy J. Blood chemical abnormalities in cattle with ruptured bladders and ruptured urethras. *Can Vet J* 1982;23:355–357.
17. Wilson D, Macwilliams P. An evaluation of the clinical pathologic findings in experimentally induced urinary bladder rupture in pre-ruminant calves. *Can J Vet Res* 1998;62:140–143.
18. Floeck M. Ultrasonography of bovine urinary tract disorders. *Vet Clin North Am Food Anim Pract* 2009;25:651–667.
19. Parrah D, Moulvi B, Malik H et al. Abdominocentesis and peritoneal fluid analysis in clinical cases of bovine obstructive urolithiasis. *Vet Practitioner* 2011;12:218–222.
20. Skarda R. Local and regional anesthesia in ruminants and swine. *Vet Clin North Am Food Anim Pract* 1996;12:579–626.
21. Scheel E, Paton I. Urinary calculi in feedlot cattle: report on treatment with amino promazine. *J Am Vet Med Assoc* 1960;137:665–667.
22. Gasthuys F, Steenhaut M, De Moor A, Sercu K. Surgical treatment of urethral obstruction due to urolithiasis in male cattle: a review of 85 cases. *Vet Rec* 1993;133:522–526.
23. Oehme F, Tillmann H. Diagnosis and treatment of ruminant urolithiasis. *J Am Vet Med Assoc* 1965;147:1331–1339.
24. Van Metre D, House J, Smith B et al. Obstructive urolithiasis in ruminants: medical treatment and urethral surgery. *Comp Cont Educ Pract Vet* 1996;18:317–327.
25. Walker D. Penile surgery in the bovine: Part I. *Mod Vet Pract* 1979;60:839–843.
26. Streeter R, Washburn K, Higbee R, Bartels K. Laser lithotripsy of a urethral calculus via ischial urethrotomy in a steer. *J Am Vet Med Assoc* 2001;219:640–643.
27. Verwilghen D, Ponthier J, Van Galen G et al. The use of radial extracorporeal shockwave therapy in the treatment of urethral urolithiasis in the horse: a preliminary study. *J Vet Intern Med* 2008;22:1449–1451.
28. Rocken M, Furst A, Kummer M, Mosel G, Tschanz T, Lischer CJ. Endoscopic-assisted electrohydraulic shockwave lithotripsy in standing sedated horses. *Vet Surg* 2012;41:620–624.
29. Kilic E, Ozba B, Atalan G. Basket catheterisation: a method for removing bovine urethral calculi. *Indian Vet J* 2003;80:43–45.
30. Oehme F. A urinary calculi retriever for nonsurgical treatment of urolithiasis in bulls. *Vet Med Small Anim Clin* 1968;63:53–57.
31. Halland S, House J, George L. Urethroscopy and laser lithotripsy for the diagnosis and treatment of obstructive urolithiasis in goats and pot-bellied pigs. *J Am Vet Med Assoc* 2002;220:1831–1834.
32. Halland S, Phelps M, House J. New methods to treat and prevent obstructive urolithiasis in small ruminants and pot-bellied pigs. In: *Proceedings of the 18th Annual Veterinary Medical Forum, American College of Veterinary Internal Medicine*. Lakewood, CO: ACVIM, 2000, pp. 268–270.
33. Haven M, Bowman K, Engelbert T, Blikslager A. Surgical management of urolithiasis in small ruminants. *Cornell Vet* 1993;83:47–55.
34. Rakestraw P, Fubini S, Gilbert R, Ward J. Tube cystostomy for treatment of obstructive urolithiasis in small ruminants. *Vet Surg* 1995;24:498–505.
35. Ewoldt J, Anderson D, Miesner M, Saville W. Short- and long-term outcome and factors predicting survival after surgical tube cystostomy for treatment of obstructive urolithiasis in small ruminants. *Vet Surg* 2006;35:417–422.
36. Van Metre D, House J, Smith B et al. Obstructive urolithiasis in ruminants: surgical management and prevention. *Comp Cont Educ Pract Vet* 1996;18:S275–S289.
37. Van Metre D, Smith B. Clinical management of urolithiasis in small ruminants. In: *Proceedings of the 9th Annual Veterinary Medical Forum, American College of Veterinary Internal Medicine, New Orleans, LA.* ACVIM, 1991, pp. 555–557.
38. Parrah J, Moulvi B, Hussain S, Bilal S, Ridwana. Comparative efficacy of tube cystostomy and cystotomy with indwelling urethral catheterization in the management of obstructive urolithiasis in bovines. *Indian J Vet Surg* 2010;31:81–85.
39. Cockcroft P. Dissolution of obstructive urethral uroliths in a ram. *Vet Rec* 1993;132:486.
40. Streeter R, Washburn K, Mccauley C. Percutaneous tube cystostomy and vesicular irrigation for treatment of obstructive urolithiasis in a goat. *J Am Vet Med Assoc* 2002;221:546–549.
41. Amarpal, Kinjavdekar P, Aithal H, Pawde A, Pratap K. Management of obstructive urolithiasis with tube cystostomy in a bullock. *Indian Vet J* 2010;87:65–67.
42. Parrah J, Moulvi B, Hussain S, Bilal S. Innovative tube cystostomy for the management of bovine clinical cases of obstructive urolithiasis. *Veterinarski Arhiv* 2011;81:321–337.
43. Van Metre D, Fubini S. Ovine and caprine urolithiasis: another piece of the puzzle. *Vet Surg* 2006;35:413–416.
44. Fazili M, Malik H, Bhattacharyya H, Buchoo B, Moulvi B, Makhdoomi D. Minimally invasive surgical tube cystostomy for treating obstructive urolithiasis in small ruminants with an intact urinary bladder. *Vet Rec* 2010;166:528–531.
45. Franz S, Dadak A, Schoffmann G et al. Laparoscopic-assisted implantation of a urinary catheter in male sheep. *J Am Vet Med Assoc* 2008;232:1857–1862.
46. Gill M, Sod G. Buccal mucosal graft urethroplasty for reversal of a perineal urethrostomy in a goat wether. *Vet Surg* 2004;33:382–385.
47. Hunter B, Huber M, Riddick T. Laparoscopic-assisted urinary bladder marsupialization in a goat that developed recurrent urethral obstruction following perineal urethrostomy. *J Am Vet Med Assoc* 2012;241:778–781.
48. May K, Moll H, Duncan R, Moon M, Pleasant R, Howard R. Experimental evaluation of urinary bladder marsupialization in male goats. *Vet Surg* 2002;31:251–258.
49. May K, Moll H, Wallace L, Pleasant R, Howard R. Urinary bladder marsupialization for treatment of obstructive urolithiasis in male goats. *Vet Surg* 1998;27:583–588.
50. Fortier L, Gregg A, Erb H, Fubini S. Caprine obstructive urolithiasis: requirement for 2nd surgical intervention and mortality after percutaneous tube cystostomy, surgical tube cystostomy, or urinary bladder marsupialization. *Vet Surg* 2004;33:661–667.
51. Canpolat I, Bulut S. Experimental evaluation of urinary bladder marsupialization in male lambs and calves. *Indian Vet J* 2005;82:398–400.
52. Hastings D. Retention catheters for treatment of steers with ruptured bladders. *J Am Vet Med Assoc* 1965;147:1329–1330.
53. Prasad B, Sharma SN, Singh J, Kohli RN. Surgical repair and management of bladder rupture in bullocks. *Indian Vet J* 1978;55:905–911.
54. Sharma S, Vikash K. Urinary bladder rupture in bovine and its management: a retrospective study of 156 cases. *Intas Polivet* 2011;12:375–378.

Chapter 20

Preparation of Teaser Bulls

Gretchen Grissett

Department of Clinical Sciences, College of Veterinary Medicine, Kansas State University, Manhattan, Kansas, USA

Introduction

Historically, a key to a successful artificial insemination (AI) program has been accurate detection of estrus. Even with the utilization of currently available estrus synchronization programs and timed AI, estrus detection is important as a tool to evaluate the efficacy of the protocol, to troubleshoot problems in real time, and identify outliers that can be bred outside the prescribed "AI window."

Several estrus detection methods exist, including tailhead paint, mount detectors, self-adhesive heat detection patches, and visual observation. All these methods depend on female cattle standing for mounting during estrus. Thus these methods might miss females with weak or short estrus behavior.[1] By far the most efficient estrus detector is the bull, with the caveat that there is the benefit of male presence.[2] Therefore, utilization of a teaser bull (intact sterilized male) is the most reliable method of estrus detection in the utilization of an AI program.

Several factors need to be considered when choosing a teaser bull procedure and each producer will have different needs and expectations. Besides the obvious need to render the bull sterile, other considerations would include herd status (open or closed herd). If an open herd, then venereal disease transmission is an important factor to consider and prevention of intromission during mounting will be an important factor when choosing a teaser bull surgical procedure. Additionally, expected longevity needs to be discussed with the producer. On average, teaser bulls will last 1–3 years within a herd.[1] Decreased libido is the most common reason for culling, with excessive size and aggression being the next most common culling reasons.[3] To summarize, the main goals of surgical preparation of teaser bulls are to render him sterile, prevent intromission and therefore the transmission of venereal disease, and avoid excessive libido reduction.[4,5]

Proper bull selection is also an important aspect of teaser bull preparation. The ideal bull needs to be moderately sized, of mild temperament, and easily handled. This bull also needs to be free of transmissible diseases. Of course, the bull also needs strong libido, but this can be difficult to assess in yearling bulls.[4] Teaser bull surgery needs to be performed well before the breeding season to allot time for healing and recovery from surgery. Ideally, the procedure should be performed on bulls less than 272 kg, primarily for ease of handling and decreased hemorrhage during surgery.[4]

Teaser bull procedures can be divided into two categories: those that block semen flow and deliver sterility (vasectomy, epididymectomy) and those that prevent penile penetration (penile–prepuce translocation, penopexy, preputial pouch). Depending on the needs and expectations of the producer, any one or combination of these procedures can be used for preparation of a teaser bull.

Vasectomy

As previously mentioned, vasectomy will render a bull sterile but does not prevent normal mating and copulation behavior. This procedure can be performed with the bull in standing restraint, recumbency, or a tilt chute if available. The typical surgical approach is an anterior approach on the neck of the scrotum. However, if standing restraint is chosen, then the approach would be the posterior aspect of the neck of the scrotum.

The neck of the scrotum should be clipped and aseptically prepared for surgery. Lidocaine 2% should be infused over the proposed incision site over each spermatic cord. A 3-cm incision should be made through the skin and tunica dartos over each spermatic cord. Then the spermatic cord is isolated by placing a hemostat underneath the entire spermatic cord. The ductus deferens is then identified via palpation. The ductus deferens is a firm structure that runs medially along the spermatic cord and is approximately 2–3 mm in diameter. Once identified, the tunica vaginalis is carefully incised, utilizing extreme caution so as not to damage the cremaster muscle or pampiniform plexus resulting in excessive hemorrhage. After the tunica vaginalis is incised, the ductus deferens is isolated with another hemostat (Figure 20.1).

Two ligatures are placed approximately 3–5 mm apart using #0 absorbable suture.[4,5] The ductus deferens is removed between the two ligatures. The skin is closed with a cruciate pattern using nonabsorbable suture. Antibiotics can be administered to prevent any postoperative infection,

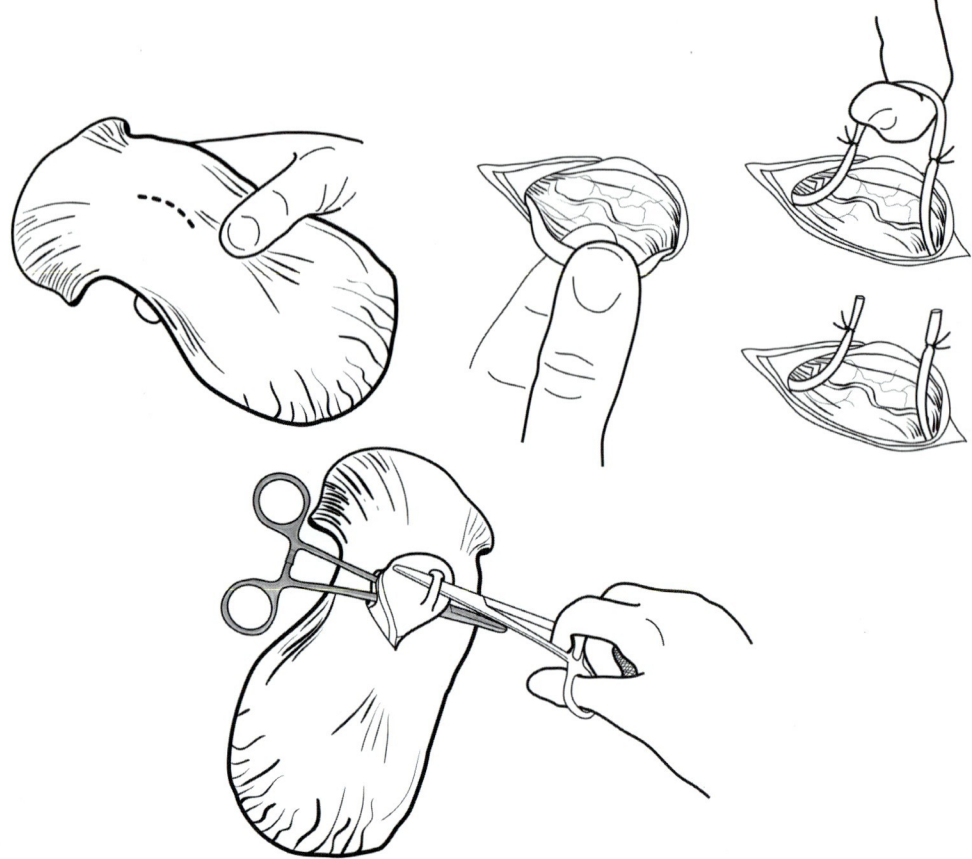

Figure 20.1 Procedure for vasectomy. Illustration by Mal Hoover.

especially if surgical contamination has occurred. It is recommended to wait 30 days prior to using the bull as a teaser animal, since sperm can be present in the reproductive tract up to 30 days postoperatively.[4,5] Additionally, it is recommended to perform yearly evaluations of the teaser animal's ejaculate to ensure sterility of the animal.

Epididymectomy

An epididymectomy is similar to a vasectomy with regard to restraint options and copulation behavior.[6] For this procedure, the base of the scrotum is clipped and aseptically prepared. Lidocaine 2% is infused over the tail of the epididymis. Once prepared, the surgeon grasps the neck of the scrotum and pushes the testicle ventrally. A 3-cm incision is made over the tail of the epididymis through the skin and common vaginal tunic until the epididymis is exteriorized. The tail of the epididymis is carefully dissected from the testicle and towel clamps or Allis tissue forceps can be used to assist in handling and manipulation of the epididymis. Then a hemostat is placed on the ductus deferens and the body of the epididymis. Ligatures with #0 absorbable suture are placed proximal to the hemostats. The tail of the epididymis is removed by transection distal to the hemostats (Figure 20.2).

The common vaginal tunic is closed using #0 absorbable suture. The skin can be closed with nonabsorbable cruciate sutures or the incisions can be left open to allow ventral drainage. Antibiotics can be administered to prevent postoperative infections. Postoperative resting recommendations and yearly ejaculate examinations are the same as previously stated for vasectomy aftercare.

Penile–prepuce translocation

Penile–prepuce translocation ("sidewinder") is the surgical transposition of the penis and prepuce from ventral midline to the right or left flank of a bull. This procedure allows normal protrusion and erection, but does not permit intromission. In general, "sidewinders" are preferred by producers due to longevity and herd retention of the teaser animal. Bulls with a penile–prepuce translocation maintain better and longer libido since this procedure allows normal protrusion and does not cause pain during erection. Some bulls are able to compensate and learn how to breed females despite the translocation of the penis and prepuce. Therefore it is recommended that a vasectomy or epididymectomy is performed to ensure sterility of the bull.

Penile–prepuce translocation is performed in lateral recumbency, so general anesthesia is the preferred method of restraint. If general anesthesia is not possible, then heavy sedation with rope restraints and local infiltration of 2% lidocaine can be used. Ideally, food should be withheld for 24 hours and water for 12 hours prior to performing the procedure.

Prior to placing the bull in recumbency, the translocation site for the preputial orifice should be identified. The translocation site should be just outside the flank fold and lateral to the original preputial orifice site.[3] The ventral abdomen from

Preparation of Teaser Bulls   183

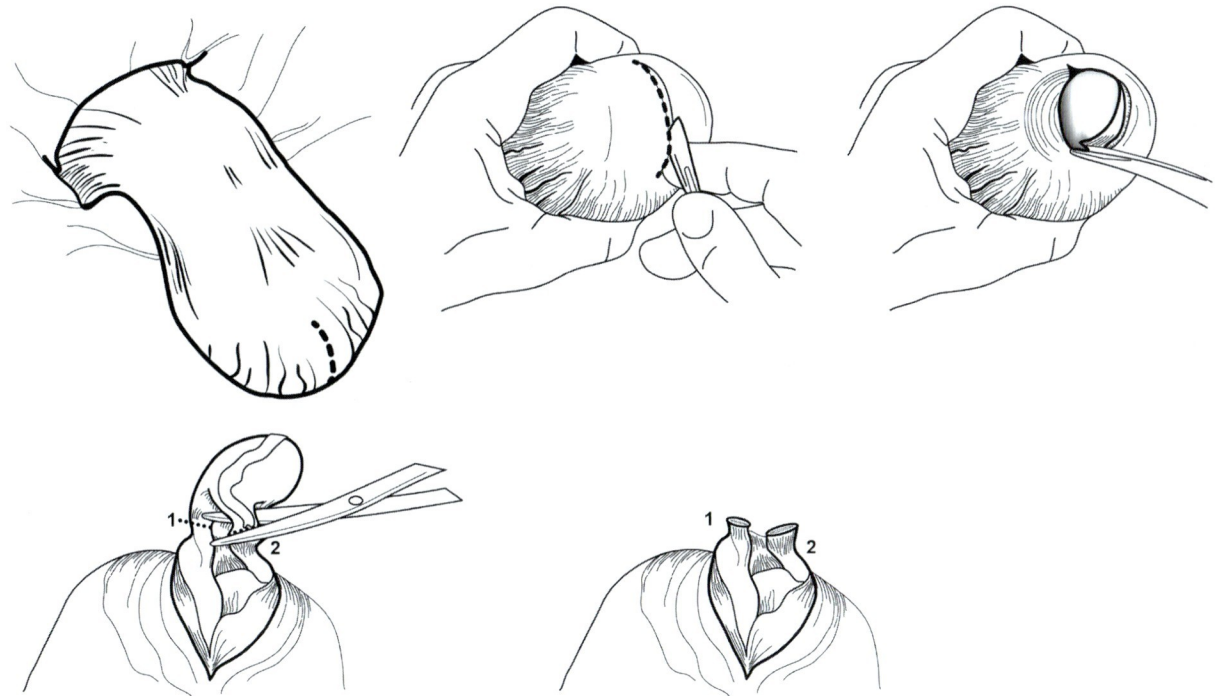

Figure 20.2 Procedure for epididymectomy. Illustration by Mal Hoover.

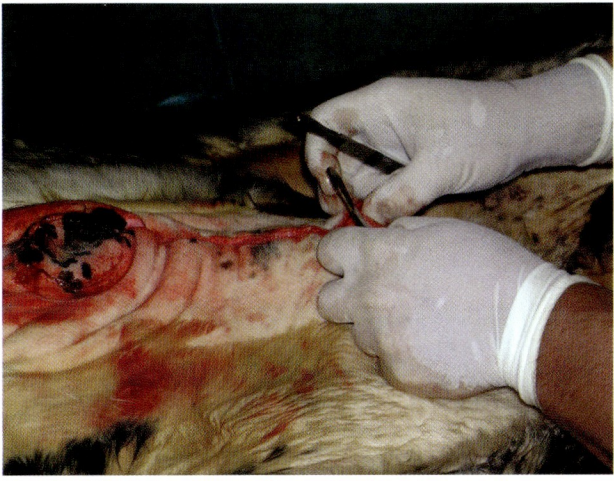

Figure 20.3 A 4-cm circumferential incision around the preputial orifice with ventral midline incision extending caudally.

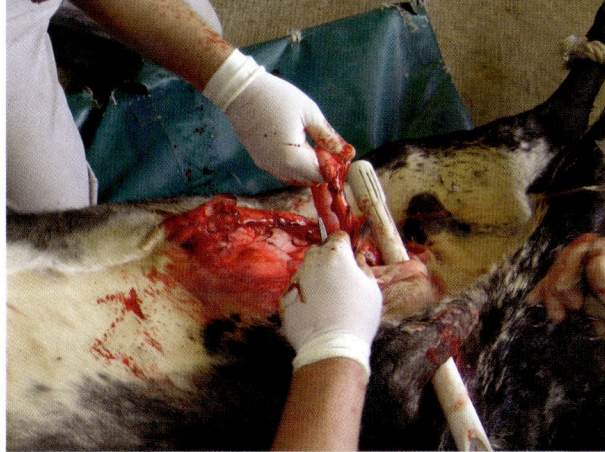

Figure 20.4 Use of a cold sterilized PVC pipe to facilitate tunneling of penile translocation and skin incision for translocation site.

the umbilicus to just cranial to the scrotum and the site of translocation of the flank should be clipped and aseptically prepared. Flush the prepuce with dilute iodine solution.

Before making the initial incision, place one simple interrupted suture at the dorsal aspect of the preputial orifice to serve as a marker and prevent twisting of the prepuce during translocation. A circumferential skin incision around the preputial orifice is made 4 cm from the orifice or a total diameter of 8–10 cm (Figure 20.3).[3,4] Extend the skin incision on the ventral midline from the preputial orifice to just cranial to the scrotum. Carefully dissect the penis and prepuce from the ventral abdomen. Avoid lacerating the prepuce; packing or tubing can be placed in the prepuce to aid with proper identification. While dissecting the penis and prepuce, avoid incising the dorsal penile vessels and control hemorrhage as it is encountered. Once the penis and prepuce are dissected, make a circular skin incision equivalent to the diameter of the preputial orifice at the desired translocation site. Use a sponge forceps to create a tunnel toward the flank incision. As the forceps is retracted, open it slightly to help facilitate penile translocation. This tunnel can also be accomplished with a cold sterilized PVC pipe (Figure 20.4).

Place a sterile glove or sleeve over the preputial orifice to minimize contamination of the subcutaneous tissues. Then run a sponge forceps from the flank incision to the ventral midline incision and grasp the preputial orifice. Manipulate the preputial orifice to the flank incision, taking care not to

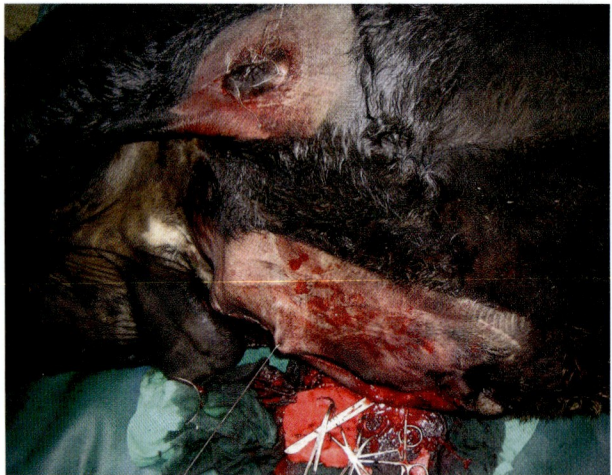

Figure 20.5 Closure of new preputial orifice with interrupted sutures and ventral midline with Ford interlocking pattern.

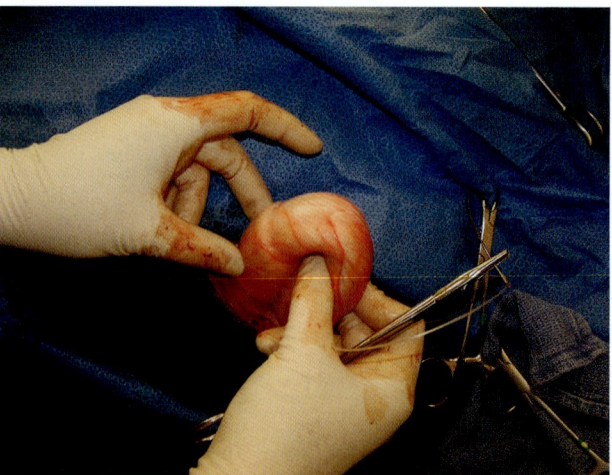

Figure 20.6 Exteriorization of the penis through the incision and identification of the caudal reflection of the penis.

twist the prepuce (use a stay suture to ensure proper alignment). Suture the skin around the preputial orifice using #3 nonabsorbable sutures with a cruciate or horizontal mattress pattern (Figure 20.5).[1] Close the subcutaneous layer of the ventral midline incision with #3 absorbable suture, closing as much dead space as possible to prevent seroma formation. Close the skin with #3 nonabsorbable suture in a Ford interlocking pattern. Place a cruciate suture at the cranial aspect of the incision to be removed for drainage if a seroma does occur.

The teaser bull should be monitored closely for 24 hours postoperatively to ensure he is able to urinate properly. Antibiotics should be administered for 3–5 days postoperatively to prevent infection. Allow 4–6 weeks of recovery time before using the teaser bull.[3–5] Penile–prepuce translocation is not a technically difficult procedure, but it is more invasive and can result in more postoperative complications. The most common complications are obviously seroma and abscess formation from the excessive dead space created. Another complication would be not translocating the preputial orifice high enough on the flank and thus the bull would still be capable of breeding a female animal. There is also one case report of a teaser bull developing paraphimosis with a penile–prepuce translocation.[7]

Penopexy

Penopexy is the iatrogenic creation of phimosis by surgically creating an adhesion of the penis to the ventral body wall. This procedure prevents protrusion of the penis, thus preventing normal intromission or copulation. Penopexy is a relatively quick procedure and can typically be performed with sedation and local infiltration of 2% lidocaine. Tilt chute restraint or general anesthesia can also be utilized. Lateral recumbency is the preferred positioning. The bull's ventral abdomen is clipped and surgically prepared from the preputial orifice to the scrotum.

A skin incision is made 2–3 cm lateral of midline and half the distance between the preputial orifice and scrotum approximately 10 cm in length. Carefully dissect the subcutaneous tissues until the penis is identified and exteriorized (Figure 20.6). Once the penis is exteriorized through the incision, identify the caudal reflection of the penis (fornix) and dissect the subcutaneous tissues on the dorsal aspect of the penis until the tunica albuginea is exposed for approximately 10 cm caudal to the fornix.[4,5] Remove the subcutaneous tissue on the linea alba in conjunction with the dorsal aspect of the penis. The tunica albuginea and corresponding linea alba are scarified to promote strong adhesion formation. After preparation of both sites, the urethral groove is identified on the ventral aspect of the penis. Beginning 6–8 cm caudal to the fornix of the penis, preplace four to six simple interrupted sutures approximately 2 cm apart using a heavy nonabsorbable suture.[4,5] The suture is placed through the dorsal third of the penis using care to not enter the urethra. The suture is then placed through a corresponding area of the linea alba (Figure 20.7).[4,5] Once all the sutures are preplaced, return the penis to the normal anatomical position and ensure it is not protruding from the preputial orifice prior to securing all the sutures (Figure 20.8). Close the subcutaneous tissue with absorbable sutures and the skin with #3 nonabsorbable suture in a Ford interlocking pattern. A vasectomy or epididymectomy is usually performed in conjunction with a penopexy to ensure sterility of the bull in case of procedure failure.

Allow 3–4 weeks of recovery to ensure proper formation of adhesions. The drawback of penopexy is the risk of entering the urethra and decreased longevity in the herd since the bull will experience pain during attempted erection, thus decreasing libido. A follow-up study of 37 bulls found that 15% of bulls maintained good libido for one breeding, 30% for 1–1.5 years, and 42% for more than 1.5 years.[5,8]

A standing perineal penopexy approach has been described using light sedation and a caudal epidural. The approach is over the distal loop of the sigmoid flexure.[4] A 4–5 cm incision is made through the skin and the tunica albuginea is exposed as mentioned previously. However, the stay sutures are placed on the lateral aspects of the penis and secured to the fibrous connective tissue in the perineal region of the bull.[4,5]

Preparation of Teaser Bulls 185

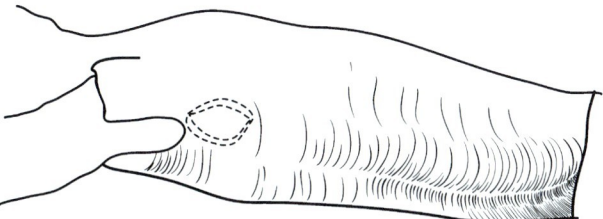

Figure 20.9 Site for incision for ventral fistula. Illustration by Mal Hoover.

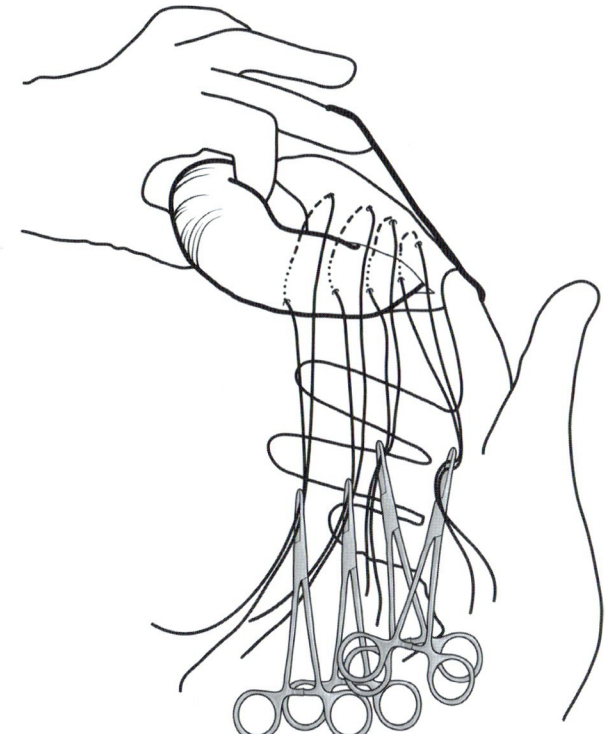

Figure 20.7 Preplacement of sutures through the dorsal third of penis and linea alba. Illustration by Mal Hoover.

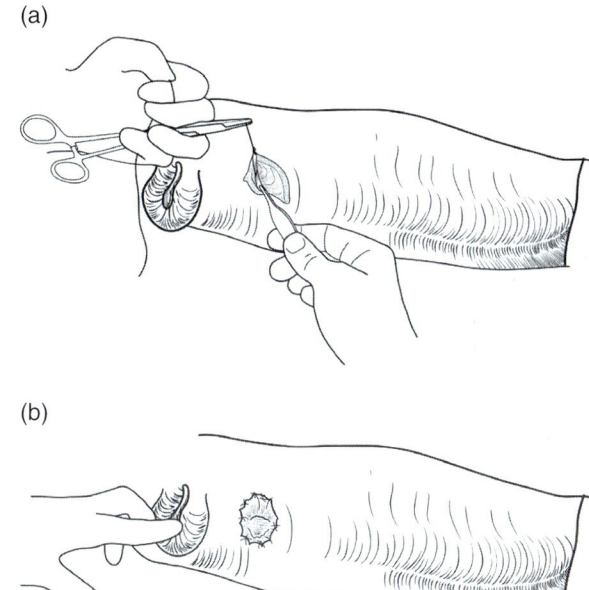

Figure 20.10 (a, b) Suturing of the preputial mucosa to the sheath skin. Illustration by Mal Hoover.

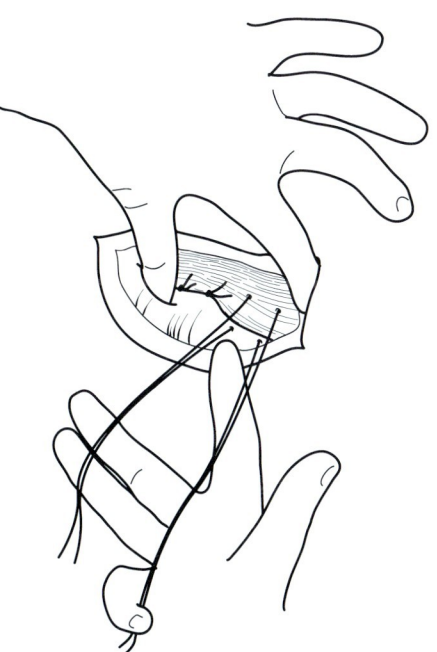

Figure 20.8 Securing the stay sutures for penopexy. Illustration by Mal Hoover.

Preputial pouch technique (ventral slot with preputial orifice obliteration)

The preputial pouch technique creates a fistula on the ventral prepuce and closes the normal preputial orifice. This technique prevents penile extension, but allows for passage of urine through the ventral fistula. Teaser bulls with a preputial pouch are typically retained in the herd longer because libido is maintained for longer due to the lack of pain during attempted breeding.[5] This procedure is performed in lateral recumbency and can be accomplished with tilt chute restraint or sedation with local infiltration of 2% lidocaine. The ventral abdomen is clipped and prepared from the umbilicus to the mid-sheath region of the bull.

Prior to initiating surgery, the penis is extended and a Penrose drain is sutured around the glans penis with 2-O PDS. An approximately 1-cm diameter elliptical incision is made through the skin 7 cm caudal to the preputial orifice (Figure 20.9). The skin incision is extended through the preputial mucosa. The excised skin and mucosa are discarded. Then the internal mucosa of the prepuce is sutured to the skin of the sheath to create the fistula. An interrupted nonabsorbable suture pattern is recommended (Figure 20.10a,b).[5] Once suturing is complete, the free end of the Penrose drain is placed through the fistula. The Penrose drain will facilitate urine divergence while the primary incision sites heal.

The preputial orifice obliteration is accomplished by removing approximately 5 mm of the sheath skin and prepuce mucosal junction around the entire preputial orifice (Figure 20.11). This incision is closed in three layers:

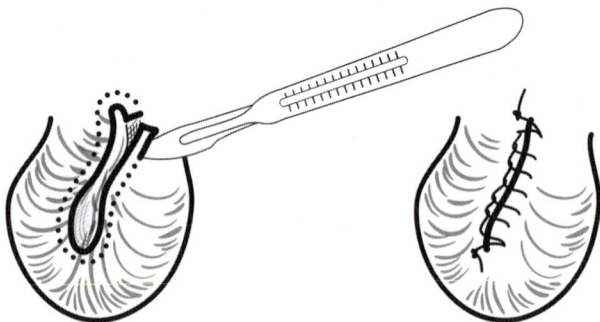

Figure 20.11 Excision of 5 mm of the preputial epithelium and sheath skin junction. Illustration by Mal Hoover.

preputial epithelium, subcutaneous layer, and skin. The Penrose drain and sutures can be removed in 2 weeks. Allow 3 weeks of postoperative recovery time prior to utilizing the teaser bull.[5]

The critical step in the preputial pouch technique is the size of the fistula. If the fistula is too small, proper urine flow is obstructed. If the fistula is too large, penile extension can occur with potential breeding. Therefore, it is recommended to perform a vasectomy or epididymectomy to ensure sterility of the bull. Some bulls may pool urine in their preputial pouch and require postoperative flushing of the pouch. Thus this technique is not recommended for *Bos indicus* breeds since their pendulous sheath would predispose them to urine pooling and calculi formation.[5]

Other procedures

Other teaser bull procedures include iatrogenic preputial stenosis, artificial corpus cavernosal thrombosis, and penectomy. These procedures are briefly discussed since their use has fallen out of favor due to high risk of failure, complication rates, or diminished libido.

Iatrogenic preputial stenosis involves a ventral mid-sheath approach to the prepuce and penis. The prepuce is identified and a stainless steel rod or Steinmann pin is secured and tightened around the prepuce. Care must be taken to tighten the ring sufficiently to prevent penile extension, but loose enough to prevent urine pooling and balanoposthitis.[5,9] Mixed success rates accompany this procedure. Complications associated with this procedure include excessive ring closure resulting in urine retention and balanoposthitis, complete stenosis resulting in subcutaneous urine accumulation, lack of stenosis resulting in penile extension and intromission, or excessive tissue reaction to the stainless steel ring.

Corpus cavernosal thrombosis involves injection of an acrylic material into the corpus cavernosum of the penis.[10] The acrylic material results in thrombus formation that prevents erection. This method is performed with standing restraint under a caudal epidural. A midline incision is made over the penis. The distal sigmoid flexure is identified by locating the retractor penis muscle, with the proximal sigmoid flexure being approximately 15 cm proximally. A 14-gauge needle is inserted at the dorsolateral aspect of the penis at the proximal sigmoid flexure and the acrylic is injected. Nonabsorbable stay sutures are placed at the lateral aspect of the penis at the level of the retractor penis muscle to prevent penile prolapse. Potential complications of this procedure include inadequate injection of acrylic into the corpus cavernosum resulting in procedure failure, or accidental injection into the corpus spongiosum or urethra resulting in urethral obstruction.[5]

Penectomy involves amputation of the penis. This can be performed by amputation of the glans penis at the fornix and suturing of the prepuce to the urethral mucosa;[11] alternatively this can be performed at the perineal region suturing urethral mucosa to the skin.[5,11] Amputation of the glans penis at the fornix results in teaser bulls that experience pain during breeding attempts, thus decreasing libido and herd retention time.[5] With penectomy via the perineal approach, bulls often lose interest and experience decreased libido due to the lack of coitus.[11] With either approach, urethral stricture is a risk factor.

Summary

Accurate heat detection is essential to any AI or embryo transfer program and teaser bulls are the best at detecting heat. There are multiple procedures for creating a teaser bull, with no single procedure being perfect. Each procedure has its advantages and disadvantages. Ultimately, the decision of which procedure to perform depends on the needs and expectations of the client (longevity of bull, postoperative complications, assured sterility). Additional factors that may impact procedural choice include facilities, veterinarian preference, cost, and herd status.

References

1. Holmann F. Economic evaluation of fourteen methods of estrous detection. *J Dairy Sci* 1987;70:186–194.
2. Hornbuckle T, Ott R, Ohl M, Zinn G, Weston P, Hixon J. Effects of bull exposure on the cyclic activity of beef cows. *Theriogenology* 1995;43:411–418.
3. Noordsy JL, Ames NK. *Food Animal Surgery*, 4th edn. Princeton, NJ: Veterinary Learning Systems, 2006, pp. 229–239.
4. Morgan G, Dawson L. Development of teaser bulls under field conditions. *Vet Clin North Am Food Anim Pract* 2008;24:443–453.
5. Gill M. Surgical techniques for preparation of teaser bulls. *Vet Clin North Am Food Anim Pract* 1995;11:123–136.
6. McCaughey W, Martin J. Preparation and use of teaser bulls. *Vet Rec* 1980;106:119–121.
7. Baird A, Wolfe D, Angel K. Paraphimosis in a teaser bull with penile translocation. *J Am Vet Med Assoc* 1992;201:325.
8. Hoffsis G, Maurer L. Preparation of detector bulls by penile retraction and fixation. In: *Proceedings of the Annual Convention of the American Association of Bovine Practitioners*, 1972, pp. 114–116.
9. Aanes W, Rupp G. Iatrogenic preputial stenosis for preparation of teaser bulls. *J Am Vet Med Assoc* 1984;184:1474–1476.
10. Wolfe D. Surgical procedures of the reproductive system of the bull. In: Morrow DA (ed.) *Current Therapy in Theriogenology*, 2nd edn. Philadelphia: WB Saunders, 1986, pp. 353–380.
11. Straub O, Kendrick J. Preparation of teaser bulls by penectomy. *J Am Vet Med Assoc* 1965;147:373–376.

SECTION II

The Cow

ANATOMY AND PHYSIOLOGY

21. Anatomy of the Reproductive System of the Cow — 191
    Ben Nabors and Robert Linford

22. Initiation of Puberty in Heifers — 195
    Charles T. Estill

23. Neuroendocrine Control of Estrus and Ovulation — 203
    Marcel Amstalden and Gary L. Williams

24. Ovarian Follicular and Luteal Dynamics in Cattle — 219
    Gregg P. Adams and Jaswant Singh

25. Maternal Recognition and Physiology of Pregnancy — 245
    Caleb O. Lemley, Leticia E. Camacho and Kimberly A. Vonnahme

Chapter 21

Anatomy of the Reproductive System of the Cow

Ben Nabors and Robert Linford

Department of Clinical Sciences, College of Veterinary Medicine, Mississippi State University, Starkville, Mississippi, USA

Introduction

The anatomy of the reproductive system in the cow is functionally grouped into the components associated with oocyte production and transport, and those involved with gestation and copulation.

Production

The cellular machinery for oogenesis and steroid production is found in the ovary (Figure 21.1). The ovary consists of a cortex and medulla. The medulla is composed of connective tissue, lymphatic vessels, blood vessels, and nerves. Surrounding the medulla is the cortex. The cortex contains the ova surrounded by follicular cells within the connective tissue stroma.[1] Exterior to the cortex, the ovary is covered by the dense fibrous tunica albuginea and a superficial epithelium.[2]

Because the ovary in the cow descends further from its embryologic origin near the kidney than other species, it is positioned closer to the pelvis. The consequence of this ovarian location and the attachment of the short mesovarium is that the uterine horns bend ventrally and caudally[3] (Figure 21.2).

Transport and gestation

The reproductive system of the cow is designed to transport spermatozoa toward the ovary and to transport an ovum toward the spermatozoa (Figure 21.3). The parts of this tubular system include the vestibule, vagina, cervix, uterine horns, and uterine tubes (oviduct).

Uterine tube

The uterine tube is arranged like a funnel near the ovary. The funnel-shaped end, or infundibulum, contains processes, the fimbriae, which collect the ovum on ovulation (Figure 21.4). The ovum is then transported through the abdominal opening of the uterine tube located at the base of the infundibulum.[4] The ampulla of the uterine tube is the region adjacent to the infundibulum where fertilization takes place. The isthmus, the continuation of the uterine tube from the ampulla toward the uterus, is relatively long due to the meandering course it takes before ending at the uterine opening where it releases the ovum into the uterine horn.[4]

Uterus

The uterus consists of a body and two horns (Figure 21.5). The body is short, beginning immediately after the cervix ends. The horns branch from the body but are joined together by peritoneum, giving the appearance that the body is longer than it truly is. As the horns progress craniad they divide at the intercornual ligaments, each turning abruptly ventrally, then proceeding caudally, and finally ending dorsal to the ovary[3] (Figure 21.6).

From external to internal, the uterus can be divided into three layers: the perimetrium, the myometrium, and the endometrium.[3] The perimetrium is the continuation of the abdominal peritoneum onto the uterus. The myometrium constitutes the muscular layers, which can undergo substantial hypertrophy.[5] The endometrium is the internal epithelial lining of the uterus and is arranged into two distinct regions, caruncular and intercaruncular.[6] The caruncles are raised mucosal regions of the endometrium that are highly vascularized (Figure 21.7). The caruncles of the uterus join with the cotyledons of the fetal placental membranes to form the placentomes of a cotyledonary placenta.[4]

Cervix

The cervix is located between the body of the uterus cranially and the vagina caudally (Figure 21.8). It is a firm, muscular, sphincter-like structure that acts as a barrier

---

Bovine Reproduction, First Edition. Edited by Richard M. Hopper.
© 2015 John Wiley & Sons, Inc. Published 2015 by John Wiley & Sons, Inc.

The Cow: Anatomy and Physiology

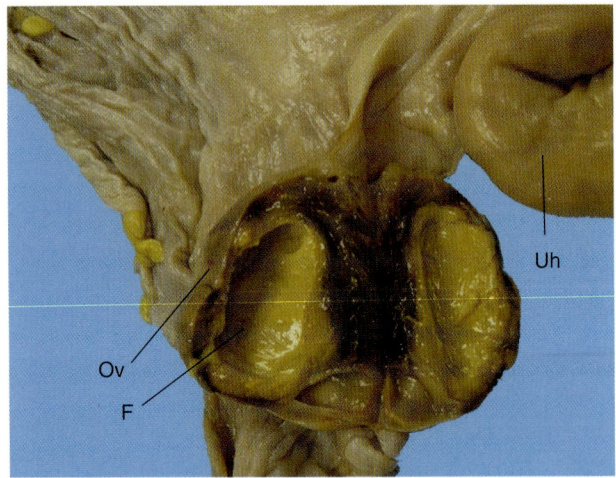

Figure 21.1 Ov, ovary; Uh, uterine horn; F, follicle.

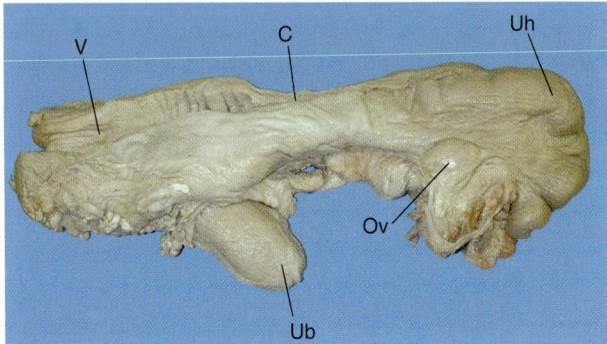

Figure 21.2 V, vagina; C, cervix; Uh, uterine horn; Ov, ovary; Ub, urinary bladder.

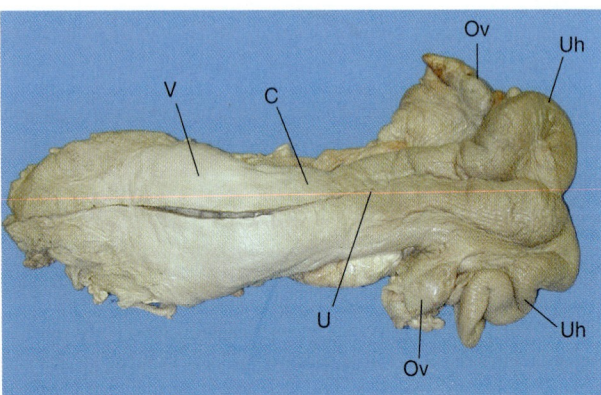

Figure 21.3 V, vagina; C, cervix; U, body of uterus; Ov, ovary; Uh, uterine horn.

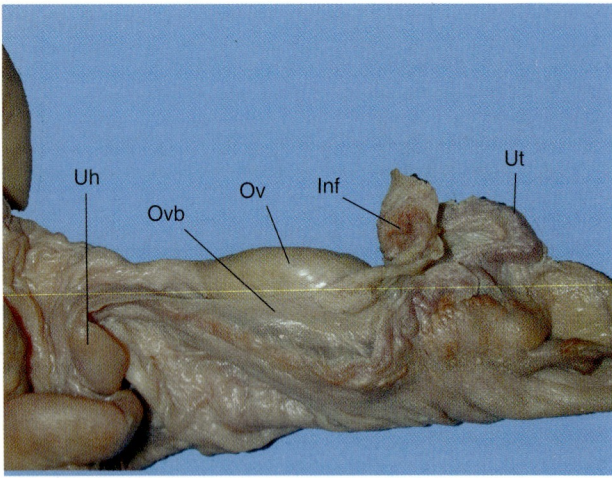

Figure 21.4 Ov, ovary; Inf, infundibulum; Ut, uterine tube; Ovb, ovarian bursa; Uh, uterine horn.

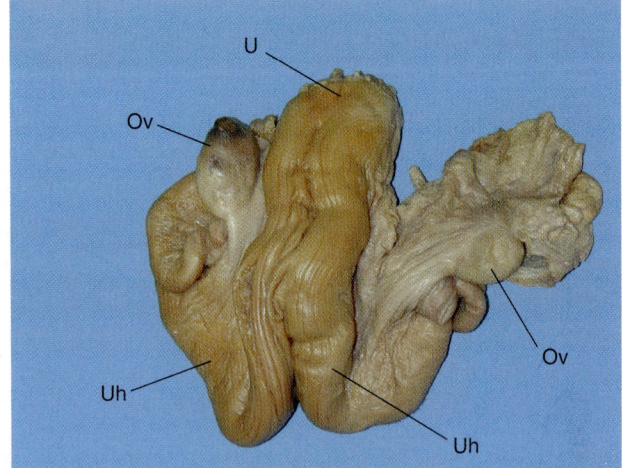

Figure 21.5 U, body of uterus; Ov, ovary; Uh, uterine horn.

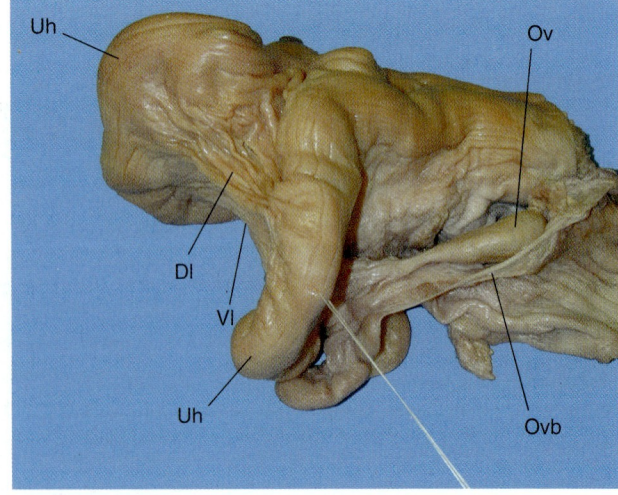

Figure 21.6 Uh, uterine horn; Dl, dorsal intercornual ligament; Vl, ventral intercornual ligament; Ov, ovary; Ovb, ovarian bursa.

separating the external genitalia from the internal genitalia.[5] Characteristic of the cervix are the three to four circular folds projecting into its lumen.

The arrangement of the cervical musculature and mucosa is responsible for this characteristic architecture (Figure 21.9). The superficial mucosa is arranged in longitudinal folds punctuated by circular folds that interdigitate to form a series of ridges and interlocking notches when the cervix is closed.[4] This arrangement effectively seals the external environment from the internal uterine environment.

Anatomy of the Reproductive System of the Cow 193

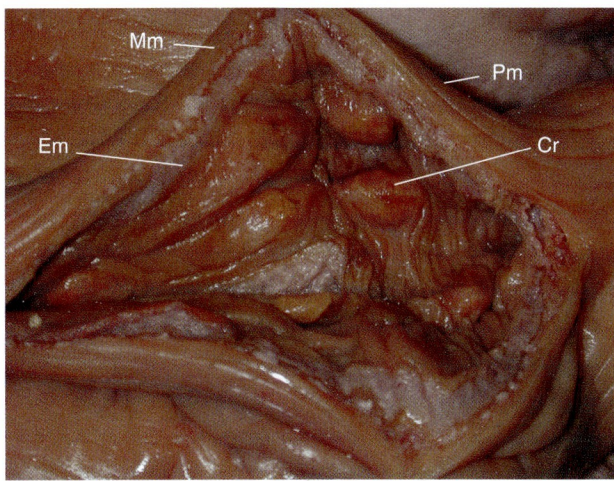

Figure 21.7 Pm, perimetrium; Mm, myometrium; Em, endometrium; Cr, caruncle.

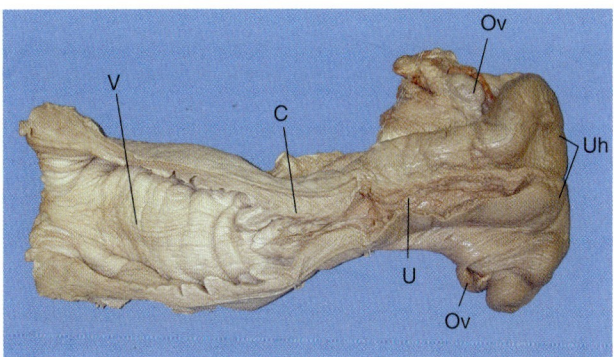

Figure 21.8 V, vagina; C, cervix; U, body of uterus; Uh, uterine horn; Ov, ovary.

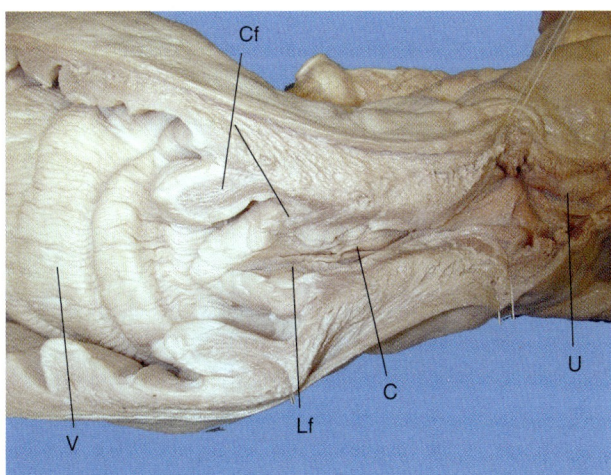

Figure 21.9 V, vagina; C, cervix; U, uterus; Cf, circular folds; Lf, Longitudinal folds.

Vagina

The vagina is positioned between the caudal extent of the cervix and border of the vestibule at the external urethral orifice. The cervix projects into the lumen of the vagina

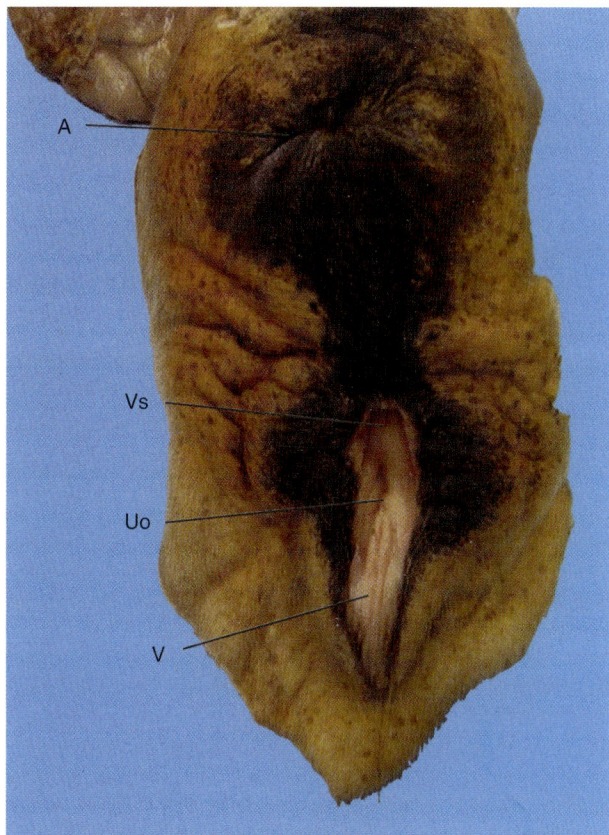

Figure 21.10 A, anus; Vs, vestibule; Uo, urethral orifice; V, vulva.

caudoventrally, causing the dorsal vaginal fornix to form a deeper recess than the ventral fornix.[4]

Vestibule

The vestibule is a small area in the cow that originates at the urethral opening and ends caudally to blend with the labia of the vulva.

Vulva

The labia of the vulva are located on either side of the labial fissure[3] (Figure 21.10). The labia meet dorsally forming the dorsal commissure and again ventrally to form the ventral commissure. The clitoris is found just cranial to the ventral commissure.[3]

Blood supply

The arterial supply to the reproductive system of the cow is provided by the ovarian, umbilical, vaginal, and internal pudendal arteries with their associated branches (Figure 21.11). The ovarian artery leaves the abdominal aorta, passing within the mesovarium to perfuse the ovary. The ovarian artery gives off branches to the uterine tube and the uterine horn.[4] The umbilical artery arises from the internal iliac artery near the pelvic inlet, and sends out a uterine branch that supplies the cervix, body, and horns

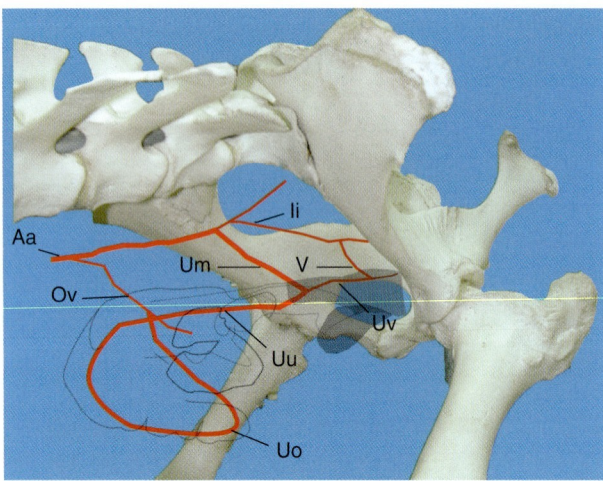

Figure 21.11 Blood supply: Aa, abdominal aorta; Ov, ovarian artery; Um, umbilical artery; Ii, internal iliac artery; V, vaginal artery; Uv, uterine branch of vaginal artery; Uu, uterine branch of the umbilical artery; Uo, uterine branch of the ovarian artery.

of the uterus. The vaginal artery arises directly from the internal iliac artery. The vaginal artery gives off a uterine branch that anastomoses with the uterine branch of the umbilical artery. The uterine artery formed by this anastomosis in turn anastomoses with the uterine branch of the ovarian artery, completing an arterial loop that supplies all the internal components of the reproductive system.[4] The internal pudendal artery is a direct continuation of the internal iliac artery. It courses caudally after the vaginal artery branches off within the ischiorectal fossa. The internal pudendal artery terminates as ventral perineal and dorsal labial branches.[4]

The innervation of the external genitalia of the cow consists of the pudendal nerve and its branches. The pudendal nerve carries motor, sensory, and parasympathetic nerve fibers.[2] The pudendal nerve passes through the pelvic cavity medial to the sacrosciatic ligament and divides as it approaches the lesser ischiatic notch of the pelvis into proximal and distal cutaneous branches that supply the skin of the caudal hip and thigh.[2,3] The pudendal nerve continues through the ischiorectal fossa and terminates as the dorsal nerve of clitoris and a mammary branch.[3] Parasympathetic components of the pudendal nerve come by way of the pelvic nerve, which originates as a coalescence of branches of the sacral spinal nerves at the sacral plexus.[4] Sympathetic components of the pudendal nerve arise from the paired hypogastric nerves, which contribute sympathetic fibers from the caudal mesenteric plexus to the genital system.[4]

Placenta

The cotyledonary bovine placenta is composed of the fetal cotyledons and the maternal caruncles that fuse and form the placentomes.[4] The placentomes are the sites for the transfer of maternal nutrients to the fetal circulation. The fetal membranes consist of two fluid-filled sacs. The innermost sac is the amnion, which surrounds the developing embryo; the outermost sac is the chorioallantois, which encircles the amnion and is composed of two separate tissues that fuse as the embryo develops. The outer layer is the chorion that develops from the trophoblast layer, which is the outer layer of the blastocyst of the embryo. As the embryo develops, the allantois, a sac that arises from the embryo hindgut, expands and eventually fuses with the chorion to form the chorioallantois.

References

1. Ross M, Kaye G, Pawlina W. *Histology: A Text and Atlas*, 4th edn. Philadelphia: Lippincott Williams & Wilkins, 2003, p. 875.
2. Schaller O, Constantinescu G. *Illustrated Veterinary Anatomical Nomenclature*. Stuttgart: F. Enke Verlag, 1992, p. 614.
3. Budras K-D. *Bovine Anatomy: An Illustrated Text*. Hannover, Germany: Schlütersche, 2003, p. 138.
4. Nickel R, Schummer A, Seiferle E, Sack WO. *The Viscera of the Domestic Mammals*. Berlin: Verlag Paul Parey; New York, Springer-Verlag, 1973, p. 401.
5. Pineda M, Dooley M. *McDonald's Veterinary Endocrinology and Reproduction*, 5th edn. Ames, IA: Iowa State Press, 2003, p. 597.
6. Mullins K, Saacke R. *Illustrated Anatomy of the Bovine Male and Female Reproductive Tracts: From Gross to Microscopic*. Blacksburg, VA: Germinal Dimensions, p. 79.

Chapter 22

Initiation of Puberty in Heifers

Charles T. Estill

Department of Clinical Sciences, College of Veterinary Medicine, Oregon State University, Corvallis, Oregon, USA

Introduction

Puberty is a critical physiological milestone in a heifer's reproductive life. In general terms, puberty can be defined as the process whereby animals become capable of reproducing themselves.[1] At puberty, plasma progesterone concentrations indicate cyclic ovarian activity before the first observed estrus.[2] Thus, puberty is defined as the first day that serum progesterone (determined in blood samples collected at weekly intervals) exceeds 1 ng/mL.[3]

Regarding heifers, puberty has been defined as the first estrus that is followed by a normal luteal phase.[4] This involves a complex series of interactions of genetic and environmental factors that direct endocrine events which culminate in puberty. In heifers, puberty is triggered when the hypothalamic–pituitary–gonadal axis first loses its sensitivity to the negative feedback effects of oestradiol-17β, allowing a surge of luteinizing hormone (LH) to occur.[4] It is now accepted that puberty and first ovulation are not necessarily coincident since in most heifers "silent" ovulations and short luteal phases may occur during the peripubertal phase.[4]

Endocrine events

Puberty encompasses the transition from the anovular state to one of regular ovulations. The mechanism of how the hypothalamus–pituitary axis loses its sensitivity to the negative feedback effects of estradiol-17β has been the subject of research efforts for many years. The classical "gonadostat" theory, originally developed in a rodent model, appears applicable to cattle.[5] It was proposed that first ovulation results when sensitivity to steroid negative feedback diminishes, allowing sufficient gonadotropin output to drive follicular maturation.[5] The hypothalamic–pituitary axis of female cattle goes through several changes during its development. *In utero*, the fetus secretes gonadotropins for the first 7 months of gestation. After this period, circulating gonadotropins are substantially reduced due to stimulation of the fetal central nervous system (CNS).[6] In sheep, it has been demonstrated that the CNS-stimulated reduction in gonadotropin release that occurs in late gestation is mediated through inhibition of N-methyl-DL-aspartate receptors, which have been demonstrated to be stimulatory to the gonadotropin-releasing hormone (GnRH) pulse generator nucleus in the fetal hypothalamus.[7] Postnatally, mean plasma LH concentrations reach a maximum around 3 months of age, then slowly decline before again rising and culminating in ovulation typically around 10–11 months of age.[8] This early transient increase in circulating concentration of LH is associated with early follicular development and is thought to regulate the timing of puberty. In an attempt to hasten the onset of sexual maturity, Madgwick *et al.*[9] noted that the early rise in LH concentration was advanced by injecting heifer calves with GnRH twice daily from 4 to 8 weeks of age. Treatment with GnRH increased mean circulating concentrations of LH at 8 weeks of age, LH pulse frequency at 4 and 8 weeks of age, and reduced the mean age at puberty by 6 weeks. Body weight gain was greater in GnRH-treated calves than in control calves and the rate of weight gain was shown to be a significant covariate with age at puberty. This early transient rise in circulating LH stimulates ovarian follicular development resulting in estradiol-17β synthesis that has a negative feedback effect on gonadotropin secretion.[10] Although an increase in circulating estradiol-17β has not been consistently demonstrated during this time period,[11,12] it is assumed the decline in LH is due to increased sensitivity to negative feedback by estradiol-17β on the hypothalamus–pituitary.[11,12] From this point until just prior to puberty, estradiol-17β continues to exert negative feedback after which sensitivity gradually declines. This period is known as the peripubertal period and begins about 50 days before puberty.[10,13] The decline in sensitivity to negative feedback by estradiol-17β has been associated with a reduction in the number of cytosolic estradiol-17β receptors in the anterior and medial-basal hypothalamus.[13] The result is that estradiol-17β becomes ineffective at suppressing LH secretion and an ovulatory surge of LH is released.[14] Progesterone levels are very low (300 pg/mL) in the peripubertal period, but there are two distinct elevations of progesterone prior to the first preovulatory peak of LH.[15] The return of the first elevation in progesterone to baseline levels is always followed by the priming peak of LH, while the second elevation in progesterone precedes the pubertal peak of LH.[15] The profile of concentrations of LH between the two major LH peaks, coincident with the second progesterone elevation, appears as a transition between prepubertal and postpubertal LH baseline concentration. This suggests that progesterone plays a key role in the changes leading to the establishment of the phasic LH release characteristic of the postpubertal heifer.[15] Coincidently, growth-related cues are monitored and regulate the activity of the GnRH pulse generator. When sufficient body size/composition is attained, the frequency of LH pulses increases because sensitivity to estradiol-17β inhibitory feedback decreases.[10] The high-frequency LH pulses stimulate follicular maturation and estradiol-17β accelerates the GnRH pulse generator resulting in the ovulatory sure of LH.[16]

The first ovulation is not synonymous with puberty and the first luteal phase is usually of short duration. Prostaglandin (PG)$F_{2\alpha}$ released from the endometrium is responsible for the reduction in luteal lifespan (premature luteolysis) following the first ovulation in heifers.[17,18] Presumably, this occurs because of an abundance of endometrial oxytocin receptors that mediate release of $PGF_{2\alpha}$.[19] Subsequently, endometrial oxytocin receptor concentration is downregulated by exposure to progesterone for 12–14 days.[20] Frequency of LH pulses increases during the 50 days preceding first ovulation and reach about one per hour around the time of first ovulation.[14] Amplitude of LH pulses also increases during this time but pulse frequency appears to be critical for initial ovulation.[14] Follicle-stimulating hormone (FSH) concentrations in blood do not fluctuate as much as LH in the peripubertal period, suggesting that FSH may play more of a permissive role in the initiation of puberty.[8] Other hormones undoubtedly play a role in initiation of puberty but LH appears to be of paramount significance. Prolactin may play a role in heifer puberty but blood concentrations do not change at puberty as in bulls. There is abundant evidence to suggest that heifers are actually capable of ovulating from early on in life but fail to do so because of insufficient gonadotropic stimulation. In fact, McLeod et al.[21] were able to induce preovulatory gonadotropin surges in GnRH-treated 5-month-old heifers. Even more impressive was the research of Seidel et al.[22] who induced ovulations in 1-month-old heifers with gonadotropin administration. Although ovulations can be artificially induced with gonadotropins, it is also notable that the hypothalamus–pituitary becomes increasingly sensitive to GnRH stimulation as the time of puberty approaches. It has been shown that the positive feedback effect of estradiol-17β on surge LH release becomes functional between 3 and 5 months of age.[23] Collectively, available data indicate heifers are capable of ovulating long before puberty but fail to do so spontaneously until the inhibitory effect of estradion-17β on GnRH release wanes.

Development of the female reproductive tract

Overall body growth and development of the reproductive tract occur in an asynchronous pattern. For example, the ovaries grow at a rate 2.7 times faster than the body until puberty, whereas the tubular reproductive tract grows at about the same rate as body growth until about 6 months of age, then enters a period of accelerated development until puberty.[24] No ovarian follicles are macroscopically visible at birth but their numbers increase to maximal at 4 months, decrease to 8 months of age, and remain relatively constant thereafter.[24] Growing and antral follicles increase in number during the first 3–4 months of age, which corresponds to the transient increase in circulating LH concentrations.[25]

Height of the luminal epithelium of the tubular reproductive tract is stimulated at birth but regresses by 1 or 2 months of age. Thereafter, increases in height of epithelia are most rapid after 6 months. From this information it is concluded, at least for Holstein heifers, that rapid peripubertal growth of the reproductive tract commences during the seventh month and is largely terminated by 10 months of age.[24]

Factors that influence age of puberty

It is generally accepted that puberty in cattle occurs around 9 or 10 months. However, there are reports of puberty occurring any time between 6 and 24 months of age,[1] with anecdotal reports of heifers calving at 13 months indicating that puberty can occur as early as 4 months of age. Age at puberty is influenced by body weight and composition, breed, nutrition, genetics, and season. Any adverse factor that decreases prepubertal growth, such as scours, pneumonia, parasitism, or harsh weather conditions, results in delay of the onset of puberty.

Weight and body composition

An early study by Sorensen[26] showed that attainment of puberty in heifers was more influenced by weight than age. He found that heifers on a higher plane of nutrition reach puberty at an earlier age than similar heifers with lower average daily gain.[26] However, heifers may have similar body weights but vary in frame size, indicating they have differing body composition. For any given frame size, heifers that are heavier reach puberty at an earlier age. Body weight at puberty has a heritability coefficient of 0.40.[27] Overall, heifers reared on higher planes of nutrition are heavier but younger than nutritionally restricted animals at puberty.[28] Puberty does not occur at similar body composition or metabolic status in all heifers[29] but is positively correlated with body fat percentage and negatively correlated with carcass moisture percentage.[29] The precise mechanisms involved in the relationship of body composition and puberty are not clearly defined. However, it is known that somatotropin and the insulin-like growth factor (IGF)-I system are involved.[30]

Breed

The process of developing heifers as replacements must begin during the cow-calf production phase. Age and weight at puberty are affected by several factors, including breed. Generally, breeds of a larger size at maturity are older and heavier when reaching puberty.[31] A classic example of the effect of breed on puberty is illustrated in the study by Laster et al.[32] They found that female progeny of a Charolais bull were 50 days older and 120 kg heavier at puberty than progeny of a Jersey bull when all dams were Angus cows. Although the Charolais × Angus heifers grew faster than the Jersey × Angus heifers, they did not reach puberty at as young an age as the Jersey × Angus heifers due to breed effect which, in this case, overrode the influence of rate of gain. Generally, European breeds reach puberty younger but at slightly heavier weights than Hereford or Angus heifers.[32] In their study on the effects of heterosis on age at puberty, Wiltbank et al.[33] found half to three-fourths of the heterosis effect on age at puberty was independent of heterosis effects on average daily gain. Thus there is a significant heterosis effect on age at puberty independent of heterosis effects on average daily gain.[33] Sire within breed also has a significant effect on age of puberty of his female offspring. The heritability coefficient for age at puberty is 0.41.[27]

Plane of nutrition

Body energy reserves and metabolic state are relevant modifiers of puberty onset and fertility. For instance, heifers in a peripubertal state may be induced to ovulate by abruptly increasing the plane of nutrition,[34] whereas heifers of adequate body weight for puberty may be rendered anestrus by severe nutritional restriction.[35] The study by Chelikani et al.[36] involving Holstein dairy heifers illustrates this point. Heifers were fed to gain 1.1, 0.8, or 0.5 kg/day from 100 kg liveweight. Age at puberty for the three groups was 9, 11, and 16 months respectively. How nutrition and metabolism are linked to reproductive cyclicity is not clearly understood but it appears several neuropeptides operating in a reciprocal manner (orexigenic and anorexigenic) are involved. A primary candidate for modulating the effects of nutrition on reproduction is kisspeptin. Much of the work with this hormone has been in a rodent model but one may speculate that the same mechanisms are operational in cattle. Kisspeptin signaling in the hypothalamus has appeared as a pivotal positive regulator of the GnRH pulse generator.[37] In the mid-1990s, the adipose hormone leptin was proven as an essential signal for transmitting metabolic information to the centers governing puberty and reproduction and kisspeptins, a family of neuropeptides encoded by the Kiss1 gene, have emerged as conduits for the metabolic regulation of reproduction and putative effectors of leptin actions on GnRH neurons.[38] Leptin is produced by adipocytes and regulated by long-term and recent nutritional status.[39] Circulating leptin concentrations increase as puberty approaches but do not change appreciably in response to dietary change when percentage of total carcass body fat is above a minimum, indicating that a certain minimum body condition is required for puberty.[40] Ghrelin and other metabolic effector hormones known to modulate the hypothalamic–pituitary–gonadal axis, such as neuropeptide Y, melanocortins, and melanocyte-concentrating hormone, are additional putative regulators of the hypothalamic kisspeptin system.[38]

Genetic markers for age at puberty

Two of the major factors that influence reproductive efficiency in cattle are age at puberty and postpartum anestrous interval. We now have the technology to evaluate genetic markers for age at puberty in cattle. For example, quantitative trait loci have been identified that predict male reproductive traits including age at puberty in cattle.[41] Similarly, random amplified polymorphic DNA markers have been used for identifying Nelore bulls with early (precocious) or late (nonprecocious) puberty.[42] In heifers, an association weight matrix (AWM) has been constructed based on 22 related traits with single nucleotide polymorphisms.[43] The AWM results recapitulated the known biology of puberty, captured experimentally validated binding sites, and identified candidate genes and gene–gene interactions for further investigation. Advances in genomic technologies will likely provide a powerful tool for selecting heifers at birth that will have a greater probability of being reproductively successful if managed correctly.

Correlation between sire scrotal circumference and age at puberty of daughters

Early studies reported a favorable genetic correlation between yearling scrotal circumference (SC) in bulls and age at puberty in half-sib heifers in populations of purebred animals.[44,45] A favorable relationship between SC in Brahman bulls and age at puberty in Brahman heifers reared under subtropical conditions has also been reported.[46] However, more recent work suggests that the correlation between genetic response in female reproductive traits (including age at puberty) and sire yearling SC may be expected to be less effective than previously reported in the literature.[47] In an Australian study, no significant relationship was found between age at puberty in heifers and the age and SC at puberty in related bulls.[48] Results of another experiment designed to investigate this correlation using Limousin bulls bred to crossbred cows indicated that selection of resulting replacement heifers based on sire SC phenotype did not significantly influence heifer age at puberty.[49] However, when sire SC estimated progeny difference was used instead, selection resulted in a significant reduction in heifer age at puberty. The authors concluded that when sires are selected for high SC estimated progeny difference within a crossbreeding system, a large percentage of heifers should reach puberty early enough to calve at 2 years of age or younger. Nonetheless, it has been reported that age of puberty in bulls, based on attainment of a critical SC, is correlated with age at puberty in female offspring.[50] One explanation of the conflicting data may be that heterosis has a significant effect on the percentage of heifers reaching puberty by 368 days, 410 days, and 452 days, with the greatest effect at the younger age.[51]

Management options to influence age at puberty

Management of replacement beef heifers should focus on factors that enhance physiological processes that promote puberty.[52] Management options that ultimately affect lifetime productivity and reproductive performance of heifers begin at birth and include decisions that involve growth-promoting implants, feeding, breed, birth date and weaning weight, social interaction, sire selection, and exogenous hormonal treatments to synchronize or induce estrus. This is especially relevant in systems where heifers are expected to calve by 2 years of age or where the breeding period is restricted.[31] Heifers that calve as 2 year olds produce more calves in their lifetime than heifers that calve as 3 year olds.[53] Breed and post-weaning rate of gain have a large influence over onset of puberty.[31] For optimal fertility, heifers should not be bred at their pubertal estrus as calving rate has been reported to be 21% less than for heifers bred at third estrus.[54] This implies that heifers should reach puberty within 1–3 months before the age at which they will be bred in order to mitigate the reduced fertility associated with breeding at pubertal estrus.

Photoperiod

Consideration should be given to time of year when calving occurs. Although cattle are not strictly seasonal breeders, there is a seasonal effect on reproduction. For example, it has been reported that fertility, cycle length, and postpartum anestrous period length varies with season. More specifically, a winter environment delays onset of puberty. Schillo et al.[55] reported that fall-born heifers were younger at puberty than those born in spring, and that heifers exposed to simulated changes in daylength from spring to fall after 6 months of age showed advanced onset of cyclic ovarian activity. The effect of season is likely mediated through photoperiod and is not a direct consequence of improved nutrient availability. As further evidence of this effect, puberty can be advanced in heifers by administration of melatonin-containing implants.[56] Treatment of 3- to 4-month-old winter-born heifers with exogenous melatonin for a period of 5 weeks at the beginning of summer significantly increased the incidence of animals attaining puberty by March 19 of the following year (58% vs. 17%).[56] Exposure of prepubertal heifers from 22 or 24 weeks of age until first ovulation to an artificially extended photoperiod will also advance first ovulation.[57]

Growth-promoting implants

Growth-promoting implants are a widely used management tool in the beef industry. However, in heifers to be retained for breeding, implants containing estradiol, zeranol, or trenbolone acetate inhibit the development of a mature reproductive endocrine system when administered to suckling beef heifers and consequently delay onset of puberty. For example, zeranol implants from birth delay the onset of puberty and decrease uterine horn diameter. Furthermore, zeranol-implanted heifers have lower pregnancy rates and higher rates of abortion than nonimplanted herdmates.[58]

Use of progestins to advance puberty

It is well established that progesterone sensitizes the hypothalamus to the positive feedback effect of estradiol-17β that results in the preovulatory LH surge. Furthermore, progesterone priming has been shown to enhance follicular estradiol-17β synthesis in ewes[59] and it has also been demonstrated that circulating estradiol-17β concentrations were higher at the first ovulatory estrus than in the preceding silent estrus.[60] Progestin administration hastens puberty by increasing LH secretion by accelerating the peripubertal decrease in estradiol-17β negative feedback effect.[61] As an example of the use of an exogenous progestin, melengestrol acetate (MGA) fed for 8 days followed by withdrawal enhanced onset of puberty by stimulating pulsatile LH secretion that accelerated follicle growth to the preovulatory stage.[62] In synchronization programs for heifers which may be peripubertal and prone to premature luteolysis, the use of a progestin improves pregnancy rates. Treatment with some progestins, but not others, before first ovulation was able to eliminate the occurrence of short-lived corpora lutea.[63–65] Progesterone via intravaginal implant for 5 days is more effective in reducing incidence of premature luteolysis after GnRH-induced ovulation than oral MGA.[64]

Nutrition optimization of puberty

Rapid growth from birth to weaning (6–10 months) and from weaning to puberty (11–16 months) are critical for beef heifer calves to be used in breeding herds. The need for rapid growth in breeding heifers is to insure that they will reach puberty in advance of the breeding season when they are yearlings. Timing of puberty is dependent on both age and body weight/composition. An animal must exceed both the age and weight "threshold" to attain puberty. Nutritional programs for breeding heifers are designed to grow animals at a rate that allows them to exceed the weight threshold before or soon after the age threshold for puberty is surpassed.[66] Across several breeds, heifers were 55–60% of mature body weight at puberty.[67] The practice of developing heifers to reach a target body weight (typically ≥65% of mature weight) prior to breeding is commonplace in the industry. However, one must bear in mind that the target body weight will vary by breed. There are multiple models for determination of feed efficiency in beef heifers. Residual feed intake (RFI) is the residual from a regression model regressing feed intake on average daily gain (ADG) and body weight $(BW)^{-0.75}$.[68] Body fat stores are greater in heifers with greater RFI than in their more efficient herdmates. A 1-unit increase in RFI resulted in a reduction of 7.54 days in age at puberty in Bos taurus beef heifers. However, Bos indicus-influenced heifers, which reach puberty at older ages, were not found to have sexual maturity influenced by election for RFI.[68] From a management standpoint, selection for low RFI results in selection of leaner heifers that reach puberty at older ages.[68] Therefore, at some point, it becomes counterproductive to select replacement heifers on RFI. This is because reproduction has been reported to be five times more important to commercial cattle producers than growth rate or milk production,[69] and heifers that calve early in their first calving season tend to calve early throughout their lives and have greater lifetime calf production.[70]

Precocious puberty (<300 days of age) in beef heifers can be induced by early weaning and continuous feeding of a high-concentrate diet. As in the case of progestin administration, puberty is preceded by increasing frequency of pulses of LH.[71] Heifers experiencing induced precocious puberty weigh significantly less at puberty than their traditionally fed counterparts.[71] Furthermore, it has been determined that feeding the high-concentrate diet from 126 days (after weaning at 112 days) through 196 days was as effective at inducing precocious puberty as continuous high-concentrate feeding.[72] High pre-weaning growth rate and heavy weaning weights were associated with early puberty and heavy weight at puberty.[73]

The concept of fetal programming was proposed by Barker[74] to explain a geographical association between poor maternal physique and health, poor fetal growth, and high death rates from cardiovascular disease in adult humans. There are numerous studies in animals which demonstrate that transient adverse events in prenatal or early postnatal life have permanent and profound effects on physiology, although such effects may remain latent until the animal is mature. One example of this comes from experiments in which the nutrition of pregnant and lactating rats was manipulated. The adult body size of the offspring was more powerfully determined by their mothers' nutrition during pregnancy and lactation than by their genetic constitution.[75] Prenatally undernourished female rats displayed delayed puberty which was associated with decreased hypothalamic kisspeptin action.[76] The mechanisms of fetal programming have only recently been explored. It is now established that maternal nutrient status can cause epigenetic alterations to the genome of the developing fetus, which can potentially impact future generations. Epigenetics is defined as heritable changes in gene expression resulting from alterations in chromatin structure but not in DNA sequence. Epigenetic programming of the genome can have lasting effects on future generations through intergenerational influences. These are described as factors, conditions, exposures, and environments in one generation that impact the health, growth, and development of subsequent generations. Three main mechanisms cause epigenetic changes to the genome: DNA methylation, histone modification, and noncoding microRNAs.[77] These processes regulate both the intensity and timing of gene expression during cell differentiation.[78,79] In cattle, nutrient restriction of gestating cows resulted in heifer offspring with reduced wet ovarian weight and decreased luteal tissue mass compared with heifers born to control-fed cows.[80] Funston et al.[81] also reported heifers born to cows that were protein-supplemented during the last one-third of pregnancy attained puberty 14 days earlier than heifers from nonsupplemented dams.

In sheep, fetal growth restriction altered pituitary LHβ expression and number of follicles in the fetal ovary.[82] Similarly, nutrient restriction for the first 110 days of gestation resulted in calves with fewer antral follicles compared with calves born to nonrestricted cows.[83] This may impact the onset of puberty in heifers as there is evidence from ultrasonographic studies that development of ovarian antral follicles and tubular genitalia occur in parallel[83] and this development is necessary for puberty.

In dairy heifers it appears that factors including colostrum intake, pre-weaning growth rate, and body composition influence age at puberty. As an example of the effect of pre-weaning (0–42 days) growth rate, heifers fed an intensive milk replacer diet were 15 days younger at conception and 14 days younger at calving than heifers fed the conventional diet.[84] The conventional diet consisted of a standard milk replacer [21.5% crude protein (CP), 21.5% fat] fed at 1.2% of BW on a dry matter basis and starter grain (19.9% CP) to attain 0.45 kg of daily gain. The intensive diet consisted of a high-protein milk replacer (30.6% CP, 16.1% fat) fed at 2.1% of BW on a dry matter basis and starter grain (24.3% CP) to achieve 0.68 kg of daily gain.[84]

Finally, one should bear in mind that consumption of certain feedstuffs may actually be deleterious to attainment of puberty. One such example is endophyte-infected tall fescue. Cattle consuming this plant are prone to decreased calving and growth rates, delayed onset of puberty, and impaired function of corpora lutea.[85]

Bull exposure

Unlike other domestic species (sheep, goats, swine), exposure to a bull has no effect on the incidence of precocious puberty.[86]

Monitoring heifers for attainment of puberty

The rearing of replacement heifers is a major financial investment for both beef and dairy cattle producers. The investment expenses do not begin to be recovered until after first calving so having heifers calve at an optimal age is paramount to enterprise profitability. For this to occur it is essential that operators know when their heifers have attained puberty and become eligible for breeding. This is most critical for herds using a restricted breeding season.

Observation of signs of estrus can predict onset of puberty but is impractical for application to larger herds. Another observational tool, reproductive tract score (RTS), is a useful predictor of heifer fertility.[87] There is a positive correlation between high reproductive tract scores in heifers and percentage conception by artificial insemination.[88] RTS is a subjective estimate of sexual maturity, based on ovarian follicular development and palpable size of the uterus. An RTS of 1 is assigned to heifers with infantile tracts, as indicated by small toneless uterine horns and small ovaries devoid of significant structures. Heifers with an RTS of 1 are likely the furthest from puberty at the time of examination. Heifers assigned an RTS of 2 are thought to be closer to puberty than those scoring 1, due primarily to larger uterine horns and ovaries. Those heifers assigned an RTS of 3 are thought to be on the verge of estrous cyclicity based on uterine tone and palpable follicles. Heifers assigned a score of 4 are considered to be estrous cycling, as indicated by uterine tone and size, coiling of the uterine horns, and presence of a preovulatory-sized follicle. Heifers assigned an RTS of 4 do not have an easily distinguished corpus luteum. Heifers with an RTS of 5 are similar to those scoring 4 except for the presence of a palpable corpus luteum.[89]

References

1. Robinson T, Shelton J. Reproduction in cattle. In: Cole HH, Cupps PT (eds) *Reproduction in Domestic Animals*, 3rd edn. New York: Academic Press, 1977, 433–454.
2. Donaldson L, Bassett J, Thorburn G. Peripheral plasma progesterone concentration of cows during puberty, oestrous cycles, pregnancy and lactation, and the effects of under-nutrition or exogenous oxytocin on progesterone concentration. *J Endocrinol* 1970;48:599–614.
3. Jones E, Armstrong J, Harvey R. Changes in metabolites, metabolic hormones, and luteinizing hormone before puberty in Angus, Braford, Charolais, and Simmental heifers. *J Anim Sci* 1991;69:1607–1615.
4. Moran C, Quirke J, Roche J. Puberty in heifers: a review. *Anim Reprod Sci* 1989;18:167–182.
5. Ramirez D, McCann S. Comparison of the regulation of luteinizing hormone (LH) secretion in immature and adult rats. *Endocrinology* 1963;72:452–464.
6. Levasseur M-C. Thoughts on puberty. The gonads. *Ann Biol Anim Biochim Biophys* 1979;19:321–335.
7. Bettendorf M, de Zegher F, Albers N, Hart C, Kaplan S, Grumbach M. Acute N-methyl-D,L-aspartate administration stimulates the luteinizing hormone releasing hormone pulse generator in the ovine fetus. *Horm Res* 1999;51:25–30.
8. Schams D, Schallenberger E, Gombe S, Karg H. Endocrine patterns associated with puberty in male and female cattle. *J Reprod Fertil Suppl* 1981;30:103–110.
9. Madgwick S, Evans A, Beard A. Treating heifers with GnRH from 4 to 8 weeks of age advanced growth and the age at puberty. *Theriogenology* 2005;63:2323–2333.
10. Day M, Imakawa K, Garcia-Winder M *et al*. Endocrine mechanisms of puberty in heifers: estradiol negative feedback regulation of luteinizing hormone secretion. *Biol Reprod* 1984;31:332–341.
11. Anderson W, Forrest D, Goff B, Shaikh A, Harms P. Ontogeny of ovarian inhibition of pulsatile luteinizing hormone secretion in postnatal Holstein heifers. *Domest Anim Endocrinol* 1986;3:107–116.
12. Moseley W, Dunn T, Kaltenbach C, Short R, Staigmiller R. Negative feedback control of luteinizing hormone secretion in prepubertal beef heifers at 60 and 200 days of age. *J Anim Sci* 1984;58:145.
13. Day M, Imakawa K, Wolfe P, Kittok R, Kinder J. Endocrine mechanisms of puberty in heifers. Role of hypothalamo-pituitary estradiol receptors in the negative feedback of estradiol on luteinizing hormone secretion. *Biol Reprod* 1987;37:1054–1065.
14. Kinder J, Day M, Kittok R. Endocrine regulation of puberty in cows and ewes. *J Reprod Fertil Suppl* 1987;34:167–186.
15. Gonzalez-Padilla E, Wiltbank J, Niswender G. Puberty in beef heifers. I. The interrelationship between pituitary, hypothalamic and ovarian hormones. *J Anim Sci* 1975;40:1091–1104.
16. Foster D, Yellon S, Olster D. Internal and external determinants of the timing of puberty in the female. *J Reprod Fertil* 1985;75:327–344.
17. Garverick H, Smith M. Mechanisms associated with subnormal luteal function. *J Anim Sci* 1986;62:92–105.
18. Garverick H, Zollers W, Smith M. Mechanisms associated with corpus luteum lifespan in animals having normal or subnormal luteal function. *Anim Reprod Sci* 1992;28:111–124.
19. Hunter M. Characteristics and causes of the inadequate corpus luteum. *J Reprod Fertil Suppl* 1991;43:91.
20. Wathes D, Lamming G. The oxytocin receptor, luteolysis and the maintenance of pregnancy. *J Reprod Fertil Suppl* 1995;49:53.
21. McLeod B, Peters A, Haresign W, Lamming G. Plasma LH and FSH responses and ovarian activity in prepubertal heifers treated with repeated injections of low doses of GnRH for 72 h. *J Reprod Fertil* 1985;74:589–596.
22. Seidel G, Larson L, Foote R. Effects of age and gonadotropin treatment on superovulation in the calf. *J Anim Sci* 1971;33:617–622.
23. Staigmiller R, Short R, Bellows R. Induction of LH surges with 17β estradiol in prepuberal beef heifers: an age dependent response. *Theriogenology* 1979;11:453–459.
24. Desjardins C, Hafs H. Maturation of bovine female genitalia from birth through puberty. *J Anim Sci* 1969;28:502–507.
25. Erickson B. Development and senescence of the postnatal bovine ovary. *J Anim Sci* 1966;25:800–805.
26. Sorensen AM. *Causes and Prevention of Reproductive Failures in Dairy Cattle. Influence of Underfeeding and Overfeeding on Growth and Development of Holstein heifers*. Vol. 936 of Bulletin of the Cornell University Agricultural Experiment Station. Ithaca, NY: Cornell University, 1959.
27. Laster D, Smith G, Cundiff L, Gregory K. Characterization of biological types of cattle (Cycle II) II. Postweaning growth and puberty of heifers. *J Anim Sci* 1979;48:500–508.
28. Short R, Bellows R. Relationships among weight gains, age at puberty and reproductive performance in heifers. *J Anim Sci* 1971;32:127–131.
29. Hall J, Staigmiller R, Bellows R, Short R, Moseley W, Bellows S. Body composition and metabolic profiles associated with puberty in beef heifers. *J Anim Sci* 1995;73:3409–3420.
30. Simpson R, Armstrong J, Harvey R, Miller D, Heimer E, Campbell R. Effect of active immunization against growth hormone-releasing factor on growth and onset of puberty in beef heifers. *J Anim Sci* 1991;69:4914–4924.
31. Ferrell C. Effects of postweaning rate of gain on onset of puberty and productive performance of heifers of different breeds. *J Anim Sci* 1982;55:1272–1283.
32. Laster D, Glimp H, Gregory K. Age and weight at puberty and conception in different breeds and breed-crosses of beef heifers. *J Anim Sci* 1972;34:1031–1036.
33. Wiltbank J, Gregory K, Swiger L, Ingalls J, Rothlisberger J, Koch R. Effects of heterosis on age and weight at puberty in beef heifers. *J Anim Sci* 1966;25:744–751.
34. Gonzalez-Padilla E, Niswender G, Wiltbank J. Puberty in beef heifers. II. Effect of injections of progesterone and estradiol-17beta on serum LH, FSH and ovarian activity. *J Anim Sci* 1975;40:1105–1109.
35. Imakawa K, Day M, Zalesky D, Clutter A, Kittok R, Kinder J. Effects of 17 beta-estradiol and diets varying in energy on secretion of luteinizing hormone in beef heifers. *J Anim Sci* 1987;64:805–15.
36. Chelikani P, Ambrose J, Kennelly J. Effect of dietary energy and protein density on body composition, attainment of puberty, and ovarian follicular dynamics in dairy heifers. *Theriogenology* 2003;60:707–725.
37. Wahab F, Atika B, Shahab M. Kisspeptin as a link between metabolism and reproduction: evidences from rodent and primate studies. *Metabolism* 2013;62:898–910.
38. Castellano J, Bentsen A, Mikkelsen J, Tena-Sempere M. Kisspeptins: bridging energy homeostasis and reproduction. *Brain Res* 2010;1364:129–138.
39. Chilliard Y, Delavaud C, Bonnet M. Leptin expression in ruminants: nutritional and physiological regulations in relation with energy metabolism. *Domest Anim Endocrinol* 2005;29:3–22.
40. Garcia M, Amstalden M, Williams S *et al*. Serum leptin and its adipose gene expression during pubertal development, the estrous cycle, and different seasons in cattle. *J Anim Sci* 2002;80:2158–2167.
41. Casas E, Lunstra D, Stone R. Quantitative trait loci for male reproductive traits in beef cattle. *Anim Genet* 2004;35:451–453.
42. Alves B, Unanian M, Silva E, Oliveira M, Moreira-Filho C. Use of RAPD markers for identifying Nelore bulls with early reproductive maturation onset. *Anim Reprod Sci* 2005;85:183–191.

43. Fortes M, Reverter A, Zhang Y et al. Association weight matrix for the genetic dissection of puberty in beef cattle. *Proc Natl Acad Sci USA* 2010;107:13642–13647.
44. Brinks J, McInerney M, Chenoweth P. Relationship of age at puberty in heifers to reproductive traits in young bulls. In: *Proceedings, Annual Meeting, Western Section, American Society of Animal Science*, Vols 29–30. Colorado State University, 1978, pp. 28–30.
45. Toelle V, Robison O. Estimates of genetic correlations between testicular measurements and female reproductive traits in cattle. *J Anim Sci* 1985;60:89.
46. Vargas C, Elzo M, Chase C, Chenoweth P, Olson T. Estimation of genetic parameters for scrotal circumference, age at puberty in heifers, and hip height in Brahman cattle. *J Anim Sci* 1998;76:2536–2541.
47. Martinez-Velazquez G, Gregory K, Bennett G, Van Vleck L. Genetic relationships between scrotal circumference and female reproductive traits. *J Anim Sci* 2003;81:395–401.
48. Perry V, Munro R, Chenoweth P, Bodero D, Post T. Relationships among bovine male and female reproductive traits. *Aust Vet J* 1990;67:4–5.
49. Moser D, Bertrand J, Benyshek L, McCann M, Kiser T. Effects of selection for scrotal circumference in Limousin bulls on reproductive and growth traits of progeny. *J Anim Sci* 1996;74:2052–2057.
50. Fortes M, Lehnert S, Bolormaa S et al. Finding genes for economically important traits: Brahman cattle puberty. *Anim Prod Sci* 2012;52:143–150.
51. Gregory K, Lunstra D, Cundiff L, Koch R. Breed effects and heterosis in advanced generations of composite populations for puberty and scrotal traits of beef cattle. *J Anim Sci* 1991;69:2795–2807.
52. Patterson D, Perry R, Kiracofe G, Bellows R, Staigmiller R, Corah L. Management considerations in heifer development and puberty. *J Anim Sci* 1992;70:4018–4035
53. Nunez-Dominquez R, Cundiff L, Dickerson G, Gregory K, Koch R. Effects of managing heifers to calve first at two vs three years of age on longevity and lifetime production of beef cows. Roman L. Hruska US Meat Animal Research Center, Paper 42, 1985. Available at http://digitalcommons.unl.edu/hruskareports/42
54. Byerley D, Staigmiller R, Berardinelli J, Short R. Pregnancy rates of beef heifers bred either on puberal or third estrus. *J Anim Sci* 1987;65:645–650.
55. Schillo K, Hall J, Hileman S. Effects of nutrition and season on the onset of puberty in the beef heifer. *J Anim Sci* 1992;70:3994–4005.
56. Tortonese D, Inskeep E. Effects of melatonin treatment on the attainment of puberty in heifers. *J Anim Sci* 1992;70:2822–2827.
57. Hansen P, Kamwanja L, Hauser E. Photoperiod influences age at puberty of heifers. *J Anim Sci* 1983;57:985–992.
58. King B, Kirkwood R, Cohen R, Bo G, Lulai C, Mapletoft R. Effect of zeranol implants on age at onset of puberty, fertility and embryo fetal mortality in beef heifers. *Can J Anim Sci* 1995;75:225–230.
59. Hunter M, Southee J. Treatment with progesterone affects follicular steroidogenesis in anestrous ewes. *Anim Reprod Sci* 1987;14:273–279.
60. Glencross R. A note on the concentrations of plasma oestradiol-17 and progesterone around the time of puberty in heifers. *Anim Prod* 1984;39:137–140.
61. Anderson L, McDowell C, Day M. Progestin-induced puberty and secretion of luteinizing hormone in heifers. *Biol Reprod* 1996;54:1025–1031.
62. Imwalle D, Patterson D, Schillo K. Effects of melengestrol acetate on onset of puberty, follicular growth, and patterns of luteinizing hormone secretion in beef heifers. *Biol Reprod* 1998;58:1432–1436.
63. Ramirez-Godinez J, Kiracofe G, McKee R, Schalles R, Kittok R. Reducing the incidence of short estrous cycles in beef cows with norgestomet. *Theriogenology* 1981;15:613–623.
64. Smith V, Chenault J, McAllister J, Lauderdale J. Response of postpartum beef cows to exogenous progestogens and gonadotropin releasing hormone. *J Anim Sci* 1987;64:540.
65. Zollers W, Garverick H, Smith M, Moffatt R, Salfen B, Youngquist R. Concentrations of progesterone and oxytocin receptors in endometrium of postpartum cows expected to have a short or normal oestrous cycle. *J Reprod Fertil* 1993;97:329–337.
66. Beal W. Life cycle of beef cattle nutrition. Virginia Polytechnic Institute and State University, 1999. Available at http://128.173.64.134/faculty/beal/Publications/FAPC96.pdf
67. Freetly H, Cundiff L. Reproductive performance, calf growth, and milk production of first-calf heifers sired by seven breeds and raised on different levels of nutrition. *J Anim Sci* 1998;76:1513–1522.
68. Randel R, Welsh T. Joint Alpharma-Beef Species Symposium. Interactions of feed efficiency with beef heifer reproductive development. *J Anim Sci* 2013;91:1323–1328.
69. Trenkle A, Willham R. Beef production efficiency. *Science* 1977;198:1009–1015.
70. Lesmeister J, Burfening P, Blackwell R. Date of first calving in beef cows and subsequent calf production. *J Anim Sci* 1973;36:1–6.
71. Gasser C, Grum D, Mussard M, Fluharty F, Kinder J, Day M. Induction of precocious puberty in heifers I. Enhanced secretion of luteinizing hormone. *J Anim Sci* 2006;84:2035–2041.
72. Gasser C, Behlke E, Grum D, Day M. Effect of timing of feeding a high-concentrate diet on growth and attainment of puberty in early-weaned heifers. *J Anim Sci* 2006;84:3118–3122.
73. Arije G, Wiltbank J. Age and weight at puberty in Hereford heifers. *J Anim Sci* 1971;33:401–406.
74. Barker D. The effect of nutrition of the fetus and neonate on cardiovascular disease in adult life. *Proc Nutr Soc* 1992;51:135–144.
75. Dubos R, Savage D, Schaedler R. Biological Freudianism: lasting effects of early environmental influences. *Pediatrics* 1966;38:789–800.
76. Iwasa T, Matsuzaki T, Murakami M et al. Effects of intrauterine undernutrition on hypothalamic Kiss1 expression and the timing of puberty in female rats. *J Physiol* 2010;588:821–829.
77. Canani R, Di Costanzo M, Leone L et al. Epigenetic mechanisms elicited by nutrition in early life. *Nutr Res Rev* 2011;24:198.
78. Zeisel S. Epigenetic mechanisms for nutrition determinants of later health outcomes. *Am J Clin Nutr* 2009;89(Suppl.):1488S–1493S.
79. McKay J, Mathers J. Diet induced epigenetic changes and their implications for health. *Acta Physiol (Oxf)* 2011;202:103–118.
80. Long N, Tousley C, Underwood K et al. Effects of early- to mid-gestational undernutrition with or without protein supplementation on offspring growth, carcass characteristics, and adipocyte size in beef cattle. *J Anim Sci* 2012;90:197–206.
81. Funston R, Martin J, Adams D, Larson D. Winter grazing system and supplementation of beef cows during late gestation influence heifer progeny. *J Anim Sci* 2010;88:4094–4101.
82. Da Silva P, Aitken R, Rhind S, Racey P, Wallace J. Impact of maternal nutrition during pregnancy on pituitary gonadotrophin gene expression and ovarian development in growth-restricted and normally grown late gestation sheep fetuses. *Reproduction* 2002;123:769–777.
83. Rawlings N, Evans A, Honaramooz A, Bartlewski P. Antral follicle growth and endocrine changes in prepubertal cattle, sheep and goats. *Anim Reprod Sci* 2003;78:259–270.
84. Davis Rincker L, VandeHaar M, Wolf C, Liesman J, Chapin L, Weber Nielsen M. Effect of intensified feeding of heifer calves on growth, pubertal age, calving age, milk yield, and economics. *J Dairy Sci* 2011;94:3554–3567.

85. Porter J, Thompson F. Effects of fescue toxicosis on reproduction in livestock. *J Anim Sci* 1992;70:1594–1603.
86. Wehrman M, Kojima F, Sanchez T, Mariscal D, Kinder J. Incidence of precocious puberty in developing beef heifers. *J Anim Sci* 1996;74:2462–2467.
87. Holm D, Thompson P, Irons P. The value of reproductive tract scoring as a predictor of fertility and production outcomes in beef heifers. *J Anim Sci* 2009;87:1934–1940.
88. Pence M, BreDahl R, Thomson J. Clinical use of reproductive tract scoring to predict pregnancy outcome. Beef Research Report 1999, Paper 32, 2000. Available at http://lib.dr.iastate.edu/beefreports_1999/32
89. Patterson D, Wood S, Randle R. Procedures that support reproductive management of replacement beef heifers. In: *Proceedings of the American Society for Animal Science*, 1999. Available at http://beefrepro.unl.edu/proceedings/2005collegestation/17_tamu_support_patterson.pdf

Chapter 23

Neuroendocrine Control of Estrus and Ovulation

Marcel Amstalden and Gary L. Williams

Department of Animal Science, Texas A&M University, Texas, USA

Introduction

The neuroendocrine control of the estrous cycle involves the integration of multiple regulatory signals. These signals form an intricate network that permits the fine control of gonadotropin release required for the complex sequence of ovarian follicular development, steroidogenesis, and ovulation. Secretion of gonadotropin-releasing hormone (GnRH) into the hypothalamic–adenohypophyseal portal circulation is considered the primary endocrinological mechanism regulating gonadotropin synthesis and release. Internal factors (e.g., gonadal and metabolic hormones, growth factors and signaling molecules) and external factors (e.g., environmental chemicals, photoperiod and stressors) are perceived at the level of the central nervous system and control GnRH secretion directly or, more often, through intermediate pathways. Under stimulation by GnRH, gonadotropes synthesize and release luteinizing hormone (LH) and follicle-stimulating hormone (FSH). These two gonadotropins reach the gonads through the blood circulation and act in the ovary to stimulate follicle growth and maturation, ovulation of the preovulatory follicle, and synthesis of gonadal steroid and peptide hormones. Gonadal hormones, in turn, exert feedback actions at the central and adenohypophyseal levels to control the release of GnRH and gonadotropins. This intricate regulatory mechanism leads to dynamic changes in circulating concentrations of reproductive hormones during the estrous cycle (Figure 23.1).

The length of the bovine estrous cycle ranges from 17 to 24 days, averaging 21 days. Day 0 of the estrous cycle is considered the day on which the female exhibits estrus (standing to be mounted) and other proceptive behavior such as mounting herdmates, restlessness, and vocalization. Estrus is induced by an elevation in circulating concentrations of estradiol-17β associated with enhanced follicular steroidogenesis during follicle development.[1] Increased concentrations of estradiol reach a threshold 12–18 hours before onset of estrus, triggering a preovulatory surge of GnRH release and, consequently, a surge in gonadotropin release.[2] Onset of the preovulatory surge of LH generally occurs coincident with the onset of estrus. Ovulation is induced approximately 30 hours after the onset of estrus and is followed by an abrupt decline in circulating concentrations of estradiol. Following ovulation, follicular cells undergo functional transformation into luteal cells, which synthesize and release progesterone.[1] During corpus luteum formation and maturation, concentrations of progesterone in the circulation increase gradually to reach their peak approximately 8 days after ovulation. The period of progesterone dominance is defined as the luteal phase of the estrous cycle. During the luteal phase, mean circulating concentrations of LH are relatively low due to infrequent episodic release of LH. In contrast, concentrations of FSH and estradiol fluctuate in association with waves of follicular growth and atresia.[3,4] At approximately day 17 of the estrous cycle in the cow, luteolytic pulses of prostaglandin $(PG)F_{2\alpha}$ released by the endometrium cause luteal regression and a consequent decrease in circulating concentrations of progesterone.[1] Therefore, the progesterone restraint of GnRH/LH release that characterizes the luteal phase declines and the frequency of episodic release of GnRH/LH increases. Increased gonadotropin support promotes follicle growth and maturation, enhanced follicular steroidogenesis, elevations in the circulating concentrations of estradiol and, ultimately, estrous behavior. The period characterized by low circulating concentrations of progesterone and enhanced follicular development leading to ovulation is termed the follicular phase. The cyclic changes in gonadotropin and ovarian hormone release, and resulting follicular activity, are usually recurrent, except when pregnancy is established or pathological conditions (e.g., undernutrition, disruption of normal luteolysis) are introduced.

The dynamic changes in concentrations of circulating LH and FSH during the estrous cycle are largely representative of distinct changes in the pattern of episodic secretion of GnRH. During the luteal phase, the frequency of episodic release of GnRH, and the correspondent pulses of LH,[5] is approximately one pulse every 4–6 hours.[2] During the follicular phase, the episodic release of GnRH/LH is increased to approximately one pulse per hour.[2,5] The follicular phase pattern of GnRH/LH release is essential for final follicle development and

---

Bovine Reproduction, First Edition. Edited by Richard M. Hopper.
© 2015 John Wiley & Sons, Inc. Published 2015 by John Wiley & Sons, Inc.

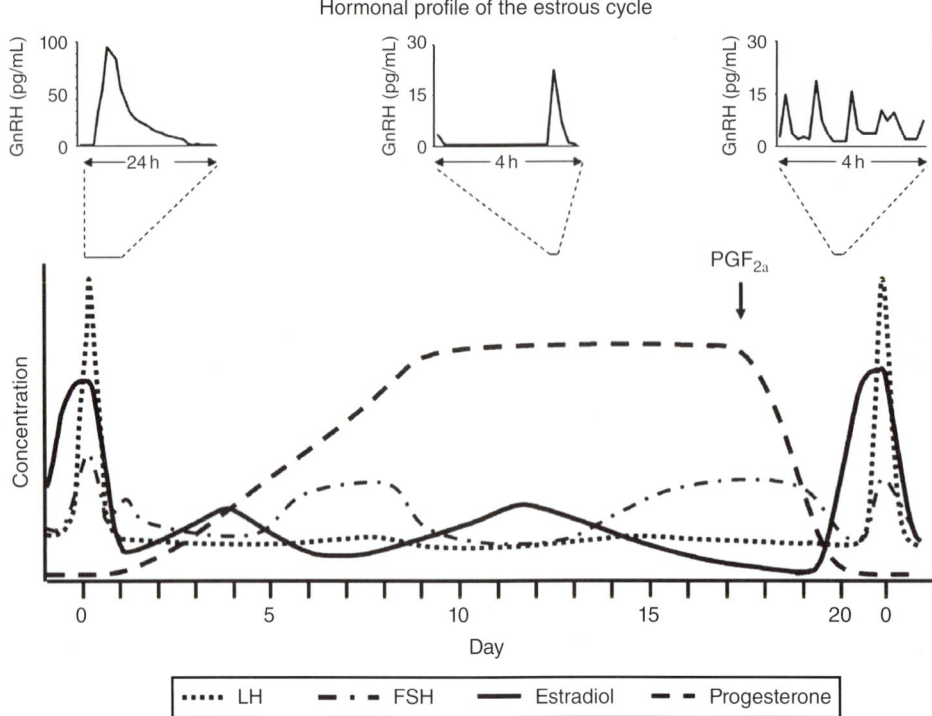

Figure 23.1 Relative changes in circulating concentrations of luteinizing hormone (LH), follicle-stimulating hormone (FSH), estradiol and progesterone during the estrous cycle in cows. The day on which estrous behavior is observed is denoted as day 0 of the estrous cycle. Circulating concentrations of progesterone increase after ovulation with formation and maturation of the corpus luteum, and remain elevated for the duration of the luteal phase. Release of prostaglandin (PG)$F_{2\alpha}$ causes luteolysis, which is followed by a decrease in circulating concentrations of progesterone. The period of rapid follicular growth and maturation, and elevated concentrations of estradiol in circulation, that follows luteolysis characterizes the follicular phase. Elevated estradiol, in the absence of progesterone, leads to estrous behavior and the preovulatory surge of gonadotropin-releasing hormone (GnRH; top, graph on left). During the luteal phase, progesterone inhibits the frequency of pulsatile release of GnRH (top, graph in middle) and consequently the pulsatile release of LH. During the follicular phase, the escape from progesterone inhibition leads to increased frequency of GnRH (top, graph on right) and LH pulses, which supports follicular maturation and enhanced ovarian steroidogenesis. Fluctuations in circulating concentrations of FSH and estradiol are observed in association with initiation and progression of follicular waves.

maturation, and enhanced steroidogenesis. The elevation in circulating concentrations of estradiol during late follicular phase leads to the preovulatory surge of GnRH. The GnRH surge is characterized by a sustained increase in concentrations of GnRH in the hypophyseal portal blood. At the onset of the surge, the pulsatile release of GnRH is of high frequency and high amplitude. With the progress of the surge, the pulsatile pattern of GnRH release switches to a continuous elevation of GnRH in the portal blood.[2] This leads in turn to a massive discharge of LH from the adenohypophysis into the peripheral circulation, similar to that of GnRH in the portal circulation.[2,5] Therefore, two modes of endogenous GnRH/LH secretion are characteristic of the estrous cycle: the tonic episodic release, and the preovulatory surge release.[6] These patterns of GnRH/LH release are regulated primarily by estradiol and progesterone, and the negative and positive feedback effects mediated by them at the hypothalamic and adenohypophyseal levels.

Functional anatomy of the hypothalamic–hypophyseal unit

The cell bodies of neurons that synthesize GnRH in cattle are located in a continuum that extends from the preoptic area to the mediobasal hypothalamic region of the diencephalon.[7,8]

GnRH neurons project to the median eminence, where GnRH is stored in secretory vesicles and released adjacent to capillaries of the hypothalamic–hypophyseal portal vasculature on stimulation. Through this specialized vasculature that connects the hypothalamus and the adenohypophysis (Figure 23.2), GnRH reaches the adenohypophysis in functionally relevant concentrations, binds to the GnRH receptor on the surface of gonadotropes, and stimulates gonadotropin release. Disruption of the hypothalamic–hypophyseal vasculature by surgical lesion leads to decreased adenohypophyseal content and circulating concentrations of LH and FSH due to the lack of gonadotrope stimulation by GnRH.[9,10] Activity of GnRH neurons and GnRH release in the median eminence are largely regulated by afferent neuronal signals and factors released by glial cells.[11,12] Evidence for structural changes in glial cells that surround the capillaries of the median eminence also exists,[13] and may contribute to the regulation of GnRH diffusion into the hypothalamic–hypophyseal circulation following a secretory episode.

Gonadal steroid control of gonadotropin secretion

Positive and negative feedback actions of estradiol and progesterone at the level of the hypothalamus and adenohypophysis explain most of the alterations in the patterns

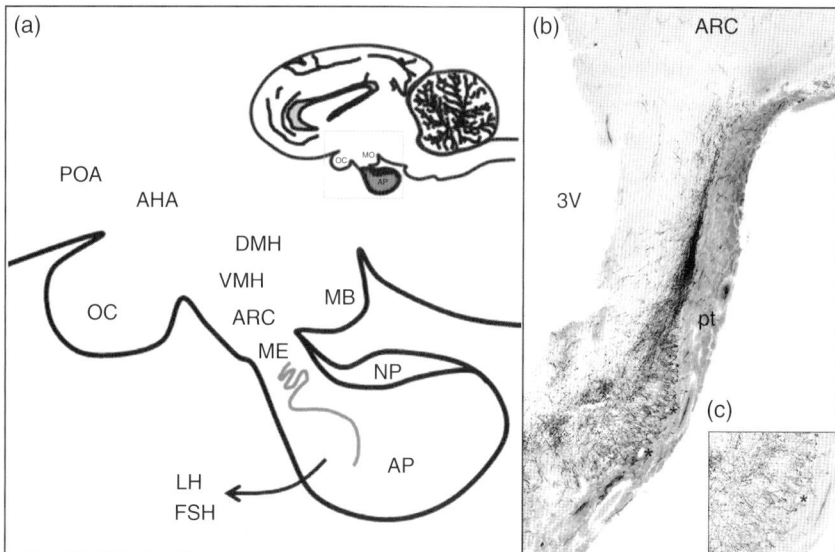

Figure 23.2 Diagram representing a sagittal view of the lower portion of the bovine brain and the hypothalamic areas involved in the neuroendocrine control of reproduction. In cattle, gonadotropin-releasing hormone (GnRH) neuronal cell bodies are distributed in a continuum that extends from the septum/preoptic area (POA) to the basal portions of the hypothalamus (a). GnRH neurons project to the median eminence (ME) where neuronal terminals are adjacent to capillaries of the hypothalamic–hypophyseal portal system. GnRH released in the ME reaches gonadotropes in the adenohypophysis (AP) and stimulates synthesis and release of luteinizing hormone (LH) and follicle-stimulating hormone (FSH). Low (b) and high (c) magnification images of GnRH-immunoreactive neuronal fibers that project basolaterally toward the median eminence and terminate adjacent to capillaries (*) of the zona externa of the ME. AHA, anterior hypothalamic area; ARC, arcuate nucleus; DMH, dorsomedial hypothalamus; MB, mammillary body; NP, neurohypophysis; OC, optic chiasm; pt, pars tuberalis of the adenohypophysis; POA, preoptic area; VMH, ventromedial hypothalamus; 3V, third ventricle.

of GnRH and LH release during the estrous cycle (Figure 23.3). The effect of progesterone in controlling the episodic release of GnRH/LH is well characterized: progesterone inhibits the frequency of LH release[14] by its action at the level of the hypothalamus.[15,16] However, the negative feedback effect of estradiol is more complex and appears largely additive to the action of progesterone.[17] Estradiol positive feedback is the primary physiological mechanism driving the development of the preovulatory surge of GnRH and LH release during the late follicular phase.[2,5,6] Interestingly, although progesterone can block the estradiol-induced LH surge in heifers and ewes,[18,19] prior exposure to endogenous (luteal phase) or exogenous (pharmacological treatment) progesterone enhances the estradiol-induced LH surge.[20]

The mechanisms by which progesterone inhibits the frequency of episodic release of LH appear to be mediated by endogenous opioid peptides, particularly dynorphin.[21] Studies in sheep have indicated that progesterone, by acting on progesterone receptor-containing dynorphin neurons located in the arcuate nucleus of the hypothalamus, inhibits the frequency of GnRH pulses via activation of endogenous opioid peptide neurotransmission. This effect appears to be mediated by the κ opioid receptor. Support for this mechanism of action is demonstrated by the observations that (i) intracerebroventricular injection of progesterone inhibits the pulsatile release of LH;[16] (ii) antagonists of endogenous opioid peptide[22] and of the κ opioid receptor,[23] injected in the mediobasal hypothalamus, increase the frequency of LH pulses; and (iii) progesterone receptor-containing dynorphin neurons, located in the arcuate nucleus of the hypothalamus, project extensively to GnRH neurons in the mediobasal hypothalamus, but to a lesser extent to GnRH neurons in the preoptic area.[23] Importantly, the population of GnRH neurons located in the mediobasal hypothalamus appears to be the major group of GnRH neurons activated during an LH pulse in sheep.[24] This provides compelling evidence that the neuronal circuitry located in this area represents a pathway by which progesterone inhibits the frequency of GnRH/LH episodic release during the luteal phase. Although the mechanisms of progesterone negative feedback on regulation of LH release have not been extensively investigated in cattle, evidence indicates that endogenous opioid peptides have a role similar to that characterized in sheep.[25]

The estradiol negative feedback actions on LH release are more complex. During periods of elevated progesterone (e.g., luteal phase), estradiol enhances progesterone inhibition of LH pulsatility.[16] This effect may involve direct actions of estradiol on endogenous opioid peptide neurons, including arcuate nucleus dynorphin neurons that are known to express estrogen receptor α and progesterone receptor in sheep.[26,27] Because estradiol increases the expression of progesterone receptor in the arcuate nucleus of ovariectomized ewes,[28] estradiol's enhancement of progesterone negative feedback may involve increased responsiveness of hypothalamic neurons to progesterone actions. Estradiol negative feedback is also demonstrated by the ability of estradiol to decrease the amplitude of LH pulses during the follicular phase in ewes.[29] However, this effect may be a consequence of the increased frequency of LH release observed during the late follicular phase[2] or by inhibition of gene expression of the α-subunit of glycoprotein hormones (common to LH, FSH and adrenocorticotropin) and the β-subunit of LH.[30,31]

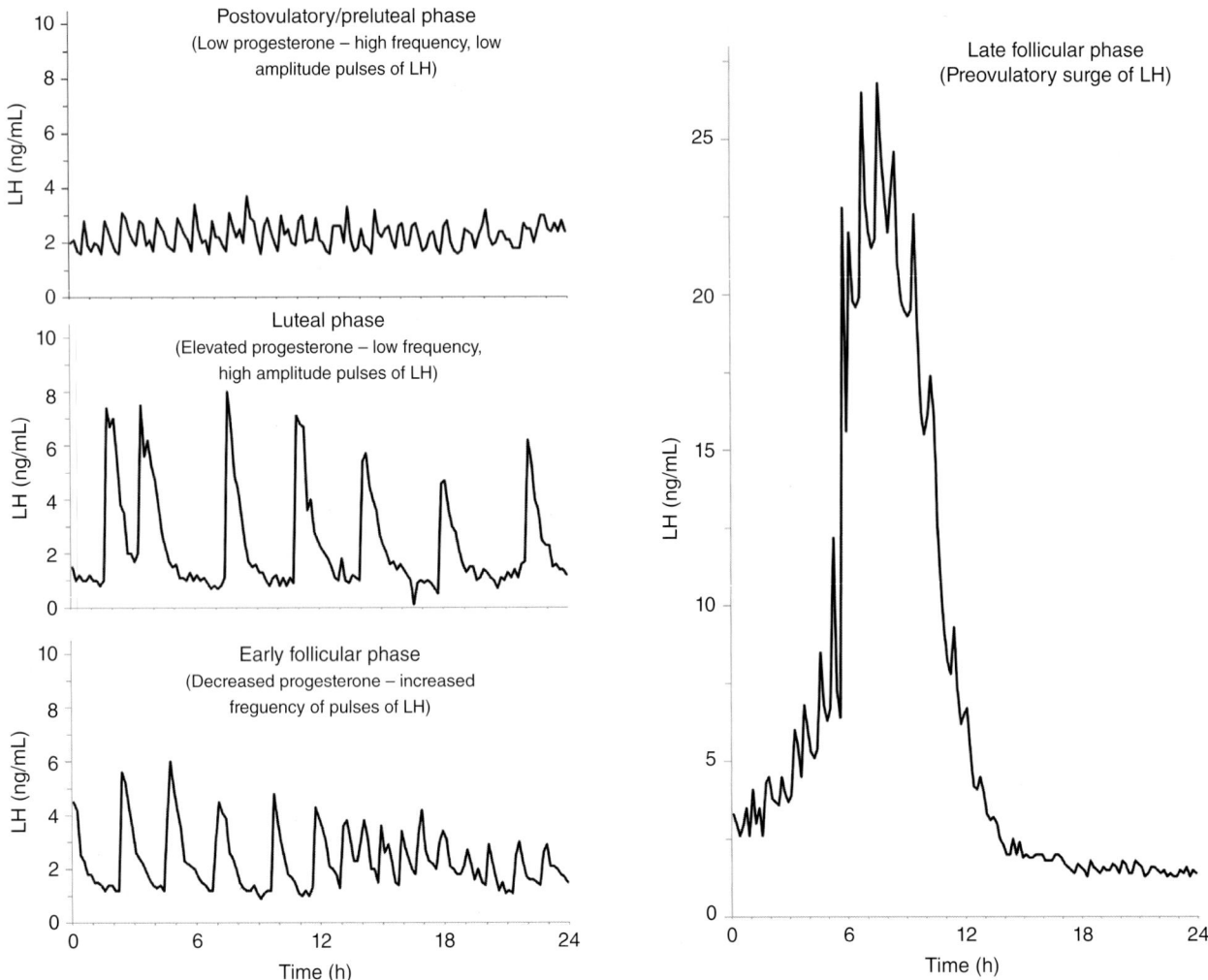

Figure 23.3 Patterns of release of luteinizing hormone (LH) during the estrous cycle in cows. Blood samples from the jugular vein were collected at 10-min intervals for 24 hours in four distinct phases of the estrous cycle in a cow. Modified with permission from Rahe et al.[2]

Estradiol appears to have a biphasic effect on secretion of LH during progression of the follicular phase. Administration of estradiol at physiological doses capable of inducing the preovulatory surge of LH leads initially to an acute reduction in the frequency of LH release.[29,32] Evidence for this effect also exists during the late follicular phase of the estrous cycle.[5] The brief period (~6 hours) of estradiol-induced inhibitory tone is followed by a considerable increase in the frequency and amplitude of GnRH/LH pulses and, subsequently, progression toward a large sustained elevation in concentrations of LH characteristic of the preovulatory surge (estradiol positive feedback; Figure 23.4).[5,6] Although the mechanisms by which estradiol exerts the positive feedback are unclear, it appears to involve activation of estradiol-receptive neurons in hypothalamus and transmission of the estradiol signal to GnRH neurons for stimulation of the surge release at the median eminence.[33] The ventromedial hypothalamus and the arcuate nucleus have been identified as critical sites for the positive effect of estradiol in ewes.[34,35] Changes in norepinephrine, β-endorphin, glutamate and γ-aminobutyric acid (GABA), neurokinin B, and kisspeptin transmission may be implicated in the transmission of estradiol positive feedback. At the level of the adenohypophysis, estradiol enhances expression of the GnRH receptor.[36] This effect is physiologically relevant because an increased response of the gonadotrope to GnRH during the late follicular phase might be critical for maintaining elevated secretion of LH during the preovulatory GnRH surge.

Prior exposure to progesterone enhances the estradiol-induced surge of LH release. This effect of progesterone appears to occur at both hypothalamic and adenohypophyseal levels. In ovariectomized ewes, the magnitude of the GnRH surge was greater in ewes pretreated with progesterone during an artificial estrous cycle than in ewes not pre-exposed to progesterone.[37] Although evidence indicates that this effect occurs by alterations in GnRH secretion, direct effects in the adenohypophysis also occur. In ovariectomized ewes with the hypothalamic–adenohypophyseal connection surgically blocked, responsiveness to estradiol-induced surge of LH was enhanced in ewes pretreated with progesterone.[38] Progesterone priming appears to influence responsiveness to estradiol feedback in cows as well.[39,40]

The control of FSH synthesis and release from gonadotropes is exquisitely regulated by GnRH and gonadal hormones, including steroid and protein hormones.

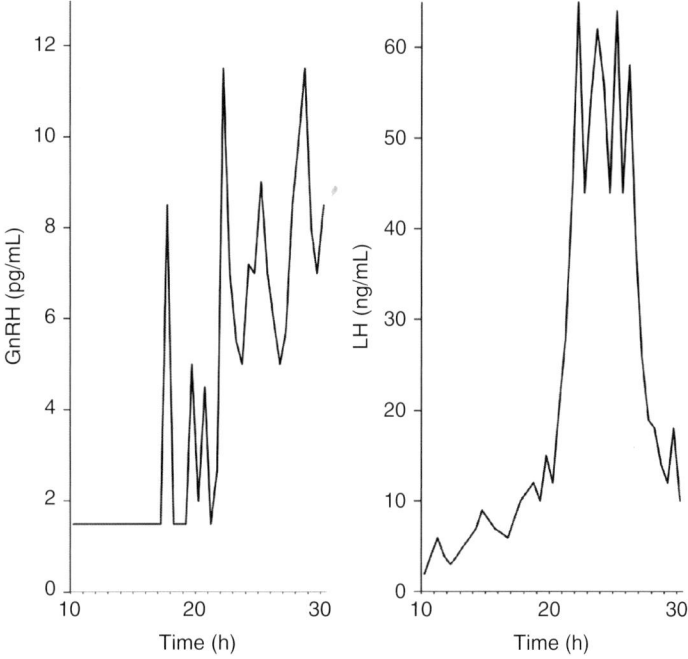

Figure 23.4 Estradiol-induced gonadotropin-releasing hormone (GnRH) and luteinizing hormone (LH) surges in cows. Estradiol (1 mg) was injected intramuscularly at Hour 0. Cerebrospinal fluid (CSF) from the third ventricle and blood from the jugular vein were obtained at 10-min intervals. Concentrations of GnRH in the CSF (top) increased considerably and remained elevated for longer than 12 hours. Concentrations of LH increased consistently with elevations in GnRH and returned to pre-surge concentrations in approximately 10 hours. Modified with permission from Gazal et al.[5]

Intra-adenohypophyseal factors may also have a role in regulating FSH secretion by paracrine actions.[41] In addition, the existence of a hypothalamic factor with FSH-releasing activity distinct from GnRH has been proposed,[42] although such a factor is yet to be identified. Similar to the release of LH, release of FSH has been shown to be pulsatile.[43,44] However, using antisera produced against the β-subunit of FSH,[45] which eliminates possible cross-reactivity with LH, yielded results in which distinct pulses of FSH release were difficult to detect in the peripheral circulation. This has been attributed in part to the long half-life of FSH in circulation.[41] In addition, fluctuations in concentrations of ovarian peptides (inhibin, follistatin) and estradiol that occur during the estrous cycle play major roles in regulating the pattern of secretion of FSH in cattle.[3,4,46] Nevertheless, studies in which concentrations of FSH were determined in blood samples collected from the hypophyseal portal vasculature in sheep have demonstrated that secretion of FSH is clearly pulsatile.[41] Interestingly, although a considerable proportion of the FSH pulses coincide with episodes of GnRH release, many non-GnRH associated episodes are also observed.

The GnRH-associated pulses of FSH observed in sheep occur coincidently with pulses of LH.[41] In sheep, LH and FSH are cosynthesized in a subpopulation of gonadotropes,[47] although the two gonadotropins are stored in distinct vesicles. Therefore, the coincident release of LH and FSH could reflect the concurrent exocytosis of readily releasable pools of vesicles containing gonadotropins. In cattle, however, there is a clearer distinction in FSH- and LH-synthesizing cells in the adenohypophysis.[48] Nevertheless, FSH is clearly secreted following GnRH administration in cattle.[39,49] In addition, the preovulatory surge of LH is accompanied by a surge in FSH.[50,51] These observations indicate that although FSH and LH might be synthesized in distinct populations of gonadotropes, both types of gonadotropes are responsive to GnRH. However, the control of secretion of each gonadotropin appears independent, and the interaction of GnRH-dependent and GnRH-independent mechanisms are important for the fine control of FSH secretion.

The mechanisms by which the non-GnRH associated pulses of FSH are generated remain unclear. Direct actions of gonadal hormones at the level of the adenohypophysis are believed to be a major regulatory mechanism. Peptide (e.g., inhibin) and steroid (e.g., estradiol) hormones act likely to control FSH synthesis and release at the level of the gonadotrope. Inhibin, a dimer of α and βA (inhibin A) or βB (inhibin B) subunits, is produced in follicular cells and acts as an endocrine factor regulating FSH release in gonadotropes. Inhibin suppresses FSH release in bovine pituitary cell cultures.[52] The role of inhibin has also been shown in immunization studies. Basal circulating concentration of FSH and GnRH-stimulated FSH release is increased in ewes immunized against bovine inhibin.[53] Similarly, mean concentrations of FSH increase in heifers immunized against inhibin, an effect associated with an increased number of small follicles present in the ovary.[54] The effect of inhibin on the regulation of gonadotropin release appears specific to the control of FSH release because there is no clear effect of inhibin on the release of LH.[55] In addition, it is unlikely that inhibin acts at the hypothalamic level for the control of GnRH secretion.[56,57]

Estradiol has been demonstrated to inhibit FSH synthesis and secretion in the adenohypophysis. The direct action of estradiol appears to occur through inhibition of FSH-β gene expression.[58] However, studies investigating the replacement of estradiol and inhibin in ovariectomized ewes indicate that a combination of estradiol and inhibin negative feedback is required for full inhibition of the post-ovariectomy rise in concentrations of FSH.[59] Additional ovarian hormones may contribute to the regulation of FSH release during the estrous cycle. Activin, a dimer of the inhibin β subunits (βAβA, βAβB and βBβB), and follistatin, a β-subunit-binding protein, are produced in follicular cells and can regulate FSH secretion. Although the endocrine actions of ovarian-derived activin and follistatin are unclear, these proteins, as well as inhibin, are synthesized in the adenohypophysis and may control FSH synthesis and release by paracrine mechanisms.[60,61] Activin stimulates FSH synthesis and release, while follistatin inhibits activin actions by averting activin binding to the receptor. Inhibin, on the other hand, disrupts activin actions by binding to the activin receptor and inducing intracellular signaling via a β-glycan-mediated pathway. Therefore, the postovulatory elevation in circulating concentrations of FSH observed shortly after the preovulatory LH and FSH surges (Figure 23.1) is largely explained by changes in feedback actions of gonadal hormones at the level of the adenohypophysis following disappearance of the dominant follicle after ovulation, and decreased circulating concentrations of estradiol and inhibin. Increases in FSH in circulation at this time contribute to recruitment of the first follicular wave observed during the early luteal phase and at subsequent waves during the remainder of the cycle.

The existence of a distinct molecule of hypothalamic origin with the ability to regulate FSH release has been proposed. Evidence in support of this hypothesis includes the observations that ablation of anterior hypothalamic neurons and their projections impair release of FSH, without clear effects on release of LH.[62,63] In addition, inhibition of LH release without notable effects on FSH release has been observed after immunization against GnRH and inhibition of GnRH actions by administration of GnRH antagonists in sheep.[41,64] Nevertheless, isolation and identification of a molecule that can be characterized as an FSH-releasing factor distinct from GnRH would be required for confirmation of this hypothesis.

Nutritional influences on hypothalamic–gonadotropic function

The nutritional control of gonadotropin release occurs primarily through the regulation of GnRH release in the median eminence. Severe nutrient restriction suppresses secretion of LH in reproductively mature and prepubertal cattle.[65,66] However, short-term restriction of feed intake has a differential impact on gonadotropin release in mature and immature female cattle. In cows, the frequency of episodes of LH release is not affected by 2–3 days of fasting.[66–68] In contrast, fasting for 48–72 hours reduces LH pulsatility in prepubertal heifers.[69,70] Gonadal steroids have a marked effect on the nutritional regulation of gonadotropin release. Pulsatile release of LH is suppressed in ovariectomized estradiol-treated rats fasted for 48 hours.[71] However, acute feed restriction does not suppress LH release in ovariectomized rats without estradiol replacement.[71] Thus, gonadal steroid hormones interact with nutritional and metabolic signals to regulate gonadotropin release.

Studies determining the concentrations of GnRH in blood samples collected from the hypothalamic–hypophyseal portal vasculature of normal-fed and feed-restricted ewes have demonstrated a considerable reduction in the frequency of GnRH pulses in response to feed restriction.[72] The reduction in the frequency of LH pulses was greater than the reduction in the episodic release of GnRH because some low-amplitude pulses of GnRH were not accompanied by a pulse of LH. Therefore, diminished amplitude of GnRH pulses may also play a role in the overall reduction in mean concentrations of LH observed during undernutrition. A major consequence of low frequency of GnRH stimulation of gonadotropes in response to feed restriction is the decrease in synthesis of LH and GnRH receptors in the adenohypophysis.[73] Pulsatile administration of GnRH in feed-restricted ewes restores circulating concentrations of LH to normal and increases pituitary expression of LH-β and LH-α genes.[74] These observations are in concordance with the hypothesis that a high frequency of GnRH stimulation is necessary for optimal expression of the common α-subunit of adenohypophyseal glycoprotein hormones and the β-subunit of LH.[75] Therefore, in conditions of low synthesis and storage of gonadotropins in the adenohypophysis and diminished numbers of GnRH receptors in gonadotropes, small-amplitude pulses of GnRH may not be sufficient to stimulate a clear episode of LH release. This may explain the observation that some small-amplitude pulses of GnRH are not followed by an episode of LH release.[72] In addition, it is possible that small-amplitude pulses of GnRH prime the adenohypophysis for enhanced responsiveness to large-amplitude pulses that induce clear episodes of gonadotropin release.[76]

Nutritional and metabolic status are perceived largely at the level of the hypothalamus, although integration with signals controlling other areas of the brain, as well as peripheral organs, are important for the precise control of gonadotropin secretion and action in target tissues. The signals that mediate the nutritional control of gonadotropin release include nutrients (e.g., glucose, fatty acids, amino acids) and metabolic hormones (e.g., leptin, insulin, ghrelin). Glucose serves as a major energy source for the central nervous system and, along with fatty acids and amino acids, serve as signaling molecules to control cellular metabolic pathways.[77–79] In sheep, insulin-induced hypoglycemia[80] and treatment with 2-deoxyglucose,[81] a competitive antagonist of glucose metabolism, inhibit secretion of LH. In cattle, hypoglycemia induced by phlorizin, an inhibitor of glucose transporter that increases renal excretion of glucose, was observed to lead to similar decrease in concentrations of LH.[82] Although it is generally accepted that glucose is essential for normal hypothalamic function, the mechanisms by which glucose deficiency impairs the release of LH are unclear. Restoration of LH pulsatility in fasted monkeys occurs not only after feeding carbohydrate-based meals, which increase circulating glucose, but also after feeding protein/fat-based meals, which do not elevate blood glucose.[83] Amino acids and fatty acids, in addition to serving as important precursor molecules for cellular biosynthesis, appear to also serve as signaling molecules for the control of reproductive neuroendocrine function.[77] Glutamate and the glutamate-derived GABA are neurotransmitters that act

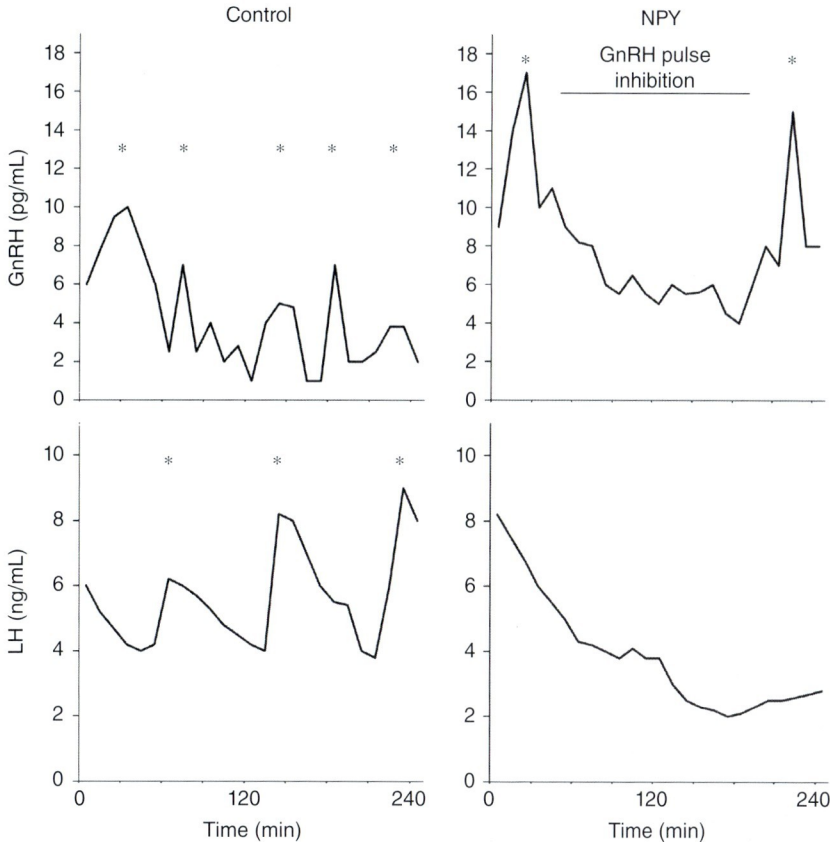

Figure 23.5 Neuropeptide Y (NPY) inhibition of release of gonadotropin-releasing hormone (GnRH) in cows. The pulsatile release of GnRH (top panel) and luteinizing hormone (LH) (lower panel) is inhibited following intracerebroventricular administration of NPY (panels on right). Asterisk denotes detected pulses of GnRH and LH. Modified with permission from Gazal et al.[5]

directly on GnRH neurons to regulate firing activity and GnRH release.[84,85]

Hormones such as insulin, insulin-like growth factor (IGF)-1, ghrelin, and leptin are reliable markers of nutritional status and appear to play important roles in mediating the metabolic control of gonadotropin secretion.[86] Leptin, a protein hormone synthesized and secreted primarily by adipocytes, has been extensively investigated as a metabolic signal for the control of reproductive neuroendocrine function in various animal species, including in cattle.[87] Circulating concentrations of leptin correlate positively with adiposity[88] and increase with elevations in body weight.[89] In contrast, undernutrition decreases concentrations of leptin in the circulation. Treatment with leptin prevents the fasting-induced decrease in LH pulsatility in peripubertal heifers.[70] In mature cows, although short-term fasting does not affect the frequency of LH release, leptin enhances LH secretion[68] by acting at both the hypothalamus to increase the amplitude of GnRH pulses[90] and at the adenohypophysis to increase sensitivity to GnRH.[91]

Mediators of leptin actions include neuropeptide Y (NPY) and proopiomelanocortin (POMC) neurons in the arcuate nucleus. These neurons contain leptin receptors.[92] NPY is a potent stimulator of feed intake and feed restriction increases expression of the NPY gene.[93] The anorexigenic actions of leptin are likely associated with the ability of leptin to decrease NPY expression.[94] NPY inhibits GnRH release in cows[5] (Figure 23.5). In contrast to its inhibitory effects on NPY gene expression, leptin enhances POMC expression.[95] One of the peptide products of the POMC gene, α-melanocyte-stimulating hormone (α-MSH), has been shown to also exert direct stimulatory effects on GnRH neurons.[96] Therefore, the actions of leptin on stimulating GnRH secretion appear to involve suppression of the inhibitory actions of NPY and stimulation of excitatory effects of α-MSH. Although NPY and POMC are believed to be major pathways involved in the control of GnRH release, it is likely that other neuronal networks contribute to the nutritional control of reproductive neuroendocrine function.

Pubertal development

Puberty is the process through which individuals acquire reproductive maturity. It involves physiological and behavioral changes that are largely driven by activation of gonadal function. In mammalian females, activation of the hypothalamic–adenohypophyseal axis, and the downstream effects of gonadotropins on gonadal function, support enhanced folliculogenesis and steroidogenesis (primarily estradiol synthesis) and leads to first ovulation.[97] The onset of puberty is characterized by an increase in the episodic release of LH, resulting from a heightened frequency of GnRH release and stimulation of gonadotropes.[98] Interestingly, exogenous estradiol induces a surge-like release of LH in prepubertal females,[99,100] indicating that estradiol positive feedback is functional before reproductive maturation is complete. Therefore, a major limiting factor for the onset of puberty is the lack of high-frequency episodic release of LH necessary for enhanced estradiol synthesis that will lead to the preovulatory surge of GnRH and LH release.

During the prepubertal period, estradiol negative feedback inhibition of episodic release of LH is enhanced. Ovariectomy in prepubertal heifers, and consequent removal of the major source of estradiol, is associated with a rapid increase in circulating concentrations of LH.[101] Replacement with physiological levels of estradiol in ovariectomized heifers prevents this increase and maintains the pulsatile release of LH at a low frequency. Around the expected time of pubertal transition, the frequency of secretory LH episodes increases to that similar to an ovariectomized heifer receiving no estradiol replacement. Thus, it is hypothesized that decreased sensitivity to estradiol negative feedback during pubertal transition leads to increased episodic release of LH characteristic of the mature state. Although the mechanisms involved in diminished sensitivity to estradiol negative feedback remain unclear, they may include changes in estradiol binding in the hypothalamus.[102] More recently, a population of neurons that synthesize the neuropeptide kisspeptin has been implicated in mediating estradiol's effect on GnRH secretion.[103–105] Kisspeptin is a potent GnRH/LH secretagogue.[106,107] In addition, estradiol regulates the expression of the gene encoding kisspeptin, likely through direct actions on kisspeptin neurons.[103]

Puberty in heifers occurs normally at approximately 10–15 months of age in Bos taurus breeds. In Bos indicus and Bos indicus-influenced breeds, puberty usually occurs later and over a wider time range (15–27 months of age).[108,109] However, age at puberty is influenced to a considerable degree by nutrition.[110] Undernutrition delays puberty in heifers.[65] In contrast, elevated body weight gain during the prepubertal period advances the onset of puberty.[111–113] The acceleration of puberty in response to increased weight gain in heifers has been associated with an earlier reduction in the sensitivity to estradiol negative feedback and hastening of the peripubertal activation of the reproductive neuroendocrine axis.[114] As discussed earlier in this chapter, nutritional signals that mediate the nutritional control of pubertal development include nutrients and metabolic hormones. Leptin, a hormone secreted primarily by adipocytes, appears to be a major permissive signal for pubertal progression. In heifers, expression of the leptin gene in the adipose tissue and circulating concentrations of leptin increase as puberty approaches.[89] In addition, a fasting-induced decrease in LH pulsatility is prevented by treatment with leptin.[70] However, chronic administration of leptin in prepubertal heifers does not advance puberty,[115] indicating that leptin may be necessary but not sufficient to induce puberty in heifers.

Neuroendocrinology of the postpartum period

The interval from parturition to resumption of ovarian cycles in the bovine female represents a physiologically distinct and economically critical period of the bovine reproductive life cycle. Therefore, it is not surprising that research to understand the biological basis of postpartum anovulation and to develop strategies for its managerial control have been subjects of intense investigation for over 50 years.[116–120]

In the cow, three main factors contribute to the evolution and maintenance of a state of ovarian quiescence after parturition: (i) a marked depletion of anterior pituitary stores of LH during late gestation, (ii) peripartum nutritional status, and (iii) suppression of high-frequency pulses of GnRH by suckling in the presence of a maternal–offspring bond. Figure 23.6 summarizes the influence of various factors that contribute to the maintenance of anovulation in the postpartum cow and the physiological events during postpartum anovulation.

Depletion of releasable pools of LH within the adenohypophysis occurs in both ewes and cows during the last third of pregnancy and is quite evident at the time of parturition.[12,121] Its basis is the ability of chronically elevated, placental-derived estradiol to suppress anterior pituitary gonadotrope synthesis of the α and β-subunits of LH and, to a certain extent, FSH.[122,123] However, this does not appear to be mediated through changes in mRNA expression for the gonadotropin subunits. In experiments with ovariectomized ewes, in which the estrogenic environment of late pregnancy was simulated, pituitary content of both LH and FSH was markedly reduced but abundance of mRNA for the α-subunit or β-subunit of LH and FSH was unchanged or increased. Moreover, treatment with exogenous GnRH in these ewes did not restore pituitary synthesis of gonadotropins. Therefore, it was concluded that the negative effects of chronically elevated estradiol are downstream of mRNA expression and represent post-translational regulation of gonadotropin synthesis.[122,123] In considering this model, it should be noted that in pregnant cows neither adenohypophyseal content[124] nor releasable pools[39] of FSH change appreciably during the postpartum period. This may be the result of a more rapid restoration of the β-subunit for FSH by gonadotropes after parturition.[125] Thus, the negative impact of high circulating concentrations of estradiol during gestation in the cow, and its carry-over into the postpartum period, is primarily mediated through effects on LH.

Although progesterone is also elevated during pregnancy, it does not have major direct effects on the pituitary.[121] The main effects of high concentrations of progesterone are on hypothalamic secretion of GnRH and this inhibition is lifted at parturition. Moreover, hypothalamic content of GnRH does not change from days 1 to 35 postpartum in cows or ewes.[121,126] Perhaps more important to postpartum recovery is the stimulatory low-level increases in progesterone observed during the resumption of ovarian cycles.[127,128] Short-lived luteal phases created by premature regression of corpora lutea during the first 30 days postpartum precede the establishment of normal-length cycles.[129] In this role, progesterone serves as a positive modulator of postpartum reproduction by priming the entire hypothalamic–pituitary–ovarian–uterine axis for normal gonadotropin secretion,[43] normal timing of uterine prostaglandin release,[130] and normal luteal function.[131–133]

Within 2–3 days following parturition, the high gestational levels of placental steroids are cleared from the circulation in both ewes and cows.[128,134] This is accompanied by large increases in the concentration of mRNA for the α- and β-subunits of LH.[121] Within 2–3 weeks postpartum, total anterior pituitary content and releasable pools of LH are replete in both dairy and beef cows.[39,124,135]

Neuroendocrine Control of Estrus and Ovulation

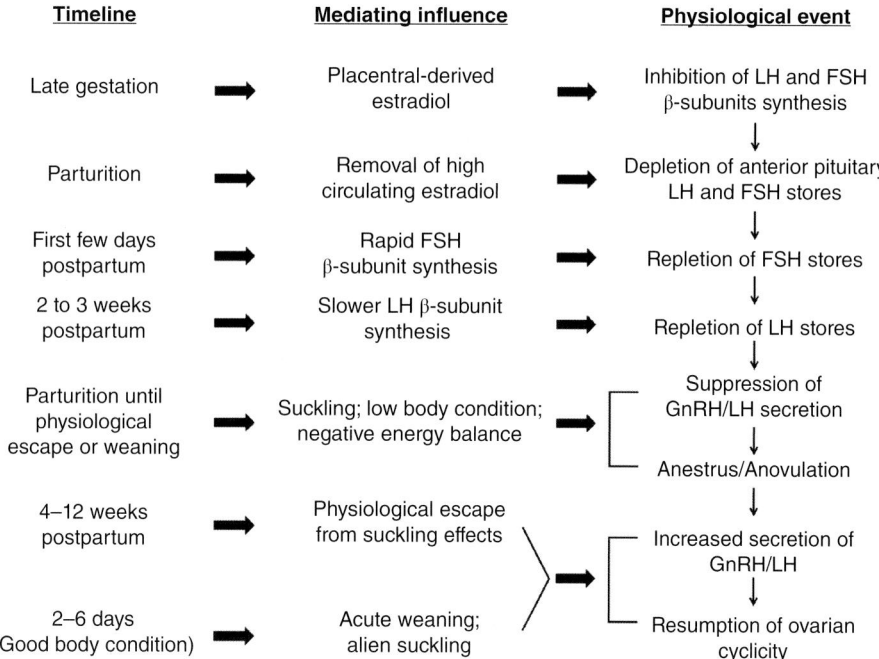

Figure 23.6 Timeline and mediating influences underlying the postpartum anovulatory period of the cow with emphasis on the suckled beef cow. In the high-producing dairy cow, negative energy balance, as opposed to suckling, is the predominant influence inhibiting re-initiation of estrous cyclicity after calving.

This process proceeds in the presence of suckling, although the amplitude and/or frequency of GnRH pulses remains low.[5] Thus, in the absence of elevated circulating concentrations of estradiol, the ruminant gonadotrope appears to require very little GnRH stimulus to synthesize and store LH.[121] By 30 days postpartum, release of LH in response to pharmacological challenge with GnRH is notably greater in suckled than in weaned cows.[39] This occurs because the pulsatile release of LH from the pituitary increases dramatically in weaned cows within the first 2 weeks after calving due to the absence of suckling.[136,137] Correspondingly, suckled cows accumulate more GnRH within the median eminence during this period than nonsuckled cows.[138] Thus, more LH is available for release in response to GnRH challenge in suckled compared with weaned cows during the first few weeks post calving.

Since the repletion of adenohypophyseal LH is essentially complete within 3 weeks after calving, it would seem reasonable to assume that a normal pattern of LH secretion and onset of ovarian cycles would also occur within this period. However, while first ovulation often does occur within 2–3 weeks postpartum in milked dairy cows and weaned beef cows, it is delayed for extended and variable periods in the presence of suckling.[118,119] Suckling frequency must be reduced to no more than once daily in order for duration of postpartum anovulation to be decreased.[139,140] Moreover, up to 6 days are required in order for the acute effects of complete weaning on secretion of LH to result in resumption of ovulatory cycles in a majority of cows.[141,142] Time of suckling during the solar day had no effect on length of the postpartum anovulatory interval.[5] However, low body condition at calving and postpartum undernutrition interacts with suckling to exacerbate its negative effects on postpartum reproduction. In some cases, the period of suckling-induced anovulation can extend for 100 days or more due to this interaction, although the average is usually 60–70 days in mature cows calving in good body condition.[139,143] Of additional interest has been the biostimulatory effect of the bull on resumption of postpartum ovarian function.[144,145] Data indicate that a pheromonal component of bull urine can hasten the recovery of a pulsatile pattern of gonadotropin secretion in postpartum cows and thus first estrus, although the neuroendocrine mechanism has not been elucidated.

How does the suckling calf influence the neuroendocrine regulation of GnRH and LH secretion? In the 1980s and 1990s, several laboratories around the world focused on this question. Based on historical perspectives in the literature, it was assumed that the suckling calf inhibited the secretion of GnRH and/or the gonadotropins through neural signals mediated by tactile stimulation of the teat.[146,147] Early endocrine-related work focused on the potential roles of neurohumoral mediators in this process, including prolactin[128,148,149] and cortisol.[149,150] It is now clear that neither of these hormones accounts for the suppressive effects of suckling on the hypothalamic–pituitary axis of cows during the postpartum period. Additional studies addressed the role of ovarian estradiol. Results indicated that enhanced sensitivity to estradiol negative feedback plays a role in mediating the suckling effect.[151,152] However, estradiol negative feedback is not required for the inhibitory effects of suckling to be manifested.[153] Concomitant with this effect is an apparent increase in endogenous opioid tone that decreases as the postpartum interval increases and in response to weaning.[154,155]

In order to define more clearly whether physical manipulation of the teat alone could account for the inhibitory effects of suckling on gonadotropin secretion, a series of experiments were conducted to examine the effects of various exteroceptive stimuli. These included chronic milking by hand as well as mechanical, electrical, and thermal stimulation of the teats of ovariectomized cows. None of these stimuli were able to recreate or simulate the suckling effect.[149,156,157] Contrasting studies attempted to simulate the "weaning effect" through mechanical masking of the udder during suckling[158] or by using either complete mastectomy or mammary denervation.[159,160] Neither approach altered gonadotropin secretion in cows suckling their own calves. Mammary denervation did impair oxytocin secretion and milk let-down following suckling, but this physiological deficit was only temporary. However, denervation had no effect on secretion of prolactin or milk production in beef cows. Collectively, the foregoing studies resulted in the conclusion that chronic tactile stimulation of mammary somatosensory pathways is not required in order for the suckling effect to be manifested,[160] and spawned a new avenue of research to assess the role of maternal behavior as a regulator of postpartum anovulation.

Initial experiments involving the role of maternal behavior demonstrated unequivocally that if a cow is forced to suckle an "alien" calf under controlled conditions, the neuroendocrine system responds as if the cow has been weaned: pulsatile secretion of LH increases, and first postpartum ovulation occurs within 2–6 days.[142] This led to experimental observations that cows must be able to identify the calf as "own," using either vision or olfaction, in order for suckling to inhibit the GnRH pulse generator and to sustain the anestrous state.[161] Elimination of both senses prevented calf identification and the negative effects of suckling by the cow's own calf on the reproductive neuroendocrine system. At the neuroendocrine level, suckling appears to suppress secretion of LH by dampening the amplitude of individual GnRH pulses (based on measurements of GnRH secretion in third ventricle cerebrospinal fluid), rather than reducing the number of detectable pulses. This has been interpreted to mean that the number of biologically effective GnRH pulses reaching the adenohypophyseal portal vessels is reduced, thus lessening the frequency of LH pulses secreted by the anterior pituitary.[5]

In addition to fundamental studies, a plethora of applied studies have been reported using temporary weaning, alien suckling, and limited suckling to hasten the timing of first postpartum estrus and ovulation, and to enhance synchronization of estrus and ovulation. A detailed consideration of this topic is beyond the scope of this chapter. The reader is referred to published reviews[119] and other recent reports that consider the topic in greater detail.

Novel neuroendocrine regulators of GnRH neuronal function

In the past decade, a number of neuropeptides have been investigated for their potential roles in controlling reproductive function. Of particular interest relative to the regulation of GnRH secretion, the peptides kisspeptin[162–164] and arginine-phenylalanine-amide related peptide (RFRP), originally named gonadotropin-inhibitory hormone, have received great attention.[165,166] These peptides are synthesized in the hypothalamus and project to GnRH neuronal cell bodies and/or terminals in the median eminence, where they likely exert direct control on GnRH neuronal activity and GnRH secretion.

Kisspeptin is a product of the *Kiss1* gene, which is expressed in two discrete populations of neurons located in the preoptic/periventricular area and in the arcuate nucleus of the hypothalamus.[167] Kisspeptin is processed from a precursor peptide to produce several isoforms that vary in length. All biologically active forms of kisspeptin contain the C-terminus sequence common to all forms. The decapeptide form of kisspeptin appears to be the shortest isoform with full capacity for stimulation of GnRH secretion. The role of kisspeptin in the regulation of reproductive function was first characterized in studies investigating mutations in the kisspeptin receptor (*Kiss1r*) gene.[168,169] Humans and mice with loss of function of kisspeptin receptors exhibit hypogonadotropic hypogonadism. Following these studies, the action of kisspeptin as a potent GnRH secretagogue was characterized in several species including cattle.[107,170] Because GnRH neurons contain the kisspeptin receptor, kisspeptin is believed to act directly on GnRH neurons at the level of the cell body and dendrites, as well as at the level of the terminals in the median eminence. Indirect effects of kisspeptin via modulation of other neuronal inputs to the GnRH neurons,[171] and direct action on adenohypophyseal cells *in vitro*,[172] has also been reported. However, the physiological relevance for a direct action of kisspeptin in the adenohypophysis is unclear because kisspeptin failed to stimulate release of LH in hypothalamic–hypophyseal disconnected sheep.[173]

Estradiol has a profound effect on the expression of the *Kiss1* gene, but estradiol actions differ between the two populations of kisspeptin neurons. In the preoptic/periventricular area, estradiol stimulates *Kiss1* gene expression. In contrast, estradiol inhibits *Kiss1* gene expression in the arcuate nucleus.[174,175] This effect of estradiol on regulation of *Kiss1* expression may be involved in the feedback actions of estradiol on gonadotropin release. In ewes, estradiol induces activation of kisspeptin neurons in the preoptic/periventricular areas, indicating that this population of neurons may be important for mediating estradiol positive feedback and onset of the preovulatory surge of LH.[176,177] Although controversial, the population of kisspeptin neurons in the arcuate nucleus may also be activated during the preovulatory surge of GnRH/LH release. Nevertheless, the observation that constant exogenous treatment with kisspeptin induces sustained release of LH in anestrous ewes and synchronizes the LH surge[178,179] demonstrates that kisspeptin, or its analogues, can be used pharmacologically to control the estrous cycle in ruminant species.

Although studies investigating the physiological actions of RFRP in cattle are limited, evidence from studies in cattle[180] and other species[166] indicate that RFRP-3 acts as an inhibitor of LH release by potential actions in the hypothalamus and at the level of the pituitary. The avian homologue of the mammalian RFRP-3 was the first peptide of the family to be isolated in vertebrates.[181] Because the avian peptide inhibited LH synthesis and secretion,[181–183] it was named GnIH. In mammals, the gene that encodes RFRP-3 can produce more than one bioactive peptide of the RFRP family.[184] Expression of the RFRP gene has been localized in the paraventricular and dorsomedial nucleus of the hypothalamus of ewes.[175] Neuronal fibers containing RFRP immunoreactivity are observed to extend to the median eminence, where they are in close proximity to GnRH-containing fibers.[182,185] Neuronal projections containing RFRP are also observed in close proximity to GnRH cell bodies and proximal dendrites.[175] Thus, RFRP-3 may control gonadotropin release by acting on GnRH neurons and/or at the level of the gonadotrope following its release into the hypothalamic–hypophyseal vasculature. The presence of GPR147, the receptor for RFRP-3, in the hypothalamus and adenohypophysis supports RFRP-3 actions at both levels. Studies in sheep[185] and pigs[186] indicate that RFRP-3 inhibits the release of LH *in vivo*. However, controversy on the effects of RFRP-3 exists because studies in sheep,[187] pigs,[188] rats,[189] and horses[190] failed to support a clear effect of RFRP. Nevertheless, the observation that RF9, an RFRP antagonist, increases gonadotropin release in laboratory rodents,[191,192] sheep,[187] and mares[190] indicates that RFRP-3 may function within a complex biological system to effect regulation of gonadotropin release.

References

1. Hansel W, Convey E. Physiology of the estrous cycle. *J Anim Sci* 1983;57(Suppl. 2):404–424.
2. Rahe C, Owens R, Fleeger J, Newton H, Harms P. Pattern of plasma luteinizing hormone in the cyclic cow: dependence upon the period of the cycle. *Endocrinology* 1980;107:498–503.
3. Fortune J, Sirois J, Turzillo A, Lavoir M. Follicle selection in domestic ruminants. *J Reprod Fertil Suppl* 1991;43:187–198.
4. Ginther O, Bergfelt D, Kulick L, Kot K. Selection of the dominant follicle in cattle: role of estradiol. *Biol Reprod* 2000;63:383–389.
5. Gazal O, Leshin L, Stanko R et al. Gonadotropin-releasing hormone secretion into third-ventricle cerebrospinal fluid of cattle: correspondence with the tonic and surge release of luteinizing hormone and its tonic inhibition by suckling and neuropeptide Y. *Biol Reprod* 1998;59:676–683.
6. Moenter S, Caraty A, Locatelli A, Karsch F. Pattern of gonadotropin-releasing hormone (GnRH) secretion leading up to ovulation in the ewe: existence of a preovulatory GnRH surge. *Endocrinology* 1991;129:1175–1182.
7. Dees W, McArthur N. Immunohistochemical localization of gonadotropin releasing hormone (GnRH) in the bovine hypothalamus and infundibulum. *Anat Rec* 1981;200:281–285.
8. Leshin L, Rund L, Crim J, Kiser T. Immunocytochemical localization of luteinizing hormone-releasing hormone and proopiomelanocortin neurons within the preoptic area and hypothalamus of the bovine brain. *Biol Reprod* 1988;39:963–975.
9. Molnár J, Köves K, Marton J, Halász B. On the origin of luteinizing hormone releasing hormone and somatostatin in the median eminence of the rat hypothalamus. *Acta Biol Acad Sci Hung* 1982;33:255–268.
10. Clarke I, Cummins J, de Kretser D. Pituitary gland function after disconnection from direct hypothalamic influences in the sheep. *Neuroendocrinology* 1983;36:376–384.
11. Herbison A. Physiology of the gonadotropin-releasing hormone neuronal network. In: Neill JD (ed.) *Knobil and Neil's Physiology of Reproduction*, 3rd edn. New York: Academic Press, 2006, pp. 1415–1482.
12. Ojeda S, Lomniczi A, Sandau U. Glial–gonadotrophin hormone (GnRH) neurone interactions in the median eminence and the control of GnRH secretion. *J Neuroendocrinol* 2008;20:732–742.
13. Prevot V. Glial–neuronal–endothelial interactions are involved in the control of GnRH secretion. *J Neuroendocrinol* 2002;14:247–255.
14. Bergfeld E, Kojima F, Cupp A et al. Changing dose of progesterone results in sudden changes in frequency of luteinizing hormone pulses and secretion of 17 beta-estradiol in bovine females. *Biol Reprod* 1996;54:546–553.
15. Karsch F, Cummins J, Thomas G, Clarke I. Steroid feedback inhibition of pulsatile secretion of gonadotropin-releasing hormone in the ewe. *Biol Reprod* 1987;36:1207–1218.
16. Skinner D, Evans N, Delaleu B, Goodman R, Bouchard P, Caraty A. The negative feedback actions of progesterone on gonadotropin-releasing hormone secretion are transduced by the classical progesterone receptor. *Proc Natl Acad Sci USA* 1998;95:10978–10983.
17. Goodman R, Bittman E, Foster D, Karsch F. The endocrine basis of the synergistic suppression of luteinizing hormone by estradiol and progesterone. *Endocrinology* 1981;109:1414–1417.
18. Kesner J, Padmanabhan V, Convey E. Estradiol induces and progesterone inhibits the preovulatory surges of luteinizing hormone and follicle-stimulating hormone in heifers. *Biol Reprod* 1982;26:571–578.
19. Kasa-Vubu J, Dahl G, Evans N et al. Progesterone blocks the estradiol-induced gonadotropin discharge in the ewe by inhibiting the surge of gonadotropin-releasing hormone. *Endocrinology* 1992;131:208–212.
20. Caraty A, Skinner D. Progesterone priming is essential for the full expression of the positive feedback effect of estradiol in inducing the preovulatory gonadotropin-releasing hormone surge in the ewe. *Endocrinology* 1999;140:165–170.
21. Goodman R, Gibson M, Skinner D, Lehman M. Neuroendocrine control of pulsatile GnRH secretion during the ovarian cycle: evidence from the ewe. *Reprod Suppl* 2002;59:41–56.
22. Whisnant C, Goodman R. Effects of an opioid antagonist on pulsatile luteinizing hormone secretion in the ewe vary with changes in steroid negative feedback. *Biol Reprod* 1988;39:1032–1038.
23. Goodman R, Coolen L, Anderson G et al. Evidence that dynorphin plays a major role in mediating progesterone negative feedback on gonadotropin-releasing hormone neurons in sheep. *Endocrinology* 2004;145:2959–2967.
24. Boukhliq R, Goodman R, Berriman S, Adrian B, Lehman M. A subset of gonadotropin-releasing hormone neurons in the ovine medial basal hypothalamus is activated during increased pulsatile luteinizing hormone secretion. *Endocrinology* 1999;140:5929–5936.
25. Stumpf T, Roberson M, Wolfe M, Hamernik D, Kittok R, Kinder J. Progesterone, 17 beta-estradiol, and opioid neuropeptides modulate pattern of luteinizing hormone in circulation of the cow. *Biol Reprod* 1993;49:1096–1101.
26. Goubillon M, Forsdike R, Robinson J, Ciofi P, Caraty A, Herbison A. Identification of neurokinin B-expressing neurons as an highly estrogen-receptive, sexually dimorphic cell group in the ovine arcuate nucleus. *Endocrinology* 2000;141:4218–4225.
27. Foradori C, Coolen L, Fitzgerald M, Skinner D, Goodman R, Lehman M. Colocalization of progesterone receptors in parvicellular dynorphin neurons of the ovine preoptic area and hypothalamus. *Endocrinology* 2002;143:4366–4374.
28. Scott C, Pereira A, Rawson J et al. The distribution of progesterone receptor immunoreactivity and mRNA in the preoptic area and hypothalamus of the ewe: upregulation of progesterone receptor mRNA in the mediobasal hypothalamus by oestrogen. *J Neuroendocrinol* 2000;12:565–575.
29. Evans N, Dahl G, Glover B, Karsch F. Central regulation of pulsatile gonadotropin-releasing hormone (GnRH) secretion by estradiol during the period leading up to the preovulatory GnRH surge in the ewe. *Endocrinology* 1994;134:1806–1811.
30. Herring R, Hamernik D, Kile J, Sousa M, Nett T. Chronic administration of estradiol produces a triphasic effect on serum concentrations of gonadotropins and messenger ribonucleic acid for gonadotropin subunits, but not on pituitary content of gonadotropins, in ovariectomized ewes. *Biol Reprod* 1991;45:151–156.
31. Cupp A, Kojima F, Roberson M et al. Increasing concentrations of 17 beta-estradiol has differential effects on secretion of luteinizing hormone and follicle-stimulating hormone and amounts of mRNA for gonadotropin subunits during the

follicular phase of the bovine estrous cycle. *Biol Reprod* 1995;52:288–296.
32. Kesner J, Convey E, Anderson C. Evidence that estradiol induces the preovulatory LH surge in cattle by increasing pituitary sensitivity to LHRH and then increasing LHRH release. *Endocrinology* 1981;108:1386–1391.
33. Evans N, Richter T, Skinner D, Robinson J. Neuroendocrine mechanisms underlying the effects of progesterone on the oestradiol-induced GnRH/LH surge. *Reprod Suppl* 2002;59:57–66.
34. Caraty A, Fabre-Nys C, Delaleu B et al. Evidence that the mediobasal hypothalamus is the primary site of action of estradiol in inducing the preovulatory gonadotropin releasing hormone surge in the ewe. *Endocrinology* 1998;139:1752–1760.
35. Clarke I, Pompolo S, Scott C et al. Cells of the arcuate nucleus and ventromedial nucleus of the ovariectomized ewe that respond to oestrogen: a study using Fos immunohistochemistry. *J Neuroendocrinol* 2001;13:934–941.
36. Schoenemann H, Humphrey W, Crowder M, Nett T, Reeves J. Pituitary luteinizing hormone-releasing hormone receptors in ovariectomized cows after challenge with ovarian steroids. *Biol Reprod* 1985;32:574–583.
37. Caraty A, Skinner D. Progesterone priming is essential for the full expression of the positive feedback effect of estradiol in inducing the preovulatory gonadotropin-releasing hormone surge in the ewe. *Endocrinology* 1999;140:165–170.
38. Clarke I, Cummins J. Direct pituitary effects of estrogen and progesterone on gonadotropin secretion in the ovariectomized ewe. *Neuroendocrinology* 1984;39:267–274.
39. Williams G, Petersen B, Tilton J. Pituitary and ovarian responses of postpartum dairy cows to progesterone priming and single or double injections of gonadotropin-releasing hormone. *Theriogenology* 1982;18:561–572.
40. Gümen A, Wiltbank M. An alteration in the hypothalamic action of estradiol due to lack of progesterone exposure can cause follicular cysts in cattle. *Biol Reprod* 2002;66:1689–1695.
41. Padmanabhan V, Brown M, Dahl G et al. Neuroendocrine control of follicle-stimulating hormone (FSH) secretion: III. Is there a gonadotropin-releasing hormone-independent component of episodic FSH secretion in ovariectomized and luteal phase ewes? *Endocrinology* 2003;144:1380–1392.
42. Dhariwal A, Nallar R, Batt M, McCann S. Separation of follicle-stimulating hormone-releasing factor from luteinizing hormone-releasing factor. *Endocrinology* 1965;76:290–294.
43. Williams G, Talavera F, Petersen B, Tilton J. Coincident secretion of follicle-stimulating hormone and luteinizing hormone in early postpartum beef cows: effects of suckling and low-level increases of systemic progesterone. *Biol Reprod* 1983;29:362–373.
44. Walters D, Schams D, Schallenberger E. Pulsatile secretion of gonadotrophins, ovarian steroids and ovarian oxytocin during the luteal phase of the oestrous cycle in the cow. *J Reprod Fertil* 1984;71:479–491.
45. Bolt D, Rollins R. Development and application of a radioimmunoassay for bovine follicle-stimulating hormone. *J Anim Sci* 1983;56:146–154.
46. Miller K, Bolt D. Treatment of ovariectomized ewes with bovine follicular fluid decreases secretion of FSH without changing secretion of LH. *Theriogenology* 1985;24:211–216.
47. Gross D, Turgeon J, Waring D. The ovine pars tuberalis: a naturally occurring source of partially purified gonadotropes which secrete luteinizing hormone in vitro. *Endocrinology* 1984;114:2084–2091.
48. Bastings E, Beckers A, Reznik M, Beckers J. Immunocytochemical evidence for production of luteinizing hormone and follicle-stimulating hormone in separate cells in the bovine. *Biol Reprod* 1991;45:788–796.
49. Amstalden M, Zieba D, Garcia M et al. Evidence that lamprey GnRH-III does not release FSH selectively in cattle. *Reproduction* 2004;127:35–43.
50. Walters D, Schallenberger E. Pulsatile secretion of gonadotrophins, ovarian steroids and ovarian oxytocin during the periovulatory phase of the oestrous cycle in the cow. *J Reprod Fertil* 1984;71:503–512.
51. Turzillo A, Fortune J. Suppression of the secondary FSH surge with bovine follicular fluid is associated with delayed ovarian follicular development in heifers. *J Reprod Fertil* 1990;89:643–653.
52. Lussier J, Carruthers T, Murphy B. Effects of bovine follicular fluid and partially purified bovine inhibin on FSH and LH release by bovine pituitary cells in culture. *Reprod Nutr Dev* 1993;33:109–119.
53. Findlay J, Doughton B, Robertson D, Forage R. Effects of immunization against recombinant bovine inhibin alpha subunit on circulating concentrations of gonadotrophins in ewes. *J Endocrinol* 1989;120:59–65.
54. Medan M, Akagi S, Kaneko H, Watanabe G, Tsonis C, Taya K. Effects of re-immunization of heifers against inhibin on hormonal profiles and ovulation rate. *Reproduction* 2004;128:475–482.
55. Kaneko H, Nakanishi Y, Taya K et al. Evidence that inhibin is an important factor in the regulation of FSH secretion during the mid-luteal phase in cows. *J Endocrinol* 1993;136:35–41.
56. Tilbrook A, de Kretser D, Clarke I. Human recombinant inhibin A and testosterone act directly at the pituitary to suppress plasma concentrations of FSH in castrated rams. *J Endocrinol* 1993;138:181–189.
57. Sharma T, Nett T, Karsch F et al. Neuroendocrine control of FSH secretion: IV. Hypothalamic control of pituitary FSH-regulatory proteins and their relationship to changes in FSH synthesis and secretion. *Biol Reprod* 2012;86:171.
58. Cupp A, Kojima F, Roberson M et al. Increasing concentrations of 17 beta-estradiol has differential effects on secretion of luteinizing hormone and follicle-stimulating hormone and amounts of mRNA for gonadotropin subunits during the follicular phase of the bovine estrous cycle. *Biol Reprod* 1995;52:288–296.
59. Martin G, Price C, Thiéry J, Webb R. Interactions between inhibin, oestradiol and progesterone in the control of gonadotrophin secretion in the ewe. *J Reprod Fertil* 1988;82:319–328.
60. Bilezikjian L, Corrigan A, Blount A, Vale W. Pituitary follistatin and inhibin subunit messenger ribonucleic acid levels are differentially regulated by local and hormonal factors. *Endocrinology* 1996;137:4277–4284.
61. Padmanabhan V, Sharma T. Neuroendocrine vs. paracrine control of follicle-stimulating hormone. *Arch Med Res* 2001;32:533–543.
62. Lumpkin M, McCann S. Effect of destruction of the dorsal anterior hypothalamus on follicle-stimulating hormone secretion in the rat. *Endocrinology* 1984;115:2473–2480.
63. Lumpkin M, McDonald J, Samson W, McCann S. Destruction of the dorsal anterior hypothalamic region suppresses pulsatile release of follicle stimulating hormone but not luteinizing hormone. *Neuroendocrinology* 1989;50:229–235.
64. Padmanabhan V, McFadden K, Mauger D, Karsch F, Midgley A. Neuroendocrine control of follicle-stimulating hormone (FSH) secretion. I. Direct evidence for separate episodic and basal components of FSH secretion. *Endocrinology* 1997;138:424–432.
65. Day M, Imakawa K, Zalesky D, Kittok R, Kinder J. Effects of restriction of dietary energy intake during the prepubertal period on secretion of luteinizing hormone and responsiveness of the pituitary to luteinizing hormone-releasing hormone in heifers. *J Anim Sci* 1986;62:1641–1648.
66. McCann J, Hansel W. Relationships between insulin and glucose metabolism and pituitary–ovarian functions in fasted heifers. *Biol Reprod* 1986;34:630–641.
67. Kadokawa H, Yamada Y. Enhancing effect of acute fasting on ethanol suppression of pulsatile luteinizing hormone release via an estrogen-dependent mechanism in Holstein heifers. *Theriogenology* 1999;51:673–680.
68. Amstalden M, Garcia M, Stanko R et al. Central infusion of recombinant ovine leptin normalizes plasma insulin and stimulates

a novel hypersecretion of luteinizing hormone after short-term fasting in mature beef cows. *Biol Reprod* 2002;66:1555–1561.
69. Amstalden M, Garcia M, Williams S et al. Leptin gene expression, circulating leptin, and luteinizing hormone pulsatility are acutely responsive to short-term fasting in prepubertal heifers: relationships to circulating insulin and insulin-like growth factor I(1). *Biol Reprod* 2000;63:127–133.
70. Maciel M, Zieba D, Amstalden M, Keisler D, Neves J, Williams G. Leptin prevents fasting-mediated reductions in pulsatile secretion of luteinizing hormone and enhances its gonadotropin-releasing hormone-mediated release in heifers. *Biol Reprod* 2004;70:229–235.
71. Cagampang F, Maeda K, Tsukamura H, Ohkura S, Ota K. Involvement of ovarian steroids and endogenous opioids in the fasting-induced suppression of pulsatile LH release in ovariectomized rats. *J Endocrinol* 1991;129:321–328.
72. I'Anson H, Manning J, Herbosa C et al. Central inhibition of gonadotropin-releasing hormone secretion in the growth-restricted hypogonadotropic female sheep. *Endocrinology* 2000;141:520–527.
73. Leonhardt S, Shahab M, Luft H, Wuttke W, Jarry H. Reduction of luteinizing hormone secretion induced by long-term feed restriction in male rats is associated with increased expression of GABA-synthesizing enzymes without alterations of GnRH gene expression. *J Neuroendocrinol* 1999;11:613–619.
74. Kile J, Alexander B, Moss G, Hallford D, Nett T. Gonadotropin-releasing hormone overrides the negative effect of reduced dietary energy on gonadotropin synthesis and secretion in ewes. *Endocrinology* 1991;128:843–849.
75. Kaiser U, Jakubowiak A, Steinberger A, Chin W. Differential effects of gonadotropin-releasing hormone (GnRH) pulse frequency on gonadotropin subunit and GnRH receptor messenger ribonucleic acid levels in vitro. *Endocrinology* 1997;138:1224–1231.
76. Clarke I, Cummins J. The significance of small pulses of gonadotrophin-releasing hormone. *J Endocrinol* 1987;113:413–418.
77. Wade G, Jones J. Neuroendocrinology of nutritional infertility. *Am J Physiol* 2004;287:R1277–R1296.
78. Levin B. Metabolic sensing neurons and the control of energy homeostasis. *Physiol Behav* 2006;89:486–489.
79. Migrenne S, Le Foll C, Levin B, Magnan C. Brain lipid sensing and nervous control of energy balance. *Diabetes Metab* 2011;37:83–88.
80. Clarke I, Horton R, Doughton B. Investigation of the mechanism by which insulin-induced hypoglycemia decreases luteinizing hormone secretion in ovariectomized ewes. *Endocrinology* 1990;127:1470–1476.
81. Bucholtz D, Vidwans N, Herbosa C, Schillo K, Foster D. Metabolic interfaces between growth and reproduction. V. Pulsatile luteinizing hormone secretion is dependent on glucose availability. *Endocrinology* 1996;137:601–607.
82. Rutter L, Manns J. Follicular phase gonadotropin secretion in cyclic postpartum beef cows with phlorizin-induced hypoglycemia. *J Anim Sci* 1988;66:1194–1200.
83. Schreihofer D, Renda F, Cameron J. Feeding-induced stimulation of luteinizing hormone secretion in male rhesus monkeys is not dependent on a rise in blood glucose concentration. *Endocrinology* 1996;137:3770–3776.
84. Smith M, Jennes L. Neural signals that regulate GnRH neurones directly during the oestrous cycle. *Reproduction* 2001;122:1–10.
85. Clarkson J, Herbison A. Development of GABA and glutamate signaling at the GnRH neuron in relation to puberty. *Mol Cell Endocrinol* 2006;254–255:32–38.
86. Crown A, Clifton D, Steiner R. Neuropeptide signaling in the integration of metabolism and reproduction. *Neuroendocrinology* 2007;86:175–182.
87. Zieba D, Amstalden M, Williams G. Regulatory roles of leptin in reproduction and metabolism: a comparative review. *Domest Anim Endocrinol* 2005;29:166–185.
88. Ehrhardt R, Slepetis R, Siegal-Willott J, Van Amburgh M, Bell A, Boisclair Y. Development of a specific radioimmunoassay to measure physiological changes of circulating leptin in cattle and sheep. *J Endocrinol* 2000;166:519–528.
89. Garcia M, Amstalden M, Williams S et al. Serum leptin and its adipose gene expression during pubertal development, the estrous cycle, and different seasons in cattle. *J Anim Sci* 2002;80:2158–2167.
90. Zieba D, Amstalden M, Morton S, Maciel M, Keisler D, Williams G. Regulatory roles of leptin at the hypothalamic–hypophyseal axis before and after sexual maturation in cattle. *Biol Reprod* 2004;71:804–812.
91. Amstalden M, Zieba D, Edwards J et al. Leptin acts at the bovine adenohypophysis to enhance basal and gonadotropin-releasing hormone-mediated release of luteinizing hormone: differential effects are dependent upon nutritional history. *Biol Reprod* 2003;69:1539–1544.
92. Cheung C, Clifton D, Steiner R. Proopiomelanocortin neurons are direct targets for leptin in the hypothalamus. *Endocrinology* 1997;138:4489–4492.
93. Schwartz M, Seeley R, Campfield L, Burn P, Baskin D. Identification of targets of leptin action in rat hypothalamus. *J Clin Invest* 1996;98:1101–1106.
94. McShane T, Petersen S, McCrone S, Keisler D. Influence of food restriction on neuropeptide-Y, proopiomelanocortin, and luteinizing hormone-releasing hormone gene expression in sheep hypothalami. *Biol Reprod* 1993;49:831–839.
95. Schwartz M, Seeley R, Woods S et al. Leptin increases hypothalamic pro-opiomelanocortin mRNA expression in the rostral arcuate nucleus. *Diabetes* 1997;46:2119–2123.
96. Roa J, Herbison A. Direct regulation of GnRH neuron excitability by arcuate nucleus POMC and NPY neuron neuropeptides in female mice. *Endocrinology* 2012;153:5587–5599.
97. Kinder J, Day M, Kittok R. Endocrine regulation of puberty in cows and ewes. *J Reprod Fertil Suppl* 1987;34:167–186.
98. Foster D, Jackson L. Puberty in sheep. In: Neill JD (ed.) *Knobil and Neil's Physiology of Reproduction*, 3rd edn. New York: Academic Press, 2006, pp. 2127–2176.
99. Foster D, Karsch F. Development of the mechanism regulating the preovulatory surge of luteinizing hormone in sheep. *Endocrinology* 1975;97:1205–1209.
100. Schillo K, Dierschke D, Hauser E. Estrogen-induced release of luteinizing hormone in prepubertal and postpubertal heifers. *Theriogenology* 1983;19:727–738.
101. Day M, Imakawa K, Garcia-Winder M et al. Endocrine mechanisms of puberty in heifers: estradiol negative feedback regulation of luteinizing hormone secretion. *Biol Reprod* 1984;31:332–341.
102. Day M, Imakawa K, Wolfe P, Kittok R, Kinder J. Endocrine mechanisms of puberty in heifers. Role of hypothalamo-pituitary estradiol receptors in the negative feedback of estradiol on luteinizing hormone secretion. *Biol Reprod* 1987;37:1054–1065.
103. Smith J. Sex steroid control of hypothalamic Kiss1 expression in sheep and rodents: comparative aspects. *Peptides* 2009;30:94–102.
104. García-Galiano D, Pinilla L, Tena-Sempere M. Sex steroids and the control of the Kiss1 system: developmental roles and major regulatory actions. *J Neuroendocrinol* 2012;24:22–33.
105. Clarkson J. Effects of estradiol on kisspeptin neurons during puberty. *Front Neuroendocrinol* 2013;34:120–131.
106. Messager S, Chatzidaki E, Ma D et al. Kisspeptin directly stimulates gonadotropin-releasing hormone release via G protein-coupled receptor 54. *Proc Natl Acad Sci USA* 2005;102:1761–1766.
107. Kadokawa H, Matsui M, Hayashi K et al. Peripheral administration of kisspeptin-10 increases plasma concentrations of GH as well as LH in prepubertal Holstein heifers. *J Endocrinol* 2008;196:331–334.
108. Chenoweth P. Aspects of reproduction in female *Bos indicus* cattle: a review. *Aust Vet J* 1994;71:422–426.

109. Nogueira G. Puberty in South American *Bos indicus* (Zebu) cattle. *Anim Reprod Sci* 2004;82–83:361–372.
110. Short R, Bellows R. Relationships among weight gains, age at puberty and reproductive performance in heifers. *J Anim Sci* 1971;32:127–131.
111. Arije G, Wiltbank J. Age and weight at puberty in Hereford heifers. *J Anim Sci* 1971;33:401–406.
112. Marston T, Lusby K, Wettemann R. Effects of postweaning diet on age and weight at puberty and mild production of heifers. *J Anim Sci* 1995;73:63–68.
113. Gasser C, Behlke E, Grum D, Day M. Effect of timing of feeding a high-concentrate diet on growth and attainment of puberty in early-weaned heifers. *J Anim Sci* 2006;84:3118–3122.
114. Gasser C, Bridges G, Mussard M, Grum D, Kinder J, Day M. Induction of precocious puberty in heifers III. Hastened reduction of estradiol negative feedback on secretion of luteinizing hormone. *J Anim Sci* 2006;84:2050–2056.
115. Maciel M, Zieba D, Amstalden M, Keisler D, Neves J, Williams G. Chronic administration of recombinant ovine leptin in growing beef heifers: effects on secretion of LH, metabolic hormones, and timing of puberty. *J Anim Sci* 2004;82:2930–2936.
116. Casida L. The postpartum interval and its relation to fertility in the cow, sow and ewe. *J Anim Sci* 1971;32(Suppl. 1):66–72.
117. McNeilly A. Suckling and the control of gonadotropin secretion. In: Knobil E, Neill JD (eds) *The Physiology of Reproduction*. New York: Raven Press, 1988, pp. 2323–2349.
118. Short R, Bellows R, Staigmiller R, Berardinelli J, Custer E. Physiological mechanisms controlling anestrus and infertility in postpartum beef cattle. *J Anim Sci* 1990;68:799–816.
119. Williams G. Suckling as a regulator of postpartum rebreeding in cattle: a review. *J Anim Sci* 1990;68:831–852.
120. Williams G, Griffith M. Sensory and behavioral control of suckling-mediated anovulation in cows. *J Reprod Fert Suppl* 1995;49:463–475.
121. Nett T. Function of the hypothalamic–hypophysial axis during the post-partum period in ewes and cows. *J Reprod Fert Suppl* 1987;34:201–213.
122. Nett T, Flores J, Carnevali F, Kile J. Evidence for a direct negative effect of estradiol at the level of the pituitary gland in sheep. *Biol Reprod* 1990;43:554–558.
123. Herring R, Hamernik D, Kile J, Sousa M, Nett T. Chronic administration of estradiol produces a triphasic effect on serum concentrations of gonadotropins and messenger ribonucleic acid for gonadotropin subunits, but not on pituitary content of gonadotropins, in ovariectomized ewes. *Biol Reprod* 1991;45:151–156.
124. Nett T, Cermak D, Braden T, Manns J, Niswender G. Pituitary receptors for GnRH and estradiol, and pituitary content of gonadotropins in beef cows. II. Changes during the postpartum period. *Domest Anim Endocrinol* 1988;5:81–89.
125. Di Gregorio G, Kile J, Herring R, Nett T. More rapid restoration of pituitary content of follicle-stimulating hormone than of luteinizing hormone after depletion by oestradiol-17 beta in ewes. *J Reprod Fertil* 1991;93:347–354.
126. Moss G, Parfet J, Marvin C, Allrich R, Diekman M. Pituitary concentrations of gonadotropins and receptors for GnRH in suckled beef cows at various intervals after calving. *J Anim Sci* 1985;60:285–293.
127. Humphrey W, Kaltenbach C, Dunn T, Koritnik D, Niswender G. Characterization of hormonal patterns in the beef cow during postpartum anestrus. *J Anim Sci* 1983;56:445–453.
128. Williams G, Ray D. Hormonal and reproductive profiles of early postpartum beef heifers after prolactin suppression or steroid-induced luteal function. *J Anim Sci* 1980;50:906–918.
129. Garverick H, Smith M. Mechanisms associated with subnormal luteal function. *J Anim Sci* 1986;62(Suppl. 2):92–105.
130. Silvia W, Lewis G, McCracken J, Thatcher W, Wilson L. Hormonal regulation of uterine secretion of prostaglandin F2 alpha during luteolysis in ruminants. *Biol Reprod* 1991;45:655–663.
131. Ramirez-Godinez J, Kiracofe J, McKee R, Schalles R, Kittock R. Reducing the incidence of short estrous cycles in beef cows with norgestomet. *Theriogenology* 1981;15:613–623.
132. Pratt B, Berardinelli J, Stevens L, Inskeep E. Induced corpora lutea in the postpartum beef cow. I. Comparison of gonadotropin releasing hormone and human chorionic gonadotropin and effects of progestogen and estrogen. *J Anim Sci* 1982;54:822–829.
133. Sheffel C, Pratt B, Ferrell W, Inskeep E. Induced corpora lutea in the postpartum beef cow. II. Effects of treatment with progestogen and gonadotropins. *J Anim Sci* 1982;54:830–836.
134. Burd L, Lemons J, Makowski E, Meschia G, Niswender G. Mammary blood flow and endocrine changes during parturition in the ewe. *Endocrinology* 1976;98:748–754.
135. Kesler D, Garverick H, Youngquist R, Elmore R, Bierschwal C. Effect of days postpartum and endogenous reproductive hormones on GNRH-induced LH release in dairy cows. *J Anim Sci* 1977;45:797–803.
136. Carruthers T, Convey E, Kesner J, Hafs H, Cheng K. The hypothalamo-pituitary gonadotrophic axis of suckled and nonsuckled dairy cows postpartum. *J Anim Sci* 1980;51:949–957.
137. Peters A, Lamming G, Fisher M. A comparison of plasma LH concentrations in milked and suckling post-partum cows. *J Reprod Fertil* 1981;62:567–573.
138. Zalesky D, Forrest D, McArthur N, Wilson J, Morris D, Harms P. Suckling inhibits release of luteinizing hormone-releasing hormone from the bovine median eminence following ovariectomy. *J Anim Sci* 1990;68:444–448.
139. Randel R. Effect of once-daily suckling on postpartum interval and cow-calf performance of first-calf Brahman × Hereford heifers. *J Anim Sci* 1981;53:755–757.
140. Lamb G, Miller B, Lynch J et al. Twice daily suckling but not milking with calf presence prolongs postpartum anovulation. *J Anim Sci* 1999;77:2207–2218.
141. Shively T, Williams G. Patterns of tonic luteinizing hormone release and ovulation frequency in suckled anestrous beef cows following varying intervals of temporary weaning. *Domest Anim Endocrinol* 1989;6:379–387.
142. Silveira P, Spoon R, Ryan D, Williams G. Maternal behavior as a requisite link in suckling-mediated anovulation in cows. *Biol Reprod* 1993;49:1338–1346.
143. Wettemann R. Postpartum endocrine function of cattle, sheep and swine. *J Anim Sci* 1980;51(Suppl. 2):2–15.
144. Tauck S, Berardinelli J, Geary T, Johnson N. Resumption of postpartum luteal function of primiparous, suckled beef cows exposed continuously to bull urine. *J Anim Sci* 2006;84:2708–2713.
145. Berardinelli J, Tauck S. Intensity of the biostimulatory effect of bulls on resumption of ovulatory activity in primiparous, suckled, beef cows. *Anim Reprod Sci* 2007;99:24–33.
146. Clapp H. A factor in breeding efficiency of dairy cattle. *J Anim Sci* 1937;(1937):259–265.
147. Short R, Bellows R, Moody E, Howland B. Effects of suckling and mastectomy on bovine postpartum reproduction. *J Anim Sci* 1972;34:70–74.
148. Forrest D, Fleeger J, Long C, Sorensen A, Harms P. Effect of exogenous prolactin on peripheral luteinizing hormone levels in ovariectomized cows. *Biol Reprod* 1980;22:197–201.
149. Williams G, Kirsch J, Post G, Tilton J, Slanger W. Evidence against chronic teat stimulation as an autonomous effector of diminished gonadotropin release in beef cows. *J Anim Sci* 1984;59:1060–1069.
150. Li P, Wagner W. In vivo and in vitro studies on the effect of adrenocorticotropic hormone or cortisol on the pituitary response to gonadotropin releasing hormone. *Biol Reprod* 1983;29:25–37.

151. Acosta B, Tarnabsky T, Platt T et al. Nursing enhances the negative effect of estrogen on LH release in the cow. *J Anim Sci* 1983;57:1530–1536.
152. Garcia-Winder M, Imakawa K, Day M, Zalesky D, Kittok R, Kinder J. Effect of suckling and ovariectomy on the control of luteinizing hormone secretion during the postpartum period in beef cows. *Biol Reprod* 1984;31:771–778.
153. Hinshelwood M, Dierschke D, Hauser E. Effect of suckling on the hypothalamic–pituitary axis in postpartum beef cows, independent of ovarian secretions. *Biol Reprod* 1985; 32:290–300.
154. Whisnant C, Kiser T, Thompson F, Barb C. Opioid inhibition of luteinizing hormone secretion during the postpartum period in suckled beef cows. *J Anim Sci* 1986;63:1445–1448.
155. Whisnant C, Kiser T, Thompson F, Barb C. Influence of calf removal on the serum luteinizing hormone response to naloxone in the postpartum beef cow. *J Anim Sci* 1986;63:561–564.
156. Williams G, Kosiorowski M, Osborn R, Kirsch J, Slanger W. The postweaning rise of tonic luteinizing hormone secretion in anestrous cows is not prevented by chronic milking or the physical presence of the calf. *Biol Reprod* 1987;36:1079–1084.
157. Cutshaw J, Hunter J, Williams G. Effects of transcutaneous thermal and electrical stimulation of the teat on pituitary luteinizing hormone, prolactin and oxytocin secretion in ovariectomized, estradiol-treated beef cows following acute weaning. *Theriogenology* 1992;39:915–934.
158. McVey W, Williams G. Mechanical masking of neurosensory pathways at the calf–teat interface: endocrine, reproductive and lactational features of the suckled anestrous cow. *Theriogenology* 1991;35:931–941.
159. Viker S, Larson R, Kiracofe G, Stewart R, Stevenson J. Prolonged postpartum anovulation in mastectomized cows requires tactile stimulation by the calf. *J Anim Sci* 1993;71:999–1003.
160. Williams G, McVey W, Hunter J. Mammary somatosensory pathways are not required for suckling-mediated inhibition of luteinizing hormone secretion and ovulation in cows. *Biol Reprod* 1993;49:1328–1346.
161. Griffith M, Williams G. Roles of maternal vision and olfaction in suckling-mediated inhibition of luteinizing hormone secretion, expression of maternal selectivity, and lactational performance in beef cows. *Biol Reprod* 1996;54:761–768.
162. Caraty A, Franceschini I. Basic aspects of the control of GnRH and LH secretions by kisspeptin: potential applications for better control of fertility in females. *Reprod Domest Anim* 2008;43(Suppl. 2):172–178.
163. Caraty A, Decourt C, Briant C, Beltramo M. Kisspeptins and the reproductive axis: potential applications to manage reproduction in farm animals. *Domest Anim Endocrinol* 2012;43:95–102.
164. Amstalden M, Alves B, Liu S, Cardoso R, Williams G. Neuroendocrine pathways mediating nutritional acceleration of puberty: insights from ruminant models. *Front Endocrinol* 2011;2:109.
165. Tsutsui K, Bentley G, Kriegsfeld L, Osugi T, Seong J, Vaudry H. Discovery and evolutionary history of gonadotrophin-inhibitory hormone and kisspeptin: new key neuropeptides controlling reproduction. *J Neuroendocrinol* 2010;22:716–727.
166. Kriegsfeld L, Gibson E, Williams W et al. The roles of RFamide-related peptide-3 in mammalian reproductive function and behaviour. *J Neuroendocrinol* 2010;22:692–700.
167. Franceschini I, Lomet D, Cateau M, Delsol G, Tillet Y, Caraty A. Kisspeptin immunoreactive cells of the ovine preoptic area and arcuate nucleus co-express estrogen receptor alpha. *Neurosci Lett* 2006;401:225–230.
168. DeRoux N, Genin E, Carel J, Matsuda F, Chaussain J, Milgrom E. Hypogonadotropic hypogonadism due to loss of function of the KiSS1-derived peptide receptor GPR54. *Proc Natl Acad Sci USA* 2003;100:10972–10976.
169. Seminara S, Messager S, Chatzidaki E et al. The GPR54 gene as a regulator of puberty. *N Engl J Med* 2003;349:1614–1627.
170. Ezzat A, Saito H, Sawada T et al. Characteristics of the stimulatory effect of kisspeptin-10 on the secretion of luteinizing hormone, follicle-stimulating hormone and growth hormone in prepubertal male and female cattle. *J Reprod Dev* 2009; 55:650–654.
171. Pielecka-Fortuna J, Moenter SM. Kisspeptin increases gamma-aminobutyric acidergic and glutamatergic transmission directly to gonadotropin-releasing hormone neurons in an estradiol-dependent manner. *Endocrinology* 2010;151:291–300.
172. Suzuki S, Kadokawa H, Hashizume T. Direct kisspeptin-10 stimulation on luteinizing hormone secretion from bovine and porcine anterior pituitary cells. *Anim Reprod Sci* 2008; 103:360–365.
173. Smith J, Rao A, Pereira A, Caraty A, Millar R, Clarke I. Kisspeptin is present in ovine hypophysial portal blood but does not increase during the preovulatory luteinizing hormone surge: evidence that gonadotropes are not direct targets of kisspeptin in vivo. *Endocrinology* 2008;149:1951–1959.
174. Smith J, Clay C, Caraty A, Clarke I. KiSS-1 messenger ribonucleic acid expression in the hypothalamus of the ewe is regulated by sex steroids and season. *Endocrinology* 2007;148:1150–1157.
175. Smith J, Coolen L, Kriegsfeld L et al. Variation in kisspeptin and RFamide-related peptide (RFRP) expression and terminal connections to gonadotropin-releasing hormone neurons in the brain: a novel medium for seasonal breeding in the sheep. *Endocrinology* 2008;149:5770–5782.
176. Hoffman G, Le W, Franceschini I, Caraty A, Advis J. Expression of fos and in vivo median eminence release of LHRH identifies an active role for preoptic area kisspeptin neurons in synchronized surges of LH and LHRH in the ewe. *Endocrinology* 2011;152:214–222.
177. Merkley C, Porter K, Coolen L et al. KNDy (kisspeptin/neurokinin B/dynorphin) neurons are activated during both pulsatile and surge secretion of LH in the ewe. *Endocrinology* 2012;153:5406–5414.
178. Caraty A, Smith J, Lomet D et al. Kisspeptin synchronizes preovulatory surges in cyclical ewes and causes ovulation in seasonally acyclic ewes. *Endocrinology* 2007;148:5258–5267.
179. Caraty A, Lomet D, Me S, Guillaume D, Beltramo M, Evans N. GnRH release into the hypophyseal portal blood of the ewe mirrors both pulsatile and continuous intravenous infusion of kisspeptin: an insight into kisspeptin's mechanism of action. *J Neuroendocrinol* 2013;25:537–546.
180. Kadokawa H, Shibata M, Tanaka Y et al. Bovine C-terminal octapeptide of RFamide-related peptide-3 suppresses luteinizing hormone (LH) secretion from the pituitary as well as pulsatile LH secretion in bovines. *Domest Anim Endocrinol* 2009;36:219–224.
181. Tsutsui K, Saigoh E, Ukena K et al. A novel avian hypothalamic peptide inhibiting gonadotropin release. *Biochem Biophys Res Commun* 2000;275:661–667.
182. Bentley G, Perfito N, Ukena K, Tsutsui K, Wingfield J. Gonadotropin-inhibitory peptide in song sparrows (*Melospiza melodia*) in different reproductive conditions, and in house sparrows (*Passer domesticus*) relative to chicken-gonadotropin-releasing hormone. *J Neuroendocrinol* 2003;15:794–802.
183. Ubuka T, Ueno M, Ukena K, Tsutsui K. Developmental changes in gonadotropin-inhibitory hormone in the Japanese quail (*Coturnix japonica*) hypothalamo-hypophysial system. *J Endocrinol* 2003;178:311–318.
184. Tsutsui K, Ubuka T, Bentley G, Kriegsfeld L. Gonadotropin-inhibitory hormone (GnIH): discovery, progress and prospect. *Gen Comp Endocrinol* 2012;177:305–314.

185. Clarke I, Sari I, Qi Y et al. Potent action of RFamide-related peptide-3 on pituitary gonadotropes indicative of a hypophysiotropic role in the negative regulation of gonadotropin secretion. *Endocrinology* 2008;149:5811–5821.
186. Li X, Su J, Fang R et al. The effects of RFRP-3, the mammalian ortholog of GnIH, on the female pig reproductive axis in vitro. *Mol Cell Endocrinol* 2013;372:65–72.
187. Caraty A, Blomenröhr M, Vogel G, Lomet D, Briant C, Beltramo M. RF9 powerfully stimulates gonadotrophin secretion in the ewe: evidence for a seasonal threshold of sensitivity. *J Neuroendocrinol* 2012;24:725–736.
188. Heidorn R, Barb C, Rogers C, Hausman G, Lents C. Effects of RFamide-related peptide-3(RFRP-3) on secretion of LH in ovariectomized prepubertal gilts. *J Anim Sci Suppl* 2010;88:566.
189. Anderson G, Relf H, Rizwan M, Evans J. Central and peripheral effects of RFamide-related peptide-3 on luteinizing hormone and prolactin secretion in rats. *Endocrinology* 2009;150:1834–1840.
190. Thorson J, Prezotto L, Caraty A, Amstalden M, Williams G. RF9, a potent antagonist of RFRP3 receptor signaling, disinhibits secretion of LH in the seasonally anovulatory mare. In: ENDO 2012 Abstracts. Endocrine Reviews 33. Communication presented at ENDO 2012, Houston, USA.
191. Pineda R, Garcia-Galiano D, Sanchez-Garrido M et al. Characterization of the potent gonadotropin-releasing activity of RF9, a selective antagonist of RF-amide-related peptides and neuropeptide FF receptors: physiological and pharmacological implications. *Endocrinology* 2010;151:1902–1913.
192. Rizwan M, Poling M, Corr M et al. RFamide-related peptide-3 receptor gene expression in GnRH and kisspeptin neurons and GnRH-dependent mechanism of action. *Endocrinology* 2012;153:3770–3779.

Chapter 24

Ovarian Follicular and Luteal Dynamics in Cattle

Gregg P. Adams and Jaswant Singh

*Veterinary Biomedical Sciences, Western College of Veterinary Medicine,
University of Saskatchewan, Saskatoon, Saskatchewan, Canada*

Introduction

The intent of this chapter is to provide an overview of our present understanding of ovarian function in cattle, with emphasis on the dynamics and control of follicular and luteal gland development. The following sections are arranged, more or less, in chronological order beginning with fetal ovarian development and follicle assembly, early folliculogenesis and oogenesis, antral follicle dynamics during the estrous cycle, ovulation, and development of the corpus luteum. A following section includes a description of ovarian dynamics during different reproductive states (i.e., prepubertal and peripubertal periods, pregnancy and the postpartum period, and during old age). Lastly, reference is made to implications of our new-found knowledge of ovarian function on clinical and breeding management practices in cattle, particularly on the design of modern protocols for ovarian synchronization and superstimulation.

Ovarian development and follicle assembly

The gross, ultrasonographic, and histologic morphology of the mature bovine ovary is illustrated in Figures 24.1 and 24.2. Similar to most mammalian species (except the horse), the bovine ovary is arranged with an outer cortex containing the transient corpus luteum and thousands of follicles at differing stages of development, and an inner medulla composed primarily of connective tissue, vessels, nerves, and the rete ovarii.

The ovarian primordium develops from an elongated gonadal ridge on the ventral aspect of the embryonic kidney (mesonephros). Recent immunohistochemical studies of bovine ovaries suggest that the ovarian primordium does not initially have a distinct surface epithelium with a basement membrane separating it from the underlying stroma, but instead is composed of a cluster of "gonadal ridge epithelial-like" (GREL) cells[1] (Figure 24.3). The GREL cells likely develop from the surface epithelium of the mesonephros and are precursors of both granulosa cells of future follicles and the surface epithelium of the future ovary. Primordial germ cells (PGC) migrate from the yolk sac/developing hindgut to the dorsal celomic wall of the embryo via the dorsal mesentery to colonize the gonadal ridge. The PGC undergo massive mitotic proliferation and intermingle with the resident GREL cells. Invading fingers of stroma ("cell streams") from the underlying mesonephros penetrate the gonadal ridge/ovarian primordium and separate the clumps of PGC and associated GREL cells into ovigerous cords that radiate from the surface epithelium toward what will become the ovarian medulla at the base of the ovary.[2,3] The developing medulla contains stroma, blood vessels, and mesonephric tubules that will persist in the adult ovary as the rete ovarii. The penetrating stroma has a basal lamina at its leading edge, hence the ovigerous cords become surrounded by a basal lamina. As the ovary develops, the ovigerous cords break down into smaller groups of PGC/GREL cells resulting in the formation of primordial follicles containing an oocyte surrounded by flattened granulosa cells separated from surrounding tissue by a basal lamina. After forming a primordial follicle, mitotic division of the oogonium ceases, it enlarges, and becomes arrested in the early stages of Meiosis I. The first follicles form close to the medulla while the outer cortex still contains shorter ovigerous cords. The gradual progression of follicle formation extending from the medulla to the surface has been observed in fetal ovaries of human, cattle and sheep, and in postnatal rat ovaries (reviewed in Ref. 1). At later stages of ovarian development, when preantral and antral follicles appear, the cells on the ovarian surface become single-layered and the stroma underlying the surface thickens to become the tunica albuginea.

Folliculogenesis and oogenesis

Folliculogenesis is the developmental process in which an activated primordial follicle develops to a preovulatory size through the growth and differentiation of the oocyte and

---

Bovine Reproduction, First Edition. Edited by Richard M. Hopper.
© 2015 John Wiley & Sons, Inc. Published 2015 by John Wiley & Sons, Inc.

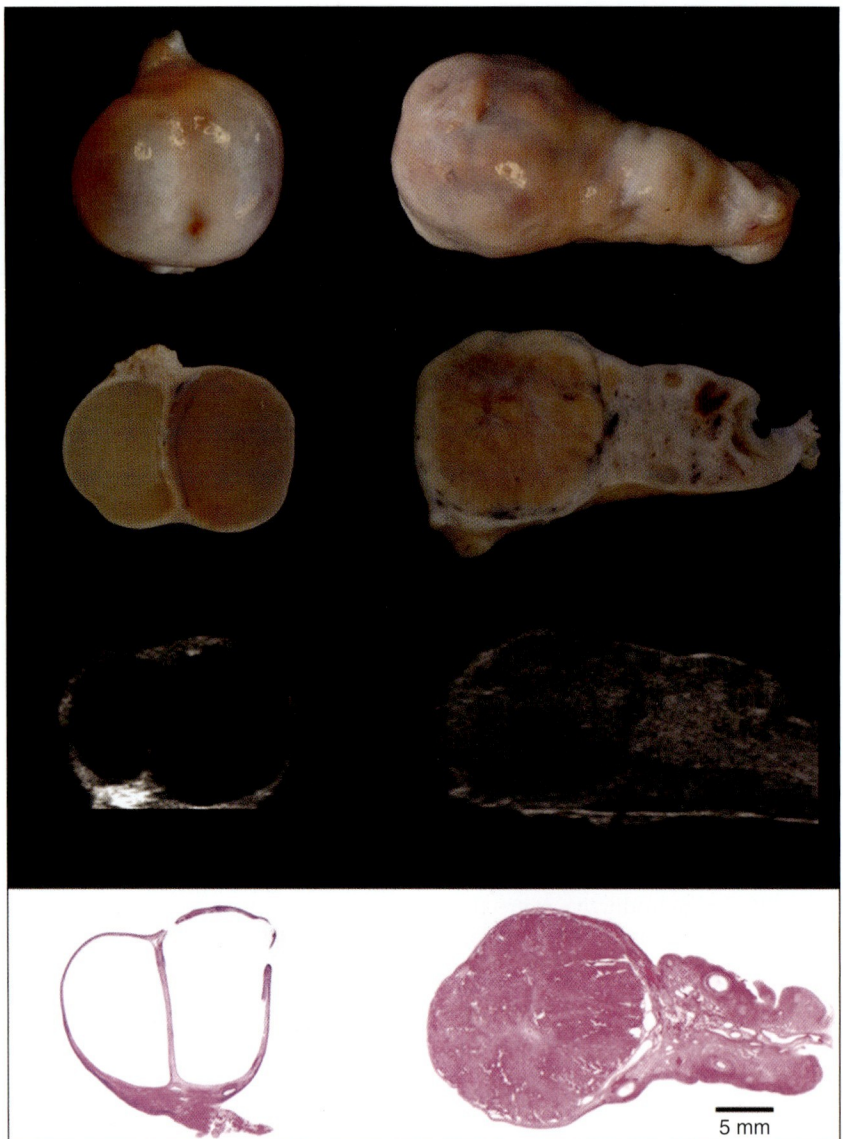

Figure 24.1 Left and right ovaries from a cow in proestrus (top to bottom: surface view, cut surface, ultrasound image, and histological section stained with H&E). In the left ovary, two large antral follicles are apparent side by side, a regressing former dominant follicle (on the left), and the preovulatory dominant follicle (on the right). In the right ovary, a regressing corpus luteum is the predominant structure. The scale bar is 5 mm for all images.

its surrounding granulosa cells.[4–6] Ovarian follicles have been classified in various ways based on morphologic characteristics, such as the number of granulosa cell layers surrounding the oocyte, the morphologic features of the granulosa cells, the diameter of the oocyte and the follicle, and the absence or presence of a fluid-filled antrum, but in general may be categorized as primordial, primary, secondary, or tertiary (antral or vesicular) (Table 24.1).[7–9]

Initiation of follicular growth, referred to as activation, begins with the transformation of the flattened pre-granulosa cells of the primordial follicle into a single layer of cuboidal granulosa (follicular) cells, at which point the follicle is referred to as a primary follicle.[8,10] Proliferation of granulosa cells results in an increase in the number of layers around the oocyte. A follicle with two to six layers of granulosa cells is referred to as a secondary follicle, and a follicle with more than six layers of granulosa cells and a fluid-filled antrum is referred to as a tertiary (or antral) follicle.[7,8] The diameter of a primordial follicle is about 0.04 mm and the diameter of the smallest antral follicle is 0.25 mm.[7] The largest antral or Graafian follicle becomes the ovulatory follicle following the preovulatory gonadotropin surge.[8] In cattle, primordial, primary, secondary and tertiary follicles first appear at days 90, 140, 210, and 250 of gestation, respectively.[11] The time required for a follicle to grow from the large preantral stage (secondary follicle) to a mature ovulatory size has been estimated to be about 42 days[7] (Table 24.2 and Figure 24.4).

Antral follicle dynamics

While early histologic studies provide insight regarding the morphologic changes that occur during folliculogenesis and an overall impression of the time required, they do not provide an understanding of day-to-day dynamics and control of antral follicle development during the estrous cycle.

Ovarian Follicular and Luteal Dynamics in Cattle

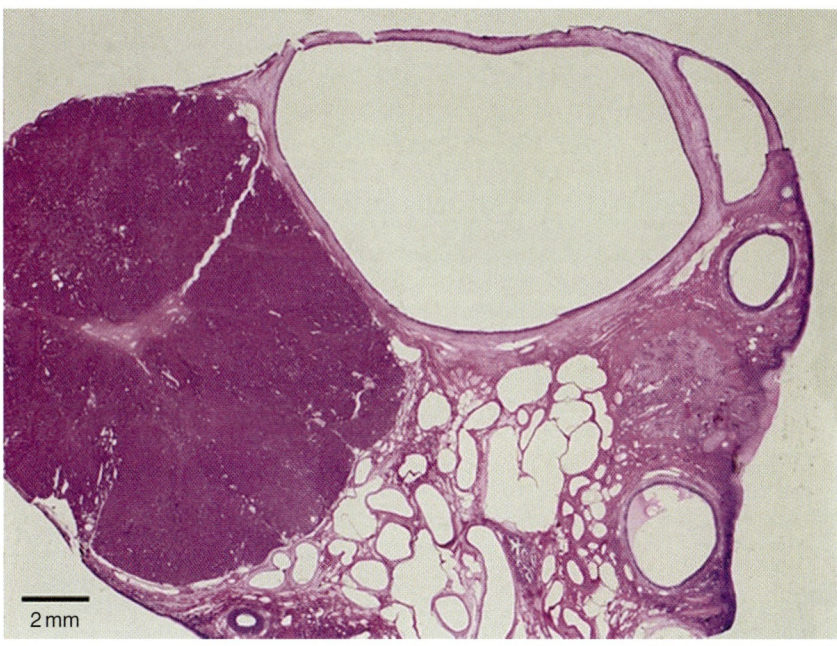

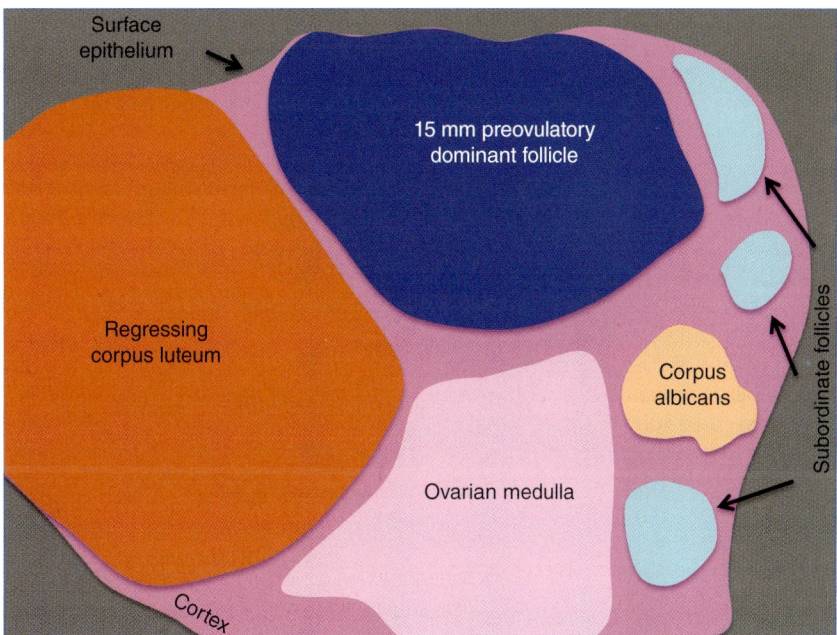

Figure 24.2 Histological section (stained with H&E) and a diagram of the ovary of a cow at day 18 of the estrous cycle (day 0 = ovulation) showing structures of the outer cortex (regressing corpus luteum, preovulatory and subordinate follicles, corpus albicans) and the inner medulla (rete ovarii, blood vessels, connective tissue).

By inference from Rajakoski's review of the early literature,[12] it appears that two schools of thought prevailed at the time (Figure 24.5): (i) follicular growth and regression is a continuous process and is independent of the phase of the reproductive cycle; or (ii) a reservoir of antral follicles remains in a resting stage, and at a particular time in each species one or several of these follicles begin to grow and reach maturity during the ensuing estrus. However, based on gross and histologic evaluation of ovaries obtained from cows slaughtered on known days of the estrous cycle ($n = 1–3$ cows/day), Rajakoski concluded that for antral follicles (>5 mm) two growth waves occurred during the estrous cycle. The first wave of growth was initiated on days 3 and 4 of the estrous cycle (estrus = day 1) and the second between days 12 and 14. His interpretation was that the follicles in each wave that did not ovulate regressed and became atretic.

Contrary to the conclusions of Rajakoski, interpretation of later histological studies supported the notion that follicular growth is continuous and independent of the phases of the cycle.[13–15] Controversy regarding the wave theory of follicular dynamics was evident in later reviews in which the two-wave theory was refuted.[16–18] Evidence was presented to support the concept that follicles are recruited continuously throughout the cycle and the follicle destined to ovulate was a result of coincidence of its stage of maturity (readiness) and the occurrence of the preovulatory gonadotropin surge,

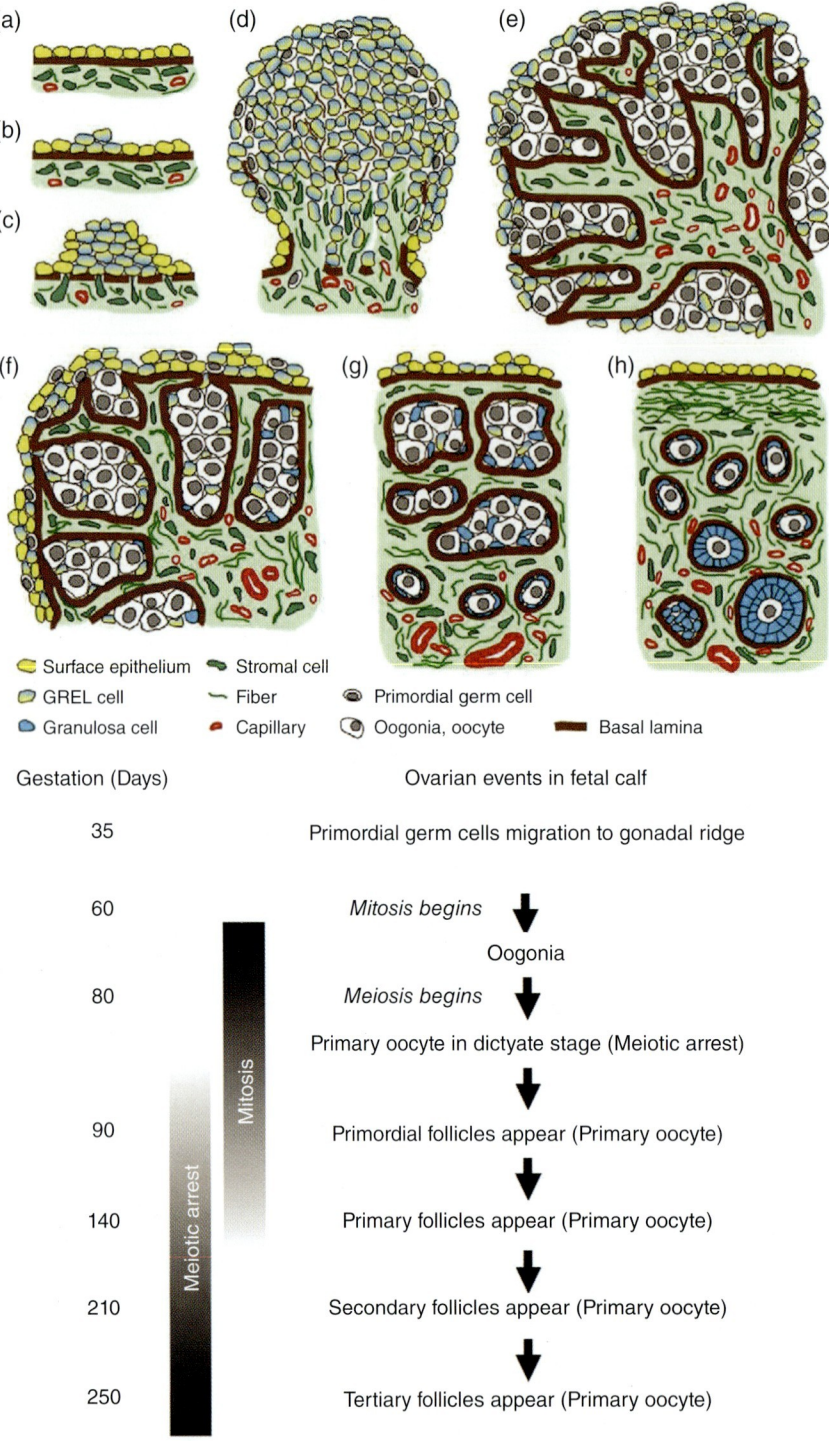

Figure 24.3 Model of bovine fetal ovarian development and follicle assembly (from Hummitzsch et al.[1]) and chronology of early folliculogenesis and oogenesis. (a) Ovary development commences at the mesonephric surface epithelium (yellow) in the location of the future gonadal ridge. (b) Some surface epithelial cells change phenotype into gonadal ridge epithelial-like (GREL) cells (yellow-blue). (c) The GREL cells proliferate and the basal lamina underlying the surface epithelium breaks down allowing stromal cells (green) to penetrate into the gonadal ridge. (d) GREL cells continue to proliferate and primordial germ cells (gray) migrate into the gonadal ridge between the GREL cells. The mesonephric stroma, including vasculature (red), continues to penetrate and expand in the ovary. (e) Oogonia proliferate and the stroma penetrates further toward the ovarian surface, thus enclosing oogonia and GREL cells into ovigerous cords. The cords are surrounded by a basal lamina at their interface with the stroma, but are open to the ovarian surface. (f) Compartmentalization into cortex and medulla becomes obvious. The cortex is characterized by alternating areas of ovigerous cords and stroma, whereas the medulla is composed of stromal cells, vasculature, and tubules originating from the mesonephros (rete ovarii). As the stroma penetrates toward the periphery, the GREL cells at the surface of the gonadal ridge are aligned by a basal lamina at their interface with the stroma and begin to differentiate into typical ovarian surface epithelium (yellow cells). (g) Ovigerous cords are partitioned into smaller cords and eventually into individual primordial follicles as the GREL cells mature to granulosa cells (blue) and surround an oogonium. (h) Finally, the surface epithelium becomes mostly single-layered and a tunica albuginea, densely packed connective tissue, develops from the stroma below the basal lamina of the surface epithelium. Some primordial follicles become activated and commence development into primary and preantral follicles.

Table 24.1 Characteristics of bovine ovarian follicles during development.

Follicle stage	Follicle type	FSH/LH receptors	Granulosa cell layers[a]	Granulosa cells per section[a]	Follicle diameter (mm)	Oocyte diameter (μm)	Zona pellucida	Theca interna
Primordial	1		1	<10 flattened	<0.04	30	Absent	Absent
Transitory	1a		Entered growing pool and surrounded by a mixture of flattened and cuboidal cells					
Primary	2	FSHr form (granulosa)	1–1.5	10–40[b] cuboidal	0.04–0.08	31	Absent	Absent
Secondary	3 (small preantral)		2–3	41–100	0.08–0.13	50	Begins to form	Begins to form
	4 (large preantral)	LHr form (theca)	4–6	101–250[c]	0.13–0.25	69	Partially formed	Partially formed
Tertiary (antral)	5 (small antral)	LHr in granulosa of dominant follicle	>6	>250	0.25–0.5	93	Fully formed	
	Large antral Graafian			40 × 10[6]	10–15	132		

[a]Largest cross-section of the follicle defined by the nucleolus of the oocyte.
[b]Oocyte commenced growth when there were at least 40 granulosa cells in the largest cross-section (fourth generation of follicle cells).
[c]Beginning of antrum formation was observed in follicles with at least 250 granulosa cells in the largest cross-section.
Source: adapted from Lussier et al.,[7] Braw-Tal and Yossefi,[8] and Xu et al.[103]

Table 24.2 Developmental rates of bovine ovarian follicles.

Follicle size range (mm)	Granulosa cell generation	Time in each size class		Percent atretic	Mitotic activity
		Hours	Days		
0.13–0.28	6	365	15.1	1.6	
0.29–0.67	8	285	11.9	6.6	
0.68–1.52	10	100	4.2	40.5	Maximum
1.53–3.67	11	83	3.5	30	
3.68–8.57	12	186	7.8	67.4	
>8.57	13			60	
Total		1019	42.5		

Source: adapted from Lussier et al.[7] and Fortune[184]

i.e., the "propitious moment" theory (Figure 24.5). However, with the introduction of ultrasonography in the late 1980s, the barrier to our understanding of the dynamic process of follicle development was suddenly removed.

The technique of detecting and monitoring ovarian structures by transrectal ultrasonography was described initially in 1984.[19] By comparing the results of ultrasonography with the results of slicing excised ovaries, the ultrasound technique was validated as a tool for detecting follicles greater than 2 mm in diameter and for monitoring the corpus luteum.[20,21] Subsequent studies using ultrasonic imaging to monitor follicle populations in different size categories[22] or to monitor individually identified follicles[23–26] convincingly documented that follicular growth in cattle occurs in a wave-like fashion and that the majority of estrous cycles in cattle comprise two or three such waves. The method by which individual follicles may be monitored throughout one interovulatory interval by ultrasonography is illustrated in Figure 24.6. During daily examinations, a sketch of each ovary is made to record the number, diameter, and relative positions of follicles and the corpora lutea. Even though the ovaries are quite moveable at the end of the mesovarium, their position remains remarkably consistent from one examination to the next, allowing accurate "mapping" of ovarian structures over time. The results of ultrasound studies are consistent with those in which three waves of follicular growth were hypothesized[27] based on the presence of one "estrogen-active" follicle during three different periods of the estrous cycle[28,29] and increases in estradiol in blood from the utero-ovarian veins during approximately the same time periods.[30]

Follicular waves during the estrous cycle

Follicular wave emergence in cattle is characterized by the sudden (within 2–3 days) growth of 8–41 small follicles that are initially detected by ultrasonography at a diameter of 3–4 mm (Figures 24.6 and 24.7).[20,25,26,31,32] The growth rate is similar among follicles of the wave for about 2 days, after which one follicle is then *selected* to continue growth (dominant follicle) while the remainder become atretic and regress (subordinate follicles). In both two- and three-wave estrous cycles, emergence of the first follicular wave occurs

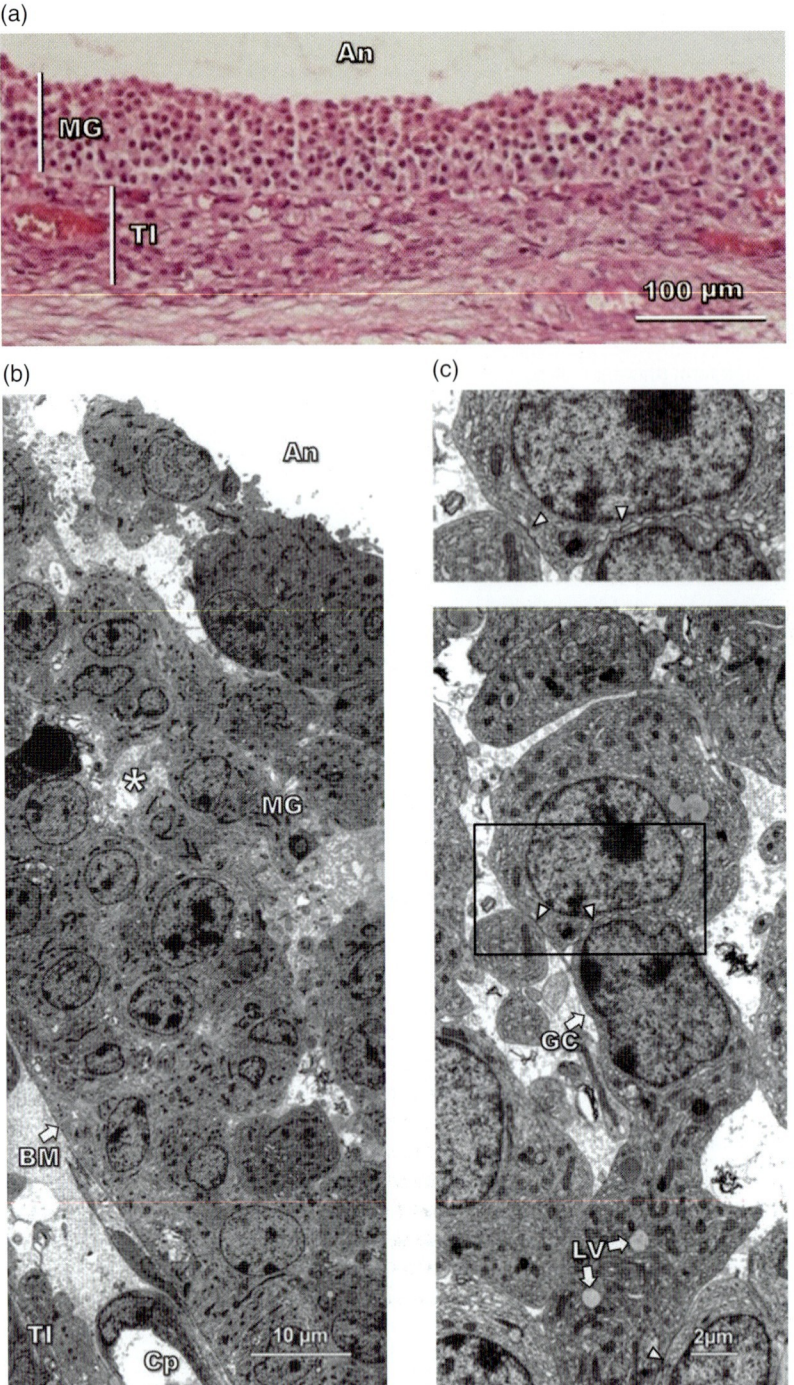

Figure 24.4 Histological section (stained with H&E) and transmission electron micrograph of the wall of the dominant follicle during (a, b) the growing phase (3 days after ovulation) and (c) the early static phase (6 days after ovulation, right) in cattle. The granulosa cells of the growing dominant follicle have many cytoplasmic processes projecting into intercellular spaces. In the early static dominant follicle, most of the granulosa cells are polyhedral (others are elongated) and the intercellular spaces are wide. The granulosa cells contain abundant mitochondria, smooth and roughendoplasmic reticulum, and occasional lipid vacuoles (LV). The area enclosed in the box is magnified in the upper right corner to show that gap junctions (arrowheads) between the cells are frequent. An, antrum; BM, basement membrane; *, intercellular spaces; MG, membrana granulosa; GC, granulosa cells; TI, theca interna; Cp, capillary.

consistently on the day of ovulation (day 0). Emergence of the second wave occurs on day 9 or 10 for two-wave cycles, and on day 8 or 9 for three-wave cycles. In three-wave cycles, a third wave emerges on day 15 or 16. Under the influence of progesterone (e.g., diestrus, pregnancy), dominant follicles of successive waves are anovulatory and undergo atresia.[33] The dominant follicle present at the onset of luteolysis becomes the ovulatory follicle, and emergence of the next wave is delayed until the day of the ensuing ovulation. The corpus luteum begins to regress earlier in two-wave cycles (day 16) than in three-wave cycles (day 19), resulting in a correspondingly shorter estrous cycle (20 days vs. 23 days, respectively). Hence, the proverbial 21-day estrous cycle of cattle exists only as an average between two- and three-wave cycles.

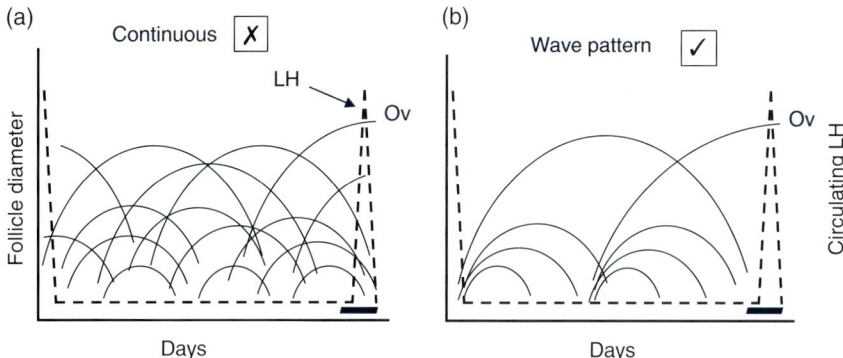

Figure 24.5 Theories of ovarian follicular dynamics during the estrous cycle in cattle, illustrated from one ovulation to the next (i.e., one interovulatory interval). Estrus is indicated by the solid bar along the *x*-axis and plasma LH concentrations by the dashed line. (a) The theory of continuous recruitment or "propitious moment" is that antral follicles grow and regress continuously throughout the interovulatory interval and the ovulatory follicle is the one that by happenstance is at the appropriate stage of development to respond to the preovulatory surge of LH. (b) The wave theory of follicle recruitment is that two or more cohorts (i.e., waves) of antral follicles are recruited (i.e., emerge) during the ovarian cycle, regardless of the presence of a corpus luteum (follicular phase or luteal phase). The dominant follicle that develops in the final wave of the interovulatory interval ovulates while preceding waves are anovulatory. Modified from Adams G, Singh J, Baerwald A. Large animal models for the study of ovarian follicular dynamics in women. *Theriogenology* 2012;78:1733–1748.

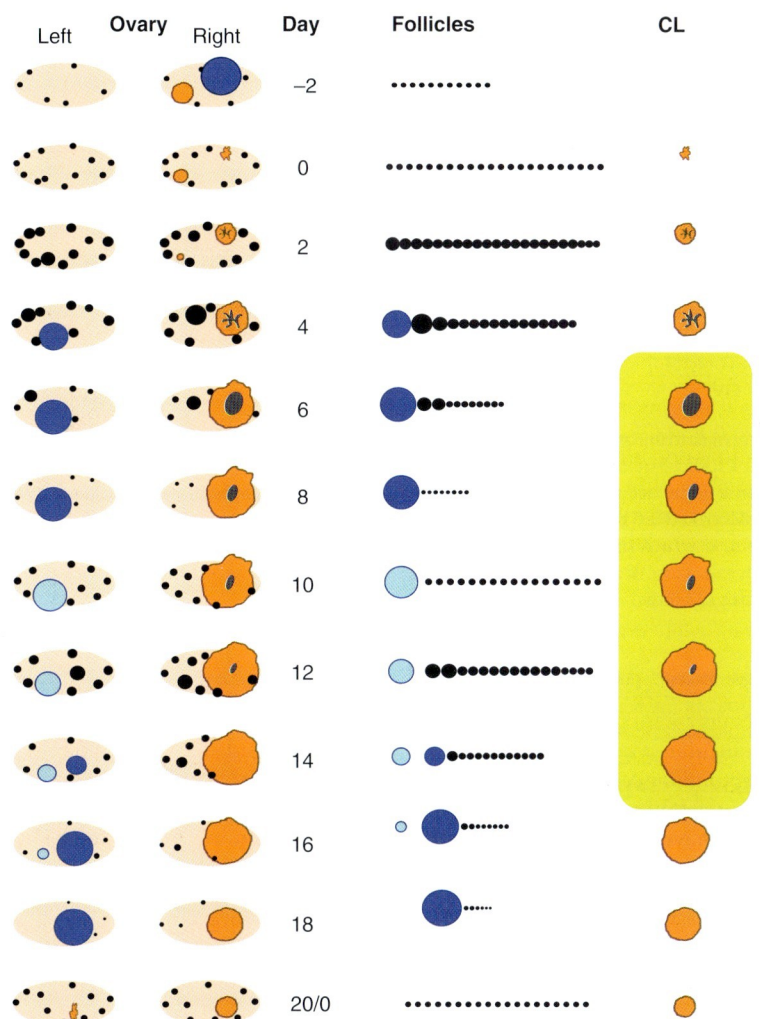

Figure 24.6 Mapping ovarian follicular and luteal dynamics in cattle by serial ultrasonography. During daily examinations (day 0=ovulation), a sketch of each ovary is made to record the number, diameter, and relative positions of follicles and the corpus luteum. The data are used to identify the number of follicular waves during the interovulatory interval (two waves in this example), follicular wave emergence (days 0 and 10), the dominant follicle of each wave (growing, dark blue; regressing, pale blue), the preselection and subordinate follicles of each wave (black), the occurrence of ovulation (days 0 and 20), and the growth and regression of the corpus luteum (orange). Note that both ovaries contribute to a given follicular wave, and successive dominant follicles may be in either ovary (random). The size and number of ovarian structures in each sketch are tabulated under the respective follicle and corpus luteum columns. The high progesterone (luteal) phase is outlined in the yellow box.

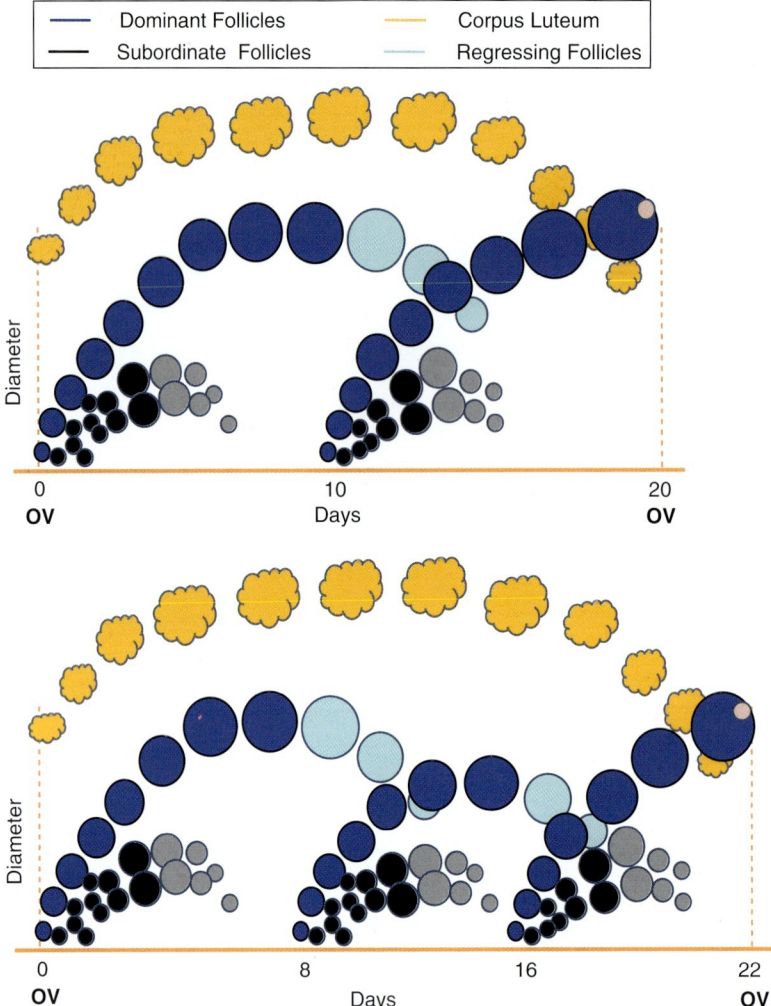

Figure 24.7 Follicular dynamics during two-wave (upper) and three-wave (lower) estrous cycles in cattle. Data from both ovaries are used to identify the dominant follicle of each wave (growing, dark blue; regressing, pale blue), the subordinate follicles of each wave (growing, black; regressing, gray), the occurrence of ovulation, and the growth and regression of the corpus luteum (dark yellow). A follicular wave in cattle is characterized by the sudden emergence (within 2–3 days) of a group of small follicles ($n = 8–41$) that grow at a similar rate for about 2–3 days, at which time one follicle is selected to continue growth (dominant follicle) while the rest become atretic and regress (subordinate follicles). The often referred to "21-day" estrous cycle of cattle exists only as an average between two-wave cycles (20 days) and three-wave cycles (22 days).

Two-wave versus three-wave patterns

The majority of bovine estrous cycles (>95%) are composed of either two or three follicular waves (Figure 24.7).[34,35] Some have reported a preponderance (>80%) of either the two-wave or the three-wave pattern during an interovulatory interval, while others have reported a more even distribution. There appears to be no breed- or age-specific predilection for a given wave pattern in *Bos taurus*. However, an increase in the proportion of three-wave patterns has been associated with a low plane of nutrition[36,37] and heat stress.[38,39] In *Bos indicus*, no seasonal effect on wave pattern was detected,[40] but the pattern was influenced by parity. The majority of Nelore heifers (65%) exhibited a three-wave pattern, whereas the majority of cows (83%) exhibited a two-wave pattern.[41] Others have reported that up to 27% of estrous cycles in *Bos indicus* cows consist of four waves of follicular development, compared with only 7% in *Bos indicus* heifers (reviewed in Ref. 42).

Predictive factors associated with a two- versus three-wave pattern may provide insight into mechanisms controlling the pattern, and have important implications for breeding management and the development of effective protocols for ovarian synchronization, superstimulation, and fixed-time artificial insemination. Pregnancy rates in cattle with two- versus three-wave patterns were compared[43–45] based on the notion that the preovulatory follicle in the two-wave pattern grows for a longer period[31] and may therefore contain a relatively aged oocyte. However, results have been contradictory; pregnancy rates did not differ between two- versus three-wave cycles in some studies,[43,45] whereas a lower pregnancy rate following ovulation at the end of the two-wave pattern was reported in another study.[44]

In a more recent study involving ultrasonographic data from 91 interovulatory intervals,[35] two- and three-wave patterns of follicular development were compared to determine the repeatability and predictive characteristics of a given wave pattern. Two-wave cycles were 3 days shorter than three-wave cycles (19.8 ± 0.2 vs. 22.5 ± 0.3; $P<0.01$). The majority of cycles of 21 days or less were of the two-wave pattern (88%; $P<0.05$), while the majority of cycles 22 days or longer were of the three-wave pattern (78%; $P<0.05$). The proportion of successive cycles in which the pattern remained the same (i.e., repeatability) was more than twofold greater than the proportion of cycles that changed patterns (70% vs. 30%; $P<0.01$). The repeatability of wave pattern, and the proportion of two- versus three-wave patterns within the herd were not affected by the season of the year; in other words not affected by pasture-based (spring and summer) versus non-pasture-based (fall and winter) seasons. Interestingly, the strongest correlate to the number of waves in the interovulatory interval was the duration of follicular dominance of Wave 1. The duration of dominance of the Wave 1 dominant follicle was 3 days longer, and the onset of regression was later in two-wave patterns than in three-wave patterns ($P<0.01$). Dominance of Wave 1 was associated with a subsequent delay in the attainment of maximum diameter by the dominant follicle of Wave 2, as well as early onset of luteolysis (Figure 24.7). Results supported the hypothesis that factors which influence the development of the dominant follicle of Wave 1 are responsible for regulating the wave pattern.

Two ovaries, one wave

The two ovaries act as a single unit – one ovary does not have follicular waves independent of the other. The ovaries respond in unison to factors regulating follicular wave dynamics, hence a follicular wave is the result of multiple follicle growth in both ovaries (Figures 24.6 and 24.7). In a critical study of the intraovarian relationships among follicles and the corpus luteum, Ginther et al.[46] concluded that the dominant follicle suppresses subordinates and new wave emergence via systemic rather than local channels. The presence of a dominant follicle in one ovary had no effect that could not be seen equally in the other ovary. Hence, no intraovarian relationships were found in the location of successive dominant follicles during the estrous cycle; in other words, the side of anovulatory and ovulatory dominant follicles was random. These observations are consistent with the findings of early studies on the effects of unilateral ovariectomy. Removal of the ovary with the largest follicle resulted in an increase in follicle development in the remaining ovary, but removal of the ovary without the largest follicle resulted in no such compensatory effect (cited in Ref. 46). While intrafollicular (autocrine and paracrine) factors are important for growth, health, and demise of an individual follicle, there is no convincing *in vivo* documentation of one follicle affecting the health/regression status of its neighbors directly.

Asymmetry in follicle dynamics in the left and right ovaries has been used to elucidate local versus systemic mechanisms of control of ovarian function, and some have reported greater follicular activity and proportion of ovulations in the right ovary in cattle (~60%),[12,22] whereas others report no such differences.[26,31] Ovarian follicular asymmetry has also been attributed to the presence of a corpus luteum. A positive intraovarian effect of the corpus luteum on the development of small antral follicles (≤3 mm) has been doc-umented in sheep[47] and cattle,[21] but the positive effect did not extend to large dominant follicles.[21,46] Conversely, the corpus luteum of pregnancy has been associated with a negative intraovarian effect on the dominant follicle.[46,48] Dominant follicles of successive waves were more frequently (75–80%) found in the ovary contralateral to the corpus luteum during pregnancy. The cause of the negative local association between the corpus luteum and the follicles is unknown but it may be directly related to the conceptus rather than to the corpus luteum. Support for this notion is found in the observation that there were no ovarian asymmetries during the first two follicular waves in pregnant cattle or in 10 successive follicular waves in unmated progesterone-treated heifers.[33] The latter observation, however, is confounded by the fact that exogenous progesterone did not forestall luteolysis in treated heifers; luteal regression occurred at the expected time for a nonpregnant animal regardless of progesterone treatment. However, an experimental design involving hysterectomy during mid-diestrus permitted direct comparison of the effects of a persistent corpus luteum due to hysterectomy or pregnancy.[49] Results demonstrated that unilateral attenuation of follicle development was due to the conceptus or the ipsilateral gravid uterine horn, and not the corpus luteum.

Hormonal interplay controlling follicular wave dynamics

Results of early studies of follicle dynamics gave rise to the hypothesis that the dominant follicle suppresses the growth of the subordinates in the existing wave, and suppresses the emergence of the next follicular wave.[31,32] Support for this hypothesis was provided in a series of studies involving systemic treatment with the proteinaceous fraction of follicular fluid and by electrocautery of the dominant follicle.[50–52] The applied implications of these findings were immediate and far-reaching, and marked a new era for ovarian synchronization and superstimulation in cattle.[53–55]

Emergence of a follicular wave and selection of the dominant follicle are temporally and causally associated with a rise and fall in circulating concentrations of follicle-stimulating hormone (FSH)[51] (Figure 24.8). Emergence of a follicular wave is preceded by a surge in plasma FSH concentrations in both spontaneous waves and induced waves.[51,56] Surges in FSH preceding each wave are of similar magnitude: mid-cycle surges are similar in amplitude and breadth as those of the preovulatory gonadotrophin surge.[51,56] Follicular products, especially those from the dominant follicle, are responsible for suppressing FSH release and therefore the emergence of the next follicular wave[57] (Figure 24.9). A direct relationship between the number of follicles emerging in response to the FSH surge and the subsequent degree of FSH suppression[58–60] documents that all follicles of the new wave contribute to FSH suppression.

Estradiol and inhibin-A and -B are the principal follicular products responsible for suppressing FSH.[57,61–65] Inhibin-A is produced by all the small growing follicles of the wave and appears to be the most important suppressor of FSH during the first 2 days of wave emergence; estradiol secreted from the dominant follicle is the most important FSH suppressor thereafter.[66–69] The nadir in FSH is reached 4 days after wave emer-

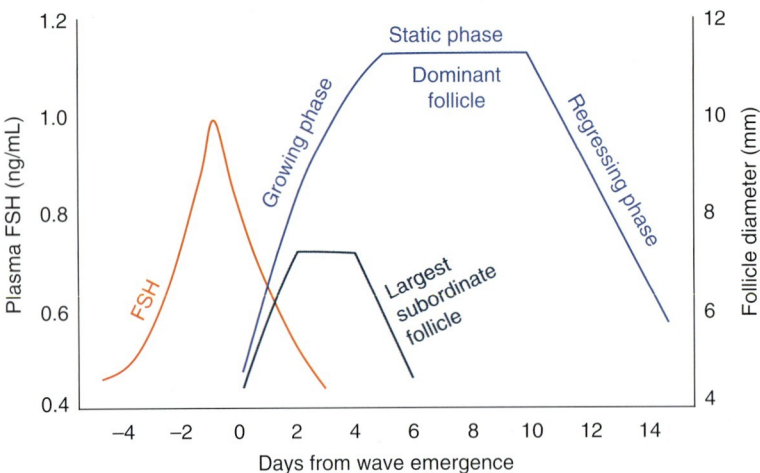

Figure 24.8 Temporal and causal relationship between surges in circulating concentrations of FSH and follicular wave emergence in cattle. Peak concentrations of FSH occur on the day before follicular wave emergence (at 4 mm) and decline to nadir by 3 days after wave emergence at the time the dominant follicle can be distinguished from the subordinates. Development of a dominant or subordinate follicle may be divided into a growing phase (best characterized by a quadratic equation), a static phase, and a regressing phase (best characterized by simple linear equations). Data from Adams et al.[51, 53]

gence and levels remain low for next 2–3 days. At the end of the period of dominance (i.e., at ovulation, or the mid-static phase of an anovulatory dominant follicle), circulating concentrations of FSH begin to rise over the next 2 days and peak about 12–24 hours before emergence of the next follicular wave, when the future dominant follicle is 4–5 mm in diameter. If an existing dominant follicle is removed (i.e., follicular ablation), a surge in FSH begins within the next 12 hours and results in emergence of a new follicular wave within the next 24 hours.[52,70] Interestingly, FSH only rises again when inhibin-A concentrations reach nadir, despite the fact that estradiol declines 2 days earlier.[63]

An inverse relationship between circulating progesterone concentrations and dominant follicle growth was suggested by the observations that (i) in three-wave estrous cycles, the maximum diameter of the dominant anovulatory follicle was greater for Wave 1 (under minimal luteal dominance during its growing phase) than for Wave 2 (under maximum luteal dominance); (ii) the maximum diameter attained by the dominant anovulatory follicle was greater for Wave 1 than for all subsequent waves in pregnant and nonmated progesterone-treated heifers; and (iii) high concentrations of progesterone (endogenous plus exogenous) resulted in a smaller maximum diameter of the dominant follicle of Wave 2 compared with subsequent waves (exogenous progesterone only) in nonmated progesterone-treated heifers (reviewed in Ref. 71). A cause-and-effect relationship was established in studies in which progesterone was shown to suppress the dominant follicle during its growing phase in a dose-dependent manner[71] through the negative feedback effect of progesterone on luteinizing hormone (LH) pulsatility.[72–74] The suppressive effects on the dominant follicle were not mediated by decreased circulating FSH since progesterone treatment did not suppress FSH.[71] High plasma progesterone during follicular growth resulted in a smaller shorter-lived dominant follicle, an earlier surge in FSH preceding the next follicular wave, and hastened emergence of the next follicular wave (Figure 24.10). Conversely, low concentrations of progesterone resulted in an oversized persistent dominant follicle, a delayed surge in FSH, and delayed emergence of the next wave.[71] The oversized dominant follicle in the low-dose situation would have been considered a pathological follicular cyst because of its size and persistence for more than 10 days in the absence of a corpus luteum.[75,76] Low-level progesterone exposure (and therefore less inhibition of LH), at the time the dominant follicle normally stops growing, appears to be a fundamental component in the etiology of ovarian follicular cysts.[71]

Selection of the dominant follicle

In monovular species, the term "selection" refers to a process by which a single follicle becomes functionally and morphologically dominant over other follicles of a wave. In cattle, the mechanism of selection of the dominant follicle is based on relative reliance of follicles within a wave on FSH, and their relative responsiveness to LH. The transient rise in FSH permits sufficient follicular growth so that some (not all) follicles acquire LH responsiveness. The ability to respond to LH imbues the follicle with the ability to survive without FSH.

It remains unclear precisely when the process begins, but selection of the dominant follicle is associated with declining levels of FSH in circulation during the first 3 days of the wave.[51,77] Small follicles of the emerging wave (<6 mm) are dependent on elevated circulating concentrations of FSH for continued development. In the face of the post-surge decline in FSH, the growth of most of the follicles of a wave stops and they begin to regress within 2–5 days of emergence. Conversely, the follicle destined to become dominant can maintain cell proliferation and estradiol production despite declining FSH concentrations, an ability that appears to be brought about by maintaining high FSH receptor mRNA expression and FSH-binding affinity.[28,29,78,79]

Divergence in the growth profiles (defined as a significant difference in diameter) of the largest and second largest follicle is a manifestation of the selection process and is

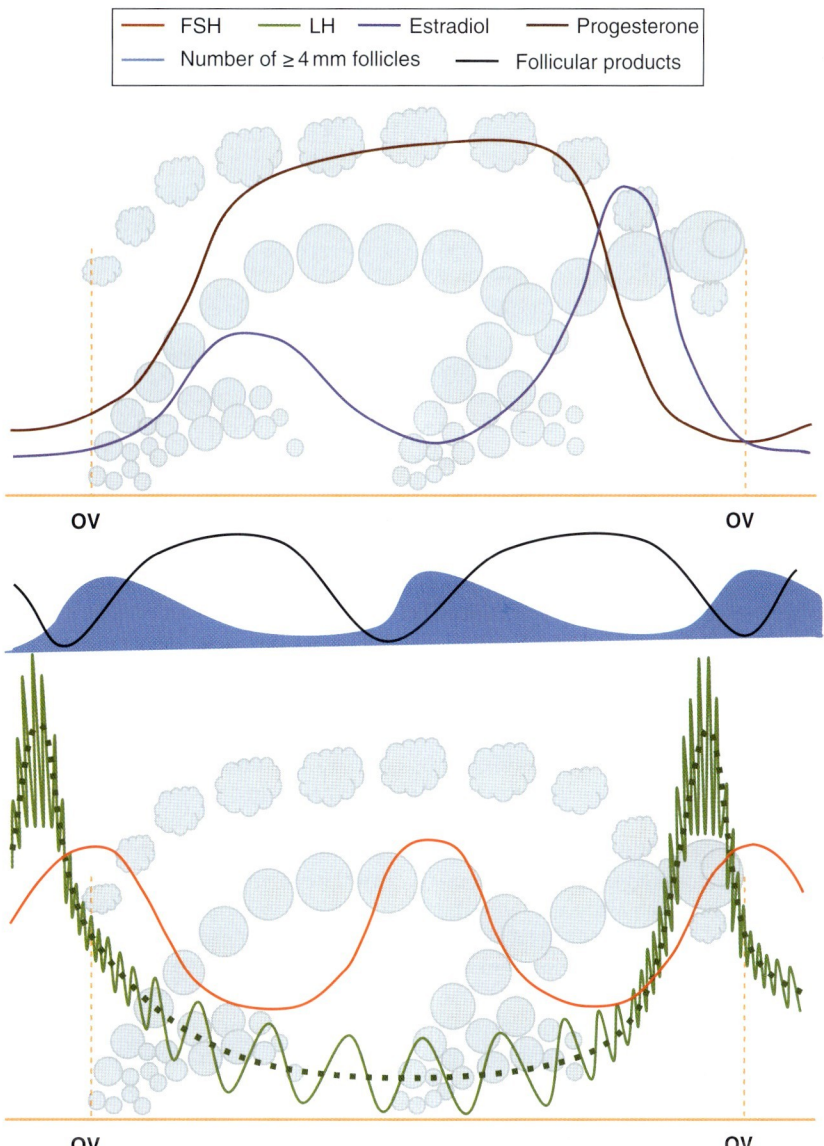

Figure 24.9 Hormonal interplay controlling follicular wave dynamics during a two-wave interovulatory interval in cattle. Dominant and subordinate follicles of each wave are indicated as gray-shaded circles, and the corpus luteum is the gray-shaded irregular shape drawn above the follicles. A surge in circulating concentration of FSH precedes emergence of each wave. A surge in circulating concentration of LH precedes ovulation. The LH surge is preceded and succeeded by a period of high LH pulse frequency as a result of low circulating concentrations of progesterone (i.e., period of luteolysis and luteogenesis, respectively). The number of follicles ≥4 mm fluctuates in a wave-like fashion in response to surges of FSH. Products of the follicles of each wave (primarily inhibin and estradiol) feed back on the hypothalamus and pituitary to suppress FSH. Concentrations of FSH decline at the time that selection of divergence in growth rate of the dominant versus subordinate follicles, and follicle numbers decline. The dominant follicle of each wave is the primary source of estradiol, and production is twice as high in the ovulatory versus anovulatory dominant follicle. Upon the demise or ovulation of the dominant follicle, the follicular products responsible for suppressing FSH drop, and circulating concentrations of FSH are again allowed to surge resulting in new wave emergence. Adapted from Adams G, Jaiswal R, Singh J, Malhi P. Progress in understanding ovarian follicular dynamics in cattle. *Theriogenology* 2008;69:72–80.

apparent by 2.5 days after wave emergence (Figures 24.8 and 24.9).[77] Similarly, follicle "deviation" (defined as a difference in growth rate between the dominant and subordinate follicles) is manifest at 2.5 days after wave emergence, when the future dominant follicle reaches a mean diameter of 8.5 mm in Holstein heifers (reviewed in Ref. 80). Continued development of the dominant follicle, beyond the 8-mm threshold (point at which divergence in the growth rate of the largest two follicles becomes evident), is associated with a transition from FSH to LH dependence.[66,68,81] Granulosa cells from the dominant follicle acquire an enhanced ability to bind LH compared with that of the subordinates, and have greater LH receptor mRNA expression.[29,78,79,82,83] Although a slight increase in LH receptor mRNA expression was detected in granulosa cells from the largest follicle a few hours before the onset of deviation,[84] it is not yet clear whether the observed increase in LH receptor expression in the dominant follicle is the cause of dominance or a consequence of the selection process. Perhaps the increased LH receptor expression is merely a reflection of a follicle with a developmental advantage before such an advantage becomes ultrasonographically apparent. In this

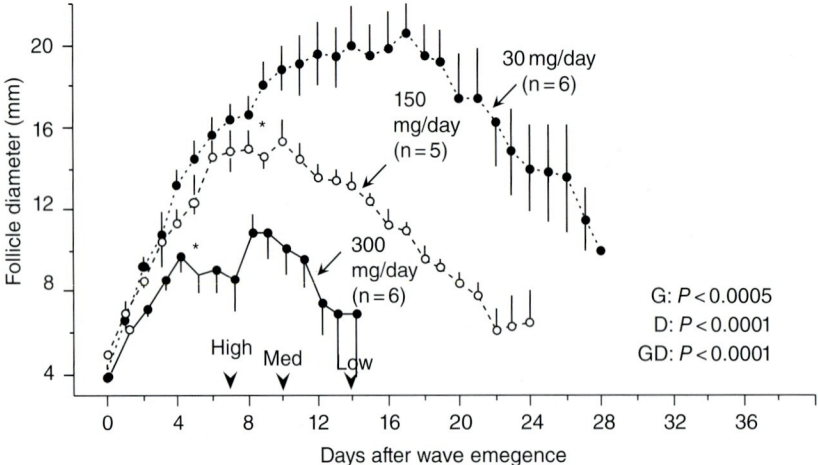

Figure 24.10 Suppressive effect of progesterone on the growth of the dominant follicle in cows. Exogenous progesterone administered in low (30 mg/day), medium (150 mg/day), and high (300 mg/day) doses for 14 days resulted in progressively greater suppression of the growth of the dominant follicle and progressively shorter interwave intervals (arrowheads on x-axis). Maximum plasma progesterone concentrations were 2, 9, and 17 ng/mL in the low-, medium- and high-dose groups, respectively. The suppressive effect on the growing phase of the dominant follicle is attributed to progressively more suppression on LH pulse frequency. Adapted from Adams G, Matteri R, Ginther O. The effect of progesterone on growth of ovarian follicles, emergence of follicular waves and circulating FSH in heifers. *J Reprod Fertil* 1992;96:627–640.

regard, results of a critical study of the development of small follicles support the hypothesis that the follicle destined to become dominant has a size advantage over the others in the wave from its earliest detection at 1 mm (Figure 24.11) (see section Small antral follicles).

Differential production of follicular-fluid factors of the dominant versus subordinate follicles has been associated with the development of greater sensitivity to LH and FSH in the developing dominant follicle, and close temporal coupling between changes in FSH and follicle diameter likely accounts for the increase in growth rate of the dominant follicle before the subordinate follicles can reach a similar stage of development.[66] In this regard, intrafollicular concentrations of insulin-like growth factor (IGF)-1 remain elevated and estradiol begins to increase in the incipient dominant follicle while both begin to decrease in the incipient subordinate follicles.[85] Higher concentrations of free IGF-1 in the largest follicle near the beginning of deviation are temporally associated with greater proteolytic degradation of IGF-binding protein (IGFBP)-4 and -5 in the largest follicle.[86] Concomitantly, the concentrations of IGFBP-4, IGFBP-5,[86] and IGFBP-2[84] are greater in the second-largest follicle. Not surprisingly, more recent transcriptomic studies have confirmed that major functional pathways in the growth and selection of dominant follicles in cattle involve genes regulating cell differentiation/proliferation and apoptosis/cell death.[82,87–89]

Codominance

Subordinate follicles can achieve dominance if the original dominant follicle is removed[58,77] or if exogenous FSH is supplied.[52] Similarly, codominance (more than one dominant follicle in a wave) is associated with a smaller diameter profile,[52,90] suggesting competition over limited availability of LH. Double ovulations result from codominance during the ovulatory wave. The incidence of codominance and double ovulation was low or nil in studies of nonlactating heifers,[26,31,91–93] but codominance ranged from 21 to 36% of follicular waves in lactating dairy cows.[93] A high incidence (10–39%) of multiple ovulations in high-producing dairy cows has been described in several studies.[90,94,95] In cows that ovulated multiple follicles, all ovulatory follicles emerged from the same follicular wave.[93] In cows that developed multiple dominant follicles, the process of deviation was associated with lower circulating concentrations of inhibin and progesterone compared with cows that developed a single dominant follicle.[90] These effects are consistent with higher circulating concentrations of FSH and LH at the expected time of deviation, continued growth of the second and third largest follicles of the wave, and development of multiple dominant follicles.

Small antral follicles: when do follicles join a wave?

Previous ultrasonographic studies have detailed the developmental dynamics of follicles 4 mm or larger (described above), but with the availability of newer ultrasound scanners capable of resolving structures as small as 1 mm, the developmental pattern of 1–3 mm follicles in cattle has been examined in detail.[96] Results revealed a change over days in the number of 1–3 mm follicles, with a maximum 1 or 2 days before conventionally defined wave emergence (i.e., when the dominant follicle is first detected at 4 mm), followed 3–4 days later by a maximum in the number of 4-mm or larger follicles. The future dominant follicle was first identified at a diameter of 1 mm and emerged 6–12 hours earlier than the first subordinate follicle ($P<0.01$; Figure 24.11). After detection of the dominant follicle at 1 mm (0 hours), its diameter was greater than that of the first and second subordinate follicles at 24 hours and 12 hours, respectively, when the dominant follicle was 2.4 ± 0.17 mm and 1.7 ± 0.14 mm (Figure 24.11). The growth rate of the dominant follicle was greater than that of the first and second subordinate follicles at 120 and 108 hours, respectively, when the dominant

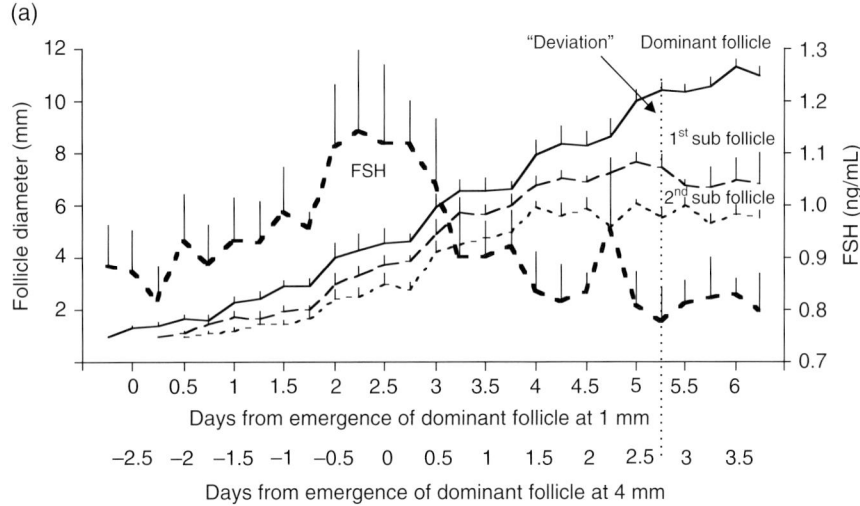

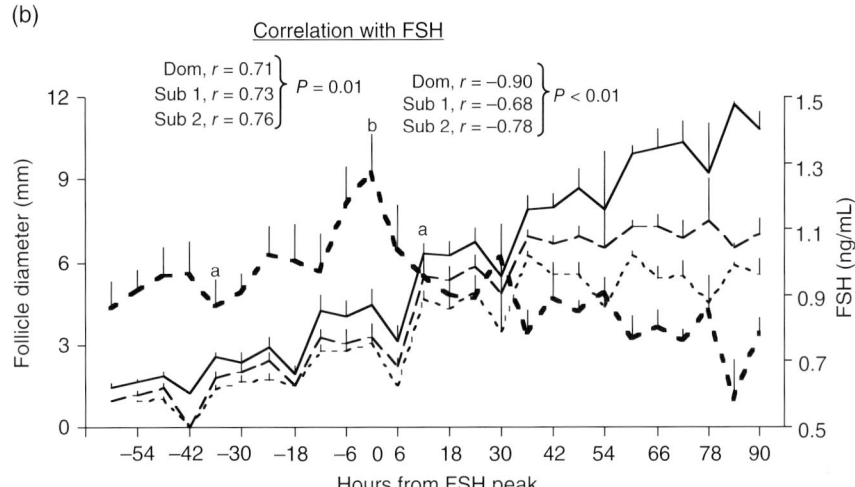

Figure 24.11 Growth (mean ± SEM) of dominant and subordinate follicles in cattle ($n=9$) relative to (a) wave emergence (defined as the day on which the dominant follicle was detected at 1 or 4 mm in diameter; double x-axis), and (b) the peak in plasma FSH concentrations (mean ± SEM). (a) The future dominant follicle had a size advantage over all other follicles of the wave from its earliest detection at 1 mm (much earlier than previously documented). Successive slowing of the growth rate of the third, second, and first largest follicles of a wave during the decline in the FSH surge suggests that the selection mechanism involves sequential suppression of progressively larger follicles over a period of 72 hours (i.e., selection hierarchy). "Deviation" is defined as the first day on which the growth rate of the largest and second largest follicles differed ($P \leq 0.05$). (b) Plasma FSH concentrations were positively correlated ($P=0.01$) with follicle diameter from the time of follicle detection at 1 mm to the time at which FSH concentrations peaked. Transient elevations in plasma FSH concentration were followed within 6 hours by an increase in the growth rate of 1–3 mm follicles. For FSH, values with no common superscript (a, b) are different ($P \leq 0.05$). Adapted from Jaiswal R, Singh J, Adams G. Developmental pattern of small antral follicles in the bovine ovary. *Biol Reprod* 2004;71:1244–1251.

follicle was 9.5 ± 0.30 mm and 8.8 ± 0.49 mm. The authors concluded that:

- 1–3 mm follicles develop in a wave-like manner in association with surges in plasma concentrations of FSH;
- 1–3 mm follicles are exquisitely responsive to transient elevations in FSH (i.e., within 6 hours);
- selection of the dominant follicle is manifest much earlier than previously documented;
- follicular dominance is characterized by a hierarchical progression over a period encompassing the entire FSH surge (5 days) (Figure 24.11);
- divergence in growth rates (referred to as "deviation")[80] of the dominant and largest subordinate follicles is apparent when the largest follicle reaches a diameter of 8–9 mm.

While the developmental dynamics of follicles 1 mm or larger have now been well characterized, the dynamics of smaller follicles remain a mystery. An early study demonstrated the binding of FSH to the granulosa of follicles with only a single layer of granulosa cells[97] but it has been argued that they may not be coupled to the adenylate cyclase second messenger system during early stages of folliculogenesis, and may be nonfunctional.[98] However, the growth-promoting effect of FSH on preantral and small antral follicles in

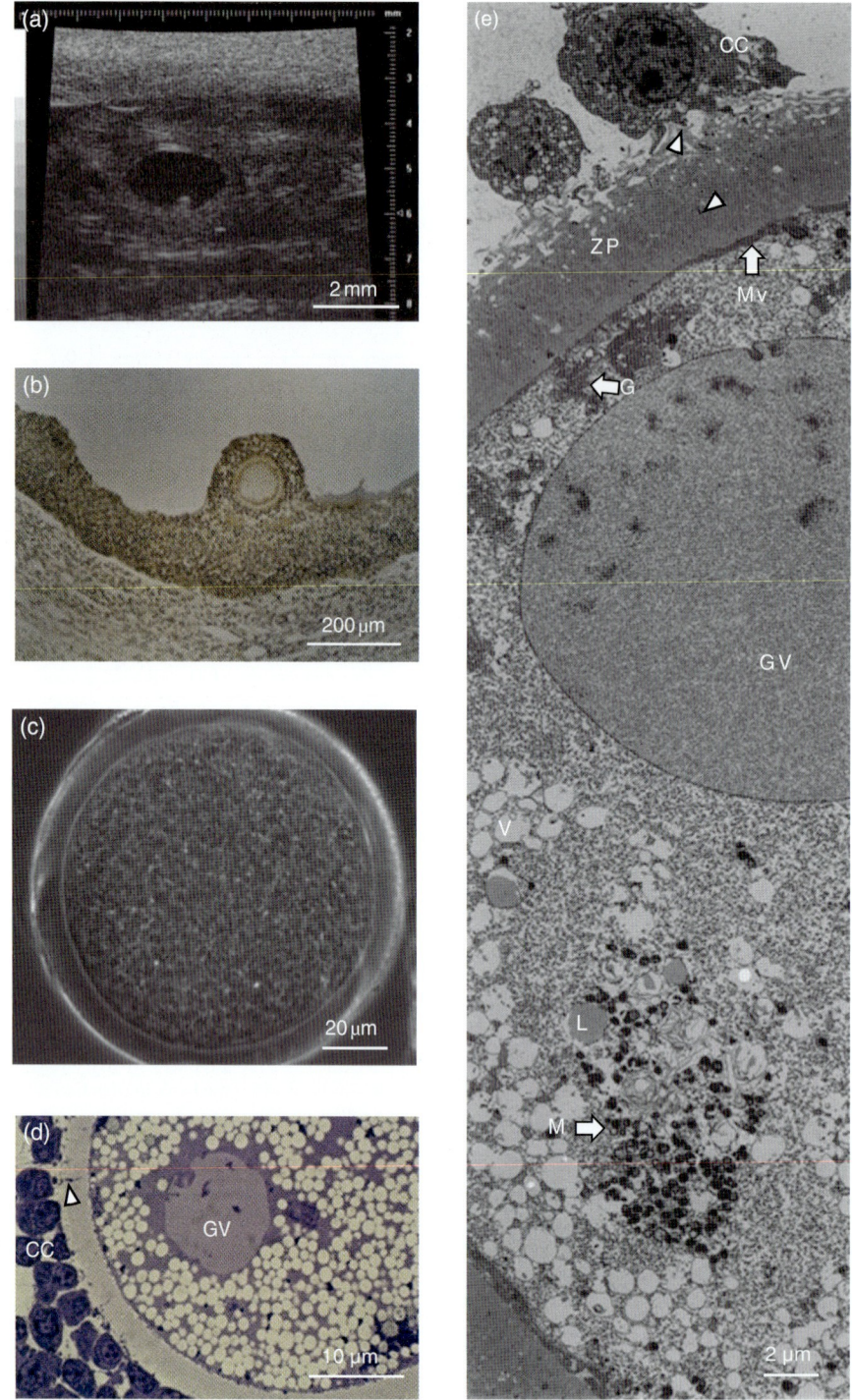

Figure 24.12 Images of the bovine cumulus–oocyte complex. (a) Ultrasound biomicroscopy using a 40-MHz transducer. (b) Histology (5-μm section) stained with Masson's trichrome. (c) Phase contrast microscopy of a denuded oocyte (40×). (d) Semi-thin section (1-μm section) of the cumulus–oocyte complex from a growing phase dominant follicle stained with toluidine blue. (e) Transmission electron microscopy of (d). CC, cumulus cells; ZP, zona pellucida; V, oocyte vesicles; L, lipid droplets; M, hooded and normal mitochondria; arrowheads, cumulus cell projections; Mv, microvilli on ooplasm surface; G, Golgi; GV, germinal vesicle.

cattle produced *in vitro*[99] and *in vivo*[100,101] suggests a role for FSH in the development of early-stage preantral and antral follicles. These observations challenge the old dogma that early follicle growth is gonadotropin-independent.

If small follicles are responsive to FSH, it is logical to postulate that their developmental dynamics follow a wave-like pattern in response to periodic endogenous surges of FSH.

Until recently, reference to a follicular wave was limited to follicles of 4 mm or greater, based simply on the limit of resolution of existing ultrasound equipment. At the microscopic level, there is no morphologic distinction between mid- to late-stage antral follicles of less than 4 mm and those 4 mm and larger.[7,8] At the cellular level, both size categories of follicles not only express FSH receptors but have a similar level of expression on a per-granulosa cell basis (reviewed in Ref. 78). The periodic emergence of waves of follicles 4 mm and larger in response to periodic surges in the circulating concentrations of FSH,[51] and the consistency in the number of follicles of 2 mm or more[102] or 3 mm or more[59] recruited into successive waves, indicates that the follicles may become progressively entrained to waves from the earliest stages of development.

Ovulation and oocyte maturation

Ovulation involves a series of events that culminates in the evacuation of the antral contents of a follicle, including its mature cumulus–oocyte complex (Figure 24.12), and the initiation of development of a corpus luteum. Several events happen in succession and are tightly coordinated at systemic, cellular, and molecular levels. The process begins with endocrine changes triggered by luteolysis. A decrease in circulating concentrations of progesterone at the time of luteolysis removes the negative feedback effects on the pituitary and hypothalamus, resulting in an increase in LH pulse frequency and amplitude. Availability of LH results in final maturation of the dominant follicle with an increase in production of estradiol and upregulation of LH receptors on granulosa cells. The preovulatory increase in LH results in extensive ovarian matrix remodeling (vascular changes, disintegration of the basement membrane, and rupture of the follicle wall), granulosa and theca cell responses (cumulus cell expansion, luteinization), and resumption of meiosis by the oocyte. At estrus, the high circulating concentration of estradiol in the absence of progesterone is responsible for a surge release of LH, and estrous behavior (increased locomotion, phonation, mounting, and standing to be mounted).

Although LH receptors are present on both granulosa and theca cells,[103] within the granulosa cells their distribution is confined to the mural granulosa cells. In addition, LH receptors are highly expressed in those layers closer to the basement membrane compared with those closer to the antrum.[104] Consequently, LH needs second messengers to be able to affect the entire follicle and the oocyte. The effect of second messengers is facilitated by the intimate relationship among granulosa cells and cumulus cells and the oocyte through gap junctions (see Figure 24.4).[105] The two main second messengers for LH are cyclic AMP[106,107] and intracellular calcium.[108] Binding of LH to the G-protein receptors on the granulosa cell surface results in activation of adenyl cyclase, an increase in cyclic AMP, and stimulation of protein kinase A and C pathways. Stimulation of protein kinase A and protein kinase C causes phosphorylation and activation of transcription factors (Egr-1, C/EBPβ, CREB, Fos, Jun, Fra2, JunD) within 4 hours of the LH surge. Expression of some of these genes is transient while others are sustained until ovulation (reviewed in Ref. 109). C/EBPβ controls expression of *COX2*, a key gene on the prostaglandin pathway. Granulosa cell products of epidermal-like growth factor (EGF) expression (epiregulin, amphiregulin, and betacellulin) are elevated 1–3 hours after the LH surge and are involved in transmitting the ovulatory signal to cumulus cells.[110] The LH surge also indirectly induces activation of the progesterone and prostaglandin pathways[111] that modify fibroblast function to produce chemokines that recruit inflammatory cells and growth factors that initiate angiogenesis (e.g., VEGF). The LH-induced extracellular protease ADAMTS1 is synthesized in granulosa cells but the protein is translocated to the extracellular matrix to induce the production of versican (chondroitin sulfate proteoglycan 2; reviewed in Ref. 112), and is implicated in the activation of tissue proteases such as metalloproteinases required for follicle rupture.

The extracellular matrix that surrounds and supports cells is a major component of connective tissue (see Figure 24.4). It is composed mainly of structural proteins (elastin and collagens), specialized proteins (fibrillin, fibronectin, laminin), and proteoglycans. It plays a prominent role in ovarian function by participating in processes such as cell migration, proliferation, growth, and demise during follicle development, ovulation, and corpus luteum formation. Two families of enzymes govern degradation of the ovarian epithelium during ovulation: metalloproteinases and plasminogen activators. Expression of both of these enzyme families is induced by the LH surge. Changes in gene expression of metalloproteinases and metalloproteinase inhibitors throughout the bovine estrous cycle have not been reported, but one factor that can induce expression is tumor necrosis factor (TNF)-α, which is secreted by preovulatory bovine follicles.[113] Heparanase (encoded by *HPSE*) cleaves heparan sulfate glycosaminoglycans during matrix remodeling and may also play a major role in follicle rupture during ovulation in cattle.[114] LH-induced ovulation is also associated with increase in the expression of yet another group of proteases called kallikreins.[115] It appears that many paracrine factors, proinflammatory cytokines, and families of protease enzymes act in concert with each other leading to degradation of extracellular connective tissue and ultimately follicle rupture, on average, 30 hours after the LH surge.[116]

Meiotic progression of the oocyte is arrested at prophase I (first meiotic arrest) during the fetal development of primordial follicles. The process of resumption of meiosis, release of half the set of chromosomes as a polar body, and progression to metaphase II is referred to as nuclear maturation (Figure 24.13). Shortly after the LH surge, the gap junctions between cumulus cells and the oocyte deteriorate.[117,118] The loss in communication between cells removes inhibitory signaling, and the oocyte can then resume meiosis. However, meiosis is stopped again at metaphase II (second meiosis arrest) until fertilization, when meiosis is finally completed. Mammalian oocytes have the propensity to resume meiosis if they are removed from the antral follicle in which they are housed,[119–121] or if they are within preovulatory follicles under the influence of the LH surge.[117,122] In cattle, a major proportion of oocytes reached metaphase II at 20 hours of *in vitro* maturation[117] or 20 hours after a spontaneous or induced LH surge.[122–124] Cyclic AMP and GMP are considered the primary inhibitors of meiotic resumption in the oocyte.[118] Cyclic GMP produced by granulosa cells is thought to enter the oocyte via gap junctions between the oocyte and cumulus cells to maintain low levels of cAMP phosphodiesterase 3A (PDE3A) in the oocyte.[125] Low PDE3A is thought to maintain high levels of cAMP produced within the oocyte[126] and thus maintain meiotic arrest. Exposure to LH decreases cGMP production by granulosa cells through as yet unknown mechanisms, and uncouples the gap junctions (by phosphorylation of connexin-43) via mitogen-activated protein kinase (MAPK).[125,127,128] Reduced production and entrance of cyclic GMP into the oocyte in

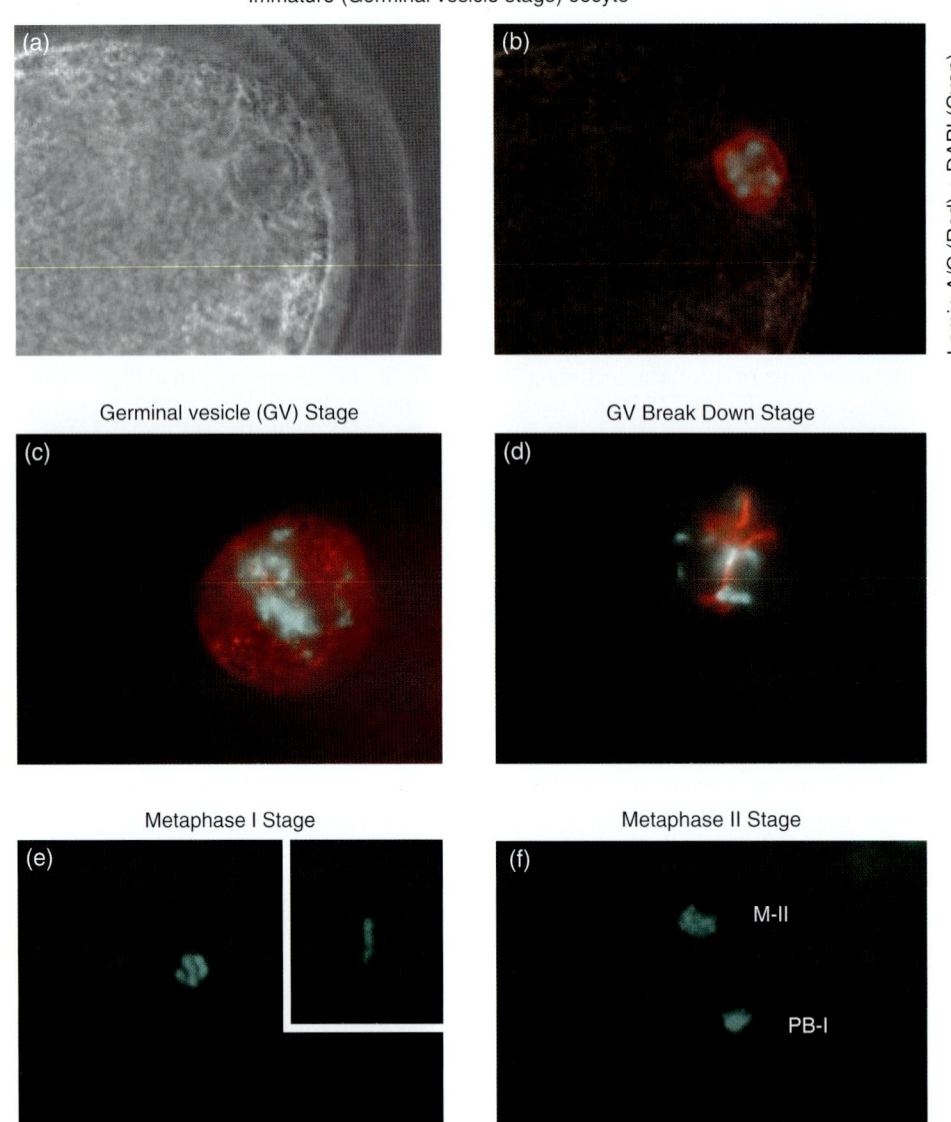

Figure 24.13 Stages of nuclear maturation of bovine oocytes. (a) Phase-contrast image of an immature oocyte with an intact nuclear membrane (germinal vesicle or GV stage) obtained from a 4-mm follicle. (b) Anti-lamin A/C (stains nuclear membrane red) and DAPI (stains DNA cyan) stained images merged with the phase-contrast image in (a). (c) Nucleus of an immature oocyte with an intact nuclear membrane (GV stage). (d) Maturing oocyte showing germinal vesicle breakdown (GVBD stage). (e) Metaphase I; inset shows metaphase I plate at a right-angle. (f) Metaphase II and the first polar body (PB-I). Adapted from Prentice-Biensch J, Singh J, Alfoteisy B, Anzar M. A simple and high-throughput method to assess maturation status of bovine oocytes: comparison of anti-lamin A/C-DAPI with an aceto-orcein staining technique. *Theriogenology* 2012;78:1633–1638.

turn increases PDE3A, which ultimately hydrolyzes cyclic AMP and enables resumption of meiosis.

Luteal dynamics

Based on the status of the corpus luteum, the estrous cycle may be divided into four successive phases: (i) proestrus, when the corpus luteum is regressing and an ovulatory follicle develops; (ii) estrus, the absence of a functional corpus luteum and the period of sexual receptivity when final follicular maturation and ovulation occur; (iii) metestrus, the period of corpus luteum formation; and (iv) diestrus, when the corpus luteum is mature and actively produces progesterone.[129] Proestrus and estrus have been referred to as the follicular phase and metestrus and diestrus as the luteal phase, though "luteal phase" should not be taken to imply the lack of follicle development. To the contrary, follicular waves continue to develop throughout follicular and luteal phases of the estrous cycle, as described above.

Origin of the corpus luteum

The growth and demise of the corpus luteum during the estrous cycle represents one of the most rapid dynamic processes in the body. At the time of ovulation, antral contents are evacuated and the wall of the ovulatory follicle

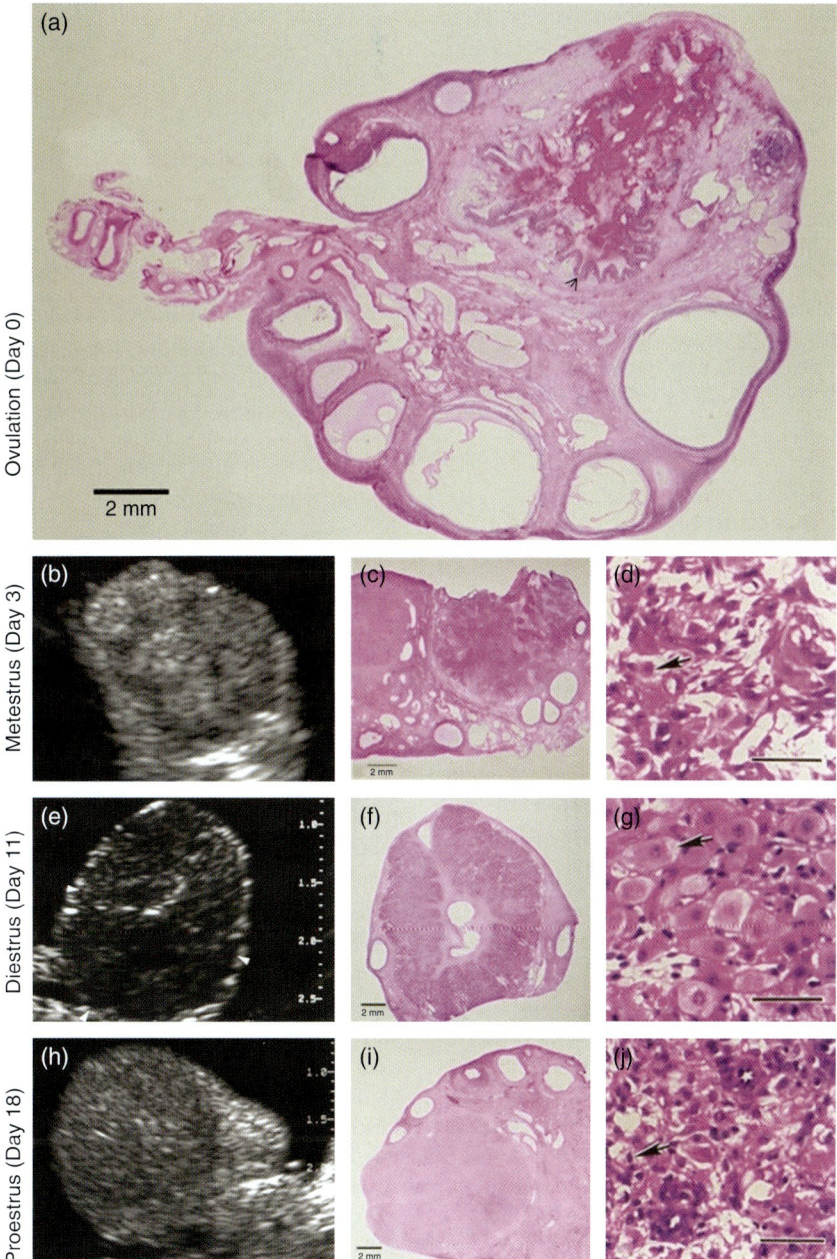

Figure 24.14 Development of the corpus luteum in cattle. (a) Histological section of an ovary taken within 1 hour of ovulation. Note the collapsed follicle with extensive crenulation of the follicle wall (arrowhead) and blood clot in the antrum. Note also numerous small antral follicles in the ovary, typical of new wave emergence at the time of ovulation. Ultrasound morphology (b, e, h) and low-magnification histology (c, f, i) and tissue details (d, g, j) of bovine corpora lutea during metestrus (b, c, d; day 3 after ovulation), mid-diestrus (e, f, g; day 11 after ovulation) and proestrus (h, i, j; day 18 after ovulation). Arrows indicate luteal cells. Greater angiogenesis during metestrus and hyalinization of blood vessels during proestrus (luteal regression) resulted in brighter echotexture of ultrasound images at these stages compared with diestrus. Histological images were taken from 5-μm thick paraffin sections stained with hematoxylin and eosin. Scale on the ultrasound images (b, d, f, h) is in centimeters. Scale bar is 2 mm for a, c, f, i and 10 μm for d, g, j. Adapted from Singh J, Pierson R, Adams G. Ultrasound image attributes of the bovine corpus luteum: structural and functional correlates. *J Reprod Fertil* 1997;109:35–44.

collapses into undulations within the vacated antrum (Figure 24.14). The granulosa and theca cells of the former follicle then expand beyond the volume of the former follicular antrum through a dramatic process of vascularization and luteinization to form a functional corpus luteum. This temporary steroid-producing gland undergoes marked structural and functional changes in a short timespan, and has been the subject of extensive morphological studies (Figure 24.14) (reviewed in Ref. 130). Plasma progesterone concentration is highly correlated with corpus luteum weight, volume,[131] histomorphology,[132] and ultrasound morphology.[130]

During metestrus, blood vascular components occupy the greatest percent volume of the corpus luteum. The luteal cells undergo hypertrophy at mid-diestrus, during peak progesterone production, in association with their greatest volume density. An increase in luteal cell size during the formation phase and a decrease during the regression phase parallel progesterone secretion. During proestrus, the regressing corpus luteum decreases in weight, in volume densities of luteal cells and blood vascular components, and in progesterone production. There is a concurrent increase in connective tissue and hyalinization of blood vessels. Individual luteal cells undergo degenerative changes reflected by decreased cell size and accumulation of coarse lipid droplets in the cytoplasm. These morphologic changes are associated with progressive changes in plasma progesterone concentration from less than 2 ng/mL at 3 days after ovulation to about 4 ng/mL by 6 days and peak concentrations of greater than 6 ng/mL between 10 and 14 days after ovulation, followed by decreasing levels after 16 days due to luteolysis.[71]

Ultrasonographic detection and morphology of the corpus luteum

The reliability and accuracy of ultrasonography for detecting the corpus luteum, and the relationship between ultrasound image attributes and corpus luteum function, have important implications in clinical settings for assessing normalcy and stage of the cycle. In a comparison between ultrasonographic detection and gross anatomic dissection, there was 83% agreement in inexperienced operators and 97% agreement in experienced operators.[133] The growing corpus luteum was detected by day 2 (day 0 = ovulation) in more than 95% of 80 estrous cycles, and the regressing corpus luteum was be detected, on average, 1.4 days into the next estrous cycle.[133]

A large proportion (40–70%) of normal corpora lutea have a central fluid-filled cavity during their development; the remainder are solid structures (reviewed in Ref. 130). The cavity of the corpus luteum is filled with a serous transudate, and remains anechoic during its entire existence as opposed to the organized clot and appearance of a network of echogenic lines within the central cavity of the equine corpus luteum. The central cavity becomes smaller and often disappears as the corpus luteum matures. Plasma progesterone concentrations are not different between corpus luteum with or without a central cavity. The profile of luteal tissue area (minus area of a cavity if present) detected by ultrasound paralleled the profile of plasma progesterone concentration, with the exception that during luteolysis progesterone began to decrease 1–2 days before ultrasonographic regression was apparent.[134]

Changes in ultrasound image attributes (echotexture) of the corpus luteum are also reflective of corpus luteum function in cattle (Figure 24.14).[130] Higher pixel values obtained by computer-assisted quantitative echotexture analysis during luteal gland formation (metestrus) and regression (proestrus) were similar to the pattern observed in horse and pony mares by subjective scoring of echotexture[135] and pixel analysis[136] of corpora lutea recorded *in vivo* by transrectal ultrasonography. In a test of the hypothesis that ultrasound images may be digitally "stained," computer algorithms were developed to produce a "virtual histology" through subtle differences in echotexture resulting from different ovarian structures.[137] Quantitative analysis showed that the "sticks" algorithms increased the statistical differences in the echotexture of ovarian stroma versus the corpus luteum, representing a first step toward virtual histology of ultrasound images and understanding dynamic changes in the ovary at the microscopic level in a safe, repeatable, and noninvasive way.[138]

Vascular flow dynamics in the corpus luteum

Major changes in the microvasculature of the ovary occur during growth and regression of large follicles and the corpus luteum.[130,139,140] Three variations of the Doppler technique have been used to study vascular flow: (i) spectral Doppler is used to display vascular flow over time in a waveform; (ii) color-flow Doppler permits superimposition of information about vascular flow direction on the B-mode gray-scale image; and (iii) power-flow Doppler is used to superimpose low-velocity vascular flow information (without directional information) on the B-mode gray-scale image. Doppler studies have provided visual evidence for time-related changes in blood flow within the preovulatory follicle wall of cows in relation to the LH surge.[140,141] Initial results indicate that a biphasic blood flow response to LH may occur in the bovine preovulatory follicle. An initial increase in blood flow was accompanied by serration of the follicular granulosa layer within 3 hours of LH treatment, and a second increase in blood flow was detected around 20 hours after LH.[141] Distinct changes in vascular flow patterns during luteogenesis, luteolysis and early pregnancy, and in response to gonadotropin-releasing hormone (GnRH), prostaglandin (PG)F, PGE, PGI, oxytocin and nitric oxide synthase have been reported in cattle.[140,142–146]

Luteolysis

In the early 1970s, $PGF_{2\alpha}$ was found to be the natural luteolysin in cattle,[147,148] and since that time $PGF_{2\alpha}$ has become the most commonly used treatment for elective induction and synchronization of estrus in cattle.[149–151] Pulsatile release of $PGF_{2\alpha}$ from the uterus initiates luteolysis beginning 4–6 days before ovulation.[152,153] Early studies showed that the maturity of the corpus luteum at the time of $PGF_{2\alpha}$ treatment influenced the luteolytic response;[154] $PGF_{2\alpha}$ did not effectively induce luteolysis during the first 5 or 6 days following estrus. Lack of responsiveness was thought to be a result of a lack of $PGF_{2\alpha}$ receptors in immature corpus luteum, but more recent work has refuted this hypothesis by demonstrating the presence of $PGF_{2\alpha}$ receptors in corpus luteum as early as 2 days after ovulation.[155] The reason for the resilience of the immature corpus luteum to $PGF_{2\alpha}$ remains unclear, but some suggest that the mature corpus luteum may possess a positive feedback loop involving luteal oxytocin and TNF that elicits endometrial release of $PGF_{2\alpha}$ and sustains the process of luteolysis.[156,157] If this notion is correct, and immature corpus luteum do not possess the positive feedback loop, then multiple injections of $PGF_{2\alpha}$ may be expected to complete the luteolytic process in early-stage corpus luteum. In this regard, luteolysis and ovulation was induced in 56 of 60 heifers (93%) given $PGF_{2\alpha}$ in the morning and evening of either day 4 or 5 after estrus in two studies involving ovarian superstimulation relative to follicle wave emergence.[158,159] Recent studies show a distinct cause-and-effect balance between the luteotrophic effect of endogenous LH pulses and the luteolytic effect of

Ovarian Follicular and Luteal Dynamics in Cattle

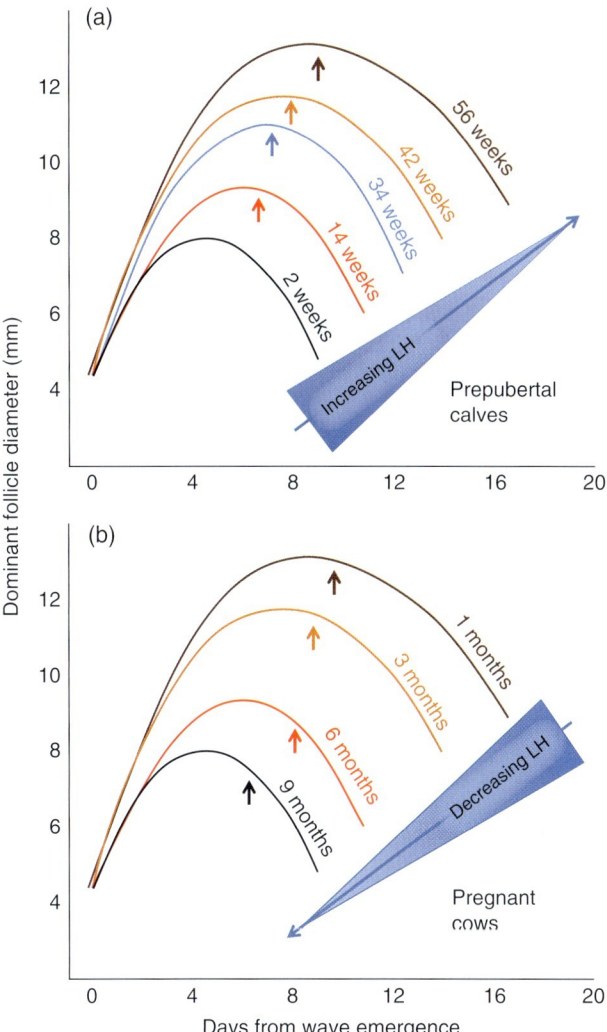

Figure 24.15 Diameter profiles of the dominant follicle of follicular waves during prepubertal development in (a) calves and (b) during gestation in pregnant cows as detected by daily transrectal ultrasonography. (a) The profile represented at 56 weeks of age is of the wave immediately preceding puberty (first ovulation; data from Adams et al.[158]; Evans et al.[162]). From 14 to 56 weeks of age, the interwave interval (indicated by arrows) increased from 6.8 to 9.2 days. (b) In pregnant cows from 1 to 9 months of gestation, the interwave interval (indicated by arrows) decreased from 9.6 to 6.4 days. Data from Ginther et al.[48]; Bergfelt et al.[33]; Ginther et al.[164]). Changes in dominant follicle profiles are associated with changes in circulating concentrations of LH.

pulses of $PGF_{2\alpha}$; repeated $PGF_{2\alpha}$ pulses from the uterus ultimately overcome the resuscitating effects of LH to complete luteolysis (reviewed in Ref. 160).

Ovarian dynamics during different reproductive states

Prepubertal and peripubertal periods

In the only study on ovarian follicle dynamics during the prepubertal and peripubertal periods, transrectal ultrasonography was used to monitor daily changes in follicle development in calves as young as 2 weeks of age; calves were monitored at regular intervals during the first year of life.[161–163] Puberty was defined as the time of the first ovulation, and was determined to be 52–56 weeks (12–13 months) in the Hereford-cross heifers used in the study. Remarkably, follicular development occurred in a wave-like fashion, similar to that in adults, at all ages examined. Individual follicle development was characterized by growing, static, and regressing phases, and periodic surges in serum concentrations of FSH were associated with each follicular wave emergence (see Figure 24.8). FSH surges spanned a mean of 3 days and were maximal 1 day before wave emergence. It was concluded that the mechanisms controlling the well-ordered phenomena of wave emergence, follicle selection, and follicle regression in mature cattle were also extant in prepubertal heifers.

The diameter profile of the dominant as well as the largest subordinate follicle in prepubertal calves increased with age (Figure 24.15). The rate of increase was greatest from 2 to 8 weeks of age and again between 24 and 40 weeks of age, in temporal association with increases in mean concentrations of LH.[162] The number of follicles detected (≥3 mm in diameter) also increased dramatically between 8 and 14 weeks of age, and then appeared to decrease as puberty approached, in temporal association with the early rise and subsequent tapering of mean FSH concentrations. The developmental reason for the early rise in circulating concentrations of LH and FSH between 4 and 14 weeks of age is not clear, but it may reflect initial maturation of the hypothalamic–pituitary axis and subsequent sensitivity to negative feedback by ovarian steroids. This notion is supported by data illustrated in Figure 24.15, and in observed changes in the interval between the emergence of dominant follicles of successive waves (interwave interval). The interwave interval was relatively long (8–9 days) at 2 and 8 weeks of age, but was significantly shorter (6.8 days) at 14 weeks. The interwave interval gradually increased thereafter until puberty (7–9 days).

Examination of data from the period immediately preceding first ovulation and encompassing the subsequent two interovulatory intervals revealed unique characteristics at the onset of ovulatory cyclicity in cattle.[163] The first ovulatory cycle was short (7.7 ± 0.2 days) and the first ovulation occurred after the dominant follicle had entered the static phase; that is, the dominant follicle was older at the time of ovulation than its counterpart of later cycles. The first corpus luteum (from the aged ovulatory follicle) was smaller and shorter-lived than corpora lutea of subsequent cycles, thus resulting in an abbreviated interovulatory interval. The second interovulatory interval was of normal duration (20.3 ± 0.5 days) and was composed of two or three follicular waves. Mean serum estradiol and LH concentrations, and LH pulse frequency increased as first ovulation approached, but FSH concentrations did not. Surges in FSH were associated with follicular wave emergence, as seen previously. These results demonstrate a striking similarity between the onset of puberty in young heifers and recrudescence of ovulatory cyclicity in postpartum cows (see below).

Follicular development during pregnancy

Results of three separate studies are consistent in that the wave phenomenon occurs throughout pregnancy, except for the last 21 days when no follicles of 6 mm or larger were detected.[33,48,164] In the presence of progesterone (endogenous or exogenous),[33] anovulatory follicular waves emerge at regular intervals. However, the maximum diameter of the dominant follicle of successive waves decreased, and was associated with a successive decrease in the interwave interval (i.e., the period of dominance became shorter) (Figure 24.15). An initial decrease occurred immediately after the first follicular wave of pregnancy, and a subsequent sharp decrease occurred after the fourth month of pregnancy. Progesterone suppresses growth of the dominant follicle in a dose-dependent manner,[71] and the progressive decrease in follicular dominance seen in pregnant cattle was attributed to the suppressive effects on LH of rising progesterone levels during mid-pregnancy, and rising estrogen levels near term.[164]

As in nonpregnant cows, periodic surges of FSH were associated with periodic emergence of follicular waves in pregnant cows. In addition, the magnitude and frequency of FSH surges were influenced by the size of the dominant follicle. Waves in which the dominant follicle reached a diameter of 10 mm or more were associated with longer intervals between successive FSH surges and lower concentrations of FSH. Conversely, waves with smaller dominant follicles (6–9 mm) were associated with shorter intervals and higher concentrations of FSH. A similar interrelationship between FSH and follicular wave dynamics was observed in prepubertal calves.

Follicular development during the postpartum period

In all species in which it has been studied, lactation and suckling have been shown to have a suppressive effect on ovarian follicular development. The degree of suppression in cattle has been associated primarily with energy balance (nutritional status, body condition, milk production, parity) and periparturient disorders (metritis, endometritis, mastitis, lameness) (reviewed in Refs 165–167). The interval from calving to first ovulation in dairy cows was a mean of 21 days (range 10–55 days),[168] but was significantly longer in primiparous cows than in multiparous cows (31.8 ± 8.3 days vs. 17.3 ± 6.3 days).[169] Emergence of the first follicular wave ranged from 2 to 7 days (mean 4.0 days) after calving in primiparous Holsteins,[164] and the dominant follicle of the first postpartum wave ovulated in 14 of 19 (74%) multiparous Friesian cows[170] on mean day 27 (range 12–58).[171] Recrudescence of follicular waves was observed within 10 days of calving in beef cattle,[172] but the first ovulation occurred somewhat later than in dairy cattle (mean 30.6 days) and only rarely from the dominant follicle of the first postpartum wave (11%).[172]

The first postpartum ovulation is frequently associated with an absence of estrous behavior and is often followed by a luteal phase of short duration (reviewed in Ref. 173). The first ovulation was not accompanied by estrus in 17 of 18 (94%) postpartum dairy cows,[170] and the length of the first postpartum interovulatory interval varied depending on when the first follicle destined to ovulate emerged. The first postpartum interovulatory interval was short (mean 11.2 days) in approximately 25% of dairy cows, was of normal duration (mean 20.6 days) in another 25%, and was long (mean 30.0 days) in the remaining 50% of cows.[170] Short cycles were associated with later detection of the first follicle destined to ovulate (≥10 days after calving), whereas normal and long cycles were associated with earlier detection of the first follicle destined to ovulate (usually ≤10 days). These findings are remarkably consistent with those of another study[168] in which short postpartum anovulatory periods (around 14 days) were followed by a normal-length cycle (18–21 days), and longer postpartum anovulatory periods (21–25 days) were followed by a short cycle (<14 days). In the majority of beef cows (78%), ovulation occurred from the second, third or fourth postpartum follicular wave and, as in dairy cattle, the first ovulations occurring after 20 days (16 of 18 cows) were followed by a short cycle (14 of 16 cows).[172] Short cycles were associated with a short luteal phase, a smaller corpus luteum, and lower circulating progesterone concentrations. The short luteal phase following the first postpartum ovulation has been attributed to premature release of $PGF_{2\alpha}$ by the uterus as a result of lower numbers of progesterone receptors and greater numbers of oxytocin receptors in endometrial cells;[173] however, luteal inadequacy as a result of altered preovulatory granulosa/theca cell differentiation in a low LH environment remains to be tested.

Follicular development during old age

Decreased fertility with maternal aging has been well documented in cattle and humans.[174–176] Endocrine and ovarian characteristics of reproductive aging were characterized in a series of studies in which old cows (≥15 years) were compared with their young daughters (≤5 years).[177–179] Mean circulating FSH concentrations were consistently higher in old cows than in their daughters, but the expected pattern of FSH secretion and wave emergence was maintained in old cows, i.e., each ovarian follicular wave was preceded by a surge in circulating FSH. Despite elevated FSH, fewer 4–5 mm follicles were recruited into each follicular wave in old cows compared with their daughters. This interesting inverse relationship between the number of follicles recruited into a wave and the peak concentrations of FSH has also been reported in studies documenting the repeatability of follicle numbers within individuals.[59,102] The duration of the interovulatory and interwave intervals did not change with age, but the ovulatory follicle of old cows with a two-wave pattern was smaller at the time of ovulation than that of young cows. Similarly, the diameter of the corpus luteum was smaller, and the plasma concentration of progesterone tended to be lower in old versus young cows. There was no age effect on circulating LH concentrations or LH pulse frequency.

The responsiveness of the hypothalamic–pituitary axis in young and old cows was examined in a study involving estradiol/progesterone-based ovarian synchronization.[178] Steroid treatment suppressed circulating FSH in both age groups, and the intervals from treatment to subsequent FSH peak and wave emergence were not different between old and young cows. In a study of the ovarian response to superstimulatory treatment,[179] fewer small (<5 mm) follicles were recruited into the follicular wave, and fewer follicles in all size categories developed after ovarian superstimulation in old cows compared with their young daughters. On average, the young daughters had eight more ovulations per cow than their mothers. Lastly, data from a study in which

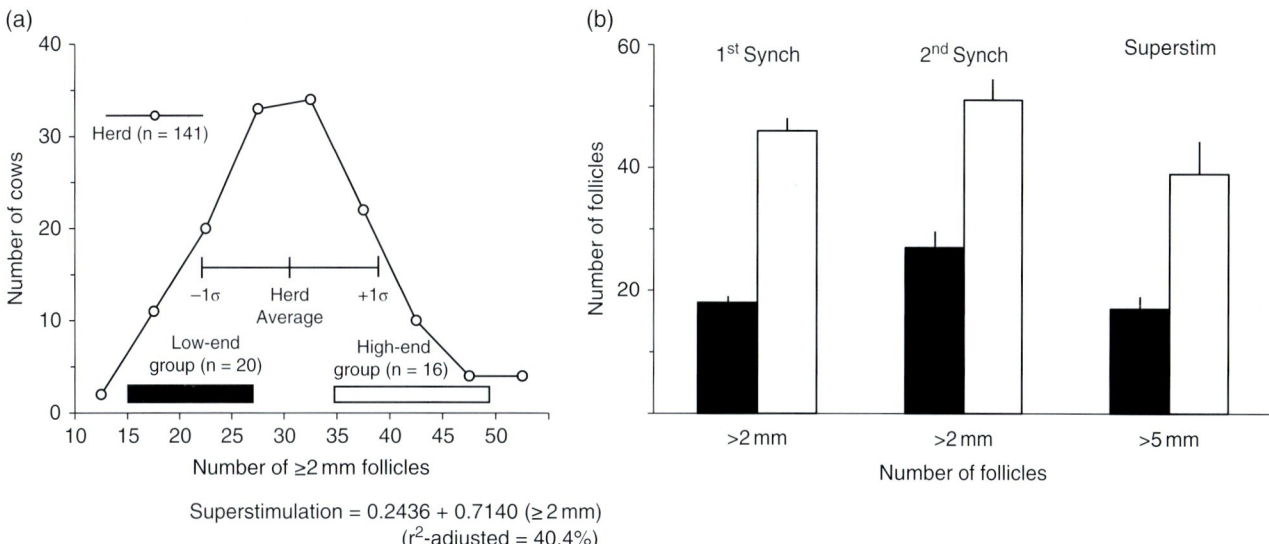

Figure 24.16 Frequency distribution of Hereford-cross beef cows in a herd ($n=141$, solid line) based on the number follicles detected in both ovaries by ultrasonography at the time of follicular wave emergence. (a) Based on antral follicle count, a sample of cows from the low end ($n=20$, black bar) and high end ($n=16$, white bar) were selected for successive synchronization and superstimulation to determine the repeatability and impact of antral follicle count. (b) Number of follicles detected at emergence of the first and second synchronized waves, and after ovarian superstimulatory treatment in the low-end group (black bars) and high-end group (white bars). The number of follicles detected was significantly greater in the high-end group after both synchronizations and after superstimulation. The superstimulatory response was related in a simple linear fashion to the number of follicles detected at wave emergence [superstimulatory response = $0.2436 + 0.7140$ (no. of follicles >2 mm); adjusted $r^2 = 40.4\%$]. Adapted from Singh J, Dominguez M, Jaiswal R, Adams G. A simple ultrasound test to predict superovulatory response in cattle. *Theriogenology* 2004;62:227–243.

embryos from old cows and their daughters were transferred to age-matched recipient cows[179] revealed that the oocytes from old cows were markedly less competent and produced only half as many embryos as their daughters.

Conclusions and implications

Knowledge of normal follicular and luteal dynamics, and the mechanisms controlling these dynamics, have had a direct impact on clinical and breeding management practices in cattle. The concepts represented in Figure 24.9 have been used as a basis for developing protocols for ovarian synchronization for fixed-time insemination and for ovarian superstimulation for the purposes of *in vivo* and *in vitro* embryo production and embryo transfer, and for controlled breeding management programs in dairy and beef herds. The concepts of the wave pattern of follicle development and selection imply predictable variations in the number and maturity of specific classes of follicles (e.g., growing, static or regressing, dominant and subordinate follicles) that may or may not be capable of responding to certain treatments. Importantly, the dynamics of follicular development are overlain by distinct influences of the corpus luteum. Some examples of the impact of endogenous ovarian rhythms on practice are offered below.

Synchronization

In cows in which luteolysis is effectively induced by $PGF_{2\alpha}$ treatment, the ensuing estrus is distributed over a 6-day period.[180] The most consistent response (i.e., the least variability in interval to estrus) resulted from $PGF_{2\alpha}$ treatment during early diestrus (7 and 8 days after estrus) and later diestrus (15 days after estrus) than from treatment during mid-diestrus (10–13 days after estrus).[180] Subsequent studies have documented that much of this variation is due to follicular status at the time of treatment. Treatment when the dominant follicle of a wave is in its late growing (≥9 mm) or early static phase will result in ovulation of that follicle within 2–3 days, whereas treatment after the mid to late static phase (i.e., when it is no longer viable) will result in ovulation of the dominant follicle of the next wave 4–5 days after treatment. That is, if $PGF_{2\alpha}$ is given at the end of the period of functional dominance of one follicle, several days are required for a new dominant follicle to grow sufficiently to ovulate.[153,181,182]

Antral follicle count

The number of follicles detectable by ultrasonography varies widely among cows and according to follicular wave status.[35,59,96,102] However, differences in follicle numbers are not associated with wave sequence (i.e., similar between Wave 1, Wave 2, and Wave 3) or the number of waves per estrous cycle (i.e., two- and three-wave patterns; reviewed in Ref. 35). Although individuals vary greatly, the intrinsic number of follicles at wave emergence is highly repeatable within individuals.[59,102] In a direct comparison of high-follicle count versus low-follicle count cows,[102] the number of follicles recruited into a wave was highly correlated with the number of follicles recruited into successive waves, and the superstimulatory response was highly correlated with the number of follicles recruited into a follicular wave for a given individual (Figure 24.16). This characteristic now forms the basis of a simple ultrasound test to predict the superstimulatory response in cattle.

In a recent study of the effect of restricting maternal nutrition to 60% of maintenance requirements during the first third of gestation, calves born to nutritionally restricted mothers had 60% lower antral follicle counts compared with calves born to mothers fed control diets.[183] Furthermore, results suggest that fertility may be compromised in animals with low antral follicle counts due to effects on oocytes, progesterone, and the endometrium compared to animals with high antral follicle counts; postpartum dairy cows with high antral follicle counts had higher pregnancy rates, shorter calving to conception intervals, and received fewer services during the breeding season compared to cows with low counts.[183]

Ovarian superstimulation

The greatest sources of variability in response to ovarian stimulation are the status of the follicular wave at the time stimulatory treatment is initiated and the intrinsic number of follicles present at the time of wave emergence in a given individual. Results of studies in cattle have documented that the superstimulatory response is (i) inferior when treatment is initiated after selection of the dominant follicle, (ii) greater when treatment is initiated at the time of wave emergence rather than 1 or 2 days later, and (iii) similar between Wave 1 (rising progesterone) and Wave 2 (high endogenous progesterone (reviewed in Ref. 102). Only approximately 20% (4 or 5 days) of the estrous cycle is available for initiating treatment at the time of follicular wave emergence. Therefore, the majority (80%) of the estrous cycle is not conducive to an optimal superovulatory response. The superstimulatory response may be predicted to be approximately 70% of the number of follicles 3 mm or larger at the time of wave emergence.[102] As a result of these findings, superstimulatory protocols in cattle now incorporate synchronization of follicular wave emergence.

References

1. Hummitzsch K, Irving-Rodgers H, Hatzirodos N *et al*. A new model of development of the mammalian ovary and follicles. *PLOS ONE* 2013;8:e55578.
2. McNatty K, Fidler A, Juengel J *et al*. Growth and paracrine factors regulating follicular formation and cellular function. *Mol Cell Endocrinol* 2000;163:11–20.
3. Juengel J, Sawyer H, Smith P, Quirke L, Heath D. Origins of follicular cells and ontogeny of steroidogenesis in ovine fetal ovaries. *Mol Cell Endocrinol* 2002;191:1–10.
4. Gougeon A. Regulation of ovarian follicular development in primates: facts and hypotheses. *Endocr Rev* 1996;17:121–156.
5. Senger P. Embryogenesis of the pituitary gland and the male or female reproductive system. In: *Pathways to Pregnancy and Parturition*, 2nd edn. Redmond, OR: Current Conceptions Inc., 2003, pp. 44–79.
6. Knight P, Glister C. Potential local regulatory functions of inhibins, activins and follistatin in the ovary. *Reproduction* 2001;121:503–512.
7. Lussier J, Matton P, Dufour J. Growth rates of follicles in the ovary of the cow. *J Reprod Fertil* 1987;81:301–307.
8. Braw-Tal R, Yossefi S. Studies in vivo and in vitro on the initiation of follicle growth in the bovine ovary. *J Reprod Fertil* 1997;109:165–171.
9. Lundy T, Smith P, O'Connell A, Hudson N, McNatty K. Populations of granulosa cells in small follicles of the sheep ovary. *J Reprod Fertil* 1999;115:251–262.
10. Eppig J. Oocyte control of ovarian follicular development and function in mammals. *Reproduction* 2001;122:829–838.
11. Rüsse I. Oogenesis in cattle and sheep. *Bibl Anat* 1983;24:77–92.
12. Rajakoski E. The ovarian follicular system in sexually mature heifers with special reference to seasonal, cyclical, and left-right variations. *Acta Endocrinol Suppl* 1960;52:1–68.
13. Choudary J, Gier H, Marion G. Cyclic changes in bovine vesicular follicles. *J Anim Sci* 1968;27:468–471.
14. Donaldson L, Hansel W. Cystic corpora lutea and normal and cystic graafian follicles in the cow. *Aust Vet J* 1968;44:304–308.
15. Marion G, Gier H, Choudary J. Micromorphology of the bovine ovarian follicular system. *J Anim Sci* 1968;27:451–465.
16. Hansel W, Convey E. Physiology of the estrous cycle. *J Anim Sci* 1983;57(Suppl. 2):404–424.
17. Spicer L, Echternkamp S. Ovarian follicular growth, function and turnover in cattle: a review. *J Anim Sci* 1986;62:428–451.
18. Staigmiller R, England B. Folliculogenesis in the bovine. *Theriogenology* 1982;17:43–52.
19. Pierson R, Ginther O. Ultrasonography of the bovine ovary. *Theriogenology* 1984;21:495–504.
20. Pierson R, Ginther O. Reliability of diagnostic ultrasonography for identification and measurement of follicles and detecting the corpus luteum in heifers. *Theriogenology* 1987;28:929–936.
21. Pierson R, Ginther O. Follicle populations during oestrous cycle in heifers: II. Influence of right and left sides and intraovarian effect of the corpus luteum. *Anim Reprod Sci* 1987;14:177–186.
22. Pierson R, Ginther O. Follicular populations during the estrous cycle in heifers: I. Influence of day. *Anim Reprod Sci* 1987;124:165–176.
23. Knopf L, Kastelic J, Schallenberger E, Ginther O. Ovarian follicular dynamics in heifers: test of two-wave hypothesis by ultrasonically monitoring individual follicles. *Domest Anim Endocrinol* 1989;6:111–119.
24. Pierson R, Ginther O. Ultrasonic imaging of the ovaries and uterus in cattle. *Theriogenology* 1988;29:21–37.
25. Savio J, Keenan L, Boland M, Roche J. Pattern of growth of dominant follicles during the estrous cycle in heifers. *J Reprod Fertil* 1988;83:663–671.
26. Sirois J, Fortune J. Ovarian follicular dynamics during the estrous cycle in heifers monitored by real time ultrasonography. *Biol Reprod* 1988;39:308–317.
27. Ireland J, Roche J. Hypotheses regarding development of dominant follicles during a bovine estrous cycle. In: Roche JF, O'Callaghan D (eds) *Follicular Growth and Ovulation Rate in Farm Animals*. Dordrecht: Martinus Nijhoff, 1987, pp. 1–18.
28. Ireland J, Roche J. Development of nonovulatory antral follicles in heifers: changes in steroids in follicular fluid and receptors for gonadotropins. *Endocrinology* 1983;112:150–156.
29. Ireland J, Roche J. Growth and differentiation of large antral follicles after spontaneous luteolysis in heifers: changes in concentration of hormones in follicular fluid and specific binding of gonadotropins to follicles. *J Anim Sci* 1983;57:157–167.
30. Ireland J, Fogwell R, Oxender W, Ames K, Cowley J. Production of estradiol by each ovary during the estrous cycle of cows. *J Anim Sci* 1985;59:674–771.
31. Ginther O, Kastelic J, Knopf L. Composition and characteristics of follicular waves during the bovine estrous cycle. *Anim Reprod Sci* 1989;20:187–200.
32. Ginther O, Knopf L, Kastelic J. Temporal associations among ovarian events in cattle during oestrus cycles with two and three follicular waves. *J Reprod Fertil* 1989;87:223–230.
33. Bergfelt D, Kastelic J, Ginther O. Continued periodic emergence of follicular waves in nonbred progesterone-treated heifers. *Anim Reprod Sci* 1991;24:193–204.
34. Adams G. Comparative patterns of follicle development and selection in ruminants. *J Reprod Fertil Suppl* 1999;54:17–32.

35. Jaiswal R, Singh J, Marshall L, Adams G. Repeatability of 2-wave and 3-wave patterns of ovarian follicular development during the bovine estrous cycle. *Theriogenology* 2009;72:81–90.
36. Murphy M, Enright W, Crowe M et al. Effect of dietary intake on pattern of growth of dominant follicles during the estrous cycle in beef heifers. *J Reprod Fertil* 1991;92:333–338.
37. Rhodes F, Fitzpatrick L, Entwistle K, Death G. Sequential changes in ovarian follicular dynamics in *Bos indicus* heifer before and after nutritional anoestrus. *J Reprod Fertil* 1995;104:41–49.
38. Badinga L, Thatcher W, Diaz T, Drost M, Wolfenson D. Effect of environmental heat stress on follicular development in lactating cows. *Theriogenology* 1993;39:797–810.
39. Wolfenson D, Thatcher W, Badinga L et al. Effects of heat stress on follicular development during the oestrous cycle in lactating dairy cattle. *Biol Reprod* 1995;52:1106–1113.
40. Zeitoun M, Rodriguez H, Randel R. Effect of season on ovarian follicular dynamics in Brahman cows. *Theriogenology* 1996;45:1577–1581.
41. Figueiredo R, Barros C, Pinheiro I, Sole J. Ovarian follicular dynamics in Nelore breed (*Bos indicus*) cattle. *Theriogenology* 1997;47:1489–1505.
42. Bo G, Baruselli P, Martinez M. Pattern and manipulation of follicular development in *Bos indicus* cattle. *Anim Reprod Sci* 2003;78:307–326.
43. Ahmad N, Townsend E, Dailey R, Inskeep E. Relationship of hormonal patterns and fertility to occurrence of two or three waves of ovarian follicles, before and after breeding, in beef cows and heifers. *Anim Reprod Sci* 1997;49:13–28.
44. Townson D, Tsang P, Butler W et al. Relationship of fertility to ovarian follicular waves before breeding in dairy cows. *J Anim Sci* 2002;80:1053–1058.
45. Bleach E, Glencross R, Knight P. Association between ovarian follicle development and pregnancy rates in dairy cows undergoing spontaneous oestrous cycles. *Reproduction* 2004;127:621–629.
46. Ginther O, Kastelic J, Knopf L. Intraovarian relationships among dominant and subordinate follicles and the corpus luteum in heifers. *Theriogenology* 1989;32:787–795.
47. Dufour J, Ginther O, Casida L. Intraovarian relationship between corpora lutea and ovarian follicles in ewes. *Am J Vet Res* 1972;33:1445–1446.
48. Ginther O, Knopf L, Kastelic J. Ovarian follicular dynamics in heifers during early pregnancy *Biol Reprod* 1989;41:247–254.
49. Thatcher W, Driancourt M, Terqui M, Badinga L. Dynamics of ovarian follicular development in cattle following hysterectomy and during early pregnancy. *Domest Anim Endocrinol* 1991;8:223–234.
50. Kastelic J, Ko J, Ginther O. Suppression of dominant and subordinate ovarian follicles by a proteinaceous fraction of follicular fluid in heifers. *Theriogenology* 1990;34:499–509.
51. Adams G, Matteri R, Kastelic J, Ko J, Ginther O. Association between surges of follicle stimulating hormone and the emergence of follicular waves in heifers. *J Reprod Fertil* 1992;94:177–188.
52. Adams G, Kot K, Smith C, Ginther O. Effect of the dominant follicle on regression of its subordinates in heifers. *Can J Anim Sci* 1993;73:267–275.
53. Adams G. Control of ovarian follicular wave dynamics in cattle: implication for synchronization and superstimulation. *Theriogenology* 1994;41:25–30.
54. Bo G, Adams G, Pierson R, Mapletoft R. Exogenous control of follicular wave emergence in cattle. *Theriogenology* 1995;43:31–40.
55. Adams G. Control of ovarian follicular wave dynamics in mature and prepubertal cattle for synchronization and superstimulation. In: *Proceedings of the XX Congress of the World Association for Buiatrics, Sydney, 6–10 July, 1998.* Australian Association of Cattle Veterinarians, 1998, pp. 595–605.
56. Sunderland S, Crowe M, Boland M, Roche J, Ireland J. Selection, dominance and atresia of follicles during the oestrous cycle of heifers. *J Reprod Fertil* 1994;101:547–555.
57. Singh J, Pierson R, Adams G. Ultrasound image attributes of bovine ovarian follicles: endocrine and functional correlates. *J Reprod Fertil* 1998;112:19–29.
58. Gibbons J, Wiltbank M, Ginther O. Functional interrelationships between follicles greater than 4 mm and the follicle-stimulating hormone surge in heifers. *Biol Reprod* 1997;57:1066–1073.
59. Burns DS, Jimenez-Krassel F, Ireland J, Knight P, Ireland J. Numbers of antral follicles during follicular waves in cattle: evidence for high variation among animals, very high repeatability in individuals, and an inverse association with serum follicle-stimulating hormone concentrations. *Biol Reprod* 2005;73:54–62.
60. Ginther O, Beg M, Gastal E, Gastal M, Baerwald A, Pierson R. Systemic concentrations of hormones during the development of follicular waves in mares and women: a comparative study. *Reproduction* 2005;130:379–388.
61. Sunderland S, Knight P, Boland M, Roche J, Ireland J. Alterations in intrafollicular levels of different molecular mass forms of inhibin during development of follicular- and luteal- phase dominant follicles in heifers. *Biol Reprod* 1996;54:453–462.
62. Mihm M, Good T, Ireland J, Ireland J, Knight P, Roche J. The decline in serum FSH concentrations alters key intrafollicular growth factors involved in selection of the dominant follicle in heifers. *Biol Reprod* 1997;57:1328–1337.
63. Bleach E, Glencross R, Feist S, Groome N, Knight P. Plasma inhibin A in heifers: relationship with follicle dynamics, gonadotropins, and steroids during the estrous cycle and after treatment with bovine follicular fluid. *Biol Reprod* 2001; 64:743–752.
64. Beg M, Bergfelt D, Kot K, Ginther O. Follicle selection in cattle: dynamics of follicular fluid factors during development of follicle dominance. *Biol Reprod* 2002;66:120–126.
65. Laven J, Fauser B. Inhibins and adult ovarian function. *Mol Cell Endocrinol* 2004;225:37–44.
66. Ginther O, Bergfelt D, Kulick L, Kot K. Selection of the dominant follicle in cattle: role of two-way functional coupling between follicle-stimulating hormone and the follicles. *Biol Reprod* 2000;62:920–927.
67. Ginther O, Bergfelt D, Kulick L, Kot K. Selection of the dominant follicle in cattle: role of estradiol. *Biol Reprod* 2000;63:383–389.
68. Ginther O, Beg M, Bergfelt D, Donadeu F, Kot K. Follicle selection in monovular species. *Biol Reprod* 2001;65:638–647.
69. Mihm M, Bleach E. Endocrine regulation of ovarian antral follicle development in cattle. *Anim Reprod Sci* 2003;78:217–237.
70. Bergfelt D, Lightfoot K, Adams G. Ovarian synchronization following ultrasound-guided transvaginal follicle ablation in heifers. *Theriogenology* 1994;42:895–907.
71. Adams G, Matteri R, Ginther O. The effect of progesterone on growth of ovarian follicles, emergence of follicular waves and circulating FSH in heifers. *J Reprod Fertil* 1992;96:627–640.
72. Savio J, Thatcher W, Badinga L, de la Sota R, Wolfenson D. Regulation of dominant follicle turnover during the oestrous cycle in cows. *J Reprod Fertil* 1993;97:197–203.
73. Savio J, Thatcher W, Morris G, Entwistle K, Drost M, Mattiacci M. Effects of induction of low plasma progesterone concentrations with a progesterone-releasing intravaginal device on follicular turnover and fertility in cattle. *J Reprod Fertil* 1993;98:77–84.
74. Stock A, Fortune J. Ovarian follicular dominance in cattle: relationship between prolonged growth of the ovulatory follicle and endocrine parameters. *Endocrinology* 1993;132:1108–1114.
75. Kesler D, Garverick H. Ovarian cysts in dairy cattle: a review. *J Anim Sci* 1982;55:1147–1159.
76. Cook D, Smith C, Parfet J, Youngquist R, Brown E, Garverick H. Fate and turnover rate of ovarian follicular cysts in dairy cattle. *J Reprod Fertil* 1990;90:37–46.
77. Adams G, Kot K, Smith C, Ginther O. Selection of a dominant follicle and suppression of follicular growth in heifers. *Anim Reprod Sci* 1993;30:259–271.

78. Bao B, Garverick H. Expression of steroidogenic enzyme and gonadotropin receptor genes in bovine follicles during ovarian follicular waves: a review. *J Anim Sci* 1998;76:1903–1921.
79. Evans A, Fortune J. Selection of the dominant follicle in cattle occurs in the absence of differences in the expression of messenger ribonucleic acid for gonadotropin receptors. *Endocrinology* 1997; 138:2963–2971.
80. Ginther O, Beg M, Donadeu F, Bergfelt D. Mechanism of follicle deviation in monovular farm species. *Anim Reprod Sci* 2003; 78:239–257.
81. Ginther O, Bergfelt D, Beg M, Kot K. Follicle selection in cattle: relationships among growth rate, diameter ranking, and capacity for dominance. *Biol Reprod* 2001;65:345–350.
82. Evans A, Ireland J, Winn M et al. Identification of genes involved in apoptosis and dominant follicle development during follicular waves in cattle. *Biol Reprod* 2004;70:1475–1484.
83. Mihm M, Baker P, Ireland J et al. Molecular evidence that growth of dominant follicles involves a reduction in follicle stimulating hormone dependence and an increase in luteinizing hormone dependence in cattle. *Biol Reprod* 2006;74:1051–1059.
84. Beg M, Bergfelt D, Kot K, Wiltbank M, Ginther O. Follicular-fluid factors and granulosa-cell gene expression associated with follicle deviation in cattle. *Biol Reprod* 2001;64:432–441.
85. Ginther O, Bergfelt D, Beg M, Meira C, Kot K. In vivo effects of an intrafollicular injection of insulin-like growth factor 1 on the mechanism of follicle deviation in heifers and mares. *Biol Reprod* 2004;70:99–105.
86. Rivera G, Fortune J. Proteolysis of insulin-like growth factor binding proteins-4 and -5 in bovine follicular fluid: implications for ovarian follicular selection and dominance. *Endocrinology* 2003;144:2977–2987.
87. Liu Z, Youngquist R, Garverick A, Antoniou E. Molecular mechanisms regulating bovine ovarian follicular selection. *Mol Reprod Dev* 2009;76:351–366.
88. Mihm M, Baker P, Fleming L, Monteiro A, O'Shaughnessy P. Differentiation of the bovine dominant follicle from the cohort upregulates mRNA expression for new tissue development genes. *Reproduction* 2008;135:253–265.
89. Skinner M, Schmidt M, Savenkova M, Sadler-Riggleman I, Nilsson E. Regulation of granulosa and theca cell transcriptomes during ovarian antral follicle development. *Mol Reprod Dev* 2008;75:1457–1472.
90. Lopez H, Sartori R, Wiltbank M. Reproductive hormones and follicular growth during development of one or multiple dominant follicles in cattle. *Biol Reprod* 2005;72:788–795.
91. Quirk S, Hickey G, Fortune J. Growth and regression of ovarian follicles during the follicular phase of the oestrous cycle in heifers undergoing spontaneous and PGF-2a-induced luteolysis. *J Reprod Fertil* 1986;77:211–219.
92. Ko J, Kastelic J, Del Campo M, Ginther O. Effects of a dominant follicle on ovarian follicular dynamics during the oestrous cycle in heifers. *J Reprod Fertil* 1991;91:511–519.
93. Sartori R, Haughian J, Shaver R, Rosa G, Wiltbank M. Comparison of ovarian function and circulating steroids in estrous cycles of Holstein heifers and lactating cows. *J Dairy Sci* 2004;87:905–920.
94. Fricke P, Wiltbank M. Effect of milk production on the incidence of double ovulation in dairy cows. *Theriogenology* 1999;52:1133–1143.
95. Vasconcelos J, Sartori R, Oliveira H, Guenther J, Wiltbank M. Reduction in size of the ovulatory follicle reduces subsequent luteal size and pregnancy rate. *Theriogenology* 2001;56:307–314.
96. Jaiswal R, Singh J, Adams G. Developmental pattern of small antral follicles in the bovine ovary. *Biol Reprod* 2004; 71:1244–1251.
97. Richards J, Midgley A. Protein hormone action: a key to understanding ovarian follicular and luteal cell development. *Endocrinology* 1976;98:929–934.
98. Wandji S, Fortier M, Sirard M. Differential response to gonadotrophins and prostaglandin E2 in ovarian tissue during prenatal and postnatal development in cattle. *Biol Reprod* 1992; 46:1034–1041.
99. Itoh T, Kacchi M, Abe H, Sendai Y, Hoshi H. Growth, antrum formation, and estradiol production of bovine preanatral follicles cultured in a serum-free medium. *Biol Reprod* 2002; 67:1099–1105.
100. Fricke P, Al-Hassan M, Roberts A, Reynolds L, Redmer D, Ford J. Effect of gonadotropin treatment on size, number, and cell proliferation of antral follicles in cows. *Domest Anim Endocrinol* 1997;14:171–180.
101. Tanaka Y, Nakada K, Moriyoshi M, Sawamukai Y. Appearance and number of follicles and change in the concentration of serum FSH in female bovine fetuses. *Reproduction* 2001; 121:777–782.
102. Singh J, Dominguez M, Jaiswal R, Adams G. A simple ultrasound test to predict superovulatory response in cattle. *Theriogenology* 2004;62:227–243.
103. Xu Z, Garverick H, Smith G, Smith M, Hamilton S, Youngquist R. Expression of follicle-stimulating hormone and luteinizing hormone receptor messenger ribonucleic acids in bovine follicles during the first follicular wave. *Biol Reprod* 1995;53:951.
104. Peng X, Hsueh A, Lapolt P, Bjersing L, Ny T. Localization of luteinizing hormone receptor messenger ribonucleic acid expression in ovarian cell types during follicle development and ovulation. *Endocrinology* 1991;129:3200–3207.
105. Lawrence T, Beers W, Gilula N. Transmission of hormonal stimulation by cell-to-cell communication. *Nature* 1978;272: 501–506.
106. Marsh J. The stimulatory effect of luteinizing hormone on adenylyl-cyclase in the bovine corpus luteum. *J Biol Chem* 1970;245: 1596–1603.
107. Cooke B. Signal transduction involving cyclic AMP-dependent and cyclic AMP-independent mechanisms in the control of steroidogenesis. *Mol Cell Endocrinol* 1999;151:25–35.
108. Kosugi S, Van Dop C, Geffner M, Rabl W, Carel J, Chaussain J. Characterization of heterogeneous mutation causing constitutive activation of the luteinizing hormone receptor familial male precocious puberty. *Hum Mol Genet* 1995;4:183–188.
109. Russell D, Robker R. Molecular mechanisms of ovulation: co-ordination through the cumulus complex. *Hum Reprod Update* 2007;13:289–312.
110. Park J, Su Y, Ariga M, Law E, Jin S, Conti M. EGF-like growth factors as mediators of LH action in the ovulatory follicle. *Science* 2004;303:682–684.
111. Murdoch W, Gottsch M. Proteolytic mechanisms in the ovulatory folliculo-luteal transformation. *Connect Tissue Res* 2003;44:50–57.
112. Richards J, Russell D, Robker R, Dajee M, Alliston T. Molecular mechanisms of ovulation and luteinization. *Mol Cell Endocrinol* 1998;145:47–54.
113. Zolti M, Meirom R, Shemesh M et al. Granulosa cells as a source and target organ for tumor necrosis factor-alpha. *FEBS Lett* 1990;261:253–255.
114. Klipper E, Tatz E, Kisliouk T et al. Induction of heparanase in bovine granulosa cells by luteinizing hormone: possible role during the ovulatory process. *Endocrinology* 2009;150:413–421.
115. Clements J, Mukhtar A, Holland A, Ehrlich A, Fuller P. Kallikrein gene family expression in the rat ovary: localization to the granulosa cell. *Endocrinology* 1995;136:1137–1144.
116. Tribulo P. *Effect of OIF in seminal plasma on ovarian function in cattle.* MSc thesis, University of Saskatchewan, Saskatoon, 2012, p. 76.
117. Hyttel P, Xu K, Smith S, Greve T. Ultrastructure of in-vitro oocyte maturation in cattle. *J Reprod Fertil* 1986;78:615–625.
118. Tornell J, Billig H, Hillensjo T. Resumption of rat oocyte meiosis is paralleled by a decrease in guanosine 3′,5′-cyclic

monophosphate (cGMP) and is inhibited by microinjection of cGMP. *Acta Physiol Scand* 1990;139:511–517.
119. Pincus G, Enzmann E. The comparative behavior of mammalian eggs in vivo and in vitro: I. The activation of ovarian eggs. *J Exp Med* 1935;62:665–675.
120. Edwards R. Maturation in vitro of mouse, sheep, cow, pig, rhesus monkey and human ovarian oocytes. *Nature* 1965:208;349–351.
121. Palma G, Arganaraz M, Barrera A, Rodler D, Mutto A, Sinowatz F. Biology and biotechnology of follicle development. *Scientific World Journal* 2012;2012:938138.
122. Kruip T, Cran D, van Beneden T, Dieleman S. Structural changes in bovine oocytes during final maturation in vivo. *Gamete Res* 1983;8:29–47.
123. Hyttel P, Callesen H, Greve T. Ultrastructural features of preovulatory oocyte maturation in superovulated cattle. *J Reprod Fertil* 1986;76:645–656.
124. Hyttel P, Greve T, Callesen H. Ultrastructure of oocyte maturation and fertilization in superovulated cattle. *Prog Clin Biol Res* 1989;296:287–297.
125. Norris R, Ratzan W, Freudzon M et al. Cyclic GMP from the surrounding somatic cells regulates cyclic AMP and meiosis in the mouse oocyte. *Development* 2009;136:1869–1878.
126. Mehlmann L. Oocyte-specific expression of Gpr3 is required for the maintenance of meiotic arrest in mouse oocytes. *Dev Biol* 2005;288:397–801.
127. Norris R, Freudzon M, Mehlmann L et al. Luteinizing hormone causes MAP kinase-dependent phosphorylation and closure of connexin 43 gap junctions in mouse ovarian follicles: one of two paths to meiotic resumption. *Development* 2008;135:3229–3267.
128. Conti M, Hsieh M, Zamah A, Oh J. Novel signaling mechanisms in the ovary during oocyte maturation and ovulation. *Mol Cell Endocrinol* 2012;356:65–73.
129. Peter A, Levine H, Drost M, Bergfelt D. Compilation of classical and contemporary terminology used to describe morphological aspects of ovarian dynamics in cattle. *Theriogenology* 2009;71:1343–1357.
130. Singh J, Pierson R, Adams G. Ultrasound image attributes of the bovine corpus luteum: structural and functional correlates. *J Reprod Fertil* 1997;109:35–44.
131. Marciel M, Rodriguez Martinez H, Gustafsson H. Fine structure of corpora lutea in superovulated heifers *Zbl Vet Med A* 1992;39:89–97.
132. Gasse H, Peukert Adam I, Schwarz R, Grunert E. Die Stellung der Follikel-lutein-zyste im Zyklusgeschehen des Rindes: histologische, zytologische und hormonanalytische Untersuchungen. *Zbl Vet Med A* 1984;31:548–556.
133. Ginther O. In: *Ultrasonic Imaging and Animal Reproduction: Cattle.* Cross Plains, WI: Equiservices Publishing, 1998, p. 304.
134. Kastelic J, Bergfelt D, Ginther O. Relationship between ultrasonic assessment of the corpus luteum and plasma progesterone concentration in heifers. *Theriogenology* 1990; 33:1269–1278.
135. Pierson R, Ginther O. Ultrasonic evaluation of the corpus luteum of the mare. *Theriogenology* 1985;23:795–806.
136. Townson D, Ginther O. Ultrasonic echogenicity of developing corpora lutea in pony mares. *Anim Reprod Sci* 1989;20:143–153.
137. Eramian M, Adams G, Pierson R. Enhancing ultrasound texture differences for developing an in vivo "virutal histology" approach to bovine ovarian imaging. *Reprod Fertil Dev* 2007;19:910–924.
138. Maldonado-Castillo I, Eramian M, Pierson R, Singh J, Adams G. Classification of reproductive cycle phase using ultrasound-detected features. In: *Proceedings of the 4th Canadian Conference on Computer and Robot Vision.* IEEE Computer Society, 2007, pp. 258–265. Available at http://www.computer.org/csdl/proceedings/crv/2007/2786/00/27860258-abs.html

139. Singh J, Adams G. Histomorphometry of dominant and subordinate bovine ovarian follicles. *Anat Rec* 2000;258:58–70.
140. Acosta T, Miyamoto A. Vascular control of ovarian function: ovulation, corpus luteum formation and regression. *Anim Reprod Sci* 2004;82–83:127–140.
141. Siddiqui M, Ferreira J, Gastal E, Beg M, Cooper D, Ginther O. Temporal relationships of the LH surge and ovulation to echotexture and power Doppler signals of blood flow in the wall of the preovulatory follicle in heifers. *Reprod Fertil Dev* 2010;22:1110–1117.
142. Honnens A, Niemann H, Herzog K, Paul V, Meyer H, Bollwein H. Relationships between ovarian blood flow and ovarian response to eCG-treatment of dairy cows. *Anim Reprod Sci* 2009;113:1–10.
143. Siddiqui M, Almamun M, Ginther O. Blood flow in the wall of the preovulatory follicle and its relationship to pregnancy establishment in heifers. *Anim Reprod Sci* 2009;113:287–292.
144. Herzog K, Voss C, Kastelic J et al. Luteal blood flow increases during the first three weeks of pregnancy in lactating dairy cows. *Theriogenology* 2011;75:549–554.
145. Brozos C, Pancarci M, Valencia J et al. Effect of oxytocin infusion on luteal blood flow and progesterone secretion in dairy cattle. *J Vet Sci* 2012;13:67–71.
146. Garcia-Ispierto I, Lopez-Gatius F. Effects of GnRH or progesterone treatment on day 5 post-AI on plasma progesterone, luteal blood flow and leucocyte counts during the luteal phase in dairy cows. *Reprod Domest Anim* 2012;47:224–229.
147. Manns J, Hafs H. Controlled breeding in cattle: a review. *Can J Anim Sci* 1076;56:121–131.
148. McCracken J. Prostaglandins and luteal regression: a review. *Prostaglandins* 1972;1:1–4.
149. Inskeep E. Potential uses of prostaglandins in control of reproductive cycles of domestic animals. *J Anim Sci* 1973; 36:1149–1157.
150. Odde K. A review of synchronization of estrus in postpartum cattle. *J Anim Sci* 1990;68:817–830.
151. Larson L, Ball P. Regulation of estrous cycles in dairy cattle: a review. *Theriogenology* 1992;38:255–267.
152. Kotwica J, Skarzynski D, Jaroszewski J, Williams GL, Bogacki M. Uterine secretion of prostaglandin F2a stimulated by different doses of oxytocin and released spontaneously during luteolysis in cattle. *Reprod Nutr Dev* 1998;38:217–226.
153. Kastelic J, Knopf L, Ginther O. Effect of day of prostaglandin F2a treatment on selection and development of the ovulatory follicle in heifers. *Anim Reprod Sci* 1990;23:169–180.
154. Momont H, Seguin B. Influence of the day of estrous cycle on response to PGF2a products: implication for AI programs for dairy cattle. In: *Proceedings of the 10th International Congress on Animal Reproduction and Artificial Insemination.* University of Illinois Urbana-Champaign, 1984, Vol. 3, p. 336.
155. Wiltbank M, Shiao T, Bergfelt D, Ginther O. Prostaglandin F2a receptors in the early bovine corpus luteum. *Biol Reprod* 1995;52:74–78.
156. Kotwica J, Skarzynski D, Miszkiel G, Melin P, Okuda K. Oxytocin modulates the pulsatile secretion of prostaglandin F2alpha in initiated luteolysis in cattle. *Res Vet Sci* 1999;66:1–5.
157. Skarzynski D, Ferriera-Dias G, Okuda K. Regulation of luteal function and corpus luteum regression in cows: hormonal control, immune mechanisms and intercellular communication. *Reprod Domest Anim* 2008;43:57–65.
158. Adams G, Evans A, Rawlings N. Follicular waves and circulating gonadotropins in 8-month old prepubertal heifers. *J Reprod Fertil* 1994;100:27–33.
159. Nasser L, Adams G, Bo G, Mapletoft R. Ovarian superstimulatory response relative to follicular wave emergence in heifers. *Theriogenology* 1993;40:713–724.
160. Ginther O, Beg M. Dynamics of circulating progesterone concentrations before and during luteolysis: a comparison between cattle and horses. *Biol Reprod* 2012;86:170.

161. Adams G, Nasser L, Bo G, Mapletoft R, Garcia A, Del Campo M. Superstimulatory response of ovarian follicles of wave 1 versus wave 2 in heifers. *Theriogenology* 1994;42:1103–1113.
162. Evans A, Adams G, Rawlings N. Follicular and hormonal development in prepubertal heifers from 2 to 36 weeks of age. *J Reprod Fertil* 1994;102:463–470.
163. Evans A, Adams G, Rawlings N. Endocrine and ovarian follicular changes leading up to the first ovulation in prepubertal heifers. *J Reprod Fertil* 1994;100:187–194.
164. Ginther O, Kot K, Kulick L, Martin S, Wiltbank M. Relationships between FSH and ovarian follicular waves during the last six months of pregnancy in cattle. *J Reprod Fertil* 1996;108:271–279.
165. Crowe M. Resumption of ovarian cyclicity in post-partum beef and dairy cows. *Reprod Domest Anim* 2008;43(Suppl. 5):20–28.
166. Santos J, Rutigliano H, Sá Filho M. Risk factors for resumption of postpartum estrous cycles and embryonic survival in lactating dairy cows. *Anim Reprod Sci* 2009;10:207–221.
167. Walsh S, Williams E, Evans A. A review of the causes of poor fertility in high milk producing dairy cows. *Anim Reprod Sci* 2011;123:127–138.
168. Rajamahendran R, Taylor C. Characterization of ovarian activity in postpartum dairy cows using ultrasound imaging and progesterone profiles. *Anim Reprod Sci* 1990;22:171–180.
169. Tanaka T, Arai M, Ohtani S et al. Influence of parity on follicular dynamics and resumption of ovarian cycle in postpartum dairy cows. *Anim Reprod Sci* 2008;108:134–143.
170. Savio J, Boland M, Hynes N, Roche J. Resumption of follicular activity in the early postpartum period of dairy cows. *J Reprod Fertil* 1990;88:569–579.
171. Savio J, Boland M, Roche J. Development of dominant follicles and length of ovarian cycles in post-partum dairy cows. *J Reprod Fertil* 1990;88:581–591.
172. Murphy M, Boland M, Roche J. Pattern of follicular growth and resumption of ovarian activity in post-partum beef suckler cows. *J Reprod Fertil* 1990;90:523–533.
173. Rhodes F, McDougall S, Burke C, Verkerk G, Macmillan K. Invited review: Treatment of cows with an extended postpartum anestrous interval. *J Dairy Sci* 2003;86:1876–1894.
174. Erickson B, Reynolds R, Murphree R. Ovarian characteristics and reproductive performance of the aged cow *Biol Reprod* 1976;15:555–560.
175. Klein J, Sauer M. Assessing fertility in women of advanced reproductive age. *Am J Obstet Gynecol* 2001;185:758–770.
176. Stensen M, Tanbo T, Storeng R, Byholm T, Fedorcsak P. Routine morphological scoring systems in assisted reproduction treatment fail to reflect age-related impairment of oocyte and embryo quality. *Reprod Biomed Online* 2010;21:118–125.
177. Malhi P, Adams G, Singh J. Bovine model for the study of reproductive aging in women: follicular, luteal and endocrine characteristics. *Biol Reprod* 2005;73:45–53.
178. Malhi P, Adams G, Pierson R, Singh J. Bovine model of reproductive aging: response to ovarian synchronization and superstimulation. *Theriogenology* 2006;66:1257–1266.
179. Malhi P, Adams G, Mapletoft R, Singh J. Oocyte developmental competence in a bovine model of reproductive aging. *Reproduction* 2007;134:233–239.
180. Seguin B. Control of the reproductive cycle in dairy cattle. In: *Proceedings of the Annual Meeting of the Society for Theriogenology*, 1987, pp. 300–308.
181. Kastelic J, Ginther O. Factors affecting the origin of the ovulatory follicle in heifers with induced luteolysis. *Anim Reprod Sci* 1991;26:13–24.
182. Savio J, Boland M, Hynes N, Mattiacci M, Roche J. Will the first dominant follicle of the estrous cycle of heifers ovulate following luteolysis on Day 7? *Theriogenology* 1990;33:677–687.
183. Evans A, Mossa F, Fair T et al. Causes and consequences of the variation in the number of ovarian follicles in cattle. *Soc Reprod Fertil Suppl* 2010;67:421–429.
184. Fortune J. Ovarian follicular growth and development in mammals. *Biol Reprod* 1994;50:225–232.

Chapter 25

Maternal Recognition and Physiology of Pregnancy

Caleb O. Lemley[1], Leticia E. Camacho[2] and Kimberly A. Vonnahme[2]

[1]*Department of Animal and Dairy Sciences, Mississippi State University, Starkville, Mississippi, USA*
[2]*Department of Animal Sciences, North Dakota State University, Fargo, North Dakota, USA*

Maternal recognition of pregnancy

Maternal recognition of pregnancy encompasses the process whereby a biochemical signal is generated to prevent luteal regression, allowing the corpus luteum to persist and continue to secret adequate progesterone. The requirement of progesterone for the successful maintenance of pregnancy exceeds the length of the bovine estrous cycle and therefore the pregnant female must quickly recognize that a conceptus is present. In cattle, this signal involves conceptus secretion of interferon (IFN)-tau which further stimulates luteotropic and antiluteolytic signals. Although some studies use these words interchangeably, we will further define a luteotropin as a signal that stimulates luteal secretion of progesterone, while antiluteolytic substances act as a signal to block luteolysis by inhibiting or masking endogenous luteolytic signals. This maternal recognition of pregnancy signal is also occurring well in advance of conceptus implantation, which begins during the third week of pregnancy, and is classified as a superficial attachment in cattle.

Several biological assays have focused on elucidating substances involved in prolonging the lifespan of the bovine corpus luteum. The majority of these earlier assays focused on examining estrous cycle length in nonpregnant cattle injected with pituitary extracts. In 1965, researchers at Cornell University reported that luteinizing hormone (LH) was the luteotropin for cattle and indeed injections of pituitary extracts or purified bovine LH extended the bovine estrous cycle by 11 or 16 days, respectively.[1] This study confirmed the luteotropic actions of LH, which could serve as the endogenous biochemical signal from the conceptus; however, apart from primates and mares, no other mammals appear to possess both genetic subunits for a chorionic gonadotropin. Therefore, a chorionic gonadotropin possessing LH properties is not the signal for maternal recognition of pregnancy in cattle. However, these early studies did contribute to the foundational groundwork examining mammalian species differences in prolonging the lifespan of the corpus luteum, which is an important component of the conceptus–maternal cross-talk during maternal recognition of pregnancy.

These early results led researchers to examine other potential substances originating from the conceptus that may act in a luteotropic and/or antiluteolytic nature to prevent luteolysis during pulsatile release of prostaglandin (PG)$F_{2\alpha}$. Intrauterine infusions of estradiol-17β and PGE_2 prolong the lifespan of the bovine corpus luteum as well as progesterone secretion by 5–6 days compared with non-pregnant controls.[2] The luteotropic actions of estradiol-17β are more than likely indirect because estrogens have been show to inhibit LH-induced secretion of progesterone by cultured bovine luteal cells.[3] The indirect actions of intrauterine estradiol-17β on prolonging luteal lifespan may involve an increase in uterine and/or ovarian blood flow. A local increase in blood flow to the gravid uterine horn occurs coincidental with maternal recognition of pregnancy in the sow, ewe, and cow.[4] In addition, this local increase in uterine blood flow may help to redirect prostaglandin transport from the uterine horn to the adjacent ovary via the utero-ovarian plexus. PGE_2 directly stimulates progesterone secretion and cyclic AMP production of bovine luteal cells *in vitro* and in a similar manner to that of LH.[5] Moreover, both the bovine endometrium and corpus luteum express the complete prostaglandin biosynthesis pathway and the differential expression of these components during the critical period of maternal recognition of pregnancy (days 15–17 of the estrous cycle) can exert luteotropic or antiluteolytic actions.[6]

Interferon-tau, first identified as bovine trophoblast protein 1,[7] is recognized as the maternal recognition of pregnancy signal in cattle.[8] Interferon-tau secretion by embryonic trophoblast cells is highest between days 15 and 17, and is observed up to day 28 of pregnancy; moreover, an inadequate response of the endometrium to IFN-tau is likely to be one of the major reasons for improper production and action of $PGF_{2\alpha}$ and PGE_2 leading to pregnancy failure. *In vitro* cultures of bovine endometrial cells with increasing concentrations of IFN-tau show preferential production of PGE$_2$ compared with PGF$_{2\alpha}$.[9] Moreover, the spatial and temporal relationship of prostaglandin metabolism, transport, and receptor density changes substantially in the uterine endometrium and myometrium following IFN-tau exposure.[6] Elevated physiological concentrations of IFN-tau from the conceptus will selectively increase the ratio of PGE$_2$ to PGF$_{2\alpha}$. This relationship sets up an antiluteolytic effect of IFN-tau combined with a luteotropic effect of PGE$_2$ which eventually leads to prolonged corpus luteum lifespan and proper establishment of pregnancy. In addition, embryonic secretion of PGE$_2$ may produce a synergistic effect with the increased endometrial production of PGE$_2$ that was stimulated by conceptus secretion of IFN-tau. This PGE$_2$ can be competitively transported through the utero-ovarian plexus to the ovary, where it increases production of luteal PGE$_2$, thereby favoring luteal maintenance as a luteotropin.[6] In contrast, an elevated ratio of PGF$_{2\alpha}$ to PGE$_2$ will stimulate uterine contractility and competitive transport of PGF$_{2\alpha}$ through the utero-ovarian plexus toward the corpus luteum, thereby initiating luteal regression. Together, uterine contractility and luteal regression can lead to pregnancy failure and/or a return to a new estrous cycle following a normal lifespan of the corpus luteum.

Prevention of corpus luteum regression and continual secretion of adequate progesterone is vital for successful maintenance of pregnancy to term. The primary targets for progesterone are the reproductive tract. The amplitude of progesterone secretion can influence the early uterine environment, including the inhibition of mitotic division of the endometrium, the induction of stromal differentiation, and the stimulation of additional glandular secretions of the uterine epithelium or histotroph.[10] This increase in glandular uterine secretions of nutrients and growth factors is essential for early conceptus development.[11] In 1956, researchers proposed the progesterone block hypothesis, which states that progesterone maintains pregnancy by directly blocking labor and removal of the progesterone source will initiate parturition.[12] Later studies revealed that progesterone supplementation prevented the insertion of gap junction proteins between uterine smooth muscle cells by decreasing the expression of connexin-43.[13] Progesterone has also been shown to downregulate uterine contractility-associated genes.[14] In general cows with less than 0.5 ng/mL of progesterone have elevated contractility in circular smooth muscle cells of the myometrium compared with that in diestrus cows (>4.0 ng/mL progesterone).[15] These differences in myometrial excitation–contraction are brought about by the genomic actions of progesterone, which lead to a relaxed and quiescent uterus able to maintain a successful pregnancy until progesterone withdrawal.[14] In addition, progesterone synthesis depends on luteal production for the first 200 days of gestation; however, the extragonadal source of progesterone, primarily the bovine placenta, does not secret adequate amounts of progesterone until the third trimester of pregnancy. This placental progesterone secretion is dependent on luteal ovarian input. The luteal placental shift in progesterone secretion, which may contribute to the maintenance of pregnancy, only occurs after PGF$_{2\alpha}$-induced luteolysis or in ovariectomized animals.[16] Moreover, in the cow, the corpus luteum may exert control over placental steroidogenesis, which does not take over progesterone secretion until ovarian luteal deficiencies become apparent during late gestation.

Conceptus elongation and extraembryonic membranes

The early elongation of the bovine conceptus involves exponential increases in length and width of the trophectoderm and requires differentiation of extraembryonic membranes as well as maternal endometrial influences. This elongation begins on day 14, following the hatching of the ovoid blastocyst at approximately day 9–10, and forms a filamentous conceptus by day 19–20 after insemination.[17] On day 14, the majority of ovoid or tubular embryos of approximately 5 mm in length will start to form an epiblast, defining the embryonic disk, which is discernible from the epithelium of the trophoblast. Both of these compartments are internally lined by the continuous hypoblast of which the majority stains positive for α-fetoprotein;[18] however, the histological appearance of the hypoblast varies in relation to its spatial orientation to the epiblast versus the trophoblast cell layers. The spatial orientation results in two types of hypoblast cells: squamous cell types if located next to the trophoblast layer, or cuboidal to columnar cell types underneath the epiblast or embryonic disk.[18] At this point a distinctive basal lamina is formed between the trophoblast or epiblast and the hypoblast cell layers. In addition, adjacent epiblast cells are connected via tight junctions, as well as peripheral epiblast cells next to the trophoblast cells. Presumptive signs of early embryonic mortality in the tubular conceptus consist of apoptotic epiblast cells or a loosening of the hypoblast from the trophoblast. During this elongation phase the initial folds of the amnion will form as an outgrowth of the trophoblast cells adjacent to the epiblast, while the hypoblast is not included in the formation of this extraembryonic membrane. The most discernible difference during this time-frame is the hypertrophic and hyperplastic trophectoderm growth of the differentiating chorionic sac, which may grow from a few centimeters up to 30 cm or more.[19]

The early attachment period of days 20–33 post insemination proceeds with sequential stages of pre-contact, apposition, and adhesion between the uterus and the chorion, and ends in a microvillous interdigitation of the uterine epithelium and trophoblast. This time period coincides with several reports of limited success in flushing intact conceptuses for experimental use.[20] During the early attachment phase the development of the allantois begins, and on day 21 after insemination a strong positive association between embryo length and width of the allantois can be observed.[18] Some researchers have observed abnormalities or underdevelopment of the extraembryonic membranes following *in vitro* fertilization or somatic cell nuclear transfer; however, the underlying mechanisms for development failure are not clearly understood. Therefore, embryonic mortality during these time points may be caused by a difficulty in nutrient mobilization brought about by the allantois showing deficient vascularization, growth retardation, or its absence.[21] In relation to the dramatic elongation of the trophectoderm of the chorionic sac, the allantois undergoes gradual growth from days 21 to 25 post insemination, when it will start to occupy both uterine horns and begin fixation to the chorion to form the chorioallantois placenta; however, this fusion remains incomplete until days 60–70 of pregnancy.[21]

In addition to the morphological change from a spherical conceptus into a filament-like tube during conceptus elongation, several molecular markers have been characterized in relation to growth and cellular differentiation. The trophoblast mononucleate cells of the trophectoderm will begin to proliferate and differentiate, which may be induced by specific transcription factors such as POU-domain class 5 transcription factor, ERG, and CDX2.[22] These transcription factors have been shown to regulate genes such as IFN-tau and trophoblast domain proteins, which are upregulated during maternal recognition of pregnancy. In addition, the differentiation of trophoblast mononucleate cells into trophoblast giant cells coincides with an increase in pregnancy-associated glycoproteins, prolactin-related proteins, and placental lactogen. These proteins are associated with cellular differentiation of the trophectoderm; however, they may not be associated with the proliferation of the trophectoderm during conceptus elongation into the filament-like tube.[22] Transcriptome profiling using a bovine-specific microarray has revealed the upregulation of approximately 500 genes and the downregulation of only 26 genes during transition from a blastocyst (day 7) to an ovoid tubular conceptus (day 14). Of these genes, IFN-tau was positively correlated with conceptus size; therefore, it may be a key factor in trophectoderm proliferation during the elongation phase. After further elongation of the bovine conceptus (gestational days 17–19), specific genes involved in trophoblast cell differentiation, migration, and adhesion are upregulated, all of which play a role in fusion of the trophoblast cells to the endometrium of the uterus. Of the 80 genes upregulated during this phase, the *CD9* gene, which encodes for cell-surface glycoprotein belonging to the tetraspanin family, of the differentiated trophoblast giant cells more than likely plays a role in cellular migration and adhesion of the conceptus to the maternal endometrium during early attachment (gestational day 20).

Uterine histotroph

The growth, development, and welfare of the bovine conceptus depend on uterine histotrophic nutrition derived from the secretions of uterine endometrial glands. The uterine endometrium is a complex tissue composed of a luminal epithelial cell layer, superficial glandular epithelium, deep glandular epithelium, as well as stromal cells with a fibroblast-like morphology. In addition to autonomous regulation of trophectoderm proliferation and differentiation during conceptus elongation, uterine input via endometrial secretions will also drive early conceptus development. As previously mentioned, the bovine conceptus must secrete sufficient quantities of IFN-tau by day 16 of pregnancy in order to inhibit the production of luteolytic pulses of $PGF_{2\alpha}$ by the endometrium to allow for maternal recognition of pregnancy. In addition to conceptus secretions that alter endometrial pathways in a paracrine manner, ovarian hormones have also been shown to stimulate or inhibit endometrial secretions and glandular output. A number of studies have demonstrated that peripheral concentrations of progesterone during the early post-conception period are associated with advanced conceptus elongation and higher pregnancy rates in cattle and sheep. In contrast, lower concentrations of progesterone in dairy cows are associated with decreased conceptus development. Ovarian progesterone can influence the early uterine environment including endometrial sections of nutrients and growth factors that are essential for early conceptus development. Moreover, supplementing exogenous progesterone from day 5 to 9 after insemination resulted in a fourfold increase in trophoblast length and a sixfold increase in uterine concentrations of IFN-tau.[23] In addition to the concentrations of progesterone in peripheral circulation, progesterone receptor expression in the endometrium (both luminal epithelium and superficial glandular epithelium) is altered in cows with decreased versus elevated concentrations of progesterone during early post-conception. In cows with low post-conception concentrations of progesterone not only are specific nutrient transporters decreased in the endometrium but also progesterone receptor expression is maintained for a longer period of time in the luminal and superficial glandular epithelium. In contrast, elevated concentrations of progesterone, brought about by exogenous treatment, increase nutrient transporter expression and decrease progesterone receptor expression in the luminal and superficial glandular epithelium.[24] Moreover, this downregulation in endometrial progesterone receptors is required to establish uterine and endometrial receptivity to conceptus attachment.

Placental vascularization

The placenta plays a major role in the regulation of fetal growth. The ruminant placenta is morphologically classified as cotyledonary and histologically as syndesmochorial. On the uterine wall there are specialized areas of endometrium called caruncles, with a button-like appearance.[25] In nonpregnant ruminants, the caruncles are organized in two dorsal and two ventral rows that run lengthwise along the uterine horns. There are approximately 100 caruncles and they can be seen even in uteri from fetal calves.[26] The chorioallantois, which has a flat surface, becomes irregular when it starts to cover the caruncles; this is caused by the growth and expansion of the conceptus on the lumen of the uterus. This process is followed by the recognition of the cotyledons.[26] The caruncular–cotyledonary unit is called a placentome and is formed from the growth and interdigitation of fetal villi and caruncular crypts adopting a convex shape.[25] Contact surface area is enhanced because the cotyledon finger-like projections enter the crypts formed in the caruncles. Placentomes vary in size, but they are bigger in the horn and decrease in size close to the tip of the horns.[26] The placentome is the primary functional area of physiological exchanges between mother and fetus.

The efficiency of placental nutrient transport is directly related to uteroplacental blood flow.[27] All nutrients and wastes that are exchanged between the maternal and fetal systems are transported via the uteroplacental unit.[27,28] Establishment of functional fetal and uteroplacental circulations is one of the earliest events during embryonic/placental development.[29,30] In order to support the exponential increase in fetal growth during the last half of gestation,[31]

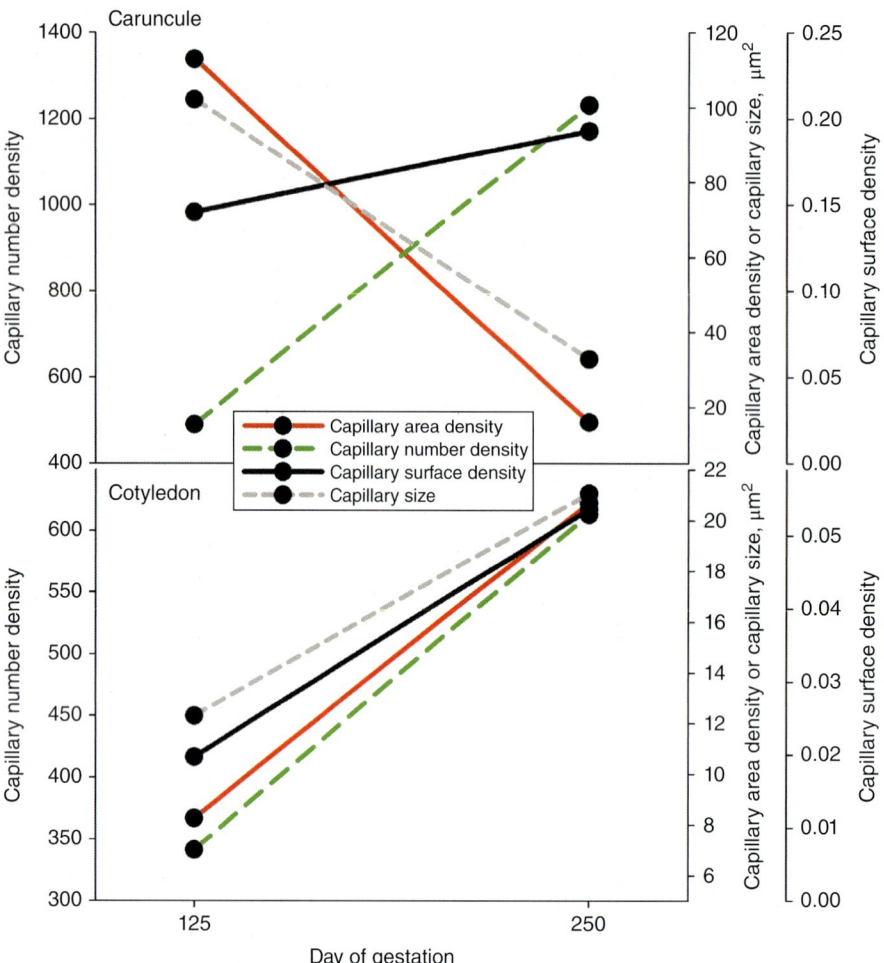

Figure 25.1 Capillary number density, area density, surface density, and size in the maternal (caruncular) and fetal (cotyledonary) portion of the bovine placentome from mid (day 125) to late (day 250) of gestation.

proper growth and development of the uteroplacental vascular bed must occur during the first half of pregnancy.[27,32] Understanding the factors that impact uteroplacental blood flow will directly impact placental efficiency and thus fetal growth. However, despite intensive research in the area of placental–fetal interactions, the regulators of placental growth and vascularization as well as uteroplacental blood flow are still largely unknown, particularly in cattle.

Limited data on development of the bovine placentomal capillary bed are available. In the cow, cotyledonary growth progressively increases throughout gestation.[33,34] Histological analysis of capillary bed development in the cow has been performed at mid and late gestation.[34] From mid to late gestation, capillary area density, a measure related to blood flow, decreases by about 30% in caruncular tissue, while it increases about 186% in cotyledonary tissue (Figure 25.1). The numbers of capillaries increase by about 150% and 80% in caruncular and cotyledonary tissue, respectively, during this time. Capillary surface density, a measurement related to nutrient exchange, is also increased in both caruncular and cotyledonary tissues (32% and 172%, respectively), while capillary size decreases 67% in caruncular tissue but increases 71% in cotyledonary tissue from mid to late gestation (Figure 25.1). The pattern of capillary development is very different between the maternal and fetal portions of the placentome possibly due to the energetic demands of these independent tissues that share a similar function – delivery of nutrients to the developing calf.

It is clear that the placenta plays a fundamental role in providing for the metabolic demands of the fetus. Although placental growth slows during the last half of gestation, placental function increases dramatically to support the exponential rate of fetal growth.[35,36] For example, in sheep and cattle, uterine blood flow increases approximately threefold to fourfold from mid to late gestation[27,37,38] (see below). The relationship between uteroplacental blood flow and conceptus size throughout gestation is further defined in the next section.

Maternal cardiovascular function

During pregnancy the physiologic state of the dam is associated with significant but reversible alterations to metabolic demand as well as alterations to the endocrine and cardiovascular systems. As previously mentioned, during early gestation (e.g., maternal recognition of pregnancy and conceptus elongation) endocrine secretions of the ovary and conceptus help establish a healthy pregnancy. Moreover, these secretions serve as biochemical cross-talk between the maternal endometrium and fetal membranes of the placenta during mid to late gestation and help assure proper nutrient and waste exchange during exponential growth of the fetus. Maternal cardiovascular functional capacity changes dramatically during pregnancy, whereby systemic arterial blood pressure and vascular resistance decrease and cardiac output, heart rate, stroke volume, and blood volume increase.[39] Although not all these variables have been determined during bovine pregnancy, several mammalian species including the ewe have shown a decrease in mean arterial pressure in early pregnancy that persists throughout gestation. Moreover, the decrease in arterial pressure (estimated as 5–10%) is minor compared with the approximate 20–30% decrease in total peripheral vascular resistance. Maternal cardiac output has been shown to increase as much as 30–40% in pregnant versus nonpregnant ruminants. Therefore, the increase in cardiac output is associated with a dramatic fall in systemic vascular resistance, allowing researchers to characterize pregnancy as a state of systemic vasodilation resulting in profound increases in total systemic flows to all vascular beds. The majority of these studies characterize this relationship by the equation:

$$\text{Systemic vascular resistance} = \frac{\text{Mean arterial pressure}}{\text{Cardiac output}}$$

Using this equation we can see that the decrease in systemic vascular resistance during pregnancy and the increase in cardiac output (total systemic blood flow) help to maintain arterial pressure. In addition, apart from this relationship, increases in blood volume and activation of the renin–angiotensin system may also contribute to maintaining blood pressure during this physiological state of substantial vasodilation; however, limited data exist in cattle. To determine specific contributions to the increase in cardiac output occurring during pregnancy we must first examine the cardiac output equation, which states that cardiac output = heart rate × stroke volume, where stroke volume equals the volume of blood pumped from one ventricle of the heart with each beat. In the majority of mammalian species heart rate increases by approximately 15% during pregnancy, which does not fully explain the 30–40% increase in cardiac output. Therefore, stroke volume may increase as much as 30–35% during pregnancy and is one of the major contributors to this increase in cardiac output. In sheep treated chronically with estrogen, researchers have shown a similar pregnancy-induced increase in left ventricular heart dimensions and enlargement; therefore the changes to the endocrine system during pregnancy may help mediate the temporal changes to maternal cardiovascular function.[39]

The rise in maternal cardiac output during pregnancy is also associated with an increase in plasma and blood volume in cows.[40] This increase in blood volume varies between species and will also depend on the nutritional status of the dam as well as singleton versus twin pregnancies. In cattle, blood volume expands by only 10–20% during pregnancy, while in litter-bearing species blood volume may expand by 30–50% during pregnancy. With the increase in plasma volume, the dam must maintain a proper balance of water and electrolyte retention; therefore, similar to the alterations in maternal arterial pressure, this increase in plasma volume will be integrated with the renin–angiotensin system which can serve an additional purpose as an extrinsic modulator of kidney function and urinary secretion. In addition to the dramatic changes in the maternal cardiovascular system during pregnancy, it is even more noteworthy to point out that the majority of mammals will return to the nonpregnant levels of cardiovascular function within 2–5 weeks after parturition. Although lactating, high-producing, nonpregnant dairy cattle will show a substantial increase in cardiac output compared with their nonpregnant nonlactating counterparts, this redistribution of blood flow during the transition period from the uteroplacental vasculature toward the mammary gland is still a phenomenal physiological feat to allow for peak lactation shortly after parturition.

Maternal nutrient partitioning and endocrinology

Interactions between the maternal endocrine system and the conceptus, and vice versa, is vital to assure continual development of the bovine fetus and to assure healthy offspring fit for life outside the uterus. One example of this interaction is the ability of the bovine conceptus to stimulate mammary growth and development during gestation. This controlled programmed development of the mammary gland allows for synchronization of milk production and secretion with the delivery of a mature calf.[41] Hormonal production by the conceptus may act in a paracrine manner to stimulate its own growth; however, this hormonal production may also interact with the maternal endocrine system to regulate mammary growth, development, and lactogenesis. Secretion of bovine placental lactogen peaks at around 160–200 days of gestation and then remains high until term. Dairy cows have elevated concentrations of bovine placental lactogen compared with beef cows; furthermore, within a subpopulation of dairy cows a significant positive association has been observed between placental lactogen concentrations during gestation and subsequent milk yield. In addition, production of estrone, conjugated estrone sulfate, and estradiol is observed to increase between days 70 and 110 of gestation. This increase in estrogens should allow for an increased exposure of maternal tissues to estrogenic activity; however, considerable conjugation of estrogen by the uterus and conceptus before transfer into peripheral circulation will decrease maternal tissue exposure. This increase in production should correspond to an increase in aromatase activity as well as estrogen sulfotransferases, generating an estrone sulfate metabolite with low biological activity but a longer peripheral half-life. However, irrespective of the lowered biological activity of conjugated estrogens, the concentration of total estrogen (conjugated and unconjugated) is highly correlated with birthweights and subsequent pre-weaning growth rates in beef calves, which may be associated with subsequent milk production of the dam. Apart from these associations between conceptus hormone production and subsequent milk production of the dam, it is still important to note that several hormones of pregnancy (progesterone, estrogens, and placental lactogens) regulate cell numbers and vascularity of the mammary gland.

Regulation of nutrient partitioning during pregnancy is an orchestrated event helping to maintain the metabolism of body tissues necessary to support the physiological state.[42] In dairy cows, pregnancy imposes a substantial cost to the animal, so that nutritional plane at the end of gestation is about 75% greater than in a nonpregnant cow of the same weight.[42] After 7 months of gestation the fetus will have acquired approximately 40% of its birthweight, while during the last 2 months of gestation fetal demands for glucose and amino acids are equal to mammary use of nutrients equivalent to about 5 kg of milk daily.[42] Therefore, in the dairy cow, drying off cows at approximately 60 days before expected calving helps ease the competition between nutrient use for lactation and the major phase of fetal growth. There is limited information on the specific nutrient requirements of the bovine fetus compared with the extensive amount of research accomplished in the sheep fetus. Gross carbon uptake of the sheep fetus is approximately 8 g/day per kilogram of fetal weight, of which 40% is retained and the remainder is shuttled back to the dam as either carbon dioxide or urea carbon. Apart from glucose, amino acids provide a large portion of the metabolic fuel, estimated at a gross nitrogen accretion rate of 0.65 g nitrogen per kilogram of fetus per day, which is in excess of the average fetal uptake of 1.5 g of nitrogen per kilogram of fetus per day. As mentioned with the carbon excretion via urea, nearly 0.3 g of nitrogen per kilogram of fetus per day is excreted as urea, indicating amino acid catabolism, presumably via gluconeogenic pathways of the fetus and placenta. It is important to note that the influxes of nutrient substrates to meet the needs of fetal oxidative metabolism are also used for fetal anabolism and it has been proposed that specific nutrient substrate fate (catabolism vs. anabolism) is controlled autonomously by the fetus. This has led some researchers to propose that the fetus should not be regarded as a "passenger" but rather that it has some self-control or autonomy in its own fate. This is more apparent when we examine some of the changes in the fetus and placenta that allow the fetus to adapt to specific environmental insults bestowed upon the dam.

Embryogenesis, fetal growth, and organ development

The embryonic period in cattle is defined as the time from conception (i.e., single-cell embryo, referred to as a zygote) to the completion of organogenesis. This period typically runs from day 1 to day 42 of gestation and the largest percentage of pregnancy wastage occurs during the embryonic period. The fetal period is defined as the remainder of gestation from day 42 until day 280 when the completion of organ differentiation occurs (Figure 25.2).

Embryogenesis

During early embryogenesis and prior to conceptus attachment, differentiation of the germ cell layers are becoming apparent. The three germ cell layers are referred to as the endoderm, mesoderm, and ectoderm. The endoderm, the innermost layer, will differentiate into the entire alimentary tract except for part of the mouth and terminal part of the rectum, which are lined by protective involutions of the ectoderm. In addition, the endoderm will develop into the respiratory tract (trachea, bronchi, and lung alveoli), the majority of the endocrine glands, and the urinary systems. The ectoderm is the most exterior germ cell layer and will give rise to the nervous system and the integumentary system. The middle germ cell layer, the mesoderm, will give rise to cardiac, skeletal, and smooth muscle, the circulatory system, as well as the majority of the reproductive tract, excluding the external portions of the genitalia which develop from the ectoderm.

During embryogenesis an invagination of the oral ectoderm toward the neural epithelium results in a diverticulum referred to as Rathke's pouch. During the formation of this diverticulum, it will lose contact with the oral ectoderm, forming the anterior portion of the pituitary. This anterior lobe is sometimes referred to as the adenohypophysis, because of the glandular derivation of the tissue. Few studies have characterized the formation of Rathke's pouch during early bovine embryogenesis; however, a common disorder known as a residual cleft cyst of the pituitary gland has been recently examined using postmortem magnetic resonance imaging of cattle.[43] Residual cleft cyst of the pituitary was observed in 26–35% of slaughtered cattle, which showed several morphological similarities to the human disorder known as Rathke's cleft cysts. These structural abnormalities of the pituitary may allow gradual accumulation of endocrine secretions over time and Holstein cattle may be more at risk for developing fluid-filled lesions compared with other breeds. More than likely these residual pituitary cysts are formed early during embryogenesis and early formation of Rathke's pouch; however, pathological conditions and/or productivity have not been associated with the appearance or relative size of the pituitary cysts in cattle.[43]

Sexual differentiation and the freemartin

In the realm of bovine fetal organogenesis, no other organ system has garnered as much attention as the development of the reproductive tract. From a sexual differentiation standpoint, the genetic sex of the offspring is established at the time of fertilization when the sperm delivers either an X (female) or Y (male) chromosome to the homogametic oocyte. In cattle, following a few weeks of embryogenesis the process of gonadal sexual differentiation will begin. One of the first steps in sexual differentiation of the gonad involves migration, proliferation, and invasion of primordial germ cells. It is now accepted that the first primordial germ cells also represent hematopoietic stem cells originating from the wall of the allantois as well as the yolk sac of the mesoderm.[44] In addition, hematopoietic stem cells also originate in other parts of the mesoderm and more recent findings suggest that the final proliferative primordial germ cells of the gonad originate exclusively in the aorta–gonad–mesoderm region, while short-lived hematopoietic stem cells of the yolk sac have a restricted differentiation capacity. It is important to note the early common lineage of hematopoietic stem cells (undifferentiated blood cells) and primordial germ cells (undifferentiated gametes). The apparent differences between these cell types in the adult may be due to the presence of

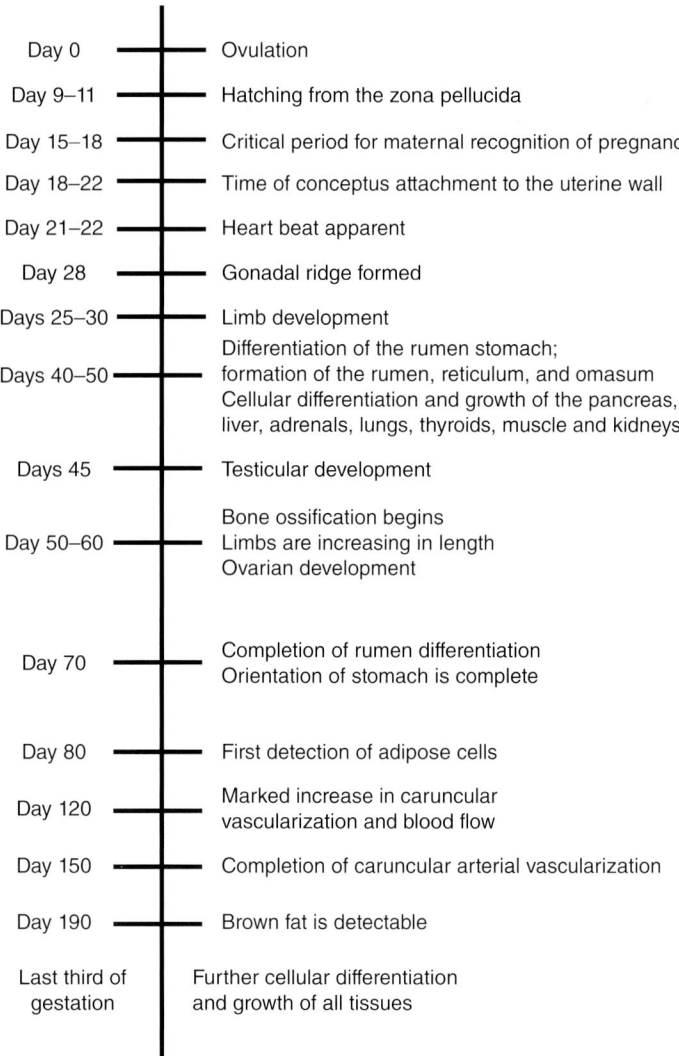

Figure 25.2 Time line of bovine fetal development. See text for references.

surrounding interstitial and stromal cells of the gonad, which help to further differentiate these early hematopoietic stem cells into germ cells. By 23–25 days of embryonic age in cattle, the primordial germ cells are located in the midgut and hindgut in close proximity to the undifferentiated gonad. By 27–31 days of embryonic age these primordial germ cells start to populate the undifferentiated gonad, which rests on the dorsal body wall known as the gonadal ridge. This accumulation of primordial germ cells at the gonadal ridge will continue until day 40 of embryonic age.

Two potential pathways for the transportation of primordial germ cells to the gonadal ridge have been proposed. First, most mammalian species are thought to have active migration of primordial germ cells via a chemoattractant produced by the gonadal ridge. The second pathway involves passive transportation via the bloodstream toward the gonadal ridge, which occurs in avian species and some reptiles. Recent advances in localizing the temporal–spatial distribution of bovine primordial germ cells during early embryogenesis have argued against both of these pathways.[45] The active migration of primordial germ cells may not be necessary because early bovine primordial germ cells (day 18 of embryonic age) have been localized to the proximal yolk sac at a distance of less than 100 μm from the indifferent germinal disk. Therefore, focus on the transportation system may be redundant due to the close proximity of the primordial germ cells to the developing gonadal ridge during early development. However, keep in mind from that day 18 to day 25 the growth and elongation of the embryo will continue to move the primordial germ cells further from their point of invasion; therefore, some form of transportation and proper signaling from the gonadal ridge of the indifferent gonad would be necessary as embryo length increases.

The embryo develops three separate renal systems: pronephros, mesonephros, and metanephros. These systems develop in pairs. The pronephros is the most basic of the three excretory organs of the bovine embryo and it will degenerate via apoptosis shortly after formation of the mesonephros (the embryonic kidney). In lower animals, the mesonephros will remain as the adult kidney; however, in the cow the final differentiation of the metanephros will form the functional adult kidney. The mesonephros is also referred to as the Wolffian ducts. At the same time as mesonephric development another pair of ducts will develop next to the mesonephros, referred to as the paramesonephric ducts or Müllerian ducts. At this stage of development the embryo is still uncommitted to sexual differentiation of the reproductive tract.[46] If the genetic sex is male (XY), the paramesonephric ducts will regress and the mesonephric ducts will differentiate to form the epididymis, vas deferens, and seminal vesicles of the male reproductive tract. If the genetic sex is female (XX), the mesonephric ducts will regress and the paramesonephric ducts will differentiate to form the oviducts, uterus, and anterior portion of the vagina. This process of differentiation starts at the level of the indifferent gonad, where mesothelial cells start a chain reaction of events based on the expression of a transcription factor known as SRY (sex-determining region of the short arm of the Y chromosome).[47] These mesothelial cells or surface epithelial cells of the indifferent gonad infiltrate the gonad during invasion and proliferation of the primordial germ cells, and become the gonadal ridge. Furthermore, in the male these mesothelial cells will differentiate into Sertoli cells, and following activation of the SRY gene these Sertoli cells begin to secrete anti-Müllerian hormone. This anti-Müllerian hormone causes the regression of the paramesonephric ducts. The interstitial cells of the differentiating testis, also known as Leydig cells, begin to secrete testosterone that helps to further differentiate the mesonephric ducts as well as the accessory sex glands of the male. In the absence of the Y chromosome (female genetic sex, XX), the ovaries will begin to differentiate and the lack of anti-Müllerian hormone leads to differentiation of the paramesonephric ducts, while the lack of testosterone leads to the regression of the mesonephric ducts. Several pioneering experiments have further elucidated the mechanisms behind sex differentiation and sex reversal.[46] For example, bilateral castration of the indifferent gonad of either a genetically sexed male (XY) or female (XX) leads to regression of the mesonephric ducts and development of the paramesonephric ducts (phenotypically female irrespective of genetic sex).

The focus on sex determination and sex reversal in the bovine is partly due to the phenomenon known as the freemartin. A freemartin is a sterile heifer born as a co-twin to a bull, and it is explained by the anastomosis that occurs between the placenta of the female and male fetuses. Because of the shared blood supply the sexually differentiating female tract (mesonephric and paramesonephric ducts) is exposed to anti-Müllerian hormone and testosterone from the male co-twin. One of the problems with studying the freemartin is reproducing the syndrome efficiently during an experimental study. For example, the presence of testosterone during critical periods of development will produce a masculinized female fetus (by helping to differentiate a portion of the mesonephric ducts); however, this experimental model is not truly a freemartin because the female fetus produced is not a chimera (containing both XX and XY cell types). This chimerism, presumably from the invasion of male (XY) hematopoietic stem cells or white blood cells from the anastomosis of the placenta, leads to an intersexed gonad, which sometimes contains a mixture of seminiferous tubules of the medulla and follicular pools in the cortex of the gonad. Interestingly, the hematopoietic stem cells may function similar to the primordial germ cells of the developing gonad because of the similarities in the lineage of these cell types. Although the regression and/or rescue of the duct system is primarily under the control of hormonal input, further research is needed to elucidate the formation of the intersexed gonad, which is presumably a function of invasion of XY blood cells of the male twin.

Fetal growth and organ development

The characterization of bovine fetal growth throughout gestation has allowed researchers to determine specific environmental insults that may perturb normal development *in utero*. In addition to the extrinsic environmental effects altering fetal growth, several inherent intrinsic mechanisms are associated with fetal growth and development. During days 70–100 of gestation, fetal weight across several breeds of cattle has been shown to increase at approximately 10 g/day.[48] Further along in gestation, from days 200 to 250 of pregnancy, the rate of fetal growth increases to approximately 200–300 g/day; however, the absolute growth of the late term fetus declined to 100 g/day, which may be due to an inherent function of the fetus exceeding the capabilities of uteroplacental exchange near term. Alternatively, the fetus may be secreting or altering its own hormone profiles near term, favoring proper maturation and differentiation of organs over growth.

During these characterization studies, several researchers noted that bovine fetal growth typically lags behind allantoic fluid volume and expansion of the chorioallantoic membrane.[48,49] As mentioned above, nutrient partitioning by the uteroplacental unit and a continuous supply of nutrients is a prerequisite for fetal growth. The chorioallantoic expansion over the endometrium allows maximal surface area for nutrient exchange during development, which should support the acceleration in fetal growth. A significant positive correlation was observed for total amniotic fluid volume and fetal weight; however, no correlation was observed for total allantoic fluid volume and fetal weight.[48] The cloning of cattle by somatic cell nuclear transfer has allowed researchers to further define this association between fetal fluid homeostasis and fetal and placental development.[50] Cloned bovine fetuses that show abnormal placentation are usually associated with excessive fluid accumulation in the fetal sacs, known as hydrops syndrome. In this experimental model, cloned embryos with hydrops syndrome surviving until mid-gestation show overgrowth of the placentomes, fetal liver, and fetal kidney compared with normal pregnancies.[50] In addition, in several studies utilizing different mammalian species, organomegaly of the offspring's liver, kidney, and heart is typically accompanied by polyhydramnios. Moreover, the underlying mechanisms of fluid sac homeostasis are not fully understood in the pregnant cow and further research is needed to elucidate the proper cause and effect relationship.

Failure of proper organ development occurs in cloned calves that die shortly after birth. Using this experimental model, researchers have examined eight developmentally important genes in six different organs (heart, liver, spleen, lung, kidney, and brain). Of these developmental genes, the kidney was the least affected organ associated with calf mortality, while the heart was the organ most affected by gene dysregulation compared with normal offspring.[51] Of these developmental genes, vascular endothelial growth factor (VEGF) was upregulated in cloned offspring associated with early neonatal mortality. VEGF has been implicated in stimulating vasculogenesis and angiogenesis to restore oxygen supply to tissues when blood circulation is inadequate. The proper concentrations of VEGF during organogenesis and early organ development are vital for proper establishment of the fetal cardiovascular system. Downregulation of VEGF could lead to a decrease in fetal organ angiogenesis, with fatal consequences for the fetus (pregnancy wastage) or neonate (calf mortality); however, upregulation may result in organomegaly or fetal cardiac dysfunction with similar fatal consequences to the newborn.

Uterine blood flow

Several techniques (electromagnetic probes, heavy water, Doppler) have been used to characterize uterine blood flow in pregnant cows. During early pregnancy and around the time of maternal recognition of pregnancy, uterine blood flow relative to the side of pregnancy (ipsilateral or contralateral) responds differently, which may be due to conceptus secretions during early establishment of pregnancy.[52] Comparing electromagnetic blood flow probes with recent observations using Doppler ultrasonography, we can see a similar increase in uterine artery blood flow during the first trimester of pregnancy (Table 25.1). Moreover, the diameter of uterine arteries remains relatively constant during this period, while the arterial resistance index decreases.[53] During the second trimester of pregnancy uterine artery blood flow increases from approximately 1–2 L/min to approximately 4–5 L/min (Table 25.2). Moreover, the first signs of increased maternal cardiac output may become apparent during this stage of pregnancy. In addition, uterine blood flow relative to the side of pregnancy (ipsilateral or contralateral) begins to diverge, allowing an increase in oxygen and nutrient delivery as well as waste exchange across the gravid pregnant uterine horn (Table 25.2). In examining Holstein cows over 147 days of gestation, Herzog et al.[54] reported similar total uterine artery blood flow in cows carrying heavy versus light calves; however, at day 175 of gestation, cows carrying heavy calves had an increase in uterine artery blood flow compared with cows carrying light calves. This characterization of uterine blood flow versus relative size of the offspring may allow researchers to determine specific time points in gestation where supplements or insults may be used to alter offspring growth and outcome.

The greatest increase in uterine blood flow has been shown in late gestation during the third trimester of pregnancy (Table 25.3). Moreover, this exponential increase in uterine blood flow during the third trimester of pregnancy also corresponds to the greatest increase in fetal weight deposition. The variation in reported blood flow during late gestation is a result of laboratory technique as well as individual animal variation. Using electromagnetic blood flow probes, Ferrell and Ford[55] reported uterine blood flow between days 202 and 224 as 4.0 ± 0.5 L/min and between days 230 and 258 as 3.1 ± 0.2 L/min. In a later study using a different technique the same laboratory reported late-gestation ipsilateral uterine artery blood flow at day 226 as 2.9 L/min and at day 250 as 13.1 L/min. In addition, there was a 4.5-fold increase in ipsilateral blood flow from days 137 to 250 and maternal heart rate was also increased.[56,57] The next experiment used Charolais cows carrying either Charolais or Brahman fetuses and Brahman cows carrying either Charolais or Brahman fetuses. At day 220 of gestation ipsilateral uterine artery blood flow of Brahman cows carrying Charolais or Brahman fetuses was similar but lower than in Charolais cows carrying Charolais or Brahman fetuses. However, with regard to the Charolais cows, those carrying Charolais fetuses had increased blood flow compared with those carrying Brahman fetuses. This study shows that late-gestation uterine blood flow and function of the uteroplacental unit may be the limiting factors for fetal growth.[38] The third study used two different breeds (Charolais and Hereford) but also looked at single versus twin fetuses. At day 190 of gestation, there was a decrease in ipsilateral uterine artery blood flow for Hereford versus Charolais cows. Another observation from this experiment was that the cows carrying twins had decreased blood flow compared with cows carrying singletons regardless of cow breed. Therefore twin pregnancies with reduced birthweight compared with singletons may be due to the reduced blood flow per fetus that was observed.[58]

More recently two other studies were conducted utilizing Doppler ultrasonography during late gestation. Bollwein et al.[53] reported ipsilateral and contralateral uterine

Table 25.1 Bovine uterine blood flow (UBF) during the first trimester of pregnancy.

Breed	Stage of pregnancy	Method	Response	Additional response	Reference
Hereford	First 30 days	Electromagnetic blood flow probes	↔ UBF until day 13 Two- to three-fold ↑ ipsilateral UBF days 14–18 ↔ contralateral UBF	Days 19–25 ↓ UBF ipsilateral UBF ↑ until day 30 Contralateral UBF ↓	Ford et al.[52]
Hereford	Approximately day 30 and 80	Electromagnetic blood flow probes	Days 31–45 0.12 ± 0.02 L/min Days 77–92 0.20 L/min		Ferrell & Ford[55]
Simmental Brown Swiss	Days 30, 60, and 90	Doppler	↔ both arteries Day 30 average 0.10 L/min Day 60 average 0.14 L/min Day 90 average 0.30 L/min	↓ RI by day 60 ↔ diameter	Bollwein et al.[53]

↔, steady change; ↑, increased; ↓, decreased; RI, resistance index.

Table 25.2 Bovine uterine blood flow (UBF) during the second trimester of pregnancy.

Breed	Stage of pregnancy	Method	Response	Additional Response	Reference
Hereford	Approximately days 140 and 180	Electromagnetic blood flow probes	Days 139–155, 2.0 ± 0.6 L/min Days 178–199, 3.2 ± 0.4 L/min		Ferrell & Ford[55]
Hereford	Days 163–166 Days 173–176	Steady-state diffusion; antipyrin solution	Mean 5.9 ± 0.5 mL/min	Mean HR 68.6 ± 2.6 bpm	Ferrell et al.[60]
Simmental Brown Swiss	Days 120, 150, 180	Doppler	Ipsilateral day 120, 0.7 L/min Contralateral day 120, 0.4 L/min Ipsilateral day 150, 1.7 L/min Contralateral day 150, 0.9 L/min Ipsilateral day 180, 4.0 L/min Contralateral day 180, 1.3 L/min		Bollwein et al.[53]
Holsteins	Days 147 and 175	Doppler	↑ Total UBF in heavy vs. light cows: Light cow day 147, 2.6 L/min Heavy cow day 147, 3.4 L/min Light cow day 175, 4.2 L/min Heavy cow day 175, 5.4 L/min	At day 147, cows had similar total UBF when carrying heavy or light calves but at day 175 cows carrying heavy calves had ↑ blood flow vs. light calves	Herzog et al.[54]

↑, increased; HR, heart rate.

Table 25.3 Bovine uterine blood flow (UBF) during the third trimester of pregnancy.

Breed	Stage of pregnancy	Method	Response	Additional Response	Reference
Hereford	Approximately days 210 and 240	Electromagnetic blood flow probes	Days 202–224, 4.0 ± 0.5 L/min Days 230–258, 3.1 ± 0.2 L/min		Ferrell & Ford[55]
Brahman Charolais	Day 220 ± 0.4	D_2O (ipsilateral) infusion	↑ ipsilateral UBF in Charolais vs. Brahman	Charolais carrying Charolais fetuses ↑ blood flow vs. Charolais carrying Brahman fetuses	Ferrell[38]
Charolais Hereford	Day 190 ± 0.5	D_2O (ipsilateral) infusion	↓ UBF Hereford 4.8 vs. Charolais 7.1 L/min	↓ UBF in cows with twins 5.2 vs. single 6.7 L/min	Ferrell & Reynolds[58]
Simmental Brown Swiss	Days 210, 240, 270, and 282	Doppler	Ipsilateral day 210, 5.0 L/min Contralateral day 210, 1.3 L/min Ipsilateral day 240, 6.0 L/min Contralateral day 240, 2.8 L/min Ipsilateral day 270, 10.5 L/min Contralateral day 270, 3.9 L/min Ipsilateral day 282, 13.1 L/min Contralateral day 282, 4.5 L/min	↑ UBF and diameter during gestation ↓ RI during first 8 months and ↔ after	Bollwein et al.[53]
Holsteins	Days 203, 231, 259, and 273	Doppler	Light cow day 203, 6.1 L/min Heavy cow day 203, 8.7 L/min Light cow day 231, 10.5 L/min Heavy cow day 231, 13.1 L/min Light cow day 259, 13.6 L/min Heavy cow day 259, 16.9 L/min Light cow day 273, 14.1 L/min Heavy cow day 273, 19.2 L/min	Linear ↑ in total UBF 3.0 to 16.9 L/min ↑ total UBF in heavy cows vs. light cows Similar-weight cows with heavy calves had ↑ total UBF vs. light calves	Herzog et al.[54]

↔, steady change; ↑, increased; ↓, decreased; RI, resistance index; D_2O, deuterium oxide.

artery blood flow on days 210, 240, 270, and 284 of gestation. They observed an increase in blood flow and arterial diameter during gestation and decreased resistance of the uterine artery during the first 8 months of gestation while the last month arterial resistance was similar. The second experiment by Herzog et al.[54] measured blood flow on days 203, 231, 259, and 273 of gestation. Throughout this experiment they observed a linear increase in total uterine artery blood flow. Also, heavy cows had a greater increase in blood flow compared with light cows; however, similar-weight cows carrying heavy fetuses had greater blood flow than those carrying light fetuses.

Developmental programming

Irrespective of the animal's genotype, an environmental stimulus or insult during critical periods of development can program the wanted phenotype of an animal. Moreover, this exposure to a stimulus or insult may establish a permanent phenotype throughout the remainder of an animal's life, which can have adverse consequences for milk production, carcass yield grade, feed efficiency, or reproductive function. This process of permanently altering an animal's phenotype through environmental stimuli has been referred to as the developmental programming hypothesis. In addition, the critical period of development, where an environmental insult or stimulus can have lasting consequences on the offspring, may occur during early embryogenesis or fetal organ differentiation. Understanding the impacts of the maternal environment on placental growth and development is especially relevant as the majority of mammalian livestock raised for red meat production spend 35–40% of their life within the uterus being nourished solely by the placenta.[59] Therefore, an understanding of factors that impact uteroplacental blood flow will directly impact placental efficiency and thus fetal growth. However, despite intensive research in the area of placental–fetal interactions, the regulators of placental growth and vascularization as well as uteroplacental blood flow in cattle are still largely unknown. In addition, the responses to different environmental stimuli in sheep, which have been thoroughly examined in the last decade, appear to be different compared with the limited amount of data in cattle. Future developmental programming studies are needed to help characterize pregnancy and offspring outcome during specific periods of gestation and following controlled environmental alterations to the dam.

References

1. Donaldson L, Hansel W. Prolongation of life span of the bovine corpus luteum by single injections of bovine luteinizing hormone. *J Dairy Sci* 1965;48:903–904.
2. Reynolds L, Robertson D, Ford S. Effects of intrauterine infusion of oestradiol-17 beta and prostaglandin E-2 on luteal function in non-pregnant heifers. *J Reprod Fertil* 1983;69:703–709.
3. Williams M, Marsh J. Estradiol inhibition of luteinizing hormone-stimulated progesterone synthesis in isolated bovine luteal cells. *Endocrinology* 1978;103:1611–1618.
4. Ford S. Control of uterine and ovarian blood flow throughout the estrous cycle and pregnancy of ewes, sows and cows. *J Anim Sci* 1982;55(Suppl. 2):32–42.
5. Marsh J. The effect of prostaglandins on the adenyl cyclase of the bovine corpus luteum. *Ann NY Acad Sci* 1971;180:416–425.
6. Arosh J, Banu S, Kimmins S, Chapdelaine P, Maclaren L, Fortier M. Effect of interferon-tau on prostaglandin biosynthesis, transport, and signaling at the time of maternal recognition of pregnancy in cattle: evidence of polycrine actions of prostaglandin E2. *Endocrinology* 2004;145:5280–5293.
7. Helmer S, Hansen P, Thatcher W, Johnson J, Bazer F. Intrauterine infusion of highly enriched bovine trophoblast protein-1 complex exerts an antiluteolytic effect to extend corpus luteum lifespan in cyclic cattle. *J Reprod Fertil* 1989;87:89–101.
8. Roberts R. Interferon-tau, a Type 1 interferon involved in maternal recognition of pregnancy. *Cytokine Growth Factor Rev* 2007;18:403–408.
9. Asselin E, Lacroix D, Fortier M. IFN-tau increases PGE2 production and COX-2 gene expression in the bovine endometrium in vitro. *Mol Cell Endocrinol* 1997;132:117–126.
10. Niswender G, Juengel J, Silva P, Rollyson M, McIntush E. Mechanisms controlling the function and life span of the corpus luteum. *Physiol Rev* 2000;80:1–29.
11. Graham J, Clarke C. Physiological action of progesterone in target tissues. *Endocr Rev* 1997;18:502–519.
12. Csapo A. Progesterone block. *Am J Anat* 1956;98:273–291.
13. Garfield R, Sims S, Kannan M, Daniel E. Possible role of gap junctions in activation of myometrium during parturition. *Am J Physiol* 1978;235:C168–C179.
14. Mesiano S. Myometrial progesterone responsiveness. *Semin Reprod Med* 2007;25:5–13.
15. Hirsbrunner G, Knutti B, Liu I, Küpfer U, Scholtysik G, Steiner A. An in vitro study on spontaneous myometrial contractility in the cow during estrus and diestrus. *Anim Reprod Sci* 2002;70:171–180.
16. Conley A, Ford S. Effect of prostaglandin F2 alpha-induced luteolysis on in vivo and in vitro progesterone production by individual placentomes of cows. *J Anim Sci* 1987;65:500–507.
17. Betteridge K, Flechon J. The anatomy and physiology of pre-attachment bovine embryos. *Theriogenology* 1988;29:155–187.
18. Maddox-Hyttel P, Alexopoulos N, Vajta G et al. Immunohistochemical and ultrastructural characterization of the initial post-hatching development of bovine embryos. *Reproduction* 2003;125:607–623.
19. Chang M. Development of bovine blastocyst with a note on implantation. *Anat Rec* 1952;113:143–161.
20. Leiser R, Wille K. Alkaline phosphatase in the bovine endometrium and trophoblast during the early phase of implantation. *Anat Embryol* 1975;148:145–157.
21. Assis Neto A, Pereira F, Santos T, Ambrosio C, Leiser R, Miglino M. Morpho-physical recording of bovine conceptus (*Bos indicus*) and placenta from days 20 to 70 of pregnancy. *Reprod Domest Anim* 2010;45:760–772.
22. Blomberg L, Hashizume K, Viebahn C. Blastocyst elongation, trophoblastic differentiation, and embryonic pattern formation. *Reproduction* 2008;135:181–195.
23. Mann G, Fray M, Lamming G. Effects of time of progesterone supplementation on embryo development and interferon-tau production in the cow. *Vet J* 2006;171:500–503.
24. Forde N, Lonergan P. Transcriptomic analysis of the bovine endometrium: what is required to establish uterine receptivity to implantation in cattle? *J Reprod Dev* 2012;58:189–195.
25. Silver M, Steven DH, Comline RS. Placental exchange and morphology in ruminants and the mare. In: Comline RS, Cross KW, Dawes GS, Nathanielz PW (eds) *Foetal and Neonatal Physiology*. Cambridge: Cambridge University Press, 1973, pp. 245–271.
26. Schlafer D, Fisher P, Davies C. The bovine placenta before and after birth: placental development and function in health and disease. *Anim Reprod Sci* 2000;60–61:145–160.
27. Reynolds L, Redmer D. Utero-placental vascular development and placental function. *J Anim Sci* 1995;73:1839–1851.
28. Reynolds L, Redmer D. Angiogenesis in the placenta. *Biol Reprod* 2001;64:1033–1040.
29. Patten B. *Foundations of Embryology*. New York: McGraw-Hill, 1964.
30. Ramsey E. *The Placenta, Human and Animal*. New York: Praeger, 1982.
31. Prior R, Laster D. Development of the bovine fetus. *J Anim Sci* 1979;48:1546–1553.
32. Meschia G. Circulation to female reproductive organs. In: Shephard JT, Abboud FM (eds) *Handbook of Physiology. The Cardiovascular System, Peripheral Circulation and Organ Blood Flow*. Bethesda, MD: American Physiological Society, 1983, pp. 241–269.
33. Reynolds L, Millaway D, Kirsch J, Infeld J, Redmer D. Growth and in-vitro metabolism of placental tissues of cows from day 100 to day 250 of gestation. *J Reprod Fertil* 1990;89:213–222.

34. Vonnahme K, Zhu M, Borowicz P et al. Effect of early gestational undernutrition on angiogenic factor expression and vascularity in the bovine placentome. *J Anim Sci* 2007;85: 2464–2472.
35. Metcalfe J, Stock M, Barron D. Maternal physiology during gestation. In: Knobil E, Neill JD, Ewing LL, Market CL, Greenwald GS, Pfaff DW (eds) *The Physiology of Reproduction*. New York: Raven Press, 1988, p. 2145.
36. Ferrell CL. Placental regulation of fetal growth. In: Campion DR, Hausman GJ, Martin RJ (eds) *Animal Growth Regulation*. New York: Springer, 1989, pp. 1–19.
37. Rosenfeld C, Morriss F Jr, Makowski E, Meschia G, Battaglia F. Circulatory changes in the reproductive tissues of ewes during pregnancy. *Gynecol Invest* 1974;5:252–268.
38. Ferrell C. Maternal and fetal influences on uterine and conceptus development in the cow: II. Blood flow and nutrient flux. *J Anim Sci* 1991;69:1954–1965.
39. Magness RR. Maternal cardiovascular and other physiologic responses to the endocrinology of pregnancy. In: Bazer FW (ed.) *Endocrinology of Pregnancy*. Totowa, NJ: Humana Press, 1998, pp. 507–539.
40. Reynolds M. Measurement of bovine plasma and blood volume during pregnancy and lactation. *Am J Physiol* 1953;175:118–122.
41. Thatcher W, Wilcox C, Collier R, Eley D, Head H. Bovine conceptus–maternal interactions during the pre- and postpartum periods. *J Dairy Sci* 1980;63:1530–1540.
42. Bauman D, Currie W. Partitioning of nutrients during pregnancy and lactation: a review of mechanisms involving homeostasis and homeorhesis. *J Dairy Sci* 1980;63:1514–1529.
43. Tsuka T, Hasegawa K, Morimoto M et al. Quantitative investigation and classification by MRI of residual cleft cysts in the pituitary glands of cows. *Vet Rec* 2009;164:588–591.
44. Rich I. Primordial germ cells are capable of producing cells of the hematopoietic system in vitro. *Blood* 1995;86:463–472.
45. Wrobel K, Suss F. Identification and temporospatial distribution of bovine primordial germ cells prior to gonadal sexual differentiation. *Anat Embryol* 1998;197:451–467.
46. Jost A. Hormonal factors in the sex differentiation of the mammalian foetus. *Phil Trans R Soc Lond B* 1970;259:119–130.
47. Gutierrez-Adan A, Behboodi E, Murray J, Anderson G. Early transcription of the SRY gene by bovine preimplantation embryos. *Mol Reprod Dev* 1997;48:246–250.
48. Eley R, Thatcher W, Bazer F et al. Development of the conceptus in the bovine. *J Dairy Sci* 1978;61:467–473.
49. Ferrell C, Garrett W, Hinman N. Growth, development and composition of the udder and gravid uterus of beef heifers during pregnancy. *J Anim Sci* 1976;42:1477–1489.
50. Lee R, Peterson A, Donnison M et al. Cloned cattle fetuses with the same nuclear genetics are more variable than contemporary half-siblings resulting from artificial insemination and exhibit fetal and placental growth deregulation even in the first trimester. *Biol Reprod* 2004;70:1–11.
51. Li S, Li Y, Du W et al. Aberrant gene expression in organs of bovine clones that die within two days after birth. *Biol Reprod* 2005;72:258–265.
52. Ford S, Chenault J, Echternkamp S. Uterine blood flow of cows during the oestrous cycle and early pregnancy: effect of the conceptus on the uterine blood supply. *J Reprod Fertil* 1979;56:53–62.
53. Bollwein H, Baumgartner U, Stolla R. Transrectal Doppler sonography of uterine blood flow in cows during pregnancy. *Theriogenology* 2002;57:2053–2061.
54. Herzog K, Koerte J, Flachowsky G, Bollwein H. Variability of uterine blood flow in lactating cows during the second half of gestation. *Theriogenology* 2011;75:1688–1694.
55. Ferrell CL, Ford SP. Blood flow steroid secretion and nutrient uptake of the gravid bovine uterus. *J Anim Sci* 1980;50: 1113–1121.
56. Reynolds L. Utero-ovarian interactions during early pregnancy: role of conceptus-induced vasodilation. *J Anim Sci* 1986;62(Suppl. 2):47–61.
57. Reynolds L, Ferrell C. Transplacental clearance and blood flows of bovine gravid uterus at several stages of gestation. *Am J Physiol* 1987;253:R735–R739.
58. Ferrell C, Reynolds L. Uterine and umbilical blood flows and net nutrient uptake by fetuses and uteroplacental tissues of cows gravid with either single or twin fetuses. *J Anim Sci* 1992;70:426–433.
59. Vonnahme K, Lemley C. Programming the offspring through altered uteroplacental hemodynamics: how maternal environment impacts uterine and umbilical blood flow in cattle, sheep and pigs. *Reprod Fertil Dev* 2011;24:97–104.
60. Ferrell C, Ford S, Prior R, Christenson R. Blood flow, steroid secretion and nutrient uptake of the gravid bovine uterus and fetus. *J Anim Sci* 1983;56:656–667.

BREEDING AND HEALTH MANAGEMENT

26. Biosecurity and Biocontainment for Reproductive Pathogens — 259
    Carla Huston

27. Beef Replacement Heifer Development — 267
    Terry J. Engelken

28. Heifer Development: From Weaning to Calving — 272
    Ricardo Stockler

29. Interaction of Nutrition and Reproduction in the Beef Cow — 276
    William S. Swecker Jr

30. Interaction of Nutrition and Reproduction in the Dairy Cow — 283
    Butch Cargile and Dan Tracy

31. Estrus Detection — 290
    Rhonda C. Vann

32. Artificial Insemination — 295
    Ram Kasimanickam

33. Pharmacological Intervention of Estrous Cycles — 304
    Ram Kasimanickam

34. Pregnancy Diagnosis: Rectal Palpation — 314
    David Christiansen

35. Biochemical Pregnancy Diagnosis — 320
    Amanda J. Cain and David Christiansen

36. Reproductive Ultrasound of Female Cattle — 326
    Jill Colloton

37. Beef Herd Health for Optimum Reproduction — 347
    Terry J. Engelken and Tyler M. Dohlman

38. Dairy Herd Health for Optimal Reproduction — 353
    Carlos A. Risco

39. Herd Diagnostic Testing Strategies — 359
    Robert L. Larson

40. Beef Herd Record Analysis: Reproductive Profiling — 364
    Brad J. White

41. Evaluating Reproductive Performance on Dairy Farms — 370
    James A. Brett and Richard W. Meiring

42. Marketing the Bovine Reproductive Practice: Devising a Plan or Tolerating a System — 374
    John L. Myers

Chapter 26

Biosecurity and Biocontainment for Reproductive Pathogens

Carla Huston

*Department of Pathobiology and Population Medicine, College of Veterinary Medicine,
Mississippi State University, Starkville, Mississippi, USA*

Introduction

The goal of dairy and beef breeding operations is to produce offspring for future production in the forms of meat, milk, or genetics. Subsequently, reproductive diseases can have detrimental effects on the reproductive efficiency of dairy and beef breeding herds. There are many infectious diseases that affect the reproductive performance in the bovine, including viral, bacterial, and parasitic. Specific pathogens may cause a range of clinical manifestations from metritis and infertility to abortion and perinatal loss.

Biosecurity measures for a livestock operation are those taken to prevent the introduction or reintroduction of infectious disease agents into susceptible populations. While many biosecurity recommendations are good general management practices that can be applied across most operations, other recommendations should be more specifically tailored to an individual premises or group of animals.

Biosecurity measures are undertaken at national, state, and local levels. At the national level, biosecurity plans are mandated by regulatory agencies such as the United States Department of Agriculture (USDA) and the United States Department of Homeland Security (USDHS) to protect livestock resources and consumer interests. Such plans are aimed at preventing the introduction of foreign animal diseases and other diseases of high economic consequence. The entry of animals, animal byproducts, germplasm (semen, embryos), and feed and feed products are carefully controlled at the national border. Producers and veterinarians participate at this level by adhering to recommendations following international travel and by reporting unusual or suspicious conditions to their state or federal animal health official.

Biosecurity plans at the state level may focus on diseases targeted for eradication and control across state lines, such as brucellosis. State plans can also target diseases of regional significance, such as trichomoniasis, and are regulated through the state animal health official, most commonly the state veterinarian. Individual states may have different regulations for interstate animal imports, and both producers and veterinarians are responsible for knowing and adhering to a state's specific animal health regulations. The issuance of certificates of veterinary inspection is one example of an important tool used to prevent the spread of livestock diseases across state lines.

While disease control measures at the national and state levels are critical to the protection of the overall US livestock population, individual producers and veterinarians have the greatest influence over the implementation of local biosecurity measures. It is ultimately individual livestock producers who are responsible for protecting their own herd against infectious reproductive diseases. Local biosecurity plans focus on preventing or controlling the spread of disease between operations and take into account endemic diseases and conditions unique to that population of animals. Local plans should consist of both biosecurity measures as well as biocontainment measures, those taken to prevent or decrease the spread of a disease among groups of animals within an operation. Biological risk management (BRM) is the term used to describe the overall awareness and management of the risk of a pathogen or disease entering and spreading throughout a population. BRM often involves the evaluation, management, and reduction of pathogens that are already present on an operation.

Some challenges to recognizing reproductive inefficiencies in a breeding herd include the lack of good reproductive records, the imperfection of diagnostic tests, and failure to identify the etiologies of abortions and perinatal losses. Furthermore, the intensive management of cattle in the United States, whether it is beef on pasture or dairy in confinement, creates challenges to ideal disease control conditions. Such challenges reinforce the importance for cattle producers to have a herd-specific biosecurity plan in place to protect their herd from reproductive diseases.

Developing a biosecurity and biocontainment plan

A properly performed biosecurity risk analysis will help determine which biosecurity and biocontainment measures are appropriate for a specific operation. While general biosecurity measures are directed at nonspecific pathogen threats, a biosecurity risk analysis can be agent specific, industry specific, or event specific.[1] Risk analyses can also be qualitative or quantitative, depending on the herd information available and desired outcome of the assessment. Many epidemiological tools are available to estimate the risk of specific disease transmission in a population, including statistical simulations and mathematical modeling.[2,3] Such risk assessments should be made using multiple sources of information, such as producer surveys and diagnostic testing, to predict probabilities of disease transmission or to monitor progress following changes in management.[4]

A biosecurity risk analysis approach involves risk assessment, risk management, and risk communication, considerations which are all key to comprehensive and successful biosecurity planning. The risk analysis approach should center on specific epidemiological characteristics of the agent and host and environmental factors affecting reproductive health in the herd (Figure 26.1). A risk analysis for disease control is similar in concept to a hazard analysis and critical control point (HAACCP) system, which helps identify critical control points throughout the production chain where disease incursions may happen and incorporates control measures at each point. Whether a risk analysis or HAACCP system approach is used for protection against reproductive diseases, considerations should involve seeking input from not only producers and veterinarians but possibly also farm employees, family members, and economists.

Risk assessment

It is commonly recognized in livestock production that every animal cannot be protected against every pathogen affecting reproduction. The purpose of a risk assessment in biosecurity planning for reproductive diseases is to evaluate the true likelihood, impacts, and consequences of an outbreak occurring in the herd. This involves hazard identification, exposure assessment, dose–response assessment, and risk characterization. The identification of hazards involves an examination of potential agents involved in reproductive diseases along with the population at risk, or susceptible hosts for disease. A risk assessment should contain estimates of reproductive disease threats, using past and current herd records for production (conception rate, calving rate, calving interval, etc.), and medical history, including vaccination practices and diagnostic testing, where available.

Several infectious agents affecting reproduction should be considered in both beef and dairy herds in the United States, including viruses such as bovine herpesvirus type 1 (BHV-1) and bovine viral diarrhea virus (BVDV), bacteria

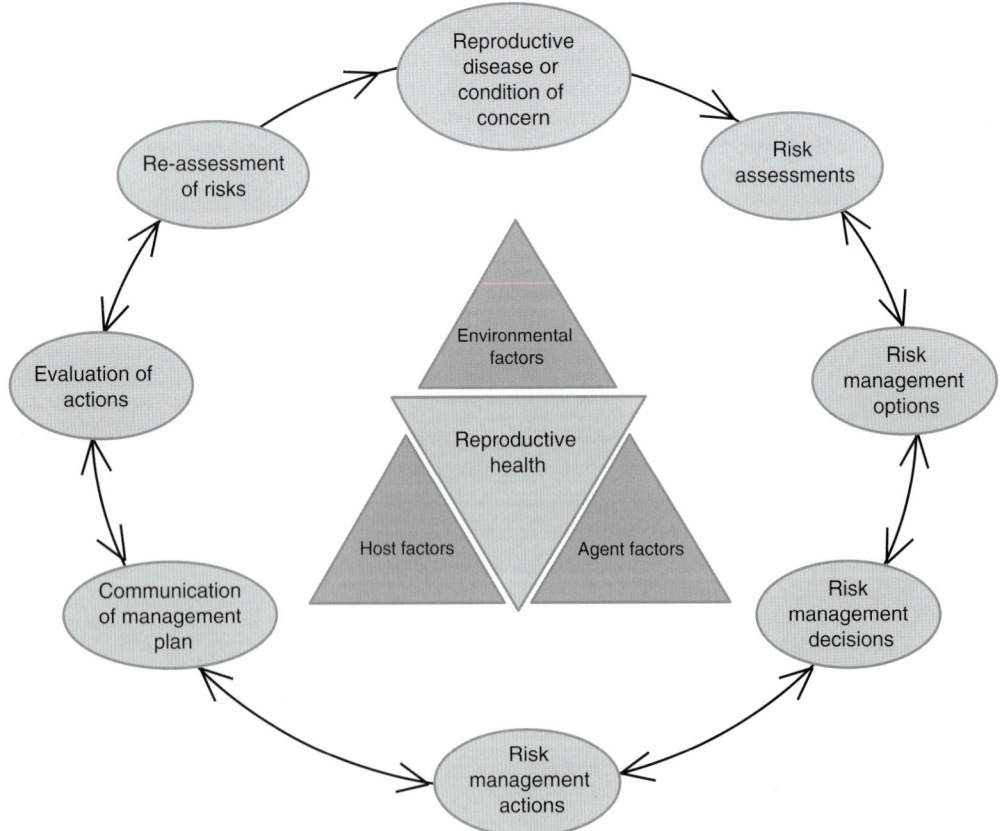

Figure 26.1 The risk analysis approach should center on specific epidemiological characteristics of the agent, host and environmental factors affecting reproductive health in the herd.

such as *Brucella abortus* and *Leptospira* spp., and parasites such as *Tritrichomonas foetus* and *Neospora caninum*. Additional agents of concern are dependent on regional and environmental exposures. Host factors for susceptibility to reproductive diseases may be influenced by animal use, age, vaccination status, nutrition, body condition, and stress. Animals most at risk for reproductive diseases – the breeding females, breeding bulls, and replacement females – can be managed in ways to reduce reproductive disease risks. For example, the use of virgin animals will eliminate the risk of pathogens transmitted venereally such as *Tritrichomonas foetus* and *Campylobacter fetus*.

It is useful to create a map of the farm environment during the initial biosecurity risk assessment to visualize the flow of animal and people traffic throughout the operation. The farm layout should include any buildings, pastures, working areas, feed storage areas, and water sources. Documenting the slope of the ground as well as the location of natural structures such as trees, streams and ponds can also help identify natural barriers for disease control. Nose-to-nose contact points should be identified in indoor as well as outdoor areas, paying special attention to sick pens, calving pens, and breeding areas. Visitor access areas should be clearly identified. Visualization of these areas and knowledge of associated risks will allow the creation of a flowchart to aid in identifying potential critical control points within an operation as well as identify potential entry points for new diseases.[5]

Exposure and dose–response assessments determine if the animals at risk are actually exposed to the identified pathogens, how often such exposure occurs, and how much exposure would cause a reproductive health risk to the animals. Environmental factors such as pasture management, animal contacts, and housing will have a great impact on pathogen exposure. Risk characterization examines the magnitude of the health risks, and how control of those risks should be prioritized. The end result of a properly performed biosecurity risk assessment is a risk determination where animal–pathogen–environment interactions are characterized as high, medium, or low priority for disease.

Risk management

The risk management stage of a risk analysis is often referred to as the "action stage." The purpose of risk management is to establish measurable goals from the risk assessment, determine risk reduction strategies using the biosecurity tools that are available, implement biosecurity practices, and evaluate the implemented strategies. Available resources (personnel, equipment, financial), current biosecurity practices, and short- and long-term herd goals must be considered when developing risk management strategies. Goal-setting should include the level of aggressiveness desired and relies on good communication between all parties involved. For example, a less aggressive goal may be desired by the producer, such as cessation of abortions, while a more aggressive goal of a veterinarian might be to establish a herd free from persistent infection by BVD (BVD-PI).

Management techniques to control reproductive diseases include a combination of tools such as movement control, diagnostic test and removal programs, vaccinations, and enhanced biosecurity measures discussed further in this chapter. Care must be taken to not implement too many changes in an operation at once such that an evaluation of impact cannot be accurately made. Furthermore, producers and employees may become overwhelmed with so many changes that none of them actually become implemented.

Risk communication

The purpose of risk communication is to translate the results of the risk assessments and summarize risk reduction strategies to an easily understandable form. Proper communication requires interaction between all parties involved in the initial herd assessment. Additional education of team members and employee training is often indicated following risk assessment and management recommendations. Standard reproductive herd health protocols such as animal addition protocols, vaccination strategies, and visitor policies should be reviewed and provided in both oral dialogue and written form to have available for future reference.

Communication methods can be very simple, such as a sign hanging in the calving barn, or very complex, such as the development of a statistical process control chart of reproductive efficiency.[6] Regardless of the communication output chosen, it should be readily understandable to the end-user and updated when changes in management practices are made. Written biosecurity risk analyses or HAACCP plans, including treatment protocols and standard operating procedures, will also provide legal documentation if needed.

While a myriad of biosecurity educational publications are available, most are nonspecific and contain wide variations in recommendations.[7] Therefore it is important that biosecurity plans are tailored to farm-specific concerns. Furthermore, it is important to note that a proper risk analysis should be a flexible system, allowing for continuous monitoring, reevaluation, and open communication of all parties involved.

Tools for intervention
Reducing effective animal contacts

The basic concept behind biosecurity and biocontainment recommendations is that diseases can be prevented or controlled by preventing effective animal contacts. This concept is even truer for reproductive diseases. Effective animal contacts are those contacts made by an uninfected animal that result in infection and disease. Effective animal contacts can be reduced by utilizing a combination of the following concepts:

- controlling exposure to pathogens;
- reducing animal contacts;
- optimizing an animal or herd's immunity.

Controlling exposure to pathogens can be accomplished by implementing traffic control and proper cleaning and disinfection measures. The use of vaccinations and prophylactic medications can complement other pathogen reduction measures by reducing pathogen shedding and exposure. Reducing animal contacts may also involve traffic control and preventing fence-line contacts with outside herds. The use of artificial insemination (AI) and embryo transfer (ET) are two ways to reduce animal contacts in a breeding herd and also prevent exposure to many of the reproductive pathogens of concern. Optimizing animal immunity is accomplished by a variety of methods but should include management strategies such as providing good nutrition, reducing stress, and delivering an appropriate vaccination scheme.

An evaluation of herd production records (including reproductive history, disease incidence and disease screening tests) will help set goals, identify needs, document actions, and benchmark progress in disease control. However, one of the challenges involved when evaluating the impact of risk reduction strategies in a breeding herd is that results are not often observed until the next calving season. This makes timely record-keeping and monitoring of all health aspects of the herd, such as body condition scores, crucial to the overall health of the herd.

The use of biosecurity practices in beef and dairy herds has been studied and described for various regions and populations across the United States.[8,9] Management factors such as confinement and pasture utilization are not easily changed, but an awareness of factors predisposing cattle to reproductive diseases under these situations can help make decisions regarding reproductive health more practical. Most biosecurity and biocontainment practices aimed at reducing effective animal contacts will focus on the utilization of movement control, barriers, and cleaning and disinfection.

Movement control

Animal additions

The biggest risk to reproductive health on a beef or dairy operation is the addition of animals from outside the farm. Animal additions include newly purchased animals as well as animals returning from outside premises such as heifer-rearing facilities and exhibitions. Maintaining a closed herd is an ideal situation where no new animals enter the herd. Unfortunately, this is not always a realistic expectation given the advances in genetics and herd expansions that often need to be made. In general, producers accept that sources of new animal additions, from perceived lowest to highest risk of disease introduction, include private sales, cattle dealers, cattle auctions, and sale barns. Breeding animal additions should come from farms with effective herd health plans as demonstrated through herd records, vaccination history, and disease surveillance programs to minimize the potential for disease introduction. Performing breeding soundness examinations on all bulls prior to arrival will also help to identify potentially infectious processes prior to the breeding season.

Diagnostic disease testing

Diagnostic disease testing is a tool that can be used to identify infectious disease risks associated with a single animal or group of animals. Given herd production parameters, statistical probabilities can also be calculated on the likelihood of disease introduction under specific herd conditions.[3]

Diagnostic testing is commonly used to diagnose cases of illness, to augment test and removal programs, to monitor progress in disease control, and to document freedom from disease in pre-purchase or post-purchase examinations. Pre-purchase testing strategies can be performed prior to bringing new animals onto a premises, and diagnostic disease testing is valuable in detecting subclinically or persistently infected animals as in the case of trichomoniasis, neosporosis, or BVD-PI infection. However, in a herd setting it is not always practical to test all incoming animals for all potential reproductive pathogens, and therefore the testing of subsets of animals or pooling of samples may be indicated. Furthermore, diagnostic tests are not perfect and should always be interpreted with test sensitivity and specificity in mind. Sensitivity of a diagnostic test is its ability to correctly identify a positive or diseased animal, and specificity is the ability to correctly identify a negative or nondiseased animal.

While the sensitivity and specificity are characteristics of the diagnostic test itself, predictive values are based on the estimated prevalence of a pathogen or disease along with sensitivity and specificity to provide a better prediction of the risk of disease in a given population. A positive predictive value estimates the probability of an animal having the pathogen or disease given that the test was positive, and the negative predictive value estimates the probability of the animal being disease-free given that the test was negative. The negative predictive value will have the most importance in pre-purchase testing as new owners wish to ensure that imported animals are free from disease. Diagnostic testing schemes are an important part of a herd biosecurity plan but should not be the sole basis for decision-making. The overall health program of the source herd is often more important to consider than a diagnostic test from a single animal. A more complete description of diagnostic testing schemes for the breeding herd can be found in Chapter 39.

Quarantine and isolation

Quarantine procedures should be followed whenever animal additions occur. The length of quarantine should cover the incubation period of the most important infectious diseases, generally a minimum of 3 weeks. Quarantine can be effective in preventing new cases of diseases transmitted by direct contact, such as BVDV, infectious bovine rhinotracheitis (BHV-1), and leptospirosis. However, many diseases, such as neosporosis or BVD-PI, can be present in the subclinical or persistent form and not easily identifiable by visual inspection during this time. In these cases, diagnostic disease testing and vaccinations specific to the operation's herd health program should be performed before the animals are allowed to commingle with the resident herd. This time will also allow the animal to acclimate to the new environment and feed sources. Ideal quarantine facilities should be separate from the rest of the operation and downwind. Water and manure runoff should not be allowed to flow into common areas or near the resident herd. Separate equipment and protective wear (boots, clothing) should be utilized in quarantine and isolation areas and clearly identified to avoid their use in other animal areas.

Dairy herd expansion efforts may pose unique challenges given space and equipment limitations.[10] The use of a modified receiving pen, with solid-sided walls or double walls, can be used as an alternative quarantine area when separate facilities are not available,[5] but strict traffic control and cleaning and disinfection procedures must be followed. If not performed prior to arrival, bull breeding soundness examinations including trichomoniasis and BVD-PI testing should also be performed under quarantine in order to detect carriers and also to allow ample time to find replacements if needed.

Sick animals should be placed under isolation for effective biocontainment. It is important that the producer understands that the medical treatment of an animal does not equate to cessation of pathogen shedding, and that convalescent carriers of disease are possible. Animals that have experienced an abortion or pregnancy loss should also be separated from the herd to allow investigation into the cause of the loss, which may include diagnostic testing of the dam or the aborted fetus. Within an operation, animal and personnel movements should occur from the least susceptible animal populations to the most susceptible animals. Sick or high-risk animals, such as those under quarantine, should be handled last whenever possible. In larger facilities, separate personnel may be designated to care for high-risk populations.

Visitor access

Both farm personnel and visitors can serve as mechanical vectors for reproductive pathogen transmission. This includes neighbors, salespersons, hoof trimmers, AI and ET technicians, and veterinarians. Such persons should be categorized into low-risk, medium-risk, or high-risk individuals based on the number and extent of animal contacts and farm visits made per day. For example, persons who have contact with animals on multiple farm visits in one day, such as veterinarians and AI technicians, are considered high risk and would be expected to follow more stringent biosecurity protocols. Employees who have their own cattle or have contact with outside animals on a regular basis should also be considered high risk. At a minimum these persons should be expected to have clean clothing and footwear when on the premises and perform stringent cleaning and disinfection before arrival. Lower-risk individuals would be those such as salespersons or equipment mechanics, who should have no contact with animals or animal areas. Visitors should have limited access to the farm, with designated entry points. Signage should be posted restricting visitor access to other areas. A visitor log should be kept and all visitors should be made aware of the farm's biosecurity protocols for visitors, including the required reporting of recent international travel.

Barriers

Effective barriers will help decrease unwanted animal contacts by providing an extra layer of protection between infected and susceptible cattle.

Farm environment

Operations may be more at risk to certain pathogens due to their geographical location. Geographical barriers such as streams, wooded areas, tree lines, and ridges can be strategically utilized to create barriers between operations or groups of animals. Perimeter control by maintaining good fences is important not only to keep unwanted animals from having contact with the resident herd, but also to keep the resident herd from moving outside the protected area. In addition to maintaining good fences, preventing fence-line contact of the resident herd with other herds is important to prevent nose-to-nose contact and potential transmission of pathogens. Within an operation, high-risk animals should also not be allowed to have nose-to-nose contact with the resident herd.

The environment in which cattle are housed has a tremendous impact on the risk of reproductive disease development. Beef cattle spend much of their lives on pasture, making it difficult to create effective barriers to other animals and wildlife. An example of proper pasture utilization would be to ensure that weaned or stocker cattle have no fence-line contact with pregnant heifers to reduce the potential for BVDV transmission. Cattle raised mainly in confinement, as many dairy cattle are, may be more susceptible to pathogens that contaminate feed and water supplies, such as *N. caninum* transmitted by canines or *Salmonella* spp. carried by bird and rodent droppings. Inclement weather, such as extreme moisture, heat or drought, can create added stress on a breeding herd, potentially affecting nutrition, decreasing immunity, and increasing susceptibility to disease.

Breeding management

Good reproductive management should include having a defined breeding season and performing pregnancy testing in a timely manner. Most beef cattle are managed according to a seasonally defined breeding season, and an outbreak of a reproductive disease may not be recognized until pregnancy checking, often several months after breeding. Reproductive diseases could potentially affect an entire calf crop, and a delay in pregnancy testing could be even more costly. Having all bred females at the same or similar stage of pregnancy can also increase the risk of diseases whose incidence is dependent on the time of infection, such as BVD-PI caused by exposure of the dam from approximately 45 to 125 days of gestation. On the other hand, dairy cattle tend to be bred year around, and changes in reproductive status are likely to be detected sooner.

The purchase, lease, or borrowing of nonvirgin bulls for breeding purposes constitutes a significant risk to the reproductive performance of a herd.[11] Trichomoniasis and campylobacteriosis cause decreased reproductive performance in the breeding females, but do not cause clinical signs in infected males. Breeding soundness examinations of nonvirgin bulls should be performed prior to purchase and should incorporate trichomoniasis testing. Breeding animals should also be current on vaccinations, including vaccination against BVDV, campylobacteriosis, and leptospirosis. States have increasingly been implementing mandatory testing for trichomoniasis for interstate and intrastate movement of animals, and producers and veterinarians should familiarize themselves with their state's regulations.

AI and ET are effective barriers to the transmission of venereal and other reproductive diseases, while allowing for advancement in genetics and herd expansion.

While the production and distribution of semen and embryos pose some risks for pathogen transmission, these are considered to be safe procedures when internationally accepted standards such as those recommended by the Certified Semen Services (CSS) and International Embryo Transfer Society (IETS) are followed.[12] Pathogens such as BVDV and *N. caninum* can cross the placental barrier, so it is important that both donors and recipients are tested free of the disease and properly vaccinated if indicated prior to embryo implantation. Disposable equipment should be used when possible; any items used on multiple animals should be thoroughly cleaned and disinfected between use. Additional information on pathogen control through AI and ET technologies are described in Chapters 73 and 81, respectively.

Personal protective equipment

Personal protective equipment is essential when working with the breeding herd. Obstetric sleeves, gloves and protective clothing should be used when working with breeding animals, especially when working with potentially infected uterine fluids, placental tissues, and aborted fetuses. A change of clothing is recommended between farms for persons who make multiple farm visits daily, such as veterinarians and AI technicians. However, this may not always be possible or practical, and guidelines are often made following an appropriate assessment of the true risk of pathogen transmission.[13] For operations with multiple farm locations, separate clothing can be kept at each premises to prevent carrying pathogens from site to site. Protective gear such as disposable coveralls and boots should be provided to visitors who are given access to animal areas if they do not have their own. Whenever possible, used or soiled protective gear should be left on the farm for disposal or cleaning.

Wildlife

Wildlife commonly share ranges and forages with breeding herds and have been shown to be efficient carriers of reproductive pathogens. However, effective physical barriers do not exist to prevent all livestock contact with wildlife. Alternative barriers should be considered to reduce the risks of disease transmission from these populations, which usually occur through the contamination of water and feed.

Wildlife can serve as direct and indirect sources of infectious reproductive pathogens and should be discouraged from feeding and cohabitating with cattle herds. BVDV has been detected in wild ungulates, and deer have experimentally transmitted the virus to naive calves.[14] While the natural transmission of BVDV between deer and cattle has not been demonstrated, the potential should not be overlooked in areas where common feeding occurs. *Brucella abortus*, a pathogen that has eluded national eradication efforts for many years due in part to its persistence in wildlife, has been shown to be transmissible from infected bison to cattle.[15,16] Wild swine have also become increasingly prevalent in pasture environments. While specific cattle pathogens have yet to be isolated from these swine herds, cases of *Brucella suis* have been reported and remain a concern to the US pork industry.[17,18] In addition to the potential risks to the reproductive herd, contact with wild ungulates can pose a significant risk to the US livestock population in the event of a foreign animal disease outbreak, such as foot-and-mouth disease. Population control of deer and swine herds may be complicated and should involve collaboration between wildlife management professionals and livestock operators.

Feed and water sources

Clean feed and water sources are important for maintaining the health of the breeding herd. *Leptospira* spp. are shed through the urine of infected domestic and wild animals such as deer and other small mammals, and are known contaminants of natural water supplies. In operations where multiple water sources are available, pregnant heifers and cows should have access to the cleanest, avoiding natural streams and ponds. When possible, areas of heavy standing water should not be used for grazing. Vaccinations schedules may be adjusted to more fully protect the breeding herd when they are exposed to outside conditions and aversion of natural water supplies is not possible.[19]

Storing commodities and feed in an enclosed or protected area helps to prevent fecal contamination with potentially harmful pathogens. Wild canids are the definitive host for *N. caninum*, a parasite causing mid- to late-term abortions in cows (see Chapter 62). Cattle become infected with the parasite when they ingest feed contaminated with feces of infected canids. The parasite has been detected more frequently in dairy cattle populations where stored feeds are more likely to be ingested. Birds and rodents have also been shown to contaminate feeds by shedding *Salmonella* spp. in their droppings. Birds should be kept from roosting in feeding areas or feed storage areas. Population control of birds and rodents by roost and nesting removal or hazing should be considered. Pesticides should only be used with care. Domestic pets should also be prevented from contaminating livestock feed and water sources.

Vaccination

Vaccination is one of the most commonly used biosecurity practices by producers and veterinarians. While vaccines rarely prevent infection, proper vaccination strategies may help reduce the incidence or clinical severity of reproductive diseases. Common pathogens targeted by vaccine manufacturers are covered elsewhere in this book and include BHV-1 (infectious bovine rhinotracheitis), BVDV, multiple *Leptospira* spp., and *Campylobacter fetus*.[20] Uncommonly used vaccines include those against *B. abortus* in females, *N. caninum*, and *Tritrichomonas foetus*. The use of these and other vaccines should be considered only after a thorough biosecurity risk assessment is performed by the producer and the veterinarian to avoid any unnecessary costs. The safety and efficacy of vaccines, including a comparison of the use of modified-live versus killed vaccines, should be determined specific to the animals on the operation.

However, the use of vaccines should not be the sole source of protection for a breeding herd and should be used to complement other biosecurity practices.

Cleaning and disinfection

The farm environment is conducive to the survival of many potential bovine pathogens. While the elimination of all debris and manure is impossible, measures can be taken to reduce the pathogen load in the farm environment. Fortunately many reproductive pathogens are susceptible to common cleaning and disinfection practices. Many excellent resources are available on cleaning and disinfection, and veterinarians and producers should be familiar with the characteristics and limitations of commonly used disinfectants in a farm environment.[21]

Equipment and tools used on individual animals should be cleaned and disinfected after each use. Special attention should be given to equipment exposed to animal body fluids such as obstetric equipment and nose leads. Tools such as rakes, shovels, buckets, and halters are ideally not shared between high-risk animals and the resident herd.

Separate machinery or equipment should be used under the following situations: feeding high-risk groups of animals, handling feed and manure, and handling carcasses. However, this is not always possible or practical, and common equipment should be cleaned and disinfected after use in any of these situations. Equipment used off-farm or shared with other operations should also be thoroughly cleaned and disinfected upon return.

The physical removal of debris such as manure and mud from equipment is important to inhibiting pathogen survival. Many viruses and bacteria are susceptible to desiccation and ultraviolet lighting, so the use of chemical disinfectants on large flooring areas or large equipment such as trailers may not always be indicated in low-risk situations such as daily farm use in a single area. Vehicles of all visitors should be clean on arrival. A designated vehicle washing area, which could consist simply of running water, detergents and disinfectants, should be placed in an area that can be used prior to entry to the farm or animal area. Runoff from the washout area should not have contact with any animal areas.

Quarantine and isolation areas should be thoroughly cleaned and disinfected between groups to avoid cross-contamination in all-in, all-out situations. Calving and other common areas should be cleaned frequently to avoid manure build-up and the potential for increased pathogen exposure. Placentas should be removed promptly in confinement situations to avoid ingestion by other livestock. Any animal carcasses, including aborted fetuses and tissues, should also be removed promptly under any conditions to avoid contact and ingestion by other cattle, wildlife, or domestic pets. Carcass disposal should also be performed in compliance with local regulations. In cases of suspected or confirmed reportable animal diseases, specific cleaning and disinfectant measures may be mandated.[22]

Feed bunks and water troughs should be cleaned out frequently. Special attention should be paid to waterers that contain lids or floats that may trap debris. BVDV has been shown to survive outside the host for extended periods of time in aqueous materials and in synthetic mucus,[23] potentially increasing the risk of transmission through feed and water sources.

Special considerations

Zoonoses

Several reproductive pathogens encountered in beef and dairy operations can be zoonotic. All persons involved in animal handling during the reproductive management of the herd should take personal protective measures when handling reproductive tissues, fluids, or pharmaceutical products such as hormones. While safer strains of modified-live *B. abortus* vaccines are now used, personal protection against iatrogenic infection is still warranted. Reproductive hormones, while not containing pathogens, should also be handled with care.

Disaster preparedness

Beef and dairy cattle producers should also include emergency provisions in their biosecurity plan. Unexpected events such as tornadoes, fires or floodings may require animals to be quickly moved to a different location. Prior to an event, producers are encouraged to identify other producers, sometimes called "sister farms," with similar health programs and biosecurity plans that their animals could be sent to with little risk of disease transmission. Quarantine procedures should be followed on arrival at the sister farm as well as on return to the home farm in order to protect both farms. Animals to be moved should be identified according to a risk assessment, with special consideration given to protecting highly susceptible animals such as pregnant heifers.

In addition to identifying alternate housing, producers should ensure that vaccinations are always current, animals are properly identified, and health records are kept in more than one location. Additional protection against waterborne pathogens such as *Leptospira* spp. may be indicated in flooding situations or events resulting in large amounts of standing water. In emergency situations, the sharing of trailers, loaders, and feed troughs may be unavoidable. If equipment must be shared, cleaning and disinfecting after use and between herds is warranted. Provisions for emergency transportation should also be included in preparedness planning. A shelter-in-place system may be most appropriate following certain emergency situations when it may be more harmful to try to move cattle out of a disaster zone. Producers can contact the local county emergency management agency or the state's animal emergency preparedness office for additional disaster planning resources or assistance with emergency sheltering or confinement.

Summary

There are many tools available to protect the beef and dairy breeding herd against reproductive pathogens. Practical and efficient biosecurity plans must be made so that they are acceptable and likely to be followed. Biosecurity and biocontainment plans should be flexible and include disease monitoring and surveillance programs to continually identify operation needs, document management plans, and benchmark progress. A combination of environmental management practices, movement control, barriers, and cleaning and disinfecting protocols can reduce the incidence of reproductive diseases by reducing effective animal contacts. Ultimately, a properly performed risk analysis or HAACCP system will assist the veterinarian and livestock producer in making the most effective biosecurity recommendations for reproductive diseases.

References

1. Wenzel J, Nusbaum K. Veterinary expertise in biosecurity and biological risk assessment. *J Am Vet Med Assoc* 2007;230: 1476–1480.
2. Lindberg A, Houe H. Characteristics in the epidemiology of bovine viral diarrhea virus (BVDV) of relevance to control. *Prev Vet Med* 2005;72:55–73.
3. Smith D. Epidemiologic tools for biosecurity and biocontainment. *Vet Clin North Am Food Anim Pract* 2002;18:157–175.
4. Luzzago C, Frigerio M, Piccinini R, Daprà V, Zecconi A. A scoring system for risk assessment of the introduction and spread of bovine viral diarrhoea virus in dairy herds in Northern Italy. *Vet J* 2008;177:236–241.
5. Villarroel A, Dargatz D, Lane V, McCluskey B, Salman M. Suggested outline of potential critical control points for biosecurity and biocontainment on large dairy farms. *J Am Vet Med Assoc* 2007;230:808–819.
6. Reneau J, Lukas J. Using statistical process control methods to improve herd performance. *Vet Clin North Am Food Anim Pract* 2006;22:171–193.
7. Moore D, Merryman M, Hartman M, Klingborg D. Comparison of published recommendations regarding biosecurity practices for various production animal species and classes. *J Am Vet Med Assoc* 2008;233:249–256.
8. National Animal Health Monitoring System. *Beef 2007–08, Part 2 Reference of Beef Cow-calf Management Practices in the United States.* Fort Collins, CO: USDA, 2009.
9. National Animal Health Monitoring System. *Dairy 2007, Heifer Calf and Mangment Practices on US Dairy Operations, 2007.* Fort Collins, CO: USDA, 2010.
10. Faust M, Kinsel M, Kirkpatrick M. Characterizing biosecurity, health, and culling during dairy herd expansions. *J Dairy Sci* 2001;84:955–965.
11. Sanderson M, Dargatz D, Garry F. Biosecurity practices of beef cow-calf producers. *J Am Vet Med Assoc* 2000;217:185–189.
12. Givens M, Marley S. Approaches to biosecurity in bovine embryo transfer programs. *Theriogenology* 2008;69:129–136.
13. Anderson D. Survey of biosecurity practices utilized by veterinarians working with farm animal species. *The Online Journal of Rural Research and Policy* 2010;5:1–13.
14. Negrón M, Pogranichniy R, Alstine W, Hilton M, Lévy M, Raizman E. Evaluation of horizontal transmission of bovine viral diarrhea virus type 1a from experimentally infected white-tailed deer fawns (*Odocoileus virginianus*) to colostrum-deprived calves. *Am J Vet Res* 2012;73:257–262.
15. Davis D, Templeton J, Ficht T et al. Brucella abortus in captive bison. I. Serology, bacteriology, pathogenesis, and transmission to cattle. *J Wildlife Dis* 1990;26:360–371.
16. Kilpatrick A, Gillin C, Daszak P. Wildlife–livestock conflict: the risk of pathogen transmission from bison to cattle outside Yellowstone National Park. *J Appl Ecol* 2009;46:476–485.
17. Sandfoss M, DePerno C, Betsill C et al. A serosurvey for *Brucella suis*, classical swine fever virus, porcine circovirus type 2, and pseudorabies virus in feral swine (*Sus scrofa*) of eastern North Carolina. *J Wildlife Dis* 2012;48:462–466.
18. Wyckoff A, Henke S, Campbell T, Hewitt D, Vercauteren K. Feral swine contact with domestic swine: a serologic survey and assessment of potential for disease transmission. *J Wildlife Dis* 2009;45:422–429.
19. Van De Weyer L, Hendrick S, Rosengren L, Waldner C. Leptospirosis in beef herds from western Canada: serum antibody titers and vaccination practices. *Can Vet J* 2011;52: 619–626.
20. Givens M. A clinical, evidence-based approach to infectious causes of infertility in beef cattle. *Theriogenology* 2006;66: 648–654.
21. Dvorak G. *Disinfection 101.* Ames, IA: Center for Food Security and Public Health, 2005. http://www.cfsph.iastate.edu/Disinfection/Assets/Disinfection101.pdf. Accessed 2/11/2013.
22. Fotheringham V. Disinfection of livestock production premises. *Rev Sci Tech (OIE)* 1995;14:191–205.
23. Stevens E, Thomson D, Wileman B, O'Dell S, Chase C. The survival of bovine viral diarrhea virus on materials associated with livestock production. *Bov Pract* 2011;45:118–123.

Chapter 27

Beef Replacement Heifer Development

Terry J. Engelken

Department of Veterinary Diagnostics and Production Animal Medicine, Lloyd Veterinary Medicine Center, College of Veterinary Medicine, Ames, Iowa, USA

Introduction

The replacement heifer represents the next generation of genetic progress for the cow herd.[1] Because of the required capital outlay that beef producers invest in these females and the time period required to recoup this investment, it is critical that reproductive efficiency is optimized. In order to get a return on this investment, replacement heifers must become pregnant early in the first breeding season, calve with a minimum of dystocia at 2 years of age, breed back in a timely fashion, and then continue to be productive for a number of years. Therefore it is critical that replacement females be selected, grown, and managed with these goals in mind. Practitioners need to be involved with heifer development programs to ensure that these females are meeting a set of performance benchmarks that include both growth and reproductive targets. This will help ensure optimum reproductive performance, increased female longevity, and increased cow herd profitability.

Replacement heifer economics

Table 27.1 summarizes the cost structure of producing a heifer that will join the mature cow herd when she is confirmed pregnant.[2] This budget assumes that the heifer is born in the spring and weaned the following November. She will be bred the following summer and then subsequently calve as a 2 year old. The single biggest expense associated with the heifer development program is the opportunity cost of the market value of the heifer at weaning. This $825.46 cost (Period 1: conception to weaning) is incurred when the producer decides to keep her in the replacement pool as opposed to selling her at market value.

Costs associated from heifer weaning until the start of the breeding season are primarily associated with the winter feeding program and summer grazing (Table 27.1). Over 85% of this $321 cost is accounted for by feed, labor, yardage, and interest. Drylot rations or supplementation programs for heifers on grass need to balance feed costs with adequate growth performance. This emphasizes the need for sound nutritional consultation during this time, especially when feed ingredient prices are high. The animal health program (including death loss) and grazing costs (pasture rent equivalent) make up the remainder of the inputs during this time frame.

The costs associated with developing replacements through the end of the breeding season are also illustrated in Table 27.1. This $129 estimated cost is primarily associated with grazing and breeding costs. The grazing season is estimated to last 4 months at $13.60 per month. This budget assumes the purchase of a $4000 bull that will breed 20 heifers per year for a total of four breeding seasons. This breeding cost also considers the annual cost of feeding the bull, interest on the initial investment, and the bull's salvage value when culled. This gives a total investment of $1275 to carry a replacement heifer to approximately 18–20 months of age (pregnancy examination).

Finally, the cost of heifer development must take into account the level of reproductive performance (percent pregnant) and cash value of the open heifer. In this example an 85% pregnancy rate increases the cost of production to $1500 for each pregnant heifer. The income from the sale of open heifers is then credited to the heifer enterprise, bringing the final cost to $1354 per pregnant heifer. This total represents a $529 developmental cash cost being added to the $825.46 opportunity cost incurred at the time of initial selection. With this magnitude of capital outlay, it is critical that these heifers be adequately developed to ensure optimum reproductive performance and longevity.

Selection at weaning

Recent surveys indicate that approximately 85% of beef replacement heifers were raised on the operation where they calved.[3] Although the majority of herds in the United States utilize some form of individual cow and calf identification, this information is underutilized when it

---

Bovine Reproduction, First Edition. Edited by Richard M. Hopper.
© 2015 John Wiley & Sons, Inc. Published 2015 by John Wiley & Sons, Inc.

Table 27.1 Cost summary for developing a replacement heifer.

Period 1: Conception to weaning	$852.46
Period 2: Weaning to breeding	$321.00
Period 3: Weaning to pregnancy examination	$129.00
Subtotal	$1275.46
Adjust for 85% heifer conception rate	$1500.00
Adjust for cull heifer credit[a]	$1354.00
Calculated heifer development costs (born 2011)	$529.00

[a] Assumes 369 kg heifer at $2.644/kg = $975.60 × 0.15 = $146.34.
Source: Hughes H. Cost of raised replacements. *Beef Magazine*, July 2012. Used with permission.

Table 27.2 Effects of dry-lot development or grazing a combination of corn residue and winter range on heifer reproduction.

Trait	DL	EXT	SEM	P-value
n, first season	3	3	—	—
Puberty by AI (%)	88	46	0.04	0.01
Conceived to AI, first season (%)	69	58	0.08	0.23
Pregnant to AI, first season (%)	64	54	0.06	0.08
Pregnant, first season (%)	94	92	0.04	0.38
n, second season	2	2	—	—
Pregnant to AI, second season (%)	62	66	0.08	0.61
Final pregnancy rate, second season (%)	93	84	0.14	0.56

DL, heifers consuming a diet in a dry lot post weaning; EXT, heifers supplemented three times per week with the equivalent of 0.45 kg/day 31% CP cube (DM basis) post weaning while grazing corn residue and dormant winter range and subsequently moved to dry lot.
Source: Funston FN, Larson DM. Heifer development systems: dry-lot feeding compared with grazing dormant winter forage. *J Anim Sci* 2011;89:1595–1602. Used with permission.

comes to making management decisions. Therefore it is critical that the practitioner help the client understand the importance of accurate data collection as it relates to the heifer development enterprise. Calf birth date, dystocia score or birthweight, female temperament, and calf weaning weight can all be used to make inferences about individual cow productivity over time. This information can then be used to make post-weaning selection decisions easier.

The process of heifer development normally begins at the time of weaning with the initial selection of potential replacements. Females should be selected based on age, weight, temperament, and the productivity of their dam. Musculoskeletal soundness must also be evaluated as poor feet and leg structure can have a negative impact on productive longevity.[4] Selecting replacements from calves born in the first half of the calving season will increase the chance of the female reaching puberty by the time the breeding season starts.[5,6] Since the age and breed of the potential replacement is already established by the producer, weaning weight becomes the primary selection parameter and the basis for the developmental program.

Regardless of their birth date, the entire calf crop is typically weaned on the same day, so uniformity of the group of replacements may be an issue. Sorting heifers by weight and frame score is normally the first step in narrowing the pool of replacements at weaning.[5] Economic considerations should be evaluated when there is a wide disparity in heifer weights. Heifers which are too large (frame score) or too light (weight) may be targeted for culling in order to improve the economic efficiency of the replacement program. Beef operations that have a sound replacement development program should only need to retain an additional 10–20% more weaned heifers than they need to become pregnant. This will enable them to maintain current herd size and avoid the opportunity cost of retaining more heifers than needed.

Pre-breeding nutritional management

Based on a heifer's genetics and expected mature size, a target weight can be selected and the feeding program tailored to meet the needed average daily gain.[5–7] Target weight is simply an estimation of the weight at which individual heifers will reach puberty. Provided we know the weaning weight, target weight, and number of calendar days before the start of the breeding season, rations can be constructed to ensure that heifers reach puberty in a timely fashion. Typically, the rate of gain needed will fall into the range 0.5–0.8 kg/day and represent a total gain of approximately 91 kg for heifers of British breeding.

Traditional intensive developmental programs have emphasized the need for heifers to reach approximately 60–65% of their expected mature weight prior to breeding.[5–7] This practice is utilized to ensure that heifers reach puberty prior to the start of the breeding season. This ensures that heifers are cycling at the beginning of the breeding season and increases the chance of early pregnancies. Recent Nebraska research has looked at the reproductive performance of heifers that were targeted to reach only 50–55% of their expected mature weight.[8–10] These heifers are overwintered on relatively poor forage (winter range or corn stalks) with supplemental feed. In these studies, reproductive performance was not different between heifers developed to reach 55% versus 60% of their expected mature weight (Table 27.2). However, heifers developed to only 50% of their expected mature weight run the risk of conceiving later in the first breeding season and weaning lighter calves.[10] Reproductive performance in subsequent years was not different among these different heifer groups. Any potential decrease in reproductive performance in these heifers must be weighed against lower feed costs. These cost savings have been documented to range from $22 to $45 per heifer compared with dry lot development.

Feeding ionophores to replacement heifers increased feed efficiency, average daily gain, and hastened the onset of puberty.[11–13] An individual daily dose of 200 mg of monensin has been used effectively in these trials. Also, heifers breeding on their third estrus have a higher first-service conception rate (21–36%) than those bred at puberty.[14,15] Therefore, gains should be managed so that heifers reach their target weight at least 3 weeks prior to the beginning of breeding. Selection of the target weight may be based on the average weight of the heifer group, a percentage of the expected mature weight, or extrapolated from the average frame score.

Pre-breeding selection and management

Approximately 30–60 days prior to breeding, heifers should undergo a pre-breeding evaluation.[4,6,16] This is the last step in determining whether or not a heifer will actually enter the breeding program. Heifers should be evaluated on their body weight, reproductive tract score, pelvic dimensions, body condition score, and structural correctness. There is a growing body of evidence which shows that female temperament is highly correlated with that of her calf, and that feeding performance of the offspring will be negatively affected.[17,18] There is anecdotal evidence that temperamental heifers have lower conception rates to artificial insemination (AI). At any rate, overly aggressive or easily frightened females should be considered for culling.

Other heifers that fail to meet specific selection targets should be culled at this time as well. This evaluation should take place far enough in advance of the breeding season to allow adequate time for needed ration changes if accelerated growth is needed. This handling also represents a good opportunity to deliver vaccines and dewormers. This information should be summarized and reported in a format that will justify program decisions to the client.

The health program for the replacement heifer should be targeted against reproductive disease and helps to optimize growth rate. This program should complement the suckling calf vaccination protocol that was administered so that protective immunity is maximized. The heifer's first breeding season, first pregnancy, and subsequent reproductive performance should all be considered when defining the timing, delivery, and antigens to include in the vaccination schedule.[19] The history of disease on the ranch, traffic patterns on and off the operation, and previous diagnostic information should also be evaluated when attempting to manage the disease risk.

Vaccines that provide fetal protection against common viral and bacterial diseases should be utilized in these schedules and Beef Quality Assurance (BQA) guidelines should be followed. Deworming schedules can vary greatly depending on rainfall, temperature variation, and forage availability. The class of anthelmintic, level of pasture contamination with infective larvae, and the relative cost of the products will also impact strategic deworming programs.[20–22] Veterinary involvement in these decisions is critical for the successful application of current vaccine and parasite control programs.

Breeding season management

Planning for the breeding season should begin early enough to allow time for bull selection (AI or natural service), as well as acquisition and implementation of the estrus synchronization protocol. Evaluation of sires should be based on their ability to pass a breeding soundness examination and their expected progeny difference (EPD) values. Bulls with large scrotal circumferences and pre-breeding bull exposure should be utilized, since this will hasten the onset of puberty in their progeny.[5,6] Progesterone-based synchronization programs do offer the advantage of inducing puberty in some heifers.[23] There are multiple protocols that can be utilized for estrus synchronization; these programs should consider cost, animal handling, and breeding options.[24,25]

Early pregnancy detection may be a useful management tool to better define the timing of pregnancy, the sire of the calf, and the overall success of the breeding season. This may occur 30–45 days after the end of the AI period or following bull removal.[26] This practice is especially helpful, in that it gives an early indication of reproductive failure so that the practitioner can begin the diagnostic work-up much earlier. Early examination of pregnancy may also be used to hasten the culling of nonpregnant or late-bred heifers in the event of forage shortages.

Pregnancy examination also provides an opportunity to provide the next step of the health program. Heifers should be dewormed between pregnancy examination and calving, and the health program should center on antigens that cause losses in mid and late gestation.[19] Vaccines for calf scours should also be delivered according to label directions as needed. Herd health procedures should follow BQA guidelines and be documented by recording the name, dose, and serial number of the vaccine and the date that it was administered.

Information concerning individual pregnancy status, body condition score, overall pregnancy rate of the group, pregnancy distribution, and culling information should all be reported to the client in the form of a written summary.[27,28] This allows the practitioner and producer to evaluate the strengths and weaknesses of the heifer development enterprise and make needed changes. This type of communication enhances opportunities for proactive planning and fine-tuning of the program. This report will also serve as the planning document to ensure that the pregnant heifers are well managed all the way through the subsequent calving season.

Post-breeding through calving

Once heifers are determined to be pregnant, they need to be managed to calve in moderate body condition (average condition score 5.5–6.5). Heifers calving in moderate condition usually have less dystocia, a shorter postpartum interval, and higher pregnancy rates after their second breeding season.

Following pregnancy examination, weight gains need to focus on having the heifers reach approximately 85% of their expected mature weight by the time they calve. Moderate daily gains in the range of 0.5 kg/day will usually be adequate to ensure that heifers reach their first calving with adequate frame, pelvis, and body condition. It is useful to remember that approximately 70–75% of fetal growth occurs during the last trimester of pregnancy and rations should be adjusted accordingly.[7,11] These adjustments need to include adequate protein, energy, and trace minerals. This is especially important for the first 3 months after calving, if heifers are expected to be able to lactate, grow, and rebreed in a timely fashion.[29,30]

It could be argued that future replacement heifer development begins while that individual heifer is *in utero*. The beef maternal environment can have a great impact on subsequent productivity even after the calf is born. Studies have reported instances of compromised maternal nutrition during gestation resulting in neonatal mortality, intestinal and respiratory dysfunction, metabolic disorders, decreased postnatal growth rates, and reduced meat quality.[31] This phenomenon has been referred to as fetal programming. Nutritional shortages may have a negative impact on the developing fetus at all stages of pregnancy but this can be especially severe during the last trimester.

Calves born to protein-restricted cows will show reduced feeding and reproductive performance.[31] Heifer calves from these cows had lower body weights at weaning, pre-breeding, and at pregnancy examination. Contemporary heifers from protein-supplemented cows reached puberty at an earlier age and had higher pregnancy rates after their first breeding season. This resulted in a 28% increase in the proportion of heifers calving in the first 21 days of the calving season. The steer mates to these heifers also showed an improvement in weaning weight, feedyard dry matter intake, percent grading choice, and marbling score.[31] Although the mechanisms by which placental and fetal programming occur are not clear, managing resources to ensure proper female nutrition during critical points of gestation can improve calf performance, calf health, and future reproductive potential.

Summary

Replacement heifers represent a relatively large capital investment on the part of the producer. Therefore, they need to be managed intensively to ensure that they breed, calve, and rebreed in a timely fashion. Events crucial to heifer development center on proper selection, nutritional management, breeding season management, and program evaluation. Careful monitoring of heifer growth and performance at these critical times is an important component of successful development. Record systems should be created and utilized so that replacement heifer program can be evaluated and benchmarked objectively over time. Practitioners are in a unique position to help their clients assess this part of their operation and make needed changes.

References

1. Engelken T. Developing replacement beef heifers. *Theriogenology* 2008;70:569–572.
2. Hughes H. Cost of raised replacements. *Beef Magazine* 2012;48(11):12–15.
3. National Animal Health Monitoring System. *Beef 2007–08, Part 1 Reference of Beef Cow-calf Management Practices in the United States*. Fort Collins, CO: USDA, 2009.
4. Rohrer G, Baker J, Long C, Cartwright T. Productive longevity of first-cross cows produced in a five-breed diallel: I. Reasons for removal. *J Anim Sci* 1988;66:2826–2835.
5. Spire M. Managing replacement heifers from weaning to breeding. *Vet Med* 1997;92:182–192.
6. Corah L, Hixon D. Replacement heifer development. In: *Beef Cattle Handbook*. Ames, IA: Midwest Plan Service, June 1999, BCH Publication 2100.
7. Larson R, Randle R. Heifer development: nutrition, health and reproduction. In: Youngquist RS, Threlfall WR (eds) *Current Therapy in Large Animal Theriogenology*, 2nd edn. St Louis, MO: Saunders Elsevier, 2007, pp. 457–463.
8. Funston R, Deutscher G. Comparison of target breeding weight and breeding date for replacement beef heifers and effects on subsequent reproduction and calf performance. *J Anim Sci* 2004;82:3094–3099.
9. Funston R, Larson D. Heifer development systems: dry-lot feeding compared with grazing dormant winter forage. *J Anim Sci* 2011;89:1595–1602.
10. Martin J, Creighton K, Musgrave J et al. Effect of prebreeding body weight or progestin exposure before breeding on beef heifer performance through the second breeding season. *J Anim Sci* 2008;86:451–459.
11. Moseley W, McCartor M, Randel R. Effects of monensin on growth and reproductive performance of beef heifers. *J Anim Sci* 1977;45:961–968.
12. Lalman D, Petersen M, Ansotegui R, Tess M, Clark C, Wiley J. The effects of rumenally undegradable protein, propionic acid, and monensin on puberty and pregnancy in beef heifers. *J Anim Sci* 1993;71:2843–2852.
13. Floyd C, Purvis H, Lusby K, Wettemann R. Effects of monensin and 4-Plex on growth and puberty in beef heifers. Oklahoma State University Animal Science Research Report, P-943, 1995. Available at www.ansi.okstate.edu/research/research-reports-1/1995/1995-1%20Floyd.pdf
14. Byerley D, Staigmiller R, Berardinelli J, Short R. Pregnancy rates of beef heifers bred either on puberal or third estrus. *J Anim Sci* 1987;65:645–650.
15. Perry R, Corah L, Cochran R, Brethour J, Olson K, Higgins J. Effects of hay quality, breed, and ovarian development on onset of puberty and reproductive performance of beef heifers. *J Prod Agric* 1991;4:13–18.
16. Hardin D. The benefits of palpation before synchronization. *Agri-Practice* 1984;5:29–32.
17. Vann R, Baker J, Randel R. Relationship between measures of cow and calf temperament and live animal body composition traits in calves at weaning. *J Anim Sci* 2004;82(Suppl. 2):24 (abstract).
18. Busby W, Strohbehn D, Beedle P, King M. Effect of disposition on feedlot gain and quality grade. Iowa State University Animal Industry Report, Extension Number ASL R2070, 2006. Available at http://lib.dr.iastate.edu/ans_air/vol652/iss1/16/
19. Spire M. Immunization of the beef breeding herd. *Comp Cont Educ Pract Vet* 1988;10:1111–1117.
20. Larson R, Corah L, Spire M, Cochran R. Effect of treatment with ivermectin on reproductive performance of yearling beef heifers. *Theriogenology* 1995;44:189–197.
21. Williams J, Loyacano A. Internal parasites of cattle in Louisiana and other southern states. LSU Research and Extension Research Information Sheet #104. Baton Rouge, LA: Louisiana Experimental Research Station, 2001. Available at http://www.lsuagcenter.com/nr/rdonlyres/5cef16ce-3571-489c-bcb8-fe318d6635a9/4101/ris104cattleparasites.pdf
22. Engelken T, Kimber M, Day T, O'Connor A, Ballweber L, Wang C. A survey to evaluate the level of anthelmintic resistance of gastrointestinal strongyles in Iowa cow/calf operations. In: *45th Annual Conference of the American Association of Bovine Practitioners, Montreal, Canada, September 20–22, 2012*.
23. Engelken T. Estrus synchronization programs for beef herds. *Vet Med* 1995;90(Suppl.):19–28.
24. Geary T. Synchronizing estrus in beef cattle. In: *Cattle Producer's Library, Reproduction Section*, CL405, 2nd edn. Western Beef Resource Committee, University of Idaho Department of Animal and Veterinary Sciences, 2002.
25. Kesler D. Review of estrous synchronization systems: CIDR inserts. In: *Proceedings, Applied Reproductive Strategies in Beef Cattle, September 5–6, 2002, Manhattan, Kansas*, pp. 47–59. Available at http://beefrepro.unl.edu/proceedings/2002manhattan/05_ksu_cidr_kesler.pdf

26. Rice L. Development of replacement heifers. In: Howard JL, Smith RA (eds) *Current Veterinary Therapy 4. Food Animal Practice*. St Louis, MO: Saunders, 1999, pp. 109–113.
27. Traffas V, Engelken T. Breeding season evaluation of beef herds. In: Howard JL, Smith RA (eds) *Current Veterinary Therapy 4. Food Animal Practice*. St Louis, MO: Saunders, 1999, pp. 98–105.
28. Engelken T, Trejo C, Voss K. Reproductive health programs for beef herds: analysis of records for assessment of reproductive performance. In: Youngquist RS, Threlfall WR (eds) *Current Therapy in Large Animal Theriogenology*, 2nd edn. St Louis, MO: Saunders Elsevier, 2007, pp. 490–496.
29. Marston T, Lusby K, Wettemann R, Purvis H. Effects of feeding energy or protein supplements before or after calving on performance of spring-calving cows grazing native range. *J Anim Sci* 1995;73:657–664.
30. Lalman D, Williams J, Hess B, Thomas K, Keisler D. Effect of dietary energy on milk production and metabolic hormones in thin, primiparous beef heifers. *J Anim Sci* 2000;78:530–538.
31. Summers AF, Funston RN. Fetal programming: implications for beef cattle production. In: *Proceedings, Applied Reproductive Strategies in Beef Cattle, September 30–October 1, 2011*, Boise, Idaho, pp. 49–59. Available at http://beefrepro.unl.edu/proceedings/2011northwest/10_nw_fetalprogramming_funston.pdf

Chapter 28

Heifer Development: From Weaning to Calving

Ricardo Stockler

Veterinary Medicine Teaching and Research Center (VMTRC),
University of California Davis, Tulare, California, USA

Performance during first lactation is the best indicator that a successful rearing program has been implemented. The typical guidelines to consider are (i) optimum age at first calving should be around 22–24 months of age; (ii) excessive rate of weight gain prior to puberty directly affects performance, reproductively and productively; and (iii) never underestimate the costs associated with rearing programs.

There are many alternatives with regard to housing and delivery of feed to recently weaned heifers. From pasture to confinement, it is imperative to avoid overcrowding, and to move heifers out of the hutches in small groups of 5–10 head at a time.[1,2] This is the beginning of a system where the heifers experience the most stressful time as a group. The main advantage of having a small group is that this allows the heifers the opportunity to adjust to diet change, pathogen challenges, and proper growth. Ultimately, the goal is to raise healthy heifers that reach the breeding group as homogeneous and balanced as possible; however it is imperative to keep in mind that when calves are grouped, the risk of diseases such as respiratory conditions and diarrhea, specifically coccidiosis in this age group (>30 days old), increases.

A palatable concentrate diet (same as that being fed in the hutch) should continue to be offered during the first 2–3 weeks after grouping, followed by a change to a heifer grower ration with inclusion of a coccidiostat product, either an ionophore or a sulfonamide. In addition to the heifer grower ration, early cut forages are preferred for the recently weaned younger heifers, as they are higher in energy and protein and lower in fiber. High-fiber products should be avoided in young calves as the rumen–reticulum is still under development, and unable to fully digest fiber-rich diets.[1] The major goal when it comes to feeding is to sustain uniform weight gain and growth, which will allow the heifer calf to attain puberty and thus breeding at the recommended age. The National Research Council[3] suggests an average daily gain of 0.7–0.8 kg, assuming that a healthy calf is being raised in a thermoneutral environment. In addition,

Drackley[4] clearly states that a growth rate of 0.9 kg/day is achievable and does not affect mammary development, but needs to be adjusted to improve animal health and farm productivity. Therefore dry matter intake (DMI) must be managed properly based on environmental conditions. DMI should be increased when temperatures are below 15 °C; failure to provide extra energy can result in decreases of up to 0.2 kg/day of average daily gain.[3] If forage availability is not an issue, heifers raised on pasture should not have problems reaching the goal suggested by the National Research Council, despite the extreme variability in quantity and quality. Special attention should be paid to the trace and major minerals, as supplementation may be warranted depending on the region and season.

Body weight is directly correlated to puberty as well as first-lactation milk yield. An easy rule of thumb to remember is that target body weight at initiation of breeding should be around 55% of mature body weight and 82% of mature body weight at calving. It is important to stress that growth rate is by far the major contributory factor to a successful breeding heifer program.[5] Monitoring body condition (weight gain) provides valuable field information with regard to weight changes. One body condition equals 54 kg of body weight.[6] Table 28.1 summarizes the ideal body condition score at different production stages.

Heifers express their first estrus at 9–12 months of age and weight of approximately 280–300 kg.[2] The identification of heifers that are suitable to be bred is essential. Individuals not reproductively sound, underdeveloped, or freemartins should be removed as soon as possible. Therefore it is recommended to move heifers into the breeding pool 2–3 months prior to expected initiation of breeding, so estrus and potential problems can be identified. The diet offered should provide a stable rate of gain, since excessive growth at this point is not necessary and wasteful. Springing heifers should be fed ingredients similar to the lactating group, with good-quality forages and a limited amount of concentrate (1% body weight), with the intention of a

---

Bovine Reproduction, First Edition. Edited by Richard M. Hopper.
© 2015 John Wiley & Sons, Inc. Published 2015 by John Wiley & Sons, Inc.

Table 28.1 Target body condition score and production cycle.

	Body condition score (1–5 scoring system)
Cows	
Calving	3.5
Peak milk	2.75
Mid lactation	3.0
Dry-off	3.5
Replacement heifers	
6 months old	2.75
Breeding age	3.0
Calving	3.5

Table 28.2 Summary of target body weight (BW), height (H), and average daily gain (ADG) for different age groups of Holstein and Jersey heifers.[2,5]

	Weight (kg)		Height (cm)		
Age (months)	Holstein	Jersey	Holstein	Jersey	ADG
6	160	113	100	90	0.7–0.8 kg/day
12	295	204	119	107	0.7–0.8 kg/day
18	408	272	127	114	(post puberty) 0.6–0.7 kg/day
24	544	363	132	122	(post puberty) 0.6–0.7 kg/day

smooth transition, which in turn supports a high DMI immediately after calving and optimum peak milk. Close attention should be paid to the sodium and potassium concentration in the diet as it is known to be one of the causes of udder edema after calving.

A summary of the optimum target body weight and height as well as average daily gain for different age groups of Holsteins and Jerseys is summarized in Table 28.2. It should be noted that these are raw guidelines and the numbers may vary according to region and type of system being utilized to raise replacement heifers.

Health events are undoubtedly a major driving factor that increases raising costs due to poor performance, extra labor associated with treatment, medication usage, premature culling, or death. As mentioned before, protozoal (*Eimeria* sp.) diarrhea followed by respiratory diseases are the main causes of post-weaning sickness. The use of feed additives is recommended, ionophores being the most common. Ionophore antibiotics do not currently require a prescription; however, it is the veterinarian and/or nutritionist's responsibility to always advise the producer to follow label directions. In this case ionophores not only aid in preventing protozoal diarrhea (coccidiosis), they also increase feed efficiency by reducing methane losses and increasing propionate production in the rumen. Propionate is essential for gluconeogenesis in the liver and therefore energy produced. This class of antibiotics is widely used in beef-feedlot cattle and in the early 2000s a label for lactating dairy cattle was established. Monensin and lasalocid are available for use in the United States as Rumensin and Bovatec, respectively. Duffield and Bagg[6] have published a review that addresses the use of monensin in lactating dairy cattle. Diets containing monensin when offered before calving have positive effects on energy metabolism and conversion in early lactation, and its use is associated with a reduction in subclinical ketosis and postpartum diseases. Despite the fact that 1 kg/day of milk production may be gained when monensin is used in early lactation, depression of milk fat components may occur with the use of ionophores. This is most likely when lasalocid is used due to its effect on volatile fatty acid production in the rumen, although the effect may be dose dependent.[7] As per the label, the dose of monensin in calves to aid in the prevention and control of coccidiosis, as well as improve feed efficiency and therefore increase the rate of weight gain, must not exceed 200 mg/day per head.[7] Rumensin is not approved for use in veal calves. Sulfonamides are also used to treat episodes of diarrhea. Sulfadimethoxine (Albon) and sulfamethazine (Sustain III) are the only sulfa products labeled for use in food animals; all others are prohibited under the Animal Medicinal Drug Use Clarification Act (AMDUCA) 1994.[8] Sulfamethazine is available in sustained-release bolus for oral administration in nonlactating dairy cattle, and when used according to label directions provides therapeutic levels for up to 72 hours. If the intent is to treat lactating dairy cows, sulfadimethoxine is the only sulfa drug available. Intravenous and oral routes exist for this antimicrobial and label recommendations should be followed with respect to indication, dosage, and frequency of administration. Most importantly, withdrawal period of 7 days for meat and withholding period of 60 hours for milk must be observed.

Another important health condition concerns the incidence of external and/or internal parasites. Parasites directly affect development due to decreases in rate of gain. They also lower production indirectly through aggravation and irritation/pain. Outbreaks tend to be seasonal and occurrence depends on the region, so treatment and control methods must be implemented accordingly. Additionally, lice and mites are a significant problem due to their potential capability as vectors, through vertical and horizontal transmission.[9] Flies are potentially one of the contributors to the increase in bacterial counts in milk due to a nonhygienic environment and are a major vector in the transmission of infectious bovine keratoconjunctivitis and blood-borne diseases.[10]

According to the 2001 National Animal Health Monitoring System, respiratory disease accounts for 50.4% of weaned heifer deaths.[11] McGuirk[12] summarizes the important points to be analyzed when investigating a respiratory outbreak. Morbidity and mortality data, vaccination and treatment protocol, nutrition, age, housing condition, and case definition are some of the records needed. These criteria facilitate the development of protocols designed to improve early detection of disease and, if indicated, the initiation of an effective treatment. A plethora of antimicrobials labeled for treatment of respiratory disease are available in the United States (Table 28.3). Dosage, frequency, route of administration, spectrum of activity, and meat withdrawal periods vary. One must consider that most of them have specific label directions with respect to age group allowed to

Table 28.3 Summary of most common labeled antimicrobials available in the United States for treatment of respiratory disease in dairy cattle.

Drug	Product	Dose	Route of administration	Duration	Dairy older than 20 months of age	Milk/meat withdrawal
Oxytetracycline	200 mg/mL LA200®	4.5 mL/45 kg	Subcutaneously	48–72 hours	Yes	96 hours/28 days
Ceftiofur (one treatment)	Exceed®	1.5 mL/45 kg	Subcutaneously in base of ear	7 days	Yes	Zero/13 days
Tulathromycin	Draxxin®	1.1 mL/45 kg	Subcutaneously	10–14 days	No (ELDU)	—/18 days
Oxytetracycline	300 mg/mL Tetradure®	3–4.5 mL/45 kg	Subcutaneously	7 days	No (ELDU)	—/28 days
Tilmicosin	Micotil®	1.5 mL/45 kg	Subcutaneously	48 hours	No (ELDU)	—/42 days
Flofenicol	NuFlor®	3 mL/45 kg 6 mL/45 kg (single TX)	Intramuscularly Subcutaneously (one-time dose)	48 hours 4 days	No (ELDU)	—/28 days (i.m.) —/38 days (s.c.)
Gamithromycin	Zactran®	2 mL/50 kg	Subcutaneously	10 days	No (ELDU)	—/35 days
Sulfamethazine	Sustain III®	2 bolus/45 kg	Orally	72 hours	Prohibited by FDA	—/12 days
Enrofloxin	Baytril®	1.1–2.3 mL/45 kg or 3.4–5.7 mL/45 kg	Subcutaneously	24 hours Single dose	Prohibited by FDA	—/28 days
Danofloxacin	A 180®	1.5 mL/45 kg	Subcutaneously	48 hours	Prohibited by FDA	—/4 days

ELDU, extra-label drug use; FDA, Food and Drug Administration.

be treated; if extra-label drug use is practiced, some of these products may or may not fall under the prohibited drug category of AMDUCA (e.g., fluroquinolones and certain sulfonamides).[8]

Intramammary antimicrobial products are often used in springing heifers with the objective of decreasing the prevalence of intramammary infection after parturition, reducing somatic cell counts, and potentially increasing milk production. Because this is considered extra-label drug use, the treatment of heifers before calving should only be recommended by a veterinarian and in herds that have demonstrated high incidence of heifer mastitis.[13,14] In 2012, Vliegher et al.[14] published a comprehensive review about mastitis in dairy heifers. The group concluded that often nonlactating heifers are infected with major mastitis-causing pathogens and there is potentially horizontal transmission. Prevention is the key; overall hygiene, especially at calving, fly control, and the prevention of cross-suckling are some of the recommendations. These authors also reported that, in general, coagulase-negative staphylococci are known to be the principal etiologic agent, followed by coagulase-positive staphylococci and environmental pathogens. Therefore, when working with herds with a high incidence of heifer mastitis the cause must be identified and the specific risk factors for such analyzed. The intramammary use of antimicrobials in nonlactating heifers has not always proven efficacious when it comes to improvements in overall udder health and milk yield.

Vaccination is a broad and variable topic. If respiratory disease is a problem in pre-weaned calves, it is necessary to evaluate the colostrum management program and the lactating/dry cow vaccination protocols. Although there is variability among farms breeding populations at risk, the general consensus is that replacement heifers must receive a modified-live respiratory five-way vaccine (IBR, PI3, BRSV, BVDV type 1 and 2) before weaning (before commingling) and at breeding time in order to acquire optimum protection.

For welfare and biosecurity reasons, prompt identification of a sick calf is crucial; however, I strongly advise the importance of addressing the potential problem in a herd health manner. Treating individual animals is important but does not solve the problem. On-farm monitoring and preventative methods are vital. This information allows the analysis of how efficacious the management program is in preventing the occurrence and transmission of the disease or condition in question.

Summary

A successful heifer rearing program consists of a consecutive series of systems employed to ultimately raise a heifer with minimum expense to be reproductively efficient and to deliver an excellent productive first lactation. Nutrition is known to be one of the most important factors at this stage of life, as it is directly correlated to reproduction, growth, and overall health. This chapter should be read in conjunction with Chapter 69 (Management to decrease neonatal loss of dairy heifers) and emphasizes the guidelines and practical approaches to developing a successful rearing program.

References

1. James R, Collins W. Heifer feeding and management systems. In: Van Horn HH, Wilcox CJ (eds) *Large Dairy Herd Management*. Champaign, IL: American Dairy Science Association, 1992, pp. 411–421.

2. Retamal P. Nutrition management of dairy heifers. In: Risco C, Melendez Retamal P (eds) *Dairy Production Medicine*. Ames, IA: Wiley-Blackwell, 2011, pp. 195–198.
3. National Research Council. *Nutrient Requirements of Dairy Cattle*, 7th edn. Washington, DC: National Academy Press, 2001.
4. Drackley J. Calf nutrition from birth to breeding. *Vet Clin North Am Food Anim Pract* 2008;24:55–86.
5. Hutchens M. *Feeding Guide Book*, 3rd edn. Fort Atkinson, WI: Hoards & Sons Company, 2008.
6. Duffield T, Bagg R. Use of ionophores in lactating dairy cattle: a review. *Can Vet J* 2000;41:388–394.
7. Rumensin 90. Elanco Animal Health, Division of Eli Lilly and Company. NADA: 95-735, approved by FDA. Available at http://elms.xh1.lilly.com/rumensin_90_label.pdf
8. Animal Medicinal Drug Use Clarification Act of 1994. Available at http://www.fda.gov/AnimalVeterinary/GuidanceCompliance Enforcement/ActsRulesRegulations/ucm085377.htm
9. Stromberg B, Moon R. Parasite control in calves and growing heifers. *Vet Clin North Am Food Anim Pract* 2008;24:105–116.
10. Butler J. External parasite control. In: Van Horn HH, Wilcox CJ (eds) *Large Dairy Herd Management*. Champaign, IL: American Dairy Science Association, 1992, pp. 568–584.
11. National Animal Health Monitoring System. *Dairy 2007. Part I: Reference of Dairy Health and Management in the United States*. Fort Collins, CO: USDA, 2002.
12. McGuirk S. Disease management of dairy calves and heifers. *Vet Clin North Am Fod Anim Pract* 2008;24:139–153.
13. Boerum A, Fox L, Leslie K *et al*. Effects of prepartum intramammary antibiotic therapy on udder health, milk production, and reproductive performance in dairy heifers. *J Dairy Sci* 2006;89: 2090–2098.
14. De Vliegher S, Fox L, Piepers S, McDougall S, Barkema H. Invited review: Mastitis in dairy heifers: nature of the disease, potential impact, prevention, and control. *J Dairy Sci* 2012;95:1025–1040.

Chapter 29

Interaction of Nutrition and Reproduction in the Beef Cow

William S. Swecker Jr

Large Animal Clinical Sciences, Virginia-Maryland Regional College of Veterinary Medicine, Blacksburg, Virginia, USA

Introduction

The relationship between nutrition and reproduction is bidirectional: reproductive status alters nutrient requirements but the nutrients assimilated by the cow also alter reproductive function. A beef cow must conceive, carry a fetus to term, give birth, and wean a calf each year, so she is either pregnant or lactating at any given time. The fetus becomes a high priority for nutrient repartitioning in the dam during late gestation, which can increase nutrient requirements by 75% at the end of pregnancy to support a weight gain of up to 70 kg for the fetus, placenta, and fetal fluids.[1] Likewise, nutrients are repartitioned again at calving to support lactation, which can increase requirements by 50–100% above maintenance.[2] Hormonal changes associated with pregnancy and lactation aid in the prioritization of nutrients; however, a negative energy balance during early lactation modifies the signals that initiate ovarian cyclic activity.

The beef cow utilizes tissue energy, which results in loss of body condition, to counter the negative energy balance of early lactation. The challenge to meet nutrient requirements becomes more difficult with the addition of environmental influences such as cold, rain, heat, and mud. The objectives of this chapter are to discuss the requirements of beef cattle relative to their reproductive status and define abnormalities of reproduction and related nutrient risk factors in beef cows. Nutrition and reproduction in the bull are discussed briefly.

Body condition scoring

Body condition scoring (BCS) is a visual and/or tactile appraisal of muscle and adipose tissue. Practically, cow BCS at critical periods such as calving, breeding, and weaning are utilized to monitor and manage feeding programs. Body condition may more accurately estimate tissue reserves than body weight. For example, weight loss during the periparturient period is associated with loss of fetus, fetal fluids, and placenta, whereas increases in weight are associated with digestive tract fill needed to support lactation and an increase in udder size. Thus weight at calving may be extremely variable, whereas BCS should be more stable.

Most beef condition scoring systems utilize a 9-point scale, with BCS 1 being extremely thin and BCS 9 being extremely obese.[3] Condition scores and body weights were evaluated in 11 301 Angus cows at weaning and a positive correlation was found between BCS and body weight. However, BCS accounted for only 16% of the total variation in weight.[4] Herd, year-month, and cow age also influenced body weight. Body weight changes among condition scores were not constant; 23–32 kg differences were found between condition scores lower than 5 and 39–45 kg differences were found between condition scores greater than 5. Apple et al.[5] evaluated various carcass characteristics of beef cows that ranged from BCS 2 to BCS 8. Of interest, backfat increased rapidy as BCS increased from 5 to 8, whereas rib eye area exhibited a linear increase from BCS 2 to 8 (Figure 29.1).

In the field, evaluation of BCS at scores of 4 or less involves changes in muscle mass as external fat is minimal, whereas increases in backfat depth contributes to the visual appraisal of scores from 5 to 9. A description of BCS 3 through 6 from several published studies is included in Table 29.1.

In conclusion, BCS scores are widely used in both field and research settings to evaluate the nutrition program of beef cows.

Nutrition and female reproduction

Detection of interactions between nutrition and reproduction

Veterinarians or producers commonly recognize reproductive abnormalities when cows do not calve or a large percentage of cows are found to be open when examined for pregnancy. However, the nutritional events that result in failure of cows to become pregnant may have occurred 4–12 months previously. One can evaluate historically or

---

Bovine Reproduction, First Edition. Edited by Richard M. Hopper.
© 2015 John Wiley & Sons, Inc. Published 2015 by John Wiley & Sons, Inc.

Interaction of Nutrition and Reproduction in the Beef Cow

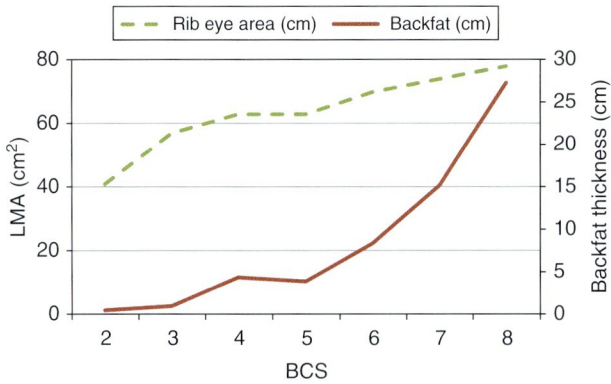

Figure 29.1 Relationship of rib eye area and backfat across condition scores in beef cows. LMA, longissiumus muscle area.

prospectively the nutrient demands of the cow relative to the feed supply available. Although there will be overlap among the groups, the practitioner can attempt to categorize the nonpregnant cows as those that failed to reestablish normal estrous cycles after calving, those that failed to conceive, or those that conceived but did not remain pregnant. Postpartum anestrus is the primary challenge that must be overcome to maximize fertility in most beef production settings. A thorough review of the diagnosis of nutritional infertility has been presented.[6]

Nutrient requirements of the beef cow

The nutritional requirements of beef cows vary with their reproductive status.[7]

1. Calving to conception (82 days): reproductive function during this period includes uterine involution, resumption of normal estrous cycles, and lactation. Nutrient requirements for a cow are greatest during this period and most cows utilize tissue energy to meet their requirements as they cannot consume enough nutrients. Clinically, this is represented by loss of body condition.
2. Pregnant and lactating (123 days): the primary demand for nutrients during this stage is lactation, and cows should maintain and potentially gain weight during this period. Increased nutrient intake supports lactation, which results in calf weight gain.
3. Mid gestation and nonlactating (140 days): nutritional requirements are lowest during this period and cows can

Table 29.1 Descriptions of body condition scores in beef cattle.

Reference	BCS 3	BCS 4	BCS 5	BCS 6
Richards et al.[3]	Thin: ribs are still individually identifiable but not quite as sharp to the touch. There is obvious palpable fat along spine and over tailhead with some tissue cover over dorsal portion of ribs	Borderline: individual ribs are no longer visually obvious. The spinous processes can be identified individually on palpation but feel rounded rather than sharp. Some fat cover over ribs, transverse processes and hip bones	Moderate: cow has generally good overall appearance. Upon palpation, fat cover over ribs feels spongy and areas on either side of the tailhead now have fat cover	High moderate: firm pressure now needs to be applied to feel spinous processes. A high degree of fat is palpable over ribs and around tailhead
Westendorf et al.[51]	Thin: ribs still identifiable but not as sharp to the touch. Some fat along the spine and over the tailhead	Borderline: individual ribs no longer obvious. The spine is still prominent but feels rounded rather than sharp. There is some fat cover over the ribs and hip bones	Moderate: good overall appearance. Fat cover over the ribs feels spongy and areas on either side of the tailhead have fat cover	High moderate: firm pressure must be applied to feel the spine. A high amount of fat is present over the ribs and around the tailhead
Corah[52]	Beginning of fat cover over the loin, back, and foreribs. Backbone still highly visible. Process of the spine can be identified individually by touch and may still be visible. Spaces between the processes are less pronounced	Foreribs not noticeable, 12th and 13th ribs still noticeable to the eye, particularly in cattle with a large spring of rib and ribs wide apart. The transverse spinous processes can be identified only by palpation (with slight pressure) to feel rounded rather than sharp. Full but straightness of muscling in hindquarters	12th and 13th ribs not visible to the eye unless the animal has been shrunk. The transverse spinous processes are rounded, though this can be felt only with firm pressure; the roundness is not visually apparent. Spaces between the processes not visible and distinguishable only with firm pressure. Areas on each side of the tailhead are fairly well filled but not rounded	Ribs fully covered, not noticeable to the eye. Hindquarters plump and full. Noticeable sponginess to covering of foreribs and on each side of the tailhead. Firm pressure required to feel transverse processes
Wiltbank[53]	Thin: fat along backbone and slight amount of fat cover over ribs	Borderline: some fat cover over ribs	Moderate: fat cover over ribs feels spongy	Moderate to good: spongy fat cover over ribs, and fat beginning to be palpable around tailhead

Source: data adapted from Apple et al.[5]

Table 29.2 Diet nutrient density requirements for a 635-kg beef cow based on reproductive cycle.

Period	Dry matter intake (kg)	NEm (Mcal/kg)	TDN (%)	Crude protein (%)
Calving to conception (85 days)	13.8	1.28	58	9.8
Conception to weaning (125 days)	13.4	1.14	54	8.3
Weaning to late gestation (110 days)	12.3	0.90	47	6.4
Late gestation (50 days)	12.4	1.16	55	8.3

Source: National Research Council. *Nutrient Requirements of Beef Cattle*, 7th revised edn. Washington, DC: National Academies Press, 1996.

be maintained on low-quality feeds during this period. This is an excellent time to increase the body weight and condition of thin cows.

4. Pre-calving (50 days): nutrient requirements rapidly increase during this phase and body condition should be maintained in addition to the increase in the weight of fetal tissues, placenta, uterine fluid, and mammary gland.

The energy and protein requirements during each of these four periods are listed in Table 29.2.

Body condition score at calving

Breed, suckling stimulus, and plane of nutrition affect return to estrus in postpartum cows.[8] BCS at calving and nutrient supply during the early postpartum period affect the return of ovarian cyclic activity and subsequent pregnancy rates (Table 29.3).

Body condition at calving alters the time between calving and conception and the pregnancy rate. Pregnancy rates for cows with BCS of 4 and 5 at calving were lower than pregnancy rates for cows with BCS of 6 and 7 at calving (64.9 and

Table 29.3 Body condition at calving and return to estrus after calving.

| Body condition score at calving | No. of cows | Per cent cycling | |
		60 days after calving	90 days after calving
Thin (1–4)	272	46	66
Moderate (5–6)	364	61	92
Good (7–9)	50	91	100

Source: Whitman R. *Weight change, body condition and beef cow reproduction*. PhD thesis, Colorado State University, 1975.

71.4% vs. 87.0 and 90.7%, respectively).[9] In a similar study, cows with BCS of 4 at calving had a pregnancy rate of 50%, while cows that calved with BCS of 5, 6, and 7 had pregnancy rates of 81, 88, and 90%, respectively.[10] BCS at calving also alters calving interval. Cows with BCS greater than 5 had 10–18 fewer days open than cows with BCS of 4.[9] Investigators using a 5-point condition score scale demonstrated a decrease in calving interval of 11.2 days for each unit increase in body condition at calving.[11] This is approximately equivalent to a change of 5–6 days in calving interval for a unit change in condition score when using a 9-point scale. Cows with condition scores greater than 5 displayed estrus 12 days earlier after calving and had 14 fewer days open than cows calving with BCS less than 4.[12] A cubic relationship between condition score at calving and pregnancy rate has been described which indicates that the effect of a 1-unit change in BCS on pregnancy rate is greater for cows with BCS between 4 and 6 than in fatter or thinner cows.[13] Crowe et al.[14] reported that the first postpartum ovulation occurs approximately 30 days after calving after three follicular waves in beef cows in good body condition, while cows in poor condition do not ovulate until 70–100 days after calving or 11 follicular waves.

Alterations in plane of nutrition and BCS at calving alter hormonal status. Flores et al.[15] reported that cows in low body condition at calving (BCS 4.3) had lower serum insulin-like growth factor (IGF)-1 concentrations than cows that calved in moderate body condition (BCS 6.2). Likewise, pulse frequency of luteinizing hormone (LH) was increased 3 weeks after parturition in cows that calved in moderate body condition (2.79 on a 5-point scale) compared with cows that calved in poor body condition (1.97 on a 5-point scale).[16] Postpartum LH secretion is less responsive to dietary protein. LH release at 20, 40, and 60 days after calving did not differ between cows fed adequate crude protein before calving (0.96 kg/day) and cows fed restricted crude protein (0.32 kg/day) starting at 90, 60, and 30 days before calving.[12]

Second-level body condition score at breeding

Body condition scores at the beginning of the breeding season influenced pregnancy rate and calving interval; however, changes in liveweight from calving to the beginning of the breeding season and BCS at the end of the breeding season had no effect on reproductive performance.[11] Kasimanickam et al.[17] evaluated the relationship of BCS at breeding to pregnancy rate from a single synchronized artificial insemination (AI) pregnancy rate and overall pregnancy rate in a group of 5500 cows over five breeding seasons. AI pregnancy rate was lower in cows with BCS 3 at breeding compared to cows with BCS 4–8. Overall pregnancy rate was higher in cows with BCS 6 and 7 at breeding compared to cows with BCS 3 and 4 at breeding. Cows with BCS 5 and 8 were intermediate in pregnancy rate.

Postpartum nutrient supplementation appears to benefit thin cows (BCS ≤4) more than adequately nourished cows. Pregnancy rates were evaluated in cows whose calves were removed for 48 hours at initiation of breeding. Cows with BCS less than 4 at calving that were fed to lose weight after calving had decreased pregnancy rates at 40 and 60 days after breeding compared with cows fed to gain weight after calving, cows fed to maintain their weight after calving, or cows fed to lose weight but with 2 weeks of increased energy intake prior to breeding (flushing). The ration fed after calving did not alter the pregnancy rate in cows that calved with BCS greater than 5.[3] Likewise, Mulliniks et al.[18] reported an increased pregancy rate in young cows (2–3 years of age) in range conditions that were supplemented twice a week with propionate salts and rumen undegradable protein (RUP) compared with cows supplemented with protein only. Supplementation was intiated 10 days after calving and continued for 65–120 days, depending on the year. Cow BCS was 4.6 at initiaton of supplementation period. Conversely, primiparous cows fed to gain 0.9 kg/day after calving had higher first estrous pregnancy rates, serum IGF-1, leptin, and glucose concentrations after calving than cows fed to gain 0.5 kg/day. Condition score at calving (group means 4.4 vs. 5.1) did not influence hormonal status after after calving.[19]

Once cows resume cycling, cessation of cycling (anestrus) requires a loss of 25% of body weight in mature cows, which indicates severe or prolonged nutrient restriction.[20] The timing of the calving season may modify the influence of nutrition on reproductive performance. BCS at calving may be more critical in spring-calving herds, whereas changes in BCS from calving to breeding may be more important in fall-calving cows due to differences in availability of feed.[21] All increases in nutrient supply before or after calving should be evaluated relative to cost of added feed versus potential increase in reproductive performance.

Protein supplementation

The effect of protein on reproductive function is less clear than is the effect of energy. An association between increased protein intake, especially rumen degradable protein, and increased early embryonic death has been reported. However, energy deficiency or protein–energy imbalance may be as critical as the protein excess.[6] Supplementation with 50% rumen undegradable or bypass protein after calving did not alter pregnancy rates in cows compared with pregnancy rates in cows fed 25% bypass protein (88 and 86%, respectively).[22] Supplementation with 250 g bypass protein (blood meal, corn gluten meal, and soybean meal) after calving increased the percentage of first-calf heifers that conceived during the first estrous cycle compared with heifers supplemented with 250 g degradable protein (wheat mill run, soybean meal, and urea). However, supplementation with bypass protein did not affect the pregnancy rate for the breeding season.[23] Pregnant beef heifers on winter range and hay were supplemented to meet either metabolizable protein requirements or crude protein requirements during gestation. Supplementation to meet metabolizable protein improved 2-year-old pregnancy rates (91% vs. 86%) over the 2-year study. The authors suggested that the response to metabolizable protein supplementation on reproduction was not by improved weight or BCS change prior to calving and the response may involve endocrine mechanisms.[24]

Fat supplementation

Fat supplements provide an energy-dense source of calories or may provide specific fatty acids that influence reproductive function. Williams and Stanko[25] reported multiple reproductive benefits of fat supplementation of beef cows. Supplemental fat fed during the early postpartum period enhanced luteal function by reducing the incidence of short cycles and prevented a postpartum decline in growth hormone. Higher dietary fat levels may increase blood concentrations of cholesterol, the precursor for progesterone.

Vitamin and mineral supplementation

Deficiencies of almost all nutrients have been thought to cause infertility, yet few have been proven to do so in cattle.[6] Dietary deficiencies of cobalt, copper, iodine, manganese, phosphorus, and selenium as well as excesses of molybdenum have been reported to cause infertility by one of three mechanisms: (i) decreased activity of rumen microorganisms with depression in digestibility; (ii) alteration of enzymatic action, which involves energy or protein metabolism or alteration of hormone synthesis; or (iii) inability to maintain the integrity of the cells of the reproductive system.[6] In addition, trace element deficiencies decrease immunocompetence, which could increase the risk of infectious causes of infertility.

Deficiencies of copper, selenium, and manganese have been reported to reduce fertility through altered embryonic survival and to delay estrus or puberty.[26] Allan et al.[27] reported an increased pregnancy rate (87%) in cows supplemented with trace element and vitamin boluses (2 mg Se, 2 mg Co, 138 mg Cu, 113 mg Zn, 71 mg Mn, 2.1 mg I, 4644 IU vitamin A, 929 IU vitamin D, and 9 IU vitamin E per day) compared with a pregancy rate of 64% in unsupplemented cows. In a second herd, length of calving season was reduced from 105 days to 49 days in supplemented cows. Ahola et al.[28] observed increased pregnancy rates after synchronized AI in cows after 2 years of supplementation with Cu, Zn, and Mn in a free choice mineral mix offered from approximately 80 days prior to calving to approximately 120 days after calving. A difference in AI pregnancy rates was not observed in the first year of the study. Trace element deficiencies often present as nonspecific disorders with impaired reproduction as one component. Thorough analysis of feeding programs and analysis of biological samples for trace elements should allow the practitioner to evaluate the role of trace elements in infertility.

Endophyte-infected tall fescue and fertility

Pregnancy rates reported for cows grazing endophyte-infected (*Neotyphodium coenophialum*, formerly *Acremonium coenophialum*) tall fescue are 67, 55, 33, 80, 55.4, and 39%, while pregnancy rates for control cows in these studies grazing uninfected or low-endophyte fescue pastures were 86, 96, 93, 90, 94.6, and 65%, respectively.[29] Plasma prolactin concentrations are decreased in cows that consume endophyte-infected fescue, yet LH concentrations are normal. The endophyte generates toxins such as ergovaline, which may affect the cow's ability to maintain pregnancy.[29] Caldwell et al.[30] reported cow productivity for spring and fall calving systems on endophyte-infected tall fescue pasture systems in Arkansas. Calving rates were higher for fall-calving than spring-calving cows.

Fetal programming

The nutrient status of the dam during gestation may also affect the reproductive productivity of the offspring; this concept is referred to as fetal programming or metabolic imprinting. Funston et al.[31] reported increased pregnancy rates in heifers from dams supplemented with protein in late gestation.

Puberty and breeding success in heifers

Most production systems require that heifers calve at 23–24 months of age. Furthermore, heifers that calve early in the calving season wean more and heavier calves during their lives.[32] Heifers should therefore reach puberty at 12–13 months of age to allow one or two estrous cycles before breeding at 14 months of age. Age at puberty is influenced by breed, season, and plane of nutrition.[32,33] Growth rates during the pre-weaning and post-weaning periods are inversely related to age at puberty, and the post-weaning growth rate is associated with plane of nutrition.[32]

The onset of puberty appears to be associated with increased frequency of LH secretion. Increased pulsatile LH secretion is associated with increased energy intake, whereas reduced energy intake suppresses LH secretion in heifers.[34] Delayed puberty is also associated with low concentrations of IGF-1. Steroidogenic capacity of thecal[35] and/or granulosa cells[36] is associated with increased concentrations of bovine somatotropin and IGF-1, respectively. Leptin may also act as a regulator for reproductive function and particularly as a mediator for nutritional cues. Short-term fasting of peripubertal heifers decreased leptin gene expression and circulating leptin concentrations were coincident with reductions in circulating concentrations of insulin and IGF-1 and LH pulse frequency resulting in delayed puberty.[37] Conversely, serum leptin concentrations were positively correlated with increasing body weight in prepubertal heifers.[38] Addition of ionophores to the ration increases growth rate but also tends to decrease age at puberty, independent of weight.[39,40]

Once puberty is reached, cessation of cycling takes moderate to severe nutrient restriction. Cassady et al.[41] restricted intake of cycling heifers to 30% of net energy for maintenance (NEm). Heifers who started the restriction period at a BCS of 5 stopped cycling in 61 days, whereas heifers who started restriction at BCS 7 stopped cycling in 148 days. The heifers lost 18 and 16% of body weight, respectively.

Historical recommendations have been that heifers should be fed to reach a target weight of 65–70% of their mature weight at breeding. Brahma-cross heifers fed to a target weight of 318 kg showed estrus earlier, had an increased rate of pregnancy after 20 days of breeding (39 vs. 9%), and had an increased rate of pregnancy at the end of the breeding season (82 vs. 66%) compared with heifers fed to a target weight of 272 kg.[42] Conversely, spring-born weaned beef heifers (213 kg) were fed over winter to achieve 55 or 60% of mature weight at breeding. Heifers fed for 60% mature weight were heavier at breeding (313 vs. 289 kg), had increased condition scores (6.0 vs. 5.6), and more were cycling before the breeding season (85 vs. 74%). However, pregnancy rate at the end of the 45-day breeding season was similar between groups (88 vs. 92%). The heifers were then followed through three calvings and production did not differ between groups except for higher 205-day adjusted weaning weights after the second calving for the heifers fed to achieve 55% of mature weight at breeding.[43]

Nutrition and male reproduction

Rate of gain after weaning and level of nutrition influence the weight and age at which bulls reach puberty. Young bulls maintained on low planes of nutrition had reduced testicular growth, ejaculate volume, sperm production, and seminal vesicle development. Conversely, overfeeding energy to young bulls can decrease reproductive performance due to increased fat deposition in the scrotum or the pampiniform plexus. Severe reduction of protein in rations fed young bulls decreased sperm production capacity. Refeeding of nutritionally stressed growing bulls improved reproductive performance in some trials but not in others. Mature bulls appear more resistant to dietary stressors; however, overfeeding and severe protein restriction have resulted in decreased libido.[44]

Barth et al.[45] conducted a series of experiments in feeding high and low levels of energy prior to traditional weaning time (approximately 26 weeks of age) or from traditional weaning to puberty. Restriction of nutrients prior to 26 weeks of age resulted in delayed puberty and reduced testicular size as yearlings. The authors concluded that nutrient intake prior to 26 weeks of age had a more profound influence on age of puberty and testicular size than post-weaning rations.

Zinc deficiency in ruminants impairs fertility in males, as evidenced by reduced testis size in zinc-deficient calves.[46] Selenium is present as glutathione peroxidase in seminal plasma. Semen from bulls with low selenium concentrations had lower motility after thawing than did semen from bulls with higher selenium concentrations.[47] Yearling beef bulls treated with sustained-release selenium boluses had increased sperm motility when compared with untreated bulls.[48]

Gossypol, a compound found in cottonseed that reduces fertility in male rats and humans, immobilizes bull spermatozoa in vitro.[40] Gossypol fed at the rate of 6 and 30 mg/kg body weight for 60 and 42 days, respectively, did not affect seminal quality or quantity or spermatogenesis in yearling Holstein bulls when compared with control animals fed soybean meal rations.[49] Rations containing gossypol 60 mg/kg (whole cottonseed), 6 mg/kg (cottonseed meal), and 0 mg/kg (soybean meal) were fed to Brahma bulls from weaning to puberty. Age at puberty was increased and 196-day weight gain was decreased in bulls fed whole cottonseed compared with bulls fed cottonseed meal. The age at puberty and 196-day weight gain was intermediate in the group fed soybean meal; therefore, gossypol did not express its effects in a dose-dependent manner. Electroejaculated semen quality and quantity at puberty did not differ among treatment groups; however, luminal diameters of the seminiferous tubules were larger, germinal epithelium was thinner, and germ cell layers were fewer in bulls fed gossypol.[50] Gossypol therefore alters the microanatomy of the testes but seems to have minimal influence on the quantity and quality of semen from young Holstein or Brahma bulls.

Summary

Research trials that investigate the relationship between nutrition and reproduction may be inadvertently biased to false-negative results because of the number of cows needed to detect small differences in pregnancy rates. An investigator needs 335 cows in each group to detect a difference between an 85% and a 90% pregnancy rate at $P<0.05$. Conversely, a trial with 50 cows per treatment group can only detect a difference between a 75% pregnancy rate and a 90% pregnancy rate at $P<0.05$. Studies with small numbers per group should therefore be evaluated carefully.

In simple terms, the developing fetus has a high priority for nutrients, lactation is a secondary priority that will decrease with decreased intake, and conception has a low priority and tends not to occur until other requirements such as maintenance, growth, and lactation are satisfied. In addition, suckling inhibits resumption of the estrous cycle. Nutrient supply to the cow prior to parturition and the resulting BCS at calving appear to be the most important factors in determining reproductive performance. Postpartum energy intake above maintenance can increase reproductive success in thin cows. In addition, the key period to increase condition in thin cows is between weaning and late gestation. Early weaning of calves expands the window of opportunity for increasing condition in cows. Thus, the management decisions of when to breed the cows and when to wean the calves are critical to maintaining reproductive success and should match the available nutrient supplies to the demands of the cows.

References

1. Bauman D, Currie W. Partitioning of nutrients during pregnancy and lactation: a review of mechanisms involving homeostasis and homeorhesis. *J Dairy Sci* 1980;63:1514–1529.
2. National Research Council. *Nutrient Requirements of Beef Cattle*, 7th revised edn. Washington, DC: National Academies Press, 1996.
3. Richards M, Spitzer J, Warner M. Effect of varying levels of postpartum nutrition and body condition at calving on subsequent reproductive performance in beef cattle. *J Anim Sci* 1986;62:300–306.
4. Northcutt S, Wilson D, Willham R. Adjusting weight for body condition score in Angus cows. *J Anim Sci* 1992;70:1342–1345.
5. Apple J, Davis J, Stephenson J, Hankis J, Davis J, Beaty S. Influence of body condition score on carcass characteristics and subprimal yield from cull beef cows. *J Anim Sci* 1999;77:2660–2669.
6. McClure T. *Nutritional and Metabolic Infertility in the Cow*. Wallingford, UK: CAB International, 1994.
7. Corah L. Nutrition of beef cows for optimizing reproductive efficiency. *Comp Cont Educ Pract Vet* 1988;10:659.
8. Masilo B, Stevenson J, Schalles R, Shiley J. Influence of genotype and yield and composition of milk on interval to first postpartum ovulation in milked beef and dairy cows. *J Anim Sci* 1992;70:379–385.
9. DeRouen S, Franke D, Morrison D et al. Prepartum body condition and weight influences on reproductive performance of first-calf beef cows. *J Anim Sci* 1994;72:1119–1125.
10. Selk GE, Wettemann RP, Lusby KS, Rasby RJ. The importance of body condition at calving on reproduction in beef cows. *OSU Agric Exp Sta Publ* 1986;118:3163–3169.
11. Osoro K, Wright I. The effect of body condition, live weight, breed, age, calf performance, and calving date on reproductive performance of spring-calving beef cows. *J Anim Sci* 1992;70:1661–1666.
12. Nolan C, Bull R, Sasser R et al. Postpartum reproduction in protein restricted beef cows: effect on the hypothalamic–pituitary–ovarian axis. *J Anim Sci* 1988;66:3208–3217.
13. Selk G, Wettemann R, Lusby K et al. Relationship among weight change, body condition and reproductive performance of range beef cows. *J Anim Sci* 1988;66:3153–3159.
14. Crowe M. Resumption of ovarian cyclicity in post-partum beef and dairy cows. *Reprod Domest Anim* 2008;43(Suppl. 5):20–28.
15. Flores R, Looper M, Rorie R, Hallford D, Rosenkrans C. Endocrine factors and ovarian follicles are influenced by body condition and somatotropin in postpartum beef cows. *J Anim Sci* 2008;86:1335–1344.
16. Wright I, Rhind S, Smith A et al. Effects of body condition and estradiol on luteinizing hormone secretion in post-partum beef cows. *Domest Anim Endocrinol* 1992;9:305–312.
17. Kasimanickam R, Whittier D, Currin J et al. Effect of body condition at initiation of synchronization on estrus expression, pregnancy rates to AI and breeding season in beef cows. *Clin Theriogenol* 2011;3:29–41.
18. Mulliniks J, Cox S, Kemp M et al. Protein and glucogenic precursor supplementation: a nutritional strategy to increase reproductive and economic output. *J Anim Sci* 2011;89:3334–3343.
19. Ciccioli N, Wettemann R, Spicer L, Lents C, White F, Keisler D. Influence of body condition at calving and postpartum nutrition on endocrine function and reproductive performance of primiparous beef cows. *J Anim Sci* 2003;81:3107–3120.
20. Burns P, Spitzer J, Henricks D. Effect of dietary energy restriction on follicular development and luteal function in nonlactating beef cows. *J Anim Sci* 1997;75:1078–1086.
21. Wettemann R. Management of nutritional factors affecting the prepartum and postpartum cow. In: Fields M (ed.) *Factors Affecting Calf Crop*. Boca Raton, FL: CRC Press, 1994, pp. 155–165.
22. Dhuyvetter D, Petersen M, Ansotegui R et al. Reproductive efficiency of range beef cows fed different quantities of ruminally undegradable protein before breeding. *J Anim Sci* 1993;71:2586–2593.
23. Wiley J, Petersen M, Ansotegui R, Bellows R. Production from first-calf beef heifers fed a maintenance or low level of prepartum nutrition and ruminally undegradable or degradable protein postpartum. *J Anim Sci* 1991;69:4279–4293.
24. Patterson H, Adams D, Klopfenstein T, Clark R, Teichert B. Supplementation to meet metabolizable protein requirements of primiparous beef heifers: II. Pregnancy and economics. *J Anim Sci* 2003;81:563–570.
25. Williams G, Stanko R. Dietary fats as reproductive nutraceuticals in beef cattle. *J Anim Sci* 2000;77:1–12.
26. Corah L. Nutritional factors affecting reproduction and practical nutrition for beef cattle ranchers. In: *25th Annual Conference of the American Association of Bovine Practitioners*, 1992, p. 244.
27. Allan C, Hemingway R, Parkins J. Improved reproductive performance in cattle dosed with trace-element vitamin boluses. *Vet Rec* 1993;132:463–464.
28. Ahola J, Baker D, Burns P et al. Effect of copper, zinc, and manganese supplementation and source on reproduction, mineral status, and performance in grazing beef cattle over a two-year period. *J Anim Sci* 2004;82:2375–2383.
29. Porter J, Thompson F Jr. Effects of fescue toxicosis on reproduction in livestock. *J Anim Sci* 1992;70:1594–1603.
30. Caldwell J, Coffey K, Jennings J et al. Performance by spring and fall-calving cows grazing with full, limited, or no access to toxic *Neotyphodium coenophialum*-infected tall fescue. *J Anim Sci* 2012;91:465–476.
31. Funston R, Summers A, Roberts A. Alpharma Beef Cattle Nutrition Symposium: Implications of nutritional management for beef cow-calf systems. *J Anim Sci* 2012;90:2301–2307.

32. Schillo K, Hall J, Hileman S. Effects of nutrition and season on the onset of puberty in the beef heifer. *J Anim Sci* 1992;70:3994–4005.
33. Martin L, Brinks J, Bourdon R et al. Genetic effects on beef heifer puberty and subsequent reproduction. *J Anim Sci* 1992;70:4006–4017.
34. Schillo K. Effects of dietary energy on control of luteinizing hormone secretion in cattle and sheep. *J Anim Sci* 1992;70:1271–1282.
35. Spicer L, Stewart R. Interaction among bovine somatotropin, insulin, and gonadotropins on steroid production by bovine granulosa and thecal cells. *J Dairy Sci* 1996;79:813–821.
36. Spicer L, Alpizar E, Echternkamp S. Effects of insulin, insulin-like growth factor I, and gonadotropins on bovine granulosa cell proliferation, progesterone production, estradiol production, and(or) insulin-like growth factor I production in vitro. *J Anim Sci* 1993;71:1232–1241.
37. Amstalden M, Garcia M, Williams S et al. Leptin gene expression, circulating leptin, and luteinizing hormone pulsatility are acutely responsive to short-term fasting in prepubertal heifers: relationships to circulating insulin and insulin-like growth factor I(1). *Biol Reprod* 2000;63:127–133.
38. Garcia M, Amstalden M, Williams S et al. Serum leptin and its adipose gene expression during pubertal development, the estrous cycle, and different seasons in cattle. *J Anim Sci* 2002;80:2158–2167.
39. Lalman D, Petersen M, Ansotegui R, Tess M, Clark C, Wiley J. The effects of ruminally undegradable protein, propionic acid, and monensin on puberty and pregnancy in beef heifers. *J Anim Sci* 1993;71:2843–2852.
40. Bagley C. Nutritional management of replacement beef heifers: a review. *J Anim Sci* 1993;71:3155–3163.
41. Cassady J, Maddock T, DiCostanzo A, Lamb G. Body composition and estrous cyclicity responses of heifers of distinct body conditions to energy restriction and repletion. *J Anim Sci* 2009;87:2255–2261.
42. Wiltbank J, Roberts S, Nix J, Rowden L. Reproductive performance and profitability of heifers fed to weigh 272 or 318 kg at the start of the first breeding season. *J Anim Sci* 1985;60:25–34.
43. Funston R, Deutscher G. Comparison of target breeding weight and breeding date for replacement beef heifers and effects on subsequent reproduction and calf performance. *J Anim Sci* 2004;82:3094–3099.
44. Brown B. A review of nutritional influences on reproduction in boars, bulls and rams. *Reprod Nutr Dev* 1994;34:89–114.
45. Barth A, Brito L, Kastelic J. The effect of nutrition on sexual development of bulls. *Theriogenology* 2008;70:485–494.
46. Hidiroglou M. Trace element deficiencies and fertility in ruminants: a review. *J Dairy Sci* 1979;62:1195–1206.
47. Slaweta R, Wasowicz W, Laskowska T. Selenium content, glutathione peroxidase activity, and lipid peroxide level in fresh bull semen and its relationship to motility of spermatozoa after freezing-thawing. *Zbl Vet Med A* 1988;35:455–460.
48. Swecker W, Kasimanickam R. Effects of nutrition on reproductive performance of beef cattle. In: Youngquist RS, Threlfall WR (eds) *Current Therapy in Large Animal Theriogenology 2*. St Louis, MO: Saunders, 2007, pp. 450–456.
49. Jimenez D, Chandler J, Adkinson R et al. Effect of feeding gossypol in cottonseed meal on growth, semen quality, and spermatogenesis of yearling Holstein bulls. *J Dairy Sci* 1989;72:1866–1875.
50. Chase C Jr, Bastidas P, Ruttle J, Long C, Randel R. Growth and reproductive development in Brahman bulls fed diets containing gossypol. *J Anim Sci* 1994;72:445–452.
51. Westendorf M, Absher C, Burris R. Scoring beef cow condition. Kentucky Extension Service, ASC-110, 1988.
52. Corah L. Body condition: an indicator of the nutritional status. *Agri-Practice* 1989;10:25.
53. Wiltbank J. Body condition scoring in beef cattle. In: Naylor JM, Ralston SL (eds) *Large Animal Clinical Nutrition*. St Louis, MO: Mosby, 1991, p. 170.

Chapter 30

Interaction of Nutrition and Reproduction in the Dairy Cow

Butch Cargile[1] and Dan Tracy[2]

[1]*Progressive Dairy Solutions, Twin Falls, Idaho, USA*
[2]*Multimin USA, Auburn, Kentucky, USA*

Introduction

Successful reproductive management of the modern dairy cow results from the implementation of best management practices. In the absence of widespread infectious disease, the largest effector of reproductive performance is prior and current nutritional status. Nutritional status of the dairy cow involves a complex interaction between macronutrients and micronutrients and herd-level management. In this chapter we discuss several of these interactions and their effects on reproductive performance. From a veterinary perspective, the primary tenet of animal nutrition is preventative medicine. Cows are programmed to succeed or fail, reproductively, based on the degree of metabolic stress experienced during the transition period from the nonlactating late-gestation stage to the lactating postpartum stage.

Reproductive goals

The goal of any commercial dairy's reproductive program is the establishment and maintenance of a subsequent pregnancy as soon as physiologically possible following parturition. This pregnancy leads to another parturition and new lactation cycle. Unfortunately, parturition is a biological event associated with considerable risk. The risks associated with several metabolic–immunological disorders are correlated with the transition period of a dairy cow, in addition to the possibility of dystocia which itself has a complex interaction with each of the metabolic disorders. Given the array of possible complications and negative outcomes associated with parturition, profitability is the only impetus strong enough to cause dairy farmers to expose their cattle to this degree of risk. Figure 30.1 displays the relative periods of the lactation cycle with regard to profitability. Over a cow's productive life, the more cumulative time she spends under the area of the curve associated with profitability, the more profitable she is for the operation. The only way to gain access to this phase of the lactation cycle is through additional parturitions.

Transition period

The transition period is widely accepted as the 3 weeks immediately before parturition through 3 weeks immediately following parturition. Disorders associated with this period of the reproductive cycle of the dairy cow include, but are not limited to, periparturient paresis (milk fever), retained fetal membranes (retained placenta), metritis, hepatic lipidosis (fatty liver syndrome), ketosis, and displaced abomasum. These disorders exert both direct and indirect effects on reproductive performance via follicular memory, immune function, and endometrial status. Table 30.1 displays incidence rates of the different metabolic disorders.

Retained placenta, metritis and endometritis

Placental detachment is an immunologically driven event. In a study by Kimura *et al.*[1] neutrophils isolated from the blood of cows with retained placenta had significantly lower function before and after parturition, extending up to 2 weeks after calving. Further evaluation in this experiment revealed lower plasma interleukin (IL)-8 concentrations in cows ultimately developing retained placenta. IL-8 functions as a chemoattractant for neutrophils and induces phagocytosis. Addition of anti-IL-8 antibodies to cotyledon preparations led to a 41% reduction in neutrophil chemotaxis *in vitro*.

Metritis is the commonly accepted vernacular for an infection contained within the uterus of the postparturient cow. These are typically Gram-negative anaerobic bacterial infections caused by invasion (growth) of microbes from the vagina

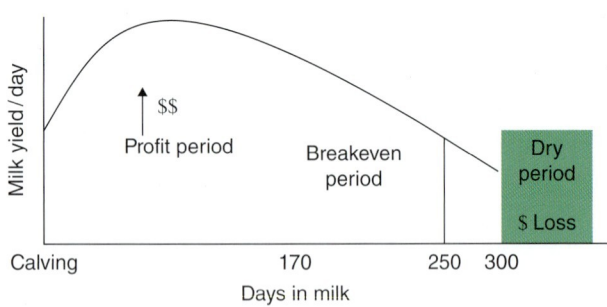

Figure 30.1 Periods of the lactation cycle with regard to profitability.

Table 30.1 Incidence of the different metabolic disorders.

Disorder	Incidence (%)
Milk fever	5–10
Retained placenta	5–10
Clinical ketosis	5–10
Metritis	8 to >40
Clinical endometritis (4–6 weeks after calving)	15–20

or delivered to the uterus iatrogenically. Immunosuppression during the periparturient period has been shown to predispose the cow to metritis. This immune suppression has been described as starting 1–2 weeks prior to parturition.[2–5] Cows experiencing dystocia have been identified as being 2.58 and 4.32 times more likely to develop metritis than cows classified as having a normal parturition.[6,7]

One method being studied in an attempt to ameliorate this prepartum immunosuppression is alteration of the ratio of omega-6 (*n*-6) to omega-3 (*n*-3) fatty acids in the duodenum.[8] It appears that by increasing the *n*-6 : *n*-3 ratio in multiparous cows before calving, functional properties of mononuclear cells may be improved. It has been suggested that a dietary ratio of *n*-6 to *n*-3 of 15 : 1 or more could be beneficial to reproductive performance during the lactation to follow.

Ketosis and hepatic lipidosis

Negative energy balance (NEB) is defined here as insufficient consumption of the calories necessary to meet metabolic requirements. This is the rule rather than the exception for late prepartum and early postpartum dairy cows. NEB begins in the prepartum period and progresses as dry matter intake (DMI) declines sharply several days before calving. This sharp decline in DMI is now thought to be the result, rather than cause, of excessive lipolysis.[9]

Ketosis is clinically classified as elevated plasma level of the ketone β-hydroxybutyrate (BHB). This occurs when the body's tissues, primarily muscular and neural, utilize less ketone than is being produced by hepatocytes. Increased production of BHB is a result of excessive mobilization of nonesterified fatty acids (NEFA). Ketosis, by itself, is a mammalian survival mechanism. However, animals of economic importance should not be approaching the survival threshold as this is well outside the zone of profitability and efficient production. Cows first testing positive for subclinical ketosis (BHB concentration 1.2–2.9 mmol/L) from 3 to 7 days in milk were 0.7 times as likely to conceive to first service compared with cows first testing positive at 8 days in milk or later,[10] intimating effect(s) based on temporal occurrence as well as incidence.

Hepatocyte uptake of NEFA depends on the circulating NEFA plasma concentration. Once internalized, NEFA can be utilized, through esterification, back into triglyceride (TG) for internal storage. TG can be repackaged into very low density lipoproteins (VLDL) and exported from the liver for use by the mammary gland and other tissues. Internalized NEFA can also be oxidized partially, which results in ketone production, or oxidized completely for adenosine triphosphate (ATP) production. This process is termed β-oxidation. Intrahepatic repackaging of TG involves the inclusion of TG into VLDL. The bovine liver has a limited capacity to produce VLDL. Therefore, during periods of excessive lipolysis, the cellular machinery of the bovine hepatocyte becomes overwhelmed with TG and is physically impaired by the amount of TG being stored intracellularly. The upper end of the spectrum of this condition is termed hepatic lipidosis and typically occurs when TG constitutes greater than 5% of liver tissue on a wet basis.

The B vitamins niacin and choline are effectively used to alter the adipose–circulatory–hepatic axis. Niacin effectively limits lipolysis through binding and stimulation of the G-protein-coupled receptor GPR109A in humans.[11] Similar effects have been seen in dairy cattle fed 12 g/day per head of rumen-protected niacin.[12] While effective at reducing lipolysis, this method of control leads to a significant reduction in energy-corrected milk production and is unacceptable in commercial settings. Lipolysis and the subsequent metabolism of NEFA by the bovine liver is an evolutionary adaptation to conserve lactation through provision of VLDL as an alternate energy substrate and provide the substrate necessary for milk fat production in colostrum and early lactation milk. When this process is limited, production is limited. Although potentially appealing from a reproductive standpoint, ultimately this control strategy is financially infeasible from a commercial perspective.

Supplementation with rumen-protected choline, on the other hand, addresses the rate-limiting step involving the inclusion of TG into VLDL, thus facilitating a net increase in hepatic energy production (translation) without increasing hepatic charge. Hepatic charge has been theorized to be a satiety initiator in the bovine.[9] Choline is integral in the production of phosphatidylcholine and critical for lipid transport and lipoprotein synthesis.[13] Feeding of rumen-protected choline in a study by Lima *et al.*[14] led to a reduction in mastitis approaching statistical significance. The link between postpartum mastitis and early reproductive performance is discussed later in this chapter. Both choline and the amino acid methionine are methyl donors and can spare each other in this process (Figure 30.2).

Another effective nutritional strategy addressing clinical and subclinical ketosis is the use of the only ionophore currently approved for use in lactating dairy cattle, monensin sodium. In Canada and other countries, controlled-release indwelling capsules that reside in the rumen and provide continuous release of monensin are approved for use. This delivery system is not currently approved in the United

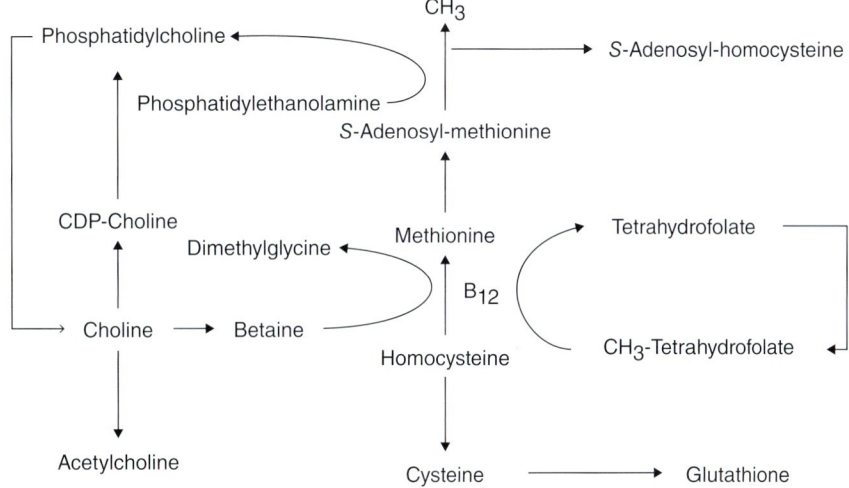

Figure 30.2 Methylation pathways (DNA, proteins, lipids).

States but feeding of monensin is approved. Monensin has been shown in a meta-analysis to reduce the incidence of clinical ketosis and severity of subclinical ketosis.[15] Monensin exerts its effects by increasing ruminal propionic acid production as a result of bacterial selection. This effect seems counter to the hepatic oxidation theory proposed by Allen[16] as the production of propionate is involved in satiety signaling in ruminants. However, this can be reconciled because the additional production of propionate is spread over a longer time-frame than the spikes from the rapid fermentation of starch following meals.[17]

β-Glucans

In its embryonic state in dairy cattle immunology, and infancy in human immunology, immune modulation by β-glucans is addressed crudely from a nutritional standpoint presently. The source and, more importantly, form of the specific β-glucan appears to have significant impact on its functionality with regard to gene expression within target immune system cells. Extensive studies are needed to elucidate the exact proinflammatory and pro-resolving effects of β-glucans and their subsequent effect(s) on reproduction.

Milk urea nitrogen

Milk urea nitrogen (MUN) is closely associated with blood urea nitrogen (BUN). BUN is an indirect measure of "excess" dietary protein or, more specifically, ammonia. Within the normal range of commercial ration formulation, BUN concentration increases as the ratio of soluble protein to nonfiber carbohydrates increases. Absolute amounts of the different protein fractions must be matched with their corresponding carbohydrate fractions when considering their respective ruminal degradation rates. When matched, efficient use of dietary protein and carbohydrate results in efficient production of microbial protein with minimal ammonia needing to be detoxified by inclusion into urea by the liver. The biochemical process of converting ammonia to urea exerts an energy cost on the cow. A study by Rajala-Schultz et al.[18] showed that cows with MUN levels between 10.0 and 12.7 mg/dL were 1.4 times more likely to be confirmed pregnant than cows with MUN levels above 15.4 mg/dL. In this same study, cows with MUN levels below 10.0 mg/dL were 2.4 times more likely to be confirmed pregnant than cows with MUN levels above 15.4 mg/dL. Studies suggest multifactorial negative fertility effects as a result of elevated BUN, including altered oocyte development,[19,20] embryonic maturation, and intrauterine environment.[21,22]

The use of metabolizable protein and amino acid balancing nutritional models has allowed crude protein and average MUN levels to be reduced. Rumen-protected forms of lysine and methionine are commercially available and appear to be economically viable. Use of these ingredients allows balancing of rations at considerably lower crude protein levels while still maintaining predicted production levels on a metabolizable protein basis since lysine and methionine are the first and second limiting amino acids in normal dairy rations.

NEFA concentration and oocyte formation

The functional micro-unit of reproduction from the female perspective is the ovum. Follicular memory is a carry-over effect that delays the influence of metabolic status on reproductive performance by as much as 60 days.[23–25] As the investigation and understanding of the influence that individual fatty acids exert on reproductive performance increases, NEB becomes a more complex proposition as it relates to fertility in the dairy cow. The negative effects of oleic acid, and a combination of oleic, palmitic and stearic fatty acids, have been studied with regard to their direct effect(s) on granulosa cells and cumulus–oocyte complexes[26] and theca cells.[27] In vivo, these negatively correlated effects are secondary to elevated circulating plasma levels of these individual and/or combinations of NEFA. Fatty acid composition of adipose tissue is influenced greatly by diet and degree of ruminal biohydrogenation of unsaturated fatty acids. Since NEFA result from the liberation of TG from adipose tissue, nutritional composition of absorbed fatty acids has a direct influence on the relative and absolute concentrations of NEFA.

Setting aside the direct influence that individual circulating fatty acids may exert on the ultimate fertility of the ovum, elevated circulating NEFA may be linked to runaway inflammatory response and insulin resistance, the latter being part of a vicious positive feedback loop that results in an ever-increasing NEFA concentration. Feeding of the omega-3 fatty acids α-linolenic acid and docosahexaenoic acid may directly dampen this hyperinflammatory state as a result of binding proliferator-activated receptors.[28] Studies completed recently have shown a decrease in embryonic loss and concomitant increase in milk production in cows supplemented with the essential fatty acids eicosapentaenoic acid and docosahexaenoic acid. Current reasoning for these results is a significant decrease in the caloric requirements of the cellular immune system, due to modulation, in these postpartum cows so that glucose is nonselectively spared, as glucose is the primary energy source for macrophages and neutrophils. As the bovine liver has a finite capacity for glucose production, primarily via gluconeogenesis, this sparing effect should return the postpartum cow to a positive energy balance at lower days in milk. In the bovine, resolution of NEB leads to reestablishment of normal ovarian cyclicity. On a molecular level, an intracellular shift from the production of proinflammatory eicosanoids to more pro-resolving eicosanoids likely explains the responses seen in these studies. This shift would break the positive feedback loop of proinflammatory eicosanoids leading to radical oxygen species production, tissue destruction, and further proinflammatory eicosanoids.

Aa alluded to above, tumor necrosis factor (TNF)-α production by adipocytes, and macrophages residing within adipose tissue, could enhance insulin resistance. Studies using exogenous TNF-α in cattle have led researchers to this conclusion.[29-31] This is especially significant in a transition cow that already has lower circulating insulin levels and decreased insulin sensitivity compared to that of a normal mid-lactation animal. One can see the possibility for a positive feedback loop here as well. It is possible this model explains obesity-induced diabetes in humans and the progression from normal periparturient lipolysis to a hyperketotic/hepatic lipidosis state in dairy cattle.

An indirect association between ketosis and reproductive efficiency, via clinical mastitis, can be created by a systemic inflammatory state. Cows experiencing ketosis have been shown to have twice the risk of developing clinical mastitis[32] and cows that experienced a case of clinical mastitis within 30 days after insemination were 16% less likely to conceive on first-service insemination. Cows experiencing subclinical or clinical mastitis prior to first-service insemination show increased days to first service, days open, and reduced first-service conception rate.[33] It is not difficult to visualize an impaired corpus luteum, hampered by proinflammatory prostaglandins, unable to produce progesterone at a level sufficient to maintain an early pregnancy.

Nutritional heat stress abatement

Heat stress, a result of the cow's physiological response to elevated temperature humidity index (THI), has been shown to lower conception rates[34] and pregnancy rates.[35] Lactating dairy cattle typically begin to experience heat stress around a THI of 70. This stress is exacerbated by dietary fiber digestion. Per unit of dry matter, the ruminal degradation of neutral detergent fiber (NDF) carries the highest heat increment of any of the macronutrients found in dairy rations. Ether extract (EE) carries the lowest heat increment of digestion. Therefore it is common practice to adjust lactating rations to reduce NDF levels and increase EE levels in response to times of intense heat stress. This serves two purposes: it increases the caloric density of the ration at a time when the cow is expending additional calories in an attempt to maintain core body temperature, and it decreases the heat increment that would have been created by digestion of the marginal fiber. Care must always be taken to avoid ration extremes that would result in fiber levels that are too low and/or EE levels that are too high. These extremes would put cows at risk of ruminal acidosis and decreases fiber digestibility/depressed DMI, respectively.

An additional method of nutritionally cooling cows has been studied[36] in beef cattle and could have application in a dairy setting. It involves feeding an osmotically active ingredient during times of heat stress and has been shown in at least one controlled study to minimize the degree of body temperature elevation. More work is needed on this concept to determine its efficacy and feasibility.

Body condition score

Probably the most complex phenotypic expression, body condition score (BCS) of the dairy cow is a combination of genetic predisposition for both reproduction and production, and numerous variables associated with management decisions. Elevated BCS is usually associated with either prolonged days to conception or abortion, extending length of lactation in both cases. This allows these cows to remain on diets that are higher in energy density than needed for maintenance due to decreasing milk production, as the lactation curve progresses at approximately a −9% monthly slope. Cows that are obese (BCS >3.75 on the 5-point scale) at calving tend to have higher risk of metabolic disorder,[37] extending the period of NEB and delaying conception. Current recommendations are to have cows at a BCS of 3.0–3.25 at the time of parturition. This axis between nutrition, reproduction, and body condition can either be coupled and provide a basis for continuous improvement in efficiency or decoupled, leading to challenges and economic loss.

Minerals and vitamins

Proper nutrition and management during the transition period are crucial for the prevention of postpartum reproductive disorders. The role of proper energy and protein supplementation in the prevention of uterine disorders is well defined, but mineral and vitamin balance and supplementation should not be ignored.[1]

Cows suffering from clinical or subclinical periparturient hypocalcemia are more at risk of experiencing secondary disorders such as dystocia, uterine prolapse, retained placenta, and metritis.[38-41] The pathogenesis and prevention of parturient hypocalcemia or milk fever continues to be a topic of debate and research. However, current management practices to prevent parturient hypocalcemia attempt to reduce the physiological lag time between the potentially hypocalcemic demands of lactation and recovery back to a functional normocalcemic state. Factors contributing to hypocalcemia include excessive prepartum dietary levels of calcium (1.35% DM), excessive levels of dietary phosphorus, low dietary levels of magnesium, increasing age, and breed with Jersey cattle being at increased risk.[38,42] The most commonly recognized presentation of milk fever (clinical hypocalcemia, parturient hypocalcemia) is during stage 2 and is diagnosed in the bovine by the clinical signs of recumbency, cold extremities, and a weak rapid pulse. Total serum calcium less than 8.0 mg/dL is considered diagnostic of hypocalcemia in the bovine.[43] The traditional nutritional method of limiting the incidence of clinical and subclinical hypocalcemia involved limiting calcium intake before calving so that the total absorbed calcium per head was less than 20 g/day. This stimulated bovine parathyroid hormone (PTH) secretion and increased bone resorption of calcium. This method has largely been replaced with the advent of dietary cation–anion difference (DCAD) diets. The equation for DCAD is (Na + K) − (S + Cl). Since sodium intake is held at a minimum during the immediate prepartum period as a means to limit udder edema and feeding excessive sulfur has negative implications, this equation can, practically, be simplified to (K − Cl). Anionic salts such as ammonium chloride, calcium chloride, and magnesium sulfate (used primarily as a magnesium source but also as an acidifier) are used to create a state of compensated metabolic acidosis. Newer commercial ingredients, usually a protein carrier treated with either hydrochloric acid or an organic acid, or combination of organic acids, serve the same function as the anionic salts. This method of hypocalcemia prevention has been described by Zeisel and Holmes-McNary.[13] Excessive acidification is undesirable as it increases the cow's caloric requirements in an effort to compensate for the greater degree of metabolic acidosis. This usually comes in the form of increased respiratory rate in an effort to reduce acid load. Decreased DMI is also associated with compensated metabolic acidosis and worsens with the degree of acidosis. Cows are typically fed these acidifying diets during the 3 weeks immediately before calving, when they are already mobilizing adipose tissue and declining in DMI. Diligent monitoring of urine pH as a measure of the degree of compensated metabolic acidosis is necessary in order for this program to be successful.

Magnesium plays an essential role in calcium homeostasis and as a result may contribute to hypocalcemia when dietary levels are insufficient.[42,44] Hypomagnesemia can blunt PTH secretion in response to low calcium as well as interfere with the interaction of PTH and its target tissues.[44] High dietary potassium will reduce the absorption of magnesium in the rumen as well as reduce the effectiveness of an anionic salt program. It is recommended that the magnesium content of the close-up ration should be between 0.35 and 0.4% of dry matter to insure adequate absorption.[44]

Trace minerals are present in the tissues at very low concentrations and their requirements are expressed in milligram or microgram amounts. Trace minerals considered to be of importance in dairy cattle include cobalt, copper, iodine, iron, manganese, molybdenum, selenium, zinc, and possibly chromium and fluorine.[38] Trace minerals are components of metalloenzymes, hormones and cofactors of functional proteins, or enzymes that are involved in many aspects of energy and protein metabolism, hormone synthesis and immune function.[38,45] Infertility in cattle, including decreased ovarian activity, reduced conception rates, increase in dystocia, abortions, and delayed onset of puberty, is linked to specific enzymatic dysfunction resulting from one or more trace mineral deficiencies.[46] In the bull, trace mineral deficiency is associated with a decrease in spermatogenesis, decrease in libido, increase in abnormal sperm, and a reduction in testicular size.[39,45] Trace minerals affect reproduction directly or indirectly through multiple functional proteins and enzyme systems. For example, selenium (Se) is well known for its role as a cofactor in the enzyme glutathione peroxidase, an important antioxidant that is present throughout bodily fluids, organs and within cells including sperm and cells of the ovary.[45] Such antioxidant activity helps maintain cellular function in the face of oxidative stress, improves cellular immunity and, when supplemented with other antioxidants such as vitamin E, can improve fertilization, reduce retained fetal membranes, and reduce the incidence of cystic ovarian disease.[47,48]

Vitamins like trace minerals are cofactors. Vitamins required by dairy cattle are classified as being fat-soluble (A, D, E and K) or water-soluble (B vitamins and vitamin C).[38] Water-soluble vitamins are produced by rumen microbes and true deficiencies in cattle with functioning rumens are rare.[38] Vitamins A and E are the only ones with a true dietary requirement because vitamin D is produced through ultraviolet (UV) radiation of the skin and vitamin K by bacteria of the rumen and intestines.[38] However, a reasonable practice includes supplementation of dry cows with 20 000–30 000 IU vitamin D daily to ensure proper calcium absorption[44] and compensate for animals not receiving sufficient UV radiation for *de novo* vitamin D synthesis. These conditions are common in dairy cattle residing in extreme northern and southern latitudes. Vitamins and trace minerals also share a synergistic relationship. Vitamin E and selenium have a sparing action on the requirement of the other and share the same characteristics.[49] Both contribute to antioxidant activity and maintain cellular health and function. Selenium contributes to intercellular, intracellular, and extracellular activity, while vitamin E preserves cellular integrity by protecting the cellular membrane.[45]

Trace mineral and vitamin supplementation challenges

As discussed earlier, trace minerals and vitamins are essential for proper immune function, cellular health and hormone synthesis, which are essential for postpartum involution, health of the reproductive tract, and resumption of normal estrous cyclicity after calving. It has been established previously that fresh cows suffering from postpartum diseases such as retained fetal membranes and metritis experience poor reproductive performance and increased days in milk.[6] However, compounding this issue is that approximately 1 month prior to calving, there is maternal transfer of stored trace minerals from the dam to the fetal tissue. Studies have demonstrated this transfer with copper, manganese, zinc, and selenium, which are the most limiting trace minerals with regard to fetal and neonatal survival and development.[50–52] In addition, transfer of vitamins A and E in colostrum occurs at calving and is a source of these vitamins for the calf after birth.[52,53] Maternal transfer of trace minerals is essential for calf health, but this transfer occurs at the cow's expense during a time of decreased DMI and demands on the immune system that occur prior to calving.[54] Further, contributing to the risk of deficiency and disease is the issue of mineral antagonism or negative mineral interactions, which reduces absorption in the gastrointestinal tract as well as antagonistic activity systemically. Excessive sulfur found in dried distillers grains, molasses and water reduce copper absorption in the gastrointestinal tract through formation of copper sulfide.[55,56] Selenium and sulfur are physically and chemically similar and when excessive sulfur is present in the ration, competition for absorption can result in reduction of selenium absorption.[57] In addition to sulfur, calcium levels greater than 1% of ration dry matter reduces the absorption of selenium from natural sources.[58] Systemic antagonism of the enzymatic function of copper can occur if excessive consumption of sulfur is combined with excessive consumption of molybdenum, resulting in the formation of thiomolybdate in the rumen that reduces copper absorption in the gastrointestinal tract.[59,60] In addition, thiomolybdate is absorbed and blocks systemic function of copper and increases biliary excretion of copper from liver stores resulting in symptoms of copper deficiency.[55] Subclinical deficiencies of trace minerals that occur due to antagonism, maternal transfer, or reduction in DMI prior to calving can persist into lactation, potentially compromising reproductive performance. Nutritionists address the risk of trace mineral deficiencies by providing proper amounts of trace minerals in the ration, reducing potential negative mineral interactions or antagonism, as well as including trace minerals with improved bioavailability such as chelates or organic trace minerals.[55] The study of trace mineral form, function, and supplementation continues to be an active area of research. Recent studies have shown a benefit to health when trace minerals are provided as an additional supplement or parenterally administered to cows that are provided with a supplemented ration. Such benefits to a cow's reproductive and physical health have been demonstrated with supplementary selenium and vitamin E. In addition, recent studies have shown an improvement in uterine health and reduction in the population of bacteria causing metritis when an injectable trace mineral product containing zinc, manganese, selenium and copper was administered during the dry period and during early lactation.[47,48,61,62]

Conclusion

Through genetic selection, economic pressure for higher milk production, and maintenance of profitability, today's dairy cow has become a physiologic athlete. However, development of today's dairy cow has resulted in many challenges to reproduction and health. Continually addressing those challenges through research has allowed better understanding of how to preemptively design rations and herd health protocols to allow for a successful lactation.

References

1. Kimura K, Goff J, Kehrli M, Reinhardt T. Decreased neutrophil function as a cause of retained placenta in dairy cattle. *J Dairy Sci* 2002;85:544–550.
2. Ishikawa H. Observation of lymphocyte function in perinatal cows and neonatal calves. *Jpn J Vet Sci* 1987;49:469.
3. Kashiwazaki Y, Maede Y, Namioka S. Transformation of bovine peripheral blood lymphocytes in the perinatal period. *Jpn J Vet Sci* 1985;47:337.
4. Kehrli M, Nonnecke B, Roth J. Alterations in bovine neutrophil function during the periparturient period. *Am J Vet Res* 1989;50:207–214.
5. Kehrli M, Nonnecke B, Roth J. Alterations in bovine lymphocyte function during the periparturient period. *Am J Vet Res* 1989;50:215–220.
6. Giuliodori M, Magnasco R, Becu-Villalobos D, Lacau-Mengido I, Risco C, de la Sota R. Metritis in dairy cows: risk factors and reproductive performance. *J Dairy Sci* 2013;96:3621–3631.
7. Ghavi H, Ardalan M. Cow-specific risk factors for retained placenta, metritis and clinical mastitis in Holstein cows. *Vet Res Commun* 2011;3:345–354.
8. Lessard M, Gagnon N, Godson D, Petit H. Influence of parturition and diets enriched in n-3 or n-6 polyunsaturated fatty acids on immune response of dairy cows during the transition period. *J Dairy Sci* 2004;87:2197–2210.
9. Allen M, Bradford B, Oba M. Board invited review: The hepatic oxidation theory of the control of feed intake and its application to ruminants. *J Anim Sci* 2009;87:3317–3334.
10. McArt J, Nydam D, Oetzel G. Epidemiology of subclinical ketosis in early lactation dairy cattle. *J Dairy Sci* 2012;95:5056–5066.
11. Gille A, Bodor ET, Ahmed K, Offermanns S. Nicotinic acid: pharmacological effects and mechanism of action. *Annu Rev Pharmacol Toxicol* 2008;48:79–106.
12. Yuan K, Shaver R, Bertics S, Espineira M, Grummer R. Effect of rumen-protected niacin on lipid metabolism, oxidative stress, and performance of transition dairy cows. *J Dairy Sci* 2012;95:2673–2679.
13. Zeisel S, Holmes-McNary M. Choline. In: Rucker RB, Suttie JW, McCormick DB (eds) *Handbook of Vitamins*, 3rd edn. New York: Marcel Dekker, 2001, pp. 513–528.
14. Lima F, Sa Filho M, Greco L et al. Effects of feeding rumen-protected choline (RPC) on lactation and metabolism. *J Dairy Sci* 2007;90(Suppl. 1):174.
15. Duffield T, Rabiee A, Lean I. A meta-analysis of the impact of monensin in lactating dairy cattle. Part 3. Health and reproduction. *J Dairy Sci* 2008;91:2328–2341.
16. Allen M. Effects of diet on short-term regulation of feed intake by lactating dairy cattle. *J Dairy Sci* 2000;83:1598–1624.
17. Allen M, Piantoni P. Metabolic control of feed intake: implications for metabolic disease of fresh cows. *Vet Clin North Am Food Anim Pract* 2013;29:279–297.
18. Rajala-Schultz P, Saville W, Frazer G, Wittum T. Association between milk urea nitrogen and fertility in Ohio dairy cows. *J Dairy Sci* 2001;84:482–489.
19. De Wit A, Cesar M, Kruip T. Effect of urea during in vitro maturation on nuclear maturation and embryo development of bovine cumulus–oocyte-complexes. *J Dairy Sci* 2001;84:1800–1804.
20. Ocon O, Hansen P. Disruption of bovine oocytes and preimplantation embryos by urea and acidic pH. *J Dairy Sci* 2003;86:1194–1200.
21. Hammon D, Wang S, Holyoak G. Effects of ammonia on development and viability of preimplantation bovine embryos. *Anim Reprod Sci* 2000;59:23–30.
22. Hammon D, Holyoak G, Dhiman T. Association between blood plasma urea nitrogen levels and reproductive fluid urea nitrogen and ammonia concentrations in early lactation dairy cows. *Anim Reprod Sci* 2005;86:195–204.

23. Britt J. Impacts of early postpartum metabolism on follicular development and fertility. In: *Proceedings of the 24th Annual Convention of the American Association of Bovine Practitioners, September 18–21, 1991, Orlando*. Auburn, AL: American Association of Bovine Practitioners, 1991, pp. 39–43.
24. Lucy M, Thatcher W, Staples C. Postpartum function: nutritional and physiological interactions. In: Van Horn HH, Wilcox CJ (eds) *Large Dairy Herd Management*. Champaign, IL: American Dairy Science Association, 1992, pp. 135–145.
25. Sartori RR, Sartor-Bergfelt R, Mertens S, Guenther J, Parrish J, Wiltbank M. Fertilization and early embryonic development in heifers and lactating cows in summer and lactating and dry cows in winter. *J Dairy Sci* 2002;85:2803–2812
26. Jorritsma R, César M, Hermans J, Kruitagen C, Kruip T. Effects of non-esterified fatty acids on bovine granulosa cells and developmental potential of oocytes in vitro. *Anim Reprod Sci* 2004;81:225–235.
27. Vanholder T, Lmr Leroy J, Van Soom A et al. Effect of non-esterified fatty acids on bovine theca cell steroidogenesis and proliferation in vitro. *Anim Reprod Sci* 2006;92:51–63.
28. de Heredia F, Gomez-Martinez S, Marcos A. Obesity, inflammation and the immune system. *Proc Nutr Soc* 2012;71:332–338.
29. Kushibiki S, Hodate K, Shingu H et al. Effects of long-term administration of recombinant bovine tumor necrosis factor-alpha on glucose metabolism and growth hormone secretion in steers. *Am J Vet Res* 2001;62:794–798.
30. Kushibiki S, Hodate K, Shingu H et al. Insulin resistance induced in dairy steers by tumor necrosis factor alpha is partially reversed by 2,4-thiazolidinedione. *Domest Anim Endocrinol* 2001;21:25–37.
31. Bradford B, Mamedova L, Minton J, Drouillard J, Johnson B. Daily injection of tumor necrosis factor-alpha increases hepatic triglycerides and alters transcript abundance of metabolic genes in lactating dairy cattle. *J Nutr* 2009;139:1451–1456.
32. Oltenacu P, Ekesbo I. Epidemiological study of clinical mastitis in dairy cattle. *Vet Res* 1994;25:208–212.
33. Kelton D, Petersson C, Leslie K, Hansen D. Associations between clinical mastitis and pregnancy on Ontario dairy farms. In: *Proceedings of the 2nd International Symposium on Mastitis and Milk Quality*, 2001, pp. 200–202. Available at http://www.nmconline.org/articles/preg.pdf
34. Gwazdauskas F, Thatcher W, Wilcox C. Physiological, environmental, and hormonal factors at insemination which may affect conception. *J Dairy Sci* 1973;56:873.
35. Jordan E. Effects of heat stress on reproduction. *J Dairy Sci* 2003;86(E Suppl.):E104–E114.
36. Perkins T, Dew R, Chestnut A, McCorkill A, Cantrell S, Watkins L. Ingestion of an osmolite included in a free choice mineral and its effect on body condition score, hair retention and temperature of beef cattle grazing fescue pastures. In: *American Society of Animal Science Southern Section*, 2009, p. 9 (Abstract).
37. Gillund P, Reksen O, Gröhn Y, Karlberg K. Body condition related to ketosis and reproductive performance in Norwegian dairy cows. *J Dairy Sci* 2001;84:1390–1396.
38. National Research Council. *Nutrient Requirements of Dairy Cattle*, 7th revised edn. Washington, DC: National Academies Press, 2001.
39. Hurley W, Doane R. Recent developments in the roles of vitamins and minerals in reproduction. *J Dairy Sci* 1989;72:784–804.
40. Risco C, Reynolds J, D. Hird D. Uterine prolapse and hypocalcemia in dairy cows. *J Am Vet Med Assoc* 1984;185:1517.
41. Oetzel G, Olson J, Curtis C, Fettmann M. Ammonium chloride and ammonium sulfate for the prevention of parturient paresis in dairy cows. *J Dairy Sci* 1988;71:3302–3309.
42. Lean I, DeGaris P, McNeil D, Block E. Hypocalcemia in dairy cows: meta-analysis and dietary cation anion difference theory revisited. *J Dairy Sci* 2006;89:669–684.
43. Oetzel G. Effect of calcium chloride gel treatment in dairy cows on incidence of periparturient diseases. *J Am Vet Med Assoc* 1996;209:958–961.
44. Goff J. Pathophysiology of calcium and phosphorous disorders. *Vet Clin of North Am Food Anim Pract* 2000;16:319–337.
45. Suttle NF. *Mineral Nutrition of Livestock*, 4th edn. Wallingford, UK: CABI, 2010.
46. Hidiroglou M. Trace element deficiencies and fertility in ruminants: a review. *J Dairy Sci* 1979;62:1195–1206.
47. Serguson E, Murray F, Moxon A, Redman D, Conrad H. Selenuim/vit E: role in fertilization of bovine ova. *J Dairy Sci* 1976;60:1001–1003.
48. Harrison J, Hancock D, Conrad H. Vit. E and Sel. for reproduction of the dairy cow. *J Dairy Sci* 1984;67:123–132.
49. Miller W. Mineral and vitamin nutrition of dairy cattle. *J Dairy Sci* 1981;64:1196–1206.
50. Gooneratne S, Christensen D. A survey of maternal copper status and fetal tissue copper concentration in Saskatchewan bovine. *Can J Anim Sci* 1989;69:141–150.
51. Abdelrahman M, Kincaid R. Deposition of copper, manganese, zinc, and selenium in bovine fetal tissue at different stages of gestation. *J Dairy Sci* 1993;76:3588–3593.
52. Kincaid R, Hodgson A. Relationship of selenium concentrations in blood of calves to blood selenium of the dam and supplemental selenium. *J Dairy Sci* 1989;72:259–263.
53. Meglia G, Holtenius K, Petersson L, Ohagan P, Persson Waller K. Prediction of vitamin A, vitamin E, selenium and zinc status of periparturient dairy cows using blood sampling during the mid dry period. *Acta Vet Scand* 2004;45,119–128.
54. Hayirli A, Grummer R, Nordheim E, Crump P. Animal and dietary factors affecting feed intake during the pre-Fresh transition period in Holsteins. *J Dairy Sci* 2001;86:3430–3443.
55. Ledoux D, Shannon M. Bioavailability and antagonists of trace minerals in ruminant metabolism. In: *16th Annual Florida Ruminant Nutrition Symposium*, 2005, pp. 23–37.
56. Arthington J, Pate F. Effect of corn- vs molasses-based supplements on trace mineral status in beef heifers. *J Anim Sci* 2002;80: 2787–2791.
57. Ivancic J, Weiss W. Effect of dietary sulfur and selenium concentrations on selenium balance of lactating Holstein cows. *J Dairy Sci* 2001;84:225–232.
58. Harrison J, Conrad H. Effect of dietary calcium on selenium absorption by the nonlactating dairy cow. *J Dairy Sci* 1984;67: 1860–1864.
59. Mason J. Thiomolybdates: mediators of molybdenum toxicity and enzyme inhibitors. *Toxicology* 1986;42:99–109.
60. Mason J. The biochemical pathogenesis of molybdenum-induced copper deficiency syndromes in ruminants: towards the final chapter. *Irish Vet J* 1990;43:18–22.
61. Machado V, Oikonomou G, Bicalho M, Knauer W, Gilbert R, Bicalho R. Investigation of postpartum dairy cows' uterine microbial diversity using metagenomic pyrosequencing of the 16S rRNA gene. *Vet Microbiol* 2012;159:460–469.
62. Machado V, Bicalho M, Pereira R et al. Effects of an injectable trace mineral supplement containing selenium, copper, zinc, and manganese on the health and production of lactating Holstein cows. *Vet J* 2013;197:451–456.

Chapter 31

Estrus Detection

Rhonda C. Vann

Brown-Loan Experiment Station, Raymond, Mississippi, USA

Introduction

Many beef and dairy producers utilize artificial insemination (AI) in their herds each year. However, one drawback to this technology is adequate estrus (heat) detection. In addition, another factor that contributes to this is lack of expression of estrus or actual mounting behavior in cattle. This can sometimes be influenced by the breed of cattle intended for AI as well as the time of year and environmental conditions at the time of the procedure. A trained observer is able to detect heat efficiently by observing the herd for 30–60 min in the morning and in the evening; however, with some breeds and with young heifers it is a good idea to carry out additional checks during the day but to space these checks evenly throughout the 24-hour period. Heat detection observation skills are developed over time and some people are just more naturally skilled at this than others. The observer must be taught to recognize the signs of estrus and must be aware that some cows show abbreviated signs of estrus while some do not show any signs of estrous behavior. If these heat periods are missed or incorrectly identified, then increased calving intervals will result, which can reduce profits.[1] It is extremely important that cows be inseminated at the proper time in relationship to ovulation. Synchronization of estrus can be used to help reduce the time spent on heat detection. However, there are many estrus synchronization programs and the choice will depend on which one suits requirements and time constraints; some protocols work better in cows while some protocols work better in heifers.

Estrus (heat) is a well-defined period that occurs in nonpregnant cows once every 18–24 days. This period is characterized by increased sexual activity and acceptance of the bull by the postpubertal cow or heifer. This period begins with the first acceptance of the bull and ends with the last acceptance of the bull. British-type breeds of cattle (*Bos taurus*) usually have 8- to 14-hour heat periods, while Brahman (Zebu)-type breeds of cattle (*Bos indicus*) usually have 6- to 12-hour intervals.

Estrus detection aids can also be used as an additional tool to supplement the visual observations but should not be the only methods of estrus detection.[2] These estrus detection aids include, but are not limited to, heat mount detectors or patches, tailhead markers (paint, chalk, crayon or paste), teaser or gomer bulls equipped with a chin-ball marker, as well as more sophisticated electronic heat detection devices (i.e., HeatWatch® system, Bovine Beacon®, or other estrus alert patches).[2]

There are a number of factors that affect the expression of heat. Adverse weather conditions or sudden changes in the weather can suppress the display of estrus. Hunger or thirst or inadequate nutritional levels for extended periods can also influence the expression of estrus. Many types of stress can also negatively impact the expression of estrus and ovulation, including rough handling, transportation, noise, overcrowding, and heat or cold stress.[1]

Visual heat detection

Generally, cattle can be observed during normal activity or as they move to and from a housing, feeding, or pasture area. In the case of dairy operations, they can be observed as they are moved to and from the milking parlor.[1] In beef operations heat detection often requires observation in the pasture or feeding area. It is important to try and not disturb the cows when making your observations so as not to interrupt mounting behavior, and so binoculars can be a useful tool in identification of the animals exhibiting estrus.

Step-by-step procedure

The onset of behavioral estrus usually occurs over a period of 4–24 hours. Animals can be characterized as restless and nervous. Some animals will tend to isolate themselves from the rest of the herd and pace fences. These females will tend to twitch their tails nervously, walk with their tails elevated, and show a clear thick mucous discharge from the vulva (an animal with a cloudy, milk-white, or pus discharge may have a vaginal or uterine infection that could reduce conception rates).[1]

The most obvious behavioral estrus activity is mounting. When a cow in heat tries to mount another cow not in heat, the mounted cow will not stand to be mounted. A cow in heat will also follow others, stand beside them, and rest her head on their backs or rump. If there is uncertainty that the cow is in heat, the observer should watch patiently a little while longer until the cow actually allows others to mount her; if she stands still and allows others to mount her, she is in heat. There may be several cows in heat at the same time and it can be difficult to sort out which ones are in heat and which ones may be coming into heat.

Cows in standing estrus or heat generally exhibit restless behavior and often remain standing throughout the day; however, this can be variable depending on environmental conditions and breed type. The cow may bellow frequently and may exhibit signs of reduced appetite. Some cows may become friendlier to other cows during the heat period. The cow may smell or nose other cows or even butt heads or rub on other cows.

A cow in estrus often has a raised tail, and may have a long string of clear mucus hanging from her vulva or on her tail or hindquarters. She may even have a reddened, loose, or slightly swollen or relaxed vulva. A roughened tailhead or mud on the back or sides of the cow indicates that the cow has been ridden.[1]

In dairy cattle, a decrease in milk production for no apparent reason may indicate that the cow is in heat or coming into heat.[1] Beef cows may also reduce milk production during estrus and bawling calves may be seen in the pasture. In addition, if there are bull calves in the pasture with the cow and multiple calves are observed following the cow and trying to mount her, it is a very good sign that the cow is in heat.

Blood on the vulva, tail, or hindquarters of a cow or heifer usually indicates that she was in heat 2–3 days previously. It is generally too late to AI this animal at this point in the cycle. This date can be recorded and used as a guide to when the animal will be cycling again. Bleeding that occurs after natural service or AI does not mean that conception has occurred and could possibly reduce conception rates.

Rule of thumb

If the cow shows estrus activity in the morning, inseminate in the evening; if the cow shows estrus activity in the evening, inseminate the next morning. However, many estrus synchronization protocols indicate a specified time for insemination and these guidelines should be followed when utilizing these protocols. In addition, breed of animal can also impact timing of insemination relative to standing heat.

A large percentage of cattle will cycle overnight when no one is available to see this activity. Therefore, the best recommendation is to observe cows as late as the light will allow in the evening and as early as possible in the morning. In addition to this visual observation, utilization of some additional heat detection aids (estrus alert patches, teaser bulls with chin-ball markers, paint, etc.)[2] can be combined to assist in catching these animals that cycle during this period.

Some secondary signs of heat can be observed in cows once the primary or standing heat signs have subsided. These include a roughened tailhead, mud on flanks or sides, pacing, licking, sniffing, or resting head on other animals, and mounting other cows but not standing to be mounted. When you see a bloody mucous discharge, estrus has occurred 2–3 days previously.

An abrupt change in the weather, particularly an extreme change in temperature, can alter normal estrus activity. When this occurs, the normal signs of estrus can be diminished and heat detection will drop sharply. Additionally, other stressors can impact this as well so keep this in mind when observing cattle for estrus.

Heat detection aids

There are multiple heat detection aids on the market, although these are meant to supplement visual heat detection and not to replace it.

Heat patch

The heat patch can be a plastic heat-detecting device that is usually glued onto the top of the tailhead of the animal. These patches usually contain a small vial of fluid with a smaller vial of color that breaks when pressure is applied by mounting activity. However, these patches can be accidentally ruptured and so also look for other signs of estrus to verify that the animal was in standing heat. Two examples of these heat patches are Kamar® and Bovine Beacon®. In addition, there are self-adhesive heat detection patches. Each patch has a silver scratch-off surface; once the animal is mounted the bright signal layer underneath is shown for easy observation (e.g., Estrotect® heat detection patches) (Figure 31.1).

Color wax marker (tailhead marking)

A thick layer of marking wax, crayon, paste or paint is applied to the very top of the tailhead of the cow. When the animal is mounted, this paint is spread down the sides of the tailhead onto the rump. This indicates the cow has been ridden by another cow. Also, look for additional signs of estrus when utilizing this method (Figure 31.2).

Electronic heat-detection systems

The HeatWatch® System is based on measurement of the pressure applied by a riding cow to a small transmitter embedded in a cloth tag that is affixed to the tailhead region of cycling cows. The transmitter is approximately 2 × 60 × 7.5 cm in size and the signal transmitted by the tag containing the transmitter is picked up by a receiver that is direct wired into the farm computer where the data are stored for later analysis. The transmitters are coded so that each mounted animal sends a unique animal-identifying signal to the computer. This unique code and the clock built into the computer enable the manger to know exactly when standing heat began and which cow is ready to be bred. This can be a useful tool if visual heat detection is limited; however, it is a more costly investment and there can be some issues with signal if animals are located under sheds or there are hills and valleys that may inhibit signal reception (Figure 31.3).

Chin-ball marker

The chin-ball marker is simply a metal paint reservoir with a large ball bearing held by a leather (or poly) harness under the chin of the marker animal. As the marker animal mounts and dismounts the cow in heat, a characteristic mark is left

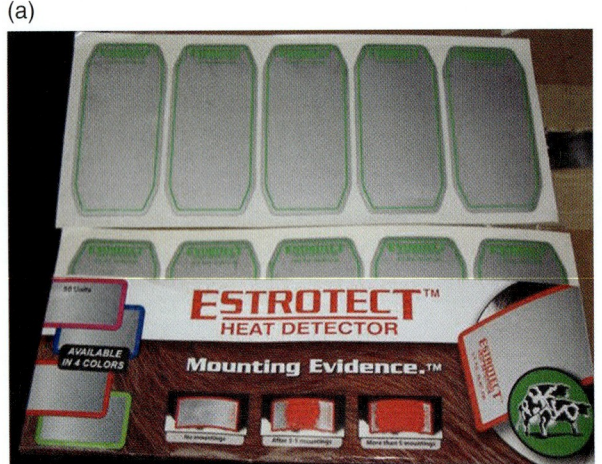 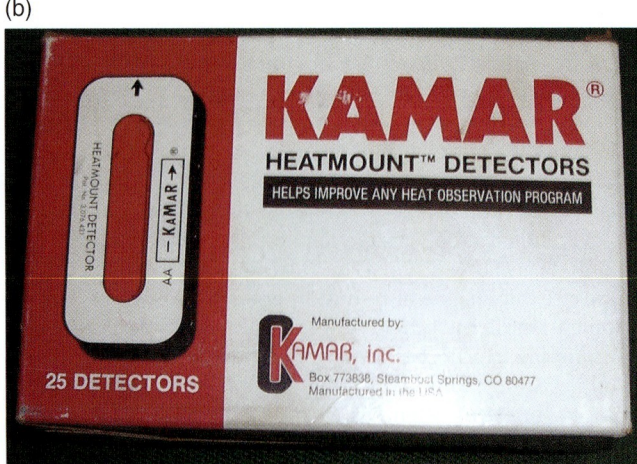

Figure 31.1 (a) Estrotect® heat detector patches and (b) Kamar® heatmount™ detector patches. Both can be used as aids in visual heat detection.

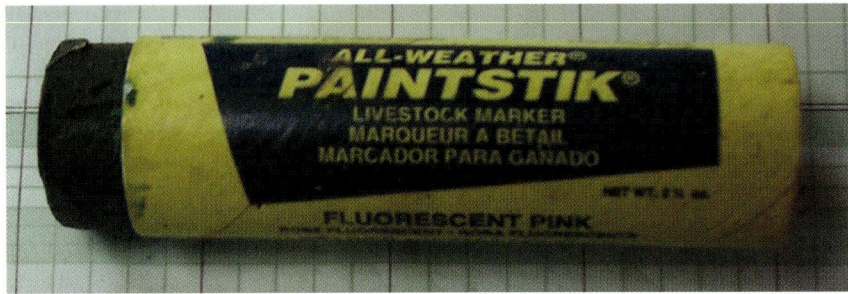

Figure 31.2 All-weather Paintstik® livestock marker can be used to place a thick layer of crayon on top of the tailhead of the cow for detection of mounting behavior.

Figure 31.3 Heifers wearing the HeatWatch® patches (in orange) over the tailhead. These heifers are in a dry-lot area so that there is close proximity to the receiver for the electronic transmitter.

on the cow's back and rump (Figure 31.4). The paint reservoir usually holds enough ink for about 25 cows, depending on mating behavior. Usually this marker animal is a bull that has been altered in some way so as to not penetrate the cow as he mounts her. These animals are called teaser (gomer) bulls; however, hormone treated animals can also be utilized. A "gomer" bull is a bull that has been surgically or nonsurgically altered so that he cannot breed the cow naturally. In some occasions, these teaser bulls will "fall in love" with one cow and can miss other cows that are in heat. In addition, some teaser bulls will lose their sex drive and interest over time once they discover that they cannot breed a cow when they mount. It is also important to pick teaser bulls that are not overly aggressive toward humans as these animals will need sorting from the cows in order to inseminate them at the appropriate time. Additionally, other heat detection aids can be utilized; for example, the heat detection patch can assist in identifying cows in heat if the teaser bull is one that likes to become attached to one cow. The other option is to remove the cow in heat from the group and force the teaser bull to go and look for new animals in heat.

Teaser animals

Several types of teaser animal can be utilized in this process, including vasectomized bulls, deviated penis and prepuce, Pen-O-Block, penectomy, and hormone-treated cows. Also, there are some combinations that can be used, such as vasectomy plus a deviated penis.

Estrus Detection

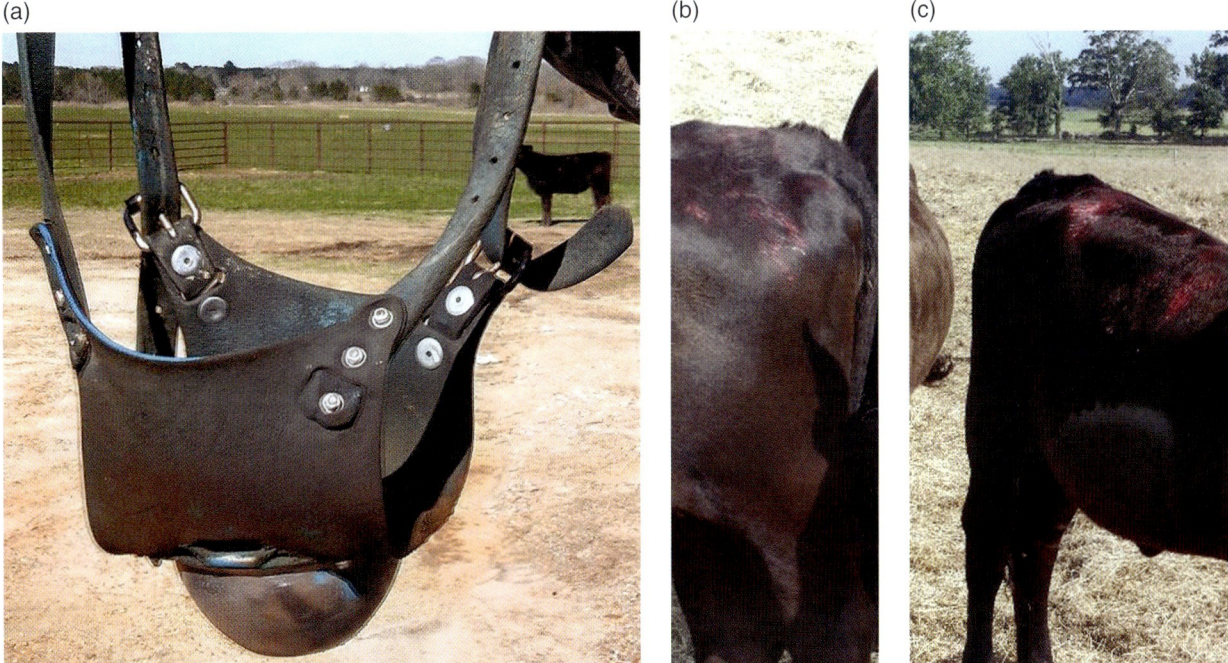

Figure 31.4 (a) Leather chin-ball marker referred to as a KowBull Marker® that can be adjusted to fit the head of the marker bull. The metal reservoir holds the paint used to mark the animals and comes in a variety of colors; it is important to use a color that shows up well on the hide of the animal being marked by the teaser bull. (b) Chin-ball marks over the rump and tailhead of a heifer indicating she has been mounted (note red or pink marks on the black animal). (c) Chin-ball marks in red or pink on a black animal located over the back and tailhead indicates that the heifer has been mounted by the teaser bull.

Vasectomized bulls

The operation involves removing part of the vas deferens from the spermatic cord. The blood and nerve supply of the spermatic cord are left intact. The testes and penis function normally, but transport of spermatozoa to the urethra is blocked. The sexual activity of the bull remains unaltered, and he will detect heat as well as an intact bull. Another form is a caudal-epididymectomized bull,[2] which involves removal of the epididymis tail to prevent sperm from reaching the penis. The disadvantages of these operations are that the bull continues to enter cows and can transmit venereal disease from one cow to another.

Deviated penis and prepuce (sidewinder)

This type of teaser (gomer) bull is altered by a surgical technique that transplants the penis and sheath from their natural position to the fold of the flank. This method allows normal erection but does not allow the bull to enter the cow. The disadvantage of this system is that the bull can become frustrated and lose his sex drive. The advantage is that the possibility of transmission of a venereal disease is usually very low unless the bull becomes skilled and discovers a way to breed the cow (Figure 31.5).

Pen-O-Block

This method involves the insertion of a commercially available plastic tube inside the sheath.[1] The device blocks the preputial opening, preventing erection and penetration. The Pen-O-Block is secured by making a small opening through the skin on each side of the sheath, inserting the plastic tube, placing a pin through the openings in the sheath and tube, and securing the pin with washers and cotter pins. The disadvantage of this system is that bulls lose their sex drive rather quickly. The advantage is no venereal disease transmission and that the device can be removed.

Penectomy

This procedure involves the removal of part of the penis.[1] Some techniques remove approximately two-thirds of the penis and exteriorize the remaining end between the anus and scrotum. The disadvantage is that sex drive is usually reduced; however, venereal disease transmission is not possible.

Hormone-treated cows

This method involves a cull cow, a hormone preparation available from the veterinarian, and a systematic injection schedule. The cost of this procedure is relatively low and many producers have cull animals available. Animals injected with a testosterone-based solution are usually sexually aggressive and mount other cows in heat within a few days after the last injection. The injections are administered periodically during the breeding season to maintain sexual activity. The advantage of this technique is that the hormone-treated cow can be prepared cheaply, the cow cannot spread venereal disease, maintenance cost of marker animals is reduced, and the cow can be sold for harvest later

Figure 31.5 Teaser bulls that have undergone surgery which deviates penis and prepuce to the fold of the flank; these bulls are referred to as "sidewinders."

after the breeding season. Selection criteria for a marker cow should include sound feet and legs, large size, and sufficiently aggressive to compete with the herd. However, it is important not to select a cow that is pregnant or nursing or is a member of the milking herd; additionally, the cow will not be able to breed after this treatment because she will be subfertile, and response to treatment can be variable.

Summary

The effectiveness of detecting heat or estrus in cattle varies depending on the method used and the time spent visually observing cattle. It is important to consider the cost, labor, and management system in deciding which method or combination of methods work best with regard to the farm operation and time constraints. Additionally, if estrus detection is used for AI programs, then the level of heat detection accuracy is more important in order to increase the success of the AI program. Heat detection is a skill and the more time spent observing cattle, the greater the accuracy of heat detection.

References

1. Battaglia RA. Heat detection. In: *Handbook of Livestock Management*, 3rd edn. Upper Saddle River, NJ: Prentice Hall, 2000, pp. 119–126.
2. Parish J, Larson J, Vann R. Estrus (heat) detection in cattle. Extension Publication 2610, Mississippi State University, 2010. Available at http://msucares.com/pubs/publications/p2610.pdf

Chapter 32

Artificial Insemination

Ram Kasimanickam

*Department of Veterinary Clinical Sciences, College of Veterinary Medicine,
Washington State University, Pullman, Washington, USA*

Introduction

Artificial insemination (AI), the introduction of sperm into the female reproductive tract by means of an instrument, is the oldest assisted reproductive technology.[1] The use of AI in domestic animal reproduction was originally pioneered for sanitary reasons. However, once frozen semen became readily available, the economic advantages of improved fertility and accelerated genetic progress became evident. Since then, AI has become the method of choice for rapidly spreading preferred animal genetics. Numerous bulls have produced hundreds of thousands of insemination doses and offspring. Furthermore, reduced numbers of sperm per insemination dose without compromising fertility has greatly increased the number of inseminations and offspring from genetically superior sires. Several important developments, including novel sperm diluents, new AI techniques, protocols for synchronizing estrus and ovulation, and the availability of gender-selected semen, have further increased the use and importance of AI as an assisted reproduction technique.

Achieving high fertility with AI requires excellent management of all phases of the AI program; it truly requires a team approach. In that regard, personnel responsible for semen collection, processing and delivery must all correctly perform their tasks, as any deficit will reduce fertility success. Furthermore, the innate fertility of bulls and cows is also of vital importance.

Large commercial AI centers generally have protocols and quality control procedures in place to ensure that good-quality semen is marketed. However, once the semen has been sold, fertility depends on the ability of others to correctly handle and thaw the semen and to correctly inseminate cows at the proper time. The primary objective of semen handling is to conserve fertilizing competency, which is accomplished by minimizing exposure of sperm to deleterious conditions. Damage to sperm during cryopreservation, storage, and thawing has been attributed to cold shock, ice crystal formation, oxidative stress, membrane alteration, cryoprotectant toxicity, and osmotic changes.[2] Knowledge of semen tank management, proper thawing and semen handling techniques, and sanitary insemination in the correct location at the correct time are critical and should be periodically reviewed.

Semen tank management

Frozen semen is stored in a specialized tank containing liquid nitrogen. These tanks have a rugged outer jacket (aluminum or stainless steel) and an inner compartment that contains liquid nitrogen. The space between the inner and outer jackets is insulated and under extreme vacuum. The lid is attached to a hard foam cylinder that protrudes into the neck of the tank to insulate liquid nitrogen and frozen semen from outside temperatures, thereby minimizing evaporation. However, the tank is not airtight; the liquid nitrogen releases gas as the temperature fluctuates (if the tank was tightly sealed, it might explode). Recent technical progress in design and construction has resulted in user-friendly good-quality semen tanks, including those with extended holding intervals (6–12 months). It is noteworthy that holding intervals vary according to tank model and the frequency with which it is opened. It is essential that the nitrogen content is routinely monitored and additional nitrogen added as required.

With newer semen tanks, maintenance of very low liquid nitrogen temperatures in the inner chamber is possible due to high-quality solid insulation material and vacuum. Regardless, all tanks are susceptible to damage from mishandling. The inner chamber containing liquid nitrogen is actually suspended from the outer shell by the neck tube. Consequently, any abnormal stress on the neck tube caused by substantial force or an excessive swinging motion can crack the tube, resulting in vacuum loss. Puncture of the outer shell will also lead to vacuum loss. Welding the tank exterior should be avoided, as this could also cause vacuum loss. Since vacuum is the major insulating component of the tank, vacuum loss causes an increase in temperature within the inner chamber and rapid evaporation of nitrogen. Accumulation of heavy frost at the top of the tank indicates

---

Bovine Reproduction, First Edition. Edited by Richard M. Hopper.
© 2015 John Wiley & Sons, Inc. Published 2015 by John Wiley & Sons, Inc.

rapid evaporation of liquid nitrogen and tank failure. Although semen can be stored at –196 °C indefinitely with minimum liquid nitrogen, a minimum depth of 5 cm should be maintained.

Semen storage tanks should be stored away from direct sunlight in a cool, clean, dry, dust-free, and well-ventilated environment. Tanks should be elevated on a wooden pallet and never stored directly on a concrete floor (to prevent corrosion on the bottom of the tank). Furthermore, tanks should not be stacked on top of each other. The tank should be placed in an area where it can be observed frequently to detect excessive evaporation of liquid nitrogen. The tank should not be laid horizontally and rolled. Storing the tank on a wooden or plastic dolly with wheels facilitates movement of the tank. Tanks should be properly secured during transportation in a vehicle to minimize damage or spillage.

Frozen semen storage

Most semen is packaged in 0.25- or 0.5-mL straws (sex-sorted semen is consistently in 0.25-mL straws). The straw is placed in a goblet with four other straws from the same bull. Two goblets are packaged onto a metal cane that has the bull's code number printed on top. Canes are stored in canisters in a semen tank. Storage and handling of 0.5- and 0.25-mL straws are similar.

Maintenance of low temperatures is the key to successful storage of frozen semen. During cooling and freezing, microenvironments are created within the semen straw. Each chemical component of extended semen freezes or solidifies at a different temperature. Water begins to freeze as temperatures are decreased below 0 °C, forming ice crystals that remain somewhat unstable at temperatures above –80 °C.[3] This instability is thought to be caused by recrystallization of the ice. Also, as water is converted to ice, sperm are exposed to the remaining concentrated solution of salts and other components of the extender which freeze at temperatures considerably below the freezing point of water. Instability of ice and concentrated solutions are harmful to sperm. Dehydration during cryopreservation has an important impact on sperm; dehydration decreases the risk of intracellular ice crystal formation, but excessive dehydration is also detrimental.[2,4] Adjusting the osmolality of the diluents is critical, as it influences water fluxes during cryopreservation.[2,4] Fortunately, cryoprotective agents and optimized freezing programs help minimize sperm damage. However, semen must be kept well below critical temperatures where the recrystallization of ice begins to occur (–100 to –80 °C). Temperatures fluctuate dangerously from the low to upper third of the neck of the tank[3] (Table 32.1). Furthermore, stored semen can be exposed to adverse high temperatures when removed from the tank for thawing, when transferring semen from tank to tank, and when handling semen within the neck when trying to locate and thaw a specific straw. In addition, other semen straws in goblets and canisters are also exposed to high temperatures. The thermal response of semen in 0.5-mL straws when raised to 5 cm from top of the tank (exposed to temperature of –22 °C) and 2.5 cm (exposed to 5 °C) is shown in Table 32.2.[3] The time to reach critical ice recrystallization temperature (–100 to –80 °C) is approximately 10–20 s for both temperatures. Thermal injury to sperm is permanent and cannot be corrected by returning semen to liquid nitrogen. To assure maintenance of sperm viability, canes and canisters should be raised into the neck of the tank and kept below the frost line only for 5–8 s. If necessary, the canister should be lowered back into the tank for cooling, and subsequently brought back to the neck of the tank.

The number of sperm per straw is variable, with some reduction in number common for high-demand elite bulls, and substantially lower numbers for sex-sorted semen. Higher dilutions (2 million sperm per 0.25-mL straw) did not affect the proportions of linearly motile spermatozoa, membrane integrity or stability, nor chromatin integrity immediately after thawing compared with a low dilution rate (15 million sperm per 0.25-mL straw). Further, there was no difference in pregnancy rate between dilution rates.[5] Regardless, it is well

Table 32.1 Fluctuating temperatures in the neck of a typical semen storage tank.

Location in neck of storage tank	Temperature (°C)
Top of neck	2.2 to 12.2
2.54 cm from top	–15 to –22.2
5.08 cm from top	–40 to –46
7.62 cm inch from top	–75 to –82
10.16 cm from top	–100 to –120
12.7 cm from top	–140 to –160
15.24 cm from top	–180 to –192

Source: adapted from Saacke R, Lineweaver J, Aalseth E. Procedures for handling frozen semen. In: *Proceedings of the 12th Conference on AI in Beef Cattle of the NAAB*, 1997, p. 49.

Table 32.2 Thermal response of semen (0.5-mL French straws) exposed to 5 and –22 °C (temperature observed in the upper portion of the semen storage tank).

Time (s)	5 °C	–22 °C
5	–180	–180
10	–115	–118
20	–85	–90
30	–62	–77
40	–50	–60
50	–39	–53
60	–30	–48
70	–24	–42
80	–21	–40
90	–19	—
110	—	–31
130	—	–25

Source: adapted from Saacke R, Lineweaver J, Aalseth E. Procedures for handling frozen semen. In: *Proceedings of the 12th Conference on AI in Beef Cattle of the NAAB*, 1997, p. 49.

known that some bulls require more sperm per insemination dose to maintain fertility. Furthermore, due to very low numbers per insemination dose, the fertility of sex-sorted sperm is almost inevitably decreased.

Tips to minimize thermal injury

- Develop a semen inventory system and mount it on a wall near the tank. Try to keep semen from one bull on each rack. Label canister and goblets with bull number/name. Such systems prevent unnecessary searching and exposure of semen to dangerously high temperatures in the neck of the tank.
- Organize rapid transfer of semen between tanks. Involve two people and arrange the tanks side by side. If possible, fill the tanks with nitrogen before transfer. Raise canisters only to a level necessary to locate the rack of semen to be transferred.
- When preparing to thaw semen, raise the canister into the lower portion of the neck where the specific goblet of semen can be grasped, then lower the canister further into the neck. Secure the goblet as low as possible in the neck, thus protecting other straws from thermal injury. If straws cannot be easily removed, bend the top tab of the goblet (45° angle) to improve access.
- Use forceps to remove the straw and to transfer it to the thaw bath. As soon as the straw is retrieved, quickly lower the rack of semen and canister into the body of the tank.

Semen thawing method

Various methods for thawing semen in straws have been recommended, including a variety of water bath temperatures and thawing periods, shirt pocket thawing, air-thawing, and thawing in the cow.[6] Regardless, it is recommended that all persons on a farm use the same method to thaw semen. The National Association of Animal Breeders recommends a semen straw be immersed in a water bath at 30–35 °C for 40 s. The thawing time for straws should be a minimum of 30 s.[7] Thawed semen should be inseminated within 15 min after thawing and drastic decreases in post-thaw temperature must be avoided.[8]

In pocket thaw methods, the initial pocket temperature was 30 ± 1 °C. However, 1 min after pocketing, the temperature decreased to approximately 19 ± 1 °C, which could have adverse effects on sperm parameters. Semen in straws began to liquefy approximately 3 min after pocketing, with thawing completed by 5 min.[8] The results of semen viability assays after pocket thaw differed widely among experiments. The rate of thawing in a shirt pocket could be quite variable depending on the ambient temperature and the insulating effect of different types and thicknesses of clothing. Furthermore, the thaw rate was much slower for pocket-thaw compared with a water bath.[8] More rapid thaw rates result in higher post-thaw viability compared with the pocket-thaw method.[8] However, a recent study claimed that irrespective of improved *in vitro* semen quality with a fast thaw rate, these semen quality measures did not increase *in vivo* fertility compared with a pocket-thaw method.[9] If frozen semen is prepared to permit flexible thawing, the thaw method used, whether pocket or warm water thaw, should not affect conception under commercial conditions.[9] Although some sire and extender (egg-yolk citrate vs. nonheated whole milk extender) combinations seem to be tolerant to thaw procedures, other combinations are more sensitive, resulting in reduced post-thaw sperm survival, conception rates, or both, in response to thaw methods.[10]

Concurrently thawing multiple straws of semen can facilitate efficiently breeding a large number of cows without compromising semen quality or conception rates. Despite several reports that more than one straw can be safely thawed, caution must be employed while preparing multiple guns, because it is paramount to follow proper procedures for straw preparation and thermal protection and to always work within skill and time constraints. An experienced inseminator can thaw multiple straws of semen and prepare insemination guns to breed up to four cows within 20 min, without an adverse effect on conception.[11] Another study concluded that, on average, conception rates differed between professional AI technicians and herd inseminators. However, elapsed time from initial thaw to completion of fourth AI and sequence of insemination (first, second, third, or fourth) had no effect on conception rate within inseminator group.[12] Oliveira *et al.*[13] concluded that sequence of insemination after simultaneous thawing of 10 semen straws can differently affect conception rates at timed AI, depending on the sire used. Regardless of number of straws thawed simultaneously, it is vital that straws are agitated immediately after immersion to prevent straws from clumping together, resulting in a decreased rate of thaw.

Semen handling after thawing

Mixing water and semen will cause irreversible sperm injury. Hence, on removal of the straw from the water bath, the straw should be dried (typically with a paper towel) before it is cut. If the straw is defective (e.g., perforated), it should be discarded.

Prevention of cold shock is critical for post-thaw semen handling. Cold shock inflicts irreversible injury to sperm and is caused by rapid post-thaw decrease in temperature; this occurs when semen is thawed and then subjected to cold environmental temperatures before being inseminated. Cold shock decreases motility and fertilizing competency due to irreversible changes in the sperm plasma membrane. The severity depends on the rate and duration of temperature drop. The most obvious sign of cold shock is loss of motility which is not regained on warming the semen. There is also a decrease in the rate of fructose breakdown by the sperm and a decrease in oxygen uptake and in ATP concentrations, which can now be no longer synthesized and used to supply energy. Cold shock caused an influx of sodium (186.2 ± 9.3 vs. 140.7 ± 10.2 mg/100 g wet weight) into sperm, as well as an efflux of potassium (101.6 ± 6.8 vs. 155.4 ± 9.6 mg/100 g wet weight) and magnesium (8.1 ± 0.2 vs. 11.8 ± 0.5 mg/100 g wet weight) from the sperm.[14] The higher total calcium concentrations in bovine seminal plasma than in sperm are primarily due to a much higher concentration of complex calcium in seminal plasma. The concentration of protein-bound calcium was approximately the same in spermatozoa and seminal plasma, whereas concentrations of ionized calcium were lower in seminal plasma than in spermatozoa. Cold shock significantly increased the concentration of total calcium in spermatozoa, with a corresponding decrease of total calcium in the seminal plasma, due to a decreased content of ionized calcium in the seminal plasma and a simultaneously increased content of complex and protein-bound calcium in spermatozoa. In contrast, there was no increase of calcium in spermatozoa following slow cooling.

Cold shock occurs most frequently when breeding is performed in cold weather and especially when a warm water bath is used. The high surface to volume ratio of a straw makes it vulnerable to cold shock. Saacke,[15] using the 0.5-mL French straw, measured the effect of static ambient temperatures (21, 4 and 16°C) on the rate of temperature drop in semen after thawing during preparation of the inseminating rod (Table 32.3). The temperature drop was only 3–6°C for an ambient temperature of 21°C; furthermore, warming the rod was effective in countering temperature drop. However, during preparation at 4 and 16°C ambient temperatures, the drops were 15 and 20°C, respectively, and preparation of the AI gun for insemination at these two temperatures only postponed the temperature drop. Clearly, precautions against cold shock must be implemented during preparation of inseminating guns in a cool environment.

Tips to prevent cold shock in cold weather

- Perform semen thawing and loading of the AI gun in a sheltered heated area located close to the animal to be bred. Then insulate the AI gun by tucking within clothing or in an AI gun warmer and carry it to the breeding chute.
- Perform preparation of AI gun and breeding in sheltered heated area.
- Utilize an AI gun warmer. An AI gun warmer is a portable, heated, temperature-controlled carrying case that maintains multiple loaded AI guns at 35°C.

Semen straw retrieval, thawing and handling tips

In order to obtain optimal results, frozen semen should be handled and thawed carefully. Insemination equipment should be kept clean and dry (including straw trimmers, which often become heavily soiled). A thermometer should be routinely utilized to monitor the temperature of a water bath, especially when several straws are thawed concurrently. The thermometer should be periodically checked for accuracy (at least every 6 months).

On retrieving straw from the semen tank, shake the straw gently to remove excess liquid nitrogen that may be retained in the cotton plug and promptly place in 35°C water to thaw for at least 45s. The thawing should be timed with a watch. While the semen is thawing, pre-warm the barrel by stroking it vigorously with a clean paper towel or by placing close to your body for several minutes in advance of loading. After thawing is completed, remove the straw from the thaw water and wipe it completely dry with a paper towel, and protect the straw from sunlight and rapid cooling. The straw should be held with the crimp upright and gently tapped so that the small air space is adjacent to the crimp. Place the cotton plug end of the straw into the AI gun. Only clean sharp scissors or a straw cutter should be used to cut the straw, perpendicular to the long axis, approximately below the crimped seal to attain a good seal with the AI gun sheath. Slide the plastic sheath over the straw and gun and firmly attach the sheath and gun together. Inspect the straw end of the gun to ensure a proper seal between sheath and straw. The loaded insemination gun is wrapped in a clean paper towel or in a plastic sheath cover to provide both thermal and hygienic protection and tuck the gun inside clothing or gun warmer for transport to the animal. The interval from removing the straw from the tank to AI should not exceed 15 min.

Insemination

Higher fertility requires proper insemination methods. Several insemination methods have been developed for AI in cattle: uterine body insemination, bilateral cornual

Table 32.3 Temperature of semen (°C) in 0.5-mL French straws after water-bath thaw at 35°C during preparation of AI gun at 21, 16 and 4°C.

Stage of preparation[a]	Preparation at 21°C		Preparation at 16°C		Preparation at 4°C	
	Warmed[b]	Unwarmed[b]	Warmed	Unwarmed	Warmed	Unwarmed
Thaw straw	35	35	35	35	35	35
Dry with paper towel	34.5	34	31.5	29.5	31.5	31
Load in gun	34.5	32	28.5	22	29	24.5
Trim the straw	34.5	31	27.5	18.5	28.5	22.5
Apply sheath	33	29.5	22	15.5	23.5	21.5
Wrap with paper towel	32.75	29.5	19.5	13.5	22	20
Tuck under shirt close to body	32.5	29.5	19	13	21	19.75

[a] Preparation of AI gun from thaw to tuck took approximately 1 min.
[b] The AI gun was warmed by rubbing rapidly with the hand (warmed) or was maintained at ambient temperature (unwarmed).
Source: adapted from Saacke R. Concepts in semen packaging and use. In: *Proceedings of the 8th Conference on AI in Beef Cattle of the NAAB*, 1994, p. 15.

insemination, deep uterine horn insemination, and intrafollicular insemination. The most common approach is to use a rectovaginal technique to deposit semen in the uterine body, immediately anterior to the internal os of the cervix. Uterine body and cornual insemination require skill to pass the insemination rod through the cervix. Various training programs are available. For example, many AI studs offer intensive 3–5 day training schools or provide instructions for individuals on the farm. Most agricultural colleges devote a whole course or part of a course to the technique of AI. However, the intensity of training and specific recommendations given to the participants varies considerably among training programs.

Developing the skill to pass the insemination rod through the cervix should not be the only objective of an AI training program. Along with mastering cervical penetration, the importance of sanitation must be emphasized and skills perfected to consistently identify the proper site of semen deposition and accurately deposit the semen. In addition, trainees should obtain a working knowledge of reproductive anatomy and appreciate the essentials of a sound reproductive management program. Failure to understand the anatomical and functional relationships of various tissues and organs of a cow's reproductive system may lead to insemination errors.

The loaded AI gun is placed in a sanitary plastic cover for protection from vaginal contamination (Figure 32.1). The cow should be adequately restrained, with care to minimize stress (as increased blood cortisol concentrations will supress a luteinizing hormone surge). The palpating arm is introduced into the rectum, and the cervix located and firmly grasped (Figure 32.2). The vulva is cleaned with a paper towel, the covered AI gun is inserted into the vagina without contacting the lips of the vulva, and the AI gun passed through the vagina to the external cervical os (Figure 32.3). Care should be taken to avoid the urethral orifice, vaginal fold, and vaginal fornix. To avoid the possibility of entering the urethral opening (on the floor of the vagina), insert the inseminating rod into the vulva at an angle of 30–40°. The fornix of vagina tends to stretch rather easily when the AI gun is pushed forward and beyond the cervix. This may give the false impression that progress is being made in advancing the rod through the cervix when really it is above, below, or to either side of the cervix (Figure 32.4). However, one should be able to feel the rod within the vaginal fold (Figure 32.5). When the rod is within

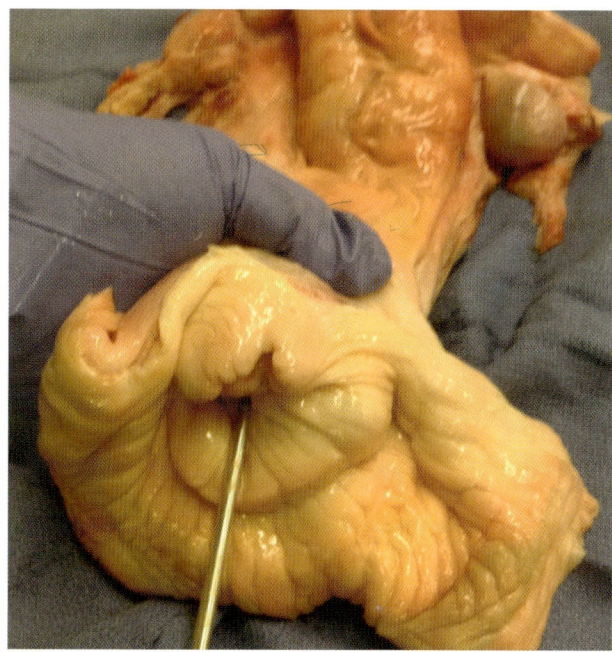

Figure 32.3 Locating external os of the cervix.

Figure 32.1 AI gun with protective sheath.

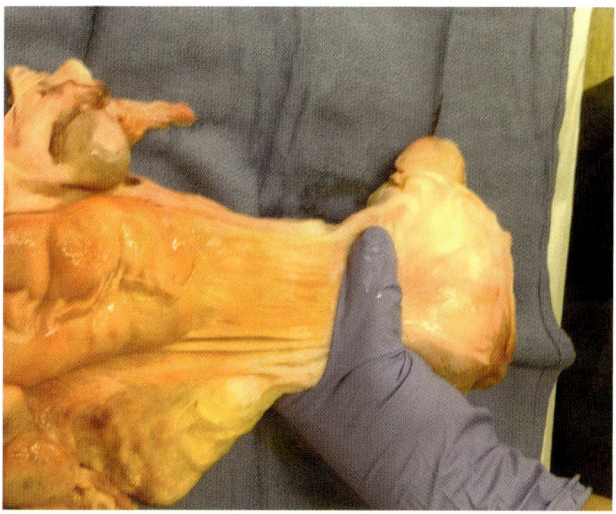

Figure 32.2 Locating and grasping cervix.

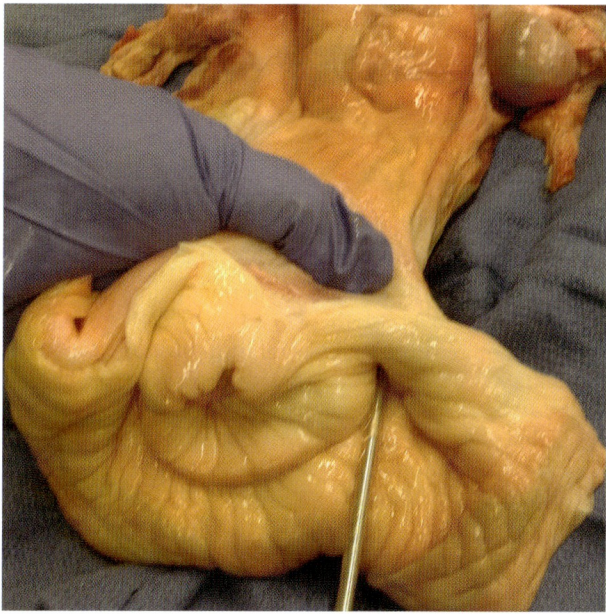

Figure 32.4 Avoiding vaginal fornix.

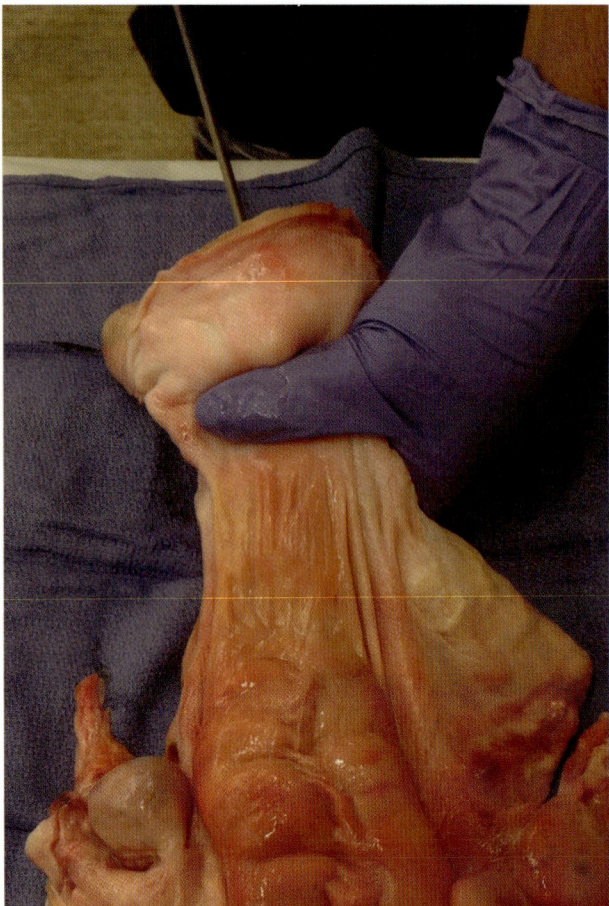

Figure 32.5 Locating tip of AI gun in the anterior vagina.

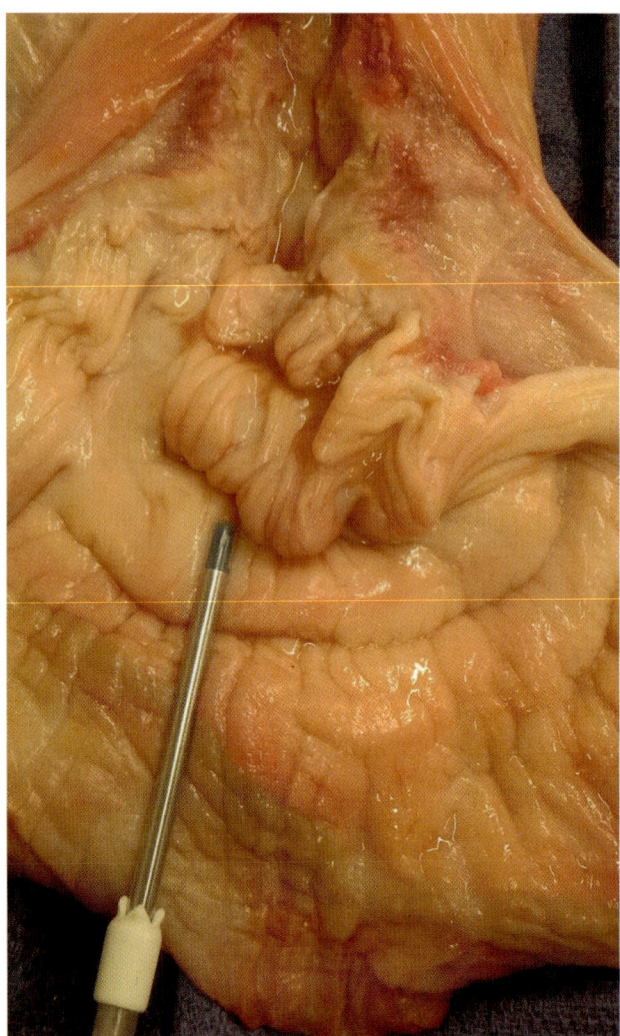

Figure 32.6 Puncturing AI gun protective sheath at external os of the cervix.

the cervix, its tip cannot be felt. The sanitary cover is punctured (Figure 32.6) and the instrument is advanced through the cervix into the base of the uterine body. There are approximately three misaligned cervical rings through which the rod must pass; the rod is advanced by manipulating the cervix and keeping consistent firm pressure on the instrument (Figure 32.7). Care should be taken to avoid catching the tip of the rod on cervical folds. The goal for semen deposition is within the uterine body. Since the uterine body is quite small, accurate AI gun placement is paramount. Generally, this site is identified by feeling the internal os of the cervix and the tip of the AI gun as it passes through the internal os and enters the uterine body (Figure 32.8). The tip of the gun should be positioned so that it is approximately 1 cm anterior to the internal os (Figures 32.9 and 32.10); this can be verified with the tip of the index finger of the hand within the rectum. Once the inseminating rod is correctly positioned, the cervix should be held parallel to the long axis of the cow and semen is deposited over a period of approximately 5 s. Slow delivery maximizes the amount of semen delivered from the straw and minimizes the flow of semen unequally into one uterine horn. During the process of semen deposition, one must insure that the fingers of the palpating hand are not inadvertently blocking a uterine horn or misdirecting the flow of semen in some manner. Care should be taken to avoid pulling of the AI gun back through the cervix while depositing the semen. If the cow moves during semen deposition or if the rod slips, reposition the rod to assure proper placement.

Even though the standard recommendation is to deposit semen in the uterine body, bicornual insemination results in no change in fertility[16,17] or improved fertility.[18] Andersson et al.[19] reported similar pregnancy rates for ipsilateral (to the ovary with the preovulatory follicle) cornual insemination with a lower insemination dose (2 million sperm) compared with uterine body insemination with normal dose (15 million sperm); however, fertility was reduced for lower doses inseminated in the uterine body or bilateral cornual insemination compared with higher dose insemination. It should be noted that anterior cornual insemination did not increase sperm retention rate compared with uterine body insemination. Although factors controlling sperm transport and retention are complex, it is clear that retrograde transport of sperm occurs regardless of site of deposition.

Verberckmoes et al.[20] in Belgium developed a novel AI device (Ghent device) for semen deposition near the uterotubal junction in cattle. The site of semen deposition on fertility was tested with insemination of 12 million sperm

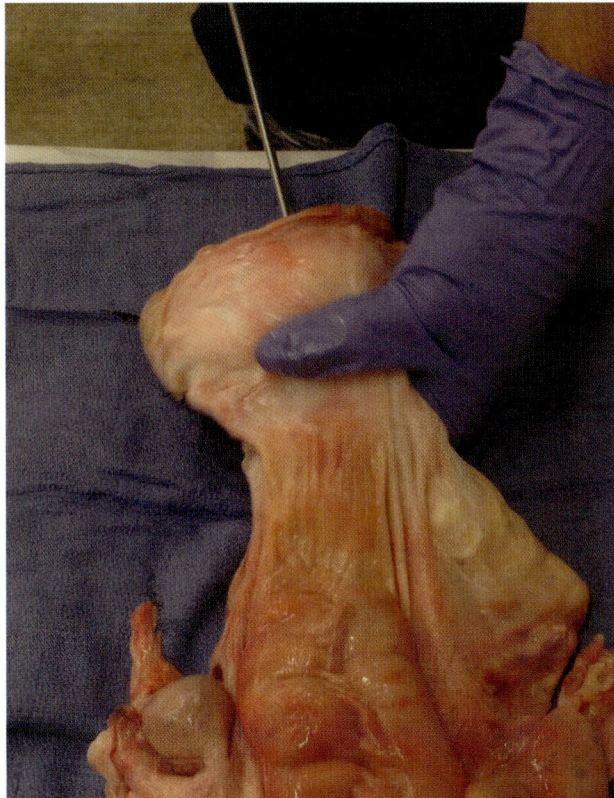

Figure 32.7 Passing AI gun through cervix.

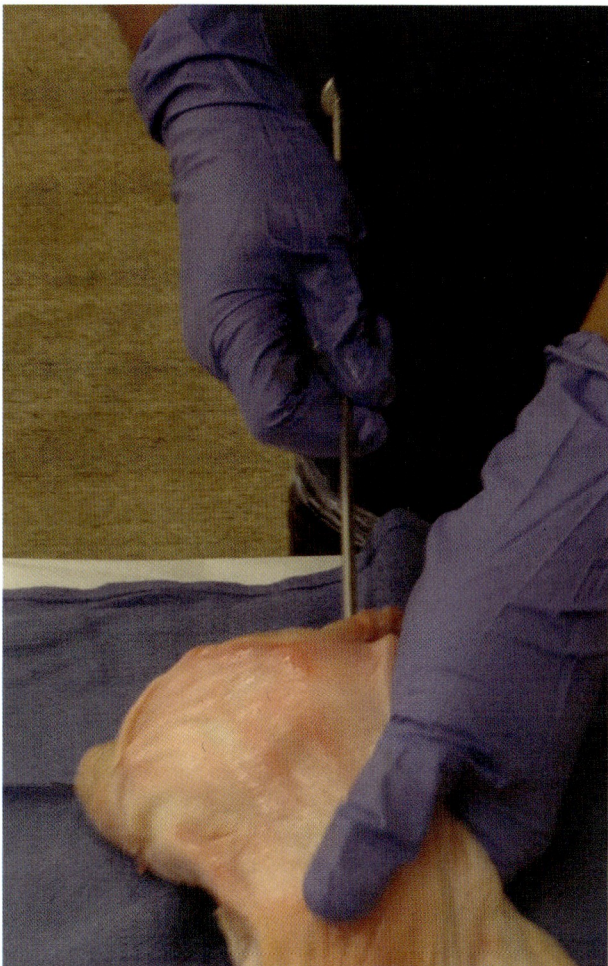

Figure 32.8 Feeling AI gun in the uterine body.

deposited in the uterine body using conventional techniques (control) or insemination with a reduced sperm dose (4 million sperm) deposited in the same manner as the control or bilateral deposition near the uterotubal junction using the Ghent device. The authors concluded that neither sperm dosage nor site of semen deposition influenced pregnancy rates.

Intrafollicular insemination is defined as direct introduction of a low volume (0.06 mL) sperm suspension into a preovulatory follicle. This technique has been recommended as an alternative procedure to the usual deposition of semen into the uterus in cows of low fertility.[21] Even though the results are comparable to intrauterine insemination in a small group of cows, further study is needed to determine its application.

Several methods have been explored to enhance sperm transport following insemination, including administration of prostaglandin $(PG)F_{2\alpha}$ at the time of insemination and clitoral massage immediately after insemination. Administration of a $PGF_{2\alpha}$ analog at the time of insemination improved fertility in heifers compared with nontreated controls, but not in cows.[22] Cooper et al.[23] reported no improvement in pregnancy rates when a 5-s clitoral massage was done after insemination.

Timing of insemination

Proper detection of estrus is critical to the success of AI. Standing estrus, or "heat," is the most reliable indication that a cow is in estrus. Studies have shown that a dairy cow displays an average of seven episodes of estrus (stands to be mounted by other cows).[24] In addition, cows in estrus show estrus activity at any time of the day and hence it is vital that visual observation should be performed several times per day. Increased frequency of visual observation increases the proportion of cows detected in estrus. Several heat detection aids have been developed to assist producers in detection of estrus, including tail paint, Estrus Alert® and Kamar® patches, pedometers, and radio-telemetric systems. A combination of visual observation and one or more of the detection aids increases the effectiveness of estrus detection compared with visual observation or detection aids alone.

Ovulation usually occurs approximately 28–32 hours after the onset of estrus in dairy cows.[25] Optimal fertility of ova is projected to be between 6 and 12 hours after ovulation.[26] The viable lifespan of sperm in the reproductive tract is estimated at 24–30 hours.[27,28] Given these limitations, the window of opportunity is very short for a successful fertilization.

For several years, researchers have investigated the optimal time to inseminate cows relative to the stage of estrus. Early work led to the establishment of the "a.m.–p.m." recommendation and based on this guideline cows in estrus during morning hours should be inseminated during the afternoon hours, and cows in estrus in the afternoon

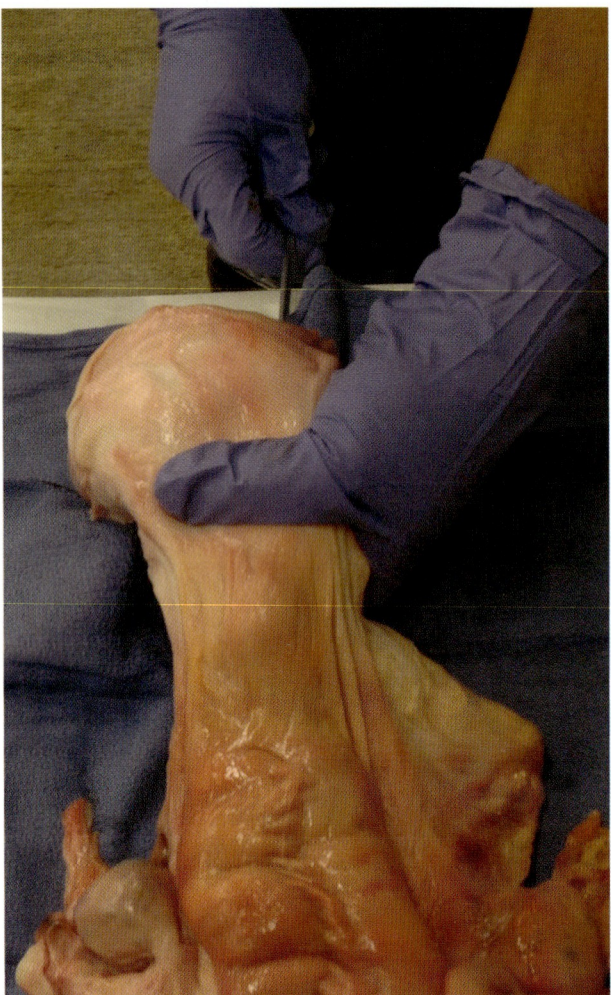

Figure 32.9 Tip of AI gun in the uterine body.

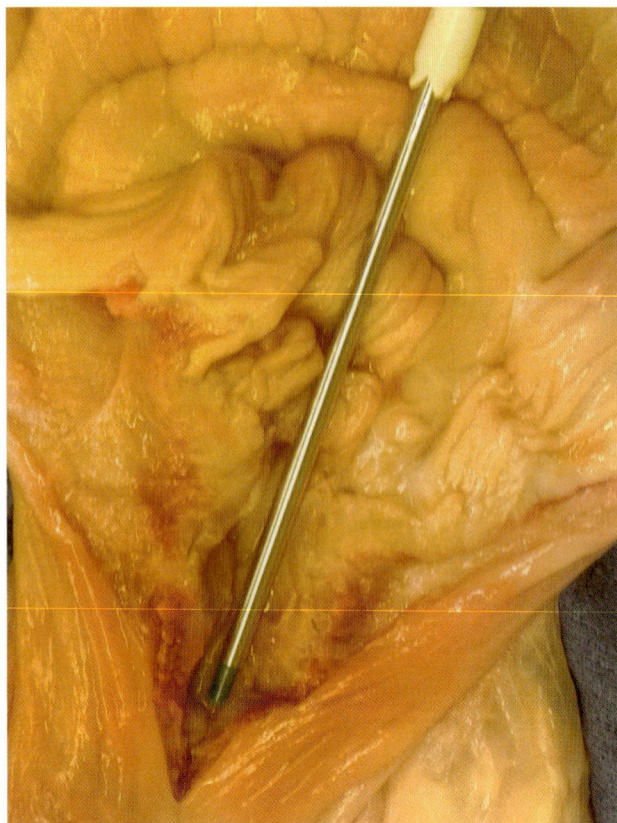

Figure 32.10 Semen deposition site (uterine body).

should be bred the following morning. However, large field trials indicate that maximum conception rates may not be achieved using the a.m.–p.m. rule. The decreased frequency of estrus detection may affect success when AI is performed based on the a.m.–p.m. rule. Highest conception rates for AI occurred between 4 and 12 hours after the onset of estrus in dairy cows.[29]

Conclusion

Accurate insemination technique requires technical and management skills. The level of success ranges from excellent to frustratingly low. Technical skill includes concentration, attention to detail, a clear understanding of reproductive anatomy, and the ability to identify the site of deposition and proper position of the insemination rod. Management skill includes proper semen storage and semen tank management. With scrupulous efforts and attention to detail, the AI program can be successful with farm inseminators. However, successful AI programs depend not only on proper skill and management but appreciable knowledge regarding risks and pitfalls. Periodic evaluation of AI parameters, such as number of services per conception and conception rate, should be done, with corrective measures taken as required, which may include seeking professional help. Herd inseminators should periodically attend a retraining course to review their techniques and be updated regarding new developments and recommendations regarding AI technique.

References

1. Foote R. The history of artificial insemination: selected notes and notables. *J Anim Sci* 2002;80(E Suppl.):E22–E32.
2. Watson P. Recent developments and concepts in the cryopreservation of spermatozoa and the assessment of their post thawing function. *Reprod Fertil Dev* 1995;7:871–891.
3. Saacke R, Lineweaver J, Aalseth E. Procedures for handling frozen semen. In: *Proceedings of the 12th Conference on AI in Beef Cattle of the NAAB*, 1997, p. 49.
4. Hammerstedt R, Graham J, Nolan J. Cryopreservation of mammalian sperm: what we ask them to survive. *J Androl* 1990;11:73–88.
5. Ballester J, Johannisson A, Saravia F *et al.* Post-thaw viability of bull AI-doses with low-sperm numbers. *Theriogenology* 2007;68:934–943.
6. Pickett B, Berndtson W. Preservation of bovine spermatozoa by freezing in straws: a review. *J Dairy Sci* 1974;57:1287–1301.
7. Almquist J, Rosenberg J, Branas R. Effect of thawing time in warm water on fertility of bovine spermatozoa in plastic straws. *J Dairy Sci* 1979;62:772–775.
8. Barth A, Bowman P. Determination of the best practical method of thawing bovine semen. *Can Vet J* 1988;29:366–369.
9. Kaproth M, Rycroft H, Gilbert G *et al.* Effect of semen thaw method on conception rate in four large commercial dairy heifer herds. *Theriogenology* 2005;63:2535–2549.

10. DeJarnette J, Marshall C. Straw-thawing method interacts with sire and extender to influence sperm motility and conception rates of dairy cows. *J Dairy Sci* 2005;88:3868–3875.
11. Kaproth M, Parks J, Grambo G, Rycroft H, Hertl J, Gröhn Y. Effect of preparing and loading multiple insemination guns on conception rate in two large commercial dairy herds. *Theriogenology* 2002;57:909–921.
12. Dalton J, Ahmadzadeh A, Shafii B, Price W, DeJarnette J. Effect of simultaneous thawing of multiple 0.5-mL straws of semen and sequence of insemination on conception rate in dairy cattle. *J Dairy Sci* 2004;87:972–975.
13. Oliveira L, Arruda R, de Andrade A et al. Effect of sequence of insemination after simultaneous thawing of multiple semen straws on conception rate to timed AI in suckled multiparous Nelore cows. *Theriogenology* 2012;78:1800–1813.
14. Karagiannidis A. The distribution of calcium in bovine spermatozoa and seminal plasma in relation to cold shock. *J Reprod Fertil* 1976;46:83–90.
15. Saacke R. Concepts in semen packaging and use. In: *Proceedings of the 8th Conference on AI in Beef Cattle of the NAAB*, 1994, p. 15.
16. Marshall C, Graves W, Meador J, Swain J, Anderson J. A fertility comparison of uterine body and bicornual semen deposition procedures in dairy cattle. *J Dairy Sci* 1989;72(Suppl. 1):455 (Abstract).
17. Graves W, Dowlen H, Kiess G, Riley T. Evaluation of uterine body or bilateral uterine horn insemination techniques. *J Dairy Sci* 1991;74:3454–3456.
18. Senger P, Becker W, Davidge S, Hillers J, Reeves J. Influence of cornual insemination on conception rate in dairy cattle. *J Anim Sci* 1988;66:3010–3016.
19. Andersson M, Taponen J, Koskinen E, Dahlbom M. Effect of insemination with doses of 2 or 15 million frozen-thawed spermatozoa and semen deposition site on pregnancy rate in dairy cows. *Theriogenology* 2004;61:1583–1588.
20. Verberckmoes S, Van Soom A, Dewulf J, Thys M, de Kruif A. Low dose insemination in cattle with the Ghent device. *Theriogenology* 2005;64:1716–1728.
21. López-Gatius F, Hunter R. Intrafollicular insemination for the treatment of infertility in the dairy cow. *Theriogenology* 2011;75:1695–1698.
22. López-Gatius F, Yániz J, Santolaria P et al. Reproductive performance of lactating dairy cows treated with cloprostenol at the time of insemination. *Theriogenology* 2004;62:677–689.
23. Cooper M, Newman S, Schermerhorn E, Foote R. Uterine contractions and fertility following clitoral massage of dairy cattle in estrus. *J Dairy Sci* 1985;68:703–708.
24. Dransfield M, Nebel R, Pearson R, Warnick L. Timing of insemination for dairy cows identified in estrus by a radiotelemetric estrus detection system. *J Dairy Sci* 1998;81;1874–1882.
25. Walker W, Nebel R, McGilliard M. Time of ovulation relative to mounting activity in dairy cattle. *J Dairy Sci* 1996;79:1555–1561.
26. Brackett B, Oh Y, Evans J, Donawick W. Fertilization and early development of cow ova. *Biol Reprod* 1980;23:189–205.
27. Senger P. The estrus detection problem: new concepts, technologies, and possibilities. *J Dairy Sci* 1994;77:2745–2753.
28. Foote RH. Time of artificial insemination and fertility in dairy cattle. *J Dairy Sci* 1979;62:355–358.
29. Trimberger G. *Breeding Efficiency in Dairy Cattle from Artificial Insemination at Various Intervals Before and after Ovulation*. Nebraska Agricultural Experiment Station Research Bulletin, Vol. 153. College of Agriculture, University of Nebraska: 1948.

Chapter 33

Pharmacological Intervention of Estrous Cycles

Ram Kasimanickam

Department of Veterinary Clinical Sciences, College of Veterinary Medicine, Washington State University, Pullman, Washington, USA

Introduction

Reproductive competence is a major aspect affecting the production and economic success of dairy and beef cow herds. For herds using artificial insemination (AI), detection of estrus (submission rate) and calving rate are the two major determinants of inter-calving interval. The economic consequences of low efficiency and poor accuracy in the detection of estrus are the main reasons why cattle reproduction research programs focus on developing practical breeding protocols. The important requirements for any effective estrous synchronization protocol are predictable and high estrus and ovulation responses during a specified interval, followed by a higher pregnancy rate to a single insemination carried out in a predetermined time.[1] To achieve better control of the estrous cycle in cattle it is necessary to synchronize emergence of new follicular waves, ensure a luteal phase, terminate the luteal phase, and synchronize ovulation.

Ovarian follicular dynamics

Growth of ovarian follicles in cattle occurs in distinct wave-like patterns during the estrous cycle, during pregnancy, and in certain anestrous conditions.[2] Each follicle wave has a lifespan of 7–10 days (range 6–12 days, depending on two vs. three waves per cycle and 18 vs. 24 days length of cycle) as the follicle progresses through the different stages of development, namely emergence, selection or atresia, and dominance and atresia or dominance and ovulation. Emergence of a new follicular wave is supported by a transient (1–2 days) increase in follicle-stimulating hormone (FSH) coinciding with the emergence of a follicular wave. Peak FSH concentrations occur just before wave emergence, and then subsequently decline over several days; concurrently, follicles grow and reach approximately 8.0 mm in diameter.[3,4] On average, these follicles grow at a constant and comparable rate and then this group of growing follicles is divided into a single dominant and several subordinate follicles. Typically, deviation begins when the diameter of the largest follicles reaches 8.0 mm.[3,4] The time difference is approximately 8 hours in cattle.[5] Once functional dominance is established, the selected dominant follicle continues to grow while the remaining follicles that emerged during the same follicular wave cease growth and become atretic. Selection of the dominant follicle occurs during declining FSH concentrations; however, the dominant follicle continues to grow (despite FSH at nadir concentrations) until it either ovulates or undergoes atresia.[6] During the final stages of selection of the dominant follicle, there appears to be a transition from mainly FSH to luteinizing hormone (LH) dependency.[6] The selected dominant follicle uniquely expresses mRNA for LH receptors in granulosa cells, which are likely the key to allowing its continued growth under LH stimulation.[6,7] Estrogen-active dominant follicles are largely LH-dependent; positive feedback of estrogen releases a large amount of gonadotropin-releasing hormone (GnRH), resulting in an LH surge and ovulation of the dominant follicle.[6,7] Conversely, if the dominant follicle fails to ovulate, the consequences are continued growth of the dominant follicle dependent on increased LH pulse frequency which, if prolonged, can lead to persistence of the dominant follicle.[8] Therefore, hormonal treatments that modify both FSH and LH clearly affect the fate of a follicular wave. Manipulation of the follicle wave, in turn, may alter systemic hormonal concentrations, the intrafollicular environment, and the oocyte. Thus, a dominant follicle capable of ovulation is present only at specific times during each wave. Therefore the interval from exogenously induced luteolysis and/or withdrawal of a progestagen treatment to estrus and ovulation depends on the stage of the follicle wave at luteolysis and/or at progestagen withdrawal (whichever occurs last if this does not occur at the same time); cattle with a selected dominant follicle will be in estrus within 2–3 days, whereas those in which follicular growth is at the preselection stage, estrus will occur 3–7 days later.[9–12]

Bovine Reproduction, First Edition. Edited by Richard M. Hopper.
© 2015 John Wiley & Sons, Inc. Published 2015 by John Wiley & Sons, Inc.

To achieve maximum synchrony of onset of estrus in a synchronization program, it is necessary to have a recently selected dominant follicle present at the end of treatment. Therefore, new wave emergence must be synchronized, because both the stage of the follicular wave and the duration of dominance affect the duration of the follicular phase and the interval from treatment to estrus. Synchrony of estrus is optimal when the duration of dominance is either less than 4 days (short) or more than 10–12 days (very long). Further, duration of dominance of the preovulatory follicle can also affect fertility. When the duration of dominance of the ovulatory follicle exceeds 10 days, there is a dramatic decline in pregnancy rate,[10–12] primarily due to aged oocytes with premature activation. Therefore it is desirable to precisely control follicular dynamics (and in particular new wave emergence) to minimize both the variation in timing of onset of estrus and to ensure high pregnancy rates.

Pharmacological control of new wave emergence

The method for controlling follicular wave emergence is based on two principles (Figure 33.1): (i) emergence of a new follicular wave is primarily FSH-dependent; and (ii) an estrogen-active dominant follicle is largely LH-dependent. Thus, the negative feedback effects of progestagens and estrogens can be used to synchronize follicle wave emergence by suppression of FSH and/or LH. However, the extent to which follicles are dependent on gonadotropins during various stages of follicular wave development is poorly defined. Further, details of local mechanisms that play key roles in regulating the sequential progression of follicles through sequential physiological stages of the wave are not well defined. This makes it difficult to develop a simple exogenous hormonal treatment that gives predictable new wave emergence in all animals treated, irrespective of the stage of the follicle wave and stage of estrus at treatment.[9,13]

The induction of new follicle wave emergence using exogenous hormonal treatments requires (i) consistent termination of an existing follicle wave, (ii) predictable induction of a transient increase in FSH to induce emergence of a new wave, and (iii) normal growth of the dominant follicle after selection. The primary dependence of the follicle wave progression on gonadotropin support has resulted in the use of steroids to suppress FSH and LH and thus terminate the existing wave. However, it should be realized that FSH and LH are differentially regulated.[6] In the case of FSH, the dominant follicle is the key regulator of the recurrent increases or decreases that occur during the wave. In contrast, although progesterone concentrations are important in regulation of LH pulse frequency, there are changes in LH pulse frequency that occur during the luteal phase which are difficult to clarify by changes in peripheral progesterone or estradiol concentrations. An alternative nonsteroidal approach is the administration of GnRH to induce endogenous LH and FSH surges that cause luteinization or ovulation of an existing dominant follicle, thereby inducing emergence of a new wave.[14]

Use of GnRH

Functional removal of the dominant follicle or cohort follicles by inducing either ovulation or luteinization using exogenously induced gonadotropin release is a practical approach for induction of a new follicular wave. Administration of GnRH induced an immediate LH and FSH surge, the magnitude of which was independent of progesterone concentration or stage of the follicle wave.[15] However, the effect of GnRH on the existing follicle wave is dependent on the presence or absence of a dominant follicle. GnRH administered after dominant follicle selection caused it to ovulate, with emergence of a new wave 1.5–2.0 days later.[14] However, when GnRH was administered before selection, it had no effect on progression of the existing follicular wave. In all cows treated after dominant follicle selection, the induced gonadotropin surge was followed by a transient increase in FSH (but not LH) which was associated with new wave emergence.[15] Therefore GnRH synchronizes new wave emergence only when administered in the presence of a functional dominant follicle, whereas if given

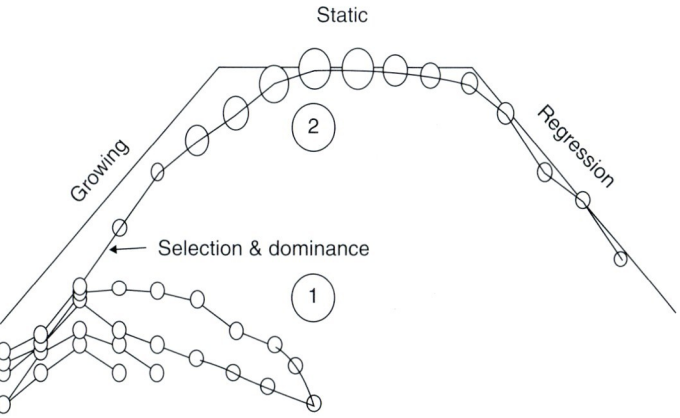

Figure 33.1 Phases of follicular wave. A follicular wave consists of growing, static and regression phases. Growing phase includes recruitment, and selection and dominance and atresia of subordinate follicles; static phase includes preparation of ovulation or regression; and regression phase includes regression of dominant follicle and preparation for subsequent wave. The method for controlling follicular wave emergence is based on two principles: (i) emergence of a new follicular wave is primarily FSH-dependent; and (ii) an estrogen-active dominant follicle is largely LH-dependent.

before dominance it would appear not to affect the subsequent progress of that wave, presumably because of lack of LH receptors on the granulosa cells of such growing follicles. The dichotomy of GnRH effects on progression of the follicle wave is a constraint that needs to be considered when using it as a treatment to synchronize new wave emergence at the start of either progesterone or prostaglandin (PG)$F_{2\alpha}$ synchronization regimens.[16] The subsequent use of $PGF_{2\alpha}$ to cause regression of the induced corpus luteum (CL) is mandatory when GnRH is used to synchronize follicle waves. In addition, not only the stage of development of the dominant follicle,[17] but also stage of the estrous cycle[18] at the time that GnRH is administered affects results. If GnRH is administered when the dominant follicle is pre- or post-dominance, ovulation may not occur and a new follicular wave will not emerge.[17] An alternative is to ensure that a viable growing dominant follicle is present at the time of GnRH treatment. In that regard, cattle will respond most consistently when GnRH is administered between days 5 and 12 of the estrous cycle.[18,19]

Use of estradiol

Even though combined treatment with progestin and estradiol has been known for decades, knowledge of the physiological effects and mechanisms of action of exogenous estradiol and progesterone on follicles is a prerequisite to the development of improved methods of controlling the estrous cycle. Estradiol suppresses the growing phase of the dominant follicle; suppression is more profound when given in combination with progesterone.[20] The mechanism responsible for estrogen-induced suppression of follicle growth appears to involve suppression of both FSH and LH.[21] Further, it exerts its negative effect on both the hypothalamus and pituitary. Studies have focused on the effects of short-term estradiol and progesterone treatments, either alone or in combination at the time of pre-follicle wave emergence, on LH and FSH, progression of the follicle wave, and changes during follicle wave development.[1] The period of elevated FSH associated with wave emergence was shortened by approximately 10 hours following estradiol, progesterone, or a combination of estradiol and progesterone treatments, compared with control. Frequency of LH pulses was reduced following progesterone treatment and was further reduced in response to a combination of estradiol and progesterone treatments. This steroid combination also significantly reduced pulse amplitude. The emergence of the follicle wave and selection of the first dominant follicle of the cycle were delayed approximately 1.5 and 2.5 days, respectively, by the combination of estradiol and progesterone treatment. Intrafollicular estradiol concentrations were suppressed in the largest follicle following combination estradiol and progesterone treatment, whereas under estradiol treatment they were unchanged. However, they were suppressed in smaller follicles following both estradiol and combination estradiol and progesterone treatments. Therefore, suppression of LH pulses by the combination estradiol and progesterone treatment is important for suppressing follicular estradiol in the largest follicle, whereas suppression of FSH alone may be sufficient to reduce follicular estradiol concentrations in smaller follicles.

In estrous synchronization protocols, estradiol is usually injected at the initiation of the protocol, concurrent with insertion of a progestin-releasing device. The combination of estradiol and progestin was used to determine if suppression of follicle growth would induce new wave emergence at a consistent interval after treatment regardless of the phases (i.e., growing, static, or regressing phases) of follicle development at which treatment was initiated. The use of estradiol-17β in progestin-implanted cattle was followed consistently by the emergence of a new wave, on average, 4.5 days later.[9,16] Once the estradiol was metabolized, there was an FSH surge and a new follicular wave emerged.[22,23] The administration of 2.5 or 5 mg estradiol-17β (or 2 mg of estradiol benzoate or estradiol valerate) in progestin-implanted cattle at random stages of the cycle was followed by the emergence of a new follicular wave approximately 4 days later, with little variability.[24–26]

The mechanism of action of estradiol would appear to be through FSH and LH, causing follicle atresia followed by synchronous release of FSH and emergence of a new follicle wave. Hence, in general, estradiol has been the hormone treatment used most successfully to synchronize follicle wave emergence in cattle. However, the use of estradiol is being prohibited in an increasing number of countries, so alternatives must be sought. Similarly, the use of GnRH or LH to induce ovulation, and to remove the suppressive effects of a dominant follicle on FSH, result in a surge in FSH and a new follicle wave at a predictable time thereafter. Therefore, synchronization of follicle wave emergence is likely to be through FSH release, which may involve manipulation of ovarian (follicular) metabolic pathways or feedback mechanisms to the hypothalamic–pituitary axis.

Controlling the lifespan of the corpus luteum

Shortening the luteal phase (premature luteolysis)

Luteolysis is initiated by estradiol from the developing preovulatory follicle, which triggers the release of hypophyseal oxytocin, which in turn stimulates release of a small quantity of uterine $PGF_{2\alpha}$. $PGF_{2\alpha}$ then initiates a positive feedback loop involving release of additional luteal oxytocin and $PGF_{2\alpha}$ of both luteal and uterine origin.[27] Oxytocin stimulates synthesis and secretion of $PGF_{2\alpha}$ from the uterus. It has recently been proposed that release of luteal $PGF_{2\alpha}$ amplifies the luteolytic signal in an autocrine or paracrine manner.[27,28] $PGF_{2\alpha}$ causes decreased blood flow to luteal parenchyma and morphological changes that reduce both the number of small steroidogenic luteal cells and the size of large luteal cells. In addition, it is involved in intracellular signaling that facilitates apoptosis of large luteal cells and stimulates intraluteal production of $PGF_{2\alpha}$, thereby resulting in the demise of the CL.[29,30]

Since the discovery of $PGF_{2\alpha}$ as the luteolytic agent in cattle,[31,32] it has been the most commonly used treatment for elective induction of luteal regression and/or synchronization of estrus.[33] The injection of $PGF_{2\alpha}$ causes immediate regression of the CL after approximately day 5 of the estrous cycle: progesterone concentrations decline rapidly to basal

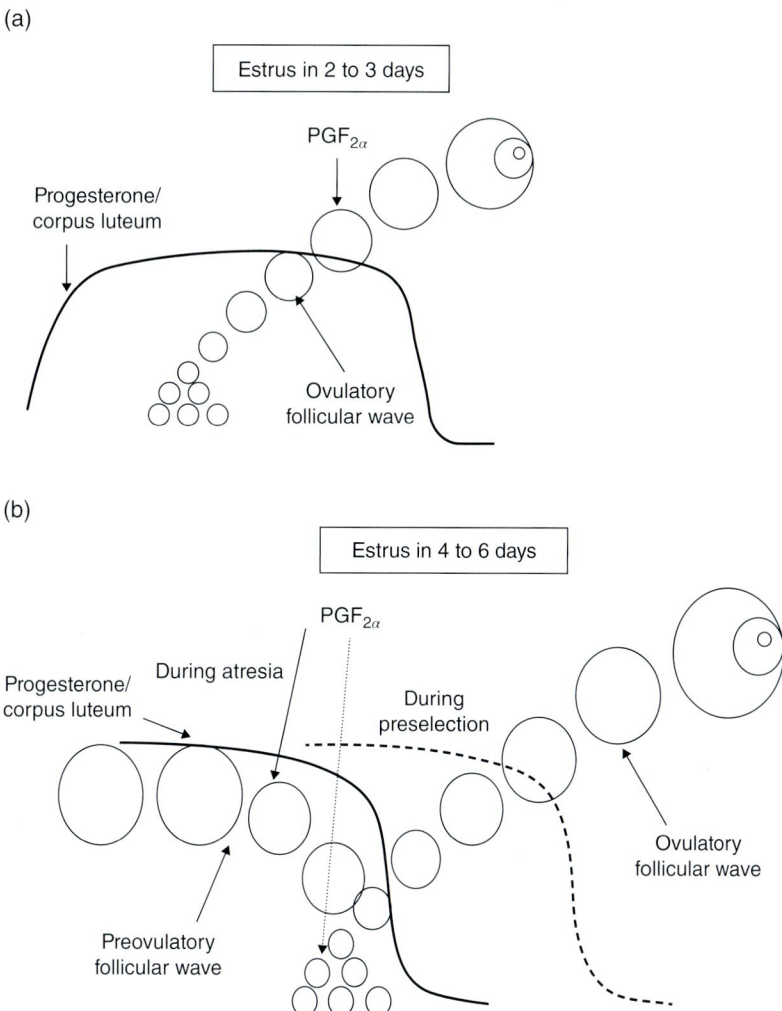

Figure 33.2 Interval from PGF$_{2\alpha}$ administration to estrus. (a) Animals with a functional dominant follicle are in estrus within 2–3 days because the dominant follicle at the time of induced luteolysis ovulates soon thereafter. (b) Animals at the pre-dominance or regression phase of the follicular wave will require 4–6 days for a dominant follicle to reach preovulatory diameters and hence have a longer and more variable interval to onset of estrus.

levels within 24 hours, and LH pulse frequency increases, causing a significant increase in estradiol from the dominant follicle and the induction of estrus and ovulation. Despite rapid luteolysis, the interval from treatment to the onset of estrus is variable and dependent on the stage of the follicle wave at treatment. Animals with a functional dominant follicle are typically in estrus within 2–3 days, because the dominant follicle at the time of induced luteolysis ovulates shortly (Figure 33.2a); however, animals at the pre-dominance or regression phase of the wave will require 4–6 days to form a dominant follicle and hence have a longer and more variable interval to onset of estrus (Figure 33.2b).

PGF$_{2\alpha}$ failed to effectively induce luteolysis during the first 5 or 6 days following estrus. Lack of responsiveness was due to lack of PGF$_{2\alpha}$ receptors in immature CL, but recent work has demonstrated the presence of PGF$_{2\alpha}$ receptors in the CL as early as 2 days after ovulation.[34] However, it appears that the mature CL may possess a positive feedback loop that results in intraluteal PGF$_{2\alpha}$ production that may continue the process of luteolysis initiated by a single treatment of exogenous PGF.[35] It should be noted that although luteolytic response was variable when a single dose of PGF$_{2\alpha}$ was given on days 5 or 6 of the estrous cycle, two doses of prostaglandin analog (12 hours apart) consistently induced luteal regression (95–100%) at 2–5 days after estrus.[36,37]

Extending/simulating a luteal phase with exogenous progesterone

Development of intravaginal progesterone delivery devices, such as the progesterone-releasing intravaginal device (PRID, 1.55 g), controlled intravaginal releasing device [CIDR, 1.38 g (US) or 1.9 g (Canada)], and Cue-Mate (1.56 g) intravaginal progesterone device, has facilitated the use of progestagens for synchronization in cattle. In general, for all progesterone-based synchronization treatments, a high proportion (up to 85%) of cows exhibit estrus 36–72 hours after removal. This higher degree of estrous synchrony allows for fixed-time AI between 48 and 84 hours after removal with hormonal induction of ovulation using GnRH or estradiol. Alternatively, cows may be observed for estrus and inseminated once after detection of estrus.

Estrous synchronization systems that include progesterone in the protocol have three primary advantages. First, maintaining the concentration of progesterone at greater than threshold level (>1 ng/mL) suppresses the LH surge and estrous behavior. Treatment with progesterone for 14 or 21 days resulted in a high estrus response within 3 days after progesterone removal.[38] However, conception rates following prolonged progesterone treatment are 10–15% below those of cows treated with progesterone for shorter duration. The reason for low conception is that intravaginal progesterone releasing devices result in progesterone concentrations that are lower than those of the luteal phase, and LH pulse frequency and peripheral estradiol concentrations are elevated. Consequently, the dominant follicle continues to receive gonadotropin support beyond the time where either ovulation or atresia would have occurred. Ultrastructural changes in the oocyte within the persistent dominant follicle are believed to reduce fertility of treated cows.[39,40] Fertility concerns for progesterone-treated cows have led to development of short-term protocols (7–9 days) for progesterone use. Progesterone is also typically combined with follicular synchronization so that a newly developing dominant follicle is present during progesterone supplementation. Progesterone supplementation is an effective method for treating anestrus, which affects approximately 20% of beef cows at the start of the breeding season.[41,42] Furthermore, in US dairies, 28% of primiparous cows and 15% of multiparous cows had not ovulated by 60 days after calving.[42] Thus the second primary advantage for progesterone in estrous synchronization protocols is that anestrous cows are responsive when progesterone is used. Progesterone supplementation in an anestrous cow decreases LH initially, but then increases LH pulsatility.[43,44] The increase in LH pulsatility increases dominant follicle development. Progesterone also primes the hypothalamic center for estrous expression and the LH surge. Removal of the progesterone is followed by estrus, ovulation, and a normal-length luteal phase in a large percentage of treated cows.[43] Treatment of cystic ovarian disease is the third advantage of progesterone supplementation. Approximately 10% of North American dairy cows are affected.[45] Abnormally high LH pulsatility and absence of an LH surge are characteristic endocrine features of the disease.[45–47] Progesterone supplementation decreases LH pulsatility in cystic cows. The decrease in LH pulsatility after progesterone treatment is followed by turnover of the cystic (dominant) follicle. Development of a new dominant follicle is normal and ovulation occurs after progesterone withdrawal.[46–48]

Induction of ovulation

Prior to the preovulatory LH surge in cyclic cows, the frequency of LH pulses gradually increases to one pulse per hour in response to endogenous GnRH pulses. In cyclic cows a single dose of exogenous GnRH induces LH release. Therefore a single injection of GnRH induces an LH surge and ovulation when given at approximately days 10–18 after calving in dairy cows and days 21–31 after calving in beef cows.[42] A single injection of GnRH combined with 48-hour calf removal also induces ovulation in beef cows. A single injection of human chorionic gonadotropin (hCG) 500–3000 IU induces ovulation in dairy and beef cows. Further, because of its positive feedback effect on the surges of FSH and LH, exogenous estradiol has been used to induce ovulation in cows. It should be noted that the GnRH-induced CL in postpartum cows usually has a shorter lifespan than a spontaneously formed CL. Further, administration of GnRH prior to complete uterine involution results in pyometra in postpartum dairy cows. Therefore, a single injection of GnRH during the early postpartum period is not very useful in practice.

Synchronization of estrus

Prostaglandin protocol

The mechanism of action of prostaglandin is luteolysis, and thus the application of a prostaglandin program is restricted only to cycling cows with a functional mature CL. The administration of a luteolytic dose of $PGF_{2\alpha}$ (25 mg) or its synthetic analogs (e.g., 500 µg cloprostenol) causes regression of the CL in most stages of diestrus, thereby inducing estrus in over half (55–65%) of a cycling herd (Figure 33.3a). A second injection given 11–14 days later should result in synchronized estrus of the herd because the cows that responded to the first injection will be on day 6–15 of diestrus with the majority being on days 7–9 (Figure 33.3b). A 100% response is not realistic, even if every cow in the group was known to have a mature CL at the time of the second injection. This is partly because not all mature corpora lutea, at least in lactating dairy cattle, are lysed by $PGF_{2\alpha}$. For example, approximately 25% of the cows that were on day 6 will respond, approximately 33% of those that were on day 7 respond, and fewer than 100% of those that were on days 8–16 of the cycle respond. Many prostaglandin-controlled synchronization programs have been developed (Figure 33.3). The selection of a program best suited to a herd depends on a variety of factors including efficiency of estrus detection on the farm, palpation skills of the attending veterinarian (Figure 33.3c), cost of treatment, availability of labor, and the objectives of the program. Further, these programs involve intramuscular injections of a single luteolytic dose of $PGF_{2\alpha}$. Various doses and injection sites have since been compared: ischiorectal fossa (one luteolytic dose) and vaginal submucosa (half a standard luteolytic dose). Administration of half a standard luteolytic dose submucosally in the vagina or injection of a single luteolytic dose of $PGF_{2\alpha}$ into the ischiorectal fossa was as efficacious as intramuscular administration.[49]

Progesterone protocols

14-day treatment

In this protocol, cows are supplemented for 14 days usually using an intravaginal progesterone device (IVPD) or oral progestin (melengestrol acetate, MGA) (Figure 33.4). Since ovulation is inhibited for up to 14 days, the follicle that ovulates following this protocol is aged (from a persistent follicle) and hence has reduced fertility. Therefore, $PGF_{2\alpha}$ is

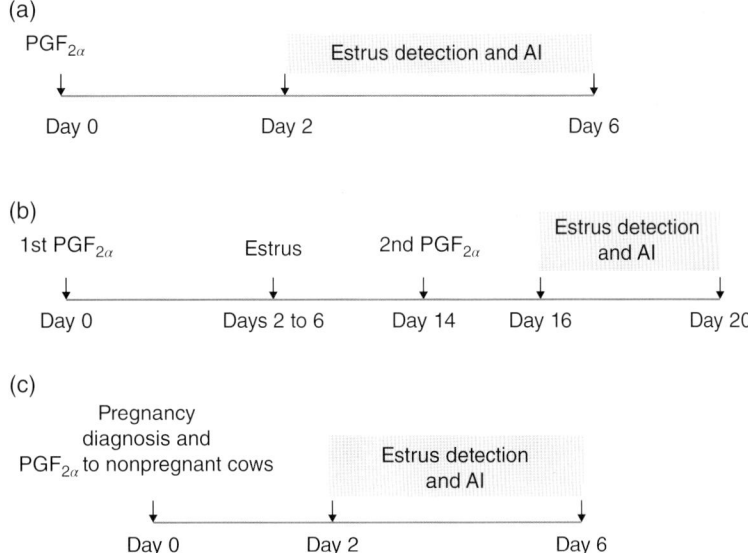

Figure 33.3 Synchronization of estrus using $PGF_{2\alpha}$. (a) Single $PGF_{2\alpha}$ program. (b) Double $PGF_{2\alpha}$ program. (c) Administration of $PGF_{2\alpha}$ to nonpregnant cows at pregnancy diagnosis. Note that under field conditions, this protocol is routinely done by a veterinarian to prevent iatrogenic pregnancy loss.

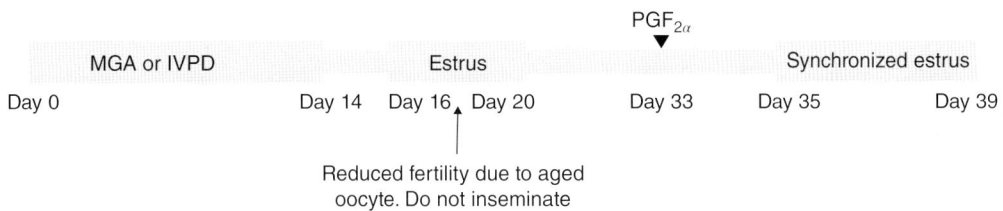

Figure 33.4 Estrus synchronization protocol with progesterone supplementation for 14 days using an intravaginal progesterone device or melengestrol acetate and $PGF_{2\alpha}$.

given 19 days following progesterone/progestin removal. The animals are observed for 2–4 days following treatment with $PGF_{2\alpha}$ and inseminated at observed estrus using the a.m.–p.m. rule. This injection of $PGF_{2\alpha}$ regresses the CL that forms after the ovulation of a persistent follicle, and an estrus with good fertility will occur. This type of synchronization protocol with longer duration of progesterone treatment (14–17 days[50] and 14–19 days[51,52]) has become a commonly used method of synchronizing estrus in replacement beef heifers. MGA, an orally active progestin,[53] is capable of inhibiting estrus and ovulation in heifers when consumed at 0.5 mg/head daily for 14 days in a supplement carrier. Usually MGA is mixed with grain or a protein supplement and fed to cattle. The MGA program is usually a cost-effective synchronization program. It is important to ensure that all animals enrolled receive the recommended dose. Recently, a 14–16 day IVPD–$PGF_{2\alpha}$ protocol was evaluated for synchronization of estrus in replacement beef heifers in an attempt to replace the oral MGA program.[54,55] This has the potential to improve synchrony of estrus compared with other protocols currently available. Progesterone has been shown to effectively presynchronize estrus[50,55,56] in replacement beef heifers and induce ovulation in heifers that did not reach puberty before administration of treatments.[57]

7-day exposure

In this synchronization program, progestin, usually an IVPD, is used for 7 days (Figure 33.5). Progesterone from the device mimics the progesterone produced by the CL and inhibits estrus and ovulation. When ovulation is inhibited for 7 days, all cycling animals are expected to have a CL that is at least 7 days old at $PGF_{2\alpha}$ treatment. Therefore all animals with a CL will respond to the $PGF_{2\alpha}$. Animals in which a CL had regressed during the 7-day period will show standing estrus following removal of the progestin. Estrus is observed 2–6 days after removal of progesterone; those detected in estrus are inseminated using the a.m.–p.m. rule (cows first detected in estrus in the morning will be inseminated the same evening and cows first detected in estrus in the evening will be inseminated the next morning).

Use of GnRH in a 7-day protocol

Progesterone exposure will not synchronize follicular waves. However, exogenous GnRH at the start of progesterone treatment will synchronize follicular waves. The normal duration of a follicular wave is 7–10 days; therefore if a mature follicle is present at the initiation of the protocol, that follicle will persist (and ovulate an oocyte with reduced

310 The Cow: Breeding and Health Management

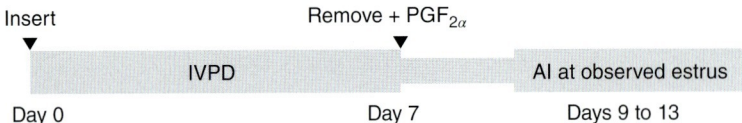

Figure 33.5 Estrus synchronization protocol with progesterone supplementation for 7 days using an intravaginal progesterone device and $PGF_{2\alpha}$.

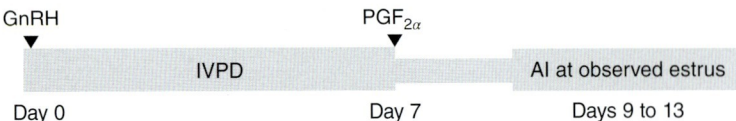

Figure 33.6 Estrus synchronization protocol with progesterone supplementation for 7 days using an intravaginal progesterone device, GnRH and $PGF_{2\alpha}$. Note that the addition of GnRH is to initiate a new follicular wave. GnRH is replaced with estradiol-17β or estradiol benzoate plus progesterone.

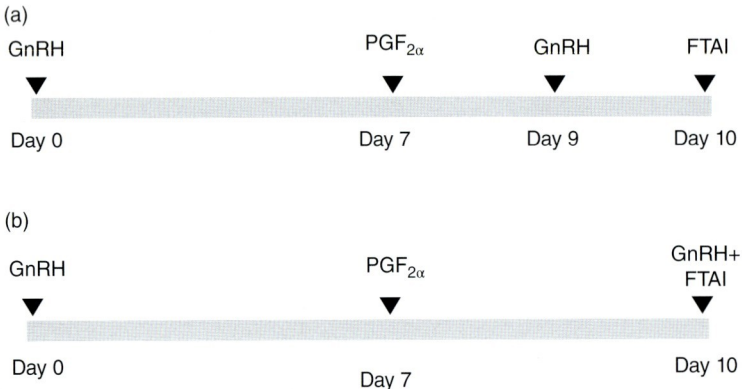

Figure 33.7 Ovulation synchronization protocol using GnRH–$PGF_{2\alpha}$–GnRH for a predetermined timed insemination (FTAI). (a) Ovsynch protocol. (b) CO-Synch protocol. Note that the first GnRH is replaced with estradiol-17β or estradiol benzoate plus progesterone and the second GnRH with estradiol-17β or estradiol benzoate. The concurrent administration of GnRH at the time of insemination in the CO-Synch protocol is preferred in beef farms as it reduces number of handlings.

fertility). Initiation of a new follicular wave at the beginning of the protocol by administering GnRH will reduce the occurrence of any persistent follicle forming. Furthermore, the progestin that is present between the GnRH and $PGF_{2\alpha}$ injections will virtually eliminate the chance of animals exhibiting signs of standing estrus before the $PGF_{2\alpha}$ injection (Figure 33.6). The animals will be inseminated at observed estrus following progesterone removal.

Synchronization of ovulation

Synchronization of ovulation involves a combination of GnRH and $PGF_{2\alpha}$, with or without progesterone supplementation during the course of the treatment. These protocols facilitate insemination at a predetermined (fixed) time without the need for detection of estrus. Several protocols have been successfully employed with acceptable pregnancy rates.

GnRH–$PGF_{2\alpha}$–GnRH protocol

The protocol involves an injection of GnRH at a random stage of the estrous cycle, followed by an injection of $PGF_{2\alpha}$ 7 days later. A second injection of GnRH follows the $PGF_{2\alpha}$ injection by 48 hours. The method is designed to be used with timed insemination 16 (range 8–24) hours after the last GnRH injection (Figure 33.7a; Ovsynch).[14] This synchronization system results in a tight synchrony of estrus, allowing breeding at an appointed time without detection of estrus. The first injection of GnRH causes either ovulation or luteinization of all dominant or large growing follicles. As a result, a new follicular wave is initiated in all cows by approximately 3 days after the injection. Therefore, all the females in the group have growing follicles of about the same stage of development. In addition, GnRH stimulates development of luteal tissue from the cells that were previously the dominant follicle. The $PGF_{2\alpha}$ injection lyses the CL resulting from the initial GnRH injection. Furthermore, the final GnRH injection serves to increase the synchrony of ovulation within a group of cows. The success of the first injection of GnRH to synchronize follicular growth is very good in cows, but less so in heifers. The CO-Synch protocol is similar to the Ovsynch protocol, except that the second injection of GnRH is given concurrent with AI (Figure 33.7b). This has a huge advantage in beef synchronization regimens because the CO-Synch regimen requires one less handling.

Presynch protocol

It was evident that stage of estrous cycle at initiation of the Ovsynch protocol influenced the success of synchrony of the follicular wave. Initiation of the protocol too early (e.g., 1–4 days after ovulation) or too late (e.g., 13–20 days after ovulation) reduced synchrony. Furthermore, if the first GnRH treatment was given when the dominant follicle was pre- or post-dominance, ovulation may not occur and synchronous emergence of a new follicular wave will not occur.[17] An alternative is to ensure that a viable growing dominant follicle is present at the time of GnRH treatment. Cattle will respond most consistently when GnRH is given between days 5 and 12 of the estrous cycle.[18,19] Consequently, several protocols have been employed to align cows at days 5–12 of the estrous cycle at initiation of an Ovsynch or CO-Synch protocol, including presynchronizing cows before Ovsynch using two injections of $PGF_{2\alpha}$ administered 14 days apart with initiation of the timed AI protocol 11 days,[58] 12 days,[19] or 14 days[59] later (Presynch-Ovsynch). Another protocol, designated G6G starts with an injection of $PGF_{2\alpha}$, which is intended to induce luteolysis of functional corpora lutea, and a GnRH injection 2 days later, intended to induce an ovulation. Collectively, the two injections of G6G are intended to initiate a new estrous cycle. The Ovsynch is scheduled to start 6 days after the GnRH injection in order for the cow to be on day 6 of a new estrous cycle and very likely to have a functional dominant follicle, capable of ovulating in response to the first GnRH of Ovsynch. Further, use of two injections of $PGF_{2\alpha}$ to presynchronize estrous cycles was not effective for anovular cows.[19] Presynchronization protocols that induce production of a CL before initiation of the Ovsynch protocol are likely to improve fertility in anovular cows. A protocol that involves induction of cyclicity in anovular dairy cows is the use of an Ovsynch protocol to presynchronize and induce cyclicity in cows before the use of a breeding Ovsynch protocol, termed Double-Ovsynch.[60]

Addition of progesterone in GnRH–$PGF_{2\alpha}$–GnRH protocol

Progesterone supplementation between the first GnRH and $PGF_{2\alpha}$ for 7 days in Ovsynch (Figure 33.8a) and CO-Synch (Figure 33.8b) protocols enhances estrus/ovulation synchrony and conception in cows and heifers. These protocols are advantageous in anestrus cows and heifers and in cows with cystic ovaries. Several protocols are available for use, modified from the above-mentioned protocol. Recently, progesterone supplementation between the first GnRH and $PGF_{2\alpha}$ for 5 days in CO-Synch protocol (Figure 33.8c) has been tested. The objective is to shorten progesterone treatment to 5 days and to increase proestrus phase (the interval from $PGF_{2\alpha}$ to the second GnRH) to 72 hours.[61,62] This approach increased AI pregnancy rates up to 15% in beef cows.[41,42] Because of the shortened interval from the initial GnRH to $PGF_{2\alpha}$ in the 5-day IVPD protocol, and variability in luteolysis in some cows when a single dose of $PGF_{2\alpha}$ is given on day 5 of the estrous cycle, all cows should receive a second dose of $PGF_{2\alpha}$

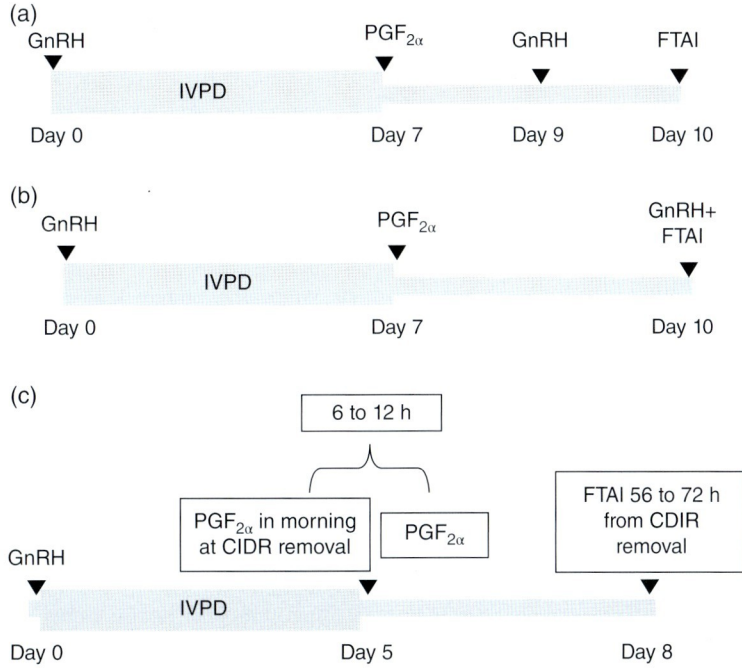

Figure 33.8 Ovulation synchronization protocol using GnRH–$PGF_{2\alpha}$–GnRH with addition of progesterone for a predetermined timed insemination (FTAI). (a) Ovsynch with IVPD. (b) CO-Synch with IVPD. (c) 5-day CO-Synch with IVPD. Note that the first GnRH is replaced with estradiol-17β or estradiol benzoate plus progesterone and the second GnRH with estradiol-17β or estradiol benzoate. The addition of progesterone is considered advantageous in anestrous cows and prepubertal and peripubertal heifers. The concurrent administration of GnRH at the time of insemination in CO-Synch + IVPD protocol is preferred on beef farms as it reduces number of handlings. Because of the shortened interval from the initial GnRH to $PGF_{2\alpha}$ in the 5-day IVPD protocol, and variability in luteolysis in some cows when a single dose of $PGF_{2\alpha}$ is given on day 5 of the estrous cycle, all cows should receive a second dose of $PGF_{2\alpha}$ approximately 6–12 hours after the first $PGF_{2\alpha}$ given at CIDR withdrawal.

treatment approximately 6–12 hours after the first PGF$_{2\alpha}$ given at CIDR withdrawal.[63,64]

Addition of equine chorionic gonadotrophin

Equine chorionic gonadotropin (eCG) is added to the synchronization treatment protocols as a means of increasing follicular development. The utilization of eCG in breeding protocols improves reproductive efficiency in seasonally calving, anestrous cattle. Addition of 4000 IU eCG at progesterone device removal resulted in 74% pregnancy in a 49-day breeding season.[65] Inclusion of eCG in 6 or 7 days of GPG/P4 protocol increased pregnancy in a 28-day breeding (58.6 vs. 52.3%) and decreased median days to conception in anestrous dairy cows.[66] In South American pasture-based lactating beef and dairy lactating anestrous cows, addition of eCG increased pregnancy rates by 10–15% after fixed-time AI compared with the same protocol without eCG.[67–69]

Conclusion

The estrous and ovulation synchronization protocols now exist to successfully achieve acceptable fertility after applying insemination at observed estrus or at a predetermined time without estrus detection. Further, success is achieved in females that may not be cycling at the onset of treatment. Initiation of fertile ovulation in peripubertal heifers and acyclic cows is likely the primary manner in which producers may improve fertility in response to estrous synchronization and timed AI protocols. Use of IVPD-based synchronization before insemination of cows and heifers enhances fertility more effectively than other hormonal synchronization interventions. Development of synchronization protocols that reduce the factors associated with reducing synchrony of estrus and/or ovulation will provide cattle producers with efficient and effective tools for improving genetics in their operations. Since location, pasture difference, diet, breed composition, body condition, postpartum interval, climate, and geographic location may affect the success of synchronization protocols, these variables should be kept in mind to enhance the success of synchronization.

References

1. Diskin M, Austin E, Roche J. Exogenous hormonal manipulation of ovarian activity in cattle. *Domest Anim Endocrinol* 2002;23:211–228.
2. Ginther O. Selection of the dominant follicle in cattle and horses. *Anim Reprod Sci* 2000;60/61:61–79.
3. Adams G, Matteri R, Ginther O. Effect of progesterone on ovarian follicles, emergence of follicular waves and circulating follicle-stimulating hormone in heifers. *J Reprod Fertil* 1992;95:627–640.
4. Ginther O, Kot K, Kulick L, Wiltbank M. Emergence and deviation of follicles during the development of follicular waves in cattle. *Theriogenology* 1997;48:75–87.
5. Ginther O, Bergfelt D, Kulick L, Kot K. Selection of the dominant follicle in cattle: establishment of follicle deviation in less than 8 hours through depression of FSH concentrations. *Theriogenology* 1999;52:1079–1093.
6. Ginther O, Bergfelt D, Kulick L, Kot K. Pulsatility of systemic FSH and LH concentrations during follicular-wave development in cattle. *Theriogenology* 1998;50:507–519.
7. Ginther O, Bergfelt D, Beg M, Kot K. Follicle selection in cattle: role of luteinizing hormone. *Biol Reprod* 2001;64:197–205.
8. Savio J, Thatcher W, Morris G, Entwhistle K, Drost M, Mattiacci M. Effects of induction of low plasma progesterone concentrations with progesterone-releasing intravaginal device on follicular turnover and fertility in cattle. *J Reprod Fertil* 1993;98:77–84.
9. Bo G, Adams G, Pierson R, Mapletoft R. Exogenous control of follicular wave emergence in cattle. *Theriogenology* 1995;43:31–40.
10. Austin J, Mihm M, Ryan M, Williams D, Roche J. Effect of duration of dominance of the ovulatory follicle on onset of estrus and fertility in heifers. *J Anim Sci* 1999;77:2219–2226.
11. Kastelic J, Knopf L, Ginther, O. Effect of day of prostaglandin treatment on selection and development of the ovulatory follicle in heifers. *Anim Reprod Sci* 1990;23:169–180.
12. Kastelic J, Ginther O. Factors affecting the origin of the ovulatory follicle in heifers with induced luteolysis. *Anim Reprod Sci* 1991;26:13–24.
13. Roche J, Mihm M, Diskin M, Ireland J. A review of regulation of follicle growth in cattle. *J Anim Sci* 1998;76:16–23.
14. Twagiramungu H, Guilbault L, Dufour J. Synchronization of ovarian follicular waves with gonadotrophin-releasing hormone agonist to increase the precision of estrus in cattle: a review. *J Anim Sci* 1995;73:3141–3151.
15. Ryan M, Mihm M, Roche J. Effect of GnRH given before or after dominance on gonadotrophin response and the fate of that follicle wave in postpartum dairy cows. *J Reprod Fertil Abstr Ser* 1998;21:61.
16. Bo G, Adams G, Caccia M, Martinez M, Pierson R, Mapletoft R. Ovarian follicular wave emergence after treatment with progestogen and estradiol in cattle. *Anim Reprod Sci* 1995;39:193–204.
17. Martinez M, Adams G, Bergfelt D, Kastelic J, Mapletoft R. Effect of LH or GnRH on the dominant follicle of the first follicular wave in heifers. *Anim Reprod Sci* 1999;57:23–33.
18. Vasconcelos J, Silcox R, Rosa G, Pursley J, Wiltbank M. Synchronization rate, size of the ovulatory follicle, and pregnancy rate after synchronization of ovulation beginning on different days of the estrous cycle in lactating dairy cows. *Theriogenology* 1999;52:1067–1078.
19. Moreira F, Orlandi C, Risco CA, Mattos R, Lopes F, Thatcher W. Effects of presynchronization and bovine somatotropin on pregnancy rates to timed artificial insemination protocol in lactating dairy cows. *J Dairy Sci* 2001;84:1646–1659.
20. Wiltbank J, Sturges J, Wideman D, LeFever D, Faulkner L. Control of estrus and ovulation using subcutaneous implants and estrogens in beef cattle. *J Anim Sci* 1971;33:600–606.
21. Bó G, Adams G, Pierson R, Tribulo H, Caccia M, Mapletoft R. Follicular wave dynamics after estradiol-17β treatment of heifers with or without a progestogen implant. *Theriogenology* 1994;41:1555–1569.
22. Martínez M, Kastelic J, Bó G, Caccia M, Mapletoft R. Effects of estradiol and some of its esters on gonadotrophin release and ovarian follicular dynamics in CIDR-treated beef cattle. *Anim Reprod Sci* 2005;86:37–52.
23. Martínez M, Kastelic J, Mapletoft R. Effects of estradiol on gonadotrophin release, estrus and ovulation in CIDR-treated beef cattle. *Domest Anim Endocrinol* 2007;33:77–90.
24. Bó G, Baruselli P, Moreno D et al. The control of follicular wave development for self-appointed embryo transfer programs in cattle. *Theriogenology* 2002;57:53–72.
25. Caccia M, Bó G. Follicle wave emergence following treatment of CIDR-B implanted beef heifers with estradiol benzoate and progesterone. *Theriogenology* 1998;49:341 (Abstract).
26. Colazo M, Martínez M, Small J et al. Effects of estradiol valerate on ovarian follicle dynamics and superovulatory response in progestin-treated cattle. *Theriogenology* 2005;63:1454–1468.

27. Milvae R. Role of luteal prostaglandins in the control of bovine corpus luteum functions. *J Anim Sci* 1986;62(Suppl. 2):72–78.
28. Niswender G, Davis T, Griffith R et al. Judge, jury and executioner: the auto-regulation of luteal function. *Soc Reprod Fertil Suppl* 2007;64:191–206.
29. Hansel W, Alila H, Dowd J, Milvae R. Differential origin and control mechanisms in small and large bovine luteal cells. *J Reprod Fertil Suppl* 1991;43:77–89.
30. Hansel W, Alila H, Dowd J, Yang X. Control of steroidogenesis in small and large bovine luteal cells. *Aust J Biol Sci* 1987;40:331–347.
31. McCracken JA, Baird DT, Carlson JC, Goding JR, Barcikowski B. The role of prostaglandins in luteal regression. *J Reprod Fertil Suppl* 1973;18:133–142.
32. Lauderdale J, Seguin B, Stellflug J et al. Fertility of cattle following PGF2a injection. *J Anim Sci* 1974;38:964–967.
33. Odde K. A review of synchronization of estrus in postpartum cattle. *J Anim Sci* 1990;68:817–830.
34. Wiltbank M, Shiao T, Bergfelt D, Ginther O. Prostaglandin F2$_\alpha$ receptors in the early bovine corpus luteum. *Biol Reprod* 1995;52:74–78.
35. Wiltbank M. How information of hormonal regulation of the ovary has improved understanding of timed breeding programs. In: *Proceedings of the Society for Theriogenology*, 1997, pp. 83–97.
36. Nasser L, Adams G, Bó G, Mapletoft R. Ovarian superstimulatory response relative to follicular wave emergence in heifers. *Theriogenology* 1993;40:713–724.
37. Adams G, Nasser L, Bó G, Garcia A, Del Campo M, Mapletoft R. Superovulatory response of ovarian follicles of Wave 1 versus Wave 2 in heifers. *Theriogenology* 1994;42:1103–1113.
38. Macmillan K, Peterson A. A new intravaginal progesterone releasing device in cattle (CIDR-B) for oestrus synchronisation, increasing pregnancy rates and the treatment of anoestrum. *Anim Reprod Sci* 1993;33:1–25.
39. Mihm M, Baguisi A, Boland M, Roche J. Association between the duration of dominance of the ovulatory follicle and pregnancy rate in beef heifers. *J Reprod Fertil* 1994;102:123–130.
40. Austin E, Mihm M, Ryan M, Williams D, Roche J. Effect of duration of dominance of the ovulatory follicle on onset of estrus and fertility in heifers. *J Anim Sci* 1999;77:2219–2226.
41. Rhodes F, McDougall S, Burke C, Verkerk G, Macmillan K. Invited review: Treatment of cows with an extended postpartum anestrous interval. *J Dairy Sci* 2003;86:1876–1894.
42. Yavas Y, Walton J. Postpartum acyclicity in suckled beef cows: a review. *Theriogenology* 2000;54:25–55.
43. Gümen A, Guenther J, Wiltbank M. Follicular size and response to Ovsynch versus detection of estrus in anovular and ovular lactating dairy cows. *J Dairy Sci* 2003;86:3184–3194.
44. Nation D, Morton J, Cavalieri J, Macmillan K. Factors associated with the incidence of "Phantom cows" in Australian dairy herds. *Proc NZ Soc Anim Prod* 2001;61:180–183.
45. Garverick H. Ovarian follicular cysts in dairy cows. *J Dairy Sci* 1997;80:995–1004.
46. Peter A. An update on cystic ovarian degeneration in cattle. *Reprod Domest Anim* 2004;39:1–7.
47. Ambrose D, Schmitt E, Lopes F, Mattos R, Thatcher W. Ovarian and endocrine responses associated with the treatment of cystic ovarian follicles in dairy cows with gonadotropin releasing hormone and prostaglandin F2alpha, with or without exogenous progesterone. *Can Vet J* 2004;45:931–937.
48. Gümen A, Wiltbank M. An alteration in the hypothalamic action of estradiol due to lack of progesterone exposure can cause follicular cysts in cattle. *Biol Reprod* 2002;66:1689–1695.
49. Colazo M, Martinez M, Kastelic J, Mapletoft R, Carruthers T. The ischiorectal fossa: an alternative route for the administration of prostaglandin in cattle. *Can Vet J* 2002;43:535–541.
50. Brown L, Odde K, King M, LeFever D, Neubauer C. Comparison of melengestrol acetate–prostaglandin F2$_\alpha$ to Syncro-Mate B for estrus synchronization in beef heifers. *Theriogenology* 1988;30:1–12.
51. Deutscher G. Extending interval from seventeen to nineteen days in the melengestrol acetate–prostaglandin estrous synchronization program for heifers. *Prof Anim Sci* 2000;16:164–168.
52. Lamb G, Nix D, Stevenson J, Corah L. Prolonging the MGA–prostaglandin F2$_\alpha$ interval from 17 to 19 days in an estrus synchronization system for heifers. *Theriogenology* 2000;53:691–698.
53. Zimbelman R, Smith L. Control of ovulation in cattle with melengestrol acetate. I. Effect of dosage and route of administration. *J Reprod Fertil* 1966;11:185–191.
54. Kojima F, Bader J, Stegner J et al. Substituting EAZI-BREED CIDR inserts (CIDR) for melengestrol acetate (MGA) in the MGA Select protocol in beef heifers. *J Anim Sci* 2004;82(Suppl. 1):225 (Abstract).
55. Mallory D, Wilson D, Busch D, Ellersieck M, Smith M, Patterson D. Comparison of long-term progestin-based estrus synchronization protocols in beef heifers. *J Anim Sci* 2010;88:3568–3578.
56. Leitman N, Busch D, Mallory D et al. Comparison of long-term CIDR-based protocols to synchronize estrus in beef heifers. *Anim Reprod Sci* 2009;114:345–355.
57. Patterson D, Corah L, Brethour J. Response of prepubertal *Bos taurus* and *Bos indicus* × *Bos taurus* heifers to melengestrol acetate with or without gonadotrophin-releasing hormone. *Theriogenology* 1990;33:661–668.
58. Galvão K, Sá Filho M, Santos J. Reducing the interval from presynchronization to initiation of timed artificial insemination improves fertility in dairy cows. *J Dairy Sci* 2007;90:4212–4218.
59. Navanukraw C, Redmer D, Reynolds L, Kirsch J, Grazul-Bilska A, Fricke P. A modified presynchronization protocol improves fertility to timed artificial insemination in lactating dairy cows. *J Dairy Sci* 2004;87:1551–1557.
60. Herlihy M, Giordano J, Souza A et al. Presynchronization with Double-Ovsynch improves fertility at first postpartum artificial insemination in lactating dairy cows. *J Dairy Sci* 2012;95:7003–7014.
61. Bridges G, Helser L, Grum D, Mussard M, Day M. Decreasing the interval between GnRH and PGF2a from 7 to 5 days and lengthening proestrus increases timed-AI pregnancy rates in beef cows. *Theriogenology* 2008;69:843–851.
62. Kasimanickam R, Day M, Rudolph J, Hall J, Whittier W. Two doses of prostaglandin improve pregnancy rates to timed-AI in a 5-day progesterone-based synchronization protocol in beef cows. *Theriogenology* 2009;71:762–767.
63. Whittier W, Kasimanickam R, Currin J, Schramm H, Vlcek M. Effect of timing of second prostaglandin F 2 alpha administration in a 5-day, progesterone-based CO-Synch protocol on AI pregnancy rates in beef cows. *Theriogenology* 2010;74:1002–1009.
64. Bridges G, Ahola J, Brauner C et al. Determination of the appropriate delivery of PGF2$_\alpha$ in the 5-day CO-Synch + CIDR protocol in suckled beef cows. *J Anim Sci* 2012;90:4814–4822.
65. Xu Z, Burton L, Macmillan K. Treatment of postpartum anestrous dairy cows with progesterone, oestradiol and equine chorionic gonadotrophin. *NZ Vet J* 1997;45:205–207.
66. Bryan M, Bó G, Mapletoft R, Emslie F. The use of equine chorionic gonadotropin in the treatment of anestrous dairy cows in gonadotropin-releasing hormone/progesterone protocols of 6 or 7 days. *J Dairy Sci* 2013;96:122–131.
67. Baruselli P, Reis E, Marques M, Nasser L, Bó G. The use of hormonal treatments to improve reproductive performance of anestrous beef cattle in tropical climates. *Anim Reprod Sci* 2004;82–83:479–486.
68. Cutaia L, Tribulo R, Moreno D, Bó G. Pregnancy rate in lactating beef cows treated with progesterone-releasing devices, estradiol benzoate and equine chorionic gonadotropin (eCG). *Theriogenology* 52003;9:216 (Abstract).
69. Veneranda G, Filippi L, Racca D, Cutaia L, Bó G. Pregnancy rates in dairy cows treated with intravaginal progesterone devices and GnRH or estradiol benzoate and eCG. *Reprod Fertil Dev* 2008;20:91 (Abstract).

Chapter 34

Pregnancy Diagnosis: Rectal Palpation

David Christiansen

Department of Pathobiology and Population Medicine, College of Veterinary Medicine, Mississippi State University, Starkville, Mississippi, USA

Introduction

Pregnancy diagnosis is an invaluable tool to the cattle producer who is trying to maintain maximum profitability. In the case of the beef cattle herd, a significant portion of the cost of producing calves is related to the feeding and care of the pregnant dam particularly during the winter months. Cows that are open (not pregnant) will add to the cost of maintaining the herd while not providing income in support of the producer. Similarly in dairy cattle, a cow that fails to become pregnant in a timely fashion will ultimately produce less milk, thus adding cost but not enhancing income. A pregnancy test that would identify pregnant and open cows before they return to estrus, thus allowing intervention and rebreeding by the subsequent estrus, would help minimize the calving interval and thereby maximize profit especially in a program utilizing artificial insemination (AI). In addition, the ideal pregnancy test would be minimally invasive, simple to perform requiring little investment in equipment, supplies and time, yield quick results allowing cowside decisions to be made by the producer, and have high sensitivity and specificity. Finally, in many herds the ability to estimate the stage of gestation, the gender of the fetus, and the presence of pathologic conditions of the reproductive organs would be very helpful.

Currently there are three reliable methods of diagnosing the pregnancy status of a cow: transrectal palpation, transrectal ultrasound examination, and endocrine testing. While none of these meet all the criteria for an ideal test, each has its advantages and disadvantages. The producer and veterinarian must carefully consider each test and determine which provides the most benefit for the conditions at hand. Of these methods, transrectal palpation is the most commonly utilized and meets the most criteria under the most circumstances (Table 34.1).

The diagnosis of pregnancy by transrectal palpation is a skill that can readily be acquired and honed to great sensitivity and specificity with good understanding of normal anatomy and physiology of the cow. A good understanding of pregnancy and fetal development also allows the palpator, with experience, to accurately estimate the stage of pregnancy. A thorough discussion of the normal anatomy and physiology and of pregnancy can be found in Chapters 21 and 25. With this understanding and practice, a palpator can achieve very consistent results approaching 100% accuracy as early as day 30–35 of pregnancy.

Transrectal examination of the reproductive tract begins with identification of the cervix, which is fibrocartilagenous and firm (sometimes described as feeling somewhat like a chicken neck) and serves as the landmark for the reproductive organs. In the nongravid or early pregnant uterus, the cervix is typically found within or just at the brim of the pelvis. In multiparous cows and as pregnancy progresses, it may be found pulled (by the weight of the uterus and contents) more cranial and ventral. It is important to remember that the cervix is relatively mobile and one may have to sweep the pelvic canal and work over the brim of the pelvis until it is found. Once identified, the cervix may be grasped and used to retract the body, horns and uterus caudally to allow better access to, and examination of, these structures. In addition, the palpator can follow the cervix cranially to the body of the uterus and identify the bifurcation of the horns. At this location, the ventral intercornual ligament can be used to retract the horns and lay them along the floor of the pelvis to facilitate thorough palpation of each horn. Just beyond the tip of each horn, the ovaries can be palpated. As pregnancy progresses, the size and weight of the fetus and fetal fluids will make retraction of the cervix and uterus impossible and they will have to be examined in place.

The definitive diagnosis of pregnancy is based on the identification of structures which can only occur if the cow is indeed pregnant; if none of these can be identified, the cow should be considered not pregnant or the diagnosis deferred. Four such structures can be readily palpated and

Table 34.1 A comparison of different methods of pregnancy diagnosis.

Criteria	Endocrine testing	Ultrasound examination	Transrectal palpation
Allows intervention prior to first estrus after breeding	Yes	No	No
Minimally invasive	Minimal to moderate	Yes	Yes
Simple to perform, little investment/cost	Moderate to high	Moderate	Simple, minimal cost
Results available quickly	Minutes to days	Instant	Instant
Estimate stage of gestation	No	Yes	Yes
Gender determination	No	Yes	No
Identify pathologic conditions of reproductive tract	No	Yes	Yes
Sensitivity/specificity	Varies (Chapter 35)	Excellent	Excellent

are considered the cardinal signs of pregnancy: chorioallantoic (membrane) slip, the amniotic vesicle, placentomes, and the fetus.

In addition, other changes in the reproductive tract can be palpated that are considered supportive of a positive diagnosis of pregnancy. These include the inability to retract the cervix, fluid within or fluctuance of the uterus, enlarged diameter and asymmetry of the uterine horn(s), and enlargement and fremitus (a buzzing sensation) of the middle uterine artery.

Cardinal signs of pregnancy

Membrane slip

Membrane slip can be palpated routinely by 30–32 days of pregnancy and sometimes as early as 28 days in heifers in the gravid horn and in both horns by days 50–60 of gestation. Following identification of the cervix and retraction of the uterus, each horn can be grasped gently between the fingers and thumb and lifted slightly. If pregnant, a distinct popping sensation can be felt as the membrane slips from the grasp within the uterine walls. At 32 days of pregnancy it gives the impression of a thread and by day 45 a small string slipping from the fingers in the gravid horn. By day 60, it is palpable in both horns as a somewhat larger string and by day 75 is approaching the feel of a piece of yarn slipping from the fingers. Membrane slip is most commonly used up to day 90 of gestation.

Amniotic vesicle

The amniotic vesicle is a very turgid, fluid-filled sac surrounding and protecting the embryo proper during early gestation. It can be palpated as a small knot or bump within the uterine horn approximately 6–7 mm in diameter by 32–35 days (5 weeks) of gestation. By 6 weeks of gestation it will be 1.5 cm in diameter and at 7 weeks it will be 3.5–5 cm in diameter. By 8 weeks, the vesicle will be 6–7 cm, will be losing its turgidity, generally allowing palpation of the fetus around day 60. As this occurs, the vesicle itself will become more difficult to identify and will generally no longer be used to determine pregnancy.

Fetus

As the fetus becomes palpable around days 55–60, it will be approximately 5–6 cm in length and can be felt floating in the fluid of the gravid horn by carefully cupping the horn with the fingers and sliding them along the length of the pregnant horn. Alternatively, the wall of the uterus may be lightly tapped by the fingers creating a wave of fluid, and causing the fetus to bump back against the fingers (referred to as ballottement). As the pregnancy advances, fetal parts may more readily be identified such as the head, feet and legs allowing the beginner to be more comfortable with the diagnosis. By 3 months of pregnancy, the uterus and fetus will reach sufficient size and weight to cause it to drop over the brim of the pelvis and into the abdominal cavity and by 4–5 months of pregnancy the fetus may descend into the abdomen far enough to make palpation of the fetus itself difficult or impossible, particularly in large deep-bodied cows. Usually by 7 months, the fetus has grown large enough that it will be readily accessible and can once again be used to determine pregnancy status.

Placentomes

Placentomes (occasionally referred to as "buttons" by lay personnel) are composed of the maternal caruncle and fetal cotyledon. With experience they may be palpated as early as 75–80 days of gestation as soft pea-sized swellings or bumps on the internal uterine wall, but are a bit more firm and readily palpated by day 90. At this time they are about the size of a dime and will reach about the size of a half dollar by the time the cow is pregnant 5 months. During mid gestation as the fetus becomes more difficult to reach, placentomes may become the best method to definitively diagnose pregnancy. At least three should be identified, ensuring that the ovaries are not mistaken for placentomes.

Supportive signs of pregnancy

As a pregnancy advances to 60 days and beyond, the weight of the fetus and fetal fluids pulls the cervix cranially over the edge of the pelvis and into the abdomen. It becomes increasingly difficult at this time to retract the uterus into the pelvic canal; thus if the cervix feels "heavy" and relatively immobile, pregnancy can immediately be suspected and the palpator can search for the cardinal signs found in more advanced pregnancy such as the fetus and placentomes. This can save time and effort that would be required to find the more subtle membrane slip or amniotic vesicle.

Fluid within or fluctuance of the uterus is sometimes used in early pregnancy (as early as 28 days) to make a presumptive diagnosis of pregnancy but the owner must clearly understand that it is presumptive only and should be confirmed at a later date. The wall of the pregnant uterus at this time thins significantly, resulting in a structure which feels much like a partially filled water balloon. This fluid, along with the developing fetus, results in an enlarged diameter of the gravid uterine horn and later, as the pregnancy progresses, both uterine horns. This asymmetry early in gestation and the enlarged uterine horn(s) are highly suggestive of a pregnant cow.

With the development of the fetus and growth of the fetoplacental unit, there must be a corresponding increase in blood delivery to the uterus. This requires considerable enlargement of the uterine arteries and by 4–5 months of gestation not only can the increased diameter of the artery supplying the gravid horn be appreciated but fremitus (a buzzing or vibrating sensation) as well. The uterine arteries originate from the aorta just caudal to the internal iliac artery and are freely movable as they traverse the broad ligament, allowing them to be readily differentiated from the internal iliac arteries (again the reader is referred to Chapter 21).

Misdiagnosis and differential diagnoses

Failure to properly determine the pregnancy status of a cow is most commonly the result of an incomplete examination or inexperience. This can be exacerbated by either the desire for speed or fatigue. Therefore it is important, especially when first learning to diagnose pregnancy, to develop a systematic routine to be followed at all times and to allot adequate time for the procedure, with additional time for rest if needed to prevent fatigue from becoming a deleterious factor. It should be reiterated that in the absence of one of the four cardinal signs of pregnancy, a definitive diagnosis should not be made and the temptation to base a decision on supportive signs of pregnancy must be resisted. Indeed, these signs are not considered definitive indications of pregnancy because they may be induced by other conditions. Adhesions from a number of disease processes may fix the cervix and uterus in location. Fluid in the uterus may be the result of a pyometra or mucometra, rather than fetal fluid. Along with causing fluctuance of the uterus, this fluid may make the cervix and uterus difficult to retract in addition to causing asymmetry and/or enlargement of the uterine horns. Physometra (air within the uterus) may also cause asymmetry and enlargement of the uterus. Typically, however, in these cases the uterine walls do not thin as they would in pregnancy, yielding somewhat more tone to the feel of the uterus. Tumors may also make the cervix difficult to retract and cause asymmetry of the horns. These pathologic conditions may also cause enlargement and, albeit rarely, fremitus of the middle uterine arteries. Fremitus and uterine artery enlargement may also persist for a short time following delivery of a calf.

The uterus is a site of predilection for tumors caused by bovine leukemia virus and, in addition to altering the supporting signs of pregnancy, these and other tumors may be difficult to differentiate from a fetus. With careful palpation, it can be noted that this mass has no identifiable fetal parts such as the head or legs and is not floating within the normal fetal fluids but is in intimate contact with the endometrium. Similarly, a mummified or macerated fetus can be palpated as a mass within the uterus and although fetal parts may be identified, once again there is no amniotic or allantoic fluid surrounding the fetus. The uterine walls are contracted down around the remnant of the fetus and it is not freely movable with the uterine lumen. In addition, a macerated fetus often has a characteristic grinding or crepitus on palpation believed to be the body parts and bones rubbing against one another during manipulation. Cows having an enlarged cervix may occasionally be misdiagnosed as pregnant by the inexperienced examiner as the cervix may approach softball size and be very firm, feeling a bit like the fetal head or pelvis. The postpartum uterus is also enlarged, may be asymmetric, and may lead one to suspect pregnancy. However, the uterine walls will be very thick, the cardinal signs cannot be identified, and often rugae (ridges or folds of the surface of the uterus as it begins to shrink and involute following parturition) may be palpated.

Various structures not associated with the uterus may also occasionally be mistaken for pregnancy. The dorsal sac of the rumen can extend near or into the pelvic canal, particularly in a cow restrained in a hydraulic squeeze chute and is sometimes mistaken for a large fetus by the beginner. It should be noted that it is much more "doughy" than the fetus. The right kidney may be bumped and may feel somewhat similar to ballottement of a fetus; however, it is located more dorsally than a calf would normally be. The bladder is also sometimes mistaken for a fluid-filled uterus and must be differentiated from a pregnancy. Finally, tumors or other masses may be misdiagnosed as a fetus. With care and patience, a thorough examination of all of these will allow the clinician to determine that these are not associated with the reproductive tract and that the cardinal signs of pregnancy are not associated with these structures.

Determining the stage of gestation

Once a cow has been determined to be pregnant, both the cardinal and supportive signs of pregnancy can be utilized to make a determination as to the stage of her gestation. This determination can be very accurate (within a few days) during the first trimester of pregnancy as fetal development through this time is relatively consistent regardless of breed or other genetic and environmental factors. Fetal development begins to vary somewhat among individuals through mid gestation and relatively more during the final trimester. Indeed, in the final trimester as the fetus goes through rapid growth, the examiner often has no precise way of predicting the size of the calf at birth. As birthweight may vary from less than 25 to more than 50 kg depending on breed, maternal and paternal factors, as well as environmental factors, estimation of gestation length at this time becomes somewhat less precise than earlier in gestation. However, if the cattle are from a purebred operation or dairy farm, the examiner can have a reliable frame of reference from which to work. In a commercial herd, the size and type of the dam and any information about the type and breed of bulls being used will be helpful in achieving the highest degree of accuracy possible. The fingers and hand can be measured and used to evaluate the size of the structures being examined to further enhance the degree of accuracy.

Early in gestation, the size of the amniotic vesicle, the magnitude of membrane slip, and the diameter of the horn can be used to determine the stage of gestation. As the pregnancy approaches 60 days and the fetus becomes palpable, it becomes another tool which can be used along with horn diameter. The size of the fetus as well as the length of the head (nose to crown) may both be utilized. By 4.5–5 months of pregnancy the uterus is pulled deep into the abdomen of the cow and the fetus itself often may not be palpated. The development of placentomes by around days 75–80 provides another aid. While the size of the placentomes is very useful, just their presence combined with the inability to reach the fetus places the cow in mid gestation. Placentome size is most consistent just anterior to the cervix. The diameter of the uterine artery(ies) may also be measured reliably at this time and the advent of fremitus ipsilateral to the pregnant horn indicates a pregnancy of about 4 months. Bilateral fremitus is usually present by about 6 months of pregnancy. As the fetus reaches 7 months of age, it will begin ascending once again allowing feet, legs and head to be examined. It is important to note that at this time the placentomes become more variable in size and less reliable indicators of gestation length. Table 34.2 lists organs and structures, where they may be located at different stages of gestation, and their approximate sizes as pregnancy develops.

Risks associated with transrectal palpation

Although palpation per rectum is a relatively noninvasive procedure, it is not without some concerns. These include the possibility of disease transmission or causing damage to the embryo or fetus especially during early gestation. The diseases most frequently mentioned are bovine leukosis virus (BLV) and bovine viral diarrhea virus (BVDV). Studies of both have met with mixed results. BLV has been shown experimentally to be spread when blood from infected cattle was repeatedly introduced rectally to noninfected cattle.[1] However, the same researchers in another study showed no transmission of the virus in a commercial dairy when the same glove was used on multiple cows compared with using a new glove for every cow. In this same report, they demonstrated transmission of BLV in a university-owned teaching herd apparently from the use of common gloves. In this instance, it seems likely that more blood and trauma from inexperienced palpators and repeated palpations contributed to the seroconversion of these cows.[2] Conversely, in a small study, Kohara et al.[3] demonstrated seroconversion in three of four steers palpated a single time with a common glove following palpation of a positive cow. In addition, BLV-negative cows palpated twice weekly for 9 weeks with obstetric sleeves that were first used for aggressive palpation (to the point there was mucus and blood on the sleeves) of two BLV-positive cows resulted in all the negative cows developing BLV antibodies.[4] Thus it may be prudent to change or wash a sleeve if there is blood on the palpation sleeve prior to examining the subsequent cow in herds in which BLV is a concern.

Lang-Ree et al.[5] tied 10 BVDV-negative heifers side by side with no partition between them. They then palpated a persistently infected (PI) heifer and using the same sleeve palpated the first three heifers in the line. With a new sleeve, they again palpated a PI heifer and examined heifers 5 to 7. Again changing to a new sleeve, heifers 9 and 10 were palpated following the same procedure. All eight of the palpated heifers showed mild clinical signs of BVDV and seroconverted within 2 weeks of the examinations. In addition, the virus was isolated from five of the eight palpated heifers. Blood was not observed at any time on the sleeve following palpation of the PI heifer. Heifers 4 and 8 (not palpated) remained asymptomatic and seronegative even though they were kept in close proximity with the others.[5] Consequently, in herds to be examined for pregnancy which are at high risk for having a PI cow, the clinician may want to take extra precautions to reduce the chance of transmitting the virus and inducing clinical signs or, more importantly, a PI calf.

Early embryonic death resulting from transrectal palpation has also been studied with variable results. Older studies seemed to indicate some level of iatrogenic abortion during early pregnancy diagnosis. Females palpated at or before 45 days of gestation had higher incidence of embryonic mortality than females examined later in gestation.[6] Similarly, fetal loss was greater in females palpated 35–51 days after insemination than in females palpated 52–70 days after insemination.[7] Decreased pregnancy rates were observed in females palpated twice between 42 and 46 days after breeding compared with females initially diagnosed pregnant by progesterone testing and first palpated at 90 days after breeding.[8] On the other hand, Vaillancourt et al.[9] found that though the rate of embryonic death was greater in females examined before 50 days of gestation, there was no conclusive indication that the increased embryonic loss was due to pregnancy diagnosis by fetal membrane slip. Conversely, Thurmond and Picanso[10] found that females palpated early in the gestational period from 28 to 42 days had a significantly lower probability ($P < 0.0001$) of early embryonic death compared with females palpated later in the 28–42 day interval. In two other studies, there was no significant difference ($P > 0.05$) in the incidence of embryonic death in females that underwent transrectal palpation and those that were not palpated.[11,12] A number of confounding factors have likely contributed to these differences. These include natural embryonic and fetal loss and the lack of a reliable noninvasive test to make the original pregnancy diagnosis. As these tests have improved, the degree of normal embryonic and fetal loss has become better understood, thus allowing for better design and evaluation of the more recent studies. In addition, one must take into consideration the number of examinations and stress placed on cattle in studies compared with normal production scenarios. As these factors may contribute to the varying

Table 34.2 Size of various reproductive organs and fetal structures which can be used to help determine the stage of gestation.

Gestation length (days)	Amniotic vesicle size (cm)	Membrane slip	Fetal body length (cm)	Fetal size	Placentome size	Fetal nose to crown length	Horn diameter	Uterine artery findings and diameter
30	0.6–0.7	Thread in gravid horn					Slightly asymmetric	3–4 mm
40	1.2–1.5	Small string in gravid horn					Slightly enlarged, asymmetric	
50	5–6						6 cm	
60	8–8.5	String in both horns	5.5–7	Mouse			7 cm	3–4 mm
70			10			1.5 cm		
80			14		Pea	3.5 cm		
90		Large string in both horns	15–16	Rat	Dime	5.5 cm	8–10 cm	
100			18–20			7.5 cm		
110			24–25			9 cm		
120			27–28	Small cat	Nickel–quarter	10.5 cm	15 cm	Fremitis on side of pregnant horn, 6 mm
150			36–38	Big cat	Half dollar		18 cm	Fremitis on side of pregnant horn, 9 mm
180			47–50	Beagle	Variable			Fremitus both horns, 10–12 mm
210			62–64	Springer spaniel				
240			73–80	Doberman				
270			85–105	German shepherd				10–15 mm

results of these studies,[13,14] it is likely that the more recent studies are more reliable and that routine transrectal palpation does not significantly contribute to early embryonic death or fetal demise when performed within normal management programs.

Conclusion

The cost of carrying open cows is significant to the beef producer and for dairy producers it is important to minimize the nonproductive time (dry period) for each cow to maximize profitability. Rectal palpation is currently the most commonly utilized method of diagnosing pregnancy in cattle. With experience, it can be performed rapidly, with very good accuracy and relatively inexpensively, not only to determine pregnancy status but stage of gestation as well. Indeed, Kasimanickam *et al.*[15] showed a very small error rate when transrectal palpation was used to separate cows that were 70 days pregnant by fixed-time artificial insemination (FTAI) from cows which were bred by a bull turned in 2 weeks following the FTAI and demonstrated increasing accuracy with the experience of the individual performing the examination. This capability allows the producer the opportunity to more accurately make crucial management decisions such as culling or knowing when to dry a cow off. However, recent data gathered by the National Animal Health Monitoring System indicate that pregnancy diagnosis by palpation was utilized in only 10.8% of beef cattle operations with less than 50 cows, 25.8% of operations with 50–99 head, 41.2% of operations with 100–199 head, and 58.3% of farms with 200 or more cows.[16] On the other hand, about two-thirds of all dairy operations (67.0%) performed pregnancy examinations monthly or more frequently. The majority of large operations (75.0%) performed pregnancy examinations weekly or every 2 weeks, while 50.2% of small operations performed examinations on a monthly basis and 69.3% of mid-size operations performed examinations once or twice a month.[17] Thus there is a significant opportunity for veterinarians with expertise in pregnancy diagnosis and pregnancy staging by rectal palpation to provide a valuable profit-building service to their clients and the effort to develop this skill is well worth the investment.

References

1. Hopkins S, Evermann J, DiGiacomo R et al. Experimental transmission of bovine leukosis virus by simulated rectal palpation. *Vet Rec* 1988;122:389–391.
2. Hopkins S, DiGiacomo R, Evermann J, Christensen J, Deitelhoff D, Mickelsen W. Rectal palpation and transmission of bovine leukemia virus in dairy cattle. *J Am Vet Med Assoc* 1991;199:1035–1038.

3. Kohara J, Konnai S, Onuma M. Experimental transmission of bovine leukemia virus in cattle via rectal palpation. *Jpn J Vet Res* 2006;54:25–30.
4. Wentink G, van Oirschot J, Pelgrim W, Wensing T, Gruys E. Experimental transmission of bovine leukosis virus by rectal palpation. *Vet Rec* 1993;132:135–136.
5. Lang-Ree J, Vatn T, Kommisrud, Loken T. Transmission of bovine viral diarrhoea virus by rectal examination. *Vet Rec* 1994;135:412–413.
6. Paisley L, Mickelson W, Frost O. A survey of the incidence of prenatal mortality in cattle following pregnancy diagnosis by rectal palpation. *Theriogenology* 1978;9:481–491.
7. Abbitt B, Ball L, Kitto G et al. Effect of three methods of palpation for pregnancy diagnosis per rectum on embryonic and fetal attrition in cows. *J Am Vet Med Assoc* 1978;173:973–977.
8. Franco O, Drost M, Thatcher M, Shille V, Thatcher W. Fetal survival in the cow after pregnancy diagnosis by palpation per rectum. *Theriogenology* 1987;27:631–644.
9. Vaillancourt D, Bierschwal C, Ogwu D et al. Correlation between pregnancy diagnosis by membrane slip and embryonic mortality. *J Am Vet Assoc* 1979;175:466–468.
10. Thurmond M, Picanso J. Fetal loss associated with palpation per rectum to diagnose pregnancy in cows. *J Am Vet Med Assoc* 1993;203:432–435.
11. Alexander B, Johnson M, Guardia R, Van de Graaf W, Senger P, Sasser R. Embryonic loss from 30 to 60 days post breeding and the effect of palpation per rectum on pregnancy. *Theriogenology* 1995;43:551–556.
12. Romano J, Thompson J, Forrest D, Westhusin M, Tomaszweski M, Kraemer D. Early pregnancy diagnosis by transrectal ultrasonography in dairy cattle. *Theriogenology* 2006;66:1034–1041.
13. Romano J, Thompson J, Kraemer D, Forrest D, Westhusin M, Tomaszweski M. Early pregnancy diagnosis by palpation per rectum: influence on embryo/fetal viability in dairy cattle. *Theriogenology* 2007;67:486–493.
14. Thompson J, Marsh W, Calvin J, Etherington W, Momont H, Kinsel M. Pregnancy attrition associated with pregnancy testing by rectal palpation. *J Dairy Sci* 1994;77:3382–3387.
15. Kasimanickam R, Whittier W, Tibary A, Inman B. Error in pregnancy diagnosis by per-rectal palpation in beef cows. *Clin Theriogenol* 2011;3:43–47.
16. National Animal Health Monitoring System. Bull management. In: *Beef 2007–08, Part 2. Reference of Beef Cow-Calf Management Practices in the United States*. Fort Collins, CO: USDA, 2009.
17. National Animal Health Monitoring System. Reproductive Practices on U.S. Dairy Operations, 2007. www.aphis.usda.gov/animal_health/nahms/dairy/downloads/dairy07/Dairy07_is_ReprodPrac.pdf

Chapter 35

Biochemical Pregnancy Diagnosis

Amanda J. Cain and David Christiansen

Department of Pathobiology and Population Medicine, College of Veterinary Medicine, Mississippi State University, Starkville, Mississippi, USA

Introduction

Prompt accurate diagnosis of pregnancy or nonpregnancy is critical for efficient livestock management, especially in the dairy industry where profitability relies largely on the reproductive efficiency of the cow herd.[1] Early identification of nonpregnant cows, particularly if they can be identified before the next expected estrus, is vital for maintenance of a desirable calving interval in both beef and dairy herds.

Traditionally, producers have relied heavily on transrectal palpation and/or transrectal ultrasound to diagnose pregnancy. Both of these methods require specialized training and technical skill and are usually performed by a veterinarian. In addition, neither can diagnose pregnancy prior to the estrus immediately following breeding. Additionally, many geographical areas are experiencing a shortage of large animal practitioners. This, coupled with the desire to identify nonpregnant animals as early after insemination as possible, has made the pursuit of reliable biochemical methods of pregnancy testing more attractive.[2] A number of biochemical tests for pregnancy in the bovine are currently available. These tests detect pregnancy-associated hormones or the presence of pregnancy-specific molecules in maternal circulation and are marketed as viable alternatives to transrectal palpation and transrectal ultrasonography. However, unlike transrectal palpation and transrectal ultrasound these tests by themselves cannot currently be used to closely estimate the stage of gestation.

When diagnosing pregnancy via transrectal palpation, reproductive structures are felt for positive signs of pregnancy. Most large animal veterinarians can accurately diagnose pregnancy as early as 30–35 days after insemination, but the level of accuracy is highly dependent on the expertise of the veterinarian and generally increases when examination is delayed to 40–50 days after breeding.[3] Concerns over early embryonic mortality as a result of transrectal palpation, particularly via fetal membrane slip, have been expressed and has been the subject of multiple studies. The risks associated with transrectal palpation for diagnosis of pregnancy are fully addressed in Chapter 34.

Through transrectal ultrasound, pregnancy can be consistently and accurately diagnosed as early as 25 days after insemination. Maximum sensitivity in diagnosing a pregnancy is reached at day 26 and day 29 in heifers and cows respectively.[4,5] For this reason, transrectal ultrasound is often used as the "gold standard" when evaluating other means of pregnancy diagnosis. Although 25 days after insemination provides a diagnosis earlier than transrectal palpation, transrectal ultrasound still does not allow diagnosis of nonpregnancy before the next expected estrus and, like transrectal palpation, requires that the veterinarian has a great deal of technical skill in order to yield consistently accurate diagnoses. Furthermore, this method of pregnancy determination requires substantial inputs to purchase and maintain necessary equipment, and these costs must be passed on to the producer.

The ideal pregnancy test would accurately diagnose both pregnancy and nonpregnancy a short time after fertilization, have excellent sensitivity (the female that tests pregnant is indeed pregnant), specificity (the female that tests nonpregnant is indeed open), and accuracy (the proportion of true results). Furthermore, the ideal pregnancy test would be noninvasive in that there would be no risk of causing embryonic or fetal death through testing. In addition, it would be minimally expensive, require little or no technical skill, and could be quickly and easily performed chute-side. An ideal pregnancy test for the bovine would be very similar to the human chorionic gonadotropin (hCG) test for pregnancy in women.

The hCG test is a simple ELISA (enzyme-linked immunosorbent assay) that detects the presence of hCG protein in the urine of pregnant women. It can easily and inexpensively be performed in the home, requiring virtually no technical skill. Commercially available hCG tests boast extremely high accuracy (>99%), and some can be performed as early as

---

Bovine Reproduction, First Edition. Edited by Richard M. Hopper.
© 2015 John Wiley & Sons, Inc. Published 2015 by John Wiley & Sons, Inc.

8–10 days after conception.[6] Unfortunately, cows and heifers do not produce hCG or any analogous molecule that could be tested in a comparable fashion. Instead, tests have been developed to ascertain the presence of certain pregnancy-associated molecules such as elevated progesterone concentration, pregnancy-associated glycoproteins (PAGs), early pregnancy factor (EPF), and interferon-stimulated genes (ISG).[2] Additionally, as technology and scientific knowledge of early pregnancy advance, there are a number of other molecules that may one day be used as pregnancy biomarkers in the bovine.

An important distinction that producers should be mindful of when opting to utilize biochemical methods to determine pregnancy status is that progesterone, PAG, EPF, and ISG assays are more accurate in detecting nonpregnant females than in correctly identifying pregnancy. Consequently, endocrine-based pregnancy tests often have excellent specificity (>90%) but slightly less desirable sensitivity. This is of significant economic importance because females in intensive production scenarios that are falsely identified as open will often be administered prostaglandin in an attempt to initiate a new cycle and reduce time to next insemination. This injection of prostaglandin will of course induce abortion in pregnant females and since the average cost of a pregnancy loss in a dairy female in the United States is estimated at $555, there is significant incentive to minimize the incidence of iatrogenic embryonic and fetal loss.[7]

Similarly, another confounding factor when attempting to identify pregnancy soon after fertilization is natural pregnancy loss. Bovine fertilization rates are usually high (>90%), and therefore the main contributor to reproductive wastage is embryonic or fetal death.[8,9] Early embryonic death, occurring prior to day 17 of pregnancy, accounts for 20.5–43.6% of loss when coupled with frequency of failure of fertilization.[10] However, late embryonic death, which represents loss from approximately 17 to 42 days of gestation, is estimated to be, on average, 12.8% in dairy cows and approximately 6.5% in beef females.[11–13] It has been estimated that pregnancy losses from fertilization to term in high-producing dairy cows may approach 60%.[13] When critically evaluating the reliability of endocrine tests performed early in gestation, it is customary to confirm pregnancy diagnosis by transrectal palpation or ultrasound days or weeks after the initial test was performed. This creates a window in which embryonic or fetal loss can occur, thus skewing the data by incorrectly labeling some results as false positives as the female was indeed pregnant when first tested but lost the pregnancy before the confirmatory pregnancy evaluation.

Progesterone concentration

Serum and milk progesterone levels steadily increase following ovulation as a functional corpus luteum (CL) grows and develops. In nonpregnant females, progesterone declines as the CL regresses and the cow returns to estrus. In pregnant females, however, progesterone levels should remain elevated during the period of the next expected estrus as the CL continues to produce progesterone in order to maintain the pregnancy.

Progesterone concentration testing functions on the premise that serum and/or milk progesterone concentration will remain elevated at the time of the next expected estrus if the female were pregnant. If the female is open, progesterone concentration will decrease to baseline as the CL regresses and a new dominant follicle develops[3] (see Chapters 23 and 24).

Discrepancies in progesterone testing may be largely caused by natural variation between individuals in luteal lifespan and estrous cycle length as progesterone levels remain high while a CL is present during diestrus.[14] Furthermore, embryonic mortality between the time of progesterone testing and pregnancy confirmation via transrectal palpation or transrectal ultrasound examination may further contribute to perceived test error.[3]

A potential drawback to the use of progesterone testing in production settings is that the breeding or insemination date must be known in order to obtain reliable results from a single sample, and therefore the applicability of progesterone testing is somewhat limited.[2,3] Typically, a milk or serum sample is taken 21–24 days after breeding and is analyzed using either radioimmunoassay (RIA), ELISA or, less frequently, latex agglutination (LA). A number of progesterone ELISAs for serum and milk are available to producers and are marketed as cowside tests.

Progesterone testing is very reliable for identifying nonpregnant females, as cows with blood progesterone levels less than 1 ng/mL 21 days after insemination cannot be pregnant, but it is slightly less reliable for positively diagnosing pregnancy because of natural variations in estrous cycle lengths between individual animals.[2] Detection rates of 95–100% have been reported in detecting nonpregnancy. Pregnancy was correctly identified 68–97% of the time, with study designs varying in the day of gestation the test was performed, the standard it was compared against, and the type of assay used (ELISA, RIA, or LA).[15–20]

Pregnancy-specific protein B

Ruminant trophoblasts release proteins as a part of maternal recognition of a viable conceptus. These proteins, known as PAGs, are aspartic peptidases and have not been found to be proteolytically active. The first PAG to be discovered was pregnancy-specific protein B (PSPB), formerly known as pregnancy-associated glycoprotein 1. This protein is produced by trophoblast binucleate cells in the bovine placenta around the time of implantation, though its exact role in pregnancy maintenance is not yet fully understood. It can be found in maternal circulation from the time of implantation throughout gestation, often persisting well into the postpartum interval.[21,22]

As its name suggests, PSPB is a pregnancy-specific molecule and is not found in the serum of virgin heifers or open cows, thus demonstrating that the protein is produced solely as a response to a conceptus and not in response to luteal activity during the estrous cycle.[23] Unlike equine chorionic gonadotropin (eCG) and hCG, PSPB is not detectable in urine, and it is only found in milk during intervals when PSPB is at very high concentration in plasma, for example shortly after parturition.[21]

An RIA capable of reliably detecting PSPB in maternal serum as early as 24 days after conception was developed by Sasser et al.[23] in 1986. A then unique advantage of PSPB testing was that, unlike progesterone assays, insemination dates are not needed to confirm pregnancy in PSPB testing as the protein is pregnancy specific and present throughout gestation, peaking around the time of parturition.[23]

In addition to the RIA test, there is also a commercially available ELISA for PSPB (BioPRYN™; Biotracking, Moscow, ID, USA). This assay is marketed in both the United States and abroad as a convenient and inexpensive pregnancy test for use by both producers and veterinarians, and furthermore is marketed as being as accurate as transrectal ultrasound examination. It can be used as early as 30 days after insemination in cows and 28 days after insemination in heifers and uses identical concentrations for the determination of pregnancy or nonpregnancy as the RIA.[24,25] When verified contemporaneously with transrectal ultrasound near the end of the late embryonic period, the PSPB ELISA yielded accuracies greater than 93%.[26,27]

The usefulness of PSPB in pregnancy diagnosis is detrimentally affected by the relatively long half-life of PSPB. PSPB can be detected in maternal blood up to 80–100 days after calving; therefore, producers must be cautious when using PSPB assays to detect pregnancy in cows bred early in the postpartum period.[23,28] In addition, this long half-life may result in false-positive tests for pregnancy following pregnancy loss. Finally, this test does not have the advantage of diagnosing nonpregnancy prior to the cow's first estrus following breeding.

Early pregnancy factor

EPF is an immunosuppressive hormone that can first be found at detectable levels in the maternal system 24–48 hours after successful fertilization and persists for at least the first half of gestation.[29] EPF acts as an immunomodulator protecting the antigenically foreign conceptus from maternal rejection.[30] It is produced by maternal tissues during the preimplantation stage in response to the presence of a conceptus. After implantation has occurred, however, the embryonic tissue assumes responsibility for EPF production. Only a viable conceptus that has reached the blastocyst stage is capable of producing EPF. Furthermore, EPF dissipates rapidly after the death or removal of the conceptus. Unlike some PAGs, including PSPB, EPF has never been detected after parturition.[29] The combination of these attributes makes EPF potentially very useful for pregnancy determination and recognition of embryonic fetal death.

Before the discovery of EPF and development of a test to determine its presence, it was impossible to detect a pregnancy prior to implantation except through invasive surgery. Although diagnosing pregnancy so soon after insemination seems appealing, the high frequency of early embryonic mortality makes this test less useful. Producers opting to use EPF to diagnose pregnancy early in gestation should retest all positive females at a later date to confirm the presence of a viable conceptus.

A producer-oriented assay for EPF would be a very valuable tool for the detection of fertilization and pregnancy in cows during the first 4 weeks of gestation as the unique properties of EPF make it particularly useful in intensive production scenarios. It is detectable within 48 hours of fertilization and, unlike PSPB, disappears rapidly on removal of the embryo or fetus, and therefore can be assayed accurately even in cows inseminated early in the postpartum interval that may have residual PSPB in circulation. Furthermore, an EPF assay would be far less likely than a PSPB assay to yield a false-positive result due to previous pregnancy loss.[31] Most importantly, however, an EPF test can accurately determine the pregnancy status of a female well before her next expected estrus.[29]

The first and currently the only reliable test for EPF is the rosette inhibition test (RIT) in which EPF, in the presence of a complement, works to cause an increase in the effects and titers of antilymphocyte serum in decreasing the formation of rosettes between lymphocytes and heterologous red blood cells.[29] The RIT for EPF is very accurate and extremely useful in identifying nonpregnant females. Unfortunately, the RIT test is time-consuming and difficult to maintain, and therefore it is not practical for regular use on the majority of commercial operations.[3]

A producer-oriented test to detect the presence of EPF has recently been developed (EDP Biotech Company, Knoxville, TN, USA). This test, the early conception factor (ECF) lateral flow assay, has been marketed as a cowside test for nonpregnancy and reportedly can detect EPF in milk or serum 48 hours to 20 days after fertilization.[32] Concerns regarding the accuracy of the ECF test performed early in gestation have been expressed. Ambrose et al.[33] reported that ECF specificity on 7, 14, and 21 days after insemination ranged from 38 to 51%. Data from other studies corroborate these findings and conclude that the ECF lateral flow assay cannot correctly identify open cows with the degree of precision needed by most commercial dairy operations.[34–36] However, the development of a more reliable assay for EPF could revolutionize early bovine pregnancy diagnosis, especially in the dairy industry.

Interferon-stimulated genes

In cattle, the molecule primarily responsible for the maternal recognition of pregnancy is interferon (IFN)-tau. It is secreted by conceptus trophectoderm cells and acts in a paracrine manner on uterine endometrium to inhibit pulsatile secretion of prostaglandin $(PG)F_{2\alpha}$, thereby effectively inhibiting luteolysis and successfully maintaining the pregnancy.[37] Unlike the molecules of maternal pregnancy recognition in some other mammalian species (i.e., hCG in humans and eCG in mares), IFN-tau is present only in very low concentrations in the maternal periphery and therefore cannot currently be directly tested for in serum or milk. Thus, maternal IFN-tau concentration cannot be used in diagnosing early pregnancy in the bovine.[38]

In addition to its role in interrupting the signal transduction pathway that regulates the release of $PGF_{2\alpha}$, IFN-tau also induces the upregulation of multiple differentially expressed ISGs on maternal leukocytes. Among these ISGs are myxovirus resistance gene 1 and 2 (Mx1 and Mx2) and interferon-stimulated gene 15-kDa protein (ISG15), a ubiquitin homolog.[38–41] ISG15 mRNA expression on maternal white blood cells follows a similar pattern to IFN-tau release, first being expressed around day 14 after conception, peaking between day 17 and 18, and declining from day 21 to day 26.[42] However, unlike IFN-tau, ISG15 is detectable in the maternal periphery in early pregnancy, and could potentially be useful in diagnosing nonpregnancy before the next expected return to estrus. Because of variability between individual females' white blood cell composition and differences in IFN-tau concentrations between conceptuses, ISG15, like most other molecules of biochemical pregnancy diagnosis, is more useful in detecting open cows than in affirmatively diagnosing pregnancy.

Though results generated in laboratories for the use of ISG15 in identifying nonpregnant females, especially heifers, have been promising, thus far a consumer-oriented test for ISG15 mRNA upregulation has not been developed, and real-time polymerase chain reaction (PCR) must be used to detect ISG15 mRNA in maternal serum.[38,43] Furthermore, serial blood tests are necessary to achieve maximum accuracy (>99.9%) when using ISG15 mRNA to identify nonpregnant cows and heifers. Accuracy in detecting nonpregnant cows from a single blood collection (day 18) was 89% in a 2006 study conducted by Han et al.[43] Further complicating the potential application of ISG15 as a means to early pregnancy determination is ISG15 activation as a response to interferon throughout the course of a viral infection. If relying solely on ISG mRNA expression levels for pregnancy detection, virally infected open females may be falsely diagnosed pregnant.[43] Though ISG15 mRNA is currently not feasible for use in early pregnancy determination in commercial operations, with refinement and further scientific advances this innovative new technology may one day be adapted for widespread use in diagnosing pregnancy and predicting embryonic mortality.

Emerging research

In addition to the four previously described pregnancy-associated molecules, which have been well tested and reviewed in scientific literature, there are a number of other peripheral indicators of pregnancy that may one day be useful for commercial pregnancy diagnosis in the bovine. As our knowledge of early gestation and early maternal responses to conception increases, more biomolecules with potential applications in pregnancy determination will be uncovered. As with existing biochemical pregnancy tests, these new and emerging tests will endeavor to detect the presence of a conceptus prior to the female's next expected estrus so that she may be rebred, if necessary, as quickly as possible.

Among the most promising molecules for potential future use in early bovine pregnancy determination are circulating nucleic acids. Circulating nucleic acids (CNAs) are cell-free DNA/RNA molecules circulating in the blood and have been used as biomarkers in both diseased and healthy humans and animals for a number of years. In humans, CNAs are primarily used in the diagnosis of certain tumor diseases and other disorders. During human pregnancy, CNAs are often used to screen for genetic abnormalities such as chromosomal aneuploidy.[44] Unlike human fetuses, bovine trophoblasts have no direct contact with maternal blood, and therefore exchange of fetal genetic material occurs less readily than it does in humans. However, it has been demonstrated that "transplacental leakage" of bovine fetal DNA into the maternal circulation can, and often does, occur in spite of the epitheliochorial placenta structure, though the mechanisms responsible for this exchange have not yet been elucidated.[45,46] Other studies have demonstrated that CNAs can be used to accurately (100% for male calves, 91% for female calves) determine calf sex after day 30 of gestation.[46,47] Mayer et al.[44,48] have recently characterized the differences in CNA composition between pregnant and open cows at day 20 of gestation, finding multiple sequences present exclusively in the serum of pregnant females. Although this particular application of CNA technology is still in its infancy, these findings indicate that it may one day be feasible to use CNAs to identify pregnant cows and heifers as early as 20 days after breeding.

Summary

Intensive reproductive management systems make early and accurate pregnancy determination a necessity in the cattle industry, especially on commercial dairy operations. Though producers have traditionally been limited to transrectal palpation and transrectal ultrasound for pregnancy diagnosis, a variety of biochemical pregnancy tests are rapidly becoming available. A summary of the tests discussed in this chapter and their attributes is provided in Table 35.1. All these tests have the potential to diagnose pregnancy or nonpregnancy earlier than traditional methods (transrectal palpation and transrectal ultrasound) while still maintaining a comparable level of accuracy. However, on their own, they do not have the capability to closely stage the pregnancy and may require some knowledge of the breeding or AI date.

Table 35.1 Summary of biochemical indicators presently used for the diagnosis of pregnancy in the bovine.

Molecule	Type of test	Producer test available?	Days of gestation effective	Significant drawbacks
Progesterone	RIA, ELISA, LA	Yes	21–285	Must know breeding date to use reliably in early gestation
PSPB	RIA, ELISA	Yes	28–285	Long half-life may skew results in females bred before 90 days post partum
EPF	RIT, lateral flow assay	Lateral flow assay not reliable	2–140	No reliable producer test
ISG15	RT-PCR	No	14–26	No reliable producer test; virally infected females may have false positives

Currently, only EPF or ISG15 would allow producer intervention (i.e., prostaglandin administration) prior to the first estrus after insemination. Unfortunately, RIT and real-time PCR, respectively, are required to accurately detect these molecules in maternal peripheral circulation. Both of these tests require expensive equipment and are difficult to run outside of a laboratory environment. Thus, the practicality of using these compounds for bovine pregnancy determination is currently limited. Producers and veterinarians should work together to make decisions regarding which pregnancy testing method to use based on the individual needs of each operation.

References

1. Plaizer J, King G, Dekkers J, Lissemore K. Estimation of economic values of indices for reproductive performance in dairy herds using computer simulation. *J Dairy Sci* 1997;80: 2775–2783.
2. Lucy M, Green J, Poock S. Pregnancy determination in cattle: a review of available alternatives. In: *Proceedings of Applied Reproductive Strategies in Beef Cattle, August 31 to September 1, 2011, Joplin, MO*. http://www.appliedreprostrategies.com/2011/Joplin/schedule.html
3. Sasser R, Ruder C. Detection of early pregnancy in domestic ruminants. *J Reprod Fert Suppl* 1987;34:261–271.
4. Fricke P. Scanning the future: ultrasonography as a reproductive management tool for dairy cattle. *J Dairy Sci* 2002;85: 1918–1926.
5. Romano J, Thompson J, Forrest D, Westhusin M, Tomaszweski M, Kraemer D. Early pregnancy diagnosis by transrectal ultrasonography in dairy cattle. *Theriogenology* 2006;66:1034–1041.
6. Marshall J, Hammond C, Ross G, Jacobson A, Rayford P, Odell W. Plasma and urinary chorionic gonadotropin during early human pregnancy. *Obstet Gynecol* 1986;32:760–764.
7. De Vries A. Economic value of pregnancy in dairy cattle. *J Dairy Sci* 2006;89:3876–3885.
8. Ayalon N. A review of embryonic mortality in cattle. *J Reprod Fertil* 1978;54:483–493.
9. Diskin M, Sreenan J. Fertilization and embryonic mortality rates in beef heifers after artificial insemination. *J Reprod Fertil* 1980;59:463–468.
10. Humblot P. Use of pregnancy specific proteins and progesterone assays to monitor pregnancy and determine the timing, frequencies and sources of embryonic mortality in ruminants. *Theriogenology* 2001;56:1417–1433.
11. Committee on Bovine Reproductive Nomenclature. Recommendations for standardizing bovine reproductive terms. *Cornell Vet* 1972;62:216–237.
12. Beal W, Perry R, Corah L. The use of ultrasound in monitoring reproductive physiology of beef cattle. *J Anim Sci* 1992;70:924–929.
13. Santos JE, Thatcher WW, Chebel RC, Cerri RL, Galvão KN. The effect of embryonic death rates in cattle on the efficacy of estrus synchronization programs. *Anim Reprod Sci* 2004;82–83:513–535.
14. Pennington J, Schultz L, Hoffman W. Comparison of pregnancy diagnosis by milk progesterone on day 21 and day 24 postbreeding: field study in dairy cattle. *J Dairy Sci* 1985;68:2740–2745.
15. Chang C, Estergreen V. Development of a direct enzyme immunoassay for milk progesterone and its application to pregnancy diagnosis in cows. *Steroids* 1983;41:173–195.
16. Laing J, Gibbs H, Eastman S. A herd test for pregnancy in cattle based on progesterone levels in milk. *Br Vet J* 1976;132:204–209.
17. Romangolo D, Nebel R. The accuracy of enzyme-linked immunosorbent assay and latex agglutination progesterone test for the validation of estrus bred early pregnancy diagnosis in dairy cattle. *Theriogenology* 1993;39:1121–1128.
18. Van de Wiel D, Koops W. Development and validation of an enzyme immunoassay for progesterone in bovine milk or blood plasma. *Anim Reprod Sci* 1986;10:201–213.
19. Waldman A. Enzyme immunoassay (EIA) for milk progesterone using a monoclonal antibody. *Anim Reprod Sci* 1993;34:19–30.
20. Zaied A, Bierschwal C, Elmore R, Youngquist R, Sharp A, Garverick H. Concentrations of progesterone in milk as a monitor of early pregnancy diagnosis in dairy cows. *Theriogenology* 1979;12:3–11.
21. Sasser R, Crock J, Ruder-Montgomery C. Characteristics of pregnancy-specific protein B in cattle. *J Reprod Fert Suppl* 1989;37:109–113.
22. Xie S, Low B, Nagel R et al. Identification of the major pregnancy specific antigens of cattle and sheep as inactive members of the aspartic proteinase family. *Proc Natl Acad Sci USA* 1991;88:10247–10251.
23. Sasser R, Ruder C, Ivani K, Butler J, Hamilton W. Detection of pregnancy by radioimmunoassay of a novel pregnancy-specific protein in serum of cows and a profile of serum concentrations during gestation. *Biol Reprod* 1986;35:936–942.
24. Howard J, Gabor G, Gray T et al. BioPRYN, a blood-based pregnancy test for managing breeding and pregnancy in cattle. In: *Proceedings, Annual Meeting, Western Section, American Society of Animal Science*, Vol. 58. Champaign, IL: American Society of Animal Science, 2007, pp. 295–298.
25. Sousa N, Ayad A, Beckers J, Gajewski Z. Pregnancy-associated glycoproteins (PAG) as pregnancy markers in the ruminants. *J Physiol Pharmacol* 2006;57(Suppl. 8):153–171.
26. Romano J, Larson J. Accuracy of pregnancy specific protein-B test for early pregnancy diagnosis in dairy cattle. *Theriogenology* 2010;74:932–939.
27. Silva E, Sterry R, Kolb D et al. Accuracy of a pregnancy-associated glycoprotein ELISA to determine pregnancy status of lactating dairy cows twenty-seven days after timed artificial insemination. *J Dairy Sci* 2007;90:4612–4622.
28. Kiracofe G, Wright J, Schalles R, Ruder C, Parish S, Sasser R. Pregnancy-specific protein B in serum of postpartum beef cows. *J Anim Sci* 1993;71:2199–2205.
29. Morton H. Early pregnancy factor (EPF): a link between fertilization and immunomodulation. *Aust J Biol Sci* 1984;37:393–407.
30. Morton H, Nancarrow C, Scaramuzzi R, Evison B, Clunie G. Detection of early pregnancy in sheep by the rosette inhibition test. *J Reprod Fertil* 1979;56:75–80.
31. Nancarrow C, Wallace A, Grewal A. The early pregnancy factor of sheep and cattle. *J Reprod Fert Suppl* 1981;30:191–199.
32. Whisnant C, Pagels L, Daves M. Case study: effectiveness of a commercial early conception factor test for use in cattle. *Prof Anim Sci* 2001;17:51–53.
33. Ambrose D, Radke B, Pitney P, Goonewardene L. Evaluation of early conception factor lateral flow test to determine nonpregnancy in dairy cattle. *Can Vet J* 2007;48:831–835.
34. Cordoba M, Sartori R, Fricke P. Assessment of a commercially available early conception factor (ECF) test for determining pregnancy status of dairy cattle. *J Dairy Sci* 2001;84:1884–1889.
35. DesCôteaux L, Carrière P, Bigras-Poulin M. Evaluation of the early conception factor (ECF™) dip stick test in dairy cows between days 11 and 15 post-breeding. *Bov Pract* 2000;34:87–91.
36. Gandy B, Tucker W, Ryan P et al. Evaluation of the early conception factor (ECF™) test for the detection of nonpregnancy in dairy cattle. *Theriogenology* 2001;56:637–647.
37. Bazer F, Spencer T, Ott, T. Interferon tau: a novel pregnancy recognition signal. *Am J Reprod Immunol* 1997;37:412–420.
38. Green J, Okamura C, Poock S, Lucy M. Measurement of interferon-tau (IFN-τ) stimulated gene expression in blood leukocytes for pregnancy diagnosis within 18–20 d after insemination in dairy cattle. *Anim Reprod Sci* 2010;121:24–33.
39. Austin K, Ward S, Teixeira M, Dean V, Moore D, Hansen T. Ubiquitin cross-reactive protein is released by the bovine uterus

in response to interferon during early pregnancy. *Biol Reprod* 1996;54:600–606.
40. Austin K, Carr A, Pru J et al. Localization of ISG15 and conjugated proteins in bovine endometrium using immunohistochemistry and electron microscopy. *Endocrinology* 2004;145:967–975.
41. Gifford C, Racicot K, Clark D et al. Regulation of interferon-stimulated genes in peripheral blood leukocytes in pregnant and bred, nonpregnant dairy cows. *J Dairy Sci* 2007;90:274–280.
42. Hansen T, Austin K, Johnson G. Transient ubiquitin cross-reactive protein gene expression in the bovine endometrium. *Endocrinology* 1997;138:5079–5082.
43. Han H, Austin K, Rempel L, Hansen T. Low blood ISG15 mRNA and progesterone levels are predictive of non-pregnant dairy cows. *J Endocrinol* 2006;191:505–512.
44. Mayer J, Soller J, Beck J et al. Early pregnancy diagnosis in dairy cows using circulating nucleic acids. *Theriogenology* 2013;79:173–179.
45. Turin L, Invernizzi P, Woodcock M et al. Bovine fetal microchimerism in normal and embryo transfer pregnancies and its implications for biotechnology applications in cattle. *Biotechnol J* 2007;2:486–491.
46. Lemos D, Takeuchi P, Rios A, Araújo A, Lemos H, Ramos E. Bovine fetal DNA in the maternal circulation: applications and implications. *Placenta* 2011;32:912–913.
47. Wang G, Cui Q, Cheng K, Zhang X, Xing G, Wu S. Prediction of fetal sex by amplification of fetal DNA present in cow plasma. *J Reprod Dev* 2010;56:639–642.
48. Mayer J, Beck J, Soller J, Wemheuer W, Schutz E, Brenig B. Analysis of circulating DNA distribution in pregnant and non-pregnant dairy cows. *Biol Reprod* 2013;88:29.

Chapter 36

Reproductive Ultrasound of Female Cattle

Jill Colloton

Bovine Services LLC, Edgar, Wisconsin, USA

Introduction

When O.J. Ginther[1] published *Ultrasonic Imaging and Animal Reproduction: Fundamentals* in 1995 field veterinarians began using ultrasound technology for reproductive examinations in the bovine. With the advent of small machines designed to be carried on the operator's person, its use has become mainstream. Many texts and other resources have been published describing the use of ultrasound for bovine reproductive examinations.[1-6] The goal of this chapter is not to reiterate those resources, but to expand on them.

Pregnancy diagnosis

It is theoretically possible to identify a conceptus by ultrasound as early as 21 days after breeding,[7] but that is usually not practical in field conditions. One proficient ultrasonographer tested his ability to accurately diagnose open versus pregnant cows beginning at 24 days after breeding.[8] The negative predictive value (diagnosis of not pregnant) was 89% at 24 days in adult cows. In other words, 11% of cows diagnosed open were found to be pregnant when rechecked at 32 days. By 26 days this ultrasonographer was able to identify open cows correctly in 99% of cows examined. In heifers he was able to achieve the same level of competency at 24 days after breeding.

Bovine ultrasonographers are advised to perform a similar analysis of their skills to identify their competence at various stages after insemination. The time it takes to perform examinations and the number of rechecks required at various stages should also be taken into consideration. Note the large difference in the amount of fluid and size of the embryo between 26 days (Figure 36.1a) and 28 days (Figure 36.1b). Note that the 26-day embryo is difficult to visualize against the uterine wall. The 28-day embryo, on the other hand, is 1 cm in length and easily seen within the uterine fluid. Embryonic development occurs very rapidly so waiting just a couple of days can improve accuracy and speed of pregnancy examinations. It is of no value to the producer if early pregnancy examinations do not lead to immediate management decisions.

Ovarian structures are helpful aids in pregnancy diagnosis. It can hasten the examination to know if a corpus luteum (CL) is present and, if so, on which side. Two corpora lutea indicate the potential for twins. A pregnancy with the CL contralateral to the embryo or fetus is very high risk for pregnancy loss. A small or poor-quality CL may indicate low progesterone levels that could lead to pregnancy loss. A cavitary CL is of no concern unless the cavity is so large and the rim of luteal tissue so small that total luteal volume is low. In cases of poor-quality or contralateral corpora lutea, treatment with gonadotropin-releasing hormone (GnRH) may produce an accessory CL to improve progesterone levels. Examination of ovarian structures is discussed further in later sections of this chapter.

The producer should understand that early diagnosis of open cows is critical for quick rebreeding of those animals, but that pregnant cows diagnosed before day 60 should be checked again after placentation is complete (see section Embryonic and fetal viability). This is especially important for lactating dairy animals due to their higher risk of late embryonic and early fetal death.

Embryonic and fetal aging

Ultrasound is far superior to manual palpation for staging pregnancies up to about 120 days in gestation. With ultrasound, embryonic or fetal size is measured rather than the size of the amniotic sac or total amount of pregnancy fluid. Embryonic and fetal growth is very consistent regardless of ambient temperature or breed, whereas the amount of fluid can vary.

Two methods can be used to determine the gestational age of the embryo or fetus. First, direct measurements of the embryo or fetal structures can be taken and correlated to age using a gestational aging chart.[9,10] Trunk diameter is the widest part of the ribcage at the level of the umbilicus. Braincase diameter is the maximum diameter of the skull just

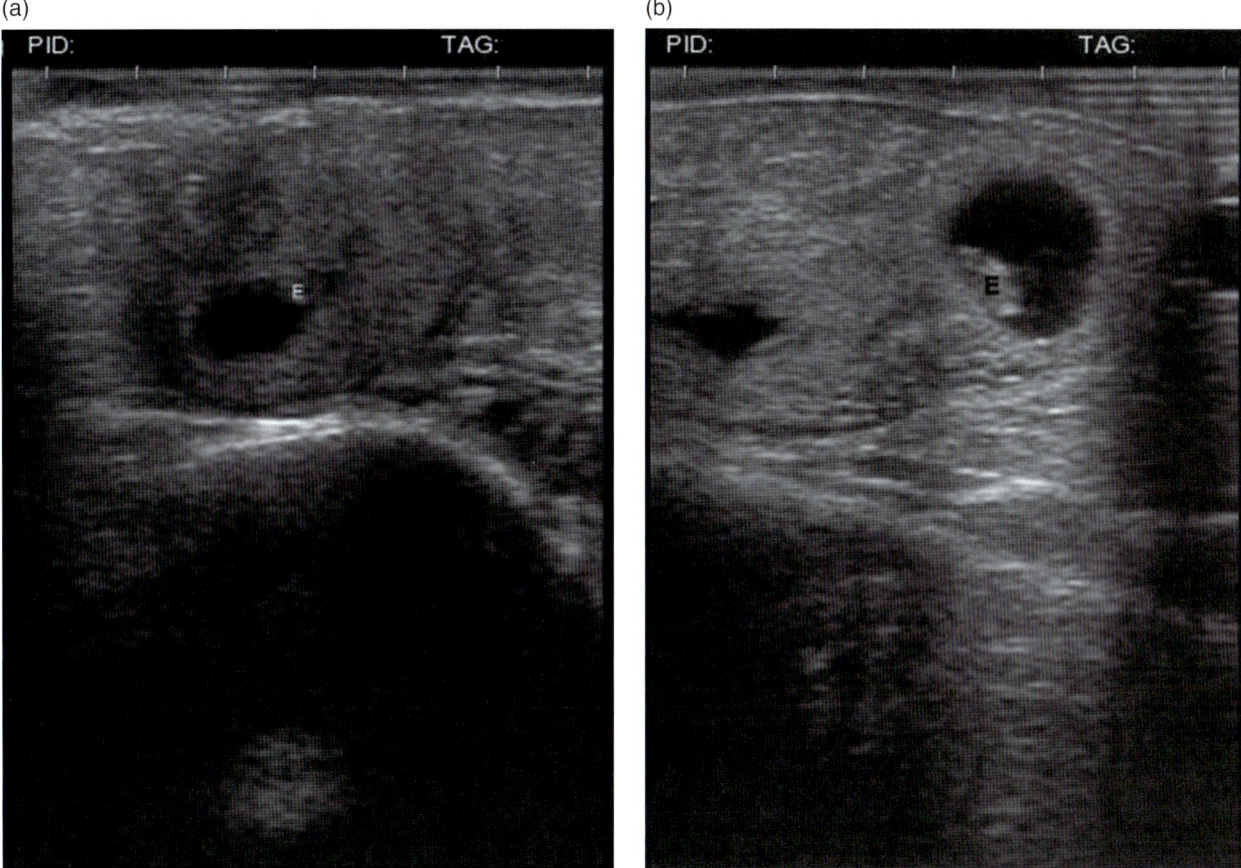

Figure 36.1 (a) A 26-day pregnancy and embryo (E). (b) A 28-day pregnancy and embryo (E).

caudal to the eyes. Head length is measured from the top of the cranium to the tip of the nose. Crown–rump length is measured from the tailhead to the crown of the skull. Of these measurements, only crown–rump length has a simple formula for a portion of gestation. Until approximately 50 days of gestation embryonic or fetal age can be calculated thus:

$$\text{Length of the embryo/fetus in millimeters} + 18 = \text{embryonic or fetal age in days}$$

For measurements of other structures, or for fetal ages greater than 50 days, one must use fetal aging charts or an ultrasound unit that is equipped with software to measure and calculate gestational age.

The second method is to visually evaluate embryonic/fetal development. Sandra Curran and colleagues have studied when various structures first appear (Table 36.1).[7] These authors find that in field conditions these structures are most easily seen at the high end of Dr Curran's range. Using embryonic/fetal size and knowledge of development, experienced ultrasonographers learn to "eyeball" age quite accurately without referring to charts or software.

After approximately 120 days of gestation fetal aging is less accurate. It is difficult to reach and measure fetal structures. Also, fetal size begins to vary among animals with the same breeding dates. Other measurements, such as fetal eye diameter, placentome size, or uterine artery diameter, are too variable among animals to provide accurate estimates of fetal age.

Table 36.1 Ultrasonic appearance of embryonic and fetal structures by days in gestation.

Characteristic	Earliest mean	Range
Embryo proper	20.3	19–24
Heartbeat	20.9	19–24
Allantois	23.2	22–25
Spinal cord	29.1	26–33
Forelimb buds	29.1	28–31
Amnion	29.5	28–33
Eye orbit	30.2	29–33
Hindlimb buds	31.2	30–33
Placentomes	35.2	33–38
Split hooves	44.6	42–49
Fetal movement	44.8	42–50
Ribs	52.8	51–55

Source: adapted from Curran S, Pierson R, Ginther O. Ultrasonographic appearance of the bovine conceptus from days 10 through 20. J Am Vet Med Assoc 1986;189:1289–1302.

Embryonic and fetal viability

Lactating dairy cattle experience considerable pregnancy loss, particularly before 60 days of gestation when placentation is not yet complete. Losses in lactating dairy cattle are

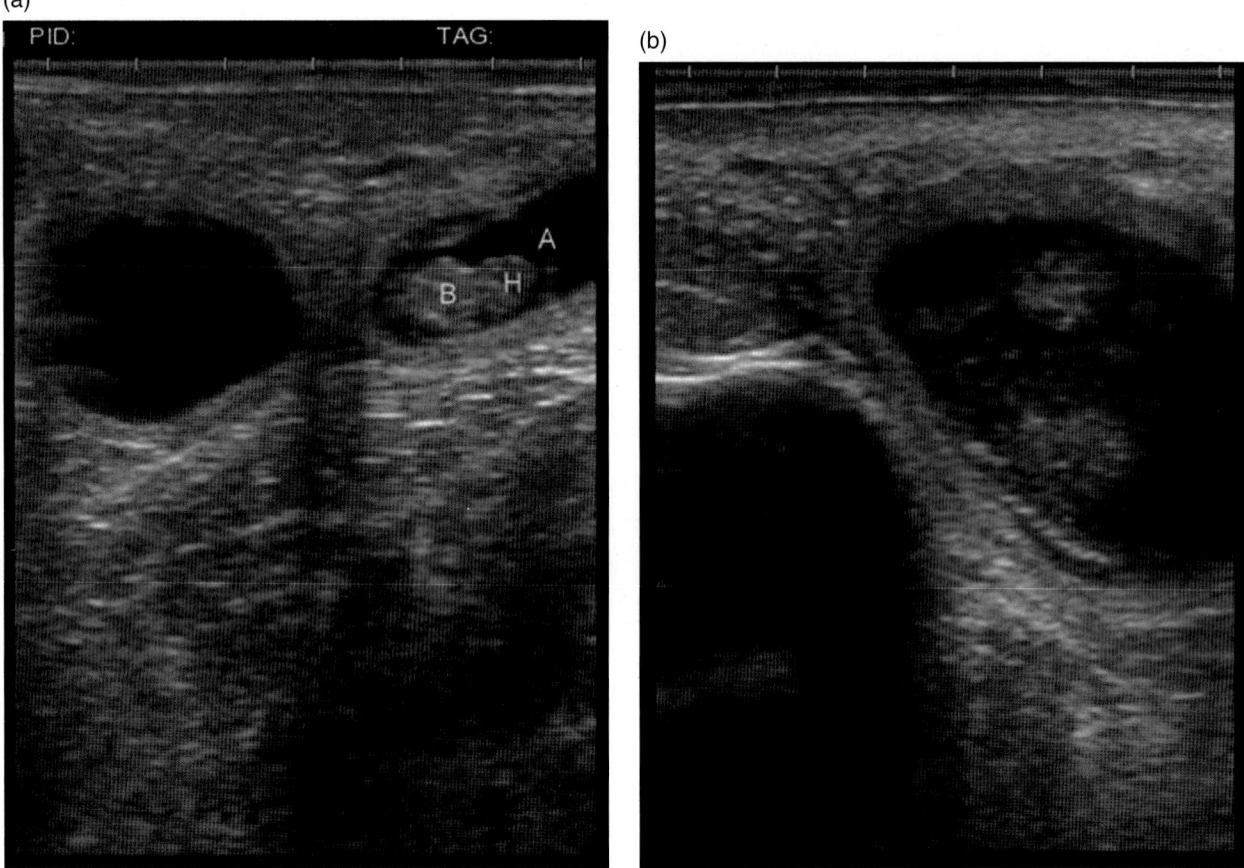

Figure 36.2 (a) Normal 35-day pregnancy: amniotic vesicle (A), head (H), body (B). (b) Dead 40-day twins.

approximately 3.6 times greater than in pregnant heifers[11] and often are 10–12% from day of initial pregnancy diagnosis to 90 days.[12] Hypotheses for this high rate of loss include reduced progesterone levels due to high milk production,[13] a history of postpartum disorders,[14] insemination by certain bulls,[15] twinning,[16] and heat stress.[17] Regardless of cause, it is critical for producers to understand pregnancy loss and the need to reconfirm pregnancies diagnosed prior to about 60 days of gestation.

Fetal viability is easily assessed with ultrasound. Beginning at about 24 days of gestation the fetal heartbeat can be seen.[7] The heart rate should be at least 130 bpm in early gestation, increasing to about 190 bpm by 60 days.[2] It can be difficult to identify the beating heart if the cow or the fetus is moving. For this reason, it is advisable to recheck pregnancies when no heartbeat is seen but all else appears normal.

The amniotic and chorioallantoic fluids should be clear and anechoic until at least 70 days of gestation. If the fluid is cloudy before 70 days and there is no heartbeat, the embryo or fetus can safely be assumed to be dead. Separated chorioallantoic membranes may also be seen in cases of late embryonic or early fetal death, often appearing to be floating in the pregnancy fluids. Compare the normal 35-day pregnancy in Figure 36.2a with the dead twin pregnancy in Figure 36.2b. Note the clear fluid and normal fetal development in the live fetus versus the flocculent fluid and unformed fetuses in the dead twins. Movement, other than a beating heart, is difficult to detect before 50 days because the limbs are not well developed yet.

After 70 days the fluids of pregnancy may begin to appear cloudy in normal pregnancies, but by that stage fetal movement can easily be appreciated. Separation of the chorioallantoic membrane is difficult to appreciate at this stage due to elongation and folding of all the placental tissues. Dead fetuses often are very heterogeneous, with many areas of high echogenicity that do not correspond to normal bony anatomy. There may also be fluid in the peritoneal cavity. Compare the dead 65-day fetus in Figure 36.3a to the normal 60-day fetus in Figure 36.3b. Fetal mummies (Figure 36.4) are easily identified by echogenic bony remnants, nearly complete lack of intrauterine fluid, lack of fetal movement, and lack of heartbeat.

In the absence of prostaglandin treatment and CL regression, dead embryos and fetuses can persist for weeks.[18] In these cases there are usually still palpable cardinal signs of pregnancy such as membrane slip or amniotic vesicle. With ultrasound, however, they can be quickly identified without additional examinations and treated with prostaglandin. In most cases expulsion of the dead embryo/fetus and uterine contents will be complete within 48 hours after prostaglandin administration.

In some cases, fetal distress can be identified before death. A heart rate of less than 130 bpm is cause for reevaluation at the next visit. Mild separation of chorioallantoic membranes or abnormal fluid in the presence of a live fetus is also cause for follow-up. A live fetus smaller than expected may be cause for concern, but is usually due to incorrect recording of breeding date. Twin fetuses are not smaller

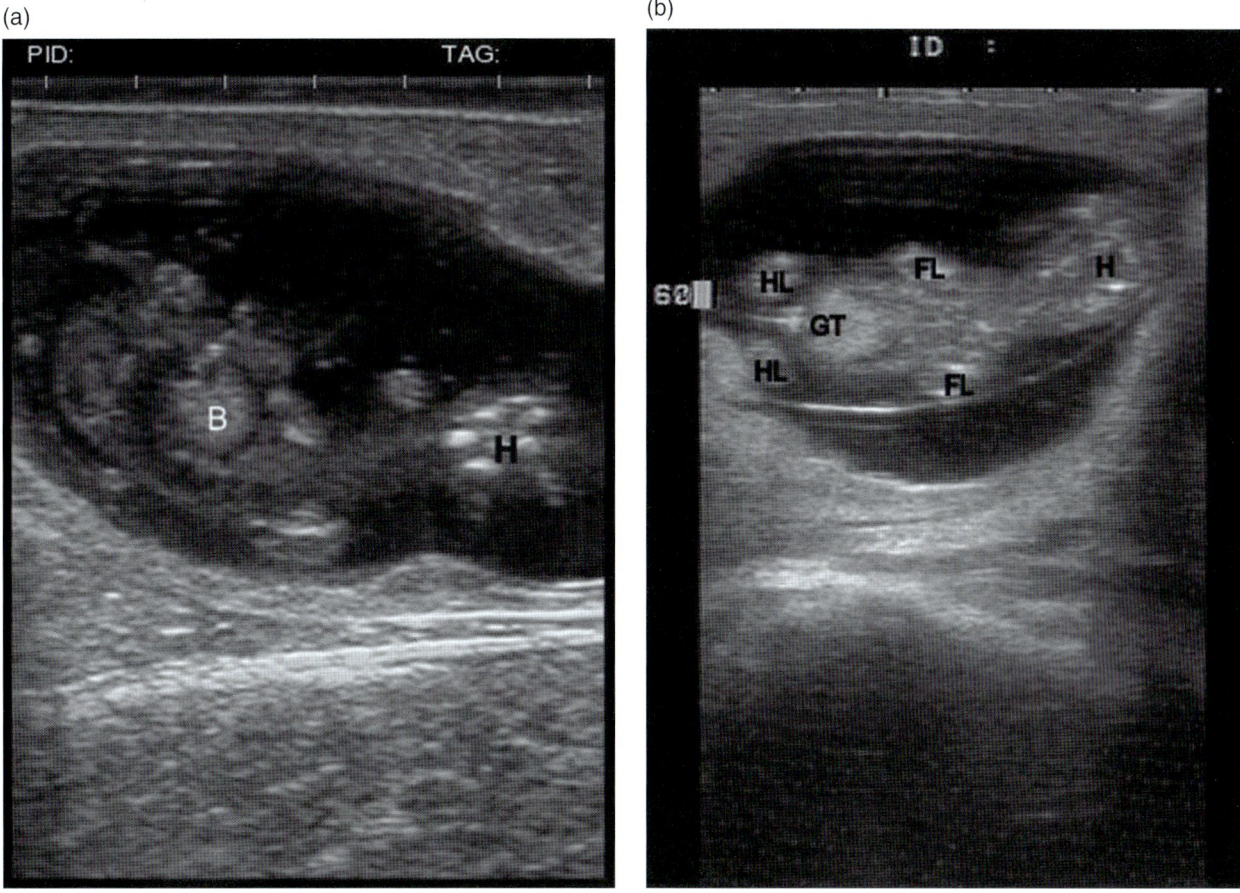

Figure 36.3 (a) Dead 65-day fetus: head (H), body (B). (b) A 60-day male fetus: forelimbs (FL), hindlimbs (HL), genital tubercle (GT).

than singletons until late in gestation. Fetal anomalies are at higher risk of loss throughout gestation and, in many cases, it is desirable to abort these abnormal pregnancies intentionally (see section Fetal anomalies). As discussed in the section on pregnancy diagnosis, a contralateral or poor-quality CL is also reason to recheck the pregnancy at the next opportunity.

Twins

High twinning rates are often seen in dairy herds and, less commonly, in beef herds. The primary mechanisms for twinning appear to be correlated to high milk production in dairy cattle[19] and to genetic propensity in certain strains of beef cattle.[20] In dairy cattle there also appears to be a higher risk of twinning in cows with a history of cystic ovarian disease.[16] In any case, twinning can lead to more pregnancy loss,[16,21] more dystocia, more periparturient disease, reduced future reproductive performance,[22] freemartinism, and a higher risk of early culling for the dam.[23]

Despite these risks it is generally not recommended to abort twin pregnancies. The possibility of failing to produce another pregnancy or of producing another set of twins makes it preferable to monitor and manage the cow carrying twins. This can include additional examinations to identify pregnancy loss, drying the cow off early to compensate for probable early calving, feeding a higher energy ration to improve body condition, providing extra monitoring and assistance at calving, and being alert for periparturient metabolic disease. In order to provide these management techniques it is important to the producer to know which animals are carrying twins. Manual palpation is not an accurate method for diagnosis of twins.[24] However, a careful ultrasonographic examination can identify most cases.

Because over 95% of twins in cattle are dizygous,[25] it is helpful to examine the ovaries for two or, occasionally, more corpora lutea. Approximately 60% of pregnant cows with two corpora lutea will have twins.[26] When twins are suspected a thorough examination of the entire uterus should be performed. It is imperative to follow the tract in a systematic fashion to avoid inadvertently counting a single fetus twice. Triplets are rare, but should be considered when three or more corpora lutea are present (Figure 36.5).

During examination of the uterus a twin line may be seen (Figure 36.6). This line represents the confluence of the chorioallantoic membranes of each fetus. It can be distinguished from the amniotic vesicle because it moves away from the fetus rather than encircling it. It will also not be confused with the umbilicus, which is not readily visible prior to 45 days and which is much thicker after that. A twin line may be seen with either ipsilateral or contralateral twins. In some cases it is not seen due to the angle of examination. In later pregnancy the twin line is difficult to identify because all the placental membranes are elongated and folded within the uterus. Twin pregnancies also tend to have more fluid than single pregnancies and often extend beyond the pelvic

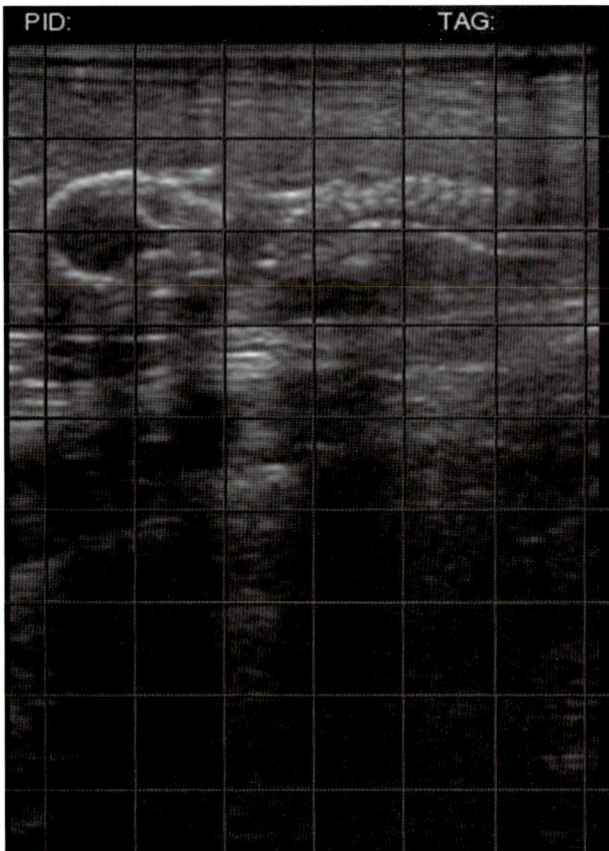

Figure 36.4 Fetal mummy.

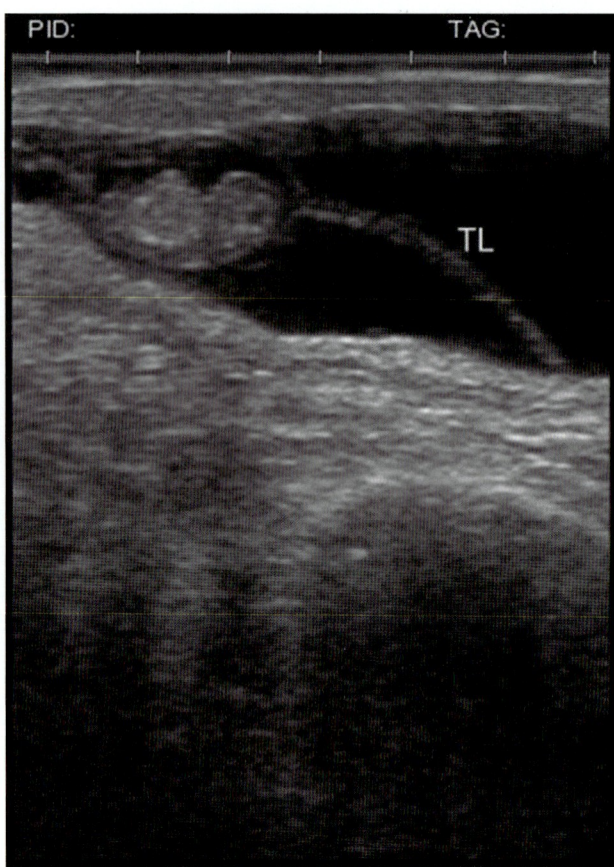

Figure 36.6 Twin line (TL).

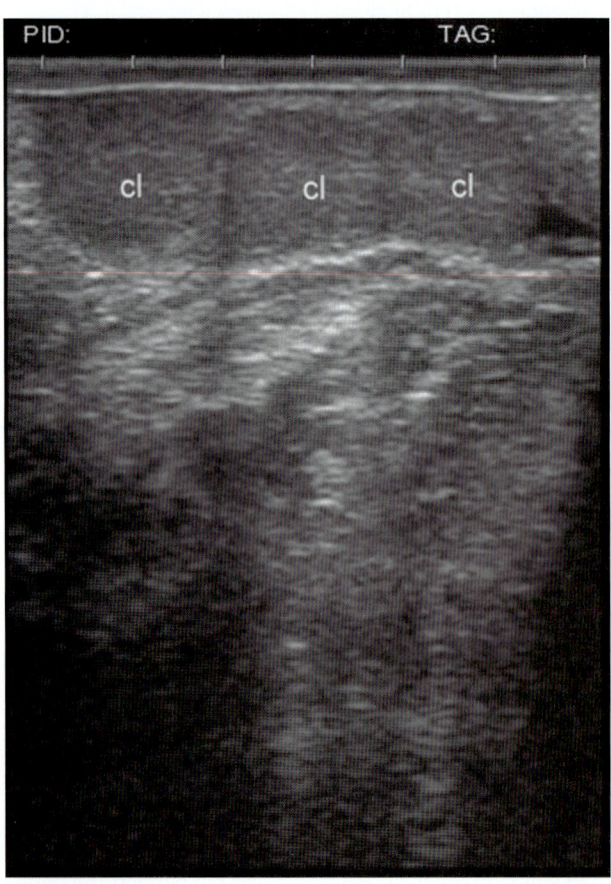

Figure 36.5 Three corpora lutea (cl) on one ovary.

brim sooner. However, the only cardinal sign of twin pregnancy is the presence of two fetuses.

Twin pregnancies, particularly if ipsilateral, have higher rates of loss (Figure 36.2b) throughout gestation, but especially in the first 60 days before placentation is complete. This loss may be 3.1–6.9 times higher than that for singleton pregnancies.[15,27] However, one unpublished study (A. Scheidegger, personal communication, 2005) indicates that loss of contralateral twins may be more likely in mid gestation. In either case, recheck examinations are warranted to identify cases of loss.

Spontaneous reduction of twin embryos has been documented,[16,21] most commonly in the late embryonic or early fetal stages. Depending on the study, 8–24% of twins reduced to a single fetus. The varying incidence is probably due at least partly to different days after insemination for initial and follow-up examinations. CL reduction often occurs in the ovary ipsilateral to the uterine horn suffering embryo reduction. Attempts to manually reduce twins in cattle have been, at best, only marginally successful.[28] In the future these results may be improved upon by ultrasound guided needle aspiration of earlier embryos as is done successfully in mares.

Fetal gender determination

Ultrasonographic gender determination of the bovine fetus was first described by Curran et al.[29] in 1989. It was found that the relative location of the genital tubercle, which will become the penis in the male and the clitoris in the female, could be used to determine fetal gender beginning at day 55 of gestation.

Reproductive Ultrasound of Female Cattle 331

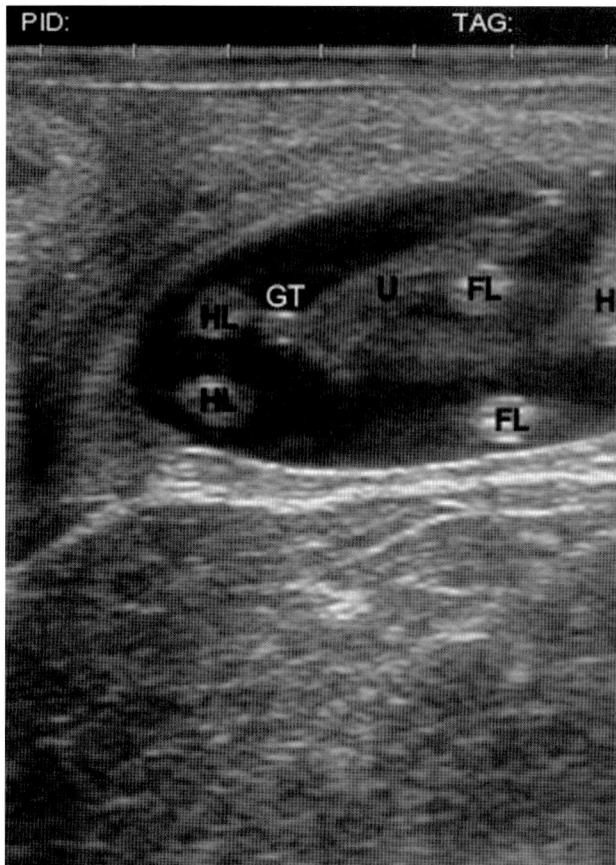

Figure 36.7 A 73-day male fetus: head (H), forelimbs (FL), hindlimbs (HL), genital tubercle (GT), umbilicus (U).

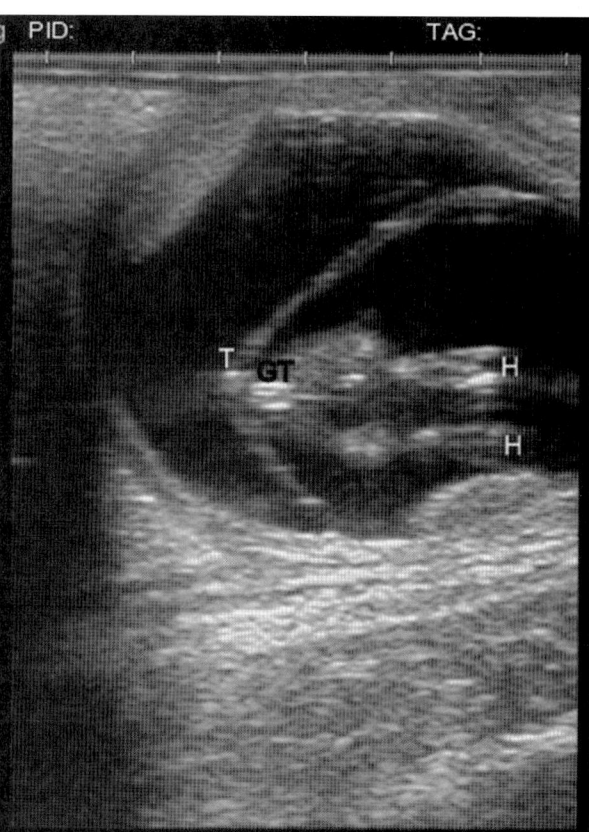

Figure 36.8 A 65-day female fetus: tail (T), genital tubercle (GT), hooves (H).

By this stage the genital tubercle has reached its final location immediately behind the umbilicus in the male (Figure 36.3b and Figure 36.7) or immediately under the tail in the female (Figure 36.8). Until about 80 days of gestation the genital tubercle usually appears as a highly echogenic bilobed structure in both genders, but may appear mono- or tri-lobed.

After 70 days of gestation the skin of the prepuce or vulva begins to cover the genital tubercle.[2] At this point the adult term for the male or female external genitalia replaces the term "genital tubercle." By about 80 days the skin of the prepuce or vulva has developed enough to reduce the bright bilobed appearance of the genital structure, as shown in Figure 36.9. Also by this time the scrotum and mammary glands can be detected with most field ultrasound units. The scrotum may appear as three bright lines representing the lateral walls and median raphe or it may appear as a soft tissue sac. Teats appear as bright dots and may resemble the scrotum when it presents as three bright lines. High-quality ultrasound units can detect teats on bull fetuses, especially after about 90 days. Figure 36.10 shows teats on the same 98-day bull fetus shown in Figure 36.9. For these reasons, it is not advisable to diagnose fetal gender based on presence of a scrotum or teats without visualizing the genital tubercle, penis, or vulvar structures.

After 90 days the uterus may be over the brim of the pelvis and difficult to reach for gender determination in some animals. In others it may be possible to reach the fetus at 120 days or even later; 60–80 days is the ideal time for

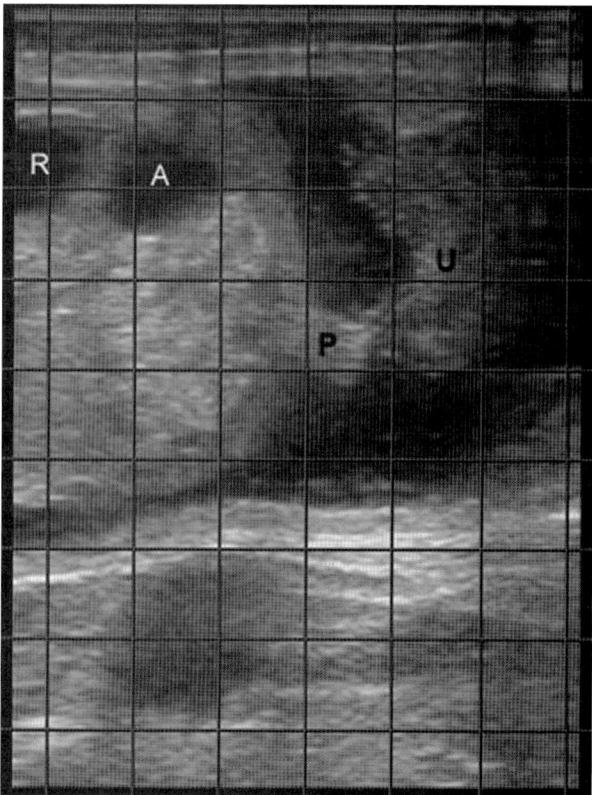

Figure 36.9 A 98-day male fetus demonstrating penis covered by prepuce: rumen (R), abomasum (A), penis (P), umbilicus (U).

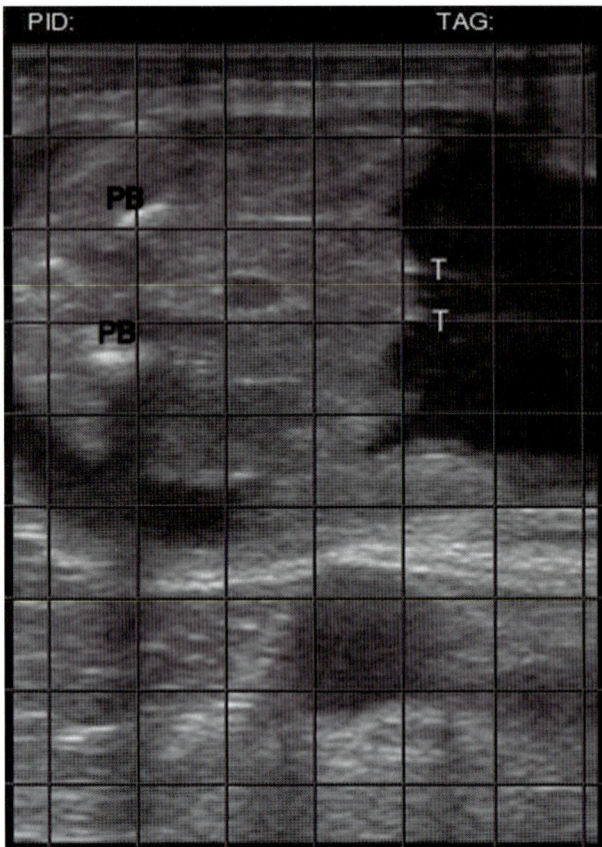

Figure 36.10 A 98-day male fetus demonstrating teats: teats (T), pelvic bones (PB).

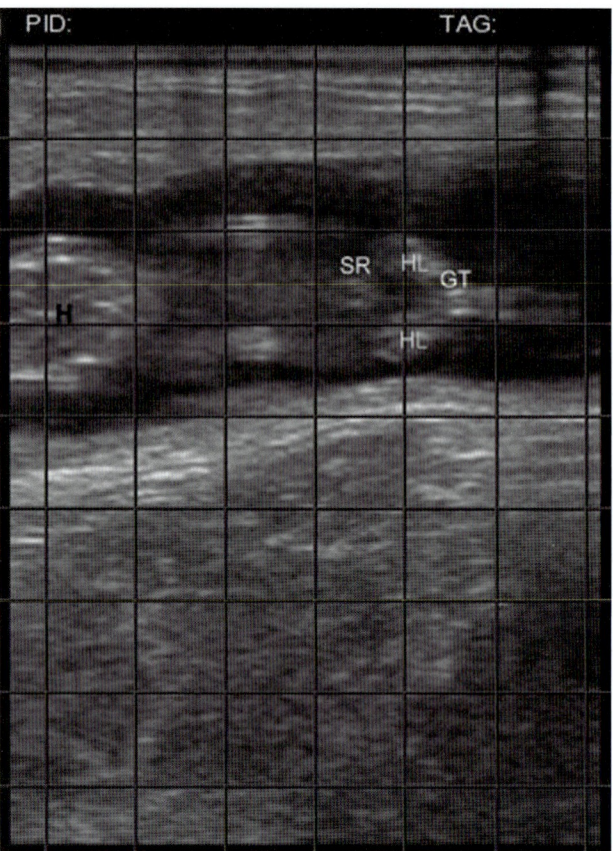

Figure 36.11 A 60-day female fetus with specular reflection artifact: head (H), specular reflection artifact (SR), hindlimbs (HL), genital tubercle (GT).

gender determination. At this stage the fetus is easily reached, small enough to orient the location of structures, and the genital tubercle is not obscured by skin.

Knowledge of fetal anatomic landmarks helps to locate the genital tubercle. The large tortuous umbilicus is easily identified (Figure 36.7) and the area immediately behind it can be examined for presence of a male genital tubercle. The fluid-filled anechoic rumen and abomasum (Figure 36.9) are also in the region of the male genital tubercle. The "V" shape of the ribcage permits easy identification of cranial and caudal aspects of the fetus. The head is easily identified by its echogenic bony structure.

The most common error in fetal gender diagnosis occurs because the operator does not appreciate how much the genital structures protrude from the body wall. For example, in a precise cross-sectional examination of a female fetus only the mono-lobed tailhead and female genital tubercle will be seen because they protrude away from other pelvic and hindlimb structures. Similarly, in the case of a male in precise longitudinal plane, only the male genital structure and the umbilicus will be seen. Therefore it is important not to stop short when scanning the fetus for gender determination.

It is also easy to move too quickly when scanning for fetal gender. There is no gap between the umbilicus and male genital tubercle by 60 days of gestation. Likewise, there is very little space between the female genital tubercle and tailhead by that stage. Very slow, small movements are critical to avoid passing through these structures before the eye can appreciate them.

Other errors occur when another structure or artifact is mistaken for the genital tubercle. Legs sometimes occlude the umbilical area and the bones in cross-section can resemble the genital tubercle. Specular reflections in the umbilical cord can be mistaken for a male genital tubercle. Figure 36.11 shows the female genital tubercle to the right of the image, along with two less distinct bright lines in the abdominal region where the male genital tubercle would be found. These lines are a specular reflection artifact between the fluid vessels and the walls of the umbilicus (see section Artifacts for further explanation). A cross-section of the tail can be mistaken for a female genital tubercle. When in doubt it is wise to examine both the umbilical region and the perineal region to confirm gender.

Finally, one must develop a quick eye to accurately identify landmarks and the genital tubercle. Because the cow and the fetus are moving it is usually not possible to isolate the structures for prolonged viewing. Training videos such as Brad Stroud's *Bovine Fetal Sexing Unedited* are valuable for training the eye to detect fetal gender quickly and accurately.[30]

Normal uterine and ovarian structures throughout the estrous cycle

Proestrus

In proestrus the uterus becomes more heterogeneous as blood flow increases.[31] Anechoic mucus begins to accumulate in the lumen about 3 days before ovulation. One, or sometimes more,

Reproductive Ultrasound of Female Cattle

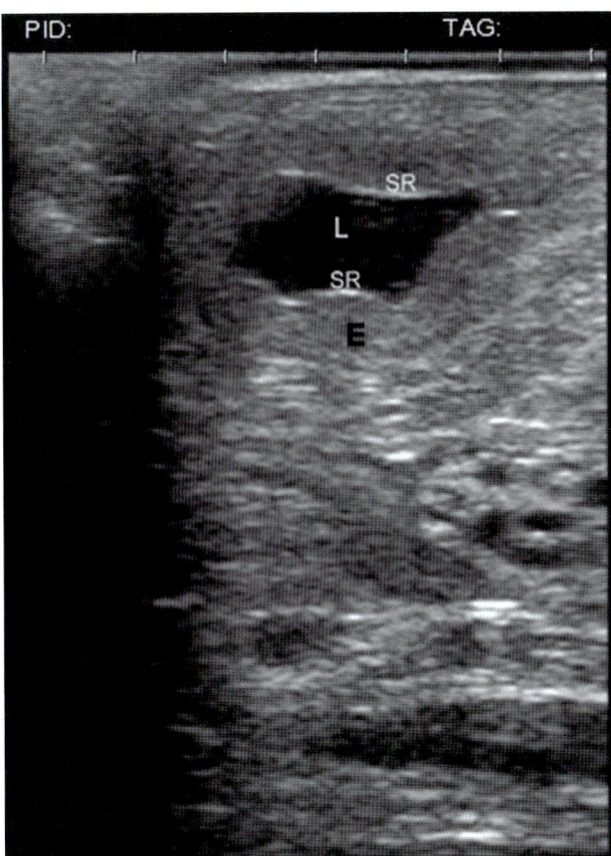

Figure 36.12 Estrous uterus in cross-section: lumen (L), endometrium (E), specular reflection artifacts (SR).

large (16–20 mm) follicles will be present. The CL is usually still visible, but is becoming smaller and more heterogeneous.

Estrus

Figure 36.12 shows a normal estrus uterus in cross-section. The uterus is most heterogeneous at this point. Varying amounts of anechoic mucus are present in the uterine lumen. The large difference in acoustic impedance between the dense endometrium and the fluid mucus often creates specular reflection artifacts on the inner wall of the endometrium. The endometrium thickens to about 2.5 times its normal volume under the influence of estrogen. When measured in cross-section at about 2 cm from the internal uterine body bifurcation, the endometrial thickness has been shown to increase to about 9.4 mm in estrus from 6.9 mm in diestrus.[32] A mature follicle is still present until ovulation at approximately 12 hours after the end of standing heat. Sometimes a portion of the follicular fluid is not expelled during ovulation. In these cases it is difficult to determine if ovulation has occurred. A small heterogeneous CL may still be seen throughout proestrus and estrus.

Metestrus

After ovulation a new wave of small follicles (<8 mm) develops. In the first day or two of the cycle these small follicles are all that is visible on the ovaries. By day 2 or 3 a

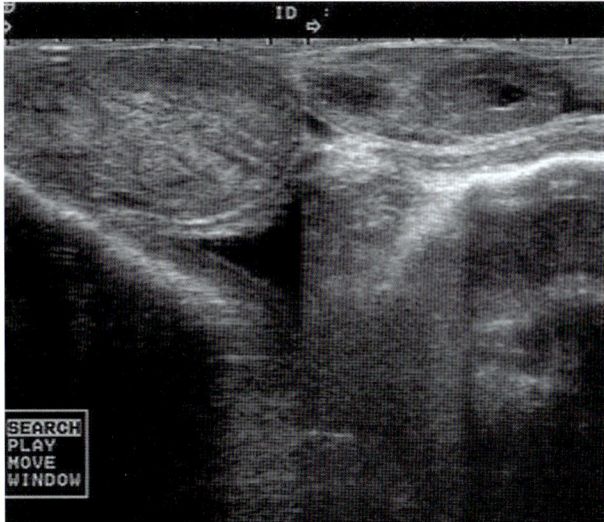

Figure 36.13 Split screen image of diestrous uterus in longitudinal section on the left and ovary with corpus luteum on the right.

small heterogeneous CL becomes visible, often with an anechoic fluid-filled cavity. Occasionally the cavity is filled with very hyperechoic material, probably clotted blood.

Diestrus

Figure 36.13 shows a normal diestrus uterus and ovary. During diestrus the uterus is homogeneous and a medium shade of gray. There should be little or no fluid in the lumen. The CL will also be homogeneous and a medium shade of gray. The CL may or may not have a fluid cavity. Figure 36.14 shows three corpora lutea in heifers 7 days after standing estrus. Despite their different appearances these are all normal corpora lutea for that stage. About 80% of corpora lutea have a cavity at some stage in development, usually early.[33] The cavity may be filled with clear fluid (Figure 36.14b), probably due to incomplete expulsion of follicular fluid during ovulation. Dense material in the cavity (Figure 36.14c) may be clotted blood remaining after ovulation. The fluid cavity tends to fill in over time, but may persist into pregnancy. A mature CL with a fluid cavity may also have bright specular reflection artifacts where the dense, fully developed luteal tissue meets the clear fluid cavity (Figure 36.15). These artifacts are less evident in less mature cavitary corpora lutea (Figure 36.14b) because there is not yet adequate acoustic mismatch for them to occur. Specular reflection artifacts are discussed further in the Artifacts section. Follicles ranging in size from 2 to 20 mm may be present depending on the stage of follicular wave cycles.

Pregnancy

Before 21 days of gestation the uterus and ovaries look much as they do in diestrus. A small amount of fluid may be visible in the uterine lumen at 21–23 days of gestation. The embryo is difficult to see under field conditions until at least 24 days. The gravid uterus does not have a thickened endometrium as in estrus. Specular reflections are either not present or much less pronounced than in estrus. The CL is usually solid at this point, but may be cavitary.

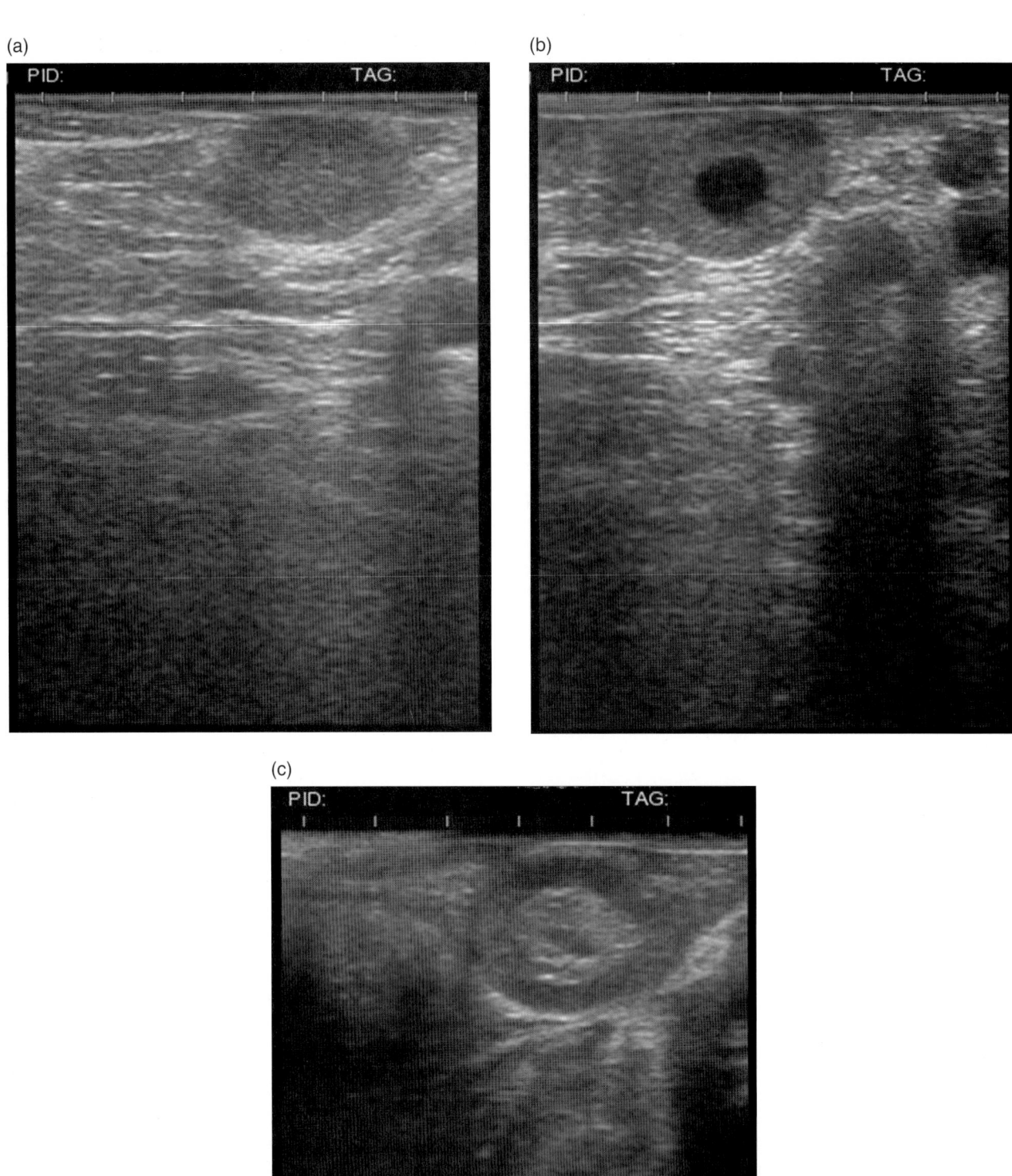

Figure 36.14 (a) Solid corpus luteum 7 days after estrus. (b) Cavitary corpus luteum 7 days after estrus. (c) Cavitary corpus luteum with dense center 7 days after estrus.

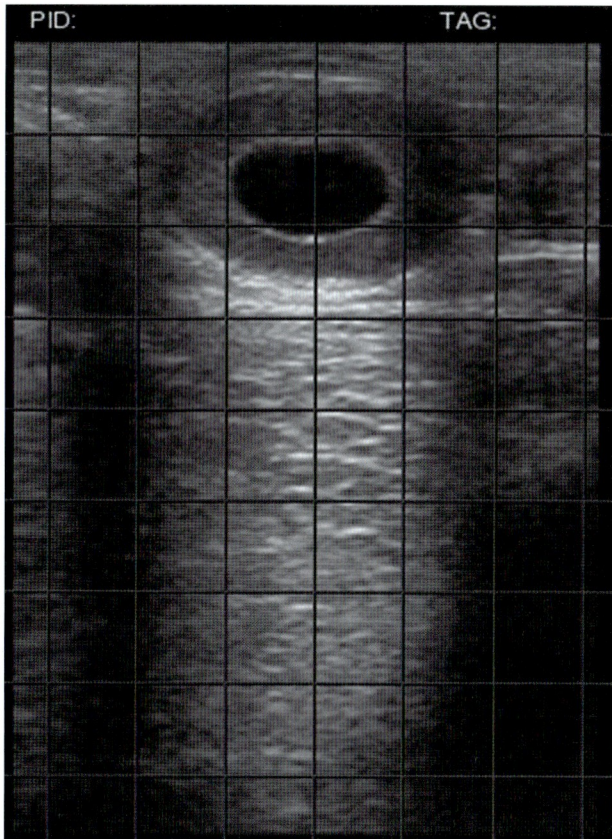

Figure 36.15 Mature cavitary corpus luteum with specular reflections.

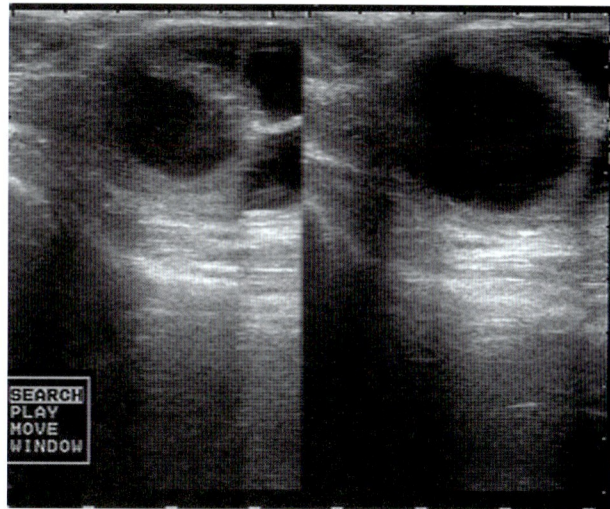

Figure 36.16 Multiple follicular cysts.

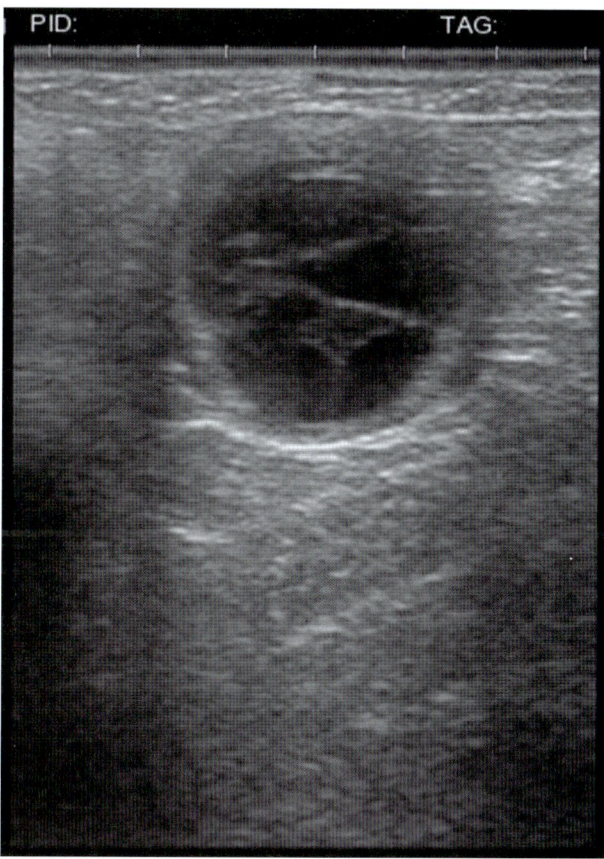

Figure 36.17 Luteinizing follicular cyst.

Ovarian pathology

Anovulation

Wiltbank et al.[34] have classified ovulation failure into four basic categories. Ultrasound can assist in determining the category of ovulation failure.

Anovulation with follicle growth only to emergence

The term "static ovaries" has been used to describe conditions in which follicles grow to emergence (about 4 mm) but do not mature further to ovulatory size of 8 mm or more. On ultrasound examination this condition resembles early metestrus with many small follicles, no follicles over 4 mm, and no visible CL. A second examination a few days later is required to differentiate this condition from metestrus. According to Wiltbank, this type of anovular condition is rare except in extreme malnutrition and possibly in Zebu cattle.

Anovulation with follicle growth to selection but not ovulatory size

This type of follicular growth pattern is typical in the postpartum period. There is no CL and the ovaries are small, but frequent ultrasound examinations reveal that follicles grow up to 8.5 mm in a dynamic wave pattern.

Anovulation with follicle growth to ovulatory or larger size (follicular cysts)

Follicular cysts have been defined as thin-walled fluid-filled follicular structures greater than 22 mm in diameter.[35] A follicular cyst (Figure 36.16) is differentiated from a cavitary CL, luteal cyst, or luteinizing follicular cyst (Figure 36.17) by the absence of a visible luteal rim. Follicular cysts are benign in most cases, with luteal structures and follicular waves developing

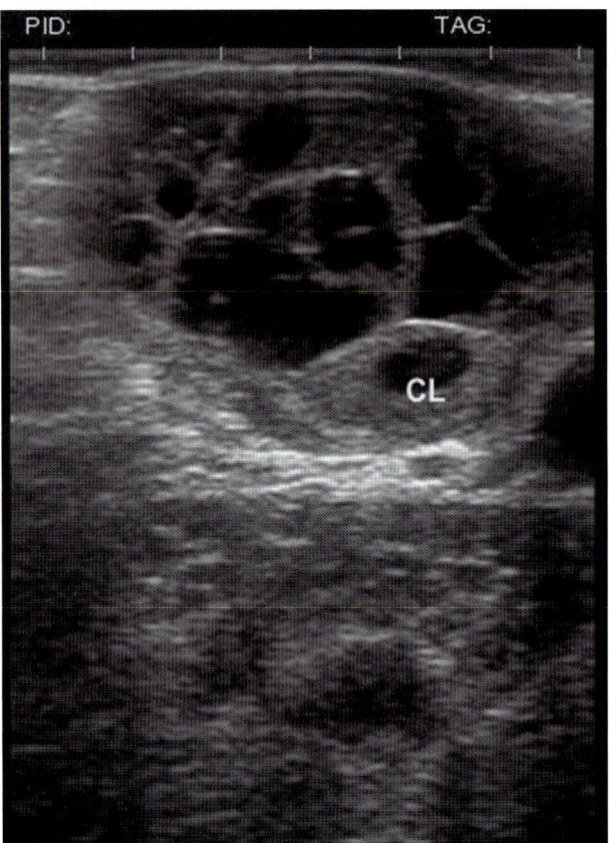

Figure 36.18 Benign follicular cysts with corpus luteum (CL).

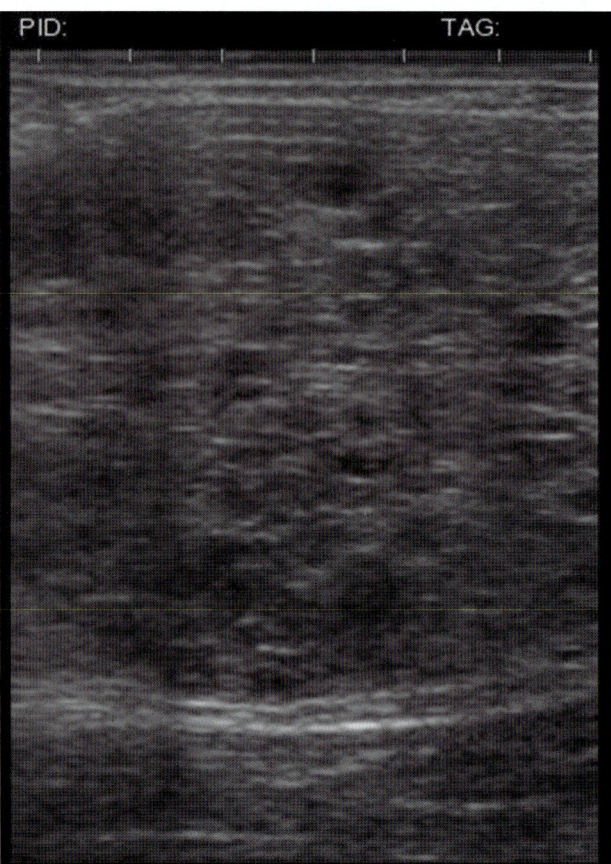

Figure 36.19 Granulosa cell tumor.

normally on the same or contralateral ovary. Benign cysts can persist throughout several cycles and into pregnancy. Ultrasound is far superior to manual palpation for diagnosis of benign versus pathologic follicular cysts because it is more capable of identifying luteal tissue within the fluid-filled structure or the presence of a small CL elsewhere.[36,37]

Thorough ultrasonographic examination of both ovaries is essential to properly classify follicular cysts. If luteal tissue is detected the cyst is benign. Figure 36.18 shows a large ovary with several follicular cysts and a normal CL at 5 o'clock. If no luteal tissue is seen, a second examination must be performed in about 5–14 days to determine if a CL has developed. If not, the cyst is likely preventing normal cyclicity.

Follicular cysts have traditionally been treated with GnRH. Following treatment, most follicular cysts luteinize rather than ovulate. In these cases, examination 7 days later will reveal that the original thin-walled structure has developed a thicker rim of luteal tissue (Figure 36.17). This large soft structure will still feel like a follicular cyst on manual palpation, but the luteal tissue is clearly seen with ultrasound. Alternatively, the cyst may remain but another, smaller follicle may be induced to ovulate. In this case the original follicular cyst will still be seen after 7 days, but a small CL will be evident elsewhere. It has been demonstrated that the presence of progesterone improves conception rates to the Ovsynch synchronization program.[38] Hence animals that respond to GnRH treatment by producing luteal tissue may be started on this protocol with good success.

Another common treatment for follicular cysts is progesterone supplementation in conjunction with the Ovsynch synchronization protocol at the time of diagnosis. Progesterone supplementation adds expense and inconvenience. In some countries it is not permitted or requires substantial documentation. For these reasons, ultrasound diagnosis is very helpful to limit progesterone supplementation only to cystic cows that do not produce luteal tissue in response to GnRH treatment.

Anovulation in the presence of luteal tissue

Prolonged luteal function prevents ovulation at normal intervals. Various papers define prolonged luteal function differently. Proposed definitions include, but are not limited to, the persistence of the CL with milk progesterone above 3 ng/mL for more than 30 days,[39] or milk progesterone above 3 ng/mL for at least 19 days.[40] Frequent multiple ultrasound examinations or progesterone tests are required to confirm persistent luteal function. Because the Ovsynch protocol works well in the presence of progesterone, it is more important to identify the presence of a CL rather than attempting to identify if it is persistent.

Other ovarian pathology

Ovarian pathologies other than the above anovular defects are rare in cattle. Granulosa cell tumors are sometimes seen (Figure 36.19). The ovary is very large. The hyperechoic areas are specular reflections between the vasculature of the tumor and the denser surrounding tissue.

Synchronization protocols

When cattle are presynchronized with prostaglandin (Presynch) before beginning Ovsynch, all normal cycling animals should have a CL when the first GnRH injection (G1) is scheduled. An examination of ovaries at this time will reveal if presynchronization was successful. If many cows do not have corpora lutea at this time, the protocol has failed and the reason must be found. Poor compliance, poor cow health, low body condition scores, and nutrition problems are all possible reasons for failure.

It has been shown that the presence of a CL and the availability of an ovulatory size (≥8 mm) follicle at G1 improve conception rates for Presynch/Ovsynch.[38] This study is summarized below.

- For cows with a CL and ovulation to G1, the conception rate was 42.2% when pregnancy examination was performed at 38 days.
- For cows with a CL and no ovulation to G1, the conception rate was 37.7%.
- For cows with no CL and ovulation to G1, the conception rate was 27.6%.
- For cows with no CL and no ovulation to G1, the conception rate was 15.4%.

Ultrasound is very sensitive for identifying the presence of a CL, including very early metestrus corpora lutea and regressing proestrus corpora lutea. These very early and late corpora lutea are probably not producing progesterone in adequate levels for optimal conception rates to Ovsynch. Therefore it has been suggested that a cutoff size be used to determine if a CL is functional. Bicalho et al.[37] found that ultrasound was nearly 90% sensitive and specific for identifying a CL producing more than 1 ng/mL blood progesterone if a 2.2-cm cutoff size was used. Palpation, by contrast, was 33–60% sensitive and 77–93% specific for identifying blood progesterone greater than 1 ng/mL, varying among palpators.

Ovsynch works best from days 5 to 10 of the cycle when some early corpora lutea may not yet be 2.2 cm in diameter. Therefore, I use a minimum cutoff size of 1.5 cm if the CL appears to be a typical early diestrus CL (see section above on ovarian structures). If the CL appears to be regressing during proestrus, it is more desirable to wait until 5–10 days after ovulation to begin the Ovsynch protocol.

Based on the findings in the study by Galvão et al.,[38] the presence of a functional CL was more important than ovulation to G1. However, some ultrasonographers[41] feel it is also of value to identify the presence of a potentially ovulatory follicle greater than 8 mm. Given that it is difficult to assess if follicles of this size are normal, persistent, or atretic, I rely more on the presence of a CL of appropriate size and stage.

Uterine, cervical and vaginal pathology

Metritis, endometritis and pyometra

Clinical and most subclinical uterine infections can be diagnosed with ultrasound. Field ultrasound units can visualize 2 mm of fluid, far less than detected by manual palpation. Ultrasound can also evaluate the character of the fluid based

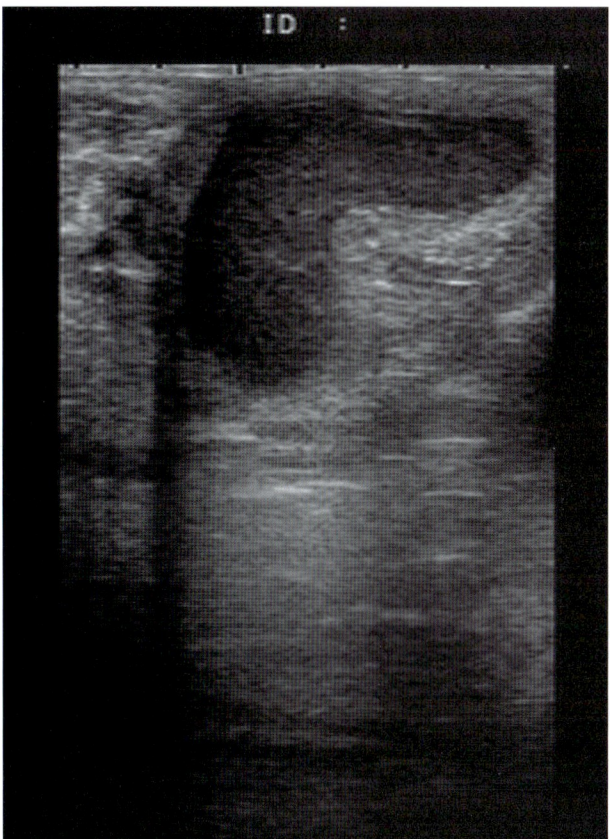

Figure 36.20 Pyometra.

on echogenicity. In most cases of metritis/endometritis, echogenic purulent material can be seen in the uterine lumen. This fluid may be any shade from light gray to very bright white depending on its density. Intraluminal fluid in cases of pyometra is usually voluminous and of medium echogenicity (Figure 36.20). Cases of chronic subclinical metritis/endometritis usually have little intraluminal fluid of higher echogenicity (Figure 36.21). The ultrasonographer should be careful not to confuse small volumes of echogenic intraluminal fluid with specular reflections due to estrus (Figure 36.12). When fluid is not detected increased endometrial and cervical thickness may be used to evaluate endometritis in the early postpartum period.[42] The ultrasonographer should keep in mind that endometrial thickness is also increased in estrus and must perform a thorough examination of both ovaries and uterus to make a differential diagnosis.

Mucometra

Mucometra is an accumulation of sterile anechoic or slightly echoic fluid in the uterine lumen not associated with pregnancy. Specular reflections may or may not be present. The endometrium is not thickened as it is in estrus. Mucometra does not appear to be related to ovarian structures. The ovaries may have follicular cysts, normal corpora lutea, or no significant ovarian structures. It is critical to confirm the absence of an embryo or fetus in cases of mucometra. When in doubt the ultrasonographer should perform a second examination a few days later to confirm the diagnosis. Mucometra

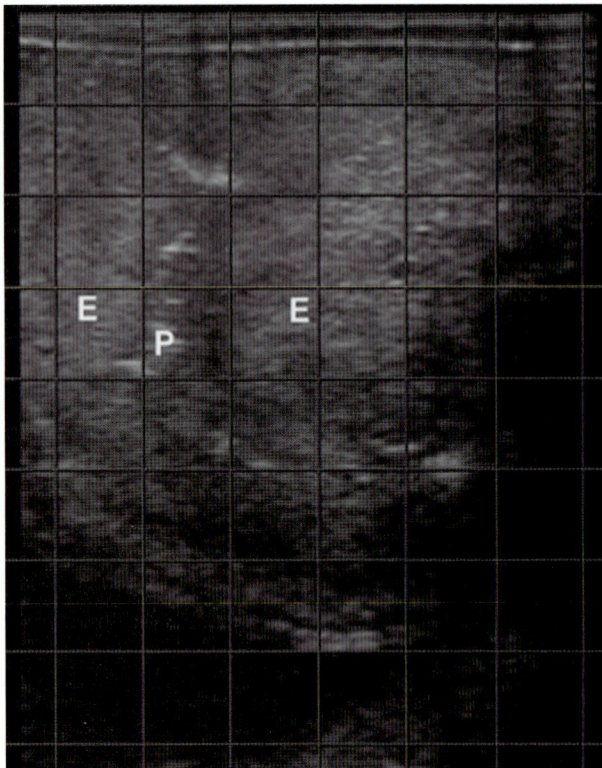

Figure 36.21 Subclinical metritis: endometrium (E), purulent material (P) in lumen.

in heifers may be related to the presence of a complete hymen or other congenital blockage of the tubal portions of the reproductive tract. The cause in adult cows is unknown.

Lymphosarcoma

Uterine lymphosarcoma is identified by very firm thickening of the uterine wall that may be localized or throughout the entire uterus. The affected uterine tissue (Figure 36.22a) looks like the tissue of a lymph node (Figure 36.22b).

Abscesses

Uterine, cervical, or vaginal abscesses can be caused by difficult parturition, rough assistance during parturition, or trauma from artificial insemination. They can be distinguished from metritis, cervicitis, or vaginitis by the presence of a capsule around the purulent material (Figure 36.23). Uterine abscesses have a poor prognosis for future fertility. Cervical abscesses may or may not be of concern. Careful ultrasound examination can determine if the abscess extends into the cervical lumen or is entirely outside it. If the latter the prognosis for future fertility and parturition is good. Vaginal abscesses generally do not cause difficulty unless they are large enough to interfere with breeding or parturition.

Vaginitis

The normal vagina presents as a thin white line representing the confluence of opposite sides of the vaginal wall. In cases of vaginitis echogenic material is visible in the vaginal vault

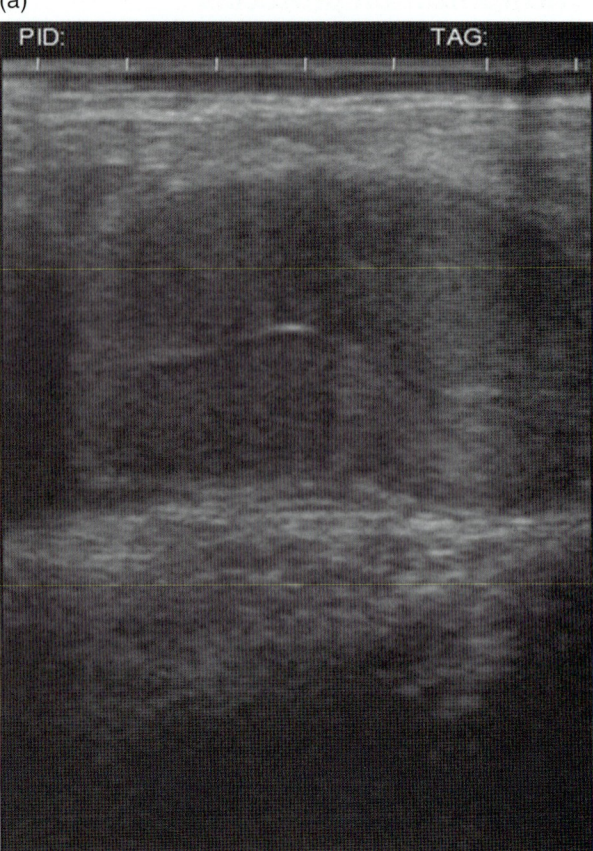

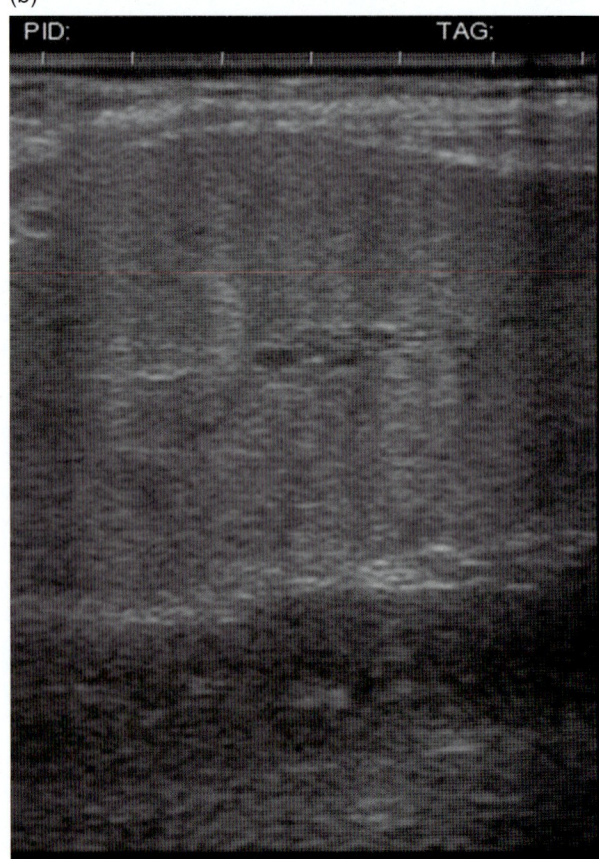

Figure 36.22 (a) Uterine lymphoma. (b) Pelvic lymph node.

Reproductive Ultrasound of Female Cattle

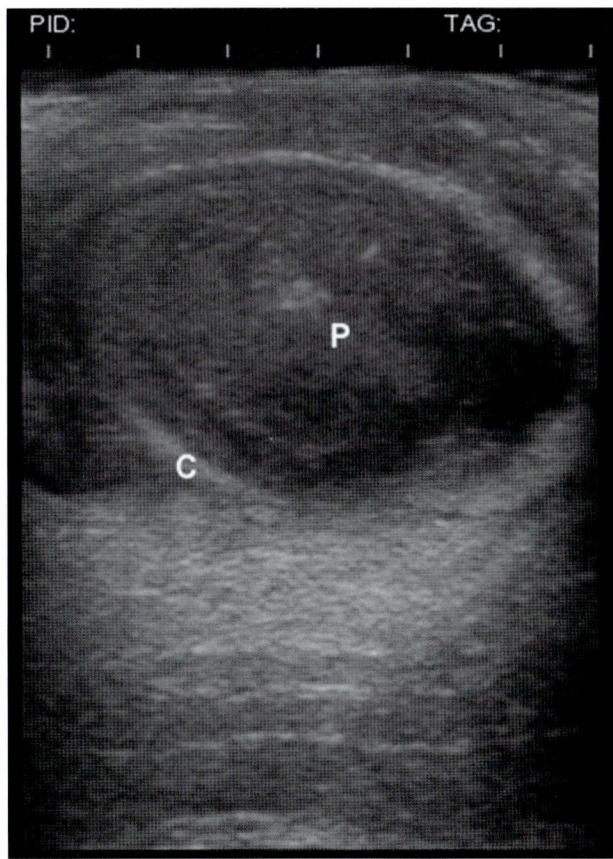

Figure 36.23 Uterine abscess: capsule (C), purulent material (P).

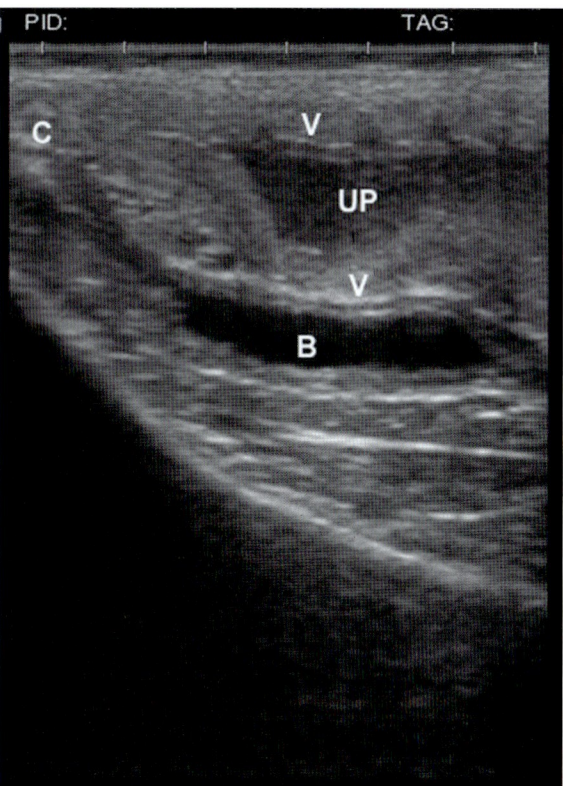

Figure 36.24 Vaginitis due to urine pooling: vaginal walls (V), urine pooling (UP), bladder (B).

and the vaginal mucosa is thickened. Unless echogenic material is also present in the uterus, the prognosis for fertility is fair to good if a protective sheath is used on the insemination gun at breeding. Vaginitis may be primary or caused by urine pooling (Figure 36.24).

Vaginal urine pooling

Urine is usually less echogenic than purulent material, but is often mixed with purulent material when accumulated in the vagina (Figure 36.24). As with vaginitis, prognosis for fertility is fair to good if there is no abnormal fluid in the uterine lumen and a protective sheath is used on the insemination gun at breeding. Urine pooling may be accompanied by cystitis.

Pneumovagina

Gas in the vagina or elsewhere presents as a reverberation artifact (Figure 36.25) (see section Artifacts).

Embryo transfer

Donor evaluation

Donors should be assessed for uterine health and ovarian function before beginning the superovulation protocol. If the cow is examined during follicular recruitment, it may be possible to predict the response to future superovulation. Cows that recruit large numbers of follicles at the beginning of each

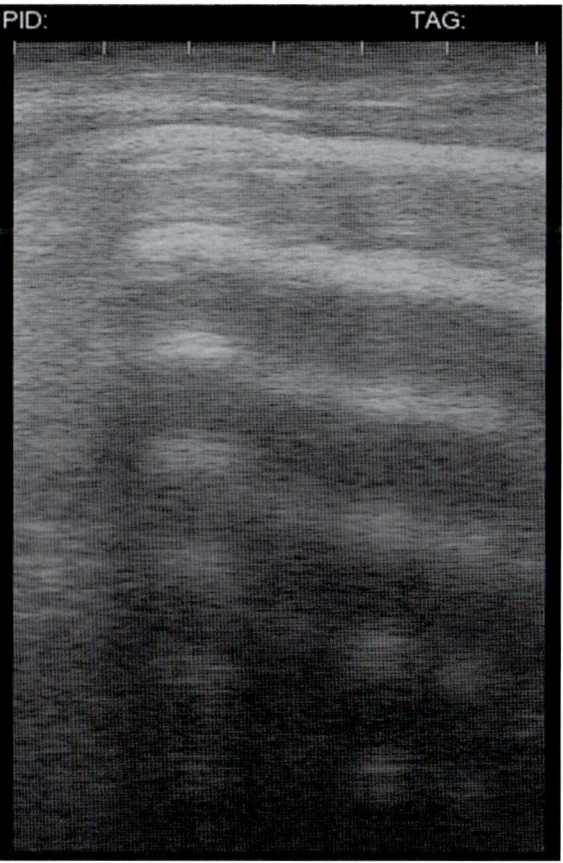

Figure 36.25 Reverberation artifact due to pneumovagina.

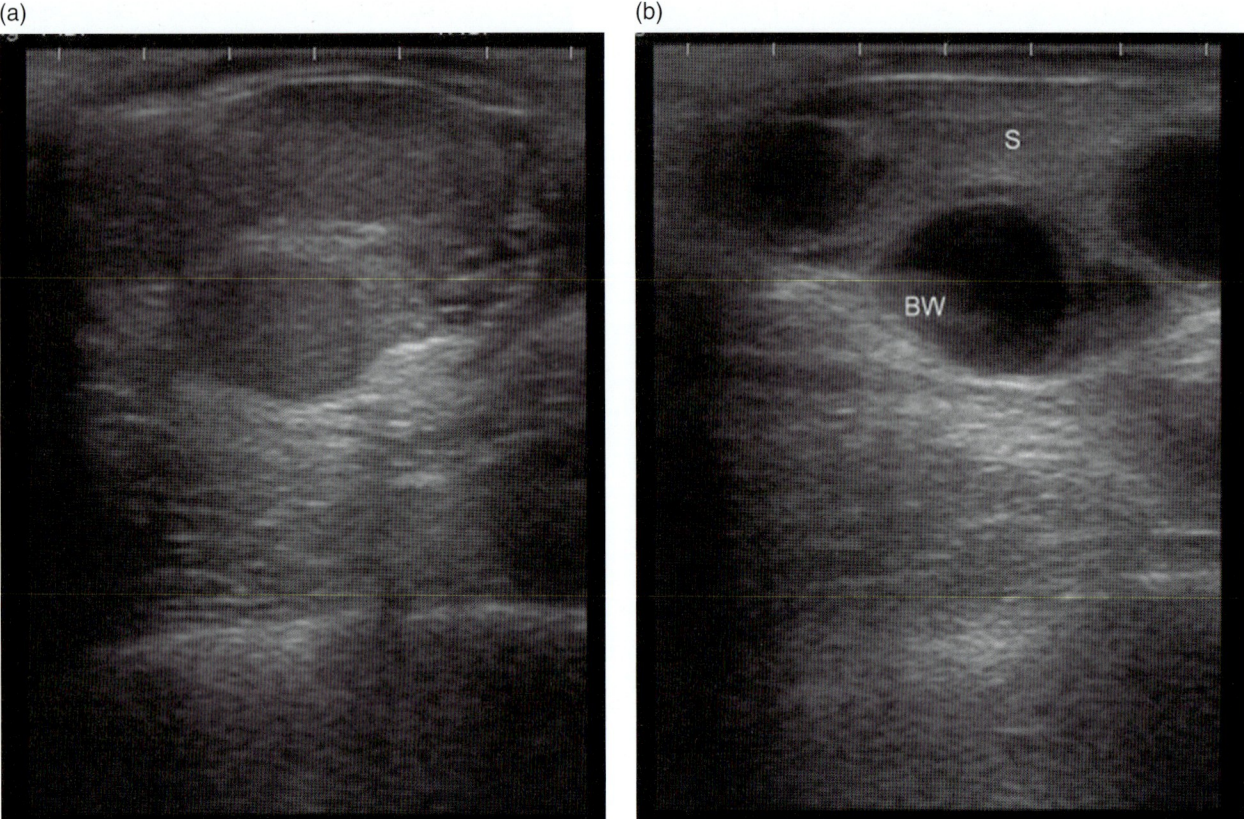

Figure 36.26 (a) Multiple corpora lutea on embryo flush day. (b) Cavitary corpora lutea or unovulated follicles on embryo flush day: ovarian stroma (S), beam width artifact (BW).

follicular wave can be expected to produce more follicles after treatment with follicle-stimulating hormone (FSH).[43,44]

It is usually not necessary to examine the donor cow on the day of insemination unless very rare or expensive semen is to be used. In those cases it is helpful to know if enough ovulatory size (≥8 mm) follicles are present to warrant insemination. Care must be taken to scan gently over the ovaries without manipulation or pressure. It is possible to rupture small follicles or even remove the ovary from the ovarian bursa with rough handling.

On collection day, ultrasound is much more useful than manual palpation for counting the number of corpora lutea.[38] Small corpora lutea, cavitary corpora lutea, and corpora lutea imbedded deep in the ovary can be seen. Figure 36.26 shows two sections of the same ovary. Note the solid corpora lutea in the first section and the cavitary corpora lutea or unovulated follicles in the second section. The ultrasonographer should keep in mind that some cavitary corpora lutea will have very thin walls and may be difficult to differentiate from unovulated follicles.[45] When these fluid-filled structures are seen the cow should still be flushed. It is very possible that more embryos than expected will be retrieved.

Recipient evaluation

Recipients should be evaluated for uterine health and ovarian function prior to beginning estrous synchronization. This is more important for adult cows than for heifers due to increased risk of uterine or ovarian pathology in parous animals.

It is critical that ultrasound is used to identify corpora lutea before embryo transfer (Figure 36.14). Because ultrasound is much more accurate than manual palpation for identifying small or fluid-filled corpora lutea,[37,45] more animals will be used. Almost 100% of corpora lutea in recipients are diagnosed with ultrasound, but only 80% with manual palpation. Maintaining open animals is costly so it is imperative to identify all appropriate recipients. Examination for uterine health should also be performed at this time, especially if not evaluated prior to synchronization.

Pregnancy examination of recipients can take place at any stage comfortable for the ultrasonographer. I have noted that some frozen embryo pregnancies appear 1 day smaller than fresh embryo pregnancies. Also, some *in vitro* fertilized (IVF) embryos appear 1 day larger. This is of no concern as long as the embryo is viable and large enough to be detected on the day post-transfer chosen for examination. Twins are rare in embryo transfer pregnancies, but are at high risk for loss when they occur. Twins may be due to splitting of the embryo, intentional transfer of two embryos, or to insemination of the recipient during estrus prior to embryo transfer.

An ideal time to perform pregnancy examination is at 21 days after transfer (28 days in embryonic development). At this stage open animals with no pathology can immediately receive an embryo if the CL is appropriate. In dairy animals, this is probably more effective in heifers than adult cows. Adult dairy cows often do not have typical 21-day estrous cycles,[46] so may not be at the correct stage of luteal development.

Recheck pregnancy examinations after 60 days of gestation should always be performed due to the high value of most embryo transfer and IVF pregnancies. Fetal gender determination and evaluation for fetal anomalies can also be done at this time.

Ovum pick-up

Collection of ova for IVF is done with a guided needle attached to a transvaginal ultrasound transducer. A curvilinear transducer is usually used, but a standard linear rectal transducer can be adapted for the purpose if attached to a special introducer containing the guided needle.

Embryonic and fetal anomalies

Embryonic and fetal anomalies are rare, but of great importance. Depending on the anomaly it may be desirable to monitor the pregnancy, abort the pregnancy, or take special precautions at parturition. Most anomalies are best seen at about 50–90 days when the fetus is developed enough to resemble a small version of an adult bovine, but not too large to reach for a thorough examination. A few abnormalities can be detected earlier.

Abembryonic vesicle

Figure 36.27 shows a 35-day embryo that is outside the amniotic vesicle. Such embryos rarely survive even if they have a heartbeat on the day of first examination. Recheck examination usually reveals an empty amniotic sac or complete lack of uterine contents.

Schistosomus reflexus

Note the fluid-filled fetal stomach outside the body cavity in Figure 36.28. In most cases the spinal column is severely bent, sometimes into a "V" shape.

Fetal ascites

Note the fluid within the peritoneal cavity in Figure 36.29. In a normal fetus the abdominal organs should completely fill the cavity with no visible fluid. In some cases there is also fluid around the heart or lungs. Careful assessment should be made to differentiate fetal ascites from a dead fetus with fluid accumulation (see Figure 36.3a).

Fetal anasarca

Compare the fluid under the skin in fetal anasarca (Figure 36.30) with the fluid within the abdominal cavity as in fetal ascites (Figure 36.29). Follow-up of fetal anasarca sometimes reveals that the fluid is no longer visibly present. In other cases there is no change or fetal death occurs.

Umbilical hernia

Figure 36.31 shows the fetal chest to the left of the screen and a large umbilicus to the right. The umbilicus contains tissues of various densities probably representing abdominal structures. Not shown in this view is the absence of the fluid-filled fetal stomach within the abdomen.

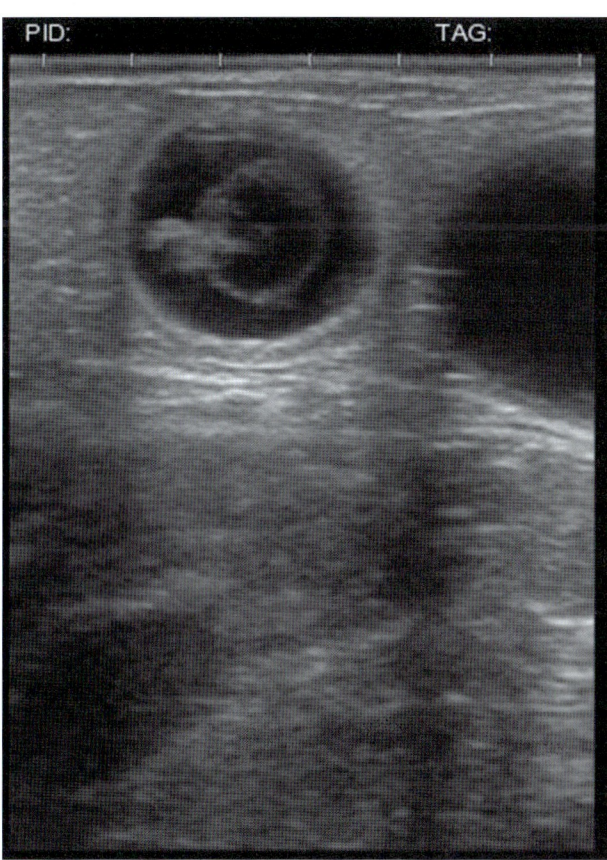

Figure 36.27 Abembryonic vesicle.

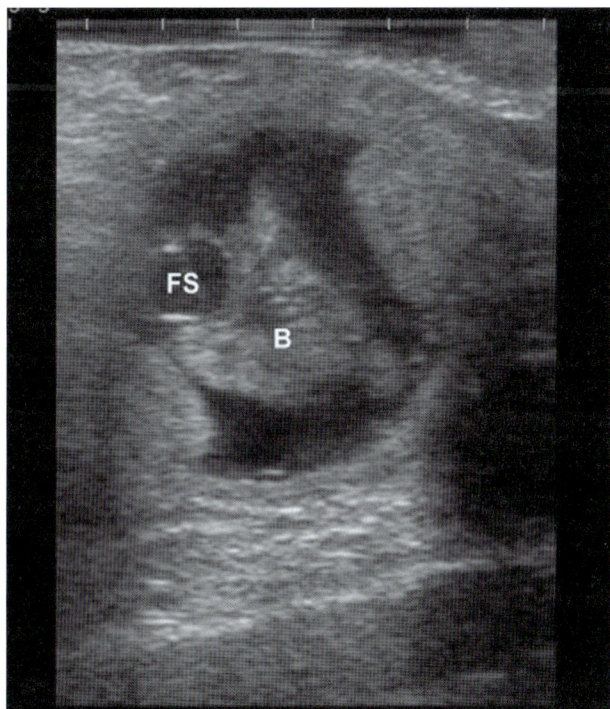

Figure 36.28 Schistosomus reflexus: fetal stomach (FS), fetal body (B).

342 The Cow: Breeding and Health Management

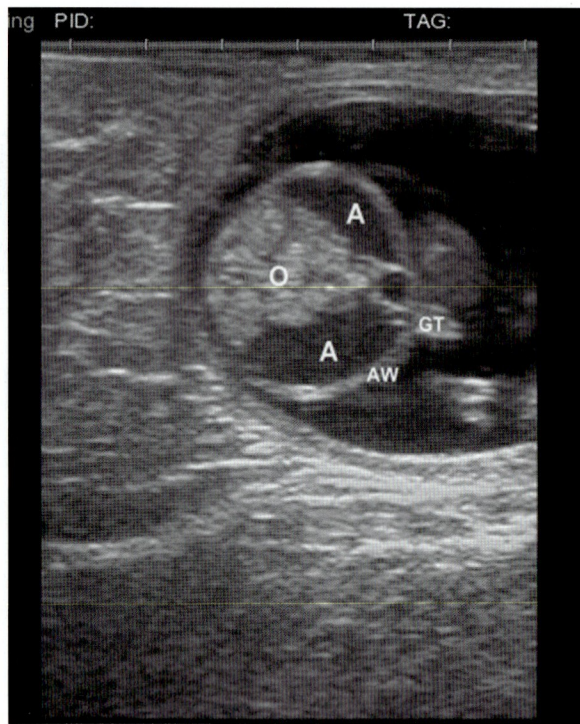

Figure 36.29 Fetal ascites: abdominal organs (O), ascites (S), abdominal wall (AW), male genital tubercle (GT).

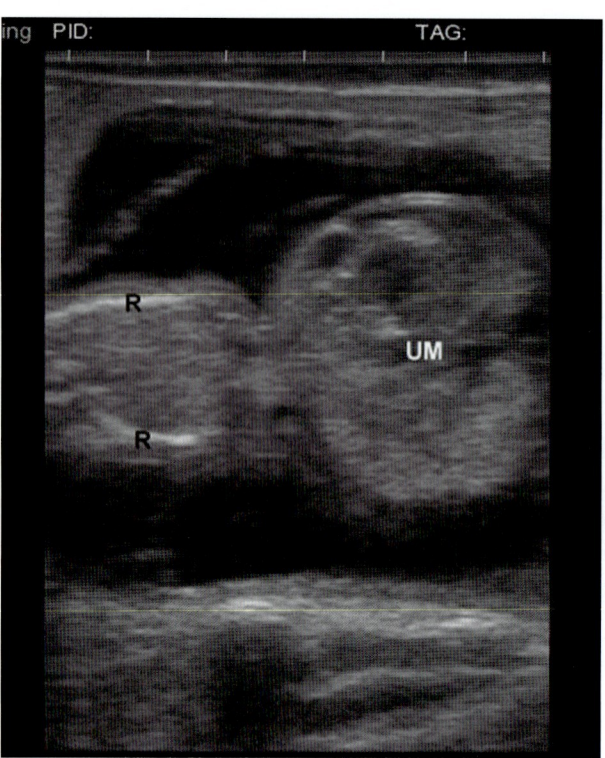

Figure 36.31 Umbilical hernia: herniated umbilicus (UM), ribs (R).

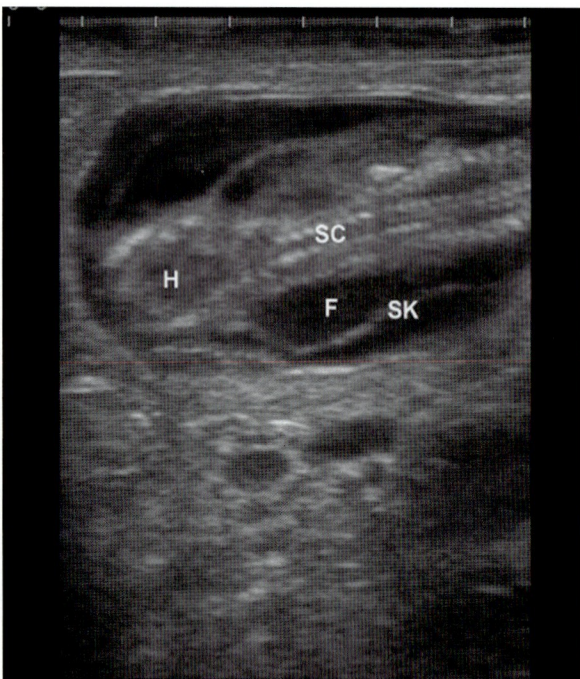

Figure 36.30 Fetal anasarca: skin (SK), fluid (F), head (H), spinal column (SC).

Amorphous globosus or acardiac twin

The round structure to the right of the normal fetus in Figure 36.32 is likely a ball of fetal tissue sustained by the circulatory system of its live twin. These anomalies usually do not grow large so are rarely a problem at parturition. They have no gender so are not of concern for freemartinism.

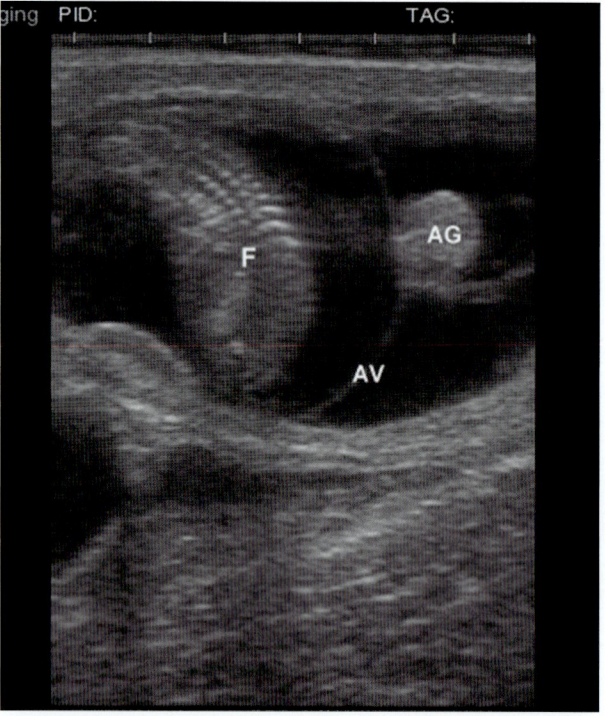

Figure 36.32 Amorphous globosus: normal fetus (F), amorphous globosus (AG), amniotic vesicle (AV).

Conjoined twins

Figure 36.33 shows a fetus with two heads. Conjoined twins are very rare and must be differentiated from normal twins lying close together.

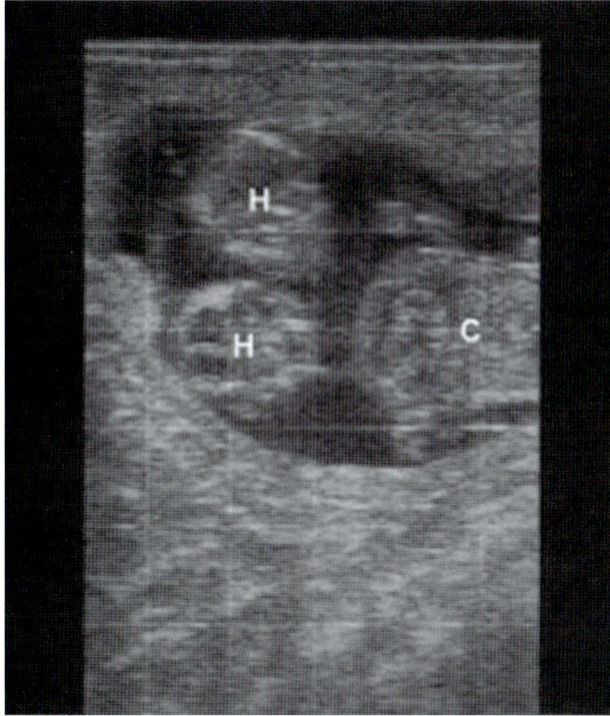

Figure 36.33 Conjoined fetus with two heads: heads (H), chest (C).

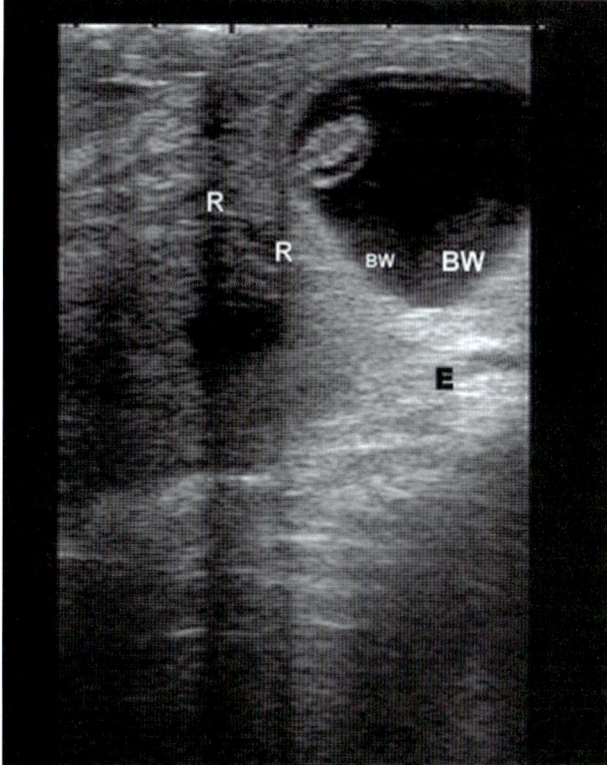

Figure 36.34 Reflection/refraction artifacts (R), beam width artifacts (BW), enhancement artifact (E).

Artifacts

It is important to recognize common artifacts in order to avoid misdiagnosing them as normal structures or pathology. When identified properly artifacts are often helpful in evaluating the reproductive tract. For a more complete description of artifacts and their interpretation refer to Ginther.[1]

Specular reflection artifacts

These common artifacts occur when two interfaces of very different densities are aligned horizontally to each other and perpendicular to the ultrasound pulse. They are most often seen when a fluid-filled structure is adjacent to denser tissue. They appear as thin bright horizontal lines of varying lengths. Figure 36.12 shows specular reflections between the turgid uterine tissue of a cow in estrus and the clear fluid in the lumen. Figure 36.15 shows specular reflections in a cavitary CL. Figure 36.19 shows many small bright specular reflections representing the confluence of the tissue and the fluid-filled vasculature of the granulosa tumor.

Reflection and refraction shadows

Fluid reflects and refracts ultrasound waves just as the surface of a lake reflects and refracts light waves. When the ultrasound beam is either refracted at an angle into the fluid or reflected away from the fluid, it is not echoed back to the transducer. Hence no image will be seen below the point of reflection/refraction. This artifact is commonly seen at the lateral edges of fluid-filled structures (Figure 36.34).

Beam width artifacts

When one ultrasound beam passes through two tissues of different densities at the same depth, a beam width artifact is produced (Figure 36.34). Two echoes reach the transducer at the same time, but only one is read. These artifacts are most common on the lateral walls of fluid-filled structures and cause the sides of the structure to appear more echogenic than the contents. Figure 36.34 shows beam width artifacts between the endometrium and fluid contents of a gravid uterus. Beam width artifacts are most obvious when using ultrasound machines with poor lateral resolution.

Beam width artifacts are also common on the edges of follicles or follicular cysts (Figure 36.26a). In these cases the artifact may be mistaken for luteal tissue. In the uterus they may appear to be abnormal contents in the lumen. In either case a slow scan throughout the structure will help differentiate the artifact from actual tissue.

Reverberation artifacts

Repeated echoing of ultrasound waves between two strong interfaces creates reverberation artifacts. One interface is usually the rectal wall/transducer. The second interface is usually gas or bone. Each time the wave is echoed it is reduced by attenuation. Hence each reverberated echo is weaker than the previous one. Also, the time delay between echoes results in each later echo being seen farther down on the ultrasound screen. The resulting image on the screen consists of equidistant, increasingly smaller, parallel artifacts. Figure 36.25 shows the reverberation artifact caused by pneumovagina.

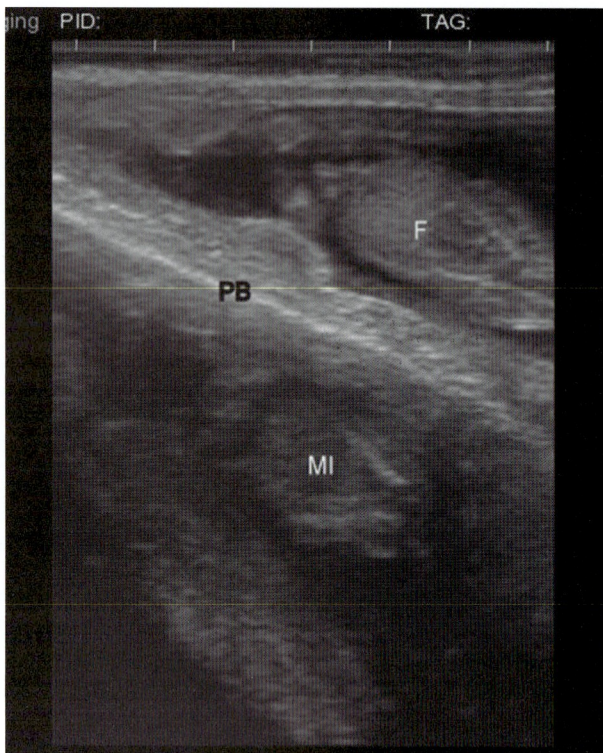

Figure 36.35 Mirror image artifact: fetus (F), pelvic bone (PB), mirror image (MI).

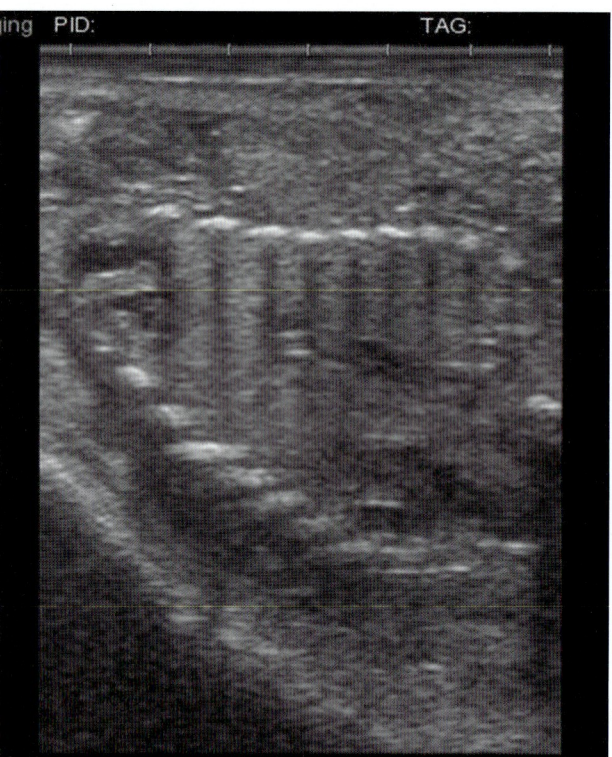

Figure 36.36 Prison bar artifacts.

Mirror image artifacts

Mirror image artifacts are a type of reverberation artifact. They are most common when the pelvic bone lies below the structure being examined. Figure 36.35 shows a 60-day pregnancy with a false "twin" in mirror image below the pelvic bone. Understanding mirror image artifacts helps avoid mistaking one structure for two.

Blockage shadows

Material dense enough to block or attenuate the ultrasound beam causes blockage shadows. Most obvious are shadows caused by bone as in the "prison bar" artifacts shown in Figure 36.36. Less dramatic shadows can also be caused by manure between the transducer and rectal wall leading to a poor-quality dark image.

Enhancement artifacts

Enhancement artifacts are a function of machine settings. Because ultrasound beams are attenuated over time and distance, it is necessary to increase the gain (volume) at deeper levels so that similar tissues at various depths from the transducer appear the same. However, when less echogenic material such as fluid is near the face of the transducer the beam is not attenuated and structures below the fluid appear brighter than they should. Figure 36.34 shows enhancement below the fluid-filled uterus. Gain controls can be manipulated to reduce enhancement artifacts if desired by the ultrasonographer.

Figure 36.37 Crystal defects.

Crystal defects

Crystal defects are seen as black vertical lines extending vertically from the transducer face (Figure 36.37). They can be differentiated from shadow artifacts or reflection/refraction artifacts because they always stay in the same location on the monitor and always begin at the face of the transducer. They may be caused by broken crystals in the transducer or by broken wires connecting the crystals to the power source in the unit. Loss of a few crystals may not be noticed, but loss of many will greatly reduce image quality.

The future

Technologies used in human medicine are being adapted to veterinary use. Three-dimensional ultrasound has been described in mares[47] and may have advantages over two-dimensional imaging for identifying fetal anomalies and for fetal aging. Color Doppler ultrasound has been studied extensively in horses and, to a lesser extent, in cattle.[48] It is currently feasible to use color Doppler ultrasound to assess fetal viability in cattle. In the future it may be possible to use blood flow changes in ovarian structures and the uterus to more accurately assess stage of the estrous cycle and predict ovulation. Some field ultrasound units already come equipped with color Doppler capabilities in addition to standard B-mode ultrasound.

Other uses

Ultrasound has many uses outside the scope of this chapter. Ultrasound of the male reproductive organs has been well described.[49] Other organ systems can be assessed with ultrasound for diagnosis or prognosis of disease.[50]

References

1. Ginther O. *Ultrasonic Imaging and Animal Reproduction: Fundamentals*. Cross Plains, WI: Equiservices Publishing, 1995.
2. Ginther O. *Ultrasonic Imaging and Animal Reproduction: Cattle*. Cross Plains, WI: Equiservices Publishing, 1998.
3. DesCôteaux L, Gnemmi G, Colloton J. *Practical Atlas of Ruminant and Camelid Reproductive Ultrasonography*. Ames, IA: Wiley-Blackwell, 2010.
4. Colloton J. Applications of ultrasonography in dairy cattle reproductive management. In: Risco CA, Melendez Retamal P (eds) *Dairy Production Medicine*. Ames, IA: Wiley-Blackwell, 2011, pp. 99–116.
5. Stroud B. *Bovine Reproductive Ultrasonography* (DVD). Redwood City, CA: BioTech Productions, 1996.
6. Carriere P, DesCôteaux L, Durocher J. *Ultrasonography of the Reproductive System of the Cow* (CD-ROM). Montreal: University of Montreal, 2005.
7. Curran S, Pierson R, Ginther O. Ultrasonographic appearance of the bovine conceptus from days 10 through 20. *J Am Vet Med Assoc* 1986;189:1289–1302.
8. Romano J. Early pregnancy diagnosis by transrectal ultrasonography in dairy cattle. *Theriogenology* 2006;66:1034–1041.
9. Kahn W. Sonographic fetometry in the bovine. *Theriogenology* 1989;31:1105–1121.
10. White I. Real-time ultrasonic scanning in the diagnosis of pregnancy and the estimation of gestational age in cattle. *Vet Rec* 1985;117:5–8.
11. Làbernia J, López-Gatius F, Santolaria P, López-Béjar M, Rutllant J. Influence of management factors on pregnancy attrition in dairy cattle. *Theriogenology* 1996;45:1247–1253.
12. Santos J, Thatcher W, Chebel R, Cerri R, Galvão K. The effect of embryonic death rates in cattle on the efficacy of estrus synchronization programs. *Anim Reprod Sci* 2004;82–82:513–535.
13. Bech-Sabàt G, López-Gatius F, Yániz JL et al. Factors affecting plasma progesterone in the early fetal period in high producing cows. *Theriogenology* 2008;69:426–432.
14. López-Gatius F, Làbernia J, Santolaria P, López-Béjar M, Rutllant J. Effect of reproductive disorders previous to conception on pregnancy attrition in dairy cows. *Theriogenology* 1996;46:643–648.
15. López-Gatius F, Santolaria P, Yániz J, Rutllant J, López-Béjar M. Factors affecting pregnancy loss from gestation Day 38 to 90 in lactating dairy cows from a single herd. *Theriogenology* 2002;57:1251–1261.
16. Silva-del-Río N, Colloton J, Fricke P. Factors affecting pregnancy loss for single and twin pregnancies in a high-producing dairy herd. *Theriogenology* 2009;71:1462–1471.
17. García-Ispierto I, López-Gatius F, Santolaria P et al. Relationship between heat stress during the peri-implantation period and early fetal loss in dairy cattle. *Theriogenology* 2006;65:799–807.
18. Kastelic J, Ginther O. Fate of conceptus and corpus luteum after induced embryonic loss in heifers. *J Am Vet Med Assoc* 1989;194:922–928.
19. Lopez H, Caraviello D, Satter L, Fricke P, Wiltbank M. Relationship between level of milk production and multiple ovulations in lactating dairy cows. *J Dairy Sci* 2005;88:2783–2793.
20. Van Vleck LD, Gregory KE. Genetic trend and environmental effects in a population of cattle selected for twinning. J Anim Sci 1996;74:522–528.
21. López-Gatius F, Hunter R. Spontaneous reduction of advanced twin embryos: its occurrence and clinical relevance in dairy cattle. *Theriogenology* 2005;63:118–125.
22. Van Saun R. Comparison of pre- and postpartum performance of Holstein cows having either a single or twin pregnancy. In: *Proceedings of the 34th Annual Conference of the American Association of Bovine Practitioners*. Auburn, AL: AABP, 2001, p. 204.
23. Andreu-Vázquez C, García-Ispierto I, Ganau S, Fricke P, López-Gatius F. Effects of twinning on the subsequent reproductive performance and productive lifespan of high-producing dairy cows. *Theriogenology* 2012;78:2061–2070.
24. Callahan C, Horstman L. The accuracy of predicting twins by rectal palpation in dairy cows. *Theriogenology* 1990;1:322–324.
25. Silva-del-Río N, Kirkpatrick J, Fricke P. Observed frequency of monozygotic twinning in Holstein dairy cattle. *Theriogenology* 2006;66:1292–1299.
26. López-Gatius F, García-Ispierto I, Hunter R. Factors affecting spontaneous reduction of corpora lutea and twin embryos during the late embryonic/early fetal period in multiple-ovulating dairy cows. *Theriogenology* 2010;73:293–299.
27. López-Gatius F, Szenci O, Bech-Sabat G et al. Factors of non-infectious nature affecting late embryonic and early foetal loss in high producing dairy herds in north-eastern Spain. *Magyar Allatorvosok Lapja* 2009;131:515–531.
28. Andreu-Vázquez C, Garcia Ispierto I, López-Bejár M, de Sousa N, Beckers JF, López-Gatius F. Clinical implications of induced twin reduction in dairy cattle. *Theriogenology* 2011;76:512–521.
29. Curran S, Kastelic J, Ginther O. Determining the sex of the bovine fetus by ultrasonic assessment of the relative location of the genital tubercle. *Anim Reprod Sci* 1989;19:217–227.
30. Stroud B. *Bovine Fetal Sexing Unedited* (DVD). Redwood City, CA: BioTech Productions, 1996.
31. Singh J, Pierson R, Adams G. Ultrasound image attributes of the bovine corpus luteum: structural and functional correlates. *J Reprod Fertil* 1998;112:19–29.
32. Souza A, Silva E, Cunha A et al. Ultrasonographic evaluation of endometrial thickness near timed AI as a predictor of fertility in high-producing dairy cows. *Theriogenology* 2011;75:722–733.

33. Kastelic J, Pierson R, Ginther O. Ultrasonic morphology of corpora lutea and central luteal cavities. *J Am Vet Med Assoc* 1990;194:922–928.
34. Wiltbank M, Gümen A, Sartori R. Physiological classification of anovulatory conditions in cattle. *Theriogenology* 2002;57:21–52.
35. Braw-Tal R, Pen, S, Roth Z. Ovarian cysts in high-yielding dairy cows. *Theriogenology* 2009;72:690–698.
36. Stevenson J. What's all the fuss about cysts? *Hoard's Dairyman* 2006 (October), p. 682.
37. Bicalho R, Galvão K, Guard C, Santos J. Optimizing the accuracy of detecting a functional corpus luteum in dairy cows. *Theriogenology* 2008;70:199–207.
38. Galvão K, Sá Fihlo M, Santos J. Reducing the interval from presynchronization to initiation of timed artificial insemination improves fertility in dairy cows. *J Dairy Sci* 2007;90:4212–4218.
39. Bulman D, Lamming G. Cases of prolonged luteal activity in the non-pregnant dairy cow. *Vet Rec* 1977;100:550–552.
40. Lamming G, Darwash A. The use of milk progesterone profiles to characterize components of subfertility in milked dairy cows. *Anim Reprod Sci* 1998;52:175–190.
41. Gnemmi G. Bovine ovary. In: DesCôteaux L, Gnemmi G, Colloton J (eds) *Practical Atlas of Ruminant and Camelid Reproductive Ultrasonography*. Ames, IA: Wiley-Blackwell, 2010, pp. 55–56.
42. López-Helguera I, López-Gatius F, Garcia-Ispierto I. Influence of genital tract status in postpartum period on the subsequent reproductive performance in high producing dairy cows. *Theriogenology* 2012;77:1334–1342.
43. Durocher J, Morin N, Blondin P. Effect of hormonal stimulation on bovine follicular response and oocyte developmental competence in a commercial operation. *Theriogenology* 2006;65:102–115.
44. Singh J, Dominguez M, Jaiswal R, Adams G. A simple ultrasound test to predict the superstimulatory response in cattle. *Theriogenology* 2004;62:227–243.
45. Stroud B. Bovine embryo transfer, in vitro fertilization, special procedures, and cloning. In: DesCôteaux L, Gnemmi G, Colloton J (eds) *Practical Atlas of Ruminant and Camelid Reproductive Ultrasonography*. Ames, IA: Wiley-Blackwell, 2010, pp. 131–132.
46. Savio J, Boland M, Roche J. Development of dominant follicles and length of ovarian cycles in post-partum dairy cows. *J Reprod Fertil* 1990;88:581–591.
47. Kotoyori Y, Yokoo N, Ito K et al. Three-dimensional ultrasound imaging of the equine fetus. *Theriogenology* 2012;77:1480–1486.
48. Ginther O. *Ultrasonic Imaging and Animal reproduction: Color-Doppler Ultrasonography*. Cross Plains, WI: Equiservices Publishing, 2007.
49. Gnemmi G, Lefebvre R. Bull anatomy and ultrasonography of the reproductive tract. In: DesCôteaux L, Gnemmi G, Colloton J (eds) *Practical Atlas of Ruminant and Camelid Reproductive Ultrasonography*. Ames, IA: Wiley-Blackwell, 2010, pp. 143–162.
50. Buczinski S (ed.) Bovine ultrasound. *Vet Clin North Am Food Anim Pract* 2009;25:553–822.

Chapter 37

Beef Herd Health for Optimum Reproduction

Terry J. Engelken and Tyler M. Dohlman

Department of Veterinary Diagnostics and Production Animal Medicine, Lloyd Veterinary Medicine Center, College of Veterinary Medicine, Ames, Iowa, USA

Introduction

Reproductive efficiency is the most important output parameter affecting the profitability of the beef cow/calf enterprise.[1] While there are many reasons for suboptimal reproductive performance and calf survival, infectious disease is a major contributor and often plays a pivotal role. Reproductive disease may manifest itself in a number of ways, depending on the pathogen involved. Early embryonic death, late-term abortion, "weak calf syndrome," and delayed conception are all common clinical scenarios. However, the end result is that the operation will have fewer kilograms to market from weaned calves. It is critical that the veterinary practitioner be able to understand the relationship of these infectious agents with the risk of exposure, timing of gestational loss, herd diagnostic information, and herd productivity. Only then can comprehensive immunization programs be constructed for the entire operation that will minimize losses associated with reproductive pathogens.

Utilization of diagnostics

It is critical that diagnostic information be utilized in the construction of the herd health program. The selection of vaccines to be used in a program and an individual producer's cattle working schedules depends on the presence of a particular pathogen in the herd or geographical area and on the risk of introduction.[2] It is important to determine the category or period of reproductive loss since reproductive pathogens have a tendency to occur within specific stages of gestation. Classically these categories have been defined as early gestational, mid gestational, late gestational, and peri-parturient (Table 37.1). Determining when gestational losses occur is the first step in understanding the etiology of these losses and represents the starting point for building the vaccination program.

Utilizing diagnostic tools in cases of abortion can be very helpful and can be very frustrating due to diagnostic limitations and challenges. In cases of abortion an etiologic diagnosis is identified less than 50% of the time.[3] However, several diagnostic advances, through conscious efforts by diagnostic laboratories, have been made to improve detection of certain pathogens.[4] Even though an etiologic diagnosis can be challenging, infectious agents causing abortions can be identified. Table 37.2 highlights the multiple pathogens that have been associated with bovine abortions, the gestational period they normally affect, and common diagnostic procedures used to identify their presence.[2,5–9] Some pathogens are considered common reproductive agents and some are rarely identified. In addition, certain pathogens can be more prevalent in certain geographical areas.[3]

Diagnostic laboratories have different capabilities in handling an abortion case. As a practitioner, it is prudent to have a close working relationship with your laboratory. Diagnostician consultations are sometimes necessary to develop a systematic approach for abortions. Appropriate tissues and bodily fluids samples are more useful than others and comprehensive submissions are valuable.

Adequate information, including herd history, with complete submission to a diagnostic laboratory is the most important step to a definitive diagnosis.[4] An abortion work-up can be costly and sometimes unrewarding. It is in the best interest to provide the diagnostic laboratory with the required information to minimize cost and maximize diagnostic success. Even though infectious agents comprise most of the reported etiology, practitioners, producers, and diagnosticians have to be aware that abortions are not limited to infectious agents.[9] Noninfectious causes, such as environment, toxins, nutrition, and genetics, have to be considered as possible sources of reproductive failures.

Serologic sampling can be utilized in diagnostic work-ups for abortions but the results should always be interpreted with caution. Single serum samples have little to no value when diagnosing causes for abortions. It is difficult to differentiate the titer values based on vaccine and natural

---

Bovine Reproduction, First Edition. Edited by Richard M. Hopper.
© 2015 John Wiley & Sons, Inc. Published 2015 by John Wiley & Sons, Inc.

exposure to certain agents/antigens. However, a lack of titer in abortions may rule out certain causes.

Paired serum samples have also demonstrated limited value. Many of the bacterial and viral pathogens that cause abortion may infect the fetus or placenta long before the abortive event occurs. This lag time between infection and abortion may prevent the practitioner from detecting the rising or falling titers associated with the initial infection. This leads to the collection of two "convalescent" serum samples that will fail to detect the increase in antibody titer, if it indeed occurred. This is especially true when only affected females are sampled at the time when the abortion is noted. Overall, the time of seroconversion is dependent on the exposure of the agent and the amount of immunity established prior to the breeding and throughout gestation. Paired sera are much more useful when it is used as part of a complete diagnostic work-up that includes samples from the placenta, fetus, and fetal fluids.

Serologic profiling is one option to optimize the use of serologic testing. The basis of serologic profiling is analyzing titers from affected/aborted and nonaffected dams over the same time period.[4] It is unclear how many samples are needed, but some suggest that the same number of affected and nonaffected animals, preferably at same stage of gestation and age, is adequate.[5] In herds with chronic gestational losses serum may be collected and frozen from a statistically relevant number of cows for future testing as needed. These frozen samples may be collected as the females are processed prior to breeding and/or at the time of pregnancy examination. Then, as fetal loss is detected, banked serum samples can be submitted along with acute and convalescent samples to provide a clearer serologic picture of the affected animals and their normal cohorts. This should give a more complete picture of when seroconversion occurred and what pathogens were involved.

Table 37.1 Pathogens associated with reproductive wastage in beef cows by gestational period.

Pathogen	Gestational period			
	Early	Mid	Late	Periparturient
Histophilus somni	Low		High	Moderate
Brucella abortus	Low		High	Low
IBR virus	Low	Moderate	High	Low
Bluetongue virus	Low		Moderate	Low
BVDV	Low	High	Low	Low
Leptospiral serovars	Low		High	Low
Campylobacter fetus subsp. *venerealis*	High	Moderate		
Tritrichomonas foetus	High	Low		
Aspergillus fumigatus			Low	
Ureaplasma spp.	Low			
Listeria monocytogenes			Low	
Chlamydia psittaci			Low	
Parainfluenza 3 virus			Low	Low
Mixed bacteria			Low	Low

Source: Spire M. Immunization of the beef breeding herd. *Comp Cont Educ Pract Vet* 1988;10:1111–1117. Used with permission.

Table 37.2 Agent-related abortions categorized by period of gestation and diagnostic procedures.

Pathogen	Gestational period	Common diagnostic procedures
Viral		
IBR	Mid to late	FA, IHC, VI
BVDV	Mid	FA, IHC, VI, PCR
Bluetongue virus	Late	PCR, VI
Bacterial		
Brucella abortus	Mid to late	Bacterial culture
Leptospira spp.	Late	Culture, IHC, PCR
Campylobacter fetus subsp. *venerealis*	Any	Culture, MAT, FA, IHC
Ureaplasma spp.	Late	Culture, PCR
Sporadic/opportunistic bacteria[a]	Any	Bacterial culture
Listeria	Late	Bacterial culture, IHC
Chlamydophila spp.	Late	PCR, IHC, FA
Protozoal		
Tritrichomonas	Early	Culture, IHC, silver stain, H+E, PCR
Neospora spp.	Mid to late	IHC, PCR
Sarcocystis spp.	Mid to late	IHC
Mycotic/fungal		
Aspergillus fumigatus	Any	Fungal culture, H+E
Mucor/Candida/Rhizopus spp.[b]	Any	Fungal culture, H+E
Unknown agent		
Epizootic/enzootic bovine abortion	Late	IHC, silver stain

[a] *Arcanobacterium pyogenes*, *Bacillus* spp., *Escherichia coli*, *Mannheimia haemolytica*, *Streptococcus* spp., *Pasteurella multocida*, *Salmonella* spp., etc.
[b] Many genus and species associated with bovine abortions.
FA, fluorescent antibody; IHC, immunohistochemistry; VI, virus isolation; PCR, polymerase chain reaction; H+E, hematoxylin and eosin; MAT, microagglutination test.

In some cases, all samples will have elevated titers due to endemic infections of specific agents. For example, in herds endemically infected with bovine viral diarrhea virus (BVDV), all animals may have seroconverted yet have no clinical evidence of etiologic diagnosis of abortions.[10] On rare occasions, fetal/precolostral serology may be beneficial. Fetuses must be immunocompetent for specific agents (*Toxoplasma gondii*/*Neospora caninum*/BVDV/infectious bovine rhinotracheitis/*Brucella*) to produce serologic evidence in fetal fluids.[3,4,11] While serology may be valuable in certain circumstances, the interpretation of these results should be carefully assessed as it relates to the entire diagnostic work-up and clinical signs in the herd.

Designing herd vaccination protocols

While vaccines represent an important tool in protecting reproductive performance, they tend to be somewhat underutilized in beef herds (Table 37.3). When designing protocols to immunize the beef breeding herd against reproductive pathogens, there are several other important factors to consider. The potential at-risk level of the herd should be considered not only from the entry of potential pathogens, but also from the standpoint of the current disease level in the resident herd, different management groups on the ranch, breeding animal movement, and the potential side effects of the immunizing agents.[2] While complete protection against every pathogen in every individual is not realistic, the goal would be to minimize the number of susceptible animals in the population. This should prevent epidemic outbreaks of reproductive disease as well as the establishment of chronic endemic losses in the cow herd.

While veterinarians and producers often think of individual vaccination protocols for different management groups on the ranch, it is our belief that vaccination programs should be viewed as a continuum. For example, if producers are developing their own replacement heifers, the suckling calf vaccination program should be viewed beyond the summer grazing season and fall weaning events. This vaccination program should be constructed to take into account the probability that these young heifer calves will join the replacement pool, become pregnant, and eventually become a productive member of the mature herd. The suckling calf protocol should be designed to prepare the calf for post-weaning disease challenges and increase the calf's response to subsequent reproductive vaccination. Research has clearly shown that calves vaccinated at an early age will mount a cell-mediated immune response that will enhance the calf's ability to respond to subsequent vaccination or disease challenge.[12,13] This approach will maximize protection against reproductive pathogens and minimize the potential for any negative vaccine side effects associated with the pre-breeding vaccination of seronegative females.[14-16] These side effects may include multifocal areas of ovarian necrosis, hemorrhage and inflammatory cell infiltrate in the ovary, as well as the development of cysts in the corpus luteum. These lesions are transitory in nature, but can result in decreased reproductive performance in the short term.

Other factors to consider in vaccine selection include fetal protection and duration of immunity. Recent advances in vaccine technology and diagnostic testing have allowed vaccine manufacturers to document the ability of their products to prevent disease organisms from spreading to the placenta and fetus following maternal infection. Challenge studies using virulent BVDV, infectious bovine rhinotracheitis (IBR), and *Leptospira borgpetersenii* (serovar *hardjo*) have shown that fetal protection against pregnancy wastage, BVDV persistent infection (PI), and leptospiral renal colonization and urine shedding is possible following vaccination.[17-21] Studies have also shown that this protection can last for 1 year or longer following vaccination of animals of various ages.[17,19,22] The concepts of fetal protection and duration of immunity are especially important for beef operations as they are more likely to come in contact with adjacent herds and may only be handled for vaccination once per year.

Before constructing any vaccination program for a cow/calf operation, the potential risk for exposure of the herd to a particular pathogen through herd additions or herd contact with clinical or inapparent carriers of a pathogen should be evaluated. The epidemiological terms "open," "closed," and "modified open" have been used to describe the potential risk level of a given herd.[2] When assessing the need for vaccination, factors such as risk-level management, the magnitude and etiology of previous reproductive losses, herd working patterns and animal management, and the producer's long-term goals should all be considered. Once this information is collected and evaluated, recommendations concerning the use of specific vaccine antigens, the type of vaccine needed, and the frequency of vaccination can be constructed to fit within the confines of the total ranch management plan.[2]

Table 37.3 Use of reproductive vaccines by cow/calf operations.

Type of vaccine	Use in weaned heifers (%)	Use in pregnant heifers (%)	Use in cows (%)
Infectious bovine rhinotracheitis	19.4	11.9	24.6
BVDV	25.1	13.7	28.1
Campylobacter	12.6	10	19
Leptospira	19.9	15.1	37.1
Brucella abortus	14.8	2.8	1
Histophilus somni	9.3	5.3	7.9
Neospora	—	0.4	0.3
Tritrichomonas	0.7	0.9	1

Source: National Animal Health Monitoring System. *Beef 2007–08, Part IV. Reference of Beef Cow-Calf Management Practices in the United States, 2007–08*. Fort Collins, CO: USDA, 2009.

First level: closed herd vaccination

Closed herds run a relatively low risk of disease introduction due to the fact that there is limited movement of breeding animals on and off the operation. These operations develop their own replacement females and have little to no contact with adjacent herds. The primary potential source for disease introduction would be in the form of venereal disease from the addition of breeding bulls. There is also the potential of exposure to leptospiral serovars from domestic livestock and wildlife populations in the area. This type of management will create a herd that is highly susceptible to viral and bacterial reproductive diseases but does so at a very low level of risk for disease introduction.[2]

Adequate quarantine protocols and facilities are important for all operations, but this is especially true for closed herds. While the risk of introduction of viral reproductive pathogens is very low in these herds, it can be devastating if it occurs. Therefore quarantine procedures must prevent nose-to-nose contact between new arrivals and the resident cow herd for a minimum of 30 days. This time period should allow animals that are shedding IBR virus following recrudescence or undergoing acute BVD to clear the viremia and stop shedding prior to introduction with the resident herd.[23] Individual animals can also be vaccinated appropriately as they enter and again as they leave the quarantine pens to ensure that herd immunity is maintained and the resident herd protected. This time frame also ensures that all diagnostic screening performed on arrival, such as testing for BVD-PI or *Tritrichomonas foetus*, will be completed and test results returned to the practitioner prior to the animal being released from quarantine. An additional step to enhance biosecurity would be to screen culled animals annually to gain insight as to the level of seroconversion that is occurring in the breeding herd.[2] Effective quarantine procedures not only help prevent disease introduction but will also enhance the effectiveness of the vaccination program.

Vaccination of closed herds should center on pathogens that are difficult to control through biosecurity measures alone. Annual boosters with multivalent products containing *Leptospira* and *Campylobacter fetus* subsp. *venerealis* should be utilized in all breeding females on the operation.[2] Depending on the normal working patterns on the operation and the anticipated timing of disease exposure, these vaccine antigens may be given before breeding prior to bull turnout or in mid gestation at the time of pregnancy examination (Table 37.4). While *C. fetus* vaccination would be most effective if given before breeding, it is commonly given in concert with the *Leptospira* serovars in mid gestation since it is unlikely the herd would be handled an additional time. Herd bull vaccination schedules should match the antigens and timing of the female group that they will be turned in with.

Replacement heifers represent the highest at-risk population for reproductive pathogens, both viral and bacterial.[2] The antigens given to these heifers during the suckling calf and weaning phases should prepare these females to respond adequately to the reproductive vaccines given just prior to the first breeding season. This pre-breeding vaccination can be administered 30–45 days prior to breeding as the final selections for the breeding pool takes place (Table 37.4). Since these heifers will have had previous vaccine exposure, are relatively stress-free, and are not pregnant, they make the ideal candidates to utilize modified-live IBR and BVDV vaccines. This will enable the herd to maximize protection and minimize potential vaccine side effects.[2,16]

Second level: modified open herd vaccination

Modified-open herds represent a moderate risk level for disease introduction and should be considered partially susceptible to viral diseases at all times. These herds are best described as those in which the addition of new animals takes place on a limited basis (herd expansion or replacement with purchased additions), individual animals move in and out of the herd (to livestock shows or exhibitions), or there may be adult animal-to-animal contact with adjoining herds.[2] It is imperative that newly introduced replacements, whether raised or purchased, carry a high level of protective immunity to the most common viral pathogens. Addition of replacements of this type will lessen the need to rely on annual viral vaccination of the resident breeding herd. Well-designed quarantine protocols are also important for modified-open herds as well.

Yearling replacement heifers represent the main concern for building an immune base within modified-open herds. The pre-breeding vaccination should include IBR, BVDV, the appropriate *Leptospira* serovars, and *C. fetus* (Table 37.4). These heifers are not pregnant so it would be preferable to take advantage of the longer duration of immunity provided by modified-live viral vaccines. BVDV types 1 and 2 should be included in the vaccine to ensure a higher level of heterologous protection against strain variation.[24] Whether these heifers are home-raised or purchased, it is important to emphasize the use of vaccines that have been proven to provide fetal protection.

If producers are purchasing pregnant replacements or additions, killed or intranasal combinations of IBR and BVDV provide the best alternative to avoiding the negative side effects of modified-live virus vaccination.[2] This is especially important if the vaccination history of these animals is undocumented prior to their arrival (Table 37.4). Following calving, modified-live viral vaccines may be used before breeding in subsequent years. There are modified-live vaccines that are labeled to be used in pregnant females. However, it is critical that the label directions on these vaccines are followed with regard to the timing and the number of doses given before breeding while the females are not pregnant. This is essential to avoid vaccine-induced pregnancy wastage. As in closed herds, annual vaccination for *Leptospira* and *Campylobacter* are conducted in all adult replacement females, and bulls.[2]

Third level: open herd vaccination

Open herds represent the greatest challenge to a vaccination program since these operations have the highest risk for the introduction of both viral and bacterial reproductive pathogens.[2] These pathogens are typically introduced via animal movement on and off the ranch. This movement typically occurs via the routine purchase of replacements, direct contact or commingling of the breeding herd with stressed high-risk stocker or feeder calves, or through the salebarn purchase of orphan calves for the purpose of grafting onto a lactating female in the resident herd. These purchased replacements may be heifers or mature cows and can be pregnant or open. The main issue with these purchased

Table 37.4 Example vaccination program based on risk management concept.

Closed herd

Adult (female and male)
Campylobacter: pre-breeding
Leptospira: pre-breeding or mid-gestational period

Replacements (pre-breeding)
IBR and BVDV: MLV s.c.
Campylobacter: pre-breeding
Leptospira: pre-breeding or mid-gestational period
Brucella (optional): 4–10 months

Modified open herd

Adult (female and male)
Campylobacter: pre-breeding
Leptospira: pre-breeding or mid-gestational period

Replacements (nonpregnant, pre-breeding)
Brucella: 4–10 months
IBR and BVDV: MLV s.c.
Campylobacter
Leptospira

Replacements (pregnant before introduction)
IBR: killed virus s.c. or i.n.
BVDV: killed virus s.c.
Campylobacter
Leptospira

Open herd

Adult (nonpregnant females, 30–45 days pre-breeding)
IBR and BVDV: MLV s.c.

Pregnant females and bulls
IBR: killed virus s.c. or i.n.
BVDV: killed virus s.c.

All adult animals
Campylobacter: pre-breeding
Leptospira: pre-breeding or mid-gestational period

Replacements (nonpregnant, pre-breeding)
Brucella: 4–10 months
IBR and BVDV: MLV s.c.
Campylobacter
Leptospira

Replacements (pregnant before introduction)
IBR: killed virus s.c. or i.n.
BVDV: killed virus s.c.
Campylobacter
Leptospira

IBR, infectious bovine rhinotracheitis; BVDV, bovine viral diarrhea virus; MLV, modified live virus; s.c., subcutaneous; i.n., intranasal.
Source: adapted from Spire M. Immunization of the beef breeding herd. *Comp Cont Educ Pract Vet* 1988;10:1111–1117. Used with permission.

females is that the practitioner and producer typically know nothing about their previous vaccination or disease history. Proper quarantine protocols and disease screening at arrival are absolutely critical components to minimize disease risk in this type of herd.

The primary emphasis of the vaccination program for open herds centers on the addition of viral components to maintain a high level of collective herd immunity to these pathogens. This will require annual viral vaccinations in the breeding herd, while continuing annual vaccination against the common bacterial pathogens.[2] A modified-live viral vaccine would be most commonly used 30–45 days prior to the start of the breeding season in open cows. In those situations where the breeding herd is only vaccinated at pregnancy examination (mid gestation) or the herd has a prolonged calving season, the use of an annual killed virus vaccine may be the only option (Table 37.4).

The use of viral vaccines in replacement females should follow the same guidelines as those for the modified-open herd. This will minimize the risk of introducing susceptible females into the herd, help establish an immune base for the operation, and ensure that the subsequent pregnancy is protected.[2] Depending on the source of breeding bulls in these herds and the risk of venereal disease introduction, *T. foetus* vaccine could be considered in these heifers and cows. However, a recent critical review and meta-analysis of the efficacy of *T. foetus* vaccine indicated that while there may be a small positive effect on decreasing abortions in vaccinated heifers, there was no effect on decreasing the infection rate or the risk of being open.[25]

The delivery of vaccines for *Leptospira* and *Campylobacter* will be the same in open herds as was described for closed herds (Table 37.4). All adult females, replacement females, and herd bulls will receive these antigens at least annually.[2] These may be given before breeding or at mid gestation, depending on the normal working patterns associated with the ranch management.

Summary

The process of designing immunization programs for beef cattle operations must take into consideration factors such as traffic patterns on and off the ranch, normal handling times, the pregnancy status of the animals to be vaccinated, historical disease patterns, and the relative risk of disease introduction. The goal of the immunization program should be to increase the level of collective herd immunity by minimizing the number of animals that are susceptible to reproductive disease. This will prevent not only epizootic outbreaks of pregnancy wastage, but should also control chronic endemic disease. Our ability to better understand the relationship between reproductive pathogens and the bovine reproductive tract has enabled vaccine manufacturers to provide products that ensure fetal protection and a long duration of immunity, while minimizing negative vaccine side effects. The end result is that the practitioner can provide the client with cost-effective vaccine options to help insure optimum reproductive performance.

References

1. Dhuyvetter K, Herbel K. Differences between high-, medium-, and low-profit producers: an analysis of the Kansas Farm Management Association Beef Cow-Calf Enterprise. Department of Agricultural Economics, Kansas State University, November, 2012. Available at http://www.agmanager.info
2. Spire M. Immunization of the beef breeding herd. *Comp Cont Educ Prac Vet* 1988;10:1111–1117.
3. Anderson M. Infectious causes of bovine abortion during mid- to late-gestation. *Theriogenology* 2007;68:474–486.

4. Holler L. Ruminant abortion diagnostics. *Vet Clin North Am Food Anim Pract* 2012;28:407–418.
5. Givens M. A clinical, evidence-based approach to infectious causes of infertility in beef cattle. *Theriogenology* 2006;66: 648–654.
6. Kirkbride C. Appendix B. Summary tables of diagnostic tests for infectious causes of abortion and neonatal loss: In: Njaa BL (ed.) *Kirkbride's Diagnosis of Abortion and Neonatal Loss in Animals*, 4th edn. Ames, IA: Wiley-Blackwell, 2012, pp. 225–227.
7. Kirkbride C. Viral agents and associated lesions detected in a 10-year study of bovine abortions and stillbirths. *J Vet Diagn Invest* 1992;4:374–379.
8. Kirkbride C. Bacterial agents detected in a 10-year study of bovine abortions and stillbirths. *J Vet Diagn Invest* 1993;5:64–68.
9. Kirkbride C. Etiologic agents detected in a 10-year study of bovine abortions and stillbirths. *J Vet Diagn Invest* 1992;4: 175–180.
10. Rodning S. Reproductive and economic impact following controlled introduction of cattle persistently infected with bovine viral diarrhea virus into a naive group of heifers. *Theriogenology* 2012;78:1508–1516.
11. Iowa State University College of Veterinary Medicine, Veterinary Diagnostic Laboratory. Abortion serology. Available at http://vetmed.iastate.edu/diagnostic-lab/user-guide/pathology#abortion_serology
12. Ellis J, Hassard L, Cortese V, Morley P. Effects of perinatal vaccination on humoral and cellular immune responses in cows and young calves. *J Am Vet Med Assoc* 1996;208:393–400.
13. Platt R, Widel P, Kesl L, Roth J. Comparison of humoral and cellular immune responses to a pentavalent modified live virus vaccine in three age groups of calves with maternal antibodies, before and after BVDV type 2 challenge. *Vaccine* 2009;27: 4508–4519.
14. Chiang B, Smith P, Nusbaum K, Stringfellow D. The effect of infectious bovine rhinotracheitis vaccine on reproductive efficiency in cattle vaccinated during estrus. *Theriogenology* 1990;33:1113–1119.
15. Perry G, Zimmerman A, Russell F *et al.* The effects of vaccination on serum hormone concentrations and conception rates in synchronized naïve beef heifers. *Theriogenology* 2013;79:200–205.
16. Spire M, Edwards J, Leipold H, Cortese V. Absence of ovarian lesions in IBR seropositive heifers subsequently vaccinated with a modified live IBR virus vaccine. *Agri-Practice* 1995;16:33–38.
17. Ficken M, Ellsworth M, Tucker C. Duration of the efficacy of a modified-live combination vaccine against abortion caused by virulent bovine herpesvirus 1 in a one-year duration-of-immunity study. *Vet Therapeut* 2006;7:275–282.
18. Fairbanks K, Reinhart C, Ohnesorge W, Loughin M, Chase C. Evaluation of fetal protection against experimental infection with type 1 and type 2 bovine viral diarrhea virus after vaccination of the dam with a bivalent modified-live virus vaccine. *J Am Vet Med Assoc* 2004;225:1898–1904.
19. Fricken M, Ellsworth M, Tucker C. Evaluation of the efficacy of a modified-live combination vaccine against bovine viral diarrhea virus types 1 and 2 challenge exposures in a one-year duration-of-immunity fetal protection study. *Vet Therapeut* 2006;7:283–294.
20. Bolin C, Alt D. Use of a monovalent leptospiral vaccine to prevent renal colonization and urinary shedding in cattle exposed to *Leptospira borgpetersenii* serovar hardjo. *Am J Vet Res* 2001;62:995–1000.
21. Rinehart C, Zimmerman A, Buterbaugh R, Jolie R, Chase C. Efficacy of vaccination of cattle with the *Leptospira interrogans* serovar *hardjo* type hardjoprajitno component of a pentavalent *Leptospira* bacterin against experimental challenge with *Leptospira borgpetersenii* serovar *hardjo* type hardjo-bovis. *Am J Vet Res* 2012;73:735–740.
22. Zimmerman A, Springer E, Barling K *et al*. Immunity in heifers 12 months after vaccination with a multivalent vaccine containing a United States *Leptospira borgpetersenii* serovar hardjo isolate. *J Am Vet Med Assoc* 2013;242:1573–1577.
23. Kelling C. Viral diseases of the fetus. In: Youngquist RS, Threlfall WR (eds) *Current Therapy in Large Animal Theriogenology*, 2nd edn. St Louis, MO: Saunders Elsevier, 2007, pp. 399–408.
24. Brock K, Cortese V. Experimental fetal challenge using type II bovine viral diarrhea virus in cattle vaccinated with modified-live virus vaccine. *Vet Therapeut* 2001;2:354–360.
25. Baltzell P, Newton H, O'Connor A. A critical review and meta-analysis of the efficacy of whole-cell killed *Tritrichomonas foetus* vaccines in beef cattle. *J Vet Intern Med* 2013;27:760–770.

Chapter 38

Dairy Herd Health for Optimal Reproduction

Carlos A. Risco

Large Animal Clinical Sciences, College of Veterinary Medicine, University of Florida, Gainesville, Florida, USA

Introduction

The profitable operation of a dairy herd is influenced by the reproductive performance of the lactating cows. The level of reproductive performance that optimizes profit is a result of the combination of the days that a cow spends in the most efficient time of the lactation curve and the cull rate due to reproductive failure.[1] The main factors that determine reproductive performance in dairy herds are the voluntary waiting period (VWP), insemination rate, pregnancy per artificial insemination (AI), and pregnancy loss.[2] The outcome of these factors depends on the reproductive herd health program of the herd.[3]

Typically, reproductive programs on dairy herds are focused on hormonal manipulation of the estrous cycle to obtain a pregnancy in a timely manner at the end of the VWP. However, due to the relationship of biological and management factors that determine when cows become pregnant at the end of the VWP and whether or not pregnancy is maintained, a reproductive program needs to integrate the principles of theriogenology and herd health in order to optimize reproductive performance.

This chapter evaluates the importance of the outcome of the factors that affect reproductive performance considering a herd health approach and discusses management strategies to improve herd fertility.

Impact of postpartum diseases on the VWP

The VWP is the set time postpartum that producers choose not to breed cows to allow the uterus to recover from infection and for cows to achieve a positive energy balance and resume estrous cyclicity. These conditions can be viewed as physiological requirements for an optimal time to pregnancy at the end of the VWP. The challenge for the dairy cow is that following parturition they experience some degree of hypocalcemia, and due to increased nutrient demands for milk yield early postpartum, a state of negative energy balance (NEB) occurs, which affect uterine health and estrous cyclity.

Hypocalcemia and NEB appear to prevent cows from mounting an effective immune response to the microbial challenge to the uterus after calving. Ribeiro et al.[4] reported that low calcium concentration in the first 7 days postpartum was associated with an increased incidence of metritis. Martinez et al.[5] reported that dairy cows that developed metritis had decreased calcium concentrations (<8.59 mg/dL) in the first 3 days postpartum, and was associated with decreased neutrophil function. Further, the relative risk of developing metritis decreased by 22% for every 1 mg/dL increase in serum calcium in the first 3 days postpartum.[5] In a study by Hammon et al.,[6] cows that were in NEB because of low dry matter intake before and after calving had suppressed neutrophil function and developed subsequent uterine infections postpartum. Dubuc et al.[7] reported that risk factors for metritis among dystocia and retained fetal membrane included increased nonesterified free fatty acids (≥0.6 mmol/L), a marker for NEB in the first week postpartum. Energy status has been linked with a delayed resumption of estrous cyclicity and ovulation postpartum.[8] NEB reduces the frequency of luteinizing hormone (LH) pulses, thereby impairing follicle maturation, and inhibits estrous expression in part due to a reduction of estrogen receptor α in the brain.[9]

There has been much discourse in the literature regarding the association between high milk yield and NEB in lowering fertility in dairy cows over the past 30 years. Nonetheless, little or no association has been observed between milk production in early lactation and the risk of anovulation, pregnancy, and pregnancy loss in high-producing dairy cows.[10] However, of greater concern to producers is the high incidence of health disorders postpartum, particularly those that affect the reproductive tract and those of metabolic origin, on reproductive performance. Santos et al.[11] evaluated data from 5719 postpartum dairy cows daily for health disorders. All cows were evaluated for

---

Bovine Reproduction, First Edition. Edited by Richard M. Hopper.
© 2015 John Wiley & Sons, Inc. Published 2015 by John Wiley & Sons, Inc.

cyclicity at 65 days postpartum and were subjected to presynchronized timed AI programs to their first service. Only 55% of cows were considered healthy and did not develop clinical disease in the first 60 days postpartum. Incidence of diseases observed were as follows: calving-related problems, 14.6%; metritis, 16.1%; clinical endometritis, 20.8%; fever, 21.0%; mastitis, 12.2%; ketosis, 10.4%; lameness, 6.8%; digestive problems, 2.8%; and pneumonia, 2.0%. Twenty seven per cent of the cows were diagnosed with a single disease event, whereas 17.2% had at least two disease events in the first 60 days postpartum. Although these health disorders did not influence milk yield (10 919 kg, 11 041 kg, and 10 858 kg for cows considered healthy, those with a single disease, and those with multiple diseases, respectively), they depressed cyclicity, reduced pregnancy at the end of the VWP, and increased risk for pregnancy loss. Moreover, those cows that experienced two or more health disorders had lower pregnancies to first AI and higher pregnancy loss than cows with one disorder. In contrast, healthy cows achieved a high (51.4%) pregnancy per AI at the end of the WVP and had lower odds for pregnancy loss.

These studies indicate that improving postpartum health by preventing periparturient diseases has the potential to improve fertility by increasing pregnancy per AI and lowering pregnancy loss. That is, cows that transition well from parturition to lactation are more likely to have good reproductive performance in their lactation. Below is a checklist that veterinarians can follow to determine whether or not a transition cow program is amenable to lowering the incidence of periparturient disorders.

1. Is feed intake maximized?
    a. Is there feed available at all times?
    b. What is the frequency of feed delivery?
    c. What is the number of cow movements during the prepartum and postpartum transition period?
    d. What is the feed bunk space (60–76 cm/cow)?
2. Are diets formulated to prevent postparturient diseases?
    a. What is the fiber content?
    b. Prepartum cows should consume daily about 15–18 Mcal of NEL and 1.1 kg of metabolizable protein.
    c. What is the mineral composition of the diet to avoid hypocalcemia?
3. What is the level of employee competency in providing calving assistance?
4. What is the incidence of hypocalcemia and ketosis? Has it changed over time?
5. Is there a postpartum health program for prompt diagnosis and treatment of sick cows (metritis, ketosis, mastitis)?

What is the ideal VWP?

The ideal VWP for an individual herd is partly determined by the pregnancy rate (PR = number of cows pregnant/number of cows eligible to breed in a 21-day cycle) of the herd. Ribeiro et al.[12] modeled the desired VWP for a herd according to the ideal day postpartum at pregnancy and the PR of the herd. Two calculations were considered, one for an ideal median day at pregnancy of 110 and another of 130 days postpartum. Herds with poor PR (<13%), even if they begin to breed immediately after calving, cannot achieve any of those days to pregnancy. In those herds with pregnancy rates above 15%, these days for pregnancy can be achieved but the VWP has to be around 30 days, which is not ideal. In contrast, those herds with excellent PR (>22.5%), days to pregnancy of 110 and 130 days can be achieved with a VWP of 60–98 days. A survey by the National Animal Health Monitoring System[13] reported that the average VWP is 55 days postpartum regardless of herd size. This range in days for the VWP is likely related to the typical PR of 16–17% found in dairy herds in the United States.[12]

Parity, milk production, and persistency are cow factors that should also be considered when establishing the length of the VWP. De Vries[14] reported on the optimal VWP and days to pregnancy as a function of these factors in Holstein herds in the United States. The optimal VWP or first insemination for the average first-parity cow is 77 days and 70 days for the average second- and third-parity cow. Low-producing cows could start their breeding period sooner and first insemination for higher-producing cows could be delayed by 1–2 weeks.

Strategies to increase pregnancy rate at the end of the VWP and decrease the interval between inseminations

The use of prostaglandin $(PG)F_{2\alpha}$ promotes estral events that are concentrated within a 7-day period, which helps to improve estrous detection. However, whether or not cows are inseminated depends on the efficiency of estrous detection, which is low (<50%) in most herds.[12] Consequently, it is now well accepted that an alternative for increasing insemination rate and increasing the proportion of pregnant cows at the end of the VWP is the incorporation of timed AI programs.[15] Timed AI programs are particularly beneficial on farms with low detection of estrus, particularly in herds with 21-day cycle insemination rates below 55%.[2] Application of timed AI increased the yearly profit by $30 per cow compared with estrous detection alone.[16] Similarly, $80–148 lower cost per pregnancy for timed AI compared with estrous detection programs was reported by De Vries.[14] Timed AI programs were considered to improve reproductive performance because AI submission rates increase; however, increased knowledge of the reproductive biology of the cow and how hormonal manipulation influences oocyte and embryo quality have now led to the development of methods that also optimize pregnancy per AI.[4] Indeed, pregnancy rates per AI above 45% have been achieved with timed AI programs in high-producing cows and 55% in grazing dairy cows.[17,18] Within an AI program, those herds that incorporate timed AI for first insemination followed by detection of estrus experience the lowest median days to pregnancy and more profit per cow.[4] In a simulation of reproductive performance and economics of various breeding programs by Galvão et al.,[19] the highest 21-day cycle PR was obtained when cows were subjected to timed AI for first AI, with 95% compliance of treatments followed by efficiency of estrous detection at 60% with 95% accuracy. This program also resulted in the shortest median days to pregnancy and increased profit from increased milk production. Therefore, incorporation of timed AI with high compliance to first service, followed by an efficient estrous detection program, improves pregnancy rate and increases profit. Further, in order to optimize reproductive performance dairy herds cannot abandon estrous detection practices altogether.

Use of bulls in a natural service (NS) breeding program is an option that some dairy producers use to avoid the need for detection of estrus and to improve PR. A study conducted in the southeast region of the United States which compared NS with AI after detected estrus demonstrated that reproductive performance was not different for NS herds.[20] In contrast, Overton and Sischo[21] (using data from 10 large California dairy herds that utilized a combination of AI and NS) reported that cows were more likely to become pregnant if they were bred by AI. Further, in a direct comparison between timed AI and NS, two breeding systems without estrous detection showed similar reproductive performance.[22] However, NS was more expensive ($10.00–60.00 per cow annually) than timed AI, related to the cost of feeding bulls, milk price, genetic merit, and whether or not to replace bulls in the lactating pen by cows.[23]

When to identify nonpregnant cows?

The value to dairy producers of pregnancy diagnostics by veterinarians is to find a nonpregnant cow early followed by successful rebreeding to reduce days not pregnant. The decision as to which "test" to use for early diagnosis of nonpregnant cows should be based on cost, practicality, sensitivity (i.e., correctly identifying pregnant animals), specificity (i.e., correctly identify nonpregnant animals), and practitioner comfort level. Palpation per rectum is the most common method used by veterinarians to diagnose pregnancy status in cattle and can be effective as early as 33–35 days of gestation depending on clinician experience. In contrast, the use of ultrasonography is useful by days 26–28 of gestation.[24] More recently, pregnancy-specific protein B (PSPB) and pregnancy-associated glycoproteins (PAGs), which are produced by the ruminant placenta and secreted into the maternal circulation throughout most of pregnancy, have been reported to be reliable in identifying pregnant animals by 30 days after breeding.[25] Pregnancy diagnosis in dairy cattle was evaluated by comparing a commercial (BioPRYN®) PSPB enzyme-linked immunosorbent assay (ELISA) with palpation per rectum, with good agreement between the two tests.[26]

Pregnancy loss

Pregnancy loss remains a significant cause of frustration and economic loss to dairy producers. De Vries calculated a weighted average cost of loss of pregnancy at $555 per case, depending on the proportion of cows in each stage of lactation, stage of gestation, lactation number, and level of milk production. Factors that contribute to pregnancy loss can be arbitrarily divided into embryonic abnormalities, inadequacies of the maternal environment, and external factors.[1] Common terms used to describe pregnancy loss include early and late embryonic death and abortion.[27] Early embryonic death occurs prior to day 17 and accounts for the largest proportion of pregnancy loss. Late embryonic losses occur from day 17 through day 42 and result in a prolonged interestrus interval due to the time beyond pregnancy recognition. Abortion refers to those losses after day 42 of gestation. Pregnancy loss at 32–60 days of gestation has been reported as 10–15%.[11] A study conducted in California showed that 39% of cows pregnant on day 23 lost their embryo by day 27, while 18% of cows that were pregnant on day 27 or 28 were not pregnant on days 35–41.[28] Risk factors identified for both time periods included insemination of pregnant cows, low progesterone concentration, and cows with mastitis (linear somatic cell count ≥ 4.5). As pointed out by the authors, veterinarians can have an impact on pregnancy loss by implementing management strategies that lower these risk factors, as listed below.

- Improve estrous detection accuracy of AI technicians to reduce breeding pregnant cows that have been incorrectly identified as being in estrus. In addition, breeding only those cows that have been identified nonpregnant would mitigate the risk of breeding pregnant animals.
- Improve milk quality by lowering the incidence of clinical and subclinical mastitis of the herd. Has it improved or worsened over time? In addition to pregnancy loss, clinical[29] or subclinical[30] mastitis during the VWP has been associated with lower odds to conception to first service and consequently a prolonged interval to pregnancy.
- Avoid prolonged postpartum NEB. Low progesterone concentration in the cycle preceding first service has been associated with NEB,[31,32] which compromises oocyte maturation and the ability of the embryo to continue normal development after fertilization.[32]

There is abundant evidence from the literature that infectious diseases such as campylobacteriosis, trichomoniasis (natural service), infectious bovine rhinotracheitis, bovine viral diarrhea, brucellosis, leptospirosis, and neosporosis contribute to pregnancy loss and infertility. According to Corteze,[33] the key for controlling these infectious agents is by a vaccination program that minimizes the amount and duration of the viremia and septicemia or which prevents the infectious agent from moving through the cervix. One should consider the following.

- Presence and degree of challenge of the particular diseases on the dairy.
- Management practices on the facility that lend themselves to, or which hinder, the implementation of vaccination programs.
- At what times or ages are the disease problems occurring and are they associated with any stresses?
- What is the status of the herd? Is it open or closed? Are they purchasing animals and at what age? Are the calves home raised or grown by others? What age are they returning?
- What immune system components are necessary to afford protection against the various diseases?
- The information that is available on products being considered and the source and quality of the information.
- Label indications for duration of immunity and maternal antibody interference.
- Warnings or restrictions on the use of the particular vaccine.
- What is the breeding program? Are clean-up bulls used? Source of the bulls and age of the bulls at purchase?

In dairy herds where NS is used, bulls should undergo the same vaccination practices as cows (except for brucellosis and *Tritrichomonas foetus*). Control of venereal diseases is essential to the success of NS breeding programs. Cows should be vaccinated against vibriosis at least 3 weeks prior to being exposed to bulls and receive a booster at 6-month intervals. Bulls can also be vaccinated for vibriosis, with some success, using twice the recommended dose for a cow.[34] Vaccination is also available for *Tritrichomonas foetus* in cows.

Evaluation of reproductive performance

In general, the two reasons for monitoring reproductive performance in dairy herds are (i) to determine if the reproductive management of the herd optimizes the number of pregnant animals in a timely manner at the end of the VWP; and (ii) to identify and correct flawed areas in management that contribute to poor reproductive performance. Typically, veterinarians evaluate an outcome from graphs, figures and reports generated from commercially available computer programs in order to evaluate the reproductive status of the herd and make a recommendation. This approach to performance evaluation without considering the population at risk, the time at risk, or if the metric reported reveals the true condition of the system could lead to wrong conclusions. Overton[35] suggests that a more prudent approach would be to use metrics that consider not only time and the population at risk but which also question the reproductive management process of the herd rather than the outcome.

When are cows inseminated at the end of the VWP?

Assuming a 70% insemination risk and 21-day cycles, Overton[35] indicates that a reasonable goal is to have 90% or more of all first inseminations occurring within the first 45 days of the VWP if cows are inseminated at detected estrus. In a timed AI breeding to first service, a reasonable goal would be to have 90–95% of all cows inseminated within 7 days of the VWP if the program is performed weekly or within 14 days of the VWP if the program is performed using 2-week calving cohorts. If a combination of estrous detection is followed by a timed AI for cows not detected in estrus, and weekly calving cohorts, a reasonable goal is to have 90–95% inseminated within 30 days of the VWP.

What is the efficiency at which eligible cows become pregnant at the end of the VWP?

Evaluation of the herd PR should consider a historical rate at which eligible cows became pregnant.[35] Pregnancy rate by a 21-day calendar period and days postpartum can be used to determine whether or not changes are occurring in specified periods of time. In addition, PR can be sorted by AI technician to determine if the technician is affecting pregnancy risk. The evaluation of PR should be able to answer these questions.

- Is first and subsequent conception risk acceptable?
- Does it change across time?
- Are there differences by parity, technicians and sires?
- Is there a seasonal effect?

Are pregnant cows remaining pregnant?

The level of pregnancy loss in a herd is influenced by the day of gestation when pregnancy diagnosis is performed because of the difference in late embryonic losses according to gestation length. Pregnancy loss in herds that use ultrasonography around day 28 after AI may be higher compared with herds that diagnose pregnancy later (>40 days), simply as a result of an earlier pregnancy diagnosis and not due to any damage caused by the earlier evaluation.[36] Consequently, cows diagnosed pregnant early in gestation (28–32 days) regardless of the method used should be reconfirmed later in gestation (42–60 days and 7 months). Reconfirmation of pregnant cows would provide an opportunity to identify when during gestation pregnancy loss occurs and determine if cows found not pregnant should be rebred and remain in lactation or culled. The evaluation of pregnancy loss should be able to answer these questions.

- Is it acceptable?
- When during gestation does it occur?
- Does it change across time?
- Are there differences by parity?
- Is there a seasonal effect?

What is the distribution of days dry in prepartum cows?

Variation in days dry (pregnant nonlactating) from the conventional 50–60 days has been associated with fertility in the subsequent lactation.[37] Longer dry periods (>100 days) were associated with increased number of days from calving to first breeding and days to conception. Cows with a conventional dry period (53–76 days) had increased odds of conception at the first service. These findings indicate that deviation from the conventional 60-day dry period negatively affects reproductive performance in the subsequent lactation. Errors in record keeping, cows that become pregnant late in lactation that require going dry at less than 7 months of gestation, and cows bred by NS are common causes for varying dry period lengths on dairy farms.

Reproductive programs for heifers

To maintain an economic advantage of higher milk production, dairy producers should breed replacement heifers to AI sires. Overton and Sischo[21] showed that milk yield in daughters of AI-proven bulls was 366–444 kg greater compared with daughters bred to NS sires. However, despite the advantage of AI, 33% of dairy farms in the United States still use NS as the main method for breeding heifers.[13] In those herds that use AI, the most common method is to administer $PGF_{2\alpha}$ when heifers reach the appropriate size and age to promote estrous expression in fewer days that improves PR.[38] However, similar to lactating cows, the success of a $PGF_{2\alpha}$ program depends on the efficiency and accuracy of estrous detection practices. On some dairy herds efficiency

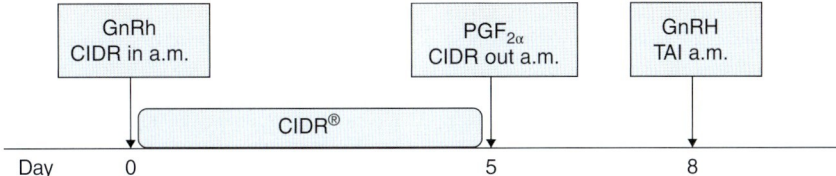

Figure 38.1 The 5-day CIDR-CoSynch 72 hours protocol (Rabaglino et al.[41]).

of estrous detection is suboptimal and is the main reason why producers choose to breed heifers to NS.

Similar to lactating cows, the use of ovulation synchronization protocols that allow timed AI to avoid the challenges of estrous detection can be used in heifers. However, less than 4% of dairy herds used timed AI for breeding heifers.[13] This low application can be attributed to reduced pregnancy per AI (30–40%) obtained with standard Ovsynch protocols compared with AI at detected estrus (50–60%).[12] Because of differences in follicular dynamics between cows and heifers,[39] dairy heifers can express estrus close to the $PGF_{2\alpha}$ injection, thereby causing asynchrony at timed AI. Nonetheless, application of a controlled internal drug release (CIDR) intravaginal device containing progesterone during the Ovsynch protocol improves ovulation synchrony at timed AI.[40] Modifications to the standard Ovsynch program by injection of gonadotropin-releasing hormone (GnRH) and insertion of a CIDR, followed 5 days later by CIDR removal and injection of $PGF_{2\alpha}$, and then 72 hours later AI concurrent with a second GnRH injection (5-day CIDR-CoSynch 72 hours) (Figure 38.1) resulted in adequate pregnancy per AI (50–60%).[41,42]

Profit of dairy replacement heifers is maximized when heifers calve between 23 and 25 months of age due to higher lifetime milk production and lower rearing costs.[43] Thus to maintain genetic progress and maximize profitability, heifer breeding programs should include AI and strive for calving to occur around 24 months. Considering the adequate fertility observed from the 5-day CIDR-CoSynch 72 hours protocol, the major economic advantage would be an increase in the number of pregnant heifers after they become eligible for breeding, which should result in younger age at calving that reduces feeding costs.[44] Similar to lactating cows, incorporating detection of estrus with timed AI would increase the number of pregnant heifers and lower cost per pregnancy.

Summary

Successful reproductive performance is critical to the economic viability of a dairy operation. The development and application of timed AI programs now allow producers to increase AI submission rates to obtain more pregnant cows on a timely manner at the end of the VWP. Recognition of the relationship between healthy cows and reproductive efficiency is essential. Because postpartum health has a major impact on fertility of dairy cows, it is essential that nutrition and health programs be designed to minimize the risk and reduce the prevalence of periparturient diseases. An effective vaccination program is needed to control infectious causes of infertility. Procedures to collect and evaluate data pertinent to the reproductive management of the herd must be in place to determine if changes are needed to improve fertility. Reproductive failure is more often related to human error than of cow origin. Therefore, dairies that have a successful reproductive program have well-trained and responsible employees who are compliant with the indicated herd management operational procedures.

Editor's note

For a more specific and in-depth coverage of dairy production management, the reader is referred to the excellent text edited by the author of this chapter, *Dairy Production Medicine*, which is also available from Wiley-Blackwell.

References

1. De Vries A. Economic value of pregnancy in dairy cattle. *J Dairy Sci* 2006;89:3876–3885.
2. Santos J. Implementation of reproductive programs in dairy herds. *Cattle Practice* 2008;16:5-1.
3. Giordano J, Fricke P, Wiltbank C, Cabrera E. An economic decision-making support system for selection of reproductive management programs on dairy farms. *J Dairy Sci* 2011;94:6216–6232.
4. Ribeiro E, Lima F, Ayres H et al. Effect of postpartum diseases on reproduction of grazing dairy cows. *J Dairy Sci* 2011;94(Suppl. 1):63 (Abstract).
5. Martinez N, Risco C, Lima F et al. Evaluation of peripartum calcium status, energetic profile, and neutrophil function in dairy cows at low or high risk of developing uterine disease. *J Dairy Sci* 2012;95:7158–7172.
6. Hammon D, Evjen I, Dhiman T, Goff J, Walters J. Neutrophil function and energy status in Holstein cows with uterine health disorders. *Vet Immunol Immunopathol* 2006;113:21–29.
7. Dubuc T, Duffield T, Leslie K, Walton J, LeBlanc S. Risk factors for postpartum uterine diseases in dairy cows. *J Dairy Sci* 2010; 93:5764–5771.
8. Butler S, Marr L, Pelton H, Radcliff P, Butler C, Butler R. Insulin restores GH responsiveness during lactation-induced negative energy balance in dairy cattle: effects on expression of IGF-I and GH receptor 1A. *J Endocrinol* 2003;176:205–217.
9. Hileman S, Lubbers L, Jansen T, Lehman M. Changes in hypothalamic estrogen receptor-containing cell numbers in response to feed restriction in the female lamb. *Neuroendocrinology* 1999; 69:430–437.
10. Santos J, Rutigliano H, Sá Filho M. Risk factors for resumption of postpartum cyclicity and embryonic survival in lactating dairy cows. *Anim Reprod Sci* 2009;110:207–221.
11. Santos J, Bisinotto R, Ribeiro E et al. Applying nutrition and physiology to improve reproduction in dairy cattle. *Soc Reprod Fertil Suppl* 2010;67:387–403.
12. Ribeiro E, Galvão K, Thatcher W, Santos JEP. Economic aspects of applying reproductive technologies to dairy herds. *Anim Reprod* 2012;9:370–387.
13. National Animal Health Monitoring System. *Dairy 2007. Part IV. Reference of Dairy Cattle Health and Management Practices in the United States, 2007*. Fort Collins, CO: USDA, 2009.

14. De Vries A. Economics of reproductive performance. In: Risco CA, Melendez Retamal P (eds) *Dairy Production Medicine*. Ames, IA: Wiley-Blackwell, 2011, pp. 139–151.
15. Caraviello D, Weigel A, Fricke P et al. Survey of management practices on reproductive performance of dairy cattle on large US commercial farms. *J Dairy Sci* 2006;89:4723–4735.
16. LeBlanc S. Economics of improving reproductive performance in dairy herds. *WCDS Adv Dairy Technol* 2007;19:201–214.
17. Bisinotto R, Santos J. The use of endocrine treatments to improve pregnancy rates in cattle. *Reprod Fertil Dev* 2011;24:258–266.
18. Wiltbank M, Sartori R, Herlihy M et al. Managing the dominant follicle in lactating dairy cows. *Theriogenology* 2011;76:1568–1582.
19. Galvão N, Federico P, De Vries A, Schuenemann M. Economic comparison of reproductive programs for dairy herds using estrus detection, Ovsynch, or a combination of both. *J Dairy Sci* 2013;96:2681–2693.
20. De Vries A, Steenholdt C, Risco C. Pregnancy rates and milk production in natural service and artificially inseminated dairy herds in Florida and Georgia. *J Dairy Sci* 2005;88:948–956.
21. Overton M, Sischo W. Comparison of reproductive performance by artificial insemination versus natural service sires in California dairies. *Theriogenology* 2005;64:603–613.
22. Lima F, Risco C, Thatcher M et al. Comparison of reproductive performance in lactating dairy cows bred by natural service or timed artificial insemination. *J Dairy Sci* 2009;92:5456–5466.
23. Lima F, De Vries A, Risco C, Santos J, Thatcher W. Economic comparison of natural service and timed artificial insemination breeding programs in dairy cattle. *J Dairy Sci* 2010;93:4404–4413.
24. Colloton J. Application of ultrasonography in dairy cattle reproductive management. In: Risco CA, Melendez Retamal P (eds) *Dairy Production Medicine*. Ames, IA: Wiley-Blackwell, 2011, pp. 99–116.
25. Humblot F, Camous S, Martal J. Pregnancy-specific protein B, progesterone concentrations and embryonic mortality during early pregnancy in dairy cows. *J Reprod Fertil* 1988;83:215–223.
26. Breed M, Guard C, White M, Smith M, Warnick L. Comparison of pregnancy diagnosis in dairy cattle by use of a commercial ELISA and palpation per rectum. *J Am Vet Med Assoc* 2009;235:292–297.
27. Roche J. Early embryo loss in cattle. In: Morrow DA (ed.) *Current Therapy in Theriogenology*, 2nd edn. Philadelphia: WB Saunders, 1986, pp. 200–204.
28. Moore D, Overton M, Chebel R, Truscott ML, BonDurant RH. Evaluation of factors that affect embryonic loss in dairy cattle. *J Am Vet Med Assoc* 2005;226:1112–1118.
29. Barker R, Schrick F, Lewis M, Dowlen H, Oliver S. Influence of clinical mastitis during early lactation on reproductive performance of Jersey cows. *J Dairy Sci* 1988;81:1285–1290.
30. Pinedo P, Melendez P, Villagomez-Cortez, J, Risco C. Effect of high somatic cell counts on reproductive performance of Chilean dairy cattle. *J Dairy Sci* 2009;92:1575–1580.
31. Beam S, Butler W. Energy balance and ovarian follicle development prior to first ovulation postpartum in dairy cows receiving three levels of dietary fat. *Biol Reprod* 1997;56:133–142.
32. Patton J, Kenny D, McNamara S et al. Relationships between milk production, energy balance, plasma analytes and reproduction in Holstein-Friesian cows. *J Dairy Sci* 2007;90:649–658.
33. Corteze V. Immunology and vaccination of dairy cattle. In: Risco CA, Melendez Retamal P (eds) *Dairy Production Medicine*. Ames, IA: Wiley-Blackwell, 2011, pp. 165–174.
34. Overton M, Risco C, Dalton J. Selection and management of natural service sires in dairy herds. In: *Proceedings of the Annual Conference of the American Association of Bovine Practitioners*, 2003, pp. 57–62.
35. Overton M. Dairy records analysis and evaluation of performance. In: Risco CA, Melendez Retamal P (eds) *Dairy Production Medicine*. Ames, IA: Wiley-Blackwell, 2011, pp. 271–302.
36. Santos J, Thatcher W, Chebel R, Cerri R, Galvão K. The effect of embryonic death rates in cattle on the efficacy of estrus synchronization programs. *Anim Reprod Sci* 2004;83:513–535.
37. Pinedo P, Risco C, Melendez P. A retrospective study on the association of the dry period and subclinical mastitis, milk yield, reproductive performance, and culling in Chilean dairy cows. *J Dairy Sci* 2011;94:106–115.
38. Stevenson, J, Dalton J, Santos J, Sartori R, Ahmadzadeh A, Chebel R. Effect of synchronization protocols on follicular development and estradiol and progesterone concentrations of dairy heifers. *J Dairy Sci* 2008;91:3045–3056.
39. Savio D, Keenan M, Boland P, Roche J. Pattern of growth of dominant follicles during the estrous cycle of heifers. *J Reprod Fertil* 1988;83:663–671.
40. Rivera H, Lopez H, Fricke P. Use of intravaginal progesterone-releasing inserts in a synchronization protocol before timed AI and for synchronizing return to estrus in Holstein heifers. *J Dairy Sci* 2005;88:957–968.
41. Rabaglino M, Risco C, Thatcher M, Kim I, Santos J, Thatcher W. Application of one injection of prostaglandin $F_{2\alpha}$ in the five-day CoSynch + CIDR protocol for estrous synchronization and resynchronization of dairy heifers. *J Dairy Sci* 2010;93:1050–1058.
42. Lima F, Ayres H, Favoreto M et al. Effects of the GnRH at initiation of the 5-d timed AI program and timing of induction of ovulation relative to AI on ovarian dynamics and fertility of dairy heifers. *J Dairy Sci* 2011;94:4997–5004.
43. Head H. Heifer performance standards: rearing systems, growth rates and lactation. In: Van Horn HH, Wilcox CJ, DeLorenzo DA (eds) *Large Dairy Herd Health Management*. Champaign, IL: American Dairy Science Association, 1992, pp. 422–433.
44. Ettema J, Santos J. Impact of age at calving on lactation, reproduction, health, and income in first parity Holsteins on commercial farms. *J Dairy Sci* 2004;87:2730–2742.

Chapter 39

Herd Diagnostic Testing Strategies

Robert L. Larson

Clinical Sciences, College of Veterinary Medicine, Kansas State University, Manhattan, Kansas, USA

Introduction

Veterinary practitioners help food animal clients meet a number of specific herd reproduction goals. These include having a high percentage of females exposed to bulls becoming pregnant (i.e., enhancing fertility or minimizing infertility); minimizing the effects of infectious, metabolic, and other disease processes that can cause pregnancy loss or other disease loss; and enhancing the genetic value of the herd through selection and multiplication of economically superior parent animals. To aid their evaluation of an operation's reproductive efficiency, practitioners have a number of tests available to them for use in screening, monitoring, and diagnosing populations and individuals. Commonly utilized diagnostic procedures include use of laboratory tests for infectious disease agents, such as serology, immunohistochemistry, polymerase chain reaction (PCR), virus isolation, and bacterial culture. These tests can be used to screen for infectious diseases in apparently healthy animals and to investigate disease outbreaks. In addition, veterinarians use diagnostic procedures to examine individual livestock and entire herds to find and change avoidable risks, for breeding soundness examination of bulls and heifers, for body condition scoring, to diagnose pregnancy via uterine palpation or ultrasonic examination per rectum, and for feed and ration evaluation (nutritional and toxicological).

Determining diagnostic test usefulness

A valid question confronting veterinary practitioners is whether to use available diagnostic tests to screen a particular herd for a specific condition.[1] The input that one needs in order to arrive at a logical conclusion includes prevalence data about the condition or disease, diagnostic test sensitivity and specificity data, disease or condition epidemiology, and economic costs of the condition and its treatment or prevention.[2] Literature review and mathematic aids such as computer spreadsheets are the tools used to calculate the post-test predictive values of diagnostic tests, economic value of testing, sensitivity of the decision to the individual inputs, and the importance of individual inputs to the decision. These calculations can then be used to evaluate alternate diagnostic testing strategies and to identify the control points that will be monitored for change that can trigger a reevaluation of the decision.

Sensitivity and specificity of diagnostic tests

Sensitivity and specificity are properties of a diagnostic test that are determined by comparing the test to a "gold standard." The gold standard is considered the true diagnosis and may be made using a variety of information such as clinical examination, laboratory results, or postmortem findings. Sensitivity is the proportion of true positive (gold standard-positive) samples that the test in question identifies as positive. Specificity is the proportion of true negative samples that the test identifies as negative. In other words, sensitivity answers the question "How effective is the test at identifying animals with the condition?" and specificity answers the question "How effective is the test at identifying animals without the condition?"

Diagnostic tests try to separate two populations. One is abnormal, diseased, or has an undesired condition and the other population is normal or has a desired condition. In most conditions of interest to veterinarians, the affected and unaffected populations overlap based on available diagnostic tests; therefore, both laboratory and clinical examination tests must use an arbitrary cutoff to separate test-positive and test-negative populations. Where one places cutoffs for diagnostic tests is very important when deciding into which of two distributions of outcomes a particular animal or herd falls. Because diagnostic measurements of affected and unaffected populations overlap to some extent, sensitivity and specificity are inversely related, and placing the cutoff is always a trade-off between the impacts of false-negative and false-positive results (Figure 39.1). For test cutoffs not set by the practitioner (i.e., set by a test manufacturer or at a

(a) In this figure, the cutoff is placed to maximize sensitivity so that there are few false-negative results, but false-positive results occur

(b) In this figure, the cutoff is placed to maximize specificity so that there are few false-positive results, but false-negative results occur

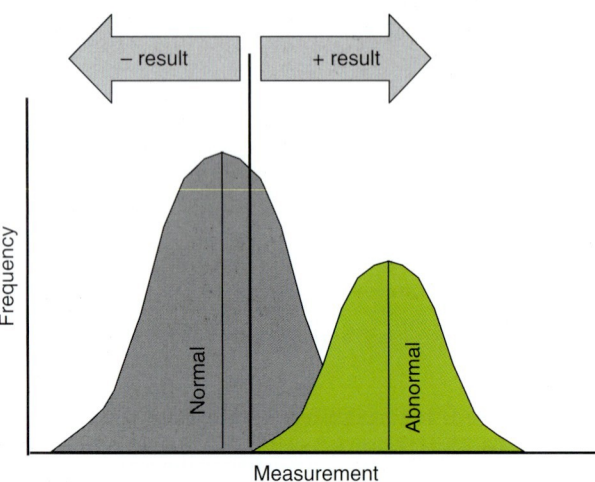
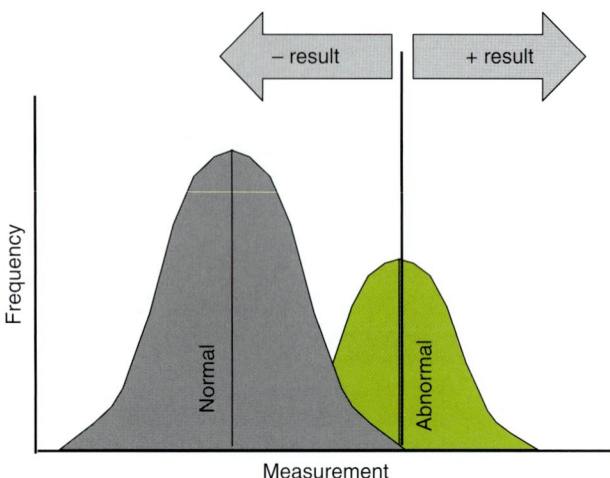

Figure 39.1 When diagnostic measurements of affected and unaffected populations overlap to some extent, sensitivity and specificity are inversely related, and placing the cutoff is always a trade-off between the impacts of false-negative and false-positive results.

diagnostic laboratory), it is important for the practitioner to be informed of test sensitivity and specificity.

Where one places a diagnostic cutoff is always a trade-off between false-negative and false-positive results because of the overlap between normal and abnormal populations.[3] Using an example of tachypnea as evidence of respiratory disease in calves requires that the veterinarian define tachypnea with a cutoff number of breaths per minute. Using a cutoff of 25 will likely be more sensitive for detecting calves with pneumonia, but is likely to result in the false-positive classification of many nonpneumonic calves. Placing the cutoff at 40 will increase specificity and will reduce the number of nonpneumonic calves classified as tachypneic, but may also fail to identify some truly pneumonic calves (increased false-negative classifications).

Prevalence

Prevalence is the proportion of animals that meet a particular case definition at a given time to the size of the population at that time. The prevalence is the probability of the condition being present in a randomly selected individual from that population. Unfortunately, there is often either limited published prevalence information or the published prevalence estimates are so broad as to be of limited usefulness for most of the infectious diseases and reproductive conditions of interest to veterinary medicine. Each practitioner's judgment, based on history and clinical examination of both individuals and the population, aided by what prevalence information is available, is often all that is available to establish the probability for both infectious diseases and reproductive conditions.

Knowing or estimating prevalence is important when interpreting diagnostic tests because for tests with imperfect specificity, an increasing proportion of the animals that test positive will be false positives as prevalence or disease probability decreases. For biosecurity reasons, veterinarian often test cattle from very low prevalence populations so that even a highly accurate test will render inaccurate positive test results when applied to low-risk populations as indicated by the test's low positive predictive value in these types of populations.

Similarly, when a test has imperfect sensitivity, an increasing proportion of the test-negative animals will be false negatives as prevalence or disease probability increases. Therefore, even a highly accurate test can result in inaccurate negative test results (low negative predictive value) when applied to a population with high prevalence or probability for the condition.

Post-test predictive value

The post-test predictive values of a test are determined not in the laboratory but in the field with the populations where the test is applied; they tell a veterinarian if a valid test is useful in specific populations. The positive predictive value is the proportion of animals with a positive test result that are actually positive for the disease or condition in question, and in most situations is influenced more heavily by test specificity than sensitivity. The negative predictive value is the proportion of animals with a negative test result that are truly negative and in most situations is influenced more heavily by test sensitivity than specificity. Both the positive and negative predictive values of a test are affected by the prevalence of the condition in a population (or probability of the condition in individual animals). As the prevalence of the condition is raised, more animals with the condition are present in the population and one has greater confidence that a positive test result is correct. With increasing prevalence, when test sensitivity and specificity are held constant, the positive predictive value of the test is increased and the negative predictive value is decreased, while the reverse is true as the prevalence of the condition is decreasing (Figure 39.2).

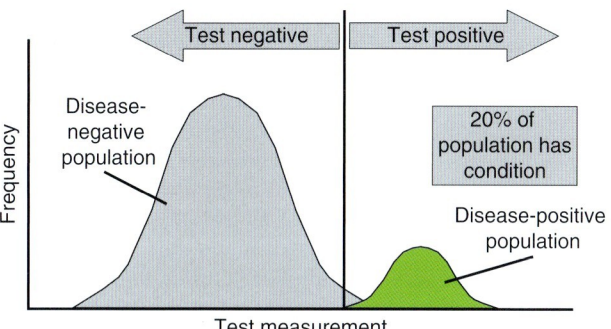
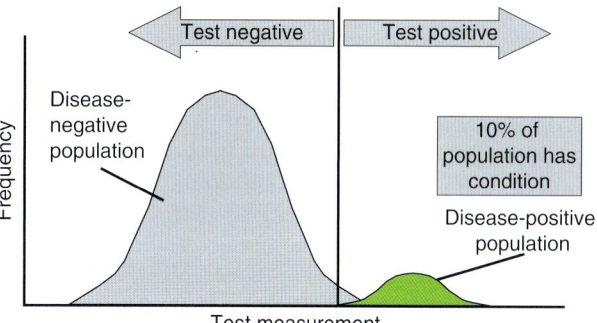
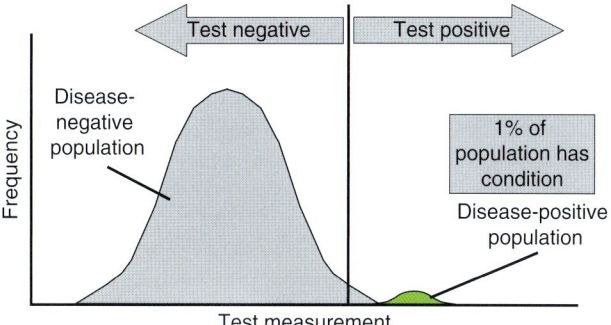
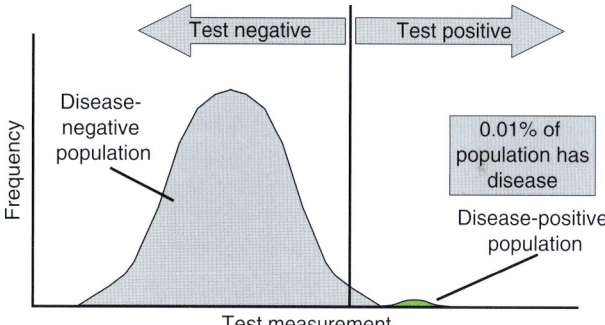

Figure 39.2 If the test has imperfect test specificity (i.e., <100%), as the prevalence of the condition decreases, the positive predictive value of a positive test (PPV) decreases. In these examples, the test sensitivity and the test specificity are consistently 99.99% and 95% respectively, and the prevalence of the condition changes.

The probability that an animal that tests positive is truly positive (positive predictive value or predictive value of a positive test) is computed as:

$$\text{Positive predictive value} = (Se \times P) / \{(Se \times P) + [(1-P) \times (1-Sp)]\}$$

The probability that an animal that tests negative is truly negative (negative predictive value or predictive value of a negative test) is computed as:

$$\text{Negative predictive value} = (1-P) / \{(1-P) + [(1-Se)/(Sp)]\}$$

where Se is test sensitivity, P is prevalence, and Sp is test specificity.

An example of how the probability of a condition in certain classes of cattle or in herds with particular previous diagnoses or other history characteristics affects test interpretation is the different certainty a veterinarian will place on a positive test result for bovine viral diarrhea persistent infection (BVD-PI) status depending on the animal signalment and herd history. Using a BVD-PI test that has a sensitivity of 99.7% and a specificity of 99.3% in different situations will result in different positive predictive values and hence the certainty of the diagnosis, which will influence the follow-up steps by the veterinarian. If the BVD-PI test is applied to a calf that died following a short disease course of pneumonia and diarrhea in a herd with confirmed cases of BVD in the past 2 months, the veterinarian may estimate that the probability of the calf being BVD-PI is about 70% before he or she requests the diagnostic test. With a 70% probability of BVD-PI status and test sensitivity of 99.7% and specificity of 99.3%, the positive predictive value is 99.7% and the veterinarian will likely believe a positive test result. In contrast, the veterinarian for a seedstock client who routinely tests all calves on the ranch for BVD-PI status as part of a rigorous biosecurity protocol and as a marketing tool to promote their high-health calves may estimate the probability of finding a truly BVD-PI animal to be very low (<0.1% probability). Using the same tests (99.7% sensitivity and 99.3% specificity) on this low-risk herd results in a positive predictive value of 12.5% and the veterinarian will likely be very skeptical that the calf is truly BVD-PI.

Diagnostic testing strategy

After a single test it is very unlikely to have both a near-perfect predictive value of a positive test result and a near-perfect predictive value of a negative test result; therefore, veterinarians must recognize that they are likely to be placed in the situation where one can only believe one test result direction, either positive or negative. The veterinarian must then decide what to do with the animals who return a test result that is not trusted: either retest it or knowingly accept the possibility of a false positive (situation with low positive predictive value) or a false negative (situation with low negative predictive value) without further diagnostic effort. That is, you accept that you may cull or treat some false-positive animals if the cost of further diagnostics is greater than the cost of false-positive intervention; or you accept that you may keep in the herd or fail to treat some false-negative animals if the cost of further diagnostics is greater than the cost of false-negative intervention.

There are two general categories for testing strategies that utilize more than one diagnostic test to determine test-positive or test-negative status. The first of these is described as "testing in series" and is usually used to retest animals with positive test results when the predictive value of a positive test is low after the initial test. The reasons a test would have a low positive predictive value are either low specificity of the test or low prevalence in the population. Interpretation of two tests in series requires positive test results in both tests in order to meet the case definition for a positive conclusion. The net results of "testing in series" are a loss of sensitivity compared with a single test, but a gain in specificity. Therefore, the positive predictive value is improved and negative predictive value is reduced when interpreting tests in series versus using a single test.

There are several reasons to test in series. In some instances the second test is more expensive, time-consuming, or difficult than the initial test; therefore, we only want to use that test in higher prevalence populations. In other situations, the effect of false positives after a single test is more detrimental than the benefit of finding all instances of disease. If a calf tests positive for BVD-PI status from a herd with low probability of BVD, the positive predictive value will be low and the value of the calf should cause the veterinarian to run a second test interpreted in series and the calf would only be classified as truly BVD-PI if it has a positive result to both the initial test and the follow-up test.

The other category of multiple testing strategies is to test "in parallel" and is usually used to retest animals with negative test results when the predictive value of a negative test is low after the initial test. The reasons that a test would have a low negative predictive value are either that the test has low sensitivity or the population has a very high prevalence for the condition.

The current tests for *Tritrichomonas foetus* have very good specificity but far less than 100% sensitivity, with estimates of 53–77% under field conditions.[4–6] If a veterinarian is testing bulls in a situation where the possibility of trichomoniasis is at least 50%, the negative predictive value is only about 77%, meaning that many test-negative bulls are truly infected. When testing a herd with known *Tritrichomonas*-positive bulls, the veterinarian will likely estimate the probability of infection in other bulls in the herd to be very high, which decreases ones confidence in negative test results

(a) In this diagram, the test has very good sensitivity and negative predictive value, but the specificity is less than 100% and the positive predictive value is poor. Therefore, the test-positive animals should be re-tested and interpreted "in series".

Positive test results are not trusted:
- Retest positive animals
- Interpret tests in series
- Second test must be positive for the animal to be considered test-positive

(b) In this diagram, the test has very good specificity and positive predictive value, but the sensitivity is less than 100% and the negative predictive value is poor. Therefore, the test-negative animals should be retested with a second test interpreted "in parallel".

Negative test results are not trusted:
- Retest negative animals
- Interpret tests in parallel
- Second test must be negative for the animal to be considered test-negative

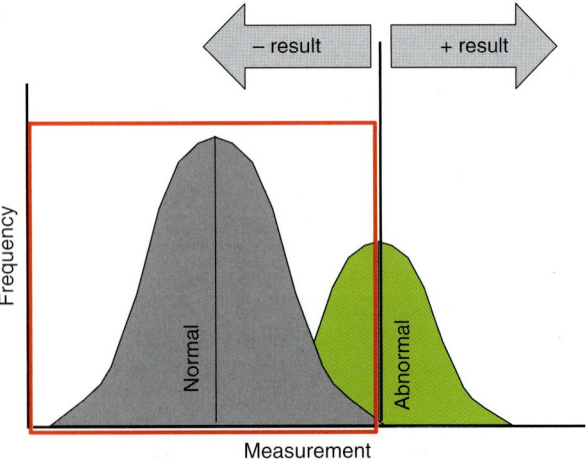

Figure 39.3 After an initial diagnostic test, one test result direction (test-positive or test-negative) should be very good (near-perfect predictive value), but the other test result direction is likely to result in a much lower post-test predictive value. For conditions affecting bovine reproduction, additional diagnostic tests are interpreted "in series" for situations where the veterinarian does not trust initial positive test results and test-positive animals are retested. Additional tests are interpreted "in parallel" for situations where the veterinarian does not trust initial negative test results and test-negative animals are retested.

even further. When testing in a herd situation with a disease probability of 75%, the negative predictive value of a single test would be around 50%.

When testing in parallel, two or more tests are applied to the entire population, or to the initial test-negative population. Parallel interpretation of multiple diagnostic tests is used when the veterinarian is concerned about making false-negative mistakes, so that a positive result in any of the tests results in a positive conclusion. The net result of testing in parallel is a gain in sensitivity and a net loss in specificity. Therefore, the positive predictive value is reduced but negative predictive value is improved. Testing and interpreting in parallel is commonly done if tests are fairly easy and inexpensive, or if it is more important to identify all animals with the condition than to avoid false positives (Figure 39.3).

Summary

Interpretation of diagnostic tests is an important component of clinical veterinary practice. Veterinarians use a variety of testing methods including physical examination, laboratory testing, diagnostic imaging, and performance record evaluation. The diagnostic endpoint may be a physical diagnosis of pregnancy, attainment of puberty, or adequate quality and quantity of sperm; or a medical diagnosis of reproductive tract pathology, presence of an infectious pathogen, or abnormal hormonal status. Proper interpretation of test results requires an understanding of how sensitivity and specificity as measures of test accuracy and prevalence of the condition affect the interpretation of an individual result. For many diagnostic questions, the proper use of more than one test interpreted either in series or in parallel allows veterinarians to optimize diagnostic accuracy and the economic return for the testing strategy.

References

1. Rothman K, Greenland S. Causation and causal inference. In Rothman KJ, Greenland S (eds) *Modern Epidemiology*, 2nd edn. Philadelphia: Lippincott Williams & Wilkins, 1998, pp. 7–28.
2. Clemen R. Modeling uncertainty: Monte Carlo simulation. In: *Making Hard Decisions: An Introduction to Decision Analysis*. Belmont, CA: Duxbury Press, 1991, pp. 167–195.
3. Romatowski J. Problems in interpretation of clinical lab tests. *J Am Vet Med Assoc* 1994;205:1186–1188.
4. Schonmann M, BonDurant R, Gardner I, Van-Hoosear K, Baltzer W, Kachulis C. Comparison of sampling and culture methods for the diagnosis of *Tritrichomonas foetus* infection in bulls. *Vet Rec* 1994;134:620–622.
5. Peter D, Fales W, Miller R. et al. *Tritrichomonas foetus* infection in a herd of Missouri cattle. *J Vet Diagn Invest* 1995;7:278–280.
6. Mukhufhi N, Irons P, Michel A, Peta F. Evaluation of a PCR test for the diagnosis of *Tritrichomonas foetus* infection in bulls: effects of sample collection method, storage and transport medium on the test. *Theriogenology* 2003;60:1269–1278.

Chapter 40

Beef Herd Record Analysis: Reproductive Profiling

Brad J. White

Department of Clinical Sciences, College of Veterinary Medicine, Kansas State University, Manhattan, Kansas, USA

Introduction

Reproductive efficiency is related to overall herd health, and both high reproductive success and low disease losses are paramount to beef cow-calf herd economic sustainability. A standard measurement of reproductive outcomes is pregnancy or calving percentage, which is often used to evaluate herd reproductive efficiency status. In addition to this standard performance assessment, the reproductive profile of the herd can also be used as a methodology to create a focused herd reproductive health program, assist in the diagnostic strategy for suboptimal reproductive efficiency, and to guide the design of intervention strategies. Utilizing a breeding season evaluation to assess the overall reproductive status of the herd is not a new concept,[1] and the evaluation can be expanded to generate specific target areas to improve overall success. Managing a herd to optimize the reproductive profile serves as a tool to optimize herd health parameters, and comparison of the current herd profile with profiles of the same breeding group in previous years facilitates early diagnosis of potential problems. The objective of this chapter is to describe the cow-calf herd reproductive profile and how to use this combination of measurements to better customize a reproductive management program based on information generated through the profile.

The herd reproductive profile

Weaned calves are the primary source of income in the cow-calf herd, and fixed costs associated with maintaining cows is an important expense category.[2] Therefore high reproductive efficiency that results in many weaned calves per cows exposed for breeding is strongly associated with financial success of the cow-calf herd.[3] One variable that encompasses both conception percentage and successful pregnancy maintenance is the number of calves born in the calving season relative to the number of cows exposed during the breeding season. Lower than expected reproductive success could be attributed to several potential causes and in order to narrow the differential diagnosis list and create focus for the preventive health program, a herd reproductive profile should be generated.

The herd reproductive profile consists of three main components.

1. The length of the calving season.
   a. Number of days from the birth of the first calf to the birth of final calf.
2. The number of calves born in the calving season divided by the number of cows exposed during the previous breeding season.
   a. Total number of calves born in a breeding group, including mortalities.
   b. Total number of cows exposed in a breeding group in the previous season.
3. The distribution of births in the calving season.
   a. Timing of births throughout the calving season aggregated to 21-day intervals.

The length of the calving season is influenced by a variety of factors including production goals, calving season, resource availability, and owner preferences. The goal is to produce one calf per cow per year, and herds typically sell the calves born within one season at a similar time point. The length of the calving season will influence the uniformity of the calf crop at the time of sale and longer calving seasons result in calves with more weaning weight variability due to the length of growing time from birth to weaning. Although a 60-day calving season is often recommended, producers with a slightly longer calving season can still maintain "front-end loaded" calving seasons each year if forage production and nutrition are adequate for short postpartum anestrous intervals. The effect of breeding season length on herd net returns is related to the estimated length of the postpartum interval and conception rate in the subsequent breeding season.[4] Most (66%) producers report that their

---

Bovine Reproduction, First Edition. Edited by Richard M. Hopper.
© 2015 John Wiley & Sons, Inc. Published 2015 by John Wiley & Sons, Inc.

calving season is 4 months or less.[5] The calving season length can be modified by either controlling the length of the breeding season (i.e., the time period that bulls are exposed to cows) or by utilizing gestational aging to select cows that will calve in the desired time period.

When evaluating potential reproductive problems, an important aspect is to compare the length of the breeding season from the current to the previous year. Breeding season length, along with calving percentage, have been associated with overall profitability of cow-calf operations.[3] If the breeding season length is the same and a disease causes reproductive losses, then the problem will be manifested as either a decreased pregnancy percentage at mid-gestation pregnancy evaluation or a higher than expected percentage of cows that were confirmed pregnant but which did not calve during the calving season. If the breeding season is unrestricted, the problem may best be evaluated by monitoring the change in the pattern of calving during the season (described below).

The number of calves born in the calving season relative to the number of cows exposed to breeding is based on both successful conception and the maintenance of pregnancy to parturition. The target mid-gestation pregnancy percentage after a 65-day or longer breeding season is 95%, but some pregnancies may not be maintained until parturition due to normal reproductive loss.[6] Pregnancy diagnosis by rectal palpation may be used to provide an evaluation of the previous breeding season and offers information to the herd manager relative to the success of the program. Pregnancy percentage may vary slightly from year to year and from herd to herd, but decreases below 85–90% should be considered problems that merit further investigation. A low percent pregnancy at a mid-gestation evaluation identifies that a herd problem is present, but does not provide insight into the root cause of the problem; therefore, the pregnancy (or calving) distribution should be evaluated.

The distribution of calving within the herd can be combined with calving season length and the pregnancy rate to create the herd reproductive profile. The calving distribution is created by aggregating the number of calves (or percent of the herd) born in each 21-day period of the calving season. The time interval used in the distribution is based on the average period between estrus for cycling cows and the fact that a mating between a cycling cow and a fertile bull should result in a pregnancy that can be detected 50 days later 60–70% of the time.[7] Therefore, if all the females are cycling at the start of the breeding season and they are exposed to an appropriate number of fertile bulls, about 65% of the herd should be bred in the first 21-day cycle. This pattern of calving is often designated as "front-end loaded" as the majority of the herd calves early in the breeding season.

For example, if the herd has 100 cows, then after the first cycle, 65 cows would be bred leaving 35 open. In the second cycle, the 35 open cows continue to be exposed to fertile bulls and if 65% get bred that would result in 23 pregnancies, leaving 12 cows open. In the third cycle, 65% of the 12 cows (8 head) would be bred leaving four open and a 96% overall pregnancy percentage. This pattern would be recognized as an ideal distribution and is similar to ideal distributions reported elsewhere.[8] The pregnancy percentage at the end of multiple bull exposures over the breeding season is predicated by the fact that the cows are cycling the entire period

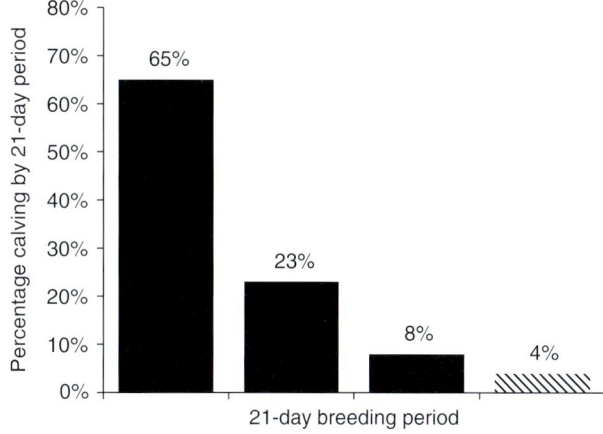

Figure 40.1 Ideal calving pattern in a 64-day breeding season cow herd with 96% calving rate and front-end loaded calving season.

until bred and the bulls maintained fertility throughout the breeding season. Therefore, the overall pregnancy percentage at the end of the breeding period can be influenced by any factor that decreases bull fertility or results in females not cycling during the breeding season.

All three aspects of the reproductive profile (calving season length, number of calves born per cow exposed, and expected or actual calving distribution) can be illustrated in a single figure created from herd data (Figure 40.1). The data necessary to create a reproductive profile is available on most cow-calf farms and profiles can be generated from data collected at the time of pregnancy diagnosis or the previous year's calving or both. If cows have pregnancy status determined at a time when pregnancies can be categorized into 21-day gestational age groups, then the results of the pregnancy diagnosis can be used to generate the reproductive profile. If pregnancy diagnosis was not performed during this time period (or at all), the reproductive profile can still be generated by plotting the number of calves born by 21-day intervals. This does not require individual animal birth dates to be recorded by the producer, but can be done with a minimum of information such as the number of calves born weekly or even a cumulative count of the number of calves at 21-day increments throughout the calving season. One caveat with both pregnancy diagnosis and birth-date records is to insure that the appropriate number of cows exposed to the bull is included in the denominator when determining the percentage of the herd. As both pregnancy diagnosis and calving occur several months after the breeding season, simply counting the number of cows present at the time of the assessment may introduce bias by missing cows that dropped out of the herd or were culled due to not becoming pregnant during the breeding season. Often the number of cows exposed can be determined by asking the producer for sales records on cull cows and the number of cows that have died in the interim since the breeding season.

Importance of the reproductive profile

The reproductive profile is an important metric for determining herd reproductive efficiency and to institute management strategies to ensure cow-calf reproductive success. The overall number of calves born has a dramatic influence on total kilograms weaned and therefore the overall farm income. Whether calves are born early or late in the calving season also influences weaning weight as days of age at weaning is a primary driver of weaning weight when calves within a herd are weaned at the same date.

The potential difference and value of weaning weight based on calving distribution is best evaluated through an example comparing herds with similar reproductive traits except the calving distribution. Both herds will be assumed to have a 95% pregnancy percentage, 2% gestational loss after pregnancy determination, 2% pre-weaning calf mortality, a 228-day period from the birth of the first calf to weaning of the calf crop, and an expected calf average daily gain from birth to weaning of 1.1 kg/head per day. If herd A follows the ideal calving distribution (Figure 40.1), the expected average calf weaning weight is 216 kg per head and a total weaning weight of 21 638 kg for the 91 weaned calves. If herd B has a uniform calving distribution over the first four cycles (approximately 24% in each 21-day period for four periods), the expected average calf weight at weaning is 193 kg per head with a total weaning weight of 19 280 kg. The large discrepancy between the two herds is primarily driven by the difference in average calf age at weaning (herd A, 210 days; herd B, 187 days) created by the calving pattern in each herd. The 23-kg difference in average weaning weight between the operations would have significant implications related to herd income.

The calving distribution influences production even after the calves leave the farm at weaning. A study that followed steer calves through the finishing phase illustrated that pre-weaning or feedlot average daily gain did not differ among calves based on when they were born in the calving season, as would be expected in cohorts from similar genetic and management backgrounds.[9] However, when the calves were harvested, calves born in the first 21 days had higher hot carcass weights (370 kg) and percent grading choice (79%) compared with calves born in the third 21-day period (352 kg hot carcass weight, 65% grading choice).

The impacts of the calving distribution are also displayed beyond the current production year in replacement animals maintained in the herd. In a long-term evaluation of calving season distribution impacts on subsequent performance, Funston et al.[9] illustrated that heifers born in the first 21 days of the calving season were more likely (70%) to be cycling at the start of their first breeding season compared with heifers born in the second 21 days (58%) or third 21 days of the calving season (39%). These effects carried through the second breeding season and heifers born in the first 21 days had higher overall pregnancy success at approximately 4 months of gestation (90%) and a higher percentage calving in the first 21 days of the season (81%) compared with heifers born in the third 21 day period (78% pregnancy percentage, 65% calving in first 21 days of season). For producers raising replacement heifers, the reproductive profile from the current year creates production momentum which carries through to subsequent breeding and calving seasons. The importance of modifying the distribution of calving dates within a calving season to maintain front-end loading should not be underestimated.

Using the reproductive profile to manage herd health

Typically, reproductive problems can be divided into three major areas: female problems (cows not cycling due to normal postpartum anestrus extending later than the onset of breeding, or prolonged postpartum anestrus due to poor body condition), male problems (bulls with traumatic injuries, bull infertility), or infectious or toxicologic diseases causing early embryonic loss or abortion. To assist the herd in overcoming reproductive issues, it is important to utilize the herd reproductive profile to narrow down the potential areas to focus the reproductive health program.

Annual momentum

In some cases the reproductive profile from the previous year will be available for comparison with the year of interest. One of the primary drivers of this year's calving distribution is the calving distribution from the previous year. Postpartum anestrus presents a biological barrier influencing the interval between calving and subsequent rebreeding. Postpartum anestrous periods in mature beef cows in moderate body condition receiving adequate nutrition averages 50–60 days,[10,11] while the average postpartum interval is typically longer (80–100 days) in heifers in moderate body condition following their first calving.[12,13] Combined with a 283-day gestation period, these limitations prevent cows from calving much before the same date as the calving date in the current season in subsequent years. However, any disruption of return to estrus can cause a cow to calve later in the calving season in subsequent years. As most beef operations calve once per year, the herd develops a momentum with cows maintaining a relatively tight annual calving interval (or calving at approximately the same time every year). The number of cows cycling at the start of the breeding season is influenced by the timing of calving relative to breeding, which was influenced by the previous breeding pattern.

Management of first-calf heifers plays a critical role in both creating and maintaining a front-end loaded calving season. If the goal is to maintain a 365-day calving interval for heifers, this only allows 82 days for heifers to get bred following their initial calving. Postpartum anestrus is 80–100 days in heifers, and therefore it is unlikely the majority of the heifers will have a 365-day interval following their first calf. This problem may manifest in subsequent years by a breeding shift, or more first-calf heifers being bred in the second and third 21-day period of the breeding season instead of the first 21-day period (Figure 40.2). This problem is commonly attributed to nutrition (and nutritional status of the heifers may play a role in the length of the postpartum anestrous period); however, the biological limitations of the system should be considered when evaluating problems with first-calf heifers. The issue could be related to a shortened time period between calving and subsequent rebreeding, poor nutrition, or a combination of the two factors. When observed in the mature cow herd, this pattern may be indicative of calving pattern in previous years and not necessarily representative of a problem in the current year. The current calving histogram and departure

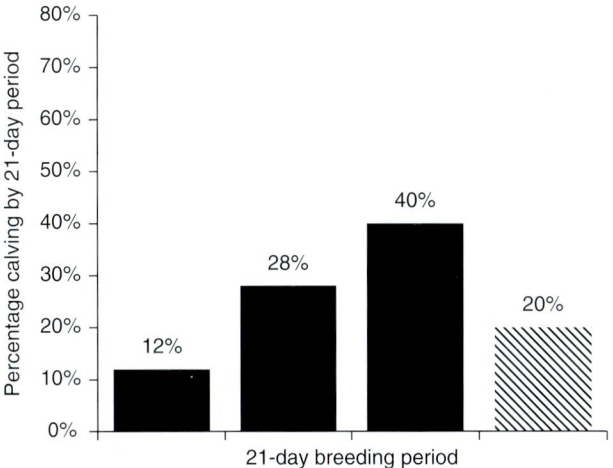

Figure 40.2 Calving pattern in a 64-day breeding season for first-calf heifers with limited time between calving and start of subsequent breeding season.

from front-end loaded breeding season represents an accumulation of errors as once cows shift to the right (or into the second and third 21-day periods of the calving season), it is very difficult to move them back toward the early breeding season. This illustrates the importance of generating and maintaining positive momentum by calving heifers 2–3 weeks prior to the mature cows and encouraging appropriate annual calving intervals.

Distinguishing male/female issues

Modifications to this ideal calving pattern (either increases in length or changes in distribution) may result from changes in management, nutritional problems, or disease problems. When confronted with an abnormal reproductive profile (higher than expected percentage of cows open and/or non-ideal calving distribution), one of the first objectives is determining if the root cause is related to the male or female side of the equation. The reproductive profile often will not provide definitive information on the cause of the reproductive insufficiency; however, evaluation of the calving pattern facilitates refining the differential list and focusing the herd health program in a specific area. The reproductive profile should be evaluated on a breeding group basis when possible, as assessment of aggregate data from the entire herd may be misleading or hide the problem entirely (Figure 40.3). In Figure 40.3a, the reproductive profile of a single herd with four equally sized separate breeding groups is represented, while Figure 40.3b represents the profile from only a single breeding group within the entire herd. Evaluation of only the entire herd (Figure 40.3a) does not reveal the reproductive insufficiency; however, the problem becomes obvious on further evaluation of the subgroup (Figure 40.3b). The breeding groups do not have to be single-sire units to enable an evaluation, but they should consist of animals managed as a unit through the breeding season.

Evaluation of the reproductive profile in Figure 40.3b provides several clues toward potential causes of the problem. In the first 21-day period of the breeding period the pregnancy rate was near the goal (62%), indicating that nearly all the cows were cycling at the initiation of the breeding season. In the subsequent 21-day period a very low percentage of the cows became bred. Assuming this herd had a normal distribution in the previous year (approximately 20% in the second period), this is a dramatic drop to 5% bred in the same period this year. This is likely a male-related issue due to the near truncation of the normal calving pattern. The first-period pregnancy success of 62% indicates that nearly all the cows were cycling during the first 21 days of breeding, and although there are some issues that may cause an individual cow to stop cycling, there are few factors that would cause an entire population of cows to stop cycling. Conversely, if this were a single-sire breeding unit, a traumatic bull injury could result in this pattern. Even in a multi-sire pasture, this pattern could be presented if the dominant bull was injured and prevented the subordinate bull from breeding as many cows as possible.

The pattern in Figure 40.3b is unlikely to be representative of an infectious disease except for pathogens with venereal transmission. If the herd was exposed to a transmissible infectious pathogen causing abortion and decreased conception rates (e.g., bovine viral diarrhea virus) during the second 21 days of the breeding season, there would be no reason to expect the pregnancies conceived in the first period would be spared from abortion. However, if the herd were exposed to a venereally transmitted disease in the second period (e.g., a bull infected with *Tritrichomonas foetus* entered the pasture), the pregnancies conceived in the first period would be unaffected. Thus, this pattern is likely representative of a problem requiring further evaluation of the bulls (e.g., traumatic injury, venereal disease carrier).

Patterns attributed to the female side often present with a slow start to the breeding season and an example of this pattern is shown in Figure 40.2. If the bulls have been fertility tested prior to the breeding season and adequate bull power is present in the breeding pasture, the percentage bred during the first 21-day period of the breeding season is indicative of the number of cows cycling at the start of the period. Mating fertile bulls to cycling cows should result in a pregnancy 65% of the time, so the percentage of cows cycling can be estimated by dividing the percentage of cows actually pregnant by 0.65. In Figure 40.2, 12% of the herd conceived during the first 21 days of the breeding season; therefore, if the bulls were fertile, it was likely that only 18% of the herd was cycling at the start of breeding. The specific interventions necessary to rectify the reproductive issues will still need to be based on the root cause of the problem, but distinguishing male from female causes of reproductive issues allows the generation of a more focused reproductive health program.

Infectious reproductive losses

Multiple etiologic agents can contribute to reproductive inefficiency in the cow-calf herd. Some pathogens may cause early embryonic death (which may be indistinguishable from a failure to conceive at the time of pregnancy diagnosis). Other infectious agents may cause abortion at different gestational stages. The reproductive profile can be useful to distinguish infectious causes of reproductive loss

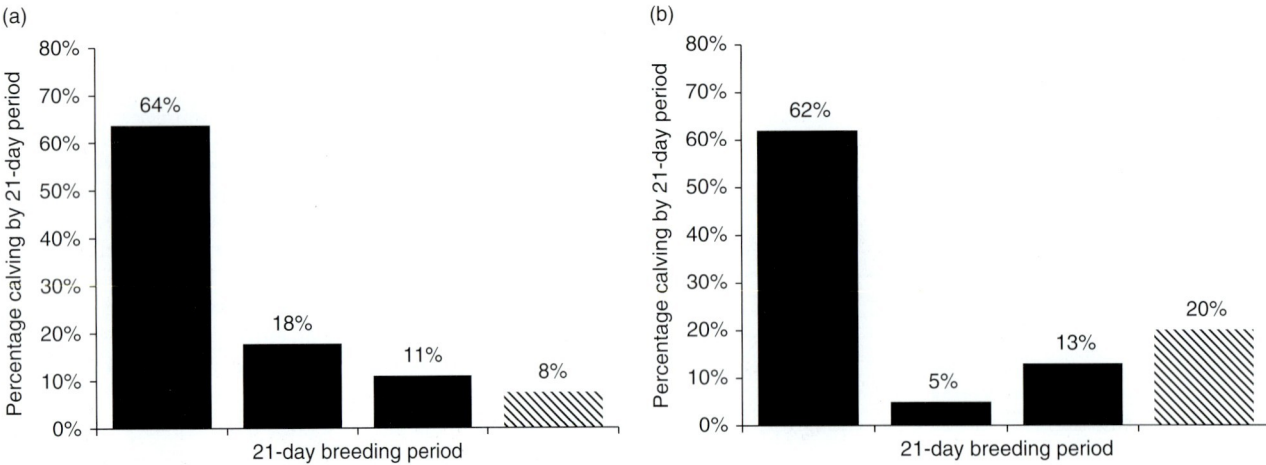

Figure 40.3 Evaluation of the reproductive profile by 21-day period for the entire herd consisting of four equally sized breeding groups (a), and for only one breeding group within the herd (b).

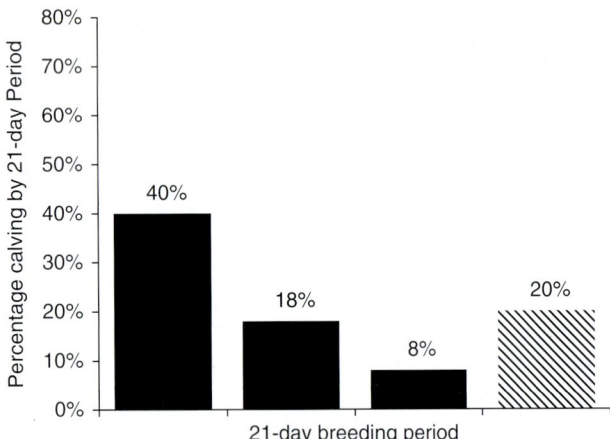

Figure 40.4 Evaluation of the reproductive profile by 21-day period for a herd with exposure to a pathogen leading to early embryonic death and abortion.

from the management and male or female issues described above. This distinction is important as the diagnostic and herd health interventions differ greatly based on the specific cause of the reproductive problem.

After introduction of an infectious agent causing reproductive loss, the reproductive profile may be similar to the previous year, yet with lower conception rates in each 21-day period. Figure 40.4 represents an example of an infectious agent pattern in a herd with a near-ideal calving pattern in the previous year. After early embryonic death or abortion, some cows may be able to rebreed (depending on the pathogen), leading to an increased percentage of cows bred late in the breeding season. This is rarely observed in herds with a confined breeding season (60 or 90 days) as there is limited time for a cow to become infected, abort, and begin cycling again prior to removal of the bulls from the pasture. Thus, the most common profile for a herd exposed to an infectious abortifacent is a similar pattern to the previous year with a reduction in the conception rate within each 21-day period. The magnitude of reduction in pregnancy rate is related to the specific pathogen.

Conclusions

The goal of a complete herd health program is management based on areas of need or opportunity, but the specific reproductive challenges differ among herds. The reproductive profile consists of the number of calves born per female exposed in the previous season, the length of the calving season, and the distribution of calves born throughout the season. The profile can be created from generally available information, and using pregnancy diagnosis information can assist in adding value to the herd and improve the value of pregnancy diagnosis at an appropriate time. Reproductive success in the cow-calf herd can be optimized by using the reproductive profile to identify specific target areas when creating and maintaining the herd health program. The reproductive herd profile can also be a useful component of a diagnostic strategy for investigating reproductive insufficiency as many diseases and management conditions manifest in relatively unique calving patterns.

Summary

- Preventing pregnancy losses and promoting optimal reproductive efficiency are critical for high beef cattle herd productivity and economic sustainability, and overall herd health can be evaluated with herd reproductive profiles.
- The herd reproductive profile is a series of key indicators related to reproductive success, including the pregnancy percentage, length of the calving season, and the distribution of calving dates within the season.
- The reproductive profile is a valuable tool for identifying specific areas for opportunity to improve reproductive efficiency, as well as highlighting potential problem areas.
- Assessment of changes in the reproductive profile can be a useful diagnostic tool, and a method to monitor success of disease and management interventions.
- Many factors (e.g., pathogen exposure, herd nutritional status) can negatively impact the reproductive profile, and an evaluation of the profile provides insight to refine the differential diagnosis list and prioritize specific health interventions for the herd.

References

1. Spire M. Breeding season evaluation of beef herds. In: Howard JL (ed.) *Current Veterinary Therapy 2: Food Animal Practice*. Philadelphia: WB Saunders, 1986, pp. 808–811.
2. Fuez DM, Umberger WJ. Beef cow-calf production. *Vet Clin North Am Food Anim Pract* 2003;19:339–363.
3. Ramsey R, Doye D, Ward C, McGrann J, Falconer L, Bevers S. Factors affecting beef cow-herd costs, production, and profits. *J Agr Appl Econ* 2005;37:91–99.
4. Werth L, Azzam S, Nielsen M, Kinder J. Use of a simulation model to evaluate the influence of reproductive performance and management decisions on net income in beef production. *J Anim Sci* 1991;69:4710–4721.
5. National Animal Health Monitoring System. *Beef 2007–08. Part II: Reference of Beef-Cow Calf Management Practices in the United States, 2007–2008*. Fort Collins, CO: USDA, 2009.
6. Alexander B, Johnson M, Guardia W, Van de Graaf W, Senger R, Sasser R. Embryonic loss from 30 to 60 days post breeding and the effect of palpation per rectum on pregnancy. *Theriogenology* 1995;43:551–556.
7. BonDurant R. Selected diseases and conditions associated with bovine conceptus loss in the first trimester. *Theriogenology* 2007;68:461–473.
8. Chenoweth PJ. Cow/calf production principles. In: Chenoweth PJ, Sanderson MW (eds) *Beef Practice: Cow-Calf Production Medicine*. Ames, IA: Wiley-Blackwell, 2005, pp. 9–25.
9. Funston R, Musgrave J, Meyer T, Larson D. Effect of calving distribution on beef cattle progeny performance. *J Anim Sci* 2012;90:5118–5121.
10. Lents C, White F, Ciccioli N, Wettemann R, Spicer L, Lalman D. Effects of body condition score at parturition and postpartum protein supplementation on estrous behavior and size of the dominant follicle in beef cows. *J Anim Sci* 2008;86:2549–2556.
11. Cushman R, Allan M, Thallman R, Cundiff L. Characterization of biological types of cattle (Cycle VII): influence of postpartum interval and estrous cycle length on fertility. *J Anim Sci* 2007;85:2156–2162.
12. Ciccioli N, Wettemann R, Spicer L, Lents C, White F, Keisler D. Influence of body condition at calving and postpartum nutrition on endocrine function and reproductive performance of primiparous beef cows. *J Anim Sci* 2003;81:3107–3120.
13. Berardinelle J, Joshi P. Introduction of bulls at different days. *J Anim Sci* 2005;83:2106–2110.

Chapter 41

Evaluating Reproductive Performance on Dairy Farms

James A. Brett and Richard W. Meiring

*Department of Pathobiology and Population Medicine, College of Veterinary Medicine,
Mississippi State University, Starkville, Mississippi, USA*

Introduction

There is no area on a dairy farm that does not affect the reproductive performance of the herd. The success or failure of a herd's reproduction program has a dramatic effect on the longevity of the cow in the herd. Whether beef or dairy, all cows that do not become pregnant will eventually be culled and culling has a major economic impact on the farm's financial health. Since all the programs and areas on the farm can affect reproduction, monitoring the farm's reproductive parameters is vital to the evaluation of those programs' success or failure. This chapter will describe some of those evaluation procedures.

Overview

"If you can measure it, you can manage it."
　　　　　　　　　　Jenks Britt DVM, Diplomate ABVP – Dairy.

Reproductive services and protocols have historically constituted the foundation of the services that veterinarians provide for dairies. Developing vaccination programs and regular herd visits to rectally palpate or ultrasound cows are commonplace. These routine visits to a dairy should be viewed as "wellness exams" of the farm. We examine and evaluate individual animals but the farm's examination is not complete until we review the data from the visit and the farm's records.

Recording information on pregnancy status still occurs on these regular veterinary visits, but the incorporation of blood and milk tests for pregnancy now provides dairies additional opportunities to gather this information. These new technologies should not limit a veterinarian's role on a dairy but may alter the services and time allocation. Progressive dairies often need outside, professional assistance to serve as independent evaluators of their operations. Competent reviewers and evaluators of a farm's records can provide a valuable service.

Records

A farm's records are only as valid as the consistency and accuracy of the data that are entered. A cow's complete history including calving date, estrus dates, body condition scores, breeding dates, technician or breeders code, and pregnancy status must be accurate to correctly access a farm's program. Traditional measures of reproductive performance are useful for comparing a herd over time or for comparing two or more herds. However, these measures do not provide information necessary to monitor current reproductive performance. The traditional measures of reproductive performance are usually found on the herd summary sheet. Most of the data, such as calving interval, is retrospective. This lag time limits its usefulness in assessing current performance. The value of the information can be enhanced by use of cohort analysis.

The most common source of data is through the Dairy Herd Improvement Association (DHIA). Data are collected and managed by several processing centers in the United States, but it is important to recognize that differences between processing centers exist. Various DHIA processing centers report the reproductive indices in different ways and the indices are not calculated in the same way by all centers. It is important to determine how an index is calculated before the numbers are analyzed. A processing center should be contacted for the definitions used in its reports before data analysis is attempted. In the southeastern United States, many dairies utilize Dairy Records Management System (DRMS). DRMS processes dairy records in Raleigh, North Carolina and has staff in Raleigh and Ames, Iowa to provide technical support.

On-farm software systems are available to dairy producers and veterinarians to assist in the evaluation of the data. The two most common systems used by dairy producers and consultants are PC Dart and Dairy Comp 305. There is regional bias to these systems, with PC Dart almost exclusively used in the southeastern United States

---

Bovine Reproduction, First Edition. Edited by Richard M. Hopper.
© 2015 John Wiley & Sons, Inc. Published 2015 by John Wiley & Sons, Inc.

and is the one discussed here. Once understood, either system can effectively help "massage" the numbers to create reports, graphs and charts to evaluate the different areas of a dairy's performance.

Guidelines to abnormal findings and parameters can be developed using the DairyMetrics© portion of DRMS. DairyMetrics© is a benchmarking tool for dairy farm performance evaluation. This service allows the veterinarian to compare a herd with other selected herds, also known as cohort herds. These cohort herds can be selected based on their individual parameters within five classifications (general, production, udder health, reproduction, and genetic information). Within these five classifications, there are 76 variables from which to choose. The DairyMetrics© program allows comparison of the means, standard deviations, minimums, and maximums of selected parameters for a herd and the cohort herds. Percentiles for each herd performance parameter in comparison to the cohort group are also given. The veterinarian has the option of graphing a herd and its cohorts for the previous 3 months and from data 1 year ago. The value of this program is the opportunity to choose the cohorts from the over 15 000 herds processed by DRMS. The database is updated nightly to maintain current herd information so data is relevant and compares a herd to those in your area, state, or region.

When reviewing the farm's DHIA records, the DRMS 202 Herd Summary Sheet gives an overview of the dairy and contains relevant information including milk production, reproduction, genetics, udder health, and feed cost information.[1] Data are normally broken down by parity (first lactation, second lactation, third and greater lactation, and all lactations). The reproduction section of the 202 Herd Summary Sheet has summary data for the current breeding herd, the total herd, and a yearly reproduction summary. The current breeding herd designation includes all cows that are past the farm's voluntary waiting period (VWP). The VWP represents the time period from calving to first breeding. Pregnant cows included in this report should be given a "P" code and are defined as those diagnosed as pregnant or those bred 65 days or more before the current test date or before they have left the herd.

The DRMS Total Herd Reproduction Summary contains breeding and pregnancy data on all cows in the herd (including those below the VWP) and is also listed for the entire herd by parity. The only cows not included in this report are those recorded as reproductive cull cows ("C" code). In this summary, various aspects of herd reproductive management and herd performance can be seen. The report contains an area that will calculate the average days open at first service; it will show the percentage of cows bred in relation to the VWP and the calculated average days to first breeding. Delaying the first breeding will increase this number and can have a dramatic effect on the herd's calving interval (CI). Measured in months, the CI is the period from one calving to the next. The actual CI is retrospective and better analysis may be made by comparing the actual versus predicted CI.

The other area that will affect the CI is the conception rate. In the Total Herd Reproduction Summary, it will be displayed as the number of services per pregnancy for both pregnant cows and for all cows. Services per pregnancy for pregnant cows includes the data for all pregnant cows currently in the herd and those that have left the herd within the last 9 months if recorded as pregnant. It is calculated by the number of services recorded to those cows divided by the number of pregnant cows. Services per pregnancy for all cows includes all cows with diagnosed pregnancies and/or cows not returning to heat in 65 days after breeding and also includes all cows that have left the herd within the last 9 months. This figure contains all services in this evaluation period divided by the number of pregnant cows. Services per pregnancy should be evaluated by parity and by days in milk at first breeding. It is also useful to compare services per pregnancy for pregnant cows with services per pregnancy for all cows. As with other parameters, it is important to note that the mean may not represent the herd's reproductive success because cows with numerous services may misrepresent the actual pregnancy data.

Service or heat intervals are also included in this report. These figures vary in accuracy since all breeding and heat dates must be reported. The figures are based on all intervals for each cow and the dates of the recorded estrus or breeding. Four categories are listed in this section: less than 18 days, 18–24 days, 36–48 days, and all other intervals. Expected estrous cycles will be 18–24 days. Intervals less than 18 days can indicate reproductive problems such as cystic ovaries, inadequate heat detection (false heats), or use of prostaglandin. Intervals of 36–48 days indicate that one heat was not observed. Service or heat intervals in the "other" category are likely to be associated with two or more missed heats, anestrous cows, or those with abnormal heat cycles.

The number of services and the percentage of successful services can also be reviewed. These are reported by first, second, and third and greater services to cows and includes current breeding herd, all breedings on cows removed from the herd in the last 9 months, and all breedings on cows coded as reproductive culls. It also calculates the "service sire merit $" which is the "average merit $" for all services to proven sires. Service number and percent successful are used with the service or heat interval to see how aggressive and successful the farm is with its breeding program. The percent successful by service number can help evaluate successful programs such as timed AI protocols or pinpoint times and areas where the program may be failing.

The Yearly Reproductive Summary is a current view of the previous year's reproductive data and is calculated using the most recent six test intervals up to 200 days from the current test date. Services on cows that have left the herd and the reproductive cull cows are also included. The chart will often include information on the month prior to the 12-month period or "month dropped." This information is for historical review of the herd 1 year ago but is not included in the current yearly figures.

Detection of estrus is critical in herds not using a timed AI program. Success or failure of estrus observation is found in the percent heats observed data. The formula for estrus detection efficiency is shown below:

$$\text{Percent heats detected} = \frac{\text{Heats observed}}{(\text{Total cows days in period} \div 21)} \times 100$$

Eligible cows are defined as all cows past the VWP which are contributing to estrous cycle days in the test period.

Percent successful services reports the successful services (pregnancies) in each test period divided by the total number of services for that test period. Percent successful services entry for the two most recent test periods is not used since pregnancy status on most of these breedings has not been determined. For herds on routine pregnancy examinations, a reported pregnancy diagnosis is used to determine percent successful services. For herds not reporting pregnancy diagnoses, the recorded estrus is used.

In recent years, veterinary practitioners and producers have recognized the need for measurements of reproductive performance that are not influenced by momentum and lag time. One parameter that is commonly used is the herd's pregnancy rate:

Pregnancy rate = percent heat detection × conception rate

Simply put, it is the percentage of cows eligible to become pregnant that are reported pregnant in a specific period of time. Some systems measure every month, while others measure 3-week periods (the typical estrous cycle). The number of cows becoming pregnant is a function of both the efficiency of heat detection and the conception rate. Both contribute equally to the percentage of eligible cows becoming pregnant. Successful reproductive programs in well-managed herds will report a pregnancy rate of over 20% but the average in most regions remains at 13–15%. Factors that negatively affect conception rates, such as poor AI technique, heat stress, loss of body condition, and inadequate heat detection, negatively affect this number. The yearly summary also reports total services, confirmed pregnancies, number of calvings, and total number of pregnant cows in the herd for that test period.

Reproductive figures can also be found in other areas of the DHIA records. The summary area includes the yearly summary of cows entered and left the herd, which measures the herd turnover and herd replacement. In the "number of cows left the herd" section, the cows that left the herd for reproduction reasons are given. Reproduction culls are considered "involuntary culls." Care should be taken since this number may not be accurate. "Milk" or "Dairy" classifications for culling are "voluntary culls." However, if the cow does not become pregnant, her production will eventually drop so she will leave the herd. If the producer codes her culling for production reasons, the accuracy of the number culled for reproduction may be falsely low.

The DHIA program[2] offers several reports specifically to aid in reproductive management:

- DHIA Form 208: a barn wall chart for recording breeding and calving records.
- DHIA Form 211: a 21-day reproduction record, which is used to signal dates when cows should be observed for heat.
- DHIA Form 212: lists management options such as cows to calve, cows to pregnancy check, cows to breed as well as dry off.
- DHIA Form 217: contains information on breeding and offspring.

After using the DairyMetrics© program, parameters should be established for intervention or investigation into the causes of substandard performance. An example for such parameters compiled by Heersche[3] at the University of Kentucky is seen in Table 41.1. The University of Nebraska-Lincoln[4] published a more comprehensive chart in 1986 (adapted from Dairy Profit Series, Cooperative Extension Service Iowa State University, DyS-2852/December 1985) that can be adapted for herds today (Table 41.2).

Caution should be used when reviewing a herd's records and data based on the number of animals or entries in each category. This offers a challenge with smaller farms since the percentages may or may not reach the farm's goal but inadequate numbers are available to accurately access the area. Other aspects are farm management practices that will artificially affect the data. In the southeastern United States, herds may choose not to breed during the early fall so cows will not calve in the summer months.

In addition to standard DRMS reports, other measurements of reproductive performance are available. These will help monitor the herd's reproductive status as well as identify areas of opportunity. Several of these measurements are listed below.

1. The percentage of the herd that is pregnant at a given time. A goal for this parameter is to have 53–55% of the herd pregnant. A herd can have acceptable reproductive performance with less than 50% of the herd pregnant if the calving pattern is seasonal.
2. Average days-in-milk for lactating cows. This is an indirect monitor of reproductive performance, but is closely related to reproductive efficiency. A goal for average days-in-milk is 160–165 days.
3. The percentage of cows open beyond 150 days-in-milk. A goal for well-managed herds is to have less than 10–12% of cows open beyond 150 days-in-milk.
4. Evaluation of method of heat detection or synchronization program on conception. Compare methods to determine deficiencies in heat detection method or type of synchronization programs used by the farm.
5. Plot of days-in-milk at first breeding by days-in-milk. This allows visualization of the patterns for first breeding. Look for signs of visual heat detection or synchronization.

Table 41.1 Economically important reproductive metrics and levels at which management interventions should be made.

Parameter	Goal	Intervention
Days open	115	160
Calving interval (months)	13	14.5
Days to first service	75	100
Conception rate, first service (%)	55	30
Conception rate, all services (%)	50	30
Heat detection rate (%)	70	40
Reproductive culls per lactation (%)	<8	15
Abortions (%)	<5	10

Table 41.2 Template for the specific client-determined on-farm goals for reproductive management.

	Goal	My herd	Need help
Calving interval (days)	365–380		
Average days to first observed heat	<40		
Percent cows in heat by 60 days post calving	≥90		
Average days open to first breeding	50–60		
Average days open to conception	85–100		
Services per conception	1.5–1.7		
First service conception rate (%)			
Replacements	65–70		
Producing females	55–60		
Percent breeding intervals between 18 and 24 days	>85		
Percent cows open >120 days	<10		
Dry period length (days)	45–60		
Average age at first freshening (months)	24		
Average age at first breeding (months)	15		
Percent cows pregnant ≤3 AI services	90		
Percent cows pregnant on examination	80–85		
Abortion rate (%)	<5		
Cull rate for infertility (%)	<10		

6. Cows not serviced by 75 days-in-milk. Cows should be inseminated by 75 days-in-milk. The cause of delayed insemination should be determined. A goal is to have 90% of cows serviced by 75 days-in-milk.

References

1. Dairy Record Management System. *DHIA handbook*. University of Pennsylvania, Computer Aided Learning Center, 2007.
2. National DHI. Boost milk quality, repro efficiency with DHI tools. *Holstein World* 2009;(February):66–67.
3. Heersche G. Benchmarks for evaluating the reproductive performance of the dairy herd. University of Kentucky College of Agriculture Cooperative Extension Service. Kentucky Vet Notes (not archived). http://afsdairy.ca.uky.edu/files/extension/reproduction/Benchmarks_for_Evaluating_the_Reproductive_Performance_of_the_Dairy_Herd.pdf
4. Keown JF. How to estimate a dairy herd's reproductive losses. University of Nebraska–Lincoln Extension, 1986. Available at http://digitalcommons.unl.edu/extensionhist/538/

Web resources

Dairy Herd Improvement Association: www.dhia.org
Dairy Records Management Systems: www.drms.org
DairyComp 305 (Valley Ag Software): www.vas.com

Chapter 42

Marketing the Bovine Reproductive Practice: Devising a Plan or Tolerating a System

John L. Myers

Pecan Drive Veterinary Services, Vinita, Oklahoma, USA

Introduction

Nothing is more common than to hear a veterinarian complain about how little income results from the practice of veterinary medicine. While such a complaint is not limited to the veterinary profession, somewhere in each of these discussions a remedy appears as if it were as simple as turning on a light or repairing a dystocia: marketing is the remedy. When it surfaces as a cure to veterinary poverty there are as many interpretations of what marketing is or does as there are individuals engaged in the discussion.

The goal in these marketing discussions is always profit, although that also has different connotations, when in fact the term "marketing" implies aspects of a business many of which do not necessarily have a direct effect on profitability. A definition of marketing as adopted in 2007 by the American Marketing Association is as follows: Marketing is the activity, set of institutions, and processes for creating, communicating, delivering, and exchanging offerings that have value for customers, clients, partners, and society at large.[1] Perhaps another definition would be that marketing involves obtaining the client, satisfying the client, and retaining the client.

Principles in marketing

Pursuit of literature about marketing frequently exposes one to the four Ps of marketing:

1. Product
2. Price
3. Promotion
4. Place.

In the context of a bovine reproductive practice, the product to be marketed would be either a service offered or some specific item such as a drug, a feed ingredient, or an instrument. The price is a fee that represents a reasonable profit and would be competitive with other businesses offering the same or similar services or items. Promotion would be those efforts taken to let prospective clients know that the practice is actively pursuing delivering what is for sale. Finally, place would be how the service or item is distributed so that clients can easily obtain the service or item at a time and location convenient to the client.

For this discussion, the assumption will be that the bovine reproductive practice would be offering a service rather than an item, and as such three more Ps come into play:

1. People
2. Physical evidence
3. Process.

There is an intangibility of service. A bull having undergone a breeding soundness examination (BSE) looks much the same after the examination as he did before. A person buying a pickup, however, has a tangible object that was not there before its purchase. Clients therefore make judgments about the service by looking for clues about quality of the service as it is accomplished. This is done by assessing the skill, appearance, and efficiency of all of the people involved in the examination from the office employees, to the assistants and the veterinarian. The physical evidence of the examination would be the professional quality of any documents such as a certificate, billing charges, and information materials or handouts that are presented after the service is accomplished. The process involves the entirety of the examination from reception of the client, the handling of the bull, the efficiency, availability and convenience of instruments or supplies, and the adequate presentation of the results resulting from the examination.

Bovine Reproduction, First Edition. Edited by Richard M. Hopper.
© 2015 John Wiley & Sons, Inc. Published 2015 by John Wiley & Sons, Inc.

A marketing system or a marketing plan

Regardless of the efforts, conscious or not, of any active business, a marketing system exists. Services are delivered, processes and people are in place, and promotion, pricing and distribution of those services occur whether this system is efficient, reasonable, profitable or successful. If one attempts to modify the system, efforts toward improvement in one aspect of the business may be detrimental to other aspects.[2] The veterinarian may be quite successful, for instance, in satisfying a client by doing excellent work in a limited number of services, but the efforts do not result in obtaining more clients. If that veterinarian embarks on a strategy of promotion that restricts the dedication to performing the service, there will be no resulting improvement in the marketing system. Likewise, if the veterinarian channels efforts into obtaining and satisfying the client but because of other competitor's superior promotion efforts the veterinarian is unable to retain existing clients, then no progress is made.

A marketing plan should therefore try to address all aspects of the system instead of merely propping up deficiencies that already exist. The resulting balance should result in a more successful, profitable, and enjoyable practice.

Three pillars of the marketing plan

Before one embarks on a plan, three prerequisites must be firmly in place with commitment dedicated to each.

1. *Competence*. It is one thing to desire to operate a practice dedicated to bovine reproduction and quite another to have available the expertise needed to make it successful. Competence cannot be acquired and then left unattended. Dedication to constant study, attendance at quality seminars and conventions, interaction with knowledgeable colleagues, and familiarity and openness to new ideas and techniques are essential for a successful practice and career. This dedication and effort is expensive in both a monetary and temporal significance.
2. *Dedication of resources*. While no program will succeed if adequate funding or resources are not dedicated to it, the amount of resources requires careful consideration. The veterinarian's time is an available resource, although calculating the value of that time is difficult but necessary. Regardless, if a well-balanced and carefully considered plan is in place, the appropriate numbers should be budgeted and documented and, like all budgets, adhered to or altered as necessary.
3. *Comfort with, or at least toleration of, the plan*. Many times marketing plans and schemes will fail because there is not complete acceptance of the proposition. The competence to perform may be in place and adequate financing is available but there is revolt at the features of the marketing plan simply because they do not lie within the comfort zone of the personality of the veterinarian. Advertising may seem too gauche, speaking to producer groups may be terrifying, and promoting oneself may strike one as arrogance. Toleration of discomfort in the pursuit of a beneficial outcome is the price one pays for personal growth. Abandoning a program because the personal discomfort over-matches any possible benefit dooms the best plans and propositions.

Devising the marketing plan

For purposes of discussion, the assumption will be made that a budgetary consideration of 1% of anticipated annual revenues will be dedicated to the marketing plan. This would not be enough in an entity that depends heavily on product sales (such as automobiles or hamburgers) but the 1% figure does fit well within budgets of large companies that provide services and products such as a rural electric cooperative or a pipeline engineering concern. Therefore, if the annual revenue of the practice was $500 000 per year, then $5000 would be budgeted toward the marketing plan. In this budget it is necessary to place a value on the time the veterinarian or other individual spends on the project.

Obtaining the client

The best, but perhaps not the easiest to attain, promotion one can do is to benefit from positive word of mouth. When first embarking on a practice or when trying to urge the practice on to greater levels of revenue, the word of mouth technique may not be sufficient to satisfy the aggressive practitioner. While there may be overlap between the following categories, four divisions will be addressed.

1. *Expanding the service repertoire*. There may be requests from current or possible clients that desire a specific service, such as fetal sexing by ultrasound or embryo transfer, which is not currently available at the practice. An investment in instruments and techniques to satisfy the requests makes possible not only client satisfaction but also brings in more clients looking for the same service.
2. *Promotion of existing services*. Many times clients are unaware that certain services already exist in the practice. Brochures should be prominently displayed explaining the services, details of the services readable on invoices or receipts, and the physical evidence of these services displayed in a place of congregation such as the reception room or laboratory. Examples of physical evidence might be ultrasound images or breeding soundness examination certificates.
3. *Development of associations that lead to programs to be given by the veterinarian*. Extension agencies, drug and pharmaceutical companies, breed associations, and livestock organizations are always in need of programs. Volunteering for these types of programs offers an ideal opportunity to display one's expertise. Further, writing articles in trade magazines is an excellent method of displaying one's proficiency and knowledge.
4. *Advertisements*. Although once controversial, other medical professions have wholeheartedly embraced advertising on billboards, magazines, clothing, and other items that are given away. This is an expensive proposition and

care should be taken to ensure maximum exposure to potential clients while still operating within budget and ethical restraints. A bovine reproductive practice may need to expand beyond normal practice boundaries and so advertisements in regional livestock publications or breed association magazines that the veterinarian could still service may be appropriate.

Satisfying the client

Client satisfaction relies on much more than veterinary competence. A wise marketing plan makes use of multiple possibilities to satisfy the client and encourage the client to return for further services.

Process

One must beautify the process. The appearance of the practice, including the appearance and demeanor of the people employed, will give the client an important clue to the satisfaction possible in the service about to be received. An investment in training of the staff to this end is important, but many times it is the veterinarian who falls short in appearance and demeanor. A holistic approach embraced by all will keep everyone, including the veterinarian, working toward an efficient and professional process in providing service that is done in an attractive and pleasant environment.

Physical evidence

As stated before, providing a service frequently leaves nothing to behold after completion of the project. It is imperative, therefore, that eye-catching, professional, easily readable and memorable physical evidence be provided for the satisfaction of the client. In the case of a BSE, the Society for Theriogenology (SFT) has for several years provided to its members a triplicate paper record that adheres to the standards for the BSE as established by the SFT (see Chapter 6). Because the form was illicitly, and in violation of the copyright, copied by nonmembers and even nonveterinarians who were not following SFT guidelines, the SFT has recently produced a computer-generated online BSE form that not only embraces the best standards on BSEs, but also provides a professional-looking document that also compiles statistics on all bulls entered into the program in a national and international database (see Chapter 83).

The possibilities of in-house produced documents should be carefully considered in the bovine reproductive practice. At the end of a pregnancy examination experience, even if individual records are not kept on individual animals, a recap of the satisfaction, or lack thereof, of the pregnancy status of the herd as discussed between the veterinarian and the manager should be documented on a form expressly for that purpose and a copy kept as clinic records and one left at the ranch. Placing the forms in folders, or mailing a copy back to the ranch or to the owner, ensures that the recommendations and evaluations are studied by all concerned and keeps that veterinary clinic's name in front of the client. Professional-looking forms and handouts should be developed for all procedures offered by the practice that would benefit from such a distribution.

Place

An important aspect of client satisfaction would be how the service offered would be placed, or distributed, for the convenience of the client. Consideration of this distribution of services encompasses facilities, scheduling, and availability of the veterinarian. There are successful bovine practices that are strictly based on offering the service at a haul-in facility, others that are strictly on-farm practices, and others that fall somewhere in between. Many times it may be more convenient for the client to have the veterinarian at the farm to perform services, but on a limited number of animals a much better job of control and execution of the service can be done at a quality facility managed by the veterinarian. Construction of such a facility is expensive and cannot be justified unless the veterinarian is available at the appropriate times and not elsewhere trying to perform veterinary procedures on animals inadequately captured or unsatisfactorily restrained. If change is made between one type of philosophy to another, client education becomes paramount.

Scheduling is frequently a primary source of dissatisfaction for clients. Unfortunately, conflicts will always exist whereby the client either has to wait longer than would be convenient for placement on the schedule or is postponed longer than anticipated to receive the service scheduled in the first place. Hopefully, the bovine reproductive practice will develop into a practice more amenable to scheduling and less dependent on servicing emergencies, but a great deal of patience and client education must take place for the satisfaction of the client and the veterinarian. Experience and a knowledgeable person at the clinic (it may be the veterinarian) should smooth out some of the rough edges that always occur when several clients need services rendered at much the same time.

Finally, distribution of services works much better when the veterinarian concentrates on fewer services and elevates the quality of the services provided. A bovine reproductive practice implies specialization, and clients tolerate delays or scheduling conflict better when the veterinarian is working hard on the business and species on which the client makes his or her living.

Price

No more difficult decision is made in the practice than how to set pricing, and price is the most sensitive and difficult issue the practice will face. With adequate and fair pricing the practice will thrive but without it the practice will stumble or sit idle, so while it is obvious that the client will usually be more satisfied with a lower price, at the same time that client may respond negatively to the corners that were cut to deliver the service with a corresponding decrease in results.

There will be advocates for either higher or lower pricing based on the benefits either side has to gain, but to the individual practice owner the ideal price means work is cheerfully accepted but at any lower level would be gladly refused and is an extremely difficult accomplishment. This is a difficult problem and therefore needs serious and thoughtful consideration but it is nonetheless possible to solve. Flexibility is sometimes beneficial in price-setting, but there are perils.

It behooves the practitioner to consider the type of client the practice intends to attract but it should realize that attracting as many clients as possible would be desirable even if the veterinarian decides later that in some cases neither the veterinarian nor the client are destined for a beneficial relationship. Some, perhaps the best, clients respond well to education. For these clients, time spent in thoughtful conversation about the benefits the bovine reproductive practice can provide and at what price is the most productive part of the association and the veterinarian should encourage, plan and efficiently execute these sessions. Other clients will not respond to education and merely want a brief explanation of what is to be done and for what price; these clients have already decided what the benefits will be and are not anxious to hear the veterinarian's viewpoint. Thus, there are three types of clients: those that respond to education, those that will respond to education in the future, and those that will never respond to education. However, it should be realized that one cannot predict which group will be most profitable, but it can be predicted which group will be most enjoyable and rewarding.

While pricing is a difficult issue, the practitioner should rejoice in the prospect that many clients do respond to flexibility and creativity rather than responding only to competitive pricing. Flexibility and creativity can be perilous, however, so once again the practitioner should carefully weigh a new or different price structure regardless how elegant it may seem.

Hourly versus price per task In a dairy practice or perhaps certain beef practices there may be an imperative to offer services by the hour and for many this is advantageous to both the veterinarian and the client. In other practices, however, there is seasonality due to set breeding seasons that results in fluctuation between being very busy to being relatively idle. In such cases bidding services by the hour necessitates a rate that must cover those times when the clinic is idle and may result in a rate that is unattractive to the client. Obviously, the hourly rate of, say, a dairy practice considers the same prospect, but the dairy practitioner may have a more level production thereby having less idle time for which an accounting must take place. An answer might be to encourage elective work during the non-busy time, even at an hourly rate.

Guarantees Many veterinarians can successfully work by a promise of outcomes. There are ethical issues at play here, but taking on a project and being rewarded by a positive outcome (higher conception rates, higher weaning weights, etc.) can be profitable if the veterinarian is confident that the strategies and procedures implemented should produce the expected positive results, especially to a reluctant prospective client. This should probably be a short-term proposition as the culture the veterinarian wants to cultivate is for clients to use the practice's services because they are already confident that outcomes will be beneficial.

Consulting It is hard to envision a bovine reproductive practice that does not provide consulting in some fashion, but it is equally as difficult to envision that practice dependent strictly on consulting. It would seem that by definition of the bovine reproductive practice there would be some skill (in addition to consulting) that would be brought into play, for example surgery, palpation, embryo transfer, BSE. During some of the procedures mentioned, there is usually time devoted to consulting in the hope of developing protocols that would improve the outcome of the procedure being executed, or at least defining what it is that has produced such a positive outcome. Consulting at this time hopefully will be covered by the charge of the procedure, but should result in positive results in the future as well as tremendous goodwill in the present. Care must be taken, however, in consulting at no charge when no other procedure or prospect of a procedure is anticipated.

Dispensing Traditionally, many veterinary clinics have made a sizeable portion of their profit from dispensing of drugs, biologicals, and other products. The profit made in these activities covered other areas of the practice that were not priced at an adequate rate (such as consulting). While a dispensing practice is entirely possible, and many times quite profitable, competition from larger corporations has complicated the issue. Some clients frequently prefer to obtain pharmaceuticals from the veterinarian, thinking this will ensure that procedures related to product administration will be done correctly, but the trend seems to be the contrary and toward requesting prescriptions for those items necessary and purchasing them online or from other large outlets. If one is to compete in the dispensing business, care should be taken not to let it interfere with other nondispensing procedures and one must develop skills in inventory control, purchasing and promotion that may not already be possessed within the practice.

Retention of client

Attention to client retention is more than admitting that existing customers are more valuable and profitable than prospective ones; it is the realization that high-quality service and exceeding expectations should be emphasized more than obtaining profits.[3] Those clients who are loyal and engaged are the best promoters of the bovine reproductive practice, and attention to these clients is imperative.

The bovine reproductive practice would in theory thrive on repeat business from existing customers, and therefore management procedures should be instigated that would keep the practitioner and the clinic in the mind of the client without seeming excessive. Follow-up phone calls from the veterinarian or staff on outcomes from previous work is warranted but hopefully a line of communication is established some time during the client–veterinarian relationship that lends itself readily to a steady realistic stream of information between the two parties. The veterinarian must decide which of several possibilities most meets these needs as well as fits within the abilities of the practice.

Newsletters

Newsletters or other means of updating clients on available procedures, current or changing issues that apply to the client's production, and diseases or syndromes that have been recently seen in the practice are usually well received but difficult to produce on a consistent basis. There are companies that will produce such newsletters that can be tailored to the practice, but many times these do not display a personal touch that resonates with the client. If there is a desire to do this and there is a corresponding willingness and capital to get it done, a newsletter can prove beneficial to client retention.

Email or social media

Information and personal contact can be easily distributed by email or one of the social media avenues (Facebook, LinkedIn), although not all clients readily embrace such communication. As the sophistication of these outlets improves, using such devices will be a convenient and obvious method of client correspondence and feedback.

Personal contact

Many veterinary clinics of different or general specialties already have reminder systems in place for vaccinations or other procedures. In theory, the bovine reproductive practice would not have the volume of clients that need reminding of, for instance, rabies vaccinations that exist in a small animal practice. This leaves in place the opportunity for a personal phone call on a calculated basis for business to be lined up. Herd palpations, BSEs in bulls, and other procedures that have been done in the past and should be accomplished in the future are excellent, although there may be preferences on the client's part of how any particular information would most effectively be received. The important part is to have a system in place that accomplishes this as a matter of procedure rather than when the spirit moves. Every service provided is an opportunity to make contact in the future.

Evaluation of the marketing plan

Regular evaluation of the marketing plan should be part of the overall design. One should attempt to measure both objectively and subjectively to see what effect the plan has made on the practice. While revenue or repeat business is easily determined, other determinations such as client satisfaction and staff morale are equally important but perhaps more difficult to assess. Building a consensus about what the marketing plan should do and how to measure the success of the plan may be the most important elements in the marketing concept.

Discussion

Every veterinary practice has a marketing system in place. The methods for attracting, satisfying and retaining clients or customers is present from the moment the business begins, but whether that system is effective enough for a prosperous venture depends on several factors, and conditions that are favorable for success at the present may not be present in the future.

The veterinarian launching a bovine reproductive practice is in a more favorable situation than many counterparts in other specialties of veterinary practice. Assuming the veterinarian is competent and in a location that is accessible to clients desiring bovine reproductive services, and further possessing no problems with health, personality or other distractions, no other individual possesses the attributes needed to deliver those services than a well-trained veterinarian. The veterinarian is the only individual who can legally provide the pharmaceuticals, techniques, and procedures needed to accomplish some of the reproductive services, no other profession can issue the health inspection papers that are important in the transport of animals, semen and embryos, no other person has the training to grasp and treat the correlation between sickness, health and reproductive performance and, finally, there is a reported shortage of individuals capable of supplying all these services.

There is, however, a marked incursion into bovine reproduction being made by those outside the veterinary profession. As this results sometimes in fair or unfair competition between those with a veterinary degree and those without, many veterinarians or groups of veterinarians have as their response objections about legislation unfavorable to the profession, grievances about the lack of enforcement of existing laws to protect veterinarians, criticisms of the public for failing to recognize the supremacy of veterinarians and, finally, protests about the profession itself for not providing its members with the skills necessary to make an adequate living.

Concurrently, however, the veterinary profession, especially the bovine practitioner, is populated with talented and creative individuals capable of providing efficient and quality medical and reproductive services under less than ideal conditions. Collectively and individually there is no reason the bovine veterinary reproductive specialist should be impeded from establishing a specialty practice of sufficient profitability if the veterinarian applies as much thought, creativity, courage, and cunning into marketing as has already been made in other parts of the business.

The success rate of businesses is fairly consistent across all types of industry and failures are not always related to poor marketing. The most common complaints about the impediment to success in bovine veterinary practice, however, can be very consistently affected by a well-conceived marketing plan even though its execution will require courage, creativity and the ever-present hard work.

In his book, *Lessons From a Desperado Poet*, author and large animal veterinarian Baxter Black demonstrates methods of marketing that have successfully ushered him into the difficult world of entertainment, publishing, and performance.[4] He is obviously a talented individual, but remarkable in this book is the relentless variety of creative strategies and hard work that preceded his success. While many of his methods would not be appropriate in a marketing plan for a bovine reproductive practice, his example as one who pursues all avenues in his quest for markets should be an inspiration to all veterinarians.

If the bovine reproductive practice is to succeed, two viewpoints must be addressed. First, the profession needs to claim what is rightfully and legally mandated from the federal level concerning issues surrounding pharmaceuticals, drugs, their use and prescription coinciding with a veterinary client–patient relationship. Notifying authorities of nonveterinarians or veterinarians not in compliance with these edicts is the responsibility of each ethical veterinarian regardless of the personal discomfort involved. The second viewpoint is that there are valuable contributions to be made by those outside the veterinary profession and a solid, pleasant and profitable relationship with these individuals can enhance the bovine reproductive practice.

Equipped with the competence necessary to deliver reproductive services to the bovine and a marketing plan that promotes and communicates the values associated with those services, the veterinarian in this type of practice should anticipate an excellent possibility of an exciting, satisfying and rewarding career. Helping clients engage to provide the nation and the world with high-quality, safe and reasonably priced food is a noble occupation. Resources and support are readily available, and while success is not assured, the prospects should be considered excellent.

References

1. American Marketing Association. Definition of marketing. Approved October 2007. Available at www.marketingpower.com/aboutama/pages/definitionofmarketing.aspx
2. A Marketing Blog by the Marketing Journal. November 14, 2005, www.marketingjournalblog.com/
3. Reichheld FF. *The Loyalty Effect: The Hidden Force Behind Growth, Profits and Lasting Value*. Boston: Bain & Company Inc. and Harvard Business Press, 1996.
4. Black B. *Lessons of a Desperado Poet*. Guilford, CT: Globe Pequot Press, 2011.

OBSTETRICS AND REPRODUCTIVE SURGERY

43. Vaginal, Cervical, and Uterine Prolapse — 383
    Augustine T. Peter

44. Inducing Parturition or Abortion in Cattle — 396
    Albert Barth

45. Management to Prevent Dystocia — 404
    W. Mark Hilton and Danielle Glynn

46. Dystocia and Accidents of Gestation — 409
    Maarten Drost

47. Obstetrics: Mutation, Forced Extraction, Fetotomy — 416
    Kevin Walters

48. Obstetrics: Cesarean Section — 424
    Cathleen Mochal-King

49. Retained Fetal Membranes — 431
    Augustine T. Peter

50. Postpartum Uterine Infection — 440
    Colin Palmer

51. Cystic Ovarian Follicles — 449
    Jack D. Smith

52. Postpartum Anestrus and its Management in Dairy Cattle — 456
    Divakar J. Ambrose

53. Surgery to Restore Fertility — 471
    Richard M. Hopper

Chapter 43

Vaginal, Cervical, and Uterine Prolapse

Augustine T. Peter

Veterinary Clinical Sciences, College of Veterinary Medicine, Purdue University, West Lafayette, Indiana, USA

Introduction

Eversion of the vagina or both uterine horns from the vulva is commonly referred to as vaginal and uterine prolapse (from *prolabi*, to fall out), respectively. It is not unusual to encounter these cases in cattle practice, particularly in the developing world. Vaginal and cervical eversions usually occur before calving, and uterine eversions after calving. There is no relationship between the occurrence of these conditions. A recent review provided an excellent description of the management of these conditions from the perspective of replacing, repairing and, if need be, amputating the uterus.[1] Noteworthy is a detailed description on manual eversion of the uterus to correct a uterine tear, a technique that involves the use of different pharmaceuticals.[2,3] This chapter will address this condition from three perspectives. Primarily, it will address these maladies and provide remedies with representative images, both real and created. Secondarily, it will provide a summary of information reported in the global literature, including practical solutions provided by practitioners. Finally, a discussion will address these clinical situations from a productive, fertility and, most importantly, an animal welfare perspective.

Vaginal and cervical eversions

Eversion of the vagina (Figure 43.1) is a relatively common occurrence, particularly in cattle compared with other species.[4] There appears to be a genetic component to its occurrence in beef cows compared with dairy cows. Among the beef breeds, Hereford,[5] heavier breeds such as Charolais and Limousin,[6] and Shorthorns are more affected than others. It has been known that *Bos indicus* breeds are predisposed to this condition, along with cervical eversion. Many factors have been suggested to contribute to this condition,[7–10] including increased intra-abdominal pressure in late pregnancy, extreme cold weather, excess perivaginal fat, prior injury to the perivaginal tissues, intake of large volumes of poorly digestible roughage, poor vaginal conformation, persistence of the medial walls of the Müllerian ducts,[11] and changes in hormonal secretions, particularly increased estrogen as observed in late pregnancy and in estrus. It has been suggested that the higher concentrations of estrogens found during pregnancy can contribute to excessive relaxation and edema of pelvic ligaments that support the vagina. Besides the above, incompetence of the constrictor vestibule and constrictor vulvae muscles may have a role in the etiology.[7] The condition is observed in embryo donor cows that are exposed to hormonal stimulation of the ovaries. Imbalance in calcium and phosphorus, and hypocalcemia has been considered to be a contributing factor by some workers.[12] It should be pointed out that the association with hypocalcemia was suggested merely based on the response to calcium therapy rather than on the clinical signs of hypocalcemia. However, as eloquently suggested,[12] it can be tentatively postulated that subclinical hypocalcemia could be derogatory to the internal environment of the physiologically stressed female because of pregnancy and parturition.

The majority of vaginal and cervical eversions occur in the last few weeks of pregnancy; however, it may occur several months before calving. In some cows, it may occur during estrus. Vaginal eversions are classified into four grades according to the severity of eversion, the extent of injury, and the exposure of cervix in such eversions.

- Grade 1 eversion: a small intermittent protrusion of the vaginal mucosa through the vulval lips is noticed (Figure 43.1) when the animal lies down. Often this tissue becomes dehydrated and traumatized and leads to a grade 2. A grade 1 eversion is readily repaired by applying any one of the retention sutures described later.
- Grade 2 eversion: vaginal mucosa protrudes through the vulval lips continuously (Figure 43.2) with the possibility

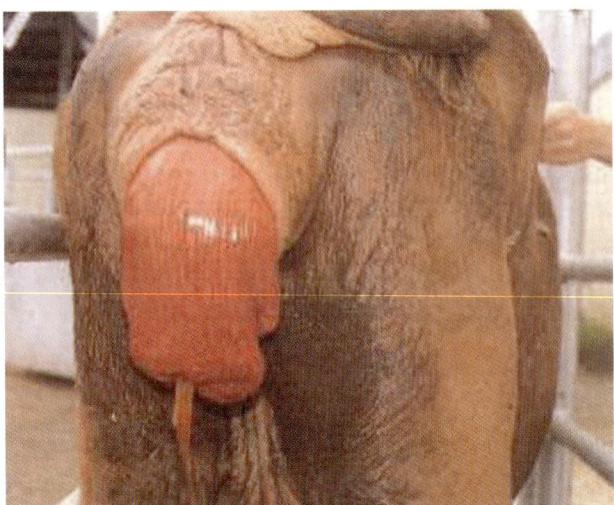

Figure 43.1 Grade 1 eversion: vaginal eversion occurs intermittently, particularly when the animal lies down. Reprinted from *Color Atlas of Diseases and Disorders of Cattle* by R. Blowey and A. Weaver with permission from Elsevier.

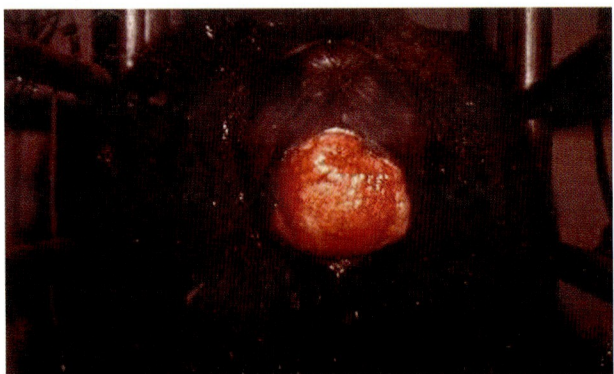

Figure 43.2 Grade 2 eversion: floor of the vagina everts and stays everted. Reprinted from *Clinical Theriogenology* with permission from the Editor Dr R. Youngquist.

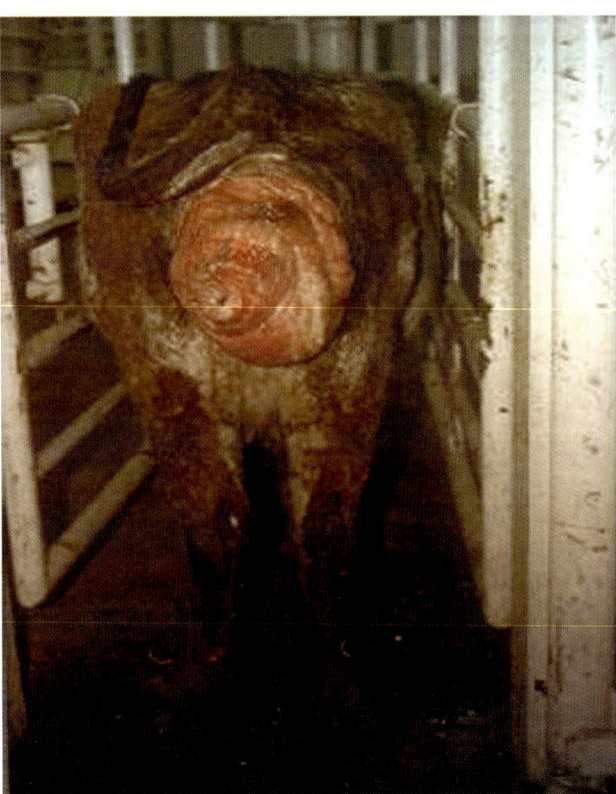

Figure 43.3 Grade 3 eversion: floor and cervix evert and they stay everted. Reprinted from *Clinical Theriogenology* with permission from the Editor Dr R. Youngquist.

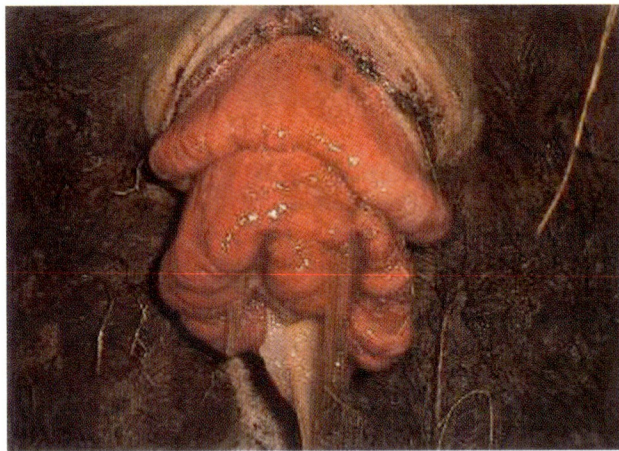

Figure 43.4 Grade 4 eversion: either a grade 2 or grade 3 that has become necrotic and/or fibrotic. Reprinted from *Color Atlas of Diseases and Disorders of Cattle* by R. Blowey and A. Weaver with permission from Elsevier.

of the urinary bladder getting trapped in the everted organ. If left untreated, grade 2 quickly progresses to a grade 3 eversion.
- Grade 3 eversion: the entire vaginal mucosa and cervix protrude continuously through the vulval lips (Figure 43.3) with a trapped urinary bladder and an exposed cervical mucous plug. The cervical seal may liquefy allowing bacterial contamination of the uterus leading to placentitis and fetal death. Often these fetuses become emphysematous and cervical dilation is insufficient for a vaginal delivery. This condition is often referred to as a cervical eversion but more correctly should be referred to as a cervico-vaginal eversion. This distinguishes it from eversion of the cervix alone as sometimes observed in *Bos indicus* breeds. In the latter case, only the cervix everts as a pedunculated mass through the vulva.
- Grade 4 eversion: an eversion of such a duration that the entire vaginal mucosa appears necrotic and fibrosed (Figure 43.4). Infection is so extensive that the urinary bladder may become involved in it and peritonitis is a possible result.

The objective of treatment in the case of pregnant animals is to replace and retain the vagina and cervix within the pelvic canal and to deliver and wean a live offspring. Many management and treatment options are described for vaginal and cervical eversions.[1,4,7,9–11,13–33] These can be placed in two general categories: those aimed at permanent reduction and others that afford a temporary solution. It should be mentioned that one approach may not fit all grades. The severity of the eversion, the time to expected

Vaginal, Cervical, and Uterine Prolapse

delivery, the veterinarian's preference, and the owner's ability or willingness to manage the patient after treatment will dictate the treatment option. The elected initial treatment option may need to be changed after the response of the patient has been assessed. This is particularly relevant if tenesmus becomes an issue. If the patient is close to parturition, induction is recommended to prevent recurrence prior to parturition.

Short-term treatment options

Caslick's suture

This technique (Figures 43.5 and 43.6) should be reserved only for a nonirritated grade 1 vaginal eversion that occurs close to parturition and when tenesmus is not expected to be a concern. Unfortunately, a simple Caslick's suture in many cases is insufficient. As intra-abdominal pressure increases, tenesmus begins and the suture is ripped out or the vagina begins to evert below the sutures. Further, the vulval softening and stretching that occur closer to parturition contributes to this phenomenon. Caslick's suture may suffice in patients that are not pregnant and have a grade 1 eversion during estrus. This is often the case with embryo donor animals whose ovaries are frequently superstimulated with hormones for the sole purpose of increasing the number of ovulations.

Buhner stitch

This is a very effective treatment for more advanced grades and that are chronic in nature.[17,18] To have a successful outcome the stab incisions have to be placed deeply such that the retention line is as cranial to the vulval lips as possible (Figure 43.7). The disadvantage is that it is unforgiving and if assistance is not available when the patient begins to calve, severe trauma to the vulva may occur or the calf is unable to be delivered and dies *in utero* or the patient may die as a result of uterine rupture and fatal hemorrhage. Following epidural anesthesia, the vulva is thoroughly washed with detergent and a 1-cm stab incision is made on the midline below the vulva and as far forward as practical. A second incision is made midway between the anus and dorsal commissure of the vulva (Figure 43.7). The Buhner needle is then passed deeply from the ventral incision up one side of the vulva and out through the dorsal incision. The Buhner tape (nonwicking) is passed through the eye of the needle and the needle retracted through the ventral incision along with the tape.

The procedure is repeated on the opposite vulval lip and the two ends tied leaving about a two-finger opening in the ventral vulva for urination. The preference is to tie a bow knot and then tie the two loops of the bow in a single square knot. This allows the knot to be loosened but to retain the suture so that if the patient should try to evert again, it can be replaced and the knot retightened. This can also be done in the evenings, if observation of the patient for parturition is not going to occur overnight.

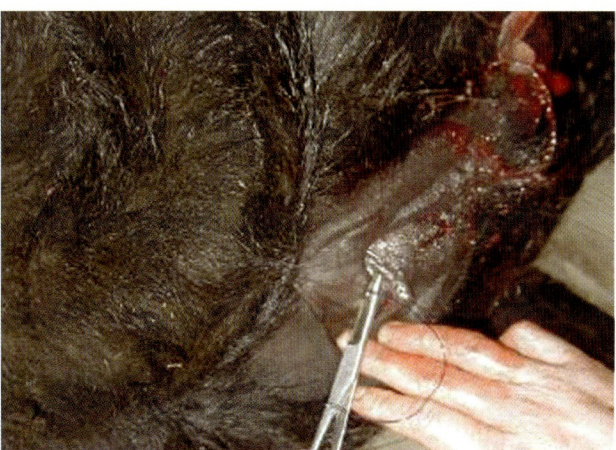

Figure 43.5 Caslick's suture being applied to a nonpregnant cow with grade 1 vaginal eversion. Reprinted from *Clinical Theriogenology* with permission from the Editor Dr R. Youngquist.

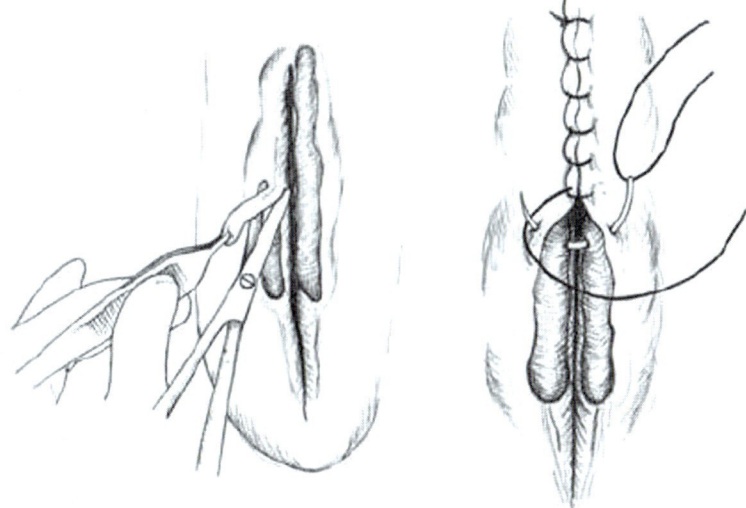

Figure 43.6 Diagrammatic representation of the Caslick's suture procedure. Illustration by Dr T. Barkley.

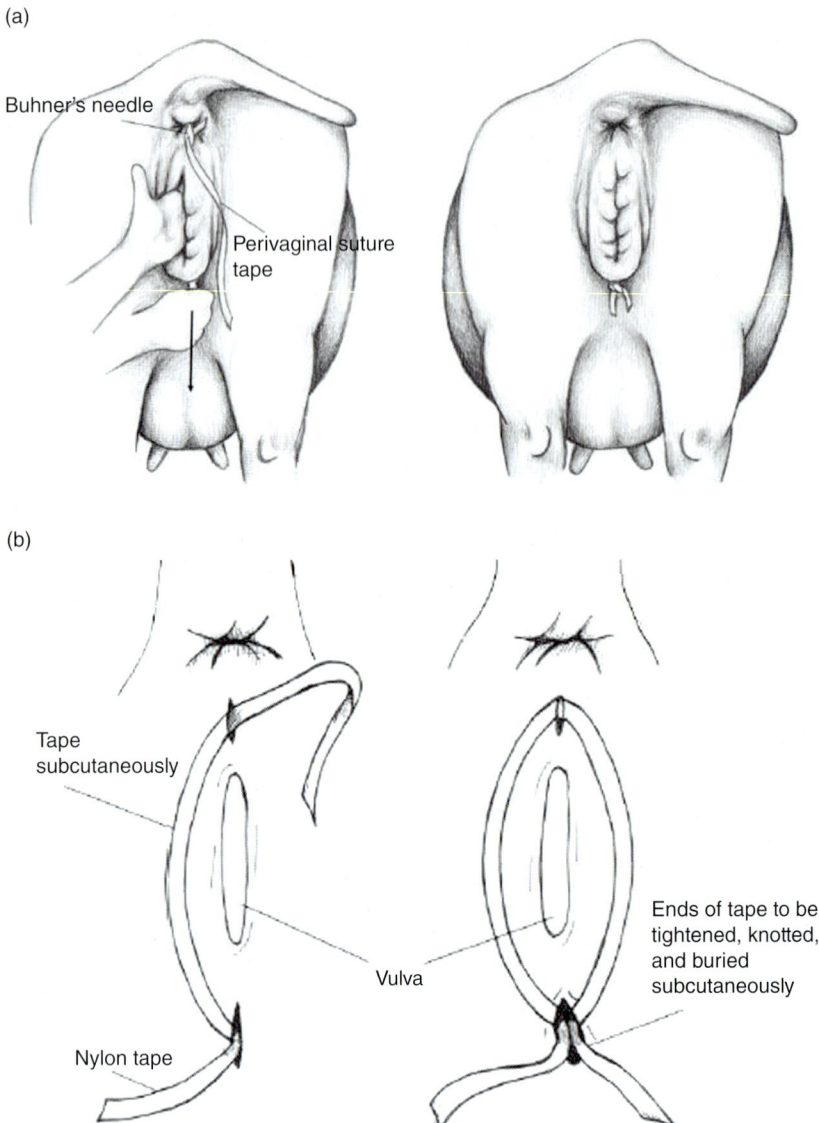

Figure 43.7 Diagrammatic representations of Buhner stich procedure. Note the locations of dorsal and ventral retention lines. Illustration by Dr T. Barkley.

If the procedure is performed properly, the tape should not be visible at the dorsal incision and this helps to prevent feces tracking down the needle paths. It should be stressed to the caretakers that their presence at the onset of calving is crucial for a successful outcome. If observation is a problem, medium-sized catgut can be doubled and used as a suture material; this helps in patients that are within 10 days of calving as the gut will break during parturition.

Bootlace technique

This is not a highly preferred technique as the vulval lips are inverted and the suture placement is tedious. However, it is stronger than a Caslick's suture. Following an epidural anesthetic procedure and disinfecting of the vulva and surrounding area, four to five small eyelets are made with umbilical tape or hog nose rings[11] on either side of the vulva in a dorsal to ventral line at the hair to hairless junction on the vulval lips. After placing the eyelets, umbilical or Buhner tape is then used to "lace up" the vulva much like a boot (Figure 43.8). As the pattern is tightened, the vulval lips invert. Again, the bootlace must be loosened prior to calving to avoid serious trauma to the vagina.

Horizontal mattress (Halstead) technique

In this technique, a needle is passed through the vulval lip on the right, beginning deep at the junction of the labia with the skin of the rump. The needle is continued across the vulval opening and through the left vulval lip at the same depth as the right. The needle is then passed back from left to right at the same plane as the first passage but beginning approximately 2–3 cm ventral to the first passage. This procedure can be done using a Buhner or large cutting needle. Place as many sutures as needed to close the vulva down to about two fingers wide to allow for urination.

The disadvantage of the technique is that even if the sutures are not pulled really tight, vulval edema frequently develops and can be quite severe. This can be lessened if the suture tension is dispersed by "stents." Polyvinyl chloride tubing (1.3 cm diameter) can be cut the length of the vulva and holes drilled 2–3 cm apart; similarly wooden dowels[24] or latex tubing can be utilized. A stent is placed either side of the vulva and the horizontal mattress suture passed through the holes and tied (Figure 43.9). This approach has good deep retention and more evenly disperses the pressure.

Deep vertical mattress technique

This is similar to the horizontal mattress pattern described above. The difference is that a vertical mattress suture pattern is used and stents are recommended to disperse tension on the vulval lips (Figure 43.10). This replacement method can also have serious consequences if assistance is not available at the onset of parturition.

Permanent treatment options

Minchev vaginopexy

There are two Minchev techniques described that differ only in location of needle insertion. In the original Minchev technique the needle is passed through the lesser sciatic foramen.

The technique is an excellent prepartum method that aims to retain the vagina in the correct pelvic location by adhesion formation between the submucosa of the vagina and surrounding fascia. An epidural is given and the gluteal area shaved and disinfected. In the original Minchev technique, a large S-shaped needle is threaded with 0.95-cm umbilical tape and a gauze bandage or large plastic or metal plate is attached at the end of the tape; the needle is then taken in vaginally and the lesser sciatic foramen is located on the dorsolateral wall of the vagina. The needle is then pushed through the dorsal area of the foramen in order to avoid the pudendal nerve. The needle is pushed through the skin in the gluteal area and the suture pulled up snug (do not tie too tightly) and another button, plate or gauze is tied into the suture. This is repeated on the other side.

Modified Minchev vaginopexy

This technique is similar except the stay sutures are placed anterior to the lesser sciatic foramen 5 cm lateral to the midline and just posterior to the shaft of the ilium, providing more cranial fixation of the vaginal wall (Figure 43.11). Care must be taken to avoid the sciatic nerve, pudendal artery, and rectum when placing the needle. These structures can be identified cranial and slightly dorsal to the lesser sciatic

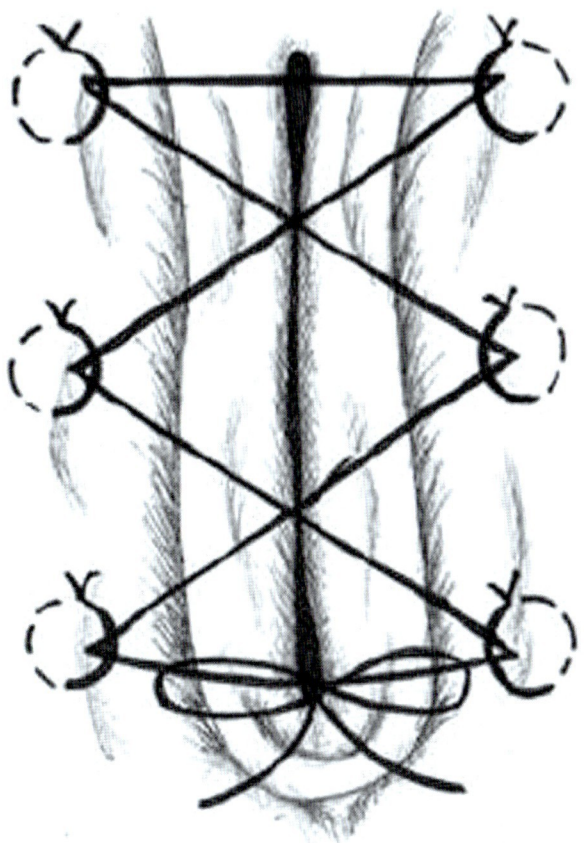

Figure 43.8 Diagrammatic representation of a bootlace technique. Illustration by Dr T. Barkley.

(a) 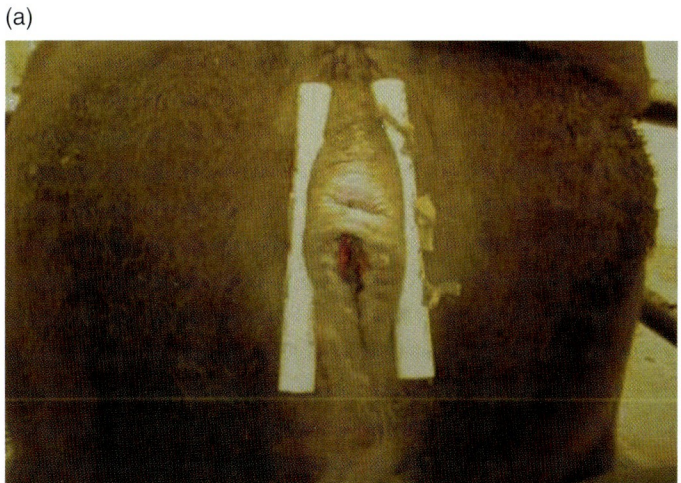 (b)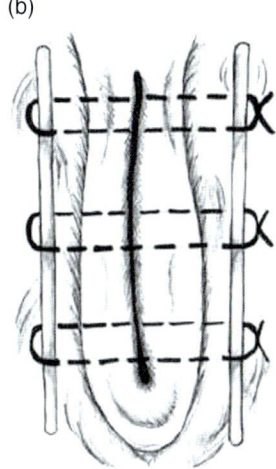

Figure 43.9 An example of a horizontal mattress technique. Note the location of the stents for reduction of vulval edema and for dispersion of pressure. Illustration (b) by Dr T. Barkley.

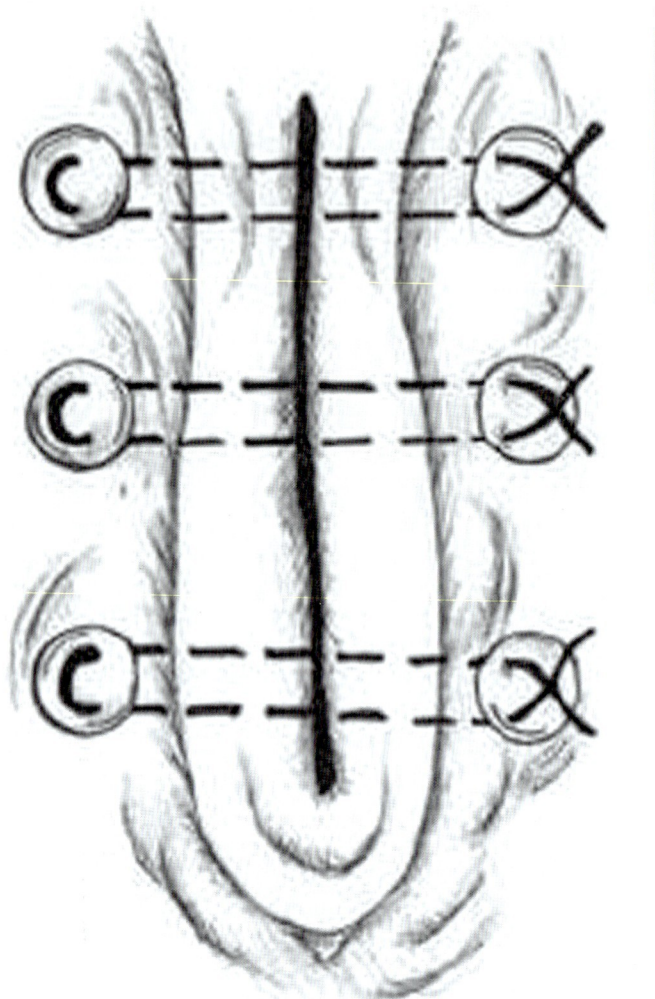

Figure 43.10 An example of a deep vertical mattress technique applied where buttons are used as stents. Illustration by Dr T. Barkley.

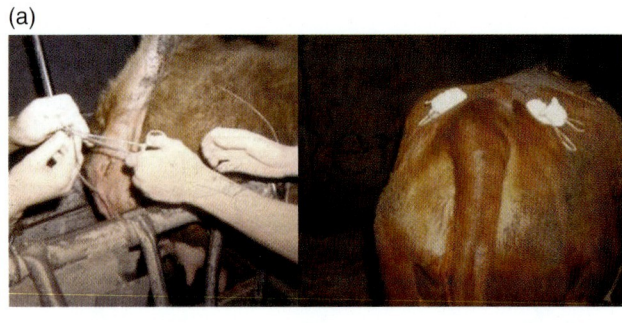

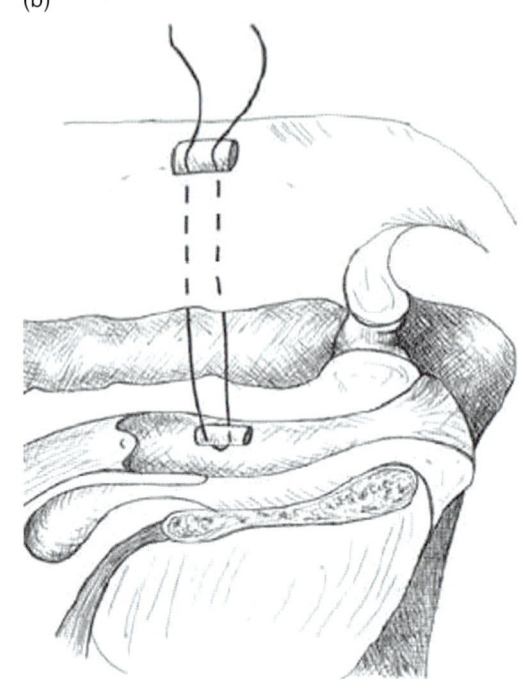

Figure 43.11 An example of a modified Minchev technique (a) with a schematic illustration (b). Illustration (b) by Dr T. Barkley.

foramen. The needle is passed through the sacrosciatic ligament approximately 4–5 cm lateral to the sacrum and just posterior to the shaft of the ilium. There are advantages to this technique, including the fact that cows rarely show tenesmus when the vagina is replaced and the cow can calve unassisted. However, it is recommended that the sutures are removed in 2–4 weeks. A commercial prolapse repair kit is available (Figure 43.12; Pro-Button Prolapse, Jorgensen Laboratories, Loveland, CO, USA). The assembled kit is soaked in a disinfectant. The stainless steel pin with the 15-cm plastic trocar passed through the 7.5-cm plastic button is carried vaginally. The trocar is pushed dorsal or anterior to the lesser sciatic foramen pointing dorsolaterally to avoid the rectum, and then through the sacrosciatic ligament and the biceps femoris to the skin outside. The stainless steel pin is removed, and the 6-cm plastic button is placed on the protruding trocar. Pushing down on the 6-cm button enables the 7.5-cm button to be held firmly between the vaginal wall and the sacrosciatic ligament. The cotter pin is inserted through the hole in the protruding trocar. The plastic trocar is cut about 2.5 cm above the cotter pin. In severe cases, it

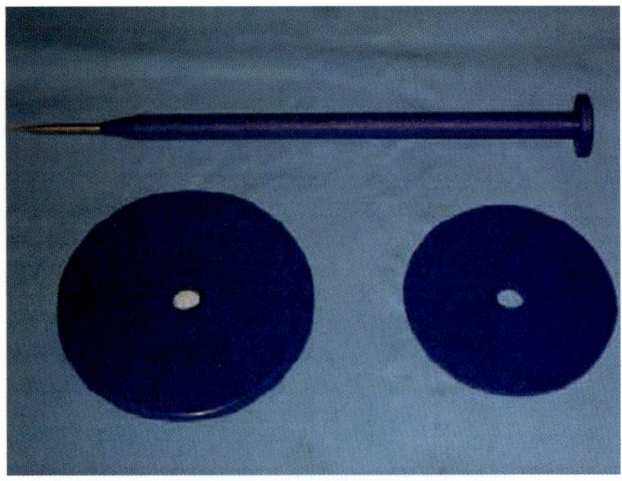

Figure 43.12 Example of trocar and plastic buttons used for a modified Minchev technique. Reprinted from *Clinical Theriogenology* with permission from the Editor Dr R. Youngquist.

may be necessary to place buttons on both sides of the anterior vagina. However, it should be recognized that recurrence of eversion is possible since only the dorsum of the vagina is held. Further, when the internal placement is too far caudal, a partial eversion may occur eliciting straining that results in total eversion.

Winkler technique cervicopexy

This technique[33] is a very effective method of vaginal retention because it fixes the cervix in place, making it difficult for the vagina to prolapse. The basic procedure is aimed at suturing the cervix to the prepubic tendon. There are numerous descriptions of placing the suture in the cervix and how to suture through the tendon. The two approaches to suture placement are either through a left paralumbar approach[26] or through the vagina. The former is more time-consuming but is safer while the vaginal approach is quicker but can result in broken needles if one is not careful. The vaginal approach could be done in two ways by either including the lower half of the cervix or by just including the external os of the cervix on one side (Figure 43.13).

Before placing the suture vaginally, an epidural is given and a mild antiseptic douching of the vagina performed. The prepubic tendon is then palpated at its attachment to the pelvic brim. As the fingers are moved laterally about 5 cm on both sides of the midline, a small triangular-shaped area free of tendon is felt. This is formed by a short band of tendon extending posterolateral to the iliopectineal eminence of the pelvis. A large curved cutting needle is bent into either a U-shape (Figure 43.14a) or other shape (Figure 43.14b) and about 150 cm of nonabsorbable suture material is attached. The needle is then carried into the vagina and passed from the inner lumen of the external cervical ring out peripherally. The needle is grasped again and passed down in a medial to lateral direction through the prepubic tendon and carefully curved around to come up through the triangular area. The needle is grasped again, passed through the external cervical ring from the periphery to the center and bought on out of the vagina. Some clinicians prefer to pass the needle through the vaginal wall initially, then pass the needle through the

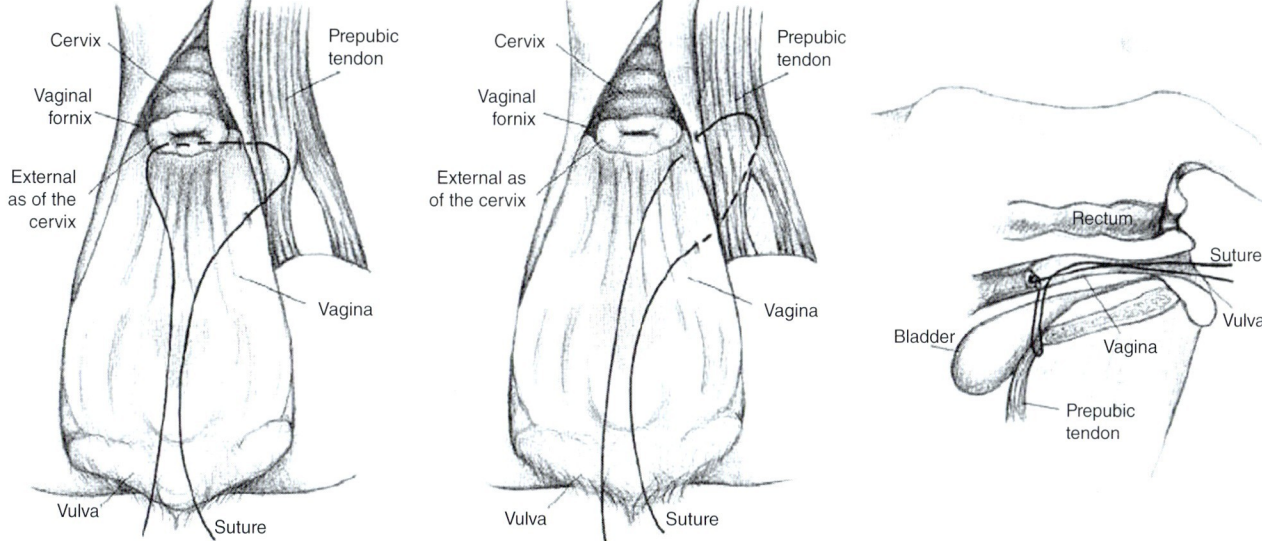

Figure 43.13 Schematic representation of a Winkler technique for cervicopexy. Redrawn from *Techniques in Large Animal Surgery* by D. Hendrickson with permission from John Wiley & Sons.

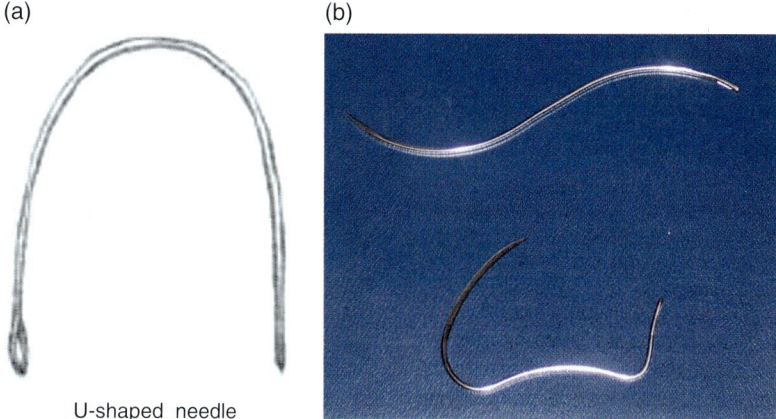

Figure 43.14 Special needle used for Winkler technique for cervicopexy. Illustration (a) by Dr T. Barkley.

prepubic tendon medial to lateral direction, then pull the needle back through toward the intravaginal part of the cervix; the needle is then directed across the lower half of the cervix and brought outside to tie the knot. This latter procedure may prevent breaking of the needle. A catheter is then passed through the urethra so that as the suture is tied the urethra is not trapped in the suture.

To tie the suture, the knots are started on the outside of the vulva and slid forward. The suture is tied tight enough to prevent the posterior movement of the cervix. To assist in placing the cervical sutures, Knowles cervical forceps can be used to either fix the cervix cranially making passing of the needle through the cervix easier or in some cows the cervix can be retracted sufficiently to be presented at the vulva again, making placing of the cervical sutures easier.

Other descriptions of this procedure do not advocate passing the needle through the cervical opening but rather through the side of the cervix without penetrating the lumen. This approach is greatly aided by fixation of the cervix with forceps.[3,26] The advantages of this approach are that the cow can calve on its own, post-repair tenesmus is rare, and the cow can still be examined per vaginum. Occasionally the suture in the cervix can break resulting in recurrence of eversion. This procedure is recommended for chronic cases which the owner wishes to keep, especially embryo donor cows in which the problem is likely hormonally induced rather then there being any genetic predisposition. However, if the condition is likely to occur again or if it is probably hereditary in nature, the affected animal should be culled.

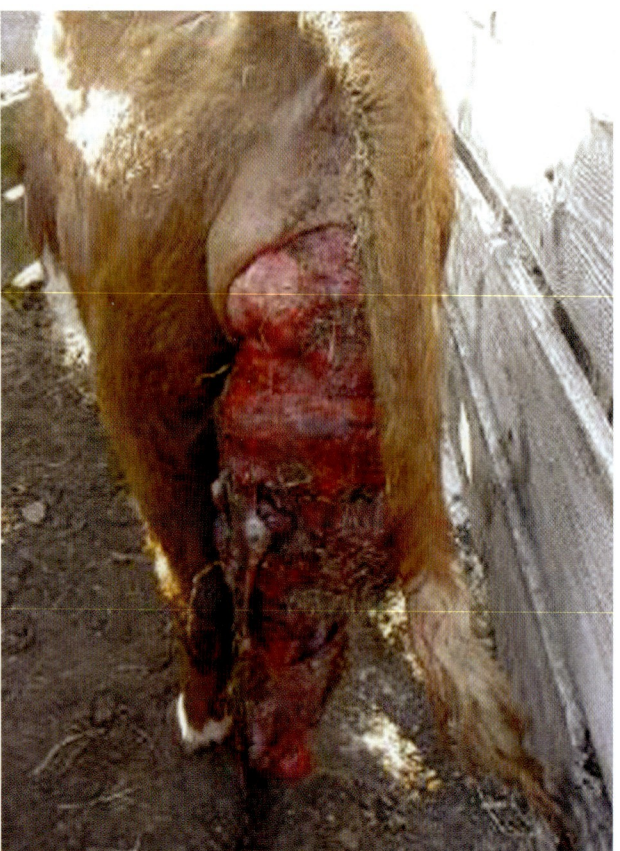

Figure 43.15 An everted uterus with evidence of contamination. Reprinted from *Clinical Theriogenology* with permission from the Editor Dr R. Youngquist.

Uterine eversion

Uterine eversion occurs mostly within 24 hours after fetal delivery, with the majority of cases occurring shortly after parturition. While relatively uncommon (<1% of parturitions),[34] it is nevertheless one of the true emergency situations one encounters in practice. In many instances the everted organ is readily contaminated (Figure 43.15), traumatized physically or by extreme weather (heat/cold), becomes edematous, may hemorrhage, and evert severely enough to rupture the uterine vessels leading to rapid shock and death.

Owners should be instructed not to transport the animal and to place the everted organ inside a plastic bag to reduce contamination. Fortunately, many of these cows are unable to stand up and run due to hypocalcemia, partial paralysis, or exhaustion. However, on occasions, the cow is capable of standing up and may become fractious as she attempts to outrun the uterus or knocks the uterus between her hocks (Figure 43.16). Since these cases often result in rupture of the uterine vessels and rapid death due to internal hemorrhage, one should approach them very cautiously.

Many factors[8,35] are believed to contribute to uterine eversion, including reduced uterine contractility due to decreased uterine activity following primary or secondary uterine inertia. Hypocalcemia and reduced prostaglandin and oxytocin receptors and myometrial defects due to overstretching contribute to primary uterine inertia. Among others, mere physical exhaustion and dystocia are responsible for secondary uterine inertia. Apart from uterine inertia, situations in which there is continuing tenesmus as noted in

Figure 43.16 Uterus is further traumatized by knocking between the hocks. Reprinted from *Clinical Theriogenology* with permission from the Editor Dr R. Youngquist.

dystocia, manual extraction of calf, retained fetal membranes, delayed cervical involution, and laceration of the reproductive tract can lead to eversions. Prolonged recumbency that follows paralysis or hypocalcemia can lead to increases in intra-abdominal pressure, a potential factor for eversion. Hypocalcemia is also believed to make the uterus atonic delaying the involution of cervix, a predisposing factor for uterine eversion.[36] However, it has been suggested that hypocalcemia can be the result of uterine eversion rather than a cause of it.[37] The role of parity[38] and hypocalcemia still remain an enigma in the etiology of uterine eversions.[39,40] Overconditioned animals may be more prone to increased intra-abdominal pressure due to increased fat reserves in the abdomen. Other nonanimal factors can contribute to this condition. For example, outbreaks can occasionally occur due to extreme weather conditions and pasture composition. Further it has been suggested that alterations in the availability of dietary calcium, phosphorus, and magnesium may have a role in creating uterine inertia and predisposing animals to this condition. The uterine eversion outbreak noticed in animals kept on a cool-season grass pasture supports this hypothesis. It has also been suggested that high estrogen concentrations in pasture may play a role. The foregoing undoubtedly attests to the multifactorial etiology of this condition.

Many methods of reduction and practical tips have been described.[1,35,36,41–59] Replacing the uterus can be done in either in a standing or a recumbent position depending on circumstances and the clinician's preference. The animal may have to be cast, rolled, and pushed to lie on its abdomen to facilitate pulling the hindlegs straight behind to facilitate reduction.[35] Obviously, if the cow is in the pasture with partial paralysis, the recumbent position is dictated. Replacing the uterus in the recumbent position is greatly facilitated by positioning the cow in the sternal position with both hindlegs extended out behind (Figure 43.17) with the head facing downhill ("Okie position"). This position tilts the pelvis forward and allows the clinician to get behind the cow and cradle the uterus in the lap as it is fed back they into the cow. Others prefer the standing position (Figure 43.18) if the cow can be adequately restrained and help is available to place a towel or bed sheet (or something similar) under the

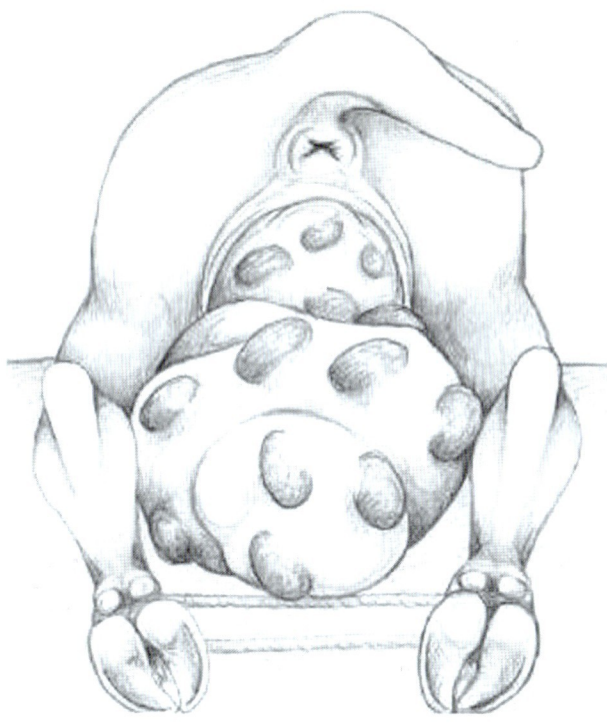

Figure 43.17 "Okie position": note the extension and elevation of hindlegs. Illustration by Dr T. Barkley.

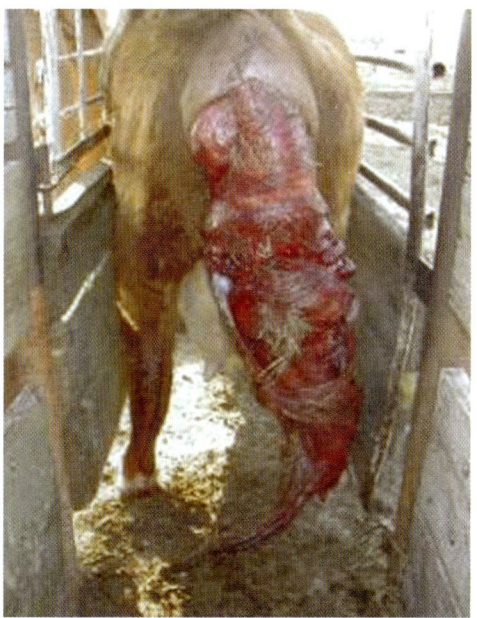
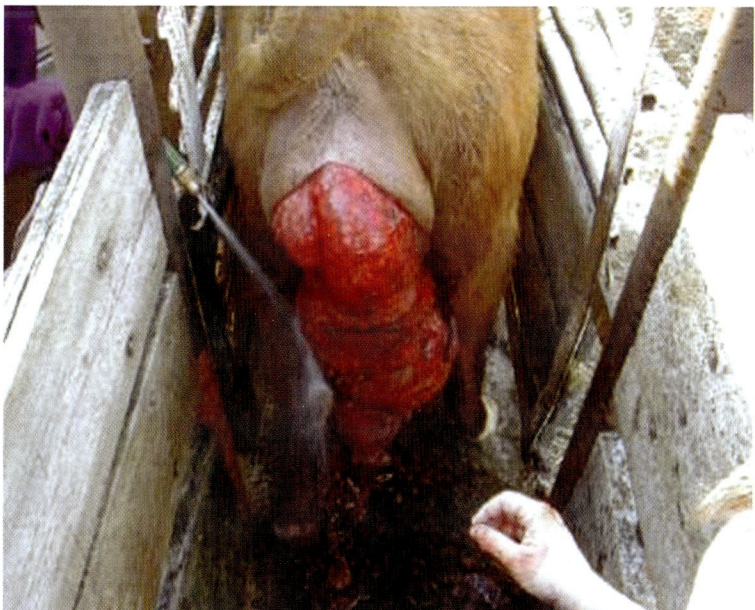

Figure 43.18 Cow is placed in a chute (a squeeze chute is preferred) for the standing reduction. Reprinted from *Clinical Theriogenology* with permission from the Editor Dr R. Youngquist.

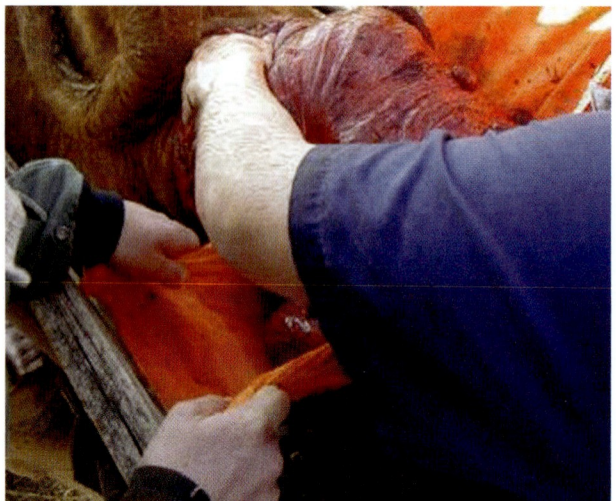

Figure 43.19 Elevation of the uterus by a sheet helps in reduction. Reprinted from *Clinical Theriogenology* with permission from the Editor Dr R. Youngquist.

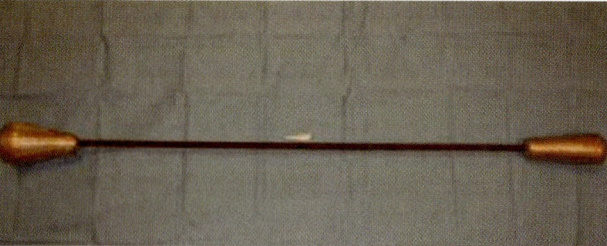

Figure 43.20 A probang is used to fully invert the uterine horns. Note that the ends are rounded. Reprinted from *Clinical Theriogenology* with permission from the Editor Dr R. Youngquist.

uterus and lift it up as the clinician replaces the uterus (Figure 43.19) back into the cow.

Replacement guidelines

- Give epidural anesthesia.
- Remove as much fetal membranes as possible.
- Clean the endometrium.
- Apply chlorhexidine ointment over the entire uterus and massage.
- Apply a pressure bandage for 10–15 min tightened from the tip of the horn. Sugar may be applied sparingly. However, care needs to be taken because sugar will dry the endometrial layer quickly, causing the tissue to become friable. This may lead to tears when applying pressure to push in the horns.
- Depending on dryness the operator may have to apply an ointment that has lubricant and emollient properties.
- Identify the nongravid horn (strong intercornual ligament prevents its eversion) by observing for the presence of an oval or slit-like opening bordered by a ring of small cotyledons and reduce it first.
- Begin applying steady pressure at the cervix/uterine body area using both hands with a clenched fist to reduce the everted uterus into the vagina. Once the uterine body is reduced into the vagina, the uterine horns are gradually massaged into the birth canal or vagina.
- Fully invert the horns. If the operator's arms are not long enough, the rounded bottom of a 1-L soda bottle or probang can be used to completely invert the uterus.
- Give nonsteroidal anti-inflammatory drugs and parenteral antibiotics.
- Oxytocin should not be given until the uterus has been reduced or replaced.
- Apply a retention suture.

A probang (Figure 43.20) is commonly used to make sure that the tip of the uterine horn is fully inverted. The rounded edges of a probang prevent possible tears that may occur to the friable uterine wall during the procedure.

Failure to fully straighten the horns can result in re-eversion or ischemic necrosis of a portion of the uterine horn. Placement of a suture in the vulva after uterine replacement is controversial. If the uterus responds to oxytocin and the cow is able to stand, then re-eversion is highly unlikely and a vulval stitch is not needed. On the other hand, if the cow is going to be unable to rise and tenesmus recommences, a vulval stitch may be indicated for 24–48 hours. After reduction, 20–30 IU oxytocin should be injected and repeated two more times at 30-min intervals and calcium therapy instituted if necessary. Intrauterine antibiotics are probably warranted in the case of a severely injured and contaminated uterus but are controversial in the case of a relatively fresh and cleanly reduced eversion.

Uterine amputation

There are two approaches to uterine amputation, open and closed.[1,2,60,61] The open approach involves incising the uterine body and location and ligation of major uterine vessels in the stretched broad ligament prior to amputation of the uterus. The advantage of this approach is that any viscera contained within the prolapse can be returned to the abdomen before the uterus is actually removed. With the surgical approach, the wall of the uterine body is opened from the cranial cervix to the bifurcation of the uterine horns to expose the contents within the everted uterus and the major vessels in the broad ligament (mesometrium). The latter are identified by "fanning out" the mesometrium and the vessels double ligated, preferably when they are not under tension. The vessels are severed between ligatures and the mesometrium incised and allowed to retract into the abdominal cavity. A series of adjacent slightly overlapping mattress sutures are then placed across the uterine body just proximal to the cervix. The uterus is then amputated 2–3 cm below this line of sutures. The stump may then be sutured with a simple continuous pattern. Postoperative treatment should include administration of tetanus toxoid, nonsteroidal anti-inflammatory drugs, and appropriate antibiotics.

In the closed approach (Figure 43.21) the entire organ is ligated and left to slough in 10–14 days. This is a salvage operation applied in cases of severe ischemic necrosis and/or lacerations. With this approach, viscera could be unintentionally trapped in the everted uterus with fatal consequences. Ligatures for this latter approach can be made of surgical tubing, bungee cord, or the bloodless castrating unit used for castrating large bulls works well.

Vaginal, Cervical, and Uterine Prolapse

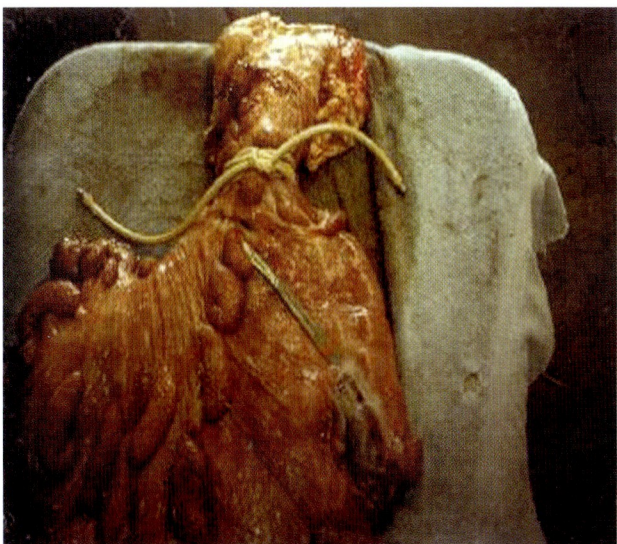

Figure 43.21 A closed approach of uterine amputation. Note the location of the ligation. Reprinted from *Clinical Theriogenology* with permission from the Editor Dr R. Youngquist.

Discussion

Vaginal and cervical eversions

Prepartum vaginal and cervical eversions can be perplexing and disturbing problems both to the producer and veterinarian. The expense and effort can be costly at times. These everted structures should be promptly replaced within the vaginal vault and secured such that contamination, laceration, sepsis, fibrosis, or necrosis does not occur. This approach will prevent compromise of cervical and vaginal relaxation at the time of parturition and most importantly will enhance animal well-being by preventing unnecessary injury at the time of parturition.[62] Some attempts made by producers to correct the situation can have many unintended consequences and can inflict unnecessary pain and injury to the animal. In this regard, client education is paramount. Although there may be temporary relief, the animal may evert again at the next parturition if the primary cause for an eversion is not corrected.[11] It is very important to realize that satisfactory reduction and retention alone are not sufficient since the recurrence rate is high if the primary or inciting cause is not removed. This not only forces subsequent repair but more importantly, in some instances, can result in mutilation of vulval and perivulval tissue, and become an important animal-welfare issue. Many techniques have been developed and stood the test of time. Each one is unique and the technique chosen depends on the condition and expertise of the clinician. What was said a few decades ago still stand true: better techniques for handling chronic vaginal eversions are needed.[28] New knowledge in this area by developing alternate techniques and improving current techniques will keep our discipline vibrant.

Not many studies have characterized the future fertility of affected animals. However, we ought to consider the possibility of future occurrence of the condition in light of animal well-being rather than fertility. In this regard, it is pertinent to consider some of the treatments administered. These are aimed at preventing recurrence by narrowing the birth canal[6, 21] by creating adhesions between retroperitoneal surfaces of the vaginal and adjacent or subjacent structures. These methods can prevent recurrence but may result in injury to the animal at the time of parturition. Further, should these animals conceive, subsequent pregnancies are frequently complicated by recurrence. Hence consideration should be given to culling affected animals after their calves are weaned.[9] Animal health advisors have an important role to play to help producers to make a decision to eliminate these animals from the breeding herd. Considering the injury the animal may endure at the present and subsequent reductions, it behooves us to strongly suggest removing such animals from the herd.[62]

Uterine eversion

The appearance of a heavy mass of tissue outside the body of an animal can be intimidating and at times the process of reduction can be overwhelming. Truly the prospect of performing a successful reduction and the thought of recurrence can place one in mental anguish. We always look for ease of the operation and reduced physical effort on our part to reduce the eversion. In an attempt to facilitate the process of reduction mechanical aids have been used successfully.[48,50,56,57] However, raising the rear end of a live cow may appear inhumane.[46] Next, the application of agents such as sugar[45,58] or salt[52] to the everted tissue to reduce the edema spectacularly before reduction is not supported uniformly. Although I prefer a careful application as supported by others in practice,[45] it remains controversial[8] since it can amplify endometrial trauma. This technique may have certain academic skepticism despite its benefits.

The chance of immediate recurrence can be eliminated if the reposition is done completely, without allowing invagination of the uterine horn to occur at the ovarian pole. Overall, the mortality rate ranges from 18 to 20%.[8,36] The prognosis for survival and future reproductive performance depends on several factors, but time from the occurrence to treatment is critical for a positive outcome. Despite the extensive trauma that may occur before replacement, those that are replaced correctly and survive have an acceptable prognosis for future fertility.[36,63] Occasional bladder eversion and intestinal entrapment that follow into the everted uterus may result in a poor prognosis. Timely intervention by a professional can minimize animal discomfort and economic loss to the producer. Treatment given by the uninitiated nonprofessional has to be discouraged since severe damage can be inflicted to the animal.[44] A unique retention technique developed to prevent recurrence[59] should be considered to obviate situations such as necrosis, gangrene, and severe trauma, necessitating amputation.

The chance of recurrence at the next parturition has been discussed[2,64] and the heritability or additive susceptibility with subsequent pregnancies for this condition is not apparent.[1] Since no known heritable component is recorded there is no increased probability of occurrence at subsequent parturition. Therefore one should not be too hasty in recommending salvage of affected cows. The chance of recurrence is minimal and this has to be discussed with the producer. It should be remembered that in the developing world, cattle are an economic asset and the cost of therapy must be balanced against the value of the animal. In addition to the value of the animal, its fertility will dictate whether the animal has to be culled or not. The major factor that can influence this decision is the degree of trauma caused to the uterus and the time it has taken to repair.[36] Future fertility of cows is reduced and there is a slight increase in risk at the subsequent calving, but the increased risk is not sufficient to justify culling.[2]

As the profession moves more and more into animal health and production than restorative therapy, it is incumbent on us to prevent the occurrence of these conditions by careful selection of breeding stock, observing quality animal husbandry methods, providing adequate care to periparturient animals, and finally eliminating potential recurring cases. All the above will enhance animal well-being and place the profession as animal-welfare contributors.

Acknowledgments

The impetus for this work came from Dr L. Dawson, an expert in ruminant theriogenology, whose help is acknowledged. The illustrations were created by Dr T. Barkley of Woodward, Oklahoma and her assistance is much appreciated. Figures 43.1 and 43.4 are reprinted from the *Color Atlas of Diseases and Disorders of Cattle* by R. Blowey and A. Weaver with permission from Elsevier. Figure 43.13 is redrawn from the *Techniques in Large Animal Surgery* by D. Hendrickson with permission from John Wiley & Sons. The remaining figures and content are reprinted from the journal *Clinical Theriogenology* with permission from the Editor Dr R. Youngquist.

References

1. Miesner M, Anderson D. Management of uterine and vaginal prolapse in the bovine. *Vet Clin North Am Food Anim Pract* 2008;24:409–419.
2. Youngquist RS. Surgical correction of abnormalities of genital organs of cows: uterine prolapse. In: Youngquist RS (ed.) *Current Therapy in Large Animal Theriogenology*. Philadelphia: WB Saunders, 1997, pp. 433–434.
3. Hopper R. Surgical correction of abnormalities of genital organs of cows: management of uterine prolapse. In: Youngquist RS, Threlfall WR (eds) *Current Therapy in Large Animal Theriogenology*, 2nd edn. Philadelphia: WB Saunders, 2004, pp. 382–390.
4. Roberts S. Vaginal prolapse. In: *Veterinary Obstetrics and Genital Diseases*, 2nd edn. Ithaca, NY: published by the author, 1971, pp. 189–196.
5. Woodward R, Quesenberry J. A study of vaginal and uterine prolapse in Hereford cattle. *J Anim Sci* 1956;15:119–124.
6. Coulthard H. Treatment of bovine vaginal prolapse. *Vet Rec* 1991;129:151.
7. Hudson R. Genital surgery of the cow. In: Morrow DA (ed.) *Current Therapy in Theriogenology*, 2nd edn. Philadelphia: WB Saunders, 1986, pp. 346–350.
8. Roberts S. Uterine prolapse. In: *Veterinary Obstetrics and Genital Diseases*, 2nd edn. Ithaca, NY: published by the author, 1971, pp. 308–313.
9. Youngquist RS. Surgical correction of abnormalities of genital organs of cows: vaginal prolapse. In: Youngquist RS (ed.) *Current Therapy in Large Animal Theriogenology*. Philadelphia: WB Saunders, 1997, pp. 436–437.
10. Wolfe D, Carson R. Surgery of the vestibule, vagina, and cervix. In: Wolfe DF, Moll HD (eds) *Large Animal Urogenital Surgery*, 2nd edn. Baltimore: Williams and Wilkins, 1998, pp. 398–410.
11. Johansen R. Repair of prolapsed vagina in the cow. *Vet Med Small Anim Clin* 1968;63:252–256.
12. Dhanotiya R, Srivastava R, Pandit R. A note on postpartum utero vaginal prolapse in Gir cows: estimation of serum calcium, phosphorus, proteins, and cholesterol. *Arch Exp Vet Med* 1989;43:79–80.
13. Arthur G. Recent advances in bovine obstetrics. *Vet Rec* 1966;79:638–639.
14. Baker G. Advances in large animal surgery, part 2. *Vet Rec* 1967;81:II.
15. Minchev P. The use of a new surgical method in eversion and prolapse of vagina in animals. *Veterinariya* 1956;33:58–60. Abstracted in *J Am Vet Med Assoc* 1957;130:344.
16. Bouckaret J, Oyaert W, Wijverkens H, van Meirhaeghe E. Prolapsus vaginae bij het Rund. *Vlaams Diergeneesk Tijdschr* 1956;25:119–132. Abstracted in *J Am Vet Med Assoc* 1957;130:344.
17. Buhner F. Simple surgical treatment of uterine and vaginal prolapse [translated from German]. *Tierdraztl Umsch* 1958;13:183–188.
18. Bierschwal C, deBois C. The Buhner method for control of chronic vaginal prolapse in the cow (review and evaluation). *Vet Med Small Anim Clin* 1971;66:230–236.
19. Dhillon K, Singh B, Kumar H, Bal M, Singh J. Treatment of vaginal prolapse in cows and buffaloes. *Vet Rec* 2006;158:312.
20. Habel R. Prevention of vaginal prolapse in the cow. *J Am Vet Med Assoc* 1957;130:344–345.
21. Hattangady S, Deshphande K. Case reports on "through and through" stay suture technique for retention of vaginal prolapse. *Indian Vet J* 1967;44:528–530.
22. Hentschl A. The button technique for correction of prolapse of the vagina in cattle. *J Am Vet Med Assoc* 1961;139:1319–1320.
23. Kerz P. Correction of vaginal prolapse in the bovine. *Vet Med Small Anim Clin* 1966;61:888–889.
24. Lamp J, Lamp T. A method for correcting vaginal prolapse in a cow. *Vet Med Small Anim Clin* 1981;76:395–396.
25. Milne F. Efocaine: an aid in the treatment of bovine vaginal prolapse. *J Am Vet Med Assoc* 1954;124:108–111.
26. Noordsy J. Cervopexy as a treatment for chronic vaginal prolapse in the cow. *Vet Med Small Anim Clin* 1981;76:383–385.
27. Norton E. External fixation of the bovine vagina after reduction of a prolapse. *J Am Vet Med Assoc* 1969;154:1179–1181.
28. Pierson R. A review of surgical procedures for correction of vaginal prolapses in cattle. *J Am Vet Med Assoc* 1961;139:352–356.
29. Sah S, Nakao T. Some characteristics of vaginal prolapse in Nepali buffaloes. *J Vet Med Sci* 2003;65:1213–1215.
30. Pittman T. Practice tips. *Can Vet J* 2010;51:1347–1348.
31. Rautenbach G. The retention of vaginal prolapse in the cow using a purse-string suture. *J South Afr Vet Assoc* 1984;55:203–204.
32. Narasimhan K, Quayam S, Gera K. A method of retention of recurrent prolapse of the vagina in cows. *Indian Vet J* 1975;52:311–313.
33. Winkler J. Repair of bovine vaginal prolapse by cervical fixation. *J Am Vet Med Assoc* 1966;149:768–771.
34. Roine K, Saloniemi H. Incidence of some diseases in connection with parturition in dairy cows. *Acta Vet Scand* 1978;19:341–353.
35. Plenderleith B. Prolapse of the uterus in the cow. *In Pract* 1986;1:14–15.
36. Odegaard S. Uterine prolapse in dairy cows. A clinical study with special reference to incidence, recovery and subsequent fertility. *Acta Vet Scand Suppl* 1977;63:1–124.
37. Richardson G, Klemmer A, Knudsen D. Observations on uterine prolapse in beef cattle. *Can Vet J* 1981;22:189–191.
38. Markusfeld O. Periparturient traits in seven high dairy herds, incidence rates, association with parity, and interrelationships among traits. *J Dairy Sci* 1987;70:158–166.

39. Risco C, Reynolds J, Hird D. Uterine prolapse and hypocalcemia in dairy cows. *J Am Vet Med Assoc* 1984;185:1517–1519.
40. Patterson D, Bellows R, Burfening P. Effects of caesarean section, retained placenta and vaginal or uterine prolapse on subsequent fertility in beef cattle. *J Anim Sci* 1981;53:916–921.
41. Caldwell H. Eversion of uterus in a maiden heifer. *Vet Rec* 1933;29:688–689.
42. Baxter K. Replacing the prolapsed bovine uterus. *Vet Rec* 2004;155:344.
43. Biggs A, Osborne R. Uterine prolapse and mid-pregnancy uterine torsion in cows. *Vet Rec* 2003;152:91–92.
44. Bullard J. A case of uterine prolapse in a Hereford with unusual lay treatment *J Am Vet Med Assoc* 1946;109:462.
45. Douch J. Uterine prolapse in the cow. *Vet Rec* 1986;118:310.
46. Parker C. Uterine prolapse in the cow. *Vet Rec* 1986;118:310.
47. Formston C. Uterine prolapse in the cow. *Vet Rec* 1986;118:310.
48. Foster S. A pneumatic appliance for the replacement of the prolapsed uterus of the cow. *Vet Rec* 1972;91:418.
49. Hibberd R. Replacing the prolapsed bovine uterus. *Vet Rec* 2004;155:96.
50. Johnston R. Uterine prolapse in the cow. *Vet Rec* 1986;118:252.
51. Levine H. Partial uterine prolapse associated with uterine foreign body in a cow. *J Am Vet Med Assoc* 1990;197:759–760.
52. Lyons A. Uterine prolapse in the cow. *Vet Rec* 1986;43:79–80.
53. Garcia-Saco E, Gill M, Paccamonti D. Theriogenology question of the month: laparoscopy to assist replacement of the uterus. *J Am Vet Med Assoc* 2001;219:443–444.
54. Abdullahi U, Kumi-Diaka J. Prolapse of the nongravid horn in a cow with a seven-month pregnancy: a case report. *Theriogenology* 1986;26:353–356.
55. Gardner I, Reynolds J, Risco C, Hird D. Patterns of uterine prolapse in dairy cows and prognosis after treatment. *J Am Vet Med Assoc* 1990;197:1021–1024.
56. White A. Uterine prolapse in the cow. *Livestock* 2007;12:21–23.
57. Wilson P. A pneumatic appliance for the replacement of the prolapsed uterus of the cow. *Vet Rec* 1972;90:729–730.
58. Munro I. Replacing the prolapsed bovine uterus. *Vet Rec* 2004;155:344.
59. Narasimhan K, Thangaraj T, Subramanyam R. A method of retention of recurrent prolapse of the uterus in bovines. *Indian Vet J* 1967;44:67–73.
60. Wenzel J, Baird A, Wolfe D, Carson R. Surgery of the uterus In: Wolfe DF, Moll HD (eds) *Large Animal Urogenital Surgery*, 2nd edn. Baltimore: Williams and Wilkins, 1998, p. 423.
61. Roberts S. Amputation of the prolapsed uterus. *Cornell Vet* 1949;39:438–439.
62. Wolfe D. Theriogenology to enhance animal well-being. *Clin Theriogenol* 2011;3:180–184.
63. Jubb T, Malmo J, Brightling P, Davis G. Survival and fertility after uterine prolapse in dairy cows. *Aust Vet J* 1990;67:22–24.
64. Murphy A, Dobson H. Predisposition, subsequent fertility, and mortality of cows with uterine prolapse. *Vet Rec* 2002;151:733–735.

Chapter 44

Inducing Parturition or Abortion in Cattle

Albert Barth

Large Animal Clinical Sciences, Western College of Veterinary Medicine University of Saskatchewan, Saskatoon, Saskatchewan, Canada

Introduction

Since the early work of Liggins[1,2] that provided insight into the mechanisms controlling pregnancy and parturition, a great deal of research has followed resulting in successful and widely accepted methods for inducing abortion or parturition in cattle. Research is still needed as the problem of retained placentas after induced parturition is still not completely solved and if a live calf is desired, known breeding dates are still a prerequisite for induction of parturition. Solutions should be possible as our understanding of the physiology and endocrinology of gestation and parturition increases and with further testing of new procedures in the field.

Endocrinological aspects

Progesterone is essential for the establishment and maintenance of pregnancy in all mammalian species that have been studied.[3] In cattle, the corpus luteum (CL) is the primary source of progesterone throughout gestation and luteal regression is necessary for parturition to occur.[4] Progesterone is also produced by the binucleate cells in the cotyledon of bovine placental tissue.[5,6] After 120 days of gestation the placenta begins to secrete progesterone and between 120–150 days and approximately 240 days the placental source of progesterone alone is sufficient to maintain pregnancy. During the last 4–6 weeks of gestation, placental progesterone gradually declines and the maintenance of pregnancy once again becomes dependent on the CL.[7,8] Thus if luteolysis occurred prior to the eighth month of gestation, pregnancy would be maintained for a time but parturition would likely occur several weeks before the normal due date.

As the fetal hypothalamus, pituitary, and adrenal glands mature, increasing amounts of cortisol are produced by the fetus. Rising fetal cortisol in the last 4–6 weeks of gestation appears to gradually reduce uteroplacental synthesis of progesterone.[9] This decline in progesterone secretion is due to cortisol induction of enzymes that convert placental progesterone to estrogen. Cortisol induces 17α-hyroxylase and 17,20-lyase activities in the placenta, catalyzing the formation of androgens from progesterone.[10] Androgens are rapidly converted to estrogens by the placental P450 aromatase system.

During pregnancy, high concentrations of progesterone maintain uterine quiescence mainly by hyperpolarizing the myometrial cells. At the end of gestation, falling progesterone and increasing estrogen produces depolarization of the myometrial cells and stimulates the formation of myometrial cell gap junctions enhancing the sensitivity of the myometrium to stimulatory agonists.[11] Declining progesterone and rising estrogen results in increased expression of oxytocin receptors in the myometrium.[12] It appears that increased estrogen concentrations also stimulate the production and release of prostaglandins.[13] Recent data from ovine studies suggest that fetal cortisol induces prostaglandin G/H synthetase-2 isozyme in placental cotyledons, thus favoring the prepartum increase in prostaglandin formation.[14] Prostaglandin production results in the demise of the CL and a precipitous drop in serum progesterone concentrations. Smooth muscle activation is associated with endometrial synthesis of prostaglandin $(PG)F_{2\alpha}$ which augments the force of contraction. Prostaglandins are also involved in the mechanisms of cervical ripening, which is essential for successful parturition.

Fetal endocrine control of parturition was demonstrated by Liggins in his studies on parturition in sheep. He showed that electrocoagulation of the fetal pituitary prevented normal term parturition and went on to demonstrate that the hypothalamus–pituitary–adrenal (HPA) axis was responsible for initiating parturition as infusion of glucocorticoid or adrenocorticotropin (ACTH) initiated premature parturition.[15]

The trigger for the spontaneous onset of labor is the greatly increased cortisol production by the adrenal gland of

---

Bovine Reproduction, First Edition. Edited by Richard M. Hopper.
© 2015 John Wiley & Sons, Inc. Published 2015 by John Wiley & Sons, Inc.

the maturing fetus. The increase in activity of the fetal HPA axis at the end of gestation is likely caused by programmed maturation in the fetal hypothalamus, or elsewhere in the fetal nervous system, rather than a response of the fetal pituitary to a chronic stress such as might be caused by a developing inadequacy of space and nourishment within the uterus. Some other external influence on the maturing hypothalamus, such as a progressive increase in the secretion of the placenta, is another possibility.[16]

Throughout the last 2–3 weeks of development *in utero* the fetal adrenal increases in size relative to body weight and the cellular sensitivity to ACTH increases.[3] The increase in sensitivity to ACTH is partially the result of increased adrenal cortical mass and partially the result of increased cellular responsiveness to ACTH. The increase in size and sensitivity of the adrenal gland combined with increasing circulation of ACTH account for the increased cortisol secretion that triggers parturition. Recent evidence suggests that at least some of the activity of the fetal HPA axis is due to positive feedback from placental estrogen. Increasing estrogen in fetal plasma greatly increases ACTH concentrations in fetal plasma.[17]

Since fetal cortisol levels gradually rise in the last month of gestation, particularly in the last week,[9] placental maturation may require exposure to elevated cortisol levels for a period of time prior to calving. This hypothesis is supported by a reduction in the incidence of retained placenta when long-acting corticosteroids are used to induce parturition.[18,19]

Induction of parturition

Indications

The use of induction of parturition in commercial beef herds is usually limited to the treatment of uterine hydrops, cardiac failure, or other health-related matters in which salvage of the fetus, or the life of the cow, is considered. The main factors preventing the use of induction of parturition as a management tool in commercial herds is the lack of known breeding dates with natural mating and elevated rates of placental retention leading to reduced first-service pregnancy rates.

The fetus can gain 0.45–0.68 kg/day in the final weeks *in utero* in normal gestations.[20,21] It appears that when gestations are 1–2 weeks overdue, as is more commonly the case in some of the large European breeds of cattle, fetal weight gains could be as high as 1 kg/day. In such cases, calf birthweight might easily become too large to allow normal parturition. Producers of purebred cattle often employ artificial insemination and thus breeding records are available, gestation lengths can be determined, and overdue gestations could be prevented by induction of parturition. It has been reported that when a short-acting glucocorticoid was used to terminate prolonged gestation in continental breed heifers, there was dystocia due to poor birth canal relaxation and ineffective labor.[22] However, too early intervention in the calving process might have been partially to blame for this finding. It has become a common practice by producers to induce parturition after 285 days of gestation rather than to allow gestations to become prolonged. In most cases, the stages of labor and pelvic relaxation are normal. Although no data are available, clinical experience from common usage of the procedure indicates that calf viability and colostral transfer are good. An additional benefit of induction of parturition in these cases might be to allow the female an extra 10–14 days to resume cycling for rebreeding in the subsequent breeding season. However, despite gestations having progressed for a normal duration, placental retention rates appear to be elevated. Thus the subsequent problem of endometritis may cancel any benefits of extra time gained through induction of parturition for the cow to resume cycling since endometritis may delay the first postpartum estrus.

In dairy herds, parturition may be induced 1–2 weeks early to prevent excessive udder edema and distension which may predispose to mastitis and difficulty in milking. Induction of parturition with long-acting corticosteroids has gained widespread acceptance with dairy producers in New Zealand and Australia as an important management tool for initiation and synchronization of lactation with the grazing season.[23,24] Calving is concentrated at a time of year when grazing is optimal for milk production. In the limited breeding season that follows it is inevitable that some cows will fail to become pregnant or will become pregnant too late to calve on time the following year. These cows would ordinarily be culled. Induced calving can be applied in association with an extended mating period to reduce herd wastage. Cows that are not pregnant at the end of a defined mating season can be mated and identified by bulls fitted with a chin-ball marking harness. In the following calving season these cows can be induced to calve 1–3 months prematurely. Calves born more than 1 month prematurely will be lost; however, milk yield can be expected to be near normal.

Precautions

At present one of the main obstacles to widespread use of induction of parturition is the need for accurate knowledge of breeding dates to prevent the birth of nonviable calves and an increased incidence of retained placentas. Calves born up to 2 weeks preterm have good vitality and are able to attain good maternal immunity derived from colostrum.[25,26] The incidence of placental retention is related to the degree of prematurity of induced parturition. Cows induced 1–2 weeks prematurely usually retain the placenta in over 75% of cases, whereas cows induced within a few days of term, or at term, have a 10–50% placental retention rate. In most reports cows with induced parturition and retained placentas did not have reduced pregnancy rates;[27,28] however, first-service pregnancy rates were significantly lower in cows that had retained placentas than cows that did not retain their placentas.[29,30]

The effect of induction of parturition on milk yield in the following lactation and on the concentration and yield of colostral immunoglobulins most likely depends on how early parturition is induced. In general, when parturition is induced within 2 weeks of normal term, the onset of milk production would be a few days slower than in natural calvings, but overall production levels for the entire lactation period are within expected limits.[9,18] When parturition was induced up to 22 days before the normally expected calving date there was a significant decrease in the yield, but not the concentration, of colostral IgG.[31] When birth occurred

10–15 days or more before due date, calves had lower levels of colostral immunoglobulin.[19,31] This may be due to an impaired ability to absorb immunoglobulin[32,33] and reduced intake due to calf weakness.[19]

When short-acting corticosteroids are used to induce calving close to term, the immunoglobulin concentration of colostrum is not different from that in naturally calving cows; however, the total immunoglobulin content may be lower due to a reduced volume of colostral secretion.[31,32] The ability of these calves to absorb immunoglobulin is not reduced as serum γ-globulin concentrations after suckling are similar to those of naturally born calves.[22,33]

The response and effects of inducing parturition on day 274 of gestation in dairy cows has been investigated in Australia. Cows were injected with dexamethasone trimethylacetate with herdmates as controls. Treated cows calved an average of 2.61 days before their due date. There were no differences in the proportion of cows displaying symptoms of milk fever, mastitis, paralysis, or acute metritis. There were no differences between groups for cow or calf mortality, or for any parameter of milk yield. The incidence of retained fetal membranes was significantly higher in cows with induced parturition but, interestingly, control cows received more assistance at parturition.[34]

Methods of induction of parturition

Various types and combinations of hormone treatments have been studied for efficacy and safety of induction of parturition, including corticosteroids or prostaglandins in combination with various estrogen preparations[27,28,35–40] and oxytocin[41] without appreciably reducing the incidence of placental retention. Dimenhydrinate[42] and relaxin[43] in combination with dexamethasone have been reported to reduce placental retention; however, the number of cows treated with dimenhydrinate was small and relaxin is not commercially available. RU-486, a potent antiprogesterone, used alone or in combination with relaxin and administered to beef cows on day 277 or 278 of gestation, resulted in basal levels of progesterone (P4) by 48 hours after injection and calving at 53–55 hours after treatment. There were no complications at calving and no retained placentas; however, there were only two animals per treatment group and inductions were done very close to normal term, so a great deal more study is needed before this method can be recommended.[44]

Short-acting corticosteroids

The most commonly used corticosteroids for inducing parturition are dexamethasone (20–30 mg) and flumethasone (8–10 mg) given as a single intramuscular injection. Parturition is induced with 80–90% efficacy when the injection is given within 2 weeks of normal term. The interval from injection to parturition is 24–72 hours, with an average of 48 hours. In cows that have not calved by 72 hours after treatment the induction is considered to have failed. Retreatment in such cases is often successful in inducing parturition. Relaxation of the pelvic ligaments, cervical dilation and filling of the udder occur rapidly, and labor and parturition are normal. Calving difficulty scores that have been recorded for induced parturition have generally not been different from those of natural calvings; however, frequently a higher incidence of minor assistance is given. This is probably because personnel are readily available during induction trials and assistance is more likely to be given even when not absolutely necessary.

Calves born less than 2 weeks prematurely are vigorous and calf mortality is not increased. The actual secretion of milk at the onset may not be plentiful in induced cows; however, colostral immunoglobulin levels and total milk production for the lactation period are very close to normal. Induced calves attain similar blood levels of immunoglobulins as calves born naturally.

Long-acting corticosteroids

Long-acting corticosteroids are not used when calf viability is of primary importance; however, they have gained wide acceptance where seasonal milk production is of primary importance and calving (lactation) is synchronized with the grazing season. Dexamethasone trimethylacetate (25 mg) or triamcinolone acetonide (4–8 mg) may be used and appear to provide similar outcomes.[18,34,35] An intramuscular injection is given once 2–4 weeks prior to the due date for calving and parturition occurs over a wide range of 4–26 days. Usually the farther a cow is from her due date, the longer it takes for a response.[24] Despite prolonged elevated systemic corticosteroid levels, cow health is generally good; however, pre-existing diseases, particularly subclinical infections, may be exacerbated by the treatment and there is a potential increase in cow mortality.

The udders of treated cows are consistently engorged with milk about 1 week after injection, although it may be another week before they actually calve. It has been suggested that these cows should be milked before calving if the udder is obviously full to prevent regression of secretory tissue. Total milk production per lactation can be expected to be reduced by 4–7%.

The incidence of retained placentas with the use of long-acting corticosteroids is quite low (9–22%) compared with short-acting corticosteroids. However, there is a high incidence of calf mortality (7–45%), which appears to be due to premature placental separation, an increased frequency of uterine inertia, and calf prematurity. Calf mortality may be reduced somewhat by careful observations and provision of prompt assistance.

The variability in response to treatment with a long-acting corticosteroid can be reduced by treating with a short-acting corticosteroid or prostaglandin about a week after the long-acting corticosteroid treatment.[19] Most cows will calve 2–3 days after the the second injection. The interval from injection to calving tends to be shorter and more predictable after prostaglandin than after a short-acting corticosteroid, and fewer repeat injections are required with prostaglandins. Calf mortality and the incidence of retained placentas are not reduced by giving a second injection.

Prostaglandins

Induction of parturition with prostaglandins intramuscularly gives very similar results to induction with short-acting corticosteroids, with a range of 24–72 (mean 44.9) hours from treatment to calving. As with short-acting corticosteroid-induced parturitions, there is a high incidence of retained placentas and a 10–20% rate of induction failure when treatments are given within 2 weeks before normal term.[45,46] The later the stage of gestation at which prostaglandins are administered, the greater the efficacy with which a single injection induces calving.

Some investigators have reported that cows with prostaglandin-induced parturition, particularly those induced more than 2 weeks prematurely, suffer a higher incidence of dystocia due to uterine inertia or malpresentation of the fetus than naturally calving cows.[47] However, others indicated no difference in the frequency of dystocia in cows induced to calve 1–2 weeks prematurely with either corticosteroids or prostaglandins. Moreover, the induced parturitions had no greater frequency of dystocia than spontaneous parturitions. In one experiment it was observed that cows induced with cloprostenol had a longer interval from appearance of the placental membranes (onset of stage 2 of parturition) until delivery of the fetus than cows treated with dexamethasone or a combination of dexamethasone and cloprostenol. Cows receiving only cloprostenol had higher concentrations of progesterone at parturition, which may have suppressed uterine contractions compared with cows receiving dexamethasone. Those receiving only cloprostenol may have had a placental source of progesterone that was eliminated in cows receiving dexamethasone.[45]

Corticosteroids and prostaglandins in combination

Hormones used to induce parturition initiate endocrine events normally triggered by fetal cortisol. Corticosteroid injections appear to efficiently remove the placental source of progesterone by induction of enzymes that convert placental progesterone to estrogen. Failure of corticoids to induce parturition may be due to failure to remove the ovarian source of progesterone. On the other hand, prostaglandin injections are efficient removers of the ovarian source of progesterone, but may fail to induce parturition due remaining placental progesterone. However, a combination of the two hormones, would act in concert to remove both sources of progesterone, resulting in fewer induction failures and less variability in the time from treatment to parturition.

Experiments have been conducted to compare parturition induced by a combination of dexamethasone and cloprostenol, dexamethasone alone, and cloprostenol alone, with saline-treated controls.[45] Cows receiving a combination of prostaglandin and dexamethasone calved earlier (range 25–42 hours from treatment to calving), and the interval from injection to calving was less variable than with dexamethasone alone (29–65 hours) or cloprostenol alone (37–57 hours). In addition, with the combination treatment there were no induction failures, whereas dexamethasone and cloprostenol treatments resulted in induction failure in 10.5–16.6%. The calving process for the combination treatment was considered to be normal. The rate of placental retention is significantly higher with all methods of induction of parturition than in control cows, but this was expected since with induction of parturition calving occurred before the due date and thus placental maturation would not have been completed.

Long-acting corticosteroids in combination with dexamethasone and cloprostenol for daylight calving and a low incidence of retained placentas

At present, the main impediments to the use of induction of parturition are the need for known breeding dates, lack of control of the time of calving, and an elevated incidence of retained placentas. Several workers have made significant progress toward alleviating these impediments. Their work is presented in greater detail in this chapter to provide readers with a clear understanding of the progress made to date and to provide direction for future research.

It has been determined that calves that are born within the last 2 weeks of gestation are usually very vital and adjust well to extrauterine life.[20] Therefore there exists a large window of opportunity to develop methods for detecting this safe 2-week period for induction of parturition, to adapt methods of induction that result in early maturation of the placenta and normal placental shedding, and for controlling the time of parturition. The ability to predict the last 2 weeks of gestation would need to be based on physical signs in the dam and these could perhaps be coupled with an endocrine indicator to more precisely determine the nearness to the end of gestation. In a study done by myself and my coworkers (unpublished results), a scoring system for teat, udder and vulval changes in late gestation in 80 beef cows and heifers had a positive predictive value of 72 for predicting the last 2 weeks of gestation. The predictive value of the scoring system was considered to be too low to be used to determine a safe induction time. Nevertheless, induction of parturition in 70 cows and heifers by two commercial beef producers, after application of the same scoring system, without known breeding dates, resulted in the loss of only three calves due to prematurity at birth. Future efforts should likely be directed to developing an adjunct test to the physical scoring system based on blood endocrine parameters. A combination of such tests should increase the predictive value of tests for fetal maturity.

Long-acting corticosteroids for induction of parturition resulted in the lowest incidence of retained placentas compared with other induction methods. It would appear that the long-acting corticosteroids more closely mimicked the natural endocrine events of gradually rising fetal cortisol resulting in placental maturation. Cows receiving a combination of dexamethasone and cloprostenol had a less variable interval from injection to calving and there were no induction failures. Thus an induction regimen using a long-acting corticosteroid pretreatment followed by a combined dexamethasone and cloprostenol treatment might be capable of reliable induction of parturition for calving in daylight hours and a low incidence of retained placentas.

Several experiments were carried out to determine the optimum dosages of two different long-acting corticosteroids and the optimum time interval from pretreatment with the long-acting corticosteroid and induction with combined dexamethasone–cloprostenol treatment.[29,30] Experiments utilized Hereford and Hereford–Angus cows with known breeding dates. This important work is summarized here. In the first experiment,[29] on day 270 of gestation 121 cows were placed in five groups. Groups I, II and III received 25 mg of Opticortenol (dexamethasone trimethylacetate, a long-acting corticosteroid). Cows that had not calved by day 277 received 25 mg dexamethasone (Dex, Group II) or a combination of DEX and 500 µg cloprostenol (DEX+CLO, Group III). Cows in Group IV received only DEX+CLO on day 277. Cows in Group 5 were untreated controls.

Pretreatment with Opticortenol followed by induction of parturition with DEX+CLO on day 277 resulted in calving during a very short time interval (28 ± 0.8 hours after induction) and the incidence of retained placentas was no different from that of the control group (13% vs. 6%). The length of labor, birthweights, and calving difficulty scores were no different between groups. There was no difference in the number of premature calves between groups and the only calf that died was in the control group.

In a second experiment,[29] cows were assigned, in a 2 × 2 factorial design, to one of four treatment groups and a control group. On day 270 of gestation cows in treatment groups received either a high dose (1 mg per 25 kg) or a low dose (1 mg per 50 kg) of Opticortenol. The cows were further subdivided to be induced with DEX+CLO on day 274 or day 276. Induction treatments were performed at 9 a.m. in order to achieve daylight calvings. All cows in treated groups calved between 7 a.m. and 11 p.m. about 30 hours after the induction treatment; however, the group receiving a high dose of Opticortenol and the induction treatment on day 276 had a lower incidence of retained placentas (29%) than other treatment groups. The incidence of retained placentas was 5% in the control group.

The length of stage 2 of labor, birthweights, calving difficulty, and calf viability were not different between groups and no premature calves were born in this experiment. Calf IgG concentrations at 48 hours of age were not different, whether the calves were born to induced cows or to untreated control cows and there was no increase in neonatal illness in calves born to induced cows. In the breeding seasons following these experiments, the first-service conception rates and pregnancy rates did not differ between groups. However, first-service conception rate and overall pregnancy rate was lower in cows that retained their placentas than in cows that did not.

Two further experiments[30] were designed to determine whether pretreatment with a different long-acting corticosteroid, triamcinolone acetonide (TRI), before induction of parturition with combined dexamethasone and cloprostenol (DEX+CLO) could be used to reduce the incidence of retained placenta and to produce a predictable calving time. The experiments were also designed to determine the optimum dosage of TRI and the optimum interval from pretreatment with TRI to induction treatment with DEX+CLO. It was determined that when cows are treated with TRI on day 270 of gestation, the optimum dosage of TRI was 1 mg per 60 kg and the optimum interval from TRI to DEX+CLO was day 277. Cows receiving this combination of treatments had no retained placentas compared with 5% retained placentas in the control group. All induced cows calved between 24 and 48 hours after DEX+CLO, and 94% began to calve between 7 a.m. and 7 p.m., whereas only 58% of control cows began to calve during the same period of time.

The length of stage 2 of labor, birthweights, calving difficulty, and calf viability did not differ among groups.

The results of these studies support the hypothesis that exposure to elevated blood corticosteroid levels prior to induction with DEX+CLO will result in a low incidence of retained placenta compared to induction with DEX+CLO alone. The use of long-acting corticosteroids to induce placental maturation followed by DEX+CLO to ensure parturition at a predictable time was quite successful in meeting the goals of calving in daylight, with a low incidence of retained placenta. Work needs to be done to combine this concept of induction of parturition with a test to predict the final 2 weeks of gestation in cows without known breeding dates. The level of estrogens at the time of induction treatment with DEX has been shown to be negatively correlated with the incidence of placental retention and induction failures.[48] Therefore circulating estrogen concentrations near term may serve as a useful indicator of placental maturity and the temporal proximity to parturition for a subsequent induction of parturition experiments.

Induction of abortion

Induction of abortion in cattle is a common procedure and is done for a variety of reasons. Heifer calves may become pregnant at a very early age, predisposing them to dystocia as well as retarded growth during pregnancy and during the subsequent lactation. Pregnancy in feedlot heifers usually results in reduced economic efficiency and calving management problems that feedlots are ill-suited to handle. Pathological conditions of pregnancy, including hydramnios, hydroallantois, fetal maceration, and fetal mummifications, must be terminated in order to save the life or breeding value of the affected animals.

Endocrinological aspects

Pregnancy maintenance depends on continuous adequate concentrations of circulating blood progesterone whether pregnancy is normal or abnormal. All treatment methods intended to terminate pregnancy, except surgical removal of the fetus, must directly or indirectly eliminate the sources of progesterone. The CL is the main source of progesterone in cattle throughout pregnancy; however, the placenta and to a minor extent the adrenal glands contribute to circulating levels of progesterone. A functional CL is essential for the maintenance of pregnancy during the first 120–150 days of gestation and during the final month of gestation. From approximately 150 to 250 days of gestation placental progesterone alone is sufficient to maintain pregnancy; therefore, lysis of the CL during this period will not terminate pregnancy.

Treatment of the pregnant bovine with prostaglandin results in luteolysis in both normal and abnormal gestations at any stage. In some cases, luteolysis appears to be incomplete and sufficient progesterone remains to maintain the pregnancy. In such cases, the female may undergo partial cervical dilation and may experience some abdominal straining, but then returns to the normal course of gestation and maintains the pregnancy to term. Ovulation and formation of a new CL may occur in these cases.[49]

Glucocorticoids such as dexamethasone may reduce placental secretion of progesterone as early as the fifth month of gestation. However, luteolysis does not occur as a result of glucocorticoid treatment until the final month of gestation when these steroids are able to induce abortion. In order for glucocorticoids to play a role in initiation of parturition, the fetoplacental unit must be functional. Thus in cases of fetal mummification or maceration, glucocorticoids will fail to induce abortion.

Methods of inducing abortion

Some of the older methods for the induction of abortions include per rectum enucleation of the CL, rupture of the fetal heart or fetal membranes and fetal decapitation, or the use of estrogenic hormones such as diethylstilbestrol, esters of estradiol. All of these methods have some significant disadvantages compared with prostaglandins.[39,50-54] The luteolytic effect of prostaglandins will induce abortion during the first 4–5 months of gestation, which are dependent on the CL. During months 5–8 of gestation, either $PGF_{2\alpha}$ or glucocorticoid treatment alone will fail to induce parturition, although treatment with a combination of $PGF_{2\alpha}$ and dexamethasone is highly reliable for inducing abortion during this period.[55]

Situations requiring induced abortion

Mismating

Veterinarians are sometimes asked to prevent pregnancy in a mismated female. The treatment of choice in these cases is to cause luteolysis. A luteolytic dose of prostaglandin may be administered as early as 7 days after mismating since the bovine CL becomes responsive to prostaglandins by 5–7 days after ovulation. Estrus will occur approximately 3 days after the prostaglandin injection and females can be rebred on that estus if desired. First-service pregnancy rates will be normal on the estrus concurring with abortion for pregnancies advanced as far as 60 days at the time of prostaglandin-induced abortion.

Feedlot heifers

The most common reason for inducing abortion in cattle is to terminate pregnancy in feedlot heifers. Precocious puberty in suckling heifer calves coupled with long breeding seasons frequently results in pregnancy before weaning. Ovarian cyclicity may occur in calves for a short period of time followed by a period of anestrus. As many as 25% of beef heifers have transient luteal function before 300 days of age.[56] In one study, the age of precocious puberty in beef heifers was 194 ± 12.4 days, the duration of cyclic luteal function was 65 ± 10.5 days, with resumption of anestrus after precocious puberty at 260 ± 15.3 days. Consequently, pregnancy rates in feedlot heifers of about 17% are common and may reach as high as 64%.[57]

Heifers that calve or are aborted in the feedlot generally show poor performance in comparison with nonpregnant heifers. Problems associated with pregnant feedlot heifers include a lower feed efficiency and growth rate, lower dressing percentage and quality grades at slaughter, and management problems such as dystocia and retained placentas should calving occur in the feedlot. It has been estimated that feedlot pregnancy results in losses of $44–115 per head.[58] The stage of gestation of the feedlot heifer will affect the degree of economic loss. In the early stages of pregnancy the anabolic effect of the luteal progesterone may actually be beneficial. Daily gains were higher in heifers pregnant up to 120 days and dressing percentage and quality grades at slaughter of heifers pregnant up to 120 days were equal to those of open heifers. Therefore heifers in early pregnancy and undergoing a relatively short feeding period would be expected to perform similarly to open heifers; however, this is seldom the case, with most pregnant heifers requiring 4–6 months on feed before slaughter.

All females that have recently entered a feedlot should be examined by transrectal palpation for pregnancy. Abortion can be induced at the time of examination. Animals induced to abort should be in good health and well adjusted to the feedlot since induction of abortion can be sufficiently stressful to increase the incidence of common feedlot diseases, particularly respiratory diseases. It is usually recommended that heifers be allowed a 2–3 week adjustment period to the feedlot before inducing abortion.

Growth-promoting agents containing progestins may interfere with induction of abortion and should be withheld until after abortion occurs. Heifers that are up to 150 days pregnant can be aborted with approximately 90% efficacy with a single luteolytic dose of prostaglandin.[57,59,60] Those that are greater than 150 days pregnant must be treated with a combination of a luteolytic dose of prostaglandin and 25 mg of dexamethasone. Abortion will occur in 95% of these cases within 3–9 days.

Behavioral estrus occurs in 75–80% of aborting females, but since physical signs of abortion may go unnoticed and some fail to abort, all abortions should be recorded and the remainder of treated animals must be reexamined about 10–14 days after the abortion treatment. Lack of luteolysis is the most likely cause of abortion failure, and retreatment will usually induce abortion. Fetal mummification occurs in 2–4% of aborted females and this incidence is likely at least twice as high as that recorded in normal populations of cattle. Cattle that are more than 4 months pregnant at the time of abortion will retain the placenta in more than 80% of cases.[13] The placenta is usually passed in about 1 week and treatment is rarely necessary. Postpartum endometritis and purulent discharges occur in many heifers, but in most cases resolve spontaneously.

Ovariectomy as a means of inducing abortion and control of estrus in feedlot heifers has been suggested. In one experiment, ovariectomy was compared with manual or chemical abortion and incorporation of melengestrol acetate (MGA) in the diet.[61] Ovariectomy reduced average daily gain and feed efficiency in the first 24 days. Prostaglandin-treated heifers reached slaughter condition earlier than ovariectomized or MGA-treated heifers. Only with $PGF_{2\alpha}$ were all pregnancies terminated.

Hydropic conditions of the uterus

Hydropic conditions of the uterus are a serious threat to the life of the affected animal. When affected animals in still in good condition, slaughter is usually the best method of salvaging some value in the animal. In more severely affected cows with dehydration and electrolyte imbalance, treatment may be attempted but the prognosis for life is poor. When abdominal distension is severe, fluid drainage by abdominal catheterization should be considered as a first step in treatment. Large volumes of intravenous fluids will be required for several days to maintain hydration. Treatment to induce abortion with a luteolytic dose of prostaglandin in combination with 25 mg of dexamethasone usually will result in partial cervical dilation and accompanying loss of fetal fluid by 30 hours after treatment. Parturition is usually abnormal, with incomplete cervical dilation, primary uterine inertia, and lack of a strong abdominal press. However, since the fetus is usually small, with most cases developing at 7–9 months of gestation, fetal removal by repositioning and traction is usually successful. Postpartum metritis is a common sequela and frequently causes death. The prognosis for cesarean section is equally unfavorable because the uterus is atonic and friable and the fetal membranes are nearly always retained.[62]

Fetal mummification and maceration

Fetal mummification or maceration occurs when a fetus dies without concomitant luteolysis and adequate cervical dilation to allow expulsion. A functional CL is often present and therefore the first step in treatment is the administration of a luteolytic dose of prostaglandin. Expulsion of the fetus usually begins within 2–4 days after treatment. A second treatment is usually not necessary. Estrogens have also been used; 50–80 mg of diethylstilbestrol or 5–10 mg of estradiol intramuscularly will usually result in fetal expulsion in 24–72 hours. A second treatment may be necessary, and in rare cases three or four treatments at 48-hour intervals are required.[62] Whether prostaglandins or estrogens are used, expulsion of the fetus may not be complete because of poor cervical dilation and dryness of the cervix and birth canal. Lubrication and manual assistance are often necessary to achieve fetal delivery and postpartum medical care may be necessary.

References

1. Liggins G. Premature parturition after infusion of corticotrophin or cortisol into fetal lambs. *J Endocrinol* 1968;42:323–329.
2. Liggins G. Premature delivery of foetal lambs infused with glucocorticoids. *J Endocrinol* 1968;45:515–523.
3. Wood C. Development of adrenal cortical function. In: Meisami E, Timiras PS (eds) *Handbook of Human Growth and Developmental Biology*, Vol. II, Part A. Boca Raton: CRC Press, 1989, pp. 81–94.
4. Robertson H. Sequential changes in plasma progesterone in the cow during the estrus cycle, pregnancy, at parturition and postpartum. *Can J Anim Sci* 1972;52:645–658.
5. Wilden S, Day M. Placental progesterone production in mid- to late-gestation in the beef cow. Beef Cattle Research Report, Ohio Agricultural Research and Development Center, 1991, No. 91-2, pp. 73–83.
6. Reimers T, Ulman M, Hansel W. Progesterone and prostanoid production by bovine binucleate trophoblastic cells. *Biol Reprod* 1985;33:1227–1236.
7. Pimmentel S, Pimentel M, Weston P, Hixon J, Wagner W. Progesterone secretion by the bovine fetoplacental unit and responsiveness of corpora lutea to steroidogenic stimuli at two stages of gestation. *Am J Vet Res* 1986;47:1967–1971.
8. Johnson W, Manns J, Adams W, Mapletoft R. Termination of pregnancy with cloprostenol and dexamethasone intact or ovariectomized cows. *Can Vet J* 1981;22:288–290.
9. Hunter J, Fairclough R, Peterson A, Welch R. Foetal and maternal hormonal changes preceding normal bovine parturition. *Acta Endocrinol* 1977;84:653–662.
10. Anderson A, Flint A, Turnbull A. Mechanism of action of glucocorticoids in induction of ovine parturition: effect on placental steroid metabolism. *J Endocrinol* 1975;66:61–70.
11. Ou C, Orsino A, Lye S. Expression of connexin-43 and connexin-26 in the rat myometrium during pregnancy and labor is differentially regulated by mechanical and hormonal signals. *Endocrinology* 1997;138:5398–5407.
12. Murata T, Murata E, Liu X, Narita K, Honda K, Higuchi T. Oxytocin receptor gene expression in rat uterus regulation by ovarian steroids. *J Endocrinol* 2000;166:45–52.
13. Lindell J, Kindahl H, Edqvist L. Prostaglandin induced early abortions in the bovine. Clinical outcome and endogenous release of prostaglandin F2 alpha and progesterone. *Anim Reprod Sci* 1981;3:289–299.
14. McLaren W, Young I, Rice G. Localization and temporal changes in prostaglandin G/H synthetase-1 and -2 content in ovine intrauterine tissues in relation to glucocorticoid-induced and spontaneous labour. *J Endocrinol* 2000;165:399–410.
15. Liggins G. Endocrinology of parturition. In: Novy MJ, Resko JA (eds) *Fetal Endocrinology*. New York: Academic Press, 1981, pp. 211–237.
16. Wood C. Control of parturition in ruminants. *J Reprod Fertil Suppl* 1999;54:115–126.
17. Saoud C, Wood C. Modulation of ovine fetal adrenocorticotrophin secretion by androstenedione and 17β- estradiol. *Am J Physiol* 1997;272:R1128–R1134.
18. Welch R, Newling P, Anderson D. Induction of parturition in cattle with corticosteroids: an analysis of field trials. *NZ Vet J* 1973;21:103–108.
19. Welch R, Day A, Duganzich D, Featherstone P. Induced calving: a comparison of treatment regimes. *NZ Vet J* 1979;27:190–194.
20. Wagner W, Willham R, Evans L. Controlled parturition in cattle. *J Anim Sci* 1974;38:485–489.
21. Muller L, Beardsley G, Ellis R, Reed D, Owens M. Calf response to the initiation of parturition in dairy cows with dexamethasone or dexamethasone with estradiol benzoate. *J Anim Sci* 1975;41:1711–1716.
22. Diskin M, Sreenan J. Induction of parturition in the cow: the use of a short-acting corticosteroid alone or at five days after an injection of a long-acting corticosteroid. *Irish Vet J* 1984;38:6–13.
23. MacDiarmid S. Induction of parturition in cattle using corticosteroids: a review. Part 1. Reasons for induction, mechanisms of induction, and preparations used. *Animal Breeding Abstracts* 1983;51:403–419.
24. MacDiarmid S. Induction of parturition in cattle using corticosteroids: a review. Part 2. Effects of induced calving on the calf and cow. *Animal Breeding Abstracts* 1983;51:499–508.
25. Wagner W. Parturition induction in cattle. In: Morrow DA (ed.) *Current Therapy in Theriogenology*. Philadelphia: WB Saunders, 1980, pp. 236–238.
26. O'Farell K, Crowley J. Some observations on the use of two corticosteroid preparations for the induction of premature calving. *Vet Rec* 1974;94:364–366.
27. Barth A, Adams W, Manns J, Rawlings N. Induction of parturition in beef cattle using estrogens in conjunction with dexamethasone. *Can Vet J* 1978;19:175–180.
28. Beardsley G, Muller L, Owens M, Ludens F, Tucker W. Initiation of parturition in dairy cows with dexamethasone. I. Cow response and performance. *J Dairy Sci* 1974;57:1061–1066.
29. Bo G, Fernandez M, Barth A, Mapletoft R. Reduced incidence of retained placenta with induction of parturition in the cow. *Theriogenology* 1992;38:45–61.

30. Nasser L, Bo G, Barth A, Mapletoft R. Induction of parturition in cattle: effect of triamcinolone pretreatment on the incidence of retained placenta. *Can Vet J* 1994;35:491–496.
31. Hoerlein A, Jones D. Bovine immunoglobulins following induced parturition. *J Am Vet Med Assoc* 1977;170:325–326.
32. Field R, Bretzlaff K, Elmore R, Rupp G. Effect of induction of parturition on immunoglobulin content of colostrum and calf serum. *Theriogenology* 1989;32:501–506.
33. Husband A, Brandon M, Lascelles A. Absorption and endogenous production of immunoglobulins in calves. *Aust J Exp Biol Med Sci* 1973;51:707–721.
34. Bailey L, McLennan M, McLean D, Harford P, Munro G. The use of dexamethasone trimethyl acetate to advance parturition in dairy cows. *Aust Vet J* 1973;19:175–179.
35. Davis K, Macmillan K. Controlled calving with induction of parturition on day 274 of gestation in dairy cows. *Proc NZ Soc Anim Prod* 2001;61:184–186.
36. Lavoie V, Moody E. Estrogen pretreatment of corticoid-induced parturition in cattle. *J Anim Sci* 1973;37:770–774.
37. Grunert E, Ahlers D, Jockle W. Effects of a high dose of diethylstilbestrol on the delivery of the placenta after corticoid-induced parturition in cattle. *Theriogenology* 1975;3:249–258.
38. Kordts E, Joechle W. Induced parturition in dairy cattle: a comparison of a corticoid (flumethasone) and prostaglandin (PGF2α) in different age groups. *Theriogenology* 1975;3:171–178.
39. Brand A, Debois C, Kommery R, Dejong M. Induction of abortion in cattle with prostaglandin F2α and oestradiol valerate. *Tijdschr Diergeneeskd* 1975;100:432–435.
40. Beardsley G, Muller L, Garverick H, Ludens F, Tucker W. Initiation of parturition in dairy cows with dexamethasone. II. Response to dexamethasone in combination with estradiol benzoate. *J Dairy Sci* 1976;59:241–247.
41. Vesnick Z, Holub A, Zraly Z et al. Regulation of bovine labor with a long-acting caba-analog of oxytocin: a preliminary report. *Am J Vet Res* 1979;40:425–429.
42. Morrison D, Humes P, Godke R. The use of dimenhydrinate in conjunction with dexamethasone for induction of parturition in beef cattle. *Theriogenology* 1983;19:221–233.
43. Musah A, Schwabe C, Willham R, Anderson L. Induction of parturition, progesterone secretions and delivery of the placenta in beef heifers given relaxin with cloprostenol or dexamethasone. *Biol Reprod* 1978;39:797–803.
44. Li YF, Perezgrovas R, Gazal OS, Schwabe C, Anderson LL. Antiprogesterone, RU 486, facilitates parturition in cattle. *Endocrinology* 1991;129:765–770.
45. Lewing F, Proulx J, Mapletoft R. Induction of parturition in the cow using cloprostenol and dexamethasone in combination. *Can Vet J* 1985;26:317–322.
46. Day A. Cloprostenol for termination of pregnancy in cattle. A) The induction of parturition. *NZ Vet J* 1977;25:136–139.
47. Henricks D, Rawlings N, Ellicott A, Dickey J, Hill J. Use of prostaglandin F2α to induce parturition in beef heifers. *J Anim Sci* 1977;44:438–441.
48. Chew B, Erb R, Randel R, Rouquette F Jr. Effect of corticosteroid induced parturition on lactation and on prepartum profiles of serum progesterone and estrogens among cows retaining and not retaining fetal membranes. *Theriogenology* 1978;10:13–25.
49. Day A. Cloprostenol for termination of pregnancy in cattle. B) The induction of abortion. *NZ Vet J* 1977;25:139–144.
50. Dawson F. Methods for early termination of pregnancy in the cow. *Vet Rec* 1974;94:542–548.
51. Lauderdale J. Effects of PGF2A on pregnancy and estrous cycles of cattle. *J Anim Sci* 1972;35:246.
52. Miller P. Methods for early termination of pregnancy in the cow. *Vet Rec* 1974;94:626.
53. Copeland D, Schultz R, Kemptrup M. Induction of abortion in feedlot heifers with cloprostenol (a synthetic analogue of prostaglandin F2A): a dose response study. *Can Vet J* 1978;19:29–32.
54. Parmigiani E, Ball L, Lefever D, Rupp G, Seidel G Jr. Elective termination of pregnancy in cattle by manual abortion. *Theriogenology* 1978;10:283–290.
55. Barth A, Adams W, Manns J, Kennedy K, Sydenham R, Mapletoft R. Induction of abortion on feedlot heifers with a combination of cloprostenol and dexamethasone. *Can Vet J* 198;22:62–64.
56. Wehrman M, Kojima F, Sanchez T, Mariscal D, Kinder J. Incidence of precocious puberty in developing beef heifers. *J Anim Sci* 1996;74:2462–2467.
57. Fernandes C, Viana J, Ferreira A, Sa W. Fertilidade de novilhas apos aborto induzido com cloprostenol. *Arq Bras Med Vet Zootec* 2002;54:279–282.
58. MacGregor S, Falkner T. Managing pregnant heifers in the feedlot. *Comp Cont Educ Pract Vet* 1977;19:1389–1391.
59. Edwards A, Laudert S. Economic evaluation of the use of feedlot abortifacients. *Bov Pract* 1984;19:148–150.
60. Coulson A. Early termination of pregnancy in cattle with dinoprost. *Vet Rec* 1979;105:553–554.
61. Horstman L, Callahan C, Morter R, Amstutz H. Ovariectomy as a means of abortion and control of estrus in feedlot heifers. *Theriogenology* 1982;17:273–292.
62. Roberts S. In: *Veterinary Obstetrics and Genital Diseases (Theriogenology)*, 3rd edn. Woodstock, VT: published by the author, 1986, pp. 216–220.

Chapter 45

Management to Prevent Dystocia

W. Mark Hilton and Danielle Glynn

College of Veterinary Medicine, Purdue University West Lafayette, Indiana, USA

Introduction

Dystocia, defined as difficult birth, is an important economic issue in the beef cattle industry. Consequences of dystocia include increased calf morbidity and mortality, increased cow morbidity and mortality, decreased animal welfare, decreased subsequent fertility in cows, and increased labor. If dystocia rate exceeds herd goals, implementation of a plan to decrease that rate should reduce the problems listed above, which will improve overall herd health, welfare, and profitability.

The focus of dystocia prevention is primarily directed at first-calf heifers, and the most common cause of dystocia in heifers is fetal–dam disparity. The incidence of calving difficulty is significantly less for mature cows compared with heifers. Table 45.1 depicts the results of a study in 1978 in Montana where no selection pressure was made for calving ease. The incidence of dystocia among females by age is outlined.[1] In another study of 386 calvings with primiparous dams and 1805 with multiparous dams, the level of dystocia was 17% and 4%, respectively.[2] This corroborates with the previous study to show the significance of age of dam with respect to dystocia, but with a much lower overall dystocia rate.

Genetics

Birthweight is a highly heritable trait, with more variation between breeds than within breeds according to research from 4639 calves and 290 sires of 14 *Bos taurus* breeds.[3] This study at the US Meat Animal Research Center (MARC) in Nebraska which concluded in 1976 demonstrated that certain breeds tend to produce calves with lower birthweight than other breeds.[3] Each year the Beef Improvement Federation (BIF) publishes a summary of numerous production traits for the most common breeds of beef cattle in the United States. Table 45.2 reiterates the conclusion from the earlier study and demonstrates that some breeds tend to produce lower birthweight calves compared with the other breeds. While the research showed more variation between breeds on average birthweight, there can still be considerable variation within breeds with regard to birthweight. Just because a bull is of a breed that normally produces lower birthweight calves does not mean every bull in that breed will do the same.

The use of expected progeny differences (EPDs) in the beef industry has allowed for a more scientifically based path to improvement in traits where EPDs are available. Bulls with birthweight EPDs well below breed average that are from breeds that produce lower birthweight calves have been selected to mate with virgin heifers and this has been quite successful in reducing dystocia rates in beef herds. Birthweight EPD is an indicator trait of calving ease and, until the newer calving ease EPDs were developed, selecting bulls with low birthweight EPD was the primary selection tool for calving ease.

Calves from 2-year-old dams that experienced difficult births were significantly heavier at birth (39.6 vs. 35.4 kg) and had significantly lower survival at 72 hours (87.1 vs. 92.9%) and at weaning (77.4 vs. 85.1%) than calves from 2-year-old dams that did not experience difficult births.[4] Two studies that examined the correlation between birthweight and dystocia showed that each kilogram increase in birthweight resulted in a 1.6–4.2% increase in calves requiring assistance.[4,5] Calf mortality was also increased with calves that were significantly lighter at birth than the mean, so selecting for extremely low birthweight does not appear to be a sound production practice.[6]

The BIF publishes an across-breed (AB) EPD table each year and data are obtained from calves born at the MARC in Nebraska. The Angus breed is used as the "base" and is assigned a value of zero for each EPD adjustment. The AB EPD allows a direct comparison of two different breeds on an equal EPD basis. For example, to compare a Hereford bull to an Angus bull on birthweight EPD, the first step is to identify the adjustment factor for the Hereford bull.

Table 45.1

Age of dam (years)	Number of calvings	Dystocia (%)
2	437	29.7
3	475	10.5
4	427	7.2
5–10	1478	2.7

Table 45.2 Breed of sire means (in pounds) for 2011-born animals under conditions similar to US MARC.

Breed	Birthweight	Weaning weight	Yearling weight
Angus	87.3	577.0	1045.3
Hereford	91.7	571.5	1009.7
Red Angus	88.1	561.5	1013.0
Shorthorn	93.7	556.5	1022.9
South Devon	91.4	566.0	1030.0
Beefmaster	92.1	575.6	1002.9
Brahman	98.3	587.7	989.3
Brangus	90.8	568.2	1008.4
Santa Gertrudis	92.6	570.5	1013.9
Braunvieh	89.9	549.4	981.8
Charolais	94.7	592.4	1047.7
Chiangus	90.9	546.2	987.0
Gelbvieh	89.6	575.4	1027.1
Limousin	90.8	574.7	1007.7
Maine-Anjou	91.8	554.1	1000.8
Salers	89.0	566.4	1019.5
Simmental	91.5	586.1	1038.8
Tarentaise	89.1	576.2	1008.2

Note: Because of the prevalent use of the English system of weights and measures in the calculation of EPDs, pounds was used in this table.
Source: courtesy of Larry Kuehn, ARS, USDA.

In Table 45.3, the adjustment is listed as +2.7 for birthweight for Hereford. So, a Hereford bull with EPD of 0 for birthweight EPD is equal to an Angus bull with birthweight EPD of +2.7 and an Angus bull with birthweight EPD of 0 is equal to a Hereford with birthweight EPD of −2.7 for birthweight. To compare a Hereford bull with a birthweight EPD of +2.0 to an Angus bull with a birthweight EPD of +4.0, add the adjustment factor of 2.7 for Hereford birthweight EPD to the Hereford bull's actual birthweight EPD of +2.0. This gives the Hereford an AB EPD +4.7 for birthweight. The Angus bull's AB EPD is still +4.0 (no adjustment needed). AB EPDs are used to compare bulls of two different breeds on an equal genetic basis, so on average this Hereford will sire calves that are 0.3 kg heavier at birth than the Angus bull. Since the AB EPD research is published each year, it is important for readers to consult the most current data to use in calculating AB EPDs. The effects of heterosis on birthweight are significantly positive (2.3 kg) while the effects on calving difficulty are generally not significant (−0.5 to +1.5%).

Table 45.3 Adjustment factors (in pounds) to add to EPDs of 18 different breeds to estimate across-breed EPDs.

Breed	Birthweight	Weaning weight	Yearling weight
Angus	0.0	0.0	0.0
Hereford	2.7	−3.5	−23.6
Red Angus	3.4	−23.2	−27.9
Shorthorn	5.8	11.3	38.8
South Devon	3.2	4.8	−6.6
Beefmaster	6.3	35.7	39.5
Brahman	11.0	42.8	5.9
Brangus	4.5	14.6	6.0
Santa Gertrudis	6.6	36.2	48.3
Braunvieh	1.9	−21.6	−42.3
Charolais	8.6	38.1	45.3
Chiangus	2.2	−20.5	−40.2
Gelbvieh	2.7	−18.2	−25.6
Limousin	3.8	−1.8	−35.9
Maine-Anjou	4.2	−15.3	−36.7
Salers	1.8	−4.8	−19.5
Simmental	3.7	−5.9	−10.9
Tarentaise	1.7	30.3	20.3

Note: Because of the prevalent use of the English system of weights and measures in the calculation of EPDs, pounds was used in this table.
Source: courtesy of Larry Kuehn, ARS, USDA.

Table 45.4 BIF calving ease scoring system.

Score	Description
1	No difficulty, no assistance
2	Minor difficulty, some assistance
3	Major difficulty, assistance or puller
4	Cesarean birth
5	Abnormal presentation

Even with a slight increase in birthweight, survival to weaning tended to be higher (1.7–2.1%) in crossbred calves.[6] Calving ease EPDs are a more recent development in the expanded EPD repertoire. In many seedstock and research herds, not only is birthweight of calves recorded but so is actual calving ease score in heifers. The BIF scoring system for calving ease is depicted in Table 45.4. The terminology for calving ease EPDs is inconsistent between breeds, which can lead to confusion. Some breeds use the simple term "calving ease" EPD (or CE EPD) while others use "calving ease direct" EPD (or CED EPD). Both of these terms refer to the relative calving ease of the calves sired by the bull in question.

There are also EPDs for calving ease of the daughters of the sire in question and these are listed as either CEM EPD for "calving ease maternal" or CED EPD for "calving ease of daughters." Each breed sire summary will have an explanation of their terminology. Since Angus and Red Angus are the two breeds most frequently used on yearling heifers for calving ease, it is convenient that they both use the same terminology: CED EPD for calving ease direct and CEM EPD for calving ease maternal. I will use these terms when discussing calving ease EPDs in this chapter.

The advantage of using CED EPD to decrease dystocia is that it is a direct measurement of differences in percentage of unassisted births between bulls. If sire A has a CED EPD of 12 and he is compared with sire B with CED EPD of 5, sire A should sire calves with 7% more unassisted births than sire B. So, if sire B produced 80% unassisted births in a group of first-calf heifers, sire A would be expected to produce 87% unassisted births in the exact same group of heifers. Currently, the BIF does not publish AB EPD adjustment factors for calving ease, so it is not possible to compare the EPDs between breeds at this time.

It is difficult to give a recommendation on what level of CED EPD or birthweight EPD is acceptable when selecting sires to be used on yearling heifers. Heifers that would be considered to be at highest risk for dystocia would have any of the following genetics:

- have low CEM;
- have high AB birthweight EPD'
- sired by bulls with low CED;
- sired by bulls with high AB birthweight EPD;
- sired by bulls of breeds that tend to produce higher birthweight calves.

In this case, a bull more extreme in high CED EPD or low birthweight EPD should be selected for these heifers. We must not forget that the dam contributes half of the genetics to the calves and low CEM or high birthweight EPD females are genetically more likely to have increased calf birthweight and calving difficulty compared with high CEM or low birthweight EPD females.

While one herd may need a bull above 12 CED EPD or below 0 AB birthweight EPD, another may tolerate a CED EPD of more than 9 or an AB birthweight EPD of less than 2.0. If a group of potential sires have CED EPDs, apply those numbers to select the ideal sire to use on the group of heifers. Do not also use the birthweight EPD as that is already factored into the CED EPD. Only use the AB birthweight EPD if there is no CED EPD.

CEM is similar in use and calculation to CED, but it measures the difference in percentage of unassisted births between sires for their first-calf daughters. As with CED, a higher number is more favorable for CEM. Even though the genetic correlation between CED and CEM is negative (−0.36), there are bulls that defy the antagonism and excel in both traits.[7] If the producer intends to retain daughters of a calving ease bull as replacement females, looking at both CED and CEM EPD is a wise choice.

Another factor to examine before selecting a bull to use on heifers is his EPD accuracy. It is ideal to breed heifers to an artificial insemination (AI) sire with high accuracy for CED or birthweight EPD as there is a greater chance that the high accuracy sire's EPDs will change less over time. A low accuracy bull will not produce calves with more variability than a high accuracy sire. There are tables that list the possible change in EPD based on accuracy. For example, a yearling Angus bull with a CED EPD of 10 and an accuracy of 0.05 has a potential change of 7.8 (American Angus Association). This means that this sire will likely have a CED of between 2.2 and 17.8 as his accuracy improves. In contrast, a bull with a CED EPD of 10 and accuracy of 0.80 has a potential change of only 1.6 as he moves toward a higher CED EPD accuracy. This sire will likely have a "true" CED EPD of 8.4–11.6. This shows that it is ideal to use a highly accurate sire on first-calf heifers so that the "true" EPD for calving ease is much less likely to change over time. An accuracy of more than 0.80 would be considered highly accurate by most advisors.

The actual birthweight of a sire is incorporated into the EPD calculation if the bull is born into a herd with a contemporary group. A young sire will have a low accuracy for both CED and birthweight EPD until he has sired numerous calves and that data submitted to a breed association. If a bull is DNA tested and has his genomic EPDs incorporated into his EPD, his accuracy will improve significantly over a bull without this technology. Adding this technology is equivalent to the data from eight calves born and recorded.[8]

If actual birthweight is the only parameter available, it would be difficult to predict if a bull has the potential to be a true calving ease sire. Many nongenetic factors, such as age of dam, season of year, and single versus twin birth, can influence birthweight. When EPDs for CED or birthweight are available, looking at actual birthweight is of almost no value.

Male calves are heavier at birth, have increased gestation, and subsequently have an increased rate of dystocia (28.4 vs. 16.98%) compared with female calves.[9] With the commercialization of techniques to produce sexed-semen, the use of heifer-specific semen in first-calf heifers is another management practice that could be used to decrease the incidence of dystocia in these females.

Gestation length is a highly heritable trait.[3] Cows with shorter gestation produce calves with reduced birthweight, so selecting for shorter gestation length can be an indirect way to select for reduced dystocia. As a primary selection tool, reducing gestation length would not be successful due to the fact that there is very little variation from short to average gestation. For example, in the Gelbvieh breed in 2012 the average gestation length EPD is −0.8 days, while a bull ranked in the top 5% of the breed for short gestation is −3.1 days. With variation of only 2.3 days between these sires, it is unlikely that gestation could be reduced enough to be of major benefit to reducing dystocia.

Abnormal presentation and twinning

In a study of 3878 calvings over 21 years, 4.1% of calvings (excluding twin births) were malpresentations. Because malpresentations are so infrequent, extensive analysis of contributing factors is difficult. In this study, 72.8% of all malpresentations were in the posterior dorsal position and birthweight, sex, year, and breed of sire of calf were all significant factors.[10] In another study that included 2088 calvings, abnormal presentation occurred in 2.7% of cows giving birth to single calves, but in 17.4% in cases of multiple births.[11] In this same study, if assistance was needed at calving, single-born calves had a 95% survival rate while twins had a 73% survival rate.

In dairy herds, cows carrying twins are a significant economic detriment to the herd. Dairy cows with twins exhibit increases in abortion, dystocia, stillbirth, retained placenta, subsequent service to conception and culling.[12] In a study with 12 576 calvings over 11 years with a twinning rate of 5.6%, 53.6% of twin calvings resulted in dystocia while 38.1% of singleton calvings required assistance.[12] There is disagreement about the management of dairy cows shown to be carrying twins after ultrasound reproductive examination, with options for selective abortion, special prepartum management, culling, or no management changes (Brent Cousin, personal communication). Scheduling a dry-off time 14 days earlier than normal combined with increased nutritional support has been an effective tool in the hands of some dairy veterinarians (Brent Cousin, personal communication).

Nutrition

Numerous trials have been conducted to examine the relationship between pre-calving nutrition, calf birthweight, and incidence of dystocia. In nine trials where pre-calving nutrition was manipulated to cause heifers to calve at body condition score (BCS) 4–6, calf birthweight increased from 0 to 3 kg as BCS increased.[6] In these same studies, the incidence of dystocia was unchanged in seven trials and increased in two. When protein levels were increased similar results were produced. While increasing pre-calving nutrition generally had no effect on dystocia rate, it had a profound effect on subsequent reproduction. In one study 2-year-old heifers nursing their first calf that calved at BCS 4 had a pregnancy percentage of 56% in a 60-day breeding season, whereas heifers calving at BCS 6 had a 96% pregnancy percentage in the same time frame.[13] It is clear that restricting feed intake to below National Research Council guidelines is not an effective way to reduce dystocia rates in heifers. Another study reported a negative correlation between heifer BCS before calving and dystocia rate, indicating that heifers with better BCS had reduced dystocia, possibly due to these heifers having more energy, resulting in a more intense labor effort.[14]

Heifer development

Traditional recommendations indicate that heifers should reach 65% of mature body weight (MBW) just prior to the first breeding season. While resulting in a maximum percent cycling, these intensive programs may not be the most cost-effective. A concern in developing heifers at less than 65% of MBW is a subsequent decrease in body weight at calving with an increased incidence of dystocia, along with reduced fertility in the subsequent breeding season. Studies in Nebraska have shown that heifers developed to 50 or 55% of MBW had no differences in calf birthweight (33 vs. 33 kg), percent dystocia (31.3 vs. 24.7%), initial conception rates in a 45-day breeding season (87.2 vs. 89.8%), and second calf conception rates (92.4 vs. 93.8%).[15] Some caution should be exercised in developing heifers if the target is 50% of MBW. If environmental stress or a change in feed availability occurs during development and heifers are actually below 50% of MBW at breeding, a reduction in fertility may be encountered.

Since the primary cause of dystocia in beef cattle is due to fetal–dam disparity, numerous researchers have looked at pelvic measurements of the yearling female to predict subsequent risk of dystocia. Unfortunately, heifers with larger pelvic openings tend to be larger in frame score, and also have calves with heavier birthweights. So selecting for larger pelvic openings does not lead to a reduction in rate of dystocia.[16] There may be some merit in identifying heifers with abnormally shaped pelvises before breeding, but because the incidence of this problem is rare the cost-effectiveness of evaluating the pelvic area and/or structure simply to reduce dystocia is questionable.

Techniques to improve calf survival with dystocia

Breeding yearling heifers prior to the mature cow herd is a proven method to increase rebreeding percentage in nursing 2 year olds. This technique can also provide a boost in calf survival from these heifers due to the potential of additional time spent observing heifers at calving time. The first weeks of the calving season come with much anticipation as it has been many months since the previous calving season. These heifers calving at the start of the calving season are more likely to be examined multiple times during the day and night, so dystocia should be caught earlier which should lead to more live calves being born.

Studies show that early intervention in cases of dystocia yield more live healthy calves and more live healthy heifers. An added benefit to early intervention is improved reproductive rate in these heifers compared with heifers where intervention was later.[17] Instead of having beef producers try to memorize how long each stage of labor lasts, our private veterinary practice had the rule of "progress every hour." Every heifer or cow needs to make progress on delivering the calf every hour and if she is not, she needs to be examined. This rule greatly increased the percentage of live calves delivered from our clients' farms.

Environmental effects on birthweight and dystocia

A study on genotype by environment interaction in Hereford cattle was conducted for 11 years where cattle of similar genetics were raised in Montana and Florida. Birthweight of calves born in Montana averaged 6.5 kg heavier than those born in Florida even though genetics were the same.[18] Research like this substantiates observations from practicing veterinarians in northern US climates that late summer and fall-born calves in the United States are lighter at birth and experience less dystocia compared with winter and early spring-born calves. Moving from a winter or early spring calving season to one later in the spring would be a management strategy that could decrease percent dystocia in a herd where cold weather stress could be a contributor to increased birthweight and subsequent dystocia.

Summary

Risk factors for dystocia in a herd can be enumerated after a thorough herd investigation. The most common population at risk for dystocia in a beef herd are the first-calf heifers and the most common reason for dystocia is due to fetal–dam disparity. Selecting high-accuracy AI sires that are well above breed average for CED EPD in the breeds known for calving ease is the most successful way to prevent dystocia in heifers. If CED EPDs are not available, then selecting highly accurate bulls well below breed average for AB birthweight EPD is a reasonable approach.

References

1. Price T, Wiltbank J. Dystocia in cattle. A review and implications. *Theriogenology* 1978;9:195–219.
2. Nix J, Spitzer J, Grimes L, Burns G, Plyler B. A retrospective analysis of factors contributing to calf mortality and dystocia in beef cattle. *Theriogenology* 1998;49:1515–1523.
3. Cundiff L, MacNeil M, Gregory K, Koch R. Between- and within-breed genetic analysis of calving traits and survival to weaning in beef cattle. *J Anim Sci* 1986;63:27–33.
4. Gregory K, Cundiff L, Koch R. Breed effects and heterosis in advanced generations of composite populations for birth weight, birth date, dystocia, and survival as traits of dam in beef cattle. *J Anim Sci* 1991;69:3574–3589.
5. Smith G, Laster D, Gregory K. Characterization of biological types of cattle. I. Dystocia and preweaning growth. *J Anim Sci* 1976;43:27–36.
6. Funston R. Nutrition and reproduction interactions. In: *Proceedings, Applied Reproductive Strategies in Beef Cattle*, 2010, pp. 175–191. Available at http://www.beefusa.org/Udocs/PR101-Nutrition.pdf
7. Phocas F, Laloë D. Evaluation models and genetic parameters for calving difficulty in beef cattle. *J Anim Sci* 2003;81:933–938.
8. GE-EPDs deliver progeny-proof to commercial bull buyers. Pfizer Animal Health web site. www.zoetisus.com
9. Laster D, Glimp H, Cundiff L, Gregory K. Factors affecting dystocia and the effects of dystocia on subsequent reproduction in beef cattle. *J Anim Sci* 1973;36:695–705.
10. Holland M, Speer N, Lefever D, Taylor R, Field T, Odde K. Factors contributing to dystocia due to fetal malpresentation in beef cattle. *Theriogenology* 1993;39:899–908.
11. Gregory K, Echternkamp S, Dickerson G, Cundiff L, Koch R, Van Vleck L. Twinning in cattle: III. Effects of twinning on dystocia, reproductive traits, calf survival, calf growth and cow productivity. *J Anim Sci* 1990;68:3133–3144.
12. Andreu-Vázquez C, Garcia-Ispierto I, Ganau S, Fricke P, López-Gatius F. Effects of twinning on the subsequent reproductive performance and productive lifespan of high-producing dairy cows. *Theriogenology* 2012;78:2061–2070.
13. Spitzer J, Morrison D, Wettemann R, Faulkner L. Reproductive responses and calf birth and weaning weights as affected by body condition at parturition and postpartum weight gain in primiparous beef cows. *J Anim Sci* 1995;73:1251–1257.
14. Arthur P, Archer J, Melville G. Factors influencing dystocia and prediction of dystocia in Angus heifers selected for yearling growth rate. *Aust J Agr Res* 2000;51:147–153.
15. Martin J, Creighton K, Musgrave J et al. Effect of prebreeding body weight or progestin exposure before breeding on beef heifer performance through the second breeding season. *J Anim Sci* 2008;86:451–459.
16. Laster D. Factors affecting pelvic size and dystocia in beef cattle. *J Anim Sci* 1974;38:496–503.
17. Doornbos D, Bellows R, Burfening P, Knapp B. Effects of dam age, prepartum nutrition and duration of labor on productivity and postpartum reproduction in beef females. *J Anim Sci* 1984;59:1–10.
18. Burns W, Koger M, Butts W, Pahnish O, Blackwell R. Genotype by environment interaction in Hereford cattle: II. Birth and weaning traits. *J Anim Sci* 1979;49:403–409.

Chapter 46

Dystocia and Accidents of Gestation

Maarten Drost

College of Veterinary Medicine, University of Florida, Gainesville, Florida, USA

Introduction

The term "dystocia" means difficult birth. The economic impact of calving difficulties is significantly associated with increased morbidity and mortality of newborn calves, as well as the subsequent fertility of the dam. The etiology can be fetal or maternal. Early recognition of the delay in the calving process and intervention can ameliorate the economic impact.

In a study conducted on three Colorado dairy farms, it was reported that 51.2% of first-calf heifers required assistance compared with 29.4% of pluriparous cows.[1] Risks were greater for bull calves than for heifer calves, and for twins than singletons. Images of the various causes, anatomic relationships, and procedures are shown in this chapter and can also be viewed in the *Bovine Reproduction Guide*.[2]

Fetal causes

The most common cause of dystocia is fetopelvic disproportion, when the fetus is too large and/or the maternal pelvis is too small. Other fetal causes are abnormal presentation, position or posture of the fetus, multiple offspring, and occasionally fetal monsters.

Nearly all calves (95%) are delivered in anterior presentation (head first) (Figure 46.1). Retention of the head (Figure 46.2) and/or a leg increases the diameter of the calf. This abnormal posture is frequently the result of a weak fetus or dead fetus that fails to participate in the delivery process by keeping its head and neck extended.

While only 5% of calves are born in posterior presentation (tail first), only about half of these are born spontaneously. The blunt conformation of the hindquarters with the thin legs (Figure 46.3) is not as effective in dilating the birth canal and in eliciting the Ferguson reflex (straining, as the cone-shaped head rests on the forelimbs). An added risk of a posterior presentation is the impingement of the umbilical cord on the pubic brim, cutting off the oxygen supply to the fetus.

Twins or triplets, while not too large individually, can create a problem by presenting extremities of two calves at one time, with or without the presence of a head.

Abnormal offspring syndrome

Conceptus development following transfer of embryos from *in vitro* production (IVP) or somatic cell nucleus transfer (SCNT) in cattle may lead to early embryonic death, fetal death, or overgrown calves (type IV abnormal offspring syndrome). The latter has also been referred to as large offspring syndrome. Abnormal phenotypes resulting from IVP and SCNT embryos are stochastic in occurrence. Therefore IVF recipient cows should be monitored closely near term, and a cesarean section should be anticipated.[3]

Fetal monsters and anomalies

Fetal monsters and anomalies, while rare, are frequently disproportionate. Incomplete twins like a bicephalic calf (Figure 46.4), or conjoined calves, a schistosomus reflexus calf (Figure 46.5), or calves with ankylosed limbs do not conform to the birth canal and will require a partial fetotomy to be delivered per vaginam. In the case of fetal anomalies such as hydrocephaly (Figure 46.6) the body of the calf may be small but the fluid-filled skull is too large to enter the pelvic canal. Diagnosis can be a challenge, and the simple solution is to drain the thin cartilaginous skull (Figure 46.7). Hydrocephalus results from an accumulation of excessive fluid in the ventricular system of the brain.

Maternal causes

Torsion of the uterus

The cause of uterine torsion remains open to speculation. Review of the hospital records of 164 cases from 24 North American veterinary schools[4] showed that there was no effect of season. It was found that large fetuses appear to predispose a cow to uterine torsion. Brown Swiss cows were at significantly higher risk ($P<0.01$), while Hereford, Angus, and Jersey cows were at lower risk compared with Holstein-Friesian cows, the largest breed population. Most cows

Bovine Reproduction, First Edition. Edited by Richard M. Hopper.
© 2015 John Wiley & Sons, Inc. Published 2015 by John Wiley & Sons, Inc.

410 The Cow: Obstetrics and Reproductive Surgery

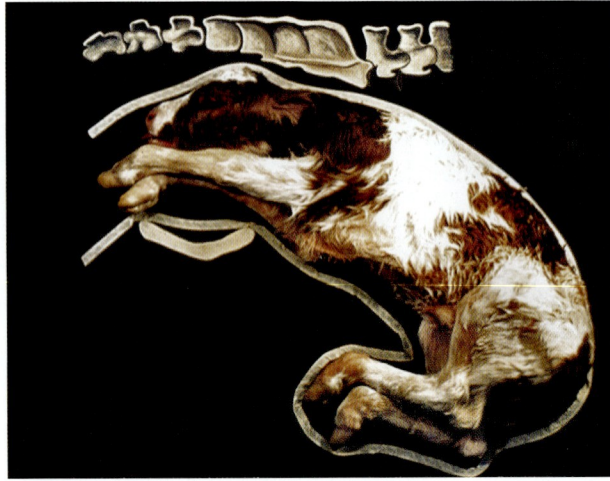

Figure 46.1 The calf is in anterior longitudinal presentation (head first), dorsosacral position (right side up), and has normal posture (both limbs and head and neck extended).

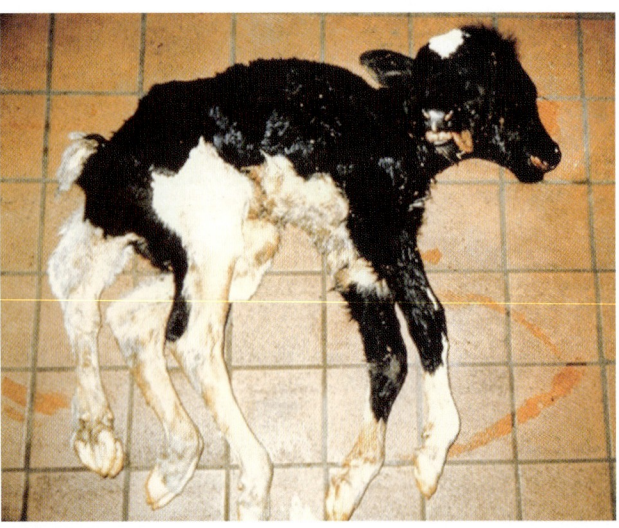

Figure 46.4 Term fetal monster with two asymmetrical heads and a parasitic (extra) pelvic limb.

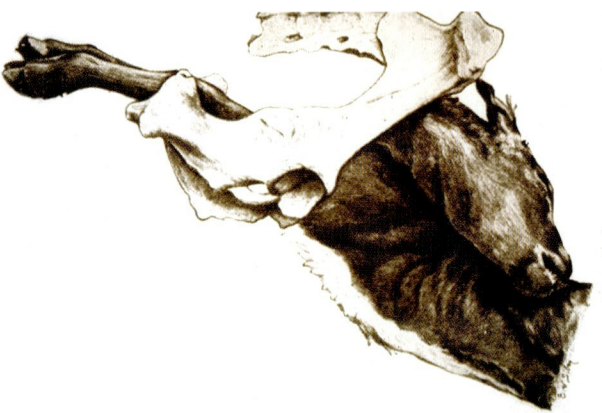

Figure 46.2 The calf is in anterior longitudinal presentation, dorsosacral position, with its head deviated to the left (of its own body).

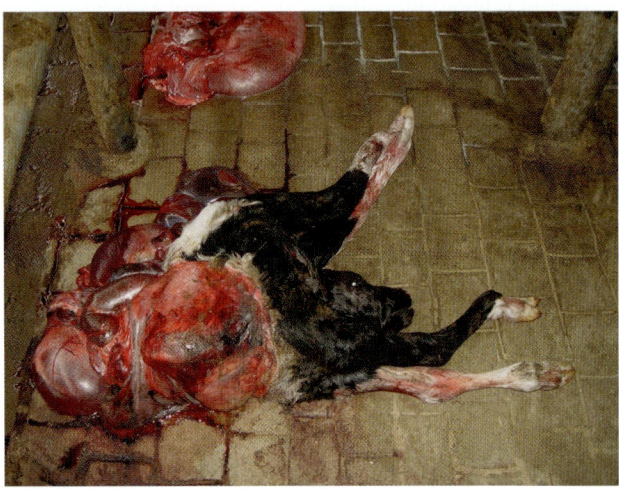

Figure 46.5 Schistosomus reflexus in a newborn Holstein calf. The vertebral column has doubled back on itself and is ankylosed. Recommended method of delivery is by fetotomy.

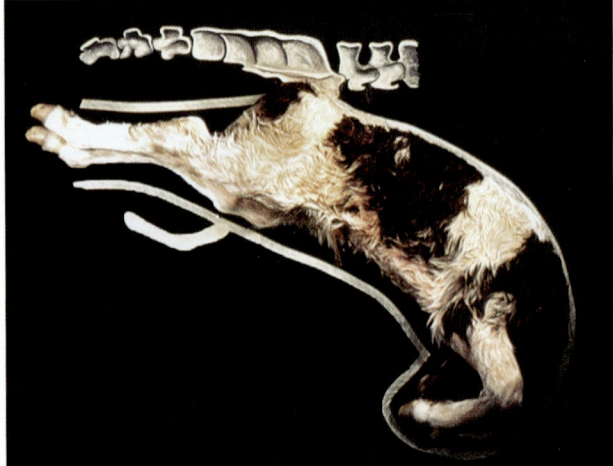

Figure 46.3 This calf is in a posterior longitudinal presentation (backwards), dorsosacral position, and normal posture (both legs extended).

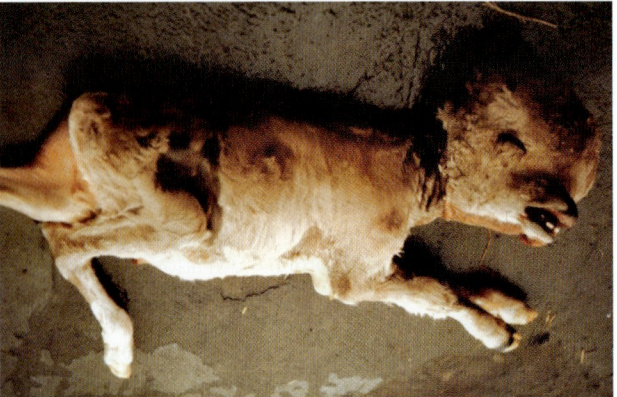

Figure 46.6 Hydrocephalus in a Jersey calf. This is uncommon in calves and can be inherited and congenital. This calf also did not have a tail nor an anus, and had a septal defect of the heart.

Dystocia and Accidents of Gestation 411

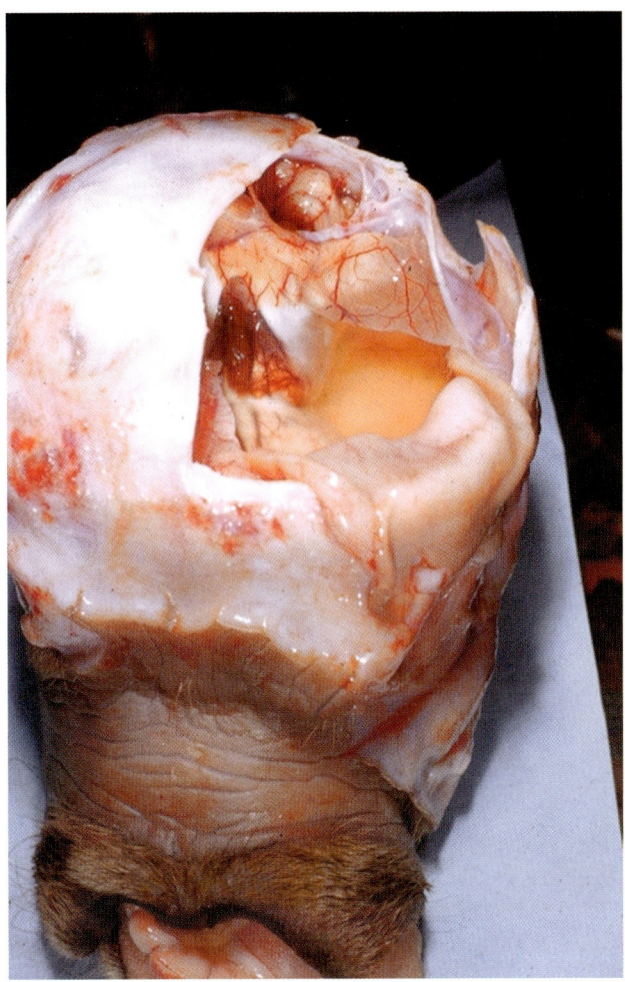

Figure 46.7 Severe hydrocephalus in a Guernsey calf. Note the membrane-like calvarium.

(81%) were at term. Vaginal delivery was possible after manual correction (20%) or rolling of the cow (18%). Cesarean section was performed immediately in 35% of the cases, after failed detorsion attempts in 7%, and due to failure of the cervix to dilate following successful correction of the torsion in 20%. Calf birthweight exceeded breed means in 89% of the cases, and a significantly greater proportion (63%) was male. Fetal survival rate was 24% (14% of dead fetuses were emphysematous), and the cow survival rate was 78% (10 were euthanized).

Predisposition

The ventral attachment of the broad ligament is along the lesser curvature of the uterus, leaving the greater curvature free, and predisposing the cow to torsion of the uterus during the third trimester. In *Bos indicus* cattle, the ventral attachment changes from ventral at the body to dorsal at the tip of the horn. As cows get up on their hindlegs first, the (gravid) uterus is temporarily suspended. The broad ligament is looser and longer in pluriparous cows. The abdomen is capacious, especially when the rumen is relatively empty. Strong fetal movements and poor maternal muscle tone further contribute to torsion of the uterus.

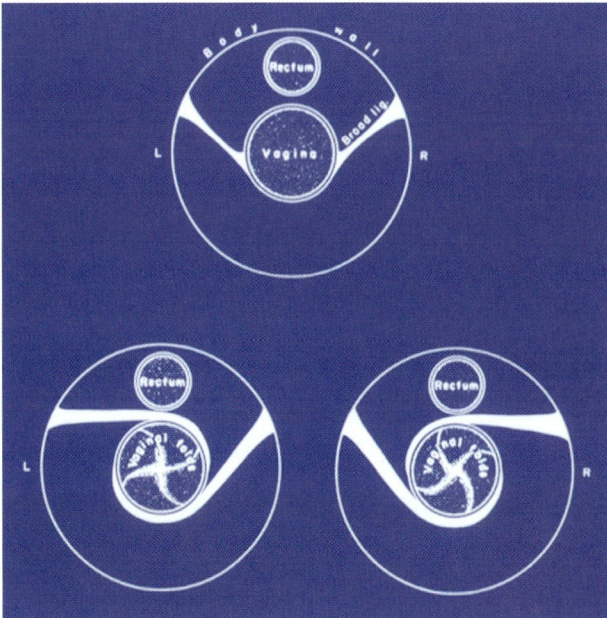

Figure 46.8 The relative position of the broad ligaments is diagrammed for a 180° right (clockwise) uterine torsion (lower left) and a 180° left (counterclockwise) uterine torsion (lower right).

Clinical signs

There is evidence of abdominal pain and discomfort due to stretching of the broad ligament. Other signs, such as anorexia, rumen stasis, constipation, and increased pulse and respiration, are usually present.

Diagnosis

The diagnosis is based on a history of advanced pregnancy. On transrectal palpation the orientation of the broad ligaments is distinctly altered; depending on whether the torsion is to the left or to the right, the respective broad ligament is pulled tightly across the uterus (Figure 46.8). Spiral folds can be palpated per vaginam. Most torsions are to the left (counterclockwise); in general, the uterus rolls toward and over the nongravid horn (approximately 60% of all pregnancies in the cow are in the right horn). Uterine torsion occurs anterior to the cervix (no vaginal involvement) in 34% of cases. Torsions of 45–90° are uncommon; 20% are 90–180°, 57% are 180–270°, and 22% are 270–360°.[4] Depending on the degree of torsion, the fetus may be in dorso-pubic position. With severe torsion, circulatory embarrassment occurs.

Treatment

Correction depends on the degree of the torsion. With rotations of 90° or less, the fetus can frequently be manually rocked into a normal dorsosacral position. Greater rotations can be corrected by rolling the cow around the fetus, which is held in place by a plank in the flank (Figure 46.9).[5] Briefly, the cow is cast with ropes to lie on the side of the direction of the torsion. A long plank is placed in the paralumbar fossa and pressure is applied by one person standing on the plank above the abdomen of the cow. Next, the front legs of the

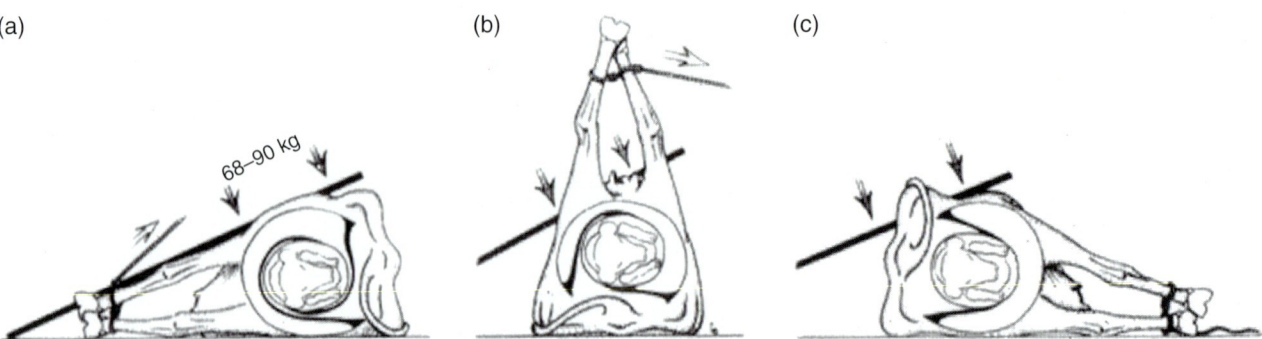

Figure 46.9 "Plank in the flank" or Schaffer method of correction of torsion of the uterus by rolling the cow around the fetus, which is kept in place by applying pressure to her abdomen by a person standing on a plank. Note that the head of the fetus does not change position. The cow is rolled around the fetus to correct the torsion.

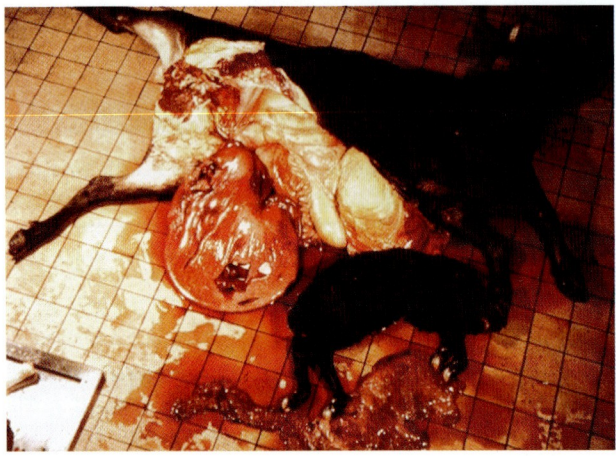

Figure 46.10 Uterine torsion at term. Severe vascular compromise rendered the uterine wall congested and fragile, and the fetus dropped through the uterine wall.

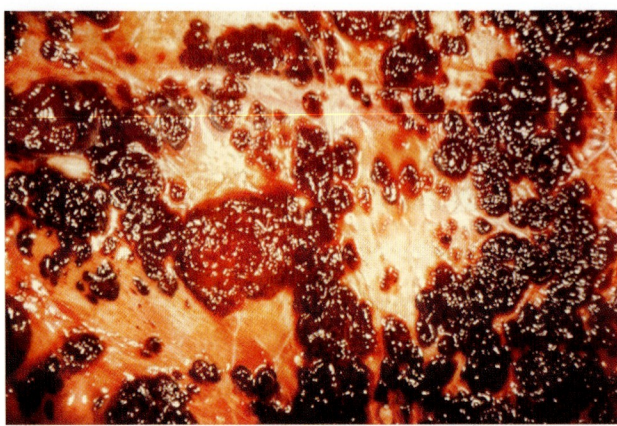

Figure 46.11 Multiple random islets of cotyledonary tissue surround a normal cotyledon. Adventitious placentation can lead to placental dysfunction and result in hydrallantois.

cow are tied together, as are the hindlegs, and they are pulled up and over the body of the recumbent cow. In intractable cases, a cesarean section must be done to deliver the fetus, suture the uterus, and untwist the uterus. Severe vascular compromise will render the uterine wall edematous, congested and fragile, hence difficult to suture.

Prognosis

The prognosis depends on the degree of severity and largely on the extent of vascular compromise (Figure 46.10); the latter may render the uterus friable, predisposing it to rupture.

Hydrops allantois (hydrallantois)

Hydrops allantois is seen sporadically in dairy as well as beef cattle. Excessive accumulation of allantoic fluid may occur progressively after mid gestation. As much as a 10-fold increase in allantoic fluid volume, up to 200 L, has been reported;[6] normal volume of allantoic fluid is 8–15 L. The prevalence of hydrops allantois is significantly higher in IVP pregnancies (1 case per 200 pregnancies) than in normal pregnancies (1 case per 7500 pregnancies).[6]

Placental dysfunction is the result of adventitious placentation characterized by a reduced number of placentomes (normal 75–120) and the development of a more primitive villous placentation (Figure 46.11). Nutritional deficiencies have also been reported to cause this condition, with a prevalence of up to 30% in Criollo cattle owned by smallholders in Zacatecas, Mexico, and raised on poor soil during the long severe dry season.[7]

Clinical signs

Affected cows show bilateral distension of the abdomen (Figure 46.12). They are distressed, anorectic, and have no rumen activity due to compression. Dehydration and constipation follow and eventually the otherwise thin cows become recumbent. During transrectal examination, the uterine wall feels too tight to distinguish fetal parts or to ballot the fetus (Figure 46.13).

Treatment

Salvage is generally recommended because of the underlying adventitious placentation. Parturition can be induced with corticosteroids and prostaglandins provided the cow is within 2 months of term. Oral fluid therapy is essential upon

Dystocia and Accidents of Gestation 413

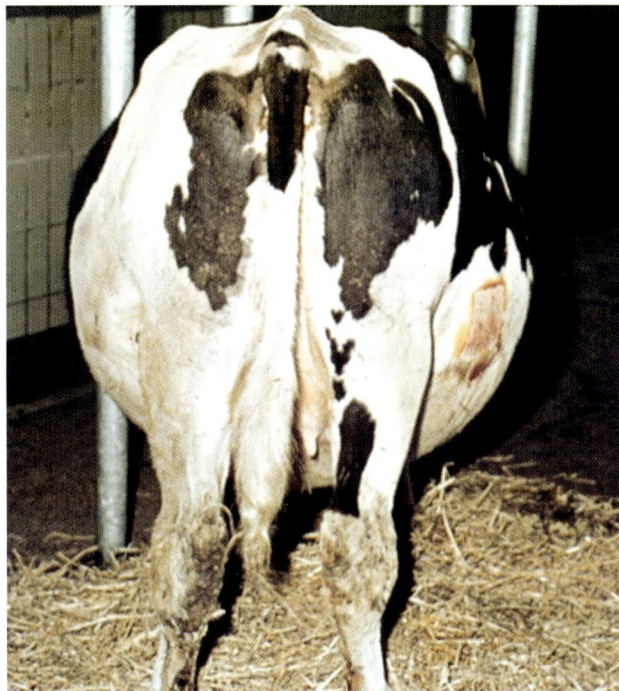

Figure 46.12 Bilateral distension of the abdomen due to hydrops allantois.

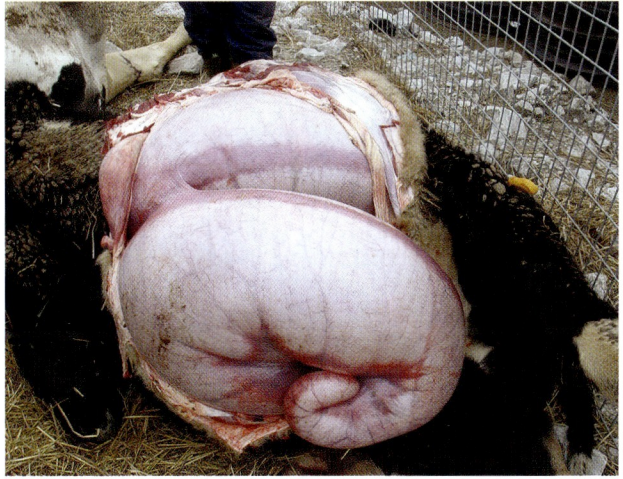

Figure 46.13 This cow was initially misdiagnosed as suffering from bloat, despite the fact that her abdomen was bilaterally distended. It was not possible to ballot a fetus through the tightly distended uterine wall.

the sudden loss of large volumes of body fluids at the time of delivery. The caruncles and adventitious caruncular tissue will be permanently deformed and lead to recurrence of hydrallantois during a subsequent pregnancy. When due to nutritional deficiency, improved nutrition will help if the condition is diagnosed early.

Hydrops amnii (hydramnios)

Hydramnios is a rare condition. The accumulation of excessive amniotic fluid is gradual and can be the result of fetal anomalies such as impaired deglutition or renal dysgenesis or agenesis. It may affect only one of twin fetuses. Hybrids produced

Table 46.1 Differential diagnosis of bovine hydrallantois versus hydramnios.

Characteristic	Hydrallantois	Hydramnios
Prevalence	85–95%	5–15%
Rate of development	Rapid within 1 month	Slow over several months
Shape of abdomen	Round and tense	Piriform, not tense
Transrectal findings of placentomes and fetus	Cannot be palpated (tight uterus)	Can be palpated
Gross characteristics of fluid	Watery, clear, amber-colored transudate	Viscid, may contain meconium
Fetus	Small, normal	Malformations present
Placenta	Adventitious	Normal
Refilling after trocharization	Rapid	Does not occur
Occurrence of complications	Common	Uncommon
Outcome	Abortion or maternal death common	Parturition at about full term

Source: adapted with permission from Roberts S (ed.) Veterinary Obstetrics and Genital Diseases (Theriogenology). Woodstock, VT: published by the author, 1986; and Jennings D (ed.) Buffalo History and Husbandry. Freeman, SD: Pine Hill Press, 1978, p. 94.

by the mating of an American bison bull with a domestic cow or heifer have an increased incidence of hydrops amnii.[8]

Because of the gradual nature of the increase in amniotic fluid, the cow does not show any overt clinical signs, except when viewed from the rear she shows a pear-shaped abdomen. The prognosis for the future breeding of the dam, preferably to a different sire, is good. The fetus is invariably defective and nonviable.

Table 46.1 shows the differential diagnosis of bovine hydrallantois and hydramnios.

Vaginal prolapse

While prolapse of the vagina is not a direct cause of dystocia, the sequelae can be depending on corrective procedures. The primary predisposition to cervico-vaginal prolapse in cattle is elevated estrogen concentrations during the late third trimester. Pluriparity, *Bos indicus* breeding, and obesity are contributing factors. The pathogenesis of the prolapse is progressive, starting with the exposure of the vaginal mucosa when the cow lies down. The mass disappears when the cow gets up. The exposed vaginal mucosa dries out and becomes inflamed. The prolapsed tissues become edematous, leading to circulatory impairment and more swelling, and the cervix may become exposed with accompanying increase in straining.

Diagnosis

The symptoms are obvious by the appearance of an angry-looking soiled mass protruding from the vulva (Figure 46.14), and the diagnosis is frequently made by the owner. However, there are several conditions that can be mistaken

414 The Cow: Obstetrics and Reproductive Surgery

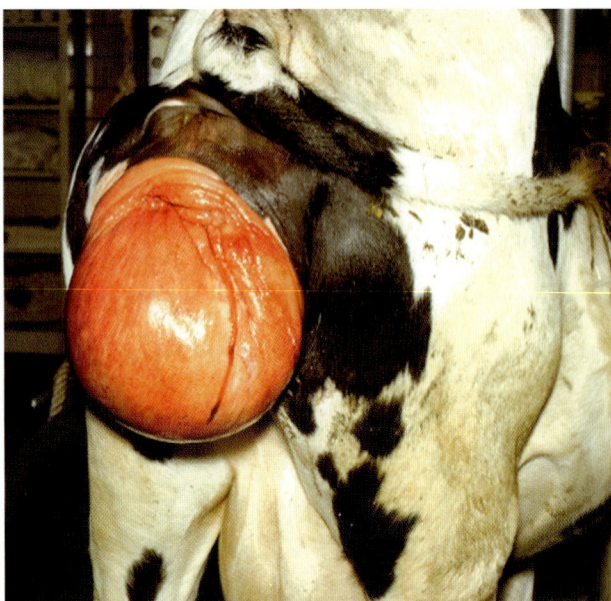

Figure 46.14 This prolapsed mass was thoroughly cleaned up for the picture. Manure and bedding typically cling to the sticky mucous membrane of the exposed vagina. Trauma is common.

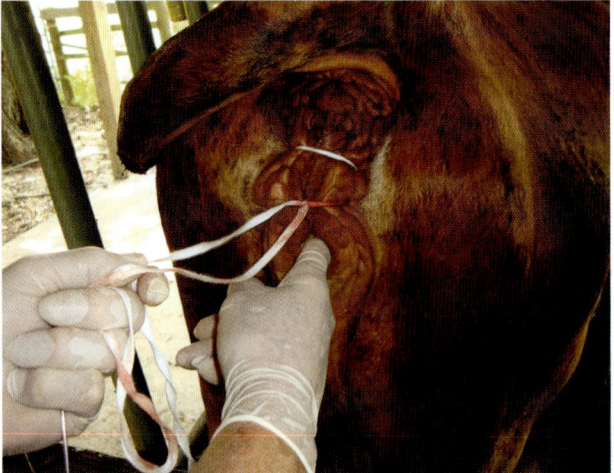

Figure 46.15 Insertion of a finger to test the patency of the vestibule for urination prior to tying the final knot.

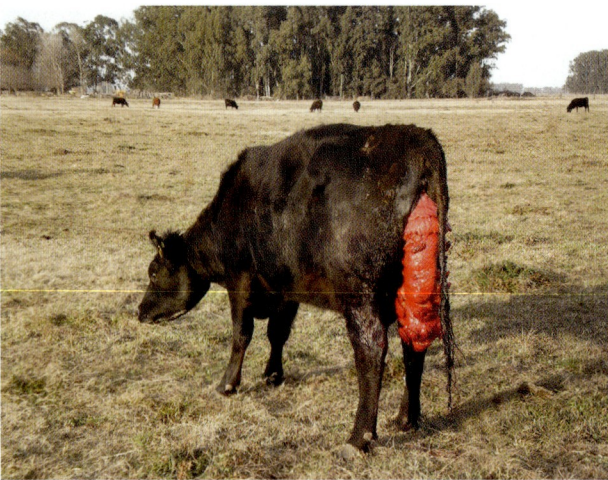

Figure 46.16 Total, fresh, prolapsed uterus in an Angus cow. Rings of the fully dilated cervix are exposed.

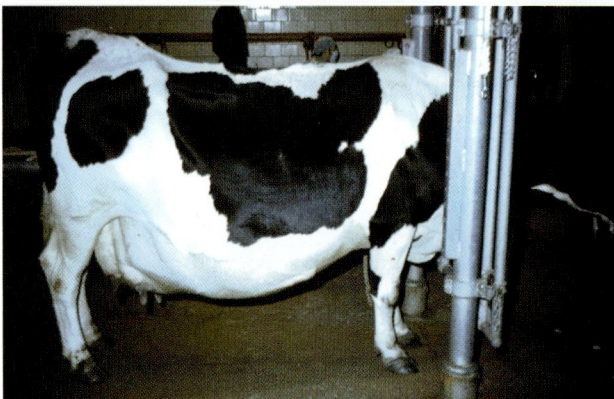

Figure 46.17 This cow suffered a rupture of the prepubic tendon due to the excessive weight of term quadruplets (14–16 kg each). There was no abdominal press, hence a cesarean section was performed and four live calves were delivered.

for a prolapsed vagina, including fetal membranes filled with bloody fluids, hematoma of the vagina, cystic vestibular glands (Bartholin glands), and tumors.

Treatment

Depending on the severity, correction varies from simply elevating the hindquarters of the cow with a platform in a tie-stall to the placement of retention sutures (Figure 46.15) or prolapse pins, which are easily removed when the animal shows early signs of impending parturition.

The prognosis depends on the severity. There is no correlation with prolapse of the uterus after parturition. Vaginal prolapse is likely to recur obsting the next pregnancy, whereas uterine prolapse (Figure 46.16) usually does not recur.

Note that for nonpregnant animals a deep pursestring suture, such as a Buhner pattern, is recommended and placed under epidural anesthesia. The animal should be culled.

Prepubic tendon rupture

The prepubic tendon is the tendon of the rectus abdominis muscle that attaches to the cranial border of the pubis. In cattle, a subpubic tendon provides added strength to the prepubic tendon. Horses do not have the added support of a subpubic tendon, hence rupture of the prepubic tendon, while still rare, is more common in mares than in cows.[4] Rupture of the prepubic tendon, or prepubic desmorrhexis, may be caused by trauma, multiple fetuses (Figure 46.17), hydrops allantois, or fetal giants in prolonged gestation.

Prolonged gestation

On occasion, a cow calculated to calve on a particular date fails to show signs of impending parturition and go into labor. Generally, improper record keeping is at fault. A cow may have returned to estrus and was reinseminated or bred by a bull.

Prolonged gestation is a syndrome in cattle that is characterized by failure of the cow to calve at the normal time. For different breeds the mean length of gestation ranges from 278 to 290 days; for individual cows it ranges from 270 to 300 days. True prolonged gestation far exceeds the upper limits because there is interference with the mechanism of the initiation of labor. The syndrome occurs as an inherited manifestation when the fetus is homozygous recessive for an autosomal gene. Heterozygous cows and bulls are of the normal phenotype, and when mated have a one-in-four chance of producing a homozygous recessive fetus that fails to trigger parturition.

In the Holstein-Friesian breed, for which prolonged gestation has been well described, the calf continues to grow *in utero* for 1–2 months past term, likely until it outgrows its placental blood supply, at which times it dies. The Holstein calves show no obvious deformities on delivery by cesarean section, but do show definite signs of postmaturity. They have a large skeletal framework and weigh 50–100% more than a normal term calf. They have long hair and overgrown hooves. The muscles of the legs are poorly developed. The umbilical cord is large and friable. Careful studies further revealed that these calves have small nonfunctional adrenal glands.

A second type of calf associated with prolonged gestation has been described in the Guernsey breed. In contrast with the Holstein, it is small, mature, or immature in appearance, with varying degrees of facial, cranial, and central nervous system anomalies. Body development resembles that of a 7-month fetus both on gross examination and by radiographic evidence. These calves appear to vegetate *in utero* for months past normal term; cases in excess of 500 days have been reported. Postmortem examination of these calves failed to demonstrate the presence of the anterior pituitary gland.

In humans, fetal anencephaly has long been recognized as being associated with abnormally prolonged gestation. Mummification of the fetus may also lead to an apparently long gestation.

Stress

Optimally, cows should be moved to the calving area in cohorts or groups to minimize stress, 2–3 weeks prior to term. Cows need time to establish social rank. Cows with early signs of impending labor should not be segregated too soon. Separation from herdmates leads to stress, which will temporarily delay labor and increase the incidence of stillbirth. It is best to wait until the amniotic membranes and/or a leg show or after 30 min of nonproductive labor, at which time a vaginal examination is indicated.

Prevention of dystocia

Heifers bred too young or at too small a body weight are more likely to experience dystocia. Pelvimetry can be a helpful selection practice. Selection of bulls with a high index of calving ease is preferable. Excess body weight leads to excess perivaginal fat constricting the birth canal. The body condition score of dairy cows (on a scale of 1–5) should not equal 4 or more, and that of beef cows (on a scale of 1–10) should not equal 8 or more. At the stage of impending parturition cows should be observed at least twice per day, and heifers four times per day.

References

1. Lombard J, Garry F, Tomlinson F, Garber L. Impacts of dystocia on health and survival of dairy calves. *J Dairy Sci* 2007;90:1751–1760.
2. Drost M. *The Visual Guide to Bovine Reproduction.* The Drost Project, http://drostproject.org.
3. Farin P, Piedrahita J, Farin E. Errors in development of fetuses and placentas from in vitro-produced bovine embryos. *Theriogenology* 2006;65:178–191.
4. Frazer G, Perkins N, Constable P. Bovine uterine torsion: 164 hospital referral cases. *Theriogenology* 1966;46:739–758.
5. Roberts S (ed.) *Veterinary Obstetrics and Genital Diseases (Theriogenology).* Woodstock, VT: published by the author, 1986.
6. Sloss V, Dufty J (eds) *Handbook of Bovine Obstetrics.* Baltimore, MD: Williams & Wilkins, 1980.
7. Flores F, Valencia J, Fernandez Baca S, Rosiles R, Ruiz H. Hidropesia de las membranas fetales de los bovinos. *Proc Congr Mundial de Buiatrica, Mexico DF, Mexico* 1977;185:1510–1513.
8. Jennings D (ed.) *Buffalo History and Husbandry.* Freeman, SD: Pine Hill Press, 1978, p. 94.

Chapter 47

Obstetrics: Mutation, Forced Extraction, Fetotomy

Kevin Walters

*Department of Pathobiology and Population Medicine, College of Veterinary Medicine,
Mississippi State University, Starkville, Mississippi, USA*

Introduction

The provision of obstetric services continues to be an important aspect of the large animal practice. Previous chapters have covered prevention and described and defined dystocia. A subsequent chapter will cover the surgical management of dystocia (cesarean section). The purpose of this chapter is to describe the nonsurgical obstetric management of a dystocia as well as the decision parameters utilized to abandon one technique and proceed to another. This begins with a basic list of equipment, restraint options, and then progresses through the management of progressively more difficult obstetric cases.

Obstetric terminology

Terminology that describes both normal and abnormal dystocia as well as techniques employed is valuable for effective case management discussion among both veterinarians and veterinary students. The following are accepted definitions of presentation, position, and posture, which serve to describe fetal orientation with respect to the dam.

- *Presentation*: describes fetal orientation by the relationship of the spinal axis of the fetus to that of the dam: thus either longitudinal or transverse (sideways), and if longitudinal either cranial or caudal (cranial–longitudinal is normal), and if transverse either dorsal or ventral. A transverse presentation is very rare in cattle.
- *Position*: further describes fetal orientation based on the relationship of the dorsum of the fetus to the quadrants of the maternal pelvis, these being sacrum (dorsosacral is normal), right and left ilium, and pubis.[1] Note that while the dorsosacral position is considered to be the normal position for a fetus, slight rotation to a position midway between dorsosacral and right or left dorso-ilial will often facilitate delivery. In fact this is what typically occurs naturally as a cow shifts position from lateral to sternal recumbency during normal parturition.
- *Posture*: describes the relationship of extremities with respect to the body of the fetus. For example, for a forelimb this could be fetlock, carpal, or shoulder flexion.

Definitions for various obstetric techniques are as follow.

- *Mutation*: defined as those procedures by which a fetus is returned to its normal orientation with respect to presentation, position, and posture. Procedures utilized include repulsion (retropulsion), rotation, version, and extension of extremities.[1,2]
- *Forced extraction*: the manual or mechanically assisted removal of the calf through traction.
- *Fetotomy*: the reduction and removal of the fetus by division and removal of extremities and sections. In a percutaneous fetotomy, the dissection is made through the skin and this is the classic fetotomy procedure as described by Bierschwal and DeBois.[3] In a subcutaneous fetotomy, a "freeing" incision is made through the skin to allow a limb's bulk to be removed. This is typically a partial fetotomy in which only one or two extremities are removed.
- *Cesarean section*: the surgical removal of a fetus via laparohysterotomy.
- *Repulsion (retropulsion)*: the pushing of the fetus back out the pelvis to facilitate other corrective steps.
- *Rotation*: the turning of the fetus on its long axis.
- *Version*: the rotation of the fetus on its transverse axis.
- *Extension of extremities*: the correction of flexural deformities of the extremities.

Basic equipment

This is an area in which the opinions of obstetricians might vary. There are instruments that one individual might consider unnecessary that another would not approach an obstetric case without. Therefore the approach taken will be to first list obstetric tools that most would deem absolutely necessary and then add others and their potential uses.

Basic obstetric kit needs include at least two obstetric chains, two handles, a fetal extractor, a Krey hook, and basic surgery pack. Obstetric chains can be obtained in lengths of 53, 76, 114, and 152 cm. Chains are preferred over straps made for obstetrics for at least two reasons. The chains are more easily cleaned and despite the appearance are less likely to cause damage to the fetus or dam. With respect to handle selection, the old Muir obstetric handles are time tested but with the caveat that the heavier models are best. The type or model of fetal extractor is usually based on operator preference and its use is not controversial from my perspective. The Krey hook, which is routinely used in fetotomies, is listed among the basics as it is often useful in the vaginal delivery of a fetal abnormality or dead calf in which only a partial fetotomy is performed. A basic surgery pack is needed for such procedures as an episiotomy, a single appendage removal (partial fetotomy), or cesarean section (Figure 47.1). Additional obstetric tools that are useful include the calf-saver snare, which is easier to place around the head than using chains, a de-torsion bar, and a double blunt eye hook (Figure 47.2). Although limbs as well as the head and neck can be removed without a fetotome, to perform the classic fetotomy the following tools are necessary: a fetotome (model selected based on operator preference), saw wire, a wire guide, an "embryotomy" knife, and if not already in the obstetric kit a Krey hook (Figure 47.3).

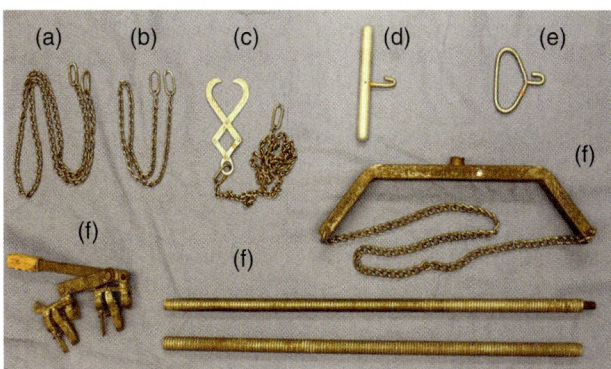

Figure 47.1 Basic obstetric equipment: (a) 152-cm obstetric chain; (b) 76-cm obstetric chain; (c) Krey hook with attached chain; (d) T-bar obstetric chain handle; (e) Moore's obstetric handle; (f) components of disassembled fetal extractor.

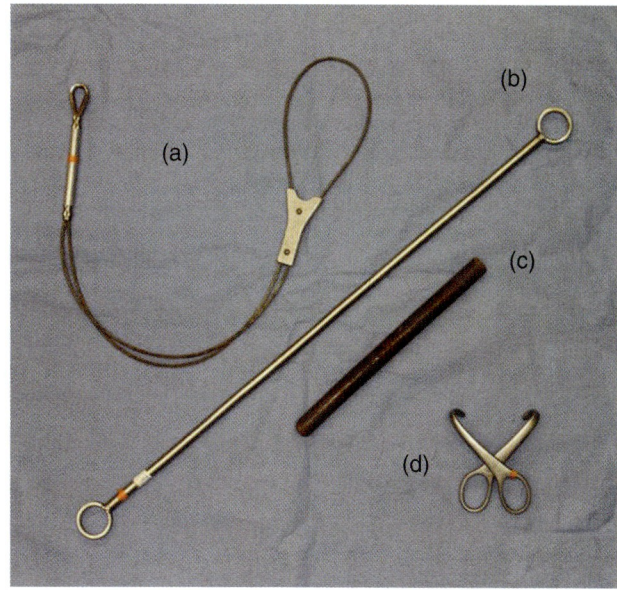

Figure 47.2 Additional obstetric equipment: (a) fetal head snare; (b) de-torsion bar or rod; (c) wooden dowel for applying torque to de-torsion bar; (d) double blunt eye hooks or Vienna scissor eye hooks.

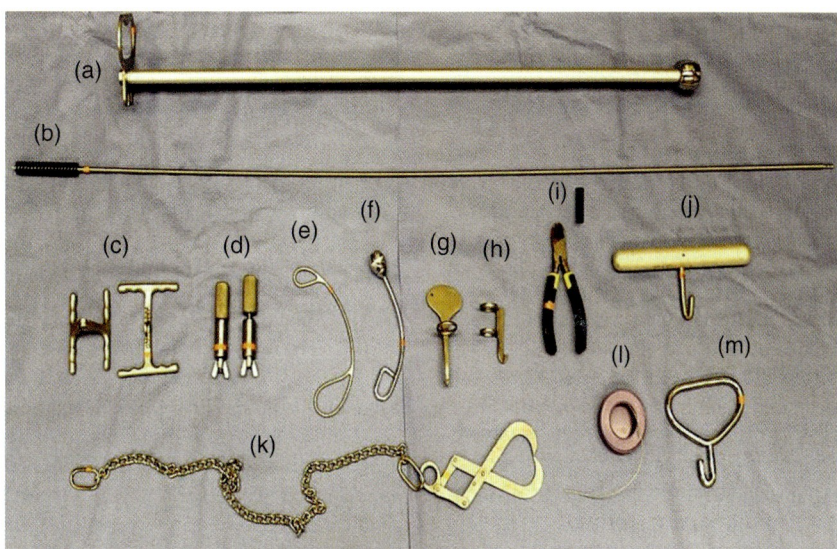

Figure 47.3 Fetotomy equipment: (a) Utrecht fetotome; (b) threader or insertion coil and wire brush; (c) Lyss wire saw handles; (d) wire saw handles; (e) Shriever wire saw introducer or passer; (f) Hauptner wire saw introducer or passer; (g) Linde's fetotomy palm knife; (h) Geunther's fetotomy finger knife; (i) side cutters to sever wire saw; (j) T-bar obstetric handle; (k) Krey hook with obstetric chain attached; (l) wire saw; (m) Moore's obstetric handle.

Case management

Restraint

The restraint approach utilized must be safe for obstetrician and patient. Both facilities available and patient temperament dictate the approach taken. Haul-in facilities are almost always superior to those found on farms and indeed if a significant percentage of a practice is devoted to bovine obstetric work, suitable facilities are a good investment. Although on-farm dystocia cases are often recumbent, most obstetric procedures and specifically mutation are easier on the well-restrained standing cow. Often a cow that has previously refused to stand will do so after the administration of an epidural, presumably due to the abatement from the pain associated with labor.[1]

A cattle chute is adequate for examination, preparation, and the administration of an epidural, but you must have the ability to open a side or otherwise get the cow out in case she goes down during the obstetric procedure. In virtually all cases, with the exception of a purpose-designed dystocia chute or stanchion, you will be better off with the cow haltered and tied in the corner of a stall/pen/lot.

History

As the large animal practitioner must often utilize to some extent a form of triage in determining which client's request or patient must be attended to first, the obstetric case begins with the perfunctory question of duration of labor. This may be of marginal value without an understanding on the part of the client of the stages of labor. For example, the time period that an owner relates is almost always from the time they first noticed the cow, without any respect to the stage of parturition. The stages of labor are as follows.

- Stage 1: early labor, with a typical duration of 4–12 hours. Its conclusion is marked by rupture of the allantois.
- Stage 2: delivery of the calf. This is also referred to as "true labor" and typically lasts 30 min to 4 hours, but the calf can often live 8–10 hours.[4]
- Stage 3: passage of the fetal membranes

Thus the appropriate question to ask the owner/attendant is whether or not fetal membranes or an extremity is visible, followed by queries to ascertain whether allantoic rupture has occurred ("Have the waters broke?"). Simply determining this provides the answer with respect to continued observation or immediate attention. Additionally, the knowledgeable stockman might be able to describe the presentation based on observation or their preliminary examination. Just as the knowledge that labor has progressed to stage 2 dictates prompt attention, so does a posterior presentation.

Other questions relate to parity, past calving history, and sire. The young veterinarian must realize that extensive questioning is often off-putting to the cattleman and one may be wise to simply expedite matters and obtain necessary information during the course of the management. The relative value of the cow versus the calf may also be important and will often be a factor in the management approach.

Examination

The examination begins as the cow is approached and as the animal is restrained. The temperament and physical condition are assessed. Prior to the obstetric examination the tail should be tied to her neck (or held by an assistant) and the perineum cleaned with soap and water. The obstetric examination can be performed before or after the administration of an epidural, but at any rate should be done quickly and with the purpose of guiding the direction of obstetric management. Thus fetal viability, the presentation, position, posture, and degree of fetus–dam disproportion are evaluated.

Fetal viability and even how long a calf may have been dead are important determinants with respect to management. If the presentation and position are normal, but an extremity is malpostured, the fetus can be repulsed and the posture corrected, followed by traction for example. A very large and viable calf in a heifer or relatively small cow directs one to cesarean section, as in turn a large dead fetus may dictate a fetotomy.

The presentation is easily assessed if the obstetrician can identify the head or conversely the tail/tailhead. When a determination must be made based on palpation of a limb, one should simply remember that the first two joints of the forelimb, the metacarpal joint (fetlock) and the carpus (knee) will both flex in the same direction, while the fetlock and hock flex in opposite directions.

The examination must therefore be performed in an efficient thoughtful manner with the goal of directing the first or next obstetric procedure and avoidance of the creation of a potential complication. It is intertwined with management, and reassessment of the case may result following an unexpected finding or the inability to complete a prescribed plan of action.

Obstetric procedures and decision-making

As the reader would hopefully surmise from the section on examination, obstetric management is fluid, but should follow a reasonable course that results in a successful outcome, which in turn is dependent on the initial findings.

A common dystocia presentation is a fetus in anterior presentation with normal posture (for this and other dystocia images, see Chapter 46). If traction results in a severe and immediate crossing of the forelimbs as the shoulders engage the maternal pelvis, the calf is likely too large to be delivered vaginally. Thus the next step would be to move to a cesarean section or fetotomy. In the case that the fetus is of a size in relation to the dam that vaginal delivery should be possible, examination to determine the cause of dystocia should follow. Was there adequate cervical dilation? Is the heifer or cow exhausted or suffering uterine atony? Are there concurrent issues such as hypocalcemia? Is there inadequate dilation of the vulva? Depending on the answer, your next step could be to simply apply traction, but these issues should all be considered.

An anterior presentation in which the fetus is rotated in position based on head or limb orientation, findings from your vaginal examination, or findings of a rectal examination should lead to a diagnosis or suspicion of torsion. Diagnosis and obstetric management of uterine torsion is discussed in Chapter 46 and surgical management is discussed in Chapter 48.

For an anterior presentation with lateral deviation of the head, the fetus can be repulsed. This is obviously easier when the patient is standing and following epidural analgesia. The head can then be pulled into position (into the pelvis) by grasping the orbits. If the operator lacks the hand strength to do this, the double blunt eye hook can be employed. In the case in which they cannot be reached, first pull at the neck and this will pull the head closer. These attempts may be alternated with repulsion. Alternatively, or in cases when the head cannot be pulled into position, the calf-saver snare or an obstetric chain may be used. Correct application of the calf-saver snare and obstetric chains is important to avoid injury to the calf. They should be placed behind the ears and through the mouth. When utilizing a chain the loop should be placed in a manner so that it will not tighten when traction is employed. Also, traction can be utilized to move the head into position, but should not be used to effect delivery.

For an anterior presentation with one or more forelimbs back, the fetus is again repulsed and the limb(s) brought into position. Depending on the severity of the malpostured limb, correction may be stepwise, first reaching for and pulling on the upper aspect of the limb below the shoulder, then the carpus, and finally the fetlock. Affixing an obstetric chain around the fetlock and providing traction alternating with repulsion of the fetus is often necessary. Often times only an elbow back is hindering delivery and is corrected by selective traction.

During the application of traction, once the thorax has cleared the pelvis, stop briefly and attend to the calf. Move any fetal membranes from over the calf's nose and mouth. Stimulate the calf if necessary by slapping the head or pulling on the tongue. Then before resuming traction, rotate the calf about 30–45°. This will likely prevent a "hip-lock" as can occur when the hips of the calf and dam engage. In the situation when a hip-lock occurs the following tips are useful. If the cow is standing, orient the direction of traction almost straight down. If the cow is recumbent (lateral) at this time, first try the same change in orientation of traction. If a fetal extractor is employed, simply take the far end and direct it toward and between the cow's hindlegs. Alternatively, or if the aforementioned technique does not work, rock the cow on her back, grabbing a hind foot, and roll her over (from right lateral recumbency to left or vice versa). Regardless, continued traction on a calf when there is a hip-lock without changing the direction of traction usually results in injury to calf and/or dam.

A posterior presentation in which the hindlimbs are flexed and back is the classical "breech." Correction requires repulsion of the fetus and in a stepwise fashion, first flexing and retracting the hock, followed by the fetlock, and finally bringing each hind foot into position for delivery. This is extremely difficult in the recumbent animal, so all attempts to restrain the patient standing should be made and obviously, as whenever repulsion is necessary, epidural analgesia is employed. Care during repulsion is also crucial as uterine tearing can occur (etiology and repair is discussed in Chapter 53). Once corrected, traction can be administered; however, with this, or in a normal posterior presentation, care must be taken to keep the calf's tail down during traction and delivery. Two possible deleterious sequelae can follow an upwardly fixed tail. First, due to the tension the sharp ventral coccygeal processes can abrade or lacerate the dam's dorsal vagina. Secondly, and more commonly, the calf's tail will break, resulting in a neurologic deficit. With this injury the calf may still be able to walk, but lacks proprioception and additionally usually lacks bladder tone.

The goal for the management of dystocia is always to avoid injury first to the dam and then the calf. Prolonged obstetric efforts rarely end well so regardless of the mutation techniques employed, a time limit of 10–15 min without progress should lead to reevaluation of technique and specifically in the case of a viable fetus to a cesarean section.

Forced extraction

The term "forced extraction" refers to both manual and mechanically assisted extraction of the fetus. Since it is rare that problems (specifically injury) occur as a result of manual traction, the focus of this section is on traction in excess of that provided manually, thus referring specifically to the use of a fetal extractor. Because a fetal extractor can provide a level of traction that can be deleterious to either or both the dam and fetus, its use has been controversial. From a practical aspect its use is safe when used judicially.

Injuries that can result from forced extraction are significant and their consideration is important. Obturator or calving paralysis, which is more likely due to injury to or around the sciatic nerve rather than the obturator nerve, can occur as a result of prolonged dystocia, but is most commonly a result of excessive traction, as is tearing of the uterus, cervix, or vagina. Metacarpal and metatarsal fractures can result from incorrect placement of obstetric chains, but likewise are more likely with excessive traction.

This said, prevention of injury can virtually always be assured with adequate lubrication and careful application of traction. Methylcellulose-based lubricants can be pumped into the uterus and around the fetus, and then added during extraction. (A note with respect to the selection of commercially available obstetric lubricants: the product J-Lube when exposed to the abdomen of cattle has resulted in death.[5] Thus if your obstetric management plan may potentially lead to a cesarean section and as abdominal contamination during surgery is often unavoidable, its use would not be recommended.) Traction should be applied slowly, carefully, and only in conjunction with the dam's contractions, and should be stopped when no progress results.

As stated previously the model of fetal extractor utilized is based on operator preference. Remembering the old adage that no more traction should be applied than that of two strong men, the "power" requirement of a fetal extractor need not be excessive. Thus many of the time-tested older models are very adequate. A model developed several years ago features the ability to provide alternate traction to each of the two forelimbs. While it seems intuitive that this provides an advantage, in a well-constructed experimental model Becker et al.[6] concluded that simultaneous traction was optimal.

In summary, providing adequate lubrication, correctly applied obstetric chains, rotation of the fetus 30–45° during extraction, and applying traction with the dam's contraction efforts are all key to safe and judicious mechanical extraction.

Fetotomy

Fetotomy by definition is the dissection of the fetus into two or more parts while still within the uterus or vagina. Subcutaneous fetotomy was developed in the eighteenth century in which instruments were used to dismember a fetus while leaving the fetal skin intact to protect the uterus and provide points of traction.[7] This technique has been abandoned for the percutaneous fetotomy technique. The percutaneous fetotomy utilizes a braided wire as a saw to cut through skin, muscles, bone, and connective tissue to dissect or dismember the fetus into small portions for assisted vaginal delivery.[3]

The standard Utrecht technique for a cranial or anterior presentation fetotomy is removal of the head, one forelimb, the other forelimb, the trunk (potentially two cuts), and bisection of the fetal pelvis. This summarizes the complete fetotomy used to dismember an oversized fetus that has resulted in fetopelvic disproportion. This procedure will encompass five to six cuts to complete. A complete fetotomy is not always necessary for delivery of the fetus. Frequently a partial fetotomy, which may include one to two cuts, will provide the relief necessary to deliver the fetus. The standard cuts have been modified to incorporate more anatomical parts but which are still manageable portions easily removed from the genital tract.[8]

Preparation for this procedure will depend on the environment in which the cow is confined and restrained, whether in a hospital, corral, or pasture setting. Deposition of obstetric lubricant or water mixed with lubricant can be infused into the uterus by the obstetrician based on their appraisal of its need to provide lubricating protection to the genital tract at any point during these obstetric procedures.

Anterior presentation

If manipulative efforts to relieve the dystocia have not been successful and fetotomy is pursued, results of the manipulative process will determine the initial cut of the fetotomy. Malposture and malposition will have an impact on the initial cut as well. The fetotome should be threaded with a section of saw wire that is at least five times the length of the fetotome. This length of wire should be sufficient to allow a loop at the head of the fetotome that is large enough to be manipulated over the fetal head, as well as extension beyond the ends of both barrels of the fetotome at the handle to allow fixation of handles to the wire for sawing. This allows tension to be applied on the wire handles in an alternating manner to perform the cut. Tension applied to the wire handles will allow placement of the wire to be confirmed. The obstetrician should evaluate the wire saw as it exits the barrels of the fetotome to ensure that the wire is not crossed just prior to entering each barrel. If so, this would lead to premature failure or breakage of the wire saw. The crossing of the saw wire can be alleviated by rotating the fetotome on its long axis. The appropriate direction is determined by the obstetrician after this evaluation. Initiation of the cutting action of the wire saw is conducted by exerting tension on the wire handles in an alternating manner. The amount of tension is increased. Initially the assistant makes short smooth strokes with the wire handles until the saw wire is fully engaged in the fetal tissue. The assistant can now extend the length of stroke and increase the amount of tension applied to the handles to perform the cut. The assistant or wire saw operator can determine when the cut is complete due to a dramatic decrease in resistance of the wire saw in tissue.

If the initial cut is intended to remove the fetal head, the threaded fetotome should be introduced into the genital tract. The loop of wire should be manipulated by hand to encompass the head. The head of the fetotome should be advanced into the tract so that it lies caudal to the ears. Tension may be placed on the wire after fetotome placement and the obstetrician can feel for placement of the wire around the neck. The standard Utrecht procedure describes placement of the fetotome head caudal to ramus of the mandible and potentially ventral as well. I prefer to advance the fetotome head to a location on the dorsum of the fetus at approximately the withers or craniomedial to both scapulae (Figures 47.4 and 47.5).

Apply tension to the wire saw handles and perform the cut through the fetal neck. The obstetrician can place a hand over the head of the fetotome during this cut to provide protection to the uterus dorsal to the instrument head. The obstetrician can provide traction to the fetal head for removal. It may be necessary to apply the fetal head snare or an obstetric chain if manual traction is unsuccessful at removal of the fetal part. This cut eliminates the neck as a potential impingement if traction is applied to the fetus at this point for delivery. This cut may also provide more medial excursion of the shoulders as they engage the maternal pelvis if traction is attempted at this time.[9]

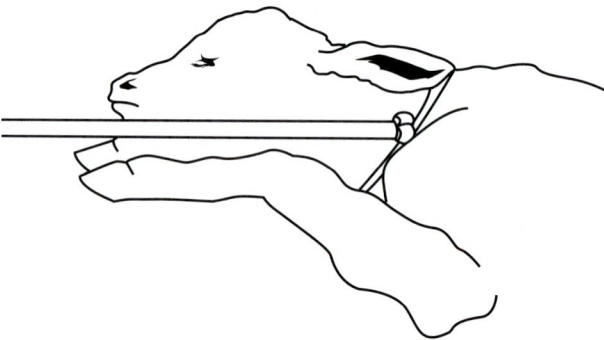

Figure 47.4 The author's recommended placement of fetotome head for removal of the fetal head and neck.

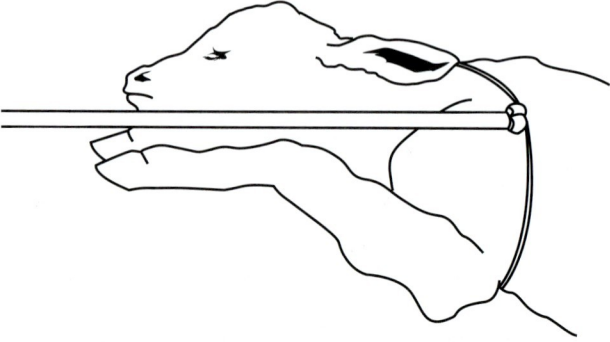

Figure 47.5 An alternate "cut" that incorporates the head and one shoulder allowing removal of both with a single section.

If removal of the fetal head does not provide satisfactory conditions for fetal extraction, the obstetrician can then proceed to removal of one of the thoracic limbs of the fetus. This can be accomplished by utilizing the previously threaded fetotome. A wire loop can be manipulated around the limb chosen for removal as the instrument is introduced into the genital tract. The instrument should be advanced along the lateral aspect of the limb to a point that the head is dorsocaudal to the caudal extent of the scapula.[1] An obstetric chain may have been previously placed on the limb. Be careful not to entangle the chain in the saw wire. If a chain was not already in place, one needs to be placed in order to provide traction during the cut. After instrument placement, tension on the wire will allow the obstetrician to determine if wire placement is correct. The wire should follow a path from the axillary region dorsally and ventrally to the upper limb and enter the instrument head dorsocaudal to the scapula. The obstetric chain can now have tension applied to it, and the chain introduced into the locking plate at the handle of the fetotome. This will maintain the leg in full extension during the sawing action. The obstetrician may place a hand over the head of the fetotome during this cut to provide protection of the uterus. Once the cut is complete, the obstetrician can disengage the chain from the fetotome and apply traction to the fetal limb for removal. The instrument can be removed at this time as well. Palpation of the caudal region of the cut should be performed to ensure no bone fragments were left as a result of the cut that could potentially traumatize the genital tract. If fragments of the scapula are discovered, they can be removed by blunt dissection with the hand to prevent trauma to the genital if traction is attempted at this time.[3]

Removal of the second thoracic limb is conducted in the same manner as the first. The same precautions should be taken to ensure that trauma to the genital tract from sharp bony fragments are minimized. Additional lubrication may need to be deposited at this time due to multiple introductions and removals of equipment from the genital tract.[10]

If an attempt is made at this time to extract the remaining portion of the fetus, evisceration of the thoracic cavity through the thoracic inlet can be conducted by hand. If the fetus is emphysematous, manual penetration through the diaphragm and evisceration of as much of the abdominal contents as possible manually through the thoracic inlet will accomplish a reduction in size of the remaining fetal portion. The Krey hook could be applied to the cranial end of the vertebral column as an appropriate place for tension application. To continue with the fetotomy, the head of the fetotome with loop of wire can be introduced into the genital tract, and the loop manipulated around the cranial end of the thoracic cavity. The loop is manipulated caudally on the trunk of the body to a point caudal to the last rib and cranial to the wings of the ilium in the lumbar region. The fetotome should be advanced into the genital tract so that the head rests in a dorsolateral position to the trunk of the fetus. It is sometimes difficult to manipulate the wire along the trunk of the fetus. The saw wire can be attached to a wire introducer or passer that is introduced into the genital tract. It is passed over the dorsal part of the trunk and then is retrieved by reaching under the ventral aspect of the fetal trunk. The wire introducer/passer is withdrawn from the genital tract. The saw wire is detached from the introducer/passer and threaded into the fetotome using a wire threader. The fetotome head should be positioned in the location previously described for this cut. The obstetrician should apply the Krey hook to the cranial end of the vertebral column. With the fetotome head in place, the obstetric chain attached to the Krey hook can have tension applied and engaged into the locking plate on the fetotome. The cut can be performed. If the thorax is of too large a diameter to pass through the genital tract, it can be reduced in size further by introducing the wire director/passer attached to the wire saw through the thoracic inlet and exiting the caudal opening created by the cut through the abdomen. The wire director/passer is retrieved from the caudal aspect of the trunk and withdrawn from the genital tract in order to thread wire saw into the fetotome. The wire should be placed close to the attachment of the ribs to the thoracic vertebra. Performing the cut at this site will allow collapse of the thoracic cavity, which reduces its size for removal from the genital tract. Remain cognizant of sharp bony edges that can potentially traumatize the genital tract during manipulation. The assistant can provide traction to the chain while the obstetrician navigates the fetal thorax through the genital tract (Figure 47.6).

Positioning of the fetal pelvis for bisection can be accomplished by attaching the Krey hook to the lumbar vertebral column and applying traction to maintain the fetal portion in the opening of the maternal pelvic inlet. Saw wire from a half-threaded fetotome with wire introducer/passer attached can be introduced into the genital tract and passed over the dorsal aspect of the fetal pelvis to about the level of the inguinal region. The obstetrician can then retrieve the introducer by passing their hand between the cranial margin of the maternal pubis and fetal pelvis in order to grasp the introducer between the fetal limbs and withdraw it from the genital tract. The wire introducer can be removed and the wire saw can be threaded into the fetotome. The fetotome can then be introduced into the genital tract while tension is maintained on the wire saw to keep it in place. The head of the fetotome should be positioned so that it rests against the previously cut lumbar region of the fetus. It should be ipsilateral to the tail groove chosen for wire placement. It should also be slightly lateral to the vertebral column, so that this cut will bisect the fetal pelvis, but not in symmetrical pieces (Figure 47.7). With the fetotome in position, the chain attached to the Krey hook should have tension applied to it and then engaged into the locking plate on the handle of the fetotome. The obstetrician can evaluate the placement of the wire saw; it should traverse a path from one barrel of the fetotome dorsally along the fetus, pass

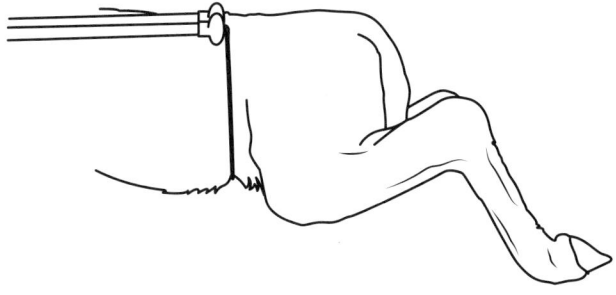

Figure 47.6 Wire placement for mid-abdomen cut.

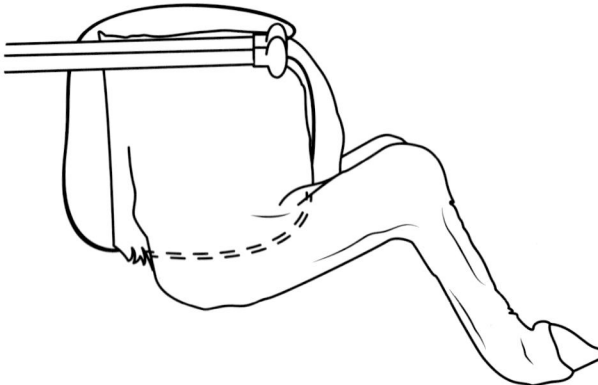

Figure 47.7 Wire placement for bisection of fetal pelvis.

through the groove between the fetal tailhead and ischial tuberosity, continue running ventrally through the inguinal region, and along the ventrum of the fetus to enter into the opposite barrel of the fetotome.[11] If all is in order, the cut can be performed. The fetotome can be withdrawn from the genital tract when the cut is complete. The portion of the fetal pelvis that is grasped by the Krey hook can be guided out from the genital tract. The remaining final portion of the fetal pelvis can then be removed.

Posterior presentation

If the calf was determined to be in posterior presentation during the genital examination, the following describes the procedures necessary to conduct a fetotomy in this presentation. Preparation of instrumentation and self will be the same as that for an anterior presentation. A fetotome threaded with the appropriate length wire saw and attached wire saw handles should be available to initiate this procedure. The prepared fetotome can be introduced into the genital tract of the cow by the obstetrician. A loop of the wire saw protruding from the head of the fetotome can be manipulated over the claw of the selected pelvic limb to encircle it. The fetotome will be advanced similarly as for thoracic limb removal. It should be advanced along the lateral aspect of the limb so that the head of the fetotome comes to be positioned on the dorsal aspect of the pelvis in the area of the greater trochanter.[1] If a chain has not already been placed on the selected fetal limb, it should be placed at this time to provide traction during wire sawing. The chain should have tension applied to it and then engaged into the locking plate on the fetotome handle. The cut can be performed to dissect the selected limb. When the cut is complete, the chain can be disengaged from the locking plate. The fetotome should be withdrawn, and the fetal pelvic limb can be removed from the genital tract.

A chain can be applied to the remaining pelvic limb, and traction applied to the chain to determine if the fetus can be delivered or additional dissection is necessary.[3] If the remaining fetal portion cannot be delivered, the obstetrician must determine whether the opposite pelvic limb requires amputation or dissection through the lumbar vertebra to allow for delivery of the remaining pelvic limb and pelvis. If the opposite pelvic limb requires amputation, it can be removed using the same technique as the first.

If the obstetrician chooses to dissect the fetus at the lumbar vertebra, there are two methods of wire placement. The obstetrician can choose to introduce a fully threaded fetotome into the genital tract and manipulate the wire loop around the claw to encircle it. The wire loop is manipulated proximally around the limb as the fetotome is advanced into the genital tract. The wire loop is manipulated cranially to a point that is just caudal to the last rib. The fetotome is advanced so that the head lies dorsolateral to cranial lumbar vertebra. The second method uses a half-threaded fetotome. The wire saw exiting the barrel from the fetotome should have the wire introducer/passer attached. This instrument is introduced into the genital tract and advanced cranially to the region of the fetal lumbar area. The wire introducer is advanced ventrally on one side of the fetus and deposited. The obstetrician can retrieve the wire introducer by withdrawing the hand to the dorsal aspect of the fetus and then advancing it ventrally on the opposite side of the fetus to the area where the wire introducer was deposited. When the introducer is identified by the obstetrician's hand, it can be grasped and withdrawn from the genital tract so that the wire may be threaded into the fetotome. An obstetric chain should be applied to the pelvic limb if one has not already been placed. Once the fetotome is advanced into the proper position and wire placement is verified, traction must be applied to the obstetric chain, and it should be engaged into the locking plate of the fetotome handle with the limb in full extension. The cut can be performed at this time and when the cut is complete, the obstetric chain should be disengaged from the fetotome. The fetotome can be removed from the genital tract as well as the pelvis and attached limb.

Manual evisceration can be performed after the fetal pelvis has been removed which will reduce bulk in the abdomen of the fetus. A Krey hook and attached obstetric chain can be introduced into the genital tract and applied to the vertebral column so that controlled traction may be applied to retain the fetus in the maternal pelvic inlet while the fetotome is positioned for the next cut. To perform the next cut, a fully threaded fetotome can be introduced into the genital tract, and a loose wire saw loop should have the obstetric chain passed through it. The fetotome should be advanced cranially to the region of the caudal aspect of the scapulae, and the wire loop should be manipulated cranially to encircle the thorax so that it lies just caudal to the scapulae. Traction can be applied to the chain, and it can then be engaged into the locking plate of the fetotome. Once fetotome and wire position are verified, the cut can be performed. The chain can be disengaged from the locking plate to allow removal of the fetotome. If this portion of the thorax is too large in diameter to be removed, it can be cut longitudinally along the rib attachment to the vertebral column as described for the anterior presentation.

After removal of the thorax, the remaining fetal forepart must be dissected. The Krey hook should be introduced to the genital tract and applied to the fetus just lateral to the vertebral column. With moderate traction applied to the fetus, the obstetrician can proceed with the division of the fetal forepart of choice. This can be accomplished by removing either both forelimbs or diagonal longitudinal division of the forepart. I choose to perform diagonal longitudinal division of the forepart. To accomplish this, a half-threaded fetotome with wire introducer attached is utilized. The wire introducer is taken into the genital tract and is manipulated dorsally to the fetus and then ventrally between the neck and one forelimb to be deposited.

The obstetrician should withdraw the hand partially. To retrieve the wire introducer the hand should then be advanced ventral to the fetus to the area of deposition of the wire introducer. Once identified and grasped, the introducer should be withdrawn diagonally so that the wire comes to pass medially to the opposite elbow. The fetotome is threaded and advanced into the genital tract. The fetotome head should be positioned just lateral to the vertebral column on the opposite side from wire passage between the neck and forelimb. The Krey hook should be in position contralateral to the fetotome head, and should not interfere with the wire saw. This cut will divide the forepart diagonally into unequal portions.[3] The fetotome can be removed after the cut is complete. Traction can be applied to the obstetric chain attached to the Krey hook to remove the fetal forepart it is grasping. The remaining fetal forepart should now be removed from the genital tract.

Fetal malpostures and monstrosities

The obstetrician can be presented with a variety of fetal malpostures that result in dystocia. In each case, the obstetrician should use clinical judgment to determine if fetal manipulation and mutation are potentially hazardous to the genital tract of the dam. If this is determined to be true, then the obstetrician can approach the dystocia by performing custom fetotomy cuts to relieve the malpostures and allow for fetal delivery. The approaches to the cuts will be determined by the variety of malpostures. Fetal monsters are another cause of dystocia where fetotomy should be given primary consideration. Because of variation in size, disfigurement, and flexibility of these fetal monsters, the obstetrician will have to determine the plan and approach for removal of such anomalies. The same principle of fetal size reduction should be applied to alleviate dystocia from such causes.[3]

References

1. Norman S, Youngquist RS. Parturition and dystocia. In: Youngquist RS, Threlfall WR (eds) *Current Therapy in Large Animal Theriogenology*, 2nd edn. St Louis, MO: WB Saunders, 2007, pp. 310–335.
2. Roberts S. *Veterinary Obstetrics and Genital Diseases (Theriogenology)*, 3rd edn. Woodstock, VT: published by the author, 1986.
3. Bierschwal CJ, DeBois CHW. *The Technique of Fetotomy in Large Animals*. Bonner Springs, KS: VM Publishing, 1972.
4. Drost M. Calving management: the team approach. In: Risco CA, Melendez Retamal P (eds) *Dairy Production Medicine*. Ames, IA: Wiley-Blackwell, 2011, pp. 19–26.
5. Frazer G. Systemic effects of peritoneal instillation of a polyethylene polymer based obstetrical lubricant in horses. In: *Proceedings of the Annual Conference of the Society for Theriogenology*, 2004, pp. 93–97.
6. Becker A, Tsousis G, Lüpke M et al. Extraction forces in bovine obstetrics: an in vitro study investigating alternate and simultaneous traction modes. *Theriogenology* 2009;73:1044–1050.
7. Vermunt J. Fetotomy in cattle: a review of the basics. In: *Proceedings from the 30th Seminar of the Society of Sheep and Cattle Veterinarians NZVA*, 1999, pp. 111–121.
8. Mortimer R, Ball L, Olson J. A modified method for complete bovine fetotomy. *J Am Vet Med Assoc* 1984;185:524–526.
9. Mortimer R, Toombs R. Abnormal bovine parturition: obstetrics and fetotomy. *Vet Clin North Am Food Anim Pract* 1993;9:323–341.
10. Jackson PGG. Fetotomy. In: Jackson PGG (ed.) *Handbook of Veterinary Obstetrics*. Philadelphia: WB Saunders, 1995, pp. 163–171.
11. Arthur G, Bee D. Dystocia due to fetal oversize: treatment. In: Arthur GH, Noakes DE, Pearson H, Parkinson TJ (eds) *Veterinary Reproduction and Obstetrics*, 7th edn. Philadelphia: WB Saunders, 1996, pp. 239–249.

Chapter 48

Obstetrics: Cesarean Section

Cathleen Mochal-King

Department of Clinical Sciences, College of Veterinary Medicine, Mississippi State University, Starkville, Mississippi, USA

Introduction

A cesarean section is a surgical procedure that is utilized to relieve, or in some cases prevent, dystocia.[1–3] Causes of dystocia are categorized as maternal or fetal and are covered in Chapter 46. Indications for cesarean sections include dystocia cases in which the calves are alive and valuable that cannot be delivered vaginally, and/or dystocia cases in which the calf is dead but delivery by another means jeopardizes the life or future fertility of the cow.

A cesarean section (C-section) is rarely completed under strict asepsis, but this is not due to the fact that it is so often performed in field settings. This is even the case when a C-section is performed in a hospital setting in a surgery room. A C-section is classified as a clean contaminated procedure. This classification describes a surgery where a nonsterile body cavity is entered under a controlled condition without unusual contamination; in this case the uterus is the nonsterile cavity.[4] Furthermore, Mijten *et al*.[4] demonstrated that uterine cultures collected during C-sections often resulted in positive bacterial cultures, with 83% of these samples heavily contaminated. This leads to the conclusion that endogenous bacterial contamination of the surgical site from the hysterotomy is likely inevitable.[4] The fact that bacterial contamination is probable does not excuse inattention to surgical preparation or to cleanliness. Therefore it is important to minimize exogenous bacterial contamination through an efficient and quick performance of the procedure. Additionally, my recommendation is to administer antimicrobials preoperatively to diminish postoperative complications as a result of the endogenous contamination.

Success of a C-section is defined as a live cow with a live calf, and the ability to rebreed the cow. These outcomes are dependent on a number of variables in addition to the surgeon's skill. Many of these variables are beyond the surgeon's control. A realistic approach to this surgery is to evaluate the case, the surgical theater, the cow's size, her condition and attitude, the owner's expectations, and the prognosis. Then tailor your plan to the circumstances and conditions presented. For example, there are several surgical approaches that have been described. It is typical for a surgeon to identify more with one approach than another due to preference and comfort level. However, there are specific case presentations for which a particular approach offers definite advantages. Newman[5] discusses the importance of case selection when determining the surgical approach. To aid a surgeon's treatment plan, Newman categorizes the C-section procedure as elective, emergency non-emphysematous, and/or emergency emphysematous. This categorization of the procedure is relative to the condition of the calf and its effects on prognosis and case management.

The elective category allows the surgeon to choose to some extent the time and the place that the surgery will be performed as well as the ability to make other preparations. Thus this category carries the best prognosis for success, the exception being when the timing of the procedure might be elective but is performed due to an infirmity of the dam (fractured limb, etc.). The emergency obstetric case in which a decision to proceed to C-section is made early in the course of the dystocia without excessive manipulation and particularly while the calf is still alive likewise carries a very good prognosis. For both of these case categories only basic postoperative management is required. The situation in which the decision to proceed to C-section to deliver an emphysematous calf should carry with it the expectation or at least readiness for complications, significant postoperative care, and a guarded prognosis. Sometimes the decision to proceed to C-section occurs after extensive obstetric manipulation or even partial fetotomy efforts. This, in my opinion, carries a less favorable prognosis than the emphysematous category.

Elective C-section

Indications for an elective C-section are a valuable fetus, complicating factors such as a dam with a prepartum vaginal prolapse, a small or malformed pelvis, a recent life-threatening condition, and/or a prolonged gestation.[1–3] In any case, a history that includes breeding date or embryo transfer date should help assess the expected calving date. Calves born less than 2 weeks prematurely should not be unduly at risk for complications or fatalities (see Chapter 44).

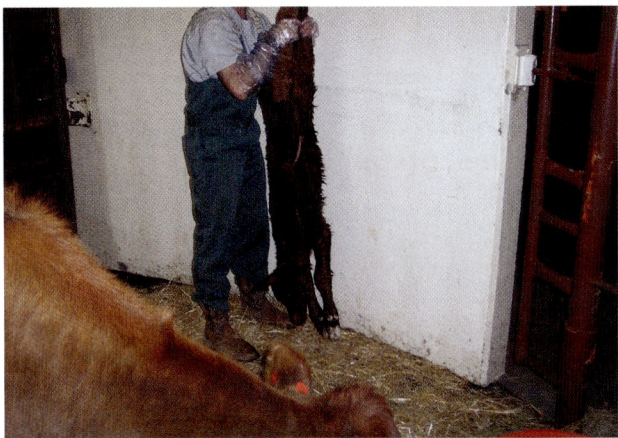

Figure 48.1 Lifting newborn by rear limbs will duplicate aspects of natural vaginal delivery. *Do not shake or swing calf.*

A C-section can be scheduled without respect to the initiation of parturition. Research performed comparing the timing of an elective C-section with respect to the stage of parturition revealed that the presence of full cervical dilatation before performing the C-section allowed better postnatal respiratory and metabolic adaptation in full-term calves.[6] When comparing elective C-section with vaginal delivery it is believed that aspects of vaginal delivery aid calves in transitioning from the womb. Specifically, in a study in which the hematologic profiles of calves delivered by elective C-section were compared with those of calves delivered by unassisted vaginal delivery, C-section calves were uniformly more hypoxic and had a greater likelihood of experiencing respiratory distress.[7] Because of the experimental design, all the deliveries (both vaginal and C-section) were uncomplicated. The hypoxia in the C-section calves was not due to, or associated with, respiratory or metabolic acidosis. The authors hypothesized that the unassisted vaginal delivery calves benefited from both the compression of the fetal thorax that occurs during transit through the pelvis and the temporary suspension by the hindlimbs improving post-delivery oxygenation.[7] Both of these events occur during spontaneous vaginal delivery.

To ameliorate these disadvantages, one option is to administer dexamethasone (20 mg) with or without a prostaglandin to induce parturition (see Chapter 44). This of course makes the C-section less "elective." Certain techniques can be performed on the neonate after delivery to improve respiratory function. These are likely beneficial following the delivery of any calf. Carefully lifting the newborn by their rear legs, as illustrated in Figure 48.1, for a few seconds should correspond with the stage of normal vaginal delivery in which the calf is momentarily suspended. This must be performed without swinging or shaking the calf, which although a common practice can be injurious to the calf. Following suspension, place the newborn in lateral recumbency. Grasp the thoracic limb, elevating and lowering the limb. This limb motion lifts the calf's chest off the ground and facilitates thoracic excursions. Repeat this motion three to four times. Additionally or alternatively respiratory efforts can be stimulated by "tickling" the nasal openings of the newborn with a straw or twig inducing a sneeze, further aiding airway clearance (Figure 48.2).

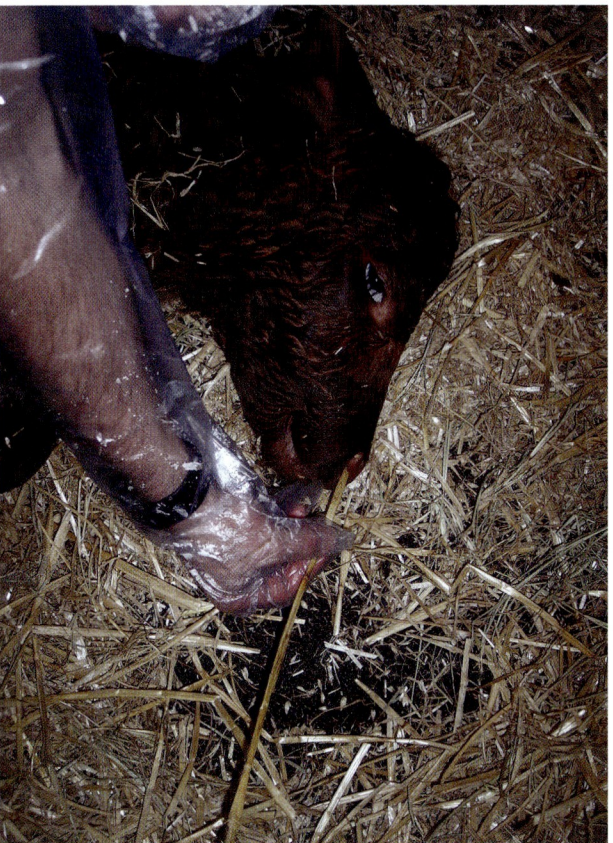

Figure 48.2 Respiratory efforts can be stimulated by "tickling" the nasal openings of the newborn with a straw or twig, inducing a sneeze and further aiding airway clearance. Technique demonstrated here with an enteral feeding tube.

Emergency C-section

For prognostic reasons the decision to perform a C-section must occur in a timely manner. Several clinical signs or circumstances can direct a quick decision to proceed to surgery. The first clinical assessment indicating a C-section is the obviously small heifer with a small or abnormally shaped pelvis.[1–3] Next, as obstetric manipulations are made and extraction of the calf is being attempted, pressure on the proximal aspect of the fetal forelimbs by the pelvis resulting in crossing of the forelimbs is an indication of fetal–maternal pelvic size incompatibility. Additionally, failure to progress in delivery of the calf is an indication for surgery. Clinicians should have a preset metric (time frame) for progress during obstetric procedures. For example, if obstetric manipulation is being made with no progress for a period of 10 min, then one must proceed to surgery. This is not to say that the calf must be delivered in that or another designated time frame, but that "progress" must be made.

In the case where the fetus is dead, and especially if it is emphysematous, the prognosis is less favorable. Case management and treatment plans must be discussed with the owner with regard to outcome prior to proceeding. Although a fetotomy is an option when the calf is dead, certain clinical presentations favor C-section. C-section is the treatment of choice for the emphysematous breech, the unreducible uterine torsion, an undilated cervix, preexisting vaginal or cervical trauma from lay delivery attempts, and uncooperative patient temperament.[1-3,5] These cases, as stated previously, dictate aggressive postoperative care and an increased risk of complications

Surgical approaches

Surgeon preference, patient temperament, environment, and case presentation dictate the surgical approach selected. C-sections may be performed under standing sedation or cast in recumbency. The surgical approach determines the type of restraint employed and the method of anesthesia, whether regional or general. Surgical approaches that have been utilized for C-section are the flank (left or right), the ventral midline, and the paramedian. There are no advantages for a paramedian over the ventral midline[8] so that approach is not covered here.

Standing left flank

The standing left flank approach is the most common surgical technique for C-section in cattle. The left flank approach is superior in that the rumen prevents prolapse of intestinal loops from the incision. This standing procedure is not recommended in animals that are likely to become recumbent during the surgery due to exhaustion, hypocalcemia, obturator nerve paralysis, and/or highly fractious nature.[2]

Standing right flank

The standing right flank approach has limited indications. It can be utilized when there is rumenal distension or adhesions on the left side from a previous left flank surgery.[9] It has also been advocated as the best approach for a C-section performed due to hydrops allantois or hydrops amnii. In these cases an attempt is made to remove the fetus and close the uterus without the massive fluid loss that will result in hypovolemic shock.[8] The right flank incision may also be utilized in the management of an uncorrected uterine torsion. The surgical technique for a right flank C-section is similar to a left flank approach, with the exception that on entering the abdomen extra care must be taken to avoid lacerating small intestine as the rumen is not available to serve as a shield.[1-3] Additionally, it is recommended that the surgical incision does not extend too far ventrally so as to prevent small intestine from escaping the incision.

Ventral midline approach

The ventral midline is preferred by some clinicians as the primary approach utilized for C-sections. This approach has several advantages and additionally some specific indications even for those who favor a standing flank method. The ventral midline approach is the preferred surgical technique for removal of an emphysematous fetus.[1-3] This technique allows greater access to the uterus and minimizes abdominal contamination during delivery. The ventral midline also offers the benefits of closure of the linea alba, allowing greater suture holding strength and less incisional hemorrhage compared with the paramedian approach.

The greatest disadvantage of the ventral midline technique is that it requires the cow to be cast in dorsal recumbency.

C-section for the treatment of uterine torsion

A C-section must be performed to deliver the calf in cases when a uterine torsion cannot be corrected by other methods. Surgically assisted correction of the torsion via a flank laparotomy followed by vaginal delivery of the calf is preferred over a hysterotomy.[8] However, Frazer et al.[10] demonstrated that one-third of all cases of uterine torsion presenting to veterinary school referral hospitals resulted in C-sections. Although either a standing flank laparotomy or a ventral midline approach can be utilized, the standing left flank approach will be described. The incision for the flank laparotomy should be made about 25% longer than ordinarily done for a routine section. Once in the abdomen an attempt is made to correct the torsion. A second operator utilizing a de-torsion bar transvaginally may assist this effort, but typically intra-abdominal manipulation is all that is necessary. Following reduction of the torsion, delivery of the calf is dependent on several factors. Fetal viability, uterine health, and cervical dilation dictate the manner of the delivery. If cervical dilation is sufficient, a vaginal delivery may be made at this time and the flank laparotomy closed. If dilation is not adequate or calf viability is questionable, proceed with a hysterotomy.

In the situation where the torsion cannot be corrected or the uterus ruptures during manipulation, realize that access to the hysterotomy and/or the uterine tear will become difficult as the uterus returns to its normal position. At this time I recommend a second approach from the right flank to assist in the closure of the hysterotomy and/or management of the tear. In both of these scenarios, either the torsion that cannot not be corrected or the uterine rupture, the fetus is likely dead and the uterus has already undergone extensive damage. Thus case management reflects heroic measures that carry a guarded prognosis. However, considering the value of some animals and the expectations of the modern owner, practicality is often dismissed.

Restraint and anesthesia

Flank laparotomies can be performed with the patient standing or in right lateral recumbency. Restraint when performed standing can range from haltered and tethered to the almost complete immobility provided by a cattle squeeze chute. In fact some cattle chutes with adjustable side panels can be set so narrow at the floor level as to prevent a cow from lying down.

An anesthetic technique described by Abrahamsen[11] that provides analgesia and enhances patient cooperation while maintaining recumbency is the "Ketamine stun," which involves the administration of ketamine in combination with butorphanol and xylazine. Specifically, it can be administered intramuscularly at the following doses: butorphanol 0.01 mg/kg, xylazine 0.02 mg/kg, and ketamine 0.04 mg/kg. This will provide duration of effect for 45 min. Abrahamsen also reported a simplified protocol utilized originally by

Dr David Anderson while at Kansas State, referred to simply as the "5–10–20" technique. This refers to an approximate dose calculated for a 500-kg cow, but reported to be safe in cattle weighing 340–660 kg. The combination dose is 5 mg butorphanol, 10 mg xylazine, and 20 mg ketamine administered intramuscularly to the patient.[11] The timing to onset of the clinical effect for this combination when administered prior to clipping and preparing the surgical site is typically perfect. Regional anesthesia options are the proximal paravertebral block or the inverted L block, both described in Chapter 15.

Surgical technique

The surgical site should be clipped and thoroughly scrubbed with chlorhexidine. I recommend using cloth or paper drapes to cover the chute to reduce contamination of the exposed uterus (Figure 48.3).

Standing flank approach

A vertical skin incision is made in the middle of the paralumbar fossa beginning 10 cm ventral to the transverse vertebral processes extending distally for a distance of 25 cm.[1–3] The incision is extended to a length sufficient to allow extraction of the calf. Alternatively, an oblique celiotomy incision can be made.[12] This skin incision is begun 8–10 cm cranial and 8–10 cm ventral to the cranial aspect of the tuber coxae and extended cranioventrally at a 45° angle, ending 3 cm caudal to the last rib. The oblique approach allows for better access to the gravid uterus.

Incision of the skin and the subcutaneous tissue exposes the fibers of the external abdominal oblique muscle. This muscle is incised vertically the entire length of the skin incision. Next the internal abdominal oblique muscle is incised in a similar fashion. On completion of this incision the aponeurosis of the transverse abdominal muscle is revealed. Care must be taken in this step to prevent inadvertent entrance into the rumen. To avoid perforating the rumen tent the transverse abdominal muscle with tissue forceps to make a small incision through the muscle and the peritoneum with a scalpel blade or Metzenbaum scissors. Once a small opening is made, elongate the incision using the tissue forceps as a tissue guard. Incise the remaining transverse abdominal muscle and the peritoneum with the blade or the Metzenbaum scissors. When the peritoneum is incised and the abdominal cavity opened, the sound of air entering the potential space is heard. The rumen is identified and pushed cranially to allow access to the uterus (Figure 48.4). In the case of a fetus in a normal anterior position, a hindlimb within the gravid horn will be both the most recognizable and accessible part of the calf. The distal hindlimb is used as a handle to exteriorize the uterus through the abdominal incision (Figure 48.5). When the part of the uterus that encases the distal limb is exteriorized, the hock is then identified and "hooked" outside the incision. The uterus is incised along the greater curvature, but care must be taken to incise only the gravid horn and avoid the uterine body. The incision begins over the hock and extends to the claw. Care must be taken to avoid incising placentomes and the fetus itself. Be forewarned that incisions that are too short will result in tearing of the uterine wall as the calf is delivered. These tears are always ragged and somewhat perpendicular to the incision line, complicating subsequent closure. After the incision is made, obstetric chains can be attached to facilitate delivery. Following delivery of the calf, the uterus is examined for the presence of another fetus, uterine trauma, and tears of the uterine wall.

Uterine closure

The placenta is not removed unless it is easily detached. Care must be taken not to include the placenta in the uterine closure. Often it is helpful to trim away some of the membranes from the incision prior to closure to prevent incorporating

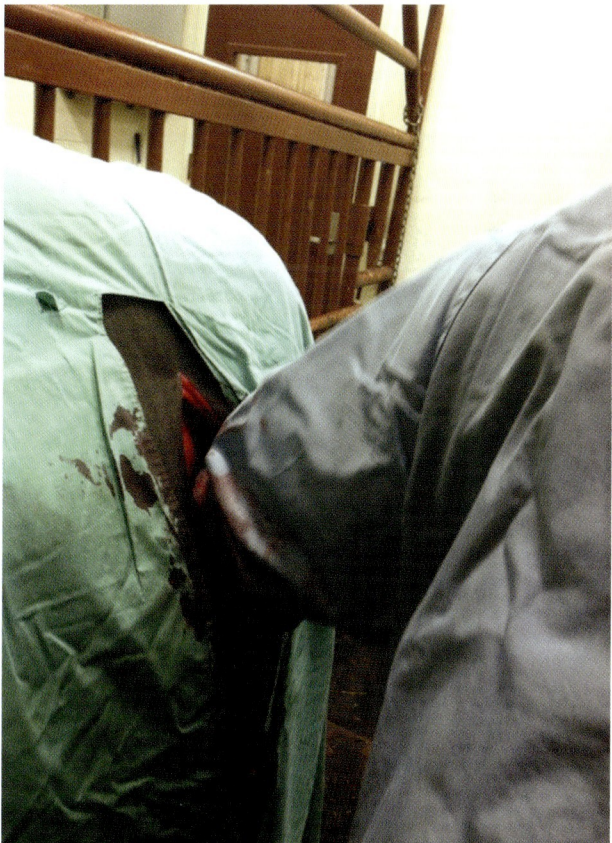

Figure 48.4 The rumen is identified and pushed cranially to allow access to the uterus.

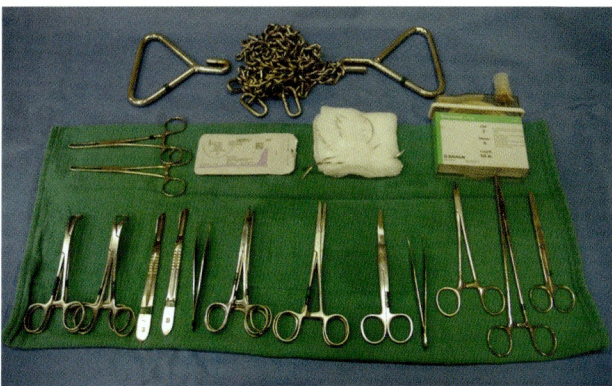

Figure 48.3 Surgical pack.

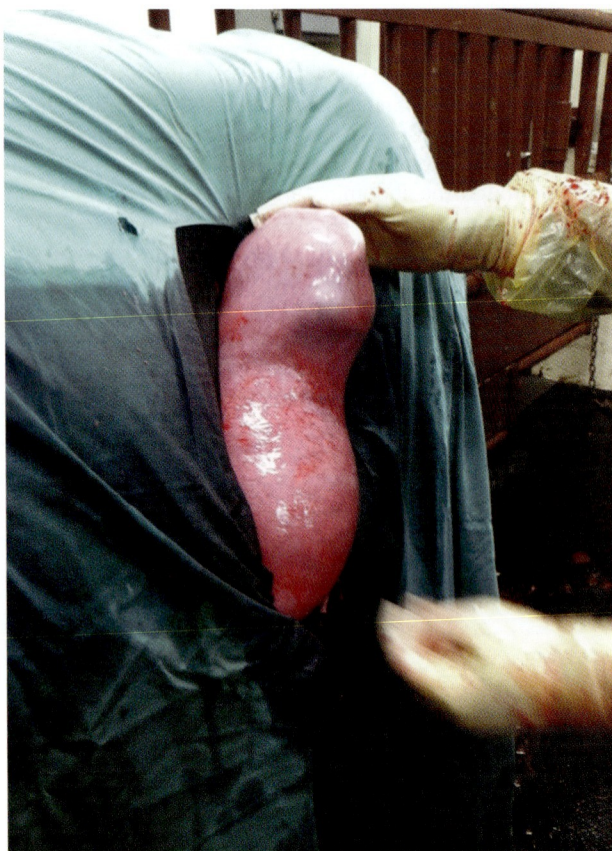

Figure 48.5 In the case of a fetus in a normal anterior position, a hindlimb within the gravid horn will be both the most recognizable and accessible part of the calf. The distal hindlimb is used as a handle to exteriorize the uterus through the abdominal incision.

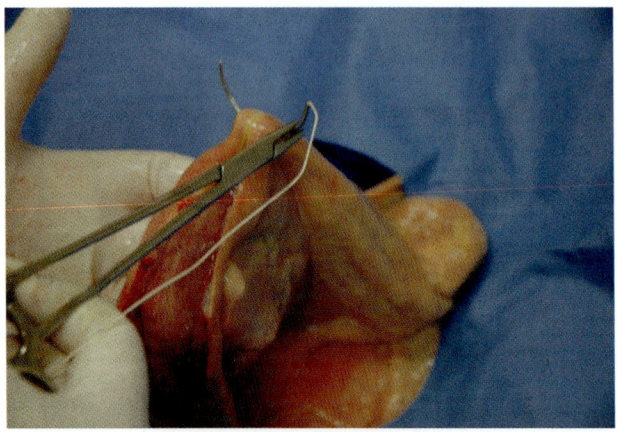

Figure 48.6 Initiating the Utrecht suture pattern: begin the suture line proximal to the uterine incision and do not enter the lumen of the uterus.

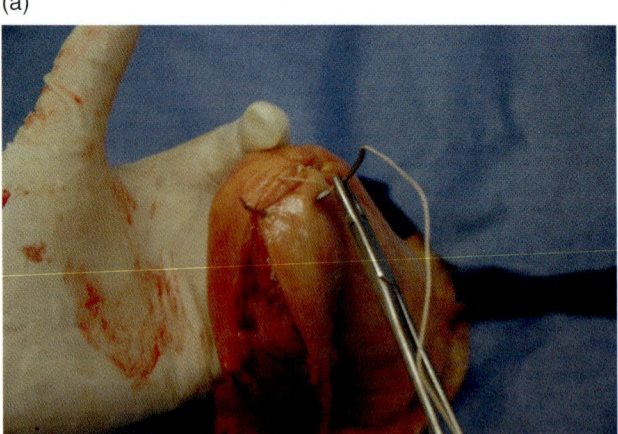

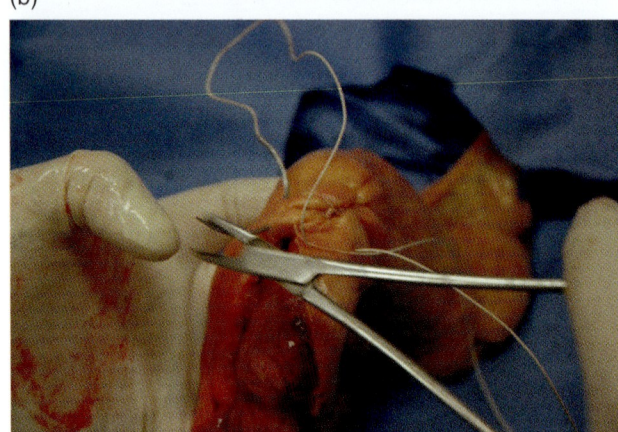

Figure 48.7 Once the knot is secure, place each bite at a 45° angle oblique to the incision proximal to the exit of the previous bite.

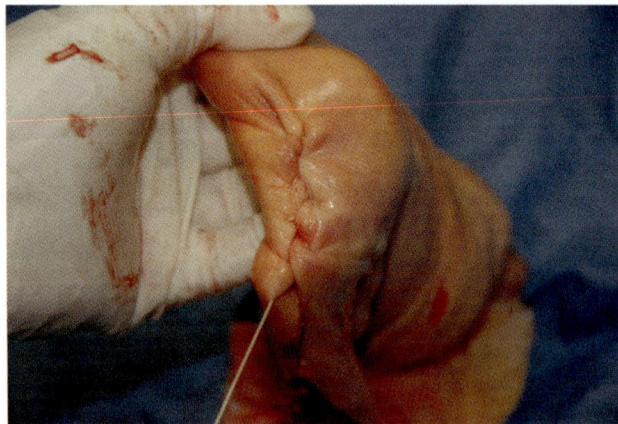

Figure 48.8 Tighten the incision line by pushing the uterus away from the suture line, so giving the closure a bunched or pursed appearance.

it into the suture line. The uterus is then closed with an inverting continuous suture pattern. The Cushing pattern or a modified Cushing pattern, the Utrecht uterine suture, is recommended (Figures 48.6, 48.7, 48.8, 48.9 and 48.10). The advantage of the Utrecht pattern is reduced postoperative adhesions because the suture and knots are buried.[3] A tapered needle and chromic or plain catgut suture (#2 or #3) is recommended as at least one study comparing plain catgut to polyglactin 910 revealed no significant advantage to the use of the synthetic product.[13] A single layer closure utilizing the Utrecht pattern is adequate unless there is concern with regard to the competency of the first suture line or the condition of the uterus. In cases of severely contaminated fluids a two-layer closure is indicated. A simple continuous

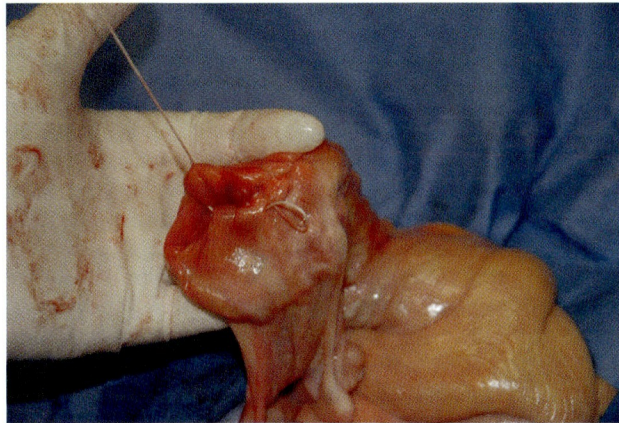

Figure 48.9 Complete the Utrecht suture line to bury the knot.

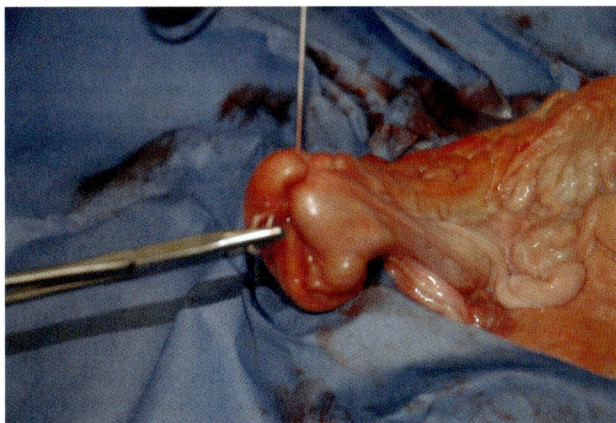

Figure 48.10 Note that as the Utrecht knot is tightened, the remaining suture is covered.

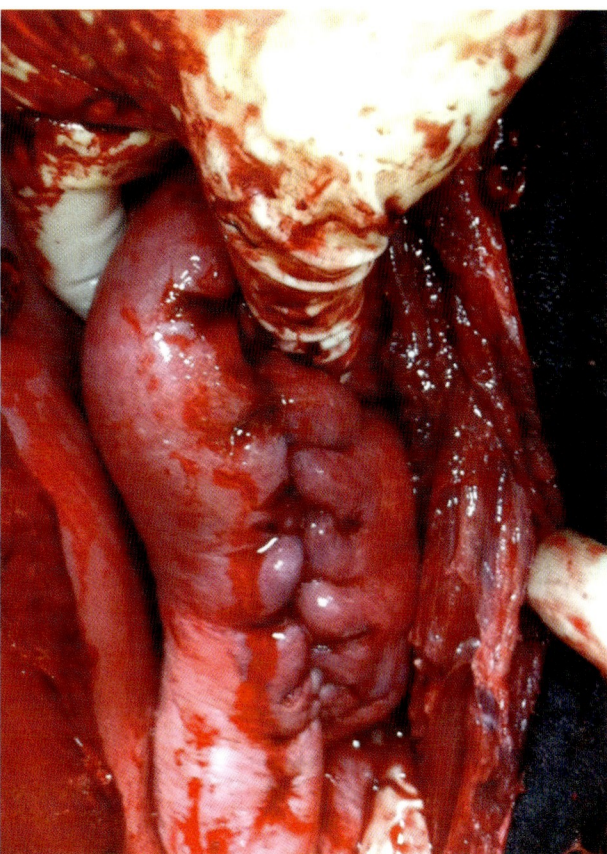

Figure 48.11 The accumulation of blood clots along the incision incite the formation of adhesions that reduce fertility. Note how the uterus is withdrawn into the incision. A surgeon must work quickly due to the rapid involution of the uterus.

followed by a Cushing pattern improves the competency of the closure in a contaminated environment. Regardless of closure it is important not to delay suturing as the uterus begins to involute quickly.[3]

Once closure of the uterus is complete, removal of blood clots and contamination from the serosal surface is imperative in reducing postoperative adhesions (Figure 48.11). Following closure of the uterus the administration of 10–20 IU of oxytocin into the uterine wall and 30–40 IU intravenously will hasten uterine involution and aid in passage of the placenta (Figure 48.12).

Then depending on the level of abdominal contamination it may be advisable to lavage the abdomen with large volumes of saline to remove clotted blood and debris. If there is gross severe contamination, large volumes can be used and egress of lavage fluids can be facilitated by placement of a rumen trocar through the ventral abdomen as described in Chapter 53. In addition to lavage, carboxymethylcellulose, heparin, potassium penicillin, flunixin meglumine, glycerol, and tetracycline have all been advocated as components of postsurgical abdominal lavage.[5,14–16]

When the exteriorized portion of the uterus has been maintained outside the abdomen for the delivery and uterine closure, it can be assumed that there is no abdominal contamination. Contamination must be assumed in cases such as an unresolved uterine torsion, a fetus in the "breech" posterior

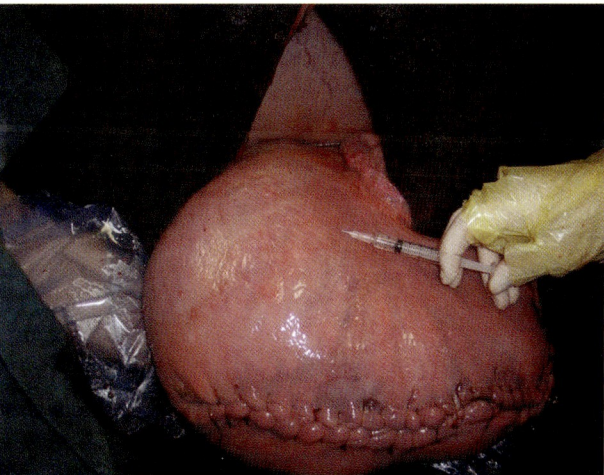

Figure 48.12 Following completion of the uterine closure and lavage of the serosal surface, the administration of 10–20 IU oxytocin into the uterine wall and 30–40 IU i.v. will hasten uterine involution and aid in passage of the placenta.

presentation, or large fetal size preventing the complete exteriorization of the uterus.[1] This raises a couple of issues. First, a cautionary note with respect to the use of a popular obstetric lubricant, J-Lube, which has been shown to be fatal to cattle, horses, and mice following intra-abdominal exposure.[14] Thus because the avoidance of abdominal contamination by the uterine contents is almost impossible, it is recommended to avoid the use of J-Lube in obstetric cases that ultimately result in C-section. Conversely, sodium carboxymethylcellulose, which is a very good obstetric lubricant, is not only safe but has been used in abdominal lavage fluids to minimize adhesions.[15]

Body wall closure

Abdominal wall closure is achieved in three layers. The peritoneum and the transverse abdominal muscle are closed together with simple continuous suture pattern using #2 or #3 chromic catgut or synthetic #3 polyglactin 910. The internal and external abdominal oblique muscles as well as the subcutaneous fascia are closed with a simple continuous using #3 polyglactin 910. I recommend closure beginning distally and working proximally to help contain viscera within the abdominal cavity. Skin closure can be performed with 0.6-mm nylon (Braunamid™) utilizing a pattern of the surgeon's choice.[1-3]

Ventral midline approach

The cow is cast in dorsal recumbency with both the head and the forelimbs tied in extension. The abdomen is clipped and aseptically prepared. A midline incision is initiated immediately cranial to the udder extending cranially 5–7 cm beyond the umbilicus.[2] Once the skin and subcutaneous tissues are incised, a small stab incision is made through the linea alba. This stab incision is large enough to pass a finger or forceps to elevate the body wall to allow extension of the incision along the linea while minimizing risk of abdominal visceral penetration. Once the abdomen is entered, downward pressure on the body wall will help to push the gravid horn of the uterus toward the incision. Manipulation of the uterus, delivery of the calf, and closure of the uterus are as previously described.

Closure of the linea alba and peritoneum is obtained by using either a simple continuous or near-far-far-near pattern with #3 polyglactin 910. To facilitate closure, placement of towel clamps along the incision or several interrupted cruciate sutures will help alleviate tension as the incision is closed. The subcutaneous tissues are apposed with absorbable suture in a simple continuous suture pattern. The skin should be closed with 0.6-mm nylon (Braunamid) in a simple interrupted cruciate pattern; continuous patterns are difficult to remove from a ventral midline incision.

Postoperative management and complications

Complications following C-section include endotoxemia, peritonitis, metritis, adhesions, and reduced fertility.[1-3,5] Mostly these complications are the result of bacterial contamination. As previously stated, all C-sections should be considered clean-contaminated surgeries even in the best of circumstances. Therefore, broad-spectrum antibiotics and nonsteroidal anti-inflammatory agents should be administered prior to surgery. Systemic antibiotics should be maintained for 3–5 days following surgery depending on the degree of surgical contamination. Severely endotoxic animals by definition should be treated with antimicrobials several days beyond the resolution of the endotoxemia. Endotoxic animals will require additional cardiovascular support of intravenous fluid therapy. Clinical signs of peritonitis, pyrexia, inappetence, and ileus usually become apparent by 3 days after surgery. In the absence of these clinical signs antibiotics may be discontinued.

Intra-abdominal adhesions can reduce future fertility. The chapter has previously stated that utilizing the Utrecht suture pattern and removing blood clots from the uterine serosa will remarkably reduce postoperative adhesions. I recommend using 1 L of carboxymethylceluose prior to body wall closure to further decrease the risk of postoperative adhesions. Postoperative fertility is reduced approximately 15% from fertility rates in cows that did not have C-sections. Furthermore, postoperatively pregnancy rates can be expected to be between 60 and 80%.[1,2]

References

1. Fubini S. Surgery of the uterus. In: Fubini SL, Ducharme N (eds) *Farm Animal Surgery*. St Louis, MO: Saunders, 2004, pp. 382–390.
2. Frazer G. Cesarean section. *Vet Clin North Am Food Anim Pract* 1995;11:19–35.
3. Turner A, McIlwraith C. Cesarean section in the cow. In: Turner A, McIlwraith C (eds) *Techniques in Large Animal Surgery*. Philadelphia: Lea & Febiger, 1989, pp. 318–316.
4. Mijten P, van den Bogard A, Hazen M, de Kruif A. Bacterial contamination of fetal fluids at the time of Cesarean section in the cow. *Theriogenology* 1997;48:513–521.
5. Newman K. Bovine Cesarean section in the field. *Vet Clin North Am Food Anim Pract* 2008;24:273–293.
6. Uysterpruyst C, Coghe J, Dorts T *et al*. Optimal timing of elective caesarean section in Belgian White and Blue breed of cattle: the calf's point of view. *Vet J* 2002;163:267–282.
7. Probo M, Giordano A, Moretti P, Opsomer G, Fiems L, Veronesi M. Mode of delivery is associated with different hematological profiles in the newborn calf. *Theriogenology* 2012;77:865–872.
8. Wenzel J, Baird A, Wolfe D, Carson R. Surgery of the uterus. In: Wolfe DF, Moll HD (eds) *Large Animal Urogenital Surgery*. Baltimore: Williams & Wilkins, 1999, pp. 413–433.
9. Youngquist R. Surgical correction of abnormalities of genital organs of cows. In: Youngquist RS (ed.) *Current Therapy in Large Animal Theriogenology*. Philadelphia: WB Saunders, 1997, pp. 429–440.
10. Frazer G, Perkins N, Constable P. Bovine uterine torsion: 164 hospital referral cases. *Theriogenology* 1996;46:739–758.
11. Abrahamsen E. Ruminant field anesthesia. *Vet Clin North Am Food Anim Pract* 2008;24:429–441.
12. Parish S, Tyler J, Ginsky J. Left oblique celiotomy approach for cesarean section in standing cows. *J Am Vet Med Assoc* 1995;207: 751–752.
13. Mijten P, de Kruif A, Van der Weyden G, Deluyker H. Comparison of catgut and polyglactin 910 for uterine sutures during bovine caesarean sections. *Vet Rec* 1997;140:458–459.
14. Frazer G. Systemic effects of peritoneal instillation of a polyethylene polymer based obstetrical lubricant in horses. In: *Proceedings of the Annual Conference of the Society for Theriogenology*, 2004, pp. 93–97.
15. Moll H, Wolfe D, Schumacher J, Wright J. Evaluation of sodium carboxymethylcellulose for prevention of adhesions after uterine trauma in ewes. *Am J Vet Res* 1992;53:1454–1456.
16. Newman K. Bovine cesarean sections: risk factors and outcomes. In: Anderson DE, Rings M (eds) *Current Veterinary Therapy: Food Animal Practice 5*. St Louis, MO: WB Saunders, 2009, pp. 372–382.

Chapter 49

Retained Fetal Membranes

Augustine T. Peter

Veterinary Clinical Sciences, College of Veterinary Medicine, Purdue University,
West Lafayette, Indiana, USA

Introduction

The placenta, an arrangement of transporting epithelia between the fetal and maternal circulations, created and owned by the fetus, protected by the dam, and escaping antigenic rejection, loses its purpose immediately after the rupture of the umbilical cord and is expelled after the delivery of the fetus. Bovine placenta fascinated the early farmers, excited many scholars to create more than 1500 documents since 1910,[1] and continues to provide remuneration for veterinarians all over the globe because of its retention. Failure of placental expulsion constitutes retained placenta. Much has been documented about retained placenta in theriogenology textbooks[2-7] based on these authors' own vast clinical experience and information gleaned from scientific publications. Excellent review articles by leading veterinary clinical investigators have more vividly encapsulated the various aspects of this clinical condition.[8-26] What is known about its clinical signs and negative influence on postpartum reproductive performance has not changed in the last two centuries. However, over the years, many hypotheses with regard to its occurrence (particularly at the tissue and cellular levels), risk factors, preventive measures, and treatment plans have been proposed. The headings for this topic include definitions, developmental morphology of placenta to better understand the process of detachment and etiology, pathogenesis, treatment, and prevention. It will become very apparent on reading this chapter that there are many instructive similarities among these topics.

Definitions

The fetal part of the placenta, namely the fetal membrane (allantochorion), normally separates from its maternal moorings during the process of parturition. The process of parturition is marked by vascular, contractile, and mechanical dynamics designed to expel the fetus and its membranes (allantochorion and amnion). The fetal and maternal parts of the placenta have equal regulatory roles in the timely separation of the fetal part of the placenta. Hence use of the term "placenta" in this chapter will point to a collective structure created by the dam and the fetus. Retention is primarily due to failure of the villi of the fetal cotyledon to detach themselves from the maternal crypts of the caruncle. The process of separation and inversion of allantochorion (allantoic surface faces outside) is completed on average within 8 hours of initiation of parturition, depending on the parity and age of the patient. In general, all parts of the allantochorion and amnion need to be expelled within 0.5–12 hours after calving. Thus retention of fetal membranes is referred to as retained placenta or, more correctly, as retained fetal membranes (RFM), with the term "fetal membranes" referring to both the allantochorion and amnion. While the use of "fetal membrane" (singular) for allantochorion is accepted, to avoid confusion the term "allantochorion" will be used here.

Developmental morphology

Presently, bovine placenta is described as cotyledonary synepitheliochorial based on the morphology that is established around 40–50 days into pregnancy. Earlier classification as syndesmochorial was based on a misunderstanding of the number and form of the layers intervening between the fetal and maternal circulation. This classification needed correction because of new evidence. The earlier assumption that the uterine epithelium is lost in the process of placentation resulting in direct apposition of trophectoderm to the maternal connective tissue is no longer tenable. Now it is known that the uterine epithelium persists although initially modified to a variable degree into patches of a hybrid fetomaternal syncytium formed by the migration and fusion of a particular type of trophectodermal cells with uterine epithelial (UE) cells. These trophoblast binucleate/giant cells (TGC) migrate and modify the uterine epithelium by apical fusion to form fetomaternal hybrid syncytial plaques with up to eight nuclei at the junction of the fetal and maternal tissue. These syncytial plaques are replaced by regrowth of uterine epithelial cells by day 40, and subsequently TGC–UE fusion produces only transient trinucleate minisyncytia throughout the remainder of pregnancy.

Research findings have necessitated a change in the morphological terminology of placentation and defined the bovine placenta as synepitheliochorial. The prefix "syn" indicates the contribution of TGC to the fetomaternal syncytium and contrasts with the simple microvillar interdigitation between trophoblast and uterine epithelium (epitheliochorial) over the rest of the placenta. Cotyledonary refers to the presence of localized areas of trophectodermal proliferation forming "cotyledons" in the placenta and each cotyledon is the fetal part of a placentome. The placentome is formed by the tuft of branching chorionic villi from the cotyledon that grow and enmesh with corresponding maternal caruncular crypts, providing a finger-in-glove arrangement. These crypts develop from the preformed flat endometrial caruncles (Figure 49.1), present in the nonpregnant cow, which are aligned along the uterine horns. Normally, the placenta consists of 80–90 of these placentomes. The placentomes are linked together by areas of flat apposition between trophoblast and caruncular epithelium (CE). The CE cells are similar to UE cells and are homogeneous in population unlike the trophoblast. The placentomes guarantee a firm anchorage by the complementary interdigitation of fetal villous trees with maternal crypts and by the interdigitation of the apical microvilli from uninucleate trophoblast cells (UTC) and CE directly by cell–cell contact or indirectly by cell–matrix contact. Secondly, placentome formation with a synepitheliochorial interhemal barrier provides the vast increase in surface area necessary for the continuously increasing substance exchange between the dam and the fetus. It is important to note that placentomal gross morphology and the pattern of fetomaternal interdigitation (villus/crypt architecture of the mid to late pregnant placenta) differ considerably between bovid species but the detailed cellular structure of the maternofetal interface is the same throughout pregnancy. It is pertinent to point out that no maternal tissue is shed in the afterbirth (secundines) of cattle. On the other hand, a few fetal villi may be caught and left in the maternal crypts.

Etiology

Placentomal maturation coupled with appropriate structural, endocrinological, and immunological changes result in easy separation and expulsion of the fetal membranes. Furthermore, vascular, contractile, and involution changes that occur in the uterus before and during parturition also help in the separation and expulsion of the fetal membranes. It should be very well recognized that the process of maturation and timely separation and expulsion of the fetal membranes involves well-orchestrated regulation by both the maternal and fetal components of the placenta.

Placentome maturation

It has been established that a physiological parturition occurs on the face of mature placental structures. The fact that induction of parturition is followed by RFM in most cases supports the observation that immaturity of placentome may be one of the contributing factors for RFM since parturition is initiated before term in such cases. Similarly, that RFM has been an issue in most abortion cases lends credence to the notion that a mature placentome is a prerequisite for the separation of the allantochorion.

Endocrine changes

Maturation of placenta is followed by specific endocrine changes. These changes are important to bring about contractile and apoptotic events, and expression of genes for matrix metalloproteinases and their inhibitors. Although a mature placenta is capable of initiating favorable steroidogenic and arachidonic cascades, altered expression of antioxidative defense mechanisms at this juncture can interfere with steroidogenesis at the cellular metabolic level. Poor expression of the antioxidative defense mechanisms against reactive oxygen species (ROS) can result in an imbalance between production and neutralization of ROS, preventing proper detachment of the allantochorion.

Closer to parturition, increases in peripheral estradiol concentrations and less perceptible changes in progesterone are observed. Progesterone perfuses into placentomes which may contribute to its imbibition within the placentome, enabling structural and functional changes. These changes facilitate the contraction of myometrium by estradiol along with other hormones including oxytocin and specific prostaglandins. It is important to note that these structural changes can possibly occur only in a mature placentome that has been exposed to adequate duration and quantity of estradiol.

Regarding prostaglandins, the type and quantity of prostaglandins produced can influence the contractility of the myometrium. For example, a lower ratio of prostaglandin (PG)E_2 to PGF$_{2\alpha}$ within the fetomaternal compartments of the placentome and an elevated steroid hormone receptor status possibly reduces the rate of apoptosis occurring in the chorionic

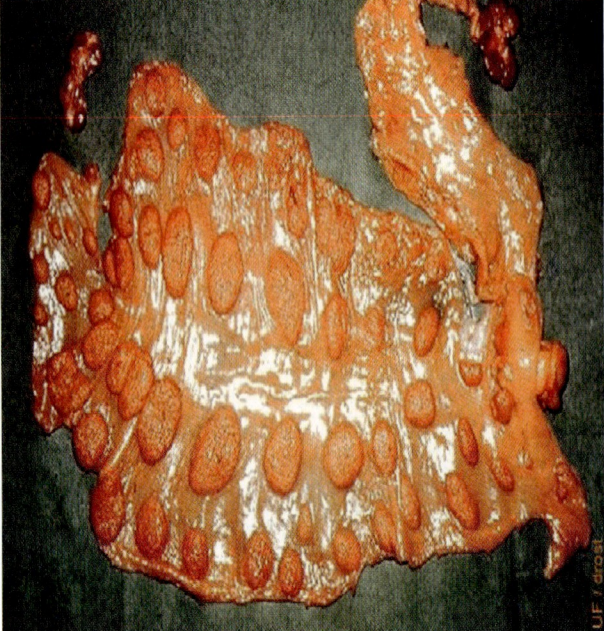

Figure 49.1 Caruncles are arranged in four rows with approximately 15 in each row. Courtesy of Maarten Drost.

epithelium before calving. There may be other unknown factors involved in the regulation of apoptosis; nevertheless, it is recognized that dysfunctional expression of apoptotic regulating factors can contribute to failure of separation.

Expression of matrix metalloproteinases and counteracting tissue inhibitors of metalloproteinases in the fetal compartment are essential to balance extracellular matrix formation and degradation, critical in the separation of the allantochorion. Specifically, they are responsible in breaking down collagen within the villi and assisting in the separation of villi from the crypts.

Finally, it should be remembered that despite the complexity of various prostaglandins and their production and our incomplete understanding of their specific roles, the use of prostaglandins to prevent the occurrence of RF in induced parturition has yielded inconclusive results.

Structural changes

Although the tight connection within the placentome is essential during pregnancy, it has to cease to allow allantochorion separation. Indeed, distinct remodeling and loosening of adherence occurs within the placentome during late pregnancy. Transformations that occur in the placental epithelium and connective tissue during the last month of pregnancy prepare the placentome for the critical loosening of the tissues involved in the fetomaternal interdigitation. This process has to proceed undisturbed and has to culminate in separation of the allantochorion at the right time. The changes include progressive collagenization of maternal and fetal connective tissues, flattening of the maternal epithelium lining of the crypts nearest to the caruncular stalk, and the appearance of TGC that become polynuclear shortly before the detachment process.

Immunological changes

Expression of the major histocompatibility complex (MHC) class I in bovine trophoblast cells and their tight regulation are biologically relevant. Early in pregnancy, a complete shutdown of major MHC class I expression by trophoblast cells appears to be critical for normal placental development and fetal survival. This immunological camouflage is vital to avoid recognition by the multifactorial array of cellular and hormonal mechanisms that mediate rejection. In a mature placenta, maternal immunological recognition of fetal MHC class I proteins triggers immune and inflammatory responses that contribute to rejection of the allantochorion at parturition. It is interesting to note that in these situations there is a clear adaptation of the immune system for a function distinct from protection against pathogens. In summary, MHC compatibility has significant molecular consequences and can result in an increased incidence of RFM.

The inflammatory response is followed by the maternal immune response. Two characteristic actions are evident in the inflammatory process, namely production of neutrophil-activating factors and the chemotactic activity to neutrophils. These are important for successful immune-assisted detachment of the allantochorion. It has been suggested that the function of peripheral leukocytes may be reduced, which may be a cause for, or an effect of, retention leading to other complications such as mastitis in RFM cases.

Vascular changes

During contractions of the uterus, uterine pressure is constantly changing and this leads to alternating anemic and hyperemic conditions and temporary changes on the surface of the fetal chorionic villi. As a result, the attachment of the chorionic epithelium in the maternal crypts becomes impaired. Further, the subsequent rupture of the umbilical cord shuts down the blood supply to the fetal villi and helps in the separation process by reducing the size of the villi. The anemic changes that occur to the fetal villi after calving because of rupture of the umbilical cord are essential and these mechanical processes of detachment should not be underestimated. Identical changes occur in the caruncles due to uterine contractions. The uterine contractions that continue reduce the amount of blood directed to the uterus. Consequently, the caruncles become smaller in size due to reduced blood supply, resulting in the dilation of maternal crypts.

Contractile changes

Abdominal contractions noticed during expulsion of the fetus result in changes to the shape of the placentome due to abdominal pressure, an essential step in the process of separation. Concurrent and intermittent uterine wall contractions change the shape of the caruncles further from oval to round due to pressure on them by the fetus, facilitating detachment of cotyledons from the caruncle. The characteristic sharp contractions of the circular myometrial muscle layers that closely follow each other and which are superimposed on contractions of the longitudinal myometrial muscle layers noticed in the parturition process help not only in expelling the fetus but also allantochorion separation. These rhythmic waves of contraction (every 3–4 min) cause the cotyledons to open up in a fan-like manner as they pass over them and enable separation of the allantochorion. Hence defective myometrial contraction can lead to retention. Progesterone diffusing from the placenta to the myometrium and estradiol synthesized locally at this time may have a role in the initiation and suppression of these contractions. The presence of uterine contractions even after the delivery of the calf helps in the process of separation. This is also necessary to prevent hemorrhage and to aid in the expulsion of the fetal membranes. These peristaltic and contraction waves, besides reducing the size of the uterus and aiding in forcing the fetal membranes into the birth canal, can also markedly reduce the amount of blood circulating in the endometrium. Further the influence of oxytocin and ongoing uterine involution in these processes cannot be overlooked, as noted where injection of oxytocin and suckling had beneficial effects in shedding of fetal membranes.

Pathogenesis

Considering the essential changes that facilitate the process of separation and expulsion of fetal membranes, it is obvious that retarded or unsuccessful separation and expulsion can be due to interference with these changes. Paying attention to interfering factors and eliminating them, if possible, will reduce the incidence. Some of these factors are discussed in the section on prevention. In general, the interfering factors can be classified as direct or indirect.

Direct factors

These include immature placentomes, noninflammatory edema of the chorionic villi that results from uterine torsion and cesarean sections, dystocia, necrosis between the crypts and villi following a possible antepartum allergic reaction, premature involution of the placentomes, hyperemia of the placentomes, inflammation, mechanical prevention of expulsion, and inflammation of fetal membrane.

Although inflammation is attributed as a direct interference, the role of prepartum uterine infection is questioned and it is more likely that uterine infection follows after retention due to either poor uterine involution or the fact that the membrane is retained. However, the incidence of retention in infectious abortion cases suggests that infection may be a factor. The most likely reason for retention in such cases may be immaturity of the placentome rather than infection. It should be remembered that poor calving hygiene can lead to parturient infection and retention, although the mechanism is not clearly understood.

In most induced parturition cases, retention of the allantochorion can be attributed to possible immature placentomes. However, retention does not occur in all induced cases and furthermore it is influenced by the type of glucocorticoid administered. Besides immaturity, administration of these agents may result in dysfunctional apoptotic processes or interfere with the immune-assisted separation process discussed earlier or specific actions such as the expression of cyclooxygenase by $PGF_{2\alpha}$. Furthermore, these treatments can also inhibit the normal loss of TGC that occurs closer to parturition and modulate the action of other hormones.

Indirect factors

These include intensive stress, duration of pregnancy, extensive distension, season, sex of the fetus, stillbirth, deficiency of trace minerals and vitamins, and situations where the uterus becomes atonic and unable to contract in cases such as dropsy, twinning, fetal gigantism, subclinical hypocalcemia, and other pathological conditions. Other indirect factors that may have a role can be broadly categorized into hormonal, hereditary, nutritional, and circulatory causes; incidentally, these may also contribute to unexplained cases of atonic uterus. The required changes for separation and expulsion of the allantochorion and the interfering factors are schematically summarized in Figure 49.2.

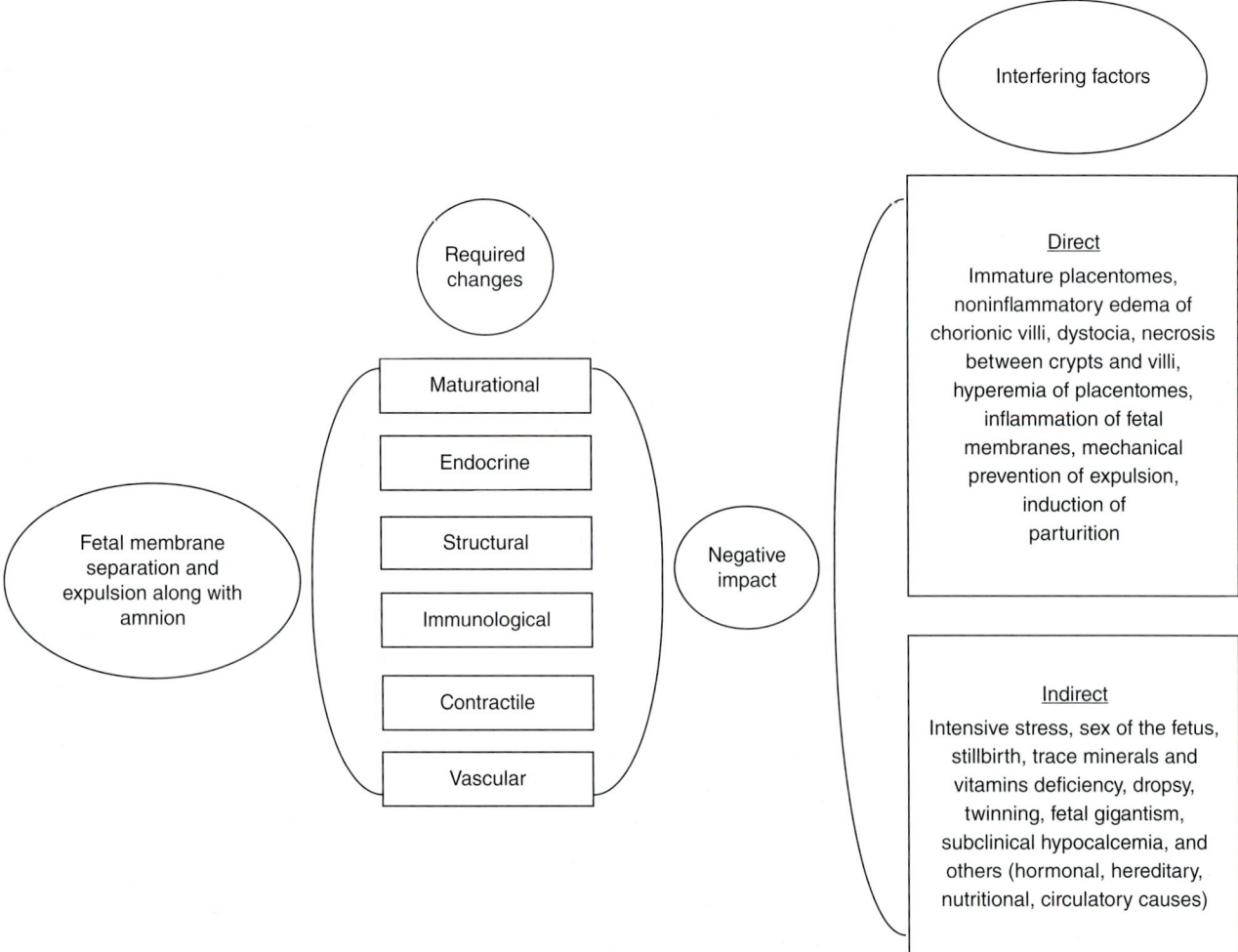

Figure 49.2 A schematic depicting the changes required for normal fetal membrane separation and expulsion. Refer to the text for details of these changes. The interfering factors that have a negative impact on these processes are listed.

It is not known whether apoptotic, degenerative, and necrotic changes are noticed in the placentomal and interplacentomal regions because of allantochorion retention. As mentioned earlier, apoptosis that started prior to separation will continue during retention. It should be emphasized that degenerative changes might have been responsible for retention; however, in almost all cases these changes will occur after retention. Similarly, in certain cases small portions of necrotic epithelium present between the chorionic villi and the cryptal wall might contribute to retention but again the process occurs after retention depending on the period of retention. In general, the lesion may vary from slight evidence of inflammation to severe necrotic or leathery or hemorrhagic inflammation of the allantochorion. Depending on the etiology and its severity, the lesions may extend to the areas of membrane that occupied the non-gravid uterine horn.

Clinical signs and complications

Normally, if the fetal membranes are not passed within 36 hours, and in the absence of manual intervention, they tend to persist for 6–10 days as a putrefying tissue and drop away as a putrescent mass in uncomplicated retention. The progress of putrefactive liquefaction of the villi allows separation, and uterine involution aids in completing the separation. Some of the cotyledons toward the cervical end would have separated by the time the calf is born. Involution of the uterus to restore its spiral configuration aids in separation. The mild contractions of uterus and cervix are accompanied by, and correlated with, transient dilations of the caudal cervix that facilitate evacuation of normal uterine exudate. In protracted cases there is a large amount of uterine exudate with its characteristic foul-smelling odor. Vaginal/uterine examination of a cow in such cases can evoke forceful straining that results in the expulsion of fetid uterine exudate along with feces.

In general, depending on the duration of retention, a case can have fresh fetal membranes hanging from the vulvar lips (Figure 49.3) without any obvious smell, or it might be one that has undergone degenerative changes as reflected by the appearance of the fetal membranes and the presence of malodorous stench. There may not be elevation of body temperature with or without marked illness in all cases even if retention has continued beyond 72 hours. The immediate complication may be the reduction in milk yield, which may not always be noticeable and the impact may be negligible on the herd basis. Some animals exhibit outward signs of ongoing uterine changes in response to spontaneous uterine infections or uterine infections that followed due to manual removal. Along with a rise in body temperature, other signs may be obvious such as anorexia, depression, increase in pulse rate, and loss of body weight. These are symptoms of either endometritis or puerperal metritis and the signs may depend on the intensity of disease process and the number of uterine layers involved in the inflammatory/infectious process. Further, the foregoing clinical signs may be severe if other conditions such as peritonitis, mastitis, necrotic vaginitis, parturient paresis, and acetonemia occur simultaneously with the uterine changes.

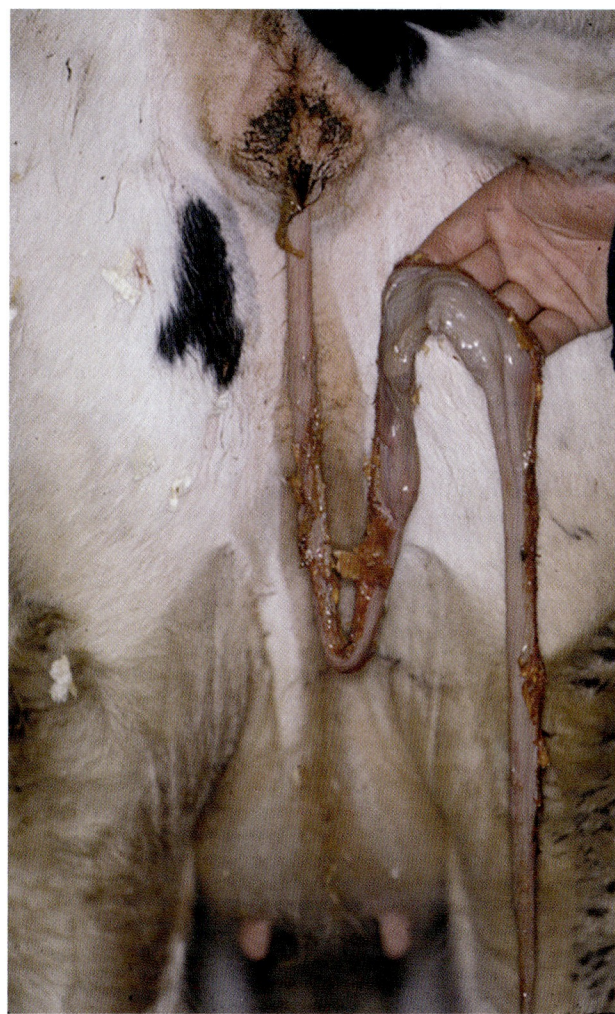

Figure 49.3 A fresh case of retained fetal membrane with minimal changes. Courtesy of Robert Lofstedt.

Apart from the immediate complications, the long-term complications include delayed uterine involution, minor to major effect on fertility, inherent loss of milk production or loss due to discarded milk because of antibiotic treatment. The immediate effect on fertility is mainly due to impaired uterine and ovarian activities. There will be an increase in days from calving to first estrus and the occurrence of subsequent estruses. In essence the resumption of normal cyclic ovarian activities is prevented, possibly due to delay in uterine involution and elimination of bacterial infection. The estrous cycles may be irregular leading to increases in the number of breedings and finally leading to increases in days open. It is difficult to determine the overall losses; however, considering the ramifications of the condition, it can have a very serious impact depending on its prevalence and incidence.

Treatment

Considering the variety of causes and mechanisms, it is not surprising that treatments have met with limited success. However, more than treatment, a case of RFM, like other clinical conditions of the immediate postpartum period, enables the veterinarian to become an educator more than a curer of all animal ailments. The basic goals of any therapeutic regimen should be return of the cow to reproductive usefulness as soon as possible and the prevention of primary and secondary complications that can lead to economic losses from lowered milk production and delayed conception. The primary complication is a normal uterus becoming a diseased or pathologic uterus and the goal of the treatment is to prevent its occurrence.

The aim of treatment is to ensure that retained fetal membranes are shed within a reasonable time with minimal intervention. It should never be the goal to manually remove by force, except in cases where the allantochorion has completely detached but the membranes are not expelled for any unexplainable reason. In other words, the only time that manual removal should be practiced is when the attending veterinarian fortuitously finds the entire fetal membranes free and expulsion only hindered by a partially closed cervix. In this case the fetal membranes are gently teased through the cervix, with care being taken not to traumatize the cervix or vagina during the procedure. Removal should be completed within a few minutes and carried out gently, paying attention to hygienic measures. Aggressive intrauterine manipulations probably increase endotoxin production and its absorption into the circulation, resulting in elevated body temperature. Intrauterine or systemic antibiotic treatment is not necessary. However, lack of secretion of thick tenacious mucus by the cervix in RFM cases allows infection to gain entrance to the uterus and the process of removal can further favor endometrial infection. Hence some cases may need systemic antibiotic treatment depending on the severity of the problem and condition of the animal.

In general, manual separation of the caruncular and cotyledonary attachment is not a preferred method. Although it appears to have many harmful effects on the animal and on future fertility, it is believed to be practiced for aesthetic reasons. This practice seems to give a "conservative trust to an old routine" and is accepted in many parts of the world and taught in the veterinary curriculum. If it is practiced, the cow's tail should be secured by tying it to the opposite front leg. The vulvar area is then prepared with a suitable antiseptic scrub, and tetracycline boluses or capsules are introduced into the cervix and into the uterine horns with a clean plastic sleeve after the allantochorion is separated carefully by releasing the attachments between the caruncle and cotyledon. The systemic administration of ceftiofur hydrochloride is beneficial for prevention of metritis. Although some veterinarians prefer a predominantly intrauterine approach for antibiotic therapy to combat uterine infection, a systemic approach has certain advantages. For example, the absorption of antibiotics from the uterus is not impaired as in local treatment, antibiotic concentrations attained in the uterine tissue are similar to blood concentrations, and uterine infection is seldom localized since it involves more layers than the uterine mucosa.

Dilating the cervix after a couple of days in an attempt to expel the fetal membranes is not recommended. This causes more trauma to the cervix and uterus, predisposing to further complications notwithstanding the pain inflicted to animals. The complications include immediate septicemia, drop in milk and appetite, elevation of temperature, loss of weight and possible death, or else a prolonged period of convalescence during which time shreds of membrane and exudate are expelled, pyometra or chronic metritis develops, or protracted or even permanent infertility may result from uterine abscess, perimetritis, salpingitis, or severe endometritis. It is a good idea not to examine the cow until 72 hours after parturition unless the cow develops clinical signs of anorexia, an elevated temperature, and other symptoms of septicemia. It should be noted that in some cases it is impossible to insert the hand into the uterus after 48 hours and cervical closure occurs in normal cases earlier than this timeframe. Delaying for 72 hours allows the endometrium to construct a protective barrier by walling the fetal membranes off and thereby lowering the chance of septicemia or toxemia. The autolysis that will eventually occur enables separation and expulsion. In such cases, treat as a large abscess in which the necrotic core has not been separated from the surrounding tissues. At no time fetal membranes should be tied to an object or cut closer to the vulva. The surgical procedures governing the handling of an abscess should also govern the handing of an infected uterus. Conservative treatment is indicated and essential. Removing fetal membranes in an incomplete, rough, and insanitary manner is not advisable. In certain cases there may be a collection of uterine exudate; gentle vaginal/uterine examination allows the fluid to drain out by the position a standing animal would take in response to such an examination.

Uncomplicated cases require no treatment and thereby the impact on future fertility is minimized. A "nondrug approach" to the problem should not be forgotten. However, the malodorous stench generated by retention cannot be ignored and it requires courage to prescribe no treatment, so it would be prudent to adopt a rigid attitude of noninterference. A conservative approach (intrauterine antibiotics, maybe including systemic drugs, and leaving the fetal membranes) is better than manual removal at the behest of the client, since clients would prefer that the unsightly fetid mass hanging from the genital tract is removed earlier.

If fetal membranes are retained for 24–48 hours and the cow is running a fever (rectal temperature >39.5 °C) or anorectic, the animal should receive immediate attention and treatment. In addition to monitoring rectal temperature, close observation of the animal is critical. Nothing replaces close observation that recognizes an animal's changing appearance and behavior. The minimum treatment would be parenteral administration of antibiotics. Systemic treatment alone is effective and intrauterine antibiotic treatment is not recommended in all cases since more harm can be done by the manipulation. For systemic administration, ceftiofur is preferred over oxytetracycline: the latter is approved for beef and nonlactating dairy cattle and has a 22-day pre-slaughter withdrawal period, while ceftiofur is appealing because of injection-site tolerance, no milk discard, and 3-day pre-slaughter withdrawal period. Although environmental fragility and residue chemistry of cephalosporin compounds have not been explored completely, these may be a concern with regard to animal and human food safety. If oxytetracycline is used, 4 days of milk discard is required. Penicillin can also be used provided a similar milk discard time is followed. It should be remembered that there may not be better reproductive performance due to antibiotic treatment, but in general proper antibiotic treatment may reduce the chance of the treated animal being culled from the herd.

In the United States, it is required that all these treatments should be carried out with the knowledge of a practitioner who is aware of the Animal Medicine Drug Use Clarification Act. It is incumbent on the practitioner to ensure that the right dosage, route of administration, and duration of treatment is strictly followed. Clients also need to be clearly instructed about the withdrawal time for milk and pre-slaughter withdrawal period. Cows that appear depressed and which are dehydrated need further examination and uterine exudate should be drained by flushing. No single course of treatment is safe and effective in every circumstance. Considering the perceived and real effects of the use of antibiotics in animals on human heath, intrauterine agents such as chlorhexidine, hydrogen peroxide, and recently ozone have been tried.

The use of ecbolic drugs has been suggested as a way to improve uterine and myometrial contractions to aid in separation and expulsion. The foremost is $PGF_{2\alpha}$ but its use is questioned since the endometrium produces this compound adequately during the immediate peripartum period. The inconclusive nature of the findings supports this notion. Some studies have advocated its use at later times in the postpartum period to improve fertility by regulating the uterine involution process and facilitating early resumption of ovarian activity. However, the findings are not uniform, real effectiveness remains undetermined, and most results seem to deny any positive effect. However, there may be instances where it may have some benefit, particularly in the presence of a corpus luteum that hinders the restoration of uterine health in the face of uterine infection.

Administration of compounds prophylactically or as a treatment option based on their ability either to facilitate separation of the allantochorion or to contract myometrium have been developed over the years and continue to be used. Many other agents continue to be used to reduce the deleterious effects on fertility. The prophylactic use of agents such as antibiotics and estrogens to prevent endometritis and improve fertility is a reasonable goal. Estradiol compounds are used as an adjunct treatment option to aid in the restoration of the endometrium and prevent metritis. The method has empirical appeal; however, paucity of beneficial information, possible side effects, lack of conclusive scientific evidence, its effect on humans as a food residue, and marginal return considering the time and cost involved limit its universal application. Lack of availability in some parts of the world due to regulatory measures also precludes its use.

Other compounds such as gonadotropin-releasing hormone (GnRH) have been used to improve fertility in the postpartum period with variable results. However, GnRH has benefit when the ovary and uterus are structurally and functionally ready for the initiation of follicular activity and the animal is fit for breeding. In certain herds that practice earlier breeding, GnRH may be beneficial. A small dose and repeated use of oxytocin has more scientific validity for treatment because of its ability to cause tubocervical contraction of the uterus and thereby expel fetal membranes and uterine exudate. Furthermore, for the same reason, its use in cesarean section is warranted to facilitate separation of the allantochorion. The amount used is important since it can enhance milk ejection prematurely. In general, lack of unified opinion has been magnified by endless clinical reports and testimonials concerning the ever-growing list of hormones and antimicrobial solutions.

Administration of collagenase into the umbilical arteries has been studied extensively and is found to aid in separation of the allantochorion. Although the technique is promising, the time, the cumbersome nature of the procedure, and the cost limits its use. Further, the effect on future fertility remains undetermined.

It is recommended that all cows affected with RFM be examined 20–30 days after removal of the fetal membranes to determine the progress of uterine involution. A good transrectal examination of the reproductive tract not only helps to identify problems, if any, but also stimulates the uterus and enables expulsion of uterine contents. Transrectal uterine massage around this time, starting from the uterine horns to the cervix, for 3–5 min may stimulate the uterine mucosa and restore the dynamic function of the myometrium. The procedure is contraindicated in cases of fibrinous, hemorrhagic, acute purulent endometritis, and closed cervix cases. The complications of unresolved cases are many and are aggravated by the animal's body condition, age, and management practices. It should be noted that permanent sterility may result due to pyometra and metritis if proper treatment is not instituted. Aged animals may need a longer time to recover, so breeding should be delayed. All cows with persistent mucopurulent discharge from the vulva following RFM should be reexamined and treated.

In summary, manual removal of retained membranes is not recommended and each case does not require treatment during or after expulsion of the fetal membranes. Systemic antibiotic treatment can be reserved for animals that have fever, while some animals may in addition need fluid therapy. Oxytocin can have beneficial effects if the correct dose and regimen is used.

Prevention

Proper treatment of the affected animal will reduce the economic losses due to increased calving interval, increased culling rate, discarded milk due to antibiotic treatment, and costs of veterinary treatment and drugs. However, in the modern cattle industry, prevention is more important than treatment in order to augment profitability. It is hard to define the percentage incidence that can be allowed or a percentage that calls for drastic preventive measures. Paying attention to the direct and indirect factors discussed earlier and eliminating these can reduce the incidence, and possibly prevent the occurrence, of RFM. In the interests of animal health and welfare, it is prudent to examine the calving cow and nutrition management, particularly dry-cow nutrition, if the incidence exceeds 10%.

It appears that there is no genetic breed predisposition to this condition; however, repeatability is possible, if problems are not addressed. No simple reason can be attributed to the occurrence. Some of the factors involved may be lack of ideal body condition at the time of calving, stress during pregnancy and calving, lack of trace minerals, and lack of sound vaccination program. The function of the placenta should be maintained during pregnancy and it should preserve its plasticity throughout pregnancy for easy separation. Selenium supplementation may have a beneficial effect in a herd with a regular history of the condition.

One factor that can be addressed is to reduce calving problems. In this regard, selection of bulls based on birth-weight expected progeny differences and calf birthweight will reduce calving problems. It should be recognized that these parameters vary in breeds. It should also be remembered that the influence of the dam on the calf's body weight is greater than that of the bull. Pelvic measurements and judicial application to herds can reduce calving problems and will reduce the incidence of dystocia and other complications of parturition including RFM. However, pelvic measurements and their application may have lost their perceived impact due to its feeble effect in reducing the incidence of dystocia.

During the course of pregnancy and parturition, the adverse effects on reproduction should be minimized. In this regard, separation of the periparturient cow and use of straw as bedding appear to be beneficial. One of the preventive measures adopted was to treat every case of retention with antibiotics. This may not be beneficial except in cases that have higher body temperatures, as discussed before. Some of the prepartum peripheral hormonal changes have been suggested to mirror the changes taking place in the placenta that most likely lead to RFM. Despite much research in prepartum endocrine/enzyme parameters to predict the occurrence, there is no current method available to apply in practice. Even if some of the hormonal measurements can help to identify these cases, it is a futile exercise to obtain a sample from each animal and, furthermore, with no clear method available to prevent its occurrence after the prediction, it is not economically justifiable.

Conclusion

While certain aspects of the condition, such as clinical manifestation and remedial measures, have remained the same over the years, the focus has been on prevention of the condition. This should be the direction in which the dairy industry should move, considering herd health, production, and animal and human welfare issues. It is incumbent on veterinary and allied professionals to improve the health of the cattle population by providing adequate housing and maintaining their body condition during the different aspects of reproduction to prevent this clinical problem.

Dedication

Dedicated to the memory of the late Professor S.J. Roberts, a founding father of theriogenology, who drew the author to theriogenology by his monumental and exhaustive textbook that he created during the noncomputerized age and nurtured this author during his academic career.

Acknowledgments

The author thanks Professor Jane Yatcilla, Veterinary Medical Librarian of Purdue University College of Veterinary Medicine for the literature search. The creation of this chapter would not have been possible without the information gleaned from earlier books and review articles on this and related topics. Only the following are listed in the bibliography and the author acknowledges the creativity of these scholars.

References

1. CAB Abstracts, OvidSP database, date of search 10/12/2012.
2. Roberts S. Parturition, and injuries and diseases of the puerperal period. In: Roberts S (ed.) *Veterinary Obstetrics and Genital Diseases (Theriogenology)*, 2nd edn. Woodstock, VT: published by the author, 1971, pp. 208–209 and 317–325.
3. Squire A. Therapy for retained placenta. In: Morrow DA (ed.) *Current Therapy in Theriogenology 1*. Philadelphia: WB Saunders, 1980, pp. 186–189.
4. Grunert E. Etiology of retained bovine placenta. In: Morrow DA (ed.) *Current Therapy in Theriogenology 1*. Philadelphia: WB Saunders, 1980, pp. 182–186.
5. Grunert E. Etiology and pathogenesis of retained bovine placenta. In: Morrow DA (ed.) *Current Therapy in Theriogenology 2*. Philadelphia: WB Saunders, 1986, pp. 237–242.
6. Eiler H. Retained placenta. In: Youngquist RS (ed.) *Current Therapy in Large Animal Theriogenology*. Philadelphia: WB Saunders, 1997, pp. 340–348.
7. Eiler H, Fecteau K. Retained placenta. In: Youngquist RS, Threlfall WR (eds) *Current Therapy in Large Animal Theriogenology*, 2nd edn. St Louis, MO: Saunders Elsevier, 2007, pp. 345–354.
8. Wetherill G. Retained placenta in the bovine. A brief review. *Can Vet J* 1965;6:290–294.
9. Arthur G. Retention of the afterbirth in cattle: a review and commentary. *Vet Annu* 1979;19:26–36.
10. Gustaffson B, Ott R. Current trends in the treatment of genital infections in large animals. *Comp Cont Educ Pract Vet* 1981;3: S147–S152.
11. Paisley L, Mickelson W, Anderson P. Mechanism and therapy for retained fetal membranes and uterine infections of cows: a review. *Theriogenology* 1986;25:353–381.
12. Momont H. Therapy for reproductive problems in the postpartum cow. *Bov Pract* 1986;21:157–158.
13. Barnouin J, Chassagne M. An aetiological hypothesis for the nutrition-induced association between retained placenta and milk fever in the dairy cow. *Ann Rech Vet* 1991;22:331–343.
14. Joosten I, Hensen E. Retained placenta: an immunological approach. *Anim Reprod Sci* 1992;28:451–461.
15. Laven R, Peters A. Bovine retained placenta: aetiology, pathogenesis and economic loss. *Vet Rec* 1996;139:465–471.
16. Peters A, Laven R. Treatment of bovine retained placenta and its effects. *Vet Rec* 1996;139:535–539.
17. Schlafer D, Fisher P, Davies C. The bovine placenta before and after birth: placental development and function in health and disease. *Anim Reprod Sci* 2000;60–61:145–160.
18. Kankofer M. Placental release/retention in cows and its relation to peroxidative damage of macromolecules. *Reprod Domest Anim* 2002;37:27–30.
19. Kindahl H, Kornmatitsuk B, Gustafsson H. The cow in endocrine focus before and after calving. *Reprod Domest Anim* 2004;39:217–221.
20. Davies C, Hill J, Edwards J *et al.* Major histocompatibility antigen expression on the bovine placenta: its relationship to abnormal pregnancies and retained placenta. *Anim Reprod Sci* 2004;82–83:267–280.
21. Frazer G. A rational basis for therapy in the sick postpartum cow. *Vet Clin North Am Food Anim Pract* 2005;21:523–568.
22. McNaughton A, Murray R. Structure and function of the bovine fetomaternal unit in relation to the causes of retained fetal membranes. *Vet Rec* 2009;165:615–622.

23. Beagley J, Whitman K, Baptiste K, Scherzer J. Physiology and treatment of retained fetal membranes in cattle. *J Vet Intern Med* 2010;24:261–268.
24. Gumen A, Keskin A, Yilmazbas-Mecitoglu G, Karakaya E, Wiltbank M. Dry period management and optimization of post-partum reproductive management in dairy cattle. *Reprod Domest Anim* 2011;46(Suppl. 3):11–17.
25. Modric T, Momcilovic D, Gwin W, Peter A. Hormonal and antimicrobial therapy in theriogenology practice: currently approved drugs in the USA and possible future directions. *Theriogenology* 2011;76:393–408.
26. Rizzo A, Roscino M, Binetti F, Sciorsci R. Roles of reactive oxygen species in female reproduction. *Reprod Domest Anim* 2012;47:344–352.

Chapter 50

Postpartum Uterine Infection

Colin Palmer

Department of Large Animal Clinical Sciences, Western College of Veterinary Medicine,
University of Saskatchewan, Saskatoon, Saskatchewan, Canada

Introduction

The postpartum period is one of the most important periods in a cow's reproductive life and is a key area for veterinarians to address in terms of prevention, diagnosis, and appropriate treatment of uterine diseases and their sequelae. Most reproductive problems in cattle occur during this period, although the ultimate goal must be to have the cow calve normally and proceed through the postpartum period with as few problems as possible.

Uterine infections are a frequent disorder during the postpartum period. Dairy cattle seem to be more susceptible than beef cattle, or are more likely to be presented for examination on an individual or herd basis. The reason for this may be that dairy cattle are under more production stress; are more intensively housed and therefore exposure to pathogenic strains of bacteria is more likely; are observed more closely, facilitating diagnosis; and resume ovarian cyclicity sooner than beef cattle, which alters local uterine immune function.

The proportion of cattle developing uterine disease varies greatly, even between what outwardly appear to be similar farms in the same geographical region. Milk production, hygiene, and other management factors probably explain most of the variability. Overall, 20–40%, or more of dairy cattle may develop one or more uterine diseases during lactation,[1] resulting in economic loss due to decreased milk production, impaired fertility, premature culling, and death. Uterine disease is not new yet the incidence seems to be increasing. A reasonable explanation may be heightened awareness and improved diagnostic capabilities facilitated by advanced statistical and herd management software that only now enables us to realize the true economic impact of uterine disease.

Predisposing factors

Prolonged calving, twins, dystocia, and retained fetal membranes (RFM) are often implicated as predisposing factors for uterine infection.[2,3] The three principal postpartum uterine infections, metritis, endometritis and pyometra, are generally considered separately in the literature and have been shown to have additional distinct risk factors which support the conjecture that they are separate conditions.[4] To put this in perspective, not all cows that develop metritis will later be diagnosed with endometritis. However, a relationship still exists as cows afflicted with metritis are approximately twice as likely to develop endometritis.[5,6] Additional risk factors for endometritis include parity, season and body condition score, ketosis, and milk production.[3,5]

Cows that developed metritis ate 2–6 kg less feed than their healthy contemporaries 2–3 weeks prior to showing signs of metritis[7] and spent significantly less time feeding during the last 3 weeks before calving.[8] It is plausible that a relationship exists between appetite and impaired immune function that begins weeks before calving.[9] Decreased dry matter intake, increased serum nonesterified fatty acid, and β-hydroxybutyrate concentrations impair neutrophil function enabling the development of metritis.[10]

Retained placenta or RFM is usually implicated as a cause of uterine infection and with good reason. The odds of a cow with RFM developing metritis are six times greater than that of her unaffected herd mate, far greater than other conditions.[11] Calving areas, bedding, and overall farm hygiene should also be considered, but are surprisingly far less important than other risk factors for metritis.[12]

Pathophysiology

During the first 6 weeks after calving the uterus undergoes the normal process of sloughing and regeneration of the endometrium known as involution.[13] Typically, lochia is passed until approximately 23 days postpartum and progresses from thick, mucoid and bloody to purulent discharge.[14] Practically all cows have bacterial infection of the uterus during the first 2 weeks postpartum;[15] however, the perpetuation of infection depends on the presence of substrate, the degree of bacterial contamination, and the various uterine defense mechanisms.[16] Pathogenic organisms must adhere to the mucosa, multiply on the surface, or penetrate the epithelium. Some bacteria produce compounds that directly promote disease or the growth of other bacteria.[1] Elevated

---

Bovine Reproduction, First Edition. Edited by Richard M. Hopper.
© 2015 John Wiley & Sons, Inc. Published 2015 by John Wiley & Sons, Inc.

progesterone and estrogen concentrations in late gestation effectively suppress the immune system of the dairy cow and this carries onward through the first 3 weeks of the postpartum period.[9,17] Both hormones are capable of modifying the expression of hormone receptors in immune cells and both, especially progesterone, reduce the secretion of prostaglandins from stromal cells stimulated by Gram-negative cell wall lipopolysaccharide (LPS). Furthermore, metabolic substances such as insulin-like growth factor (IGF)-1 and numerous other proteins have been shown to have immunomodulatory properties in the endometrium.[17] During and around the time of parturition the physical barriers to genital infection – the vulva, vagina and cervix – are compromised enabling entry of bacteria into the uterus. Throughout the immediate postpartum period the pattern of infection, clearance, and reinfection occurs repeatedly.[13,14] A variety of bacterial species may be cultured from the uterus during the postpartum period; however, virulent strains of *Trueperella pyogenes*, capable of causing disease alone or in combination with the Gram-negative anaerobes *Fusobacterium necrophorum* and *Prevotella melaninogenicus*, are most frequently encountered.[14] Until recently the microbial flora of the postpartum uterus has been characterized exclusively by bacterial culture. Metagenomic DNA extracted from the uterine fluid of postpartum dairy cows and amplified using polymerase chain reaction (PCR) technology has provided evidence of a more diverse and complex microbiota than previously described. Many clone sequences belonged to groups of bacteria that had not yet been cultured and there were distinct differences in the types of bacteria between healthy cows and those diagnosed with metritis. Fusobacteria, Bacteroidetes and Proteobacteria represented the majority of clones from metritic cows; however, *Escherichia coli* and *T. pyogenes* were not detected, probably because sampling was performed 10 days after calving.[18]

Escherichia coli is a common culture isolate from the uterus, but tends to be more prevalent earlier in the postpartum period;[13,14] *T. pyogenes* is prevalent more than 2 weeks after calving.[18] Infection with *E. coli* during the puerperal period may predispose to infection with other bacterial species.[1,17,19–21] Until recently the true significance of *E. coli* beyond that of an opportunistic pathogen was debatable; however, six *E. coli* virulence factors were significantly associated with a higher risk of infected cows developing metritis, providing strong evidence that *E. coli* should be considered a primary uterine pathogen.[22]

Trueperella pyogenes is responsible for the most severe damage to the endometrium, facilitated by a cytotoxin pyolysin capable of destroying endometrial cells.[17] Although not consistently isolated together, a synergistic relationship exists between *T. pyogenes*, *F. necrophorum*, and *P. melaninogenicus* facilitated by the production of growth-promoting factors and substances shown to limit neutrophil chemotaxis and phagocytosis.[14] Damage to the endometrium interferes with the establishment and maintenance of early pregnancy and inflammatory mediators will reduce the number of trophectoderm cells around the embryo.[1,23] Uterine infections also impair the growth of dominant follicles, impair ovulation, and are associated with lowered plasma estradiol and progesterone concentrations,[21,24–26] and have been associated with prolonged luteal phases and cystic ovarian disease.[27]

Defining uterine infections

Uterine infections primarily are named for the severity, extent and characteristics of the inflammatory reaction. However, since inflammation is associated with normal involution, it is important to consider the number of days since calving to determine if the degree and severity of signs are consistent with disease.

Metritis

Metritis is defined as inflammation of the uterine wall including the endometrium, muscular layers, and serosa. Metritis differs greatly from endometritis both histologically and clinically. Unfortunately, the term "metritis" is commonly misused when referring to both of these conditions. Researchers and clinicians should make an effort to use the correct terminology to mitigate confusion.

Most cases of metritis occur within 1–2 weeks of calving. The most common and severe presentation, referred to as puerperal or toxic metritis, occurs within the first 2 weeks after calving and is characterized by a fetid, watery, brown uterine discharge and severe systemic illness. Affected animals are usually depressed, anorexic, dehydrated, and febrile (body temperature >39.5°C). The uterus is palpably enlarged, flaccid, and lacks longitudinal rugae or ridges typical of a normal postpartum uterus. If metritis occurs in the absence of systemic illness, then the condition may be referred to as simply metritis[28] or as acute postpartum or acute puerperal metritis.[29,30]

Multiparous cows with puerperal metritis consume less feed and produce less milk and are more likely to be culled than their healthy herdmates, but these effects are apparently not realized in primiparous cows.[31,32] Uterine inflammatory conditions with purulent exudate occurring beyond 2 weeks after calving are difficult to distinguish from normal uterine involution. The current belief is that the severity of metritis decreases with time and that a proportion of these cases will persist as endometritis.[33]

Endometritis

Endometritis is defined as inflammation limited to the endometrium. Histologically, evidence of inflammation should not extend beyond the stratum spongiosum.[34] Endometrial inflammation involves disruption of the epithelium, increased blood flow, edema, and an influx of inflammatory cells, mostly neutrophils and lymphocytes.[9]

Endometritis has been further subcategorized as clinical and subclinical disease. Diagnostic signs associated with clinical endometritis are obvious to the eye and include purulent or mucopurulent (50% pus, 50% mucus) discharge visible at the vulva or present within the cranial vagina.[35] With respect to subclinical endometritis, purulent or mucopurulent uterine discharge is not evident, yet infection and inflammation persist. Examination of cellular material harvested directly from the uterine lumen and endometrium is used to assess the level of inflammation.

Neutrophils are the first and most significant inflammatory cell involved in endometritis, but are also foremost during normal uterine involution. The inflammatory cell response in cases of subclinical endometritis is widely believed to be quantifiably more severe than that associated with normal involution yet milder than clinical endometritis. However, despite supposed differences in volume of neutrophils, fluid, mucus and other debris, it is doubtful that the negative impact on reproductive performance is any less severe. Because it is so difficult to differentiate real disease having a negative impact on reproductive performance from normal uterine involution, diagnostic indicators must be evaluated relative to the number of days after calving. The current dogma among many of those who study endometritis is that attempts to diagnose endometritis should be delayed until after the first month post calving.

Dubuc et al.[4] have provided evidence that clinical and cytological endometritis may be distinct manifestations of reproductive tract disease. When endometrial cytology was assessed in cows presenting with vaginal discharge typical of clinical endometritis, only 38% and 36% of cows examined at, respectively, days 35 and 56 postpartum also had evidence of cytological endometritis. Cows having each of these conditions experienced a negative effect on reproductive performance; moreover, when they were present in the same cow the detrimental effect on reproductive performance was additive. The source of the vaginal discharge was speculated to be the cervix, specifically cervicitis. Another plausible reason for the disparity in findings was that cytological preparations prepared from material harvested from the uterine body might not be representative of inflammatory cell accumulations in the uterine horns. These authors proposed that purulent vaginal discharge was a more descriptive and accurate term for vaginal discharge previously considered diagnostic for clinical endometritis. Furthermore, cows with purulent vaginal discharge responded to antibiotic treatment deposited within the uterus. This finding suggests that there is much more to learn regarding the inflammatory response of the postpartum uterus, cervix, and vagina.

Pyometra

Pyometra is defined as the accumulation of pus within the uterine lumen facilitated by a closed cervix and the presence of a corpus luteum.[28,36] The persistence of a corpus luteum is the key feature distinguishing pyometra from closely related endometritis and ensures that the cervix remains closed and there is absence of myometrial contractions to expel the pus into the vagina. Furthermore, the persistence of a progesterone-dominant state ensures that immune defenses remain inactive. Prolongation of the luteal phase may be attributed to increased concentrations of luteotrophic prostaglandin (PG)E_2 associated with endometrial bacterial infection.[37] Endometrial cells exposed to E. coli LPS, T. pyogenes exotoxin, and bovine herpesvirus 4 preferentially secrete PGE_2 rather than $PGF_{2\alpha}$.

However, extension of the luteal phase is not exclusive to pyometra and may occur under the right conditions in cases of endometritis.[1] Pyometra is relatively rare in comparison to metritis and endometritis,[33] but will persist indefinitely until luteolysis occurs. Cows that ovulate when pathogenic bacteria are still present in the uterus are predisposed to the development of pyometra.[38] An example of iatrogenically induced pyometra is the use of gonadotropin-releasing hormone (GnRH) to advance ovarian cyclicity in postpartum cows without a follow-up injection of $PGF_{2\alpha}$.

Diagnosis

Metritis

A complete physical examination is the most effective way to diagnose puerperal metritis. Typically, the individual cow is presented because she has failed to pass her placenta, is off her feed, is "sickly," or is experiencing decreased milk production, or all four. Systemic signs such as depression, dehydration, anorexia, and body temperature above 39.5 °C in any cow occurring within 1–2 weeks of calving should warrant further examination of the uterus and its contents. Not all sick cows with metritis will have a fever and not all cows with a fever in the immediate postpartum period will have metritis, so herdspersons should be instructed to observe postpartum cows carefully rather than simply flagging cows with elevated body temperatures for further examination. Transrectal palpation of the uterus is a useful tool for evaluating uterine size, tone and texture, and to facilitate the expulsion of fluid by gentle massage. The presence of a fetid, reddish brown, watery discharge in addition to the aforementioned systemic signs can be considered diagnostic for metritis. Normal lochia is reddish brown to white and lacks a significant odor. In cases of metritis the uterus is also very large and flaccid without longitudinal ridges (also known as rugae) indicative of normal postpartum uterine contraction. However, assessment of uterine size and tone is subjective. Clinicians should strive to gather experience by palpating otherwise normal postpartum cows so the that abnormal becomes apparent when presented. Other common clinical findings are leukocytosis with or without a left shift and ketosis. Displacement of the abomasum may be diagnosed simultaneously and will present additional clinical signs specific to this condition.

Endometritis

The diagnosis and treatment of endometritis has been controversial. Researchers and practitioners constantly struggle with appropriate case definitions and ultimately with determination of treatment effectiveness. A perfect diagnostic test must identify all cows at risk of impaired reproductive performance due to both clinical and subclinical endometritis and must offer an immediate answer.

Diagnosis was once based solely on transrectal uterine palpation. Enlarged uterine horn(s), asymmetry of the uterine horns, thickness of the endometrium, and the presence of a palpable uterine lumen and/or palpable fluid within the lumen were attributed to this condition. One of the most significant obstacles to overcome when diagnosing endometritis has been to establish the most appropriate time to examine cows for evidence of disease. In recent years there has been a concerted effort to establish diagnostic protocols to facilitate an accurate diagnosis. Measures of reproductive performance including pregnancy rate, calving to first service interval, first-service conception rate, service per pregnancy, and overall conception rate have been used to retrospectively determine diagnostic thresholds. Understanding that uterine involution involves sepsis and an inflammatory process yet is a normal physiologic process is critical. Realizing that impairment of reproductive performance caused by complicated or delayed uterine involution explains the majority of cases of endometritis is paramount. Even in cows experiencing a normal postpartum period, the uterine horn diameter does not reach the pregravid size of 4–5 cm until 25–30 days after calving.[39] LeBlanc et al.[35] reported the results of a large-scale study involving 1865 dairy cows examined between 27 and 33 days in milk. One of the objectives was to investigate and validate diagnostic criteria for endometritis. Uterine characteristics including location of the uterus, symmetry of the uterine horns, uterine horn diameter, evidence of a palpable uterine lumen, and thickness of the uterine horn wall were assessed by transrectal palpation. None of the palpable uterine characteristics were associated with the relative pregnancy rate. Use of transrectal palpation of the uterus as the principal method for diagnosing endometritis should be avoided as it is far too subjective and imprecise. From a pragmatic point of view, far too many cases of endometritis will be missed, potentially leaving a number of diseased animals at risk for impaired reproductive performance if left untreated.

Endometrial tissue biopsy and uterine bacteriologic culture could perhaps be considered the most direct, objective, and probably reliable diagnostic tests for uterine inflammation and infection, respectively. Neither technique is used, however, as uterine biopsy has been associated with decreased first-service conception rates and the vast majority of infections beyond 3 weeks after calving are associated with *T. pyogenes*. Furthermore, a single uterine biopsy is not considered representative of the entire endometrium in the cow.[34]

Vaginoscopy is a rapid and simple technique for the diagnosis of purulent vaginal discharge. The use of vaginoscopy for the diagnosis of clinical endometritis is based on the premise that purulent exudate present in the cranial vagina is probably the result of drainage from the uterus. The nature of the discharge is important. Clear mucus is normal, whereas purulent (>50% pus) and mucopurulent (approximately 50% pus and 50% mucus) and foul-smelling discharge are indicative of disease.[4,35] *Trueperella pyogenes*, *Fusobacterium necrophorum*, and *Proteus* sp. uterine infections have been associated with purulent and mucopurulent vaginal discharge, whereas fetid discharge was associated with *T. pyogenes*, *E. coli*, and *Mannheimia haemolytica* infections. Other known uterine pathogens and opportunistic bacteria were not associated with vaginal mucus character or odor.[40] By utilizing vaginoscopy after 26 days postpartum, 44% of cows with an abnormal discharge were identified that would have otherwise gone unnoticed if only palpation and external examination were used. By delaying the vaginal examination until approximately 1 month after calving, false positives (i.e., cows undergoing normal involution) were less likely.[35] Other ways of detecting uterine discharge have been studied, including the gloved hand[41] and the Metricheck device (Simcrotech, Hamilton, New Zealand) (Figure 50.2).[42] These alternatives are at least as efficacious as vaginoscopy and may offer the advantage of detecting

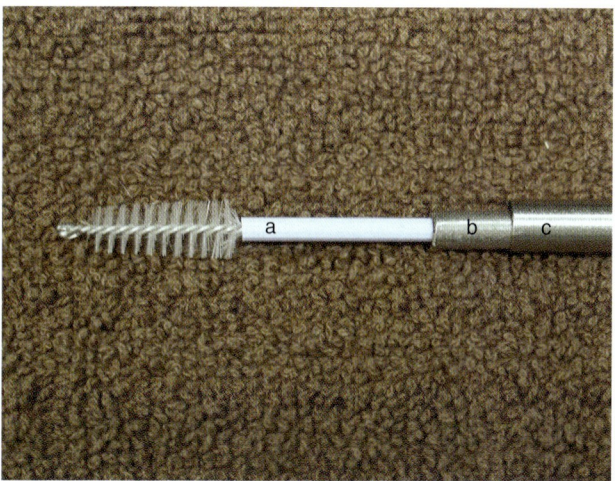

Figure 50.1 Cytobrush and modified handle (a) threaded into steel rod (b) and outer guard (c).

exudate that would otherwise go unnoticed, especially in cases where the cranial vagina slopes ventrally. Another practical advantage is that it is much easier for the examiner to avoid being soiled. Those with larger hands and arms may find the gloved hand technique difficult to employ, whereas the Metricheck device is easy to insert and easy to clean between cows.

Vaginoscopy, or a similar procedure, offers an immediate result, but fails to identify all cows at risk of poor reproductive performance due to endometritis. Subclinical endometritis cannot be diagnosed by inspection of vaginal exudate; however, if no other screening tests are being used, routine vaginal examination to detect mucopurulent or purulent exudate is a simple, reliable, and cost-effective way to identify cows at risk of impaired reproductive performance.

Endometrial cytology, based on the presence of cellular evidence of inflammation, is currently considered to be the most accurate way to diagnose endometritis in cattle, both clinical and subclinical. Inflammatory cells may be recovered by either of two techniques: uterine lavage or cytobrush. Uterine lavage has the advantage of utilizing readily available materials: 20–60 mL of 0.9% saline is flushed into the uterine lumen using a sterile infusion pipette followed by gentle massage of the uterus before aspiration of the fluid back into the syringe attached to the infusion pipette. Fluid samples are centrifuged, cellular debris is recovered and smeared onto a slide, and microscopic analysis is performed after staining with modified Wright Giemsa stain.[43] The cytobrush (Figure 50.1) technique employs a small brush and handle combination designed for human cervical cytological examination that has been modified for use in the cow.[44] The handle is cut to a length of 3 cm or less and is threaded into a 65-cm long, 4-mm diameter stainless steel rod. A 5–6 mm diameter stainless steel tube 50-cm long is used to guard the modified cytobrush and rod for passage through the cervix. A sanitary plastic sleeve is placed over the tube containing the rod and cytobrush to protect the instrument from contamination during passage through the vagina. Commercially available cytobrush devices designed specifically for harvesting endometrial cellular material from large animals are an alternative offering many obvious advantages.

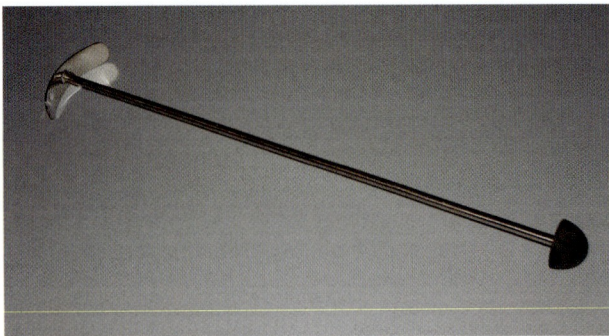

Figure 50.2 Metricheck® device.

Table 50.1 Published threshold neutrophil percentages indicative of subclinical endometritis at differing points in the postpartum period.

Neutrophil percentage	Days postpartum	Reference
≥4	56	Dubuc et al.[4]
>5	40–60	Gilbert et al.[43]
≥6	35	Dubuc et al.[4]
>8	28–41	Barlund et al.[45]
>10	34–47	Kasimanickam et al.[46]
>18	20–33	Kasimanickam et al.[46]

The guard and cytobrush are passed through the vagina to the cervix where the outer plastic sheath is perforated; the guard while still protecting the brush is gently passed through the cervix and then the brush is advanced into the uterine body. Most operators rotate the brush one-quarter to half a turn, but squeezing of the uterine walls should be avoided to prevent damage to the endometrium. The brush is withdrawn into the guard prior to removal from the reproductive tract.[45] Cellular material adhering to the brush is rolled onto a clean microscope slide and air dried prior to staining with Wright Giemsa stain. The cytobrush technique does not require centrifugation and collection of cellular material; however, cellular material is only collected from the area where the cytobrush contacts the endometrial surface. The lavage technique offers the potential to gather cellular debris from deeper within the uterus. Despite this difference, Barlund et al.[45] reported a high level of agreement between lavage and cytobrush cytology. Difficulties with fluid recovery[44] and a loss of cell definition[45] in cytosmears prepared from lavage fluid have been reported disadvantages.

The cytological criteria for the diagnosis of subclinical endometritis continue to be refined, with the postpartum interval for sampling being a key variable. Assessments of the severity of inflammation are made by determining the percentage of polymorphonuclear (PMN) cells per 100 cells (PMNs plus endometrial cells) at 400× magnification.[4,43,45,46] Barlund et al.[45] reported that a threshold of more than 8% PMNs was the lowest proportion of PMNs significantly affecting pregnancy status at 150 days postpartum. Despite the specificity of this threshold at 89.9%, the sensitivity was poor at 12.9%, indicating that there are many reasons for nonpregnancy apart from cytological evidence of endometritis. Other investigators have defined subclinical endometritis based on similar proportions of PMNs, but with a decreasing threshold as the postpartum period advanced (Table 50.1).

Ghasemi et al.[47] used expression of the proinflammatory cytokines tumor necrosis factor (TNF)-α, interleukin (IL)-6, and IL-8 by endometrial cells harvested using the cytobrush technique to determine a PMN percentage threshold indicative of endometritis in postpartum cows. A threshold above 18% PMNs was the lowest percentage significantly associated with the elevation of all three cytokines in cows sampled 31 ± 1 days after calving. This threshold is much higher than those based on survival analysis of days to pregnancy.

Ultrasonographic evidence of intrauterine fluid has been shown to be associated with bacterial growth and impaired uterine involution.[27] Uterine fluid detected in cows examined between 20 and 47 days postpartum was associated with a 62–63 day increase in median days open compared with those having no fluid.[46] Both Kasimanickam et al.[46] and Barlund et al.[45] reported that the presence of fluid in the uterine lumen and positive endometrial cytology were effective for diagnosing endometritis based on increased days to pregnancy; however, the tests did not identify the same cows. Evidence to date suggests that endometrial cytology identifies cows that have a cellular inflammatory response, whereas the presence of intrauterine fluid may be due to decreased uterine clearance mechanisms.

Attempts to improve on the diagnostic capabilities of ultrasonography by measuring uterine luminal diameter and quantifying endometrial thickness have been made. Carefully selected thresholds of >1 and >3 mm uterine luminal diameter, and >7 and >8 mm of endometrial thickness were not reliable for predicting pregnancy status at 150 days and compared poorly with cytobrush cytology.[45] Ultrasound technology is useful for identifying some but not all cows with endometritis. This is unfortunate as ultrasound has the advantage of offering an immediate diagnosis.

Novel techniques are being studied to improve our diagnostic abilities. Optical density measurements (wavelength 620 nm) of processed uterine lavage samples were highly associated with an accepted threshold proportion of neutrophils considered diagnostic for endometritis, and the presence of T. pyogenes in the sample was associated with higher mean optical densities. Uterine lavage fluid optical density may be a much less laborious yet equally accurate alternative to conventional cytobrush cytology.[48]

Pyometra

Pyometra is often diagnosed during routine herd visits to palpate cows presented for anestrus. The presence of palpable fluid in a uterine horn and a palpable corpus luteum in the absence of a fetus or fetal membrane slip is diagnostic. Ultrasonography is also an excellent diagnostic tool, especially for the less experienced clinician. A distinct advantage of ultrasonography is the ability to visualize the uterine fluid, which differs greatly from amniotic or allantoic fluid. In cases of pyometra the uterine lumen of one or both horns is filled with variable quantities of thick fluid of mixed echodensity owing to the high cellular content of pus. The presence of a corpus luteum supports the diagnosis.

Treatment

Metritis

Metritis has traditionally been treated with antibiotics delivered systemically, or infused directly into the uterine lumen. The value of supportive care should never be underestimated; in severely affected animals, anti-inflammatory agents and intravenous fluid therapy should also be employed. In a retrospective analysis of the management of 78 cases of postpartum metritis it appeared that all that was needed was to administer therapy for life-threatening changes while the uterus healed itself.[49]

Systemic antibiotic therapy appears to offer many advantages over intrauterine infusion. Withdrawal times are known, distribution to all layers of the uterus is possible, and systemic antibiotic administration is less harmful to the uterine environment.[50] Penicillin was once the preferred antibiotic for treating postpartum metritis as it penetrates all layers of the uterus, is inexpensive, and most of the bacteria penetrating the endometrium and causing septicemia were sensitive to penicillin.[50,51] A typical dose is procaine penicillin G 22 000 IU/kg i.m. once a day for 3–5 days. Milk should be withheld for at least 96 hours and the animal should not be slaughtered for use as food for 10 days after the last treatment. Alternatively, ceftiofur sodium 1 mg/kg i.m. or s.c. may be administered for 3–5 days with the advantage of no milk withdrawal at this dosage. Ceftiofur sodium and ceftiofur hydrochloride have been found to concentrate in uterine tissues at levels exceeding the mean inhibitory concentration for *T. pyogenes*, *F. necrophorum* and *E. coli*.[50,52] Ceftiofur hydrochloride sterile suspension was also shown to be effective for the treatment of acute postpartum metritis at a dosage of 1.1 or 2.2 mg/kg i.m. or s.c. for 5 days with the higher dose being more beneficial.[29] At 1.1 mg/kg, labeled meat and milk withdrawal instructions may be followed; however, 2.2 mg/kg represents an off-label usage.

Oxytetracycline has also been a popular drug for treating postpartum metritis. Intravenous dosages of 11 mg/kg oxytetracycline administered twice daily only maintained a mean tissue concentration above 5 µg/g in the uterine wall for the first 4 hours after treatment, reaching a maximum of 9 hours by the fifth day of treatment. Slightly higher levels were found in the caruncles and endometrium for longer periods of time and were approximately equivalent to a serum concentration of 5 µg/mL. The minimum inhibitory dose for *T. pyogenes* in uterine isolates has been reported to be 20.4 µg/mL. This information suggests that oxytetracycline is a poor treatment choice for postpartum metritis.[53] The only value may be for treating cases of metritis having only mild to moderate systemic involvement (e.g., slight depression).

A variety of compounds have been infused into the uterus in an attempt to destroy bacteria, enhance uterine defense mechanisms, or increase uterine tone and blood flow. The ideal treatment should remove harmful bacteria from the uterus but at the same time should not be damaging to the uterus or impair its defense mechanisms. Perhaps the most common infusion is a water or saline solution containing iodine. Not surprisingly, there is no evidence to support the use of iodine-based uterine infusions. In fact, as little as 50–100 mL of 2% polyvinylpyrrolidone iodine (povidone-iodine) solution as a routine therapy 30 days postpartum had a detrimental effect on fertility compared with nontreated animals.[54] Others have supported dextrose infusions; the outcome of infusion of 1 L of 50% dextrose into the uterus of cows experiencing an abnormal postpartum period has never been critically evaluated and probably does little more than cause an increase in uterine tone.

As a rule, intrauterine antibiotic or antiseptic infusion should be avoided as a treatment for postpartum metritis. When antibiotics are infused into the uterus we are often uncertain whether the drug is distributed throughout all layers of the uterus. Moreover, many agents injected systemically or absorbed from the uterus raise concerns about appropriate meat and milk withdrawal periods as there is at least some absorption of these compounds.[51,55] Most common drugs are not registered for intrauterine use and many are rendered ineffective in the postpartum uterus.

The penicillin family of drugs performs poorly when infused early in the postpartum period as there are a number of organisms producing inactivating (β-lactamase) enzymes.[51] Tetracyclines are very irritating to the bovine uterus and most formulations should not be used for intrauterine therapy. All intrauterine antibacterials have been found to have a negative effect on leukocyte function and placement risks iatrogenic contamination or further injury to the uterus.[52] Oxytetracycline-resistant strains of *T. pyogenes* are reportedly common, providing further evidence that intrauterine oxytetracycline infusion should be avoided.[56]

Prostaglandin therapy in the immediate postpartum period has been directed toward increasing uterine tone and expulsion of uterine fluid and bacteria rather than changing the hormonal influence through luteolysis. For the most part, studies extolling the benefits of $PGF_{2\alpha}$ for the treatment of metritis have largely been refuted.[57] A notable exception was a study showing that two doses of $PGF_{2\alpha}$ intramuscularly 8 hours on day 8 postpartum in cows with acute puerperal metritis increased first-service conception rates by 17%, but only in primiparous cows; there was no benefit in multiparous cows.[30] Such a disparity in findings coupled with relatively small group sizes suggests that this finding may not have been real. Similarly, oxytocin 25–30 IU at least three times per day during the first week postpartum in cows with metritis was reported to be the most physiologic means of promoting fluid expulsion, but to date there have been no large-scale scientific studies to support its use.[57]

Estrogens have long been used in the immediate postpartum period for the treatment or prevention of metritis in cows based on the rationale that they increase uterine tone and stimulate phagocytosis by inducing an estrogenic state, and they also enhance the myometrial response to oxytocin.[58] Critical evaluation using dosages of 4 and 5 mg of estradiol cypionate within 36 hours of calving failed to substantiate these beliefs.[58,59] Unsubstantiated use of steroid hormones in food animals will only contribute to perceived food safety issues and should be avoided.

Despite the importance of RFM in the pathogenesis of metritis, it must be understood that RFM alone only has mild effects on future fertility. Problems occur when cases are complicated by metritis. Any therapy for RFM should be directed toward the treatment of metritis. Manual removal of RFM has been shown to be detrimental to future fertility, especially in cows with metritis[60] and intrauterine antibiotics impair putrefactive processes necessary for dissolution of the membranes.[51] Ecbolic agents such as $PGF_{2\alpha}$ and its synthetic analogs, ergot derivative, and oxytocin have not been shown to affect the expulsion of the placenta.[51,61] The best advice is to disregard the membranes and focus efforts on the treatment of metritis.

Endometritis

Treatment of endometritis is currently based on two different regimens, the intramuscular injection of $PGF_{2\alpha}$ and intrauterine infusion of antibiotics. Numerous other treatment protocols have been popularized, but have not withstood scientific scrutiny. Adequate sample sizes, improved understanding of uterine involution, and using measures of reproductive performance in treatment outcomes assessment are requirements of valid studies.

The rationale for treatment of endometritis with $PGF_{2\alpha}$, or a synthetic analog, is to stimulate uterine defense mechanisms by destroying the corpus luteum and removing the progesterone source.[62,63] Uterotonic effects in the absence of a corpus luteum are believed to be negligible.[9] Treatment of subclinical endometritis with cloprostenol between 20 and 33 days in milk resulted in a 70% improvement in the risk of those cows becoming pregnant compared with their untreated counterparts.[64] Presence or absence of a functioning corpus luteum was not reported; however, it is believed that $PGF_{2\alpha}$ treatment is more effective when luteal tissue is present,[65,66] hence the recommendation that the use of $PGF_{2\alpha}$ for the treatment of uterine disease be reserved until after the first month postpartum.[9] Estrous synchronization programs, particularly those utilizing multiple injections of $PGF_{2\alpha}$, showed a beneficial effect in cows diagnosed with endometritis;[62] however, in two studies treatment with $PGF_{2\alpha}$ at 35 days and 49 days postpartum showed no benefit for resolving uterine disease.[67,68] Although it appears that $PGF_{2\alpha}$ does not have a direct benefit for the treatment of uterine disease, a positive effect of $PGF_{2\alpha}$ treatment on reproductive performance was demonstrated in cows with poor body condition.[68]

Intrauterine infusion of antimicrobials is aimed at obtaining greater concentrations of drug at the site of infection than would be achieved by systemic administration. Oxytetracycline was once the most popular agent infused into the uterus of cows, but is relatively ineffective as it is poorly absorbed into the deeper layers of the uterus.[53,65] Despite this, some absorption does occur and therefore milk and meat residues must also be a concern.[69] Some preparations may also cause coagulation necrosis of the endometrium.[70] Post-breeding uterine infusions on the day of, or the day following, insemination based on the premise that therapeutic efficacy is greater during the follicular phase have been shown to have a detrimental effect on pregnancy in the majority of cows, which was speculated to be caused by endometrial inflammation.[71]

Where available, benzathine cephapirin (Metricure, Merck Animal Health), a first-generation cephalosporin with no meat or milk withdrawal is the drug of choice for uterine infusion. In a large field study, 316 cows between 27 and 33 days in milk with clinical endometritis received cephapirin infusion, $PGF_{2\alpha}$ intramuscularly or were untreated controls. Cephapirin-treated animals became pregnant 60% faster than controls; however, time to pregnancy in the $PGF_{2\alpha}$-treated animals did not differ significantly from that in control animals.[66] It appears that not all cephalosporin preparations are alike; intrauterine infusion with 125 mg of ceftiofur hydrochloride (Spectramast LC, Pfizer Animal Health) at 44 days postpartum in cows with clinical endometritis reduced the prevalence of bacterial infections, but did not affect the pregnancy rate.[72]

Pyometra

Undisputedly, the best treatment for pyometra is a luteolytic (labeled dosage) injection of $PGF_{2\alpha}$ or its synthetic analog cloprostenol. Recommendations have not changed for decades, with the majority of animals responding after a single injection. Uterine evacuation begins as soon as 24 hours after treatment and estrous behaviour will often be displayed in 3–4 days. Insemination should be withheld until the first or even second estrus following the induced estrus, with more than two-thirds of the animals reportedly becoming pregnant.[73] Rare cases may involve some damage to the cervix, impairing uterine clearance. Gentle lavage with saline solution may be beneficial in these instances.

References

1. Sheldon I, Price S, Cronin J, Gilbert R, Gladsby J. Mechanisms of infertility associated with clinical and subclinical endometritis in high producing dairy cattle. *Reprod Domest Anim* 2009; 44(Suppl. 3):1–9.
2. Markusfeld O. Periparturient traits in seven high dairy herds. Incidence rates, association with parity, and interrelationships among traits. *J Dairy Sci* 1987;70:158–166.
3. Gröhn Y, Erb H, McCulloch C, Saloniemi H. Epidemiology of reproductive disorders in dairy cattle: associations among host characteristics, disease and production. *Prev Vet Med* 1990; 8:25–39.
4. Dubuc J, Duffield T, Leslie K, Walton J, LeBlanc S. Definitions and diagnosis of postpartum endometritis in dairy cows. *J Dairy Sci* 2010;93:5225–5233.
5. Cheong S, Nydam D, Galvão K, Crosier B, Gilbert R. Cow-level and herd-level risk factors for subclinical endometritis in lactating Holstein cows. *J Dairy Sci* 2011;94:762–770.
6. Rutigliano H, Lima F, Cerri R et al. Effects of method of presynchronization and source of selenium on uterine health and reproduction in dairy cows. *J Dairy Sci* 2008;91:3323–3336.
7. Huzzey J, Veira D, Weary D, von Kerserlingk M. Prepartum behaviour and dry matter intake identify cows at risk for metritis. *J Dairy Sci* 2007;90:3220–3233.
8. Urton G, von Keyserlingk M, Weary D. Feeding behaviour identifies dairy cows at risk for metritis. *J Dairy Sci* 2005;88: 2843–2849.
9. LeBlanc S. Postpartum uterine disease and dairy herd reproductive performance: a review. *Vet J* 2008;176:102–114.
10. Hammon D, Evjen I, Dhiman T, Goff J, Walters J. Neutrophil function and energy status in Holstein cows with uterine health disorders. *Vet Immunol Immunopathol* 2006;113:21–29.
11. Smith B, Risco C. Predisposing factors and potential causes of postpartum metritis in dairy cattle. *Comp Cont Educ Pract Vet* 2002;24:S74–S80.

12. Noakes D, Wallace L, Smith G. Bacterial flora of the uterus of cows after calving on two hygienically contrasting farms. *Vet Rec* 1991;128:440–442.
13. Herath S, Dobson H, Bryant C, Sheldon I. Use of the cow as a large animal model of uterine infection and immunity. *J Reprod Immunol* 2006;69:13–22.
14. Lewis G. Health problems of the postpartum cow, uterine health and disorders. *J Dairy Sci* 1997;80:984–994.
15. Sheldon I, Rycroft A, Zhou C. Association between postpartum pyrexia and uterine bacterial infection in dairy cattle. *Vet Rec* 2004;154:289–293.
16. Noakes D, Parkinson T, England G, Arthur G. *Arthur's Veterinary Reproduction and Obstetrics*, 8th edn. Elsevier Science, 2002, pp. 399–408.
17. Sheldon I, Cronin J, Goetze L, Donofrio G, Schuberth H. Defining postpartum uterine disease and the mechanisms of infection and immunity in the female reproductive tract in cattle. *Biol Reprod* 2009;81:1025–1032.
18. Santos T, Gilbert R, Bicalho R. Metagenomic analysis of the uterine bacterial microbiota in healthy and metritic postpartum dairy cows. *J Dairy Sci* 2011;94:291–302.
19. Dohmen M, Joop K, Sturk A, Bols P, Lohuis J. Relationship between intra-uterine bacterial contamination, endotoxin levels and the development of endometritis in postpartum cows with dystocia or retained placenta. *Theriogenology* 2000;54:1019–1032.
20. Gilbert R, Santos N, Galvão K, Brittin S, Roman H. The relationship between postpartum uterine bacteria and subclinical endometritis. *J Dairy Sci* 2007;90(Suppl. 1):469 (Abstract).
21. Williams E, Fischer D, Noakes D *et al*. The relationship between uterine pathogen growth density and ovarian function in the postpartum dairy cow. *Theriogenology* 2007;68:549–559.
22. Bicalho R, Machado V, Bicalho M *et al*. Molecular and epidemiological characterization of bovine intrauterine *Escherichia coli*. *J Dairy Sci* 2010;93:5818–5830.
23. Hill J, Gilbert R. Reduced quality of bovine embryos cultured in media conditioned by exposure to an inflamed endometrium. *Aust Vet J* 2008;86:312–316.
24. Peter A, Bosu W. Effects of intrauterine infection on the function of the corpora lutea formed after first postpartum ovulations in dairy cows. *Theriogenology* 1987;27:593–609.
25. Peter A, Bosu W. Relationship of uterine infections and folliculogenesis in dairy cows during early puerperium. *Theriogenology* 1988;30:1045–1051.
26. Sheldon I, Noakes D, Rycroft A, Pfeiffer D, Dobson H. Influence of uterine bacterial contamination after parturition on ovarian dominant follicle selection and follicle growth and function in cattle. *Reproduction* 2002;123:837–845.
27. Mateus L, Lopes da Costa L, Bernardo F, Robalo Silva J. Influence of puerperal uterine infection on uterine involution and postpartum ovarian activity in dairy cows. *Reprod Domest Anim* 2002;37:31–35.
28. Sheldon I, Lewis G, LeBlanc S, Gilbert R. Defining postpartum uterine disease in cattle. *Theriogenology* 2006;65:1516–1530.
29. Chenault J, McAllister J, Chester T, Dame K. Efficacy of ceftiofur hydrochloride sterile suspension administered parenterally for the treatment of acute postpartum metritis in dairy cows. *J Am Vet Med Assoc* 2004;224:1634–1639.
30. Melendez P, McHale J, Bartolome J, Archbald L, Donovan G. Uterine involution and fertility of Holstein cows subsequent to early postpartum $PGF_{2\alpha}$ treatment for acute puerperal metritis. *J Dairy Sci* 2004;87:3238–3246.
31. Østergaard S, Gröhn Y. Effect of diseases on test day milk yield and body weight of dairy cows from Danish research herds. *J Dairy Sci* 1999;82:1188–1201.
32. Wittrock J, Proudfoot K, Weary D, von Keyserlingk M. Metritis affects milk production and cull rate of Holstein multiparous and primiparous dairy cows differently. *J Dairy Sci* 2011;94:2408–2412.
33. Sheldon I, Williams E, Miller A, Nash D, Herath S. Uterine diseases in cattle after parturition. *Vet J* 2008;176:115–121.
34. BonDurant R. Inflammation in the bovine reproductive tract. *J Dairy Sci* 1999;82(Suppl. 2):101–110.
35. LeBlanc S, Duffield T, Leslie K *et al*. Defining and diagnosing postpartum uterine clinical endometritis and its impact on reproductive performance in dairy cows. *J Dairy Sci* 2002;85:2223–2236.
36. Ott R, Gustafsson B. Use of prostaglandins for the treatment of bovine pyometra and postpartum infections: a review. *Comp Cont Educ Pract Vet* 1981;3:S184–S187.
37. Manns J, Nkuuhe J, Bristol F. Prostaglandin concentrations in uterine fluid of cows with pyometra. *Can J Comp Med* 1985;49:436–438.
38. Noakes D, Wallace L, Smith G. Pyometra in a Friesian heifer: bacteriological and endometrial changes. *Vet Rec* 1990;126:509.
39. Mortimer R, Farin P, Stevens R. Reproductive examination of the non-pregnant cow. In: Youngquist RS (ed.) *Current Therapy in Large Animal Theriogenology*. Philadelphia: WB Saunders, 1997, pp. 268–275.
40. Williams E, Fischer D, Pfeiffer D *et al*. Clinical evaluation of postpartum vaginal mucus reflects uterine bacterial infection and the immune response in cattle. *Theriogenology* 2005;63:102–117.
41. Sheldon I, Noakes D, Rycroft A, Dobson H. Effect of postpartum manual examination of the vagina on bacterial contamination in cows. *Vet Rec* 2002;15:531–534.
42. MacDougall S, Macaulay R, Compton C. Association between endometritis diagnosis using a novel intravaginal device and reproductive performance in dairy cattle. *Anim Reprod Sci* 2007;99:9–23.
43. Gilbert R, Shin S, Guard C, Erb H, Frajblat M. Prevalence of endometritis and its effects on reproductive performance of dairy cows. *Theriogenology* 2005;64:1879–1888.
44. Kasimanickam R, Duffield T, Foster R *et al*. A comparison of the cytobrush and uterine lavage techniques to evaluate endometrial cytology in clinically normal postpartum dairy cows. *Can Vet J* 2005;46:255–259.
45. Barlund C, Carruthers T, Waldner C, Palmer C. A comparison of diagnostic techniques for postpartum endometritis in dairy cattle. *Theriogenology* 2008;69:1516–1530.
46. Kasimanickam R, Duffield T, Foster R *et al*. Endometrial cytology and ultrasonography for the detection of subclinical endometritis in postpartum dairy cows. *Theriogenology* 2004;62:9–23.
47. Ghasemi F, Gonzalez-Cano P, Griebel P, Palmer C. Proinflammatory cytokine gene expression in endometrial cytobrush samples harvested from cows with and without subclinical endometritis. *Theriogenology* 2012;78:1538–1547.
48. Machado V, Knauer W, Bicalho M, Oikonomou G, Gilbert R, Bicalho R. A novel diagnostic technique to determine uterine health of Holstein cows at 35 days postpartum. *J Dairy Sci* 2012;95:1349–1357.
49. Pugh D, Lowder M, Wenzel, J. Retrospective analysis of the management of 78 cases of postpartum metritis in the cow. *Theriogenology* 1994;42:455–463.
50. Smith B, Risco C. Therapeutic and management options for postpartum metritis in dairy cattle. *Comp Cont Educ Pract Vet* 2002;24:S92–S100.
51. Paisley L, Mickelsen W, Anderson P. Mechanisms and therapy for retained membranes and uterine infections of cows: a review. *Theriogenology* 1986;25:353–381.
52. Okker H, Schmitt E, Vos P, Scherpenisse P, Bergwerff A, Jonker F. Pharmacokinetics of ceftiofur in plasma and uterine secretions and tissues after subcutaneous postpartum administration in lactating dairy cows. *J Vet Pharmacol Ther* 2002;25:33–38.
53. Bretzlaff K, Ott R, Koritz G, Bevill R, Gustafsson B, Davis L. Distribution of oxytetracycline in genital tract tissues of postpartum cows given the drug by intravenous and intrauterine routes. *Am J Vet Res* 1983;44:764–769.

54. Nakao T, Moriyoshi M, Kawata K. Effect of postpartum intrauterine treatment with 2% polyvinylpyrrolidone-iodine solution on reproductive efficiency in cows. *Theriogenology* 1988;30:1033–1043.
55. Risco C, Youngquist R, Shore M. Postpartum uterine infections. In: Youngquist RS, Threlfall WR (eds) *Current Therapy in Large Animal Theriogenology*, 2nd edn. Philadelphia: WB Saunders, 2007, pp. 339–344.
56. Cohen R, Bernstein M, Ziv G. Isolation and antimicrobial susceptibility of *Actinomyces pyogenes* recovered from the uterus of dairy cows with retained fetal membranes and postparturient endometritis. *Theriogenology* 1995;43:1389–1397.
57. Frazer G. Hormonal therapy in the postpartum cow days 1 to 10. Fact or fiction? In: *Proceedings of the Annual Meeting of the Society for Theriogenology*, 2001, pp. 161–183.
58. Burton M, Dziuk H, Fahning M, Zemjanis R. Effects of oestradiol cypionate on spontaneous and oxytocin-stimulated postpartum myometrial activity in the cow. *Br Vet J* 1990;146:309–315.
59. Overton M, Sischo W, Reynolds J. Evaluation of effect of estradiol cypionate administered prophylactically to postparturient dairy cows at high risk for metritis. *J Am Vet Med Assoc* 2003;223:846–851.
60. Olson J, Bretzlaff K, Mortimer R, Ball L. The metritis–pyometra complex. In: Morrow DA (ed.) *Current Therapy in Theriogenology*, 2nd edn. Philadelphia: WB Saunders, 1986, pp. 227–236.
61. Peters A, Laven R. Treatment of bovine retained placenta and its effects. *Vet Rec* 1996;139:535–539.
62. Kasimanickam R, Cornwell J, Nebel R. Effect of presence of clinical and subclinical endometritis at the initiation of Presynch–Ovsynch program on the first service pregnancy in dairy cows. *Anim Reprod Sci* 2006;95:214–223.
63. Hendricks K, Bartolome J, Melendez P, Risco C, Archbald L. Effect of repeated administration of $PGF_{2\alpha}$ in the early postpartum period on the prevalence of clinical endometritis and probability of pregnancy at first insemination in lactating dairy cows. *Theriogenology* 2006;65:1454–1464.
64. Kasimanickam R, Duffield T, Foster R *et al*. The effect of a single administration of cephapirin or cloprostenol on the reproductive performance of dairy cows with subclinical endometritis. *Theriogenology* 2005;63:818–830.
65. Sheldon I, Noakes D. Comparison of three treatments for bovine endometritis. *Vet Rec* 1998;142:575–591.
66. LeBlanc S, Duffield T, Leslie K *et al*. The effect of treatment of clinical endometritis on reproductive performance in dairy cows. *J Dairy Sci* 2002;85:2237–2249.
67. Dubuc J, Duffield T, Leslie K, Walton J, LeBlanc S. Randomized clinical trial of antibiotic and prostaglandin treatments for uterine health and reproductive performance in dairy cows. *J Dairy Sci* 2011;94:1325–1338.
68. Galvão K, Frajblat M, Brittin S, Butler W, Guard C, Gilbert R. Effect of prostaglandin F2α on subclinical endometritis and fertility in dairy cows. *J Dairy Sci* 2009;92:4906–4913.
69. Black W, MacKay A, Doig P, Claxton M. A study of drug residues in milk following intrauterine infusion of antibacterial drugs in lactating cows. *Can Vet J* 1979;20:354–357.
70. Gilbert R, Schwark W. Pharmacologic considerations in the management of peripartum conditions in the cow. *Vet Clin North Am Food Anim Pract* 1992;8:29–56.
71. Dohoo I. A retrospective evaluation of postbreeding infusions in dairy cattle. *Can J Comp Med* 1984;48:6–9.
72. Galvão K, Greco L, Vilela J, Sá Filho M, Santos J. Effect of intrauterine infusion of ceftiofur on uterine health and fertility in dairy cows. *J Dairy Sci* 2009;92:1532–1542.
73. Gustafsson B, Ott R. Current trends in the treatment of genital infections in large animals. *Comp Cont Educ Pract Vet* 1981;3:S147–S151.

Chapter 51

Cystic Ovarian Follicles

Jack D. Smith

Department of Pathobiology and Population Medicine, College of Veterinary Medicine, Mississippi State University, Starkville, Mississippi, USA

Introduction

Cystic ovarian follicles (COF) have been recognized as a frequent cause of subfertility and reduced reproductive performance in cattle for over 100 years. This condition of anovulation is noted to occur in a number of mammalian species. The condition in cattle results in a major cause of economic loss worldwide, primarily to the dairy industry but its economic effects are seen in certain sectors of the beef industry as well. The major causes of economic loss involve increased days open in the postpartum period leading to extended calving intervals, drug costs associated with treatment, increased semen costs associated with the increase in servicers per conception, and higher culling rates of affected animals.[1,2] It has been shown that dairy cows with COF take 6–11 days longer to reach first service and 20–30 more days to conception than control cows.[3]

Terminology

Numerous terms and definitions have been used to describe the anovulatory condition of cattle, which is most commonly termed cystic ovarian disease (COD). Terms which have been used in the past to describe the same clinical condition include adrenal virilism, nymphomania, ovarian cysts, cystic ovarian degeneration, and cystic ovaries, among others.[2] The classic definition proposed by Roberts[4] was the presence of an anovulatory structure on the ovary more than 2.5 cm in diameter and which has persisted for at least 10 days in the absence of a corpus luteum (CL). Several aspects of the former definitions are relatively arbitrary and do not reflect our current understanding of the condition. Cattle typically ovulate a dominant follicle with an average follicular diameter of 1.6–1.9 cm. The use of a size limit of 2.5 cm thereby excludes all cystic follicles which are smaller than this arbitrary size.[5] Also, many previous definitions stated the requirement that the cystic structure be present for a minimum of 10 days. This requirement should also be questioned as it is now known that these structures are dynamic and can change significantly over this period and even be replaced by newer cystic structures. Additionally, current management systems would often not allow the cow to be reexamined to fulfill the 10-day duration requirement. Older definitions have also required the absence of a CL. The necessity for the absence of a CL is also not universally fulfilled as it is clearly evident that some cysts are not steroidogenic and thus are hormonally inactive and would not necessarily influence the estrous cycle. Such cystic structures can be found in the presence of a CL.

A more recent term put forth by Vanholder *et al.*[5] to more accurately define and describe this anovulatory condition of cattle and its effects on reproduction is "cystic ovarian follicles." COF is defined as the presence of a follicle(s) with a diameter of at least 20 mm, present on one or both ovaries in the absence of any active luteal tissue, and which clearly interferes with normal ovarian cyclicity.[5]

Further subclassification of the COF condition into follicular or luteal cyst is based on the degree of luteinization and on progesterone levels in blood or milk. Follicular cysts are typically thin-walled structures that secrete varying amounts of estradiol. The thickness of the layers of granulosa cells dictates the amount of estradiol present in the intrafollicular fluid. Both cyst types are considered to be different forms of the same condition, with luteal cysts being a follicular cyst in which theca and granulosa cells have undergone some luteinization and are producing progesterone.[2] Follicular cysts do not secrete progesterone, whereas luteal cysts secrete varying amounts of progesterone depending on the degree of luteinization. A single universal threshold level of progesterone used to distinguish follicular cysts from luteal cysts has not been consistently used among studies of this condition.[5] Thus the ability to accurately classify each cyst as a follicular or luteal cyst remains subject to personal interpretation of the clinical examination and history in addition to ovarian ultrasound findings and serum progesterone levels.

Incidence

The incidence of COF in cattle is very dependent on their intended use and their respective management system. The incidence of COF is much greater in dairy cows compared

---

Bovine Reproduction, First Edition. Edited by Richard M. Hopper.
© 2015 John Wiley & Sons, Inc. Published 2015 by John Wiley & Sons, Inc.

with beef cattle and dairy heifers, where the incidence is relatively low. Utilizing data from a number of studies, the average incidence of COF in lactating dairy cattle is estimated to be near 10–12%, with various studies reporting incidences ranging from 3 to 32%.[2] There are numerous factors that can impact the apparent incidence of COF. The time of diagnosis can easily influence the apparent incidence of COF as it has been shown that as many as 60% of dairy cattle that develop COF recover spontaneously prior to their first postpartum ovulation and may never be diagnosed.

In beef cattle the incidence of COF is much lower when the cattle are under routine management. However, it is when the reproduction of beef cattle is intensively managed that the incidence of COF begins to rise above that of the general population. Beef cattle used for embryo donation after superovulation appear to develop COF at a much higher incidence than the general population and may approach the incidence rate of dairy cattle.

Physiology of COF formation

A dysfunction or neuroendocrine imbalance involving the normal hypothalamic–pituitary–gonadal axis resulting in ovulation failure is the basic accepted mechanism of COF formation.[1,2,6–13] However, it is the precise or primary mechanism(s) leading to ovulation failure that is the focus of extensive study. To date, the aforementioned mechanism(s) has remained elusive and is still not yet clearly identified. A complicating factor in this search is that it is clear that there are multiple factors involved in the pathogenesis, with genetic, phenotypic, environmental, cellular, and management factors all appearing to be involved in some capacity.[5]

The most widely accepted hypothesis involves the altered release of luteinizing hormone (LH) from the pituitary gland. The preovulatory surge of LH is absent, insufficient in magnitude, or improperly timed leading to failure of the dominant follicle to ovulate. The dominant follicle continues to grow and becomes large and anovulatory.[1,2,5,14] There does not appear to be a reduction in gonadotropin releasing hormone (GnRH) content in the hypothalamus or a reduction in GnRH receptors in the pituitary.[5] LH content in the pituitary also does not appear to be reduced in cows with COF.[5,15] Injections of exogenous estradiol to cows with naturally occurring or induced follicular cysts fail to induce an LH surge, although an injection of GnRH in affected cows can induce the release of LH.[16–19] These findings support the belief that there is a functional breakdown in the positive estradiol feedback loop which controls LH secretion. Under normal conditions, preovulatory follicles secrete estrogen that has a positive feedback effect on the hypothalamic–pituitary–ovarian axis, inducing an LH surge which is responsible for ovulation of the dominant follicle.[16] Because of the functional abnormality in the regulation of LH secretion, cystic cows lack an LH surge which leads to ovulation failure.

In the anovulatory condition of COF, there is also loss of the negative feedback effects of progesterone leading to relatively high levels of LH pulse secretion. These LH pulses have been observed to be higher than those in cattle with a normal luteal phase. High LH pulses promote continued and excessive growth of the dominant follicle, ultimately leading to the abnormally large preovulatory follicle.[16,20,21] Additionally, the period of estradiol and inhibin production by the dominant follicle is also protracted, which helps establish long-term dominance of the cystic follicle via suppression of follicle-stimulating hormone (FSH) production by the anterior pituitary.[16,22–25] When the COF ultimately undergoes regression and no longer produces sufficient estradiol and inhibin at levels sufficient to suppress FSH, a new follicular wave emerges. Unless the abnormality in the estradiol feedback loop is corrected, the new dominant follicle of the subsequent wave will also become cystic and anovulatory thereby sustaining or perpetuating the anovulatory condition.[16] This apparent turnover and replacement of the COF condition can be responsible for the protracted anovulation noted in some cows.

There also appears to be an intraovarian component in the pathogenesis of COF. It is known that cows with follicular cysts are often concurrently or were previously exposed to various kinds of insults or stress such as oxidative stress, negative energy balance, reduced or poor liver function, and low circulating insulin-like growth factor (IGF)-1.[26–28] There has been an association with these stressors and an increase in heat shock protein (HSP) in the ovaries of cows with COF. It is speculated that the altered expression of HSP genes decreases apoptosis in the follicular wall and leads to the delayed regression of cystic follicles.[26] Although the precise relationship between the aberrant amounts of HSP in the ovaries of cattle under stress and the COF condition is not completely understood, it is speculated that there is an association with HSP and an intraovarian component of COF pathophysiology.[26]

Clinical signs

The clinical signs associated with COF are generally related to altered reproductive behavior and, less frequently, changes in physical characteristics. Most commonly cows with COF show prolonged periods of anestrus, which is seen in over 80% of affected cows. This is most evident in cattle that develop COF early in the postpartum period.[4] Nymphomania with prolonged periods of persistent and frequent estrus is seen in approximately 10% of cows with COF. It is common for the affected cow to frequently participate in riding activity of estral cows but not stand for mating herself.[4] The presence of an extremely masculine phenotype has also been associated with cows with COF who show prolonged nymphomania.[29] It appears that as the interval from calving to diagnosis of COF increases, the likelihood of observing nymphomania behavior increases.[4] Estrous cycles with sporadic and irregular patterns can also be the primary clinical sign in cattle with COF.

Diagnosis

Transrectal palpation of the reproductive tract has been the primary means of diagnosing COF for many years. However the accuracy with which one can determine the specific type of cyst present is relatively poor. However, transrectal ultrasound can be very useful in determining the specific type of cyst present. Follicular cysts typically have a thin wall (≤3mm) whereas luteal cysts typically have a thicker and more echogenic wall (≥3mm). The follicular fluid is often hypoechoic in follicular cysts, whereas with luteal cysts echogenic strands may be present creating a cobweb-like appearance.[30] The collective findings of a rectal examination of the reproductive tract including ultrasonography, blood progesterone levels, and the clinical history of the cow will allow the most accurate diagnosis regarding the type of cyst present. It is uncommon for blood progesterone levels to be determined in routine cases in clinical practice.

The accuracy with which a skilled palpator can identify the type of cysts present based on palpation alone is relatively poor.[31] The dynamic nature of both cysts and developing corpora lutea can complicate the diagnosis when palpation alone is used. Farin et al.[32] showed that 10% of cows diagnosed as having cysts based on rectal examination were found to have a structure consistent with a normal CL when the ovaries were subsequently examined with ultrasound. When ultrasound technology was used the accuracy of a correct diagnosis of cyst type was 74% for follicular cysts and almost 90% for luteal cysts.[33] Progesterone concentrations have been shown to correlate very well with cyst wall thickness, with 3 mm often being the threshold between follicular and luteal cysts.[1,34] When blood progesterone concentration is combined with both palpation and ultrasound findings, the diagnosis of cyst type approaches 100%, although this too is rarely done due to economic considerations and impracticality in farm situations. Knowledge of progesterone concentrations could certainly aid in treatment decisions, although most of the currently recommended treatments effectively treat both luteal and follicular cysts, thereby negating the need for collecting this information. The added cost of labor and expense to gather this information is not economically justifiable.

Treatment strategies

Many techniques, drugs, and strategies have been used in the treatment of COF in cattle over the last 100 years. Currently the most effective and most widely used treatments all involve hormonal therapy(s) of some kind. Older therapies that are no longer recommended include ovariectomy, uterine infusions of antibiotics or antiseptics, and crushing of the COF. Studies in the 1950s showed that when compared with hormonal therapies of the time, manual crushing showed no advantage and there may even be some adverse conditions such as trauma and adhesions associated with its use.[35] Fatal hemorrhage associated with crushing or manual rupture of COFs has been reported.[36] This form of treatment, although still found in several texts, can no longer be recommended due to the potential side effects and availability of safer more effective treatments.

GnRH

Hormone therapy aimed at either causing (GnRH) or mimicking (human chorionic gonadotropin) an LH surge can be used to treat follicular cysts. Of these two, GnRH is generally chosen first due to its low molecular weight and size, which reduces the likelihood of an adverse immune reaction.[1,7] The standard dose of 100 μg was shown in a recent pharmacokinetic study to be more than adequate for the production of an LH surge of 5 ng/mL in cows with COF.[37] GnRH was first reported in the early 1970s as an effective treatment for follicular cysts.[38] In this small early study, all treated cows (n = 5) were observed in standing estrus within 20–24 days following GnRH treatment.[38] Many additional subsequent studies have shown the value of treating cows with COF with GnRH products; however, depending on the study design and data collected, there have been wide variations in the results of GnRH treatment alone. It is clear that GnRH has a place in current treatments due to our current understanding of the pathophysiology involved in COF.

After an injection of GnRH a surge of LH from the anterior pituitary occurs within 2 hours.[17] This LH surge can cause luteinization of the follicular cyst, which will generally undergo spontaneous luteolysis in 18–20 days followed by a normal estrous cycle. However, it is also common for the injection of GnRH to cause ovulation of a dominant follicle also present on one of the ovaries at the time of treatment. With this situation, ovulation will occur as well as initiation of a new follicular wave followed by a normal luteal phase.[39] Subsequent increase in progesterone concentrations from the luteinized cyst lasts for 15–18 days and allows the resetting of the normal hypothalamic–pituitary–ovarian axis and resumption of normal cyclicity in most cows. These cows typically return to estrus in 3 weeks, although the interval can be shortened with the use of prostaglandins administered 7–10 days after GnRH treatment.

In one study looking at the effectiveness of a single injection of GnRH for treatment of cows with COF, 72% of cows resumed normal cyclicity within 20 days of treatment compared with 16% of control cows.[17] However, as mentioned earlier, findings have not been consistent among studies. There was no difference in the period of time between treatment with GnRH alone and resolution of COF or in the period of time between treatment and a CL being observed.[40] These studies highlight the difficulty in evaluating the effectiveness of GnRH alone as therapy for COF because of the high number of cows which appear to recover spontaneously and the variability and confounding factors present in the numerous study designs.

No differences have been observed when comparing synthetic GnRH to GnRH analogs or among different concentrations of GnRH, indicating that treatment of cows with any of the commercially available GnRH products should assure similar results.[41,42]

Human chorionic gonadotropin

Human chorionic gonadotropin (hCG) is a protein hormone that has been used to treat refractory cysts that fail to respond to GnRH. hCG has high LH-like properties that will often cause the cyst or other follicles present to luteinize and subsequently begin producing progesterone.[43] Once luteinized, the cyst can then be treated with prostaglandins to induce regression of the functional luteal structures and reinitiate normal estrous cycles. Its use has occasionally been noted to stimulate an immune reaction leading to refractoriness, although the importance of this reaction is poorly understood.[1] It is a more costly treatment than GnRH and it has been known to occasionally cause anaphylactic reaction following treatment. For these reasons, it is generally only used when GnRH or other therapies have failed to render a cure.

Progesterone

Use of progesterone as a treatment for COF was proposed over 40 years ago.[44] With the approval of progesterone-impregnated controlled intravaginal drug-releasing devices (CIDRs) for use in lactating dairy cattle there have been continuing projects evaluating their effectiveness in the treatment of COF. Progesterone administration has been shown to reestablish the normal feedback mechanisms involving the hypothalamic–pituitary–ovarian axis and allow cows with COF to resume normal cyclicity. The precise mechanism for how that occurs remains somewhat elusive, but the following model proposed by Todoroki et al. is the current hypothesis. Progesterone released from the CIDR raises circulating progesterone levels to those comparable with levels during the normal luteal phase. This heightened level of progesterone induces atresia of the cystic follicles by lowering the LH pulse frequency. This then allows for relief from the inhibitory effects of elevated estradiol and inhibin produced by the cystic follicle so that emergence of new follicular wave can occur. Additionally, progesterone prevents excessive growth of the emerging dominant follicle and allows for turnover of follicles at normal intervals by suppressing LH secretion. Progesterone is believed to be able to restore the hypothalamic–pituitary–ovarian axis to produce an adequate LH surge in response to elevated estradiol production by the dominant follicle and release of the ovum on removal of the progesterone source.[16,45–47] The duration of progesterone treatment which is sufficient to reestablish the normal feedback mechanism involving the hypothalamic–pituitary–ovarian axis appears to be as short as 3 days but exogenous progesterone is typically used for 7–14 days.[48]

Progesterone can be used as the sole treatment for COF. There were comparable pregnancy rates in cows with COF who were treated with the Ovsynch protocol and bred via artificial insemination (AI) versus cows who received a CIDR for 7 days with a prostaglandin injection on removal and bred via AI after heat detection, suggesting that CIDRs alone can be effective in treating cows with COF.[49]

Todoroki et al.[45] studied the effects of using CIDRs for the treatment of superovulated beef cow donors who were diagnosed with follicular cysts and found favorable results. Over a period of 2 years, 28 cows were diagnosed with COF and treated with a CIDR for 14 days. After removal of the CIDR all cows exhibited estrus behavior and ovulated between 2 and 10 days following CIDR removal and subsequently formed corpora lutea. In 18 of 28 (64%) treated cows that showed initial recovery, the COF condition did not recur after repeated embryo recovery attempts. However, the COF condition did recur in 10 of 28 (36%) cows after the next embryo recover.

The use of CIDRs in place of traditional prostaglandin for synchronization of estrus after embryo recovery has also shown promise in reducing the incidence of follicular cysts in superovulated beef cow donors. Immediately after embryo recovery, a CIDR was placed in the vagina of each cow and left for 14 days. Todoroki et al.[45] showed that when CIDRs were used in place of prostaglandins for synchronization of estrus after embryo recovery, there was a marked reduction in the number of cows subsequently diagnosed with COF. In year one of the study of 61 donors, of which 22 had a history of COF, only two developed COF. In year 2 of the study, among 54 donors, of which 17 had a history of COF, no cows developed COF. CIDRs proved effective in both the treatment of COF as well as lowering the incidence of COF in beef cow donors where it was used in place of prostaglandin for synchronization of estrus after embryo recovery.

Progesterone clearly appears to have some healing effects on the functions of the hypothalamic–pituitary–ovarian axis in cows with COF.[45–47] Further study of progesterone's effects on the COF state certainly warrants continuing research.

Prostaglandins

Prostaglandins are effective in the treatment of all luteal cysts and those follicular cysts which have undergone luteinization due to prior GnRH treatment.[1,2,7,39] Prostaglandins have no effect on follicular cysts, highlighting the importance of accurate diagnosis prior to using prostaglandins as a single modality treatment option. After prostaglandin administration luteal cysts regress, with estrus occurring in 90% of cows by day 8 after treatment.[2] Regarding current treatment options for COF, prostaglandins are most commonly used as part of the Ovsynch or similar protocol or at the conclusion of a 7–14 day CIDR protocol.

Ovsynch protocol

Numerous protocols have been reported involving a series of hormonal injections aimed at treating COF and restoring the cow to normal cyclicity.[1,8,49–52] It has been shown that progesterone levels in cows with COF who were treated with a single injection of GnRH are elevated at 5 days following treatment. This indicates that the Ovsynch protocol can also be used with good results.[50] The classic Ovsynch protocol has been used for several years in the treatment of COF with acceptable pregnancy rates varying from 17 to 25%.[52,53] The rationale for using the Ovsynch protocol is to effectively treat COF regardless of the specific cyst type as well as breed the cow without the need for heat detection using a timed AI protocol.[8,49–52] This has allowed the Ovsynch protocol to gain popularity because it is familiar to many producers and its "one size fits all" usefulness regarding treatment of cows with COF regardless of the specific cyst type.

The Ovsynch plus CIDR protocol for treating cows with COF is very effective in altering the endocrine state to allow recruitment of a new follicular wave, causing COF turnover and resetting the hypothalamic–pituitary axis while allowing for timed AI. The combination of the additional progesterone should assure all cows are exposed to the beneficial effects of progesterone. Results have shown an increase in pregnancy rates in cows with COF that were treated with Ovsynch plus CIDR (37.5% pregnancy rate) versus Ovsynch alone (16.7% pregnancy rate).[51]

Additional therapies

A recent report evaluating the effectiveness of the opioid antagonist naloxone as a treatment for COF was published. It has been shown that stress may be a contributing factor in the pathogenesis of COF in cattle. Endogenous opioid peptides are involved in many responses to stress including the regulation of various endocrine systems.[54] Endogenous opioid peptides are believed to block the release of GnRH from the hypothalamus as well as the estrogen-induced LH surge.[55] It has been shown that administration of the opioid antagonist naloxone results in elevated LH release in cattle under various stressful states.[56] When naloxone was combined with an injection of the GnRH agonist buserelin in treatment of cows with COF, 77.5% of the treated cows had begun cystic regression as viewed with ultrasonography or had begun cycling normally within 2 weeks after treatment.[54] Additional studies evaluating this drug as a sole treatment are warranted due to the conflicting response likely associated with the buserelin administration.

Associations

Numerous risk factors have been associated with the formation of COF. The association between high milk production and the incidence of COF has been noted for over 50 years. Most studies show that there is either a positive association with increased milk production and the incidence of COF or that there is no association with increased milk production and the incidence of COF. However, it should be noted that no study to date has shown that as milk production increases, the incidence of COF goes down. Considering that the definition of "high milk production" has certainly changed over time and the fact that there can be such individual variability, it is understandable that a universal association between milk production and COF has not been discovered. However, it does seem that there likely is an association between COF and increases in milk yield. As cows are increasingly selected for high milk production and are pushed metabolically to meet their lactation potential, it seems obvious that the stressors associated with these demands can contribute to the formation of COF. Whether the association between milk production and COF formation is a cause-and-effect relationship has not been yet been clearly identified.

Use of bovine somatotropin does not appear to be associated with any increased incidence of COF in dairy cattle.[57,58] Dairy cows that deliver twins were 2.0–2.7 times more likely to develop COF in the postpartum period compared with cows who did not deliver twins, suggesting that delivery of twins may predispose the cow to COF in the subsequent lactation.[59,60]

A study attempting to identify risk factors associated with postpartum COF formation showed that the condition was most commonly seen in cows calving in the summer months, higher-producing dairy cattle, older cattle, and in cows which increased in body condition immediately preceding calving.[9] The presence of COF in the early postpartum period was the single greatest risk factor for the presence of COF at the time of insemination in dairy cattle. The risk of having COF at the time of insemination was 36.6% higher for cows which had developed COF early in the postpartum period.[9] These findings would seem to suggest that treatment of multiparous cows with COF early in the postpartum period would be beneficial whereas younger cows could be delayed until the end of the voluntary waiting period to allow for spontaneous resolution.[9] Dairy cows who were identified as having had an abnormal puerperium were 1.9 times more likely to develop COF in the postpartum period. Also cattle that had a 1 unit increase in body condition score in the prepartum period were 8.4 times as likely to develop COF.[8]

Metabolic factors that have been shown to be associated with increase in the incidence of COF include overconditioned cattle, reduction in insulin and IGF-1, and increased nonesterified fatty acids.[5,7] Additional factors that have been shown to be associated with COF formation but without a clear causal relationship include heat stress, utilization of bulls with high genetic merit for fat and protein, high stocking rates in barns, and melengestrol acetate use in beef heifers.[61,62]

There is a genetic predisposition for COF in dairy cattle, although the heritability is low at 0.07–0.12.[63,64] Certain cow families in the dairy breeds appear to have a higher incidence of COF than their contemporaries. Swedish workers showed that when sires were removed from the breeding pool after producing daughters that developed COF, they could significantly reduce the incidence of COF from 10% to 3%.[65] If one chooses to select against sires that produce daughters with COF, one may also be selecting against sires whose progeny also have high milk production. Genetic selection as a means of prevention of COF can be effective; however it will likely be a lengthy endeavor due to the low heritability as well as the multifactorial causation of this condition.

Conclusions/questions

Since most of the research regarding COF in cattle has been performed on cattle already exhibiting the condition, it makes it difficult to evaluate the likely causes that led up to the formation of the COF. We currently have no good way of inducing COF formation that clearly mimics that seen in naturally occurring cases, which limits our access to study models. Efforts moving forward should be focused on minimizing the contributing factors that have been associated with COF formation and treating cows in a prompt time frame to assure they can be available for breeding in a timely manner to avoid the economic losses associated with this condition.

References

1. Peter A. An update on cystic ovarian degeneration in cattle. *Reprod Domest Anim* 2004;39:1–7.
2. Garverick H. Ovarian follicular cysts in dairy cattle. *J Dairy Sci* 1997;80:995–1004.
3. Fourichon C, Seegers H, Malher X. Effect of disease on reproduction in the dairy cow: a meta-analysis. *Theriogenology* 2000;53:1729–1759.
4. Roberts SJ. *Veterinary Obstetrics and Genital Diseases (Theriogenology)*, 2nd edn. Woodstock, VT: published by the author, 1971, pp. 421–432.

5. Vanholder T, Opsomer G, Kruif A. Aetiology and pathogenesis of cystic ovarian follicles in dairy cattle: a review. *Reprod Nutr Dev* 2006;46:105–119.
6. Woolums A, Peter A. Cystic ovarian condition in cattle. Part I. Folliculogenesis and ovulation. *Comp Cont Educ Pract Vet* 1994;16:935–942.
7. Woolums A, Peter A. Cystic ovarian condition in cattle. Part II. Pathogenesis and treatment. *Comp Cont Educ Pract Vet* 1994; 16: 1247–1253.
8. Bartolome J, Thatcher W, Melendez P, Risco C, Archbald L. Strategies for the diagnosis and treatment of ovarian cysts in dairy cattle. *J Am Vet Med Assoc* 2005;227:1409–1414.
9. Lopez-Gatius F, Santolaria P, Yaniz J, Fenech M, Lopez-Bejar M. Risk factors for postpartum ovarian cysts and their spontaneous recovery or persistence in lactating dairy cows, *Theriogenology* 2002;58:1623–1632.
10. Farin P, Estill C. Infertility due to abnormalities of the ovaries in cattle. *Vet Clin North Am Food Anim Pract* 1993;9:291–308.
11. Ijaz A, Fahning M, Zemjanis R. Treatment and control of cystic ovarian disease in dairy cattle: a review. *Br Vet J* 1987;143: 226–237.
12. Lopez-Diaz M, Bosu W. A review and update of cystic ovarian degeneration in ruminants. *Theriogenology* 1992;37:1163–1183.
13. Peter A. Infertility due to abnormalities of the ovaries. In: Youngquist RS (ed.) *Current Therapy in Large Animal Theriogenology*. Philadelphia: WB Saunders, 1997, pp. 349–354.
14. Roche J. The effect of nutrition management of the dairy cow on reproductive efficiency. *Anim Reprod Sci* 2006;96:282–296.
15. Brown J, Schoenemann H, Reeves J. Effect of treatment on LH and FSH receptors in chronic cystic-ovarian diseased dairy cows. *J Anim Sci* 1986;62:1063–1071.
16. Todoroki J, Kaneko H. Formation of follicular cysts in cattle and therapeutic effects of controlled internal drug release. *J Reprod Dev* 2006;52:1–11.
17. Cantley T, Garverick H, Bierschwal C, Martin C, Youngquist R. Hormonal response of dairy cows with ovarian cysts to GnRH. *J Anim Sci* 1975;41:1666–1673.
18. Garverick H, Kesler D, Cantley T, Elmore R, Youngquist R, Bierschwal CJ. Hormonal response of dairy cows with ovarian cysts after treatment with hCG or GnRH. *Theriogenology* 1976: 6:413–425.
19. Kesler D, Garverick H, Elmore R, Youngquist R, Bierschwal C. Reproductive hormones associated with ovarian cysts response to GnRH. *Theriogenology* 1979;12:109–114.
20. Savio J, Thatcher W, Badinga L, de la Sota R, Wolfenson D. Regulation of dominant follicle turnover during the oestrous cycle in cows. *J Reprod Fertil* 1993;97:197–203.
21. Stock A, Fortune J. Ovarian follicular dominance in cattle: relationship between prolonged growth of the anovulatory follicle and endocrine parameters. *Endocrinology* 1993;132:1108–1114.
22. Kaneko H, Nakanishi Y, Akagi S et al. Immunoneutralization of inhibin and estradiol during the follicular phase of the estrous cycle in cows. *Biol Reprod* 1995;53:931–939.
23. Kaneko H, Nakanishi Y, Taya K et al. Evidence that inhibin is an important factor in the regulation of FSH secretion during the mid-luteal phase in cows. *J Endocrinol* 1993;136:35–41.
24. Glencross R, Bleach E, Wood S, Knight P. Active immunization of heifers against inhibin: effects on plasma concentrations of gonadotropins, steroids and ovarian follicular dynamics during prostaglandin-synchronized cycles. *J Reprod Fertil* 1994;100: 599–605.
25. Price C. The control of FSH secretion in the larger domestic species. *J Endocrinol* 1992;31:177–184.
26. Velazquez M, Alfaro N, Dupuy C, Salvetti N, Rey F, Ortega H. Heat shock protein patterns in the bovine ovary and relation with cystic ovarian disease. *Anim Reprod Sci* 2010;118:201–209.
27. Ortega H, Palomar M, Acosta J et al. Insulin-like growth factor I in ovarian follicles and follicular fluid of cows with spontaneous and induced cystic ovarian disease. *Res Vet Sci* 2008;84: 419–427.
28. Silvia W, Alter T, Nugent A, Naranja da Foseca L. Ovarian follicular cysts in dairy cows and abnormality in folliculogenesis. *Domest Anim Endocrinol* 2002;23:167–177.
29. Seguin B. Ovarian cysts in dairy cows. In: Morrow DA (ed.) *Current Therapy in Theriogenology*. Philadelphia: WB Saunders, 1980, pp. 199–204.
30. Jeffcoate I, Ayliffe T. An ultrasonographic study of bovine cystic ovarian disease and its treatment. *Vet Rec* 1995;136:406–410.
31. Farin P, Youngquist R, Parfet J, Garverick H. Diagnosis of luteal and follicular ovarian cysts by palpation per rectum and linear-array ultrasonography in dairy cows. *J Am Vet Med Assoc* 1992; 200:1085–1089.
32. Farin P, Youngquist R, Parfet J, Garverick H. Diagnosis of follicular cysts in dairy cows by sector scan ultrasonography. *Theriogenology* 1990;34:633–642.
33. Hanzen C, Pieterse M, Scenczi O, Drost M. Relative accuracy of the identification of ovarian structures in the cow by ultrasonography and palpation per rectum. *Vet J* 2000;159: 161–170.
34. Dobson H, Ribadu A, Noble K, Tebble J, Ward W. Ultrasonography and hormone profiles of adrenocorticotrophic hormone (ACTH)-induced persistent ovarian follicles (cysts) in cattle. *J Reprod Fertil* 2000;120:405–410.
35. Roberts S. Clinical observations on cystic ovaries in dairy cattle. *Cornell Vet* 1955;45:497–508.
36. Dobson H, Rankin J, Ward W. Bovine cystic ovarian disease: plasma hormone concentrations and treatment. *Vet Rec* 1977; 101:459–461.
37. Monnoyer S, Guyonnet J, Toutain P. A preclinical pharmacokinetic/pharmacodynamic approach to determine a dose of GnRH for treatment of ovarian follicular cysts in cattle. *J Vet Pharmacol Ther* 2004;27:527–535.
38. Kittok R, Britt J, Convey E. Endocrine response after GnRH in luteal phase cows and cows with ovarian follicular cysts. *J Anim Sci* 1973;37:985–989.
39. Kesler D, Elmore R, Brown E, Garverick H. Gonadotropin releasing hormone treatment of dairy cows with ovarian cysts. I. Gross ovarian morphology and endocrinology. *Theriogenology* 1981;16:207–217.
40. Jou P, Buckell BC, Liptrap R, Summerlee A, Johnson W. Evaluation of the effect of GnRH on follicular ovarian cysts in dairy cows using trans-rectal ultrasonography. *Theriogenology* 1999;52:923–937.
41. Dinsmore R, White M, Guard C et al. A randomized double blind clinical trial of two GnRH analogs for the treatment of cystic ovaries in dairy cows. *Cornell Vet* 1987;77:235–243.
42. Osawa T, Nakao T, Kimura M et al. Fertirelin and buserelin compared by LH release, milk progesterone and subsequent reproductive performance in dairy cows treated for follicular cysts. *Theriogenology* 1995;44:835–847.
43. Yamauchi M. Studies on the ovarian cysts in the cow. IV. The course of recovery from ovarian follicle cyst by the chorionic gonadotropin therapy. *Jpn J Vet Sci* 1955;13:47–55.
44. Johnson A, Ulberg L. Influence of exogenous progesterone on follicular cysts in dairy cattle. *J Dairy Sci* 1967;50:758–761.
45. Todoroki J, Yamakuchi H, Mizoshita K et al. Restoring ovulation in beef donor cows with ovarian cysts by progesterone-releasing intravaginal silastic devices. *Theriogenology* 2001;55: 1919–1932.
46. Calder M, Salfen B, Bao B, Youngquist R, Garverick H. Administration of progesterone to cows with ovarian follicular cysts results in a reduction in mean LH and LH pulse frequency and initiates ovulatory follicular growth. *J Anim Sci* 1999;77: 3037–3042.
47. Nanda A, Ward W, Williams P, Dobson H. Retrospective analysis of the efficacy of different hormone treatments of cystic ovarian disease in cattle. *Vet Rec* 1988;122:155–158.

48. Gumen A, Wiltbank M. Length of progesterone exposure needed to resolve large follicle anovular condition in dairy cows. *Theriogenology* 2005;63:202–218.
49. Crane M, Bartolome J, Melendez P, deVries A, Risco C, Archbald L. Comparison of synchronization of ovulation with timed insemination and exogenous progesterone as therapeutic strategies for ovarian cysts in lactating dairy cattle. *Theriogenology* 2006;65:1563–1574.
50. Bartolome J, Sozzi A, McHale J et al. Resynchronization of ovulation and timed insemination in lactating dairy cows. II. Assigning protocols according to stages of the estrous cycle, or presence of ovarian cysts or anestrus. *Theriogenology* 2005;63: 1628–1642.
51. Bartolome J, Sozzi A, McHale J et al. Resynchronization of ovulation and timed insemination in lactating dairy cows. III. Administration of GnRH 23 days post AI and ultrasonography for nonpregnancy diagnosis on day 30. *Theriogenology* 2005;63: 1643–1658.
52. Bartolome J, Archbald L, Morresey P et al. Comparison of synchronization of ovulation and induction of estrus as therapeutic strategies for bovine ovarian cysts in the dairy cow. *Theriogenology* 2000;53:815–825.
53. Ambrose D, Schmitt E, Lopes F, Mattos R, Thatcher W. Ovarian and endocrine responses associated with the treatment of cystic ovarian follicles in dairy cows with gonadotropin releasing hormone and prostaglandin F2 alpha, with or without exogenous progesterone. *Can Vet J* 2004;45:931–937.
54. Palomar M, Acosta J, Salvetti N et al. Treatment of cystic ovarian disease with naloxone in high production dairy cows. *J Vet Pharmacol Ther* 2008;31:184–186.
55. Malven P. Inhibition of pituitary LH release resulting from endogenous opioid peptides. *Domest Anim Endocrinol* 1986;3: 135–144.
56. Byerley D, Kiser T, Bedrand J, Kraeling R. Release of luteinizing hormone after administration of naloxone in pre- and peripubertal heifers. *J Anim Sci* 1992;70:2794–2800.
57. Cole W, Eppard P, Boysen B et al. Response of dairy cows to high doses of a sustained release bovine somatotropin administered during two lactations. 2. Health and reproduction. *J Dairy Sci* 1992;75:111–123.
58. Judge L, Bartlett J, Lloyd J, Erskine R. Recombinant bovine somatotropin: association with reproductive performance in dairy cows. *Theriogenology* 1999;52:481–496.
59. Bendixen P, Oltenacu P, Anderson L. Case-referent study of cystic ovaries as a risk indicator for twin calvings in dairy cows. *Theriogenology* 1989;31:1059–1066.
60. Emanuelson U, Bendixen P. Occurrence of cystic ovaries in dairy cows in Sweden. *Prev Vet Med* 1991;10:261–271.
61. Zulu V, Penny C. Risk factors of cystic ovarian disease in dairy cattle. *J Reprod Dev* 1998;44:191–195.
62. Johnson C. *Cystic ovarian disease in cattle on dairies in central and western Ohio: ultrasonic, hormonal, histologic, and metabolic assessments*. PhD thesis, Ohio State University, 2004.
63. Cole W, Bierschwal C, Youngquist R, Braun W. Cystic ovarian disease in a herd of Holstein cows: a hereditary correlation. *Theriogenology* 1986;25:813–820.
64. Uribe H, Kennedy B, Martin S, Kelton D. Genetic parameters for common health disorders of Holstein cows. *J Dairy Sci* 1995; 78: 421–430.
65. Garverick H. Ovarian follicular cysts. In: Youngquist RS, Threlfall WR (eds) *Current Therapy in Large Animal Theriogenology*. Philadelphia: WB Saunders, 2007, pp. 379–383.

Chapter 52

Postpartum Anestrus and its Management in Dairy Cattle

Divakar J. Ambrose

Livestock Research Branch, Alberta Agriculture and Rural Development and University of Alberta, Edmonton, Alberta, Canada

Introduction

Postpartum anestrus refers to a condition where cows have not been observed or reported in estrus for several weeks after calving, often to the end of the voluntary (elective) waiting period, in dairy cattle. While a short period of ovarian inactivity during the immediate postpartum period is normal, extended anestrus in nonsuckled cattle such as dairy cows could have negative ramifications on the timely reestablishment of pregnancy.[1–4] Postpartum anestrus has been recognized as a problem by animal scientists for several decades. According to one source,[5] anestrus in cattle was reported first as a reproductive problem back in the 1920s.[6] There has been significant interest in the past two to three decades in the understanding and management of postpartum anestrus, both in beef and dairy cows.[5–13] Nevertheless, postpartum anestrus continues to be a challenge to dairy practitioners and farmers alike. This chapter will focus primarily on the problem of postpartum anestrus and its management in dairy cattle, although references to the condition in beef cattle will be made where appropriate.

Fertility of the high-producing dairy cow

The modern dairy cow is an important livestock species with a remarkable ability to produce tens of thousands of liters of milk during her lifetime. Understandably, the profitability of dairy farming is closely integrated with the reproductive success of the dairy cow. However, suboptimal fertility of the modern dairy cow is a problem that afflicts the dairy industry worldwide,[14–18] with reproductive failure often being the primary reason for culling dairy cows.[19–21] While the reasons for declining fertility in dairy cattle will continue to be debated through the years to come, it is important to bear in mind that the modern dairy cow is physiologically quite different from her counterparts of past decades,[16] calling for new and improved strategies of reproductive management to maintain or enhance reproductive success. An extended period of postpartum anestrus is one among the many factors contributing to poor reproductive efficiency in dairy cattle. Rajala-Schultz and Gröhn[22] showed that if a cow had not been inseminated at all, her risk of culling was ten times higher than if she was inseminated at least once. While a deliberate managerial decision to not breed a cow, due to past history, may have directly contributed to this increased culling risk, anestrus could also have increased the risk, emphasizing the importance of detecting cows in estrus and having them bred in a timely manner.

Prevalence and implications of postpartum anestrus

The prevalence of postpartum anestrus in dairy cattle is herd-specific and varies widely from one herd to another. However, typical prevalence rates at the end of the elective waiting period (60–80 days after calving) are in the range 10–30%,[23–28] whereas higher rates of up to 59% have been reported in individual herds.[16,25,27] In a Canadian study involving 23 herds, milk progesterone concentrations were measured twice weekly in 637 cows to determine the onset of cyclicity, starting approximately 7 days after calving up to 120 days after calving.[28] The overall interval from calving to first rise in progesterone, indicative of an ovulation, was 32.0±0.7 days. The cumulative percentages of cows initiating cyclicity by 3, 6, 9 and more than 9 weeks after calving are shown in Figure 52.1. Although these findings may suggest that postpartum anestrus is not necessarily a huge problem because 90% of the cows had initiated cyclicity by 9 weeks, which is the typical end of the elective waiting period, it is important to note that cows that initiated cyclicity during the early postpartum period had higher conception rates at first service (Figure 52.2). Similar findings have been reported by others.[1,2,4,24,28]

The biggest impact of postpartum anestrus is the extension of the interval from calving to conception, which

---

Bovine Reproduction, First Edition. Edited by Richard M. Hopper.
© 2015 John Wiley & Sons, Inc. Published 2015 by John Wiley & Sons, Inc.

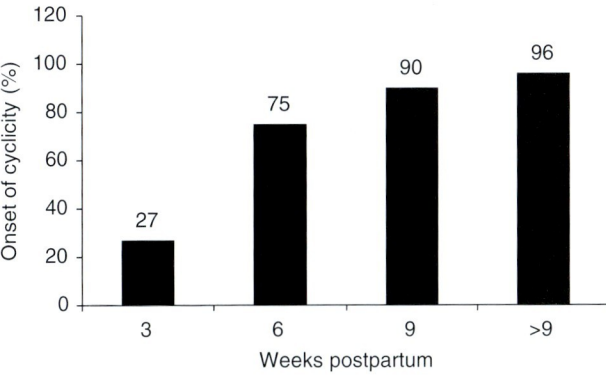

Figure 52.1 Onset of cyclicity in dairy cows by 3, 6, 9 and >9 weeks postpartum determined based on milk progesterone concentrations (data presented as cumulative percentages). Milk samples were collected twice weekly from 7 to 120 days postpartum, or until first service from 637 cows (23 herds). The mean interval from calving to first rise in progesterone (indicative of first ovulation) was 32.0 ± 0.7 days (range 5–113 days). Reproduced with permission from Ambrose D, Colazo M. Reproductive status of dairy herds in Alberta: a closer look. In: *Advances in Dairy Technology, Proceedings of Western Canadian Dairy Seminar*, 2007, pp. 227–244.

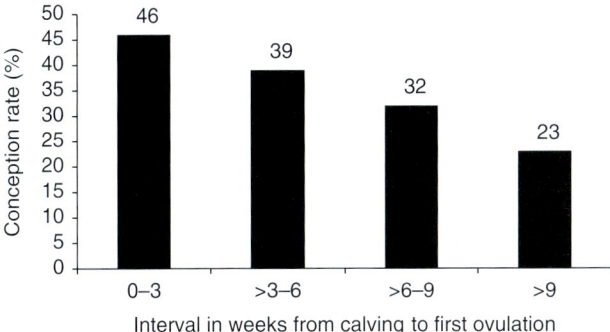

Figure 52.2 Conception rate to the first service by interval from calving to first ovulation determined based on rise in milk progesterone concentrations. Milk samples were collected from 7 to 120 days postpartum or until first service from 637 cows (23 herds). The first rise in progesterone was indicative of ovulation. Cows that had the first rise in progesterone within 3 weeks after calving had the highest ($P < 0.03$) conception rate to first service. Reproduced with permission from Ambrose D, Colazo M. Reproductive status of dairy herds in Alberta: a closer look. In: *Advances in Dairy Technology, Proceedings of Western Canadian Dairy Seminar*, 2007, pp. 227–244.

consequently impacts the calving interval. Whereas high-producing dairy cows that fail to return to estrus and conceive soon after the end of voluntary waiting period may continue to remain profitable for a period of time, cows with lower production that remain acyclic for an extended period, resulting in delayed conception, are often unprofitable and become potential targets for culling.

Physiological basis for postpartum anestrus and resumption of cyclicity

Endocrine events leading to anestrus

During gestation, circulating progesterone and estrogen concentrations remain high, exerting a suppressive effect on the hypothalamic–pituitary axis.[29] Although peripheral progesterone concentrations in late gestation are not much higher than that of mid-luteal or early gestational periods,[30,31] circulating estrogens progressively increase starting at about 60 days of gestation reaching 50 to 1000-fold higher concentrations in the days immediately preceding parturition.[30,32] The sustained long exposure to high concentrations of estradiol, either by itself or in tandem with progesterone, inhibits hypothalamic gonadotropin-releasing hormone (GnRH) secretion, which negatively affects the synthesis of luteinizing hormone (LH) leading to a gradual depletion of pituitary LH reserves. Although the mechanisms for release of LH from the anterior pituitary gland remain intact and functional throughout pregnancy,[29] the absence of pulsatile GnRH secretion from the hypothalamus is what appears to negatively impact LH production.

Unlike LH, changes in the concentrations of follicle stimulating hormone (FSH) during late gestation are less remarkable. While mean FSH concentrations do not differ between prepartum and postpartum periods,[30,33] the recurrent rise and fall (surge) of FSH that typically occurs every few days in cyclic and pregnant cows is largely absent[30] or slowed down[33] during the last weeks preceding parturition. Consequently, the associated follicular wave emergence is also absent during the last 30 days before parturition.[33]

Whereas it is possible to restore the pulsatile release of LH under experimental conditions by administering low-dose GnRH injections at frequent intervals (1–2 hours), not all cows respond to such treatment. Because of the inconsistent response to such an approach, it is believed that the inhibition of GnRH secretion is perhaps not the only mechanism involved in anestrus.[5]

Endocrine events leading to the resumption of cyclicity

As parturition approaches there is a gradual decrease in progesterone concentrations. In contrast, estrogen concentrations keep rising and peak by doubling during the last 24 hours prior to parturition. Since almost all circulating estrogens are of placental origin, their concentrations plummet sharply after expulsion of the conceptus reaching basal levels within 1 day after calving.[32] In the absence of progesterone and estradiol in the immediate postpartum period, the hypothalamic–pituitary axis is "liberated" from the suppressive effects of these steroids, resulting in the restoration of normal function of the GnRH pulse generator and the resultant pulsatile release of GnRH, which acts on the pituitary to restore a normal pattern of FSH release. Normal FSH surges resume within 3 days after calving, triggering the emergence of a follicular wave.[30] Whereas early stages of follicle development are gonadotropin independent, FSH support is required for the growth of follicles from 4 to 9 mm and LH pulses are indispensable for continued growth of follicles beyond 9 mm in diameter.[34,35]

During the early postpartum period the LH pulse frequency remains very low, generally less than one pulse every 4 hours, which later increases to about one to two pulses per hour as GnRH pulsatility increases. The restoration of pituitary function and reaccumulation of LH could take up to 3 weeks or longer depending on several factors. Thus, when conditions are optimal for pituitary LH reserves to be replenished, normal LH pulsatility is restored leading to the growth and establishment of dominant follicles capable of producing greater quantities of estradiol. A high concentration of estradiol from the dominant follicle is essential to trigger an LH surge, which is critical for the first ovulation to occur. Although the first ovulation signals resumption of cyclicity, not all cows would resume normal ovarian cycles following the first ovulation and could relapse back into anestrus.[36]

If cows are challenged with either exogenous GnRH or estradiol during the early postpartum period, not only is LH release greatly diminished due to lack of pituitary LH reserves, but also its bioactivity is lowered.[5]

Fate of the first dominant follicle

The fate of the first dominant follicle in the postpartum cow is determined by the pulsatile LH support it receives and by its own steroidogenic capacity to produce estradiol. Depending on these factors, the first dominant follicle may (i) ovulate and form a corpus luteum (CL); (ii) become anovulatory, possibly developing into a cystic follicle; or (iii) regress, making way for a new dominant follicle.[37,38]

Classification of postpartum anestrus

Postpartum anestrus can be classified into different types based on either ovarian follicle dynamics or progesterone profiles.

Categorizing anestrus based on ovarian follicular dynamics

The classification of anovulatory conditions in cattle based on three functionally critical follicular diameters relating to emergence (~4 mm), deviation (~9 mm), and ovulation (10–20 mm) has been proposed.[39] With the above classification as the basis, four types of anestrus, types I to IV, were described by Peter et al.[13] (Figure 52.3).

In type I anestrus, there is growth of follicles up to emergence but no deviation occurs resulting in lack of selection of a dominant follicle. This type of anestrus is presumed due to extreme malnutrition, which could exert a negative effect on FSH production suppressing follicular growth, although other factors could also be involved. Ovaries under this type may be described as either "inactive" or "smooth" as a reference to the lack of palpable ovarian structures during reproductive examinations per rectum. An Israeli study[40] from the 1980s reported an incidence of 8.5% inactive ovaries from 7751 lactations. While type I anestrus is not widely seen in dairy cattle of developed countries, this type of anestrus could be more prevalent in regions of the world where balanced energy-dense rations are unavailable to dairy cattle.

In type II anestrus both follicular deviation and growth occur, followed by regression, in some cases, after a follicle attained dominance. Regression of a dominant follicle is usually followed by the emergence of a new follicular wave, 2–3 days later. In this type of anestrus, there may be sequential emergence of follicular waves prior to the eventual occurrence of the first ovulation. Up to nine waves of follicular growth have been reported in one study.[41]

In type III anestrus, deviation, growth, and establishment of a dominant follicle takes place, but the dominant follicle fails to ovulate, becoming a persistent structure which may either linger as an anovular follicle or continue to grow and develop into a cystic follicle. Anovular follicles are considered different from cystic follicles because in cows with the latter condition there is disruption of the feedback mechanisms in the hypothalamic–pituitary axis.[42] The secretory pattern of LH in cows with cystic ovarian follicle is quite different from that in normal cyclic cows in that mean concentrations of LH, frequency of LH pulses, and amplitude of LH pulses are all higher;[43] in addition, in cows with an actively growing cystic follicle, the estradiol concentration is also significantly higher.[44] A cystic follicle that is growing actively usually exerts dominance, suppressing the growth of other follicles. Retrospective analysis of data from twice-weekly

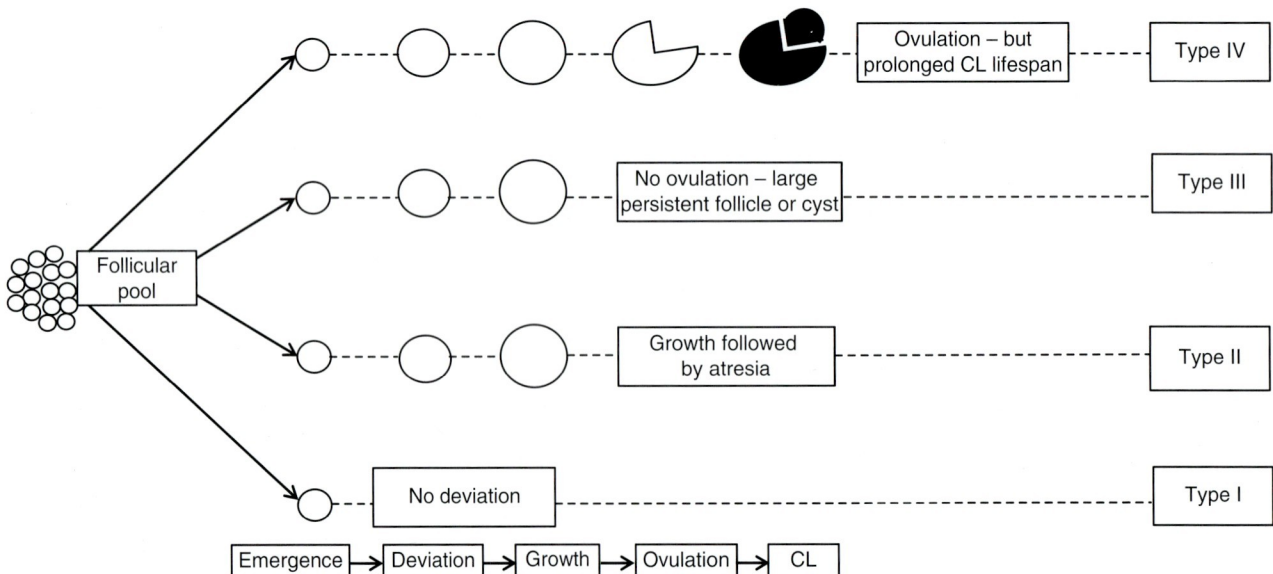

Figure 52.3 Schematic representation of types of anestrous conditions based on ovarian follicular dynamics. Reproduced with permission from Peter A, Vos P, Ambrose D. Postpartum anestrus in dairy cattle. *Theriogenology* 2009;71:1333–1342.

ultrasonography, from 7 to 35 days after calving, on eight lactating dairy cows that developed cystic ovarian follicles, a first-wave dominant follicle became cystic in seven cows, and a second-wave dominant follicle became cystic in one cow. No other follicle greater than about 5–6 mm diameter was detected on either ovary of cows until the cystic follicle turned over. In five cows, the cyst persisted until (possibly beyond) day 35; in two cows, the initial cystic follicle was replaced by another one; in one cow two follicles from the second follicular wave became dominant around day 30 at which time the original cyst had regressed to below 20 mm diameter giving way for both dominant follicles to ovulate (Divakar J. Ambrose, personal observations).

In type IV anestrus, a dominant follicle ovulates and forms a CL, but the luteal phase is prolonged due to the absence of timely luteolysis followed by CL regression. Aberrant follicular growth patterns resulting in the absence of an estrogenic dominant follicle at the ideal time to trigger luteolysis could be a reason for this condition of persistent CL.[39]

Categorizing anestrus based on progesterone profiles

Cows can also be classified into different categories of anestrus based on progesterone measured in milk or blood at least on a twice-weekly basis. Using twice-weekly measurement of progesterone in milk fat on 334 dairy cows from six herds, Opsomer et al.[45] reported that 51% of the cows had a normal progesterone profile with the first rise in progesterone occurring before 50 days after calving, followed by regular cyclicity. Among the remaining population, 21.5% of the cows had delayed cyclicity (anestrus for the first 50 days after calving, characterized by consistently low progesterone during that entire period), 21.5% had prolonged luteal phase (progesterone remained elevated for >20 days without a preceding insemination), 4% had cessation of cyclicity (normal cycles interrupted by at least 14 days of consistently low progesterone concentrations), 0.5% had short luteal phases (characterized by more than one luteal phase, excluding the first luteal phase, that was shorter than 10 days' duration), and 1.5% of the cows had irregular progesterone profiles (atypical profiles, that did not belong to any of the previously described categories).

Factors contributing to postpartum anestrus

The causes of postpartum anestrus in cattle are often multifactorial,[5,8,10,12,13,38,46–50] but can be broadly categorized into physiological, nutritional, managerial, environmental, and pathological as detailed in Table 52.1. Among the numerous potential contributors, suckling and nutrition (negative energy balance) are considered the most important.

Suckling

Suckling is a major contributor to anestrus in beef cattle,[46] but it is generally not considered a cause for anestrus in dairy cattle because of the prevailing industry practice of separating dairy calves from the dam soon after birth. However, under experimental conditions when identical twin pairs of dairy cows were assigned to either suckling by multiple calves or twice-daily machine milking, the suckled cows had a 5–9 week longer interval to first estrus after calving than their machine-milked twins. Basal plasma concentrations of LH did not differ between suckled and machine-milked twins, yet GnRH-induced LH release increased progressively in machine-milked cows from weeks 0 to 6 after calving, and the LH response at 2 weeks after calving was significantly higher in machine-milked cows than in their suckled twins.[73] Estradiol-17β-induced LH release within 24 hours after treatment was delayed by up to 6 weeks in suckled cows compared with their machine-milked twins, indicating that the response of the hypothalamic–pituitary axis to both GnRH and estradiol-17β is impaired by suckling. In another study,[74] suckled dairy cows had lower mean LH concentrations than nonsuckled controls on day 13 after calving. This decrease resulted from a 60% reduction in the frequency and a 40% reduction in the amplitude of LH pulses. These authors concluded that decreased frequency and amplitude of episodic LH secretion and reduced capacity of pituitaries to respond to GnRH may account for suckling-induced inhibition of postpartum ovulation. For further reading on the effects of suckling on postpartum anestrus in cattle, please refer to reviews on the topic[46,49,50] and Chapter 23.

Nutrition

Inadequate consumption of dry matter (i.e., energy) resulting in negative energy balance is an important contributing factor to anestrus in dairy cattle.[11,38,48] Soon after calving, the energy needs of the newly lactating cow increase dramatically due to the higher energy requirements associated with milk production, pushing the cow into a catabolic metabolism. Energy balance is the net result of energy intake minus energy expended for maintenance, growth, and milk yield. Although dairy cows rarely enter into a state of negative energy balance during dry period, except in the immediate prepartum period when dry matter intake declines, most lactating dairy cows are unable to meet their energy demands after calving and invariably become energy deficient. Physical fill in the rumen is one of the factors limiting maximum dry matter intake in early postpartum dairy cows. Even if dry matter intake is maximized, cows selected for high milk production (e.g., North American Holstein) selectively use all available nutrients for milk production at the expense of body condition.[65] Thus nutrient partitioning prevents cows from ending the catabolic state because all the available glucose is preferentially used up for milk synthesis.[75] Despite the advances made in the understanding of dairy cow nutrition and the availability of high-precision ration-balancing software, meeting all the energy needs of a lactating dairy cow continues to remain a challenge, placing cows in a state of negative energy balance that extends for 10–12 weeks into the postpartum period.[11]

Dairy cows that are in negative energy balance have lower blood concentrations of insulin and insulin-like growth factor (IGF)-1. Low circulating concentrations of IGF-1 reduce the negative feedback on growth hormone (GH), resulting in an increase in GH concentrations.

Table 52.1 Direct and indirect causes of anestrus in cattle.

Causal factor	Comment
Physiological	
Suckling	Suckling suppresses episodic LH secretion, thereby delaying ovulation.[46,49,51] Suckling is a major contributor to postpartum anestrus in beef cattle, but not so in dairy cattle because dairy calves are usually separated from the dam soon after birth
Parity	Parity as a risk factor for anestrus has been reported in numerous studies.[45,47,52] Most reports suggest that first-parity cows are at a higher risk of anestrus due to delayed first ovulation. Multiparous cows, particularly those with four or more calvings, are reportedly at a significantly higher risk for anestrus due to prolonged luteal phases. The reason may be that with increasing parity the uterus takes more time to return to its pregravid size, thereby increasing the risk of pyometra and persistent CL[45]
Milk yield	Although high milk production has been identified as a risk factor for anestrus by some researchers,[27,52,53] others have not found such an association. In one study,[27] the probability of anovulation decreased by 2% for every 100-kg increase in the first 305-day mature equivalent milk projection between 5000 and 9500 kg, and the risk of anovulation did not change as first 305-day mature-equivalent milk increased above 9500 kg. Similarly, another large study[52] found that cows with lower daily average milk yield in the first 90 days postpartum (32.1 vs. 39.1, 43.6, 50.0 kg/day) were at higher risk for anestrus. It is unlikely that high milk production itself is a cause for anestrus; however, the inability to meet energy requirements leading to negative energy balance, a common problem in lactating dairy cows, is the most likely reason
Ratio of increase in milk yield	The ratio of increase in milk yield from the first week postpartum to the week when milk yield reached its peak is reportedly smaller in ovular than in anovular cows[54]
Social or chemical cues	In beef cattle, the interval from calving to resumption of cyclicity is significantly reduced by the presence of bulls.[55] In one study the exposure of bulls in the early postpartum period had an acute effect on LH release in anestrous dairy cows.[56] Likewise, the presence of other estrous females or exposure to cervical mucus collected from estrous cows can also shorten the duration of postpartum anestrus,[57] particularly in cows with extended anestrous periods
Breed	The influence of breed or type (genetic) differences on the interval from calving to commencement of luteal activity has been reported, mainly in beef cattle and crossbreds[47,58]
Nutritional	
Malnutrition	Severe feed restriction will induce anestrus in cattle. In one study,[59] nonlactating beef cows placed on a restricted diet formulated for 1% of body weight loss each week became anestrous after 26 weeks and lost 24% of their initial body weight. Estrous cycles resumed about 9 weeks after cows were returned to a diet that was 160% of the maintenance diet given to control cows. Anestrus was associated with a decrease in the frequency of LH pulses
Negative energy balance	Cows in negative energy balance are at a very high risk of anestrus and delayed onset of cyclicity.[11] Elevated circulating concentrations of nonesterified fatty acids are associated with adipose tissue mobilization and indicative of negative energy balance
Body condition and rate of loss of body condition	A body condition score (BCS) of less than 3.0 on a 1–5 scale is commonly an indicator of negative energy balance.[45,52] Cows with low BCS are more likely to be anestrous and it has been reported that the influence of body condition on the duration of postpartum anestrous period is mediated through differences in LH pulse frequency.[60] Cyclic cows with low BCS tend to have poor conception rates and higher embryonic losses.[52,61] A higher rate of body condition loss has been associated with extended intervals from calving to resumption of cyclicity or first service
Managerial	
Length of previous dry period	Cows having a dry period longer than 77 days were almost three times more at risk of anestrus and delayed ovulation.[45] In contrast, the interval from calving to first ovulation in cows given no dry period (fed a high-energy diet continuously) compared to those given a 56-day dry period (fed a low-energy diet from 56 to 29 days prepartum followed by a moderate-energy diet for the last 28 days) was significantly shorter (13.2 vs. 31.9 days)[62,63]
Increased milking frequency	Increasing the frequency of milking from twice to three or four times daily lengthens the interval from calving to first ovulation by less than a week. When milking frequency is increased from three to six times daily, the interval to first ovulation is lengthened, but still not to the same extent as that of frequent suckling.[49] Therefore, increased milking frequency is only a minor contributor to anestrus
Efficiency of estrus detection	Unobserved estrus due to poor estrus detection efficiency can frequently contribute to anestrus,[28] exacerbating the severity of the condition
Feed and feeding management	Inadequate dry matter intake is one of major reasons for cows entering a state of negative energy balance.[38,64,65] Maximizing dry matter intake through particle size management, frequency of feed delivery (more than one feed), grouping cows by parity/production levels/social order, feed bunk management (e.g., frequent push-ups), discouraging sorting behavior by adding water, avoiding overstocking, etc. are some examples of strategies to improve dry matter intake, thereby reducing negative energy balance

(Continued)

Table 52.1 (Continued)

Causal factor	Comment
Environmental	
Calving season	Cows calving in the winter have delayed resumption of estrous cycles compared with those calving in the summer.[7,27,45] While some of the variations attributed to season are likely due to seasonal differences in management, photoperiod does seem to play a role in the regulation of postpartum cyclicity
Weather	Under extreme hot and humid conditions, cows have reduced expression of estrus,[7] which contributes to an increase in anestrus due to sub-estrus and silent estrus
Social hierarchy	Social hierarchy (e.g., presence of dominant cows) could be a factor that has detrimental effects on feed intake, particularly in overstocked barns. It is not uncommon to see "bullying" among cows. The subordinate or the "bullied" cows frequently end up eating leftover feed which may be of poor quality as it is likely to have been sorted through. Such cows are at higher risk for negative energy balance, thus anestrus
Housing type	Cows under intensive housing systems compared with those on pasture have a higher risk of anestrus. Tie-stalled cows are more likely to be in anestrus than other types of housing;[47] one explanation is that tie-stalled cows have few opportunities to form sexually active groups, limiting overt behavioral signs of estrus
Stocking density	Stocking density could have a significant impact on access to feed bunks, negatively affecting feed intake, particularly where cows of different pecking order are mingled in the same feeding group
Stall design (cow comfort)	Poorly designed stalls (e.g., lack of lunge space) would limit the frequency of accessing feed bunk/manger, negatively affecting dry matter intake, predisposing cows to negative energy balance and consequently anestrus
Stray voltage	Few studies have examined the effects of stray voltage on reproductive function in cattle. Although no direct evidence of stray voltage on impairment of ovarian function could be found, it is a potential contributor to the problem of anestrus as bothersome stray voltage in and around the feeding area could be a deterrent to feed intake
Pathological	
Uterine inflammation	Cows with acute forms of metritis and those with severe bacterial contamination of the uterus have delayed resumption of cyclic ovarian activity.[45,66] The mean interval from calving to first ovulation was longer (45 vs. 32 days) in cows that were determined to have uterine inflammation (based on higher proportion of polymorphonuclear, or PMN, cells at endometrial cytology performed on day 25 postpartum) compared with cows with no uterine inflammation.[67] The cumulative pregnancy (%) at 270 days postpartum tended to be lower in cows with high PMN (58 vs. 80) than in those with low PMN counts
Persistent corpus luteum	Cows with uterine infections accompanied by abnormal vaginal discharge are at much higher risk for anestrus due to prolonged luteal cycles.[45] However, prolonged luteal phases of >30 days' duration have been reported in cows that were not previously inseminated, even in the absence of any uterine infections[15,68]
Cystic ovarian follicle (COF)	Though not strictly a pathological condition, cows diagnosed with COF are often anestrous and anovulatory.[69] The precise etiology of COF is still not fully understood, but an aberration in LH secretory pattern is believed to trigger the formation of COF; cows with active COF have increased LH pulsatility, which is essential for continued growth of the cyst[42,70]
Calving disorders	Dystocia, twinning, and retained placenta could predispose cows to anestrus[27,45]
Metabolic disorders	Ketosis and displaced abomasum have been associated with delayed cyclicity and anovulation,[27,45] particularly in cows with subclinical ketosis during the first week postpartum[27]
Lameness	The increased risk of anestrus associated with lameness has been reported by many studies[45,71,72]
Mastitis	Mastitis by itself has been identified as a risk factor for anestrus.[45] Cows that suffered lameness and mastitis, or lameness and another severe stressor, were at much greater risk for delayed onset of cyclicity[72] than cows with mastitis alone
Other disease	Cows that developed pneumonia during the first month after calving were 5.4 times more at risk for delayed resumption of ovarian activity.[45] Numerous other types of systemic illnesses can contribute to the condition as sick cows will have reduced dry matter intake, increasing their risk to enter a state of negative energy balance, thereby anestrus

Note: This list should not be deemed complete and the references are representative citations only.

Increased GH increases liver gluconeogenesis and promotes lipolysis (mobilization of adipose tissue) resulting in the release of nonesterified fatty acids (NEFAs). High concentrations of GH and NEFA antagonize insulin action, creating a state of insulin resistance in postpartum dairy cows,[64,75] which diminishes glucose utilization by nonmammary tissues, conserving glucose for milk synthesis. Negative energy balance is strongly associated with the postpartum anovulatory period through decreased LH pulse frequency and low concentrations of blood glucose, insulin, and IGF-1 that collectively limit estrogen production by dominant follicles[11] which is essential for the LH surge, thereby delaying ovulations.

High stocking density that increases competition for feed, poor stall design that discourage cows from getting up frequently to eat, physical barriers that limit free and continuous access to feed, infrequent feed push-ups, excessive sorting by cows that get to the feed first, and inadequate hoof care leading to high prevalence of lameness are all indirect causes that can reduce dry matter intake. Furthermore, social hierarchy situations where younger or subordinate animals are intimidated by herdmates, stray voltage, slippery footing, and other deterrents can all dissuade cows from comfortably accessing feed, limiting dry matter intake and predisposing cows to enter a state of negative energy balance.

Unobserved estrus or silent estrus

Unobserved estrus is often a major contributor to anestrus as many of the cows considered anestrous could in fact be cycling normally, yet never seen or reported in estrus. There are two categories of animals.

1. Sub-estrus: cows that express poor estrous behavior (e.g., short duration, low-intensity estrus) and get missed during routine estrus detection.
2. Silent estrus: cows that fail to express any overt signs of estrus.

Both categories of cows may ovulate spontaneously and develop a CL; however, because they were never seen in estrus, they are falsely categorized as "anestrus." Performing more frequent visual detection of estrus (say three to four daily periods of dedicated surveillance) and observing for secondary signs of estrus such as chin-resting, vulva sniffing, licking, tailgating, and partial attempts to mount, rather than strictly for "standing estrus," will increase the probability of detecting estrus in these cows. The first ovulation after calving is rarely preceded by overt estrous behavior; only 10–13% of dairy cows reportedly exhibit standing estrus prior to the first ovulation,[36,76] although this proportion increases significantly in the second and third ovulatory cycles (Figure 52.4). The level of milk production could also influence the proportion of cows detected in standing estrus at first ovulation,[36,76,77] with a higher proportion of low-producing cows detected in standing estrus at first ovulation (Figure 52.5).

As previously stated, cows that initiate cyclicity earlier in the postpartum period have better reproductive performance.[1,2,4,24,28] Therefore, it becomes extremely important to assess the ovarian status of postpartum cows at regular intervals to differentiate the cases of "anestrus due to unobserved estrus" from those that are "true anestrus." This is discussed in more detail in the section Managing anestrus.

Pregnancy

The primary physiological reason for cows being anestrus is the interruption of cyclicity by pregnancy. Although most dairy farmers keep excellent breeding records, it is not uncommon for errors to occur at the time of updating breeding records. The inadvertent transposition of numeric characters during data entry is a common error that can falsely identify inseminated cows as not inseminated, resulting in the occasional case of anestrus due to a "surprise pregnancy." During reproductive examinations, practitioners should always bear this in mind and rule out pregnancy first.

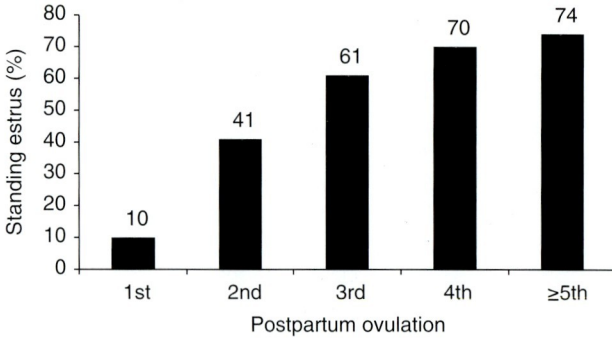

Figure 52.4 Frequency distribution of standing estrus behavior associated with first through fifth or greater postpartum ovulation in dairy cows. Standing estrus was associated with only 10% of first ovulations, but increased progressively at each subsequent ovulation. Redrawn with permission from Sakaguchi M. Oestrous expression and relapse back into anoestrus at early partpartum ovulations in fertile dairy cows. *Vet Rec* 2010;167:446–450.

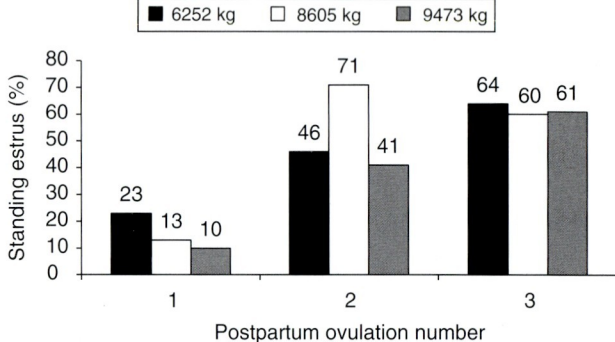

Figure 52.5 Frequency of occurrence of standing estrus and its association with level of milk production at first, second, and third postpartum ovulations. The percentage of standing estrus accompanying first ovulation was higher in cows with lowest annual production level of 6252 kg compared to those with higher production levels. The incidence of standing estrus at second and third ovulations increased substantially in higher-producing cows. Data from Refs 36, 76 and 77 for production levels of 9473, 8605 and 6252 kg, respectively.

Managing anestrus

Before planning a course of action to manage anestrus, it is important to assess the prevalence of anestrus at herd level. Anestrous postpartum beef cows, if treated, can recover from anestrus with normal fertility,[78] whereas dairy cows treated for anestrus have much lower fertility,[79] possibly due to an associated negative energy balance. Therefore, every effort should be taken to return the cows to positive energy balance before first insemination.

General strategies to diagnose and minimize the impact of anestrus

Some strategies to consider for assessing prevalence of anestrus at herd level are listed here.

1. *Perform body condition scoring (BCS).* Assessing BCS of cows will provide an indirect indication of whether cows are

meeting their energy needs adequately. Performing BCS of the whole herd, paying particular attention to close-up dry cows, fresh cows and cows in the breeding group, would be of great value. Close-up dry cows that are in high body condition (BCS >3.5/5) should be considered high risk for metabolic disorders and anestrus. A disproportionately large number of cows beyond 60 days in milk with poor body condition (BCS <3.0/5) should also be considered a problem as the prevalence of anestrus will be higher in those herds. Herdsmen should be educated on the importance of BCS, properly trained in the technique, and encouraged to perform condition scoring every 2 weeks, or at least every month, and keep a good record of the scores. Inseminating cows in poor body condition must be avoided and breeding delayed until a threshold BCS is attained, as low BCS cows have poor conception rates and a much higher risk of pregnancy loss.

2. *Perform gait (locomotion) scoring.* Herds where gait scoring is not performed should be encouraged to do it, routinely, to identify cows that are showing early signs of lameness. Periodic gait scoring, regular hoof trimming, and the use of foot baths must be encouraged to keep lameness in check. Lame cows will often be the last ones to reach the feed bunks and they are less motivated to leave the comfort of their stall to access feed frequently. As a result, they consume less feed predisposing them to an increased risk of negative energy balance and anestrus. Grouping cows with even early signs of lameness and feeding them separately would be one approach to improve dry matter intake in these cows, mitigating negative energy balance.

3. *Improve the efficiency of estrus detection.* The use of tail chalk or tail paint to improve estrus detection efficiency is one of the simplest strategies that could be implemented in herds of any size. Using three different colors to mark fresh cows (to determine cyclicity), cows at the end of the elective waiting period (to be flagged for insemination), and previously inseminated cows (for return service) would provide an efficient yet simple system for systematically detecting cows in estrus. In the breeding group, tail chalk scores should be recorded daily for this system to be effective and fresh chalk applied at least once weekly. Applying tail chalk or tail paint on fresh cows 30–40 days before the end of the elective waiting period is highly recommended. If tail chalk remains largely intact until the end of the elective waiting period, the cow is probably anestrous. This is because "standing estrus," which is essential for tail chalk to be removed completely, is only exhibited by 50% of lactating dairy cows in estrus[36] at 9–12 weeks after calving (Figure 52.4).

4. *Use of ultrasonography.* Transrectal ultrasonography is an excellent diagnostic tool[80,81] for the objective assessment of ovaries, and must be used wherever practical during routine reproductive examinations. Examining the ovaries by ultrasonography at approximately 5 weeks after calving provides great diagnostic value. Diagnosis of the anestrous condition based on ovarian ultrasonography must involve at least two consecutive examinations performed at an interval of 7–14 days.[82] Although two examinations at a 7–14 day interval are preferred, a single ultrasound examination has fairly good diagnostic value.[83]

5. *Monitoring progesterone profiles.* Albeit somewhat expensive, monitoring milk progesterone concentrations can be a reliable approach for assessing the prevalence of anestrus in dairy herds. During the early postpartum period progesterone concentrations will be very low and will continue to remain so until first ovulation occurs. Progesterone will rise following the first ovulation but will decline after a few days as the first luteal phase is typically short-lived. The second and later luteal phases will be of normal duration. Dairy cows relapsing back into anestrus after the first postpartum ovulation has been documented;[36] regular measurement of progesterone can assist in determining the patterns of cyclicity in dairy herds. The recently introduced Herd Navigator® system (DeLaval International) has the capability to measure in-line milk progesterone concentrations at the milking parlor. It has been claimed that Herd Navigator detects consistently above 95% of all estrus events including silent heats and is very useful in the detection of postpartum anestrus, pregnancy, and ovarian cysts.[84] While this is a welcome addition to the reproductive management tool-kit, its efficacy and reliability remains to be proven.

6. *Measuring plasma metabolites.* Measuring plasma NEFAs, an indicator of negative energy balance[64,75] and the early diagnosis of subclinical ketosis, particularly during the first week[27] can assist in the early identification of cows at risk.

While clinical management of individual cases of anestrus has its place, a whole herd approach is often more beneficial. For a practitioner to focus only on anestrous cows presented for reproductive examinations during herd health visits is not always the best approach. Instead, taking a proactive approach of whole herd assessment is highly recommended, particularly in herds with a history or suspicion of high prevalence of anestrus.

Walking through the barns without disturbing the cows to assess stall comfort and stall design and how those might influence cow behavior would be an essential component of the exercise. A poorly designed stall in a free-stall barn, such as one with limited lunge space, may discourage cows from getting up to access feed several times a day because of the discomforts associated with frequent getting up and lying down in a poorly designed stall.

It is equally, if not more, important for a dairy practitioner to be fully aware of reproductive management practices of the herd and to have regular conversations with the person in charge of reproductive management and the herdsmen/technicians performing estrus detection and artificial insemination (AI). In herds that have frequent staff turnover, it would be prudent to ensure that standard operating protocols are available for chores such as estrus detection, estrus synchronization treatments, semen handling and AI prominently posted, and strictly followed. The veterinarian must know who is in charge of estrus detection, and how frequently and for how long estrus detection is performed each day. In herds where electronic estrus detection systems are used, the system must be optimized for the herd to minimize false positives (wasted semen and labor) and false negatives (missed breeding opportunity). Complete reliance on electronic estrus detection systems must be discouraged, especially in herds where the system is newly installed.

Nutritional intervention

Interactions between nutrition and reproductive processes are complex[75,85] and it is often difficult to precisely study effects of specific nutrients on ovarian function. Excessive energy intake during the close-up dry period can result in changes in metabolism that may predispose cows to decreased dry matter intake and higher circulating NEFAs during the immediate peripartal period;[64] therefore monitoring feed intake in close-up dry cows is also quite important.

Specific nutritional or nutraceutical interventions to reduce the incidence of postpartum anestrus may include feeding diets high in starch[86,87] or daily oral administration of monopropylene glycol[88] during the postpartum period. Adding starch[89] or oilseeds enriched in long-chain fatty acids such as linoleic or linolenic acid[90] during the prepartum period have also reduced the interval from calving to first ovulation. One study[91] reported that cows that consumed an energy-restricted (80% of calculated requirement) diet during the entire dry period had lower NEFA concentrations and higher dry matter intake during the postpartum period than cows consuming 160% of predicted energy requirements during the dry period. Thus, several options exist for dietary manipulations that could mitigate negative energy balance and reduce anestrus. However, results are variable and research findings may not always translate into practical or economic solutions under commercial herd settings.

Therapeutic intervention

Hormonal treatment to induce estrus or ovulation is a more tangible approach that is well within the realm of a veterinary practitioner. Specific hormone therapy can be administered to either an individual case or a group of cattle as required. The commonly used hormones for treatment of anestrus are progesterone, estradiol, and GnRH, although equine chorionic gonadotropin (eCG), porcine LH, and human chorionic gonadotropin (hCG) have also been used. In herds where estrus detection efficiency is poor and undetected estrus (sub-estrus or silent estrus) is the primary suspected cause of anestrus, the use of prostaglandin (PG)$F_{2\alpha}$ followed by intensive estrus detection could be quite effective.[92,93] Of course, $PGF_{2\alpha}$ could also be used in combination with most of the above hormones in protocols that induce or synchronize estrus and ovulation.

Progesterone

A short period of exposure to high progesterone during the postpartum period is important for expression of estrus and for the establishment of normal luteal function. Treating anestrous cows with progesterone increased LH pulsatility, estradiol concentrations, and LH receptors in granulosa and theca cells of the preovulatory follicle.[94] Thus intravaginal progesterone devices can be quite effective in treating true anestrus. When progesterone is used alone, a treatment regimen for true anestrous cases would constitute insertion of a progesterone device for 7–9 days, removal of the device, and insemination at detected estrus. The insertion of a progesterone-releasing intravaginal device (PRID; 1.55 g progesterone) for 7 days and its removal at the end of that period induced cyclicity in 64 of 90 (71%) of treated cows in contrast to only 33 of 83 (40%) of untreated controls;[82] cyclicity was confirmed by the presence of a CL at ovarian ultrasonography 11 days after PRID removal. For therapy of anestrus at herd level, a 7-day treatment followed by $PGF_{2\alpha}$ injection at device removal is often recommended to ensure luteolysis in cows that were "silent ovulators" (sub-estrus) and in those that were falsely categorized as anestrous due to poor estrus detection efficiency (unobserved estrus).

The two brands of intravaginal progesterone devices available in the North American market are Eazi-Breed CIDR® (Zoetis, Florham Park, NJ, USA; 1.38 g progesterone) and PRID® (Vétoquinol, Lavaltrie, Quebec, Canada; 1.55 g progesterone). While PRID® is approved in Canada specifically for treatment of anestrus in dairy cattle, Eazi-Breed CIDR® has general approval for use in estrus synchronization. To my knowledge neither product has milk or meat withdrawal requirements. The use of oral progestagens (e.g., melengestrol acetate) is not discussed in this chapter because it is (i) registered as a product to improve growth rate and feed efficiency and suppress estrus, and (ii) it is not approved for use in any animal other than beef heifers intended for slaughter (Veterinary Drug Directorate, Ottawa, Canada).

Estradiol

Estradiol alone may be used for inducing estrus in anestrous cattle, but it is not recommended because the ovulatory response to estradiol is often variable. Furthermore, unless anestrous cows were previously exposed to progesterone, even if estradiol treatment induces estrus and ovulation, the ensuing luteal phase will usually be of short duration. However, using estradiol to induce estrus after a 7-day exposure to progesterone has been found effective in both beef and dairy cattle and has been widely used in New Zealand for treating anestrous dairy cattle. For example, the use of a CIDR device for 6 days, followed by an injection of 1 mg estradiol benzoate 24 hours after CIDR removal resulted in 87% of dairy cows being detected in estrus within 7 days of estradiol treatment; 79% of cows detected in estrus ovulated and a conception rate of 42% was attained.[94] Estradiol is available in different forms: estradiol-17β, estradiol benzoate, estradiol cypionate, and estradiol valerate. To my knowledge no injectable estradiol preparation is commercially available in the United States with approval for use in cattle. Estradiol cypionate (ECP®), which was previously marketed by Pfizer Animal Health in the United States and Canada for use in cattle, was withdrawn in 2005. Prepackaged estradiol cypionate is still available in the Canadian market through small pharmaceutical companies. Estradiol products may be legally dispensed by a compounding pharmacist under veterinary prescription in Canada. However, the Center for Veterinary Medicine has cautioned pharmacies and veterinarians about the illegality of using compounded estradiol cypionate in dairy and beef cattle in the United States.[95] Considering prevailing concerns about the use of estrogens in food animals, it would be prudent to choose alternative products for management of anestrus.

GnRH

Exogenous administration of GnRH induces LH release as early as the first week after calving in dairy cows, although the response is significantly higher by the second week (Figure 52.6). If a dominant LH-responsive follicle is present at the time of GnRH injection, there is a high probability of it ovulating, especially when the follicle has not been exposed to progesterone. In our laboratory, four of five lactating dairy cows given 100 µg GnRH in the second week after calving, ovulated (R. Salehi, M. Colazo and D. Ambrose, unpublished data, 2013). Inducing an ovulation in anestrous cows with no prior exposure to progesterone could result in short luteal phases, increasing the chances for cows to revert back into anestrus. In one study,[82] 93 of 108 (86%) anestrous cows ovulated in response to a GnRH injection, but 20% of these cows reverted to anestrus. Thus, although GnRH can be used independently to induce ovulations in anestrous dairy cows, one must recognize that short luteal phases and reversion to anestrus are drawbacks associated with this approach.

Using GnRH in combination with $PGF_{2\alpha}$ in a protocol such as Ovsynch[96] followed by timed insemination has been effective in the management of anestrous cows. Although the conception rates in anestrous cows following this approach are generally lower than in cyclic controls, several studies[82,83,97–99] have reported acceptable conception rates (Table 52.2).

LH or hCG can be used in lieu of GnRH in the Ovsynch protocol but these products are significantly more expensive than GnRH. Furthermore, there is limited research on the use of these products in the management of anestrous cattle. In one study,[98] replacing the second GnRH injection of the Ovsynch protocol with a commercial preparation of porcine LH, in predominantly cyclic lactating dairy cows, significantly improved pregnancy per AI by 14 percentile points (42% vs. 28%; $n = 308$). In the same study, when porcine LH was used in lieu of the first GnRH injection of the Ovsynch protocol, on a different subset of cows, the ovulatory response to porcine LH treatment was significantly higher than to GnRH treatment (62% vs. 44%). Since several studies have reported that cows ovulating in response to the first GnRH treatment of the Ovsynch protocol have higher pregnancy rates after timed AI,[79,83,99] using porcine LH in the treatment of individual cases of anestrus has potential value.

In this regard, pregnancy per AI was greatly enhanced in anestrous cows that ovulated in response to the first GnRH treatment of the Ovsynch protocol (46% vs. 6%) than in anestrous cows that failed to ovulate.[99] These results strongly indicate that determining the ovulation response to the first GnRH treatment in anestrous cows, by ultrasonography, would have significant value in predicting the probability of pregnancy.

Combined use of progesterone, GnRH, and $PGF_{2\alpha}$

The incorporation of an intravaginal progesterone device into a synchronized ovulation program such as Ovsynch is one approach to improve treatment outcomes (pregnancy) in anestrous cows. A second approach is to use either progesterone, GnRH and $PGF_{2\alpha}$, or $PGF_{2\alpha}$ alone, as a presynchronization treatment before the initiation of the

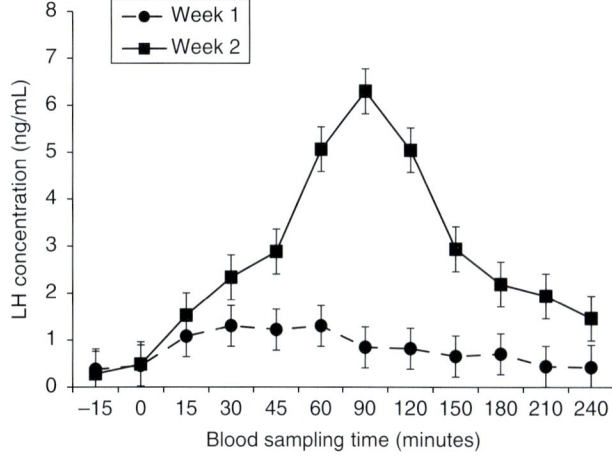

Figure 52.6 GnRH-induced LH release in early postpartum dairy cows when challenged during the first (circles) or second (squares) week postpartum. Cows ($n = 10$, 5 each week) were given 100 µg gonadorelin acetate i.m. and LH measured in bloods samples collected over 4 hours. Mean and peak LH concentrations differed between weeks 1 and 2. Unpublished results by R. Salehi, M. Colazo and D. Ambrose.

Table 52.2 Pregnancy per AI after Ovsynch/timed AI in dairy cows classified as anestrous or cyclic.

Per cent pregnancy per AI			
Anestrous (n)	Cyclic (n)	P	Reference
18.2 (24)	39.0 (303)	<0.02	Moreira et al.[97]
24.1 (116)	38.0 (799)	<0.05	Stevenson et al.[83]
24.4 (45)	30.8 (321)	>0.05	Colazo et al.[98]
30.1 (83)	n/a	n/a	Yilmazbas-Mecitoglu et al.[82]
21.8 (55)	32.0 (259)	>0.05	Colazo et al.[99]
23.1 (26)	32.8 (58)	>0.05	Öztürk et al.[101]

n/a, not applicable; only anestrous cows were assigned to the study.

Ovsynch protocol. Figure 52.7 illustrates some of the potential hormonal treatment options for postpartum anestrus in dairy cattle.[82,95,99–102]

Studies show that anestrous cows subjected to Ovsynch plus timed AI, with the insertion of a progesterone device for the first 7 days of the protocol, have higher pregnancy per AI than that of anestrous cows subjected to the same protocol without progesterone. Pregnancies per AI are comparable to that of cyclic cows subjected to Ovsynch plus timed AI with no progesterone treatment (Table 52.3).[83,98,101] A large field trial (12 herds, 1662 cows) conducted in New Zealand on anestrous dairy cows concluded that therapeutic intervention is the most economic option compared with no treatment, and that including an intravaginal progesterone device in the Ovsynch protocol was most cost-effective, attaining the highest net benefit of NZ$80.40 per cow.[103] The Ovsynch/timed AI protocol has been used successfully in cows with cystic ovarian follicles,[44] although the addition of a progesterone device to this protocol did not improve pregnancy.[104] Ovarian responses documented by ultrasonography, and pregnancy

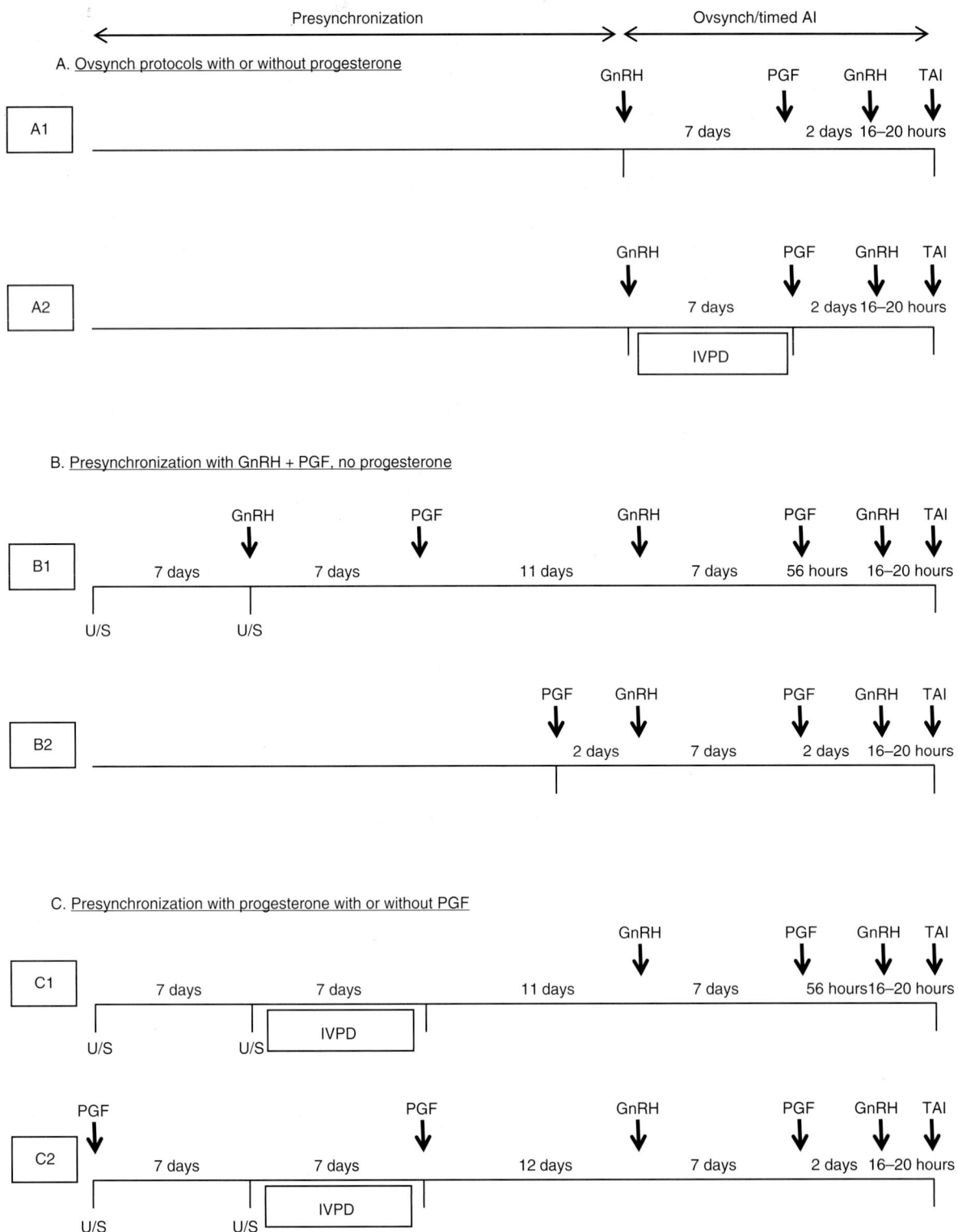

Figure 52.7 Fixed timed insemination (TAI) protocols for anestrous dairy cows. Protocol A1 is a standard Ovsynch protocol.[96] A2 is a modified Ovsynch protocol that includes an intravaginal progesterone device (IVPD) during the first 7 days with the device inserted at GnRH injection and removed at $PGF_{2\alpha}$ (PGF) administration. Protocols B1 and B2 include a presynchronization step with GnRH and PGF; in B1 the presynchronizing GnRH and PGF, in that order, are given 7 days apart and the Ovsynch protocol is stared after 11 days.[82] In B2, PGF is given 2 days prior to initiation of the Ovsynch protocol (Doublesynch);[101] a 72% pregnancy per AI was obtained (18 of 25 anestrous cows pregnant) with this protocol, and the authors speculated that the PGF pretreatment may have augmented LH response to GnRH resulting in higher ovulation rates and pregnancy per AI. For more details, please see Öztürk et al.[101] Protocol C1 involves the insertion of an IVPD for 7 days followed by initiation of Ovsynch protocol 11 days later.[82] Protocol C2[100] is identical to C1 except that two PGF injections are given 14 days apart with the IVPD inserted 7 days after the initial PGF injection; the Ovsynch protocol is started 12 days later. Although only protocols B1, C1, and C2 indicate ultrasonography (U/S), I recommend the use of ultrasonography (twice, 7–14 days apart) to assess cyclicity prior to assigning cows to any of these protocols. Furthermore, using ultrasonography to confirm ovulation following the first GnRH injection of the Ovsynch protocol in anestrous cows will be of great value, because anestrous cows that fail to ovulate in response to the first GnRH injection rarely conceive.[99]

Table 52.3 Pregnancy per AI in anestrous dairy cows subjected to Ovsynch/timed AI with or without insertion of an intravaginal progesterone device (IVPD) for the first 7 days of the Ovsynch protocol.

Anestrous		Cyclic		
No IVPD	Plus IVPD	No IVPD	Plus IVPD	Reference
24.1[a] (28/116)	32.3[b] (50/155)	38.0[b] (303/799)	n/a	Stevenson et al.[83]
33.9[a] (139/410)	45.7[b] (178/389)	34.3[a] (142/415)	n/a	McDougall[102]
21.8[a] (12/55)	34.6[b] (18/52)	32.0[b] (83/259)	43.0 (104/242)	Colazo et al.[99]

[a,b]Data with no common superscript within row, differ ($P<0.05$).
n/a, not applicable; cows confirmed cyclic did not receive an IVPD in this study.

Table 52.4 Ovarian responses to GnRH treatments, based on ultrasonography, and pregnancy rates after Ovsynch/timed AI in cows with cystic ovarian follicles in two studies.

Observation	Ambrose et al.[44]	Ambrose & Colazo[104]
Number of cows	18	109
Mean diameter of cyst (mm)	33.9 ± 1.5	33.6 ± 6.0
Ovulation of cystic follicle after first GnRH	0/18 (0)	12/109 (11.0)[a]
Ovulation of an existing follicle after first GnRH	8/18 (44.4)	68/109 (62.4)
New follicle developed after first GnRH	18/18 (100)	102/109 (93.6)
Ovulation of new follicle after second GnRH	15/18 (83.3)	91/109 (83.4)
Cows pregnant 32 days after timed AI	7/17 (41.2)[b]	45/109 (41.3)

[a]Some of the smaller cystic follicles (25–30 mm range) ovulated.
[b]Only 17 cows were available for pregnancy diagnosis.

rates after timed AI, in cows with cystic ovarian follicles in the above studies[44,104] are summarized in Table 52.4.

Equine chorionic gonadotropin

New reports[105,106] have found that including eCG into a progesterone-based protocol in treatment of anestrous cows is beneficial. Anestrous cows subjected to GnRH plus 6–7 days of an intravaginal progesterone device followed by PGF$_2\alpha$ at device withdrawal plus 400 IU eCG increased first-service conception rates by 5.4 percentile points.[105] More studies are needed to corroborate these findings.

Post-breeding treatments

Because of the higher risk of pregnancy loss in anestrous dairy cows, post-breeding interventions aimed to improve progesterone concentrations during the early phases of embryonic development have been attempted. Insertion of a PRID for 7 days starting at 4.5 days after AI in anestrous dairy cows greatly reduced pregnancy losses (5.6% vs. 33.3%) compared with anestrous cows that did not receive a PRID after breeding,[99] although the difference only tended to be significant ($P=0.10$). In another study giving a GnRH injection 5 days after timed AI increased pregnancy per AI in anestrous dairy cows (45.5 vs. 31.1%) compared with cyclic cows.[107]

Summary

The underlying reasons for postpartum anestrus in dairy cattle are often multifactorial, but negative energy balance is recognized as a major contributor. To maximize reproductive success, cows that are considered to be at higher risk for anestrus should be identified early and managed for increased dry matter intake. Assessment of ovarian status by ultrasonography offers great diagnostic value. Improving estrus detection efficiency would greatly diminish anestrus due to undetected estrus. Where therapeutic intervention is chosen, a progesterone-based Ovsynch protocol, or a progesterone-based presynchronization protocol followed by Ovsynch/timed AI provide the best options. Treatment of anestrous cows is the cost-effective alternative to no treatment.

References

1. Thatcher W, Wilcox C. Postpartum estrus as an indicator of reproductive status in the dairy cow. *J Dairy Sci* 1973;56:608–610.
2. Kawashima C, Kaneko E, Amaya Montoya C et al. Relationship between the first ovulation within three weeks postpartum and subsequent ovarian cycles and fertility in high producing dairy cows. *J Reprod Dev* 2006;52:479–486.
3. Gautam G, Nakao T, Yamad K, Yoshida C. Defining delayed resumption of ovarian activity postpartum and its impact on subsequent reproductive performance in Holstein cows. *Theriogenology* 2010;73:180–189.
4. Kim I, Jeong J, Kang H. Field investigation of whether corpus luteum formation during weeks 3–5 postpartum is related to subsequent reproductive performance of dairy cows. *J Reprod Dev* 2012;58:552–556.
5. Short R, Bellows R, Staigmiller R, Berardinelli J, Custer E. Physiological mechanisms controlling anestrus and infertility in postpartum beef cattle. *J Anim Sci* 1990;68:799–816.
6. Hammond J. *The Physiology of Reproduction in the Cow*. University Press, Cambridge, 1927. Cited in Short R, Bellows R, Staigmiller R, Berardinelli J, Custer E. Physiological mechanisms controlling anestrus and infertility in postpartum beef cattle. *J Anim Sci* 1990;68:799–816.
7. Hansen P. Seasonal modulation of puberty and the postpartum anestrus in cattle: a review. *Livestock Prod Sci* 1985;12:309–327.
8. Opsomer G, Mijten P, Coryn M, de Kruif A. Postpartum anoestrus in dairy cows: a review. *Vet Quart* 1996;18:68–75.
9. Mihm M. Delayed resumption of cyclicity in postpartum dairy and beef cows. *Reprod Domest Anim* 1999;34:277–284.
10. Mwaanga E, Janowski T. Anoestrus in dairy cows: causes, prevalence and clinical forms. *Reprod Domest Anim* 2000;35:193–200.
11. Butler W. Energy balance relationships with follicular development, ovulation and fertility in postpartum dairy cows. *Livestock Prod Sci* 2003;83:211–218.
12. Crowe M. Resumption of ovarian cyclicity in postpartum beef and dairy cows. *Reprod Domest Anim* 2008;43(Suppl. 5):20–28.
13. Peter A, Vos P, Ambrose D. Postpartum anestrus in dairy cattle. *Theriogenology* 2009;71:1333–1342.

14. Macmillan K, Lean I, Westwood C. The effects of lactation on the fertility of dairy cows. *Aust Vet J* 1996;73:141–147.
15. Royal M, Darwash A, Flint A, Webb R, Woolliams J, Lamming G. Declining fertility in dairy cattle: changes in traditional and endocrine parameters on fertility. *Anim Sci* 2000;70:487–501.
16. Lucy M. Reproductive loss in high-producing dairy cattle: where will it end? *J Dairy Sci* 2001;84:1277–1293.
17. Westwood C, Lean I, Garvin J. Factors influencing fertility of Holstein cows: a multivariate description. *J Dairy Sci* 2002;85:3225–3237.
18. López-Gatius F. Is fertility declining in dairy cattle? A retrospective study in northeastern Spain. *Theriogenology* 2003;60:89–99.
19. Ahlman T, Berglund B, Rydhmer L, Strandberg E. Culling reasons in organic and conventional dairy herds and genotype by environment interaction for longevity. *J Dairy Sci* 2011;94:1568–1575.
20. CanWest DHI. 2011 Western Herd Improvement Report. CanWest DHI 2012, p. 30.
21. National Animal Health Monitoring Sysytem. *Dairy 2007. Part I. Reference of Dairy Cattle Health and Management Practices in the United States, 2007.* Fort Collins, CO: USDA, 2007.
22. Rajala-Schultz P, Gröhn Y. Culling of dairy cows. Part II. Effects of diseases and reproductive performance on culling in Finnish Ayrshire cows. *Prev Vet Med* 1999;41:279–294.
23. Francos G, Mayer E. Analysis of fertility indices of cows with extended postpartum anestrus and other reproductive disorders compared to normal cows. *Theriogenology* 1988;29:399–412.
24. Darwash A, Lamming G, Woolliams J. The phenotypic association between the interval to postpartum ovulation and traditional measures of fertility in dairy cattle. *Anim Reprod Sci* 1997;65:9–16.
25. Stevenson J, Pursley J, Garverick H et al. Treatment of cycling and noncycling lactating dairy cows with progesterone during Ovsynch. *J Dairy Sci* 2006;89:2567–2578.
26. Ambrose D, Kastelic J, Corbett R et al. Lower pregnancy losses in lactating dairy cows fed a diet enriched in α-linolenic acid. *J Dairy Sci* 2006;89:3066–3079.
27. Walsh R, Kelton D, Duffield T, Leslie K, Walton J, LeBlanc S. Prevalence and risk factors for postpartum anovulatory condition in dairy cows. *J Dairy Sci* 2007;90:315–324.
28. Ambrose D, Colazo M. Reproductive status of dairy herds in Alberta: a closer look. In: *Advances in Dairy Technology, Proceedings of Western Canadian Dairy Seminar*, 2007, pp. 227–244. Available at http://www.wcds.ca/proc/2007/Manuscripts/Divakar.pdf
29. Nett T. Function of the hypothalamic–hypophysial axis during the postpartum period in ewes and cows. *J Reprod Fertil Suppl* 1987;34:201–213.
30. Crowe M, Padmanabhan V, Mihm M, Beitins Z, Roche F. Resumption of follicular waves in beef cows is not associated with periparturient changes in follicle-stimulating hormone heterogeneity despite major changes in steroid and luteinizing hormone concentrations. *Biol Reprod* 1998;58:1445–1450.
31. Mann G, Lamming G. Relationship between maternal endocrine environment, early embryo development and inhibition of the luteolytic mechanism in cows. *Reproduction* 2001;121:175–180.
32. Patel O, Takenouchi N, Takahashi T, Hirako M, Sasaki N, Domeki I. Plasma oestrone and oestradiol concentrations throughout gestation in cattle: relationship to stage of gestation and fetal number. *Res Vet Sci* 1999;66:129–133.
33. Ginther O, Kot K, Kulick L, Martin S, Wiltbank M. Relationships between FSH and ovarian follicular waves during the last six months of pregnancy in cattle. *J Reprod Fertil* 1996;108:271–278.
34. Savio J, Thatcher W, Badinga L, De la Sota R, Wolfeusun D. Regulation of dominant follicle turnover during the estrous cycle in cows. *J Reprod Fertil* 1993;97:197–203.
35. Gong J, Campbell B, Bramley T, Gutierrez C, Peters A, Webb R. Suppression in the secretion of follicle stimulating hormone and luteinizing hormone, and ovarian follicle development in heifers continuously infused with a gonadotropin-releasing hormone agonist. *Biol Reprod* 1996;55:68–74.
36. Sakaguchi M. Oestrous expression and relapse back into anoestrus at early partpartum ovulations in fertile dairy cows. *Vet Rec* 2010;167:446–450.
37. Savio J, Boland M, Hynes N, Roche J. Resumption of follicular activity in the early post-partum period of dairy cows. *J Reprod Fertil* 1990;88:569–579.
38. Beam S, Butler W. Effects of energy balance on follicular development and first ovulation in postpartum dairy cows. *J Reprod Fertil* 1999;54:411–424.
39. Wiltbank M, Gümen A, Sartori R. Physiological classification of anovulatory conditions in cattle. *Theriogenology* 2002;57:21–52.
40. Markusfeld O. Inactive ovaries in high-yielding dairy cows before service: aetiology and effect on conception. *Vet Rec* 1987;121:149–153.
41. McDougall S, Burke C, MacMillan K, Williamson N. Patterns of follicular development during periods of anovulation in pasture-fed dairy cows after calving. *Res Vet Sci* 1995;58:212–216.
42. Vanholder T, Opsomer G, de Kruif A. Aetiology and pathogenesis of cystic ovarian follicles in dairy cattle: a review. *Reprod Nutr Dev* 2006;46:105–119.
43. Cook D, Parfet J, Smith C, Moss G, Youngquist R, Garverick H. Secretory patterns of LH and FSH during development and hypothalamic and hypophysial characteristics following development of steroid-induced ovarian follicular cysts in dairy cattle. *J Reprod Fertil* 1991;91:19–28.
44. Ambrose D, Schmitt E, Lopes F, Mattos R, Thatcher W. Ovarian and endocrine responses associated with the treatment of cystic ovarian follicles in dairy cows with gonadotropin releasing hormone and prostaglandin F2α, with or without exogenous progesterone. *Can Vet J* 2004;45:931–937.
45. Opsomer G, Gröhn Y, Hertl J, Coryn M, Deluyker H, de Kruif A. Risk factors for postpartum ovarian dysfunction in high producing dairy cows in Belgium: a field study. *Theriogenology* 2000;53:841–857.
46. Williams G. Suckling as a regulator of postpartum rebreeding in cattle: a review. *J Anim Sci* 1990;68:831–852.
47. Ducrot C, Gröhn YT, Humblot P, Bugnard F, Sulpice P, Gilbert R. Postpartum anestrus in French beef cattle: an epidemiological study. *Theriogenology* 1994;42:753–764.
48. Jolly P, McDougall S, Fitzpatrick L, MacMillan K, Entwistle K. Physiological effects of undernutrition on postpartum anoestrus in cows. *J Reprod Fertil Suppl* 1995;49:477–492.
49. Stevenson J, Lamb G, Hoffman D, Minton J. Interrelationships of lactation and postpartum anovulation in suckled and milked cows. *Livestock Prod Sci* 1997;50:57–74.
50. Montiel F, Ahuja C. Body condition and suckling as factors influencing the duration of postpartum anestrus in cattle: a review. *Anim Reprod Sci* 2005;85:1–26.
51. Carruthers T, Hafs H. Suckling and four-times daily milking: influence on ovulation, estrus and serum luteinizing hormone, glucocorticoids and prolactin in postpartum Holsteins. *J Anim Sci* 1980;50:919–925.
52. Santos J, Rutigliano H, Sá Filho M. Risk factors for resumption of postpartum estrous cycles and embryonic survival in lactating dairy cows. *Anim Reprod Sci* 2009;110:207–221.
53. Rajala P, Gröhn Y. Disease occurrence and risk factor analysis in Finnish Ayrshire cows. *Acta Vet Scand* 1998;39:1–13.
54. Kawashima C, Amaya Montoya C, Masuda Y et al. A positive relationship between the first ovulation postpartum and the increasing ratio of milk yield in the first part of lactation in dairy cows. *J Dairy Sci* 2007;90:2279–2282.

55. Berardinelli J, Joshi P. Introduction of bulls at different days postpartum on resumption of ovarian cycling activity in primiparous beef cows. *J Anim Sci* 2005;83:2106–2110.
56. Roelofs J, Soede N, Dieleman S, Voskamp-Harkema W, Kemp B. The acute effect of bull presence on plasma profiles of luteinizing hormone in postpartum, anestrous dairy cows. *Theriogenology* 2007;68:902–907.
57. Wright I, Rhind S, Smith A, Whyte T. Female–female influences on the duration of the post-partum anoestrous period in beef cows. *Anim Prod* 1994;59:49–53.
58. Pleasants A, McCall D. Relationships among post-calving anoestrous interval, oestrous cycles, conception rates and calving date in Angus and Hereford × Friesian cows calving in six successive years. *Anim Prod* 1993;56:187–192.
59. Richards M, Wettemann R, Schoenemann H. Nutritional anestrus in beef cows: body weight change, body condition, luteinizing hormone in serum and ovarian activity. *J Anim Sci* 1989;67:1520–1526.
60. Wright I, Rhind S, Russel A, Whyte T, McBean A, McMillen S. Effects of body condition, food intake and temporary calf separation on the duration of the post-partum anoestrous period and associated LH, FSH and prolactin concentrations in beef cows. *Anim Prod* 1987;45:395–402.
61. Moreira F, Orlandi C, Risco C, Mattos R, Lopes F, Thatcher W. Effects of presynchronization and bovine somatotropin on pregnancy rates to a timed artificial insemination protocol in lactating dairy cows. *J Dairy Sci* 2001;84:1646–1659.
62. Kim I, Suh G. Effect of the amount of body condition loss from the dry to near calving periods on the subsequent body condition change, occurrence of postpartum diseases, metabolic parameters and reproductive performance in Holstein dairy cows. *Theriogenology* 2003;60:1445–1456.
63. Gümen A, Rastani R, Grummer R, Wiltbank M. Reduced dry periods and varying prepartum diets alter postpartum ovulation and reproductive measures. *J Dairy Sci* 2005;88:2401–2411.
64. Overton T. Feeding and managing dairy cows for improved early cyclicity. In: *Proceedings of the Dairy Cattle Reproduction Conference, Sacramento, CA, 8–9 November, 2012.* Champaign, IL: Dairy Cattle Reproduction Council, 2012, pp. 32–40. Available at http://www.dcrcouncil.org/resources/proceedings-presentations/2012-november-8-9,-sacramento,-california.aspx
65. Bauman D, Currie W. Partitioning of nutrients during pregnancy and lactation: a review of mechanisms involving homeostasis and homeorhesis. *J Dairy Sci* 1980;63:1514–1529.
66. Huszenicza G, Fodor M, Gacs M et al. Uterine bacteriology, resumption of cyclic ovarian activity and fertility in postpartum cows kept in large-scale dairy herds. *Reprod Domest Anim* 1999;34:237–245.
67. Dourey A, Colazo M, Patricio P, Ambrose D. Relationships between endometrial cytology and interval to first ovulation, and pregnancy in postpartum dairy cows in a single herd. *Res Vet Sci* 2011;91:e149–e153.
68. Bulman D, Lamming G. Cases of prolonged luteal activity in the non-pregnant dairy cow. *Vet Rec* 1977;100:550–552.
69. Kinsel M, Etherington W. Factors affecting reproductive performance in Ontario dairy herds. *Theriogenology* 1998;50:1221–1238.
70. Hamilton S, Garverick H, Keisler D et al. Characterization of ovarian follicular cysts and associated endocrine profiles in dairy cows. *Biol Reprod* 1995;53:890–898.
71. Garbarino E, Hernandez J, Shearer J, Risco C, Thatcher W. Effect of lameness on ovarian activity in postpartum Holstein cows. *J Dairy Sci* 2004;87:4123–4131.
72. Peake K, Biggs A, Argo C et al. Effects of lameness, subclinical mastitis and loss of body condition on reproductive performance of dairy cows. *Vet Rec* 2011;168:301–306.
73. Smith J, Payne E, Tervit H et al. The effect of suckling upon the endocrine changes associated with anoestrus in identical twin dairy cows. *J Reprod Fert Suppl* 1981;30:241–249.
74. Carruthers T, Convey E, Kesner J, Hafs H, Cheng K. The hypothalamo-pituitary gonadotrophic axis of suckled and nonsuckled dairy cows postpartum. *J Anim Sci* 1980;51:949–957.
75. Lucy M. Fertility in high-producing dairy cows: reasons for decline and corrective strategies for sustainable improvement. *Soc Reprod Fertil Suppl* 2007;64:237–254.
76. Kyle S, Callahan C, Allrich R. Effect of progesterone on the expression of estrus at the first postpartum ovulation in dairy cattle. *J Dairy Sci* 1992;75:1456–1460.
77. Morrow D, Roberts S, McEntee K, Gray H. Postpartum ovarian activity and uterine involution in dairy cattle. *J Am Vet Med Assoc* 1966;149:1596–1609.
78. Lucy M, Billings H, Butler W et al. Efficacy of an intravaginal progesterone insert and an injection of PGF2α for synchronizing estrus and shortening the interval to pregnancy in postpartum beef cows, peripubertal beef heifers, and dairy heifers. *J Anim Sci* 2001;79:982–995.
79. Gümen A, Guenther J, Wiltbank M. Follicular size and response to Ovsynch versus detection of estrus in anovular and ovular lactating dairy cows. *J Dairy Sci* 2003;86:3184–3194.
80. Rajamahendran R, Ambrose D, Burton B. Clinical and research applications of real time ultrasonography in bovine reproduction: a review. *Can Vet J* 1994;35:563–572.
81. Colazo M, Ambrose D, Kastelic J. Practical uses for transrectal ultrasonography in reproductive management of cattle. In: Wittwer F et al. (eds) *Updates on Ruminant Production and Medicine: XXVI World Buiatrics Congress, 14–18 November, 2010, Santiago, Chile.* Budapest: World Association for Buiatrics, 2010, pp. 146–156.
82. Yilmazbas-Mecitoglu G, Karakaya E, Keskin A, Alkan A, Okut H, Gümen A. Effects of presynchronization with gonadotropin-releasing hormone-prostaglandin F2α or progesterone before Ovsynch in noncyclic dairy cows. *J Dairy Sci* 2012;95:7186–7194.
83. Stevenson J, Tenhouse D, Krisher R et al. Detection of anovulation by heatmount detectors and transrectal ultrasonography before treatment with progesterone in a timed insemination protocol. *J Dairy Sci* 2008;91:2901–2915.
84. Mazeris F. DeLaval Herd Navigator® Proactive Herd Management. In: *The First North American Conference on Precision Dairy Management*, 2010, pp. 25–26.
85. Chagas L, Bass J, Blache D et al. New perspectives on the roles of nutrition and metabolic priorities in the subfertility of high-producing dairy cows. *J Dairy Sci* 2007;90:4022–4032.
86. Gong J, Lee W, Garnsworthy P, Webb R. Effect of dietary-induced increases in circulating insulin concentrations during the early postpartum period on reproductive function in dairy cows. *Reproduction* 2002;123:419–427.
87. Dyck B, Colazo M, Ambrose D, Dyck M, Doepel L. Starch source and content in postpartum dairy cow diets: effects on plasma metabolites and reproductive processes. *J Dairy Sci* 2007;94:4636–4646.
88. Chagas L, Gore P, Meier S, Macdonald K, Verkerk G. Effect of monopropylene glycol on luteinizing hormone, metabolites, and postpartum anovulatory intervals in primiparous dairy cows. *J Dairy Sci* 2007;90:1168–1175.
89. Gardner N, Reynolds C, Phipps R, Jones A, Beever D. Effect of different diet supplements in the pre- and post-partum period on reproductive performance in the dairy cow. In: *Fertility in the High Producing Dairy Cow.* British Society of Animal Science Occasional Publication no. 26, 2001, pp. 313–321.
90. Colazo M, Hayirli A, Doepel L, Ambrose D. Reproductive performance of dairy cows is influenced by prepartum feed restriction and dietary fatty acid source. *J Dairy Sci* 2009;92:2562–2571.

91. Douglas G, Overton T, Bateman II H, Dann H, Drackley J. Prepartal plane of nutrition, regardless of dietary energy source, affects periparturient metabolism and dry matter intake in Holstein cows. *J Dairy Sci* 2006;89:2141–2157.
92. Grimard B, Laumonnier G, Ponsart C et al. Postpartum suboestrus in dairy cows: comparison of treatment with prostaglandin F2α or GnRH + prostaglandin F2α + GnRH. In: *Fertility in the High Producing Dairy Cow*. British Society of Animal Science Occasional Publication no. 26, 2001, pp. 347–352.
93. López-Gatius F, Mirzaei A, Santolaria P et al. Factors affecting the response to the specific treatment of several forms of clinical anestrus in high producing dairy cows. *Theriogenology* 2008;69:1095–1103.
94. Rhodes F, McDougall S, Burke C, Verkerk G, Macmillan K. Treatment of cows with an extended postpartum anestrous interval. *J Dairy Sci* 2003;86:1876–1894.
95. Center for Veterinary Medicine. Cattle Reproductive Drug Estradiol Cypionate (ECP) Illegal, CVM Reminds Industry. FDA Veterinarian Newsletter November/December 2005, Vol. XX, No. VI. http://www.fda.gov/AnimalVeterinary/NewsEvents/FDAVeterinarianNewsletter/ucm092836.htm
96. Pursley R, Mee M, Wiltbank M. Synchronization of ovulation in dairy cows using PGF2α and GnRH. *Theriogenology* 1995;44:915–923.
97. Moreira F, Risco C, Pires M, Ambrose D, Drost M, Thatcher W. Use of bovine somatotropin in lactating dairy cows receiving timed artificial insemination. *J Dairy Sci* 2000;83:1237–1247.
98. Colazo M, Gordon M, Rajamahendran R, Mapletoft R, Ambrose D. Pregnancy rates to timed artificial insemination in dairy cows treated with gonadotropin-releasing hormone or porcine luteinizing hormone. *Theriogenology* 2009;72:262–270.
99. Colazo M, Dourey A, Rajamahendran R, Ambrose D. Progesterone supplementation before timed AI increased ovulation synchrony and pregnancy per AI, and supplementation after timed AI reduced pregnancy losses in lactating dairy cows. *Theriogenology* 2013;79:833–841.
100. Cerri R, Rutigliano H, Bruno R, Santos J. Progesterone concentration, follicular development and induction of cyclicity in dairy cows receiving intravaginal progesterone inserts. *Anim Reprod Sci* 2009;110:56–70.
101. Öztürk Ö, Cirit Ü, Baran A, Ak K. Is Doublesynch protocol a new alternative for timed artificial insemination in anestrous dairy cows? *Theriogenology* 2010;73:568–576.
102. McDougall S. Effects of treatment of anestrous dairy cows with gonadotropin-releasing hormone, prostaglandin, and progesterone. *J Dairy Sci* 2010;93:1944–1959.
103. McDougall S. Comparison of diagnostic approaches, and a cost–benefit analysis of different diagnostic approaches and treatments of anoestrous dairy cows. *NZ Vet J* 2010;58:81–89.
104. Ambrose D, Colazo M. Cows diagnosed with ovarian cysts derive no benefit from progesterone supplementation but conceive normally to a GnRH-based timed A.I. protocol. *Reprod Domest Anim* 2012;47(Suppl. 4):424–425.
105. Bryan M, Bo G, Mapletoft R, Emslie F. The use of equine chorionic gonadotropin in the treatment of anestrous dairy cows in the gonadotropin-releasing hormone/progesterone protocols of 6 or 7 days. *J Dairy Sci* 2013;96:122–131.
106. Garcia-Ispierto I, López-Helguera I, Martino A, López-Gatius F. Reproductive performance of anoestrous high-producing dairy cows improved by adding equine chorionic gonadotrophin to a progesterone-based oestrous synchronizing protocol. *Reprod Domest Anim* 2012;47:752–758.
107. Sterry R, Welle M, Fricke P. Treatment with gonadotropin-releasing hormone after first timed artificial insemination improves fertility in noncycling lactating dairy cows. *J Dairy Sci* 2006;89:4237–4245.

Chapter 53

Surgery to Restore Fertility

Richard M. Hopper

*Department of Pathobiology and Population Medicine, College of Veterinary Medicine,
Mississippi State University, Starkville, Mississippi, USA*

Introduction

For cows of genetic merit or perceived to be of high value, urogenital procedures are typically performed with the goal of restoring fertility. However, a desire to extend the life of the cow so that a calf can be reared, a lactation completed, or simply to add condition prior to sale often justifies attempts at surgical correction as well. The procedures that will be described can be performed on the standing animal with regional analgesia and sedation.

Repair of injuries occurring from dystocia

Parturition is clearly one of the most dangerous events in the life of an animal, whether the dam or the offspring. Common injuries include trauma and tears of the uterus and caudal urogenital system. Tearing of the uterus, whether spontaneous or iatrogenic, can be life-threatening. At the very least, the resultant fibrosis and structural incompetence from perineal body, cervical, and vaginal tears typically results in infertility. Identification of the severity and the provision of a realistic prognosis are critical to the producer, so that an informed decision can be made with respect to treatment versus culling or even immediate salvage.

Repair of uterine tears

Most injures to the uterus are iatrogenic and occur during efforts to relieve dystocia. With few exceptions they occur on the dorsal aspect of the uterus (cranial to the cervix) when the obstetrician is attempting to repel a breech presentation or during the forced extraction of a large fetus. In the first case, repair is secondary to removal of the calf. If a cesarean section is performed, the uterine defect may be approached and repaired after the calf is delivered. This is almost always the case when dealing with a tear that occurred during repulsion of a breech position fetus. If the calf can be delivered vaginally following this type of tear or, as in the second case, in which the tear occurs during the delivery, a "blind" suture repair through the vagina can be performed.

I will first examine how these tears occur. In the case of the tear caused by "excessive" force in the obstetrician's attempt to repel the calf, the uterine wall was probably weakened by some level of pressure (ischemic) necrosis. This occurs when the tailhead of the fetus in the breech position is pressed for hours against the dorsal uterine wall. In the second situation, a large calf, a "dry" birth canal, and forced extraction combine to create an overfolding and subsequent shearing of the uterine wall. These tears are usually 5–30 cm in length, although some seem almost circumferential.

Regardless of the tear's etiology, repair may not be necessary if it is small and dorsal. Because of rapid uterine involution, tears smaller than the width of a hand will usually be satisfactory without repair. Treatment for these can be limited to repeated oxytocin injections for the first 24–48 hours and 7–10 days of antibiotics. If the tear is larger, it can be closed with a "blind" suture technique.[1]

To perform a one-handed blind closure of a uterine rent, first thread an atraumatic needle (size and shape determined by preference) with a 150 cm of #2 or #3 catgut. With the needle at the halfway point of the suture (doubled), make a knot at the end (knot 1) and about 25 cm from the end (knot 2). Next introduce the needle into and through the vagina, guarding the point with the fingers until the tear is reached. Begin closure at one end of the tear. After piercing the uterine wall (about 1 cm lateral to the tear) run the needle between the doubled suture and cinch at knot 2 (the only purpose of knot 1 is to keep the two loose ends together). Then close the tear with a continuous pattern until the last bite is made 1 cm lateral to the other end. It is important to pull the suture tight with each throw. Then reverse and continue the closure back to the original site and tie to the suture tail. Uterine involution serves to cover small appositional problems. Recommended treatment includes antibiotics, nonsteroidal anti-inflammatory drugs (NSAIDs), and oxytocin (Figures 53.1, 53.2, 53.3 and 53.4).

Alternatively, the uterus can be prolapsed to facilitate suture repair. To prolapse a uterus, the cervix must still be

---

Bovine Reproduction, First Edition. Edited by Richard M. Hopper.
© 2015 John Wiley & Sons, Inc. Published 2015 by John Wiley & Sons, Inc.

Figure 53.1 With the suture "doubled" the first bite is made at the lateral edge of the wound most easily accessed.

Figure 53.2 After the first suture bite is taken, pull the needle through the loop created in the suture (this is not a knot, the suture is cinched).

Figure 53.3 Continuous suture bites.

Figure 53.4 After reaching the end of the wound, return. This would look similar to a cruciate pattern and serves to correct the appositional deficiencies of the first row of sutures. When the pattern is complete, the suture is tied to the "tail."

completely dilated. Therefore this procedure must be performed within 6–12 hours of parturition. Administer 10 mL of 1 in 1000 epinephrine in 250–500 mL of saline intravenously very slowly. Do not perform an epidural or administer oxytocin. The cow must be able to strain and the uterus must be flaccid. If possible, an assistant should administer the epinephrine solution, so that the surgeon is free to perform the procedure. As the solution is administered, enter the vagina with an arm, reaching as far cranial as possible to grasp a placentome and pull slowly. This will invert the tip of a uterine horn and the presence of the arm along with the invaginated uterus will stimulate straining. Pull with the cow's straining, grasping other placetomes if necessary. The uterus will usually come out within minutes. At this time if the rent or tear can be visualized, it can easily be repaired. The uterus is then cleansed and replaced.

It should be noted that this procedure has limited application in that many of the iatrogenic tears encountered as a result of, or during the management of, dystocia are very close to the cervix; thus after the uterus is prolapsed, the area of the tear will usually still be within the vagina, so not allowing the accessibility desired. Additionally, many of these tears can be near circumferential, and manual prolapse may be synonymous with uterine extraction – because the uterus is held in place with only a small amount of tissue, upon prolapse it simply tears and falls off. Thus careful evaluation of the tear and its location and severity is warranted prior to prolapse for repair. As with the "blind suture" repair, antibiotics, NSAIDS, and oxytocin should be given postoperatively.

Occasionally a cow will present two or more days following parturition with signs of peritonitis, depression, and anorexia. The cow may or may not have a history of obstetric intervention or even suspected dystocia. Examination may reveal a distal uterine tear which is likely the result of a spontaneous tear or, of course, an injury similar to the iatrogenic injury previously described. Additionally, the peritonitis could be a complication following cesarean section. In any case because of the duration of injury and peritonitis, surgical exploration of the abdomen via a flank approach should be performed. Treatment of the peritonitis begins with thorough abdominal lavage, which in turn is facilitated by the ventral placement of a drain.

Utilization of a rumen trocar placed from within the abdominal cavity at the ventralmost aspect will provide very efficient egress for the volume of fluid that is required (Figures 53.5 and 53.6). Additionally, because of the "siphon" effect that is created, debris (fibrin clots, necrotic material, and placental remnants) will be "pulled" to the drainage site and can be more efficiently retrieved. The rent in the uterus can then be repaired, the uterus lavaged, and finally the flank incision closed. This of course is accompanied by a guarded prognosis, so this level of effort is typically reserved for the cow of genetic merit or the involvement of a highly motivated owner.

Surgery to Restore Fertility 473

Figure 53.5 A rumen trocar placed from within the abdomen and directed outward at the ventralmost location.

Figure 53.6 Fluid flow through the trocar cannula from the abdominal lavage.

Injury to the vulva, vagina, and cervix

Tearing of the vulva, vagina, and cervix occur almost exclusively during dystocia. Tearing and bruising can often occur during parturition in the absence of obstetric intervention; however injury to these structures typically occurs during obstetric management of dystocia.

Tearing of the vulva during the management of a dystocia can be prevented by performing an episiotomy. This procedure is discussed in Chapter 46, but briefly it is an incision typically made at the dorsolateral (10 o'clock or 2 o'clock) aspect of the vulva. This incision will serve to either avoid or, at the very least, "guide" the tearing that occurs with the extreme stretching of the vulva during a dystocia due to an oversized fetus. If an episiotomy is not performed, the tear will most likely occur at the most dorsal aspect of the vulva toward the anus and this in fact is the mechanism for grade 1, 2, and 3 perineal lacerations, a grade 3 being a rectovaginal tear.

Lacerations that involve only the vulva and even episiotomy incisions can be left unrepaired especially if the cow is to be culled. However, this often results in compromised vulvar competence so these tears should be sutured immediately. A modified vertical mattress pattern can be used. Take a deep bite that includes the skin, fibrous tissue, and mucosal layers, and then the skin only for the superficial bite.[2] Although nonabsorbable suture material is often recommended, #1 chromic gut can be used.

Vaginal tearing usually occurs in overfit heifers during the forced extraction of a calf. In these heifers vaginal fat serves to decrease the cross-sectional area of the birth canal, thus serving as a contributing factor in the dystocia. During parturition, and specifically forced extraction, the vaginal wall will split allowing the protrusion of this fat. This fat can and should be removed, but suturing the laceration is neither necessary nor recommended.[2]

Occasionally the bruising that occurs during parturition will be so severe that the vaginal tissues will become necrotic. Depending on the severity systemic illness will follow, but at the very least the cow will display tenesmus, a vaginal discharge, discomfort at urination, and vaginal prolapse. Treatment depends on severity and ranges from systemic antibiotics and vaginal lavage to surgical debridement of the affected tissue.

Trauma to the vagina can also result in severe and extensive adhesions. These can occur following dystocia or as a complication of vaginal or cervico-vaginal prolapse. If the adhesions completely occlude the vagina, fluid of a mucoid consistency will accumulate proximally. If this fluid has filled the uterus (mucometra), this is indicative of likely damage to the endometrium and therefore a poor prognosis for future fertility. If the condition is of short duration or if the uterus is not distended, attempts at treatment are warranted. Utilizing epidural anesthesia, insert a gloved hand into the vagina and attempt to break down the adhesions manually. The difficulty with which this is done and the depth of the lesions are prognostic indicators (deeper or more cranial lesions indicate poorer prognosis).

Insuring that the vagina remains patent after the adhesions are broken down is the key to successful resolution. An inflatable rubber beach ball (25 cm long deflated) with a grommet on one end has been recommended as a pessary.[2] Using #3 Braunamid suture, fix the grommet of the well-lubricated (petrolatum jelly and antibiotic ointment) deflated ball to the cervix. Then slightly inflate the ball. To further insure retention the vulva is closed with vertical mattress sutures or prolapse pins. Remove the ball in 7–10 days, but continue the application of topical antibiotic ointment until healing. When a ball as described is unavailable, I have used a reusable latex palpation sleeve. The sleeve, closed at one end, is placed into the vagina, filled with loose cotton, and fixed in place.

Injuries to the cervix occur during dystocia and most are iatrogenic. If this is identified immediately, as in the case of a iatrogenic uterine tear or an extension of a uterine tear, it is managed basically as the uterine tear (one-handed "blind" suture). However, many injuries to the cervix are not noticed until artificial insemination (AI) is attempted. These injuries are for the most part not as amenable to correction as those that occur in mares. However, I have examined several cows with cervical injuries or scarring that did not conceive via AI, but became pregnant after natural mating. Interestingly, cervical structure following the subsequent calving and involution appeared normal and the cows were again able to conceive following AI.

Rectovaginal tears

Also termed third-degree perineal laceration, this exclusively occurs as a sequela to a dystocia. It can be iatrogenic, but most occur naturally due to fetal oversize. In the cow the constrictor vestibule muscle and cervix are able to provide an adequate anatomic barrier to prevent uterine contamination and infection.[3] Indeed, most cattle with rectovaginal tears or fistulas will continue to conceive and carry pregnancies to term.

Cases in which surgical repair is advisable would be the cow that has been presented as open after attempts at breeding or cows of high genetic merit, specifically the embryo donor cow, for which the owner/manager does not wish to leave the results of future use up to chance. In any case, allow at least 4–6 weeks following the injury before attempting repair.[4]

Several techniques have been described for use in the mare and any of those would be appropriate. The procedure I have used and will describe involves the separate closure of the rectal mucosa, the perineal body, and the roof of the vagina much the same as a technique recently described for the mare.[5] Unlike the mare, the cow has typically soft manure so dietary alterations prior to surgery are unnecessary. Likewise, it is a matter of choice whether to use a "rectal tampon," a stockinette (diameter 7.5 cm) filled with cotton and placed in the rectum cranially to the surgery site. Regardless, catheterize and empty the bladder, evacuate the feces from the caudal rectum, and administer epidural anesthesia. Alternatively, it is possible to utilize one of the regional blocks described in Chapter 15. Sedation depends on the cow's behavior. The restraint utilized can be either a chute or stanchion. The tail is tied to the side and the surgical site prepared. After cleaning the area, lavage with dilute Betadine solution and dry. I prefer the Betadine scrub and solution for this area and do not use chlorhexidine-based products. The use of a chlorhexidine product is discouraged as these products are irritating to the vaginal mucosa and can result in luminal adhesions.

A transverse incision is made along the scar and deepened to a depth of 1–2 cm carefully dissecting the newly formed "flaps" away from the lateral aspects of the vaginal wall. With careful attention to suture placement in the crani-almost aspect, begin suturing the roof of the vagina. Various suture patterns and suture materials have been recommended for this procedure in mares, but a simple continuous pattern utilizing #1 chromic gut is adequate. The rectal mucosa should be closed with an interrupted horizontal mattress pattern in such a way as to avoid suture exposure in the rectal lumen. As the closure of the rectal mucosa is advanced caudally, stop every 3–5 cm to place a suture in the perineal body. This effectively closes the "dead" space as you go. The skin is closed with a vertical mattress suture pattern and a nonabsorbable (Braunamid) suture.

The advantage of this technique over others, specifically the "6-bite" closure of all layers with one suture, is that it seems to create a thicker, stronger shelf. While this procedure is tedious, the results are usually very rewarding and, as alluded to earlier, it is a procedure reserved for an animal of high value. Postoperative care consists of antibiotics and NSAIDs for 5–7 days, as well as care during transrectal palpation for the next 2–3 months and perhaps attention at the subsequent calving.

Management of iatrogenic injury from artificial insemination or embryo transfer

Because of the physical structure of the vagina and cervix, iatrogenic injury to the vagina and cervix during the course of AI or embryo transfer (ET) is unlikely or at least rare. However, the uterus and specifically the endometrial lining can be damaged easily. The most common injury related to AI is probably uterine puncture following passage of the AI rod completely through the cervix and aggressively continuing anterior passage. This injury in typically related to lack of experience of the technician, but is likely on the increase due to the utilization of "deep-horn" AI when utilizing semen with low sperm numbers, specifically sex-sorted semen or semen that has been evaluated and determined to have low number or low post-thaw motility. Another problem occurs when a straw is broken and the straw or portions of the straw are left in the uterus. Injuries from ET are also typically associated with the novice and include overextension of the Foley balloon, which results in rupture of the endometrium, or stylet-related uterine punctures which would be similar to those created by AI pipettes.

Because it is hard to ascertain how often punctures occur, little is known about the incidence of resultant infertility. Although these injuries likely heal successfully without intervention, a reasonable management tactic when presented with such a case would be a low-volume antibiotic infusion following the injury and then to "short-cycle" the cow (administration of a prostaglandin). This would be done 7–10 days later following the AI attempt or immediately if occurring during ET. Cows managed such have conceived on the resultant estrus (note that it is not recommended to breed a cow on the next estrus following an ET flush, so this is applicable only for AI-associated injury).

A fragment of or an entire straw can inadvertently be left in the uterus. This can be the result of a prolonged and aggressive AI attempt by a novice, but can also occur when the subject is unruly and/or poorly restrained. These "foreign bodies" serve to function as excellent intrauterine devices and thus result in infertility. Depending on the duration of retention, it can be removed transcervically, which should be attempted during estrus or surgically. Transcervical removal can be attempted with any long forceps, including an equine biopsy forceps. Often the straw becomes "embedded" in the endometrium. Removal in these cases can be facilitated by the use of a Foley catheter and the introduction of 80–150 mL of sterile lavage fluid. This serves to expand the uterus and help with removal. Surgical removal can be accomplished via a flank approach. I have utilized a colpotomy approach as well. In either case if the straw appears intact it can simply be manipulated and forced out manually without an incision. If there are straw fragments, an incision must be made and the uterus lavaged following removal. Closure is with a disposable suture and pattern of individual preference.

Endometrial rupture during an ET flush is not uncommon and will occur following overextension of the Foley balloon, but also from unexpected and dramatic movement by the subject. In any case it results in some level of hemorrhage, but also the inability to recover lavage fluid. Immediately stop flush attempts, administer 5–10 mg prostaglandin (PG)

$F_{2\alpha}$ (Lutalyse) intravenously via the tail vein, and decompress the Foley balloon. After 5–10 min have elapsed the flush can be resumed, this time replacing the catheter in a position anterior to the previous location.

Urovagina (urine pooling)

Urovagina is recognized as one of the most common reproductive disorders in the mare. Although there have been various reports that describe surgical correction of this condition in the cow,[6–8] only relatively recently has work been published that describes risk factors, its effect on reproductive performance, and prevalence.[9] This work revealed that based on the evaluation of Holstein cows over a total of 344 lactations, this condition occurred at a prevalence of 26.7%. Urovagina was identified by vaginal speculum examination and classified as mild, moderate, or severe based on the amount of urine pooled within the vagina. Only moderate or severe categories of the condition resulted in a decrease in fertility and these in turn represented about half of the cases. Risk factors included poor body condition, a history of endometritis, and poor vulvar conformation.

This condition is therefore worthy of our attention and attempts at correction. Although risk factors have been identified, a specific etiology is less clear. A common case presentation at our referral clinic is that of the aged donor cow with the etiology in this case likely to be iatrogenic, specifically due to repeated superovulatory treatments. The reason I believe this is that most of the time these cows do not present with a torn or injured urethra or possess any of the identified risk factors, but instead a large vagina in which the cranial aspect falls over the floor of the pelvis. This condition of course can also occur following injury, usually associated with dystocia or is manifest as splanchtosis from multiple calvings.

One of the urethral extension procedures described for the mare can be utilized, although I have typically employed one of three easier alternatives. Regardless of etiology, if the case presented involves a cow to be used for ET and the client's desire is to simply get one more embryo flush, a long Foley catheter (20–24 French, 30-mL balloon) can be placed in the urethra at the initiation of the superovulatory regime and maintained through embryo flush/recovery. A one-way valve (finger of a latex glove with a small slit) is placed over the egress of the catheter to allow urine out, but precludes air from entering. This will temporarily prevent urine from accumulating in the cranial vagina, but irritation to the bladder and urethra from the presence of the catheter limits its value as a long-term solution.

For a longer semi-permanent correction, a circumferential suture placed in the vestibulovaginal junction[6] will allow several embryo recovery flushes or the maintenance of pregnancy (note that the suture should be removed prior to parturition). To accomplish this procedure the cow is restrained, epidural analgesia employed, the tail tied to the side, and the vagina and perineum prepared for surgery. Next place a catheter into the bladder to prevent the inadvertent restriction of the urethra by the subsequent suture pattern. Introduce a hand into the vagina, utilizing either a hand-held curved cutting edge needle or a Deschamps needle to place a subcutaneous suture cranial to the urethral orifice. The needle is introduced at the 4 o'clock position and

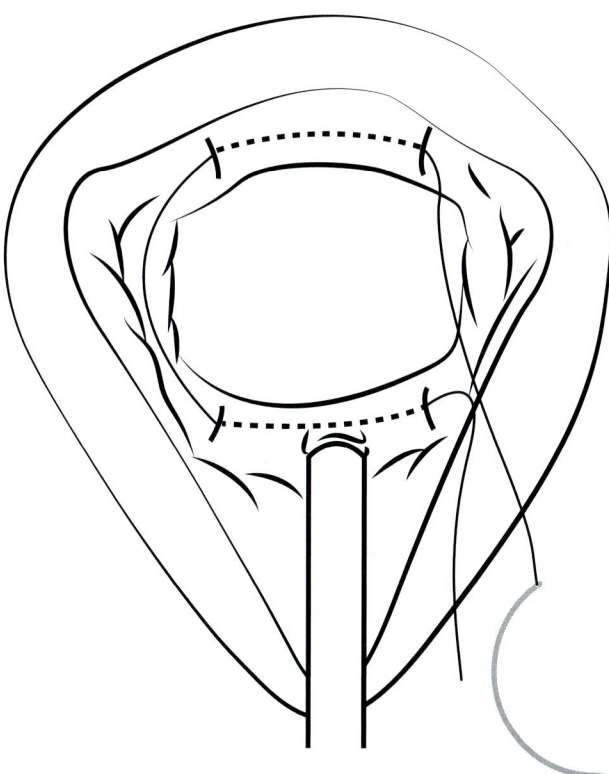

Figure 53.7 Suture placement for circumferential suture in the vestibule of the vagina. A catheter is placed into the urethra to avoid incorporating it into suture. Note that this is a modification of the procedure as described by González-Martín et al.[6]

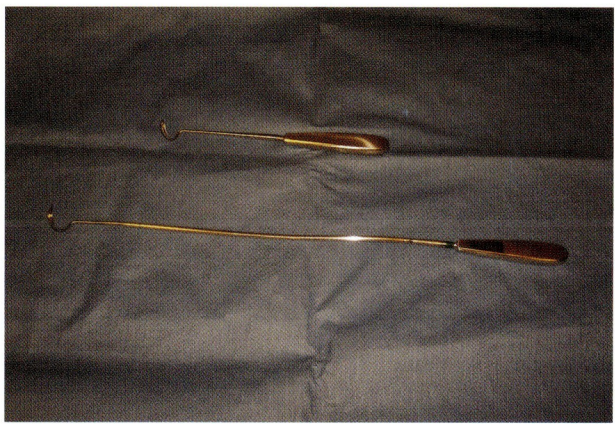

Figure 53.8 Deschamps needles. One of the instruments has been made longer. This may not be an improvement as this instrument is less stable and will bend under pressure.

exits at the 8 o'clock position. The needle is reintroduced at 10 o'clock and exits at the 2 o'clock position. Tighten the suture and tie. There should be ample opening for passage of an AI pipette or ET catheter (Figures 53.7 and 53.8).

Additionally, the Minchev procedure or a modified Minchev utilizing the commercially available Johnson button can be utilized as well. This procedure is performed exactly as it is when utilized for vaginal or cervico-vaginal prolapse retention. It is effective for the resolution of urine pooling because it has the effect of both raising the vagina and pulling it caudally. This typically facilitates urine passage.

Removal of a mummified fetus via colpotomy

A mummified fetus is an infrequent finding during routine herd pregnancy checks. Additionally, this condition is discovered when a cow is presented for signs of anestrus or had been previously diagnosed pregnant and is overdue. Diagnosis can be made by per rectal palpation. An enlarged uterine horn with a palpably hard intrauterine mass, accompanied by a lack of fluid, is the most common finding, as well as the presence of a corpus luteum on the ipsilateral ovary. Utilization of ultrasonography is also of value (see Chapter 36).

Mummification of a fetus can occur any time after fetal ossification,[10] typically between the third and eighth month of gestation. The condition is dependent on the maintenance of the corpus luteum and a closed cervix. Several etiologies for this condition are possible. Trichomoniasis and bovine viral diarrhea virus are the most common of the many infectious agents that cause early fetal death. Chromosomal abnormalities have been implicated as well as genetic anomalies. One genetic cause, an autosomal recessive gene for deficiency of uridine monophosphate synthesis (DUMPS), is known to cause embryonic and fetal death and was identified in the DNA extracted from two mummified fetuses.[10]

Depending on the age of the cow and/or economic considerations the cow may be immediately culled. If an attempt to treat the cow is chosen, conservative medical treatment should always be the first approach, this being the administration of a native or synthetic prostaglandin (Lutalyse, Estrumate).[11] Depending on whether estrogen products are legally available for use in cattle, a single dose of estrogen, either estradiol cypionate or estradiol benzoate (2–5 mg), can be given with or 24–48 hours after prostaglandin treatment with the goal of facilitating response by softening or "priming" the cervix. PGE available as a suppository used in human obstetrics, a compounded form, or Cytotec (misoprostol) tablets can also be used intracervically to aid in cervical dilatation. It is worthwhile repeating medical treatment several times before resorting to either surgery or culling.

In cases in which there is no response to these treatments, a surgical remedy may be effective. A hysterotomy via either a flank approach or a colpotomy is recommended. In a retrospective study of cases not responsive to prostaglandin treatment,[10] some mummified fetuses were removed by a combination of medical treatment (estradiol-17β, $PGF_{2\alpha}$, and oxytocin) followed by manual cervical dilation and manual extraction and some by a flank approach hysterotomy. In this study 36% of cows became pregnant following a hysterotomy removal, while none conceived following the manual extraction.

A flank approach or a midline is recommended in cases in which the mummy is greater than 30 cm.[12] For those that are smaller a hysterotomy via a colpotomy[13] approach can be performed easier and in less time. The cow is restrained in a chute. It is recommended but not absolutely necessary to fast the cow, removing feed for 24 hours and water overnight. Sedation or tranquilization is dependent on the patient's personality. Administer an epidural and evacuate the rectum as effectively as possible. Then place a length of 7.5-cm stockinette packed with cotton within the rectum. Clean the vulva and prepare the vagina by lavaging with a dilute disinfectant (Betadine). Using dampened cotton pledgets dry or at least remove water from the vaginal vault. Creating a pneumovagina by introducing or allowing air in will facilitate the procedure. With a blade carefully guarded between finger and thumb, introduce a hand into the vagina and at a location in the anterior vagina that is dorsolateral (10 o'clock) to the cervix make a small stab incision. Alternatively, a specially made trocar can be utilized to make this stab incision. After an incision is made, the vagina is reentered and the incision is enlarged bluntly, first with the fingers and then the hand until a hand can be introduced into the abdomen. This should be performed aggressively and not gingerly, as overly cautious manual entrance serves to "tent" the peritoneum rather than push through it and this is actually more painful to the patient. Once the hand is inside the abdomen, the uterus is easily palpated; when the horn with the mummy is located, grasp and retract it out through the rent created in the vagina (Figures 53.9 and 53.10). Once it is exposed, make an incision into the uterus and extract the mummy. The lumen of the uterus is then lavaged to make sure that all foreign material is removed. Then suture the uterus with a continuous inverting suture pattern as you would a cesarean section hysterotomy. Then

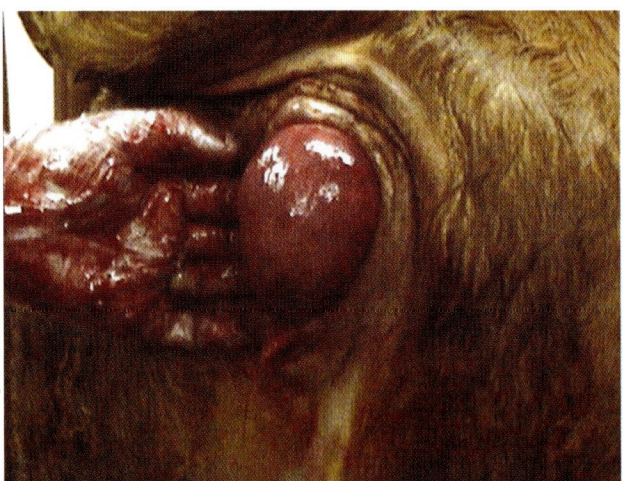

Figure 53.9 The uterus (uterine horn containing a mummified fetus) exposed through a colpotomy incision.

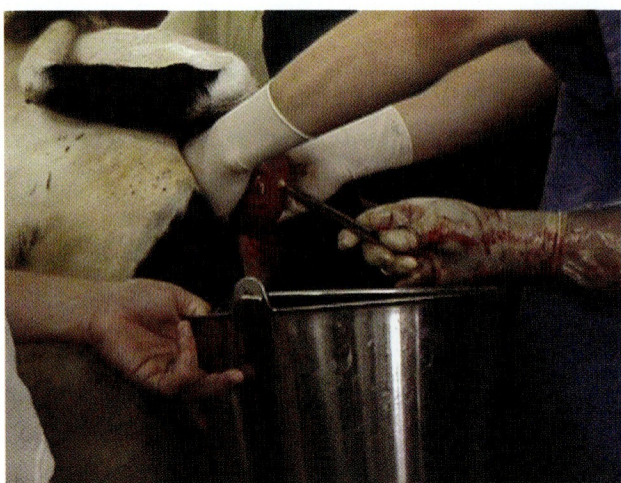

Figure 53.10 Incision into the exposed uterine horn.

rinse the uterus removing any debris or fibrin tags. Addition of carboxymethylcellulose to the saline used to both lavage the uterine lumen and cleanse the uterus before replacement is likely beneficial in reducing the possibility of adhesion formation. Then replace the uterus. Although suturing the rent created in the vagina is considered optional, I do not attempt to close colpotomy incisions on mares but I close it in cows. This can be done with a continuous pattern (the suture technique described for uterine tears can be utilized). If the decision is not to close the colpotomy incision, a Caslick should be performed.

Unilateral ovariectomy

The indications for ovariectomy include neoplasia or the presence of severe adhesions, either of which affect the function of the ovary. There should be reasonable evidence that the contralateral ovary is normal and that fertility will be restored following removal of the abnormal ovary. An ovary that is small enough to be easily grasped during per rectal palpation can be removed via colpotomy. Larger ovaries or those with extensive adhesions should be removed via flank celiotomy.

Ovariectomy via colpotomy

The colpotomy approach was described in the previous section. Once the abdomen has been entered the chain loop of an écraseur is placed around the ovary. The chain loop is slowly tightened, making sure that the placement does not include bowel. The loop is tightened further until the ovary's attachment is severed and the ovary can be removed. Although it is recommended that only smaller ovaries (those that can easily be held in the operator's hand) are to be removed by this technique, larger ovaries can be removed by carrying a sterile plastic bag into the abdomen, dropping the ovary into the bag, and then removing the bag with ovary.

Ovariectomy via celiotomy

The left flank celiotomy approach is described in Chapter 48 and the only difference is that the incision for this should be made slightly more dorsal and caudal. This approach is preferred over the colpotomy when the ovary to be removed is large, adhesions are severe, or there is concern relating to hemostasis.

References

1. Hopper R. Surgical correction of abnormalities of genital organs of cows. In: Youngquist RS, Threlfall WR (eds) *Current Therapy in Large Animal Theriogenology*, 2nd edn. St Louis, MO: WB Saunders, 2007, pp. 463–472.
2. Hudson R. Genital surgery of the cow. In: Morrow DA (ed.) *Current Therapy in Theriogenology 2*. Philadelphia: WB Saunders, 1986, pp. 341–352.
3. Fubini S. Surgery of the perineum. In: Fubini SL, Duscharme NG (eds) *Farm Animal Surgery*. St Louis, MO: WB Saunders, 2004, pp. 399–403.
4. Hudson R. Surgical procedures of the reproductive system of the cow. In: Morrow DA (ed.) *Current Therapy in Theriogenology*. Philadelphia: WB Saunders, 1980, pp. 257–271.
5. Mosbah E. A modified one-stage repair of third-degree rectovestibular lacerations in mares. *J Equine Vet Sci* 2012 ;32: 211–215.
6. González-Martín J, Astiz S, Elvira L, Lopez-Gatius F. New surgical technique to correct urovagina improves the fertility of dairy cows. *Theriogenology* 2008;69:360–365.
7. Prado T, Schumacher J, Hayden S, Donnell R, Rohrbach B. Evaluation of a modified surgical technique to correct urine pooling in cows. *Theriogenology* 2007;67:1512–1517.
8. Wenzel J, Baird A. Surgery of the female urinary tract. In: Wolfe DF, Moll HD (eds) *Large Animal Urogenital Surgery*. Baltimore: Williams & Wilkins, 1998, pp. 447–449.
9. Gautam G, Nakao T. Prevalence of urovagina and its effects on reproductive performance in Holstein cows. *Theriogenology* 2009;71:1451–1461.
10. Lefebvre R, Saint-Hilaire E, Morin I, Couto G, Francoz D, Babkine M. Retrospective case study of fetal mummification in cows that did not respond to prostaglandin F2α treatment. *Can Vet J* 2009;50:71–76.
11. Wenkoff M, Manns J. Prostaglandin-induced expulsion of bovine fetal mummies. *Can Vet J* 1977:18:44–45.
12. Hopper R, Hostetler D, Smith J, Christiansen D. Surgical removal of a mummified fetus via colpotomy. *Bov Pract* 2006;40:57–58.
13. Irons P. Hysterotomy by a colpotomy approach for treatment of foetal mummification in a cow. *J South Afr Vet Assoc* 1999; 70:127–129.

PREGNANCY WASTAGE

54. Fetal Disease and Abortion: Diagnosis and Causes 481
    Wes Baumgartner

55. Infectious Agents: *Campylobacter* 518
    Misty A. Edmondson

56. Infectious Agents: *Trichomonas* 524
    Mike Thompson

57. Infectious Agents: Leptospirosis 529
    Daniel L. Grooms

58. Infectious Agents: Brucellosis 533
    Sue D. Hagius, Quinesha P. Morgan and Philip H. Elzer

59. Infectious Agents: Infectious Bovine Rhinotracheitis 541
    Ahmed Tibary

60. Infectious Agents: Bovine Viral Diarrhea Virus 545
    Thomas Passler

61. Infectious Agents: Epizootic Bovine Abortion (Foothill Abortion) 562
    Jeffrey L. Stott, Myra T. Blanchard and Mark L. Anderson

62. Infectious Agents: *Neospora* 567
    Charles T. Estill and Clare M. Scully

63. Infectious Agents: Mycotic Abortion 575
    Frank W. Austin

64. Early Embryonic Loss Due to Heat Stress 580
    Peter J. Hansen

65. Bovine Abortifacient and Teratogenic Toxins 589
    Brittany Baughman

66. Heritable Congenital Defects in Cattle 609
    Brian K. Whitlock and Elizabeth A. Coffman

67. Abnormal Offspring Syndrome 620
    Charlotte E. Farin, Callie V. Barnwell and William T. Farmer

68. Strategies to Decrease Neonatal Calf Loss in Beef Herds 639
    David R. Smith

69. Management to Decrease Neonatal Loss of Dairy Heifers 646
    Ricardo Stockler

Chapter 54

Fetal Disease and Abortion: Diagnosis and Causes

Wes Baumgartner

*Department of Pathobiology and Population Medicine, College of Veterinary Medicine,
Mississippi State University, Starkville, Mississippi, USA*

Introduction

Reproductive failure is a significant problem in breeding management that arises from physiological, anatomical, inherited, and infectious causes. The inability to achieve or maintain a pregnancy may be divided into four main stages: failure to ovulate after estrus, failure of fertilization, embryonic death (prior to gestation day 42), and fetal death.[1] Pregnancy wastage, which probably comprises the largest portion of these losses, encompasses the combined embryonic, fetal, and neonatal deaths that achieve nothing. It has been estimated that about 75% of pregnancy wastage occurs in the embryonic stage.[2,3] The majority of reproductive failure occurs in the first 2 weeks, at the time of development from a morula to blastocyst where the conceptus begins enhanced protein synthesis and placental development.[3,4] Despite the importance of wastage at this time of development, the mechanisms contributing to embryonic death are poorly understood and in most cases the cause is never established.[1] Factors typically thought to contribute to embryonic death include heat stress, infections of the uterus/gametes/embryo, local trauma, genetic factors, heavy lactation in dairy cattle causing energy imbalances, maternal illness, aged oocytes from persistent follicles, small follicles, fetal–maternal incompatibilities, twinning, postpartum breeding intervals, and abnormal progesterone levels due to estrus synchronization.[1,3–5] Investigation into the causes of early loss are hampered by the rarity of available early conceptuses that are either expelled or resorbed and go unnoticed, only to result in a cow that returns to estrus or fails to deliver a calf.

Abortion typically refers to pregnancy loss in the fetal stage, between days 42 and 260. It is in this stage of gestation that the tissues of conception cannot be resorbed and, when expelled, are more easily noticed. During the fetal period there is a progressive decrease over time in the risk of abortion, with a slight increase in the last month.[6] Although losses during this time are a minority of overall wastage, the cost of investment by the cow and the manager are substantial. A mid-term abortion represents a loss of at least US$600–1000.[6] For practical purposes, unavoidable losses of 3% after pregnancy confirmation (6 weeks' gestation), with 1–2% loss in the periparturient period, may be considered acceptable or typical, although opinion in this area varies.[6–8]

This discussion will concentrate on fetal death and the approach to diagnosis, but will necessarily include some aspects of embryonic death, congenital abnormalities, and common gross findings. Additionally, a brief overview of anatomy and development are included to familiarize the reader with general concepts.

The investigation of abortion is a vital part of herd management and its importance is hard to overemphasize. Accurate diagnosis is the only path to the effective control of disease. It should go without saying that a focused, efficient, and thorough investigation technique with a definite purpose by prepared veterinarians is the only relevant way to identify and understand the etiologies of abortion. The necropsy itself contributes to this in several ways: by establishing definitive causes of death, identifying unsuspected findings, providing information concerning zoonotic disease, contributing to discovering new diseases and pathogenic mechanisms, and providing a means to test new diagnostic and treatment techniques.[9]

Pathology

The expression of disease in the conceptus is remarkably varied in a general sense, in that it gives rise to bizarre developmental defects of potentially every sort. This is due to the intricate physiology and morphology of fertilization and gestation that is shared between the dam and fetus. The sequential stages of organ system development *in utero* provide unique opportunities for infectious and noninfectious etiologies to manifest themselves at the gross level, if pregnancy is maintained. In addition the placenta, a unique organ of gestational necessity, is a target of many disease-causing agents and can itself develop anomalous defects that in turn directly affect the fetus.

Bovine Reproduction, First Edition. Edited by Richard M. Hopper.
© 2015 John Wiley & Sons, Inc. Published 2015 by John Wiley & Sons, Inc.

Despite this, in many cases aborted tissues from infectious and noninfectious causes exhibit little if any recognizable changes of significance. This is partly due to autolysis *in utero* masking subtle changes; the relative ease and rapidity by which fetuses succumb to disease, allowing only a brief window for gross changes to manifest; and a rudimentary inflammatory response to injury.[10] Not only does the conceptus vary in its susceptibility to particular insults across the gestational period, but the degree and duration of insult also determines the outcome, whether it be life, death, malformation, or inflammation.[7]

The physiology of pregnancy is well studied and dealt with in detail in preceding chapters. Despite our understanding of normal pregnancy, the physiology of abortion and the mechanisms involved are poorly understood. In many cases abortion may be mediated through the same pathways that occur in normal parturition, but different mechanisms and pathways are also possible.[6]

Categories of reproductive loss

The cause of abortion in many cases is not known, which is a consistent source of frustration for manager and clinician alike. In all species, abortion may be caused by infectious and noninfectious etiologies. Specific causes and agents are dealt with in detail later in this and subsequent chapters. Of these, infectious causes of abortion (bacteria, viruses, fungi, protozoa) are probably the best understood and characterized. It is common to see reports where the percentage of cases with a specific diagnosis is less than 50%.[11] Of these, at least half are due to infectious agents, with the majority of these due to bacterial infection.[6]

Bacteria involved in abortions can be broadly grouped into those that are contagious and those that are opportunistic. The majority of bacterial abortions are caused by opportunists, and these may be further divided into those that are part of the natural flora, such as *Trueperella pyogenes* (formerly *Arcanobacterium pyogenes*) and *Histophilus somni*, and those from the environment, such as *Bacillus* spp. and *Escherichia coli*.[12] The significance attached to abortion by opportunists depends on the situation in the herd. If these bacteria are only seen in isolated cases, then the significance to the herd is minimal.[12] However, if these organisms are consistently associated with abortions, then further investigation is warranted.

Noninfectious causes of abortion are similar to those known to cause embryonic loss, including nutritional imbalances, malnutrition, stress, environmental toxins, teratogenic compounds, hormone imbalances, and genetic abnormalities. An expanded list of etiologic agents is shown in Table 54.1 and in the following chapters.

Pathophysiology of injury to the conceptus

Maintenance of the first half of gestation (the first 200 days) in cattle requires a persistent corpus luteum (CL), which is maintained by the fetus.[7,13] Luteolysis may occur in pregnant cows due to excess prostaglandins (exogenous administration or secondary to heat stress) or Gram-negative bacterial septicemia.[14] If the CL is destroyed at an early stage, death and rapid loss of the embryo with minimal

Table 54.1 Disease-causing agents associated with endemic fetal losses.

Histophilus somni, *Listeria monocytogenes*, *Trueperella pyogenes*, *Leptospira* spp., *Ureaplasma diversum*
Bluetongue virus, BVDV
Neospora caninum, fungi, epizootic bovine abortion
Inbreeding, sire-derived lethal traits, chromosomal abnormalities
Feed estrogens (silage, poultry litter), progesterone aberrations (high pasture protein)
Protein, vitamin A, iodine, selenium deficiency
Protein/urea, copper, iodine excess
Endotoxins due to Gram-negative bacterial sepsis in dam
Endotoxin in Gram-negative bacterial vaccines, especially given during first or last 2 months
Pine needle, broomweed, locoweed, narrow leaf sumpweed toxicosis
High plant estrogens
Aflatoxin, ergotamine, fusarium (zearalenone), nitrate fertilizer, organophosphate toxicosis

Source: Whittier W. Investigation of abortions and fetal loss in the beef herd. In: Anderson DE, Rings M (eds) *Current Veterinary Therapy: Food Animal Practice*, 5th edn. St Louis, MO: Saunders, 2009, pp. 613–618.

degeneration may be seen.[15] If there is fetal death prior to luteolysis, the CL may regress with eventual expulsion of autolyzed fetal tissues. However, in some cases the CL is maintained after fetal death, which can result in expulsion, resorption, or mummification.[16] Early abortions are typically not recognized; most are occultly expelled, severely autolyzed, or mummified.

The maintenance of late-stage pregnancy requires both the fetus and placenta. At this stage in development, sufficient fetal stress can induce parturition through natural endocrine mechanisms. In this way, chronically diseased fetuses can initiate their own premature delivery.

Fetal death often leads to abortion within a few days rather than immediately, allowing sufficient time for autolytic changes to occur. The mechanism for expulsion of dead fetuses is unknown, but may share many similarities with normal parturition.[16] Autolytic changes include generalized edema, pallor, and hemoglobin-stained fluids in the body cavities. Visceral tissues become pulpy or semi-liquid.[16] In some cases, the time between fetal death and expulsion may be characteristic for a pathogen, and may therefore be useful to the clinician. Fetal infections due to *Listeria monocytogenes*, *Trueperella pyogenes*, nonseptate fungi (*Absidia* spp., *Mucor* spp., *Rhizopus* spp.), and bovine herpesvirus (BHV)-1 may lead to fetal death with expulsion days later. Fetuses infected by *Campylobacter fetus* and *Aspergillus* spp. may be delivered alive.[17]

It is important to recognize that infection of the conceptus does not necessarily lead to fetal death. The ability of a pathogen to injure the conceptus is influenced by the dam (general health, previous exposure), the stage of fetal development, and the virulence of the infectious agent. Fetal development is a continuum of organogenesis, physiological development, and immune development. Early stages of gestation are more prone to infection and severe disease, as the capabilities of the fetus are underdeveloped. The closer the fetus is to parturition, the stronger and more capable it is of defending itself from pathogens. Thus infection at different stages of development will produce different outcomes in the fetus.

Bovine viral diarrhea virus (BVDV) infection is a good example of this. Infection in the first trimester often leads to fetal death and resorption, while infection in the second trimester may lead to developmental anomalies. The character of the developmental defects depends on which cells or tissues are susceptible to the virus at the time of infection. In the second trimester, cerebellar growth is maximal and BVDV infection at this time may lead to cerebellar necrosis and cerebellar hypoplasia. Also at this time, hair growth is highly active and BVDV infection may damage follicles, resulting in hypotrichia. By the third trimester, the immunocompetent fetus may react to BVDV infection by mounting a sufficient immune response with antibody production and virus elimination. The only evidence for such an infection would be fetal specific antibody titers.[18]

Other viruses, such as BHV-1, cause extensive cell necrosis and hemorrhage, leading to rapid fetal death. *Neospora caninum* infection in the third trimester may cause only mild inflammation, with the birth of asymptomatic calves.

Effects of maternal disease

Infectious pathogens may cause abortion either by directly infecting the conceptus, or by affecting the dam such that severe physiological disturbance leads to abortion. The dam is predisposed to infection through nonspecific immunosuppression that occurs during pregnancy.[19] Fever (mastitis, pneumonia), circulatory disease (myocarditis, severe anemia), hypoxia, and endotoxemia (Gram-negative sepsis) are possible conditions that may cause abortion indirectly.[18] Prostaglandins may be elevated in febrile states, which can lead to luteolysis and abortion.[20] The likelihood of abortion probably increases with the number of pregnancies or previous abortions by the dam.[6]

Routes of infection to the conceptus

In cases where infectious organisms invade the conceptus, four routes are likely.[18] Hematogenous spread from the dam to the placenta is commonly suspected in cases of infectious abortion, particularly due to *Listeria monocytogenes*, *Leptospira interrogans*, *Salmonella enterica*, *Brucella abortus*, fungus, BHV-1, or BVDV infection. Ascending infection from the vagina through the cervix can occur from primary vaginitis (characteristic of *Tritrichomonas foetus*) or contamination at insemination. The presence of pathogens within the uterus (endometritis due to *Campylobacter fetus*, *Trueperella pyogenes*) and descent from the abdomen via the uterine tubes (mycobacteriosis) are also possible pathways for organisms to reach the conceptus.

Route from the placenta to the fetus

Once a pathogen reaches the placenta, it may proceed to the fetus through the blood, via the umbilical veins, or by contamination of the amniotic fluid[18] (Figure 54.1). For example, BVDV primarily infects the fetus hematogenously whereas BHV-1 may first infect the placenta, and then proceed to the fetus. Primary placental infection with subsequent amniotic fluid contamination is characteristic of fungal infections. In some cases a pathogen need not infect the fetus, such as with *Salmonella enterica* infection, where severe infection of the placenta may lead to generalized fetal hypoxia and subsequent death.

Lesions in placenta

Placentitis most commonly develops by one of three ways: hematogenous spread, extension from a diseased uterus, or an ascending vaginal infection. Gross placental lesions are characteristic of chronic bacterial and fungal disease, while viral infections generally produce no lesions.[21] Changes of significance may occur in both the cotyledons and/or intercotyledonary spaces. The presence of fibrin (yellow, stringy, friable material) on cotyledons is an indication of inflammation (Figure 54.2). Fibrin must be distinguished from necrotic inflammatory exudate or entrapped caruncle fragments

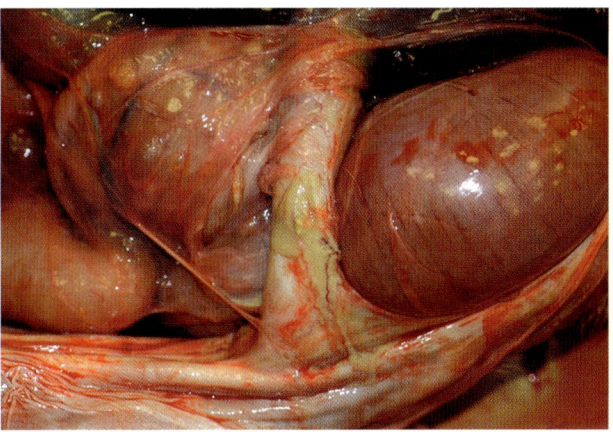

Figure 54.1 Fetus, omphalitis, ascending bacterial infection from the placenta. Purulent material is easily expressed from the cord. Photo courtesy of J. Edwards.

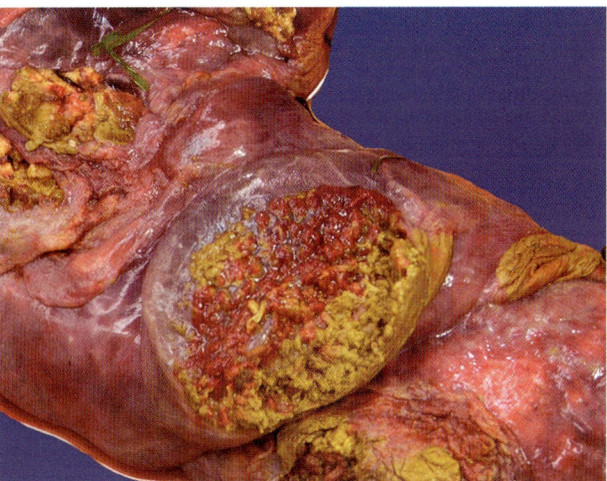

Figure 54.2 Chorioallantois, placentitis. The cotyledons are thickened with yellow discoloration due to fibrin and necrosis. Intercotyledonary tissues are relatively unaffected. Photo courtesy of J. Cooley.

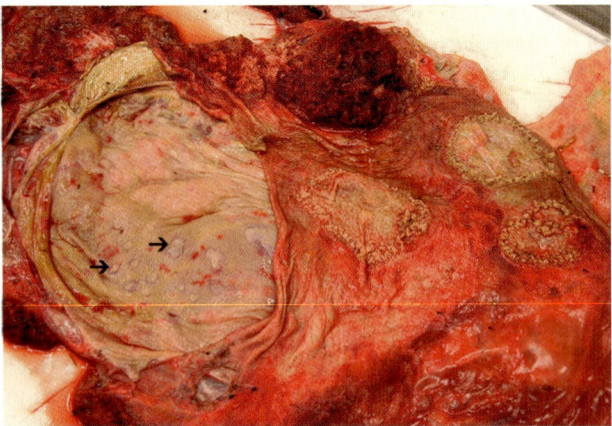

Figure 54.3 Chorioallantois, fungal placentitis; maternal aspect on right, fetal aspect on left. Cotyledons are thick, irregular, and tan to hemorrhagic. Intercotyledonary areas are thickened. The fetal side shows yellow discoloration and multiple discrete infarcts (arrows) due to vascular thrombosis. Photo courtesy of J. Edwards.

(necrotic uninfected endometrial tissue). The presence of ecchymosis and extension of the changes to the intercotyledonary space are suggestive of an infectious inflammatory process. With chronicity, stereotypical changes of placentitis include (i) placental fibrosis and edema; (ii) cupping of cotyledons; (iii) exudate on chorionic surfaces; and (iv) necrosis of the cotyledons.[10]

Fungal and bacterial inflammation of the cotyledons is often associated with necrosis, which imparts a firm to friable, tan, irregular or nodular consistency to the tissues with interspersed ecchymoses. Fungal organisms prefer to grow along and within blood vessels; this often leads to thrombosis, hemorrhage, and necrosis (Figure 54.3). Fungal organisms may be particularly abundant at the periphery of lesions rather than in the centers. A useful feature for interpretation is the extension of the pathologic process into the pericotyledonary chorioallantois. Such infiltrates may bridge cotyledons and often impart a leathery thickened texture. Another differential for such a change would be adventitial placentation (discussed in the section on pathology). Leptospirosis and BHV-1 infections may also produce a placentitis and appear similar to one another.[22]

Ureaplasma diversum is known to produce a rather characteristic hemorrhagic amnionitis, where the membranes are thickened, opaque, and ecchymotic with fibrin, necrosis, and fibrosis.[23]

Lesions in fetus

Gross lesions of the fetus, aside from malformations, are generally uncommonly noticed. They may be inapparent for four reasons: (i) the fetal immune system is not of sufficient robustness to mount a response that is easily observed at the gross level; (ii) fetuses often die before lesions appear; (iii) lesions are masked by autolysis, and (iv) fetal pathology may manifest itself in ways that are difficult to recognize.[24] Table 54.2 lists some causes for commonly encountered lesions in aborted fetuses.

The most common finding in abortion wastage is autolysis, the degree of which depends on the cause of death, the time from death to abortion, and the time from abortion to examination.[22] In experimentally induced, sterile, dead ovine fetuses, changes associated strictly with autolysis included (i) lack of odor, (ii) subcutaneous blood-tinged gelatinous edema, (iii) blood-tinged fluids in body cavities, (iv) renal cortex softening, (v) liver softening, (vi) abomasal content that was cloudy yellow to red, and (vii) uniform color (pink/red) of tissues[25] (Figures 54.4 and 54.5). By 12 hours after death, the fetal corneas were cloudy, the liver and kidneys were friable, and abomasal content became cloudy with brown flecks. After 36 hours, the subcutis contained edema and the skin would slough. By 144 hours, progressive dehydration was obvious (mummification).[25] Signs of fetal infection include fibrinous exudates in body cavities, white to tan foci in the liver or lungs, and evidence of abnormal development[22] (Figure 54.6).

Pathogens that invade the fetus via umbilical veins may produce lesions in the liver, as it is the first organ

Table 54.2 Possible causes for gross and microscopic lesions in bovine fetuses.

Gross lesion

Mummification	BVDV, *Neospora*
Ascites/anasarca	Congenital heart defect, BVDV, *Neospora*
Arthrogryposis, musculoskeletal deformities	Akabane/bunyaviruses, reduced *in utero* motility
Fibrinous peritonitis/pleuritis	Bacteria (*Trueperella*, *Campylobacter*, *Bacillus*, *Brucella*) and fungi
Fibrinous pericarditis	*Bacillus*, *Campylobacter*
Icterus	*Leptospira*
Cerebellar hypoplasia	BVDV
Hydrocephalus, hydranencephaly, porencephaly	BTV, BVDV
Microphthalmia	BVDV
Pulmonary/renal hypoplasia	BVDV
Dermatitis/hyperkeratosis	Fungi, EBA
Multifocal liver necrosis	*Listeria*, BHV-1, *Yersinia*, *Salmonella*
Splenomegaly/lymphadenopathy	EBA
Placentitis, infarctions	Fungi

Microscopic lesion

Encephalitis	*Neospora*, BHV-1, BVDV
Meningitis	EBA, *Leptospira*, *Brucella*, other bacteria
Myocarditis	*Neospora*, BVDV
Suppurative bronchopneumonia	Bacteria
Bronchointerstitial pneumonia	*Ureaplasma*, *Brucella*
Multifocal hepatic necrosis	BHV-1, *Listeria*, *Salmonella*, *Yersinia*
Interstitial nephritis	*Neospora*, *Leptospira*
Abomasitis/enterocolitis	Bacteria, fungi
Conjunctivitis	Bacteria, *Ureaplasma*, fungi
Placentitis	Bacteria, fungi

BVDV, bovine viral diarrhea virus; BTV, bluetongue virus; EBA, epizootic bovine abortion; BHV-1, bovine herpesvirus 1.
Source: Njaa BL. *Kirkbride's Diagnosis of Abortion and Neonatal Loss in Animals*, 4th edn, 2012.

Fetal Disease and Abortion: Diagnosis and Causes

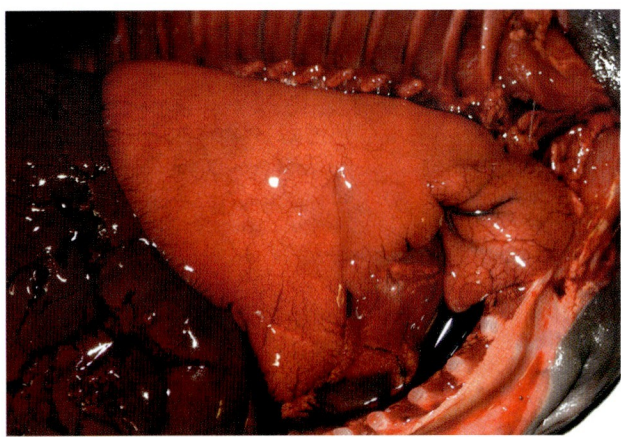

Figure 54.4 Aborted fetus, opened chest and abdomen. The lungs are uninflated (congenital atelectasis) and red with a shiny smooth pleura. Autolytic change is present (generalized reddening of tissues, friable liver). Photo courtesy of J. Edwards.

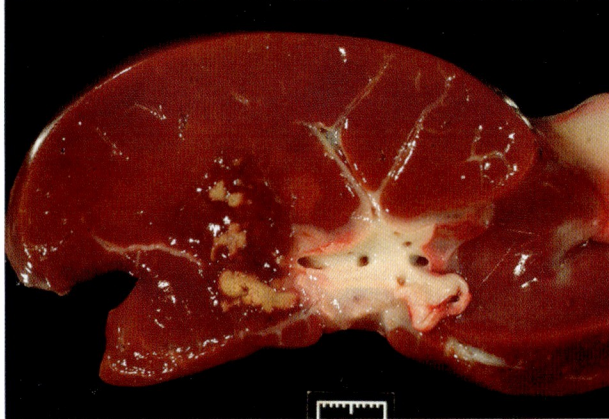

Figure 54.7 Fetus, liver section, hepatitis. Multiple tan foci of necrosis are surrounded by hyperemic parenchyma. Photo courtesy of J. Cooley.

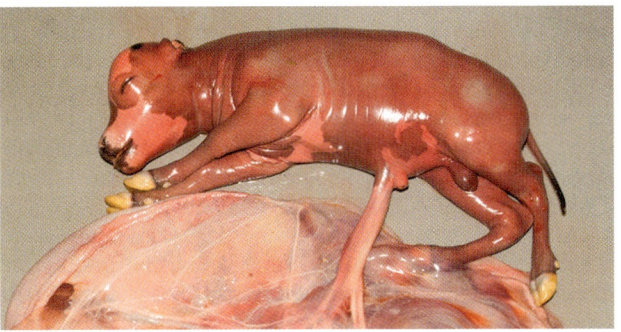

Figure 54.5 Fetus and placenta. The fetus exhibits generalized red discoloration due to autolysis. Photo courtesy of J. Cooley.

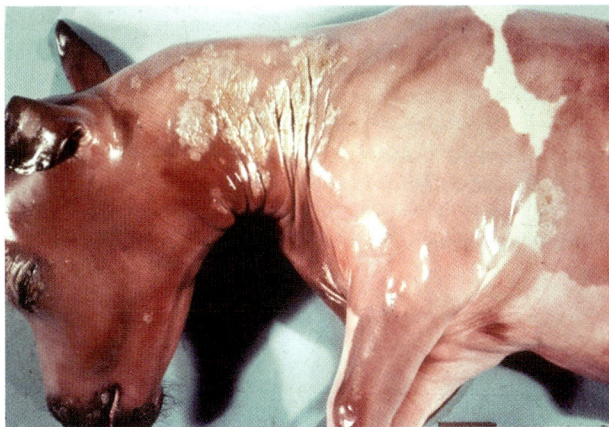

Figure 54.8 Fetus, fungal dermatitis. Coalescing, mildly bulging, gray/tan foci are present in the skin of the neck. Photo courtesy of J. Edwards.

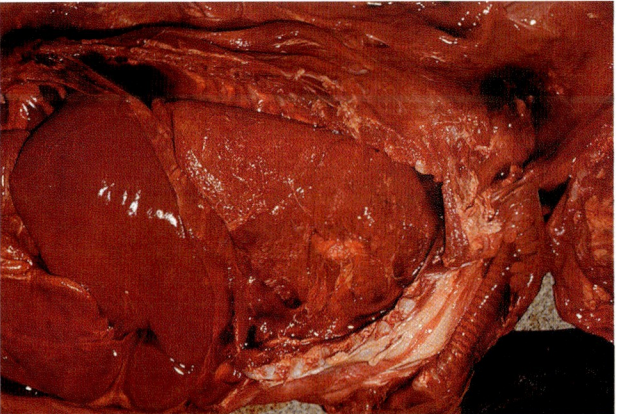

Figure 54.6 Aborted fetus, *Trueperella pyogenes* infection; opened chest and abdomen. Fibrin strands are present on the pleural surfaces. Autolytic change (generalized reddening of tissues) is evident. Photo courtesy of J. Cooley.

encountered (Figure 54.7). *Listeria monocytogenes*, BHV-1, *Yersinia pseudotuberculosis*, and *Salmonella enterica* infections are often associated with liver necrosis. For pathogens that infect the amniotic fluid, exposure of the skin, lung, and intestines may occur. Skin lesions are particularly well known in cases of fungal infection, where pale patches of thickened skin are evident (Figure 54.8). Lung and pleural disease may be manifestations of hematogenous systemic spread or inhalation of infected amniotic fluid during fetal distress.[18]

Epizootic bovine abortion causes distinctive lesions in the fetal liver, lymph nodes, and spleen. The liver is irregular and nodular; however this change may also be seen in cases of congenital heart disease where chronic passive congestion occurs (Figure 54.9). The spleen and lymph nodes are characteristically enlarged due to lymphoid and mononuclear cell hyperplasia.[16]

Outcome of exposure to abortifacient agents

Fetuses exposed to abortifacient agents or conditions may die (resorption/abortion/stillbirth), they may be infected (with or without disease), they may be malformed (live/dead), or they may be normal.[19]

Death in the embryonic stage leads to resorption. In the fetal stage the presence of skin and musculoskeletal structures, in addition to placental endocrine activity, prevents resorption. Instead fetal death leads to autolysis and expulsion, or retention with mummification/maceration. In late gestation, a sick or stressed fetus may precipitate its own

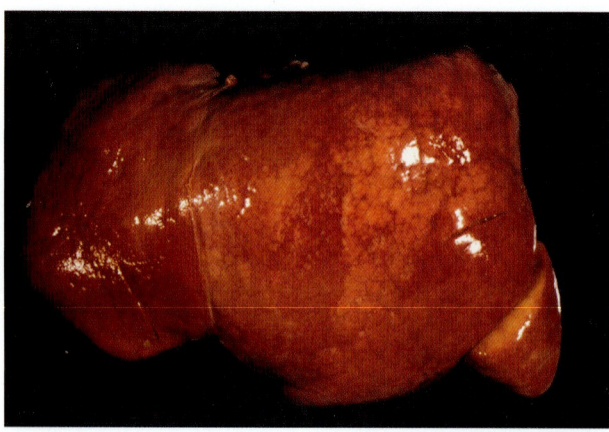

Figure 54.9 Fetal liver. The surface is tan and coarsely irregular. This change may be seen with chronic passive congestion (heart anomalies) or epizootic bovine abortion. Photo courtesy of J. Edwards.

premature delivery with one of three results: the fetus may be born and then die, it may be born and fail to thrive, or it may be born and thrive even if infected.[18]

Disease may impair fetal motion, a necessary requirement for proper musculoskeletal development. In such cases, the calves may exhibit severe changes, including arthrogryposis, ankylosis, scoliosis, torticollis, and similar defects that lead to dystocia, stillbirth, or death in the perinatal period.

Teratology

Abnormal development (congenital defects) in the fetus may be caused by teratogens: toxic substances, infectious disease agents (particularly viruses), nutritional imbalances, endocrine imbalances, hypoxia, extreme temperatures, and inherited or spontaneous genetic mutations. The manifestation of these defects includes resorption/abortion, malformation, alteration of growth, and functional deficits. The late embryonic period and early fetal periods are the times of greatest susceptibility for the conceptus. During this time, each organ system is established in an orderly fashion with morphogenesis occurring in "critical periods." During these periods organs are at their most sensitive to teratogens.

In many cases congenital defects have no defined cause. Seasonal occurrence, history of stress, or maternal disease are often noted, with or without a familial component.[26] The genetic composition of the fetus as well as the nature and degree of insult figure largely in the outcome.[11] Genetic considerations are discussed in greater detail in Chapter 66.

In cattle, viral infections are well-known causes of defects. BVDV, infectious bovine rhinotracheitis (BHV-1), Wesselsbron, bluetongue, Akabane (bunyaviruses), and Rift Valley fever viruses are commonly associated with embryonic loss and neuromuscular defects.[27,28] In addition, toxic plants such as lupines, *Conium maculatum*, and locoweeds are associated with defects.[26] Further discussion of such toxins is presented in Chapter 65.

Calves with congenital defects often die during parturition or soon thereafter. In many cases the anomalies are grossly obvious, often affecting the neural and musculoskeletal systems. A wide range of malformations have been described, some of which have been intensively studied and a genetic component has been found. For the purposes of this chapter, relevant anomalies are generally discussed for those which are caused by infectious agents (BVDV, Akabane virus) or those that are not known to be inherited but are particularly puzzling when encountered (acardiac twins, moles). Reviews are available.[11,29]

Neoplasia

Congenital tumors and neoplasms are rare in calves and are most often diagnosed as lymphomas, mesotheliomas, embryonic tumors, and hamartomas. They have been reviewed.[30]

Anatomy

The placenta is a transient metabolic organ derived from the fetal chorion and maternal endometrium to provide nutrition and metabolic exchange between mother and calf. Gestation can be divided into the embryonic stage with organogenesis, and the fetal stage with fetal growth. The embryonic stage extends to around day 45; the fetal stage continues until delivery. Together, the chorioallantois and amnion form the extraembryonic fetal membranes.

After fertilization, the zygote undergoes cleavage to form a ball of cells (morula). With further division the morula develops into the blastocyst, which is an inner cell mass (the embryo proper) and a blastocele (fluid-filled cavity) lined by the trophoblast cell layer. The blastocyst, now about 1.5 mm wide, hatches from the zona pellucida at 9–11 days after ovulation in order to attach to the endometrium of the uterus.[31] By 19 days, the bovine blastocyst trophoblast enlarges as a threadlike emanation/sac along the entire length of the uterine horn.[28,32] At this time, defective embryos will die and be resorbed or expelled. Also, the CL is developing and producing progesterone.[11]

The primitive endoderm from the inner cell mass migrates along the inner aspect of the trophoblast, lining the cavity that will become the yolk sac. The yolk sac provides nourishment to the conceptus early in development, but will quickly regress to an inapparent rudiment.[32,33] At this time, prior to uterine attachment, the conceptus derives nourishment from the secretions of the uterine mucosal glands, known as histotrophe. This material is a yellow/white, opaque, thick secretion sometimes mistaken for purulent exudate.[11]. Once the fetomaternal placental circulation develops, fetal nutrition is supplied by diffusion of nutrients from the blood across the placentome (hemotrophe).[28]

The primitive mesoderm migrates between the endoderm and trophoblast (now a trophectoderm) to form the chorion. The chorion then rapidly expands as a transparent sac to fill the uterine lumen.[34] Around this time, dorsomedially migrating folds composed of mesoderm/trophectoderm surround the embryo between days 13 and 16, providing complete envelopment of the embryo in an amniotic cavity. The site where these folds meet, the raphe or mesamnion, persists as a broad attachment of the dorsum of the amnion to the chorioallantois in Bovidae. The amnion is therefore composed of an inner trophectoderm and subjacent mesoderm with extensive attachment to the allantois.

Between days 14 and 21, a diverticulum of the hindgut extends from the embryo into the mesoderm to form the allantois and, along with it, the vasculature supplying the chorion and amnion.[21] The allantois continues to expand to fill both uterine horns, with widespread apposition and eventual fusion to the chorion (chorioallantois). It is thus T-shaped, with the stem of the T as the allantoic stalk of the umbilicus and the other ends of the T as extensions of the chorioallantois into both uterine horns.[28] The chorioallantois attaches at 4 weeks' gestation to the uterus, forming irregular villous projections over the uterine caruncles, with eventual development into placentomes.

By days 30–35, three to four fragile attachments (early placentomes) are present in the pregnant horn; by day 40 attachments are present in both horns. At day 70, 40–50 placentomes are present, which will number 75–140 by mid gestation. Placentome size may vary widely, from 5 to 15 cm in diameter. The fetal cotyledon has a velvety red surface (Figure 54.10). Both the weight and size of placentomes increase during gestation, particularly in the pregnant horn.[35] Between placentomes, there is minimal chorioallantoic villous proliferation, which may be seen as fine red villous proliferation occurring diffusely or in multiple, 5–10 mm wide, regularly spaced foci. Such proliferation may cover nongravid horns that contain no cotyledons.

Placentomes are usually arranged in two dorsal and two ventral rows along both horns, although an extra row or reduced rows may be seen (see Figure 54.14). They become progressively larger the closer they are to the fetus and the majority of cotyledons are present in the gravid horn. The largest placentomes are closest to the main arteries and become progressively smaller toward the periphery. In some cases all cotyledons are present in the gravid horn with few to no cotyledons on the nongravid side. The cotyledonary rows may be interrupted by a circular bare area associated with the junction of the horns, near the cervix.[36]

As development progresses, there is fusion of membranes to one another in the areas where they come into contact. These attachments may be transient. In early gestation, the amnion is relatively large and compresses the allantois laterally. As the amnion grows and contacts the chorion, the amniochorion is formed; on the opposite side of the calf where the amnion contacts the allantois, the allantoamnion is formed. With maturity, the allantois increases in volume to almost entirely surround the amnion.

Thus the fetus is immediately surrounded by the amniotic fluid, which is produced by transamniotic fluid fluxation, fluid from the fetal lungs, oral glands, and urination. Initially the amniotic fluid is watery, slightly yellow and clear; later in development it becomes viscous, translucent, opaque, and white or yellow. Urine throughout early development is excreted through the urachus and umbilicus to the allantois, but it may also be excreted through the urethra. The contribution of urine to the amniotic fluid after 240 days' gestation in cattle has not been conclusively decided; both increasing and decreasing late-term transurethral urination has been reported.[21,28] Amniotic fluid in full-term cattle varies from 2 to 8 L, with reported averages at 2.2 or 5–6 L. Allantoic fluid at term ranges from 4 to 15 L, with an average of 9.5–10 L.[11,37]

The umbilicus is relatively short, being about one-quarter the length of the fetus, or about 30–40 cm.[11] As such, rupture of the cord occurs during parturition when the fetal pelvis passes through the dam's pelvis.[38] The umbilicus is exclusively within the amnion, and is composed of two arteries, two veins (which merge into one prior to entering the fetus), the allantoic stalk, and rarely yolk sac remnants. The umbilical arteries and veins are sheathed in smooth muscle, which contracts during parturition due to stretching, stopping blood flow after birth.[28] The umbilical arteries arise from the caudal aortae, running cranioventrally along the urachus, exiting with the allantoic stalk to supply the chorioallantois. In necropsied neonates, it is common to see swollen, purple, thrombosed umbilical arteries at the level of the urachus; this is a normal inflammatory reaction to the severing of fetomaternal circulation (Figure 54.11).

By day 45, fetal organogenesis is complete and the fetal growth stage then begins. Average weights, crown–rump lengths, and external characteristics achieved during growth are summarized in Table 54.3.

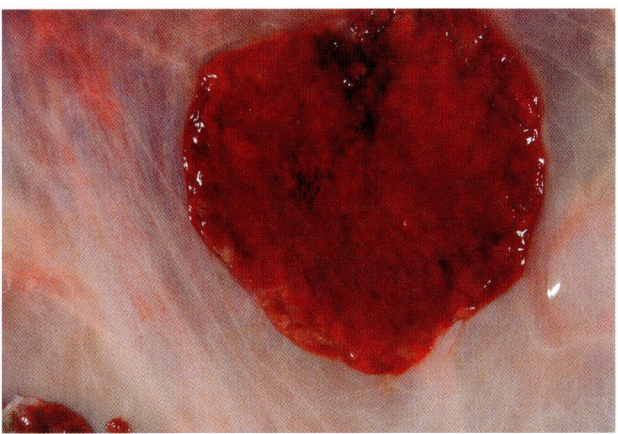

Figure 54.10 Normal placenta. A cotyledon has a red velvety surface. The intercotyledonary spaces are smooth, shiny, opaque, and white.

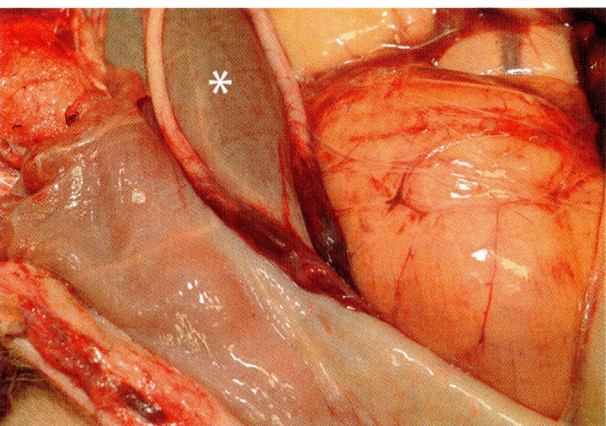

Figure 54.11 Calf, umbilical artery thrombosis. Bilaterally the arteries alongside the urinary bladder (asterisk) are focally swollen and red due to the tearing and thrombosis associated with parturition.

Table 54.3 Gestational age estimates of bovine fetuses.

Age (months)	Relative size	Crown–rump length (cm)	Weight	External characteristics
2	Mouse	6–8	8–30 g	Claw buds and scrotum present
3	Rat	13–17	200–400 g	Hair on lips, chin, and eyelids
4	Small cat	22–32	1–2 kg	Fine hair on eyebrows, claws developed
5	Large cat	30–45	3–4 kg	Hair on eyebrows and lips, testes in scrotum, teats developing
6	Small dog, beagle	40–60	5–10 kg	Hair on inside of ear and around horn pits, tip of tail and muzzle
7	Dog	55–75	8–18 kg	Hair of metatarsal, metacarpal, and phalangeal region of extremities and beginning on back, long hair on tail tip
8	Large dog	60–85	15–25 kg	Fine short hair all over body, incisor teeth not erupted

Source: Njaa BL. Kirkbride's Diagnosis of Abortion and Neonatal Loss in Animals, 4th edn, 2012. This material is reproduced with permission of John Wiley & Sons, Inc. Original content from Roberts S. Veterinary Obstetrics and Genital Diseases (Theriogenology), 3rd edn, 1986, p. 19.

Approach to the problem

Fetal loss in a herd may be divided into four epidemiologic presentations, which may simplify and better direct investigations. These four include (i) baseline losses, (ii) endemic losses that exceed baseline and occur consistently and chronically, (iii) epidemic losses characterized by high losses in a specific time frame, and (iv) fetal losses that are confused with conception failure or neonatal losses.[24]

Baseline losses, as mentioned in the introduction, are considered largely unavoidable and may be due to illness in the dam, lethal genetic make-up of the fetus, trauma, or physiological abnormalities. It is important to monitor such losses from a management standpoint as it will aid the discovery of endemic problems. While it is prudent to thoroughly investigate all abortions, it is also costly. However, with continued monitoring and testing, the manager and veterinarian have a clearer picture of herd health and are better prepared to identify endemic problems. Furthermore, upcoming epidemics may be identified at a stage where losses can be minimized.[24]

Endemic abortions significantly affect herd productivity due to chronic excessive losses. Not only may it be difficult to realize that there is a problem, but it may be equally difficult to identify the source of the losses. Accurate record keeping and routine pregnancy diagnosis is therefore vitally important in order to recognize these issues.[24] Table 54.1 lists many disease-causing agents that are associated with endemic losses in cattle. BVDV, Neospora caninum, and Leptospira spp. are of particular importance.

Epidemic abortion storms may be due to many of the same agents that cause endemic losses. BHV-1, *Brucella abortus*, and *Leptospira* spp. are well-known causes.[19] Herd health status, vaccination history, pathogen virulence, access to toxic agents, and clustering of pregnancies all have relevance in determining the likelihood of epidemic loss. Herds with tight pregnancy patterns are more susceptible to this type of outbreak. Investigations should make note of the particular time/season and area where the losses are occurring as this will aid in determining differentials.[24]

Finally, managers should be encouraged to document and report all abortions, stillbirths, premature births, dysmature calves, and abnormal calves.

History/background investigation

The importance of a thorough history cannot be overstated; essential information is found in a systematic query of farm details. Abortion is both an individual and potential herd problem, and lack of investigation in both will severely hamper efforts to alleviate unacceptably high losses. It is important to keep in mind that more than one agent may be involved in abortion. Similarly, although more than one infectious agent may be present in an aborted fetus, only one may be the cause of death. Organisms regarded as low virulence may be the cause of abortion, but are also commonly present as contaminants.[17]

Sporadic abortions may be a manifestation of an ongoing herd problem or they may be the beginning of an outbreak; in either case investigation is warranted. Standard forms are used to reliably collect important cow and herd information (see Appendix); systematic evaluation of management practices will help direct investigations. Pertinent questions are listed below.[39] These questions and other information are incorporated into the investigation form provided. An expanded questionnaire has been published.[40]

Aborted calf

- When was it due?
- Was it born alive?

Aborting cow

- Breed, age, lactation/parity number, source, dam/sire, bred naturally or AI, conception date/last service, date of last normal parturition, previous abortions and any work-up done.
- Vaccination history, vaccine brand/lot number, vaccine handling and storage.
- Deworming history.
- Any clinical signs?

Herd

- Number of animals, number that are immature, number purchased in last year, health problems/body condition.

- Herd abortion history (sporadic/recurrent/recent, how many in last week/month/year, which trimester affected, affects older/younger/all cows, any particular time of year), problems with term calves (congenital defects).
- Are postpartum examinations performed routinely?

Nutrition

- Feeding regime, types of forage, concentrates, minerals, selenium.
- Any changes in management, housing, pasture?
- What is the quality of the forage?
- Water source/quality?

Environment

- Access of herd to other cattle, animals, predators.
- Nature and quality of pasture.
- Any toxic plants on pasture?
- Nutrient deficiencies in area pasture.

Examination of the cow

In some cases, the health condition of the cow has direct implications as to the cause of abortion (plane of nutrition, respiratory/intestinal disease, *Anaplasma* sp. infection). Any vaginal discharge or cervical mucus should be collected and sent to the diagnostic laboratory for microbiological analysis. Blood samples for serology should be taken from the dam as well as unaffected cows and those that have recently aborted. Serum should be acquired 2–3 weeks later in order to provide paired samples. Collecting serum from 10 other herdmates (or 10% of the herd) will make serological assessment more meaningful.[6] Whole blood in EDTA is useful for blood smear and serum chemistry analysis of the dam. Gloves are to be worn during all examinations, as zoonotic agents may be present.

Examination of tissues

A necropsy is an examination with a definite purpose in order to yield the maximum amount of information. As such, it is vitally important to be systematic so that a full complement of tissues is taken during each examination. Optimally one should send the entire fetus and membranes to a diagnostic laboratory for evaluation. It is of primary importance to send fetal membranes as these may be diagnostic. If this is not convenient, then a standard approach should be adopted, as described here.

Examinations should be directed toward suspected etiologies, but keep in mind that more than one pathogen may be present. Typical findings in aborted calves that died *in utero* include edematous red tissues that lack normal tinctorial distinction, clear red fluids in body cavities, pale soft livers, and friable autolytic kidneys. Digital photographs are an excellent way to document lesions and may be useful in communications with diagnostic laboratories.

Examination of the fetus

It is important to verify the approximate time of death, which can be done based on gross findings. In antepartum fetal death, tissues are autolyzed or mummified and include the findings described above. Partum death is characterized by signs of life (localized head or limb edema, partial lung aeration, limb/rib fractures, subcutaneous shoulder hemorrhage, liver fractures with hemorrhage) in addition to proper degrees of fetal maturation/formation. Neonatal deaths are characterized by aeration of the lungs, umbilical artery thrombosis (swollen purple/red nodules along the arteries that run lateral to the urinary bladder), loss of the eponychium (soft fetal hoof keratin), and milk in the stomachs.[41]

The method of necropsy for aborted fetuses is basically the same as for any animal. The degree of autolysis and weight of the fetus and placenta are noted. Even if severely autolyzed, fetal tissues may be of use for polymerase chain reaction (PCR) screening for pathogens. Estimation of gestational age can be accomplished by the following formula: $x = 2.5(y + 21)$, where y is crown–rump length in centimeters and x is gestation age in days.[7] The crown is a point midway between the orbits on a line traversing the frontal eminence; the rump is the base of the tail or first coccygeal vertebra.[42] In embryos, the greatest total length may be used instead of crown to rump. For stillbirths, fetal weight is useful in the diagnosis of dystocia due to fetal–maternal disproportion. Also, comparison of calf weight to thyroid weight in cases of suspect hyperthyroidism (goiter) may be useful. Reports of normal thyroid weight in calves vary from 6.5 to over 18 g.[43] Further discussion on this subject can be found in the section on noninfectious causes of abortion.

Most aborted calves have no gross lesions. Small white/tan foci in the lungs, liver, and kidneys may be present and indicate necrosis or inflammation. The presence of fibrin on organ surfaces is helpful for confirming inflammation. Meconium on the perineum or in airways/stomach indicates fetal stress and anoxia, typically associated with placentitis or dystocia (Figure 54.12). Hemorrhage and fractures of shoulder and hip joints are most often seen as a result of dystocia.

Slightly raised, white skin plaques that are firmly adherent, particularly around the eyes and face indicate fungal infection. Sometimes calves will have scattered white mineral on the hair, which is easily removed; this finding is

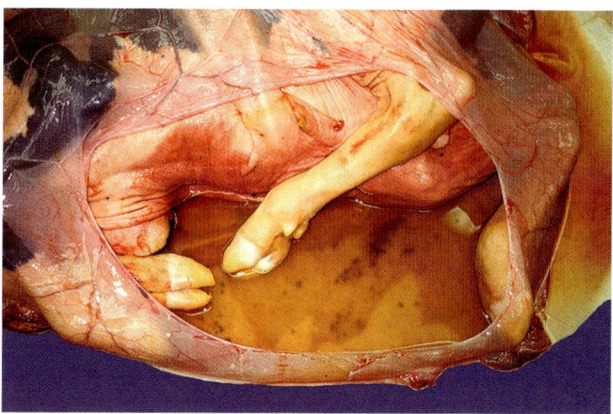

Figure 54.12 Fetus and amnion (opened). The amniotic fluid and skin are stained yellow/brown due to meconium. Photo courtesy of J. Cooley.

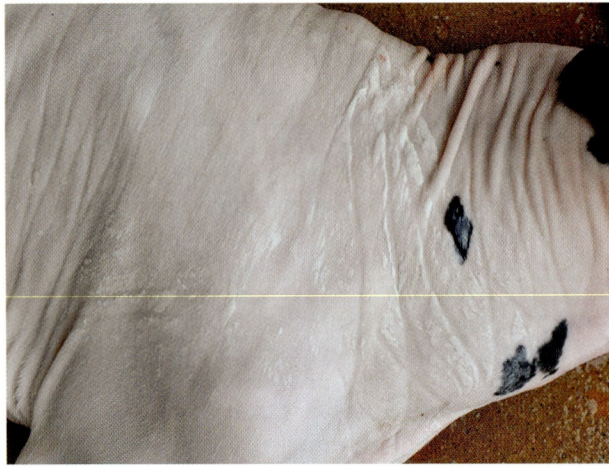

Figure 54.13 Fetus with white mineral grit present over the shoulder. Not to be confused with fungal dermatitis. Photo courtesy of J. Edwards.

not significant but must be distinguished from adherent fungal plaques (Figure 54.13). In calves born dead, the lungs are uninflated; they are uniformly dull red/plum, heavy, and sink in formalin (see Figure 54.4). In cases of dystocia where fetuses may have taken a partial breath, there may be incompletely inflated lungs; assess lung aeration in sections of cranial and caudal lobes by degree of the flotation in fixative/fluid. Make note of any milk in the stomachs, as this indicates colostrum consumption, which will interfere with bacteriological and serological findings.[7]

Necropsy method

It is best to have an abortion kit prepared beforehand; Table 54.4 enumerates materials to have in a kit. Zoonotic pathogens may be present, so always wear gloves, protective clothing, eye/face protection, and use careful technique when examining dams and fetal tissues. In addition, materials used in standard field necropsies are needed: clean water, bucket, brush, and disinfectant. Make sure to disinfect contaminated sites and equipment. If there are questions as to which samples are best to take, contact the diagnostic laboratory to which samples will be submitted for advice. Taking photographs of lesions is a relatively simple and excellent practice which will aid diagnosis.

1. Measure crown–rump length along the vertebral ridge (see above) and weigh fetus and placenta separately. Examine skin for exudate and abnormalities, around the eyes and face in particular for signs of fungal infection.
2. Examine the body for congenital abnormalities, umbilical hernias, cleft palates, facial anomalies.
3. Wash the surface of fetus with water to remove gross contaminants.
4. With the left side down, reflect right legs with a knife and continue the incision from cranial to caudal, reflecting the skin to the dorsal and ventral midlines. Collect any pooled blood between incised tissue planes with a syringe if available (for serology).
5. Open abdomen at the highest point along the caudal aspect of the ribs, being careful not to contaminate any peritoneal fluid or tissues. Aspirate any abdominal fluids with a sterile syringe, place in sterile tube, label, refrigerate. Continue the cut along the caudal aspect of the ribs, along the lumbar area, and cranial to the pelvis to expose the abdominal viscera. Note any lesions, and volume, character and color of fluids.
6. Incise the diaphragm (noting if negative pressure is present) and cut the ribs at the costal angles and along the sternum; remove the chest wall. Aspirate any thoracic fluids with a sterile syringe, place in sterile tube, label, refrigerate. Note any lesions, and volume, character and color of fluids. The pleural surface of the removed chest wall may be used as a makeshift cutting board/sterile field for dissection.
7. Aspirate fluids from the unopened abomasum with sterile syringe, place in sterile tube, label, refrigerate. Note the character of the abomasal fluid. Take samples of any other suspicious fluids at this time.
8. Examine the body cavities and organs *in situ*, making note of fibrin, hemorrhages, or abnormally situated organs.
9. Remove thoracic viscera from tongue to lung (pluck). Examine the mouth; one may split the mandibular symphysis with a knife. Incise the esophagus along the entire length; incise the dorsal tracheal membrane from the larynx to the bronchi. Examine the chest for any fractures. Slice multiple lung lobes, examining the cut sections for changes in the lobules and airways. Examine the thyroid gland. Collect tissues, especially lesions (see numbers 16 and 17).
10. Examine the heart, making sure to inspect all four main valves and four chambers. Look for ventricular septal defects or other anomalies.
11. Collect visceral organs, examining gastrointestinal tract last. Identify the kidneys, ureters, and urinary bladder. Remove the gastrointestinal tract by incising the root of the mesentery and continuing the cut to the diaphragm along the dorsum, avoiding the urinary tract. The tract may then be lifted out of the abdomen, leaving the colon attached and uncut. Collect tissues, especially lesions (see numbers 16 and 17).

Table 54.4 Contents for abortion kit.

Knife (15-cm boning) and steel
Nitrile/rubber gloves
String and ruler for measurements
Shears (small tree limb variety works well)
Culturette swabs
Overalls, washable apron, rubber boots
Alcohol swabs, syringes, needles to draw blood/fluids
Scalpel blades and handle
Red-top glass vacutainer tubes, purple-top EDTA tubes
Hatchet/cleaver and hacksaw
Sterile rat-tooth forceps and scissors
Styrofoam box/cooler for temporary specimen storage
Sterile containers/whirltop bags/ziplock bags, permanent pen
Fixative (10% neutral buffered formalin), at least 500 mL in plastic sealable container/jar

12. Remove the adrenal glands.
13. Remove the urinary tract, making multiple slices through the kidneys. Identify genital organs.
14. Serially section the liver and spleen.
15. Open the forestomachs and abomasum, examine for content. Continue into the intestine, making several incisions along each major section (duodenum, jejunum, ileum, cecum, colon). Examine content and the mucosa.
16. Collect organs (kidney, liver, lung, spleen, thymus, adrenal) in sterile bags (placenta kept in separate bag), refrigerate.
17. Collect in formalin for histopathology: lung, liver, kidney, spleen, heart, adrenal glands, skeletal muscle (three to four sections including tongue, diaphragm), thyroid, ileum, skin (including eyelid), thymus, and mesenteric lymph node in 5-mm sections.
18. Remove the head from the body at the atlanto-occipital joint. Harvest entire brain, even if severely autolyzed, and place whole in formalin. If the fetus is severely autolyzed, take a swab culture of the brain tissue prior to fixation. The skull is relatively soft and can be easily opened with a saw, making two cuts along the inner aspects of the occipital condyles, cranially (and somewhat laterally along the curve of the calvaria) toward the inner aspects of the orbital rims. Make a third incision across the orbital ridge, connecting the first two cuts. Pry the cranium from the skull in one piece. Place the brain in the palm of your hand and with a scalpel gently cut the nervous attachments of the base of the brain, allowing the brain to fall into your hand.
19. Collect ocular fluid with a needle/syringe for nitrate/nitrite levels if needed, freeze.
20. Open several joints, looking for purulent exudate, fibrin, and hemorrhage.
21. Examine fetal membranes for exudates, lesions, abnormalities (see next section for detailed instruction). Remove dirt/debris from the membranes prior to sampling.
22. Place two to three sections with cotyledons in a sterile bag, refrigerate; make impression smears of any lesions.
23. Harvest the placenta for histopathology (at least three to four sections including cotyledons and intercotyledonary areas).
24. Collect amniotic fluid if available, refrigerate.
25. Collect dam blood in red top tube (herdmate blood samples may be collected as well; 10 cows or 10% of the herd has been suggested).

Examination of the membranes

Collect as much of the membranes as possible and carefully examine the chorioallantois and amnion, as lesions may be subtle, focal, and easily overlooked. In many cases, little if any placenta can be examined. It is important to try to verify if all membranes have been expelled. Because of the attachments of the amnion to the chorion, as well as the short umbilicus, calves will rupture and escape both the amnion and chorioallantois at delivery. Typically membranes are found with cotyledons on the outside.

The fetal membranes may be laid out lengthwise to identify the umbilicus and with it the internal aspect of the amnion. Examination of amniotic membranes and fluids is best done at this time. The umbilicus is composed of

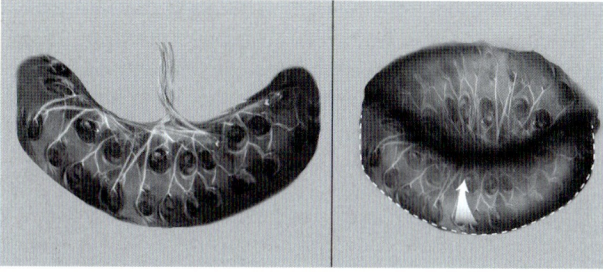

Figure 54.14 Placental membrane examination method. (*Left*) The long axis of the chorioallantois is arranged perpendicular to the umbilical stalk. (*Right*) An incision is made along the dotted white line, allowing the chorioallantoic sac to be opened (direction of white arrow) and laid flat. In this way, the vascular axis, cotyledons, and membranes can be thoroughly examined. Courtesy of Rachael Fishman.

four muscular turgid vessels (arteries and veins) loosely enmeshed in slippery membranes. The gravid horn is easily identified as the larger half of the tissue, and should have larger placentomes. Avascular chorionic tips should be identified and inspected carefully, as true placental lesions may be present in the gravid horn near the tip.[16] The umbilicus is continuous with the umbilical arteries and veins, which form a useful longitudinal axis by which to orient the placenta. From there the internal aspect of the allantoic cavity may be examined, with attention to the nature of any fluids present, thickness and color of membranes, and peculiar odors. Allantoic calculi are typically found at this time.

By placing the umbilical vessels at one side of the tissue, a lengthwise incision along the opposite side of the placenta will allow the placenta to be opened and laid flat, with the umbilical vessels forming the central axis (Figure 54.14). The horns are unequal in size, the gravid horn being much larger. In this way, vessels and cotyledon rows (normally four) may be examined thoroughly. Fresh cotyledons have a red velvety texture with a mildly irregular surface. The size variation, shape, color, consistency, and degree of cotyledon mottling should be noted, as well as degree of autolysis. Occasional small tags of attached caruncular tissue may be seen. Cotyledon margins are typically sharply defined, and the intercotyledonary chorion is white, translucent, and smooth. Adventitial placentation is a frequent finding that appears as small cotyledons or villous tuft formations adjacent to the placentomes that may cover the entire placenta.

Sample collection

A complete set of tissues should be collected in every case.[44] Biological specimens must be properly collected, prepared, stored, and transported otherwise diagnostic accuracy will suffer. It is important to maintain aseptic technique despite apparent tissue contamination; this is necessary to maximize chances for diagnosis as well as for prevention of zoonotic infection. Microbiology samples are to be kept chilled and sent to the diagnostic laboratory as soon as possible in a Styrofoam cooler with ice packs for overnight courier shipping. If fresh tissue samples cannot be sent within 2–3 days, then freezing at −70 °C and delivering on dry ice is ideal.[44] If necessary, specimens can be placed at −20 °C until dry ice can be obtained. Conventionally frozen tissues (−20 °C) are often acceptable for PCR testing of many pathogens. Abomasal and thoracic/pericardial fluid samples should be sent in sterile tubes without additives. Frozen fetal liver, lung, and kidney (2.5–5 cm chunks, −20 °C) should be kept for testing in case nutritional or toxicologic disease is suspected after initial examinations.[39]

Chilled or frozen tissues should be placed in sealed bags (whirlpack-type bags often leak), preferably with a secondary bag around them. Fluids shipped in glass tubes require ample padding in tight-fitting containers. Absorbent materials should also be included in the package to prevent soaking of the exterior, which complicates and may delay the shipping process. Disposable diapers are an effective option.[39] Contact your diagnostic laboratory for proper packaging and shipping requirements for abortion samples as they may contain infectious and/or zoonotic agents.

Viruses may have very specific tissue tropisms and therefore require certain tissues for isolation. Specimens need to be collected as soon as possible as autolysis can be very damaging to virions. Virology samples for abortions are typically pooled, although placing tissues in separate bags is ideal. Placental tissues should be bagged separately. Do not pool samples from multiple abortions. Tissues should be kept cool but not frozen until arrival at the diagnostic laboratory. If tissues will not be immediately used, freezing at −70 °C is better than at −20 °C.[39] If live viruses are sent packed in dry ice, be sure to place samples in airtight containers as the gaseous phase of dry ice is carbon dioxide, which can lower sample pH and inactivate some viruses. Certain laboratories may have preferences for virological samples, particularly if a certain pathogen is highly suspected.

Bacteriology tissue samples should be 2.5–5 cm per side in order that surfaces may be heat seared for proper culture. External and gut samples should not be bagged with internal organs. If samples are contaminated, sending a swab culture (Culturette) in addition to tissues may be helpful. For severely decomposed specimens, aseptically collected brain tissue may be useful for isolating pertinent organisms.[22]

Fungal sampling should at least include affected cotyledons; however it is best to send the entire placenta. Fetal infection is inconsistent, but lung, skin, and abomasal content are useful.

For histopathology, sections approximately 5 mm wide in 10% neutral buffered formalin are best. The brain, however, should be immersed whole in formalin, without dissection. If the brain is liquefied, fix separately. For small lesions, several foci should be submitted. For large lesions, a few sections including the margin with normal and abnormal tissues should be collected. Identify clearly any lesions seen in order to alert the diagnostician.

At least 10 volumes of formalin per volume of tissue is needed for proper fixation. Tissues should be allowed to fix for at least 24 hours, preferably 3 days. The brain in particular requires ample fixative over 3–4 days. Addition of 1 part ethanol to 9 parts formalin (10% ethanol, 90% formalin) may be used to prevent freezing in cold climates.[45] If formalin is not available, 70% ethanol may be used. Tissues may be sent in formalin, otherwise fix tissues for 24 hours, drain, and send in bags with formalin-moistened gauze. Be sure to double-bag samples as containers may rupture in transit. Contact your diagnostic laboratory for instructions.

For blood samples, blood from the dam as well as the fetus is ideal. Fresh blood smears are particularly useful if anaplasmosis is suspected. Blood in EDTA is preferred for PCR testing. For serum samples, draw 8–10 mL of serum into tubes without additives, allow to sit for 1–2 hours until the clot begins to retract, then place in the refrigerator overnight. Centrifuge at $1000 \times g$ for 10 min and decant serum; send in a separate sterile tube. One may freeze the serum (do not freeze whole blood) until needed.

Table 54.5 lists typical sample storage conditions.

Diagnosis

Arrival at an accurate diagnosis is the result of the cooperative efforts of the attending veterinarian and the diagnosticians at the laboratory. Even with the submission of proper samples, the establishment of a definitive diagnosis is problematic. Expulsion of a dead fetus often occurs hours to days after death; the resulting autolysis makes the difficult task of fetal lesion identification even more so. Fetal membranes, which may have the only significant lesions to be found, are often not present for examination. Also, in many cases examined tissues are severely contaminated, whether due to incomplete expulsion or from the environment. Finally, diagnostic laboratories run routine tests to identify well-known causes of abortion in a timely manner. As such, investigation of unusual or poorly understood etiologies is beyond their purview.[17]

Keep in mind that the majority of pregnancy wastage occurs at earlier stages of development, when expelled tissues are least likely to be noticed. Early gestation abortions are often detected only weeks after the fact, making investigation and diagnosis all but impossible. The majority of

Table 54.5 Fresh specimens collected for fetal and neonatal diagnostics.

Test	Storage	Specimens	Notes
Bacteriology/mycology (culture, PCR)	Keep refrigerated *not* frozen unless PCR only	Stomach contents, pericardial fluid, liver, lung, brain, placenta	Collect stomach contents or pericardial fluid in a syringe with a large-gauge needle Package each specimen in separate containers
Virology (virus isolation, fluorescent antibody test, PCR)	May be frozen (at −70 °C if necessary)	Lung, liver, kidney, heart, blood, placenta	Package each specimen in separate containers
Nutrition, toxicology	May be frozen	Liver, kidney, ocular fluid	
Fetal/maternal fluids	Keep refrigerated	Dam serum/fetal fluids	Paired (acute and convalescent) if possible

tissues examined by clinicians and diagnostic laboratories are of large fetuses, which may bias the interpretation of a herd problem where losses occur at various gestation stages. Late-stage abortions may be the result of injury many weeks prior to expulsion; they may not exhibit important or telling changes that could be present in earlier aborted fetuses, preventing accurate diagnosis. Worse yet, in an effort to find a cause, other coincident agents may be erroneously blamed, depending on what is found in the submitted fetus.[6]

It is important to realize that more than one disease or management problem may be at work in a herd, complicating the resolution of endemic problems. Multiple pathogens may be present; some may not be obvious while those that are readily evident may not be directly contributing to abortion. In these cases, determining the actual cause of abortion is challenging.

Serology results must be cautiously interpreted. A positive sample from a cow indicates exposure, which may or may not be relevant in light of vaccination status. Twofold to fourfold increases in titer over 2 weeks may be significant, particularly with herdmate samples for comparison.[6] However, it is likely that dam serum antibody levels have reached their peak by the time abortion occurs, so it is not unexpected to see a lack of rise in titer 2–3 weeks later.[24] Serology may be particularly useful in diagnosing BVDV, leptospirosis, anaplasmosis, and abortions where the agent is difficult to demonstrate by other methods.[17] The presence of a specific antibody in fetal serum does not mean that particular agent is the cause of abortion; BVDV and *Neospora caninum* are two examples. Serology results must be carefully weighed with microbiology, histopathology, and herd history data. Table 54.6 lists common diagnostic tests for specific pathogens.

Table 54.6 Summary of diagnostic tests for bovine abortifacient pathogens.

Agent	Preferred fetal tissues	Preferred fetal diagnostic test	Additional diagnostics
Bovine herpesvirus 1	Kidney, adrenal, liver, lung	FA (frozen tissue), IHC, VI	IHC of placenta if placentitis is present
Bovine viral diarrhea virus	Lung, heart, kidney, placenta, skin	FA, IHC, VI, PCR	Antigen-capture ELISA, RT-PCR
Bluetongue virus	Brain, spleen	PCR, VI	Fetal serology
Bunyaviruses	None	None	Fetal/precolostral serology; congenital abnormalities
Brucella	Placenta, lung, abomasal content, uterine fluid	Bacterial culture	Culture of milk from dam; dam and fetal serology
Listeria	Placenta, lung, brain, abomasal contents	Bacterial culture, IHC	Gram's stain of freshly aborted tissues, PCR
Salmonella	Placenta, liver, lung, abomasal contents	Bacterial culture	
Yersinia pseudotuberculosis	Placenta, liver, lung, abomasal contents, intestines	Bacterial culture	
Leptospira	Kidney, placenta	FA (fetal kidney smear), IHC, PCR	PCR of fetal urine; fetal and maternal serology
Ureaplasma	Lung, abomasal content, placenta	*Ureaplasma* culture, PCR	
Campylobacter	Lung, abomasal contents, placenta	*Campylobacter* culture, IHC, darkfield microscopy	Silver staining of histology tissues, PCR
Epizootic bovine abortion	Thymus, spleen, lymph nodes	Modified Steiner's silver stain of histology slides, IHC	Elevated fetal serum immunoglobulins
Chlamydophila	Placenta	PCR, IHC, FA	Macchiavello's, Gimenez, or modified acid-fast stains of histology slides
Coxiella burnetii	Placenta	PCR, IHC	Macchiavello's, Gimenez, or modified acid-fast stains of histology slides
Fungi	Placenta, abomasal contents, lung, skin lesions	Fungal culture, H&E	Direct identification by KOH wet mounts of skin or placenta or GMS and PAS stained histology slides
Tritrichomonas	Placenta, abomasal contents, lung	*Tritrichomonas* culture, Bodian's silver stain of histology slides, IHC, H&E	Darkfield microscopy of abomasal contents, PCR
Neospora	Brain, lung, kidney, skeletal muscle, liver, placenta	IHC, PCR	Fetal serology (IFA, microagglutination titer, ELISA)
Sarcocystis	Brain, lung, liver, kidney, skeletal muscle, placenta	IHC	Genomic probe; ribosomal RNA assays

ELISA, enzyme-linked immunosorbent assay; FA, fluorescent antibody test; GMS, Gomori–methenamine silver stain; H&E, standard histopathology stains; IFA, indirect fluorescence test; IHC, immunohistochemistry (usually formalin-fixed tissues); KOH, potassium hydroxide test; PAS, periodic acid–Schiff stain; PCR, polymerase chain reaction; RT-PCR, reverse transcription PCR; VI, virus isolation.
Source: Njaa BL. *Kirkbride's Diagnosis of Abortion and Neonatal Loss in Animals*, 4th edn, 2012.

Specific manifestations of the conceptus relating to pathology

Mummification

Mummification is an uncommon event in which there is often no certain cause. It is the process of progressive dehydration and compaction of a sufficiently mature dead fetus *in utero*. In order for this to occur, bacterial infection (tissue lytic organisms) of the dead fetus must be absent, no air can be present in the uterus (closed cervix), and a functional CL must be present. The time course, which may take months, depends on the size of the fetus. This process generally occurs in the second and third trimesters, when fetal bones are sufficiently developed to resist resorption. As fetal fluids are absorbed, tissues become compressed, shrunken, red/brown, and sticky to dry, without odor. As the membranes atrophy and the uterus involutes, hemorrhage occurs between the uterus and fetal tissues, imparting the red "hematic" sticky material onto the tissues (Figures 54.15 and 54.16). This hematic type of mummification apparently is distinctly bovine.[11,27]

The uterus tightly wraps around the fetal materials, compacting the mummy. The longer this proceeds, the drier and firmer the fetus becomes, remaining *in utero* for as much as 2 years. In the cow, mummification is most common at the end of the first and beginning of the second trimesters.[46] Various infectious and noninfectious etiologies have been implicated in their occurrence (BVDV, tritrichomoniasis), but the cause is usually unknown.[10,16] Mummification may be associated with certain breeds (Guernsey, Jersey) or certain breeding/genetic combinations.[11] In cases of uncomplicated mummification, prognosis is good for return to fertility.[27]

Maceration

Maceration describes the effect of dead fetal tissue softening due to fluid soaking and saprophytic digestion *in utero* (Figure 54.17). In cattle this is due to exposure of the retained fetal tissues to bacteria. This occurs when there is failure of fetal expulsion after death with retention of the carcass partially or entirely within the uterus. The bacteria may be those that caused fetal death or those that arrive through an open cervix.[10] Maceration is distinct from autolysis, which is digestion of tissues by endogenous enzymes and requires no bacteria. When bacteria gain access to a dead fetus in the reproductive tract, they are able to multiply rapidly at body temperature and are relatively insulated from the dam's immune system. Bacterial digestion of tissues and gas formation (emphysema) ensues. The uterus surrounds the fetus and there is an intense metritis.

In early gestation, macerated embryos are resorbed or expelled with little exudate. After the first trimester, the fetal skeleton is sufficiently formed to resist disintegration, and skeletal fragments may be retained. Bone fragments may become embedded in the uterus or may perforate the wall and remain free in the abdomen.[46] CL regression and cervical dilation may be found on palpation. Cows may display vague signs of illness due to a persistent and severe uterine infection.[16]

Emphysema usually occurs when incomplete abortion or near-term dystocia allow putrefactive vaginal flora to invade the dead fetus. Incomplete abortion occurs when there is partial cervical dilation and insufficient expulsion of the fetus, which is partly or totally retained in the uterus.[16]

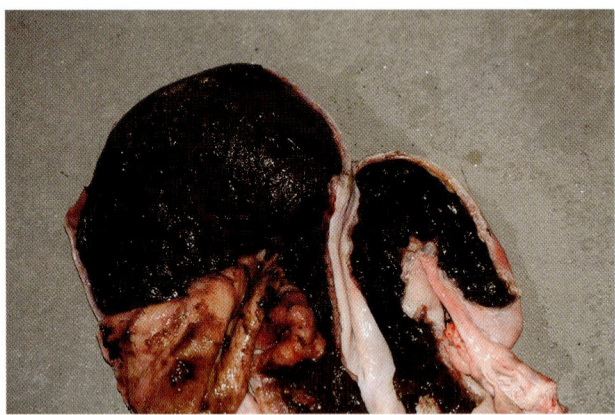

Figure 54.15 Opened uterus with mummified fetuses. Friable, non-odorous, hematic material obscures the fetuses. Photo courtesy of J. Edwards.

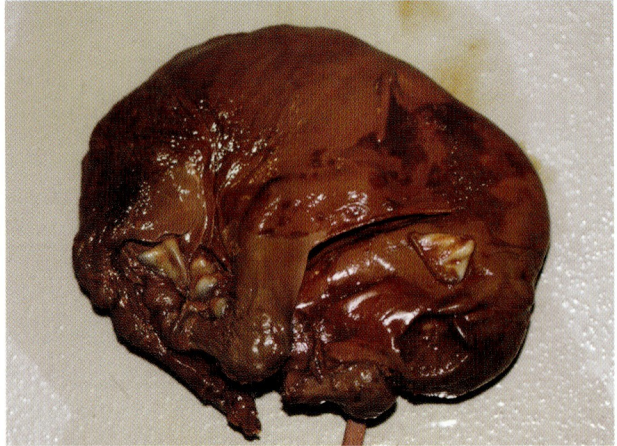

Figure 54.16 Fetal mummy. The carcass is shrunken and contracted with diffuse red/brown discoloration, dry tissues, and sunken eyes. Tissues lack odor and emphysema. Photo courtesy of J. Edwards.

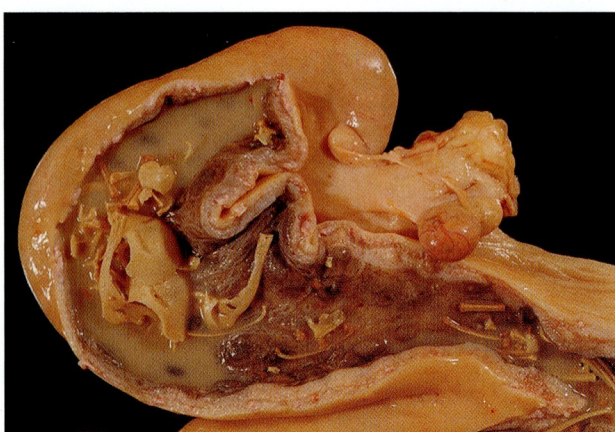

Figure 54.17 Opened uterine horn with a macerated fetus. Bone fragments in purulent fluid remain. The ovary has a retained corpus luteum. Photo courtesy of J. Cooley.

Embryonic death with persistent membranes (moles)

Moles are not neoplasms, but are masses of hyperplastic or edematous immature placental tissue.[47] Once fertilization occurs, the blastocyst divides into embryonic and extraembryonic tissues; the latter becomes the trophoblast, which is the forerunner of the placenta. Later in development, endodermal and mesodermal tissues from the embryo migrate along the inner aspect of the trophoblast to form the chorion, which is the outermost fetal tissue that is in contact with the uterus. Thus, embryonic mesenchyme contributes to the stroma of the placenta. This is an important point, for in order to form placental tissues, embryonic tissue must have been present.[47,48] Thus it is not surprising to find fetal tissues in moles, even though it is rare.

Hydatidiform moles are defined in humans by macroscopic features, and are termed (complete and partial) based on the degree of fluid accumulation. They are named hydatidiform for their abundant, bulbous, botryoid to polypoid villi, which are edematous with watery fluid-filled cisterns. Villous edema and trophoblastic hyperplasia are constant and characteristic features.[47,49] The hydropic swelling seen in moles may also occur in unrelated gestational disorders, such as maldevelopment of the placental vasculature or edema following early fetal death.[50]

In humans, complete moles are derived from an anucleate egg fertilized by one or two sperm, leading to a paternally derived (androgenetic) tissue, most often 46XX. Conversely, partial moles arise from the fertilization of an ovum by two sperm, generating a triploid conceptus.[47] Such anomalies lead to fetal loss, but membranes survive and proliferate (mole).

Moles in cattle are rare and have been reported as hydatidiform or cystic.[16,51–53] Bovine placental masses similar to hydatidiform moles in humans have been reported to be between 40 and 60 cm wide and weighing up to 48 kg.[52] The masses are smooth, soft/firm, ovoid/irregular, and pale grey/tan with multiple pendulous nodules that appear fatty. They are often associated with twinning.[52,53] Microscopic descriptions resemble descriptions in the human literature.[53,54] Recently, a mole of androgenetic origin (60XX) was reported in association with a stillborn male calf.[52]

More commonly, moles in cattle have been termed "cystic placental moles." They are thought to arise after fetal death, with or without resorption, where membranes persist as an empty cystic structure of a size similar to normal membranes at 3-4 months' gestation.[16,51] The cyst is filled with clear gel-fluid and stroma. In some cases, the tissues become infected, leading to pyometra and expulsion.[16]

Acardiac twins

Acardiac twins (acardiac monster, amorphous globosus, holoacardius acephalus, holoacardius amorphous) are the most bizarre and severely malformed fetuses imaginable. They represent a twin whose development is completely disturbed and, unlike the name, may indeed have some cardiac tissue remnants. Despite this, a functional heart is absent. They develop in association with a twin that has sufficiently normal cardiovascular function to support the monster through placental vascular anastomoses.[55] In cattle, such twins often manifest as globoid, hair-covered masses of soft tissue attached to an umbilicus (Figure 54.18). Other monsters have rudimentary limbs, faces, and teeth. In rare cases multiple monstrous anomalies may be present.[11]

Figure 54.18 Acardiac monster with umbilicus. A hair-covered soft tissue globe. Photo courtesy of J. Cooley.

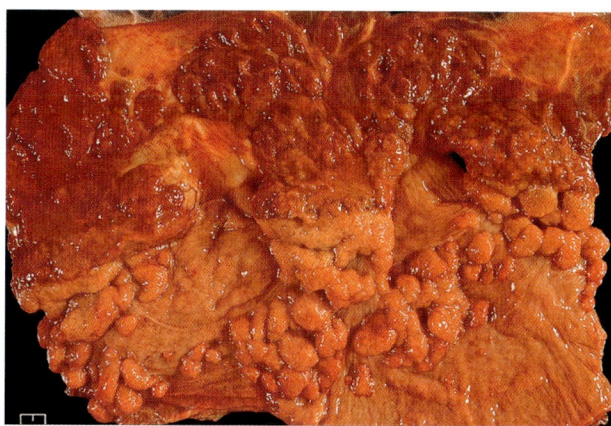

Figure 54.19 Chorioallantois, maternal aspect, adventitial placentation. Cotyledon proliferation is irregular but tissues are shiny, without fibrin, intercotyledonary hemorrhage, or necrosis. Photo courtesy of J. Cooley.

Adventitial placentation

In some cases, particularly in older multiparous cows and those with a history of metritis, there is inadequate placentomal development due to uterine damage. To compensate, the intercotyledonary chorion develops bright-red to tan, fine, villous structures and small accessory cotyledons (adventitial placentation) (Figures 54.19 and 54.20). In some cases proliferation may be abundant.[11] Rarely, villous proliferation may appear as a diffuse process with few or no obvious placentomes due to destruction of caruncles.[56,57] In some cases this process is incidental, but excessive adventitial placentation may be associated with pathologic conditions such as hydrallantois.[16,58]

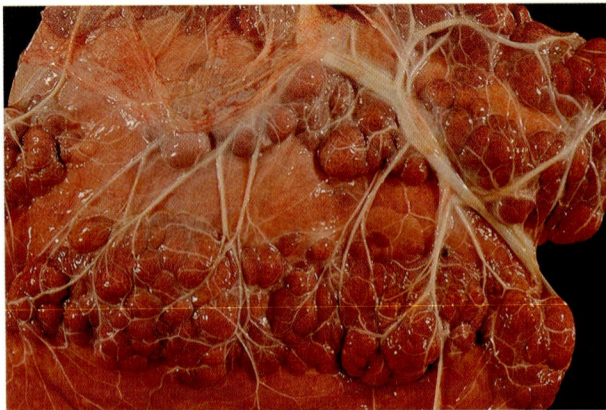

Figure 54.20 Chorioallantois, fetal aspect, adventitial placentation. Arteries and veins (white) arborize onto the coalescing placentomes. Photo courtesy of J. Cooley.

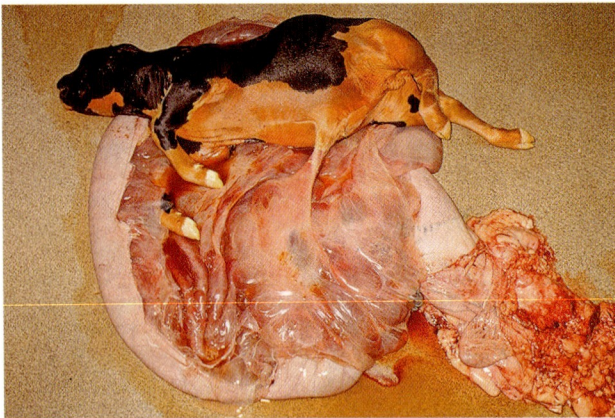

Figure 54.21 Fetus and placenta. The fetus is edematous (anasarca) with yellow/brown skin due to meconium staining. The allantois is markedly expanded by edema (hydrallantois). Photo courtesy of J. Cooley.

Dropsy of the fetal sacs (hydramnios/hydrallantois)

These conditions refer to excessive accumulation of fluids in the amnion and allantois cavities respectively; edema of the chorioallantoic membrane may also be present. In most cases, aside from Dexter bulldog calves, fluid accumulation occurs in the last trimester.[27] Fetal swallowing partly regulates amniotic fluid volume in the second half of gestation; conditions that prevent or slow fetal swallowing may lead to fluid retention.[36] Allantoic fluid is a watery fluid generated by the fetal kidney. It is excreted through the urachus, reaching 8–20 L at term. Amniotic fluid is more viscous and glairy (albuminous), due in part to fetal salivary secretions.

Hydrallantois is associated with placental dysfunction and is seen with uterine or placental disease. It is generally considered a maternal placentation problem, and adventitial placentation is commonly present.[59] This condition is seen sporadically in dairy and beef cattle. It is more common in twin pregnancies. Membranes may be more fibrous than normal but the character of the fluid is grossly normal. Fluid accumulation may reach 250 L.[59] Fetuses are usually born dead or are small, and may have ascites or anasarca[16,58] (Figure 54.21).

Hydramnios is much less common than hydrallantois. It is associated with fetal malformations that may be congenital or inherited. Congenital defects include any severe anomaly, such as schistosomus reflexus, that impairs normal swallowing. Inherited defects include Dexter bulldog fetuses, hydrocephalic Hereford fetuses, Angus fetuses with osteopetrosis, and Guernsey fetuses with pituitary hypoplasia.[11]

Fetal dropsy

Fetal dropsy encompasses three main presentations: anasarca, ascites, and hydrocephalus. Fetal anasarca may be seen with dropsy of fetal sacs (Figures 54.21 and 54.22). It is sometimes associated with genetic defects such as pulmonary hypoplasia. Fluid accumulates in soft tissues, often enlarging the fetus such that cesarean section is required.[60] Fetal ascites may be commonly seen in cases of

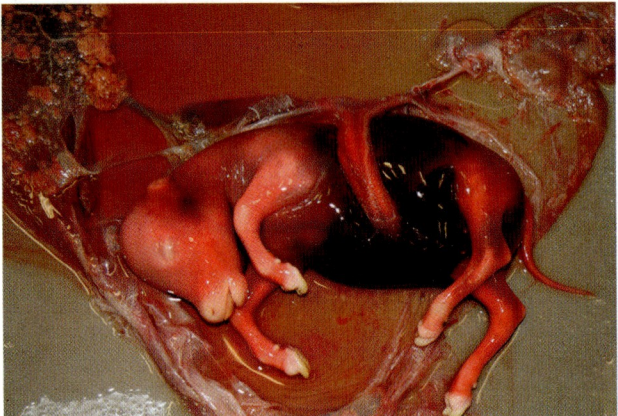

Figure 54.22 Fetus, anasarca. Photo courtesy of J. Edwards.

infectious abortion or developmental defects. Ascites or placental edema may occur as a rare accompaniment to *Brucella abortus* infection.[11,27]

Hydrocephalus

Hydrocephalus refers to an absolute increase in cerebrospinal fluid (CSF) within the brain ventricular system, within the meninges, or both. Congenital hydrocephalus occurs in two main forms, obstructive (hypertensive) and compensatory (normotensive). Obstructive hydrocephalus is associated with an inherited malformation, the most common of which is stenosis of the mesencephalic aqueduct.[61] Six such heritable forms have been described in Holstein, Jersey, Hereford, Charolais, Ayrshire, Dexter, and white Shorthorn breeds. Affected calves often have many other obvious defects.[29] Characteristically, there is marked doming of the skull with malformation of the bones of the cranium.[62]

Compensatory hydrocephalus, on the other hand, arises as a result of CSF accumulation after severe brain necrosis. It is often associated with *in utero* infections with BVDV and Akabane, Rift Valley fever, Wesselsbron, Schmallenberg, and bluetongue viruses. Hydranencephaly is a form of compensatory hydrocephalus where the neopallium (outer

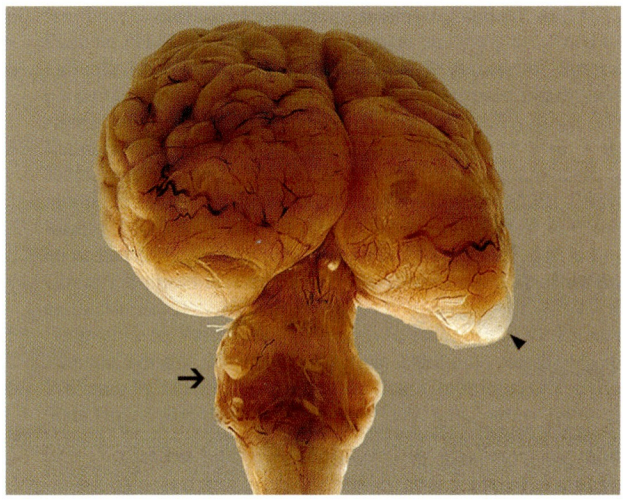

Figure 54.23 Brain, dorsolateral aspect. Hydranencephaly and cerebellar hypoplasia, BVDV *in utero* infection. The caudal aspect of the cerebrum lacks gyri and is thin to the point of translucency (arrowhead). The cerebellum is vestigial (arrow). Photo courtesy of J. Cooley.

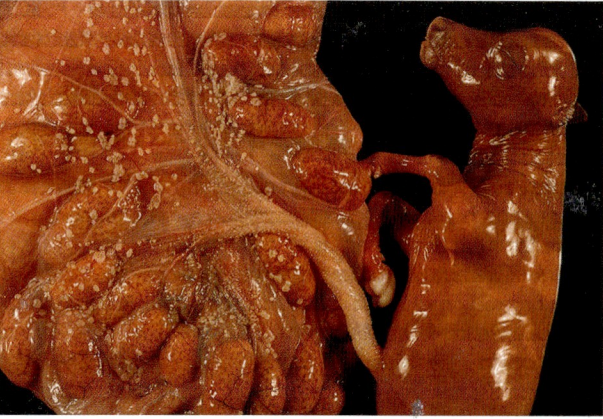

Figure 54.24 Fetus and placenta (amniotic surface exposed). The umbilicus and amnion are covered by amniotic plaques. Photo courtesy of J. Cooley.

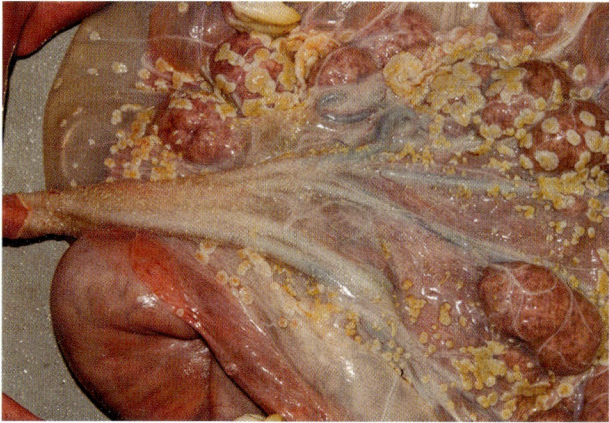

Figure 54.25 Placenta, umbilicus and fetal aspect of amnion. Numerous coalescing amniotic plaques cover the amnion. Photo courtesy of J. Cooley.

cerebrum) has been reduced to a very thin, almost translucent membrane with little to no parenchyma (Figure 54.23). Such lesions result from the aplastic and necrotizing effects of viral infection. The brainstem, cerebellum, hippocampus, and olfactory regions may be spared. The bones of the skull cavity are normal.[61] Other causes of congenital hydrocephalus include *Veratrum californicum* ingestion by pregnant cows, and vitamin A deficiency.[63] The cause of hydrocephalus may be difficult or impossible to determine in some cases.

Amniotic plaques

Amniotic plaques are normal, discrete, flat, white, 2–20 mm wide, round to irregular foci of squamous epithelium on the internal surface of the amnion (Figures 54.24 and 54.25). Immediately around the umbilicus they may become papillary or villous. They often resolve by 6 months' gestation, occur most often around the umbilicus, and are not found on the fetus. Plaques are not considered to be of pathologic significance and are not associated with signs of inflammation. They are to be distinguished from fungal plaques. Their presence is useful for orientation of tissues for examination as they occur on the internal (fetal) side of the amnion.[16]

Mineralization

Mineralization of the allantois or amnion appears as a network of white spots or streaks, typically associated with small vessels. The material is not gritty but is soft, occurring most often in the second trimester.[16] After this time the mineral disappears and is probably incorporated into fetal bone.[11] Mineral may also be present on the coat of the calf which may be easily wiped from the hair (see Figure 54.13). The origin and significance of this finding is uncertain, but it must be differentiated from fungal dermatitis.

Necrotic tips

Necrotic tips (Figure 54.26), or avascular chorion, is a normal area of necrosis occurring at the extreme tips of the chorioallantois of both horns. These areas are pale tan, shrunken, up to a few centimeters long, and clearly demarcated from the villous chorioallantois. Necrotic tips may be unilateral, particularly in the nongravid horn. In some cases, necrosis of the nongravid horn may involve as much as the entire horn.[36]

Allantoic calculi (hippomanes in horses)

Known since the time of Aristotle (*hippomane* means "horse madness"), these allantoic aggregates are composed of mineral and mucoprotein that deposit in progressive laminations around a nidus of cell debris[64] (Figure 54.27). Allantoic calculi are irregular, discoid, rubbery, tan masses with tapering margins, are 3–15 cm wide and up to 4 cm thick.[11] They have been described in horses, donkeys, zebras, cattle, deer, goats, and ewes, among others.[65] In cattle they are common and may be multiple. They are not known to be of significance.

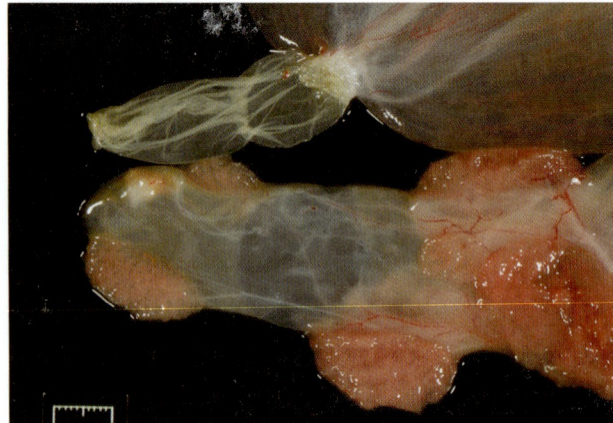

Figure 54.26 Chorioallantois, uterine horn tips. The horn at the top is avascular and shrunken. The lower horn tip is relatively normal. Photo courtesy of J. Edwards.

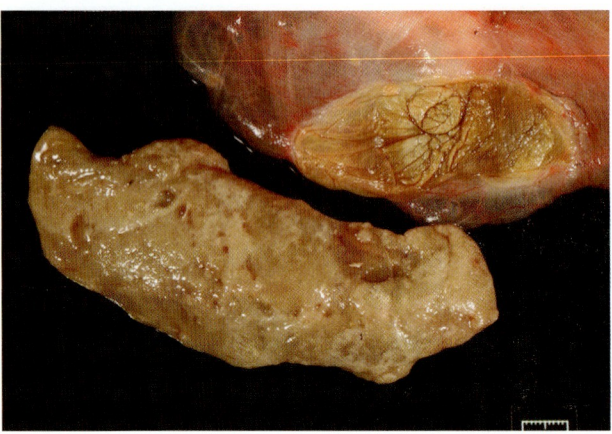

Figure 54.27 Hippomane and avascular chorion tip. Photo courtesy of J. Edwards.

Meconium

Meconium is a combination of sloughed intestinal and epidermal cells, mucus, bile acids, and various other metabolites of fetal development. These contributed materials collect in the fetal colon, where water and electrolyte absorption forms thick, brown, tenacious feces that may develop into pellets. Calves will also shed hair into the amniotic fluid, which is then swallowed by the fetus and collected in the colon to be expelled as part of the meconium. As such, meconium pellets may be composed of abundant matted hair, which may be expelled into the amniotic fluid or may be ingested and found in the alimentary tract of the fetus.[32,66] In cases of fetal stress/hypoxia, meconium may be prematurely expelled, causing discolored amniotic fluid or smearing of meconium on the fetus (see Figures 54.12 and 54.21).

Arthrogryposis

Arthrogryposis describes a condition where congenital joint contractures occur in one or many limbs and may be associated with scoliosis, kyphosis, lordosis, and torticollis. The limbs are rigidly fixed and variably bent in every sort of unnatural position. These changes occur as a result of decreased limb movement due to myogenic or neurologic abnormalities. As a result of neurologic disease *in utero*, muscles are underdeveloped and articular surfaces are misshapen due to lack of proper stimuli. It may occur sporadically, in association with viral infection (bluetongue, Akabane, BVDV, Schmallenberg), or due to maternal toxin ingestion (lupines, poison hemlock, some types of tobacco). Breed-associated syndromes in cattle have been described. Arthrogrypotic fetuses are associated with stillbirth, hydramnios, and dystocia for obvious reasons.[67]

Causes of abortion

Embryonic death

The early part of gestation is a time when fetal organ systems and extraembryonic membranes develop to achieve placentation, which is the point at which the chorion and endometrium attach. Prior to this, the uterus must be presented with an embryo by days 15–17 in order to prevent luteolysis and maintain pregnancy.[68] Failure to do so leads to early embryonic death. Late embryonic death occurs after maternal recognition but prior to the fetal stage (days 42–260).

A variety of factors contribute to embryonic death, including heat stress, metritis, maternal endocrine abnormalities (progesterone), aged gametes, trauma, genetic abnormalities, nutritional deficiencies, and maternal–fetal incompatibilities.[1] Infectious disease is able to cause embryonic loss in three major ways: (i) the uterine environment of a febrile cow is also hyperthermic and may lead to loss; (ii) infectious organisms may directly infect the embryo or may interfere with the uterine environment; and (iii) endotoxin from Gram-negative bacterial infection may lead to premature luteolysis.[1]

Fetal death

Noninfectious causes of abortion

Twinning is associated with higher rates (three to seven times normal) of stillbirth and abortion as well as retained membranes.[69] Abortion may occur at any time during gestation; rates of 30–40% have been reported for twin pregnancies.[11] In bovine twins, placental vessel anastomoses are almost always present, leading to fetal chimerism and freemartins.

Umbilical cord constriction is a rarely reported cause of abortion. This is largely due to three factors. Firstly, in the calf the umbilicus remains entirely within the amnion and is relatively short near term, such that rotation about the long axis (rolling over) is rarely possible. Secondly, the amnion is extensively fused to the chorioallantois, preventing rotation of the amnion within the chorioallantois, unlike the mare. Finally, at 5–6 months' gestation, the length of the calf exceeds the amnion width. By this stage, 95% of calves adopt an anterior presentation and cannot rotate around their transverse axes, preventing umbilical twisting and compression. In the mare, fetal anterior presentation is not firmly adopted until the ninth month.[70] If these assumptions are valid, then any constriction, twist, or entanglement should have occurred before the third trimester. Development would then proceed with relative normalcy until near term, when progressive umbilical constriction could impair venous and allantoic outflow. The result would be placental edema and possible abortion. Such a situation was suggested in a report of term abortion with placental edema in a Belgian White Blue calf, a breed in which it may be more common. Calves have had their cords wrapped around the abdomen, legs, or neck.[71]

Abortion may be observed in association with severe abdominal trauma or stress, starvation, fetal hyperthermia or hypoxia, genetic abnormalities, environmental causes (toxic plants, phytotoxins, mycotoxins, pharmaceuticals), and from iatrogenic intervention.

Toxic causes of abortion may be particularly difficult to diagnose, particularly if they are not one of the better-known varieties. Specific lesions associated with certain toxins may be poorly described and therefore not easily recognized. Subclinical toxin exposure may lead to fetal death and abortion in an otherwise normal-appearing cow. Diagnosis in such cases is unlikely to be recognized by routine examination.[17] For specifics the reader is referred to the following chapter on the subject.

Therapeutic estrogenic compounds, prostaglandins, or glucocorticoids may induce abortion.[11] Vitamin A is teratogenic, whether in deficiency or excess, and is associated with defective musculoskeletal and neural development which may lead to abortion.[72]

Iodine deficiency has been reported in cases of fetal hyperplastic goiter in association with weak, alopecic, stillborn, and aborted calves.[73] The thyroid gland in such cases is large and may weigh more than 30 g. Both iodine-deficient diets and high calcium intake have been implicated.[74] For suspected cases, removal of the entire thyroid gland is advised. Weigh the entire gland, fix half in formalin, and submit the rest fresh for iodine analysis. Gland weight above 30 g in addition to thyroid iodine concentrations below 1000 mg/kg and consistent histological changes are necessary for diagnosis.[74]

Excessive dietary selenium has been associated with abortion or weak calves, with accumulation in the liver.[75] Conversely, selenium/vitamin E deficiency has been associated with poor fertility and abortion.[76]

Finally, allergies or assumed anaphylactic reactions due to vaccines have been associated with abortion.[11]

Infectious causes of abortion

The remainder of this chapter deals with contagious and noncontagious abortion-related pathogens. In Table 54.7, the approximate gestational stage of abortion is listed for various agents.[77] For each broad category of pathogen (bacteria, virus, protozoa, fungi), sequential preference is often given in order of perceived importance. Many pathogens are included in this section, partly due to their customary inclusion in other references concerning abortion, but also for the sake of completeness and utility as a reference. Several of these organisms are discussed in other chapters and detailed information can be found therein. For those pathogens that are only mentioned in this chapter, further discussion is included.

Zoonotic pathogens include *Brucella abortus*, *Campylobacter* spp., *Mycobacterium bovis*, *Listeria monocytogenes*, *Coxiella burnetii*, *Leptospira* spp., and *Salmonella* spp.

Table 54.7 Approximate stages of development when abortion occurs for specific pathogens.

Agent	Embryo	Early	Mid	Late
Anaplasma marginale				L
Trueperella pyogenes			L	L
Aspergillus fumigatus			L	L
Bluetongue virus		L		
Bovine herpesvirus 1 (IBR)	V		V	V
Bovine viral diarrhea virus	V	V	V	
Brucella abortus				H
Campylobacter fetus	L		L	L
Chlamydophila spp.				L
Coxiella burnetii				L
Epizootic bovine abortion				V
Histophilus somni				L
Leptospira spp.				V
Listeria monocytogenes				L
Mortierella wolfii			L	L
Mucor spp.			L	L
Neospora caninum			V	
Salmonella spp.			V	V
Tritrichomonas foetus	L	L		
Ureaplasma spp.				L

L, low herd abortion rate; V, variable herd abortion rate; H, high herd abortion rate.
Source: after Givens D, Marley M. Infectious causes of embryonic and fetal mortality. *Theriogenology* 2008;70:270–285, with a few additions by the author of this chapter.

Bacteria

Campylobacter fetus subsp. *venerealis* infection
Campylobacter fetus is the cause of widespread venereal disease in cattle associated with infertility, embryonic mortality and, uncommonly, abortion.[78] *Campylobacter fetus* is an obligate pathogen of the female reproductive tract, causing a mild transient endometritis.[12] After coitus, bacteria spread from the vagina to the uterus within 2 weeks. However, infection does not seem to impair fertilization and early embryonic development. Subsequent rapid bacterial replication in the uterus incites inflammation, leading to embryonic death between 15 and 80 days' gestation. This often occurs after maternal recognition of pregnancy, leading to prolonged estrus in cows. If bacterial replication in the uterus is slower, mid-gestation abortions are seen.[12] Nonvenereally transmitted *C. fetus* subsp. *fetus* or *C. jejuni* infections are associated with sporadic abortion due to ingestion of infected material/tissues.[78]

In infected herds, low fertility rates and abortion (<10%) are typical. Abortions are uncommon and occur most often at 4–6 months' gestation, but may occur in the last trimester; fetuses may be autolyzed, freshly dead, or even delivered alive.[17,23] Gross fetal lesions are nonspecific and include fibrinous pleuritis, pericarditis, and peritonitis.[23] Membranes are not typically retained and may have lesions that resemble *Brucella* sp. infection, such as thickened leathery intercotyledonary spaces and variably necrotic cotyledons.[16] However, placental lesions are variable and often subtle or inapparent.

Darkfield examination of abomasal contents will reveal motile bacteria with rapid darting movements; this is characteristic of *C. fetus* or *C. jejuni*.[12] Organisms can be cultured from aborted tissues (lung, placenta, abomasal fluid) but isolation may be problematic as drying is highly detrimental to viability. Amies transport medium with or without charcoal is particularly useful for cultures.[23,78]

Leptospirosis Leptospirosis is a worldwide infectious disease of many species (including humans), with highest prevalence in tropical areas.[79] Antigenically distinct serovars of *Leptospira* spp. (Icterohaemorrhagiae, Hardjo, Pomona, etc.) may belong to more than one genomic species (*L. interrogans sensu stricto* and *L. borgpetersenii*), since bacteria are classified into serogroups as well as genomic species.[79,80] In cattle, *L. interrogans* serovars Hardjo and Pomona are the most important causes of abortion, stillbirth, and reproductive failure.[16] Two types of Hardjo are of particular importance in cattle; both are adapted/maintained in cattle. Such a relationship is characterized by efficient transmission, high incidence of infection, chronic disease, persistence in the urinary tract, poor antibody response, and few organisms in tissues.[12] The Hardjo strain important in North America and found throughout the world is *L. borgpetersenii* type hardjobovis. The other, *L. interrogans* Hardjo type hardjoprajitno, is a highly virulent organism found primarily in Europe.[16] The Pomona serovar is maintained in swine; cattle are incidental hosts. Serovar Pomona infections are associated with severe acute disease in adults, with hemolytic anemia, hemoglobinuria, mastitis, and abortion during the septicemic stage.[81]

Bacteria are effectively shed in urine and may spread transplacentally, venereally, orally, via inhalation, and across conjunctiva.[80] Bacteremia follows a 4–10 day incubation phase, resulting in localization to the kidney and reproductive tracts. This in turn results in persistent shedding in urine and reproductive fluids. Maintenance of *Leptospira* spp. in the uterus may lead to fetal infection and abortion.[12]

Pregnant cows generally abort in the last trimester or deliver weak infected calves; often abortion may be the only sign of infection in the herd.[23,80] They may be sporadic or epidemic, occurring most often in the second and third trimesters.[23,82] Abortion rates vary according to serotype; serovar Pomona can approach 50%, while serovar Hardjo typically ranges from 3 to 10%.[16] Fetuses are often autolyzed and may be icteric.[23] In some cases the liver may have a green tinge due to cholestasis.[83] Membranes may have intercotyledonary edema, and less commonly will have yellow/brown viscous fluid with diffuse cotyledonary necrosis (tan, uniformly affected).[22] These changes are similar to those seen in BHV-1 infection.

Fresh specimens are essential for isolating the bacteria, as bacterial competition, desiccation, and acid pH are detrimental to leptospiral survival.[80] Pericardial fluid, abomasal fluid, and kidney are particularly good samples for finding bacteria. Bacterial lability and autolysis of samples makes isolation difficult and inefficient.[23,84] Opportunistic bacteria (*Trueperella pyogenes*, *Bacillus* spp.) may also be present in aborted specimens, obscuring leptospiral involvement.[16]

One of the best methods for diagnosis is by serology using the microscopic agglutination test (MAT). Immunohistochemical tests are used for identifying bacteria in fixed tissues and fluorescent antibody tests may be used on frozen material. The kidney, lung, placenta, and adrenal glands are most useful for finding antigen.[12,16] Refrigerated samples should be sent to the diagnostic laboratory promptly. PCR testing is available. For detailed discussion concerning diagnosis, see Chapter 57.

Brucellosis *Brucella abortus* is a highly contagious, zoonotic, intracellular bacterium that is an important cause of abortion and infertility in cattle. Cattle are the primary host of *B. abortus*, which is widespread in Africa, Asia, Europe, and Latin America.[85] In areas where cattle are housed with sheep or goats (western Asia, southern Europe), *B. melitensis* may infect cattle and cause abortions.[86]

The cow is the source of infection, and the bacterium is mainly spread by placental tissue and uterine discharge, but spread via milk, contaminated feed, and *in utero* does occur. Uterine discharge may contain *B. abortus* for 2 weeks prior to, and 2–3 weeks after, calving or abortion.[12] Acute infection is characterized by abortion, delivery of weak or premature calves, retained fetal membranes, metritis, infertility, and sterility in cows and bulls.[85] Carpal hygromas can be common in certain areas and may be the only obvious sign of infection.[86] In managed vaccinated herds infection may be subclinical, whereas in fully susceptible herds abortion rates may be as high as 70%.[87]

Bacteria penetrate the nasal/oral mucosa and incite a local or regional lymphadenitis, with enlarged lymph nodes that may be hemorrhagic. In time, bacteria spread hematogenously, with a bacteremia that may be recurrent and persistent. *Brucella abortus* has an affinity for the placental trophoblasts, in which they grow to large numbers, and is associated with persistent inflammation and necrosis. Infection then spreads to the fetus.[16]

Abortion is the principal sign, occurring in the second half of gestation, after the fifth month. Infected cows may abort only once, thereafter delivering term calves that may or may not be infected.[87]

Placentitis is a consistent feature. Fetal membranes are often retained and contain yellow slimy exudate with grey/yellow flocculent debris between and especially surrounding the cotyledons.[16] Cotyledons are variably necrotic with thickening and fissures, or are soft to leathery, sometimes with sticky, brown, odorless exudate.[16] Intercotyledonary tissues may be similarly affected.

Fetuses are expelled 1–3 days after death; they are often autolyzed and may not have gross lesions.[23,87] They may be edematous with serosanguineous cavity fluids and fibrinous pleuritis/pericarditis/peritonitis, pneumonia, and multiorgan hemorrhage.[16,85,88] Abomasal fluids may have yellow flocculent material.[87] Pneumonia in aborted calves is an important and outstanding feature; tissues are firm, sometimes presenting with a cobblestoned surface or small white foci.[16,87]

The gold standard for diagnosis is the isolation of *B. abortus* from the fetus or placenta (abomasal contents, spleen, lung), but this may be difficult. Serology is the best method to detect infection, using the milk ring test and card agglutination test. PCR methods are used to differentiate vaccine strains from wild-type bacteria. Organisms are plentiful in aborted tissues. Vaccination of pregnant cattle with strain RB51 is associated with abortion and should be avoided.[86] For detailed discussion refer to Chapter 58.

Listeria infection   *Listeria monocytogenes* and *Listeria ivanovii* are associated with disease in cattle, which takes the form of meningoencephalitis, neonatal septicemia, and abortion.[89] *Listeria* spp. are common in the environment, particularly in colder temperate climates. Disease in cattle is often associated with feeding spoiled silage where bacteria grow to large numbers. Hematogenous spread to the placenta 5–12 days later with subsequent fetal sepsis occurs. If infection occurs early in the last trimester, rapid fetal septicemia and death occurs. Abortion occurs a few days later, with partial masking of lesions due to autolysis.[16] Infections near term are associated with significant metritis and septicemia in the dam, with dystocia and retained membranes. In these cases, autolysis may be less severe, allowing better visualization of fetal lesions.[16]

Abortions are sporadic and rarely exceed 15%.[89] They typically occur in the third trimester but may occur in the second. Fetuses die *in utero* and exhibit varying degrees of autolysis. Even if fetuses are well preserved, lesions may be few; the most commonly seen are pinpoint white/tan foci of liver necrosis that may mimic herpesvirus infection.[90] Small abomasal erosions, fibrinous serositis, pneumonia, and polyarthritis may also be seen.[17,41] Placentas are often retained; changes include multifocal cotyledonary necrosis and intercotyledonary placentitis with grey/white to red/brown exudate.[16,89] Aborting cows exhibit disease before, during, and after abortion, including weight loss, metritis, and septicemia.[12,16] Recurrent abortion from year to year and in the same cow may be seen.

Organisms can be isolated from numerous fetal tissues; in some cases refrigerating specimens can aid in recovery of *Listeria* spp.[16] Highly selective media are available. Gram staining of fetal fluids or tissue impressions reveals Gram-positive short rod bacteria in large numbers.[12]

Histophilus somni (formerly *Haemophilus somnus*) infection   *Histophilus somni* is a Gram-negative bacterial pathogen in cattle best known for causing pneumonia, arthritis, myocarditis, and thrombotic meningoencephalitis. It is a normal inhabitant of the male and female bovine genital tracts and is associated with infertility, abortion, and endometritis. Orchitis and epididymitis have been reported in bulls. Natural breeding spreads the bacterium. Transmission by artificial insemination is unlikely, as many strains are sensitive to the antimicrobials used to preserve frozen semen.[16]

The association of *H. somni* with infertility is controversial; this bacterium is a normal inhabitant of the reproductive tract and research results are mixed.[12] As a cause of abortion, *H. somni* is not common; abortions are sporadic and infection rarely results in outbreaks. Typically abortions occur in the second half of gestation. Placental lesions concentrate on cotyledons, where there is necrosis due to vasculitis and thrombosis. Such lesions are characteristic of this organism and indicate hematogenous spread from the vagina or respiratory tract. Fetal death is rapid and autolyzed carcasses are characteristic. Gross fetal lesions are uncommon; fibrinous bronchopneumonia may be seen.[16]

Organisms may be cultured from abomasal contents and membranes. However, light bacterial growth from the placenta may be due to contamination from the birth canal. Culture must be interpreted in conjunction with necropsy findings.[12]

Salmonella infection   *Salmonella* spp. cause infectious and contagious disease in humans and livestock around the world. *Salmonella enterica* subsp. *enterica* is divided into hundreds of serovars of which Dublin is best associated with cattle abortion, although other serotypes may be involved, like Typhimurium.[91] *Salmonella* Dublin is regionally distributed around the world and is known to persist on farms due to periodic fecal shedding, particularly around the time of parturition.[91] It is a cattle-adapted pathogen that causes enteric disease/dysentery, pneumonia, or polyarthritis in adults, but may also present with abortion as the principal sign.[91] Organisms survive in the environment for months to years and may be transported from farm to farm.[92] Bacterial infection in the dam leads to hematogenous spread to the placenta and subsequent placental failure and abortion, with or without fetal infection.[16] Abortion may be due to infection of the fetus, but as abortion may also occur in association with stressful events, pyrexia and maternal enteritis may precipitate expulsion.[16] Abortions are typically sporadic to epizootic, occurring most often in the second half of gestation, often with retained membranes.[91]

Aborted calves are often autolyzed and may be emphysematous. Pale foci of necrosis may be seen in the liver. Fetal fluids are fibrinous and amber colored; membranes are thickened and red/grey with yellow exudate and caruncular fragments on the cotyledons.[16,23] Organisms can be recovered from placental tissues and abomasal contents.[16]

Yersinia pseudotuberculosis infection   This Gram-negative bacterium is associated with enterocolitis, caseous mesenteric lymphadenitis, and septicemia in calves. Apparently healthy cattle may carry the bacteria subclinically.[93] Sporadic abortion (often in the second half of gestation) has been reported, possibly due to uterine infection via transient bacteremia from the gut flora.[94]

Fetuses typically have minimal autolysis; body cavities contain abundant fluid with fibrin tags and livers may have multiple pale tan foci of necrosis.[16,23,94] Lungs may be firm due to bronchopneumonia.[94,95] Microscopic lesions are characterized by abundant necrosis, vasculitis, and large bacterial colonies. Cotyledons are thickened, red/tan, with little exudate and caruncular tissue tags; intercotyledonary areas have fibrosis.[16,23] Bacteria may be cultured from various tissues using Culturettes.

Trueperella (Arcanobacterium) pyogenes infection   A common cause of suppurative inflammation in cattle that may act as a primary pathogen or a secondary invader. It is a Gram-positive commensal organism in the respiratory, gastrointestinal, and urogenital tracts in healthy cattle. Tissue invasion leads to abscessation in any organ, causing a wide variety of clinical signs.[96] *Trueperella pyogenes* is often isolated from aborted tissues and is considered to be a primary pathogen in cattle abortion.[16]

Uterine infection occurs hematogenously or via ascending infection, which is often acquired in the postpartum period, in part due to retained membranes. Fetal infections are not consistently present. The ensuing endometritis is associated with discharge and pyometra, although the cow may appear normal.[96]

Abortion can occur at any time throughout gestation, but is often in the second half of gestation. They are generally sporadic, but may occur in clusters. Retained membranes are common; cotyledons are covered with yellow/brown exudate and intercotyledonary tissues are thick, opaque, and covered with similar exudate.[23,96] Fetuses exhibit variable degrees of autolysis with fibrinous exudates on serosal surfaces; sometimes lungs are dark red with yellow foci or there is a characteristic hemorrhagic tracheal cast consistent with suppurative bronchopneumonia[16,41] (see Figure 54.6).

Trueperella pyogenes is readily isolated from fetal lung, abomasal content, or membranes.

Various opportunistic bacterial infections Sporadic abortions in cattle may be caused by environmental and endogenous opportunistic bacteria, including *Bacillus* spp., *Escherichia coli*, *Pasteurella* spp., *Pseudomonas* spp., *Staphylococcus* spp., *Streptococcus* spp., and *Serratia marcescens* among others.[11,97] Almost any bacterium that can gain access to the dam's bloodstream has to potential to infect the conceptus.

Abortions occur most often in the second half of gestation. Fetal membrane retention and autolysis vary. Membranes are thickened with yellow/brown exudate (Figure 54.28). Fibrinous exudates may be found in body cavities. Fetal bronchopneumonia (white spotted lungs) is sometimes present.[23]

Bacillus licheniformis is considered an important isolate associated with abortion, particularly in Europe. Outbreaks may be common. *Bacillus licheniformis* apparently has little inherent pathogenic potential, and reports of disease often implicate coinfection, such as with *Neospora caninum* and BVDV.[7,98] Fetal edema, fibrinous pericarditis, granulomatous hepatitis, bronchopneumonia, and placentitis have been reported.[98] Infected silage is implicated as a cause of cow infection, with a resulting bacteremia that may infect the cotyledons and spread to the fetus.

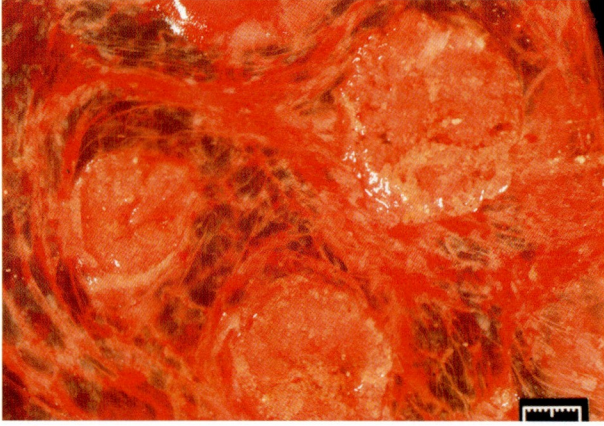

Figure 54.28 Chorioallantois, bacterial placentitis. Hemorrhage and fibrinous exudate covers the cotyledons and intercotyledonary spaces. Photo courtesy of J. Edwards.

Mollicute infection The Mollicutes are bacteria that lack cell walls and have a preference for mucosal and serosal surfaces. They are known causes of chronic respiratory, urogenital, mammary, musculoskeletal, and ocular disease in ruminants. Coinfection with other pathogens is common. *Mycoplasma*, *Ureaplasma*, and *Acholeplasma* spp. have been associated with male and female reproductive disease in cattle, including abortion.[99] These species, including those known to cause disease, are often found in the mucous membranes and reproductive tracts of normal animals.

Endometritis, granular vulvovaginitis, salpingitis, infertility, seminal vesiculitis, epididymitis, and granular balanoposthitis occur with infection. Transmission may occur venereally, mechanically (including artificial insemination as they can withstand freezing), and by respiratory routes.[99] Disease is multifactorial and associated with intensive husbandry.

Mycoplasma bovis, *M. bovigenitalium*, *M. canadense*, *M. leachii* (type 7) and *Ureaplasma diversum* have been reported causes of abortion in cattle, but *M. bovis* and *U. diversum* abortions are the best known and characterized.[99–103] The association of these organisms with abortion in general is controversial and is not well understood. However, *M. bovis* is not a normal inhabitant of the reproductive tract and thus its presence in an abortion is significant. Mycoplasmal abortion is uncommon. Reported lesions include placentitis, fetal bronchopneumonia, and myocarditis.[12]

Ureaplasma diversum is the most important abortion-related pathogen in this group, causing embryonic death, abortion (often in the last trimester), stillbirth, weak calves, and neonatal pneumonia.[16,99] Outbreaks of *U. diversum* infertility have been reported, lasting up to 6 months with conception rates as low as 20%.[12] Fetal membrane retention is common. Since few diagnostic laboratories attempt to culture this organism, there is wide regional variability in diagnosis frequency. In North America many cases are reported from Canada.[23]

Grossly, the amnion is often severely affected with thickening/fibrosis, mineralization, necrosis, hemorrhage, fibrin, and meconium staining.[16,104] The chorioallantois may be similarly affected with white to brown exudate. Aborted fetuses are commonly fresh.[90] Fetuses and newborn calves may have firm poorly aerated lungs. The constellation of amnion, chorioallantois, and lung lesions is characteristic of *U. diversum* infection.[16]

Definitive diagnosis requires culture of organisms from stomach contents, placental tissues, or lung in addition to consistent lesions. These organisms preferentially localize to the fetal–maternal interface (placentome). Specific media are required for culture; consult the diagnostic laboratory for preferred methods. Samples should be refrigerated and transported immediately. Deliver samples in frozen media (Stuart broth, Amies without charcoal), preferably in liquid nitrogen or dry ice rather than wet ice.[105] PCR methods for *M. bovis* and general *Mycoplasma* spp. are available.

***Chlamydia*-like organism infection** Bacteria of the order Chlamydiales are obligate intracellular parasites, ubiquitous, infect a wide range of vertebrate species, and cause a wide variety of diseases.[106] *Chlamydophila abortus*, *C. pecorum*, *C. psittaci*, *Waddlia* spp., and *Parachlamydia* spp. are of interest in cattle production. Abortion in cattle is usually caused by *C. abortus*, but *C. psittaci* abortion has been reported.[107] *Chlamydophila* spp. have also been associated with mastitis and infertility.[12] *Chlamydophila pecorum* is associated with enteritis, pneumonia, neurologic disease, polyarthritis, and conjunctivitis. Infectious forms may be shed in feces, urine, semen, uterine fluids, and ocular/nasal discharges.[108] Ingestion or inhalation of the organism leads to infection.

Parachlamydia spp. and *Waddlia chondrophila* have been isolated from aborted calves in Europe and in the United States.[109–111] The overall contribution of *Waddlia* and *Parachlamydia* spp. to cattle reproductive disease is uncertain.

Chlamydophila abortus is an uncommonly diagnosed cause of abortion in cattle.[23,112,113] Fetal loss is more sporadic than in sheep, but abortion rates of up to 20% can occur.[112] Although the prevalence of occult chlamydial infections may be high in some areas, currently the contribution of such infections to fetal losses is unclear.[112,114,115] The method of natural transmission of infection in cattle has not been verified, but is probably through contact with contaminated fetal fluids and tissues, as in sheep.[116]

Affected cows may show little sign of disease, delivering aborted, stillborn, or weak calves, sometimes with retained membranes.[117,118] *Chlamydophila abortus* abortions typically occur during 6–8 months of gestation, often presenting with necrotizing placentitis, with or without vasculitis.[111] Tissues have leathery, reddish, opaque intercotyledonary patches and multifocal cotyledonary necrosis. Fetuses may have pink/red subcutaneous edema, ascites, thymic and subcuticular petechiae, and serofibrinous pleuritis/peritonitis, which in some cases may be severe.[117,118] Chronically infected calves may have nodular mottled livers and enlarged lymph nodes.[118] *Parachlamydia* and *Waddlia* related abortions also exhibit necrotizing placentitis, but vasculitis is uncommon.[111]

Diagnosis of chlamydial abortion requires examination of the placenta, where organisms multiply in cotyledons.[116] Cotyledons and uterine discharges contain abundant organisms.[12] Isolation is the gold standard for diagnosis, which is necessary to identify species and strains/serotypes/subtypes; such methods require specialized media and special expertise.[119] It is important to recognize that many cattle harbor chlamydiae in their intestinal tracts, which may contaminate aborted tissues and confound complement fixation tests and PCR, leading to false-positive results.[116] Exposure of cows may be common and single titers are of questionable worth, although a rise in titer is suggestive of significant infection as delivery of an infected fetus leads to a rise in titer 2–3 weeks after parturition.[12] Specific antibody in fetal thoracic fluid or tissues is confirmative.[16] The presence of placentitis with identification of organisms within lesions by immunohistochemistry as well as PCR is diagnostic. Placental smears stained with Giemsa, Gimenez, or modified Ziehl–Neelsen acid-fast methods are useful for diagnosis.[16,118] Chlamydial organisms should be considered as serious zoonotic agents and may cause abortion in humans.[120]

Epizootic bovine abortion   Epizootic bovine abortion is a regionally important and specific disease confined to the western United States. It is associated with certain pine ranges that sustain the argasid tick *Ornithodoros coriaceus*.[121] The proposed etiologic agent is a deltaproteobacterium that is as yet unculturable and is transmitted by the tick.[121] Disease is typified by premature calving and abortion in otherwise normal cattle.[23] The disease in the fetus is a chronic one, developing over 3 months or more prior to abortion. Abortions occur sporadically or in severe outbreaks, often in the last trimester. Fetuses are expelled fresh; retained membranes are not typical. Lesions include petechiae in the mouth (particularly the ventral tongue), enlarged peripheral lymph nodes, ascites, splenomegaly, dermatitis, large and coarsely nodular livers, and small thymuses surrounded by edema and hemorrhages.[23,83,121] The presentation and lesions are characteristic. Diagnosis requires compatible lesions and herd history in addition to elimination of other causes.[121]

Coxiella burnetii infection   The cause of Q fever in humans, *Coxiella burnetii* is a Gram-negative, obligate intracellular rod bacterium that replicates within macrophages.[122] Organisms are hardy and resist desiccation, pH change, ultraviolet radiation, and disinfectants. Found worldwide except New Zealand, *C. burnetii* is present in a wide variety of rodents, birds, wild and domestic animals (including reptiles, dogs, cats, pigs, horses), ticks, and domestic ruminants; the latter serving as reservoirs for human infections.[122,123] Prevalence is variable worldwide; in the United States prevalence in 316 bulk milk samples was greater than 90%.[124] Routine pasteurization kills the organism.[81]

In endemic areas infected calves may clear the infection or become latently infected. During late pregnancy hormonal changes are thought to trigger *C. burnetii* replication in the uterus, fetus, and mammary gland.[122] Infected sheep, goats, and cattle are typically asymptomatic when they present for late-term abortion, stillbirths, and weak neonates.[122,125] Abortion in cattle is uncommon to rare. The organisms are shed in feces, milk, urine, fetal fluids, and fetal membranes in large numbers.[122] Infection usually occurs through inhalation of infected aerosols, but spread through infected equipment, manure, and ticks is possible.

Approximately half of seropositive heifers shed *C. burnetii* at calving; there is reduced shedding with age. Chronically infected cows may shed organisms in milk, feces, and vaginal mucus intermittently for several months to years.[122,123] Gross lesions are confined to the placenta, where the intercotyledonary zones are covered by abundant, white, inspissated exudate, giving the tissue a thick leathery appearance.[16] Multifocal mineralization and pericotyledonary necrosis may develop.[122]

Placental exudate smears stained with Gimenez or Ziehl–Neelsen acid-fast staining methods show numerous organisms, but these may be easily confused with *Chlamydophila* spp. PCR and enzyme-linked immunosorbent assay (ELISA) are effective in diagnosis. The mere presence of *C. burnetii* in aborted tissues is not sufficient evidence for diagnosis; cows may carry the organism long term and they may be in placentas of subsequent pregnancies.[16,123]

Tuberculosis (*Mycobacterium bovis* and *M. caprae*)   Although largely eliminated from many countries around the world, bovine tuberculosis is present enzootically or in limited geographic areas in Central and South America, North America, Africa, Europe, Asia, and Oceania. *Mycobacterium caprae*, previously a subspecies of *M. bovis* found in Europe, is now considered a separate species.[126] Disease caused by *M. caprae* is not significantly different from that caused by *M. bovis*. *Mycobacterium bovis* infection occurs in a wide range of hosts, including domestic livestock, cats, and dogs. The bacterium is a major economic problem in terms of trade and industry as well as being a zoonotic pathogen. One of the major problems encountered in terms of eradication is the prevalence of *M. bovis* in wildlife reservoirs such as farmed and wild cervids, European badgers, buffalo, bison, brushtail possums, and European wild boars.[127,128]

Mycobacterium bovis is typically spread by inhalation or ingestion. With reference to theriogenology, bull carriers may spread infection through coitus.[11] An initial site of infection may develop into a persistent focus of bacterial replication, with subsequent hematogenous spread to various organs. Disease is characterized by the presence of tubercles, which are discrete, pale tan granulomas of various sizes in virtually any organ. They may resemble abscesses, with abundant caseous necrosis. In adult cattle, tubercles are most common in the lungs and the lymph nodes of the mesentery, pharynx, and thorax; generalized lesions are uncommon. *Mycobacterium bovis* reaches the uterus hematogenously or by peritoneal spread to the uterine tubes. Uterine infection may primarily affect the serosa/muscularis, or may concentrate in the endometrium. Often the uterine tubes are severely affected.[129] Abortion due to endometritis often occurs late in gestation.[130] Retained placentas and purulent to caseous uterine exudate may be seen.[11]

Congenital tuberculosis occurs in about 1% of at-risk calves.[130] Transplacental transmission may occur hematogenously via tuberculous endometritis with umbilical spread to the fetal liver (hepatic primary complex) and lymph nodes.[128] Alternatively, fetal swallowing or inhalation of tuberculous amniotic fluid may result in enteric or pulmonary lesions, respectively.[130] *Mycobacterium avium* has been reported as a cause of abortion in cattle housed with infected chickens.[11]

At least 10 g of fat-free affected tissue should be submitted for culture.

Anaplasma infection There are currently three species of anaplasmal bacteria that infect cattle: *Anaplasma marginale*, *A. centrale*, and *A. phagocytophilum* (previously *Ehrlichia phagocytophila*, *Ehrlichia equi*, and human granulocytic ehrlichiosis agent).[131] *Anaplasma marginale* is the cause of anaplasmosis in cattle and can infect buffalo, bison, deer, and other hoofstock. This infectious but noncontagious agent is spread by arthropod bites, mechanical means, or transplacentally.[131] Anaplasmosis occurs in tropical and subtropical areas worldwide. *Anaplasma marginale* infects erythrocytes, leading to extravascular hemolysis, anemia, icterus, fever, and weakness.[132] Initial infections are the most severe, leading to chronic persistent infection that is often subclinical. *Anaplasma centrale* is a milder pathogen, usually employed as an immunizing agent against *A. marginale*.[133]

Anaplasma marginale is rarely implicated as a cause of abortion. Suggested causes of fetal loss include pyrexia, severe anemia, or stress in the dam. Aborted calves show variable liver and lung petechiation with splenic enlargement.[134]

Anaplasma phagocytophilum is the cause of pasture fever in cattle and tick-borne fever of sheep and goats in northern Europe, although the bacterium is distributed worldwide. While currently considered to be one species, *A. phagocytophilum* strains causing ruminant disease differ in distribution, disease, severity, and target hosts from those causing human granulocytic ehrlichiosis.[135] The organism infects circulating granulocytes and large lymphocytes, which are best seen during clinical illness.[133] Signs include fever, cough, nasal discharge, and leukopenia.[132] Although present in the United States, cases of clinical illness in cattle have not been reported. *Anaplasma phagocytophilum* has been known to cause abortions (and storms) during first exposure to infected ticks during late pregnancy.[7,135]

Diagnosis of *Anaplasma* spp. can be made on Giemsa-stained blood smears during the acute phase in cattle, but it is not useful for carrier states. Blood from live cattle should be collected in anticoagulant and refrigerated to be sent to a diagnostic laboratory. Smears from the liver, kidney, heart, lungs, or peripheral blood vessels can be made from dead animals and air dried for staining.[136] Serologic tests and PCR methods are available.

Viruses

Herpesvirus

Bovine herpesvirus 1 (BHV-1, infectious bovine rhinotracheitis) infection. This virus is an important worldwide alphaherpesvirus pathogen of cattle, known to cause abortion, genital disease, respiratory disease, encephalomyelitis, and severe systemic disease in neonates.[23] Other ruminants, including deer, pigs, goats, wildebeest, and buffalo are susceptible.[16,137] Infected cattle shed virus from oculonasal and reproductive fluids. Like other herpesviruses, latency is established after infection through the genital or respiratory tract.[137] Respiratory infection in naive pregnant females is likely to lead to fetal infection and abortion.[138] Clinical signs in adults are generally mild, with conjunctivitis or respiratory tract disease. However signs may be absent at the time of abortion, which typically occurs several weeks after initial infection.[23,137] Abortion is typically seen in the second half of gestation, anywhere from 2 weeks to 2–3 months after clinical disease; fetuses are expelled 3–7 days after death.[11,16,137] Abortions may be sporadic in herds with previous exposure, which includes modified live vaccines.[138] If naive pregnant cattle are exposed, storms affecting 25–60% may occur.[23,139]

The virus spreads hematogenously in the dam to the caruncle, where it invades the placenta and then the fetus. Herpesviruses induce widespread cell necrosis in multiple organs, leading to rapid fetal death.[16] Fetuses show advanced autolysis with few lesions; tissues are red due to hemoglobin staining and abundant reddish fluid may be present in body cavities.[137] Pinpoint white foci of hepatic necrosis may be present.[23] Less commonly, pulmonary and renal hemorrhage with necrosis are seen.[16,140] Renal necrosis may be pronounced.[141] Diffuse placentitis similar to that seen in leptospirosis may be found and yellow/brown amniotic fluid may be evident.

Fetal lung, liver, spleen, kidney, adrenal, and placenta are good sources of virus; samples should be kept cold. Freezing at −70°C is acceptable, but standard freezing (−20°C) will inactivate the virus.[137] Histopathology may give a presumptive diagnosis, which can be confirmed by immunohistochemistry, PCR, fluorescent antibody detection, immunoperoxidase tests, or virus isolation. Kidney and adrenal gland are useful for fluorescent antibody testing.[23] Cows that abort typically have low titers at the time of abortion and fetuses die so rapidly that fetal antibody production may be poor or nonexistent.[16]

Bovine herpesvirus 4 (BHV-4) infection. A widespread gammaherpesvirus in cattle considered minimally virulent or avirulent in adults, BHV-4 may be an important cause of abortion.[16,139,142] Presence of the virus in abortions is often associated with other pathogens.[16] Associated disease includes pneumonia, keratoconjunctivitis, mastitis, encephalitis, diarrhea, vulvovaginitis, postpartum metritis, and abortion.[139,142] The contribution of BHV-4 as a cattle pathogen at this point is mostly hypothetical, with the exception of the latter three conditions.[142] The role of BHV-4 in abortion is incompletely understood.[142,143] Abortion is reported to occur between 5 and 9 months' gestation. Descriptions of lesions in the literature are few, with no gross fetal or placental lesions reported. Cytomegalic intranuclear inclusions are seen microscopically in multiple organs.[145]

Bovine herpesvirus 5 (BHV-5) infection. A pathogen typically associated with severe meningoencephalitis in young calves, BHV-5 outbreaks are seen in South America and sporadically in Australia, Europe, and the United States.[146] Transmission may occur through the genital tract, with evidence suggesting possible negative embryological effects.[147–149] The virus has been isolated from an aborted fetus.[150]

Bovine viral diarrhea virus infection BVDV is one of the most important infectious causes of reproductive loss in cattle worldwide. It is a pestivirus that uses the reproductive system for spread within herds. Negative effects on herd reproduction range from genital tract inflammation, reduced reproductive efficiency, and embryonic death to abortion storms, delivery of small calves, persistently infected calves, and congenital defects.[16,82]

Transplacental infection is likely in infected cows cattle.[138] Viremia leads to fetal infection as it is a primary target for the virus.[151] The manifestations of disease in pregnancy vary according to virus strain and stage of fetal development at the time of infection.[138,151] Fetal infection in the first and second trimesters often leads to embryonic death, resorption, mummification, or abortion.[23] Fetal calves infected with noncytopathic viral strains between 1 and 4 months become persistently infected, immunotolerant calves that produce no antibody to the virus.[82] Those infected after 4 months may successfully eliminate the virus, but in some cases those infected between 100 and 150 days develop congenital anomalies.[23,82] Such anomalies are quite variable but include central nervous system anomalies (hydranencephaly, hydrocephalus, cerebellar hypoplasia, microphthalmia, retinal dysplasia, cataracts), thymic hypoplasia, hypotrichosis, bone defects, brachygnathism, arthrogryposis, and pulmonary or renal hypoplasia/dysplasia (see Figure 54.23). Necrotizing myocarditis leading to chronic passive hepatic congestion with ascites as well as congenital diabetes mellitus have been reported[23,82,152,153] (see Figure 54.9). Individual herds may have only a few calves with congenital defects; consecutive calves may have the same defects.[154] More often, fully developed but small (50–66% normal size) calves are delivered.[155] Infections in the last trimester do not typically produce disease as calves are immunocompetent and produce anti-BVDV antibodies.[151]

Fetal death and abortion typically occur in the first trimester, but can occur at any time. The abortion rate varies but is usually low.[151] There is a delay between fetal death and expulsion (as much as 50 days), which often produces autolyzed specimens from which it is difficult to detect the virus. Additionally, dams often seroconvert during this delay, making serum titers of little value.[84] Lesions aside from teratogenic changes are therefore difficult to appreciate and rarely noted. Enlarged spleens, lymph nodes, and nodular mottled livers may be seen.[151] Placental lesions are not seen.[41]

The virus is fairly resistant and can survive in moist tissues for several days; preferred samples for virus isolation include lymphoid organs.[151] Various other diagnostic methods are used including immunohistochemistry, fluorescent antibody, and ELISA.[23] BVDV is widespread, and molecular evidence of infection in an aborted fetus does not prove it to be the etiologic agent. Herd history and fetal lesions consistent with BVDV infection are necessary for diagnosis.[23] A history of ataxic, small, or blind calves in the herd is suggestive. Fresh specimens and often multiple fetuses are required before the virus can be identified.[84] Evidence for transplacental transmission includes detection of BVDV in fetal tissues or pre-colostral antibodies in fetal serum or fetal fluids. However, a persistently infected (noncytopathic BVDV infection) fetus will not produce antibody. Infected dams that carry persistently infected calves may have high titers that continue to increase until delivery.[156] Contaminated modified live BVDV vaccines have been reported to cause disease and congenital defects.[138]

Recently another pestivirus (HoBi-like, BVDV-3) has been associated with abortion and respiratory disease in cattle in Brazil, Southeast Asia, and Italy.[157–159] The origins of this virus are poorly understood; hypotheses include viral emergence from South America or transmission from water buffalo to cattle. Natural infection has not been reported in North America, Europe, Australia, India, and Africa so far.[159]

In natural cases, seromucous oculonasal discharge, tracheitis, bronchopneumonia, and gastroenteritis have been reported in infected calves. Abortions were seen in the second trimester. Experimentally, hyperthermia, oculonasal discharge, and lymphopenia were seen in calves. Lambs also exhibit nasal discharge while pigs show no signs but will seroconvert. Peak seroconversion occurs in calves by 21 days after infection; nasal and fecal shedding occur.[159]

HoBi-like virus has been isolated from bull semen. *In utero* infection may produce persistently infected calves, similar to BVDV. Current testing is complicated by the similarity of the HoBi-like viruses to BVDV.

Orbivirus

Bluetongue virus (BTV) infection. BTV is the cause of a widespread arthropod-borne orbiviral disease of domestic and wild ruminants, particularly of sheep.[160] Virus distribution is confined to areas where susceptible hosts (wild game, sheep, and cattle) are found and the *Culicoides* spp. biting midges are present for transmission.[160] BTV is generally considered noncontagious and infected animals shed little virus, except possibly in the semen of viremic bulls.[16,160] BTV infection may be seasonal, depending on the climate and numbers of *Culicoides*; in temperate zones late summer/ early autumn is common.[161] After infection, animals may be viremic for several weeks, but this does not lead to immunotolerance or persistent infection.[138,162] Clinical signs in adult cattle are rare and mild, particularly in enzootic areas.[139] When clinical disease occurs, signs include the formation of ulcers in the mouth and tongue, as well as on the muzzle. The coronary bands develop hyperemia, ulceration, and hoofs may slough.[154]

After introduction into the skin, the virus is transported to local lymph nodes, where it replicates. Subsequent hematogenous spread to secondary organs then occurs. BTV replicates in macrophages, lymphocytes, and vascular endothelial cells, causing host cell death. This results in widespread injury to small blood vessels which leads to vascular leakage, edema, subsequent thrombosis, and tissue infarction.[163]

Infections during pregnancy cause congenital abnormalities more commonly than abortion; early embryonic death, mummification, and stillbirth also occur.[16,164] Congenital abnormalities due to BTV are sporadic and are known to occur principally in the United States and South Africa, where modified live vaccines are used. These vaccine strains were considered largely responsible for fetal malformations.[164] Until recently, transplacental transmission was thought to occur only in laboratory modified strains, although now serotype 8 (BTV8) in Europe has been shown to frequently cross the placenta.[23,165,166] BTV8 is unusual not only for its virulence in cattle, but also for its wild-type transplacental transmission and oral transmission in field and experimental reports.[167–169]

Like BVDV, fetal disease is associated with gestational age at infection. Fetal infection prior to day 70 may lead to fetal death and absorption, while infection between 70 and 130 days can cause stillbirth, weak calves, hydranencephaly, or abortion. If infection is just prior to fetal immunocompetence, hydrocephalus or porencephaly may result. Fetal infection after 150 days can cause encephalitis or premature birth, but malformations do not occur.[16,23,164] Abortions occurring in the last trimester have been associated with hydranencephaly.[170]

Isolation of BTV is difficult and the virus is inactivated by freezing at −20°C.[138] Virus may be isolated from blood (in anticoagulant: EDTA, heparin), spleen, or lymph nodes if tissues are sent refrigerated to an appropriate laboratory.[162] PCR techniques use whole blood in EDTA. Live attenuated and killed vaccines are available. Live vaccines are effective, but are associated with abortion and congenital anomalies when used in pregnant females in the first period of pregnancy.[162] Fetal BTV antibody from serum or cavity fluids indicates *in utero* exposure.[23] Complementary lesions are needed for diagnosis.

Epizootic hemorrhagic disease virus (EHDV) infection. EHDV is an orbivirus transmitted by *Culicoides* midges that affects wild and domestic ruminants, particularly white-tailed deer.[171] Of the many serotypes found worldwide, Ibaraki virus, an Asian strain of EHDV serotype 2, has long been considered the only strain of significance to cattle. However, recent outbreaks of severe disease in cattle have been seen with serotypes 6 and 7 in Turkey and Israel.[172,173] Disease in cattle mimics BTV; mortality is often low and morbidity varies.[171] Abortions and stillbirths have been reported in outbreaks of Ibaraki virus infection in Japan.[174,175]

Palyam serogroup orbivirus infection. This group includes at least 15 viruses that are primarily found in Africa, Asia, and Australia and are associated with abortion and teratology in cattle.[176] *Culicoides* midges are natural vectors, but the virus has also been isolated from mosquitoes and ticks. Of these, the Chuzan (Kasba) virus is best known, causing hydranencephaly and cerebellar hypoplasia in calves.[177,178] The presence of infection in herds is usually found after abortion and malformations are seen. Abortions may occur at any stage of gestation, being sporadic or epidemic.[176] Antibodies in pre-colostral serum are helpful in establishing a diagnosis.[16,176]

Bunyaviridae
Rift Valley fever phlebovirus (RVF) infection. RVF is the cause of severe disease in domestic ruminants and humans; the virus is largely confined to sub-Saharan Africa and Madagascar.[179] Mosquitoes are the main route of transmission, with many capable species, including some found in North America.[16,179] Disease is most severe in young ruminants. In calves, severe hemorrhagic diarrhea and liver necrosis (with icterus) develop over a rapid course, with mortalities ranging from 10 to 70%.[179] Adult cattle may show only abortion, which can occur at any gestational stage and may reach 80–90%.[139,179] Abortion is due to fetal death; specimens are autolyzed and exhibit liver necrosis.[180] Fetal livers are enlarged, soft, friable, and yellow/red with hemorrhages or subcapsular hematomas.[181] Placental lesions have not been described. Aborted tissues contain large amounts of virus.[16] Hepatic lesions are considered diagnostic; differentials include BTV and Wesselsbron virus. The virus may be isolated from the liver, spleen, and brain; keep refrigerated or freeze and send on dry ice. Serological tests are available.[182]

Akabane and Simbu serogroup orthobunyavirus infection. This group includes about 25 viruses in the Bunyavirus genus that occur worldwide, including the Akabane, Aino, Tinaroo, and Schmallenberg viruses.[183–185] The Akabane virus, which occurs in Australia, Japan, China, Korea, the Middle East, and Kenya, is possibly the most important pathogen of this group, although Schmallenberg virus (SBV) is now on the rise in Europe.[16,186] Transmission is via *Culicoides* gnats and probably mosquitoes.[184,187] Clinical disease in adult cattle may be mild or inapparent, with reduced milk yield, diarrhea, or inappetence.[184,188] Clinical disease is more characteristic of SBV than in other viruses of this group.[189] Certain variants (genogroup Ia, Iriki strain) of Akabane virus are known to cause neurologic disease outbreaks (bovine epizootic encephalomyelitis) in calves and adult cattle in East Asia.[190]

Infection during pregnancy may produce abortions, malformations, mummies, and stillbirths.[16] Abortions occur in the third trimester and can occur as clusters or epidemics.[184,187] Viruses in this group are neurotropic and are known to cause arthrogryposis and hydranencephaly defects in domestic ruminants.[188,191,192] Hydrocephalus, cerebellar hypoplasia, porencephaly, scoliosis and cardiopulmonary defects are also reported.[139,188] The gestational stage at which infection occurs determines the severity and outcome of infection, with decreased severity seen in later infections.[184,187]

SBV was first noticed in the summer of 2011 in northern Germany and the Netherlands as a febrile syndrome in adult cattle; signs included fever, decreased milk production, and diarrhea.[189] SBV is the first outbreak of a Simbu serogroup virus in Europe and has now spread across much of Western Europe.[193] The virus is associated with abortion and congenital malformations in cattle, sheep, and goats. The losses in cattle due to decreased milk production and return to service may be more costly than losses due to malformed calves, which in the Netherlands outbreak occurred in 1–3% of farms.[193]

SBV antibody seroprevalence was 72% in cattle from the Netherlands between November 2011 and February 2012; a high seroprevalence in red deer, roe deer, and mouflon was also found. Current evidence suggests that biting midges (*Culicoides* spp.) are involved in horizontal transmission, but the details of horizontal transmission and its significance are poorly understood.[193,194] SBV has been found to be shed in semen, unlike Akabane virus.[195] The susceptibility of nonruminant species to SBV is unclear.[193] SBV is unlikely to pose a risk to humans.[196]

Virus can be isolated from blood and various fetal tissues and fluids.[16] PCR and immunochemical methods are available for detection in tissues. For SBV, brainstem tissue for PCR and pleural fluids for antibody testing have been reported as an optimal combination for diagnosis.[197] Differentials for arthrogryposis/hydranencephaly include BTV, BVDV, Chuzan, Wesselsbron, SBV, genetic causes, and toxins such as lupines.

Wesselsbron virus infection Wesselsbron virus is a mosquito-borne flavivirus primarily found in Africa. Disease is most severe in young small ruminants. Infection in adult cattle is typically subclinical.[198] Pregnant cattle may sporadically abort, deliver weak calves, or deliver apparently healthy calves. Certain viral strains can produce encephalitis, porencephaly, and cerebellar hypoplasia.[16]

Bovine parvovirus infection Bovine parvovirus (BPV) is widespread in the United States and worldwide. Often associated with diarrhea in calves, it is an uncommonly diagnosed cause of abortion.[16,155,199] Virus can be transplacentally transmitted to developing fetuses.[200] Experimental data indicate first and second trimester fetuses are most susceptible.[200] If infected between days 107 and 150, fetal cerebellar hypoplasia may be seen.[16] Fetuses infected in the last trimester can respond immunologically, preventing death. BPV should be considered in herds with abortion and repeat breeding problems.[199]

BPV is present in multiple fetal visceral tissues and in membranes.[16,200] BPV can be grown in cell cultures, but care should be used that supplemental sera do not contain BPV antibodies. Serum from fetuses infected in the third trimester will have specific antibody to BPV.[201]

Miscellaneous viruses Bovine enterovirus (picornavirus), adenovirus, pseudorabies virus (Suid herpesvirus 1), parainfluenza virus 3 (PI3), lumpy skin capripoxvirus, malignant catarrhal fever (gammaherpesvirus), foot and mouth disease virus (picornavirus), and bovine leukosis virus have been associated with abortion.[11,155]

Bovine PI3, generally considered a common mild respiratory pathogen, has been diagnosed uncommonly in abortions. Prevalence in herds is high, and association with abortion may be incidental or contributory with other pathogens. Experimental infection may cause bronchiolitis and interstitial pneumonia. Abortion due to PI3 is rare.[23,202]

Fetal antibodies to bovine enterovirus have been detected in high percentages of aborted fetuses tested in two studies.[203,204] This virus has been associated rarely with significant enteric disease in adult cattle, and experimental reproduction of illness in calves has been difficult.[205,206]

Protozoa

Neospora caninum **infection** *Neospora caninum* is a common worldwide infectious apicomplexan parasite of domesticated livestock, dogs, and some species of deer.[207] Dogs (including coyotes, dingoes, and gray wolves) are definitive and intermediate hosts. Currently there is no evidence for infection in foxes.[208] Livestock are intermediate hosts that harbor tachyzoites that form tissue cysts.[209–211] Tachyzoites may be found in a variety of cell types, but tissue cysts are almost exclusively found in the nervous tissues, including brain, spinal cord, eye, and nerves. Infection occurs in both beef and dairy herds and it is a major cause of dairy cattle abortion in the United States and the United Kingdom.[212–214] Vertical transmission is a common and highly efficient route of infection in ruminants and can progress through multiple generations in congenitally infected heifers.[209,215] Cattle may also become infected by ingestion of sporulated oocysts from canid feces.

Maternal seroprevalence is highly variable; seropositive cattle are more likely to abort than seronegative cattle.[23,216] Antibody titers in infected cows may fluctuate or decline over time and many may be persistently infected, making interpretation of sample results difficult.[23] Serum samples from pre-colostral calves are useful for determining infection in suspect cows, due to the efficiency of transplacental transmission.[217] For abortions, specific antibodies in fetal serum (after 100 days' gestation) or brain/spinal cord tissue are indicative of infection but do not prove *N. caninum* as the abortigenic agent. Similarly, negative titers do not rule out neosporosis since fetal antibody formation depends on exposure level, gestational age, and duration of infection.[23,207,218,219]

Abortion is the only clinical sign in cattle, occurring from 3 months' gestation to term, but predominating at 5–6 months, a hallmark pattern of this pathogen.[209,212,220,221] Abortions may be sporadic, endemic, or epidemic.[222] Calves may be stillborn, mummified, malformed, or born alive; calves may rarely show neurologic signs.[58,207] They may also be small for their age, ataxic, or have flexion/hyperextension of the limbs.[16] In most cases of infection, asymptomatic congenitally infected calves are produced.[220]

For definitive diagnosis an examination of the fetus is necessary. Fetuses are typically autolyzed with serosanguineous fluid accumulation. No consistent gross lesions are seen, although pale streaking in skeletal or cardiac muscle may be present.[220] Cotyledons may be necrotic and areas of malacia (soft, pitted, darker areas) may be seen in the brain or brainstem.[16] Submitting the entire fetus with placenta is optimal. Fetal brain is the most consistently affected tissue for finding lesions. Immunohistochemistry is highly useful for diagnosis since organisms may be few and difficult to find in autolyzed tissues.[207] The results of such methods should be viewed cautiously, as fetuses infected later in gestation may only have a mild *N. caninum* infection, but die from a secondary pathogen.[23] Congenitally infected calves often have high pre-colostral titers, but antibody levels vary and may even be undetectable.[7,16]

Tritrichomoniasis This is a worldwide venereal disease of bull-bred cattle due to *Tritrichomonas foetus* parasitism. Commercial AI services in the United States often screen for this pathogen, although custom-collected semen may be contaminated. This flagellated protozoan is an obligatory inhabitant of male and female genital tracts, occurring in both beef and dairy herds. Concurrent infection with *Campylobacter fetus* subsp. *venerealis* may be common. Females of all ages exhibit equal susceptibility and present with a history of bull-bred infertility due to embryonic death.

Infection typically causes vaginitis that may progress to endometritis; such infections may cause embryonic death and abortion (5% to greater than 20%) in the first half of gestation, with uncommon later stage abortions. The majority of pregnancies are probably lost by day 17. Fetuses aborted in mid pregnancy may be macerated and expelled with membranes. Cows may develop pyometra due to retained fetuses, which may present as multiple pyometras in a bull-bred herd. Uterine fluids are copious with little odor and can vary from watery and flocculent to red/brown and inspissated. Embryonic death after recognition of pregnancy (17 days) can lead to irregular cycling and return to estrus 60–90 days after breeding.

Placentas show little evidence of pathology aside from edema or the presence of white flocculent material and mild cotyledonary hemorrhage. Fetuses present in varying stages of autolysis, from fresh to macerated, and in many cases show no gross pathologic changes. In late-stage abortions bronchopneumonia and necrotizing enteritis have been seen.

Diagnosis depends on finding organisms in fresh fetal fluids/abomasal contents, and pyometra fluids (directly from the uterus) where they are often plentiful. Tissues older than 48 hours are poor specimens. Commercial media is preferred for transport and culture (In-Pouch TF Bovine, Biomed Diagnostics, OR, USA) and samples should be maintained between 22 and 37 °C. Buffered saline solutions (physiologic saline, lactated Ringer's) can be used, but samples should be kept cool (not frozen) and delivered as soon as possible (<48 hours) to a diagnostic laboratory. Positive identification of *T. foetus* may be complicated by other flagellated protozoan species. Immunohistochemical methods are also available to detect organisms in fixed tissues.

Tritrichomonas foetus infections may mimic *C. fetus* subsp. *venerealis* infections.[7,11,16,23,222–225]

Sarcocystis infection *Sarcocystis* spp. (*S. cruzi, S. hirsuta/S. bovifelis, S. hominis*) are coccidian protozoal parasites that commonly infect cattle. Carnivores are the definitive hosts while herbivores are the intermediate hosts. Cattle are infected by ingesting sporulated oocysts in carnivore feces. Once ingested, sporozoites are released into the gut, where they penetrate the mucosa and develop in arteries as meronts. The parasites are released from vessels as merozoites that then infect capillaries throughout the body. After a second round of replication, merozoites infect mononuclear cells and then enter muscle and neural tissues to encyst. Clinical disease is uncommon in adult cattle, but is associated with hemorrhage and edema due to the effects of intravascular maturation.[226] Signs include fever, ptyalism, lymphadenopathy, anemia, and abortion with experimental oral infection.[227]

Abortion is rare and has been reported in the third trimester; fetal hypoxia due to dam anemia, premature parturition due to prostaglandin (PG)F$_2\alpha$ release, pyrexia, or direct infection of the fetus are suggested etiologies.[11,16] Parasitic cysts may be present in the placenta or the fetus, but this is not typical. Gross lesions may resemble *Neospora caninum* infection or be inapparent.[16,23] Microscopic lesions are typically plentiful, consisting of necrosis in soft tissues and tissue cysts.[23]

Fungal infections

The incidence of mycotic abortion varies widely.[228] In some areas fungal abortion is the most common abortion diagnosis.[229] Losses are sporadic as these organisms are not contagious. The degree of environmental contamination may increase the risk of fetal infection.[23,230]

A variety of ubiquitous environmental fungi cause abortions in cattle worldwide. Prevalence of fungal types and commonality of fungal abortion varies geographically. *Aspergillus* spp., Zygomycetes (*Absidia, Mucor, Rhizopus, Mortierella*), *Candida* spp., and *Pseudallescheria boydii* are among the more frequently reported.[16] *Aspergillus fumigatus* and other *Aspergillus* species are reported as the most common causes of mycotic abortion, with Zygomycetes the second most common.[228] It is thought that primary maternal disease in the respiratory or gastrointestinal tract leads to hematogenous spread to the placentomes, as lesions are first seen there and are most severe. Clinical signs in the dam are typically not seen. However, fetal abortion with *Mortierella wolfii* is associated with a severe post-abortion pneumonia in the dam.[23,230]

Abortions occur in the last trimester and they generally do not exceed 10% incidence in a herd.[230] They may be seasonal in association with increased environmental exposure. In the northern hemisphere, the winter and spring are most often associated with abortion.[230] Retained membranes are common. Infection causes severe thickening of cotyledons with extension into the intercotyledonary areas, imparting a cracked or leathery appearance to tissues. Cotyledons may have entrapped shreds of maternal caruncular tissue. Fetuses may have minimal autolysis.

Placentitis develops slowly, interfering with fetal nutrition and often leading to fetal death. However, normal live calves may be born in cases of mild or even severe placentitis.[11] Fungal infection then extends to the fetus by infection of amniotic fluid. Fetuses may have numerous irregular tan epidermal plaques, particularly on the face, back and sides.[16] Infections due to septate fungi (*Aspergillus* spp.) often cause raised, dry, and wrinkled skin lesions, while infections due to nonseptate fungi (*Mucor* spp., *Rhizopus* spp.) may cause flat, moist, white/red foci.[17] Because of the chronicity and severity of infection in placentomes, incarceration of cotyledons may prevent membrane detachment for up to 10 days. At this stage, complete placentomal necrosis may ensue, with retention of necrotic tissues, maceration and/or pyometra.[11]

Gross lesions are distinctive. Placental lesions may mimic *Brucella abortus* or *Campylobacter* sp. infection.[230] Organisms may be cultured from the placenta, abomasal fluid, or lung. Placentas or, less commonly, abomasal fluids may have incidental fungal contamination. Diagnosis requires compatible lesions in addition to the presence of fungi.

Definitions

Asterisk denotes definitions from an authoritative source.[42]

Abortion Denotes the expulsion of a conceptus incapable of independent life.* Calves born after day 260 are generally thought to be able to survive independently. The expulsion of the entire conceptus (complete abortion) or portions (incomplete abortion) may occur.
Allantois A secondary outgrowth of the extraembryonic splanchnopleure from the ventral hindgut. The mesoderm is vascular.*
Amnion Extraembryonic somatopleure that is inverted around the embryo such that the ectoderm lines a cavity filled with amniotic liquor. The mesoderm is avascular.*
Birth Expulsion of a live fetus.
Blastocyst The stage in development after the morula, where the cells form a hollow sphere with an inner cell mass.*
BVDV Bovine viral diarrhea virus (bovine pestivirus).
Calf A live-born bovine.
Caruncle (plural caruncae) Discrete, monomorphic, regularly placed fleshy elevations of the inner uterine wall.
Chorion Extraembryonic somatopleure that is composed of an outer ectodermal trophoblast and is not part of the amnion. The mesoderm is avascular.*
Conceptus An inclusive term denoting the embryo or fetus and membranes.
Congenital Present at birth.*
Cotyledon Discrete, monomorphic, regularly placed elevations of the chorioallantois that are directly apposed and adhered to the uterine caruncles.
Embryo An animal in early development that does not have a form recognizable as a particular species.
Embryonic death Death of the conceptus prior to day 42.
Endemic abortion Abortion in a continuous pattern, at a rate greater than 5% throughout the year.
Epizootic/epidemic abortion A storm of abortions where greater than 10% of at-risk cattle abort in a short period of time, usually within 2–3 months.
Fetal death Death of a fetus prior to complete expulsion from the dam, subdivided into antepartum and intrapartum.*
Fetal period The stage of development between days 42 and 260, divided into early (days 42–120), middle (days 120–180), and late (days 180–260).*
Fetus An animal in development that has identifiable features of a given species.
Gestation The period of pregnancy; breeds typically vary in length, from 278 to 293 days.
Hamartoma A focally excessive overgrowth of mature normal cells in an organ that typically contains those cells.
IBR Infectious bovine rhinotracheitis virus (bovine herpesvirus 1).
Inner cell mass The group of cells in the blastocyst at one pole that develops into the embryo.*
Morula The early stage when eight or more cells are arranged in a solid mass without organization.*
Periparturient (peripartum) With reference to the dam, the time span occurring during the last 10% (30 days) of gestation to the first few weeks of calf life.
Placenta The fetal and maternal specialized organs of transfer between mother and offspring.
Placentome The union of chorionic villi (cotyledon) and caruncle.
Premature parturition Delivery of an immature viable calf prior to completion of normal gestation (about 260 days).
Somatopleure The lateral and ventral body wall of the embryo composed of an outer ectoderm and an underlying mesoderm.
Splanchnopleure The embryonic layer formed by the fusion of the visceral mesoderm to the endoderm.
Sporadic abortion Individual or small numbers of abortions occur at irregular intervals and typically do not exceed 3–5% annually.
Stillbirth Delivery of a dead fetus within the time of expected viability (after day 260).
Teratogen Anything, infectious or otherwise, that causes abnormal development in a conceptus.
Trophoblast The outer cellular layer composed of extraembryonic ectoderm through which all physiological exchange takes place.*
Zona pellucida The noncellular layer surrounding the ovum.
Zygote A diploid cell formed from the fusion of male and female pronuclei lasting until the first cleavage is complete.

Acknowledgments

Dr F. Austin contributed to portions of this chapter regarding microbiology sampling of specific pathogens. The author wishes to thank Drs J. Cooley and J. Edwards for their figure contributions and editorial comments.

Appendix

Abortion case submission form (Fig. A1).

Abortion Diagnosis

Owner
Address
Phone/Fax/Email:
Veterinarian Name/Phone/Address:

Aborted calf: date aborted:_____ Estimated due date: _____ Crown-rump length _____ cm

Assistance needed for delivery? Y/N   Vet involved? Y/N   Calf alive when problem first noticed? Y/N

Gross fetal lesions:

Gross placental lesions:

Formalin fixed tissues: 10 parts fixative for each part tissue: Brain (whole) ☐	Fresh/chilled tissues: DO NOT FREEZE	Chilled samples: not in syringes
Liver ☐ Ileum ☐	Placenta (3 cotyledons) ☐ **bag separately**	Abomasum contents (3ml)- sterile tube ☐
Heart ventricle ☐ Adrenal ☐	The following **can be pooled if necessary**:	Thoracic/abdomen fluid/heart blood (3ml)-sterile tube ☐
Thymus (w/ trachea & esophagus) ☐	Kidney (whole organ) ☐	Maternal serum, red top tube;
Lung (2 sections) ☐ Eyelid ☐	Liver (2"x2") ☐	at abortion ☐ 10-14 days later ☐
Placenta (3-4 cotyledons) ☐ Thyroid ☐	Lung (2"x2") ☐	Herdmate serum, red top tube ☐
Lymph Node (mesenteric) ☐ Kidney ☐	Spleen (1/2 organ) ☐	(several cows are preferable)
Spleen ☐ Colon ☐	Thymus (1cm slice) ☐	10-14 days later ☐
Sk.muscle (limb,tongue & diaphragm) ☐		
T.foetus: TF pouch culture fetal fluids keep at rec. temp.	Adrenal (whole) ☐	Fetal liver, kidney (frozen); bagged separately ☐

Cow/Dam

Breed:	Age/Tag#:	Lactation#:	Source/dam/sire:
AI ○ Bull ○ Hired bulls Y/N	Conception Date/Last Service:		Date of last normal parturition:

Figure A1

Vaccines/ date given/ brand/ lot #/booster intervals	Previous Abortions: Y / N When? Workup Done? Y/N Results?	
IBR: _____ BVD: _____ Neospora: _____ Lepto 5way: _____ Lepto hardjobovis: _____ Tritrichomonas: _____ C.fetus: _____ Chlamydophila: _____	(DairyFarm) Avg. days to 1st service:_____ Calving interval _____ Breedings per conception: _____	Previous clinical disease in cow: Y/ N (circle all that apply): Present in last Week Month 3 mo. 6 mo. 12 mo. 24 mo.

Dam disease signs (circle all that apply):   GI   Respiratory   Fever   Mastitis
Retained placenta   Anemia, dyspnea, emaciation (Anaplasmosis)

Changes in cow management/ pasture/ feeding regime:
Housing:
Forage: Grass hay _____ Legume Hay _____ Pasture _____ Corn Silage _____ Other _____
Forage quality:
Concentrates/Minerals/Commodities/Selenium:
Any testing of feed/water for nutrients/toxins?
Feeding Program: Prepared or made on-farm   Water source

# adults in herd/ body condition:	#immature in herd/ body condition:	# purchased in last: 6 mo._____ 12 mo. _____	
Number of herd abortions in last: Week_____ Month_____ 6mo._____ Associated with cold/hot weather? Affects which trimester: 1st ○ 2nd ○ 3rd ○ Affects which cows: old young mixed Problems with term calves: Weak Ataxic Stillborn Malformed Describe/how many:		Source: Health problems:	
		FarmAbortion Frequency: Sporadic ○ Recurrent ○ Recent problem ○	Pregnancy exams routinely done: Y / N Postpartum exams routinely done: Y / N _____ days after calving

Herd reproductive problems? Y / N What %?:_____ Circle all that apply: Anestrus   Metritis
Repeat Breedings   Irregular Cycles   Retained Placentas   Cystic Ovaries
Length of Breeding/calving season(beef):

Access to other cattle/ animals/ predators?	Toxic plants on pasture?

References

1. Barrett D, Boyd H, Mihm M. Failure to conceive and embryonic loss. In: Andrews AH, Blowey RW, Boyd H, Eddy RG (eds) *Bovine Medicine: Diseases and Husbandry of Cattle*, 2nd edn. Ames, IA: Wiley-Blackwell, 2004, pp. 552–576.
2. Diskin M, Parr M, Morris D. Embryo death in cattle: an update. *Reprod Fert Dev* 2012;24:244–251.
3. Inskeep E, Dailey R. Embryonic death in cattle. *Vet Clin North Am Food Anim Pract* 2005;21:437–461.
4. BonDurant R. Selected diseases and conditions associated with bovine conceptus loss in the first trimester. *Theriogenology* 2007; 68:461–473.
5. Maillo V, Rizos D, Besenfelder U et al. Influence of lactation on metabolic characteristics and embryo development in postpartum Holstein dairy cows. *J Dairy Sci* 2012;95:3865–3876.
6. Peter A. Abortions in dairy cows: new insights and economic impact. In: *Western Canadian Dairy Seminar*, 2000, pp. 233–244.
7. Caldow G, Gray D. Fetal loss. In: Andrews AH, Blowey RW, Boyd H, Eddy RG (eds) *Bovine Medicine: Diseases and Husbandry of Cattle*, 2nd edn. Ames, IA: Wiley-Blackwell, 2004, pp. 577–593.
8. Jonker F. Fetal death: comparative aspects in large domestic animals. *Anim Reprod Sci* 2004;82–83:415–430.
9. Cooper E, Laing I. The clinicians' view of fetal and neonatal necropsy. In: Keeling JW, Yee Khong T (eds) *Fetal and Neonatal Pathology*, 4th edn. London: Springer-Verlag, 2007, pp. 1–19.
10. Foster R. Female reproductive system and mammary gland. In: Zachary JF, McGavin MD (eds) *Pathologic Basis of Veterinary Disease*, 5th edn. St Louis, MO: Mosby, 2012, pp. 1085–1126.
11. Roberts S. *Veterinary Obstetrics and Genital Diseases (Theriogenology)*, 3rd edn. Woodstock, VT: published by the author, 1986.
12. Yaeger M, Holler L. Bacterial causes of bovine infertility and abortion. In: Youngquist RS, Threlfall WR (eds) *Current Therapy in Large Animal Theriogenology 2*. Philadelphia: WB Saunders, 2007, pp. 389–399.
13. Niswender G, Juengel J, Silva P, Rollyson M, McIntush E. Mechanisms controlling the function and life span of the corpus luteum. *Physiol Rev* 2000;80:1–29.
14. Foley G. Pathology of the corpus luteum of cows. *Theriogenology* 1996;45:1413–1428.
15. Kastelic J, Ginther O. Fate of conceptus and corpus luteum after induced embryonic loss in heifers. *J Am Vet Med Assoc* 1989;194:922–928.
16. Schlafer D, Miller R. Female genital system. In: Maxie MG (ed.) *Jubb, Kennedy, and Palmer's Pathology of Domestic Animals*, 5th edn. Philadelphia: WB Saunders, 2007, pp. 429–564.
17. Kirkbride C. Diagnostic approach to abortions in cattle. *Comp Cont Educ Pract Vet* 1982;4:S341–S346.
18. Miller R. A summary of some of the pathogenetic mechanisms involved in bovine abortion. *Can Vet J* 1977;18:87–95.
19. Mickelsen W, Evermann J. In utero infections responsible for abortion, stillbirth, and birth of weak calves in beef cows. *Vet Clin North Am Food Anim Pract* 1994;10:1–14.
20. Vanroose G, De Kruif A, Van Soom A. Embryonic mortality and embryo-pathogen interactions. *Anim Reprod Sci* 2000;60–61: 131–143.
21. Schlafer D, Fisher P, Davies C. The bovine placenta before and after birth: placental development and function in health and disease. *Anim Reprod Sci* 2000;60–61:145–160.
22. Dennis S. Pregnancy wastage in domestic animals. *Comp Cont Educ Pract Vet* 1981;3:S62–S69.
23. Anderson M. Disorders of cattle. In: Njaa BL (ed.) *Kirkbride's Diagnosis of Abortion and Neonatal Loss in Animals*, 4th edn. Singapore: John Wiley & Sons, 2012, pp. 13–48.
24. Whittier W. Investigation of abortions and fetal loss in the beef herd. In: Anderson DE, Rings M (eds) *Current Veterinary Therapy: Food Animal Practice*, 5th edn. St Louis, MO: Saunders, 2009, pp. 613–618.
25. Dillman R, Dennis S. Sequential sterile autolysis in the ovine fetus: macroscopic changes. *Am J Vet Res* 1976;37:403–407.
26. Leipold H, Cole D. Defective fetuses, abortion, and prematurity. In: Kirkbride CA (ed.) *Laboratory Diagnosis of Livestock Abortion*, 3rd edn. Ames, IA: Iowa State University Press, 1990, pp. 7–16.
27. Long S. Abnormal development of the conceptus and its consequences. In: Noakes DE, Parkinson TJ, England GCW (eds) *Arthur's Veterinary Reproduction and Obstetrics*, 8th edn. Philadelphia: WB Saunders, 2001, pp. 119–143.
28. Noden DN, De Lahunta A. *The Embryology of Domestic Animals*. Baltimore: Williams & Wilkins, 1985.
29. Washburn K, Streeter R. Congenital defects of the ruminant nervous system. *Vet Clin North Am Food Anim Pract* 2004;20:413–434.
30. Misdorp W. Congenital tumours and tumour-like lesions in domestic animals. 1. Cattle. A review. *Vet Quart* 2002;24:1–11.
31. Senger P. *Pathways to Pregnancy and Parturition*, 2nd edn. Ephrata, PA: Cadmus Professional Communications Science Press Division, 2003.
32. Williams W. *Veterinary Obstetrics*, 4th edn. Ithaca, NY: WL Williams, 1943.
33. Witschi E. *Development of Vertebrates*. Philadelphia: WB Saunders, 1956.
34. Assis Neto A, Pereira F, Santos T, Ambrosio C, Leiser R, Miglino M. Morpho-physical recording of bovine conceptus (*Bos indicus*) and placenta from days 20 to 70 of pregnancy. *Reprod Domest Anim* 2010;45:760–772.
35. Laven R, Peters A. Gross morphometry of the bovine placentome during gestation. *Reprod Domest Anim* 2001;36:289–296.
36. Śloss V, Dufty J. *Handbook of Bovine Obstetrics*. Baltimore: Williams & Wilkins, 1980.
37. Noakes DE. *Fertility and Obstetrics in Cattle*, 2nd edn. Oxford: Blackwell Science, 1997.
38. Bleul U, Lejeune B, Schwantag S, Kähn W. Ultrasonic transit-time measurement of blood flow in the umbilical arteries and veins in the bovine fetus during stage II of labor. *Theriogenology* 2007;67:1123–1133.
39. Taylor R, Njaa B. General approach to fetal and neonatal loss. In: Njaa BL (ed.) *Kirkbride's Diagnosis of Abortion and Neonatal Loss in Animals*, 4th edn. Singapore: John Wiley & Sons, 2012, pp. 1–12.
40. Kinsel M. An epidemiologic approach to investigating abortion problems in dairy herds. In: *Proceedings of the 32nd Annual Convention of the American Association of Bovine Practitioners*, 1999, pp. 152–160.
41. Dennis S. Infectious bovine abortion: a practitioner's approach to diagnosis. *Vet Med Small Anim Clin* 1980;75:459–466.
42. Anon. Recommendations for standardizing bovine reproductive terms. *Cornell Vet* 1972;62:216–237.
43. Mee J. Goitre in stillborn calves. *Vet Rec* 1993;133:404.
44. Nietfeld J. Field necropsy techniques and proper specimen submission for investigation of emerging infectious diseases of food animals. *Vet Clin North Am Food Anim Pract* 2010;26:1–13.
45. Blanchard P. Sampling techniques for the diagnosis of digestive disease. *Vet Clin North Am Food Anim Pract* 2000;16:23–36.
46. Drost M. Complications during gestation in the cow. *Theriogenology* 2007;68:487–491.
47. Baergen RN. *Manual of Benirschke and Kaufmann's Pathology of the Human Placenta*. New York: Springer Science, 2005.
48. Telugu B, Green J. Comparative placentation. In: Schatten H, Constantinescu GM (eds) *Comparative Reproductive Biology*. Ames, IA: Blackwell Publishing, 2007, pp. 271–320.
49. Shi I, Mazur M, Kurman R. Gestational trophoblastic disease and related lesions. In: Kurman RJ (ed.) *Blaustein's Pathology of the Female Genital Tract*, 5th edn. New York: Springer-Verlag, 2002, pp. 1193–1247.

50. Jauniaux E. Partial moles: from postnatal to prenatal diagnosis. *Placenta* 1999;20:379–388.
51. Fathalla M, Williamson N, Parkinson T. A case of bovine placental mole associated with twin embryonic death and resorption. *NZ Vet J* 2001;49:119–120.
52. Meinecke B, Kuiper H, Drögemüller C, Leeb T, Meinecke-Tillmann S. A mola hydatidosa coexistent with a foetus in a bovine freemartin pregnancy. *Placenta* 2003;24:107–112.
53. Morris F, Kerr S, Laven R, Collett M. Large hydatidiform mole: an unusual finding in a calving cow. *NZ Vet J* 2008;56:243–246.
54. Gopal T, Leipold H, Dennis S. Hydatidiform moles in Holstein cattle. *Vet Rec* 1980;107:395–397.
55. Benirschke K, Kaufmann P, Baergen R. *Pathology of the Human Placenta*, 5th edn. New York: Springer, 2006.
56. Williams W. *Retained Placenta in the Cow*. Columbus, OH: Ohio State University, 1914.
57. Wild A. Untersuchungen über den aufbau der placenta fetalis des rindes und ihre auswirkung auf die gesundheit des kalbes. *Zbl Vet Med B* 1964;11:60–89.
58. Buergelt C. *Color Atlas of Reproductive Pathology of Domestic Animals*. St Louis, MO: Mosby, 1997.
59. Peek S. Dropsical conditions affecting pregnancy. In: Youngquist RS, Threlfall WR (eds) *Current Therapy in Large Animal Theriogenology 2*. Philadelphia: WB Saunders, 2007, pp. 428–431.
60. Whitlock B, Kaiser L, Maxwell H. Heritable bovine fetal abnormalities. *Theriogenology* 2008;70:535–549.
61. De Lahunta A, Glass E. *Veterinary Neuroanatomy and Clinical Neurology*, 3rd edn. St Louis, MO: Saunders Elsevier, 2009.
62. Maxie M. Nervous system. In: Maxie MG (ed.) *Jubb, Kennedy, and Palmer's Pathology of Domestic Animals*, 5th edn. Philadelphia: WB Saunders, 2007, pp. 281–457.
63. Nietfeld J. Neuropathology and diagnostics in food animals. *Vet Clin North Am Food Anim Pract* 2012;28:515–534.
64. Dickerson J, Southgate D, King J. The origin and development of the hippomanes in the horse and zebra. II. The chemical composition of the foetal fluids and hippomanes. *J Anat* 1967;101:285–293.
65. King J. The origin and development of the hippomanes in the horse and zebra. I. The location, morphology and histology of the hippomanes. *J Anat* 1967;101:277–284.
66. Sherman D, Ervin M, Ross M. Fetal gastrointestinal composition: implications for water and electrolyte absorption. *Reprod Fert Dev* 1996;8:323–326.
67. Van Vleet J, Valentine B. Muscle and tendon. In: Maxie MG (ed.) *Jubb, Kennedy, and Palmer's Pathology of Domestic Animals*, 5th edn. Philadelphia: WB Saunders, 2007, pp. 185–280.
68. Bearden HJ, Fuquay JW. *Applied Animal Reproduction*, 4th edn. Upper Saddle River, NJ: Prentice Hall, 1997.
69. Andreu-Vázquez C, Garcia-Ispierto I, López-Béjar M, de Sousa N, Beckers J, López-Gatius F. Clinical implications of induced twin reduction in dairy cattle. *Theriogenology* 2011;76:512–521.
70. Noakes D. Development of the conceptus. In: Noakes DE, Parkinson TJ, England GCW (eds) *Arthur's Veterinary Reproduction and Obstetrics*, 8th edn. Philadelphia: WB Saunders, 2001, pp. 57–68.
71. Van Aert M, Piepers S, De Vhegher S et al. Late abortion and placental edema associated with umbilical cord constriction in a cow. *Vlaams Diergen Tijds* 2009;78:261–265.
72. Thompson K. Bone and joints. In: Maxie MG (ed.) *Jubb, Kennedy, and Palmer's Pathology of Domestic Animals*, 5th edn. Philadelphia: WB Saunders, 2007, pp. 1–184.
73. Seimiya Y, Ohshima K, Itoh H, Ogasawara N, Matsukida Y, Yuita K. Epidemiological and pathological studies on congenital diffuse hyperplastic goiter in calves. *J Vet Med Sci* 1991;53:989–994.
74. Cabell E. Bovine abortion: aetiology and investigations. *In Practice* 2007;29:455–463.
75. Yaeger M, Neiger R, Holler L, Fraser T, Hurley D, Palmer I. The effect of subclinical selenium toxicosis on pregnant beef cattle. *J Vet Diagn Invest* 1998;10:268–273.
76. Parkinson T. Infertility in the cow: structural and functional abnormalities, management deficiencies and non-specific infections. In: Noakes DE, Parkinson TJ, England GCW (eds) *Arthur's Veterinary Reproduction and Obstetrics*, 8th edn. Philadelphia: WB Saunders, 2001, pp. 383–472.
77. Givens D, Marley M. Infectious causes of embryonic and fetal mortality. *Theriogenology* 2008;70:270–285.
78. Irons P, Schutte A, Van Der Walt M, Bishop G. Genital campylobacteriosis in cattle. In: Coetzer JAW, Tustin RC (eds) *Infectious Diseases of Livestock*, 2nd edn. Cape Town: Oxford University Press Southern Africa, 2004, pp. 1459–1468.
79. André-Fontaine G. Leptospirosis. In: Lefèvre PC, Blancou J, Chermette R, Uilenberg G (eds) *Infectious and Parasitic Diseases of Livestock*. Paris: Lavoisier, 2010, pp. 1139–1149.
80. Hunter P. Leptospirosis. In: Coetzer JAW, Tustin RC (eds) *Infectious Diseases of Livestock*, 2nd edn. Cape Town: Oxford University Press Southern Africa, 2004, p. 1445–1456.
81. Garry F. Miscellaneous infectious diseases. In: Divers TJ, Peek S (eds) *Rebhun's Diseases of Dairy Cattle*, 2nd edn. St Louis, MO: WB Saunders, 2008, pp. 606–639.
82. Grooms D. Reproductive losses caused by bovine viral diarrhea virus and leptospirosis. *Theriogenology* 2006;66:624–628.
83. Anderson M. Bovine abortion diseases and diagnostics. Lecture presentation, American Association of Veterinary Laboratory Diagnosticians Workshop, Greensboro, NC, 2008.
84. Grooms D, Bolin C. Diagnosis of fetal loss caused by bovine viral diarrhea virus and *Leptospira* spp. *Vet Clin North Am Food Anim Pract* 2005;21:463–472.
85. Saegerman C, Berkvens D, Godfroid J, Walravens K. Bovine brucellosis. In: Lefèvre PC, Blancou J, Chermette R, Uilenberg G (eds) *Infectious and Parasitic Diseases of Livestock*. Paris: Lavoisier, 2010, pp. 991–1021.
86. Nielsen K, Ewalt D. Bovine brucellosis. In: Edwards S, Caporale V, Schmitt B et al. (eds) *OIE Manual of Diagnostic Tests and Vaccines for Terrestrial Animals*, 7th edn. Paris: OIE, 2012, pp. 616–650.
87. Godfroid J, Bosman P, Herr S, Bishop G. Bovine brucellosis. In: Coetzer JAW, Tustin RC (eds) *Infectious Diseases of Livestock*, 2nd edn. Cape Town: Oxford University Press Southern Africa, 2004, pp. 1510–1527.
88. Xavier M, Paixão T, Poester F, Lage A, Santos R. Pathological, immunohistochemical and bacteriological study of tissues and milk of cows and fetuses experimentally infected with *Brucella abortus*. *J Comp Pathol* 2009;140:149–157.
89. Schneider D. Listeriosis. In: Coetzer JAW, Tustin RC (eds) *Infectious Diseases of Livestock*, 2nd edn. Cape Town: Oxford University Press Southern Africa, 2004, pp. 1904–1907.
90. Anderson M. Infectious causes of bovine abortion during mid- to late-gestation. *Theriogenology* 2007;68:474–486.
91. Millemann Y, Evans S, Cook A, Sischo B, Chazel M, Buret Y. Salmonellosis. In: Lefèvre PC, Blancou J, Chermette R, Uilenberg G (eds) *Infectious and Parasitic Diseases of Livestock*. Paris: Lavoisier, 2010, pp. 947–984.
92. Fenwick S, Collett M. Bovine salmonellosis. In: Coetzer JAW, Tustin RC (eds) *Infectious Diseases of Livestock*, 2nd edn. Cape Town: Oxford University Press Southern Africa, 2004, pp. 1582–1593.
93. Fenwick S, Collett M. *Yersinia* spp. infections. In: Coetzer JAW, Tustin RC (eds) *Infectious Diseases of Livestock*, 2nd edn. Cape Town: Oxford University Press Southern Africa, 2004, pp. 1617–1627.
94. Jerrett I, Slee K. Bovine abortion associated with *Yersinia pseudotuberculosis* infection. *Vet Pathol* 1989;26:181–183.
95. Welsh R, Stair E. *Yersinia pseudotuberculosis* bovine abortion. *J Vet Diagn Invest* 1993;5:109–111.
96. Collett M, Bath G. *Arcanobacterium pyogenes* infections. In: Coetzer JAW, Tustin RC (eds) *Infectious Diseases of Livestock*, 2nd edn. Cape Town: Oxford University Press Southern Africa, 2004, pp. 1946–1957.

97. Kirkbride C. Bacterial agents detected in a 10-year study of bovine abortions and stillbirths. *J Vet Diagn Invest* 1993;5:64–68.
98. Müller M, Mölle G, Ewringmann T. Abortion in cattle caused by *Bacillus licheniformis*. *Tierarztl Umschau* 2005;60:258–262.
99. Irons P, Trichard C, Schutte A. Bovine genital mycoplasmosis. In: Coetzer JAW, Tustin RC (eds) *Infectious Diseases of Livestock*, 2nd edn. Cape Town: Oxford University Press Southern Africa, 2004, pp. 2076–2082.
100. Doig P. Bovine genital mycoplasmosis. *Can Vet J* 1981;22:339–343.
101. Hum S, Kessell A, Djordjevic S et al. Mastitis, polyarthritis and abortion caused by *Mycoplasma* species bovine group 7 in dairy cattle. *Aust Vet J* 2000;78:744–750.
102. Manso-Silván L, Vilei E, Sachse K, Djordjevic S, Thiaucourt F, Frey J. *Mycoplasma leachii* sp. nov. as a new species designation for *Mycoplasma* sp. bovine group 7 of Leach, and reclassification of *Mycoplasma mycoides* subsp. *mycoides* LC as a serovar of *Mycoplasma mycoides* subsp. *capri*. *Int J Syst Evol Microbiol* 2009;59:1353–1358.
103. Hermeyer K, Peters M, Brügmann M, Jacobsen B, Hewicker-Trautwein M. Demonstration of *Mycoplasma bovis* by immunohistochemistry and in situ hybridization in an aborted bovine fetus and neonatal calf. *J Vet Diagn Invest* 2012;24:364–369.
104. Miller R, Ruhnke H, Doig P, Poitras B, Palmer N. The effects of *Ureplasma diversum* inoculated into the amniotic cavity in cows. *Theriogenology* 1983;20:367–374.
105. Sanderson M, Chenoweth P. The role of *Ureaplasma diversum* in bovine reproduction. *Comp Cont Educ Pract Vet* 1999;21:S98–S101.
106. Kaltenboeck B, Hehnen H, Vaglenov A. Bovine *Chlamydophila* spp. infection: do we underestimate the impact on fertility? *Vet Res Commun* 2005;29(Suppl. 1):1–15.
107. Cox H, Hoyt P, Poston R, Snider Iii T, Lemarchand T, O'Reilly K. Isolation of an avian serovar of *Chlamydia psittaci* from a case of bovine abortion. *J Vet Diagn Invest* 1998;10:280–282.
108. Perez-Martinez J, Storz J. Chlamydial infections in cattle. Part 1. *Mod Vet Pract* 1985;66:517–522.
109. Dilbeck P, Evermann J, Crawford T et al. Isolation of a previously undescribed rickettsia from an aborted bovine fetus. *J Clin Microbiol* 1990;28:814–816.
110. Henning K, Schares G, Granzow H et al. *Neospora caninum* and *Waddlia chondrophila* strain 2032/99 in a septic stillborn calf. *Vet Microbiol* 2002;85:285–292.
111. Blumer S, Greub G, Waldvogel A et al. *Waddlia*, *Parachlamydia* and Chlamydiaceae in bovine abortion. *Vet Microbiol* 2011;152:385–393.
112. Livingstone M, Longbottom D. What is the prevalence and economic impact of chlamydial infections in cattle? The need to validate and harmonise existing methods of detection. *Vet J* 2006;172:3–5.
113. Storz J, Call J, Jones R, Miner M. Epizootic bovine abortion in the intermountain region. Some recent clinical, epidemiologic, and pathologic findings. *Cornell Vet* 1967;57:21–27.
114. DeGraves F, Gao D, Hehnen H, Schlapp T, Kaltenboeck B. Quantitative detection of *Chlamydia psittaci* and *C. pecorum* by high-sensitivity real-time PCR reveals high prevalence of vaginal infection in cattle. *J Clin Microbiol* 2003;41:1726–1729.
115. Barr B, Anderson M. Infectious diseases causing bovine abortion and fetal loss. *Vet Clin North Am Food Anim Pract* 1993;9:343–368.
116. Borel N, Thoma R, Spaeni P et al. *Chlamydia*-related abortions in cattle from Graubunden, Switzerland. *Vet Pathol* 2006;43:702–708.
117. Anderson A. Chlamydiosis. In: Coetzer JAW, Tustin RC (eds) *Infectious Diseases of Livestock*, 2nd edn. Cape Town: Oxford University Press Southern Africa, 2004, pp. 550–564.
118. Perez-Martinez J, Storz J. Chlamydial infections in cattle. Part 2. *Mod Vet Pract* 1985;66:603–608.
119. Sachse K, Vretou E, Livingstone M, Borel N, Pospischil A, Longbottom D. Recent developments in the laboratory diagnosis of chlamydial infections. *Vet Microbiol* 2009;135:2–21.
120. Longbottom D, Coulter L. Animal chlamydioses and zoonotic implications. *J Comp Pathol* 2003;128:217–244.
121. BonDurant R, Anderson M, Stott J, Kennedy P. Epizootic bovine abortion (foothill abortion). In: Youngquist RS, Threlfall WR (eds) *Current Therapy in Large Animal Theriogenology 2*. Philadelphia: WB Saunders, 2007, pp. 413–416.
122. Kelly P. Q fever. In: Coetzer JAW, Tustin RC (eds) *Infectious Diseases of Livestock*, 2nd edn. Cape Town: Oxford University Press Southern Africa, 2004, pp. 565–572.
123. Rodolakis A. Q fever. In: Lefèvre PC, Blancou J, Chermette R, Uilenberg G (eds) *Infectious and Parasitic Diseases of Livestock*. Paris: Lavoisier, 2010, pp. 1239–1246.
124. Kim S, Kim E, Lafferty C, Dubovi E. *Coxiella burnetti* in bulk tank milk samples, United States. *Emerg Infect Dis* 2005;11:619–621.
125. Moeller R. Disorders of sheep and goats. In: Njaa BL (ed.) *Kirkbride's Diagnosis of Abortion and Neonatal Loss in Animals*, 4th edn. Singapore: John Wiley & Sons, 2012, pp. 49–87.
126. Good M, Duignan A. Perspectives on the history of bovine TB and the role of tuberculin in bovine TB eradication. *Vet Med Int* 2011; Article ID 410470.
127. Palmer MV, Thacker TC, Waters WR, Gortázar C, Corner LAL. *Mycobacterium bovis*: a model pathogen at the interface of livestock, wildlife, and humans. *Vet Med Int* 2012; Article ID 236205.
128. Caswell J, Williams K. Respiratory system. In: Maxie MG (ed.) *Jubb, Kennedy, and Palmer's Pathology of Domestic Animals*, 5th edn. Philadelphia: WB Saunders, 2007, pp. 523–653.
129. Parkinson T. Specific infectious diseases causing infertility in cattle. In: Noakes DE, Parkinson TJ, England GCW (eds) *Arthur's Veterinary Reproduction and Obstetrics*, 8th edn. WB Saunders, 2001, pp. 473–509.
130. Boschiroli M, Thorel M. Tuberculosis. In: Lefevre PC, Blancou J, Chermette R, Uilenberg G (eds) *Infectious and Parasitic Diseases of Livestock*. Paris: Lavoisier, 2010, pp. 1075–1107.
131. Aubry P, Geale D. A review of bovine anaplasmosis. *Transbound Emerg Dis* 2011;58:1–30.
132. Hoar B, Nieto N, Rhodes D, Foley J. Evaluation of sequential coinfection with *Anaplasma phagocytophilum* and *Anaplasma marginale* in cattle. *Am J Vet Res* 2008;69:1171–1178.
133. Valli V. Hematopoietic system. In: Maxie MG (ed.) *Jubb, Kennedy, and Palmer's Pathology of Domestic Animals*, 5th edn. Philadelphia: WB Saunders, 2007, pp. 107–324.
134. Correa W, Correa C, Gottschalk A. Bovine abortion associated with *Anaplasma marginale*. *Can J Comp Med* 1978;42:227–228.
135. Woldehiwet Z. The natural history of *Anaplasma phagocytophilum*. *Vet Parasitol* 2010;167:108–122.
136. McElwain T. Bovine anaplasmosis. In: Edwards S, Caporale V, Schmitt B et al. (eds) *OIE Manual of Diagnostic Tests and Vaccines for Terrestrial Animals*, 7th edn. Paris: OIE, 2012, pp. 589–600.
137. Babiuk L, Van Drunen Little-Van Den Hurk S, Tikoo S. Infectious bovine rhinotracheitis/infectious pustular vulvovaginitis and infectious pustular balanoposthitis. In: Coetzer JAW, Tustin RC (eds) *Infectious Diseases of Livestock*, 2nd edn. Cape Town: Oxford University Press Southern Africa, 2004, pp. 875–886.
138. Kelling C. Viral diseases of the fetus. In: Youngquist RS, Threlfall WR (eds) *Current Therapy in Large Animal Theriogenology 2*. Philadelphia: WB Saunders, 2007, pp. 399–408.
139. Ali H, Ali A, Atta M, Cepica A. Common, emerging, vector-borne and infrequent abortogenic virus infections of cattle. *Transbound Emerg Dis* 2012;59:11–25.

140. Rodger S, Murray J, Underwood C, Buxton D. Microscopical lesions and antigen distribution in bovine fetal tissues and placentae following experimental infection with bovine herpesvirus-1 during pregnancy. *J Comp Pathol* 2007;137:94–101.
141. Muylkens B, Thiry J, Thiry É. Infectious bovine rhinotracheitis. In: Lefèvre PC, Blancou J, Chermette R, Uilenberg G (eds) *Infectious and Parasitic Diseases of Livestock*. Paris: Lavoisier, 2010, pp. 449–460.
142. Thiry É. Bovine herpesvirus 4 infection. In: Lefèvre PC, Blancou J, Chermette R, Uilenberg G (eds) *Infectious and Parasitic Diseases of Livestock*. Paris: Lavoisier, 2010, pp. 489–492.
143. Deim Z, Szeredi L, Egyed L. Detection of bovine herpesvirus 4 DNA in aborted bovie fetuses. *Can J Vet Res* 2007;71:226–229.
144. Czaplicki G, Thiry E. An association exists between bovine herpesvirus-4 seropositivity and abortion in cows. *Prev Vet Med* 1998;33:235–240.
145. Schiefer B. Bovine abortion associated with cytomegalovirus infection. *Zbl Vet Med B* 1974;21:145–151.
146. Meyer G, Thiry J, Thiry É. Bovine herpesvirus 5 infection. In: Lefèvre PC, Blancou J, Chermette R, Uilenberg G (eds) *Infectious and Parasitic Diseases of Livestock*. Paris: Lavoisier, 2010, pp. 493–498.
147. Oliveira M, Campos F, Dias M et al. Detection of bovine herpesvirus 1 and 5 in semen from Brazilian bulls. *Theriogenology* 2011;75:1139–1145.
148. Rodríguez M, Barrera M, Sánchez O et al. First report of bovine herpesvirus 5 in bull semen. *Arch Virol* 2012;157:1775–1778.
149. Silva-Frade C, Martins Jr A, Borsanelli A, Cardoso T. Effects of bovine herpesvirus type 5 on development of in vitro-produced bovine embryos. *Theriogenology* 2010;73:324–331.
150. Salvador S, Lemos R, Riet-Correa F, Roehe P, Osório A. Meningoencephalitis in cattle caused by bovine herpesvirus-5 in Mato Grosso do Sul and São Paulo. *Pesquisa Vet Brasil* 1998;18:76–83.
151. Potgieter L. Bovine viral diarrhoea and mucosal disease. In: Coetzer JAW, Tustin RC (eds) *Infectious Diseases of Livestock*, 2nd edn. Cape Town: Oxford University Press Southern Africa, 2004, pp. 946–969.
152. Tajima M, Yuasa M, Kawanabe M, Taniyama H, Yamato O, Maede Y. Possible causes of diabetes mellitus in cattle infected with bovine viral diarrhoea virus. *J Vet Med B* 1999;46:207–215.
153. Taniyama H, Hirayama K, Kagawa Y et al. Immunohistochemical demonstration of bovine viral diarrhoea virus antigen in the pancreatic islet cells of cattle with insulin-dependent diabetes mellitus. *J Comp Pathol* 1999;121:149–157.
154. Van Metre D, Tennant B, Whitlock R. Infectious diseases of the gastrointestinal tract. In: Divers TJ, Peek S (eds) *Rebhun's Diseases of Dairy Cattle*, 2nd edn. St Louis, MO: WB Saunders, 2008, pp. 200–294.
155. Kirkbride C. Viral agents and associated lesions detected in a 10-year study of bovine abortions and stillbirths. *J Vet Diagn Invest* 1992;4:374–379.
156. Lindberg A, Groenendaal H, Alenius S, Emanuelson U. Validation of a test for dams carrying foetuses persistently infected with bovine viral-diarrhoea virus based on determination of antibody levels in late pregnancy. *Prev Vet Med* 2001;51:199–214.
157. Decaro N, Lucente M, Mari V et al. Hobi-like pestivirus in aborted bovine fetuses. *J Clin Microbiol* 2012;50:509–512.
158. Decaro N, Lucente M, Mari V et al. Atypical pestivirus and severe respiratory disease in calves, Europe. *Emerg Infect Dis* 2011;17:1549–1552.
159. Bauermann F, Ridpath J, Weiblen R, Flores E. HoBi-like viruses: an emerging group of pestiviruses. *J Vet Diagn Invest* 2013;25:6–15.
160. Verwoerd D, Erasmus B. Bluetongue. In: Coetzer JAW, Tustin RC (eds) *Infectious Diseases of Livestock*, 2nd edn. Cape Town: Oxford University Press Southern Africa, 2004, pp. 1201–1226.
161. Maclachlan N, Drew C, Darpel K, Worwa G. The pathology and pathogenesis of bluetongue. *J Comp Pathol* 2009;141:1–16.
162. Daniels P, Oura C. Bluetongue and epizootic haemorrhagic disease. In: Edwards S, Caporale V, Schmitt B et al. (eds) *OIE Manual of Diagnostic Tests and Vaccines for Terrestrial Animals*, 7th edn. Paris: OIE, 2012, pp. 112–129.
163. Sperlova A, Zendulkova D. Bluetongue: a review. *Vet Med-Czech* 2011;56:430–452.
164. MacLachlan N, Conley A, Kennedy P. Bluetongue and equine viral arteritis viruses as models of virus-induced fetal injury and abortion. *Anim Reprod Sci* 2000;60–61:643–651.
165. Lefèvre P, Mellor P, Saegerman C. Bluetongue. In: Lefèvre PC, Blancou J, Chermette R, Uilenberg G (eds) *Infectious and Parasitic Diseases of Livestock*. Paris: Lavoisier, 2010, pp. 663–688.
166. Maclachlan N. Bluetongue: history, global epidemiology, and pathogenesis. *Prev Vet Med* 2011;102:107–111.
167. Backx A, Heutink R, van Rooij E, van Rijn P. Transplacental and oral transmission of wild-type bluetongue virus serotype 8 in cattle after experimental infection. *Vet Microbiol* 2009;138:235–243.
168. Menzies F, McCullough S, McKeown I et al. Evidence for transplacental and contact transmission of bluetongue virus in cattle. *Vet Rec* 2008;163:203–209.
169. Mayo C, Crossley B, Hietala S, Gardner I, Breitmeyer R, MacLachlan N. Colostral transmission of bluetongue virus nucleic acid among newborn dairy calves in California. *Transbound Emerg Dis* 2010;57:277–281.
170. Wouda W, Roumen M, Peperkamp N, Vos J, Van Garderen E, Muskens J. Hydranencephaly in calves following the bluetongue serotype 8 epidemic in the Netherlands. *Vet Rec* 2008;162:422–423.
171. Savini G, Afonso A, Mellor P et al. Epizootic hemorragic disease. *Res Vet Sci* 2011;91:1–17.
172. Temizel E, Yesilbag K, Batten C et al. Epizootic hemorrhagic disease in cattle, Western Turkey. *Emerg Inf Dis* 2009;15:317–319.
173. Yadin H, Brenner J, Bumbrov V et al. Epizootic haemorrhagic disease virus type 7 infection in cattle in Israel. *Vet Rec* 2008;162:53–56.
174. Ohashi S, Yoshida K, Watanabe Y, Tsuda T. Identification and PCR-restriction fragment length polymorphism analysis of a variant of the Ibaraki virus from naturally infected cattle and aborted fetuses in Japan. *J Clin Microbiol* 1999;37:3800–3803.
175. Kitano Y. Ibaraki disease in cattle. In: Coetzer JAW, Tustin RC (eds) *Infectious Diseases of Livestock*, 2nd edn. Cape Town: Oxford University Press Southern Africa, 2004, pp. 1221–1226.
176. Swanepoel R. Palyam serogroup orbivirus infections. In: Coetzer JAW, Tustin RC (eds) *Infectious Diseases of Livestock*, 2nd edn. Cape Town: Oxford University Press Southern Africa, 2004, pp. 1252–1255.
177. Miura Y, Kubo M, Goto Y, Kono Y. Hydranencephaly–cerebellar hypoplasia in a newborn calf after infection of its dam with Chuzan virus. *Jpn J Vet Sci* 1990;52:689–694.
178. Goto Y, Miura Y, Kono Y. Epidemiological survey of an epidemic of congenital abnormalities with hydranencephaly–cerebellar hypoplasia syndrome of calves occurring in 1985/86 and seroepidemiological investigations on Chuzan virus, a putative causal agent of the disease, in Japan. *Jpn J Vet Sci* 1988;50:405–413.
179. Lefèvre P, Shimshony A. Rift Valley fever. In: Lefèvre PC, Blancou J, Chermette R, Uilenberg G (eds) *Infectious and Parasitic Diseases of Livestock*. Paris: Lavoisier, 2010, pp. 633–648.

180. Gerdes G. Rift Valley fever. *Vet Clin North Am Food Anim Pract* 2002;18:549–555.
181. Swanepoel R, Coetzer J. Rift Valley fever. In: Coetzer JAW, Tustin RC (eds) *Infectious Diseases of Livestock*, 2nd edn. Cape Town: Oxford University Press Southern Africa, 2004, pp. 1037–1070.
182. Gerdes G. Rift Valley fever. In: Edwards S, Caporale V, Schmitt B et al. (eds) *OIE Manual of Diagnostic Tests and Vaccines for Terrestrial Animals*, 7th edn. Paris: OIE, 2012, pp. 283–293.
183. Saeed M, Li L, Wang H, Weaver S, Barrett A. Phylogeny of the Simbu serogroup of the genus Bunyavirus. *J Gen Virol* 2001; 82:2173–2181.
184. St George T, Kirkland P. Diseases caused by Akabane and related Simbu-group viruses. In: Coetzer JAW, Tustin RC (eds) *Infectious Diseases of Livestock*, 2nd edn. Cape Town: Oxford University Press Southern Africa, 2004, pp. 1029–1036.
185. Hoffmann B, Scheuch M, Höper D et al. Novel orthobunyavirus in cattle, Europe, 2011. *Emerg Inf Dis* 2012;18:469–472.
186. Gibbens N. Schmallenberg virus: a novel viral disease in Northern Europe. *Vet Rec* 2012;170:58.
187. Lefèvre P. Akabane disease. In: Lefèvre PC, Blancou J, Chermette R, Uilenberg G (eds) *Infectious and Parasitic Diseases of Livestock*. Paris: Lavoisier, 2010, pp. 657–662.
188. Herder V, Wohlsein P, Peters M, Hansmann F, Baumgärtner W. Salient lesions in domestic ruminants infected with the emerging so-called Schmallenberg virus in Germany. *Vet Pathol* 2012;49:588–591.
189. Garigliany M, Bayrou C, Kleijnen D et al. Schmallenberg virus: a new Shamonda/Sathuperi-like virus on the rise in Europe. *Antivir Res* 2012;95:82–87.
190. Oem J, Lee K, Kim H et al. Bovine epizootic encephalomyelitis caused by Akabane virus infection in Korea. *J Comp Pathol* 2012;147:101–105.
191. Konno S, Nakagawa M. Akabane disease in cattle: congenital abnormalities caused by viral infection. Experimental disease. *Vet Pathol* 1982;19:267–279.
192. Tsuda T, Yoshida K, Ohashi S et al. Arthrogryposis, hydranencephaly and cerebellar hypoplasia syndrome in neonatal calves resulting from intrauterine infection with Aino virus. *Vet Res* 2004;35:531–538.
193. Beer M, Conraths F, Van Der Poel W. "Schmallenberg virus": a novel orthobunyavirus emerging in Europe. *Epidemiol Infect* 2013;141:1–8.
194. Veronesi E, Henstock M, Gubbins S et al. Implicating *Culicoides* biting midges as vectors of Schmallenberg virus using semi-quantitative RT-PCR. *PLoS ONE* 2013;8:e57747.
195. ProMED-mail. Schmallenberg virus - Europe (76) virus RNA in bovine semen. http://wwwpromedmailorg/ Archive Number: 201212201460864. Accessed February 1, 2013.
196. ProMED-mail. Schmallenberg virus - Europe (40) no risk to humans, EU. http://wwwpromedmailorg/ Archive Number: 201205111130349. Accessed February 1, 2013.
197. De Regge N, Van Den Berg T, Georges L, Cay B. Diagnosis of Schmallenberg virus infection in malformed lambs and calves and first indications for virus clearance in the fetus. *Vet Microbiol* 2013;162:595–600.
198. Swanepoel R, Coetzer J. Wesselsbron disease. In: Coetzer JAW, Tustin RC (eds) *Infectious Diseases of Livestock*, 2nd edn. Cape Town: Oxford University Press Southern Africa, 2004, pp. 987–994.
199. Thomson G. Bovine parvovirus infection. In: Coetzer JAW, Tustin RC (eds) *Infectious Diseases of Livestock*, 2nd edn. Cape Town: Oxford University Press Southern Africa, 2004, pp. 815–816.
200. Storz J, Young S, Carroll E, Bates R, Bowen R, Keney D. Parvovirus infection of the bovine fetus: distribution of infection, antibody response, and age-related susceptibility. *Am J Vet Res* 1978;39:1099–1102.
201. Manteufel J, Truyen U. Animal bocaviruses: a brief review. *Intervirology* 2009;51:328–334.
202. Swift B, Trueblood M. The present status of the role of the parainfluenza-3 (PI-3) virus in fetal disease of cattle and sheep. *Theriogenology* 1974;2:101–107.
203. Dunne H, Ajinkya S, Bubash G, Griel C. Parainfluenza 3 and bovine enteroviruses as possible important causative factors in bovine abortion. *Am J Vet Res* 1973;34:1121–1126.
204. Dunne H, Huang C, Lin W. Bovine enteroviruses in the calf: an attempt at serologic, biologic, and pathologic classification. *J Am Vet Med Assoc* 1974;164:290–294.
205. Blas-Machado U, Saliki J, Boileau M et al. Fatal ulcerative and hemorrhagic typhlocolitis in a pregnant heifer associated with natural bovine enterovirus type-1 infection. *Vet Pathol* 2007;44:110–115.
206. Blas-Machado U, Saliki J, Sánchez S et al. Pathogenesis of a bovine enterovirus-1 isolate in experimentally infected calves. *Vet Pathol* 2011;48:1075–1084.
207. Dubey J. Neosporosis. In: Coetzer JAW, Tustin RC (eds) *Infectious Diseases of Livestock*, 2nd edn. Cape Town: Oxford University Press Southern Africa, 2004, pp. 382–393.
208. Dubey J, Schares G. Neosporosis in animals: the last five years. *Vet Parasitol* 2011;180:90–108.
209. Dubey J. Recent advances in *Neospora* and neosporosis. *Vet Parasitol* 1999;84:349–367.
210. Dubey J, Carpenter J, Speer C, Topper M, Uggla A. Newly recognized fatal protozoan disease of dogs. *J Am Vet Med Assoc* 1988;192:1269–1285.
211. Dubey J, Jenkins M, Rajendran C et al. Gray wolf (*Canis lupus*) is a natural definitive host for *Neospora caninum*. *Vet Parasitol* 2011;181:382–387.
212. Anderson M, Blanchard P, Barr B, Dubey J, Hoffman R, Conrad P. *Neospora*-like protozoan infection as a major cause of abortion in California dairy cattle. *J Am Vet Med Assoc* 1991;198:241–244.
213. Thilsted J, Dubey J. Neosporosis-like abortions in a herd of dairy cattle. *J Vet Diagn Invest* 1989;1:205–209.
214. Davison H, Otter A, Trees A. Significance of *Neospora caninum* in British dairy cattle determined by estimation of seroprevalence in normally calving cattle and aborting cattle. *Int J Parasitol* 1999;29:1189–1194.
215. Anderson M, Reynolds J, Rowe J et al. Evidence of vertical transmission of *Neospora* sp infection in dairy cattle. *J Am Vet Med Assoc* 1997;210:1169–1172.
216. Moen A, Wouda W, Mul M, Graat E, Van Werven T. Increased risk of abortion following Neospora caninum abortion outbreaks: a retrospective and prospective cohort study in four dairy herds. *Theriogenology* 1998;9:1301–1309.
217. Williams D. Bovine neosporosis. In: Lefèvre PC, Blancou J, Chermette R, Uilenberg G (eds) *Infectious and Parasitic Diseases of Livestock*. Paris: Lavoisier, 2010, pp. 1795–1806.
218. Barr B, Anderson M, Sverlow K, Conrad P. Diagnosis of bovine fetal *Neospora* infection with an indirect fluorescent antibody test. *Vet Rec* 1995;137:611–613.
219. Wouda W, Brinkhof J, Van Maanen C, De Gee A, Moen A. Serodiagnosis of neosporosis in individual cows and dairy herds: a comparative study of three enzyme-linked immunosorbent assays. *Clin Diagn Lab Immunol* 1998;5: 711–716.
220. Anderson M, Andrianarivo A, Conrad P. Neosporosis in cattle. *Anim Reprod Sci* 2000;60–61:417–431.
221. Dubey J, Schares G. Diagnosis of bovine neosporosis. *Vet Parasitol* 2006;140:1–34.
222. Hillman R, Gilbert R. Reproductive diseases. In: Divers TJ, Peek S (eds) *Rebhun's Diseases of Dairy Cattle*, 2nd edn. Philadelphia: WB Saunders, 2008, pp. 395–446.
223. Irons P, Schutte A, Herr S, Kitching J. Trichomonosis. In: Coetzer JAW, Tustin RC (eds) *Infectious Diseases of Livestock*,

2nd edn. Cape Town: Oxford University Press Southern Africa, 2004, pp. 305–315.
224. Rhyan J, Blanchard P, Kvasnicka W, Hall M, Hanks D. Tissue-invasive *Tritrichomonas foetus* in four aborted bovine fetuses. *J Vet Diagn Invest* 1995;7:409–412.
225. Rhyan J, Stackhouse L, Quinn W. Fetal and placental lesions in bovine abortion due to *Tritrichomonas foetus*. *Vet Pathol* 1988;25:350–355.
226. Gardiner C, Fayer R, Dubey J. *An Atlas of Protozoan Parasites in Animal Tissues*. Washington, DC: American Registry of Pathology, 1998.
227. Abbitt B, Rae D. Protozoal abortion in cattle. In: Youngquist RS, Threlfall WR (eds) *Current Therapy in Large Animal Theriogenology 2*. Philadelphia: WB Saunders, 2007, pp. 409–413.
228. Knudtson W, Kirkbride C. Fungi associated with bovine abortion in the northern plains states (USA). *J Vet Diagn Invest* 1992;4:181–185.
229. McCausland I, Slee K, Hirst F. Mycotic abortion in cattle. *Aust Vet J* 1987;64:129–132.
230. Walker R. Mycotic bovine abortion. In: Youngquist RS, Threlfall WR (eds) *Current Therapy in Large Animal Theriogenology 2*. Philadelphia: WB Saunders, 2007, pp. 417–419.

Chapter 55

Infectious Agents: *Campylobacter*

Misty A. Edmondson

Department of Clinical Sciences, College of Veterinary Medicine, Auburn University, Auburn, Alabama, USA

Introduction

Bovine venereal campylobacteriosis is caused by *Campylobacter fetus* subsp. *venerealis* and is an important cause of infertility, death of the late embryo or early fetus, and sporadic abortion in cattle. This organism was first recognized in the early 1900s as a cause of abortion in cattle and was initially called *Vibrio fetus*.[1] Based on differences in clinical presentation the agent was further classified as *Vibrio fetus* subsp. *intestinalis* and *Vibrio fetus* subsp. *venerealis*. *Vibrio fetus* subsp. *intestinalis* was associated with sporadic abortions due to bacteremia from the intestinal tract of the pregnant cow. *Vibrio fetus* subsp. *venerealis* was also associated with occasional abortions, but the primary importance of infection was a significant decrease in herd fertility. However in the 1960s there was evidence that many of the *Vibrio* species were sufficiently different to warrant classification under a separate genus, *Campylobacter*.[2] Consequently *Vibrio fetus* subsp. *intestinalis* was renamed *Campylobacter fetus* subsp. *fetus* and *Vibrio fetus* subsp. *venerealis* was renamed *Campylobacter fetus* subsp. *venerealis*. The term "vibrio" is still commonly used to refer to the venereal disease caused by *Campylobacter fetus* subsp. *venerealis*.

Etiology

Campylobacter fetus subsp. *venerealis*, a slender, motile, Gram-negative, comma-shaped or S-shaped rod that appears as "seagull wings" in silhouette, requires a microaerophilic environment for growth.[2,3] *Campylobacter* spp. require an atmosphere of 10–20% CO_2 with reduced oxygen concentration at 5% or less and optimal growth occurring at 37 °C.[2] There are two biotypes within the subspecies, biotype *venerealis* and biotype *intermedius*. These biotypes are differentiated by the production of hydrogen sulfide with lead acetate strips when organisms are grown in media containing cysteine. Biotype *intermedius* produces hydrogen sulfide, whereas biotype *venerealis* does not. Both biotypes are capable of causing disease; however, biotype *intermedius* appears to be less pathogenic than biotype *venerealis*.[4]

Pathogenesis

Campylobacter fetus subsp. *venerealis* is primarily transmitted by the venereal route. Bulls serve as carriers and are the major mode of transmitting the organism to naive cows. The rate of transmission from infected bulls to naive cows is quite high, approaching 100%. The bacterium may be spread to cows and heifers at the time of artificial insemination (AI) with infected semen. Direct transmission from cow to cow is unlikely, although contaminated insemination equipment may allow for transmission of the organism.[5] Campylobacteriosis may be spread from bull to bull if contaminated semen collection equipment is not sufficiently cleaned between bulls.[4] In addition, transmission has been reported to occur from bull to bull due to contaminated bedding.[6]

Bull

It is uncertain if the age of the bull has a role in the course and duration of infection. One study demonstrated no difference in duration of infection between bulls 5.5–6 years of age and 3.5–4 years of age.[7] However, another study revealed a higher incidence of infection in bulls greater than 6 years of age.[8] It has been speculated that younger bulls may be more resistant to infection, more likely to clear infection, and clear infections within a few weeks. In addition, it has also been speculated that older bulls are more likely to become persistent carriers due to an increase in the number and depth of crypts in the preputial epithelium.[4] However, more recent work evaluating the surface architectural anatomy of bulls revealed no differences in penile and preputial epithelial anatomy between young and old bulls.[9]

The prepuce of bulls is the natural reservoir of infection of naive cows through natural mating.[10] The addition of carrier bulls poses a significant risk to a naive herd, although the disease is most commonly seen in newly introduced cows and heifers in herds where bovine campylobacteriosis is endemic.[11]

Cow/fetus

Venereal campylobacteriosis causes low reproductive efficiency with severe economic losses.[12] This is associated with temporary infertility with mild endometritis, death of the early embryo or early fetus (30–70 days of gestation), and sporadic abortions.[13,14] After exposure the bacteria colonize in the mucosa of the anterior vagina and cervix of cows.[10] The infection then spreads to the uterus and oviducts under the influence of progesterone. Fertilization and development of the embryo are not directly affected by *C. fetus* subsp. *venerealis*. The resulting inflammatory response in the uterus and/or oviducts results in early embryonic death. The majority of pregnancy loss occurs between 30 and 70 days of gestation.[13,15] Reactions associated with heat-stabile endotoxins produced by *C. fetus* subsp. *venerealis* are thought to be responsible for the occasional mid-term abortion.[4] In addition, *C. fetus* subsp. *venerealis* can reach the placenta and the amniotic fluid and cause severe placentitis and fetal hypoxia.[16] Cows and heifers are normally capable of mounting an immune response, clearing the infection, and returning to normal fertility 4–8 months after the initial infection.[13]

Immune response

Infection with *C. fetus* subsp. *venerealis* in cows and bulls does not produce significant systemic antibodies during or after infections.[17] Unlike bulls, cows and heifers infected with *C. fetus* subsp. *venerealis* produce a local antibody response in the vagina and uterus. Heifers challenged with *C. fetus* subsp. *venerealis* had transient agglutinating and immobilizing IgM antibodies followed by persistent IgA and IgG antibodies in vaginal secretions and predominantly IgG antibodies in uterine secretions.[17,18] Both IgA and IgG antibodies to *C. fetus* subsp. *venerealis* surface antigens immobilize and mediate neutrophil and mononuclear cell killing of the bacteria.[19] The production of IgG in the uterus directs clearance through opsonization and phagocytosis. In the lower reproductive tract, IgA is also produced and helps immobilize the bacteria and block adherence. Conversely, IgA may also be responsible for blocking opsonization and promoting persistent colonization of the lower genital tract. Because of the stronger antibody response in cows and heifers, prolonged infections in females are not as persistent as they are in bulls.[18,20] However, in the case of the chronic carrier cow, the bacterium may be able to elude the immune response and be maintained in the cervix through a full gestation, although this is quite unusual (<1%).[6] Persistence of infection in the cervix and vagina in the presence of an immune response may be partly due to antigenic changes in the organism that allow for avoidance of the host immune response. Antigenic changes in the S-layer, a regular array of high-molecular-mass surface proteins that compose the outermost part of the cell envelope, result in the expression of different immunodominant epitopes and persistence of the organism.[4]

Clinical signs

Herd

A herd history of poor conception rates, infertility, or second trimester abortions may be one of the first signs that bovine venereal campylobacteriosis could be a problem. Trichomoniasis, poor detection of estrus, and bull subfertility or infertility should also be considered as possible differential diagnoses. A confirmatory diagnosis of bovine venereal campylobacteriosis should be attempted through laboratory evaluation of samples obtained from the herd. Samples for diagnosis include tissues from aborted fetuses, vaginal mucus samples, vaginal washings, or preputial scrapings.

When beginning a diagnostic evaluation of a herd, it seems advantageous to begin with the breeding males in the herd due to the fewer number of males and the self-limiting type of infection associated with females. Thus preputial scrapings should be obtained from herd bulls. If bulls are not available, then vaginal mucus samples should be obtained from a representative sample of cows and heifers and submitted to the diagnostic laboratory.

Cow

In females the main clinical signs are related to infertility. Venereal campylobacteriosis is characterized by repeat breeding, irregular estrous cycles, prolonged calving seasons, and reduced calf crops.[13,21] The average number of services per conception is extended to 2.5–3.5. When infection is introduced into a naive herd, the initial pregnancy rates are quite low. Over time the conception rate of the herd improves as animals clear the infection. However, the resulting calving season is extended as more cows conceived later in the season. Infected cows show few signs of infection. Most infected cows do not have any vaginal discharge. However, a mucopurulent secretion may be detected in the occasional cow.[4] As previously stated, the majority of pregnancy loss occurs between 30 and 70 days of gestation.[13,15] Although uncommon, abortion due to *C. fetus* subsp. *venerealis* has been reported between 4 and 8 months of gestation.[13] Abortions appear to be more commonly associated with biotype *intermedius* than with biotype *venerealis*. When mid-term abortions due to *C. fetus* subsp. *venerealis* are reported, there is typically a recent history of infertility in the herd.[4]

Bull

Infection with *C. fetus* subsp. *venerealis* in bulls is not apparent as bulls show no clinical signs of infection. Although genital infections in mature bulls may persist for months or years, there are no alterations in semen quality.[22–24]

Pathology

The pathology associated with infection of *C. fetus* subsp. *venerealis* is limited to the reproductive tract. Endometritis, mild cervicitis, and salpingitis are usually the only lesions seen in cows and heifers.[25] Grossly the cervix may appear reddened and the uterus may have a small amount of mucopurulent exudate that can extend through the cervix into the vagina. Histologically, the lesions have mild infiltration of inflammatory cells (plasma cells and foci of lymphocytes in the stroma) with slight desquamation of the superficial epithelium with no significant vascular changes.[25] Involvement of the endometrial glands is minimal, although cystic glands with slight periglandular fibrosis have been noted on clearance of infection.[25]

No gross abnormalities are seen on the penis and prepuce of infected bulls. Histologically, there is only a diffuse infiltration of mononuclear cells within the lamina propria.[22] Plasma cells are located in clusters at the apex of the dermal papillae.[7] However, the infiltration of subepithelial lymphocytes and plasma cells has been found in both infected and uninfected bulls.[22,24,26]

The most common fetal lesions associated with *C. fetus* subsp. *venerealis* are neutrophilic bronchopneumonia and interstitial pneumonia.[16] Other commonly observed lesions include nonsuppurative interstitial enteritis, hepatitis, pericarditis, myositis, myocarditis, serositis, occasional abomasitis, and meningitis.[16,27,28]

Diagnosis

It is important to be able to differentiate *C. fetus* subsp. *venerealis* from *C. fetus* subsp. *fetus*. The natural habitat for *C. fetus* subsp. *fetus* is the intestinal tract of cattle, but it can also cause sporadic abortions.[29] Differences in the epidemiology and clinical relevance of these bacteria make accurate detection imperative. The traditional means for differentiating between both species, and the recommended international standard, is based on tolerance to 1% glycine;[30] *C. fetus* subsp. *venerealis* is sensitive to glycine whereas *C. fetus* subsp. *fetus* is tolerant to glycine. However, the glycine tolerance characteristic may not be completely reliable for differentiation of the two subspecies as it has been demonstrated that glycine tolerance can be acquired by transduction or mutation.[31] In addition, glycine-tolerant variants of *C. fetus* subsp. *venerealis* have also been described and designated as *C. fetus* subsp. *venerealis* biotype *intermedius*.[30,32] Glycine-sensitive *C. fetus* subsp. *fetus* isolates have also been detected as well.[33]

Sample collection
Bull

A simple method for collecting preputial smegma is to collect a preputial scraping. The most common method for collecting a preputial scraping involves using a plastic AI pipette, which is fast, easy, and inexpensive.[14] The tip of the AI pipette is placed into the proximal preputial fornix. Scraping of the prepuce should be somewhat vigorous in an effort to loosen epithelial cells from the penile and preputial surfaces and include 10–15 strokes. A 12-mL syringe should be attached to the distal end of the AI pipette, and the clinician should apply negative pressure to the pipette throughout the scraping procedure.[14] A small amount of hemorrhage may be seen following preputial scraping but usually does not indicate any harm to the bull. As the pipette is being withdrawn from the preputial cavity, the negative pressure on the pipette should be released gently. An adequate sample consists of a 2–5cm column of cloudy smegma within the pipette. A plastic sheath may be used to minimize contamination prior to introducing the pipette into the preputial cavity.

Cow

A cranial vaginal mucus sample can be obtained from the cow by using an AI pipette with a plastic sheath to minimize contamination. A 12-mL syringe is placed on the distal end of the AI pipette, which is then placed into the cranial vaginal fornix of the cow. Negative pressure is placed on the syringe while the pipette is gently moved within the vaginal fornix. Collection of vaginal mucus samples should be performed around the time of estrus. Culturing of vaginal mucus samples is considered most useful diagnostically when samples are obtained early in the course of infection before immunity develops.[4]

Fetus

Samples that should be collected from an aborted fetus include abomasal fluid, lung, liver, heart, intestine, and placenta. Abomasal fluid from aborted fetuses serves as the best diagnostic sample. Fetal membranes may also be useful, but diagnosis is often complicated due to the presence of contaminating bacteria.[4]

Sample transport

The handling of preputial scrapings, vaginal mucus samples, or samples from aborted fetuses is critical for accurate diagnosis of *C. fetus* subsp. *venerealis*. In order to optimize recovery of the organism, it is best to have samples at the laboratory within 6 hours of sample collection. *Campylobacter* spp. require a complex support medium, antimicrobials to suppress growth of commensals and contaminants, and near anaerobic conditions.[34] The success of culture of *Campylobacter* spp. is directly related to the choice and availability of media. Several different transport enrichment media (TEM) are available for isolation and culture of samples for *C. fetus* subsp. *venerealis*. These TEM include Weybridge medium, Cary–Blair medium, Clark's selective medium, Australian TEM, Blaser–Wang medium, Lander's transport medium, Campy-Bac, and Thomann transport and enrichment medium.[3,14,35,36] It is best for clinicians to contact their local diagnostic laboratory for preferred media and sample handling prior to collection and submission of samples. In many cases the laboratory will supply the transport media and instructions for its use. Once in the TEM, the sample is safe, to some degree, from small changes in temperatures. If an inoculated TEM is to be in transit more than 4 hours, it is recommended that the sample not be incubated above room temperature until it reaches the diagnostic laboratory.[14,37]

Diagnostic techniques
Antigen or antibody detection

Antigen detection techniques are most often used to identify an active carrier of venereal campylobacteriosis in bulls, although antibody detection may be more useful for detection of the organism in cows. The detection of antigen includes the fluorescent antibody test, which is most commonly used for preputial scrapings from bulls.[35] The sensitivity and specificity are reported to be 93% and 88.9%, respectively.[38] An immunohiostochemical technique has also been developed, with a sensitivity of 94% for detecting *C. fetus* in aborted fetal tissues.[39]

Systemic antibody responses are not helpful because they are often due to nonpathogenic *Campylobacter* spp. One specific antibody detection method is the vaginal mucus agglutination test to detect antibodies in preputial scrapings, vaginal mucus, or aborted fetal fluids.[40] However, due to the low sensitivity of the test, it is best used for herd evaluation utilizing vaginal mucus samples collected between 37 and 70 days after infection.[3] In addition, the specificity of this test may be negatively affected by the presence of blood or an unrelated inflammation.[4] A vaginal mucus sample that causes complete agglutination at a dilution of 1 : 25 is considered a positive diagnosis.[4]

Specific IgG antibodies have been detected via enzyme-linked immunosorbent assay (ELISA) in order to measure the humoral immune response in cattle following vaccination against *C. fetus*.[41,42] An ELISA has also been used to detect IgA antibodies in vaginal mucus from cows following abortion.[43] In addition, a monoclonal antibody-based ELISA has been developed as a screening test for detecting *C. fetus* subsp. *venerealis* in preputial scrapings and vaginal mucus samples.[44,45]

Isolation in culture

Campylobacter spp. are not easily cultured and isolated because they are fastidious and have special requirements for growth in media.[3,46] Once at the laboratory the TEM is streaked onto selective media and incubated in a microaerophilic atmosphere (6% O_2, 7% CO_2, 7% H_2, 80% N_2) at 37°C. The media plates should be incubated for at least 6 days before being called negative. In many cases colonies of *C. fetus* subsp. *venerealis* will be visible within 3 days of inoculation. Colonies are round, raised, and 1–3 mm in diameter. These colonies may initially appear clear but become opaque on further incubation. Phenotypic characteristics are most commonly used for identification. *Campylobacter fetus* subsp. *venerealis* isolates are oxidase positive, catalase positive, resistant to nalidixic acid but sensitive to cephalothin, and grow at 25 °C but not at 42 °C. It is imperative to differentiate between *C. fetus* subsp. *venerealis* and *C. fetus* subsp. *fetus* due to potential sample contamination. These two subspecies of *C. fetus* only differ in that *C. fetus* subsp. *venerealis* does not grow in the presence of 1% glycine whereas *C. fetus* subsp. *fetus* does grow in 1% glycine.[4] As previously mentioned, the glycine tolerance characteristic may not be completely reliable for differentiation of the two subspecies as it has been demonstrated that glycine tolerance can be acquired by transduction or mutation.[31]

Direct examination via darkfield microscopy and Gram's staining of abomasal contents from aborted fetuses may be a useful additional tool for diagnosing abortions due to *C. fetus* subsp. *venerealis*. Darkfield microscopy may allow observation of the morphologic traits and darting motility of *C. fetus*. Gram staining reveals a Gram-negative, comma-shaped or S-shaped rod that appears as "seagull wings."[2,3] *Campylobacter* spp. are more easily visualized by Gram's stain when counterstained with carbolfuchsin rather than safranin, the standard counterstain.[2–4] Darkfield microscopy and Gram staining may not be as beneficial for other samples.

PCR

Molecular methods for differentiating between *C. fetus* subsp. *venerealis* and *C. fetus* subsp. *fetus* have been developed to validate the results of phenotypic differentiation in order to avoid misidentification.[29] Most of these PCR assays use the colonies grown from the cultured vaginal mucus or preputial scrapings.[14] However, Thomann transfer and enrichment medium was developed for the enrichment of *C. fetus* subsp. *venerealis* and to allow direct PCR amplification from media if overgrowth with contaminants occurs.[36] Differentiation of *C. fetus* spp. is achieved by a specific PCR assay.[47] This assay utilizes a species-specific 764-bp amplicon that is produced with primers MG3F and MG4R with the DNA of both species of *C. fetus*. Subsequently, primers VenSF and VenSR are used for differentiation. The identification of *C. fetus* subsp. *venerealis* is based on the presence of a 142-bp amplicon that is not formed by *C. fetus* subsp. *fetus*.[47] However, the reliability of this PCR assay has been called into question[30] due to misidentification in up to 10% of strains.[48] Therefore this PCR assay may best be used to confirm traditional phenotypic characterization of *C. fetus*.

Other studies have focused on the use of a multiplex PCR for differentiation of *C. fetus* spp.[48] Multiplex PCR consists of multiple primer sets within a single PCR mixture. These primer sets produce amplicons of varying sizes that are specific to different DNA sequences, which are observed as distinct bands when visualized by gel electrophoresis. However, utilization of multiplex PCR will require further validation with regard to a larger pool of clonal strains.[49,50]

Cloning and nucleotide sequencing, with and without PCR, are becoming more useful methods for genotyping bacterial isolates.[51] In fact a novel 5' *Taq* nuclease PCR assay using a 3' minor groove binder DNA probe has been developed for the direct detection of *C. fetus* subsp. *venerealis* in clinical samples from cattle.[52] Molecular methods to discriminate between the two subspecies of *C. fetus* have also utilized the discovery of an 80-kb genomic sequence (consisting of type IV secretion system [T4SS] components, putative plasmid genes, other proteins) and an insertion element (ISCfe1) unique to *C. fetus* subsp. *venerealis*.[53,54]

Prevention, control, and treatment
Prevention

Vaccination programs are the most effective means of preventing campylobacteriosis in a cattle herd and should include vaccination of bulls as well as cows and heifers.[13] Initial immunization of females requires two injections of bacterin, with the second injection coming within 3 weeks of breeding. Annual boosters should also be administered within 3 weeks prior to breeding.[14] This pre-breeding schedule of vaccination increases antibody levels (IgG_1 and IgG_2)

at the time when exposure is most likely to occur and helps protect against uterine infection.[14]

Many different *C. fetus* vaccines are available for use in cattle, including both oil-adjuvanted and aluminum hydroxide-absorbed vaccines.[13] Bacterins in oil adjuvant are more effective and provide a longer duration of immunity and protection after a single dose.[55] Two commercial aluminum hydroxide-absorbed vaccines were evaluated in herds in Argentina and were found to be ineffective at preventing reproductive loss due to *C. fetus*.[13,56] Subcutaneous administration of a single dose of an oil-adjuvanted *Campylobacter* vaccine demonstrated protection in heifers from reproductive losses. However, oil-adjuvanted vaccines cause localized granulomas and granulomas at the injection site.[13]

Vibrin (Pfizer Animal Heath) is the only oil-adjuvanted *C. fetus* bacterin available in the United States that has been evaluated in bulls.[13] Two 5-mL doses, which is 2.5 times the dosage recommended for vaccination of cows, should be administered subcutaneously to breeding bulls at 4-week intervals beginning 8 weeks prior to the breeding season. Annual revaccination at the same dose is recommended 4 weeks prior to the breeding season.[13,57]

Control

Elimination of natural breeding and the use of AI is the best means of controlling bovine venereal campylobacteriosis. However, producers often find this a difficult recommendation to follow. If AI is used the semen should be adequately screened for *C. fetus* subsp. *venerealis* and have antibiotics added to the semen extender to ensure that contaminated semen does not allow reintroduction of the bacterium into the herd. If AI is not possible, the previous battery of bulls should be replaced with virgin bulls.[14]

Treatment

Because of the lack of availability and effectiveness of certain antibiotics (streptomycin), the treatment of animals infected with *C. fetus* subsp. *venerealis* has focused on immunotherapy to clear the infection, which involves immunizing animals with an appropriately adjuvanted antigen. Systemic immunization of cows that have been actively or passively immunized with *C. fetus* subsp. *venerealis* has cured these animals of the vaginal carrier state.[24,58,59] In addition, systemic immunization may also cure, or clear infection for 9 weeks or more, in some bulls infected with *C. fetus* subsp. *venerealis*.[23,60,61] Immunization induces both local and systemic IgG antibody production, which prevents microbial adherence, immobilizes the bacteria, activates complement, and opsonizes bacteria for phagocytosis.[24] The oil-adjuvanted vaccine preparations are most often used for therapeutic immunotherapy.

References

1. Karmali M, Skirrow M. Taxonomy of the genus *Campylobacter*. In: Butzler JP (ed.) *Campylobacter Infection of Man and Animals*. Boca Raton, FL: CRC Press, 1984, p. 1.
2. Timoney JF, Gillespie JH, Scott FW, Barlough JE. The genus *Campylobacter*. In: *Hagan and Bruner's Microbiology of Infectious Diseases of Domestic Animals*, 8th edn. Ithaca, NY: Cornell University Press, 1988, pp. 153–160.
3. Mshelia G, Amin J, Woldehiwet R, Murray R, Egwu G. Epidemiology of bovine venereal campylobacteriosis: geographic distribution and recent advances in molecular diagnostic techniques. *Reprod Domest Anim* 2010;45:e221–e230.
4. Walker R. Bovine venereal campylobacteriosis. In: Howard JL, Smith RA (eds) *Current Veterinary Therapy 4: Food Animal Practice*. Philadelphia: WB Saunders, 1999, pp. 323–326.
5. Office International des Epizooties, 2005. Campylobacteriosis. Center for Food Security and Public Health. Institute for International Cooperation in Animal Biologics, Iowa State University College of Veterinary Medicine, Ames, Iowa. Available at http://www.cfsph.iastate.edu/ (December 2012).
6. Dekeyser P. Bovine genital campylobacteriosis. In: Morrow DA (ed.) *Current Therapy in Theriogenology*, 2nd edn. Philadelphia: WB Saunders, 1986, pp. 263–266.
7. Bier P, Hall C, Duncan J, Winter A. Experimental infections with *Campylobacter fetus* in bulls of different ages. *Vet Microbiol* 1977;2:13–27.
8. Wagner W, Dunn H, Vanvleck L. Incidence of vibriosis in an AI stud. *Cornell Vet* 1965;55:209–220.
9. Strickland L, Wolfe D, Carson R et al. Surface architectural anatomy of the penile and preputial epithelia of bulls. *Clin Theriogenol* 2011;3:362 (Abstract).
10. Mshelia M, Singh J, Amin J, Woldehiwet Z, Egwu G, Murray R. Bovine venereal campylobacteriosis: an overview. *CAB Rev Perspect Agr Vet Sci Nutr Nat Resources* 2007;2:080. Available at http://www.cabi.org/cabreviews/?page=4051&reviewid=80000&site=167
11. Roberts J. Infertility in the cow. In: Roberts S (ed.) *Veterinary Obstetrics and Genital Diseases (Theriogenology)*, 2nd edn. Woodstock, VT: published by the author, 1971, pp. 376–506.
12. Campero C. Las enfermedades reproductivas de los bovinos: ayer y hoy. *Acad Nac Agron Vet An* 2000;53:88–112.
13. Givens M. A clinical, evidence-based approach to infectious causes of infertility in beef cattle. *Theriogenology* 2006;66:648–654.
14. BonDurant R. Venereal diseases of cattle: natural history, diagnosis, and the role of vaccines in their control. *Vet Clin North Am Food Anim Pract* 2005;21:383–408.
15. Peter D. Bovine venereal diseases. In: Youngquist RS (ed.) *Current Therapy in Large Animal Theriogenology*. Philadelphia: WB Saunders, 1997, pp. 355–363.
16. Morrell E, Barbeito C, Odeon C, Gimeno E, Campero C. Histopathological, immunological, lectinhistochemical and molecular findings in spontaneous bovine abortions by *Campylobacter fetus*. *Reprod Domest Anim* 2011;46:309–315.
17. Corbeil L, Schurig G, Duncan J, Wilkie B, Winter A. Immunity in the female bovine reproductive tract based on the response to *Campylobacter fetus*. *Adv Exp Med Biol* 1981;137:729–743.
18. Corbeil L, Schurig G, Duncan J, Corbeil R, Winter A. Immunoglobulin classes and biological functions of *Campylobacter (Vibrio) fetus* antibodies in serum and cervicovaginal mucus. *Infect Immun* 1974;10:422–429.
19. Corbeil L, Corbeil R, Winter A. Bovine venereal vibriosis: activity of inflammatory cells in protective immunity. *Am J Vet Res* 1975;36:403–406.
20. Schurig G, Hall C, Corbell L, Duncan J, Winter A. Bovine venereal vibriosis: cure of genital infection in females by systemic immunization. *Infect Immun* 1975;11:245–251.
21. Zemjanis R. Vaccination for reproductive efficiency in cattle. *J Am Vet Med Assoc* 1974;165:689–692.
22. Bier P, Hall C, Duncan J, Winter A. Measurement of immunoglobulin in reproductive tract fluid of bulls. *Vet Microbiol* 1977;2:1–12.

23. Vasquez L, Ball L, Bennett B *et al.* Bovine genital campylobacteriosis (vibriosis): vaccination of experimentally infected bulls. *Am J Vet Res* 1983;44:1553–1557.
24. Cobo E, Corbeil L, BonDurant R. Immunity to infections in the lower genital tract of bulls. *J Reprod Immunol* 2011;89:55–61.
25. Hoffer M. Bovine campylobacteriosis: a review. *Can Vet J* 1981;22:327–330.
26. Samuelson J, Winter A. Bovine vibriosis: the nature of the carrier state in the bull. *J Infect Dis* 1966;116:581–592.
27. Campero C, Moore D, Odeon A, Cipolla A, Odriozola E. Aetiology of bovine abortion in Argentina. *Vet Res Commun* 2003;27:359–369.
28. Anderson M. Infectious causes of bovine abortion during mid- to late gestation. *Theriogenology* 2007;68:474–486.
29. Schulze F, Bagon A, Mueller W, Hotzel H. Identification of *Campylobacter fetus* subspecies by phenotypic differentiation and PCR. *J Clin Microbiol* 2006;44:2019–2024.
30. Office International des Epizooties. Bovine genital campylobacteriosis. In: *OIE Manual of Diagnostic Tests and Vaccines for Terrestrial Animals*, 5th edn. Paris: OIE, 2004, pp. 439–450. Available at http://www.oie.int/ (December 2012).
31. Chang W, Ogg J. Transduction and mutation to glycine tolerance in *Vibrio fetus*. *Am J Vet Res* 1971;32:649–653.
32. Salama M, Garcia M, Taylor D. Differentiation of the subspecies of *Campylobacter fetus* by genomic sizing. *Int J Syst Bacteriol* 1992;42:446–450.
33. On S, Harrington C. Evaluation of numerical analysis of PFGE DNA profiles for differentiating *Campylobacter fetus* subsp. by comparison with phenotypic, PCR and 16S rDNA sequencing methods. *J Appl Microbiol* 2001;90:285–293.
34. Bryner J. Bovine abortion caused by *Campylobacter fetus*. In: Kirkbride C (ed.) *Laboratory Diagnosis of Livestock Abortion*. Ames, IA: Iowa State University Press, 1990, pp. 70–81.
35. Redwood D, Lander K, Gill K. Diagnostic techniques for campylobacteriosis in animals. Central Veterinary Laboratory, Weybridge, 1989.
36. Harwood L, Thomann A, Brodard I, Makaya P, Perreten V. *Campylobacter fetus* subspecies *venerealis* transport medium for enrichment and PCR. *Vet Rec* 2009;165:507–508.
37. Garcia M, Stewart R, Ruckerbauer G. Quantitative evaluation of a transport enrichment medium for *Campylobacter fetus*. *Vet Rec* 1984;115:434–436.
38. Figueiredo J, Pellegrin A, Foscolo C, Machada R, Miranda K, Lage A. Evaluation of the direct fluorescent antibody test for the diagnosis of bovine genital campylobacteriosis. *Rev Latinoam Microbiol* 2002;44:118–123.
39. Campero C, Anderson M, Walker R *et al.* Immunohistochemical identification of *Campylobacter fetus* in natural cases of bovine and ovine abortions. *J Vet Med B* 2005;52:138–141.
40. Lander K. The development of a transport and enrichment medium for *Campylobacter fetus*. *Br Vet J* 1990;146:327–333.
41. Hewson P, Lander K, Gill K. An enzyme-linked immunosorbent assay for the detection of antibodies to *C. fetus* in bovine vaginal mucus. *Res Vet Sci* 1985;38:41–45.
42. Repiso M, Baraibar M, Olivera M, Silveyra S, Battistoni J. Development and evaluation of an enzyme-linked immunosorbent assay for the quantification of humeral response of cattle vaccinated against *Campylobacter fetus*. *Am J Vet Res* 2002;63:586–590.
43. Hum S, Stephens L, Quinn C. Diagnosis by ELISA of bovine abortion due to *Campylobacter fetus*. *Aust Vet J* 1991;68:272–275.
44. Brooks B, Davenish J, Lutze-Wallace C, Milnes D, Robertson R, Berlie-Surujballi G. Evaluation of a monoclonal antibody-based enzyme-linked immunosorbent assay for the detection of *Campylobacter fetus* in bovine preputial washing and vaginal mucus samples. *Vet Microbiol* 2004;103:77–84.
45. Devenish J, Brooks B, Perry K *et al.* Validation of a monoclonal antibody-based capture enzyme-linked immunosorbent assay for detection of *Campylobacter* fetus. *Clin Diagn Lab Immunol* 2005;12:1261–1268.
46. Penner J. The genus campylobacter: a decade of progress. *Clin Microbiol Rev* 1988;1:157–172.
47. Mueller W, Hotzel H, Schulze F. Identifizierung und differenzierung der *Campylobacter fetus* subspeziesmittels PCR. *Tierarztl Wochenschr* 2003;110:55–59.
48. Hum S, Quinn K, Brunner J, On S. Evaluation of a PCR assay for identification and differentiation of *Campylobacter fetus* subspecies. *Aust Vet J* 1997;75:827–831.
49. Willoughby K, Nettleton P, Quirie M *et al.* A multiplex polymerase chain reaction to detect and differentiate *Campylobacter fetus* subsp *fetus* and *Campylobacter fetus* subsp *venerealis*: use on UK isolates of *C. fetus* and other *Campylobacter* spp. *J Appl Microbiol* 2006;99:756–766.
50. Mshelia G. *Bovine venereal campylobacteriosis and reproductive efficiency in the Lake Chad Basin of Nigeria*. PhD dissertation, University of Maiduguri, Nigeria, 2008.
51. Wassenaar T, Newell D. Genotyping of *Campylobacter* spp. *Appl Environ Microbiol* 2000;66:1–9.
52. McMillan L, Fordyce G, Doogan V, Lew A. Comparison of culture and a novel 5′ Taq nuclease assay for direct detection of *Campylobacter fetus* subsp. *venerealis* in clinical specimens from cattle. *J Clin Microbiol* 2006;44:938–945.
53. Moulhuijzen P, Lew-Tabor A, Wlodek B *et al.* Genomic analysis of *Campylobacter fetus* subspecies: identification of candidate virulence determinants and diagnostic assay targets. *BMC Microbiol* 2009;9:86.
54. Abril C, Vilei E, Brodard I, Burnens A, Frey J, Miserez R. Discovery of insertion element ISCfe1: a new tool for *Campylobacter fetus* subspecies differentiation. *Clin Microbiol Infect* 2007;13:993–1000.
55. Hoerlein A, Carroll E. Duration of immunity to bovine genital vibriosis. *J Am Vet Med Assoc* 1970;156:775–778.
56. Cobo E, Cipolla A, Morsella C, Cano D, Campero C. Effect of two commercial vaccines to *Campylobacter fetus* subspecies on heifers naturally challenged. *J Vet Med B* 2003;50:75–80.
57. Cortese V. Bovine reproductive disease vaccines. In: Smith BP (ed.) *Large Animal Internal Medicine*, 2nd edn. St Louis, MO: Mosby, 2002, pp. 1421–1423.
58. Berg R, Firehammer B, Border M, Myers L. Effects of passively and actively acquired antibody on bovine campylobacteriosis (vibriosis). *Am J Vet Res* 1979;40:21–25.
59. Schurig G, Hall C, Burda K, Corbeil L, Duncan J, Winter A. Infection patterns in heifers following cervicovaginal or intrauterine instillation of *Campylobacter* (*Vibrio*) *fetus venerealis*. *Cornell Vet* 1974;64:533–548.
60. Ball L, Dargatz D, Cheney J, Mortimer R. Control of venereal disease in infected herds. *Vet Clin North Am Food Anim Pract* 1987;3:561–574.
61. Bouters R, De Keyser J, Vandeplassche M *et al. Vibrio fetus* infection in bulls: curative and preventative vaccination. *Br Vet J* 1973;129:52–57.

Chapter 56

Infectious Agents: *Trichomonas*

Mike Thompson

Willow Bend Animal Clinic, Holly Springs, Mississippi, USA

Introduction

Trichomoniasis is an insidious venereal disease of cattle caused by the protozoan *Tritrichomonas foetus*. Reproductive failure due to *T. foetus* has been recognized worldwide during the past century. Bulls usually carry the organism asymptomatically, while females bear the brunt of the disease, most often suffering early embryonic loss. While females possess an adequate ability to rid themselves of the infection, bulls are likely to become chronic inapparent carriers, and this likelihood increases as the bull's age increases. Testing for *T. foetus* is problematic in both males and females, with multiple tests performed serially to attain satisfactory sensitivity and specificity. There is no legal effective treatment for trichomoniasis in the United States, leading to state regulations requiring removal of positive bulls from the herd. A vaccine is available, though the efficacy has been questioned, particularly in older bulls. Biosecurity plays an important role in prevention of reproductive failure in cattle herds. The terms "trichomoniasis" and "trichomonosis" are both used for *T. foetus* infections in cattle.

Characteristics of the organism

Tritrichomonas foetus is a motile flagellated protozoan measuring 10–15 × 5–10 μm. The trophozoite form has three anterior and one posterior flagellae and one undulating membrane along the long axis of the organism.[1,2] The pseudocyst form usually occurs when trophozoites are subjected to stressful conditions. During the transformation from trophozoite to pseudocyst, the organism loses its teardrop shape, becoming rounder, and internalizes its flagella. While existing as a pseudocyst, internal cellular structures continue to replicate resulting in a multinucleated organism.[1,2] The organism is microaerophilic, surviving best in areas of low oxygen tension. When observed in wet mount preparations, *T. foetus* moves in a rolling, jerky pattern.[1] Although the appearance and movement of the organism in culture media is quite characteristic, diagnosticians should be aware that trichomonads other than *T. foetus* may be present in cultures of preputial scrapings or cervical mucus. These organisms can be a cause of false-positive test results[1-7] (Figure 56.1).

Clinical signs

Clinical signs of trichomoniasis in the male are rare to nonexistent, although mild balanoposthitis has been associated with *T. foetus*. In the female, infections are characterized by early embryonic loss, post-coital pyometra, and occasionally abortion.[1,3,6,8] Trichomoniasis should be suspected in herds with low overall pregnancy rates and widely distributed gestational ages. History of exposure to nonvirgin bulls of unknown trichomoniasis status, along with signs of early embryonic loss and post-coital pyometra, should raise suspicion of *T. foetus* in herds with poor reproductive performance.[1,4,8]

Following infection, bulls may become inapparent carriers and fail to eliminate infection. It has been long-standing dogma that older bulls have different histological architecture of the penile and preputial epithelium compared with younger bulls, with crypts and folds deepening and enlarging as bulls age.[1] Although this has been related to older bulls more commonly becoming chronic carriers of *T. foetus*, Strickland et al.[9] showed no statistical difference in histological samples from younger bulls compared with those of older bulls with respect to preputial and penile epithelium and number of and area contained within epithelial folds. Infected cows and heifers mount an immune response and are often able to eliminate the organism 6–12 weeks following infection. Females less frequently develop a carrier state that persists throughout pregnancy and into the next breeding season or longer.[1,3,4,6,8]

Diagnosis

Diagnostic tests commonly utilized to detect *T. foetus* include culture with microscopic identification, and polymerase chain reaction (PCR).[1,3] The organism may occasionally be detected with microscopic examination of direct samples, but this method lacks sensitivity. For both culture and PCR, the sample is incubated in Diamond's media or, more

---

Bovine Reproduction, First Edition. Edited by Richard M. Hopper.
© 2015 John Wiley & Sons, Inc. Published 2015 by John Wiley & Sons, Inc.

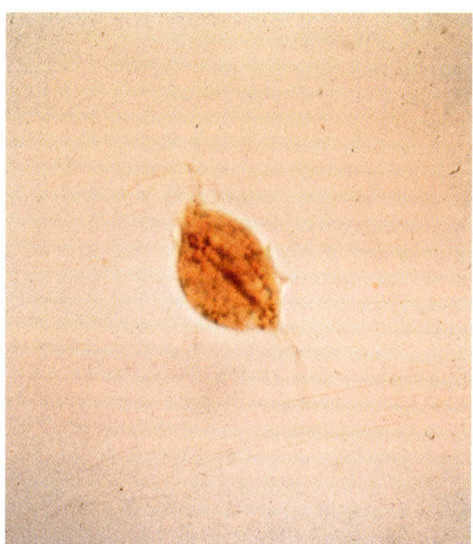

Figure 56.1 *Tritrichomonas foetus*. Courtesy of Maarten Drost.

commonly, in a commercially available, two-chambered, media-filled pouch (InPouch™).[4] For culture, the sample is incubated for up to 6 days at approximately 36 °C. The sample is inspected under a light microscope daily to check for the presence of *T. foetus*.[5] Identification of *T. foetus* in culture is based on the morphological and movement characteristics of the organism.[1]

Test sensitivity can be defined as the proportion of actual positive animals which are correctly identified. Test specificity can be defined as the proportion of actual negative animals which are correctly identified. Sensitivity and specificity of culture and PCR for *T. foetus* can be influenced by factors occurring at every level of the testing process, including sample acquisition, handling and shipping, as well as laboratory procedures and the type of PCR test utilized by the laboratory.[10] Reported sensitivity for a single culture ranges from 84 to 96% under experimental conditions, but may be as low as 65% in suboptimal conditions in the field.[7] Specificity of culture can be impaired by the presence of morphologically similar trichomonads.[1,3,5–7,10]

PCR allows the detection of small quantities of the genetic material of the target organism by amplification of specific DNA fragments unique to the organism.[10] PCR may also be used to confirm the identity of organisms recovered by culture to avoid false-positive results from morphologically similar trichomonads.[1,3,4,7,10,11] With PCR being able to detect smaller numbers of *T. foetus* than culture, samples may be handled in a different manner than culture.[10,11] These options include, but are not limited to, immediate media sample testing with PCR, incubation for 1–2 days in culture media followed by PCR, incubation for 1–2 days in culture media and then freezing the sample prior to shipping, and direct sample PCR testing without any type of culture media utilized.[5,7,10,12] Because the sensitivity and specificity of these different methods vary, the receiving laboratory should be consulted prior to shipment to attain the specific requirements of that laboratory.

Sensitivity and specificity of a single PCR test have been shown to be greater than those for a single culture, although both parameters can be improved by using serial testing, as well as combining PCR and culture tests on the same sample.[5] Comparison of different testing regimens utilizing different combinations of PCR and culture, as well as different field conditions and sample handling, reveals a wide range of sensitivities and specificities achieved.[3,5,7,12]

As with most testing regimens, increasing sensitivity can have a negative impact on specificity. Although all types of PCR tests offer specificities of greater than 90%, veterinarians and producers alike should be aware that false positives can occur, though this has been shown to be diminished by serial and/or combination testing.[5,7] Because of the possibility of false positives, veterinary, laboratory, and regulatory personnel should use proper discretion regarding animals that test positive, particularly until the test results have been confirmed.

Trichomoniasis is reportable to state boards of animal health and several states require nonvirgin bulls to have preputial samples tested either by one PCR or three weekly cultures. These regulations often change and they differ from state to state. Data from field studies utilizing common methods of testing, as well as from experimental epidemiological models, do not support the practice of equating one PCR test with culture of three weekly preputial samples.[5,7] Cobo *et al.*[7] showed sensitivity of a single culture to be 67.8%, while sensitivity of a single PCR was 65.9%. Combining both tests on a single sample increased the sensitivity to 78.3%, which was equivalent to two PCR tests taken 1 week apart. Ondrak *et al.*[5] reported real-time PCR to be less sensitive and specific than gel PCR when results were compared with samples taken from bulls which were either positive or negative on three cultures taken 1 week apart. Montilla *et al.*[12] showed excellent sensitivity utilizing a single gPCR when the samples were incubated for 24 hours and then frozen for shipment. Unpublished experimental epidemiological data formulated with models devised by Sanderson show a significant decrease in sensitivity and specificity when a single PCR is compared to three weekly cultures (M. Sanderson, personal communication). Because of the complexities involved in the studies cited, testing regimens should be tailored to the prevalence of trichomoniasis in the herd and surrounding area, the age and history of the bull being tested, individual state regulations, and the laboratory utilized.

Producers should be made aware of the limitations of these protocols and how the sensitivity and specificity can be improved by using more intensive protocols than the regulations may require.[5,7] Since one test protocol does not fit all situations, these protocols should be based on the risk involved, with increased levels of testing in areas of higher trichomoniasis incidence, as well as in individual bulls of higher risk such as older bulls known not to be virgin.[8,11] The practice of 6 weekly cultures has historically been considered the gold standard in the AI industry.[7] Any number of weekly cultures, PCR tests, or combinations of both should be employed to satisfy the risk level of the bull being tested, regardless of the minimum testing requirements to satisfy state regulations.

Proper sampling of a bull by a veterinarian requires training, attention to detail, and basic knowledge of the organism and how it behaves in the transport/culture media.[13] Materials required for sampling include an infusion pipette, 12-mL syringe, marking pen, receptacle to maintain trich pouches in an upright position after inoculation, scissors, disinfectant, paper towels, examination gloves,

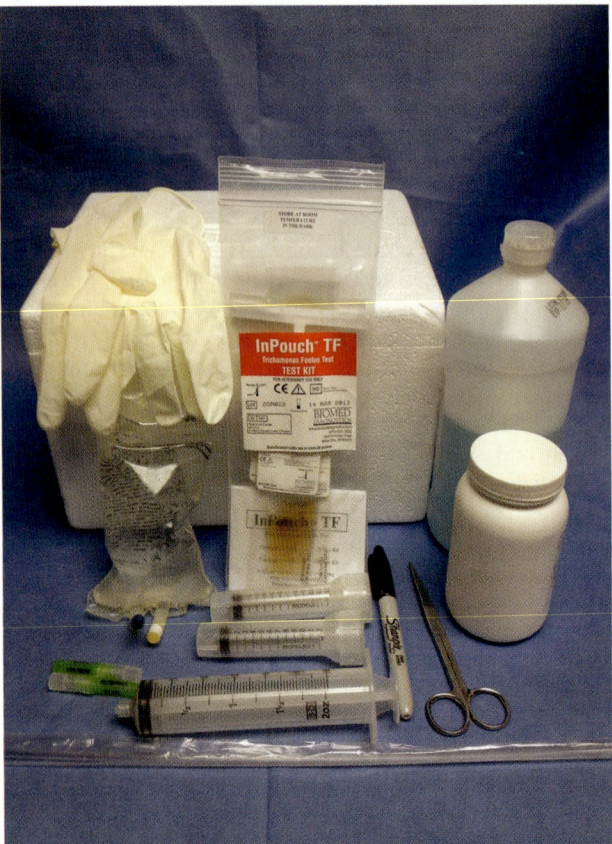

Figure 56.2 Testing supplies. Courtesy of Josh Thompson.

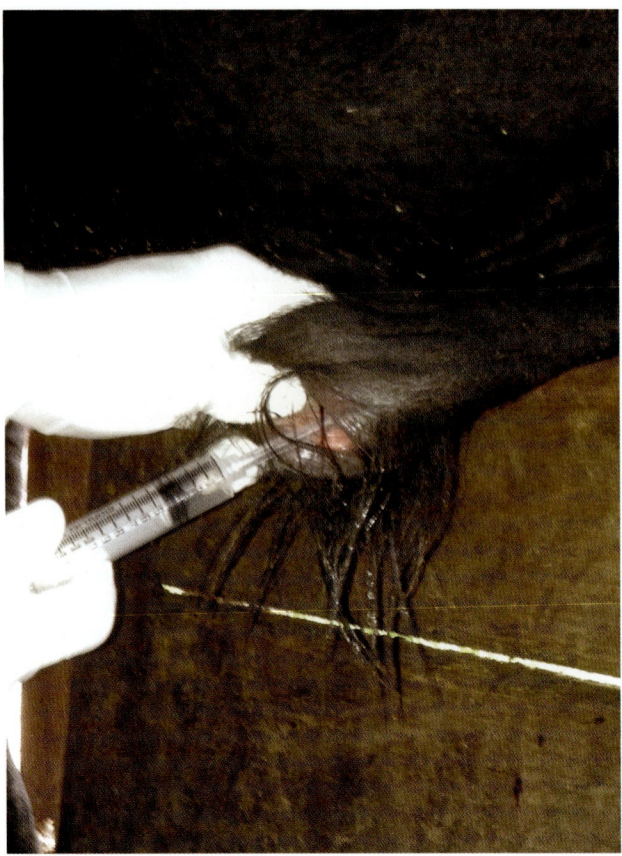

Figure 56.3 Sample collection. Courtesy of David Childers.

60-mL syringe, saline, InPouch™, secondary shipping containers, and an insulated shipping box[5,7,13] (Figure 56.2).

Bulls should be sexually rested for 1 week prior to testing. Sample collection begins with assimilating the necessary materials and restraining the bull in a manner to allow the sample to be taken correctly and safely. The hindleg nearest the practitioner can be tied if necessary for safety. The veterinarian dons a clean pair of latex gloves to reduce the possibility of cross-contamination between bulls. The hair is trimmed from the sheath and the preputial opening is dried with a paper towel. The prepuce may now be lavaged with a small amount of saline in bulls that are excessively dirty or dry. An infusion pipette is attached to a new 12-mL syringe. The pipette is preferably guarded with a plastic sleeve. The pipette is advanced into the prepuce approximately to the level of the fornix, and the prepuce is aggressively scraped while applying negative pressure to the attached syringe (Figure 56.3). Scraping should be aggressive enough to capture at least 1.5–2 cm of smegma within the pipette, but not so aggressive that the sample is contaminated with blood. The sample is then inoculated into a properly identified InPouch™. The pipette is inserted into the media in the top chamber of the InPouch™, aspirating and expelling the fluid several times to effectively wash the sample into the media. The pouch is then folded over several times according to manufacturer's recommendations to express the fluid and sample from the upper to the lower compartment of the pouch. The pouch is then closed with the wire ties and is placed in an upright position. The pouch should now be protected from direct sunlight and kept at temperatures between 20 and 37 °C until it arrives at the laboratory by hand delivery or overnight shipping. As the shipping requirements can vary between laboratories, the receiving laboratory should be consulted prior to testing.[4,5,7,10,13]

For shipping, the pouches should be placed in a secondary shipping container that is not easily collapsed, and this is placed into an insulated container that is properly labeled for biological products to be shipped overnight.[7,13] These samples should be shipped to a veterinary diagnostic laboratory that adheres to guidelines that match or exceed those set forth by the American Association of Veterinary Laboratory Diagnosticians. These laboratories are encouraged to utilize the best testing methods available, and utilize outside laboratories for confirmation as needed.

Recently, new devices for sampling bulls have become available. A device known as a bull rasper or Tricamper™ marketed in Australia, and a device known as Trichit™ which is marketed in the United States are both introduced into the preputial cavity as with the pipette and used to scrape the tissues. These devices are then rinsed into the culture media to extract the sample. These devices may allow better sampling with less blood contamination. Females may also be sampled with a pipette or either of these devices to collect vaginal mucus for culture or PCR[14] (Figure 56.4).

Since proper sampling and handling techniques are directly related to the labile nature of the organism, the Society for Theriogenology encourages veterinarians to receive proper training before submitting T. foetus samples.[15] Several states require or are considering training and certification in T. foetus sampling before accepting samples from veterinarians.

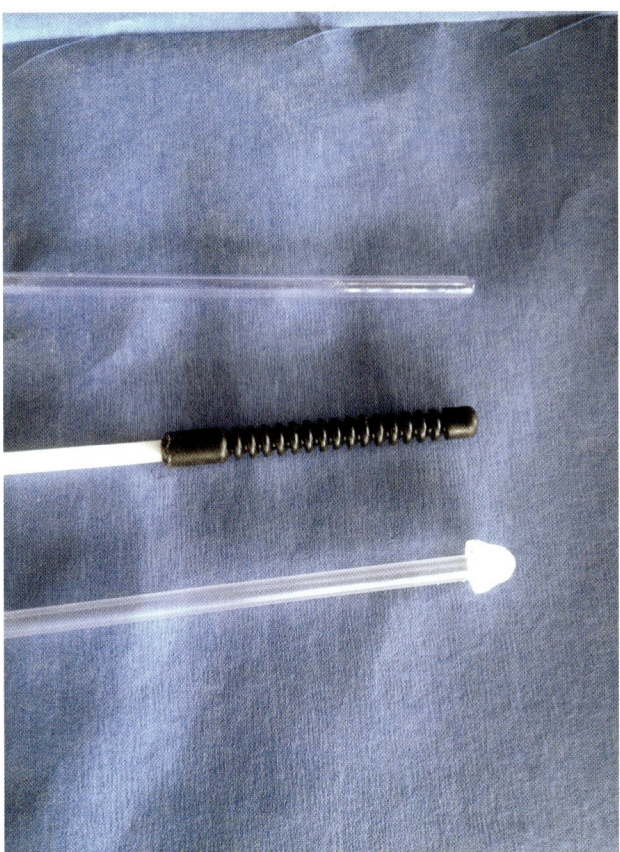

Figure 56.4 Sample collection devices. Courtesy of David Childers.

In herds known to be infected with *T. foetus*, serial testing coupled with management changes should be utilized to establish a disease-free herd.[5] As there is no legal or effective treatment for bulls infected with *T. foetus*, these bulls should be slaughtered.[1,5] Disposition of infected animals should be coordinated with regulatory officials. There is no legal or effective treatment for cows infected, but since females commonly clear the organism, options other than slaughter may be considered.[1,3,6] Positive cows can be maintained in a separate herd and bred by AI. In herds of known positive incidence, open cows should be culled, aborted fetuses should be tested for *T. foetus*, and vaccination can be used to limit the impact of the disease.[1,16]

Conclusion

In 1947, Dr Dave Bartlett summarized practical trichomoniasis control measures for naturally bred cattle into two basic principles: keep from breeding susceptible cows to infected bulls, and keep from breeding susceptible bulls to infected cows.[17] These principles form the basis for trichomoniasis control today and can be accomplished through attention to detail, knowledge of the disease, and common sense. Commingling of cattle from different herds should be avoided. A limited breeding season with timely pregnancy diagnosis can increase the chances of identifying a problem earlier. Older bulls should be replaced with younger bulls, and all bulls should be tested before introduction to the herd and prior to each breeding season.

Cows found open at pregnancy diagnosis should be culled. Cows that abort or have a noticeable discharge should be culled and aborted fetuses or any discharge should be tested for *T. foetus*. There is a vaccine for *T. foetus*, and although the efficacy of the vaccine has historically been regarded as limited, it has been shown to reduce the impact of the disease in females. Recent work by Cobo et al.[16] indicates vaccination of bulls may be of benefit as well. AI using semen from bulls tested negative for trichomoniasis obviously limits exposure to the disease, though AI personnel should adhere to rigid disease prevention procedures. Vaccination for other reproductive diseases will make it easier to discount a trichomoniasis problem in a herd with reproductive failure.

As with many other herd health problems, a team approach including producers, veterinarians, and regulatory personnel is the best way to develop and implement epidemiologically sound biosecurity protocols to prevent introduction of venereal trichomoniasis. Proper testing is paramount to success in preventing introduction and also clearing infected herds. The value of testing is directly related to sample quality, which in turn is directly related to proper sample acquisition, handling, and shipping.

References

1. BonDurant R. Venereal diseases of cattle: natural history, diagnosis, and the role of vaccines in their control. *Vet Clin North Am Food Anim Pract* 2005;21:383–408.
2. Edmondson M. Managing bovine trichomoniasis in the female. *Clin Theriogenol* 2013;5:225–230.
3. Corbeil B, Campero M, Hoosear V, BonDurant H. Detection of trichomonad species in the reproductive tracts of breeding and virgin bulls. *Vet Parasitol* 2008;154:226–232.
4. Peter D, Fales W, Miller R et al. *Tritrichomonas foetus* infection in a herd of Missouri cattle. *J Vet Diagn Invest* 1995;7:278–280.
5. Ondrak J, Keen J, Rupp G, Kennedy J, McVey D, Baker W. Repeated testing by use of culture and PCR assay to detect *Tritrichomonas foetus* carrier bulls in an infected Nebraska herd. *J Am Vet Med Assoc* 2010;237:1068–1073.
6. Ho M, Conrad P, Conrad P, LeFebvre R, Perez E, BonDurant R. Detection of bovine trichomoniasis with a specific DNA probe and PCR amplication system. *J Clin Microbiol* 1994;32:98–104.
7. Cobo E, Favetto P, Lane V et al. Sensitivity and specificity of culture and PCR of smegma samples of bulls experimentally infected with *Tritrichomonas foetus*. *Theriogenology* 2007;68:853–860.
8. Rae D, Crews J, Greiner E, Donovan G. Epidemiology of *Tritrichomonas foetus* in beef bull populations in Florida. *Theriogenology* 2004;61:605–618.
9. Strickland L, Wolfe D, Carson R et al. Surface architectural anatomy of the penile and preputial epithelia of bulls. *Clin Theriogenol* 2011;3:362 (Abstract).
10. Mukhufhi N, Irons P, Michel A, Peta F. Evaluation of a PCR test for the diagnosis of *Tritrichomonas foetus* infection in bulls: effects of sample collection method, storage and transport medium on the test. *Theriogenology* 2003;60:1269–1278.
11. BonDurant R, Campero C, Anderson M, Van Hoosear K. Detection of *Tritrichomonas foetus* by polymerase chain reaction in cultured isolates, cervicovaginal mucus, and formalin-fixed tissues from infected heifers and fetuses. *J Vet Diagn Invest* 2003;15:579–584.
12. Montilla H, Estill C, Villarroel A, Bounpheng M. Proceeding of the American Association of American Veterinary Diagnosticians,

52nd Annual Conference, October 7–14, 2009, San Diego, CA. Retrieved January 1, 2011 from www.aavld.org/assets/documents/2009%20Proceedings%20Book%20_9-23-09.pdf

13. Parker S, Campbell J, Ribble C, Gajadhar A. Sample collection factors affect the sensitivity of the diagnostic test for *Tritrichomonas foetus* in bulls. *Can J Vet Res* 2003;67:138–141.

14. McMillan L, Lew A. Improved detection of *Tritrichomonas foetus* in bovine diagnostic specimens using a novel probe-based real time PCR assay. *Vet Parasitol* 2006;141:204–215.

15. Position Statement on *Tritrichomonas foetus* Testing. Society for Theriogenology. Available at http://www.therio.org/?page=PositionStatement#TRitri

16. Cobo E, Corbiel L, Gershwin C, BonDurant R. Preputial cellular and antibody responses of bulls vaccinated and/or challenged with *Tritrichomonas foetus*. *Vaccine* 2010;28:361–370.

17. Bartlett D, Hanson E, Teeter K. Occurrence of *Tritrichomonas foetus* in preputial samples from infected bulls. *J Am Vet Med Assoc* 1947;110(839):114–120.

Chapter 57

Infectious Agents: Leptospirosis

Daniel L. Grooms

*Department of Large Animal Clinical Sciences, College of Veterinary Medicine,
Michigan State University, East Lansing, Michigan, USA*

Introduction

Bovine leptospirosis is recognized as a cause of significant reproductive losses including infertility, early embryonic death, abortions, stillbirths, and birth of weak calves. In addition, several other nonreproductive clinical manifestations of leptospirosis have been described.[1] Leptospirosis occurs worldwide and is caused by infection with the spirochete bacteria belonging to the genus *Leptospira*. *Leptospira* can infect all mammalian species including humans and is a significant zoonotic pathogen.[2]

Leptospira can be classified on several different levels. The most definitive classification is based on genomospecies which is established by genetic sequence analysis. *Leptospira* are also classified into serovars based on antigenic relatedness. Over 200 different serovars of pathogenic *Leptospira* have been identified.[3] Serovars can further be defined by genotypic analysis, for example there are two genotypes of serovar Hardjo that are important in cattle, serovar Hardjo type hardjoprajitno and type hardjo-bovis.

Epidemiology of leptospirosis

Prevalence

Leptospira are found worldwide, though different serovars may only be endemic in certain geographic regions. The most common serovars in the United States are Hardjo, Grippotyphosa, and Pomona; however, other serovars can also be found and have been associated with reproductive losses.

Serologic surveys have repeatedly shown that the most common cause of leptospirosis in cattle is infection with serovar Hardjo.[4-7] As mentioned earlier, two genetically distinct types of serovar Hardjo have been identified: serovar Hardjo type hardjo-bovis and serovar Hardjo type hardjoprajitno. Serovar Hardjo type hardjo-bovis is common in cattle populations throughout the world, including North America,[5] while type hardjoprajitno has been found primarily in the United Kingdom.

Limited studies in the United States estimate herd prevalence of serovar Hardjo to be near 60% in dairies[6] and 40% in beef herds.[7] In Ontario, Canada, 8% of dairy herds and 44% of beef herds surveyed that were not vaccinated against leptospirosis were found to have cows serologically positive for serovar Hardjo,[8] whereas in western Canada seroprevalence appears to be much less in the beef herds surveyed there.[9] Herd prevalence may vary regionally with higher rates more likely in temperate climates.[7,10] In a large survey of cull cows sampled at slaughter in 49 states, the percentage of cattle serologically positive for a *Leptospira* serovar was much higher in southern tier and west coast states than other areas of the United States.[4] In studies of individual adult cows, seroprevalence to serovar Hardjo ranges from 10 to 30% in both the United States and Canada. Higher seroprevalence rates can be expected in individual herds that are endemically infected.

Accurate data for the frequency of abortions attributable to leptospirosis are lacking in North America. In Northern Ireland, leptospirosis was recognized as being responsible for over half of all bovine abortions, with serovar Hardjo being identified as the causative agent in 97% of those in one study.[11] In the previously cited study performed in Ontario, Canada, serovar Hardjo was associated with 6% of abortions while Pomona abortions were not recognized.[8] In the United States, multiple diagnostic laboratory surveys of cases in which there was a definitive diagnosis found that *Leptospira* was identified in less than 10% of aborted fetuses.[12] Because of the overall difficulty in making a definitive diagnosis in bovine abortions cases in general, it is likely these surveys underestimate the true prevalence of *Leptospira*-related abortions.

Maintenance versus incidental hosts

Animals can be divided into maintenance hosts and incidental hosts of *Leptospira*. Different *Leptospira* serovars are associated with one or more maintenance host(s), which serve as transmission reservoirs. Animals may be maintenance hosts of some serovars but incidental or accidental hosts

Table 57.1 Important maintenance hosts for different *Leptospira* serovars that commonly cause infection in cattle in North America.

Leptospira serovar	Maintenance host
Bratislava	Pigs, mice, horses
Canicola	Dogs
Grippotyphosa	Raccoons, opossums, skunks
Hardjo	Cattle, deer
Icterohaemorrhagiae	Rats
Pomona	Pigs, raccoons, opossums, skunks

Table 57.2 Summary of clinical differences between infections with Hardjo and non-Hardjo serovars.

	Serovar Hardjo	Non-Hardjo serovars (e.g., Pomona and Grippotyphosa)
Presentation in adult cows	Subclinical to mild signs	Subclinical to mild signs
Persistence of infection	6–18 months	<6 months
Infertility	Common	Uncommon
Abortion	Occurs with initial infection	Occurs with initial infection
Time to abortion after infection	Months after infection	Weeks after infection
Herd presentation of abortion	Sporadic abortions	Sporadic abortions or abortion storms
Stillbirths	Yes	Yes
Weak calves at birth	Yes	Yes
Persistent infection in newborn calves	Yes	No

of others. In a maintenance host, the disease is maintained by chronic infection of the renal tubules. Transmission of the infection among maintenance hosts is efficient and the incidence of infection is relatively high at a population (or herd) level. Incidental hosts may become infected by direct or indirect contact with the maintenance host. Incidental hosts are not important reservoirs of infection and the rate of transmission is low. Transmission of the infection from one incidental host to another is relatively uncommon. Examples of important maintenance hosts for different *Leptospira* serovars that commonly cause infection in cattle in North America are listed in Table 57.1.

Transmission

Transmission of *Leptospira* can occur either directly or indirectly. Transmission among maintenance hosts, such as serovar Hardjo among cattle, can be both direct and indirect and involves contact with infected urine, placental fluids, or milk. In addition, the infection can be transmitted venereally or transplacentally. Infection of incidental hosts, such as cattle becoming infected with serovar Grippotyphosa originating from raccoons, occurs more commonly by indirect contact with environments contaminated with urine of maintenance hosts.

Leptospira can survive in the environment for days to weeks outside of the animal under warm moist conditions. The organism survives only briefly in hot dry conditions or at freezing temperatures. Because of these characteristics, leptospirosis occurs more commonly during the spring and fall and in temperate climates.[6]

Pathogenesis

Leptospira gains entrance to the body through mucous membranes and water-softened or abraded skin. After a 3–20 day incubation period, leptospiremia occurs and the organism is disseminated to many organs including the renal system and genital tract. During this period, acute clinical disease may be seen. Antibodies can be detected in serum soon after leptospiremia begins and coincides with the clearance of *Leptospira* from blood and most organs. However, organisms can persist in the kidneys for several weeks in an incidental host, or 6 months or more in a maintenance host.[5,6] *Leptospira* can also persist in the genital tract of a maintenance host for similar periods of time.[6] *Leptospira* excretion in urine is variable and may last weeks in incidental hosts and months in maintenance hosts.

In cattle, serovar Pomona shedding has been observed for up to 15 weeks under experimental conditions,[13] while serovar Hardjo shedding typically is for 6–8 months[14] but has been reported to be as long as 18 months.[15]

Clinical disease

Clinical outcomes of *Leptospira* infection in cattle will depend on the infecting serovar, level of immunity, and physiological state. A summary of clinical reproductive differences between infections with Hardjo and non-Hardjo serovars can be found in Table 57.2.

Infections by serovar Hardjo typically result in subclinical or mild clinical signs. Infection of cattle with serovar Hardjo commonly leads to a renal carrier state associated with long-term urinary shedding; urinary shedding is the major mode of transmission between cattle. Persistent infection of the male and female genital tract with serovar Hardjo also occurs. The most significant outcome of serovar Hardjo infection is decreased reproductive efficiency. Initial infections with serovar Hardjo during pregnancy can result in abortion, stillbirth, or birth of weak calves. Abortions may occur many weeks after infection of the dam and are usually not associated with any obvious illness in the cow. Infected but apparently healthy calves may also be born.[16] Abortions due to serovar Hardjo infection tend to occur sporadically as opposed to abortion "storms" which may occur as a result of infection with other non-Hardjo serovars such as Pomona or Grippotyphosa.

Persistent infection of the reproductive tract with serovar Hardjo can lead to infertility and early embryonic death. In infected herds, increased services per conception and prolonged calving interval have been described.[17–20]

Infections with non-Hardjo *Leptospira* serovars are typically subclinical. If infection occurs during pregnancy, it can result in abortion, stillbirth, or birth of weak calves. Abortions are more common with non-Hardjo serovar infections. Abortion of autolyzed fetuses typically occurs 1–4 weeks after infection and without any overt clinical signs in the dam.

Diagnosis of *Leptospira* infections

Diagnosis of leptospirosis-related reproductive losses is dependent on putting together history, clinical signs, and results of diagnostic testing. Important information to consider include history of vaccination (type and timing of vaccine), risk of exposure to wildlife and other domestic animals, recent herd additions, and clinical signs. Because *Leptospira* are known to cause reproductive sequelae in cattle, identification of the infection by detection of the organism or significant levels of antibody in the dam or in the herd is considered indicative that leptospirosis may be involved in the reproductive problems seen. Unfortunately, because of the lability of the organism and poor condition of many fetuses when expelled, identification of the organism in the placenta or fetal tissues can be challenging. Persistent diagnostic efforts are often warranted to get a definitive diagnosis of leptospirosis.

Diagnostic tests for leptospirosis include those designed to detect antibodies against the organism and those designed to detect the organism or its DNA in tissues or body fluids (urine) of cattle. Each of the diagnostic procedures – for detection of the organism or for antibodies directed against the organism – has a number of advantages and disadvantages. Use of a combination of tests allows maximum sensitivity and specificity in establishing the diagnosis.

Serologic assays are the most commonly used technique for diagnosing leptospirosis in animals. Serology is inexpensive, reasonably sensitive, and widely available. The microscopic agglutination test (MAT) is the most common serologic assay conducted and involves mixing appropriate dilutions of serum with live *Leptospira* of relevant serovars.[21] Detection of high titers of antibody in animals with a clinical disease and history consistent with leptospirosis may be sufficient to establish the diagnosis. This is particularly true in the investigation of abortions caused by incidental host serovars (i.e., serovars Pomona, Grippotyphosa, Canicola, and Icterohaemorrhagiae) in which the dam's agglutinating antibody titer is often 1600 or greater. In contrast, serovar Hardjo-infected cattle often have a poor antibody response to infection; anti-Hardjo antibody titers may be quite low or negative at the time of abortion. It should also be noted that antibody titers are often at peak levels at the time of abortion and persist for months following infection, making paired serology around the time of abortion unlikely to demonstrate significant changes in titers.

Interpretation of *Leptospira* serologic results is complicated by a number of factors including cross-reactivity of antibodies and antibody titers induced by vaccination. Antibodies produced in an animal in response to infection with a given serovar of *Leptospira* often cross-react with other serovars. In general, the infecting serovar is assumed to be the serovar to which that animal develops the highest antibody titer. However, with serovar Hardjo, a significant percentage of cattle that are actively infected and shedding *Leptospira* have anti-Hardjo antibody titers of 100 or less.[22] Therefore, a low antibody titer does not necessarily rule out a diagnosis of leptospirosis, especially infections caused by serovar Hardjo.

Widespread vaccination of cattle for leptospirosis also complicates the interpretation of *Leptospira* serology. In general, cattle develop relatively low antibody titers (100–400) in response to vaccination and these titers persist for 1–3 months after vaccination, whereas titers after natural infection will typically be much higher.[22]

Leptospira organisms or DNA in tissues or body fluids can be detected by immunofluorescence, immunohistochemistry, culture, and polymerase chain reaction (PCR) assays. Immunofluorescence can be used to identify *Leptospira* in tissues (fetal liver, lung, kidney, liver, or placenta) or urine sediment. The test is rapid, has reasonable sensitivity, and can be used on frozen samples. The use of special stains on formalin-fixed tissues can also be effective for identification of *Leptospira* histologically. Tissues to be examined include kidney in adults, and fetal placenta, lung, liver, and kidney in the case of abortions. Bacteriologic culture of urine or tissue specimens is the definitive method for the diagnosis of leptospirosis. Leptospires are usually present in the urine of cattle for 10 days after the onset of clinical signs. Culture of *Leptospira* is difficult and is not practiced widely. The use of PCR is widely available, although procedures may vary between laboratories.[23] PCR is a sensitive and specific technique for the diagnosis of leptospirosis when appropriate samples are collected. In general, PCR testing of urine is more reliable than testing of tissues. Collecting urine midstream following the use of furosemide increases the sensitivity of both culture and PCR.[24] Furosemide increases the glomerular filtration rate thereby flushing more organisms into the urine. In addition, urine dilution enhances survival of *Leptospira* for culture purposes.

Control of leptospirosis

Given the widespread exposure to wildlife and domestic animals that may be carriers of leptospirosis and the high prevalence of serovar Hardjo infection in cattle, prevention of all exposure to leptospirosis is virtually impossible in most dairy and beef operations. Therefore, control of leptospirosis in cattle requires a comprehensive approach that includes reducing the risk of *Leptospira* exposure, vaccination, and selective antimicrobial treatment to clear carriers of serovar Hardjo when indicated.

Limiting direct and indirect contact between cattle and carriers of *Leptospira* should be a priority. Strategies would include controlling wildlife and domestic species known to be maintenance hosts of serovars that commonly infect cattle (Table 57.1). Strategies should focus on reducing access of both wild and domestic animals to cattle feeds, especially those in storage such as hay, grain, and other commodity feedstuffs. Control measures should also attempt to limit cattle access to standing surface water such as swamps or ponds to which wildlife commonly have access. Quarantine procedures can be implemented to reduce the risk of introducing serovar Hardjo into a herd through purchase of infected cattle. During quarantine, steps can be taken to eliminate carrier states of serovar Hardjo in new cattle including antimicrobial treatment and vaccination.

Vaccination is commonly practiced in North America to prevent leptospirosis. Whole cell bacterins containing serovars Pomona, Canicola, Icterohaemorrhagiae, Grippotyphosa, and Hardjo ("5-way" vaccines) are widely available and commonly used. In general, these vaccines provide good protection against disease induced by each of the serovars, except serovar Hardjo.[16,25] New-generation leptospirosis vaccines specifically targeting serovar Hardjo have demonstrated superior ability to prevent renal infection, urinary shedding, and fetal infection.[26–28] These new vaccines induce a strong and long-lasting cell-mediated immune response compared with cattle naturally infected with serovar Hardjo and cattle vaccinated with nonprotective serovar Hardjo vaccines.[29,30]

A general vaccination strategy for leptospirosis would include a whole cell bacterin directed against serovars Pomona, Canicola, Icterohaemorrhagiae, Grippotyphosa and a new-generation vaccine directed against serovar Hardjo that has been shown to be effective in reducing the risk of renal infection, urinary shedding, and fetal infection. Immunity should be induced in replacement breeding cattle following the manufacturer's recommendations and then revaccination is recommended every 6–12 months.

Antimicrobials can be used to treat individual cattle including those harboring persistent serovar Hardjo infections. Antibiotic treatment stops urinary shedding and is likely to improve clinical signs associated with persistent colonization of the reproductive tract with serovar Hardjo.[31,32] Long-acting oxytetracycline, tilmicosin, tulathromycin, and ceftiofur have been shown to be effective in the treatment of serovar Hardjo infections. Antimicrobial treatment should be combined with vaccination to prevent reinfection from occurring.

References

1. Callan R. Leptospirosis. In: Smith (ed.) *Large Animal Internal Medicine*, 4th edn. St Louis, MO: Mosby, 2009, pp. 967–971.
2. Guerra M. Zoonosis "Update: Leptospirosis." *J Am Vet Med Assoc* 2009;234:472–478.
3. Levett P. Leptospirosis. *Clin Microbiol Rev* 2001;14:296–326.
4. Miller D, Wilson M, Beran G. Survey to estimate prevalence of *Leptospira interrogans* infection in mature cattle in the United States. *Am J Vet Res* 1991;52:1761–1765.
5. Wikse S. Update on *Leptospira hardjo-bovis* control in beef herds. In: *Proceedings of the Annual Meeting of the American Association of Bovine Practitioners*, 2006, pp. 79–87.
6. Bolin C. Diagnosis and control of bovine leptospirosis. In: *Proceedings of the 6th Western Dairy Management Conference*, 2003, pp. 155–159.
7. Wikse S, Rogers G, Ramachandran S *et al*. Herd prevalence and risk factors of *Leptospira* infection in beef cow/calf operations in the United States: *Leptospira borgpetersenii* serovar Hardjo. *Bov Pract* 2007;41:15–23.
8. Prescott J, Miller R, Nicholson V, Martin S, Lesnick T. Seroprevalence and association with abortion of leptospirosis in cattle in Ontario. *Can J Vet Res* 1988;52:210–215.
9. Van De Weyer L, Hendrick S, Rosengren L, Waldner C. Leptospirosis in beef herds from western Canada: serum antibody titers and vaccination practices. *Can Vet J* 2011;52:619–626.
10. Miller D, Wilson M, Beran G. Relationships between prevalence of *Leptospira interrogans* in cattle, and regional, climatic, and seasonal factors. *Am J Vet Res* 1991;52:1766–1768.
11. Ellis, W, O'Brien J, Bryson D, Mackie D. Bovine leptospirosis: some clinical features of serovar Hardjo infection. *Vet Rec* 1985;117:101–104.
12. Anderson M. Infectious causes of bovine abortion during mid- to late-gestation. *Theriogenology* 2007;68:474–486.
13. Ferguson L, Ramge J, Sanger V. Experimental bovine leptospirosis. *Am J Vet Res* 1957;18:43.
14. Leonard F, Quinn P, Ellis W, O'Farrell K. Duration of urinary excretion of leptospires by cattle naturally or experimentally infected with *Leptospira interrogans* serovar hardjo. *Vet Rec* 1992;131:435–439.
15. Thiermann A. Experimental Leptospiral infections in pregnant cattle with organisms of the Hebdomadis serogroup. *Am J Vet Res* 1982;43:780–784.
16. Bolin C, Thiermann A, Handsaker A, Foley J. Effect of vaccination with a pentavalent leptospiral vaccine on *Leptospira interrogans* serovar hardjo type hardjo-bovis infection of pregnant cattle. *Am J Vet Res* 1989;50:161–165.
17. Dhaliwal G, Murray R, Ellis W. Reproductive performance of dairy herds infected with *Leptospira interrogans* serovar hardjo relative to the year of diagnosis. *Vet Rec* 1996;138:272–276.
18. Dhaliwal G, Murray R, Dobson H, Montgomery J, Ellis W. Reduced conception rates in dairy cattle associated with serological evidence of *Leptospira interrogans* serovar Hardjo infection. *Vet Rec* 1996;139:110–114.
19. Dhaliwal G, Murray R, Dobson H, Montgomery J, Ellis W. Effect of vaccination against *Leptospira interrogans* serovar Hardjo on milk production and fertility in dairy cattle. *Vet Rec* 1996;138:334–335.
20. Ellis W. Effects of leptospirosis on bovine reproduction. In: Morrow DA (ed.) *Current Therapy in Theriogenology*, 2nd edn. Philadelphia: WB Saunders, 1986, pp. 267–271.
21. Bolin C. Leptospirosis. In: *OIE Manual of Diagnostic Tests and Vaccines for Terrestrial Animals*, 5th edn. Geneva, Switzerland: Office Internationale des Epizooties, 2004, pp. 316–327.
22. Grooms D, Bolin C. Diagnosis of fetal loss caused by bovine viral diarrhea virus and *Leptospira* spp. *Vet Clin North Am Food Anim Pract* 2005;21:463–472.
23. Wagenaar J, Zuerner R, Alt D, Bolin C. Comparison of PCR assays with culture, immunofluorescence, and nucleic acid hybridization for detection of *Leptospira borgpetersenii* serovar hardjo in bovine urine. *Am J Vet Res* 1998;61:316–320.
24. Nervig R, Garrett L. Use of furosemide to obtain urine samples for leptospiral isolation. *Am J Vet Res* 1979;40:1197–1200.
25. Bolin CA, Zuerner RL, Trueba G. Effect of vaccination with a pentavalent leptospiral vaccine containing *Leptospira interrogans* serovar hardjo type hardjo-bovis on type hardjo-bovis infection of cattle. *Am J Vet Res* 1989;50:2004–2008.
26. Ellis W, McDowell S, Mackie D, Pollock J, Taylor M. Immunity to bovine leptospirosis. In: *Proceedings of the 21st World Buiatrics Congress, Punta del Este, Uruguay*, 2000, pp. 601–611.
27. Bolin C, Alt D. Use of a monovalent leptospiral vaccine to prevent renal colonization and urinary shedding in cattle exposed to *Leptospira borgpetersenii* serovar Hardjo. *Am J Vet Res* 2001;62:995–1000.
28. Guard C, Nydam D, Eicker S. Field trial of vaccination against *Leptospira borgpetersenii* serovar hardjo bovis in a single New York dairy herd. *Bov Pract* 2006;39:160–161.
29. Brown R, Blumerman S, Gay C, Bolin C, Duby R, Baldwin C. Comparison of three different leptospiral vaccines for induction of a type 1 immune response to *Leptospira borgpetersenii* serovar Hardjo. *Vaccine* 2003;21:4448–4458.
30. Zuerner R, Alt D, Palmer M, Thacker T, Olsen S. A *Leptospira borgpetersenii* serovar Hardjo vaccine induces a Th1 response, activates NK cells, and reduces renal colonization. *Clin Vaccine Immunol* 2011;18:684–691.
31. Alt D, Zuerner R, Bolin C. Evaluation of antibiotics for treatment of cattle infected with *Leptospira borgpetersenii* serovar hardjo. *J Am Vet Med Assoc* 2001;219:636–639.
32. Cortese V, Behan S, Galvin J *et al*. Evaluation of two antimicrobial therapies in the treatment of *Leptospira borgpetersenii* serovar hardjo infection in experimentally infected cattle. *Vet Ther* 2007;8:201–208.

Chapter 58

Infectious Agents: Brucellosis

Sue D. Hagius, Quinesha P. Morgan and Philip H. Elzer

Department of Veterinary Science, Louisiana State University, Baton Rouge, Louisiana, USA

Introduction

Brucellosis is an infectious bacterial disease of humans and ungulates caused by *Brucella* species. It is a worldwide zoonosis that has significant economic and human health impacts. The disease has been afflicting humans and animals since ancient times as anthropological evidence suggests the presence of the organism in carbonized cheese and the typical bone lesions in human remains from areas affected by the eruption of Mount Vesuvius in AD 79.[1] Human infection is generally due to contact with animals or animal products contaminated with the bacteria. Animal disease in the United States has been controlled through vaccination and an eradication program, although the bacteria still persists in certain wildlife populations. Brucellosis remains a problem in countries where animal disease control programs have not reduced the infection among animals. Management and control of animal brucellosis is the key to prevention of human disease.

The genus *Brucella* consists of six classical species which are generally associated with their host of preference: *Brucella abortus* (cattle); *B. melitensis* (goats and sheep); *B. suis* (pigs); *B. canis* (dogs); *B. ovis* (sheep); and *B. neotomae* (desert wood rat). Relatively new species have been isolated from marine mammals: *B. pinnipedialis* (seals, sea lions and walruses) and *B. ceti* (whales, porpoises and dolphins) are of unknown significance and have the potential to infect humans.[2,3] *Brucella microti*, found in the common vole, red foxes, and the soil in some areas of Europe, is of unknown pathogenicity for livestock and humans.[4] Because of the advent of molecular technologies, several new atypical species have been proposed due to their genetic similarities with classical brucellae.[5-8] Although they have a distinct preference for their chosen host, the brucellae are capable of infecting other vertebrates also. Bovine brucellosis is primarily caused by *B. abortus* and results in abortion and infertility, but *B. melitensis* and *B. suis* can infect cattle and cause a range of symptoms.

Brucella spp. are facultative intracellular pathogens that can be grown in the laboratory environment in nutrient-rich media selected for fastidious organisms in an aerobic atmosphere at 37°C. Laboratory strains usually produce individual colonies in 2–5 days, but animal isolates from infected tissues may take up to 4 weeks to grow on selective media. The organisms are small Gram-negative coccobacilli that are catalase-positive, nonmotile, and have variable oxidase and urease activity. Species and biovars identification is achieved through the use of dye sensitivities, phage typing, and molecular analysis. The species are classified as "smooth" or "rough" based on the composition of their lipopolysaccharide (LPS) outer membrane. Smooth species (*B. abortus*, *B. melitensis*, *B. suis*) have a complete O-polysaccharide group (OPS) while rough species (*B. canis* and *B. ovis*) lack this polysaccharide chain or possess only a portion of it.[9] This affects colonial morphology and may be used both microbiologically and serologically in species and strain determination.

Animal brucellosis is classically a reproductive disease, resulting in abortion, stillbirth, and weak offspring. It may affect fertility and milk production. Chronically infected animals which exhibit no clinical signs pose a threat to naive animals and their handlers. The presence of the disease in a region or country can result in international regulations that restrict animal movement and trade, adversely affecting exports and the economy.

Brucella melitensis

Brucella melitensis, the first species in the genus *Brucella* to be described, causes abortions in female goats and sheep, unilateral orchitis in males, and Malta fever in humans.[10] Sir David Bruce, a British army surgeon, discovered the organism in 1887 as the causative agent of Mediterranean or Malta fever.[11] The preferred hosts may be goats and sheep, but this organism is the least animal species-specific of the brucellae,[10] and in countries with a high prevalence of this infection in small ruminants it is not uncommon to isolate it from cattle.[12]

Although sheep and goats and their products are the main source of human infection, *B. melitensis* in cattle has emerged as an important problem in some southern European countries, Israel, Kuwait, and Saudi Arabia. The disease in goats closely resembles the disease observed in *B. abortus*-infected cattle.[13] *Brucella melitensis* infection is particularly problematic because *B. abortus* vaccines do not

Bovine Reproduction, First Edition. Edited by Richard M. Hopper.
© 2015 John Wiley & Sons, Inc. Published 2015 by John Wiley & Sons, Inc.

effectively protect against this species. Thus bovine *B. melitensis* infection is emerging as an increasingly serious public health problem in some countries. The United States was thought to be free of this pathogen; however, in October 1999, four cows and some goats and sheep in South Texas were found to be infected with *B. melitensis*.[14] The likely source of the outbreak was goats from northern Mexico that may have crossed the border and mixed with this herd.

Brucella suis

Brucellosis caused by *B. suis* was first described by J. Traum in 1914 in swine herds in Indiana. Although originally considered a pathogenic *B. abortus*, it was later named *B. suis* by I.F. Huddleson.[15] Domestic and feral swine are natural hosts of *B. suis*.[16–18] In sows, abortion is the primary indicator of disease, which may occur at any stage of the pregnancy. An infected sow may deliver healthy live piglets as well as dead or weak ones. In boars, there may be brucellae present in the semen without any visual indications of disease. Boars may exhibit unilateral swelling and atrophy of the epididymides and testes usually resulting in infertility. Reports of lameness; swollen joints, bursae, and tendons; and paralysis because of abscess formation near the spine have also been documented. Brucellosis caused by *B. suis* is considered to be a venereal disease with the infected boar passing the disease on to uninfected sows.[15]

Since pigs do not produce dairy products, contraction of the human disease is primarily limited to feral pig hunters and as an occupational hazard of farmers, veterinarians, and abattoir workers. However, *B. suis* can infect cattle that have had direct contact with feral swine, leading to the eventual consumption of unpasteurized dairy products and the perpetuation of this zoonotic pathogen.[19,20]

Brucella abortus

Brucella abortus, initially isolated as *Bacillus abortus* by Bang in 1897 and eventually renamed in 1920, is the etiological agent of bovine brucellosis, an infection that leads to spontaneous abortion, premature calving, and infertility in cattle. At parturition, the fetus, placenta, uterine fluid, and milk contain large quantities of infectious bacteria, which are shed into the environment, potentially infecting naive animals or human handlers.[13,21]

In cattle, brucellosis is primarily a disease of the female. Bulls can be infected, but they do not readily spread the disease. The organism localizes in the testicles of the bull resulting in orchitis. In the female, the organism localizes in the udder, uterus, and lymph nodes adjacent to the uterus. The infected cows exhibit symptoms that may include abortion during the last trimester of pregnancy, retained afterbirth, and weak calves at birth.[13] Infected cows typically abort only once. Subsequent calves may be born weak or healthy and normal. Some infected cows will not exhibit any clinical symptoms of the disease and give birth to normal calves. Millions of organisms are shed in the afterbirth and fluids associated with calving and abortion. The disease is spread when cattle ingest contaminated feed or lick calves or aborted fetuses from infected cattle. Once lactation begins, the organism migrates to the lymphatics associated with the mammary gland, thus making milk a potential source of human infection.

This species is able to cross the species barrier affecting both livestock and humans.[22] In livestock, it causes billions of dollars in losses due to the reproductive consequences in cattle. It is also listed as a civilian, military, and agricultural bioterrorism agent.

Pathogenicity

Pathogenic *Brucella* spp. are intracellular pathogens that are capable of survival and replication inside host phagocytic and nonphagocytic mammalian cells, which is essential for virulence.[23] Following penetration of the mucosal epithelium, the bacteria are transported to the regional lymph nodes. The spread and multiplication of *Brucella* in lymph nodes, spleen, liver, bone marrow, mammary glands, and sex organs occurs via macrophages. The increase of brucellae in the host is mainly due to the organism's ability to avoid the killing mechanisms of the macrophages. Virulent *Brucella* species not only resist killing by neutrophils following phagocytosis,[24–26] but they also replicate inside macrophages[27] and nonprofessional phagocytes.[28] Brucellae are capable of establishing themselves in replicative phagosomes inside host macrophages for extended intracellular survival. They also appear to be capable of withstanding exposure to reactive oxygen intermediates, acidic pH, and nutrient deprivation during their time inside a variety of host macrophages, including in mice, humans, and cattle.[23,29–33] There is evidence that smooth LPS probably plays a vital role in intracellular survival since smooth organisms tend to survive much more effectively than rough ones.[34,35] Macrophages containing the bacteria journey to the draining lymph node via the lymphatics.[13,36] Within 2–3 weeks, the lymph node becomes hemorrhagic due to destruction of the vasculature within the node. Some macrophages within the lymph node are lysed, and the brucellae are released to enter the bloodstream which results in a subsequent bacteremia.[13,36,37] The organisms migrate throughout the host, localizing in the reticuloendothelial system.

During pregnancy, *Brucella* spp. replicate to high numbers inside the trophoblastic cells of the ungulate placenta.[13,36,38,39] The mechanism of invasion of nonphagocytic cells, such as placental trophoblasts, is not clearly established. Within nonphagocytic cells, brucellae tend to localize in the rough endoplasmic reticulum. Placental trophoblasts are a part of the epithelial layers of the placenta of the natural host. They serve as an important interface between the maternal and fetal circulation. During late gestation, *Brucella* are known to replicate within the placental trophoblasts of their natural ruminant host, causing the degradation of placental integrity, infection of the fetus, and possibly abortion or the birth of weak or infected animals.[40] Erythritol, which may serve as a growth stimulant for brucellae, is produced in large amounts by ruminant placental trophoblasts.[13] Further experiments utilizing microscopic analysis of placental tissues from *B. abortus*-infected cows and goats also revealed that brucellae replicate in intracellular compartments associated with the rough endoplasmic reticulum of trophoblasts, suggesting a similar intracellular environment to that inside host macrophages.[38,41,42] The exact mechanism by which fetal infection causes abortions is unclear. The interference with fetal circulation due to placentitis has been suggested as a cause of fetal abortion.[43]

Diagnostics

To diagnose brucellosis in cattle one could use three methodologies: (i) serology, the detection of antibodies to the bacteria; (ii) bacteriology, the isolation of the organism from animal tissues; and (iii) molecular biology, using molecular techniques to isolate and identify bacterial DNA. Serology is commonly used to test animals because results can be obtained in a short period of time. All the serologic tests are based on bovine antibodies binding to *Brucella* antigens. Typically a country would follow the suggested OIE or USDA serological methods.[44,45] Conventional serologic tests include the Rose Bengal test (RBT), the milk ring test, the Card test, the Rivanol test, the standard tube test (STT), buffered acidified plate antigen test, and the complement fixation test. Primary binding tests include particle concentration fluorescence immunoassays (PCFIA), indirect and competitive enzyme-linked immunosorbent assays (ELISA), and fluorescence polarization assay (FPA). Current US surveillance involves the milk ring test for dairies and serologic testing of market cattle. Ideally, one would use serology in conjunction with either *Brucella* culture or molecular typing for a definitive diagnosis.[46,47]

Identification of a serologic positive animal will trigger an accepted regulatory process that may involve test and slaughter, quarantine, vaccination, and herd management. Human disease detection initiates the appropriate antibiotic therapy once the bacterial genus is elucidated. Epidemiologic surveillance generally requires speciation so all three methodologies may be utilized.

Immune response

The host immune response to a typical virulent strain of *B. abortus* includes both cell-mediated and humoral reactions. The humoral facet involves the production of antibodies that target the surface components of the cell's outer membrane. Complement has been shown to be ineffective in combating *B. abortus* directly but may aid in phagocytosis.[48] The primary immunogenic element of the smooth outer membrane is the OPS portion of the LPS. The host's antibody response to the OPS is insufficient in providing protection against a challenge infection. The efficacy of the rough RB51 vaccine strain, which lacks the OPS, also demonstrates that the humoral response is not of primary importance. The IgG_1 antibodies predominantly produced in response to *B. abortus* do not correlate with elimination of the pathogen.[49] Antibodies can promote intracellular killing through opsonization and slow the reproduction rate of intracellular *B. abortus*, but antibodies alone cannot protect the host from the bacteria.[48–50] In terms of bovine vaccination, the cell-mediated immune response must be evoked for protective immunity.

The host's cell-mediated immune reaction to *B. abortus* is primarily accomplished by activation of the T-cells of the adaptive immune response. The function of cell-mediated immunity is to seek out and destroy host cells containing intracellular pathogens, such as viruses and facultative or obligate intracellular bacteria.[51] The macrophage is primarily responsible for ingestion and clearance of brucellae from the extracellular environment. Antigen presentation to both CD4 and CD8 T-cells is a crucial step in the host's protective immunity. Macrophages that are not activated prior to ingestion of *B. abortus* differ in their response compared with previously activated macrophages, which are considerably more effective in killing the brucellae.[52,53] *Brucella abortus* ingested by inactivated macrophages can inhibit phagolysosome development or neutralize the acidic environment, allowing them to avoid killing and replicate within the cell relatively undeterred.[52] In these nonactivated phagocytes, the brucellae cause the macrophages to suppress their reaction to a strong CD4+ Th1 response.[54] Activated macrophages degrade *B. abortus* on ingestion and phagolysosomal fusion, where the bacteria can be recognized and presented to CD4+ Th1 cells by the major histocompatibility complex II receptor. An effector CD4+ Th1 cell secretes interferon (IFN)-γ which further activates macrophages, causing an increase in antimicrobial activities against the pathogen. *Brucella abortus* triggers a strong CD8+ T-cell response that mediates cytotoxic activity to lyse infected cells. If a macrophage containing replicating *B. abortus* is lysed, the bacteria are released into the extracellular environment where they can either infect other cells or be exposed to the immune system and ingested and destroyed by activated macrophages.[27,53,55–57]

Vaccines

Development of efficacious vaccines against brucellosis is at the forefront of prevention of the disease. Many experiments have been conducted using killed and live cultures of possible vaccine candidates. Killed vaccine candidates usually confer poor immunity, whereas live vaccines of virulent strains typically provide good immunity against abortion but frequently lead to release of the pathogenic organisms and possible exposure of susceptible animals to infection.[58] *Brucella* vaccines should be characterized by attenuation, induction of a protective cell-mediated immune response, and stability on multiple passages *in vivo*.[58–60] Ideal qualities of a vaccine candidate include (i) long duration of protection; (ii) minimum interference with diagnostic tests; (iii) easy production and storage with extended stability; and (iv) minimum adverse effects in vaccinated animals with no danger to humans in the event of exposure.[59,61] No vaccination program is 100% effective, and brucellosis vaccines do not prevent infection. The goal of brucellosis vaccination is to increase herd immunity and to prevent abortion and shedding. Control programs are multifaceted, involving vaccination, surveillance, and herd management.

Vaccination against bovine brucellosis is a powerful tool used in the eradication program in the United States. The first brucellosis vaccine approved by the federal government was smooth *B. abortus* Strain 19 (S19), which was implemented in 1941.[58] After many years of successful use, conflicts with S19 arose due to interference with diagnostic tests, and a new rough vaccine *B. abortus* RB51 (RB51) was approved by the USDA to replace S19 as the official bovine brucellosis vaccine in 1996.[62] Although RB51 resolved the S19 issues with diagnostic testing, some feel that RB51 may provide a less efficacious immunization than S19.[63] The immunogenic differences between S19 and RB51 are principally their variability in structure, function, and survivability in the host.

The three primary vaccines used to prevent brucellosis in cattle as well as in goats and sheep include *B. abortus* strain 19, *B. abortus* strain RB51, and *B. melitensis* strain Rev. 1.[64] There are no effective commercial vaccines available for swine or humans in the United States.

B. abortus Strain 19

Strain 19 was the original vaccine used in the brucellosis control programs for cattle in the United States. After it became the official vaccine candidate in the United States, it was subsequently used throughout the world. Strain 19 is the most widely used vaccine for the prevention of bovine brucellosis.[58,59,65] It originated from a virulent *B. abortus* animal isolate that was serially passaged and stored at room temperature for several months and appeared to be attenuated upon replication.[66] Protection in pregnant heifers induced by S19 is reported to be 70–90% against abortion and/or infection.[66–68] The S19 vaccine was and still is an effective tool in brucellosis control; however, the vaccine has its advantages and disadvantages. It is a live vaccine that stimulates both the humoral and cell-mediated immune responses of vaccinated animals to protect them from virulent *Brucella* spp. for an extended period of time (years), and the detectable antibodies usually disappear in a few months. Some disadvantages include abortion if administered during late gestation, orchitis, pyrexia, anorexia, chronic low titers, and the occasional persistent udder infection.[58]

The host's immunologic response to S19 is similar to the exposure to an infectious strain of *B. abortus* but does not result in a clinical infection.[58] As a smooth strain of *B. abortus*, S19 expresses a fully intact and functional LPS with the OPS on the surface of the outer membrane. It is capable of inducing OPS-specific antibodies in the host as it stimulates both cell-mediated and humoral immunity. However, this aspect of S19 may be problematic since standard diagnostic tests detect the presence of anti-OPS immunoglobulins, and one cannot distinguish a vaccinated animal from a field-strain-infected one. Normally, a vaccinated animal retains resistance to disease for a protracted period of time (years), but the serologic detection of antibodies disappears in a few months. A few animals may become persistently infected with S19 following vaccination, continuing to produce anti-OPS antibodies for years following exposure.[69] For these chronic shedders, diagnostic tests will result in false-positive results for vaccinated animals that are not infected with a field strain.

Because of this issue, researchers sought a vaccine that would not interfere with the eradication program, and a new vaccine named RB51 was licensed in 1996.[62] A rough strain of *B. abortus* 2308, RB51 lacks the OPS found on virulent strains as well as on the S19 vaccine so the problem with diagnostic tests was solved. RB51 does not promote the production of anti-OPS antibodies, yet is effective in providing adequate protection against challenge infections.[70]

B. abortus RB51

The current US vaccine used to protect against bovine brucellosis is *B. abortus* RB51, which was developed through serial passages of a smooth strain of *B. abortus* on rifampicin and penicillin-supplemented tryptic soy agar plates.[71] The resulting rough strain was rifampicin-resistant and lacked the OPS of the parental strain due to a change in the LPS biosynthesis loci.[64,72] Although there was some hesitation concerning the release of an antibiotic-resistant strain of *B. abortus*, the United States designated RB51 as the vaccine of choice in place of S19 for the eradication program in 1996.[62,70] RB51 does not stimulate production of smooth anti-OPS antibodies that might interfere with serologic testing so vaccinated animals do not react on routine brucellosis diagnostic tests. This rough vaccine is attenuated and does not revert to a smooth phenotype after administration. RB51 is as efficacious as S19 but is less abortogenic and does not produce disease symptoms after vaccination.[70,73] Administration of the vaccine elicits a cell-mediated response that is necessary to protect the animal from a full infection.[70] Studies have shown that the protection conferred by RB51 is equal to S19, being 70–90% effective in preventing abortions and disease.[62,74,75] RB51 remains the current bovine brucellosis vaccine and continues to be an important component of regulatory programs monitoring brucellosis. Domestic heifers receive RB51 at 4–12 months of age at a dosage of $1.0–3.4 \times 10^{10}$ CFU and are tagged as vaccinates.[62,70] Following immunization, RB51 is usually undetectable in the bloodstream 3 days after vaccination, is quickly cleared from the draining lymph nodes, and cannot be cultured from the animal after 6–8 weeks.[62,76,77] The rough vaccine strain RB51 provides protection against bovine brucellosis without the undesirable side effects associated with S19.

B. melitensis Rev. 1

Rev. 1 vaccine was developed by Elberg and Herzberg in 1957 as a live vaccine against *B. melitensis* in goats and sheep.[66] Rev. 1 is a smooth bacterium with a complete LPS which induces similar antibody responses as those caused by field strains and cannot be easily differentiated by conventional serology. The vaccine is virulent to humans and is streptomycin-resistant, which makes prophylactic treatment difficult.[64,77]

Subcutaneous vaccination with Rev. 1 is recommended for goats and sheep between 4 and 6 months of age. It induces a powerful and long-lasting serologic response that frequently interferes with diagnostic tests.[78] If given during pregnancy, the vaccine may cause abortions in sheep and goats. In endemic areas with widespread infection or the likelihood of reinfection, it is recommended that adult animals receive reduced doses of the vaccine.[79] Rev. 1 is an effective vaccine to control brucellosis in small ruminants, but although tested in cattle, it is not generally used against bovine disease.

Human infection, treatment, and disease prevention

Human brucellosis (undulant fever) manifests itself as a febrile illness that has a range of clinical symptoms.[22] Individuals infected with the microbes may suffer from cycling fever, malaise, and headaches, as well as joint and back pain. Typically the acute phase of the disease causes flu-like symptoms, whereas the chronic phase manifests itself as crippling arthritis or, in rare cases, meningitis, endocarditis and psychoneurosis.[80–82] Although not generally a fatal disease, untreated human infections can result in mortality.[82,83] Human brucellosis is diagnosed by obtaining a complete case history, physical examination, serologic testing, and by culturing bacteria from blood, lymph, or cerebrospinal fluid.[65]

Generally, humans are infected in one of three ways: ingestion of contaminated animal products; inhalation of *Brucella* organisms via the nasal, oral, and pharyngeal cavities; or having the bacteria enter the body through open skin wounds.[84] The most common way to be infected is by eating or drinking contaminated milk products. Inhalation of *Brucella* organisms is not a common route of infection, but it can be a significant hazard for people in certain occupations, including those working in laboratories where the organism is cultured, veterinarians assisting with births or administering brucellosis vaccines, and abattoir employees. Contamination of skin wounds may be a problem for those mentioned above; hunters may also be infected through contact with the bacteria in the carcasses of infected bison, elk, moose, or wild pigs.

Patients suffering from brucellosis are routinely treated with a combination of antibiotics, such as rifampicin and doxycycline or streptomycin and doxycycline.[85] The World Health Organization recommends treatment with 600–900 mg rifampicin and 200 mg doxycycline daily for a minimum of 6 weeks for acute brucellosis in adults.[86] Because of the zoonotic aspects of this infectious disease, the control of brucellosis in livestock and wildlife is crucial in order to eradicate human brucellosis.

General precautions should be taken when dealing with animals suspected of having brucellosis. Personal hygiene and protective clothing or equipment should be used to prevent occupational exposure when assisting an animal birth, performing a necropsy, or butchering potential infected animals. Ideally gloves would be worn at all times to avoid direct contact between the skin and fetal or maternal fluids/membranes, and some type of respirator should be used to prevent inhalation of infectious particles. Hunters, veterinarians, and abattoir workers should follow these practices, and all exposed body parts should be washed with soap and water to minimize infection. Veterinarians should take extra precautions when clearing the air bubbles out of syringes since this leads to aerosolization of live vaccines, and needlestick injuries should be avoided since the vaccines can cause infection in humans. Milk and cheese products should always be made with pasteurized milk since the organism is sensitive to heating. Travelers to endemic areas need to be aware of the potential infectious nature of dairy products. The majority of human brucellosis cases (about 100 annually) in the United States occurs in Texas and California and is likely due to the consumption of illegally imported unpasteurized milk products. Worldwide about 500 000 people are infected every year.[87]

Good laboratory practices with strict adherence to biosafety protocols and the proper personnel protective equipment will prevent most laboratory infections. Potentially infected tissues and bacterial cultures should be worked with under a Class II biosafety cabinet, and procedures minimized that produce aerosols. Diagnostic laboratories may work with routine clinical samples using biosafety level 2 equipment and practices. The Centers for Disease Control and USDA consider three *Brucella* species as select agents and require laboratories to receive special licensing to work with the bacteria.[88]

New technologies and research tools

The genomes of *B. abortus*, *B. melitensis*, and *B. suis* are available to researchers searching for new virulence factors and potential vaccine candidates. Molecular technologies have aided in species and biovar identification. Traditionally, serologic and microbiologic methods have been used for species typing of *Brucella* isolates. In more recent times, serotyping results are confirmed by a molecular genotyping method, a variety of which are available for this purpose: PCR restriction fragment length polymorphism (PCR-RFLP),[89] cytoplasmic protein-specific gene probe analysis,[90] multiple locus variable number tandem repeat analysis (MLVA),[91,92] and typing with the *rpoB* gene coding the DNA-dependent RNA polymerase (RNAP) β subunit.[93,94]

An important step in understanding the molecular basis of pathogenesis is identification of genes causing disease. Opportunities to determine virulence genes in *Brucella* spp. increased with the characterization of the species' genomes. Various techniques are used to evaluate potential virulence genes and survival genes.[95–101]

The ideal goal is to develop a new generation of vaccines which provide 100% immunity across bacterial species and hosts. These vaccines should not interfere with serologic tests; they should be cold-chain independent; they should be safe to all animals, regardless of age or pregnancy status; and they should not cause disease in humans.[61] The elimination of human infection depends on the eradication of animal disease, which will require a united effort on the parts of governmental agencies, veterinarians, and producers worldwide.

References

1. Capasso, L. Bacteria in two-millennia-old cheese, and related epizoonoses in Roman populations. *J Infect* 2002;2:122–127.
2. The Center for Food Security and Public Health, Iowa State University. Brucellosis in marine mammals. July 2009. Available at www.cfsph.iastate.edu/Factsheets/pdfs/brucellosis_marine.pdf
3. Carvalho Neta A, Mol J, Xavier M, Paixao T, Lage A, Santos R. Pathogenesis of bovine brucellosis. *Vet J* 2010;184:146–155.
4. Jiménez de Bagüés M, Ouahrani-Bettache S, Quintana J et al. The new species *Brucella microti* replicates in macrophages and causes death in murine models of infection. *J Infect Dis* 2010;202:3–10.

5. Scholz H, Nöckler K, Göllner C et al. *Brucella inopinata* sp. nov. isolated from a breast implant infection. *Int J Syst Evol Microbiol* 2010;60:801–808.
6. Tiller R, Gee J, Lonsway D et al. Identification of an unusual *Brucella* strain (BO2) from a lung biopsy in a 52 year-old patient with chronic destructive pneumonia. *BMC Microbiol* 2010;10:23.
7. Tiller R, Gee J, Frace M et al. Characterization of novel *Brucella* strains originating from wild native rodent species in North Queensland, Australia. *Appl Environ Microbiol* 2010;76:5837–5845.
8. Cook I. Campbell R, Barrow G. Brucellosis in North Queensland rodents. *Aust Vet J* 1966;42:5–8.
9. Schurig G, Sriranganathan N, Corbel M. Brucellosis vaccines: past present and future. *Vet Microbiol* 2002;90:479–491
10. Alton G. *Brucella melitensis*. In: Nielsen K, Duncan JR (eds) *Animal Brucellosis*. Boca Raton, FL: CRC Press, 1990, pp. 383–409.
11. Moreno E, Moriyon I. *Brucella melitensis*: a nasty bug with hidden credentials for virulence. *Proc Natl Acad Sci USA* 2002;99:1–3.
12. Verger J. B. *melitensis* infection in cattle. In: Verger JM, Plommet M (eds) *Brucella melitensis*. Dordrecht: Martinus Nijhoff, 1985, pp. 197–203.
13. Enright F. The pathogenesis and pathobiology of *Brucella* infection in domestic animals. In: Nielsen K, Duncan JR (eds) *Animal Brucellosis*. Boca Raton, FL: CRC Press, 1990, pp. 301–320.
14. Kahler S. *Brucella melitensis* infection discovered in cattle for first time, goats also infected. *J Am Vet Med Assoc* 2000;216:648.
15. Alton G. *Brucella suis*. In: Nielsen K, Duncan JR (eds) *Animal Brucellosis*. Boca Raton, FL: CRC Press, 1990, pp. 411–422.
16. Norton J, Thomas A. *Brucella suis* in feral pigs. *Aust Vet J* 1976;52:293–294.
17. Becker H, Belden R, Breault T, Burridge M, Frankenberger W, Nicoletti P. Brucellosis in feral swine in Florida. *J Am Vet Med Assoc* 1978;173:1181–1182.
18. Zygmont S, Nettles V, Shotts E, Carmen W, Blackburn B. Brucellosis in wild swine: a serologic and bacteriologic survey in the South Eastern United States and Hawaii. *J Am Vet Med Assoc* 1982;181:1285–1287.
19. Borts IH, Harris DM, Joynt MF, Jennings JR, Jordan CF. A Milk-borne epidemic of brucellosis, caused by the porcine type of brucella (*Brucella suis*) in a raw milk supply. *JAMA* 1943;121:319–322.
20. Ewalt D, Payeur J, Rhyan J, Geer P. *Brucella suis* biovar 1 in naturally infected cattle: a bacteriological, serological, and histological study. *J Vet Diagn Invest* 1997;9:417–420.
21. Perry QL. *Brucella melitensis*: the evaluation of a putative hemagglutinin gene's effect on virulence in the caprine model. Dissertation, Louisiana State University, 2007.
22. Young E. An overview of human brucellosis. *Clin Infect Dis* 1995;21:283–290.
23. Celli J. Surviving inside a macrophage: the many ways of *Brucella*. *Res Microbiol* 2006;157:93–98.
24. Riley L, Robertson D. Brucellacidal activity of human and bovine polymorphonuclear leukocyte granule extracts against smooth and rough strains of *Brucella abortus*. *Infect Immun* 1984;46:231–236.
25. Riley L, Robertson D. Ingestion and intracellular survival of *Brucella abortus* in human and bovine polymorphonuclear leukocytes. *Infect Immun* 1984;46:224–230.
26. Canning P, Roth J, Deyoe B. Release of 5′-guanosine monophosphate and adenine by *Brucella abortus* and their role in the intracellular survival of the bacteria. *J Infect Dis* 1986;154:464–470.
27. Jones S, Winter A. Survival of virulent and attenuated strains of *Brucella abortus* in normal and gamma interferon-activated murine peritoneal macrophages. *Infect Immun* 1992;60:3010–3014.
28. Detilleux P, Deyoe B, Cheville N. Penetration and intracellular growth of *Brucella abortus* in nonphagocytic cells in vitro. *Infect Immun* 1990;58:2320–2328.
29. Harmon B, Adams L, Frey M. Survival of rough and smooth strains of *Brucella abortus* in bovine mammary gland macrophages. *Am J Vet Res* 1988;49:1092–1097.
30. Price R, Templeton J, Smith R, Adams L. Ability of mononuclear phagocytes from cattle naturally resistant or susceptible to brucellosis to control in vitro intracellular survival of *Brucella abortus*. *Infect Immun* 1990;58:879–886.
31. Rittig M, Alvarez-Martinez M, Porte F, Liautard J, Rouot B. Intracellular survival of *Brucella* spp. in human monocytes involves conventional uptake but special phagosomes. *Infect Immun* 2001;69:3995–4006.
32. Watarai M, Makino S, Fujii Y, Okamoto K, Shirahata T. Modulation of *Brucella*-induced macropinocytosis by lipid rafts mediates intracellular replication. *Cell Microbiol* 2002;4:341–355.
33. Celli J, de Chastellier C, Franchini D, Pizarro-Cerda J, Moreno E, Gorvel J. *Brucella* evades macrophage killing via VirB-dependent sustained interactions with the endoplasmic reticulum. *J Exp Med* 2003;198:545–556.
34. Zhan Y, Cheers C. Endogenous interleukin-12 is involved in resistance to *Brucella abortus* infection. *Infect Immun* 1995;63:1387–1390.
35. Caron E, Peyrard T, Kohler S, Cabane S, Liautard J, Dornand J. Live *Brucella* spp. fail to induce tumor necrosis factor alpha excretion upon infection of U937-derived phagocytes. *Infect Immun* 1994;2:5267–5274.
36. Thoen C, Enright F, Cheville N. *Brucella*. In: Gyles CL, Thoen CO (eds) *Pathogenesis of Bacterial Infections in Animals*, 2nd edn. Ames, IA: Iowa State University Press, 1993, pp. 236–247.
37. Edmonds M. *Creation and characterization of 25 kDa outer membrane protein (Omp25) deletion mutants in Brucella species*. Dissertation, Louisiana State University, 2000.
38. Anderson T, Cheville N. Ultrastructural morphometric analysis of *Brucella abortus*-infected trophoblasts in experimental placentitis. Bacterial replication occurs in rough endoplasmic reticulum. *Am J Pathol* 1986;124:226–237.
39. Anderson T, Meador V, Cheville N. Pathogenesis of placentitis in the goat inoculated with *Brucella abortus*. I. Gross and histologic lesions. *Vet Pathol* 1986;23:219–226.
40. Roop R, Bellaire B, Valderas M, Cardelli J. Adaptation of the brucellae to their intracellular niche. *Mol Microbiol* 2004;52:621–630.
41. Anderson T, Meador V, Cheville N. Pathogenesis of placentitis in the goat inoculated with *Brucella abortus*. II. Ultrastructural studies. *Vet Pathol* 1986;23:227–236.
42. Meador V, Deyoe B. Intracellular localization of *Brucella abortus* in bovine placenta. *Vet Pathol* 1989;26:513–515.
43. Payne J. The pathogenesis of experimental brucellosis in the pregnant cow. *J Pathol Bacteriol* 1959;78:447–463.
44. Bovine brucellosis. In: Edwards S, Caporale V, Schmitt B et al. (eds) *OIE Manual of Diagnostic Tests and Vaccines for Terrestrial Animals*, 7th edn. Paris: OIE, 2012, pp. 616–650. Available at http://www.oie.int/fileadmin/Home/eng/Health_standards/tahm/2.04.03_BOVINE_BRUCELL.pdf
45. Animal and Plant Health Inspection Service, United States Department of Agriculture. Brucellosis Eradication: Uniform Methods and Rules, Effective October 1, 2003. Available at http://www.aphis.usda.gov/animal_health/animal_diseases/brucellosis/downloads/umr_bovine_bruc.pdf
46. Gall D, Nielsen K. Serological diagnosis of bovine brucellosis: a review of test performance and cost comparison. *Rev Sci Tech Off Int Epiz* 2004;23:989–1002.
47. Poester F, Nielsen K, Samartino L, Yu W. Diagnosis of brucellosis. *Open Vet Sci J* 2010;4:46–60.

48. Timoney J, Gillespie J, Scott F, Barlough J. The genus *Brucella*. In: *Hagan and Bruner's Microbiology and Infectious Diseases of Domestic Animals*, 8th edn. New York: Cornell University Press, 1988, pp. 135–152.
49. Bellaire B, Roop R, Cardelli J. Opsonized virulent *Brucella abortus* replicates within nonacidic, endoplasmic reticulum-negative, LAMP-1-positive phagosomes in human monocytes. *Infect Immun* 2005;73:3702–3710.
50. Arenas G, Staskevich A, Aballay A, Mayorga L. Intracellular trafficking of *Brucella abortus* in J774 macrophages. *Infect Immun* 2000;68:4255–4263.
51. Abbas AK, Lichtman AHH, Pillai S. Immunity to microbes. In: *Cellular and Molecular Immunology*, 7th edn. Philadelphia: Elsevier Saunders, 2012, pp. 345–365.
52. Barquero-Calvo E, Chaves-Olarte E, Weiss D et al. *Brucella abortus* uses a stealthy strategy to avoid activation of the innate immune system during the onset of infection. *PLoS One* 2007;2:e631.
53. Duhon LE. *In vitro and in vivo evaluation of a Brucella putative hemagglutinin*. Dissertation, Louisiana State University, 2010.
54. Forestier C, Deleuil F, Lapaque N, Moreno E, Gorvel J. *Brucella abortus* lipopolysaccharide in murine peritoneal macrophages acts as a down-regulator of T cell activation. *J Immunol* 2000;165:5202–5210.
55. Araya L, Elzer P, Rowe G, Enright F, Winter A. Temporal development of protective cell-mediated and humoral immunity in BALB/c mice infected with *Brucella abortus*. *J Immunol* 1989;143:3330–3337.
56. Baldwin CL, Winter AJ. Macrophages and *Brucella*. *Immunol Ser* 1994;60:363–380.
57. Atluri V, Xavier M, de Jong M, den Hartigh A, Tsolis R. Interactions of the human pathogenic *Brucella* species with their hosts. *Annu Rev Microbiol* 2011;65:523–541.
58. Nicoletti P. Vaccination. In: Nielsen K, Duncan JR (eds) *Animal Brucellosis*. Boca Raton, FL: CRC Press, 1990, pp. 283–299.
59. Nicoletti P. Vaccination against *Brucella*. *Adv Biotechnol Processes* 1990;13:147–168.
60. Adams L. Development of live *Brucella* vaccines. In: Adams L (ed.) *Advances in Brucellosis Research*. College Station, TX: Texas A&M University Press, 1990, pp. 250–276.
61. Costa Oliveira S, Costa Macedo G, de Almeida L et al. Recent advances in understanding immunity against brucellosis: application for vaccine development. *Open Vet Sci J* 2010;4:102–108.
62. Animal and Plant Health Inspection Service, United States Department of Agriculture. *Brucella abortus* strain RB51 vaccine licensed for use in cattle. Available at http://www.aphis.usda.gov/animal_health/animal_dis_spec/cattle/downloads/rb51_vaccine.pdf
63. Moriyon I, Grillo M, Monreal D et al. Rough vaccines in animal brucellosis: structural and genetic basis and present status. *Vet Res* 2004;35:1–38.
64. Avila-Calderón E, Lopez-Merino A, Sriranganathan N, Boyle S, Contreras-Rodríguez A. A history of the development of *Brucella* vaccines. *Biomed Res Int* 2013;2013:743509.
65. Seleem M, Boyle S, Sriranganathan N. *Brucella*: a reemerging zoonosis. *Vet Microbiol* 2010;140:392–398.
66. Samartino L. Brucellosis vaccines. In: *The 58th Brucellosis Conference, Merida*, Yucatan, Mexico, 2005, pp. 31–41.
67. Confer A, Hall S, Faulkner C et al. Effects of challenge dose on the clinical immune responses of cattle vaccinated with reduced doses of *Brucella abortus* strain 19. *Vet Microbiol* 1985;10:561–575.
68. Nicoletti P. Vaccination of cattle with *Brucella abortus* strain 19 administered by differing routes and doses. *Vaccine* 1984;2:133–135.
69. Jacob J, Hort G, Overhoff P, Mielke M. *In vitro* and *in vivo* characterization of smooth small colony variants of *Brucella abortus* S19. *Microbes Infect* 2006;8:363–371.
70. Poester F, Goncalves V, Paixao T et al. Efficacy of strain RB51 vaccine in heifers against experimental brucellosis. *Vaccine* 2006;24:5327–5334.
71. Schurig GG, Roop II RM, Bagchi T, Boyle S, Buhrman D, Sriranganathan N. Biological properties of RB51, a stable rough strain of *Brucella abortus*. *Vet Microbiol* 1991;8:171–188.
72. Vemulapalli R, Contreras A, Sanakkayala N, Sriranganathan N, Boyle S, Schurig G. Enhanced efficacy of recombinant *Brucella abortus* RB51 vaccines against *B. melitensis* infection in mice. *Vet Microbiol* 2004;10:237–245.
73. Stevens M, Olsen S, Cheville N. Comparative analysis of immune responses in cattle vaccinated with *Brucella abortus* strain 19 or strain RB51. *Infect Immun* 1995;44:223–235.
74. Olsen S. Immune responses and efficacy after administration of a commercial *Brucella abortus* strain RB51 vaccine to cattle. *Vet Ther* 2000;1:183–191.
75. Cheville N, Jensen A, Halling S et al. Bacterial survival, lymph node changes, and immunologic responses of cattle vaccinated with standard and mutant strains of *Brucella abortus*. *Am J Vet Res* 1992;53:1881–1888.
76. Cheville N, Stevens M, Jensen A, Tatum F, Halling S. Immune responses and protection against infection and abortion in cattle experimentally vaccinated with mutant strains of *Brucella abortus*. *Am J Vet Res* 1993;57:1153–1156.
77. Blasco J, Diaz R. *Brucella melitensis* Rev-1 vaccine as a cause of human brucellosis. *Lancet* 1993;342:805.
78. Blasco J. Existing and future vaccines against brucellosis in small ruminants. *Small Ruminant Res* 2006;62:33–37.
79. Blasco J. A review of the use of *B. melitensis* Rev 1 vaccine in adult sheep and goats. *Prev Vet Med* 1997;31:275–283.
80. Harris H, Kemple C. Chronic brucellosis and psychoneurosis. *Psychosom Med* 1954;16:414–425.
81. Alapin B. Psychosomatic and somato-psychic aspects of brucellosis. *J Psychosom Res* 1976;20:339–350.
82. Franco M, Mulder M, Gilman R, Smits H. Human brucellosis. *Lancet Infect Dis* 2007;7:775–786.
83. Park K, Kim D, Park C et al. Fatal systemic infection with multifocal liver and lung nodules caused by *Brucella abortus*. *Am J Trop Med Hyg* 2007;77:1120–1123.
84. Centers for Disease Control and Prevention. Brucellosis transmission. Available at http://www.cdc.gov/brucellosis/transmission/index.html
85. Solera J, Martinez-Alfaro E, Espinosa A. Recognition and optimum treatment of brucellosis. *Drugs* 1997;53:245–256.
86. Joint FAO/WHO Expert Committee on Brucellosis. Sixth Report. World Health Organization Technical Report Series No. 740. Geneva: World Health Organization, 1986.
87. Pappas G, Papadimitriou P, Akritidis N, Christou L, Tsianos EV. The new global map of human brucellosis. *Lancet Infect Dis* 2006;6:91–99.
88. Chosewood LC, Wilson DE (eds) *Biosafety in Microbiological and Biomedical Laboratories*, 5th edn. Washington, DC: US Government Printing Office, 2007. Available at http://www.cdc.gov/biosafety/publications/bmbl5/index.htm
89. Al Dahouk S, Tomaso H, Prenger-Berninghoff E, Splettstoesser W, Scholz H, Neubauer H. Identification of *Brucella* species and biotypes using polymerase chain reaction-restriction fragment length polymorphism (PCR-RFLP). *Crit Rev Microbiol* 2005;31:191–196.
90. Verger J, Grayon M, Tibor A, Wansard V, Letesson J, Cloeckaert A. Differentiation of *B. melitensis*, *B. ovis* and *B. suis* biovar 2 strains by use of membrane protein- or cytoplasmic protein-specific gene probes. *Res Microbiol* 1998;149:509–517.
91. Bricker B, Ewalt D, Halling S. *Brucella* HOOF-prints strains typing by multilocus analysis of variable number tandem repeats (VNTR). *BMC Microbiol* 2003;3:1–15

92. Le Flèche P, Jacques I, Grayon M et al. Evaluation and selection of tandem repeat loci for a *Brucella* MLVA typing assay. *BMC Microbiol* 2006;6:1–14.
93. Marianelli C, Ciuchini F, Tarantino M, Pasquali P, Adone R. Molecular characterization of the rpoB gene in *Brucella* species: new potential molecular markers for genotyping. *Microbes Infect* 2006;8:860–865.
94. Sayan M, Yumuk Z, Bilenoglu O, Erdenlig S, Willke A. Genotyping of *Brucella melitensis* by rpoB gene analysis and re-evaluation of conventional serotyping method. *Jpn J Infect Dis* 2009;62:160–163.
95. Hensel M, Holden D. Molecular genetic approaches for the study of virulence in both pathogenic bacteria and fungi. *Microbiology* 1996;142:1049–1058.
96. O'Callaghan D, Cazevieille C, Allardet-Serven A et al. A homologue of the *Agrobacterium tumefaciens* VirB and *Bordetella pertussis* Ptl type IV secretion systems is essential for intracellular survival of Brucella suis. *Mol Microbiol* 1999;33:1210–1220.
97. Hong P, Tsolis R, Ficht T. Identification of genes requires for chronic persistence of *Brucella abortus* in mice. *Infect Immun* 2000;68:4102–4107.
98. Lestrate P, Delrue RM, Danese I et al. Identification and characterization of *in vivo* attenuated mutants of Brucella melitensis. *Mol Microbiol* 2000;38:543–551.
99. Kahl-McDonagh M, Ficht T. Evaluation of protection afforded by *Brucella abortus* and *Brucella melitensis* unmarked deletion mutants exhibiting different rates of clearance in BALB/c mice. *Infect Immun* 2006;74:4048–4057.
100. Wu Q, Pei J, Turse C, Ficht T. Mariner mutagenesis of *Brucella melitensis* reveals genes with previously uncharacterized roles in virulence and survival. *BMC Microbiol* 2006;6:102.
101. Zygmunt M, Hagius S, Walker J, Elzer P. Identification of *Brucella melitensis* 16 M genes required for bacterial survival in the caprine host. *Microbes Infect* 2006;8:2849–2854.

Chapter 59

Infectious Agents: Infectious Bovine Rhinotracheitis

Ahmed Tibary

Department of Clinical Sciences, College of Veterinary Medicine, Washington State University, Pullman, Washington, USA

Introduction

Infectious bovine rhinotracheitis (IBR) due to bovine herpesvirus type 1 (BHV-1 or BoHV-1) is a common disease of cattle which is responsible for significant economic loss worldwide. Infection with BHV-1 is associated with mild to severe respiratory disease and represents a high risk for bovine respiratory disease complex. Infections with this agent can also manifest as ocular, neonatal, gastrointestinal, and neurologic disease as well as reproductive failure due to abortion and other genital symptoms (infectious pustular vulvovaginitis or IPV and infectious pustular balanoposthitis or IPB).[1,2] BHV-1 is one of the most important pathogens involved in the respiratory disease syndrome of shipping fever.

Etiology

The disease is caused by bovine herpesvirus 1 (BHV-1) which belongs to the order *Herpesvirales*, family *Herpesviridae*, subfamily *Alphaherpesvirinae*, genus *Varicellovirus*. The viral genome consists of a double-stranded DNA. Three subtypes (genotypes) of BHV-1 (BHV-1.1, BHV-1.2 subdivided into BHV-1.2a and BHV-1.2b, and BHV-1.3) have been described and characterized.[3] BHV-1.1 and BHV-1.2a are responsible primarily for the respiratory syndrome and abortion; BHV-1.2b causes primarily a genital syndrome and BHV-1.3 is a neuropathogenic subtype causing encephalitis. Isolates of BHV-1.1 are more virulent than are isolates of BHV-1.2b. Recently, BHV-1.3, the neuropathogenic agent, has been renamed bovine herpesvirus type 5 (BHV-5). BHV-5 was first isolated from calves and adult cattle with encephalitis as well as from aborted fetuses. The virus has been isolated from semen of a healthy bull.[4,5] Recently, BHV-5 was isolated from semen used for artificial insemination (AI) of a small group of cows that experienced an outbreak of venereally transmitted genital syndrome (vulvovaginitis and reduced conception rates).[6] BHV-4 has been associated with mastitis.

BHV-1 is very sensitive in the environment. Inactivation of BHV-1 depends on several factors, including temperature, pH, light, humidity, and tissue. The virus can survive for 1 month at 4°C and up to 10 days at body temperature. The virus is sensitive to organic solvents and many disinfectants.

Epidemiology

The disease caused by BHV-1 has long been recognized in cattle-raising countries throughout the world. Cattle are more sensitive to the virus than buffaloes and there seems to be some variation in sensitivity among breeds or genotypes of cattle. However, this genetic variation in sensitivity has not been fully investigated. Seroprevalence of BHV-1 ranges from 10 to 70% depending on the country. In Africa seroprevalence varies from 14 to 60%, while in Central and South American the prevalence is 36–50%. Most countries with significant cattle production systems have national control programs. Countries in the European Union have been aggressive in the development of control and eradication programs. BHV-1 has been successfully eradicated in Austria, Denmark, Finland, Italy (Province of Bolzano-Bozen), Sweden, and Switzerland.[1,7]

Latency and shedding are the primary factors in the spread of the disease. The genital carrier state is important in the maintenance of venereal BHV and the occurrence of sporadic IPV and IBP. Although these clinical manifestations have decreased substantially with vaccination and use of AI, they remain an important clinical syndrome in some countries. In Australia up to 96% of bulls and 52% of cows are seropositive, probably due to venereal transmission. Seroprevalence probably reflects vaccination strategies and degree of commingling with other herds. An outbreak of IPV and IBP in dairy herds reported up to 80% of cows with respiratory and genital symptoms.[8] Latency can result from primary infection with a field strain or vaccination with attenuated or modified live virus (MLV) strain. Vaccination

---

Bovine Reproduction, First Edition. Edited by Richard M. Hopper.
© 2015 John Wiley & Sons, Inc. Published 2015 by John Wiley & Sons, Inc.

does not inhibit shedding of a wild strain that was already present. Presence of colostral antibody does not prevent initial virus replication and latency can persist after decline in colostral immunity. Cattle from herds where BHV is endemic must always be considered a potential source of the virus.

The main sources of infection are nasal exudates and cough droplets, preputial secretions, semen, and fetal fluids and tissues. Transmission via shedding from the vagina or prepuce (IPV or IBP) is possible but less efficient.[9] Cattle infected with BHV-1.1 excrete much higher titers of virus in nasal fluids than do cattle infected with BHV-1.2b. Venereal transmission through semen during natural service or AI is possible.[10-15] Semen is most likely contaminated during ejaculation from the prepuce and not through viral replication in the testis or accessory sex glands. Bulls can start shedding BHV-1 from the prepuce 2–7 days after primary preputial infection. Bulls may become intermittent shedders for several months to years. Latently infected bulls may resume shedding after stress.[16] Mechanical transmission by vectors such as ticks has been suspected. Vaccinated cattle can also excrete the BHV-1 virus after exposure to infection but at a lower level and for a shorter period of time than do unvaccinated animals. Outbreaks usually reach their maximum impact 2–3 weeks after introduction of shedders and ends by the sixth week. The impact of the disease on a herd depends on the level of natural or acquired immunity.

The BHV-1 virus can become latent following a primary infection with a field isolate or vaccination with an attenuated strain and remains localized near the site of its first multiplication. The virus is usually detectable in the ganglia of the trigeminal nerve in IBR and in sacral spinal ganglia in IPV/IPB cases.[17] Latently infected animals excrete BHV-1 in nasal, vaginal, or preputial secretions during the time of recrudescence. Transportation of cattle with latent infection can reactivate the virus, resulting in re-excretion of virus and a rise in neutralizing antibodies. Attenuated vaccine strains can remain in a latent state in the body and vaccination does not provide protection against establishment of a latent infection with a wild strain.[18,19] Vaccination in latently infected animals does not prevent re-excretion of a wild strain.

Clinical presentation

Clinical symptoms of BHV-1 often appear 10–20 days following natural infection and can take one of several forms, including respiratory (IBR), genital (IPV, IBP), ocular (conjunctivitis), or encephalomyelitis form.

IBR may occur as a subclinical, mild, or severe disease. Morbidity and mortality rate may approach 100% and 10%, respectively. Affected animals show fever (40–42 °C), inappetence, increased respiratory rate, salivation, dyspnea, persistent harsh cough, conjunctivitis, depression, and severe drop in milk production. Nasal discharge is initially serous and becomes mucopurulent as the disease progresses. Nasal lesions ("red nose") progress from pustular necrosis to large hemorrhagic and ulcerated areas covered by diphtheritic membrane. Secondary bacterial or viral infection may lead to an increase in the low mortality rate. Animals may show signs of bronchitis and pneumonitis. Some animals recover rapidly, within a few days, and become carriers and may shed the virus for a long period. In mild cases, clinical signs may be limited to a serous nasal and ocular discharge. Respiratory signs become more severe (open mouth breathing) in cases with secondary bacterial infection (*Mannheimia haemolytica*, *Pasteurella multocida*, *Haemophilus somnus* or *Mycoplasma bovis*) and there is a concurrent increased risk of mortality.

Affected animals may show varying degrees of unilateral or bilateral conjunctivitis with profuse lacrimation, photophobia, and epiphora. Secondary bacterial infection is common, resulting in purulent discharge and keratitis and corneal ulceration. Pregnant seronegative cows experience abortion in the last trimester of pregnancy following BHV-1 infection, viremia, and lethal infection of the fetus.

The encephalitic form caused by BHV-5 is characterized by neurologic signs including nervousness, muscle tremor, incoordination leading to ataxia and coma. Affected animals usually die within 4 days from the onset of the neurologic signs. Blindness is a common sequel in recovering animals.

Genital form of the disease

In addition to abortion, IBR virus is known to cause IPV in cows and IBP in bulls. The acute form of the infection in naive animals is characterized by the presence of fever, depression, and anorexia within 1–3 days of exposure. In the female, a severe and painful inflammation of the vagina and vulva is present. Affected animals show frequent posturing for urination and constant tail twitching or raising due to pain. The mucosa surface of the vulva and vestibular area is hyperemic and shows small (2–3 mm) raised pustules. As the disease progresses, the pustules coalesce and form ulcers. Mucopurulent discharge may develop due to secondary bacterial infection. The lesions usually heal 10–14 days after the onset of the disease but purulent vaginal discharge may persist for several weeks.[13,20] Insemination of animals with BHV-1-contaminated semen can cause IPV, cervicitis with copious mucopurulent discharge, endometritis, and poor conception rate in cows. Experimental inoculation of heifers results in oophoritis, decreased luteal function, and early pregnancy loss.[15,21]

In bulls, IPB is characterized by similar lesions on the preputial mucosa and surface of the penis. Some bulls may show severe edema and prolapse of the preputial mucosa. This often leads to complication by secondary bacterial infection followed by development of adhesions. Recovery may occur 2 weeks after of development of clinical signs. Some bulls may experience decreased libido for several weeks. Reactivation and shedding of the virus may occur in carrier bulls at the time of mating. BHV-1 has also been associated with epididymitis and poor semen quality. Mild or subclinical forms of IPV/IPB are common.[20]

Abortions due to BHV-1 usually occur between 4 and 8 months of gestation after natural infection or vaccination. The cotyledons are usually blanched and degenerate without any gross lesions on the aborted fetus. The abortion is due to death of the fetus. The placenta may harbor the virus in a latent state for up to 3 months without transmitting the virus to the fetus. Abortion may occur up to 90 days following vaccination with MLV vaccine.

Pathophysiology

The respiratory syndrome is caused by multiplication of the virus in the upper respiratory tract, resulting in rhinitis, laryngitis, and tracheitis. The syndrome is complicated by lower resistance to secondary bacterial infection due to *M. haemolytica*, *P. multocida*, and *H. somnus*. Conjunctivitis results from contamination through the nasolacrimal duct. Meningoencephalitis results from brain infection through the trigeminal nerves. In pregnant nonvaccinated cows, the virus is thought to be transported by peripheral leukocytes to the placental barrier and transferred to the fetus, resulting in a peracute fatal infection and abortion. Infections in the last trimester of pregnancy may result in mummification, abortion, stillbirth, or birth of weak calves. Calves infected *in utero* often show severe necrotizing enteritis, hepatitis, nephritis, and ulceration of the upper digestive tract.

The virus can access the reproductive system by the intranasal or venereal route. Natural respiratory infection has been shown to result in shedding of BHV-1 in semen. The virus has also been isolated from preputial washing 2–10 days following experimental intranasal or genital infection of bulls. However, the virus has not been isolated from the proximal urethra, accessory glands, epididymis, or testicles of experimentally infected bulls.[22]

Histopathologic changes in uncomplicated cases include acute catarrhal inflammation. A neutrophilic infiltration of the mucosa is present. The submucosa often shows lymphocytic and plasmacytic infiltration with presence of macrophages. Intranuclear (Cowdry type A) inclusions may be present in the epithelial cells during the first few day of infection. Aborted fetuses show necrosis in a variety of tissues and especially the liver (focal necrotizing hepatitis). Hemorrhagic lesions and foci of necrosis surrounded by leukocytes are visible on a variety of fetal tissues (liver, brain, kidney, adrenal cortex, and lymph nodes). Occasionally, intranuclear inclusion bodies are observed. Viral antigen can be demonstrated in sections of the lung, liver, spleen, kidney, adrenal gland, placenta, and in mummified fetuses using the avidin–biotin complex system.

Both cell-mediated and humoral immune response are activated following infection or vaccination. However, serum neutralizing antibody titers are not a reliable indicator of resistance or protection.

Diagnosis

The virus can be detected by viral isolation on specimens collected during the febrile acute phase of the infection. BHV-1 can be readily isolated in cell culture from nasal swabs, conjunctival swabs, vaginal swabs, preputial washings, placental cotyledons of aborted fetus, fetal liver, lung, spleen, kidney, lymph node, mucous membrane of respiratory tract, tonsils and lungs collected in virus transport medium. Primary cell culture is more sensitive.

Histologically, intranuclear viral inclusions (Cowdry type A) can occasionally be identified in the epithelial cells of vaginal biopsy tissues collected in the early stage of IPV. These inclusions are also present in tissues of aborted fetuses. However, absence of inclusions does not rule out BHV-1 as they are usually transitory.[13]

The most sensitive and rapid diagnostic technique is detection of viral particles using polymerase chain reaction (PCR) on nasal swabs, bovine fetal serum, and semen samples.[23] Real-time PCR provides high specificity and sensitivity for detection of BHV-1 in semen.[24,25]

Serology of paired serum samples can detect recent infection. Both the virus neutralization test (VNT) and enzyme-linked immunosorbent assay (ELISA) have been used for the detection of antibodies against BHV-1 infection. The ELISA is a specific, sensitive, and practical test for the detection of antibody and has advantages over the VNT. The detection of latent BHV-1 infection in cattle is important in control programs.

Neutralizing antibodies, mainly of the IgM followed by the IgG class, are usually detected around 7 days after infection.[13] Bulk tank milk testing for BHV-1 antibodies may be useful in eradication and monitoring programs.[26]

Diagnosis of BHV-1-associated abortions relies on gross and histopathologic lesions in fetal and placental tissues. Diagnoses are supported by the detection of viral antigen by fluorescent antibody test, immunohistochemistry, or viral DNA by PCR methods. A recent retrospective study on 19 459 bovine abortions in five diagnostic laboratories in the United States showed that only 35.7% of submissions were tested for BHV-1. Of the samples tested, 3.8% were positive as a proportion of the total abortions; BHV-1 diagnosis was rare.[2]

Prevention and control

Several vaccines are available and include MLV vaccines, inactivated vaccines, subunit vaccines, and marker vaccines.[27] None of these vaccines has been shown to prevent viral latency. MLV can be administered parenterally (bovine fetal kidney tissue culture origin) or intranasally (rabbit tissue culture origin or bovine tissue culture origin containing a temperature-sensitive mutant of BHV-1). The live vaccine strains can be differentiated from the field strains by digestion with restriction endonucleases. MLV vaccines induce a rapid immune response, long-lasting immunity, and result in local and mucosal immunity. Intranasal MLV BHV-1 may be transiently shed for up to 10 days after vaccination.[28] Parenteral MLV vaccines can cause abortion and should not be used in nonvaccinated pregnant cows.[29] MLV vaccines can also induce latency and subsequent shedding of the vaccine virus. Intranasal vaccines are safer for pregnant cows and highly effective for the prevention of abortion. They usually produce rapid protection and can be used in the face of an outbreak. Inactivated vaccines do not cause immunosuppression, abortion, or latency. However, they are not as efficacious and do not prevent the development of latency following exposure to field virus.

Vaccination programs are helpful in disease prevention but not prevalence of BHV-1 infection within a herd. Marker vaccines can be used in conjunction with appropriate testing to differentiate between infected and vaccinated animals. Several European countries have adopted a test and slaughter strategy to eradicate the disease. Effective monitoring and biosecurity measures are required to avoid the risk of reintroducing BHV-1 into BHV-1 free herds.

Conclusion

Although IBR has been identified as an important cause of economic loss due to respiratory and genital disease, its eradication is not simple because of its presence in latent infections and peculiar adaptation. The main manifestation remains the respiratory syndrome and abortion. However, occasional outbreaks of IPV and IPB may occur. Genital manifestations seem to be more frequently reported in Europe. Several European countries have been successful in eradicating the disease through a program combining vaccination and testing.

Although vaccines are available for BHV-1, vaccination may not provide complete protection or efficacy may be reduced by factors such as poor vaccination technique and handling. In addition, the National Animal Health Survey of 2007 shows that only 68.9% of beef cow-calf herds vaccinate for BHV-1. Consequently, BHV-1 infection remains a potential differential diagnosis for abortion of beef cow-calf herds. The isolation of BHV-5 from semen and its association with genital disease merit further investigation.

References

1. Cowley D, Clegg T, Doherty M, More S. Aspects of bovine herpesvirus-1 infection in dairy and beef herds in the Republic of Ireland. *Acta Vet Scand* 2011;53:40.
2. Gould S, Cooper V, Reichardt N, O'Connor A. An evaluation of the prevalence of Bovine herpesvirus 1 abortions based on diagnostic submissions to five U.S.-based veterinary diagnostic laboratories. *J Vet Diagn Invest* 2013;25:243–247.
3. Muylkens B, Thiry J, Kirten P, Schynts F, Thiry E. Bovine herpesvirus 1 infection and infectious bovine rhinotracheitis. *Vet Res* 2007;38:181–209.
4. Esteves P, Aspilki F, Franco D et al. Bovine herpesvirus type 5 in the semen of a bull not exhibiting clinical signs. *Vet Rec* 2003;152:658–659.
5. Gomes L, Rocha M, Souza J, Costa E, Barbosa-Stancioli E. Bovine herpesvirus 5 (BoHV-5) in bull semen: amplification and sequence analysis of the US4 gene. *Vet Res Commun* 2003;27:495–504.
6. Kirkland PD, Poynting AJ, Gu X, Davis RJ. Infertility and venereal disease in cattle inseminated with semen containing bovine herpesvirus type 5. *Vet Rec* 2009;165:111–1113.
7. Memeteau S. The regulatory situation of I.B.R. in Europe. *Le Nouveau Praticien Vétérinaire Élevages et Santé* 2011:29–30.
8. Pritchard GC, Banks M, Vernon R. Subclinical breakdown with infectious bovine rhinotracheitis virus infection in dairy herd of high health status. *Vet Rec* 2003;153:113–117.
9. Edwards S, Newman R, White H. The virulence of British isolates of bovid herpesvirus 1 in relationship to viral genotype. *Br Vet J* 1991;147:216–231.
10. Givens MD. Bull biosecurity: diagnosing pathogens that cause infertility of bulls or transmission via semen. *Clin Theriogenol* 2012;4:302–307.
11. Pozzi N, Guérin B. Eradication and mastery of the health risk, IBR in the field of "animal insemination" for the bovine species. *Le Nouveau Praticien Vétérinaire Élevages et Santé* 2011:39–43.
12. Wrathall A, Simmons H, Soom A. Evaluation of risks of viral transmission to recipients of bovine embryos arising from fertilisation with virus-infected semen. *Theriogenology* 2006;65:247–274.
13. Turin L, Russo S. BHV-1 infection in cattle: an update. *Vet Bull* 2003;73:15R–21R.
14. Oirschot J. Bovine herpesvirus 1 in semen of bulls and the risk of transmission: a brief review. *Vet Quart* 1995;17:29–33.
15. Miller J, Maaten M. Reproductive tract lesions in heifers after intrauterine inoculation with infectious bovine rhinotracheitis virus. *Am J Vet Res* 1984;45:790–794.
16. Schynts F, Meurens F, Detry B, Vanderplasschen A, Thiry E. Rise and survival of bovine herpesvirus 1 recombinants after primary infection and reactivation from latency. *J Virol* 2003;77:12535–12542.
17. Ackermann M, Wyler R. The DNA of an IPV strain of bovid herpesvirus 1 in sacral ganglia during latency after intravaginal infection. *Vet Microbiol* 1984;9:53–63.
18. Jones C. Herpes simplex virus type 1 and bovine herpesvirus 1 latency. *Clin Microbiol Rev* 2003;16:79–95.
19. Jones C, Chowdhury S. A review of the biology of bovine herpesvirus type 1 (BHV-1), its role as a cofactor in the bovine respiratory disease complex and development of improved vaccines. *Anim Health Res* 2007;8:187–205.
20. Tikoo S, Campos M, Babiuk L. Bovine herpesvirus 1 (BHV-1): biology, pathogenesis, and control. *Adv Virus Res* 1995;45:191–223.
21. Miller J, Maaten M. Experimentally induced infectious bovine rhinotracheitis virus infection during early pregnancy: effect on the bovine corpus luteum and conceptus. *Am J Vet Res* 1986;47:223–228.
22. Engels M, Ackermann M. Pathogenesis of ruminant herpesvirus infections. *Vet Microbiol* 1996;53:3–15.
23. De-Giuli L, Magnino S, Vigo PG, Labalestra I, Fabbi M. Development of a polymerase chain reaction and restriction typing assay for the diagnosis of bovine herpesvirus 1, bovine herpesvirus 2, and bovine herpesvirus 4 infections. *J Vet Diagn Invest* 2002;14:353–356.
24. Wang J, O'Keefe J, Orr D et al. An international inter-laboratory ring trial to evaluate a real-time PCR assay for the detection of bovine herpesvirus 1 in extended bovine semen. *Vet Microbiol* 2008;126:11–19.
25. Wang J, O'Keefe J, Orr D et al. Validation of a real-time PCR assay for the detection of bovine herpesvirus 1 in bovine semen. *J Virol Methods* 2007;144:103–108.
26. Yan B, Chao Y, Chen Z et al. Serological survey of bovine herpesvirus type 1 infection in China. *Vet Microbiol* 2008;127:136–141.
27. Fulton R, d'Offay J, Eberle R. Bovine herpesvirus-1: comparison and differentiation of vaccine and field strains based on genomic sequence variation. *Vaccine* 2013;31:1471–1419.
28. Baker J, Rust S, Walker R. Transmission of a vaccinal strain of infectious bovine rhinotracheitis virus from intranasally vaccinated steers commingled with nonvaccinated steers. *Am J Vet Res* 1989;50:814–816.
29. Kennedy P, Richards W. The pathology of abortion caused by the virus of infectious bovine rhinotracheitis. *Pathol Vet* 1964;1:7–17.

Chapter 60

Infectious Agents: Bovine Viral Diarrhea Virus

Thomas Passler

*Departments of Clinical Sciences and Pathobiology, College of Veterinary Medicine,
Auburn University, Auburn, Alabama, USA*

Introduction

Bovine viral diarrhea virus (BVDV) is a pathogen of cattle and other artiodactyls with substantial impact on the health and well-being of affected livestock. Originally described by two North American research groups in 1946,[1,2] infections with BVDV are still cause for enormous financial losses to the cattle industries worldwide. While the two earliest reports on BVDV described a disease that was mostly of severe acute nature, less severe subacute cases were also observed.[1,2] In the decades following their original descriptions, BVDV infections in immunocompetent animals were reported to be mainly subclinical, and it was estimated that 70–90% of acutely infected animals did not display clinical signs.[3] The perception of BVDV as a mainly apathogenic bovine virus changed in the early to mid 1990s with the description of two new BVDV-associated syndromes, hemorrhagic syndrome and severe acute BVD that were responsible for severe clinical signs with mortality rates of up to 100%.[4,5] Genetic sequencing of BVDV isolates from severe outbreaks led to the description of a second genotype of BVDV (BVDV 2) based on its genetic dissimilarity of greater than 30% from classic BVDV strains.[5,6] Subsequent retrospective genetic typing revealed that BVDV 2 had been present in Ontario as early as 1981 without causing severe disease symptoms.[4] Variation of the severity and observed clinical signs among BVDV-associated disease outbreaks emphasizes the influence of the tremendous genetic variability of BVDV isolates.

BVDV is an RNA virus and is the prototypic member of the genus *Pestivirus* within the family *Flaviviridae*. Currently, the genus *Pestivirus* contains four recognized species: BVDV 1 and 2, border disease virus, and classical swine fever virus (CSFV).[6] Additional genetically distinct genotypes such as Pronghorn pestivirus, Giraffe pestivirus, strain V60 (Reindeer-1), and HoBi strain have been described, but have not yet been officially classified.[7–11] The lack of strict host specificity and antigenic relatedness of pestiviruses has caused difficulties with the traditional method of classification based on the host from which a strain was isolated, and more accurate taxonomic information is now based on monoclonal antibody binding assays and phylogenetic analysis of genomic sequences.[12–16] Although sequence variations are located throughout the entire genome, the 5′ untranslated region (UTR) is most commonly used for the differentiation of pestiviral species.[12,13,15,17,18]

Pestiviruses are enveloped viruses with a single-stranded RNA genome that is approximately 12.5 kb in length and contains a single large open reading frame flanked by UTRs at the 5′ and 3′ termini.[19] During replication of BVDV in the host cell cytoplasm, mutation and variation occurs readily with an approximate mutation frequency of 10^{-4} base substitution per base site.[20] At this high mutation frequency, at least one point mutation occurs per replication cycle of the viral RNA.[21] Frequent mutation is the consequence of error-prone viral RNA polymerases and results in variation of each progeny virus from the infecting parent virus.[20] Although millions of varying progeny virions are produced daily during the peak of infection, only copies with a selective advantage and without deleterious mutations will progress in the course of an infection.[20,22] Consequently, a swarm of mutant virions exists in the host during a BVDV infection.[20] The virions of this mutant swarm are similar in base sequence and as adapted to the host's organism and are referred to as a quasispecies.[20] The existence of BVDV as a quasispecies with frequent mutation events offers a significant advantage to viral survivability, as viral mutants are readily available that are capable of adapting to adverse conditions such as the host's immunity. With respect to virulence, adaptation to either lower or higher virulence may benefit viral survivability. While low virulence and host survival, as are common for host-adapted viruses, enable prolonged shedding, viral mutations that increase rates of replication and virulence allow for a burst of viral shedding.[20,23]

Bovine Reproduction, First Edition. Edited by Richard M. Hopper.
© 2015 John Wiley & Sons, Inc. Published 2015 by John Wiley & Sons, Inc.

BVDV isolates can be classified as noncytopathic or cytopathic, and this classification is based on the effects of an isolate on cultured cells. Cytopathic isolates cause cytoplasmic vacuolation and death of cultured cells, but infections with noncytopathic isolates are inapparent.[24,25] The categorization into biotype is not associated with the virulence of a BVDV isolate, and mild or severe disease can be observed with either biotype. The noncytopathic biotypes are more prevalent in nature and 60–90% of isolates from specimens at diagnostic laboratories are noncytopathic.[26,27] Only noncytopathic BVDV strains can induce persistent infections, and these noncytopathic strains are believed to be the most common source for cytopathic BVDV strains following homologous or heterologous recombination in the NS2–3 genomic region.[20] A third biotype of BVDV, the lymphocytopathic biotype, has been proposed, and refers to BVDV isolates that have cytopathic effects only on cultured lymphoid but not epithelial cells.[28] Lymphocytopathic isolates are BVDV 2 strains that were associated with severe clinical disease.[28]

Prevalence in cattle populations

Following the initial descriptions in North America, BVDV has been detected in cattle populations worldwide. The prevalence of seropositive animals depends largely on the management of cattle, including type of husbandry, use of vaccination protocols, addition of new herdmates, and presence of persistently infected (PI) animals.[29,30] Cattle herds that do not use vaccination programs or only use killed vaccines and do not have PI animals have fewer numbers of seropositive cattle than herds containing PI cattle.[29] In North America, serosurveys demonstrated seropositive rates between 40 and 90%.[31,32] The individual seroprevalence rate among unvaccinated dairy heifers in Canada for BVDV 1 and BVDV 2 was 28.4% and 8.9%, respectively, and the herd-level prevalence was 53.4% for BVDV 1 and 19.7% for BVDV 2.[33] In feedlot cattle, seroconversion to respiratory pathogens including *Pasteurella haemolytica*, *Mycoplasma* spp., and BVDV following arrival occurred in 40% of cattle,[34,35] but varied between 0 and 100% of cattle in 11 pens.[34] Similar rates of seroprevalence have been detected in cattle herds worldwide, but the observed range varied widely among countries (13–90%) and was largely influenced by the examined population, potentially reflecting global differences in cattle management.[36–52]

In contrast to seroprevalence rates, the prevalence of PI cattle is considerably lower and is generally believed to be less than 1% of all cattle. Persistently infected cattle may be found in clusters within groups of cattle, elevating the prevalence within populations.[53] The prevalence of PI calves arriving at feedlots has been demonstrated to be between 0.1 and 0.3%, and this rate is similar to the reported rate of 0.17% for beef cow-calf operations in the United States.[34,54–56]

Host range

Pestiviral infections have been identified in various species of the mammalian order Artiodactyla, and lack of host specificity may play an important role in the pestiviral survival strategy.[57] Among pestiviruses, BVDV has been the species most intensely studied as an infectious agent of heterologous hosts. Evidence of BVDV infection exists in over 50 species of artiodactyls in the families Antilocapridae, Bovidae, Camelidae, Cervidae, Giraffidae, Suidae, and Tragulidae.[58] The implications of heterologous pestivirus infections are twofold. First, heterologous BVDV infections may hamper the success of control strategies, as are in progress for BVDV in cattle and CSFV in swine. Second, heterologous BVDV infections may threaten the health and well-being of other mammalian hosts, such as free-ranging and captive wildlife. The role of heterologous infections in the epidemiology of BVDV is incompletely understood, but transmission and maintenance of BVDV within three host clusters –domestic small ruminants and swine, free-ranging and captive wildlife, and camelids – has been postulated.[59]

BVDV in small ruminants

The close phylogenetic relatedness of small ruminants to cattle and resulting greater possibility of permissive infection may promote a unique role of sheep and goats in the ecology of BVDV. Sheep and goats often have closer contact with cattle compared with other artiodactyls, especially in certain management conditions such as subsistence agriculture, alpine farming, or hobby farming. While one report did not identify a positive correlation between BVDV seropositivity of sheep flocks and presence of cattle,[60] other studies have reported significantly greater seroprevalence rates in sheep on farms with cattle, especially where communal alpine pasturing of sheep, goats, and cattle was practiced.[61]

Several serological surveys have demonstrated the occurrence of BVDV in small ruminants; however, cross-reactivity of pestiviral antibodies warrants caution when evaluating the results of such surveys. In recent reports on pestiviral seroprevalence in sheep, cross-neutralization assays identified highest reactivity against BVDV 1 rather than border disease virus.[60,61] When antibodies against pestivirus are detected in small ruminants, the distinction of pestiviral species may hold critical information on epidemiology, prevention, and control.[62]

Infections with BVDV in domestic small ruminants result in clinical signs of border disease. Postnatal infections commonly cause mild clinical signs, including pyrexia and leukopenia.[63] Experimental inoculation of juvenile sheep resulted in anorexia, tachycardia, pyrexia, and lung lesions especially associated with the pulmonary vasculature.[64] Infections with BVDV in pregnant small ruminants may result in uteroplacental pathology, pregnancy loss by fetal resorption or abortion, and birth of mummified or stillborn fetuses.[65] As in cattle, congenital BVDV infection in sheep and goats can result in damage to the central nervous system, and neuropathogenicity may be more severe with cytopathic biotypes.[66] Although fetal death and nonviability of lambs are common sequelae of transplacental BVDV infections in sheep, reports of viable PI offspring exist.[67,68] In contrast to cattle and sheep, viable PI goat kids appear to be rare and reproductive failure is the more common outcome of BVDV infection in goats.[65,69]

BVDV in swine

The close antigenic relationship between ruminant pestiviruses and CSFV has important implications for diagnostic testing and control of CSFV. Currently, examination of blood samples for CSFV antibodies relies on a combination of enzyme-linked immunosorbent assay (ELISA) and virus neutralization techniques, and considerable efforts are necessary in establishing a CSFV-specific diagnosis.[70] During the CSFV outbreak in the Netherlands in 1997/1998, 26.5% of CSFV ELISA-positive samples were caused by the presence of antibodies to ruminant pestiviruses.[70] Only a limited number of BVDV serosurveys have been performed in domestic swine, and reported seroprevalence rates are relatively low compared with other domestic livestock. Risk factors associated with seropositivity of domestic swine include the presence of cattle on the same farm, high density of small ruminants near swine populations, vaccination with BVDV-contaminated vaccines, and age of tested swine.[71–73] Postnatal BVDV infections in nonpregnant swine rarely result in clinical signs.[74,75] Experimental infection of 8-week-old pigs with either BVDV 1 or BVDV 2 did not result in clinical signs or increases in body temperature in any group, despite the use of a BVDV 2 isolate that was demonstrated to be virulent for cattle.[75] In contrast, BVDV infection of pregnant gilts resulted in reproductive disease, including intrauterine infection, pregnancy loss, and reduction in litter size.[76] While transplacental BVDV infection in pigs appears to less common than in ruminants, the birth of PI piglets has been reported.[77]

BVDV in New and Old World camelids

BVDV infections were identified through detection of antibodies as early as 1975 in Old World camelids (OWC) and 1983 in New World camelids (NWC).[78,79] However, most seroepidemiologic and experimental infection studies suggested that BVDV may cause infections without presenting a serious risk to NWC. Reports of BVDV isolation and identification of persistent infections in alpacas have prompted increased interest in BVDV infections in camelids, and BVDV is now considered an emerging pathogen of NWC.[80] The first description of a PI alpaca was made in Canada following natural exposure of a pregnant alpaca to a chronically ill cria.[81] From the PI cria, a BVDV 1b strain was isolated from buffy coat cells on different occasions before euthanasia of this animal. Several cases of PI alpacas have since been reported in North America and the United Kingdom.[80,82,83] PI alpacas may survive for several months but are affected by low birthweight, failure to thrive, inappetence, lethargy, chronic diarrhea, and chronic recurrent infections especially of the respiratory tract. In OWC, reproductive disease including abortion, stillbirth, weak calves, early neonatal death, and neonatal hemorrhagic disease have also been reported.[84] Noncytopathic BVDV has also been isolated from tissues of camels in which histopathologic changes were consistent with a BVDV infection.[85]

From NWC in Chile, both BVDV 1 and BVDV 2 have been isolated.[86] In contrast, in reports from North America and the United Kingdom in which the subgenotype of the infecting isolate was determined, all isolates belonged to BVDV 1b.[80–82] A study analyzed 46 BVDV isolates from over 12 000 North American alpacas and classified all isolates as noncytopathic strains of subgenotype 1b with 99% or more nucleotide homology in the 5′-UTR.[83] Two explanations for this apparent predominance of BVDV 1b in NWC have been proposed: first, exposures of alpacas to BVDV are rare and spread of the existing BVDV 1b strain is by extensive movement of a few PI animals; or, second, only unique 1b subgenotypes are able to establish transplacental infections in alpacas.[83] Movement of alpacas (including dams with cria by foot) between farms, mainly for breeding purposes, is common practice and have been described in reports of reproductive disease and birth of PI offspring, highlighting the importance of sound biosecurity practices.[81,82,87]

BVDV in wildlife

Although sources of BVDV infections in free-ranging wildlife are unknown, a likely source is contact with cattle. This is supported by the absence of antibodies in deer that were without contact with cattle for over 50 years.[88] Significantly greater seroprevalence rates were detected in white-tailed deer on ranches where cattle were present compared with ranches without cattle.[89] In contrast, an association between BVDV seroprevalence and cattle density was not detected in another study.[90] Similar findings have been described in a recent report from Minnesota where a greater percentage of deer were seropositive in a region with a lower cattle density.[91] The authors of this report concluded that cattle use and management (i.e., dairy or beef) may have an important impact on interspecific BVDV transmission, as there is likely less wildlife contact with housed dairy cattle compared with pastured beef cattle.[91] Various factors likely influence the transmission of BVDV between cattle and wildlife as has been described for other pathogens such as bovine tuberculosis.[92] With presence of suitable environmental and management factors, transmission of BVDV from cattle to deer is possible as has been demonstrated in a cohabitation experiment.[93] In this study, BVDV was efficiently transmitted from PI cattle to white-tailed deer and resulted in seroconversion in all adult deer and birth of PI fawns from four of seven pregnant does.[93]

Maintenance of BVDV within a cervid population without presence of cattle was suggested by the presence of antibodies in over 60% of caribou that had not been in contact with domestic ruminants for over 25 years.[94] The presence of endemic BVDV infections as indicated by high seroprevalence rates was also suggested in reindeer in Norway and US cervid populations.[95–97] This is supported by identification of high seroprevalence rates in eland populations of Zimbabwe, in which a PI and a virus-isolation-positive eland were identified.[98]

Only a limited number of studies have surveyed free-ranging wildlife populations for the presence of BVDV or BVDV antigen. Anderson and Rowe[98] utilized an antigen-capture ELISA to detect BVDV in a subset of 303 seronegative animals during a serosurvey in Zimbabwe and detected two BVDV-positive eland. In Germany, cytopathic BVDV was detected in spleen samples from two of 203 deer and both animals were seronegative roe deer.[99] Three surveys utilizing immunohistochemistry (IHC) or ELISA investigated the occurrence of BVDV in free-ranging cervids in the United States and results suggest that PI cervids exist in wildlife populations.[100–102] In Alabama, one of 406 skin samples from white-tailed deer was positive on IHC and the antigen distribution resembled that of PI cattle.[101] The skin sample of one of 5597 deer in Colorado was positive on IHC and this result was confirmed by detection of viral RNA in skin and lymph node samples.[100] In Indiana, two of 745 white-tailed deer were positive for BVDV by antigen-capture ELISA with subsequent isolation of cytopathic and noncytopathic BVDV.[102] To date, validation of BVDV assays for use in wildlife has not been performed, and this may be critical as considerable variations were observed among the IHC and antigen-capture ELISA on skin samples of white-tailed deer (T. Passler and P. Walz, unpublished observations). In addition to surveys, isolation of BVDV was successful in free-ranging roe deer in Hungary, a mule deer in Wyoming, and two white-tailed deer in South Dakota.[97,103,104] In these reports, clinical illness including emaciation, weakness, and death prompted further investigations leading to the isolation of BVDV from tissues.

Few reports on the clinical outcome of BVDV infection in wildlife exist, but reported clinical signs are similar to BVDV infections in cattle. Experimental inoculation with BVDV NY-1 did not result in clinical signs in four mule deer and one white-tailed deer fawns, despite evidence of viremia and nasal shedding.[105] Similar findings were made in yearling elk that were inoculated with either BVDV 1 Singer or BVDV 2 24515.[106] In two young reindeer, loose, bloody, and mucoid feces, transient laminitis, or coronitis were observed after inoculation with BVDV Singer and mild lesions were detected at necropsy.[107] Naive white-tailed deer fawns developed moderate pyrexia and marked to moderate decreases in lymphocyte populations in response to either BVDV 1 or BVDV 2; and lethargy or coughing was observed in individual fawns.[108,109] Similar to other species, BVDV infections in wildlife may have the most important implications for reproductive health. Experimental inoculation of pregnant white-tailed deer may result in reproductive failure, including fetal resorption, fetal mummification, stillbirth, and abortion.[110,111]

Clinical disease syndromes

Infections with BVDV may result in a wide variety of clinical manifestations as the outcome of infection depends on the triad of host-associated factors, environmental stressors, and viral characteristics.[112] Host-associated factors include (i) immune status and immunocompetence, (ii) pregnancy status, and (iii) the gestational age of the fetus at the time of infection. Viral characteristics that influence the outcome of an infection include the biotype of the infecting strain and genotypic variation that determine virulence and antigenicity. Infection with both species of BVDV may result in a broad spectrum of clinical manifestations, but severe acute BVD, thrombocytopenia, and hemorrhagic syndrome, as reported in North America, were associated only with BVDV 2.[4,5,113]

Postnatal infections in immunocompetent cattle

Postnatal infections in cattle that are able to immunologically respond to the BVDV infection are referred to as "acute" or "transient" infections, and the severity of clinical signs depends on the infecting BVDV strain. After infection of susceptible cattle, the tonsils and respiratory tract are the first sites of BVDV replication. Subsequently the pathogen disseminates to many other epithelial and lymphoid tissues, resulting in infection of gastrointestinal, integumentary, and respiratory tissues.[114,115]

The majority of BVDV infections in immunocompetent seronegative cattle proceed subclinically.[3] However, close observation of infected animals usually reveals mild signs including hyperthermia, leukopenia, and decreased milk production. Affected animals develop neutralizing antibodies.[116] Symptomatic infections are most commonly observed in 6–24-month-old cattle following waning of maternal immunity, in colostrum-deprived calves, or in seropositive cattle as a result of infection with a heterologous BVDV strain.[117] Clinical signs in affected animals include pyrexia, lethargy, leukopenia, ocular and nasal discharge, oral erosions and ulcers, blunting of oral papillae, diarrhea, and decreased milk production. More virulent BVDV strains can also cause epithelial erosions at the interdigital spaces, coronary bands, teats, or vulva. Although BVDV-associated immunosuppression can result in pneumonia, the tachypnea observed in acutely infected animals is likely the result of hyperthermia and other nonpulmonary factors.[117]

In the 1990s, severe acute BVD was described in cattle herds in the United States and Canada.[5,113] The outbreaks were characterized by a peracute course and caused unusually high rates of morbidity and mortality in all ages of cattle. The described clinical signs and postmortem findings were principally those of mucosal disease, with severe diarrhea, fever, and ulcers and erosions of the upper alimentary tract.[4] Histopathologic findings included a dramatic lymphocytolysis and lymphoid depletion of Peyer's patches, necrosis of intestinal crypt epithelium, and diffuse ulcerative lesions in the upper alimentary tract. Sudden death without premonitory clinical signs, abortions, and pneumonia were also prominent findings in some affected herds. Interestingly, the gross and histopathologic postmortem changes were more pronounced and more likely to be observed in older animals, which is in contrast to most previous descriptions of transient BVDV infections.[4] The viral isolates from all reports of severe acute BVD were determined to be different from traditional BVDV isolates in genotype and monoclonal antibody binding patterns, prompting the designation BVDV 2.[5,15,118]

Another form of severe acute BVD associated with BVDV 2 is the hemorrhagic syndrome, which is characterized by marked thrombocytopenia. Hemorrhagic syndrome is observed only in a minority of cattle with severe acute BVD, contributing to a diverse clinical picture in an outbreak.[117] Clinical signs in affected cattle include petechiation and ecchymoses of mucosal surfaces, epistaxis, bloody diarrhea, bleeding from injection sites or trauma, fever, and death.[113] The marked thrombocytopenia and leukopenia of affected animals is accompanied by altered function of platelets, and thus quantitative and qualitative platelet defects contribute to the observed hemorrhagic diathesis.[119]

In addition to direct effects on the host organism, BVDV suppresses the number and function of various innate and adaptive immune components, further damaging tissues by enabling infections with secondary pathogens. BVDV-induced immunosuppression is especially important in polymicrobial diseases such as the bovine respiratory disease complex of feedlot cattle and dairy calves. Infection of cells of the innate immune system may result in impairment of function and decreases in the number of circulating leukocytes. The microbicidal, chemotactic, and antibody-dependent cell-mediated cytotoxicity was impaired in neutrophils that were infected with BVDV.[120] Infected monocytes may undergo apoptosis, and a reduction of 30–70% in the number of monocytes has been observed after BVDV inoculation of calves.[121–123] Infection with BVDV diminishes the ability of antigen-presenting cells to present antigen to T-helper cells by reduction of Fc and C3 receptor expression and downregulation of major histocompatibility complex (MHC)II and B7 molecules.[121,124–126] A strain-dependent inhibition of function with reduced ability to kill bacterial and fungal pathogens was observed following BVDV infection of cultured macrophages; however, significant functional inhibition was induced only by virulent strains of BVDV.[125]

Similar to innate immunity, adaptive immune responses are affected quantitatively and qualitatively by BVDV infections. On infection with BVDV, circulating lymphocytes are reduced and lymphocytes in lymphoid tissues are depleted, and this depletion is strain-dependent.[121,127] Of the T-lymphocyte subpopulations, CD8$^+$ cytotoxic T-lymphocytes are most dramatically depleted, followed by CD4$^+$ T-helper cells, with little reduction of circulating γ/δ T-cells.[128] In addition to lymphocyte depletion, functional alterations of CD4$^+$ cells are pronounced, and noncytopathic BVDV infections result in a shift toward the Th2 response, with reduced cell-mediated but high levels of humoral immunity.[129–131] Virulent BVDV isolates can cause reductions in MHCI expression of infected cells, thus affecting cytotoxic CD8$^+$ cell function; however, this reduction appears to be strain-dependent.[125]

The humoral immunity to BVDV infection depends largely on the quantity of neutralizing antibodies and the antigenic relatedness of the infecting strain with that of the original exposure. Although an amount of cross-reactive protection is conferred by vaccination against heterologous strains, and even between BVDV 1 and 2, superior protection from infection appears to result from homologous exposure.[132,133]

Infections of the reproductive organs

While BVDV-associated suppression of the host immune response and respiratory disease are cause for the greatest losses in feedlot and stocker operations, susceptible cow-calf and dairy herds suffer considerable economic damage from BVDV-induced reproductive failure. Reproductive losses as a result of BVDV infection can occur throughout gestation, and BVDV infection at the time of breeding can result in reduced conception rates and embryonic losses.

In heifers experimentally infected with BVDV 9 days prior to insemination, a conception rate of 44% was detected compared with 79% in control animals.[134] In the same study, other heifers were infected 4 days after artificial insemination (AI) resulting in a conception rate of 60% by day 20; however, significant embryonic loss was detected by day 79 when only 33% of animals were still pregnant. Detrimental effects of BVDV infection on fertility were also reported in field studies, in which poor conception rates, early embryonic losses, and pregnancy losses were detected.[135–138] In contrast, heifers exposed to PI calves for 220 days beginning at 50 days prior to the breeding season had the same pregnancy rates as unchallenged control heifers, resulting from development of protective immune responses prior to breeding.[139] Highlighting the benefits of a controlled breeding season, this study demonstrated that BVDV infections can be self-limiting in certain herds, when all animals are protected from BVDV during gestation and PI calves are not born.[139]

Both male and female gonads and the entire female reproductive system are a refuge for BVDV replication.[140] In postpubertal heifers experimentally infected with a noncytopathic BVDV, the virus was isolated from the corpora lutea of two animals and BVDV antigen staining was detected predominantly in the stroma of ovarian cortex. In this study, prolonged antigen staining and associated chronic lymphocytic oophoritis were detected from 6 to 60 days following intranasal inoculation.[141] Similarly, experimental inoculation of susceptible heifers with cytopathic BVDV resulted in chronic oophoritis for days 5–61 after infection, and BVDV was detected in the ovaries of a subset of cattle immunized with a commercial vaccine containing cytopathic BVDV.[142,143] Ovarian inflammation may alter ovarian cytokine concentrations or disrupt tissue macrophages, thus interfering with normal ovarian dynamics, and while the effects of BVDV on ovarian function remain incompletely understood, alteration of follicular formation and inhibition of ovulation were demonstrated following BVDV infection.[144] A decrease in the number of subordinate follicles and retarded growth of dominant anovulatory and ovulatory follicles were observed in heifers experimentally infected with BVDV, but the infection did not result in alterations of peripheral estradiol or progesterone patterns.[144] In contrast, experimental infection with noncytopathic BVDV 2 days prior to estrus resulted in significantly reduced plasma estradiol concentrations but did not affect progesterone secretion or plasma prostaglandin (PG)F$_{2\alpha}$ concentrations.[145] While the average number of follicles in each size category was not affected in BVDV-infected superovulated heifers, infection resulted in the absence or delay of the preovulatory surge of luteinizing hormone, reduced number of corpora lutea and recovered ova/embryos, and oophoritis with necrosis of granulosa cells.[146–148] Alteration of sex hormone concentrations with potential for disruption of ovulatory follicles to form a competent corpus luteum was also detected in heifers experimentally inoculated 9 days before synchronized ovulation.[149]

Collectively, these studies suggest that transient BVDV infections can interfere with normal ovarian dynamics and result in transiently or prolonged reduced reproductive ability.[148,150] Other causes for reproductive losses associated with BVDV infection at the time of breeding have been proposed and include disruption of the hypothalamic–pituitary–gonadal axis, deficiency of gonadal leukocytes, and incompatibility of the oviductal and uterine environment for embryonic development.[140,150] Infection with BVDV was demonstrated in oviductal cells following intravenous inoculation with noncytopathic BVDV and intrauterine infection with cytopathic BVDV resulting in salpingitis for 21 days.[151,152] In uterine flushes from a subset of experimentally infected heifers, BVDV was isolated on days 8 and 16 following infection with BVDV, demonstrating a potential risk for the developing embryo in the uterine environment.[152]

Widespread BVDV infection of the reproductive system and reduced reproductive performance are described in PI cattle. Extensive BVDV immunostaining and pathoanatomical abnormalities including significantly reduced follicular structures and ovarian and follicular hypoplasia were described in PI cows.[153] In addition to ovarian structures, including follicular epithelium, theca cells, cumulus oophorus, stromal macrophages, and walls of small arteries, BVDV antigen is present in oviductal epithelial cells, uterine glandular and luminal epithelia, arterial walls, and smooth muscle of PI cattle.[154,155] While oocytes in primordial and secondary follicles did not contain BVDV antigen in one study, approximately 20% of oocytes in all stages of follicular development contained BVDV antigen in another report.[145,155]

Infections of the conceptus

Infections with BVDV can have deleterious effects on the bovine pregnancy, and infections before the third trimester are most harmful. BVDV is highly efficient in crossing the placental barrier, and fetal infections are common with either biotype.[156] Increasing rates of fetal infection of 33, 33, 86, and 100% were detected in cows experimentally infected at 4, 11, 18, and 30–34 days after insemination, respectively.[157] In ewes infected with BVDV at 18 days of gestation, the virus was detected by reverse transcription polymerase chain reaction (RT-PCR) and IHC in fetal tissues approximately 100 hours following infection, and BVDV-positive cells were more numerous in fetal than maternal tissues.[158] Fetal infections occur in an orchestrated fashion, beginning with infection of fetal membranes by 72 hours, followed by infection of bulbus cordis and fetal aorta by 7 days, and subsequent spread to many fetal tissues by days 10–14.[158,159] Further dissemination through fetal tissues can be observed as the infection proceeds, and a primary target organ of fetal BVDV infection is the brain.[160] While viral antigen can be detected in most tissues from PI fetuses, histopathologic changes are primarily detected in the brain, liver, and spleen.[161] Invasion of the fetal brain may occur in a "Trojan horse" mode of invasion, in which BVDV enters the central nervous system (CNS) within infected microglial precursors.[162]

To date, the mechanism of BVDV entry into the fetus is poorly understood. Infection of pregnant cows with noncytopathic BVDV results in pronounced upregulation of interferon (IFN)-stimulated genes such as *ISG15* in the first week after infection, and blood concentrations return to baseline by 45 days after infection.[163,164] In heifers carrying persistently or transiently infected fetuses, BVDV infection results in measurable serum concentrations of type I IFN, but this antiviral activity is not capable of preventing fetal infection.[164] This may be due to failure of maternal type I IFN and antibodies to enter the fetal bloodstream as a result of the blood–placental barrier.[164] Likewise, BVDV infection results in measurable IFN responses in fetal blood, which is pronounced in fetuses infected after the susceptible window for development of persistent infection.[165] While IFN responses are also present in PI fetuses and resulting PI calves, these responses are of a more modest and chronic nature and are not able to eliminate the infecting BVDV in the absence of specific adaptive immune responses.[165,166]

The outcome of fetal infection depends on the viral virulence and biotype, and the gestational age. Fetal death and abortions can occur at any gestational age, but are most common during the first trimester.[150] Fetal death during early pregnancy may result in fetal resorption or mummification (Figure 60.1) and expulsion of the fetus may occur up to 50

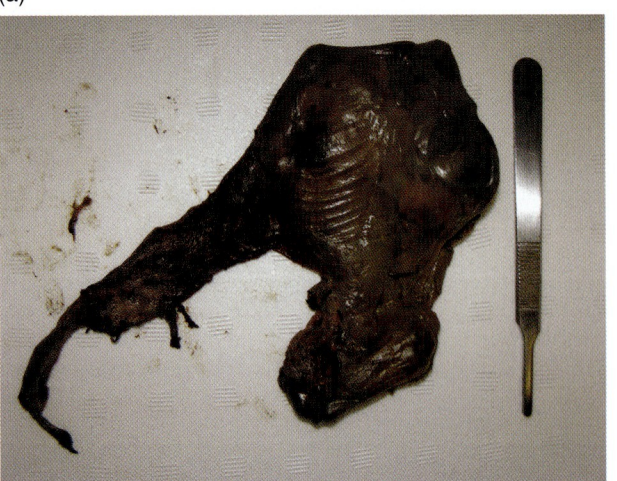
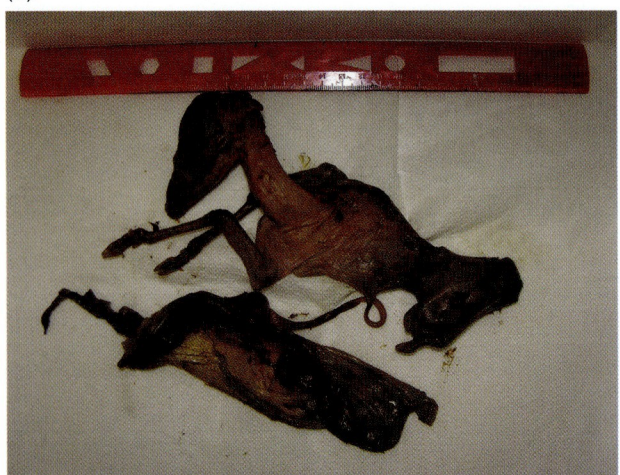

Figure 60.1 Fetal mummification in white-tailed deer. Fetal death was estimated to have occurred at 94 days of gestation, approximately 50 days after BVDV infection of the dam. The mummified fetus was born together with a viable persistently infected fawn after completion of the full gestational period. (a) Mummified fetus as found in birthing nest; (b) fetal membranes removed.

days following exposure to BVDV.[167] Of four heifers infected between 29 and 41 days of pregnancy, two abortions and two fetal resorptions were observed.[168] While infections in the preimplantation phase (before day 40 of pregnancy) result in large numbers of pregnancy losses, fetuses that survive infection with noncytopathic BVDV of low to moderated virulence from approximately 40 to 125 days become PI.[169,170] Persistent infections can only occur with noncytopathic BVDV strains, and experimental infections with cytopathic isolates do not result in the development of PI calves.[170,171] Experimental inoculation of seronegative cattle with noncytopathic BVDV at 75 days of gestation results in persistent infection in up to 100% of fetuses, and this gestational age is commonly used to induce persistent infection in research studies.[164,166,172,173] The pathophysiology of persistent infection is incompletely understood, but this phenomenon occurs in fetuses before completion of immune system development, when viral antigens are recognized as self antigens, resulting in negative selection of BVDV-specific lymphocytes.[150] The negative selection of reacting CD4+ lymphocytes is highly specific and the exchange of a single amino acid in the BVDV antigen may be sufficient to stimulate immune responses in CD4+ lymphocytes from PI cattle.[174] The lack of immune responses results in widespread distribution of BVDV throughout the host's tissues, and PI cattle shed BVDV in multiple secretions and excretions, including nasal discharge, urine, semen, colostrum, milk, and feces.[175]

Fetal infections during the period of organogenesis (approximately 100–150 days of gestation) may result in congenital malformations, many of which are associated with abnormal development of nervous tissues. Defects of the CNS include cerebellar hypoplasia (Figure 60.2), hydranencephaly, hydrocephalus, porencephaly, microencephaly, spinal cord hypoplasia, and a variety of ocular changes such as cataracts, optic neuritis, retinal degeneration, and microphthalmia.[150,176,177] Less commonly, congenital BVDV infection can cause CNS hypomyelination, which has been reported with both species of BVDV.[178,179] In addition to CNS abnormalities, congenital BVDV infections can result in brachygnathism, osteopetrosis, thymic aplasia, hypotrichosis, hair coat abnormalities, pulmonary hypoplasia, and growth retardation.[150,175,176,180]

While abortions are still possible in later stages of gestation, congenital infections following the completion of organogenesis and immunocompetence result in effective immune responses and clearance of the virus. These calves are born with precolostral serum antibody titers and usually appear normal at birth.[150] However, one study suggested that calves born with precolostral antibodies against BVDV are more likely to develop severe illness during the first months of life, indicating that late gestational infections with BVDV may not be benign.[181]

Infections in persistently infected animals (mucosal disease)

Although PI calves may survive to reproductive age, many are born weak, stunted, and die shortly after birth.[112] PI cattle have a 50% greater risk of leaving the herd before the first year of age, dying, or being slaughtered due to unthriftiness as compared to normal cohorts, which may be partly due to a poor immune response and susceptibility to opportunistic infections.[182,183] An additional risk for PI cattle is the development of fatal mucosal disease, which develops when cattle that are PI with a noncytopathic BVDV become superinfected with a cytopathic strain. The severity of the developing clinical signs depends on the homology between both strains, and mucosal disease can be classified as acute, chronic, or delayed onset.[184]

Acute mucosal disease occurs when the superinfecting cytopathic strain is in close antigenic homology with the PI strain, which most commonly is the result of mutational events that alter the PI strain's biotype.[185] Because of the close homology of both strains, acute mucosal disease results in case fatality rates that approach 100%.[117] Because acute mucosal disease affects only PI cattle, the disease is characterized by a low morbidity but very high mortality.[117] However, in groups of cattle with multiple PI cattle that are infected with similar BVDV strains, outbreaks of mucosal disease have been reported.[186] Clinical signs of acute mucosal disease include fever, anorexia, tachycardia, polypnea, nasal discharges, and profuse foul-smelling diarrhea with mucosal shreds, fibrin casts, and blood. In addition, erosion and ulcers are commonly present on the alimentary epithelia (Figure 60.3), coronary bands, prepuce, and vulva.[117]

Superinfections with heterologous cytopathic strains may result in chronic forms of mucosal disease that are not fatal but result in chronic clinical signs. The source for heterologous cytopathic strains is commonly vaccine strains in modified-live vaccines.[184] Clinical signs include anorexia, weight loss, diarrhea, bloat, lameness, and erosive and ulcerative lesions on epithelial surfaces.[184] Sufficient heterology between the PI strain and the cytopathic strain may result in an immune response that clears the superinfection, enabling the animal's recovery from mucosal disease, but not persistent infection.[184]

Under experimental conditions, the incubation period of acute fatal mucosal disease is 7–14 days.[187] In contrast, delayed-onset mucosal disease occurs in PI cattle several

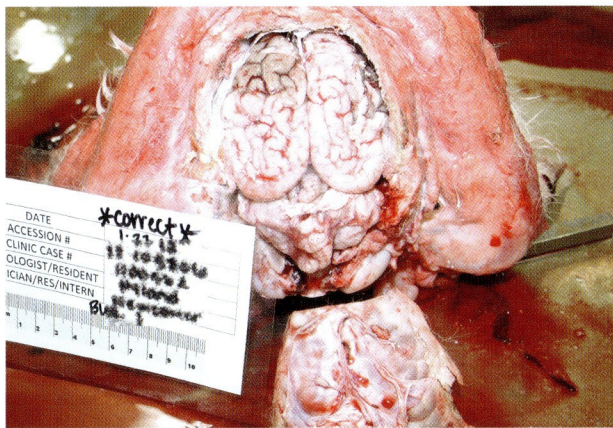

Figure 60.2 Cerebellar hypoplasia in an 11-day-old bull calf. Calf had clinical signs of cerebellar disease including intention tremors, ataxia, and hypermetria. Histopathologically, diffuse cerebellar hypoplasia characterized by shortened and blunted cerebellar folia with marked decrease in cellularity of the granular cell layer, paucity of Purkinje cells, and a markedly thin molecular layer were identified. Image courtesy of Dr Joanna Hyland, Auburn, Alabama.

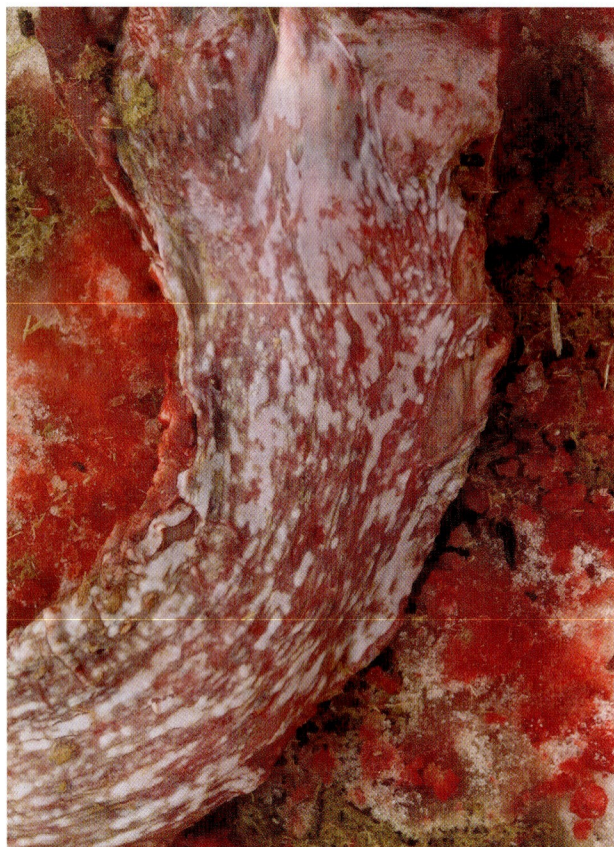

Figure 60.3 Severe esophageal erosions and ulcers in a steer with acute mucosal disease. Image courtesy of Dr Jenna Bayne, Auburn, Alabama.

weeks to months after inoculation with the cytopathic strain.[188] As for chronic mucosal disease, delayed-onset mucosal disease is caused by a heterologous cytopathic strain from an external source, but clinical sings of mucosal disease occur after RNA recombination events between the PI strain and the heterologous cytopathic strain. This RNA recombination creates a homologous cytopathic strain and fatal mucosal disease.[189]

Infections of bulls

Both persistently and acutely infected bulls shed BVDV in semen, but the duration and level of shedding is much greater in PI bulls. In experimentally infected bulls, viral titers in semen ranged from 5 to 75 $CCID_{50}/mL$, and virus isolation was similarly effective from raw and extended semen.[190] In a group of heifers experimentally inseminated with contaminated semen from an acutely infected bull, approximately 5% became systemically infected.[191] While this initial rate of infection was low, it was sufficient for further transmission to susceptible heifers and maintenance of BVDV in the herd.[191] Under field conditions, AI with semen from two acutely infected bulls resulted in infections rates in exposed cows of 15.4% and 1.8%, respectively.[192]

Semen from PI bulls can contain large quantities of BVDV (10^3–10^7 $CCID_{50}/mL$).[193,194] Insemination of susceptible heifers with semen from PI bulls consistently results in infection.[193-195] Semen quality from PI bulls ranges from normal on the basis of concentration, motility, and morphology of spermatozoa to abnormal, with morphological abnormalities of spermatozoa and low concentration and motility.[193,196] The use of semen from PI bulls usually affects conception rates negatively, which is likely due to a combination of factors, including lower quality semen, ill-thrift of affected bulls, and effects of BVDV on the reproductive tract of exposed cows.[150,193,197]

In acutely infected bulls, formation of neutralizing antibodies results in viral clearance from most organ systems, but chronic infections of testicular tissues despite formation of serum antibodies are described.[198] This phenomenon was first described in a seropositive nonviremic bull that consistently shed BVDV in semen, thought to be the result of an acute BVDV infection prior to completion of a functional blood–testes barrier.[199] Insemination of seronegative heifers with the bull's semen containing less than 2×10^3 $CCID_{50}/mL$ of BVDV resulted in infection and seroconversion.[200] Localized and prolonged testicular infections have also been experimentally produced following intranasal inoculation and subcutaneous vaccination of peripubertal bulls with BVDV.[201,202] Viral RNA was detected in semen for 2.75 years following BVDV exposure, and infectious BVDV was isolated from testicular tissue for up to 12.5 months after BVDV exposure.[203] Semen from a bull with prolonged testicular infection resulted in BVDV transmission when given intravenously to a seronegative calf.[201] However, this finding could not be repeated in other trials, and artificial or natural insemination did not result in transmission of BVDV, indicating that the chronic presence of BVDV in testicular tissues could be of limited epidemiologic importance as a result of neutralization of infectious virus by humoral immunity in the ejaculate.[198,203]

BVDV and *in vivo* and *in vitro* derived embryos

Potential introduction of BVDV to *in vivo* and *in vitro* derived embryo production systems can result from three sources, including donor cows, recipient animals, and materials of animal origin such as fetal bovine serum, and all three have to be free from BVDV to prevented transmission.[204] The World Organization of Animal Health (OIE) states that "BVDV is a particularly important hazard to embryo transfer and the manufacture of biological products for other diseases due to the high frequency of contamination of batches of fetal calf serum."[205] *In vivo*-derived embryos could be exposed to BVDV if they were collected from acutely infected or PI cattle, or if contaminated sera were used in embryo collection, culture, or cryopreservation media.[204] However, BVDV appears to present a more serious threat in the production of embryos via *in vitro* production than via *in vivo*-derived embryos. The virus can be introduced via serum or abattoir-origin oocytes and somatic cells used in *in vitro* embryo production. Hence unrecognized infections with noncytopathic isolates could lead to amplification of the virus throughout *in vitro* maturation of oocytes, *in vitro* fertilization, and *in vitro* culture of embryos.[206] Furthermore, the virus readily adheres to the zona pellucida of *in vitro*-produced embryos rendering the embryo washing and trypsin treatment procedures even less effective than they are for *in vivo*-derived embryos.[207] Production of embryos by somatic cell nuclear transfer presents even more unique hazards than that of *in vivo*-derived or *in vitro*-produced embryos. Introduction of BVDV via serum used to culture both cytoplast-donor cell lines and the resulting embryos can readily occur. The cultured embryos are especially vulnerable because the zona pellucida has been either broken or removed to facilitate the process of cell fusion. Moreover, cumulus–oocyte complexes are usually collected at abattoirs from donors of unknown BVDV status and cell lines are often of fetal origin, creating enhanced opportunities for introduction and amplification of BVDV as embryos are produced and develop.[206] Therefore, BVDV still continues to be a concern in embryo technologies and a more detailed discussion of BVDV and embryo technologies is included in Chapter 80.

Diagnosis, prevention, and control

As for other infectious pathogens, the main principles of BVDV control are the elimination of pathogen reservoirs and the prevention of transmission from infected to susceptible animals. Any BVDV control program should focus on three major principles: (i) identification and elimination of PI animals, (ii) enhancement of immunity through vaccination, and (iii) improvement of biosecurity to prevent exposure of susceptible cattle.[208] PI cattle are considered the most important source for BVDV and are central in the epidemiology of the virus. Ideally, PI cattle are detected before entry into a herd, and all new herd additions including breeding bulls should be screened. In herds using a controlled breeding season, all calves, replacement heifers, open cows, and breeding bulls should be tested simultaneously before the beginning of the breeding season.[208] If pregnant dams are present, they should be segregated and their calves tested as soon as possible. All newborn calves should be tested as soon as possible to prevent exposure of susceptible animals. Because PI calves can be born to acutely infected or PI cows, the dams of PI calves must be retested.

When attempting BVDV control on a herd level, diagnostic assays should have a high sensitivity to detect BVDV, use an easily obtained sample, and be highly specific for PI status. For this purpose, antigen-capture ELISA or IHC assays utilizing skin biopsy samples (ear notches) are well suited. While various diagnostic tests are available to detect infection with BVDV, not all available tests are applicable to each clinical problem. Absence of rational use hampers efforts in solving production problems caused by BVDV.[209] Principally, all available diagnostic assays can be grouped into methods that culture BVDV in tissue cells, detect antigen in diagnostic specimens, amplify the RNA genome of BVDV, or detect the presence of antibodies in serum or milk.[209]

Virus isolation

The isolation of BVDV from various specimens, including serum, buffy coat, semen, nasal swabs, and tissue samples remains the "gold standard" of diagnosis.[201,209,210] For BVDV diagnostics on live animals, white blood cells are the preferred sample for virus isolation as there is no interference by serum neutralizing antibodies and the virus is highly cell-associated. Tissues of the lymphoid system including spleen, thymus, and lymph nodes are the preferred sample from postmortem examinations or aborted fetuses.[209] BVDV grows readily in various cell lines, including those of other species, and this may result in difficulties when using BVDV-contaminated cell cultures.[211–213] Successful culture of cytopathic BVDV strains is indicated by vacuolation and cell death within a few days of virus inoculation. For noncytopathic strains, immunofluorescence or immunoperoxidase staining techniques are utilized to visualize cellular infection with BVDV.[214,215] Virus isolation may also be used as a herd screening test in the form of the immunoperoxidase monolayer assay.[216] Although economic, this test uses serum as the diagnostic sample and is not ideal for detection of acute infections or testing of cattle below 3 months of age that have acquired colostral immunity. Failure to detect PI cattle with the immunoperoxidase monolayer assay has been reported, although at an extremely low rate of detection error.[217]

Antigen detection

Direct detection of BVDV antigen allows more rapid and economic testing of samples compared with virus isolation. Antigen detection can be performed by immunofluorescence, IHC, and ELISA techniques. In general, the analytical sensitivity and specificity of antigen detection assays are less than those of virus isolation, which accommodates the detection of PI cattle rather than of acute infections. In contrast to the traditional laboratory identification of PI cattle using two positive virus isolations performed 3 weeks apart, the commercially available ELISA kits and IHC are used to identify PI cattle with a single sample.[218] Several studies have demonstrated excellent results for antigen-capture ELISA and IHC for use as screening tests for PI cattle.[219–222] False-positive results due to acute infections are rare, and the antigen-capture ELISA and IHC tests can be utilized to test calves before waning of colostral immunity, which interferes with other assays.[209,223] Antigen detection is most frequently performed on ear-skin biopsies, but various samples including tissues, whole blood, serum, milk, and nasal swabs are suitable after appropriate sample preparation.[209] In addition to laboratory-based ELISA tests, a chute-side individual sample ELISA test has been developed, which can be used on ear-notch or serum samples to PI animals in 20 min.[224]

RT-PCR

The description of the genomic sequence of BVDV and the development of suitable probes and commercially available kits has enabled the widespread use of RT-PCR as a routine diagnostic method.[225] The technique is highly sensitive and can be utilized for various diagnostic samples including milk, urine, tissues, serum, whole blood, semen, and embryos.[226–231] Although RT-PCR techniques can be used on diagnostic specimens that are fixed in formalin, RNA fragmentation occurs and reduces assay sensitivity.[232]

The high sensitivity of RT-PCR exceeds that of virus isolation, and the assay can detect the viral RNA of a single viremic sample in a pool of 100 negative sera.[233] The ability to pool samples such as milk, serum, or whole blood offsets the initial higher cost of the assay and allows economic screening for PI animals. Pooling strategies using skin biopsies or blood samples rather than bulk-tank milk do not abolish the need to collect a sample from each animal in the herd.[229] In milk, BVDV was detected by RT-PCR in samples that were serially diluted to 1 : 640.[234] The number of samples that is optimal for pooling has been controversial, and pools of 30 serum samples to 100 samples have been recommended.[235,236] The economic benefits of sample pooling decrease when greater prevalence rates exist in populations, and the competitive benefit of pooling is diminished when prevalence rates exceed 3%.[237]

Serology

The detection of serum antibodies directed against BVDV can be performed by various assays but the ELISA and serum neutralization tests predominate and have replaced indirect immunoperoxidase or indirect immunofluorescence techniques.[238] With appropriate application, serologic tests allow the assessment of vaccine efficacy and compliance with established vaccination protocols, the detection of herd exposures to BVDV, and the association of clinical signs with BVDV infection. Difficulties exist in determining the etiology of detected antibodies, as seroconversion may be the result of acute infections, vaccination, or transfer of maternal immunity through colostrum. Commercial blocking ELISA assays that detect antibodies directed against the NS3 protein of BVDV are unable to differentiate antibodies derived from natural infection or vaccination, and marker vaccines are currently unavailable for BVDV.[239] Serologic testing of previously unexposed sentinel animals can be utilized to detect the presence of PI cattle in herds.[240] However, the analysis and comparison of serologic titers do not allow a clear distinction between PI-free herds and PI cattle.[241]

Vaccination

In the United States, BVDV modified-live and inactivated vaccines are widely available, most commonly in the form of multivalent vaccines containing other respiratory and reproductive pathogens. While modified-live vaccines stimulate a broader immune response in vaccinated animals, are less expensive, and provide some protection after administration of a single dose, inactivated vaccines are utilized when safety concerns exist. However, failure to administer the necessary second booster dose at initial immunization can result in breakdown of BVDV control programs using inactivated vaccines.[242]

Because of antigenic diversity of BVDV strains and failure of some animals to adequately respond to immunization, vaccination efficacy cannot be 100%. Therefore, vaccination programs must be part of a control strategy that includes removal of PI animals and measures of biosecurity.[243] Traditionally, only BVDV 1 strains were used in vaccine production, but the antigenic diversity between both BVDV species has prompted the inclusion of BVDV 2 in most vaccines. Studies have documented protection from BVDV 2-induced disease in cattle vaccinated with BVDV 1-containing modified-live or attenuated vaccines,[244–246] but antigenic diversity of BVDV and resulting failure of cross-protection continues to be of concern.

While protection of each individual animal from BVDV-induced disease is the desirable goal of vaccination programs, the increase in herd immunity and prevention of viral spread following accidental introduction are more reasonable objectives in BVDV control. Vaccination against BVDV-induced disease is principally aimed at two goals: (i) protection from clinical disease or (ii) prevention of fetal infection and birth of PI offspring. While identical immune mechanisms are stimulated to achieve either goal, a much larger level of immune protection is required to prevent fetal infection, which requires T- and B-cell responses that approach sterilizing immunity.[243] Vaccination programs should be designed with a focus on either goal, dictating the optimal timing of vaccination and choice of vaccine type.

Protection from clinical disease is the desired goal of BVDV vaccination in calves, and the greatest humoral and cell-mediated immunity levels should be present at the time of highest risk, such as at commingling, arrival at a stocker operation, or shipping to a feedyard.[247] In calves arriving at the feedyard with low antibody titers against BVDV and other respiratory pathogens, decreased economic returns, increased morbidity rates, and increased treatment costs were observed.[247,248] Increased morbidity rates and treatment costs were associated with inappropriate timing of vaccination close to delivery of calves or lack of a booster dose in programs using inactivated vaccines, emphasizing the value of adequate preconditioning programs.[248] Commonly, precondition programs as implemented in cow-calf herds require weaning of calves 45 days prior to shipment, and include dehorning, castration, bunk-training, and the administration of anthelmintics and vaccines. Calves vaccinated at 30 and 17 days before shipment to a feedlot using a multivalent modified-live vaccine did not become viremic or shed BVDV when exposed to PI pen mates.[249] While modified-live vaccines are able to rapidly confer clinical protection, and protected calves from experimental challenge as early as 3–5 days after vaccination under field conditions, complete protection cannot be expected this promptly.[250,251]

When planning a BVDV vaccination protocol for calves, blocking of protective immune responses by the presence of maternal antibodies is of concern, especially in early-weaned calves.[252] Two-week-old calves vaccinated with a modified-life vaccine in the presence of high titers of colostral antibodies were not protected from BVDV-induced disease at 4 month of age, in contrast to calves BVDV-seronegative at vaccination.[253] Additionally, large variation between individual maternal antibody titers exists, causing difficulty in estimating the ideal time of initial vaccination in calves.[254,255] However, different investigators have recently demonstrated that vaccination of young calves in the face of maternal antibodies can result in cell-mediated anamnestic immune responses and protection from BVDV-induced disease, especially when two doses of modified-live vaccines are used.[256,257] While a decline in antibody titers following vaccination with a single dose of modified-live vaccine was observed in three age groups of calves, all groups were clinically protected from BVDV-induced disease and underwent T-cell anamnestic responses, indicating that vaccination in the face of maternal immunity can be efficacious when early exposure to BVDV is expected.[258]

In cow-calf herds, vaccination against BVDV should prevent fetal infection and birth of PI offspring, thus interrupting the endemic cycle of BVDV. Protective immunity should be achieved at least 30 days before breeding and vaccinations should not be administered in the immediate pre-breeding season to avoid adverse effects of vaccine strains on fertility.[142] In various studies, experimental intranasal challenge of vaccinated dams resulted in protection from fetal infection of up to 100%; however, complete protection was often not demonstrated.[259-263] Lesser degrees of protection were observed in other studies when vaccinated pregnant cattle were exposed to PI cattle, which is considered a more rigorous and realistic challenge model.[264-267] While prolonged and complete protection from development of PI calves was demonstrated in a study using PI cattle as the challenge model,[264] under field conditions, success of vaccination to prevent fetal infection depends on various factors including timing of vaccination, nutritional and immune status of animals, cross-protection of vaccine and field strains, and number of vaccinations administered before breeding.

References

1. Childs T. X disease in cattle: Saskatchewan. *Can J Comp Med* 1946;10:316–319.
2. Olafson P, MacCallum A, Fox F. An apparently new transmissible disease of cattle. *Cornell Vet* 1946;36:205–213.
3. Ames T. The causative agent of BVD: its epidemiology and pathogenesis. *Vet Med* 1986;81:848–869.
4. Carman S, van Dreumel T, Ridpath J et al. Severe acute bovine viral diarrhea in Ontario, 1993–1995. *J Vet Diagn Invest* 1998;10:27–35.
5. Pellerin C, van den Hurk J, Lecomte J, Tussen P. Identification of a new group of bovine viral diarrhea virus strains associated with severe outbreaks and high mortalities. *Virology* 1994;203:260–268.
6. Ridpath J. Classification and molecular biology. In: Goyal S, Ridpath J (eds) *Bovine Viral Diarrhea Virus: Diagnosis, Management and Control*. Ames, IA: Blackwell Publishing, 2005, pp. 65–89.
7. Avalos-Ramirez R, Orlich M, Thiel H, Becher P. Evidence for the presence of two novel pestivirus species. *Virology* 2001;286:456–465.
8. Becher P, Avalos Ramirez R, Orlich M et al. Genetic and antigenic characterization of novel pestivirus genotypes: implications for classification. *Virology* 2003;311:96–104.
9. Dekker A, Wensvoort G, Terpstra C. Six antigenic groups within the genus pestivirus as identified by cross neutralization assays. *Vet Microbiol* 1995;47:317–329.
10. Schirrmeier H, Strebelow G, Depner K, Hoffmann B, Beer M. Genetic and antigenic characterization of an atypical pestivirus isolate, a putative member of a novel pestivirus species. *J Gen Virol* 2004;85:3647–3652.
11. Vilcek S. Discrepancies in genetic typing of bovine viral diarrhoea virus isolates. *Vet Microbiol* 2005;106:153–154.
12. Giangaspero M, Harasawa R. Numerical taxonomy of the genus Pestivirus based on palindromic nucleotide substitutions in the 5′ untranslated region. *J Virol Methods* 2007;146:375–388.
13. Harasawa R. Phylogenetic analysis of pestivirus based on the 5′-untranslated region. *Acta Virol* 1996;40:49–54.
14. Hofmann M, Brechtbuhl K, Stauber N. Rapid characterization of new pestivirus strains by direct sequencing of PCR-amplified cDNA from the 5′ noncoding region. *Arch Virol* 1994;139:217–229.
15. Ridpath J, Bolin S, Dubovi E. Segregation of bovine viral diarrhea virus into genotypes. *Virology* 1994;205:66–74.
16. Vilcek S, Durkovic B, Kolesarova M, Paton D. Genetic diversity of BVDV: consequences for classification and molecular epidemiology. *Prev Vet Med* 2005;72:31–35; discussion 215–219.
17. Harasawa R, Mizusawa H. Demonstration and genotyping of pestivirus RNA from mammalian cell lines. *Microbiol Immunol* 1995;39:979–985.
18. Ridpath J, Bolin S. Comparison of the complete genomic sequence of the border disease virus, BD31, to other pestiviruses. *Virus Res* 1997;50:237–243.
19. Colett M, Larson R, Gold C, Strick D, Anderson D, Purchio A. Molecular cloning and nucleotide sequence of the pestivirus bovine viral diarrhea virus. *Virology* 1988;165:191–199.
20. Bolin S, Grooms D. Origination and consequences of bovine viral diarrhea virus diversity. *Vet Clin North Am Food Anim Pract* 2004;20:51–68.
21. Drake JW, Holland JJ. Mutation rates among RNA viruses. *Proc Natl Acad Sci USA* 1999;96:13910–13913.
22. Figlerowicz M, Alejska M, Kurzynska-Kokorniak A, Figlerowicz M. Genetic variability: the key problem in the prevention and therapy of RNA-based virus infections. *Med Res Rev* 2003;23:488–518.
23. Saiz J, Domingo E. Virulence as a positive trait in viral persistence. *J Virol* 1996;70:6410–6413.
24. Lee K, Gillespie J. Propagation of virus diarrhea virus of cattle in tissue culture. *Am J Vet Res* 1957;18:952–953.
25. Underdahl N, Grace O, Hoerlein A. Cultivation in tissue-culture of cytopathogenic agent from bovine mucosal disease. *Proc Soc Exp Biol Med* 1957;94:795–797.
26. Fulton R, Ridpath J, Ore S et al. Bovine viral diarrhoea virus (BVDV) subgenotypes in diagnostic laboratory accessions: distribution of BVDV1a, 1b, and 2a subgenotypes. *Vet Microbiol* 2005;111:35–40.
27. Fulton R, Saliki J, Confer A et al. Bovine viral diarrhea virus cytopathic and noncytopathic biotypes and type 1 and 2 genotypes in diagnostic laboratory accessions: clinical and necropsy samples from cattle. *J Vet Diagn Invest* 2000;12:33–38.
28. Ridpath J, Bendfeldt S, Neill J, Liebler-Tenorio E. Lymphocytopathogenic activity in vitro correlates with high virulence in vivo for BVDV type 2 strains: criteria for a third biotype of BVDV. *Virus Res* 2006;118:62–69.
29. Houe H, Baker J, Maes R et al. Prevalence of cattle persistently infected with bovine viral diarrhea virus in 20 dairy herds in two counties in central Michigan and comparison of prevalence of antibody-positive cattle among herds with different infection and vaccination status. *J Vet Diagn Invest* 1995;7:321–326.
30. Houe H. Epidemiology of bovine viral diarrhea virus. *Vet Clin North Am Food Anim Pract* 1995;11:521–547.
31. Durham P, Hassard L. Prevalence of antibodies to infectious bovine rhinotracheitis, parainfluenza 3, bovine respiratory syncytial, and bovine viral diarrhea viruses in cattle in Saskatchewan and Alberta. *Can Vet J* 1990;31:815–820.
32. Bolin S, McClurkin A, Coria M. Frequency of persistent bovine viral diarrhea virus infection in selected cattle herds. *Am J Vet Res* 1985;46:2385–2387.
33. Scott H, Sorensen O, Wu J, Chow E, Manninen K, VanLeeuwen JA. Seroprevalence of *Mycobacterium avium* subspecies *paratuberculosis*, *Neospora caninum*, bovine leukemia virus, and bovine viral diarrhea virus infection among dairy cattle and herds in Alberta and agroecological risk factors associated with seropositivity. *Can Vet J* 2006;47:981–991.
34. Taylor L, Van Donkersgoed J, Dubovi E et al. The prevalence of bovine viral diarrhea virus infection in a population of feedlot calves in western Canada. *Can J Vet Res* 1995;59:87–93.
35. Martin S, Bateman K, Shewen P, Rosendal S, Bohac J. The frequency, distribution and effects of antibodies, to seven putative respiratory pathogens, on respiratory disease and weight gain in feedlot calves in Ontario. *Can J Vet Res* 1989;53:355–362.

36. Brulisauer F, Lewis F, Ganser A, McKendrick I, Gunn G. The prevalence of bovine viral diarrhoea virus infection in beef suckler herds in Scotland. *Vet J* 2010;186:226–231.
37. Duong M, Alenius S, Huong L, Bjorkman C. Prevalence of *Neospora caninum* and bovine viral diarrhoea virus in dairy cows in Southern Vietnam. *Vet J* 2008;175:390–394.
38. Niza-Ribeiro J, Pereira A, Souza J, Madeira H, Barbosa A, Afonso C. Estimated BVDV-prevalence, contact and vaccine use in dairy herds in Northern Portugal. *Prev Vet Med* 2005;72:81–85; discussion 215–219.
39. Solis-Calderon J, Segura-Correa V, Segura-Correa J. Bovine viral diarrhoea virus in beef cattle herds of Yucatan, Mexico: seroprevalence and risk factors. *Prev Vet Med* 2005;72:253–262.
40. Kampa J, Stahl K, Moreno-Lopez J, Chanlun A, Aiumlamai S, Alenius S. BVDV and BHV-1 infections in dairy herds in northern and northeastern Thailand. *Acta Vet Scand* 2004;45:181–192.
41. Vilcek S, Mojzisova J, Bajova V et al. A survey for BVDV antibodies in cattle farms in Slovakia and genetic typing of BVDV isolates from imported animals. *Acta Vet Hung* 2003;51:229–236.
42. Stahl K, Rivera H, Vagsholm I, Moreno-Lopez J. Bulk milk testing for antibody seroprevalences to BVDV and BHV-1 in a rural region of Peru. *Prev Vet Med* 2002;56:193–202.
43. Mainar-Jaime R, Berzal-Herranz B, Arias P, Rojo-Vazquez F. Epidemiological pattern and risk factors associated with bovine viral-diarrhoea virus (BVDV) infection in a non-vaccinated dairy-cattle population from the Asturias region of Spain. *Prev Vet Med* 2001;52:63–73.
44. Obando C, Baule C, Pedrique C et al. Serological and molecular diagnosis of bovine viral diarrhoea virus and evidence of other viral infections in dairy calves with respiratory disease in Venezuela. *Acta Vet Scand* 1999;40:253–262.
45. Sudharshana K, Suresh K, Rajasekhar M. Prevalence of bovine viral diarrhoea virus antibodies in India. *Rev Sci Tech* 1999;18:667–671.
46. Obando R, Hidalgo M, Merza M, Montoya A, Klingeborn B, Moreno-Lopez J. Seroprevalence to bovine virus diarrhoea virus and other viruses of the bovine respiratory complex in Venezuela (Apure State). *Prev Vet Med* 1999;41:271–278.
47. Grom J, Barlic-Maganja D. Bovine viral diarrhoea (BVD) infections: control and eradication programme in breeding herds in Slovenia. *Vet Microbiol* 1999;64:259–264.
48. Luzzago C, Piccinini R, Zepponi A, Zecconi A. Study on prevalence of bovine viral diarrhoea virus (BVDV) antibodies in 29 Italian dairy herds with reproductive problems. *Vet Microbiol* 1999;64:247–252.
49. Ferrari G, Scicluna MT, Bonvicini D et al. Bovine virus diarrhoea (BVD) control programme in an area in the Rome province (Italy). *Vet Microbiol* 1999;64:237–245.
50. Nuotio L, Juvonen M, Neuvonen E, Sihvonen L, Husu-Kallio J. Prevalence and geographic distribution of bovine viral diarrhoea (BVD) infection in Finland 1993–1997. *Vet Microbiol* 1999;64:231–235.
51. Houe H. Epidemiological features and economical importance of bovine virus diarrhoea virus (BVDV) infections. *Vet Microbiol* 1999;64:89–107.
52. Rossmanith W, Deinhofer M. The occurrence of BVD virus infections in lower Austrian dairy farms [In German]. *Dtsch Tierarztl Wochenschr* 1998;105:346–349.
53. Houe H. Age distribution of animals persistently infected with bovine virus diarrhea virus in twenty-two Danish dairy herds. *Can J Vet Res* 1992;56:194–198.
54. Loneragan G, Thomson D, Montgomery D, Mason G, Larson R. Prevalence, outcome, and health consequences associated with persistent infection with bovine viral diarrhea virus in feedlot cattle. *J Am Vet Med Assoc* 2005;226:595–601.
55. O'Connor A, Sorden S, Apley M. Association between the existence of calves persistently infected with bovine viral diarrhea virus and commingling on pen morbidity in feedlot cattle. *Am J Vet Res* 2005;66:2130–2134.
56. Wittum T, Grotelueschen D, Brock K et al. Persistent bovine viral diarrhoea virus infection in US beef herds. *Prev Vet Med* 2001;49:83–94.
57. Vilcek S, Nettleton P. Pestiviruses in wild animals. *Vet Microbiol* 2006;116:1–12.
58. Passler T, Walz P. Bovine viral diarrhea virus infections in heterologous species. *Anim Health Res Rev* 2010;11:191–205.
59. Evermann J. Pestiviral infection of llamas and alpacas. *Small Ruminant Res* 2006;61:201–206.
60. Graham D, Calvert V, German A, McCullough S. Pestiviral infections in sheep and pigs in Northern Ireland. *Vet Rec* 2001;148:69–72.
61. Krametter-Frotscher R, Loitsch A, Kohler H et al. Serological survey for antibodies against pestiviruses in sheep in Austria. *Vet Rec* 2007;160:726–730.
62. Konig M, Cedillo Rosales S, Becher P, Thiel H. Heterogeneity of ruminant pestiviruses: academic interest or important basis for the development of vaccines and diagnostics? [In German] *Berl Munch Tierarztl Wochenschr* 2003;116:216–221.
63. Taylor W, Okeke A, Shidali N. Experimental infection of Nigerian sheep and goats with bovine virus diarrhoea virus. *Trop Anim Health Prod* 1977;9:249–251.
64. Meehan J, Lehmkuhl H, Cutlip R, Bolin S. Acute pulmonary lesions in sheep experimentally infected with bovine viral diarrhoea virus. *J Comp Pathol* 1998;119:277–292.
65. Loken T, Bjerkas I. Experimental pestivirus infections in pregnant goats. *J Comp Pathol* 1991;105:123–140.
66. Hewicker-Trautwein M, Trautwein G, Frey H, Liess B. Variation in neuropathogenicity in sheep fetuses transplacentally infected with non-cytopathogenic and cytopathogenic biotypes of bovine-virus diarrhoea virus. *Zbl Vet Med B* 1995;42:557–567.
67. Carlsson U. Border disease in sheep caused by transmission of virus from cattle persistently infected with bovine virus diarrhoea virus. *Vet Rec* 1991;128:145–147.
68. Scherer C, Flores E, Weiblen R et al. Experimental infection of pregnant ewes with bovine viral diarrhea virus type-2 (BVDV-2): effects on the pregnancy and fetus. *Vet Microbiol* 2001;79:285–299.
69. Broaddus C, Lamm C, Kapil S, Dawson L, Holyoak G. Bovine viral diarrhea virus abortion in goats housed with persistently infected cattle. *Vet Pathol* 2009;46:45–53.
70. de Smit A, Eble P, de Kluijver E, Bloemraad M, Bouma A. Laboratory decision-making during the classical swine fever epidemic of 1997–1998 in The Netherlands. *Prev Vet Med* 1999;42:185–199.
71. Liess B, Moennig V. Ruminant pestivirus infection in pigs. *Rev Sci Tech* 1990;9:151–161.
72. Loeffen W, van Beuningen A, Quak S, Elbers A. Seroprevalence and risk factors for the presence of ruminant pestiviruses in the Dutch swine population. *Vet Microbiol* 2009;136:240–245.
73. Vannier P, Leforban Y, Carnero R, Cariolet R. Contamination of a live virus vaccine against pseudorabies (Aujeszky's disease) by an ovine pestivirus pathogen for the pig. *Ann Rech Vet* 1988;19:283–290.
74. Beckenhauer W, Brown A, Lidolph A, Norden C. Immunization of swine against hog cholera with a bovine enterovirus. *Vet Med* 1961;56:108–112.
75. Walz P, Baker J, Mullaney T, Kaneene J, Maes R. Comparison of type I and type II bovine viral diarrhea virus infection in swine. *Can J Vet Res* 1999;63:119–123.
76. Stewart W, Miller L, Kresse J, Snyder M. Bovine viral diarrhea infection in pregnant swine. *Am J Vet Res* 1980;41:459–462.
77. Terpstra C, Wensvoort G. A congenital persistent infection of bovine virus diarrhoea virus in pigs: clinical, virological and immunological observations. *Vet Quart* 1997;19:97–101.

78. Burgemeister R, Leyk W, Goessler R. Untersuchungen über vorkommen von parasitosen, bakteriellen und viralen infektionskrankheiten bei dromedaren in südtunesien. *Dtsch Tierarztl Wochenschr* 1975;82:352–354.
79. Doyle L, Heuschele W. Bovine viral diarrhea virus infection in captive exotic ruminants. *J Am Vet Med Assoc* 1983;183:1257–1259.
80. Byers S, Snekvik K, Righter D et al. Disseminated bovine viral diarrhea virus in a persistently infected alpaca (*Vicugna pacos*) cria. *J Vet Diagn Invest* 2009;21:145–148.
81. Carman S, Carr N, DeLay J, Baxi M, Deregt D, Hazlett M. Bovine viral diarrhea virus in alpaca: abortion and persistent infection. *J Vet Diagn Invest* 2005;17:589–593.
82. Foster A, Houlihan M, Holmes J et al. Bovine viral diarrhoea virus infection of alpacas (*Vicugna pacos*) in the UK. *Vet Rec* 2007;161:94–99.
83. Kim S, Anderson R, Yu J et al. Genotyping and phylogenetic analysis of bovine viral diarrhea virus isolates from BVDV infected alpacas in North America. *Vet Microbiol* 2009;136:209–216.
84. Hegazy A, Lotifa S, Saber M, Aboellail T, Yousif A, Chase CCL. Bovine viral diarrhea virus infection causes reproductive failure and neonatal deaths in dromedary camel. In: *International Symposium, Bovine Viral Diarrhea Virus: a 50 Year Review, Ithaca, New York.* Cornell University College of Veterinary Medicine, 1996, p. 205.
85. Wahab A, Hussein M, Asfowr H, Shalaby M. Detection of BVDV associated with mortalities in camels. *Int J Virol* 2005;1:45.
86. Celedon M, Osorio J, Pizarro J. Isolation and identification of pestiviruses in alpacas (*Lama pacos*) and llamas (*Lama glama*) introduced to the Region Metropolitana, Chile. *Archivos De Medicina Veterinaria* 2006;38:247–252.
87. Topliff C, Becker E, Smith D, Clowser S, Steffen D. Prevalence of bovine viral diarrhea virus infections in US alpacas. In: *American Association of Veterinary Laboratory Diagnosticians Annual Conference, Reno, Nevada*, 2007, p. 196.
88. Sadi L, Joyal R, St-Georges M, Lamontagne L. Serologic survey of white-tailed deer on Anticosti Island, Quebec for bovine herpesvirus 1, bovine viral diarrhea, and parainfluenza 3. *J Wildl Dis* 1991;27:569–577.
89. Cantu A, Ortega S, Mosqueda J, Garcia-Vazquez Z, Henke S, George J. Prevalence of infectious agents in free-ranging white-tailed deer in northeastern Mexico. *J Wildl Dis* 2008;44:1002–1007.
90. Frolich K. Bovine virus diarrhea and mucosal disease in free-ranging and captive deer (Cervidae) in Germany. *J Wildl Dis* 1995;31:247–250.
91. Wolf K, Deperno C, Jenks J et al. Selenium status and antibodies to selected pathogens in white-tailed deer (*Odocoileus virginianus*) in southern Minnesota. *J Wildl Dis* 2008;44:181–187.
92. Schmitt S, O'Brien D, Bruning-Fann C, Fitzgerald S. Bovine tuberculosis in Michigan wildlife and livestock. *Ann NY Acad Sci* 2002;969:262–268.
93. Passler T, Walz P, Ditchkoff S et al. Cohabitation of pregnant white-tailed deer and cattle persistently infected with bovine viral diarrhea virus results in persistently infected fawns. *Vet Microbiol* 2009;134:362–367.
94. ElAzhary M, Frechette J, Silim A, Roy R. Serological evidence of some bovine viruses in the caribou (*Rangifer tarandus caribou*) in Quebec. *J Wildl Dis* 1981;17:609–612.
95. Aguirre A, Hansen D, Starkey E, McLean R. Serologic survey of wild cervids for potential disease agents in selected national parks in the United States. *Prev Vet Med* 1995;21:313–322.
96. Lillehaug A, Vikoren T, Larsen I, Akerstedt J, Tharaldsen J, Handeland K. Antibodies to ruminant alpha-herpesviruses and pestiviruses in Norwegian cervids. *J Wildl Dis* 2003;39:779–786.
97. Van Campen H, Ridpath J, Williams E et al. Isolation of bovine viral diarrhea virus from a free-ranging mule deer in Wyoming. *J Wildl Dis* 2001;37:306–311.
98. Anderson E, Rowe L. The prevalence of antibody to the viruses of bovine virus diarrhoea, bovine herpes virus 1, rift valley fever, ephemeral fever and bluetongue and to *Leptospira* sp. in free-ranging wildlife in Zimbabwe. *Epidemiol Infect* 1998;121:441–449.
99. Frolich K, Hofmann M. Isolation of bovine viral diarrhea virus-like pestiviruses from roe deer (*Capreolus capreolus*). *J Wildl Dis* 1995;31:243–246.
100. Duncan C, Van Campen H, Soto S, Levan I, Baeten L, Miller M. Persistent bovine viral diarrhea virus infection in wild cervids of Colorado. *J Vet Diagn Invest* 2008;20:650–653.
101. Passler T, Walz P, Ditchkoff S, Walz H, Givens M, Brock K. Evaluation of hunter-harvested white-tailed deer for evidence of bovine viral diarrhea virus infection in Alabama. *J Vet Diagn Invest* 2008;20:79–82.
102. Pogranichniy R, Raizman E, Thacker H, Stevenson G. Prevalence and characterization of bovine viral diarrhea virus in the white-tailed deer population in Indiana. *J Vet Diagn Invest* 2008;20:71–74.
103. Chase C, Braun L, Leslie-Steen P, Graham T, Miskimins D, Ridpath J. Bovine viral diarrhea virus multiorgan infection in two white-tailed deer in southeastern South Dakota. *J Wildl Dis* 2008;44:753–759.
104. Romvary J. Incidence of virus diarrhoea among roes. *Acta Vet Acad Sci Hung* 1965;15:451–455.
105. VanCampen H, Williams E, Edwards J, Cook W, Stout G. Experimental infection of deer with bovine viral diarrhea virus. *J Wildl Dis* 1997;33:567–573.
106. Tessaro S, Carman P, Deregt D. Viremia and virus shedding in elk infected with type 1 and virulent type 2 bovine viral diarrhea virus. *J Wildl Dis* 1999;35:671–677.
107. Morton J, Evermann J, Dieterich R. Experimental infection of reindeer with bovine viral diarrhea virus. *Rangifer* 1990;10:75–77.
108. Mark C, Ridpath J, Chase C. Exposure of white tailed deer to bovine diarrhea virus. In: *Annual Meeting of the Conference of Research Workers in Animal Diseases, St Louis, MO*, 2005, p. 141.
109. Ridpath J, Mark C, Chase C, Ridpath A, Neill J. Febrile response and decrease in circulating lymphocytes following acute infection of white-tailed deer fawns with either a BVDV1 or a BVDV2 strain. *J Wildl Dis* 2007;43:653–659.
110. Passler T, Walz P, Ditchkoff S, Givens M, Maxwell H, Brock K. Experimental persistent infection with bovine viral diarrhea virus in white-tailed deer. *Vet Microbiol* 2007;122:350–356.
111. Ridpath J, Driskell E, Chase C, Neill J, Palmer M, Brodersen B. Reproductive tract disease associated with inoculation of pregnant white-tailed deer with bovine viral diarrhea virus. *Am J Vet Res* 2008;69:1630–1636.
112. Baker J. The clinical manifestations of bovine viral diarrhea infection. *Vet Clin North Am Food Anim Pract* 1995;11:425–445.
113. Corapi W, Elliott R, French T, Arthur D, Bezek D, Dubovi E. Thrombocytopenia and hemorrhages in veal calves infected with bovine viral diarrhea virus. *J Am Vet Med Assoc* 1990;196:590–596.
114. Ohmann H. Experimental fetal infection with bovine viral diarrhea virus. II. Morphological reactions and distribution of viral antigen. *Can J Comp Med* 1982;46:363–369.
115. Ohmann H. Pathogenesis of bovine viral diarrhoea-mucosal disease: distribution and significance of BVDV antigen in diseased calves. *Res Vet Sci* 1983;34:5–10.
116. Moerman A, Straver P, de Jong M, Quak J, Baanvinger T, van Oirschot J. Clinical consequences of a bovine virus diarrhoea virus infection in a dairy herd: a longitudinal study. *Vet Quart* 1994;16:115–119.
117. Evermann J, Barrington G. Clinical features. In: Goyal SM, Ridpath JF (eds) *Bovine Viral Diarrhea Virus: Diagnosis, Management, and Control.* Ames, IA. Blackwell Publishing, 2005, pp. 105–119.

118. Ridpath J, Bolin S. Differentiation of types 1a, 1b and 2 bovine viral diarrhoea virus (BVDV) by PCR. *Mol Cell Probes* 1998;12:101–106.
119. Walz P, Steficek B, Baker J, Kaiser L, Bell T. Effect of experimentally induced type II bovine viral diarrhea virus infection on platelet function in calves. *Am J Vet Res* 1999;60:1396–1401.
120. Potgieter L. Immunology of bovine viral diarrhea virus. *Vet Clin North Am Food Anim Pract* 1995;11:501–520.
121. Archambault D, Beliveau C, Couture Y, Carman S. Clinical response and immunomodulation following experimental challenge of calves with type 2 noncytopathogenic bovine viral diarrhea virus. *Vet Res* 2000;31:215–227.
122. Glew E, Carr B, Brackenbury L, Hope J, Charleston B, Howard CJ. Differential effects of bovine viral diarrhoea virus on monocytes and dendritic cells. *J Gen Virol* 2003;84:1771–1780.
123. Lambot M, Douart A, Joris E, Letesson J, Pastoret P. Characterization of the immune response of cattle against non-cytopathic and cytopathic biotypes of bovine viral diarrhoea virus. *J Gen Virol* 1997;78:1041–1047.
124. Adler B, Adler H, Pfister H, Jungi TW, Peterhans E. Macrophages infected with cytopathic bovine viral diarrhea virus release a factor(s) capable of priming uninfected macrophages for activation-induced apoptosis. *J Virol* 1997;71:3255–3258.
125. Chase C, Elmowalid G, Yousif A. The immune response to bovine viral diarrhea virus: a constantly changing picture. *Vet Clin North Am Food Anim Pract* 2004;20:95–114.
126. Welsh M, Adair B, Foster J. Effect of BVD virus infection on alveolar macrophage functions. *Vet Immunol Immunopathol* 1995;46:195–210.
127. Brodersen B, Kelling C. Alteration of leukocyte populations in calves concurrently infected with bovine respiratory syncytial virus and bovine viral diarrhea virus. *Viral Immunol* 1999;12:323–334.
128. Ellis J, Davis W, Belden E, Pratt D. Flow cytofluorimetric analysis of lymphocyte subset alterations in cattle infected with bovine viral diarrhea virus. *Vet Pathol* 1988;25:231–236.
129. Collen T, Carr V, Parsons K, Charleston B, Morrison W. Analysis of the repertoire of cattle CD4(+) T cells reactive with bovine viral diarrhea virus. *Vet Immunol Immunopathol* 2002;87:235–238.
130. Collen T, Morrison W I. CD4(+) T-cell responses to bovine viral diarrhoea virus in cattle. *Virus Res* 2000;67:67–80.
131. Waldvogel A, Hediger-Weithaler B, Eicher R et al. Interferon-gamma and interleukin-4 mRNA expression by peripheral blood mononuclear cells from pregnant and non-pregnant cattle seropositive for bovine viral diarrhea virus. *Vet Immunol Immunopathol* 2000;77:201–212.
132. Fulton R, Burge L. Bovine viral diarrhea virus types 1 and 2 antibody response in calves receiving modified live virus or inactivated vaccines. *Vaccine* 2000;19:264–274.
133. Fulton R, Ridpath J, Confer A, Saliki J, Burge L, Payton M. Bovine viral diarrhoea virus antigenic diversity: impact on disease and vaccination programmes. *Biologicals* 2003;31:89–95.
134. McGowan M, Kirkland P, Richards S, Littlejohns I. Increased reproductive losses in cattle infected with bovine pestivirus around the time of insemination. *Vet Rec* 1993;133:39–43.
135. Houe H, Pedersen K, Meyling A. The effect of bovine virus diarrhea virus-infection on conception rate. *Prev Vet Med* 1993;15:117–123.
136. Kale M, Yavru S, Ata A, Kocamuftuoglu M, Yaplcl O, Haslrcloglu S. Bovine viral diarrhea virus (BVDV) infection in relation to fertility in heifers. *J Vet Med Sci* 2011;73:331–336.
137. Munoz-Zanzi C, Thurmond M, Hietala S. Effect of bovine viral diarrhea virus infection on fertility of dairy heifers. *Theriogenology* 2004;61:1085–1099.
138. Robert A, Beaudeau F, Seegers H, Joly A, Philipot J. Large scale assessment of the effect associated with bovine viral diarrhoea virus infection on fertility of dairy cows in 6149 dairy herds in Brittany (Western France). *Theriogenology* 2004;61:117–127.
139. Rodning S, Givens M, Marley M et al. Reproductive and economic impact following controlled introduction of cattle persistently infected with bovine viral diarrhea virus into a naive group of heifers. *Theriogenology* 2012;78:1508–1516.
140. Fray M, Paton D, Alenius S. The effects of bovine viral diarrhoea virus on cattle reproduction in relation to disease control. *Anim Reprod Sci* 2000;60–61:615–627.
141. Grooms D, Brock K, Ward L. Detection of bovine viral diarrhea virus in the ovaries of cattle acutely infected with bovine viral diarrhea virus. *J Vet Diagn Invest* 1998;10:125–129.
142. Grooms D, Brock K, Ward L. Detection of cytopathic bovine viral diarrhea virus in the ovaries of cattle following immunization with a modified live bovine viral diarrhea virus vaccine. *J Vet Diagn Invest* 1998;10:130–134.
143. Ssentongo Y, Johnson R, Smith J. Association of bovine viral diarrhoea-mucosal disease virus with ovaritis in cattle. *Aust Vet J* 1980;56:272–273.
144. Grooms D, Brock K, Pate J, Day M. Changes in ovarian follicles following acute infection with bovine viral diarrhea virus. *Theriogenology* 1998;49:595–605.
145. Fray M, Mann G, Clarke M, Charleston B. Bovine viral diarrhoea virus: its effects on ovarian function in the cow. *Vet Microbiol* 2000;77:185–194.
146. Kafi M, McGowan M, Jillella D, Davies F, Johnston S, Kirkland P. The effect of bovine viral diarrhoea virus (BVDV) during follicular development on the superovulatory response of cattle. *Theriogenology* 1994;41:223.
147. Kafi M, McGowan M, Kirkland P, Jillella D. The effect of bovine pestivirus infection on the superovulatory response of Friesian heifers. *Theriogenology* 1997;48:985–996.
148. McGowan M, Kafi M, Kirkland P et al. Studies of the pathogenesis of bovine pestivirus-induced ovarian dysfunction in superovulated dairy cattle. *Theriogenology* 2003;59:1051–1066.
149. Fray M, Mann G, Bleach E, Knight P, Clarke M, Charleston B. Modulation of sex hormone secretion in cows by acute infection with bovine viral diarrhoea virus. *Reproduction* 2002;123:281–289.
150. Grooms D. Reproductive consequences of infection with bovine viral diarrhea virus. *Vet Clin North Am Food Anim Pract* 2004;20:5–19.
151. Archbald L, Gibson C, Schultz R, Fahning M, Zemjanis R. Effects of intrauterine inoculation of bovine viral diarrhea-mucosal disease virus on uterine tubes and uterus of nonpregnant cows. *Am J Vet Res* 1973;34:1133–1137.
152. Bielanski A, Sapp T, Lutze-Wallace C. Association of bovine embryos produced by in vitro fertilization with a noncytopathic strain of bovine viral diarrhea virus type II. *Theriogenology* 1998;49:1231–1238.
153. Grooms D, Ward L, Brock K. Morphologic changes and immunohistochemical detection of viral antigen in ovaries from cattle persistently infected with bovine viral diarrhea virus. *Am J Vet Res* 1996;57:830–833.
154. Booth P, Stevens D, Collins M, Brownlie J. Detection of bovine viral diarrhoea virus antigen and RNA in oviduct and granulosa cells of persistently infected cattle. *J Reprod Fertil* 1995;105:17–24.
155. Shin T, Acland H. Tissue distribution of bovine viral diarrhea virus antigens in persistently infected cattle. *J Vet Sci* 2001;2:81–84.
156. Duffell S, Harkness J. Bovine virus diarrhoea-mucosal disease infection in cattle. *Vet Rec* 1985;117:240–245.
157. Kirkland P, McGowan M, Mackintosh S. Factors influencing the development of persistent infection of cattle with pestiviruses. In: *Proceedings of the Second Symposium on Pestiviruses, Lyon, France*, 1993, pp. 117–121.

158. Swasdipan S, Bielefeldt-Ohmann H, Phillips N, Kirkland P, McGowan M. Rapid transplacental infection with bovine pestivirus following intranasal inoculation of ewes in early pregnancy. *Vet Pathol* 2001;38:275–280.
159. Fredriksen B, Press C, Loken T, Odegaard S. Distribution of viral antigen in uterus, placenta and foetus of cattle persistently infected with bovine virus diarrhoea virus. *Vet Microbiol* 1999;64:109–122.
160. Montgomery D, Van Olphen A, Van Campen H, Hansen T. The fetal brain in bovine viral diarrhea virus-infected calves: lesions, distribution, and cellular heterogeneity of viral antigen at 190 days gestation. *Vet Pathol* 2008;45:288–296.
161. Bielefeldt-Ohmann H, Tolnay A, Reisenhauer C, Hansen T, Smirnova N, Van Campen H. Transplacental infection with non-cytopathic bovine viral diarrhoea virus types 1b and 2: viral spread and molecular neuropathology. *J Comp Pathol* 2008;138:72–85.
162. Bielefeldt-Ohmann H, Smirnova N, Tolnay A et al. Neuroinvasion by a "Trojan Horse" strategy and vasculopathy during intrauterine flavivirus infection. *Int J Exp Pathol* 2012;93:24–33.
163. Charleston B, Brackenbury L, Carr B et al. Alpha/beta and gamma interferons are induced by infection with noncytopathic bovine viral diarrhea virus in vivo. *J Virol* 2002;76:923–927.
164. Smirnova N, Bielefeldt-Ohmann H, Van Campen H et al. Acute non-cytopathic bovine viral diarrhea virus infection induces pronounced type I interferon response in pregnant cows and fetuses. *Virus Res* 2008;132:49–58.
165. Shoemaker M, Smirnova N, Bielefeldt-Ohmann H et al. Differential expression of the type I interferon pathway during persistent and transient bovine viral diarrhea virus infection. *J Interferon Cytokine Res* 2009;29:23–35.
166. Smirnova N, Webb B, Bielefeldt-Ohmann H et al. Development of fetal and placental innate immune responses during establishment of persistent infection with bovine viral diarrhea virus. *Virus Res* 2012;167:329–336.
167. Murray R. Lesions in aborted bovine fetuses and placenta associated with bovine viral diarrhoea virus infection. *Arch Virol Suppl* 1991;3:217–224.
168. Carlsson U, Fredriksson G, Alenius S, Kindahl H. Bovine virus diarrhea virus, a cause of early-pregnancy failure in the cow. *J Vet Med A* 1989;36:15–23.
169. Liess B, Orban S, Frey H, Trautwein G, Wiefel W, Blindow H. Studies on transplacental transmissibility of a bovine virus diarrhoea (BVD) vaccine virus in cattle. II. Inoculation of pregnant cows without detectable neutralizing antibodies to BVD virus 90–229 days before parturition (51st to 190th day of gestation). *Zbl Vet Med B* 1984;31:669–681.
170. McClurkin A, Littledike E, Cutlip R, Frank G, Coria M, Bolin S. Production of cattle immunotolerant to bovine viral diarrhea virus. *Can J Comp Med* 1984;48:156–161.
171. Brownlie J, Clarke M, Howard C. Experimental infection of cattle in early pregnancy with a cytopathic strain of bovine virus diarrhoea virus. *Res Vet Sci* 1989;46:307–311.
172. Brock K, Chase C. Development of a fetal challenge method for the evaluation of bovine viral diarrhea virus vaccines. *Vet Microbiol* 2000;77:209–214.
173. Brock K, Cortese V. Experimental fetal challenge using type II bovine viral diarrhea virus in cattle vaccinated with modified-live virus vaccine. *Vet Therapeut* 2001;2:354–360.
174. Collen T, Douglas A, Paton D, Zhang G, Morrison W. Single amino acid differences are sufficient for CD4(+) T-cell recognition of a heterologous virus by cattle persistently infected with bovine viral diarrhea virus. *Virology* 2000;276:70–82.
175. Baker J. Bovine viral diarrhea virus: a review. *J Am Vet Med Assoc* 1987;190:1449–1458.
176. Done J, Terlecki S, Richardson C et al. Bovine virus diarrhoea-mucosal disease virus: pathogenicity for the fetal calf following maternal infection. *Vet Rec* 1980;106:473–479.
177. Hewicker-Trautwein M, Liess B, Trautwein G. Brain lesions in calves following transplacental infection with bovine-virus diarrhoea virus. *Zbl Vet Med B* 1995;42:65–77.
178. Otter A, Welchman D de B, Sandvik T et al. Congenital tremor and hypomyelination associated with bovine viral diarrhoea virus in 23 British cattle herds. *Vet Rec* 2009;164:771–778.
179. Porter B, Ridpath J, Calise D et al. Hypomyelination associated with bovine viral diarrhea virus type 2 infection in a longhorn calf. *Vet Pathol* 2010;47:658–663.
180. Webb B, Norrdin R, Smirnova N et al. Bovine viral diarrhea virus cyclically impairs long bone trabecular modeling in experimental persistently infected fetuses. *Vet Pathol* 2012;49:930–940.
181. Munoz-Zanzi C, Hietala S, Thurmond M, Johnson W. Quantification, risk factors, and health impact of natural congenital infection with bovine viral diarrhea virus in dairy calves. *Am J Vet Res* 2003;64:358–365.
182. Houe H. Survivorship of animals persistently infected with bovine virus diarrhoea virus (BVDV). *Prev Vet Med* 1993;15:275–283.
183. Roberts D, Lucas M, Wibberley G, Westcott D. Response of cattle persistently infected with bovine virus diarrhoea virus to bovine leukosis virus. *Vet Rec* 1988;122:293–296.
184. Bolin S. The pathogenesis of mucosal disease. *Vet Clin North Am Food Anim Pract* 1995;11:489–500.
185. Tautz N, Thiel H, Dubovi E, Meyers G. Pathogenesis of mucosal disease: a cytopathogenic pestivirus generated by an internal deletion. *J Virol* 1994;68:3289–3297.
186. Odeón A, Risatti G, Kaiser G et al. Bovine viral diarrhea virus genomic associations in mucosal disease, enteritis and generalized dermatitis outbreaks in Argentina. *Vet Microbiol* 2003;96:133–144.
187. Fritzemeier J, Haas L, Liebler E, Moennig V, Greiser-Wilke I. The development of early vs. late onset mucosal disease is a consequence of two different pathogenic mechanisms. *Arch Virol* 1997;142:1335–1350.
188. Westenbrink F, Straver P, Kimman T, de Leeuw P. Development of a neutralising antibody response to an inoculated cytopathic strain of bovine virus diarrhoea virus. *Vet Rec* 1989;125:262–265.
189. Ridpath J, Bolin S. Delayed onset postvaccinal mucosal disease as a result of genetic recombination between genotype 1 and genotype 2 BVDV. *Virology* 1995;212:259–262.
190. Kirkland P, Richards S, Rothwell J, Stanley D. Replication of bovine viral diarrhoea virus in the bovine reproductive tract and excretion of virus in semen during acute and chronic infections. *Vet Rec* 1991;128:587–590.
191. Kirkland P, McGowan M, Mackintosh S, Moyle A. Insemination of cattle with semen from a bull transiently infected with pestivirus. *Vet Rec* 1997;140:124–127.
192. Rikula U, Nuotio L, Laamanen U, Sihvonen L. Transmission of bovine viral diarrhoea virus through the semen of acutely infected bulls under field conditions. *Vet Rec* 2008;162:79–82.
193. Kirkland P, Mackintosh S, Moyle A. The outcome of widespread use of semen from a bull persistently infected with pestivirus. *Vet Rec* 1994;135:527–529.
194. Meyling A, Jensen A. Transmission of bovine virus diarrhoea virus (BVDV) by artificial insemination (AI) with semen from a persistently-infected bull. *Vet Microbiol* 1988;17:97–105.
195. McClurkin A, Coria M, Cutlip R. Reproductive performance of apparently healthy cattle persistently infected with bovine viral diarrhea virus. *J Am Vet Med Assoc* 1979;174:1116–1119.
196. Revell S, Chasey D, Drew T, Edwards S. Some observations on the semen of bulls persistently infected with bovine virus diarrhoea virus. *Vet Rec* 1988;123:122–125.
197. Barlow R, Nettleton P, Gardiner A, Greig A, Campbell J, Bonn J. Persistent bovine virus diarrhoea virus infection in a bull. *Vet Rec* 1986;118:321–324.

198. Givens M, Marley M. Immunology of chronic BVDV infections. Biologicals 2013;41:26–30.
199. Voges H, Horner G, Rowe S, Wellenberg G. Persistent bovine pestivirus infection localized in the testes of an immuno-competent, non-viraemic bull. Vet Microbiol 1998;61:165–175.
200. Niskanen R, Alenius S, Belak K et al. Insemination of susceptible heifers with semen from a non-viraemic bull with persistent bovine virus diarrhoea virus infection localized in the testes. Reprod Domest Anim 2002;37:171–175.
201. Givens M, Heath A, Brock K, Brodersen B, Carson R, Stringfellow DA. Detection of bovine viral diarrhea virus in semen obtained after inoculation of seronegative postpubertal bulls. Am J Vet Res 2003;64:428–434.
202. Givens M, Riddell K, Walz P et al. Noncytopathic bovine viral diarrhea virus can persist in testicular tissue after vaccination of peri-pubertal bulls but prevents subsequent infection. Vaccine 2007;25:867–876.
203. Givens M, Riddell K, Edmondson M et al. Epidemiology of prolonged testicular infections with bovine viral diarrhea virus. Vet Microbiol 2009;139:42–51.
204. Gard J, Givens M, Stringfellow D. Bovine viral diarrhea virus (BVDV): epidemiologic concerns relative to semen and embryos. Theriogenology 2007;68:434–442.
205. Office International des Epizooties. Bovine viral diarrhea. In: OIE Manual of Diagnostic Tests and Vaccines for Terrestrial Animals, 6th edn. Paris: OIE, 2008, pp. 698–711.
206. Givens M, Waldrop J. Bovine viral diarrhea virus in embryo and semen production systems. Vet Clin North Am Food Anim Pract 2004;20:21–38.
207. Vanroose G, Nauwynck H, Soom A et al. Structural aspects of the zona pellucida of in vitro-produced bovine embryos: a scanning electron and confocal laser scanning microscopic study. Biol Reprod 2000;62:463–469.
208. Walz P, Grooms D, Passler T et al. Control of bovine viral diarrhea virus in ruminants. J Vet Intern Med 2010;24:476–486.
209. Saliki J, Dubovi E. Laboratory diagnosis of bovine viral diarrhea virus infections. Vet Clin North Am Food Anim Pract 2004;20:69–83.
210. Bruschke C, Weerdmeester K, Van Oirschot J, Van Rijn P. Distribution of bovine virus diarrhoea virus in tissues and white blood cells of cattle during acute infection. Vet Microbiol 1998;64:23–32.
211. Makoschey B, van Gelder P, Keijsers V, Goovaerts D. Bovine viral diarrhoea virus antigen in foetal calf serum batches and consequences of such contamination for vaccine production. Biologicals 2003;31:203–208.
212. Rossi C, Bridgman C, Kiesel G. Viral contamination of bovine fetal lung cultures and bovine fetal serum. Am J Vet Res 1980;41:1680–1681.
213. Wessman S, Levings R. Benefits and risks due to animal serum used in cell culture production. Dev Biol Stand 1999;99:3–8.
214. Saliki J, Fulton R, Hull S, Dubovi E. Microtiter virus isolation and enzyme immunoassays for detection of bovine viral diarrhea virus in cattle serum. J Clin Microbiol 1997;35:803–807.
215. Smith G, Collins J, Carman J, Minocha H. Detection of cytopathic and noncytopathic bovine viral diarrhea virus in cell culture with an immunoperoxidase test. J Virol Methods 1988;19:319–324.
216. Deregt D, Prins S. A monoclonal antibody-based immunoperoxidase monolayer (micro-isolation) assay for detection of type 1 and type 2 bovine viral diarrhea viruses. Can J Vet Res 1998;62:152–155.
217. Grooms D, Kaiser L, Walz P, Baker J. Study of cattle persistently infected with bovine viral diarrhea virus that lack detectable virus in serum. J Am Vet Med Assoc 2001;219:629–631.
218. Njaa B, Clark E, Janzen E, Ellis J, Haines D. Diagnosis of persistent bovine viral diarrhea virus infection by immunohistochemical staining of formalin-fixed skin biopsy specimens. J Vet Diagn Invest 2000;12:393–399.
219. Edmondson M, Givens M, Walz P, Gard J, Stringfellow D, Carson R. Comparison of tests for detection of bovine viral diarrhea virus in diagnostic samples. J Vet Diagn Invest 2007;19:376–381.
220. Fulton R, Hessman B, Ridpath J et al. Multiple diagnostic tests to identify cattle with bovine viral diarrhea virus and duration of positive test results in persistently infected cattle. Can J Vet Res 2009;73:117–124.
221. Hilbe M, Arquint A, Schaller P et al. Immunohistochemical diagnosis of persistent infection with bovine viral diarrhea virus (BVDV) on skin biopsies. Schweiz Arch Tierheilkd 2007;149:337–344.
222. Thur B, Zlinszky K, Ehrensperger F. Immunohistochemical detection of bovine viral diarrhea virus in skin biopsies: a reliable and fast diagnostic tool. Zbl Vet Med B 1996;43:163–166.
223. Brodersen B. Immunohistochemistry used as a screening method for persistent bovine viral diarrhea virus infection. Vet Clin North Am Food Anim Pract 2004;20:85–93.
224. Anonymous. SNAP BVDV Test, 2012. http://www.idexx.com/view/xhtml/en_us/livestock-poultry/ruminant/bvdv.jsf
225. Driskell E, Ridpath J. A survey of bovine viral diarrhea virus testing in diagnostic laboratories in the United States from 2004 to 2005. J Vet Diagn Invest 2006;18:600–605.
226. Drew T, Yapp F, Paton D. The detection of bovine viral diarrhoea virus in bulk milk samples by the use of a single-tube RT-PCR. Vet Microbiol 1999;64:145–154.
227. Givens M, Galik P, Riddell K et al. Validation of a reverse transcription nested polymerase chain reaction (RT-nPCR) to detect bovine viral diarrhea virus (BVDV) associated with in vitro-derived bovine embryos and co-cultured cells. Theriogenology 2001;56:787–799.
228. Hamel A, Wasylyshen M, Nayar G. Rapid detection of bovine viral diarrhea virus by using RNA extracted directly from assorted specimens and a one-tube reverse transcription PCR assay. J Clin Microbiol 1995;33:287–291.
229. Kennedy J, Mortimer R, Powers B. Reverse transcription-polymerase chain reaction on pooled samples to detect bovine viral diarrhea virus by using fresh ear-notch-sample supernatants. J Vet Diagn Invest 2006;18:89–93.
230. Kim S, Dubovi E. A novel simple one-step single-tube RT-duplex PCR method with an internal control for detection of bovine viral diarrhoea virus in bulk milk, blood, and follicular fluid samples. Biologicals 2003;31:103–106.
231. Tajima M, Ohsaki T, Okazawa M, Yasutomi I. Availability of oral swab sample for the detection of bovine viral diarrhea virus (BVDV) gene from the cattle persistently infected with BVDV. Jpn J Vet Res 2008;56:3–8.
232. Finke J, Fritzen R, Ternes P, Lange W, Dolken G. An improved strategy and a useful housekeeping gene for RNA analysis from formalin-fixed, paraffin-embedded tissues by PCR. Biotechniques 1993;14:448–453.
233. Weinstock D, Bhudevi B, Castro A. Single-tube single-enzyme reverse transcriptase PCR assay for detection of bovine viral diarrhea virus in pooled bovine serum. J Clin Microbiol 2001;39:343–346.
234. Radwan G, Brock K, Hogan J, Smith K. Development of a PCR amplification assay as a screening test using bulk milk samples for identifying dairy herds infected with bovine viral diarrhea virus. Vet Microbiol 1995;44:77–91.
235. Kennedy J. Diagnostic efficacy of a reverse transcriptase-polymerase chain reaction assay to screen cattle for persistent bovine viral diarrhea virus infection. J Am Vet Med Assoc 2006;229:1472–1474.

236. Smith R, Sanderson M, Walz P, Givens M. Sensitivity of polymerase chain reaction for detection of bovine viral diarrhea virus in pooled serum samples and use of pooled polymerase chain reaction to determine prevalence of bovine viral diarrhea virus in auction market cattle. *J Vet Diagn Invest* 2008;20:75–78.
237. Munoz-Zanzi C, Johnson W, Thurmond M, Hietala S. Pooled-sample testing as a herd-screening tool for detection of bovine viral diarrhea virus persistently infected cattle. *J Vet Diagn Invest* 2000;12:195–203.
238. Muvavarirwa P, Mudenge D, Moyo D, Javangwe S. Detection of bovine-virus-diarrhoea-virus antibodies in cattle with an enzyme-linked immunosorbent assay. *Onderstepoort J Vet Res* 1995;62:241–244.
239. Raue R, Harmeyer S, Nanjiani I. Antibody responses to inactivated vaccines and natural infection in cattle using bovine viral diarrhoea virus ELISA kits: assessment of potential to differentiate infected and vaccinated animals. *Vet J* 2011;187:330–334.
240. Pillars R, Grooms D. Serologic evaluation of five unvaccinated heifers to detect herds that have cattle persistently infected with bovine viral diarrhea virus. *Am J Vet Res* 2002;63:499–505.
241. Waldner C, Campbell J. Use of serologic evaluation for antibodies against bovine viral diarrhea virus for detection of persistently infected calves in beef herds. *Am J Vet Res* 2005;66:825–834.
242. Rauff Y, Moore D, Sischo W. Evaluation of the results of a survey of dairy producers on dairy herd biosecurity and vaccination against bovine viral diarrhea. *J Am Vet Med Assoc* 1996;209:1618–1622.
243. Ridpath J. Immunology of BVDV vaccines. *Biologicals* 2013;41:14–19.
244. Dean H, Leyh R. Cross-protective efficacy of a bovine viral diarrhea virus (BVDV) type 1 vaccine against BVDV type 2 challenge. *Vaccine* 1999;17:1117–1124.
245. Fairbanks K, Schnackel J, Chase C. Evaluation of a modified live virus type-1a bovine viral diarrhea virus vaccine (Singer strain) against a type-2 (strain 890) challenge. *Vet Therapeut* 2003;4:24–34.
246. Hamers C, Couvreur B, Dehan P et al. Assessment of the clinical and virological protection provided by a commercial inactivated bovine viral diarrhoea virus genotype 1 vaccine against a BVDV genotype 2 challenge. *Vet Rec* 2003;153:236–240.
247. Fulton R, Cook B, Blood K et al. Immune response to bovine respiratory disease vaccine immunogens in calves at entry to feedlot and impact on feedlot performance. *Bov Pract* 2011;45:1–12.
248. Fulton R, Cook B, Step D et al. Evaluation of health status of calves and the impact on feedlot performance: assessment of a retained ownership program for postweaning calves. *Can J Vet Res* 2002;66:173–180.
249. Fulton R, Johnson B, Briggs R et al. Challenge with bovine viral diarrhea virus by exposure to persistently infected calves: protection by vaccination and negative results of antigen testing in nonvaccinated acutely infected calves. *Can J Vet Res* 2006;70:121–127.
250. Brock K, Widel P, Walz P, Walz H. Onset of protection from experimental infection with type 2 bovine viral diarrhea virus following vaccination with a modified-live vaccine. *Vet Therapeut* 2007;8:88–96.
251. Fulton R, Briggs R, Ridpath J et al. Transmission of bovine viral diarrhea virus 1b to susceptible and vaccinated calves by exposure to persistently infected calves. *Can J Vet Res* 2005;69:161–169.
252. Kelling C. Evolution of bovine viral diarrhea virus vaccines. *Vet Clin North Am Food Anim Pract* 2004;20:115–129.
253. Ellis J, West K, Cortese V, Konoby C, Weigel D. Effect of maternal antibodies on induction and persistence of vaccine-induced immune responses against bovine viral diarrhea virus type II in young calves. *J Am Vet Med Assoc* 2001;219:351–356.
254. Fulton R, Briggs R, Payton M et al. Maternally derived humoral immunity to bovine viral diarrhea virus (BVDV) 1a, BVDV1b, BVDV2, bovine herpesvirus-1, parainfluenza-3 virus, bovine respiratory syncytial virus, *Mannheimia haemolytica* and *Pasteurella multocida* in beef calves, antibody decline by half-life studies and effect on response to vaccination. *Vaccine* 2004;22:643–649.
255. Munoz-Zanzi C, Thurmond M, Johnson W, Hietala S. Predicted ages of dairy calves when colostrum-derived bovine viral diarrhea virus antibodies would no longer offer protection against disease or interfere with vaccination. *J Am Vet Med Assoc* 2002;221:678–685.
256. Endsley J, Roth J, Ridpath J, Neill J. Maternal antibody blocks humoral but not T cell responses to BVDV. *Biologicals* 2003;31:123–125.
257. Woolums A. Vaccinating calves: new information on the effects of maternal immunity. In: *40th Annual Convention of the American Association of Bovine Practitioners, Vancouver, BC*, 2007, pp. 10–17.
258. Platt R, Widel P, Kesl L, Roth J. Comparison of humoral and cellular immune responses to a pentavalent modified live virus vaccine in three age groups of calves with maternal antibodies, before and after BVDV type 2 challenge. *Vaccine* 2009;27:4508–4519.
259. Cortese V, Grooms D, Ellis J, Bolin S, Ridpath J, Brock K. Protection of pregnant cattle and their fetuses against infection with bovine viral diarrhea virus type 1 by use of a modified-live virus vaccine. *Am J Vet Res* 1998;59:1409–1413.
260. Fairbanks K, Rinehart C, Ohnesorge W, Loughin M, Chase C. Evaluation of fetal protection against experimental infection with type 1 and type 2 bovine viral diarrhea virus after vaccination of the dam with a bivalent modified-live virus vaccine. *J Am Vet Med Assoc* 2004;225:1898–1904.
261. Ficken M, Ellsworth M, Tucker C. Evaluation of the efficacy of a modified-live combination vaccine against bovine viral diarrhea virus types 1 and 2 challenge exposures in a one-year duration-of-immunity fetal protection study. *Vet Therapeut* 2006;7:283–294.
262. Meyer G, Deplanche M, Roux D et al. Fetal protection against bovine viral diarrhoea type 1 virus infection after one administration of a live-attenuated vaccine. *Vet J* 2012;192:242–245.
263. Xue W, Mattick D, Smith L, Maxwell J. Fetal protection against bovine viral diarrhea virus types 1 and 2 after the use of a modified-live virus vaccine. *Can J Vet Res* 2009;73:292–297.
264. Ellsworth M, Fairbanks K, Behan S et al. Fetal protection following exposure to calves persistently infected with bovine viral diarrhea virus type 2 sixteen months after primary vaccination of the dams. *Vet Therapeut* 2006;7:295–304.
265. Givens M, Marley M, Jones C et al. Protective effects against abortion and fetal infection following exposure to bovine viral diarrhea virus and bovine herpesvirus 1 during pregnancy in beef heifers that received two doses of a multivalent modified-live virus vaccine prior to breeding. *J Am Vet Med Assoc* 2012;241:484–495.
266. Grooms D, Bolin S, Coe P, Borges R, Coutu C. Fetal protection against continual exposure to bovine viral diarrhea virus following administration of a vaccine containing an inactivated bovine viral diarrhea virus fraction to cattle. *Am J Vet Res* 2007;68:1417–1422.
267. Leyh R, Fulton R, Stegner J et al. Fetal protection in heifers vaccinated with a modified-live virus vaccine containing bovine viral diarrhea virus subtypes 1a and 2a and exposed during gestation to cattle persistently infected with bovine viral diarrhea virus subtype 1b. *Am J Vet Res* 2011;72:367–375.

Chapter 61

Infectious Agents: Epizootic Bovine Abortion (Foothill Abortion)

Jeffrey L. Stott, Myra T. Blanchard and Mark L. Anderson

School of Veterinary Medicine, University of California, Davis, California, USA

Introduction

Epizootic bovine abortion (EBA), commonly referred to as "foothill abortion," was first recognized in the early 1950s in California. EBA is characterized by late-term abortions, or birth of weak calves, delivered by dams grazed in foothill regions of California, Nevada, and southern Oregon[1] during their first and second trimesters of pregnancy. The disease has been, and continues to be, the leading cause of abortion in beef cattle in the state of California; estimates are not available for Oregon and Nevada.[1-4]

Etiologic agent

Multiple quests to identify the etiologic agent of EBA span 50 years. These efforts produced a number of candidate microbes as being the causative agent including a chlamydia, an uncharacterized virus (C-type-like particles), Borrelia coriaceae, and an uncharacterized spirochete.[5] Development of a reproducible method for experimental disease transmission in 2000 facilitated efforts to identify the etiologic agent. Inoculation of naive pregnant heifers (~100 days' gestation) with cryopreserved fetal thymus homogenates, derived from select EBA-diseased fetuses, provided a reliable mechanism to predictably transmit the pathogen to the developing fetus with concomitant development of classical EBA. Antibiotic susceptibility studies were employed using disease-transmission experiments in pregnant heifers; aggressive treatment at the time of heifer infection with penicillin and/or tetracycline abrogated subsequent development of fetal disease, providing strong evidence the etiologic agent was a prokaryote.[5] Application of molecular biology techniques to diseased fetal necropsy tissues ultimately identified the agent in 2005 as being a deltaproteobacterium. 16S rRNA sequences derived from the pathogen established an 89.4% homology to that of *Polyangium cellulosum*, a member of the order Myxococcales.[6] The only other mammalian pathogen currently recognized in this class of prokaryotes is *Lawsonia intracellularis*, a deltaproteobacterium primarily associated with enteritis in pigs. The bacterial agent of EBA (aoEBA) is pending assignment of genus and species until further characterization studies are complete.

Diagnosis

EBA was historically diagnosed with a combination of unique fetal pathology, elevated levels of immunoglobulin in fetal serum, and history of the dam being grazed in areas considered endemic for the disease during pregnancy.[2,3] Gross and microscopic pathology associated with EBA is typically dramatic. Aborted fetuses are usually in excellent condition (not autolyzed) and present with a variety of gross abnormalities that can include extensive ascites with excessive fibrin (Figure 61.1a) resulting in a distended abdomen, swollen liver, enlarged spleen and lymph nodes (Figure 61.1b), skin lesions (Figure 61.1c), and petechial hemorrhages on mucosal membranes (oral cavity, tongue and eyelids) (Figure 61.1d) and in the thymus (Figure 61.2a). Microscopically, lesions are widespread and have a vascular orientation characterized by infiltration with lymphocytes and mononuclear phagocytes.[2,3] Focal necrotizing lesions may be present in organized lymphatic tissues and pathognomonic lesions occur in the thymus, characterized by depletion of cortical thymocytes and infiltration with macrophages[3] (Figure 61.2b).

Bovine Reproduction, First Edition. Edited by Richard M. Hopper.
© 2015 John Wiley & Sons, Inc. Published 2015 by John Wiley & Sons, Inc.

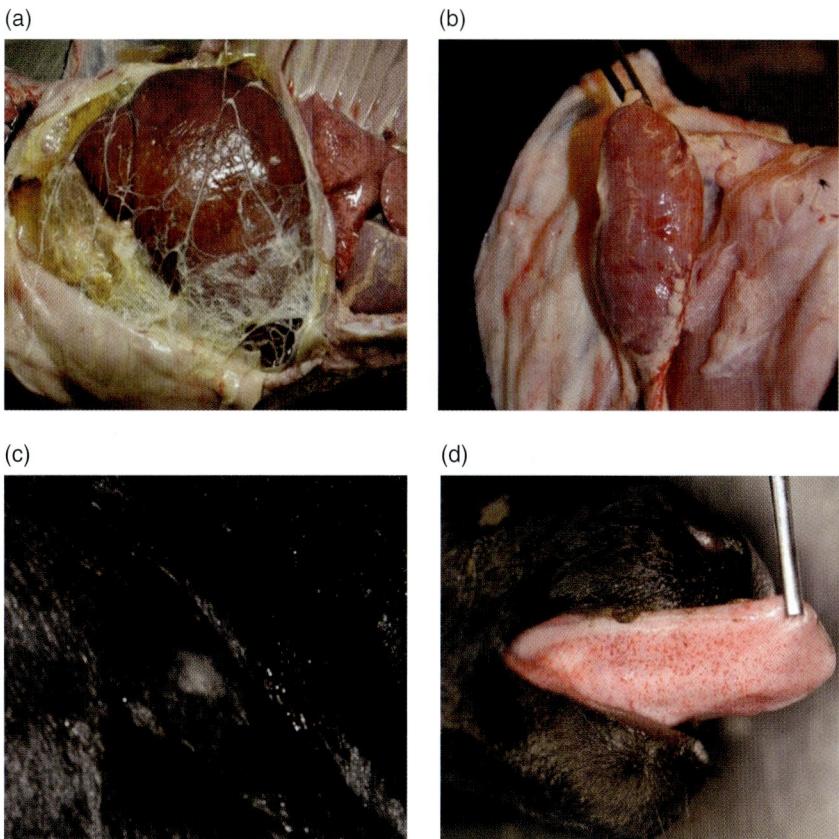

Figure 61.1 Gross changes in aoEBA-infected fetuses: (a) fibrinous peritoneal effusion and enlarged nodular liver; (b) enlarged prescapular lymph node; (c) skin lesion; (d) tongue with extensive petechiation of ventral ingual mucosa.

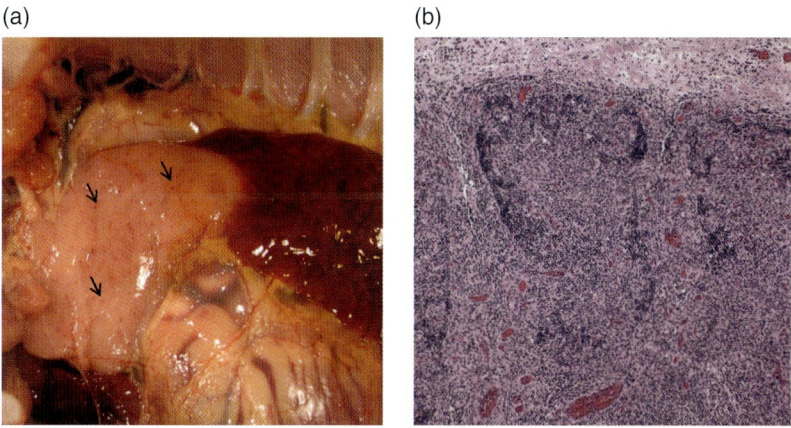

Figure 61.2 Gross and histologic changes of thymus in aoEBA-infected fetuses. (a) Thoracic thymus with petechiation. Note the petechial hemorrhages indicated by arrows. (b) Thymus exhibiting lobular atrophy with depletion of cortical thymocytes and extensive lymphohistiocytic infiltration of lobule medullary regions and the interlobular septae. Hematoxylin and eosin stain.

Microbiological diagnosis of EBA continues to be hampered by an inability to culture the causative agent. While limited short-term replication of aoEBA has been observed in primary cultures of macrophages derived from fetal lymphoid necropsy tissues, substantive propagation of the bacterium in either primary or established cell lines has not been possible; all attempts to culture aoEBA on synthetic media have also failed.[5] Polymerase chain reaction (PCR) assays, both traditional and TaqMan-based, have been developed and target the 16S bacterial ribosomal gene.[6–8] Both assays can be used to identify the presence of aoEBA in necropsy tissues of infected fetuses presenting with classical EBA pathology. The TaqMan is the assay of choice when identifying low levels of bacteria in necropsy tissues derived from fetuses suspected of being infected, such as fetuses presenting with equivocal or nonclassical EBA pathology and weak calves that may or may not suckle. The TaqMan assay is also preferable for identifying aoEBA in the vector as only the occasional infected tick will have sufficient levels of bacteria detectable by a standard PCR.[7]

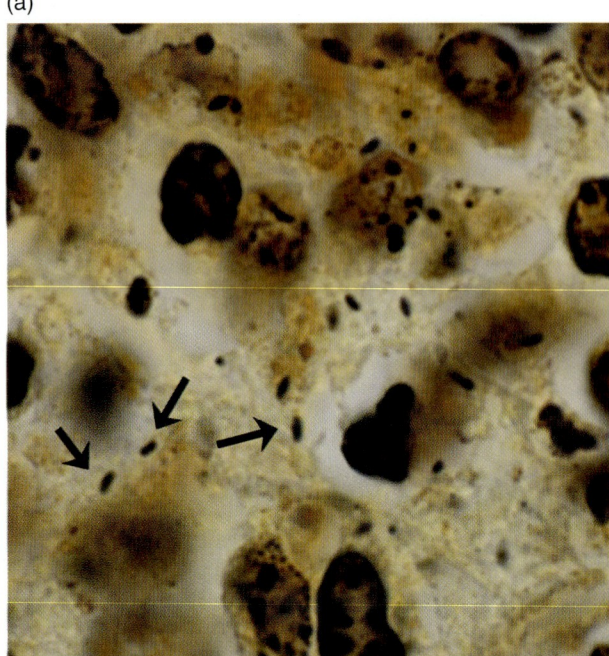

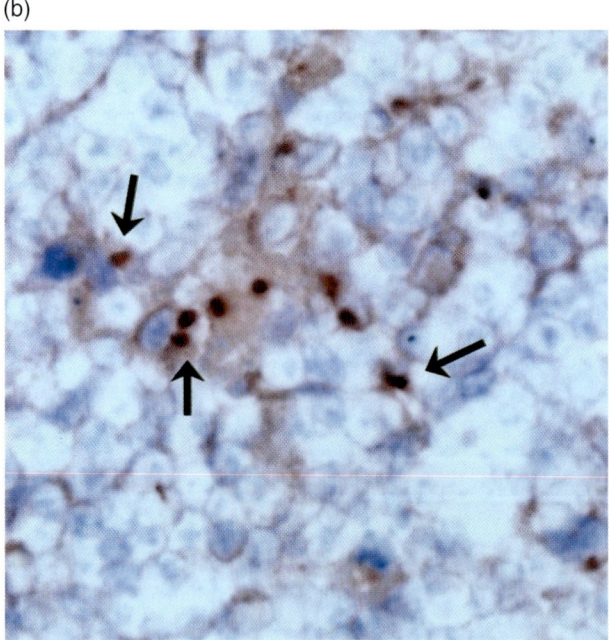

Figure 61.3 (a) aoEBA-infected thymus, modified Steiner silver stain. Note the numerous intracytoplasmic 2-μm bacterial rods indicated by the arrows. (b) aoEBA-infected thymus, immunohistochemistry. Note the numerous bacterial rods indicated by the arrows.

Routine bacterial histologic stains do not demonstrate the aoEBA agent in fetal tissues so a modified Steiner silver staining technique applied to the distantly related deltaproteobacterium *Lawsonia intracellularis* was used to microscopically visualize aoEBA in sections of formalin-fixed necropsy tissues. Steiner stain and immunohistochemistry applied to thin sections of formalin-fixed necropsy tissues (Figure 61.3a,b), and immunofluorescence applied to impression smears and/or homogenates of fresh tissue, are the current methods of choice for microscopic diagnosis of EBA.[9,10] A fluorescent antibody-based serologic assay is currently being validated. The assay appears both specific and sensitive and should prove to be a useful diagnostic when the majority of aoEBA has been eliminated by the fetal immune system. However, interpretation of fetal serology will be difficult in calves that have suckled due to the presence of maternal antibody.

Pathogenesis

The aoEBA apparently gains entrance to the developing fetus across the placenta from the infected dam. The dam, whether infected by tick bite or experimental inoculation, presents with no overt clinical signs. Reactions to experimental tick bites recorded in historical communications were most likely a response to contaminating microbes and/or tick saliva. The fetus has a "window" in development in which it is susceptible to an infection that will ultimately lead to fetal disease. This window is poorly defined but susceptibility is generally believed to be 60–140 days of gestation. Susceptibility early in gestation (<60 days) is not known. While the pathogen does not appear to impact conception, the possibility that it may induce early embryonic death in a small percentage of infected animals is under investigation. Infection of the fetus is a slow process, with definable lesions not becoming evident for 2–3 months.[3] The bacterium appears to replicate intracellularly in histiocytes as determined by immunohistochemistry (Figure 61.2); replication is extremely slow (~1 day) when observed in short-term cultures of macrophages derived from fetal necropsy tissues. The intracellular nature of the infection, vascular orientation of lesions throughout the fetus, and slow and persistent development of disease would all be consistent with the hypothesis that lesions are immunologically mediated. Slow growth of the bacteria is undoubtedly responsible for slow dissemination throughout the body and lesion development coinciding with development of immunologic maturation of the fetal immune system.

Pregnant sheep do not appear to be susceptible to foothill abortion; unpublished studies in which pregnant ewes were infected in the first trimester resulted in the birth of healthy lambs. These latter studies would be consistent with the hypothesis that the bacteria replicate and disseminate slowly, and in the case of sheep the gestational period is too short to develop immunologically mediated lesions and abortion. Mule deer (*Odocoileus hemionus*) serve as an important host to the tick vector, but there is no evidence of EBA occurring in this species. Their gestation period is intermediate between sheep and cattle (6–7 months) and may be too short to allow for disease development. Alternatively, regular exposure to the organism through tick bites may confer immunity and, in those instances when abortions occur, fetuses would be rapidly scavenged leaving no evidence. Multiple attempts to produce disease in immunocompetent laboratory animals, such as mice and rabbits, have failed.

Mice with severe combined immunodeficiency disease (SCID) are susceptible to infection with aoEBA. C3H-*scid* mice develop evidence of wasting at 7–8 weeks following infection and require euthanasia 2–3 weeks later.[10] The only obvious gross lesion is an enlarged spleen that harbors large numbers of infectious bacteria. Cryopreserved aoEBA-infected SCID mouse spleen cells are an excellent source of bacteria to predictably transmit infection to pregnant heifers. Unlike bacteria derived from the bovine fetus, recovered organisms are not opsonized and appear to have excellent viability. The absence of EBA-like lesions in immunodeficient infected mice would support the hypothesis that the fetal bovine lesions are immunologically mediated as suggested above.[10]

Transmission

Observations that the distribution of the *Ornithodoros coriaceus* tick could be superimposed upon the distribution of foothill abortion precipitated studies to establish them as potential biological vectors.[4,7,8] This unusual soft-shelled tick (commonly called the Pajaroello tick) carries a colorful past as Native Americans in California feared its bite over that of a rattlesnake. They believed a second or third exposure to this tick bite would result in death. The tremendous pain and swelling of appendages that sometimes occurs following sequential Pajaroello tick bites is assumed to be an extreme allergic reaction to the tick saliva and/or associated contaminating microbes. An unusual and classical experiment was conducted in which naive pregnant heifers were divided into two groups in an area considered endemic for the disease. One half of the heifers were maintained in tree houses with the tree bases tightly wrapped with metal to interfere with the potential ascent of crawling insects, and the other half were maintained on the ground below. All tree-housed heifers produced healthy calves while foothill abortion occurred in the land-based animals. Studies were subsequently conducted in which wild-trapped ticks were experimentally fed on pregnant naive heifers, further demonstrating disease transmission.[4,5] The *O. coriaceus* tick has been collected in Mexico, California, Nevada, and Oregon (Figure 61.4), but foothill abortion has only been described in the above-mentioned three US states.[7,8,11] Cattle ranching practices in Mexico may well be responsible for a lack of diagnosis of foothill abortion in this country; aborted fetuses can be rapidly scavenged in free-ranging cattle herds, making diagnosis unlikely.

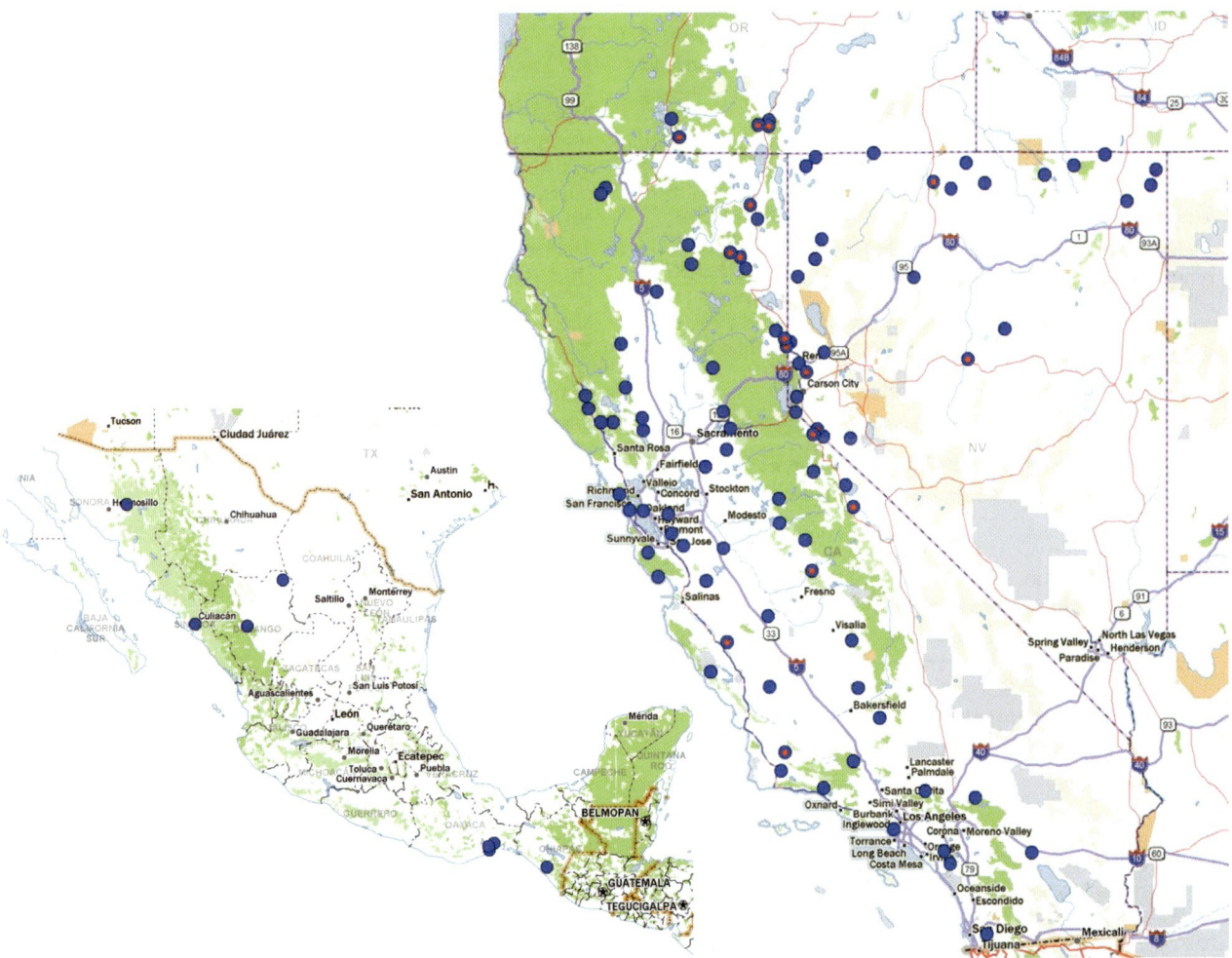

Figure 61.4 Published reports of trapped *Ornithodoros coriaceus* (blue circles). Recent select collections of ticks were subjected to TaqMan-based detection of aoEBA; blue circles with red centers represent positive results. The majority of tick collection sites were sampled prior to the identification of aoEBA and associated PCR-based diagnostics. Tick collections in Mexico are based on historical literature and their locations roughly approximated. © DeLorme. Topo North America™ 10.

Recent published studies have been directed at better understanding the vector–pathogen relationship using field-trapped ticks and PCR/TaqMan diagnostics to identify the presence of aoEBA. *Ornithodoros coriaceus* can be readily collected in the field using dry-ice traps in the duff beneath brush and trees that have previously served as bedding areas for wildlife and/or livestock. These studies confirmed the extensive distribution of *O. coriaceus* in foothill regions of southern Oregon, central and western Nevada, and California.[7,8] Application of diagnostics to DNA extracted from ticks demonstrated a highly variable rate of infection with aoEBA. A positive result obtained by standard PCR was rare.[8] Data obtained using the more sensitive TaqMan suggest that infection rates are typically less than 20% but may exceed 50% (examples of high infection rates may be a mathematical artifact due to low tick numbers). The authors suggest that those ticks that are positive by TaqMan but not standard PCR may not be capable of biological transmission of the bacteria to cattle.[8] This would be supported by studies where experimental feeding of hundreds of ticks per heifer results in definitive disease transmission approximately 50% of the time.[4,5] Efforts to better characterize the tick as a vector of aoEBA demonstrated that the bacterium was preferentially located to the salivary gland of field-collected ticks.[8] The source of this bacterium is currently unknown; possibilities being considered include soil or the presence of a mammalian/avian reservoir. A blood meal is required prior to each molt (progressive instar stages result in increased tick size) and maturation into either a male or female. The authors hypothesize that if ticks are infected while feeding, the multiple blood meals required during the process of maturation would increase the incidence of infection in larger (more mature) ticks. Studies suggest little difference in infection rate in nymphs, males and females and therefore appear to contradict the concept of infection though a blood meal.[8]

Control

Control of foothill abortion has largely been limited to management practices. One approach is minimizing exposure of naive pregnant cattle to tick-infested habitat during the window of gestation in which the fetus is susceptible to infection. Another approach is to intentionally expose naive cattle to the tick prior to breeding such that immunity can be established; historically, EBA was not as prevalent prior to the current practice of breeding young heifers (<2 years of age). Such practices can sometimes modestly reduce losses; however, the reduction in foothill abortion is often offset by inefficient use of rangelands and increased costs associated with altered breeding and grazing strategies. An experimental vaccine is currently being field-tested with encouraging results. The vaccine is cryopreserved live virulent bacteria derived from spleen cells of aoEBA-infected SCID mice; the vaccine cannot be safely administered to pregnant cattle. The duration of natural- and/or vaccine-induced immunity beyond 1 year is unknown. Anecdotal evidence would suggest that natural immunity can begin to wane after 2–3 years.

References

1. Howarth J, Moulton J, Frazier L. Epizootic bovine abortion characterized by fetal hepatopathy. *J Am Vet Med Assoc* 1956; 128:441–449.
2. Kennedy P, Olander H, Howarth J. Pathology of epizootic bovine abortion. *Cornell Vet* 1960;50:417–429.
3. Kennedy P, Casaro A, Kimsey P, BonDurant R, Bushnell R, Mitchell G. Epizootic bovine abortion: histogenesis of the fetal lesions. *Am J Vet Res* 1983;44:1040–1048.
4. Schmidtmann E, Bushnell R, Loomis E, Oliver M, Theis J. Experimental and epizootiologic evidence associating *Ornithodoros coriaceus* Koch with the exposure of cattle to epizootic bovine abortion in California. *J Med Entomol* 1976; 13:292–299.
5. Stott J, Blanchard M, Anderson M et al. Experimental transmission of epizootic bovine abortion (foothill abortion). *Vet Microbiol* 2002;88:161–173.
6. King D, Chen C, Blanchard M et al. Molecular identification of a novel deltaproteobacterium as the etiologic agent of epizootic bovine abortion (foothill abortion). *J Clin Microbiol* 2005;43: 604–609.
7. Teglas M, Drazenovich N., Stott J, Foley J. The geographic distribution of the putative agent of epizootic bovine abortion in the tick vector, *Ornithodoros coriaceus*. *Vet Parasitol* 2006;140: 327–333.
8. Chen C, King D, Blanchard M et al. Identification of the etiologic agent of epizootic bovine abortion in field-collected *Ornithodoros coriaceus* Koch ticks. *Vet Microbiol* 2007;120:320–327.
9. Anderson M, Kennedy P, Blanchard M et al. Histochemical and immunohistochemical evidence of a bacterium associated with lesions of epizootic bovine abortion. *J Vet Diagn Invest* 2006; 18:76–80.
10. Blanchard M, Chen C, Anderson M, Hall M, Barthold S, Stott J. Serial passage of the etioloigic agent of epizootic bovine aboriton in immunodeficient mice. *Vet Microbiol* 2010;144:177–182.
11. Loomis E, Schmidtmann E, Oliver M. A summary review of the distribution of *Ornithodoros coriaceus* Koch in California (Aracina: Argasidae). *California Vector News* 1974;21:57–62.

Chapter 62

Infectious Agents: *Neospora*

Charles T. Estill and Clare M. Scully

Department of Clinical Sciences, College of Veterinary Medicine, Oregon State University, Corvallis, Oregon, USA

Introduction

Neosporosis is caused by the protozoan parasite *Neospora caninum*, an obligate intracellular parasite found in dogs and cattle as well as other species. True prevalence is difficult to determine since published reports tend to be specific for a particular region. However, it is estimated that *Neospora* infects 10–20% of cattle and is responsible for 20% of abortions worldwide.[1] This parasite was first described in dogs having encephalomyelitis in Norway[2] in 1984 and was previously misidentified as *Toxoplasma gondii* until 1988 when the new species, *N. caninum*, was described.[3] The first report of bovine abortion associated with *Neospora* infection was from a New Mexico dairy in the United States in 1989.[4] Both domesticated[5,6] and wild[7–10] canids can serve as definitive hosts while cattle,[1] deer,[11] and chickens[12] are intermediate hosts. However, there is some doubt that the red fox is truly a definitive host.[13] Although transmission can occur both horizontally[14] and vertically, vertical transmission is the primary source of infection in cattle.[1] The economic impact of *N. caninum* is the result of numerous factors including abortions, culling, decreased milk production, increased calving interval, and increased veterinary costs.[15] Global losses are estimated at $1.3 billion and in excess of $600 million in the United States annually.[16]

Epidemiology

Neospora caninum has been identified worldwide and is considered a major cause of abortion in cattle.[17] Globally, 15–20% of dairy cattle and 10–20% of beef cattle have been found seropositive for *N. caninum*.[1] Surveys in California,[17] the Netherlands,[18] and New Zealand[19] indicate that approximately 20% of all aborted bovine fetuses submitted to diagnostic laboratories tested positive for this infection. The estimates of *Neospora* infection in US dairy cattle, based on serology, are variable but appear to be in the range 10–20% and ranges from 5 to 98% of cows in individual dairy herds. Many animal species test seropositive for *N. caninum* but do not exhibit clinical signs.[1] The gold standard for identifying infection is recognition of tissue cysts on histological examination, finding oocysts in feces,[6] or identification of the parasite by immunohistochemical staining[20] or polymerase chain reaction (PCR).[21]

Life cycle

The life cycle of *N. caninum* is depicted in Figure 62.1. *Neospora caninum* has a facultative heteroxenous life cycle, meaning that the organism may use more than one host during its life cycle.[22,23] Unsporulated oocytes (10–14 μm) are shed in the feces of definitive hosts (i.e., dog or wild canid).[24] The oocytes sporulate within 24 hours to the infective form. Each sporulated oocyte contains two sporocysts, and each sporocyst contains four sporozoites. When consumed by cattle or a wide range of other warm-blooded animals which can serve as intermediate hosts, eight sporozoites are released into the gastrointestinal tract for each oocyte consumed.[1] The sporozoites differentiate into tachyzoites (5–7 μm), which subsequently invade the epithelial cells of the gastrointestinal tract. The tachyzoites replicate rapidly via asexual endodyogeny.[25] Endodyogeny is a form of asexual reproduction that involves an unusual process where two daughter cells are produced inside a mother cell, which is then consumed by the offspring prior to their separation.[26] Replication occurs rapidly within cells and tachyzoites may spread hematogenously within mononuclear phagocytes[27] to infect many cell types including neural cells, vascular endothelial cells, myocytes, hepatocytes, renal cells, alveolar macrophages, and placental trophoblasts.[28–30] Tachyzoites differentiate into bradyzoites which are the slowly replicating encysted stage of the parasite. Each cyst may contain hundreds of bradyzoites.[23] In congenitally infected bovine fetuses and calves, tissue cysts are found in the brain and spinal cord.[31] A few thin-walled cysts have been found in skeletal muscle of two naturally infected 2-day-old calves.[32] Bradyzoites are thought to persist for the duration of the host's life and be responsible for persistent infection, although tissue cysts have not been observed in histological sections of naturally infected cattle older than 2 months.[23,28] Bradyzoites are believed to reactivate and

Bovine Reproduction, First Edition. Edited by Richard M. Hopper.
© 2015 John Wiley & Sons, Inc. Published 2015 by John Wiley & Sons, Inc.

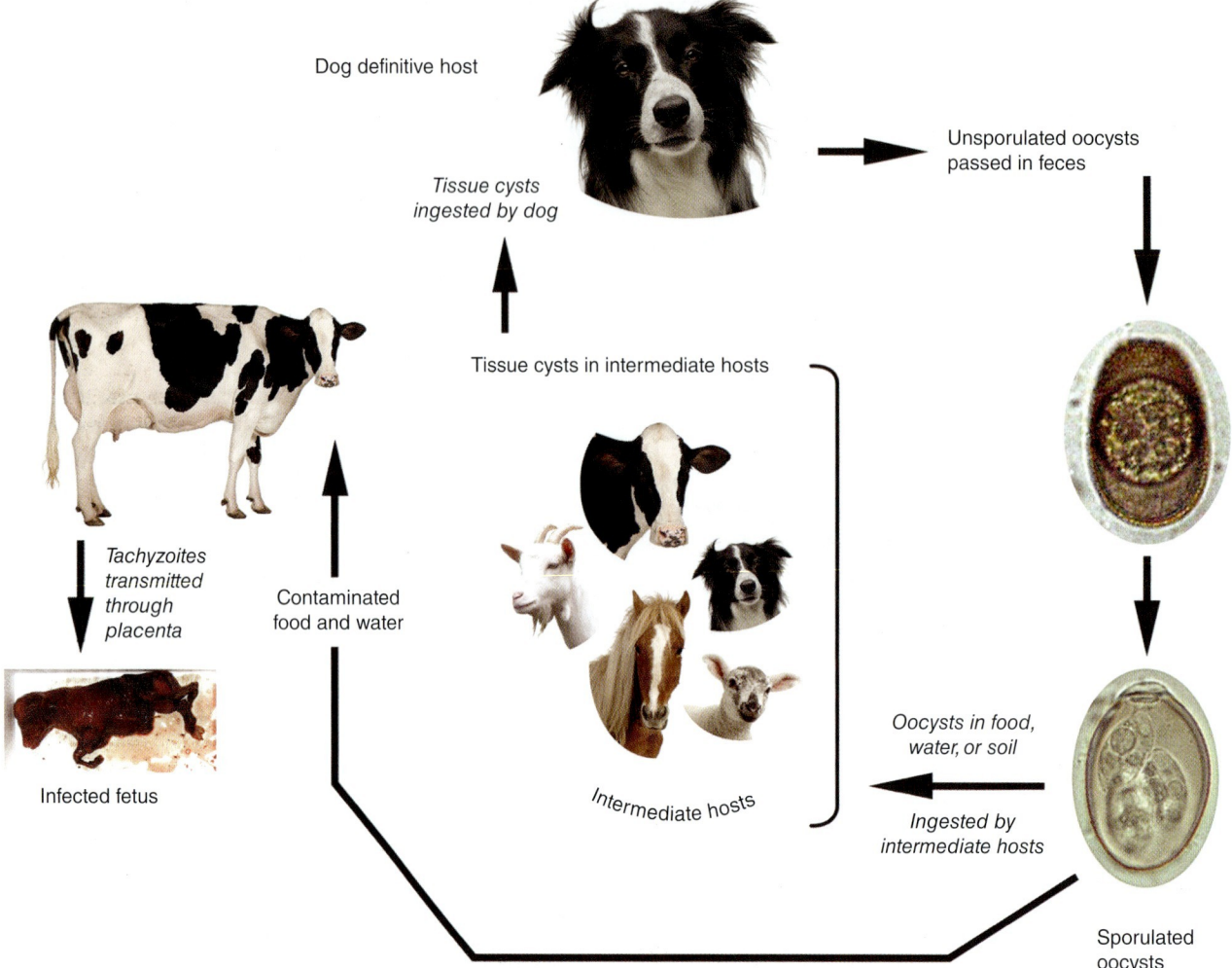

Figure 62.1 Life cycle of *Neospora caninum*. Reproduced with permission from Almeria S, Ferrer D, Pabón M, Castella J, Manas S. Red foxes (*Vulpes vulpes*) are a natural intermediate host of *Neospora caninum*. Vet Parasitol 2002;107:287–294.

differentiate back into tachyzoites, thereby establishing persistent infection.[23] Bradyzoites can be differentiated in tissue sections from tachyzoites by labeling with a specific antibody.[33] The cycle continues when tissue cysts shed by the intermediate host (cattle) are consumed by the definitive host (canids).[1]

Transmission

Neospora can be transmitted vertically or horizontally with both routes being vital to long-term survival of the parasite.[23] Vertical transmission takes place when tachyzoites migrate transplacentally from dam to fetus during pregnancy.[1,24] *Neospora caninum* is one of the most efficiently transplacentally transmitted parasites among all known microbes in cattle.[1] The terms "endogenous transplacental transmission" and "exogenous transplacental transmission" have been used to more precisely describe the origin and route of fetal infection (Figure 62.2).[22] Endogenous transplacental transmission occurs in a persistently infected dam when the infection crosses the placenta and enters the fetus.[22] The endogenous form has a higher rate of associated abortion.[34] Exogenous transplacental transmission occurs when a previously noninfected dam ingests infective oocysts while pregnant and her fetus subsequently also becomes infected *in utero*.

Vertical transmission can result in abortion but in most cases the calf is congenitally infected but asymptomatic with no evidence of deleterious effects on subsequent calf health.[35] If the infection occurs during the second or third trimester when immunocompetence of the fetus is greater, the most likely outcome will be a liveborn but persistently infected calf.[23] Once infected, cattle are presumed to be infected for life[36] and females can transmit the infection to successive generations.[37,38] Although not every pregnancy results in transmission of the disease, it has been reported that the rate of transmission is as high as 75–100%.[16] In the case of exogenous transplacental transmission, the number of oocysts ingested by the dam and stage of gestation influence pregnancy outcome.[24,39] There is a report indicating that in persistently infected cattle, vertical transmission is more efficient in younger than older cows.[40] Also, transplacental infection may be more likely to occur in dams that were themselves prenatally infected compared with postnatally infected dams.[41] The risk of abortion was positively correlated with *N. caninum*-specific antibodies

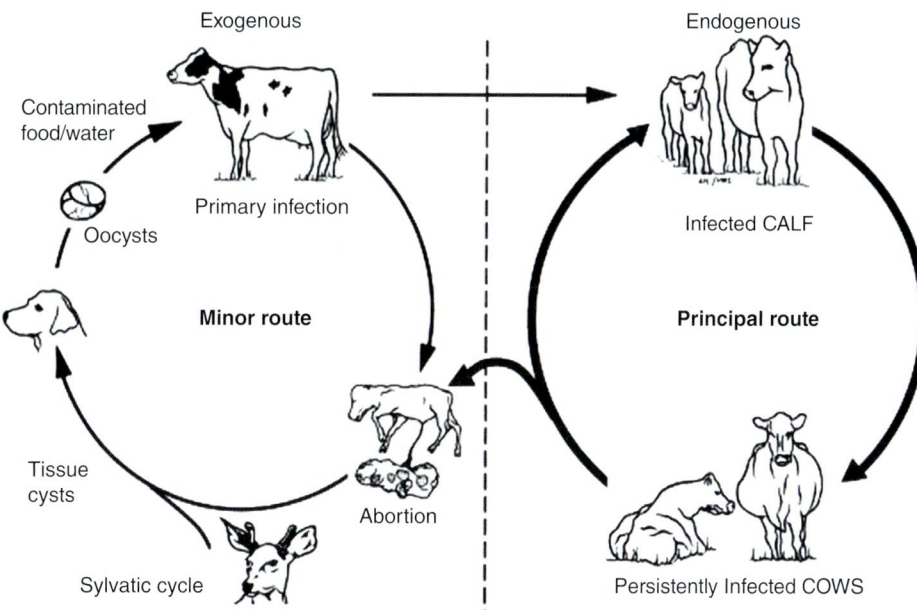

Figure 62.2 Transmission of bovine neosporosis. Oocysts are produced by the canine definitive host and their subsequent ingestion by a susceptible pregnant cow leads to infection of the fetus (exogenous transplacental transmission). Liveborn infected heifer calves would be expected to remain infected into adulthood when they, in turn, may pass infection to their fetus (endogenous transplacental transmission). Spread of N. caninum in this second way is the principal route whereby the parasite is propagated in a herd. Adapted with permission from Dubey JP. Neosporosis in cattle. Vet Clin North Am Food Anim Pract 2005;21:473–483.

in individual animals[42] and severity of fetal *Neospora*-associated lesions also increased with higher maternal seropositivity.[43] Cows with high antibody titers (≥400) showed higher vertical transmission frequency (94.8%) than cows with low antibody titers of 25–200 (14.8%).[44] Occasionally, a seronegative dam will give birth to a seropositive calf.[45,46] This can occur when the dam has a long-standing infection with low antibody titer and she was wrongly diagnosed as uninfected.[1]

Horizontal transmission is less common than vertical transmission.[16] The ingestion of sporulated oocysts from the environment is the only demonstrated natural mode of postnatal infection in cattle (Figure 62.2).[47,48] Both domestic and wild dogs are frequently implicated in horizontal transmission. This is attributed to canids consuming infected bovine placentas or fetuses and subsequently defecating in the environment where the oocytes are ingested by cattle.[49] There has been no horizontal cow-to-cow transmission reported at this time.[37] Experimentally, calves may become infected through ingestion of milk contaminated with tachyzoites[50,51] and *N. caninum* DNA has been detected in colostrum of seropositive cows.[52] However, it remains doubtful that lactogenic transmission occurs naturally.[49] Since *N. caninum* DNA was reported in fresh and frozen semen from naturally infected bulls,[53] the possibility of venereal transmission was investigated. Intrauterine inoculation of nine heifers with 10^7 *N. caninum* tachyzoites reacted with seroconversion and a specific interferon (IFN)-γ response. Also, *N. caninum* DNA was demonstrated in the blood and tissues of all nine heifers.[54] However, the numbers of tachyzoites used to test the venereal route of infection are higher than those previously found in semen of naturally infected bulls, which range from 1 to 10 organisms per milliliter associated with the cell fraction of semen.[55,56] Therefore, it seems unlikely that venereal transmission is significant under natural conditions.

Pathogenesis of abortion

Neosporosis is primarily a disorder of the fetoplacental unit that occurs subsequent to maternal parasitemia resulting either from exogenous or endogenous infection.[23] In the case of exogenous infection, the ingested oocytes excyst in the small intestine and presumably release sporozoites which parasitize the maternal intestinal epithelium and transform into tachyzoites which, in turn, multiply in the mesenteric lymph nodes.[23] The tachyzoites are released into the blood and have been detected in the leukocyte fraction[27] of naturally infected cattle. The parasitemia results in dissemination throughout the body including the gravid uterus. The endogenous (vertical) route of transmission is most common.[37,57] It is suspected that latent infections are reactivated due to the immunosuppression associated with mid-gestation pregnancy.[58] It is also reported that endogenous transplacental transmission is more likely in cattle that were themselves infected *in utero*.[59,60]

The mechanism by which *N. caninum* actually causes abortion is a topic of ongoing investigation and multiple hypotheses have been proposed. First, since the parasite is able to establish itself in the caruncular septum before crossing to the fetal placental villi,[61,62] the resultant placental damage may result in fetal death due to placental insufficiency.[23] Placental insufficiency, because it causes fetal stress and release of adrenocorticotropin, could also cause premature delivery of either viable or compromised calves. Second, infection by *N. caninum* stimulates cell-mediated immune responses associated with cytokine release from the damaged placenta, resulting in fetal rejection by the maternal immune system culminating in fetal expulsion.[23,63,64] Finally, the placental damage may result in endometrial prostaglandin $(PG)F_{2\alpha}$ release causing luteolysis and abortion as has been demonstrated in goats with toxoplasmosis.[65] However, cows with high *Neospora* antibody titers have higher plasma progesterone levels throughout gestation than seronegative cows or cows with low antibody titers;[66] in a study in which exogenous progesterone was administered during the mid-gestation period, the risk of abortion increased dramatically in cows with high levels of antibodies against *N. caninum*.[67] It seems most likely that all three of the proposed mechanisms play a part in inducing *Neospora*-associated abortion in cattle.

The first-trimester fetus seems particularly susceptible to damage by invading *N. caninum* since it is not immunocompetent[68] and unlikely to survive infection. In the middle trimester the fetus may be able to mount an immune response to *N. caninum* challenge, although the response may not be robust enough to preclude fetal death.[58,69,70] By the third trimester the fetus is capable of mounting an immune response of sufficient magnitude that it is likely to survive *N. caninum* infection, leading to birth of clinically normal but infected calves. In beef cows experimentally infected with *N. caninum* at various times during gestation by either subcutaneous or intravenous routes, transplacental infection rate increased with gestational age and cows infected during later gestation generally delivered normal but infected calves.[39]

In cows experimentally infected with *N. caninum* the most severe lesions are found in the placenta and the fetal brain.[39,69] Lesions consist of a nonsuppurative inflammatory response.[62] The associated inflammatory cells have been shown capable of producing INF-γ,[58] which suggests fetal death may be more a consequence of the maternal immune response rather than direct effects of the parasite.[58] At the time of placental invasion the parasite also enters the fetal bloodstream with a predilection for the central nervous system,[62] where it is located perivascularly,[29] and in younger fetuses causes widespread destruction of neuropil with very little inflammatory response.[71] When fetuses later in gestation become infected with *N. caninum* the tissue reaction is more moderate, with areas of necrosis restricted to small foci surrounded by intense inflammatory infiltrate in brain tissue[18,72] and sometimes, mild meningitis.[73] *Neospora caninum* fetal infection also causes characteristic lesions of inflammation and necrosis in tissues such as liver and heart.[18]

Clinical signs

Neospora caninum infection causes abortion in both beef and dairy cattle. Abortion may occur from 3 months of gestation to term but typically occurs at 5–6 months of gestation.[4,18,74,75] Possible outcomes of infection include fetal death followed by resorption,[76] mummification,[77] autolysis,[78] stillbirth,[79] born with clinical signs,[78] or born clinically normal[35] but persistently infected. Fetuses dying *in utero* between 3 and 8 months are usually expelled with evidence of moderate autolysis while those dying before 5 months of gestation are more likely to mummify.[23] Experimental infection of pregnant cows at various points during gestation showed that parasitemia during the first 10 weeks resulted in fetopathy and resorption of the fetal tissue 3 weeks after infection. Infection at 30 weeks of gestation resulted in birth of asymptomatic congenitally infected calves.[59] The period in between appears as most likely for *N. caninum*-induced abortion.

In *Neospora*-infected calves clinical signs have been reported only in animals less than 2 months of age.[80] Infected calves may be born underweight, unable to rise, and with neurologic signs.[80] Hindlimbs and/or forelimbs may be flexed or hyperextended and the calf may be ataxic, have diminished patellar reflexes, and loss of conscious proprioception.[80] Exophthalmia has been observed in some cases.[80]

Within herds, *Neospora*-associated abortions may be sporadic, enzootic, or endemic.[4,17,29,81] In comparison with endemic abortion, epidemic abortion storms are less common.[82] It appears that most sporadic and endemic abortions in cattle due to neosporosis are the result of reactivation of chronic infections because *Neospora*-seropositive cows are at twofold to threefold higher risk of abortion than are seronegative cows.[75,77] Reactivation of a chronic *Neospora* infection may occur as a result of factors that cause immunosuppression. We have observed a *Neospora*-associated abortion storm in a dairy herd concomitant with an outbreak of coliform mastitis.[45] In herds with endemic *Neospora*-associated abortions there is often a positive correlation between serostatus of the dam and her daughters, indicating that the major route of transmission is vertical.[83,84]

Epidemic abortion outbreaks (abortion storms) have been defined as those in which over 10–15% of at-risk cows abort within a 4–8 week period.[77,83,85] There is epidemiological evidence that epidemic *Neospora*-associated abortions may be caused by the postnatal infection of naive cattle via exposure to oocyst-contaminated feed or water, with up to 57% of pregnant dairy cows aborting over a few weeks to months.[83,86,87] However, in an epidemiological investigation of abortion storms in the Netherlands, the authors concluded that the abortion storms appeared to be induced by factors causing recrudescence of *N. caninum* infection in chronically infected animals rather than being the result of a recent introduction.[85]

Diagnosis

Definitive diagnosis of neosporosis is challenging and relies on numerous diagnostic techniques. There are serologic assays available to test the maternal serum and fetal fluids (serum, fluid from body cavities) for *Neospora*-specific antibodies.[88] Serology of fetal fluids is useful after 5 months of gestation as the fetus becomes immunocompetent but sensitivities reported varied from 50 to 84%.[89] Immature fetal immunocompetence and a short interval between infection and death can all lead to a false-negative result.[89] Finding the antibodies does not definitively confirm *Neospora* as the cause of abortion since clinically normal calves may have congenital antibodies.[75]

Serology of the dam can aid in diagnosis of *Neospora*-associated abortion. Sera obtained within 2 weeks of abortion generally have the highest titers and then titers decline and may be undetectable 60 days after abortion.[88,90] Seropositive cows do not display detectable clinical signs of infection and most deliver healthy congenitally infected calves. Therefore, seropositivity has a high predictive value for congenital infection in cows but not for abortion[91] (although there is an increased probability of abortion of twofold to threefold compared with seronegative cows[75]). High antibody levels persist for prolonged periods in some cows[90] or may increase again in subsequent gestations.[89] Seronegativity in a mother cow or heifer does not exclude N. caninum-associated abortions.[92]

PCR has also been utilized to detect parasite DNA,[93] but again it is not always found.[28] In a Swiss study, N. caninum was detected by PCR in the brains of 21% of all aborted fetuses. Microscopic lesions indicative of cerebral protozoal infection were detected in 84% of PCR-positive fetal brains.[92]

Histopathology of fetal tissues can be examined to identify characteristic lesions and parasites.[94] Multifocal inflammation and necrosis of the heart, brain, liver, and skeletal muscle are common lesions and tachyzoites are found primarily in the brain.[18,89] Lesions can be found in the placenta but are not particularly valuable.[89] Even in grossly autolyzed or mummified fetuses it is possible to find suspect lesions, especially in the brain.[89] On necropsy, gross lesions are not frequently seen.

To positively identify parasites associated with lesions, immunohistochemistry is commonly employed.[17,31] In one study, in which N. caninum tachyzoites were identified by immunohistochemistry performed on tissues taken from calves with confirmed neosporosis, 85% of the brains were positive, 14% of the hearts, and 26% of the livers.[18] Significant differences between epizootic and sporadic abortion cases with regard to positivity on immunohistochemical examination were found only in the liver, where tachyzoites were more frequently found and in higher numbers in epizootic cases than from sporadic cases.[18] One must bear in mind that positive results of any diagnostic tests without the presence of life-threatening lesions are insufficient for a definitive diagnosis.[28]

In investigations of herds in which multiple abortions occurred, a tentative diagnosis can be supported by cross-sectional or herd serology in which the N. caninum status of aborting and nonaborting cows are compared to determine if there is an association between serologic status and abortions.[90,95] In the case of an abortion storm, a suggested protocol would be to immediately take blood samples from all animals at risk, then take post-abortion samples to determine if seroconversion occurred.[89]

Control/prevention

A number of approaches have been employed to control neosporosis in cattle herds. These include improving farm biosecurity, test and cull programs, test and exclude from breeding, and artificially inseminating with beef semen. Biosecurity may prove beneficial to avoid introducing the parasite into a closed herd not already infected with Neospora.[96] Farm management practices to reduce N. caninum infection can include (i) minimizing fecal contamination of cattle feed or water by canids,[1] (ii) prompt removal of aborted bovine fetuses and fetal membranes; and (iii) limiting the introduction of infected cattle into the herd and culling infected animals. If employing the test and cull method of control, every animal must be serologically tested and those that test positive removed from the herd. Test-negative animals should be periodically retested since titers tend to wax and wane and there is a possibility of horizontal transmission resulting in new infections. This approach can be devastating depending on how many in the herd test positive. Notwithstanding the risk of horizontal transmission, a more economic approach would be to test each animal and exclude the daughters of seropositive cows from the replacement pool. If a cow is very valuable there is always the opportunity to preserve the genetics by embryo transfer into a seronegative dam.[97]

It is possible to reduce the risk of Neospora abortions in dairy cattle by inseminating seropositive dams with beef breed semen. This is effective because crossbreed pregnancies have a more robust placentation with higher levels of peripartum pregnancy-associated glycoprotein, which may have protective value.[98] Beef breeds in general, and the Limousin breed in particular, are more resistant to Neospora infection than are dairy breeds.[98] Beef cow-calf herds that manage their cows on range for summer grazing have lower seroprevalence than those that do not, while increased seroprevalence is associated with higher winter stocking density.[99]

These control methods might not be economically or practically feasible on dairies or beef cattle operations.[100] Buying replacements originating from seronegative herds can mitigate the risk of introducing infected replacements into a negative herd.[101] General management efforts to reduce stress from concurrent disease, environmental and social stress, and providing an adequate wholesome ration may reduce immunosuppression and hence abortion rates.

There is accumulating evidence that some N. caninum-infected cows can develop a degree of protective immunity against abortion and/or congenital transmission, indicating that immunoprophylaxis to prevent abortion or congenital transmission is a feasible goal.[65] A commercial vaccine is not currently available. In the future, vaccine development is likely to depend on identification of specific Neospora genes that may enable the production of genetically engineered vaccines.

Treatment

Currently, there are no approved practical chemotherapeutic compounds useful for treatment of bovine neosporosis.[16] Various antimicrobial agents have been tested in vitro[102–104] or in vivo in mice,[105] but there is no known drug that can be used to clear N. caninum infection in adult cattle. However, when toltrazuril was administered to congenitally infected newborn calves the infection was eliminated, suggesting that vertical transmission in subsequent generations may be reduced by treatment of calves soon after birth.[106]

References

1. Dubey J, Schares G, Ortega-Mora L. Epidemiology and control of neosporosis and *Neospora caninum*. *Clin Microbiol Rev* 2007;20: 323–367.
2. Bjerkås I, Mohn S, Presthus J. Unidentified cyst-forming sporozoon causing encephalomyelitis and myositis in dogs. *Parasitol Res* 1984;70:271–274.

3. Dubey J, Carpenter J, Speer C, Topper M, Uggla A. Newly recognized fatal protozoan disease of dogs. *J Am Vet Med Assoc* 1988;192:1269–1285.
4. Thilsted J, Dubey J. Neosporosis-like abortions in a herd of dairy cattle. *J Vet Diagn Invest* 1989;1:205–209.
5. McAllister M, Dubey J, Lindsay D, Jolley W, Wills R, McGuire A. Dogs are definitive hosts of *Neospora caninum*. *Int J Parasitol* 1998;28:1473–1479.
6. Lindsay D, Dubey J, Duncan R. Confirmation that the dog is a definitive host for *Neospora caninum*. *Vet Parasitol* 1999;82:327–333.
7. Gondim L, McAllister M, Pitt W, Zemlicka D. Coyotes (*Canis latrans*) are definitive hosts of *Neospora caninum*. *Int J Parasitol* 2004;34:159–161.
8. Almería S, Ferrer D, Pabón M, Castella J, Manas S. Red foxes (*Vulpes vulpes*) are a natural intermediate host of *Neospora caninum*. *Vet Parasitol* 2002;107:287–294.
9. Dubey J, Jenkins M, Rajendran C et al. Gray wolf *Canis lupus* is a natural definitive host for *Neospora caninum*. *Vet Parasitol* 2011;181:382–387.
10. King J, Šlapeta J, Jenkins D, Al-Qassab S, Ellis J, Windsor P. Australian dingoes are definitive hosts of *Neospora caninum*. *Int J Parasitol* 2010;40:945–950.
11. Gondim L, McAllister M, Mateus-Pinilla N, Pitt W, Mech L, Nelson M. Transmission of *Neospora caninum* between wild and domestic animals. *J Parasitol* 2004;90:1361–1365.
12. Costa K, Santos S, Uzeda R et al. Chickens *Gallus domesticus* are natural intermediate hosts of *Neospora caninum*. *Int J Parasitol* 2008;38:157–159.
13. Schares G, Heydorn A, Cüppers A et al. In contrast to dogs, red foxes (*Vulpes vulpes*) did not shed *Neospora caninum* upon feeding of intermediate host tissues. *Parasitol Res* 2002;88:44–52.
14. Davison H, Otter A, Trees A. Estimation of vertical and horizontal transmission parameters of *Neospora caninum* infections in dairy cattle. *Int J Parasitol* 1999;29:1683–1689.
15. Trees A, Davison H, Innes E, Wastling J. Towards evaluating the economic impact of bovine neosporosis. *Int J Parasitol* 1999;29:1195–1200.
16. Reichel M, Alejandra Ayanegui-Alcérreca M, Gondim L, Ellis J. What is the global economic impact of *Neospora caninum* in cattle: the billion dollar question. *Int J Parasitol* 2013;43:133–142.
17. Anderson M, Blanchard P, Barr B, Dubey J, Hoffman R, Conrad P. *Neospora*-like protozoan infection as a major cause of abortion in California dairy cattle. *J Am Vet Med Assoc* 1991;198:241–244.
18. Wouda W, Moen A, Visser I, Van Knapen F. Bovine fetal neosporosis: a comparison of epizootic and sporadic abortion cases and different age classes with regard to lesion severity and immunohistochemical identification of organisms in brain, heart, and liver. *J Vet Diagn Invest* 1997;9:180–185.
19. Thornton R, Thompson E, Dubey J. *Neospora* abortion in New Zealand cattle. *NZ Vet J* 1991;39:129–133.
20. Jenkins M, Baszler T, Björkman C, Schares G, Williams D. Diagnosis and seroepidemiology of *Neospora caninum* associated bovine abortion. *Int J Parasitol* 2002;32:631–636.
21. Ho M, Barr B, Marsh A et al. Identification of bovine *Neospora* parasites by PCR amplification and specific small-subunit rRNA sequence probe hybridization. *J Clin Microbiol* 1996;34:1203–1208.
22. Trees A, Williams D. Endogenous and exogenous transplacental infection in *Neospora caninum* and *Toxoplasma gondii*. *Trends Parasitol* 2005;21:558–561.
23. Dubey J, Buxton D, Wouda W. Pathogenesis of bovine neosporosis. *J Comp Pathol* 2006;134:267–289.
24. McCann C, McAllister M, Gondim L et al. *Neospora caninum* in cattle: experimental infection with oocysts can result in exogenous transplacental infection, but not endogenous transplacental infection in the subsequent pregnancy. *Int J Parasitol* 2007;37:1631–1639.
25. Dubey J. Neosporosis in cattle. *Vet Clin North Am Food Anim Pract* 2005;21:473.
26. Robert-Gangneux F, Dardé M. Epidemiology of and diagnostic strategies for toxoplasmosis. *Clin Microbiol Rev* 2012;25:264–296.
27. Okeoma C, Williamson N, Pomroy W, Stowell K, Gillespie L. The use of PCR to detect *Neospora caninum* DNA in the blood of naturally infected cows. *Vet Parasitol* 2004;122:307–315.
28. Dubey J, Schares G. Diagnosis of bovine neosporosis. *Vet Parasitol* 2006;140:1–34.
29. Barr B, Anderson M, Dubey J, Conrad P. *Neospora*-like protozoal infections associated with bovine abortions. *Vet Pathol Online* 1991;28:110–116.
30. Barr B, Conrad P, Dubey J, Anderson M. *Neospora*-like encephalomyelitis in a calf: pathology, ultrastructure, and immunoreactivity. *J Vet Diagn Invest* 1991;3:39–46.
31. Dubey J, Leathers C, Lindsay D. *Neospora caninum*-like protozoon associated with fatal myelitis in newborn calves. *J Parasitol* 1989;75:146–148.
32. Peters M, Lütkefels E, Heckeroth A, Schares G. Immunohistochemical and ultrastructural evidence for *Neospora caninum* tissue cysts in skeletal muscles of naturally infected dogs and cattle. *Int J Parasitol* 2001;31:1144–1148.
33. McAllister M, Huffman E, Hietala S, Conrad P, Anderson M, Salman M. Evidence suggesting a point source exposure in an outbreak of bovine abortion due to neosporosis. *J Vet Diagn Invest* 1996;8:355–357.
34. Williams D, Hartley C, Björkman C, Trees A. Endogenous and exogenous transplacental transmission of *Neospora caninum*: how the route of transmission impacts on epidemiology and control of disease. *Parasitology* 2009;136:1895–1900.
35. Paré J, Thurmond MC, Hietala SK. Congenital *Neospora caninum* infection in dairy cattle and associated calfhood mortality. *Can J Vet Res* 1996;60:133–139.
36. Piergili Fioretti D, Pasquali P, Diaferia M, Mangili V, Rosignoli L. *Neospora caninum* infection and congenital transmission: serological and parasitological study of cows up to the fourth gestation. *J Vet Med B* 2003;50:399–404.
37. Björkman C, Johansson O, Stenlund S, Holmdahl O, Uggla A. *Neospora* species infection in a herd of dairy cattle. *J Am Vet Med Assoc* 1996;208:1441–1444.
38. Wouda W, Moen A, Schukken Y. Abortion risk in progeny of cows after a *Neospora caninum* epidemic. *Theriogenology* 1998;49:1311–1316.
39. Gondim L, McAllister M, Anderson-Sprecher R et al. Transplacental transmission and abortion in cows administered *Neospora caninum* oocysts. *J Parasitol* 2004;90:1394–1400.
40. Dijkstra T. *Horizontal and vertical transmission of Neospora caninum*. PhD dissertation, Utrecht University, 2002.
41. Dijkstra T, Lam T, Bartels C, Eysker M, Wouda W. Natural postnatal *Neospora caninum* infection in cattle can persist and lead to endogenous transplacental infection. *Vet Parasitol* 2008;152:220–225.
42. Quintanilla-Gozalo A, Pereira-Bueno J, Seijas-Carballedo A, Costas E, Ortega-Mora L. Observational studies in *Neospora caninum* infected dairy cattle: relationship infection–abortion and gestational antibody fluctuations. *Int J Parasitol* 2000;30:900–906.
43. De Meerschman F, Speybroeck N, Berkvens D et al. Fetal infection with *Neospora caninum* in dairy and beef cattle in Belgium. *Theriogenology* 2002;58:933–945.
44. Moré G, Bacigalupe D, Basso W et al. Frequency of horizontal and vertical transmission for *Sarcocystis cruzi* and *Neospora caninum* in dairy cattle. *Vet Parasitol* 2009;160:51–54.
45. Estill C. *Neospora*-associated abortion and field experience with a commercial vaccine in a dairy herd. In: *23rd World Buiatrics Congress*, Quebec City, Canada, July 11–16, 2004. Available at http://www.ivis.org/proceedings/wbc/wbc2004/abstr_301_400.htm

46. Paré J, Thurmond MC, Hietala SK. Congenital *Neospora* infection in dairy cattle. *Vet Rec* 1994;134:531–532.
47. Trees A, McAllister M, Guy C, McGarry J, Smith R, Williams D. *Neospora caninum*: oocyst challenge of pregnant cows. *Vet Parasitol* 2002;109:147–154.
48. De Marez T, Liddell S, Dubey J, Jenkins M, Gasbarre L. Oral infection of calves with *Neospora caninum* oocysts from dogs: humoral and cellular immune responses. *Int J Parasitol* 1999;29: 1647–1657.
49. Dijkstra T, Eysker M, Schares G, Conraths F, Wouda W, Barkema H. Dogs shed *Neospora caninum* oocysts after ingestion of naturally infected bovine placenta but not after ingestion of colostrum spiked with *Neospora caninum* tachyzoites. *Int J Parasitol* 2001; 31:747–752.
50. Davison H, Guy C, McGarry J et al. Experimental studies on the transmission of *Neospora caninum* between cattle. *Res Vet Sci* 2001;70:163–168.
51. Uggla A, Stenlund S, Holmdahl O et al. Oral *Neospora caninum* inoculation of neonatal calves. *Int J Parasitol* 1998;28:1467–1472.
52. Moskwa B, Pastusiak K, Bien J, Cabaj W. The first detection of *Neospora caninum* DNA in the colostrum of infected cows. *Parasitol Res* 2007;100:633–636.
53. Ortega-Mora L, Ferre I, del-Pozo I et al. Detection of *Neospora caninum* in semen of bulls. *Vet Parasitol* 2003;117:301–308.
54. Serrano E, Ferre I, Osoro K et al. Intrauterine *Neospora caninum* inoculation of heifers. *Vet Parasitol* 2006;135:197–203.
55. Serrano-Martínez E, Ferre I, Martínez A et al. Experimental neosporosis in bulls: parasite detection in semen and blood and specific antibody and interferon-gamma responses. *Theriogenology* 2007;67:1175–1184.
56. Ferre I, Aduriz G, del-Pozo I et al. Detection of *Neospora caninum* in the semen and blood of naturally infected bulls. *Theriogenology* 2005;63:1504–1518.
57. Anderson M, Reynolds J, Rowe J et al. Evidence of vertical transmission of *Neospora* sp. infection in dairy cattle. *J Am Vet Med Assoc* 1997;210:1169–1172.
58. Innes E, Wright S, Bartley P et al. The host–parasite relationship in bovine neosporosis. *Vet Immunol Immunopathol* 2005;108: 29–36.
59. Williams D, Guy C, McGarry J et al. *Neospora caninum*-associated abortion in cattle: the time of experimentally-induced parasitaemia during gestation determines foetal survival. *Parasitology* 2000; 121:347–358.
60. Innes E, Wright S, Maley S et al. Protection against vertical transmission in bovine neosporosis. *Int J Parasitol* 2001;31:1523–1534.
61. Maley S, Buxton D, Rae A et al. The pathogenesis of neosporosis in pregnant cattle: inoculation at mid-gestation. *J Comp Pathol* 2003;129:186–195.
62. Macaldowie C, Maley S, Wright S et al. Placental pathology associated with fetal death in cattle inoculated with *Neospora caninum* by two different routes in early pregnancy. *J Comp Pathol* 2004; 131:142–156.
63. Innes E, Andrianarivo A, Björkman C, Williams D, Conrad P. Immune responses to *Neospora caninum* and prospects for vaccination. *Trends Parasitol* 2002;18:497–504.
64. Quinn H, Ellis J, Smith N. *Neospora caninum*: a cause of immune-mediated failure of pregnancy? *Trends Parasitol* 2002;18:391–394.
65. Engeland I, Waldeland H, Kindahl H, Ropstad E, Andresen Ø. Effect of *Toxoplasma gondii* infection on the development of pregnancy and on endocrine foetal–placental function in the goat. *Vet Parasitol* 1996;67:61–74.
66. García-Ispierto I, Nogareda C, Yániz J et al. *Neospora caninum* and *Coxiella burnetii* seropositivity are related to endocrine pattern changes during gestation in lactating dairy cows. *Theriogenology* 2010;74:212–220.
67. Bech-Sabat G, López-Gatius F, Santolaria P et al. Progesterone supplementation during mid-gestation increases the risk of abortion in *Neospora*-infected dairy cows with high antibody titres. *Vet Parasitol* 2007;145:164–167.
68. Osburn BI. Ontogeny of immune responses in cattle. In: Morrison WI (ed.) *The Ruminant Immune System in Health and Disease*. Cambridge: Cambridge University Press, 1986, pp. 252–260.
69. Andrianarivo A, Barr B, Anderson M et al. Immune responses in pregnant cattle and bovine fetuses following experimental infection with *Neospora caninum*. *Parasitol Res* 2001;87:817–825.
70. Bartley P, Kirvar E, Wright S et al. Maternal and fetal immune responses of cattle inoculated with *Neospora caninum* at mid-gestation. *J Comp Pathol* 2004;130:81–91.
71. Buxton D, McAllister M, Dubey J. The comparative pathogenesis of neosporosis. *Trends Parasitol* 2002;18:546–552.
72. Barr B, Rowe J, Sverlow K et al. Experimental reproduction of bovine fetal *Neospora* infection and death with a bovine *Neospora* isolate. *J Vet Diagn Invest* 1994;6:207–215.
73. Jardine J, Last R. *Neospora caninum* in aborted twin calves. *J South Afr Vet Assoc* 1993;64:101.
74. Dubey J. Neosporosis: the first decade of research. *Int J Parasitol* 1999;29:1485–1488.
75. Paré J, Thurmond MC, Hietala SK. *Neospora caninum* antibodies in cows during pregnancy as a predictor of congenital infection and abortion. *J Parasitol* 1997;83:82–87.
76. Dubey J, Lindsay D. Neosporosis. *Parasitol Today* 1993;9:452–458.
77. Moen A, Wouda W, Mul M, Graat E, Van Werven T. Increased risk of abortion following *Neospora caninum* abortion outbreaks: a retrospective and prospective cohort study in four dairy herds. *Theriogenology* 1998;49:1301–1309.
78. Barr B, Conrad P, Breitmeyer R et al. Congenital *Neospora* infection in calves born from cows that had previously aborted *Neospora*-infected fetuses: four cases (1990–1992). *J Am Vet Med Assoc* 1993;202:113–117.
79. Otter A, Jeffrey M, Griffiths I, Dubey J. A survey of the incidence of *Neospora caninum* infection in aborted and stillborn bovine fetuses in England and Wales. *Vet Rec* 1995;136:602–606.
80. Dubey J. Recent advances in *Neospora* and neosporosis. *Vet Parasitol* 1999;84:349–367.
81. Yaeger M, Shawd-Wessels S, Leslie-Steen P. *Neospora* abortion storm in a midwestern dairy. *J Vet Diagn Invest* 1994;6:506–508.
82. Anderson M, Andrianarivo A, Conrad P. Neosporosis in cattle. *Anim Reprod Sci* 2000;60:417–431.
83. Schares G, Bärwald A, Staubach C et al. p38-avidity-ELISA. Examination of herds experiencing epidemic or endemic *Neospora caninum*-associated bovine abortion. *Vet Parasitol* 2002;106:293–305.
84. Thurmond M, Hietala S. Effect of congenitally acquired *Neospora caninum* infection on risk of abortion and subsequent abortions in dairy cattle. *Am J Vet Res* 1997;58:1381–1385.
85. Wouda W, Bartels C, Moen A. Characteristics of *Neospora caninum*-associated abortion storms in dairy herds in The Netherlands (1995 to 1997). *Theriogenology* 1999;52:233–245.
86. Jenkins M, Caver J, Björkman C et al. Serological investigation of an outbreak of *Neospora caninum*-associated abortion in a dairy herd in southeastern United States. *Vet Parasitol* 2000;94:17–26.
87. Sager H, Hüssy D, Kuffer A, Schreve F, Gottstein B. First documentation of a *Neospora*-induced "abortion storm" (exogenous transplacental transmission of *Neospora caninum*) in a Swiss dairy farm. *Schweiz Arch Tierheilkd* 2005;147:113–120.
88. Otter A, Jeffrey M, Scholes S, Helmick B, Wilesmith J, Trees A. Comparison of histology with maternal and fetal serology for the diagnosis of abortion due to bovine neosporosis. *Vet Rec* 1997;141:487–489.
89. Wouda W. Diagnosis and epidemiology of bovine neosporosis: a review. *Vet Quart* 2000;22:71–74.
90. Wouda W, Brinkhof J, Van Maanen C, De Gee A, Moen A. Serodiagnosis of neosporosis in individual cows and dairy

herds: a comparative study of three enzyme-linked immunosorbent assays. *Clin Diagn Lab Immunol* 1998;5:711–716.
91. Atkinson R, Harper P, Reichel M, Ellis J. Progress in the serodiagnosis of *Neospora caninum* infections of cattle. *Parasitol Today* 2000;16:110–114.
92. Sager H, Fischer I, Furrer K *et al.* A Swiss case–control study to assess *Neospora caninum*-associated bovine abortions by PCR, histopathology and serology. *Vet Parasitol* 2001;102:1–15.
93. Ellis J. Polymerase chain reaction approaches for the detection of *Neospora caninum* and *Toxoplasma gondii*. *Int J Parasitol* 1998;28:1053–1060.
94. Anderson M, Blanchard P, Barr B, Dubey J, Hoffman R, Conrad P. *Neospora*-like protozoan infection of a major cause of abortion in California dairy cattle. *J Am Vet Med Assoc* 1991;198:241–244.
95. Thurmond M, Hietala S. Strategies to control *Neospora* infection in cattle. *Bov Pract* 1995;29:60–63.
96. Noall D, Kasimanickam R, Memon M, Gay J. Neosporosis in cattle. *Clin Theriogenol* 2013;5:109–119.
97. Larson R, Hardin D, Pierce V. Economic considerations for diagnostic and control options for *Neospora caninum*-induced abortions in endemically infected herds of beef cattle. *J Am Vet Med Assoc* 2004;224:1597–1604.
98. Almería S, López-Gatius F, García-Ispierto I *et al.* Effects of crossbreed pregnancies on the abortion risk of *Neospora caninum*-infected dairy cows. *Vet Parasitol* 2009;163:323–329.
99. Sanderson M, Gay J, Baszler T. *Neospora caninum* seroprevalence and associated risk factors in beef cattle in the northwestern United States. *Vet Parasitol* 2000;90:15–24.
100. Barr B, Dubey J, Lindsay D, Reynolds J, Wells S. Neosporosis: its prevalence and economic impact. *Comp Cont Educ Pract Vet* 1998;20:1–16.
101. Haddad JPA, Dohoo IR, VanLeewen JA. A review of *Neospora caninum* in dairy and beef cattle: a Canadian perspective. *Can Vet J* 2005;46:230–243.
102. Dubey J, Lindsay D. A review of *Neospora caninum* and neosporosis. *Vet Parasitol* 1996;67:1–59.
103. Lindsay D, Butler J, Blagburn B. Efficacy of decoquinate against *Neospora caninum* tachyzoites in cell cultures. *Vet Parasitol* 1997;68:35–40.
104. Kim J, Park J, Seo H *et al.* In vitro antiprotozoal effects of artemisinin on *Neospora caninum*. *Vet Parasitol* 2002;103:53–63.
105. Gottstein B, Eperon S, Dai W, Cannas A, Hemphill A, Greif G. Efficacy of toltrazuril and ponazuril against experimental *Neospora caninum* infection in mice. *Parasitol Res* 2001;87:43–48.
106. Haerdi C, Haessig M, Sager H, Greif G, Staubli D, Gottstein B. Humoral immune reaction of newborn calves congenitally infected with *Neospora caninum* and experimentally treated with toltrazuril. *Parasitol Res* 2006;99:534–540.

Chapter 63

Infectious Agents: Mycotic Abortion

Frank W. Austin

*Department of Pathobiology and Population Medicine, College of Veterinary Medicine,
Mississippi State University, Starkville, Mississippi, USA*

Introduction

Fungi are a diverse group of eukaryotes that includes yeasts, molds, mushrooms, rusts, smuts, and the recently reclassified microsporidia. They constitute one of the five kingdoms of life (Prokaryotae, Fungi, Protista, Plantae, and Animalia) in the current classification system. Like other eukaryotic organisms, fungi have membrane-bound nuclei, membrane-bound cytoplasmic organelles, and 80S ribosomes contained within a plasma membrane. Their membranes contain sterols, namely ergosterol, which determines cell membrane fluidity and acts as boundary lipid. Fungi lack chloroplasts, found in plants and some protists, and have cell walls that contain chitin, unlike other eukaryotes. Many can reproduce both sexually (teleomorphic state) and asexually (anamorphic state). There are several valid and recognized classification schemes for the kingdom Fungi. Recent phylogenetic analysis, based on ribosomal RNA and associated genes, divides the kingdom Fungi into one subkingdom, the Dikarya including the phyla Ascomycota and Basidiomycota, and five additional phyla composed of Chytridiomycota, Neocallimastigomycota, Blastocladiomycota, Glomeromycota, and the Microsporidia.[1,2] Taxonomy based on molecular detection of genetic sequences has had little effect on the commonly known genus and species names of fungi, but did substantially change their higher taxonomic relationships. From a classical medical perspective, pathogenic fungi can also be classified into the dermatophytes, yeasts, dimorphic fungi (having two anamorphic forms), hyaline hyphomycetes (e.g., *Aspergillis* spp.), dematiaceous hyphomycetes (a disparate group of melanin-pigmented fungi), coelomycetes (anamorphic fungi that form conidia in a cavity called pycnidia), zygomycetes (aseptate fungi), and the basidiomycetes (a large group of fungi including puffballs, shelf fungi, rusts, smuts, and mushrooms that bear sexually produced spores on specialized cells called basidia).[3] Additionally, based on morphology and reproductive features fungi have been historically classified into the Ascomycota (sac fungi), Basidiomycota (club fungi), Zygomycota (conjugation fungi, aseptate), Deuteromycota (fungi imperfecti, with no known sexual state), and the Mycophycophyta (lichens and symbiotic fungi).[4] At a very basic level, fungi are easily separated into two major morphologic categories: the filamentous or true fungi and yeasts. Most fungi are ubiquitous saprophytes in the environment, which feed on and recycle decaying organic matter. They prefer dark moist environments in which to grow. Filamentous saprophytic fungi, represented by the hyaline hyphomycetes, dematiaceous hyphomycetes and the zygomycetes, are by far the most commonly encountered agents producing bovine mycotic abortion.

Mycotic agents of abortion

A wide variety of filamentous fungi and several yeasts have been reported to cause bovine abortion.[5] Mycotic abortion is a sporadic event. The incidence of mycotic abortion varies from 2 to 20% depending on the environment, location, and time of the year and generally less than 10% of the herd is affected.[6] The most common cause of bovine mycotic abortion is *Aspergillus* spp., responsible for approximately 84% of the cases reported. Of these, *A. fumigatus* was cited as the cause in 64% of 369 cases in which only one fungal isolate was obtained in culture.[7] In mixed fungal infections, occurring in about 11% of the cases, *A. fumigatus* was found in association with other Zygomycetes in 87% of cases examined. In approximately 14% of the 369 cases examined, Zygomycetes were confirmed as the sole causal agents, constituting the second most common cause of bovine mycotic abortion. Zygomycetes were responsible for 21% of the abortions when considered with mixed infections. Dematiaceous ascomycetes as a group were associated with 7.5% of these cases. Yeasts of the genus *Candida* caused 2.4% of bovine abortions[7] (Table 63.1).

Table 63.1 Fungi isolated in culture from 369 North American cases in which a single isolate was obtained.

Fungus	Classification/features	Prevalence (%) in 369 cases
Filamentous fungi		
Aspergillus fumigatus	Ascomycota, hyaline hyphomycetes	64
Aspergillus terreus	Ascomycota, hyaline hyphomycetes	7.3
Aspergillus flavus	Ascomycota, hyaline hyphomycetes	2.7
Aspergillus nidulans	Ascomycota, hyaline hyphomycetes	3.8
Aspergillus rugulosus	Ascomycota, hyaline hyphomycetes	<1.0
Penicillium thermophilus	Ascomycota, hyaline hyphomycetes	<1.0
Penicillium vermiculatus	Ascomycota, hyaline hyphomycetes	<1.0
Penicillium flavus var. flavus	Ascomycota, hyaline hyphomycetes	<1.0
Scedosporium boydii	Ascomycota, hyaline hyphomycetes	2.4
Phialophora mutabilis	Ascomycota, dematiaceous hyphomycetes	<1.0
Curvularia geniculata	Ascomycota, dematiaceous hyphomycetes	<1.0
Exophilia jeanselmei	Ascomycota, dematiaceous hyphomycetes	<1.0
Scytalidium dimidiatum	Ascomycota, dematiaceous coelomycete	<1.0
Exophiala dermatitidis	Ascomycota, dematiaceous hyphomycetes	<1.0
Absidia corymbifera	Zygomycetes, aseptate	6.5
Rhizomucor pusillus	Zygomycetes, aseptate	3.0
Rhizopus arrhizus	Zygomycetes, aseptate	3.8
Rhizopus rhizopodoformis	Zygomycetes, aseptate	<1.0
Mortierella wolffi	Zygomycetes, aseptate	<1.0
Yeasts		<1.0
Candida krusei	Ascomycota, yeast	1.0
Candida pseudotropicalis	Ascomycota, yeast	<1.0
Candida tropicalis	Ascomycota, yeast	<1.0
Candida lusitaniae	Ascomycota, yeast	<1.0
Candida glabrata	Ascomycota, yeast	<1.0

Source: adapted from Knudtson WU, Kirkbride CA. Fungi associated with bovine abortion in the northern plains states (USA). *J Vet Diagn Invest* 1992;4:181–185.

Pathogenesis and pathology

The most likely routes for exposure to fungi causing mycotic placentitis and abortion are the respiratory and gastrointestinal tracts.[8] Intrauterine inoculation of *Aspergillus* conidia has been unsuccessful in experimental induction of mycotic abortion and ascending reproductive tract infections therefore seem unlikely.[9] Moreover, mycotic abortions have been experimentally induced by intravenous administration of *Aspergillus* microconidia and the hematogenous route is thought to be the path because lesions develop initially in the placentomes.[8] Fungi may enter the circulation through the alveolar septa or through gastrointestinal lesions resulting from penetration of mucosal barriers by rumen, reticular, or omasal infections or ulcers. Once the infection is established in the placentomes, it advances laterally to the intercotyledonary spaces. The placenta becomes grossly thickened, with a leather-like appearance of the intercotyledonary spaces. There is necrosis of the cotyledons and caruncular tissue and a thickening at the margins of the caruncle, imparting a concave or dished appearance. The placenta may have dry, thick, yellow plaques covering the intercotyledonary areas and most cotyledons when infected by *Aspergillus* spp.[10] (Figure 63.1). A severe placental necrotizing vasculitis with thrombosis is a hallmark lesion associated with mycotic abortions. However, placental lesions with similar appearance can be produced by brucellosis and genital campylobacteriosis. Subsequently, fetal infection may occur involving the skin, lungs, brain, or liver. Fetal death is usually attributed to placental insufficiency. Skin lesions typically appear as dry, wrinkled, raised plaques with *Aspergillus* spp.; contrastingly, the fetus may appear wet and moist when infected by the Zygomycetes. The fetus is aborted soon after death, which occurs most commonly between 6 and 8 months of gestation. Frequently the placenta is retained and the cotyledons remain firmly attached to the maternal caruncles, which may rupture at the peduncle. These latter events are most commonly observed with zygomycotic abortions and may lead to subsequent secondary infections.[5] Without severe uterine damage, most cows recover and have normal pregnancies later in life.

In approximately 25% of abortions caused by *Mortierella wolfii*, a zygomycete, severe post-abortion pneumonia is recognized to occur.[7] Following abortion, the caruncles and uterus can release large numbers of infective elements, resulting in an acute fungal embolic pneumonia. Cows usually

Infectious Agents: Mycotic Abortion

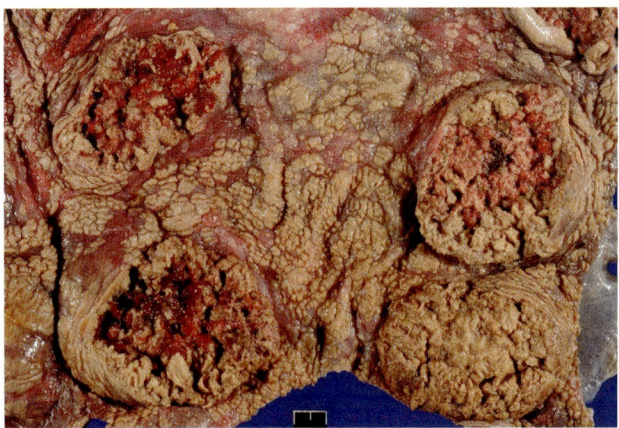

Figure 63.1 Bovine mycotic placentitis produced by *Aspergillus fumigatus*. The cotyledons are necrotic and there are thick yellow plaques covering the intercotyledonary areas. Photograph courtesy of Dr J. Cooley.

die within 72 hours following the onset of clinical signs that include rapid shallow respiration with forced expiration.[11] In *M. wolfii* fetal infections, the liver and brain are commonly involved, unlike infections caused by other fungi.[12]

Diagnosis

Diagnosis of mycotic abortion can be aided by gross, microscopic, histopathologic, or cultural methods of examination. Additionally, molecular detection of genetic sequences has been employed to identify specific fungal agents.[13] Definitive diagnosis is generally achieved through agreement of many of these methods. The visual demonstration of fungal hyphae and elements by microscopic methods in placental or fetal tissues and fluids combined with compatible gross lesions may confer a presumptive diagnosis. Histopathology combined with cultural or molecular methods allows for definitive confirmation of the etiologic agent responsible for abortion.

Specimens

Specimens for mycotic abortion can be used for gross examination, direct microscopic and histopathologic examination, fungal culture, and molecular detection. Specimens can include the placenta, fetus, fetal skin, fetal lung, and stomach fluid of the fetus. Because a definitive diagnosis of mycotic abortion frequently depends on agreement using multiple methods, the comprehensive specimen list presented is generally required. Specimens should be carefully selected and independently collected for each of the methods used. Extreme care should be taken to avoid contamination of specimens used for culture or molecular detection, which can easily lead to false-positive results.

The placenta is the most common and useful specimen for diagnosis of mycotic abortions. Because the lesions of mycotic placentitis may be focal and not involve some areas, the entire placenta should be submitted. The cotyledons with caruncular tissue contain the highest number of organisms and are considered to be the best specimen. The placenta will always be involved in mycotic abortion; however, the fetus or fetal tissues are not consistently affected.

Following abortion, the fetus can become easily contaminated by contact with the environment. In approximately 25% of mycotic abortions the fetal skin is infected and lesions appear as circumscribed, slightly raised, cutaneous, gray to yellow plaques. These lesions may coalesce into larger areas along the dorsum, the top of the head, and base of the tail. Although amniotic fluid is not commonly available following abortion, it will contain fungal elements when the skin is affected and transport them to other fetal areas and tissues that are not as easily contaminated.[10,14]

Stomach fluid is a commonly used specimen for culture and molecular detection when the fetus has been expelled and contaminated on its surface. Stomach fluid can also be used as an excellent specimen for direct microscopy. Likewise, the fetal lung is exposed to the amniotic fluid and commonly used to circumvent contamination of the fetal surface when diagnostic amplification techniques are employed.

Microscopic and histopathologic examination

Direct microscopic examination of the placenta, fetal skin, and abomasal fluid following digestion in 10–15% KOH can reveal fungal hyphae. Digestion of host debris is time and temperature dependent and can be quickened by the application of gentle heat. Direct visualization of fungal elements can be enhanced with the inclusion of calcofluor white in 10% KOH solution, which specifically binds cellulose and chitin of the fungal cell wall. Calcofluor white (or Uvitex B, Blankophor) fluoresces blue-white or apple green when exposed to ultraviolet light.[15–17] Additionally, Parker permanent blue ink (10%) can be added to the KOH solution to impart contrast and aid visualization of hyaline hyphal elements. Lactophenol cotton blue dye can be used as a simple positive stain with stomach fluid specimens that do not necessarily need alkaline digestion.

The microscopic morphology and characteristics of the fungi can aid in identification and confirmation of the cultural isolate. The hyphae of the Zygomycetes are relatively broad (5–20 μm), thin-walled and coenocytic (aseptate).[18] Budding yeasts with narrow bases and broad pseudohyphae having constricted septa can indicate pathogenic *Candida* species.[19] The dematiaceous fungi causing abortions have brown to black intracellular pigment (melanin) that is easily seen. *Aspergillus* species have relatively narrow (2–6 μm) hyphae and frequent septae spanning parallel walls.[10] It is important to note that identification to the species level is not attainable through direct microscopic methods, but these methods can be quickly used to confirm a presumptive diagnosis of mycotic abortion, suggest a group of responsible fungi, and rule out other causes of abortion.

Adequate areas and portions of the placenta and fetal skin and lung should be collected into 10% buffered neutral formalin for histopathology if the whole fetus is not submitted to the laboratory. In routine hematoxylin/eosin-stained histopathology slide sections, hyaline fungal elements are poorly visualized. Therefore requests for special staining of slides using periodic acid–Schiff or Gomori's methenamine silver stain should be made if mycotic abortion is suspected.

Fungal culture methods and molecular identification

Because of the ubiquity of fungi in the natural environment, culture results should be interpreted with caution. Contamination of specimens from abortion cases is frequent. For this reason, specimens of the fetal lung and abomasal fluid or contents are commonly sought. The morphology and characteristics of the fungi recovered in culture should be in agreement with the results of direct microscopy and histopathology.

A common medium for the culture and isolation of fungi is Sabouraud's dextrose agar (SDA) with and without the addition of antibiotics. Antibiotics such as chloramphenicol and cycloheximide are commonly included in media, such as Mycobiotic Agar (Difco), to suppress the growth of bacteria and saprophytic fungi respectively. Care must be taken in choosing media as cycloheximide will inhibit the Zygomycetes. It can also partially inhibit some *Aspergillus* spp. and *Candida* spp. involved in mycotic abortions. For this reason, SDA with only chloramphenicol is recommended for the culture of fungal agents from mycotic abortions. Furthermore, the Zygomycetes can be difficult to culture as they do not remain viable in tissue for extended periods of time. Failure to recover these fungi when observed in direct microscopy or histopathology can be due to the toxic effects of placental or fetal autolysis.[11] Agar plates are incubated at 25–30°C for several weeks prior to morphologic identification.

Following isolation and maturation of growth, filamentous fungi are classically identified on the basis of colonial and microscopic morphology (Figure 63.2). Speciation of the filamentous fungi can be problematic using these techniques and advanced molecular methods may be necessary. Yeasts are commonly identified on the basis of carbohydrate assimilation and biochemical testing using commercially available systems.[19]

Molecular detection of specific genetic sequences obtained through polymerase chain reaction (PCR) methods is becoming commonly used in mycology reference laboratories for fungal identification. These methods can be applied to fresh tissues, tissues fixed for histopathology, or cultural isolates.[13] Identification of fungal species using unique genetic sequences can save considerable time and offer unparalleled specificity. However, because of the ubiquity of fungi in the environment and the extreme sensitivity of molecular methods, results should be used in conjunction with other methods to assure a conclusive diagnosis.

Prevention and control

Since mycotic abortions occur sporadically and are produced by ubiquitous environmental fungi that are not contagious, the best means of control are preventative. Avoidance of moldy hay and feedstuffs and poor-quality silage can decrease the occurrence of mycotic abortion. Likewise, dark and moist environments with a high organic component favoring the growth of fungi should be eliminated. Decreasing environmental exposure to fungi by providing adequate ventilation, avoiding animal confinement, and decreasing their density can further aid in preventing mycotic abortions. Therapy should be aimed at preventing uterine damage from retained placenta and associated secondary infections.

Figure 63.2 Photomicrograph of a lactophenol cotton blue-stained slide of *Aspergillus fumigatus* from a Sabouraud's dextrose agar plate culture. Note the conidiophores with large terminal vesicles supporting the numerous sterigmata that bear long chains of conidia. Photomicrograph (20×) courtesy of Dr F.W. Austin.

References

1. James TY, Kauff F, Schoch CL *et al.* Reconstructing the early evolution of the fungi using a six gene phylogeny. *Nature* 2006; 443:818–822.
2. Hibbett DS, Binder M, Bischoff JF *et al.* A higher-level phylogenetic classification of the fungi. *Mycol Res* 2007;111:509–547.
3. Mycology Online, School of Molecular and Biomedical Science, University of Adelaide, Australia. Available at http://www.mycology.adelaide.edu.au/, accessed December 2012.
4. Guarro J, Gene J, Stchigel A. Developments in fungal taxonomy. *Clin Micobiol Rev* 1999;12:454–500.
5. Kirkbride C. Etiologic agents detected in a 10-year study of bovine abortions and stillbirths. *J Vet Diagn Invest* 1992;4: 175–180.
6. Williams B, Shreeve B, Hergert C, Swire P. Bovine mycotic abortion: epidemiologic aspects. *Vet Rec* 1977;100:382–385.
7. Knudtson WU, Kirkbride CA. Fungi associated with bovine abortion in the northern plains states (USA). *J Vet Diagn Invest* 1992;4:181–185.
8. Kennedy P, Miller R. The female genital system. In: Jubb K, Kennedy P, Palmer N (eds) *Pathology of Domestic Animals*, 4th edn, Vol. 3. San Diego, CA: Academic Press, 1993, pp. 420–421.
9. Hill M, Whiteman C, Benjamin M *et al.* Pathogenesis of experimental bovine mycotic placentitis produced by *Aspergillus fumigatus*. *Vet Pathol* 1971;8:190–193.
10. Glover A, Rech R, Howerth E. Pathology in practice. Mycotic abortion. *J Am Vet Med Assoc* 2011;239:319–321.
11. Carter M, Cordes D, Di Menna M, Hunter R. Fungi isolated from bovine mycotic abortion and pneumonia with special reference to Mortierella wolfii. *Res Vet Sci* 1973;14:201–206.

12. Cordes D, Carter M, Di Menna M. Mycotic pneumonia caused by *Mortierella wolfii*. II. Pathology of experimental infection in cattle. *Vet Pathol* 1972;9:190–193.
13. Dongyou Liu (ed.) *Molecular Detection of Human Fungal Pathogens*. Boca Raton, FL: CRC Press, 2011.
14. Miller R. A summary of the pathogenic mechanisms involved in bovine abortion. *Can Vet J* 1977;18:87–95.
15. McGinnis MR. *Laboratory Handbook of Medical Mycology*. New York: Academic Press, 1980.
16. Koneman E, Roberts G. *Practical Laboratory Mycology*. Baltimore: Williams and Wilkins, 1985.
17. Hageage G, Harrington B. Use of calcofluor white in clinical mycology. *Lab Med* 1984;15:109–112.
18. Ribes JA, Vanover-Sams CL, Baker DJ. Zygomycetes in human disease. *Clin Microbiol Rev* 2000;13:236–301.
19. Chengappa M, Maddux R, Greer S, Pincus D, Geist L. Isolation and identification of yeasts and yeastlike organisms from clinical veterinary sources. *J Clin Microbiol* 1984;19:427–428.

Chapter 64

Early Embryonic Loss Due to Heat Stress

Peter J. Hansen

Department of Animal Sciences, University of Florida, Gainesville Florida, USA

Nature of the problem

Adverse changes in the maternal environment can perturb embryonic development and lead to embryonic loss. Among the most consequential of these changes is hyperthermia, which occurs most typically because of heat stress but also via pyrogens generated during inflammation. Even a small increase in body temperature can compromise embryonic survival. In lactating dairy cows,[1] conception rate declined 6.9–12.8% for each 0.5 °C increase in uterine temperature above the mean temperature of 38.3–38.6 °C. In this study, rectal temperature was about 0.2 °C lower than uterine temperature[1] and therefore it can be inferred that fertility declines when rectal temperature reaches about 39 °C. In an earlier study, fertility of lactating dairy cows also declined at rectal temperatures of 39 °C and higher.[2]

Embryonic loss caused by maternal hyperthermia is largely caused by direct and indirect actions of elevated body temperature on oocyte function and embryonic development. The window in the reproductive process when hyperthermia can cause embryonic loss is a wide one. It extends from as early as 15–19 weeks before ovulation, because exposure to a 28-day heat stress reduces the competence of the oocyte to become a blastocyst 15 weeks after the termination of heat stress,[3] and continues until 2–3 days after ovulation when the preimplantation embryo becomes resistant to disruption by elevated temperature.[4–8]

This chapter will focus on the mechanisms by which heat stress compromises the oocyte and embryo and the implications for management strategies to improve fertility during heat stress. Throughout the chapter, the term "heat stress" will refer to the combination of environmental characteristics that reduce the capacity of an animal to regulate its body temperature while "heat shock" will refer to the effect of elevated temperature on cellular function.

Contribution of the oocyte to embryonic loss

Heat stress can affect oocyte function during two distinct periods of oocyte development. The first is the period of follicular growth, when the oocyte experiences activation and then silencing of the genome, establishes communication with cumulus cells, and synthesizes cortical granules.[9] The second period is following ovulation when the oocyte undergoes nuclear and cytoplasmic maturation in preparation for fertilization and embryonic development.

Oocyte damage during follicular growth

Damage to the oocyte during follicular growth has been demonstrated in experiments to evaluate seasonal variation in the competence of oocytes (harvested at slaughter or via ultrasound-guided aspiration) to be fertilized *in vitro* and develop into blastocysts. In these experiments, which used lactating Holstein cows, the proportion of oocytes that became blastocysts after *in vitro* fertilization[10–12] or chemical activation[13] was lower in summer than winter (see Figure 64.1 for an example).

The seasonal effects on blastocyst formation largely represented decreased competence of cleaved embryos to develop rather than competence of the oocyte to cleave. One reason is that heat stress reduces accumulation of molecules in the oocyte important for early embryonic development. Actually, there are differences in transcript abundance between embryos produced by *in vitro* fertilization of oocytes collected in summer versus winter.[14] Also, perhaps, oocytes in the summer are more likely to undergo abnormal chromosomal segregation so that subsequent embryos are chromosomally abnormal. Occurrence of chromosomal abnormalities can affect the rate of embryonic development.[15]

Bovine Reproduction, First Edition. Edited by Richard M. Hopper.
© 2015 John Wiley & Sons, Inc. Published 2015 by John Wiley & Sons, Inc.

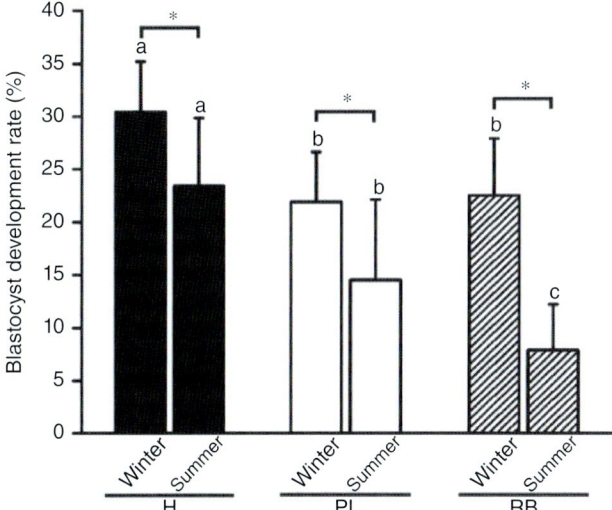

Figure 64.1 Seasonal variation in Brazil for oocyte competence for development in Holstein heifers (H), lactating cows at the peak of lactation (PL), and lactating cows classified as repeat breeders (RB). Oocytes were harvested by transvaginal aspiration and subjected to *in vitro* maturation and fertilization. Resultant embryos were cultured to the blastocyst stage. Asterisks represent differences within animal type between seasons. Letters above each bar represent differences between animal types within an individual season. Data are reproduced from Ferreira R, Ayres H, Chiaratti M *et al.* The low fertility of repeat-breeder cows during summer heat stress is related to a low oocyte competence to develop into blastocysts. *J Dairy Sci* 2011;94:2383–2392 with permission.

Appropriate follicular development is critical to the ovulation of an oocyte competent to be fertilized and able to transform into an embryo capable of successful development.[16,17] Heat stress can alter follicular dynamics by reducing growth of the dominant follicle and increasing the number of smaller-sized follicles.[3,18–20] Furthermore, heat stress can reduce follicular concentrations of estradiol-17β[21] and insulin-like growth factor (IGF)-1,[20] while increasing follicular concentrations of androstenedione.[21]

The mechanism by which heat stress alters follicular development is not well understood. Experimentally, heat stress can reduce luteinizing hormone (LH) secretion.[22,23] There may also be direct effects of elevated body temperature on steroid synthesis. Culture of cumulus cells at 40.5 °C reduced estradiol-17β production in the fall but not in the summer or winter.[21] In the same study, incubation at 40.5 °C reduced forskolin-stimulated androstenedione production by thecal cells when cells were collected in winter but not when cells were collected in summer or fall. Heat shock of 41 °C reduced basal and gonadotropin-stimulated synthesis of androstenedione and estradiol-17β by cultured follicular walls from the dominant follicle.[24]

In some cases, heat stress can cause a reduction in circulating progesterone concentrations[25] and such a decline could conceivably contribute to a reduction in oocyte competence. This idea is based on the observation that fertility following ovulation of a first-wave dominant follicle, when circulating progesterone concentrations are low, is less than fertility following ovulation of a second-wave dominant follicle, when progesterone concentrations are high.[26,27] Moreover, supplementation with progesterone during the first follicular wave improved fertility.[27]

Summer heat stress is followed by a gradual restoration of oocyte competence for fertility[28,29] and oocyte competence for chemically activated development.[30] These observations are symptomatic that heat stress can affect follicular development early in the 16-week period that is required for a primordial follicle to reach dominance.[31] Indeed, there is experimental evidence for delayed effects of heat stress on function of the follicle and oocyte. In one study,[32] exposure of lactating Holsteins to heat stress affected follicular steroid production 20–26 days later. An experiment in Gir cattle (*Bos indicus*) indicated that carry-over effects of heat stress can persist for up to 133 days after initiation of heat stress (i.e., 105 days after the end of heat stress).[3] Cows were exposed to heat stress for 28 days and oocytes aspirated and subjected to *in vitro* maturation and fertilization at weekly intervals for up to 133 days after the beginning of heat stress. Heat stress did not affect follicular dynamics during the 28-day heat stress period but increased the diameter of the first and second largest follicles, as well as the number of follicles over 9 mm, from days 28 to 49 after initiation of heat stress. Moreover, the percent of oocytes that became a blastocyst was reduced by heat stress even at the end of the experiment at 133 days after initiation of heat stress (Figure 64.2).

Oocyte maturation

The oocyte remains susceptible to disruption by heat stress during the process of nuclear and cytoplasmic maturation coincident with ovulation. Heat stress of superovulated heifers for a 10-hour period beginning at the onset of estrus and before insemination at 15–20 hours after the beginning of estrus decreased the proportion of embryos recovered at day 7 after estrus that were classified as normal.[33] There was no effect of heat stress on fertilization rate so damage to the oocyte gave rise to an embryo with reduced competence for development.

Some of the effects of heat stress on oocyte function during maturation could be due to alterations in hormone secretion because heat stress lowered the magnitude of the preovulatory surge in LH and estradiol-17β.[23,34] In addition, exposure to elevated temperature can induce errors in oocyte maturation. Culture at elevated temperatures during maturation reduced the proportion of oocytes that completed nuclear maturation, increased the proportion with abnormal spindle formation, cortical granule distribution, depolarized mitochondria and apoptotic pronuclei, and decreased the proportion that cleaved and became blastocysts after insemination.[35–40]

Apoptosis has been implicated in disruption of oocyte function during maturation. Inhibition of apoptosis with a caspase inhibitor,[36] sphingosine 1-phosphate,[41] or inhibitor of ceramide synthesis[42] reduced the effects of heat shock on developmental competence of the oocyte. There are also indications that the effects of heat shock involve acceleration of maturation and oocyte aging. Exposure of oocytes to 41 °C for the first 12 hours of maturation hastened both nuclear and cytoplasmic maturation.[43] Moreover, the deleterious effects of heat shock on development could be avoided if oocytes were fertilized at 19 hours of maturation instead of 24 hours[43] or if chemical activation was used to trigger development.[44]

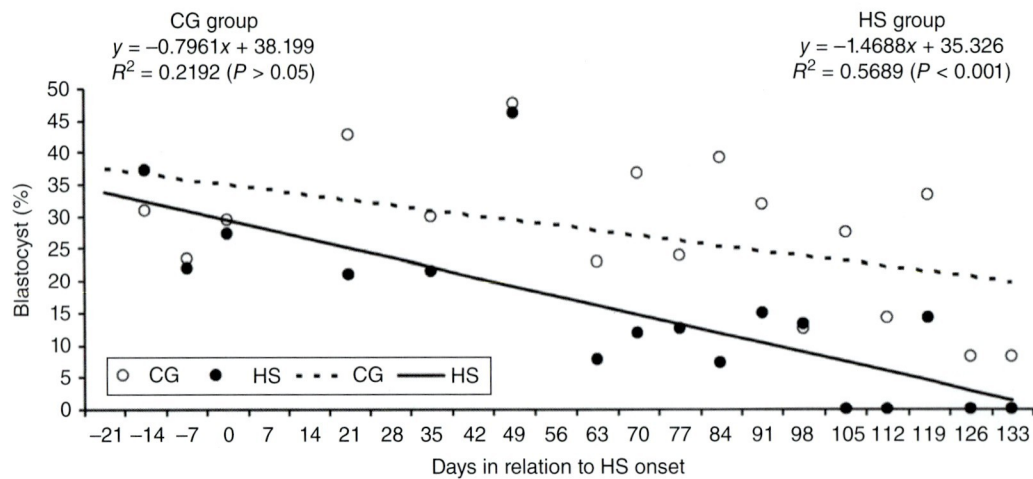

Figure 64.2 Changes in developmental competence of oocytes harvested from Gir cows that were either experimentally heat-stressed for 28 days (HS) or were maintained in a control environment (control group, CG). Oocytes were subjected to *in vitro* maturation and fertilization. Resultant embryos were cultured to the blastocyst stage. Note that the competence of oocytes to become blastocysts remained lower for the HS group for up to 133 days after the initiation of heat stress (i.e., 105 days after the end of heat stress). Data are reproduced from Torres-Júnior JR de S, de FA Pires M, de Sá WF et al. Effect of maternal heat-stress on follicular growth and oocyte competence in *Bos indicus* cattle. *Theriogenology* 2008;69:155–166 with permission.

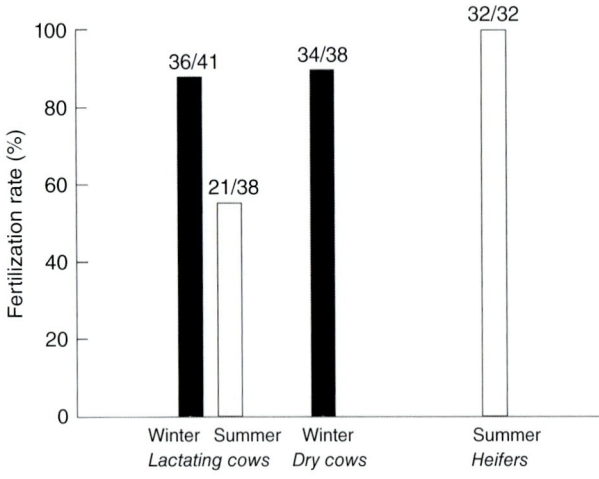

Figure 64.3 Seasonal changes in fertilization rate in Holsteins in Wisconsin. The reproductive tract of inseminated females was flushed 5 days after ovulation to recover oocytes and embryos. Fertilization rate was based on the number of recovered embryos (two-cell or greater) divided by the number of recovered oocytes and embryos (see numbers above each bar). For lactating cows, there was a decrease in fertilization rate in the summer compared with winter. For nonlactating animals, however, there was no difference in fertilization rate between seasons (compare dry cows in winter with heifers in summer). Data are taken from Sartori R, Sartor-Bergfelt R, Mertens S, Guenther J, Parrish J, Wiltbank M. Fertilization and early embryonic development in heifers and lactating cows in summer and lactating and dry cows in winter. *J Dairy Sci* 2002;85:2803–2812.

Fertilization failure: oocyte versus sperm

As shown in Figure 64.3, there are differences in fertilization rate between lactating cows in summer and winter.[45] Reduced fertilization rate is probably largely caused by damage to the oocyte. The other cell important for fertilization, the spermatozoon, is much more resistant to elevated temperature. Formation of spermatozoa in the testis can be compromised by heat stress,[46] but function of the spermatozoa deposited in the female reproductive tract is probably not disrupted by maternal hyperthermia. Heat shock did not reduce fertilizing capacity of sperm *in vitro* or subsequent development of embryos to the blastocyst stage.[47,48] There are reports in the rabbit that exposure of sperm to elevated temperature *in vitro*[49] or *in vivo*[50] can result in abnormalities in development, leading to a reduction in the number of implantation sites despite a lack of reduction in fertilization rate. More work is necessary to determine if damage to the spermatozoa by heat shock can lead to developmental abnormalities later in development in cattle.

Damage to the preimplantation embryo

When first formed, the preimplantation embryo is very susceptible to elevated temperature. A heat shock as low as 40 °C for 12 hours can reduce development of zygote-stage embryos to the blastocyst stage.[51] Heat shock can also reduce developmental potential of embryos at the 2-cell stage of development,[7,52] but thereafter the embryo becomes more resistant to heat shock. Development of 4- to 8-cell embryos can be compromised by heat shock but to a lesser extent than for 2-cell embryos.[52] By the morula stage of development, the embryo is substantially resistant to deleterious effects of heat shock.[7,8,51,52] The same phenomenon is observed *in vivo*. Exposure of superovulated cows to heat stress reduced blastocyst yield at day 8 after estrus when occurring at day 1 after estrus but not when occurring at day 3, 5, or 7 after estrus.[4]

Mechanism for embryonic mortality caused by heat stress

The 2-cell embryo exposed to 41 °C experiences alterations in ultrastructure characterized by swelling of mitochondria and movement of organelles toward the center of the blastomere[53] (Figure 64.4). Organellar movement is caused by

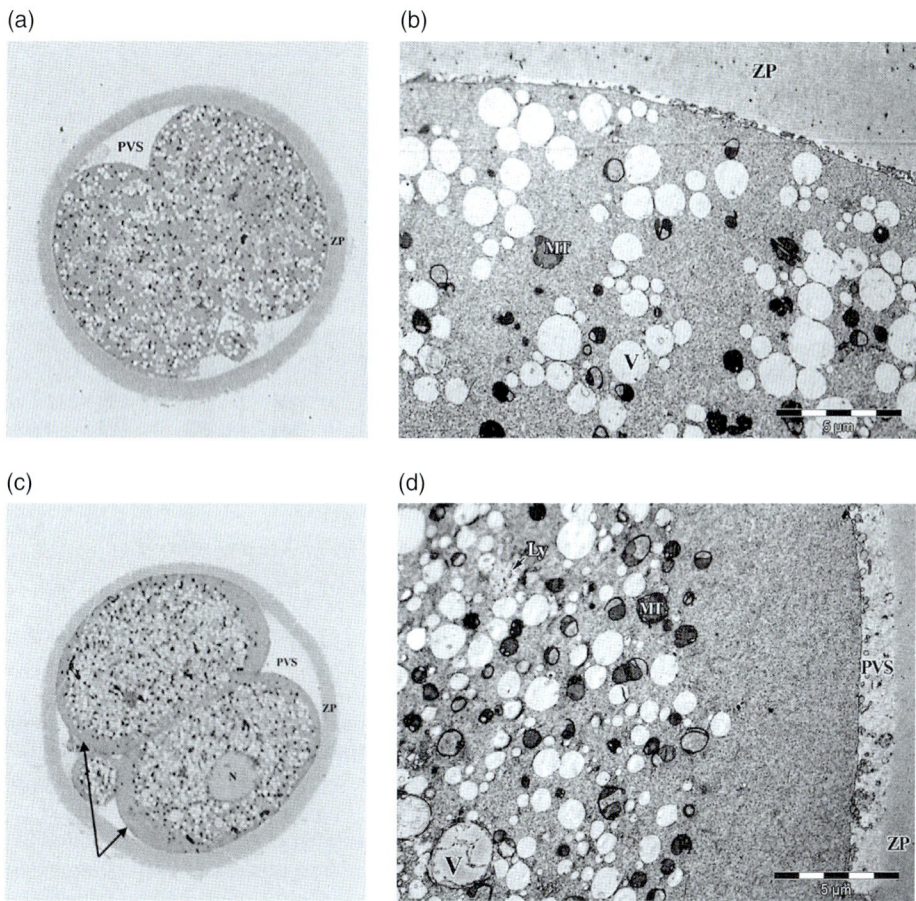

Figure 64.4 Effects of heat shock (41°C for 6 hours) on the ultrastructure of the two-cell bovine embryo produced *in vivo*. (a, c) Images taken using light microscopy after staining with toluidine blue O; (b, d) images from transmission electron microscopy. (a, b) Embryos cultured at 38.5°C; (c, d) heat-shocked embryos. Arrows show an area of the cytoplasm devoid of organelles. ZP, zona pellucida; PVS, perivitelline space; N, nucleus; MT, mitochondrion; Ly, lysosome. Images are reproduced from Rivera R, Kelley K, Erdos G, Hansen P. Alterations in ultrastructural morphology of two-cell bovine embryos produced in vitro and in vivo following a physiologically-relevant heat shock. *Biol Reprod* 2003;69:2068–2077 with permission.

disruption of the cytoskeleton.[54] The early embryo exposed to heat shock also experiences a decrease in protein synthesis.[5]

Some of the damage caused by heat shock could be the result of increased production of free radicals as has been reported for embryos at days 0 and 2 after insemination.[55] Also, effects of heat shock were greater in 2-cell embryos cultured in a high oxygen atmosphere that would promote oxidative stress.[56] In contrast, the effects of heat shock were independent of oxygen tension in other studies.[51,57] Similarly, the antioxidant anthocyanin[58] and dithiothreitol[56] have been reported to mitigate the actions of heat shock whereas other antioxidants including vitamin E,[59] glutathione,[60] and glutathione ester[60] were reported to be without embryoprotective properties.

Effects of heat stress on embryonic survival *in vivo* could include actions on the reproductive tract as well as on the embryo directly. Heat stress can reduce the magnitude of the preovulatory surge in estradiol-17β[23] and this could conceivably change function of the oviduct or endometrium. The direct effects of heat shock on oviductal or endometrial function have not been examined at temperatures characteristic of those experienced by heat-stressed cows. However, it is likely that these tissues are resistant to heat shock because even a temperature as high as 43°C did not have major effects on protein synthesis by endometrium or oviduct.[61]

Developmental acquisition of thermotolerance

Several hypotheses have arisen to explain the process by which the embryo transitions from a state in which it is very susceptible to heat shock, at the 1-cell to 2-cell stage, to one where it has acquired resistance to heat shock, by the 16-cell stage. One possibility is that the early embryo is very susceptible to heat shock because processes required for embryonic genome activation are easily disrupted by heat shock. If this was the case, embryos exposed to heat shock at the 2- or 4-cell stage should block in development between the 8-cell and 16-cell stages; embryonic genome activation is required for development past this stage.[62] Such a preferential block at the 8-cell to 16-cell transition was seen in one experiment with 2-cell embryos[57] but not in another where embryos were heat-shocked at the 1-cell stage.[51]

Another possibility is that the acquisition of capacity for transcription after the 8-cell to 16-cell stage allows the embryo to engage a wider range of stress responses to stabilize cellular function during heat shock. However, steady-stage amounts of mRNA for some important stress proteins are higher in the 1-cell or 2-cell embryo than the morula[51,62] (Figure 64.5), probably because transcripts are inherited from the oocyte. Heat shock can induce synthesis

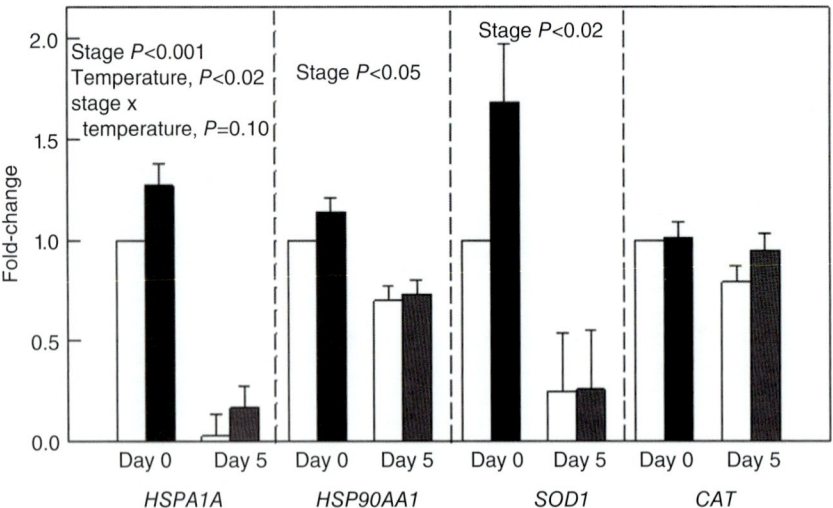

Figure 64.5 Differences in steady-stage amounts of mRNA for *HSPA1A* (heat-shock protein 70), *HSP90AA1* (heat-shock protein 90), *SOD1* (superoxide dismutase), and *CAT* (catalase) between embryos harvested at the zygote (day 0) or morula stage (day 5) and cultured for 24 hours at 38.5°C (open bar) or 40°C (black bar). Data are reproduced from Sakatani M, Alvarez N, Takahashi M, Hansen P. Consequences of physiological heat shock beginning at the zygote stage on embryonic development and expression of stress response genes in cattle. *J Dairy Sci* 2012;95:3080–3091 with permission.

of heat-shock protein 70 as early as the 2-cell stage.[5,63] Moreover, the morula-stage embryo undergoes a muted transcriptional response to heat shock, with only 4 of 68 genes associated with the heat-shock response showing an increase in response to elevated temperature.[64] Thus it might be that the damage to cellular macromolecules caused by heat shock that trigger stress protein responses is reduced in embryos at resistant stages of development. One reason might be a change in the balance between free radical generation and antioxidant protection. Heat shock caused increased free radical production at days 0 and 2 but not at days 4 and 6.[55] In addition, the cytoplasmic antioxidant glutathione is at its lowest during the 2- to 8-cell stages.[65]

One mechanism that protects embryos from heat shock is the capacity for apoptosis. It is not until after embryonic genome activation that pro-apoptotic signals can activate the apoptosis response; the process is inhibited in 2-cell embryos because of mitochondrial resistance to depolarization[66] and inaccessibility of caspase-activated deoxyribonuclease to the internucleosomal regions of DNA.[67] Exposure of the morula to 41°C for 9 hours results in about 10–25% of cells becoming apoptotic.[68,69] Prevention of apoptosis exacerbates the deleterious effects of heat shock on development,[68,69] probably by eliminating those cells most damaged by heat shock.

Genetic variation in embryonic resistance to heat shock

Certain breeds of cattle that evolved in hot climates have acquired genes that protect embryos from elevated temperature. At later cleavage stages (day 4 or 5 after insemination), Brahman, Nelore and Romosinuano embryos are more resistant to elevated temperature than Angus, Holstein, or Jersey embryos.[8,70–73] There was also a tendency for pregnancy rates after transfer into recipients to be reduced for Angus blastocysts that were previously exposed to heat shock but not for Nelore embryos previously exposed to heat shock.[73]

The factors responsible for breed differences in thermotolerance appear to be inherited through the oocyte. In experiments in which crossbred embryos were produced *in vitro*, it was the breed of oocyte and not breed of spermatozoa that determined whether embryos exhibited increased thermotolerance.[72,74] One possibility is that the genes conferring thermotolerance are imprinted, with only the female allele expressed. Alternatively, transcripts that accumulate in the oocyte are responsible for thermotolerance later in development. The oocyte and newly fertilized embryo have higher transcript abundance for specific stress response genes than embryos at day 5 of development.[51,62] Also, oocytes collected during the warm season are more resistant to heat shock applied *in vitro* than oocytes collected during the cool season.[75] Accordingly, the characteristics of transcripts accumulated in the oocyte during follicular development could be an important determinant of subsequent resistance to heat shock.

An important unanswered question is whether breed differences exist in resistance of the oocyte or early embryo to heat shock. The silencing of transcription in these cells may not allow for expression of genetic differences in thermotolerance. If so, genotype is likely to be of little importance *in vivo* because it is the oocyte and very early embryo that are most susceptible to elevated temperature.

Consequences of heat stress later in pregnancy

While abundant evidence indicates embryos become substantially resistant to heat stress by the morula stage of development, heat stress can compromise some aspects of pregnancy later in gestation.

Results are conflicting as to whether late embryonic and fetal loss is affected by heat stress. A controlled experiment to test this hypothesis has not been carried out. Epidemiological analyses indicated an association between warm weather and increased pregnancy loss between days

35–45 and day 90 of gestation for cows in Spain.[76] In another study from Spain, increased number of warm events on days 11–20 or 0–40 of gestation was associated with pregnancy loss between days 28–34 and 80–86 of gestation.[77] In contrast, there was no relationship between occurrence of heat stress before or after insemination on pregnancy loss between days 31 and 45 of pregnancy in California.[78] Similarly, there was an absence of seasonal variation in pregnancy loss between days 40–50 and 70–80 or between days 70–80 and term among dairy cattle in Florida.[79]

During late gestation, heat stress can compromise placental function and fetal growth.[80-83] As a result, there is a reduction in subsequent milk yield of the dam[80,81,83,84] and postnatal growth rate and immune function of the calf.[82]

Consequences for management

The best method for reducing the effects of heat stress is to cool cows by altering animal housing. The level of fertility in Israeli dairy farms in the summer, for example, is directly related to the magnitude of cooling that cows receive.[85] In many cases, however, summer depression in fertility occurs despite efforts to provide cooling.[85,86] Understanding the cellular and physiological basis for embryonic loss during heat stress has resulted in some possible strategies for managing reproductive processes that could be used in conjunction with cooling.

Embryo transfer

The only reproductive management method that has been shown consistently to improve fertility in cattle during heat stress is embryo transfer. Improvements in fertility using embryo transfer during heat stress have been seen using embryos produced by superovulation[87-89] or by in vitro fertilization.[90-92] In one commercial dairy in Brazil, seasonal variation in fertility was eliminated by use of embryo transfer.[89]

Follicle removal

Work in Israel has shown that hastening the removal of follicles from the ovary in the fall can hasten the restoration of fertility following heat stress. Beneficial effects on oocyte quality have been observed by turning over the follicular pool by either follicular aspiration[30] or administration of follicle-stimulating hormone[93] or bovine somatotropin.[93] Such treatments would not be expected to be effective in the summer because follicles emerging into the growing pool would also be damaged by heat stress. Treatment of lactating cows with three rounds of a gonadotropin-releasing hormone and prostaglandin $(PG)F_{2\alpha}$ protocol to hasten removal of follicles in the summer and fall did not improve pregnancy rate at first AI in multiparous cows, although it tended to improve pregnancy rate in primiparous cows.[94] Perhaps greater responses would have been seen if the study was limited to the late summer or fall.

Antioxidants

Treatments to improve fertility in summer through antioxidant administration have yielded mixed results. There was no improvement in fertility when cows were injected with vitamin E at insemination,[95] given multiple injections of vitamin E and selenium before and after calving,[96] or administered injections of β-carotene at –6, –3 and 0 days relative to insemination.[97] In contrast, feeding supplemental β-carotene for at least 90 days beginning at about 15 days after calving increased the proportion of cows that were pregnant at 120 days postpartum.[98] In addition, treatment of cows with melatonin implants in the summer reduced days open.[99] Interpretation of the melatonin study is difficult because melatonin has many biological actions including as an antioxidant[100] but also as an important mediator of photoperiodic regulation.[101]

Genetic control of embryonic resistance to elevated temperature

A study in Brazil with lactating Holsteins in the summer indicated that pregnancy rate per insemination was greater when cows were inseminated with Gir semen than when inseminated with Holstein semen.[102] Such an effect could represent genetic differences in embryonic resistance to heat shock or embryo heterosis.

References

1. Gwazdauskas F, Thatcher W, Wilcox C. Physiological, environmental, and hormonal factors at insemination which may affect conception. *J Dairy Sci* 1973;56:873–877.
2. Ulberg L, Burfening P. Embryo death resulting from adverse environment on spermatozoa or ova. *J Anim Sci* 1967;26:571–577.
3. Torres-Júnior JR de S, de FA Pires M, de Sá WF et al. Effect of maternal heat-stress on follicular growth and oocyte competence in Bos indicus cattle. *Theriogenology* 2008;69:155–166.
4. Ealy A, Drost M, Hansen P. Developmental changes in embryonic resistance to adverse effects of maternal heat stress in cows. *J Dairy Sci* 1993;76:2899–2905.
5. Edwards J, Ealy A, Monterroso V, Hansen P. Ontogeny of temperature-regulated heat shock protein 70 synthesis in preimplantation bovine embryos. *Mol Reprod Dev* 1997;48:25–33.
6. Ju J, Parks J, Yang X. Thermotolerance of IVM-derived bovine oocytes and embryos after short-term heat shock. *Mol Reprod Dev* 1999;53:336–340.
7. Sakatani M, Kobayashi S, Takahashi M. Effects of heat shock on in vitro development and intracellular oxidative state of bovine preimplantation embryos. *Mol Reprod Dev* 2004;67:77–82.
8. Eberhardt B, Satrapa R, Capinzaiki C, Trinca L, Barros C. Influence of the breed of bull (Bos taurus indicus vs. Bos taurus taurus) and the breed of cow (Bos taurus indicus, Bos taurus taurus and crossbred) on the resistance of bovine embryos to heat. *Anim Reprod Sci* 2009;114:54–61.
9. Fair T. Mammalian oocyte development: checkpoints for competence. *Reprod Fertil Dev* 2010;22:13–20.
10. Rocha A, Randel R, Broussard J et al. High environmental temperature and humidity decrease oocyte quality in Bos taurus but not in Bos indicus cows. *Theriogenology* 1998;49:657–665.
11. Al-Katanani Y, Paula-Lopes F, Hansen P. Effect of season and exposure to heat stress on oocyte competence in Holstein cows. *J Dairy Sci* 2002;85:390–396.
12. Ferreira R, Ayres H, Chiaratti M et al. The low fertility of repeat-breeder cows during summer heat stress is related to a low oocyte competence to develop into blastocysts. *J Dairy Sci* 2011;94:2383–2392.
13. Zeron Y, Ocheretny A, Kedar O, Borochov A, Sklan D, Arav A. Seasonal changes in bovine fertility: relation to developmental

competence of oocytes, membrane properties and fatty acid composition of follicles. *Reproduction* 2001;121:447–454.
14. Gendelman M, Roth Z. Seasonal effect on germinal vesicle-stage bovine oocytes is further expressed by alterations in transcript levels in the developing embryos associated with reduced developmental competence. *Biol Reprod* 2012;86:1–9.
15. Kawarsky S, Basrur P, Stubbings R, Hansen P, King W. Chromosomal abnormalities in bovine embryos and their influence on development. *Biol Reprod* 1996;54:53–59.
16. Sirard M. Follicle environment and quality of in vitro matured oocytes. *J Assist Reprod Genet* 2011;28:483–488.
17. Pohler K, Geary T, Atkins J, Perry G, Jinks E, Smith M. Follicular determinants of pregnancy establishment and maintenance. *Cell Tissue Res* 2012;349:649–664.
18. Badinga L, Thatcher W, Diaz T, Drost M, Wolfenson D. Effect of environmental heat stress on follicular development and steroidogenesis in lactating Holstein cows. *Theriogenology* 1993; 39:797–810.
19. Wolfenson D, Thatcher W, Badinga L et al. Effect of heat stress on follicular development during the estrous cycle in lactating dairy cattle. *Biol Reprod* 1995;52:1106–1113.
20. Shehab-El-Deen M, Leroy J, Fadel M, Saleh S, Maes D, Van Soom A. Biochemical changes in the follicular fluid of the dominant follicle of high producing dairy cows exposed to heat stress early post-partum. *Anim Reprod Sci* 2010;117:189–200.
21. Wolfenson D, Lew B, Thatcher W, Graber Y, Meidan R. Seasonal and acute heat stress effects on steroid production by dominant follicles in cows. *Anim Reprod Sci* 1997;47:9–19.
22. Wise M, Armstrong D, Huber J, Hunter R, Wiersma F. Hormonal alterations in the lactating dairy cow in response to thermal stress. *J Dairy Sci* 1988;71:2480–2485.
23. Gilad E, Meidan R, Berman A, Graber Y, Wolfenson D. Effect of tonic and GnRH-induced gonadotrophin secretion in relation to concentration of oestradiol in plasma of cyclic cows. *J Reprod Fertil* 1993;99:315–321.
24. Bridges P, Brusie M, Fortune J. Elevated temperature (heat stress) in vitro reduces androstenedione and estradiol and increases progesterone secretion by follicular cells from bovine dominant follicles. *Domest Anim Endocrinol* 2005;29:508–522.
25. Wolfenson D, Roth Z, Meidan R. Impaired reproduction in heat-stressed cattle: basic and applied aspects. *Anim Reprod Sci* 2000;60–61:535–547.
26. Bisinotto R, Chebel R, Santos J. Follicular wave of the ovulatory follicle and not cyclic status influences fertility of dairy cows. *J Dairy Sci* 2010;93:3578–3587.
27. Denicol A, Lopes G Jr, Mendonça L et al. Low progesterone concentration during the development of the first follicular wave reduces pregnancy per insemination of lactating dairy cows. *J Dairy Sci* 2012;95:1794–1806.
28. Al-Katanani Y, Webb D, Hansen P. Factors affecting seasonal variation in 90 day non-return rate to first service in lactating Holstein cows in a hot climate. *J Dairy Sci* 1999;82:2611–2615.
29. Huang C, Tsuruta S, Bertrand J, Misztal I, Lawlor T, Clay J. Environmental effects on conception rates of Holsteins in New York and Georgia. *J Dairy Sci* 2008;91:818–825.
30. Roth Z, Arav A, Bor A, Zeron Y, Braw-Tal R, Wolfenson D. Improvement of quality of oocytes collected in the autumn by enhanced removal of impaired follicles from previously heat-stressed cows. *Reproduction* 2001;122:737–744.
31. Webb R, Campbell B. Development of the dominant follicle: mechanisms of selection and maintenance of oocyte quality. *Soc Reprod Fertil Suppl* 2007;64:141–163.
32. Roth Z, Meidan R, Shaham-Albalancy A, Braw-Tal R, Wolfenson D. Delayed effect of heat stress on steroid production in medium-sized and preovulatory bovine follicles. *Reproduction* 2001;121: 745–751.
33. Putney D, Mullins S, Thatcher W, Drost M, Gross T. Embryonic development in superovulated dairy cattle exposed to elevated ambient temperatures between the onset of estrus and insemination. *Anim Reprod Sci* 1989;19:37–51.
34. Gwazdauskas F, Thatcher W, Kiddy C, Paape M, Wilcox C. Hormonal patterns during heat stress following $PGF_{2\alpha}$-tham salt induced luteal regression in heifers. *Theriogenology* 1981; 16:271–285.
35. Payton R, Romar R, Coy P, Saxton A, Lawrence J, Edwards J. Susceptibility of bovine germinal vesicle-stage oocytes from antral follicles to direct effects of heat stress in vitro. *Biol Reprod* 2004;71:1303–1308.
36. Roth Z, Hansen PJ. Involvement of apoptosis in disruption of oocyte competence by heat shock in cattle. *Biol Reprod* 2004; 71:1898–1906.
37. Roth Z, Hansen PJ. Disruption of nuclear maturation and rearrangement of cytoskeletal elements in bovine oocytes exposed to heat shock during maturation. *Reproduction* 2005;129:235–244.
38. Ju J, Jiang S, Tseng J, Parks J, Yang X. Heat shock reduces developmental competence and alters spindle configuration of bovine oocytes. *Theriogenology* 2005;64:1677–1689.
39. Soto P, Smith L. BH4 peptide derived from Bcl-xL and Bax-inhibitor peptide suppresses apoptotic mitochondrial changes in heat stressed bovine oocytes. *Mol Reprod Dev* 2009;76:637–646.
40. Andreu-Vázquez C, López-Gatius F, García-Ispierto I, Maya-Soriano MJ, Hunter RH, López-Béjar M. Does heat stress provoke the loss of a continuous layer of cortical granules beneath the plasma membrane during oocyte maturation? *Zygote* 2010;18:293–299.
41. Roth Z, Hansen P. Sphingosine 1-phosphate protects bovine oocytes from heat shock during maturation. *Biol Reprod* 2004;71:2072–2078.
42. Kalo D, Roth Z. Involvement of the sphingolipid ceramide in heat-shock-induced apoptosis of bovine oocytes. *Reprod Fertil Dev* 2011;23:876–888.
43. Edwards J, Saxton A, Lawrence J, Payton R, Dunlap J. Exposure to a physiologically relevant elevated temperature hastens in vitro maturation in bovine oocytes. *J Dairy Sci* 2005;88: 4326–4333.
44. Rispoli L, Lawrence J, Payton R et al. Disparate consequences of heat stress exposure during meiotic maturation: embryo development after chemical activation vs fertilization of bovine oocytes. *Reproduction* 2011;142:831–843.
45. Sartori R, Sartor-Bergfelt R, Mertens S, Guenther J, Parrish J, Wiltbank M. Fertilization and early embryonic development in heifers and lactating cows in summer and lactating and dry cows in winter. *J Dairy Sci* 2002;85:2803–2812.
46. Meyerhoeffer D, Wettemann R, Coleman S, Wells M. Reproductive criteria of beef bulls during and after exposure to increased ambient temperature. *J Anim Sci* 1985;60:352–357.
47. Monterroso V, Drury K, Ealy A, Howell J, Hansen P. Effect of heat shock on function of frozen/thawed bull spermatozoa. *Theriogenology* 1995;44:947–961.
48. Hendricks K, Martins L, Hansen P. Consequences for the bovine embryo of being derived from a spermatozoon subjected to post-ejaculatory aging and heat shock: development to the blastocyst stage and sex ratio. *J Reprod Dev* 2009;55:69–74.
49. Burfening P, Ulberg L. Embryonic survival subsequent to culture of rabbit spermatozoa at 38° and 40°C. *J Reprod Fertil* 1968;15:87–92.
50. Howarth B Jr, Alliston C, Ulberg L. Importance of uterine environment on rabbit sperm prior to fertilization. *J Anim Sci* 1965;24:1027–1032.
51. Sakatani M, Alvarez N, Takahashi M, Hansen P. Consequences of physiological heat shock beginning at the zygote stage on embryonic development and expression of stress response genes in cattle. *J Dairy Sci* 2012;95:3080–3091.
52. Edwards J, Hansen P. Differential responses of bovine oocytes and preimplantation embryos to heat shock. *Mol Reprod Dev* 1997;46:138–145.

53. Rivera R, Kelley K, Erdos G, Hansen P. Alterations in ultrastructural morphology of two-cell bovine embryos produced in vitro and in vivo following a physiologically-relevant heat shock. *Biol Reprod* 2003;69:2068–2077.
54. Rivera R, Kelley K, Erdos G, Hansen P. Reorganization of microfilaments and microtubules by thermal stress in two-cell bovine embryos. *Biol Reprod* 2004;70:1852–1862.
55. Sakatani M, Kobayashi S, Takahashi M. Effects of heat shock on in vitro development and intracellular oxidative state of bovine preimplantation embryos. *Mol Reprod Dev* 2004;67:77–82.
56. de Castro e Paula L, Hansen P. Modification of actions of heat shock on development and apoptosis of cultured preimplantation bovine embryos by oxygen concentration and dithiothreitol. *Mol Reprod Dev* 2008;75:1338–1350.
57. Rivera R, Dahlgren G, De Castro E, Paula L, Kennedy R, Hansen P. Actions of thermal stress in two-cell bovine embryos: oxygen metabolism, glutathione and ATP content, and the time-course of development. *Reproduction* 2004;128:33–42.
58. Sakatani M, Suda I, Oki T, Kobayashi S, Kobayashi S, Takahashi M. Effects of purple sweet potato anthocyanins on development and intracellular redox status of bovine preimplantation embryos exposed to heat shock. *J Reprod Dev* 2007;53:605–614.
59. Paula-Lopes F, Al-Katanani Y, Majewski A, McDowell L, Hansen P. Manipulation of antioxidant status fails to improve fertility of lactating cows or survival of heat-shocked embryos. *J Dairy Sci* 2003;86:2343–2351.
60. Ealy A, Howell J, Monterroso V, Aréchiga C, Hansen PJ. Developmental changes in sensitivity of bovine embryos to heat shock and use of antioxidants as thermoprotectants. *J Anim Sci* 1995;73:1401–1407.
61. Malayer J, Hansen P. Differences between Brahman and Holstein cows in heat-shock induced alterations of protein synthesis and secretion by oviducts and uterine endometrium. *J Anim Sci* 1990;68:266–280.
62. Fear J, Hansen P. Developmental changes in expression of genes involved in regulation of apoptosis in the bovine preimplantation embryo. *Biol Reprod* 2011;84:43–51.
63. Chandolia R, Peltier M, Tian W, Hansen P. Transcriptional control of development, protein synthesis and heat-induced heat shock protein 70 synthesis in 2-cell bovine embryos. *Biol Reprod* 1999;61:1644–1648.
64. Sakatani M, Bonilla L, Dobbs K et al. Changes in the transcriptome of morula-stage bovine embryos caused by heat shock: relationship to developmental acquisition of thermotolerance. *Reprod Biol Endocrinol* 2013;11:3.
65. Lim J, Liou S, Hansel W. Intracytoplasmic glutathione concentration and the role of β-mercaptoethanol in preimplantation development of bovine embryos. *Theriogenology* 1996;46:429–439.
66. Brad A, Hendricks K, Hansen P. The block to apoptosis in bovine two-cell embryos involves inhibition of caspase-9 activation and caspase-mediated DNA damage. *Reproduction* 2007;134:789–797.
67. Carambula S, Oliveira L, Hansen P. Repression of induced apoptosis in the 2-cell bovine embryo involves DNA methylation and histone deacetylation. *Biochem Biophys Res Commun* 2009;388:418–421.
68. Paula-Lopes F, Hansen P. Apoptosis is an adaptive response in bovine preimplantation embryos that facilitates survival after heat shock. *Biochem Biophys Res Commun* 2002;295:37–42.
69. Jousan F, Hansen P. Insulin-like growth factor-I promotes resistance of bovine preimplantation embryos to heat shock through actions independent of its anti-apoptotic actions requiring PI3K signaling. *Mol Reprod Dev* 2007;74:189–196.
70. Paula-Lopes F, Chase C Jr, Al-Katanani Y et al. Genetic divergence in cellular resistance to heat shock in cattle: differences between breeds developed in temperate versus hot climates in responses of preimplantation embryos, reproductive tract tissues and lymphocytes to increased culture temperatures. *Reproduction* 2003;125:285–294.
71. Hernández-Cerón J, Chase C Jr, Hansen P. Differences in heat tolerance between preimplantation embryos from Brahman, Romosinuano, and Angus Breeds. *J Dairy Sci* 2004;87:53–58.
72. Satrapa R, Nabhan T, Silva C et al. Influence of sire breed (*Bos indicus* versus *Bos taurus*) and interval from slaughter to oocyte aspiration on heat stress tolerance of in vitro-produced bovine embryos. *Theriogenology* 2011;76:1162–1167.
73. Silva C, Sartorelli E, Castilho A et al. Effects of heat stress on development, quality and survival of *Bos indicus* and *Bos taurus* embryos produced in vitro. *Theriogenology* 2013;79:351–357.
74. Block J, Chase C Jr, Hansen P. Inheritance of resistance of bovine preimplantation embryos to heat shock: relative importance of the maternal vs paternal contribution. *Mol Reprod Dev* 2002;63:32–37.
75. Maya-Soriano M, López-Gatius F, Andreu-Vázquez C, López-Béjar M. Bovine oocytes show a higher tolerance to heat shock in the warm compared with the cold season of the year. *Theriogenology* 2013;79:299–305.
76. García-Ispierto I, López-Gatius F, Santolaria P et al. Relationship between heat stress during the peri-implantation period and early fetal loss in dairy cattle. *Theriogenology* 2006;65:799–807.
77. Santolaria P, López-Gatius F, García-Ispierto I et al. Effects of cumulative stressful and acute variation episodes of farm climate conditions on late embryo/early fetal loss in high producing dairy cows. *Int J Biometeorol* 2010;54:93–98.
78. Chebel R, Santos J, Reynolds J, Cerri R, Juchem S, Overton M. Factors affecting conception rate after artificial insemination and pregnancy loss in lactating dairy cows. *Anim Reprod Sci* 2004;84:239–255.
79. Jousan F, Drost M, Hansen P. Factors associated with early and mid-to-late fetal loss in lactating and nonlactating Holstein cattle in a hot climate. *J Anim Sci* 2005;83:1017–1022.
80. Collier R, Doelger S, Head H, Thatcher W, Wilcox C. Effects of heat stress during pregnancy on maternal hormone concentrations, calf birth weight and postpartum milk yield of Holstein cows. *J Anim Sci* 1982;54:309–319.
81. Wolfenson D, Flamenbaum I, Berman A. Dry period heat stress relief effects on prepartum progesterone, calf birth weight, and milk production. *J Dairy Sci* 1988;71:809–818.
82. Tao S, Monteiro A, Thompson I, Hayen M, Dahl G. Effect of late-gestation maternal heat stress on growth and immune function of dairy calves. *J Dairy Sci* 2012;95:7128–7136.
83. Thompson I, Tao S, Branen J, Ealy A, Dahl G. Environmental regulation of pregnancy-specific protein B concentrations during late pregnancy in dairy cattle. *J Anim Sci* 2013;91:168–173.
84. do Amaral B, Connor E, Tao S, Hayen J, Bubolz J, Dahl G. Heat-stress abatement during the dry period: does cooling improve transition into lactation? *J Dairy Sci* 2009;92:5988–5999.
85. Flamenbaum I, Galon N. Management of heat stress to improve fertility in dairy cows in Israel. *J Reprod Dev* 2010;56(Suppl.):S36–S41.
86. Hansen P, Aréchiga C. Strategies for managing reproduction in the heat-stressed dairy cow. *J Anim Sci* 1999;77(Suppl. 2):36–50.
87. Putney D, Drost M, Thatcher W. Influence of summer heat stress on pregnancy rates of lactating dairy cattle following embryo transfer or artificial insemination. *Theriogenology* 1989;31:765–778.
88. Drost M, Ambrose J, Thatcher M et al. Conception rates after artificial insemination or embryo transfer in lactating dairy cows during summer in Florida. *Theriogenology* 1999;52:1161–1167.
89. Rodriques C, Ayres H, Reis E, Nichi M, Bo G, Baruselli P. Artificial insemination and embryo transfer pregnancy rates in high production Holstein breedings under tropical conditions. In: *Proceedings of the 15th International Congress on Animal Reproduction*, 2004, Vol. 2, p. 396 (Abstract).

90. Ambrose J, Drost M, Monson R et al. Efficacy of timed embryo transfer with fresh and frozen in vitro produced embryos to increase pregnancy rates in heat-stressed dairy cattle. *J Dairy Sci* 1999;82:2369–2376.
91. Al-Katanani Y, Drost M, Monson R et al. Pregnancy rates following timed embryo transfer with fresh or vitrified in vitro produced embryos in lactating dairy cows under heat stress conditions. *Theriogenology* 2002;58:171–182.
92. Stewart B, Block J, Morelli P et al. Efficacy of embryo transfer in lactating dairy cows during summer using fresh or vitrified embryos produced in vitro with sex-sorted semen. *J Dairy Sci* 2011;94:3437–3445.
93. Roth Z, Arav A, Braw-Tai R, Bor A, Wolfenson D. Effect of treatment with follicle-stimulating hormone or bovine somatotropin on the quality of oocytes aspirated in the autumn from previously heat-stressed cows. *J Dairy Sci* 2002;85:1398–1405.
94. Friedman E, Voet H, Reznikov D, Dagoni I, Roth Z. Induction of successive follicular waves by gonadotropin-releasing hormone and prostaglandin $F_{2\alpha}$ to improve fertility of high-producing cows during the summer and autumn. *J Dairy Sci* 2011;94:2393–2402.
95. Ealy A, Aréchiga C, Bray D, Risco C, Hansen P. Effectiveness of short-term cooling and vitamin E for alleviation of infertility induced by heat stress in dairy cows. *J Dairy Sci* 1994;77:3601–3607.
96. Paula-Lopes F, Al-Katanani Y, Majewski A, McDowell L, Hansen P. Manipulation of antioxidant status fails to improve fertility of lactating cows or survival of heat-shocked embryos. *J Dairy Sci* 2003;86:2343–2351.
97. Aréchiga C, Vázquez-Flores S, Ortíz O et al. Effect of injection of β-carotene or vitamin E and selenium on fertility of lactating dairy cows. *Theriogenology* 1998;50:65–76.
98. Aréchiga C, Staples C, McDowell L, Hansen P. Effects of timed insemination and supplemental β-carotene on reproduction and milk yield of dairy cows under heat stress. *J Dairy Sci* 1998;81:390–402.
99. Garcia-Ispierto I, Abdelfatah A, López-Gatius F. Melatonin treatment at dry-off improves reproductive performance postpartum in high-producing dairy cows under heat stress conditions. *Reprod Domest Anim* 2013;48:577–583.
100. Carpentieri A, Díaz de Barboza G, Areco V, Peralta López M, Tolosa de Talamoni N. New perspectives in melatonin uses. *Pharmacol Res* 2012;65:437–444.
101. Hansen P. Seasonal modulation of puberty and the post-partum anestrus in cattle: a review. *Livestock Prod Sci* 1985;12:309–327.
102. Pegorer M, Vasconcelos J, Trinca L, Hansen P, Barros C. Influence of sire and sire breed (Gyr vs Holstein) on establishment of pregnancy and embryonic loss in lactating Holstein cows during summer heat stress. *Theriogenology* 2007;67:692–697.

Chapter 65

Bovine Abortifacient and Teratogenic Toxins

Brittany Baughman

Mississippi Veterinary Diagnostic Laboratory, Jackson, Mississippi, USA

Introduction

A toxicant is a material that interferes with normal biological processes and causes adverse effects when it enters the body via ingestion, inhalation, dermal contact, or injection.[1] Reproductive toxicants are those capable of causing infertility, subfertility, early embryonic death, fetal developmental deformities (teratogens), fetal death, abortion, and stillbirths. Reproductive toxicants can cause deleterious effects by acting alone or by interacting with environmental, nutritional, or infectious stressors.[2] The effect of naturally occurring toxicants as well as anthropogenic compounds on reproduction in cattle while often sporadic can have considerable economic impacts.[3]

Nitrate/nitrite-accumulating plants

Ruminants, particularly cattle, are at increased risk for developing acute nitrate/nitrite poisoning from the ingestion of excessive amounts of nitrate-accumulating forages, weeds, and crops. Plants attain nitrogen from the soil in the form of nitrate and nitrite. Excessive nitrogen accumulation may occur in a variety of grasses and weeds under specific environmental and artificial conditions including increased fertilization, drought, hail, frost, herbicide damage, decreased sunlight, increased soil acidity, and in soils deficient in sulfur, phosphorus, and molybdenum. Commonly incriminated nitrate-concentrating plants include cereal grains, forage crops, and common weeds.[2,4,5] Nitrate accumulates greatest in the roots and lower portions of the plant including the stalks and stems due to decreased photosynthesis-dependent nitrate reductase activity at these sites.[2] Cattle may also ingest excessive nitrite and nitrate compounds directly from nitrate-based fertilizers (storage of fertilizers in areas accessible to cattle) and from water contaminated with manure or fertilizer runoff.[4,6] A few of the more commonly implicated nitrate-accumulating weeds are Johnson grass (*Sorghum halepense*), pigweed (*Amaranthus retroflexus*), jimsonweed (*Datura stramonium*), and lamb's quarters (*Chenopodium* spp.).[6,7] However, a variety of other plants species have also been implicated in nitrate poisoning[4] (Table 65.1).

Pathogenesis/mechanism of action

Normally dietary nitrate is converted to nitrite within the reticulorumen and is slowly metabolized to ammonia. However, when present in levels that overwhelm the microbial organisms' ability to reduce them, nitrate and nitrite are rapidly absorbed into the circulation. Once in the circulation nitrite ions readily interact with hemoglobin molecules of blood, promoting the oxidation of hemoglobin iron from the ferrous (Fe^{2+}) to the ferric (Fe^{3+}) state and forming methemoglobin. The methemoglobin molecule is incapable of binding and transporting oxygen molecules. Once sufficient hemoglobin is oxidized to methemoglobin, clinical signs of hypoxia develop including exercise intolerance, dyspnea, muscle tremors, and cyanotic and occasionally brown-colored membranes and blood. Although maternal methemoglobin molecules are incapable of crossing the placenta, nitrite and nitrate readily cross and can induce fetal erythrocyte methemoglobinemia and anoxia. Typically, reproductive repercussions including abortions, stillbirths, and the birth of weak calves have been reported approximately 3–7 days or more following maternal toxic exposure. In addition to the effects of methemoglobinemia, nitrite and nitrates also act as potent vasoactive compounds resulting in vasodilation and hypotension.[2,8]

Diagnosis of nitrate/nitrite poisoning

In adult cattle nitrate/nitrite poisoning may be suspected based on clinical signs including brownish discoloration of the blood and mucous membranes and a history of exposure to possible nitrate-accumulating forages.[2,4] However, the characteristic chocolate-colored mucous membranes and blood are not always present and can become less obvious as autolysis proceeds.[9]

Table 65.1 Nitrate-accumulating plants.

Sorghum sudan (*Sorghum vulgare*)
Pigweed (*Amaranthus* spp.)
Button grass (*Dactyloctenium radulans*)
Sugar beet (*Beta vulgaris*)
Corn (*Zea mays*)
Soybean (*Glycine max*)
Pearl millet (*Pennisetum typhoides*)
Flax (*Linum* spp.)
Rye (*Secale cereale*)
Sunflower (*Helianthus annuus*)
Oats (*Avena* spp.)
Lamb's-quarter (*Chenopodium* spp.)
Turnips and rape (*Brassica* spp.)
Canada thistle (*Cirsium arvense*)
Wheat (*Triticum aestivum*)
Goldenrod (*Solidago* spp.)
Nightshades (*Solanum* spp.)
Barley (*Hordeum* spp.)
Sorrel, curly leaf dock (*Rumex*)
Alfalfa (*Medicago sativa*)
Russian thistle (*Salsola kali*)
Johnson grass (*Sorghum halepense*)
Sweet clover (*Melilotus* spp.)
Tall fescue (*Lolium arundinaceum*)
Kochia weed (*Kochia scoparia*)
Timothy grasses (*Phleum* spp.)
Smart weed (*Polygonum* spp.)
Ragweed (*Ambrosia* spp.)
Jimsonweed (*Datura stramonium*)
Fireweed (*Kochia* spp.)
Field bindweed (*Convolvulus arvense*)
Docks (*Rumex*)
Barnyard grass (*Echinochloa crus-galli*)
Cheese weed (*Malva* spp.)

Source: adapted from Knight A, Walter R. *A Guide to Plant Poisoning of Animals in North America*. Jackson, WY: Teton New Media, 2001.

In cases of suspect nitrate toxicosis-induced abortion, a full necropsy should be performed on the fetus to rule out other potential causes. Ocular fluid from the fetus or cow can be collected for postmortem nitrate/nitrite analysis by ion chromatography with a conductivity detector. This test can be quite reliable when fluid is collected within hours of death and refrigerated. Standard nitrate dipsticks are also available and can be used to make a presumptive diagnosis of nitrate poisoning. Ideally, nitrate levels should approach 30 ppm or greater to be highly suggestive of nitrate/nitrite-associated abortion. However, a compatible history including maternal grazing on potential nitrate-accumulating forages, or some other form of excessive maternal ingestion of nitrates, is also needed.[6] Forage nitrate analysis should also be conducted for confirmation. Forages containing greater than 0.6–1.0% nitrates (6000–10 000 ppm on a dry-matter basis) have been associated with reproductive losses (abortion, stillbirths, and weak calves).[2]

Treatment and management

Treatment of individual animals with intravenous methylene blue in a 1% or 2% aqueous solution at a rate of 1–2 mg/kg body weight with up to 10 mg/kg in severe cases can be performed. However, due to withdrawal times, cattle should not be sold for slaughter for up to 180 days.[9] Although methylene blue treatment can be effective in individual animals, it is likely cost prohibitive in herd outbreaks. When nitrate/nitrite poisoning is suspected cattle should be removed from pasture. *Propionibacterium acidipropionici* strain P5 (a bacterial feed additive) may be supplemented if exposure to high nitrate concentrations in feed or water is unavoidable. Water should be tested when nitrate contamination is suspected from manure or fertilizer runoff. Water nitrate concentrations should be below 440 mg/L, although acute toxicity is typically not appreciated until water nitrate levels exceed 1300 mg/L.[6]

Ponderosa pine

Ponderosa pine or western yellow (*Pinus ponderosa*) are common large coniferous trees of the pine tree family (Pinacea) found in all states west of the Great Plains and throughout western Canada. *Pinus ponderosa* is characterized by 18–28 cm long, yellow-green needles with two to three needles per fascicle.[4,10] Ponderosa pine contains a number of compounds that have been associated with toxicity including phytotoxins, mycotoxins, resins, and ligands.[4] Overt toxicosis and death in cattle have occasionally been reported; however, the primary toxicological effects of ponderosa pine needle ingestion in cattle are abortion, retained fetal membranes, and metritis. Open cows, steers, and bulls appear to be unaffected by pine needle ingestion.[11] Isocupressic acid, a labdane resin acid, is the principal abortifacient compound found in pine needles and the bark of *P. ponderosa*.[12] Two related derivatives of isocupressic acid, acetylisocupressic acid and succinylisocupressic acid, have also been found to be abortifacient following hydrolytic conversion to isocupressic acid in the rumen.[13] Other potentially abortifacient acids present in lesser amounts in ponderosa pine needles include agathic acid, imbricatolic acid, and dihydroagathic acid.[14]

Cattle most often ingest early green ponderosa pine needles as well as dry pine needles from the ground in the late fall, winter, and early spring when normal forages are scarce or when they are stressed.[4,10,11] Ponderosa pine needle toxicity can result in late-term abortion in cows when ingested primarily during the last trimester. However, abortions have been reported as early as 3 months of gestation. Cattle may undergo premature parturition anywhere from 2 to 21 days following ingestion of pine needles, although occasionally immediate abortions have been reported.[4,11] Premature parturition has been induced in cattle fed pine needles 2.2–2.7 kg/day for a period of at least 3 days.[4,15] Calves may be born stillborn or premature (small and weak) depending on the stage of the gestation at the time of ingestion.[16] Affected pregnant cows develop edematous swelling of the vulva and mucoid vaginal discharge followed by premature parturition or abortion.[4] Abortions are typically characterized by weak uterine contractions, uterine bleeding, incomplete cervical dilation, dystocia, birth of weak calves, agalactia, and retained placentas. The mechanism of action of isocupressic acid on pregnancy is not completely understood.[16] However, decreased blood flow to the uterus and increased caruncular arterial tone resulting in chronic vasoconstriction of the vascular bed with subsequent fetal parturition are reported. Histologic lesions of placental tissues may include necrosis and hemorrhage secondary to vasoconstriction.[11,16]

Management and treatment

There are no methods to completely prevent pine needle-induced abortions, although the potential impacts may be minimized. Ideally, pregnant cows should be denied access to pine needles when they are most susceptible to the abortifacient effects of isocupressic acid (third trimester). Pregnant cows should be maintained in adequate body condition when grazing ranges with ponderosa pines to reduce consumption. Adequate food and shelter should also be provided for cattle to minimize stress and consumption of pine needles. Calving schedules may also be changed to late spring or fall when winter weather conditions are less likely to increase pine needle ingestion.[10]

There is no effective treatment for pine needle toxicosis. Supportive care including antibiotic treatment and uterine infusion in cases of placental retention are recommended for cows that abort. Supplemental colostrum and milk may be required for live premature calves.[11]

Abortions similar to those caused by ponderosa pine in cattle have also been associated with accidental ingestion of Monterey cypress (*Cupressus macrocarpa*) in New Zealand. Monterey cypress is a native of California that has been widely planted throughout Europe and New Zealand. Analysis of *Cupressus macrocarpa* in New Zealand detected elevated concentrations of isocupressic acid. Isocupressic acid is also the primary diterpene acid in *Pinus contorta* (lodgepole pine), *Juniperus communis* (common juniper), Korean pine (*Pinus koraiensis*), and California juniper (*Juniperus californica*), all of which have been associated with late-term abortions in cattle. Cattle abortions have also been attributed to the ingestion of bark from the Utah juniper (*Juniperus osteosperma*) and western juniper (*Juniperus occidentalis*), both of which contain increased concentrations of agathic acid but minimal isocupressic acid.[13]

Ateleia glazioviana, Tetrapterys acutifolia, and Tetrapterys multiglandulosa

Abortions and neonatal mortality in cattle have been attributed to *Ateleia glazioviana*, *Tetrapterys acutifolia*, and *Tetrapterys multiglandulosa* poisoning in Brazil. *Ateleia glazioviana* is a deciduous tree found in western Santa Catarina and northwestern Rio Grande do Sul. Cattle may browse the leaves of this tree when other forages are scarce due to overgrazing or drought. Ingestion of leaves can cause abortions at any time during pregnancy but most occur when green leaves are ingested during the fall and winter months.

Tetrapterys acutifolia and *Tetrapterys multiglandulosa* are toxic shrubs that can cause significant outbreaks of abortion and deliveries of weak calves. *Ateleia glazioviana*, *T. acutifolia*, and *T. multiglandulosa* can cause similar clinical signs and lesions. Although abortions can occur at any stage of pregnancy, reports are most often described between 6 and 9 months of gestation. Pregnant cows become lethargic 1–3 days prior to abortion and occasionally exhibit blindness. Fetal lesions are similar to those observed in adult intoxications. Aborted fetuses may demonstrate subcutaneous edema of the limbs, yellow clear thoracic and abdominal effusion, and myocardial pallor with increased thickness of the right ventricular wall and interventricular septum. The toxin or toxins responsible for abortions and clinical disease in these plants has not yet been identified.[17]

Enterolobium spp. and Stryphnodendron spp.

Stryphnodendron spp. and *Enterolobium* spp. are trees in the family Fabaceae that produce pods that when ingested have been linked to gastrointestinal signs, photosensitization, and abortions in cattle and other ruminant species in Brazil.[18] *Stryphnodendron obovatum* trees grow in the central-western region of Brazil and are considered to be one of the most important causes of cattle abortions in this region. Cattle readily ingest the fallen pods during the dry season. Abortions may occur during all stages of gestation but are often observed during late pregnancy. The abortive effects of *S. obovatum* were experimentally reproduced by feeding mature pods to seven pregnant cows that were between 3 and 7 months of gestation. Three of seven cows aborted between 20 and 30 days following the start of the study. A fourth expelled a mummified fetus 7 months following the experiment. The underlying toxic properties of *S. obovatum* pods were not identified.[19] The abortive properties of *Enterolobium contortisiliquum* have been demonstrated experimentally in guinea pigs but not in cattle.[18] *Enterolobium contortisiliquum* and *Dimorphandra mollis* (also of the Fabaceae family) are commonly blamed for abortions in Brazil although this has not yet been proven experimentally.[19]

Hairy vetch (*Vicia villosa*)

Hairy vetch (*Vicia villosa* Roth) (Figure 65.1), originally introduced from Europe as a forage legume and cover crop, can be found throughout much of the United States and in many other countries including Argentina, Australia, and South Africa.[4,20,21] Hairy vetch poisoning in cattle is most often associated with the development of multiorgan, granulomatous and eosinophilic perivascular inflammation most notably the kidney, heart, adrenal gland, and skin. Clinical findings in affected cattle most often include pruritic dermatitis, conjunctivitis, weight loss, hematuria, and diarrhea. However, reports of sporadic abortion, neurologic dysfunction, subcutaneous swelling of the head, neck, and body, multifocal ulceration of the oral mucous membranes, purulent nasal discharge, rales, cough, and pulmonary congestion have also been reported.[21]

Figure 65.1 Hairy vetch (*Vicia villosa*). USDA-NRCS PLANTS Database.

Figure 65.2 Broom snakeweed (*Gutierrezia sarothrae*). Photograph courtesy of Clarence A. Rechenthin, USDA-NRCS PLANTS Database.

Diagnosis, mechanism of action, and clinical signs

Hairy vetch poisoning in cattle is most often a diagnosis of exclusion based on clinical signs, histologic findings, and grazing history.[20] Clinical manifestations of hairy vetch toxicity typically develop weeks following ingestion of the plant. The toxic principle of hairy vetch has not been definitively determined, but the pattern and composition of the inflammatory response is highly suggestive of a type IV hypersensitivity reaction. It has been postulated that the inflammatory reaction may be caused by one or more ingested plant constituents (absorbed as haptens or complete antigens) sensitizing antigen-specific lymphocytes that then become distributed throughout the body. Subsequent exposure to these antigens presumably evokes a type IV hypersensitivity response. Another proposed mechanism of toxicity is that vetch lectins bind directly to T-lymphocytes inducing T-lymphocyte activation, lymphokine production, cytotoxicity, and the classic granulomatous inflammatory response.[21] Once clinical signs develop recovery is not promising as there is no effective treatment. Cattle do not appear to be uniformly susceptible to hairy vetch poisoning. Cattle under the age of 3 years appear to be less prone to clinical disease. Vetch toxicosis has been reported in multiple breeds of cattle, although Holstein and Angus breeds appear to be more commonly affected.[4]

Broom snakeweed

Broom snakeweed (*Gutierrezia sarothrae*) (Figure 65.2) is an invasive, herbaceous, short-lived, woody perennial shrub native to the western rangelands of North America. The plant can be found from Saskatchewan and Alberta, Canada south through western Nebraska and Kansas into Texas and central Mexico and extending to Washington, Oregon, California, and Baja California. Heavy growth of broom snakeweed in rangelands is often indicative of poor range management (such as overgrazing), or may be seen in times of drought or following fires.[4,22] There are two major species of broom snakeweed, *Gutierrezia sarothrae* (turpentine weed, slinkweed, perennial snakeweed) and *G. microcephala* (threadleaf broomweed). *Gutierrezia sarothrae* grows up to 70 cm tall and has a woody base and branching stems with alternate linear leaves and clusters of small yellow flowers. *Gutierrezia microcephala* is similar in appearance to *G. sarothrae* but has smaller flowers.[4]

Broom snakeweeds are toxic and abortifacient to cattle and other livestock species (sheep and goats).[11] Potentially toxic and abortifacient compounds in broom snakeweed include saponins, furanoditerpene acids, terpenoids, steroids, and flavones.[22] Furanoditerpene acids and flavones found in the leaves of broom snakeweed are similar to isocupressic acid, the abortifacient compound in ponderosa pine. In a study by Molyneux in 1980, saponin extracts from threadleaf snakeweed induced abortion in cattle at low concentrations and death in cattle at high concentrations.[11] Abortions caused by broom snakeweed occur mainly when cattle graze during the last trimester of pregnancy.[23]

Reported clinical signs include anorexia, weight loss, poor hair coat, listlessness, mucopurulent nasal discharge, abortion, diarrhea, constipation, jaundice, rumen stasis, and death. Abortions can occur throughout gestation and are often accompanied by retained fetal membranes that may lead to uterine infection. Signs of abortion include weak uterine contractions, occasional incomplete cervical dilation, and excessive discharge of vaginal mucus. Low nutrition exacerbates fertility problems caused by broom snakeweed. Supplemental nutrition may enhance degradation and elimination of terpenes. Treatment of clinically affected animals is symptomatic and may include antibiotic use in cows with retained placentas to prevent infection.[4,11]

Snakeweed is typically not very palatable to cattle and is usually only grazed when there is significant grazing pressure due to depleted vegetation. The overall palatability of snakeweed is dependent on the concentrations of plant resins (including terpenes) which accumulate over the growing season. Higher resin concentrations decrease palatability.[22] Proper range management, including providing adequate nutrition for cattle by maintaining the range in good condition, decreases the likelihood of snakeweed toxicosis in cattle. In range regions where snakeweed dominates management strategies to control the shrub may include the use of herbicides such as Tordon® 22K (active ingredient: picloram) and Escort® XP (active ingredient: metsulfuron methyl) or prescribed burning. Low nutrition exacerbates fertility problems caused by broom snakeweed, whereas supplemental protein and energy may enhance degradation and elimination of terpenes.[11]

Sumpweed (*Iva angustifolia*)

Marshelder, also known as narrowleaf sumpweed, is a native warm-season annual found throughout the south central United States and is a member of the sunflower family (Asteraceae). Reportedly cattle consuming large amounts of the plant may display premature mammary development, premature milk let-down, and abortion. Sumpweed is seldom eaten by cattle; however, ingestion of young plants during the two- to eight-leaf stage of development has been associated with cattle abortions during 4 to 8 months of gestation and premature calves. The toxic agent/compound is not known.[24]

Moldy sweet clovers

White sweet clover (*Melilotus alba*) and yellow sweet clover (*Melilotus officinalis*) are biennial legumes (family Fabaceae) commonly grown as forages in the northwestern United States and western Canada. Both species have compound leaflets with serrated edges and pea-like flowers produced on axillary racemes.[4] Ingestion of hay or silage containing spoiled sweet clover (yellow or white) can induce toxicity. Sweet clover naturally contains coumarin which may be converted to the toxin dicoumarol, a potent vitamin K antagonist, by dimerization and oxidation through the actions of fungi such as *Aspergillus*, *Penicillium*, and *Mucor*.[25,26] Dicoumarol interferes with synthesis of vitamin K-dependent coagulation factors II, VII, IX, and X. These factors are necessary for prothrombin conversion to thrombin, which polymerizes fibrinogen to fibrin. Impaired fibrin production causes poor stabilization of platelet plugs, predisposing affected animals to hemorrhage.[26]

Clinical signs of moldy sweet clover toxicity are often nonspecific and vary from excessive hemorrhage following trauma to spontaneous hemorrhages including epistaxis, melena, hemarthrosis, hematuria, and anemia. Hemorrhages typically occur several days to weeks following ingestion of moldy sweet clover. Dicoumarol levels of 20–30 mg/kg in hay are usually required before clinical disease develops.[27] Dicoumarol can cross the placenta causing fetal hemorrhage and has been associated with sporadic abortions in cattle.[4,25,28]

Clinical signs of dicoumarol toxicity secondary to moldy sweet clover ingestion are indistinguishable from anticoagulant rodenticide toxicosis. A diagnosis of moldy sweet clover poisoning should be suspected when cattle have a history of prolonged consumption of sweet clover combined with compatible clinical signs, and prolonged blood clotting times. Dicoumarol analysis of blood from affected animals (live or dead) and analysis of silage or hay can confirm the diagnosis. In cases of clinical disease affected silage or hay should be removed. Animals demonstrating clinical signs should receive an intramuscular or subcutaneous injection of vitamin K_1 (phytonadione) at a dose of 1.1–3.3 mg/kg body weight every 12 hours for 5 days or until prothrombin time and activated partial thromboplastin time return to normal.[26] In severe cases cattle may be treated with whole blood transfusions (2–4 L of blood per 450 kg) from a donor animal not being fed moldy clover.[28] Blood transfusions are typically indicated when packed cell volumes fall below 12%.[26] Other plants containing coumarin include sweet vernal grass (*Anthoxanthum odoratum*), *Lespedeza stipulacea*, many species of Umbelliferae, and other species of *Melilotus* (melilot).[4,26] Herd outbreaks of dicoumarol toxicity have also been reported in cattle ingesting moldy *A. odoratum* hay or silage.[26]

Teratogenic plants

Teratogens are compounds that are capable of inducing developmental malformations or birth defects via insult to a developing fetus. The vast majority of scientific research and investigation in the field of veterinary and human teratology has involved laboratory animals due to their short gestation periods compared with livestock and due to the significant differences in cost involved with project design and the care and maintenance of large animal subjects.[29] Many factors are involved in the development of teratogen-induced congenital defects.

Teratogen susceptibility can vary greatly among genotypes within the same species which may result in variations in incidence and severity of malformations within the same herd. In order for a teratogenic compound to exert its effect on a fetus, it and/or its metabolites must first reach the conceptus by crossing the placenta. Teratogens, like many other toxins, are dose dependent. Sufficient amounts of the toxic compounds must be ingested by the dam, absorbed into the maternal circulation, and eventually pass through the placenta to the embryo or fetus. Teratogenic compounds in sufficient quantity may result in early fetal loss (abortion or resorption) rather than congenital malformation. In order for a teratogen to cause a specific defect, it must first exert its effect at the appropriate period of time during gestation. For example, for cyclopia to develop in a fetus exposed to *Veratrum californicum*, the toxic plant must be ingested on day 14 of gestation by sheep during the stage of primitive neural plate development.[29]

Astragalus and *Oxytropis* (locoweeds)

Locoweeds, or milk vetches (Figure 65.3), include several species of plants in the genera *Astragalus* and *Oxytropis*.[23] *Astragalus* and *Oxytropis* are members of the Fabaceae (Leguminosae) or pea family. These plants are characterized by their butterfly-like flowers displaying a single large banner petal, two side petals, and two lower fused petals forming a keel. *Astragalus* and *Oxytropis* species have a worldwide distribution. Collectively, locoweeds are considered to be the most

Figure 65.3 Milk vetch (*Astragalus* sp.). USDA-NRCS PLANTS Database.

detrimental poisonous plant of cattle in the western United States and are responsible for significant economic losses.[4,30]

Clinical signs of locoweed poisoning are caused by the toxic effects of the indolizidine alkaloid swainsonine found in multiple species of *Astragalus* and *Oxytropis*. Swainsonine is produced by the fungal endophyte *Undifilum oxytropis* (formerly *Embellisia oxytropis*) which can be found in the parenchymal layers of the infected plant seeds.[11] Locoweed poisoning is the chronic condition that develops in ruminants after weeks of ingesting swainsonine-containing plants. Common clinical signs of locoism include depression, incoordination, tremors, weight loss, emaciation, ataxia (staggering), altered mentation, belligerence, aggression, proprioceptive deficits, and eventually death. Reproductive failure, embryo loss, abortion, and occasional birth defects can be associated with locoweed poisoning and are often seen prior to more severe clinical signs in the cow.[31]

Mechanism of action

Swainsonine inhibits the lysosomal enzyme α-mannosidase, which is required for normal hydrolysis of mannose-rich oligosaccharides. In the absence of hydrolysis incompletely processed oligosaccharides accumulate within the cells of multiple organs and tissues resulting in a loss of cellular function and cell death.[30,31] α-Mannosidosis is the most important acquired lysosomal storage disease in grazing livestock.[32] Microscopic lesions of locoism can be found in many organs and tissues throughout the body; however, cellular vacuolation due to accumulated oligosaccharides is reportedly most pronounced within the brain, liver, kidney, pancreas, and thyroid gland.[31]

Teratogenic effects

Sporadic outbreaks of abortion and congenital limb deformity have been reported in a number of pregnant domestic animal species including cattle, sheep, goats, and horses grazing on species of *Astragalus* and *Oxytropis*. Fetal malformations may include brachygnathia, contracture or overextension of joints, limb rotations, osteoporosis, and bone fragility. The toxic agent responsible for the fetal malformations has not been determined but is likely unrelated to those responsible for the development of lysosomal storage disease.[33]

Management and treatment

Because of the transient nature of locoism, adult animals may recover if they are promptly removed from the swainsonine-containing plants early in the course of their intoxication. Swainsonine is readily cleared from the body making recovery rapid unless significant tissue damage is present. Although many locoweeds are not highly palatable to cattle, once forced to graze it (usually due to extreme grazing conditions) cattle may develop a preference for the plant that can lead to intoxication.[11]

Other swainsonine-containing plants

A number of other swainsonine-containing plant species cause livestock poisonings similar to locoweed poisoning. Some of these include *Ipomea* spp. in Brazil and Africa, *Swainsona* spp. in Australia (swainsonine was first identified in *Swainsona canescens*), and toxic species of *Ipomea*, *Sida*, *Solanum*, *Physalis*, and *Convolvulus*.[11,31] Swainsonine-containing plants in Brazil include *Sida carpinifolia*, *Ipomea carnea* subsp. *fistulosa*, and *Turbinata cordata*. Fetal malformations and abortions with these plants are similar to those seen with swainsonine-containing locoweed.[17,32]

Other toxic effects of *Astragalus* spp.

Over 260 North American species of *Astragalus* have been identified as nitrogen accumulators. Nitrogen-containing *Astragalus* spp. can cause both an acute form and a chronic form of nitrate poisoning in cattle. In the acute form cattle develop methemoglobinemia and respiratory distress. Chronic intoxication manifests as generalized weakness, incoordination, and other central nervous system signs (knuckling of the fetlocks, goose-stepping, heel-clicking paralysis) and death. Methylene blue treatment can reverse methemoglobinemia in affected cattle but does not prevent mortality, indicating that the cause of death is likely attributed to other factors. Over 20 *Astragalus* species have also been identified as selenium accumulators and may induce selenium toxicosis ("alkali disease") with prolonged grazing.[11]

Lupines

Lupines (*Lupinus* spp.), commonly known as blue bonnet, belong to the Fabaceae (legume) family (Figure 65.4). These perennial herbaceous plants are indigenous to North America and range throughout much of the United States. Over 100 species of lupines have been identified. Plants grow up to 1 m high. Leaves are alternate and palmately divided with 5–17 leaflets, which can be smooth or hairy depending on the species. Flowers of *Lupinus* spp. are bonnet-shaped and are present on racemes that are

Figure 65.4 Silver lupine (*Lupinus albifrons* Benth). Photograph by Pete Veilleux, East Bay Wilds.

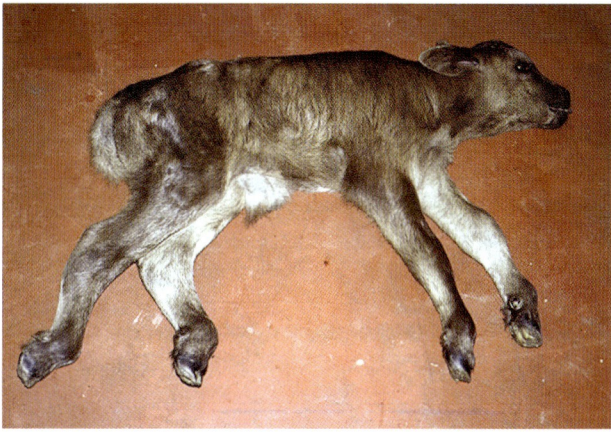

Figure 65.5 Arthrogryposis of hindlimbs (crooked calf).

located on a single central stalk. Poisonous varieties of lupine responsible for cattle toxicosis in the United States are predominantly found in the Rocky Mountain region and in states located to the west of the Rockies. Lupines are typically not very palatable to cattle and are most often grazed only when other forages are not readily available.[4,11]

Three specific syndromes in livestock have been attributed to lupine toxicosis in North America: (i) "crooked calf disease," a teratogenic condition caused by pregnant cows grazing lupines; (ii) acute fatal neurologic disease in adult cattle; and (iii) lupinosis. Lupinosis is not an alkaloid-induced toxicity but is instead caused by cattle grazing lupines infected with the fungus *Phomopsis leptostromiformis* that produces the mycotoxin phomopsin. Clinical signs of phomopsin poisoning may include severe hepatic, renal, and muscle disease in cattle and sheep.[4] Clinical signs of the neurologic syndrome in adult cattle include muscle weakness, agitation, frequent urination and defecation, depression, frothing at the mouth, relaxation of the third eyelid, muscle fasciculations, ataxia, lethargy, collapse, and sternal recumbency followed by lateral recumbency, respiratory failure, and death.[11]

Maternal ingestion of toxic lupines during days 40–70 of gestation results in significant fetal malformations depending on the concentration of teratogenic compounds present in the plant.[29] Alkaloids responsible for fetal teratogenic effects (crooked calf syndrome) include anagyrine, ammodendrine, and N-methyl-ammodendrine.[11]

The most commonly observed lupine-induced fetal malformation is arthrogryposis of the forelimbs (Figure 65.5), which may be accompanied by scoliosis, torticollis, lordosis, kyphosis, or palatoschisis (cleft palate).[11,34] Muscles from affected limbs are often hypoplastic or atrophied. Muscle atrophy is attributed to disuse of the affected limb(s) *in utero*.[34] The mechanism of action for teratogenic fetal effects is an alkaloid-induced decrease in fetal movement secondary to desensitization of skeletal muscle-type nicotinic acetylcholine receptors. Decreased or inhibited fetal movement results in skeletal malformations including skeletal muscle contractions and cleft palate formation.[11]

Management and treatment

Prevention and management of lupine poisoning can be achieved by coordinating grazing periods of susceptible pregnant cows with growth stages of the plant, ensuring that susceptible pregnant cows do not graze during the flower and pod stages of growth when anagyrine concentrations are at their highest. Other effective measures may include changing the calving periods from spring to fall and by instituting intermittent grazing between lupine-free pasture and lupine-infested pasture to break the cycle of lupine ingestion. Broadleaf herbicides such as 2,4-D can also be utilized to reduce lupine numbers in heavily infested pastures, although lupine populations commonly reestablish due to the presence of resident seeds within the soil.

Treatment of clinically ill cattle involves supportive and symptomatic care. Animals displaying signs of intoxication should not be stressed and should be allowed to rest. Because the toxic alkaloids are rapidly excreted in the urine, affected adult animals will often make a full recovery within 24 hours. No treatment is available for severe fetal malformations. However, calves with mild contracture defects, of the forelimbs primarily, may recover if the elbow joint can be locked into place within 1 week of birth.[11]

Poison hemlock (*Conium maculatum*)

Poison hemlock, also known as European hemlock, spotted hemlock, and California fern, is a biennial perennial member of the parsley/carrot (Apiaceae, formerly Umbelliferae) family found all across North America (Figure 65.6). *Conium maculatum* is one of the most toxic members of the plant kingdom.[35] It was originally introduced to North America from Europe as an ornamental plant and is also native to Asia but can be found throughout North and South America, North Africa, Australia, and New Zealand. Poison hemlock is a tall branched plant with distinctive purple spots along its stems, multiple small white five-petaled flowers, a single taproot, and a distinctive mousy odor.[4,31,36]

Poison hemlock contains eight known piperidine alkaloids.[36] The two major toxic alkaloids are γ-coniceine found in immature growing plants and coniine found in

Figure 65.6 Poison hemlock (*Conium maculatum*). USDA-NRCS PLANTS Database.

mature plants and seeds.[4,31,36] However, all vegetative organs, flowers, and fruits contain alkaloids.[35] Alkaloid levels vary based on environmental factors (such as temperature and moisture), stage of growth, and geographic location. Fresh drying the plant in the sun for 7 days results in significantly decreased levels of the toxic piperidine alkaloids.[31]

Poison hemlock is highly toxic to cattle and in acute poisoning cases clinical signs may develop rapidly following ingestion (30 min to 1 hour). Reported clinical manifestations include depression, abdominal pain, mydriasis (dilated pupils), frequent urination and defecation, increased salivation (ptyalism), lacrimation, muscle tremors and fasciculations, incoordination and ataxia, respiratory distress, recumbency, collapse, and death. Animals surviving acute poisoning usually recover but abortions may result. The teratogenic effects of *C. maculatum* are most often appreciated in cases of chronic ingestion by pregnant cattle. However, fetal malformations have also been described in piglets and lambs.[36] Ingestion of the plant between days 40 and 70 of gestation can result in a wide array of skeletal malformations in the neonate.[4] Reported teratogenic malformations are indistinguishable from those caused by lupines and tobacco species. Fetal malformations may include torticollis (twisted neck), scoliosis (vertebral column curvature), palatoschisis (cleft palate), arthrogryposis (multiple joint contractures), and excessive flexure of the carpal joints (crooked calf disease).[35,36]

Mechanism of toxicity in acute poisoning

Conium alkaloids, specifically γ-coniceine, coniine, and N-methylconiine, act as nondepolarizing blockers (akin to curare) resulting in paralysis of motor nerves through their action on the medulla. It has been proposed that these alkaloids initially stimulate the skeletal muscles and produce subsequent neuromuscular blockade through their action on nicotinic receptors.[4,35,36] Death occurs when the phrenic nerve is affected and respiratory muscles become paralyzed.[36]

Mechanism of teratogenicity

Poison hemlock-related fetal malformations, such as joint contractures, are believed to result from fetal immobility induced by neuromuscular blockade of nerves. Palatoschisis in calves may be secondary to the loss of fetal mandibular and tongue movements, which are both required for palatal shelves to become closer in the developing fetus. In the absence of jaw and tongue movements, the tongue permanently occupies the site where palatal shelves should fuse.[36,37]

Management and treatment

Treatment of poison hemlock toxicosis relies largely on supportive care. In cases of acute toxicity gastric lavage and activated charcoal treatment may be performed on individual animals. Control of the weed can be managed through the use of broadleaf herbicides, although populations will reestablish via seed reserves in the soil.[11]

Tobacco poisoning

Nicotiana glauca (tree tobacco), *Nicotiana tabacum* (cultivated tobacco), *Nicotiana trigonophylla* (wild or desert tobacco), and *Nicotiana attenuata* (wild or coyote tobacco) are members of the nightshade (Solanaceae) family. Wild tobacco varieties are indigenous to the southwestern United States while cultivated tobacco grown in the United States is predominantly raised in the southeastern states. Tobacco is a herbaceous branching annual that grows between 0.3 and 1.2 m in height, with hairy stems and leaves and fragrant tubular flowers with five parts. Tree tobacco (*N. glauca*) is an evergreen shrub or small tree growing 2–6 m tall with alternate bluish-green hairless leaves. Flowers are 5 cm long, tubular, and produced on the leafless branches.[4]

Tobacco contains numerous alkaloids (over 40) most of which are neurotoxic pyridine alkaloids including nicotine. Animals may become intoxicated by grazing the plant or ingesting cut stalks.[38] In adult cattle clinical signs of acute toxicity may include excitation, followed by depression, respiratory failure, and death in severe cases. Treatment of affected animals is symptomatic. The principal toxins responsible for tobacco toxicosis in cattle are nicotine and anabasine. *Nicotiana glauca* produces the piperidine alkaloid anabasine, a potent nicotinic acetylcholine receptor agonist. Anabasine is acutely toxic as well as teratogenic.[38] Ruminants are relatively more tolerant of the neurotoxic alkaloid effects of nicotine. However, the teratogenic anabasine can induce fetal skeletal malformations including arthrogryposis, palatoschisis, torticollis, lordosis, and kyphosis. Fetal defects resulting from tobacco intoxication occur when cattle ingest the plant between days 30 and 60 of gestation.

The fetal malformations caused by tobacco toxicosis are indistinguishable from those induced by poison hemlock, locoweeds, and lupine poisoning. This similarity of teratogenic fetal malformations is attributed to the analogous molecular structures of anabasine and coniine (principal toxin of *C. maculatum* and lupines) and its analogs.[4]

At low doses anabasine stimulates depolarization of postsynaptic membranes of ganglia and at high does causes neuromuscular blockade. Like poison hemlock and lupine toxicosis, the mechanism of teratogenicity is believed to be caused by the neuromuscular blockade of nerves and subsequent decrease in intrauterine fetal movement resulting in the development of joint contractures, palatoschisis formation, and other skeletal malformations.[39,40] In acute toxicosis, cattle should be removed from contaminated fields or pastures. In severe cases supportive and symptomatic treatment is recommended for affected animals and may include gastric lavage.[28]

Veratrum californicum (false hellebore)

Veratrum californicum grows in dense sharply defined stands in high moist meadows or hills.[29] Individual plants grow approximately 2–3 m tall and have short rootstalks and alternate smooth, broad, parallel-veined oval leaves that are present in three ranks. Numerous toxic alkaloids have been identified in V. californicum; however, the jervine alkaloids including cyclopamine, jervine, and cycloposine are associated with the plant's teratgenic effects.[4,39] Among the toxic alkaloids, cyclopamine is present at the highest levels and is the principal teratogenic alkaloid.[29]

Teratogenic effects are predominantly seen in sheep but fetal malformations have also been reported in goats and cattle.[33] The teratogenic alkaloids exert their effects by interfering with the sonic hedgehog (SHH) signal transduction pathway, by inhibition of neuroepithelial cell mitosis and migration, and by decreasing chondrocyte proliferation. In pregnant ewes, ingestion of V. californicum on days 12–14 of gestation (most notably day 14) results in cyclopia and prolonged gestation. Embryo death can result from maternal exposure on days 19–21. Cleft palate results from maternal exposure on gestational days 24–30 and metacarpal and metatarsal defects occur when ewes are exposed on days 28–31 of gestation.[39]

Ingestion of V. californicum in pregnant cows between days 12 and 30 of gestation may cause fetal malformations, including cleft palate, harelip, brachygnathia, hypermobility of the hocks, syndactyly, and decreased numbers of coccygeal vertebrae. When pregnant cows ingest the plant in later stages of pregnancy (days 30–36), teratogenic alkaloids inhibit growth in length of metacarpal and metatarsal bones.[33]

There is no specific treatment for V. californicum-induced fetal malformations. However, malformations can be prevented by removing pregnant animals from V. californicum-infested pastures and ranges during the first trimester of pregnancy.[4]

Mimosa spp.

Mimosa tenuiflora and Mimosa ophthalmocentra are perennial trees or small shrubs native to Brazil. Mimosa tenuiflora and possibly M. ophthalmocentra (Fabaceae, Mimosoideae) can induce fetal malformations and embryonic death in goats, sheep, and less frequently cattle. Affected calves, lambs, and kids may be born with an array of malformations, which can include arthrogryposis, micrognathia, cheiloschisis (harelip), palatoschisis, kyphosis, lordosis, torticollis, blindness, microphthalmia, corneal opacity, ocular dermoids, acephaly, bicephaly, hydranencephaly, glossal hypoplasia, meningocele, and syringocele. The toxic principle of M. tenuiflora is unknown, but tryptamine-derived alkaloids have been isolated from the leaves and seeds.[17]

Mycotoxins

Mycotoxins are naturally occurring compounds or secondary metabolites produced by fungi that cause deleterious effects (mycotoxicoses) when ingested or exposed to susceptible animal species. Mycotoxisosis in livestock occurs when animals graze on infected grasses or crops or are fed infected grains. Chemical analysis via thin-layer chromatography, gas and high performance liquid chromatography, and mass spectrometry can accurately detect the concentrations of various mycotoxins in properly collected and prepared samples.[41,42]

Ergot alkaloids

Ergot alkaloids include more than 80 indole compounds and are associated with two mycotoxicoses in livestock species: fescue toxicosis and ergotism.[2] Both conditions may share similar clinical manifestations and in many cases may be indistinguishable from one another.[43] Ergot alkaloids are produced by the mold Claviceps purpurea on wheat, rye, and other cereal grains and by the endophyte Neotyphodium coenophialum growing on tall fescue (Lolium arundinaceum [Shreb]). Ergot alkaloids have similar molecular structures to the biogenic amines norepinephrine, serotonin, and dopamine and act as agonists at various serotonin receptors.[44,45] This agonistic effect is thought to be responsible for clinical manifestations of ergot alkaloid toxicosis. In both fescue toxicosis and ergotism the principal ergot alkaloid believed to be most responsible for clinical disease is ergovaline. Ergovaline is a potent α_2-adrenergic agonist on arterioles and other blood vessels causing vasoconstriction.[42]

Tall fescue

Fescue toxicosis is one of the most costly grass-related intoxications of livestock in the United States and throughout many other countries. Fescue toxicosis is a descriptive term used for several clinical syndromes (summer slump, fescue foot, and fat necrosis) attributed to ingestion of endophyte-infected tall fescue.[42]

Tall fescue (L. arundinaceum [Shreb]) is a cool-season perennial grass grown in temperate climates and is one of the most abundant forage crops in the continental United States.[46] In cases of fescue toxicosis, livestock ingest the fungal endophyte Neotyphodium coenophialum (formerly Acremonium coenophialum and Epichloe typhina) that grows within the intercellular spaces of the leaf sheaths, stems, and seeds of tall fescue grass.[2] Tall fescue and its endophyte N. coenophialum share a mutually beneficial relationship. The endophyte produces toxins (ergot alkaloids, loline alkaloids, and peramine) that are distributed throughout the plant making it more resistant to drought, parasitism, and fungal infection.[42] The endophyte of tall fescue produces an

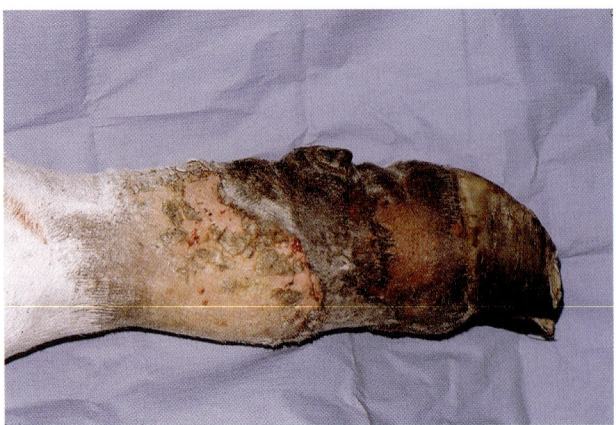

Figure 65.7 Fescue foot: dry gangrenous dermatitis secondary to "fescue toxicosis." Photograph courtesy of Dr Jim Cooley.

array of ergot alkaloids but ergovaline is the primary ergopeptine alkaloid believed to be responsible for inducing toxicosis. Ergot alkaloids are highly concentrated within the seeds of tall fescue and levels are greatest in the summer and early fall when seed heads are present.[46] Drought and rainy conditions as well as fertilization with nitrogen- and phosphorus-based fertilizers tend to increase ergovaline concentrations.[42]

Several syndromes including summer slump, fescue foot (Figure 65.7), and fat necrosis (lipomatosis) are associated with fescue toxicosis in cattle. The most costly of the three is summer slump. As the name implies clinical manifestations of this condition are most often appreciated in summer months when the ambient temperature is above 31 °C.[42] Affected cattle typically have a rough hair coat, unthrifty appearance, reduced feed intake, decreased milk production with lower weaning weights of calves, reduced conception rates, agitation/nervousness, ptyalism, and increased respiratory rates.[42,47]

Mechanism of action

Ergot alkaloids are dopamine D_2 receptor agonists and mimic the endogenous tonic inhibition of pituitary lactotropes by dopamine, thereby inhibiting prolactin secretion from the anterior pituitary gland.[42] Decreased serum prolactin is a relatively consistent finding in livestock feeding studies involving ergot alkaloid-infected forage. Therefore serum prolactin concentrations may be utilized as a diagnostic indicator of endophyte toxicosis.[48]

Reproductive effects

The hormone prolactin plays an important role in mammary gland growth, milk production, corpus luteal function, gonadotropin secretion, lipogenesis, copper homeostasis, and immune modulation. Decreased prolactin production caused by the D_2 dopaminergic effects of ergovaline and other ergot alkaloids can cause agalactia at parturition, most notably in horses. Cattle and other ruminants produce a placental lactogen that can overcome the initial lack of prolactin stimulation at the time of parturition. However, lower concentrations of prolactin coupled with decreased feed intake can cause decreased milk production in these species.[42] Prolactin also plays an important role in gonadotropin secretion and corpus luteal function and development. Decreased concentrations of luteinizing hormone and prostaglandins have been documented in animals exposed to ergot alkaloids.[48] Ingestion of endophyte-positive fescue and ergopeptide exposure have been associated with low progesterone production in cattle and horses. Decreased progesterone, as well as other imbalances in reproductive hormones, can result in difficulty maintaining early pregnancy in cattle and in late pregnancy problems in mares.[42]

In addition to their dopaminergic effects, ergot alkaloids also act as potent serotonergic agonists and α_2-adrenergic agonists on blood vessels, particularly arterioles. Peripheral vasoconstriction of vessels is believed to contribute to the hyperthermia exhibited in some cases of summer slump in the warm summer months along with dysregulation of the hypothalamic thermoregulatory center due to decreased prolactin secretion. In winter months, toxicosis may result in fescue foot, a condition caused by peripheral vasoconstriction of vessels of the distal limbs secondary to the combined effects of ergovaline and cold environmental temperatures. The tail switch and ear tips as well as the distal limbs (most often the hindlimbs between the coronary band and fetlock) may also be affected.[42]

Management and treatment

Cattle experiencing clinical signs of summer slump should be removed from endophyte-contaminated pasture or their diet should be diluted by feeding supplemental grain or nonfescue forages.[42] Cattle with gangrenous dermatitis (fescue foot) can be treated with antibiotics to prevent secondary infections. Shelter or wind breaks may also be beneficial in minimizing the effects of cold weather. Domperidone, a D_2 dopamine antagonist, is used to treat fescue toxicosis in mares with agalactia and prolonged gestation. Experimentally, domperidone has been shown to ameliorate some clinical signs of fescue toxicosis in heifers, including reversal of decreased weight gain and increased circulating concentrations of progesterone. These results suggest that treatment of cattle with fescue toxicosis may have clinical and economic merit.[49]

Ergotism

Ergot is a general term referring to a group of parasitic fungi belonging to the genus *Claviceps*, most notably *Claviceps purpurea*, that infect rye, barley, wheat, oats, and other cereal grains.[44] Infection of grains with *C. purpurea* results in colonization of the grain ovary wall or base with fungal mycelia. The mycelia produce a sticky exudate called honeydew containing conidia (asexually produced spores). Following germination within the hardened honeydew (sclerotia), conidia can be transferred to and infect other seeds by insect vectors or direct contact.[44,45] Signs of ergotism – the various toxic effect(s) of ergot alkaloids produced by *Claviceps* spp. – develop secondary to the ingestion of toxic levels of ergot-infected grains.[44]

Clinical signs of ergotism are primarily caused by the ergopeptine class of ergot alkaloids, which include ergotamine, ergocristine, ergosine, ergocornine, ergocryptine, and ergovaline. Ergot alkaloids act as serotonin receptor agonists due to the similarities of their molecular structures with the biogenic amines dopamine, serotonin, and histamine.[44,45] Vasoconstriction results from the agonistic activity of ergot alkaloids. Other potential effects include uterine stimulation and dopaminergic activity at D_2 receptors resulting in decreased prolactin secretion.[45] Documented cases of ergotism in the United States are uncommon because milling and cleaning processes are successful in removing ergot-infected grains.[41]

Ergot alkaloid toxicity is known to cause four distinct disease forms in livestock: (i) cutaneous and gangrenous lesions on the tail and extremities, (ii) hyperthermia and production loss, (iii) reproductive failure, and (iv) the less common and frequently debated neurologic form.[2,45] The gangrenous form of ergot poisoning represents the chronic form of intoxication by ergot-producing fungi. Clinical signs of gangrenous ergotism include acute lameness with swelling and hyperemia. Lesions present most often in the rear limbs, although all four limbs may be affected in some cases; however, necrosis is usually restricted to below the fetlock but occasionally extend to the metatarsals.[20] Vasoconstriction of small arteries and arterioles due to the potent α_2-adrenergic agonistic effects of ergovaline causes ischemia and necrosis of the distal limbs, ear tips, and the distal third of the tail. This gangrenous form is most often associated with winter months due to exacerbation of the vasoconstrictive effects by cold weather.[45] The hyperthermic form of the disease is the most commonly reported form of ergotism and is seen most often in summer months when weather is warm.[44] The neurologic form of ergotism appears to be rare and may be associated with acute intoxication with large doses of ergot alkaloids.[45]

Management and treatment

When ergotism is suspected, infected grains should be removed. No effective treatment is available for ergotism in cattle, although animals with the hyperthermic form should recover given time.[45]

Aflatoxins

Aflatoxins (B_1, B_2, G_1, and G_2, named for their respective blue or green fluorescence under ultraviolet light) are a group of structurally related difuranocoumarin compounds produced primarily by *Aspergillus flavus* and *Apergillus parasiticus*. Aflatoxins can be found throughout the world and can commonly contaminate animal feed and crops in warm subtropical and tropical climates under appropriate conditions.[40,41,43] Grains stored under high moisture (>14%), at warm temperatures (>20 °C), or those improperly dried are at risk of becoming contaminated.[50] *Aspergillus flavus* strains invade damaged plant tissue and predominantly produce aflatoxin B_1 which is considered to be the most toxic and carcinogenic aflatoxin.[43]

All animals are susceptible to the toxic effects of aflatoxins, although differences in species sensitivities do exist. Susceptibility can vary with breed, age, dose, nutritional status, and length of exposure.[50] Cattle and other ruminants are considered to be relatively resistant in comparison with monogastric species. However, young calves are at greater risk of developing clinical toxicosis compared with adult animals.[43]

Aflatoxins are primarily hepatotoxic but can also have immunosuppressive, carcinogenic, mutagenic, and teratogenic effects.[43] Some aflatoxins have been shown to cross the placenta in laboratory animals and humans. Decreased male fertility via decreased spermatozoa production, embryo loss, fetal death, and possible teratogenic effects have been reported with aflatoxicosis, most notably with aflatoxin B_1 and G_1 in animals.[51] Abortions in cattle are uncommonly reported with aflatoxicosis, but sporadic herd outbreaks have been attributed to ingestion of feeds contaminated with aflatoxin B_1, B_2, G_1, and G_2.[43] However, the negative reproductive impacts do not appear to be due to any direct activity of aflatoxins in the oocyte or embryo but instead are attributed to altered maternal homeostasis.[41]

Treatment

There is no specific treatment for aflatoxicosis. Instead treatment of affected cattle involves supportive and symptomatic care. Contaminated feed should be removed and replaced with aflatoxin-free feeds.[51]

Zearalenone

Zearalenone is a potent estrogenic mycotoxin produced by the fungus *Fusarium graminearum* and multiple other species of *Fusarium*. Zearalenone contaminates cereal grains including corn and to a lesser extent wheat, barley, sorghum, millet, rice, and oats under specific environmental and storage conditions.[39] Zearalenone can have profound effects on reproductive function due to its estrogenic actions. Clinical signs attributed to zearalenone toxicity are most often reported in swine, although signs may also be observed in ruminant species as well. Clinical signs of zearalenone toxicity mimic those of estrogenic stimulation and can include decreased fertility, abnormal estrous cycles, vulvar swelling, vaginitis, reduced milk production, and mammary gland enlargement. Heifers appear to be more susceptible to the effects of zearalenone than adult cows.[41,52] However, in pregnant cows the potential for abnormal embryonic or fetal development with possible fetal death, abortion, and dystocia exists.[2]

Zearalenone and its derivatives (α- and β-zearalenol) are believed to exert their toxic effects by competitively binding to specific estrogen receptors and by modification of steroid metabolites. Zearalenone and α- and β-zearalenol have chemical structures closely resembling naturally occurring estrogen.[53] Zearalenone and its metabolites can directly bind with cytoplasmic estradiol-17β receptors and translocate receptor sites to the nucleus where stimulation of RNA leads to protein synthesis and clinical signs of estrogenism.[52] β-Zearalenol appears to be the predominant zearalenone metabolite in cattle as opposed to α-zearalenol which predominates in most other species including swine.[2]

Treatment

Treatment of zearalenone intoxication relies on removal of contaminated feedstuff and replacement with high-quality clean feeds. Clinical signs will typically resolve within 3–7 weeks. Prevention of *Fusarium* spp. growth and subsequent zearalenone production can be achieved by maintaining low moisture concentrations (15–16%).[52]

Phytoestrogens

Phytoestrogens are biologically active nonsteroidal plant compounds that are similar in structure or function to mammalian estrogen and can compete for estrogen receptors on cells.[4,54] Plant estrogens can have broad effects on both human and animal populations. Phytoestrogens are found in a number of legumes including alfalfa (*Medicago sativa*), soybeans (*Glycine max* L.), red clover (*Trifolium pratense*), subterranean clover (*Trifolium subterraneum*), white clover (*Trifolium repens*), aslike clover (*Trifolium hybridium*), and burr medic (*Medicago* spp.) among others.[4,39,55]

Concentrations of phytoestrogens in plant material can vary significantly. Estrogenic compounds are often produced in response to environmental or physiologic plant stressors such as fungal infection, insect infestation, and animal predation. Phytoestrogen concentrations have also been shown to increase in cool wet conditions.[54]

Over 100 phytoestrogen molecules have been identified. Phytoestrogens are subdivided into broad categories based on their chemical structure. Categories of phytoestrogens include isoflavones, flavones, coumestans, stilbenes, and lignans.[54,56] Isoflavone compounds can be found in high concentrations in soybean, red clover, and white clover. Coumestrol (potent estrogenic phytoestrogen) can be found in alfalfa as well as white clover and soybean sprouts.[54] Stilbenes are found in cocoa- and grape-containing products and lignans are prevalent in flax seed.[56,57]

Once ingested, phytoestrogens are absorbed in the rumen by resident microbial activity. Because of structural similarity with endogenous hormones, phytoestrogens can bind to nuclear receptors.[56] Although phytoestrogens may mimic estradiol-17β (E2), their effects may not always be similar.[58] The affinity phytoestrogens have with ERα receptors and ERβ receptors is relatively weak in comparison with endogenous E2.[56] However, phytoestrogens can be present in much higher concentrations in the body (as much as 100-fold higher) than endogenous estrogens, allowing for potentially profound effects. Phytoestrogens may exert agonistic, partial agonistic, or antagonistic effects depending on the concentration of E2.[56,57]

The toxic effects of phytoestrogens are dependent on dose, route of exposure, duration of exposure, and timing of exposure.[57] Organ systems in young animals appear to be more sensitive to phytoestrogens than adults. Cattle infertility has been reported in herds fed red clover silage containing isoflavones and alfalfa-containing coumestans.[58] However, cattle appear to be less sensitive than sheep to the phytoestrogenic effects of clover forages.[54] The effects of high concentrations of phytoestrogens may include clinical signs of hyperestrogenemia (nymphomania, cystic ovaries, swollen genitalia, infertility, impaired pregnancy maintenance) or antiestrogenic effects such as gonadal hypoplasia and signs of anestrus depending on the action of the phytoestrogen on estrogenic receptors.

Uterine endometrial release of prostaglandin (PG)$F_{2\alpha}$ is regulated by oxytocin, progesterone, and estradiol (E2) in ruminants and induces luteolysis and corpus luteum regression. In a study conducted by Mlynarczuk *et al.* the phytoestrogens coumestrol, daidzein, and genistein stimulated genes responsible for synthesis of oxytocin precursors (neurophysin-I/oxytocin), and post-translational synthesis of oxytocin, PGA (peptidyl-glycine-α-amidating monooxygenase) in granulosa cells and luteal cells in cattle. These phytoestrogens stimulated oxytocin secretion from the ovarian follicle and corpus luteum. It is speculated that phytotoxin-induced oxytocin release during early pregnancy or implantation may result in early luteolysis, formation of persistent corpus luteum, and subsequent impaired pregnancy maintenance and early embryonic death.[54,55]

Management and treatment

In order to prevent or minimize potential reproductive problems, forages can be analyzed for isoflavones and coumestrol concentrations and feeds containing low or no levels can be substituted for high phytoestrogen-containing forages. Once phytoestrogen-affected forages have been eliminated from cattle rations, animals will typically resume normal reproductive cycling within a period of 4–6 weeks.[54]

Trace minerals and heavy metals

Heavy metals and trace minerals are common environmental contaminants and when present in sufficient levels can cause significant health and reproductive impacts in domestic livestock. Heavy metals include elements which under appropriate conditions play a necessary role for normal body function, such as iron, copper, zinc, and chromium, as well as toxic elements that have no biological functions. Many minerals are crucial for animal health, survival, and production due to the essential roles they play in the physiological, structural, and regulatory functions of the body. However, ingestion of excessive amounts of minerals may lead to acute poisoning or chronic poisoning, when animals ingest lower concentrations over prolonged periods of time.[59]

Potentially toxic heavy metals include lead, mercury, cadmium, chromium, arsenic, and nickel. Environmental pollution with trace elements, toxic metals, and metalloids is often attributed to mining, industrial wastes, urbanization, and intensification of farming practices (pesticide and herbicide activity). Transfer of heavy metals from contaminated soils into plants and successive levels of the food chain can be a significant health risk in both animals and humans. Clinical manifestations of trace mineral and heavy metal intoxication can vary depending on the element involved, health of the animal, and degree of exposure. Several heavy metals have been identified as potentially teratogenic and abortifacient in animals.[60] However, in cattle the most commonly reported heavy metal toxicities involve lead and arsenic.

Arsenic

Arsenic (As) is a ubiquitous metalloid element present in multiple forms, the most common oxidative states being the As^{3+} and As^{5+} valence forms.[61] Arsenic can be found in both organic and inorganic forms although inorganic arsenite is the most toxic form of the element.

Arsenic is predominantly used in the manufacture of pesticides, wood preservatives, glass, alloys, semiconductors, and pharmaceuticals. Environmental contamination of arsenic has been attributed to smelting of lead, copper, and nickel, combustion of fossil fuels, manufacturing of arsenic-containing compounds, and overuse or improper storage of arsenic-containing herbicides and insecticides.[62]

Peracute exposure to large doses of inorganic arsenic is often fatal and may result in death within minutes to hours depending on dose. Clinical signs of acute toxicosis may be profound and include colic, bloody diarrhea, neurologic signs such as ataxia and incoordination, weakness, and collapse followed by death. Gross postmortem lesions in cattle may include abomasal hyperemia, increased intraluminal intestinal fluid, edema of the rumen, omasum, reticulum, and abomasum, rumenal mucosal necrosis, and gastrointestinal hemorrhage.[61]

Mechanism of action

Absorption of arsenic compounds depends on the particle size and its aqueous solubility. Soluble forms of arsenic are readily absorbed via both passive and active means.[61,62] In pregnant animals inorganic arsenic crosses the placenta and accumulates in the neuroepithelium of the developing fetus. Organic arsenic accumulates in the placenta but does not appear to cross the membrane. Laboratory animal studies have demonstrated the embryotoxic effects of high doses of inorganic arsenic, with development of teratogenic defects such as cleft palate, encephalocele, microphthalmia in hamsters and failure of limb bud development, prosencephalon hypoplasia, and somite abnormalities in rats and mice exposed during early gestation. Arsenic at lower concentrations can also have significant placental and fetal toxic effects.[40,62] However, reports of abortion in cattle with subclinical or chronic arsenic toxicosis are sporadic.

Diagnosis, treatment, and prevention

A presumptive diagnosis of arsenic toxicosis is often based on an appropriate clinical history of exposure and suspicious clinical signs. Necropsy lesions can also be suggestive. However, definitive diagnosis requires demonstration of elevated liver and kidney arsenic concentrations. Arsenic levels of approximately 8–10 ppm in fresh liver and kidney and 2–4 ppm in liver and kidney from autolyzed carcasses (several days old) are diagnostic.

Early recognition and medical intervention is imperative for successful treatment of arsenic intoxication cases. In cattle not showing clinical signs, gastric lavage with a saline purgative and 20–30 g sodium thiosulfate administered orally or intravenously in a 10–20% solution may be effective. Intramuscular injection of dimercaprol 1.5–5 mg/kg i.m. two to four times daily for 10 consecutive days should also be administered if over 4 hours have elapsed since time of exposure. Additional treatment with sodium thiosulfate 30–40 mg/kg i.v. two to four times daily until recovery (typically 2–4 days) is also recommended.[61]

Cadmium

Cadmium (Cd) is a divalent transition metal with chemical properties similar to those of zinc.[63] Unlike many minerals, cadmium has no essential physiologic or biochemical function in the body and is highly toxic to animals and humans.[59] Cadmium compounds are widely utilized in industrial applications. They are used in iron and steel plating and can be found in batteries, semiconductors, solders, plastic stabilizers, and solar cells. Environmental contamination of water and soil is often secondary to industrial pollution. Cadmium may enter the environment from steel and iron production, zinc refining and smelting, mine wastes, coal combustion, improper battery disposal, and the use of sewage sludge and rock phosphate as fertilizers.[63] Batteries are considered to be one of the most common sources of cadmium intoxications in cattle in the United States. However, phosphate fertilizers and sewage sludge are also important sources of cadmium toxicosis in cattle.[64] Some plants readily accumulate cadmium from contaminated soil making it available for consumption by grazing animals.[59,63]

Cadmium, once ingested, is primarily absorbed through the respiratory system and digestive tract and accumulates mainly within the liver and kidneys. Gastrointestinal absorption of cadmium is much less than other divalent cations (zinc and iron). However, intestinal absorption can be increased in animals with iron deficiency. Following absorption, the metal circulates either in red blood cells or bound to albumin in the plasma. Cadmium interacts with the metabolism of essential minerals such as calcium, zinc, iron, copper, and selenium. Cadmium ions can displace divalent metals such as zinc from their binding sites on metallothionein (major metal-binding protein).[63,64] Metallothionein functions in cadmium detoxification through high-affinity binding with the mineral that results in a toxicologically inert compound.[59,64] Cadmium along with other heavy metals (Cu, Zn, and toxic mercury) induces metallothionein synthesis in multiple organs including the liver and kidneys. However, because metallothionein's action is limited, excessive amounts of cadmium ingested acutely or over time can exceed metallothionein's ability to bind it and significant multiorgan damage can develop.[59] In the liver, cadmium toxicity results in hepatocellular apoptosis and necrosis.[63] High levels of dietary cadmium can cause decreased feed intake, poor weight gain, anemia, decreased bone absorption, and abortions.[64] There is no specific treatment for cadmium poisoning in livestock and therefore exposure should be minimized or prevented in the environment and in feeds.[63]

Lead

Lead poisoning in cattle often results from environmental contamination. Cattle may become acutely intoxicated from the ingestion of large quantities of lead-containing materials, licking old storage batteries, licking or ingesting lead-containing paint chips on outbuildings or from discarded paint cans, eating lead-containing lubricants, and drinking water from lead-containing water pipes.[6] Chronic exposure most often occurs from ingesting forages grown on polluted pastures in the vicinity of mines or industrial sites.

Only about 2–10% of ingested lead is absorbed by the digestive tract and distributed throughout the body.[6] Absorbed lead is slowly excreted in bile, milk, and urine and is deposited in tissues, predominantly in the kidney, liver, brain, and bone.

Lead induces toxic effects via multiple mechanisms including combining with calcium- and zinc-binding proteins, random hydrolysis of nucleic acids, and by inducing RNA catalysis via activation of ribosomal 5S RNA.[65] Lead interferes with protein/hemoprotein biosynthesis, inhibits mitochondrial and membrane enzymes, and causes deficits in cholinergic, dopaminergic, and glutamatergic functions.[40] Lead disrupts calcium homeostasis, causing the accumulation of calcium in lead-exposed cells and mitochondrial release of calcium inducing apoptotic cell death.

In adult cattle acute toxicity typically causes death within 12–24 hours. When poisoning is less acute animals may survive 4–5 days. Clinical signs may include cortical blindness, delayed menace response, delayed withdrawal reflex, decreased glossal tone, ptyalism, frothing at the mouth, champing of the jaws, respiratory difficulty, tachycardia, ataxia, head pressing, convulsions, circling, and recumbency. Signs of subacute poisoning may include depression, anorexia, bruxism, colic, diarrhea, and rumenal atony, and central nervous system (CNS) signs and abortion.[40,65]

Lead readily crosses the placenta and can pass through the immature blood–brain barrier of developing fetuses where it concentrates in the CNS.[6,40,65] Lead can also accumulate in the placenta during times of fetal stress. The toxicity of lead is primarily attributed to its ability to mimic and substitute calcium. Effects of lead on the developing fetus have been demonstrated in laboratory animals and may include growth retardation, reduced fertility, neural tube defects, brain defects, urogenital defects, and tail defects.[39,65]

Diagnosis

Antemortem diagnosis of acute lead poisoning should be based on a history of exposure, clinical signs, and blood lead concentrations greater than 0.35 μg/mL. Hematologic abnormalities may also be indicative of lead toxicity and may include anemia, anisocytosis, poikilocytosis, polychromasia, basophilic stippling, metarubricytes, and hypochromia. Postmortem gross and histologic lesions in cattle are nonspecific and therefore definitive diagnosis relies on the clinical history and liver and kidney lead concentrations.[2] Acid-fast positive lead inclusion bodies within nuclei of proximal renal tubular epithelial cells may be observed in some cases, although the presence of these inclusions is variable. Brain lesions are usually absent but mild to moderate brain swelling may be appreciated along with vascular congestion. In select cases of subacute lead poisoning laminar cortical necrosis may be observed.[66]

Treatment

In cattle, chelation with calcium disodium edetate (Ca-EDTA) is the preferred option. It may be administered intravenously or subcutaneously at a dose of 110 mg/kg daily divided into two doses for 2–3 days and then repeated 2 days later. Thiamine may also be utilized as an adjunct to Ca-EDTA treatment in ruminants; in adult cattle a dose of 250–2500 mg/day is recommended and for calves 2 mg/kg daily. Cathartics such as magnesium sulfate (400 mg/kg p.o.) may be administered or a rumenotomy may be performed to remove larger ingested lead sources but is rarely successful in cases of particulate lead ingestion. Convulsions may be controlled by barbiturates or tranquilizers. Withdrawal times in cattle may be greater than 1 year and can be estimated by periodically monitoring blood lead concentrations.[28]

Selenium toxicosis

Selenium poisoning is most commonly attributed to (i) ingestion of selenium-concentrating plants grown in soils high in selenium content; (ii) accidental overdoses of selenium from ingestion of improperly mixed rations or injections; and (iii) from environmental contamination due to either plant accumulation or water contamination.[67] Selenium is a nonmetallic element that is an important antioxidant but can be toxic in excessive amounts. Naturally occurring selenium toxicosis is most often a regional problem in the western United States between the Rocky Mountains and the Mississippi river. Grains and forages grown in high selenium regions may passively accumulate toxic levels of selenium causing chronic toxicosis in herbivores.[68] The toxic dose of selenium in ruminants is unclear. References range from 2.2 to greater than 20 mg/kg body weight across species.[67] Table 65.2 lists common obligate selenium-accumulating plants and potential secondary selenium-accumulating plants.

In adult cattle clinical signs of acute selenium toxicosis may include respiratory distress, lethargy, restlessness, anorexia, watery diarrhea, pyrexia, teeth grinding, ptyalism, stilted gait, tetanic spasms, and death. Signs have been reported as early as 8–10 hours following exposure or as late as 36 hours. In cases of chronic selenium toxicosis ("alkali disease") resulting from long-term ingestion of seleniferous forages, clinical signs including depression, weakness, anorexia, weight loss, emaciation, diarrhea, anemia, hair loss, hoof deformities, and lameness may be observed.[67,68] Reproductive abnormalities secondary to clinical and subclinical selenium toxicosis have been reported in multiple domestic species including cattle. However, a clear association between selenium toxicosis and abortion is not well documented. Reproductive abnormalities including decreased conception rates and increased fetal resorption rates were documented in cattle, sheep, and horses fed diets containing selenium 20–50 mg/kg.[67]

Selenium is efficiently transferred across the placenta into fetal tissues, most notably the liver, during gestation. Excessive selenium concentration in the fetus may be teratogenic and abortifacient. Selenium levels in the fetal liver have been reported to reach as high as eight times that of the dam. Normal fetal hepatic selenium concentrations are

Table 65.2 Selenium-accumulating plants of North America.

Obligate selenium-accumulating plants

Milk vetches (*Astragalus* spp.)
Golden weeds (*Conopsis* spp.)
Woody aster (*Xylorhiza* spp.)
Princes plume (*Stanleya pinnata*)

Secondary selenium-accumulating plants

Acacia (*Acacia* spp.)
Sages (*Artemisia* spp.)
Asters (*Aster* spp.)
Saltbrush (*Atriplex* spp.)
Paintbrush (*Castilleja* spp.)
Bear tongue (*Penstemon* spp.)
Gumweed (*Grindelia* spp.)

Source: adapted from Knight A, Walter R. *A Guide to Plant Poisoning of Animals in North America*. Jackson, WY: Teton New Media, 2001.

reported to be 0.3–1.2 ppm. In a study conducted by Yaeger et al.[68] high fetal liver levels (3.47–5.25 ppm) were present in clinically healthy calves so concentrations above 3.0 ppm are not necessarily indicative of toxicosis. Subclinical selenium may influence the immune system and pregnancy outcome.[68]

Management and treatment

Avoidance of excessive selenium exposure or ingestion is recommended for prevention of clinical disease. Once diagnosed, sources of excess selenium (forages, rations) should be removed. Effective selenium chelators are not available and therefore medical treatment relies largely on supportive care, which can be lengthy especially in cases of hoof lesions.[67]

Vitamin A

Vitamin A is essential for vision, growth differentiation and proliferation of epithelial tissues, bone growth, reproduction, and fetal and embryonic development. Deficiencies in pregnant animals can result in abnormal development including disrupted organogenesis, failure of embryo growth, and resorption. Conversely, excessive vitamin A, through oversupplementation (or injection), has been linked to reproductive failure in cattle. Though uncommon, hypervitaminosis A has been associated with fetal malformations, including neurologic and craniofacial defects with preimplantation exposure.[69] Teratogenicity appears to be increased by protein or protein-energy malnutrition.[33] Hypervitaminosis A may disrupt the retinoic acid receptor, impairing normal homeobox gene expression. Alterations in specific homeobox genes have been associated with retinoic acid-induced teratogenesis.[70] In laboratory studies using the rat as the animal model, anomalies were stage and dose dependent. When exposure occurred early following implantation, limb and genitourinary abnormalities developed.[69]

Endocrine-disrupting compounds

Endocrine-disrupting compounds are a diverse group of compounds, natural and synthetic, that can potentially modulate or disrupt hormone homeostasis.[71,72] Common synthetic endocrine disrupters include pesticides (DDT and organophosphorus pesticides), herbicides, fungicides, plasticizers, polystyrenes, polybrominated biphenyls and polychlorinated biphenyls (PBBs and PCBs), polychlorinated dibenzodioxins, and alkylphenolic compounds.[71,73] These compounds are typically present in the environment at low concentrations but are persistent and have the potential to exert adverse effects on humans and animals. Exposure to endocrine disrupters may in some cases contribute to, or directly cause, infertility, pregnancy loss, intrauterine fetal death, and fetal teratogenic effects. The effects of endocrine-disrupting compounds can have impacts on developing fetuses and postnatal offspring by affecting different biochemical pathways.[71] Synthetic endocrine disrupters are suspected to have potential reproductive effects in a number of animal species but clinical studies are limited.

Polycyclic aromatic hydrocarbons

Polycyclic aromatic hydrocarbons (PAHs) are a family of organic compounds with two or more fused aromatic rings. PAHs are formed from incomplete combustion of organic materials and are produced from industrial and automobile emissions, forest fires and wildfires, coal burning, volcanic eruptions, and hazardous waste sites.[74,75] Livestock are exposed to PAHs through contaminated water, air, soil, and forages. PAH compounds are ubiquitous in the environment and are of significant concern due to their carcinogenic and other detrimental health effects demonstrated in laboratory animals and their suspected health effects in humans and domestic animals. The reproductive and developmental impacts of PAHs in domestic animals have not been thoroughly studied. PAH compounds have a steric resemblance to steroid molecules. In studies using the rat as a model, the PAH benzo(a)pyrene affected serum concentrations of progesterone, androgens, estrogens, prolactin and, indirectly, luteinizing hormone. PAHs have also been shown to have toxic effects on rodent fetal development, resulting in fetal death, absorption, and teratogenic effects. The effect of these compounds on the reproductive health of cattle has not yet been clearly elucidated.[74]

Polychlorinated biphenyls

PCBs are chlorinated hydrocarbons that have been extensively utilized in industry in hydraulic fluids, transformer oils, plastic material, and lubricants. PCBs were widely synthesized from the 1920s to the 1970s in the United States and are still being produced in many developing countries. The chemical stability of PCBs allows these chemicals to accumulate in the environment and their lipophilic nature permits them to concentrate in fatty tissues.[76] Huge concentrations of PCBs have been found in sewage sludge from industrial and agricultural origins that is utilized as pasture and forage fertilizer. Cattle may become exposed from ingesting contaminated water and food. Although studies in large animals are scarce, PCBs have been shown to be embryotoxic in a number of species. PCBs either alone or in combination can impair normal oocyte maturation.[71] PCBs act as synthetic endocrine disrupters that affect steroid synthesis by increasing the activity of cytochrome P450 via their high affinity for estradiol and/or arylhydrocarbon receptors. Select PCBs have been shown to increase the secretion of oxytocin from the bovine ovary and $PGF_{2\alpha}$ from uterine endometrial cells. Alterations in progesterone and oxytocin induced by PCBs and other endocrine disrupters can impair normal ovulation via either luteal regression or prolonging the lifespan of the corpus luteum, but also potentially impair blastocyst implantation and recognition of pregnancy due to disturbed ratios of $PGF_{2\alpha}$ to PGE_2.[77] In a study using the rabbit as an animal model, direct exposure of the pregnant uterus to PCBs resulted in high embryo PCB concentrations and in some cases induced embryo toxicity and pregnancy loss.[76]

Organochlorine insecticides

Organochlorine insecticides encompass a wide array of compounds that can be divided into three distinct groups including the dichlorodiphenylethanes (e.g., DDT, dicofol, perthane, methocychlor), the hexachlorocyclohexanes (e.g., benzene hexachloride, chlordane), and the chlorinated cyclodienes (e.g., aldrin, endrin, dieldrin).[40] The most infamous organochlorine insecticide is dichlorodiphenyltrichloroethane or DDT.

DDT is a chlorinated hydrocarbon that was widely used in the United States as an insecticide following World War II until it was banned in 1972 due to potential human and animal health hazards and environmental impacts. DDT is highly resistant to degradation and thereby is able to persist in the environment for long periods. DDT and other organochlorines can be absorbed orally or topically. In some developing countries persistent chlorinated hydrocarbons such as DDT, aldrin, and dieldrin (all of which are banned in the United States and Europe) are still used for insect control and can potentially contaminate cattle pastures and forages leading to intoxication. Organochlorine insecticides exert their effects on the peripheral nerves and CNS by slowing Na^+ influx and inhibiting K^+ efflux, which causes partial depolarization of the neuron and thereby decreases the threshold for another action potential resulting in premature depolarization of the neuron. DDT can also induce necrosis of the zona fasciculata and the zona reticularis layers of the adrenal gland, resulting in mineralocorticoid and glucocorticoid deficiency. The aryl hydrocarbons and cyclodienes decrease action potentials and also inhibit postsynaptic binding of γ-aminobutyric acid (GABA).[78]

In laboratory animal studies, organochlorines such as DDT have demonstrated estrogenic-like effects resulting in impaired fertility and pregnancy losses.[40] In a study by Mlynarczuk et al.[77] DDT along with hexachlorocyclohexanes and PCBs appeared to stimulate oxytocin synthesis in the corpus luteum of pregnant cows, decreasing the ratio of progesterone to oxytocin which presumably affected the force of myometrial contractions and increased the risk of abortions.

Clinical signs in cases of acute toxicosis in large animals may include apprehension, hypersensitivity, belligerence, muscle fasciculations of the face, neck and limbs, and convulsions. There is no specific antidote for organochlorine poisoning; instead treatment is symptomatic. In cases of dermal exposure the animal should be washed with soap and water. Activated charcoal may be administered in cases of acute toxicity.[78]

Organophosphate and carbamate insecticides

Livestock intoxication with most insecticides is often accidental and due to topical or oral dosage miscalculations for parasite control and formulation errors. However, intoxication may also result from the improper storage of insecticides that allows cattle access to the chemicals.[79]

Organophosphates and carbamates both inhibit the activity of acetylcholinesterase by binding the enzyme and preventing it from hydrolyzing acetylcholine.[80] Inhibition of acetylcholinesterase results in the accumulation of acetylcholine, which causes overstimulation of muscarinic and nicotinic acetylcholine receptors. Toxicity is typically acute in nature with rapid onset of clinical signs (15 min to 1 hour). Common clinical signs in animals may include ptyalism, lacrimation, respiratory distress, muscle fasciculations and tremors, agitation, ataxia, and death.[80] Anticholinesterase compounds are potentially embryotoxic, fetotoxic, and placentotoxic due to the high concentration of acetylcholinesterase in these placental and fetal tissues. Organophosphates and carbamates readily cross the placenta and act on cholinergic and noncholinergic components of the developing nervous system and other organs.[40] Diagnosis of suspected organophosphate or carbamate poisoning relies on quantifying the levels of acetylcholinesterase inhibition in the blood from live animals and in brain tissue (cerebral cortex preferably) from deceased animals. Inhibition of more than 70% acetylcholinesterase is considered positive for intoxication.[80]

Treatment

In cases of cattle intoxication secondary to topical exposure the animal should be bathed with soap and water to remove any residual insecticide. Medical treatment is restricted to the use of atropine, activated charcoal, and supportive care. Atropine competitively inhibits acetylcholine at muscarinic receptors thereby reducing muscarinic effects. The recommended dose of atropine is 0.25–0.5 mg/kg body weight. One-fourth of the initial dose may be given intravenously and the rest subcutaneously or intramuscularly. The dose may be repeated if clinical signs recur. Dosages of activated charcoal range from 0.45 to 0.9 kg in adult cattle orally.[80]

Exogenous hormones
Therapeutic use

Birth in the cow is triggered by an abrupt rise in fetal plasma cortisol, which induces increased synthesis of estrogens and prostaglandins. The elevation of estrogens and prostaglandins leads to regression of the corpus luteum and the rapid decline of maternal progesterone levels. The prepartum withdrawal of progesterone resulting from luteolysis appears to be essential in the calving process. Induction and manipulation of bovine parturition via the use of exogenous (synthetic) hormones is often instituted as a management tool. Exogenous hormones, including prostaglandins, glucocorticoids, estrogens, and oxytocin, are commonly utilized to induce abortion or parturition according to the stage of pregnancy.[81] Potential reasons for termination of pregnancy or early parturition in cattle may include advancing the calving date in late-conceiving cows, synchronizing the beginning of lactation, terminating prolonged gestations, controlling fetal development in cases of oversized fetuses, treating conditions such as uterine hydrops, or where termination is necessary for the health of the cow.[82]

Accidental or inappropriate use

$PGF_{2\alpha}$ and its analogs (cloprostenol and fluprostenol) are powerful luteolytic agents that induce rapid corpus luteum regression and cessation of its secretory activity, uterine smooth muscle contraction, and cervical relaxation.[83] In cases where induced abortion is desired, $PGF_{2\alpha}$ and its analogs may be administered. Administration of $PGF_{2\alpha}$ 5–7 days following ovulation and up to 150 days of gestation is typically effective for termination of pregnancy. Estrogens administered within 72 hours of ovulation can impair oviductal embryo transit thereby preventing embryo implantation. After this period estrogens induce luteolysis, thereby inducing abortion.

A combination of $PGF_{2\alpha}$ and dexamethasone/flumethasone is commonly used to induce abortion between 150 and 180 days (5–8 months) of gestation but is also effective in all stages of gestation. The combination of these two hormones allows the elimination of both ovarian corpus luteum progesterone and placental progesterone production. Following administration abortions will usually occur within 5 days. During the last month of gestation dexamethasone or prostaglandins may be used alone to induce parturition.[84]

Oxytocin is a nonapeptide hormone produced by neurons in the paraventricular or supraoptic nucleus of the hypothalamus and stored in the pars nervosa of the pituitary gland.[85] During parturition, lactation, and in response to osmotic challenge, oxytocin is released into the systemic circulation from axon terminals in the neurohypophysis and acts directly on the smooth muscle of the uterus to induce rhythmic contractions. Exogenous oxytocin is often used as a uterine contractor to precipitate and accelerate normal parturition and postpartum evacuation of retained placentas

Table 65.3 Potentially harmful drugs in pregnancy.

Acepromazine	Phenothiazines should be avoided near term due to fetal CNS depression
Aminoglycosides (amikacin, gentamicin)	Cross placenta. In rare cases amikacin has been associated with eighth cranial nerve toxicity or nephrotoxicity in the fetus
Atropine	May produce fetal tachycardia
Barbituates (pentobarbital, thiamylal, thiopental)	Cross placenta and induce respiratory depression; pentobarbital possibly increases neonatal mortality
Chloramphenicol	Synthetic bacteriostatic antibiotic that may decrease protein synthesis in the fetus particularly in the bone marrow
Corticosteroids (cortisol, dexamethasone, prednisone, prednisolone)	Associated with fetal and neonatal articular cartilage defects Associated with increased incidence of congenital defects including cleft palate Induces parturition in cattle during late gestation
Diazepam	Associated with congenital defects in humans, rats, and mice in first trimester
Dimethylsulfoxide (DMSO)	Teratogenic in laboratory animals, not advised for use in pregnant animals
Diethylstilbestrol (DES) and estradiol	Contraindicated during pregnancy, cause fetal malformations of genitourinary system, possible fetal bone marrow depression
Griseofulvin	Teratogenic in cats and rats
Lidocaine	Probably safe but may induce fetal bradycardia
Metronidazole	Teratogenic in laboratory animals
Prostaglandins ($PGF_{2\alpha}$, cloprostenol, misoprostol)	Used as abortifacient or parturition inducer in cattle
Quinolones (ciprofloxacin, enrofloxacin)	Associated with fetal cartilage defects
Sulfonamides (trimethoprim sulfa, sulfadimethoxine)	Cross the placenta, and have caused congenital defects in rats and mice but defects not reported in other species. Avoid long-acting sulfonamides due to risk of folate antagonism and bone marrow depression
Testosterone	Masculinization of female fetus
Tetracyclines (doxycycline, oxytetracycline, tetracycline)	Can induce bone and teeth malformations and dental staining in the fetus; may cause toxicity in mother
Warfarin	Causes embryotoxicity and congenital malformations (neural tube defects in laboratory animals)

Source: adapted from Evans T. Reproductive toxicity and endocrine disruption, table 19.2. In: Gupta RC (ed.) *Veterinary Toxicology*, 2nd edn. San Diego, CA: Academic Press, 2012, pp. 278–318.

or uterine debris. It also stimulates contraction of smooth muscle cells of the mammary gland for milk let-down and is used in cases of agalactia.[83]

The responsiveness of uterine musculature to oxytocin varies significantly with the stage of the reproductive cycle. Oxytocin can result in luteolysis in the cycling cow or heifer by inducing the release of $PGF_{2\alpha}$ from endometrial cells. During early phases of pregnancy the uterus is thought to be relatively insensitive to the effects of oxytocin, while in the late phases the sensitivity is markedly increased. However, in a study by Yildiz and Erisir[86] oxytocin administration to pregnant cows 4–7 days following insemination resulted in a higher risk of pregnancy loss and early embryonic mortality.

Pharmaceuticals

The administration of certain pharmaceuticals to pregnant animals can pose a significant risk to the developing embryo and/or fetus. Before any medication is administered to a pregnant animal, the drug label should be closely reviewed to ensure that the product is safe for prenatal use. Table 65.3 lists some contraindicated medications along with their potential fetal effects. However, it should be noted that many of the effects have been documented in laboratory animals.

References

1. McClellan R. Concepts in veterinary toxicology. In: Gupta RC (ed.) *Veterinary Toxicology*, 2nd edn. San Diego, CA: Academic Press, 2012, pp. 8–36.
2. Evans T. Diminished reproductive performance and selected toxicants in forages and grains. *Vet Clin North Am Food Anim Pract* 2011;27:345–371.
3. Panter K, Stegelmeier B. Effects of xenobiotics and phytotoxins on reproduction in food animals. *Vet Clin North Am Food Anim Pract* 2011;27:429–446.
4. Knight A, Walter R. *A Guide to Plant Poisoning of Animals in North America*. Jackson, WY: Teton New Media, 2001.
5. Nicholson S. Nitrate and nitrite accumulating plants. In Gupta RC (ed.) *Veterinary Toxicology*, 2nd edn. San Diego, CA: Academic Press, 2012, pp. 1117–1120.
6. Varga A, Puschner: Retrospective study of cattle poisonings in California: recognition, diagnosis, and treatment. *Veterinary Medicine: Research and Reports* 2012;3:111–127.
7. Aslani M, Vojdani M. Nitrate intoxication due to ingestion of pigweed red-root (*Amaranthus retroflexus*) in cattle. *Iranian J Vet Res* 2007;8:377–380.
8. Cockburn A, Brambilla G, Fernandez M *et al.* Nitrite in feed: from animal health to human health. *Toxicol Appl Pharmacol* 2013;270:209–217.
9. Nicholson S. Southeastern plants toxic to ruminants. *Vet Clin North Am Food Anim Pract* 2011;27:447–458.
10. Cook D, Gardner D, Pfister J *et al.* Differences in ponderosa pine isocupressic acid concentrations across space and time. *Society of Range Management* 2010;32:14–17.
11. Panter K, Gardner D, Lee S *et al.* Poisonous plants of the United States. In: Gupta RC (ed.) *Veterinary Toxicology*, 2nd edn. San Diego, CA: Academic Press, 2012, pp. 1031–1079.
12. Gardner D, Panter K, James L. Pine needle abortion in cattle: metabolism of isocupressic acid. *J Agric Food Chem* 1999;47:2891–2897.
13. Gardner D, Panter, Molyneux R, James L, Stegelmeir B. Abortifacient activity in beef cattle of acetyl- and succinylisocupressic acid from ponderosa pine. *J Agric Food Chem* 1996;44:3257–3261.
14. Welch K, Gardner D, Pfister J, Panter K, Zieglar J, Hall J. A comparison of the metabolism of the abortifacient compounds from Ponderosa pine needles in conditioned versus naive cattle. *J Anim Sci* 2012;90:4611–4617.
15. Panter K, James L, Molyneux R. Ponderosa pine needle-induced abortion in cattle. *J Anim Sci* 1992;70:1604–1608.
16. Wang S, Panter K, Gardner D, Evans R, Bunch T. Effects of the pine needle abortifacient, isocupressic acid, on bovine oocyte maturation and preimplantation embryo development. *Anim Reprod Sci* 2004;81:237–244.
17. Riet-Correa F, Medeiros M, Schild A. A review of poisonous plants that cause reproductive failure and malformations in the ruminants of Brazil. *J Appl Toxicol* 2012;32:245–254.
18. Furlan F, Colodel E, Lemos R, Castro M, Mendonca F, Riet-Correa F. Poisonous plants affecting cattle in central-western Brazil. *IJPPR* 2012;2:1–13.
19. Tokarnia C, Döbereiner J, Peixoto P. Poisonous plants affecting livestock in Brazil. *Toxicon* 2002;40:1635–1660.
20. Ginn P, Joanne, Mansell J, Rakich P. Skin and appendages. In: Maxie MG (ed.) *Jubb, Kennedy, and Palmer's Pathology of Domestic Animals*, 5th edn. Philadelphia: WB Saunders, 2007, pp. 553–781.
21. Panciera R, Mosier D, Ritchey J. Hairy vetch (*Vicia villosa* Roth) poisoning in cattle: update and experimental induction of disease. *J Vet Diagn Invest* 1992;4:318–325.
22. Ralph M, McDaniel, K. Broom snakeweed (*Gutierrezia sarothrae*): toxicology, ecology, control, and management. *Invasive Plant Sci Manag* 2011;4:125–132.
23. James L, Panter K, Nielsen D, Molyneux R. The effect of natural toxins on reproduction in livestock. *J Anim Sci* 1992;70:1573–1579.
24. Murphy M, Rowe L, Ray A, Reagor J. Bovine abortion associated with ingestion of *Iva angustifolia* (narrowleaf sumpweed). In: *Proceedings of the Annual Meeting of the American Association of Veterinary Laboratory Diagnosticians*, 1983, pp. 161–166.
25. Blakley B. Moldy sweet clover (dicoumarol) poisoning in Saskatchewan cattle. *Can Vet J* 1985;26:357–360.
26. Runciman D, Lee A, Reed K, Walsh J. Dicoumarol toxicty in cattle associated with ingestion of silage containing sweet vernal grass (*Anthoxanthum odoratum*). *Aust Vet J* 2002;80:28–32.
27. Valli T. Hematopoitic system. In: Maxie MG (ed.) *Jubb, Kennedy, and Palmer's Pathology of Domestic Animals*, 5th edn. Philadelphia: WB Saunders, 2007, pp. 107–324.
28. Aiello S. *The Merk Veterinary Manual*, 8th edn. Philadelphia: National Publishing Inc., 1998.
29. Keeler R. *Teratogens in plants*. *J Anim Sci* 1984;58:1029–1039.
30. Pfister J, Stegelmeier B, Gardner D, James L. Grazing of spotted locoweed (*Astragalus lentiginosus*) by cattle and horses in Arizona. *J Anim Sci* 2003;81:2285–2293.
31. Molyneux R, Panter K. Alkaloids toxic to livestock. In: Cordell GA (ed.) *The Alkaloids: Chemistry and Biology*, Vol. 67. San Diego, CA: Academic Press, 2009, pp. 143–216.
32. Armien A, Tokarnia C, Peixoto P, Frese K. Spontaneous and experimental glycoprotein storage disease of goats induced by *Ipomea carnea* subsp. *fistulosa* (Convoluaceae). *Vet Pathol* 2007;44:170–184.
33. Thompson K. Bones and joints. In: Maxie MG (ed.) *Jubb, Kennedy, and Palmer's Pathology of Domestic Animals*, 5th edn. Philadelphia: WB Saunders, 2007, pp. 1–184.
34. Abbott L, Finnell H, Chernoff G, Parish S, Gay C. Crooked calf disease: a histological and histochemical examination of eight affected calves. *Vet Pathol* 1986;23:734–740.

35. Vetter J. Poison hemlock (*Conium maculatum* L.). *Food Chem Toxicol* 2004;42:1373–1382.
36. Lopez T, Cid M, Bianchini M. Biochemistry of hemlock (*Conium maculatum* L.). *Toxicon* 1999;37:841–865.
37. Green B, Lee S, Panter K, Brown D. Piperidine alkaloids: human and food animal teratogens. *Food Chem Toxicol* 2012;50:2049–2055.
38. Green B, Lee S, Panter K, Welsh K, Cook D. Actions of piperidine alkaloid teratogens at fetal nicotinic acetylcholine receptors. *Neurotoxicol Teratol* 2010;32:383–390.
39. Evans T. Reproductive toxicity and endocrine disruption. In: Gupta RC (ed.) *Veterinary Toxicology*, 2nd edn. San Diego, CA: Academic Press, 2012, pp. 278–318.
40. Gupta RC. Placental toxicity. In: Gupta RC (ed.) *Veterinary Toxicology*, 2nd edn. San Diego, CA: Academic Press, 2012, pp. 319–336.
41. Diekman M, Green M. Mycotoxins and reproduction in domestic livestock. *J Anim Sci* 1991;70:1615–1627.
42. Evans T, Blodgett D, Rottinghaus G. Fescue toxicosis. In: Gupta RC (ed.) *Veterinary Toxicology*, 2nd edn. San Diego, CA: Academic Press, 2012, pp. 1166–1177.
43. Mostrom M, Jacobsen B. Ruminant mycotoxicosis. *Vet Clin North Am Food Anim Pract* 2011;27:315–344.
44. Belser-Ehrlich S, Harper A, Hussey J, Hallock R. Human and cattle ergotism since 1900: symptoms, outbreaks, and regulations. *Toxicol Ind Health* 2012;29:307–316.
45. Nicholson S. Ergot. In: Gupta RC (ed.) *Veterinary Toxicology*, 2nd edn. San Diego, CA: Academic Press, 2012, pp. 1200–1204.
46. Smith S, Caldwell J, Popp M *et al.* Tall fescue toxicosis mitigation strategies: comparisons of cow-calf returns in spring- and fall-calving herds. *J Agric Appl Econ* 2012;44:577–592.
47. Koontz AF, Bush LP, Klotz JL, McLeod KR, Schrick FN, Harmon DL. Evaluation of ruminally dosed tall fescue seed extract as a model for fescue toxicosis in steers. *J Anim Sci* 2012;90:914–921.
48. Lehner A, Duringer J, Estill C, Tobin T, Craig M. ESI-mass spectrometric and HPLC elucidation of a new ergot alkaloid from perennial ryegrass hay silage associated with bovine reproductive problems. *Toxicol Mech Methods* 2011;21:606–621.
49. Jones K, King S, Griswold, Cazac D, Cross D. Domperidone can ameliorate deleterious reproductive effects and reduced weight gain associated with fescue toxicosis in heifers. *J Anim Sci* 2003;81:2568–2574.
50. Richard J. Some major mycotoxins and their mycotoxicoses: an overview. *Int J Food Microbiol* 2007;119:3–10.
51. Coppock R, Christian R, Jacobsen B. Aflatoxins. In: Gupta RC (ed.) *Veterinary Toxicology*, 2nd edn. San Diego, CA: Academic Press, 2012, pp. 1181–1199.
52. Mostrom M. Zearalenone. In: Gupta RC (ed.) *Veterinary Toxicology*, 2nd edn. San Diego, CA: Academic Press, 2012, pp. 1266–1271.
53. Filannino A, Stout T, Gacella B *et al.* Dose–response effects of estrogenic mycotoxins (zearalenone, alpha-and beta-zearalenol) on motility, hyperactivation and the reaction of stallion sperm. *Reprod Biol Endocrinol* 2011;9:134–144.
54. Mostrom M, Evans T. Phytoestrogens. In: Gupta RC (ed.) *Veterinary Toxicology*, 2nd edn. San Diego, CA: Academic Press, 2012, pp. 1012–1028.
55. Mlynarczuk J, Wrobel M, Kotwica J. The adverse effects of phytoestrogens on the synthesis and secretion of ovarian oxytocin in cattle. *Reprod Domest Anim* 2011;46:21–28.
56. Woclawek-Potocka I, Mannelli C, Boruszewska D, Kowalczyk I, Wasniewski T, Skarynski D. Diverse effects of phytoestrogens on the reproductive performance: cows as a model. *Int J Endocrinol* 2013;2013:1–15.
57. Jefferson W, Patisaul H, Williams C. Reproductive consequences of developmental phytoestrogens exposure. *Reproduction* 2012;143:247–260.
58. Adams N. Detection of the effects of phytoestrogens on sheep and cattle. *J Anim Sci* 1995;73:1509–1515.
59. Reis L, Pardo P, Camargos A, Oba E. Mineral elements and heavy metal poisoning in animals. *J Med Med Sci* 2010;1:560–579.
60. Tomza-Marciniak A, Pilarczyk B, Bakowska M, Pilarczyk R, Wojcik J. Heavy metals and other elements in serum of cattle from organic and conventional farms. *Biol Trace Elem Res* 2011;143:863–870.
61. Garland T. Arsenic. In: Gupta RC (ed.) *Veterinary Toxicology*, 2nd edn. San Diego, CA: Academic Press, 2012, pp. 499–502.
62. DeSesso J, Jacobson C, Scialli A, Farr C, Holson J. An assessment of the developmental toxicity of inorganic arsenic. *Reprod Toxicol* 1998;12:385–433.
63. Hooser S. Cadmium. In: Gupta RC (ed.) *Veterinary Toxicology*, 2nd edn. San Diego, CA: Academic Press, 2012, pp. 503–507.
64. Lane E, Canty M. Cadmium exposure in cattle: A review. Overview appendix 7.
65. Maxie M, Youssef S. Nervous system. In: Maxie MG (ed.) *Jubb, Kennedy, and Palmer's Pathology of Domestic Animals*, 5th edn. Philadelphia: WB Saunders, 2007, pp. 281–456.
66. Zachary J. Nervous system. In: McGavin MD, Zachary JF (eds) *Pathologic Basis of Veterinary Disease*, 4th edn. St Louis, MO: Mosby Elsevier, 2007, pp. 833–972.
67. Hall J. Selenium. In: Gupta RC (ed.) *Veterinary Toxicology*, 2nd edn. San Diego, CA: Academic Press, 2012, pp. 549–557.
68. Yaeger M, Neiger R, Holler L, Fraser T, Hurley D, Palmer I. The effect of subclinical selenium toxicosis on pregnant beef cattle. *J Vet Diagn Invest* 1998;10:268–276.
69. McEvoy T, Robinson J, Ashworth C, Rooke J, Sinclair K. Feed forage toxicants affecting embryo survival and fetal development. *Theriogenology* 2001;55:113–129.
70. Ross S, McCaffery P, Drager C, De Luca L. Retinoids in embryonal development. *Physiol Rev* 2000;80:1021–1054.
71. Gandolfi F, Pocar P, Brevini T, Fischer B. Impact of endocrine disrupters on ovarian function and embryonic development. *Domest Anim Endocrinol* 2002;23:189–201.
72. Sweeney T. Is exposure to endocrine disrupting compounds during fetal/post-natal development affecting the reproductive potential of farm animals? *Domest Anim Endocrinol* 2002;23:203–209.
73. Majdic G. Endocrine disrupting chemicals and domestic animals. *Slov Vet Res* 2010;47:5–11.
74. Ramesh A, Archibong A, Huderson A *et al.* Polycyclic aromatic hydrocarbons. In: Gupta RC (ed.) *Veterinary Toxicology*, 2nd edn. San Diego, CA: Academic Press, 2012, pp. 797–809.
75. Ramesh A, Walker S, Hood D, Guillen M, Schneider K, Weyland E. Bioavailability and risk assessment of aromatic hydrocarbons. *Int J Toxicol* 2004;23:301–333.
76. Lindenau A, Fischer B. Embryotoxicity of polychlorinated biphenyls (PCBs) for preimplantation embryos. *Reprod Toxicol* 1996;10:227–230.
77. Mlynarczuk J, Wrobel M, Kotwica J. The influence of polychlorinated biphenyls (PCBs), dichlorodiphenyltrichloroethane (DDT) and its metabolite dichlorodiphenyldichloroethylene (DDE) on mRNA expression for NP-I/OT and PGA, involved in oxytocin synthesis in bovine granulosa and luteal cells. *Reprod Toxicol* 2009;28:354–358.
78. Ensley S. Organochlorines. In: Gupta RC (ed.) *Veterinary Toxicology*, 2nd edn. San Diego, CA: Academic Press, 2012, pp. 591–595.
79. Muhammad F, Riviere J. Dermal toxicity. In: Gupta RC (ed.) *Veterinary Toxicology*, 2nd edn. San Diego, CA: Academic Press, 2012, pp. 337–350.
80. Gupta R. Organophosphates and carbamates. In: Gupta RC (ed.) *Veterinary Toxicology*, 2nd edn. San Diego, CA: Academic Press, 2012, pp. 573–585.

81. Taverne M, Breeveld-Dwarkasing V, van Dissel-Emiliani F, Bevers M, de Jong R, van der Weijden G. Between prepartum luteolysis and onset of expulsion. *Domest Anim Endocrinol* 2002;23:329–337.
82. Yildiz A. Induction of parturition in cows with misoprostol. *J Anim Vet Adv* 2009;8:876–879.
83. Plumb D. *Veterinary Drug Handbook*, 3rd edn. Ames, IA: Iowa State University Press, 1999.
84. Purohit G, Shekher C, Kumar P, Solanki K. Induced termination of pregnancy in domestic farm animals. *Iranian J Appl Anim Sci* 2012;2:1–12.
85. Capen C. Endocrine glands. In: Maxie MG (ed.) *Jubb, Kennedy, and Palmer's Pathology of Domestic Animals*, 5th edn. Philadelphia: WB Saunders, 2007, pp. 325–428.
86. Yildiz A, Erisir Z. Effects of exogenous oxytocin on embryonic survival in cows. *Acta Vet Brno* 2006;75:73–78.

Chapter 66

Heritable Congenital Defects in Cattle

Brian K. Whitlock[1] and Elizabeth A. Coffman[2]

[1]*Department of Large Animal Clinical Sciences, College of Veterinary Medicine, University of Tennessee, Knoxville, Tennessee, USA*
[2]*Department of Veterinary Clinical Sciences, College of Veterinary Medicine, Ohio State University, Columbus, Ohio, USA*

Introduction

The etiologies of congenital defects in cattle can be divided into heritable, toxic, nutritional, and infectious categories. Heritable congenital defects are being recognized at an increasing rate and are most likely propagated as a result of specific but unrelated trait selection. In some cattle, the occurrence of inherited defects has become a frequent and economically important source of pregnancy loss/wastage. Veterinarians, animal scientists, and cattle breeders should be aware of heritable defects and be prepared to investigate and report animals exhibiting abnormal phenotypes.

Genetic testing has changed cattle production, not only in terms of traits beneficial to production, but also in the ability to identify and manage harmful genetic defects. As selection concentrates the genetics of certain individuals, the potential for emergence of heritable defects will increase. The surveillance of such disorders has become an important part of bovine health programs.

Fetal abnormalities are very often discounted as randomly occurring "accidents of gestation"; the defects may not be deemed reportable and appropriate samples may not be collected. Failure of identification or delay in detection of heritable defects may allow further distribution of the mutated genetics. Obvious defects such as skeletal malformations, extensive soft tissue abnormalities, severe neurological disorders, and diseases of the skin are more likely to be recognized, whereas defects involving internal organs or abortions and stillbirths may be less obvious and more easily missed. Surveillance may be further compromised by the reluctance to report potentially heritable disorders or the reluctance of breed associations to aggressively pursue potentially heritable disorders.

Several reviews on inherited disorders in cattle have been published,[1–3] and a regularly updated electronic database, Online Mendelian Inheritance in Animals (OMIA), is available at http://omia.angis.org.au/.[4]

This chapter describes the morphologic characteristics, mode of inheritance, breeding lines affected, and availability of testing for selected heritable bovine fetal abnormalities. A comprehensive summary of all known heritable congenital abnormalities would be extensive and is beyond the scope of this chapter. The purpose of this review is to discuss those heritable bovine fetal abnormalities recently described for which the mutation has been identified and a test is available.

Arthrogryposis multiplex

Arthrogryposis multiplex (AM) is a lethal autosomal recessive genetic defect that originated in Angus cattle. Beginning in 2008, researchers in collaboration with the American Angus Association (AAA) investigated abnormal calves believed to fit the description of what was then called AM and commonly referred to as "curly calf syndrome" in Angus cattle. Within 2 months, researchers obtained samples and pedigrees from affected calves and their parents, the mutation was identified, the DNA test was developed and validated, and the status of over 700 AI bulls was determined.[5]

Calves with AM are born dead or die shortly after birth. They are small-for-gestational age (15–25 kg) and have markedly diminished muscle mass, but dystocia is common as a result of the congenital arthrogryposis, scoliosis, torticollis, and possibly hydroamnion. It appears that in AM an essential protein that allows communication between nerves and muscle tissue is absent, so the calf (which fails to move *in utero*) is born with the joints of all four limbs fixed and the legs twisted. There are several characteristics of AM,

Bovine Reproduction, First Edition. Edited by Richard M. Hopper.
© 2015 John Wiley & Sons, Inc. Published 2015 by John Wiley & Sons, Inc.

Figure 66.1 Arthrogryposis multiplex in an Angus calf that was born dead. Note the contracted forelimbs and extended hindlimbs. Image courtesy of Dr Robert L. Carson of Auburn University, Alabama.

including arthrogryposis (fixed twisted joints), kyphoscoliosis (twisted spine), and decreased muscling[5] (Figure 66.1).

The mutation is a deletion that involves three genes; one of these genes is involved in the development of nerve and muscle. Affected calves are missing about 23 kb. These missing base pairs result in complete loss of function of all three genes in homozygous calves.[6]

Bovine arachnomelia syndrome

Bovine arachnomelia syndrome (AS) is an inherited monogenetic autosomal recessive trait with complete penetrance. Affected calves are usually stillborn with skeletal abnormalities including a "spidery" appearance of the limbs (dolichostenomelia) with marked thinning of the diaphysis and an abnormally shaped skull[7] (Figure 66.2). In cattle there are two virtually identical AS phenotypes in Brown Swiss cattle and German Fleckvieh/Simmental cattle.

In Brown Swiss cattle, AS is the result of a single base (G) insertion in exon 4, leading to a frameshift and a premature stop codon in the coding sequence of the bovine sulfite oxidase (*SUOX*) gene, interfering with the expression of the SUOX protein (molybdohemoprotein sulfite oxidase, a terminal enzyme in the oxidative degradation pathway of sulfur-containing amino acids).[7] Deficiencies of the SUOX enzyme in humans are characterized by major neurological abnormalities and early death.[8–11] Cattle that are heterozygous for the mutation do not exhibit any clinical signs. Of the 302 unaffected Brown Swiss cattle tested, 10 (3.3%) were identified as carriers of the mutation in the *SUOX* gene.[7]

More recently, there were more than 150 confirmed cases of AS in German Fleckvieh/Simmental cattle by 2008, with an estimated 6% prevalence in the cow population.[12] The mutation in German Fleckvieh/Simmental cattle is a 2-bp deletion in the bovine gene encoding molybdenum cofactor synthesis 1 (*MOCS1*) resulting in a frameshift and premature termination of the bovine MOCS1 protein (73 vs. 633 amino acids). When cattle were sampled randomly from the Bavarian Simmental population, excluding first-degree relatives of known carriers, 17 of 616 (2.8%) were heterozygous for the genetic mutation (2-bp deletion in *MOCS1*) for AS in Simmental cattle.[13] The MOCS1 protein is required for the synthesis of the molybdoprotein cofactor, which forms the active site in SUOX (the defective protein in Brown Swiss cattle affected with AS).[7] The involvement of SUOX and MOCS1 in a common biochemical pathway mutually supports the causality of these mutations in Brown Swiss and German Fleckvieh/Simmental cattle.

Bovine citrullinemia

Bovine citrullinemia is an autosomal recessively inherited disease of Holstein cattle that was first described in Australia.[14] Bovine citrullinemia was disseminated throughout the Australian Holstein population following importation of semen from the North American sire Linmack Kriss King.[15,16] Affected cattle are clinically normal at birth but within 24 hours become depressed, wander aimlessly, head press, appear blind, teeth grind and, within 4–5 days, develop proprioceptive deficits and ataxia, become recumbent, convulse, collapse, and die. Lesions include mild to moderate diffuse astroglial swelling in the cerebrocortical gray matter and mild to severe hepatocellular hydropic degeneration.[17] Bovine citrullinemia is caused by transition of cytosine (CGA/arginine) to thymine (TGA/Stop codon) at codon 86 of the gene coding for argininosuccinate synthase, leading to impairment of the urea cycle and extreme elevation of citrulline and ammonia in plasma of affected cattle.[17,18]

Bovine dwarfism

Dwarfism as a heritable condition has been reported in many mammals, including multiple breeds of cattle.[1,19–25] Several types of inherited chondrodysplasia have been reported in the bovine. Although the phenotypic expression is variable, all are characterized by systemic skeletal disorders, including shortness and deformity of limbs, head, and vertebrae. Dwarfism has been recognized in the Angus, Brown Swiss, Danish Red, Dexter, Hereford, Holstein, Japanese Brown, and Shorthorn breeds.[25] No single gene or mutation is responsible for all reported cases of bovine dwarfism. Here we describe two unique forms of bovine dwarfism.

Dexter bulldog dwarfism has been a major problem for the Dexter breed since its description in the early nineteenth century.[26] Homozygosity for the Dexter dwarf mutation is lethal and fetuses are generally aborted at approximately 7 months of gestation. Clinically, the defective fetuses are characterized by an extreme disproportionate dwarfism, a short vertebral column, marked micromelia, large abdominal hernia, relatively large head with a retruded muzzle, cleft palate, and protruding tongue.[27]

Bulldog dwarfism of Dexter cattle is caused by two discrete mutations in the aggrecan (*ACAN*) gene, and DNA testing is available for the mutation.[27] When defined strictly on the lethal bulldog phenotype, the inheritance pattern of Dexter dwarfism is described as autosomal recessive (i.e., the bulldog dwarf is homozygous for the genetic mutation). However, because heterozygotes have an intermediate dwarf phenotype, the trait can also be described as having

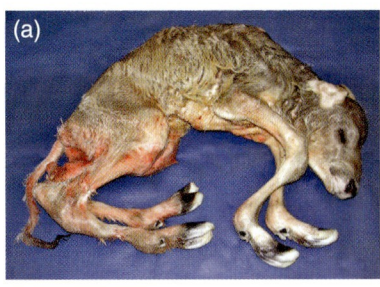 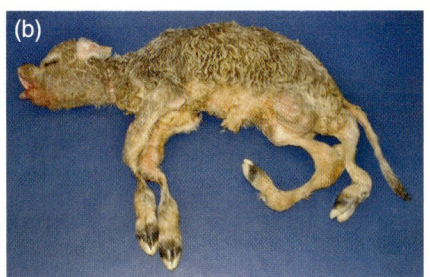 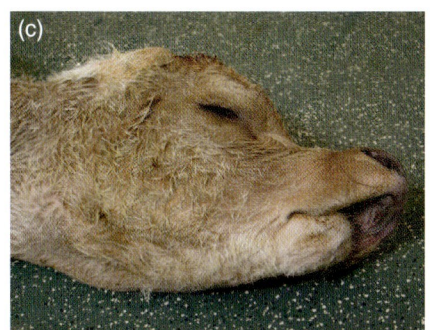
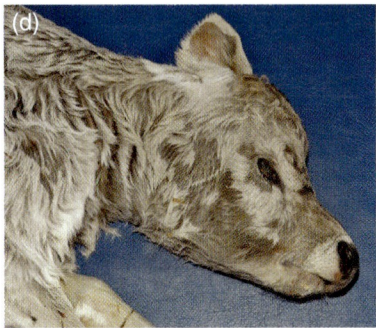 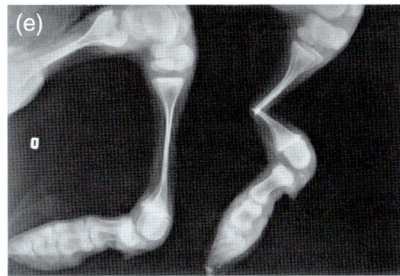 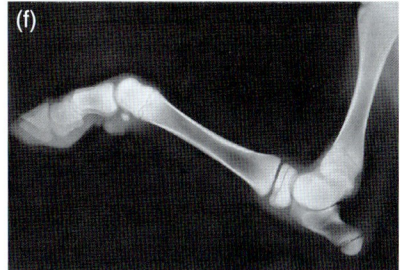

Figure 66.2 Phenotype of bovine arachnomelia in Brown Swiss cattle. (a, b) Stillborn affected calves; note the abnormal length of all legs (dolichostenomelia), the angular deformities in the distal part of the hindlegs with arthrogryposis of the distal joints, and muscular atrophy of the legs. (c, d) Typical facial deformities of affected calves; note a concave rounding of the dorsal profile of the maxilla and brachygnathia inferior. (e) Radiography of the hindlimbs of the affected calf from (c); note that the joint ends (epiphyses) of the long bones are of normal size but with marked thinning of the shafts (diaphyses) showing increased fragility. (f) Radiography of the left hindleg of a nonaffected control calf. From Drögemüller C, Tetens J, Sigurdsson S *et al*. Identification of the bovine arachnomelia mutation by massively parallel sequencing implicates sulfite oxidase (SUOX) in bone development. *PLoS Genet* 2010;6(8):e1001079.

an incomplete dominance mode of inheritance (8% expression of *ACAN* in heterozygotes compared with homozygous normal).[27] This has presented a major problem for Dexter breeders, because the favored Dexter phenotype, featuring short legs (a mild form of disproportionate dwarfism), tends to be heterozygous for the bulldog mutation. Selecting for the favored Dexter phenotype maintains a lethal allele at a high frequency.[27]

Dwarfism in Aberdeen Angus calves was described in 1951, and determined to be autosomal recessive.[21] The AAA handled the issue by the virtual annihilation of all animals associated with the primary source herd.[28] Following the efforts of the breed association to eliminate dwarfism, there were no certified reports to the AAA of dwarfism in registered Angus cattle from the 1970s until 2002. At the start of the twenty-first century, abnormal Angus calves were reported in several herds in the western United States. Unlike the previous form of dwarfism, these calves appeared normal at birth, but failed to grow, and after several months appeared to have abnormally short legs and thick bodies. Gross and histopathological examination of these calves indicated evidence for diminished endochondral ossification and other features consistent with dwarfism, including the protrusion of the alar wing of the basisphenoid bone into the cranial cavity, abnormalities of the ventral vertebral bodies, and curving of the transverse vertebral processes ("long-nosed" dwarfism).[29] The genetic defect of the long-nosed Angus dwarf has not been definitively identified but is not the same as the defect in Dexter (mutations in *ACAN*) or Japanese brown cattle (*LIMBIN* mutations).[29,30]

Brachyspina syndrome

Brachyspina syndrome is a recently reported lethal malformation in the Holstein breed.[31-33] The syndrome is characterized by calves born dead following a slightly prolonged gestation. The gross morphology of brachyspina syndrome shares many features with short spinal lethal syndrome in cattle, first described in Old Norwegian Mountain calves born in 1930.[34] Brachyspina calves are growth retarded, with severe shortening of the entire vertebral column and limbs. Most vertebral segments have some lack of organization, with irregular ossification separated by cores of cartilage that prevent identification of individual vertebra. Multiple defects of the internal organs, including renal dysplasia and intestinal atresia, are consistently present, whereas brachygnathism and caudal dislocation and compression of the brain are inconsistently present. The appendicular skeleton may or may not be affected.

Although variation in morphology caused by a common syndrome is not unusual, the six recently reported cases of brachyspina syndrome may have had more than one etiology, since they were morphologically diverse. In two reports of brachyspina syndrome, genealogical examination showed that the cases occurred in familial patterns,[32,33] a common ancestor to all parents was found, and the occurrence of the abnormal calves could potentially be explained by transmission of recessive alleles. Although these observations supported the hypothesis of a genetic basis of brachyspina syndrome, association does not prove causation. Widely used sires may occur as common ancestors in the pedigree of malformed calves without that individual being a carrier, or without the defect having an inherited etiology. If brachyspina syndrome is inherited as an autosomal recessive anomaly, it may be an emerging worldwide disease in the Holstein breed.

Complex vertebral malformation

Complex vertebral malformation (CVM) syndrome is a recessively inherited lethal disorder in the Holstein breed that increases embryonic death, abortion, and perinatal death. CVM fetuses have a composite of phenotypic abnormalities, including axial skeletal deformities (e.g., hemi- and mis-shaped vertebra, ankylosis of mainly the cervicothoracic vertebra, and scoliosis), symmetric arthrogryposis of the lower limbs, craniofacial dysmorphism, and cardiac anomalies.[35-37] Clinical heterogeneity among affected calves may make it difficult to make a diagnosis of CVM; however, a presumptive diagnosis can be made at necropsy (if pedigree information is available).

Genealogical research identified Carlin-M Ivanhoe Bell (registration number 1667366), an elite Holstein-Friesian bull born in 1974, to be the main ancestor of cattle carrying this mutation.[36] Because of the superior lactation performance of his daughters, Ivanhoe Bell was extensively used for two decades. Carriers of the CVM mutation exist in Holstein cattle populations worldwide and the frequency of CVM carriers among Holstein sires has reached an alarming level.[35,38-41]

By comparing DNA sequences of unaffected and affected calves, recent research[42] has uncovered a point mutation in the form of a transversion (G→T) in the *SLC35A3* gene, causing a valine at position 180 to be replaced with phenylalanine. The *SLC35A3* gene product is a Golgi-resident transporter critical for the formation of glycoproteins and, ultimately, axial skeleton development. A screening test (DNA) for CVM is now available and has been widely performed on Holstein sires. Pedigree information now includes the designation "TV" for tested animals free from CVM and "CV" for carriers of CVM.

Congenital contractural arachnodactyly

Congenital contractural arachnodactyly (CA), or "fawn calf syndrome," is a nonlethal autosomal recessive genetic defect of Angus cattle. CA calves are normally born alive and most can walk, suckle, and survive. The birthweight of CA calves is "normal." The phenotype is subtle, so that CA may not initially be recognized as a defect. CA is a developmental defect involving reduced elasticity of the connective tissue of muscles first identified in Victoria, Australia in 1998 but now reported in many countries.[43] Although CA is a less severe disease than lethal genetic defects of Angus calves, without human intervention up to 20% of CA calves die soon after birth because they are unable to stand and suckle.[43] CA manifests in newborn calves as elongated limbs, congenital proximal limb contracture, congenital distal limb hyperextension, and congenital kyphosis with significant postnatal improvement in these clinical signs as the calf grows and matures.[43]

Researchers have identified the genetic defect that causes CA and have partially characterized the specific mutation responsible for CA as a deletion of at least 38 kb, severely compromising function.[44] The complete sequence of the deleted DNA segment is not known, making it difficult to develop a diagnostic test that is 100% accurate.[44] Until recently the breed associations (AAA and Angus Australia) have avoided identifying any animal as a CA carrier because the current diagnostic test is less than 100% accurate.[45] However, some specifically identified animals have been named as either carriers or "highly likely" to be carriers of the CA mutation by Angus Australia.[44] The current assay generates some false positives in a number of pedigrees, creating a significant danger of misinterpretation of test results. The current test does allow an overall estimation of frequency of the AM mutation in the population. With more than 500 animals genotyped with several of the genetic markers for AM, the maximum frequency of AM in the AI sire population is approximately 3–4%.[44]

Crooked tail syndrome

The Belgian Blue breed is well known for its exceptional muscular development. This phenotype is due in part to an 11-bp loss-of-function deletion in the myostatin gene that has been fixed in the breed. Intense selection in the Belgian Blue has substantially reduced the effective population size, resulting in greater inbreeding and therefore recurrent outbreaks of recessive defects. Most recently a novel defect referred to as crooked tail syndrome (CTS) has been described in the Belgian Blue.[46,47] In addition to the deviation of the tail, cattle with CTS have general growth retardation manifested at approximately 1 month of age, abnormal skull shape manifested as a shortened broad head, and extreme muscular hypertrophy.[46] Although the defect is not lethal itself, the most severe cases (~25%) are euthanized on welfare grounds.[46] The surviving 75% nevertheless cause important economic losses as a result of growth retardation and carcass depreciation.[46]

CTS is caused by mutations in the mannose receptor C type 2 (*MRC2*) gene. *MRC2* encodes the 180-kDa endocytic transmembrane glycoprotein (Endo180), a recycling endocytic receptor that is expressed in mesenchymal cells such as stromal fibroblasts and in the chondrocytes, osteoblasts and osteocytes in developing bones, and is thought to play a role in regulating extracellular matrix degradation and remodeling. Two mutations in *MRC2* have been identified as causing CTS. The first mutation is a 2-bp deletion in the open reading frame of *MRC2* resulting in a frameshift and a premature stop codon that causes a nonsense-mediated decay of the mutant mRNA; the second identified mutation is a T→C substitution in *MRC2*. Both defects result in a virtual absence of functional Endo180.[46,47]

When 1899 healthy Belgian Blue cattle were tested, unexpectedly 24.7% were identified as carriers of the first mutation (2-bp deletion in *MRC2*).[46] The unusually high frequency of the first CTS mutation in Belgian Blue cattle suggests that it might confer heterozygotes an advantage in this highly selected population. In fact, CTS carrier animals are smaller, stockier, and more heavily muscled, and they have a thinner skeleton and more rounded ribs. Moreover, the *MRC2* genotype accounts for 3.6, 3.6, and 2.6% genetic variance of height, muscularity, and general appearance, respectively within 519 Belgian Blue pedigree bulls.[46] Enhanced muscularity of CTS carriers may contribute greatly to the rapid increase in the CTS mutation in Belgian Blue cattle. Indeed, carrier animals are approximately twice as likely to be selected as elite sires than their noncarrier siblings.[46]

Deficiency of uridine monophosphate synthase

Deficiency of uridine monophosphate synthase (DUMPS) is a hereditary lethal autosomal recessive disorder in Holstein cattle causing early embryonic mortality during implantation in the uterus and possibly fetal mummification.[48] About 2% of the Holstein cattle in the United States possess an autosomal recessive form of the gene for DUMPS.[49] DUMPS interferes with *de novo* biosynthesis of pyrimidine (a nucleotide constituent of DNA and RNA). It is inherited as a single autosomal locus with two alleles.[50,51] In mammalian cells, the last step of pyrimidine nucleotide synthesis involves the conversion of orotic acid to uridine monophosphate (UMP) and is catalyzed by UMP synthase.[52] Growth and development of homozygous recessive embryos is arrested, leading to mortality around 40 days after conception.[49] DUMPS is caused by a single point mutation (C → T) at codon 405 within exon 5.[53]

Developmental duplication

Initially observed in Australia, developmental duplication (DD) is an autosomal recessive genetic defect of Angus cattle. Affected DD calves are born with additional limbs (polymelia). Duplication of the front legs is typical, with the limbs originating from the neck or shoulder region. Variations of DD are classified according to the point of attachment to the body. The foot of the supernumerary limb may be normal or syndactyl. Affected DD calves may also present as conjoined twins. There has been one case of conjoined twins in Australia that were reported to be homozygous for the DD mutation.

The main economic impacts of DD were thought to be from losses related to dystocia and costs associated with supernumerary limb amputation. However, the frequency of reported DD cases is unexpectedly low, especially with carrier frequency among US sires being moderately high at approximately 6%. These data indicate that calves presenting with polymelia at birth are rare events that survive embryonic death. Early DD events may prevent many embryos from developing to term, resulting in embryonic death and the reduced frequency of live births that are being observed. This may be a more significant economic impact of DD than losses related to dystocia and limb amputation. Genetic testing for DD is available through Angus Genetics Inc. and Zoetis Genetics. Either association archived or newly collected samples may be used for testing.

Inherited congenital myoclonus

Inherited congenital myoclonus (ICM), also known as idiopathic epilepsy, is a seizure disorder observed in Hereford cattle and their crosses caused by an autosomal recessive genetic defect incompatible with life. ICM is characterized by spontaneous and stimulus-responsive myoclonic spasms that are prenatal and prevent calves from rising at birth. Affected calves have a "normal" phenotype when they are not experiencing seizures. Environmental stressors (heat, cold, weaning) can trigger the seizures, and the seizures can last from minutes to more than 1 hour. Bovine ICM has been attributed to a severe disturbance of glycine-mediated neurotransmission in the spinal cord.[54,55] The observed phenotype is due to a nonsense mutation in codon 24 of the glycine receptor polypeptide, which results in truncation of the α_1-subunit and subsequent loss of cell-surface expression.[56]

Microdeletion in the maternally imprinted *Peg3* domain

A deletion in the maternally imprinted *Peg3* domain has been identified in Finnish Ayrshire cattle that results in loss of paternal *MIMT1* expression and causes late-term abortion and stillbirth has been identified. The stillborn calves weigh approximately 20 kg, or half the average normal birthweight for the breed, and the lungs are not inflated. No other visible abnormalities are detected (Figure 66.3). Fetal and placental development in mammals are both affected by imprinted genes for which either the paternally or maternally inherited allele has become epigenetically inactivated, leading to monoallelic expression. Approximately half of the fetuses can be adversely affected when a male transmits the allele for the defective fully penetrant imprinted gene that is

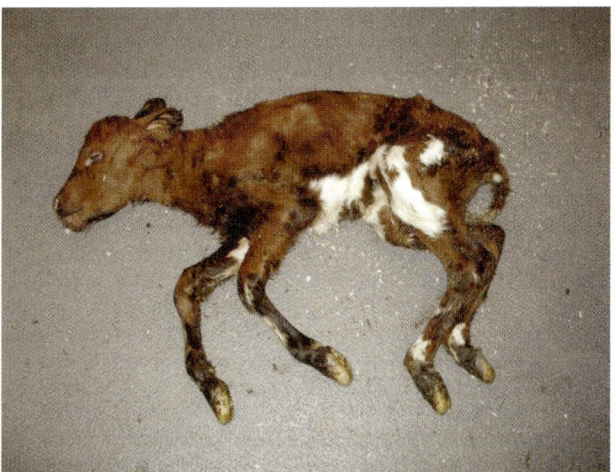

Figure 66.3 A stillborn calf resulting from a deletion in the maternally imprinted *Peg3* domain that caused loss of paternal *MIMT1* expression. The stillborn calf is approximately half the average normal birthweight for the breed. No other visible abnormalities were detected. From Flisikowski K, Venhoranta H, Nowacka-Woszuk J *et al*. A novel mutation in the maternally imprinted PEG3 domain results in a loss of MIMT1 expression and causes abortions and stillbirths in cattle (*Bos taurus*). *PLoS ONE* 2010;5:e15116.

silenced on the maternal allele. In mammalian cloning experiments, the loss of epigenetic control of imprinted genes has led to low success rates and birth of animals with health problems.[57] Defects in maternally imprinted genes will be transmitted silently from females to their progeny, whereas phenotypically normal males will transmit these mutations to their progeny as though they were heterozygous for a dominant mutation. The mutation in *Peg3* was associated with late fetal death and stillbirth in at least 42.6% of the offspring of one Finnish Ayrshire bull (YN51). The mutation, when inherited from the sire, is semi-lethal for his progeny, with an observed mortality rate of 85%. The survival of 15% is presumably due to the incomplete silencing of maternally inherited *MIMT1* alleles, which is a common phenomenon for imprinted loci.[58] Crossbreeding with Holstein heifers and cows did not impact the survival rate of fetuses, consistent with the imprinting model of inheritance. The surviving female calves with the deletion should transmit the mutation to 50% of their offspring without any impact on fetal death, and bull calves inheriting this mutation from these dams could regenerate the problem.[59] Although the biological role of *MIMT1* is unknown, "natural" and experimentally induced (mice) knockouts suggest an important regulatory mechanism of *Peg3* affecting late prenatal development. Moreover, this inherited disorder of cattle stresses the importance of defects that cause stillbirth and abortion as well as the potentially drastic effects of mutations in imprinted genes.

Neuropathic hydrocephalus

Neuropathic hydrocephalus ("water head") is a lethal autosomal recessive genetic defect of Angus cattle. At the same time that AM calves were being submitted, calves with hydrocephalus were also submitted. These calves were similar in description and pedigree to calves described by Dr Denholm in Australia.

Affected calves are born near term and weigh 10–20 kg at birth. The head is markedly enlarged (Figure 66.4). The bones of the skull are malformed and appear as loosely organized bony plates that fall apart when the head is opened. The cranium is filled with fluid, and no recognizable brain tissue is evident. The spinal canal is also dilated, and no observable spinal tissue is found.[60] Because of the hydrocephalic cranial enlargement, dystocia is almost invariable. Delivery per vagina can normally be achieved by incising the cephalic vesicle of the domed cranium to drain the fluid. Manually removing the skull bones from the collapsed vesicle before applying traction may be necessary.

The genetic mutation leads to the abnormal function of a protein involved in the development and maintenance of the central nervous system and results in neuropathic hydrocephalus.[60] Nearly 10% of AI sires representing a broad cross-section of registered Angus genetics were found to be carriers of this syndrome.[61] Given the number of calves reported, this frequency is higher than expected. This phenomenon could be explained by a relatively high percentage (50–70%) of pregnancy wastage in embryos and/or fetuses. This is consistent with what is known about mutations in this gene for other species (i.e., complete disruption of this gene in mice results in 100% fetal mortality before the halfway point of gestation).[61]

Osteopetrosis

Osteopetrosis ("marble bone") is a lethal autosomal recessive genetic defect previously identified in humans and animal species. Cattle breeds known to be affected are Black and Red Angus, Hereford, Simmental, and Holstein. The defect was most recently reported in the Red Angus breed.[62] Calves affected with osteopetrosis are born 10–30 days early. They usually have head abnormalities that consist of brachygnathia inferior, impacted molars, and a protruding tongue. The long bones are shorter than normal, and the marrow cavities are filled with unresorbed bone (primary spongiosa) but are very fragile and can be easily broken.[62]

The disease is caused by a deletion of *SLC4A2*, a gene necessary for bone remodeling during development. *SLC4A2* encodes an anion exchanger protein necessary for proper osteoclast function. Loss of *SLC4A2* function appears to induce premature cell death, and results in cytoplasmic alkalization of osteoclasts, which in turn may disrupt acidification of resorption lacunae. Histological examination of affected tissues reveals scarce, morphologically abnormal osteoclasts displaying evidence of apoptosis.[63] Genetic mutations that cause osteopetrosis in Red Angus and Black Angus breeds are not the same. Genetic testing is available through AgriGenomics, Pfizer, MMI Genomics, and Igenity. However, the osteopetrosis test is for the mutation in the Red Angus breed only.

Paunch calf syndrome

An outbreak of a lethal multiorgan developmental dysplasia known as paunch calf syndrome (PCS) has recently been reported in Romagnola cattle in Italy.[64] Affected calves are usually stillborn and have craniofacial deformities, an enlarged fluid-filled abdomen, and hepatic fibrosis (Figure 66.5).[64,65] Pedigree analysis indicates that all PCS cases trace back to a once-popular Romagnola sire born in 1969.[65]

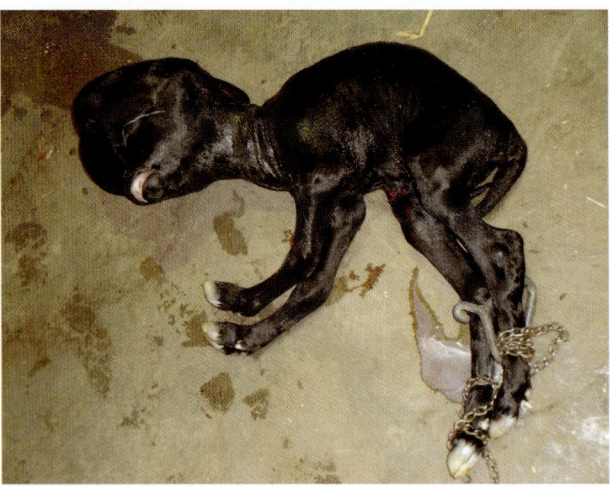

Figure 66.4 Neuropathic hydrocephalus in an Angus calf that was born dead. Note the markedly enlarged cranium. Image courtesy of Dr Kristie Steuer, Animal Health Center, Laramie, Wyoming.

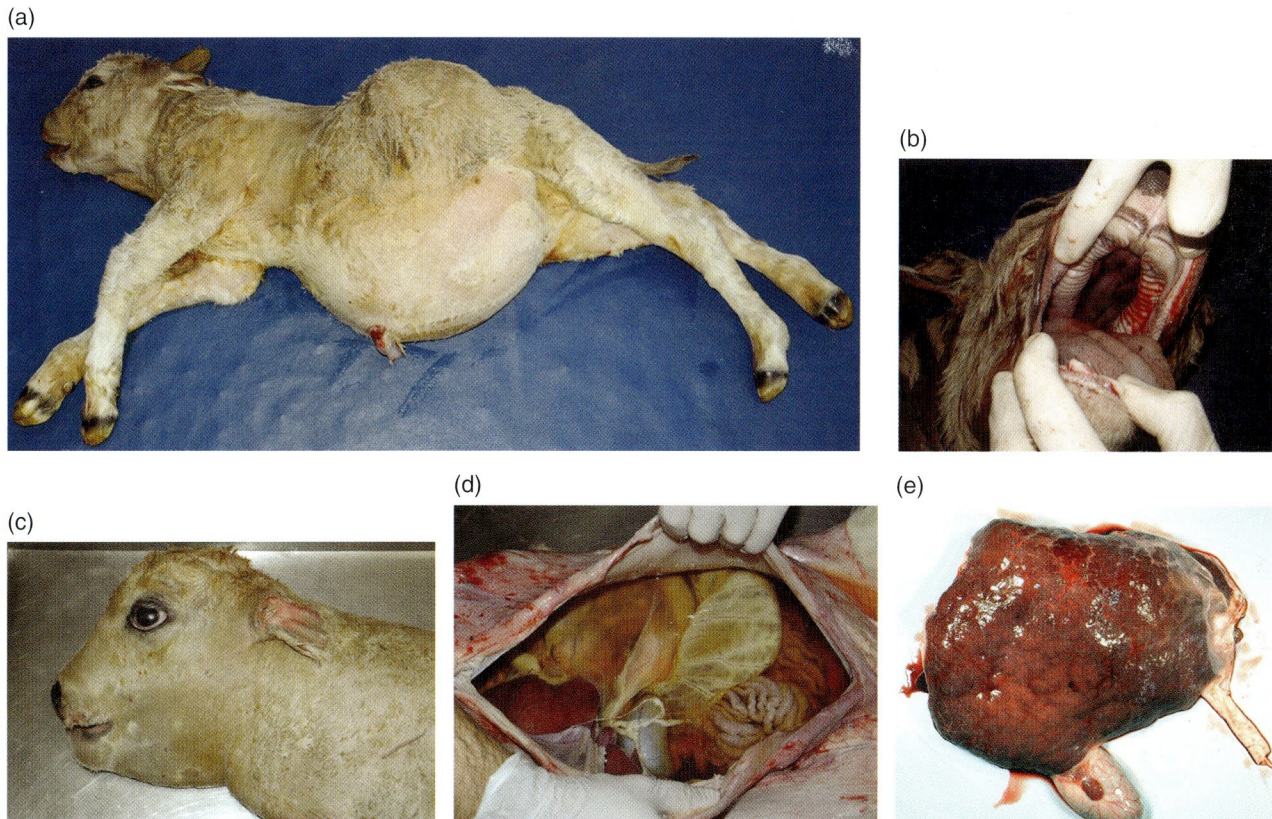

Figure 66.5 Phenotype of the paunch calf syndrome in Romagnola cattle. Note the shortened face and the abdominal distension with a considerable peritoneal liquid presence (a) and, in some cases, tongue protrusion and cleft palate (b), shortened and flattened splanchnocranium (c), accumulation of dark-yellowish turbid fluid in the abdominal cavity (d), and irregular surface of the liver and presence of a hepatic cyst with reddish fluid content (e). From Flisikowski K, Venhoranta H, Nowacka-Woszuk J et al. A novel mutation in the maternally imprinted PEG3 domain results in a loss of MIMT1 expression and causes abortions and stillbirths in cattle (Bos taurus). PLoS ONE 2010;5:e15116.

PCS is caused by a mutation (G→A) in the lysine (K)-specific demethylase 2B gene (*KDM2B*).[65] *KDM2B* encodes a histone H3 lysine 36 dimethyl-specific demethylase that may be critical in transcription regulation during embryonic development (epigenetic mark) in ectodermal (e.g., neural tube closure)[66] and endodermal- and mesodermal-derived organs like bone and liver.[65] The DNA-based test for the detection of PCS carriers has been adopted by the Italian Romagnola Breed Association. The carrier frequency was 32% within 466 healthy unaffected Italian Romagnola individuals not related to PCS cases at the parent level. This outbreak in PCS can most likely be attributed to selective breeding practices involving extensive use of a particular highly selected sire leading to an increase in consanguineous/co-ancestry breeding.

Pulmonary hypoplasia with anasarca

Pulmonary hypoplasia with anasarca (PHA) is a lethal autosomal recessive disorder of a single missense mutation characterized most notably by dystocia, profound anasarca, death of the calf, and not uncommonly death or morbidity of the cow.[67,68] Originally, PHA calves were identified as Maine-Anjou, Shorthorn, or their crosses (composite cattle); however, cases are also reported in Australian Dexter, Hereford, and belted Galloway cattle.[3,67,68]

Breeding of two PHA carriers would theoretically produce 25% affected fetuses. However, there appears to be a high percentage of pregnancy wastage from approximately 90 to 200 days of gestation. When PHA fetuses survive to near term, they invariably result in dystocia. It appears that PHA calves thought to be "bred to the AI date" are usually premature births of calves conceived due to natural service following an AI program.

The most obvious external characteristic of PHA calves is marked anasarca, resulting in large calves (Figure 66.6), frequently requiring cesarean section. Furthermore, PHA is characterized by the absence or near absence of lung tissue (pulmonary hypoplasia; Figure 66.7), lymphatic agenesis (absence of lymph duct and nodes), and athymia.[68,69]

Heterozygous PHA carriers have a "normal" phenotype and can be identified only by genetic testing. Following identification of the gene in late 2006, over 40 popular Maine-Anjou, Shorthorn, and "club calf" sires were identified as PHA carriers. Interestingly, several of the most popular Shorthorn bulls are carriers for both tibial hemimelia and PHA. A syndrome similar to PHA was identified in Dexter cattle in 2006 when Windsor et al.[67] reported three cases of PHA in the Dexter breed. The cows were related and inheritance was felt to be autosomal recessive. Although the mutation in the Dexter breed involves the same gene as the condition seen in Maine cattle, it is not identical. The

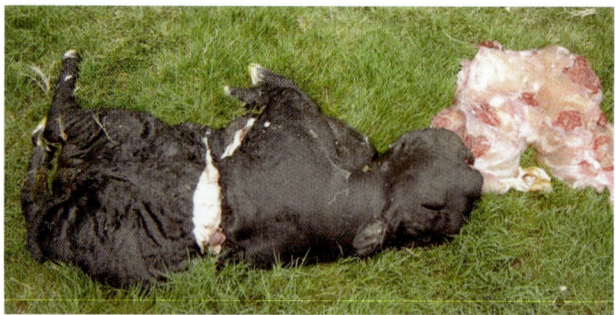

Figure 66.6 Calf with pulmonary hypoplasia with anasarca syndrome delivered by partial fetotomy. Note the severe anasarca. Photograph courtesy of Dr Brian K. Whitlock.

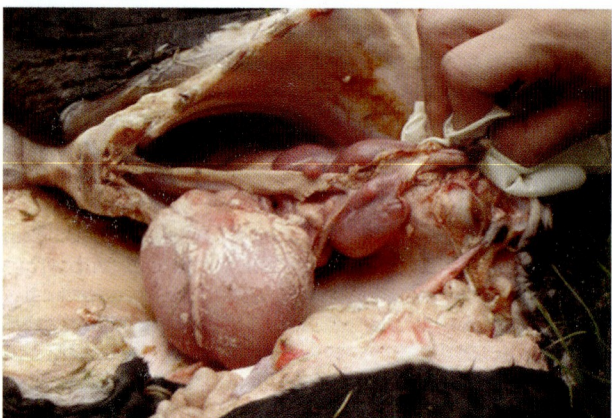

Figure 66.7 Thorax of a calf with pulmonary hypoplasia with anasarca syndrome. While the lungs of a "normal" bovine fetus at a similar stage of gestation are proportionate to those of a neonate, note the severe pulmonary hypoplasia (severely hypoplastic lungs are to the right of the heart). Photograph courtesy of Dr Brian Whitlock.

Maine-Anjou mutation was a simple point mutation, whereas the Dexter mutation was an 84-bp deletion involving the same exon as the Maine mutation (J. Beever, personal communication).

Recessive defects discovered from the absence of homozygous haplotypes

Lethal recessive defects were often discovered in the past from reports of abnormal calves and subsequent breeding trials to confirm the inheritance. Defects that affect embryo loss were sometimes discovered from occasional abnormal calves that reached late gestation or parturition[31,35] or from physiological differences between normal and heterozygous phenotypes.[70] Instead of relying on breeder reports of abnormalities, embryonic defects may more easily be discovered using fertility traits and genotypes stored in national or international databases.[71]

Genotypes and haplotypes for large numbers of individuals provide a new tool for identifying lethal recessive genes. Defects that cause embryo loss are difficult to detect without genomic data, even with very large sets of phenotypic and pedigree data, because of insufficient observations per estimated interaction.[72] With genomic data, lethal recessives may be discovered from haplotypes that are common in the population but are never homozygous in live animals. This method requires genotypes only from apparently normal individuals and not from affected embryos and therefore is the opposite of most previous strategies.[73] The new approach can discover lethal defects without phenotypic data. If the number of genotyped individuals is large, the expected number of homozygous haplotypes is sufficiently large so that their complete absence is unlikely to be attributed to chance.

Five new recessive defects have been discovered in Holstein, Jersey, and Brown Swiss breeds by examining haplotypes that had high population frequency but which were never homozygous. Genotypes from the BovineSNP50 BeadChip (Illumina, San Diego, CA) were examined for 58 453 Holsteins, 5288 Jerseys, and 1991 Brown Swiss apparently normal individuals with genotypes in the North American database. Phenotypic effects were confirmed using records for conception rate in a smaller population of cows from the affected breeds. Estimated effects on conception rate ranged from −3.0 to −3.7 percentage points, which was slightly larger than estimated effects for previously known lethal alleles of −2.5 percentage points for brachyspina and −2.9 percentage points for CVM.[74] Estimated effect of carrier interaction for stillbirth rate was small (≤1.8 percentage points) but comparable to the effect of CVM (1.4 percentage points).[74] For each of the five new defects, 2.7–20.7% of elite animals in the current population are carriers. However, selection against carriers would improve fertility only slightly. For example, conception rates would increase by less than 1 percentage point if any of the haplotypes were eliminated from the population.[74]

Tibial hemimelia

The tibial hemimelia (TH) syndrome was first described in 1951 in Scotland in the Galloway breed of cattle.[75] This lethal condition is characterized by multiple congenital skeletal deformities, most notably shortened or absent tibia, abdominal hernia, cryptorchidism, failed paramesonephric duct development, hirsutism, and improper neural tube closure resulting in meningocele. Calves are born dead or die shortly after birth. The Scottish Galloway Association used test breeding and pregnancy termination to identify carriers and eliminate the autosomal recessive defect from Galloway breeding stock.[76]

The syndrome was first described in the United States in 1974,[77] and subsequently in a case report of a female Simmental calf with arthrogryposis, ventral abdominal hernia, TH, and nonunion of the paramesonephric ducts, with the suggestion that it might be a genetic disorder.[78] A decade later, TH was identified in six genetically related registered Shorthorn calves, three from the United States and three from Canada.[79] Although the skeletal abnormalities varied, the combination of TH, abdominal hernia, and meningocele was felt to be strikingly similar to the condition previously described and determined to be inherited as an autosomal recessive trait in Galloway cattle. Pedigree analysis of the Shorthorn calves suggested that homozygosity of an autosomal recessive allele was responsible for the defect.

Identification of the defective gene mutation causing TH was made in 2005.[80] The defective gene has been shown to be aristaless-like homeobox 4 (*ALX4*), a major regulator of hindlimb formation (J. Beever, personal communication). The deletion removes approximately one-third of the *ALX4* gene, including the upstream regulatory sequence, and involves approximately 46 kb. After identification of the initial deletion, it was noted that although the parentage of some TH calves was DNA-verified, some parents did not test positive for this deletion. Additional studies revealed that other TH carriers possess a larger deletion of 450 kb that overlaps the previous/smaller deletion, and removes four genes including *ALX4*. The later and larger identified deletion is rare and affected calves sired by carriers of that deletion are compound heterozygotes (heterozygous with the smaller mutation on one chromosome and the larger mutation on the other). Although "heterozygous," these calves are affected like those homozygous for the smaller deletion. The frequency of the larger mutation is so low, and the magnitude of the deletion so much larger, that it is unlikely that there have been any TH calves homozygous for the larger deletion. Because samples from the Scottish Galloway cattle are unavailable for testing, it is unclear whether the smaller or larger deletions in Shorthorns originate from the Galloway breed or are different mutations.

Conclusion

In recent years, several heritable defects have been identified that have had significant impact on specific cattle populations. Molecular geneticists, veterinary pathologists, breeders, and breed associations have cooperatively identified genetic disorders, characterized the pathology, determined the genetic mutations responsible, and developed tests to identify carriers. In the very near future, it is likely that the genetic mutations in other heritable congenital defects of cattle will be identified through similar collaborative efforts.

Although the investigation of heritable bovine fetal anomalies has often been left to those in academia (specifically, animal scientists and veterinary pathologists), many of the currently recognized inherited disorders of cattle would have gone undiscovered without the assistance of private practitioners and producers. For any surveillance program to be successful, recognition of a potentially heritable defect is the first step. The anomaly must be reported, appropriate samples collected and preserved, and pedigree information made available. Veterinarians in the field can play a pivotal role in discovery and surveillance by recognizing and reporting inherited defects of cattle.

References

1. Leipold H, Huston K, Dennis S. Bovine congenital defects. *Adv Vet Sci Comp Med* 1983;27:197–271.
2. Huston K, Saperstein G, Steffen D *et al.* Clinical, pathological and other visible traits loci except coat colour (category 2). In: Lauvergne JJ, Dolling CHS, Millar P (eds) *Mendelian Inheritance in Cattle*. Wageningen: Wageningen Academic Publishers, 2000.
3. Whitlock B, Kaiser L, Maxwell H. Heritable bovine fetal abnormalities. *Theriogenology* 2008;70:535–549.
4. Nicholas FW. Genetic databases: online catalogues of inherited disorders. *Rev Sci Tech* 1998;17:346–350.
5. American Angus Association. Arthrogryposis multiplex information. American Angus Association, St Joseph, MO, 2010. http://www.angus.org/pub/AM/AMInfo.aspx
6. Kaiser L. The gene gurus have been busy this decade: recessive defect, mutations and DNA-based tests . http://kaisercattle.com/pdf/Deadcows.pdf.
7. Drogemuller C, Tetens J, Sigurdsson S *et al*. Identification of the bovine arachnomelia mutation by massively parallel sequencing implicates sulfite oxidase (SUOX) in bone development. *PLoS Genet* 2010;6:e1001079.
8. Kisker C, Schindelin H, Pacheco A *et al.* Molecular basis of sulfite oxidase deficiency from the structure of sulfite oxidase. *Cell* 1997;91:973–983.
9. Garrett R, Johnson J, Graf T, Feigenbaum A, Rajagopalan K. Human sulfite oxidase R160Q. Identification of the mutation in a sulfite oxidase-deficient patient and expression and characterization of the mutant enzyme. *Proc Natl Acad Sci USA* 1998;95:6394–6398.
10. Johnson J, Coyne K, Garrett R *et al.* Isolated sulfite oxidase deficiency: identification of 12 novel SUOX mutations in 10 patients. *Hum Mutat* 2002;20:74.
11. Seidahmed M, Alyamani E, Rashed M *et al.* Total truncation of the molybdopterin/dimerization domains of SUOX protein in an Arab family with isolated sulfite oxidase deficiency. *Am J Med Genet A* 2005;136:205–209.
12. Buitkamp J, Luntz B, Emmerling R *et al.* Syndrome of arachnomelia in Simmental cattle. *BMC Vet Res* 2008;4:39.
13. Buitkamp J, Semmer J, Gotz K. Arachnomelia syndrome in Simmental cattle is caused by a homozygous 2-bp deletion in the molybdenum cofactor synthesis step 1 gene (MOCS1). *BMC Genet* 2011;12:11.
14. Harper P, Healy P, Dennis J, O'Brien J, Rayward D. Citrullinemia as a cause of neurological disease in neonatal Friesian calves. *Aust Vet J* 1986;63:378–379.
15. Healy P, Dennis J, Camilleri L, Robinson J, Stell A, Shanks R. Bovine citrullinemia traced to the sire of Linmack Kriss King. *Aust Vet J* 1991;68:155.
16. Healy P. Testing for undesirable traits in cattle: an Australian perspective. *J Anim Sci* 1996;74:917–922.
17. Healy P, Harper P, Dennis J. Bovine citrullinemia: a clinical, pathological, biochemical and genetic study. *Aust Vet J* 1990;67:255–258.
18. Windsor P, Agerholm J. Inherited diseases of Australian Holstein-Friesian cattle. *Aust Vet J* 2009;87:193–199.
19. Gregory P, Rollins W, Pattengale P *et al.* A phenotypic expression of homozygous dwarfism in beef cattle. *J Anim Sci* 1951;10:922–933.
20. Johnson L, Harshfield G, McCone W. Dwarfism, a hereditary defect in beef cattle. *J Hered* 1950;41:177–181.
21. Baker M, Blunn C, Plum M. "Dwarfism" in Aberdeen-Angus cattle. *J Hered* 1951;42:141–143.
22. Wild P, Rowland A. Systemic cartilage defect in a calf. *Vet Pathol* 1978;15:332–338.
23. Jones J, Jolly R. Dwarfism in Hereford cattle: a genetic morphological and biochemical study. *NZ Vet J* 1982;30:185–189.
24. Horton W, Jayo M, Leipold H *et al.* Bovine achondrogenesis: evidence for defective chondrocyte differentiation. *Bone* 1987;8:191–197.
25. Agerholm J. Inherited disorders in Danish cattle. *APMIS* 2007;115(Suppl. 122):1–76.
26. Seligmann C. Cretinism in calves. *J Pathol Bacteriol* 1904;9:311–322.
27. Cavanagh J, Tammen I, Windsor P *et al.* Bulldog dwarfism in Dexter cattle is caused by mutations in ACAN. *Mamm Genome* 2007;18:808–814.

28. Evans K. *A Historic Angus Journey: The American Angus Association 1883–2000*. St Joseph, MO: American Angus Association, 2001.
29. Mishra B, Reecy J. Mutations in the limbin gene previously associated with dwarfism in Japanese brown cattle are not responsible for dwarfism in the American Angus breed. *Anim Genet* 2003;34:311–312.
30. Koltes J, Mishra B, Kumar D et al. A nonsense mutation in cGMP-dependent type II protein kinase (PRKG2) causes dwarfism in American Angus cattle. *Proc Natl Acad Sci USA* 2009;106:19250–19255.
31. Agerholm J, McEvoy F, Arnbjerq J. Brachyspina syndrome in a Holstein calf. *J Vet Diagn Invest* 2006;18:418–422.
32. Agerholm J, Peperkamp K. Familial occurrence of Danish and Dutch cases of the bovine brachyspina syndrome. *BMC Vet Res* 2007;3:8.
33. Testoni S, Diana A, Olzi E, Gentile A. Brachyspina syndrome in two Holstein calves. *Vet J* 2008;177:144–146.
34. Mohr O, Wriedt C. Short spine, a new recessive lethal in cattle; with a comparison of the skeletal deformities in short-spine and in amputated calves. *J Genet* 1930;22:279–297.
35. Agerholm J, Bendixen C, Andersen O, Arnbjerg J. Complex vertebral malformation in Holstein calves. *J Vet Diagn Invest* 2001;13:283–289.
36. Agerholm J, Andersen O, Almskou M et al. Evaluation of the inheritance of the complex vertebral malformation syndrome by breeding studies. *Acta Vet Scand* 2004;45:133–137.
37. Nielsen U, Aamand G, Andersen O, Bendixen C, Neilsen V, Agerholm J. Effects of complex vertebral malformation on fertility traits in Holstein cattle. *Livestock Prod Sci* 2003;79:233–238.
38. Duncan R, Carrig C, Agerholm J, Bendixen C. Complex vertebral malformation in a Holstein calf: report of a case in the USA. *J Vet Diagn Invest* 2001;13:333–336.
39. Revell S. Complex vertebral malformation in a Holstein calf in the UK. *Vet Rec* 2001;149:659–660.
40. Wouda W, Visser I, Borst G, Vos J, Zeeuwen A, Peperkamp N. Developmental anomalies in aborted and stillborn calves in The Netherlands. *Vet Rec* 2000;147:612.
41. Nagahata H, Oota H, Nitanai A et al. Complex vertebral malformation in a stillborn Holstein calf in Japan. *J Vet Med Sci* 2002;64:1107–1112.
42. Thomsen B, Horn P, Panitz F et al. A missense mutation in the bovine SLC35A3 gene, encoding a UDP-N-acetylglucosamine transporter, causes complex vertebral malformation. *Genome Res* 2006;16:97–105.
43. Denholm L. Congenital contractural arachnodactyly ("fawn calf syndrome") in Angus cattle. New South Wales Government, Department of Primary Industries, Agriculture. 2010. Available at http://www.dpi.nsw.gov.au/agriculture/livestock/health/specific/cattle/ca-angus
44. American Angus Association. Fawn calf syndrome update. American Angus Association, St Joseph, MO, 2010. https://www.angus.org/pub/CA/CAInfo.aspx
45. Steffen D, Beever J. Fawn calf syndrome update as of December 21, 2009. American Angus Association, St Joseph, MO, 2009. https://www.angus.org/Pub/CA/FC_Notice_122109.aspx
46. Fasquelle C, Sartelet A, Li W et al. Balancing selection of a frame-shift mutation in the MRC2 gene accounts for the outbreak of the crooked tail syndrome in Belgian Blue cattle. *PLoS Genet* 2009;5(9):e1000666.
47. Sartelet A, Klingbeil P, Franklin C et al. Allelic heterogeneity of crooked tail syndrome: result of balancing selection? *Anim Genet* 2012;43:604–607.
48. Ghanem M, Nakao T, Nishibori M. Deficiency of uridine monophosphate synthase (DUMPS) and X-chromosome deletion in fetal mummification in cattle. *Anim Reprod Sci* 2006;91:45–54.
49. Shanks R, Dombrowski D, Harpestad G, Robinson J. Inheritance of UMP synthase in dairy-cattle. *J Hered* 1984;75:337–340.
50. Kuhn M, Shanks R. Association of deficiency of uridine monophosphate synthase with production and reproduction. *J Dairy Sci* 1994;77:589–597.
51. Shanks R, Greiner M. Relationship between genetic merit of Holstein bulls and deficiency of uridine-5′-monophosphate synthase. *J Dairy Sci* 1992;75:2023–2029.
52. Jones M. Pyrimidine nucleotide biosynthesis in animals: genes, enzymes, and regulation of UMP biosynthesis. *Annu Rev Biochem* 1980;49:253–279.
53. Schwenger B, Tammen I, Aurich C. Detection of the homozygous recessive genotype for deficiency of uridine monophosphate synthase by DNA typing among bovine embryos produced in-vitro. *J Reprod Fertil* 1994;100:511–514.
54. Gundlach A, Dodd P, Grabara C et al. Deficit of spinal cord glycine/strychnine receptors in inherited myoclonus of Poll Hereford calves. *Science* 1988;241:1807–1810.
55. Gundlach A. Disorder of the inhibitory glycine receptor: inherited myoclonus in Poll Hereford calves. *FASEB J* 1990;4:2761–2766.
56. Pierce K, Handford C, Morris R et al. A nonsense mutation in the alpha1 subunit of the inhibitory glycine receptor associated with bovine myoclonus. *Mol Cell Neurosci* 2001;17:354–363.
57. Couldrey C, Lee R. DNA methylation patterns in tissues from mid-gestation bovine foetuses produced by somatic cell nuclear transfer show subtle abnormalities in nuclear reprogramming. *BMC Dev Biol* 2010;10:27.
58. Doornbos M, Sikkema-Raddatz B, Ruijvenkamp C et al. Nine patients with a microdeletion 15q11.2 between breakpoints 1 and 2 of the Prader–Willi critical region, possibly associated with behavioural disturbances. *Eur J Med Genet* 2009;52:108–115.
59. Flisikowski K, Venhoranta H, Bauersachs S et al. Truncation of MIMT1 gene in the PEG3 domain leads to major changes in placental gene expression and stillbirth in cattle. *Biol Reprod* 2012;87:140.
60. American Angus Association. Neuropathic hydrocephalus (NH) fact sheet. American Angus Association, St Joseph, MO, 2010. http://www.angus.org/pub/nh/nhfactsheet.pdf
61. Beever J. An update on neuropathic hydrocephalus in cattle. American Angus Association, St Joseph, MO, 2009. http://www.angus.org/nh_summary.pdf
62. Nietfeld J. Osteopetrosis in calves. Diagnostic insights, Kansas State Diagnostic Laboratory, Manhattan, KS, 2007.
63. Meyers S, McDaneld T, Swist S et al. A deletion mutation in bovine SLC4A2 is associated with osteopetrosis in Red Angus cattle. *BMC Genomics* 2010;11:337.
64. Testoni S, Militerno G, Rossi M, Gentile A. Congenital facial deformities, ascites and hepatic fibrosis in Romagnola calves. *Vet Rec* 2009;164:693–694.
65. Testoni S, Bartolone E, Rossi M et al. KDM2B is implicated in bovine lethal multi-organic developmental dysplasia. *PLoS ONE* 2012;7:e45634.
66. Fukuda T, Tokunaga A, Sakamoto R, Yoshida N. Fbxl10/Kdm2b deficiency accelerates neural progenitor cell death and leads to exencephaly. *Mol Cell Neurosci* 2011;46:614–624.
67. Windsor P, Cavanagh J, Tammen I. Hydrops fetalis associated with pulmonary hypoplasia in Dexter calves. *Aust Vet J* 2006;84:278–281.
68. Agerholm J, Arnbjerg J. Pulmonary hypoplasia and anasarca syndrome in a belted Galloway calf. *Vet Rec* 2011;168:190.
69. Steffen D. Bovine congenital malformations of current concern. In: *25th Annual American College of Veterinary Internal Medicine Forum, Seattle, Washington*, 2007, pp. 241–242.
70. Shanks R, Robinson J. Embryonic mortality attributed to inherited deficiency of uridine monophosphate synthase. *J Dairy Sci* 1989;72:3035–3039.

71. Veerkamp R, Beerda B. Genetics and genomics to improve fertility in high producing dairy cows. *Theriogenology* 2007;68:S266–S273.
72. VanRaden P, Miller R. Effects of nonadditive genetic interactions, inbreeding, and recessive defects on embryo and fetal loss by seventy days. *J Dairy Sci* 2006;89:2716–2721.
73. Charlier C, Coppieters W, Rollin F *et al.* Highly effective SNP-based association mapping and management of recessive defects in livestock. *Nature Genet* 2008;40:449–454.
74. VanRaden P, Olson K, Null D, Hutchinson J. Harmful recessive effects on fertility detected by absence of homozygous haplotypes. *J Dairy Sci* 2011;94:6153–6161.
75. Young G. A case of tibial hemimelia in cattle. *Br Vet J* 1951;107:23–28.
76. Pollock D, Fitzsimons J, Deas W, Fraser J. Pregnancy termination in the control of the tibial hemimelia syndrome in Galloway cattle. *Vet Rec* 1979;104:258–260.
77. Ojo S, Guffey M, Saperstein G, Leopold H. Tibial hemimelia in Galloway calves. *J Am Vet Med Assoc* 1974;165:548–550.
78. Ko J, Evans L, Haynes J. Multiple congenital defects in a female calf: a case report. *Theriogenology* 1990;34:181–187.
79. Lapointe J, Lachance S, Steffen D. Tibial hemimelia, meningocele, and abdominal hernia in Shorthorn cattle. *Vet Pathol* 2000;37:508–511.
80. Marron B, Thurnau G, Hannon C, Beever J. Mapping of the locus causing tibial hemimelia (TH) in Shorthorn cattle. In: *Plant and Animal Genome XIII, San Diego, California*, 2005.

Chapter 67

Abnormal Offspring Syndrome*

Charlotte E. Farin, Callie V. Barnwell and William T. Farmer

Department of Animal Science, North Carolina State University, Raleigh, North Carolina, USA

Utilization of embryo biotechnologies in cattle

Development of nonsurgical embryo transfer techniques[1-3] led to the establishment of the bovine embryo transfer industry during the 1970s and 1980s. In nonsurgical embryo transfer, valuable female cattle are superovulated by treatment with follicle-stimulating hormone (FSH)-containing preparations, bred by artificial insemination (AI), and then subjected to nonsurgical uterine lavage, usually 7 days after insemination, to recover embryos at the morula and blastocyst stages of development. On average, five to seven embryos are recovered from a donor cow and individual embryos are then nonsurgically transferred to recipient females, who then carry the pregnancy to term. In 1997 approximately 360 000 bovine embryos produced *in vivo*, primarily from superovulated donor cows, were being transferred annually. By 2001, this number had risen to 452 546,[4,5] and in the decade between 2001 and 2011 the number of commercial *in vivo* produced bovine embryo transfers roughly stabilized, averaging approximately 550 000 transfers annually worldwide.[5,6]

With the development of systems for *in vitro*, or laboratory-based, production of bovine embryos, hundreds of thousands of calves have resulted from the commercial transfer of *in vitro* produced (IVP) embryos. From 2001 through 2011, the number of commercial IVP bovine embryos transferred annually rose more than 10-fold, from a total of 30 260 transfers in 2001 to 373 869 in 2011 (Figure 67.1).[5,6] Thus, *in vitro* embryo production and transfer now represents the actively growing segment of the commercial bovine embryo transfer industry. Advantages of using *in vitro* embryo production systems include the ability to produce embryos from valuable female cattle whose oviducts or uterus may have been damaged due to disease or dystocia as well as the ability to fertilize oocytes from a single collection of a donor female to different sires. In addition, the use of transferred IVP embryos is an effective method to improve pregnancy rate in dairy cows exposed to summer heat stress compared with that following AI.[7-9]

In vitro production of bovine embryos is also a key technology supporting the application of more advanced embryo technologies such as gene transfer and somatic cell nuclear transfer (SCNT). From a commercial perspective, cloned cattle have been developed as bioreactors for producing biopharmaceuticals, critical human blood components, and modified milk components.[10-14] In addition, genetic modification via nuclear transfer has been used to study the function and importance of individual genes and their involvement in diseases, developmental origins, and other physiological processes in both cattle and swine.[15-19] SCNT has also been suggested as a potential technology that could be applied to rescue endangered or extinct species.[20,21]

Recognition of the effects of embryo manipulation on offspring development

Manipulation of bovine embryos became a reality with the development of techniques for embryo recovery and transfer. As noted above, the development of the commercial embryo transfer industry in cattle soon followed the introduction of nonsurgical embryo transfer techniques.[1-3] Bovine embryos retrieved from donor dams could then be handled in the laboratory (*in vitro*) prior to their transfer to recipient females. Initially, embryos spent little time in the laboratory between transfers but as embryo culture medium was developed, embryos could be successfully maintained in the laboratory for a number of hours or even in overnight culture prior to transfer. In commercial practice, however, bovine embryos were rarely maintained outside the animal for longer than few hours prior to transfer. Assessment of development following transfer and calving supported the conclusion that normal development of offspring was associated with production by embryo transfer.[22]

Coincident with the commercialization of embryo transfer techniques in cattle was the advancement of manipulation-intense methods to study development of mammalian embryos in species such as mice[23,24] and sheep.[25,26] The

---

*This chapter is dedicated to the memory of Peter W. Farin DVM, PhD, Diplomate ACT (1954–2011).

Bovine Reproduction, First Edition. Edited by Richard M. Hopper.
© 2015 John Wiley & Sons, Inc. Published 2015 by John Wiley & Sons, Inc.

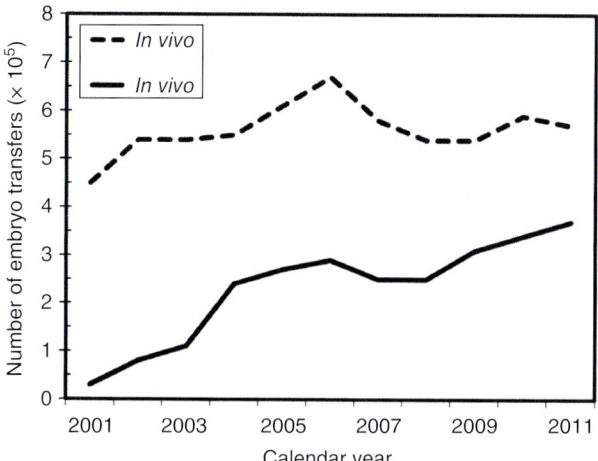

Figure 67.1 Number of commercial transfers of *in vivo*-produced or *in vitro*-produced bovine embryos worldwide between 2001 and 2011.

intersection of investigations in these areas quickly led to the application of *in vitro* manipulation of bovine embryos in an attempt to improve efficiency of commercial production, initially through the application of embryo splitting and later through embryo cloning or nuclear transfer (NT). For splitting, embryos were recovered nonsurgically, typically at the morula to early blastocyst stage of development, bisected to produce identical twins and transferred within hours into suitable recipient females using nonsurgical methods.[27–29] Splitting and other biopsy techniques were also applied for determination of embryo sex with one resulting demi-embryo transferred and the other portion used for sexing.[30] As calves from these procedures continued to be produced, there were no reports of developmental abnormalities of offspring derived from the use of these methods. In contrast to reports for bisected bovine embryos, development of cattle embryos produced by NT using embryonic blastomeres as a source of donor nuclei[26] resulted in reports of the production of oversized offspring and evidence of abnormal fetal and placental development.[31–33]

Coincident with the development of manipulation-intensive technologies such as embryo bisection and NT, techniques to successfully fertilize bovine oocytes *in vitro*[34] or successfully *in vitro* mature, *in vitro* fertilize, and *in vitro* culture bovine embryos up to the 8- to 16-cell stage of development were also established.[35,36] However, few IVP bovine embryos were transferred because they could not be cultured beyond the 8- to 16-cell stage of development *in vitro*.[37,38] Because nonsurgical embryo transfer had by then become the industry standard, only embryos at the morula (32–64 cells) or blastocyst (100–128+ cells) stages of development were considered suitable candidates for transfer. In the late 1980s, the use of oviductal cell coculture as a method to successfully overcome the developmental arrest of cultured bovine embryos at the 8- to 16-cell stage was established.[39] As the decade of the 1990s progressed, blastocyst-stage bovine embryos could be routinely produced from bovine oocytes entirely matured *in vitro*, fertilized *in vitro*, and cultured *in vitro*.[40–42]

An early report of calf development following the transfer of IVP bovine embryos,[38] although anecdotal, initially suggested a pattern of pregnancy loss during the second and third trimesters associated with the transfer of IVP bovine embryos. Similarly, a report of the development of sheep zygotes produced *in vivo* and recovered surgically from the oviduct that were then cultured *in vitro* for 5 days[43] demonstrated that prolonged exposure of embryos to *in vitro* conditions prior to embryo transfer resulted in extended gestation lengths, increased lamb birthweights, and greater mortality compared with noncultured controls. In 1995, three papers were published[44–46] documenting the occurrence of oversized fetuses or calves that resulted from the transfer of IVP bovine embryos. In addition, increased perinatal mortality was associated with fetuses and calves resulting from the transfer of embryos produced *in vitro*.[44,45] In each of these cases, *in vitro* oocyte maturation was used in conjunction with *in vitro* fertilization and *in vitro* culture (IVC) for production of the embryos, fetuses and calves studied. This confounding made impossible the assignment of the effects of individual components of the *in vitro* process on the production of oversized fetuses and calves. Thus for cattle it was not possible to clearly determine if only IVC was responsible for production of oversized individuals as suggested from the data in sheep[47] or whether other components of the *in vitro* production process also contributed. Collectively, however, these observations were the first to suggest that manipulation of the environment during bovine *in vitro* oocyte maturation, fertilization, and preimplantation culture contributed to alterations in subsequent gestational and perinatal development. Moreover, the developmental alterations were largely reminiscent of those observed in some of the offspring that resulted from blastomere cloning (NT)[31–33] and, later, SCNT[15,48,49] in cattle.

Classification system for abnormal offspring syndrome

As noted above, transfer of IVP, NT or SCNT manipulated bovine embryos results in a proportion of conceptuses, fetuses or offspring that exhibit developmental abnormalities or abnormalities in their associated placental tissues.[32,44,45,50,51] Because the occurrence of an overgrowth phenotype of the fetus or offspring was the most striking observation initially identified as resulting from the transfer of either IVP or SCNT embryos, the phenomenon initially became known as "large offspring syndrome."[31–33,44,47,50] However, as more investigators began studying the phenomenon, it became clear that not all pregnancies from IVP and SCNT embryos resulted in phenotypes that exhibited excessive fetal overgrowth.[52–58] Therefore, the term "abnormal offspring syndrome" (AOS) was used to more accurately describe the range of characteristics found associated with this syndrome.[59]

Designation of this syndrome as AOS is useful when discussing developmental anomalies associated with *in vitro* manipulation of embryos in other mammalian species. In swine, for example, an overgrowth phenotype was not associated with SCNT.[60,61] However, cloned pigs can exhibit other developmental aberrations that would still be appropriately described as AOS.[60] Because the etiology of AOS-like characteristics is associated with epigenetic alterations to the genome resulting from *in vitro* embryo manipulations, recognition that AOS occurs in a variety of mammalian species rather than only in cattle and sheep will assist in further defining all aspects of this syndrome.

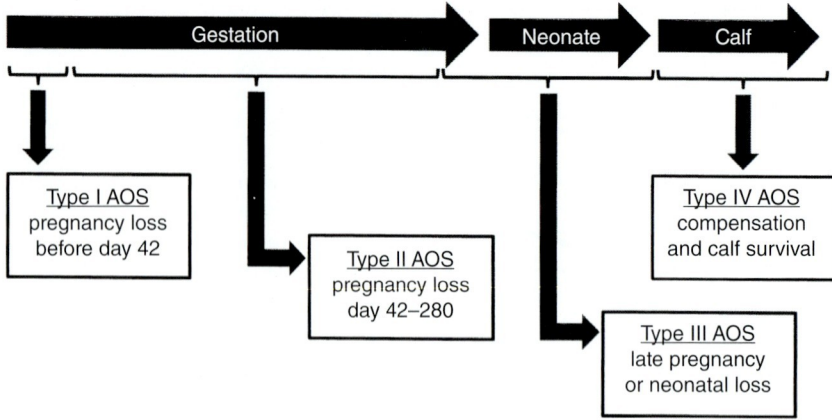

Figure 67.2 Classification of the subtypes of abnormal offspring syndrome (AOS).

Characteristics associated with AOS in cattle have been grouped into four types[59] (Figure 67.2). In type I AOS, abnormalities are exhibited in pregnancy establishment. In cattle, this is during the period between days 0 and 42 of gestation when organogenesis is occurring. In type II AOS, abnormalities occur that are associated with failure to maintain the fetoplacental unit resulting in abortion during the period following organogenesis until term, namely between days 42 and 280 of bovine gestation. In type III AOS, a range of abnormalities can occur in the fetoplacental unit that are either less severe or that permit compensation by the fetus and placenta, thus allowing the pregnancy to go to term; however, survival of the fetus during parturition or during the neonatal period does not occur, resulting in neonatal mortality. These compromised nonsurviving offspring exhibit altered clinical, hematological, or biochemical parameters. In type IV AOS, a range of abnormalities can occur in fetoplacental development that are either less severe or permit compensation by the fetus and placenta. In this case, the pregnancy succeeds in going to term and the neonatal calf survives; however, the offspring may still exhibit differences in anatomy, physiology, or metabolism compared with animals from nonmanipulated embryos.

Production of embryos *in vitro* is associated with a greater incidence of early and mid-gestational loss.[45,59] These cases represent type II AOS, with pregnancy losses typically occurring between days 40 and 90 of gestation.[45] Similarly, the majority of SCNT pregnancies are lost between days 40 and 90 of gestation[15,62] and would also be classified as type II AOS. In this case, losses are typically associated with improper development of the placenta.[62] When SCNT pregnancies do survive, 65–70% of neonates exhibit some form of AOS and their condition would be classed as either type III or IV AOS. In some instances, intensive care is required to support survival of the neonate, without which the neonate typically does not survive (type III AOS), whereas in others little or no assistance is needed (type IV AOS).[57]

Characteristics associated with AOS during bovine development

In cattle, the expression of anomalies associated with AOS can range from changes in gene expression in preimplantation-stage embryos[63] to moderate alterations with little apparent phenotypic compromise[57] to severe alteration in phenotype accompanied by abortion or neonatal death[59,62,64] (Figure 67.3). Expression of AOS is influenced by culture conditions,[58,65,66] species,[61] extent of embryo manipulation,[15,67] and even levels of maternal nutrients.[68] At any specific stage of development, every embryo, fetus or calf does not exhibit a uniform set of AOS characteristics. Therefore, we have sought to compile observations reported as altered characteristics associated with the manipulation of bovine embryos *in vitro* at various stages of development. In each case, however, a direct correspondence between the type of manipulation, altered characteristic, and occurrence of AOS has not always been established.

Gametes

Sperm

There is a limited number of reports which suggest that alterations in the *in vitro* environment of spermatozoa prior to fertilization can be associated with abnormalities in subsequent embryo development.[69,70] For example, exposure of bull sperm to polyvinylpyrrolidone (PVP), an agent commonly included in handling medium for intracytoplasmic sperm injection,[70] increased the incidence of acrosome-reacted sperm following 30 or 60 min of PVP exposure and decreased the proportion of oocytes fertilized by PVP-exposed sperm following coincubation with *in vitro* matured oocytes.[69] However, it is unclear whether exposure of sperm to PVP prior to fertilization had any effect on the subsequent development of the resulting zygotes in terms of morphology, time of cleavage, time of blastocoel formation, or inner cell mass and trophoblast cell number.

Oocytes

Oocytes at the metaphase II stage of meiosis are ovulated from preovulatory follicles that, in cattle, typically range from 18 to 22 mm in diameter. In contrast to litter-bearing species such as swine, it is technically difficult to obtain large numbers of *in vivo* matured bovine oocytes for laboratory manipulations. Therefore, bovine oocytes used for both IVP and SCNT procedures are matured *in vitro*. Bovine oocytes used for *in vitro* production systems are derived

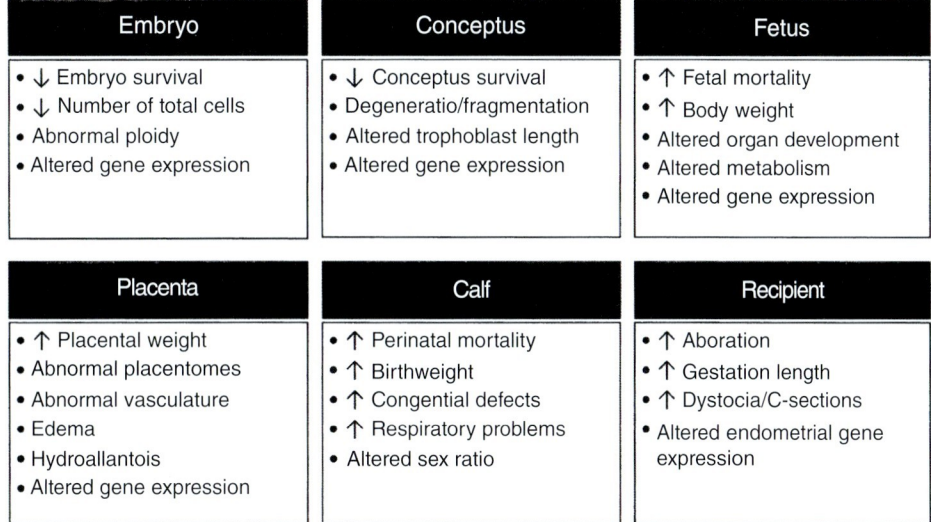

Figure 67.3 Identification of abnormalities commonly associated with abnormal offspring syndrome (AOS). In individual cases, each abnormality listed may or may not be present.

from follicles usually 2–10 mm in diameter and are recovered at the dictyate stage within prophase I of meiosis. These dictyate oocytes must be subjected to *in vitro* maturation (IVM) for 20–24 hours during which they reach the metaphase II stage of meiosis and become ready for fertilization. Whereas 50–80% of *in vivo* matured oocytes develop to the blastocyst stage following fertilization, only 15–40% of *in vitro* matured oocytes reach the blastocyst stage in standard *in vitro* production systems.[71–75] Furthermore, blastocysts from oocytes that were matured *in vitro* result in fewer numbers of cells per embryo[75] and within the inner cell mass[76–78] than blastocysts from oocytes matured *in vivo*. Similarly, development of transgenic blastocysts was greater from oocytes that were matured *in vivo* compared with oocytes that were matured *in vitro*.[79] In sheep, more pregnancies were produced from SCNT blastocysts derived from *in vivo* matured oocyte cytoplasts compared with SCNT blastocysts from *in vitro* matured oocyte cytoplasts.[80] In contrast, there was no difference in pregnancy rate following transfer of transgenic embryos at the morula or blastocyst stages based on whether they were from *in vivo* versus *in vitro* maturation conditions.[79] However, for transgenic blastocysts developed from bovine oocytes matured *in vivo* versus those matured *in vitro*, there was a difference in the sex ratio of the calves, with a higher percentage of males born in the *in vitro* matured group compared with the *in vivo* fertilized group.[79] Interestingly, the mean birthweights of male or female transgenic calves resulting from the transfer of blastocysts from oocytes matured *in vitro* were significantly greater than those for male or female calves produced from blastocysts from oocytes matured *in vivo*.[79] In addition, 48% of calves from *in vitro* matured oocytes were 45 kg or more at birth, whereas only 19% of calves from *in vivo* produced oocytes were 45 kg or more at birth. Finally, female calves produced from *in vitro* matured oocytes were significantly heavier than male calves produced from *in vivo* matured oocytes. Taken together, observations in both cattle and sheep support the conclusion that differences between *in vitro* and *in vivo* oocyte maturation conditions plays an important role in determining the occurrence of at least type IV AOS and possibly other AOS subtypes as well.

Other factors that can affect oocyte developmental competence include body condition score and diet.[81] Oocytes recovered from heifers in moderate body condition that were fed diets containing high levels of metabolizable energy showed reduced yields of IVP blastocysts whereas oocytes from heifer that were in low body condition on the same high dietary energy levels showed improved *in vitro* blastocyst development.[81] Oocytes from heifers on diets designed to generate elevated levels of plasma ammonia demonstrated reduced *in vitro* developmental competence (cleavage to 2-cell and blastocyst stages) compared with controls.[82] In sheep, embryos produced *in vitro* from oocytes exposed *in vivo* to elevated dietary urea nitrogen experienced reduced pregnancy rates following transfer.[68] Taken together, these observations provide insight into possible mechanisms accounting for early observations that diets high in protein are associated with reduced fertility in dairy cattle[83] and also provide insight into potential mechanisms contributing to the occurrence of type I AOS.

Preimplantation development

In cattle, embryos are typically transferred on day 7 of gestation at the blastocyst stage of development. The greatest degree of pregnancy loss in cattle occurs within the first 2–3 weeks of gestation, with the majority of losses occurring between days 8 and 16.[84] Compared with AI or the transfer of *in vivo* produced embryos, IVP embryos can experience even greater rates of early embryonic death; however, this can be influenced by the quality of the embryo transferred. For example, survival rates for bovine embryos of quality grade 1 were the same whether derived from *in vivo* or *in vitro* production systems; however, for embryos of quality grade 2, survival rates were lower if from *in vitro* production systems.[45] Pregnancy rates and calving rates following the transfer of IVP embryos vary both across and within laboratories, although they can be lower than those seen following

AI or the transfer of *in vivo* produced embryos.[54,55,85-88] Other factors such as embryo evaluator[89] and recipient characteristics[86] can also influence the successful establishment and maintenance of pregnancy following transfer of nonfrozen *in vivo* and IVP bovine embryos.

Preimplantation-stage bovine embryos produced *in vitro* differ in morphology compared with *in vivo* produced embryos. If produced in serum-containing medium, bovine embryos appear darker, have a less-organized inner cell mass, and exhibit a less complete compaction.[76,78] IVP embryos, regardless of whether they are produced in serum-containing or serum-free media, have greater numbers of lipid droplets, more immature mitochondria, and do not exhibit a comparable increase in mitochondrial density as they transition from the morula to blastocyst stage compared with *in vivo* embryos.[53,90-94]

Total cell number per blastocyst and allocation of blastomeres to the inner cell mass of blastocysts also differs between *in vitro* and *in vivo* production systems. IVP blastocysts demonstrate fewer total cells and a smaller proportion of cells allocated to the inner cell mass.[94,95] Differences also exist in the extent of abnormal ploidy (chromosome number) among blastomeres of embryos produced *in vivo* compared with their *in vitro* counterparts.[96] Whereas only 20–25% of blastocysts from *in vivo* production systems demonstrate mixoploidy, approximately 45–70% of IVP blastocysts were determined to be mixoploid with the incidence dependent on the method used to detect chromosome number.[97,98] Although embryos carrying increased proportions of mixoploidy can successfully establish and maintain pregnancy,[96] severe increases in the number of mixoploid blastomeres are not conducive to embryo survival.[99]

Early comparisons of gene expression for bovine preimplantation embryos produced *in vivo* and *in vitro* demonstrated that expression of both imprinted and nonimprinted genes can be affected by *in vitro* embryo production.[100-109] Additional support for the hypothesis that the competence of blastocysts to establish pregnancy is associated with expression of specific gene sets was provided by investigators who analyzed mRNA expression from biopsies of blastocysts taken prior to embryo transfer and with their subsequent post-transfer events. Blastocysts that were not destined to establish pregnancy were found to express elevated levels of the implantation inhibitor CD9, as well as the inflammatory cytokine tumor necrosis factor (TNF) and mRNAs associated with carbohydrate metabolism.[110] Conversely, blastocysts whose transfer resulted in delivery of a calf demonstrated increased expression of a number of specific mRNAs, including some associated with implantation (COX2, CDX2, PLAC8) and differentiation (BMP15). These observations were consistent with an earlier report correlating mRNA expression profiles with post-transfer pregnancy success of demi-embryo pairs produced by splitting blastocysts obtained from superovulated donor cows.[111] These investigators also noted elevated expression of PLAC8 in demi-embryos whose twinned demi-blastocyst was associated with calf delivery following transfer. To address the question of when mRNA profiles predictive of developmental competence could be determined during development, Held *et al*.[112] assessed the mRNA expression pattern from isolated blastomeres from 2-cell IVP bovine embryos with the developmental competence of their sister blastomeres. These investigators determined that sister blastomeres destined to develop to the blastocyst stage expressed elevated levels of mRNAs associated with oxidative stress responses, antioxidant activity, and oxidative phosphorylation.[112] Taken together, these observations suggest that mRNA signatures of embryos at various stages of development can be associated with both pre- and post-transfer developmental outcome. Furthermore, they are consistent with the hypothesis that, in the future, specific mRNA signatures for embryos destined to exhibit specific AOS subtypes following embryo transfer may also be identifiable. This, in turn, will support development of pre-transfer embryo screening procedures or refinement of *in vitro* embryo production methods to eliminate the occurrence of AOS. For example, inclusion of specific supplements to medium used for production of bovine embryos *in vitro*, including insulin-like growth factor (IGF)-1, colony-stimulating factor (CSF)-2 and hyaluronan, have been demonstrated to improve pregnancy rate after transfer.[113] Interestingly, addition of CSF-2 into culture medium eliminated the occurrence of embryonic loss prior to day 35 of gestation (type I AOS); however, the number of observations in this study was somewhat limited.[113]

Peri-implantation development

It is estimated that moderate-producing dairy cows experience about a 40% embryonic and fetal mortality rate; about 70–80% of this loss is sustained between days 8 and 16 of gestation.[114] Maternal recognition of pregnancy occurs around day 15–16 of gestation in cattle, and requires adequate expression of interferon (IFN)-tau from the developing conceptus to prevent the luteolytic cascade.[115,116] Higher rates of pregnancy loss occur for conceptuses around the peri-implantation period following the transfer of IVP embryos.[53,105,117,118] Similarly, significant embryonic loss has been reported in the 2-week period following the transfer of IVP and SCNT embryos, although a portion of this loss could be attributed to difficulties in recovering these conceptuses. For example, recovery rates of 51.2% at day 12 and 33–38% at day 14 were found following the transfer of IVP embryos.[105,119] At day 14, 75% of IVP embryos transferred were lost, compared with 85–93% for SCNT embryos.[118] At day 25 of gestation, only 19% and 15% of conceptuses derived from IVP and cloned embryos, respectively, were recovered compared with 67% of conceptuses produced by AI.[120]

In addition to lower recovery rates, conceptuses derived from IVP embryos also exhibited some morphological differences from those derived *in vivo*. For conceptuses derived from IVP embryos that were recovered on day 17 of gestation, there was an increased incidence of fragmentation and/or degeneration compared with *in vivo*-derived controls.[45] In addition, IVP conceptuses are also longer than *in vivo*-derived conceptuses both at day 12 and day 17 of gestation.[105,117] In contrast, other investigators found no statistical difference in the recovery rate of intact conceptuses on day 16 between IVP and *in vivo*-derived groups; however, they did report that there were more total structures or degenerated nonhatched embryos recovered in the IVP group compared with *in vivo*-derived controls.[104] In that study, conceptuses derived from IVP embryos and recovered on day 16 of gestation had smaller embryonic disks and tended to have shorter trophoblastic lengths compared with in vivo-derived controls. Differences in conceptus length between studies could be attributed to differences in in vitro conditions used to produce the embryos as well as to the timing of recovery relative to maternal recognition of pregnancy.

Compared with conceptuses derived from the transfer of IVP embryos, 20% fewer conceptuses from SCNT embryos had an embryonic disk at recovery on day 17.[121] Recovery rate is generally higher when an embryonic disk is detected with no signs of conceptus degeneration.[119] Culture conditions can influence the presence of an embryonic disk in developing conceptuses; one study reported a higher percentage of embryonic disks present when KSOM medium (72%) was utilized compared with SOF medium (46%).[119]

Peri-implantation conceptuses derived from IVP or SCNT embryos may have different developmental and ultrastructural properties.[118] Both IVP and SCNT conceptuses were developmentally delayed at day 14 and day 21 compared with the expected developmental timeline described for in vivo conceptuses.[118] When comparing conceptuses from IVP- and SCNT-derived embryos, SCNT conceptuses at both days 14 and 21 of gestation were shorter in length and displayed more ultrastructural deficiencies at both time points compared with IVP conceptuses.[118] For example, SCNT conceptuses exhibited a greater occurrence of apoptosis, more vacuoles, fewer mitochondria in the hypoblast, disintegration of the basement membrane between the hypoblast and epiblast, less defined tight junctions between trophectoderm cells, fewer desmosomes, and fewer polyribosomes.

Differences in gene expression have been observed between conceptuses derived from the transfer of IVP or in vivo-produced embryos.[104] For example, in vivo control embryos showed a temporal reduction in transcriptional activity from day 7 to day 16 that was not seen during this period in IVP embryos. Furthermore, on day 16 of gestation conceptuses from IVP-derived embryos had greater expression of IGF-1R mRNA transcript compared with in vivo controls.[104] While day 7 IVP blastocysts had a lower relative level of IFN-tau transcript compared with in vivo-produced controls, developmentally delayed day 16 IVP conceptuses had greater content of IFN-tau mRNA.[104]

Compared with IVP conceptuses, conceptuses derived from the transfer of SCNT embryos lacked expression of NANOG and FGF4, and overexpressed IFN-tau mRNA at day 17 of gestation.[121] Using a custom microarray, 47 genes were identified as differentially expressed at day 17 between conceptuses from SCNT and IVP embryos. Furthermore, expression of genes involved in trophoblast proliferation (Mash2), differentiation (Hand1), and trophoblast function (PAG-9) were altered in SCNT conceptuses at day 17 compared with conceptuses from either IVP or AI.[122] At day 25 of gestation, conceptuses derived from IVP or NT embryos had a greater expression of IGF-1 mRNA compared with AI controls. In addition, conceptuses derived from SCNT embryos had greater expression of IGF-2 mRNA than IVP-derived conceptuses.[120] Interestingly, gene expression profiles from uterine endometria of pregnancies derived from SCNT and IVP embryos differ.[123] These observations support the suggestion that during the peri-implantation period, abnormal embryo–maternal communication could be responsible for the altered phenotype in cloned animals.[123] In effect, altered expression of genes vital to normal trophoblast differentiation and subsequent placental development associated with IVP and SCNT may be factors contributing to increased pregnancy losses and other abnormalities observed later in development.

Fetal development

Development of fetuses following the transfer of IVP or SCNT embryos is associated with increased fetal mortality and a multitude of fetal abnormalities. Following the period of embryonic loss in cattle, fetuses derived from IVP or SCNT embryos experience higher rates of gestational loss compared with those produced in vivo (Figure 67.4). A large-scale study analyzing results from over 30 datasets on IVP and SCNT reported that after adjusting for effects such as season, parity, and sex of calf, fetal losses were higher for both IVP and SCNT embryos compared with AI or embryo transfer pregnancies.[55] Furthermore, IVP and SCNT calves experienced longer gestation lengths, higher birthweights, and increased incidences of dystocia and perinatal losses compared with AI or embryo transfer calves.

Higher rates of pregnancy loss following the transfer of IVP embryos are seen during the late embryonic period (day 30) and early fetal period up to day 90 of gestation compared with in vivo controls.[59] While the rate of pregnancy loss from approximately 2 months of gestation to term is less than 10% in recipients carrying in vivo-produced embryos,[124–126] this loss can be two to three times greater in recipients carrying IVP embryos.[7,54,127,128] IVP pregnancies experience embryonic and fetal losses particularly between days 30 and 65 of gestation, with the majority of the losses occurring between days 30 and 44.[128] By comparison, pregnancies derived from embryos produced in vivo had lower rates of loss that were distributed more evenly between days 30 and 90.[128] Furthermore, pregnancy rate at day 60 of gestation was 42% lower in recipients with grade 1 IVP embryos compared to those with grade 1 in vivo-produced embryos.[45] Pregnancy rate at day 50–60 of gestation following the transfer of grade 1 fresh IVP embryos (59%) was significantly lower compared

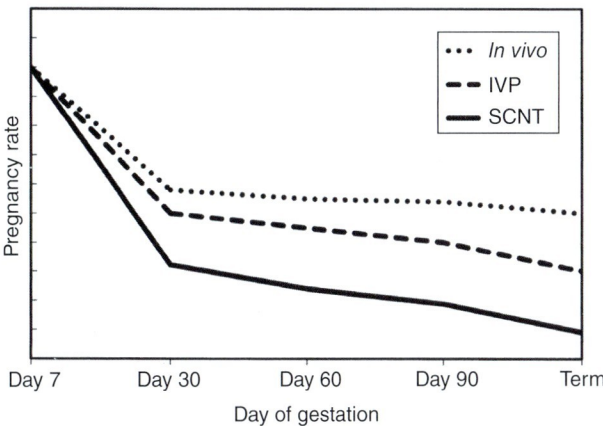

Figure 67.4 Relative timing of pregnancy losses following the transfer of bovine embryos produced in vivo, in vitro (IVP), or by somatic cell nuclear transfer (SCNT).

with that of grade 1 fresh *in vivo*-produced embryos (76%).[54] However, based on data from a large-scale study, pregnancy rate between days 90 and 150 of gestation was similar for both fresh and frozen IVP embryos (43.5%).[129]

The occurrence of early fetal mortality is even greater following the transfer of cloned (NT and SCNT) embryos. Losses of 50–100% of pregnancies from cloned embryos can occur between days 30 and 60 of gestation, compared with a 2–10% loss of pregnancies from natural service that normally occurs during this period.[15,20,62,130] A second period of high pregnancy loss occurs during the second trimester of pregnancy for cloned pregnancies.[15,62,131] In many of these cases, while few gross abnormalities may be noted in the aborted fetuses, the placentas are often grossly abnormal.[15,62,131] Cloned pregnancies may also be subjected to greater losses during the third trimester, often characterized by placental abnormalities including hydrallantois, reduced placentome number, and edema in the placental membranes.[15,62] Based on a nationwide survey in Japan on the survival of embryos and calves derived from SCNT from 1998 to 2007, the survival rate of SCNT embryos and calves had not improved during that 10-year period.[132] Pregnancy rates at day 30 following the transfer of SCNT embryos ranged from 21.6 to 32.9% and were lower compared with pregnancy rates following the transfer of *in vivo*-derived embryos (50–52%) and IVP embryos (37–46%).[132]

In addition to an increased incidence of fetal mortality, fetuses derived from IVP and SCNT embryos exhibit a variety of morphological abnormalities during early gestation. Compared with AI controls, SCNT and IVP fetuses at day 80 of gestation had increased body weight, liver weight, and thorax circumference.[133] Furthermore, these investigators found that the ratio of crown–rump length to thorax circumference was a defining morphological characteristic for the phenotype of disproportionate fetal overgrowth. In cloned fetuses, decreased crown–rump length and/or fetal heart beat were predictive ultrasound markers for fetal death that occurred in about 27% of clones that died prior to day 90 of gestation.[49] These investigators noted that, based on ultrasound assessments after day 100 of gestation, risk factors for loss of SCNT-derived fetuses included placental edema, hydrops, and increased abdominal circumference.

The frequency of congenital malformations, which include hydrallantois and abnormal limbs, is increased in calves from IVP embryos (3.2%) compared with AI controls (0.7%).[129] At day 90 of gestation, IVP pregnancies were associated with increased fetal crown–rump lengths and organ weights compared with *in vivo* controls.[134] At day 180 of gestation, IVP pregnancies were associated with greater uterine, placental and fetal weights compared with *in vivo* controls. Fetal weight, fetal liver, fetal heart, and placentome weight were all correlated and were significantly greater in IVP pregnancies.[134] At day 222 of gestation, fetuses from IVP embryos were 17% larger than *in vivo*-produced controls and exhibited significant increases in fetal body weight, heart girth, heart weight, and long-bone length.[45] Furthermore, fetuses from IVP embryos exhibited no difference in gross visceral organ weight but had smaller skeletal muscle measurements per kilogram body weight compared with fetuses from the *in vivo*-derived group, suggesting that IVP fetuses develop disproportionately to their body weight. Histologically, IVP fetuses at day 222 of gestation exhibited a ratio of secondary to primary skeletal muscle fiber number that was significantly greater than that for *in vivo*-derived controls. Similarly, the proportional volume of tissue present between myofibrils was also greater in muscle of IVP fetuses compared with controls.[135] Expression of myostatin mRNA was decreased in the skeletal muscle of fetuses from the IVP group, while expression of glyceraldehyde 3-phosphate dehydrogenase (GAPDH) tended to be increased compared with the *in vivo*-derived group. These observations are consistent with the observed changes in muscle fiber ratios since myostatin is an inhibitor of muscle fiber development and GAPDH is a key enzyme in glycolysis, a pathway that is heavily used by secondary muscle fibers.[135]

Perinatal mortality rates of 8.6–15.6% from IVP calves at birth or shortly after birth were reported in recipient heifers, and were as high as 17.9% for recipient cows.[7,54,129] In contrast, perinatal mortality rates in AI calves were approximately 6.2–9.0% for *in vivo* embryo transfer pregnancies.[22,129] Some studies have reported an increased gestation length for IVP fetuses,[129] whereas others have not.[44] The incidence of stillbirth and perinatal mortality within the first 48 hours after calving was increased in IVP calves compared with *in vivo*-derived controls but was not due to altered gestation length.[136] Stillbirth and postnatal mortality were also increased in SCNT pregnancies. Only about 70% of cloned cattle survive the perinatal period and these losses are generally sustained within the first week of postnatal life (reviewed in ref. 48). Based on the 2012 national survey by the Japanese Ministry of Agriculture, of 594 cloned calves born in Japan between 1998 and 2007, 14.8% were stillborn and 95% were dead within 24 hours after parturition.[137] The calving rate for SCNT pregnancies (5.1–12.6%) was also lower than for *in vivo* embryo transfer (23.9–34.8%) and for IVP (20.2–22.3%) pregnancies.[132] Based on data from 5 years of commercial cloning practice in the United States, Argentina and Brazil,[49] of 3374 SCNT embryos transferred, 41% were alive at day 30 of gestation but only 9% of the total embryos transferred resulted in live births. Of these 388 live SCNT calves, about 42% died between parturition and 150 days of life.[49] The efficiency of cell lines for successful production of viable SCNT pregnancies ranged from 0 to 45%, with about 24% of cell lines failing to produce live calves.[49]

The most conspicuous abnormalities seen in the perinatal period following the transfer of IVP embryos is the occurrence of high birthweight calves in association with a broader distribution of birthweight compared with *in vivo* controls. High rates of perinatal pregnancy loss associated with IVP pregnancies can be largely attributed to dystocia occurring as a result of large-sized fetuses.[44,59] Dystocia can have long-term effects on recipient animal health, production, and future reproductive performance.[59] High birthweight in association with IVP of transferred embryos has been reported for several *Bos taurus* breeds including Holstein, Angus, Simmental, and Japanese Black.[55,129,136,138,139] Transfer of IVP-derived embryos in *Bos indicus* Gyr cattle also resulted in higher birthweights than AI controls, with particular effect observed for male calves.[140] Interestingly, IVP calves demonstrated not only higher birthweights but also elevated plasma fructose concentrations after birth compared with *in vivo* controls.[134] These factors appeared to be associated with metabolic distress in the IVP calves.[134]

Common morphological abnormalities reported following the transfer of SCNT embryos included enlarged umbilical cord, respiratory problems, prolonged recumbency, and contracted flexor tendons.[49] In addition, cloned calves demonstrated other abnormalities at birth including hypothermia, hypoglycemia, severe metabolic acidosis, and hypoxia, although these parameters were not significantly correlated with birthweight. Because plasma insulin concentrations were significantly increased and plasma glucagon concentrations correlated with birthweight, aberrant energy availability and energy utilization *in utero* were proposed to induce conditions similar to that seen with human gestational diabetes.[141] At birth, cloned calves often exhibit respiratory distress syndrome[142] and, in some cases, exhibited left-sided cardiopathy with ventricular dilation and pulmonary hypertension.[142] In addition, calves that appeared abnormal at birth also exhibited an edematous placenta.[142] Other studies have reported enlarged umbilical vessels and varying degrees of edema as well as illnesses such as rumenitis, abomasitis, and coccidiosis at birth.[15,20,52,64,143] Mean placentome number was also lower and mean placentome weight higher in placentas of cloned calves compared with IVP controls.[52]

Fetuses and calves from IVP or SCNT embryos exhibit a right-skewed weight distribution compared with AI or *in vivo*-produced embryo transfer controls.[52,54,55,144] Cloned calves were 20% larger at birth compared with *in vivo*-produced embryo transfer or AI calves, and exhibited greater variation in birthweight.[33] Interestingly, cloned calves derived from oocytes matured *in vitro* had a significantly higher birthweight than those cloned calves derived from oocytes matured *in vivo*. In fact, the mean birthweight of the female *in vitro*-derived calves was significantly higher than that of the *in vivo*-derived male calves.[79] The proportion of male offspring resulting from the transfer of IVP embryos has been reported to be increased from the expected 1 : 1 ratio compared with AI controls where a normal sex ratio was maintained.[104,129,140] In contrast, other studies have not reported a difference in the proportion of males born following the transfer of IVP embryos.[45,54,145]

Fetuses derived from IVP embryos may also differ in gene and protein expression compared with those produced *in vivo*. At day 90 of gestation, IVP pregnancies were associated with lower fetal d-glucose concentrations than *in vivo* controls.[128] Expression of IGF-2 mRNA was twofold greater in day 63 liver tissue but decreased in skeletal muscle tissue of fetuses derived from IVP embryos compared with that of *in vivo*-derived controls.[146] Furthermore, skeletal muscle tissue from fetuses at day 222 of gestation derived from the transfer of IVP embryos had lower expression of myostatin mRNA and tended to have increased expression of GAPDH mRNA compared with *in vivo* controls.[147] In contrast, despite differences in fetal and placental morphology at day 90 or day 180 of gestation between pregnancies derived from IVP or *in vivo* produced embryos, no significant difference in levels of mRNA expression were detected in fetal liver and placental tissues from these pregnancies when a 13 000 oligonucleotide microarray was used for analysis.[148] Because these results contrasted with data from earlier studies, Jiang et al.[148] proposed that abnormal gene expression could be stage and tissue specific. Alternatively, the limited number of observations used within each treatment/stage subgroup, coupled with a limited size expression array, may have precluded identification of differences in mRNA expression.

Placental development

In addition to fetal abnormalities, the production of calves following the transfer of IVP or SCNT embryos is associated with defects in placental development and vasculature. Placentas of IVP or SCNT pregnancies can exhibit differences in morphology, function, gene expression, hormone production, or protein production compared with placentas from *in vivo*-derived embryos or AI pregnancies.[15,51,52,59,62,149] Perturbations in placentation and placental function could affect growth and development of those fetuses produced from *in vitro* techniques. Abnormalities in placental development are even more pronounced in pregnancies from SCNT embryos; in fact, the majority of pregnancy loss in SCNT pregnancies is attributed to placental deficiencies.[62,142] In contrast, in normal (*in vivo*-derived) pregnancies, the majority of losses are due to fetal abnormalities.[54]

Loss of IVP pregnancies during early gestation can be attributed to abnormalities and failures in placental development. Common morphological abnormalities reported in IVP placentas include enlargement (placentomegaly), increased placental fluid volume (hydrallantois), and reduced placentome surface area (reviewed in ref. 59). In an analysis of placentas at day 70 of gestation from embryos produced *in vitro* using either a semi-defined culture medium (mSOF) or an undefined serum-supplemented medium (M199) and compared with *in vivo*-produced embryo transfer controls,[51] placentas from embryos produced in mSOF medium were heavier, had decreased placental efficiency, fewer placentomes, and decreased placental fluid compared with those from the M199 or *in vivo* control groups. In addition, placentas from the mSOF group had decreased blood vessel density and levels of vascular endothelial growth factor (VEGF) mRNA in the cotyledons, indicative of impaired vascular development. Furthermore, an increased density of fetal pyknotic cells in placentomes from the mSOF group was observed compared with the M199 and *in vivo* control groups, indicative of increased cell death in the fetal villi.[51] Other placental abnormalities that were observed in IVP pregnancies during early gestation include differences in placentome size distribution, weight, and surface area at days 90 and 180 of gestation.[128,134] Furthermore, placentomes directly surrounding the fetus were observed to be longer and thinner during the first trimester of IVP pregnancies compared with controls. At day 180 of gestation, increased placentome surface area and weight was associated with increased glucose and fructose accumulation in the fetal plasma and fluids.[134] While IVP calves had placentas with significantly fewer cotyledons than placentas from *in vivo* controls, placentas from IVP pregnancies had a larger cotyledonary surface area. It was hypothesized that increased substrate utilization in the larger IVP fetuses could be responsible for the higher fetal weights observed in that treatment group.[128,134]

First-trimester placentas of SCNT pregnancies can lack proper vascularization and exhibit abnormalities including hypoplasia and reduced cotyledonary development.[62,150–153]

Similarly, pregnancy loss following transfer of SCNT embryos later in gestation was attributed to a number of factors including reduced cotyledonary or placentome number, hydrallantois, placental edema, and fetal anasarca.[15] Improper allantoic development is often the leading cause of embryo mortality after day 30 of pregnancy, as the allantois is necessary for the establishment of the chorioallantoic placenta. Placentomes are formed starting around day 30.[154] SCNT embryos can have impaired cotyledonary formation or other problems that prevent the proper development of placentome structure and number.[150] Placentome number is more variable in SCNT pregnancies compared with AI controls[62,150] and there is a high likelihood of pregnancy loss by day 90 of gestation when placentome numbers are very low.[59,62,150]

Later in gestation, placentas from either IVP or SCNT pregnancies can exhibit other anomalies in morphology or function. At day 222 of gestation, placentas recovered from IVP pregnancies had an increased proportional volume of blood vessels in the maternal caruncles compared with *in vivo* controls.[50] Furthermore, placentas from the IVP group had an increased ratio of the tissue volume occupied by blood vessels to placentome surface area, indicative of potentially increased vascularization compared with controls. In contrast, no differences were identified in the volume density of fetal villi at day 70 between IVP and *in vivo* groups, suggesting perturbations in villous formation likely occur later in gestation.[51] Similarly, there was no difference in placental weight at day 70 of gestation, whereas at day 222 of gestation placentas from IVP pregnancies were heavier than *in vivo* controls.[50] These results also support the suggestion that deviations in placental growth associated with IVP pregnancies occurs later in gestation.[51] Fetal overgrowth associated with *in vitro* production systems likely leads to an increased need for nutrients and gas exchange and placentas from IVP pregnancies may develop compensatory mechanisms within the vascular system to address the decreased surface area available for fetal and maternal contact.[50] The incidence of hydrallantois in IVP pregnancies is about 1 in 200 pregnancies and is significantly greater than the 1 in 7500 frequency reported for normal pregnancies.[54]

During the second half of gestation, SCNT placentas may be exhibit placentomegaly and hydrallantois with edema of the fetal and placental tissues as the leading cause of fetal mortality in SCNT pregnancies.[150] Furthermore, umbilical vessel diameter is generally larger and the umbilical cords have a larger luminal space due to the lack of luminal folds in SCNT pregnancies compared with AI control pregnancies.[150,151] Placental overgrowth in SCNT pregnancies could be due to inherent defects in the placenta rather than due to fetal overgrowth, and can occur in up to 50% of late-gestation SCNT pregnancies.[151] Placental overgrowth occurred prior to fetal overgrowth and the ratio of fetal to placentome weight was lower in SCNT pregnancies as compared with AI and *in vitro* fertilized controls.[151]

Placentas from IVP or SCNT pregnancies demonstrate differences in gene expression or protein production. For example, caruncular levels of PPARγ, a transcription factor involved in VEGF expression and vascular development, were increased in day 222 placentas from IVP pregnancies as compared with *in vivo* controls.[50] Differences in placentome gene expression between SCNT, IVP and AI pregnancies sampled between day 174 of gestation and term were assessed by microarray analysis.[149] A total of 162 differentially expressed genes (DEGs) were identified between IVP and SCNT placentomes with a total of 276 DEGs identified between AI and SCNT placentomes.[149] Some DEGs were shared between IVP and SCNT placentomes, suggesting an influence of culture system on the placentome transcriptome. SCNT placentas had significantly altered levels of gene expression, regardless of the presence of either normal or abnormal morphology. Furthermore, these authors suggested that failure to compensate for placental dysfunction leads to fetal death and other abnormalities associated with AOS.[149] This was also consistent with the suggestion that fetal starvation due to placental insufficiency may be a leading cause of SCNT pregnancy loss after day 50.[62] It is highly likely that deficiencies in fetal–maternal nutrient exchange and waste exchange are a major cause of pregnancy loss in mid to late gestation in association with type III AOS.

Aston *et al.*[155] used the Affymetrix GeneChip bovine genome array to examine gene expression between cotyledons of day 70 SCNT and AI placentas and reported 19 DEGs between the groups. However, none of their DEGs overlapped with those from Everts *et al.*[149] This is likely due to sampling of different tissues at different time points. Recently, the same Affymetrix GeneChip was used to investigate the global transcriptome of day 50 placentas derived from AI, IVP and SCNT pregnancies.[156] A total of 1196 DEGs were identified between placentas of the SCNT and AI groups, of which 9% were due to transcriptional reprogramming error.[156] Placentas from the IVP group had only 72 DEGs compared with the AI group, with 58 of those DEGs also common to the SCNT-derived placentas. The DEGs in both SCNT and IVP groups were involved in organ development, blood vessel development, extracellular matrix organization, and the immune system.[156] Interestingly, 96% of the DEGs in the SCNT group were not differentially regulated in the IVP groups, indicating that these placentas experience a unique perturbation of gene expression. Furthermore, the higher level of transcriptome dysregulation in SCNT placentas is consistent with the higher degree of morphological dysregulation reported in SCNT pregnancies.

Mean glucose concentrations in fetal fluids did not differ between SCNT and AI pregnancies early in pregnancy; however, fetal fluid glucose levels were lower in failed SCNT pregnancies or those with either a large placenta or reduced placentome numbers.[157] The rate of placental growth in SCNT pregnancies was not reduced as in AI pregnancies, and there is evidence that expression of the glucose transporter GLUT3 was increased in the SCNT placenta, indicating an increase in glucose uptake by the SCNT placenta.[158,159] The bovine placenta is also the main source of estrone, and placental-derived estrogens play a role in vascular permeability and extracellular fluid distribution.[150] Elevated estradiol levels have been reported in SCNT pregnancies starting at day 80 of gestation and could contribute to the placental and fetal edema frequently observed.[150] Perturbed estrogen levels in SCNT pregnancies may also be responsible for insufficient mammary gland development observed near term and for the observed lack of preparation for parturition.[131] Abnormally high concentrations of the pregnancy-associated glycoproteins (PAGs) at day 50 of gestation were correlated with fetal death in SCNT pregnancies.[131] Based on data derived using a global proteomics approach to examine differential protein expression in placentas derived from either the afterbirth of SCNT pregnancies that ended in postnatal death or the afterbirth of normal AI pregnancies,[160] 60 differentially expressed proteins were identified. These proteins were related to protein repair, cytoskeleton, signal transduction, immune system, and metabolism.[160] These investigators confirmed the abnormal expression levels of several putative differentially expressed proteins, including TIMP-2, in the SCNT-derived placentas that are known to play important roles in placental development.

Long-term effects on calf viability

Calves resulting from the transfer of IVP or SCNT embryos that survive to parturition may still experience compromised health. These calves can exhibit congenital problems such as cerebellar hypoplasia, respiratory distress, and heart enlargement.[59] Despite differences in birthweight between calves derived from the transfer of IVP and in vivo-produced embryos, postnatal growth rate and yearling weight were unaffected by embryo source or birthweight.[33,139]

Cloned calves can also present with abnormalities during the first few months of life. Erythrocyte mean cell volume was higher at birth in cloned calves and, during the first week following birth, body temperature and plasma leptin concentrations were higher and thyroxine levels lower in cloned calves compared with controls.[52] Interestingly, in that study, elevated body temperature was observed until at least 50 days postpartum. Other reported differences in cloned calves include varying telomere lengths among cloned calves associated with different donor nuclei cell types[161] and a later onset of puberty in cloned heifers compared with AI controls.[162] For bulls resulting from the transfer of IVP embryos that were oversized at birth, an increase in heart weight per kilogram body weight at slaughter was reported.[139] Thus despite appearing healthy, these data suggest that at least a portion of the calves resulting from the transfer of IVP and SCNT embryos are not physiologically normal during the first few months of life.

While AOS is observed in a high number of cloned pregnancies, some individuals experience a normal birth and require little veterinary care after birth. Despite a 24% differences in birthweight between calves derived from SCNT embryos and in vivo controls, accelerated growth and weight variability did not continue during postnatal development and was less than 5% at 1 year of age.[33] Those clones that survive to adulthood appear to have normal reproductive performance and progeny. Cloned heifers used in an embryo transfer program had embryo production rates similar to controls as well as normal breeding and calving rates.[49] Similarly, adult male clones were found to produce good-quality semen with normal fertility with AI. In addition, the progeny of the few cloned adults in that study also appeared to have normal reproductive performance and milk production.[49] A number of studies have found little to no differences in growth rate, estrous cycle length, hormonal profiles including luteinizing hormone (LH), FSH, E2, P4, IGF-1, and growth hormone as well as in lactation characteristics and overall reproductive performance for clones compared with noncloned calves of similar age.[15,33,52,163] In addition, production of both milk and meat by cloned cattle was found to be similar to that from AI cattle.[164–166] Thus, although the transfer of IVP or SCNT embryos can be associated with a wide variety of abnormalities during gestation and even after parturition, those calves that survive to adulthood can lead healthy productive lives.

Mechanisms underlying AOS

The abnormalities observed in bovine embryos, fetuses, placentas, and offspring following the transfer of IVP or SCNT embryos were initially hypothesized to be the result of epigenetic dysregulation and subsequent aberrant mRNA expression of developmentally important imprinted genes.[63,105,167,168] Imprinted genes are expressed monoallelically, in a parent-of-origin specific manner, as a result of epigenetic modifications.[169] Epigenetic modifications are chemical alterations to the DNA or to histone proteins that alter chromatin structure without altering the nucleotide sequence.[170] Embryos derived from IVP and SCNT are exposed to the IVC environment which is thought to disrupt the acquisition or maintenance of epigenetic patterns regulating imprinted gene expression.[115,171,172] In addition to insults from the IVC environment, SCNT embryos must also contend with rupture of the zona, loss of maternal cytoplasm and cytoskeleton, electrical or chemical fusion and activation of enucleated oocyte with donor cell, and epigenetic reprogramming of the donor nucleus to a totipotent state.[173] Another source of epigenetic dysregulation contributing to abnormal phenotypes associated with AOS may arise from the aspiration and IVM of bovine oocytes prior to IVP or SCNT manipulations. The acquisition of maternal epigenetic modifications that regulate imprinted gene expression occurs during the latter stages of follicular development, a time at which cumulus–oocyte complexes are aspirated and may also become altered during IVM.[174] Interestingly, many imprinted genes regulate the expression of nonimprinted genes. Therefore, abnormal phenotypes may arise from aberrant expression of nonimprinted genes as a downstream effect of epigenetic dysregulation on imprinted genes.[59] Alternatively, IVC environments and SCNT protocols may have adverse effects on the DNA modifications regulating the expression of nonimprinted genes independent of genomic imprinting.[59] Adding to the complexity of the potential mechanisms underlying AOS comes from the observation that IVC and SCNT protocols also negatively impact the expression of genes that regulate the acquisition and maintenance of epigenetic modifications necessary for the appropriate expression of imprinted genes.

Epigenetic dysregulation of imprinted genes

Abnormalities observed in embryos, fetuses, placentas, and offspring derived from the transfer of IVP or SCNT embryos are proposed to be the result of inadequate IVC conditions or improper epigenetic reprogramming during SCNT procedures leading to the incorrect establishment or maintenance of epigenetic patterns regulating imprinted gene expression during critical periods of preimplantation development.[59,168,175,176] Numerous experiments in cattle and sheep investigating aberrant expression of imprinted genes following the transfer of IVP or SCNT embryos support this hypothesis.[146,172,177–183] Recently, single nucleotide polymorphisms in a *Bos taurus* × *Bos indicus* model were used to investigate imprinted gene expression for *in vivo*, IVP and SCNT conceptuses at day 17 of gestation and for fetuses and placentas at day 45.[171] Monoallelic expression of the imprinted genes investigated was not perfectly maintained across the tissues studied and in some cases relaxation of DNA methylation imprints was associated with inappropriate biallelic expression.[171] Additional studies investigating the DNA methylation imprints of developmentally important imprinted genes also support the hypothesis that epigenetic dysregulation is a key factor that likely contributes to the occurrence of abnormalities associated with AOS.[175,177,182,184–186]

Further support for this hypothesis was provided by investigations of other known epigenetic changes such as covalent histone modifications.[187–193] Histone modifications, including acetylation and methylation, are proposed to play a role in regulating imprinted gene expression through RNA silencing[194,195] or upregulation of monoallelic expression.[196]

Currently, there are more than 140 known imprinted genes in mice and humans.[170] However, the studies in cattle thus far have focused on only a small subset of imprinted genes. It is clear that the bovine preimplantation embryo is sensitive to altered epigenetic modifications produced from IVC system or through SCNT protocols based on the resulting aberrant expression of imprinted genes.[63,101,197] The observation that some imprinted genes have the ability to regulate expression of nonimprinted genes suggests that the initial epigenetic dysregulation caused by IVC and SCNT may be exacerbated by downstream effects on the expression of nonimprinted genes.[59,198,199]

Epigenetic dysregulation of nonimprinted genes

The IVC environment and SCNT protocols may impact the expression of nonimprinted genes through downstream effects of aberrantly expressed imprinted genes and epigenetic dysregulation of the nonimprinted genes themselves. In human cancer cells, aberrant expression of the imprinted gene p73 downregulated expression of VEGF, a nonimprinted gene that is an important placental angiogenic factor.[198,199] It is reasonable to expect that similar effects on nonimprinted downstream regulators may also occur in bovine embryos and fetuses generated from conditions, such as IVC and SCNT, that induce aberrant expression of imprinted genes. Alternatively, aberrant expression of nonimprinted genes may occur as a result of hypermethylation at CpG islands surrounding the promoters of nonimprinted genes.[200] In this case, such epigenetic errors would be independent of imprinting mechanisms. For example, in the chorion of SCNT bovine pregnancies, the promoter of the epidermal cytokeratin gene and satellite 1 repeat elements were hypermethylated compared with that for *in vivo* controls.[59,201] While the exact mechanisms that regulate expression of nonimprinted genes remain unclear, it is apparent that IVC and SCNT protocols can negatively influence nonimprinted gene expression and, in turn, contribute to abnormalities associated with AOS.[51,63,135,202–204]

Altered expression of mRNAs encoded by nonimprinted genes including myostatin and GAPDH has been associated with an increased ratio of secondary-to-primary muscle fiber types in IVP-derived fetuses.[135] Similarly, altered expression of mRNAs or proteins for VEGF, PPARγ, leptin, bovine placental lactogen, transforming growth factor (TGF)-β1, TGF-β2, TGF-β3, TGF-β receptors, and major histocompatibility class I antigens have been found in placentas from IVP or SCNT pregnancies.[50,51,202–204] These more subtle alterations in fetal or placental development following IVP or SCNT procedures that do not fall within the classically defined "large offspring" syndrome may, however, still reflect the occurrence of epigenetic dysregulation characteristically associated with AOS.[59]

Examination of gene expression patterns in embryos, fetuses, and placentas of cattle, sheep and mice produced using IVC or manipulated using SCNT procedures have demonstrated altered expression of imprinted,[63,104,146,177,205] X-linked,[63,201,206] and nonimprinted[51,63,135,203,204] genes. Changes in embryo-wide methylation patterns and global methylation levels have been associated with altered preimplantation development[207,208] and adverse pregnancy outcomes for SCNT fetuses.[209] These effects may only be secondary responses that are resulting from an initial effect of the altered epigenetic status of imprinted genes early in preimplantation or peri-implantation development. Alternatively, they may indicate that IVC and manipulations may influence methylation patterns that may be involved in the regulation of expression nonimprinted genes as has been proposed for other systems[200,210] and observed in SCNT pregnancies.[201] It is plausible, then, that hypermethylation of promoters containing CpG islands in SCNT pregnancies can lead to changes in gene expression independent of imprinting. Together, these observations lend further support to the suggestion that global epigenetic modifications affecting the expression of nonimprinted genes may play an important role in contributing to the occurrence of AOS.

In studies of human placental tissues, the importance of both imprinted and nonimprinted genes in control of fetal growth and placental function has been recognized. For example, dysregulation of imprinted genes was identified in association with the occurrence of human intrauterine growth restriction, a syndrome that affects both placental function and fetal growth.[211] Interestingly while epigenetic dysregulation in species such as cattle and sheep results in fetal overgrowth, analogous dysregulation results in the opposite phenotype in humans. These observations reinforce the need for a change in nomenclature from "large offspring" syndrome to the more accurate and broadly applicable term AOS.[59]

Disruption of epigenetic regulators

During early embryonic development in mammals, DNA methylation patterns that regulate gene expression are acquired and maintained by a family of enzymes called DNA methyltransferases (DNMTs).[168,170,212] In the growing fetus, primordial germ cells emerge and migrate to the genital ridge while undergoing mitotic division. During this period their DNA undergoes complete demethylation. DNMT3A, 3B and 3L are *de novo* methyltransferases that function to establish the sex-specific methylation patterns for imprinted gene loci during gametogenesis.[168,174,212]

Maternal imprints are acquired postnatally during folliculogenesis, coinciding with an increase in global DNA methylation that reaches its maximum prior to ovulation in fully grown oocytes that are arrested at the dictyate (GV) stage.[213,214] In contrast, paternal imprints are established prior to birth in the developing male's germ cells.[176] In the zygote, the paternal genome undergoes active DNA demethylation which occurs rapidly following fertilization.[176,212] By comparison, demethylation of the maternal genome occurs passively due to exclusion of DNMT1 from the nucleus during early cleavage divisions.[176] DNMT1 is responsible for maintaining methylation patterns while DNA is replicating.[215] While global DNA demethylation is occurring in the parental genomes of the early zygote, sex-specific methylation imprints are maintained. In the mouse, the maternal and paternal imprints are believed to be maintained in the zygote by an oocyte-specific isoform of DNMT1 designated DNMT1o.[216] In cattle, however, this oocyte isoform has not been identified.[217,218] In mice, DNMT1 protein is present in early stage follicles and is then relegated to the cytoplasm during the later phases of oocyte differentiation and early embryogenesis.[219,220] Demethylation of the parental genome is thought to occur as a result of DNMT1 exclusion from the nucleus during early cleavage divisions.[175] In cattle, the newly formed embryonic genome is remethylated as it progresses to the blastocyst stage, occurring by the action of the *de novo* methyltransferases DNMT3A, 3B, and 3L.[168,174,176,212]

In embryos produced *in vivo*, active demethylation of the paternal genome and passive demethylation of the maternal genome results in a hypomethylated state during the early cleavage divisions.[175,221] As the embryo reaches the blastocyst stage, *de novo* methylation coincides with differentiation of cells contributing to the inner cell mass and the trophectoderm. Thus, tissue- and time-specific methylation patterns are established that will be carried throughout development.[212,222,223] Because DNMTs are responsible for erasure, acquisition, and maintenance of methylation imprints in gametes as well as for establishing and maintaining DNA methylation patterns during epigenetic reprogramming of the embryo, it has been suggested that abnormalities associated with AOS following the transfer of IVP or SCNT embryos may result from aberrant expression or function of the DNMTs.[224-226] Altered mRNA expression of DNMT1 and DNMT3a have been observed in IVP and SCNT preimplantation embryos,[63] further supporting the hypothesis that DNMT dysregulation leads to a cascade of aberrant expression of imprinted genes and subsequent developmental abnormalities.[63] Additional insults may occur from alterations to chromatin structure and histone modifications that can foster inappropriate access of DNMTs to the genome during development.[59]

Bovine SCNT embryos exhibit hypermethylated DNA compared with *in vivo* embryos, a situation that resembles the methylation status of the somatic donor cell.[187,224,227,228] Failure of the oocyte to demethylate and reprogram the donor nucleus during cloning procedures is often cited as the primary reason for abnormal gene expression and low cloning efficiency observed in SCNT embryos.[222,223] In SCNT bovine embryos, DNA demethylation does not occur after the 2-cell stage.[175] Thus, some developmental failures in SCNT pregnancies may arise from improper DNA demethylation with subsequent maintenance of a hypermethylated state for cells of the trophectoderm lineage resulting in aberrant gene expression and placental abnormalities.[223,224,229,230] Indeed, expression of DNMT1 mRNA and levels of DNA methylation were significantly higher at all preimplantation stages for SCNT embryos compared with IVP controls.[223] When the level of DNMT1 transcripts was reduced by insertion of siRNAs for DNMT1, methylation was reduced and the number of embryos reaching the blastocyst stage rose significantly.[223] In a similar study, SCNT embryos with siRNA-reduced levels of DNMT1 showed an increase to the blastocyst stage, although there was no difference in development to term.[217]

Oocyte-specific epigenetic abnormalities

Aberrant epigenetic modifications leading to abnormalities in embryos, fetuses, and placentas following the transfer of IVP or SCNT embryos may initially originate in the gametes. More emphasis has been placed on evaluation of potential aberrations of imprinting status of genes within the oocyte rather than in sperm because mammalian oocytes acquire their maternal imprints during folliculogenesis and are finalized during oocyte maturation occurring immediately prior to ovulation.[231] By comparison, sperm acquire paternal imprints *in vivo* during spermatogenesis and their DNA is packaged in a highly protected manner. However, sperm subjected to freezing and thawing as well as sex selection procedures may also incur epigenomic alterations based on the observation that they exhibit low blastocyst development rates during IVP embryo production and low pregnancy rates following AI.[232-236] In an analysis of two specific imprinted genes, *IGF2* and its receptor *IGF2R*, no difference in DNA methylation pattern within these genes' regulatory differentially methylated regions was found between Nelore bull sperm that were frozen-thawed, unsorted, and sex-sorted.[236] In addition, no differences in DNA methylation imprints for these genes were found between unsorted sperm and sex-selected X-bearing sperm or Y-bearing sperm.[236] These observations support the premise that DNA methylation imprints on the paternal genome are intact prior to *in vitro* fertilization. Low blastocyst and pregnancy rates following insemination or fertilization with sex-sorted semen may be the result of DNA fragmentation, altered chromatin packaging, and epigenetic defects to DNA methylation within interspersed repeats or histone modifications.[236,237]

In mammals, maternal imprints are acquired during folliculogenesis and therefore immature follicles that are aspirated and subjected to IVM may exhibit altered maternal imprints compared with oocytes fully matured and ovulated *in vivo*. Thus, use of IVM systems may be a contributing factor to aberrant expression of imprinted genes and abnormalities associated with the occurrence of AOS. In bovine oocytes there are substantial developmental and physiological differences between *in vitro* matured cumulus–oocyte complexes compared with those matured *in vivo*. For example, 50–80% of *in vivo* matured oocytes develop to the blastocyst stage, whereas only 15–40% of IVM oocytes reach the blastocyst stage.[71–74] Several early studies sought to correlate IVM of oocytes with an overgrowth phenotype. For example, ovine oocytes that were subjected to IVM and *in vitro* fertilization followed by *in vivo* culture for 6.5 days prior to subsequent transfer into recipient females resulted in lambs that were heavier at birth than those that were *in vivo* matured, fertilized, and cultured.[238] A fetal overgrowth phenotype was also observed in transgenic calves derived from IVM oocytes when compared with those derived from *in vivo* matured oocytes.[79] Interestingly, when immature follicles (1–3 mm) were used to generate SCNT conceptuses it was observed that they possessed smaller extraembryonic membranes compared with those from more mature follicles (6–12 mm).[57] However, there were no apparent differences between SCNT pregnancies following the transfer of immature or mature oocytes for fetal size, allantoic vascularization, pregnancy rates, or survival to term.[57]

Recent studies have focused on assessment of epigenetic profiles in oocytes and the effect of IVM on mRNA expression and subsequent embryonic development. For example, expression of oocyte mRNA for enzymes necessary to establish or maintain DNA methylation and histone modification, including G9A, SUV39H1 and DNMT1, were significantly affected by follicle size.[239] In addition, mRNAs for genes involved in the tricarboxylic acid cycle and oxidative phosphorylation pathways were found to be aberrantly expressed in metaphase II bovine oocytes that were matured *in vitro*.[240] Furthermore, developmentally important imprinted genes, including *IGF2R*, *PEG3* and *SNRPN*, were also aberrantly expressed, suggesting that errors in establishing or maintaining oocyte DNA methylation patterns occurred during IVM.[240]

Maternal DNA methylation imprints for *SNRPN*, *MEST*, *IGF2R*, *PEG10*, and *PLGAL1* were analyzed in bovine oocytes that were grouped based on a small (101–110 μm) or large (>120 μm) diameter.[174] Differentially methylated regions (DMRs) for these imprinted genes in small oocytes were hypomethylated. In contrast, in large oocytes these DMRs were hypermethylated. These observations support the premise that acquisition of maternal imprints is correlated with oocyte size.

Fagundes *et al.*[241] evaluated the methylation imprint for the *IGF2* gene for oocytes aspirated from small (<3 mm) and large (>8 mm) follicles and subjected to IVM. *IGF2* is paternally imprinted and becomes hypomethylated in mature preovulatory oocytes.[241] Oocytes taken from small or large follicles were further subdivided into matured or immature following IVM. Matured oocytes were identified as those that had extruded a second polar body by the termination of culture.[241] Interestingly, matured oocytes from large follicles exhibited a greater degree of hypomethylation compared with matured oocytes obtained from small follicles.[241] These results are consistent with the premise that methylation imprints for *IGF2* are established correctly during IVM for oocytes that are obtained from large follicles and matured *in vitro*. Furthermore, assessment of *IGF2* imprints may be potentially useful as a marker for oocyte developmental competence.[241] These results also support the suggestion that when oocytes are obtained from smaller follicles, maternal imprints may not be established correctly during IVM, despite the fact that the oocytes are successful in reaching metaphase II of meiosis during culture. Such incorrectly established imprints have the potential to contribute to the occurrence of abnormalities in development after fertilization.

While DNA methylation imprints appear to be established correctly in oocytes from larger-sized follicles that are matured *in vitro*, it should be noted that the studies discussed above analyzed pools of oocytes rather than oocytes as individuals. Using analysis of methylation patterns in individual matured oocytes using limited dilution bisulphate sequencing, El Hajj *et al.*[242] found that DNA methylation patterns for *H19*, *SNRPN*, and *PEG3* were correctly established in seven of eight oocytes, whereas one of eight exhibited a reversal of imprinting for *H19* and *SNRPN*.[242] Thus it is possible to examine the methylation pattern of multiple genes from single cells and, more importantly, that IVM can result in aberrant establishment of epigenetic patterns in a subset of bovine oocytes prior to fertilization.[242]

During oocyte maturation, DNA methylation of regulatory CpG islands does not always correlate with suppressed expression of mRNA from the corresponding gene. For example, DNA methylation of CpG islands within the *PEG3*, *H19*, and *SNRPN* genes was not different or was only slightly different between *in vivo* matured oocytes and *in vitro* matured oocytes.[243] However, mRNA expression for *H19* and *PEG3* were significantly greater in oocytes matured *in vitro* compared with oocytes matured *in vivo* and was similar to levels of expression found for immature oocytes.[243] Thus although DNA methylation patterns appear to be correctly established, some imprinted genes retain expression patterns characteristic of immature oocytes even after being subjected to IVM conditions.[243] It is reasonable to expect that additional levels of regulation exist beyond the methylation status of CpG islands within gene regulatory regions and are adversely affected by IVM conditions.

References

1. Elsden R, Hasler J, Seidel G. Non-surgical recovery of bovine eggs. *Theriogenology* 1976;6:523–532.
2. Bowen J, Elsden R, Seidel G. Non-surgical embryo transfer in the cow. *Theriogenology* 1978;10:89–95.
3. Seidel G. Superovulation and embryo transfer in cattle. *Science* 1981;211:351–358.
4. Thibier M. Embryo transfer statistics from around the world. *IETS Newsletter* 1998;16. http://www.iets.org/pdf/comm_data/December1998.pdf online.
5. Thibier M. A contrasted year for the world activity of the animal embryo transfer industry: a report from the IETS Data Retrieval Committee. *IETS Newsletter* 2002;20:13–19.
6. Stroud B. The year 2011 worldwide statistics of embryo transfer in domestic farm animals. *IETS Newsletter* 2012;30:16–26.
7. Block J, Drost M, Monson R *et al.* Use of insulin-like growth factor-I during embryo culture and treatment of recipients with gonadotropin-releasing hormone to increase pregnancy rates following the transfer of *in vitro*-produced embryos to heat-stressed, lactating cows. *J Anim Sci* 2003;81:1590–1602.
8. Hansen P, Block J. Towards an embryocentric world: the current and potential uses of embryo technologies in dairy production. *Reprod Fertil Dev* 2004;16:1–14.

9. Hansen P. Realizing the promise of IVF in cattle: an overview. *Theriogenology* 2006;65:119–125.
10. Wall R, Kerr D, Bondioli K. Transgenic dairy cattle: genetic engineering on a large scale. *J Dairy Sci* 1997;80: 2213–2224.
11. Keefer C. Production of bioproducts through the use of transgenic animal models. *Anim Reprod Sci* 2004;82–83:5–12.
12. van Berkel P, Welling M, Geerts M et al. Large scale production of recombinant human lactoferrin in the milk of transgenic cows. *Nat Biotechnol* 2002;20:484–487.
13. Brophy B, Smolenski G, Wheeler T, Wells D, L'Huillier P, Laible G. Cloned transgenic cattle produce milk with higher levels of beta-casein and kappa-casein. *Nat Biotechnol* 2003;21:157–162.
14. Wang J, Yang P, Tang B et al. Expression and characterization of bioactive recombinant human alpha-lactalbumin in the milk of transgenic cloned cows. *J Dairy Sci* 2008;91:4466–4476.
15. Edwards J, Schrick F, McCracken M et al. Cloning adult farm animals: a review of the possibilities and problems associated with somatic cell nuclear transfer. *Am J Reprod Immunol* 2003; 50:113–123.
16. Sommer J, Jackson L, Simpson S, Collins E, Piedrahita J, Petters R. Transgenic Stra8–EYFP pigs: a model for developing male germ cell technologies. *Transgenic Res* 2012;21:383–392.
17. Sommer J, Estrada J, Collins E et al. Production of ELOVL4 transgenic pigs: a large animal model for Stargardt-like macular degeneration. *Br J Ophthalmol* 2011;95:1749–1754.
18. Narbonne P, Miyamoto K, Gurdon J. Reprogramming and development in nuclear transfer embryos and in interspecific systems. *Curr Opin Genet Dev* 2012;22:450–458.
19. Prather R. Pig genomics for biomedicine. *Nat Biotechnol* 2013; 31:122–124.
20. Wells D, Misica P, Tervit H, Vivanco W. Adult somatic cell nuclear transfer is used to preserve the last surviving cow of the Enderby Island cattle breed. *Reprod Fertil Dev* 1998;10:369–378.
21. Gomez M, Pope C, Ricks D, Lyons J, Dumas C, Dresser B. Cloning endangered felids using heterospecific donor oocytes and interspecies embryo transfer. *Reprod Fertil Dev* 2009;21: 76–82.
22. King K, Seidel G, Elsden R. Bovine embryo transfer pregnancies. I. Abortion rates and characteristics of calves. *J Anim Sci* 1985;61:747–757.
23. Nagashima H, Matsui K, Sawasaki T, Kano Y. Production of monozygotic mouse twins from microsurgically bisected morulae. *J Reprod Fertil* 1984;70:357–362.
24. Surani M, Barton S, Norris M. Experimental reconstruction of mouse eggs and embryos: an analysis of mammalian development. *Biol Reprod* 1987;36:1–16.
25. Willadsen S. Nuclear transplantation in sheep embryos. *Nature* 1986;320:63–65.
26. Willadsen S. Cloning of sheep and cow embryos. *Genome* 1989; 31:956–962.
27. Baker R, Shea B. Commercial splitting of bovine embryos. *Theriogenology* 1985;23:3–12.
28. Heyman Y. Factors affecting the survival of whole and half-embryos transferred in cattle. *Theriogenology* 1985;23:63–75.
29. Warfield S, Seidel G, Elsden R. Transfer of bovine demi-embryos with and without the zona pellucida. *J Anim Sci* 1987;65:756–761.
30. Lopes R, Forell F, Oliveira A, Rodrigues J. Splitting and biopsy for bovine embryo sexing under field conditions. *Theriogenology* 2001;56:1383–1392.
31. Willadsen S, Janzen R, McAlister R, Shea B, Hamilton G, McDermand D. The viability of late morulae and blastocysts produced by nuclear transplantation in cattle. *Theriogenology* 1991;35:161–170.
32. Keefer C, Stice S, Matthews D. Bovine inner cell mass cells as donor nuclei in the production of nuclear transfer embryos and calves. *Biol Reprod* 1994;50:935–939.
33. Wilson J, Williams J, Bondioli K, Looney C, Westhusin M, McCalla D. Comparison of birth weight and growth characteristics of bovine calves produced by nuclear transfer (cloning), embryo transfer and natural mating. *Anim Reprod Sci* 1995;38:73–83.
34. Brackett B, Bousquet D, Boice M, Donawick W, Evans J, Dressel M. Normal development following *in vitro* fertilization in the cow. *Biol Reprod* 1982;27:147–158.
35. Brackett B. *In vitro* oocyte maturation and fertilization. *J Anim Sci* 1985;61(Suppl. 3):14–24.
36. Younis A, Brackett B, Fayrer-Hosken R. Influence of serum and hormones on bovine oocyte maturation and fertilization *in vitro*. *Gamete Res* 1989;23:189–201.
37. Wright R, Bondioli K. Aspects of *in vitro* fertilization and embryo culture in domestic animals. *J Anim Sci* 1981;53:702–729.
38. Sirard M, Parrish J, Ware C, Leibfried-Rutledge M, First N. The culture of bovine oocytes to obtain developmentally competent embryos. *Biol Reprod* 1988;39:546–552.
39. Eyestone W, First N. Co-culture of early cattle embryos to the blastocyst stage with oviductal tissue or in conditioned medium. *J Reprod Fertil* 1989;85:715–720.
40. Saeki K, Hoshi M, Leibfried-Rutledge M, First N. *In vitro* fertilization and development of bovine oocytes matured in serum-free medium. *Biol Reprod* 1991;44:256–260.
41. Keskintepe L, Burnley C, Brackett B. Production of viable bovine blastocysts in defined *in vitro* conditions. *Biol Reprod* 1995;52:1410–1417.
42. Keskintepe L, Brackett B. *In vitro* developmental competence of *in vitro*-matured bovine oocytes fertilized and cultured in completely defined media. *Biol Reprod* 1996;55:333–339.
43. Walker S, Heard T, Seamark R. *In vitro* culture of sheep embryos without co-culture: successes and perspectives. *Theriogenology* 1992;37:111–126.
44. Behboodi E, Anderson G, BonDurant R et al. Birth of large calves that developed from *in vitro*-derived bovine embryos. *Theriogenology* 1995;44:227–232.
45. Farin P, Farin C. Transfer of bovine embryos produced *in vivo* or *in vitro*: survival and fetal development. *Biol Reprod* 1995; 52:676–682.
46. Sinclair K. *In vitro* produced embryos as a means of achieving pregnancy and improving productivity in beef cows. *Anim Sci J* 1995;60:55–64.
47. Walker S, Hartwich K, Seamark R. The production of unusually large offspring following embryo manipulation: concepts and challenges. *Theriogenology* 1996;45:111–120.
48. Chavatte-Palmer P, Remy D, Cordonnier N et al. Health status of cloned cattle at different ages. *Cloning Stem Cells* 2004;6:94–100.
49. Panarace M, Aguero JI, Garrote M et al. How healthy are clones and their progeny: 5 years of field experience. *Theriogenology* 2007;67:142–151.
50. Miles J, Farin C, Rodriguez K, Alexander J, Farin P. Angiogenesis and morphometry of bovine placentas in late gestation from embryos produced *in vivo* or *in vitro*. *Biol Reprod* 2004;71: 1919–1926.
51. Miles J, Farin C, Rodriguez K, Alexander J, Farin P. Effects of embryo culture on angiogenesis and morphometry of bovine placentas during early gestation. *Biol Reprod* 2005;73:663–671.
52. Chavatte-Palmer P, Heyman Y, Richard C et al. Clinical, hormonal, and hematologic characteristics of bovine calves derived from nuclei from somatic cells. *Biol Reprod* 2002;66:1596–1603.
53. Farin P, Crosier A, Farin C. Influence of *in vitro* systems on embryo survival and fetal development in cattle. *Theriogenology* 2001;55:151–170.
54. Hasler J, Henderson W, Hurtgen P et al. Production, freezing and transfer of bovine IVF embryos and subsequent calving results. *Theriogenology* 1995;43:141–152.
55. Kruip T, den Daas J. *In vitro* produced and cloned embryos: effects on pregnancy, parturition and offspring. *Theriogenology* 1997;47:43–52.
56. van Wagtendonk-de Leeuw AM, Mullaart E, de Roos A et al. Effects of different reproduction techniques: AI, MOET or IVP,

57. Piedrahita J, Wells D, Miller A et al. Effects of follicular size of cytoplast donor on the efficiency of cloning in cattle. *Mol Reprod Dev* 2002;61:317–326.
58. Wells D, Laible G, Tucker F et al. Coordination between donor cell type and cell cycle stage improves nuclear cloning efficiency in cattle. *Theriogenology* 2003;59:45–59.
59. Farin P, Piedrahita J, Farin C. Errors in development of fetuses and placentas from *in vitro*-produced bovine embryos. *Theriogenology* 2006;65:178–191.
60. Park K, Cheong H, Lai L et al. Production of nuclear transfer-derived swine that express the enhanced green fluorescent protein. *Anim Biotechnol* 2001;12:173–181.
61. Archer G, Friend T, Piedrahita J, Nevill C, Walker S. Behavioral variation among cloned pigs. *Appl Anim Behav Sci* 2003;82:151–161.
62. Hill J, Burghardt R, Jones K et al. Evidence for placental abnormality as the major cause of mortality in first-trimester somatic cell cloned bovine fetuses. *Biol Reprod* 2000;63:1787–1794.
63. Wrenzycki C, Herrmann D, Lucas-Hahn A, Lemme E, Korsawe K, Niemann H. Gene expression patterns in *in vitro*-produced and somatic nuclear transfer-derived preimplantation bovine embryos: relationship to the large offspring syndrome? *Anim Reprod Sci* 2004;82–83:593–603.
64. Wells D, Misica P, Tervit H. Production of cloned calves following nuclear transfer with cultured adult mural granulosa cells. *Biol Reprod* 1999;60:996–1005.
65. Rooke J, McEvoy T, Ashworth C et al. Ovine fetal development is more sensitive to perturbation by the presence of serum in embryo culture before rather than after compaction. *Theriogenology* 2007;67:639–647.
66. Sinclair K, McEvoy T, Maxfield E et al. Aberrant fetal growth and development after *in vitro* culture of sheep zygotes. *J Reprod Fertil* 1999;116:177–186.
67. Lawrence J, Schrick F, Hopkins F et al. Fetal losses and pathologic findings of clones derived from serum-starved versus serum-fed bovine ovarian granulosa cells. *Reprod Biol* 2005;5:171–184.
68. Powell K, Rooke J, McEvoy T et al. Zygote donor nitrogen metabolism and *in vitro* embryo culture perturbs in utero development and IGF2R expression in ovine fetal tissues. *Theriogenology* 2006;66:1901–1912.
69. Kato Y, Nagao Y. Effect of PVP on sperm capacitation status and embryonic development in cattle. *Theriogenology* 2009;72:624–635.
70. Kato Y, Nagao Y. Effect of polyvinylpyrrolidone on sperm function and early embryonic development following intracytoplasmic sperm injection in human assisted reproduction. *Reprod Med Biol* 2012;11:165–176.
71. Blondin P, Bousquet D, Twagiramungu H, Barnes F, Sirard M. Manipulation of follicular development to produce developmentally competent bovine oocytes. *Biol Reprod* 2002;66:38–43.
72. Rizos D, Fair T, Papadopoulos S, Boland M, Lonergan P. Developmental, qualitative, and ultrastructural differences between ovine and bovine embryos produced *in vivo* or *in vitro*. *Mol Reprod Dev* 2002;62:320–327.
73. Choi Y, Carnevale E, Seidel G, Squires E. Effects of gonadotropins on bovine oocytes matured in TCM-199. *Theriogenology* 2001;56:661–670.
74. Ward F, Enright B, Rizos D, Boland M, Lonergan P. Optimization of *in vitro* bovine embryo production: effect of duration of maturation, length of gamete co-incubation, sperm concentration and sire. *Theriogenology* 2002;57:2105–2117.
75. Dieleman S, Hendriksen P, Viuff D et al. Effects of *in vivo* prematuration and *in vivo* final maturation on developmental capacity and quality of pre-implantation embryos. *Theriogenology* 2002;57:5–20.
76. Van Soom A, Boerjan M, Bols P et al. Timing of compaction and inner cell allocation in bovine embryos produced *in vivo* after superovulation. *Biol Reprod* 1997;57:1041–1049.
77. van Soom A, Ysebaert M, de Kruif A. Relationship between timing of development, morula morphology, and cell allocation to inner cell mass and trophectoderm in *in vitro*-produced bovine embryos. *Mol Reprod Dev* 1997;47:47–56.
78. Van Soom A, Boerjan M, Ysebaert M, De Kruif A. Cell allocation to the inner cell mass and the trophectoderm in bovine embryos cultured in two different media. *Mol Reprod Dev* 1996;45:171–182.
79. Behboodi E, Groen W, Destrempes M et al. Transgenic production from *in vivo*-derived embryos: effect on calf birth weight and sex ratio. *Mol Reprod Dev* 2001;60:27–37.
80. Wells D, Misica P, Day T, Tervit H. Production of cloned lambs from an established embryonic cell line: a comparison between *in vivo*- and *in vitro*-matured cytoplasts. *Biol Reprod* 1997;57:385–393.
81. Adamiak S, Mackie K, Watt R, Webb R, Sinclair K. Impact of nutrition on oocyte quality: cumulative effects of body composition and diet leading to hyperinsulinemia in cattle. *Biol Reprod* 2005;73:918–926.
82. Sinclair K, Kuran M, Gebbie F, Webb R, McEvoy T. Nitrogen metabolism and fertility in cattle: II. Development of oocytes recovered from heifers offered diets differing in their rate of nitrogen release in the rumen. *J Anim Sci* 2000;78:2670–2680.
83. Canfield R, Sniffen C, Butler W. Effects of excess degradable protein on postpartum reproduction and energy balance in dairy cattle. *J Dairy Sci* 1990;73:2342–2349.
84. Sreenan J, Diskin M. Early embryonic mortality in the cow: its relationship with progesterone concentration. *Vet Rec* 1983;112:517–521.
85. Hasler J. In-vitro production of cattle embryos: problems with pregnancies and parturition. *Hum Reprod* 2000;15(Suppl. 5):47–58.
86. Peterson A, Lee R. Improving successful pregnancies after embryo transfer. *Theriogenology* 2003;59:687–697.
87. Farin P, Slenning B, Britt J. Estimates of pregnancy outcomes based on selection of bovine embryos produced *in vivo* or *in vitro*. *Theriogenology* 1999;52:659–670.
88. Pontes J, Nonato-Junior I, Sanches B et al. Comparison of embryo yield and pregnancy rate between *in vivo* and *in vitro* methods in the same Nelore (*Bos indicus*) donor cows. *Theriogenology* 2009;71:690–697.
89. Farin P, Britt J, Shaw D, Slenning B. Agreement among evaluators of bovine embryos produced *in vivo* or *in vitro*. *Theriogenology* 1995;44:339–349.
90. Plante L, King WA. Light and electron microscopic analysis of bovine embryos derived by *in vitro* and *in vivo* fertilization. *J Assist Reprod Genet* 1994;11:515–529.
91. Abe H, Yamashita S, Itoh T, Satoh T, Hoshi H. Ultrastructure of bovine embryos developed from *in vitro*-matured and -fertilized oocytes: comparative morphological evaluation of embryos cultured either in serum-free medium or in serum-supplemented medium. *Mol Reprod Dev* 1999;53:325–335.
92. Crosier A, Farin P, Dykstra M, Alexander J, Farin C. Ultrastructural morphometry of bovine compact morulae produced *in vivo* or *in vitro*. *Biol Reprod* 2000;62:1459–1465.
93. Crosier A, Farin P, Dykstra M, Alexander J, Farin C. Ultrastructural morphometry of bovine blastocysts produced *in vivo* or *in vitro*. *Biol Reprod* 2001;64:1375–1385.
94. Van Soom A, Vanroose G, De Kruif A. Blastocyst evaluation by means of differential staining: a practical approach. *Reprod Domest Anim* 2001;36:29–35.
95. Iwasaki S, Yoshiba N, Ushijima H, Watanabe S, Nakahara T. Morphology and proportion of inner cell mass of bovine blastocysts fertilized *in vitro* and *in vivo*. *J Reprod Fertil* 1990;90:279–284.

96. Hyttel P, Viuff D, Laurincik J et al. Risks of in-vitro production of cattle and swine embryos: aberrations in chromosome numbers, ribosomal RNA gene activation and perinatal physiology. *Hum Reprod* 2000;15(Suppl. 5):87–97.
97. Iwasaki S, Hamano S, Kuwayama M et al. Developmental changes in the incidence of chromosome anomalies of bovine embryos fertilized *in vitro*. *J Exp Zool* 1992;261:79–85.
98. Viuff D, Rickords L, Offenberg H et al. A high proportion of bovine blastocysts produced *in vitro* are mixoploid. *Biol Reprod* 1999;60:1273–1278.
99. Viuff D, Greve T, Avery B, Hyttel P, Brockhoff P, Thomsen P. Chromosome aberrations in *in vitro*-produced bovine embryos at days 2–5 post-insemination. *Biol Reprod* 2000;63:1143–1148.
100. Wrenzycki C, Hermann D, Carnwath J, Niemann H. Expression of RNA from developmentally important genes in preimplantation bovine embryos produced in TCM supplemented with BSA. *J Reprod Fertil* 1998;112:387–398.
101. Niemann H, Wrenzycki C. Alterations of expression of developmentally important genes in preimplantation bovine embryos by *in vitro* culture conditions: implications for subsequent development. *Theriogenology* 2000;53:21–34.
102. Wrenzycki C, Herrmann D, Keskintepe L et al. Effects of culture system and protein supplementation on mRNA expression in pre-implantation bovine embryos. *Hum Reprod* 2001;16:893–901.
103. Yaseen M, Wrenzycki C, Herrmann D, Carnwath J, Niemann H. Changes in the relative abundance of mRNA transcripts for insulin-like growth factor (IGF-I and IGF-II) ligands and their receptors (IGF-IR/IGF-IIR) in preimplantation bovine embryos derived from different *in vitro* systems. *Reproduction* 2001;122:601–610.
104. Bertolini M, Beam S, Shim H et al. Growth, development, and gene expression by *in vivo*- and *in vitro*-produced day 7 and 16 bovine embryos. *Mol Reprod Dev* 2002;63:318–328.
105. Lazzari G, Wrenzycki C, Herrmann D et al. Cellular and molecular deviations in bovine *in vitro*-produced embryos are related to the large offspring syndrome. *Biol Reprod* 2002;67:767–775.
106. Niemann H, Wrenzycki C, Lucas-Hahn A, Brambrink T, Kues W, Carnwath J. Gene expression patterns in bovine *in vitro*-produced and nuclear transfer-derived embryos and their implications for early development. *Cloning Stem Cells* 2002;4:29–38.
107. Rizos D, Lonergan P, Boland M et al. Analysis of differential messenger RNA expression between bovine blastocysts produced in different culture systems: implications for blastocyst quality. *Biol Reprod* 2002;66:589–595.
108. Wrenzycki C, Lucas-Hahn A, Herrmann D, Lemme E, Korsawe K, Niemann H. *In vitro* production and nuclear transfer affect dosage compensation of the X-linked gene transcripts G6PD, PGK, and Xist in preimplantation bovine embryos. *Biol Reprod* 2002;66:127–134.
109. Rizos D, Gutierrez-Adan A, Perez-Garnelo S, De La Fuente J, Boland MP, Lonergan P. Bovine embryo culture in the presence or absence of serum: implications for blastocyst development, cryotolerance, and messenger RNA expression. *Biol Reprod* 2003;68:236–243.
110. El-Sayed A, Hoelker M, Rings F et al. Large-scale transcriptional analysis of bovine embryo biopsies in relation to pregnancy success after transfer to recipients. *Physiol Genomics* 2006;28:84–96.
111. Ghanem N, Salilew-Wondim D, Gad A et al. Bovine blastocysts with developmental competence to term share similar expression of developmentally important genes although derived from different culture environments. *Reproduction* 2011;142:551–564.
112. Held E, Salilew-Wondim D, Linke M et al. Transcriptome fingerprint of bovine 2-cell stage blastomeres is directly correlated with the individual developmental competence of the corresponding sister blastomere. *Biol Reprod* 2012;87:154.
113. Block J, Hansen P, Loureiro B, Bonilla L. Improving post-transfer survival of bovine embryos produced *in vitro*: actions of insulin-like growth factor-1, colony stimulating factor-2 and hyaluronan. *Theriogenology* 2011;76:1602–1609.
114. Diskin M, Parr M, Morris D. Embryo death in cattle: an update. *Reprod Fertil Dev* 2012;24:244–251.
115. Farin C, Farmer W, Farin P. Pregnancy recognition and abnormal offspring syndrome in cattle. *Reprod Fertil Dev* 2010;22:75–87.
116. Lonergan P. Influence of progesterone on oocyte quality and embryo development in cows. *Theriogenology* 2011;76: 1594–1601.
117. Farin P, Farin C, Crosier A, Blondin P, Alexander J. Effect of *in vitro* culture and maternal insulin-like growth factor-I on development of bovine conceptuses. *Theriogenology* 1999;51:238.
118. Alexopoulos N, Maddox-Hyttel P, Tveden-Nyborg P et al. Developmental disparity between *in vitro*-produced and somatic cell nuclear transfer bovine days 14 and 21 embryos: implications for embryonic loss. *Reproduction* 2008;136:433–445.
119. Fischer-Brown A, Lindsey B, Ireland F et al. Embryonic disc development and subsequent viability of cattle embryos following culture in two media under two oxygen concentrations. *Reprod Fertil Dev* 2005;16:787–793.
120. Moore K, Kramer J, Rodriguez-Sallaberry C, Yelich J, Drost M. Insulin-like growth factor (IGF) family genes are aberrantly expressed in bovine conceptuses produced *in vitro* or by nuclear transfer. *Theriogenology* 2007;68:717–727.
121. Rodriguez-Alvarez L, Sharbati J, Sharbati S, Cox J, Einspanier R, Castro F. Differential gene expression in bovine elongated (Day 17) embryos produced by somatic cell nucleus transfer and *in vitro* fertilization. *Theriogenology* 2010;74:45–59.
122. Arnold DR, Bordignon V, Lefebvre R, Murphy B, Smith L. Somatic cell nuclear transfer alters peri-implantation trophoblast differentiation in bovine embryos. *Reproduction* 2006;132:279–290.
123. Bauersachs S, Ulbrich S, Zakhartchenko V et al. The endometrium responds differently to cloned versus fertilized embryos. *Proc Natl Acad Sci USA* 2009;106:5681–5686.
124. Silke V, Diskin M, Kenny D et al. Extent, pattern and factors associated with late embryonic loss in dairy cows. *Anim Reprod Sci* 2002;71:1–12.
125. Santos JEP, Thatcher WW, Chebel RC, Cerri RLA, Galvao KN. The effect of embryonic death rates in cattle on the efficacy of estrus synchronization programs. *Anim Reprod Sci* 2004;82–83: 513–535.
126. Lopez-Gatius F, Santolaria P, Yaniz J, Rutllant J, Lopez-Bejar M. Factors affecting pregnancy loss from gestation Day 38 to 90 in lactating dairy cows from a single herd. *Theriogenology* 2002;57: 1251–1261.
127. Agca Y, Monson R, Northey D, Abas Mazni O, Schaefer D, Rutledge J. Transfer of fresh and cryopreserved IVP bovine embryos: normal calving, birth weight and gestation lengths. *Theriogenology* 1998;50:147–162.
128. Bertolini M, Anderson G. The placenta as a contributor to production of large calves. *Theriogenology* 2002;57:181–187.
129. van Wagtendonk-de Leeuw A, Aerts B, den Daas J. Abnormal offspring following *in vitro* production of bovine preimplantation embryos: a field study. *Theriogenology* 1998;49:883–894.
130. Hill J, Winger Q, Long C, Looney C, Thompson J, Westhusin M. Development rates of male bovine nuclear transfer embryos derived from adult and fetal cells. *Biol Reprod* 2000;62:1135–1140.
131. Heyman Y, Chavatte-Palmer P, LeBourhis D, Camous S, Vignon X, Renard J. Frequency and occurrence of late-gestation losses from cattle cloned embryos. *Biol Reprod* 2002;66:6–13.
132. Watanabe S, Nagai T. Survival of embryos and calves derived from somatic cell nuclear transfer in cattle: a nationwide survey in Japan. *Anim Sci J* 2011;82:360–365.

133. Hiendleder S, Mund C, Reichenbach HD et al. Tissue-specific elevated genomic cytosine methylation levels are associated with an overgrowth phenotype of bovine fetuses derived by in vitro techniques. Biol Reprod 2004;71:217–223.
134. Bertolini M, Moyer A, Mason J et al. Evidence of increased substrate availability to in vitro-derived bovine foetuses and association with accelerated conceptus growth. Reproduction 2004;128:341–354.
135. Crosier A, Farin C, Rodriguez K, Blondin P, Alexander J, Farin P. Development of skeletal muscle and expression of candidate genes in bovine fetuses from embryos produced in vivo or in vitro. Biol Reprod 2002;67:401–408.
136. Numabe T, Oikawa T, Kikuchi T, Horiuchi T. Birth weight and birth rate of heavy calves conceived by transfer of in vitro or in vivo produced bovine embryos. Anim Reprod Sci 2000;64:13–20.
137. Akagi S, Geshi M, Nagai T. Recent progress in bovine somatic cell nuclear transfer. Anim Sci J 2013;84:191–199.
138. Bertolini M, Mason J, Beam S et al. Morphology and morphometry of in vivo- and in vitro-produced bovine concepti from early pregnancy to term and association with high birth weights. Theriogenology 2002;58:973–994.
139. McEvoy T, Sinclair K, Broadbent P, Goodhand K, Robinson J. Post-natal growth and development of Simmental calves derived from in vivo or in vitro embryos. Reprod Fertil Dev 1998;10:459–464.
140. Camargo L, Freitas C, de Sa W, Ferreira A, Serapiao R, Viana J. Gestation length, birth weight and offspring gender ratio of in vitro-produced Gyr (Bos indicus) cattle embryos. Anim Reprod Sci 2010;120:10–15.
141. Garry F, Adams R, McCann J, Odde K. Postnatal characteristics of calves produced by nuclear transfer cloning. Theriogenology 1996;45:141–152.
142. Hill J, Roussel A, Cibelli J et al. Clinical and pathologic features of cloned transgenic calves and fetuses (13 case studies). Theriogenology 1999;51:1451–1465.
143. Vignon X, Chesné P, Le Bourhis D, Fléchon J, Heyman Y, Renard J. Developmental potential of bovine embryos reconstructed from enucleated matured oocytes fused with cultured somatic cells. C R Acad Sci Ser III 1998;321:735–745.
144. Yazawa S, Aoyagi Y, Konishi M, Takedomi T. Characterization and cytogenetic analysis of Japanese black calves produced by nuclear transfer. Theriogenology 1997;48:641–650.
145. Schmidt M, Greve T, Avery B, Beckers J, Sulon J, Hansen H. Pregnancies, calves and calf viability after transfer of in vitro produced bovine embryos. Theriogenology 1996;46:527–539.
146. Blondin P, Farin P, Crosier A, Alexander J, Farin C. In vitro production of embryos alters levels of insulin-like growth factor-II messenger ribonucleic acid in bovine fetuses 63 days after transfer. Biol Reprod 2000;62:384–389.
147. Crosier A, Farin C, Rodriguez K, Blondin P, Alexander J, Farin P. Development of skeletal muscle and expression of candidate genes in bovine fetuses from embryos produced in vivo or in vitro. Biol Reprod 2002;67:401–408.
148. Jiang L, Marjani S, Bertolini M, Anderson G, Yang X, Tian XC. Indistinguishable transcriptional profiles between in vitro- and in vivo-produced bovine fetuses. Mol Reprod Dev 2011;78:642–650.
149. Everts R, Chavatte-Palmer P, Razzak A et al. Aberrant gene expression patterns in placentomes are associated with phenotypically normal and abnormal cattle cloned by somatic cell nuclear transfer. Physiol Genomics 2008;33:65–77.
150. Chavatte-Palmer P, Camous S, Jammes H, Le Cleac'h N, Guillomot M, Leed R. Placental perturbations induce the developmental abnormalities often observed in bovine somatic cell nuclear transfer. Placenta 2012;33:S99–S104.
151. Constant F, Guillomot M, Heyman Y et al. Large offspring or large placenta syndrome? Morphometric analysis of late gestation bovine placentomes from somatic nuclear transfer pregnancies complicated by hydrallantois. Biol Reprod 2006;75:122–130.
152. Lee R, Peterson A, Donnison M et al. Cloned cattle fetuses with the same nuclear genetics are more variable than contemporary half-siblings resulting from artificial insemination and exhibit fetal and placental growth deregulation even in the first trimester. Biol Reprod 2004;70:1–11.
153. Hoffert K, Batchelder C, Bertolini M et al. Measures of maternal–fetal interaction in day-30 bovine pregnancies derived from nuclear transfer. Cloning Stem Cells 2005;7:289–305.
154. Ushizawa K, Herath C, Kaneyama K et al. cDNA microarray analysis of bovine embryo gene expression profiles during the pre-implantation period. Reprod Biol Endocrinol 2004;2:77.
155. Aston K, Li G, Hicks B et al. Global gene expression analysis of bovine somatic cell nuclear transfer blastocysts and cotyledons. Mol Reprod Dev 2009;76:471–482.
156. Salilew-Wondim D, Tesfaye D, Hossain M et al. Aberrant placenta gene expression pattern in bovine pregnancies established after transfer of cloned or in vitro produced embryos. Physiol Genomics 2013;45:28–46.
157. Li N, Wells D, Peterson A, Lee R. Perturbations in the biochemical composition of fetal fluids are apparent in surviving bovine somatic cell nuclear transfer pregnancies in the first half of gestation. Biol Reprod 2005;73:139–148.
158. Hirayama H, Sawai K, Hirayama M et al. Prepartum maternal plasma glucose concentrations and placental glucose transporter mRNA expression in cows carrying somatic cell clone fetuses. J Reprod Dev 2011;57:57–61.
159. Constant F, Camous S, Chavatte-Palmer P et al. Altered secretion of pregnancy-associated glycoproteins during gestation in bovine somatic clones. Theriogenology 2011;76:1006–1021.
160. Kim H, Kang J, Yoon J et al. Protein profiles of bovine placenta derived from somatic cell nuclear transfer. Proteomics 2005;5:4264–4273.
161. Miyashita N, Shiga K, Yonai M et al. Remarkable differences in telomere lengths among cloned cattle derived from different cell types. Biol Reprod 2002;66:1649–1655.
162. Enright B, Taneja M, Schreiber D et al. Reproductive characteristics of cloned heifers derived from adult somatic cells. Biol Reprod 2002;66:291–296.
163. Pace M, Augenstein M, Betthauser J et al. Ontogeny of cloned cattle to lactation. Biol Reprod 2002;67:334–339.
164. Heyman Y, Chavatte-Palmer R, Fromentin G et al. Quality and safety of bovine clones and their products. Animal 2007;1:963–972.
165. Tian XC, Kubota C, Sakashita K et al. Meat and milk compositions of bovine clones. Proc Natl Acad Sci USA 2005;102:6261–6266.
166. Walsh M, Lucey J, Govindasamy-Lucey S, Pace M, Bishop M. Comparison of milk produced by cows cloned by nuclear transfer with milk from non-cloned cows. Cloning Stem Cells 2003;5:213–219.
167. Farin C, Farin P, Piedrahita J. Development of fetuses from in vitro-produced and cloned bovine embryos. J Anim Sci 2004;82:E53–E62.
168. Reik W, Dean W, Walter J. Epigenetic reprogramming in mammalian development. Science 2001;293:1089–1093.
169. Prickett A, Oakey R. A survey of tissue-specific genomic imprinting in mammals. Mol Genet Genomics 2012;287:621–630.
170. Barlow D. Genomic imprinting: a mammalian epigenetic discovery model. Annu Rev Genet 2011;45:379–403.
171. Smith L, Suzuki J, Goff A et al. Developmental and epigenetic anomalies in cloned cattle. Reprod Domest Anim 2012;47:107–114.
172. Farin C, Alexander J, Farin P. Expression of messenger RNAs for insulin-like growth factors and their receptors in bovine fetuses at early gestation from embryos produced in vivo or in vitro. Theriogenology 2010;74:1288–1295.
173. Niemann H, Lucas-Hahn A. Somatic cell nuclear transfer cloning: practical applications and current legislation. Reprod Domest Anim 2012;47(Suppl. 5):2–10.

174. O'Doherty A, O'Shea L, Fair T. Bovine DNA methylation imprints are established in an oocyte size-specific manner, which are coordinated with the expression of the DNMT3 family proteins. *Biol Reprod* 2012;86:67.
175. Dean W, Santos F, Stojkovic M et al. Conservation of methylation reprogramming in mammalian development: aberrant reprogramming in cloned embryos. *Proc Natl Acad Sci USA* 2001;98:13734–13738.
176. Reik W, Santos F, Dean W. Mammalian epigenomics: reprogramming the genome for development and therapy. *Theriogenology* 2003;59:21–32.
177. Young L, Fernandes K, McEvoy T et al. Epigenetic change in IGF2R is associated with fetal overgrowth after sheep embryo culture. *Nat Genet* 2001;27:153–154.
178. Young L, Schnieke A, McCreath K et al. Conservation of IGF2–H19 and IGF2R imprinting in sheep: effects of somatic cell nuclear transfer. *Mech Dev* 2003;120:1433–1442.
179. Tveden-Nyborg P, Alexopoulos N, Cooney M et al. Analysis of the expression of putatively imprinted genes in bovine peri-implantation embryos. *Theriogenology* 2008;70:1119–1128.
180. Perecin F, Meo S, Yamazaki W et al. Imprinted gene expression in *in vivo*- and *in vitro*-produced bovine embryos and chorio-allantoic membranes. *Genet Mol Res* 2009;8:76–85.
181. Guillomot M, Taghouti G, Constant F et al. Abnormal expression of the imprinted gene Phlda2 in cloned bovine placenta. *Placenta* 2010;31:482–490.
182. Su J, Wang Y, Liu Q et al. Aberrant mRNA expression and DNA methylation levels of imprinted genes in cloned transgenic calves that died of large offspring syndrome. *Livestock Sci* 2011;141:24–35.
183. Hori N, Nagai M, Hirayama M et al. Aberrant CpG methylation of the imprinting control region KvDMR1 detected in assisted reproductive technology-produced calves and pathogenesis of large offspring syndrome. *Anim Reprod Sci* 2010;122:303–312.
184. Khosla S, Dean W, Brown D, Reik W, Feil R. Culture of preimplantation mouse embryos affects fetal development and the expression of imprinted genes. *Biol Reprod* 2001;64:918–926.
185. Su H, Li D, Hou X et al. Molecular structure of bovine Gtl2 gene and DNA methylation status of Dlk1–Gtl2 imprinted domain in cloned bovines. *Anim Reprod Sci* 2011;127:23–30.
186. Long J, Cai X. Igf-2r expression regulated by epigenetic modification and the locus of gene imprinting disrupted in cloned cattle. *Gene* 2007;388:125–134.
187. Wee G, Shim J, Koo D, Chae J, Lee K, Han Y. Epigenetic alteration of the donor cells does not recapitulate the reprogramming of DNA methylation in cloned embryos. *Reproduction* 2007;134:781–787.
188. Lin L, Li Q, Zhang L, Zhao D, Dai Y, Li N. Aberrant epigenetic changes and gene expression in cloned cattle dying around birth. *BMC Dev Biol* 2008;8:14.
189. Geng-Sheng C, Yu G, Kun W, Fang-Rong D, Ning L. Repressive but not activating epigenetic modifications are aberrant on the inactive X chromosome in live cloned cattle. *Dev Growth Differ* 2009;51:585–594.
190. McLean C, Wang Z, Babu K et al. Normal development following chromatin transfer correlates with donor cell initial epigenetic state. *Anim Reprod Sci* 2010;118:388–393.
191. Breton A, Le Bourhis D, Audouard C, Vignon X, Lelievre J-M. Nuclear profiles of H3 histones trimethylated on Lys27 in bovine (*Bos taurus*) embryos obtained after *in vitro* fertilization or somatic cell nuclear transfer. *J Reprod Dev* 2010;56:379–388.
192. Wu X, Li Y, Xue L et al. Multiple histone site epigenetic modifications in nuclear transfer and *in vitro* fertilized bovine embryos. *Zygote* 2011;19:1931–1945.
193. Herrmann D, Dahl J, Lucas-Hahn A, Collas P, Niemann H. Histone modifications and mRNA expression in the inner cell mass and trophectoderm of bovine blastocysts. *Epigenetics* 2013;8:281–289.
194. Regha K, Latos P, Spahn L. The imprinted mouse Igf2r/Air cluster: a model maternal imprinting system. *Cytogenet Genome Res* 2006;113:165–177.
195. Regha K, Sloane M, Huang R et al. Active and repressive chromatin are interspersed without spreading in an imprinted gene cluster in the mammalian genome. *Mol Cell* 2007;27:353–366.
196. Latos P, Stricker S, Steenpass L et al. An *in vitro* ES cell imprinting model shows that imprinted expression of the Igf2r gene arises from an allele-specific expression bias. *Development* 2009;136:437–448.
197. Lonergan P, Rizos D, Gutierrez-Adan A, Fair T, Boland M. Oocyte and embryo quality: effect of origin, culture conditions and gene expression patterns. *Reprod Domest Anim* 2003;38:259–267.
198. Matsuura T, Takahashi K, Nakayama K et al. Increased expression of vascular endothelial growth factor in placentas of p57(Kip2) null embryos. *FEBS Lett* 2002;532:283–288.
199. Salimath B, Marme D, Finkenzeller G. Expression of the vascular endothelial growth factor gene is inhibited by p73. *Oncogene* 2000;19:3470–3476.
200. Tufarelli C, Stanley J, Garrick D et al. Transcription of antisense RNA leading to gene silencing and methylation as a novel cause of human genetic disease. *Nat Genet* 2003;34:157–165.
201. Dindot S, Farin P, Farin C et al. Epigenetic and genomic imprinting analysis in nuclear transfer derived *Bos taurus*/*Bos taurus* hybrid fetuses. *Biol Reprod* 2004;71:470–478.
202. Davies C, Hill J, Edwards J et al. Major histocompatibility antigen expression on the bovine placenta: its relationship to abnormal pregnancies and retained placenta. *Anim Reprod Sci* 2004;82–83:267–280.
203. Ravelich S, Shelling A, Ramachandran A et al. Altered placental lactogen and leptin expression in placentomes from bovine nuclear transfer pregnancies. *Biol Reprod* 2004;71:1862–1869.
204. Ravelich S, Shelling A, Wells D et al. Expression of TGF-beta1, TGF-beta2, TGF-beta3 and the receptors TGF-betaRI and TGF-betaRII in placentomes of artificially inseminated and nuclear transfer derived bovine pregnancies. *Placenta* 2006;27:307–316.
205. Doherty A, Mann M, Tremblay K, Bartolomei M, Schultz R. Differential effects of culture on imprinted H19 expression in the preimplantation mouse embryo. *Biol Reprod* 2000;62:1526–1535.
206. Nolen L, Gao S, Han Z et al. X chromosome reactivation and regulation in cloned embryos. *Dev Biol* 2005;279:525–540.
207. Shi W, Haaf T. Aberrant methylation patterns at the two-cell stage as an indicator of early developmental failure. *Mol Reprod Dev* 2002;63:329–334.
208. Santos F, Zakhartchenko V, Stojkovic M et al. Epigenetic marking correlates with developmental potential in cloned bovine preimplantation embryos. *Curr Biol* 2003;13:1116–1121.
209. Cezar G, Bartolomei M, Forsberg E, First N, Bishop M, Eilertsen K. Genome-wide epigenetic alterations in cloned bovine fetuses. *Biol Reprod* 2003;68:1009–1014.
210. Jones P. The DNA methylation paradox. *Trends Genet* 1999;15:34–37.
211. McMinn J, Wei M, Schupf N et al. Unbalanced placental expression of imprinted genes in human intrauterine growth restriction. *Placenta* 2006;27:540–549.
212. Li E. Chromatin modification and epigenetic reprogramming in mammalian development. *Nat Rev Genet* 2002;3:662–673.
213. Kageyama S, Liu H, Kaneko N, Ooga M, Nagata M, Aoki F. Alterations in epigenetic modifications during oocyte growth in mice. *Reproduction* 2007;133:85–94.
214. Spinaci M, Seren E, Mattioli M. Maternal chromatin remodeling during maturation and after fertilization in mouse oocytes. *Mol Reprod Dev* 2004;69:215–221.

215. Bird A. DNA methylation patterns and epigenetic memory. *Genes Dev* 2002;16:6–21.
216. Howell CY, Bestor TH, Ding F et al. Genomic imprinting disrupted by a maternal effect mutation in the Dnmt1 gene. *Cell* 2001;104:829–838.
217. Golding M, Williamson G, Stroud T, Westhusin M, Long C. Examination of DNA methyltransferase expression in cloned embryos reveals an essential role for Dnmt1 in bovine development. *Mol Reprod Dev* 2011;78:306–317.
218. Russell D, Betts D. Alternative splicing and expression analysis of bovine DNA methyltransferase 1. *Dev Dynam* 2008;237:1051–1059.
219. Mertineit C, Yoder J, Taketo T, Laird D, Trasler J, Bestor T. Sex-specific exons control DNA methyltransferase in mammalian germ cells. *Development* 1998;125:889–897.
220. Hirasawa R, Chiba H, Kaneda M et al. Maternal and zygotic Dnmt1 are necessary and sufficient for the maintenance of DNA methylation imprints during preimplantation development. *Genes Dev* 2008;22:1607–1616.
221. Santos F, Hendrich B, Reik W, Dean W. Dynamic reprogramming of DNA methylation in the early mouse embryo. *Dev Biol* 2002;241:172–182.
222. Armstrong L, Lako M, Dean W, Stojkovic M. Epigenetic modification is central to genome reprogramming in somatic cell nuclear transfer. *Stem Cells* 2006;24:805–814.
223. Yamanaka K, Sakatani M, Kubota K, Balboula A, Sawai K, Takahashi M. Effects of downregulating DNA methyltransferase 1 transcript by RNA interference on DNA methylation status of the satellite I region and *in vitro* development of bovine somatic cell nuclear transfer embryos. *J Reprod Dev* 2011;57:393–402.
224. Kang Y, Koo D, Park J et al. Aberrant methylation of donor genome in cloned bovine embryos. *Nat Genet* 2001;28:173–177.
225. Casillas M, Lopatina N, Andrews L, Tollefsbol T. Transcriptional control of the DNA methyltransferases is altered in aging and neoplastically-transformed human fibroblasts. *Mol Cell Biochem* 2003;252:33–43.
226. Liu J, Liang X, Zhu J et al. Aberrant DNA methylation in 5' regions of DNA methyltransferase genes in aborted bovine clones. *J Genet Genomics* 2008;35:559–568.
227. Kang Y, Park J, Koo D et al. Limited demethylation leaves mosaic-type methylation states in cloned bovine pre-implantation embryos. *EMBO J* 2002;21:1092–1100.
228. Kang Y, Lee H, Shim J et al. Varied patterns of DNA methylation change between different satellite regions in bovine pre-implantation development. *Mol Reprod Dev* 2005;71:29–35.
229. Beaujean N, Taylor J, Gardner J, Wilmut I, Meehan R, Young L. Effect of limited DNA methylation reprogramming in the normal sheep embryo on somatic cell nuclear transfer. *Biol Reprod* 2004;71:185–193.
230. Young L, Beaujean N. DNA methylation in the preimplantation embryo: the differing stories of the mouse and sheep. *Anim Reprod Sci* 2004;82–83:61–78.
231. Lucifero D, Mertineit C, Clarke H, Bestor T, Trasler J. Methylation dynamics of imprinted genes in mouse germ cells. *Genomics* 2002;79:530–538.
232. Seidel G, Schenk J, Herickhoff L et al. Insemination of heifers with sexed sperm. *Theriogenology* 1999;52:1407–1420.
233. Andersson M, Taponen J, Kommeri M, Dahlbom M. Pregnancy rates in lactating Holstein-Friesian cows after artificial insemination with sexed sperm. *Reprod Domest Anim* 2006;41:95–97.
234. Peippo J, Vartia K, Kananen-Anttila K et al. Embryo production from superovulated Holstein-Friesian dairy heifers and cows after insemination with frozen-thawed sex-sorted X spermatozoa or unsorted semen. *Anim Reprod Sci* 2009;111:80–92.
235. Underwood SL, Bathgate R, Pereira DC et al. Embryo production after *in vitro* fertilization with frozen-thawed, sex-sorted, re-frozen-thawed bull sperm. *Theriogenology* 2010;73:97–102.
236. Carvalho J, Sartori R, Machado G, Mourao G, Dode M. Quality assessment of bovine cryopreserved sperm after sexing by flow cytometry and their use in *in vitro* embryo production. *Theriogenology* 2010;74:1521–1530.
237. Tavalaee M, Razavi S, Nasr-Esfahani M. Influence of sperm chromatin anomalies on assisted reproductive technology outcome. *Fertil Steril* 2009;91:1119–1126.
238. Holm P, Walker S, Seamark R. Embryo viability, duration of gestation and birth weight in sheep after transfer of *in vitro* matured and *in vitro* fertilized zygotes cultured *in vitro* or *in vivo*. *J Reprod Fertil* 1996;107:175–181.
239. Racedo S, Wrenzycki C, Lepikhov K, Salamone D, Walter J, Niemann H. Epigenetic modifications and related mRNA expression during bovine oocyte *in vitro* maturation. *Reprod Fertil Dev* 2009;21:738–748.
240. Katz-Jaffe M, McCallie B, Preis K, Filipovits J, Gardner D. Transcriptome analysis of *in vivo* and *in vitro* matured bovine MII oocytes. *Theriogenology* 2009;71:939–946.
241. Fagundes N, Michalczechen-Lacerda V, Caixeta E et al. Methylation status in the intragenic differentially methylated region of the IGF2 locus in *Bos taurus indicus* oocytes with different developmental competencies. *Mol Hum Reprod* 2011;17:85–91.
242. El Hajj N, Trapphoff T, Linke M et al. Limiting dilution bisulfite (pyro)sequencing reveals parent-specific methylation patterns in single early mouse embryos and bovine oocytes. *Epigenetics* 2011;6:1176–1188.
243. Heinzmann J, Hansmann T, Herrmann D et al. Epigenetic profile of developmentally important genes in bovine oocytes. *Mol Reprod Dev* 2011;78:188–201.

Chapter 68

Strategies to Decrease Neonatal Calf Loss in Beef Herds

David R. Smith

Department of Pathobiology and Population Medicine, College of Veterinary Medicine, Mississippi State University, Starkville, Mississippi, USA

Introduction

Calving is a highly anticipated event for the beef cow-calf producer. The first few weeks of each calf's life are critical to its future health and productivity, and collectively the health and productivity of the herd as well as the financial success of the enterprise. Successful calving seasons are planned in advance, with much consideration to minimizing the likelihood and costs of the known hazards of this phase of cattle production. Important hazards during the calving season are birthing problems, dangers from the environment, and sickness or death from disease. Understanding these hazards allows the cattle producer to make long-term and near-term plans to minimize risk. Veterinarians can help beef cattle producers take appropriate actions to prevent neonatal calf losses by conducting a herd-specific risk assessment.

Risk assessment is a process of:

1. evaluating the likelihood and costs (or benefits) of potential hazards (or opportunities), termed risk analysis;
2. determining what actions, at what relative cost, can be taken to mitigate those hazards, termed risk management; and
3. sharing the action plan with all members of the team, as well as keeping records to show what was done and whether the actions were successful, termed risk communication.

The process of risk assessment can be iterative to provide a continuum of actions toward ongoing improvement.[1] A risk assessment may be completed formally or informally, and the process may be qualitative or quantitative.[2,3] As part of the risk analysis phase it may be necessary, or at least useful, to supplement published evidence with herd-specific data from health records,[4] outbreak investigation,[5] or clinical trials.[6] It is possible to recognize important hazards and estimate their costs without quantitative data during the risk analysis stage, but it is more difficult to evaluate progress or compliance in the risk management stage without using records.

Unfortunately, few cow-calf operations collect animal health data in a format that is easily analyzed.[7] The lack of a simple record-keeping system on many farms hinders the process of recognizing important hazards and their costs, makes it difficult to document that risk management actions were implemented, or to evaluate if those actions were effective. If records exist at all they may be on paper (e.g., in pocket-sized health record books) or in a computerized digital format (e.g., spreadsheets or databases) with free text fields, none of which allow the veterinarian or the producer to quickly and easily discover animal health relationships.[7] Some proprietary database systems capture health information and may present information in standardized graphs or summaries, but do not always allow easy querying of other health relationships. Free text fields are difficult or impossible to analyze because of misspelled words, multiple names or descriptors for the same disease (e.g., diarrhea, scours, loose, enteric), and multiple pieces of information in the same field. Often it is too costly to wade through inefficient record systems to get the needed information from which a risk assessment can be performed. Therefore, it may be beneficial for the veterinarian to help the cattle producer establish a system for capturing and analyzing health and performance data.

A desirable system for evaluating calf health and performance captures data on all calves at risk, not just treated calves. Useful information includes the calf's identification, the identification of the dam, the age of the dam, the calf's birth date, date of onset of health concerns, and information describing variables of interest (e.g., birth and weaning weights, risk factors for disease, economic data).

A risk assessment evaluates the reasons for the occurrence of hazards, the likelihood of the hazard, and its cost. In the absence of farm specific information, risk assessments are often based on published information and expert opinion. For example, based on a national survey of beef cattle herds completed in 2007–2008,[7] 2.9% of calves were born dead and another 3.5% died or were lost prior to weaning. These rates were similar regardless of the herd size. In this survey, the reasons for beef calves to die in the first 3 weeks of life, in order of frequency, were (i) birth related (25.7% of deaths); (ii) weather related (25.6%); (iii) unknown causes (18.6%); (iv) digestive system related (14%); (v) respiratory disease (8.2%); and (vi) predation or injury (6.2%). Note that these are population averages and not every beef herd experiences losses according to this distribution. However, in the absence of herd-specific information, these data inform us that on average the important hazards to the survival of neonatal calves are problems occurring during calving, dangers from the environment, and contagious diseases.

Managing risks requires greater understanding of the factors associated with those hazards, how to mitigate those factors, and the associated costs.[1] Economics should not be the sole basis for making decisions about the care of animals. However, the cost of healthcare remains an important financial constraint to most cattle producers, and is therefore an important consideration. The relative costs of disease prevention and control practices are relevant to risk management decisions. Comparing the relative costs of different approaches to managing risk can help inform animal care decisions.

The cost of even highly effective risk management practices should not be overlooked. As a simplified example, assume that a vaccine is 75% effective and costs $10 per head to administer (e.g., for product, labor, and supplies). If the disease occurs at an annual incidence of 40 cases per 10 000 head, then vaccination would reduce the incidence to 10 cases per 10 000 head per year. The annual cost to prevent 30 cases in a population of 10 000 would be $100 000, or $3333 to prevent each case. In this scenario, use of the vaccine might only be economically justified to prevent extremely costly diseases (e.g., causing high mortality in herds of high value). If the annual incidence of disease were 400 cases per 10 000, the cost of preventing each case would be $333. In this scenario, if the cost of the disease (e.g., costs for labor to treat, medications, lost performance, and deaths) averaged $100 per case, then use of the vaccine would not be economically justified unless the expected incidence in the absence of vaccine was 1333 cases per 10 000.

Other factors besides direct monetary costs or benefits play into a farmer's decision to use a preventive intervention. It might be more useful to consider units of utility rather than units of money in the calculation of risk management costs and benefits.[1] For example, in addition to the animals' monetary value, a particular farmer's unit of utility might also include their concern for the health and welfare of their cattle or pride of owning healthy and productive cattle. Also, average or expected values do not consider the impact of extreme events. For example, the occurrence of infectious diseases may cluster in time and place as an outbreak. Therefore, just as with many insurance plans, the decisions to use vaccines or to take other preventive actions may be based on minimizing the risk of a catastrophic outbreak, even if the average cost of the prevention is greater than the average cost of the disease. Finally, the relative level of risk aversion of the individual producer may affect the utility breakpoint at which a risk management practice is adopted.[1]

What follows are qualitative risk assessments with generalized approaches to managing the risk of the most common hazards of the neonatal period in beef herds. The veterinary practitioner may be able to provide a more tailored approach to managing health risks within specific cattle herds using specific herd data. The emphasis of this chapter is on disease and injury prevention; however, timely and effective treatment is also important in managing the overall cost of disease or injury.

Risk assessment of calving hazards

Successful calving occurs when a live calf is born without complications to the calf or the dam. Problems with the birthing process are called dystocia. Dystocia may be due to factors of the calf or factors of the dam.[8]

Calf factors of dystocia

The factors of dystocia related to the calf are size, posture, and presentation. Large calves are more likely to experience dystocia because they have difficulty fitting through the dam's pelvis. Posture refers to the position of the calf's head and legs as the calf enters the birth canal. Presentation refers to the calf's orientation relative to the dam. The most normal presentation and posture at birth is the calf upright, facing toward the rear of the cow, with the head between extended front legs. A breech birth is an example of both abnormal posture and presentation; the calf is backwards with the rear legs folded under the body. Of these factors, calf size is the most common problem and most preventable, through genetic selection.[8]

The consequences of dystocia to the calf are metabolic or physical injury that may result in death during or following the birthing process. Lack of oxygen in the blood causes injury to cells and results in acidosis and low blood sugar. Physical injuries include congestion and swelling of the head and tongue which may prevent nursing, or broken bones due to excessive force during calving assistance.

Dam factors of dystocia

Factors of dystocia attributable to the dam are age, pelvic size, and metabolic health.[8] Dystocia is more common among heifers, and also cows with small pelvic dimensions. Common metabolic problems at calving are from muscle weakness due to protein–energy malnutrition, exhaustion during prolonged muscular contractions, and low blood levels of calcium or magnesium. The dam may experience metabolic or physical injury during or following the birthing process. The most common problems for the dam are exhaustion from muscular contractions, pressure injury to leg muscles while being recumbent, and bruises or tears to the uterus and vagina. The consequences of these problems include failure of the dam to get up after calving, prolapse of the uterus, excessive bleeding, or infection of the reproductive tract. Each may ultimately be fatal.

Managing the risk of dystocia

There are long-term and near-term strategies for preventing problems associated with dystocia.[9] Long-term strategies include genetic selection for calving ease and pelvic size, and implementing sound heifer breeding and development programs. Near-term strategies include providing exercise and balanced nutrition to heifers and cows prior to and after calving, frequent monitoring of calving progress, early and appropriate calving assistance, ready access to the appropriate tools for calving assistance, copious use of lubricants, and attention to sanitation (e.g., use of soap and water) during birthing assistance. Describing methods for calving assistance are beyond the scope of this chapter; however, an important aspect of managing dystocia risk is helping producers know when veterinary assistance should be sought. Cattle producers should seek veterinary assistance when they:

- do not know what is wrong;
- know what is wrong, but either do not know what to do or recognize it is beyond their abilities; or
- know what is wrong and what to do about it, but have been unsuccessful after 30 min of trying.[10]

Risk assessment of environmental hazards

The environmental conditions in the neonatal period are important to the health and well-being of the dam and her calf.

Environmental hazards to the calf

Common environmental factors that present a hazard to the neonatal calf are weather extremes, crowding, predators, and physical sources of injury. At birth the calf is limited in its ability to regulate its body temperature so extremely warm or cold environmental conditions present a risk for hyperthermia or hypothermia, respectively, especially when accompanied by wet and muddy or dry and dusty conditions. Crowded herd conditions increase the chance for injury from being stepped on, butted or otherwise injured by others in the herd, and increase opportunities for pathogen exposure and transmission. Predators present the greatest hazard to calves recently born, or calves weakened by illness or injury. Calves may become injured from a variety of physical hazards in the calving environment including protruding nails, broken posts, loose wire, holes, steep embankments, standing water, various sources of electricity, and toxins (e.g., from lead batteries or chemical containers).

Environmental hazards to the dam

Cows are less susceptible to weather stressors compared to their calves, but dystocia or metabolic disease increases their risk for hypothermia or hyperthermia. During the periparturient period cows may be more likely to slip and fall, and the likelihood further increases when the floor surface has a steep slope or is slippery due to snow, ice, or mud. Cows calving near ditches and streams or other low spots are at risk to fall or not be able to rise after lying down. As with calves, cows may be injured by a variety of physical hazards that may be present in the calving environment.

Risk management of environmental hazards

The risk of injury to cow or calf can be minimized by providing favorable environmental conditions. Long-term strategies include breeding in a time frame such that calving occurs during favorable weather conditions and designing and using calving facilities with minimal physical hazards. Near-term strategies for environmental safety include a pre-calving survey of the environment and facilities for potential sources of injury, and frequent surveillance of the herd during the calving season.

Risk assessment of neonatal calf diarrhea

Neonatal calf diarrhea, commonly called scours, is one of the most likely reasons for morbidity or mortality of beef calves.[7] Neonatal calf diarrhea is a detriment to calf health and well-being, and the disease is costly to cattle producers because of reduced calf growth performance, death loss, and the expense of labor and medicines to treat sick calves.[11,12] Also, restraint and treatment of calves puts veterinarians, herd owners, and their employees at risk for injury or zoonotic infection. Finally, it is disheartening to most cattle producers to see young calves sicken or eventually die following often valiant but futile treatment attempts.

Neonatal calf diarrhea is a complex disease,[13–16] with many interrelated component causes. Component causes are the factors which in combination are sufficient to cause signs of disease.[17] The component causes of neonatal diarrhea in calves include more than the etiologic agent. In fact, it may be rare that knowledge of the presence of an agent by itself leads to a solution to outbreaks of neonatal calf diarrhea. Agent, host, and environmental factors collectively explain the occurrence of clinical signs of diarrhea, and these factors interact dynamically over the course of time.[18] To make effective risk management decisions cattle producers and their veterinarians must understand the dynamic relationships occurring between agent, host, and environmental factors within the context of the specific production system.[19] For example, even if the etiologic agent(s) involved are known, understanding the various sources of the agent and routes of transmission on the farm might be useful for prevention and control.

Agent factors of neonatal diarrhea

Numerous pathogens have been recovered from cases of neonatal calf diarrhea including bacteria such as *Escherichia coli* and *Salmonella* spp., viruses such as rotavirus and coronavirus, and protozoa such as cryptosporidia.[13,14,19–26] The adult cow herd commonly serves as the source of pathogens from one year to the next.[27–32] The common pathogens associated with neonatal calf diarrhea can commonly be recovered from calves in most cattle herds including calves in herds of cattle not experiencing calf diarrhea.[19] Recovering a pathogen from a calf with diarrhea might help to explain the immediate cause of the calf's morbidity or mortality; however, that knowledge alone rarely explains the reason for the outbreak, nor does it often provide a complete approach to treatment, control, or prevention of cases.[16] Multiple agents can often be recovered from calves in herds experiencing outbreaks of calf diarrhea, suggesting that even during outbreaks more than one agent may be involved.

Host factors

After absorbing antibodies from colostrum or colostrum supplements in the first 24 hours of life, calves may obtain passively acquired maternal immunity against diarrhea-causing pathogens common to the herd.[33–35] The quantity of antibodies absorbed is determined by the quality and quantity of colostrum the calf ingests, how soon after birth it is ingested, and the calf's physiologic state.[34,36] The presence of antibodies in colostrum requires prior exposure of the dam to the wild-type agent or vaccine antigens. Vaccines are sometimes used to immunize the dam against specific agents so that the calf will receive passive antibody protection from colostrum.[16,37] Also, some commercially available colostrum supplements and replacers contain antibodies directed against specific calf diarrhea pathogens.[38,39] Unfortunately, the use of vaccines or colostrum supplements has not always prevented outbreaks of calf diarrhea in beef cattle herds.

Calves become ill or die from neonatal diarrhea within a narrow range of age, typically 1–3 weeks of age.[13,20,22,40] The narrow range of age within which calf diarrhea occurs cannot be explained by the incubation period of the agents because diarrhea is observed within a few days of experimental pathogen challenge in colostrum-deprived or gnotobiotic calves, regardless of age.[41–43] It may be that calves experience an age-specific susceptibility to diarrhea agents that occurs as maternal immunity is waning and before the calf is primed for an effective immune response.[37] Therefore, the first 7–21 days of age typically defines the age of susceptibility as well as the age calves are most likely to initially shed diarrhea-causing pathogens in their feces, becoming infective to other calves in the herd.[44–48] In some calving systems, the relative number of susceptible and infective calves can change dynamically with time.[18] Exposure to pathogens of a dose-load or duration sufficient to cause disease is called an effective contact. At times, particularly late in the calving season, the number of infective calves may greatly outnumber the number of susceptible calves, resulting in widespread opportunities for effective contacts.

The dam's age also contributes to a neonatal calf's risk for diarrhea. Often calves born to heifers have a higher risk for diarrhea and have lower maternal antibody levels than calves born to older cows.[49] Calves born to heifers are probably more susceptible to disease in the neonatal period because heifers produce a lower volume and quality of colostrum, may have poor mothering skills, and are more likely to experience calving difficulty.[50,51]

Environmental factors of neonatal calf diarrhea

Environmental conditions may contribute to both the level of pathogen exposure and the ability of the calf to resist disease. Exposure to pathogens may occur through direct contact with other cattle or via contact with contaminated environmental surfaces. Keeping the environment clean has long been recognized as important for preventing calf diarrhea,[52,53] but it may not be possible to do so effectively. Crowded conditions increase opportunities for effective contacts with infected animals or contaminated surfaces. Ambient environmental temperature (e.g., excessive heat or cold) and moisture (e.g., mud or snow) are important stressors to the neonatal calf, may impair the calf's resistance to disease, and also affects pathogen concentration in the environment, opportunity for exposure to pathogens, and dose-load of ingestion.

Temporal factors of neonatal calf diarrhea

Host susceptibility, pathogen exposure load, and pathogen transmission occur dynamically over time within the calving season.[19] Although the adult cow-herd likely serves as the source of most calf scour pathogens from year to year,[27–32] the average dose-load of pathogen exposure to calves is likely to increase over time within a calving season because calves infected earlier serve as pathogen multipliers and become the primary source of exposure to younger susceptible calves. This multiplier effect can result in high prevalence of infective calves and widespread environmental contamination with pathogens.[54] Each calf serves as growth media for pathogen production, growing the number of pathogens to which it was exposed to even greater numbers.[43–45] Therefore, calves born later in the calving season may receive larger dose-loads of pathogens and, in turn, may become relatively more infective by growing even greater numbers of agents. Eventually the dose-load of pathogens overwhelms the calf's ability to resist disease. These factors alone or in combination may explain observations that calves born later in the calving season are at greater risk for disease or death.[18]

Risk management of neonatal calf diarrhea

Biosecurity is the sum of actions taken to prevent introducing a disease-causing agent into a population (e.g., pen, herd, or region), and biocontainment describes the actions taken to control a pathogen already present in the population.[55] In the context of risk management, one might think of biosecurity as the actions that reduce the likelihood of a pathogen being present in a herd, whereas biocontainment actions reduce the likelihood or severity of disease once the pathogen is present. In theory there are three approaches to preventing outbreaks of neonatal calf diarrhea: (i) eliminate the pathogens from the population; (ii) increase calf immunity against the pathogens; or (iii) alter the production system to reduce opportunities for pathogen exposure and transmission.[56] However, the endemic nature of the common pathogens of calf diarrhea makes it difficult or impossible to make cattle populations biosecure from these agents. Colostral immunity is critical to protect neonatal calves from disease.[37] However, passive immunity against diarrhea pathogens decreases with time,[37] and managers of beef cattle herds have limited ability to improve calf ingestion and absorption of colostral antibodies beyond not interfering with maternal bonding. Unfortunately, vaccines are not available against all pathogens associated with calf diarrhea, vaccines may not provide sufficient cross-protection against all pathogen strains,[48] and pathogens may evade the protection afforded by vaccination by evolving away from the vaccine strains.[57] For these reasons, a biocontainment approach for controlling calf diarrhea primarily by reducing effective contacts may be a useful risk management strategy in many cattle production systems.[56]

Effective contacts can be prevented by physically separating animals, reducing the dose-level of exposure (e.g., through the use of sanitation or dilution over space), or minimizing contact time.[56] For example, these principles have been successfully used to prevent neonatal illnesses in dairy calves by using calf hutch systems.[58] Various biocontainment systems for beef herds have been developed to prevent calf diarrhea.[59-62] Each of these strategies is designed to manage cattle in a system that prevents calves from having effective contacts with pathogens by reducing opportunities for exposure and minimizing dose-load.

One example of a beef cattle biocontainment system for controlling neonatal calf diarrhea is the Sandhills Calving System.[62,63] The management actions defined as the Sandhills Calving System prevent effective contacts among beef calves by segregating calves in calving pastures by week of age. This is achieved through scheduled weekly movement of pregnant cows to clean calving pastures. The objective of the system is to re-create, during each subsequent week of the season, the more ideal conditions that exist at the start of the calving season, where cows are calving on ground that has been previously unoccupied by cattle (for at least some months) in the absence of older, infective calves.

The Sandhills Calving System prevents effective contacts by using clean calving pastures, preventing direct contact between younger calves and older infective calves, and preventing later-born calves from being exposed to an accumulation of pathogens in the environment. The specific actions to implement the system may differ between herds in order to meet the specific needs of each production system. Key components of the systems are age segregation of calves, the frequent movement of pregnant cows to clean calving pastures, and opportunity for maternal bonding and colostrum ingestion with little management interruption. Age segregation prevents the serial passage of pathogens from older calves to younger calves. The routine movement of pregnant cows to new calving pastures prevents the build-up of pathogens in the calving environment over the course of the calving season, and helps to prevent exposure of the latest born calves to an overwhelming dose-load of pathogens.

Ranchers using the Sandhills Calving System have observed meaningful and sustained reductions in morbidity and mortality from neonatal calf diarrhea, and greatly reduced use of medications;[63] however, there may be additional costs associated with this production system compared with other methods of calving management. For example, there may be changes in land usage that require fencing or placement of water sources. Time and labor devoted to checking cows and delivering feed may increase. Cattle calving in extensive pastures may have greater health risks due to inclement weather or unobserved dystocia. Development of a ranch-specific plan for implementing the Sandhills Calving System or other biocontainment system must take place well in advance of the calving season. Available pastures must be identified and their use coordinated with the calving schedule. Water, feed, shelter, and anticipated weather conditions must be considered. The size of the pastures should be matched to the number of calves expected to be born in a given week so as not to be damaging to later grazing.

The costs of implementing the Sandhills Calving System may be offset by benefits to labor management. For example, there may be some efficiency because cattle movement could be scheduled once a week as labor is available. Moving cows without calves to a new pasture is often easier than sorting and moving individual cow-calf pairs. Also, the workload is partitioned between pasture groups such that cows at risk for dystocia are together in one pasture while calves at risk for diarrhea are in another.

Summary

Beef cattle producers can anticipate the most likely and most costly hazards to their cows and calves at calving time, and can use this knowledge to plan for their prevention. Minimizing the risk of these hazards requires long-range and near-term planning. Calving problems may occur because of factors of the calf or the dam. Planning ahead for calving problems and close monitoring of the herd during calving can minimize the likelihood and/or cost of dystocia. Environmental conditions such as weather or physical hazards in the calving area are also important sources of injury to cows and calves. Planning to calve during favorable weather seasons and monitoring the environment for dangerous conditions minimizes the risk of these hazards.

A common cause of sickness or death of baby calves is diarrhea. Understanding the complex interactions that cause calf diarrhea is the basis for developing strategies for disease control and prevention. The common pathogens of calf diarrhea are common to most cattle herds, and it is unlikely that cattle could be made biosecure from these agents. Managers of beef cattle herds have few opportunities to improve rates of colostrum uptake and absorption, and vaccines are not always protective. Colostral immunity wanes, making calves age-susceptible and age-infective. Each calf serves as growth media for pathogen production, amplifying the dose-load of pathogen it receives and resulting in high calf infectivity and widespread environmental contamination over time in a calving season. For these reasons it is logical to apply biocontainment strategies to prevent effective transmission of the pathogens causing diarrhea. Cattle management systems based on an understanding of infectious disease dynamics have successfully reduced sickness and death due to calf diarrhea.

References

1. Moore P. The manager's struggles with uncertainty. *J R Stat Soc A* 1977;140:129–165.
2. Stark K, Regula G, Hernandez J et al. Concepts for risk-based surveillance in the field of veterinary medicine and veterinary

public health: review of current approaches. *BMC Health Service Res* 2006;6:20.
3. Hathaway S. The application of risk assessment methods in making veterinary public health and animal health decisions. *Rev Sci Tech* 1991;10:215–231.
4. Rae D. Assessing performance of cow-calf operations using epidemiology. *Vet Clin North Am Food Anim Pract* 2006;22:53–74.
5. Smith D. Field disease diagnostic investigation of neonatal calf diarrhea. *Vet Clin North Am Food Anim Pract* 2012;28:465–481.
6. Sanderson M. Designing and running clinical trials on farms. *Vet Clin North Am Food Anim Pract* 2006;22:103–123.
7. National Animal Health Monitoring System. *Beef 2007–08. Part II: Reference of Beef-Cow Calf Management Practices in the United States, 2007–2008*. Fort Collins, CO: USDA, 2009.
8. Rice L. Dystocia-related risk factors. *Vet Clin North Am Food Anim Pract* 1994;10:53–68.
9. Sanderson M. Calving and calf management in beef herds. In: Sanderson M, Chenoweth P (eds) *Beef Practice: Cow-Calf Production Medicine*. Ames, IA: Blackwell Publishing, 2005, pp. 193–213.
10. Mortimer R. Problems handling dystocia. In: *Proceedings Range Beef Cow Symposium, Cheyenne, Wyoming*, 1993. Proceedings archived at http://digitalcommons.unl.edu/rangebeefcowsymp
11. Anderson D, Kress D, Bernardini M, Davis K, Boss D, Doornbos D. The effect of scours on calf weaning weight. *Professional Animal Scientist* 2003;19:399–403.
12. Swift B, Nelms G, Coles R. The effect of neonatal diarrhea on subsequent weight gains in beef calves. *Vet Med Small Anim Clin* 1976;71:1269–1272.
13. Acres S, Laing C, Saunders J, Radostits O. Acute undifferentiated neonatal diarrhea in beef calves. I. Occurrence and distribution of infectious agents. *Can J Comp Med* 1975;39:116–132.
14. Acres S, Saunders J, Radostits O. Acute undifferentiated neonatal diarrhea of beef calves: the prevalence of enterotoxigenic *E. coli*, reo-like (rota) virus and other enteropathogens in cow-calf herds. *Can Vet J* 1977;18:274–280.
15. Saif L, Smith K. Enteric viral infections of calves and passive immunity. *J Dairy Sci* 1985;68:206–228.
16. Foster D, Smith G. Pathophysiology of diarrhea in calves. *Vet Clin North Am Food Anim Pract* 2009;25:13–36.
17. Rothman K. Causes. *Am J Epidemiol* 1976;104:587–592.
18. Smith D, Grotelueschen D, Knott T, Clowser S, Nason G. Population dynamics of undifferentiated neonatal calf diarrhea among ranch beef calves. *Bov Pract* 2008;42:1–8.
19. Barrington G, Gay J, Evermann J. Biosecurity for neonatal gastrointestinal diseases. *Vet Clin North Am Food Anim Pract* 2002;18:7–34.
20. Bulgin M, Anderson B, Ward A, Evermann J. Infectious agents associated with neonatal calf disease in southwestern Idaho and eastern Oregon. *J Am Vet Med Assoc* 1982;180:1222–1226.
21. Mebus C, Stair E, Rhodes M, Twiehaus M. Neonatal calf diarrhea: propagation, attenuation, and characteristics of coronavirus-like agents. *Am J Vet Res* 1973;34:145–150.
22. Trotz-Williams L, Jarvie B, Martin S, Leslie K, Peregrine A. Prevalence of *Cryptosporidium parvum* infection in southwestern Ontario and its association with diarrhea in neonatal dairy calves. *Can Vet J* 2005;46:349–351.
23. Athanassious R, Marsollais G, Assaf R et al. Detection of bovine coronavirus and type A rotavirus in neonatal calf diarrhea and winter dysentery of cattle in Quebec: evaluation of three diagnostic methods. *Can Vet J* 1994;35:163–169.
24. Naciri M, Lefay M, Mancassola R, Poirier P, Chermette R. Role of *Cryptosporidium parvum* as a pathogen in neonatal diarrhoea complex in suckling and dairy calves in France. *Vet Parasitol* 1999;85:245–257.
25. Morin M, Lariviere S, Lallier R. Pathological and microbiological observations made on spontaneous cases of acute neonatal calf diarrhea. *Can J Comp Med* 1976;40:228–240.
26. Lucchelli A, Lance S, Bartlett P, Miller G, Saif L. Prevalence of bovine group A rotavirus shedding among dairy calves in Ohio. *Am J Vet Res* 1992;53:169–174.
27. Crouch C, Bielefeldt Ohman H, Watts T, Babiuk L. Chronic shedding of bovine enteric coronavirus antigen–antibody complexes by clinically normal cows. *J Gen Virol* 1985;66:1489–1500.
28. Collins J, Riegel C, Olson J, Fountain A. Shedding of enteric coronavirus in adult cattle. *Am J Vet Res* 1987;48:361–365.
29. Crouch C, Acres S. Prevalence of rotavirus and coronavirus antigens in the feces of normal cows. *Can J Comp Med* 1984;48:340–342.
30. McAllister T, Olson M, Fletch A, Wetzstein M, Entz T. Prevalence of *Giardia* and *Cryptosporidium* in beef cows in southern Ontario and in beef calves in southern British Columbia. *Can Vet J* 2005;46:47–55.
31. Watanabe Y, Yang CH, Ooi H. *Cryptosporidium* infection in livestock and first identification of *Cryptosporidium parvum* genotype in cattle feces in Taiwan. *Parasitol Res* 2005;97:238–241.
32. Ralston B, McAllister T, Olson M. Prevalence and infection pattern of naturally acquired giardiasis and cryptosporidiosis in range beef calves and their dams. *Vet Parasitol* 2003;114:113–122.
33. Barrington G, Parish S. Bovine neonatal immunology. *Vet Clin North Am Food Anim Pract* 2001;17:463–476.
34. Besser T, Gay C. The importance of colostrum to the health of the neonatal calf. *Vet Clin North Am Food Anim Pract* 1994;10:107–117.
35. Besser T, Gay C, McGuire T, Evermann J. Passive immunity to bovine rotavirus infection associated with transfer of serum antibody into the intestinal lumen. *J Virol* 1988;62:2238–2242.
36. Nocek J, Braund D, Warner R. Influence of neonatal colostrum administration, immunoglobulin, and continued feeding of colostrum on calf gain, health, and serum protein. *J Dairy Sci* 1984;67:319–333.
37. Cortese VL Neonatal immunology. *Vet Clin North Am Food Anim Pract* 2009;25:221–227.
38. Swan H, Godden S, Bey R, Wells S, Fetrow J, Chester-Jones H. Passive transfer of immunoglobulin G and preweaning health in Holstein calves fed a commercial colostrum replacer. *J Dairy Sci* 2007;90:3857–3866.
39. Godden S, Haines D, Hagman D. Improving passive transfer of immunoglobulins in calves. I. Dose effect of feeding a commercial colostrum replacer. *J Dairy Sci* 2009;92:1750–1757.
40. Clement J, King M, Salman M, Wittum T, Casper H, Odde K. Use of epidemiologic principles to identify risk factors associated with the development of diarrhea in calves in five beef herds. *J Am Vet Med Assoc* 1995;207:1334–1338.
41. El-Kanawati Z, Tsunemitsu H, Smith D, Saif L. Infection and cross-protection studies of winter dysentery and calf diarrhea bovine coronavirus strains in colostrum-deprived and gnotobiotic calves. *Am J Vet Res* 1996;57:48–53.
42. Heckert R, Saif L, Mengel J, Myers G. Mucosal and systemic antibody responses to bovine coronavirus structural proteins in experimentally challenge-exposed calves fed low or high amounts of colostral antibodies. *Am J Vet Res* 1991;52:700–708.
43. Saif L, Redman D, Moorhead P, Theil K. Experimentally induced coronavirus infections in calves: viral replication in the respiratory and intestinal tracts. *Am J Vet Res* 1986;47:1426–1432.
44. Kapil S, Trent A, Goyal S. Excretion and persistence of bovine coronavirus in neonatal calves. *Arch Virol* 1990;115:127–132.
45. Uga S, Matsuo J, Kono E et al. Prevalence of *Cryptosporidium parvum* infection and pattern of oocyst shedding in calves in Japan. *Vet Parasitol* 2000;94:27–32.
46. Nydam D, Wade S, Schaaf S, Mohammed H. Number of *Cryptosporidium parvum* oocysts or *Giardia* spp cysts shed by dairy calves after natural infection. *Am J Vet Res* 2001;62:1612–1615.

47. O'Handley R, Cockwill C, McAllister T, Jelinski M, Morck D, Olsen M. Duration of naturally acquired giardiosis and cryptosporidiosis in dairy calves and their association with diarrhea. *J Am Vet Med Assoc* 1999;214:391–396.
48. Murakami Y, Nishioka N, Watanabe T, Kuniyasu C. Prolonged excretion and failure of cross-protection between distinct serotypes of bovine rotavirus. *Vet Microbiol* 1986;12:7–14.
49. Schumann F, Townsend H, Naylor J. Risk factors for mortality from diarrhea in beef calves in Alberta. *Can J Vet Res* 1990;54:366–372.
50. Odde K. Reducing neonatal calf losses through selection, nutrition and management. *Agri-Practice* 1996;17:12–15.
51. Odde K. Survival of the neonatal calf. *Vet Clin North Am Food Anim Pract* 1988;4:501–508.
52. Law J. Diseases of young calves. Special Report on Diseases of Cattle. United States Department of Agriculture, Bureau of Animal Industry, Washington, DC, 1916, pp. 245–261.
53. Van Es L. White scours. In: *The Principles of Animal Hygiene and Preventive Veterinary Medicine*. New York: John Wiley & Sons, 1932, pp. 504–513.
54. Atwill E, Johnson E, Pereira M. Association of herd composition, stocking rate, and duration of calving season with fecal shedding of *Cryptosporidium parvum* oocysts in beef herds. *J Am Vet Med Assoc* 1999;215:1833–1838.
55. Dargatz D, Garry F, Traub-Dargatz J. An introduction to biosecurity of cattle operations. *Vet Clin North Am Food Anim Pract* 2002;18:1–5.
56. Sanderson M, Smith D. Biosecurity for beef cow/calf production. In: Chenoweth P, Sanderson M (eds) *Beef Practice: Cow-Calf Production Medicine*. Ames, IA: Blackwell Publishing, 2005, pp. 81–88.
57. Lu W, Duhamel G, Benfield D, Grotelueschen D. Serological and genotypic characterization of group A rotavirus reassortants from diarrheic calves born to dams vaccinated against rotavirus. *Vet Microbiol* 1994;42:159–170.
58. Sanders D. Field management of neonatal diarrhea. *Vet Clin North Am Food Anim Pract* 1985;1:621–637.
59. Radostits O, Acres S. The control of acute undifferentiated diarrhea of newborn beef calves. *Vet Clin North Am Food Anim Pract* 1983;5:143–155.
60. Thomson J. Implementing biosecurity in beef and dairy herds. In: *Proceedings of the 30th Annual Convention of the American Association of Bovine Practitioners*, 1997, pp. 8–14.
61. Pence M, Robbe S, Thomson J. Reducing the incidence of neonatal calf diarrhea through evidence-based management. *Comp Cont Educ Pract Vet* 2001;23:S73–S75.
62. Smith D. Management of neonatal diarrhea in cow-calf herds. In: Anderson DE, Rings DM (eds) *Current Veterinary Therapy: Food Animal Practice*, 5th edn. St Louis, MO: Saunders Elsevier, 2009, pp. 599–602.
63. Smith D, Grotelueschen DM, Knott T, Ensley S. Prevention of neonatal calf diarrhea with the Sandhills calving system. In: *Proceedings of the 37th Annual Convention of the American Association of Bovine Practitioners*, 2004, pp. 166–168.

Chapter 69

Management to Decrease Neonatal Loss of Dairy Heifers

Ricardo Stockler

Veterinary Medicine Teaching and Research Center (VMTRC), University of California Davis, Tulare, California, USA

Introduction

Raising healthy and productive replacement dairy heifers is well known to be challenging to dairy farmers. This chapter emphasizes practical approaches for decreasing morbidity and mortality of dairy heifers from birth to weaning, focusing on topics such as the pregnant dam, the calving period, neonatal dietary requirements, and husbandry and health maintenance of the newborn heifer calf.

From birth to weaning

Perinatal mortality is directly associated with both genetic and nongenetic causes.[1] Calf sire and breed,[2] dam breed,[3] trait heritability,[4,5] and gestation length[3] are some examples of genetic variabilities that will in some degree influence neonatal death.[1] Nongenetic factors would include season,[6] calving environment, dry cow nutrition, dam vaccination history, and other specific risk factors associated with dystocia such as age at first calving,[2] gender and birthweight,[3,6] stress and uterine inertia as well as infectious and noninfectious causes following eutocia such as congenital defects,[7] twinning,[6,8] prematurity or dysmaturity, and abnormal placentation or premature placenta separation[9] leading to the demise of the calf during or shortly after calving. Thus, knowing the risk factors, farmers must then focus on identifying cows that are in labor (close up) and move them promptly to a designated area separate from the rest of the herd. This can be determined based on the dam's due date and/or clinical signs of labor.

Maternity area or pen should have good lighting, minimum human and farm equipment traffic, and be separate from the rest of the herd to prevent transfer of diseases between newborns and adults. A well-designed facility does not replace good management, and thus close observation is essential so that cows spend the minimum necessary amount of time in the calving area. It is imperative to note that if cows are to be calved in a designated pen or stall, providing a well-bedded dry enclosure with good air quality is essential. Larger dairies may opt to use calving paddocks or pastures, close observation of cows, and the prompt movement of close up and fresh cows.

The first hours of life to 3 months of age are crucial for the replacement heifer calf. Several variables will directly affect the health of the newborn calf and therefore its growth and potential productivity. The costs associated with raising a dairy heifer are positively related to price of feed (milk, milk replacer, waste milk and pasteurization, concentrate), type of housing, death losses, culling rate, and other fixable variable costs.[10] In an article published in 2007, Zwald *et al.*[11] reported that the average cost of raising a replacement heifer in northern United States was about $4.28 per day from birth to moving the heifers out of hutches into group pens.

Colostrum

Colostrum management is critical: for neonates to achieve successful transfer of maternal immunoglobulin, a sufficient amount of immunoglobulin must be present in the colostrum, combined with a low bacterial count (measured in colony-forming units or cfu), and be consumed by the calf shortly after birth while the large immunoglobulin molecules are still able to be absorbed through the intestinal mucosa. High-quality colostrum contains immunoglobulin at a concentration in excess of 50 g/L with a bacterial count of less than 100 000 cfu/mL.[12] Every neonate should be ingesting a minimum of 150–200 g of immunoglobulin within the first 4 hours of life.[13] As far as the volume to be offered, the rule of thumb is 3–4 L of good-quality colostrum; however, one must also consider the 10–12% of body weight rule[14] such that overfeeding should be avoided in some lighter calves, such as Jerseys and low-birthweight Holsteins. For optimum health, calves less than 1 week old that received sufficient good-quality colostrum should manifest minimum serum IgG_1 concentrations of 1 g/dL.[15] Recent work by Elizondo-Salazar and Heinrichs[16] showed that the best method to reduce bacterial counts in colostrum is to batch pasteurize at 60 °C for 60 min, known as the 60/60 rule. This treatment has been proven to reduce the concentration of pathogenic bacteria while preserving the nutrient values, and provides a superior-quality immunoglobulin to the neonate. The shelf-life with refrigeration is 8–10 days.[14]

Calf nutrition

Drackley[17] published a comprehensive report on calf nutrition in 2008. To summarize, a 45-kg calf requires 1.75 Mcal/day for maintenance in a thermoneutral environment. This is equivalent to 325 g of milk solids or 2.5 L of whole milk or 3 L of milk replacer, as the latter is lower in fat content compared with whole milk. As a guideline, producers should be feeding calves a milk diet from birth to weaning at a measured quantity of 8–10% of body weight per day, divided into two feedings; this amount may be adjusted up to 12% during the winter months.[18] With regard to solid diet, 1.5% of body weight for the first week after birth increasing to 2% until weaning is recommended. It is also important to note that a calf should not be weaned off milk completely until it is regularly consuming 1 kg of starter feed daily.[17]

Calf housing

Housing is also particularly important and it is commonly understood that all calves must be kept dry, free from drafts, and separate from each other (no nose-to-nose contact) until weaning. The reasoning behind each is well described in the literature.[18–20] Calf hutches or individual pens have been standard in the dairy industry;[21] a minimum of 3 m² is recommended and about an extra 10–15% open area between calves must be available to allow for cleaning and disinfection.[20] Different calf hutches/pens are available in the market and their use will vary depending on region of the United States. Calf barns are very popular in the northern region, whereas outdoor hutches are commonly found in other areas of the United Statres. Hutches are usually not bedded (slatted wood floors, California style hutches), or bedded with sand, straw, shavings, or other suitable material. The calf hutch should face south in the winter and north in the summer,[18] and be located in a prepared sloped area to allow water drainage away from the hutch. Dirt, gravel, concrete, or slatted wood may be used as flooring material. A rack to hold water and feed buckets and or nipple bottle should be placed outside the hutch, as this significantly reduces bedding maintenance. Common biosecurity practices should be practiced daily. Nipple bottles, water and feed bucks must be properly disinfected and rinsed, and must not be swapped between calves.

Weaning

Although age (12 weeks), weight (90–100 kg), and health condition must be considered, a calf should be weaned off milk and moved into a group pen only when consuming at least 1 kg of solids for three consecutive days.[17,18] It is advisable to keep the calf in the hutch up to 10 days with starter feed ad libitum, after complete removal of milk diet; this practice allows the calf to lose its urge to nurse, as well as encourage the consumption of dry feed on a regular basis, adapting and promoting rumen growth.

Health maintenance

As discussed previously, good-quality colostrum delivered to the neonate within the first 4–6 hours after birth is essential for successful passive transfer of maternal immunoglobulins. It is recommended practice to remove the calf to a clean and dry area, protected from severe weather conditions where basic calf care can be done. Close examination for any obvious gross abnormalities, such as cleft palate or other birth defects, followed by inspection and disinfection of the umbilicus is critical. Proper identification and maintenance of accurate records is also important. The use of a "calf card" with information such as date of birth, sex, dam number, colostrum feeding (amount and quality), calving ease and general comments is recommended. This information can be easily transferred to the dairy management software for easy access at a later date.

Chase et al.[22] recently published a review that summarized the role of vaccines and of the neonatal immune system, as well as its impact on vaccine response. They concluded that, first and foremost, one must assess the disease risks at the production site, consider the dam's immunity (i.e., colostrum quality), and age of the calf in order to determine the vaccination protocol. Despite new research being done to avoid vaccine interference with maternal antibodies, one must never undervalue the importance of delivering good-quality colostrum, nor support the extra-label drug use of vaccines.[20] Vaccination schedules are farm specific and vary according to dairy practices; however, an example of a classic timetable is summarized in Table 69.1.

Dehorning, commonly called disbudding, is a livestock procedure practiced for a number of reasons, of which the most important are on-farm safety for herdmates and employees, and easier use of milking facilities and feed bunk area. A variety of methods are available to perform this procedure. Chemical caustic paste, Barnes or scoop-type instrument, electric or gas-filled iron, or a combination of Barnes and dehorning iron are the options available.[23] The application of caustic paste is advantageous if applied correctly over the horn buds and in calves that are less than 2 days old, housed individually, and protected from a wet environment.[24] The use of an electric or butane gas-filled dehorning iron at 1–4 weeks of age is the quickest and most efficacious method for removal of the horn bud.[18,25,26] The procedure is simple, but requires precise restraint of the head. The use of an injectable sedative (xylazine 0.025–0.05 mg/kg i.v.) may be used although this is often not necessary. I strongly recommend the use of local anesthesia to numb the cornual nerve bilaterally. Administering 3–5 mL of 2% lidocaine using an 18 or 20 gauge 2.5-cm needle on each side should provide adequate anesthesia of the area within 3–5 min.[22,24]

According to the USDA survey in 2009,[21] diarrhea and respiratory disease are accountable for most of the deaths of young calves prior to weaning and, as mentioned before, this usually occurs at 12 weeks of age when replacement

Table 69.1 Typical vaccination protocol. Variation and inclusion of other pathogens may vary among farms.

Pathogen	Route	Age at first dose	Booster	Comments
IBR	i.m., s.c. or i.n.	4–6 weeks	8–10 weeks	Revaccinate at breeding age, then yearly. Immunosuppressant
PI3	i.m., s.c. or i.n.	4–6 weeks	8–10 weeks	Revaccinate at breeding age, then yearly Immunosuppressant
BRSV	i.m., s.c. or i.n.	4–6 weeks	8–10 weeks	Revaccinate at breeding age, then yearly. Immunosuppressant
BVD (type 1 and 2)	i.m. or s.c.	4–6 weeks	8–10 weeks	Revaccinate at breeding age, then yearly. Immunosuppressant
Clostridium spp.	s.c.	4–6 weeks	8–10 weeks	Revaccinate at breeding age, then yearly. Inactivated toxoid, local reaction may occur
Rotavirus Coronavirus	p.o.	1 day	N/A	May be affected by colostral antibody

Follow the vaccine label with respect to indication, dose, route, frequency, and meat withdrawal. i.n., intranasal; s.c., subcutaneous; i.m., intramuscular; p.o., orally. *Source*: adapted from Chase C, Hurley D, Reber A. Neonatal immune development in the calf and its impact on vaccine response. *Vet Clin North Am Food Anim Pract* 2008;24:87–104.

Table 69.2 The most common infectious causes of calf diarrhea and age of occurrence.

Etiology	Age of onset
Clostridium perfringens type C	<2 days old
Enterotoxigenic *E. coli*	<4–5 days old
Rotavirus	5–14 days old
Coronavirus	5–30 days old
Cryptosporidium parvum	4–14 days old
Salmonella	>5–7 days old
Nutritional	Until weaned
Giardia	>14 days old
Eimeria sp.	>25 days old

heifers are moved from hutches into a group pen. Prior to 30 days of age, diarrhea accounts for more than half of calf mortality on dairies, followed by pneumonia.[27]

It is imperative to recognize that the use of drugs and/or nutraceuticals do not replace a competent immune system. Colostrum management, good nutrition, and solid husbandry practices are often what are needed to raise healthy calves during the first weeks of life. In 2009, Foster and Smith[28] published a review that summarizes the pathophysiology of the most common infectious causes of diarrhea in calves. A herd investigation of diarrheic calves is primarily accomplished by understanding the age of onset, mortality and morbidity, as well as the prevalence and incidence of the problem.[28] The age of onset is crucial to rule out the likely etiology. Table 69.2 reviews the age and etiology of calf diarrhea. However, it should be noted that the majority of enteric diseases in calves are caused by mixed infections,[29] and definitive diagnosis is valuable when the incidence of diarrhea is increased within a population with the goal of addressing potential medical, biosecurity (zoonosis), husbandry, and dietary problems, whereas diagnosing individual sporadic cases is known to be of little value. Frequently, clinical signs and age of the neonate dictates the likely cause of the diarrhea and supportive care is usually the treatment option, as the majority of etiologies have no specific therapy. As pointed out by Constable in 2009 in a review,[30] the main goals to achieve when treating an individual is to prevent septicemia or bacteremia, decrease the bacterial load (mainly *Escherichia coli*) in the small intestine and abomasum, prevent negative energy balance, and reduce stress by providing analgesia.

In general, utilizing a combination of Gram-negative spectrum antimicrobials and a nonsteroidal anti-inflammatory drug, while continuing to feed a milk diet but in smaller volumes and more frequently, are the current recommendations.[30,31] There is a plethora of antimicrobials available to treat bacterial enteritis in diarrheic calves,[31] so it is important to consider the prudent use of these drugs when it comes to following the manufacturer's label; discretionary use must follow extra-label drug use rules if applicable. Most importantly, a protocol should be used that addresses the appropriate dosing, route and frequency that reaches and maintains an effective therapeutic concentration at the site of infection.[31]

Oral electrolytes are considered the hallmark of routine treatment to achieve rehydration. Smith[32] states that these products must supply a sufficient amount of sodium to normalize extracellular fluid, provide agents that facilitate the absorption of sodium and water from the intestines into the circulatory system, correct the metabolic acidosis, and provide energy.

References

1. Mee J. Managing the dairy cow at calving time. *Vet Clin North Am Food Anim Pract* 2004;20:521–546.
2. Heins B, Hansen L, Seykora A. Comparison of first-parity Holstein, Holstein-Jersey crossbred, and Holstein-Normande crossbred cows for dystocia and stillbirth. *J Dairy Sci* 2003;86 (Suppl. 1):130.
3. Johnanson J, Berger P. Birth weight as a predictor of calving ease and perinatal mortality in Holstein cattle. *J Dairy Sci* 2003; 86:3745–3755.

4. Wilcox C. Growth, type, and dairy beef. In: Van Horn HH, Wilcox CJ (eds) *Large Dairy Herd Management*. Champaign, IL: American Dairy Science Association, 1992, pp. 36–41.
5. Young C. Breeding dairy cattle for disease resistance. In: Van Horn HH, Wilcox CJ (eds) *Large Dairy Herd Management*. Champaign, IL: American Dairy Science Association, 1992, pp. 42–49.
6. Silva del Río N, Stewart S, Rapnicki P, Chang Y, Fricke P. An observational analysis of twin births, calf sex ratio, and calf mortality in Holstein dairy cattle. *J Dairy Sci* 2007;90:1255–1264.
7. Bellows R, Patterson D, Burfening P, Phelps D. Occurrence of neonatal and postnatal mortality in range beef cattle. II. Factors contributing to calf death. *Theriogenology* 1987;28:573–586.
8. Mee J. Factors affecting the spontaneous twinning rate and the effect of twinning on calving problems in nine Irish dairy herds. *Irish Vet J* 1991;44:14–20.
9. Mee J. Perinatal calf mortality: recent findings. *Irish Vet J* 1991;44:80–83.
10. Tozer P, Heinrichs A. What affects the costs of raising replacement dairy heifers: a multiple-component analysis. *J Dairy Sci* 2001;84:1836–1844.
11. Zwald A, Kohlman T, Gunderson S, Hoffman P, Kriegl T. Economic costs and labor efficiencies associated with raising dairy herd replacements on Wisconsin dairy farms and custom heifer raising operations. University of Wisconsin Extension and Cooperative Extension, 2007. http://www.uwex.edu/ces/heifermgmt/documents/ICPAfinalreport.pdf
12. McGuirk S, Collins M. Managing the production, storage, and delivery of colostrum. *Vet Clin North Am Food Anim Pract* 2004;20:593–603.
13. Hopkins B, Quigley J. Effects of method of colostrum feeding and colostrum supplementation on concentrations of immunoglobulin G in the serum of neonatal calves. *J Dairy Sci* 1997;80:979–983.
14. Godden S. Colostrum management for dairy calves. *Vet Clin North Am Food Anim Pract* 2008;24:19–39.
15. Weaver D, Tyler J, VanMetre D, Hostetler D, Barrington G. Passive transfer of colostral immunoglobulins in calves. *J Vet Intern Med* 2000;14:569–577.
16. Elizondo-Salazar J, Heinrichs A. Feeding heat-treated colostrum or unheated colostrum with two different bacterial concentrations to neonatal dairy calves. *J Dairy Sci* 2009;92:4565–4571.
17. Drackley J. Calf nutrition from birth to breeding. *Vet Clin North Am Food Anim Pract* 2008;24:55–86.
18. Morrill J. The calf: birth to 12 weeks. In: Van Horn HH, Wilcox CJ (eds) *Large Dairy Herd Management*. Champaign, IL: American Dairy Science Association, 1992, pp. 401–410.
19. Nordlund K. Practical considerations for ventilating calf barns in winter. *Vet Clin North Am Food Anim Pract* 2008;24:41–54.
20. McGuirk S. Management of dairy calves from birth to weaning. In: Risco CA, Melendez Retamal P (eds) *Dairy Production Medicine*. Ames, IA: Wiley-Blackwell, 2011, pp. 175–193.
21. National Animal Health Monitoring System. *Dairy 2007: Heifer Calf Health and Management Practices on U.S. Dairy Operations, 2007*. Fort Collins, CO: USDA, 2010.
22. Chase C, Hurley D, Reber A. Neonatal immune development in the calf and its impact on vaccine response. *Vet Clin North Am Food Anim Pract* 2008;24:87–104.
23. Anderson N. Dehorning of calves. Ontario Ministry of Agriculture and Food, Factsheet 87-038, 2009. http://www.omafra.gov.on.ca/english/livestock/dairy/facts/09-003.htm
24. Villarroel A. Dehorn calves early. In: *Proceedings of the 44th Annual Conference of the American Association of Bovine Practitioners*, 2011, Practice tips session
25. Duffield T. Dehorning dairy calves to minimize pain. In: *Proceedings of the Fortieth Annual Conference of the American Association of Bovine Practitioners*, 2007, pp. 200–202.
26. McGuirk S. Disease management of dairy calves and heifers. *Vet Clin North Am Food Anim Pract* 2008;24:139–153.
27. National Animal Health Monitoring System. Dairy heifer morbidity, mortality, and health management focusing on preweaned heifers. USDA survey, 1994.
28. Foster D, Smith G. Pathophysiology of diarrhea in calves. *Vet Clin North Am Food Anim Pract* 2009;25:13–36.
29. Hall G, Reynolds D, Parsons K, Bland A, Morgan J. Pathology of calves with diarrhea in southern Britain. *Res Vet Sci* 1988;45:240–250.
30. Constable P. Treatment of calf diarrhea: antimicrobial and ancillary treatments. *Vet Clin North Am Food Anim Pract* 2009;25:101–120.
31. Constable P. Antimicrobial use in the treatment of calf diarrhea. *J Vet Intern Med* 2004;18:8–17.
32. Smith G. Treatment of calf diarrhea: oral fluid therapy. *Vet Clin North Am Food Anim Pract* 2009;25:55–72.

SECTION III

Assisted and Advanced Reproductive Technologies

70. Use of Technology in Controlling Estrus in Cattle 655
Fred Lehman and Jim W. Lauderdale

71. Cryopreservation of Semen 662
Swanand Sathe and Clifford F. Shipley

72. Utilization of Sex-sorted Semen 671
Ram Kasimanickam

73. Control of Semen-borne Pathogens 679
Rory Meyer

74. Bovine Semen Quality Control in Artificial Insemination Centers 685
Patrick Vincent, Shelley L. Underwood, Catherine Dolbec, Nadine Bouchard, Tom Kroetsch and Patrick Blondin

75. Superovulation in Cattle 696
Reuben J. Mapletoft and Gabriel A. Bó

76. Embryo Collection and Transfer 703
Edwin Robertson

77. Cryopreservation of Bovine Embryos 718
Kenneth Bondioli

78. Selection and Management of the Embryo Recipient Herd for Embryo Transfer 723
G. Cliff Lamb and Vitor R.G. Mercadante

79. Evaluation of *In Vivo*-Derived Bovine Embryos 733
Marianna M. Jahnke, James K. West and Curtis R. Youngs

80. Control of Embryo-borne Pathogens 749
Julie Gard

81. *In Vitro* Fertilization 758
John F. Hasler and Jennifer P. Barfield

82. Cloning by Somatic Cell Nuclear Transfer 771
J. Lannett Edwards and F. Neal Schrick

83. The Computer-generated Bull Breeding Soundness Evaluation Form 784
John L. Myers

Chapter 70

Use of Technology in Controlling Estrus in Cattle

Fred Lehman[1] and Jim W. Lauderdale[2]

[1]Overland Park, Kansas, USA
[2]Lauderdale Enterprises, Augusta, Michigan, USA

This chapter is largely a reproduction of work which generated a centennial paper in the *Journal of Animal Science* by James (Jim) Lauderdale, PhD. Dr Lauderdale's meticulous efforts to capture and acknowledge key scientific details reveals the small but incremental progress that has provided a foundation for technologic advances and the development of tools for controlling bovine reproduction. Without these tools, many of which are embodied in commercial pharmaceuticals, veterinarians and producers would be limited in their ability to implement control of estrus and superovulation programs en masse. An appreciation for the historic accomplishments that led to progressive enhancements of our understanding of reproductive physiology and pharmacology can only augment the efforts of theriogenologists and practitioners as they seek to make the next contribution to our discipline. Our thanks go to Dr Lauderdale and the *Journal of Animal Science* for collaborating on this chapter.

Introduction

Reproductive efficiency is one of the most important factors for successful cow-calf and dairy enterprises. Certainly, in the absence of reproduction, there is no cow-calf or dairy enterprise. During the 1950s, frozen bovine semen was developed and artificial insemination (AI) with progeny-tested bulls became recognized as effective in making more rapid genetic progress for milk yield and beef production. During the 1950s through 1960s, a major detriment to AI in beef cattle was the requirement for daily detection of estrus and AI over 60–90 days or more. Therefore, with the availability of AI, further control of estrus and breeding management was of greater interest and value, especially to the beef producer. This chapter presents a review of the research, published primarily in the *Journal of Animal Science*, contributing to successful management of estrus and breeding of cattle. Additional publications, not addressed herein, provide reviews of the development of cattle estrus and breeding management.[1–11]

Managing the estrous cycle of cattle

Early studies with P4

Trimberger and Hansel[12] injected progesterone (P4) in corn oil subcutaneously daily to dairy cows ($n=30$, each cow functioning as its control). For cows receiving P4, the interval from last P4 injection to estrus was 4.6 days and pregnancy rate (PR) was 12.5%; 50% had abnormal follicles and 53% had abnormal estrus. However, the estrous cycle subsequent to the synchronized estrus for the nonpregnant cows was normal for estrous cycle length, estrus, and ovarian structures; PR was 65.2%, indicating no carry-over effect of P4 on reproduction.

Nellor and Cole[13] ground crystalline P4 in a starch emulsion and injected beef heifers once subcutaneously with 540–1120 mg P4 on various days of the estrous cycle. Estrus and corpus luteum (CL) formation were prevented. Estrus was detected in 89% of heifers 15–19 days after 540–560 mg P4 but 15–23 days after 700–1120 mg P4 (fat heifers receiving these doses did not express estrus days 15–23). In a second study, the P4 emulsion was injected once subcutaneously in 19 beef heifers at 540 mg followed by 2140 IU of equine gonadotropin 15 days after P4; estrus was detected in 84% 1–4 days after equine gonadotropin and 14% were pregnant to AI at detected estrus. PR was 67% for six controls. In a third study, the P4 emulsion was injected once subcutaneously in 35 beef heifers at 540 mg followed by 750 IU of equine gonadotropin 15 days after P4; 89% were detected in estrus during a 4-hour period 1 day after equine gonadotropin administration and all heifers were artificially inseminated 48 hours after equine gonadotropin. Unfortunately, pregnancy rates were not reported for the timed AI (TAI). This is the first report of using TAI as a component of managing estrus and breeding of cattle. An additional 20 beef heifers, 10 estrous cycling and 10 nonestrous cycling, were treated with 540 mg P4 emulsion followed by 750 IU equine gonadotropin 15 days after P4; 100% of estrous cycling and 50% of nonestrous cycling heifers were detected in estrus during 3 days, suggesting that P4 could initiate estrus in some nonestrous-cycling heifers. PR was 20% to AI at detected estrus.

The Second Brook Lodge Workshop on problems of reproductive biology, held May 1965, facilitated discussion by research leaders in reproductive biology of domestic animals to address use of estrogens, P4 and progestogens, and gonadotropins to manage estrus and breeding in cattle, the luteotrophic and luteolytic mechanisms controlling CL lifespan, and mode of action of luteinizing hormone (LH) on steroidogenesis of CL.[14] Meeting participants were encouraged to pursue existing fledgling cattle estrous synchronization research for potential commercialization. Additionally, John Babcock[14] asked if prostaglandin (PG), a new class of compounds with vasoconstrictive properties released from the uterus, might be the luteolytic factor controlling regression of the CL. Babcock's question stimulated research that led to the identification of $PGF_{2\alpha}$ as luteolytic in cattle and to $PGF_{2\alpha}$ products becoming available for commercial use in cattle.

These initial studies using P4, with and without gonadotropins, along with the data derived from studies addressing hormonal factors affecting the estrous cycle of cattle, stimulated research to find commercially viable products to manage the estrous cycle and breeding of cattle. During these early years, orally active cost-effective progestogens, fed for about 18 days to block estrus, were of greatest interest for practical estrous synchronization.

Development of progestogens for commercial use

Repromix®

Hansel's group investigated use of medroxyprogesterone acetate (MAP), an orally active synthetic progestogen, for cattle estrous synchronization.[15] Hereford cattle ($n=32$) were fed 968/500 mg MAP for 20 days, with 16 being injected with 0.5 mg estradiol-17β at time of AI. Estrus or CL formation, or both, was detected in 91% during 3–5 days after last feeding of MAP and 25% conceived to that AI. Injection of 0.5 mg estradiol-17β at time of AI had no effect on conception rate.

Based on dose–response studies, Zimbelman[16] identified the effective oral dose of MAP for cattle to be 180 mg fed daily for 18 days, with feeding beginning at random stages of the estrous cycle. In five studies with 170 beef heifers and cows, 86% of the cattle were detected in estrus during 1–6 days after last MAP feeding; 93% of those detected in estrus were detected on days 2–4, conception rate to AI at the synchronized estrus was 51% but was highly variable among the five studies, and conception rate to AI at estrus subsequent to the synchronized estrus was 76% for previously fed MAP cattle and 74% for control cattle. Gestation length and calf body weight were not different between cattle of the MAP and control groups. During MAP feeding, no new corpora lutea were formed, old corpora lutea regressed, but follicular development was not altered. These data were interpreted as showing that MAP inhibited the ovulatory surge of LH. Feeding MAP to cattle postpartum before resumption of estrous cycles resulted in a significant reduction in the variability but not average interval from calving to first post-treatment ovulation, suggesting that a progestogen could stimulate resumption of estrous cycles in postpartum cattle.

Hansel et al.[17] investigated MAP and chlormadinone acetate (CAP) for estrous synchronization in beef cattle. These orally active progestogens were fed at 240 mg MAP daily and 10 mg CAP daily, each for 18 days. Pooling the data for 1963 and 1964, estrous detection rate for days 1–9 after last feeding was 84% for MAP ($n=232$) and 87% for CAP ($n=236$); 93% of controls ($n=229$) were detected in estrus in 20 days. PR to AI was 49% for MAP and 31% for CAP at synchronized estrus days 1–9 and was 46% for control AI during 20 days. PR from AI at synchronized estrus plus subsequent estrus for MAP and CAP and AI for 40 days for controls was 74, 68, and 66%, respectively.

The research by Hansel's group at Cornell (Ithaca, NY) and Zimbelman's group at The Upjohn Company (Kalamazoo, MI) stimulated the commercial development by The Upjohn Company of MAP, which was sold as Repromix®. Repromix® was the first product for estrous synchronization of cattle. *The Repromix Story* was a 45-page booklet that provided information on the reproductive cycle of cattle, synchronization of the reproductive cycle, effectiveness and safety of Repromix as a cattle estrous synchronization product, field trial data, and good management needed for successful cattle estrous synchronization and AI.[18] Cattle were fed MAP 180 mg daily for 18 days beginning at unknown days of the estrous cycle. University ($n=9$) and commercial ($n=63$) facilities participated in the research, with 4326 cattle fed MAP and 1899 cattle as untreated controls. Pooling the data for 1962–1963 (52 studies) with 1964 (18 studies), 76% of the MAP cattle were detected in estrus days 1–6 after last feeding of MAP with a PR of 36%. Control cattle had a PR of 42% for AI at estrus detected during 20 days. PR for AI at detected estrus during 26 days was 60% for MAP and 45% for controls.

Repromix® was sold in the United States for cattle estrous synchronization from about 1965 to 1967, but was too expensive for commercial cattle producers, and sales were ceased voluntarily by The Upjohn Company in 1967.

Syncro-Mate-B®

Wiltbank et al.[19] investigated progestogens alone and in combination with estrogens for cattle estrous synchronization. In the initial study, beef heifers ($n=324$ in two trials) were injected subcutaneously with various doses of P4 in corn oil and estradiol for various durations staring on various days of the estrous cycle. Estrous synchrony during 4 days was 68–100% (control 100% over 21 days) and conception rates were 12–53% (control 50% over 21 days).

Wiltbank et al.[20] fed dihydroxyprogesterone acetophenide (DHPA) to beef heifers either alone or in combination with estradiol valerate (EV) to synchronize estrus. Beef heifers were assigned either to be fed 500 mg DHPA daily for 20 days ($n=50$) beginning at unknown days of the estrous cycle or untreated controls ($n=54$). Estrous detection over 48 hours was 96% and conception rate was 54% for DHPA heifers. AI at detected synchronized estrus and conception rate was 26% for controls. Ova were collected from a subset of the heifers at 48 hours after AI and fertilization rate was 54% for DHPA and 86% for controls. In a second study,

100 beef heifers were assigned to a treatment–control switch-back design. Treated heifers were fed 400 mg DHPA daily for 9 days and injected intramuscularly with 5 mg EV on day 2 of DHPA feeding. Estrous detection was 84% in 96 hours; conception rates were 32% for DHPA/EV and 50% for control heifers. Further studies investigated the combination of DHPA fed for 9 days and EV injected intramuscularly on day 2 of DHPA feeding. Beef heifers were fed 400 mg DHPA for 9 days and were injected with 5 mg EV on day 2 of DHPA feeding ($n=66$) or served as untreated controls ($n=33$). The DHPA feeding was started without regard to day of the estrous cycle. Percentage of treated heifers detected in estrus were 82% in 48 hours, 86% in 72 hours, and 95% in 96 hours. Conception rates were 54% for estrous-synchronized heifers and 52% for control AI during 21 days.

Based on data from the above-cited studies, Wiltbank et al.[21] investigated use of polyhydroxy polmer subcutaneous implants to deliver an estrus inhibition agent (norethandrolone, Nor) in combination with EV injected intramuscularly at implantation to regress the CL. Beef heifers were implanted with the Nor implant for 9 days and injected with 5 mg EV at implantation ($n=42$) or assigned as untreated controls ($n=62$). Percentage in estrus at 96 hours was 93% for Nor. Conception rates were 56% for Nor treated and AI at synchronized estrus and 67% for control AI at estrus during 22 days. Intramuscular injection of 2 mg estradiol-17β 24 hours after implant removal resulted in 98–100% estrous detection in a 48-hour interval and 100% ovulation in a 36-hour interval.

Three studies were published on synchronization of estrus in beef cattle using 9-day polyhydroxy polymer subcutaneous implants containing norgestomet instead of Nor and either an intramuscular injection of EV or a combination injection or EV and norgestomet.[22–24] These studies provided data that led to the final product investigated as the commercial product, Syncro-Mate-B® (SMB).

Syncro-Mate-B® is a 6-mg norgestomet polyhydroxy polymer implant inserted subcutaneously for 9 days plus an intramuscular injection of 3 mg norgestomet and 5 mg EV at time of implantation. Spitzer et al.[23] investigated use of SMB with AI either at detected estrus or at specific times (TAI) after implant removal. Beef heifers were assigned to control AI at detected estrus during 21 days ($n=276$); SMB and AI at synchronized estrus ($n=307$); SMB and TAI twice at 48 and 60 hours ($n=47$); SMB and TAI at 45, 48, or 50 hours ($n=176$); and SMB TAI at 54 or 55 hours ($n=152$). PR was 62% in 21 days for controls; 50% for SMB AI at estrus in 5 days; 45% for SMB TAI at 48 and 60 hours; 62% for SMB TAI 45, 48, or 50 hours; and 58% for SMB TAI 54 or 55 hours. PR after 21 days of AI was 62, 61, 57, 70, and 67% for the five groups, respectively.

Based on data such as presented above, SMB was approved by the Food and Drug Administration Center for Veterinary Medicine (FDA CVM) as "For synchronization of estrus/ovulation in cycling beef cattle and non-lactating dairy heifers."[25]

Melengestrol acetate

Zimbelman and Smith[26,27] reported a series of studies designed to investigate the effective oral dose of melengestrol acetate (MGA) to inhibit estrus and effect changes in ovarian follicles and corpora lutea in cattle. The effective oral daily dose for estrous inhibition and prevention of CL formation but continued follicular development was determined to be 0.25–0.50 mg. Feeding MGA for 14–18 days was equally effective in synchronizing estrus after last feeding. During these studies, Bloss's group and Zimbelman and Smith observed that heifers fed MGA appeared to increase body weight gain compared with control heifers, especially at MGA doses of 0.25–0.75 mg.[27,28] As a result of the observation of increased body weight gain in heifers fed MGA, Bloss's group also investigated MGA as a growth promotant for beef heifers.[28] Beef heifers ($n=255$) were fed MGA in a typical feedlot ration at various doses between 0.35 and 0.53 mg and physiological conditions (pubertal, immature, ovariectomized) daily for 106–119 days. Ovariectomized or immature heifers did not increase body weight gain or feed efficiency greater than or less than controls. However, pubertal heifers fed MGA had a mean increase in average daily gain of 6.2% and feed efficiency of 6.4% over controls. Postmortem examination of ovaries frequently revealed multiple follicles, suggesting that the progestogen suppressed the release of LH resulting in anovulatory but estrogen-competent follicles, confirmed by Imwalle et al.[29] Persistent low-level endogenous estrogen was theorized, and data support that interpretation, resulting in the enhanced growth rate through anabolic actions.[30] Subsequent studies in commercial feedlots led to the approval of MGA as "For increased rate of weight gain, improved feed efficiency, and suppression of estrus in heifers fed in confinement for slaughter."[31]

Research at The Upjohn Company during 1960 through 1969 was directed to achieve FDA CVM approval of MGA for estrous synchronization of beef cattle. However, the estrous synchronization label claim was delayed until 1997 due to business, political, and regulatory decisions. Because MGA was commercially available through the feedlot approval and extensive data were available on effective beef cattle estrous synchronization programs, MGA was used for beef cattle estrous synchronization beginning about 1970. Estrous synchronization was investigated in 15 trials with 556 MGA-fed and 829 untreated control cattle.[32] Estrus was detected in 70% of MGA heifers 3–8 days after last feeding and 86% were detected during 20 days; 71% of controls were detected in estrus during 20 days. The range in estrous detection among the 15 trials was 39–95% for MGA 3–8 days and 28–90% for controls 20 days. First-service conception rates were based on 24 trials with 1853 MGA and 537 control cattle and were 36% (range 11–75%) for MGA 3–8 days and 50% (range 24–91%) for controls 20 days. Second-service conception rate for MGA heifers was 61% (range 8–100%). The observed 0.72 conception rate of MGA-fed heifers AI at estrus 3–8 days compared with control conception rate over 20 days has been observed consistently for the past 40 years and the apparent increase in conception rate of MGA-fed heifers at the second estrus after MGA has been a consistent observation. The basis for reduced conception rate at the estrus immediately following use of progesterone/progestogens is due to "persistent follicles."[33–38]

MGA was approved in 1997 by FDA CVM for feeding 0.5 mg daily for up to 24 days to suppress estrus in heifers intended for breeding.[39]

Development of PGF$_{2\alpha}$

Publications identified (i) that the human uterus would either contract or relax on instillation of fresh semen; (ii) the strong smooth muscle-stimulating activity of seminal fluid from human, monkey, sheep and goat, and extracts of vesicular glands of male sheep; and (iii) lipid extracts of sheep vesicular glands, the fraction containing lipid-soluble acids, elicited strong smooth muscle stimulation.[40–42] The active factor was named "prostaglandin." Research with prostaglandin was quiet until the 1960s when scientists at the Karolinska Institute (Stockholm, Sweden) and The Upjohn Company collaborated to produce sufficient quantities for research. The question by J. Babcock during the Brook Lodge meeting initiated research on prostaglandins for their luteolytic action.[43]

PGF$_{2\alpha}$ was reported to be luteolytic in cattle.[44–46] Lauderdale reported that heifers injected subcutaneously with 30 mg PGF$_{2\alpha}$ tromethamine salt returned to estrus in 2–4 days if injected between 6 and 9 days and 13 and 16 days but not 2 and 4 days of the estrous cycle. Liehr's group reported that 6 mg PGF$_{2\alpha}$ tromethamine salt introduced into the ipsilateral uterine horn during the responsive days of the estrous cycle resulted in return to estrus in 2.4 ± 0.5 days, while Rowson's group reported that an analog of PGF$_{2\alpha}$, cloprostenol, was luteolytic in the bovine and cattle returned to estrus in about 3 days.

The following year it was reported that following intrauterine delivery of 5 mg PGF$_{2\alpha}$ tromethamine salt into the ipsilateral horn, that serum P4 decreased by 12 hours, CL diameter decreased by 24 hours, and the intervals to estrus, peak LH, and ovulation were 72 ± 5, 71 ± 4, and 95 ± 5 hours, respectively and that the subsequent estrous cycle was of a normal duration (21 ± 3 days).[47]

Introducing 1.5 or 2.0 mg PGF$_{2\alpha}$ into the ipsilateral uterine horn of 34 beef cows once between 6 and 15 days of the estrous cycle resulted in 74% in estrus 60–80 hours after PGF$_{2\alpha}$ and 52% conceived to AI at detected estrus. If 400 μg estradiol-17β was injected intramuscularly 48 hours after PGF$_{2\alpha}$, 69% were in estrus 60–80 hours after PGF$_{2\alpha}$ with a 69% conception rate to AI at detected estrus.[48]

Lauderdale's group investigated fertility of cattle at four locations in three states after intramuscular injection of 30 mg PGF$_{2\alpha}$ tromethamine salt. Cattle were palpated and those with a CL were assigned randomly in replicates to control, AI at detected estrus during 18–25 days ($n=153$), PGF$_{2\alpha}$ with AI at estrus detected within 7 days after injection ($n=119$), and PGF$_{2\alpha}$ with AI at 72 and 90 hours after injection ($n=120$). Estrous detection was 80, 58, and 72% and PR was 42, 30, and 40% for the three groups, respectively.[49]

Peters' group synchronized estrus in beef cattle with 25 mg PGF$_{2\alpha}$ injected intramuscularly either once or twice at a 12-day interval and either with or without 400 μg estradiol benzoate (EB) injected intramuscularly 48 hours after PGF$_{2\alpha}$. Cattle were artificially inseminated at detected estrus. The percentage of cattle detected in estrus 56–86 hours after PGF$_{2\alpha}$ was 72% for no EB ($n=333$) and 90% for EB ($n=272$); the calving rate to that AI was 34 and 42% for the two groups.[50]

Lauderdale's group described the dose–response for PGF$_{2\alpha}$ to synchronize estrus in cattle. The effective dose to regress the CL leading to return to estrus was identified in a dose–response study involving nine herds and 1215 beef cattle and dairy heifers; the dose identified was 25 mg PGF$_{2\alpha}$ injected intramuscularly[51] and several practical use programs for PGF$_{2\alpha}$ to synchronize estrus in cattle.[52] The earliest programs injected cattle twice at a 10–12 day interval in an attempt to synchronize all cattle because cattle will not respond to a luteolytic dose of PGF$_{2\alpha}$ injected during 0–5 days of the estrous cycle. The efficacy study was completed with 24 herds and 1844 cattle. Controls were artificially inseminated at estrus detected during 24 days (control); cattle assigned to PGF$_{2\alpha}$ were injected intramuscularly with 25 mg PGF$_{2\alpha}$ at an interval of 10–12 days and were artificially inseminated either at estrus during 5 days after second PGF$_{2\alpha}$ (PGF$_{2\alpha}$ AI estrus) or at 80 hours after second PGF$_{2\alpha}$ (PGF$_{2\alpha}$ TAI). Percentage of cattle in estrus was 66% for controls and 47% for PGF$_{2\alpha}$ AI estrus for cows and 81% for controls and 66% for PGF$_{2\alpha}$ AI estrus for heifers. Conception rate was 61% for controls and 61% for PGF$_{2\alpha}$ AI estrus for cows and 58% for controls and 55% for PGF$_{2\alpha}$ AI estrus for heifers. PR was 48, 34, and 35% for the respective groups for cows and 53, 38, and 36% for the respective groups for heifers.

PGF$_{2\alpha}$ (Lutalyse sterile solution) was approved by the FDA CVM for synchronization of estrus of cattle for double injection at 11–14 days (1979) and single injection (1981) programs.[53] Subsequently, generics and analogs of Lutalyse® (Pfizer, New York) have been approved (ProstaMate™, Teva, St Joseph, MO; Estrumate™, Intervet/Schering-Plough, Millsboro, DE; In Synch™, Pro Labs, St Joseph, MO; and EstroPLAN™, Pfizer).

Development of GnRH

Publications by Kittock et al.,[54] Mauer and Rippel,[55] and Zolman et al.[56] documented that gonadotropin-releasing hormone (GnRH) released LH in cattle. Kaltenbach et al.[57] reported that both intracarotid and intramuscular injections of GnRH released both LH and follicle-stimulating hormone (FSH) in cattle. Additionally, these authors reported that SMB-treated heifers responded with an LH surge, estrus, and ovulation to 250 μg GnRH injected intramuscularly 24 or 36 hours after implant removal.

Documentation that large or dominant, or both, ovarian follicles in cattle either ovulate or continue to regress by atresia in response to exogenous GnRH followed.[58–62] When GnRH is a component of estrous synchronization and breeding management protocols, timing (day of the estrous cycle relative to stage of follicle dominance) of GnRH injection is important for follicle turnover and ovulation management to be successful as measured by acceptable pregnancy rates, especially when TAI is the method of breeding.

A GnRH product, Cystorelin®, was approved by the FDA CVM for treatment of ovarian follicular cysts in cattle in 1986.[63] Subsequently, generics of Cystorelin® (Merial, Athens, GA) have been approved (Factryl™, Fort Dodge, IA; Fertagyl™, Intervet/Schering-Plough; and OvaCyst™, Teva).

Development of transrectal ultrasonography to identify ovarian follicular waves

In a series of papers, transrectal ultrasonic imaging was reported to allow noninvasive monitoring of ovarian follicle recruitment, selection, dominance and atresia, ovulation, and regression of the CL.[64–67] The authors identified that cattle exhibit two or three ovarian follicle waves each estrous cycle. Ultrasonography was essential to understanding stage of ovarian follicle development by day of the estrous cycle and follicle responsiveness to GnRH. This information and ultrasonography contributed significantly to understanding that time of administration of GnRH is critical, relative to the day of the estrous cycle and stage of follicle dominance at the time of GnRH injection, for follicle turnover and ovulation management to achieve acceptable pregnancy rates, especially when TAI is the method of breeding.

Understanding ovarian follicle recruitment, selection, dominance, and atresia provided understanding as to why progestogen and $PGF_{2\alpha}$-based estrous synchronization protocols resulted in estrus detected over 4–6 days and the variance in TAI pregnancy rates. Progestogen and $PGF_{2\alpha}$-based estrous synchronization protocols control CL lifespan but do not control ovarian follicles. Control of each is essential to minimize variance in return to estrus and achieve acceptable TAI pregnancy rates.

Use of progesterone/progestogens and $PGF_{2\alpha}$

Roche[68] investigated estrous synchronization and breeding in dairy and beef cattle using Silastic coils impregnated with P4 wrapped around a stainless steel core and inserted into the vagina (P4-releasing intravaginal device, PRID) for 12 days plus an intramuscular injection of 5 mg EB and 50 mg P4 in corn oil at time of PRID insertion. Control cattle were artificially inseminated at estrus and PRID cattle were artificially inseminated at estrus detected during 2–6 days after PRID removal. Estrous synchrony was 86% for dairy cows ($n=159$) and 81% for dairy heifers ($n=253$). Calving rates to first AI for controls and synchronized estrous AI for PRID were 49 and 45% for dairy cows and 45 and 53% for dairy heifers. In a second study, beef cows were assigned to control with AI at estrus ($n=14$), PRID with AI at synchronized detected estrus ($n=16$), PRID with AI at 48 hours after PRID removal ($n=14$), and PRID plus 100 μg GnRH at 30 hours with AI at 48 hours after PRID removal ($n=23$). PR was 71, 69, 21, and 52%, respectively. In a third study, beef heifers were assigned to control with AI at estrus ($n=24$), PRID with AI at 56 hours after PRID removal ($n=26$), PRID with AI at 74 hours after PRID removal ($n=25$), and PRID with AI at 56 and 74 hours after PRID removal ($n=25$). PR was 58, 65, 46, and 68%, respectively.

Smith et al.[69] investigated estrous synchronization and breeding of Holstein heifers using the PRID plus $PGF_{2\alpha}$. In this study, unlike the previous study, PRID were in place for either 6 or 7 days and 25 mg $PGF_{2\alpha}$ was injected intramuscularly on day 6. Cattle were assigned to control with AI at estrus during 25 days ($n=79$), PRID in place for 6 days with AI 1–5 days after PRID removal ($n=80$), and PRID in place for 7 days with AI 1–5 days after PRID removal ($n=83$). Estrous detection rate was 97, 99, and 99%, respectively. PR was 72, 82, and 73%, respectively. In a second study, estrous synchronization was investigated using the double injection of 25 mg $PGF_{2\alpha}$ at an 11-day interval. Cattle were assigned to control AI at estrus during 25 days ($n=91$), 25 mg $PGF_{2\alpha}$ injected twice at 11 days and TAI at 80 hours ($n=90$), and PRID in place for 7 days and TAI at 84 hours ($n=93$). Estrous detection rate was 93, 84, and 94%, respectively. PR was 73, 52, and 66%, respectively. These data documented effective estrous synchronization with a PRID plus $PGF_{2\alpha}$ or $PGF_{2\alpha}$ alone but enhanced PR with PRID plus $PGF_{2\alpha}$ compared with $PGF_{2\alpha}$ alone.

Moody et al.[70] synchronized estrus by feeding 0.5 mg MGA daily to beef heifers for 6 days (MGA6, $n=32$) or 7 days (MGA7, $n=31$) and injected 25 mg $PGF_{2\alpha}$ on day 6 to cattle of each group; controls ($n=33$) and MGA-fed cattle were artificially inseminated at detected estrus. First-service conception rate was 58% (controls), 44% (MGA6), and 61% (MGA7). PR was 18, 25, and 42% for the respective groups for 5 days of AI and was 61, 47, and 65% for the respective groups for 19 days of AI.

Lucy et al.[71] published results of an extensive field trial investigating estrous synchronization using the intravaginal P4-releasing insert containing 1.38 g P4 (controlled internal drug-releasing device, CIDR) inserted for 7 days plus 25 mg $PGF_{2\alpha}$ on day 6. Cattle were artificially inseminated at estrus during 31 days for control and 3 days for CIDR. For control and CIDR cows estrous cycling at the beginning of the study, estrous detection rate was 82 and 72%, conception rate was 64 and 63%, and PR was 58 and 46%. For control and CIDR cows not estrous cycling at the beginning of the study, estrous detection rate was 67 and 45%, conception rate was 58 and 57%, and PR was 42 and 26%. For control and CIDR heifers estrous cycling at the beginning of the study, estrous detection rate was 87 and 80%, conception rate was 61 and 61%, and PR was 64 and 49%. For control and CIDR heifers not estrous cycling at the beginning of the study, estrous detection rate was 54 and 48%, conception rate was 56 and 58%, and PR was 31 and 28%.

The Eazi-Breed CIDR® (CIDR®, Pfizer) specifically to be used with $PGF_{2\alpha}$, for estrous synchronization of beef cattle and dairy heifers was approved by FDA CVM in 1997.[72]

Progestogens, $PGF_{2\alpha}$, and GnRH

Estrus can be blocked with progestogens and estrus is synchronized after removal of the P4 block, but fertility generally was decreased when cattle were artificially inseminated at the synchronized estrus. Commercial progestogen products are now available. $PGF_{2\alpha}$ will regress the CL of cattle and estrous synchronization programs have been developed utilizing $PGF_{2\alpha}$ but its administration is not effective if cattle are not cycling at time of treatment. $PGF_{2\alpha}$ and its various analogs are commercially available. GnRH can be used in estrous synchronization and breeding management protocols to turn over follicles and induce ovulation and CL formation. Numerous estrous and breeding management protocols are available for beef and dairy cattle that incorporate progestogens, $PGF_{2\alpha}$, and GnRH in different combinations. The breeding protocols with pregnancy rates to TAI on the order of 55% and greater are from cattle studies with Bos taurus breeding; cattle with Bos indicus breeding have not responded as well.[73,74] However, Williams et al.[75] developed a breeding protocol (GnRH plus $PGF_{2\alpha}$ on day of CIDR insertion and CIDR inserted for 5 days, $PGF_{2\alpha}$ on day of CIDR removal, GnRH and TAI at 66 hours with pregnancy rates of greater than 50% consistently) in cattle with Bos indicus genetics.

Summary

Estrous synchronization and breeding management protocols can be found in a number of papers[76-78] and specific protocols and summary data can be found in Johnson.[79] Additionally, updated beef breeding management protocols are published annually in each of the AI Company Beef Sire Directories, and protocols for dairy cattle are now available from the Dairy Cattle Reproduction Council. Note that Chapters 31, 32, and 33 cover estrus detection, artificial insemination, and the pharmacological control of the estrous cycle. This chapter's primary goal was to provide a historic perspective of the advances in this area.

Conclusions

Discovery research led to applied research, which led to products for estrous synchronization and breeding management of cattle being available today. Such research contributed significantly and positively to animal agriculture and society. The estrous synchronization and breeding management cattle protocols enhance use of AI for increased genetic capability to produce meat and milk and are essential for viable commercial embryo transfer. Use of the protocols can increase efficiency for beef and dairy production, contributing both to enterprise economic viability and a positive environmental effect. The cost–benefit of the protocols is positive for most beef and dairy enterprises and protocols exist to meet the breeding management needs of most beef and dairy enterprises. The protocols are based on biology of the cow, and the hormones used in the protocols are FDA CVM approved and have been documented to be safe to the animal and environment, to be effective, and the animal products safe for human consumption.

On the consumer side, producers who use these protocols are meeting consumer wants by providing high-quality beef and dairy products at an acceptable price, are decreasing production effects on the environment through increased efficiency of production, and the hormones in use have no negative animal welfare issues.

References

1. Wiltbank J. Research needs in beef cattle reproduction. *J Anim Sci* 1970;31:755–762.
2. Wiltbank J. Management programs to increase reproductive efficiency of beef herds. *J Anim Sci* 1974;38:58–67.
3. Odde K. A review of synchronization of estrus in postpartum cattle. *J Anim Sci* 1990;68:817–830.
4. Chenault J, Boucher J, Hafs H. Synchronization of estrus in beef cows and beef and dairy heifers with intravaginal progesterone inserts and prostaglandin F2α with or without gonadotropin-releasing hormone. *Prof Anim Sci* 2003;19:116–123.
5. Kesler D. Synchronization of estrus in heifers. *Prof Anim Sci* 2003;19:96–108.
6. Kojima F. The estrous cycle in cattle: physiology, endocrinology, and follicular waves. *Prof Anim Sci* 2003;19:83–95.
7. Lamb G, Dahlen C, Brown D. Reproductive ultrasonography for monitoring ovarian structure development, fetal development, embryo survival, and twins in beef cows. *Prof Anim Sci* 2003;19:135–143.
8. Mapletoft R, Martinez M, Colazo M, Kastelic J. The use of controlled internal drug release devices for the regulation of bovine reproduction. *J Anim Sci* 2003;81(E Suppl. 2):E28–E36.
9. Patterson D, Kojima F, Smith M. A review of methods to synchronize estrus in replacement beef heifers and postpartum cows. *J Anim Sci* 2003;81(E Suppl. 2):E166–E177.
10. Patterson D, Kojima F, Smith M. Methods to synchronize estrous cycles of postpartum beef cows with melengestrol acetate. *Prof Anim Sci* 2003;19:109–115.
11. Stevenson J, Johnson S, Milliken G. Incidence of postpartum anestrus in suckled beef cattle: treatments to induce estrus, ovulation, and conception. *Prof Anim Sci* 2003;19:124–131.
12. Trimberger G, Hansel W. Conception rate and ovarian function following estrus control by progesterone injection in dairy cattle. *J Anim Sci* 1955;14:224–232.
13. Nellor J, Cole H. The hormonal control of estrus and ovulation in the beef heifer. *J Anim Sci* 1956;15:650–661.
14. Duncan G, Ericsson R, Zimbelman R. Ovarian Regulatory Mechanisms: Proceedings of the Second Brook Lodge Workshop on Problems of Reproductive Biology. *J Reprod Fertil Suppl* 1966;1:1–136.
15. Hansel W, Malven P, Black D. Estrous cycle regulation in the bovine. *J Anim Sci* 1961;20:621–625.
16. Zimbelman R. Determination of the minimal effective dose of 6α-methyl-17α-acetoxyprogesterone for control of the estrual cycle of cattle. *J Anim Sci* 1963;22:1051–1058.
17. Hansel W, Donaldson L, Wagner W, Brunner M. A comparison of estrous cycle synchronization methods in beef cattle under feedlot conditions. *J Anim Sci* 1966;25:497–503.
18. Upjohn: The Repromix® Story, P-1216. TUCO Products Company, *Division of the Upjohn Company*, 1965, Kalamazoo, MI.
19. Wiltbank J, Zimmerman D, Ingalls J, Rowden W. Use of progestational compounds alone or in combination with estrogen for synchronization of estrus. *J Anim Sci* 1965;24:990–994.
20. Wiltbank J, Shumway R, Parker W, Zimmerman D. Duration of estrus, time of ovulation and fertilization rate in beef heifers synchronized with dihydroxyprogesterone acetophenide. *J Anim Sci* 1967;26:764–767.
21. Wiltbank J, Sturges J, Wideman D, LeFever D, Faulkner L. Control of estrus and ovulation using subcutaneous implants and estrogens in cattle. *J Anim Sci* 1971;33:600–606.
22. Spitzer J, Miksch E, Wiltbank J. Synchronization following norgestomet and 5 or 6 mg EV. *J Anim Sci* 1976;43:305 (Abstract).
23. Spitzer J, Jones D, Miksch E, Wiltbank J. Synchronization of estrus in beef cattle. III. Field trial in heifers using a norgestomet implant and injections of norgestomet and estradiol valerate. *Theriogenology* 1978;10:223–229.
24. Miksch E, LeFever D, Mukembo G, Spitzer J, Wiltbank J. Synchronization of estrus in beef cattle. II. Effect of an injection of norgestomet and an estrogen in conjunction with a norgestomet implant in heifers and cows. *Theriogenology* 1978;10:201–221.
25. Food and Drug Administration. Syncro-Mate-BR. *Federal Register* 47 FR 55477, December 10, 1982.
26. Zimbelman R, Smith L. Control of ovulation in cattle with melengestrol. I. Effect of dosage and route of administration. *J Reprod Fertil* 1966;11:185–191.
27. Zimbelman R, Smith L. Control of ovulation in cattle with melengestrol. II. Effects on follicular size and activity. *J Reprod Fertil* 1966;11:193–201.
28. Bloss R, Northam J, Smith L, Zimbelman R. Effects of oral melengestrol acetate on the performance of feedlot cattle. *J Anim Sci* 1966;25:1048–1053.
29. Imwalle D, Fernandez D, Schillo K. Melengestrol acetate blocks the preovulatory surge of luteinizing hormone, the expression of behavioral estrus, and ovulation in beef heifers. *J Anim Sci* 2002;80:1280–1284.
30. Lauderdale J. Use of MGA (melengestrol acetate) in animal production. In: Meissonier E, Mitchell-Vigneron J (eds) *Anabolics in Animal Production: Public Health Aspects, Analytical Methods and Regulation*. Symposium held in Paris, February 15–17, 1983. Paris: OIE, 1983.
31. Food and Drug Administration. MGA. *Federal Register* 33 FR 2602. February 6, 1968.

32. Zimbelman R, Lauderdale J, Sokolowski J, Schalk T. Safety and pharmacological evaluations of melengestrol acetate in cattle and other animals: a review. *J Am Vet Med Assoc* 1970;157:1528–1536.
33. Siros J, Fortune J. Lengthening the bovine estrous cycle with low levels of exogenous progesterone: a model for studying ovarian follicular dominance. *Endocrinology* 1990;127:916–925.
34. Stock A, Fortune J. Ovarian follicular dominance in cattle: relationship between prolonged growth of the ovulatory follicle and endocrine parameters. *Endocrinology* 1993;132:1108–1114.
35. Savio J, Thatcher W, Badinga L, de la Sota R, Wolfenson D. Regulation of dominant follicle turnover during the estrous cycle in cows. *J Reprod Fertil* 1993;97:197–203.
36. Savio J, Thatcher W, Morris G, Entwistle G, Drost M, Mattiacci M. Effects of low plasma progesterone concentration with a progesterone-releasing intra-vaginal device on follicle turnover in cattle. *J Reprod Fertil* 1993;98:77–84.
37. Wehrman M, Roberson M, Cupp A et al. Increasing exogenous progesterone during synchronization of estrus decreases endogenous 17β-estradiol and increases conception in cows. *Biol Reprod* 1993;49:214–220.
38. Sanchez T, Wehrman M, Bergfeld E et al. Pregnancy rate is greater when the corpus luteum is present during the period of progestin treatment to synchronize time of estrus in cows and heifers. *Biol Reprod* 1993;49:1102–1107.
39. FDA CVM Freedom of Information. NADA 141–200.
40. Kurzroc R, Lieb C. Biochemical studies of human semen. II. The action of semen on the human uterus. *Proc Soc Exp Biol Med* 1930;28:268–272.
41. Goldblatt M. A depressor substance in seminal fluid. *J Soc Chem Ind* 1933;52:1056–1057.
42. von Euler U. A depressor substance in the vesicular gland. *J Physiol* 1935;84:21P.
43. Duncan G, Ericsson R, Zimbelman R. Ovarian Regulatory Mechanisms: Proceedings of the Second Brook Lodge Workshop on Problems of Reproductive Biology. *J Reprod Fertil Suppl* 1966;1:1–136.
44. Lauderdale J. Effects of PGF$_{2\alpha}$ on pregnancy and estrous cycles of cattle. *J Anim Sci* 1972;35:246 (Abstract).
45. Liehr R, Marion G, Olson H. Effects of prostaglandin on cattle estrous cycles. *J Anim Sci* 1972;35:247 (Abstract).
46. Rowson L, Tervit R, Brand A. The use of prostaglandin for synchronization of oestrus in cattle. *J Reprod Fertil* 1972;29:145 (Abstract).
47. Louis T, Hafs H, Seguin B. Progesterone, LH, estrus and ovulation after prostaglandin F2α. *Proc Soc Exp Biol Med* 1973;143:152–155.
48. Inskeep E. Potential uses of prostaglandins in control of reproductive cycles of domestic animals. *J Anim Sci* 1973;36:1149–1157.
49. Lauderdale J, Seguin B, Stellflug J et al. Fertility of cattle following PGF$_{2\alpha}$ injection. *J Anim Sci* 1974;38:964–967.
50. Peters J, Welch J, Lauderdale J, Inskeep E. Synchronization of estrus in beef cattle with PGF$_{2\alpha}$ and estradiol benzoate. *J Anim Sci* 1977;45:230–235.
51. Lauderdale J, Moody E, Kasson C. Dose effect of PGF$_{2\alpha}$ on return to estrus and pregnancy in cattle. *J Anim Sci* 1977;45(Suppl. 1):181 (Abstract).
52. Lauderdale J, McAllister J, Kratzer D, Moody E. Use of prostaglandin F2α (PGF$_{2\alpha}$) in cattle breeding. *Acta Vet Scand Suppl* 1981;77:181–191.
53. Food and Drug Administration. Lutalyse® sterile solution. 1979, 1981 FDA CVM Freedom of Information. NADA 108–901.
54. Kittock R, Brittand J, Convey E. Effect of GnRH on LH and progesterone in cystic cows. *J Anim Sci* 1972;35:1120 (Abstract).
55. Mauer R, Rippel R. Response of cattle to synthetic gonadotropin releasing hormone. *J Anim Sci* 1972;35:249 (Abstract).
56. Zolman J, Convey E, Britt J, Hafs H. Release of bovine luteinizing hormone by purified porcine and synthetic gonadotropin-releasing hormone. *Proc Soc Exp Biol Med* 1973;142:189–193.
57. Kaltenbach C, Dunn T, Kiser T, Corah L, Akbar A, Niswender G. Release of FSH and LH in beef heifers by synthetic gonadotrophin releasing hormone. *J Anim Sci* 1974;38:357–362.
58. Thatcher W, Macmillan K, Hansen P, Drost M. Concepts for regulation of corpus luteum function by the conceptus and ovarian follicles to improve fertility. *Theriogenology* 1989;31:149–164.
59. Twagiramungu H, Guilbault L, Prouix J, Dufour J. Effects of Syncro-Mate-B and prostaglandin F2α on estrus synchronization and fertility in beef cattle. *Can J Anim Sci* 1992;72:31–39.
60. Twagiramungu H, Guilbault L, Prouix J, Dufour J. Synchronization of estrus and fertility in beef cattle with two injections of buserelin and prostaglandin. *Theriogenology* 1992;38:1131–1144.
61. Twagiramungu H, Guilbault L, Prouix J, Villeneuve P, Dufour J. Influence of an agonist of gonadotropin releasing hormone (Buserelin) on estrus synchronization and fertility in beef cows. *J Anim Sci* 1992;70:1904–1910.
62. Schmitt E, Diaz T, Drost M, Thatcher W. Use of a GnRH-agonist for a timed-insemination protocol in cattle. *J Anim Sci* 1994;72(Suppl. 1):292 (Abstract).
63. Food and Drug Administration. Cystorelin®. FDA CVM Freedom of Information, 1986. NADA 098–379.
64. Pierson R, Ginther O. Ultrasonography of the bovine ovary. *Theriogenology* 1984;21:495–504.
65. Savio J, Keenan L, Boland M, Roche J. Pattern of growth of dominant follicles during the oestrous cycle in heifers. *J Reprod Fertil* 1988;83:663–671.
66. Sirois J, Fortune J. Ovarian follicular dynamics during the estrous cycle in heifers monitored by real-time ultrasonography. *Biol Reprod* 1988;39:308–317.
67. Ginther O, Knopf L, Kastelic J. Temporal associations among ovarian events in cattle during oestrous cycles with two and three follicle waves. *J Reprod Fertil* 1989;87:223–230.
68. Roche J. Calving rate of cows following insemination after a 12-day treatment with silastic coils impregnated with progesterone. *J Anim Sci* 1976;43:164–169.
69. Smith R, Pomerantz A, Beal W, McCann J, Pilbeam T, Hansel W. Insemination of Holstein heifers at a preset time after estrous cycle synchronization using progesterone and prostaglandin. *J Anim Sci* 1984;58:792–800.
70. Moody E, McAllister J, Lauderdale J. Effect of PGF$_{2\alpha}$ and MGA on control of the estrous cycle in cattle. *J Anim Sci* 1978;47(Suppl. 1):36 (Abstract).
71. Lucy M, Billings H, Butler W et al. Efficacy of an intravaginal progesterone insert and an injection of PGF$_{2\alpha}$ for synchronizing estrus and shortening the interval to pregnancy in postpartum beef cows, peripubertal beef heifers, and dairy heifers. *J Anim Sci* 2001;79:982–995.
72. Food and Drug Administration. Eazi-Breed™ CIDR® (CIDR) 1997.
73. Mikeska J, Williams G. Timing of preovulatory endocrine events, estrus and ovulation in Brahman × Hereford females synchronized with norgestomet and estradiol valerate. *J Anim Sci* 1988;66:939–946.
74. Lemaster J, Yelich J, Kempfer J et al. Effectiveness of GnRH plus prostaglandin F2α for estrus synchronization in cattle of Bos indicus breeding. *J Anim Sci* 2001;79:309–316.
75. Williams G, Sanko R, Allen C et al. Evidence that luteal regression at the onset of a 5-day Co-Synch + CIDR synchronization protocol markedly improves fixed-time AI pregnancy rates in *Bos indicus*-influenced cattle. *J Anim Sci* 2011;89:251.
76. Kesler J. Development of protocols to synchronize estrus in beef cattle. In: *Proceedings of Applied Reproductive Strategies for Beef Cattle, Billings, Montana*, 2007, pp. 47–61.
77. Lamb G, Larson J, Dahlen C. Estrus synchronization protocols for cows. In: *Proceedings of Applied Reproductive Strategies for Beef Cattle, Billings, Montana*, 2007, pp. 99–114.
78. Patterson D, Busch D, Leitman N, Wilson D, Mallory D, Smith M. Estrus synchronization protocols for heifers. In: *Proceedings of Applied Reproductive Strategies for Beef Cattle, Billings, Montana*, 2007, pp. 63–97.
79. Johnson S. Protocols for synchronization of estrus and ovulation. In: *Proceedings of Applied Reproductive Strategies for Beef Cattle, Billings, Montana*, 2007, pp. 115–126.

Chapter 71

Cryopreservation of Semen

Swanand Sathe[1] and Clifford F. Shipley[2]

[1]*Lloyd Veterinary Medical Center, College of Veterinary Medicine, Iowa State University, Ames, Iowa, USA*
[2]*Agricultural Animal Care and Use Program, College of Veterinary Medicine, University of Illinois, Urbana, Illinois, USA*

Introduction

Artificial insemination (AI) and cryopreservation of spermatozoa are probably the two great advances that have revolutionized the breeding industry. The dairy industry in particular has benefited from the extensive use of AI, which has permitted an accelerated rate of genetic selection and improvement. It is estimated that more than 60% of dairy cattle in the United States are bred with AI programs as compared with just 10% of beef cattle. The history of semen cryopreservation dates back half a century to the discoveries of the protective agents in egg yolk for cooling and glycerol for freezing fowl and bull sperm,[1] and by the birth of the first calf by AI using frozen/thawed spermatozoa.[2] Bratton *et al.*[3] in their field trials demonstrated that bovine sperm frozen to –79 °C and packed on dry ice could still yield high fertility. Since then several media formulations termed "extenders" have been investigated based on their ability to improve the survivability and post-thaw motility of cryopreserved sperm. Similarly, several cryoprotective compounds have been investigated for their protective role during cryopreservation of semen. There is little doubt of the profound effect this technology has had on the cattle industry. This has largely been possible due to the remarkable success that has been achieved with bull semen as compared with semen of other species. This has mainly been due to the higher tolerance of bovine sperm to cryoprotectants such as glycerol as compared with other species and the relatively few number of spermatozoa required for conception.[4]

Principles of cryopreservation

Mechanisms of cell injury during cryopreservation

Semen cryopreservation techniques have been in practice for the past 50 years and allow long-term storage of semen and hence its virtually unlimited availability. This facilitates the large-scale provision and dissemination of highly valuable genetic material at will. Cryopreservation of semen involves the freezing of spermatozoa to –196 °C, the boiling point of liquid nitrogen, a commonly used medium for freezing and storage purpose. This increases the viability of the individual cells by slowing down their metabolic rate, thereby reducing the rate at which substrates are used and toxins are produced. Semen may also be stored after cooling to 5–8 °C and will survive for 24–48 hours without a significant decline in motility, and even up to 96 hours without a significant drop in fertilization rates. Although this may provide an efficient and successful means of short-term storage, it has some adverse effects on the spermatozoa manifested as a decline in viability rate, structural integrity, motility, and conception rates.[5,6] Cryopreservation, on the other hand, offers the option of indefinite storage of semen in liquid nitrogen with acceptable post-thaw fertility rates. Cryopreservation can be detrimental to sperm function and fertility even with the latest advances in techniques. However, it is not the long-term storage of cells at these temperatures that is damaging, but rather the progression to these temperatures and back to normothermia which results in cryoinjury.[7] Thus the changes that occur during freezing are mainly ultrastructural, biochemical, and functional. These can impair sperm transport and survival in the female reproductive tract and reduce fertility in domestic species.[8]

The effects of cryopreservation on function and viability have been extensively studied for bovine sperm. Spermatozoa have a very limited biosynthetic activity of their own and depend mostly on catabolic processes to function.[9] Thus to halt these metabolic processes the cells need to be cooled below –130 °C. Sperm, like other cells in the body, are composed of various organelles containing water, which can form intracellular and extracellular ice crystals during the freezing process. At temperatures around –5 °C the intracellular and extracellular water remains unfrozen in a supercooled metastable state. However, between –5 and –10 °C, ice forms in the

Bovine Reproduction, First Edition. Edited by Richard M. Hopper.
© 2015 John Wiley & Sons, Inc. Published 2015 by John Wiley & Sons, Inc.

extracellular medium, while intracellular water remains supercooled. At this point, the rate of cooling must be slow enough to permit cellular dehydration to occur, avoiding the freezing of the intracellular water yet fast enough to avoid exposing the cell to a hyperosmotic condition subsequent to dehydration. Severe dehydration leads to solution-effect injury caused by denaturation of macromolecules and extreme shrinkage of the cell up to irreversible membrane collapse.[10] Another damaging effect is the mechanical stress of ice formation all around the cell, which will be constrained to a very limited space of unfrozen solutes.[10] Cell viability plotted as a function of rate of freezing presents an inverted U-shaped curve. Primary causes for cellular damage are largely dependent on the cooling rate employed during the freezing process. The most appropriate freezing rate is the fastest one that allows freezing of extracellular water without intracellular ice formation. Fast cooling between 30 °C and 0 °C results in cell injury in some sperm cells, called "cold shock," with deleterious effects on the cytoskeleton and genome-related structures and causing cytoplasmic fracture.[11] Cold shock also alters permeability of various membrane structures on the plasma membrane, mitochondria, and acrosome. During cryopreservation, ice crystals form in the extracellular medium, increasing the osmolality of the unfrozen water. Because of this difference in the osmotic gradient, the intracellular water diffuses out of the sperm thus dehydrating the cell and the plasma membrane. At thawing, this phenomenon is repeated in reverse order as extracellular ice crystals melt and water starts diffusing in and rehydrating the cell. Because of such drastic changes in the volume and osmotic stress, there can be irreversible damage and ultrastructural deformation of the plasma membrane.

Sperm membranes are composed of many phospholipids (depending on species), with each phospholipid having a precise phase transition temperature. The degree of structural damage to these membranes depends on the temperature and the lipid composition of the membrane.[12] The characteristics of membranes that affect their sensitivity include cholesterol/phospholipid ratio, content of nonbilayer-preferring lipids, degree of hydrocarbon chain saturation, and protein/phospholipid ratio.[1] Some lipids aggregate in domains of gel-like (frozen) lipid, thus excluding other lipid types that remain in the liquid crystalline (melted) state.[1,13] This ultimately leads to phase separation in which membrane proteins can become irreversibly clustered, leading to loss of function.[14] The specific phase transition temperatures for the different phospholipids in the membrane result in lateral migration with rearrangement of membrane components and lipid phase separations within the plane of the membrane. The lateral migration may create microdomains of nonbilayer-forming lipids and may modify protein surrounding environments. On thawing, these alterations predispose apposing membranes to fuse and affect protein activity, leading to overall altered membrane permeability to water and solutes.[1] This loss of membrane permeability also interferes with ion pumps and results in influx of Ca^{2+} into the sperm cell as well as loss of membrane ATPase activity. Damage to the Ca^{2+} regulatory systems by freezing and thawing may predispose spermatozoa to inaccurate timing of capacitation and the acrosome reaction, contributing to reduced fertilizing capacity of cryopreserved bull semen.[15]

Role of extenders used for cryopreservation

Extenders or diluents are routinely added to protect sperm during liquid storage or cryopreservation. Regardless of species, the role of semen extenders is to (i) provide nutrients as an energy source; (ii) buffer against harmful changes in pH; (iii) ensure appropriate physiologic osmotic pressure and concentration of electrolytes; (iv) prevent growth of bacteria; (v) protect from cold shock during cooling; and (vi) have cryoprotectant(s) to reduce the amount of freezing damage.[16] Over the past 65 years, the cryoprotective media for sperm storage have been continuously revised but the basic ingredients remain unchanged, with egg yolk and/or milk and glycerol representing the indispensable compounds of practically all media used for bull sperm preservation in the liquid or frozen states. Extenders used for diluting bull semen typically are egg yolk and/or skimmed milk-citrate/Tris-based buffers with added simple sugars and antibiotics, with or without cryoprotectants depending on whether they are one-step or two-step extenders. These components provide an acceptable buffering capacity, osmolality and energy in the form of metabolizable substrates, minimize bacterial growth, and also protect sperm from decreases in temperature.

As mentioned earlier, sperm can suffer cold shock as a result of a sudden reduction in temperature, which causes structural and biochemical damage. This can be prevented by cooling semen slowly in the presence of protective agents. Phillips[17] first reported the value of adding egg yolk to bovine semen to afford such protection. Subsequent studies have shown that egg yolk not only increases the fertilizing ability of spermatozoa at ambient temperatures but also appears to prevent sperm cell damage during cooling and freezing.[18–21] Over the years the concentration of egg yolk added to extenders has been reduced from 1 : 1 (volume/volume) to 20–25% of volume as this has been shown to improve sperm survival.[22,23] The low-density lipoproteins (LDLs) in yolk have been shown to play an important role in protecting sperm, with some studies reporting that LDL by itself is better than whole egg yolk in preserving sperm motility after freezing.[24,25] Besides egg yolk, skimmed milk has also been found to be very efficient in protecting sperm during semen storage at 4 °C or in cryopreservation,[26–28] and has the same composition as whole milk but contains less than 0.1% lipids (mostly triglycerides). The protective constituent of skimmed milk is not the lipid fraction but rather the protein constituents known as casein micelles. There are various hypotheses on how LDLs provide protection. It has been suggested that the phospholipid fraction of LDL protects sperm by forming a protective film on the sperm surface[29] or by replacing sperm membrane phospholipids that are lost or damaged during the cryopreservation process.[30,31] Vishwanath et al.[32] suggested that egg yolk lipoproteins compete with detrimental seminal plasma cationic peptides (<5 kDa) in binding to the sperm membrane and thus protect the sperm. However, recent studies and research favors the idea that LDL interacts with the major proteins of bull seminal plasma and this interaction appears to be crucial for sperm protection.[33] A family of major proteins, known as binder of sperm (BSP), found in seminal plasma binds to sperm at ejaculation and modifies the sperm membrane by removing cholesterol and phospholipids. This may adversely affect the ability of sperm to be preserved. LDL from egg yolk and casein micelles and whey proteins in skimmed milk sequester these BSPs, thereby maintaining sperm motility and viability during storage.[34]

Apart from egg yolk and skimmed milk, extenders used for diluting bovine semen also contain various buffers, with phosphate being one of the earliest to be used. However, sodium citrate has largely replaced phosphate due to its superior ability to promote sperm survival at 5°C. Citrate also improves the solubility of proteins fractions in the egg yolk due to its chelating properties. Many zwitterionic buffers such as Tris [tris(hydroxymethyl)aminomethane], TES [N-tris(hydroxymethyl)-methyl-2-aminoethane sulfonic acid], and Tris titrated with TES (TEST) have also been developed and tested over a wide pH range and have proved comparable or superior to citrate buffers. Of these, Tris-based diluents combined with egg yolk have been tested extensively and are used universally for extending bovine semen. In addition, protein components of skimmed milk extenders have also been thought to provide buffering capacity to semen diluents. Because of the potential risk of xenobiotic contamination, research has also started focusing on animal product-free extenders such as coconut and soy milk with satisfactory results. A recent study comparing soy milk tris extender (SMT) with an egg yolk tris (EYT) extender showed no significant differences between sperm in EYT extender and SMT extender with regard to post-thaw motility, viability, membrane integrity, acrosome integrity, and cryocapacitation.[35]

Spermatozoa require energy for motility and are capable of both aerobic and anaerobic metabolism.[36] Most diluents provide an energy source in the form of simple sugars such as lactose, mannose, fructose, and arabinose. Sugars also add osmotic pressure to the medium and act as cryoprotectants. The main effect of sugars and polyols such as glycerol is their ability to replace the water molecule in the normally hydrated polar groups, which helps to stabilize the sperm plasma membrane during transition through the critical temperature zones.[13] Sugars, like glycerol, also increase the viscosity of the diluent and prevent the eutectic crystallization of solutes increasing the glass-forming tendency of the medium, a property used increasingly in vitrification media.[37]

Most commercial bovine semen extenders also contain antibiotics to reduce the rate of bacterial overgrowth and its subsequent deleterious effect on semen quality. Bacteria are present in the genitalia and reproductive tract of bulls regardless of fertility status and can be difficult to screen even with the most hygienic and sanitary collection procedures and processing. Moreover, the presence of egg yolk and other extending media provide a good nutrient environment for the growth of these organisms. Historically, penicillin and streptomycin have been the preferred antibiotics used in combination with bovine semen extenders as they are relatively harmless to sperm and, when combined, inhibit a broad spectrum of microorganisms. However, these antibiotics fail to control growth of organisms such as *Campylobacter fetus* subsp. *venerealis* and have questionable efficacy against *Mycoplasma* and *Ureaplasma* spp. which can survive the freezing process. Continued use of these antibiotics has also led to development of bacterial resistance. Because of these drawbacks the current international standards for semen extenders favor protocols that include treatment of semen and extender with the antibiotics gentamicin, tylosin, lincomycin, and spectinomycin (GTLS), as they are more effective in controlling *Mycoplasma*, *Ureaplasma*, *Campylobacter fetus*, *Haemophilus somnus* and *Pseudomonas* in bovine semen.[38,39] Recent studies have shown that bacterial presence in semen can adversely affect the DNA integrity of sperm, and the rate at which this damage takes place correlates positively with the initial bacterial load and bacterial growth rate.[40] Longevity of DNA is also adversely affected in the presence of GTLS combination. Use of the quinolone class of antibiotics has been shown to increase sperm DNA longevity in semen samples containing bacteria and thus could be of interest in terms of promoting alternative cryopreservation strategies to increase reproductive outcome.[41]

Role of cryoprotectants in bovine semen freezing

In 1949, Polge *et al.*[42] made a pivotal discovery showing that the use of glycerol (a permeating solute) could provide protection to cells at low temperatures. This is often cited as the defining moment in the establishment of modern sperm cryobiology. The development of cryopreservation protocols for the bull to be used for AI in the dairy industry began in the 1950s. Bratton *et al.*[3] demonstrated in field trials that bovine sperm frozen to −79°C and packed on dry ice could still yield high fertility. The discovery of the protective properties of egg yolk lipids and glycerol further aided the development of freezing extenders for cryopreservation of bull sperm. The sum of these discoveries led to the development of the Tris–egg yolk–glycerol method for freezing bull sperm, which has now become a standard.[28,43] Cryoprotectants are included in cryopreservation medium to reduce the physical and chemical stresses derived from cooling, freezing, and thawing of sperm cells,[10,44] and are classified as either penetrating or nonpenetrating based on their ability to cross membranes. Penetrating cryoprotectants (glycerol, dimethyl sulfoxide, ethylene glycol, propylene glycol) cause membrane lipid and protein rearrangement, resulting in increased membrane fluidity, greater dehydration at lower temperatures, reduced intracellular ice formation, and increased survival to cryopreservation.[4] They act like solvents and dissolve sugars and salts in the medium.[44] Nonpenetrating cryoprotectants (egg yolk, nonfat skimmed milk, trehalose, amino acids, dextrans, sucrose) do not cross the plasma membrane and act extracellularly. They may alter the plasma membrane or act as a solute, lowering the freezing temperature of the medium and decreasing extracellular ice formation.[45,46]

Glycerol, a penetrating cryoprotective agent, is the most favored of cryoprotectants, largely because of the ability of bovine semen to withstand much higher levels of glycerol compared with other species. Numerous studies have shown that glycerol yields better post-thaw motility, lesser membrane damage, and better survival rates compared with other cryoprotective agents.[47,48] Glycerol readily enters the cell after its addition and acts to lower the freezing point of the medium to a temperature lower than that of water. This reduces the proportion of the medium which is frozen at any one time, reducing the effect of low temperature on solute concentrations and hence on osmotic pressure differences.[1,49,50] It also provides channels of unfrozen medium between ice crystals in which spermatozoa may exist while at low temperatures and acts as a salt buffering agent. Cryoprotective agents in general are believed to act by increasing the osmotic pressure of the extracellular fluid and hence draw water out of the spermatozoa, thereby decreasing the risk of formation of ice crystals and hence physical damage. However, they do not alleviate, and may even exacerbate, the problem of dehydration and increases in solute concentration. Glycerol and dimethyl sulfoxide can induce osmotic stress and toxic effects on spermatozoa, the extent of which vary according to the species and on their concentration in the extender solution.[44] The concentration of glycerol in the diluent for optimal cryosurvival for bull semen ranges between 6 and 9%, with upper levels causing damage to sperm and lowering their post-thaw survival rate. While formulating semen extenders, glycerol may be added initially or later in a separate fraction after semen refrigeration. In the first situation, also known as the one-step method, the entire semen extender along with glycerol is joined after semen collection. In the second situation, or two-step method, a fraction of extender without glycerol is joined first. This is followed by refrigeration, after which the glycerolated fraction is joined just before semen freezing.[51] The protocols used for adding cryoprotectants and freezing is ultimately a compromise between the advantages and detrimental effects of their incorporation and may vary between individual breeding bulls in order to obtain optimal results.

Steps for cryopreservation of bovine semen

Semen cryopreservation procedures have been evolving over the past 50 years. However, despite continued advances in the development of newer extenders and cryoprotectants, the post-thaw motility of bovine sperm remains at around 50%. This is because semen cryopreservation involves several general steps, of which dilution, cryoprotection, cooling and freezing, storage, and thawing can affect sperm structure and function.[52] Bovine sperm are thought to be relatively more resistant to higher concentrations of cryoprotective agents and the freezing process. However, to optimize post-thaw motility and membrane stability rates, the semen collected should be of high quality and more detailed attention should be paid to evaluation techniques than with those performed during a routine breeding soundness examination. Similarly, individual differences can be observed based on sensitivity of semen to extenders and freezing techniques among bulls. Thus, allowances have to be made by packaging straws with more spermatozoa or by adjusting freezing protocols for individual bulls.[53] Certified Semen Services Inc. (CSS), a wholly owned subsidiary of the National Association of Animal Breeders (NAAB), was established to assure standards and authenticity of semen products. It is also responsible for disseminating information concerning the handling and processing of semen. There are currently two standard protocols approved for freezing bovine semen commercially by the CSS. The standard CSS protocol or the two-step method is the most common method utilized in the United States, whereas the one-step method (also known as the alternative CSS protocol) is more popular in Europe. We routinely employ the two-step method for bovine semen freezing, which will be described later in greater detail.

Dilution, initial cooling and packaging

After collection and evaluation, the semen should be extended to an appropriate dilution/concentration using various diluents/extenders. The dilution stage is performed to minimize any potential toxic effects of seminal plasma, to suspend the spermatozoa in the freezing diluent, and to extend the semen to allow maximum usage.[54] A wide range of diluents of varying composition have been used in successful freezing protocols but the basic requirements are the same in all cases. The diluent must maintain osmolality, pH, and ionic strength, provide an energy substrate, contain a cryoprotective agent, and usually contain antimicrobial agents. A number of commercially available diluents for bovine semen seem to adequately fulfill these requirements (Table 71.1). The most commonly used diluents

Table 71.1 Some common commercially available extenders for bovine semen cryopreservation and their components.

Extender name and manufacturer	Diluents used	Cryoprotective agents	Freezing protocol	Antibiotics
Bovine Semen Extender by Agtech®	Tris buffer + citric acid + fructose	Glycerol and egg yolk[a]	Two step	GTLS[b]
Biladyl® by Minitube	Tris buffer + citric acid + sugars	Glycerol and egg yolk	Two step	GTLS
Triladyl® by Minitube	Tris buffer + citric acid + fructose	Glycerol and egg yolk	One step	GTLS
Triladyl® CSS by Minitube	Tris buffer + citric acid + sugars	Glycerol and egg yolk	One step	None[c]
AndroMed® by Minitube	Tris buffer + citric acid + fructose + antioxidants	Glycerol and soybean lecithin	Two step	GTLS
AndroMed® CSS by Minitube	Tris buffer + citric acid + fructose + antioxidants	Glycerol and phospholipids (animal origin free)	Available as one step and two step	None
BioXcell® Bovine semen preservation medium by IMV Technologies	N/A	Glycerol and soybean lecithin	Available as one step and two step	GTLS
Biociphos-Plus® by IMV, L'Aigle, France	N/A	Glycerol and soybean lecithin	One step	GTLS

[a] Egg yolk and water to be added prior to the addition of semen.
[b] Gentamicin sulfate, tylosin tartrate, lincomycin, and spectinomycin combination.
[c] None: one-step extenders require additional quantities of GTLS antibiotics usually supplied by the manufacturer.

contain Tris–egg yolk/milk–glycerol as buffers and cryoprotective agents, although there are some extenders available now in the market that are free of products of animal origin and contain phospholipids instead. These have been manufactured given current needs for disease control and therefore the avoidance of biologically derived substances in cryoprotective media. The diluents are usually available as a mixture of buffers and antibiotics to which the egg yolk is added at the time of semen extension. Similarly, depending on the process (one-step or two-step method), the glycerol fraction of the extender is added initially or later after refrigeration and cooling of the extended semen. The more commonly used two-step protocol will yield a final concentration of 20% egg yolk, 7% glycerol, 50 μg tylosin, 250 μg gentamicin, 150 μg lincomycin, and 300 μg spectinomycin in each milliliter of extended frozen semen. The one-step protocol yields an antibiotic concentration double that of the two-step procedure.

Dilution rates for bovine semen may vary depending on the concentration of spermatozoa in the ejaculate. In general, high dilution rates (typically 1 : 5 semen/diluent) are usually sufficient as the insemination doses used to achieve fertilization in bovine are surprisingly low (typically 10–15 million) compared with those required for other species such as horse and pig. However, care has to be taken to expose semen to an appropriate volume and concentration of extender, as overdilution may in fact result in loss of motility and increase in vital staining.[55] This phenomenon is known as the "dilution effect" and represents loss of cell viability, probably through leaching of structural components of the cell membrane.[56] The quantitative difference in the ratio of semen to extender between species is an important determinant of the fertility of cryopreserved semen.[4] In species where a large number of spermatozoa are required for conception, the dilution rates are correspondingly low, which means spermatozoa are more tightly packed resulting in lesser tolerance to the cryopreservation process and poor survival rates.

Semen thus diluted is then cooled slowly to 4 °C over a period of 2 hours. This helps in reducing the metabolic rate of sperm greatly and minimizes the damage due to cold shock, apparent when semen is cooled below 15 °C.[54] Metabolic rate is believed to double for every 10 °C rise in temperature; hence cooling semen from body temperature of about 39 °C to refrigerator temperature of 4 °C reduces the metabolic rate to about one-tenth that at body temperature. Though bull spermatozoa are sensitive to cooling to refrigerated temperatures, the phospholipid components present in the egg yolk and whey proteins in skimmed milk play an important role in reducing the damage and also permit faster cooling. Once the partially diluted semen attains a holding temperature of 4 °C, it is extended to its final desired concentration by addition of the glycerol fraction if the two-step protocol for cryopreservation is followed. (Note that in the one-step protocol, semen is extended to its desired concentration and includes addition of glycerol prior to the cooling process.) The step involving addition of glycerol is sometimes erroneously referred to as the glycerol equilibration period. Since glycerol has the ability to equilibrate quite rapidly across the cell membrane at 5 °C, it can realistically be added at any time during the cooling period.[21,57] The equilibration period is thought to involve a process of membrane stabilization. Semen thus extended to its appropriate

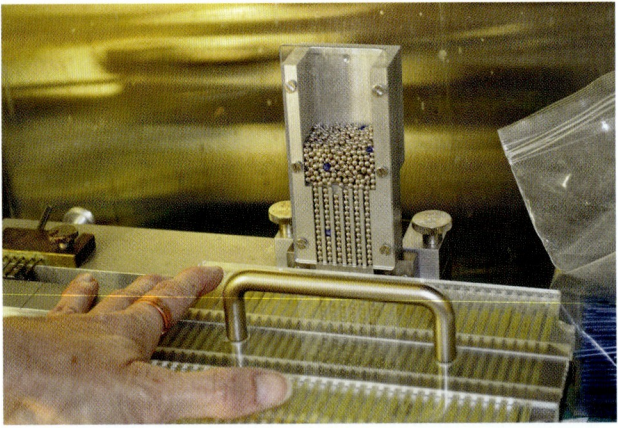

Figure 71.1 Sealing of straws using metal beads.

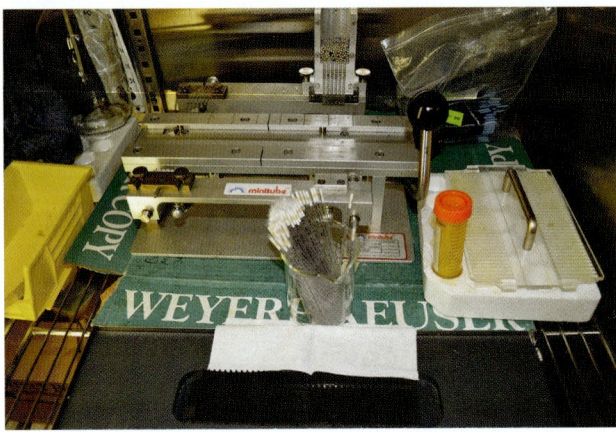

Figure 71.2 Cold cabinet used for semen cooling and packaging.

concentration contains around 7% glycerol and is further held at 4 °C for a few hours. During this holding period semen may be filled, packed, and sealed in 0.25 or 0.5 mL (standard size used in the United States) polyvinyl straws for further processing (Figure 71.1). All procedures involving the handling of cooled semen are performed in a cold handling cabinet in order to maintain the semen equilibration temperature (Figure 71.2).

Cooling rates and freezing

During the process of freezing and thawing semen, spermatozoa are prone to damage caused by the rapid and dramatic changes in the physiochemical conditions occurring during cooling and ice formation. The extent of damage brought about by these abrupt changes is strongly affected by the cooling rate and composition of the cryoprotective medium. Cooling rates for cryopreservation of semen have been determined for various domestic species of animals and humans, sometimes in combination with the cryoprotectant concentration. Optimal rates ideally should fall in ranges that are neither too fast nor too slow. During the process of cooling and freezing, the first change that spermatozoa have to cope with is reduction from body temperatures to near freezing point of water.[1,6] During semen refrigeration with slow cooling rates (0.5–1 °C/min), temperature reduction induces stresses on membranes, probably ascribed to the phase changes in lipid bilayers and altered functional state of membranes.[6] This is believed to impair the function of membrane proteins that are necessary for structural integrity (cytoskeleton) or ion metabolism (ion pumps).[6] Major changes in bull spermatozoa during this phase occur in the vicinity of 15 to 5 °C but do not occur below 0 °C.[6] A second major change in the environment is coincident with the formation of ice crystals after the solution is cooled to a temperature between −5 and −15 °C. The release of the "latent heat of fusion" then causes the sample to warm up abruptly, until the freezing/melting temperature of the solution (of the remaining unfrozen fraction) is reached.[6] This results in spontaneous ice nucleation as the solution is supercooled and ice crystals then grow rapidly in all directions. Ice formation will depend on the rate at which the medium is further cooled and what remains is the so-called unfrozen fraction in which all solutes are confined. The osmolality of this fraction increases due to the increased concentration of glycerol, sugars, and salts and causes a rapid efflux of water from the cells or cellular dehydration. Fast cooling at such a stage can in fact lead to very sudden changes in size, shape, and ultrastructure of cells and lead to structural deformation. Spermatozoa may find themselves confined to very narrow channels of unfrozen solution, squeezed between growing masses of ice, which could also lead to mechanical stress in cells. As cooling continues, these processes continue until the viscosity of the unfrozen fraction becomes too high for any further crystallization. A two-factor theory accounts for the effects of cooling rates on spermatozoa.[58,59] Slow cooling rates expose spermatozoa to "solution effects" such as increasing salt concentration, increasing osmolality, and changing pH, whereas fast rates may not allow intracellular water to pass out leading to intracellular ice nucleation. Optimal cooling rates for semen before freezing in domestic species are generally considered to be in the range 10–100 °C/min,[54] resulting in good survival rates after thawing. In bulls, cooling rates can range between 50 and 60 °C/min, although Woelders et al.[13] have suggested cooling rates as high as 100 °C/min with optimal results. One advantage of such high cooling rates could be a decrease in glycerol concentration in the medium since glycerol is thought to protect from effects of slow cooling damage. This would also help avoid the osmotic damage caused due to the high glycerol concentrations used presently for freezing bovine sperm. However, there are technical difficulties in achieving such uniform cooling rates for straws independent of their position in the freezing apparatuses.[13]

Currently there are two options available for freezing bovine sperm: controlled freezing by use of automated programmable freezers, and conventional vapor freezing (nonprogrammable) by holding semen straws over liquid nitrogen vapors at a specific time and distance. Commercial semen freezing machines (Figure 71.3) offers the option of accurately controlling the temperature inside the cooling chamber as well as programming the time course of the temperature. However, because semen straws are placed in more than one layer, this contributes to considerable variation in cooling rates in individual layers within a freezing cycle and the time course of temperature inside the straws may be different due to the generation of latent heat of

Figure 71.3 Programmable freezers for semen straws as well as embryos.

Figure 71.4 Vapor freezing by placing straws on a rack 5 cm above liquid nitrogen level in a Styrofoam box.

fusion. Also, programmable freezers are more expensive, and they do not necessarily yield more satisfactory results, especially for experienced technicians and cryobiologists. Conventional vapor freezing or nonprogrammable freezing can be performed by partially filling a nonventilated Styrofoam box with liquid nitrogen and freezing straws by arranging them on a rack placed inside the box (Figure 71.4). The height of the straws above the liquid nitrogen determines the rate of heat exchange, with the rate of decrease in temperature being between 150 and 300 °C/min. In addition, the advantage of conventional freezers is that all the straws in any given freezing cycle are subjected to the same cooling rates, because the straws are placed in one layer above the liquid nitrogen level. In nonprogrammable freezers the cooling curve (the time course of cooling and freezing) is by default the form theoretically predicted to be optimal for slow freezing.[60] This results in relatively low cooling rates directly after ice formation begins and higher cooling rates later. The overall steepness of the freezing curve can be adjusted in nonprogrammable systems by choosing the height of the straws above the liquid nitrogen, which is proportional to the temperature of the vapor around the straws.

Semen thawing and storage

Frozen spermatozoa can experience damage during the thawing process as there is a complete reversal of the cellular events that occur during freezing. During slow thawing, the recrystallization of ultra-microscopic ice crystals to form comparatively large ice crystals[6,13] can occur when spermatozoa pass through the critical temperature zone of −5 to −15 °C[61] and can be harmful to spermatozoa. Faster thawing rates (optimum: 37 °C for at least 45 s) limit the time for recrystallization and thus increase the survivability of sperm. However, when the duration of thawing is insufficient for the outflow of excess cryoprotectant from the cell, it suffers osmotic stress and the spermatozoa swell and lyse as the medium becomes abruptly diluted by the melting of extracellular ice.[62] The thawing rate can be influenced by factors such as the temperature and nature of the environment (air or water bath) and the thermal conductivity of the packaging as related to the diameter of the lumen of the packing. Faster thawing rates may be achieved by using higher temperatures for shorter periods of time (e.g., 65 °C for 5 s); however, care needs to be taken to avoid overheating the semen.

Optimal storage and handling of semen straws is a crucial factor that can contribute significantly toward the success of any AI program. Semen is usually packaged in 0.5-mL straws, although some organizations (especially in Europe) prefer packaging in 0.25-mL straws. Regardless of size the handling procedure is common and straws are usually placed in goblets which in turn are clipped onto a metal cane. The canes are thus identified properly with the bull's name or identification number and are stored in canisters of specifically designed liquid nitrogen tanks. Though a detailed description of the construction of a liquid nitrogen tank is beyond the scope of this chapter, the need for proper care in storage and handling of these tanks cannot be overemphasized. Though semen tanks are designed to be sturdy, they are more fragile than they appear and require regular care and attention to prolong their productive life. Failure to do so can lead to a lost inventory or reduced pregnancy rates because semen quality is compromised. Tanks should be stored in a dry, well-lit and well-ventilated area away from direct sunlight. They should be kept elevated above the concrete floor or other wet and poorly ventilated surfaces as corrosion of the outer shell shortens the functional life of the tank and can possibly cause tank failure. One of the key external signs of tank failure is frost build-up (especially around the neck), indicating loss of vacuum. The level of liquid nitrogen in the tanks needs to be regularly monitored and topped off when necessary. While retrieving straws from the tank, the canisters should be kept suspended as low as possible in the tank in order to maintain the temperature of the straws at a constant −196 °C. When straws cannot be retrieved from or located in canisters in less than 10 s, the canister should be lowered back into the tank and held for a period of at least 20 s to recool before trying again.

Stepwise procedure for the two-step protocol for cryopreservation of bovine semen

Note that if directions for the extender supplied by the manufacturer differ, then it is advisable to follow those.

1. Prepare Fraction A (containing buffers and added 200 g of egg yolk) of the extender and place in the incubator before semen collection. Place Fraction B (containing glycerol) of the extender in the refrigerator.
2. After examining and assessing semen quality and concentration, dilute the semen slowly with a small volume of the warmed Fraction A extender in a 50-mL dilution tube. Depending on the calculated number of straws to be processed, the ejaculate can be extended with Fraction A to 50% of the calculated final volume.
3. The extended semen can be cooled to 4 °C by first keeping at ambient room temperature and later placing in a refrigerator or a cold room. The sample should be held at 4 °C for a minimum of 2 hours. The extended semen can also be placed in a water bath having the same temperature, and then placed to be cooled to 4 °C. This helps with the cooling rate.
4. Fraction B, which has been previously cooled to 4 °C, can now be added to the extended cooled semen in a 1 : 1 ratio gradually over a period of 20 min.
5. The semen now extended to its final concentration should be placed again in the cold room or cabinet to equilibrate for a period of 4 hours. During this time straws may be labeled with the appropriate information (e.g., breed, registration, number, name, date). They are filled, sealed and placed on racks for freezing and counting. Straws should be shaken so that the air bubble is in the center after filling and before freezing. All procedures should be performed at 4 °C in a cold room or cooling cabinet.
6. Straws are frozen by placing them on a rack at a distance of 5 cm above the liquid nitrogen level in a Styrofoam airtight container. They are thus held over the liquid nitrogen vapor for about 15 min and then plunged in liquid nitrogen (Figure 71.5).
7. Straws are then loaded into canes (labeled with the bull information) and transferred to liquid nitrogen canisters. One straw should be thawed for assessing post-thaw motility as well as concentration. Ideally the concentration per straw should be equal to one insemination dose (10–15 million sperm). This may be adjusted to bull fertility data if available.

Figure 71.5 Straws plunged into liquid nitrogen.

Newer advances in bovine semen cryopreservation

Over the years several advances in the field of cryobiology have benefited the bovine AI industry greatly. These have resulted in the development of superior semen diluents, standardized optimal freezing protocols, predictable fertilizing ability of sperm through different semen viability tests, and a better understanding of the pathophysiology of cryopreservation-related sperm damage.

The current need for disease control and the avoidance of biologically derived substances of animal origin in cryoprotective media have led to development of coconut milk and soy lecithin-based semen extenders, which have been shown to have comparable protective properties as egg yolk and skimmed milk. Substances such as synthetic liposomes[63] and cholesterol-loaded cyclodextrins[64] have been shown to afford better protection to sperm during cryopreservation and have yielded increased post-thaw survival rates. Similarly, addition of substances to semen diluents such as antioxidants, to protect the sperm from reactive oxygen species and ultimately lipid peroxidation-associated damage, has been investigated. Curcumin and erythritol[65] and n-3 fatty acids combined with α-tocopherol[66] have shown promising results pertaining to better cryosurvival of spermatozoa by altering membrane lipid composition. Supplementation with antioxidants such as methionine, carnitine, and inositol prior to the cryopreservation process has been shown to protect DNA integrity against cryodamage.[67] Recent advances in freeze-drying techniques for equine spermatozoa and subsequent birth of live foals[68] have opened up a whole new perspective for cryopreservation of sperm. Though no live calves have been reported yet, lyophilization of bovine sperm may offer an attractive alternative to liquid nitrogen storage of sperm. Since lyophilized sperm can fertilize an oocyte only via procedures such as intracytoplasmic sperm injection, and require fewer numbers per dose, the number of doses obtained per ejaculate could literally be in hundreds of thousands. Lyophilization of semen may also help overcome several drawbacks associated with conventional cryopreservation such as high maintenance costs, problems associated with transportation of the frozen semen, as well as accidental loss due to improper temperature control.

References

1. Medeiros C, Forell F, Oliveira A, Rodrigues J. Current status of sperm cryopreservation: why isn't it better? *Theriogenology* 2002;57:327–344.
2. Stewart D. Storage of bull spermatozoa at low temperatures. *Vet Rec* 1951;63:65–66.
3. Bratton R, Foote R, Cruthers J. Preliminary fertility results with frozen bovine spermatozoa. *J Dairy Sci* 1955;38:40–46.
4. Holt W. Basic aspects of frozen storage semen. *Anim Reprod Sci* 2000;62:3–22.
5. Batellier F, Vidament M, Fauquant J et al. Advances in cooled semen technology. *Anim Reprod Sci* 2001;68:181–190.
6. Watson P. The causes of reduced fertility with cryopreserved semen. *Anim Reprod Sci* 2000;60:481–492.
7. Mullen S, Critser J. The science of cryobiology. *Cancer Treat Res* 2007;138:83–109.
8. Salamon S, Maxwell W. Storage of ram semen. *Anim Reprod Sci* 2000;62:77–111.
9. Hammerstedt R, Andrews J. Metabolic support of normothermia. In: Karow A, Critser JK (eds) *Reproductive Tissue Banking*. San Diego, CA: Academic Press, 1997, pp. 136–166.
10. Gao D, Mazur P, Critser J. Fundamental cryobiology of mammalian spermatozoa. In: Karow A, Critser JK (eds) *Reproductive Tissue Banking*. San Diego, CA: Academic Press, 1997, pp. 263–327.
11. Isachenko E. Vitrification of mammalian spermatozoa in the absence of cryoprotectants: from past practical difficulties to present success. *Reprod BioMed Online* 2003;6:191–200.
12. White I. Lipids and calcium uptake of sperm in relation to cold shock and preservation: a review. *Reprod Fertil Dev* 1993;5:639–658.
13. Woelders H, Mathijs A, Engel B. Effects of trehalose, and sucrose, osmolality of the freezing medium and cooling rate on viability and intactness of bull sperm after freezing and thawing. *Cryobiology* 1997;35:93–105.
14. De Leeuw F, Colenbrander B, Verkleij A. The role membrane damage plays in cold shock and freezing injury. *Reprod Domest Anim* 1991;1:95–104.
15. Bailey J, Buhr M. Cryopreservation alters the Ca^{2+} flux of bovine spermatozoa. *Can J Anim Sci* 1994;74:45–52.
16. Concannon P, Battista M. Canine semen freezing and artificial insemination. In: Kirk RW, Bonagura JD (eds) *Current Veterinary Therapy X*. Philadelphia: WB Saunders, 1989, pp. 1247–1258.
17. Phillips P. The preservation of bull semen. *J Biol Chem* 1939;130:415.
18. Dunn H, Bratton R, Collins W. Fertility and motility of bovine spermatozoa in buffered whole egg extenders. *J Dairy Sci* 1950;33:434–437.
19. Shannon P, Curson B. Effect of egg yolk levels on the fertility of diluted bovine sperm stored at ambient temperatures. *NZ J Agric Res* 1983;26:187–189.
20. Barak Y, Amit A, Lessing J, Paz G, Hommonai Z, Yogev L. Improved fertilization rate in an in vitro fertilization program by egg yolk-treated sperm. *Fertil Steril* 1992;58:197–198.
21. De Leeuw F, De Leeuw A, Den Daas J, Colenbrander B, Verkleij A. Effects of various cryoprotective agents and membrane-stabilizing compounds on bull sperm membrane integrity after cooling and freezing. *Cryobiology* 1993;30:32–44.
22. Pickett B, Berndtson W. Principles and techniques of freezing spermatozoa In: Salisbury GW, VanDemark NL, Lodge JR (eds) *Physiology of Reproduction and Artificial Insemination of Cattle*, 2nd edn. San Francisco: WH Freeman, 1978, pp. 494–554.
23. Smith R, Berndston W, Unal M, Picket B. Influence of percent egg yolk during cooling and freezing on survival of bovine spermatozoa. *J Dairy Sci* 1979;62:1297–1303.
24. Moussa M, Martinet V, Trimeche A, Tainturier D, Anton M. Low density lipoproteins extracted from hen egg yolk by an easy method: cryoprotective effect on frozen-thawed bull semen. *Theriogenology* 2002;57:1695–1706.
25. Amirat L, Tainturier D, Jeanneau L et al. Bull semen in vitro fertility after cryopreservation using egg yolk LDL. A comparison with Optidyl, a commercial egg yolk extender. *Theriogenology* 2004;61:895–907.
26. Almquist J, Flipse R, Thacker D. Diluters for bovine semen. IV. Fertility of bovine spermatozoa in heated homogenised milk and skim milk. *J Dairy Sci* 1954;37:1303–1307.
27. Chen Y, Foote RH, Tobback C, Zhang L, Hough S. Survival of bull spermatozoa seeded and frozen at different rates in egg yolk-tris and whole milk extenders. *J Dairy Sci* 1993;76:1028–1034.
28. Foote R. Fertility of bull semen at high extension rates in Tris buffered extenders. *J Dairy Sci* 1970;53:1475–1477.
29. Quinn P, Chow P, White I. Evidence that phospholipid protects ram spermatozoa from cold shock at a plasma membrane site. *J Reprod Fertil* 1980;60:403–407.
30. Foulkes J. The separation of lipoproteins from egg yolk and their effect on the motility and integrity of bovine spermatozoa. *J Reprod Fertil* 1977;49:277–284.

31. Graham J, Foote R. Effect of several lipids, fatty acyl chain length, and degree of unsaturation on the motility of bull spermatozoa after cold shock and freezing. *Cryobiology* 1987;24:42–52.
32. Vishwanath R, Shannon P, Curson B. Cationic extracts of egg yolk and their effects on motility, survival and fertilizing ability of bull sperm. *Anim Reprod Sci* 1992;29:185–194.
33. Manjunath P, Nauc V, Bergeron A, Ménard M. Major proteins of bovine seminal plasma bind to the low-density lipoprotein fraction of hen's egg yolk. *Biol Reprod* 2002;67:1250–1258.
34. Manjunath P. New insights into the understanding of the mechanism of sperm protection by extender components. *Anim Reprod Sci* 2012;9:809–815.
35. Singh V, Singh A, Kumar R, Atreja S. Development of soya milk extender for semen cryopreservation of Karan Fries (crossbreed cattle). *Cryo Letters* 2013;34:52–61.
36. Ford W. Glycolysis and sperm motility: does a spoonful of sugar help the flagellum go around? *Hum Reprod Update* 2006;12:269–274.
37. Nicolajsen H, Hvidt A. Phase behaviour of the system trehalose–NaCl–water. *Cryobiology* 1994;31:199–205.
38. Shin S, Lein D, Patten V, Ruhnke H. A new antibiotic combination for frozen bovine semen: 1. Control of mycoplasmas, ureaplasmas, *Campylobacter fetus* subsp. venerealis and *Haemophilus somnus*. *Theriogenology* 1988;29:577–591.
39. Lorton S, Sullivan J, Bean B, Kaproth M, Kellgren H, Marshall C. A new antibiotic combination for frozen bovine semen: 2. Evaluation of seminal quality. *Theriogenology* 1988;29:593–607.
40. González-Marín C, Roy R, López-Fernández C *et al.* Bacteria in bovine semen can increase sperm DNA fragmentation rates: a kinetic experimental approach. *Anim Reprod Sci* 2011;123:139–148.
41. Gonzalez-Marín C, Kjelland M, Roy R *et al.* Use of quinolones in bull semen extenders to reduce sperm deoxyribonucleic acid damage. *Am J Anim Vet Sci* 2012;7:180–185.
42. Polge C, Smith A, Parkes A. Revival of spermatozoa after vitrification and dehydration at low temperatures. *Nature* 1949;164:666.
43. Watson P. The interaction of egg yolk and ram spermatozoa studied with a fluorescent probe. *J Reprod Fertil* 1975;42:105–111.
44. Purdy P. A review on goat sperm cryopreservation. *Small Ruminant Res* 2006;63:215–225.
45. Amann R. Cryopreservation of sperm In: Knobil E, Neill JD (eds) *Encyclopedia of Reproduction*. San Diego, CA: Academic Press, 1999, pp. 773–783.
46. Kundu C, Chakraborty J, Dutta P, Bhattacharyya D, Ghosh A, Majumder G. Effects of dextrans on cryopreservation of goat cauda epididymal spermatozoa using a chemically defined medium. *Reproduction* 2002;123:907–913.
47. Jayendran R, Graham E. An evaluation of cryoprotective compounds on bovine spermatozoa. *Cryobiology* 1980;17:458–464.
48. Forero-Gonzalez R, Celeghini E, Raphael C, Andrade A, Bressan F, Arruda R. Effects of bovine sperm cryopreservation using different freezing techniques and cryoprotective agents on plasma, acrosomal and mitochondrial membranes. *Andrologia* 2012;44:154–159.
49. Amann R, Pickett B. Principles of cryopreservation and a review of cryopreservation of stallion spermatozoa. *J Equine Vet Sci* 1987;7:145–173.
50. Watson P, Duncan A. Effect of salt concentration and unfrozen water fraction on the viability of slowly frozen ram spermatozoa. *Cryobiology* 1988;25:131–142.
51. Barbas J, Mascarenhas R. Cryopreservation of domestic animal sperm cells. *Cell and Tissue Banking* 2009;10:49–62.
52. Hammerstedt R, Graham J, Nolan P. Cryopreservation of mammalian sperm: what we ask them to survive. *J Androl* 1990;11:73–88.
53. Parkinson T, Whitfield C. Optimization of freezing conditions for bovine spermatozoa *Theriogenology* 1987;27:781–797.
54. Curry M. Cryopreservation of mammalian sperm. In: Day JG, Stacey G (eds) *Cryopreservation and Freeze-Drying Protocols*, 2nd edn. Totowa, NJ: Humana Press, 2007, pp. 303–311.
55. Mann T. *The Biochemistry of Semen and the Male Reproductive Tract*, 2nd edn. London: Methuen, 1964.
56. Parkinson T. Artificial insemination. In: Noakes DE, Parkinson TJ, England GCW (eds) *Veterinary Reproduction and Obstetrics*, 9th edn. Philadelphia: Saunders Elsevier, 2009, pp. 765–806.
57. Martin I. Effects of cooling to 58 °C, storage at 58 °C, glycerol concentration, sodium chloride, fructose and glycine on the revival of deep frozen spermatozoa. *J Agric Sci* 1965;64:425–432.
58. Mazur P. Kinetics of water loss from cells at subzero temperatures and the likelihood of intracellular freezing. *J Gen Physiol* 1963;47:347–369.
59. Mazur P, Leibo S, Chu E. A two-factor hypothesis of freezing injury. *Exp Cell Res* 1972;71:345–355.
60. Woelders H, Chaveiro A. Theoretical prediction of "optimal" freezing programmes. *Cryobiology* 2004;49:258–271.
61. Kumar S, Millar J, Watson P. The effect of cooling rate on the survival of cryopreserved bull, ram and boar spermatozoa: a comparison of two controlled-rate cooling machines. *Cryobiology* 2003;46:246–253.
62. Pegg D. The history and principles of cryopreservation. *Semin Reprod Med* 2002;20:5–13.
63. Röpke T, Oldenhof H, Leiding C, Sieme H, Bollwein H, Wolkers W. Liposomes for cryopreservation of bovine sperm. *Theriogenology* 2011;76:1465–1472.
64. Purdy P, Graham J. Effect of cholesterol-loaded cyclodextrin on the cryosurvival of bull sperm. *Cryobiology* 2004;48:36–45.
65. Bucak MN, Başpınar N, Tuncer PB *et al.* Effects of curcumin and dithioerythritol on frozen-thawed bovine semen. *Andrologia* 2012;44:102–109.
66. Nasiri AH, Towhidi A, Zeinoaldini S. Combined effect of DHA and α-tocopherol supplementation during bull semen cryopreservation on sperm characteristics and fatty acid composition. *Andrologia* 2012;44:550–555.
67. Bucak MN, Tuncer PB, Sarıözkan S *et al.* Effects of antioxidants on post-thawed bovine sperm and oxidative stress parameters: antioxidants protect DNA integrity against cryodamage. *Cryobiology* 2010;61:248–253.
68. Choi Y, Varner D, Love C, Hartman D, Hinrichs K. Production of live foals via intracytoplasmic injection of lyophilized sperm and sperm extract in the horse. *Reproduction* 2011;142:529–538.

Chapter 72

Utilization of Sex-sorted Semen

Ram Kasimanickam

*Department of Veterinary Clinical Sciences, College of Veterinary Medicine,
Washington State University, Pullman, Washington, USA*

Introduction

Gender selection of offspring has considerable benefit for livestock management systems. Planning the sex of the offspring prior to conception is the most cost-effective means of achieving the desired outcome. Utilization of gender-selected semen allows production of male and female offspring to take advantage of sex-limited and sex-influenced traits.[1] To be widely used, gender preselection should be effective and efficient, result in acceptable fertility, and be reasonably inexpensive and convenient.[1] In mammals, sex is determined by gonosomes: an XX chromosome combination determines a female, whereas an XY results in a male. Since X and Y chromosomes differ significantly in length, discrimination of X and Y chromosome-bearing sperm populations is possible due to the difference in their DNA content. Several methods to separate X and Y chromosome-bearing sperm populations of ejaculates have been investigated. The only accurate and potentially cost-effective approach for achieving gender selection currently involves separating X from Y chromosome-bearing sperm by flow cytometry, followed by its use for artificial insemination (AI) for breeding or multiple embryo production systems, or for *in vitro* fertilization (IVF) with subsequent embryo transfer programs.[2,3] This method can produce populations of X or Y sperm with up to 90% accuracy and subsequent offspring whose phenotypic sex is consistent with the initial accuracy of the sex-sorted sperm population. Limitations of the technology are associated with the number of sperm required for fertilization, timing of insemination, individual bull variation in sexed-sperm sorting, and subsequent production of offspring. This technology examines each sperm separately for its DNA content, thus limiting the number of sorted X or Y sperm to approximately 5–6 million sperm per hour.[2]

Sorting sperm by DNA content for gender selection

Flow cytometric examination of the DNA content of X and Y chromosome-bearing sperm of cattle (*Bos taurus* and *Bos indicus*) indicated a difference in their DNA content. The difference in total DNA content between those from bulls and cows is approximately 4.2%.[4] This average DNA content difference is sufficient to detect by flow cytometric systems[5–7] (Table 72.1) and subsequent sorting of X and Y chromosome-bearing sperm.

The sperm sex-sorting procedure involves several steps. Briefly, sperm DNA is quantitatively stained with Hoechst 33342, and the sperm are then forced in a stream in front of a laser beam at specific wavelengths.[3,8,9] The illuminated stained sperm emit a very bright blue fluorescence. This fluorescence is rapidly measured using a photomultiplier tube as the sperm flow in single file in front of the tube. A computer is used to rapidly analyze the relative fluorescence of the X and Y sperm populations as they flow through the instrument in a fluidic stream. A crystal vibrator is used to break the stream into individual droplets, many of which contain a sperm. The fluorescently stained sperm are sorted by DNA content by placing opposite charges on droplets containing X sperm from those containing Y sperm (Figures 72.1 and 72.2). The droplets fall past positive and negative electrical fields, and since opposite charges attract, the droplets separate into two streams for collection. A third stream of uncharged droplets is discarded, as these droplets have sperm that could not accurately be sexed (over half), no sperm, rarely two sperm, and dead sperm (gating out dead sperm is a valuable additional benefit). This sperm-sorting technique is known as the Beltsville sperm sexing technology.[3,8,9]

Bovine Reproduction, First Edition. Edited by Richard M. Hopper.
© 2015 John Wiley & Sons, Inc. Published 2015 by John Wiley & Sons, Inc.

Table 72.1 Average DNA content difference (%) in X and Y chromosome-bearing sperm of different breeds.

Holstein	4.98
Jersey	4.24
Angus	4.05
Hereford	4.05
Brahman	3.73

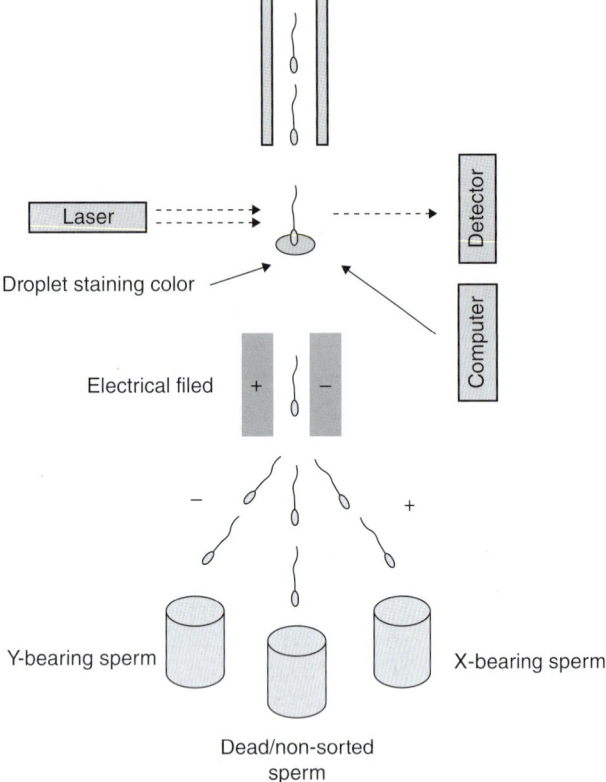

Figure 72.1 Schematic of flow cytometry for sorting X- and Y-bearing bovine sperm.

Figure 72.2 Genesis III sperm sorter for gender selection. Image courtesy of XY Inc.

Commercialization of bovine gender-selected semen

Advances in sex sorting sperm have enabled its incorporation into commercial reproductive management. Currently, gender-selected semen is more applicable on a large-scale basis to the dairy industry than to the beef industry. Sexed semen will contribute to increased profitability of dairy and beef cattle production in several ways. It could be used to produce offspring of the desired sex from a particular mating to take advantage of differences in value of males and females for specific marketing purposes. Dairy herds are more interested in producing replacement heifers, as dairy bull calves are inherently less valuable. Advantages are production of more replacement heifers, thereby enabling greater selection pressure, contributing to more rapid genetic gains.[10] Furthermore, there are opportunities for herd size increase, sale of more females, and simultaneous increase in milk production and decrease in milk price (the latter is an advantage for consumers). Furthermore, once replacement heifer needs are met, dairy producers could also use sexed semen to produce beef crossbred bulls from the remainder of their cow population. Seed-stock dairy producers gain their income from selling genetically superior animals in addition to selling milk. Their goal is to maximize genetic merit of breeding stock for economically important commercial production traits in a cost-effective manner. Sexed semen will be useful to seed-stock breeders if it allows them to increase rate of response to selection, or if it reduces the cost of achieving genetic change.

In order to produce genetically superior replacement heifers using sexed semen, commercial dairy farmers should first rank their cows according to estimated genetic and economic merit for dairy production traits and then use X chromosome-enriched semen from genetically superior bulls to inseminate a sufficient number of the highest-ranking cows to produce the needed number of replacements. By selecting only the best cows to produce replacement heifers, farmers would increase selection pressure for dairy production traits. Methods to assess the economic impact, generated through increased response to selection among dams of commercial cows, of a unit of semen with an increased proportion of X chromosome-bearing sperm have been discussed.[11] It was reported that the predicted impact on annual genetic progress for sexed semen technology is a 9%[12] or 15%[13] increase compared with a mating system using AI with conventional semen.

The commercial beef cattle farmer produces calves for eventual sale as slaughter animals and for female herd replacements. Depending on the planned role for a calf, males and females are of quite different value, so sexed semen potentially is a powerful technology to affect genetic and economic efficiency. With beef cattle herds, the value of shifting the calf crop gender ratio to favor bulls or heifers depends on herd genetics and marketing plans. For beef cattle seed-stock producers, bull sales to commercial cattle producers are often the most significant source of revenue. Replacement heifers are also important for genetic improvement. For beef cattle producers who depend on bull sales and require fewer replacement heifers, utilization of male (Y-bearing) gender-selected semen could be practical. Steers are usually heavier at weaning and more valuable than heifers. Thus, for commercial cattle producers who market feeder calves, utilization of Y-bearing semen in a commercial setting would lead to more steer calves that could offset increased semen cost. Furthermore, if heifers are retained as replacements, individual hand-selected matings could be done to produce heifers from more maternal dams and bull calves from terminal crosses. It should also be noted that in some beef cattle breeding programs, heifers are more valuable than their male counterparts.

Artificial insemination

Sexed semen technology became practical with the live birth of offspring of the predicted sex.[14] Continued research recognized the ability of flow sorting to consistently produce approximately 90% gender-selected offspring and facilitated eventual commercial adaptation.[15,16] The technology was introduced into commercial application in the United Kingdom in 2000. Currently, several American AI centers have this technology and market sexed semen from their bulls worldwide.

Conception rates to AI in cattle are related to sire, semen processing, sperm dosage, timing of insemination, and service number. Despite successes, the invasive nature of the sex-sorting procedure has detrimental effects on sperm viability and quality.[17,18] In addition, for economic reasons, the number of sperm in sex-sorted samples is reduced to approximately 2 million sperm per dose.[18] In breeding programs that use conventional semen, some factors that cause differences in fertility are amenable to increased sperm numbers in an insemination dose (compensable), whereas others are not (uncompensable).[19] An extensive insemination dosage trial concluded that the compensable component is mostly satisfied at total sperm dosages of approximately 4.0 million; however, considering differences among sires, the range is 1–11 million sperm per dose. Seidel[18] and Schenk et al.[16] predicted that commercial application of doses of 2.1 million sex-sorted sperm in Holstein heifers would yield conception rates approximately 75–80% of those obtained with conventional semen, consistent with field results in the United States.[20] Increasing sperm number in the insemination dose seems to be a plausible solution to increase fertility. A recent study compared conception rates for 2.1 or 10 million sperm dosages of sex-sorted or conventional sperm, respectively.[21] Across herds and sires, the conception rates of sperm dosage by semen type combinations were 38 and 44% for 2.1 and 10 million sex-sorted semen, and 55 and 60% for 2.1 and 10 million conventional semen dosages, respectively. A marginal improvement in conception rates was observed with insemination dose 10 million for some sires, suggesting that a portion of the decrease in conception rates with sex-sorted sperm may be compensable. However, that conception rates achieved with 10 million gender-selected sperm are not comparable to either dose of conventional semen dose in this study implies that a portion of the decrease in conception rates of sex-sorted semen is induced by the staining and/or sorting procedure. Further, although the sorting method is standardized, bull-related dilution effects may contribute to the decreased fertility independent of sorting stress,[22] but this remains unproven as insemination of high numbers of sex-sorted sperm (10 million) led to reduced pregnancy rates compared with unsorted sperm inseminated at the reduced dose (2.1 million).[21] Further, reducing the available insemination sex-sorted semen dose to 20% to achieve a 6% increase in pregnancy rate is a poor use of genetic potential.[21]

Current recommendations for use of sex-sorted semen in a commercial operation are as follows.

1. Use only in herds with AI pregnancy rates of 60% or more with conventional semen.
2. Select healthy cycling females with good body condition, preferably heifers.
3. Inseminate only animals observed in standing estrus. If using fixed-time AI, make sure a high percentage of the animals were in estrus before fixed-time AI.
4. Inseminate heifers with sex-sorted semen once and subsequently with conventional semen if they do not conceive.
5. Be extremely careful with semen thawing and handling.
6. The inseminator should have experience and be known to achieve good pregnancy rates.

At present, the recommendation for use of sex-sorted semen is restricted to females exhibiting estrus.[23] However, the use of sex-sorted semen in conjunction with estrus synchronization for a timed AI has huge commercial appeal. Besides, estrus synchronization schemes are of particular value for insemination with sex-sorted sperm, as highly accurate insemination time is extremely important. The current focus is to develop an optimal protocol for fixed-timed AI of sex-sorted sperm. Insemination with sex-sorted sperm closer to ovulation[16,24,25] and insemination of females with large follicles at the time of AI[23] increased the likelihood of pregnancy. Additionally, uterine horn insemination could improve the conception rate when using sex-sorted sperm.[15,26,27] However, placement of sex-sorted sperm deeper into the uterus did not increase pregnancy rates to timed AI.[23]

Multiple ovulation and embryo transfer

Several studies have reported *in vivo* embryo production from superovulated cows and heifers following insemination with sex-sorted sperm. In many of those studies, there was a reduced fertilization rate and increased percentage of degenerate embryos from sexed sperm, which resulted in fewer recovered transferable embryos.[28–30] Suboptimal sperm quality and improper timing of AI were regarded as major factors contributing to fewer embryos per flush from sorted sperm.[28,29] In contrast to single-ovulating cattle, multiple inseminations are believed necessary to cover the interval required for multiple ovulations.[31] Because the number of sperm per insemination dose is limited by the high cost of sexed semen, multiple inseminations with low dose (2 million) sorted sperm have been adopted.[29,30] When donors were inseminated with sexed semen 12 and 24 hours after detected estrus, the number of transferable embryos was increased compared with donors receiving only one dose of sexed semen,[29] indicating that two inseminations of sexed semen were more effective than one insemination at attaining embryo production characteristics similar to those for conventional semen. However, another study claimed that superstimulated donor cows inseminated four times had fewer Grade 1 embryos and more unfertilized ova with sexed versus conventional semen.[32] Additionally, when sexed semen was inseminated at various times after a detected estrus,[30] the number of transferable embryos was decreased compared with donors receiving conventional semen.

Higher numbers of sperm are normally required for superovulated cows. The successful use of sexed sperm in bovine multiple ovulation and embryo transfer (MOET) schemes critically depends on the AI dose. The dose of sex-sorted semen that is used for regular AI, 2 million sperm, is insufficient.[29] The success with higher doses (10–20 million sperm) seems to depend on the sire or MOET conditions. Further, increasing the dosage to 20 million sex-sorted sperm lowered fertilization rates compared with controls.[29] Superstimulated heifers inseminated with 5 million sexed (X-sorted; $n=5$) or unsexed ($n=5$) frozen-thawed sperm from one bull at 12 and 24 hours after detection of estrus showed no difference in rates of transferable embryos (53.4 vs. 68.1%).[33] Furthermore, rates of degenerate embryos and unfertilized ova between sexed and unsexed sperm were 24.8 versus 26.6% and 21.8 versus 5.3%, respectively.[33] Producing higher numbers of good-quality transferable embryos and better pregnancy rates after transfer has great potential for a significant impact on the efficiency of utilization of sex-sorted sperm.

Contract matings between elite sires and dams that use MOET greatly increase the probability of producing a desirable offspring, even without sexed semen. Suppose that a MOET procedure resulted in three embryos that produced viable offspring. With conventional semen, the probability that at least one of the calves is of the desired gender would 0.879 (1 minus the probability that all the offspring are the other gender). With 90% purity, the probability of producing at least one calf of the desired sex would be 0.999. For four viable offspring, the probabilities would be 0.934 and 0.999, respectively. Thus it should be noted that whereas sexed semen increases the likelihood of producing at least one offspring with desired gender from a planned mating, the improvement is not huge.

The goal of most embryo transfer programs is to promote genetic gain both from the dam and sire. Utilization of sexed semen should yield such genetic gain equal to or better than conventional semen. A study estimated that the genetic trend in the entire population was improved by 3.0% when X chromosome-bearing semen was used on all production heifers without use of MOET in the nucleus herd and by 2.2% when MOET was performed on all bulls and dams.[34] When sexed semen was combined with MOET in the nucleus, the effect of using sexed semen was minimal, except when marker-assisted selection was applied.[35] Despite small genetic gain, a recent study reported a beneficial effect on the rate of inbreeding by using sexed semen in the breeding scheme.[36]

In vitro fertilization

IVF requires low numbers of sperm for embryo production. Thus use of sex-sorted sperm in IVF seems advantageous. Cleavage rates were similar to unsorted sperm, but blastocyst development with sorted sperm was approximately 70% of controls produced with unsorted sperm.[37] Further, there was a 0.5–1 day delay of embryonic development when oocytes were inseminated with sex-sorted sperm compared with unsorted sperm.[37] Similarly, in other studies, cleavage rates after IVF with sex-sorted sperm were 30% below those using unsorted sperm from the same ejaculate; blastocyst formation on day 8 was 30–40% lower than for the controls and the number of cell cycles was reduced.[38,39] It was postulated that the impact of sorting on molecular aspects of sperm function, including capacitation, acrosome reaction, zona binding, and embryonic development, are reasons for decreased blastocyst production. Regardless, subsequent reports indicated negative effects of sorting on sperm integrity, but neither cleavage nor blastocyst rates were affected when sperm were centrifuged with a Percoll gradient.[40] In addition, IVF results are also affected by sperm concentration, which needs to be higher than for IVF with unsorted sperm, and by the hydrodynamic pressure used for sorting.[41] Additionally, there was a positive correlation between the percentage of progressively moving sorted sperm and blastocyst development.[42]

Regarding differences in the expression of developmentally important genes, bovine IVF embryos derived from sex-sorted sperm had a reduction in the relative abundances of glucose transporter 3 (Glut3) and glucose 6-phosphate dehydrogenase (G6PD) compared with embryos derived from unsorted sperm, which may be deleterious to the developmental competence of embryos.[43] Further, it is interesting to note that there are significant sex-related (X- vs. Y-bearing sperm) differences in day 7 blastocysts for several genes, namely glutathione S-transferase M3 (*GSTM3*), DNA (cytosine-5)-methyltransferase 3A (*DNMT3A*), and progesterone receptor membrane component 1 (*PGRMC1*), plausibly contributing to delayed onset of the first cleavage.[38,44] Additionally, blastocysts produced from sex-sorted sperm are observed with immature mitochondria and rough endoplasmic reticulum, as well as a lower percentage of intact nuclear envelope membranes than the nonsorted controls.[42] Although the sorting process has significant effects on sperm viability, and bull effects are obvious, compromised DNA does not affect reduced fertility, as most nonviable sperm, and those with fragmented DNA, are out-gated during sorting. Regardless, further research is needed to invstigate this phenomenon, as there is time-related DNA damage to sex-sorted sperm compared with nonsorted counterparts.

Reverse sorting

Reverse sorting is sex-sorting of sperm from previously frozen samples and refreezing those samples after sorting. This technique allows previously frozen sperm to be separated for sex. This technique is useful for males that are located at a greater distance from a sex-sorting site or which are deceased and thus unable to provide fresh ejaculates for sex-sorting. The "reverse sorting" technique surmounts staining problems[45] by washing frozen sperm through a density gradient, thereby eliminating dead sperm, glycerol, and egg yolk, which may interfere with staining or negatively impact the remaining viable sperm.[46,47]

The first pre-sexed offspring produced from reverse-sorted sperm were achieved in sheep following IVF and embryo transfer,[48] and a few years later using AI.[49] Reverse-sorted bull sperm have poorer motility and viability *in vitro* compared with frozen-thawed controls.[50] Although pre-sexed calves have been born following AI of reverse-sorted bull sperm, fertility is very low, with increased pregnancy loss.[51,52] Greater success has been achieved with the use of reverse-sorted bull sperm in IVF, where cleavage and blastocyst production have been comparable to those obtained with sex-sorted frozen sperm and nonsorted frozen controls.[53]

It was apparent that there was a difference among individual sheep with regard to success following reverse sorting. Although the reason for this difference between the species is the source of some disagreement,[54] a better understanding would likely improve fertility of sorted and reverse-sorted bull sperm.

Effect of sex-sorting procedure on structural and functional parameters of sperm

During the sorting procedure, sperm are subjected to several steps, including (i) high working pressures during hydrodynamic focusing and passage through the injection nozzle; (ii) DNA staining with the fluorescent dye bis-benzimide (Hoechst 33342); (iii) sperm exposure to wavelengths of the ultraviolet light spectrum; (iv) repeated exposure of sperm DNA to electrical currents and fields; and (v) subsequent passage through the electrostatic deflection field that may cause stress to sex-sorted sperm.[55,56] Further impact on sperm quality may be related to preservation of stained sperm at low temperature until insemination. In addition, impact on the motility, viability, and functional integrity of sorted sperm may be attributable to a dilution effect following removal of seminal plasma. Perhaps a major reason for a decrease in sorted sperm survival is production of reactive oxygen species (ROS) during several steps of the sorting process. Addition of sodium pyruvate and catalase to the dilution medium and freezing extenders exerted a positive effect on the motility, morphology, acrosomal status, and viability of the sex-sorted frozen-thawed sperm, confirming production of ROS in several steps during the sorting process.[57]

Hoechst 33342 (bis-benzimide) is suitable for sperm DNA labeling as it preferentially binds to the minor groove of the DNA helix (for review see ref. 58). However, this dye is known to be mutagenic to cells and may induce disturbances in embryo development. Moreover, the fate of this dye transported by the spermatozoon into the oocyte and thereby into the embryo and offspring, is an area that requires further investigation.[59]

Assessment of sperm DNA fragmentation using non-sorted and sex-sorted sperm samples during a 72-hour incubation at 37 °C detected reduced DNA longevity in sex-sorted frozen-thawed sperm, with sperm DNA damage appearing between 24 and 48 hours.[60] Although the baseline DNA fragmentation level was higher in conventional frozen-thawed than in sex-sorted frozen-thawed sperm samples initially,[60,61] after approximately 30 hours of incubation there was cross-over in the tendency of DNA fragmentation for both conventional and sex-sorted sperm samples, with higher DNA fragmentation for sex-sorted than conventional semen.[60] Furthermore, it was concluded that sperm chromatin was more resistant to external stressors affecting sperm after sorting.[60]

Cost and revenues

Comparisons of heifers that were inseminated at their first AI breeding with either sexed or conventional semen (in both groups, all services after first AI used conventional semen) were made to evaluate subsequent reproductive performance and health, production, and reproductive performance during their first lactations (refer to ref. 62 for detailed economic outcomes). Briefly, age of heifers at first insemination was 13.1 ± 0.1 and 13.8 ± 0.1 months for sex-sorted and conventional semen, respectively. Rearing cost from first AI to calving was greater for sex-sorted heifers than for conventional heifers ($\$775.3 \pm 6.7$ vs. $\$750.0 \pm 5.9$). However, calf revenue tended to be greater for sex-sorted heifers ($\$142.0 \pm 7.2$ vs. $\$126.7 \pm 6.4$) and cost per female calf produced was less for sex-sorted heifers than for conventional heifers ($\$-809.4 \pm 10.8$ vs. $\$-1249.7 \pm 10.9$). Treatment did not affect calving difficulty, proportion of heifers needing assistance, and incidence of retained fetal membranes or metritis. Among heifers that conceived to first AI, those inseminated with sex-sorted semen were more likely to be culled within 30 days in milk (3.3 vs. 1.6%) and tended to be more likely to be culled within 60 days in milk (5.5 vs. 3.4%) compared with heifers inseminated with conventional semen, but overall replacement cost was not different ($\$136.8 \pm 13.4$). Total milk yield ($9245.5 \pm 84.7$ kg) and income over feed cost ($\$554.7 \pm 5.1$) were not different. Overall economic return was greater for heifers inseminated with sex-sorted semen than conventional semen ($\$-83.7 \pm 36.7$ vs $\$-175.3 \pm 33.4$). Use of sex-sorted semen for first insemination of virgin heifers reduced the cost per female calf produced and increased the economic return during the first lactation. Female to male sex ratios were 0.86 : 0.14 and 0.48 : 0.52 for sex-sorted and conventional semen, respectively. Heifers inseminated with sex-sorted semen were more likely to deliver a dead calf (8.8 vs. 3.4%), and thus the difference in proportion of heifers delivering a live female calf was lower than reported (sex-sorted 79.1%; conventional 47.2%).

Using sex-sorted semen, dairy herds could eliminate 70% of cows in third and greater parity as dams of replacements while maintaining heifer birth rates equal to those of conventional semen.[63] Elimination of older cows as dams of replacements reduces age of dam of replacements and increases genetic merit of dams by about $23. This change doubles the genetic change per year in the dam–daughter path compared with conventional semen, primarily due to shorter generation intervals.[63] Calculated economic values of sexed semen programs for dairy and beef heifers are available.[64,65]

Incidence of stillbirth

Incidence of stillbirth was similar between heifers inseminated with conventional or sex-sorted semen.[20,66] Furthermore, there were no breed differences for incidence of stillbirth among Holstein, Jersey, and Danish Red heifers inseminated with sex-sorted or conventional semen from the same bulls.[67] In contrast, a higher incidence of stillbirth following insemination with sex-sorted semen has also been reported.[62] In 3 years of using sex-sorted semen, total incidence of stillbirths was not influenced by the use of sex-sorted semen compared with conventional semen (10.4 vs. 11.8%).[20] Stillbirth for female calves resulting from sex-sorted semen (9.2%) is similar or slightly lower than conventional semen stillbirths. However, the incidence of stillborn births for male calves from sex-sorted semen (19.9%) is much higher than for conventional semen (12.9%). Even though heifers delivering bull calves seemed to have a higher stillbirth rate, only 10% of calves born were male. However, this might be a concern where a higher sex-ratio toward males is preferred. It was suggested that increased incidence of stillbirths may occur when pregnancies result from Y chromosome-bearing sperm that are mistakenly identified as X chromosome-bearing sperm during the sorting process due to the presence of aneuploidy, which would result in greater DNA content and increased risk for malformation.[20] In this case, it should be noted that the Y-bearing sperm sorted for male selection based on sperm having lesser DNA and the Y-bearing sperm that have aneuploidism with greater DNA will not be mixed with this Y-bearing sperm population as they will be selected out. Hence the stillbirth incidence will be similar to conventional semen when selection for male sperm is occurring.

Closing remarks

It is now clear that sexed semen is a commercial reality for cattle, deer, and sheep, and is nearing commercial application in several other species, particularly pigs, horses, and dogs. Reports on sexed semen being applied for conservation purposes have also become common in zoo animals and other wildlife, some of which are of threatened or endangered status.[68]

The one aspect that has remained constant until recently and perhaps most concerning is fertility. In cattle and possibly other species it believed that fertility of sexed semen is compromised by about 10 percentage points by an unknown mechanism that appears uncompensable by increasing the number of sperm per inseminate.[69] Sexing Technologies has invested significantly in research and development in modifying processing media as well as biochemically altering the overall sorting process (Vish Vishwanath and Juan Moreno, personal communication). Recently completed trials with Select Sires indicate a significant increase in fertility, which has narrowed the gap between conventional semen and sexed semen down to 3–5 percentage points. It is not unlikely that this gap can be narrowed further, thereby making sexed semen almost as fertile as conventional semen. Some reports do show that in sheep, sexed semen outperforms conventional semen as far as fertility is concerned.[54] Reasons for this are varied but the sorting process does clear out many dead and compromised sperm and if conditions are right and timing of insemination is accurately matched with ovulation, there is a real opportunity to improve the fertilization outcome.

Conclusion

Despite key technological developments for more than three decades, most commercial operations that utilize sorted sperm are dairy herds. Research to improve sperm sexing has been largely focused on improving sample throughput by increasing the rate at which sperm are introduced to the sorter, and on improving measurement resolution, which has increased the proportion of sperm that can be reliably measured and sorted. Although the speed of sorting has greatly increased, sperm still need to be individually assessed and sorted. As a consequence, the use lower sperm dosage for AI has reduced fertility (compared with conventional, sperm) and has contributed greatly to the limited commercial use. Furthermore, more than half of the sperm are discarded. Several improvements have recently made sperm sorting more efficient and less harmful to sperm. Modification of the technology with a harmless sorting method has great promise to improve the longevity and fertilizing capacity of sorted sperm. With further improvements in speed of sorting and precision, the cost of the process will be reduced, particularly for high value samples. In the future, the focus may shift to other aspects of the overall process such as improvement in post-sort sperm quality to reduce bull-to-bull variations and increase the semen supply from high-demand bulls. Further, it is essential that, as commercialization increases, more data regarding fertility, embryonic losses, stillbirth, and heifer health be collected. Such information will help to evaluate the technology correctly for its practical use.

References

1. Foote R, Miller P. What might sex ratio control mean in the animal world? In: Kiddy CA, Hafs HD (eds) *Sex Ratio at Birth Prospects for Control*. Champaign, IL: American Society of Animal Science, 1971, pp. 1–9.
2. Johnson L, Welch G. Sex preselection: high-speed flow cytometric sorting of X- and Y-sperm for maximum efficiency. *Theriogenology* 1999;52:1134–1323.
3. Johnson L. Sexing mammalian sperm for production of offspring: the state-of-the-art. *Anim Reprod Sci* 2000;60–61:93–107.
4. Moruzzi J. Selecting a mammalian species for the determination of X- and Y-chromosome-bearing sperm. *J Reprod Fertil* 1979;57:319–323.
5. Garner D, Gledhill B, Pinkel D et al. Quantification of the X- and Y-chromosome-bearing sperm of domestic animals by flow cytometry. *Biol Reprod* 1983;28:312–321.
6. Garner D. Sex-sorting mammalian sperm: concept to application in animals. *J Androl* 2001;22:519–526.
7. Garner D. Flow cytometric sexing of mammalian sperm. *Theriogenology* 2006;65:943–957.
8. Johnson L, Flook J, Look M. Flow cytometry of X and Y chromosome-bearing sperm for DNA using an improved preparation method and staining with Hoechst 33342. *Gamete Res* 1987;17:203–212.
9. Johnson L, Welch G, Rens W. The Beltsville sperm sexing technology: high-speed sorting gives improved sperm output for in vitro fertilization and AI. *J Anim Sci* 1999;77 (Suppl. 2):213–220.

10. Weigel K, Barlass K. Results of a producer survey regarding crossbreeding on US dairy farms. *J Dairy Sci* 2003;86:4148–4154.
11. Van Vleck L, Everett R. Genetic value of sexed semen to produce dairy heifers. *J Dairy Sci* 1976;59:1802–1807.
12. Dematawewa C, Berger P. Break-even cost of cloning in genetic improvement of dairy cattle. *J Dairy Sci* 1998;81:1136–1147.
13. Van Vleck L. Potential genetic impact of artificial insemination, sex selection, embryo transfer, cloning, and selfing in dairy cattle. In: Brackett B, Seidel G, Seidel S (eds) *New Technologies in Animal Breeding*. New York: Academic Press, 1981, pp. 222–242.
14. Johnson L, Flook J, Hawk H. Sex preselection in rabbits: live births from X and Y sperm separated by DNA and cell sorting. *Biol Reprod* 1989;41:199–203.
15. Seidel G Jr, Schenk J, Herickhoff L *et al.* Insemination of heifers with sexed sperm. *Theriogenology* 1999;52:1407–1420.
16. Schenk J, Cran D, Everett R, Seidel G Jr. Pregnancy rates in heifers and cows with cryopreserved sexed sperm: effects of sperm numbers per inseminate, sorting pressure and sperm storage before sorting. *Theriogenology* 2009;71:717–728.
17. Seidel G Jr, Garner D. Current status of sexing mammalian spermatozoa. *Reproduction* 2002;124:733–743.
18. Seidel G Jr. Overview of sexing sperm. *Theriogenology* 2007;68:443–446.
19. Saacke R, Dalton J, Nadir S, Nebel R, Bame J. Relationship of seminal traits and insemination time to fertilization rate and embryo quality. *Anim Reprod Sci* 2000;60–61:663–677.
20. DeJarnette J, Nebel R, Marshall C. Evaluating the success of sex-sorted semen in US dairy herds from on farm records. *Theriogenology* 2009;71:49–58.
21. DeJarnette J, Leach M, Nebel R, Marshall C, McCleary C, Moreno J. Effects of sex-sorting and sperm dosage on conception rates of Holstein heifers: is comparable fertility of sex-sorted and conventional semen plausible? *J Dairy Sci* 2011;94:3477–3483.
22. Den Daas J, De Jong G, Lansbergen L, Van Wagtendonk-De Leeuw A. The relationship between the number of spermatozoa inseminated and the reproductive efficiency of individual dairy bulls. *J Dairy Sci* 1998;81:1714–1723.
23. Sá Filho M, Girotto R, Abe E *et al.* Optimizing the use of sex-sorted sperm in timed artificial insemination programs for suckled beef cows. *J Anim Sci* 2012;90:1816–1823.
24. Sá Filho M, Ayres H, Ferreira R *et al.* Strategies to improve pregnancy per insemination using sex-sorted semen in dairy heifers detected estrus. *Theriogenology* 2010;74:1636–1642.
25. Sales J, Neves K, Souza A *et al.* Timing of insemination and fertility in dairy and beef cattle receiving timed artificial insemination using sex-sorted sperm. *Theriogenology* 2011;76:427–435.
26. Seidel Jr G, Allen C, Johnson L, Holland M, Brink Z, Welch G. Uterine horn insemination of heifers with very low numbers of non-frozen and sexed spermatozoa. *Theriogenology* 1997;48:1255–1264.
27. Kurykin J, Jaakma U, Jalakas M, Aidnik M, Waldmann A, Majas L. Pregnancy percentage following deposition of sex-sorted sperm at different sites within the uterus in estrus synchronized heifers. *Theriogenology* 2007;67:754–759.
28. Sartori R, Souza A, Guenther J *et al.* Fertilization rate and embryo quality in superovulated Holstein heifers artificially insemination with X-sorted or unsorted sperm. *Anim Reprod* 2004;1:86–90.
29. Schenk J, Suh T, Seidel Jr G. Embryo production from superovulated cattle following insemination of sexed sperm. *Theriogenology* 2006;65:299–307.
30. Peippo J, Vartia K, Kananen-Anttila K *et al.* Embryo production from superovulated Holstein-Friesian dairy heifers and cows after insemination with frozen thawed sex-sorted X spermatozoa or unsorted semen. *Anim Reprod Sci* 2008;111:80–92.
31. Dalton J, Nadir S, Bame J, Noftsinger M, Saacke R. The effect of time of artificial insemination on fertilization status and embryo quality in superovulated cows. *J Anim Sci* 2000;78:2081–2085.
32. Larson J, Lamb G, Funnell B, Bird S, Martins A, Rodgers J. Embryo production in superovulated Angus cows inseminated four times with sexed-sorted or conventional, frozen-thawed semen. *Theriogenology* 2010;73:698–703.
33. Hayakawa H, Hirai T, Takimoto A, Ideta A, Aoyagi Y. Superovulation and embryo transfer in Holstein cattle using sexed sperm. *Theriogenology* 2009;71:68–73.
34. Sørensen M, Voergaard J, Pedersen L, Buch L, Berg P, Sørensen A. Genetic gain in dairy cattle populations is increased using sexed semen in commercial herds. *J Anim Breed Genet* 2011;128:267–275.
35. Abdel-Azim G, Schnell S. Genetic impacts of using female-sorted semen in commercial and nucleus herds. *J Dairy Sci* 2007;90:1554–1563.
36. Pedersen L, Kargo M, Berg P, Voergaard J, Buch L, Sørensen A. Genomic selection strategies in dairy cattle breeding programmes: sexed semen cannot replace multiple ovulation and embryo transfer as superior reproductive technology. *J Anim Breed Genet* 2012;129:152–163.
37. Lu K, Cran D, Seidel G Jr. In vitro fertilization with flow-cytometrically sorted bovine sperm. *Theriogenology* 1999;52:1393–1405.
38. Bermejo-Alvarez P, Rizos D, Rath D, Lonergan P, Gutierrez-Adan A. Can bovine in vitro-matured oocytes selectively process X- or Y-sorted sperm differentially? *Biol Reprod* 2008;79:594–597.
39. Beyhan Z, Johnson L, First N. Sexual dimorphism in IVM-IVF bovine embryos produced from X and Y chromosome-bearing spermatozoa sorted by high speed flow cytometry. *Theriogenology* 1999;52:35–48.
40. Carvalho J, Sartori R, Machado G, Mourão G, Dode M. Quality assessment of bovine cryopreserved sperm after sexing by flow cytometry and their use in in vitro embryo production. *Theriogenology* 2010;74:1521–1530.
41. Barcelo-Fimbres M, Campos-Chillon L, Seidel G Jr. In vitro fertilization using non-sexed and sexed bovine sperm: sperm concentration, sorter pressure, and bull effects. *Reprod Domest Anim* 2011;46:495–502.
42. Palma G, Olivier N, Neumuller C, Sinowatz F. Effects of sex-sorted 985 spermatozoa on the efficiency of in vitro fertilization and ultrastructure of 986 in vitro produced bovine blastocysts. *Anat Histol Embryol* 2008;37:67–73.
43. Morton K, Herrmann D, Sieg B *et al.* Altered mRNA expression patterns in bovine blastocysts after fertilization in vitro using flow-cytometrically sex-sorted sperm. *Mol Reprod Dev* 2007;74:931–940.
44. Bermejo-Alvarez P, Lonergan P, Rath D, Gutierrez-Adan A, Rizos D. Developmental kinetics and gene expression in male and female bovine embryos produced in vitro with sex-sorted spermatozoa. *Reprod Fertil Dev* 2010;22:426–436.
45. Stap J, Hoebe R, Merton J, Haring R, Bakker P, Aten J. Improving the resolution of cryopreserved X- and Y-sperm during DNA flow cytometric analysis with the addition of Percoll to quench the fluorescence of dead sperm. *J Anim Sci* 1998;76:1896–1902.
46. O'Brien J, Hollinshead F, Evans K, Evans G, Maxwell W. Flow cytometric sorting of frozen-thawed spermatozoa in sheep and non-human primates. *Reprod Fert Dev* 2003;15:367–375.
47. Underwood S, Bathgate R, Maxwell W, Evans G. Development of procedures for sex-sorting frozen-thawed bovine spermatozoa. *Reprod Domest Anim* 2009;44:460–466.
48. Hollinshead F, Evans G, Evans K, Catt S, Maxwell W, O'Brien J. Birth of lambs of a pre-determined sex after in vitro production of embryos using frozen-thawed sex-sorted and re-frozen-thawed ram spermatozoa. *Reproduction* 2004;127:557–568.
49. de Graaf S, Evans G, Maxwell W, Cran D, O'Brien J. Birth of offspring of pre-determined sex after artificial insemination of frozen-thawed, sex-sorted and re-frozen-thawed ram spermatozoa. *Theriogenology* 2007;67:391–398.
50. Underwood S, Bathgate R, Maxwell W, Evans G. In vitro characteristics of frozen-thawed, sex-sorted bull sperm after refreezing or incubation at 15 or 37 °C. *Theriogenology* 2009;72:1001–1008.

51. Underwood S, Bathgate R, Maxwell W, Evans G. Birth of offspring after artificial insemination of heifers with frozen-thawed, sex-sorted, re-frozen thawed bull sperm. *Anim Reprod Sci* 2010;118:171–175.
52. Underwood S, Bathgate R, Ebsworth M, Maxwell W, Evans G. Pregnancy loss in heifers after artificial insemination with frozen-thawed, sex sorted, re-frozen-thawed dairy bull sperm. *Anim Reprod Sci* 2010;118:7–12.
53. Underwood S, Bathgate R, Pereira D et al. Embryo production after in vitro fertilization with frozen-thawed, sex-sorted, re-frozen-thawed bull sperm. *Theriogenology* 2010;73:97–102.
54. De Graaf SP, Beilby K, Underwood S, Evans G, Maxwell W. Sperm sexing in sheep and cattle: the exception and the rule. *Theriogenology* 2009;71:89–97.
55. Maxwell W, Long C, Johnson L, Dobrinsky J, Welch G. The relationship between membrane status and fertility of boar spermatozoa after flow cytometric sorting in the presence or absence of seminal plasma. *Reprod Fertil Dev* 1998;10:433–440.
56. Maxwell W, Parrilla I, Caballero I et al. Retained functional integrity of bull spermatozoa after double freezing and thawing using Puresperm density gradient centrifugation. *Reprod Domest Anim* 2007;42:489–494.
57. Klinc P, Rath D. Reduction of oxidative stress in bovine spermatozoa during flow cytometric sorting. *Reprod Domest Anim* 2007;42:63–67.
58. Rath D, Johnson L. Application and commercialization of flow cytometrically 1010 sex-sorted semen. *Reprod Domest Anim* 2008;43(Suppl. 2):338–346.
59. Garner D. Hoechst 33342: the dye that enabled differentiation of living X-and Y-chromosome bearing mammalian sperm. *Theriogenology* 2009;71:11–21.
60. Gosálvez J, Ramirez M, López-Fernández C et al. Sex-sorted bovine spermatozoa and DNA damage. II. Dynamic features. *Theriogenology* 2011;75:206–211.
61. Gosálvez J, Ramirez M, López-Fernández C et al. Sex-sorted bovine spermatozoa and DNA damage. I. Static features. *Theriogenology* 2011;75:197–205.
62. Chebel R, Guagnini F, Santos J, Fetrow J, Lima J. Sex-sorted semen for dairy heifers: effects on reproductive and lactational performances. *J Dairy Sci* 2010;93:2496–2507.
63. Cassell B. Implications of sex-sorted dairy semen for genetic change. *J Anim Sci* 2010;88(E Suppl. 2):783 (Abstract).
64. Cabrera V. Economic value of sexed semen programs for dairy heifers. http://www.sexingtechnologies.com/articles/economic_aids
65. http://www.sexingtechnologies.com/articles/economic_aids
66. Tubman L, Brink Z, Suh T, Seidel G Jr. Characteristics of calves produced with sperm sexed by flow cytometry/cell sorting. *J Anim Sci* 2004;82:1029–1036.
67. Borchersen S, Peacock M. Danish A.I. field data with sexed semen. *Theriogenology* 2009;71:59–63.
68. O'Brien J, Steinman K, Robeck T. Application of sperm sorting and related technology for wildlife management and conservation. *Theriogenology* 2009;71:98–107.
69. DeJarnette J, Leach M, Nebel R, Marshall C, McCleary C, Moreno J. Effects of sex sorting and sperm dosage on conception rates of Holstein heifers: is comparable fertility of sex-sorted and conventional semen plausible? *J Dairy Sci* 2011;94:3455–3483.

Chapter 73

Control of Semen-borne Pathogens

Rory Meyer

Alta Genetics, Watertown, Wisconsin, USA

Introduction

The control of semen-borne pathogens in the bovine is an essential task in the cattle artificial insemination (AI) industry to prevent the dissemination of disease to multiple locations. Each year, millions of units of frozen bovine semen are produced and distributed both domestically in the United States and also internationally throughout the world. To reduce the risk of disease spread, standards for the production of semen from donor bulls have been established and donor bulls must complete a rigorous testing program in order to qualify for entry into an AI center (AIC) and be continually monitored while being collected for semen production. As new tests are developed and new research is conducted, testing programs to monitor donor bulls in AICs are continually assessed and improved to reduce the probability of spread of diseases through frozen bovine semen.

In the United States, Certified Semen Services, Inc. (CSS) establishes minimum testing requirements for donor bulls entering and residing in AICs for domestic distribution of semen. Additional testing requirements may have to be met if the frozen semen from the donor bull is intended for international export and these requirements are negotiated by the United States Department of Agriculture (USDA) and the importing country. Many countries establish their own requirements for testing and these are often based on the guidelines published by the World Organisation for Animal Health (OIE) for the collection and processing of bovine semen. One of the intended purposes of these OIE guidelines is to maintain the health of animals in an AIC at a level that allows the international distribution of semen with negligible risk of infecting other animals or humans with pathogens transmissible by semen.[1]

In general, donor bulls and animals intended for the use as mount animals undergo testing in three phases. Selected animals are first tested prior to entering quarantine facilities of the AIC. The second phase consists of assembling animals for a quarantine period to included testing throughout and at the end of the quarantine period. And finally, once the animal has entered the resident population in the AIC, it is tested on an annual or semi-annual basis to monitor the disease status of the AIC. Additional testing may be required depending on the requirements of different export markets.

This chapter covers the major diseases of concern that are at risk for transmission through frozen bovine semen. The pathogens involved and the testing strategies to minimize the risk of transmission to other cattle are reviewed.

Bovine viral diarrhea

Bovine viral diarrhea virus (BVDV) is a single-stranded RNA *Pestivirus* belonging to the *Flaviviridae* family. It occurs worldwide and is a significant disease with regard to economic consequences in infected herds. Clinical consequences vary and include respiratory, gastrointestinal, and reproductive effects, with the major source of infection coming from persistently infected (PI) animals present in a herd. PI animals will transmit BVDV to susceptible animals in the herd causing an acute transient infection that may lead to reduced pregnancy rates in heifers and cows. Several studies have shown BVDV can be present in semen from bulls that are persistently and acutely infected and are able to transmit BVDV to susceptible recipient cattle.[2-5] In addition, there are reports of bulls that have testicular infections that are able to transmit BVDV in the semen without being viremic.[6,7]

PI animals are a result of exposure to the virus through transplacental infection of the developing fetus prior to 125 days of gestation. Bulls that are PI can develop normally but are immunotolerant to the strain of BVDV the animal was exposed to.[8] The PI bull is capable shedding BVDV for long periods of time and in large amounts, despite producing semen with normal concentration, motility, and morphology.[2] The effect of using semen from PI bulls and exposing susceptible females is significant. One study showed that 12 of 12 seronegative and virus-negative heifers artificially inseminated with semen from a PI bull seroconverted within

2 weeks after exposure.[3] Another study showed that 161 of 162 cows that were inseminated with semen from a PI bull were seropositive compared with 95 of 143 cows that were inseminated with semen from other bulls, and concluded that semen from PI bulls has the potential to transmit BVD infection in susceptible herds.[2] This highlights the importance of properly screening bulls for PI status prior to entry into an AIC.

Testing requirements for BVD PI status of bulls prior to entry to an AIC includes two tests for the BVDV antigen in serum or whole blood at least 10 days apart. CSS requires that serum or whole blood is tested using a virus isolation (VI), polymerase chain reaction (PCR), or antigen capture enzyme-linked immunosorbent assay (ELISA) test with the first sample collected on the farm of origin within 30 days prior to entry into the AIC isolation facility with a negative result. The bull is sampled for a second BVDV antigen test on serum or whole blood during the isolation period at least 10 days following entry into the AIC isolation facility with a negative result. Because of the interference of maternal antibodies on circulating BVDV in the serum, bulls that are less than 6 months of age must be tested using whole blood while in isolation.[9]

Susceptible bulls exposed to BVDV can become acutely or transiently infected and are viremic for a period of 6–21 days following exposure. The period of viremia also coincides with a significant increase in antibody production to the infecting strain of BVDV, indicating exposure to the virus.[10] BVDV can be detected in semen from acutely infected bulls throughout the duration of viremia, although the levels of virus in the semen are significantly less than those found in bulls that are persistently infected.[2,7,11] As a result, the seroconversion rates of exposed animals inseminated using semen from acutely infected bulls that have virus present are lower than those of animals inseminated using semen from PI bulls (1.8–15.4%).[4,5] Despite the lower seroconversion rates, testing for acutely infected bulls in an AIC is important to prevent the spread of BVDV to susceptible herds.

Testing requirements for BVDV acute infections is similar to testing of PI bulls. CSS requires that the bull must be tested on-farm using VI, PCR, or antigen capture ELISA on serum or whole blood within 30 days of entry into the AIC isolation facility with a negative result. The bull is tested again at least 10 days from the beginning of the period in the AIC isolation facility with a negative result. Similar to testing for PI status, bulls less than 6 months of age require testing to be completed on whole blood. If the result is positive, the bull is isolated from the remaining animals in the isolation facility and tested again at least 21 days later, in addition to the remaining animals in the isolation group. If the second result is positive, this indicates the bull is a PI animal and is not allowed entry into the resident herd. If the second result is negative, the bulls must be tested a third time at least 10 days from the second sample with a negative result prior to entering the production herd.[9] In the event that BVDV is isolated from a bull as a result of an acute infection, all semen batches collected 30 days prior and 30 days after the positive result are to be tested with VI or PCR before they can be distributed.

Optional testing for BVDV acute infections recommended by the OIE includes a test for the detection of BVDV antibodies. Serum samples are collected prior to or at the beginning of the isolation period and at least 21 days after entry into the isolation facility. Seroconversion indicates exposure during the isolation period and the bull must be retested prior to entering the resident herd. The OIE also recommends that bulls are sampled on an annual basis and tested for BVDV antibodies to monitor for seroconversion while they are residents of the production herd of the AIC.[1]

There have been two reports of bulls with persistent testicular infection that were able to shed virus particles in the semen despite absence of a viremic state and the presence of circulating antibodies for BVDV.[6,12] This is believed possible because an antibody response was generated from the initial infection, but virus in the testicles was protected from circulating antibodies due to the blood–testes barrier. Both reports demonstrated that the bulls consistently produced semen containing virus over an extended period of time, although the viral load in the insemination dose was less than that of semen produced by PI bulls but more than that present in semen from acutely affected bulls.[7] One study reported that use of frozen semen from one of the reported bulls in three heifers resulted in seroconversion in one of the three heifers that were inseminated.[13]

There are reports of bulls with prolonged testicular infection (PTI) that had detectable BVDV virus antigen in the semen for 2.75 years following experimental acute infections. However, BVDV could only be detected in semen using VI testing during the viremic period of infection, although infectious virus could be detected in testicular tissue 12.5 months after exposure.[7,14] The positive result was due to detection of neutralized virus with PCR but subsequent investigation was unable to detect infectious virus 17 days after infection using VI of the semen. Despite this, one experiment in one of the studies demonstrated that raw semen collected from one of the positive PTI bulls at 5 months after infection resulted in the seroconversion of a 6-month-old calf when the semen was injected intravenously.[14] In the other study, a repeated trial resulted in no seroconversions of calves inoculated intravenously with raw semen, and breeding trials using frozen semen from a PTI bull to inseminate heifers also resulted in no seroconversions. It was concluded that the risk of viral transmission from prolonged testicular infections with BVDV may be negligible, but additional information is required to substantiate this conclusion.[7]

CSS requires that bulls housed in an AIC must be tested in one of two ways for PTI prior to semen release for distribution. The first method is to test all bulls in isolation with a BVDV serum neutralization test. Bulls that test positive are required to have one batch code of frozen processed semen tested using VI or PCR. The second method is to require all bulls in isolation, regardless of their antibody status, to have one batch code of frozen processed semen tested using VI or PCR. Any bull found to have persistent testicular infection for BVDV is not eligible for semen collection and is not permitted to remain in the resident production herd.[9]

There are currently no recommendations regarding testing of bulls to determine their PTI status. As more information is made available, further assessment may be required to include additional testing.

Leptospirosis

Leptospirosis is caused by a Gram-negative spirochete bacterium that occurs worldwide in cattle, sheep, goats, pigs, and horses. It is also a zoonotic disease and humans acquire the infection by contact with the urine of affected animals. *Leptospira interrogans* is the pathogenic form of the bacteria and has many serovars based on cell surface antigens. Clinical signs in cattle include hemolytic anemia, drop in milk production and changes in milk quality, stillbirths, and abortions.[15,16] In the United States, common serovars include *Leptospira interrogans* serovar Hardjo (*L. hardjo*), *L. pomona*, *L. grippotyphosa*, *L. canicola*, and *L. icterohaemorrhagiae*. Cattle are the reservoir host for *L. hardjo*.

Although the main route of transmission of leptospirosis is urine, studies have demonstrated that transmission of the organism may be possible through fresh and frozen semen.[17,18] The addition of antibiotics to extended semen has been shown to be effective in controlling the number of leptospiral organisms that may be present in the semen without affecting semen quality and reduces the risk of transmission through AI.[17]

In the United States, CSS requires bulls entering an AIC to be negative to a microscopic agglutination test (MAT) at 1 : 100 or demonstrate a low stable titer (negative at less than 1 : 400) when tested at least 2–4 weeks apart within 30 days of entering the isolation facility of the AIC for each of the following serovars: *L. hardjo*, *L. pomona*, *L. grippotyphosa*, *L. canicola*, and *L. icterohaemorrhagiae*. The same test and test results must be obtained while the bull is in the isolation facility before entering the resident production facility and the test repeated at 6-month intervals while the bull is in the resident production facility of the AIC.[9] Because of the test requirements for bulls to enter an AIC, vaccination for leptospirosis is not done as the MAT is not able to distinguish between titers that are the result of natural infections and those due to vaccines.

Brucellosis

Brucella abortus is the main causative agent of brucellosis in cattle that causes abortion, retained placenta, orchitis, epididymitis, and impaired fertility. It is mainly spread by contact with the uterine discharge of an aborted or full-term fetus but spread through milk and semen of infected cows and bulls is also possible. Bulls will shed *B. abortus* organisms in semen during the acute phase of infection, after which shedding will cease or become intermittent.[19] It as an important zoonotic disease in humans and many countries have implemented programs to eradicate *B. abortus* through vaccination and test and slaughter strategies. In the United States, the USDA has designated all 50 states brucellosis free since 2009, although infections continue to be detected in the Greater Yellowstone Area of Idaho, Montana, and Wyoming.

Several serological tests are available to detected antibodies to *B. abortus*. For bulls entering an AIC, CSS requires a negative result on a buffered *Brucella* antigen test (Card or BAPA) or a complement fixation (CF) test within 30 days prior to entry into the isolation facility of the AIC. During the isolation period, the bull is required to have a negative result on both the buffered *Brucella* antigen test (Card or BAPA) and the CF test before the bull can enter the resident production facility and the same tests are repeated at 6-month intervals while the bull is in the resident production facility. If the test results are non-negative, another approved serological test can be completed with a negative result; however, final classification of the bull is determined by the regulatory officials of the state from which the testing is completed.[9]

Tuberculosis

Mycobacterium bovis is the main causative agent of tuberculosis in cattle and is a chronic progressive disease that frequently causes progressive emaciation of the affected animal. It is spread to other animals through inhalation or ingestion of *M. bovis* organisms via exhaled air, sputum, milk, urine, and feces. Transmission of tuberculosis has been reported via natural breeding from bulls with genital lesions caused by *M. bovis* and the organism has been demonstrated in fresh and frozen semen samples using PCR.[20,21] Like brucellosis, bovine tuberculosis is a zoonotic disease and many countries have attempted to eradicate the disease through test and slaughter programs. In the United States, the USDA has designated the majority of the states as Accredited-free states, with the remainder classified as Modified Accredited Advanced states with designated prevalence rates of less than 0.01%.[22]

In the United States, surveillance testing for *M. bovis* is performed on cattle by intradermal injection of bovine tuberculin in the caudal fold area at the base of the tail and observation for a reaction is performed 72 hours later. Animals that show a palpable response are followed up with a comparative cervical test using bovine and avian tuberculin with measurement of the response 72 hours later. A serological test is available that measures gamma interferon response to bovine tuberculin but is limited for use only by regulatory officials.

For bulls entering an AIC, CSS requires a negative intradermal tuberculin test within 60 days prior to entry into the isolation facility. Another intradermal test is completed while the bull is in the isolation facility and a negative test is required to enter the resident production facility and the same test is repeated every 6 months with a negative result while the bull is a resident of the production facility of the AIC.[9]

Campylobacteriosis and trichomoniasis

Campylobacteriosis and trichomoniasis are venereal diseases in cattle that cause infertility, early embryonic death, and abortion and, in the case of trichomoniasis, pyometra in infected females. Campylobacteriosis is caused by *Campylobacter fetus* subsp. *venerealis* and is a microaerophilic Gram-negative bacterium. Trichomoniasis is caused by *Tritrichomonas foetus*, a flagellated protozoan. Both organisms reside on the penile and preputial mucosa of infected bulls, although they do not cause any clinical signs and bulls are asymptomatic and acquire either infection after breeding infected females. A carrier state exists for both diseases, almost exclusively in older bulls. Infected bulls transmit *C. fetus* subsp. *venerealis* and *T. foetus* by natural breeding of females, and transmission through the use of AI for both diseases is possible.[23,24]

Diagnostic testing for *C. fetus* subsp. *venerealis* and *T. foetus* depends on the demonstration of both organisms from preputial smegma collected by aspirating or washing of the internal prepuce and penile mucosa. The preputial smegma is inoculated immediately into transport media and shipped to the diagnostic laboratory for testing. CSS requires a culture test for both organisms for each sample with a negative result. CSS does allow initial screening for *C. fetus* subsp. *venerealis* using a fluorescent antibody test, but all positive results using this test must be confirmed using bacterial culture.[9] Other diagnostic tests on preputial smegma are available including PCR and ELISA for *C. fetus* subsp. *venerealis*[25,26] and PCR for *T. foetus*.

The frequency of testing is the same for both *C. fetus* subsp. *venerealis* and *T. foetus*. CSS requires serial testing at weekly intervals of bulls present in the isolation facility of the AIC prior to entering the resident production herd. The number of tests required in isolation depends on the age of the bull, with bulls older than 1 year of age requiring six tests done at weekly intervals on preputial smegma with negative results. Bulls less than 1 year of age require three tests on preputial smegma with negative results. While resident in the production herd, bulls are tested every 6 months for each disease with negative results.[9] The OIE recommends that bulls older than 6 months of age are tested three times at weekly intervals and bulls less than 6 months of age are tested once with a negative result while in the isolation facility and once on an annual basis when in the resident production herd.[1]

The addition of antibiotics has been shown to control *C. fetus* subsp. *venerealis* organisms present in bovine semen.[27,28] In addition to the testing requirements for *C. fetus* subsp. *venerealis* for bulls to enter and remain in an AIC, CSS requires the addition of an antibiotic combination of gentamicin, tylosin, and lincomycin–spectinomycin to be added to the raw semen and again included in the extender that is added for final dilution prior to freezing the semen.[9] Acceptable antibiotic combinations that the OIE recommends for addition to extended semen include gentamicin, tylosin, lincomycin–spectinomycin; penicillin, streptomycin, lincomycin–spectinomycin; or amikacin and dibekacin.[1] The addition of antibiotics does not control for the presence of *T. foetus* in fresh or frozen semen.

Infectious bovine rhinotracheitis

Infectious bovine rhinotracheitis (IBR) is a viral disease of cattle that causes upper respiratory disease, abortions, infertility, encephalitis, conjunctivitis, vulvovaginitis, and balanoposthitis. The disease is caused by bovine herpesvirus type 1 (BHV-1) and type 5 (BHV-5). BHV-1 can be further classified into two subtypes. BHV-1.1 is associated with the respiratory form of the disease (IBR) and BHV-1.2 is associated with the genital form of the disease causing infectious pustular vulvaginitis (IPV) in females and infectious pustular balanoposthitis (IPB) in males. Both subtypes are capable of causing abortions. BHV-1 is spread via aerosol transmission for the respiratory form and venereally for the genital form, and the virus has been found in fresh and frozen semen and transmission through AI demonstrated.[29] BHV-5 is an encephalitic form of the disease that is primarily spread through aerosol transmission and has recently been reported to cause genital tract lesions and abortions. BHV-5 has been detected in semen from infected bulls and transmission through AI is possible.[30]

Primary infection with BHV-1 in bulls can result in shedding of the virus from the prepuce, beginning 7–10 days after infection and can last for several days to several weeks. After the initial phase, BHV-1 becomes latent in the infected bull and intermittent shedding can result throughout the animal's lifetime. The virus can remain latent indefinitely, although administration of corticosteroids or a period of stress such as transport can result in the recrudescence of BHV-1 and subsequent shedding of the virus from the prepuce. Contamination of semen with BHV-1 is likely the result of virus present on the mucosa of the prepuce and not from the testicles, epididymis, or accessory sex glands.[29] Although exposure leads to antibody production, the presence of antibodies may not prevent the recrudescence and shedding of latent virus in the prepuce and levels of antibodies may be very low or absent in bulls that have remained latent for a long period.[18]

There are several recommendations for the testing and maintenance of bulls in an AIC. CSS does not require testing of bulls for IBR for the production of semen. This is based on several studies showing that semen from seropositive bulls appropriately managed in AICs did not shed BHV-1 in the semen and attributed this to the stress-free environment in which bulls reside.[31] Serological tests are available to detect antibodies to BHV-1 and include serum neutralization and ELISA and it is generally recommended that an antibody test prior to and at least 21 days after semen production will indicate that the bull was not actively infected during the collection interval. For seropositive bulls, VI and PCR tests are available to test semen batches for presence of the virus. The OIE recommends maintenance of an IBR-free herd in an AIC. Bulls are tested within 28 days prior to entry into the isolation facility with a serological test for antibodies with a negative result or the animals should come from an IBR-free herd. The bull is tested in isolation with a serological test with a negative result and all the bulls in the isolation facility must be negative on a serological test prior to entry into the production herd. The production herd is tested with a serological test on an annual basis and all resident bulls require a negative result.

Arboviruses

Cattle are susceptible to arthropod-borne diseases that have the potential to be transmitted through semen. Cattle become infected by insects that harbor the virus from feeding on viremic animals and subsequently feeding on susceptible animals, transmitting the virus in the process. Distribution of the disease is limited to the endemic areas where the insect vectors reside and the viruses are often maintained in these areas through wildlife reservoirs. New epidemic areas occur through the movements of virus-infected vectors into new areas or the movements of viremic host animals to new areas with competent vectors.

Bluetongue virus (BTV) and epizootic hemorrhagic disease virus (EHD) are orbiviruses that have worldwide distribution. There are at least 24 identified serotypes of BTV and the virus is mainly transmitted by midges (*Culicoides* species) and cattle are a major reservoir host for the virus. Infected cattle rarely show clinical signs but the disease may cause fever, facial edema, hemorrhages, and ulceration of the mucous membranes in some cattle. BTV has been found in semen but only during the viremic period of infection in bulls, likely associated with blood that has leaked from the genital tract.[8] EHD is similar to BTV and there are at least seven serotypes, with significant immunological cross-reactivity between the two viruses. EHD is also similar to BTV in that it rarely causes clinical signs in cattle and is mainly transmitted by midges (*Culicoides* species). There have been no reports of EHD in semen but, due to its similarity to BTV, international recommendations include testing for the virus in bulls used for AI.

Serological and virus identification tests are available for monitoring bulls and semen used for AI. Serological tests that are available include agar gel immunodiffusion and ELISA. The OIE recommends that seronegative bulls are tested prior to semen collection and again 21–60 days after collection with negative results. Virus identification tests include VI and PCR done on whole blood. The OIE recommends either VI or PCR prior to collection and again at least every 7 days during the collection period for VI or at least every 28 days during the collection period for PCR and a final test at the end of the collection period.[32] VI and PCR tests are also available for frozen semen.

Recently, a new arbovirus has emerged in Europe caused by Schmallenberg virus (SBV), first detected in November 2011 from dairy cattle that had reduced milk production and diarrhea; it was later found to cause congenital malformations in aborted and newborn calves. SBV is an *Orthobunyavirus* and is a member of the Simbu serogroup and is transmitted by *Culicoides* midges. The virus has been detected in bovine semen by PCR and is able to be shed in consecutive semen batches or intermittently for 43 days in one report[33] and as long as 3 months in another report.[34] It has not yet been determined if SBV detected using PCR in the semen is viable and able to transmit the disease to susceptible animals. Serological tests are available including serum neutralization and ELISA, while PCR and VI are available to detect virus antigen in blood.

Enzootic bovine leukosis

Enzootic bovine leukosis is a retroviral disease of cattle that causes persistent lymphocytosis and lymphosarcoma in affected cattle and is caused by bovine leukemia virus (BLV). BLV is found in lymphocytes and can be transmitted horizontally via contact with secretions from infected animals, iatrogenically via contaminated instruments such as tattoo pliers and dehorning gouges, and vertically via transplacental exposure.

There has been one study that reported the presence of BLV in bull semen by demonstrating seroconversion of sheep that were injected intraperitoneally with semen collected from a BLV-positive bull.[35] The semen from the bull was collected by rectal massage, which may have resulted in excess trauma and presence of lymphocytes in the semen. A follow-up study repeated the experiment using 32 seropositive bulls that were collected normally and no seroconversions were reported in inoculated sheep.[36] A study following the use of semen from BLV-seropositive bulls collected and processed in an AIC showed no evidence of seroconversion of females in a BLV-free herd over a 5-year period when bred with AI.[37] Another study tested frozen semen batches from BLV-seropositive bulls using PCR and did not find any evidence of BLV.[38] Semen collected in these studies were from bulls residing in AICs where semen was collected using an artificial vagina and routine laboratory procedures used to evaluate the semen for quality to include the presence of white blood cells.

Serological tests for BLV are available to accommodate international requirements for trade in bovine semen and include agar gel immunodiffusion and ELISA. The OIE recommends that donor bulls either reside in a BLV-free herd or serological testing of the bull at least 30 days before and at least 90 days after the collection of semen. In the United States, CSS does not require testing of donor bulls for leukosis for residence in a production herd. VI and PCR tests are available to detect virus antigen in the blood and a PCR test is available to test semen.

Control

Control of pathogens in bovine semen, especially from bulls residing in AICs, is primarily dependent on frequent testing for diseases of concern for both the pathogen antigen as well as antibody response to the disease. Several diseases such as BVDV and leptospirosis require the resident production herd to maintain freedom from diseases status, and a positive test result generally requires the removal of the positive animal and retesting of the resident herd population. Other diseases such as BTV and EHD allow seropositive animals to be present in the resident herd and tested differently to minimize the risk of the pathogen being present in the semen. Seronegative animals are tested for antibodies before collection of semen and again after the collection to show no exposure to the pathogen. Similar testing is done on seropositive animals using tests to detect the pathogen antigen during the time of collection. The testing strategy for each AIC will depend on the regulations and standards of the country where the AIC is located and the standards that are needed to export frozen semen to foreign markets.

Minimizing the introduction of diseases is an important consideration for maintaining a disease-free herd. Biosecurity protocols are necessary to reduce the probability of introduction of disease. Bulls should be housed indoors or in paddocks with perimeter fencing to prevent contact with other livestock and wildlife. Barn personnel should change into clothing dedicated to the facility and, at a minimum, should wash their boots prior to entry into the facilities and when moving between quarantine and isolation facilities. Outside vehicles should be clean and have tires disinfected prior to entering the AIC. Bulls should be maintained in a clean environment and not soiled during collection and be maintained in good health and physical status. Sick animals should be isolated and not used to collect semen. Many countries require that minimum standards for biosecurity are maintained in order for semen to be released for distribution both domestically and internationally. Biosecurity and standard operating practice manuals should be present and frequently updated and followed to reduce the introduction of disease into the AIC.

References

1. World Organisation for Animal Health (OIE). Terrestrial Animal Health Code (2012), Volume 1, Section 4, Chapter 4.6.
2. Kirkland P, Mackintosh S, Moyle A. The outcome of widespread use of semen from a bull persistently infected with pestivirus. *Vet Rec* 1994;135:527–529.
3. Meyling A, Jensen A. Transmission of bovine virus diarrhea virus (BVDV) by artificial insemination (AI) with semen form a persistently-infected bull. *Vet Microbiol* 1988;17:97–105.
4. Kirkland P, McGowan M, Mackintosh S, Moyle A. Insemination of cattle with semen from a bull transiently infected with pestivirus. *Vet Rec* 1997;140:124–127.
5. Rikula U, Nuotio L, Laamanen L, Sihvonen L. Transmission of bovine viral diarrhea virus through the semen of acutely infected bulls under field conditions. *Vet Rec* 2008;162:79–82.
6. Voges H, Horner G, Rowe S, Wellenberg G. Persistent bovine pestivirus infection localized in the testes of an immuno-competent, non-viraemic bull. *Vet Microbiol* 1998;61:165–175.
7. Givens D, Riddell K, Edmondson M et al. Epidemiology of prolonged testicular infections with bovine viral diarrhea virus. *Vet Microbiol* 2009;139:42–51.
8. Wrathall A, Simmons H, Van Soom A. Evaluation of risks of viral transmission to recipients of bovine embryos arising from fertilisation with virus-infected semen. *Theriogenology* 2006;65:247–274.
9. Certified Semen Services. CSS Minimum Requirements for Disease Control of Semen Produced for AI. 2011. Available at http://www.naab-css.org/about_css/CSSMinReqJan2011rev0312.htm.
10. Nickell J, White B, Larson R et al. Onset and duration of transient infections among antibody-diverse calves exposed to a bovine viral diarrhea virus persistently infected calf. *Int J Appl Res Vet Med* 2011;9:29–39.
11. Kirkland P, Richards S, Roghwell J, Stanley D. Replication of bovine viral diarrhea virus in the bovine reproductive tract and excretion of virus in semen during acute and chronic infections. *Vet Rec* 1991;128:587–590.
12. Givens D, Kurth K, Zhang Y et al. A rare case of peristent testicular infection causes shedding of infectious virus in semen. *Clin Theriogenol* 2012;4:426.
13. Niskanen R, Alenius S, Belák K et al. Insemination of susceptible heifers with semen from a non-viraemic bull with persistent bovine virus diarrhea virus infection localized in the tests. *Reprod Domest Anim* 2002;37:171–175.
14. Givens D, Heath A, Brock K, Brodersen B, Carson R, Stringfellow D. Detection of bovine viral diarrhea virus in semen obtained after inoculation of seronegative postpubertal bulls. *Am J Vet Res* 2003;64:428–434.
15. Radostits OM, Gay CC, Hinchcliff KW, Constable PD. *Veterinary Medicine*, 10th edn. Philadelphia: Saunders, 2007, pp. 1094–1110.
16. Johnson RC. Leptospira. In: Baron S (ed.) *Medical Microbiology*, 4th edn. Galveston, TX: University of Texas Medical Branch at Galveston, 1996, ch. 35.
17. Miraglia F, Morais Z, Cortez A et al. Comparison of four antibiotics for inactivating leptospires in bull semen diluted in egg yolk extender and experimentally inoculated with *Leptospira santarosai* serovar cuaricura. *Braz J Microbiol* 2003;34:147–151.
18. Eaglesome M, Garcia M. Disease risks to animal health from artificial insemination with bovine semen. *Rev Sci Tech Off Int Epiz* 1997;16:215–225.
19. Bercovich Z. Maintenance of *Brucella abortus*-free herds: a review with emphasis on the epidemiology and the problems in diagnosing brucellosis in areas of low prevalence. *Vet Quart* 1998;20:81–88.
20. Bartlett D. Comments on the episode of bovine tuberculosis in a population of bulls used for artificial insemination. In: *Proceedings of the Annual Meeting of the United States Animal Health Association*, 1967, pp. 166–171.
21. Niyaz Ahmed A, Khan J, Ganai N. DNA amplification assay for rapid detection of bovine tubercle bacilli in semen. *Anim Reprod Sci* 1999;57:15–21.
22. USDA APHIS. Bovine Tuberculosis Eradication. Uniform Methods and Rules, effective January 1, 2005.
23. Bondurant R. Venereal diseases of cattle: natural history, diagnosis, and the role of vaccines in their control. *Vet Clin North Am Food Anim Pract* 2005;21:383–408.
24. Thibier M, Guerin B. Hygienic aspects of storage and use of semen for artificial insemination. *Anim Reprod Sci* 2000;62:233–251
25. McMillen L, Fordyce G, Doogan V, Lew A. Comparison of culture and a novel 5′ *Taq* nuclease assay for direct detection of *Campylobacter fetus* subsp. *venerealis* in clinical specimens from cattle. *J Clin Microbiol* 2006;44:938–945.
26. Devenish J, Brooks B, Perry K et al. Validation of monoclonal antibody-based capture enzyme linked immunoabsorbant assay for detection of *Campylobacter fetus*. *Clin Diagn Lab Immunol* 2005;12:1261–1268.
27. Shin S, Lein D, Patten V, Ruhnke H. A new antibiotic combination for frozen bovine semen. *Theriogenology* 1988; 29:577–591.
28. Shin S. Comparative efficacy study of bovine semen extension: 1-step vs 2-step procedure. In: *Proceedings of the 18th Technical Conference on Artificial Insemination and Reproduction*, 2000, pp. 60–62.
29. van Oirschot J. Bovine herpesvirus 1 in semen of bulls and the risk of transmission: a brief review. *Vet Quart* 1995;17:29–33.
30. Favier P, Marin M, Pérez S. Role of bovine herpesvirus type 5 (BoHV-5) in diseases of cattle. Recent findings on BoHV-5 association with genital disease. *Open Vet J* 2012;2:46–53.
31. Monke D. IBR research update and commentary. In: *Proceedings of the 14th Technical Conference on Artificial Insemination and Reproduction*, 1992, pp. 26–29.
32. World Organisation for Animal Health (OIE). Terrestrial Animal Health Code (2012). Volume 2, Section 8, Chapter 8.3.
33. ProMED-Mail. Published date: 2012-12-20. Schmallenberg virus, Europe (76): virus RNA in bovine semen. Archive Number: 20121220.1460864. http://www.promedmail.org/direct.php?id=20121220.1460864.
34. ProMED-Mail. Published date: 2012-12-21. Schmallenberg virus, Europe (77): (Netherlands, France) virus RNA in bovine semen. Archive Number: 20121221.1462748: http://www.promedmail.org/direct.php?id=20121221.1462748.
35. Lucas M, Dawson M, Chasey D, Wibberley G, Roberts D. Enzootic bovine leucosis virus in semen. *Vet Rec* 1980;106:128.
36. Kaja R, Olson C. Non-infectivity of semen from bulls infected with bovine leukosis virus. *Theriogenology* 1982;18:107–112.
37. Monke D. Noninfectivity of semen from bulls infected with bovine leukosis virus. *J Am Vet Med Assoc* 1986;188:823–826.
38. Choi K, Monke D, Stott J. Absence of bovine leucosis virus in semen of seropositive bulls. *J Vet Diagn Invest* 2002;14:403–406.

Chapter 74

Bovine Semen Quality Control in Artificial Insemination Centers

Patrick Vincent[1], Shelley L. Underwood[1], Catherine Dolbec[1], Nadine Bouchard[1], Tom Kroetsch[2] and Patrick Blondin[1]

[1]*L'Alliance Boviteq Inc., St-Hyacinthe, Quebec, Canada*
[2]*The Semex Alliance, Guelph, Ontario, Canada*

Introduction

Fertility is a multiparametric phenomenon that relies on the use of semen of sufficient quality and quantity, accurate timing and method of insemination, and appropriate herd management. When using artificial insemination (AI), the dairy producer must manage a range of these factors, including heat detection, timing of insemination in relation to estrus, and correct handling of the frozen straws. However, it is the onus of the semen production centers (SPCs) to supply straws containing spermatozoa of good viability that produce acceptable conception rates if all other variables are managed correctly.

To ensure acceptable fertility after AI, frozen-thawed spermatozoa must be present in sufficient number in each straw (concentration), and possess a number of characteristics important for fertilization. Accordingly, spermatozoa must survive the thawing procedure with normal morphology, an intact acrosome, DNA integrity, active mitochondria, and maintain forward progressive motility to traverse the female reproductive tract. Some or many of these characteristics are measured during post-thaw quality control procedures undertaken by SPCs prior to distribution. Quality control (QC) is the assurance that each batch of straws has undergone semen analysis to verify that the sample is likely to be fertile.

Although semen analysis may seem easy to perform, meticulous attention to detail and technique is essential in order to obtain an accurate and reproducible analysis. Manual semen analysis using a light microscope has been the standard method for analysis in most SPCs. However, manual analyses can be very subjective and prone to within- and between-technician errors. Similarly, the use of fluorescence microscopy to assess spermatozoa for acrosome, membrane, and DNA integrity is markedly slow and limited due to the low number of spermatozoa analyzed from each sample and the incapacity for an extensive multiparametric analysis.

To maximize accuracy in QC, SPCs are realizing the benefit of a multiparametric approach and have increased the rigor of their semen testing, moving from time-consuming basic subjective assessment of a few hundred spermatozoa for concentration, motility and morphology using microscopy, to the use of computer-assisted tracking to assess motility, and flow cytometry to analyze thousands of cells within seconds for characteristics such as viability, mitochondrial activity, acrosome, DNA and capacitation status. The topics for discussion in this review are the various tools and assays in use in cattle SPCs to determine QC values, factors to consider when using these tools, and how the efficacy of QC procedures may be maximized in order to predict field fertility.

Objective assessment of sperm motility

Computer-assisted sperm analysis (CASA) is a powerful tool for the objective assessment of sperm motility and is hence now frequently used for evaluating semen quality. The basic components of this technology consist of a microscope to visualize the sample, a digital camera to capture images, and a computer with specialized software to analyze the movement of the spermatozoa. The essential principle behind most microscopy-based CASA systems is that a series of successive images of motile spermatozoa within a static field of view are acquired by computer software algorithms, which then scan these image sequences to identify individual spermatozoa and trace their progression across the field of view. This involves recognizing the same cell in each image by its position, and inferring its next position by estimating the likelihood that it will only have moved a certain maximum distance between frames. CASA can also provide information about sperm concentration, morphology, viability, and index of DNA fragmentation of frozen-thawed sperm. However, these more specialized techniques are not routinely applied for regular analysis of frozen-thawed semen in SPCs.

With the use of CASA, several motility parameters describing the specific movements of spermatozoa can be obtained in greater detail than possible in subjective assessment. These computerized measurements can be useful for assessing various sperm characteristics simultaneously and objectively, and are valuable for the detection of subtle changes in sperm motion that cannot be identified by conventional subjective semen analysis as reviewed elsewhere.[1-3] The parameters typically collected using CASA systems are motility, velocity, linearity, and lateral displacement of spermatozoa as they progress along their trajectories in a sample. The percentages of total and progressive motility are the most important motility parameters in the evaluation of spermatozoa. Total motility refers to the fraction of spermatozoa that display any type of movement, whereas progressively motile spermatozoa swim forward in an essentially straight line. Spermatozoa that swim but in an abnormal way, such as in tight circles, are not included in the proportion of progressively motile sperm. In addition to evaluation of sperm motility, the software calculates the kinetic values of each spermatozoon, which covers the velocity of movement, the width of the sperm head's trajectory, and frequency of the change in direction of the sperm head. The velocity values that are determined by CASA are the curvilinear velocity, straight-line velocity, and average path velocity.[2] The amplitude of lateral head displacement and beat cross frequency are two other characteristics measured with CASA instruments.[2]

Limitations of CASA instruments

Despite the power of an objective evaluation by CASA, there are some constraints associated with this technology. Many factors are known to affect CASA results. The type of specimen chamber used for analysis can affect the movement of sperm, the accuracy of the cell count number, and therefore the percentage of motile spermatozoa.[4] The temperature at which semen is analyzed is also an important factor that may affect CASA results. Independent studies showed that analyzing semen below 37°C significantly affected results.[5,6] These groups performed CASA on spermatozoa maintained at 37°C with a stage warmer and compared the results with spermatozoa analyzed at room temperature or at 30°C. The data demonstrated a decrease in the motility parameters (percentage of motile spermatozoa and track speed) when spermatozoa were not analyzed at 37°C. These experiments suggest that a simple variation introduced in the analysis of sperm motility can have a considerable effect on the results. The concentration at which semen is analyzed is an essential aspect that influences CASA results. It has been established that at low semen concentrations (<20 million/mL), an overestimation of the concentration and thus an underestimation of the percentage of motile cells can occur due to the acquisition of nonspermatic particles (debris). On the other hand, at higher concentrations (>50 million/mL), a large proportion of the fast-moving cells will be excluded from analysis because of cell collisions, spermatozoa exiting the analysis area, or excluded on the basis of nearest-neighbor effects, leading to an underestimation of the motility.[5,7]

Sampling condition is a source of error when acquiring data with CASA. Computer and video camera equipment are continuously evolving and different CASA systems use various models of video camera. Most of the CASA systems allow 30 or 60 Hz as a frame rate to analyze sperm tracks and speed. Studies have shown the importance of the frame rate for reliability of the analysis.[8-10] It is generally accepted that a higher frame rate is required to render an evaluation closer to the real path for a fast nonlinear sperm cell. To study a hyperactivated sperm cell, Mortimer and Swan[11] suggested using the highest frame rate available on the system in order to have the most accurate evaluation.

The type of extender in which semen is diluted is another aspect that should be taken into consideration when evaluating spermatozoa with CASA. Some extenders contain debris of a size similar to a sperm head, causing CASA software to include them in the analysis. Egg yolk- and milk-based diluents are examples of extenders containing such particles. In addition, when observing semen diluted with milk extender, the globular lipids mask the spermatozoa thus rendering CASA analysis impossible. To assess motility analysis in these conditions, samples could be washed to remove extender debris from semen. However, it has been established that washing the semen affects the motility of the spermatozoa,[12] making correct evaluation more difficult. To overcome this problem, fluorescence technology allows discrimination of sperm cells from particles in the extender by staining sperm heads with a DNA-binding fluorochrome. Under fluorescent light, only DNA-containing objects will be detected by the CASA software, thus omitting the need for washes. This technique improves the accuracy of the concentration[13] as well as the motility analysis[14] when working with semen diluted in these extenders. Therefore, standardizing the type of chamber, the temperature, the concentration, and the type of extender is crucial to assure repeatable standard QC at an SPC.

Motility is one of the most important characteristics believed to be associated with the fertilizing ability of spermatozoa. Several groups have reported a significant correlation between total[15-18] and progressive[19] motility of bull semen and its associated field fertility. However other groups have reported that the subjective analysis of semen motility did not correlate with fertility.[20,21] CASA instruments collect a wide range of sperm motility parameters, allowing a more detailed and accurate analysis of sperm movements and track speed. Researchers have also tried to correlate the kinetic parameters with the field fertility of semen, with some groups able to show a positive correlation between straight-line velocity of spermatozoa and field fertility.[18-22] Another study used a combination of several motility parameters to reach a very high correlation with bull fertility.[20] Taken together, these studies show the high potential of CASA for estimating the quality of the semen and therefore becoming a powerful tool to measure sperm characteristics and predict bull fertility compared with standard semen evaluation. However, as mentioned above, standardization of conditions and parameters of all CASA analyses are key to obtain repeatable and valid correlations with fertility.

Several models of CASA instruments are now available to evaluate the quality and motility of spermatozoa. Each system operates on similar principles but they differ in their parameter settings and use different algorithms to determine speed and trajectories. Parameter settings, threshold settings, video frame rate, and other variables will affect CASA results as reviewed by Davis and Katz.[23] As mentioned above, new technologies and CASA software evolve quickly. Our laboratory undertook a small study to measure the aptitude of the CEROS (Hamilton-Thorne, USA) and the Sperm Class Analyzer (SCA; Microptics, Spain) in evaluating the motility and the concentration of frozen-thawed bovine spermatozoa diluted in an egg yolk-based extender (unpublished data). A total of 18 different frozen-thawed ejaculates were analyzed with both systems and the mean total and progressive motility percentages, concentration, average path velocity, curvilinear velocity, straight-line velocity, beat cross frequency, and amplitude of lateral head displacement were compared (Figure 74.1). Among all parameters analyzed, only the percentage of total motile cells was not significantly different between the systems. The discrepancies can be explained by the better capacity of the SCA to exclude egg yolk particles from the analysis. The SCA discriminated nonspermatic particles based on size in microns while the CEROS used pixels to estimate the size of the cells. Differences in the algorithms used to calculate slow, medium, and fast spermatozoa may also explain the variation in the motility and kinetic parameters observed between each system. Overall, this mini-study indicates high variability between CEROS and SCA systems in estimating sperm motility parameters.

Analysis of sperm function by flow cytometry

Flow cytometry analyzes cells suspended in a stream of fluid passing at high velocity in front of one or several lasers. The light emitted by fluorochrome-bound cells is captured by photomultiplier tubes and converted into an electronic signal subsequently digitalized by cytometry software. Key features of flow cytometry are the acquisition and analysis of thousands of cells within seconds and the multiparametric potential of the technology. The most modern cytometers are routinely equipped with three lasers and at least 10 photomultiplier tubes, allowing cell labeling with several probes at the same time and thus enabling analysis of numerous parameters simultaneously. In the last few years, the multiparametric aspect of flow cytometry has allowed this technology to become a popular tool for evaluating sperm attributes.[24–26] A wide range of fluorochromes has been developed to assess numerous characteristics of sperm cells. Here we review some of the fluorochromes used to study sperm cells with flow cytometry.

Sperm attributes analyzed by flow cytometry
Viability/mortality

Propidium iodide is the most popular dye used to identify dead cells. This membrane-permeable fluorochrome enters spermatozoa with damaged cellular membranes and binds to DNA where it can be excited with a 488-nm laser present on most cytometers.[27–29] Propidium iodide is often used in combination with SYBR-14, another DNA-labeling probe.[30,31] SYBR-14 is also excited by the 488-nm laser and is a permeant probe staining all cells. Added to the cells simultaneously, propidium iodide displaces or quenches the SYBR-14 fluorescence in damaged cells. A new fixable dye commercialized by Invitrogen under the name Live/Dead® fixable dead cell kit is now available to evaluate the viability of cells.[32] This dye reacts with cellular amines on the surface of cells or inside the cytoplasm of cells with damaged membranes. Cell-surface staining of amines of viable cells will result in relatively dim staining compared with the bright staining of dead cells. This fixable dye belongs to a large family available in different wavelengths of excitation/emission, allowing its use on most cytometers.

Acrosome integrity

Evaluation of acrosomal status is mainly assessed by using plant lectins recognizing acrosomal ligands. *Pisum sativum* agglutinin binds mannose and galactose moieties of the acrosomal matrix. As *Pisum sativum* agglutinin cannot penetrate the intact acrosomal membrane, only spermatozoa with a reacted or damaged acrosome will be stained.[21,33,34] However, it has been shown that *Pisum sativum* agglutinin has an affinity for egg yolk and nonspecific binding sites on the sperm cell surface.[35,36] This aspect could become a problem when analyzing semen frozen in egg yolk-based extender and result in misinterpretation of the acrosomal status of sperm. *Arachis hypogaea* (peanut) agglutinin binds galactose moieties of the outer acrosome membrane and is the most popular lectin used to study the integrity of the acrosomal membrane with flow cytometry.[37–39] In addition, *Arachis hypogaea* agglutinin seems the most reliable probe for identifying spermatozoa with a damaged acrosome as it displays less nonspecific binding to other areas of spermatozoa.[40] *Pisum sativum* agglutinin and *Arachis hypogaea* agglutinin are usually labeled with FITC fluorochromes, allowing them to be used by all cytometers.

Mitochondrial activity

Mitochondria are very important organelles involved primarily in the generation of the energetic substrates for the motility of the sperm cell. Rhodamine 123 was one of the first probes to monitor mitochondrial activity.[41,42] Rhodamine 123 is sequestered in active mitochondria and washed out from the cell when the membrane potential is lost. This characteristic limits its use when quantification is needed or when fixation of spermatozoa is required. To overcome the fixation problem, Mitotracker® dye could become a solution. This fixable dye accumulates and stains active mitochondria and has the advantage of availability in different ranges of excitation and emission fluorescence.[42–44] The most popular probe for evaluating mitochondrial activity is JC-1 (5,5′,6,6′-tetrachloro-1,1′,3,3′ tetraethylbenzimidazolylcarbocyanine iodide).[45–47] In spermatozoa containing mitochondria with a high membrane potential, JC-1 enters the mitochondrial matrix where it accumulates and forms J-aggregates and become fluorescent red. In spermatozoa containing mitochondria with low membrane potential, JC-1 cannot

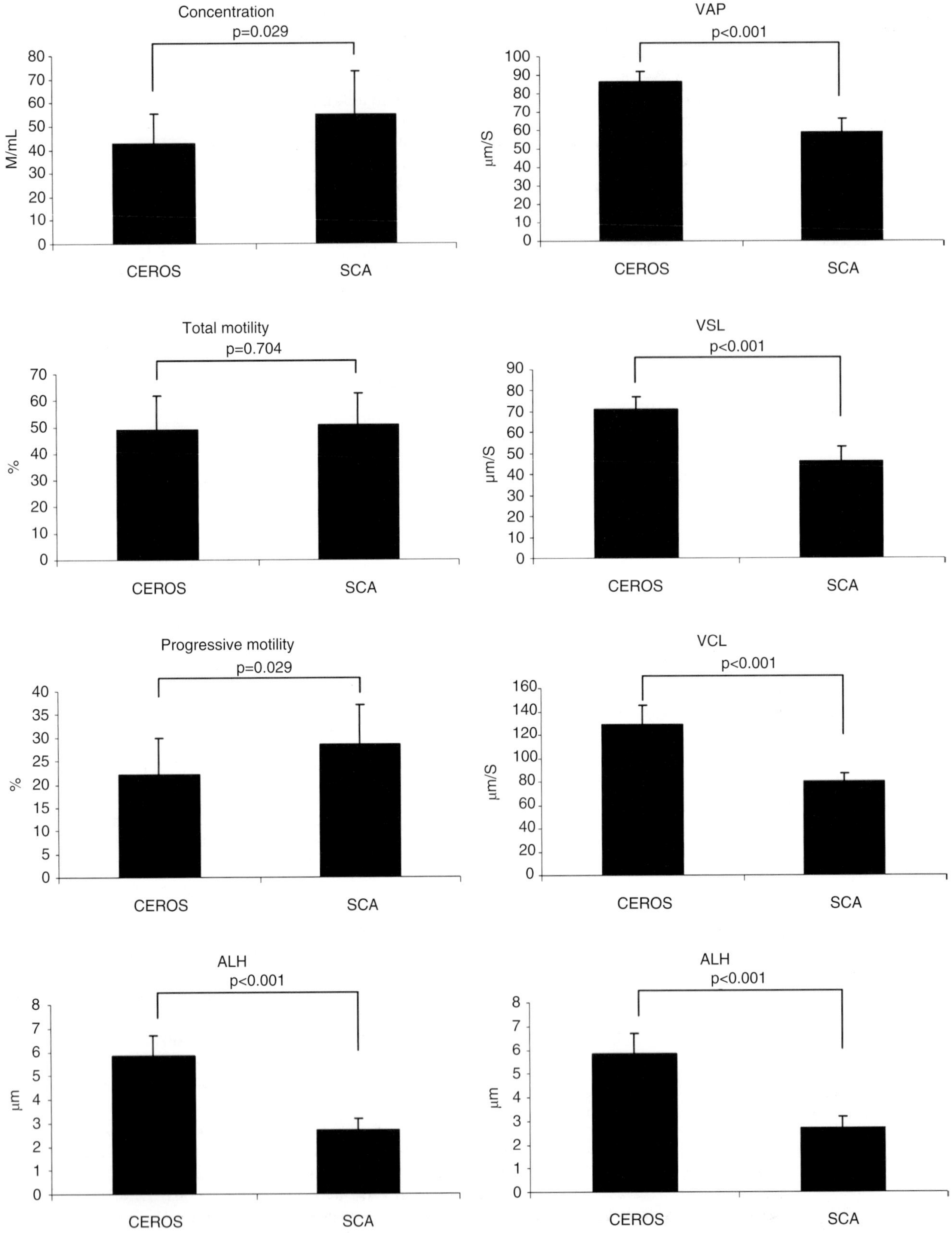

Figure 74.1 Comparison of CEROS (Hamilton-Thorne) and SCA (Sperm Class Analyzer) for determining concentration, percentage of total and progressive motility, lateral head displacement (ALH), average path velocity (VAP), straight line velocity (VSL), curvilinear velocity (VCL), and beat cross frequency (BCF) from 18 different frozen-thawed bovine ejaculates. Columns represent mean values ±SEM. *P*-value <0.05 indicates a statistically significant difference between CEROS and SCA, Student's paired *t*-test.

accumulate within the mitochondria and remains in the cytoplasm in a green fluorescent monomeric form. JC-1 has the advantage of being able to quantify the mitochondrial burst of the cell compared with Rhodamine 123 and Mitotracker. A disadvantage of the JC-1 probe is its dual fluorescence emission that limits its combination with other probes emitting at the green and red wavelngths.

DNA integrity

Assessment of chromatin status is important in determination of the fertility potential of spermatozoa. In recent years, the sperm chromatin structure assay developed by Evenson and Jost[48] is the main technique used to evaluate chromatin integrity in spermatozoa by flow cytometry.[49,50] The sperm chromatin structure assay uses the dual fluorescence emission of acridine orange depending on whether it binds to single-stranded DNA (red fluorescence) or double-stranded DNA (green fluorescence). Following a denaturation step, the sperm sample is incubated with acridine orange and then analyzed by flow cytometry. Denaturation will induce single-strand DNA formation when DNA breaks are present and generate a heterogeneous population of red and green fluorescence depending on the integrity of the chromatin. The most important result derived from the sperm chromatin structure assay is the ratio of red/green plus red fluorescence called the DNA fragmentation index, where a high DNA fragmentation index correlates with high DNA damage. The DNA fragmentation index has shown correlation with fertility in different species.[51-53] The large luminal spectrum covered by acridine orange and the denaturation step required to induce single-strand DNA are two main inconveniences of the sperm chromatin structure assay for a multiparametric analysis. Acridine orange fluoresces in the green and red spectrum; that leaves few possibilities for adding other fluorochromes in these spectral areas and the denaturation step is performed with an acid/detergent solution not compatible with all probes. Another assay to assess DNA integrity developed for flow cytometry is the TUNEL assay (terminal transferase dUTP nick end labeling), which can identify DNA strand breaks.[54-56] Transferase enzyme incorporates fluorescently modified nucleotides at the sites of DNA breakage. The resultant labeled cells can then be analyzed by flow cytometry. The TUNEL assay allows quantification of labeled nucleotides incorporated into fragmented DNA reflected by the increase in fluorescence, which gives an appreciable advantage over the sperm chromatin structure assay.

Calcium influx

Calcium influx is one of the primary steps involved in the sperm capacitation process. The rise in intracellular calcium ultimately leads to the phosphorylation of tyrosine and serine residues in proteins regulating the signaling cascade. The most popular dye used to determine the intracellular calcium concentration in sperm cells is the Fluo-3/4 family probe excited by the 488-nm laser line.[57-59] Calcium-unbound Fluo-3 is a nonfluorescent molecule but when calcium ions enter the cell and bind Fluo-3 the latter becomes fluorescent. Fluo-4 is a derivative of Fluo-3 bearing higher fluorescence intensity. Fura red is a probe also excited by the 488-nm laser whose fluorescence emission decreases on calcium binding. Dual labeling of spermatozoa with Fluo-3/Fura red allows calculation of the ratio of unbound to bound calcium. The ratio between the two mean fluorescence intensities (Fluo-3/Fura red) is proportional to the intracellular calcium concentration of the spermatozoa. This experimental approach has been used to assess dog semen by Peña et al.[60] One critical aspect when using these dyes is cell loading. Because mean fluorescence intensity is the parameter used to indicate the level of intracellular calcium, errors in pipetting of the probe will change the mean intensity and result in misinterpretation of the results. The drawback of Fluo-3/Fura red combination is the need for two different fluorescence detectors, which decreases the scope of a multiparametric approach. Indo-1 pentaacetoxymethyl ester is a membrane-permeable calcium sensor dye used to monitor changes in intracellular calcium. Once Indo-1 enters the cell, esterases cleave the acetoxymethyl group, yielding a membrane-permeable dye. Unbound Indo-1 has peak emission at 485 nm. On binding calcium, the peak emission shifts down to 410 nm. Measurement over time can be represented as a ratio of the two emission wavelengths. As Indo-1 acetoxymethyl is a ratiometric probe, cell-loading concerns (as for Fluo-3/4) are less important. One restriction with this probe is that not all laboratories are equipped with an instrument comprising the ultraviolet laser needed to excite Indo-1 acetoxymethyl. For those with this instrument, Indo-1 acetoxymethyl becomes a very good probe for a multiparametric approach because it is one among few probes using the ultraviolet laser, thus leaving the 488 and 633 nm lasers available to study other parameters.

Limitations of flow cytometry

Several factors influence the choice of cytometer for the analysis of sperm cells. The price of the instrument remains a major factor that will influence this choice. Multiparametric analysis is usually obtained with instruments containing more than one laser and many photomultiplier tubes, which increases the price of the equipment substantially. Indeed, the type of analyses to be performed is also a factor that will determine the choice of flow cytometer. Depending on the objectives of the breeding center and the experimental design, the combination of lasers (number and wavelength) and the number of photomultiplier tubes included in the instrument must be taken into account. An instrument with only one laser and three photomultiplier tubes allows detection of a maximum of three parameters on each cell, while a multiparametric analysis including four or more parameters will usually require an instrument having at least two lasers and four photomultiplier tubes. The software operating the flow cytometer is another important aspect in the choice of instrument. Most software products available are fairly easy and straightforward to operate for a novice user in flow cytometry. However, some software requires certain knowledge of flow cytometry concepts, making the instrument more difficult to operate. As an example, samples stained with a cocktail containing several probes are subjected to subpopulation gating analysis. In order to obtain representative results, gates need to be associated to the proper population in the correct hierarchy, a perspective that is difficult to handle with some software for a novice user. Moreover, some programs have gaps in export and data compilation, making it more difficult to analyze the data and these shortcomings are time-consuming for the user.

As mentioned in the section on analysis of sperm function by flow cytometry, flow cytometry is a relatively new avenue for the SPCs. The unique characteristics of spermatozoa must be considered when selecting an instrument. The paddle shape of the head and presence of the flagellum make spermatozoa very different in size and cellular complexity compared with most cells studied by flow cytometry. Indeed, the majority of cells studied with this technology have a round shape and passage in front of the laser leads to a neat forward scatter versus side scatter plot. However, when a sperm cell hits the laser, it could be on the thick or the thin side of the head. This unique feature of sperm cells will lead to a scatter plot of different size and complexity. A very important aspect we found when studying frozen-thawed spermatozoa with flow cytometry, and which has been discussed in this chapter, is the extender in which semen is diluted. As stated, different types of extenders are used to dilute semen, and some contain particles of a similar size to spermatozoa. This aspect of particle contamination of the target population is a concern when considering the purchase of a cytometer for multiparametric analyses. An apparatus powerful enough to discriminate sperm cells from particles based only on the size and complexity of the cells allows gating of the sperm cell population without using any fluorochromes. As a result, more photomultiplier tubes are left available for cell characterization with fluorochromes. Our laboratory has compared several flow cytometers, and only half of them were sensitive enough to accomplish this discrimination between foreign particles and spermatozoa without having to use any fluorochromes. A research laboratory possessing a cytometer that cannot discriminate debris from spermatozoa will have to either wash the cells to remove particles or stain the sperm cells with a fluorochrome, which leaves fewer photomultiplier tubes to study other parameters. Not all fluorochromes are suitable with semen extenders, especially egg yolk-derived extenders known to quench some fluorescent dyes. Hoechst 33342 is routinely used to stain the nucleic acid of spermatozoa when using a cytometer equipped with an ultraviolet or violet laser. This approach allows elimination of the particles by gating them out, resulting in a more accurate analysis (Figure 74.2). Unfortunately, not all systems are equipped with such lasers to use Hoechst 33342 as a cell tracking dye, but a cell permeant dye like SYBR-14 could be added to sperm cells and this fluorescent population gated for further analysis.

A multiparametric approach for standardization and QC among SPCs
Use of CASA and flow cytometry as QC tools

Subjective evaluation of semen by conventional microscopy is still used by numerous SPCs. As discussed earlier, this type of semen evaluation results in variation of the final decision regarding whether the semen lot will be accepted or rejected during QC. Even though CASA is used in some SPCs, fertility is a multiparametric phenomenon; thus motility parameters of spermatozoa are not sufficient for evaluating the global fertility potential of a semen sample. Introduction of flow cytometry in SPCs will allow better characterization of the spermatozoa because, by itself, flow cytometry has the potential for multiparametric analysis of spermatozoa. Combining CASA and cytometry will provide SPCs with a powerful multiparametric approach to evaluate the quality of the semen produced and allow the establishment of standardized procedures to make accurate and repeatable decisions on the outcome of the semen.

We have evaluated the potential of these tools to accurately estimate the quality of semen produced in SPCs across Canada. The percentage of total and progressive motility and the percentage of viability, acrosome integrity, and high mitochondrial activity were evaluated with CASA and flow cytometry. We obtained and reanalyzed 660 lots of semen diluted in egg yolk produced by different SPCs that had previously been processed by standard subjective QC, where 58% ($n=385$) were accepted and 42% ($n=275$) were rejected. Semen evaluation was performed at two different time points: immediately after thawing and 2 hours later after a thermoresistance stress. Cutoff values (Table 74.1) were established for each parameter analyzed with CASA/flow cytometry at each time point. Semen lots that did not reach our cutoff standard would have been rejected and discarded. On the other hand, if the quality of the semen was good and met our cutoff values, the lot was considered as accepted using CASA and flow cytometry and would have been distributed.

Table 74.2 highlights the percentage of lots rejected after analyzing the 660 lots with CASA and flow cytometry. CASA analyses revealed that 14% (94 of 660) unique lots failed to meet the cutoff immediately after thawing compared with 27% (178 of 660) unique lots after 2 hours incubation. The same observation could be made from flow cytometry analyses, where 24% (156 of 660) unique lots were rejected after thawing compared with 32% (214 of 660) unique lots after 2 hours incubation. This analysis demonstrates the need for a thermoresistance test to be included in QC to increase precision where the majority of lots were discarded based on the 2-hour incubation. In addition, CASA evaluation alone resulted in the discard of 28% of the lots while flow cytometry analysis alone discarded 34% (224 of 660) of the lots. When CASA and flow cytometry are used in combination, a total of 41% (268 of 660) unique lots were rejected, resulting in an increase in rejection of 13% compared with CASA alone (28%). Overall, these results demonstrate the ability of the multiparametric approach to provide very high authority to the rejection/acceptance decision.

Detailed analyses of accepted and rejected lots with CASA/flow cytometry compared with standard QC showed 77.4% agreement in the decision made using the two different methods of QC assessment, but 22.6% of the samples were discordant between them (Figure 74.3). This discrepancy would be considered the "precision impact" of the multiparametric technique over the standard subjective evaluation. This 22.6% consists of 10.8% of the accepted lots by standard methods that would have been rejected with our multiparametric tools and 11.8% of the

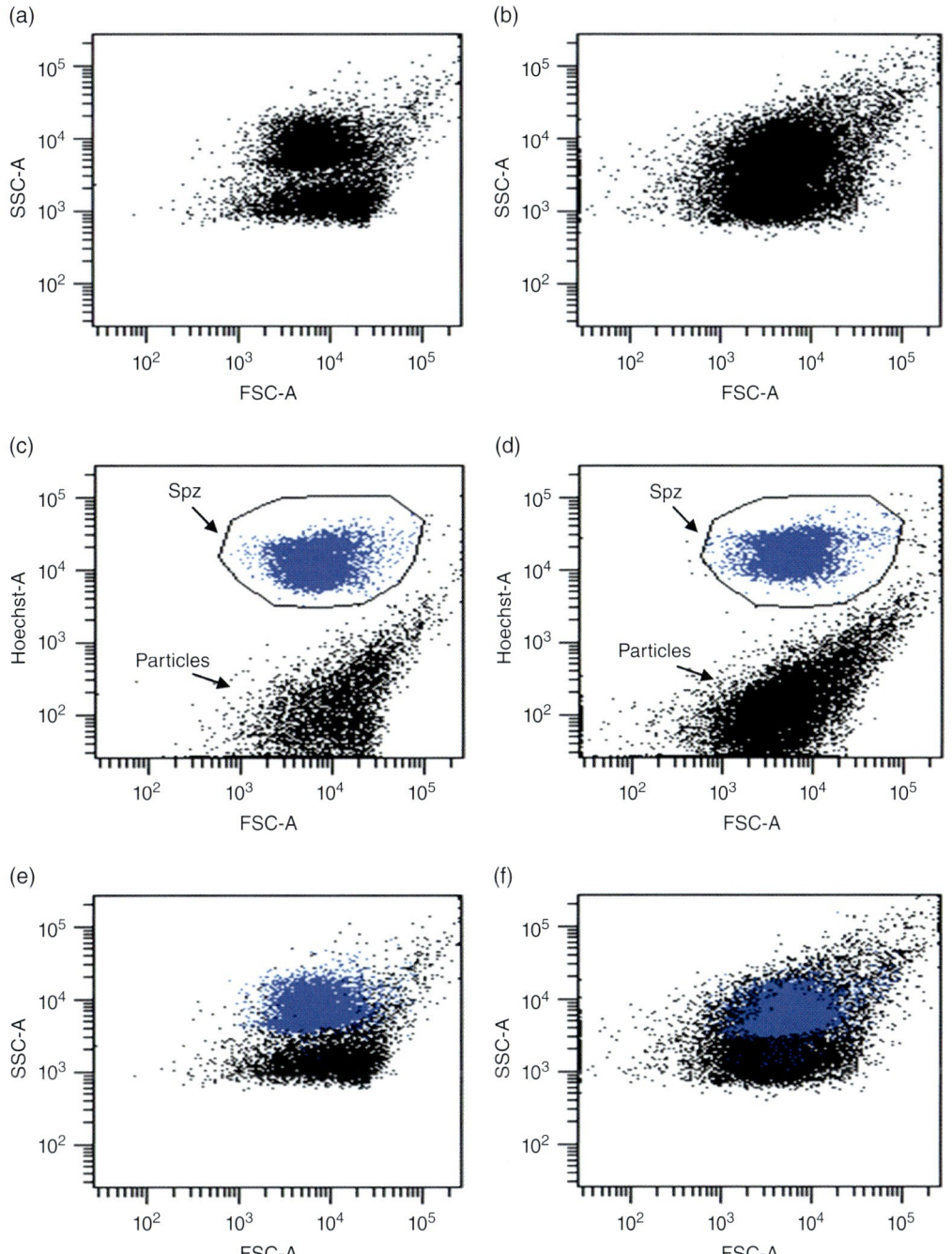

Figure 74.2 Comparison of (a) good and (b) bad resolution of spermatozoa from nonspermatic particles in frozen-thawed ejaculates. Labeling cells with Hoechst 33342 (blue cells in c, d) allows exclusive staining of spermatozoa and discrimination from cellular debris and egg yolk-derived nonspermatic particles. The positive population can be used for further characterizations in a multiparametric approach (e, f). Spz, spermatozoa; SSC, side scatter plot; FSC, forward scatter plot.

rejected lots by standard methods that would have been accepted with the multiparametric approach. At first sight, 22.6% of discrepancies between objective and subjective evaluation would appear high. However CASA/flow cytometry would have discarded 268 lots (Table 74.2) while accepting 392 lots compared with the original 385 accepted lots and 275 rejected by the SPCs. This represents a 1% difference of semen production and would be considered as "production impact" of CASA/flow cytometry over standard QC. This study demonstrates that the production impact would be negligible but the precision impact would be quite considerable.

To estimate the impact of this multiparametric approach on semen fertility in the field, we applied this analysis on semen lots of known fertility that were released in the field after using standard QC. CASA and flow cytometry were performed on 192 lots with at least 250 first inseminations. The fertility associated to each lot was obtained from the Canadian Dairy Network and was expressed as FERTSOL, which represents the 56-day nonreturn rate, adjusted for

Table 74.1 CASA/flow cytometry cutoff values used to determine pass/fail rates during quality control of frozen-thawed semen immediately after thawing and after 2 hours thermoresistant stress. Semen lots meeting or exceeding these cutoff values passed the evaluation.

Parameter	Post thaw	After 2 hours' stress
Total motility (%)	40	35
Progressive motility (%)	15	10
Intact acrosome (%)	66	61
Membrane intact cells (%)	40	40
Mitochondrial activity (%)	40	45

Table 74.2 Number (%) of samples rejected (of 660 total) for at least one QC parameter based on CASA alone, flow cytometry alone, or CASA plus flow cytometry.

Parameter	Post-thaw (0 hours)	After 2 hours' stress	0 hours + 2 hours[a]
CASA alone			
Total motility (%)	76 (12)	157 (24)	163 (25)
Progressive motility (%)	70 (11)	137 (21)	151 (23)
Total CASA	94 (14)	178 (27)	188 (28)
Flow cytometry alone			
Intact acrosome (%)	104 (16)	97 (15)	117 (18)
Membrane intact cells (%)	132 (20)	144 (22)	159 (24)
Mitochondrial activity (%)	133 (20)	208 (32)	212 (32)
Total flow cytometry	156 (24)	214 (32)	224 (34)
Total CASA + flow cytometry	176 (27)	260 (39)	268 (41)

[a]Unique lots rejected at 0 hours + unique lots rejected at 2 hours.

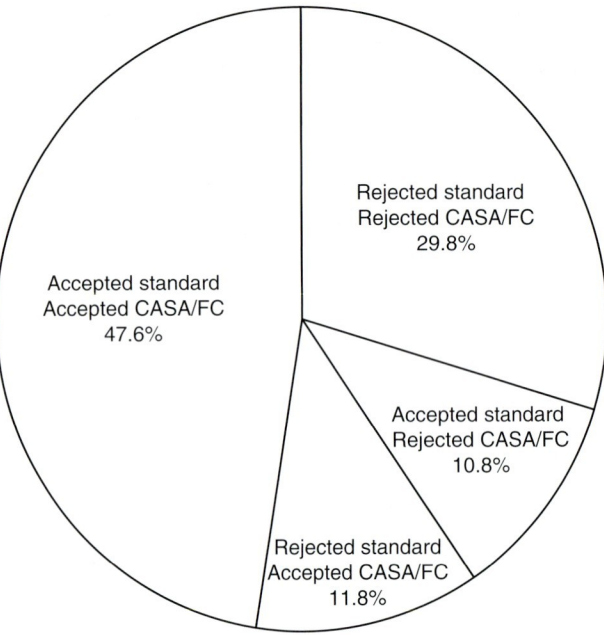

Figure 74.3 Comparative analysis of pass/fail percentage between standard QC and CASA/flow cytometry during quality control of 660 frozen-thawed semen lots.

multiple parameters including season of insemination, inseminator, number of inseminations, etc. Using our multiparametric approach 9.4% (18 of 192) of the semen lots would have been discarded before distribution. These lots corresponded to 5.3% of the low-fertility semen (<–1 FERTSOL), 3.6% of average fertility semen (between –1 and +1 FERTSOL), and 0.5% of high-fertility semen (>+1 FERTSOL). Again, this multiparametric approach would have increased the overall fertility of semen released in the field.

Specific applications of multiparametric approach in semen evaluation

The multifactorial aspect of fertility can result in misevaluation of semen quality when assessing only a few parameters. Even semen that passes rigorous QC steps may have low fertility once distributed in the field. Such lots would have passed all our CASA/flow cytometry cutoff values but still resulted in few calves despite the high number of inseminations. A more detailed analysis of this semen is then required to understand its low fertility. However, deeper analyses of semen samples are time-consuming and expensive, so they are performed only on high-value semen. Here, we report two cases where flow cytometry helped us to target a putative cause of low fertility-associated semen.

The first case involved a young bull (bull A), one of whose semen lots produced very few embryos after *in vitro* fertilization compared with the average embryo production. This lot of semen was compared (in triplicate) with one from an average bull (bull C) used as an internal control for most of our experiments. Motility and viability parameters estimated with CASA/flow cytometry showed average quality spermatozoa from both bulls (average of 36% total motility and 15% progressive motility). Evaluation of the level of intracellular calcium and the integrity of the acrosome membrane was carried out immediately after thawing and after 5 hours of post-thaw incubation with heparin at 38.5°C. At both time points, an aliquot of spermatozoa was incubated with Indo-1 acetoxymethyl and FITC-peanut agglutinin, and then challenged with thapsigargin to induce the intracellular cascade involved in the capacitation process. The percentages of live spermatozoa with high intracellular levels of calcium and reacted acrosomes were determined by flow cytometry. Analysis immediately after thawing before the heparin incubation and thapsigargin challenge showed a similar percentage of acrosome-reacted cells and cells with high intracellular calcium between bull A and bull C. Following thapsigargin stimulation, a similar percentage of spermatozoa from each bull acrosome reacted, but semen from bull A showed an increase in cells with high calcium compared with bull C (23.3±6.3% vs. 13.3±2.5%, respectively; $P=0.14$) (Table 74.3). After 5 hours of incubation with heparin at 38.5°C and without thapsigargin

Table 74.3 Comparison of acrosome integrity and calcium level of live spermatozoa from bulls A and C before and after a thapsigargin challenge, both post-thaw and after 5 hours' incubation with heparin.

	Immediately post-thaw				After 5 hours incubation with heparin			
	−Thaps		+Thaps		−Thaps		+Thaps	
Bull	AR	High Ca	AR	High Ca	AR	High Ca	AR	High Ca
A	1.8±0.7	3.1±1.2	16.3±3.8	23.3±6.3	10.5±2.3[a]	5.3±0.5	10.5±2.0[a]	25.3±2.3
C	1.5±0.2	1.1±0.2	17.8±4.0	13.3±2.5	24.1±0.9[b]	5.7±0.1	23.1±0.5[b]	36.6±8.8

Values are mean percentages from three replications±SEM. Different letters (a, b) on the same column represent significant differences, $P<0.01$, Student's paired t-test. AR, acrosome reacted; High Ca, elevated calcium level; Thaps, thapsigargin.

Table 74.4 Comparison of sperm functions based on CASA and flow cytometry analysis of bull B and C immediately after thawing.

Bull	Batch	Total motility (%)	Progressive motility (%)	Acrosome reacted (%)	Dead (%)	Mitochondrial activity (%)	DNA fragmentation index (%)
B	1	57.5±4.5	32.5±2.5	19.1±0.2	43.5±0.1*	55.3±0.2*	7.05±0.09**
B	2	44.5±3.5	24.5±3.5	19.0±0.2	51.7±1.2**	46.7±1.1**	5.76±0.12**
B	3	63.5±1.5	34.5±2.5	17.9±0.1	39.8±0.7	60.0±0.6	4.24±0.01**
B	4	56.0±0.0	32.5±0.5	16.1±0.5	38.7±1.3	60.3±2.1	4.83±0.11**
B	5	61.0±5.0	34.5±1.5	14.0±0.3*	35.0±0.3	64.6±0.4	3.85±0.03**
C	1	60.5±0.5	29.5±0.5	17.6±1.0	38.6±0.9	62.7±0.2	1.61±0.16

Values are mean percentages from two replications±SEM. Significant differences from control are represented by *, $P<0.05$ and **, $P<0.001$, Dunnett's multiple comparison test.

challenge, spermatozoa from bull A had fewer (10.5±2.3% vs. 24.1±0.9%; $P<0.01$) acrosome-reacted cells compared with bull C, but a similar percentage of cells possessing elevated calcium (5.3±0.5% vs. 5.7±0.1%, respectively; $P=0.4$) (Table 74.3). After the thapsigargin challenge, bull A still had fewer acrosome-reacted cells compared with bull C (10.5±2.0% vs. 23.1±0.5%; $P<0.01$), but also had a proportionally smaller increase in the percentage of cells having high calcium. These observations suggest that bull A spermatozoa could not capacitate in heparin-containing medium, demonstrated by fewer acrosome-reacted cells and high intracellular calcium sperm. Altogether, this analysis suggests that this specific lot of semen from bull A was not mature enough to reach its full capacitation process and fertility potential. Saacke et al.[61] described some compensable and uncompensable seminal deficiencies related to subfertility. Compensable factors include functional or molecular traits reflected in this case study by the low percentage of cells having reacted acrosomes or elevated calcium. This case could benefit from increasing the number of spermatozoa per straw of bull A to compensate its subfertility.

The second case studied was an adult bull (bull B) that exhibited extremely low fertility in the field (FERTSOL<−2) based on several different semen lots released from an SPC after standard QC tests. We firstly performed the basic CASA/flow cytometry tests on five different batches and compared the results with an average bull (bull C). All the lots studied would have passed our QC cutoffs. Surprisingly, total and progressive motility percentages were not significantly different in bull B compared with bull C (Table 74.4). On the other hand, acrosomal reaction, cell death, and mitochondrial activity were significantly different in some lots analyzed compared with bull C (Table 74.4). As these results could not explain the very low fertility of bull B, we conducted a sperm chromatin structure assay analysis as discussed previously on the same batches and compared results with those obtained by bull C. The DNA fragmentation index percentage was approximately two to four times higher in bull B than in bull C ($P<0.001$; Table 74.4). High DNA fragmentation index percentage has been correlated with subfertility, mostly in humans.[53,62] Therefore, the very low fertility observed for bull B could be explained by his high level of DNA denaturation as measured by the sperm chromatin structure assay. In opposition to the first case study, chromatin aberration is a noncompensable deficiency in subfertility.[61] Increasing the number of spermatozoa per straw in this case cannot compensate for the extremely low fertility of bull B observed in the field.

Conclusions

Multiparametric analyses of the semen produced by SPCs with CASA and flow cytometry demonstrates very high predictive potential for semen quality and fertility. As fertility has multiparametric aspects, research and development of new markers to identify high-fertility semen needs to be extensive. Incorporation of new markers in a multiparametric approach will lead to a better evaluation of semen quality and fertility. In addition, the implementation of these tools in SPCs will help to standardize the QC procedures by eliminating the subjective aspect of semen evaluation. This will standardize semen produced within a SPC and within multiple SPCs for AI companies that have semen produced in multiple laboratories. Another application of these tools in the AI industry is the characterization of semen with high genetic value. In the AI industry, most of the semen produced arises from a few high genetic value animals. These highly demanded semen samples should be very well characterized to optimize production in the number of straws produced and in the quality of the semen available for the market. Overall, the AI industry will benefit from the implementation of CASA and flow cytometry in SPCs.

Editor's note

This chapter is a reproduction of work prepared by the authors that has previously appeared first in the Proceedings Association for Applied Animal Andrology Conference (2012) and Animal Reproduction, July/September 2012, Volume, pages 153–165.

References

1. Verstegen J, Iguer-Ouada M, Onclin K. Computer assisted semen analyzers in andrology research and veterinary practice. *Theriogenology* 2002;57:149–179.
2. Mortimer S. CASA: practical aspects. *J Androl* 2000;21:515–524.
3. Kathiravan P, Kalatharan J, Karthikeya G, Rengarajan K, Kadirvel G. Objective sperm motion analysis to assess dairy bull fertility using computer-aided system: a review. *Reprod Domest Anim* 2011;46:165–172.
4. Massányi P, Chrenek P, Lukáč N et al. Comparison of different evaluation chambers for analysis of rabbit spermatozoa motility parameters using CASA system. *Slovak J Anim Sci* 2008;41:60–66.
5. Iguer-ouada M, Verstegen J. Evaluation of the "Hamilton Thorn computer-based automated system" for dog semen analysis. *Theriogenology* 2001;55:733–749.
6. Purdy PH, Tharp N, Stewart T, Spiller S, Blackburn H. Implications of the pH and temperature of diluted, cooled boar semen on fresh and frozen-thawed sperm motility characteristics. *Theriogenology* 2010;74:1304–1310.
7. Contri A, Valorz C, Faustini M, Wegher L, Carluccio A. Effect of semen preparation on CASA motility results in cryopreserved bull spermatozoa. *Theriogenology* 2010;74:424–435.
8. Mortimer S, Swan M. Effect of image sampling frequency on established and smoothing-independent kinematic values of capacitating human spermatozoa. *Hum Reprod* 1999;14:997–1004.
9. Brito L. Variations in laboratory semen evaluation procedures and testing. In: *Proceedings of the 23rd Technical Conference on Artificial Insemination and Reproduction*, 2010, pp. 61–67.
10. Castellini C, Dal Bosco A, Ruggeri S, Collodel G. What is the best frame rate for evaluation of sperm motility in different species by computer-assisted sperm analysis? *Fertil Steril* 2011;96:24–27.
11. Mortimer S, Swan M. The development of smoothing-independent kinematic measures of capacitating human sperm movement. *Hum Reprod* 1999;14:986–996.
12. Fernández-Santos M, Martínez-Pastor F, Matias D et al. Effects of long-term chilled storage of red deer epididymides on DNA integrity and motility of thawed spermatozoa. *Anim Reprod Sci* 2009;111:93–104.
13. Zinaman M, Uhler M, Vertuno E, Fisher S, Clegg E. Evaluation of computer-assisted semen analysis (CASA) with IDENT stain to determine sperm concentration. *J Androl* 1996;17:288–292.
14. Tardif A, Farrell P, Trouern-Trend V, Simkin M, Foote R. Use of Hoechst 33342 stain to evaluate live fresh and frozen bull sperm by computer-assisted analysis. *J Androl* 1998;19:201–206.
15. Wood P, Foulkes J, Shaw R, Melrose D. Semen assessment, fertility and the selection of Hereford bulls for use in AI. *J Reprod Fertil* 1986;76:783–795.
16. Kjaestad H, Ropstad E, Berg K. Evaluation of spermatological parameters used to predict the fertility of frozen bull semen. *Acta Vet Scand* 1993;34:299–303.
17. Correa J, Pace M, Zavos P. Relationships among frozen-thawed sperm characteristics assessed via the routine semen analysis, sperm functional tests and fertility of bulls in an artificial insemination program. *Theriogenology* 1997;48:721–731.
18. Gillan L, Kroetsch T, Maxwell W, Evans G. Assessment of in vitro sperm characteristics in relation to fertility in dairy bulls. *Anim Reprod Sci* 2008;103:201–214.
19. Kathiravan P, Kalatharan J, Edwin M, Veerapandian C. Computer automated motion analysis of crossbred bull spermatozoa and its relationship with in vitro fertility in zona-free hamster oocytes. *Anim Reprod Sci* 2008;104:9–17.
20. Farrell P, Presicce G, Brockett C, Foote R. Quantification of bull sperm characteristics measured by computer-assisted sperm analysis (CASA) and the relationship to fertility. *Theriogenology* 1998;49:871–879.
21. Januskauskas A, Gil J, Söderquist L et al. Effect of cooling rates on post-thaw sperm motility, membrane integrity, capacitation status and fertility of dairy bull semen used for artificial insemination in Sweden. *Theriogenology* 1999;52:641–658.
22. Budworth P, Amann R, Chapman P. Relationships between computerized measurements of motion of frozen-thawed bull spermatozoa and fertility. *J Androl* 1988;9:41–54.
23. Davis R, Katz D. Operational standards for CASA instruments. *J Androl* 1993;14:385–394.
24. Gillan L, Evans G, Maxwell W. Flow cytometric evaluation of sperm parameters in relation to fertility potential. *Theriogenology* 2005;63:445–457.
25. Martínez-Pastor F, Mata-Campuzano M, Alvarez-Rodríguez M, Alvarez M, Anel L, de Paz P. Probes and techniques for sperm evaluation by flow cytometry. *Reprod Domest Anim* 2010;45(Suppl. 2):67–78.
26. Hossain M, Johannisson A, Wallgren M, Nagy S, Siqueira A, Rodriguez-Martinez H. Flow cytometry for the assessment of animal sperm integrity and functionality: state of the art. *Asian J Androl* 2011;13:406–419.
27. Graham J, Kunze E, Hammerstedt R. Analysis of sperm cell viability, acrosomal integrity, and mitochondrial function using flow cytometry. *Biol Reprod* 1990;43:55–64.
28. Partyka A, Nizański W, Łukaszewicz E. Evaluation of fresh and frozen-thawed fowl semen by flow cytometry. *Theriogenology* 2010;74:1019–1027.
29. Oldenhof H, Blässe A, Wolkers W, Bollwein H, Sieme H. Osmotic properties of stallion sperm subpopulations determined by simultaneous assessment of cell volume and viability. *Theriogenology* 2011;76:386–391.
30. Garner D, Johnson L, Yue S, Roth B, Haugland R. Dual DNA staining assessment of bovine sperm viability using SYBR-14 and propidium iodide. *J Androl* 1994;15:620–662.
31. Garner D, Johnson L. Viability assessment of mammalian sperm using SYBR-14 and propidium iodide. *Biol Reprod* 1995;153:276–284.
32. Marchiani S, Tamburrino L, Giuliano L et al. Sumolylation of human spermatozoa and its relationship with semen quality. *Int J Androl* 2011;34:581–593.
33. Maxwell W, Welch G, Johnson L. Viability and membrane integrity of spermatozoa after dilution and flow cytometric sorting in the presence or absence of seminal plasma. *Reprod Fertil Dev* 1996;8:1165–1178.
34. Nagy S, Jansen J, Topper E, Gadella B. A triple-stain flow cytometric method to assess plasma- and acrosome-membrane integrity of cryopreserved bovine sperm immediately after thawing in presence of egg-yolk particles. *Biol Reprod* 2003;68:1828–1835.

35. Purvis K, Rui H, Schølberg A, Hesla S, Clausen O. Application of flow cytometry to studies on the human acrosome. *J Androl* 1990;11:361–366.
36. Lybaert P, Danguy A, Leleux F, Meuris S, Lebrun P. Improved methodology for the detection and quantification of the acrosome reaction in mouse spermatozoa. *Histol Histopathol* 2009;24:999–1007.
37. Carvalho J, Sartori R, Machado G, Mourão G, Dode M. Quality assessment of bovine cryopreserved sperm after sexing by flow cytometry and their use in in vitro embryo production. *Theriogenology* 2010;74:1521–1530.
38. Anzar M, Kroetsch T, Boswall L. Cryopreservation of bull semen shipped overnight and its effect on post-thaw sperm motility, plasma membrane integrity, mitochondrial membrane potential and normal acrosomes. *Anim Reprod Sci* 2011;126:23–31.
39. Yi Y, Zimmerman S, Manandhar G et al. Ubiquitin-activating enzyme (UBA1) is required for sperm capacitation, acrosomal exocytosis and sperm–egg coat penetration during porcine fertilization. *Int J Androl* 2012;35:196–210.
40. Carver-Ward J, Moran-Verbeek I, Hollanders J. Comparative flow cytometric analysis of the human sperm acrosome reaction using CD46 antibody and lectins. *J Assist Reprod Genet* 1997;14:111–119.
41. Evenson D, Darzynkiewicz Z, Melamed M. Simultaneous measurement by flow cytometry of sperm cell viability and mitochondrial membrane potential related to cell motility. *J Histochem Cytochem* 1982;30:279–280.
42. Garner D, Thomas C, Joerg H, DeJarnette J, Marshall C. Fluorometric assessments of mitochondrial function and viability in cryopreserved bovine spermatozoa. *Biol Reprod* 1997;57:1401–1406.
43. Hallap T, Nagy S, Jaakma U, Johannisson A, Rodriguez-Martinez H. Mitochondrial activity of frozen-thawed spermatozoa assessed by Mitotracker Deep Red 633. *Theriogenology* 2005;63:2311–2322.
44. Sousa A, Amaral A, Baptista M et al. Not all sperm are equal: functional mitochondria characterize a subpopulation of human sperm with better fertilization potential. *PLoS ONE* 2011;6:e18112.
45. Thomas C, Garner D, DeJarnette J, Marshall C. Effect of cryopreservation of bovine sperm organelle function and viability as determined by flow cytometry. *Biol Reprod* 1998;58:786–793.
46. Garner D, Thomas C. Organelle-specific probe JC-1 identifies membrane potential differences in the mitochondrial function of bovine sperm. *Mol Reprod Dev* 1999;5:222–229.
47. Guthrie H, Welch G. Determination of high mitochondrial membrane potential in spermatozoa loaded with the mitochondrial probe 5,5′,6,6′-tetrachloro-1,1′,3,3′-tetraethylbenzimidazolyl-carbocyanine iodide (JC-1) by using fluorescence-activated flow cytometry. *Methods Mol Biol* 2008;477:89–97.
48. Evenson D, Jost L. Sperm chromatin structure assay is useful for fertility assessment. *Methods Cell Sci* 2000;22:169–189.
49. Januskauskas A, Johannisson A, Rodriguez-Martinez H. Assessment of sperm quality through fluorometry and sperm chromatin structure assay in relation to field fertility of frozen-thawed semen from Swedish AI bulls. *Theriogenology* 2001;55:947–961.
50. Januskauskas A, Johannisson A, Rodriguez-Martinez H. Subtle membrane changes in cryopreserved bull semen in relation with sperm viability, chromatin structure, and field fertility. *Theriogenology* 2003;60:743–758.
51. Karabinus D, Evenson D, Jost L, Baer R, Kaproth M. Comparison of semen quality in young and mature Holstein bulls measured by light microscopy and flow cytometry. *J Dairy Sci* 1990;73:2364–2371.
52. Love C, Kenney R. The relationship of increased susceptibility of sperm DNA to denaturation and fertility in the stallion. *Theriogenology* 1998;50:955–972.
53. Evenson D, Jost L, Marshall D et al. Utility of the sperm chromatin structure assay as a diagnostic and prognostic tool in the human fertility clinic. *Hum Reprod* 1999;14:1039–1049.
54. Anzar M, He L, Buhr M, Kroetsch T, Pauls K. Sperm apoptosis in fresh and cryopreserved bull semen detected by flow cytometry and its relationship with fertility. *Biol Reprod* 2002;66:354–360.
55. Sutovsky P, Neuber E, Schatten G. Ubiquitin-dependent sperm quality control mechanism recognizes spermatozoa with DNA defects as revealed by dual ubiquitin-TUNEL assay. *Mol Reprod Dev* 2002;61:406–413.
56. Waterhouse K, Haugan T, Kommisrud E et al. Sperm DNA damage is related to field fertility of semen from young Norwegian Red bulls. *Reprod Fertil Dev* 2006;18:781–788.
57. Colás C, Grasa P, Casao A et al. Changes in calmodulin immunocytochemical localization associated with capacitation and acrosomal exocytosis of ram spermatozoa. *Theriogenology* 2009;71:789–800.
58. Guthrie H, Welch G, Theisen D, Woods L. Effects of hypothermic storage on intracellular calcium, reactive oxygen species formation, mitochondrial function, motility, and plasma membrane integrity in striped bass (Morone saxatilis) sperm. *Theriogenology* 2011;75:951–961.
59. Kumaresan A, Siqueira A, Hossain M, Bergqvist A. Cryopreservation-induced alterations in protein tyrosine phosphorylation of spermatozoa from different portions of the boar ejaculate. *Cryobiology* 2011;63:137–144.
60. Peña A, López-Lugilde L, Barrio M, Becerra J, Quintela L, Herradón P. Studies on the intracellular Ca^{2+} concentration of thawed dog spermatozoa: influence of Equex from different sources, two thawing diluents and post-thaw incubation in capacitating conditions. *Reprod Domest Anim* 2003;38:27–35.
61. Saacke R, Dalton J, Nadir R, Nebel R, Bame J. Relationship of seminal traits and insemination time to fertilization rate and embryo quality. *Anim Reprod Sci* 2000;60–61:663–677.
62. Evenson D, Darzynkiewicz Z, Melamed M. Relation of mammalian sperm chromatin heterogeneity to fertility. *Science* 1980;210:1131–1133.

Chapter 75

Superovulation in Cattle

Reuben J. Mapletoft[1] and Gabriel A. Bó[2]

*[1]Department of Large Animal Clinical Sciences, Western College of Veterinary Medicine,
University of Saskatchewan, Saskatoon, Saskatchewan, Canada
[2]Instituto de Reproducción Animal Córdoba (IRAC), Cno. General Paz – Paraje Pozo del Tigre- Estación
General Paz, CP 5145, Córdoba, Argentina*

Introduction

The objective of superovulation in cattle is to maximize the number of fertilized and transferable embryos with a high probability of producing pregnancies.[1] However, wide ranges in superovulatory responses and embryo yield have been reported in reviews of commercial embryo transfer records. In 2048 beef donor collections, a mean of 11.5 ova/embryos with 6.2 transferable embryos was recorded.[2] Variability was great, both in the superovulatory response and embryo quality; 24% of the collections did not yield viable embryos, 64% produced fewer than average numbers of transferable embryos, and 30% yielded 70% of the embryos. Embryo recovery from 987 dairy cows yielded slightly fewer ova/embryos but there was similar variability.[3] The high degree of unpredictability in superovulatory response creates problems that affect both the efficiency and profitability of embryo transfer programs.[4]

Variability in ovarian response has been related to differences in superstimulatory treatments, such as gonadotropin preparation, batch and total dose of gonadotropins, duration and timing of treatments, and the use of additional hormones. Additional factors, which may be more important, are inherent to the animal and its environment. These may include nutritional status, reproductive history, age, season, breed, effects of repeated superstimulation, and ovarian status at the time of treatment. While considerable progress has been made in the understanding of bovine reproductive physiology, factors inherent to the donor animal are only partially understood.

Gonadotropins and superovulation

Factors affecting superovulatory response associated with the administration of exogenous gonadotropins include source, batch, and biological activity.[5] Three different types of gonadotropin preparations have been used to induce superovulation in the cow: gonadotropins from extracts of porcine or other domestic animal pituitaries, equine chorionic gonadotropin (eCG), and human menopausal gonadotropin.[5,6] Human menopausal gonadotropin did not offer any advantages in cattle[7] and has not been used commercially; therefore it will not be discussed further.

Pituitary extracts contain follicle-stimulating hormone (FSH). The biological half-life of FSH in the cow has been estimated to be 5 hours or less,[8] so it must be injected twice daily to successfully induce superovulation.[9,10] The usual treatment regimen is twice-daily intramuscular treatments with FSH for 4 or 5 days, with a total dose of 28–50 mg (Armour) of a crude pituitary extract, or 260–400 mg NIH-FSH-PI of a partially purified pituitary extract (Folltropin®-V, Bioniche Animal Health Inc., Belleville, Ontario, Canada). At 48 or 72 hours after initiation of treatment, prostaglandin (PG)$F_2\alpha$ is administered to induce luteolysis. Estrus occurs in 36–48 hours, with ovulations beginning 24–36 hours later.[11,12]

Equine chorionic gonadotropin is a complex glycoprotein with both FSH and luteinizing hormone (LH) activity.[13] It has been shown to have a half-life of 40 hours in the cow and persists for up to 10 days in the bovine circulation; thus it is normally injected intramuscularly once followed by a $PGF_2\alpha$ injection 48 hours later.[14] The long half-life of eCG causes continued ovarian stimulation, unovulated follicles, abnormal endocrine profiles, and reduced embryo quality.[15–18] These problems have been largely overcome by the intravenous injection of antibodies to eCG at the time of the first insemination, 12–18 hours after the onset of estrus.[14,19] Recommended doses of eCG range from 1500 to 3000 IU, with 2500 IU by intramuscular injection commonly chosen.

Monniaux et al.[10] treated groups of cows with either 2500 IU eCG or 50 mg (Armour) pituitary FSH and observed that ovulation rate and the percentage of cows with more than three transferable embryos was slightly higher with FSH than eCG. Although these results were in agreement with those of Elsden et al.,[20] others have found no differences between pituitary extracts containing FSH and eCG.[21,22] Endocrine studies have revealed that eCG-treated animals more frequently had abnormal profiles of LH and progesterone,[15,23] which were associated with reductions in both ovulation and fertilization rates[24] as compared with FSH-treated cows.

Although folliculogenesis in mammals requires both FSH and LH, there is considerable variability in FSH and LH content of crude gonadotropin preparations.[5] Radioreceptor assays and *in vitro* bioassays have revealed variability in both the FSH and LH activity of eCG, not only among pregnant mares but also within the same mare at different times during gestation.[13] The effects of the FSH/LH ratio of eCG on superovulatory responses have been examined and there was a positive correlation between the ratio of FSH/LH activity and superovulatory response. Lower ratios of FSH/LH activity reduced ovulatory response in immature rats and LH added to eCG reduced superovulatory response in cows.[5,13]

Purified pituitary extracts with low LH contamination have been reported to improve superovulatory response in cattle. Chupin *et al.*[25] superstimulated three groups of dairy cows with an equivalent amount of 450 μg pure porcine (p) FSH and varying amounts of LH and showed that the mean ovulation rate and the number of total and transferable embryos increased as the dose of LH decreased. It has been suggested that high levels of LH during superstimulation cause premature activation of the oocyte.[16] Several experiments with an LH-reduced pituitary extract[26] utilizing several different total doses, ranging from 100 to 900 mg of NIH-FSH-P1, revealed no evidence of detrimental effects of dose on embryo quality.[6,27] On the other hand, doubling the dose of crude pituitary extracts containing both FSH and LH resulted in significantly reduced fertilization rates and percentages of transferable embryos.[6] Collectively, data support the hypothesis that the detrimental effects of high doses of pituitary gonadotropins on ova/embryo quality is due to an excess of LH.

Although it is generally believed that some LH is required for successful superovulation, endogenous LH levels may be adequate. Looney *et al.*[28] reported that recombinantly produced bovine (b)FSH induced high superovulatory responses without the addition of exogenous LH. These data suggest that LH is not needed in superstimulatory preparations and that embryo quality may be superior with pure FSH. The very high fertilization rates and transferable embryo rates in the absence of exogenous LH suggest that administration of LH in superstimulation protocols, at any dose, may be detrimental to embryo quality. Further, these results and more recent results with recombinant bFSH[29] indicate that future progress lies in the use of recombinantly derived gonadotropins.

The effects of LH on superovulatory response has also been demonstrated in an experiment involving Brahman-cross (*Bos indicus*) heifers superstimulated with 400 mg NIH-FSH-P1 containing 100%, 16% or 2% LH.[30] Although the more purified preparations in this experiment caused higher superovulatory responses, there were obvious seasonal effects; responses with the most purified and intermediate preparations were superior to the least purified preparation during summer months, but only the most purified preparation was highly efficacious during winter months. These results would appear to contradict the findings of Page *et al.*[31] who reported that superovulatory response and embryo quality in Holstein heifers was not affected by LH levels in cool weather, but that during heat stress a more purified preparation yielded more corpora lutea and significantly more fertilized ova and transferable embryos. It becomes apparent that stress is the common factor; *Bos taurus* breeds likely find summer heat stressful, whereas *Bos indicus* breeds likely find winter temperatures stressful. In either case, the more purified extracts resulted in greater superovulatory responses during conditions of environmental stress.

Follicle wave dynamics and superstimulation

It has been reported that superstimulatory response is greater if treatment is initiated before selection of a dominant follicle. In an early study recombinant bFSH given to heifers on day 1 of the cycle (ovulation = day 0), before the time of selection of the dominant follicle of wave 1, resulted in more ovulations than that given on day 5, after the time of selection.[32] A subsequent study was done to determine if exogenous FSH given at the expected time of the endogenous wave-eliciting FSH surge had a positive effect on the superstimulatory response.[33] As the endogenous surge of FSH was expected to peak 1 day before wave emergence,[34] superstimulatory treatments were initiated on the day before, the day of, or 1 or 2 days after wave emergence. Significantly more follicles were recruited when treatments were initiated on the day of, or the day before, follicular wave emergence and more ovulations were detected when treatment was initiated on the day of wave emergence rather than the day before or 1 or 2 days after wave emergence. A subsequent study indicated that superovulation was induced with equal efficacy when treatments were initiated during the first or second follicular waves and that the superstimulatory response was optimized when treatment was initiated at the time of follicle wave emergence.[35]

Superstimulation: the traditional approach

In the very early days of bovine embryo transfer, treatment with eCG was made to coincide with natural regression of the corpus luteum (i.e., about day 16 of the cycle).[12] With the introduction of PGF$_{2\alpha}$ in the 1970s, it became possible to initiate gonadotropin treatments at other times during the estrous cycle and most practitioners began treating with FSH during mid-cycle (i.e., 8–12 days after estrus).[36,37] Although this was initially based on anecdotal evidence, and then some experimental data,[38] it is now known that this encompasses the time of emergence of the second follicular wave in cattle exhibiting two- or three-wave cycles.[39,40]

Many practitioners prefer decreasing FSH dose schedules and treating with PGF$_{2\alpha}$ on the third day of the treatment protocol, while others prefer to treat with PGF$_{2\alpha}$ on the fourth day, and many do not treat with FSH on the day after the administration of PGF$_{2\alpha}$. Although no differences have been found between 4- and 5-day treatment protocols, recent experiments have shown that ovulation rate can be improved in some donors if FSH treatments are prolonged to 6 or 7 days.[41,42] Regardless, most superstimulation protocols have been successful in inducing superovulation under most circumstances.[12] Still others incorporate a progestin insert into the protocol which ensures that donors do not come into estrus early, especially if it is not possible to confirm the presence of a corpus luteum before initiating FSH treatments. In all cases inseminations are normally done 12 and 24 hours after the onset of estrus.[11,37]

Although the initiation of superstimulatory treatments during mid-cycle has served the embryo transfer industry over the years, conventional treatment protocols have two drawbacks: (i) the requirement to have trained personnel dedicated to the detection of estrus, both before and after initiating treatments; and (ii) the necessity to have all donors in estrus at the same time in order to begin the superstimulatory treatments at the most appropriate time in groups of cows (i.e., mid-cycle). To obviate these problems, protocols that facilitate superstimulation subsequent to elective induction of follicular wave emergence have been developed.

Synchronization of follicle wave emergence for superovulation

Estradiol and progesterone

The ability to electively induce follicular wave emergence permits initiation of superstimulation without regard to the stage of the estrous cycle and eliminates the need for estrus detection or waiting 8–12 days to initiate gonadotropin treatments.[37] In the 1990s the use of progestins and estradiol to induce synchronous emergence of a new follicular wave was reported[36] and its use in superstimulation protocols has been reviewed extensively.[37,43] The estradiol treatment causes suppression of FSH release and follicle atresia. Once the estradiol has been metabolized, FSH surges and a new follicular wave emerges on average 4 days after treatment.[36,44]

The most common protocol involves the administration of 5 mg estradiol-17β or 2.5 mg estradiol benzoate, plus 100 or 50 mg progesterone by intramuscular injections at the time of insertion of an intravaginal progestin device (day 0).[37,43] Twice-daily intramuscular FSH treatments are then initiated on day 4. On day 6, $PGF_{2\alpha}$ is injected in the morning and evening and the progestin device is removed in the evening. Estrus normally occurs on day 8 (approximately 48 hours after the first $PGF_{2\alpha}$ injection) and inseminations are done 12 and 24 hours later.

Unfortunately, estradiol cannot be used in many countries around the world because of concerns about the effects of estrogenic substances in the food chain.[45] This restriction leaves many embryo transfer practitioners with a serious dilemma and created the need to develop treatments that do not involve the use of estradiol.

Follicle ablation

An alternative to the use of estradiol in superstimulation protocols is to eliminate the suppressive effect of the dominant follicle by ultrasound-guided follicle aspiration of all follicles of 5 mm or more.[46] Follicle wave emergence occurs very consistently 24–36 hours later and superstimulatory treatments are initiated at that time. This approach to the synchronization of follicle wave emergence for superstimulation is very efficacious and results in superovulatory responses that do not differ from the use of estradiol.[46] In addition it has been found that it is necessary to ablate only the two largest follicles to effectively synchronize follicle wave emergence.[47] The protocol involves the ablation of the two largest follicles at random stages of the estrous cycle and the insertion of a progestin device; FSH treatments are initiated 24–48 hours later and the remainder of the protocol is as described above. The disadvantage of ultrasound-guided follicle aspiration is that it requires ultrasound equipment and trained personnel, making it appropriate only when donors are held in an embryo production facility; it is very difficult to apply in the field.

Gonadotropin-releasing hormone

Another alternative for the synchronization of follicle wave emergence is to induce ovulation of the dominant follicle by treatment with gonadotropin-releasing hormone (GnRH),[48,49] which is followed by wave emergence 1–2 days later.[50] However, emergence of the new follicular wave is synchronized only when treatment causes ovulation, and when administered at random stages of the estrous cycle GnRH results in ovulation in less than 60% of animals.[50,51] Not surprisingly, treatment with GnRH prior to initiating superstimulatory treatments at random stages of the estrous cycle resulted in lower superovulatory responses than treatments initiated after follicular ablation or estradiol treatment.[52]

More recently, retrospective analysis of commercial embryo transfer data has revealed no differences in the numbers of transferable embryos between donors superstimulated 4 days after treatment with estradiol and those superstimulated 2 days after treatment with GnRH (Randall Hinshaw, personal communication).[53,54] It is noteworthy that in each of these reports GnRH was administered 2 days after insertion of a progestin device. The improved responses may have been a consequence of progestin-induced development of a persistent dominant follicle that was more responsive to treatment with GnRH.[55] Obviously, controlled studies with the use of GnRH must be conducted to validate these promising results.

Improving the ovulatory response to GnRH

Most fixed-time artificial insemination (AI) protocols utilizing GnRH to synchronize follicle wave emergence employ a form of presynchronization to improve the ovulatory response to the first injection of GnRH.[55,56] Another alternative is to synchronize ovulation and then initiate FSH treatments at the time of emergence of the first follicular wave.[33] To avoid the need to detect estrus and ovulation in Nelore (*Bos indicus*) donors, Nasser et al.[57] induced synchronous ovulation with an estradiol-based protocol designed for fixed-time AI. Gonadotropin treatments were then initiated at the expected time of ovulation (and emergence of the first follicular wave). Superovulatory response did not differ from a contemporary group superstimulated 4 days after treatment with estradiol. However, the number of transferable embryos was reduced in cows superstimulated during the first follicular wave unless accompanied by the use of a progestin device. Similar results were obtained by Rivera et al.[58] In this study Holstein cows superstimulated during the first

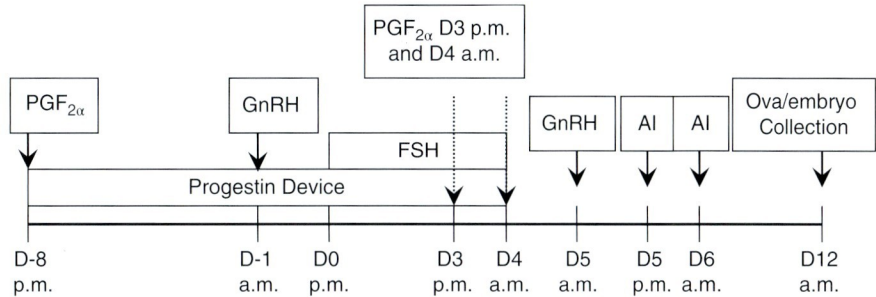

Figure 75.1 Treatment schedule for superovulation of donor cows during the first follicular wave after GnRH-induced ovulation. Donors receive a progestin device along with PGF$_{2\alpha}$ followed by GnRH 7 days later. On day (D)0 (36 hours after GnRH), superstimulation with FSH is initiated (twice-daily decreasing doses over 4 days). PGF$_{2\alpha}$ is administered with the last two FSH injections and the progestin device is removed with the last FSH injection. Ovulation is induced with GnRH 24 hours after progestin removal, donors are fixed-time inseminated (AI) 12 and 24 hours later, and ova/embryos are collected 7 days later. From Bó G, Carballo Guerrero D, Tríbulo A et al. New approaches to superovulation in the cow. *Reprod Fertil Dev* 2010;22:106–112.

follicular wave also produced a greater number of viable embryos following the addition of a progestin device.

Bó et al.[59] recently reported on a series of experiments with the overall objective of developing a protocol for superstimulation following ovulation induced synchronously by the administration of GnRH. This approach was based on a previous study in which ovulatory response was increased by causing a persistent follicle to develop with the administration of PGF$_{2\alpha}$ and the insertion of a progestin device 7–10 days before the administration of GnRH.[55] In that study, ovulation and follicle wave emergence occurred 1–2 days after the administration of GnRH, indicating that this approach could be used in groups of randomly cycling donors.

The recommended superstimulation protocol is schematically presented in Figure 75.1. It consists of the administration of PGF$_{2\alpha}$ at the time of insertion of a progestin device. Seven days later (with the progestin device still in place) GnRH is administered to induce ovulation of the persistent follicle and follicle wave emergence; FSH treatments are initiated 36 hours after the administration of GnRH. Although this protocol was designed for 4 days of FSH treatments, a 5-day superstimulation protocol can be accomplished by simply delaying the removal of the progestin device by 1 day. Overall in this series of experiments, more than 95% of animals ovulated to the first GnRH administration and superovulatory response and ova/embryo numbers and quality were similar to that obtained when estradiol was used to synchronize follicular wave emergence.[59]

Fixed-time AI of donors

Barros and Nogueira[60] have developed a superstimulatory protocol for *Bos indicus* cattle that they refer to as the P-36 protocol. In this protocol the progestin device that is inserted prior to the initiation of superstimulatory treatments is left in place for 36 hours after PGF$_{2\alpha}$ administration and ovulation is induced by the administration of pLH 12 hours after withdrawal of the progestin device (i.e., 48 hours after PGF$_{2\alpha}$ administration). Since ovulation occurs between 24 and 36 hours after pLH administration,[61] fixed-time AI is scheduled 12 and 24 hours later, eliminating the need for estrus detection.

In a series of experiments in which the timings of ovulations were monitored ultrasonically, Bó et al.[11] developed a protocol for fixed-time AI in *Bos taurus* donors without the need for estrus detection and without compromising results. Basically, the time of progestin device removal was delayed to prevent early ovulations and allow late-developing follicles to "catch up," followed by induction of ovulation with GnRH or pLH. In this protocol follicular wave emergence was synchronized with estradiol and a progestin device on day 0 and FSH treatments were initiated on day 4. On day 6, PGF$_{2\alpha}$ was administered in the morning and evening and the progestin device was removed on the morning of day 7 (24 hours after the first administration of PGF$_{2\alpha}$). On the morning of day 8 (24 hours after the removal of the progestin device), GnRH or pLH was administered and fixed-time AI was done 12 and 24 hours later. Delaying the removal of the progestin device to the morning of day 7 resulted in a higher number of ova/embryos and fertilized ova than removal on the morning of day 6 (P. Chesta, MSc thesis, National University of Cordoba, Argentina, 2010). From a practical perspective, fixed-time AI of donors has been shown to be useful in eliminating the need for estrus detection for busy embryo transfer practitioners.[62]

Studies in high-producing Holstein cows (*Bos taurus*) in Brazil have indicated that it is preferable to allow an additional 12 hours before removing the progestin device (i.e., evening day 7; P36) followed by GnRH or pLH 24 hours later (i.e., evening day 8).[63] In *Bos indicus* breeds, Baruselli et al.[64] confimed that it was preferable to remove the progestin device on the evening of day 7 (P36), followed by GnRH 12 hours later (i.e., morning day 8). Although donors are typically inseminated twice, 12 and 24 hours after administration of pLH or GnRH,[11] it is possible to use a single insemination with high-quality semen 16 hours after pLH.[64]

Reducing the need for multiple treatments with FSH

Because the half-life of pituitary FSH is short in the cow,[8] traditional superstimulatory treatment protocols consist of twice daily intramuscular injections over 4 or 5 days.[9] This requires frequent attention by farm personnel and increases

Table 75.1 Mean (±SEM) ova/embryo production in beef donors treated with Folltropin-V given by twice-daily intramuscular injections over 4 days (control) or diluted in 2% hyaluronan and given by a single intramuscular injection.

Treatment	N	Total ova/embryos	Fertilized ova	Transferable embryos	Cows with "0" transferable embryos (%)
Control (twice daily)	146	12.1 ± 0.6	8.8 ± 0.5	6.7 ± 0.4	12 (8.2%)
Single injection	146	11.2 ± 0.8	8.3 ± 0.6	6.4 ± 0.5	19 (13.0%)

Means and percentages did not differ ($P > 0.1$).
Source: Bó G, Carballo Guerrero D, Tríbulo A et al. New approaches to superovulation in the cow. Reprod Fertil Dev 2010;22:106–112.

the possibility of failures due to noncompliance. In addition, twice-daily treatments may cause undue stress in donors with a subsequent decreased superovulatory response and/or altered preovulatory LH surge.[65,66] Simplified protocols may be expected to reduce donor handling and improve response, particularly in less tractable animals.

A single subcutaneous administration of FSH has been shown to induce a superovulatory response equivalent to the traditional twice-daily treatment protocols in beef cows in high body condition (i.e., body condition score >3 on scale of 5)[67] but results were not repeatable in Holstein cows which had less adipose tissue. However, superovulatory responses were improved in Holstein cows when the single injection was split into two; 75% of the FSH dose was administered subcutaneously on the first day of treatment and the remaining 25% was administered 48 hours later when $PGF_{2\alpha}$ is normally administered.[68]

An alternative for inducing a consistent superovulatory response with a single injection of FSH is to combine the pituitary extract with agents that cause the hormone to be released slowly over several days. These agents are commonly referred to as polymers that are biodegradable and nonreactive in the tissues, facilitating use in animals.[69] In a series of experiments in which FSH diluted in a 2% hyaluronan solution was administered as a single intramuscular injection (to avoid the effects of body condition), a similar number of ova/embryos was produced as in the traditional twice-daily FSH protocol[70] (Table 75.1). However, 2% hyaluronan was viscous and difficult to mix with FSH, especially in the field. Although more dilute preparations of hyaluronan were less efficacious as a single administration, their use was improved by splitting them into two injections 48 hours apart as was done with subcutaneous injections of FSH.

The split intramuscular treatment protocol consists of diluting the FSH lyophilized powder with 10 mL of a 1% or 0.5% hyaluronan solution and the administration of two-thirds of the total dosage of FSH on the first day, followed by administration of the remaining one-third 48 hours later, when $PGF_{2\alpha}$ is normally administered.[71] When compared with the twice-daily treatment protocol (controls) in 29 beef cows superstimulated three times in a cross-over design, the numbers of transferable embryos with the split-injection protocol (1.0% hyaluronan, 5.0 ± 0.9; 0.5% hyaluronan, 6.1 ± 1.3) did not differ from controls (4.0 ± 0.8). Data were interpreted to suggest that splitting the FSH dose in either concentration of hyaluronan into two intramuscular injections 48 hours apart would result in a comparable superovulatory response to the traditional twice-daily intramuscular injection protocol in beef cattle. Furthermore, the less concentrated solutions of hyaluronan were not difficult to mix with FSH, even under field conditions. A recent report derived from commercial embryo transfer data has now confirmed these results in beef cattle in North America.[72]

Concluding remarks

Although the development of protocols that control follicular wave emergence and ovulation has not eliminated the variability in superovulatory response, these treatments have had a positive impact on the application of commercial on-farm embryo transfer by permitting the initiation of treatments at a self-appointed time. These protocols offer the convenience of being able to initiate superstimulatory treatments quickly without the necessity of estrus detection. In addition these protocols have facilitated fixed-time AI of donors, thereby avoiding the necessity of estrus detection during the superstimulatory protocol. However, estradiol, which has proven to be most useful in the field, has been withdrawn from many veterinary markets around the world, leaving only follicle ablation, which is difficult to utilize in the field, as a reliable method to synchronize follicular wave emergence for superstimulation. Although the synchrony of follicular wave emergence following the administration of GnRH or pLH at random stages of the estrous cycle has been considered to be too variable for superstimulation, several recent reports suggest that this approach should be revisited. An alternative may be to initiate FSH treatments at the time of emergence of the first follicular wave following GnRH-induced ovulation of a progestin-induced persistent follicle. Finally, a single or split-single intramuscular injection of FSH in hyaluronan can be used to induce a satisfactory superovulatory response in beef donors, simplifying the entire protocol and reducing animal handling and the stress associated with the administration of superstimulatory treatments.

References

1. Armstrong D. Recent advances in superovulation of cattle. Theriogenology 1993;39:7–24.
2. Looney C. Superovulation in beef females. In: Proceedings of the 5th Annual Meeting of the American Embryo Transfer Association, Fort Worth, Texas, 1986, pp. 16–29.
3. Lerner S, Thayne W, Baker R et al. Age, dose of FSH and other factors affecting superovulation in Holstein cows. J Anim Sci 1986;63:176–183.
4. Hasler J, McCauley A, Schermerhorn E, Foote R. Superovulatory responses of Holstein cows. Theriogenology 1983;19:83–99.
5. Murphy B, Mapletoft R, Manns J, Humphrey W. Variability in gonadotrophin preparations as a factor in the superovulatory response. Theriogenology 1984;21:117–125.

6. Alkemade S, Murphy B, Mapletoft R. Superovulation in the cow: effects of biological activity of gonadotropins. In: *Proceedings of the 12th Annual Meeting of the American Embryo Transfer Association, Portland, Maine*, 1993, pp. 1–19.
7. McGowan M, Braithwaite M, Jochle W, Mapletoft R. Superovulation of beef cattle with pergonal (HMG): a dose response trial. *Theriogenology* 1985;24:173–184.
8. Laster D. Disappearance of and uptake of I125 FSH in the rat, rabbit, ewe and cow. *J Reprod Fertil* 1972;30:407–415.
9. Looney C, Boutle B, Archibald L, Godke R. Comparison of once daily FSH and twice daily FSH injections for superovulation in beef cattle. *Theriogenology* 1981;15:13–22.
10. Monniaux D, Chupin D, Saumande J. Superovulatory responses of cattle. *Theriogenology* 1983;19:55–82.
11. Bó G, Baruselli P, Chesta P, Martins C. The timing of ovulation and insemination schedules in superstimulated cattle. *Theriogenology* 2006;65:89–101.
12. Mapletoft R, Bó G. The evolution of improved and simplified superovulation protocols in cattle. *Reprod Fertil Dev* 2012; 24: 278–283.
13. Murphy B, Martinuk S. Equine chorionic gonadotropin. *Endocrine Rev* 1991;12:27–44.
14. Dieleman S, Bevers M, Vos P, de Loos F. PMSG/anti-PMSG in cattle: a simple and efficient superovulatory treatment. *Theriogenology* 1993;39:25–42.
15. Mikel-Jenson A, Greve T, Madej A, Edqvist L. Endocrine profiles and embryo quality in the PMSG-PGF$_{2\alpha}$-treated cow. *Theriogenology* 1982;18:33–34.
16. Moor R, Kruip T, Green D. Intraovarian control of folliculogenesis: limits to superovulation? *Theriogenology* 1984;21:103–116.
17. Saumande J, Chupin D, Mariana J, Ortavant R, Mauleon P. Factors affecting the variability of ovulation rates after PMSG stimulation. In: Sreenan JM (ed.) *Control of Reproduction in the Cow*. The Hague: Martinus Nijhoff, 1978, pp. 195–224.
18. Schams D, Menzer D, Schalenberger E, Hoffman B, Hahn J, Hahn R. Some studies of the pregnant mare serum gonadotrophin (PMSG) and on endocrine responses after application for superovulation in cattle. In: Sreenan JM (ed.) *Control of Reproduction in the Cow*. The Hague: Martinus Nijhoff, 1978, pp. 122–142.
19. Gonzalez A, Wang H, Carruthers T, Murphy B, Mapletoft R. Increased ovulation rates in PMSG-stimulated beef heifers treated with a monoclonal PMSG antibody. *Theriogenology* 1994;41:1631–1642.
20. Elsden R, Nelson L, Seidel G. Superovulation of cows with follicle stimulating hormone and pregnant mare's serum gonadotrophin. *Theriogenology* 1978;9:17–26.
21. Goulding D, Williams D, Roche J, Boland M. Factors affecting superovulation in heifers treated with PMSG. *Theriogenology* 1996;45:765–773.
22. Mapletoft R, Pawlyshyn V, Garcia A *et al.* Comparison of four different gonadotropin treatments for inducing superovulation in cows with 1:29 translocation. *Theriogenology* 1990;33:282 (Abstract).
23. Greve T, Callesen H, Hyttel P. Endocrine profiles and egg quality in the superovulated cow. *Nord Vet Med* 1983; 35: 408–421.
24. Callesen H, Greve T, Hyttel P. Preovulatory endocrinology and oocyte maturation in superovulated cattle. *Theriogenology* 1986; 25:71–86.
25. Chupin D, Combarnous Y, Procureur R. Antagonistic effect of LH in commercially available gonadotrophins. *Theriogenology* 1984;25:167 (Abstract).
26. Armstrong D, Opavsky M. Biological characteristics of a pituitary FSH preparation with reduced LH activity. *Theriogenology* 1986;25:135 (Abstract).
27. Gonzalez A, Lussier J, Carruthers T, Murphy B, Mapletoft R. Superovulation of beef heifers with Folltropin. A new FSH preparation containing reduced LH activity. *Theriogenology* 1990;33:519–529.
28. Looney C, Bondioli K, Hill K, Massey J. Superovulation of donor cows with bovine follicle-stimulating hormone (bFSH) produced by recombinant DNA technology. *Theriogenology* 1988;29:271 (Abstract).
29. Rogan D, Strauss C, Mapletoft R *et al.* Opportunities for the production of recombinant gonadotropins for assisted reproduction and embryo transfer. In: *Proceedings of the Joint Annual Meeting of the Canadian and American Embryo Transfer Associations, Montreal, Quebec*, 2009, pp. 59–67.
30. Tribulo H, Bó G, Jofre F, Carcedo J, Alonso A, Mapletoft R. The effect of LH concentration in a porcine pituitary extract and season on superovulatory response of *Bos indicus* heifers. *Theriogenology* 1991;35:286 (Abstract).
31. Page R, Jordan J, Johnson S. Superovulation of Holstein heifers under heat stress with FSH-P or Folltropin. *Theriogenology* 1985;31:236 (Abstract).
32. Adams G. Control of ovarian follicular wave dynamics in cattle: implications for synchronization and superstimulation. *Theriogenology* 1994;41:19–24.
33. Nasser L, Adams G, Bó G, Mapletoft R. Ovarian superstimulatory response relative to follicular wave emergence in heifers. *Theriogenology* 1993;40:713–724.
34. Adams G, Matteri R, Kastelic J, Ko J, Ginther O. Association between surges of follicle stimulating hormone and the emergence of follicular waves in heifers. *J Reprod Fertil* 1992;94:177–188.
35. Adams G, Nasser L, Bó G, Garcia A, Del Campo M, Mapletoft R. Superovulatory response of ovarian follicles of Wave 1 versus Wave 2 in heifers. *Theriogenology* 1994;42:1103–1113.
36. Bó G, Adams G, Pierson R, Mapletoft R. Exogenous control of follicular wave emergence in cattle. *Theriogenology* 1995;43: 31–40.
37. Mapletoft R, Bennett-Steward K, Adams G. Recent advances in the superovulation of cattle. *Reprod Nutr Dev* 2002;42:1–11.
38. Lindsell C, Murphy B, Mapletoft R. Superovulatory and endocrine responses in heifers treated with FSH at different stages of the estrous cycle. *Theriogenology* 1986;26:209–219.
39. Pierson R, Ginther O. Follicular populations during the estrous cycle in heifers I. Influence of day. *Anim Reprod Sci* 1987;14: 165–176.
40. Ginther O, Kastelic J, Knopf L. Temporal associations among ovarian events in cattle during estrous cycles with two and three follicular wave. *J Reprod Fertil* 1989;87:223–230.
41. Bó G, Carballo Guerrero D, Adams G. Alternative approaches to setting up donor cows for superstimulation. *Theriogenology* 2008;69:81–87.
42. García Guerra A, Tribulo A, Yapura J, Singh J, Mapletoft R. Lengthening the superstimulatory treatment protocol increases ovarian response and number of transferable embryos in beef cows. *Theriogenology* 2012;78:353–360.
43. Bó G, Baruselli P, Moreno D *et al.* The control of follicular wave development for self-appointed embryo transfer programs in cattle. *Theriogenology* 2002;57:53–72.
44. Mapletoft R, Bó G, Baruselli P. Control of ovarian function for assisted reproductive technologies in cattle. *Anim Reprod* 2009;6:114–124.
45. Lane E, Austin E, Crowe M. Estrus synchronisation in cattle. Current options following the EU regulations restricting use of estrogenic compounds in food-producing animals: a review. *Anim Reprod Sci* 2008;109:1–16.
46. Bergfelt D, Bó G, Mapletoft R, Adams G. Superovulatory response following ablation-induced follicular wave emergence at random stages of the oestrous cycle in cattle. *Anim Reprod Sci* 1997;49:1–12.
47. Baracaldo M, Martinez M, Adams G, Mapletoft R. Superovulatory response following transvaginal follicle ablation in cattle. *Theriogenology* 2000;53:1239–1250.

48. Macmillan K, Thatcher W. Effect of an agonist of gonadotropin-releasing hormone on ovarian follicles in cattle. *Biol Reprod* 1991;45:883–889.
49. Pursley J, Mee M, Wiltbank M. Synchronization of ovulation in dairy cows using $PGF_{2\alpha}$ and GnRH. *Theriogenology* 1995;44:915–923.
50. Martinez M, Adams G, Bergfelt D, Kastelic J, Mapletoft R. Effect of LH or GnRH on the dominant follicle of the first follicular wave in heifers. *Anim Reprod Sci* 1999;57:23–33.
51. Colazo M, Gordon M, Rajamahendran R, Mapletoft R, Ambrose D. Pregnancy rates to timed artificial insemination in dairy cows treated with gonadotropin-releasing hormone or porcine luteinizing hormone. *Theriogenology* 2009;72:262–270.
52. Deyo C, Colazo M, Martinez M, Mapletoft R. The use of GnRH or LH to synchronize follicular wave emergence for superstimulation in cattle. *Theriogenology* 2001;55:513 (Abstract).
53. Wock J, Lyle L, Hockett M. Effect of gonadotropin-releasing hormone compared with estradiol-17β at the beginning of a superstimulation protocol on superovulatory response and embryo quality. *Reprod Fertil Dev* 2008;20:228 (Abstract).
54. Steel R, Hasler J. Comparison of three different protocols for superstimulation of dairy cattle. *Reprod Fertil Dev* 2009;21:246 (Abstract).
55. Small J, Colazo M, Kastelic J, Mapletoft R. Effects of progesterone presynchronization and eCG on pregnancy rates to GnRH-based, timed-AI in beef cattle. *Theriogenology* 2009;71:698–706.
56. Pursley J, Martins J. Impact of circulating concentrations of progesterone and antral age of the ovulatory follicle on fertility of high-producing lactating dairy cows. *Reprod Fertil Dev* 2012;24:267–271.
57. Nasser L, Sá Filho M, Reis E et al. Exogenous progesterone enhances ova and embryo quality following superstimulation of the first follicular wave in Nelore (*Bos indicus*) donors. *Theriogenology* 2011;76:320–327.
58. Rivera F, Mendonca L, Lopes G et al. Reduced progesterone concentration during growth of the first follicular wave affects embryo quality but has no effect on embryo survival post-transfer in lactating dairy cows. *Reproduction* 2011;141:333–342.
59. Bó G, Carballo Guerrero D, Tríbulo A et al. New approaches to superovulation in the cow. *Reprod Fertil Dev* 2010;22:106–112.
60. Barros C, Nogueira M. Superovulation in zebu cattle: Protocol P-36. *IETS Embryo Transfer Newsletter* 2005;23:5–9.
61. Nogueira M, Barros C. Timing of ovulation in Nelore cows superstimulated with P-36 protocol. *Acta Sci Vet* 2003;31:509 (Abstract).
62. Larkin S, Chesta P, Looney C, Bó G, Forrest D. Distribution of ovulation and subsequent embryo production using Lutropin and estradiol-17β for timed AI of superstimulated beef females. *Reprod Fertil Dev* 2006;18:289 (Abstract).
63. Martins C, Rodrigues C, Vieira L et al. The effect of timing of the induction of ovulation on embryo production in superstimulated lactating Holstein cows undergoing fixed-time artificial insemination. *Theriogenology* 2012;78:974–980.
64. Baruselli P, Sá Filho M, Martins C et al. Superovulation and embryo transfer in *Bos indicus* cattle. *Theriogenology* 2006;65:77–88.
65. Stoebel D, Moberg G. Repeated acute stress during the follicular phase and luteinizing hormone surge of dairy heifers. *J Dairy Sci* 1982;65:92–96.
66. Edwards L, Rahe C, Griffin J, Wolfe D, Marple D, Cummins K. Effect of transportation stress on ovarian function in superovulated Hereford heifers. *Theriogenology* 1987;28:291–299.
67. Bó G, Hockley D, Nasser L, Mapletoft R. Superovulatory response to a single subcutaneous injection of Folltropin-V in beef cattle. *Theriogenology* 1994;42:963–975.
68. Lovie M, Garcia A, Hackett A, Mapletoft R. The effect of dose schedule and route of administration on superovulatory response to Folltropin in Holstein cows. *Theriogenology* 1994;41:241 (Abstract).
69. Sutherland W. Alginate. In: Byrom D (ed.) *Biomaterials: Novel Material From Biological Sources*. New York: Stockton Press, 1991, pp. 307–333.
70. Tríbulo A, Rogan D, Tribulo H et al. Superstimulation of ovarian follicular development in beef cattle with a single intramuscular injection of Folltropin-V. *Anim Reprod Sci* 2011;129:7–13.
71. Tríbulo A, Rogan D, Tríbulo H, Tríbulo R, Mapletoft R, Bó G. Superovulation of beef cattle with a split-single intramuscular administration of Folltropin-V in two concentrations of hyaluronan. *Theriogenology* 2012;77:1679–1685.
72. Hasler J, Hockley D. Efficacy of hyaluronan as a diluent for a two-injection FSH superovulation protocol in *Bos taurus* beef cows. *Reprod Domest Anim* 2012;47:459 (Abstract).

Chapter 76

Embryo Collection and Transfer

Edwin G. Robertson

Harrogate Genetics International, Harrogate, Tennessee, USA

Introduction

Just as the Burger King slogan "Have It Your Way" applies to fast food, so too will the experienced embryo collection practitioner say it applies to the technique of collecting bovine embryos. Since Drs Robert Rowe, Peter Elsden, and Martin Drost reported the successful nonsurgical collection of bovine embryos in 1976, embryo transfer has become common practice in the cattle industry. After visiting Dr Rowe during September 1977 in Middleton, Wisconsin to observe his technique, I've spent 36 years collecting over 180 000 embryos from 30 000 cows and transferring over 80 000 embryos. During that time, my techniques have evolved. As I've taught beginners and visited other practitioners to observe their techniques, I quickly came to realize that "my way" was not "the only way." Finding a procedure that is comfortable in your hands will lead to confidence that will translate into you becoming a successful embryo transfer practitioner.

I have observed embryo transfer techniques worldwide and can honestly say that many techniques can be utilized to successfully collect and transfer embryos. However, I'm pretty much about "simple," "easy," and "quick," so through my career I have always looked for techniques that limit my exposure to error as well as those that save time. So, the techniques I will describe are "my way," the most efficient and effective techniques in my hands. I hope you will pick up a few helpful ideas that will increase your skills and productivity.

Previous chapters have discussed superovulation, site of fertilization, and transport of sperm up the oviduct and embryo/ova down the oviduct into the uterus. We will start our discussion assuming that the embryos have arrived into the uterus by day 4.5 or 5.0.

Timing

Since embryos are most versatile at day 6–7, most practitioners plan collections 6.5–7 days after the donor's standing heat is first observed. When several donors are to be collected on the same day, there will usually be at least 12–24 hours variation in the onset of standing heat among them. Since embryos may hatch by day 7.5–8.0 and become less useable, it is usually wise to collect the donors first that exhibit standing heat the earliest. Also realize that collection on or before day 6 may result in embryos being too immature to endure the freezing process very well.

Collection media

Many commercially prepared collection media are available today. I use a ready-to-go medium from Bioniche called Vigro Complete Flush™, which contains a surfactant (polyvinyl alcohol) and antibiotics (gentamicin and kanamycin). Some practitioners use lactated Ringers which they have available in their veterinary clinics and simply add 0.1% bovine serum albumin as a surfactant. For short-term holding, this appears to work well and is much more economic.

If embryos are collected for export, some countries will not allow animal byproducts such as bovine serum albumin to be added (due to the scare over bovine spongiform encephalopathy, or BSE). In these collections, use of a complete medium that uses polyvinyl alcohol as the surfactant is acceptable.

Donor preparation

In my early years, I carried a ramp with me to the farms and placed it in the chute, or I had the cattleman elevate the front of the chute 30–35 cm. Elevation shifts the viscera posteriorly and tends to lift the uterus upward for ease of accessing it for manipulation. I also learned a lesson the hard way about withholding feed from donors and recipients. Although it sounded like a good idea to reduce feed intake prior to my arrival in order to lessen feces, I quickly learned that one of an embryo transfer practitioner's best friends is a full rumen. The full rumen not only elevates the uterus upward and caudally but it also prevents negative abdominal pressure caused by an empty rumen. Negative abdominal pressure causes air to

---

Bovine Reproduction, First Edition. Edited by Richard M. Hopper.
© 2015 John Wiley & Sons, Inc. Published 2015 by John Wiley & Sons, Inc.

be sucked around your arm and into the rectum and colon during the flush, making manipulation of the uterus very difficult. Ramps and feed are two great friends.

Once the donor is in the chute or stanchion, a lidocaine (3–5 mL) epidural is administered using a 3.75 cm × 18-gauge needle. No preparation is needed for the epidural in my opinion. Little is accomplished by clipping or washing the hair unless one is willing to surgically prepare the tailhead. I never do any preparation for epidurals on donors or recipients. Immediately after the epidural, I palpate each ovary and record my estimate as to the number of corpora lutea that are on each. Unovulated follicles are noted as well as the size of ovaries if they are unusual. This information is helpful once the results are in whether good or bad.

Tranquilization is another friend in some opinions, an enemy in others. Most often, I give 10 mg (0.1 mL) Rompun (xylazine) subcutaneously or intramuscularly just before the epidural. Although it causes a slight increase in uterine tone, I've never considered that to be a problem. Even though I've collected many cows without xylazine, I like it because it takes the edge off excitable cows and it makes life easier for me because they stand still throughout the collection process.

I never use disinfectants (enemies) around the collection materials. I simply wipe the perineal region after the tail is tied off to the side. In the absence of water, I simply wipe the perineal area clean with a disposable towel. Once prepared, the cow is ready for collection.

Equipment (and supplier)

- Catheter: 52 cm silicone, 16 French with 5-mL cuff (Bioniche).
- Stylet: 60 cm stainless steel (Bioniche).
- Y tubing: 150 cm plastic tubing, one end with syringe tip and the other attaches to EZ flow filter (Pets-inc).
- Disposable three-way plastic valve (Pets-inc).
- 50 or 60 mL airtight syringe (Pets-inc).
- EZ Flow Embryo Filter (Pets-inc).
- KY jelly (nonspermicidal).
- Cervical dilator: Harrogate AI/ET gun and cervical dilator (Stone Mfg. Co.).
- Medium: Vigro Complete (Bioniche).
- 10-mL syringe to inflate cuff.

Sterilization

Most equipment is sterilized by the manufacturer using gamma radiation. Ethylene oxide was found to be embryotoxic years ago and is no longer used. If catheters and tubing are not autoclavable, simple washing with Alconox™ laboratory detergent and several rinses with distilled water is sufficient. Many practitioners discard all equipment after single use. I reuse as much as feasible depending on price.

Embryo collection

The choice of catheter is affected by the type of flush (horn or body), age of the cow, and size of the cow. For years I used a standard human Foley catheter 45 cm long and either 16 or 18 French made of latex to flush cows of all sizes. Today we have many choices available to add to our ease and comfort during the collection. I currently use a 52 cm silicone 16 or 18 French catheter with a 5-mL balloon for all flushes (Bioniche). The 5-mL balloon will inflate to 20 mL if needed and it gives a rounder, more defined shape than the more elongated effect the 30-mL balloon produces (Figure 76.1).

Catheter placement is the first critical step in nonsurgical embryo collection. I will discuss the placement of the catheter and collection procedures for both single horn and uterine body collection techniques (Figure 76.2). The following are important considerations. Is the donor a virgin heifer or cow? If virgin, how mature is her tract? Does the donor have a healthy reproductive tract or are there abnormalities? If abnormalities (e.g., adhesions, deformities, presence of infection, breed, age), what are they and how will they potentially affect the collection? What is the temperament of donor? What are the facilities available for the flush?

First, let's talk about a normal collection and later discuss special circumstances, abnormalities, and tricks of the trade. Experts train bank tellers to detect counterfeit money by learning to first identify what real money looks like, then they can spot any abnormality. That's the way I like to think of difficult collections. Once you learn to collect normal cows efficiently and with confidence, then when trouble arises you will, hopefully, not panic but will smartly navigate through the trouble and salvage a successful but difficult collection.

A metal stylet must be inserted into the catheter to make it rigid enough to be passed through the cervix. A small amount of KY jelly is applied to the tip of the stylet before entry into the catheter to aid in its withdrawal after passage. The stylet should be slightly longer than the catheter (60-cm stylet, 52-cm catheter) to allow the catheter to be stretched so the stylet will fit tightly inside it (Figure 76.3). This prevents the tip of the stylet from accidentally popping out of one of the fluid ports (holes) near the catheter tip during its passage through the cervix. Cervical damage and uterine hemorrhage are enemies. Passage of the catheter is similar to that in artificial insemination (AI) except the catheter is considerably larger. In addition, stretch must be kept on the catheter or finger pressure on the cervix may manipulate the stylet tip out of the lumen through the fluid port on the side of the catheter tip.

Eventually every practitioner runs into a crooked cervix which cannot be passed. An invaluable tool is a cervical dilator. I would never leave home without one even though I may use it only once or twice per year. Stone Mfg. Co. made one I designed and sells it as the Harrogate AI/ET gun and cervical dilator. (*Editor's note*: Product #10050, Stone Mfg, 1212 Kansas Ave, Kansas City, MO 64127. inquiries@stonemfg.net.) The dilator can be passed through the most difficult cervix. Then, as the dilator is withdrawn by an assistant, you can quickly pass the catheter through before the cervix has time to return to its abnormal shape. Dilators are especially useful in Brahman and Brahman crossbreeds which tend to get "S" curves and/or 90° bends in the cervix as they age (Figure 76.4).

Passing catheters through difficult or crooked cervices requires tenacity. Steady forward pressure must be kept on the stylet. Squeezing each cervical ring in front of the

Embryo Collection and Transfer

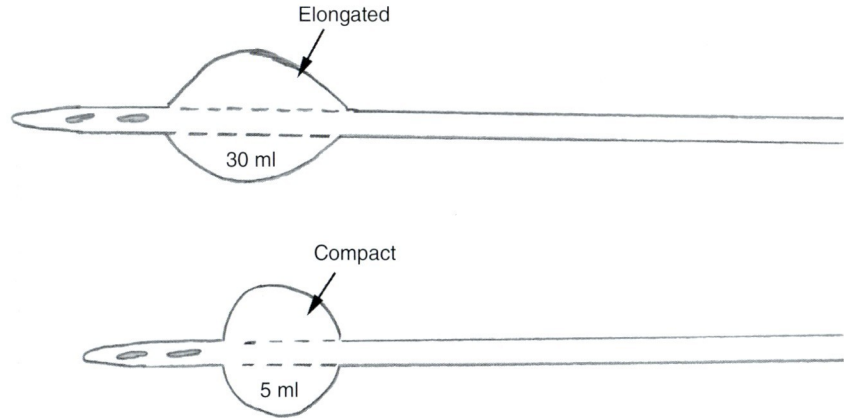

Figure 76.1 Shapes of Foley catheter balloon when inflated.

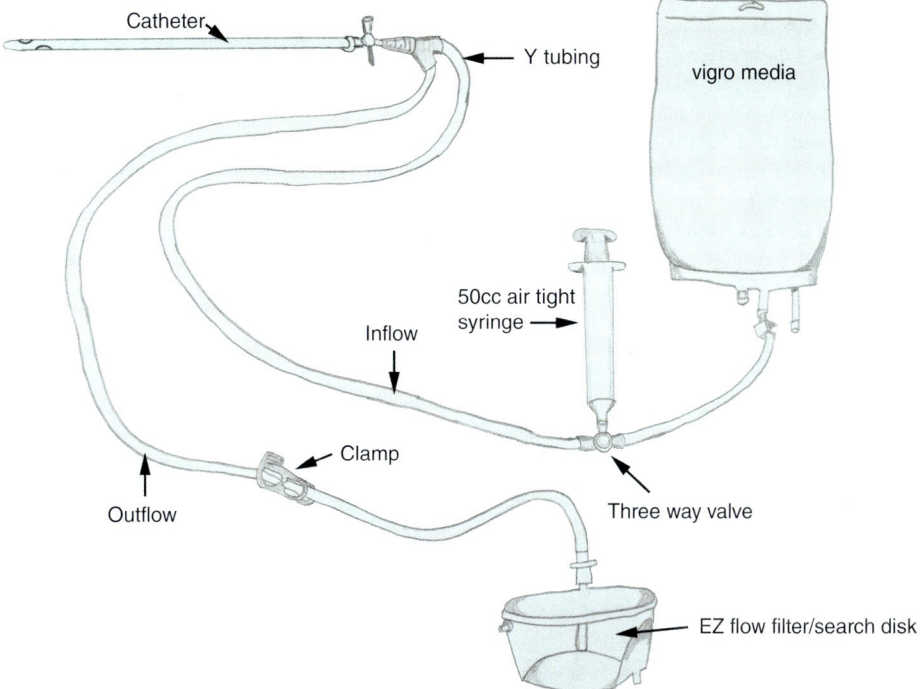

Figure 76.2 Schematic of flush equipment set-up.

Figure 76.3 Stylet protruding through catheter.

catheter often helps by "flattening" the ring, thereby making a "wide slit" out of the ring whose opening may be offset to one side or the other. The four cervical rings are usually aligned, with the opening of each ring being straight in front of the preceding one, but it is not uncommon to have one ring's opening in the center, the next at 9 o'clock, the next at 3 o'clock, and the last at 12 o'clock. Pressing holes into slits often aids passage. Dilators are diagnostic friends in discovering where each "hole" is located.

Body flush

During my first 5 years, I performed all horn flushes. For the past 30 years I have used the body flush unless I had a diagnostic reason to collect single horns or if I was performing a single embryo recovery. For the body flush, understanding the anatomy of the uterine body is necessary to prevent frustration (Figure 76.5).

In veterinary school, I was taught to use the external bifurcation to manipulate the uterus. Depending on the age, breed, and size of the cow, the external bifurcation can be 5–15 cm from the cervical opening, causing us to believe there is a lot of open space in the uterine body immediately anterior to the cervical os. Not so! In reality, there is only 1.25–5 cm of open space (heifers usually 1.25–1.9 cm while larger aged cows have 2.5–5 cm of space). Because the catheter has 2.5–3.75 cm of tip anterior to the cuff balloon, when we inflate the cuff the tip of the catheter protrudes into one

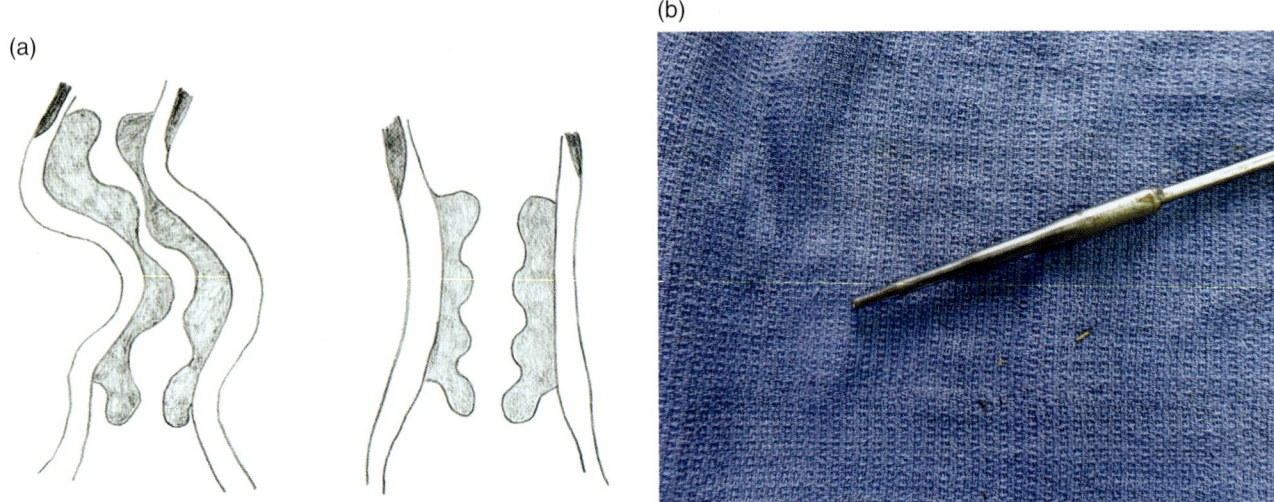

Figure 76.4 (a) Normal versus crooked cervix. (b) Cervical dilator tip.

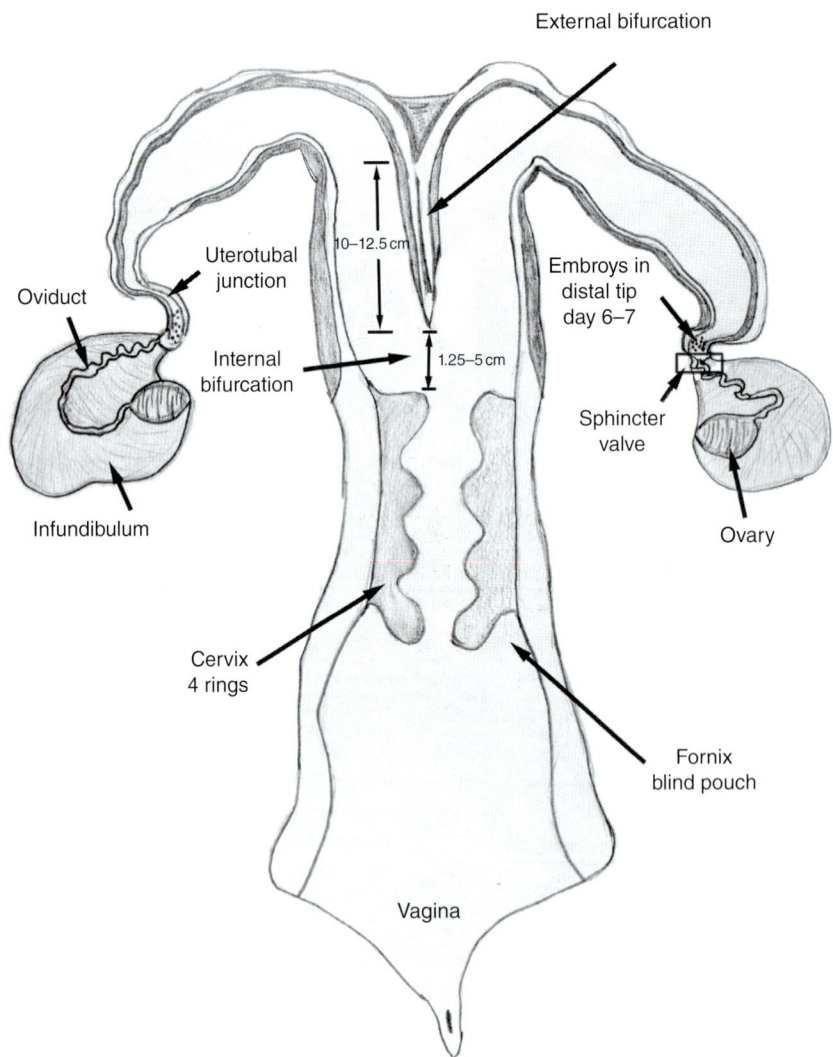

Figure 76.5 Normal anatomy.

of the horns. The balloon will occupy 1.25 cm with 2–3 mL of inflation, but will occupy 2.5–3.75 cm if 6–10 mL inflation is required to make the balloon large enough so that it won't retract into the cervix. Catheter placement is easiest if you direct the catheter tip into the uterine horn opposite whichever arm you have in the rectum. For example, I use my left arm in the rectum, so it is easiest for me to place the tip of the catheter into the right horn.

Inflation of the cuff and seating the catheter's balloon is the most important part of a successful flush. I use flush media rather than air to inflate the cuff so I can see it dripping out of the valve if it leaks. The catheter should be gently introduced into the uterine horn 5 cm or so. Then, 1–2 mL of collection fluid should be injected through the valve into the cuff until you can manually feel the cuff expand. Once you locate the slightly expanded cuff with your fingertips, retract stylet about 5 cm into catheter, and then gently retract the catheter until the cuff is in the uterine body just anterior to the cervical os. Once there, continue inflating the cuff. Gently tug the catheter caudally to see if sufficient inflation has been achieved to keep the cuff from retracting into the cervix, typically 2–3 mL being required in heifers and 4–10 mL in mature cows. Once satisfied that the inflation is correct, gently remove the stylet. (NB: Always put small amounts of KY jelly on the stylet tip when it's introduced into the catheter to prevent difficulty in withdrawing the stylet at this point.)

Overinflating and underinflating the cuff are both enemies. If underinflated, the cuff will pull back into the last cervical ring of the cervix as it relaxes. Even though the catheter tip will still protrude into the uterine body, the cervical pressure on the cuff will cause the catheter lumen to collapse and no fluid can be retrieved. If that happens, you must "pinch" behind the cuff in the cervix and squirt it back out into the uterine body where you can add more inflation. The only other option is to deflate the cuff, remove the catheter, reinsert the stylet, and start over. Usually it can be pinched forward without having to remove the catheter, which could result in loss of some embryos (Figure 76.6).

Overinflation results in only one horn being flushed unless some inflation is removed. How? See Figure 76.7. Note that if a heifer or cow has only 1.9–2.5 cm or so in her uterine body, overinflation causes the cuff to engage both the cervical os and the leaf of the internal bifurcation, preventing fluid from entering both horns. Whenever this occurs, you will only be flushing one horn and unless you figure out what's not normal (i.e., counterfeit) embryos will be left in the other horn.

Once the catheter is seated correctly, flush medium may be infused by gravity, syringe, or pump (Figure 76.8). I prefer to use a 50-mL airtight syringe connected to a disposable three-way valve so I can measure exactly how much fluid I am using. (NB: Prime the tubing to cut down on air in the uterus. Figure 76.1 displays the set-up I use today for collections.) I use a total of 400 mL of flush medium to collect both horns at the same time. Five infusions and retrievals are performed with the following amounts. Make sure to remove each infusion as completely as possible. Prior to the first infusion, the inflow side of the Y tubing should be filled (primed) with flush fluid to prevent air from being injected into the uterus, causing bubbles in the fluid.

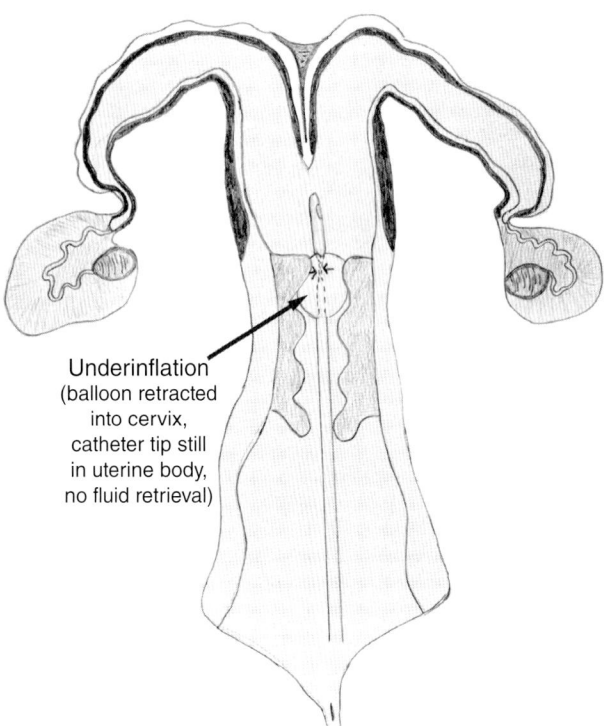

Figure 76.6 Underinflation.

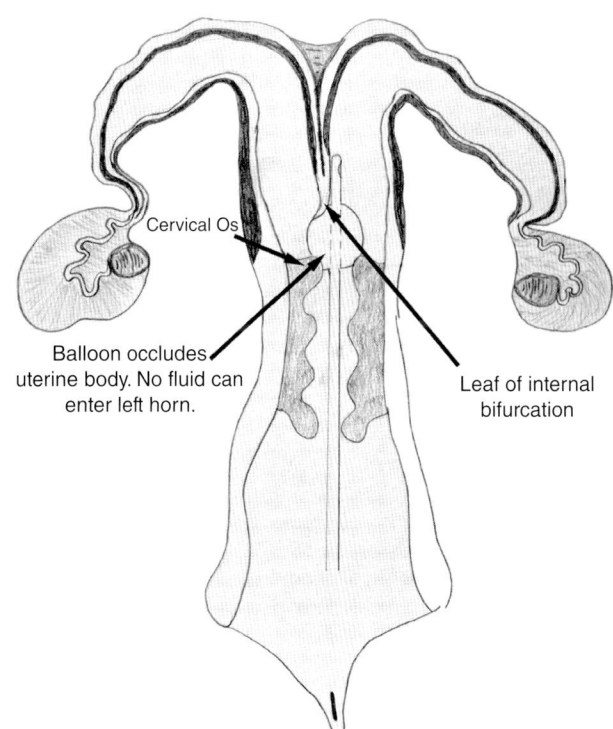

Figure 76.7 Overinflation and the occlusion of one horn.

- First infusion: 50 mL (25 mL into each horn), remove fluid.
- Second infusion: 50 mL (25 mL into each horn), remove fluid.
- Third infusion: 100 mL (50 mL into each horn), remove fluid.
- Fourth infusion: 150 mL (75 mL into each horn), remove fluid.
- Fifth infusion: 50 mL (25 mL into each horn), remove fluid.

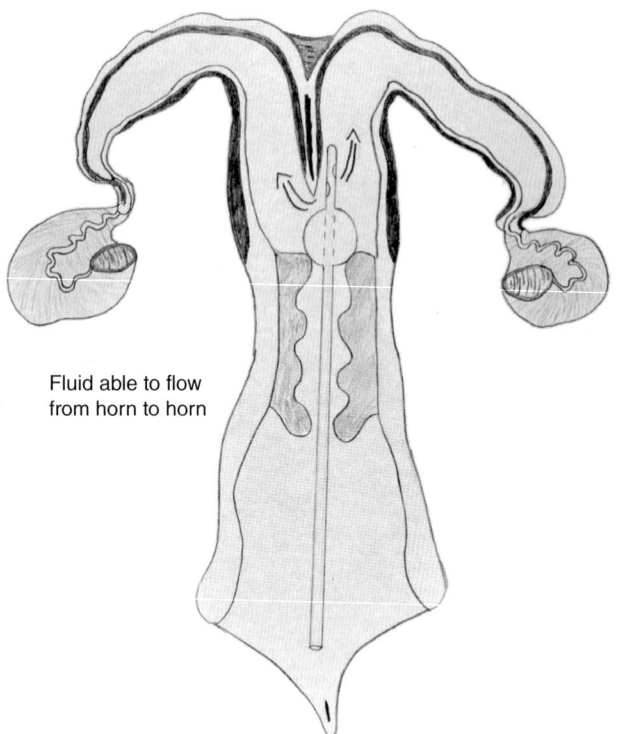

Figure 76.8 Correct catheter–balloon placement allows flush medium to flow freely throughout both horns.

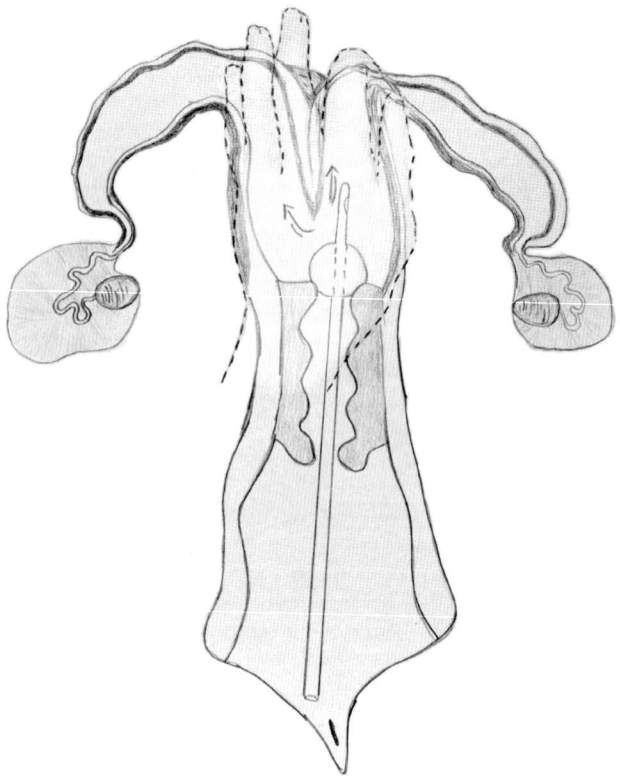

Figure 76.9 One horn can be held off to better fill the opposite horn.

Follow the flush with infusion of 25 mg Lutalyse through the catheter and into uterus. Deflate cuff with a 10-mL syringe and gently remove. To assure that I get equal amounts into each horn, I hold off the first horn just in front of the catheter tip. This forces the first half of each infusion to "dam up" and then run over into the other horn. I then allow the next half of that infusion to go straight ahead into that horn. You actually learn to "feel" the fluid running under your hand and into the opposite horn as it is injected via syringe. If the fluid continues to "dam up" between your fingers and the cuff and does not run over into the opposite horn, the cuff inflation is too large and must be reduced sufficiently to allow fluid to flow between the cuff and the leaf of the internal bifurcation and into the opposite horn. Be sure the outflow side of the tubing is clamped closed before each infusion and then opened after you've massaged the fluid throughout the tract (Figure 76.9). After each infusion, the fluid is squeezed forward into the tip of each uterine horn and vigorously massaged to suspend the embryos. This massage forces media all the way to the uterotubal junction (UTJ) where the embryos are located.

Once the massage is completed, the clamp is removed from the outflow tubing and the fluid is retrieved. About half of the fluid usually runs out by gravity while the other half must be reverse milked out. To reverse milk the fluid, I usually clamp the horn near the uterine tip between my second and third fingers and use my thumb to pull my fingers caudally with a motion similar to that of an inchworm. I alternate doing this between both horns and the fluid is easily moved back into the uterine body where it drains out the catheter. Sometimes I gently tug on the catheter, which produces a pumping action in the body area that aids in fluid retrieval. Each person develops his or her own technique for fluid retrieval. Repeat removal procedure after each infusion, trying to ensure the uterus is evacuated after each.

Before embryo filters were available, we collected all the fluid and searched it all. It was very time-consuming to search 12 bowls or Petri plates for each flush. However, that method taught me a lot about my technique and my efficiency (or lack of) in getting the embryos removed early in the flush. I still encourage beginners to do the same today to build confidence in their technique (Figure 76.10).

To run this quality assurance assessment on yourself, use 12 plastic embryo transfer search bowls, each capped with a plastic lid that fits tightly (these lids can be acquired from a fast food restaurant). Drain infusion no. 1 into dishes 1 and 2, infusion no. 2 into dishes 3 and 4, infusion no. 3 into dishes 5 and 6, infusion no. 4 into dishes 7–10, and infusion no. 5 (the last) into dishes 11 and 12. By searching those dishes quickly, you can determine at what point in the flush you are retrieving the most embryos; 90% of the embryos will be found in the first six dishes if you are doing a good job. If you find embryos in dishes 11 or 12, you may have left embryos in the donor. By repeating the flush you can determine if you've left any behind or not.

Note that using this technique, the fluid will be too deep in each dish for you to easily see the embryos on the bottom. Since the embryos sink to the bottom quickly, you can use a 20-mL syringe to "skim" the top of the fluid until you've got the depth down to 0.6–1.25 cm so you can see the debris on the bottom of the dish easily. After you've searched every dish and recorded the number of embryos found per dish, you can then pour all the flush fluid through a filter and find any embryos you missed while searching the bowls.

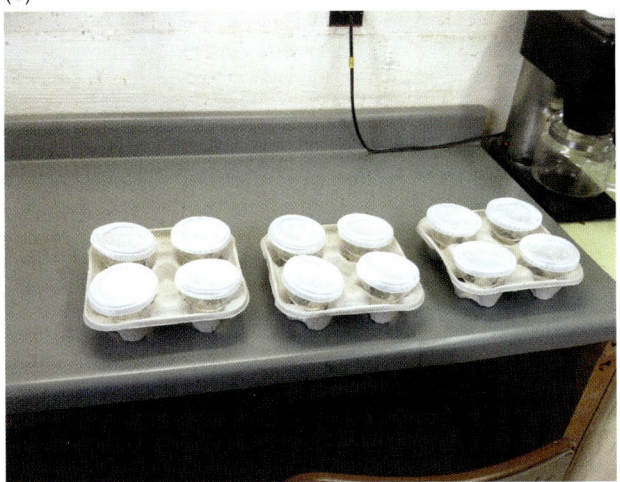

Figure 76.10 Search bowls safely held in fast food containers and covered with Solo™ cup lids.

I kept records on 200 repeat flushes performed early on in my career. Cows were reflushed whenever embryos were found in the last four bowls. The result? Twice I found one embryo and once I found 13. None were found on the other 197 flushes. I learned valuable information on the reflush which netted 13. The donor was a virgin 12-month-old Chianina heifer weighing 450 kg with a small reproductive tract. On the original flush I had found no embryos in dishes 1–8, but found 13 in dishes 9–12.

The lesson learned was that the virgin uterus, having never been expanded by a pregnancy, requires more dilatation with fluid and massage to get fluid into that distal horn near the UTJ which is where the embryos are on day 7. Although care must be taken not to overexpand the virgin uterus, causing hemorrhage or rupture, it must be expanded until full and tight and the uterine tips must be massaged to agitate fluid into that area.

My suggestion is for all novice practitioners to devise some way to evaluate their collection technique in order to develop confidence that *all* embryos are being efficiently collected. Collecting fluid into bottles and pouring the fluid through several filters is more costly but would be quicker than my method which I used in the era before filters.

Embryo Collection and Transfer

Horn flush

Horn flushes require less flush medium, but their main value to me is in the single-egg flush or when oviductal blockage is suspected. For instance, if a cow had six corpora lutea on each ovary and only six embryos were recovered with a body flush, you really can't know where they came from. With a horn flush, if the fluid is searched separately from each horn, it is possible to diagnose oviductal blockage.

In my opinion, the horn flush is quick and safe in the hands of experience but has more potential pitfalls for the beginner. The key to a successful horn flush is to have a perfectly seated cuff in the horn each time. If the balloon is overinflated, the endometrium will tear, allowing flush medium and embryos to leak into the broad ligament, causing crepitation in the uterine wall and broad ligament that resembles the feel of blackleg infected tissue. In addition, there will be a lot of hemorrhage to deal with. To salvage the flush, deflate the cuff, reinsert the stylet and go past the site of the tear, reinflate the cuff and continue the flush.

If you underinflate the cuff, the catheter will slip caudally, eventually ending up at the cervix. In that event, you may choose to reinsert the stylet and reposition the catheter in the horn or just reduce cuff inflation enough to allow fluid to freely pass from horn to horn and switch to a body flush. When properly inflated half to two-thirds down the horn, the cuff will require 8–15 mL of inflation. Fingers should sense the cuff's expansion and inflation should stop when the cuff causes the horn to bulge slightly when adequately engaged; the cuff should feel slightly flattened and elongated rather than still round (Figure 76.11).

When properly positioned, the cuff should be as far as possible down the horn. Minimal amounts of fluid will be required for each flush. I usually infuse medium five to six times into each horn tip with amounts that vary according to that necessary to achieve slight expansion of the horn. During each infusion, the uterine tip near the UTJ should be massaged before the fluid is retrieved. Most of the fluid will drain out by gravity flow, only slight manipulation being required to finish evacuation of all medium. I usually begin with 15-mL infusions, increasing to 30–50 mL if the horn is large enough. When one horn is completed, the catheter is removed, the stylet reinserted, and the catheter is then seated in the opposite horn. Experienced practitioners can rethread the stylet while the catheter is still *in utero*, but it's best not to try that at first.

The gravity flow technique of infusing medium is used by many practitioners, but I prefer to use syringes so I will have a known quantity infused each time to standardize my procedure. If gravity flow infusion is used, it is still necessary to massage the uterine tip after each infusion or embryos will be left behind. If collecting virgin heifers, vigorous massage of fluid into the UTJ area is a must. Some practitioners use injection pumps to infuse the fluid, but "simple" still works well for me.

Filter and search

Once the fluid is in the EZ Flow filter, the filter is tilted so all the remaining flush fluid drains out through the filter mesh. All the embryos should therefore be on the 75-µm filter

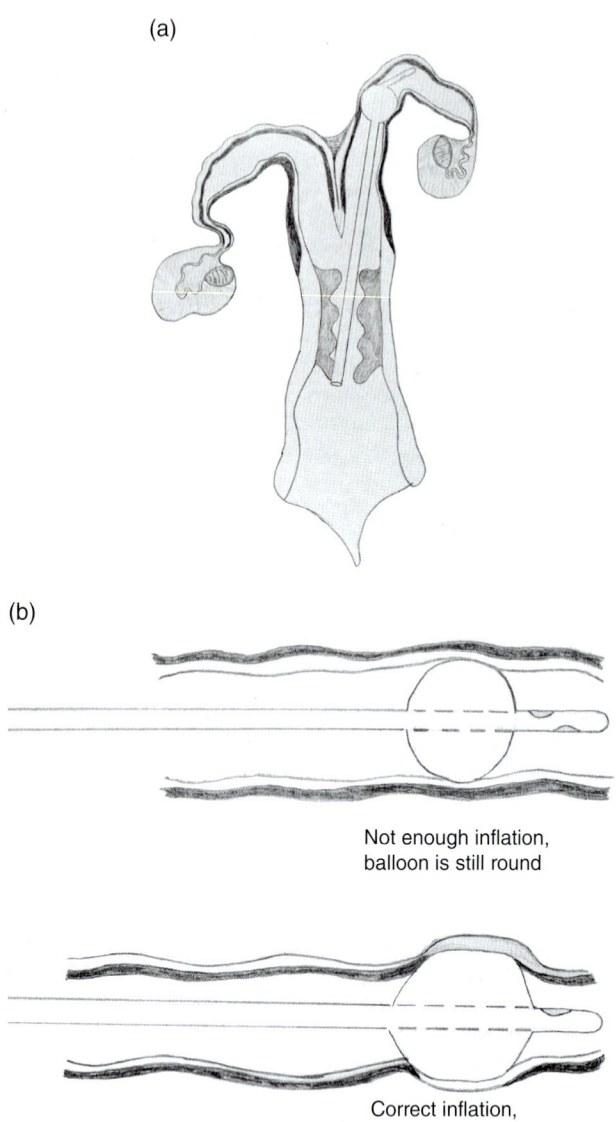

Figure 76.11 Proper position and inflation for "horn" flush.

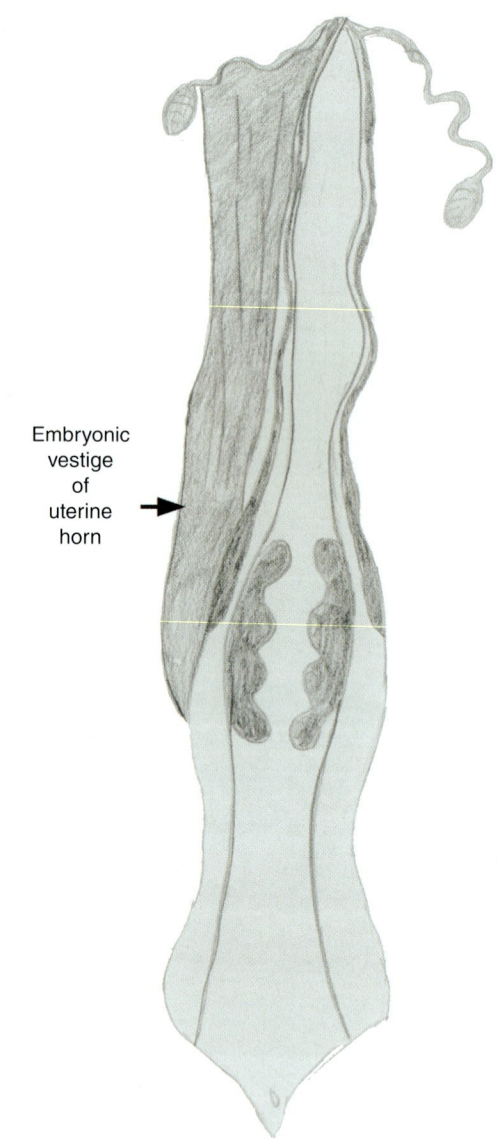

Figure 76.12 Uterus unicornis (segmental aplasia of the uterine horn).

mesh on the side of the EZ Flow. I use a 20-mL syringe filled with fresh flush medium with a 22-gauge × 2.5-cm needle attached to spray the filter mesh off, allowing the wash to collect in the bottom of the EZ Flow dish which will then be used as the search dish. I like to rinse the filter mesh off twice using 20 mL each time.

"Counterfeit money"

Now let's talk a little about those abnormalities and those not so perfect flushes. Some may be due to genetic abnormalities, AI technician damage, or self-inflicted trauma.

Genetic abnormalities encountered

Breed associations should be notified when these genetic abnormalities are found so that family lines which are carriers can be identified. However, it has been my experience that most breed associations are reluctant to "harm" a breeder by identifying one of their animals as a possible carrier.

Unicorn uterus

Yes, strangely enough, some donors make it to the flush barn with only one uterine horn. They are easy to flush because there is just one straight horn. Although I've usually found two ovaries in these animals, I only recover half the amount of embryos expected by corpus luteum count of the ovaries. The conclusion is that one ovary is connected to only a vestige of the missing horn while the other is connected to the existing horn (Figure 76.12).

Cervical deformities

In time, you will encounter all three of these cervical anomalies and unless you're aware that something is counterfeit you'll get frustrated.

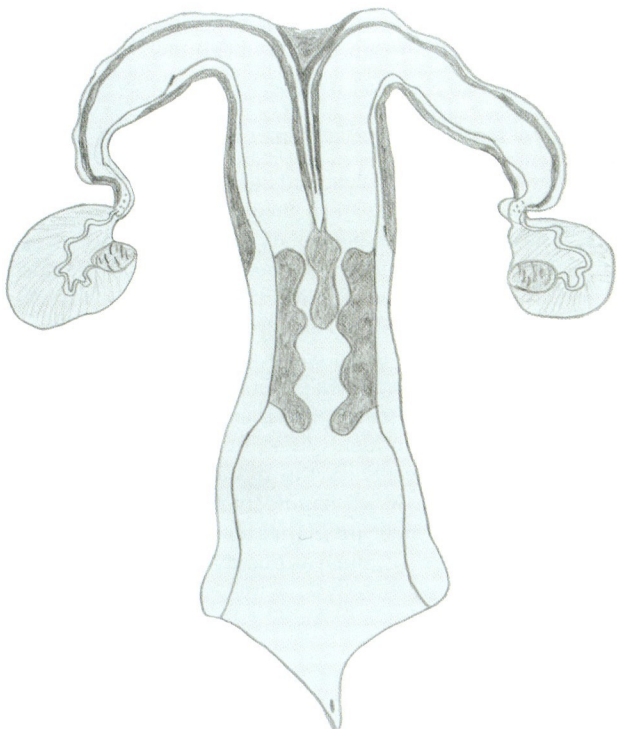

Figure 76.13 Y-shaped cervix (one external os, but a separate opening to each uterine horn).

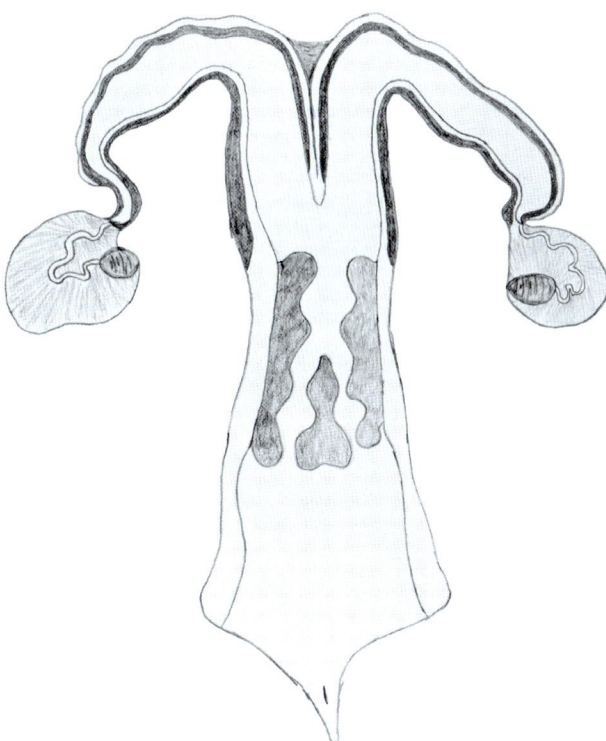

Figure 76.14 Inverted Y cervix (two external os, but only one internal os).

Y cervix

The "Y" cervix has one opening in the vagina, and then somewhere midway through the cervix the cervix bifurcates. Occasionally, there is a common uterine body but frquently each cervical opening goes to only one horn that does not communicate via a common uterine body. This means that for AI and embryo collection, each horn must be managed (both for AI and embryo collection) separately. Usually one of the cervical pathways is larger and unless the abnormality is diagnosed, all breedings will go to the same horn. Astute practitioners will notice the counterfeit cervix due to the lack of an internal uterine body. These are especially noticeable with recipients that have Y cervices. Nonprofessionals frequently artificially inseminate and even collect these abnormal uteri without knowing what was wrong (Figure 76.13).

Inverted Y cervix

The inverted Y cervix has two openings in the vaginal vault but both converge mid-cervix and only one opening is into the internal uterine body. These may be artificially inseminated and flushed, and if not an astute practitioner, they may not even be diagnosed as counterfeit but just classified as a difficult cervix to pass. These are not as critical for diagnosis since they share a common uterine body; therefore all embryos should be fertilized and collected regardless of which vaginal opening is used (Figure 76.14).

Double barrel cervix

This has two openings in the vagina and are two completely separate cervices, each going to its own horn. These are easier to diagnose by palpation. The cervix is noticeably wider than normal. With deep palpation, a groove can be felt which resembles the groove between the barrels of a double-barrel shotgun. Each "barrel" must be artificially inseminated separately and each horn collected separately since there is no common uterine body (Figure 76.15).

AI technician damage

Thickened rectal wall

When you palpate the donor for collection 7 days after AI and you feel a thickened edematous rectal wall, it's a good sign the technician had difficulty with the AI. Occasionally, the rectum is so badly traumatized that it is impossible to safely collect the cow. If collected, it is not unusual to collect nonfertile ova, indicating that the semen was never deposited in the uterus.

Brown blood clots

Finding brown blood clots in the first fluid retrieved usually indicates that the AI rod traumatized the uterine body resulting in mild to severe hemorrhage. The brown color indicates it is several days old, most likely from AI. This is another bad sign that embryos may not be fertile. Brown blood clots are enemies.

Infused fluid disappears, not retrieved

Three causes come to mind. First, there may be a perforation from the AI rod that has not yet healed. Second, you may have perforated the wall. Previous trauma usually has

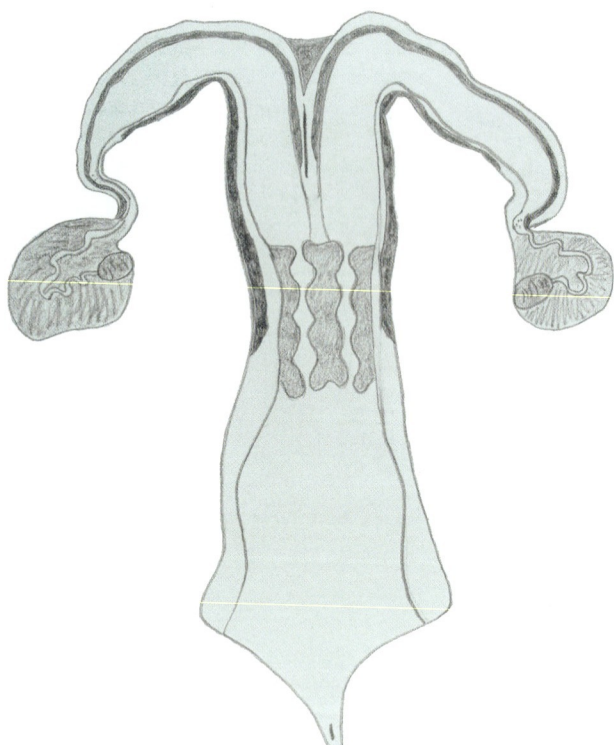

Figure 76.15 Double barrel cervix.

brown blood clots whereas fresh trauma is always red. The flush can be salvaged in either case if the perforation can be located; just go past the hole, inflate the balloon, and flush horns separately.

The third cause is serious and almost always results in all embryos being lost. When many unovulated follicles exist that are full of active estrogen, the sphincter valve at the UTJ does not close properly on day 5 or so. When fluid is infused into the uterus on day 7, it simply flows through the UTJ, into the oviduct, and out into the abdomen carrying all the embryos with it. This leaky valve is not common but I have frequently encountered them. I have had some that did not leak on early flushes of 25 mL per horn but gave way when I increased the volume to 50 or 75 mL per horn. This is reason to make the first two infusions of fluid a lesser amount and then increase volume on subsequent infusions. Consoling news is that these embryos were usually not fertilized anyway due to the increased estrogen level.

External uterine masses

A marble to golf ball-size mass found near the uterine body indicates probable trauma from AI gun or harsh palpation resulting in hematoma. These will resolve in time, but if very large cause difficulty that day during the flush.

Pelvic hematomas

Pelvic hematomas due to trauma from improper CIDR placement may prevent you from accessing the uterus. These are generally unilateral, from either 3 o'clock or 9 o'clock, and protrude into the pelvic canal. Treat with systemic antibiotics for 1 week to prevent abscessation. If an abscess occurs, drain via vaginal puncture.

Embryo transfer practitioner damage

Crepitation in the broad ligament indicates the endometrium has been torn, allowing flush medium to leak into it. The crepitation mimics the gaseous crepitation caused by blackleg in affected muscles. If encountered, stop the flush, infuse the uterus with penicillin, and give systemic penicillin for 5 days. Embryos will be lost into the ligament. A frequent cause is slippage of the cuff backwards during horn flushes or overinflation of the cuff on either horn or body flush. It is an unforgettable feel. Crepitation is an enemy.

Hematomas can suddenly arise during the flush near the uterine body, especially in the bifurcation, or along the serosa of the horn if we handle the uterus too roughly. Always use "flat finger pressure" for palpation and avoid "fingertip pressure" as much as possible.

Other problem areas

Traumatic adhesions

These have many etiologies. Cesarean section or dystocia that resulted in perforation can cause severe impingements, making uterine manipulation difficult to impossible. The presence of ovarian spider web-like adhesions is likely due to a previous ascending uterine infection or salpingitis. These are enemies and have a poor outlook. These spider-web adhesions often involve the ovarian bursa and interfere with ova pick-up during ovulation. The best long-term therapy is to try to get the cow pregnant by transplanting an embryo and hope the uterine expansion resolves enough adhesions to allow for sperm and ova transport through the oviduct. Alternatively, *in vitro* fertilization (IVF) could be attempted.

Sometimes a granular or sandpaper feeling to uterine serosa can be felt during catheter passage. This has an unknown origin but a distinctive feel when it is encountered. I suspect trauma from CIDR "wings" may cause pelvic inflammation resulting in fibrin deposition on serosa. Most respond to repeated infusion of penicillin and systemic penicillin (100 mL daily for 7 days).

Paraovarian cysts

Small nodules around the ovarian bursa usually do not cause pathology unless they become very large. Most are less than 10 cm.

Preventing unwanted or multiple pregnancies

Prostaglandin should be given to all donors post flush. I usually infuse 5 mL Lutalyse into the uterus at the end of the flush and then ask the owner to administer 5 mL (25 mg) i.m. twice a day on the third day following the flush. This quickly reduces ovarian size as well as aborting unwanted pregnancies. If cows are collected on 21–28 day intervals, it is very important to reduce ovary size quickly. However, infusing $PGF_{2\alpha}$ into the uterus before the dishes are searched for embryos negates your ability to reflush the donor if needed. The intramuscular injections twice on day 10 (3 days post flush) will be sufficient to prevent pregnancy if the owner can be trusted to remember to give the injections.

Helpful hints

Ballottement test

Even in normal passes of the catheter, occasionally a blockage occurs in the catheter. After the first fluid is infused and before any massage is done on the uterine tips, you can be sure the catheter lumen is patent with a simple test. With the fluid in the uterus before opening the exit clamp or palpating the tips, simply squeeze the bifurcation area gently two or three times. You should see movement of air bubbles from the catheter back into your clear tubing each time you squeeze, indicating patency. If no fluid or bubbles move caudally, you've got a flow problem. If it cannot be corrected by massaging the catheter tip, adding or reducing cuff inflation, or moving the cuff around in the body, now is the best and safest time to remove the catheter and reintroduce a new one. If you have not massaged the uterine tips, you're very unlikely to lose embryos.

Washout technique

When cervical passage of the catheter is difficult, quite often the catheter gets impacted with cervical mucus that enters through the side ports near the catheter tip. Excess mucus in the flush dish is an enemy. The mucus can easily be removed by a washout technique before any fluid infusion is allowed to go down the horns. With the catheter cuff inflated in the uterine body, wrap your thumb and forefinger completely around both horns just anterior to the uterine body (Figure 76.16). Infuse 15–20 mL of medium through the catheter quickly and open the exit side of the tubing immediately, allowing that fluid to drain onto the floor. Repeat this two or three times until the fluid draining out the exit tubing is mucus-free. Remember, keep both horns pinched off so fluid only enters and exits from the uterine body. Disconnect the tubing from the catheter and completely wash out the tubing before reattaching to both catheter and flush dish.

Cold weather outdoor collections

To prevent chilling of medium, we use two plastic stackable storage drawers (from Wal-Mart). Each one has a small electric heating pad on the bottom of the drawer. The top drawer has the medium bag lying on the pad. In the bottom drawer we have the filter dish sitting on the edge of the pad at the front of the drawer where a hole has been made to allow the fluid to drain from the filter cup onto the floor.

To protect the tubing from the cow to the dish, we use a 120-cm section of foam rubber water pipe insulation which has its side split open the entire length. We press the tubing inside the foam insulation section so the fluid can be kept relatively warm during the flush. In severely cold situations, an electric heat tape can be run through the piece of insulation as well.

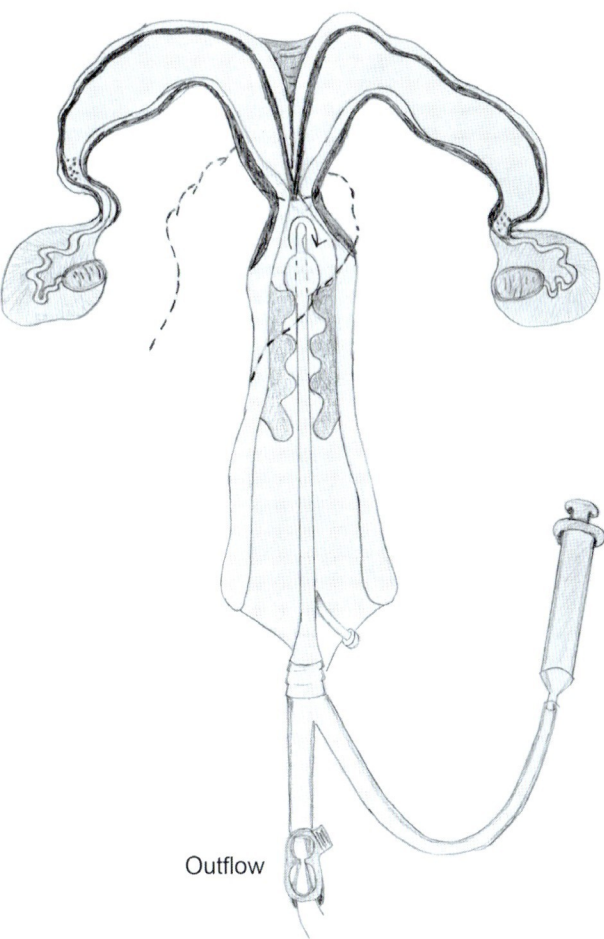

Figure 76.16 The author's "washout" technique.

Concluding remarks on embryo collection

Collecting embryos is a challenging and exciting procedure for practitioners. New techniques, instruments, and supplies are always thought of and experimented with. My hope is that you will develop a consistent technique you're confident with and that you can successfully dominate all the normal uteri and sufficiently navigate through all the counterfeit (abnormal) collections. After 35 years, I still find it amazing that this actually works and that we can have so much fun earning a living and helping our clients add to genetic progress in the cattle industry.

Now you're ready to send the dish to the microscope for searching, grading, and preparation for either fresh transfer or freezing. Since these are discussed in other chapters, let's now look at the technique for nonsurgical embryo transfer on both fresh and frozen embryos.

Transferring the embryo

Nonsurgical embryo transfer is a great example of the old saying "A chain is only as strong as its weakest link." Many factors, some of which are out of your control as the practitioner, will determine your success. I am far more keenly aware now of all external influences that bear upon my potential recipients than I was in the old days. Maybe you can pick up an idea or two from this discussion.

Selection for the recipient pool

I frequently say that "the outside of a cow looks like the way her insides feel." Cows in poor general body condition with dull hair coats and kept in mud lots generally are not the best choice. So what we are looking for includes the following. (*Editor's note*: The selection and management of the recipient herd is covered in depth in Chapter 78. The following is an overview of areas that the author feels are critical to the overall success of an embryo transfer program.)

Size

Dystocia is our enemy. Dead calves impress no one, especially our banker. Allow for the expected birthweight based on the breed of the embryo when adding cows to the pool. If using heifers, I mostly want only dairy cross or large beef heifers weighing a minimum of 360 kg when synchronization begins (actually I prefer those weighing 385–408 kg).

Milk

Sound udders are our friend. Getting a live calf is only the first skirmish. The recipient must be able to milk well and grow an exceptional calf. If transferring dairy embryos, this is not important since they will be bottle raised.

Prior history

Whenever possible, it is great to know that a recipient has calved regularly and has been an early breeder each year. Information is not available oftentimes though.

Lactation

Lactating cows in good body condition are usually great recipients. Usually, cows are usable by 45–50 days after calving *if* they are in good condition. Dry cows are excellent recipients if they have been held open, but if they are problem breeders or have already weaned a calf or been dried off, stay away from them.

Body condition score

The optimal body condition score (BCS) for fertility is 4–6 on a 1–9 scale. When ribs and spine are prominent on BCS 3 beef cattle, these are cows that are generally on the verge of going anestrus. They usually have smaller firmer corpora lutea if they are cycling at all. These usually give disappointing results. Overly fat cows and heifers have gotten a bad reputation from the belief that "fat cows are infertile." However, my experience has been that if the cow is fat because she has been held open or been overfed, she will still have decent fertility. However, if a recipient has received two embryos or been exposed to several AI attempts without conceiving (three at most), she should be turned in with the bull or culled.

Mineral deficiency

Every area of the United States has certain mineral supplementation needs. You should know those in your region.

Disposition

I have no use for crazy recipients. Each recipient is handled often, so a good disposition is a key ally. It also rubs off on the calf, which should have far more value than its mother.

Low stress

Stresses need to be eliminated. Adequate shelter from cold in winter and heat in other months is necessary. Mud lots are enemies. Grassy pastures are friends. Fresh water is a friend. Stagnant ponds or troughs are enemies. Excellent forage quality is essential to success.

Biosecurity

Several diseases will wreck an embryo transfer program and are discussed in detail in Chapter 80. Test all recipients and eliminate them from the herd if they are positive for the following diseases.

Bovine viral diarrhea virus

Animals persistently infected with bovine viral diarrhea virus (BVDV) are created when a pregnant cow is exposed to BVDV between 30 and 150 days of gestation. When born alive, these animals will shed live BVDV from birth until they die. Most die by 1 year of age but reports of some living until 7 years highlight the need to eliminate these. Ear notch test all animals under 6 months and ELISA test serum or ear notch the rest.

Neospora canis

This canine disease causes abortion in infected cows and is supposedly seen primarily west of the Mississippi River. Over the last 8 years, we have tested over 8000 beef and beef/dairy cross recipients in Tennessee, Kentucky, and Alabama and found 9% to be infected. *Neospora*-positive dams will infect 100% of their calves *in utero*. Male calves are infected but have no ill effects and cannot pass the disease. Females may abort but, if not, will infect each heifer calf they birth. Cows contract *Neospora* by eating hay, silage, or feed contaminated by canine feces containing *Neospora* oocysts (dog or coyote). *Neospora*-positive donors may safely be flushed, with embryos being transferred to negative recipients. Utilize available serology testing for *Neospora*.

Johne's disease

Testing serum is unreliable but is a good screening exercise. Fecal culture is best but time-consuming. Testing and removing positives should keep infection in the herd low. Positive dams spread Johne's disease *in utero* 26%

of the time. Most other infections occur orally when the newborn calf nurses on the cow's teats.

Bovine leukosis

Some dairies have successfully culled, divided herds, and used bovine leukosis virus (BLV)-free colostrum to get BLV-free herds. This is very difficult to do with the present level of infection in the United States. The best prevention is testing and separating positive cows into one herd or selling them if possible. Spread is via blood transfer by using the same needles, same sleeve to palpate all cows, and large blood-sucking insects. Prevention is effected by using a clean needle for every cow, changing sleeves between cows, and using colostrum from negative testing cows.

Vaccination

All recipients should be vaccinated for BVDV, infectious bovine rhinotracheitis (IBR) virus, parainfluenza 3 virus, bovine respiratory syncytial virus, and leptospirosis. In my opinion, killed virus products are safer but not as effective. Modified live virus (MLV) vaccines with fetal protection for BVDV are best on open cattle. If MLV IBR vaccine is used, recipients will have lower fertility for 30 days due to the IBR virus colonizing the corpus luteum, resulting in decreased progesterone levels. Be safe. A leptospirosis vaccine that includes *Leptospira borgpetersonii* (serovar Hardjo) must be added to the staple five-way leptospiral vaccine. This variant causes not only abortion, but infertility and early embryonic death. It has become a major problem in most of the United States. Other vaccines as needed in your locale.

Recipient synchronization

Estrus synchronization is discussed in Chapter 33. Use whatever means best fits your operation. Experience and data have taught us that natural heats are no more fertile than induced heats. Many practitioners today use timed embryo transfer protocols making heat detection unnecessary. This is a preferable method on farms with average management and poor heat detection. If a farm is well managed and has excellent personnel trained in heat detection, I prefer to use observed heats. In cases of very valuable embryos, observed heats are safer.

Timed embryo transfer results in a slightly lower pregnancy percentage but produces more total pregnancies from a group of recipients. For example, if 100 recipients are in a group synchronized and heats are observed and recorded, approximately 85 will be seen in heat in the acceptable "window" or time frame for use. On the average, 75 of those 85 will have good corpora lutea and will receive an embryo. If timed embryo transfer is used on 100 recipients, approximately 92 will have corpora lutea, so 92 embryos will be transferred. If the pregnancy rate is 60% for observed heats and 55% for timed embryo transfer, the total pregnancies produced will be 45 from the observed heat group and 50.6 from the timed heat group (plus the added advantage of no labor being spent on heat detection).

Acceptable synchrony on recipients

Generally, we use heats on recipients which are 24 hours before the donor (8 day heats) to 48 hours after the donor (5 day heats). Of course, 7-day recipients are always best. Several considerations must be taken into account.

- Fresh embryos can tolerate greater asynchrony than frozen embryos. I prefer to use day 7 (exact synchrony) to day 5.5 (36 hours after the donor) recipients for frozen embryos while 5–8 day recipients are acceptable for fresh embryos.
- When more embryos are collected than recipients are available on a given day, the Grade 2 and Grade 3 embryos are usually transferred fresh while the Grade 1 embryos are frozen (see embryo evaluation in Chapter 79).
- Frozen embryos have undergone stress and therefore need closer synchrony. I rarely use an 8-day recipient but have no hesitation using a 5.5-day recipient for frozen embryos.
- IVF embryos are readily available today, both fresh and frozen. We use the same guidelines for synchrony on IVF as we do frozen embryos.

Recipient preparation

So now finally we are getting our first recipient into the chute. The same feeding principle applies to recipients as well as to donors: full rumens are your friends, empty rumens are enemies. Never withhold feed: a full abdomen makes transferring much easier.

When the recipient enters the chute, I inject 3–5 mL of lidocaine as an epidural and quickly palpate for a good corpus luteum. If the corpus luteum is good and the uterus is acceptable on palpation, I load the embryo straw into a 52-cm IMV embryo transfer gun with blue sheath and immediately transfer the embryo. I always place an identification ear tag into each recipient, showing the mating and expected calving date.

The question always arises if it is acceptable to move or transport recipients to and from the transfer site. The only hesitation I have is if the weather conditions are very hot or very cold. Short-distance hauls should have no bearing on pregnancy rate.

Site of transfer

Embryos should be deposited into the uterine horn (ipsilateral to the corpus luteum) as deeply as possible without causing trauma (Figure 76.17). Prostaglandin is an enemy and is present in the endometrial cells. If we cause trauma as we pass the embryo transfer gun down the uterine horn, $PGF_{2\alpha}$ "leaks" from each damaged cell. Gentle flat-fingered pressure is necessary to smoothly guide the transfer gun without engaging or "bumping" the uterine wall. To do so requires the following conditions.

Restraint of recipient

Some animals stand more quietly with their head not caught in a head gate, while others must be caught and squeezed. It is very frustrating to have the embryo near the drop zone and the recipient suddenly go berserk or be spooked by

716 Assisted and Advanced Reproductive Technologies

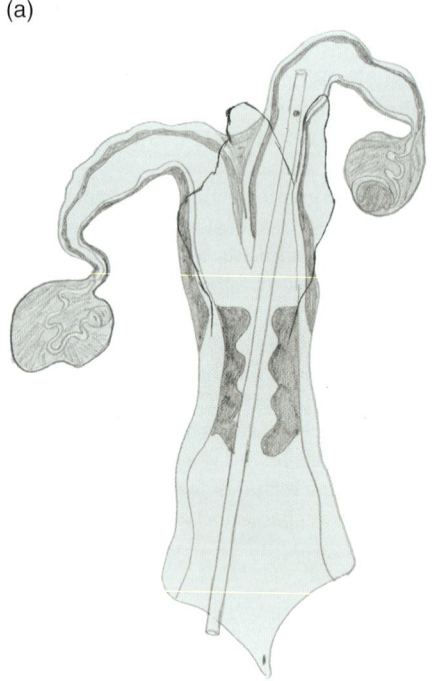

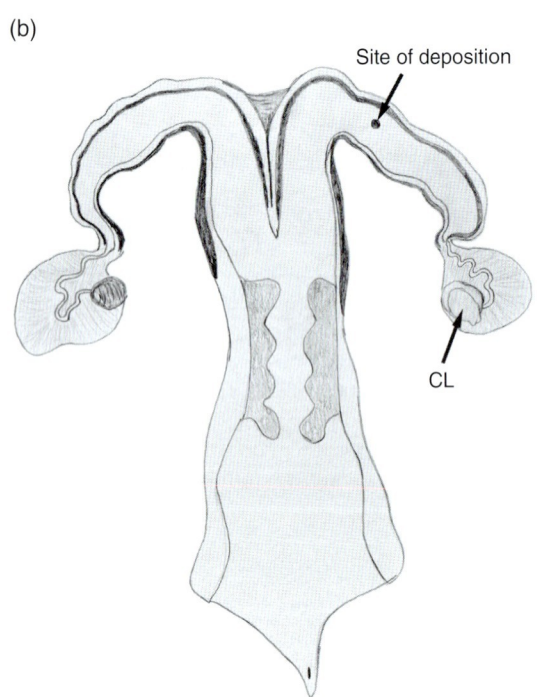

Figure 76.17 Site of embryo deposition. CL, corpus luteum.

Climate

Overheated animals yield poor results. Every effort must be made to avoid the hot part of the day or have fans and plenty of air moving. Plenty of water should be available. Minimal cattle movement should be done in the heat. Cold temperatures are generally not a problem unless extreme or if the conditions are also wet and muddy with no dry place to lie down. Recipient comfort is a friend.

Embryo deposition: gentle and fast

Your technique inside the cow will evolve as you gain experience. Gentle and fast are the rules. Data have shown that time spent in passing through the cervix has little effect on pregnancy rates, but increased time spent going down the uterine horn usually affects success negatively. To achieve gentle and fast, I will describe the uterine manipulations I use. Remember, I'm left-handed in the cow, so reverse my procedure if you use your right hand inside. Here are the steps I follow after the recipient has received an epidural and been checked for a corpus luteum.

After the epidural and before placing my hand in the rectum, I place the embryo transfer gun into the vagina and extend it to near the cervix. Once there, I push the gun through the thin plastic wrapper that is over the blue sheath. Then I place my hand into the rectum, locate cervix, and stretch it forward thereby removing any vaginal folds that may be obstructing the gun. Then I pass the embryo transfer gun to the fourth cervical ring and hold it there. If the corpus luteum is on the right, before I pass the gun into the uterine body, I retract the cervix caudally and then reach forward and pull the right uterine tip upward and caudally (I envision it as though a cobra snake is raised up to strike me in the face). Now, this horn will not stay there very long, so I quickly pass the embryo transfer gun into the right horn and begin sliding the gun down the horn, having my four fingers on the medial side and ventral to the horn, my fingers being used to straighten the horn's wall as I continue forward. When I reach the curvature of the horn, I use my thumb and forefinger to pull the horn forward and downward from its upright or sideways position, allowing me to get at least some distance around the curvature.

Each cow will be a different challenge. Some will allow passage almost to the uterine tip while others must be deposited near the beginning of the curvature due to their size, fatness, etc. The key is to be atraumatic. Use flat finger pressure under the horn. Do not use fingertips to squeeze the horn and do not attempt to pull the horn or "thread" it onto the gun. Both of these movements will usually result in trauma.

For the left horn, which is backhanded for me with my left hand in the cow, I start with my four fingers lateral and ventral to the horn with thumb on top of the horn. I push the uterus slightly to the right in the pelvic cavity to enable me to enter the left horn atraumatically. Then I continue down the horn while my fingers push the horn from left to right to allow for a straight path forward. When I reach the curvature, I move the gun with my outside hand (right hand, outside of the cow) as far to the right as I can (which moves the gun tip to a left-sided direction internally). With the gun in the lumen at the curvature, I wrap my fingers underneath

someone walking up near her head or yelling. I transfer most embryos with heads caught in a chute or stanchion, but I'm also comfortable with a low bar behind their hips with head not caught, unless they're crazy.

Footing

Solid footing is a major factor in an animal's willingness to stand still. Metal or slick concrete floors cause animals to panic and struggle. A good layer of bedding in the chute is a friend.

the curvature to the medial side and cradle the curvature, bending it around straight as far as I can and slide the gun as far as I can without bumping or scraping the wall, and then deposit the embryo.

Always visualize yourself as sliding or laying the gun down into the horn rather than threading the horn onto the gun. When depressing the stylet to expel the embryo, envision yourself as laying a bead of toothpaste on your toothbrush. As you depress the stylet, retract the gun a slight distance in order to distance the tip from the embryo as it comes out and I like to rotate the gun 360° to "wipe" the embryo off the gun tip if, by chance, it was still adhered.

Frozen-thaw transfer

The transfer of frozen embryos is identical to that for fresh transfers except that frozen-thawed embryos should be transferred immediately on thawing whereas fresh embryos can be held several hours (6–8) before transfer.

The preferred thawing process has been debated. The various options are a water thaw, air thaw, or a combination. In a water thaw, embryos from the liquid nitrogen tank (−193 °C) are plunged directly into a water bath at 35 °C. In an air thaw, embryos are held in the air until totally thawed. In combination thawing, embryos are removed from the liquid nitrogen tank, held in the air for 5–6 s, and then plunged into water at 35 °C. Putting embryos directly into water increases the possibility that the zona may crack, which may be undesirable for international work. Embryos may not be frozen for export if the zona is cracked, which could allow viruses or bacteria to gain access to the embryo. Allowing the straw to be held in the air for 5–6 s (or even a total air thaw) has been shown to prevent zona cracking. On a positive side, cracked zonas have been shown to be beneficial in the hatching process, since some embryos have been shown via microphotography to be unable to crack the zona (which enables or allows their growth), resulting in embryonic death. For this reason, I use the combination air–water thaw.

Most embryo transfer practitioners today use the IMV 52-cm embryo transfer gun with blue sheath. The sheath has a highly polished stainless steel tip that causes much less endometrial trauma than the clear or green-plugged Cassou gun AI sheaths of the 1970s through 1990s. The side port on the IMV sheath (compared with the AI sheath's open end) helps prevent cervical debris from being injected right at the spot where the embryo is deposited.

Success rates for frozen-thawed embryos are very close to those for fresh embryos, usually within 5% according to annual data published by the American Embryo Transfer Association (AETA).

Conclusion

As we end, let me encourage you to become involved in bovine embryo transfer if you think you may enjoy it. You will miss 100% of the shots you don't take. Embryo transfer was an addition to our large animal practice for 13 years (1977–1990) and has been our entire focus for the past 23 years. If you endeavor to participate, I encourage you to join the American Embryo Transfer Association, the Canadian Embryo Transfer Association, or the Embryo Transfer Association in your own country. Attend their meetings, meet colleagues who gladly will share their knowledge, and enjoy one of the great experiences available through your veterinary career.

Web resources

American Embryo Transfer Association: www.aeta.org
Canadian Embryo Transfer Association: www.ceta.org
International Embryo Transfer Society: www.iets.org
*Agtech: http://www.agtechinc.com
*Professional Embryo Transfer Supply: www.pets-inc.com
*Reproduction Resources: www.reproductionresources.com
Stone Manufacturing (source for cervical dilator described): www.stonemfg.net

*Bioniche products mentioned in the text are available from these and other sources.

Chapter 77

Cryopreservation of Bovine Embryos

Kenneth Bondioli

School of Animal Sciences, Louisiana State University, Baton Rouge, Louisiana, USA

Introduction

The utilization of embryo cryopreservation has become an integral part of the technique of embryo transfer, particularly the commercial application in the bovine embryo transfer industry. This is demonstrated by the industry statistics compiled by the International Embryo Transfer Society (IETS). Of the 572 342 *in vivo*-derived bovine embryos transferred worldwide in 2011, 324 149 or 57% were cryopreserved.[1] Embryo cryopreservation in conjunction with embryo transfer allows for efficient utilization of recipients, reduces the need to move cattle, allows an efficient means for marketing of genetics, and greatly facilitates the international movement of bovine genetics.

Cryopreservation methods

There are generally two types of procedures used in the cryopreservation of bovine embryos: (i) slow-rate or "conventional" cryopreservation and (ii) vitrification. Vitrification refers to the solidification of water as a glass-like structure without the formation of ice crystals. Slow-rate freezing employs a low concentration of cryoprotectant, induction of ice crystal formation outside of embryonic cells, and slow reduction of temperature so that these ice crystals grow and draw water from the embryonic cells without ice crystal formation inside the cells. This process is continued until the concentration of cryoprotectant within the cells becomes high enough to allow the formation of a glass-like structure. Vitrification on the other hand employs high concentrations of cryoprotectants and very rapid cooling rates so that a glass-like structure is immediately formed within the cells. This brief description of the two methods indicates that the terminology is not technically accurate since both methods ultimately result in vitrification. The terminology of "slow-rate freezing" and "vitrification" used to distinguish the different methods of arriving at this state is well accepted and commonly used.

Selection of embryos for cryopreservation

Regardless of the cryopreservation method used, the selection of embryos to be cryopreserved is always a major factor influencing the success of the procedure. The stage of embryonic development can be a factor influencing or affecting the efficiency of cryopreservation. Oocytes and early-stage (pre-compaction morula) embryos generally survive the freeze–thaw cycle of cryopreservation less efficiently than later stages. Fortunately, the compacted morula to blastocyst stage bovine embryo (normally recovered 7 days after behavioral estrus), which is the stage utilized for nonsurgical bovine embryo transfer, survives cryopreservation with a high degree of success.

The "quality" or embryo grade is a factor that varies considerably and has a major influence on the efficiency of cryopreservation. Embryo quality or embryo grading is an attempt of estimate the probability that a given embryo will result in full-term development if transferred to a suitable recipient. The morphological characteristics of bovine embryos that are the basis of embryo grading were described by Linder and Wright[2] and have become reasonably standardized within the commercial embryo transfer industry. These criteria for embryo grading, as well as the criteria for classification of embryonic stage, have been described in the *IETS Manual of Embryo Transfer*.[3] This manual has been widely accepted as a standard for procedures employed for embryo transfer throughout the world. (*Editor's note*: see Chapter 79 for complete coverage of embryo evaluation and grading.)

It is important to realize that many of the differences in morphological characteristics that constitute the embryo grading criteria are amplified by the freeze–thaw process. Many of these differences that result in embryos being assigned lower quality grades are a direct or indirect reflection of the number of viable cells within the embryo. The freeze–thaw cycle can only be expected to reduce the number of viable cells, so these characteristics that result in embryos being assigned lower grades are amplified by the freeze–thaw cycle. Thus it is common practice to limit the application of cryopreservation to embryos with higher-quality grades (grades 1 or 2).

While stage of development is a factor in embryo cryopreservation for some species (human, pig, and horse for example), embryonic stage is not a major factor for cryopreservation of bovine embryos within the context of normal embryo transfer. Differences in the ability to survive cryopreservation based on embryonic stage have been observed for bovine embryos,[4] although all stages normally recovered from a day 7 (day 0 being observed or expected estrus) nonsurgical embryo collection survive cryopreservation at an equal rate. A possible exception to this exists for hatched blastocysts, which have been observed to not survive cryopreservation well. Hatched blastocysts are not normally recovered at day 7 from cattle but are sometimes encountered when collection is delayed until day 8.

Cryoprotectants

A cryoprotectant is defined as any substance that aids in cell survival during freezing and thawing. Cryoprotectant compounds can be divided into two general classes, penetrating and nonpenetrating. Penetrating cryoprotectants are those which cross the cell membrane and act intracellularly. These compounds have a small molecular weight and a polar molecular structure that can mimic that of water. For slow-rate or conventional cryopreservation of bovine embryos one of two penetrating cryoprotectants are commonly used, glycerol or ethylene glycol (EG). For vitrification, penetrating cryoprotectant molecules frequently used include glycerol and EG as well as dimethyl sulfoxide (DMSO) and isopropyl alcohol. Nonpenetrating cryoprotectant compounds do not cross the cell membrane and work extracellularly. These compounds affect the osmolarity of a freezing solution and provide membrane-stabilizing properties. These compounds are generally sugars and include sucrose, galactose, and trehalose. These cryoprotectant compounds are not normally included in the freezing solutions utilized in slow-rate freezing but are frequently used to control the movement of water across the cell membrane after thawing (discussed below). Nonpenetrating cryoprotectant compounds are commonly included in vitrification solutions. Freezing solutions used in slow-rate freezing commonly include only one cryoprotectant compound, while those used for vitrification commonly include multiple cryoprotectant compounds.

Freezing solutions are prepared by addition of a cryoprotectant to a buffer solution. Traditionally, this buffer has been phosphate-buffered saline supplemented with either serum or bovine serum albumin (BSA). More recently, freezing solutions have become available from commercial sources. These solutions often utilize a zwitterionic buffer such as MOPS or HEPES instead of the phosphate buffer and a synthetic macromolecule such as polyvinyl alcohol instead of serum or BSA. This latter substitution is extremely important for the international movement of cryopreserved embryos. Penetrating cryoprotectants are normally added at 1.4 mol/L for glycerol or 1.5 mol/L for EG.

Slow-rate freezing

The method of cryopreservation commonly referred to as slow-rate freezing entails a reduction in temperature at a slow or controlled rate to allow extracellular ice crystals to grow and draw water from the cells without ice crystals forming within the cells. As mentioned previously the terminology is somewhat of a misnomer since the slow rate of cooling only occurs for a short period of time followed by very rapid cooling and vitrification of the remaining water. The rate of cooling must be controlled by some manner and this is typically performed by some version of an electronically controlled freezing apparatus.

Addition of cryoprotectants

The freezing solutions described above are clearly hypertonic and will cause cells to shrink on initial exposure and then slowly expand to their original size as the cryoprotectants enter the cells. Initially, it was believed that this type of osmotic stress would reduce embryo viability and embryos were exposed to the cryoprotectant solutions in a stepwise manner. Empirical studies demonstrated that stepwise addition of cryoprotectant was not necessary. This is particularly true for EG. Embryos are placed in the freezing solution and allowed to equilibrate with the cryoprotectant. Since the cryoprotectants at this concentration are not very toxic, equilibration times are not critical and equilibration will continue while the embryos are "loaded" into the appropriate "package" and during the first part of the freezing process. Embryos placed in a freezing solution will float because of the hypertonic nature of the solution and slowly sink as cryoprotectants enter the cells. Normal practice is to start loading embryos as soon as they have settled to the bottom of dish, with a minimum of 10 min of equilibration before starting the freezing process.

Packaging of embryos for cryopreservation

The container embryos are loaded into for the freezing process needs to meet several criteria. The material should transfer heat in an efficient manner and must be able to withstand the transition from body temperature to the temperature of liquid nitrogen (LN). Advantageous features include an ability to conveniently label the product and not requiring excessive room for storage in LN. Cryopreserved embryos are stored in the liquid phase of LN so space in storage is at a premium and it is important that the container exclude LN. If LN enters the container during storage, the rapid heating at thawing can cause physical damage to the container and possible loss of embryos. In the initial utilization of cryopreserved bovine embryos for transfer, glass ampoules were frequently used with multiple embryos per ampoule. The slow rate of heat transfer (particularly during thaw) and the amount of room occupied in LN for glass ampoules rapidly led to the application of semen straws for this purpose. As "direct transfer" cryopreservation methodology (discussed below) has gained popularity, the 0.25-mL semen straw has become the container of choice. With direct transfer methodology the cryopreservation container is also used for transfer and each

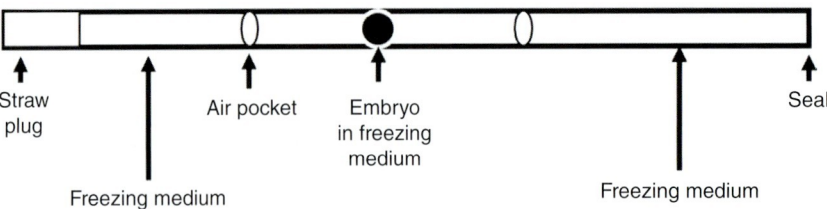

Figure 77.1 Embryos in straws of freezing medium separated by small columns of air or "air bubbles."

container will contain a single embryo. Embryos are loaded into straws by drawing several columns of freezing medium into the straw separated by small columns of air or "air bubbles." The first column does not contain an embryo and seals the plugged end of the straw. The second column contains the embryo and is separated from the first and third columns by columns of air (Figure 77.1).

One of the major aspects of packaging embryos for freezing is how the packaging allows the embryos to be labeled. The amount of information required to adequately label cryopreserved embryos will vary considerably. The IETS Manual recommends certain information be included for cryopreserved embryos. Minimal information should include sire and dam of the embryo and the date the embryo was frozen. Additional information might include the practitioner or company conducting the freezing, breed of sire and dam, registration numbers, and the developmental stage and quality grade of the embryo at the time of freezing. The IETS provides "freezing codes" to its members that can be used to identify the practitioner or company conducting the freezing. This information can be printed directly on the straw. Straw printing machines are available but not frequently used for embryos because the information could be unique for each embryo. Information can be printed by hand but this can lead to legibility problems. One approach to labeling is to heat seal a 0.5-mL straw to the 0.25-mL straw and apply labeling information on the 0.5-mL straw. Another approach is the use of plastic identification and sealing rods. These rods attach to the end of a semen straw and can be used for labeling. In both of these approaches information can be printed on adhesive labels that can be wrapped around the labeling straw or plug. It is not recommended that labels be placed on the straw containing the embryo because this could interfere with the efficient transfer of heat during freezing and thawing. Another advantage of the approaches that place labeling information separate from the straw containing the embryo is that they provide a handle such that the information can be read without removing the straw containing the embryo from LN.

The freezing process

Slow-rate freezing is often referred to as controlled-rate freezing because the rate of cooling is controlled during most of the freezing process. The processes of equilibration with cryoprotectant and packaging for freezing are normally carried out at ambient or room temperature. Once embryos are loaded into straws, properly labeled and straws sealed, the temperature is reduced to a point 1 or 2 °C below the true freezing temperature of the freezing solution. For commonly used freezing solutions this temperature will be approximately −6 °C. Since there is no ice crystal formation at this point, the rate of cooling during this phase is not important. A common approach is to place embryos in the appropriate containers in a controlled-rate freezing apparatus and lower the temperature at 1–2 °C per minute to the −6 °C target temperature. At this point ice crystal formation is induced in the freezing solution, a process known as "seeding." Seeding is generally required because solutions cooled in this manner will undergo supercooling, and remain liquid below the true freezing point. If supercooling is allowed to occur, when ice crystal formation does occur there will be a rapid decrease in temperature for a short time. This rate can easily exceed the rate that allows water to leave the cells and ice crystals will form within the cells. Ice formation is seeded at a temperature 1 or 2 °C below the true freezing point by cooling the solution in a portion of the container away from the embryo. Touching the container with a pair of forceps cooled in LN or a cotton swab dipped in LN is an effective and simple way of inducing ice crystal formation provided the temperature remains slightly below the freezing temperature of the solution. After seeding the temperature is maintained for several minutes to allow osmotic equilibration to occur between the embryonic cells and the freezing solution with ice crystals. At this point the true controlled-rate freezing process begins with a reduction of the temperature at a rate 0.3–0.5 °C per minute. It is during this phase that ice crystals grow in the solution outside the cells and draw water from the cells. The rate of cooling that allows the water to be drawn from the cells without ice crystal formation within the cells is determined by a combination of membrane permeability and cell size. For cells of the bovine morula to blastocyst stage embryo, this rate of cooling is 0.3–0.5 °C. Controlled-rate cooling continues to approximately −35 °C. At this point the amount of water has been reduced sufficiently that the concentration of cryoprotectant and salts are high enough to allow vitrification to occur if the temperature is reduced rapidly. After a short equilibration time at −35 °C the temperature is rapidly dropped by "plunging" directly into LN. Controlled-rate cooling is terminated at the temperature that allows vitrification to occur because the concentration of salts increases as well as the concentration of cryoprotectant. Continuation of controlled-rate cooling will increase cellular damage due to ionic stress and cryoprotectant toxicity. Once the embryos have been cooled to the temperature of LN they are stored in the liquid phase until thawed. A number of embryo freezing apparatuses have been developed and several are commercially available. Most of these so-called freezing machines use LN for cooling and control the rate of cooling with heat. Commercially available apparatuses utilize some sort of computer interface to control the cooling rate within a freezing program designed by the operator.

Thawing

Even though a slow rate of cooling is used in the early phase, a rapid rate of cooling is used in the final phase that results in vitrification of the remaining water in the cells. This results in a requirement for rapid thawing. If thawing is allowed to occur slowly, ice crystals will form during thaw and cellular damage will occur. Rapid warming is accomplished by transfer of the containers into warm (35–37 °C) water until all ice has thawed. It has been observed that if this rapid warming starts immediately from the liquid phase of LN there can be a high incidence of damage to the zona pellucida. Empirical studies have shown that if warming occurs in air for about 30s before transfer to warm water, the incidence of zona pellucida damage is reduced. With this approach, warming through the critical temperature phase of above approximately –60 °C is still rapid and cellular damage due to ice crystal formation during thaw is avoided.

Removal of cryoprotectant

Since the initial application of embryo cryopreservation to bovine embryo transfer, the removal of cryoprotectants after thaw has been a major issue. The challenge results from the fact that water moves across biological membranes faster than most cryoprotectants. During removal of cryoprotectants water enters the cells faster than the cryoprotectant exits and results in swelling of the cells. Cells are far less resistant to swelling than shrinkage and the degree of swelling that occurs with normal concentrations of DMSO and glycerol can be very detrimental. Initially, this problem was circumvented by removal of cryoprotectant by equilibration with stepwise lower concentrations of cryoprotectant. A number of approaches employing sucrose have been utilized, including a "one step" method using a sucrose solution with an osmotic pressure equal to that of the freezing solution.[5] With this approach cryoprotectant can leave the cells without water entering and no swelling occurs. After cryoprotectant has been removed embryos can be placed in a physiological solution and water reenters the cells. Other approaches have utilized lower concentrations of sucrose,[6] which allows cryoprotectant to be removed in two steps. This approach also prevents cellular swelling and also limits the degree of cellular shrinkage and dehydration. All these approaches have been successful in the preservation of viability of cryopreserved embryos but involve the movement of embryos between different solutions, requiring a certain degree of embryo manipulation skills and equipment between thaw and transfer.

Direct transfer

The methods for preventing cellular damage during removal of cryoprotectant require movement of embryos between different solutions. This results in the need for some degree of embryo manipulation skill and equipment at the time of transfer for cryopreserved embryos. A number of methods have been described for the cryopreservation of embryos such that no manipulation is required between thaw and transfer. This allows cryopreserved embryos to be thawed and transferred into suitable recipients in the same container they were frozen in or so-called "direct transfer." A number of approaches involving the use of sucrose have been described but in 1992 Voelkel and Hu[7] reported that the cryoprotectant EG moved across the membranes of cells of the day 7 bovine embryo very rapidly. This rate of movement is sufficiently fast that when embryos were frozen in 1.5 mol/L EG they could be transferred directly into recipients without loss of viability. At this concentration EG is an effective cryoprotectant and no manipulation is required between thaw and transfer and so avoids loss of viability due to cellular swelling. The method of cryopreservation using a freezing solution of 1.5 mol/L EG has become the standard of the bovine embryo transfer industry.

Vitrification

The cryopreservation method commonly referred to as vitrification differs from slow-rate freezing in two major aspects: (i) higher concentrations of cryoprotectant are used; and (ii) very rapid cooling is initiated at a warmer temperature. Details of the procedures used for vitrification vary widely and a "universal" method for vitrification of bovine embryos has not been established. The discussion here will be limited to general aspects of the procedure.

Vitrification solutions

In order for vitrification to occur at warm temperatures (20–25 °C), the cryoprotectant concentration must be higher than that described for slow-rate freezing. The majority of vitrification solutions used for bovine embryos consist of a combination of two or more penetrating cryoprotectants and a nonpenetrating cryoprotectant such as sucrose or trehalose. A typical vitrification solution for bovine embryos would consist of 15–20% EG, 15–20% DMSO, and 0.3–0.5 mol/L sucrose or trehalose. The combined concentration of EG and DMSO is at least three times the concentration of EG or glycerol used for slow-rate freezing. Extended exposure to these concentrations of cryoprotectants can be toxic, so embryos are exposed to stepwise higher concentrations and the exposure time to the final vitrification solution is limited to 40–45s.

Rapid cooling

In order for vitrification to occur at high temperatures the cooling rate must be very rapid, in the order of 1000 °C per minute or greater. To achieve these rates of cooling, even by transfer directly into LN, the surface area to volume ratio must be greater than that of the 0.25-mL semen straw. One method of increasing surface area to volume ratio is the "open pulled straw,"[8] which has been successfully used for the vitrification of bovine embryos. This is accomplished by gently heating the plastic semen straw and pulling it to yield a much smaller diameter and thus increasing the surface area to volume ratio. Additional methods of producing ultra-rapid cooling rates include the Cryoloop™, Cryotop™, Cryoleaf™, and others.[9] These examples consist of various platforms or carriers that the embryo can be placed on in a minimal amount of vitrification solution. These create a high surface area to volume ratio and when placed directly in LN yield ultra-rapid cooling. All these carriers result in exposure of the embryo to LN, so once cooling is complete the embryo and carrier are placed in some sort of outer container that can be sealed. The same principle of high surface area to volume ratio that results in ultra-rapid cooling will result in ultra-rapid warming when placed in warm thawing solution. Thawing solutions will typically consist of a sucrose solution to ensure rapid heat transfer and control cellular swelling during removal of the cryoprotectants.

Applications for vitrification

Vitrification has been applied to the cryopreservation of human embryos and oocytes extensively but far less for bovine embryos. At present there is no commercial application for cryopreservation of bovine oocytes and the value of vitrification for the day 7 late morula or blastocyst stage bovine embryos has not been established. Vitrification has the advantage of not requiring a controlled-rate freezer and the freezing time for a single embryo is much less. However, the precise nature of the timing required for vitrification requires that this procedure be conducted one embryo at a time. Because of this, when even a moderate number of embryos need to be processed, as much or more time will be required for vitrification compared with slow-rate freezing in which groups of embryos are frozen in a single cycle of a controlled-rate freezer. The precise nature of the timing and the handling procedures required during ultra-rapid cooling and warming during vitrification can result in variable success rates depending on the skill level of the person performing the procedure. There are very few data from controlled studies under practical conditions comparing success rates for vitrification compared with slow-rate freezing for late morula or blastocyst stage bovine embryos. One application where vitrification may have greater significance is the cryopreservation of *in vitro*-produced embryos. Bovine embryos produced from aspirated oocytes, matured, fertilized and cultured *in vitro* do not survive cryopreservation as well as the same stage embryos produced *in vivo* and collected from superovulated cattle. Vitrification may have increased application for cryopreservation of these embryos.

Transfer of cryopreserved embryos

The parameters for transfer of cryopreserved embryos to recipients are no different from those for transfer of noncryopreserved embryos. Synchrony requirements between the donor and recipient estrous cycles are the same as for noncryopreserved embryos. Of course, the concept of synchrony with the donor estrous cycle is technically lost since transfer of the cryopreserved embryo may occur years after collection. This emphasizes the importance of including information concerning the stage of development on the labeling of cryopreserved embryos. If the embryonic stage is between a late compacted morula and a blastocyst, the assumption of a day 7 embryo is appropriate. It is difficult to discuss expected pregnancy rates with any embryo transfer because there are so many variables. The same is true for the transfer of cryopreserved embryos. If high-quality day 7 bovine embryos are cryopreserved by the methods described here for slow-rate cooling, there should be an expectation of pregnancy rates very comparable to those expected for noncryopreserved embryos under the same conditions. The qualification of "high-quality embryos" is very important for cryopreserved embryos because the difference in pregnancy rates becomes significant for lower-quality embryos.[10] The same embryos vitrified by an experienced and skilled technician should produce similar pregnancy rates.

References

1. Stroud B. IETS 2012 statistics and data retrieval committee report. *Embryo Transfer Society Newsletter* 2012;30:16–26.
2. Linder G, Wright R Jr. Bovine embryo morphology and evaluation. *Theriogenology* 1983;20:407–441.
3. Stringfellow DA. Certification and identification of embryos. In: Stringfellow DA, Givens MD (eds) *Manual of the International Embryo Transfer Society*, 4th edn. Champaign, IL: IETS, 2010.
4. Looney C, Westhusin M, Bondioli K. Effect of cooling temperatures on pre-compacted bovine embryos. *Theriogenology* 1989;31:218 (Abstract).
5. Leibo S. A one-step method for direct nonsurgical transfer of frozen-thawed bovine embryos. *Theriogenology* 1986;21:767–790.
6. Niemann H, Sacher B, Schilling E, Smidt D. Improvement of survival rates of bovine blastocysts with sucrose for glycerol dilution after a fast freezing and thawing method. *Theriogenology* 1982;17:102 (Abstract).
7. Voelkel S, Hu YX. Use of ethylene glycol as a cryoprotectant for bovine embryos allowing direct transfer of frozen-thawed embryos to recipient females. *Theriogenology* 1992;37:687–697.
8. Vajta G, Holm P, Kuwayama M *et al*. Open pulled straw (OPS) vitrification: a new way to reduce cryoinjuries of bovine ova and embryos. *Mol Reprod Dev* 1998;51:53–58.
9. Saragusty J, Arav A. Current progress in oocyte and embryo cryopreservation by slow freezing and vitrification. *Reproduction* 2011;141:1–19.
10. Hasler J, McCauley A, Lathrop W, Foote R. Effect of donor–embryo–recipient interactions on pregnancy rate in a large scale bovine embryo transfer program. *Theriogenology* 1987;27:139–168.

Chapter 78

Selection and Management of the Embryo Recipient Herd for Embryo Transfer

G. Cliff Lamb and Vitor R.G. Mercadante

North Florida Research and Education Center, University of Florida, Marianna, Florida, USA

Introduction

The primary use of embryo transfer in cattle has been to amplify reproductive rates of valuable females. Ideally, embryo transfer can be used to enhance genetic improvement and to increase marketing opportunities with purebred cattle. Embryo transfer is especially useful with cattle because of their relatively low reproductive rate and long generation interval.[1] Once transferable embryos are collected from a donor cow, a decision is made as to which of the available recipients should receive embryos to achieve the greatest number of pregnancies.[2] The success of embryo transfer depends on factors associated with the embryo, the recipient, and the embryo transfer technician, or an interaction among these factors. Suitability of recipients is dependent on numerous management, nutritional, and estrous cycle control factors to ensure the presence of a functional corpus luteum (CL) at transfer. Although studies have focused on these factors, differences in techniques, sample sizes, and other elements have limited the applicability of the results of these studies. In many ways, management of the recipient is more critical to the success of an embryo transfer program than the donor, since the recipient is expected to conceive to the transferred embryo, maintain the pregnancy until full term, calve without assistance, and raise a calf of high genetic merit. This chapter focuses on recipient-related factors responsible for the success of an embryo transfer program.

Nutritional management

Body condition score as an indicator of reproductive efficiency

Insufficient intake of energy, protein, vitamins, and microminerals and macrominerals has been associated with suboptimal reproductive performance. Of these nutritional effects on reproduction, energy balance is probably the single most important nutritional factor related to poor reproductive function in cattle. The metabolic use of available energy in ruminants for each physiological state is ranked, in order of importance, as follows: (i) basal metabolism; (ii) activity; (iii) growth; (iv) energy reserves; (v) pregnancy; (vi) lactation; (vii) additional energy reserves; (viii) estrous cycles and initiation of pregnancy; and (ix) excess energy reserves.[3] Based on this list of metabolic priorities, reproductive function is compromised because available energy is directed toward meeting minimum energy reserves and milk production. Generally, beef cows do not experience a period of negative energy balance because they fail to produce the quantity of milk that dairy cows produce; however, beef cows need to be in sufficient body condition to resume estrous cycles after parturition and overcome anestrus, short estrous cycles, and uterine involution just to become pregnant every year.

Body condition score (BCS) is a reliable method for assessing the nutritional status of recipients. A visual BCS system developed for beef cattle uses a scale from 1 to 9, with 1 representing emaciated and 9 obese.[4] A linear relationship exists between body weight change and BCS, where an approximate 40-kg weight change is associated with each unit change in BCS. Managers of recipients should understand when cows can be maintained on a decreasing plane of nutrition, when they should be maintained on an increasing plane of nutrition, or when they can be kept on a maintenance diet. Understanding the production cycle of the cow and how to manipulate the diet will improve the ability of the recipients to conceive to the transferred embryo.[5,6]

BCS at calving has been shown to be a more predictable indicator of the duration of postpartum anestrus than prepartum change in either weight or BCS.[4,7] When cows were thin at calving or had a BCS of 4 or less, increased

---

Bovine Reproduction, First Edition. Edited by Richard M. Hopper.
© 2015 John Wiley & Sons, Inc. Published 2015 by John Wiley & Sons, Inc.

postpartum level of energy increased the percentage of females exhibiting estrus during the breeding season. BCS at parturition and breeding are the dominant factors influencing pregnancy success, although body weight changes during late gestation modulate this effect. However, altering poor body condition after parturition may reduce the negative impact on reproduction, but seldom overcomes or eliminates those negative effects. A recent study by Stevenson et al.[8] using blood samples at initiation of the breeding season to determine estrous cycling status demonstrated that only 47.2% of the cows were cycling at the onset of the breeding season. However, as BCS increased, the percentage of cows that were cycling also increased. It is important to note that when cows had a body condition score of less than 4 at the beginning of the breeding season, only 33.9% had resumed their estrous cycles.

Prepartum nutritional effects on reproduction

The general belief is that cows maintained on an increasing plane of nutrition prior to parturition usually have a shorter interval to their first ovulation than cows on a decreasing plane of nutrition. Energy restriction during the prepartum period results in a low BCS at calving, prolonged postpartum anestrus, and a decrease in the percentage of cows exhibiting estrus during the breeding season.[9] Pregnancy rates and intervals from parturition to pregnancy are also affected by level of prepartum energy.[9] Conversely, when prepartum nutrient restriction was followed by increased postpartum nutrient intake, the negative effect of prepartum nutrient restriction was partially overcome; however, the effectiveness of elevated postpartum nutrient intake depended on the severity of prepartum nutrient restriction.[7,9] The effect of BCS prior to calving also has implications for calf birth and weaning weights. When cows were fed to achieve a BCS of either 4 or 6 prior to calving, body weights were greater and calf birth and weaning weights (with similar genetics) were also greater for those cows at a BCS of 6.[10] Despite the greater birthweights, there was no difference in calving difficulty, demonstrating the added advantage for recipients to wean calves with greater weaning weights. In addition, there tended to be an increased number of cows calving with a medium BCS that were cycling at the beginning of the breeding season and after a 60-day breeding season than cows in poor condition, resulting in a greater proportion of cycling cows at various stages of the breeding season.[10]

Postpartum nutrition

Numerous studies document that increasing nutritional levels following parturition increases conception and pregnancy rates in beef cows.[4,11] Increasing the postpartum dietary energy density increased body weight and BCS and decreased the interval to first estrus.[7] However, suckled beef cows in relatively poor body condition gaining in excess of 1 kg/day while consuming an 85% concentrate diet did not resume cyclic ovarian activity before 70 days postpartum.[7] Therefore, although an enhanced plane of nutrition after calving may partially overcome the negative effects of poor prepartum nutrition, the added stress and negative impact of suckling and lactation must also be considered.

A major impact on postpartum fertility is the length of the breeding season. Having a restricted breeding season has many advantages, such as a more uniform, older calf crop, but most importantly a breeding season of 60 days or less increases the percentage of females cycling during the next breeding season. If the breeding season is shortened, then all cows have a higher probability for pregnancy during the next breeding season. Strategic feeding to obtain ideal BCS can be achieved by understanding the production cycle of the cow. The period of greatest nutritional need occurs shortly after calving; a cow is required to produce milk for a growing calf, regain weight lost shortly before and after parturition, and repair her reproductive tract to become pregnant within 3 months after calving. During this stage, a cow is usually consuming as much feed as she can and adjusting BCS at this time often is futile. Cows are usually grazing and tend to consume their full protein, vitamin, and mineral requirements; however, the grass is often lush with a high percentage of moisture, which occasionally can cause a deficiency in energy.[12]

Estrous cycle control

Development of estrus synchronization protocols

Factors such as nutrition, management, and inefficient detection of estrus affect the widespread use of embryo transfer in commercial beef cattle operations. The most useful alternative for increasing the number of animals receiving embryos is to utilize protocols that allow for embryo transfer without the need for estrus detection, usually called fixed-time embryo transfer (FTET) protocols. However, much of the research related to the systems currently used in embryo transfer programs was for fixed-time artificial insemination (TAI) rather than FTET. Transfer of embryos into estrus synchronized cows has been most effective when embryos were transferred 6–8 days after detected estrus or gonadotropin-releasing hormone (GnRH) injection.[13] Early estrus synchronization systems focused on altering the estrous cycle by inducing luteolysis with an injection of prostaglandin (PG)$F_{2\alpha}$ followed by estrus detection. Once systems involving a single PGF$_{2\alpha}$ treatment became successful, researchers focused on multiple injections of PGF$_{2\alpha}$ to further reduce days required for estrus detection.[14,15] The next generation of estrus synchronization systems involved the use of exogenous progestins, such as an intravaginal progesterone release insert (CIDR) or melengestrol acetate (MGA), which were used to delay the time of estrus following natural or induced luteolysis and extend the length of the estrous cycle.[16,17]

Not until the discovery that growth of follicles in cattle occurs in distinct wave-like patterns[18] were scientists able to embark on the third generation systems for estrus synchronization. Controlling follicular waves with a single injection of GnRH at random stages of the estrous cycle involves release of a surge of luteinizing hormone (LH) that causes synchronized ovulation or luteinization of dominant follicles.[19–21] Consequently, a new follicular wave is initiated in most

Selection and Management of the Embryo Recipient Herd for Embryo Transfer

(>60%) cows within 1–3 days of GnRH administration. Luteal tissue that forms after GnRH administration will undergo $PGF_{2\alpha}$-induced luteolysis 6–7 days later.[22] A drawback of this method of estrus synchronization is that approximately 5–15% of cows are detected in estrus on, or before, the day of $PGF_{2\alpha}$ treatment, reducing the proportion of females that are detected in estrus during the synchronized period.[23–25]

Advances in protocols for beef cows

Preliminary studies identified significant improvements in fertility among cows that received MGA prior to the administration of $PGF_{2\alpha}$ compared with cows that received only $PGF_{2\alpha}$.[26] When cows received a CIDR for 7 days and an injection of $PGF_{2\alpha}$ the day before CIDR removal, estrus synchrony and pregnancy rates were improved.[17] When GnRH was given 6 or 7 days prior to $PGF_{2\alpha}$, 70–83% of cows were in estrus within a 4-day period.[22]

The use of GnRH to control follicular wave emergence and ovulation and $PGF_{2\alpha}$ to induce luteolysis led to the development of the Ovsynch protocol for dairy cows.[27] Combining the second injection of GnRH with TAI (CO-synch) proved to be more practical than estrus detection for beef producers because it had no negative effects on fertility.[28] However, a disadvantage of this protocol is that approximately 5–15% of suckled beef cows exhibit estrus prior to, or immediately after, the $PGF_{2\alpha}$ treatment.[24] Unless these cows are detected in estrus and inseminated, they will fail to become pregnant to TAI. Therefore we hypothesized that the addition of a CIDR to a GnRH-based protocol would prevent the premature occurrence of estrus and result in enhanced fertility following TAI. Overall pregnancy rates were enhanced by the addition of a CIDR to a GnRH-based TAI protocol (59% vs. 48%, respectively). The CIDR delayed the onset of ovulation, resulting in more synchronous ovulation, and induced cyclicity in noncycling cows.[24] However, the efficacy of these CIDR-based TAI protocols had not been evaluated concurrently with AI protocols requiring detection of estrus in suckled beef cows. Therefore, we implemented and coordinated a multi-state, multi-location experiment to discern whether a GnRH-based plus CIDR protocol for TAI could yield pregnancy rates similar to protocols requiring detection of estrus.[29] Results demonstrated that the TAI protocol yielded pregnancy rates that were similar to the estrus detection protocol, even though 35% of the cows were in postpartum anestrus at the time of treatment.

A detailed version of current estrus synchronization and TAI protocols reviewed by the Beef Reproduction Task Force is available in Figure 78.1. Utilizing a similar protocol on recipients using FTET would be practical and effective in yielding high pregnancy rates in recipients.

Advances in protocols for beef heifers

Early studies in beef heifers demonstrated that feeding MGA for 14 days followed by $PGF_{2\alpha}$ 17 days later was an effective method of estrous cycle control in heifers.[16,30] However, when heifers were treated with $PGF_{2\alpha}$ 19 days after the 14-day MGA feeding period, there was no difference in fertility but estrus was more synchronous.[31] Following the success of this protocol, researchers began to include GnRH

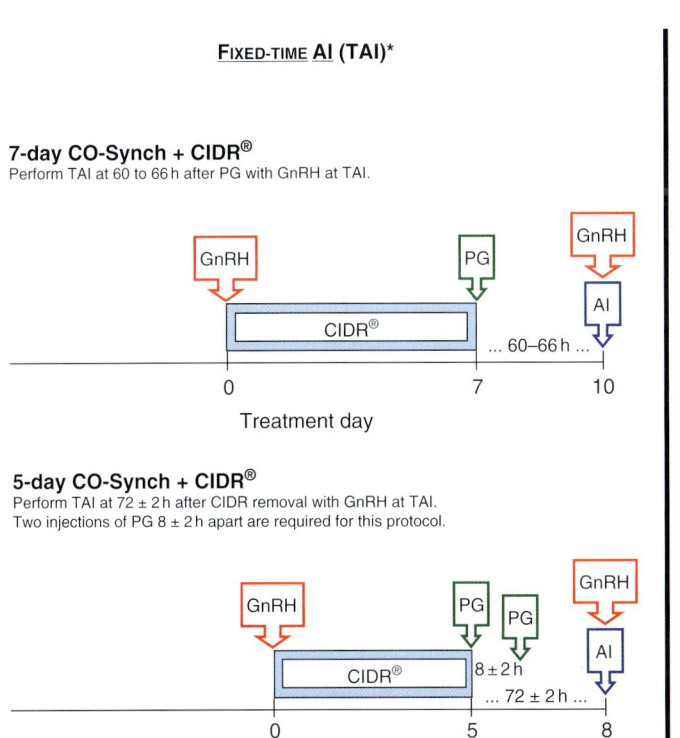

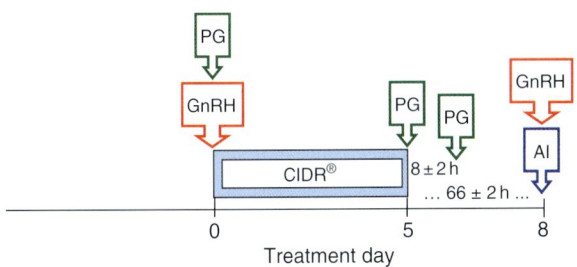

* The time listed for "fixed-time AI" should be considered as the approximate average time of insemination. This should be based on the number of cows to inseminate, labor, and facilities.

Figure 78.1 Current protocols for estrus synchronization and TAI of beef cows reviewed by the Beef Reproduction Task Force. Available at http://beefrepro.unl.edu.

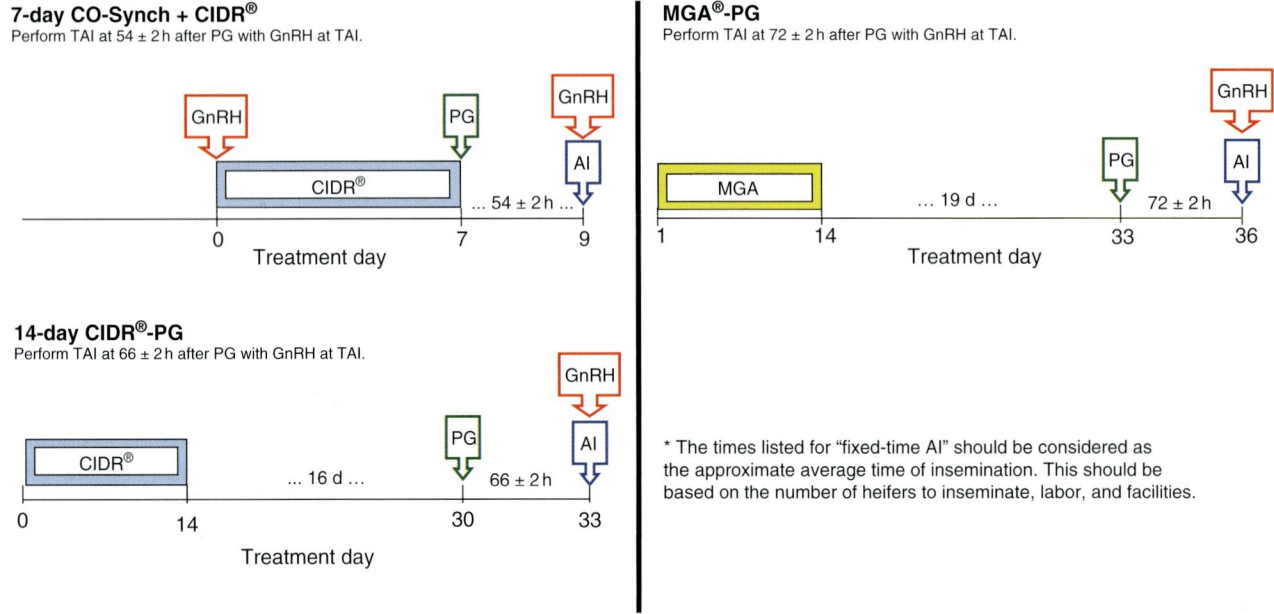

Figure 78.2 Current protocols for estrus synchronization and TAI of beef heifers reviewed by the Beef Reproduction Task Force. Available at http://beefrepro.unl.edu.

in estrus synchronization protocols for TAI. However, addition of GnRH to the above protocol failed to increase pregnancy rates following TAI in heifers.[32] Estrus synchronization using GnRH followed by $PGF_{2\alpha}$ successfully synchronized heifers, but the above MGA–$PGF_{2\alpha}$ protocol led to greater synchrony of estrus and therefore tended to be more effective.[31]

Development of a TAI protocol in beefs heifers has not been as straightforward as in cows, especially considering that at the time of estrus synchronization a majority (>85%) of heifers have attained puberty.[33] The primary reason for failure of TAI in heifers appears to be the inability to synchronize follicular waves with GnRH. After an injection of GnRH at random stages of the estrous cycle, 75–90% of postpartum beef cows ovulated,[34,35] whereas only 48–60% of beef and dairy heifers ovulated in response to the same treatment.[27,36,37] We have found no difference in synchrony of estrus or pregnancy rate in CIDR-treated heifers whether or not GnRH is administered at CIDR insertion, suggesting that response to GnRH in heifers at CIDR insertion may be of limited value.[33]

In a large multi-location study using GnRH, $PGF_{2\alpha}$, and CIDR, GnRH did not enhance pregnancy rates following estrus detection but the addition of a CIDR to a GnRH-based TAI protocol yielded similar pregnancy rates to those utilizing estrus detection.[33] Nevertheless, a bewildering fact remains that the average pregnancy rate for these protocols ranged from 53 to 58%, whereas pregnancy rates in MGA (with $PGF_{2\alpha}$ administered 19 days after MGA removal) or long-term CIDR (with $PGF_{2\alpha}$ administered 16 days after MGA removal) protocols followed by $PGF_{2\alpha}$ have been reported to range from 60 to 75%.[23,31,33,38] A detailed version of current estrus synchronization and TAI protocols reviewed by the Beef Reproduction Task Force is available in Figure 78.2. Utilizing a similar protocol on recipients using FTET would be practical and effective in yielding high pregnancy rates in heifer recipients.

Resynchronization of estrus

Reinsemination of nonpregnant cows at the first eligible estrus can be facilitated by resynchronization of the estrous cycle,[39] which would have wide application in intense embryo transfer programs. To most effectively condense the calving season, the second round of estrus synchronization needs to begin before the pregnancy status of the animals is known. Although resynchronization with a progestin increased synchronized return rates of nonpregnant females,[40,41] resynchronization with CIDR devices and estradiol cypionate or estradiol benzoate decreased subsequent conception rates to AI.[40] In contrast, further studies did not note a decrease in fertility when estrogens were utilized for resynchronization with a CIDR.[42] Furthermore, insertion of a CIDR for 13 days on the day of embryo transfer 7 days after estrus[43] or from 5 days after TAI until day 21[29] was effective in resynchronizing estrus in nonpregnant cows, but insertion of a CIDR failed to enhance fertility compared to controls.

Recipient-related factors

Embryo transfer factors

The procedure of removing an embryo from its natural uterine environment, and in many cases freezing and thawing, increases the stress experienced by those embryos resulting in a decreased survival rate following transfer. Our findings of a decrease in pregnancy rate from 83% with fresh embryos ($n=122$) to 69% with frozen-thawed embryos

($n=326$) are similar to the 10–15% decrease in pregnancy rates reported previously,[44,45] which is similar to the difference in averages reported by the American Embryo Transfer Association and the International Embryo Transfer Association (Savoy, IL). Results from two studies reveal that pregnancy rates among cows receiving a Grade 1 or Grade 2 fresh embryo[46,47] were not different. Previous reports[48–51] noted a decrease in pregnancy rate with each corresponding decrease in quality score.

Pregnancy rates have been shown to vary with the synchrony of the donor and recipient. Higher pregnancy rates were observed when recipients were in estrus coinciding with the donor or 12 hours before the donor. Pregnancy rates decreased in recipients in estrus 12 hours after the donor[50] but not until 24 hours in other reports.[45–47]

The variability in progesterone concentrations in recipients reflects a combination of different rates of CL development and the fluctuation of progesterone secretion during the early luteal phase. It has been suggested that the optimum circulating concentration of progesterone for establishing pregnancy ranges between 2.0 and 5.0 ng/mL.[52] However, a recent study has revealed that the minimum threshold progesterone concentration on the day of embryo transfer essential for the establishment and maintenance of pregnancy may be lower than previously reported; there were no differences in pregnancy rates when progesterone concentrations were as low as 0.58 ng/mL or exceeded 16.0 ng/mL ($n=448$).[47] In another study, 8 of 177 pregnant recipients had concentrations of progesterone of less than 0.5 ng/mL on days 10, 11, and 12 of the transfer cycle.[53] In addition, the diameter and volume of the CL differed among recipients that received embryos from 6.5 to 8.5 days after estrus,[47] increasing as days after estrus increased. However, pregnancy rates did not differ among recipients receiving embryos 6.5–8.5 days after estrus.

Recipient movement while the embryo transfer gun is in the uterus increases the risk of damage to the endometrium, causing the release of $PGF_{2\alpha}$. Pregnancy rates have been inversely correlated with the time spent in the uterus during embryo transfer.[54] In addition, when the prostaglandin synthesis inhibitor flunixin meglumine was administered at the time of embryo transfer, pregnancy rates were reportedly enhanced, but only when poorer-quality embryos were transferred.[43,55] However, the FDA has released a reminder that flunixin meglumine is only approved for intravenous use in cattle and extra-label drug use has resulted in increased residues found at slaughter.

General recipient considerations

Selection and identification of high-quality recipients is not simple. Many prefer the use of virgin heifers, whereas others choose cows with a known history of high fertility. When heifers are to be used for recipients, the selection criteria should be the same as for high-quality replacement heifers. Heifers need to be cycling (which can be assessed indirectly by using reproductive tract scores[56]), on a high plane of nutrition, have an adequately sized and normally shaped pelvic canal, and have no history of receiving growth implants.

Lactating recipients have an advantage of a known reproductive history. Since the health of the calf is dependent on the recipient, records should be kept of calf health and weaning performance. Recipients that carry an embryo transfer calf to term but do not raise a normal calf to weaning should be reevaluated as a recipient prospect. Similarly, open cows with an unknown reproductive history need to be carefully examined prior to being included in a recipient herd or program.[57] The reproductive tract needs to be thoroughly examined via rectal palpation or transrectal ultrasonography for pregnancy or uterine anomalies such as fluid or fetal remnants or evidence of metritis or endometritis and the ovaries examined for normal follicular or luteal structures. In addition, recipients should have good teeth and eyes and a good udder and be less than 8 years of age and structurally sound. Highest fertility occurs in herds where handling facilities are designed to ensure that cattle are handled with a minimum of stress.

It is wise to keep the new arrivals separate from the breeding herd until sufficient time has elapsed for diagnostic screening tests to return and any incubating disease to become apparent. Many purebred producers and embryo transfer companies take blood samples to test for exposure to bovine leukosis virus (BLV), *Mycobacterium paratuberculosis* (Johne's disease), bovine viral diarrhea virus (BVDV), anaplasmosis, and *Neospora caninum*. Brucellosis testing or vaccination is no longer required in many areas but it is prudent to test cattle from areas where the disease is present in wildlife. Many of these pathogens have been associated with decreased fertility, by preventing fertilization or by causing embryonic death, fetal loss or ovarian dysfunction.[58] The use of vaccinations to control livestock diseases is a common and proven practice. Conventional recommendations suggest that modified-live virus vaccines be given at least 30 days prior to breeding. However, recent research has shown that in previously vaccinated cattle, there are no detrimental effects of vaccinating cattle at the time of $PGF_{2\alpha}$ treatment or initial GnRH treatment.[58] Cattle with an unknown or questionable history of vaccination should receive primary and booster vaccinations at least 30 days prior to breeding.[59]

Management after confirmation of pregnancy

Once the recipient is confirmed pregnant, she needs to be managed to remain pregnant. There are several environmental risks that cause abortion depending on the area of the country where the cow is maintained. Some of these include *N. caninum*, locoweed, ponderosa pine needles, fescue toxicity, nitrates, and mycotoxins and/or moldy feed. Handling stress has been shown to cause heifers to abort, but no reliable data have focused on the ideal interval after embryo transfer to transport recipients. However, pregnancy rates were lower in females that were transported between 8 and 33 days after AI as compared with those transported within the first 4 days.[60] Therefore, it appears to be more desirable to transport recipients prior to embryo hatching than after hatching. Infectious agents (e.g., BVDV, infectious bovine rhinotracheitis and, rarely, bluetongue virus) are also a threat to the developing fetus. Often it is not possible to diagnose the cause of abortions but diagnostic success can be increased if the proper samples and history are submitted.[61]

Biosecurity

To protect the herd from the introduction of infectious diseases that could cause abortion or reduce the value of the calves, a well-designed biosecurity program needs to be in place. This includes a test and quarantine program for all new cattle introduced into the herd. A commonsense program of single-use needles and rectal sleeves, disinfection of equipment between cows (drench applicators, tattoo pliers, and other instruments), external parasite control, the use of clean coveralls for employees and visitors as well as footbaths or shoe coverings, and not allowing equipment or clothing to be taken from one farm to another without first thorough washing and disinfection can minimize risk of introducing pathogens onto a facility.

Pregnancy diagnosis

Knowing when cows conceive and when they will calve helps concentrate calving supervision. Ultrasonography can be used to accurately determine the presence of a conceptus as early as 28 days, but it is recommended to recheck all cows after 45 days to confirm pregnancy.[62] Through the use of ultrasonography and breeding dates to determine the estimated date of calving, cows can be sorted into calving groups and managed to save on feed, labor, and veterinary expenses. Pre-calving vaccinations can also be timed to insure the most effective response. Also, avoiding overcrowding of calving pastures and/or calving cows on "fresh" pastures that have not been grazed by cows with calves has been shown to reduce calf morbidity and mortality due to infectious calfhood scours.[63]

Management for efficient recipient utilization

Effective management of a recipient herd requires preparing the recipient to receive an embryo and identifying and preparing open cows to be resynchronized and reused or inseminated. In any group of synchronized recipients, a small percentage will not be detected in estrus and not all detected in estrus will receive an embryo, due to either an asynchronous estrus or lack of a suitable CL at the time of transfer. If 80% of the synchronized recipients are detected in estrus and 90% of those receive embryos and 60% become pregnant, then less than 45% of any group of recipients will become pregnant. Therefore, it is important to devise a strategy to resynchronize recipients as soon as possible.

In large herds with extended calving seasons, it is necessary to synchronize more than one group of recipients. Resynchronization strategies vary depending on the resources and capabilities of the ranch. With the use of ultrasonography, nonpregnant recipients may be confidently identified and resynchronized as early as 3 weeks after embryo transfer.[64] A CIDR can also be inserted at the time of transfer and removed 12–13 days later; ultrasonographic pregnancy diagnosis can then be performed before reuse.

We do not transfer an embryo to a recipient more than twice in any season. Unpublished data from Granada BioSciences Inc., Marquez, Texas demonstrated that the pregnancy rate between recipients that became pregnant following one or two transfers was not significantly different, but less than 20% of the recipients that received a third embryo transfer became pregnant. However, a single report has reported up to a 12% decrease in pregnancy rates for recipients receiving a second embryo transfer.[65]

Many purebred producers contract with commercial cow/calf producers to use cows as recipients and purchase the weaned embryo transfer calf. Normally, the cooperator synchronizes and observes cows for estrus, sorts recipients 6–8.5 days after estrus, and provides the labor to move the cows through the working facility for a technician to transfer embryos. The purebred producer pays the synchronization expenses and the embryo transfer technician, and pays the cow owner a premium for every embryo transfer calf weaned. This arrangement can be a win–win for both producers. However, there are issues that arise when first beginning this program. Not all commercial producers have the satisfactory management; a commercial producer already familiar with AI is a good candidate for such a program. It is essential that a producer has a good record-keeping system, is able to weigh and identify each calf at birth, be comfortable with estrus detection, and have a good herd health program. The purebred producer may expect the cooperator to screen cows for the aforementioned diseases for biosecurity purposes.

American Embryo Transfer Association Survey

In 2011, we conducted a survey of all members of the American Embryo Transfer Association (AETA), regarded as leaders in embryo transfer of cattle in the United States. The goal of the survey was to assess the utilization of current embryo transfer practices, the perception by AETA members of each practice, and whether perceptions were supported by research. A total of 218 professional members received the survey, of whom 63 (29%) responded to the survey; 47 (75%) of the members who responded were certified AETA members, while 27 (43%) classified their embryo transfer business as primarily beef cattle related, 13 (21%) as dairy cattle related, 21 (34%) as a mixture of beef and dairy, and one (2%) member was not cattle related. We asked the survey participants to rate their perception regarding several factors that could potentially affect the fertility of the recipients to embryo transfer, and the results are shown in Figure 78.3. Embryo quality was rated as the factor with the greatest impact on fertility. Quality grade of the embryo has a huge impact on fertility; in a study comparing pregnancy rates of embryo transfer using embryos graded from 1 to 3 (1 being excellent and 3 being poor), recipients receiving grade 1 embryos had greater pregnancy rates than recipients receiving grade 3 embryos (56.1% vs. 33.3%, respectively).[66] Embryo transfer technician was rated the second most important factor affecting fertility of embryo transfer; indeed differences in pregnancy rates can be seen across embryo transfer technicians[67] and pregnancy rates have been inversely correlated to the time spent in the uterus during embryo transfer.[54]

Recipient BCS at embryo transfer was rated of moderate impact on fertility and the effect of nutrition on the success of embryo transfer programs has been discussed in the section on nutritional management. Although rated as moderate,

Selection and Management of the Embryo Recipient Herd for Embryo Transfer

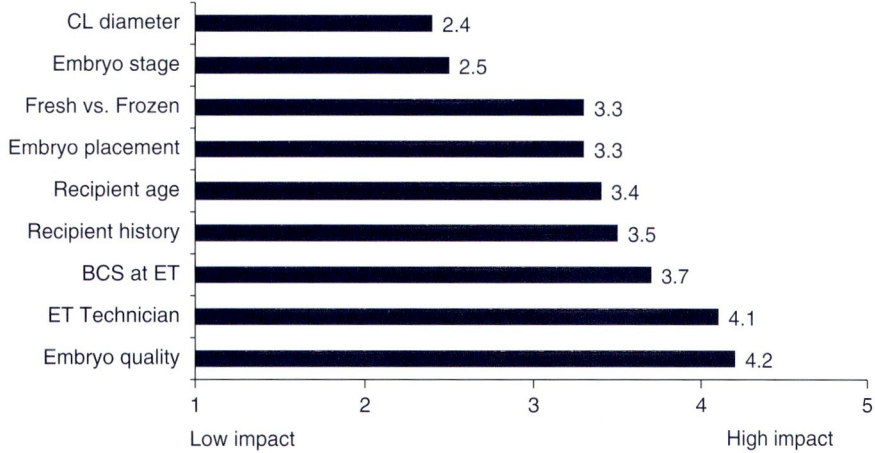

Figure 78.3 Perception of the relative impact on fertility of recipients to embryo transfer (1, least impact to 5, greatest impact). BCS, body condition score; ET, embryo transfer.

the use of fresh or frozen embryos as well as embryo stage at transfer and recipient history has a great impact on the fertility of embryo transfer programs.[67] The diameter of the CL is perceived as having a low impact on fertility of recipients of embryo transfer. Indeed, in a study evaluating luteal characteristics and plasma concentrations of progesterone in pregnant and nonpregnant recipients, no differences were found in CL diameter, luteal volume, and plasma progesterone concentrations.[47]

Surprisingly, 61% of the survey participants claimed to have no experience with the use of human chorionic gonadotropin (hCG), while 23% claimed that the use of hCG is unsuccessful, and only 16% believed that the use of hCG is successful in embryo transfer programs. Nevertheless, our recent work[68] indicated that treatment with hCG at embryo transfer increased the incidence of accessory corpora lutea formation, increased progesterone in pregnant recipients, and increased transfer pregnancy rates. At three locations, purebred and crossbred Angus, Simmental, and Hereford recipients (n = 719) were assigned alternately to receive hCG 1000 IU i.m. or 1 mL saline (control) at embryo transfer. Pregnancy rates at the first diagnosis on day 33 were 61.8 and 53.9% for hCG and controls, respectively (Figure 78.4) and were 59.0 and 51.4%, respectively at the second diagnosis on day 68. In agreement with our findings, the use of hCG at embryo transfer also increased pregnancy rates of Holstein recipient cows receiving fixed-time embryo transfer.[69]

When questioned about the effects of flunixin meglumine (FM), a prostaglandin synthesis inhibitor, on fertility of embryo transfer recipients, 53% of the survey participants claimed that the use of FM was not successful, 18% claimed to have no experience with FM, and only 29% rated the use of FM as successful in embryo transfer programs. In a study evaluating the effects of FM prior to embryo transfer, pregnancy rates were greater for recipients receiving an injection of FM. However, the results varied with location.[43] In another study, FM treatment improved overall pregnancy rates of recipients (65.3% vs. 60.2% for FM and control, respectively) and the effects of FM appeared to be even more beneficial when transferring Grade 2 embryos.[70]

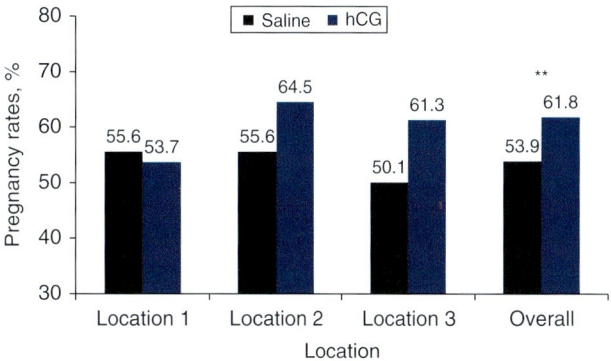

Figure 78.4 Effects of hCG 1000 IU at embryo transfer on pregnancy rates at day 33 (fresh and frozen embryos combined). **, hCG differs from saline, $P = 0.03$). Adapted from Wallace L, Breiner C, Breiner R et al. Administration of human chorionic gonadotropin at embryo transfer induced ovulation of a first wave dominant follicle, and increased progesterone and transfer pregnancy rates. *Theriogenology* 2011;75:1506–1515.

Finally, we asked participants for their perception of the success of FTET programs; 73% claimed that FTET is successful, while 14% claimed that FTET is unsuccessful and 13% had no experience with FTET. The use of synchronization protocols excludes the need for estrus detection and increases the number of animals receiving embryos. Most current FTET protocols are based on progestin-releasing devices combined with estradiol or GnRH, which control and synchronize follicular wave dynamics and ovulation. Conception rates to a single FTET have been reported to be similar to those after detection of estrus (Figure 78.5); however, pregnancy rates are higher because these treatments have increased the proportion of recipients that receive an embryo.[66]

Conclusion

For an embryo transfer program to be effective, numerous factors need to be put in place to ensure success. Nutrition, estrous cycle control, and recipient management are all

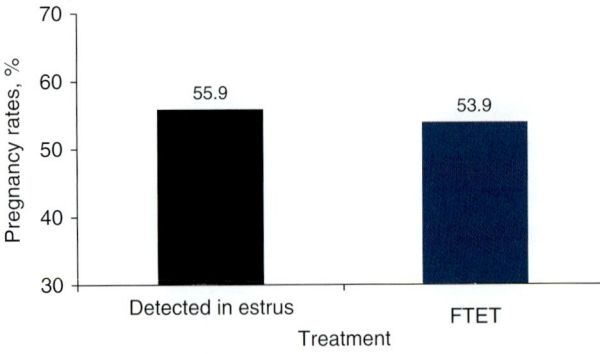

Figure 78.5 Pregnancy rates after embryo transfer (frozen embryos) to recipients detected in estrus or to FTET ($P = 0.36$). Adapted from Bó G, Peres L, Cutaia L et al. Treatments for the synchronisation of bovine recipients for fixed-time embryo transfer and improvement of pregnancy rates. *Reprod Fertil Dev* 2011;24:272–277.

responsible for the success or failure in a given program. Producers, embryologists, veterinarians, and all members of the recipient management team need to be aware of the short- and long-term factors that contribute to recipients conceiving to embryo transfer, maintaining the embryo/fetus to term, delivering the calf without assistance, and raising and weaning a healthy calf.

Summary

A commercially viable cattle embryo transfer industry was established during the early 1970s. Initially, techniques for transferring cattle embryos were exclusively surgical. However, by the early 1980s, most embryos were transferred nonsurgically. For an embryo transfer program to be effective, numerous factors need to be put in place to ensure success. Nutrition, estrous cycle control, and recipient management are all responsible for the success or failure in fertility for a given herd. Utilization of BCS is a practical method for determining nutritional status of the recipient herd. Prepartum nutrition is critical to ensure that cows calve in adequate body condition to reinitiate postpartum estrous cycles early enough to respond to synchronization protocols. Estrus synchronization for embryo transfer after detected estrus or for fixed-time embryo transfer are effective methods for increasing the number of calves produced by embryo transfer. In addition, resynchronization of nonpregnant females effectively ensures that a high percentage of recipients will return to estrus during a 72-hour interval and are eligible for subsequent embryo transfers. Numerous additional factors need to be assessed to ensure that the recipient herd achieves its reproductive potential. These factors include assessing the merits of nulliparous, primiparous, or multiparous cows, ensuring that facilities allow for minimal stress, and that the herd health program is well defined and followed. Numerous short- and long-term factors contribute to recipients conceiving to a transferred embryo, maintaining the embryo/fetus to term, delivering the calf without assistance, and raising and weaning a healthy calf.

References

1. Seidel GE Jr, Seidel SM. Applications of embryo transfer. In: *Training Manual for Embryo Transfer in Cattle*. Rome: FAO, 1991, pp. 3–13.
2. Wright J. Non-surgical embryo transfer in cattle, embryo-recipient interaction. *Theriogenology* 1981;15:43–56.
3. Short R, Adams D. Nutritional and hormonal interrelationships in beef cattle reproduction. *Can J Anim Sci* 1988;68:29–39.
4. Whitman R. *Weight change, body condition and beef-cow reproduction*. PhD Dissertation, Colorado State University, Fort Collins, 1975.
5. Mapletoft R, Lindsell C, Pawlyshyn V. Effects of clenbuterol, body condition and non-surgical embryo transfer equipment on pregnancy rates in bovine recipients. *Theriogenology* 1986; 25:172 (Abstract).
6. Beal W. Streamlining embryo transfer. In: *18th Annual Convention of the American Embryo Transfer Association*, Colorado Springs, CO, 1999, pp. 78–85.
7. Lalman D, Keisler D, Williams J, Scholljegerdes E, Mallet D. Influence of postpartum weight and body condition change on duration of anestrus by undernourished suckled beef heifers. *J Anim Sci* 1997;75:2003–2008.
8. Stevenson J, Johnson S, Milliken G. Incidence of postpartum anestrus in suckled beef cattle: treatments to induce estrus, ovulation, and conception. *Prof Anim Sci* 2003;19:124–134.
9. Perry R, Corah L, Cochran R et al. Influence of dietary energy on follicular development, serum gonadotropins, and first postpartum ovulation in suckled beef cows. *J Anim Sci* 1991;69:3762–3773.
10. Spitzer J, Morrison D, Wetteman R, Faulkner L. Reproductive responses and calf birth and weaning weights as affected by body condition at parturition and postpartum weight gain in primiparous beef cows. *J Anim Sci* 1995;73:1251–1257.
11. Wiltbank J, Rowden W, Ingalls J, Gregory K, Koch R. Effect of energy level on reproductive phenomena of mature Hereford cows. *J Anim Sci* 1962;21:219–225.
12. National Research Council. *Nutrient Requirements of Beef Cattle*, 7th edn. Washington, DC: National Academy Press, 1996, pp. 85–96.
13. Bó G, Baruselli P, Moreno D et al. The control of follicular wave development for self-appointed embryo transfer programs in cattle. *Theriogenology* 2002;57:53–72.
14. Lauderdale J, Seguin B, Stellflung J et al. Fertility of cattle following $PGF_{2\alpha}$ injection. *J Anim Sci* 1974;38:964–967.
15. Seguin B, Gustafson B, Hurtgen J et al. Use of the prostaglandin F2α analog cloprostenol (ICI 80,996) in dairy cattle with unobserved estrus. *Theriogenology* 1978;10:55–64.
16. Brown L, Odde K, LeFever D, King M, Neubauer C. Comparison of MGA- $PGF_{2\alpha}$ to Syncro-Mate B for estrous synchronization in beef heifers. *Theriogenology* 1988;30:1–12.
17. Lucy M, Billings H, Butler W et al. Efficacy of an intravaginal progesterone insert and an injection of PGF2alpha for synchronizing estrus and shortening the interval to pregnancy in postpartum beef cows, peripubertal beef heifers, and dairy heifers. *J Anim Sci* 2001;70:1904–1910.
18. Fortune J, Sirois J, Quirk S. The growth and differentiation of ovarian follicles during the bovine estrous cycle. *Theriogenology* 1988;29:95–109.
19. Garverick H, Elmore R, Vaillancourt D, Sharp A. Ovarian response to gonadotropin-releasing hormone in postpartum dairy cows. *Am J Vet Res* 1980;41:1582–1585.
20. Bao B, Garverick H. Expression of steroidogenic enzyme and gonadotropin receptor genes in bovine follicles during ovarian follicular waves: a review. *J Anim Sci* 1998;76:1903–1921.
21. Sartori R, Fricke P, Ferreira J, Ginther O, Wiltbank M. Follicular deviation and acquisition of ovulatory capacity in bovine follicles. *Biol Reprod* 2001;65:1403–1409.
22. Twagiramungu H, Guilbault A, Dufour J. Synchronization of ovarian follicular waves with a gonadotropin-releasing hormone

agonist to increase the precision of estrus in cattle: a review. *J Anim Sci* 1995;73:3141–3151.
23. Kojima F, Salfen B, Ricke W, Lucy M, Smith M, Patterson D. Development of an estrus synchronization protocol for beef cattle with short-term feeding of melengestrol acetate: 7–11 Synch. *J Anim Sci* 2000;78:2186–2191.
24. Lamb G, Stevenson J, Kesler D, Garverick H, Brown D, Salfen B. Inclusion of an intravaginal progesterone insert plus GnRH and prostaglandin F2α for ovulation control in postpartum suckled beef cows. *J Anim Sci* 2001;79:2253–2259.
25. Martinez M, Kastelic J, Adams G, Mapletoft R. The use of GnRH or estradiol to facilitate fixed-time insemination in an MGA-based synchronization regimen in beef cattle. *Anim Reprod Sci* 2001;67:221–229.
26. Patterson D, Hall J, Bradley N, Schillo K, Woods B, Kearnan J. Improved synchrony, conception rate, and fecundity in postpartum suckled beef cows fed melengestrol acetate prior to prostaglandin $F_{2\alpha}$. *J Anim Sci* 1995;73:954–959.
27. Pursley J, Mee M, Wiltbank M. Synchronization of ovulation in dairy cows using PGF2alpha and GnRH. *Theriogenology* 1995; 44:915–923.
28. Geary T, Whittier J, Hallford D, MacNeil M. Calf removal improves conception rates to the Ovsynch and CO-Synch protocols. *J Anim Sci* 2001;79:1–4.
29. Larson J, Lamb G, Stevenson J *et al.* Synchronization of estrus in suckled beef cows for detected estrus and artificial insemination and timed artificial insemination using gonadotropin-releasing hormone, prostaglandin F2alpha, and progesterone. *J Anim Sci* 2006;84:332–342.
30. Patterson D, Kiracofe G, Stevenson J, Corah L. Control of the bovine estrous cycle with melengestrol acetate (MGA): a review. *J Anim Sci* 1989;67:1895–1906.
31. Lamb G, Nix D, Stevenson J, Corah L. Prolonging the MGA-prostaglandin $F_{2\alpha}$ interval from 17 to 19 days in an estrus synchronization system for heifers. *Theriogenology* 2000; 53:691–698.
32. Wood-Follis S, Kojima F, Lucy M, Smith M, Patterson D. Estrus synchronization in beef heifers with progestin-based protocols. I. Differences in response based on pubertal status at the initiation of treatment. *Theriogenology* 2004;62:1518–1528.
33. Lamb G, Larson J, Geary T *et al.* Synchronization of estrus and artificial insemination in replacement beef heifers using GnRH, $PGF_{2\alpha}$, and progesterone. *J Anim Sci* 2006;84:3000–3009.
34. Thompson K, Stevenson J, Lamb G, Grieger D, Loest C. Follicular, hormonal, and pregnancy responses of early postpartum suckled beef cows to GnRH, norgestomet, and prostaglandin$F_{2\alpha}$. *J Anim Sci* 1999;77:1823–1832.
35. El-Zarkouny S, Cartmill J, Hensley B, Stevenson J. Pregnancy in dairy cows after synchronized ovulation regimens with or without presynchronization and progesterone. *J Dairy Sci* 2000; 87:1024–1037.
36. Macmillan K, Thatcher W. Effects of an agonist of gonadotropin-releasing hormone on ovarian follicles in cattle. *Biol Reprod* 1991;45:883–889.
37. Moreira F, de la Sota R, Diaz T, Thatcher W. Effects of day of the estrous cycle at the initiation of a timed artificial insemination protocol on reproductive responses in dairy heifers. *J Anim Sci* 2000;78:1568–1576.
38. Patterson D, Wood S, Randle R. Procedures that support reproductive management of replacement beef heifers. *J Anim Sci* 2000;77:1–15.
39. Van Cleeff J, Macmillan K, Drost M, Thatcher W. Effects of administering progesterone at selected intervals after insemination of synchronized heifers on pregnancy rates and resynchronization of returns to estrus. *Theriogenology* 1996; 4:1117–1130.
40. Stevenson J, Johnson S, Medina-Britos M, Richardson-Adams A, Lamb G. Resynchronization of estrus in cattle of unknown pregnancy status using estrogen, progesterone, or both. *J Anim Sci* 2003; 81:1681–1692.
41. Colazo M, Kastelic J, Mainar-Jaime R *et al.* Resynchronization of previously timed-inseminated beef heifers with progestins. *Theriogenology* 2006;65:557–572.
42. Cavalieri J, Hepworth G, Smart V, Ryan M, Macmillan K. Reproductive performance of lactating dairy cows and heifers resynchronized for a second insemination with an intravaginal progesterone-releasing device for 7 or 8d with estradiol benzoate injected at the time of device insertion and 24 h after removal. *Theriogenology* 2007;67:824–834.
43. Purcell S, Beal W, Gray K. Effect of a CIDR insert and flunixin meglumine, administered at the time of embryo transfer, on pregnancy rate and resynchronization of estrus in beef cattle. *Theriogenology* 2005;64:867–878.
44. Leibo S. Commercial production of pregnancies from one-step diluted frozen-thawed bovine embryos. *Theriogenology* 1986; 25:166 (Abstract).
45. Sreenan J, Diskin M. Factors affecting pregnancy rate following embryo transfer in the cow. *Theriogenology* 1987;27: 99–113.
46. Hasler J. Factors affecting frozen and fresh embryo transfer pregnancy rates in cattle. *Theriogenology* 2001;56:1401–1415.
47. Spell A, Beal W, Corah L, Lamb G. Evaluating recipient and embryo factors that affect pregnancy rates of embryo transfer in beef cattle. *Theriogenology* 2001;56:287–298.
48. Coleman D, Dailey R, Leffel R, Baker R. Estrous synchronization and establishment of pregnancy in bovine embryo transfer recipients. *J Dairy Sci* 1987;70:858–866.
49. Hasler J, McCauley A, Lathrop W, Foote R. Effect of donor–recipient interactions on pregnancy rate in a large-scale bovine embryo transfer program. *Theriogenology* 1987;27: 139–168.
50. Schneider H Jr, Castleberry R, Griffin J. Commercial aspects of bovine embryo transfer. *Theriogenology* 1980;13:73–85.
51. Wright J. Non-surgical embryo transfer in cattle: embryo–recipient interaction. *Theriogenology* 1981;15:43–56.
52. Niemann H, Sacher B, Elasaesser F. Pregnancy rates relative to recipient plasma progesterone levels on the day of non-surgical transfer of frozen/thawed bovine embryos. *Theriogenology* 1985;23:631–639.
53. Hasler J, Bowen R, Nelson L, Seidel G Jr. Serum progesterone concentrations in cows receiving embryo transfers. *J Reprod Fertil* 1980;58:71–77.
54. Rowe R, Del Campo M, Crister J, Ginther O. Embryo transfer in cattle: nonsurgical transfer. *Am J Vet Res* 1980:41:1024–1028.
55. Schrick F. Influence of a prostaglandin synthesis inhibitor administered at embryo transfer on pregnancy rates of recipient cows. *Prostaglandins Other Lipid Mediat* 2005;78:38–45.
56. Patterson D, Wood S, Randle R. Procedures that support reproductive management of replacement beef heifers. *J Anim Sci* 1999;77:1–15.
57. Stroud B, Hasler J. Dissecting why superovulation and embryo transfer usually work on some farms but not on others. *Theriogenology* 2006;65:65–76.
58. Whittier W, Baitis H. Timing of vaccinations in estrous synchronization programs. In: *Proceedings, Applied Reproductive Strategies in Beef Cattle, Lexington, KY*, 2005, pp. 147–156.
59. Daly R. Timing of vaccination in beef cattle herds. In: *Proceeding, Applied Reproductive Strategies in Beef Cattle, Rapid City, SD*, 2006, pp. 190–195.
60. Harrington T, King M, Mihura H, LeFever D, Hill R, Odde K. Effect of transportation time on pregnancy rates of synchronized yearling beef heifers. Colorado State University Beef Program Report, Fort Collins, 1995, pp. 81–86.
61. Mickelson W, Evermann J. In utero infections responsible for abortion, stillbirth and birth of weak calves in beef cows. *Vet Clin North Am Food Anim Pract* 1994;10:1–14.

62. Jones A, Beal W. Reproductive applications of ultrasound in the cow. *Bov Pract* 2003;37:1–9.
63. Smith D, Grotelueschen D, Knott T, Ensley S. Managing to alleviate calf scours: the Sandhills calving system. In: *Proceedings, The Range Beef Cow Symposium XVIII, Mitchell, NE,* 2003. http://digitalcommons.unl.edu/rangebeefcowsymp/70/
64. Jones A, Marek D, Wilson J, Looney C. The use of ultrasonography to increase recipient efficiency through early pregnancy diagnosis. *Theriogenology* 1990;33:259.
65. Looney C, Nelson J, Schneider H, Forrest D. Improving fertility in beef cow recipients. *Theriogenology* 2006;65:201–209.
66. Bó G, Peres L, Cutaia L et al. Treatments for the synchronisation of bovine recipients for fixed-time embryo transfer and improvement of pregnancy rates. *Reprod Fertil Dev* 2011;24:272–277.
67. Lamb G, Larson J, Marquenzini G, Vasconcelos J. Factors affecting pregnancy rates in an IVF embryo transfer program. In: *Proceedings of the American Embryo Transfer Association and Canadian Embryo Transfer Association Joint Convention,* 2005, pp. 31–36.
68. Wallace L, Breiner C, Breiner R et al. Administration of human chorionic gonadotropin at embryo transfer induced ovulation of a first wave dominant follicle, and increased progesterone and transfer pregnancy rates. *Theriogenology* 2011;75:1506–1515.
69. Vasconcelos J, Sá Filho O, Justolin P et al. Effects of postbreeding gonadotropin treatments on conception rates of lactating dairy cows subjected to timed artificial insemination or embryo transfer in a tropical environment. *J Dairy Sci* 2011;94:223–234.
70. Scenna F, Hockett M, Towns T et al. Influence of a prostaglandin synthesis inhibitor administered at embryo transfer on pregnancy rates in recipient cows. *Prostaglandins Other Lipid Mediat* 2005;78:38–45.

Chapter 79

Evaluation of *In Vivo*-Derived Bovine Embryos

Marianna M. Jahnke[1], James K. West[2] and Curtis R. Youngs[3]

[1]*Veterinary Diagnostic and Production Animal Medicine Department, College of Veterinary Medicine, Iowa State University, Ames, Iowa, USA*
[2]*Director of Embryo Transfer Services, Iowa State University, Ames, Iowa, USA*
[3]*Department of Animal Science, Iowa State University, Ames, Iowa, USA*

Introduction

Morphological evaluation of embryo quality is performed for three major reasons. First reason is to differentiate between embryos and unfertilized ova. Careful evaluation of harvested embryos/ova is necessary to ensure that viable embryos are not discarded or, conversely, that unfertilized ova are not transferred or cryopreserved. The second reason to perform morphological assessment of embryos harvested from a donor is to determine if the developmental stage of the embryos is consistent with the expected developmental stage based on the days since estrus that the embryos were collected. There is often a high degree of variability in observed stage of embryonic development among embryos obtained during embryo recovery from a single donor female, and this variation in stage of development should be considered when selecting a suitable recipient female. The third reason to assess embryo quality is to enable technicians to have sufficient information on which to base the decision to transfer or to cryopreserve the harvested embryos. Poorer-quality embryos that may result in pregnancies if transferred fresh usually do not survive the cryopreservation process.

Despite the importance of this topic, embryo evaluation is a difficult aspect of the overall embryo transfer process to learn, especially if attempting to self-train. Although this chapter will provide numerous helpful photographs of embryos and unfertilized ova, beginners would benefit from joining an experienced embryo transfer mentor for intensive and repetitive training sessions on morphological embryo evaluation. Proper training, proper equipment, and extensive experience are the three major factors affecting one's proficiency with embryo evaluation.

History of bovine embryo evaluation

The very earliest stages of bovine embryonic development (ovum and 2-cell embryo) were first described in 1931 by Hartman *et al*.[1] In the following decade two other research groups[2,3] described and illustrated various stages of bovine embryonic development. These very early descriptions of embryonic development served as the foundation of the morphological embryo evaluation method that subsequently evolved and became widely used in the commercial bovine embryo transfer industry.[4]

The morphological embryo evaluation method depends on the human eye to critically evaluate subtle differences among harvested embryos. During the early years of its use, the method was criticized as being too subjective, inconsistent, and prone to error. As a result, several alternative methods for assessing embryo quality were investigated for their potential to be less subjective and more consistent than morphological evaluation. Some of these methods included the ability of an embryo to continue its development during *in vitro* culture,[5,6] dye exclusion test,[7] measurement of embryonic metabolic activity,[6] glucose uptake,[8] live–dead stains,[9] computerized image analysis,[10] fluorescein diacetate vital stain,[11,12] DAPI (4′,6′-diamidino-2-phenyl-indole) staining for dead cells,[13] and *in vivo* culture in rabbit oviducts.[14]

Refinement of the morphological embryo evaluation method

Despite the numerous attempts to develop a more precise and predictable method to evaluate the viability of bovine embryos, no methods have been devised that are considered better or more accurate than morphological embryo evaluation. This method is the preferred method of embryo viability assessment under clinical conditions because it is a relatively simple and fairly rapid technique for evaluation of bovine embryos.

After the early report on bovine embryo evaluation in 1976 by Shea *et al.*,[4] three other landmark papers on embryo evaluation were published within the next 7 years.[15–17] These two groups of practitioners and scientists described the criteria that are, for the most part, still used today as a general

Bovine Reproduction, First Edition. Edited by Richard M. Hopper.
© 2015 John Wiley & Sons, Inc. Published 2015 by John Wiley & Sons, Inc.

guideline for embryo evaluation and prediction of potential pregnancy. The latest of those papers[17] described numerous parameters commonly used for embryo evaluation, including stage of embryonic development and structural features such as embryo size and shape, presence of extruded cells or degenerated cells, color characteristic, number and compactness of the cells (also called blastomeres) that comprise the early embryo, integrity of the zona pellucida, and presence or absence of vacuoles within the cytoplasm of the embryonic blastomeres.

History of the IETS grading and classification system

The International Embryo Transfer Society (IETS) is an international organization that was born in 1974 in Denver, Colorado, USA. It was created by a group of North American practitioners and scientists who recognized the need to share scientific discoveries. As the worldwide commercial embryo transfer industry grew, it became critical to develop standardized embryo nomenclature so that buyers and sellers of embryos knew precisely the stage and quality of embryos being marketed.

The IETS published in their manual[18] a recommended two-digit coding system to uniformly and systematically describe embryos and their characteristics. On completion of the morphological assessment of an embryo, it is assigned two numbers, separated by a hyphen, that correspond to the embryonic stage of development and embryo quality grade. The standardized coding system used to describe the stage of embryonic development ranges from 1 (an unfertilized oocyte) to 9 (expanding hatched blastocyst). The standardized code for embryo quality, based on morphological characteristics of embryos, is also numerical and ranges from 1 (excellent/good) to 4 (dead/degenerating). This IETS embryo grading and classification system must be used for international embryo trade, and because of this mandate most practitioners have adopted this system for everyday use. A more detailed description of the IETS embryo grading and classification system is discussed later in this chapter.

The importance of a good microscope

Before delving into the specific methodology used for morphological evaluation of embryos, it is important to remind readers that proper equipment is needed for accurate embryo evaluation. Although many beginners in embryo transfer are tempted to purchase an inexpensive microscope as a means to reduce start-up costs, doing so will likely create difficulties in finding and evaluating embryos. A poor-quality microscope can make it extremely challenging for a practitioner to tell the difference between an unfertilized ovum (UFO) and a compact morula. Similarly, a practitioner will be hampered by a poor-quality microscope when attempting to discern the various embryo quality grades.

Unlike semen evaluation, where a compound microscope is used, embryo transfer requires the use of a stereomicroscope. Stereomicroscopes, also called dissecting microscopes, are designed for use with three-dimensional specimens. Preimplantation bovine embryos, although microscopic in size, are round like a basketball and thus are best evaluated using a stereomicroscope.

There are several important features of a stereomicroscope that can directly influence one's ability to easily and accurately evaluate embryos. One important feature is the overall magnification. Overall magnification of a stereomicroscope is a function of the magnification of the eyepieces, objective lens, and zoom changer. For example, total magnification of a stereomicroscope equipped with 10× eyepieces, a 1× objective lens, and a 2× zoom changer would be approximately 20× (10 × 1 × 2 = 20). Many technicians prefer to search for embryos at a total magnification between 10× and 15×, but to effectively evaluate the fine details of an embryo and to identify any abnormalities, such as cracks in the zona pellucida, embryos should be observed under a good optical quality stereomicroscope with a magnification range that reaches at least 50×. It should be noted that the IETS-approved procedure for sanitary handling of embryos requires that embryos be inspected at a magnification of at least 50×.

Most embryo transfer technicians prefer that the eyepieces not exceed 10× in order to maintain an adequate field of view and image resolution. A microscope equipped with a diopter adjustment on one eyepiece (if not both) is very useful for compensating differences in eyesight between a technician's left and right eyes. An adjustable binocular tube will also enable multiple technicians with varying interpupillary distances to use the same stereomicroscope.

A second important feature of a stereomicroscope for embryo handling and evaluation is the working distance (distance from the embryo to the objective lens). The working distance should be long to give maximal flexibility when searching for embryos and the microscope stage should be large and free of obstacles such as stage clips, power buttons, or similar items that make sliding of Petri dishes on the stage difficult. A large and clear stage is important for providing adequate space to move Petri dishes and to operate the embryo handling device without accidently spilling the Petri dish and losing the embryos.

A third important feature of a stereomicroscope used for embryo evaluation is the illumination. The illumination should come from underneath the stage plate and it should be bright and uniform across the entire stage plate. A clear stage plate is needed; frosted stage plates are not suitable for embryo work. Brightfield illumination is standard for most stereomicroscopes, but darkfield illumination can be helpful when searching through "cloudy" fluids in a search dish.

Normal *in vivo* bovine embryonic development

In order to know whether embryos harvested from a donor have developed at the expected rate, it is important for technicians to have a good understanding of normal *in vivo* bovine embryonic development. The overall diameter of a bovine embryo is 150–190 μm (or about 15 times the size of white blood cells), which includes the zona pellucida thickness of approximately 12–15 μm.[17] The overall diameter of the zona pellucida-enclosed embryo remains virtually unchanged from the 1-cell stage of development until the blastocyst stage of development because blastomeres

Evaluation of *In Vivo*-Derived Bovine Embryos

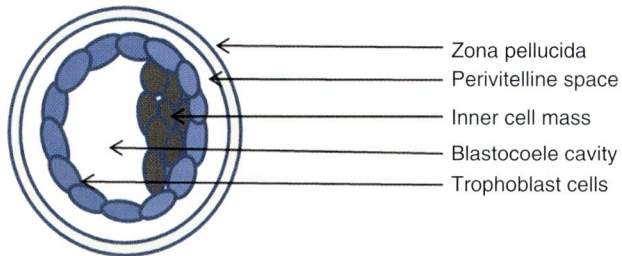

Figure 79.1 Illustration of a blastocyst-stage bovine embryo.

(individual cells of a fertilized embryo prior to cellular differentiation at the blastocyst stage of development) become progressively smaller with each cell division.

The zona pellucida is a translucent shell surrounding an embryo from the 1-cell until the expanded blastocyst stage of development. The zona pellucida functions as a physical barrier against pathogens, contains receptors for sperm, and blocks accessory sperm from reaching the ovum. Once an embryo reaches the expanded blastocyst and continues to grow, the embryo becomes so large that it ruptures the zona pellucida and "hatches."

During normal *in vivo* embryonic development, blastomeres go through a series of cleavage divisions. The ovum, after fertilization, divides and develops into a 2-cell embryo, the 2-cell develops into a 4-cell, the 4-cell into an 8-cell, and so on. When the blastomeres appear like a cluster of grapes and individual blastomeres cannot be differentiated, this stage of embryonic development is known as the morula stage. As the embryo further develops, it prepares to undergo its first differentiation event known as blastulation. Just prior to this differentiation event, cells of the morula "compact" and are allocated to the "inside" and "outside" parts of the embryo.

Once blastulation has occurred, the embryo is at the blastocyst stage of embryonic development. The blastocyst contains two populations of differentiated cells: an outer ring of trophoblast cells and an inner clump of cells known as the inner cell mass. The inner cell mass (ICM) will give rise to the fetus and most layers of the placenta, while the trophoblast cells will eventually form the outermost layer of the placenta. The blastocyst stage of embryonic development has another distinguishing feature, the presence of a blastocoele cavity. This fluid-filled cavity is surrounded by the trophoblast cells and grows progressively larger to help cause thinning of the zona pellucida prior to embryo hatching. Figure 79.1 illustrates the features observed in a blastocyst-stage bovine embryo, while Figure 79.2 shows the stages of bovine embryonic development.

IETS grading and classification system

Embryos are typically recovered 6–8 days after the onset of estrus. Harvested embryos should be classified into groups based on their stage of embryonic development and quality grade. Table 79.1 illustrates the current IETS embryo evaluation system followed by a detailed description of these developmental stages. After a technician completes the embryo evaluation process, a two-digit code is assigned to each embryo. For example, a fair-quality early blastocyst would be denoted as 5-2. A blastocyst-stage embryo of excellent or good quality would be denoted as 6-1.

Stage of embryonic development

Stage code 1: 1-cell

For any entity recovered on days 6–8 that is truly only a single cell, it is safe to assume that it is a UFO. However, a challenging task for an inexperienced technician is to distinguish a UFO from a compact morula (stage code 4, see below). Discerning a UFO from other stages of embryonic development can also be difficult because not all UFOs have the same morphological appearance. Many "normal" UFOs have a perfectly spherical zona pellucida, perfectly spherical oolemma (also called the vitelline membrane), relatively uniform granularity in the cytoplasm, and a moderate volume of perivitelline space. However, other UFOs may show varying signs of degeneration including cytoplasm fragmentation, which can lead to the illusion of a blastocoele cavity and/or appearance of cell division, and extreme condensation, which can lead to confusion with a compact morula.

Stage code 2: 2-cell to 12-cell

Cells of a preimplantation embryo prior to morphological and functional differentiation are called blastomeres. Any embryo recovered 6–8 days after the onset of estrus that contains 2–12 blastomeres is almost certainly dead or degenerate. Under normal conditions this stage of embryonic development is expected to be present in the oviduct (not the uterus) prior to day 5 of the bovine estrous cycle (day 0 = onset of estrus). Clearly, if a 2- to 12-cell embryo were somehow still viable, its development is severely delayed and it should not be considered for transfer or cryopreservation.

Stage code 3: early morula

The term "morula" has its origin in the Latin word for mulberry. Embryos at the morula stage of embryonic development consist of a group of at least 16 cells. Although some individual blastomeres are easy to visualize, others are only partially visible as they are "hidden" by other blastomeres. This stage of embryonic development, like all stages of development discussed previously, is not well suited for cryopreservation.

Stage code 4: compact morula

Individual blastomeres present in this embryo have coalesced to form a tightly compacted ball of cells. It is impossible to discern individual blastomeres at this stage of embryonic development. As a part of the compaction process, cells are allocated to the "inside" part of the ball of cells and also to the "outside" part of the ball of cells. The cells on the outside of the ball form tight junctions with one another in a contiguous ring. The cells comprising the embryo occupy approximately 60–70% of the space inside the zona pellucida (called perivitelline space). As stated earlier, it is this stage of embryonic development that can be easily confused with a UFO.

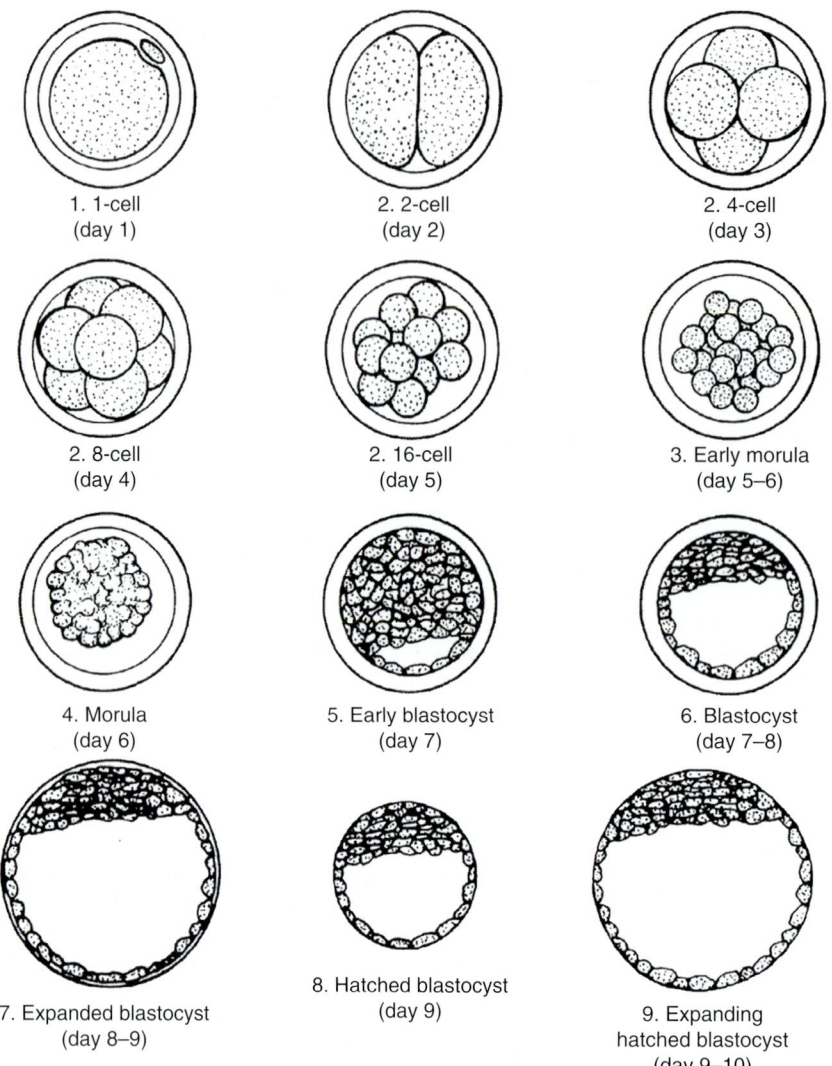

Figure 79.2 Normal embryonic development of bovine embryos. Reproduced with permission from Stringfellow D, Givens M. *Manual of the International Embryo Transfer Society*, 4th edn. Champaign, IL: International Embryo Transfer Society, 2010.

Stage code 5: early blastocyst

The most distinguishing characteristic of an early blastocyst-stage embryo is the presence of a small fluid-filled cavity known as the blastocoele. Several differentiated trophoblast cells will also be visible between the blastocoele cavity and the zona pellucida, although the ability to distinguish them from the cells of the ICM is quite limited at this stage. The cells comprising the embryo occupy approximately 70–80% of the perivitelline space. The zona pellucida is still the same thickness as the less developmentally advanced stages of embryonic development previously discussed.

Stage code 6: blastocyst

A blastocyst-stage embryo will have a clearly defined trophoblast layer, blastocoele cavity, and group of ICM cells, along with a normal-thickness zona pellucida. There will be little to no perivitelline space present except in cases where the embryo may be partially collapsed.

Stage code 7: expanded blastocyst

The most distinguishing feature of an expanded blastocyst-stage embryo is an increase in its overall diameter, 1.2–1.5 times larger than its original diameter of 150–190 μm, coupled with a thinning of the zona pellucida to approximately one-third of its original thickness of 12–15 μm. Because of the large increase in the volume of the blastocoele cavity at this stage of embryonic development, no perivitelline space is present. Interestingly, expanded blastocyst-stage embryos frequently appear collapsed. This phenomenon is characterized by a partial loss of blastocoele cavity fluid and is believed to be a normal phenomenon that contributes to embryo hatching. The zona pellucida, once thinned, never regains its original thickness.

Stage code 8: hatched blastocyst

Hatched blastocyst-stage embryos can be undergoing the process of hatching or may have completely hatched from the zona pellucida. Embryos at this stage may be spherical

Table 79.1 IETS recommended codes for embryo stage of development and embryo quality grade.

Stage of development
1. Unfertilized
2. 2-cell to 12-cell
3. Early morula
4. Morula
5. Early blastocyst
6. Blastocyst
7. Expanded blastocyst
8. Hatched blastocyst
9. Expanded hatched blastocyst

Quality of embryos
10. Excellent or good
11. Fair
12. Poor
13. Dead or degenerating

Source: Stringfellow D, Givens M. *Manual of the International Embryo Transfer Society*, 4th edn. Champaign, IL: International Embryo Transfer Society, 2010.

with a well-defined blastocele cavity or may be collapsed. Identification of this stage of embryonic development can be quite difficult if the embryo has completely hatched from the zona pellucida and has nearly totally collapsed because it can easily be mistaken for a piece of endometrial tissue. This stage of embryonic development is not eligible for international trade because the nonintact zona pellucida makes it impossible to wash the embryo in accordance with IETS procedures for sanitary handling of embryos.[19]

Stage code 9: expanded hatched blastocyst

This stage of embryonic development is identical in appearance to a completely hatched stage code 8, except for being significantly larger in diameter. It is uncommon to recover expanded hatched blastocysts from bovine donor females unless flushing is performing more than 8 days after the onset of estrus. This stage of embryonic development is not eligible for international trade.

Embryo quality grade

Embryo quality grade is determined by visual assessment of an embryo's morphological characteristics. Characteristics used to determine the quality grade of an individual embryo include uniformity of blastomeres (size, color, and shape), presence of extruded cells, presence of dead/degenerating blastomeres, degree of cytoplasmic granularity, presence of cytoplasmic vacuoles, cytoplasmic fragmentation, and structural integrity of the zona pellucida.

Quality grade code 1: excellent or good

A quality grade 1 embryo has a symmetrical and spherical embryo mass with individual blastomeres that are uniform in size, color, and density. The embryonic stage of development is consistent with that expected based on the day of the donor female's estrous cycle when the embryo was recovered. Irregularities (such as cytoplasmic vacuoles), if any, are relativity minor, and at least 85% of the embryonic cells are an intact, viable and cohesive mass (i.e., no more than 15% of embryonic cells are extruded into the perivitelline space). The zona pellucida should be spherically shaped, smooth, and with no concave or flat surfaces that might cause the embryo to adhere to a Petri dish or straw.

Quality grade code 2: fair

Moderate irregularities exist in the overall shape of the embryonic mass or in the size, color and density of individual cells comprising the embryo. Mild unevenness of cytoplasmic pigmentation within individual blastomeres (do not confuse this with translucency of the blastocoele cavity and darkness of the ICM cells of a blastocyst-stage embryo), "peppering" of cytoplasm with individual blastomeres, and/or a small number of vacuoles within the cytoplasm of blastomeres would be sufficient to warrant a quality grade 2 score. At least 50% of the embryonic cells should be an intact, viable and cohesive embryonic mass.

Quality grade code 3: poor

Major irregularities exist in the shape of the embryonic mass or in the size, color and density of individual cells. Extreme unevenness of cytoplasmic pigmentation within individual blastomeres, significant "peppering" of cytoplasm within individual blastomeres, and/or large numbers of vacuoles within individual blastomeres are sufficient to warrant a quality grade 3 score. At least 25% of the embryonic cells should be an intact, viable and cohesive embryonic mass. Multiple extruded cells of widely varying sizes, indicating a prolonged and persistent problem with the cohesiveness of embryonic blastomeres, would cause an embryo to be assigned a quality grade 3 score.

Quality grade code 4: dead or degenerating

Entities that are assigned a quality grade 4 score are nonviable and should not be transferred to recipient females or cryopreserved. Dead/degenerating embryos typically fall in this category because of extremely dark cytoplasm, nonintact cell membranes, and other significant defects. Because embryos typically are recovered on approximately day 7 of the estrous cycle, any embryo that has not advanced to at least the morula stage of embryonic development should be assigned an embryo quality grade 4 score.

Despite the strong correlation that exists between embryo quality grade and post-transfer pregnancy rate, some quality grade 1 embryos fail to produce a pregnancy after transfer whereas some quality grade 3 embryos produce a higher than expected pregnancy rate The IETS Manual[20] states:

> it should be recognized that visual evaluation of embryos is a subjective valuation of a biological system and is not an exact science. Furthermore, there are other factors such as environmental conditions, recipient quality, and technician capacity that are important in obtaining pregnancies after transfer of embryos.

Embryo evaluation procedure

During the recovery procedure, embryos are typically collected using an embryo filtration device that minimizes the volume of flushing medium through which the technician must search to locate the embryos. Some embryo filtration devices also serve as the embryo search dish, whereas other devices require that embryos be transferred from the filtration device to a Petri dish for searching. After embryos are recovered, they must be located in the recovery medium using a stereomicroscope typically under 10–15× magnification. Embryos are subsequently transferred to a Petri dish containing embryo holding medium where they will be evaluated at a minimum magnification of 50× before being assigned the two-digit IETS code for stage of development and embryo quality.

As illustrated in Figure 79.2, the compact morula, early blastocyst, and blastocyst stages of embryonic development are the ones most commonly recovered on day 7 after estrus. The expanded blastocyst is typically recovered on day 8 but can sometimes be found after a day 7 embryo recovery. The zona pellucida, because of its translucency and ability to refract light, is the landmark structure that helps technicians identify embryos and ova during the search procedure. Once an embryo progresses to the hatched blastocyst stage it may become harder to identify because it may have a collapsed appearance and, without the zona pellucida, it can easily be mistaken for endometrial cells or other cellular debris in the Petri dish.

Although each harvested embryo must be assigned its own two-digit code, there is great value in examining all harvested embryos/ova as a group. This comparative evaluation usually makes it easy for the technician to distinguish UFOs from embryos, as well as to compare embryos that are expanded and have a thinner zona pellucida with embryos that have not yet expanded.

A good predictor of an embryo's viability is its stage of development relative to what it should be on a given day after estrus. From a superovulated cow, one should expect variety in the stages of embryonic development recovered because there is variation in the time at which ovarian follicles rupture and release their ova. Similarly, one should also expect variation in embryo quality grade. It is not uncommon to obtain from the same donor female excellent-quality embryos in addition to UFOs and degenerate embryos. Based on the American Embryo Transfer Association survey report,[21] the average number of viable embryos recovered from dairy donors was 5.7 (representing 56% of the total ova/embryos recovered) and from beef donors was 6.9 (also representing 56% of total ova/embryos recovered). The average number of degenerate embryos and unfertilized ova recovered was 1.6 (14% of total embryos/ova) and 3.1 (29% of total embryos/ova), respectively, from dairy donors and 2.0 (15% of total ova/embryos) and 3.5 (27% of total ova/embryos), respectively, from beef donors.

Practical application of embryo evaluation guidelines

To help technicians learn how to properly evaluate bovine *in vivo* derived embryos, Figure 79.3 depicts (using photographs of actual bovine ova and embryos) the range in embryonic developmental stage that may be obtained after performing a day 7 embryo recovery. A more developmentally advanced embryo (expanded hatched blastocyst) is shown in Figure 79.3b, although an embryo at this developmental stage is not usually expected on day 7. Many people find the side-by-side photos of actual specimens more helpful for learning embryo grading and classification than illustrations/sketches. In addition to the composite image in Figure 79.3 depicting the range in embryonic developmental stages, the figures in the remainder of this chapter show not only the principles of embryo grading and classification but also some of the common challenges faced by embryo technicians.

One important point to make before proceeding further is that although each stage of embryonic development is assigned a specific code, embryos are continually developing and are gradually transitioning from one stage of development to the next. As a result, many embryos may actually be between two different developmental stages at the time of recovery. Any time an embryo is making the transition from one stage to the next, it increases the likelihood that it may be called one developmental stage by one technician and a different developmental stage by another technician (e.g., stage 4 by some and stage 5 by others).

The following discussion of embryo evaluation is divided into sections on nontransferable-quality embryos and transferable-quality embryos.

Nontransferable-quality embryos

During the initial phase of the embryo evaluation process, most technicians segregate nonviable (nontransferable-quality) ova/embryos from viable (transferable-quality) embryos. Obviously, nonviable embryos should not be transferred to recipient females nor should they be cryopreserved.

Unfertilized ova

Unfertilized ova collected 6–8 days after the onset of estrus can have many different physical appearances (Figure 79.4). The three major keys to identifying a UFO are (i) the vitelline membrane (also called the oolemma) that surrounds the cytoplasm of the ovum is smooth and nearly perfectly spherical unless the ovum is fragmented or degenerate; (ii) the cytoplasm is granular and not cellular in appearance; and (iii) there is a moderate amount of perivitelline space (the space between the ovum and the zona pellucida), except in cases of degeneration.

As alluded to above, it can be frustrating to definitely determine the stage of development of ova that have undergone degeneration or fragmentation. Degenerate ova can, on occasion, appear very similar to a compact morula. Fragmentation, sometimes referred to as cytoplasmic blebbing, is a process during which some of the cytoplasm (which does not contain any chromosomes or chromosomal DNA) segregates itself from the ovum, resulting in a specimen that appears to be multicellular. Typically, the structure formed as a result of fragmentation is far smaller in size than would be expected for a blastomere of a 2-cell or 4-cell embryo. However, unless a fragmented ovum is exposed to a DNA-specific stain (e.g., bis-benzimide), one cannot determine with certainty if the entity is a stage code 2

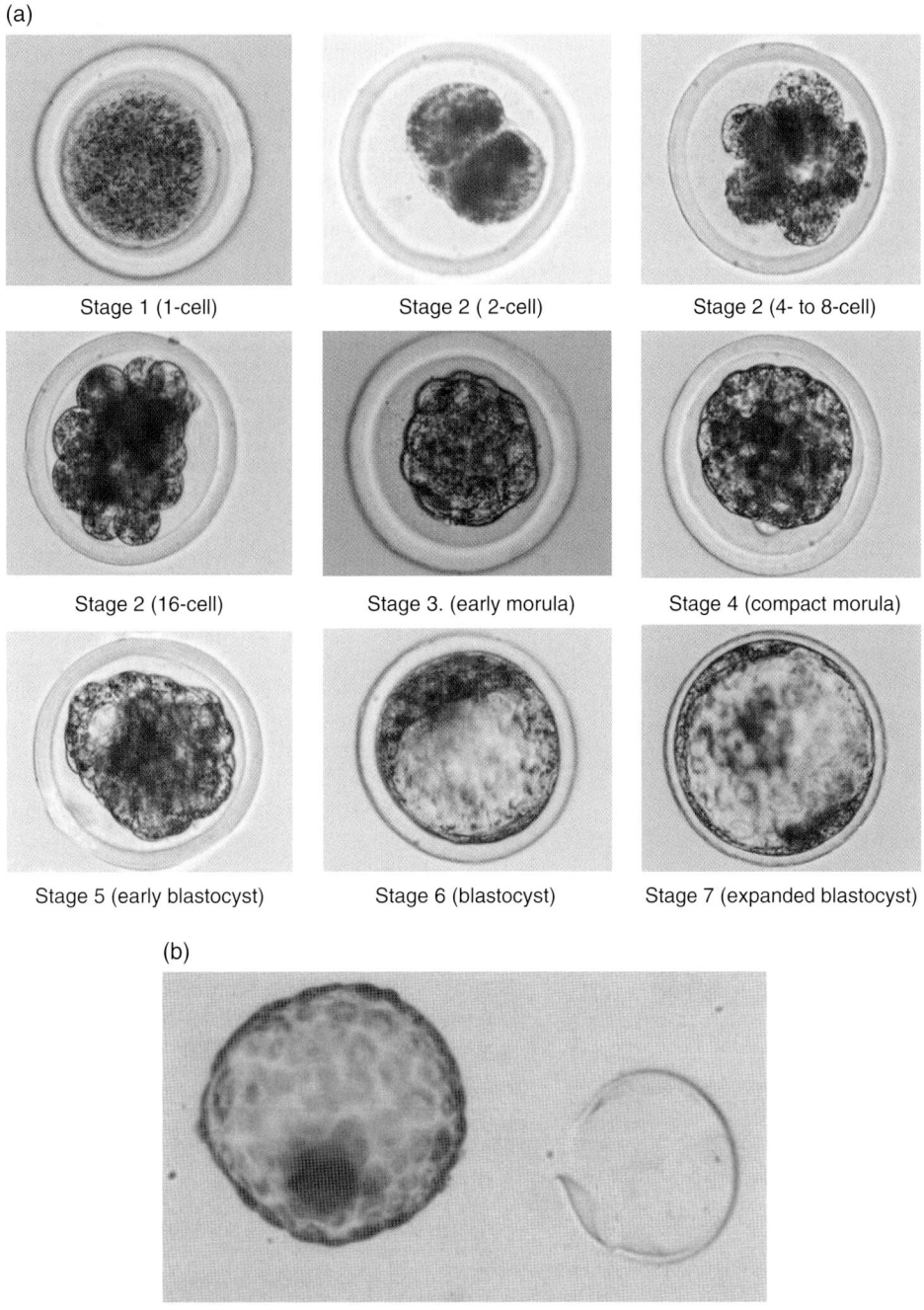

Figure 79.3 (a) Comparison of developmental stages of *in vivo*-derived bovine embryos collected on days 6–8 after estrus. At day 7, a stage code 1 embryo is considered an unfertilized ovum and a stage code 2 embryo is considered dead or degenerate. (b) Expanded hatched blastocyst typically recovered on or after day 8 after estrus; note the empty zona pellucida on the right and the expanded embryo on the left (courtesy of Dr Brad Lindsey).

embryo or a fragmented ovum. From a very practical perspective, however, the distinction between stage code 1 and stage code 2 is irrelevant because neither is suitable for transfer.

Dead or degenerating embryos

If embryos at stage code 2 are collected on day 7, the stage of embryonic development is not consistent with the expected stage of development. Therefore, these embryos are classified as degenerate (quality grade 4) because they are severely retarded in their development. Degenerate embryos are ones that have been fertilized but which died some time during the developmental process.

An embryo could also be considered dead or degenerate if its blastomeres are not fused to one another (i.e., there is lack of tight cell junction formation between the "outer" cells of the embryo). Failure of blastomeres to

Transferable-quality embryos

Once a technician has discarded UFOs and nontransferable-quality embryos, attention should be directed to the remaining embryos that are of transferable quality. In a perfect world, all transferable-quality embryos would be quality grade 1. However, this is not the case. There are a number of "defects" in transferable-quality embryos that can cause their embryo quality grade to be lowered to a grade 2 or 3. These "defects" are discussed in some detail in the following.

What are the characteristics of a quality grade 1 embryo? In essence, it is lack of the "defects" that would lead to assignment of a quality grade 1. Although a number of quality grade 1 embryos have already been shown (e.g., Figures 79.3 and 79.6), the main emphasis in the remaining discussion will be on identification of the embryo "defects" that lead to assignment of lower quality grade scores. In many photographs, however, quality grade 1 embryos will be shown side by side with quality grade 2 and 3 embryos, as well as UFOs, for easy comparison.

A number of morphological characteristics of an embryo can be considered as a deviation from normal. These abnormalities can include irregular sizes of embryonic blastomeres, large and/or multiple vacuoles within the cytoplasm of embryonic cells, degeneration of one or more cells in the embryo, some cells not adhering to the other cells comprising the embryo (i.e., extruded blastomeres/cells), and a damaged or misshapen zona pellucida. Many viable embryos have some detectable morphological abnormality such as a small number of extruded cells. Embryo quality grading must consider not only the presence or absence of these abnormalities but also the extent/degree to which these deviations are expected to influence post-transfer embryo survival. Experience has shown, for example, that an embryo which possesses a few degenerate cells but that consists mostly of normal-appearing cells has an excellent chance of establishing a pregnancy and producing a live calf.

Extruded cells

One abnormality that occurs somewhat frequently during preimplantation embryonic development is failure of all cells of an embryo to continue developing. Ordinarily, all cells of an embryo adhere to and communicate with one another. However, in certain cases some cells do not participate with their cohorts in normal embryonic development. The cells that do not fully participate in embryonic development are known as extruded cells because they are not adhered to the rest of the embryo and are found in the perivitelline space. Extruded cells often detract from the developmental potential of an embryo, particularly when the cells are extruded early in the developmental process. Thus, the size and number of extruded cells directly impacts embryo quality grade.

Extruded cells typically do not divide, and the extrusion of cells from an embryo can occur at any developmental stage beginning at the 2-cell stage. As long as a substantial number of blastomeres within an embryo remains healthy, the embryo can produce a viable pregnancy even when extruded cells are present. The rate of pregnancy depends

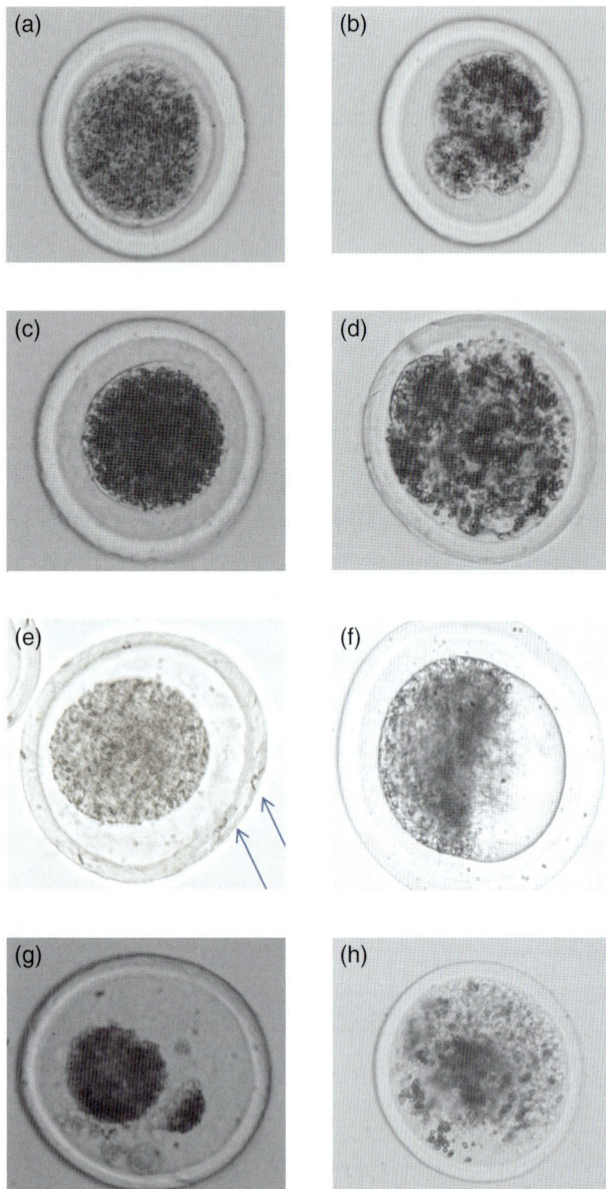

Figure 79.4 Unfertilized ova recovered nonsurgically from superovulated donor cows 6–8 days after the onset of estrus. Note the extremely wide variation in morphology which makes accurate identification difficult for inexperienced embryo technicians. (a) Unfertilized ovum with smooth vitelline membrane and granular cytoplasm that can be mistaken for a compact morula. (b) Fragmented unfertilized ovum. (c) Unfertilized ovum possessing dark cytoplasm and a smooth vitelline membrane; can be mistaken for a compact morula. (d) Degenerate unfertilized ovum. (e) Unfertilized ovum with light-colored granular cytoplasm; note sperm (arrows) attached to the zona pellucida. (f) Unfertilized ovum with smooth vitelline membrane; ovum is granular in one half and clear in the other half; can be mistaken for a blastocyst-stage embryo. (g) Fragmented unfertilized ovum. (h) Degenerate unfertilized ovum; note "loss" of vitelline membrane.

develop in a tight and cohesive group to compact and form tight cell junctions is a sign that the embryo is in the process of degenerating or dying (Figures 79.5 and 79.6). Note in Figure 79.6 a compact morula (a) next to a degenerate/dead embryo (b).

Evaluation of *In Vivo*-Derived Bovine Embryos 741

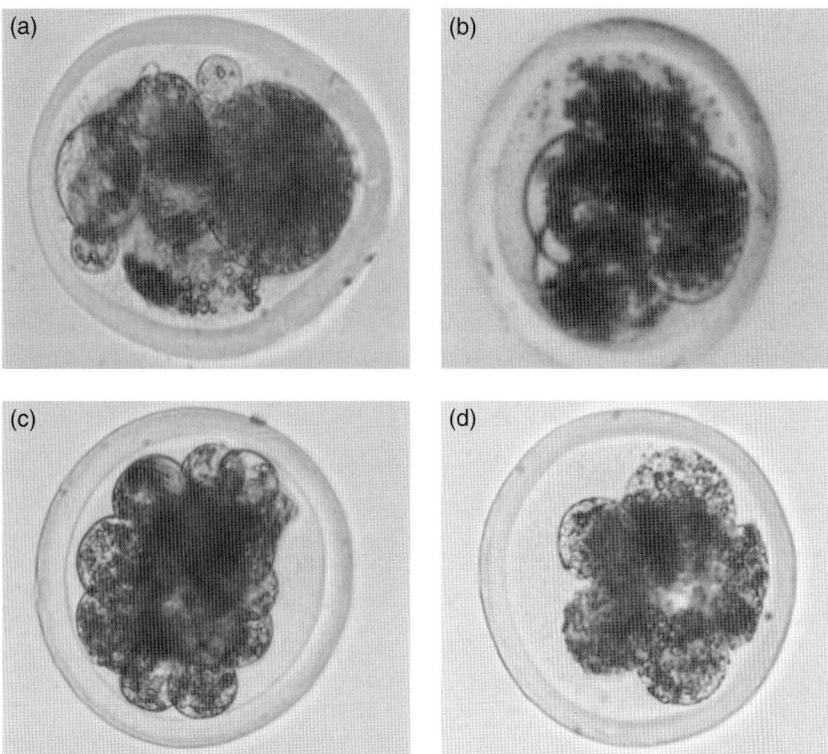

Figure 79.5 Degenerate bovine embryos collected on day 7: (a) and (b) are degenerate and are considered dead; (c) and (d) are not as developmentally advanced as expected for a day 7 embryo, and the cells are not fused to one another, so are also considered degenerate.

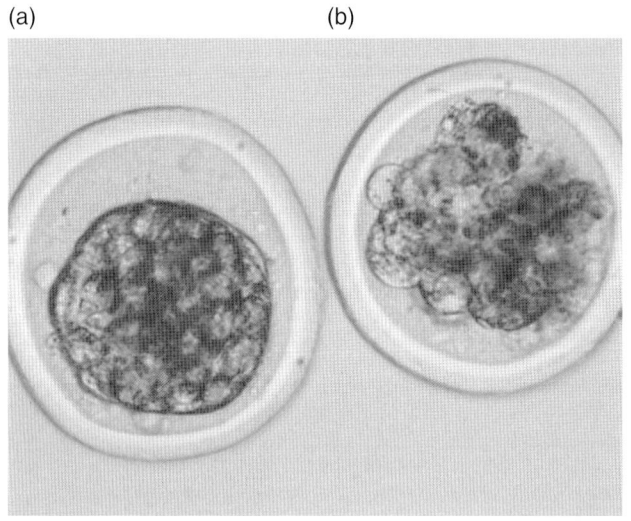

Figure 79.6 Day 7 compact morula (a) adjacent to a degenerate embryo (b).

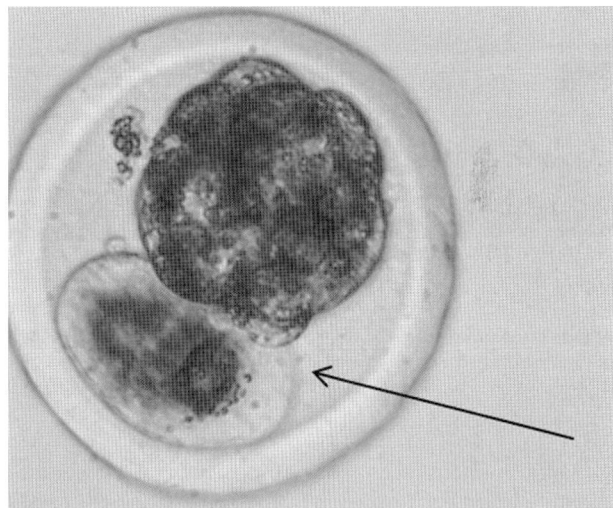

Figure 79.7 Grade 2 compact morula. Note one large blastomere (arrow) has stopped dividing while others have formed a compact morula.

on the number and size of extruded cells. Several examples of embryos possessing extruded cells are shown in Figures 79.7, 79.8, 79.9, 79.10, and 79.11. It should be noted that extruded cells in stage code 6 and 7 embryos get pressed against the zona pellucida and may not be obvious (see Figure 79.11) or may lead to assignment of a lower than deserved quality grade score (see Figure 79.9). Embryos should be rolled in the Petri dish to enable views from multiple angles.

Misshapen or flat zona pellucida

A zona pellucida is considered misshapen if it is not spherical or has a flat or concave surface that can cause the embryo to stick to a Petri dish or straw. This defect, by itself, is sufficient to cause a lowering of embryo quality grade by one grade (e.g., from quality grade 1 to quality grade 2). Figure 79.12 depicts an early blastocyst-stage bovine embryo with a misshapen zona pellucida.

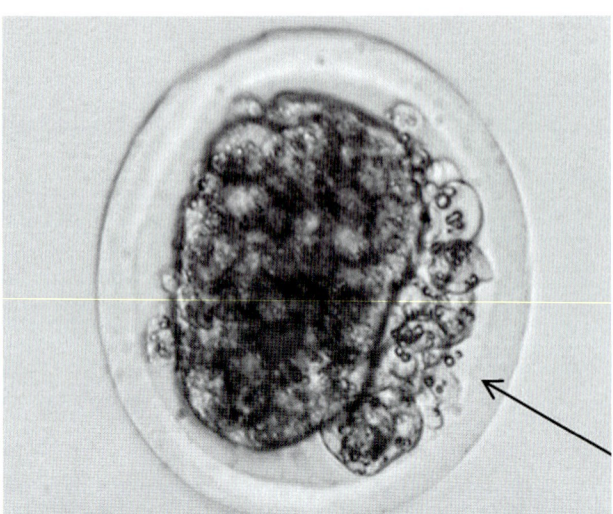

Figure 79.8 Grade 2 morula. Note several extruded blastomeres (arrow) to the right side of the embryo.

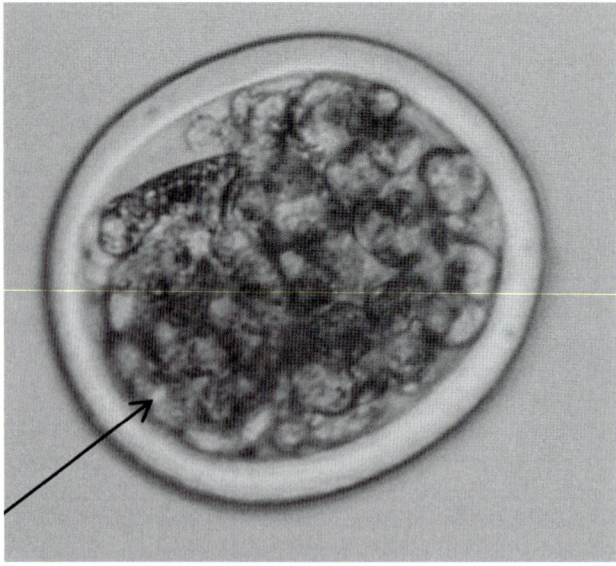

Figure 79.10 Grade 4 morula. Note a small group of cells on the left side (arrow) and a very large number of extruded cells on the right. This embryo was identified as grade 4 by our laboratory but may be called grade 3 by others.

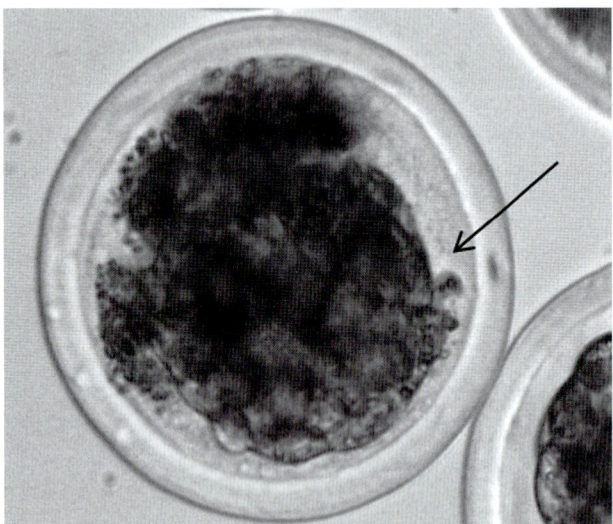

Figure 79.9 Grade 3 morula. Note the embryonic cell mass on the right side (arrow) hidden by a large number of extruded cells in the upper left corner and left side. May be called grade 4 by some without rolling this embryo.

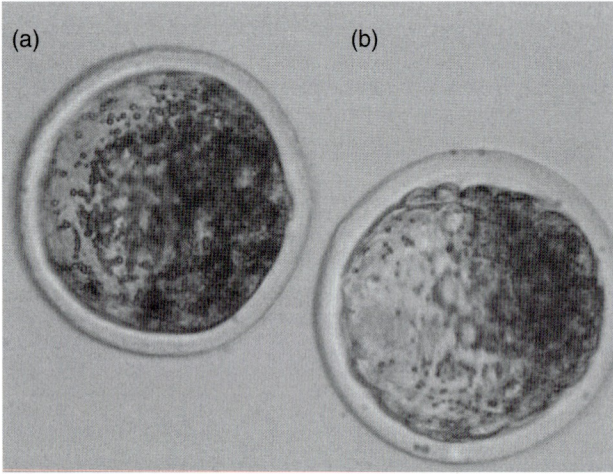

Figure 79.11 Embryos (a) and (b) are both IETS 6-1 embryos. Extruded cells in stages 6 and 7 embryos are often pressed against the zona pellucida and may not be obvious unless the embryo is collapsed. (a) It is difficult to identify this embryo as a blastocyst without rolling the embryo and having a good view of the blastocoele cavity which is obscured by the extruded cells on top. It could be easily mistaken for a degenerating unfertilized ovum.

Lipids in the cytoplasm

Although lipids are an essential component of cell membranes and must be present in viable embryos, some embryos possess an abnormal accumulation of lipid droplets (also called vacuoles) within the cytoplasm of individual cells. Most technicians will lower the quality grade of an embryo that possesses more than a few small vacuoles. A high cytoplasmic lipid content may be a cause of decreased pregnancy rates after transfer of frozen-thawed embryos, especially with embryos from the Jersey breed of dairy cattle.[22] Figures 79.13 and 79.14 show the presence of vacuoles within individual cells of bovine embryos.

Cracked zona pellucida

Typically, a cracked zona pellucida is found only in hatching blastocyst-stage embryos (stage code 8). However, some embryos can exhibit a cracked zona pellucida at earlier developmental stages, presumably due to the hydrostatic pressure placed on the embryo during the embryo recovery process (Figures 79.15, 79.16, and 79.17). Although a cracked zona pellucida does not automatically lower the embryo quality grade, only embryos possessing an intact zona pellucida can be washed properly[19] and remain eligible for export. However, it can still be frozen for domestic use.

Irregular shape of the cells comprising the embryo

An irregular shape of the group of cells comprising the embryo is commonly observed, especially in blastocyst-stage embryos when the blastocoele cavity is forming.

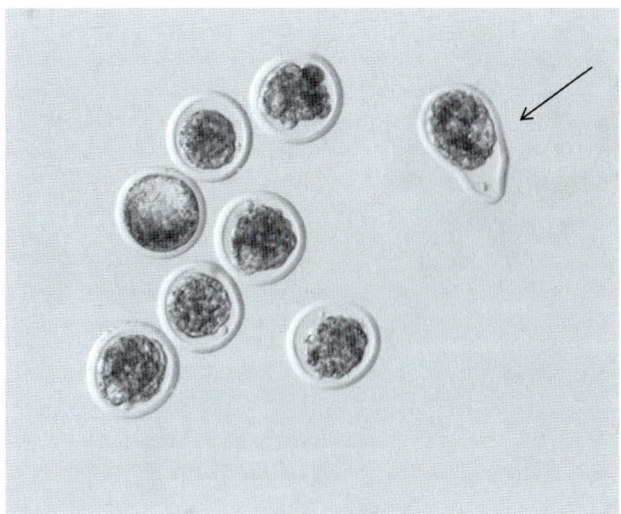

Figure 79.12 The arrow points to a day-7 grade 2 early blastocyst-stage embryo with a misshapen zona pellucida (50×).

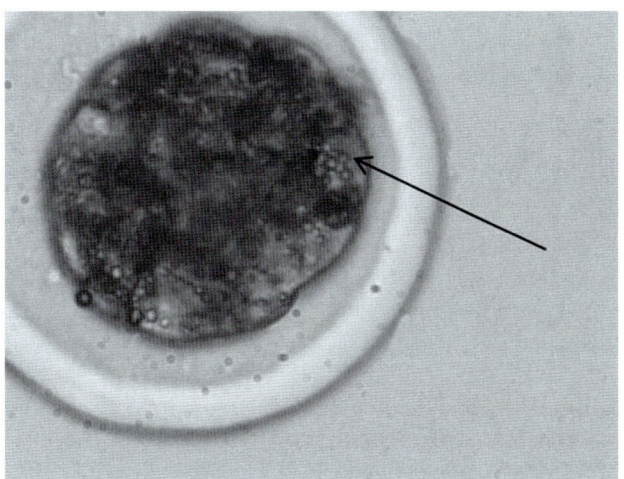

Figure 79.13 Grade 1 compact morula with vacuoles (arrow) in the cytoplasm.

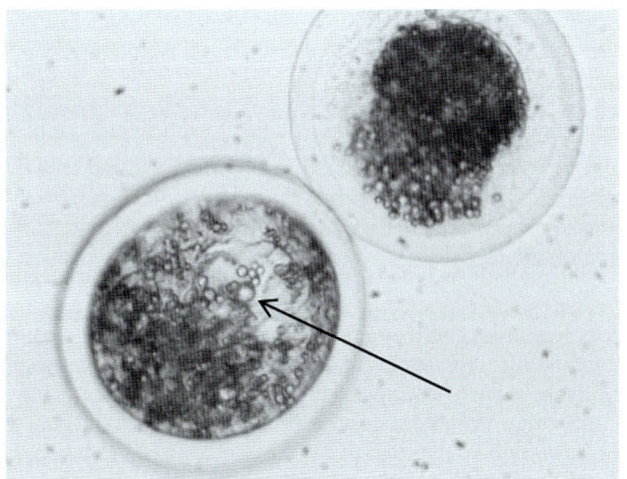

Figure 79.14 Grade 2 bovine blastocyst containing vacuoles (arrow) in the cytoplasm beside an unfertilized ovum. Note that it is difficult to identify the blastocyst embryo without rolling the embryo and having a good view of the blastocoele cavity. Could easily be mistaken for a degenerating unfertilized ovum.

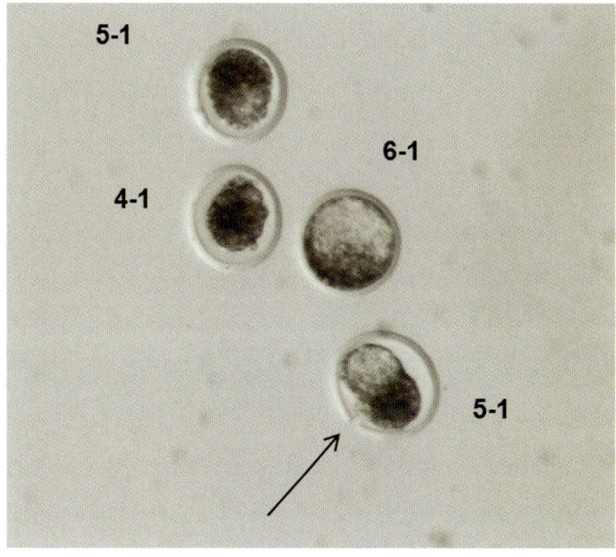

Figure 79.15 Collapsed early blastocyst with a cracked zona pellucida (arrow).

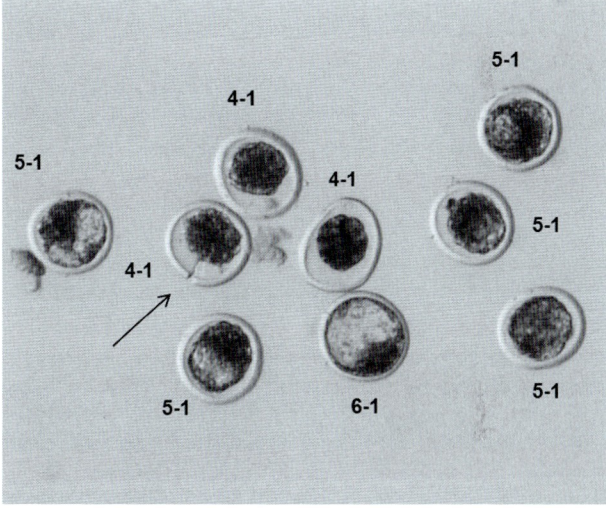

Figure 79.16 Day 7 group of embryos collected from a single donor. Note the embryo with cracked zona pellucida (arrow) that could be either a morula or a collapsed blastocyst. The thinning of the zona pellucida and slightly larger overall diameter gives supporting evidence that it is a collapsed expanded blastocyst.

Collapsing of the blastocoele cavity is considered a normal physiological process that does not automatically lower the quality grade, except in cases when the zona pellucida is also misshapen. Figure 79.18 depicts a group of embryos with normal and irregular shapes.

Material adhering to the zona pellucida

At the time of embryo recovery, it is not uncommon for some embryos to have material adhered to the outer surface of the zona pellucida. The adherent material may be mucus, individual cells, or a clump of tissue. Adherent material on the zona pellucida may be a sign of endometritis in the donor, cumulus cells that were not removed during normal gamete transport, or donor endometrial cells. Figures 79.19 and 79.20 depict bovine embryos with adherent material. The adherent

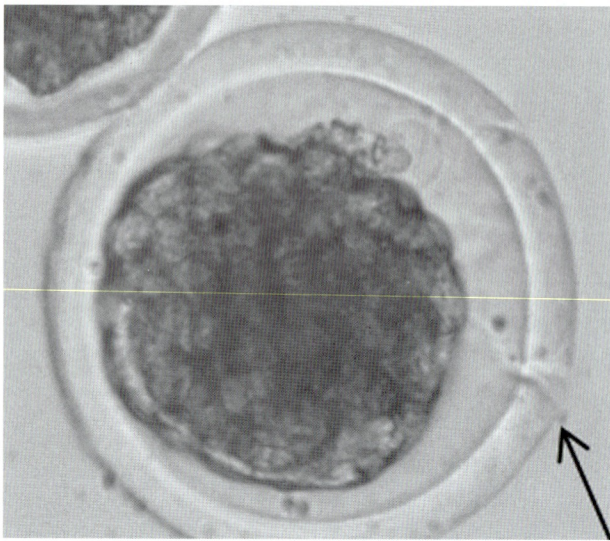

Figure 79.17 Day 7 compact morula. Note the crack in the zona pellucida (arrow).

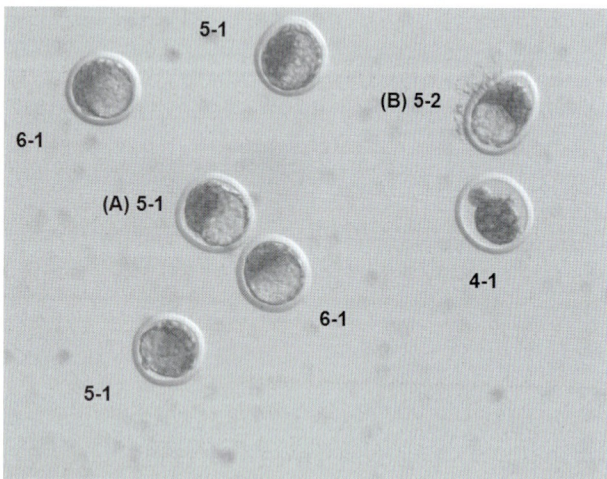

Figure 79.18 Day 7 grade 1 bovine early blastocyst (a) with slightly irregularly shaped cell mass. Day 7 grade 2 early blastocyst (b) with misshapen cell mass and zona pellucida.

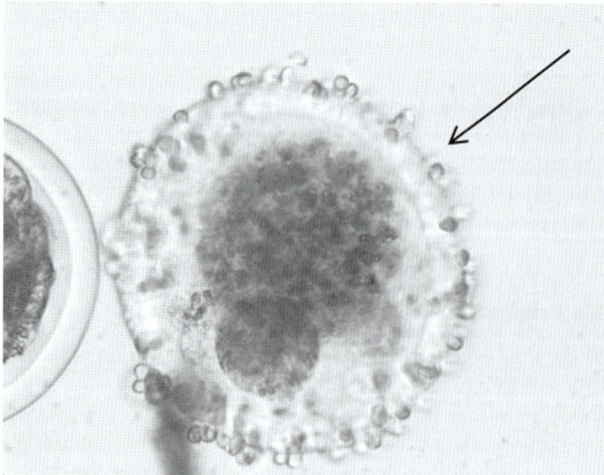

Figure 79.19 Fragmented unfertilized ovum. Note the white blood cells (arrow) attached to the zona pellucida

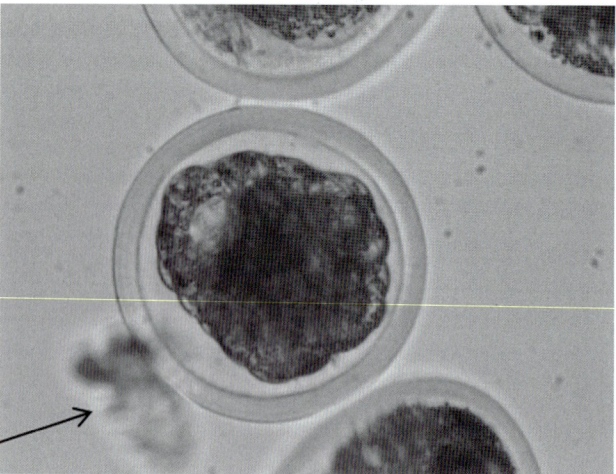

Figure 79.20 Day 7 early blastocyst (5-1). Note the endometrial cells (arrow) adhering to the zona pellucida. These cells must be removed prior to transfer or cryopreservation.

debris can frequently be washed off the zona pellucida during the IETS-approved embryo washing procedure.[19] In some cases, it may be necessary to include washing in a 0.25% solution of trypsin. Washing embryos with an intact zona pellucida in accordance with IETS washing guidelines should remove the adherent debris, as well as many pathogens.

The presence of debris attached to the outer surface of the zona pellucida does not necessarily influence the assignment of an embryo quality grade. However, any embryo with adherent material is not eligible for commercial trade, either domestically or internationally.

Challenges to accurate embryo grading and classification

One very difficult task for technicians learning to grade and classify embryos is to properly distinguish between a compact morula and a UFO, two entities which can look very similar in size, shape, and color. Figure 79.21 shows a group of compact morula and unfertilized ova side by side, and it is easy to envision how an inexperienced technician may inadvertently describe all these entities as either compact morula or unfertilized ova. This would be especially true when examining the specimens at a lower magnification or with a poor optical quality stereomicroscope. A more highly magnified view of an excellent quality compact morula and a UFO is shown in Figures 79.22 and 79.23. These are good examples of how these two stages of embryonic development are sometimes difficult to differentiate. As stated earlier, it is vitally important to be able to distinguish a compact morula from a UFO to prevent transfer (or cryopreservation) of a UFO and/or the disposal of a compact morula. With sufficient experience in grading embryos, it becomes easier to distinguish these two developmental stages.

Another common misclassification error can occur in instances when a portion of a UFO degenerates and becomes lighter, resembling the blastocoele cavity in a blastocyst-stage embryo (Figure 79.24). Experienced technicians will be able to distinguish the two by looking for the trophoblast layer which "surrounds" the blastocoele cavity in a blastocyst-stage embryo, as well as the ICM cells which are found only in a blastocyst.

Evaluation of *In Vivo*-Derived Bovine Embryos 745

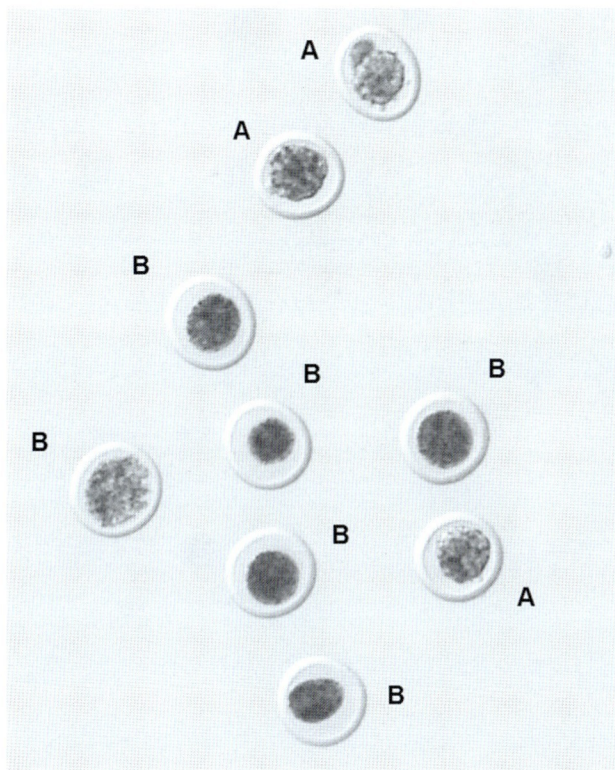

Figure 79.21 Group of compact morula (a) and unfertilized ova (b) recovered from a single donor cow on day 7. Note the similarity in diameter and sphericity between the compact morula and unfertilized ova. Unfertilized ova from this donor appear darker in color, and vitelline membrane of each ovum is extremely smooth in contrast to the wavy edge of the compact morula.

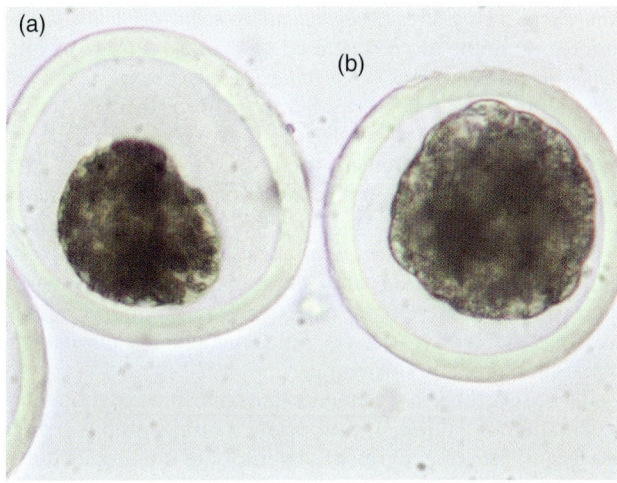

Figure 79.22 An unfertilized ovum (a) and excellent-quality compact morula stage embryo (b).

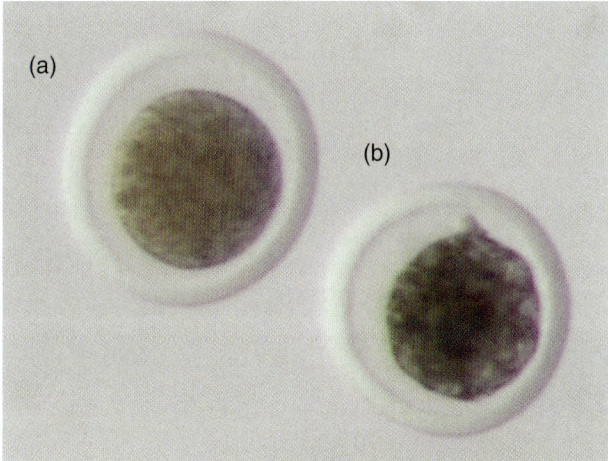

Figure 79.23 An unfertilized ovum (a) and excellent-quality compact morula stage embryo (b).

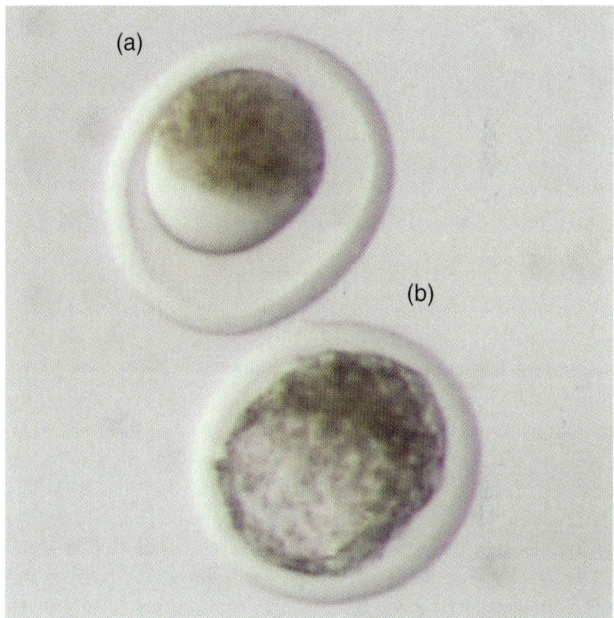

Figure 79.24 A degenerating unfertilized ovum possessing a translucent region (a) and an excellent-quality early blastocyst-stage embryo (b) with discernible trophoblast layer, blastocoele cavity, and inner cell mass cells.

Embryo evaluation tip

Although the potential exists for inexperienced technicians to make serious grading and classification errors, experienced embryo specialists seldom make such large mistakes. The most common disagreement among experienced technicians of embryo grading and classification is typically differentiating between excellent/good and fair and between fair and poor-quality embryos. This is because of the subjective basis of embryo evaluation procedure.

To avoid errors in classification and grading of embryos it is vitally important to "roll" the embryos in the Petri dish to enable the technician to thoroughly evaluate the entire surface of the zona pellucida as well as all sides of the cells that make up the embryo. This can be particularly important when donors yield a large number of ova/embryos in a wide range of embryo developmental stages and embryo quality grades (Figure 79.25). A very thorough visual appraisal will reduce the likelihood of embryo grading and classification mistakes.

Each embryo should be examined individually by focusing up and down, keeping in mind that the embryo is a three-dimensional object and that a single two-dimensional

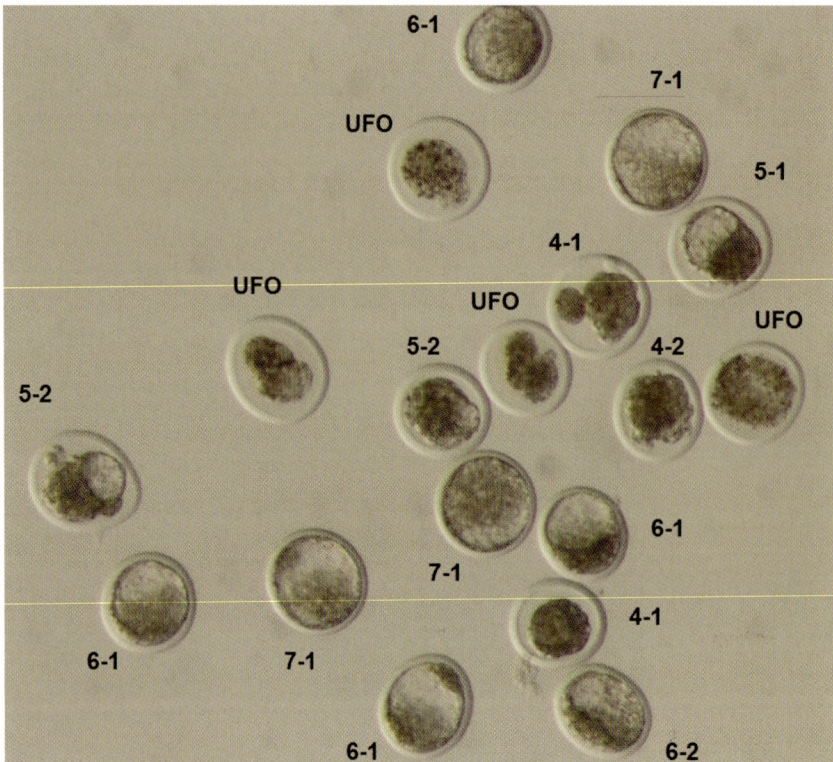

Figure 79.25 A group of 18 day 7 ova and embryos recovered from a single donor.

view is inadequate for proper evaluation. The embryo should then be rolled (with an embryo handling pipette or by gently and carefully swirling the medium in the dish) to enable the same embryo to be viewed from a different angle. Figure 79.26 consists of three panels (1, 2 and 3) which show the same group of eight embryos from three different angles. Note how different the embryos look in each picture, even though they are the same embryos. These three panels clearly help readers visualize the importance of rolling the embryos around within the Petri dish to enable correct identification of every detail of each embryo. Rolling the embryo and evaluating it from different angles is the key to proper embryo grading and classification.

The effectiveness of morphological embryo evaluation

Morphological embryo evaluation remains one of the most subjective aspects of the overall embryo transfer process. The photographs presented in this chapter were examined by several embryo specialists who, for the most part, agreed with the stage of development and grade assigned to each embryo. However, there were some mild disagreements over grades assigned to some embryos, and most of the disagreements were between quality grade 1 and 2 embryos. For a couple of the photographs one specialist called a quality grade 1 blastocyst-stage embryo a UFO. These discrepancies most likely occurred from that technician's inability to view the embryos through a stereomicroscope and to roll them around to see them from different angles.

Photographs can be great learning tools, but they are not equivalent to seeing embryos under the stereomicroscope and being able to roll them. One study[23] showed that agreement among six experienced embryo evaluators who viewed video images of embryos was higher for stage of embryo development (89.2%) than it was for embryo quality grade (68.5%). The inability to roll embryos within the Petri dish and to view them from multiple different angles undoubtedly contributed to this lack of agreement. Disagreement among evaluators on embryo stage of development may be attributed to embryos being between two developmental stages. Disagreement on embryo quality grading was higher when attempting to distinguish between good- and fair-quality embryos. This likely reflects differences in individual technician philosophies on embryo evaluation. Disagreement was lower when evaluating excellent and degenerate quality embryos. The higher rate of agreement when viewing excellent-quality embryos (possessing few, if any, "defects") or degenerate embryos perhaps should not be surprising.

There are three key elements that contribute to successful evaluation of embryos: appropriate training, ample experience, and proper equipment. Appropriate training includes learning the correct morphology of embryos at different times after estrus and the significance of any deviations from normal embryo morphology. One must also learn how to maneuver embryos within a Petri dish in order to thoroughly examine them. Ample experience is gained by examining numerous embryos at all the different stages of embryonic development and different quality grades. Ideally, hundreds of embryos should be evaluated while under the guidance of a highly experienced embryo technician. Experienced personnel can evaluate more than 95% of ova/embryos accurately provided they have a suitable stereomicroscope. Proper equipment includes a good optical

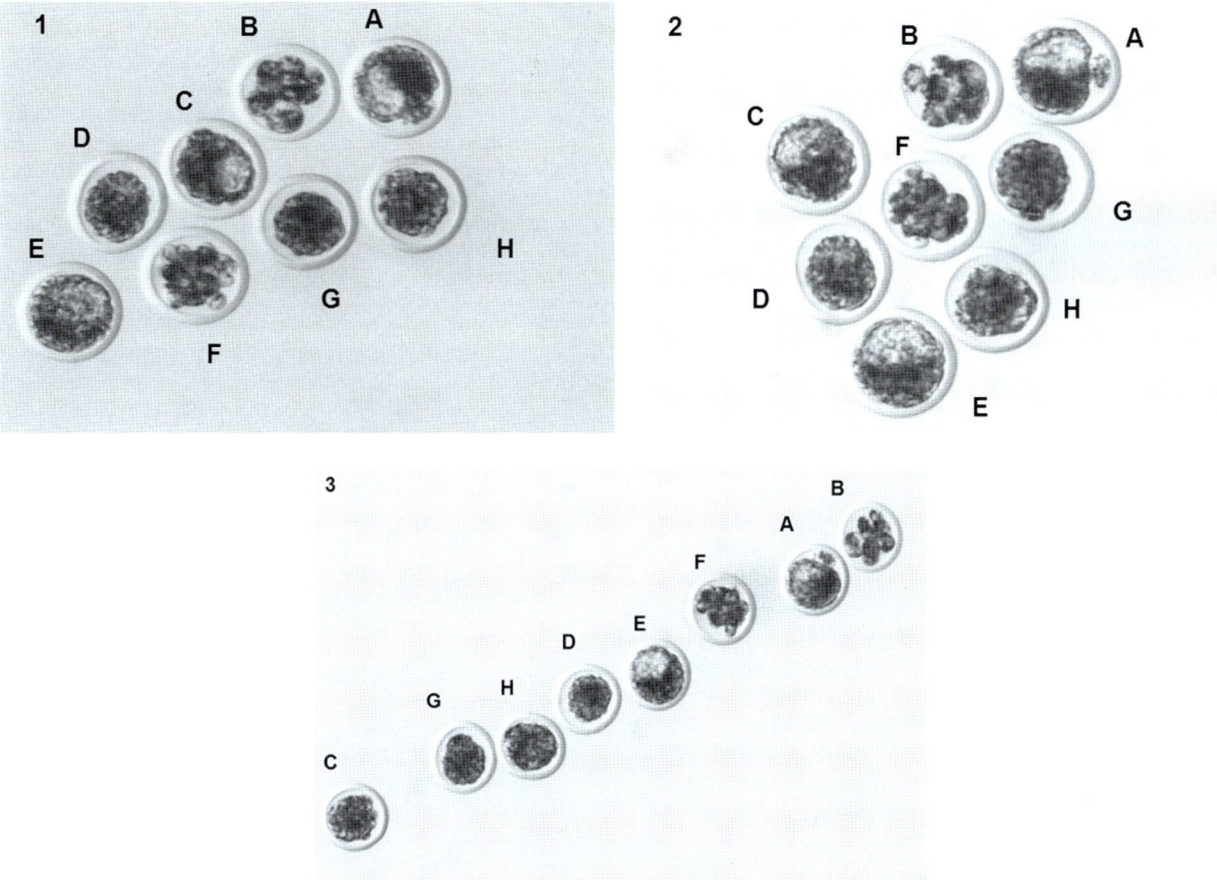

Figure 79.26 A group of eight ova/embryos collected on day 7 from a single donor cow seen from three different angles (panels 1, 2, and 3) after repositioning of embryos within the Petri dish. Note how different each embryo looks in each photo (e.g., embryo C), highlighting the importance of a thorough visual evaluation.

quality stereomicroscope of sufficient magnifying capabilities, appropriate embryo handling pipettes (or other handling devices), and satisfactory Petri dishes and embryo media.

Embryo transfer service providers routinely use the IETS embryo grading and classification system. Although this system is subjective, it has allowed practitioners and researchers to record data in a manner that lends itself to subsequent analysis. Although most large embryo transfer datasets originate from the commercial embryo transfer industry, a number of research studies have also contributed to the body of knowledge.

For example, embryonic stage of development has little influence on post-transfer pregnancy rate following transfer of fresh *in vivo*-derived embryos to synchronous recipients if the embryos are IETS stage codes 4 (compact morula) to 7 (expanded blastocyst). This result has been obtained in both smaller-scale studies[24] as well as larger-scale studies involving retrospective analysis of multiple years of data obtained from commercial embryo transfer companies.[25] However, the latter study did document a lower pregnancy rate following transfer of hatched blastocysts (stage code 8).

Similarly, IETS stage code 4, 5 and 6 *in vivo*-derived embryos survive the freezing and thawing procedures equally well (i.e., there is no difference in post-transfer pregnancy rate of embryos of the same quality grade). However, embryo quality grade is a critical factor that influences how well embryos survive the freezing and thawing process. Quality grade 1 and 2 embryos typically yield higher pregnancy rates than quality grade 3. One study[24] showed no difference in pregnancy rate among cows receiving grade 1 or 2 embryos (74.1% and 71.8%, respectively), whereas another study[16] showed that quality grade 1 embryos had a greater probability to conceive (odds ratio 1.62) compared with quality grade 2 or 3 embryos. Conception rate of quality grade 1 embryos was reported as 76% for fresh embryos and 64% for frozen embryos.[26]

Conclusion

The ability to accurately evaluate the developmental stage and quality of cattle embryos is critical to the success of an embryo transfer program. Careful examination with a stereomicroscope of ova/embryos recovered from donors is necessary to ensure that viable embryos are not discarded or that unfertilized ova are not transferred or cryopreserved. It is recommended that embryo evaluators undergo intensive and repetitive training and also that a consistent method be used for embryo evaluation. It is highly important to roll the embryos around within the Petri dish to accurately evaluate the entire surface of the zona pellucida as well as all cells comprising the embryo itself. During any given embryo recovery from a superovulated cow one should expect a range of embryo developmental stages and embryo quality grades. The best predictor of an embryo's post-transfer pregnancy rate is its embryo quality grade, with quality grade 1 embryos yielding higher pregnancy rates than either quality grade 2 or 3 embryos. Compact morula (stage code 4) through blastocyst (stage code 6) frozen-thawed embryos are expected to yield equivalent pregnancy rates when transferred to synchronous recipient females. Accurate recording keeping will allow embryo transfer service providers the opportunity to monitor their success and to identify any areas needing potential improvement.

Summary

Embryo transfer is a reproductive biotechnology that facilitates rapid genetic improvement in the dairy and beef cattle industries. An important determinant of the overall success of the embryo transfer process is the quality of embryos obtained from superovulated donor females. Although a variety of different methods have been used over the years to assess the quality of embryos, the only proven and industry-accepted method to assess embryo quality is morphological evaluation. The purpose of this chapter is to provide readers with the necessary information to enable accurate morphological assessment of bovine embryo quality. Although substantial experience is needed before one becomes proficient at this technique, it is an important skill to master because embryo quality is predictive of pregnancy rate obtained after transfer of embryos.

References

1. Hartman C, Lewis W, Miller F, Swett W. First findings of tubal ova in the cow, together with notes on oestrus. *Anat Rec* 1931;48:267–275.
2. Winters, L, Green W, Comstock R. Prenatal development of the bovine. Technical Bulletin 151, University of Minnesota Agricultural Experiment Station, 1942.
3. Hamilton W, Laing J. Development of the egg of the cow up to the stage of blastocyst formation. *J Anat* 1946;80:194–204 plus plates.
4. Shea B, Hines D, Lightfoot D, Ollis G, Olson S. The transfer of bovine embryos. In: Rowson LEA (ed.) *Egg Transfer in Cattle*. Luxembourg: EEC Publication 5491, 1976, pp. 145–152.
5. Trounson A, Willadsen S, Rowson L. The influence of in vitro culture and cooling on the survival and development of cow embryos. *J Reprod Fertil* 1976;47:321–327.
6. Renard P, Menezo Y, Saumande J, Heyman Y. Attempts to predict the viability of cattle embryos produced by superovulation. In: Streenan JR (ed.) *Control of Reproduction in the Cow*. Current Topics in Veterinary Medicine, Vol. 1. Berlin: Springer Verlag, 1977, pp. 398–417.
7. Kardymovicz O. A method of vital staining for determining the viability of fertilized sheep ova stored in vitro. In: Proceedings 7th International Congress on Animal Reproduction and AI, Munich, Germany, 1972, pp. 503–506.
8. Renard J, Philippon A, Menezo Y. In-vitro uptake of glucose by bovine blastocysts. *J Reprod Fertil* 1980;58:161–164.
9. von Schilling E, Niemann H, Cheng SP, Doepke H-H. DAPI: a further fluorescence test for diagnosing the viability of early cow and rabbit embryos. *Reprod Domest Anim* 1979;14:170–172.
10. Youngs C, Pendleton R, Rorie R, Godke R. The use of a computerized image analysis system for photometric and morphometric measurements of bovine preimplantation embryos. *Theriogenology* 1987;27:299 (Abstract).
11. von Schilling E, Döpke H. A rapid diagnostic test for the viability of early cattle and rabbit embryos using diacetyl-fluorescein. *Naturwissenschaften* 1978;65:658–659.
12. Hoppe R, Bavister B. Evaluation of the fluorescein diacetate (FDA) vital dye viability test with hamster and bovine embryos. *Anim Reprod Sci* 1984;6:323–335.
13. von Schilling E, Niemann H, Smidt D. Evaluation of fresh and frozen cattle embryos by fluorescence microscopy. In: Hafez ESE, Semm K (eds) *In Vitro Fertilization and Embryo Transfer*. New York: Alan R. Liss, 1982, pp. 349–355.
14. Boland M. Use of the rabbit oviduct as a screening tool for the viability of mammalian eggs. *Theriogenology* 1984;21:126–137.
15. Shea B. Evaluating the bovine embryo. *Theriogenology* 1981;15:31–42.
16. Scenna F, Munar C, Mujica I et al. Factors affecting pregnancy rate following timed embryo transfer program in cattle under field conditions. *Reprod Fertil Dev* 2009;21:172 (Abstract).
17. Lindner G, Wright R. Bovine embryo morphology and evaluation. *Theriogenology* 1983;20:407–416.
18. Stringfellow D, Seidel S. *Manual of the International Embryo Transfer Society*, 3rd edn. Champaign, IL: International Embryo Transfer Society, 1998.
19. Stringfellow D. Recommendations for the sanitary handling of in vivo derived embryos. In: Stringfellow DA, Givens MD (eds) *Manual of the International Embryo Transfer Society*, 4th edn. Champaign, IL: International Embryo Transfer Society, 2010, pp. 65–68.
20. Stringfellow D, Givens M. *Manual of the International Embryo Transfer Society*, 4th edn. Champaign, IL: International Embryo Transfer Society, 2010.
21. Stroud B. AETA 2009 Survey Report. http://www.aeta.org/docs/09statsreport.pdf. Downloaded January 13, 2013.
22. Steel R, Hasler J. Pregnancy rates resulting from transfer of fresh and frozen Holstein and Jersey embryos. *Reprod Fertil Dev* 2004;16:182–183 (Abstract).
23. Farin P, Britt J, Shaw D, Slenning B. Agreement among evaluators of bovine embryos produced in vivo or in vitro. *Theriogenology* 1995;44:339–349.
24. Spell A, Beal W, Corah L, Lamb C. Evaluating recipient and embryo factors that affect pregnancy rates of embryo transfer in beef cattle. *Theriogenology* 2001;56:287–297.
25. Hasler J. Bovine embryo transfer: are efficiencies improving? In: *Proceedings, Applied Reproductive Strategies in Beef Cattle*, Nashville, TN, 2010, pp. 265–282.
26. Hasler J. Current status and potential of oocyte recovery, in vitro embryo production, and embryo transfer in domestic animals with an emphasis on the bovine. *J Anim Sci* 1998;76(Suppl. 3):52–74.

Chapter 80

Control of Embryo-borne Pathogens

Julie Gard

Department of Clinical Sciences, College of Veterinary Medicine, Auburn University, Auburn, Alabama, USA

Introduction

Embryos are vulnerable to contamination from numerous sources, including the donor, semen, culture and washing media, and animal source proteins utilized in media. These embryonic pathogens can have a significant effect on the clinical proficiency of embryo technologies. It is an epidemiologic adage that agent, host, and environmental factors combine to determine whether health or disease will prevail within a population of individuals.[1] This axiom is important to keep in mind when discussing embryonic pathogens. Three distinct generations of embryo technologies have been developed and with them have surfaced critical challenges.[2] Each technology imposes environmental changes that have the potential to either enhance or diminish the spread of infectious agents among susceptible cattle via utilization of these technologies. Hence it is necessary for these technologies to be evaluated, as done yearly by the Health and Safety Advisory Committee of the International Embryo Transfer Society (IETS).[2] The World Organisation of Animal Health (OIE) adopts the committee's suggestions on standards for embryo movement to minimize disease transmission, standard protocols and treatments in order to minimize pathogenic embryonic associations, and the validity of experimental treatments currently being utilized in embryo technologies to maximize viability and minimize embryonic loss. It is important to evaluate all areas when controlling pathogens but prevention (biosecurity) is the foundation of prevention of all disease states.

The first two generations – *in vivo*- and *in vitro*-derived embryo production, respectively – are widely utilized in commercial cattle production and in basic research. The International Embryo Transfer Society Newsletter (http://www.iets.org/pdf/Newsletter/Dec12_IETS_Newsletter.pdf) reported that in 2011 there were 104 651 *in vivo*-derived embryo collections, with a resulting 732 862 total embryos collected of which 238 194 were transferred fresh and 309 806 were transferred frozen. For *in vitro*-produced embryos there were 453 471 transferrable embryos produced, with 373 836 of these embryos being transferred fresh and the remainder transferred frozen.[3] As far as current numbers for cloned embryos, there were no reported numbers for 2010 or 2011 but there were 105 fresh transfers in 2009 according to the IETS 2010 Newsletter.[4] Obviously, there has been an apprehension and this continues to be a legitimate concern due to the level of domestic and international activity which could facilitate broad distribution of any pathogen that might be even occasionally transmitted via embryos.

First generation of embryo technologies

In vivo embryo production can be considered the first generation of embryo technologies. The original techniques were developed more than 60 years ago as detailed in a review by Betteridge.[5] These fundamental procedures are widely used in cattle production today and include superovulation of donors, nonsurgical embryo collection, embryo cryopreservation, and nonsurgical transfer of embryos to synchronized recipients.[6] The expansive implementation of these reproductive techniques represented a potentially significant environmental change that has the potential to increase or decrease distribution of indigenous or foreign animal pathogens.

Over the past 30 years, numerous pathogens of cattle have been evaluated within the context of *in vivo* embryo production and transfer to determine if the techniques might diminish or enhance the normal spread of disease among populations of cattle.[7-9] The original hypotheses were that an intact zona pellucida would protect the early conceptus from infectious agents and that proper handling of zona-intact embryos was likely to reduce and often eliminate rather than enhance the spread of naturally occurring pathogens. There have been four primary strategies utilized to investigate these hypotheses.[10] Initially, zona pellucida-intact embryos were exposed *in vitro* to a variety of pathogens. These pathogens were chosen due to regulatory concerns surrounding them, such as institution of and/or

Bovine Reproduction, First Edition. Edited by Richard M. Hopper.
© 2015 John Wiley & Sons, Inc. Published 2015 by John Wiley & Sons, Inc.

current domestic disease control programs and/or international trade restrictions. For example, the OIE considers bluetongue virus (BTV) a reportable disease due to the socioeconomic concern and the major importance of the international trade in animal and animal products.[11,12] Additionally, venereal and transplacental transmission exists with most serotypes of BTV.[13–17] Hence, the spread of BTV can be heavily influenced by human interventions preventing or facilitating the transmission pathways,[18] emphasizing the necessity of these trials.

The first strategy consisted of embryos being exposed to the pathogen of concern. Following *in vitro* artificial exposure, embryos were then washed with or without trypsin treatment and assayed for the presence of virus either through implementation of an *in vitro* assay or transfer of embryos to disease-free recipients. The recipients and progeny were monitored closely for evidence of infection. The results of these experiments were promising in that the procedures employed for processing embryos seemed to provide a broad-spectrum approach to prevention of spread of several important infectious agents.[19,20] Additional trials were also conducted in order to validate these embryo processing procedures within the context of normal embryo production and transfer protocols. In these studies, embryos were collected from diseased (acute or convalescent) donors, washed or trypsin treated, and then either assayed for pathogens *in vitro* or transferred to disease-free recipients which, along with their offspring, were monitored for transmission of disease.[7–10,19–21] The results of these studies served to certify that the embryo processing procedures now endorsed by the IETS[21] and recommended by the OIE[22] have indeed been effective widespread deterrents to the spread of certain infectious diseases, which are categorized accordingly by the OIE.[8,13,22,23] These procedures include the standard processing procedures for *in vivo*-derived embryos as listed in the IETS Manual[21] and endorsed by the OIE, again as stated above.[22] The *Embryo Training Manual* of the Food and Agriculture Organization of the United Nations (FAO)[24] recommends the following.

1. Embryos should be washed at least 10 times with at least 100-fold dilutions between each wash, and a fresh pipette should be used for transferring the embryos through each wash.
2. Only embryos from the same donor should be washed together, and no more than 10 embryos should be washed at any one time.
3. Sometimes, for example when inactivation or removal of certain viruses such as bovine herpesvirus (BHV)-1 and Aujeszky's disease virus is required, the standard washing procedure should be modified to include additional washes with the enzyme trypsin, as described in the IETS Manual.[21]
4. The zona pellucida of each embryo, after washing, should be examined over its entire surface area at not less than 50× magnification to ensure that it is intact and free of adherent material.[24]

The protocol for washing and trypsin treatment for adherent viruses such as infectious bovine rhinotracheitis virus (BHV-1), Aujeszky's disease virus, and vesicular stomatitis vurus requires that the embryos be washed five times in phosphate-buffered saline to which antibiotics and 0.4% bovine serum albumin have been added, but without Ca^{2+} and Mg^{2+} (because they inhibit trypsin), then through two washes of trypsin of 30–45 s each, then again through five washes of saline containing antibiotics and 2% heat-inactivated serum (not bovine serum albumin) to inactivate the trypsin.[24] The trypsin enzyme used to prepare the washes should have an activity such that 1 g will hydrolyze 250 g of casein in 10 min at 25°C and pH 7.6. The sterile trypsin solution should contain a concentration of 0.25% trypsin in Hank's balanced salt solution without Ca^{2+} and Mg^{2+}. There are ready-to-use solutions commercially available. After embryos have been washed, they should be microscopically examined over their entire surface area at no less than 50× magnification.[24] Also, for purposes of disease control, only embryos that have an intact zona pellucida and that are free of adherent material should be transferred or cryopreserved.[21–24]

Sources of pathogens

Although some areas of embryo production seem to have received adequate attention, other important sources of pathogens in embryo production have received insufficient consideration. These sources of pathogens are contaminated materials (reagents) of animal origin and diseased recipients.[25–27] The term "materials of animal origin" embraces a variety of constituents, including supplements to media (e.g., fetal bovine serum or bovine serum albumin), hormones that might be used to treat donors (e.g., pituitary-origin follicle-stimulating hormone), and enzymes for treatment of embryos (e.g., trypsin). Detailed descriptions of the hazards associated with these products have been previously reported.[26–28] To highlight some, bovine serum albumin is an unlikely source of pathogens because production protocols seem to preclude the survival of contaminants.[29,30] On the other hand, fetal bovine serum could be the source of adventitious bacteria, fungi or viruses when adequate quality assurances are not in place.[26,31] The most common contaminant in fetal bovine serum is bovine viral diarrhea virus (BVDV).[26,31–33]

A number of pathogens are known to be transmitted from mother to fetus via the placenta. These include BVDV, *Neospora caninum*, BTV, bovine leukemia virus (BLV), and *Mycobacterium avium* subsp. *paratuberculosis* (Johne's disease).[13–17,34–36] Thus, a complete risk assessment should always give due consideration to the entirety of infectious agents capable of establishing transient, latent, or persistent infection of the donor and/or recipient which can result in fetal disease. Otherwise, collection and transfer of embryos, albeit even pathogen-free embryos, could lead to the birth of diseased offspring as a result of subsequent *in utero* infections.

As of December 2012, the official biosecurity recommendations for international movement of embryos are outlined in Chapter 4.7.5 of the Terrestrial Animal Health Code under "Risk Management."[37] These general strategies were developed to help ensure that the transfer of *in vivo*-derived embryos would not result in the transmission of infectious agents. Three general strategies are available to ensure that transfer of *in vivo*-derived embryos will not result in the transmission of infectious agents originating from donor animals.[38,39] Each method represents a contrived environmental change intended to deter pathogen transmission. The preferred method in cattle is embryo processing (i.e., washing or trypsin treatment),[21] as previously described, because it is relatively inexpensive and has been substantiated by a considerable amount of research to prevent the spread of a variety of viruses, bacteria, and bovine spongiform encephalopathy (see Table 80.1).[8,22,37–39]

Other methods have been employed in cattle, including donor testing and recipient quarantine and testing.[38] By testing donors for presence of disease at the time of embryo collection, and again weeks or months later (while embryos are held in the cryopreserved state), it can be confirmed that donors are not a possible pathogen source. The testing of recipients following transfer of embryos and testing of offspring after birth function as bioassays utilizing the recipients and offspring as sentinel animals. However, this requires a lengthy quarantine of animals. Additional disadvantages are that these methods are expensive and that multiple tests have to be conducted if there is a need to detect more than one disease and so as to determine that accidental exposure did not occur during the quarantine period. Therefore, economically, the use of embryo processing is the preferred method for certifying health of embryos. This method has greatly facilitated international trade in embryos because of the combination of animal welfare, disease control, and economic advantages.[38]

Categorization of pathogenic agents of embryos

In order to further facilitate international trade of embryos, an estimation, or categorization, of the known risk of transmission of a particular disease via embryo transfer is compiled by the OIE with the help of the Health and Safety Research Subcommittee of the IETS and is updated yearly.[22,23] The official categorization of diseases of embryos is detailed further in chapter 1.4.7.14 of the Terrestrial Animal Health Code under "Categorization of Diseases and Pathogenic Agents."[22] Category 1 diseases are defined and outlined in Table 80.1.

Terrestrial Animal Health Code 1.4.7.14 states that category 2 diseases are those for which substantial evidence has accrued to show that the risk of transmission is negligible provided that the embryos are properly handled between collection and transfer according to the IETS Manual, but for which additional transfers are required to verify existing data.[22] The following diseases are in category 2: bluetongue (sheep), caprine arthritis/encephalitis, and classical swine fever. Category 3 diseases or pathogenic agents are those for which preliminary evidence indicates that the risk of transmission is negligible provided that the embryos are properly handled between collection and transfer according to the IETS Manual, but for which additional *in vitro* and *in vivo* experimental data are required to substantiate the preliminary findings.[22] The following diseases or pathogenic agents are in category 3: bovine immunodeficiency virus, bovine spongiform encephalopathy (goats), BVDV (cattle), *Campylobacter fetus* (sheep), foot and mouth disease (pigs, sheep and goats), *Haemophilus somnus* (cattle), maedi-visna

Table 80.1 Category 1 diseases or pathogenic agents for which there is sufficient research to show that the risk of transmission via *in vivo*-derived embryos is negligible if embryos are properly handled between collection and transfer.

Donor species	Disease or agent	Embryo treatment
Cattle	Bluetongue	Washing
Cattle	Bovine spongiform encephalopathy	Washing
Cattle	*Brucella abortus*	Washing
Cattle	Enzootic bovine leukosis	Washing
Cattle	Foot and mouth disease	Washing
Cattle	Infectious bovine rhinotracheitis (BHV-1)	Trypsin treatment
Swine	Aujeszky's disease	Trypsin treatment
Sheep	Scrapie (not including atypical scrapie which is presently not categorized)	Washing

Source: chapter 4.7.14 of the Terrestrial Animal Code (http://www.oie.int/index.php?id=169&L=0&htmfile=chapitre_1.4.7.htm).

(sheep), *Mycobacterium paratuberculosis* (cattle), *Neospora caninum* (cattle), ovine pulmonary adenomatosis, porcine reproductive and respiratory disease syndrome (PRRS), rinderpest (cattle; eradicated), and swine vesicular disease.[22] Category 4 diseases and pathogenic agents as listed in Terrestrial Animal Health Code 1.4.7.14 are those for which studies have been done, or are in progress, that indicate no conclusions are yet possible with regard to the level of transmission risk; or that the risk of transmission via embryo transfer might not be negligible even if the embryos are properly handled according to the IETS Manual between collection and transfer.[22] The following diseases or pathogenic agents are in category 4: African swine fever, akabane (cattle), bovine anaplasmosis, bluetongue (goats), border disease (sheep), BHV-4, *Chlamydia psittaci* (cattle, sheep), contagious equine metritis, enterovirus (cattle, pigs), equine rhinopneumonitis, equine viral arteritis, *Escherichia coli* 09:K99 (cattle), *Leptospira borgpetersenii* serovar Hardjobovis (cattle), *Leptospira* sp. (pigs), lumpy skin disease, *Mycobacterium bovis* (cattle), *Mycoplasma* spp. (pigs), ovine epididymitis (*Brucella ovis*), parainfluenza-3 virus (cattle), parvovirus (pigs), porcine circovirus (type 2) (pigs), scrapie (goats), *Tritrichomonas foetus* (cattle), *Ureaplasma* and *Mycoplasma* spp. (cattle, goats), and vesicular stomatitis (cattle, pigs).[22]

Control of viral pathogens

Considerations of space preclude discussion of each disease listed in all four categories. Therefore, the primary focus of the remainder of this chapter will be on BVDV, which is of great importance due to the tremendous economic impact it has on both the beef and dairy industries through its ability to diminish production via decreases in feed efficiency and milk production and to cause early embryonic death, abortion, mummification, fetal abnormalities, pneumonia, and other various debilitating disease states.[28,40–47] The economic impact of BVDV was evaluated in a study by Houe[41] which estimated BVDV losses to the cattle industry to be $20–57 million per million calvings. Recently, Hessman et al.[42] investigated the economic effects and the health and performance of cattle populations after exposure to persistently infected (PI) animals. They reported that for cattle with direct exposure to a PI animal, economic losses due to fatalities were $5.26 per animal and performance losses were much more significant at $88.26 per animal. Also, due to the significant economic implications resulting from BVDV infections, a number of countries such as Denmark, Finland, Norway, Sweden, the Shetland Islands, and Slovenia have established eradication programs.[28,48–52] Several countries are utilizing control programs and are moving toward implementation of eradication programs while others, such as the United States, are leaving the individual states and specifically the producers within those states to set standards for BVDV control by the formation of voluntary control programs. Control of BVDV in the United States is further complicated by the existence of BVDV in the white tail deer population.[53,54] Hence, the focus of most control programs is appropriate biosecurity, including quarantine, testing, identification and removal of PI animals, and prevention of PI formation.

Embryo transfer as a means of controlling pathogen transmission

Historically, embryo transfer has been utilized as a way to minimize disease transmission along with maximizing genetics.[55,56] However, in the years following its inception questions regarding the ability of embryos to transmit pathogens surfaced. Numerous studies were performed which showed that embryo transfer was indeed a good way to prevent or minimize pathogen transmission if good-quality controls and appropriate processing procedures were instituted.[8,38,39,55–61] However, there were some organisms for which prevention of pathogen transmission was not refutable and in some cases embryo transfer might plausibly serve as a mechanism for introduction of disease. The original placement of BVDV in category 3 was based on a study conducted over 20 years ago.[56] This study validated the efficacy of embryo washing procedures for removal of virus after artificial exposure of embryos to a cytopathic isolate of BVDV.[56] Four later studies collectively produced five BVDV-free calves after transfer of washed or trypsin-treated embryos from PI donor cows.[57–60] While data from these four reports were limited, the results were encouraging. However, reports by Lindberg et al.[62] and Drew et al.[63] stimulated further investigations as to the possibility of transmission of BVDV via embryo transfer due to seroconversion of heifers after embryo transfer and birth of a PI calf following embryo transfer, respectively. Contaminated fetal bovine serum was thought to be the inciting cause of these infections. Hence additional studies were necessary to determine if BVDV would remain associated with embryos following standard IETS processing procedures for *in vivo*-derived embryos and whether this associated virus could be transmitted via embryo transfer. So, further studies were performed which highlighted the variation in affinity between different strains of BVDV and the affinity of these strains for embryos.[64,65] Some strains were shown to maintain association with embryos following IETS processing procedures, including when trypsin treatment was added to the processing procedures.[64,65] These high-affinity strains became more of a concern when it was determined that the embryo-associated virus was indeed infective in both *in vitro* and *in vivo* studies.[65,66] Current research by Gard et al.[67] reported that 27% of *in vivo*-derived and 42% of *in vitro*-produced embryos had embryo-associated virus (EAV) following artificial exposure to a high-affinity strain of BVDV (SD-1, type 1a) following washing procedures in accordance with IETS (without trypsin). It was also found that the amount of EAV ranged from 100 to 450 $CCID_{50}$ (cell culture infective doses to 50% endpoint) per embryo.[67] In previous studies, EAV was also determined to be infectious in an *in vitro* culture system and in an *in vivo* model through intravenous inoculation of embryos into virus-negative and seronegative recipients.[64,66,67] An additional study was recently performed which utilized twice the amount of EAV known to be previously associated with individual *in vivo*-derived and *in vitro*-produced embryos following processing procedures.[68] This amount was then doubled (878 $CCID_{50}$ per straw) to imitate what is sometimes done in the embryo transfer industry, namely transferring two embryos simultaneously into one recipient, and in keeping with the worse-case scenario. It was found that all recipients of embryos and virus became viremic, and then seroconverted. There were six pregnancies at 30 days and only one at 60 days after transfer. However, all pregnancies were lost. The 60-day pregnancy was removed via colpotomy and tested for BVDV. Sequencing found that the strain of BVDV was SD-1, the same as the infecting viral strain.[68] The finding in this study that no BVDV-positive offspring were produced is similar to the results of Bielanski et al.[69] In this study, *in vitro*-produced embryos were exposed to noncytopathic biotypes of BVDV for 1 hour, namely type 2 strain (P-131) or a type 1 strain (NY-1), and then washed in accordance with IETS guidelines (no trypsin treatment) and transferred to seronegative and BVDV-negative recipients. Following intrauterine transfer of embryos, none of the recipients exposed to the type 1 strain seroconverted; however, of the 35 recipients receiving embryos exposed to the type 2 strain, 18 seroconverted (51%) and there were 11 pregnancies at 30 days after transfer but of these only two resulted in live offspring. These two offspring were determined to be BVDV negative and seronegative.[69] Hence, the infection seemed not to be recognized by the fetus and/or the virus was destroyed prior to development of immune competence, so in fact no antibodies were formed by the fetus. The results of no detectable seroconversion from the type 1 strain and abortion due to type 2 BVDV may be the results of mutations within the virus. In a study by Meyers et al.,[70] decreases in interferon production, abortion, and presence of virus in fetal tissues did not result when pregnant cattle were injected with two different mutant strains of virus. Each of these mutant strains had mutations specifically affecting both the N-terminal protease (N(pro)) and the deletion of codon 349, which abrogates the RNAse activity of the structural glycoprotein E(rns). However, decreases in interferon production, abortion, and presence of virus in fetal tissues did occur with wild-type viruses and in viruses in which only one mutation of either N(pro) or E(rns) occurred.[70] Therefore, the establishment of persistent infections requires both N(pro) and E(rns). It is logical to ascertain that the type 1 strain utilized in Bielanski's study might have had mutations within these areas, resulting in adequate interferon production and no fetal infection. Each specific strain may affect interferon production differently and therefore may or may not result in fetal infection and that fetal infection may or may not be fatal.

Based on these studies and the variety of BVDV strains, it is still plausible that a PI animal could result via transfer of a contaminated embryo. The results from Gard's and Bielanski's studies show that some high-affinity isolates can result in viremia and seroconversion in susceptible recipients followed by early embryonic death and abortion.[68,69] However, the possibility of the development of a PI animal following transfer of contaminated embryos seems more likely to result from the viremia of the recipient (acute exposure) and shedding of virus to other gestating animals than from transfer of a contaminated embryo based on the findings of these recent studies. However, additional studies involving embryos naturally infected with various strains of BVDV and transferred into the uterus need to be completed in order to establish the ability of BVDV to undergo intrauterine transmission and the potential for development of a PI calf. A thorough evaluation is important as the ability to transmit BVDV via an embryo would necessitate reevaluation of embryo health certification procedures along with implementation of additional regulations on embryos exported from BVDV-positive countries to those countries where BVDV has been eradicated.

Second generation of embryo technologies

In vitro embryo production has been described as the second generation of embryo technologies and most of the essential components of this process in cattle have been developed and refined over the past 35 years.[28,39,71] Fundamental procedures currently used with *in vitro* production include collection of oocytes via transvaginal follicular aspiration, *in vitro* maturation of oocytes (IVM), *in vitro* capacitation of spermatozoa, *in vitro* fertilization (IVF), *in vitro* culture of embryos (IVC) to blastocysts, and embryo transfer to a recipient.[72] This technology represents multiple changes in the animal breeding environment and provides opportunities for pathogens to enter, amplify and be disseminated. Hence there are a number of logical sources of pathogens in *in vitro* embryo production. These sources include gametes and their associated fluids, somatic cells used in IVM, IVF or IVC, and the materials of animal origin used to supplement maturation, fertilization or culture media (e.g., fetal bovine serum, bovine serum albumin, or follicle-stimulating hormone).[26]

Studies surveying for contaminants in *in vitro* embryo production laboratories have been conducted in a number of different countries, such as Canada,[73,74] Denmark,[75] France,[7,76] and the United States.[71,77,78] As expected, levels of bacterial contamination can be quite high, from 13 to 68% of samples when abattoir-origin materials are utilized in *in vitro* production systems.[71] Also, low to moderate levels of BVDV and BHV-1 have been found in abattoir-origin materials, with a range of 1–12% of samples testing positive for BVDV and 0–12% of samples testing positive for BHV-1.[71,78]

Because BVDV is widely distributed among populations of cattle and is known to infect reproductive tissues, it has received the most attention by researchers.[77] In fact, all the materials (animal products, ovaries, oocytes, follicular fluid, cumulus, uterine tubal cells, and serum) of animal origin that have been integral to IVF embryo production in cattle have been shown to contain BVDV when harvested from actively infected animals.[77,79] This hazard is complicated by the fact that a proportion of animals infected by BVDV are asymptomatic.[41,80] Cattle that are persistently infected with BVDV are often asymptomatic and serve as reservoirs for large quantities of virus.[41] Also, most of the field isolates of BVDV are noncytopathic and could cause unapparent persistent infections in established laboratory cell lines[31,79] that are used in cocultures with embryos.[31,79] Finally, it is noteworthy that the cultures in each phase of IVF embryo production are ideally suited for replication of BVDV, providing adequate time and substrate as well as permissive cells for viral replication. Therefore, even a low level of contamination can be effectively amplified, leading to exposure of embryos to a high level of virus by the end of IVC.

As with *in vivo*-derived embryos, it was originally hypothesized that the zona pellucida of IVF embryos would protect the developing conceptus from infections and, indeed, artificial exposure studies have demonstrated that an intact zona is an effective barrier that is seldom penetrated.[81] Nevertheless, it is clear that the processing procedures recommended for use on *in vivo*-derived embryos are not as effective for IVF embryos, since several pathogens (e.g., BTV, BHV-1, foot-and-mouth disease virus, BVDV) have been shown to adhere more readily to the zona of IVF embryos.[77,78] Furthermore, the presence of virus has been shown to result in reduced rates of maturation, fertilization, and development.[77,82–85]

Studies have shown that BVDV remains associated with *in vitro*-produced embryos after viral exposure and washing or trypsin treatment.[86,87] However, infection of susceptible recipients after transfer of exposed embryos has not been demonstrated. Additionally, some interesting results have been achieved in attempts to determine whether the amount of BVDV associated with individual embryos would constitute an infective dose *in utero*.[88,89] The cumulative results of these studies evaluating risks of transmitting BVDV via embryo transfer of *in vitro*-produced embryos are somewhat contradictory. On the one hand, it is clear that multiple strains of BVDV will remain associated with IVF embryos after washing or trypsin treatment.[88–90] On the other hand, there appears to be innate deterrents to transmission of the embryo-associated virus to permissive cells *in vitro*.[88–90] Obviously, resolution of these issues requires further studies.

Official recommendations for international shipment of *in vitro*-produced embryos in order to control the spread of disease are outlined in chapter 4.8 of the Terrestrial Animal Health Code of the OIE.[91] Impediments to the establishment of universally recognized sanitary precautions for the production of bovine IVF embryos include the laboratory-to-laboratory variability in techniques utilized, the comparative ease with which pathogens adhere to the zona pellucida, and the fact that raw materials are often collected from abattoirs.[92] It is apparent that mechanical washing and trypsin treatment of IVF embryos[86,87] do not provide the same level of effectiveness when utilized for cleansing *in vivo*-derived embryos. However, these treatments have been shown to definitely reduce the level of contamination of IVF embryos following artificial exposure to agents such as BHV-1 and BVDV.[86,87,93] Thus, as is the case with *in vivo*-derived embryos, further studies are necessary in order to complete the risk assessment, which includes evaluating fully if the quantity of embryo-associated BVDV is sufficient to infect recipients and/or their fetuses after transfer into the uterus. Obviously, the strain of the virus seems to have a tremendous effect on the affinity of the virus for embryos and will therefore need to be taken into account when assessing the true risk.[65]

Third generation of embryo technologies

The third generation of embryo technologies includes cloning, such as somatic cell nuclear transfer, and transgenics. These technologies specifically involve maturation and enucleation of oocytes, insertion of somatic cell nuclei, activation, *in vitro* culture, possibly cryopreservation, and nonsurgical transfer. The same hazards that are associated with the second generation of embryo technologies apply directly to the third generation of embryo technologies with a few added caveats. These caveats include that the zona pellucida is either broken or removed, and that donor nuclei are harvested from additional cell lines. The cell lines utilized are often cultured for 8–12 weeks in order to allow for cellular amplification. Additionally, the amplification procedures can allow noncytopathic BVDV replication and infection in these cell lines due to the use of animal origin materials such as abattoir-origin oocytes and fetal bovine serum utilized in formation and culturing of embryos.

The presence of noncytopathic BVDV may be manifested in poor cloning efficiency and development, low pregnancy rates, early embryonic death, fetal wastage, and neonatal abnormalities. An example of the problems that might be associated with BVDV contamination in cloning systems is well documented in a report by Shin *et al.*[94] This report describes a high incidence of developmental failure in bovine fetuses that were derived by cloning using BVDV-infected cells. Specifically, a fetal fibroblast cell line was inadvertently infected with a noncytopathic strain of BVDV. As a result of the infection an extraordinary amount of embryonic death and fetal resorption occurred with the use of this cell line.[94] In an additional study by Stringfellow *et al.*,[95] screened samples sent to veterinary diagnostic laboratories revealed that 15 of 39 fetal fibroblast cell lines used in cloning research were positive for BVDV as determined by various assays including reverse transcription-polymerase chain reaction (RT-PCR). However, it was determined that only 5 of 39 cell lines were actually infected with BVDV. Furthermore, three of these five lines were determined not to be infected at the earliest cryopreserved passage, leading to the conclusion that the cell lines became infected after culture in media containing contaminated fetal bovine serum. Sequence comparison of the amplified cDNA from one lot of fetal bovine serum confirmed that it was the source of infection for one of these cell lines.[95] Since BVDV was isolated from the remaining two cell lines at the earliest available passage, the fetuses from which they were established could not be ruled out as the source of the virus. These studies highlight the additional concerns that cloning and transgenics pose and the greater potential for the production of contaminated embryos via utilization of these techniques. Therefore excellent quality control is imperative in order to prevent embryonic pathogens and transmission of pathogens via embryo transfer to recipients is of major importance with all generations of embryo technologies, and the needs increase with each subsequent generation.

The use of embryo technologies continues to increase in other species, including the pig, goat, sheep, horse, and wildlife as well. As seen in cattle, there is a variety of pathogens which pose a concern and can result in embryonic and fetal mortality. A review by Givens and Marley[96] summarizes bacterial, fungal, protozoan, and viral causes of reproductive dysgenesis in cattle, sheep, goats, pigs, horses, dogs, and cats. This study illustrates how it is imperative to make prudent use of immunization and biosecurity protocols when incorporating embryonic technologies in order to minimize reproductive losses.

Experimental treatments

The addition of antibiotics and antifungals to media significantly decreases problems with bacteria and fungi in culture. However, viral contamination poses a more complicated threat. A number of recent studies have investigated the addition of antiviral substances such as interferon-tau, which has been reported to decrease the amount of BVDV in culture but not BHV-1.[97] Additionally, other antiviral compounds such as 2-(4-[2-imidazolinyl]phenyl)-5-(4-methoxyphenyl)furan (DB606) have been shown to inhibit the replication of BVDV in bovine uterine tubal epithelial cells, Madin Darby bovine kidney cells, and fetal fibroblast cells without negatively affecting embryonic development nor future fertility of heifers derived from IVF embryos treated with DB606.[98,99] A more recent study by Newcomer *et al.*[100] evaluated the scope of antiviral activity of another similar compound, 2-(2-benzimidazolyl)-5-[4-(2-imidazolino)phenyl]furan dihydrochloride (DB772), among diverse pestiviruses. Isolates of BVDV 2, border disease virus, HoBi virus, pronghorn virus, and Bungowannah virus were all tested for *in vitro* susceptibility to DB772 by incubating infected cells in medium containing various concentrations of DB772. It was found that DB772 effectively inhibits all pestiviruses studied at concentrations above 0.20 μmol/L with no evidence of cytotoxicity.[100] These compounds, DB606 and DB772, along with interferon-tau show potential for minimization of viral pathogens, specifically pestivirus, in culture that may result in their inclusion during culture of embryos and amplification of fibroblast cell lines for cloning and transgenics.[99–101]

Summary

Continued evaluation of these new modalities and incorporation in culture systems for prevention of embryonic pathogens is necessary but quite possibly these compounds might provide a reliable methodology to curtail infectious virus associated with embryos in the future. However, proper biosecurity and quality control are always at the forefront of prevention of embryonic pathogens and should be the focus of standard operating procedures when utilizing any of the generations of embryo technologies.

References

1. Smith RD. Definitions. In: Smith RD (ed.) *Veterinary Clinical Epidemiology*, 3rd edn. Boca Raton, FL: CRC Press, 2005, pp. 10–12.
2. Thibier M, Stringfellow D. Health and Safety Advisory Committee (HASAC) of the International Embryo Transfer Society (IETS) has managed critical challenges for two decades. *Theriogenology* 2003;59:1067–1078.
3. Stroud B. IETS Data Retrieval Committee Annual Report. Embryo Transfer Newsletter December 2012, pp. 16–26. http://www.iets.org/pdf/Newsletter/Dec12_IETS_Newsletter.pdf
4. Thibier M. IETS Data Retrieval Committee Annual Report. Embryo Transfer Newsletter, December 2010, pp. 11–21. http://www.iets.org (Newsletter Archive restricted to IETS members).
5. Betteridge K. Reflections on the golden anniversary of the first embryo transfer to produce a calf. *Theriogenology* 2000;53:3–10.
6. Mapletoft R. Bovine embryo transfer. In: Morrow DA (ed.) *Current Therapy in Theriogenology*, 2nd edn. Philadelphia: Saunders, 1986, p. 54.
7. Guerin B, Nibart M, Marquant-Le Guienne B, Humblot P. Sanitary risks related to embryo transfer. *Theriogenology* 1997;47:33–42.
8. Wrathall A, Brown K, Sayers A et al. Studies of embryo transfer from cattle clinically affected by bovine spongiform encephalopathy (BSE). *Vet Rec* 2000;150:365–378.
9. Stringfellow D, Givens M. Epidemiologic concerns relative to in vivo and in vitro production of livestock embryos. *Anim Reprod Sci* 2000;60–61:629–642.
10. Bielanski A, Hare W. Procedures for design and analysis of research on transmission of infectious disease by embryo transfer. In: Stringfellow DA, Seidel SM (eds) *Manual of the International Embryo Transfer Society*, 3rd edn. Savoy, IL: International Embryo Transfer Society, 1998, pp. 143–149.
11. Backx A, Heutink R, Van Rooij E, Van Rijn P. Transplacental and oral transmission of wild-type bluetongue virus serotype 8 in cattle after experimental infection. *Vet Microbiol* 2009;138:235–243.
12. Jimmeénze-Clavero M, Aguero M, san Miguel E et al. High throughput detection of bluetongue virus by a new real-time fluorogenic reverse transcription polymerase chain reaction: application on clinical samples from current Mediterranean outbreaks. *J Vet Diagn Invest* 2006;18:7–17.
13. Wrathall A, Simmons H, Van Soom A. Evaluation of risks of viral transmission to recipients of bovine embryos arising from fertilization with virus-infected semen. *Theriogenology* 2006;65:247–274.
14. Kirschvink N, Raes M, Saegerman C. Impact of natural bluetongue serotype 8 on semen quality of Belgian rams in 2007. *Vet J* 2009;182:244–251.
15. DeClercq K, De Leeuw I, Verheyden B et al. Transplacental infection and apparently immunotolerance induced by a wild-type bluetongue virus serotype 8 natural infection. *Transbound Emerg Dis* 2008;55:352–359.
16. Maclachalan M, Drew C, Darpei K, Worwa G. The pathology and pathogenesis of bluetongue. *J Comp Pathol* 2009;141:1–16.
17. Menzies F, McCullough S, McKeown I et al. Evidence for transplacental and contact transmission of bluetongue in cattle. *Vet Rec* 2008;163:203–209.
18. Mintiens K, Meroc E, Mellor P et al. Possible routes of introduction of bluetongue virus serotype 8 into the epicentre of the 2008 epidemic in north-western Europe. *Prev Vet Med* 2008;87:131–144.
19. Hare W. Status of disease transmission studies and their relationship to the international movement of bovine embryos. *Can Vet J* 1985;27:37–45.
20. Singh E. The disease control potential of embryos. *Theriogenology* 1987;27:9–20.
21. Stringfellow D. Recommendations for the sanitary handling of in-vivo-derived embryos. In: Stringfellow DA, Seidel SM (eds) *Manual of the International Embryo Transfer Society*, 3rd edn. Savoy, IL: International Embryo Transfer Society, 1998, pp. 79–84.
22. Anonymous. Terrestrial Animal Health Code 4.7.14. World Organization for Animal Health (OIE), Paris, 2013. http://www.oie.int/index.php?id=169&L=0&htmfile=chapitre_1.4.7.htm.
23. Thibier M. Embryo transfer: a comparative biosecurity advantage in international movements of germplasm. *Rev Sci Tech* 2011;30:177–188.
24. Food and Agriculture Organization (FAO) of the United Nations. *Training Manual for Embryo Transfer in Cattle*. http://www.fao.org/docrep/004/T0117E/T0117E14.htm.
25. Mapletoft R. Bovine embryo transfer. In: Morrow DA (ed.) *Current Therapy in Theriogenology*, 2nd edn. Philadelphia: Saunders, 1986, pp. 59, 60.
26. Brock K, Grooms D, Ridpath J, Bolin S. Changes in levels of viremia in cattle persistently infected with bovine viral diarrhea virus. *J Vet Diagn Invest* 1998;10:22–26.
27. Stringfellow D. Use of materials of animal origin in embryo production schemes: continued caution is recommended. *Embryo Transfer Newsletter* 2002;20:11.
28. Gard J, Givens M, Stringfellow D. Bovine viral diarrhea virus (BVDV): epidemiologic concerns relative to semen and embryos. *Theriogenology* 2007;68:434–442.
29. Cohn E, Strong L, Hughes W Jr et al. Preparations and properties of serum and plasma proteins IV. A system for the separation into fractions of the protein and lipoprotein components of biological tissues and fluids. *J Am Chem Soc* 1946;68:459–475.
30. Schiewe M. General hygiene and quality control practices in an embryo production laboratory. In: Stringfellow DA, Seidel SM (eds) *Manual of the International Embryo Transfer Society*, 3rd edn. Savoy, IL: International Embryo Transfer Society, 1998, pp. 93–102.
31. Rossi C, Bridgman R, Kiesel G. Viral contamination of bovine fetal lung cultures and bovine fetal serum. *Am J Vet Res* 1980;41:1680–1681.
32. Bolin S, Ridpath J. Prevalence of bovine viral diarrhea virus genotypes and antibody against those viral genotypes in fetal bovine serum. *J Vet Diagn Invest* 1998;10:135–139.
33. Nettleton P, Vilcek S. Detection of pestiviruses in bovine serum. In: *Proceedings of the International Workshop organized by EDQM, Paris*, 2001, pp. 69–75.
34. Sweeney R, Whitlock R, Rosenberger A. *Mycobacterium paratuberculosis* isolated from fetuses of infected cows not manifesting signs of disease. *Am J Vet Res* 1992;53:477–480.
35. Whittington R, Windsor P. In utero infection of cattle with *Mycobacterium avium* subspecies *paratuberculosis*: a critical review and meta-analysis. *Vet J* 2009;179:60–69.
36. Fray M, Prentice H, Clarke M, Charleston B. Immunohistochemical evidence for localization of bovine viral diarrhea virus, a single-stranded RNA virus, in ovarian oocytes in the cow. *Vet Pathol* 1998;35:253–259.
37. Anonymous. Terrestrial Animal Code Chapter 4.7.5 of the Terresterial Animal Code, under "Risk Management". www.oie.int.
38. Stringfellow D. The potential of bovine embryo transfer for infectious disease control. *Rev Sci Tech Off Int Epiz* 1985;4:859–866.
39. Stringfellow D, Riddell K, Zurovac O. The potential of embryo transfer for infectious disease control in livestock. *NZ Vet J* 1991;39:8–17.
40. Baker J. Bovine viral diarrhea virus: a review. *J Am Vet Med Assoc* 1987;190:1449–1458.

41. Houe H. Epidemiological features and economical importance of bovine virus diarrhea virus (BVDV) infections. *Vet Microbiol* 1999;64:135–144.
42. Hessman B, Fulton R, Sjeklocha D. Evaluation of economic effects and the health and performance of the general cattle population after exposure to cattle persistently infected with bovine viral diarrhea virus in a starter feedlot. *Am J Vet Res* 2009;70:73–85.
43. Kirkland P, Mackintosh S, Moyle A. The outcome of widespread use of semen from a bull persistently infected with pestivirus. *Vet Rec* 1994;135:527–529.
44. McClurkin A, Coria M, Cutlip R. Reproductive performance of apparently healthy cattle persistently infected with bovine viral diarrhea virus. *J Am Vet Med Assoc* 1979;174:1116–1119.
45. Grooms D. Reproductive consequences of infection with bovine viral diarrhea virus. *Vet Clin North Am Food Anim Pract* 2004: 20:5–19.
46. Gregg K, Chen S, Guerra T et al. A sensitive and efficient detection method for bovine viral diarrhea virus (BVD) in single reimplantation bovine embryos. *Theriogenology* 2009;71:966–974.
47. Houe H. Economic impact of BVDV infection in dairies. *Biologicals* 2003;31:137–143.
48. Lindberg A, Alenius S. Principles for eradication of bovine viral diarrhea virus (BVDV) infections in cattle populations. *Vet Microbiol* 1999;64:197–222.
49. Houe H, Lindberg A, Moennig V. Test strategies in bovine viral diarrhea virus control and eradication campaigns in Europe. *J Vet Diagn Invest* 2006;18:427–436.
50. Grom J, Barlic-Maganja D. Bovine viral diarrhea (BVD) infections: control and eradication program in breeding herds in Slovenia. *Vet Microbiol* 1999;64:259–264.
51. Ferrari G, Scicluna M, Bonvicini D et al. Bovine virus diarrhea (BVD) control program in an area in the Rome province (Italy). *Vet Microbiol* 1999;64:237–245.
52. Synge B, Clark A, Moar J, Nicolson J, Nettleton P, Herring J. The control of bovine virus diarrhea virus in Sheltland. *Vet Microbiol* 1999;64:223–229.
53. Walz P, Givens M, Cochran A, Navarre C. Effect of dexamethasone administration on bulls with a localized testicular infection with bovine viral diarrhea virus. *Can J Vet Res* 2008;72:56–62.
54. Passler T, Walz P, Ditchkoff S et al. Cohabitation of pregnant white-tailed deer and cattle persistently infected with bovine viral diarrhea virus results in persistently infected fawns. *Vet Microbiol* 2009;134:362–367.
55. Singh E. The disease control potential of embryos. *Theriogenology* 1987;27:9–20.
56. Singh E, Eaglesome M, Thomas F, Papp-Vid G, Hare W. Embryo transfer as a means of controlling the transmission of viral infections. The in vitro exposure of preimplantation bovine embryos to akabane, bluetongue and bovine viral diarrhea viruses. *Theriogenology* 1982;17:437–444.
57. Wentink G, Aarts T, Mirck M, Van Exsel A. Calf from a persistently infected heifer born after embryo transfer with normal immunity to BVDV. *Vet Rec* 1991;129:449–450.
58. Bak A, Callesen H, Meyling A, Greve T. Calves born after embryo transfer from donors persistently infected with BVD virus. *Vet Rec* 1992;131:37.
59. Brock K, Lapin D, Skrade D. Embryo transfer from donor cattle persistently infected with bovine viral diarrhea virus. *Theriogenology* 1997;47:837–844.
60. Smith A, Grimmer S. Birth of a BVDV-free calf from a persistently infected embryo donor. *Vet Rec* 2000;146:49–50.
61. Guerin B. A secure health status associated with the production and trade of in vitro-derived cattle embryos. *Livestock Prod Sci* 2000;62:271–285.
62. Lindberg A, Ortman K, Alenius S. Seroconversion to bovine viral diarrhea virus (BVDV) in dairy heifers after embryo transfer. In: Proceedings 14th ICAR, 2–6 July, 2000;1:250 (Abstract).
63. Drew T, Sandvik T, Wakeley P, Jones T, Howard P. BVD virus genotype 2 detected in British cattle. *Vet Rec* 2002;51:551.
64. Waldrop J, Stringfellow D, Galik P et al. Infectivity of bovine viral diarrhea virus associated with in vivo-derived bovine embryos. *Theriogenology* 2004;62:387–397.
65. Waldrop J, Stringfellow D, Riddell K et al. Different strains of noncytopathic bovine viral diarrhea virus (BVDV) vary in their affinity for in vivo-derived bovine embryos. *Theriogenology* 2004;62:45–55.
66. Waldrop J, Stringfellow D, Galik P et al. Seroconversion of calves following intravenous injection with embryos exposed to bovine viral diarrhea virus (BVDV) in vitro. *Theriogenology* 2005;65:594–605.
67. Gard J, Givens M, Marley M et al. Bovine viral diarrhea virus (BVDV) associated with single in vivo-derived and in vitro-produced preimplantation bovine embryos following artificial exposure. *Theriogenology* 2009;71:1238–1244.
68. Gard J, Givens M, Marley M et al. Intrauterine inoculation of seronegative heifers with bovine viral diarrhea virus simultaneous to transfer of in vivo-derived bovine embryos. *Theriogenology* 2010;73:1009–1017.
69. Bielanski A, Sapp T, Lutze-Wallace C. Association of bovine embryos produced by in vitro fertilization with a noncytopathic strain of BVDV type II. *Theriogenology* 1998;49:1231–1238.
70. Meyers G, Ege A, Fetzer C et al. Bovine viral diarrhea virus: prevention of persistent fetal infection by a combination of two mutations affecting Erns RNase and Npro protease. *Virology* 2007;81:3327–3338.
71. Galik P, Givens M, Stringfellow D, Crichton E, Bishop M, Eilertsen K. Bovine viral diarrhea virus (BVDV) and anti-BVDV antibodies in pooled samples of follicular fluid. *Theriogenology* 2002;57:1219–1227.
72. Gordon I. Laboratory produced embryos. In: Gordon I (ed.) *Laboratory Production of Cattle Embryos*, 2nd edn. Wallingford, UK: CAB International, 1994, pp. 15–20.
73. Bielanski A, Loewen K, Del Campo M, Sirard M, Willadsen S. Isolation of bovine herpesvirus-1 (BVH-1) and bovine viral diarrhea virus (BVDV) in association with the in vitro production of bovine embryos. *Theriogenology* 1993;40:531–538.
74. Bielanski A, Stewart B. Ubiquitous microbes isolated from in vitro fertilization (IVF) system. *Theriogenology* 1996;45:269 (Abstract).
75. Bielanski A, Algire J, Lalonde A, Nadin-Davis S. Transmission of bovine viral diarrhea virus (BVDV) via in vitro-fertilized embryos to recipients, but not to their offspring. *Theriogenology* 2009;71:499–508.
76. Marquant-Le Guienne B, Remond M, Cosquer R et al. Exposure of in vitro-produced embryos to foot-and-mouth disease virus. *Theriogenology* 1998;50:109–116.
77. Stringfellow D, Givens M. Preventing disease transmission through the transfer of in vivo-derived bovine embryos. *Livestock Prod Sci* 2000;62:237–251.
78. Stringfellow D, Givens M, Waldrop J. Biosecurity issues associated with current and emerging embryo technologies. *Reprod Fertil Dev* 2004;16:93–102.
79. Engles M, Ackermann M. Pathogenesis of ruminant herpesvirus infections. *Vet Microbiol* 1996;53:3–15.
80. Nettleton P, Vilcek S. Detection of pestiviruses in bovine serum. In: *Proceedings of the International Workshop organized by EDQM, Paris*, 2001, pp. 69–75.
81. Vanroose G, Nauwynek H, Van Soom H et al. Structural aspects of the zona pellucida of in-vitro-produced embryos: a scanning electron and confocal laser scanning microscopic study. *Biol Reprod* 2000;62:463–469.
82. Bielanski A, Dubuc B. In vitro fertilization of ova from cows experimentally infected with a non-cytopathic strain of bovine viral diarrhea virus. *Anim Reprod Sci* 1995;38:215–221.
83. Booth P, Collins M, Jenner L et al. Noncytopathic bovine viral diarrhea virus (BVDV) reduces cleavage but increases blastocyst

yield of in vitro produced embryos. *Theriogenology* 1998; 50:769–777.
84. Vanroose G, Nanwynek H, Vansoom A, Vanopdenbosch E, DeKruif A. Effects of bovine herpesvirus-1 on bovine viral diarrhea virus on development of in vitro-produced bovine embryos. *Mol Reprod Dev* 1999;54:255–263.
85. Dinkin M, Stallknecht D, Brackett B. Reduction of infectious epizootic hemorrhagic disease virus associated with in vitro-produced bovine embryos by non-specific protease. *Anim Reprod Sci* 2001;65:205–213.
86. Bielanski A, Jordan L. Washing or washing and trypsin treatment is ineffective for removal of non-cytopathic bovine viral diarrhea virus from bovine oocytes or embryos after experimental viral contamination of an in vitro fertilization system. *Theriogenology* 1996;46:1467–1476.
87. Trachte E, Stringfellow DA, Riddell KP, Galik PK, Riddell MG, Wright J. Washing and trypsin treatment of in vitro derived bovine embryos exposed to bovine viral diarrhea virus. *Theriogenology* 1998;50:717–726.
88. Givens M, Galik P, Riddell K, Brock K, Stringfellow D. Quantity and infectivity of embryo-associated bovine viral diarrhea virus and antiviral influence of a blastocyst impede in vitro infection of uterine tubal cells. *Theriogenology* 1999;52:887–900.
89. Givens M, Galik P, Riddell K, Stringfellow D. Uterine tubal cells remain uninfected after culture with in vitro-produced embryos exposed to bovine viral diarrhea virus. *Vet Microbiol* 1999;70:7–20.
90. Givens M, Galik P, Riddell K, Brock K, Stringfellow D. Replication and persistence of different strains of bovine viral diarrhea virus in an in vitro embryo production system. *Theriogenology* 2000;54:1093–1107.
91. Anonymous. Terrestrial Animal Health Code, chapter 4.8. www.oie.int/index.php?id=169&L=0&htmfile=chapitre_1.4.8.htm.
92. Nibart M, Marquant-Le Guienne B, Humbolt P. General sanitary procedures associated with in vitro-production of embryos. In: Stringfellow DA, Seidel SM (eds) *Manual of the International Embryo Transfer Society*, 3rd edn. Savoy, IL: International Embryo Transfer Society, 1998, pp. 67–77.
93. Edens M, Galik P, Riddell K, Givens M, Stringfellow D, Loskutoff N. Bovine herpesvirus-1 associated with single, trypsin-treated embryos was not infective for uterine tubal cells. *Theriogenology* 2003;60:1495–1504.
94. Shin T, Sneed L, Hill J, Westhusin M. High incidence of developmental failure in bovine fetuses derived by cloning bovine viral diarrhea virus-infected cells. *Theriogenology* 2000;53;243.
95. Stringfellow D, Riddell K, Givens M *et al*. Bovine viral diarrhea virus (BVDV) in cell lines used for somatic cell cloning. *Theriogenology* 2005;63:1004–1013.
96. Givens M, Marley M. Infectious causes of embryonic and fetal mortality. *Theriogenology* 2008;3:270–285.
97. Galik P, Gard J, Givens M *et al*. Effects of ovine interferon-tau on replication of bovine viral diarrhea virus and bovine herpesvirus-1. *Reprod Fertil Dev* 2007;20:156–157.
98. Givens M, Stringfellow D, Riddell K *et al*. Normal calves produced after transfer of in vitro fertilized embryos cultured with an antiviral compound. *Theriogenology* 2006;65:344–355.
99. Givens M, Stringfellow D, Dykstra C *et al*. Prevention and elimination of bovine viral diarrhea virus infections in fetal fibroblast cells. *Antiviral Res* 2004;64:113–118.
100. Newcomer B, Marley M, Ridpath J *et al*. Efficacy of an antiviral compound to inhibit replication of multiple pestivirus species *Antiviral Res* 2012;96:127–129.
101. Walker A, Kimura K, Roberts R. Expression of bovine interferon-tau variants according to sex and age of conceptuses. *Theriogenology* 2009;72:44–53.

Chapter 81

In Vitro Fertilization

John F. Hasler[1] and Jennifer P. Barfield[2]

[1]Bioniche Animal Health, Inc., Laporte, Colorado, USA
*[2]Department of Biomedical Sciences, Animal Reproduction and Biotechnology Laboratory,
Colorado State University, Fort Collins, Colorado, USA*

Introduction

In Latin, *in vitro* means "in glass," and *in vitro* fertilization (IVF) is frequently used as a general term for the process of generating embryos outside the body, which also includes *in vitro* maturation (IVM) and *in vitro* culture (IVC). These procedures are conducted in the sequence IVM–IVF–IVC to produce embryos exclusively *in vitro* (IVP).

In 1959, the rabbit was the first mammalian species in which live offspring were produced by IVF,[1] followed by laboratory mice in 1968.[2] Because the mouse was not a good model for cattle, the first successful IVF with cattle was in 1977 when sperm were capacitated in the oviduct or uterus of a cow or rabbit[3] and the first live calf born in 1981 when a 4-cell embryo was transferred into the oviduct of a recipient cow.[4] The first calves produced entirely from IVM, IVF, and IVC were born in 1987.[5] Since that time, research on *in vitro* procedures has grown dramatically. In addition, *in vitro* procedures are now being used commercially in conjunction with embryo transfer in a number of countries. In this chapter we intend to describe the most current techniques for oocyte collection, IVP of embryos, and the utilization of IVP embryos in the cattle industry. Some of the applications of IVF technology in cattle are listed in Table 81.1.

Collection of oocytes

Originally, *in vitro* procedures in cattle were conducted primarily for research purposes and utilized oocytes collected from superovulated females. When collected from preovulatory follicles or oviducts 20–24 hours after onset of estrus, oocytes have already undergone maturation and are ready for IVF and IVC. Oocytes collected from slaughterhouse-procured ovaries require IVM, so research utilizing these oocytes lagged until IVM techniques were improved. Today, ovaries from cattle slaughterhouses are used extensively as a source of oocytes.

Oocyte collection from living cattle

The primary method of collecting oocytes from live cattle is aspiration of ovaries manipulated per rectum and guided by a vaginally inserted ultrasound probe and needle. Prior to the development of this technique, oocytes were obtained surgically through a flank incision[6] or laparoscopic procedures via the paralumbar fossa,[7] though these approaches were expensive, inefficient, and risked the formation of adhesions with subsequent loss of fertility. The introduction of real-time transrectal ultrasonic imaging (for review, see ref. 8) led to the development of techniques for the repeated collection of oocytes from bovine females. Ultrasound-guided collection of oocytes via the paralumbar fossa was described in 1987 by Callesen *et al.*[9] and the first repeatable efficient technique involving transvaginal ultrasound-guided aspiration was developed in 1988.[10] This technique has become widely known as ovum pick-up (OPU) or transvaginal aspiration. An ultrasound transducer and an attached needle guide of the type used in cattle are shown in Figure 81.1. A diagram of how the transducer is inserted into the vaginal fornix so that a needle can be guided through the vaginal wall and into the ovary is also shown.

The number of oocytes collected from a cow during a single session of OPU depends on a variety of factors. On the mechanical side, the vacuum pressure used to remove the oocyte from the follicle, the gauge of the needle,[12] and the length of the bevel on the needle[13] all impact the number of intact cumulus–oocyte complexes collected. Frequency of collection is another factor. OPU can be performed once or twice a week on the same cow without the use of exogenous hormones. Studies have demonstrated no advantage of interval length for the number of oocytes retrieved per collection when 3, 4, or 7 days were compared.[14–16] The ultimate goal is to generate embryos that can be transferred to recipients and from that perspective there is evidence that a 3- or 4-day interval (8 oocytes per OPU) is better than a 7-day interval (5.5 oocytes per OPU).[17]

Bovine Reproduction, First Edition. Edited by Richard M. Hopper.
© 2015 John Wiley & Sons, Inc. Published 2015 by John Wiley & Sons, Inc.

Table 81.1 Applications of IVF technology.

Commercial applications
Offspring from infertile cattle
Offspring from pregnant cattle
Offspring from young heifers prior to breeding age
Salvage of genetics from terminally ill/injured cattle
Efficient use of sexed semen
Use of resorted semen (frozen-thawed and sexed after thawing)
Use of multiple sires in a short period of time
Utilization of slaughterhouse-derived oocytes for production of research and/or inexpensive embryos

Research applications
Improvements of IVF technology
Improvement of IVC for cloning and transgenic procedures

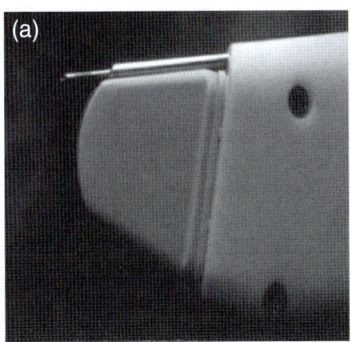
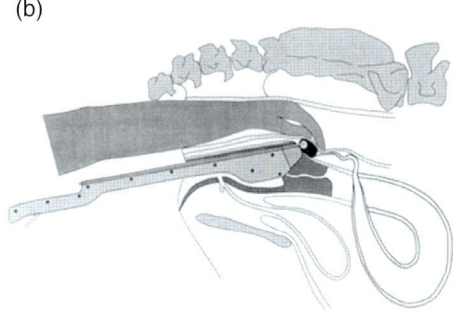

Figure 81.1 (a) Ultrasound transducer with oocyte aspiration needle protruding from needle guide. (b) Diagram showing position of the ultrasound transducer pressed against vaginal fornix, with ovary manually manipulated per rectum and held up against the vaginal wall. Reproduced with permission from Hasler.[11]

The number of oocytes collected per OPU can be increased by pretreatment of the donor with gonadotropins.[18,19] In a very comprehensive comparison of different OPU intervals with and without follicle-stimulating hormone (FSH), Chaubal et al.[20] reported that the most productive protocol in terms of oocyte and embryo production involved dominant follicle removal, FSH treatment 36 hours later, and OPU 48 hours after FSH. This treatment was alternated weekly with simple once-weekly OPU. The OPU weekly session involving FSH averaged 10.6 oocytes and 2.4 blastocysts, while the alternating weekly OPU with no FSH averaged 4.6 oocytes and 0.9 blastocysts. The interaction of factors, including the frequency of OPU and the inclusion or absence of FSH stimulation, has been recently reviewed.[21]

Breed of cattle is also an important factor. Low mean numbers of usable oocytes, ranging from 4.1 to 5.3, were reported by Hasler et al.[22] and Looney et al.[23] from combined populations of dairy and beef breeds, including older females with a variety of fertility problems. The use of FSH with a coasting period is now successfully employed in several commercial North American bovine IVP businesses and has resulted in a mean of approximately 20 oocytes collected from *Bos taurus* cattle once every 2 weeks (data courtesy of Trans Ova Genetics). In South America (primarily Brazil), 17–25 oocytes are routinely collected with no FSH priming from Nelore cattle, an indigenous breed of Brahman origin known to develop higher numbers of follicles in each follicular wave than *Bos indicus* females (data courtesy of In Vitro Brazil).

Age of the donor can also impact the number of oocytes recovered per OPU. Evidence in Holsteins suggests that a significantly higher number of oocytes can be collected in cows 6–9 years old than in cows aged 14–18 years.[16] However, these were primarily cattle that were undergoing OPU/IVP due to previously existing fertility problems. Recent unpublished data from a large commercial *in vitro* program involving reproductively healthy donors indicated little variation in recovered oocyte numbers to age 13 and a steady decline from age 14 to 18 years (data courtesy of Trans Ova Genetics). Interestingly, the developmental potential of oocytes was similar from all donor age categories in this program.

There is considerable variation among donors in terms of oocyte production over time. For example, an 8-year-old Holstein donor, previously diagnosed as infertile, produced a total of 176 embryos from 167 collections by 23 different sires over a period of 167 weeks.[16] With repeated OPU collections, the mean number of oocytes decreased; however, the developmental potential of oocytes produced by OPU did not change. This clearly shows the potential for production of large numbers of embryos by OPU–IVP methods and the advantageous opportunity to use different sires weekly or even biweekly.

Oocyte collection from excised ovaries

In addition to bovine oocyte retrieval by OPU, there are several situations in which oocytes are removed from excised ovaries. When cows develop terminal diseases or become crippled, ovaries can be removed via a flank laparotomy or through a vaginal incision and the oocytes recovered by follicular aspiration or slicing. Stringfellow et al.[24] reported collecting an average of 46 oocytes from 18 culled dairy cows by slicing the ovarian cortex. In a highly successful commercial program, Green and McGuirk[25] reported producing an average of 46 oocytes and nine embryos per donor in more than 100 cases of chronically ill, injured, and senile cows.

Oocyte collection from slaughterhouse-procured ovaries

A large source of oocytes is ovaries from cows processed in slaughterhouses. All follicles between 3 and 8 mm in diameter are aspirated and the collected oocytes used to generate IVP embryos. In a mass-production system using ovaries from the slaughterhouse, Lu and Polge[26] reported producing more than 200 000 in vitro-derived blastocysts from approximately 700 000 oocytes in 2 years. Although this program is no longer in operation, it demonstrated the potential for production of IVP embryos from cattle after slaughter. Reports indicate that as many as 100 and 200 oocytes can sometimes be harvested from one pair of *Bos taurus* ovaries.[26] In Japan, ovaries from individual Kobi cows are procured at the time of slaughter and sent to an IVP laboratory for processing. Later in the same day, carcass values of the individually identified cattle determine which oocytes are retained and processed through the entire IVP procedure. In addition to their commercial value, IVP embryos are produced in universities and laboratories around the world as a valuable source of material for bovine reproduction experiments.[27]

Media

Once oocytes are collected, they are cultured in a series of media that support IVM, IVF, and *in vitro* development. In general, there are at least six different media utilized for the production of *in vitro* embryos. Most of these media are not commercially available and must be produced in the laboratory, though some commercial media systems exist. All these media have common components, which are briefly reviewed here, followed by specifics about media used during specific stages of development.

Water

Pure water is the single most abundant component of all media used for all types of embryo procedures. In fact, water quality might be considered the single most important component of embryo media. What constitutes pure water and how best to produce it has been thoroughly reviewed.[28]

Salts

A number of salts, such as sodium chloride and potassium phosphate, are used in the preparation of virtually all media that are utilized for IVP. These components, and in fact all chemicals used in media preparation, should be reagent quality and whenever possible should be certified as having been "mouse embryo tested."

Energy sources

Glucose, fructose and/or pyruvate are often included in media as energy sources. It is well established that optimal glucose levels for bovine embryos must be low, around 0.5 mmol/L, during early embryonic development. Once the late morula stage is reached, however, higher glucose levels are desirable, and most embryo transfer media contain approximately 2.0 mmol/L glucose.

Macromolecules

Macromolecules include serum, bovine serum albumin (BSA), hyaluronic acid, and synthetic molecules such as polyvinyl alcohol. The most important reason to include a macromolecule in IVP media is the surfactant properties of these molecules, which reduces the chances that embryos will float or stick to equipment, pipette tips, straws, and Petri dishes. More specifically, serum contains a wide range of components that are beneficial for embryo development, such as proteins, growth factors, vitamins, minerals, and energy substrates, and it is well established that serum promotes blastocyst development.[29] Serum and BSA also provide a number of physical properties, including chelation, colloidal osmotic regulation, and pH regulation. Unfortunately, the inclusion of serum during IVC has been associated with a high degree of abnormal pregnancies and offspring, often referred to as large offspring syndrome.[30–33]

Polyvinyl alcohol is primarily used as a surfactant in flushing and holding media that are formulated to include no components of animal origin. These media have been used on hundreds of thousands of bovine donors with highly satisfactory results, based on extensive repeated sales and verbal expressions of satisfaction from a large number of bovine and equine embryo transfer practitioners (unpublished results courtesy of Bioniche Animal Health, Inc.).

Hyaluronic acid (HA) is a glycosaminoglycan that is present in follicular, oviductal, and uterine fluids.[34] Embryos have been shown to have surface receptors for HA, and it is involved in the regulation of gene expression, cell proliferation, and cell differentiation. Addition of HA to bovine embryo culture systems improves development to the blastocyst stage, increases hatching rates, and improves cryotolerance.[35–38] The traditional source of HA has been cock's combs, but can be produced biosynthetically in a *Streptococcus equi* fermentation that does not contain any products of animal origin.

Amino acids

The addition of all 20 essential and nonessential amino acids to embryo culture systems is almost universal. The role of certain individual amino acids and also that of all 20 combined in the development of embryos and enhancing their viability was reviewed by Gardner.[39] When serum is absent from culture media, the inclusion of amino acids substitutes for some of the benefits that serum provides.

Antioxidants

Antioxidants reduce the formation of damaging free radicals in media. Glutathione, BSA, ascorbic acid, and catalase are frequently used as antioxidants in embryo media.

Chelators

Chelators are chemicals that bind heavy metal ions. In a perfectly pure medium, there would be no need for chelators, but there is always a chance that traces of heavy metals can enter the formulation via water or reagent salts. Transferrin, BSA, and ethylenediaminetetraacetic acid (EDTA) all function as chelators.

Osmolytes

Osmolytes are organic compounds that play a role in maintaining volume of cells and organelles within cells as well as fluid balance. When cells swell due to external osmotic pressure, membrane channels open and allow efflux of osmolytes, which carry water with them, restoring normal cell volume. Glycine is an osmolyte, and the smallest of the so-called nonessential amino acids that are usually added to culture media and to some formulations of holding and cryopreservation media. The physiological basis for the osmolyte properties of glycine in embryo culture was described in detail by Steeves et al.[40]

pH and buffers

The intracellular pH (pHi) of in vitro-derived bovine embryos is 7.2 as measured with a pH-sensitive probe.[41] This study also demonstrated that bovine embryos in culture recovered after being subjected to an acid pH more much readily than from a basic pH. In fact, pHi was decreased only to approximately 7.8 when embryos were subjected to a pH above 8. In a classic study, Kane[42] showed that the highest percentage of rabbit blastocysts hatched during IVC when pH was in the range 7.2–7.4, with a precipitous decline when pH was below 7.0 or above 7.8. Most bovine IVP media are in the 7.3–7.4 range except for fertilization media which are generally 7.5–7.6.

Phosphate-buffered saline (PBS), which has been widely used for in vivo-derived embryo recoveries, contains phosphate as a buffer. Although phosphate is not a particularly effective buffer, the inclusion of either serum or BSA adds additional buffering capacity to PBS. IVP media, on the other hand, are normally maintained in an incubator with a 5% CO_2 atmosphere. The inclusion of bicarbonate in these media provides a stable pH due to the equilibrium that is established between the bicarbonate and the CO_2. However, when media based on this system are removed from the incubator, the pH rapidly decreases and will reach an unacceptable level of acidity within a few minutes in air. As a consequence, the media used for rinsing ova/zygotes between IVP steps and for holding them in air prior to transfer or cryopreservation contain MOPES or HEPES zwitterionic buffers.

Osmolarity

The optimum osmolarity for maintaining bovine embryos during IVP procedures is approximately 275 mmol/L, which is also true of other species such as the rabbit, rat, and pig.[43] On a practical basis, bovine and probably equine embryos are not adversely affected by exposure to osmolarities ranging between 250 and 300 mmol/L for relatively short periods. Research has suggested that a total osmolarity between 250 and 270 mmol/L appeared to be optimal when bovine in vitro-derived embryos were cultured in KSOM.[44]

An important consideration when adjusting the osmolarity of media is the issue of what molecules are added or reduced in order to raise or lower osmolarity. The most abundant salt in all media utilized in IVF procedures is NaCl, but it had been shown to be detrimental to bovine embryo development at concentrations above 95 mmol/L.[44] Thus the benefit of having osmolarity below 300 mmol/L in most studies is likely due to lowered Na^+ as opposed to lowered osmolarity per se (oviduct fluid is >300 mmol/L).

Maturation

Much of the success of embryo IVP depends on the quality of the starting material, the oocyte. Photomicrographs of the four basic categories of freshly collected oocytes are shown in Figure 81.2. Oocytes in these four categories have greatly different potentials for maturation and development into embryos, ranging from type 1 with the best potential to type 4 with almost zero potential.[16] Table 81.2 demonstrates the potential for these oocytes to be fertilized, develop to blastocysts, and result in pregnancies.

An oocyte that is considered competent for IVP is referred to as "matured." Maturation involves a series of events that begin in fetal life with the initiation of meiosis. At birth oocytes are arrested at the diplotene stage while continuing to grow and they do not become meiotically active again until puberty, when they are exposed to preovulatory surges of the gonadotropins luteinizing hormone (LH) and FSH, which induce changes in gene expression as well as in morphological and physiological features of the theca, granulosa, cumulus, and oocyte compartments of the ovarian follicle. A subset of oocytes is continually recruited to proceed with meiosis during each estrous cycle, only to arrest at metaphase II, the stage at which they are ovulated. The changes that occur between the diplotene and metaphase II stages of arrest are considered maturation.[45]

IVM of immature oocytes occurs by a different mechanism from that of in vivo-matured oocytes. Removal of an oocyte from the inhibitory follicular environment followed by favorable culture conditions results in spontaneous maturation without the physiological series of events that occur in vivo, though some morphological changes can be observed (e.g., expansion of cumulus cells).[46] In vitro conditions simulate some of the physiological changes with the addition of hormones, including LH, FSH, and estradiol-17β, which are typical of the preovulatory follicle, and growth hormones or epidermal growth factor.[47]

The way in which oocytes are collected and their maturation status dictate how they must be handled after collection. Immature oocytes can be collected from unstimulated cows

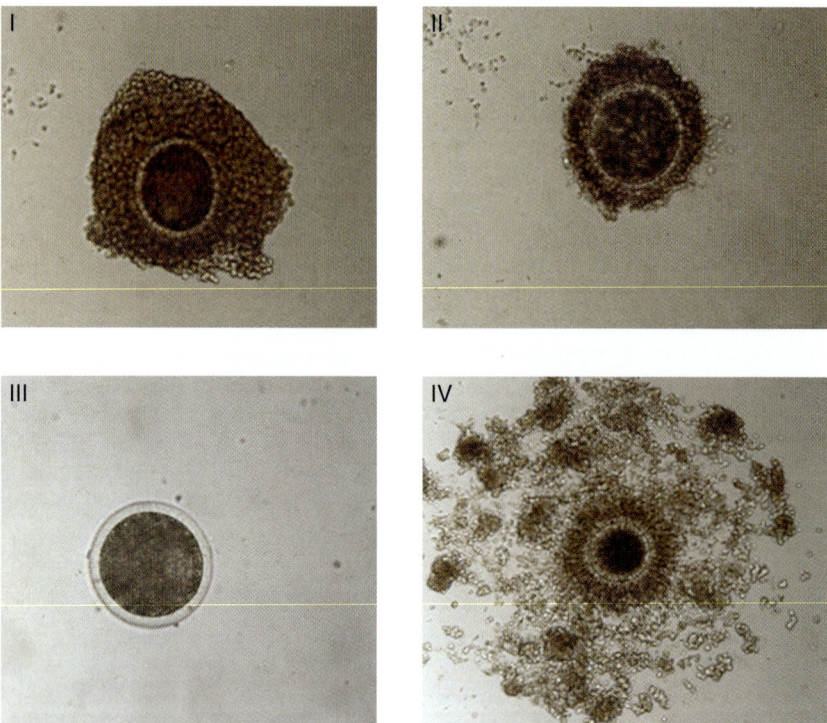

Figure 81.2 Types of bovine oocytes collected from cattle ovaries prior to IVM. Brief descriptions: I, more than four layers of cumulus; II, one to three layers of cumulus; III, nude oocyte with no cumulus; IV, expanded cumulus. Reproduced with permission from Hasler.[11]

Table 81.2 Potential for fertilization and embryonic development in different types of immature bovine oocytes.[11]

Oocyte type	Cumulus	No. oocytes	Percent cleaved	Percent blastocysts	Percent hatched
I	>4 layers	571	65[a]	29[a]	19[a]
II	1–3 layers	228	40[b]	8[b]	6[b]
III	Nude	289	38[b]	<1[c]	<1[c]
IV	Expanded	151	38[b]	7[b]	5[b]

Values within a column with different superscripts (a, b or c) differ significantly ($P<0.025$).
Source: reproduced with permission from Hasler J. Embryo transfer and in vitro fertilization. In: Schatten H, Constantinescu GM (eds) *Comparative Reproductive Biology*. Ames, IA: Blackwell Publishing, 2007, pp. 171–211.

undergoing OPU every 3–7 days, irrespective of the stage of their estrous cycle; however, it is also possible to time aspiration after ovarian stimulation and injection of LH or gonadotropin-releasing hormone (GnRH) so that one collects mature oocytes. These *in vivo*-matured oocytes do not require further maturation *in vitro* though some hours in maturation medium prior to fertilization may be beneficial. Oocytes aspirated from ovaries collected from cows processed in slaughterhouses are primarily from follicles 3–8 mm in diameter and the resulting cohorts of oocytes vary in their maturational status. Even so, the majority of these oocytes can achieve maturation during a 24-hour period of IVM.

Fertilization

Once IVM is complete, oocytes are ready to be fertilized. This involves the coincubation of oocytes with spermatozoa and is generally done in a 4-well dish, but also can be done in a microdrop under oil. Most laboratories allow for 18 hours of coincubation, even though the majority of fertilization events will be complete by 12 hours of coincubation.

On ejaculation, spermatozoa are motile but not capable of fertilizing an oocyte. The changes that a spermatozoon must go through before it is capable of fertilizing an oocyte are collectively called capacitation. Even though this process was discovered in 1951 by Chang[48] and Austin,[49] the molecular mechanisms of capacitation remain unclear. Capacitation generally occurs in the isthmus of the oviduct, although in some species it may begin in the cervix where cervical mucus facilitates the removal of sperm surface proteins. The challenge for those performing IVF is to mimic these conditions by providing spermatozoa with an environment that supports capacitation. Temperature is a critical factor for successful capacitation. Evidence in sheep and pigs suggests that capacitation is most efficient at body temperature rather than 37°C, the temperature to which many incubators are preset (reviewed in ref. 50). This may be associated with the physical state of membrane lipids outside of physiological conditions. In cattle, IVC is typically done at 38.5 °C, including the fertilization process.

Media have been developed to support capacitation *in vitro*. The medium in which sperm and oocytes are coincubated is generally slightly higher in pH (7.5 vs. 7.3 in the maturation and culture media). A basic medium that can be used is Tyrode's and Krebs–Ringer solutions supplemented with an energy source (glucose, lactate, and pyruvate) and albumin.[50] The primary capacitation agent added to fertilization media is heparin, which is thought to assist in the removal of seminal plasma components from the sperm surface by binding to proteins and stimulating the efflux of cholesterol and phospholipids.[50,51] Other capacitation agents include caffeine, an inhibitor of phosphodiesterases; bicarbonate, which induces protein kinase A-dependent changes in lipid architecture; adenosine, which interacts with membrane receptors and increases intracellular cyclic AMP; and reactive oxygen species (ROS) such as superoxide anion, hydrogen peroxide, and nitric oxide which activate membrane targets to trigger intracellular mechanisms including protein tyrosine phosphorylation.[51]

The majority of spermatozoa used in bovine IVF are frozen-thawed. The cryopreservation process induces changes in bovine spermatozoa that resemble capacitated spermatozoa. They have elevated intracellular calcium, a net loss of plasma membrane proteins, and an increase of ROS.[52] These sperm also require a shorter exposure to, and lower doses of, capacitation agents than fresh spermatozoa, which are ideal for IVF. In a well-managed IVF laboratory, fertilization rates with frozen spermatozoa can average 80% or higher.

The most common method for preparing spermatozoa for fertilization is by centrifuging them through a concentration gradient. This discontinuous gradient is prepared with a sperm TALP solution mixed with Percoll® (Sigma) in different proportions. The most common gradient is a 45% Percoll mixture layered on a 90% Percoll mixture. The semen or contents of a thawed straw of semen are layered on top of the gradient and centrifuged at $800 \times g$ for a time dependent on the volume of the column through which the sperm must travel. This isolates intact sperm at the bottom of the centrifuge tube. An alternative method for isolating intact sperm is a "swim up," which involves layering a medium over the contents of a thawed straw and allowing the spermatozoa to swim up into this medium where they are aspirated after a period of time.

The use of spermatozoa that have been sorted based on whether they carry an X or Y chromosome is an increasingly important part of the bovine IVP process, particularly in the dairy industry.[53] The accuracy of the semen sorting process is approximately 90%. In a recent study, there was no difference in pregnancy rates between dairy cows receiving IVP embryos produced with sexed semen (30%) or traditional semen (27%), although there was a significant increase in pregnancy loss of IVP sex-sorted embryos compared with cows that had been artificially inseminated with traditional semen.[54] In this study and in most commercial settings, frozen-thawed semen that was sorted prior to cryopreservation is used. In some cases, frozen semen is sorted after thawing and used to fertilize oocytes immediately after sorting. This is called reverse sorting and has resulted in *in vitro* blastocyst production and pregnancy rates that are comparable to rates obtained with fresh sexed semen and better than sexed semen that was frozen after sexing[55] (unpublished results courtesy of Trans Ova Genetics).

In vitro culture of embryos

IVC of bovine embryos is the last step in the IVP process and involves approximately 6 days of culture from the 1-cell zygote stage following IVF to a blastocyst stage endpoint.

The most common media for culturing bovine embryos are variations of Tervit *et al.*'s[56] original synthetic oviduct medium (SOF), supplemented with amino acids and, in some cases, low levels of serum.[57–60] SOF-based culture may involve one medium formulation for the duration of IVC or a two or three step "sequential" system in which the medium formulation changes at certain points in the culture period. Sequential systems are based in an attempt to mimic the physiological changes that *in vivo* zygotes encounter as they move down the oviducts and into the uterus during the first 6 or 7 days of development. The volume of medium used in IVC systems also varies, with some laboratories culturing embryos in microdrops as small as 25 µL under a layer of oil, ranging up to 500 µL with or without an oil layer. Some IVC systems are conducted in an incubator atmosphere of 5% CO_2 in air; however, there is significant evidence that in the absence of coculture, bovine embryos develop better in a 5% CO_2, 5% O_2, and 90% N_2 atmosphere.[61,62]

In the past, most commercial IVP involved the use of coculture systems that included a monolayer of somatic cells. Cell types that have been used in the successful coculture of bovine embryos include primary cells from oviducts (bovine or porcine), mouse oviductal ampullae, cumulus, granulosa, and uterus, as well as established cell lines such as Buffalo Rat Liver (BRL) and Vero cells. These cells have been combined in coculture systems with a number of different media, including TCM-199, Menezo's B2, and Ham's F-10.

Coculture systems are not widely used today, as most laboratories have largely pursued investigating IVP systems based on "defined" media formulations. Defined media are those in which all the components are known and, as a consequence, products such as serum and BSA are not included. It should theoretically be possible for a defined system to be exactly replicated in any laboratory at any time. Also, the use of defined systems is essential for understanding and optimizing the culture requirements of preimplantation embryos. However, only moderately successful rates of blastocyst production have been reported with the use of chemically defined, protein-free media.[63,64] Some very comprehensive reviews have been written on this subject.[65–68]

Commercial embryo transfer/IVP businesses are primarily concerned with efficiently and predictably producing as many viable embryos as possible. Consequently, BSA (semi-defined medium) and, in some cases, moderate levels of serum (nondefined medium) are frequently added to various media, including IVC media. In addition, nondefined systems may involve coculture, in which any of a number of different somatic cells are included in the system. These systems can be difficult to reproduce because of variations among batches of commercially available cells or a lack of "immortality" when some cell types are cultured through multiple generations in IVP laboratories.

Numerous qualitative and quantitative differences between *in vivo*- and *in vitro*-produced embryos have been described in detail for cattle and other species.[16] There is evidence that IVP embryos have fewer cells and lower proportions of inner cell mass (ICM) cells to trophoblast cells than *in vivo* embryos at the same stage of development and this variation can be related to different culture systems.[69,70] However, total cell counts do not necessarily reflect the overall integrity and viability of preimplantation IVP embryos. Van Soom *et al.*[71] used differential staining to show that although embryos produced in Menezo's B2 medium had higher cell counts than embryos of the same age produced in TCM-199, the B2 embryos had a larger range of variation in the number of ICM cells. The additional cells in the embryos produced in B2 were mainly trophoblast. Including serum can also increase cell numbers but there is evidence that this may decrease the embryo's ability to survive cryopreservation.[72] Thus one should not choose an IVC medium on the basis of increasing cell numbers in embryos.

In addition to differences in cell numbers, the rate of embryonic development varies among IVC media.[73–75] It is clear that within a given IVC medium, faster developing embryos are frequently of higher quality based on cell number,[76] end-stage of development,[77] and pregnancy rate after transfer.[22]

A study that utilized fluorescence *in situ* hybridization showed that a high proportion (72%) of bovine embryos cultured in a Menezo's B2 coculture system were mixoploid, compared with 25% of blastocysts developed *in vivo*.[78] Of the total number of *in vitro* embryos, 17% contained more than 10% polyploidy cells. In contrast, all the *in vivo* embryos contained less than 10% polyploidy cells. The authors of this study offered a hypothetical link between the polyploidy in *in vitro* embryos and the possibility of polyploidy placental cells developing from them, leading to overactive placentas that in turn are responsible for large calves (see section Pregnancies and offspring).

A summary of the IVP process compared with *in vivo* embryo production and stage of embryo development is represented in Figure 81.3.

Cryopreservation of IVP embryos

Cryopreservation of bovine embryos is routinely done in research and commercial IVF laboratories around the world. Traditional slow-freeze cryopreservation has been used extensively with *in vivo*-produced embryos, with pregnancy rates up to 50–55% in large-scale trials.[22,79] Pregnancy rates with frozen-thawed IVP embryos have been lower.

The most common protocol for slow-freeze cryopreservation of bovine oocytes involves cooling and holding the embryo at −6°C, during which time the straw containing the embryo is seeded, followed by a decrease to −32°C at 0.3–0.6°C per minute and plunging the straw into liquid nitrogen. Bovine embryos are typically frozen in a solution containing 1.5 mol/L ethylene glycol and 0.1–0.5 mol/L sucrose. The use of ethylene glycol as the cryoprotectant allows one to transfer the embryo to a recipient directly from the straw, whereas other cryoprotectants such as glycerol must be removed via dilution of the freezing medium prior to transfer.[80] Other cryoprotectants that have been used to freeze bovine embryos include dimethyl sulfoxide, propandiol, methanol, various sugars, and combinations of these.[81]

There are several factors that influence embryo survival post thaw including stage of development. Morulae and blastocysts survive the freeze-thaw process better than earlier stage embryos, but even the day of blastocyst development can impact post-thaw survival. Pregnancy rate following transfer of IVF blastocysts frozen on day 7 was 42%, but dropped to 20% for blastocysts frozen on day 8 with a higher incidence of early abortions and dystocias.[22] The quality of the embryos also impacts post-thaw results. Pregnancy rates for frozen-thawed bovine embryos by grade were reported by Hasler *et al.*[22]: grade 1, 63%; grade 2, 57%; grade 3, 44%; and grade 4, 36%.

There has been extensive research conducted on the effects of cooling and warming rates on embryos of a variety of species. In one study bovine embryos were cooled at 0.3, 0.6, 0.9, 1.2, or 1.5°C/min and warmed either rapidly in a 35°C water bath (warming rate >1000°C/min) or slowly in air at 25–28°C (<250°C/min).[82] The best treatment was cooling at 0.3°C/min with rapid warming; 42.1% of the morulae developed into hatching blastocysts. These results indicate that the survival of IVP bovine morulae and blastocysts is improved by very slow cooling, but slow warming was deleterious to morulae and blastocysts even when coupled with very slow cooling.

Vitrification is an alternative to slow-freeze cryopreservation that involves ultra-rapid cooling resulting in a glass-like state rather than ice formation. Since its introduction as an effective embryo cryopreservation method in 1985,[83] vitrification has been used primarily for research. Limitations of vitrification include toxicity from exposure to high concentrations of cryoprotectants, damage caused by osmotic shock, and the need for highly skilled technicians. However, vitrification has proven more effective for cryopreservation of embryos sensitive to cryopreservation, such as IVP embryos.[84] Another advantage of vitrification is that it does not require expensive cell freezers used for slow freezing procedures, and the equilibration and subsequent cryopreservation takes less time per embryo. In addition, studies have shown that slow-freeze cryopreservation may induce changes in gene expression levels of genes involved in development in the preimplantation bovine embryo when compared with vitrified embryos.[85] Vitrification combined with an in-straw dilution would provide an alternative to slow-freeze cryopreservation of embryos produced *in vitro* or *in vivo*.

Transfer of IVF-derived embryos

In most cases, IVP embryos are transferred in the same manner as *in vivo*-derived embryos. Once removed from the incubator, IVP embryos are usually transferred immediately, whereas *in vivo*-derived embryos can be held without loss of viability at room temperature for up to 24 hours prior to transfer.

Pregnancy rates of IVP embryos are consistently lower than those of *in vivo*-produced embryos. Significantly lower

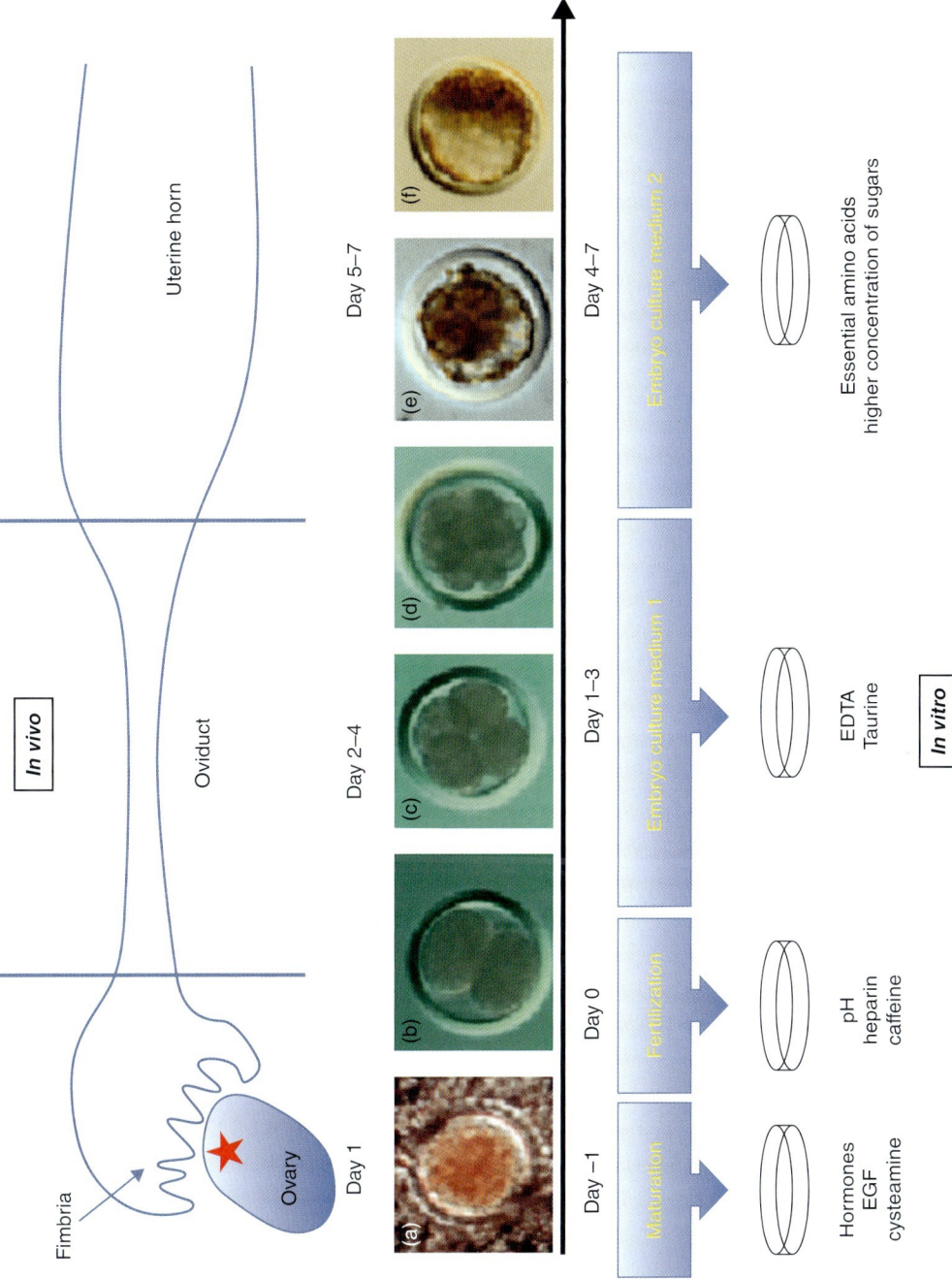

Figure 81.3 A comparison of embryonic development *in vivo* and *in vitro*. The upper portion depicts the location of each of the embryo stages during their transit through the reproductive tract. The lower portion highlights some of the differences between the media at different stages of the IVF process. Photomicrographs: (a) mature oocyte with expanded cumulus (b) 2-cell embryo (c) 4-cell embryo (d) 16-cell embryo (e) early blastocyst (f) blastocyst.

pregnancy rates were achieved, in order, by *in vivo* frozen, fresh IVP day-7, fresh IVP day-8, frozen IVP day-7, and (lowest) frozen IVP day-8 embryos.[22] With regard to stage of development independent of age and grade, the transfer of large numbers of *in vivo* embryos showed that early blastocysts (EB) and mid-blastocysts (MB) resulted in higher pregnancy rates than morulae (M), expanded blastocysts (XB), or hatched blastocysts (HB).[86] Transfer of grade 1 day-7 embryos at the EB, MB, or XB stages results in higher pregnancy rates than transfer of corresponding day-8 embryos.

Synchrony of recipient cows to embryo development is an important factor for the establishment of pregnancy. For IVP embryos, day 0 is considered the day of fertilization, while in recipients day 0 commences when the cow is in standing estrus. With regard to embryo transfer, zero synchrony refers to the transfer of an embryo into a recipient of the same day, for example transfer of a day-7 embryo into a recipient that was in estrus 7 days earlier. Minus 1 day synchrony is the transfer of an embryo to a recipient that was in estrus 1 day less than the age of the embryo, for example a day-7 embryo transferred into a cow that was in estrus 6 days prior. While this is the convention, it is misleading because ovulation in the recipient occurs approximately 12 hours after the end of estrus, which lasts on average 18 hours (day 0 begins at the onset of standing estrus), while fertilization *in vitro* occurs within hours. Consequently, if an IVP embryo is designated as day 7, it is actually 1 day older than a day-7 *in vivo*-derived embryo.[22,87] Pregnancy rates reflect this asynchrony. When day-7 embryos were transferred into day-6 recipients, pregnancy rates were 50.3% compared with 55% for day-7 and 58% for day-8 recipients. Further support is provided by recent data based on more than 10 000 transfers from Trans Ova Genetics showing pregnancy rates of 42, 50, and 49% when day-6, -7 and -8 recipients, respectively, were used for day-7 IVP embryos (unpublished results from Trans Ova Genetics). To avoid decreased pregnancy rates, the use of day-6 recipients should be avoided whenever possible.

Commercial production of bovine embryos via IVP is now successful in a number of laboratories around the world. BOVITEQ Inc. in Quebec[88] and Trans Ova Genetics[89] and Em Tran, Inc.[22] in the United States either have been or are commercial embryo transfer companies that have successfully produced OPU/IVP-derived pregnancies from infertile donors. In addition, Trans Ova Genetics, Sexing Technologies, and Ova Tech in the United States and BOVITEQ in Canada are currently producing large numbers of IVP embryos from healthy cattle. It is estimated that, as of 2012, OPU/IVP embryos represent approximately 10% of the total number of embryo transfers in the United States (data courtesy of American Embryo Transfer Association).

The trend in the international transfer of *in vitro*- and *in vivo*-derived embryos on a yearly basis between 2000 and 2011 is shown in Figure 81.4. It is clear that the number of *in vivo* transfers, involving embryos produced by superovulation and collection via "flushing," has remained relatively stable, whereas there has been a yearly increase in the numbers of IVP embryos. More IVP embryos are currently produced in Brazil than in any other country. One company, In Vitro Brazil, reported the transfer of more than 200 000 IVP embryos in 2012. It is estimated that superovulation/flushing currently represents only 9% of the embryo transfer

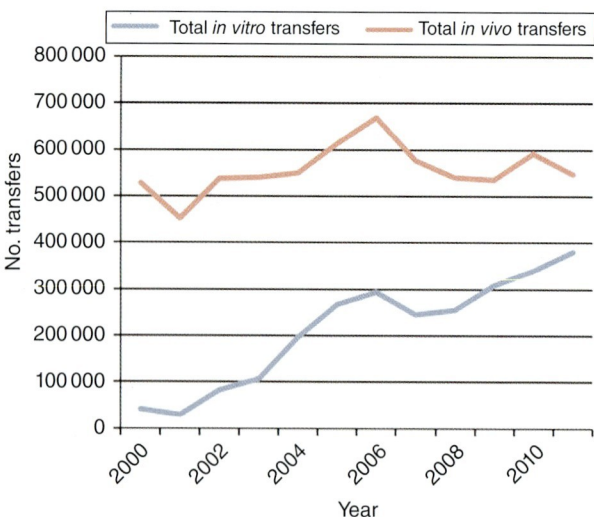

Figure 81.4 Comparison of the annual number of *in vivo*-derived and IVP embryos transferred internationally since 2000. Compliments of the International Embryo Transfer Society.

Table 81.3 Aberrant characteristics associated with the large offspring syndrome of IVP bovine pregnancies and calves.

Large calves
Increased gestation length
Decreased intensity of labor
Increased rate of abortions
Increased rate of congenital malformations
Increased rate of perinatal mortality
Increased rate of hydroallantois in recipients

work in Brazil, with OPU/IVP now providing 91% of the embryos (data courtesy of In Vitro Brazil).

In the Netherlands, IVP embryos are produced from oocytes obtained by OPU from young, reproductively healthy donors and transferred on-farm around the country.[90] In Japan, the production of IVF-derived embryos from slaughterhouse ovaries has been commercialized and, in some cases, very large numbers of embryos have been produced. Kuwayama *et al.*[91] described the production of more than 150 000 frozen IVP embryos in a 1-year period from more than 21 000 ovaries of Japanese Black Cattle.

Pregnancies and offspring

Problems with pregnancies resulting from bovine IVP embryos have been described by a number of authors.[16,32,59,90,92] One of the more obvious problems is larger-than-normal calves at birth and is referred to as large offspring syndrome. Some of the problems associated with this syndrome are shown in Table 81.3.

A number of reports indicate that the average birthweights of IVF calves are heavier than those of calves produced by artificial insemination (AI).[90,93,94] Furthermore, Farin *et al.*[92] showed that 7-month fetuses derived from IVC embryos were heavier than those from *in vivo* embryos. Other studies indicate that serum, coculture, IVM/IVF, and uterine environment are all capable of influencing fetal development in sheep. These studies have primarily been done with sheep because the high maintenance costs and longer gestation period make the bovine a more expensive model for studying the effects of *in vitro* procedures on subsequent offspring. Nevertheless, the findings that lambs from *in vitro*-cultured embryos exposed to serum can be up to 20% heavier than control embryos could be of significant consequence to the cattle industry.[95] Longer gestations have also been noted but not correlated with the weight of the offspring.[95–97]

Because serum has been implicated as a cause of higher birthweights of *in vitro*-derived lambs (for review, see ref. 30), its use in bovine IVC systems has received a good deal of attention. Deletion of serum during the first 4 days of coculture failed to decrease the incidence of problems with IVP-derived pregnancies and calves,[31] whereas Agca et al.[98] showed that birthweights and gestation lengths were normal for a relatively small number of calves resulting from embryos cultured in serum-free CR1aa.[99] Another study comparing pregnancies resulting from cocultured embryos versus those produced in SOF containing BSA showed that the use of SOF resulted in lower birthweights and easier calvings.[59] Unfortunately, comprehensive data regarding late-term pregnancies, incidence of dystocias, and calf normalcy are not available from the large commercial *in vitro* programs currently in business.

Higher-than-normal abortion rates, ranging from 8 to 47%, during the first two trimesters of pregnancy were reported for *in vitro*-derived bovine pregnancies.[22,100–102] In contrast, abortion rates of 5.3% from 2 to 7 months,[103] 4.7% from 2 to 6 months,[86] and 4% from 40 to 250 days[104] for bovine *in vivo* embryo transfer pregnancies were reported.

Perinatal and neonatal losses have also been reported as higher in recipients carrying *in vitro*-derived fetuses compared with AI or *in vivo* embryo pregnancies.[22,90,105,106] There was no difference in the incidence of calf loss at birth between calves produced in the TCM-199 and B2 coculture systems at Em Tran. However, the overall loss at birth of the IVF calves was 14.9% (205 of 1376), which is higher than the 9% loss reported for *in vivo* embryo transfer.[103]

Increased incidence of hydrallantois was diagnosed among pregnant Holstein recipients carrying IVP-derived fetuses (1%).[22,90] A frequency of 1% is substantially higher than the cited frequency of 1 per 7500 calvings reported in non-IVP pregnancies.[107]

Finally, although the data were not quantified, it was reported that labor was often not clearly pronounced in the Holstein heifers used as recipients for IVP embryos.[16] As a result, these recipients were often not observed in labor and calvings frequently went unassisted, resulting in increased perinatal losses. However, when calves were delivered by scheduled cesarean section, the survival rate was close to 98%.

A number of reports show that male bovine IVP embryos grow more rapidly than female embryos during the first 7–8 days in IVP.[108–111] Bovine IVP embryos are usually transferred on day 7 or 8 and the most advanced embryos are usually selected because they tend to be more viable.[22] The percentage of male calves born following transfer of IVP embryos has been reported as 60%,[100] 62%,[102] and 53.4%.[16] The 53.4% was based on more than 2000 IVP calves and fetuses and differed significantly ($P < 0.001$) from 50% and also from the 51.1% males reported for 1751 calves produced from transferred *in vivo*-derived embryos[103] and the 50.5% of 24 000 Holstein AI-produced calves that were males.[112]

Summary

IVF has come a long way since 1959 when the first IVF mammal was produced, and is now a growing component of the bovine embryo transfer industry. Media systems continue to be improved to more closely mimic the *in vivo* environment, resulting in better-quality embryos. This, coupled with the introduction of sexed semen to the IVP process, enhances the value of IVP embryos, particularly in the dairy industry. Additionally, as cryopreservation techniques improve for IVP embryos, embryos can be shipped greater distances and further integrated into the cattle industry.

References

1. Chang M. In vitro fertilization of mammalian eggs. *J Anim Sci* 1968;27:15–22.
2. Whittingham D. Fertilization of mouse eggs in vitro. *Nature* 1968;220:592–593.
3. Iritani A, Niwa K. Capacitation of bull spermatozoa and fertilization in vitro of cattle follicular oocytes matured in culture. *J Reprod Fertil* 1977;50:119–121.
4. Brackett B, Bousquet D, Boice M, Donawick W, Evans J, Dressel M. Normal development following in vitro fertilization in the cow. *Biol Reprod* 1982;27:147–158.
5. Fukuda Y, Ichikawa M, Naito K, Toyoda Y. Birth of normal calves resulting from bovine oocytes matured, fertilized, and cultured with cumulus cells in vitro up to the blastocyst stage. *Biol Reprod* 1990;42:114–119.
6. Bederian K, Shea B, Baker R. Fertilization of bovine follicular oocytes in bovine and porcine oviducts. *Can J Anim Sci* 1975;55:251–256.
7. Lambert R, Sirard M, Bernard C et al. In vitro fertilization of bovine oocytes matured in vitro and collected at laparoscopy. *Theriogenology* 1986;25:117–133.
8. Griffen P, Ginther O. Research applications of ultrasonic imaging in reproductive biology. *J Anim Sci* 1992;70:953–972.
9. Callesen H, Greve T, Christensen F. Ultrasonically guided aspiration of bovine follicular oocytes. *Theriogenology* 1987;27:217 (Abstract).
10. Pieterse M, Kappen K, Kruip T, Taverne M. Aspiration of bovine oocytes during transvaginal ultrasound scanning of the ovaries. *Theriogenology* 1988;30:751–762.
11. Hasler J. Embryo transfer and *in vitro* fertilization. In: Schatten H, Constantinescu GM (eds) *Comparative Reproductive Biology*. Ames, IA: Blackwell Publishing, 2007, pp. 171–211.
12. Bols P, Van Soom A, Ysebaert M, Vandenheede J, de Kruif A. Effects of aspiration vacuum and needle diameter on cumulus oocyte complex morphology and development capacity of bovine oocytes. *Theriogenology* 1996;45:1001–1014.
13. Bols P, Ysebaert M, Van Soom A, de Kruif A. Effects of needle tip bevel and aspiration procedure on the morphology and developmental capacity of bovine compact cumulus oocyte complexes. *Theriogenology* 1997;47:1221–1236.
14. Gibbons J, Beal W, Krisher R, Faber E, Pearson R, Gwazdauskas F. Effects of once- versus twice-weekly transvaginal follicular aspiration on bovine oocyte recovery and embryo development. *Theriogenology* 1994;42:405–419.
15. Broadbent P, Dolman D, Watt R, Smith A, Franklin M. Effect of frequency of follicle aspiration on oocyte yield and subsequent superovulatory response in cattle. *Theriogenology* 1997;47:1027–1040.

16. Hasler J. The current status of oocyte recovery, in vitro embryo production, and embryo transfer in domestic animals, with an emphasis on the bovine. *J Anim Sci* 1998;76:52–74.
17. Garcia A, Cherdieu J, Rademakers A, Salaheddine M. Once versus twice-weekly transvaginal follicular aspiration in the cow. In: *Proceedings of the 12th Scientific Meeting of the AETE, Lyon, France*, 1996, p. 130 (Abstract).
18. Goodhand K, Watt R, Staines M, Hutchinson J, Broadbent P. In vivo oocyte recovery and in vitro embryo production from bovine donors aspirated at different frequencies or following FSH treatment. *Theriogenology* 1999;51:951–961.
19. de Ruigh L, Mullaart E, van Wagtendonk-de Leeuw AM. The effect of FSH stimulation prior to ovum pick-up on oocyte and embryo yield. *Theriogenology* 2000;53:359 (Abstract).
20. Chaubal SA, Molina JA, Ohlrichs CL et al. Comparison of different transvaginal ovum pick-up protocols to optimize oocyte retrieval and embryo production over a 10-week period in cows. *Theriogenology* 2006;65:1631–1648.
21. De Roover R, Feugang JMN, Bols PEJ, Genicot G, Hanzen Ch. Effects of ovum pick-up frequency and FSH stimulation: a retrospective study on seven years of beef cattle in vitro embryo production. *Reprod Domest Anim* 2007;43:239–245.
22. Hasler J, Henderson W, Hurtgen P et al. Production, freezing and transfer of bovine IVF embryos and subsequent calving results. *Theriogenology* 1995;43:141–152.
23. Looney CR, Lindsey BR, Gonseth CL, Johnson DL. Commercial aspects of oocyte retrieval and in vitro fertilization (IVF) for embryo production in problem cows. *Theriogenology* 1994; 41:67–72.
24. Stringfellow D, Riddell M, Riddell K et al. Use of in vitro fertilization for production of calves from involuntary cull cows. *J Assist Reprod Genet* 1993;10:280–285.
25. Green D, McGuirk B. Embryo production from valuable individual salvage cows. In: *Proceedings of the 12th Scientific Meeting of the AETE, Lyon, France*, 1996, p. 134 (Abstract).
26. Lu K, Polge C. A summary of two years' results in large scale in vitro bovine embryo production. In: *Proceedings of the 12th International Congress on Animal Reproduction, The Hague, Netherlands*, 1992, Vol. 3, pp. 1315–1317.
27. Vajta G, Holm P, Greve T, Callesen H. Overall efficiency of in vitro embryo production and vitrification in cattle. *Theriogenology* 1996;45:683–689.
28. Wiemer KE, Anderson A, Stewart B. The importance of water quality for media preparation. *Hum Reprod* 1998;13:166–172.
29. Van Langendonckt A, Donnay I, Schuurbiers N et al. Effects of supplementation with fetal calf serum on development of bovine embryos in synthetic oviduct fluid medium. *J Reprod Fertil* 1997;109:87–93.
30. Young L, Sinclair K, Wilmut I. Large offspring syndrome in cattle and sheep. *Rev Reprod* 1998;3:155–163.
31. Hasler J. In vitro culture of bovine embryos in Ménézo's B2 medium with or without coculture and serum: the normalcy of pregnancies and calves resulting from transferred embryos. *Anim Reprod Sci* 2000;60–61:81–91.
32. Farin P, Piedrahita J, Farin C. Errors in development of fetuses and placentas from in vitro-produced bovine embryos. *Theriogenology* 2006;65:178–191.
33. Thompson J, Mitchell M, Kind K. Embryo culture and long-term consequences. *Reprod Fertil Dev* 2007;19:43–52.
34. Lee C, Ax R. Concentration and composition of glycoaminoglycans in the female reproductive tract. *J Dairy Sci* 1984; 67:2006–2009.
35. Furnus C, de Mantos D, Martinez A. Effect of hyaluronic acid on development of in vitro produced bovine embryos. *Theriogenology* 1998;49:1489–1499.
36. Lane M, Maybach J, Hooper K, Hasler J, Gardner D. Cryosurvival and development of bovine blastocysts are enhanced by culture with recombinant albumin and hyaluronan. *Mol Reprod Dev* 2003;64:70–78.
37. Block J, Bonilla L, Hansen P. Effect of addition of hyaluronan to embryo culture medium on survival of bovine embryos in vitro following vitrification and establishment of pregnancy after transfer to recipients. *Theriogenology* 2009;71:1063–1071.
38. Palasz A, Beltrán Breña P, Martinez M et al. Development, molecular composition and freeze tolerance of bovine embryos cultured in TCM-199 supplemented with hyaluronan. *Zygote* 2008;16:1–9.
39. Gardner D. Dissection of culture media for embryos: the most important and less important components and characteristics. *Reprod Fertil Dev* 2008;20:9–18.
40. Steeves C, Hammer M-A, Walker G, Rae D, Stewart N, Baltz J. The glycine neurotransmitter transporter GLYT1 is an organic osmolyte transporter regulating cell volume in cleavage-stage embryos. *Proc Natl Acad Sci USA* 2003;100:13982–13987.
41. Lane M, Bavister B. Regulation of intracellular pH in bovine oocytes and cleavage stage embryos. *Mol Reprod Dev* 1999; 54:396–401.
42. Kane M. The effects of pH on culture of one-cell rabbit ova to blastocysts in bicarbonate-buffered medium. *J Reprod Fertil* 1974;38:477–480.
43. Baltz JM. Osmoregulation and cell volume regulation in the reimplantation embryo. *Curr Top Dev Biol* 2001;52:55–106.
44. Liu Z, Foote R. Sodium chloride, osmolyte and osmolarity effects on blastocysts formation in bovine embryos produced by in vitro fertilization (IVF) and cultured in simple serum-free media. *J Assist Reprod Genet* 1996;13:562–568.
45. Wassarman P, Schultz R, Letourneau G, LaMarca M, Josefowicz W, Bleil J. Meiotic maturation of mouse oocytes in vitro. *Adv Exp Med Biol* 1979;112:251–268.
46. Silvestre F, Fissore R, Tosti E, Boni R. [Ca^{2+}]i rise at in vitro maturation in bovine cumulus–oocyte complexes. *Mol Reprod Dev* 2012;79:369–379.
47. Mtango N, Varisanga M, Dong Y, Rajamahendran R, Suzuki T. Growth factors and growth hormone enhance in vitro embryo production and post-thaw survival of vitrified bovine blastocysts. *Theriogenology* 2003;59:1393–1402.
48. Chang M. Fertilizing capacity of spermatozoa deposited into the fallopian tubes. *Nature* 1951;168:697–698.
49. Austin C. Observations on the penetration of sperm into the mammalian egg. *Aust J Sci Res B* 1951;4:581–596.
50. Yanagamachi R. Mammalian fertilization. In: Knobil E, Neill J (eds) *The Physiology of Reproduction*, 2nd edn. New York: Raven Press, 1994.
51. Breininger E, Cetica C, Beconi M. Capacitation inducers act through diverse intracellular mechanisms in cryopreserved bovine sperm. *Theriogenology* 2010;74:1036–1049.
52. Pons-Rejraji H, Bailey J, Leclerc P. Cryopreservation affects bovine sperm intracellular parameters associated with capacitation and acrosome exocytosis. *Reprod Fertil Dev* 2009;21: 525–537.
53. Norman H, Hutchison J, Miller R. Use of sexed semen and its effect on conception rate, calf sex, dystocia, and stillbirth of Holsteins in the United States. *J Dairy Sci* 2010;93:3880–3890.
54. Rasmussen S, Block J, Seidel G et al. Pregnancy rates of lactating cows after transfer of in vitro produced embryos using X-sorted sperm. *Theriogenology* 2013;79:453–461.
55. Avelino K, Rossetto M, Maraia A, Lima M, Mansano A, Garcia J. Reports of in vitro production and pregnancy rates of bovine embryos produced by post-thawing sexed semen. *Reprod Fertil Dev* 2010;22:288.
56. Tervit H, Whittingham D, Rowson L. Successful culture in vitro of sheep and cattle ova. *J Reprod Fertil* 1972;38:177–179.
57. Holm P, Booth, P, Schmidt M, Greve, T, Callesen H. High bovine blastocyst development in a static in vitro production system using SOFaa medium supplemented with sodium citrate and myo-inositol with or without serum-proteins. *Theriogenology* 1999;52:683–700.

58. Lane M, Maybach J, Hooper K, Hasler J, Gardner D. Cryosurvival and development of bovine blastocysts are enhanced by culture with recombinant albumin and hyaluronan. *Mol Reprod Dev* 2003;64:70–78.
59. van Wagtendonk-de Leeuw AM, Mullaart ER, de Roos APW *et al*. Effects of different reproduction techniques: AI, MOET, or IVP, on health and welfare of bovine offspring. *Theriogenology* 2000;53:575–597.
60. De La Torre-Sanchez J, Preis K, Seidel G. Metabolic regulation of *in vitro*-produced bovine embryos. I. Effects of metabolic regulators at different glucose concentrations with embryos produced by semen from different bulls. *Reprod Fertil Develop* 2006;18:585–596.
61. Thompson J, Simpson A, Pugh P, Donnelly P, Tervit H. Effect of oxygen concentration on in-vitro development of preimplantation sheep and cattle embryos. *J Reprod Fertil* 1990;89:573–578.
62. Fukui Y, McGowan L, James R, Pugh P, Tervit H. Factors affecting the in-vitro development to blastocysts of bovine oocytes matured and fertilized in vitro. *J Reprod Fertil* 1991;92:125–131.
63. Bavister B, Rose-Hellekant T, Pinyopummintr T. Development of in vitro matured/in vitro fertilized bovine embryos into morulae and blastocysts in defined culture media. *Theriogenology* 1992;37:127–146.
64. Kim J-H, Niwa K, Lim J-M, Okuda K. Effects of phosphate, energy substrates, and amino acids on development of in vitro-matured, in vitro-fertilized bovine oocytes in a chemically defined, protein-free culture medium. *Biol Reprod* 1993;48:1320–1325.
65. Gardner, D, Lane M. Embryo culture systems. In: Trounson AO, Gardner DK (eds) *Handbook of In Vitro Fertilization*. Boca Raton, FL: CRC Press, 1993, pp. 85–114.
66. Bavister B. Culture of preimplantation embryos: facts and artifacts. *Hum Reprod Update* 1995;1:91–148.
67. Gordon I. *Laboratory Production of Cattle Embryos*. Wallingford, UK: CAB International, 1994.
68. Thompson J. Defining the requirements for bovine embryo culture. *Theriogenology* 1996;45:27–40.
69. Goto K, Iwai N, Ide K, Takuma Y, Nakanishi Y. Viability of one-cell bovine embryos cultured in vitro: comparison of cell-free culture and co-culture. *J Reprod Fertil* 1994;100:239–243.
70. Rieger D, Grisart B, Semple E, Van Langendonckt A, Betteridge K, Dessy F. Comparison of the effects of oviductal cell co-culture and oviductal cell-conditioned medium on the development and metabolic activity of cattle embryos. *J Reprod Fertil* 1995;105:91–98.
71. Van Soom A, Boerjan M, Ysebaert M-T, De Kruif A. Cell allocation to the inner cell mass and the trophectoderm in bovine embryos cultured in two different media. *Mol Reprod Dev* 1996;45:171–182.
72. Mucci N, Aller J, Kaiser G, Hozbor F, Cabodevila J, Alberio R. Effect of estrous cow serum during bovine embryo culture on blastocyst development and cryotolerance after slow freezing or vitrification. *Theriogenology* 2006;65:1551–1562.
73. Hawk H, Wall R. Improved yields of bovine blastocysts from in vitro-produced oocytes. 11. Media and co-culture cells. *Theriogenology* 1994;41:1585–1594.
74. Donnay I, Van Langendonckt A, Auquier P *et al*. Effects of coculture and embryo number on the in vitro development of bovine embryos. *Theriogenology* 1997;47:1549–1561.
75. Farin C, Hasler J, Martus N, Stokes J. A comparison of Menezo's B2 and Tissue Culture Medium-199 for in vitro production of bovine blastocysts. *Theriogenology* 1997;48:699–709.
76. Jiang H, Wang W, Lu K, Gordon I, Polge C. Examination of cell numbers of blastocysts derived from IVM, IVF and IVC of bovine follicular oocytes. *Theriogenology* 1992;37:229 (Abstract).
77. Grisart B, Massip A, Collette L, Dessy F. The sex ratio of bovine embryos produced in vitro in serum-free oviduct cell-conditioned medium is not altered. *Theriogenology* 1995;43:1097–1106.
78. Viuff D, Rickords L, Offenberg H *et al*. A high proportion of bovine blastocysts produced in vitro are mixoploid. *Biol Reprod* 1999;60:1273–1278.
79. Nibart M, Humblot P. Pregnancy rates following direct transfer of glycerol sucrose or ethylene glycol cryopreserved bovine embryos. *Theriogenology* 1997;47:371 (Abstract).
80. Voelkel SA, Hu YX. Use of ethylene glycol as a cryoprotectant for bovine embryos allowing direct transfer of frozen-thawed embryos to recipient females. *Theriogenology* 1992;37:687–697.
81. Penfold L, Watson P. The cryopreservation of gametes and embryos of cattle, sheep, goats and pigs. In: Watson PF, Holt WV (eds) *Cryobanking the Genetic Resource: Wildlife Conservation for the Future?* London: Taylor & Francis, 2001, pp. 279–316.
82. Hochi S, Semple E, Leibo S. Effect of cooling and warming rates during cryopreservation on survival of in vitro-produced bovine embryos. *Theriogenology* 1996;46:837–847.
83. Rall W, Fahy G. Ice-free cryopreservation of mouse embryos at –196 degrees C by vitrification. *Nature* 1985;313:573–575.
84. Wurth Y, Reinders J, Rall W, Kruip T. Developmental potential of *in vitro* produced bovine embryos following cryopreservation and single-embryo transfer. *Theriogenology* 1994;42:1275–1284.
85. Stinshoff H, Wilkening S, Hanstedt A, Brüning K, Wrenzycki C. Cryopreservation affects the quality of in vitro produced bovine embryos at the molecular level. *Theriogenology* 2011;76:1433–1441.
86. Hasler J, McCauley A, Lathrop W, Foote R. Effect of donor–embryo–recipient interactions on pregnancy rate in a large-scale bovine embryo transfer program. *Theriogenology* 1987;27:139–168.
87. Avery B, Brandenhoff H, Greve T. Development of in vitro matured and fertilized bovine embryos, cultured from days 1–5 post insemination in either Menezo-B2 medium or in HECM-6 medium. *Theriogenology* 1995;44:935–945.
88. Bousquet D, Twagiramungu H, Morin N, Brisson C, Carboneau G, Durocher J. In vitro embryo production in the cow and effective alternative to the conventional embryo production approach. *Theriogenology* 1998;51:59–70.
89. Faber D, Molina J, Ohlrichs C, Vander Zwaag D, Ferré L. Commercialization of animal biotechnology. *Theriogenology* 2003;59:125–138.
90. Kruip T, den Daas J. In vitro produced and cloned embryos: effects on pregnancy, parturition and offspring. *Theriogenology* 1997;47:43–52.
91. Kuwayama M, Hamano S, Kolkeda A, Matsukawa K. Large scale in vitro production of bovine embryos. In: *Proceedings of the 13th International Congress on Animal Reproduction, Sydney, Australia*, 1996, p. 71 (Abstract).
92. Farin P, Farin C. Transfer of bovine embryos produced in vivo or in vitro: survival and fetal development. *Biol Reprod* 1995;52:676–682.
93. Behboodi E, Anderson G, BonDurant R *et al*. Birth of large calves that developed from in vitro-derived bovine embryos. *Theriogenology* 1995;44:227–232.
94. Sinclair K, Broadbent P, Dolman D. In vitro produced embryos as a means of achieving pregnancy and improving productivity in beef cows. *Anim Sci* 1995;60:55–64.
95. Thompson JG, Gardner DK, Pugh PA, McMillan WH, Tervit HR. Lamb birth weight is affected by culture system utilized during in vitro pre-elongation development of ovine embryos. *Biol Reprod* 1995;53:1385–1391.
96. Walker S, Heard T, Seamark R. In vitro culture of sheep embryos without coculture: successes and perspectives. *Theriogenology* 1992;37:111–126.
97. Holm P, Walker S, Seamark R. Embryo viability, duration of gestation and birth weight in sheep after transfer of in vitro

matured and in vitro fertilized zygotes cultured in vitro or in vivo. *J Reprod Fertil* 1996;107:175–181.
98. Agca Y, Monson R, Northey D, Schaefer D, Rutledge J. Transfer of fresh and cryopreserved IVP bovine embryos: normal calving, birth weight and gestation lengths. *Theriogenology* 1998;50:147–162.
99. Rosenkrans J, Zeng G, McNamara G, Schoff P, First N. Development of bovine embryos in vitro as affected by energy substrates. *Biol Reprod* 1993;49:459–462.
100. Reichenbach H, Liebrich J, Berg U, Brem G. Pregnancy rates and births after unilateral transfer of bovine embryos produced in vitro. *J Reprod Fertil* 1992;95:363–370.
101. Reinders J, Wurth Y, Kruip T. From embryo to calf after transfer of in vitro produced bovine embryos. *Theriogenology* 1995;43:306 (Abstract).
102. Massip A, Mermillod P, Van Longendonckt A, Touze J, Dessy F. Survival and viability of fresh and frozen-thawed in vitro bovine blastocysts. *Reprod Nutr Dev* 1995;35:3–10.
103. King K, Seidel G, Elsden R. Bovine embryo transfer pregnancies. I. Abortion rates and characteristics of calves. *J Anim Sci* 1985;61:747–762.
104. Callesen H, Bak A, Greve T. Embryo recipients: dairy cows or heifers. In: *Proceedings of the 10th Scientific Meeting of the AETE, Lyon, France*, 1994, pp. 125–135.
105. Van Soom A, Mijten P, Van Vlaenderen I, Van den Branden J, Mahmoudzadeh A, de Kruif A. Birth of double-muscled Belgian blue calves after transfer of in vitro produced embryos into dairy cattle. *Theriogenology* 1994;41:855–867.
106. Schmidt M, Greve T, Avery B, Beckers J, Sulon J, Hansen H. Pregnancies, calves, and calf viability after transfer of in vitro produced bovine embryos. *Theriogenology* 1996;46:527–539.
107. Sloss V, Dufty J. Disorders during pregnancy. In: Sloss V, Dufty J (eds) *Handbook of Bovine Obstetrics*. Baltimore: Williams & Wilkins, 1980, pp. 88–97.
108. Avery B, Madison V, Greve T. Sex and development in bovine in-vitro fertilized embryos. *Theriogenology* 1991;35:953–963.
109. Avery B, Jorgensen C, Madison V, Greve T. Morphological development and sex of bovine in vitro-fertilized embryos. *Mol Reprod Dev* 1992;32:265–270.
110. Xu K, Yadav B, King W, Betteridge KJ. Sex related differences in developmental rates of bovine embryos produced and cultured in vitro. *Mol Reprod Dev* 1992;31:249–252.
111. Carvalho R, Del Campo M, Palasz A, Plante Y, Mapletoft R. Survival rates and sex ratio of bovine IVF embryos frozen at different developmental stages on day 7. *Theriogenology* 1996;45:489–498.
112. Foote R. Sex ratios in dairy cattle under various conditions. *Theriogenology* 1977;8:349–356.

Chapter 82

Cloning by Somatic Cell Nuclear Transfer

J. Lannett Edwards and F. Neal Schrick

Department of Animal Science, University of Tennessee, Knoxville, Tennessee, USA

Introduction

In 1997 Dr Ian Wilmut and coworkers[1] startled the world by announcing the birth of Dolly, the first ever clone of an adult mammal (Figure 82.1). Because producing a clone of an adult mammal (i.e., a genetic replica of an existing individual) was thought biologically impossible, the announcement of Dolly's birth reverberated around the world, capturing the attention of scientists and conjuring up fantastical visions from the lay public. Fascination with Dolly, continuing even today, stems from her unexpected origin. Dolly originated from the transfer of a quiescent-induced mammary cell originating from the udder of a 6-year-old adult ewe into an enucleated oocyte (i.e., an oocyte whose nuclear DNA had been previously removed). A brief electrical pulse was all that was needed to initiate embryonic development, highlighting the fact that the spermatozoon is not needed for this type of "asexual" reproduction.

The notion of cloning by transferring the nucleus of a cell into an enucleated oocyte was not novel to Dolly. The first successful production of clones by nuclear transfer was in amphibians in 1952.[2] Thirty one years later McGrath and Solter[3] announced the first successful cloning of mice. Soon thereafter came the cloning of sheep,[4] cattle,[5] rabbits,[6] and pigs.[7] Interestingly, the early successes in obtaining cloned offspring were related to the use of totipotent or pluripotent cell types comprising the early embryo; any and every attempt using more differentiated cell types failed. Worldwide failures in different species gave credibility to the long and widely held dogma that it was biologically impossible to clone mammals by nuclear transfer. An initial hint that this dogma would not withstand the test of time came with the reporting of the births of Megan and Morag in 1996 by Dr Wilmut's group with Dr Keith Campbell as lead author.[8] Unlike the others that had been born up to that time, these cloned sheep were obtained by nuclear transfer using a cultured cell line that was embryonic in origin but which was passaged up to 6–13 times (i.e., number of times the cell line was subcultured for producing a larger number of cells from the initial number harvested). Because the cultured cell line was serum starved before nuclear transfer, the first ever success in obtaining live cloned offspring using a differentiated cell type was ascribed to the use of quiescent-induced cells. Subsequent successes in producing cloned sheep reinforced this notion by using established lines from mammary (i.e., Dolly), fetal and embryo cells after inducing cells into a quiescent state (Table 82.1).

Although the need for induction of quiescence before nuclear transfer was later questioned, Dolly's birth served as a reminder to all that the impossible is possible assuming a willingness to persevere while thinking outside the box. The cloning of Dolly and the other sheep using cultured cell lines reinvigorated interests of scientists worldwide as the ability to do so provided the opportunity to genetically modify livestock species. Mindful that "the impossible *was* possible," the list of adult animals cloned using cultured cell lines for nuclear transfer now includes, but is not limited to, sheep, cattle, mouse, goat, pig, gaur, mouflon, rabbit, cat, horse, rat, African wild cat, mule, banteng, deer, dog, ferret, wolf, buffalo, camel, and Spanish ibex (Figure 82.2). Realizing that what can be done in one species is likely possible in others, the list of different cloned species will continue to grow. Since Dolly, nuclear transfer clones have been constructed using a variety of different body cells such that cloning by nuclear transfer is now appropriately termed somatic cell nuclear transfer (SCNT).

Procedures involved in SCNT

The ultimate challenge of SCNT is to reprogram a somatic nucleus in a manner allowing for proper embryo development; in other words, transfer to an environment that would "force" the nucleus to forget its somatic programming and start functioning as a 1-cell zygote. The predominant cell type of choice for reprogramming a somatic nucleus is an oocyte arrested at metaphase II (MII). This is a logical choice for reprogramming a somatic nucleus because it has within it the majority, if not all, of the key critical components required for directing early embryo development. In general, SCNT is remarkably similar in the numerous species that have been cloned to date, highlighting the relevance of the procedural steps depicted in Figure 82.3.

Collection of somatic cells from the animal to be cloned (i.e., somatic cell donor) is a first important step in SCNT. In cattle and other species, adult animals have been cloned using a variety of different diploid (2n) cell types including, but not exclusive to, fibroblasts obtained from various sources, granulosa cells from antral follicles, cumulus cells, along with those originating from mammary tissue, muscle, oviduct, uterus and other sources. From a practical viewpoint, the relative ease of obtaining fibroblasts from a simple skin biopsy make this cell type a logical preference for cloning adult males and females. With regard to cows, ovarian/granulosa cells are easily obtained by the same ultrasound-guided transvaginal aspiration used for ovum pick-up (OPU). After collection somatic cells may be utilized immediately or after long-term culture. Dispersion into a single cell suspension is readily achieved using a trypsin-based solution with most somatic cell types. Interestingly, cell line rather than cell type is more influential on outcomes. In cases where outcomes are less than expected, individuals may choose to obtain another biopsy for establishing a different cell line from somatic cell donor before switching cell type. It is now known that induction of quiescence in somatic cells before nuclear transfer through serum starvation is not required for producing clones of adult animals.

The next important step in SCNT involves the collection of oocytes from donor females. Abattoir-derived ovaries provide an abundant and affordable supply of oocytes. Demonstrated ability to produce cloned offspring after *in vitro* maturation makes this the preferred source/approach for obtaining MII oocytes. In cases where it is important to control the genetic and/or maternal background of donor oocytes, OPU procedures provide a means for collecting smaller numbers of oocytes from live cows. *In vitro* maturation, while not as optimal as *in vivo* maturation occurring within Graafian follicle(s), provides a greater level of control for selecting MII oocytes after 18–20 hours.

The next and perhaps most labor-intensive step in SCNT involves the removal of maternal (nuclear) DNA from MII stage oocytes using microtools (see Figure 82.3). Addition of a microfilament inhibitor (i.e., cytochalasin B) to the holding medium relaxes the cytoplasm, allowing mechanical removal via aspiration of 5–15% of the oocyte's cytoplasm containing the maternal (nuclear) DNA. Doing so is important to minimize the incidence of lysis that would otherwise occur without the addition of a microfilament inhibitor. Addition of Hoechst stain to the holding medium allows verification that the nuclear DNA was successfully removed. Limiting exposure to ultraviolet transillumination of the specific region of the microtool containing the maternal DNA avoids harm to the remaining "bag" of oocyte cytoplasm devoid of nuclear DNA but still containing mitochondrial DNA, essential organelles, and other essential components unique to the oocyte.

The next challenge of the SCNT procedure is transferring a somatic nucleus from a cell obtained from the animal to be cloned into the enucleated bag of oocyte cytoplasm. In most species, mice being an exception, this is achieved by electrical-induced fusion of the somatic cell with the enucleated bag of oocyte cytoplasm. To this end an intact somatic cell is mechanically inserted into the perivitelline space (available

Figure 82.1 Dolly, a lamb derived from the mammary gland of a Finn Dorset ewe, with the Scottish Blackface ewe which was the recipient. Reproduced from Wilmut I, Schnieke A, McWhir J, Kind A, Campbell K. Viable offspring derived from fetal and adult mammalian cells. *Nature* 1997;385:810–813 with permission from Nature Publishing Group. Image kindly provided by the Roslin Institute, University of Edinburgh.

Table 82.1 Development of cloned sheep embryos constructed with three different cell types

Cell type	No. of fused couplets (%)	No. of morulae/blastocysts (%)	No. of pregnancies/no. of recipients (%)	No. of live lambs
Mammary epithelium	277 (63.8)	29 (11.7)	1/13 (7.7)	1 (0.36%)
Fetal fibroblast	172 (84.7)	47 (37.9)	5/16 (31.3)	3 (1.74%)
Embryo-derived	385 (82.8)	126 (54.5)	15/32 (46.9)	4 (1.04%)

Source: adapted from Wilmut I, Schnieke A, McWhir J, Kind A, Campbell K. Viable offspring derived from fetal and adult mammalian cells. *Nature* 1997;385:810–813 with permission from Nature Publishing Group.

Figure 82.2 Different animals cloned by somatic cell nuclear transfer; camel, Spanish ibex, water buffalo, and wolf are not shown. When possible, images were obtained from relevant journal articles after obtaining publisher permission (i.e., Nature Publishing Group and Elsevier). Images of a cloned gaur and banteng and permission for use were provided Dr Robert Lanza of Advanced Cell Technology. Texas A&M College of Veterinary Medicine and Biomedical Sciences provided images and permission to include cloned cat and deer. Dr Martha Gomez at the Audubon Center for Research of Endangered Species provided image and permission to include an African wildcat.

space between the zona pellucida and the enucleated bag of oocyte cytoplasm) using microtools. The resulting "couplet" (bag of enucleated oocyte cytoplasm and somatic cell) is aligned between two electrodes and briefly pulsed with an electrical current. In the case of cattle, 2.2 kV/cm for 40 s induces greater than 70% of couplets to fuse. Applying an electrical current induces pore formation in the respective membranes of the two cells. Electrofusion as a means of transferring the somatic nucleus into the bag of oocyte cytoplasm depends on continuing contact between the

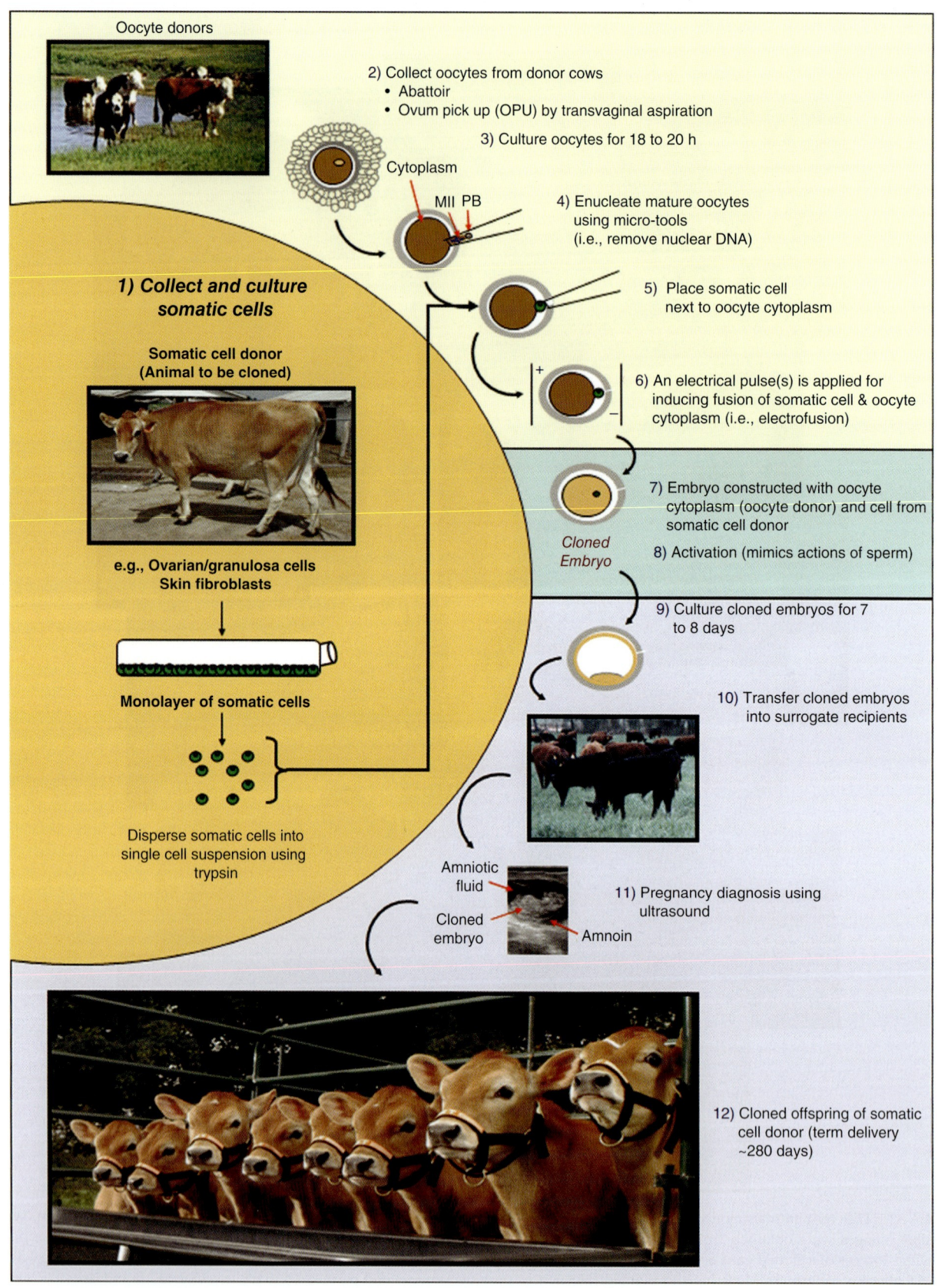

Figure 82.3 The different SCNT steps utilized for cloning adult animals. Although the schematic depicts efforts to clone a cow, SCNT procedures are in general similar across species. (1) The first step involves collection of somatic cells from the animal to be cloned (somatic cell donor). (2) Oocytes are collected from abattoir-derived cows or by transvaginal aspiration (ovum-pick up-OPU). (3) Oocytes may be matured *in vitro*. (4) DNA is then removed from an oocyte using microtools. (5) A somatic cell is then placed beside the bag of oocyte cytoplasm. (6) Exposure of the oocyte cytoplasm and somatic cell to an electrical current will induce the majority to fuse, (7) effectively constructing the equivalent of a 1-cell cloned embryo. (8) In most cases, electrical induction of fusion of the oocyte cytoplasm with somatic cell is sufficient to activate the cloned embryo to begin development. (9) Cloned embryos may be cultured in the laboratory for a defined period of time. (10) Thereafter, embryos must be transferred to surrogate recipients for continued development. (11) Ultrasound is useful for determination and monitoring of resulting pregnancies. (12) A few of the cloned embryos will develop to term resulting in the birth of live offspring.

Table 82.2 Developmental potential of SCNT cow clones (constructed with ovarian/granulosa cells and skin fibroblasts) or *in vitro*-produced embryos using *in vitro* fertilization

Embryos	No.	No. cleaved (%)	No. of day 6 and 7 morulae and blastocyst stage embryos (%)
Clones	686	ND	207 (30.2)
In vitro produced	863	705 (81.7)	235 (27.2)

ND, not determined.
Source: adapted from Edwards J, Schrick F, McCracken M *et al*. Cloning adult farm animals: a review of the possibilities and problems associated with somatic cell nuclear transfer. *Am J Reprod Immunol* 2003;50:113–123.

somatic cell and the oocyte cytoplasm. Depending on the extent to which membranes intermingle, fusion may occur. Within minutes of introducing the somatic nucleus into the bag of oocyte cytoplasm, the nuclear membrane breaks down and the chromatin condenses. Success in doing so effectively constructs the equivalent of a 1-cell embryo.

In some cases the electrical pulse utilized for fusion is also sufficient to "activate" the cloned embryo to begin development, while others choose various chemical combinations. Regardless of how one chooses to "jump start" embryo development, the ultimate challenge of activation protocols is to mimic the developmentally important processes occurring after fertilization. Note the deliberate use of the word "mimic" because the spermatozoon is *not* required for this type of asexual reproduction. After activation and depending on the species, cloned embryos may be transferred into ligated oviducts (preventing entry into the uterus of temporary recipients; this is what was done for Dolly), the uterus (pigs), or cultured in the incubator for a period of time required for development to the compact morula or blastocyst stage (cattle). Cloned bovine embryos develop to the blastocyst stage at an equivalent rate as those undergoing *in vitro* fertilization (Table 82.2).

Embryo transfer into surrogate recipients uses the same approach as for transferring *in vitro*- or *in vivo*-derived embryos, but with a few extra challenges. Depending on the level of difficulty, cloned embryos may be damaged while being transferred into the uterus of the surrogate recipients. In cattle, effort should be taken to transfer embryos before or during the early stages of blastocoele expansion, otherwise the expanding blastocoele will protrude through the holes in the zona pellucida created by earlier use of microtools (Figure 82.4). Extensive protrusions appearing as a "figure 8" could explain why twins are occasionally noted after the transfer of individual cloned embryos. Pregnancy rates after the transfer of one or two embryos do not differ. Transfer of a single embryo is helpful for minimizing complications associated with twinning. Nonetheless, earlier stage embryos may be transferred (i.e., compact morula or early blastocyst stage embryos) to avoid possible issues with the aberrant hatching of clones. When doing so, extra care should be taken to ensure that embryos are stage-matched with the uterus (i.e., compact morulae would be transferred to day-6 recipients, not day 7 or 8). Depending on the cell line, it is reasonable to expect day-28 pregnancy rates in

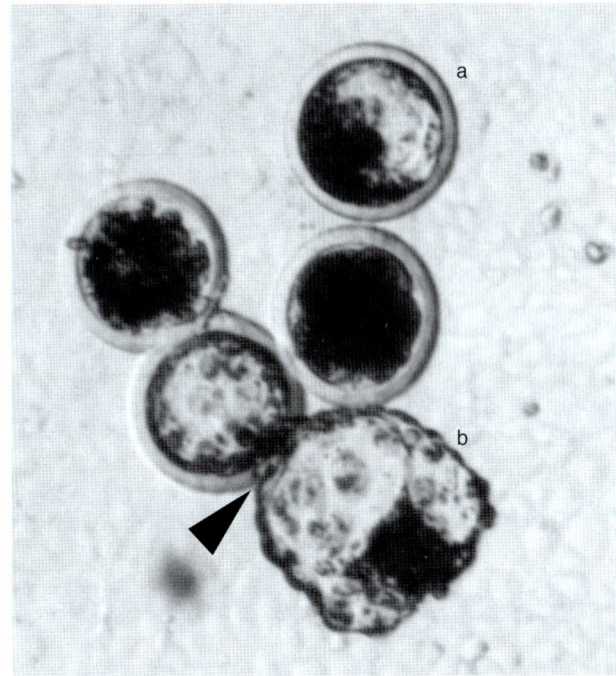

Figure 82.4 Bovine cloned blastocysts (a) before expansion and (b) expanding blastocyst where much of the blastocoele is protruding through the hole in the zona pellucida (arrowhead), appearing as a "figure 8."

cattle approaching 50% after the transfer of single embryos, as indicated by the presence of an embryo proper with heartbeat (Table 82.3). A limited number of cloned embryos resulting in a confirmed pregnancy will progress to term and result in the delivery of live offspring.

Applications of SCNT

Success in cloning sheep, cattle, pigs, and goats confirms the usefulness of SCNT for clonal expansion of agriculturally important food-producing animals. Additional efforts of producers to cross their best clonal lines to create new genetic combinations are imperative for making genetic improvements while maximizing genetic diversity. Clones produced by SCNT are merely genetic replicas of existing individuals, affirming the safety of consuming food-derived products.[9] Furthermore, the usefulness of this technology for restoring endangered or even extinct species has been realized, with the first successful efforts being to preserve the last surviving cow of the Enderby Island cattle breed[10] and the cloning of an endangered species.[11] Although SCNT has been used to clone companion animals, obvious phenotypic differences in genetically identical twins (nature's truest form of clone) should serve as a reminder that cloning in and of itself is not an effective postmortem means of restoring an animal's or individual's life.

Because somatic cells can be easily obtained, clonally expanded, and genetically modified in the laboratory before nuclear transfer, SCNT is the preferred approach for producing transgenic farm animals.[12] Taking the effort to genetically modify somatic cells before nuclear transfer effectively increases the efficiency of producing transgenic

Table 82.3 Comparison of SCNT pregnancy losses in the cow related to MHC homozygous and heterozygous SCNT cell lines

Cell line	Year	MHC type[a]	Number of embryos transferred	Number of pregnant recipients			
				Day 28[b]	Day 90[c]	Day 200[c]	Term[c]
MHC class I homozygous SCNT cell lines							
UT3388	2001	AH68/AH68	48	29 (60.4%)	21	21	13
UT3388	2002	AH68/AH68	21	6 (28.5%)	5	5	4
UT4585	2002	AH67/AH67	39	10 (25.6%)	8	6	6
		Total	108	45 (41.6%)	34	32	23
MHC class I heterozygous SCNT cell lines							
UT4381	2001	AH12/AH68	48	18 (37.5%)	0	0	0
UT4472	2002	AH12/AH68	5	3 (60%)	1	1	1
		Total	53	21 (39.6)	1	1	1

[a] MHC class I haplotypes (AH) carried by the SCNT donor cell lines were determined by MHC class I microarray typing and genomic MHC class I sequencing.
[b] Day 28 pregnancy rates following transfer of a single SCNT embryo to each recipient were not significantly different for the two groups ($P = 0.80$).
[c] Pregnancy losses for class I homozygous and heterozygous SCNT cell lines were compared for three time periods: days 28–90, days 28–200, and day 28 to term. For all three time periods pregnancy loss was significantly different at $P < 0.001$ (chi-square test).
Source: adapted from Davies C, Hill J, Edwards J et al. Major histocompatibility antigen expression on the bovine placenta: its relationship to abnormal pregnancies and retained placenta. *Anim Reprod Sci* 2004;82–83:267–280 with permission from Elsevier.

farm animals to 100%. The ability to produce transgenic farm animals to serve as bioreactors for the production of pharmaceuticals or possibly serving as organ donors for the human population has created a new form of "pharming" with unlimited potential benefit for human health. Transgenic animals also provide a means in agricultural and biomedical disciplines to study genes involved in a variety of biological and physiological systems. Nonetheless, SCNT provides a powerful tool in developmental biology for ascertaining the interdependence of nuclear and cytoplasmic components necessary for embryonic development and determination of cell fate.

Cloning for the purpose of producing genetic replicas of existing individuals is collectively referred to as "reproductive" cloning. Because SCNT could be used for cloning humans, the moral and ethical issues of its use have been, and will continue to be, debated by various governments worldwide. While the majority would agree that cloning humans for reproductive purposes should be banned, the usefulness of the technolog for therapeutic purposes remains contentious. Therapeutic cloning refers to the use of SCNT to reprogram patient-derived somatic cells to provide a source of stem cells for the regeneration of tissues afflicted by a disease or injury. Use of patient-derived stem cells would obviate concerns related to tissue or cell rejection. However, efforts to harvest stem cells resulting after SCNT destroy the human embryos created in the process.

Challenges associated with cloning by SCNT

A major limitation of SCNT for the purposes of cloning agriculturally important food-producing animals and others continues to be the inefficiency in producing live offspring. Dolly was one of 277 cloned embryos that developed to term (0.3% efficiency).[1] "Millie," a clone of an adult Jersey cow produced in our laboratory, was just one offspring that resulted after 95 attempts (1.1% efficiency). The Food and Drug Administration (FDA) have provided a detailed assessment of the potential hazards and risks to highlight the magnitude of inefficiencies that continue to be problematic in food-producing animals such as cattle, sheep, pigs, and goats.[9] Data provided by numerous academic and private laboratories around the world emphasize that, regardless of species or somatic cell type, death of cloned embryos and fetuses occur throughout pregnancy. Moreover, an unusually high proportion of cloned offspring are larger than normal and die soon after birth.

Interestingly, the development of cloned bovine embryos during the first 29–32 days parallels development of embryos produced after *in vitro* maturation, fertilization, and culture (IVMFC) to the blastocyst stage (Figure 82.5 and Table 82.3). However, similarities beyond this time period end abruptly. Depending on the cell line, between about 30 and 90 days of pregnancy embryonic deaths (defined by the absence of a heartbeat and detachment of placental membranes[13]) may occur in up to 100% of cloned pregnancies (Table 82.3 and Figure 82.6). Without ill effects on the dam, pregnancy losses in cattle of this magnitude are significantly higher than those expected after breeding by natural service (2–10%)[14] or after development using IVMFC procedures (16%).[15,16] Detailed efforts examining placentas from cloned embryos between days 40 and 50 of gestation showed that a high percentage were hypoplastic or partially developed with rudimentary cotyledons[13] (Figure 82.7). Placental insufficiency would explain fetal losses of clones occurring during this time period. Although not entirely understood, immune-mediated rejection related to inappropriate expression of trophoblast major histocompatibility complex (MHC) class I antigens may be problematic.[17] While others were not able to provide evidence to support abnormal expression of MHC class I on the trophoblast of cloned fetuses,[18] it is interesting to note that the two cell lines used extensively in our laboratory and shown to produce a substantial number of successful pregnancies were both derived

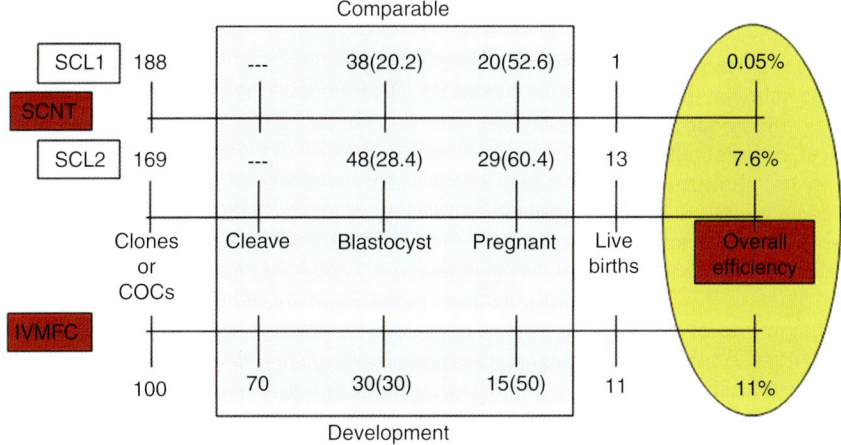

Figure 82.5 Overall efficiency of SCNT for producing offspring using different somatic cell lines (SCL1 versus SCL2) and after performing *in vitro* maturation, fertilization and culture of embryos (IVMFC). Numbers presented for SCL1 and SCL2 represent actual numbers obtained using two different cell lines. Numbers presented for IVMFC were derived from our own experience and the volumes of data presented by Dr John Hasler in the published literature.[15,16,28] Shaded areas highlight time periods with similar development for SCNT and IVMFC embryos (i.e., up to ~29 days in gestation with pregnancy rates approximating 50%).

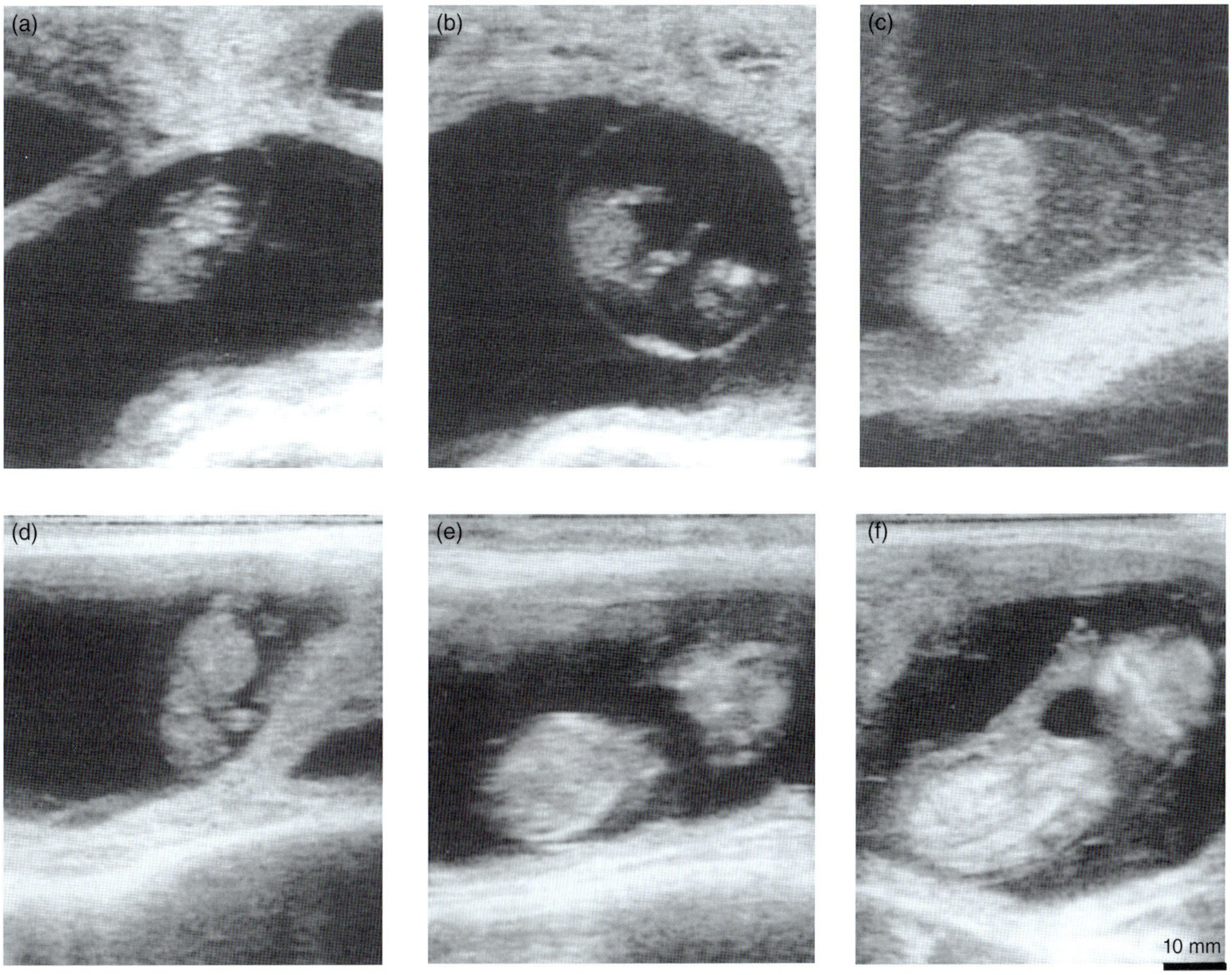

Figure 82.6 A comparison of the development of nonviable (a–c) and viable (d–f) SCNT fetuses using transrectal ultrasonography at days 45 (a, d), 55 (b, e), and 65 (c, f). The nonviable fetus was alive at day 45 (heartbeat detected) and dead at day 55 (no heartbeat). At day 65, increased echogenicity of the amniotic fluid was apparent as the fetus degenerated. The crown–rump lengths (cm) for the nonviable and viable fetuses at each stage were, respectively, 3 vs. 2.8, 3.7 vs. 5.7, and 3.7 vs. 7.5. Reproduced from Hill J, Burghardt R, Jones K *et al.* Evidence for placental abnormality as the major cause of mortality in first-trimester somatic cell cloned bovine fetuses. *Biol Reprod* 2000;63:1787–1794 with permission from Wiley-Blackwell. Image kindly provided by Dr Jonathan Hill at the University of Queensland.

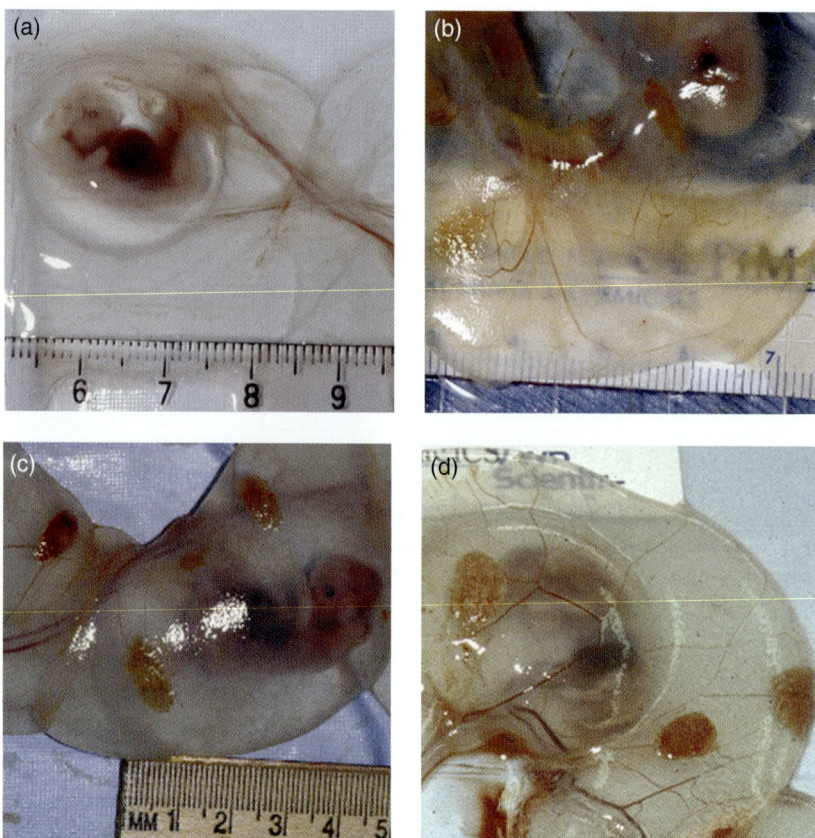

Figure 82.7 Each observed SCNT placenta could be grouped as hypoplastic (a, day 40), partially developed with a reduced number of rudimentary cotyledons (b, day 45), or essentially normal (c, day 50). A control placenta is included for reference (d, day 53 placenta from an IVMFC derived fetus). Reproduced from Hill J, Burghardt R, Jones K et al. Evidence for placental abnormality as the major cause of mortality in first-trimester somatic cell cloned bovine fetuses. *Biol Reprod* 2000;63:1787–1794 with permission from Wiley-Blackwell. Image kindly provided by Dr Jonathan Hill at the University of Queensland.

from MHC homozygous cows[19] (see Table 82.3). In contrast, the cell lines resulting in a high degree of fetal losses were from MHC heterozygous cows, suggesting some degree of incompatibility.

The incidence of spontaneous abortions increases during the second trimester of cloned pregnancies. Complete macroscopic and histopathologic examinations of aborted fetuses reveal few abnormalities. However, placentas are oftentimes grossly abnormal with a marked reduction in fetal cotyledons (occasionally fewer than 20 compared with the expected 70–120)[9] (Figure 82.8). Fetal membranes are generally thickened and edematous.

Pregnancy losses during the third trimester, especially between 200 and 265 days (term is 280 days), are more challenging. Many of the fetal losses occurring during this time relate to an increased incidence of hydrallantois (i.e., hydrops), an excessive accumulation of fluid in different parts of the placenta (e.g., hydrallantois, hydramnios)[20] and fetus (hydrops fetalis). In moderate to severe cases of hydrops, the abdomen becomes noticeably distended from the excessive accumulation of fluids (Figure 82.9). Uterine rupture and other harmful complications to the surrogate recipient may be problematic in rapidly developing cases warranting immediate euthanasia.[20] Attempts to "rescue"

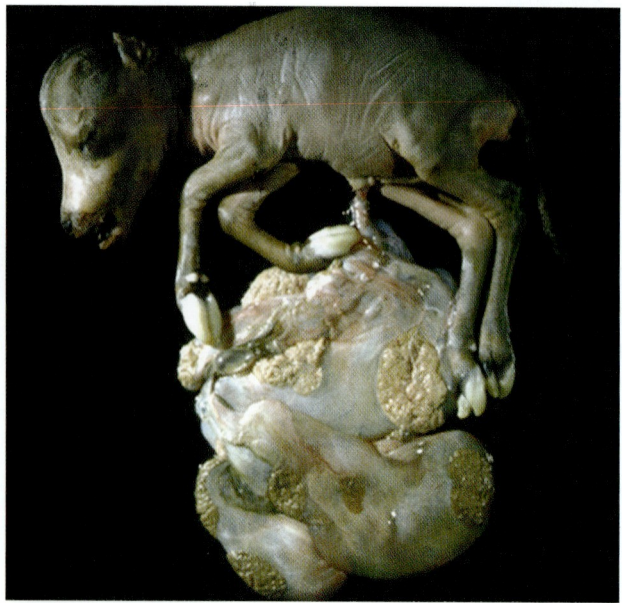

Figure 82.8 Aborted SCNT fetus aborted during second trimester. Placenta was thickened and edematous with about 14 fetal cotyledons.

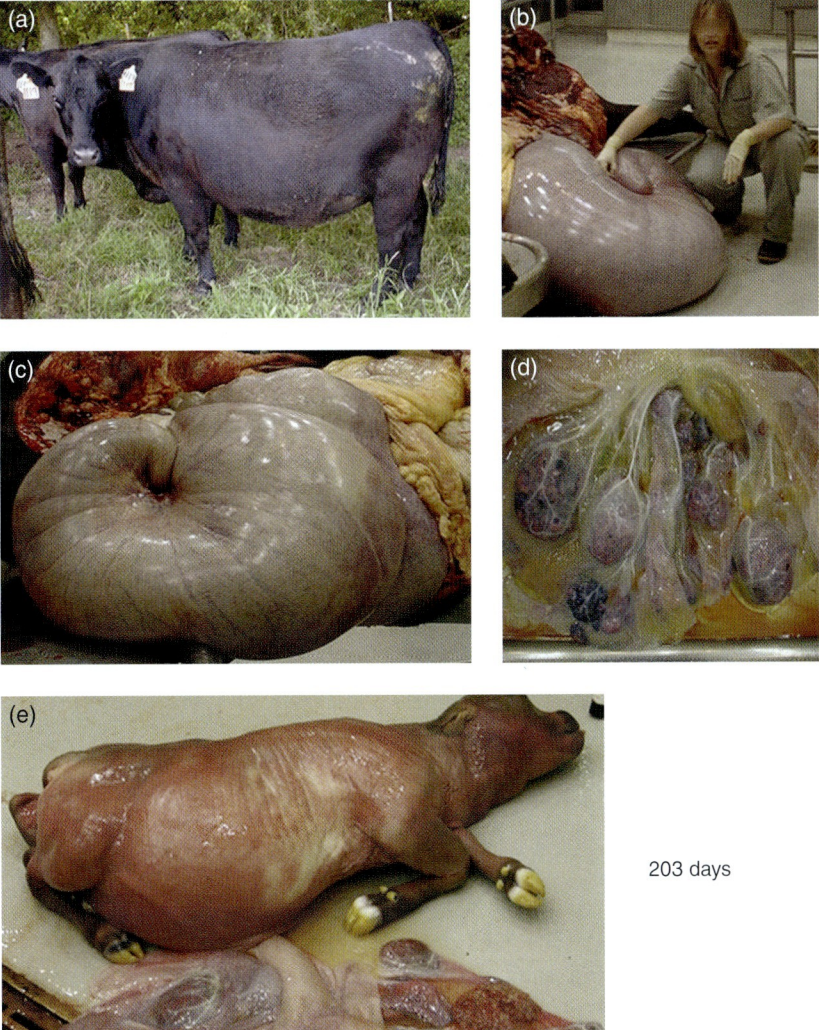

Figure 82.9 (a) Hydropic surrogate recipient at 203 days of gestation. (b) After euthanasia; ovary is located on the tip of the first uterine horn (indicated by individual's hand). (c) Clearer image of hydropic uterus; the fetus is located in the uterine horn behind the first one. (d) An abnormal cloned placenta with variations in size of placentomes (50 g to 1 kg) and edematous fetal membranes. (e) A Jersey cloned fetus weighing 30 kg with an enlarged umbilicus. Image in (d) reproduced from Edwards J, Schrick F, McCracken M *et al*. Cloning adult farm animals: a review of the possibilities and problems associated with somatic cell nuclear transfer. *Am J Reprod Immunol* 2003;50:113–123 with permission from John Wiley & Sons.

the fetus by emergency cesarean section often prove futile as the fetus is often already dead. Even if delivered early enough, neonatal care is inadequate to sustain the life of those neonates delivered at the time when most cases of hydrops are diagnosed. Marked reductions in numbers of placentomes of varying sizes, along with marked hypertrophy of many cotyledons, are evident with cloned placentas. Often, there is an increased incidence of adventitial placentation and severe edema in the intercotyledonary placental membranes (see Figure 82.9).

Hydrops fetalis is often observed in clones originating from hydropic pregnancies. Most fetal abnormalities associated with overly large clones are consistent with consequences related to hydrops and placental insufficiency. Fetal anasarca with generalized edema and marked edema of the umbilicus is usually present (see Figure 82.9). Amniotic squames and meconium are generally present in the lungs of late-term fetuses indicating some degree of stress *in utero* before death.

Although the occurrence and severity of hydrops varies with cloned pregnancies, incidences as high as 86% have been observed.[9] In our experience, 37% of the 58% of total pregnancies developing hydrops were severe enough to warrant euthanasia of the surrogate recipient to avoid further harm from complications. Some producers interviewed by the FDA as a part of their risk assessment of cloning animals monitored surrogate recipients closely for developing signs of hydrops beginning at 150 days of pregnancy using ultrasound. Because of the poor prognosis for the fetus and possible harmful complications to the surrogate recipient, termination of the pregnancy is warranted following diagnosis.

While clones may die immediately or soon after birth (within about 48 hours), survival rates up to 50–70% are obtainable. Survival is related to birthweight to some extent (Figure 82.10). In the sample data provided in Figure 82.10, the majority of Jersey cloned calves surviving (L, live)

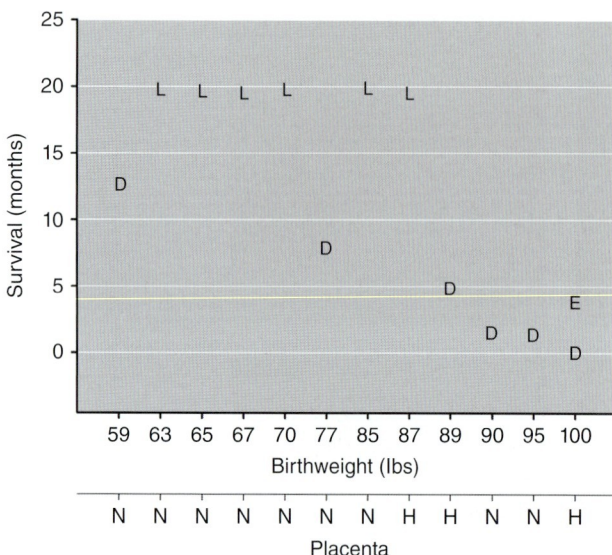

Figure 82.10 Representative image depicting survival (months) for SCNT clones of Jersey cows (L, live; D, dead; E, euthanasia) considering birthweight (lbs) and placenta (H, hydropic; N, no obvious signs of hydrops).

weighed about 39 kg or less at birth. Jersey clones weighing 40–45 kg died at birth, and at times thereafter. Of the few cloned embryos that develop to term, calves are typically larger at birth, with some exhibiting clinical signs associated with large offspring "dummy calf" syndrome. Report incidences of cloned calves exhibiting large offspring syndrome varies considerably, ranging from 8 to 100%.[9] In our experience and others,[9] large calves are slow to stand, have difficulty with regulating body temperature, have a weak sucking reflex, and their umbilicus is often enlarged and edematous with a patent urachus. Some have deformities in limbs related to tendon contracture, have disproportionate or immature organ development, and may be more susceptible to infections. Calves exhibiting clinical signs of large offspring syndrome are intensively monitored and may be provided with therapies to treat a whole plethora of complications, including but not limited to lung dysmaturity, pulmonary hypertension, respiratory distress, hypoxia, hypothermia, hypoglycemia, metabolic acidosis, enlarged umbilical veins and arteries, and/or the development of sepsis in either the umbilical structures or lungs. In severe cases euthanasia is appropriate as even after extraordinary efforts to intervene more often than not clones die.

Complications in cloned calves may not be evident for several months, with the majority of deaths occurring by 6 months of age.[9] Thereafter surviving clones may appear "normal" and "healthy" until some unexpected pathology presents itself. In our experience three clones died at 7.5, 8, and 9 months of age while two others died at 1 and 2.5 years of age. Although different pathologies were noted, several died or were euthanized after periodic episodes of scours unresponsive to treatment. A few but documented instances of sudden death in clones without any ill effects in the non-clones housed in the same environment makes for good discussion whether death was attributable to the animal being a clone or merely a victim of its circumstances.

Figure 82.11 Dolly with Bonnie, her first-born lamb after being mated to a small Welsh Mountain ram. Image and permission to reproduce were provided by the Roslin Institute, University of Edinburgh.

Fortunately, most of the healthy appearing clones from otherwise normal pregnancies exhibit normal growth and development. To date, numerous clones have been bred and have given birth. Dolly gave birth to six lambs (Bonnie was the first born in 1998, twins followed the next year, with triplets the year after that; Figure 82.11) several years before contracting the terminal contagious disease sheep pulmonary adenomatosis, likely from her contact with other herdmates. Extensive efforts examining numerous different endpoints in different species confirm the normalcy of resultant progeny.

Regardless of species, only a few offspring result after extraordinary efforts to construct cloned embryos using a somatic cell and the cytoplasm from an enucleated oocyte. Like cattle, much of the losses are attributable to early embryonic and fetal losses. Fortunately, with the majority of other species cloned to date, including goats, complications related to hydrops and large offspring syndrome are not problematic. Litter sizes ranging from zero to eight pigs are possible after the transfer of more than 100 1-cell cloned embryos into the uterus of surrogate recipients.[9] While a few abnormalities have been noted, most clones are smaller, not larger, at birth than pigs derived from natural mating or IVMFC procedures.[9] Later application of SCNT for cloning horses and companion animals such as the dog is more attributable to the limited amount of information on the optimal conditions for maturing oocytes *in vitro* and supporting early embryo development during culture. Once information was made available, success with cloning horses using SCNT was realized and numerous clones using

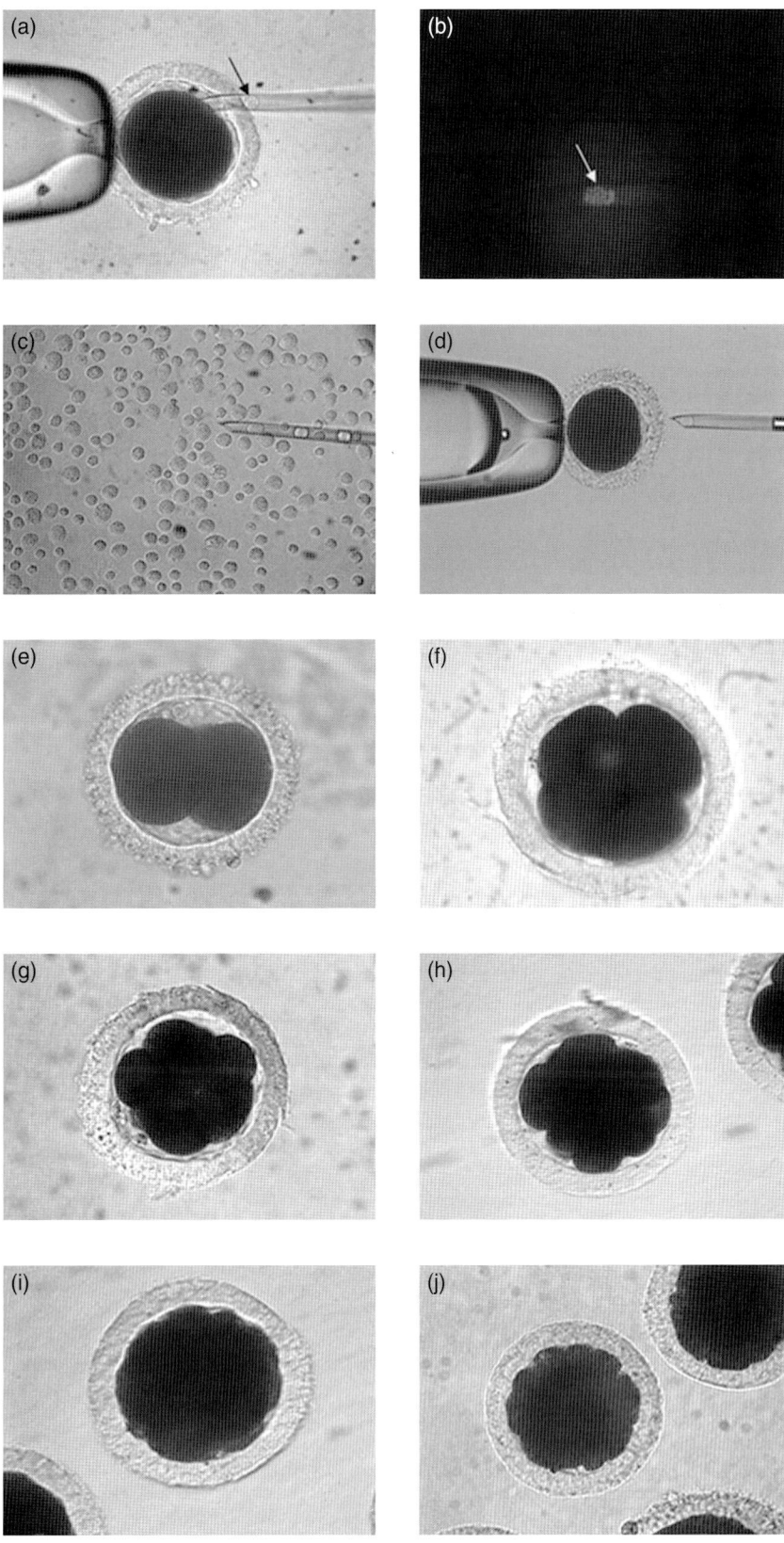

Figure 82.12 Representative image depicting SCNT procedure and *in vitro* embryo development for cloned dogs. The first polar body (a, arrow) and oocyte nucleus (b, arrow) are removed in *in vivo*-matured oocytes using an aspiration pipette. Enucleation may be confirmed by nuclear staining (b, arrow). Somatic cells are aspirated into a micropipette (c) and placed into the perivitelline space of an enucleated oocyte (d). Fused couplets are activated and cultured for 7 days. Cloned cleavage stage embryos: 2-cell (e), 4-cell (f), 8-cell (g), 16-cell (h), 32-cell or morula (i, j). Reproduced from Jang G, Oh H, Kim M *et al*. Improvement of canine somatic cell nuclear transfer procedure. *Theriogenology* 2008;69:146–154 with permission from Elsevier.

adult somatic cells from horses have been produced. Although the incidence of embryonic and fetal losses parallel those occurring in cattle (75–100% loss may be incurred depending on cell line), pregnancies and offspring in available literature are without complications related to hydrops and large offspring syndrome.[21,22] For the most part, placentas appear "normal" at term with some foals having an enlarged edematous umbilical cord.

Although the procedure of SCNT in the dog is remarkably similar to what has been done in cattle and other species (Figure 82.12), inability to mature appreciable numbers of oocytes *in vitro* warrants use of *in vivo* matured oocytes.[23] Aside from dogs being monoestrus, the fact that oocytes undergo maturation in the oviducts after ovulation, which may occur over a time period of 48–72 hours, poses a unique challenge related to controlling stage of maturation utilized for SCNT. With success after transferring cloned embryos to the oviducts, much of the inefficiency related to SCNT in dogs has been related to early embryonic and fetal losses. Abnormal offspring syndrome appears to be problematic with cloning of dogs. As observed in cattle, heavier placentas were associated with lower survival rates.[24] Different pathologies have been noted in nonsurviving cloned fetuses and offspring including increased muscle mass.[25]

Potential factors contributing to the inefficiency of cloning by SCNT

Mindful of the physical insult inherent in the SCNT procedure and the complexity of factors within the oocyte's cytoplasm that are "expected" to reprogram a somatic nucleus, it is truly remarkable that any clones develop into healthy offspring. Some of the procedural aspects related to the extensive micromanipulations required to construct cloned embryos using oocytes and somatic cells are not ideal (see Figure 82.3). Removal of nuclear DNA from the oocyte typically involves the physical removal of 5–15% or more of the ooplasm while exposing it to an otherwise "noxious" chemical. Verification of this process may expose the ooplasm to ultraviolet light, which has been shown to alter membrane integrity, increase methionine uptake, and alter protein synthesis and mitochondrial activity.[26] To insert the somatic cell nucleus into the bag of oocyte cytoplasm generally involves electrical-induced fusion or microinjection. No matter how careful, some degree of lysis occurs with either procedure. Furthermore, a variety of protocols have been described for activating cloned embryos, with the majority involving exposure to otherwise noxious chemicals.

Moreover, culture of cloned embryos from *in vitro* matured oocytes before transfer into surrogate recipients is not without ill effects. Like SCNT, complications inherent to the production of embryos using IVMFC procedures include alterations in gene expression in early embryos and fetuses,[27] increased abortion rates (~16%),[16] abnormal placentas, increased incidence of hydrallantois (~1%),[15] heavier fetuses, dystocia, large offspring syndrome, and mortality rates (~15%),[15,16,28–32] albeit at a much lower rate (see Chapter 67). The specific SCNT factors exacerbating problems inherent to IVMFC procedures are poorly understood. However, stark differences in the ability of certain cell lines to produce live offspring after nuclear transfer, approaching values derived from IVMFC procedures (see Figure 82.5), implicate improper or incomplete reprogramming of the somatic nucleus by the oocyte cytoplasm. To this end genomic methylation status in cloned blastocysts often resembles that observed in somatic cells, which would explain the aberrant expression of developmentally important genes observed in cloned embryos and fetuses.[33,34]

Summary and conclusions

Cloning by SCNT defies basic principles in developmental biology. Eventual improvements related to increasing the number and survival rates of cloned embryos, fetuses, and offspring are imperative for fully realizing SCNT applications for animal agriculture and human medicine. Because of the complexity involved with the construction of an embryo, using nothing more than a somatic nucleus and the cytoplasm of an oocyte, it is likely that many more years of effort will be required to understand the process well enough to improve outcomes.

References

1. Wilmut I, Schnieke A, McWhir J, Kind A, Campbell K. Viable offspring derived from fetal and adult mammalian cells. *Nature* 1997;385:810–813.
2. Briggs R, King T. Transplantation of living nuclei from blastula cells into enucleated frogs' eggs. *Proc Natl Acad Sci USA* 1952;38:455–463.
3. McGrath J, Solter D. Nuclear transplantation in the mouse embryo by microsurgery and cell fusion. *Science* 1983;220:1300–1302.
4. Willadsen S. Nuclear transplantation in sheep embryos. *Nature* 1986;320:63–65.
5. Prather R, Barnes F, Sims M, Robl J, Eyestone W, First N. Nuclear transplantation in the bovine embryo: assessment of donor nuclei and recipient oocyte. *Biol Reprod* 1987;37:859–866.
6. Stice S, Robl J. Nuclear reprogramming in nuclear transplant rabbit embryos. *Biol Reprod* 1988;39:657–664.
7. Prather R, Sims M, First N. Nuclear transplantation in early pig embryos. *Biol Reprod* 1989;41:414–418.
8. Campbell K, McWhir J, Ritchie W, Wilmut I. Sheep cloned by nuclear transfer from a cultured cell line. *Nature* 1996;380:64–66.
9. Food and Drug Administration. Animal Cloning: A Risk Assessment, chapter V, animal health risks, 2008. http://www.fda.gov/AnimalVeterinary/SafetyHealth/AnimalCloning/ucm124840.htm. Accessed January 15, 2013.
10. Wells D, Misica P, Tervit H, Vivanco W. Adult somatic cell nuclear transfer is used to preserve the last surviving cow of the Enderby Island cattle breed. *Reprod Fertil Dev* 1998;10:369–378.
11. Lanza R, Cibelli J, Diaz F et al. Cloning of an endangered species (*Bos gaurus*) using interspecies nuclear transfer. *Cloning* 2000;2:79–90.
12. Schnieke A, Kind A, Ritchie W et al. Human factor IX transgenic sheep produced by transfer of nuclei from transfected fetal fibroblasts. *Science* 1997;278:2130–2133.
13. Hill J, Burghardt R, Jones K et al. Evidence for placental abnormality as the major cause of mortality in first-trimester somatic cell cloned bovine fetuses. *Biol Reprod* 2000;63:1787–1794.
14. Hawk H. *Infertility in Dairy Cattle*. Montclair, NJ: Allanheld, Osmun and Co. Publishing, 1979.
15. Hasler J. The current status of oocyte recovery, *in vitro* embryo production, and embryo transfer in domestic animals with an emphasis on the bovine. *J Anim Sci* 1998;76(Suppl. 3):52–74.

16. Hasler J, Henderson W, Hurtgen P *et al.* Production, freezing and transfer of bovine IVF embryos and subsequent calving results. *Theriogenology* 1995;43:141–152.
17. Hill J, Schlafer D, Fisher P, Davies C. Abnormal expression of trophoblast major histocompatibility complex class I antigens in cloned bovine pregnancies is associated with a pronounced endometrial lymphocytic response. *Biol Reprod* 2002;67: 55–63.
18. Chavatte-Palmer P, Guillomot M, Roiz J *et al.* Placental expression of major histocompatibility complex class I in bovine somatic clones. *Cloning Stem Cells* 2007;9:346–356.
19. Davies C, Hill J, Edwards J *et al.* Major histocompatibility antigen expression on the bovine placenta: its relationship to abnormal pregnancies and retained placenta. *Anim Reprod Sci* 2004;82–83:267–280.
20. Peek S. Dropsical conditions affecting pregnancy. In: Youngquist RS, Threlfall WR (eds) *Current Therapy in Large Animal Theriogenology*. St Louis, MO: Saunders Elsevier, 2007, pp. 428–431.
21. Galli C, Lagutina I, Duchi R, Colleoni S, Lazzari G. Somatic cell nuclear transfer in horses. *Reprod Domest Anim* 2008;43(Suppl. 2):331–337.
22. Lagutina I, Lazzari G, Duchi R *et al.* Somatic cell nuclear transfer in horses: effect of oocyte morphology, embryo reconstruction method and donor cell type. *Reproduction* 2005;130:559–567.
23. Jang G, Kim M, Lee B. Current status and applications of somatic cell nuclear transfer in dogs. *Theriogenology* 2010;74: 1311–1320.
24. Oh H, Hong S, Park J *et al.* Improved efficiency of canine nucleus transfer using roscovitine-treated canine fibroblasts. *Theriogenology* 2009;72:461–470.
25. Hong I, Jeong Y, Shin T *et al.* Morphological abnormalities, impaired fetal development and decrease in myostatin expression following somatic cell nuclear transfer in dogs. *Mol Reprod Dev* 2011;78:337–346.
26. Smith L. Membrane and intracellular effects of ultraviolet irradiation with Hoechst 33342 on bovine secondary oocytes matured *in vitro*. *J Reprod Fertil* 1993;99:39–44.
27. Blondin P, Farin P, Crosier A, Alexander J, Farin C. *In vitro* production of embryos alters levels of insulin-like growth factor-II messenger ribonucleic acid in bovine fetuses 63 days after transfer. *Biol Reprod* 2000;62:384–389.
28. Hasler J. In-vitro production of cattle embryos: problems with pregnancies and parturition. *Hum Reprod* 2000;15(Suppl. 5): 47–58.
29. Farin P, Farin C. Transfer of bovine embryos produced *in vivo* or *in vitro*: survival and fetal development. *Biol Reprod* 1995;52:676–682.
30. Garry D, Bassel-Duby R, Richardson J, Grayson J, Neufer P, Williams R. Postnatal development and plasticity of specialized muscle fiber characteristics in the hindlimb. *Dev Genet* 1996;19: 146–156.
31. Walker S, Hartwich K, Seamark R. The production of unusually large offspring following embryo manipulation: concepts and challenges. *Theriogenology* 1996;45:111–120.
32. Young L, Sinclair K, Wilmut I. Large offspring syndrome in cattle and sheep. *Rev Reprod* 1998;3:155–163.
33. Niemann H, Tian X, King W, Lee R. Epigenetic reprogramming in embryonic and foetal development upon somatic cell nuclear transfer cloning. *Reproduction* 2008;135:151–163.
34. Rodriguez-Osorio N, Urrego R, Cibelli J, Eilertsen K, Memili E. Reprogramming mammalian somatic cells. *Theriogenology* 2012;78:1869–1886.

Chapter 83

The Computer-generated Bull Breeding Soundness Evaluation Form

John L. Myers

Pecan Drive Veterinary Services, Vinita, Oklahoma, USA

Marketing tool or "pretty picture"

It was apparent by the second semester of my eighth grade year that Sue Ellen Owens was the smartest student in our class. It was obvious that while our teachers admired her a great deal, she earned her high grades fairly through a combination of rapt concentration in class, innate intelligence, and attention to detail. One of these details was her consistent placement of our English assignments done at home in an attractive binder that distinguished her papers from everyone else and surely caught the attention of the teacher, the rest of the class, and especially me.

I think of Sue Ellen today as I tell the story of the Society for Theriogenology (SFT) computer-generated and online bull breeding soundness evaluation form. I am sure that the English homework she produced in the eighth grade would have scored just as high a grade had she turned them in without a binder, just as surely as my work justified its same sub-Sue Ellen result even when I imitated her technique of presenting my themes in an over-the-top folder my father used to present his engineering work. However, I wonder whether this new form will promote the practice of quality soundness examinations, establish the standards by which bulls should be measured, encourage the use of a common, efficient, recognizable and attractive form that will be embraced by a large number of veterinarians, and provide a method by which measurements can be learned or enhanced, or are we merely providing to the veterinary consuming public the same uneven and substandard work placed in an pretty plastic binder.

Directive to produce the computer-generated breeding soundness evaluation form

The machinations of committee work as told through reports or minutes surpass in boredom a thoughtful reading of the IRS code. From the perspective of the committee itself, however, embarking on modification of the existing breeding soundness evaluation (BSE) form became an adventure that changed from what we thought was a concept of a "fill-in-the-blank" spreadsheet exercise to a full-fledged emersion into the bizarre and mysterious world of program writing. Along the way the committee had long serious discussions on graphic artistry, color combinations, pathology prevalence, confidentiality, keystroke shortcuts, economic impact to the SFT, and the most efficient process in making our wishes known to the programmer.

Dr Richard Hopper, then president of the SFT, gave the directive at the winter meeting of the 2011 SFT board in Milwaukee, Wisconsin to develop a computer-generated BSE form for bulls. Assigned to the committee were Drs John L. Myers, Herris Maxwell, and Will Shultz. Dr Shultz had commissioned a company to design a program for tracking and identification of frozen canine semen and recommended the same company be retained for development of this initiative. Soon thereafter Drs Michael Thompson and Brian Keith Whitlock were added to the committee as well as the SFT executive director Dr Charles Franz.

Between January of 2011 SFT board meeting and the subsequent SFT convention in August of the same year the committee worked to develop a temporary model of the computer-generated BSE to be displayed for comments and suggestions at both the August SFT meeting and later the next month at the American Association of Bovine Practitioners (AABP) convention. To describe the product on display at the two conventions as a BSE form would be similar to saying "since a six-grader can read, he's ready for college."

Many useful comments were garnered, however, especially at the SFT convention. Perhaps the most useful response from the AABP would have been the general feeling of apathy to any sort of BSE form produced by the SFT because individual clinics and veterinarians developed their own forms separate and distinct from any other entities and not necessarily conforming to standards set by the SFT

Bovine Reproduction, First Edition. Edited by Richard M. Hopper.
© 2015 John Wiley & Sons, Inc. Published 2015 by John Wiley & Sons, Inc.

or any other organization. As an aside, the programmer developing the computer-generated BSE was given money for travel and lodging to attend the AABP convention to hear first hand any comments or suggestions.

Shortly after the AABP convention, progress on the computer-generated BSE form stopped. Through negotiations between the SFT executive director Dr Charles Franz and the programmer's superior it was decided that either the SFT and the company doing the programming would abandon the project with a full monetary refund, or the company would need a great deal more money to move forward. Through the counsel of Dr Franz, the committee decided to take the refund, sever ties with the programming company, and look elsewhere for help in completing the project. Once again, Dr Will Shultz had a prospect.

The programmer Dr Shultz suggested was David Riedle of Riedle Consulting and he is the person responsible for the form as it now exists. The relationship between Dave and those working on this project has been excellent, and while Dave seems to enjoy working with us we have no idea how he would assess our committee about cooperation, quality of direction, or understanding of the concepts underlying computer programming. From the committee side of the equation, however, we all believe that if it is possible for a computer by means of a program to do anything at all, Dave Reidle can make it happen.

This brings up the seductive sensation experienced by the committee once we began working on the project. We realized that magical feeling that movie makers must experience when they imagine Mark Ruffalo's muscles enlarging as he turns into the Great Hulk that the seams burst on his clothes as he climbs a building and throws a bus at someone evil. Similarly, we now possessed the possibility that we could design a form that could streamline the process of completion, feature accoutrements that visually enhanced the professionalism and attractiveness of the form, provide for accumulation of a large amount of data, encourage careful and complete examinations of the bulls being tested, and maintain an efficient method of continuing a revenue stream for the SFT.

The committee established priorities that insured the delivery of the features listed above while preventing the allure of making our form turn green and throw buses.

Principles guiding the development of the computer-generated BSE

1. The guidelines and standards previously set by the SFT would be maintained in the new form.
2. The computer-generated BSE would not only be easy to use, but would require less time to complete than filling out the paper form by hand. Through the use of a digital signature, a repeatable method in memory to provide information on the veterinary clinic on each form, an efficient method of quickly finding owner information that can be placed instantly into the proper fields, and several dropdown menus that provide appropriate information with the click of a mouse has produced a product that can be completed very quickly. Those familiar with the program can easily complete the form in 30 seconds or less.
3. The computer-generated BSE would be capable of producing an abbreviated form which lists the classification of the bull without revealing specific defects or motility as well as the conventional form that would list all specifics. Some members of the committee have witnessed confusion in instances when a bull buyer, unfamiliar with sperm defects, displays concern over any sperm defects regardless if those numbers are within the limits of a satisfactory classification. When the particular bull's specific data are uploaded into the centralized database, however, the specifics remain intact even if the form did not display them.
4. The computer-generated BSE would be attractive, professional, and difficult to reproduce by copying. The paper form of the BSE has a picture of Nandi, which is not conducive to use in the computer form. Because of the affection for Nandi among influential members in the SFT, the committee spent a great deal of time and discussion on the substitution of a silhouette of a generic bull into the logo. Computers can do many things, but in this case it could not keep Nandi from looking like a smudge. Colors were used to highlight two different items: the SFT logo and the classification of the bull. The committee felt that the color in the logo (SFT Green-Pantone 363 green) is good marketing for the SFT and the red, yellow and green in the classification (signifying in order "unsatisfactory," "deferred," and "satisfactory") lent not only an immediate and familiar assessment of the examination but further provided gravity for the reason the bull was tested in the first place. Finally, it is not uncommon for veterinary practitioners to fabricate their own BSE form by copying the current paper form and performing modifications such as removing the SFT logo and inserting the veterinarian's name and clinic. This is not only a violation of copyright but deprives the SFT of revenue for which it is justifiably entitled. The computer-generated BSE form makes it difficult to perform those types of sinister behaviors.
5. The computer-generated BSE would encourage careful and complete examination of the bulls. One feature of the new form is its inclusion of a dropdown menu of 14 common sperm defects as well as an illustration of a normal sperm. The committee believes this offers a great marketing tool for the veterinarian, in that he or she can, by mean of several illustrations, convey to the client what was discovered while examining the morphology stain. While it is not a requirement for completion of the form that the illustrations be included, this will be one method that will enforce perception of the differences between those veterinarians who do a complete evaluation from those who pronounce a bull satisfactory from a cursory examination and a quick look at motility. Further, we believe the illustrations not only provide a guide for those new to the BSE business to see defects that are commonly seen, but it also provides a means – limited though it may be – to look for persistence in specific defects in subsequent examinations.
6. The use of the computer-generated BSE should make economic sense for both the SFT and the users of the product. The sales of the paper form of the bull BSE produce an annual income that while not large is nonetheless important, and it is hoped that the new

computer-generated BSE will gain equal or greater acceptance. There is no plan to discontinue the paper form of the SFT BSE form. Just as paper forms are now sold in booklet fashion by the SFT office, the computer-generated BSE forms will be purchased by tokens. There will be a tab within the program that, on opening, will give access to purchase a batch of forms through use of a credit card online transaction. Additionally, there is a mechanism that will alert the user to the number of forms yet unused, and we think that this will not only streamline the access to SFT BSE forms but in fact creates a greener, paperless solution to the BSE examination. Currently, the committee has considered a price structure for the program download that includes 50 forms. The program purchase is a one-time event that includes support, and subsequently the forms will be sold in the form of tokens at $0.25 apiece. The current price to purchase one unit of the paper BSE form is $0.33 but when subtracting the cost to print the form the gross net income to the Society is greater when the computer-generated form is used rather than the paper form.

7. The records for each bull examined will stay within the program under the owner's account. Revisions can be made to any part of the computer-generated BSE form up until the record is marked as complete, and at that time the token for that form is spent and changes can no longer be made. The form (either the long or the short form) can be printed on the completed form, however, as many times as needed or desired. While our particular clinic will never be nominated for the Model of Efficiency Award (should there be one), we will examine around 1000 bulls a year and we are perpetually looking into files in pursuit of our findings on a bull we examined previously. Since we retain the yellow copy of the paper form in our files, we must thumb through several sheets of paper to find our previous examination. Many times that is successful if we can find the owner's file and if we have filed the paperwork properly. Since instituting the computer-generated BSE form we have been able to recover needed previous examinations relatively effortlessly, although we still print a copy of the examination and place it in the owner's file if we can find it. From a personal viewpoint, retaining the examinations on computer and unlimited printing of the form have been some of the more surprising and valuable features of this program.

8. Finally, within this program is the ability to accumulate a large amount of information that can then be stored in a centralized location. Once again, a tab exists that takes one through the process of uploading data, and we believe it is as simple, intuitive and safe as uploading other information (accounting, brucellosis procedures, etc.) that we already perform. We do not presume that all the data will be of high quality, but the committee does imagine the possibility of a large quantity. While I understand the repulsion those scientists among us have to large amounts of questionable information, the committee believes this is an avenue by which the quality of the BSE of bulls can be elevated. If individual practitioners can compare their culling percentage to other practitioners, their evaluation of their own methods may cause changes for the better. Comparing number of abnormal sperm or scrotal circumferences between breeds, ages and locations of bulls would be of great interest, but only in numbers large enough to have confidence that the trends seen are valid. The committee feels that from data collected from this program already and the anticipated percentage of those who would buy based on practitioners already using the paper form, information from 10 000–20 000 bulls a year would not be unreasonable.

Discussion

As of June 1, 2013, data from 1567 bulls have been uploaded into a database generated from the computer-generated BSE form. In addition to the original four veterinarians or clinics that first tested and used the form, an additional four practitioners participated in the beta testing and the uploading of information into the database. The breakdown on breeds of the 1567 bulls is as follows:

- Angus: 71.7%
- Simmental: 5.3%
- Charolais: 4.8%
- Red Angus: 3.6%
- Hereford: 3.2%
- Brangus: 3.1%
- Simmental/Angus: 2.8%
- Gelbvieh: 1.3%
- Limousin: 1.1%
- All other breeds: 3.1%

The classification of the 1567 bulls is as follows:

- Satisfactory potential breeder: 87.6%
- Classification deferred: 6.5%
- Unsatisfactory: 5.3%
- Unclassified: 6%

Within the computer-generated BSE there is the opportunity to indicate and display on the form 14 seminal defects. While the document can be completed without displaying these schematic drawings of the morphological abnormalities, the database accumulates how often a particular defect was selected. The breakdown of the most common defects is as follows:

- Detached heads: 39.6%
- Distal midpiece reflex: 30.4%
- Proximal droplet: 10.9%
- Coiled midpiece and tails: 7.8%
- Microcephalic heads: 7.0%
- Kinked tails: 4.3%

It was apparent during beta testing that one of the significant effects of the use of a common form for BSE used by different veterinarians but combined into one set of data result in outcomes and observations that, while surprising, may produce enhanced clarity and quality for practitioners participating in the future.

For instance, that 87.6% of the bulls were rated as satisfactory potential breeders may or may not fit within what

Bull Breeding Soundness Evaluation

Guidelines Established by Society for Theriogenology
P.O. Box 3007 - Montgomery, AL 36109
Phone (334) 395-4666 - Fax (334) 3399 - www.therio.org

Jeffrey Brooks
P.O. Box 199
Starkville, MS 39759
(662) 321-4567

BSE Date: 3/21/2013	BSE Case No: 13-401
Bull Name:	Breed: Angus
Bull I.D. No: 94/X135	Brand ☐ Tattoo ☐ Ear Tag ■
Bull Birth Date:	Age (Mo.) 30

PHYSICAL EXAMINATION

Body Condition Score: Beef - 6
Thin ☐ Moderate ☐ Good ☐ Obese ☐

Pelvic Height — Pelvic Width — Pelvic Area

Feet/Legs	■
Eyes	■
Vesicular Glands	■
Ampullae/Prostate	■
Inguinal Rings	■
Penis/Prepuce	■
Testes/Spermatic Cord	■
Epididymides	■
Scrotum (Shape)	■
Other	
SCROTAL CIRCUMFERENCE (CM)	42.0

This bull has been examined for physical soundness and quality of semen only. Unless otherwise noted, no diagnostic tests were undertaken for libido, mating ability, or infectious disease status of this bull.

Proximal Droplet — Detached Heads — Distal Mid Piece Reflex

Remarks and interpretation (diagnosis, prognosis, recommendations)
TRICH TESTED RESULTS PENDING

SEMEN EXAMINATION

Collection Method: EE ■ AV ☐ Massage ☐
Response: Erection ■ Protrusion ■ Ejaculation ■

Semen Characteristics	Ejaculate 1	Ejaculate 2
Motility: Gross		
Individual (%)	30	40
% Normal Cells	39	52
% Primary Abnormalities	5	2
% Secondary Abnormalities	56	46
WBC, RBC, Other		

CLASSIFICATION
Interpretation of data resulting from this examination would indicate that on this date this bull is a:

Unsatisfactory Potential Breeder

Re-examination recomended on: 5-1-2013

Signed: _____
Member - Society for Theriogenology

© Copyright - 2012/2013 - Society for Theriogenology

Figure 83.1 Printout of the computer-generated BSE long form. The "long form" includes all the sperm evaluation parameters. An advantage of this form over the original "paper" form is the ability to select images from the sperm abnormalities window and include on the form. This is an excellent client education tool. Courtesy of *Clinical Theriogenology*.

Bull Breeding Soundness Evaluation

Guidelines Established by Society for Theriogenology
P.O. Box 3007 - Montgomery, AL 36109
Phone (334) 395-4666 - Fax (334) 3399 - www.therio.org

Jeffrey Brooks
P.O. Box 199
Starkville, MS 39759
(662) 321-4567

BSE Date: 6/10/2013   BSE Case No: 13-401
Bull Name: "TEDDY"   Breed: Angus
Bull I.D. No: Brand ☐   Tattoo ☐   Ear Tag ■
Bull Birth Date: Age (Mo.) 84

PHYSICAL EXAMINATION

Body Condition Score: Beef - 6
Thin ☐   Moderate ☐   Good ■   Obese ☐

Pelvic Height   Pelvic Width   Pelvic Area

Feet/Legs	■ RR foot grown out no lameness
Eyes	■
Vesicular Glands	■
Ampullae/Prostate	■
Inguinal Rings	■
Penis/Prepuce	■
Testes/Spermatic Cord	■
Epididymides	■
Scrotum (Shape)	■
Other	
SCROTAL CIRCUMFERENCE (CM)	40.0

This bull has been examined for physical soundness and quality of semen only. Unless otherwise noted, no diagnostic tests were undertaken for libido, mating ability, or infectious disease status of this bull.

CLASSIFICATION
Interpretation of data resulting from this examination would indicate that on this date this bull is a:

Satisfactory Potential Breeder

Re-examination recomended on:

Signed: _____
Member - Society for Theriogenology

Mirocephalic Heads   Detached Heads   Coiled Mid Pieces and Tails

. Remarks and interpretation (diagnosis, prognosis, recommendations)

© Copyright - 2012/2013 - Society for Theriogenology

Figure 83.2 Printout of the computer-generated BSE form without the sperm evaluation parameters ("short form"). This avoids confusion created by potential buyers who do not understand thresholds for sperm abnormalities. Courtesy of *Clinical Theriogenology*.

would be expected from the testing of 1567 animals. The opportunity to contribute and reflect how an individual's success rate compares with others who tested a similar age, breed, and body condition at a particular time of the year would be invaluable both to veterinarians but also seedstock breeders. It is the wish of the SFT to monitor the collected data and make available the results to interested and responsible individuals who would analyze, criticize, and theorize about the trends and developments of the bulls being tested as well as the people who perform those tests.

There has been considerable discussion within the committee about the attributes and impediments to creating a national (and international) database from information collected using the computer-generated BSE form. The positive outcomes have been previously stated but there is an obvious admission that retrieving relevant quality data will be, at the least, cumbersome. In order for the database to be large enough to recognize or predict trends, there must be widespread usage of the form by many veterinarians whose primary emphasis will be to complete an incredibly full day's work rather than upload data. Additionally, reluctance to release such data because of suspicion of confidentiality issues or simply because of confusing information about the procedures to execute such an upload are real problems that will need to be addressed over the next years.

One solution to the disinclination to upload data would be a monetary incentive such as a discount on future computer-generated BSE token purchases each time information is shared. Because the tokens are so moderately priced, such an incentive may not induce all parties to participate and it may tilt the database to reflect an abnormal collection of bulls. At present the committee decision has been to focus on achieving a broad usage of the form and to concentrate on retrieving data as the form's popularity gains momentum.

There is now a version of the program compatible with the iPad™ and the program should be able to convert to other mobile devices not presently available or even barely imagined. The versatility inherent in the computer-generated BSE is such that adaptations and innovations created by its users should make for exciting possibilities in times to come.

Within the board of the SFT, its membership, and the American College of Theriogenologists the possibilities perceived from a computer-generated form for bulls has sparked an interest for similar forms that could be developed for other animals. The benefits derived from both the standardization of guidelines and the accumulation of the data influencing fertility in bulls would apply to any practitioner desiring to learn more about problems with, or enhancement of, reproduction in the particular species and breed presented.

Conclusion

There has been a serious amount of work and cooperation to produce the computer-generated BSE form presented today. While we hope the points presented above as to the principles the BSE committee used are valid and convincing, we would like to conclude with the idea that this form could be used as a marketing tool for the following:

- a private clinic or practitioner,
- a university clinic,
- the importance of BSEs,
- the SFT.

The origins of the Society began with a few interested and dedicated individuals who were involved with bull fertility. The form as we present it today represents but a blip in the progress and expansion of a field that encompasses more species, procedures, and knowledge than the gentlemen in 1954 would have ever imagined. With that history the committee not only feels gratitude but more importantly responsibility.

I began by relating my impressions of an eighth grade girl who turned in her assignments in a pretty plastic binder. Sue Ellen moved away in the ninth grade but I still remember how attractive and intelligent she was. The binders into which she placed her work were no guarantee that what was inside was of high quality, but it became clear that the binder was evidence that she took pride and care with what she produced. The computer-generated BSE form's usage also does not guarantee high-quality work, but it does represent one way to make what is done accessible, attractive, and memorable.

We feel that the form represents yet another step in presenting the knowledge and work that has gone before us and it is our hope that with its acceptance we can help in the progress and continued excellence of the SFT.

Glossary

activin A nonsteroidal regulator synthesized in the pituitary glands and gonads that stimulates the secretion of follicle-stimulating hormone (FSH) (Chapter 23).

actual calving interval Interval, expressed in months, between calving dates calculated for multiparous cows using the formula: number of days between calving/30.4 (Chapter 41).

adenohypophysis Anterior lobe of the hypophysis (pituitary gland). It secretes hormones under the control of releasing agents that arrive from the hypothalamus via a hypothalamo-hypophyseal portal system (Chapter 23).

agenesis Absence of an organ due to nonappearance of its primordium in the embryo (Chapter 45).

antiluteolytic Substance that blocks luteolysis by inhibiting or masking endogenous luteolytic signals (Chapter 23).

antral follicle Follicle with at least six layers of granulosa cells and a fluid-filled antrum; tertiary follicle (Chapter 24).

arginine-phenylalanine-amide related peptide (RFRP) Peptide that acts on GnRH neurons to alter GnRH secretion; formerly known as gonadotropin inhibitory hormone (Chapter 23).

arthrogryposis Permanent fixation of a joint in a contracted position (Chapter 54).

athymia Absence of functioning thymus tissue (Chapter 66).

atretic follicle Ovarian follicle that degenerates before fully maturing or ovulating (Chapter 24).

Aujeszky's disease A disease primarily of pigs but can occur in other secondary host species; caused by porcine herpesvirus 1 and characterized by respiratory, reproductive, and nervous signs; pseudorabies (Chapter 80).

autolysis Destruction of tissues or cells of an organism by the action of substances, such as enzymes, that are produced within the organism (Chapter 54).

average days to first service Average days in milk at first service for all cows in a group that have been bred at least once (Chapter 41).

axoneme Bundle of fibrils that constitutes the central core of a cilium or flagellum (Chapter 2).

azotemia Excess of nitrogen-containing compounds in the blood (Chapter 19).

balanoposthitis Inflammation of the penis and prepuce (Chapter 9).

ballottement Palpatory technique for detecting or examining a floating object in the body, such as the use of a finger to push against the uterus to detect the presence of a fetus by its return impact (Chapter 34).

beta-oxidation Process by which fatty acids are converted for use by the body as a source of energy (Chapter 30).

binder-of-sperm proteins (BSP) Group of proteins found in the seminal plasma that bind to sperm cells and modify their membranes (Chapter 71).

biovar Group or strain of a species of microorganism having differentiable biochemical or physiological characteristics (Chapter 58).

bougie A slender, flexible, hollow or solid, cylindrical instrument for introduction into the urethra or other tubular organs, usually for calibrating or dilating constricted areas (Chapter 19).

bougienage Examination or dilation of the interior of a canal by the passage of a bougie; balloon dilation (Chapter 19).

bovine somatotropin (bST) Protein secreted by the pituitary gland that stimulates body cell growth and milk production. A synthetic version is available for use in cattle (Chapter 51).

brachygnathism Abnormal shortness of the mandible resulting in a maxilla that is longer; overshot; parrot mouth (Chapter 54).

bradyzoite Slow-growing stage of certain parasite life cycles in which the microorganisms present in encysted clusters (Chapter 62).

bruxism Gnashing, grinding, or clenching the teeth (Chapter 19).

CD9 Transmembrane protein that plays a role in signal transduction and may also function as an implantation inhibitor (Chapter 67).

celiotomy Incision into the abdominal cavity (Chapter 53).

chemotaxis Directional movement, either toward or away, in response to chemical stimulation (Chapter 30).

cicatrix Fibrous tissue left after the healing of a wound; a scar (Chapter 14).

citrate Any salt of citric acid; plays an important role in fatty acid synthesis (Chapter 71).

cofactor A molecule such as a coenzyme with which another must associate in order to function properly (Chapter 30).

colpotomy Incision into the vagina; vaginotomy (Chapter 53).

cotylendonary placenta Distribution of the villi on the fetal chorion is localized in multiple circumscribed areas, the cotyledons (Chapter 25).

CpG island Genomic regions containing high frequencies of CpG sites, which are usually unmethylated if the gene is to be expressed (Chapter 67).

cryoprotectant Substance added to living biological material (e.g., embryo, ova, spermatozoa) that is to be preserved in a viable state by freezing (Chapter 77).

cystorrhexis Rupture of the urinary bladder (Chapter 19).

cytokine Any of a class of regulatory proteins, such as the interleukins and lymphokines, that are released by the immune system to act as intercellular mediators in the generation of an immune response (Chapter 2).

days in milk Number of days from calving to day of test; if cow is dry, days in milk is the number of days from calving date to dry date (Chapter 41).

dehiscence Bursting open or splitting along natural or sutured lines (Chapter 19).

desmorrhexis Rupture of a ligament (Chapter 46).

dictyate A stage in the development of oocytes which are arrested at the same stage of meiotic prophase (Chapter 67).

dietary cation–anion difference (DCAD) Diet balancing equation used frequently in the dairy industry to mitigate the occurrence of hypocalcemia; the DCAD equation is (Na+K) – (S+Cl) (Chapter 30)

dihydrotestosterone (DHT) An androgen derived from testosterone (Chapter 2).

diopter Unit of measurement of the refractive power of a lens (Chapter 79).

diplotene Stage of the first meiotic prophase, following the pachytene, in which the two chromosomes in each bivalent begin to repel one another and a split occurs between the chromosomes (Chapter 81).

dizygous Pertaining to or derived from two separate fertilized ova (Chapter 36).

dropsy Abnormal accumulation of serous fluid in a body cavity or in the cellular tissues; called also hydrops, edema (Chapter 54).

dynorphin Any of a family of opioid peptides found throughout the central and peripheral nervous systems; most are agonists at opioid receptor sites (Chapter 23).

dysgenesis Defective development; malformation (Chapter 45).

ecbolic agent Agent that stimulates contractions of the myometrium; oxytocic (Chapter 50).

echotexture Appearance of a tissue or organ on ultrasound (Chapter 24).

endodyogeny Form of asexual reproduction in which two daughter cells are produced inside a mother cell, which is then consumed by the offspring prior to their separation (Chapter 62).

endogenous transmission *See* vertical transmission (Chapter 62).

endometritis Inflammation limited to the endometrium of the uterus (Chapter 50).

endophyte An organism, especially a fungus or microorganism, that lives inside a plant, in a parasitic or mutualistic relationship (Chapter 65).

enzootic Prevalent among or restricted to animals of a specific geographic area (Chapter 54).

epitheliochorial placenta Type of placentation in which the uterine epithelium of the uterus and the chorion are in contact and there is no erosion of the epithelium; also known as adeciduate placenta

equine chorionic gonadotropin (eCG) Hormone secreted by endometrial cups in pregnant mares to promote the formation of secondary corpora lutea and maternal recognition of pregnancy; formerly known as pregnant mare serum gonadotropin (PMSG) (Chapter 75).

ergovaline Ergot alkaloid found in *Neotyphodium (Acremonium) coenophialum* (Chapter 29).

extravasation Discharge or escape, as of blood, from a vessel into the tissues; blood or other substance so discharged (Chapter 14).

fetotomy Reduction and removal of the fetus by division and removal of extremities and sections (Chapter 47).

follistatin Gonadal peptide hormone isolated from the ovarian follicular fluid that suppresses follicle-stimulating hormone (FSH) secretion (Chapter 23).

forced extraction Manual or mechanically assisted removal of the calf through traction (Chapter 47).

fremitus A palpable vibration felt when the hand is placed on a part of the body (Chapter 34)

gap junction Intercellular network of protein channels that facilitates the cell-to-cell passage of molecules (Chapter 24).

ghrelin Peptide hormone secreted mainly by stomach cells that is important in appetite regulation, growth hormone release, and regulation of energy homeostasis (Chapter 23).

gnotobiotic Free of germs or associated only with known or specified germs (Chapter 68).

gonocyte Primitive reproductive cell of the embryo (Chapter 4).

Graafian follicle Largest antral follicle which becomes the ovulatory follicle following the preovulatory gonadotropin surge (Chapter 24).

hamartoma Benign tumor-like nodule composed of an overgrowth of mature cells and tissues normally present in the affected part, but often with one element predominating (Chapter 54).

heat interval Time between periods of estrus; average is 21 days (Chapter 41).

heat shock protein (HSP) Any of a group of proteins (mostly molecular chaperones) synthesized in response to stress, particularly in response to stress from hyperthermia or hypoxia (Chapter 51).

hematogenous Originating or spread through the blood or characterized by producing blood (Chapter 62).

hemimelia Developmental anomaly characterized by absence of all or part of the distal half of a limb (Chapter 66).

hemotrophe The sum total of the nutritive material from the circulating blood of the maternal body, utilized by the early embryo. It is absorbed directly from the maternal blood by the allantochorion or vitellochorion (Chapter 54).

heteroxenous Requiring more than one host to complete the life cycle (Chapter 62).

horizontal transmission Lateral spread of a disease to others in the same group and over the same interval of time (Chapter 62).

human chorionic gonadotropin (hCG) Glycopeptide hormone produced by the fetal placenta that maintains the function of the corpus luteum during the first few weeks of pregnancy (Chapter 78).

hyalin Translucent albuminoid substance obtainable from the products of amyloid degeneration (Chapter 12).

hydatidiform mole Vesicular or polycystic placental mass resulting from the proliferation of the trophoblast and the hydropic degeneration and avascularity of the chorionic villi, usually indicative of an abnormal pregnancy; also called *mole* (Chapter 54).

hygroma Accumulation of fluid in a sac, cyst, or bursa (Chapter 54).

immunoflouresence Method of determining the location of antigen (or antibody) in a tissue section or smear by the pattern of fluorescence resulting when the specimen is exposed to the specific antibody (or antigen) labeled with a fluorochrome (Chapter 57).

immunohistochemistry (IHC) Microscopic localization of specific antigens in tissues by staining with antibodies labeled with fluorescent or pigmented material (Chapter 54).

inhibin Peptide hormone secreted by the follicular cells of the ovary and the Sertoli cells of the testis that inhibits secretion of follicle-stimulating hormone (FSH) from the anterior pituitary (Chapter 23).

interferon tau (IFN-t) Molecule secreted by embryonic trophoblast cells that serves as the molecule of maternal recognition of pregnancy in the bovine; formerly known as bovine trophoblast protein 1 (Chapter 21).

ketosis Elevated plasma levels of the ketone β-hydroxybutyrate occurring when the body's tissues utilize less ketone than is being produced by hepatocytes (Chapter 30).

kisspeptin Product of the *Kiss1* gene; acts on GnRH neurons to modulate GnRH secretion (Chapter 23).

kyphosis Abnormally increased convexity in the curvature of the thoracic spine as viewed from the side (Chapter 54).

lability Tendency to constantly undergo change, as a chemical compound; unstable (Chapter 57).

latex agglutination test Type of test in which antigen to a given antibody is adsorbed to latex particles and mixed with a test solution to observe for agglutination of the latex (Chapter 35).

lordosis Abnormal forward curvature of the spine in the lumbar region (Chapter 54).

low-density lipoproteins (LDL) Class of plasma lipoproteins that transport cholesterol to extrahepatic tissues (Chapter 71).

luteotropin A signal that stimulates luteal secretion of progesterone (Chapter 25).

lyase Enzyme that catalyzes the formation of double bonds by removing chemical groups without hydrolysis or catalyzes the addition of chemical groups to double bonds (Chapter 2).

lyophilization Creation of a stable preparation of a biological substance by rapid freezing and dehydration of the frozen product under high vacuum; freeze-drying (Chapter 71).

maceration Softening of a solid by soaking. In obstetrics, the degenerative changes with discoloration and softening of tissues, and eventual disintegration, of a fetus retained in the uterus after its death (Chapter 54).

major histocompatibility complex (MHC) proteins Surface antigens that aid immune cells in identifying "self" or infected cells (Chapter 82).

marsupialization Conversion of a closed cavity, such as an abscess or cyst, into an open pouch, by incising it and suturing the edges of its wall to the edges of the wound (Chapter 19).

meconium Combination of sloughed intestinal and epithelial cells, mucus, bile acids, and various other metabolites of fetal development (Chapter 54).

melanocyte-stimulating hormone alpha (α-MSH) Peptide hormone that stimulates melanin production and may have some effect on GnRH secretion (Chapter 23).

metallothionein (MT) A family of cysteine-rich proteins that are capable of binding a wide range of metals (Chapter 65).

methemoglobin Form of hemoglobin that occurs when hemoglobin is oxidized either during decomposition of the blood or by the action of various oxidizing drugs or toxic agents. It contains iron in the ferric state and cannot function as an oxygen carrier (Chapter 65).

metritis Inflammation of the uterine wall including the endometrium, muscular layers, and serosa (Chapter 50).

micelle Supermolecular colloid particle, most often a packet of chain molecules in parallel arrangement occurring in certain colloidal solutions due to the amphipathic nature of fatty acids (Chapter 71).

microaerophilic Requiring oxygen for growth but at a lower concentration than is present in the atmosphere, in conjunction with an enhanced carbon dioxide concentration; said of bacteria (Chapter 55).

MIMT1 Nonprotein-coding gene that forms part of the imprinted *Peg3* domain (Chapter 66).

mole Mass of hyperplastic or edematous immature placental tissue. See also hydatidiform mole (Chapter 54).

mollicutes Bacteria that lack cell walls and have a preference for mucosal and serosal surfaces (Chapter 54).

mutation Those procedures by which a fetus is returned to its normal orientation with respect to presentation, position, and posture, including repulsion (retropulsion), rotation, version, and extension of extremities (Chapter 47).

myoclonus Brief involuntary twitching of a muscle or group of muscles (Chapter 66).

negative energy balance (NEB) Insufficient consumption of the calories necessary to meet metabolic requirements (Chapter 30).

neuropeptide Y (NPY) Neurotransmitter involved in the regulation of a number of functions including, but not limited to, feeding behavior, circadian rhythms, and reproductive functions (Chapter 23).

neutralization test Test for the neutralization power of an antiserum or other substance by testing its action on the pathogenic properties of a microorganism, toxin, virus, bacteriophage, or toxic substance (Chapter 59).

nidus Point of origin or focus of a morbid process (Chapter 19).

nymphomania Behavioral state in which the female is in estrus continually or for longer periods at shorter intervals than is normal (Chapter 51).

oolemma Zona pellucida; the transparent noncellular secreted layer surrounding an ovum (Chapter 79).

opsonization The rendering of bacteria and other foreign substance subject to phagocytosis (Chapter 61).

osmolarity Concentration of a solution in terms of osmoles of solutes per liter of solution (Chapter 81).

P450 aromatase Enzyme that catalyzes the conversion of androgens to estrogens (Chapter 44).

Peg3 Paternally expressed gene 3; may play a role in cell proliferation and p53-mediated apoptosis as well as late prenatal development (Chapter 66).

phlegmon Diffuse inflammation of soft or connective tissue due to infection (Chapter 14).

phlorizin (phloridzin) A dihydrochalcone extracted from the root bark of the apple tree that blocks the renal tubular reabsorption of glucose, hence promoting the development of glycosuria (Chapter 23).

position Further describes fetal orientation based on the relationship of the dorsum of the fetus to the quadrants of the maternal pelvis (Chapter 47).

posture Describes the relationship of extremities with respect to body of the fetus (Chapter 47).

predicted calving interval Calculated for pregnant cows as the period between the most recent calving date and the due date; for nonpregnant females, predicted calving interval may be computed as average days open plus gestation length (Chapter 41).

pregnenolone Cholesterol-derived intermediate in the synthesis of steroid hormones (Chapter 2).

presentation Describes fetal orientation by the relationship of the spinal axis of the fetus to that of the dam (Chapter 47).

primary follicle Ovarian follicle characterized by a single layer of cuboidal granulosa cells (Chapter 24).

primordium The first beginnings of an organ or structure in the developing embryo; anlage (Chapter 24).

probang A long, slender, flexible rod usually having a tuft or sponge at the end (Chapter 43).

proopiomelanocortin (POMC) Polypeptide that is a precursor to several important hormones, including ACTH (Chapter 23).

propidium iodide Intercalating agent and fluorescent stain commonly used to identify dead or damaged cells (Chapter 74).

ptyalism Excessive secretion of saliva (Chapter 54).

pyelonephritis Inflammation of the kidney and renal pelvis due to bacterial infection (Chapter 19).

pyometra Accumulation of pus within the uterine lumen facilitated by a closed cervix and the presence of a corpus luteum (Chapter 50).

quasispecies Group of viruses with similar genetic structure that share a host with other quasispecies, usually descended from a single ancestor strain (Chapter 60).

quiescent At rest; latent; the G_0 stage of the cell cycle (Chapter 82).

random amplified polymorphic DNA (RAPD) Type of polymerase chain reaction (PCR) in which random segments of DNA are amplified (Chapter 22).

raphe Seam; line of union of the halves of various symmetrical parts (Chapter 1).

repulsion (retropulsion) Pushing the fetus back out the pelvis to facilitate other corrective steps (Chapter 47).

reticuloendothelial system (RES) Group of cells having the ability to take up and sequester inert particles, including macrophages and macrophage precursors, specialized endothelial cells, and reticular cells of lymphatic tissue and bone marrow (Chapter 58).

rotation Turning of the fetus on its long axis (Chapter 47).

ruminal acidosis Accumulation of acid or depletion of bases characterized by an increase in hydrogen ion concentration and a corresponding decrease in the pH of rumen fluid caused by an altered metabolic state (Chapter 30).

saprophytic Relating to any organism living on dead or decaying organic matter (Chapter 54).

secondary follicle Ovarian follicle with two to six layers of granulosa cells (Chapter 24).

second messenger Any of several classes of intracellular signals that translate electrical or chemical messages from the environment into cellular responses (Chapter 24).

secretagogue Causing flow of secretion; an agent that stimulates secretion (Chapter 23).

secundines Afterbirth; the placenta and membranes expelled after parturition (Chapter 49).

sensitivity Proportion of individuals in a population that will be correctly identified when administered a test designed to detect a particular disease, calculated as the number of true positive results divided by the number of true positive and false-negative results (Chapter 39).

serovar Group of closely related microorganisms distinguished by a characteristic set of antigens (Chapter 57).

service interval *See* heat interval.

service number Number of times a cow or heifer was bred (Chapter 41).

services per pregnancy Number of services required to establish a pregnancy (Chapter 41).

somatopleure Part of the embryo which forms the lateral and ventral walls of the fetus and consists of somatic mesoderm and ectoderm (Chapter 54).

sonic hedgehog (SHH) signal transduction pathway Plays a role in the patterning of many systems during development, including the central nervous system (Chapter 65).

specificity Statistical probability that an individual which does not have the particular disease being tested for will be correctly identified as negative, expressed as the proportion of true negative results to the total of true negative and false-positive results (Chapter 39).

splanchnopleure Layer of embryonic cells formed by association of part of the mesoderm with the endoderm and developing into the wall of the viscera (Chapter 54).

sporocyst Sac-like structure or cysts that contains spores or other reproductive cells (Chapter 62).

sporozoite Any of the minute undeveloped sporozoans produced by multiple fission of a zygote or spore; the infective stage of a sporozoan organism (Chapter 62).

stereoscopic Giving objects a solid or three-dimensional appearance (Chapter 79).

substance P Short-chain polypeptide that functions as a neurotransmitter especially in the transmission of pain impulses from peripheral receptors to the central nervous system (Chapter 8).

superstimulation Planned production of a number of ova from one female in the same ovulation period; an essential part of the technique of bovine embryo transfer and is typically achieved via injection of eCG or FSH and LH; commonly referred to as superovulation (Chapter 24).

SYBR-14 Sperm cell membrane permeable stain that binds DNA and is commonly used in computer-assisted sperm analysis (Chapter 74).

syndesmochorial placenta Type of placentation characterized by an endometrial attachment to the chorion with a limited amount of destruction of the endometrial epithelium (Chapter 25).

tachyzoite Motile form of cyst-causing microorganisms which rapidly grow and asexually replicate via endodyogeny (Chapter 62).

TaqMan Hydrolysis probes designed to increase the specificity of real-time PCR assays (Chapter 61).

teratogenic Of, relating to, or causing malformations of an embryo or a fetus (Chapter 54).

torticollis Wryneck; a contracted state of the cervical muscles, producing torsion of the neck (Chapter 54).

totipotent Able to differentiate along any line or into any type of cell (Chapter 67).

transcription factor Protein that binds to a sequence of DNA and controls the transcription of information from DNA to mRNA (Chapter 25).

tropism Growth response in a nonmotile organism elicited by an external stimulus; either toward (positive tropism) or away from (negative tropism) the stimulus (Chapter 54).

version Rotation of the fetus on its transverse axis (Chapter 47).

vertical transmission Spread of a disease from one generation to the next either congenitally or genetically (Chapter 62).

vitrification Process of changing or making into glass or a glassy substance, especially through heat fusion (Chapter 81).

voluntary waiting period (VWP) Desired amount of time between calving and first service (Chapter 41).

xenobiotic Any substance, harmful or not, that is foreign to the animal's biological system (Chapter 71).

zwitterion Ion with both positive and negative regions of charge; dipolar ion (Chapter 77).

Index

Abaxial placement, 74
Abnormal midpiece, 74
Abnormal offspring syndrome
   classifications, 621, 622
   development, 22–7
   dystocia, 409
   long term effects, 629
   mechanisms, 629–32
Abnormal sperm morphology
   classification, 71–3
   due to temperature, 27–9, 76
Abortion
   causes
      infectious, 499–509
         bacterial, 499–504
         fungal, 508, 509, 575–8
         protozoal, 507, 508
         viral, 504–7
      iatrogenic, 317, 318
      non-infectious, 498, 499
   diagnosis, 492, 493
   induction, 401
   pathogenesis, 482, 483
   pathology, 481, 482
Abscess
   cervical, 338
   preputial, 142
   retropreputial, 117, 118, 142, 143
   uterine, 338, 339
   vaginal, 338
Acardiac twins, 342, 495
Accessory organs, 9, 51
Aflatoxins, 599
American Embryo Transfer Association (AETA), 728
Amorphous globosus, 342
Anaplasmosis, 504
Anasarca, 341, 615, 616
Anatomy
   bull, 5–9
   cow, 191–4, 332–4
Anesthesia (local), 131–5
Anovulation, 211, 335, 336
Antiluteolytic, 245
Antral follicle count, 239, 240
Arboviruses, 682, 683

Arcanobacterium (Trueperella) pyogenes infection, 501, 502
Arginine-phenylalanine-amide related peptide (RFRP), 212–13
Arsenic, 601
Arthrogryposis multiplex (AM), 609
Artifacts, 343–5
Artificial insemination (AI), 295–302
Artificial insemination center (AIC), 97–102, 679–83
Ateleia glazioviana, 591

Balanoposthitis, 144
Bang's disease *see* Brucellosis
Basket catheter, 176
Beam width artifacts, 343
Beltsville sperm sexing technology, 671
Bent tail defect, 77
Biocontainment plan, 260
Biosecurity, 95, 97–9, 261–5, 642, 714, 715, 728
Bladder rupture, 179
Blockage shadows, 340
Blom sperm defect classification, 71
Bluetongue virus (BTV), 505, 506, 682, 683
Body condition scoring (BCS)
   beef, 276–9
   dairy, 286
Bootlace technique, 386
Bovine arachnomelia syndrome (AS), 610
Bovine citrullinemia (BC), 610
Bovine herpesvirus-1 *see* Infectious bovine rhinotracheitis
Bovine herpesvirus-4, 505
Bovine herpesvirus-5, 505
Bovine leukosis virus (BLV), 683, 715
Bovine parvovirus, 507
Bovine viral diarrhea virus (BVDV)
   characteristics, 545, 546
   control, 93, 554, 555
   diagnosis, 98, 505, 553, 554, 680
   infection of conceptus, 505, 550, 551
   persistently infected animals, 551, 552
   prevalence, 546–8
   transmission, 551, 552, 679, 680, 752–4
Brachyspina syndrome, 611, 612

Breeding injury, 102, 122
Breeding soundness examination (BSE)
   classifications, 59, 92
   electronic records, 784–6, 789
   evaluation protocol, 59, 67
   form, 59, 60, 787, 788
   history, 58, 59, 784, 785
   limitations, 59, 61
   physical examination, 64–6, 137, 142, 143
   semen evaluation, 68
   standards, 62, 66, 137
Broom snakeweed, 592
Brucellosis
   abortion, 500–501
   diagnosis, 98, 535, 681
   pathogenicity, 534, 535, 681
   strains, 533, 534
   vaccination, 535, 536
   zoonoses, 537
Buhner stitch, 385, 386
Bull to cow ratio, 94, 95

Cadmium, 601
Calf diarrhea
   causes, 641, 642, 648
   management, 642, 643, 647, 648
Calf management
   following dystocia, 407
   health maintenance, 639–43, 647, 648
   housing, 647
   neonates, 639–43, 646, 647
   nutrition, 646, 647
   weaning, 647
Calf presentation, 406, 418, 419
Calving ease, 405, 406
Campylobacteriosis
   abortion, 499–500
   diagnosis, 520, 521
   etiology, 518
   pathogenesis, 518, 519
   pathology, 519, 520
   prevention, 93, 521
   testing, 98, 520, 521, 682
   transmission, 518, 519, 681
   treatment, 522

Carbamate insecticide toxicity, 604
Cardiovascular function (during pregnancy), 248, 249
Caslick's suture, 385
Caudal epidural anesthesia, 133
Cavernosography, 159, 160
   double barrel cervix, 711
   inverted Y cervix, 711
   Y cervix, 711
Cervical eversion, 383–90, 393
Cervical prolapse see Cervical eversion
Cervical tear, 473
Cervicopexy, 389, 390
Cervix anatomy, 191, 192
Cesarean section
   indications, 424–6
   surgical approaches, 426–30
Chlamydia, 502, 503
Circulating nucleic acids, 323
Circumcision, 148–50
Client relations, 375–8
Cloning, 771–82
Closed herd, 349, 350
Clover (moldy), 593
Codominance, 230
Coiled tail, 74, 76, 77
Coitus, 113, 114
Cold shock, 297, 298, 663
Colostrum, 646, 647
Colpotomy, 476, 477
Compensable defects, 71, 72
Complex vertebral malformation (CVM), 612
Computer assisted sperm analysis (CASA), 685–7, 690–693
Congenital contractural arachnodactyly (CA), 612
Conjoined twins, 342
Corpus luteum (CL), 234–6
Corticosteroids (pharmacological use to induce parturition), 398–400
Coxiella burnetii infection, 503
Crooked tail syndrome (CTS), 612, 613
Cryopreservation
   cell injury, 662
   embryos, 718–22, 764
   freezing, 666, 667, 720
   handling, 297, 298
   packaging, 666, 719, 720
   pathogen control, 683, 749–54
   protocol, 665–8
   storage, 296, 297, 668
   thawing, 297, 298, 668, 721
   vitrification, 721, 722
Cryoprotectants, 664, 665, 719
Crystal defects, 344
Cystic ovarian disease (cystic ovarian follicles)
   classification, 449
   diagnosis, 334–6, 450, 451
   incidence, 449, 450
   physiology, 450
   risk factors, 453
   treatment, 451–3
Cystotomy, 176
Cytokines, 15
Cytoplasmic droplet, 73

Dag defect, 72, 74
Deficiency of uridine monophosphate synthase (DUMPS), 476, 613
Dematiaceous hyphomycetes, 575
Detached head, 76
Developmental duplication (DD), 613
Developmental programming, 199, 240, 255, 269, 270, 280, 629–32
Diagnostic testing, 262, 347–9, 359–63
Dietary cation–anion difference (DCAD), 287
Diphallus, 120
Disaster preparedness, 265
Disinfection, 265
Distal midpiece reflex, 73, 74
Dominant follicle, 228–30
Donor evaluation, 339, 340
Double barrel cervix, 711
Dropsy, 496
Ductus deferens, 6
Dwarfism, 610, 611
Dystocia
   causes, 409–15 see also Calf presentation
   diagnosis, 411–13, 418
   genetics, 404–6
   injuries, 471–4
   prevention, 404, 415
   risk factors, 640, 641
   treatment, 411, 412, 419–23

Early pregnancy factor (EPF), 322
Electroejaculation, 68–70
Electronic bull breeding soundness evaluation see Breeding soundness examination
Elongation (of the conceptus), 246
Embryo collection
   biosecurity, 750, 751
   collection media, 703
   collection procedure, 704–10
   equipment, 704
   horn flush, 709
   issues, 711–13
   timing, 703
Embryo grading, 718, 733–8, 744–7
Embryo transfer
   effects of heat stress, 585
   effects on offspring development, 620, 621
   embryo deposition, 715–17
   embryonic pathogens, 749–54
   frozen embryos
      thawing, 717
      transfer, 717
   IVF derived embryos, 766
   prevalence, 620
   procedure, 715–17
Embryonic development, 246, 247, 250–252, 622–5, 734–7, 765
Endocrine disrupting compounds, 603, 604
Endocrinology
   abortion, 400, 401
   bull puberty, 30–37
   estrous cycle, 203–9
   heifer puberty, 195, 196, 209, 210
   parturition, 396, 397
   placenta, 432, 433
   postpartum, 210–212, 457, 458

   pregnancy, 249–50
   testes, 14–19
Endometritis
   characterization, 441
   diagnosis, 337, 442–4
   treatment, 446
Enhancement artifacts, 344
Enterolobium spp., 591
Epididymectomy, 182
Epigenetics see Developmental programming
Epizootic bovine abortion (EBA)
   abortion, 503
   control, 566
   diagnosis, 562
   etiology, 562–4
   pathogenesis, 564, 565
   transmission, 565, 566
Epizootic hemorrhagic disease virus (EHDV), 506, 682
Equine chorionic gonadotropin (eCG), 312, 467, 696, 697
Erection, 113, 157, 158
Ergot alkaloids, 597–9
Ergotism, 598, 599
Estradiol
   estrous cycle, 203–8
   estrus synchronization, 306
   follicular dynamics, 227–30
   heifer sexual development
      infantile period, 195
      peripubertal period, 195, 196, 209, 210
      prenatal period, 195
   postpartum, 210–212, 464
   superovulation, 698
Estrous cycle, 203–8
Estrus detection
   chin ball marker, 291–3
   effectiveness, 371
   electronic systems, 291, 292
   patches, 291, 292
   tail paint, 291, 292
   teaser animals, 181, 291–4
   visual observation, 290, 291
Estrus synchronization
   CO-synch, 357
   dairy heifers, 356, 357
   eCG, 312
   estradiol, 306
   evaluation via ultrasound, 337, 658, 659
   GnRH, 305, 306, 309–11, 658, 659
   history, 655, 656
   melengestrol acetate, 657
   Presynch, 311
   progestins, 307–12, 655–7, 659
   prostaglandins, 239, 309–12, 658, 659
   recipients, 715, 724–6
Expected progeny differences (EPDs), 404–6
Extenders, 663–5
Extraembryonic membranes, 246, 247
Extruded cells, 740, 741

Fascia lata graft, 167–70
Fat (dietary), 279, 285, 286
Fawn calf syndrome see Congenital contractural arachnodactyly

Fescue toxicosis, 279, 597, 598
Fetal aging, 326
Fetal anomalies, 341, 342
Fetal development, 250, 252, 253, 625–7
Fetal lesions, 484, 485
Fetal maceration, 402, 494
Fetal programming *see* Developmental programming
Fetal sexing, 330–332
Fetotomy, 420–423
Fibropapilloma, 155, 156
Fibrotic lesions of the testes, 81, 82, 105–7
Flow cytometry, 671, 687–93
Follicle ablation, 698
Follicle stimulating hormone (FSH)
  bull, 14, 19, 34–5
  estrous cycle, 203, 204, 206–8
  follicular dynamics, 227–30, 304
  postpartum, 210–212, 457, 458
  superovulation, 696–700
Follicular cysts *see* Cystic ovarian disease
Follicular dynamics
  normal estrous cycle, 220–223, 304, 305
  old age 238, 239
  postpartum, 238
  pregnancy, 237
  prepubertal, 237
  superovulation, 697
Follicular waves, 223–8, 230, 231, 696, 697
Folliculogenesis, 219, 220
Foothill abortion *see* Epizootic bovine abortion
Forced extraction, 419
Freemartinism, 250–252
Frostbite, 144, 145

Glial cell-derived neurotrophic factor (GDNF), 19
Gonadotropin releasing hormone (GnRH)
  bull sexual development
    infantile period, 30–32
    prepubertal period, 32–5
    puberty, 35
  cystic follicles, 451
  estrous cycle, 203–8
  estrus synchronization, 305, 306, 309–11, 658, 659
  heifer sexual development
    infantile period, 195
    peripubertal period, 195, 196, 209, 210
    prenatal period, 195
  postpartum, 210–212, 457, 458
  resumption of cyclicity, 465–7
  superovulation, 696, 698, 699
Gossypol, 104, 280
Growth implants, 199

Hairy vetch, 591
Halstead technique, 386, 387
Head defects, 74–6
Heat *see* Estrus detection; Estrus synchronization
Heat stress, 286, 580–585
Heavy metal toxicity, 600
Hemlock, 595, 596
Hepatic charge, 284
Hepatic lipidosis, 284, 285

Herd health
  beef, 347–51
  dairy, 273, 274, 353–7
Histophilus somni infection, 501
Histotroph, 247
Hormone teratogenicity, 604, 605
Human chorionic gonadotropin (hCG), 320, 451, 452
Hydrallantois, 412, 413, 496
Hydramnios, 413, 496
Hydrocephalus, 409, 410, 496, 497
Hypocalcemia, 286
Hypospadias, 119
Hypothalamus-pituitary-gonadal axis, 14, 30, 51, 195

Induction of abortion, 401
Induction of parturition
  indications, 397
  methods, 398–401
  precautions, 397, 398
Infectious bovine rhinotracheitis (IBR)
  clinical signs, 542
  control, 93, 543
  diagnosis, 543, 682
  epidemiology, 541, 542
  etiology, 541
  infection of conceptus, 504, 505
  pathophysiology, 144, 145, 543
Infrared thermography, 27, 138
Inherited congenital myoclonus (ICM), 613
Inhibin, 19, 227
Insemination, 298–302
Insulin-like growth factor 1 (IGF-1), 36, 37, 209, 230
Insulin-like peptide 3 (INSL3), 18
Interferon stimulated genes, 322, 323
Interferon-tau, 245, 246, 322, 323
International Embryo Transfer Society (IETS), 734
Intromission, 7, 119
Inverted L block, 132
Inverted Y cervix, 711
In vitro fertilization
  cryopreservation, 764
  culture, 763, 764
  fertilization, 762, 763
  issues, 766, 767
  media, 760, 761
  oocyte collection, 758–60
  oocyte maturation, 761, 762
  transfer, 764, 766

Johne's disease, 714, 715

Ketosis, 284–6
Kisspeptin, 212
Knobbed acrosome, 74, 75

Lameness, 101, 102
Large offspring syndrome *see* Abnormal offspring syndrome
Lead, 601, 602
Leptospirosis
  abortion, 500
  control, 93, 531, 532
  diagnosis, 98, 500, 531, 681
  epidemiology, 529, 530

  pathogenesis, 530, 531
  transmission, 530, 681
Leydig cell, 13–18
Libido, 51, 59, 61
Line block, 131
Listeriosis, 501
Lithotripsy, 175
Locoweeds 104, 593, 594
Lupines, 594, 595
Luteal dynamics, 234–6
Luteinizing hormone (LH)
  bull, 14, 30–35
  follicular dynamics, 228–30, 304
  heifer sexual development
    infantile period, 195
    peripubertal period, 195, 196
    prenatal period, 195
  luteolysis, 236, 237
  ovulation, 233
  postpartum, 457, 458
  superovulation, 697
Luteolysis, 236, 237, 306–8
Luteotropin, 245
Lymphosarcoma (uterine), 338

Major sperm defect, 71
Management
  breeding management (to minimize disease), 263
  bulls
    breeding season, 94, 95
    bull stud, 97–102
    post-breeding season, 95
    pre-breeding, 92–4
  dairy calves, 646–8
  donor cows, 703
  dystocia prevention, 404–7
  effects on heifer age at puberty, 197–9
  heat stress, 585
  neonates, 639–43, 646, 647
  recipients, 715, 727, 728
  replacement beef heifers
    breeding season, 269
    post-breeding, 269, 270, 407
    pre-breeding, 268, 269
Marketing the practice, 374–8
Marsupialization, 178
Maternal recognition of pregnancy, 245–50
Maternally imprinted PEG3 domain, 613, 614
Meiosis, 20
Melengestrol acetate (MGA), 657
Membrane slip, 315
Metabolic disorders, 284–7
Metritis
  causes, 283, 284
  characterization, 441, 442
  diagnosis, 337, 442
  treatment, 445, 446
Midpiece defects, 73, 74
Milk fever, 286
Milk urea nitrogen, 285
Mimosa spp., 597
Minchev vaginopexy, 387–9
Minor sperm defect, 71
Mirror image artifacts, 344
Mismating, 401
Moles, 495

Mollicute infection, 502
Morphology (sperm), 70–77, 687
Motility (sperm), 70, 685, 686
Mucometra, 337, 338
Mummified fetus, 330, 476, 477, 494
Mycotic abortion, 575–8
Mycotoxins, 597

Necropsy (of aborted fetus), 488–93
Negative energy balance, 284
Neonatal death
    causes, 641, 642
    prevention, 407, 643, 647, 648
    risk assessment, 640, 641
Neosporosis
    abortion, 507, 508
    clinical signs, 570
    control, 571
    diagnosis, 570, 571
    epidemiology, 567
    Neospora life cycle, 567, 568
    pathogenesis, 569, 570
    transmission, 568, 569
    treatment, 571
Neuropathic hydrocephalus (NH), 614
Neuropeptide Y (NP-Y), 209
Nitrate toxicity, 589, 590
Nuclear vacuole, 75, 76
Nutrient partitioning, 249, 250
Nutrition
    beef cows, 277–9
    bulls, 280
    dairy cows, 283–8, 459
    dairy heifers, 272, 273, 646, 647
    effects on bull sexual development, 51–3
    effects on heifer sexual development, 197, 280
    effects on hypothalamic-gonadal function, 208, 209
    management of dystocia, 407
    recipients, 723, 724
    replacement beef heifers
        post-breeding, 269, 270
        pre-breeding, 198, 268, 280

Oocyte damage, 580–582
Open herd, 350, 351
Orbivirus infection, 505, 506
Organochlorine insecticide toxicity, 604
Organogenesis, 252, 253, 487
Organophosphate insecticide toxicity, 604
Orthobunyavirus infection, 506, 507
Osteopetrosis, 614
Ovariectomy, 477
Ovary
    anatomy, 191, 333, 334
    cystic follicles *see* Cystic ovarian disease
    early development, 219, 222
    pathology, 335, 336
    physiology, 191
Ovulation
    first, 196
    induction, 308
    process, 233, 234
Ovum pickup, 341
Oxytocin, 18

Paraphimosis, 122, 152
Paravertebral block
    distal, 133
    proximal, 132, 133
    sacral, 134
Paunch calf syndrome (PCS), 614, 615
Pelvic hematoma, 712
Pen-o-Block, 293
Penectomy, 293
Penile adhesions, 166
Penile deviations
    S-shaped, 121
    spiral, 121, 166, 167
    ventral, 121, 167–70
Penile fibropapilloma
    diagnosis, 155
    etiology, 155
    treatment, 155–7
Penile hair rings, 157
Penile hematoma
    abscessation, 166
    cause, 123, 160
    diagnosis, 123, 142, 143, 162–5
    treatment, 123, 162–6
Penile lacerations, 166
Penile prepuce translocation, 182–4, 293
Penis
    abnormalities, 119–22, 155–67
    anatomy, 7, 8, 113
    anesthesia, 134, 135
    denervation, 125, 126
    examination, 155
    injury, 122, 123, 125
    innervation, 9
    vasculature, 8
Penopexy, 184
Persistent frenulum, 114, 144, 157
Pharmaceutical teratogenicity, 606
Phimosis, 118, 119, 152
Photoperiod, 198
Phytoestrogens, 600
Placenta
    anatomy, 194, 486, 487
    development, 431–3, 495, 627–9
    expulsion, 433
    morphology, 247, 431, 432, 497
    retained placenta *see* Retained fetal membranes
    vasculature, 247, 248
Placentitis, 483, 484
Pneumovagina, 339
Polychlorinated biphenyls (PCB), 603, 604
Polycyclic aromatic hydrocarbons (PAH), 603
Ponderosa pine, 590
Postpartum anestrus
    classification, 458, 459
    contributing factors, 211, 459–62
    endocrinology, 210–212, 457, 458
    management, 462–7
Precocious puberty, 199
Predictive value, 360, 361
Pregnancy associated glycoproteins, 321
Pregnancy diagnosis
    biochemical, 321–3
    transrectal palpation, 314, 315, 320
    ultrasound, 320, 326

Pregnancy loss
    embryonic loss, 355, 498, 580, 582–4
    fetal loss *see* Abortion
    ultrasound, 327–9
Pregnancy rate
    calculation, 372
    strategies to increase, 354, 355
Pregnancy specific protein B (PSPB), 321, 322
Pregnancy staging, 316–18, 326, 327
Prepuce
    amputation, 150, 151
    anatomy, 114, 142
    examination. 142
    injury, 114–19, 148
    reconstruction, 151, 152
Preputial avulsion, 119, 148
Preputial laceration, 102, 115–17, 145, 146
Preputial pouch, 185
Preputial prolapse, 145, 146
Preputial stenosis, 114, 152–4
Preputial stoma, 152–4
Prevalence, 360
Primary sperm abnormalities, 71, 72
Progestins
    estrous cycle, 204, 206
    follicular dynamics, 228–30
    parturition, 396
    pharmacological use
        attainment of puberty, 198
        cystic follicles, 452
        estrus synchronization, 307–12, 655–7, 659
        postpartum anestrus, 464
        superovulation, 698
        pregnancy diagnosis, 321
        pregnancy maintenance, 246
Prolactin, 14, 15
Prostaglandins
    cystic follicles, 452
    heifer sexual development, 196
    luteolysis, 236, 237
    parturition, 398, 399
    pharmacological use
        estrus synchronization, 239, 309–12, 658, 659
        induction of parturition, 398–401
        postpartum anestrus, 465–7
        superovulation, 696–9
    placenta, 432, 433
    pregnancy recognition, 245, 246
Prolonged gestation, 414
Propitious moment theory, 221, 223, 225
Protein (dietary), 279, 285
Pseudodroplet, 74
Puberty *see* Sexual development
Pulmonary hypoplasia with anasarca (PHA), 615, 616
Pyometra
    characterization, 442
    diagnosis, 337, 444
    treatment, 446
Pyriform head, 75

Recipient evaluation, 340, 341
Record keeping, 364–6, 370, 639
Rectal wall thickening, 711
Rectovaginal tear, 474
Reflection artifacts/shadows, 343

Relaxin, 18
Reproductive efficiency
  dairy herd, 356, 357, 370–373
  evaluation, 356, 357, 364–7
  identification of problems, 356, 367, 368
Reproductive profile, 364–6
Retained fetal membranes
  clinical signs, 435
  pathogenesis 283, 284, 433–5
  prevention, 437
  treatment 435–7
Reverberation artifacts, 343
Rift Valley fever, 506
Risk assessment
  calving hazards, 640, 641
  environmental hazards, 641
  herd health, 349
Risk management, 260, 261, 349, 639, 640

Salmonella infection, 501
Sandhills calving system, 643
Sarcocystis infection, 508
Schistosomus reflex, 341, 409
Schmallenberg virus (SBV), 683
Scrotal circumference, 41–3, 62, 66, 137
Scrotal insulation, 28
Scrotum
  anatomy, 12, 13, 26–9, 136
  vasculature see Thermoregulation
Secondary sperm abnormalities, 71, 72
Selection
  bulls, 42
  heifers, 267, 268
  recipients, 340, 341, 714, 727
Selenium toxicosis, 602, 603
Semen-borne pathogen control, 683
Semen collection, 68–70, 100, 101
Semen evaluation, 62, 68, 70–73, 685–94
Semen handling, 297, 298
Semen sex sorting
  artificial insemination, 673
  cell injury, 675
  cost, 675
  embryo transfer, 673, 674
  in vitro fertilization, 674
  reverse sorting, 674, 675
Semen storage, 296, 297, 668
Semen tanks, 295, 296, 668
Semen thawing, 297, 298, 668
Seminiferous epithelium, 18, 22
Sensitivity, 359, 360, 362
Sertoli cell, 18, 19
Serving capacity test, 59, 61, 94
Sexual behavior
  bulls, 51
  cows, 203
Sexual development
  bulls
    infantile period, 30–32
    peripubertal period, 35–7, 44–9
    prepubertal period, 32–5
  heifers
    factors affecting, 196–9
    infantile period, 195
    ovarian dynamics, 237
    peripubertal period, 195, 196
    prenatal period, 195
Sexual differentiation, 250–252

Short penis, 120
Sidewinder see Penile prepuce translocation
Silent estrus, 462
Society for Theriogenology (SFT), 59–62, 784–6
Somatic cell nuclear transfer (SCNT)
  applications, 775, 776
  challenges, 625–9, 776–82
  history, 771
  procedure, 771–5
Specificity, 359, 360, 362
Specular reflection artifacts, 343
Sperm morphology, 70–77, 687
Sperm motility, 70, 685, 686
Spermatic cord, 5
Spermatocytogenesis, 20
Spermatogenesis
  initiation, 35, 36, 44, 47
  physiology, 5
  stages, 19, 20
  waves, 22, 23
Spermiation, 20
Spermiogenesis, 20
Spermogram see Semen evaluation
Steroidogenesis, 15–18
Stocking rate see Bull to cow ratio
Stryphnodendron spp., 591
Stump tail defect, 76
Sumpweed, 593
Superovulation
  induction, 697–700
  management of donors, 698–700
  protocols, 699, 700
  role of follicular wave dynamics, 240, 697
Superstimulation, 240, 697
Swainsonine-containing plants, 594
Synchronization of follicular wave, 697
Synchronization of recipients see Estrus synchronization

Teratogen, 486, 589–606
Terminally coiled tail, 76
Test mating, 158
Testes
  anatomy, 5, 12
  function, 12
  vasculature, 8, 26, 27
Testicular degeneration
  diagnosis, 104–5
  pathogenesis, 28, 103, 104
  prognosis, 105
Testicular development, 37, 41–3, 49
Testicular injury
  diagnosis, 136, 137
  surgical repair, 138–41
Testicular hypoplasia, 104
Testicular vascular cone, 26, 49
Testing in parallel, 362, 363
Testing in series, 362
Tetraptery spp., 591
Thermoregulation, 12, 13, 26–9, 136
Tibial hemimelia (TH), 616, 617
Tobacco poisoning, 596, 597
Trace minerals, 103, 104, 279, 280, 287, 288, 600
Transition period, 283

Transport of spermatozoa, 5, 6
Transrectal palpation, 65, 314, 315
Trichomoniasis
  abortion, 508
  characteristics, 524
  control, 94
  diagnosis, 98, 524–7, 682
  transmission, 524, 681
Tuberculosis
  abortion, 503, 504
  testing, 98, 681
  transmission, 681
Twins, 329, 330, 406

Ultrasound
  bulls
    accessory glands, 87–9
    penis, 89, 90
    scrotum, 80–87, 137, 138
  females
    antral follicle count, 239, 240
    corpus luteum, 236, 333
    ovaries, 333–6
    pregnancy, 326–32
    uterus, 332, 333, 337–40
    vagina, 338, 339
Umbilical hernia, 341
Uncompensable defects, 71, 72
Unicorn uterus, 710
Urethral fistula, 122
Urethral obstruction, 172–9
Urethral rupture, 178, 179
Urethral stricture, 179
Urethrostomy
  ischial, 174
  perineal, 174
Urethrotomy, 175, 177, 178
Urine pooling see Urovagina
Urolithiasis
  diagnosis, 173
  pathophysiology, 172, 173
  prevention, 172, 179
  treatment, 173–9
Urovagina, 339, 475
Uterine amputation, 392
Uterine blood flow, 253, 254
Uterine body flush, 704–8
Uterine eversion, 390–394
Uterine histotroph, 247
Uterine horn flush, 709
Uterine infection
  pathophysiology, 440, 441
  predisposing factors, 440
Uterine mass, 712
Uterine prolapse see Uterine eversion
Uterine tears, 471, 472
Uterine torsion, 409–12, 426
Uterus
  anatomy, 191, 332, 333
  innervation, 194
  pathology, 337
  vasculature, 193, 194

Vaccination, 92, 264, 265, 269, 274, 349–51, 647, 648, 715
Vagina
  anatomy, 193
  vasculature, 194

Vaginal eversion
  dystocia, 413
  grade, 383–5
  treatment, 385–90, 414
Vaginal prolapse *see* Vaginal eversion
Vaginal tear, 473
Vaginitis, 338, 339
Vaginopexy, 387–9
Vascular cone, 26, 49
Vascular shunts
  cavernosography, 124, 125, 159, 160
  congenital, 120
  repair, 165, 166

Vasectomy, 181, 182, 293
Veratrum californicum, 597
Vesicular adenitis
  diagnosis, 110
  pathogenesis, 109
  prevention, 111
  treatment, 110, 111
Vesiculectomy, 111
Vestibule, 193
Vitamins, 279, 286–8
  Vitamin A, 103, 104, 497, 603
  Vitamin E, 287, 288
Vitrification
  applications, 722

  protocols, 721, 722
  solutions, 721
Voluntary waiting period, 353
Vulva, 193

Wesselsbron virus, 507
Winkler cervicopexy, 389, 390

Y cervix, 711
Yersinia pseudotuberculosis infection, 501

Zearalenone, 104, 599, 600
Zoonoses, 265
Zygomycetes, 575